S0-AFH-020

Make Reform a Reality—with HRW MATHEMATICS!

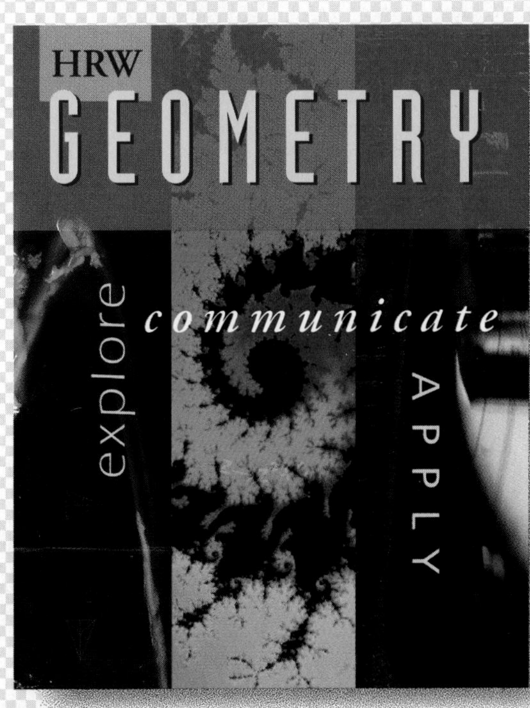

INTEGRATING

- **MATHEMATICS**
- **TECHNOLOGY**
- **EXPLORATIONS**
- **APPLICATIONS**
- **ASSESSMENT**

EXPLORE • COMMUNICATE • APPLY

The Sign of Progress—SATISFIED TEACHERS!

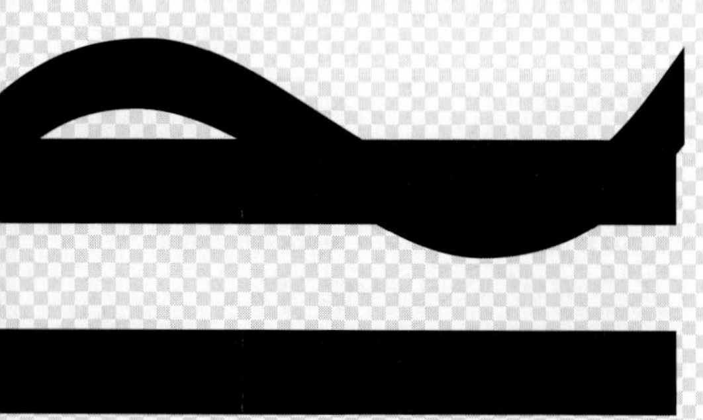

"“Nicely done! The chapters, concepts, and materials were introduced in a real-world reform framework, and the transitions from informal to formal proofs were smooth."

Nanci Takagi White
San Diego, CA

"Your integration of the discovery method for proving triangles congruent is outstanding!"

Joyce Haney
Russellville, AR

"I am impressed by the connection between mathematics and the real world. The examples are interesting and informative, and they motivated students to apply concepts."

Audrey Beres
Bridgeport, CT

"It's exciting to see the integration of portfolio activities in the textbook! I like the way discovery is carried throughout an entire chapter. Your activities foster excellent problem-solving skills!"

Betty Mayberry
Gallatin, TN

"The real-life examples are meaningful to the students—so much, in fact, that many have even taken it upon themselves to discuss their work with their parents."

Melanie Gasperec
Olympia Fields, IL

"I really like the way you integrate material, especially cooperative- learning activities."

Dianne Hershey
Jonesburg, GA

Making Progress– The Authors of HRW MATHEMATICS!

KATHLEEN A. HOLLOWELL

"Mathematics classrooms should become laboratories of learning where excited students collect data, look for patterns, make and test conjectures, and explain their reasoning."

Dr. Hollowell's keen understanding of what takes place in the mathematics classroom recently helped her win a major NSF research grant to enhance mathematics teaching methods. She is particularly well-versed in the special challenge of motivating students and making the classroom a more dynamic place to learn.

JAMES E. SCHULTZ

"Technology has the capability of opening new worlds of mathematics to more students than ever."

Dr. Schultz is a co-author of the ***NCTM Curriculum and Evaluation Standards for Mathematics*** and ***A Core Curriculum: Making Mathematics Count for Everyone.*** He is especially well regarded for his inventive and skillful integration of mathematics and technology. Jim's dynamic vision of classroom reform recently earned him the prestigious Morton Chair at Ohio University.

WADE ELLIS, JR.

"Integration can cultivate an appreciation for the relevance of mathematics—provided you meaningfully unify material with an intuitive, common-sense approach. Otherwise, diversity fosters confusion and becomes a liability."

Professor Ellis has co-authored numerous books and articles on how to integrate technology realistically and meaningfully into the mathematics curriculum. He was a key contributor to the landmark study, ***Everybody Counts: A Report to the Nation on the Future of Mathematics Education***.

Reform You Can Relate

"HRW's approach makes a real world of difference!

"I'm a big believer in teaching concepts by building from a student's base of knowledge rather than my own. I've seen too many students become frustrated because a book imposes rules that just don't relate to them and their world.

Fortunately, the intuitive approach you'll find in *HRW Geometry* makes a world of difference for me and my class. That's because HRW seamlessly connects the geometry students encounter every day in the world around them with the formal expression of geometry they find in Euclidean geometry."

To — HRW GEOMETRY!

"Students get their hands on a convincing approach!"

"In Chapter One, my students start on solid ground with very specific and convincing explanations of how to build a portfolio and why that's so important. Then they move on to discover significant properties and concepts as they construct, measure, and explore geometry through hands-on activities, real-world applications, and technology-based explorations.

HRW Geometry creatively uses tables for collecting data, which my students can use to construct their own mathematical models. From there, they can make and test conjectures about their findings, which allows them to make smooth transitions to proof."

"Here's the proof!"

"If the bottom line is getting students to understand geometry, then I'm all for employing any way possible to gain access to "big ideas" and concepts. It doesn't matter to me whether it's through informal or formal proofs, and fortunately, HRW takes the same practical approach.

I love the way my students are primed to make smooth and confident transitions from informal to formal proofs. Throughout the book, ***HRW Geometry*** does a great job of integrating proofs by synthetic, coordinate, and transformational methods. It's a whatever-works approach, and best of all, it works!"

"It works—just the way you want it to!"

"H***RW Geometry*** effectively uses a balanced approach to proofs throughout Chapters 5-9, which feature the topics you'll find at the heart of any geometry course. The program is even more flexible in Chapters 10-12, which independently treat key topics so you can explore your own areas of interest in whatever order you'd like.

You'll find a unique and comprehensive coverage of trigonometry in Chapter 10, from the unit circle to the laws of sine and cosine. Chapter 11 offers you the option of exploring fun topics—fractals, topology and so on. And in Chapter 12, you can formalize the study of logic with court cases, T.V. ads—there's something for everyone!"

KNOCK DOWN THE WALLS!

"No sooner do I open a new chapter in *HRW Geometry* than its applications take my students beyond the walls of the classroom—to the places where math really matters.

***Every chapter* begins with a credible connection to the real world, including a lively *portfolio activity*. HRW really helps me to knock down the walls that some of my students build around themselves when it comes to learning math."**

Look for Some

"Turn snoring to exploring with Exploratory Lessons!"

"When I really want to get my students off to a good start, I give them a job to do. That's exactly what HRW does with its *Exploratory Lessons*, which present concepts using a discovery approach.

For example, I have my students pretend they're paraskiing, and if they want to get off the ground, they'll need to explore the trigonometric ratios of the rope's height and length."

"Set an example with Expository Lessons"

"Another lesson format, *Expository Lessons,* provides step-by-step examples in a relevant, applied, or hands-on context. My students begin with applications or activities that involve them in what they're doing. Sometimes they continue on to 'mini explorations' when a discovery approach is particularly useful."

Sure Signs of Progress!

"But first ask WHY?"

"***HRW Geometry*** begins each lesson from a student's point of view, asking *"Why should I learn this?"* This spirit of inquiry is kept alive throughout HRW's concept development with a series of strategically positioned questions.

My students are prompted to reflect upon and clarify their understandings as they move from explorations and examples to *Try This* practice, integrated *Critical Thinking*, and open-ended assessment questions. And how can you miss them—they're all highlighted in yellow ?"

"It's great exercise!"

"***HRW Geometry*** pumps up math comprehension in four sessions of the best, no-nonsense math workout I've ever come across. In *Communicate*, my students discuss, explain, or write about math to exercise one of the most powerful and underdeveloped problem-solving muscles—the logic of language.

In the next session, they break out in a healthy sweat with some robust *Practice and Apply* problems and applications. And when it's time to wind down, my students can *Look Back* and review what they've learned, and *Look Beyond* to prepare them for future workouts."

"Expect to see some healthy changes."

"When it's time to take a deep breath and measure individual or group progress, my students stretch their minds with *Portfolio Activities*, long-term *Chapter Projects*, and *Eyewitness Math* activities. And they can further examine their progress with *Chapter Reviews* and *Chapter* and *Cumulative Assessments*.

Plug Into Math with

PUT SOME ELECTRICITY IN THE AIR!

"My classroom really comes to life whenever we plug into technology. With HRW, technology is more than a computational toy—it's one of the most serious instructional advances ever to hit mathematics education.

HRW seamlessly integrates calculators and software into the text at the right place and the right time with all the help you need. My students are motivated to *explore, communicate,* and *apply* technology in a reasonable and balanced progression.

Although the book only requires a scientific calculator, my students have the opportunity to explore concepts much further if other technologies are available to them."

"Explore!"

Reduced from actual size

Exploration 1 *Parallelograms*

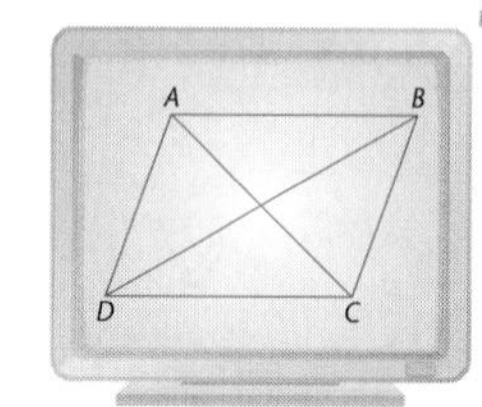

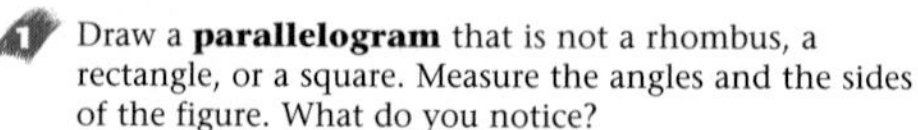

You will need
Geometry technology or
Ruler and protractor

Geometry Graphics

1. Draw a **parallelogram** that is not a rhombus, a rectangle, or a square. Measure the angles and the sides of the figure. What do you notice?
2. Draw diagonals connecting the opposite vertices. List any parts of your figure that appear to be congruent. You may want to measure these parts to verify that they are congruent. If you are using geometry technology, vary the shape of your figure by dragging one of the vertices.
3. What conjectures can you make about sides, angles, or triangles in the figure? (One was given just before this exploration. Try to find at least five conjectures in all.) ❖

Exploration 2 *Rhombuses*

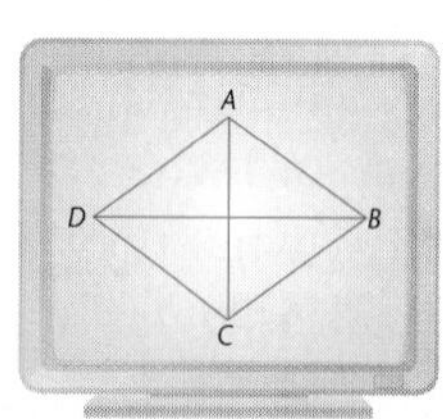

You will need
Geometry technology or
Ruler and protractor

Geometry Graphics

1. Draw a rhombus that is not a square. Draw diagonals connecting the opposite vertices. Make measurements as in the previous exploration.
2. Do the conjectures you made about parallelograms in Exploration 1 seem to be true of your rhombus? Discuss why they should or should not be true.
3. What new conjectures can you make about rhombuses that are not true of general parallelograms? (Try to find two new conjectures.) ❖

"HRW's use of technology allows me to go far beyond the surface of a math problem. Right now my class is using drawing software to explore properties of quadrilaterals."

"The book asked us to combine our algebra skills with what we've learned about quadrilaterals in geometry in order to find the measure of the angles of a rhombus, and then the measure of a rectangle's diagonal length."

TECHNOLOGY!

"Communicate!"

Reduced from actual size

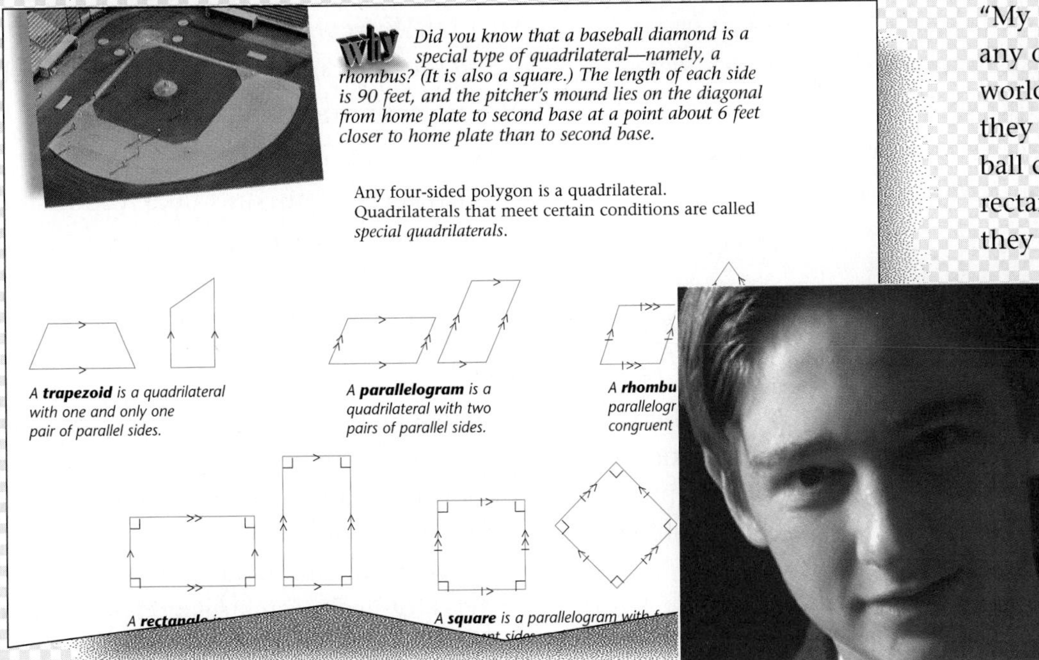

Why Did you know that a baseball diamond is a special type of quadrilateral—namely, a rhombus? (It is also a square.) The length of each side is 90 feet, and the pitcher's mound lies on the diagonal from home plate to second base at a point about 6 feet closer to home plate than to second base.

Any four-sided polygon is a quadrilateral. Quadrilaterals that meet certain conditions are called *special quadrilaterals.*

A **trapezoid** is a quadrilateral with one and only one pair of parallel sides.

A **parallelogram** is a quadrilateral with two pairs of parallel sides.

A **rhombu**[...] parallelogr[...] congruent

A **rectangle** i[...]

A **square** is a parallelogram with f[...]

"My students get a real kick out of any opportunity to redefine their world—even in small ways. When they recently discovered that a baseball diamond is also a parallelogram, rectangle, square, and a rhombus, they talked for days about playing nine innings on the ol' rhombus."

By the time class was over, we had a real debate brewing over the way fields and stadiums are designed for baseball, football, tennis—you name it."

"Apply!"

"Technology can be simple, useful, and even awe-inspiring—something my class always discovers when they use software to generate tessellation figures. Not only do they surprise themselves with how beautiful the patterns are, they also learn about the properties of quadrilaterals. And they remember what they learn!"

"Some of the drawings were so good that we actually printed and framed them, and then put them up on the classroom wall!"

Reduced from actual size

Draw your figures, as indicated below, on large-grid graph paper. The grid will help you copy the parts of your drawing. Or, if you like, you can use tracing paper. You can also do the steps using computer tessellation software.

a. Start with a square, rectangle, or other parallelogram. Replace one side of the parallelogram, such as $\overline{AB}$, by a broken line or curve.

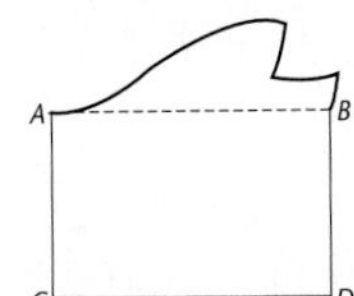
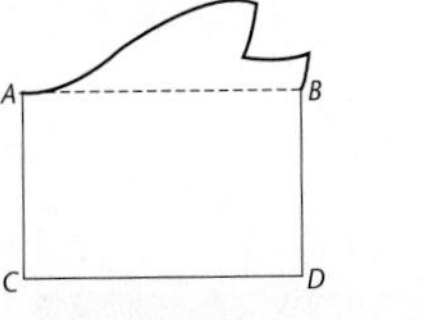

b. Translate the broken line or curve to the other side of the parallelogram. (If you are using a tracing of the curve, place it under your drawing and trace it again.)

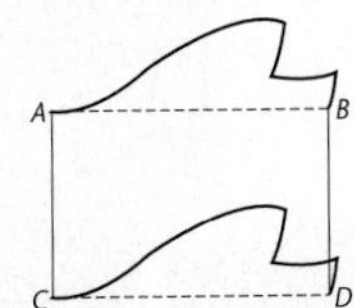

c. Repeat Steps A and B for the other two sides of your parallelogram.

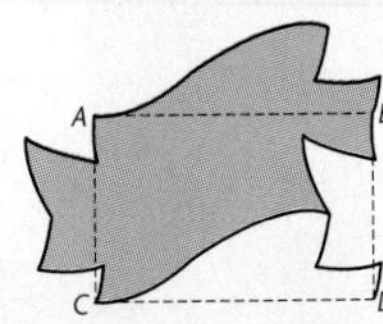

d. Your figure will now fit together with itself on all sides. Make repeated tracings of your figure in interlocking positions. You may want to add pictorial details.

Look for exciting signs

"Don't miss the action!"

"It seems strange to me that geometry books tend to be so stationary, because I've always thought of geometry as an active and practical discipline—something you do.

That's probably why I'm so comfortable with ***HRW Geometry.*** Students learn by doing with thought-provoking materials, explorations and hands-on activities. This approach really helps my students to get interested in, and take responsibility for, their own work."

"Explore!"

"When I want my students to make discoveries, HRW gives them something worth exploring, like using paper-folding activities for a visual representation of how to measure a dihedral angle.

My students discovered there's only one correct way to measure a dihedral angle. That came in handy when describing the dihedral angles formed by an airplane's plane of flight as it leaves the runway."

Reduced from actual size

Exploration 3 *Measuring Angles Formed by Planes*

You will need
Stiff folding paper or an index card
Scissors

1. Some of the faces of a cube form right angles (are perpendicular to each other). Each face of the cube is perpendicular to how many other faces of the cube?
2. Draw a horizontal line l on a piece of paper. Mark and label points A, B, and C, with B between A and C. Make a crease through B so that line l folds onto itself. What is the relationship between l and the line of the crease?
3. Open the paper slightly. The angle formed by the sides of the paper is known as a **dihedral angle**. The measure of the dihedral angle is the measure of $\angle ABC$.
4. Write your own definition of the measure of a dihedral angle by completing the following sentence: "The measure of a dihedral angle is the measure of the angle formed by . . ."

The paper-folding activity cleared things up because I could see for myself that defining the flight path really depends on the way you measure the angle of two planes."

# of life in your class!

"Communicate!"

Reduced from actual size

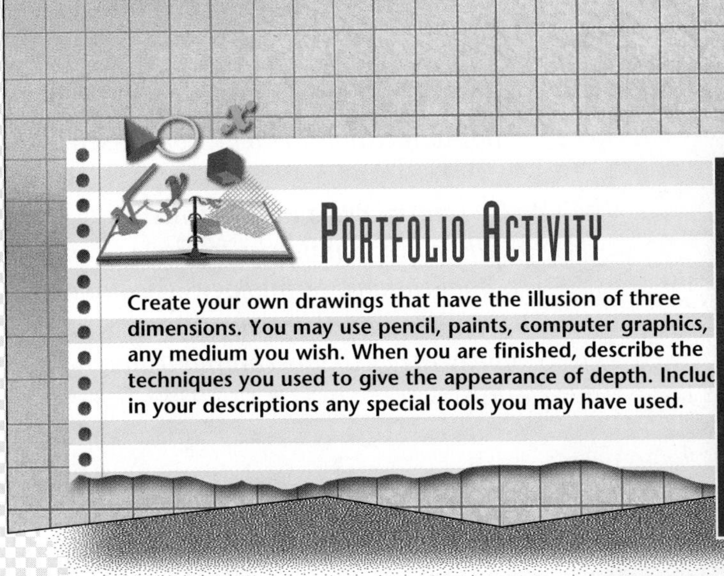

PORTFOLIO ACTIVITY
Create your own drawings that have the illusion of three dimensions. You may use pencil, paints, computer graphics, any medium you wish. When you are finished, describe the techniques you used to give the appearance of depth. Incluc in your descriptions any special tools you may have used.

"There are three things that students everywhere love to use—their hands, their minds, and their mouths. HRW puts all three to work by asking students to create, consider, and discuss their own three-dimensional drawings."

"Some people were impressed with my drawing, but when I told them how I did it, they realized they could do the same thing—it's really just a matter of technique and knowing the geometry."

"Apply!"

"My students always understand the purpose and relevance of what they're doing with ***HRW Geometry***, and I think that's the secret to success in motivating students. They're especially great at coming up with unique and thought-provoking perspectives on things they encounter every day, like music, movies, or sports.

For example, my class recently studied the concept of a parallax perspective and how it affects the way a T.V. viewer sees a pitch come over the plate in a baseball game."

Reduced from actual size

'Parallax Conspiracy' Has Angry Umpires in a Tizzy

From the *Albuquerque Journal*, October 21, 1993
By Bill Conlin

"Parallax", the Major League Umpires Association charged Tuesday night in a sharply worded letter of outrage.

The unsigned statement said this:

"The 'overhead' camera creates what experts have termed a parallax (the apparent change in the position of an object resulting from the change in the direction or position from which it is viewed.)

"I've found myself complaining a lot less about the calls I see umpires make on T.V., because I know that things don't always appear as they are when you've got a parallax perspective."

Put it All Together With

"Students climb out of the zip lock™ bag!"

"Sometimes it seems as if geometry lives in two worlds. In the real world, geometry explores the thousands of shapes, patterns, and figures we encounter every day. But in text-books, geometry concepts are all too often isolated from one another, not to mention the real world itself.

***HRW Geometry** helps my students bring those worlds together with seamlessly connected concepts, activities, applications, technology, disciplines and cultures. My students make smooth, intuitive transitions as they explore, communicate, and apply mathematics."*

"Explore!"

"I like to take my students to the places where math happens. With HRW, that means climbing into a taxicab and using geometric concepts to figure out the best route to take to a given destination.

This sort of challenge motivates my students to see how time and distance relate to integer operations on a coordinate grid. It's unbelievable just how much more my students get out of class when concepts are connected to real-world challenges."

Reduced from actual size

Exploring Taxidistances

Transportation

You will need
Graph paper (large grid)

Part I: The taxidistance from Central Dispatch

Assume that all taxis leave for their destinations from a central terminal at point $O(0, 0)$.

1. Draw the six destination points on a taxicab grid as shown in the diagram. Label the points A through F and their coordinates.
2. Find the taxidistances from (0, 0) to each of the six destination points. (Make sure that you have found the shortest taxidistance in each case.) Arrange your information in a chart.
3. Write a conjecture about the taxidistance between the point (0, 0) and a point (x, y) on a taxicab grid.

Destination point coordinates	Taxi distance from O
A (?, ?)	
B	
C	
D	
E	
F	

Part II: The taxidistance between any two points

1. Using the diagram from Part 1, find the taxidistance between at ...

x_1, y_1	x_2, y_2	x_1	x_2	y_1	y_2	Taxi distance
A (1, 1)	B (3, 4)	1				

It's a good thing our route saved us plenty of time. Otherwise, the heavy traffic would have made us late for the movie!"

SEAMLESS INTEGRATION!

"Communicate!"

"Sometimes the most important connection students make is with one another, but you've got to have an interesting and relevant context to make that a rewarding experience. HRW comes up with connections you can build on—things you can honestly say students would talk about.

For example, I have my students pretend they're parking police, and they've got to diagram the most efficient route for collecting money from parking meters."

Reduced from actual size

Law Enforcement A member of the police force is collecting money from the parking meters on the streets diagramed below. The dots represent the meters. If meters occur on both sides of the street, the police officer must go down the street twice. Use this diagram to complete Exercises 17–19.

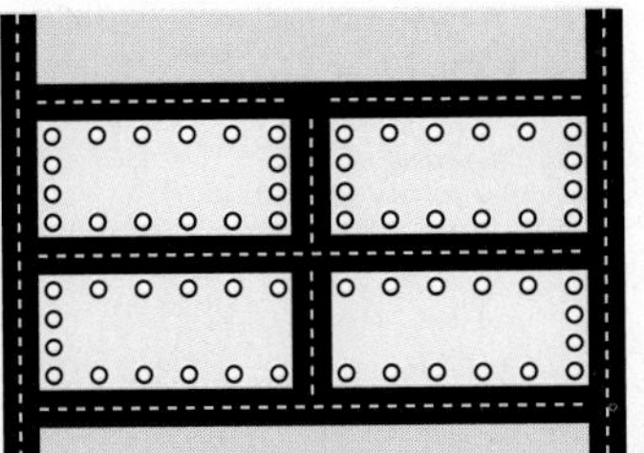

17. Draw a network that represents the parking meter problem.

18. Is this network traversable? Is it an Euler circuit? Explain your reasoning.

19. Where should the officer park in order to start and end at the same point, and to take the most efficient route possible? Explain your reasoning.

"Everyone had their own ideas about the best route, but it turned out that a combination of ideas worked best."

"Apply!"

"Integration works best when you have applications that stand at the intersection of concepts, disciplines, technology, or cultures, but most importantly, student interest. HRW does exactly that, like when students apply their knowledge of fractals to create a kolam design."

"The people of India who first made these designs probably never heard of words like algorithm, but they sure understood the concept. It makes you realize how much geometry is a part of everyday life, and you start looking at things differently."

Reduced from actual size

Cultural Connection: Asia People of India have been using fractals as art for many centuries. They are taught to draw a *Kolam* very quickly. A Kolam has an algorithm very similar to a curve known as the Hilbert Curve, which you will construct in Exercises 12–15.

12. Start with a square with no bottom side. Divide the to[p] three congruent segments, and construct a square inw[ard] middle segment. Erase the middle segment of the top [of] square. Check student constructions.

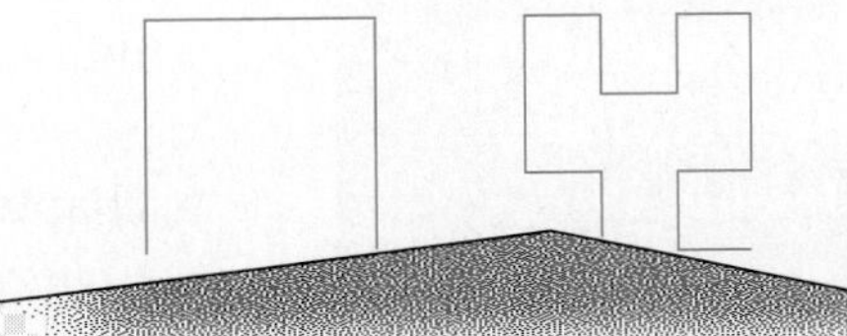

Integrated Instruction

Reduced from actual size

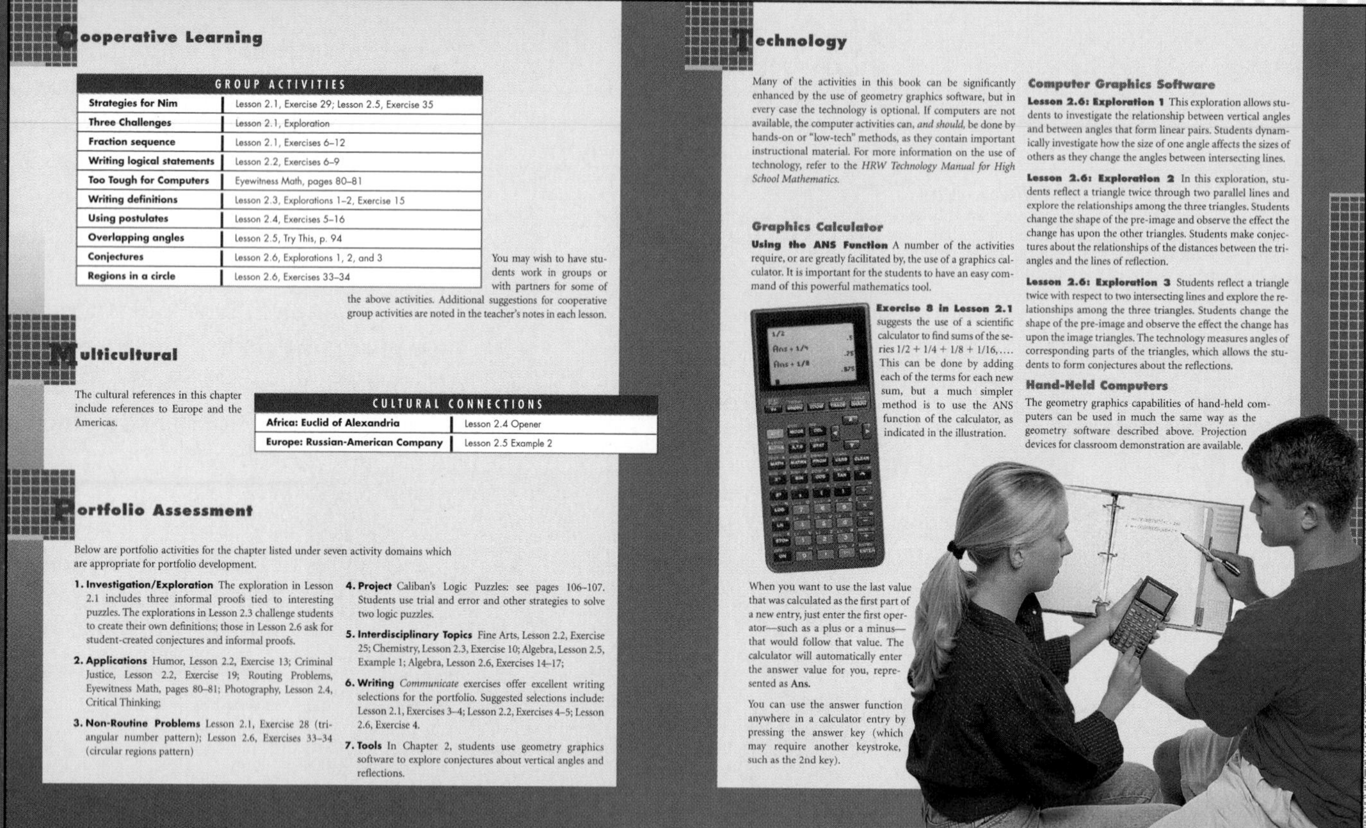

Cooperative Learning

GROUP ACTIVITIES	
Strategies for Nim	Lesson 2.1, Exercise 29; Lesson 2.5, Exercise 35
Three Challenges	Lesson 2.1, Exploration
Fraction sequence	Lesson 2.1, Exercises 6–12
Writing logical statements	Lesson 2.2, Exercises 6–9
Too Tough for Computers	Eyewitness Math, pages 80–81
Writing definitions	Lesson 2.3, Explorations 1–2, Exercise 15
Using postulates	Lesson 2.4, Exercises 5–16
Overlapping angles	Lesson 2.5, Try This, p. 94
Conjectures	Lesson 2.6, Explorations 1, 2, and 3
Regions in a circle	Lesson 2.6, Exercises 33–34

You may wish to have students work in groups or with partners for some of the above activities. Additional suggestions for cooperative group activities are noted in the teacher's notes in each lesson.

Multicultural

The cultural references in this chapter include references to Europe and the Americas.

CULTURAL CONNECTIONS	
Africa: Euclid of Alexandria	Lesson 2.4 Opener
Europe: Russian-American Company	Lesson 2.5 Example 2

Portfolio Assessment

Below are portfolio activities for the chapter listed under seven activity domains which are appropriate for portfolio development.

1. **Investigation/Exploration** The exploration in Lesson 2.1 includes three informal proofs tied to interesting puzzles. The explorations in Lesson 2.3 challenge students to create their own definitions; those in Lesson 2.6 ask for student-created conjectures and informal proofs.
2. **Applications** Humor, Lesson 2.2, Exercise 13; Criminal Justice, Lesson 2.2, Exercise 19; Routing Problems, Eyewitness Math, pages 80–81; Photography, Lesson 2.4, Critical Thinking;
3. **Non-Routine Problems** Lesson 2.1, Exercise 28 (triangular number pattern); Lesson 2.6, Exercises 33–34 (circular regions pattern)
4. **Project** Caliban's Logic Puzzles: see pages 106–107. Students use trial and error and other strategies to solve two logic puzzles.
5. **Interdisciplinary Topics** Fine Arts, Lesson 2.2, Exercise 25; Chemistry, Lesson 2.3, Exercise 10; Algebra, Lesson 2.5, Example 1; Algebra, Lesson 2.6, Exercises 14–17;
6. **Writing** *Communicate* exercises offer excellent writing selections for the portfolio. Suggested selections include: Lesson 2.1, Exercises 3–4; Lesson 2.2, Exercises 4–5; Lesson 2.6, Exercise 4.
7. **Tools** In Chapter 2, students use geometry graphics software to explore conjectures about vertical angles and reflections.

Technology

Many of the activities in this book can be significantly enhanced by the use of geometry graphics software, but in every case the technology is optional. If computers are not available, the computer activities can, *and should*, be done by hands-on or "low-tech" methods, as they contain important instructional material. For more information on the use of technology, refer to the *HRW Technology Manual for High School Mathematics.*

Graphics Calculator

Using the ANS Function A number of the activities require, or are greatly facilitated by, the use of a graphics calculator. It is important for the students to have an easy command of this powerful mathematics tool.

Exercise 8 in Lesson 2.1 suggests the use of a scientific calculator to find sums of the series 1/2 + 1/4 + 1/8 + 1/16, …. This can be done by adding each of the terms for each new sum, but a much simpler method is to use the ANS function of the calculator, as indicated in the illustration.

When you want to use the last value that was calculated as the first part of a new entry, just enter the first operator—such as a plus or a minus—that would follow that value. The calculator will automatically enter the answer value for you, represented as **Ans**.

You can use the answer function anywhere in a calculator entry by pressing the answer key (which may require another keystroke, such as the 2nd key).

Computer Graphics Software

Lesson 2.6: Exploration 1 This exploration allows students to investigate the relationship between vertical angles and between angles that form linear pairs. Students dynamically investigate how the size of one angle affects the sizes of others as they change the angles between intersecting lines.

Lesson 2.6: Exploration 2 In this exploration, students reflect a triangle twice through two parallel lines and explore the relationships among the three triangles. Students change the shape of the pre-image and observe the effect the change has upon the other triangles. Students make conjectures about the relationships of the distances between the triangles and the lines of reflection.

Lesson 2.6: Exploration 3 Students reflect a triangle twice with respect to two intersecting lines and explore the relationships among the three triangles. Students change the shape of the pre-image and observe the effect the change has upon the image triangles. The technology measures angles of corresponding parts of the triangles, which allows the students to form conjectures about the reflections.

Hand-Held Computers

The geometry graphics capabilities of hand-held computers can be used in much the same way as the geometry software described above. Projection devices for classroom demonstration are available.

"Get full instructional support!"

"As a teacher, nothing is more important than the transfer of knowledge between me and my students. *HRW Geometry* gives teachers a full stream of instructional options and suggestions, so I can maintain a seamless relationship between me, the text, and my students.

HRW really supports instruction today rather than someone's vision of math instruction in the year 2010. I get useful information, like the *Alternative Teaching Strategies* you'll find in every lesson. Plus, I really appreciate HRW's easy-to-follow layout and organization."

"Before..."

"Before you begin a chapter, HRW's opening *Side Columns* on the *Chapter Openers* help you prepare to introduce new material by providing *Background Information, Chapter Objectives*, a list of available *Resources*, and more.

Then there's a series of *Interleaves*, which I find particularly useful. I read them before starting a chapter, then refer back to them whenever necessary. Interleaves include everything from a *Planning Guide* to *Reading, Visual*, and *Hands-on Strategies* for helping individual students.

You also get a quick-look reference to the chapter's *Cooperative Learning Activities, Cultural Connections*, and *Portfolio Assessment*, as well as a full page dedicated to *Technology* instruction."

begins with You!

Reduced from actual size

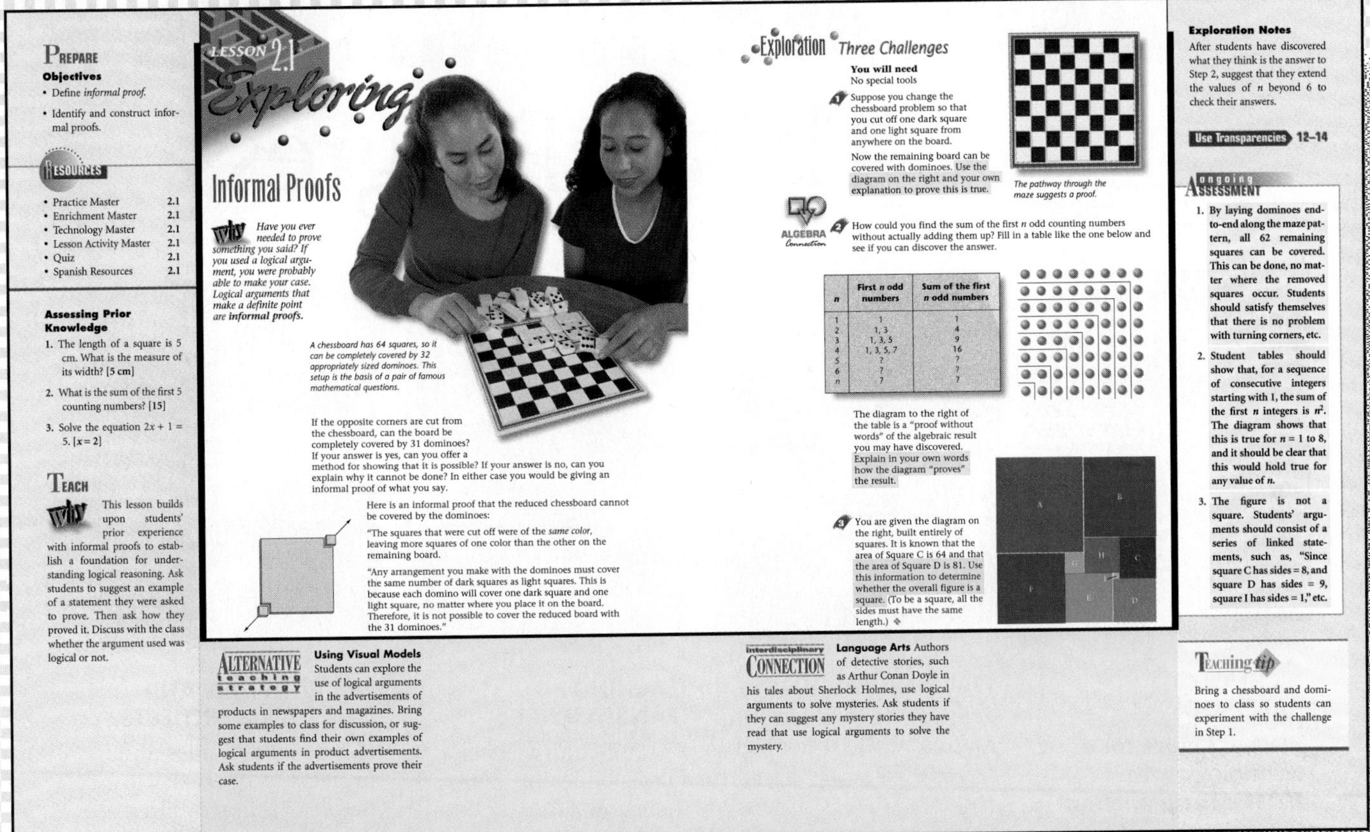

PREPARE

Objectives

- Define *informal proof.*
- Identify and construct informal proofs.

RESOURCES

• Practice Master	2.1
• Enrichment Master	2.1
• Technology Master	2.1
• Lesson Activity Master	2.1
• Quiz	2.1
• Spanish Resources	2.1

Assessing Prior Knowledge

1. The length of a square is 5 cm. What is the measure of its width? [5 cm]
2. What is the sum of the first 5 counting numbers? [15]
3. Solve the equation $2x + 1 = 5$. [$x = 2$]

TEACH

WHY This lesson builds upon students' prior experience with informal proofs to establish a foundation for understanding logical reasoning. Ask students to suggest an example of a statement they were asked to prove. Then ask how they proved it. Discuss with the class whether the argument used was logical or not.

LESSON 2.1 *Exploring* Informal Proofs

WHY *Have you ever needed to prove something you said? If you used a logical argument, you were probably able to make your case. Logical arguments that make a definite point are* ***informal proofs.***

A chessboard has 64 squares, so it can be completely covered by 32 appropriately sized dominoes. This setup is the basis of a pair of famous mathematical questions.

If the opposite corners are cut from the chessboard, can the board be completely covered by 31 dominoes? If your answer is yes, can you offer a method for showing that it is possible? If your answer is no, can you explain why it cannot be done? In either case you would be giving an informal proof of what you say.

Here is an informal proof that the reduced chessboard cannot be covered by the dominoes:

"The squares that were cut off were of the *same color,* leaving more squares of one color than the other on the remaining board.

"Any arrangement you make with the dominoes must cover the same number of dark squares as light squares. This is because each domino will cover one dark square and one light square, no matter where you place it on the board. Therefore, it is not possible to cover the reduced board with the 31 dominoes."

Exploration *Three Challenges*

You will need
No special tools

1. Suppose you change the chessboard problem so that you cut off one dark square and one light square from anywhere on the board.
Now the remaining board can be covered with dominoes. Use the diagram on the right and your own explanation to prove this is true.

The pathway through the maze suggests a proof.

ALGEBRA Connection

2. How could you find the sum of the first n odd counting numbers without actually adding them up? Fill in a table like the one below and see if you can discover the answer.

n	First n odd numbers	Sum of the first n odd numbers
1	1	1
2	1, 3	4
3	1, 3, 5	9
4	1, 3, 5, 7	16
5	?	?
6	?	?
n	?	?

The diagram to the right of the table is a "proof without words" of the algebraic result you may have discovered. Explain in your own words how the diagram "proves" the result.

3. You are given the diagram on the right, built entirely of squares. It is known that the area of Square C is 64 and that the area of Square D is 81. Use this information to determine whether the overall figure is a square. (To be a square, all the sides must have the same length.) ❖

Exploration Notes
After students have discovered what they think is the answer to Step 2, suggest that they extend the values of n beyond 6 to check their answers.

Use Transparencies 12–14

Ongoing ASSESSMENT

1. By laying dominoes end-to-end along the maze pattern, all 62 remaining squares can be covered. This can be done, no matter where the removed squares occur. Students should satisfy themselves that there is no problem with turning corners, etc.
2. Student tables should show that, for a sequence of consecutive integers starting with 1, the sum of the first n integers is n^2. The diagram shows that this is true for $n = 1$ to 8, and it should be clear that this would hold true for any value of n.
3. The figure is not a square. Students' arguments should consist of a series of linked statements, such as, "Since square C has sides = 8, and square D has sides = 9, square I has sides = 1," etc.

ALTERNATIVE teaching strategy

Using Visual Models Students can explore the use of logical arguments in the advertisements of products in newspapers and magazines. Bring some examples to class for discussion, or suggest that students find their own examples of logical arguments in product advertisements. Ask students if the advertisements prove their case.

Interdisciplinary CONNECTION

Language Arts Authors of detective stories, such as Arthur Conan Doyle in his tales about Sherlock Holmes, use logical arguments to solve mysteries. Ask students if they can suggest any mystery stories they have read that use logical arguments to solve the mystery.

TEACHING tip

Bring a chessboard and dominoes to class so students can experiment with the challenge in Step 1.

"During..."

"As you move through a lesson, you'll notice side column support is designed to lead you through content in a timely and logical progression.

Key sections include *PREPARE*, with *Objectives, Resources*, and *Prior Knowledge Assessing*, and *TEACH,* which features some substantial teaching strategies, notes, and tips, from *Critical Thinking* to *Alternative Teaching Strategies* to *Interdisciplinary Connections."*

"And After Your Lessons!"

"***HRW Geometry*** gives you plenty of ways to measure the progress of your students with *Ongoing* and *Alternative Assessment*, and *Practice* and *Technology Masters*.

You'll also get lots of practical tips for the *Chapter Project*, which applies skills presented throughout the chapter. And finally, you'll find some great ideas for *Eyewitness Math*, a fun cooperative-learning activity that springs from today's headlines and stories."

Reform That's Based on What You Need!

HRW MATHEMATICS... This *is* Progress

TECHNOLOGY HANDBOOK
Teacher's guide for using technology with ***HRW Mathematics.***

TEACHING RESOURCES
Support for each lesson in four convenient booklets.

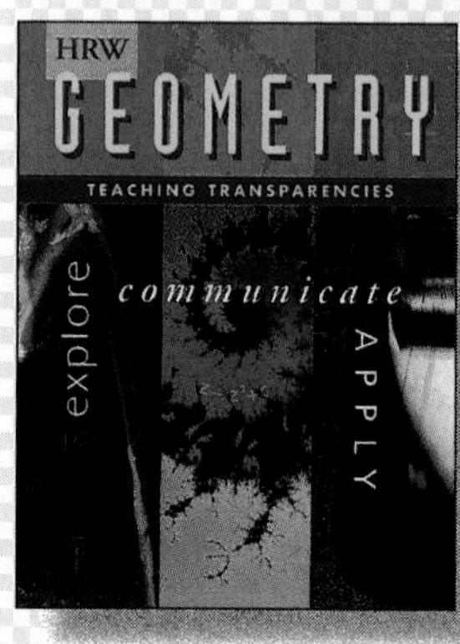

TEACHING TRANSPARENCIES
150 full-color visuals with suggested lesson plans for their use.

LAB ACTIVITIES AND LONG-TERM PROJECTS
Hands-on activities and projects to be used before, during, or after each chapter.

TEST GENERATOR
A variety of assessment types delivered on user-friendly software.
(Two versions available–Macintosh® and IBM®PC and compatibles.)

TEST GENERATOR: ASSESSMENT ITEM LISTING
Printout of all items included on the Test Generator.

SOLUTION KEY
Worked-out solutions for all *Exercises & Problems.*

SPANISH RESOURCES
Spanish translation of objectives, main ideas of lessons, and terminology.

EXPLORE • COMMUNICATE • APPLY

HRW

GEOMETRY

TEACHER'S EDITION

explore *communicate* APPLY

Integrating

MATHEMATICS
TECHNOLOGY
EXPLORATIONS
APPLICATIONS
ASSESSMENT

HOLT, RINEHART AND WINSTON
Harcourt Brace & Company
Austin • New York • Orlando • Atlanta • San Francisco • Boston • Dallas • Toronto • London

AUTHORS

Kathleen A. Hollowell

Dr. Hollowell is widely respected for her keen understanding of what takes place in the mathematics classroom. Her impressive credentials feature extensive experience as a high school mathematics and computer science teacher, making her particularly well-versed in the special challenges associated with integrating math and technology. She currently serves as Associate Director of the Secondary Mathematics Inservice Program, Department of Mathematical Sciences, University of Delaware and is a past-president of the Delaware Council of Teachers of Mathematics.

James E. Schultz

Dr. Schultz is one of the math education community's most renowned mathematics educators and authors. He is especially well regarded for his inventive and skillful integration of mathematics and technology. He helped establish standards for mathematics instruction as a co-author of the NCTM "Curriculum and Evaluation Standards for Mathematics" and "A Core Curriculum, Making Mathematics Count for Everyone." Following over 25 years of successful experience teaching at the high school and college levels, his dynamic vision recently earned him the prestigious Robert L. Morton Mathematics Education Professorship at Ohio University.

Wade Ellis, Jr.

Professor Ellis has gained tremendous recognition for his reform-minded and visionary math publications. He has made invaluable contributions to teacher inservice training through a continual stream of hands-on workshops, practical tutorials, instructional videotapes, and a host of other insightful presentations, many focusing on how technology should be implemented in the classroom. He has been a member of the National Research Council's Mathematical Sciences Education Board, the MAA Committee on the Mathematical Education of teachers, and is a former Visiting Professor of Mathematics at West Point.

(Acknowledgements appear on pages 725–726, which are extensions of the copyright page.)

Printed in the United States of America
1 2 3 4 5 6 7 041 00 99 98 97 96

ISBN: 0-03-097776-2

CONTRIBUTING AUTHORS

Larry Hatfield Dr. Hatfield is Department Head and Professor of Mathematics Education at the University of Georgia. He is recipient of the Josiah T. Meigs Award for Excellence in Teaching, his university's highest recognition for teaching. He has served at the National Science Foundation and is Director of the NSF funded Project LITMUS.

Bonnie Litwiller Professor of Mathematics, University of Northern Iowa, Cedar Falls, Iowa, Dr. Litwiller has been co-director of NSF institutes, project coordinator for the NCTM Addenda Project, and co-author of three books and 650 articles.

Martin Engelbrecht A mathematics teacher at Culver Academy, Culver, Indiana, Mr. Engelbrecht also teaches statistics at Purdue University—North Central. An innovative teacher and writer, he integrates applied mathematics with technology to make mathematics more accessible to all students.

Kenneth Rutkowski A mathematics teacher at James Bowie High School, Austin, Texas, Mr. Rutkowski is an innovative geometry teacher, who sponsors his school's mathematics honor society, serves on various professional committees and conducts creative teacher workshops.

Editorial Director of Math
Richard Monnard

Executive Editor
Gary Standafer

Senior Editor
Ronald Huenerfauth

Project Editors
Charles McClelland
Joel Riemer
Michelle Dowell
Michael Funderburk

Design and Photo
Pun Nio
Diane Motz
Lori Male
Julie Ray
Rhonda Holcomb
Robin Bouvette
Candice Moore
Ophelia Wong
Sam Dudgeon
Victoria Smith
Mavournea Hay
Cindy Verheyden
Michael Obershan
Alicia Sullivan
Katie Kwun
Monotype Editorial Services

Production and Manufacturing
Donna Lewis
Amber Martin
Jenine Street
Shirley Cantrell

Editorial Staff
Steve Oelenberger
Richard Zelade
Andrew Roberts
Pam Garner
Jane Gallion
Desktop Systems Assistant
Jill Lawson
Department Secretary

CONTENT CONSULTANT

Kenneth Rutkowski A mathematics teacher at James Bowie High School, Austin, Texas, Mr. Rutkowski is an innovative geometry teacher, who sponsors his school's mathematics honor society, serves on various professional committees and conducts creative teacher workshops.

MULTICULTURAL CONSULTANT

Beatrice Lumpkin A former high school teacher and associate professor of mathematics at Malcolm X College in Chicago, Professor Lumpkin is a consultant for many public schools for the enrichment of mathematics education through its multicultural connections. She served as a principal teacher-writer for the *Chicago Public Schools Algebra Framework* and has served as a contributing author to many other mathematics and science publications that include multicultural curriculum.

REVIEWERS

James A. Bade
Adlai Stevenson High School
Sterling Heights, Michigan

Tom Fitzgerald
Cocoa High School
Cocoa, Florida

Ona Lea Lentz
North High School
Minneapolis, Minnesota

Karen M. Lesko
Pacific High School
San Bernardino, California

Jean Mariner
St. Steven's Episcopal School
Austin, Texas

Gregory Massarelli
Watkins Memorial High School
Pataskala, Ohio

Susan May
Science Academy of Austin at LBJ
Austin, Texas

John S. Nesladek
Ozaukee High School
Fredonia, Wisconsin

Gary Nowitzke
Jefferson High School
Monroe, Michigan

Roger O'Brien
Polk County School District
Bartow, Florida

Ruth R. Price
Lee County High School
Sanford, North Carolina

Robert J. Russell
West Roxbury High School
West Roxbury, Massachusetts

Sandra Seymour
Science Academy of Austin at LBJ
Austin, Texas

Rosalind Taylor
W. C. Overfelt High School
San Jose, California

Jean D. Watson
High School of Commerce
Springfield, Massachusetts

Nanci Takagi White
University City High School
San Diego, California

A Message from the Authors

It is certainly a challenge in our rapidly changing world to capture in a textbook the essence of what students need and teachers can provide. Frequent visits to schools confirm that mathematics programs often continue to focus on skills - many of which are diminishing in importance - even though there is an increasing need for students to be able to understand concepts and to apply them to real-world problems using appropriate technology. To make matters worse, limited school budgets make it difficult for teachers to implement desired changes while they are faced with significant challenges that compete for their time and energy.

The authors are dedicated to the idea that mathematics programs should help all students gain mathematical power in a technological society. Based on careful examination of current recommendations and school mathematics programs, the authors have developed a program that strikes a balance in maintaining the strengths of former approaches while moving to mathematics content and methods of learning that are up-to-date and relevant to the present and future lives of students. In education, as in so many other areas, even well-intended change should not be so rapid that students, teachers, and parents cannot cope with it. This program makes carefully chosen strides in the most vital areas, while staying within the comfort zones of students, teachers, and parents.

Our textbooks reflect a vision of mathematics instruction that includes three components:

- active students
- solving interesting and relevant problems
- using appropriate technology

For example, in these books students use readily available technology (ordinary calculators). Problems that 40 percent of high school juniors traditionally attempt using advanced techniques can now be solved by almost 100 percent of high school freshmen using simple, inexpensive technology. This advanced study of important topics makes mathematics more interesting and relevant.

This program successfully solves a long-standing dilemma: textbooks don't feature the use of technology because schools don't have the equipment. And, schools don't have the equipment because it's not used in the textbooks. Thus, we have seen too many textbooks that do not reflect the needs of the students. In this series, technology is highly profiled, but only a limited amount is required. Courses that are strongly rooted in mathematical content can be enhanced by including the technology as it becomes available.

We wish you well in pursuing this timely, balanced approach!

TABLE OF CONTENTS

CHAPTER 3 Parallels and Polygons 114

MATH *Connections*

CHAPTER 4 Congruent Triangles 172

APPLICATIONS

APPLICATIONS

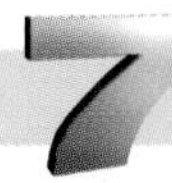

CHAPTER 7 Surface Area and Volume 352

MATH *Connections*

CHAPTER 8 Similar Shapes 410

APPLICATIONS

CHAPTER  Circles 468

MATH Connections

CHAPTER  Trigonometry 522

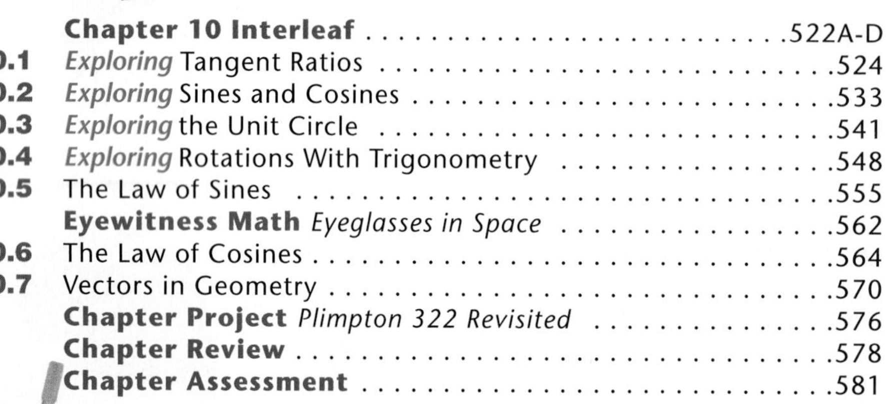

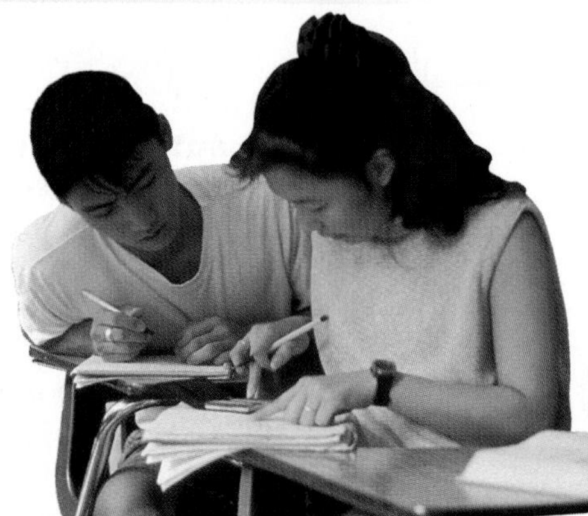

APPLICATIONS

MATH *Connections*

Algebra 588, 631, 632
Coordinate Geometry 594–606, 679–684

APPLICATIONS

Science
Environmental Science 678
Genetics 607

Social Studies
History 630

Language Arts
Communicate 590, 597, 604, 611, 618, 626, 633, 647, 653, 661, 668, 675, 683
Eyewitness Math 656

Business and Economics
Materials Handling 612

Sports and Leisure
Candy Making 635
Nine Coin Puzzle 628

Other
Communications 632
Klein Bottle 613
Law Enforcement 605
Transportation 595

Interactive Learning

Students learn best if they are active participants in the learning process. Technology transforms today's mathematics classroom into a laboratory where students explore and experiment with mathematical concepts rather than just memorize isolated skills. Students make generalizations and reach conclusions about mathematical concepts and relationships and apply them to real-life situations. *HRW Geometry* encourages instruction that utilizes technology with numerous explorations and examples. Where appropriate, students explore, work examples, solve exercises, and confirm mathematical ideas for themselves.

To learn geometry actively, students need to be able to make and test conjectures about geometric figures and their properties. The availability of geometry software today provides a unique opportunity for students to learn geometry in an interactive environment that employs the full power of computer technology.

One of the most fundamental tenets of constructivist learning theory is that initial learning by a student should come from physical or mental experiments or explorations, that is, from constructivist approaches and methods and not from memorizing facts and definitions. Such an approach to learning geometry requires that students create many geometric figures of various sizes and shapes in order to explore them and make conjectures about their properties. The availability of geometry software programs that focus on the basic concepts of high school geometry has made it possible to implement constructivist approaches to geometry in new and exciting ways. Tasks that were formerly time-consuming and tedious are now fast and trouble-free. More time may be devoted to explorations, making conjectures, and solving original problems.

Historically, geometry first developed as a large collection of discrete facts that were arrived at by making observations and measurements. Only later did Euclid organize these facts into a coherent whole by using the process of deductive reasoning. The use of computer technology supports both a sound psychological development of learning and the historical development. In other words, computers can help students to first discover relationships and properties of figures which can then be stated in formal terms as theorems and proved deductively.

Role of the Teacher

Technology usage is changing the role of the teacher in the mathematics classroom. The National Council of Teacher's of Mathematics' *Curriculum and Evaluation Standards for School Mathematics* describes "the emergence of a new classroom dynamic in which teachers and students become natural partners in developing mathematical ideas and solving mathematical problems."[1]

As a facilitator of learning, the teacher becomes a guide, leading students to their own mathematical discoveries and generalizations. Mathematics is no longer a static subject, a group of abstract symbols. Mathematics becomes a dynamic field of related concepts that can be explored and experimented with in the same way as science concepts.

In a dynamic classroom setting with students actively working on computers, the role of the teacher changes from lecturer to facilitator of learning. As students work individually or in small groups, the teacher avoids directing the process, but instead asks questions to help students think creatively to make discoveries and solve problems. The teacher is there to guide students along the path of creating or constructing their own understanding and knowledge of geometry. This is an exciting, motivating, and powerfully effective way to teach.

Changes in the Classroom

Teachers are finding that technology encourages cooperative learning. Cooperative-learning groups allow students to compare results, brainstorm, and reach conclusions based on group results. As in real life where scientists or financial analysts often consult each other, students in the technology-oriented classroom learn to

[1] *National Council of Teachers of Mathematics.* Curriculum and Evaluation Standards for School Mathematics. *Reston, VA: The Council, 1989. (p. 128)*

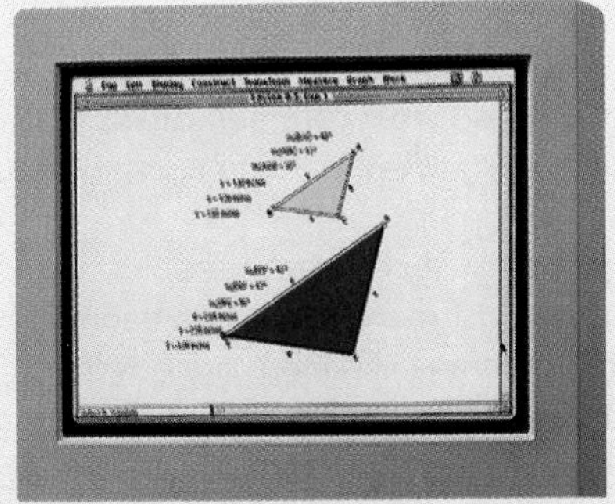

communicate and consult with each other. Students learn that such consultations are not "cheating," but rather are a method of sharing information that will be used to solve a problem.

Another change in the classroom will be the role of the teacher from a dispenser of facts and formulas to a guide along the exploration trail. Teachers will become monitors to keep students headed in the right directions. As students gain confidence in their ability, they will ask more questions, including many higher-level questions. Teachers no longer always need to have the "right" answer—teachers can suggest further exploration or research. In this way, teachers model real-life situations where experts with the right answer are not always available or where there is no right answer.

Software Methodology

Exploratory lessons occur throughout *HRW Geometry.* Many of these exploratory lessons utilize the power of technology as a tool of exploration. Geometry software employs various methods of involving students in the learning process. Students can draw (construct) basic geometric figures such as points, lines, triangles, polygons, and circles. They can also make and test conjectures, measure segments and angles, vary the size and shape of figures, check certain properties, such as lines being parallel or perpendicular, and they can do all of these things rather quickly. This basic aspect of the software empowers students to visualize and examine many figures in a relatively short period of time.

Geometry software can also engage students in specific investigations, explorations, and problems that lead them to discover relationships and make conjectures. Having discovered a relationship they think is true, students can then test it by examining different cases. If the relationship continues to be valid, students can generalize their results and then try to prove them by writing a proof.

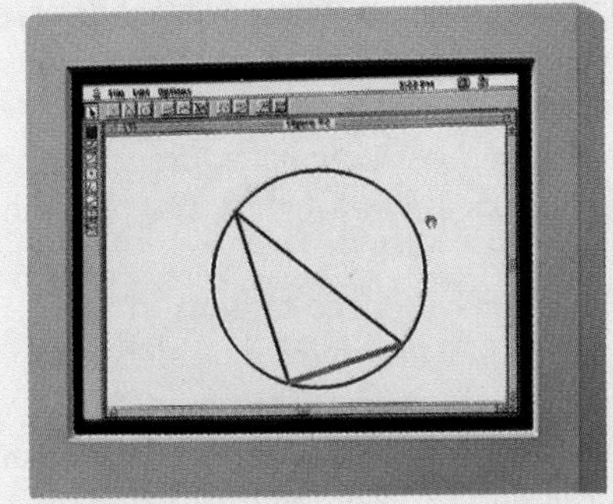

This aspect of the software actively involves students in discovering geometric facts and constructing their own understanding of geometry.

Graphics Calculators

The scope of capabilities of this graphics tool is wide, including traditional arithmetic calculations, table setups, statistics mode, function graphing, parametric graphing, polar graphing, sequence graphing, and more. With the comprehensive integration of algebra into the geometry instruction, the graphics calculator becomes a viable tool for performing many specialized and complex mathematical actions.

HRW Geometry

Geometry graphics software programs available today can involve students in the major concepts of this course. These concepts include the study of lines and angles, triangles, quadrilaterals and other

polygons, circles, congruence and similarity, area, the Pythagorean theorem, transformations, symmetry, tessellations, coordinates, trigonometry, and vectors.

Throughout the student textbook, references are made to those activities that can be done using computer software. This approach begins in Chapter 1, Lesson 1.3, where students explore some special points in triangles and then construct circles having these points as centers.

Although the use of computer technology is integrated throughout the student text, all computer activities can also be done using compass-and-straightedge or paper-folding strategies. Specific suggestions for using geometry software also appear frequently in the side-column notes of the Teacher's Edition. Sample activities are given on the Technology page of the introductory material that precedes each chapter.

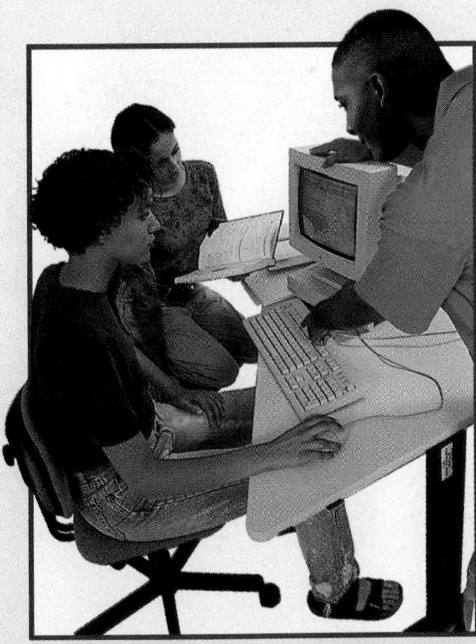

Explorations

Facilitator of Learning

An increasing number of educators are realizing that more learning takes place when students construct knowledge for themselves. This approach, sometimes called *constructivism,* necessitates changes in both the form of instructional materials and the vehicles used to deliver instruction.

Instead of presenting students with rules, theorems, principles, and worked-out examples, a constructivist approach calls for students to be given questions to investigate, problems to explore, and conjectures to verify or disprove. During the implementation of constructivism, a teacher is no longer the source of all information. Instead, the teacher acts a a facilitator, presenting the questions to explore and pointing to areas that need further discussion and clarification.

An activity-oriented approach gives teachers several advantages. Perhaps the most important of these is the ability to provide for learning styles other than verbal and visual. Students who need hands-on tactile experiences—the kinesthetic learners—are no longer penalized by materials and instructional approaches that emphasize primarily listening and recording information.

Students who need to work in collaboration with others to understand geometric concepts—the interpersonal learners—now have the opportunities that occur during small-group instruction. In fact, all students benefit from the increased written and verbal communication that occurs when active learning is implemented in the classroom.

So, what specific form should constructivist materials take in geometry? Although many variations are possible, materials suitable for active learning usually involve identification of patterns, making and testing conjectures, and exploring alternative approaches to problems. Students should be encouraged to try any approach that occurs as they work on a problem. Concrete models, graphic representations, tabular methods such as those facilitated by spreadsheets—all of these have an appropriate place in today's geometry classroom.

In many classroom situations, a key teacher decision will be how much constructivism to use and when to use it. Some topics and situations will continue to require direct instruction; others will open opportunities for both students and teachers to experience the richer knowledge formation that results from students' active and personal involvement in their own learning.

Uses of Technology

Rapid and continuing changes in technology provide teachers and students in the geometry classroom with more choices for strategies and procedures than in past years. Geometry software is an excellent tool for use in student explorations. Using geometry graphics software, students can now explore multiple geometric relationships without tedious manipulations. Activities and problems that in the past could only be done with compass and straightedge can now be approached using geometry graphics software on a computer. Applications in analytic geometry and integrated algebra problems can be done using spreadsheets or graphics calculators.

Teachers and students will continue to use hands-on materials, paper folding, and compass-straightedge approaches. But they will also have available various interactive technologies to make the learning experiences more exciting and more varied.

HRW Geometry

In the past, many geometry textbooks included little, if any, material appropriate for active learning situations. Thus, teachers who realized a need for this type of approach had to search out or create special materials.

In marked contrast, this geometry program has been structured to include the frequent use of activity-style lessons and activities. In Lessons titled with the word "Exploring," the entire instructional focus is centered around exploration and discovery. Also many of the "non-exploration" lessons contain explorations which involve students in active learning experiences. Most are appropriate for small-group work, many allow for use of alternative technologies, and all give students a chance to discover new ideas on their own.

As students begin a set of exercises at the end of a lesson, they may notice that they do not start with a set of simple-minded practice or drill problems. Instead, each exercise set starts with questions called *Communicate.* Here, students can talk or write about key ideas as a check on what they have learned from the lesson. Students may be asked to compare and contrast two similar concepts, explain how they would solve a particular problem, or discuss why a particular answer is not reasonable for a given problem.

Students for whom active learning is a new approach may at first be confused by the way in which material is presented. They may have come to expect that they are simply to repeat whatever the teacher or the textbook has explained. But in this geometry program, the textbook doesn't "give it all away." Students must explore and think a bit to get at the central concepts. With a few experiences, students will come to enjoy doing geometry in their own way, rather than being tied to someone else's thinking processes.

Real-World Applications

Teachers who implement activity-style learning can expect to see gradual improvement in long-term recall of information, clearer understanding of connections among concepts, more creative approaches to problem solving, and more sophisticated ways of using and communicating geometric ideas.

In addition, teachers will be preparing students for future "real-world" problem-solving situations. Few, if any, of the problems and decisions that face people in their personal and professional lives come in tidy, textbook packaging. "Real" problems are vaguely defined and messy, with missing information or too much information, and difficult compromises to be made. Students who have learned to tackle geometric problems by actively exploring approaches, collaborating with others in groups or team approaches, and comparing results to look for alternative strategies will be equipped to transfer these methodologies to personal and professional challenges later in their lives.

Activity-style instruction that makes frequent use of explorations and investigations gives students more than just a knowledge of geometric concepts and skills. Careful and consistent implementation of these techniques will prepare all students for facing and solving future challenges.

ASSESSMENT

Assessment Goals

An essential aspect of any learning environment, such as a geometry classroom, is the process of assessing or evaluating what students have learned. Informally this has been done using paper-and-pencil tests given by the teacher on a regular basis to measure students' performance against the material being studied. Formal evaluations using standardized tests are generally conducted over a period of years to establish performance records for both individuals and groups of students within a school or school district. Both types of tests are very good at measuring the ability of a student to use a particular mathematical skill or to recall a specific fact. They fall short, however, in evaluating other key goals of learning mathematics, such as being able to solve problems, to reason critically, to understand concepts, and to communicate mathematically, both verbally and in writing. Other techniques, usually referred to as alternative assessment, are needed to evaluate students' performance on these *process* goals of instruction.

The goals of an alternative assessment program are to provide a means of evaluating students' progress in non-skill areas of learning mathematics. Thus, the design and structure of alternative assessment techniques must be quite different from the skill-oriented, paper-and-pencil tests of the past.

Types of Alternative Assessment

In the world outside of school, a person's work is evaluated by what that person can do, that is, by the results the person achieves, and not by taking a test. For example, a musician demonstrates skill by making music, a pilot by flying an airplane, a writer by writing a book, and a surgeon by performing an operation. Students learn to think mathematically and to solve problems on a continual basis over a long period of time as they study mathematics at many grade levels. Students, too, can demonstrate what they have learned and understand by collecting a representative sample of their best work in a **portfolio.** A portfolio should illustrate achievements in problem solving, critical thinking, writing mathematics, mathematical applications or connections, and any other activity that demonstrates an understanding of both concepts and skills.

Specific examples of the kinds of work that students can include in their portfolios are solutions to nonroutine problems, proofs of geometric theorems, graphs, tables or charts, computer printouts, group reports or reports of individual research, simulations, and examples of artwork or models. Each entry should be dated and should be chosen to show the student's growth in mathematical competence and maturity.

A portfolio is just one way for students to demonstrate their performance on a mathematical task. Performance assessment can also be achieved in other ways, such as by asking students questions and evaluating their answers, by observing their work in cooperative-learning groups, by having students give verbal presentations, by having students work on extended projects and investigations, and by having students keep journals.

Peer assessment and self-evaluation are also valuable methods of assessing students' performance. Students should be able to critique their classmates' work and their own work against standards set by the teacher. In order to evaluate their own work, students need to know the teacher's goals of instruction and the criteria that have been established (scoring rubrics) for evaluating performance against the goals. Students can help to design their own self-assessment forms that they then fill out on a regular basis and give to the teacher. They can also help to construct test items that are incorporated into tests given to their classmates. This work is ideally done in small groups of four students. The teacher can then choose items from each group to construct the test for the entire class. Another alternative testing technique is to have students work on *take-home* tests that pose more open-ended and non-routine questions and problems. Students can devote more time to such tests and, in so doing, demonstrate their understanding of concepts and skills and their ability to do mathematics independently.

Scoring

The use of alternative assessment techniques implies the need to have a set of standards against which students' work is judged. Numerical grades are no longer sufficient because growth in understanding and problem solving cannot be measured by a single number or letter grade. Instead, scoring rubrics or criteria can be devised that allow the teacher more flexibility to recognize and comment on all aspects of a student's work, pointing out both strengths and weaknesses that need to be corrected.

A scoring rubric can be created for each major instructional goal, such as being able to solve problems or communicate mathematically. A rubric generally consists of four or five short descriptive paragraphs that can be used to evaluate a piece of work. For example, if a five-point paragraph scale is used, a rating of 5 may denote that the student has completed all aspects of the assignment and has a comprehensive understanding of problems. The content of paragraph 5 specifies the details of what constitutes the rating of highly satisfactory. On the other hand, a rating of 1 designates essentially an unsatisfactory performance, and paragraph 1 would detail what is unsatisfactory. The other three paragraphs provide an opportunity for the teacher to recognize significant accomplishments by the student and also aspects of the work that need improvement. Thus, scoring rubrics are a far more realistic and educationally substantive way to evaluate a student's performance than a single grade, which is usually determined by an answer being right or wrong.

The guide pictured, the Kentucky Mathematics Portfolio, was developed by the Kentucky Department of Education for use by school districts throughout that state. This guide is illustrated to show an example of an excellent and effective holistic scoring guide currently in use by math teachers who are practicing performance assessment in their classrooms. The scorer uses the Workspace/Annotations section of the guide to gather evidence about a student's mathematical ability. The top of the guide is then used to assign a single performance rating, based on an overall view of the full contents of the student's portfolio. The lower right-hand corner of the Holistic Scoring Guide lists the Types and Tools for

Portfolio Holistic Scoring Guide

An individual portfolio is likely to be characterized by some, but not all, of the descriptors for a particular level. Therefore, the overall score should be the level at which the appropriate descriptors for a portfolio are clustered.

		NOVICE	APPRENTICE	PROFICIENT	DISTINGUISHED
PROBLEM SOLVING	Undestanding/ Strategies	• Indicates a basic understanding of problems and uses strategies	• Indicates an understanding of problems and selects appropriate strategies	• Indicates a broad understanding of problems with alternate strategies	• Indicates a comprehensive understanding of problems with efficient, sophisticated strategies
	Execution/ Extensions	• Implements strategies with minor mathematical errors in the solution without observations or extentions	• Accurately implements strategies with solutions, with limited of observations or extentions	• Accurately and efficiently implementes and analyzes strategies with correct solutions with extensions	• Accurately and efficiently implements and evaluates sophisticated strategies with correct solutions and includes analysis, justifications and extensions
REASONING		• Uses mathematical reasoning	• Uses appropriate mathematical reasoning	• Uses perceptive mathematical reasoning	• Uses perceptive, creative, and complex mathematical reasoning
MATHEMATICAL COMMUNICATION	Language	• Uses appropriate mathematical time	• Uses appropriate mathematical reasoning	• Uses percise and appropriate methematical language most of the time	• Uses sophisicated, precise, and appropriated mathematical language
	Representations	• Uses few mathematical representations		• Uses a wide variety of mathematical representations accurately and appropriately; uses multiple representations with some entries	• Uses a wide variety of mathematical representations accurately and appropriately uses multiple representations within entries and states their connections
UNDERSTANDING/ CONNECTING CORE CONCEPTS		• Indicates a basic understanding of core concepts	• Indicates an understanding of core concepts with limited connections	• Indicates a broad understanding of some core concepts with corrections	• Indicates a comprehensive understanding of core concepts with connections throughout
TYPES AND TOOLS		• Indudes few types; uses few tools	• Indudes a variety of types; uses tools appropriately	• Includes a wide variety of types; uses a wide variety of tools appropriately	• Includes all types; uses a wide variety of tools appropriately and insightfully

PERFORMANCE DESCRIPTORS	WORKSPACE/ANNOTATIONS
PROBLEM SOLVING • Understands the features of a problem (understands the question, restates the problem in own words) • Explores (draws a diagram, constructs a model and/orchart, records data, looks for patterns) • Selects an appropriate strategy (guesses and checks, makes an exhaustive list, solves a simpler but simular problem, works backward, estimates a solution) • Solves (implements a strategy with an accurate solution) • Reveiws, revises, and extends (verifies, explores, analyzes, evaluates strategies/ solutions; formulates a rule)	
REASONING • Observes data, records and recognizes pattern, makes mathematical conjectures (indictive reasoning) • Validates mathematical conjectures through logical arguments or counter-examples; constructs valid arguments (deductive reasoning)	
MATHEMATICAL COMMUNICATION • Provides quality explanations and expresses concepts, ideas, and reflections clearly • Uses appropriate mathematical notation and terminology • Provides various mathematical representations (model, graphs, charts, diagrams, words, pictures, numerals, symbols, equations)	

PORTFOLIO CONTENTS
- Table of Contents
- Letter to Reviewer
- 5–7 Best Entries

BREADTH OF ENTRIES

TYPES
0 INVESTIGATIONS/DISCOVERY
0 APPLICATIONS
0 NON-ROUTINE PROBLEMS
0 PROJECTS
0 INTERDISCIPLINARY
0 WRITING

TOOLS
0 CALCULATORS
0 COMPUTER AND OTHER TECHNOLOGY
0 MODELS MANIPULATIVES
0 MEASUREMENT INSTRUMENTS
0 OTHERS
0 GROUP ENTRY

UNDERSTANDING/CONNECTING CORE CONCEPTS
- Demonstrates an understanding of core concepts
- Recognizes, makes, or applies the connections among the mathematical core concepts to other disciplines, and to the real world

Place an X on each continuum to indicate the degree of understanding demonstrated for each core concept.

DEGREE OF UNDERSTANDING OF CORE CONCEPTS

	Basic							
NUMBER								
MATHEMATICAL PROCEDURES								
SPACE & DIMENSIONALITY								
MEASUREMENT								
CHANGE								
MATHEMATICAL STRUCTURE								
DATA: STATISTICS AND PROBABILITY								

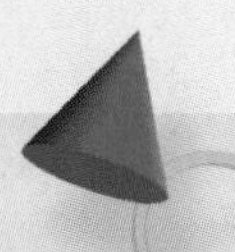

the Breadth of Entries that are appropriate for students to place in their portfolio. Within the interleaf pages that precede each chapter of this Annotated Teachers Edition is a list of seven activity domains that correspond to the Breadth of Entries. Each of these activity domains is correlated to specific examples of activities in the pupil's book that are appropriate for portfolio development.

Assessment and HRW Geometry

HRW Geometry provides many opportunities to employ alternative assessment techniques to evaluate students' performance. These opportunities are an integral part of the textbook itself and can be found in the Explorations, Try This, interactive questions (which are highlighted), Chapter Projects, Critical Thinking questions, and exercise and problem sets.

Throughout the textbook, students are asked to explain their work; to describe what they are doing; to compare and contrast different approaches; to analyze a problem; to make sketches, graphs, tables, and other models; to hypothesize, conjecture, and look for counterexamples; and to make and prove generalizations.

All of these activities, including the more traditional responses to routine problems, provide the teacher with a wealth of assessment opportunities to see how well students are progressing in their understanding and knowledge of geometry. The assessment task can be aligned with the major process goals of instruction and scoring rubrics established for each goal. For example, a teacher may decide to organize his or her assessment tasks in the following general areas of doing mathematics.

- Problem solving
- Reasoning
- Communication
- Connections

Within each of these areas, specific goals can be written and shared with students. In this way, the assessment process becomes an integral part, not only of evaluating students' progress, but also of the instructional process itself. The results of assessment can and should be used to modify the instructional approach to enhance learning for all students.

In addition to the many opportunities for performance assessment found in HRW Geometry, a variety of assessment types are integrated into the chapter-end material. The Chapter Review, the Chapter Assessment, and the Cumulative Assessment include both traditional and alternative assessment. All Chapter Tests include both multiple-choice and open-ended type questions. The Cumulative Reviews are formatted in the style of college preparatory exams. In addition to multiple-choice and free-response questions, each Cumulative Assessment contains quantitative-comparison questions that emphasize concepts of equalities, inequalities, and estimation. Another type of college-entrance-exam question found in the Cumulative Review is student-produced response questions with gridded solutions. These Cumulative Assessments expose students to the new types of assessment that they will encounter when they take the latest form of college entrance examinations.

Eyewitness Math

Special two-page features called Eyewitness Math appear in almost every other chapter. These feature pages provide students with opportunities to read about current developments in mathematics and to solve real-life problems by working together in cooperative groups. Students' performance on Eyewitness Math can be assessed through group reports in writing or orally.

	Column A	Column B	Answers
1.	Graph: $D(-7,7)$, $E(-3,3)$; Slope of $\overline{DE}$	Graph: $A(5,6)$, $B(1,-5)$; Slope of $\overline{AB}$	(A) (B) (C) (D) **[Lesson 3.8]**
2.	$\triangle MNO \cong \triangle PQR$; $\triangle MNO$ with angles 64° at N, 60° at M, 56° at O; x	$\triangle PQR$ with angles $(2x + y) - 1$ at P, $2y + 14$ at Q, $3x + 2$ at R; y	(A) (B) (C) (D) **[Lesson 3.5]**
3.	Triangle XYZ; $\angle XYZ$	Triangle TUV; $\angle TUV$	(A) (B) (C) (D) **[Lesson 4.2, 4.3]**

HANDS-ON to Proof

> *"In geometry, the learner should [first] be instructed in the demonstrations of theorems which are at once startling and easily verifiable by actual drawing, such as those in which it is shown that three or more lines meet in a point. . . ."*

Hands-On in HRW Geometry

These words of Bertrand Russell embody the philosophy of *HRW Geometry.* This book provides opportunities and experiences for students to construct, measure, and explore geometry through visualization, pictorial representations, and hands-on manipulation of geometric shapes. In the first three chapters, students get hands-on experience with geometric relationships as they use paper folding to explore figures, make discoveries, and formulate conjectures. Paper folding makes it possible for students to make mathematically precise constructions without special drawing instruments.

At first students use paper folding to explore perpendicular lines and parallel lines. As they move on to other explorations that involve paper folding, they learn how the discoveries they are making lead to conjectures. For example, during a paper folding activity in which students explore segment and angle bisectors, they are asked to write a conjecture about the points on the perpendicular bisector of a segment. They are then asked to make a conjecture about the distance from a point in the angle bisector of an angle to the sides of the angle.

Students also use paper folding activities to explore transformations. They are first asked to formulate conjectures about a segment and its reflection through a line and are then asked to formulate conjectures about triangles and their reflection through a line. In Chapter 2, paper folding is used to explore reflections over parallel lines.

> *". . . It is desirable also that the figure illustrating a theorem should be drawn in all possible cases and shapes. . ."*

Another tool that lets students construct and explore geometry concepts is geometry graphics software. These programs offer construction opportunities with points, lines, triangles, polygons, circles, and other basic objects. Objects can be translated, dilated, and reflected. Some programs offer easily constructed conics, including ellipses and hyperbolas. Geometry graphics software is interactive and provides many opportunities for geometric discoveries and resultant conjectures. The user can drag points, lines, and circles to manipulate the figures and then observe the changes and constants. Figures can be manipulated to display all possible cases of a theorem. Many of the explorations in *HRW Geometry* are written so that the student can take advantage of this powerful tool. Students who have access to a computer graphics tool begin making discoveries and conjectures about topics that traditionally could not be made until many prerequisite skills and concepts had been covered.

> *". . . the abstract demonstrations should form but a small part of the instruction, and should be given when, by familiarity with concrete illustrations, they have come to be felt as the natural embodiment of visible fact. . . ."*

Proof in HRW Geometry

Reasoning and proof are themes infused throughout the book. Formal proof is not a filter through which students must pass to gain access to the "big ideas" in this textbook. Although the theoretical framework for the textbook is based on postulates, the book does not focus on Euclidean geometry as a complete axiomatic system. Students gradually build competence in proof by using it throughout the various chapters in many different forms and contexts.

In the first three chapters, students begin by exploring figures, creating definitions, and looking for geometric relationships. As students formulate different conjectures, a genuine need for proof arises. In Chapter 2, students work with if-then statements and converses. By building chains of if-then statements, proof naturally evolves. As the text moves into Chapter 4, Congruent Triangles, students will have gained enough of an understanding of geometric relationships and logical reasoning to use a formal two-column proof. From this point on, a balance is provided between paragraph proofs and two-column proofs, the latter being used primarily as a means of organizing and sequencing several geometric relationships.

Tables are often used in *HRW Geometry* to represent various components of a figure. Students are expected to look for patterns and generalize relationships using algebraic expressions and functions. The if-then statements required in a formal proof are often suggested as students move from column to column and discover the underlying relationships among the quantities involved. This unique use of and emphasis on "Table Proofs" is a special feature of *HRW Geometry.*

> *". . . In this way belief is generated; it is seen that reasoning may lead to startling conclusions, which nevertheless the facts will verify; and thus the instinctive distrust of whatever is abstract or rational is gradually overcome."*

CHAPTER 1

Exploring Geometry

Meeting Individual Needs

1.1 Modeling the World with Geometry

Core Resources

Inclusion Strategies, p. 11
Reteaching the Lesson, p. 12
Practice Master 1.1
Enrichment Master 1.1
Technology Master 1.1
Lesson Activity Master 1.1

[2 days]

Core Plus Resources

Practice Master 1.1
Enrichment, p. 11
Technology Master 1.1
Interdisciplinary Connection, p. 10

[1 day]

1.2 Exploring Geometry Using Paper Folding

Core Resources

Inclusion Strategies, p. 18
Reteaching the Lesson, p. 19
Practice Master 1.2
Enrichment Master 1.2
Lesson Activity Master 1.2
Interdisciplinary Connection, p. 17

[2 days]

Core Plus Resources

Practice Master 1.2
Enrichment, p. 18
Technology Master 1.2

[2 days]

1.3 Exploring Geometry with a Computer

Core Resources

Inclusion Strategies, p. 24
Reteaching the Lesson, p. 25
Practice Master 1.3
Enrichment Master 1.3
Lesson Activity Master 1.3

[2 days]

Core Plus Resources

Practice Master 1.3
Enrichment, p. 24
Technology Master 1.3

[2 days]

1.4 Measuring Length

Core Resources

Inclusion Strategies, p. 31
Reteaching the Lesson, p. 32
Practice Master 1.4
Enrichment Master 1.4
Lesson Activity Master 1.4
Mid-Chapter Assessment Master

[2 days]

Core Plus Resources

Practice Master 1.4
Enrichment, p. 24
Technology Master 1.4
Interdisciplinary Connection, p. 30
Mid-Chapter Assessment Master

[1 day]

1.5 Measuring Angles

Core Resources

Inclusion Strategies, p. 39
Reteaching the Lesson, p. 40
Practice Master 1.5
Enrichment Master 1.5
Lesson Activity Master 1.5
Interdisciplinary Connection, p. 38

[2 days]

Core Plus Resources

Practice Master 1.5
Enrichment, p. 39
Technology Master 1.5

[1 day]

1.6 Motions in Geometry

Core Resources

Inclusion Strategies, p. 46
Reteaching the Lesson, p. 47
Practice Master 1.6
Enrichment Master 1.6
Technology Master 1.6
Lesson Activity Master

[2 days]

Core Plus Resources

Practice Master 1.6
Enrichment, p. 46
Technology Master 1.6
Interdisciplinary Connection, p. 45

[2 days]

1.7 Exploring Motion in the Coordinate Plane

Core Resources	Core Plus Resources
Inclusion Strategies, p. 46 Reteaching the Lesson, p. 47 Practice Master 1.7 Enrichment Master 1.7 Lesson Activity Master 1.7 Interdisciplinary Connection, p. 52	Practice Master 1.7 Enrichment, p. 46 Technology Master 1.7
[2 days]	**[1 day]**

Chapter Summary

Core Resources	Core Plus Resources
Chapter 1 Project, pp. 58–59 Lab Activity Long-Term Project Chapter Review, pp. 60–62 Chapter Assessment, p. 63 Chapter Assessment, A/B Alternative Assessment	Chapter 1 Project, pp. 58–59 Lab Activity Long-Term Project Chapter Review, pp. 60–62 Chapter Assessment, p. 63 Chapter Assessment, A/B Alternative Assessment
[3 days]	**[2 days]**

Hands-On Strategies

Lesson 1.2 introduces students to the use of paper folding for making geometric figures. Students learn to make perpendicular lines, parallel lines, angle bisectors, perpendicular bisectors, and other figures. Models made using wax paper are particularly useful to students—the lines are easy to see, and fairly precise measurements can be made for both line segments and angles. With the wax paper, students can check for congruency by laying one sheet of paper on top of another. They can create translations, reflections, and rotations by tracing over figures. Paper folding can be used instead of, or along with, compass and straightedge work.

In Lessons 1.4 and 1.5, students review their ruler and protractor measurement skills. Lesson 1.4 includes the first use of a compass as students use this tool to mark off equal segments on a line segment. Lesson 1.7 involves constructing figures on a coordinate grid system. Students should have ready access to the correct tools for geometry. These include sharp pencils, metric and customary rulers, a compass, and a protractor. Pencils in different colors will also be useful. Help students learn to distinguish between situations in which they need a careful and precise drawing, and those in which a rough sketch is sufficient.

Cooperative Learning

GROUP ACTIVITIES	
Perpendiculars, parallels, bisectors	Lesson 1.2, Explorations 1 and 2
Triangles and circles	Lesson 1.3, Explorations 1 and 2
Center of mass	Lesson 1.3, Look Beyond
Reflections of points, segments, triangles	Lesson 1.6, Explorations 1–2
Translations, reflections, rotations	Lesson 1.7, Explorations 1–3

You may wish to have students work in groups or with partners for some of the above activities. Additional suggestions for cooperative group activities are noted in the teacher's notes in each lesson.

Multicultural

The cultural references in this chapter include references to Asia, the Americas, and Africa.

CULTURAL CONNECTIONS	
Asia: Islamic art	Lesson 1.0
Americas: Aztec calendar	Lesson 1.0
Asia: Origami	Lesson 1.2
Africa: Egyptian royal cubit	Lesson 1.4, Exercise 24
Asia: Babylonian measurement	Lesson 1.5, Exercises 47–48
Africa: Egyptian vault plan	Lesson 1.7, Exercise 47

Portfolio Assessment

Below are portfolio activities for the chapter listed under seven activity domains which are appropriate for portfolio development.

1. **Explorations** Lesson 1.1 (models of points, lines, and planes); Lesson 1.2 (paper folding); Lesson 1.3 (special points in triangles, special circles); Lesson 1.5 (linear pairs); Lesson 1.6 (transformations); Lesson 1.7 (coordinate plane transformations)

2. **Applications** Origami, p. 19; Construction, p. 26; Slide rule, p. 36; Navigation, p. 42; Scuba Diving, p. 42.

3. **Nonroutine Problems** Card exchange p. 15; Construction, p. 27; Unit fractions (Ex. 25), p. 35.

4. **Project** Origami: pp. 58–59. Classic paper crane.

5. **Interdisciplinary Topics** Algebra, pp. 32, 41–42, 50, 56; Architecture, p. 5; Fine Arts, pp. 4, 5, 7, 8, 27 (Ex. 20); Geology, p. 35; Physics, p. 28; Statistics, p. 35

6. **Writing** *Communicate* exercises on the following pages are suggested: p. 20, pp. 25–26, p. 55.

7. **Tools** Geometry software: Lesson 1.3 exploration (special points in triangles, special circles)

Technology

Many of the activities in this book can be significantly enhanced by the use of geometry graphics software, but in every case the technology is optional. If computers are not available, the computer activities can, *and should,* be done by hands-on or "low-tech" methods, as they contain important instructional material. For more information on the use of technology, refer to the HRW Technology Manual for High School Mathematics.

When instructing students in the use of the software, you may find that the best approach is to give students a few well-chosen hints about the different tools and let them discover their use on their own. Today's computer-savvy students can be quite impressive in their ability to discover the uses of computer software! Once the students are familiar with the geometry software, they can design and create their own computer sketches for the explorations.

A second approach is for you, the teacher, to demonstrate the creation of the exploration sketches "from scratch" using a projection device. The students will find this exciting to watch, and they will be eager to learn to use the software on their own.

In the interest of convenience or time-saving, you may want to use the Geometry Investigations Software available from HRW. Each of the files on the disks contains a "sketch," custom-designed for an HRW Geometry Exploration. The sketches are to be used with *Geometer's Sketchpad*™ (Key Curriculum Press) or *Cabri Geometry II*™ (Texas Instruments).

Computer Graphics Software

Lesson 1.3: Explorations 1 & 2 These explorations have students make a discovery about the angle bisectors, perpendicular bisectors, medians, and altitudes of triangles. Students drag vertices of triangles in which each of these sets of lines are drawn, and observe dynamically that each set of three lines intersects in a single point. Then students draw a circle at each special point and make conjectures about inscribed and circumscribed circles.

Lesson 1.6: Exploration 1 In this exploration students explore the reflection of a point through a line. They drag the point to different positions and observe certain angle and segment measures as the point moves. Then they make a conjecture about the properties of reflections and, using this conjecture, write a definition of a reflection.

Lesson 1.6: Exploration 2 Students study the reflections of a segment and a triangle and observe their behavior as the pre-images are dragged into different positions. They then make a conjecture about the properties of these reflections.

Lesson 1.7: Exploration 2 Students observe the reflections of triangles through the x- or y-axis. After dragging the pre-image and observing changes in coordinates, they are asked to state a rule for reflecting figures through one axis in the coordinate plane.

Lesson 1.7: Exploration 3 Students observe a rotation of a figure about the origin in a coordinate plane. After dragging the pre-image and observing changes in coordinates, they are asked to state a rule for rotating a figure 180 degrees about the origin.

Hand-Held Computers

The geometry graphics capabilities of hand-held computers can be used in much the same way as the geometry software described above. Projection devices for classroom demonstration are available.

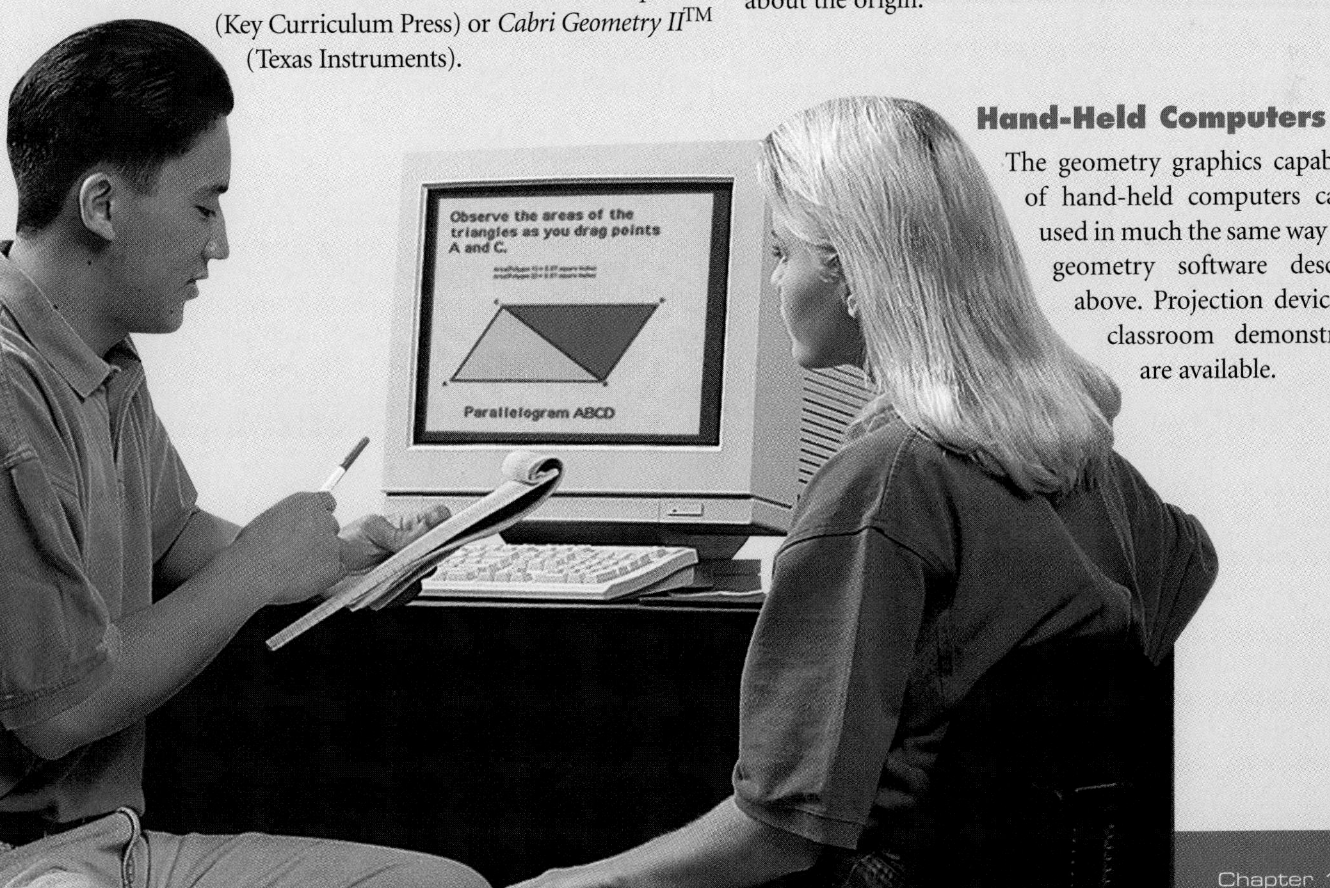

About the Chapter

Background Information

This chapter introduces students to basic notions of geometry and geometric objects. The chapter provides immediate experience with geometry through paper folding, computer graphics, and exploratory lessons.

- Practice Masters
- Enrichment Masters
- Technology Masters
- Lesson Activity Masters
- Lab Activity Masters
- Long-Term Project Masters
- Assessment Masters
  Chapter Assessments, A/B
  Mid-Chapter Assessment
  Alternative Assessments, A/B
- Teaching Transparencies
- Spanish Resources

Chapter Objectives

- Explain how idealized geometric figures differ from real-world objects.
- Define and identify *point, line, segment, plane, collinear, noncollinear, ray,* and *angle.*
- Construct perpendicular lines, parallel lines, segment bisectors, and angle bisectors using folding paper.
- Define and make geometric conjectures.
- Define *altitude, median, angle bisector,* and *perpendicular bisector* of a triangle.
- Construct a triangle altitude, median, angle bisector, and perpendicular bisector.

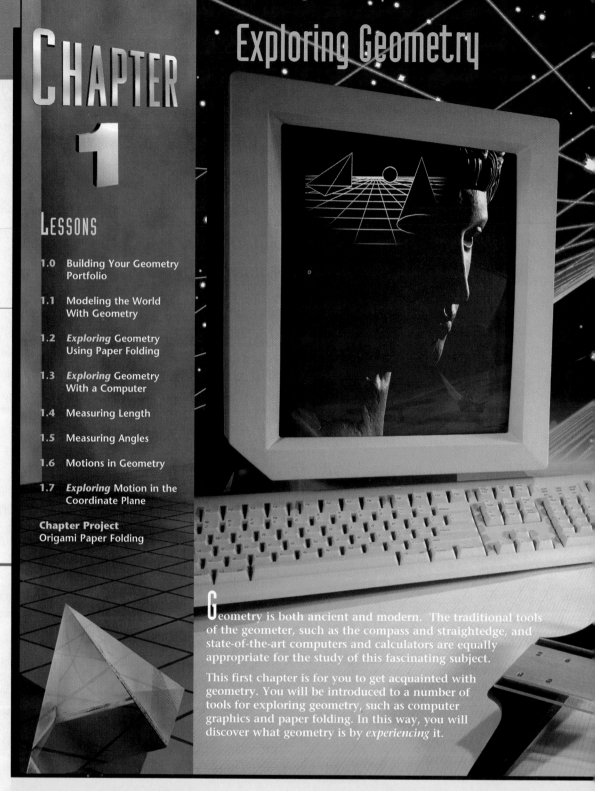

Chapter 1

Exploring Geometry

Lessons

Geometry is both ancient and modern. The traditional tools of the geometer, such as the compass and straightedge, and state-of-the-art computers and calculators are equally appropriate for the study of this fascinating subject.

This first chapter is for you to get acquainted with geometry. You will be introduced to a number of tools for exploring geometry, such as computer graphics and paper folding. In this way, you will discover what geometry is by *experiencing* it.

About the Photo

The photo features tools of geometry. Students in ancient cultures contemplated the elegance of straightedge and compass constructions. Today's geometry student can contemplate geometry using the power of a computer.

PORTFOLIO ACTIVITY
As you begin your study of geometry, set up your own portfolio. The first lesson of this book, Lesson 1.0, tells you how to do this. Throughout this book you will find many suggestions of things to include in your portfolio.
Start your portfolio right away. Look for examples of geometry that can be copied from or cut out of magazines or books. You can also use your own drawings or photographs.

- Construct a geometry ruler.
- Define *congruence.*
- Identify and use the Segment Addition Postulate.
- Measure angles using a protractor.
- Identify and use the Angle Addition Postulate.
- Identify three basic transformations.
- Construct reflections of points, segments, and triangles with a ruler and protractor.
- Define *coordinate plane, origin, x-* and *y-coordinates, ordered pairs, image,* and *preimage.*
- Construct a translation, a reflection about an axis, and a rotation about the origin on the coordinate plane.

PORTFOLIO ACTIVITY

Students will collect pictures of objects that show geometric properties. Lesson 1.0 gives suggestions for setting up and starting the portfolio. After students have collected a set of pictures, they can work in groups to develop descriptions for each picture. Each group can pick five to ten interesting pictures and sketch the underlying geometric objects related to the pictures. Each group can present its pictures to the class. Students should look for patterns and categories of shapes. For example, buildings are usually rectangular solids, and evergreen trees have cone shapes.

Additional Pupil's Edition portfolio activities can be found in the exercises for Lessons 1.3 and 1.6.

ABOUT THE CHAPTER PROJECT

In the Chapter 1 Project, on pages 58–59, students will create an origami crane. Origami is a precise geometric art. Origami figures consist of various geometric shapes created by folding paper according to a set of instructions.

PREPARE

Lesson 1.0 is a unique lesson in that it does not follow the same format as the other lessons in the text. This lesson does not contain instructional examples, nor does the lesson include any exercises to assign. The main objective of the lesson is to introduce teachers and students to the nature and possible contents of a student's portfolio. The portfolio is presented as both a tool for assessment and as a means for students to utilize the geometry of the real world.

Students using this text are not required to keep portfolios. However, there are many reasons why a teacher might use portfolio assessment. The main purpose of this type of assessment is to document learning over a period of time. Other advantages include better communication between student and teacher, positive reinforcement for students, and an increase in student awareness of evaluation standards.

LESSON 1.0 Building Your Geometry Portfolio

Why *Artists and other professionals often keep portfolios of their work. Although you are probably not yet a professional in any area, the work you do in school may help you decide on your future work and career.*

Building a portfolio will help you organize and display your work. Design it to show your work in a way that reflects your interests and your strengths. You should concentrate on the things you enjoy; these will probably be the things you do best. You might want to create geometric constructions at the computer, study the geometry of beehives and spider webs, or explore the geometry found in works of art.

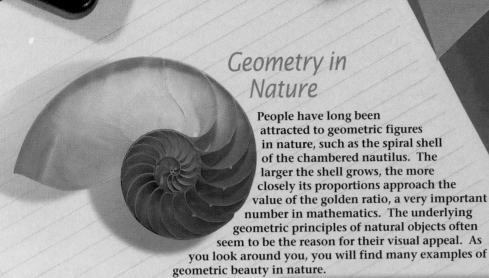

Geometry in Nature

People have long been attracted to geometric figures in nature, such as the spiral shell of the chambered nautilus. The larger the shell grows, the more closely its proportions approach the value of the golden ratio, a very important number in mathematics. The underlying geometric principles of natural objects often seem to be the reason for their visual appeal. As you look around you, you will find many examples of geometric beauty in nature.

Geometry in Art

The artist Piet Mondrian (1872-1944) had a very deep interest in geometry. His desire to break art down to its purest forms led him to create solid shapes based solely on right angles which echo the rectangular shape of the canvas itself.

If you enjoy doing art, you should include works of your own in your portfolio. Even if you think you have little talent or interest in producing works of your own, try your hand at it. You might surprise yourself! The topics presented in this book will suggest different kinds of art with which to experiment.

Piet Mondrian. *Composition with Red, Blue and Yellow.* 1930. Oil on Canvas. 20" x 20".

Geometry in Architecture

The dimensions of the Parthenon reveal the ancient Greek fascination with geometry ideas. The ratio of the height of the original structure to its width is very close to the exact value of the golden ratio. Geometry ideas are still in use in architecture. From the principles of geometry and physics, an architect knows how to design structures that will be both strong and beautiful.

Teach

You might begin a class discussion of the lesson by introducing the portfolio as a tool used by artists, architects, engineers, and other professionals. From this general definition, students should be asked how a portfolio might be used by a geometry student.

As the "Why" suggests, the portfolio should reflect a student's interests and strengths. A portfolio must include what the audience most wants to see. Since a teacher or parent would like to see what a student has accomplished over a period of time, the contents could include endless possibilities.

Introduce students to the geometry of the real world as pictured and discussed on page 5. If you plan to use portfolio assessment in your geometry course, discuss how a portfolio is put together using the activities on pages 7 and 8.

Your Notebook and Journal

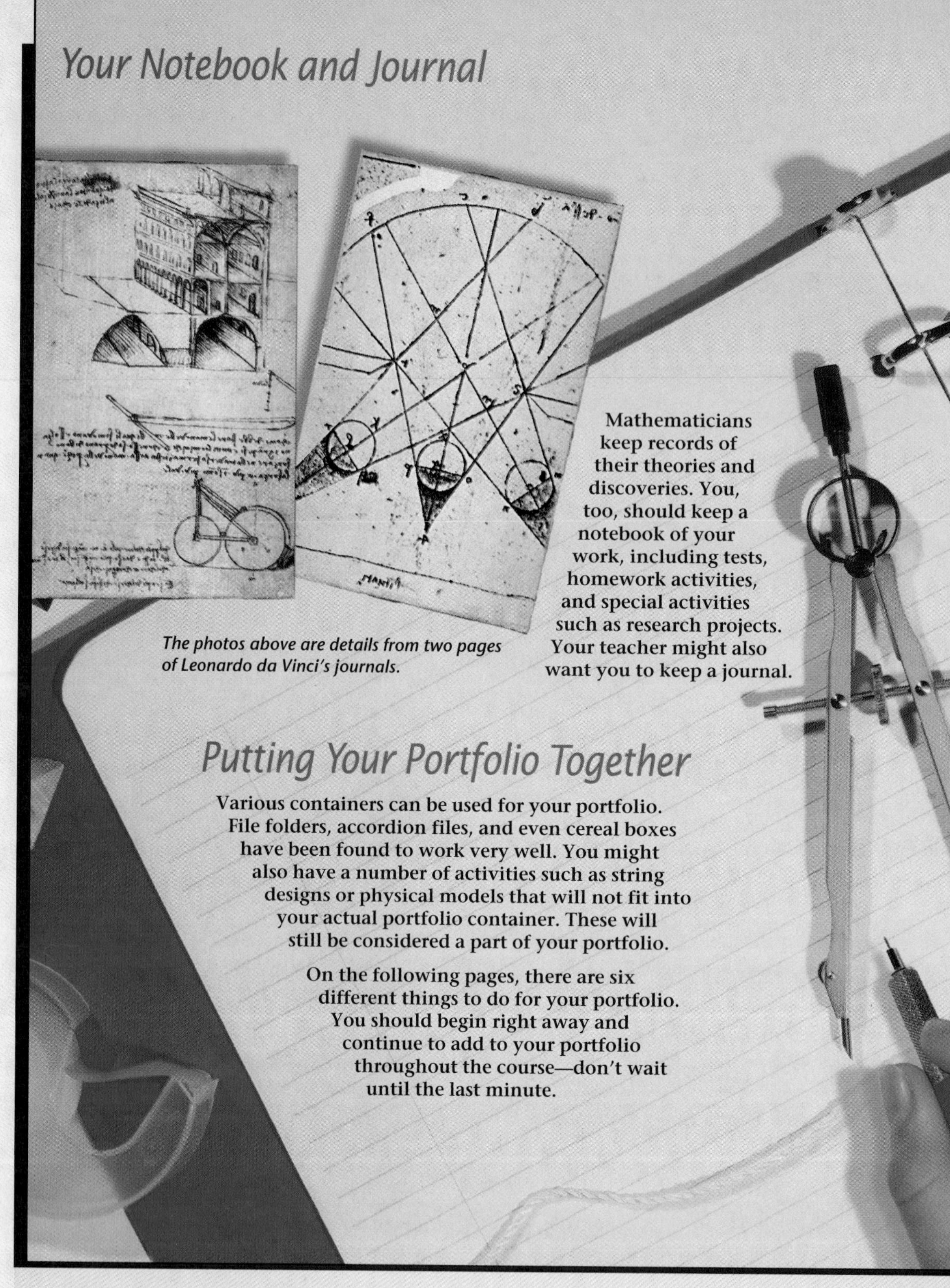

The photos above are details from two pages of Leonardo da Vinci's journals.

Mathematicians keep records of their theories and discoveries. You, too, should keep a notebook of your work, including tests, homework activities, and special activities such as research projects. Your teacher might also want you to keep a journal.

Putting Your Portfolio Together

Various containers can be used for your portfolio. File folders, accordion files, and even cereal boxes have been found to work very well. You might also have a number of activities such as string designs or physical models that will not fit into your actual portfolio container. These will still be considered a part of your portfolio.

On the following pages, there are six different things to do for your portfolio. You should begin right away and continue to add to your portfolio throughout the course—don't wait until the last minute.

You Can Begin Now . . .

1. Collect illustrations of geometry in nature, art, and engineering. Include drawings or photographs of your own if you wish.

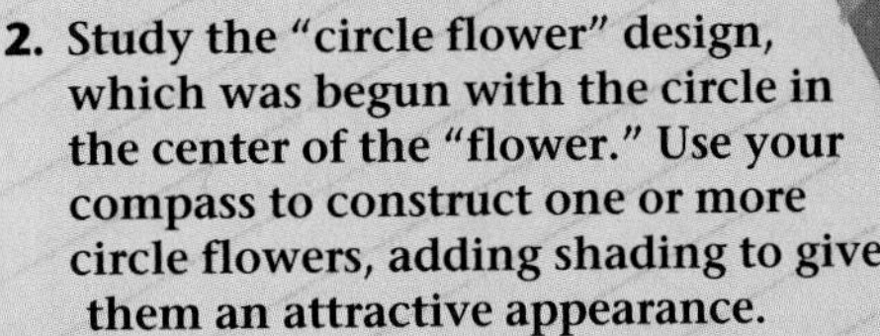

2. Study the "circle flower" design, which was begun with the circle in the center of the "flower." Use your compass to construct one or more circle flowers, adding shading to give them an attractive appearance.

You can continue adding circles to a circle flower to make a more elaborate design. Experiment by using ideas of your own.

3. Interesting designs can be created using only straight lines. One type of line design is made from string and is known as "string art." Make your own design, using either string or pencil, paper, and a straightedge.

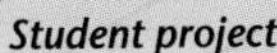

Student project

4. Figures such as the ones shown on the right are known as **mandalas**, from the Sanskrit word for "circle" or "center." Try creating mandalas of your own. Write a report on mandalas and their history, including illustrations.

Aztec "calendar"

5. A special kind of art is the creation of "knot" designs such as those shown here. Experiment with your own designs. A good rule is to first create the overall pattern and then erase lines as necessary for the parts that go behind other parts in the figure.

6. Islamic culture, which at times has forbidden direct representations of real-world objects, became extremely rich in geometric and calligraphic art. Collect examples of Islamic or other geometric art for your portfolio, and perhaps try creating some designs in the style of Islamic art.

LESSON 1.1 Modeling the World With Geometry

why *After years of research, scientists have discovered the most fundamental elements of the physical world. The ideal world of geometry also has its fundamental elements.*

Physicists recognize particles by their electronic signatures. In this computer-generated view, particle tracks emerge from the center of a collision.

Mathematically Perfect Figures

The points, lines, and planes of geometry can be used to create models of things in the physical world. However, geometric figures, unlike atoms and quarks, are not physical. **Lines** and **planes** in geometry have no thickness, and **points** have no size at all. Do you think such things can actually exist?

This book contains many drawings of geometric figures. But remember, the drawings themselves are not the same thing as the geometric figures they represent. Geometric figures exist "only in the mind."

ALTERNATIVE teaching strategy

Technology Have students work in pairs using a computer graphics program to draw a variety of lines, line segments, rays, angles, and planes. Each figure should be labeled with appropriate letters. Students can use words they like when naming the figures.

PREPARE

Objectives

- Explain how idealized geometric figures differ from real-world objects.
- Define and identify *point, line, segment, plane, collinear, noncollinear, ray,* and *angle.*

RESOURCES

- Practice Master 1.1
- Enrichment Master 1.1
- Technology Master 1.1
- Lesson Activity Master 1.1
- Quiz 1.1
- Spanish Resources 1.1

Assessing Prior Knowledge

Find three examples of each of these in the classroom.

1. point
2. line segment
3. angle
4. line
5. plane

[**Answers will vary.**]

TEACH

For many students, the distinction between a real-world object and a geometric figure is unclear. Point out that geometric figures are "perfect." Geometric figures are abstract ideas—they are idealized versions of objects students see every day.

TEACHING tip

As students read the explanations of how to name lines and segments, point out that the two letters can be written in either order. In contrast, the order of the letters for a ray is fixed—the letter for the endpoint must come first.

Alternate Example

Name a line, a line segment, a ray, a plane, and an angle in this figure.

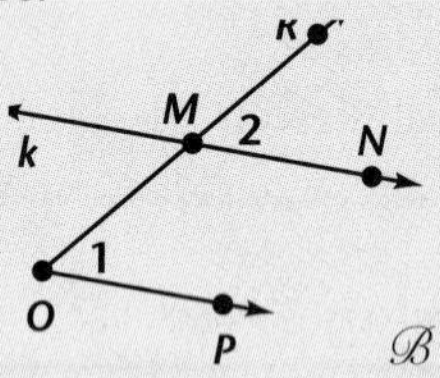

[line: k or $\overleftrightarrow{MN}$; line segment: $\overline{MR}$, $\overline{MN}$, $\overline{MO}$, $\overline{RO}$, or $\overline{OP}$; ray: $\overrightarrow{OR}$, $\overrightarrow{MR}$, $\overrightarrow{MN}$, or $\overrightarrow{OP}$; plane: $\mathscr{B}$; angle: $\angle 1$, $\angle 2$, $\angle RMN$, $\angle NMO$, or $\angle MOP$]

Use Transparency 1

Points Many familiar objects can be used to illustrate geometric points. When you look at the night sky and see the stars, the tiny dots of light seem like points. What are some other examples that could illustrate geometric points?

Points are often shown as dots, but unlike physical dots, geometric points have no size. In geometry, points are named with capital letters such as A or X.

Lines A geometric **line** has no thickness. Unlike real-world "lines," it goes forever. A line is often named using two points on the line, such as line $\overleftrightarrow{AB}$, or just $\overleftrightarrow{AB}$. You can also name a line using a lowercase letter, such as line m.

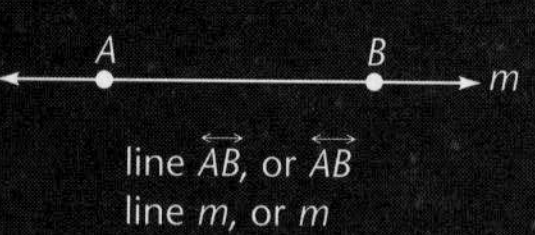

line $\overleftrightarrow{AB}$, or $\overleftrightarrow{AB}$
line m, or m

Segments In drawings, geometric lines are usually represented by physical lines with arrowheads. The arrowheads are used to suggest that lines go on infinitely. **Segments,** on the other hand, have definite beginnings and endings. A segment is a portion of a line from one endpoint to another. It is named using its endpoints.

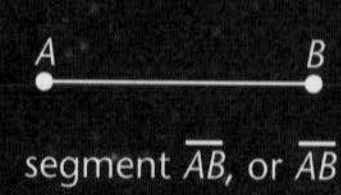

segment $\overline{AB}$, or $\overline{AB}$

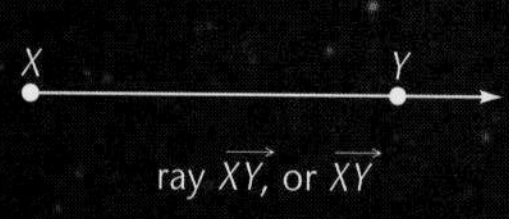

ray $\overrightarrow{XY}$, or $\overrightarrow{XY}$

Rays A **ray** is a "half-line" that starts at a point and goes on forever. How is a geometric ray like a laser beam? A ray is named using its endpoint and one other point that lies on the ray. The endpoint is named first.

EXAMPLE 1

Name each of the figures below. Use the shorter form of the notation in each case.

A 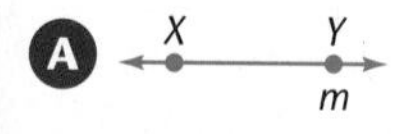

B 

C 

D 

Solution ➤

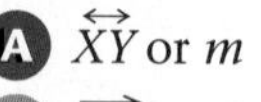

A $\overleftrightarrow{XY}$ or m **B** $\overline{PQ}$
C $\overrightarrow{MN}$ **D** Y ❖

interdisciplinary CONNECTION **Architecture** Have students collect photographs of various bridges. Ask them to consider why triangular shapes are found so often in these structures. A triangle is the only polygon that forms a rigid figure. The triangular sections are used to brace or support the structure.

Angles An **angle** is formed by two rays that have the same endpoint. When there is no danger of confusion, an angle can be named by an angle symbol (∠) and just a single letter or number. Otherwise, an angle is named using three capital letters, with the **vertex** of the angle (the common endpoint of the rays) placed in the middle.

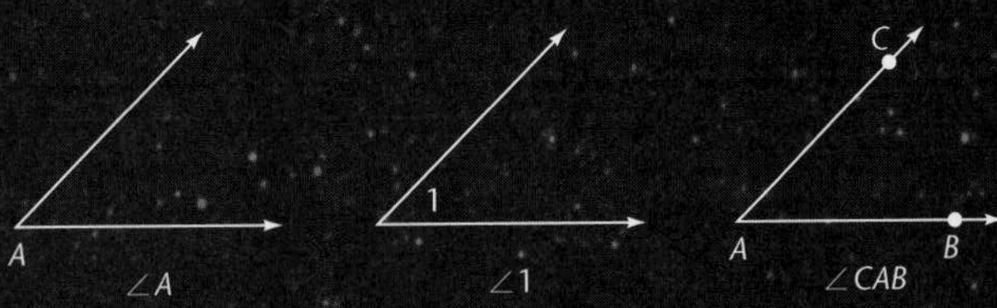

Planes A geometric **plane** extends infinitely in all directions. It is flat and has no thickness. You can think of any flat surface, such as the top of your desk or the front of this book, as representing a portion of a plane.

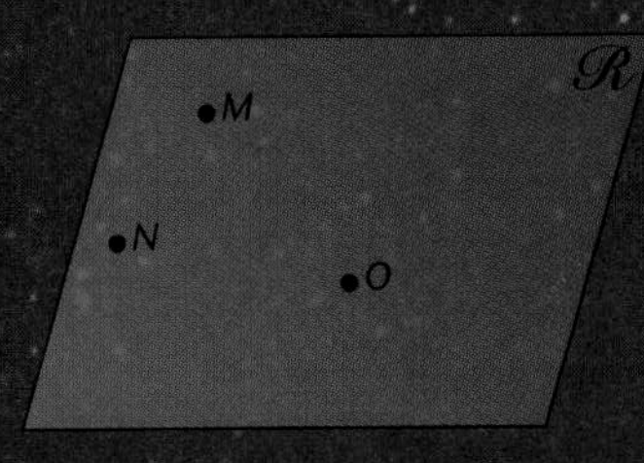

Plane *NMO*, or $\mathscr{R}$

In the figure on the left, points *M*, *N*, and *O* lie on a flat surface. This flat surface represents a plane. A plane can be named by three points that lie on the plane (such as *M*, *N*, and *O*) or by a special script capital letter, such as $\mathscr{R}$.

In naming a plane, the three points you use must be **noncollinear**. That is, they must not lie in a straight line. (**Collinear** points lie on the same line.) If the points M, N, and O were collinear, could there be more than one plane that they would name? Make a sketch to illustrate your answer.

EXAMPLE 2

Name each of the figures below. Use the shorter form of the notation in each case.

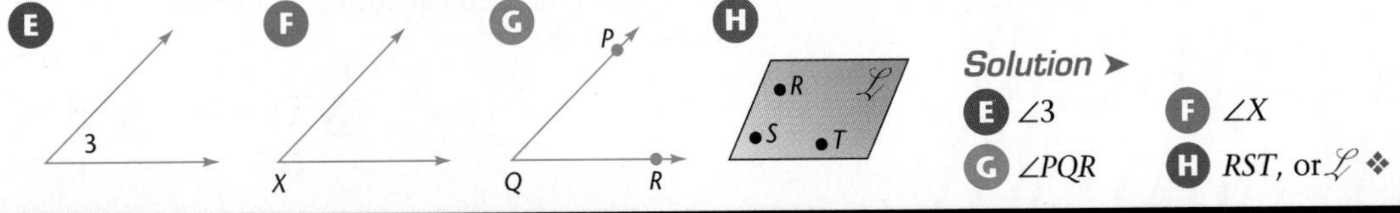

Solution ➤

E ∠3 **F** ∠*X*

G ∠*PQR* **H** *RST*, or $\mathscr{L}$ ❖

TEACHING tip

Emphasize that when naming an angle with three letters, the middle letter must represent the vertex. The other two letters can be in either order. Thus, ∠*ABC* is the same as ∠*CBA*.

Critical Thinking questions provide students an opportunity to analyze, assimilate, and expand their understanding of lesson concepts. The answers are provided in the Teacher's Edition side copy.

CRITICAL Thinking

No. Three collinear points do not determine a plane. Three collinear points can be contained in an infinite number of planes.

ENRICHMENT Explain that there are many kinds of geometries. In the type presented in this book, a plane is a flat surface. Every line is contained in one or more flat planes. However, a geometry can be created on a sphere like the Earth. In this type of geometry, the great circles are the "lines."

INCLUSION strategies **Using Visual Models** Some students may need help interpreting geometric drawings and visualizing geometric ideas. Starting with this lesson, have these students keep a list of term definitions. They should use each term in a sentence and include several sketches to link the verbal concept to the visual images.

Use Transparency 2

Cooperative Learning

In groups, have students build the figure on this page using index cards and tape, and have them use this model to answer the exploration questions. Then have each group label points, lines, planes, etc., within the classroom using construction paper, markers, and tape.

Exploration Notes

Model Steps 3 and 4 using pencils and a piece of cardboard. Place the cardboard on the point of one pencil and move it through various angles to show that an infinite number of planes pass through the point. Repeat with two pencils for two points, three pencils in a line for three collinear points, and three noncollinear points.

Throughout the text, answers to exploration steps can be found in Additional Answers beginning on page 727.

Ongoing Assessment

1. lines, point
2. planes, line
4. . . . one and only one line passes; . . . one and only one plane passes.
5. . . . is in the plane.

Throughout the Pupil's Edition, text highlighted in yellow is designed to be used for ongoing assessment. The answers are provided in the Teacher's Edition side copy.

Exploration Discovering Geometry Ideas in a Model

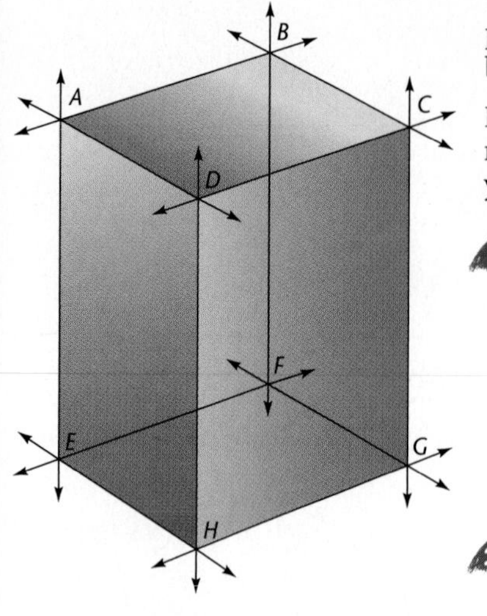

The drawing on the left illustrates how points, lines, and planes can be used to model a real-world object such as a box or perhaps the room you are in as you read this.

Each of the following questions represents an important mathematical idea. Record your answers to the questions in your notebook for future reference.

1. Examine the drawing. Identify the places where the lines pass through, or **intersect**, each other. What kind of geometric figure is suggested where lines intersect?

   Complete the following statement: The intersection of two _?_ is a _?_.

   How many lines intersect at each corner of the drawing? Do you think there is a limit to the number of lines that can intersect at a single point?

2. Identify the places in the drawing where the planes pass through, or intersect, each other. What are the intersections called?

   Complete the following statement: The intersection of two _?_ is a _?_.

3. Look at points A and B in the drawing. How many lines pass through *both* of these points? Do you think it is possible for there to be another line, different from the one shown, that passes through both points A and B?

   Complete the following statement: Through any two points _?_.

4. Look at points A, B, and C in the drawing. How many planes pass through these three noncollinear points? Do you think it is possible for there to be another plane, different from the one shown, that passes through all three points A, B, and C?

   Complete the following statement: Through any three noncollinear points _?_.

5. Pick any plane in the drawing. Then pick two points that are on the plane. Name the line that passes through them. Is the line in the plane that you picked?

   Complete the following statement: If two points are in a plane, then the line containing them _?_. ❖

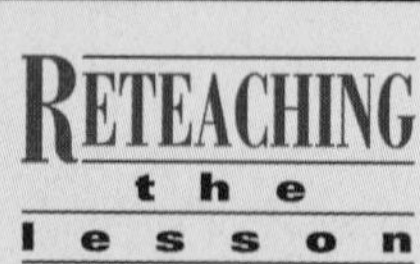

Reteaching the Lesson

Hands-On Strategies Have students make abstract artworks by creating drawings that include three lines, three line segments, three rays, and three angles. Each figure should be labeled with letters. The regions created by the various line segments should be colored in different colors or patterns. Display the art so that students can see how different the results are even though everyone started with the same directions.

EXERCISES & PROBLEMS

Communicate

1. Explain how geometric figures are different from real-world objects.
2. Discuss why it is useful to have more than one way to name a line.
3. Tell how to count the number of segments that can be named in the figure below. Is there more than one way? Explain.

4. Tell how to count the number of angles that can be named in the figure below. Is there more than one way? Explain.

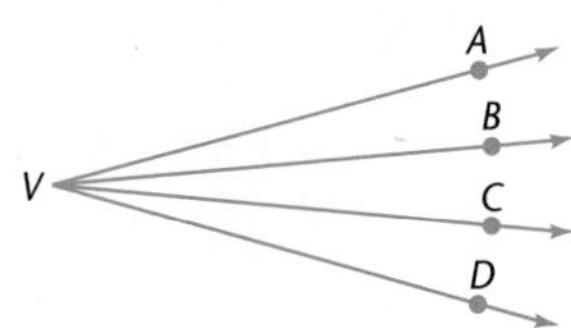

How are geometric figures important in the design of this car?

5. Tell how to count the number of rays that can be named in the figure on the right using just the given information. Why is the order of letters important in naming a ray?
6. Discuss why it is useful to have more than one way to name an angle.

Practice & Apply

In Exercises 7–10, refer to the triangle at right.

7. Name all the segments in the triangle. $\overline{AB}$, $\overline{BC}$, $\overline{AC}$
8. Name each of the angles in the triangle using three different methods. ∠A, ∠1, ∠BAC; ∠B, ∠2, ∠ABC; ∠C, ∠3, ∠ACB
9. Name the rays that form each of the angles of the triangle.
10. Name the plane that contains the triangle. plane *ABC*

∠A: $\overrightarrow{AB}$, $\overrightarrow{AC}$
∠B: $\overrightarrow{BA}$, $\overrightarrow{BC}$
∠C: $\overrightarrow{CA}$, $\overrightarrow{CB}$

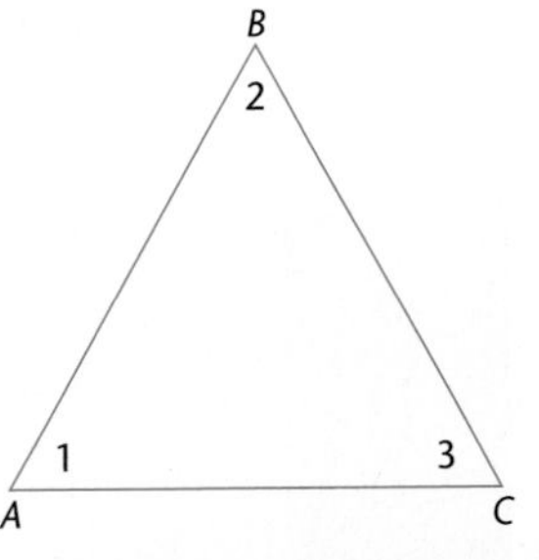

ASSESS

Selected Answers

Odd-numbered Exercises 7–49

"Communicate" Exercises provide an opportunity for students to discuss their discoveries, conjectures, and the lesson concepts. Throughout the text, the answers to all Communicate Exercises can be found in Additonal Answers beginning on page 727.

Assignment Guide

Core 1–20, 24–34, 42–50

Core Plus 1–6, 11–41, 51–52

Technology

Students can practice using geometry graphics software for Exercises 1–10. Students can sketch the various figures on the screen and change the shapes by dragging the points.

Error Analysis

For Exercises 3 and 4, some students will see only three segments or angles at first. If necessary, provide a hint by pointing out $\overline{AC}$ and $\overline{BD}$ in Exercise 3.

alternative ASSESSMENT

Portfolio Assessment

Have students find all the geometric terms used in this lesson and then group the terms into three sets as nouns, verbs, or adjectives. Students should use each term in a sentence and illustrate it with a sketch.

17. False. Lines go on indefinitely in both directions.

18. False. Planes extend indefinitely in all directions and have no thickness.

19. False. Suppose two lines intersect and another line is perpendicular to the lines at the intersection point. The third line does not lie in the same plane as the first two.

Aquarium In Exercises 11–16, identify each as being best modeled by a point, line, or plane.

11. An edge of the aquarium line
12. A grain of sand point
13. A side of the aquarium plane
14. The bottom plane
15. A corner of the aquarium point
16. Floating algae many points

In Exercises 17–23, tell whether each statement is true or false, and explain your reasoning.

17. Lines have endpoints.
18. Planes have edges.
19. Three lines that intersect in the same point must all be in the same plane.
20. Two planes may each intersect a third plane without intersecting each other.
21. Three planes may all intersect each other in exactly one point.
22. Any three points are contained in some plane.
23. Any four points are contained in some plane.

For Exercises 24–33, specify whether each statement is true in the figures that apply.

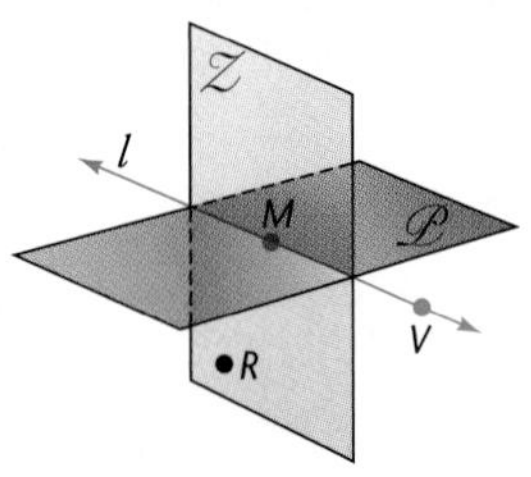

Figure 1

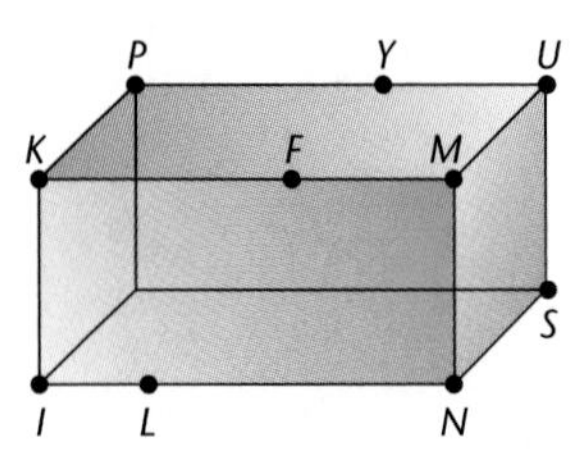

Figure 2

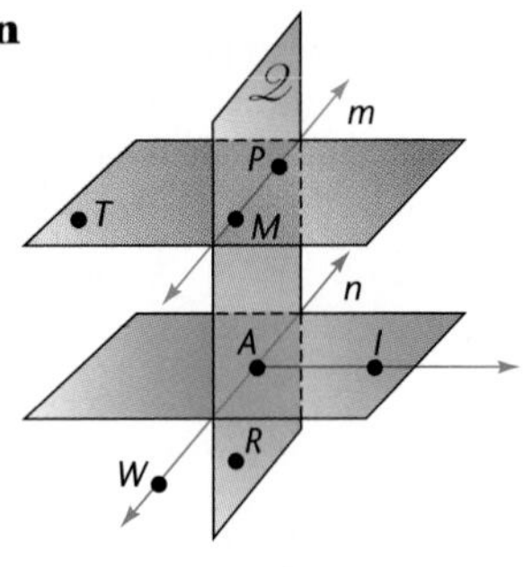

Figure 3

24. $\overline{KM}$ contains F. True
25. Line n is the same as $\overline{WA}$. False
26. $\overline{MU}$ and $\overline{MN}$ intersect at M. True
27. Line l is on plane $\mathscr{P}$. True
28. Line n intersects $\overline{AI}$ at point A. True
29. Points I, N, and M are coplanar. True
30. Points K, F, and M are collinear. True
31. Points F, Y, and N are coplanar. True
32. Plane $\mathscr{P}$ and plane $\mathscr{Z}$ intersect at $\overleftrightarrow{MV}$. True
33. T, M, and P are coplanar. True

Draw each of the figures for Exercises 34–36. Be sure to label each part you draw.

34. Plane $\mathscr{F}$ intersecting plane $\mathscr{L}$ in line p
35. Vertical planes $\mathscr{R}$ and $\mathscr{T}$ intersecting at line m
36. Three planes that never intersect each other

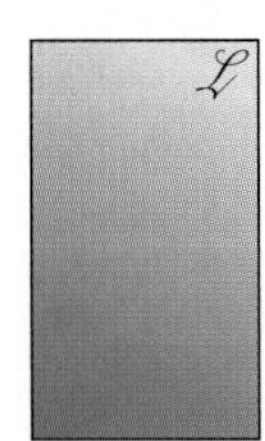

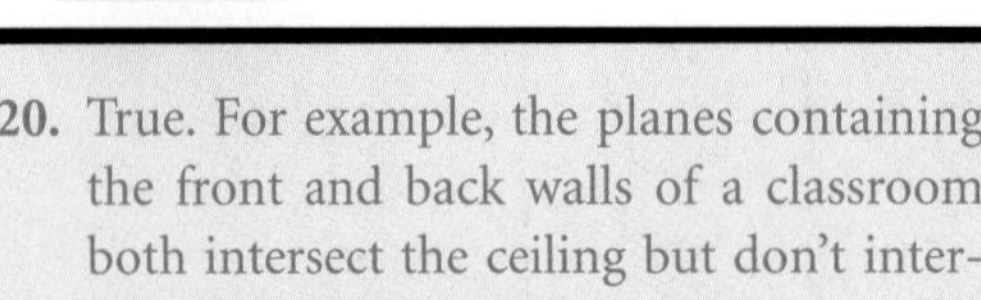

20. True. For example, the planes containing the front and back walls of a classroom both intersect the ceiling but don't intersect each other.

21. True. For example, the planes containing the front wall, a side wall, and the floor of a classroom intersect at a point, the corner of the room.

22. True. Through any three noncollinear points, there is exactly one plane. If the points are collinear, an infinite number of planes contain them.

23. False. Any three points are contained in a unique plane, so a point not on the plane cannot be coplanar with them.

34.

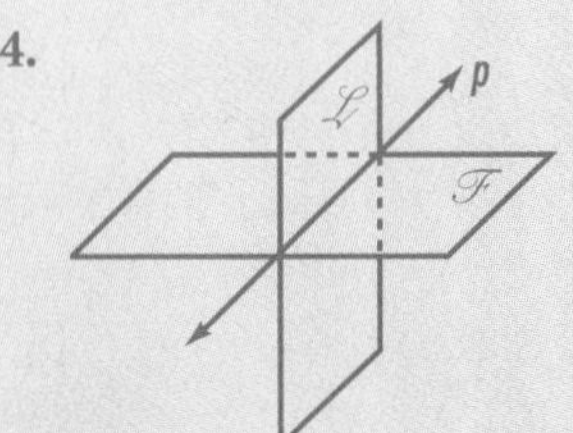

Practice Master

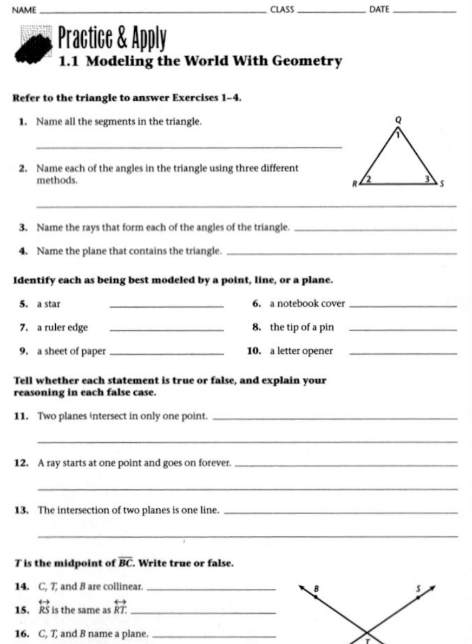

NAME ______ CLASS ______ DATE ______

Practice & Apply
1.1 Modeling the World With Geometry

Refer to the triangle to answer Exercises 1–4.

1. Name all the segments in the triangle.
2. Name each of the angles in the triangle using three different methods.
3. Name the rays that form each of the angles of the triangle.
4. Name the plane that contains the triangle.

Identify each as being best modeled by a point, line, or a plane.

5. a star
6. a notebook cover
7. a ruler edge
8. the tip of a pin
9. a sheet of paper
10. a letter opener

Tell whether each statement is true or false, and explain your reasoning in each false case.

11. Two planes intersect in only one point.
12. A ray starts at one point and goes on forever.
13. The intersection of two planes is one line.

***T* is the midpoint of $\overline{BC}$. Write true or false.**

14. C, T, and B are collinear.
15. $\overleftrightarrow{RS}$ is the same as $\overleftrightarrow{RT}$.
16. C, T, and B name a plane.
17. R, T, and C are collinear.
18. Four rays start at T.

HRW Geometry Practice & Apply 1

How many segments can be named in the figures below?

37.  A B — 1

38. 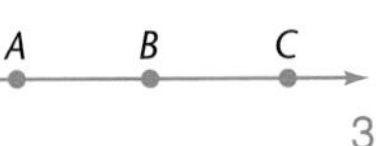

3

39. 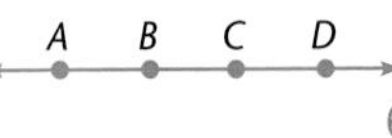

6

40. **Algebra** Write a general rule or a formula for finding the number of segments that are determined by a given number of points on a line. Can you explain why the rule works? Does your explanation prove that the rule will always work?

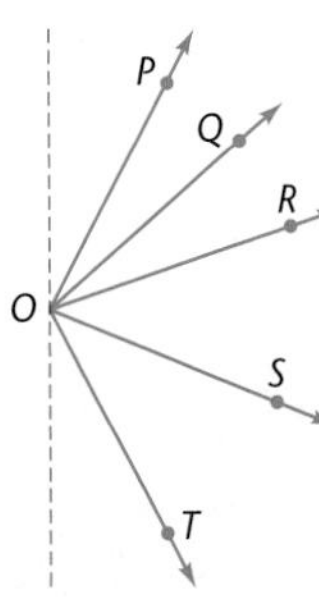

41. How many angles can be named in the figure. Write a general rule or a formula for finding the number of angles formed by a given number of rays radiating from a given point. (Assume all the rays lie to one side of a dotted line, as shown.)

10; $\frac{N(N-1)}{2}$ angles for N points.

Look Back

Complete each of the following. Draw a number line if you find it helpful.

42. $22 + (-6) = 16$

43. $7 + 15 = 22$

44. $11 - (-4) = 11 + 4 = 15$

45. $-81 - (-30) = -81 + 30 = -51$

46. $|-14 + (-35)| = |-49| = 49$

47. $|13 - 10| = |3| = 3$

48. $-123 - 41 = -164$

49. $|21 + (-35)| = |-14| = 14$

50. $|-54 + (-20)| = |-74| = 74$

Look Beyond

51. **Algebra** You can use geometry to help you answer the following question. Suppose that 4 people are exchanging cards. Each person gives 1 card to each of the other 3 people. How many exchanges are made? (**Hint:** Start by drawing 4 noncollinear points. These points represent the people. Now draw lines between the points to represent exchanges of cards. What is the total number of lines that you can draw?) 6 exchanges

52. What if 5 people exchanged cards? How many exchanges would there be? Explain how to determine the number of exchanges when n people exchange cards. 10 exchanges

40. If there are n points, there are $\frac{n(n-1)}{2}$ segments. From each of the n points, $n-1$ segments may be drawn, giving $n(n-1)$ segments. But each segment is then counted twice, so divide by 2 to get the formula.

"Look Back" exercises regularly review important concepts and skills. It is important that they be assigned to all students.

"Look Beyond" questions are of three types. They foreshadow upcoming concepts, extend concepts introduced in the lesson, or involve nonroutine problem solving.

Look Beyond

Nonroutine problems. Although these exercises can be solved using logical thinking without the help of a diagram, part of the goal is to find a visual representation of the problem.

Technology Master

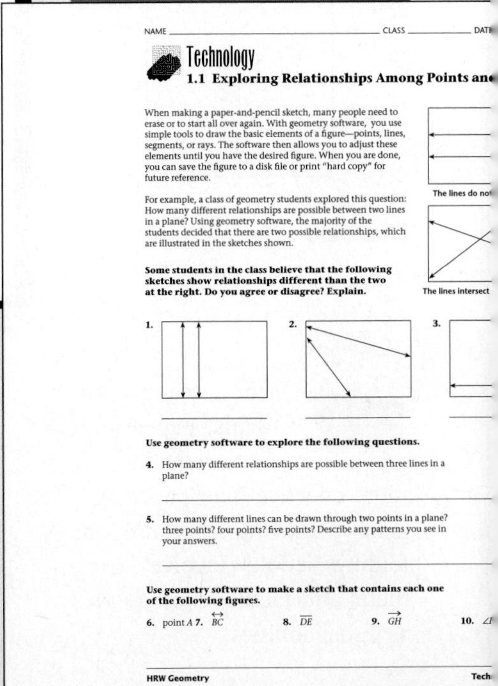

NAME ______ CLASS ______ DAT

Technology
1.1 Exploring Relationships Among Points an

When making a paper-and-pencil sketch, many people need to erase or to start all over again. With geometry software, you use simple tools to draw the basic elements of a figure—points, lines, segments, or rays. The software then allows you to adjust these elements until you have the desired figure. When you are done, you can save the figure to a disk file or print "hard copy" for future reference.

For example, a class of geometry students explored this question: How many different relationships are possible between two lines in a plane? Using geometry software, the majority of the students decided that there are two possible relationships, which are illustrated in the sketches shown.

The lines do no

The lines intersect

Some students in the class believe that the following sketches show relationships different than the two at the right. Do you agree or disagree? Explain.

1. 2. 3.

Use geometry software to explore the following questions.

4. How many different relationships are possible between three lines in a plane?

5. How many different lines can be drawn through two points in a plane? three points? four points? five points? Describe any patterns you see in your answers.

Use geometry software to make a sketch that contains each one of the following figures.

6. point A 7. $\overleftrightarrow{BC}$ 8. $\overline{DE}$ 9. $\overrightarrow{GH}$ 10. ∠

HRW Geometry Tech

35.

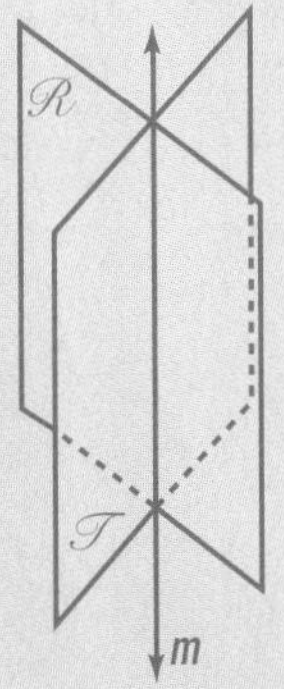

36.

R S T

PREPARE

Objectives

- Use paper folding to construct perpendicular lines, parallel lines, segment bisectors, and angle bisectors.
- Define and make geometry conjectures.

- Practice Master 1.2
- Enrichment Master 1.2
- Technology Master 1.2
- Lesson Activity Master 1.2
- Quiz 1.2
- Spanish Resources 1.2

Assessing Prior Knowledge

Identify the following in the figure below:

1. perpendicular line segments
2. a right angle
3. a 45° angle
4. a line that divides a right angle in half

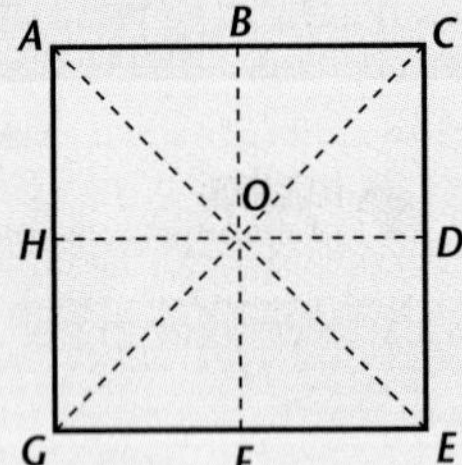

[**Answers will vary.**]

TEACH

For many students, paper folding activities are more effective than paper-and-pencil drawings because the folding actions involve kinesthetic and tactile senses as well as visual abilities.

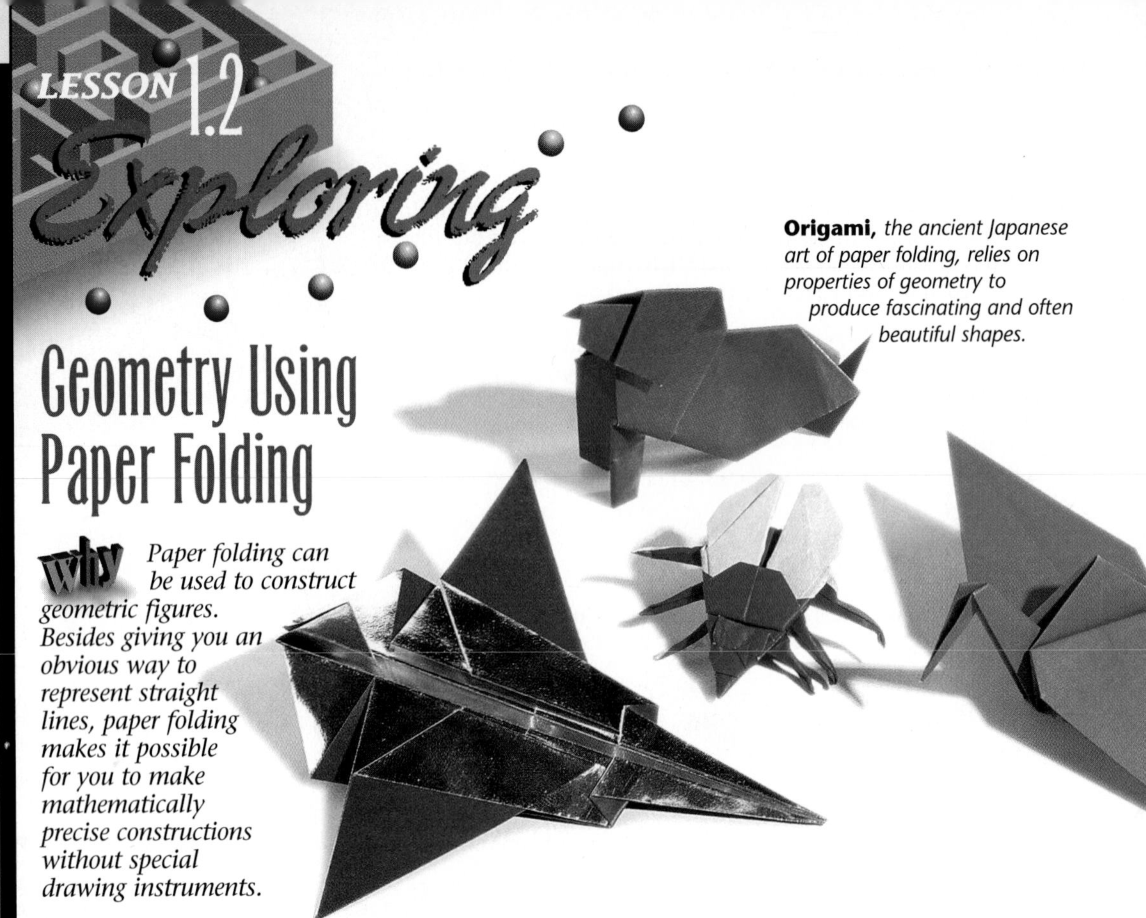

Geometry Using Paper Folding

why *Paper folding can be used to construct geometric figures. Besides giving you an obvious way to represent straight lines, paper folding makes it possible for you to make mathematically precise constructions without special drawing instruments.*

***Origami,** the ancient Japanese art of paper folding, relies on properties of geometry to produce fascinating and often beautiful shapes.*

Paper Folding: The Basics

Wax paper is especially good for folding geometric figures. For one thing, it is transparent, which allows you to match figures precisely when you fold. Another advantage of wax paper is that creases made in it are easy to see, especially against a dark background.

If you use wax paper, experiment with different tools for marking on paper. Also consider using "patty paper," which has the advantage of being easy to write on.

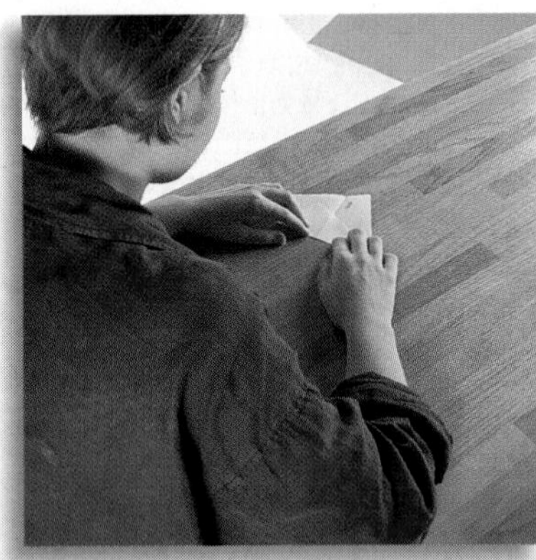

In the following explorations, work with a partner. One person can read the instructions while the other person does the paper folding. Discuss with your partner *why* you think the constructions work. The idea of knowing why things work is one of the most important things you should get from this course.

Technology After the activities and exercises in this lesson have been completed using paper-folding techniques, students can use geometry graphics software to explore the same topics.

All of the terms in the explorations are probably already familiar to you. But to refresh your memory, consider the following three important geometry terms.

A **right angle** is a "square" angle, like the corner of a piece of paper. **Perpendicular lines** are lines that form right angles where they meet. **Parallel lines** are lines that are in the same plane but never meet, no matter how far they are extended.

Exploration 1 *Perpendicular Lines and Parallel Lines*

You will need
Folding paper
Marker that will write on folding paper

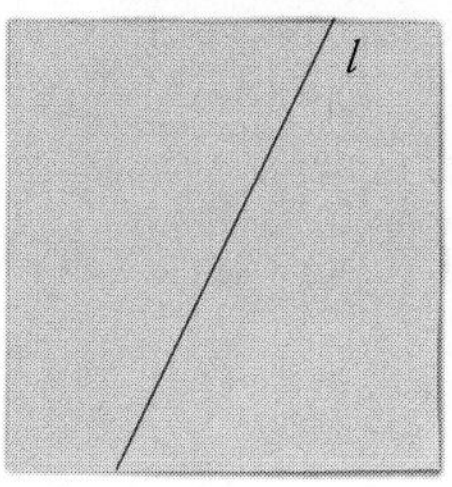

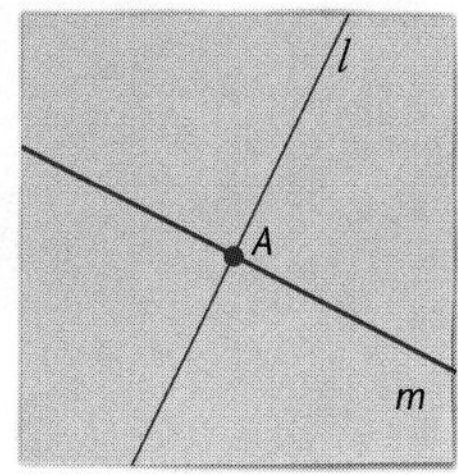

1. Fold the paper once to make a line. Label the line l.

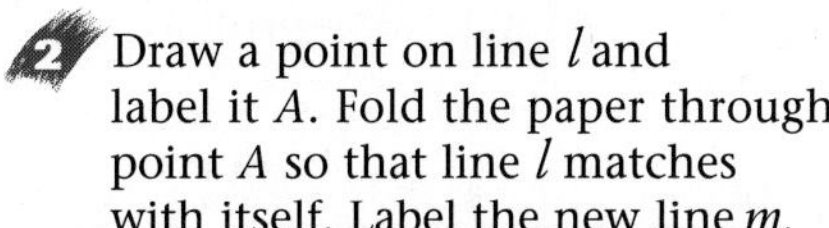

2. Draw a point on line l and label it A. Fold the paper through point A so that line l matches with itself. Label the new line m.

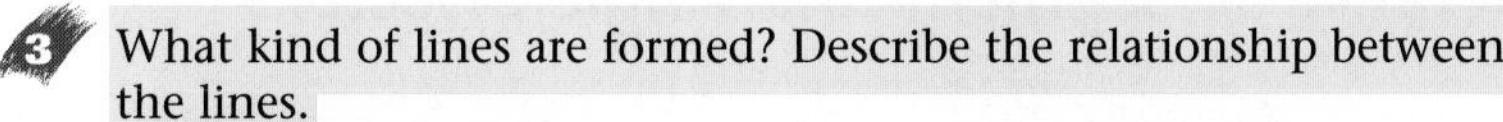

3. What kind of lines are formed? Describe the relationship between the lines.

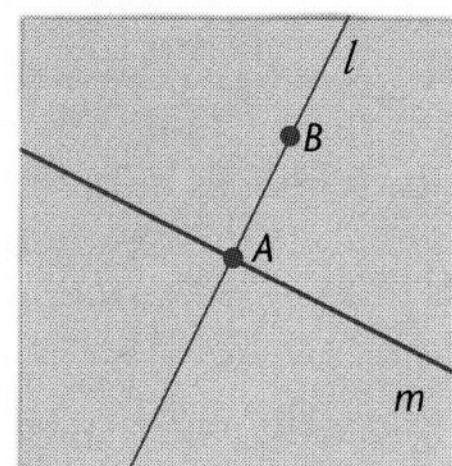

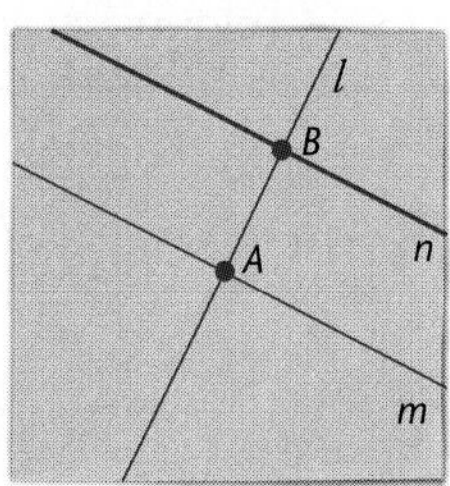

4. Start with perpendicular lines l and m. Mark a new point on line l, and label it as B.

5. Fold the paper through point B so that line l matches with itself as before. Label the new crease as line n.

6. What is the relationship between lines n and m? ❖

interdisciplinary CONNECTION **Physics** Airplanes made by paper folding are sometimes entered in contests. Have students use library resources to investigate what types of rules are used for the contests and what factors affect the speeds and distances flown.

Cooperative Learning

Students working in small groups can generate examples for each exploration. Each group should make several different examples of each figure to provide a basis for conjectures and conclusions.

TEACHING tip

Students can use the corner of an index card, a student ID card, or any square sheet of paper as a right angle "tester."

Exploration 1 Notes

Students investigate a special case of a more general theorem: If two lines are cut by a transversal in such a way that corresponding angles are congruent, then the two lines are parallel.

ongoing ASSESSMENT

3. **Lines l and m are perpendicular lines.**
6. **Lines m and n are parallel lines.**

CRITICAL *Thinking*

The perpendicular segment is the shortest distance between a point and a line. Using this definition, any two students will find the same distance between a line and a point.

Exploration 2 Notes

When constructing segment and angle bisectors with paper, students should fold the paper carefully. In Step 1, encourage students to line up points *A* and *B* precisely.

TEACHING *tip*

For Step 2 in the exploration, have students connect the various points on line *m* to the points *A* and *B* creating a set of "nested" isosceles triangles. This may help students come up with an appropriate conjecture.

ongoing ASSESSMENT

2. **The points on the perpendicular bisector of a segment are equidistant from the endpoints of the segment.**

Making Conjectures in Geometry

A conjecture is a statement we think is true. It is an "educated guess" based on observations.

Mathematical discoveries often start out as conjectures. In the explorations that follow, you will make conjectures about lines that divide angles and segments into congruent parts.

How would you measure the distance from the tree to the fence? Would you choose *a*, *b*, or *c*?

In geometry, the distance from a point to a line is the length of the perpendicular segment from the point to the line. Thus, the distance from the tree (point *X*) to the fence (line $\overleftrightarrow{AC}$) is the length of segment $\overline{XB}$, ($m\overline{XB}$, or *XB*).

CRITICAL *Thinking* 

Why do you think that the distance from a point to a line is defined along a perpendicular segment? What are the advantages of this definition?

Exploration 2 Segment and Angle Bisectors

You will need
Folding paper and marker
Ruler (for measuring only)

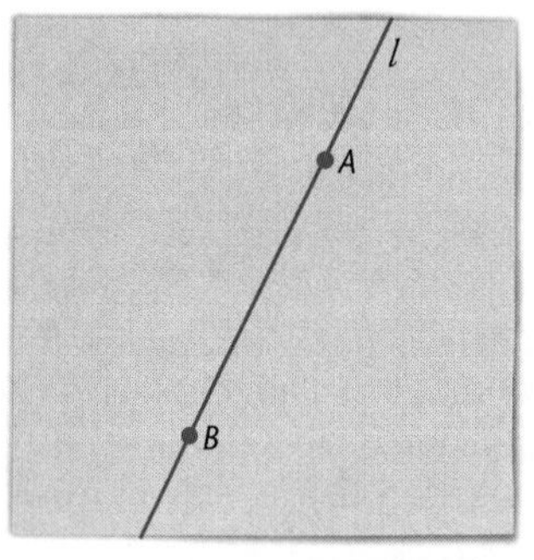

A **segment bisector** is a line that divides a segment into two equal parts. If that segment is perpendicular to the line, it is the **perpendicular bisector**. An **angle bisector** is a line that divides the angle into two equal angles.

1. Start with line *l* containing points *A* and *B*. Fold the line so *B* falls on *A*. Label the crease line *m*. Describe line *m* in relation to segment $\overline{AB}$.

2. Choose several points on line *m*. Measure the distance from each point to *A* and to *B*. Write a conjecture about the points on the perpendicular bisector of a segment.

ENRICHMENT Constructing the perpendicular bisector of a segment automatically gives the *midpoint*. Have students use paper folding to justify *this* statement: If point *M* is the midpoint of $\overline{AB}$, then $AM = 1/2\ AB$ and $MB = 1/2\ AB$.

INCLUSION strategies  **English Language Development** Although the paper-folding activities in this lesson are quite simple, students who have difficulty reading will need to work with a partner or a group as they follow the instructions.

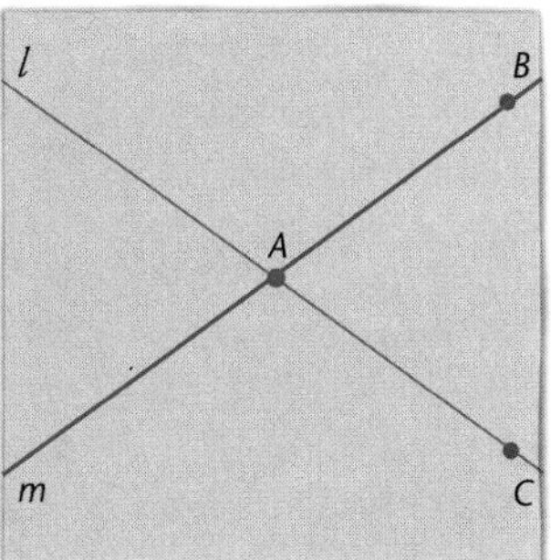

3. Fold two non-perpendicular lines l and m. Label their intersection point A. Label a point B on line m and a point C on line l. Fold the paper through A so that segment $\overline{AB}$ falls on segment $\overline{AC}$. Label the crease as line n. Describe line n in relation to $\angle BAC$.

4. Label point X on the part of line n that is in the interior of $\angle BAC$.

5. Fold the paper to make a line through X perpendicular to line l. Where the perpendicular intersects l, label the point S. Measure the segment $\overline{XS}$.

6. Fold the paper to make a line through X perpendicular to line m. Where the perpendicular intersects m, label the point T. Measure segment $\overline{XT}$.

7. Make a conjecture about the distance from a point in the angle bisector of an angle to the sides of the angle.

8. Test your conjecture with a point in each of the bisectors of the other angles formed by l and m. ❖

APPLICATION

Cultural Connection: Asia The *crane base* is the starting point for many of the traditional origami figures, including the popular paper crane. The folds below are one way of beginning a crane base.

1. Fold a square piece of paper into quarters. Then crease it along the diagonals as shown.

2. Fold the outside edges to align with the diagonal creases.

Now unfold the paper. If you have made your diagonal creases correctly, you should see a four-pointed star in the folds of your paper. How many perpendicular bisectors can you find? How many angle bisectors? ❖

TEACHING tip

Before students try to make conjectures for Step 7, remind them that "distance" means the distance along a perpendicular segment. Thus, the segments that are measured should be perpendicular to the sides of the angle.

ongoing ASSESSMENT

7. The points on the angle bisector are equidistant from the sides of the angle.

TEACHING tip

Books for making origami figures make a distinction between *mountain folds* and *valley folds.* A mountain fold is a concave crease in which the paper is folded up. A valley fold is a convex crease in which the paper is folded down. The two types of folds are represented by different types of line segments, for example, a series of dashes for valley folds and dots and dashes for mountain folds.

RETEACHING the lesson

Hands-On Strategies On a piece of paper, have each student create a four-inch square design for a quilt. Each square must include one or more examples of perpendicular lines, parallel lines, segment bisectors, and angle bisectors. Display the class "quilt" on a bulletin board for discussion and comparison.

Assess

Selected Answers

Odd-numbered Exercises 5–29

The answers to Communicate Exercises can be found in Additional Answers beginning on page 727.

Assignment Guide

Core 1–3, 5–10, 14–30

Core Plus 1–4, 7–23, and 30

Technology

Exercises 5–13 can be done with geometry graphics software. You may wish to have students do the exercises twice, once by folding paper and once with the computer.

Error Analysis

For Exercises 5 and 6, make sure students do not create three lines intersecting in a single point. Explain that the three lines should create a triangle.

Practice Master

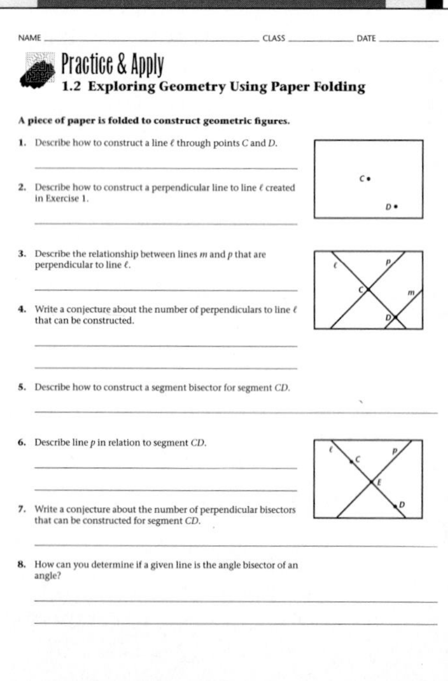

NAME ______ CLASS ______ DATE ______

Practice & Apply
1.2 Exploring Geometry Using Paper Folding

A piece of paper is folded to construct geometric figures.

1. Describe how to construct a line ℓ through points *C* and *D*.
2. Describe how to construct a perpendicular line to line ℓ created in Exercise 1.
3. Describe the relationship between lines *m* and *p* that are perpendicular to line ℓ.
4. Write a conjecture about the number of perpendiculars to line ℓ that can be constructed.
5. Describe how to construct a segment bisector for segment *CD*.
6. Describe line *p* in relation to segment *CD*.
7. Write a conjecture about the number of perpendicular bisectors that can be constructed for segment *CD*.
8. How can you determine if a given line is the angle bisector of an angle?

2 Practice & Apply — HRW Geometry

Exercises & Problems

Communicate

1. When you fold line *l* onto itself in Exploration 1, which pairs of angles match up? What conclusion can you draw about the angles from the fact that they match up? In view of this, how can you define perpendicular lines?
2. When you construct parallel lines *m* and *n* in Exploration 1, how many right angles are formed? Make a conjecture about how you can determine if two lines are parallel.
3. When you fold *A* onto *B* in Exploration 2, how do you know (without measuring) that the new line divides $\overline{AB}$ into equal parts?
4. How can you use the results of Exploration 2 to determine if a given line is a perpendicular bisector of a segment? an angle bisector? How many measurements would you need to make? What would they be? Discuss.

Practice & Apply

Use pieces of unlined paper to do each of the following Exercises. Tracing paper or other paper that you can see through works best. You will also need a ruler.

5. Fold a piece of paper to make three lines, so that each line intersects the other two at different points. Label the lines *l*, *m*, and *n*. Use a marker or pencil to emphasize the three segments connecting the intersection points of the lines. What kind of geometric shape have you drawn? A triangle
6. Label the intersection of lines *l* and *m* as point *A*, of lines *l* and *n* as point *B*, and of lines *m* and *n* as point *C*. Then label the segment connecting *A* and *B* as *c*, the one connecting *A* and *C* as *b*, and the one connecting *B* and *C* as *a*.

   Describe the relationship of the names of the points to the names of the sides. How do you think this arrangement of names might be useful?

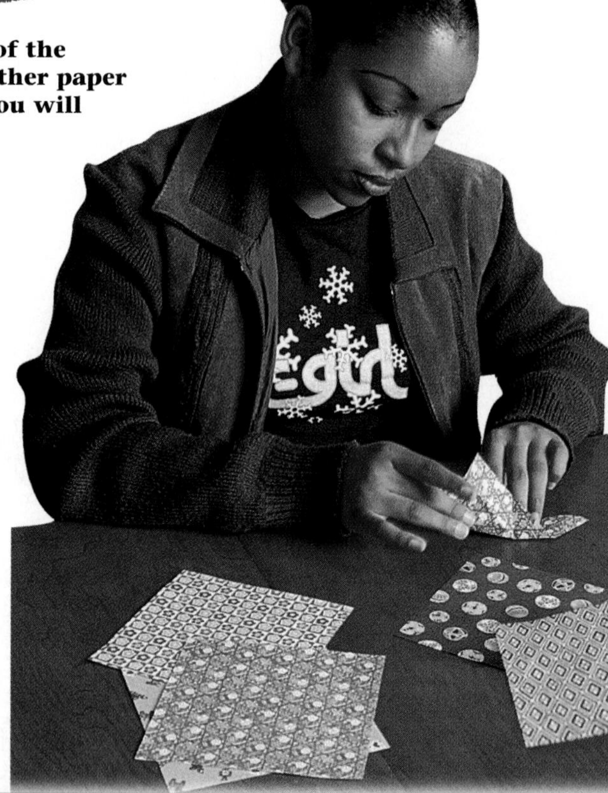

6. The corner points are labeled with capital letters, while the segments opposite them are labeled with the same letter set in lowercase. This provides a standard way of labeling a triangle and makes it easy to refer to.

7. Recall how you constructed perpendicular lines in Exploration 1. Using a new piece of paper, construct a triangle with two sides perpendicular to each other. What kind of triangle have you formed? right triangle

8. Fold a new piece of paper and label two points on the fold line as A and B. Use these points to construct the perpendicular bisector of $\overline{AB}$. Select a point C on the perpendicular bisector. Draw $\overline{CA}$ and $\overline{CB}$. Using a conjecture you made in Exploration 2, what can you conclude about the lengths of $\overline{CA}$ and $\overline{CB}$? They are equal

9. Write a conjecture about triangles that have a vertex located on the perpendicular bisector of one of the sides.

10. Construct a rectangle using the techniques for folding perpendicular lines. (Recall that all four angles of a rectangle are right angles.) Do any of the sides of your rectangle seem to be equal? parallel? Write your own conjectures about the sides of a rectangle.

11. Fold a new piece of paper twice to make two intersecting lines. Construct the angle bisector of the smaller angle between the two lines. Label the angle bisector l. Construct the angle bisector of the larger angle between your original two lines. Label this angle bisector m. What do you observe about the lines l and m? Write a conjecture about the relationship between the bisectors of angles formed by intersecting lines.

12. Draw segment $\overline{AB}$ on a piece of waxed paper; then construct its perpendicular bisector. Trace over the crease. Next, fold the paper once through the intersection of the two lines so that A and B each align with the perpendicular bisector. Poke holes through the paper to mark the point where A lines up on the perpendicular as A' and the point where B lines up on the perpendicular as B'. Connect A, A', B, and B'. What kind of shape does this appear to be? Name the segments that form the *diagonals* of the shape.

13. Write your own conjectures about the diagonals of the geometric figure you constructed in Exercise 12.

9. Such a triangle would have two equal sides.

10. Opposite sides of a rectangle are both parallel and equal in length.

11. Lines l and m are perpendicular. The bisectors of angles formed by intersecting lines are perpendicular.

12. A square. The diagonals are $\overline{AB}$ and $\overline{A'B'}$.

13. The square diagonals are perpendicular bisectors of each other and are equal in length.

Alternative Assessment

Performance Assessment

Have students work in pairs. Each student should write a set of directions for folding paper to make a triangle that has one right angle and two equal sides. Students should exchange and carry out their partner's instructions.

Technology Master

NAME ______ CLASS ______ DATE ______

Technology
1.2 "Virtual Folding"

When you draw a geometric figure on a computer screen, it is impossible to actually fold the figure. However, most geometry software has features that allow you to fold a figure *virtually*.

"Fold" a line segment along its perpendicular bisector by using geometry software to complete Exercises 1–3.

1. Using the *segment* tool of the software, draw a line segment. Label its endpoints A and B.
2. Using the *midpoint* feature, locate the midpoint of $\overline{AB}$. Label this point C.
3. Using the *perpendicular line* feature, draw a line through point C perpendicular to $\overline{AB}$. The sketch should look like the one shown.

A C B

4. Devise a method for adjusting your sketch to "fold" $\overline{AB}$ into four equal parts. Write a brief description of your method.

______

"Fold" an angle along its bisector by using geometry software to complete Exercises 5–6.

5. Using the *line* tool of the software, draw two intersecting lines, $\overleftrightarrow{FG}$ and $\overleftrightarrow{HJ}$. Place a point where they intersect and label it K, as shown in the diagram.
6. Using the *angle bisector* feature, draw the bisectors of $\angle GKJ$ $\angle JKF$, $\angle FKH$, and $\angle HKG$.

H G K F J

7. Use your sketch from Exercise 6. Using the *select* tool, grab point G and move it to several different positions. Use your observations to make a conjecture about the angle bisectors.

______

8. Make two sketches of the figure shown. Draw the first sketch using just the *line* tool, placing the lines and points into the correct positions visually. Draw a second sketch using the *parallel line* and *perpendicular line* features of the software.
9. Use your sketches from Exercise 8. Use the *select* tool to grab point M and move it to several different positions. Then move point O. Describe the difference between the two sketches.

L P M N Q O

______

16 **Technology** **HRW Geometry**

Look Back

Name the geometric figure that each item suggests. [Lesson 1.1]

14. The edge of a table line or segment

15. The wall of your classroom plane

16. The place where two walls meet line or segment

17. The place where two walls meet the ceiling point

Name each figure, using more than one method when possible. [Lesson 1.1]

18. 

line $\overleftrightarrow{XY}$, $\overleftrightarrow{YX}$, l

19. 

Segment $\overline{MN}$, $\overline{NM}$

20. 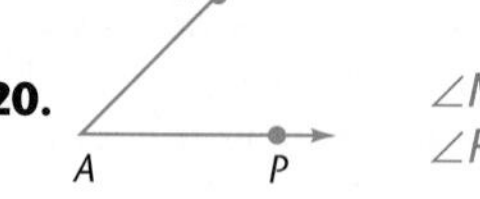

$\angle MAP$, $\angle A$, $\angle PAM$,

21. 

$\overrightarrow{PS}$

22. 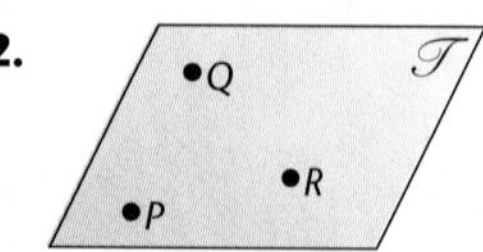

23. A (point) A

Evaluate each of the following expressions.

24. $13 - (-13)$ $13 + 13 = 26$

25. $-13 + (-13)$ -26

26. $-13 - 13$ -26

27. $|19 + 27|$ 46

28. $|19 - 27|$ 8

29. $|-19 - 27|$ 46

Look Beyond

Extension. The activity in Exercise 30 is just one example of how geometric ideas can be used to make art objects. Other examples of art applications will occur throughout the text.

Look Beyond

30. Cultural Connection: Asia Using a few simple folds, origami artists are able to suggest the shapes of real-world objects. The results, which range from humorous to elegant, are often quite striking.

There seems to be no limit to the powers of imagination of origami artists through the years. A recent creation is this sleek rendition of a fighter plane.

Experiment with creations of your own. You can find origami instruction books and special paper in hobby stores.

22. Plane PQR, PQR, or $\mathcal{T}$.

30. Check student work.

LESSON 1.3

Exploring Geometry With a Computer

Why *Powerful computer software has been created to solve problems in design. There is also computer software available to explore geometry ideas.*

If you do not have access to a computer graphics tool, you should complete the activities in this lesson using paper folding or pencil and paper. The ideas presented here will be used in later lessons.

Special Parts of Triangles

Before you begin the explorations, you will need to know some special parts of triangles.

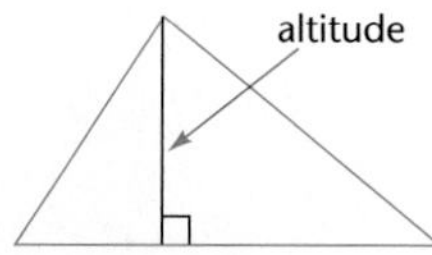

An ***altitude*** *is a line segment from a vertex drawn perpendicular to the line containing the opposite side.*

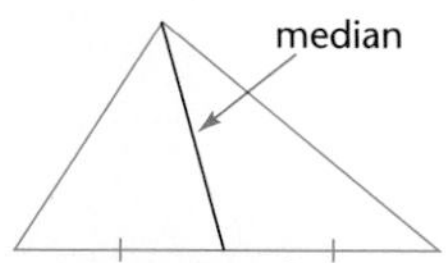

A ***median*** *is a line segment from a vertex to the midpoint of the opposite side.*

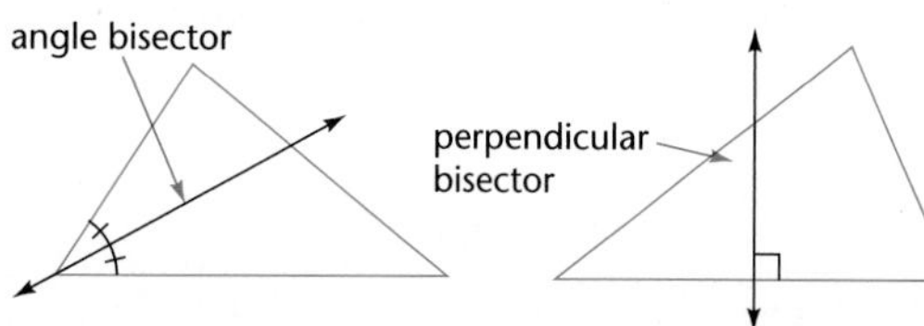

An ***angle bisector*** *is a line, a segment, or a ray that bisects an angle of the triangle.*

A ***perpendicular bisector*** *is a line or a segment that bisects and is perpendicular to a side of the triangle.*

For any given triangle, how many different altitudes can you draw? How many medians? angle bisectors? perpendicular bisectors?

Hands-On Strategies The instructions in this lesson are written so that paper-folding or paper-and-pencil methods can be used with all activities. Some students can use computer techniques while other students use folding or paper and pencil. Students can then compare their procedures and results.

PREPARE

Objectives

- Define *altitude, median, angle bisector,* and *perpendicular bisector* of a triangle.
- Construct a triangle altitude, median, angle bisector, and perpendicular bisector.

RESOURCES

- Practice Master 1.3
- Enrichment Master 1.3
- Technology Master 1.3
- Lesson Activity Master 1.3
- Quiz 1.3
- Spanish Resources 1.3

Assessing Prior Knowledge

Use paper and pencil to draw an example of each kind of triangle—right triangle, equilateral triangle, isosceles triangle, acute triangle, obtuse triangle, and scalene triangle.

[Answers will vary.]

Use Transparency 3

TEACH

Using geometry graphics software is helpful to students because they can make accurate drawings more quickly than with paper-and-pencil methods.

ongoing ASSESSMENT

You can draw three different altitudes, medians, and angle bisectors—one each from each vertex; three different perpendicular bisectors, one through each side.

Exploration 1 Some Special Points in Triangles

You will need
Geometry technology or
Folding paper and a compass

In each of the following activities, you should discover a special point related to the triangles you create. Save your figures for the second exploration. If you are using computer software, you may be able to "drag" each of the vertices of the triangles to explore different triangle shapes.

1. Draw or fold a triangle. Then construct the angle bisector of each of the angles.
2. Draw or fold a triangle. Then construct the perpendicular bisector of each of the sides.
3. Draw or fold a triangle. Then construct the three medians of a triangle.
4. Draw or fold a triangle. Then construct the three altitudes of the triangle.
5. Share your results with other members of your class. What do you notice? Write down your observations in the form of four separate conjectures. ❖

Special Circles Related to Triangles

For any triangle, you can draw an **inscribed circle** and a **circumscribed circle**. An inscribed circle, as the name suggests, is *inside* the triangle and touches its three sides. A circumscribed circle is *outside* a triangle (*circum* means "around") and contains each of its three vertices.

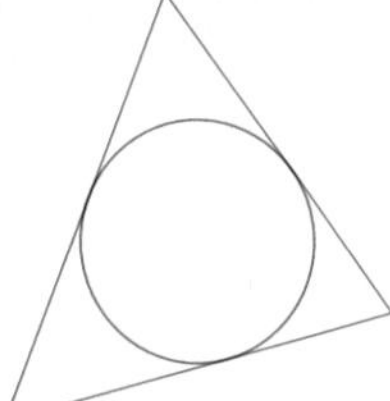
inscribed circle

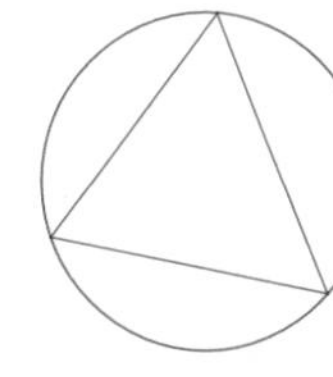
circumscribed circle

Before you do the following exploration, think: How could you find the center points for drawing the inscribed and circumscribed circles of a given triangle?

THE METROPOLITAN MUSEUM OF ART, ROGERS

Archaeology *An archaeologist wants to find the original diameter of a broken plate. She makes an outline of it, then draws a triangle with all three of its vertices on the circumference. The outer edge of the plate is thus part of a circle that circumscribes the triangle. If she can find the center of the circle she can determine the plate's original diameter.*

TEACHING tip

Although this lesson emphasizes the use of geometry graphics technology, it is not technology dependent. You may want to use paper folding or compass-and-straightedge constructions to teach the lesson.

Exploration 1 Notes

For Exploration 1, make sure students do not use equilateral or isosceles triangles as their only cases. Emphasize that their conjectures should apply to triangles of *any* shape.

Use Transparency 4

ongoing ASSESSMENT

In each of the four exercises, the lines containing the three segments meet in one point. Four conjectures: The angle bisectors of a triangle all meet at one point. The perpendicular bisectors of the sides of a triangle all meet at one point. The medians of a triangle all meet at one point. The altitudes of a triangle all meet at one point.

ENRICHMENT

A quadrilateral is *cyclic* if its four vertices lie on a circle. Challenge students to discover a property of the perpendicular bisectors of the four sides of a cyclic quadrilateral. (The bisectors are *concurrent;* that is, they intersect at a single point.)

INCLUSION strategies

Using Visual Models Students may gain more benefit from using geometry graphics or drawing software by making the drawings larger or bolder than usual.

Exploration 2 Constructing Special Circles

You will need
Geometry technology or
Folding paper and a compass

1. Use the triangles you created in Exploration 1. Draw circles with their centers at the special points you discovered in each activity. If you are using paper and pencil, vary the size of the circles by changing your compass settings. If you are using a computer, vary the size of the circles by using the pointer tool.

2. What do you discover about each special point? Write down your discoveries as conjectures about special points connected with triangles. ❖

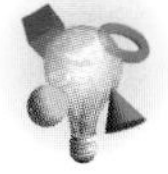

Why do you think that points at the intersection of certain lines (or segments or rays) work as centers for circumscribed or inscribed circles? (**Hint:** Recall the conjecture you made about segment and angle bisectors in Lesson 1.2.)

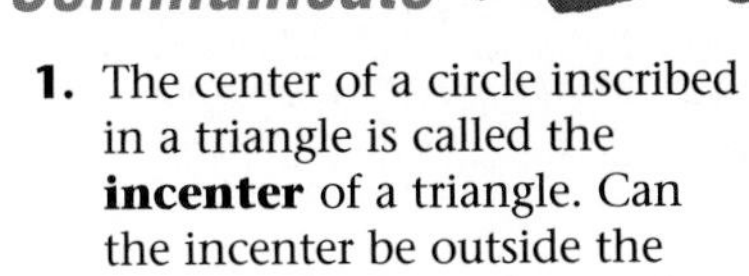

Exercises & Problems

Communicate

1. The center of a circle inscribed in a triangle is called the **incenter** of a triangle. Can the incenter be outside the triangle? Explain why or why not.
2. The center of a circle circumscribed around a triangle is called the **circumcenter** of the triangle. Can the circumcenter be outside the triangle? Explain why or why not.
3. The intersection of the medians of a triangle is called the **centroid** or **center of mass** of the triangle. Can the centroid be outside the triangle? Explain why or why not.
4. The intersection of the altitudes of a triangle is called the **orthocenter** of the triangle. Can the orthocenter be outside the triangle? Explain why or why not.

Reteaching the lesson

Cooperative Learning
Have students work in pairs. Each student should create one drawing of a triangle showing the altitudes, the medians, the angle bisectors, and the perpendicular bisectors of the sides. Segments should be labeled with letters. Students should then exchange drawings and identify the four different types of segments.

Exploration 2 Notes

In the previous exploration, students found that all four types of lines are concurrent. Here, they test whether each meeting point is the center of an inscribed or circumscribed circle.

Ongoing Assessment

2. **A circle that circumscribes the triangle can be drawn using the point where the perpendicular bisectors of the sides of the triangle meet. A circle that is inscribed in a triangle can be drawn using the point where the angle bisectors of the triangle meet.**

Critical Thinking

The point where the three perpendicular bisectors intersect will be the same distance from the three vertices of the triangle, and the point where the three angle bisectors intersect will be the same distance from the three sides of the triangle. Therefore, these points of intersection become the centers of circles.

Cooperative Learning

Have students use geometry graphics software in groups to explore the relationship between the center of a circle and a triangle inscribed in the circle. To do this, students start with any circle. They mark three points on the circumference and connect them to form an inscribed triangle. Now students can look for special segments in the triangle—angle bisectors, medians, etc.—to see if they can find segments that intersect at the center of the circle.

ASSESS

Selected Answers

Odd-numbered Problems 7–27

The answers to Communicate Exercises can be found in Additional Answers beginning on page 727.

Assignment Guide

Core 1–4, 6–14, 20–31

Core Plus 1–5, 10–20, 29–31

Error Analysis

Students may confuse the terms *incenter, circumcenter, centroid,* and *orthocenter.*

Technology

Geometry graphics software can be used for Exercises 6–19.

Practice Master

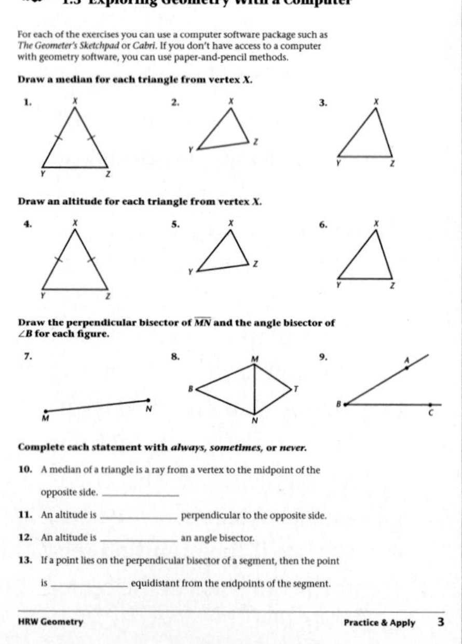

NAME ______________ CLASS ________ DATE ________

Practice & Apply
1.3 Exploring Geometry With a Computer

For each of the exercises you can use a computer software package such as *The Geometer's Sketchpad* or *Cabri.* If you don't have access to a computer with geometry software, you can use paper-and-pencil methods.

Draw a median for each triangle from vertex *X*.

1. 2. 3.

Draw an altitude for each triangle from vertex *X*.

4. 5. 6.

Draw the perpendicular bisector of $\overline{MN}$ and the angle bisector of $\angle B$ for each figure.

7. 8. 9.

Complete each statement with *always, sometimes,* or *never.*

10. A median of a triangle is a ray from a vertex to the midpoint of the opposite side. ________
11. An altitude is ________ perpendicular to the opposite side.
12. An altitude is ________ an angle bisector.
13. If a point lies on the perpendicular bisector of a segment, then the point is ________ equidistant from the endpoints of the segment.

HRW Geometry — Practice & Apply 3

5. Which of the special circles can you draw using the intersections of the perpendicular bisectors of a triangle? Explain why you can do this. To help you form an explanation, discuss the following questions.

Every point on the perpendicular bisector of $\overline{AB}$ is the same distance from what two points on the triangle? (Recall your conjectures from Lesson 1.2.) What can you say about the perpendicular bisectors of the other two sides?

If a point is on all three perpendicular bisectors of a triangle, what can you conclude about its distance from the vertices of the triangle? Explain why.

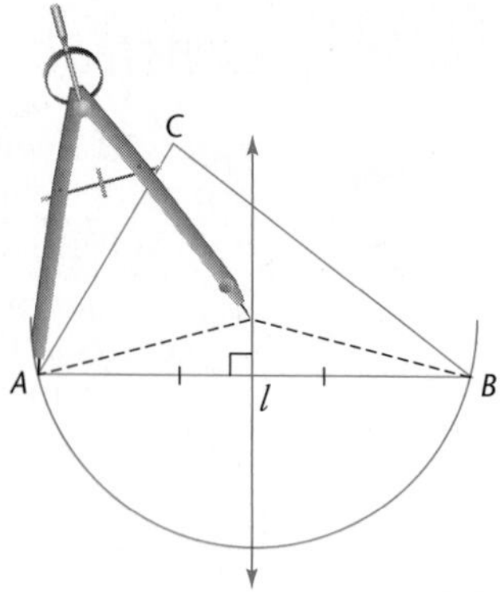

Try this with more than one point on *l*.

Practice & Apply

Technology For each of the Exercises below, you can use a computer software package such as *The Geometer's Sketchpad* ™ or *Cabri* ™. If you don't have access to geometry technology, you can use paper-and-pencil methods.

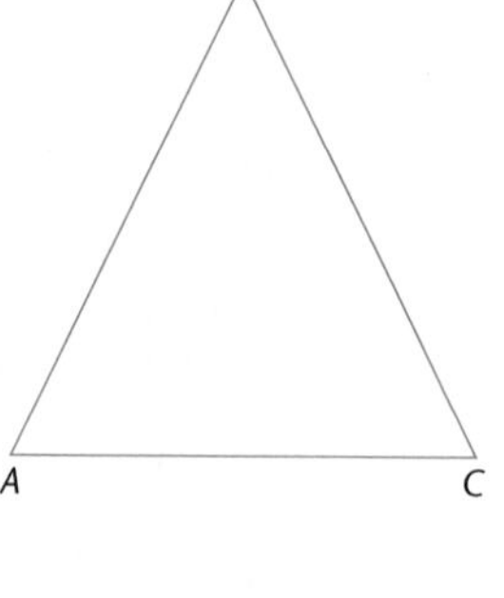

Draw a triangle like the one on the right. Draw altitude $\overline{AD}$, angle bisector $\overline{AE}$, and median $\overline{AF}$. Then draw a perpendicular bisector *m* of side $\overline{BC}$.

6. What is the order of points *D*, *E*, and *F*?

7. Experiment with different shapes of triangles. What do you observe about the order of points *D*, *F*, and *E*?

8. See if you can find a triangle in which $\overline{AF}$, $\overline{AE}$, $\overline{AD}$, and *m* are collinear. What seems to be special about the triangle? Sides $\overline{AB}$ and $\overline{AC}$ have the same length.

9. If you drew all possible angle bisectors, medians, altitudes, and perpendicular bisectors of a triangle and found that they all intersected in a single point, what would be true of the triangle? Illustrate.

In Exercises 10–14 you will discover an important geometry idea.

10. Draw a triangle with angles all less than 90°. Find its circumcenter (the intersection of its perpendicular bisectors). Is the circumcenter inside or outside the triangle? Inside

11. Draw a triangle that has one of its angles greater than a right angle. Find its circumcenter. Is the circumcenter inside or outside the triangle? Outside

12. If the pattern of Exercises 10 and 11 holds, where do you think the intersection of the perpendicular bisectors would be for a right triangle—inside, outside, or somewhere else? On the triangle.

Draw a right triangle. Test your conjecture by finding the intersection of the perpendicular bisectors. Where is it located? On the midpoint of the largest side of the triangle.

13. Use the intersection point in Exercise 12 (the circumcenter) to draw a circle through each of the vertices of the triangle.

6. *F*, *E*, then *D*, moving from *B* to *C*

7. *E* is between *F* and *D* unless all three points coincide.

9. The sides of the triangle must all have the same length.

How does the longest side of the triangle divide the circle? Are the parts of the circle equal or unequal? It divides the circle into two equal parts.

14. Express your result from Exercise 13 as a conjecture about certain triangles and half-circles.

15. **Construction** A mechanical contractor is installing the air conditioning system in a large commercial building. He needs to run a round duct through a triangular opening in the structural steel above the ceiling of the top floor of the building. The dimensions of the opening are 38 inches, 68 inches, and 92 inches. What is the diameter of the largest duct that can pass through the opening? Use geometry software or paper folding to model the problem. About 23 in.

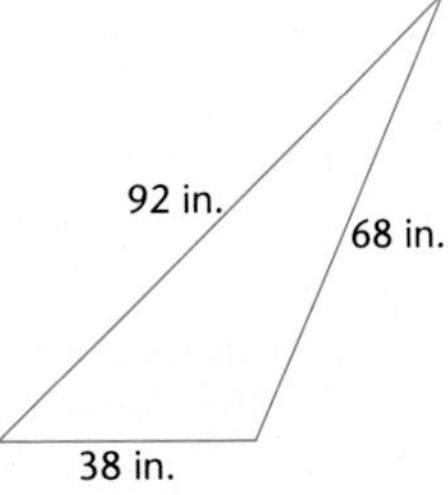

For Exercises 16 and 17, use triangles with all their angles smaller than right angles—acute triangles. You will discover three different results.

16. Find the midpoint of each side of an acute triangle. Then connect them to form another triangle. Your drawing should now contain four small triangles inside the larger one. Cut out the four triangles and compare them. What do you observe? The four small triangles are exactly the same.

17. Construct a triangle *ABC*, with medians. For each median, measure the distance from the intersection or centroid to the side it touches, such as $\overline{OX}$, as well as to the vertex it touches, such as $\overline{OC}$. Record your measurements in a table like the one shown.

What do you notice about the relationship between the figures in the first row and those below in the second row? Express your answer in the form of a conjecture about the centroid of a triangle.

Triangle ABC. Triangles are named using their "corners," or vertices.

OA = ?	OB = ?	CO = ?
OY = ?	OZ = ?	OX = ?

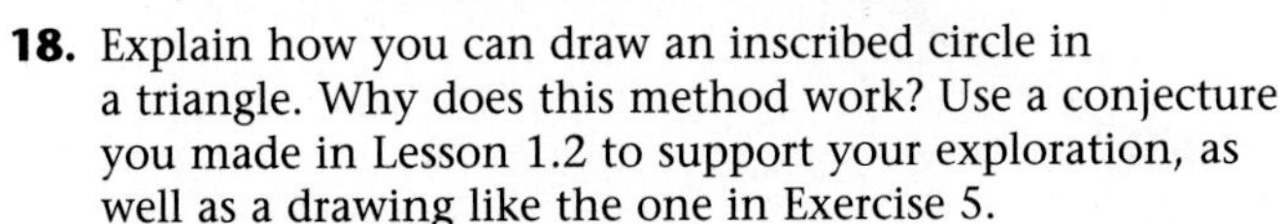

18. Explain how you can draw an inscribed circle in a triangle. Why does this method work? Use a conjecture you made in Lesson 1.2 to support your exploration, as well as a drawing like the one in Exercise 5.

Euler Lines The following exercise is especially suited for computer software, but it can also be done by using paper-and-pencil methods.

19. In a single triangle, find the circumcenter, the orthocenter, and the centroid. Connect each of the points with segments. What do you notice? Repeat with different triangles, or compare your results with those of your classmates. Write a conjecture about these three points.

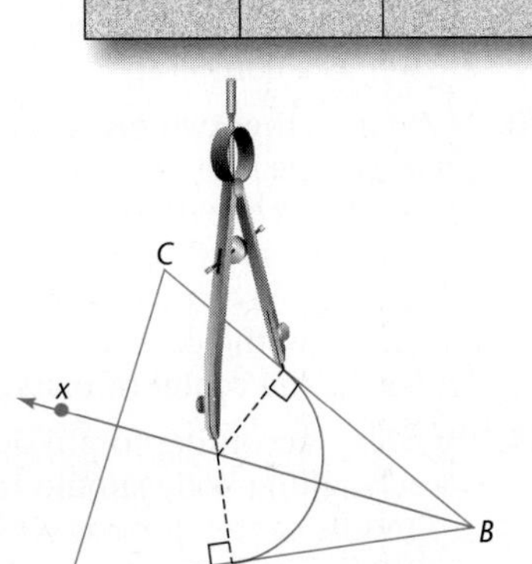

Try this with more than one point on $\overrightarrow{BX}$.

20. **Portfolio Activity** Collect samples of computer-generated art. Do some research to find out about different ways computers are used in art and publishing. Check student samples.

14. A right triangle can be inscribed in a half circle.

17. The distance from the vertex to the centroid is $\frac{2}{3}$ of the distance from the vertex to the midpoint opposite the vertex.

18. Locate the point where the angle bisectors meet. Draw a segment perpendicular to one of the sides from the intersection point. Because the angle bisectors are equidistant from their sides, a circle of this radius drawn from the intersection point is inscribed in the circle.

alternative ASSESSMENT

Authentic Assessment

Have the whole class agree on a sign that might be needed in your school or community. Possible ideas include Walk Your Bike, No Skating, and Slippery When Wet. Students should work alone or in groups to design the needed sign. Their signs should have the shape of an equilateral triangle with a circle inscribed in the triangle.

19. The orthocenter, centroid, and circumcenter of any triangle are on the same line.

Technology Master

NAME ______ CLASS ______ DATE ______

Technology
1.3 Special Lines and Segments in Other Figures

In Lesson 1.3, you explored the behavior of altitudes, medians, perpendicular bisectors, and angle bisectors in triangles. Did you find yourself wondering how these special lines and segments behave in other figures—or whether they even exist in other figures?

Use geometry software for Exercises 1–7.

1. Using the *segment* tool of the software, draw four segments joined at their endpoints. Adjust their positions to form a rectangle like the one shown. Label the vertices *A, B, C,* and *D,* as shown.
2. Use the angle bisector feature to bisect each angle. The figure should look much like the second one shown.
3. Unlike the angle bisectors of a triangle, these angle bisectors do not intersect at one point. What is the relationship among the angle bisectors in this figure?
4. Using the *select* tool, grab $\overline{CD}$ and drag it closer to $\overline{AB}$, then farther away. Be careful that *ABCD* remains a rectangle. Do the angle bisectors ever intersect at one point? Use your observations to write a conjecture.
5. Now grab $\overline{CD}$ and drag it so that *ABCD* no longer is a rectangle. Do the angle bisectors ever intersect at one point? Use your observations to write a conjecture.
6. Position $\overline{CD}$ so that *ABCD* again is a rectangle. Grab point *B* and drag it to different positions. Do the angle bisectors ever intersect at one point? Use your observations to write a conjecture.
7. Make a new sketch of rectangle *ABCD*. Use the *midpoint* and *perpendicular* features to construct the perpendicular bisector of each side. Using Exercises 4–6 as a guide, explore the behavior of the perpendicular bisectors as you change *ABCD*. Write your observations as one or more conjectures.

HRW Geometry Technology 17

24. The absolute values are the same. The absolute value of the difference between the numbers is the same, regardless of the order in which they are subtracted.

Look Back

Choose any two different positive numbers.

21. Add them together. Do you get a positive or a negative number? positive

22. Subtract the smaller number from the larger number. What kind of number do you get? Now take the absolute value of this number. positive, positive

23. Subtract the larger number from the smaller number. What kind of number do you get? Now take the absolute value of this number. negative, positive

24. Compare your answers from Exercises 22 and 23 above. Explain why they are different or the same.

Choose any two different negative numbers.

25. Add them together. Do you get a positive or a negative number? negative

26. Subtract the smaller number from the larger number. What kind of number do you get? Now take the absolute value of this number. positive, positive

27. Subtract the larger number from the smaller number. What kind of number do you get? Now take the absolute value of this number. negative, positive

28. Compare your answers from Exercises 26 and 27 above. Explain why they are different or the same.

Look Beyond

Extension. In an object of uniform composition, the center of mass is the average position of all the particles of mass that make up an object. For a baseball the center of mass is in the center of the ball, for a baseball bat the center of mass is closer to the thick end, and for a boomerang the center of mass is outside the object completely. Finding the center of mass for groups of objects such as the earth and moon or the entire solar system can help astronomers predict paths of objects in the sky.

Look Beyond

29. Cut out a triangle from a stiff piece of cardboard. Draw a median of the triangle. See if you can balance the triangle along the line you drew. What can you conclude about the area of the triangle on each side of a median? Express your answer as a conjecture.

30. Draw the other two medians of your triangle. See if you can now balance the triangle on a single point placed at the centroid. In your own words, explain why the centroid is known as the center of mass.

31. **Physics** According to physics, a freely falling body should rotate around its center of mass. Test this theory by giving your triangle a toss like a Frisbee. What do you observe?

28. The absolute values are the same. The absolute value of the difference between the numbers is the same, regardless of the order in which they are subtracted.

29. A median of a triangle divides it into two triangles with equal area.

30. The weight of the triangle is evenly distributed around the centroid.

31. The triangle appears to spin around the centroid.

LESSON 1.4 Measuring Length

We use rulers and other measuring devices all the time. We ask each other questions such as these: How tall are you? How far do you live from here? Rulers and other methods of measurement work because of some basic agreements and assumptions.

A ruler in the eyepiece of the microscope makes it possible to measure in units of one-millionth of a meter, which are known as microns.

The Measure of a Segment

In defining the measure or length of a segment, we will use a ruler.

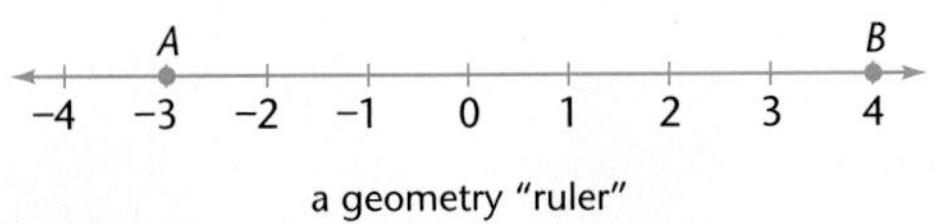

a geometry "ruler"

In geometry, a "ruler" is a **number line**—that is, a line that has been set up to correspond with the real numbers. The coordinate of a point on a number line is the real number that matches the point. In the illustration, -3 is the coordinate of A and 4 is the coordinate of B.

How would you find the distance between points A and B? Try $|-3 - 4|$. Also try $|4 - (-3)|$. What do you notice about the absolute values? This leads to the definition of the measure of a segment.

Technology A geometry graphics program can be used to create rulers with unit lengths of different sizes. Have students create one ruler. Then the ruler can be made larger or smaller using the appropriate tool in the computer program.

PREPARE

Objectives

- Construct a geometric ruler.
- Define *congruence.*
- Identify and use the Segment Addition Postulate.

RESOURCES

- Practice Master 1.4
- Enrichment Master 1.4
- Technology Master 1.4
- Lesson Activity Master 1.4
- Quiz 1.4
- Spanish Resources 1.4

Assessing Prior Knowledge

Draw a number line showing the first four members of each set.

1. positive multiples of 5
2. positive multiples of 10
3. quarter inches
4. half centimeters
5. thirds of yards

[**Answers will vary**]

5

TEACH

Although many aspects of geometry can be studied without using either length or angle measures, most real-world applications involve quantities. The text introduces ideas of measurement early in the book so that students will be able to apply what they learn to real-world problems.

Make certain every student clearly understands the meaning of *absolute value* and is familiar with the symbols used to show the absolute value of a number or expression.

Alternate Example 1

Find the measures (lengths) of $\overline{DO}$, $\overline{OG}$, and $\overline{DG}$ on the number line.

D O G
−3 −2 −1 0 1 2

[2, 3, 5]

"Try This" questions provide opportunities for ongoing assessment. The answers are provided in the Teacher's Edition side copy.

ongoing ASSESSMENT

Try This $ST = |3-(-2)| = |-2-3| = 5$

MEASURE OF SEGMENT $\overline{AB}$

Let A and B be points on a number line, with coordinates a and b. Then the measure of segment $\overline{AB}$, which is called the length, is $|a-b|$ or $|b-a|$.

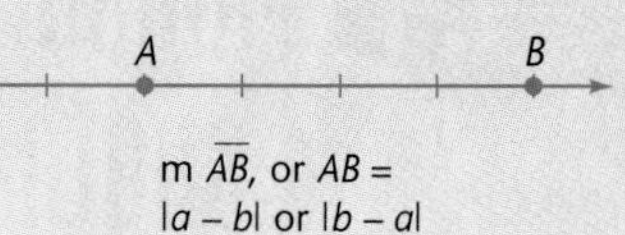

m $\overline{AB}$, or AB = $|a-b|$ or $|b-a|$

The measure of segment $\overline{AB}$ is written as m$\overline{AB}$ or simply as AB.

1.4.1

EXAMPLE 1

Find the measures (lengths) of $\overline{AB}$, $\overline{AX}$, and $\overline{XB}$ on the number line.

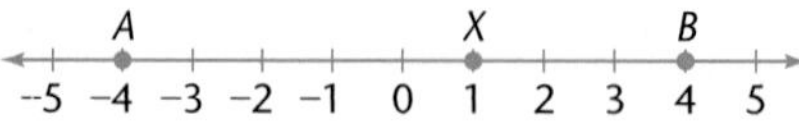

Solution ➤

AB (or m$\overline{AB}$) $= |-4-4| = |-8| = 8$ or $|4-(-4)| = |8| = 8$

AX (or m$\overline{AX}$) $= |-4-1| = |-5| = 5$ or $|1-(-4)| = |5| = 5$

XB (or m$\overline{XB}$) $= |1-4| = |-3| = 3$ or $|4-1| = |3| = 3$ ❖

Compare these results

Try This Find ST.

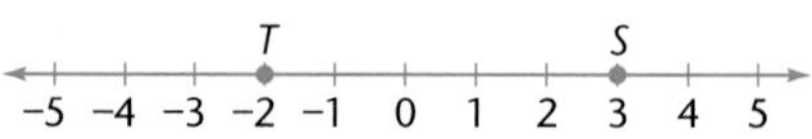

Constructing a Ruler

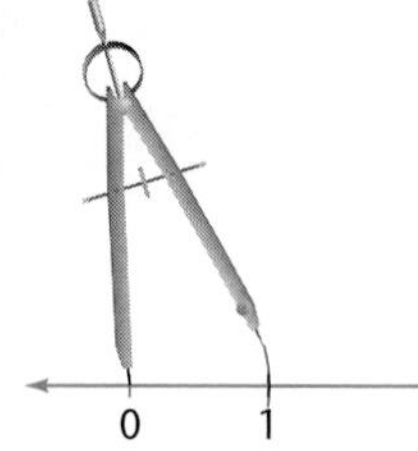

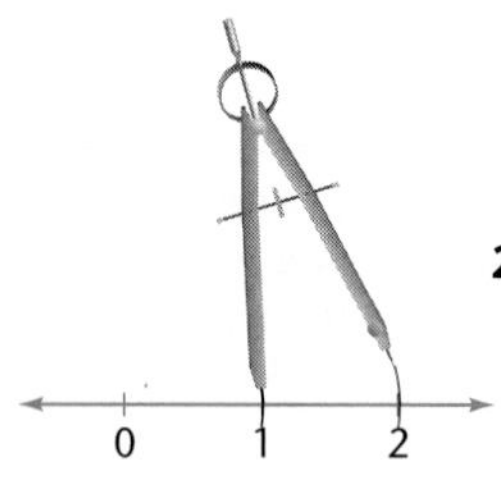

The geometry rulers that you will use in this book—with a few interesting exceptions—have evenly spaced divisions. A compass can be used to make a geometry ruler on a number line.

1. Choose any point on a line and label it as 0. Adjust your compass to an appropriate spacing and set the point of the compass on the zero point of the line.

Use the pencil part of the compass to draw a short mark that crosses the number line on one side of zero. Label the point of intersection as 1.

2. Set the point of the compass on the newly labeled point and draw another mark that crosses the line in a new place. Label the new intersection with an integer that is 1 larger than the coordinate of the previous point.

interdisciplinary CONNECTION

Language Arts A *pedometer* is an instrument for measuring distance walked. Have students list as many different types of measuring instruments as they can. Many of the words end with the suffix *-meter.* Students can use a dictionary to find out why *meter* is used for instruments that don't necessarily measure meters.

3. Repeat the previous step as many times as desired. ❖

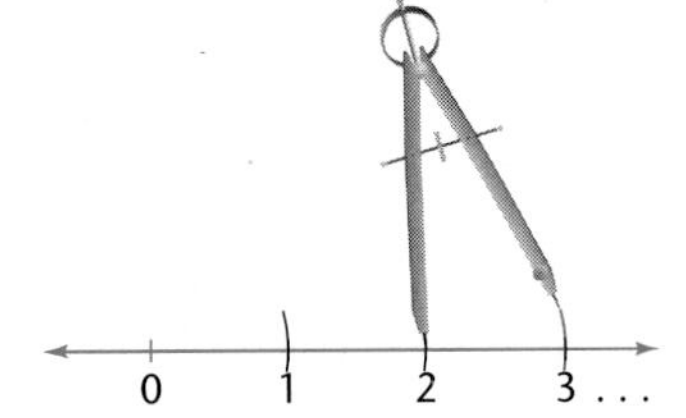

CRITICAL Thinking How would you construct the negative numbers on the number line? Do you think it matters on which side of the zero point you place the negative numbers?

Once the integers are placed on the number line, a standard is usually applied to name the **unit length**. Two standards commonly used are the inch and the meter.

Congruence

Segments that have the same length are **congruent**. That is, they match exactly. If you move one of them onto the other, they fit together perfectly.

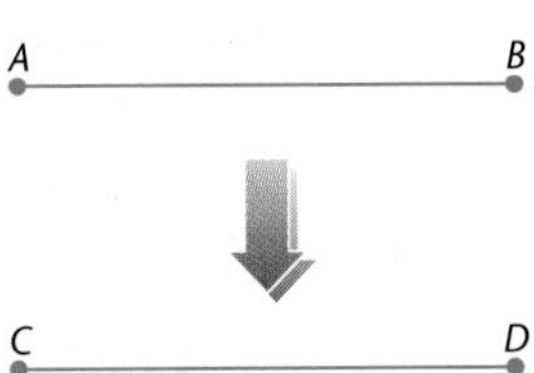

The segments on the ruler that was constructed above were all congruent because the same compass setting was used for each one.

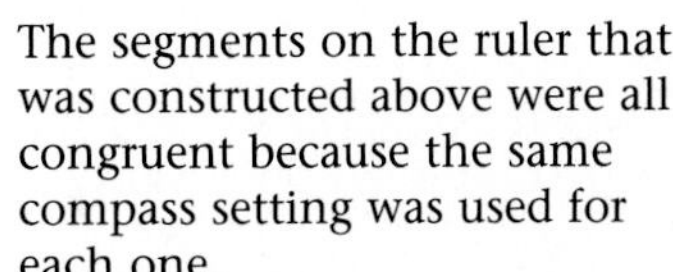

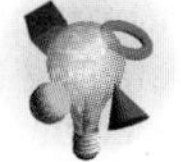

What do you think would happen if the divisions on a ruler were not evenly spaced? Could you be sure that segments with the same measure would be congruent, or that congruent segments would have equal measures? Explain your answer.

In this book, "tick marks" show that segments are congruent. Segments that have a single tick mark are congruent. Similarly, all segments with two tick marks are congruent to each other, and so on.

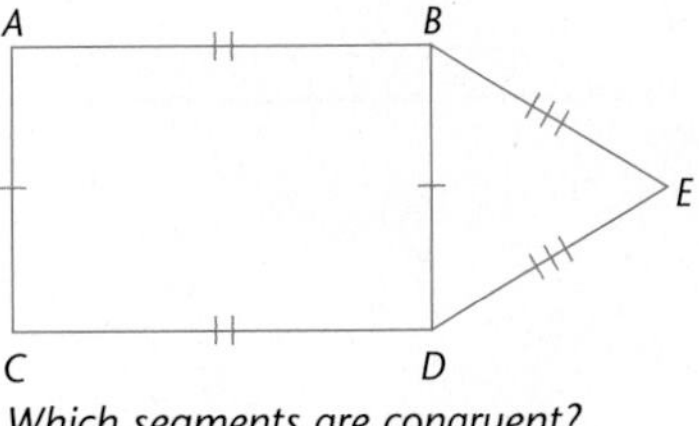

Which segments are congruent?

Segment Addition

Look again at the number line in Example 1. Notice that X is between A and B. The relationship among the distances AX, XB, and AB leads to an important assumption for geometry, known as the Segment Addition Postulate.

CRITICAL Thinking

You should do the same thing as when making positive numbers, just go the opposite direction from the point marked O. No, it does not matter as long as the negative numbers are on the opposite side of the number line from the positive numbers.

Cooperative Learning

After each student has finished constructing a ruler, have students work in small groups. The unit divisions on the rulers will be of different lengths, so students can discuss the results of measuring with different rulers. Discuss which rulers would be better tools to measure objects of different sizes and how the size of the unit length affects precision of measurement.

CRITICAL Thinking

The measured length of a segment would be different, depending on which part of the ruler you used.

ENRICHMENT In the Segment Addition Postulate, the three points must be collinear. Have students look for a relationship among points R, P, and Q, where R is between P and Q but the points are not collinear. (Students will discover a version of the Triangle Inequality: $PR + RQ > PQ$.)

INCLUSION strategies **English Language Development** Students new to the English language may have difficulty with terms such as *congruence* and *unit length.* These students can benefit from keeping a vocabulary list in their portfolios. Students can illustrate each definition to help learn the term.

TEACHING tip

The key idea in the Segment Addition Postulate is that point *R* must be *between* the other two points. Have students investigate what happens when point *R* is not between points *P* and *Q*.

CRITICAL Thinking

Measures of segments are being added. $AB + CD = 5$ and $m\overline{AB} + m\overline{CD} = 5$ make sense.

Alternate Example 2

It is 60 miles from Eastlake to Fielding, and it is half that distance from Eastlake to Danville. Find the distance from Danville to Eastlake and from Danville to Fielding.

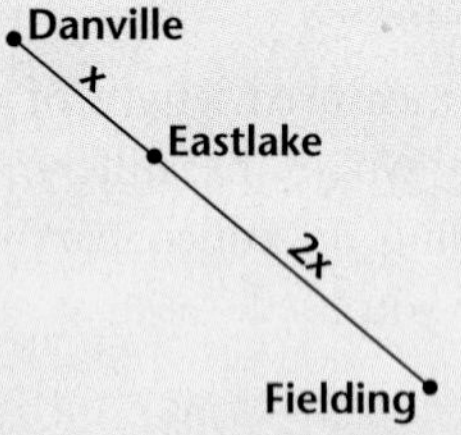

[Danville to Eastlake: 30 miles; Danville to Fielding: 90 miles]

SEGMENT ADDITION POSTULATE

If point *R* is between points *P* and *Q*, then $PR + RQ = PQ$.

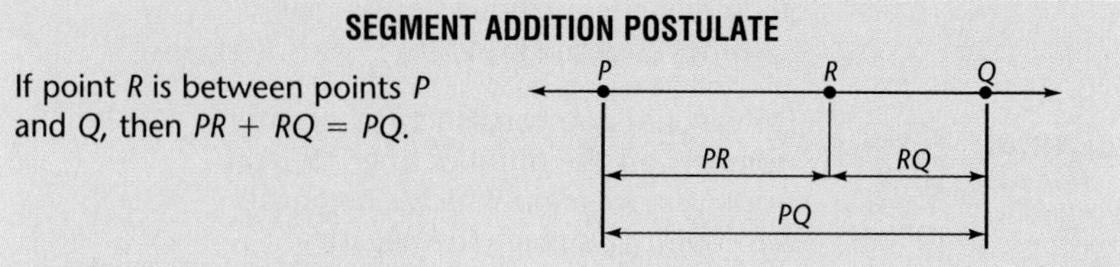

1.4.2

CRITICAL Thinking

Are *segments* really being added in this postulate? or *measures* of segments? In this course, addition and other arithmetic operations are defined for numbers, not geometric figures.

Which ones make sense?

$$\begin{cases} AB + CD = 5 \\ \overline{AB} + \overline{CD} = 5 \\ m\overline{AB} + m\overline{CD} = 5 \end{cases}$$

The Segment Addition Postulate makes an important connection between ideas in algebra and geometry.

EXAMPLE 2

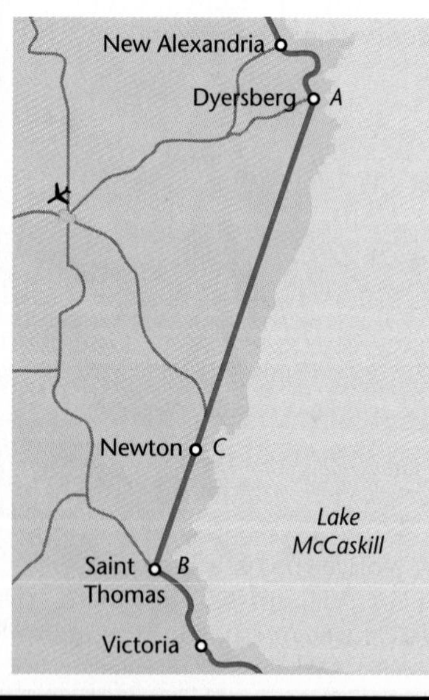

Algebra The towns of Dyersberg, Newton, and Saint Thomas are located along a straight portion of Ventura Highway. The distance from Dyersberg to Saint Thomas is 25 miles. The distance from Dyersberg to Newton is one mile more than three times the distance from Newton to Saint Thomas. Find the distances from Dyersberg to Newton and from Newton to Saint Thomas.

Solution ➤

First represent each town as a point on a line segment. Dyersberg = *A*, Saint Thomas = *B*, and Newton = *C*. Let *x* be the distance in miles from *B* to *C*. Then the distance from *A* to *C* will be $3x + 1$.

Since *C* is between *A* and *B*, $AC + CB = AB$. So,

$$(3x + 1) + x = 25$$
$$4x + 1 = 25$$
$$4x = 24$$
$$x = 6 = BC$$

So $AC = 3x + 1 = 3(6) + 1 = 19$ miles. ❖

RETEACHING the lesson

Using Visual Models Have students create floor plans of simple houses. All congruent segments should be marked with the appropriate number of tick marks. Students can exchange drawings with a partner and use a ruler to verify that congruent segments are actually congruent.

Communicate

1. "Congruent segments are segments that can be moved so that they match exactly." Explain how this idea is used in paper folding to find the perpendicular bisector of a segment.
2. Explain why it is important for a ruler to have equally spaced divisions.
3. The distance from 0 to 1 on a ruler, known as the **unit length**, can be any desired size. Give as many examples as you can of units of different length.
4. If the centimeter were the only unit of measure for length, what problems would this create? Discuss why it is convenient to have different units of measure for measuring length.
5. Explain why each of the following statements does or does not make sense.
   $\overline{MN} + \overline{OP} = 30$ cm $MN + OP = 30$ cm
   $m\overline{MN} + m\overline{OP} = 30$ cm
6. Once you have constructed the integers on your ruler, why would you want to subdivide the unit length into smaller divisions?

Geometry graphics software often allows you to select your units of measure.

Practice & Apply

7. Find the measures of the segments determined by the points on the number line. Show that it does not matter which point you subtract from which.

$AB = |-3 - (-1)| = |-1 - (-3)| = 2$
$AC = |-3 - 3| = |3 - (-3)| = 6$
$BC = |-1 - 3| = |3 - (-1)| = 4$

In Exercises 8 and 9, name all the congruent segments.

8.

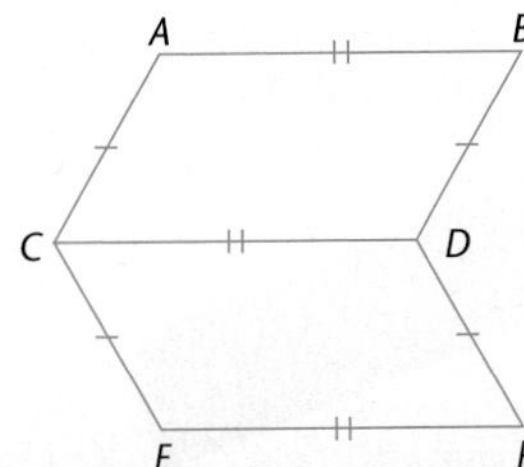

9.

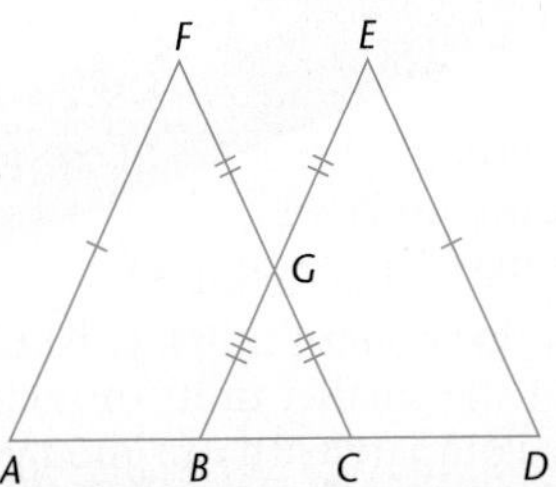

Assess

Selected Answers

Odd-numbered Exercises 7–35

The answers to Communicate Exercises can be found in Additional Answers beginning on page 727.

Assignment Guide

Core 1–17, 24, 29–35

Core Plus 8–28, 36–37

Technology

In Exercises 36 and 37 a scientific or graphics calculator can be used to supplement the investigation of logarithms. For example, does log 2 + log 3 = log 6? And does log 3.3 − log 1.1 = log 3? The students can verify these results to as many decimal places as their calculators give.

Error Analysis

Some students will need help with the algebra required in Exercises 13–15. Suggest that they make a three-column table. In the first column they write the three segment names $\overline{XY}$, $\overline{YZ}$, and $\overline{XZ}$. The second column has lengths in terms of x. The third column has numerical lengths.

8. $\overline{AC}$, $\overline{BD}$, $\overline{DF}$, and $\overline{CE}$ are congruent; $\overline{AB}$, $\overline{CD}$, and $\overline{EF}$ are congruent

9. Congruent segment pairs are $\overline{AF}$ and $\overline{ED}$, $\overline{FG}$ and $\overline{EG}$, and $\overline{GB}$ and $\overline{GC}$.

alternative ASSESSMENT

Performance Assessment

Each student should draw a number line marked from −10 to 10. Four points should be chosen at random and labeled with the points *A*, *B*, *C*, and *D*. Students find the lengths of all the various segments. They then write at least five statements using addition and subtraction of segments. Example statements include $AB + CD = 10$ units or $AD - AB = BD$.

In Exercises 10–12 you are given that *A* is between points *M* and *B*. Sketch each figure and find the missing measures.

10. $MA = 30$ $AB = 15$ $MB =$? 45

11. $MA = 15$ $AB =$? 85 $MB = 100$

12. $MA =$? 16.3 $AB = 13.3$ $MB = 29.6$

Find the indicated value in Exercises 13–15.

13. $XZ = 25$ $x =$? 5

14. $XY = 25$ $XZ =$? 40

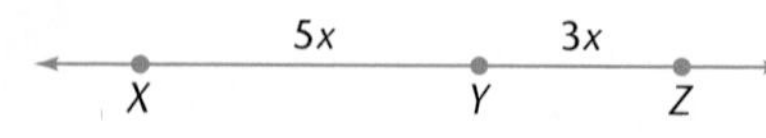

15. $YZ = 25$ $XY =$? 51

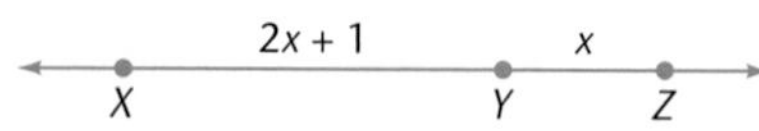

Algebra Towns *A*, *B*, and *C* are situated along a straight highway. *B* is between *A* and *C*. The distance from *A* to *C* is 41 miles. The distance from *B* to *C* is 2 miles more than twice the distance from *A* to *B*.

16. Write an equation for the distances, and then solve the equation to find the distances between the towns. $x + (2 + 2x) = 41$; $AB = 13$; $BC = 28$

17. Town *X* is between *A* and *B*, 6 miles from *A*. What is the distance from *X* to *C*. 35

For each of the statements in Exercises 18–23, write S for "Sense" and N for "Nonsense."

18. $XY = 5{,}000$ yd S **19.** $\overline{PQ} = 32$ in. N **20.** $m\overline{ST} = 6$ cm N

21. $XY + XZ = 32$ cm S **22.** $m\overline{PR} = 46$ cm S **23.** $XY - XZ = 12$ cm S

24. Cultural Connection: Africa The Egyptian Royal Cubit is subdivided into 28 units known as digits or fingers. From this basic unit, a number of others were created:

4 digits = 1 palm	12 digits = 1 small span
5 digits = 1 hand's breadth	14 digits = 1 great span
6 digits = 1 fist	16 digits = 1 foot (t'eser)
8 digits = 1 double palm	24 digits = 1 short cubit

How does the modern metric system of length measure resemble the ancient Egyptian system in subdividing larger units into smaller ones and in building up larger units of length from smaller ones? Give examples.

The length of the Royal Cubit, in modern units, is 20.67 inches. Calculate the length of the smaller units in inches. How does the Egyptian foot (t'eser) compare with our modern foot?

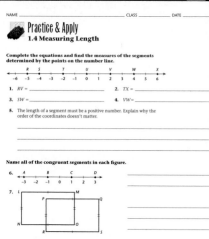

24. Meters are subdivided into decimeters, centimeters, millimeters, and so on. Decameters, kilometers, and so on are built up from the meter.

1 digit = .74 in.
1 small span = 8.86 in.
1 palm = 2.95 in.
1 great span = 10.34 in.
1 hand's breadth = 3.69 in.
1 foot (t'eser) = 11.81 in.
1 fist = 4.42 in.
1 short cubit = 17.72 in.
1 double palm = 5.91 in.
The t'eser is slightly shorter than our 12-inch foot.

Practice Master

NAME ______ CLASS ______ DATE ______

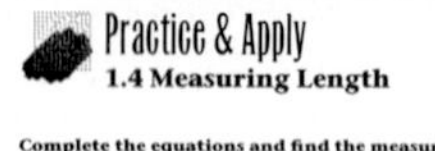

Practice & Apply
1.4 Measuring Length

Complete the equations and find the measures of the segments determined by the points on the number line.

1. RV = ______ 2. TX = ______
3. SW = ______ 4. VW = ______
5. The length of a segment must be a positive number. Explain why the order of the coordinates doesn't matter.

Name all of the congruent segments in each figure.

6.
7.

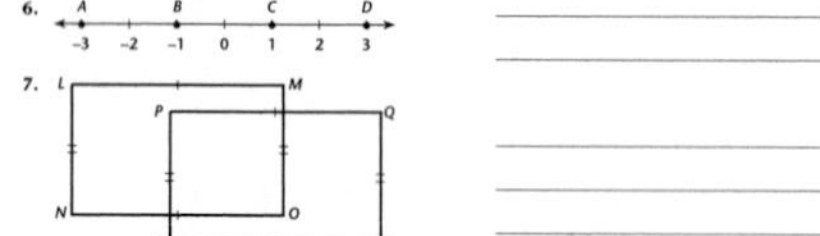

B is between points Q and R. Sketch each figure and find the missing measures.

8. QB = 20; BR = 10 QR = ______
9. QB = 50; BR = ______; QR = 110
10. QB = ______; BR = 16.9; QR = 51.5

Find the indicated value.

11. AC = 45; x = ______
12. DF = 135; x = ______

4 Practice & Apply HRW Geometry

25. Think how you could use paper folding to subdivide a given unit into smaller parts. Suppose you first bisected the interval, then bisected each half, and so on. Write the next few terms of the sequence below, which gives the size of the units you could obtain by this process.

$\frac{1}{2}, \frac{1}{4}, \ldots$ $\frac{1}{8}, \frac{1}{16}, \frac{1}{32}$

To find a distance like three-fourths, you can add three intervals of a fourth together. Name at least three distances that you *cannot* find by this process. Sample answers: $\frac{4}{11}, \frac{1}{5}, \frac{2}{7}$

There is more than one way of creating a measuring scale, and some of them might seem very strange. Do you think that "alternative rulers" might have real-world uses?

26. What can happen if the divisions of a ruler are not appropriately spaced? Draw your own "alternative ruler" to illustrate your answer. Show how segments of equal "length" might not be congruent.

27. Geology The **Richter scale** for indicating the intensity of earthquakes has an interesting feature. An earthquake of magnitude 2 has 10 times the ground movement of an earthquake of magnitude 1. Magnitude 3 is 10 times magnitude 2. In fact, each magnitude level has 10 times the ground movement of the next lower magnitude. What is the relationship between a magnitude 1 and a magnitude 3 earthquake? between a magnitude 1 and a magnitude 8 earthquake?

Two of the earth's continental plates meet at the San Andreas fault. When the plates slip, the resulting shock waves may be felt all over the world.

28. Statistics On a standardized test, Susan scored higher than anyone else in the class. Her **raw score** (number of correct answers) was 1 answer higher than James', who came in second. Compare the raw scores and the **percentile** ratings of the four students below to see how the percentile measuring scale is like an alternative ruler. Discuss, using calculations to emphasize the point.

Name	Raw Score	Percentile Ranking
Susan	63	99.0
James	62	98.6
Stewart	43	57.3
Myrna	42	55.3

26. The "length" of a segment might be different depending on which part of the ruler is used to measure it. Two segments that are not congruent could have the same "length" when measured with different parts of the ruler. Check student drawings.

27. An earthquake with magnitude 3 is 100 times more powerful than an earthquake with magnitude 1. An earthquake of magnitude 8 has 10,000,000 times the energy of an earthquake with magnitude 1.

28. There is a 1-point difference between 63 and 62, which corresponds to a difference of 0.4 in percentile ranking. There is also a 1-point difference between 43 and 42, but the difference in percentile ranking is 2, showing that percentile rank is an alternative ruler.

Technology Master

NAME ________________ CLASS ________ DATE ________

Technology
1.4 Modeling the Segment Addition Postulate

According to the Segment Addition Postulate:

If point *R* is between points *P* and *Q*, then $PR + RQ = PQ$.

A powerful aspect of geometry software is its ability to give you an instantaneous reading of the length of a segment. You can use this feature to create a dynamic model of the Segment Addition Postulate.

Model the Segment Addition Postulate by using geometry software to complete Exercises 1–3.

1. Use the *segment* tool to draw a horizontal segment. Label the endpoints of the segment *P* and *R*. Using *R* as one endpoint, draw a second horizontal segment. Label the other endpoint *Q*.

P R Q

2. Use the measurement feature to display:
   distance from *P* to *R*
   distance from *R* to *Q*
   distance from *P* to *Q*

   Use the calculation feature to display:
   distance from *P* to *R* + distance from *R* to *Q*

3. Drag point *R* closer to point *P*, then closer to point *Q*. (Be careful not to "bend" $\overline{PQ}$.) While you are doing this, observe what happens to the numbers that are displayed. Write a brief description of your observations. What geometric principle do you think this illustrates?

4. Now drag point *R* so that you do bend $\overline{PQ}$. While you are doing this, observe what happens to the numbers that are displayed. Write a brief description of your observations. What geometric principle do you think this illustrates?

5. Choose a measure greater than *PQ*. Drag point *R* to a location such that the distance from *P* to *R* plus the distance from *R* to *Q* is equal to the measure you chose. How many such locations are there? Suppose you drew a point at each of these locations. Make a conjecture about the type of figure the points would form.

18 **Technology** **HRW Geometry**

29. Circumcenter, can be outside the triangle

30. Incenter, cannot be outside the triangle

31. Centroid, can't be outside the triangle

32. Orthocenter, can be outside the triangle

Look Back

In Exercises 29–32, give the name for the intersection points of each of the special objects related to a triangle. Tell whether each can be outside of the triangle or not. [Lesson 1.3]

29. the perpendicular bisectors

30. the angle bisectors

31. the medians

32. the altitudes

In Exercises 33 and 34, tell which of the above intersection points can be used to do the following. [Lesson 1.3]

33. Draw a circle "around" a triangle. circumcenter

34. Draw a circle "inside" a triangle. incenter

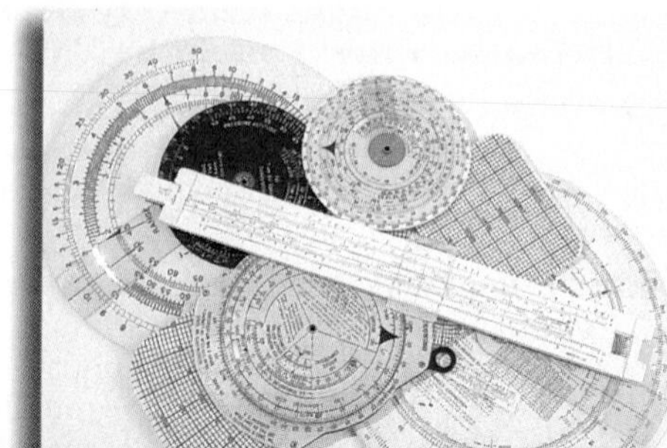

In Exercise 35, recall the famous result you found when exploring the perpendicular bisectors of a triangle. [Lesson 1.3]

35. State a conjecture about the relationship between a right angle and a part of a circle. A right angle can be inscribed in a half circle.

Look Beyond

Extension. Exercises 36 and 37 explore the use of slide rules. The locations of the numbers on the slide rule scales are based on exponents with a base of 10. For example, if the full scale is 1 unit long, 4 is placed at a point 0.602 of this distance from the starting point of the scale, because 10 raised to the power 0.602 is about 4.

Using logarithms is a way to turn multiplication into addition. To illustrate this, ask students to think of the product 100 • 10,000. If we write the factors as powers of ten, we get $10^2 \cdot 10^4$. The product is 10^6, or 1,000,000.

The two logarithmic scales can also be used to do division. Students may recall that division can be converted to subtraction through the use of logarithms.

Look Beyond

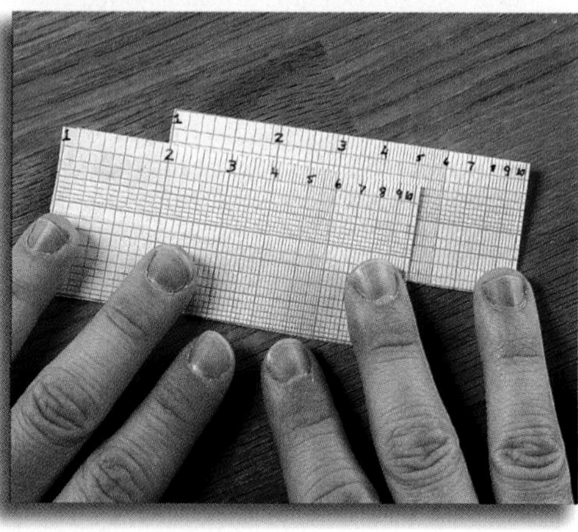

36. **Algebra** Slide rules, which were once used for doing approximate calculations, depended upon the properties of logarithms. You can make a simple slide rule from logarithm paper.

Cut out two strips of paper and place them together as shown. To multiply the numbers 2 and 3, move the left end of the top strip (the index) to the number 2 on the lower strip. Notice that the number 3 is just above the number 6 on the lower strip, which is the product of the 2 and 3.

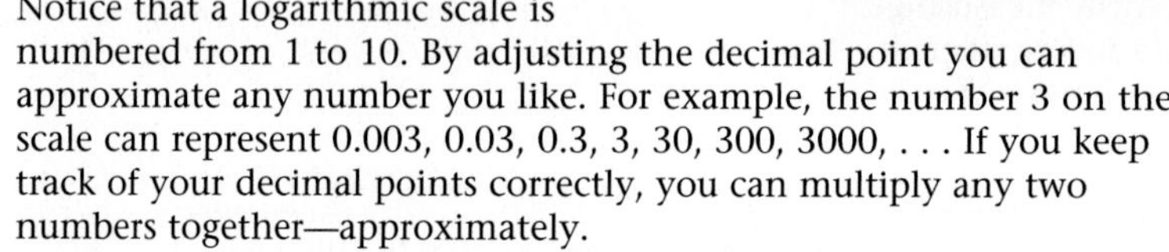

Notice that a logarithmic scale is numbered from 1 to 10. By adjusting the decimal point you can approximate any number you like. For example, the number 3 on the scale can represent 0.003, 0.03, 0.3, 3, 30, 300, 3000, . . . If you keep track of your decimal points correctly, you can multiply any two numbers together—approximately.

Multiply at least five pairs of numbers of your own choosing using logarithmic scales. Compare the answer you get with the answers you get from your calculator. How accurate were your answers?

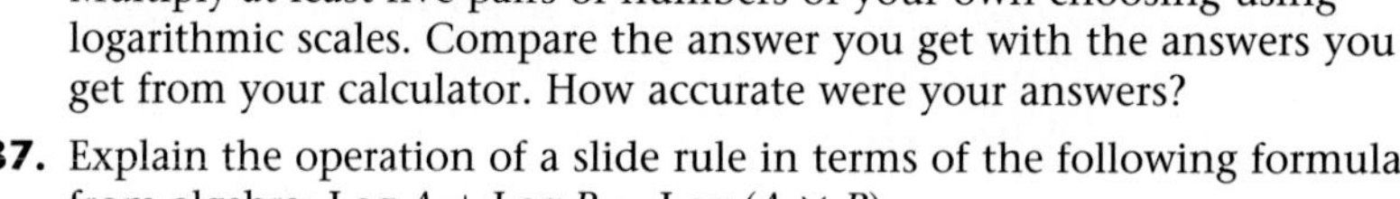

37. Explain the operation of a slide rule in terms of the following formula from algebra: Log A + Log B = Log ($A \times B$)

36. The numbers should be accurate to one decimal place if the scale is lined up correctly.

37. Logarithms are exponents. To find the exponent of the product of two numbers, add the exponents of the two numbers when the bases are the same.

LESSON 1.5 Measuring Angles

Why *Angle measure is used in many professions. For example, professional pilots use an angle measure known as the "heading" of a plane to navigate safely through the skies.*

Defining Angle Measure

A **protractor** is used to measure angles on a flat surface. As with a ruler for measuring length, the divisions of a protractor must be evenly spaced. Then you can be sure that if two angles have the same measure they are congruent, and *vice versa*.

To be sure you understand how a protractor is used, study the following example.

EXAMPLE 1

Use a protractor to find the measure of $\angle CAB$.

Solution ➤

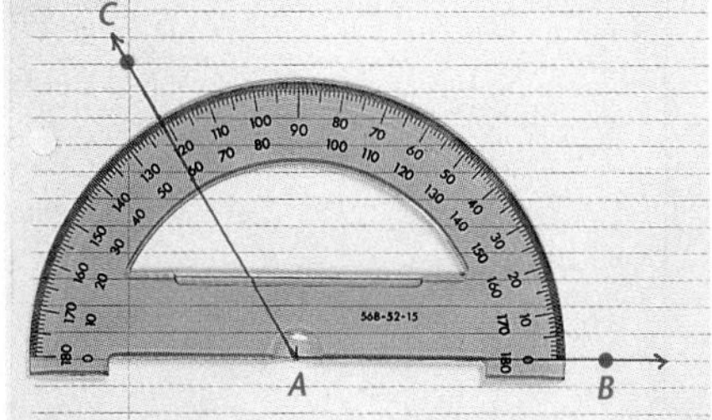

1. Put the center of the protractor at the vertex.
2. Align the 0 point of the protractor scale with ray $\overrightarrow{AB}$.
3. Read the measure (in degrees) where ray $\overrightarrow{AC}$ intersects the protractor.

The measure of $\angle CAB$ is 121°, or $m\angle CAB = 121°$. ❖

Why isn't the measure of $\angle CAB$ 59° instead of 121°?

ALTERNATIVE teaching strategy

Technology Geometry graphics software can be used to explore the Angle Addition Postulate. Have students draw any three rays, all with the same initial point. They should use the "measure" command to find the measurements of the three angles created in the figure. Their measurements should be recorded in a table. Then they can drag on the rays to change the angle sizes. They will see that the Angle Addition Postulate holds true regardless of the angle sizes.

PREPARE

Objectives

- Measure angles using a protractor.
- Identify and use the Angle Addition Postulate.

RESOURCES

- Practice Master 1.5
- Enrichment Master 1.5
- Technology Master 1.5
- Lesson Activity Master 1.5
- Quiz 1.5
- Spanish Resources 1.5

Assessing Prior Knowledge

Draw four angles of obviously different measures on the chalkboard or overhead.

1. Order the figures from largest to smallest.
2. Which angles are larger than a right angle? Which are smaller?

[Answers will vary.]

TEACH

Why Students are likely to use angle measures more often than they think. Every time they straighten a picture or move an object so that it "lines up" with something else, they are probably adjusting angles. Many angles in everyday life are 90°. Angles that aren't 90° often look "crooked."

Alternate Example 1

Have students work in pairs to practice measuring angles with a protractor. One student provides an angle and the other measures.

Cooperative Learning

Have students work in small groups on the Critical Thinking question. Each student should use a half circle of a different size. By comparing and discussing results, students can see that the size of the half circle does not change the angle measure.

CRITICAL *Thinking*

No. The measure of an angle is based on a fraction of a circle; any size circle would give the same circle fraction.

TEACHING *tip*

The protractors students will be using will probably have double scales, and students may become confused about which number gives the measure of the angle. Tell them that if the angle is **acute** (less than 90°), they should use the smaller number, and if the angle is **obtuse** (more than 90°), they should use the larger number.

Use Transparency 6

Alternate Example 2

Have students draw any angle that measures between 90° and 180°. Students should draw a ray in the middle of this angle so that the two new, smaller angles have equal measure. Students should measure all three angles and record their results.

A protractor may be thought of as another type of geometry ruler. This "ruler" is a half-circle with coordinates from 0 to 180. You can use it to define the measure of an angle such as $\angle AVB$.

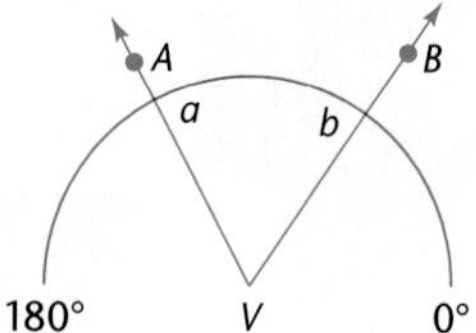

MEASURE OF ANGLE $\angle AVB$

Suppose the vertex V of $\angle AVB$ is placed on the center point of a half-circle with coordinates from 0 to 180. Let a and b be the coordinates of the points where $\overrightarrow{VA}$ and $\overrightarrow{VB}$ cross the half-circle.

Then the **measure of $\angle AVB$,** written as $m\angle AVB$, is $|a - b|$ or $|b - a|$. **1.5.1**

Angles, like segments, are measured in standard units. The unit of angle measure is the degree. This is the measure of the angle that results when a half-circle is divided into 180 equal parts.

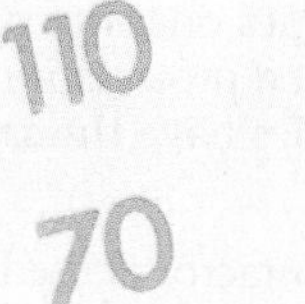

CRITICAL *Thinking* In creating a protractor, does the size of the half-circle make a difference? Why or why not?

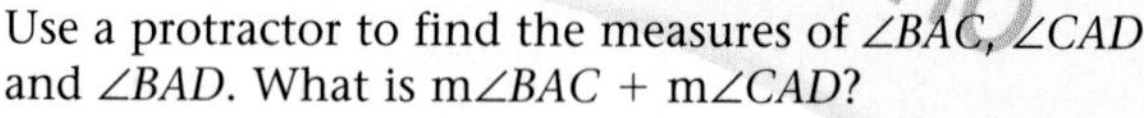

EXAMPLE 2

Use a protractor to find the measures of $\angle BAC$, $\angle CAD$, and $\angle BAD$. What is $m\angle BAC + m\angle CAD$?

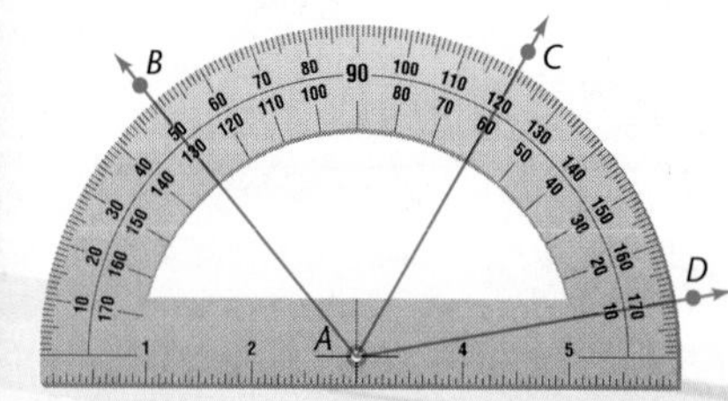

Solution ➤

1. To measure $\angle BAC$, notice the points where $\overrightarrow{AB}$ and $\overrightarrow{AC}$ pass through the half-circle scale of the protractor (50 and 120). Using these coordinates, $m\angle BAC = |50 - 120| = |-70| = 70°$.
2. The coordinate of $\overrightarrow{AD}$ is 170. So, $m\angle CAD = |120 - 170| = |-50| = 50°$.
3. Similarly, $m\angle BAD = |50 - 170| = |-120| = 120°$.
4. Using the answers from Steps 1 and 2, $m\angle BAC + m\angle CAD = 70 + 50 = 120°$. Notice that this agrees with the answer to Step 3. ❖

interdisciplinary CONNECTION **Sports** Have students research responsibilities of the navigator in boat racing. What types of tools does the navigator use? What factors affect the accuracy of his or her predictions?

Congruent Angles

Angles, like segments, are congruent if one can be moved onto the other so that they match exactly. Tick marks are also used to show that angles are congruent.

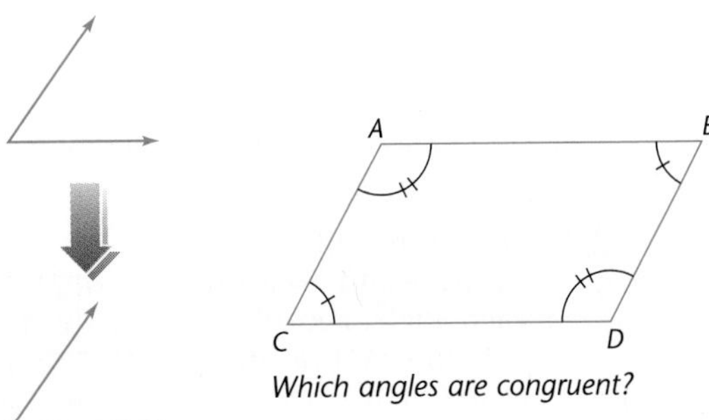

Which angles are congruent?

Pairs of Angles

The result in Step 4 of Example 2 suggests that $m\angle BAC + m\angle CAD = m\angle BAD$. This leads to the following postulate.

ANGLE ADDITION POSTULATE

If point S is in the interior of $\angle PQR$, then $m\angle PQS + m\angle SQR = m\angle PQR$.

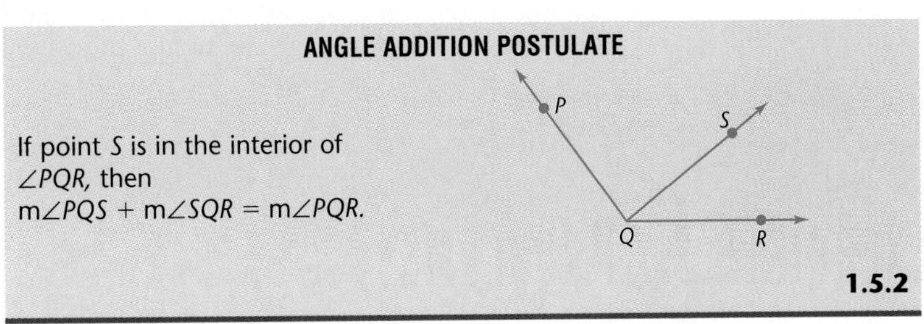

1.5.2

As in the Segment Addition Postulate, you should ask yourself: What is really being added? Is it angles or measures of angles?

Which statement makes sense? $\begin{cases} \angle A + \angle B = 180° \\ m\angle A + m\angle B = 180° \end{cases}$

SPECIAL ANGLE SUMS

If the sum of the measures of two angles is 90°, then the angles are **complementary.**

If the sum of the measures of two angles is 180°, then the angles are **supplementary.** **1.5.3**

TEACHING tip

To prevent students from reading the wrong degree mark on a protractor, encourage estimation. Angles smaller than right angles should measure less than 90°, and angles larger than right angles should measure more than 90°.

CRITICAL Thinking

It is the measures that are added. $m\angle A + m\angle B = 180°$ makes sense.

ENRICHMENT Real numbers, segments, and angles satisfy the Trichotomy Property. For every x and y, one and only one of the following statements can be true: $x < y$, $x > y$, or $x = y$. Have students list some real-world relationships or situations that are trichotomous and some that are not.

Hands-On Strategies Learning to measure angles with a protractor is difficult for many students. Show them how to extend the rays by using index cards or stick-on notes. Also, some students will have more success with larger or smaller protractors.

Use Transparency 7

Exploration Notes

Students should look for the relationship between the angles in a linear pair. There are two cases they can investigate: (1) the ray is perpendicular to the line, and (2) the ray is not perpendicular to the line. Ask students what might happen if they only considered case 1. (They might incorrectly conclude that the angles in a linear pair are always congruent.)

Ongoing Assessment

3. **Yes. Yes. The sum of the angle measures must be the angle measure of a half circle, or 180°.**

Teaching tip

Ask students to consider whether supplementary angles always form a linear pair.

Assess

Selected Answers

Odd-numbered Exercises 7–45

Answers to all Communicate exercises can be found in Additional Answers beginning on page 727.

Assignment Guide

Core 1–5, 7–34, 40–47

Core Plus 1–6, 13–39, 47–48

Exploration *Linear Pairs*

You will need
Protractor

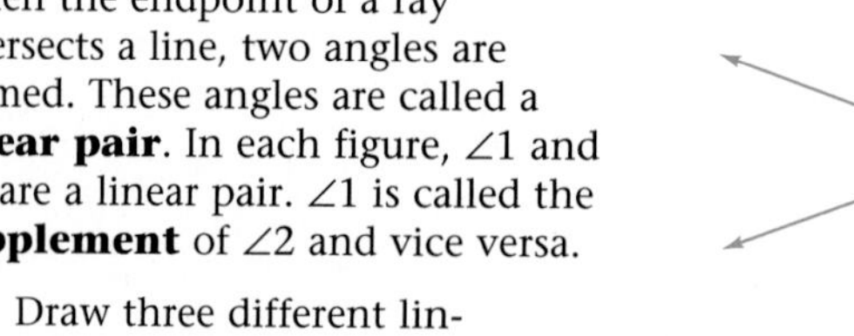

When the endpoint of a ray intersects a line, two angles are formed. These angles are called a **linear pair**. In each figure, ∠1 and ∠2 are a linear pair. ∠1 is called the **supplement** of ∠2 and vice versa.

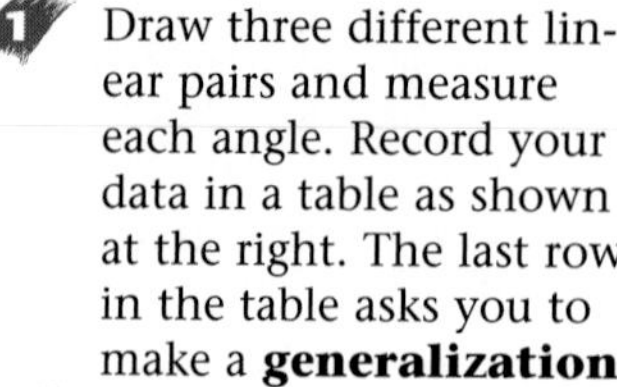

1. Draw three different linear pairs and measure each angle. Record your data in a table as shown at the right. The last row in the table asks you to make a **generalization**.

m∠1	m∠2	m∠1 + m∠2
?	?	?
?	?	?
?	?	?
x	?	?

2. Make a conjecture about linear pairs based on the information in your table.

3. If two angles form a linear pair, must they be supplementary? If two angles are supplementary, will they form a linear pair? Explain your reasoning.

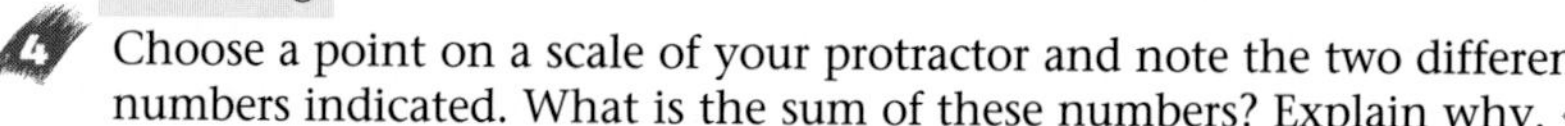

4. Choose a point on a scale of your protractor and note the two different numbers indicated. What is the sum of these numbers? Explain why. ❖

Exercises & Problems

Communicate

1. When you travel on land, you do not ordinarily use compass directions, or headings, to find your way. What kinds of information do you use instead?
2. If an angle is 43°, what is its complement? What is its supplement?
3. How is angle measure like length measure? How is it different? Discuss.
4. Explain why one of the following statements makes sense and the other does not.

$m\angle A + m\angle 2 = 190°$
$\angle X + \angle Y = 150°$

Name some other activities in which you would use a compass.

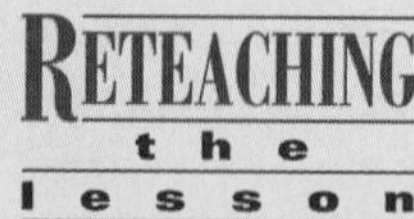

Reteaching the lesson

Cooperative Learning Each student should create a simple dissection puzzle by dividing a square into 6 regions using straight line segments. The student should measure all the angles. Then each student writes several hints for a puzzle solver. For example, a hint might be that one of the square's corners is made up of a 15° angle and a 75° angle. Students then should exchange puzzles with a partner and try to reassemble the original square. If necessary, more hints can be provided.

5. If two supplementary angles are congruent, what kind of angles are they? Why must each angle measure 90°?
6. Explain why folding a line onto itself produces perpendicular lines.

Practice & Apply

Find the measure of each angle in the diagram.

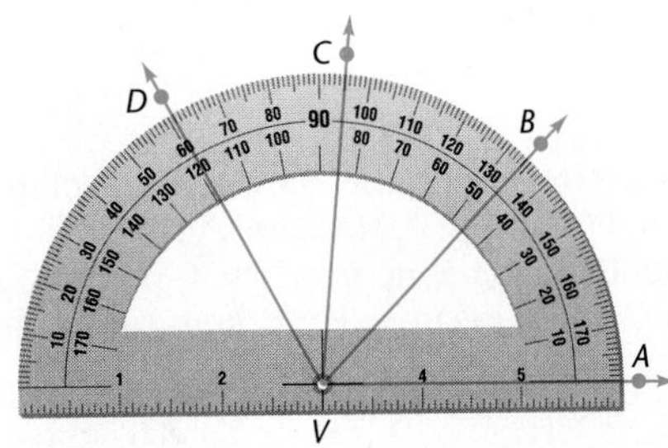

7. m∠AVB 46°
8. m∠AVC 85°
9. m∠AVD 119°
10. m∠BVC 39°
11. m∠DVB 73°
12. m∠CVD 34°

13. Name all the congruent angle pairs in the figure below. ∠BVO and ∠MVT: 20°
∠BVM, ∠OVT and ∠TVR: 45°
∠BVT and ∠MVR: 65°

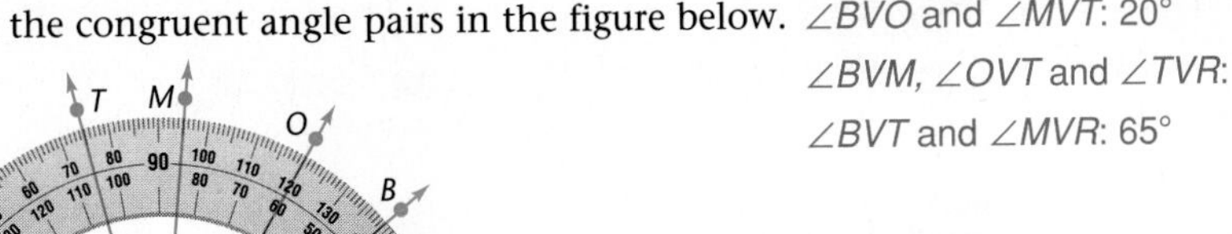

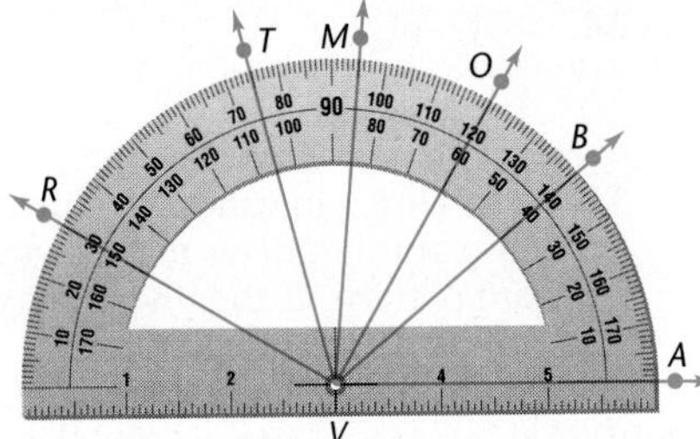

14. What is the angle between the hands of a clock if the time is 3:00? 90°
15. Through what angle does the minute hand move between 3:00 and 3:10. Explain your answer.
16. In the diagram, m∠ALE = 31° and m∠SLE = 59°. What is the measure of ∠SLA? Which postulate allows you to draw this conclusion? What is the relationship between ∠SLE and ∠ALE?

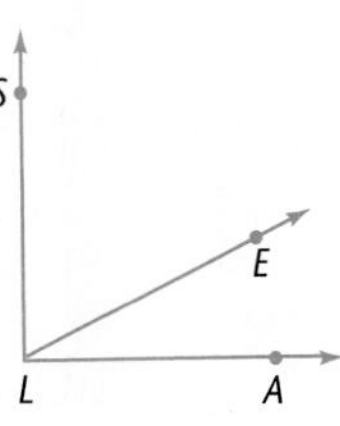

Find the missing measures in Exercises 17–19.

17. m∠SML = 80°, m∠LMI = 34°, m∠SMI = ? 114°
18. m∠SML = 21°, m∠LMI = ?, m∠SMI = 43° 22°
19. m∠SML = ?, m∠LMI = 34°, m∠SMI = 52° 18°
20. Suppose that three rays each have their endpoint at the same point V but are not coplanar. Would the Angle Addition Postulate apply to the angles that are formed? Illustrate your answer with a sketch. (You can use rays starting from the corner of a box.)

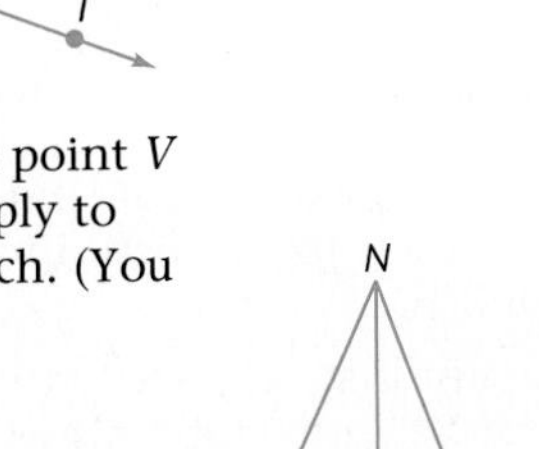

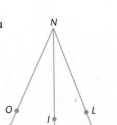

Algebra In the diagram, m∠LNO = 87°, m∠LNI = (2x − 8)°, and m∠INO = (x + 50)°.

21. What is the value of x? 15°
22. What is m∠INO? 65°

Technology

Students can use geometry graphics software with Exercises 26–34 to construct the appropriate diagram on the computer and use the computer's tools to find the angle measures. Note, however, that some computer programs will not measure angles greater than 180°.

Error Analysis

For Exercises 7–12, remind students to check the degree mark at both rays of each angle. Unless one ray of the angle is lined up with the 0° mark, students will need to subtract to find the measure of the angle.

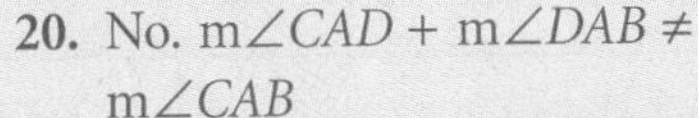

20. No. m∠CAD + m∠DAB ≠ m∠CAB

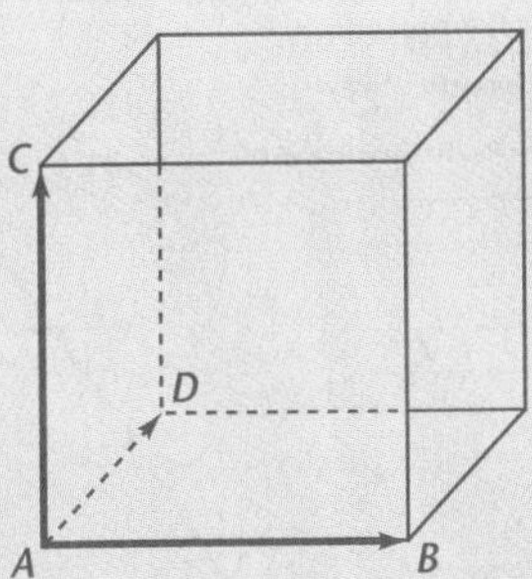

15. 60°. The 12 numbers on a clock divide the circle into 12 angles, each of measure 30°.

16. m∠SLA = 90°, by the Angle Addition Postulate. They are complementary.

alternative ASSESSMENT

Performance Assessment

Have students approximate a 90° angle, a 45° angle, and a 60° angle in two different ways: (1) using a pencil and straightedge to draw the angles, and (2) using folding paper to make the angles. Students should measure their estimated and folded angles with a protractor to see how close they came to the goal measures.

Use Transparency 8

Practice Master

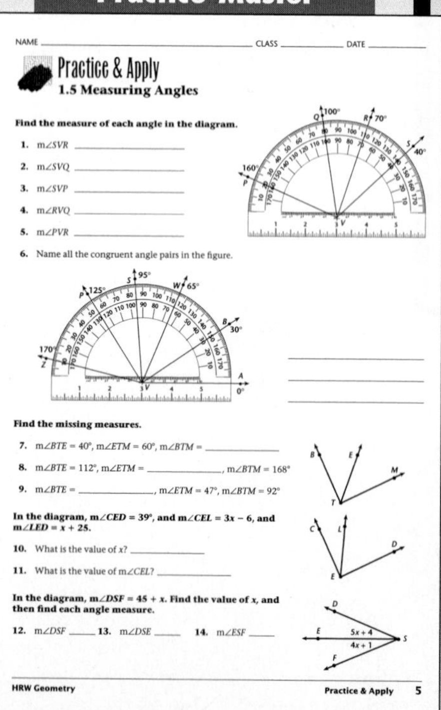
NAME ______ CLASS ______ DATE ______

Practice & Apply

1.5 Measuring Angles

Find the measure of each angle in the diagram.

1. m∠SVR ______
2. m∠SVQ ______
3. m∠SVP ______
4. m∠RVQ ______
5. m∠PVR ______
6. Name all the congruent angle pairs in the figure.

Find the missing measures.

7. m∠BTE = 40°, m∠ETM = 60°, m∠BTM = ______
8. m∠BTE = 112°, m∠ETM = ______, m∠BTM = 168°
9. m∠BTE = ______, m∠ETM = 47°, m∠BTM = 92°

In the diagram, m∠CED = 39°, and m∠CEL = 3x − 6, and m∠LED = x + 25.

10. What is the value of x? ______
11. What is the value of m∠CEL? ______

In the diagram, m∠DSF = 45 + x. Find the value of x, and then find each angle measure.

12. m∠DSF ____ 13. m∠DSE ____ 14. m∠ESF ____

HRW Geometry — Practice & Apply 5

Algebra In the diagram, m∠ARB = 65 + x. Find the value of x, and then find each indicated angle measure.

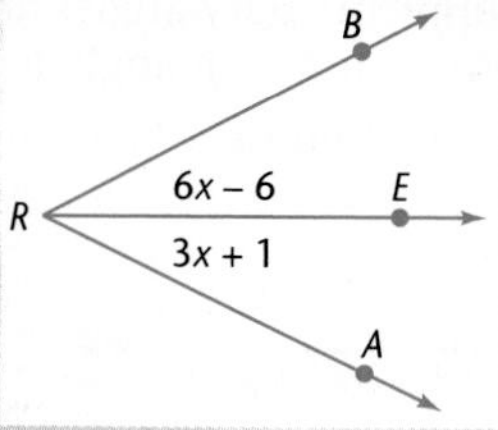

23. m∠ARB X = 8.75 73.75°

24. m∠BRE 46.5°

25. m∠ERA 27.25°

Navigation The heading of an airplane is defined in terms of a 360° circle, or "protractor." North is defined as 0°, and the numbering is done from the 0 point along a circle in the clockwise direction. Find the heading for each of the following compass directions.

26. N 0° **27.** E 90° **28.** S 180°

29. W 270° **30.** NE 45° **31.** NW 315°

32. NNE 22.5° **33.** ENE 67.5° **34.** SSW 202.5°

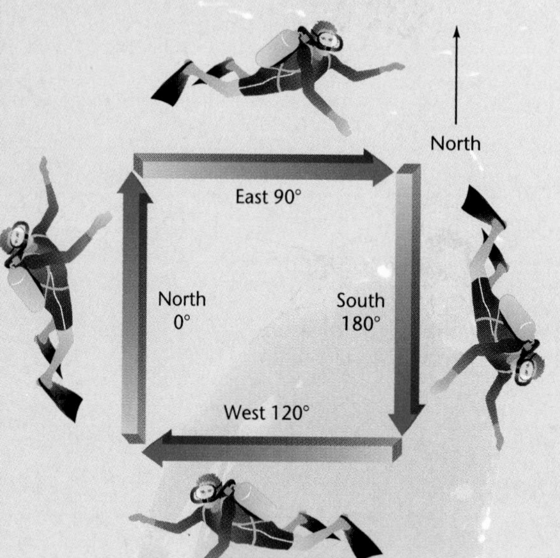

35. Scuba Diving In order to return to the same point, a diver may navigate a square pattern. If the diver plans to swim NE first, then SE, what compass headings must he or she use to navigate a square? Besides getting each angle correct, what else must a diver make sure of in order to move in a square?
45°, 135°, 225°, 315°
The diver must swim the same distance in each direction.

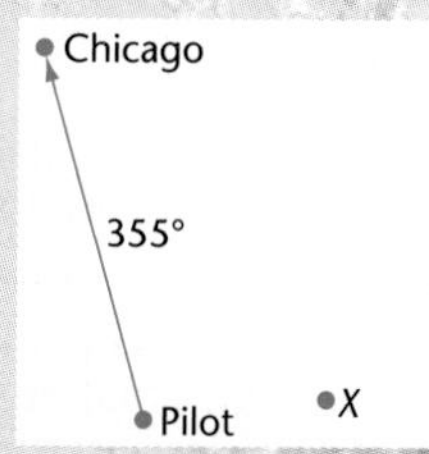

Navigation In Exercises 36 and 37, a pilot is flying to Chicago. In order to fly there in a straight line, he finds that he must keep his headng at 355°.

36. Using the diagram, would the pilot's heading be greater or less than 355° if he started further east at point x? How can you tell? Less than 355°. The pilot must fly more to the west.

37. On the way, the pilot encounters a thunderstorm along his planned route to Chicago. He finds he must adjust his heading to 15° to make the detour past the storm. Through how many degrees does the nose of his plane need to swing before the plane is on its new course? 20°

38. In creating a ruler, the part of the ruler between 0 and 1 can be made any desired size. Can the part of a 180° protractor between 0 and 1 be made of any desired size? Explain your answer.

39. Angles may be measured using units other than degrees. One way of measuring angles is in gradians. There are 100 gradians in 90 degrees. Is 1 gradian smaller or larger than 1 degree? Why?

Look Back

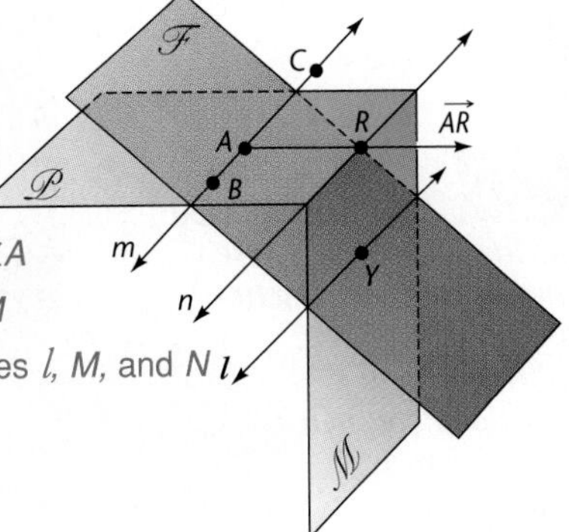

Answer questions 40–43 using the diagram shown. [Lesson 1.1]

40. Name an angle in two different ways. ∠*RAB* or ∠*BAR* or ∠*A*

41. Identify a line that is coplanar to the angle. Lines *N* and *M*

42. Name all the lines that are intersections of two planes. Lines *l*, *M*, and *N*

43. Identify three points that are collinear. Delete or ∠*A*

Find the measure of the segments below using two different methods. [Lesson 1.4]

44.

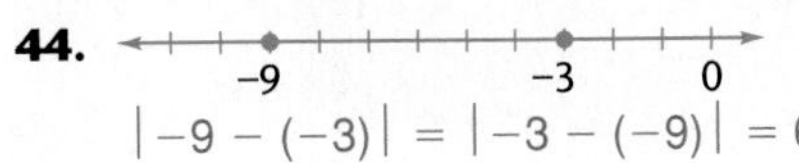

$|-9-(-3)| = |-3-(-9)| = 6$

45.

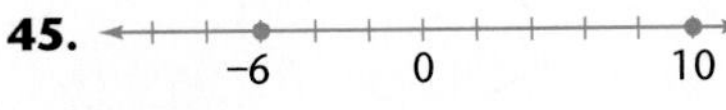

46. If $AB = 27$, find AC and BC.

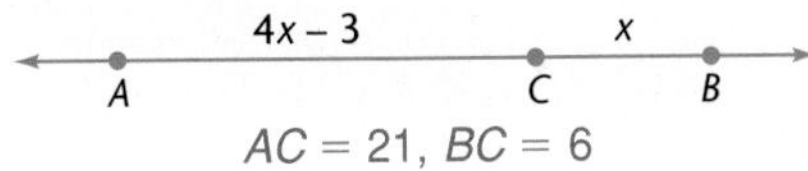

$AC = 21$, $BC = 6$

Look Beyond

47. Cultural Connection: Asia Our system of degrees comes from the Babylonians. The Babylonians considered a circle to be composed of a set of 360 congruent angles with each vertex at the center of a circle. Why do you think they chose 360? (Hint: How many numbers divide 360 evenly?)

48. Each degree measures a small angle; however, people who measure angles to navigate, to find angles between stars, or to build houses may need even more precise measurements. Often, instead of using decimals to divide the degree, they use minutes and seconds. There are 60 minutes in a degree and 60 seconds in a minute.

How many seconds are there in each degree? 3600

How many seconds are there in 1.5 minutes? 90

How many minutes are there in 1.75 degrees? 105

The ancient Babylonians were sophisticated mathematicians.

38. The protractor part between 0 and 1 can be changed by changing the size of the circle used to make the protractor.

39. One gradian is smaller, since one gradian is $\frac{9}{10}$ of a degree.

47. Probably, because the number 360 has a large number of divisors. Angles derived from these are easily measured.

Use Transparency 9

Look Beyond

Extension. Exercises 47 and 48 explore the system of marking degrees on a circle. The mathematical advantages of a 360° circle can be illustrated by having students divide circles into 2, 3, 4, 5, 6, 8, 9, 10, and 12 equal parts. The number of degrees in each sector is a whole number. Ask the students how many other numbers of equal divisions of a circle give whole numbers of degrees in each sector. (Thirteen more—15, 18, 20, 24, 30, 36, 40, 45, 60, 72, 90, 120, 180.)

If there were 100° in a circle, there would be fewer whole-number possibilities—just 2, 4, 5, and 10 for numbers less than 15.

Technology Master

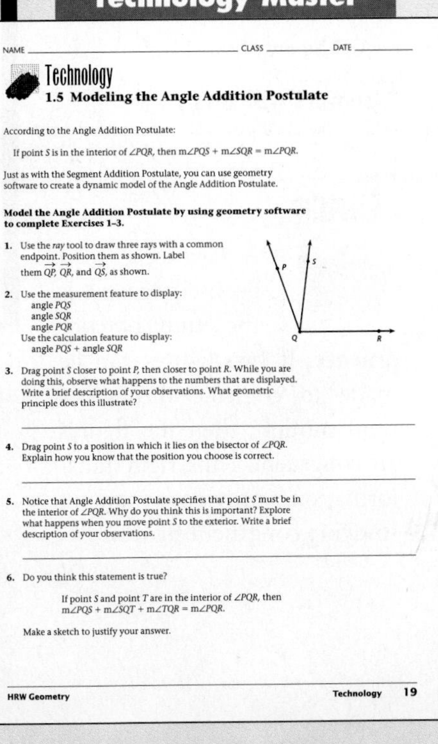

NAME ______ CLASS ______ DATE ______

Technology

1.5 Modeling the Angle Addition Postulate

According to the Angle Addition Postulate:

If point *S* is in the interior of ∠*PQR*, then m∠*PQS* + m∠*SQR* = m∠*PQR*.

Just as with the Segment Addition Postulate, you can use geometry software to create a dynamic model of the Angle Addition Postulate.

Model the Angle Addition Postulate by using geometry software to complete Exercises 1–3.

1. Use the *ray* tool to draw three rays with a common endpoint. Position them as shown. Label them $\overrightarrow{QP}$, $\overrightarrow{QR}$, and $\overrightarrow{QS}$, as shown.
2. Use the measurement feature to display:
   angle *PQS*
   angle *SQR*
   angle *PQR*
   Use the calculation feature to display:
   angle *PQS* + angle *SQR*
3. Drag point *S* closer to point *P*, then closer to point *R*. While you are doing this, observe what happens to the numbers that are displayed. Write a brief description of your observations. What geometric principle does this illustrate?
4. Drag point *S* to a position in which it lies on the bisector of ∠*PQR*. Explain how you know that the position you choose is correct.
5. Notice that Angle Addition Postulate specifies that point *S* must be in the interior of ∠*PQR*. Why do you think this is important? Explore what happens when you move point *S* to the exterior. Write a brief description of your observations.
6. Do you think this statement is true?

   If point *S* and point *T* are in the interior of ∠*PQR*, then m∠*PQS* + m∠*SQT* + m∠*TQR* = m∠*PQR*.

   Make a sketch to justify your answer.

HRW Geometry — Technology 19

PREPARE

Objectives

- Identify the three basic rigid transformations: translation, rotation, and reflection.
- Construct reflections of points, segments, and triangles with a ruler and protractor.

RESOURCES

- Practice Master 1.6
- Enrichment Master 1.6
- Technology Master 1.6
- Lesson Activity Master 1.6
- Quiz 1.6
- Spanish Resources 1.6

Assessing Prior Knowledge

Have each student cut out a cardboard template of an irregular figure. Create a pair of drawings to show each of these motions.

1. a translation or "slide"
2. a reflection across a "mirror" line
3. a rotation about one vertex of the cardboard figure

[**Answers will vary.**]

TEACH

A key topic in a geometry course is the study of congruence. If two figures can be made to coincide through a rigid motion, then the figures are congruent. Thus, rigid transformations provide a way of studying congruent figures.

LESSON 1.6 Motions in Geometry

Why *Every motion can be modeled mathematically. These three photos represent the three basic mathematical "motions," which are known as transformations.*

In mathematics, three basic **transformations** are used to model motion. They are **translations, rotations,** and **reflections**. Any real-world motion can be analyzed in terms of one or more of these transformations.

Translations

In geometry, where we deal with mathematical images, we speak of the **image** and **pre-image** of an object that undergoes a motion or transformation. Compare the two positions or "snapshots" of the same skier in motion.

The pre-image of the skier has been **translated**. Notice that the pre-image and the image are exactly the same size and shape—that is, they are congruent. In geometry, a motion or transformation that does not change the size and shape of a figure is known as **rigid**.

Technology A computer graphics or drawing program is an excellent tool for exploring rigid motions. Have students use the reflection and rotation tools to create transformed copies of freely drawn figures. Results can be printed and used for discussion during the lesson.

EXAMPLE 1

Describe how the points have been translated in the figure. Identify the pre-image and the image. How far has point A moved? How far have you moved points B and C? What do you notice about the *direction* of the motion of points?

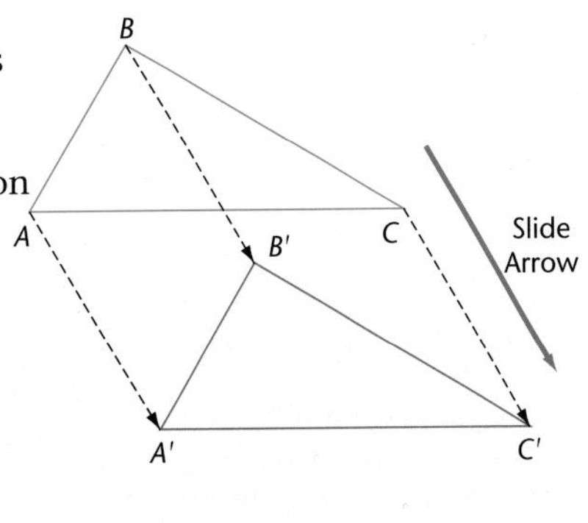

Solution ➤

The pre-image is $\triangle ABC$ and the image is $\triangle A'B'C'$. Point A has moved 2 centimeters in the direction of the arrow. In fact, every point in $\triangle A'B'C'$ has moved 2 centimeters in the same direction. ❖

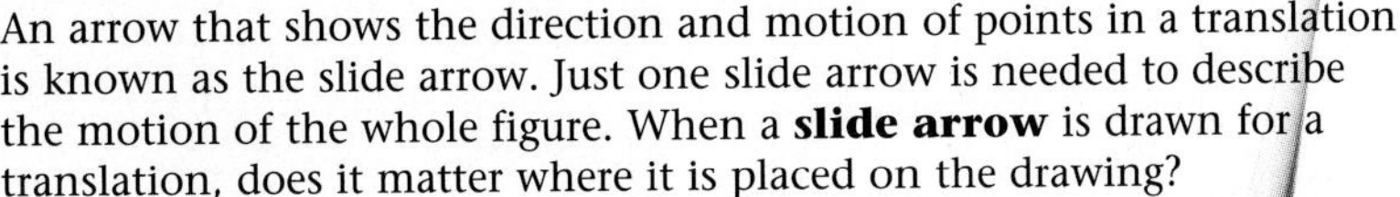

CRITICAL *Thinking* An arrow that shows the direction and motion of points in a translation is known as the slide arrow. Just one slide arrow is needed to describe the motion of the whole figure. When a **slide arrow** is drawn for a translation, does it matter where it is placed on the drawing?

Rotations

A **rotation** is a rigid motion that moves a geometric figure about a point known as the **turn center**. Probably the most obvious example of a rotation is a wheel turning in place. In this case the turn center is at the center of the rotating object. But this is not true of all rotations.

EXAMPLE 2

Describe the rotation in the figure. Identify the pre-image, the image, and the rotation center. Where is the rotation center of the figure? Describe the motion of the different points of the figure.

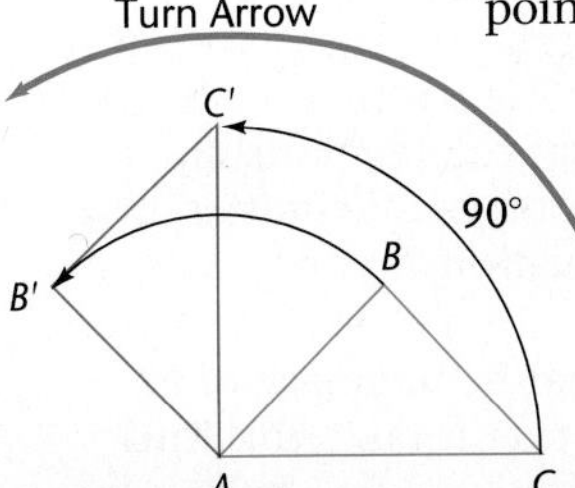

Solution ➤

The pre-image is $\triangle ABC$ and the image is $\triangle AB'C'$. The rotation center is the point A. Every point in $\triangle ABC$ has been moved 90° in a counterclockwise direction about point A. ❖

interdisciplinary CONNECTION **Physics** The study of what happens to images reflected by mirrors or other surfaces can provide interesting experiments for students. Have them investigate what happens to an image when it is reflected in a convex or concave surface.

Use Transparency 10

Alternate Example 1

Identify the pre-image and the image of the translated figure. How far have points X, Y, and Z been moved? Describe the three dotted line segments that show the movement of points X, Y, and Z.

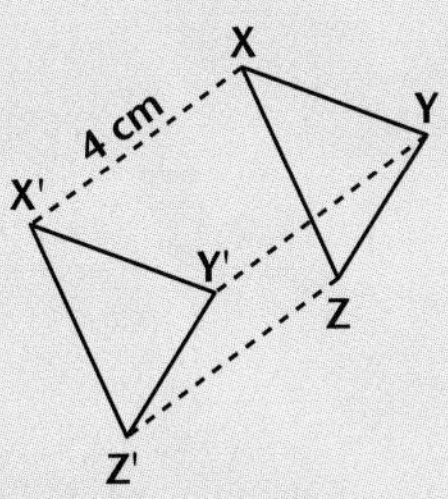

[pre-image: ΔXYZ; image: $\Delta X'Y'Z'$; Each point is moved 4 cm; The dotted movement lines are parallel]

CRITICAL *Thinking*

No. Every point in the drawing is moving by the same amount.

Alternate Example 2

Describe the rotation in the figure. Identify the pre-image, the image, and the rotation center. Describe the motion of the points in the figure.

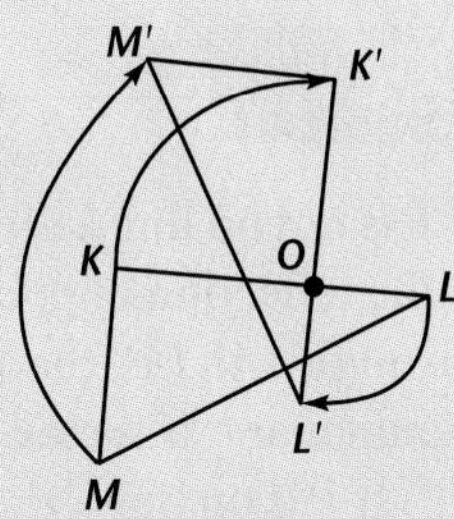

[pre-image: ΔKLM; image: $\Delta K'L'M'$; rotation center: point O; Each point is moved 90° clockwise about point O]

Cooperative Learning

Have students work in pairs. One student mimics the actions of the other. Have students discuss whether or not they are performing identical motions.

Exploration 1 Notes

In this exploration students look for some of the key ideas of a geometric reflection. Students may have intuitive notions about reflections. The exploration helps formalize those notions using geometric language and symbols.

ongoing ASSESSMENT

4. If P is not on line l, then P' is the reflection of point P through l if l is the perpendicular bisector of PP'. If P is on line l, then P is its own reflection through l.

CRITICAL Thinking

If P is on l, P is its own image.

Reflections

Hold a pencil in front of a mirror and focus on the pencil point P. The reflection P', called the image of the point, will appear to be on the opposite side of the mirror at the same distance from the mirror as P.

In a reflection, a line plays the role of the mirror, and geometric figures are "flipped" over the line. In the explorations that follow, paper folding is used to produce the appropriate motions.

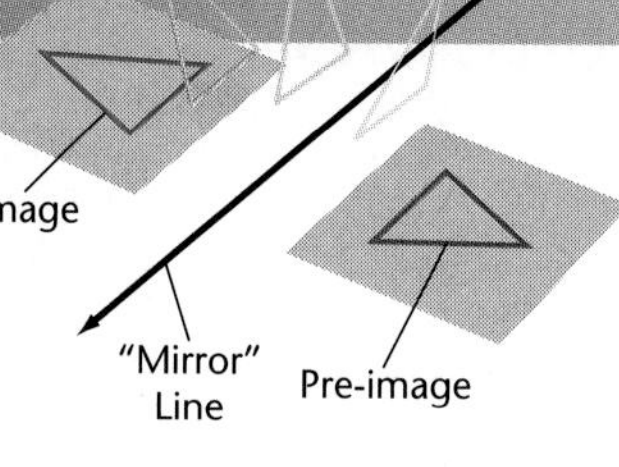

Exploration 1 The Reflection of a Point

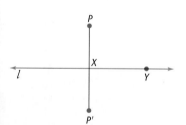

You will need
Pointed object
Ruler
Protractor

1. Draw line l and a point P not on line l.
2. Reflect point P through l as follows. Fold your paper along l. Use a pointed object such as a sharp pencil or a compass point to punch a hole through the paper at point P. Label the new point P'.
3. Draw $\overline{PP'}$ and label the intersection of $\overline{PP'}$ and l as point X. Add a new point Y on l.
   a. Measure the length of $\overline{XP}$ and $\overline{XP'}$.
   b. Measure $\angle PXY$ and $\angle P'XY$.
4. Formulate a conjecture about the relationship between l and $\overline{PP'}$. Write your own definition of a reflection in terms of the relationship. ❖

CRITICAL Thinking

What happens if you place point P on line l? What is the image of P? Make sure your conjecture in Step 4 is written so that it takes this into consideration.

ENRICHMENT A single-plane mirror can be used to demonstrate a reflection. Interesting effects are produced by using two mirrors hinged together with tape. By varying the angle between the mirrors, different numbers of reflected objects are produced. Challenge students to explore the relationship between the hinge angle and the number of images.

INCLUSION strategies

English Language Development The terms in this lesson may be difficult for some students. Illustrate the ideas of *pre-image, image,* and *reflection* using a mirror. *Translation* and *rotation* can each be acted out by having a student physically perform the movement.

Exploration 2 *Reflections of Segments and Triangles*

You will need
Pointed object
Ruler
Protractor

1 Draw line l and segment $\overline{AB}$ that does not intersect l. Fold your paper along l and punch holes through A and B to obtain image points A' and B'. Connect the image points to form segment $\overline{A'B'}$.

2 Measure $\overline{AB}$ and $\overline{A'B'}$.

Formulate a conjecture about a segment and its reflection through a line.

3 Draw line l with $\triangle ABC$ on one side.

Fold your paper along l and create the reflection $\triangle A'B'C'$ of $\triangle ABC$.

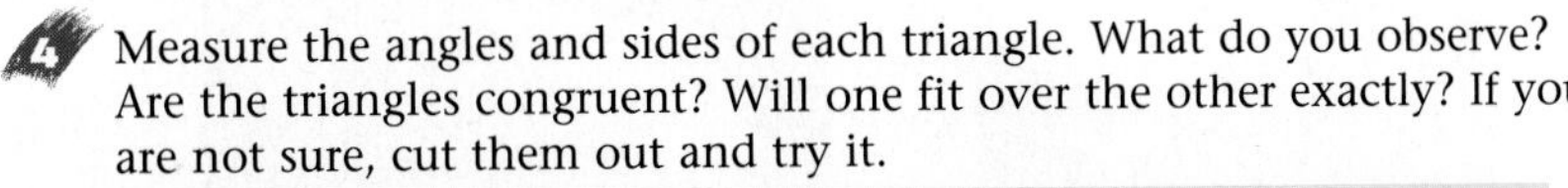

4 Measure the angles and sides of each triangle. What do you observe? Are the triangles congruent? Will one fit over the other exactly? If you are not sure, cut them out and try it.

Formulate a conjecture about triangles and their reflection through a line. ❖

Try This Reflect each of the figures through the reflection line l. Does your conjecture about reflections of triangles and segments seem to hold when the figures touch the reflection line?

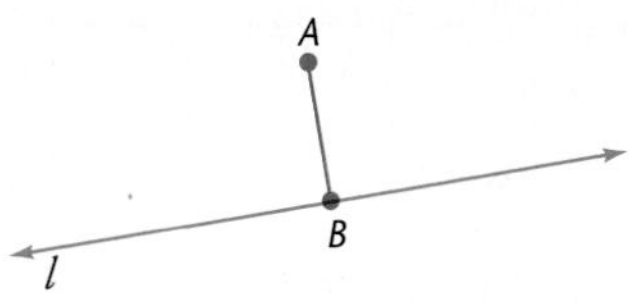

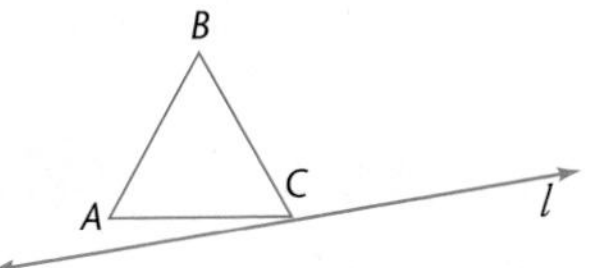

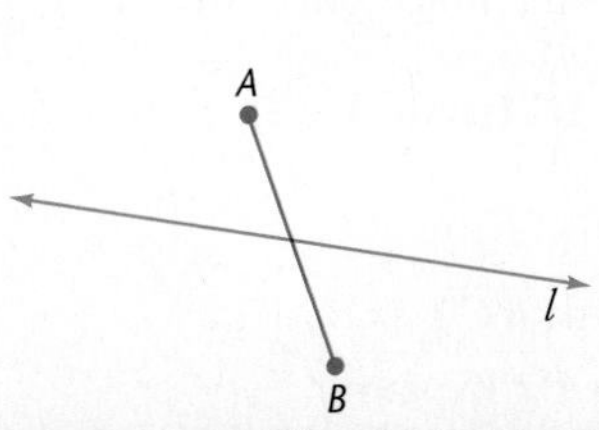

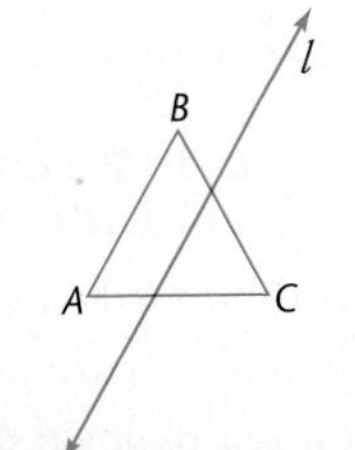

Exploration 2 Notes

Students investigate what happens to the points on a triangle under a reflection.

Ongoing Assessment

2. **A segment and its image under a reflection are congruent.**

4. **A triangle and its image under a reflection are congruent.**

Teaching tip

Point out that Exploration 2 deals with just one case. In both drawings, the figure being reflected is completely on one side of the reflection line. The Try This problems involve two other cases: one point of the pre-image is on the reflection line, and the pre-image crosses over the reflection line. Ask students to predict whether their conjecture from Step 4 in Exploration 2 will hold true for the other cases.

Ongoing Assessment

Try This Yes

Reteaching the lesson

Cooperative Learning
Some students may find the ideas involved in rotations difficult. Provide each student with a cardboard square, equilateral triangle, and regular hexagon. Have them create designs by rotating the figures about center points and vertex points. Students can work in groups and choose the number of degrees used in the various rotations.

Assess

Selected Answers

Odd-numbered Problems 9–33

The answers to Communicate Exercises can be found in Additional Answers beginning on page 727.

Assignment Guide

Core 1–6, 8–20, 25–36

Core Plus 5–7, 10–24, 35–36

Technology

Geometry graphics software will be helpful for the rotation and reflection activities in Exercises 10–12 and 16–17.

Error Analysis

Before students do the Look Back exercises, you may wish to have them write definitions for the terms *collinear, noncollinear, coplanar,* and *noncoplanar.*

Practice Master

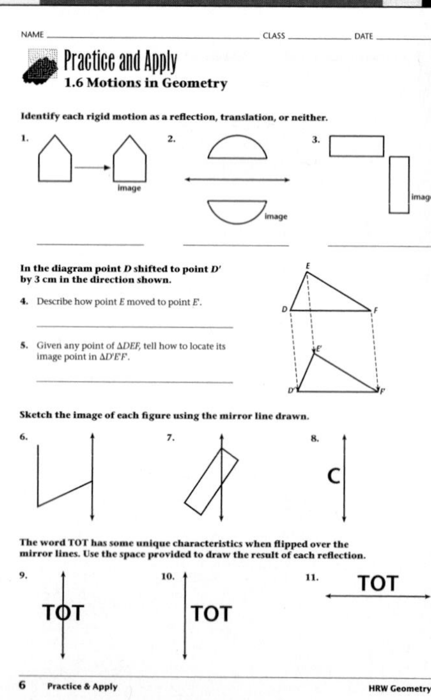

NAME ______ CLASS ______ DATE ______

Practice and Apply

1.6 Motions in Geometry

Identify each rigid motion as a reflection, translation, or neither.

1. image 2. image 3. image

In the diagram point D shifted to point D' by 3 cm in the direction shown.

4. Describe how point E moved to point E'.

5. Given any point of ΔDEF, tell how to locate its image point in $\Delta D'E'F'$.

Sketch the image of each figure using the mirror line drawn.

6. 7. 8. C

The word TOT has some unique characteristics when flipped over the mirror lines. Use the space provided to draw the result of each reflection.

9. TOT 10. TOT 11. TOT

6 Practice & Apply HRW Geometry

Exercises & Problems

Communicate

In Exercises 1–4, determine whether each of the following is a translation or a rotation. Explain why.

1. A canoe floating with the current
2. A canoe that capsizes in still water
3. The movement of the Earth's center about the sun in 6 months
4. Migrating birds flying in formation overhead
5. Explain how the pre-image and the image of a wheel rolling straight ahead is a combination of two different transformations.
6. Of the two figures below, which one best illustrates a reflection of a figure through a line?

7. When you project the image of a slide onto a screen, the pre-image (on the slide) is expanded to fill the screen for viewing. Explain how this transformation is different from those studied in this lesson.

Practice & Apply

In the diagram in Example 1 on page 45, point A shifted to A' by 2 cm in the direction shown.

8. Describe how point B moved to point B'.
9. Given any point of $\triangle ABC$, tell how to locate its image point in $\triangle A'B'C'$.

In the diagram in Example 2 on page 45, $\overline{AC}$ turned 90 degrees to coincide with $\overline{AC'}$.

10. $\overline{AB}$ turned how many degrees to coincide with $\overline{AB'}$? 90°
11. $\overline{BC}$ turned how many degrees to coincide with $\overline{B'C'}$? 90°
12. Which point traveled farther, B to B' or C to C'? Explain.

8. It shifted 2 cm in the same direction as A.
9. Shift it by 2 cm in the same direction that A moved.
12. The distance from C to C' is larger because C is farther away from A.

Refer to items 3 and 4 in Exploration 2 on page 47. Sketch the diagram.

13. Draw $\overline{BB'}$. What line appears to be equidistant from B and B′? Line l

14. Repeat Exercise 13 for $\overline{CC'}$ and $\overline{AA'}$. Line l is equidistant from both C and C' and D and D'.

15. Given any point in $\triangle ABC$, use the results of Exercises 13 and 14 to describe the location of its image point in $\triangle A'B'C'$. Each image point is the same distance from the reflection line, but on the opposite side of the line.

Copy the pictures shown and reflect them through the mirror line drawn. What is the result?

16.

The word *MOM* has some unique characteristics when flipped over the mirror lines. Show the result on your own paper.

17.

A. B.

While visiting the beach, Theresa saw footprints like those shown.

18. Which pair of footprints is a reflection?

19. Which pair of footprints is a translation?

20. The bird footprints illustrate a fourth type of transformation known as a **glide**. A glide is a combination of a reflection and a translation. Using your own illustrations, explain how a glide is a combination of these two transformations.
Check students' drawings.

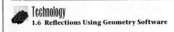

The shape below, a *net*, can be drawn, cut, and folded on the dotted lines to construct a cube. Use this net to help you answer Exercises 21–23.

21. How can you draw this net using one square and performing translations.

22. How can you draw this net using one square and performing reflections.

23. How can you draw this net using one square and performing rotations.

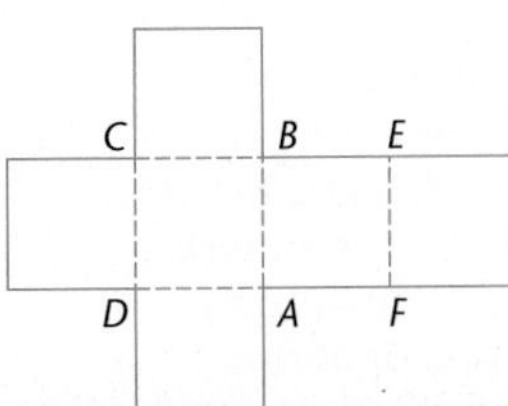

18. A

19. B

alternative ASSESSMENT

Authentic Assessment

Have students use scissors and cardboard to make a template for a simple figure such as a flower or star. Their task is then to design a wallpaper pattern using the figure. The design must include translations, rotations, and reflections. Students also write descriptions of how the transformations are used in their design.

The answers to Exercises 16, 17, 21, 22, and 23 can be found in Additional; Answers beginning on page 727.

Technology Master

NAME ______________ CLASS ________ DATE ________

Technology

1.6 Reflections Using Geometry Software

Most geometry software has a built-in reflection feature. If you select an object and identify a line of reflection, the software will automatically reflect the object across the line. Although you should develop your own good sense of the process of reflection, using this feature of the geometry software may make it easier for you to make and test conjectures about reflections.

Explore the reflection of a triangle over a line by using geometry software to complete Exercises 1–2.

1. Use the *segment* or *triangle* tool to draw a triangle. Label the vertices A, B, and C, as shown. Use the *line* tool to draw a line and label it t.

2. Use the *reflection* feature to identify t as the line of reflection. Reflect the triangle over it. Label the vertices of the new triangle A', B', and C', as shown.

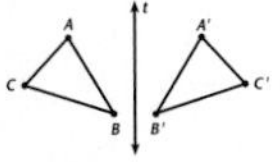

Use the figure from Exercises 1 and 2 to explore each conjecture. Do you think it is *always*, *sometimes*, or *never* true?

3. If the vertices of a triangle are points A, B, and C in clockwise order, then the vertices of its image are points A', B', and C', in clockwise order.

____

4. Under a reflection, a triangle and its image have no points in common.

____

5. Suppose that there is a "bug" in your software, and the reflection feature is not working. However, all the other tools and features are working properly. Devise a method that you can use to reflect a triangle over a line. Write a brief description of your method.

____

6. Just for Fun Using geometry software, create a sketch like the one shown. Identify the vertical line as a line of reflection. Then reflect all the segments over the line. Explore ways to change the expression on the face.

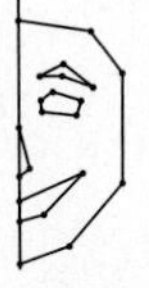

20 Technology HRW Geometry

24. What transformation describes the relationship between the picture carved into the wood and the picture imprinted on the paper? Reflection

Wood block

Print

Look Back

In Exercises 25–26, complete the sentences. [Lesson 1.1]

25. Points that lie on the same line are said to be ___?___. Collinear

26. Points or lines that lie on the same plane are said to be ___?___. Coplanar

In Exercises 27–29, explain your answer using illustrations when helpful. [Lesson 1.1]

27. Is it possible for two points to be collinear? Why or why not?

28. Is it possible for three points to be noncoplanar? Why or why not?

29. Is it possible for two lines to be noncoplanar? Why or why not?

30. According to a conjecture you made in an earlier lesson, the diagonals of a square are ___?___ and ___?___ to each other. **[Lesson 1.2]** Congruent Perpendicular

Algebra For Exercises 31–34, find the indicated measures. WN = 48, ∠CAR ≅ ∠EAR.

31. IN **[Lesson 1.4]** 23

32. WI **[Lesson 1.4]** 25

33. $m\angle CAR$ **[Lesson 1.5]** 13°

34. $m\angle EAR$ **[Lesson 1.5]** 13°

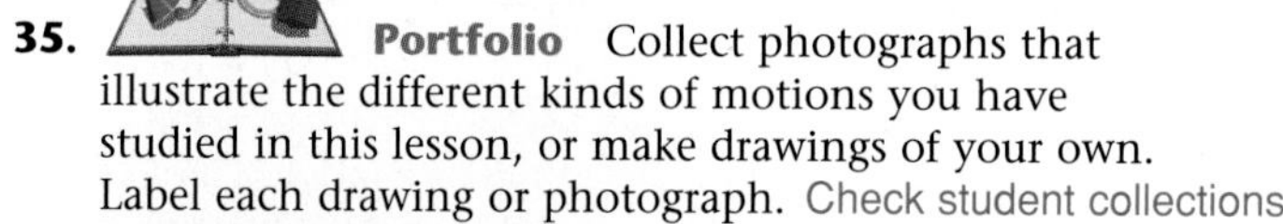

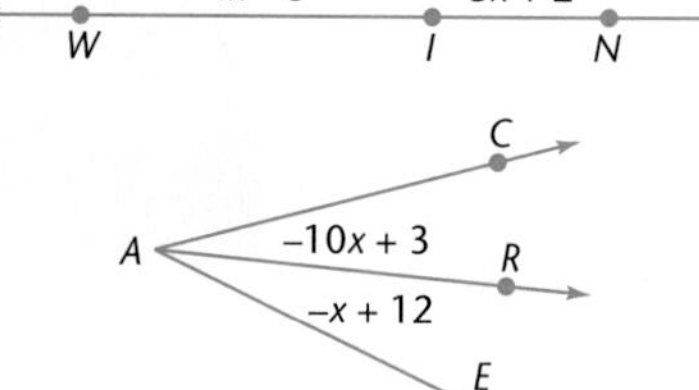

Look Beyond

35. **Portfolio** Collect photographs that illustrate the different kinds of motions you have studied in this lesson, or make drawings of your own. Label each drawing or photograph. Check student collections

36. **Visual Arts** Trace the shape shown onto your paper. To the right side of your picture, draw two parallel lines about an inch and a half apart. Reflect the picture over the closest parallel line; then reflect the result over the other parallel line. How might you describe the final picture in terms of another transformation discussed in this lesson? Translation

Look Beyond

Extension. Books of wallpaper patterns and books of tile designs are good sources that students can use for their collections of transformations for Exercises 35 and 36.

27. Yes. Through any two points there is exactly one line.

28. No. Through any three noncollinear points there is exactly one plane. If the points are collinear, they lie in an infinite number of planes.

29. Yes. The line containing the intersection of the front wall and the ceiling of a classroom, and the line containing the intersection of a side wall and the floor of the classroom are noncoplanar lines.

LESSON 1.7 *Exploring*

Motion in the Coordinate Plane

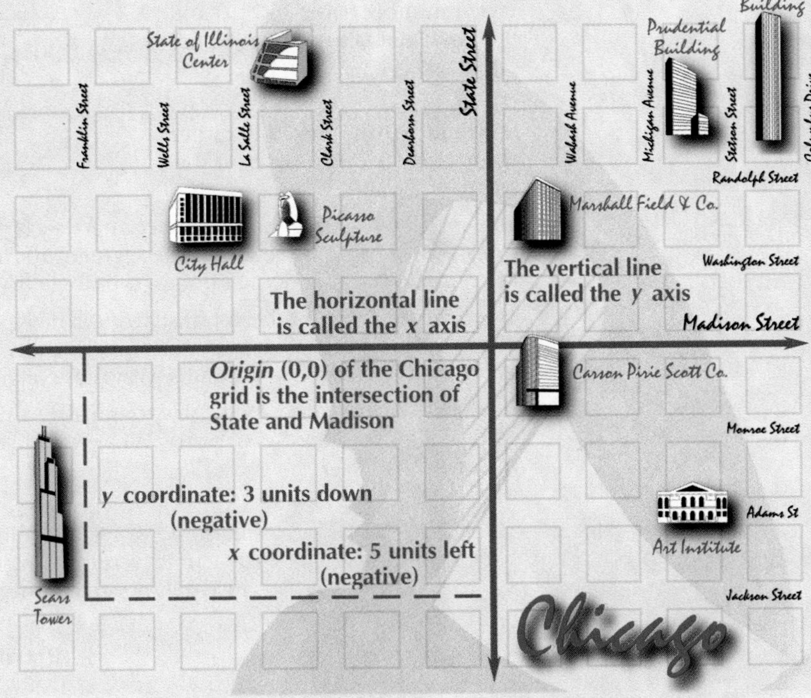

The city of Chicago is laid out like a coordinate system. The Sears Tower, once the tallest building in the world, is located at (–5, –3).

WHY *A coordinate plane enables you to describe geometry ideas using algebra. For example, you can use algebra to find the transformation of a geometric figure without using special construction tools.*

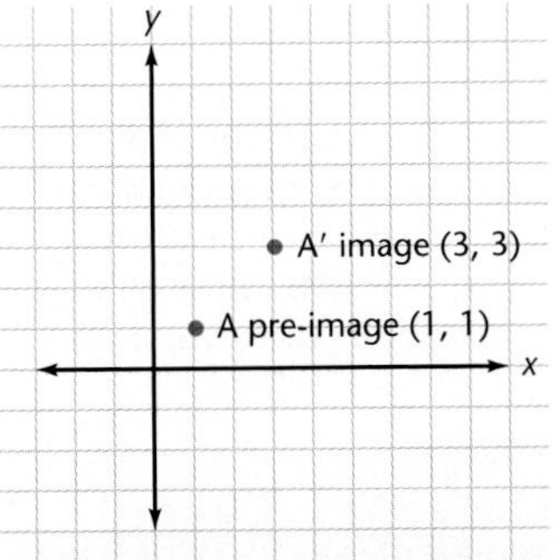

Algebra By doing algebraic operations on the coordinates of a point, you can relocate it on the coordinate plane. For example, if you multiply the x- and y-coordinates of the point (1, 1) by 3, the result is the new point (3, 3):

Pre-image	$\rightarrow$	**Image**
(1, 1)	$\rightarrow$	(3, 3)

This operation, which is known as a transformation, can be expressed as a rule using the following notation:

$$(x, y) \rightarrow (3x, 3y)$$

Technology The activities and exercises in this lesson can be done with a computer program such as *f(g) Scholar*™ or with graphics calculators. Set the range for x and y at -10 to 10.

PREPARE

Objectives

- Define *coordinate plane, origin, x- and y-coordinates, ordered pairs, image,* and *pre-image.*
- Construct a translation, a reflection about an axis, and a rotation about the origin on the coordinate plane.

RESOURCES

• Practice Master	1.7
• Enrichment Master	1.7
• Technology Master	1.7
• Lesson Activity Master	1.7
• Quiz	1.7
• Spanish Resources	1.7

Assessing Prior Knowledge

Provide students with blank coordinate grids with each axis numbered from –10 to 10.

1. Find and draw the set of all ordered pairs in which the first number is –3.

   [a vertical line crossing the *x*-axis at –3]

2. Find three different squares with (2, 3) as one vertex.

   [Answers will vary.]

TEACH

In this lesson, a most important concept from algebra is brought to bear on geometry. Transformations are treated as **functions;** that is, as rules that apply to a set of points on a plane to produce a unique new point for every point in the original set.

Cooperative Learning

Students working in small groups on the three explorations will be able to create cases that cover all four quadrants on the coordinate plane. Be sure that each group includes cases in which the pre-image triangle crosses one or both of the axes.

Exploration 1 Notes

Students look for a way to manipulate the coordinates of ordered pairs so that the result is a translation.

ongoing ASSESSMENT

5. **To translate a figure horizontally, add the same number to the x-coordinate of each point in the figure. To translate a figure vertically, add the same number to the y-coordinate of each point in the figure.**

6. **Horizontal movement: $(x, y) \rightarrow (x + h, y)$; vertical movement: $(x, y) \rightarrow (x, y + v)$**

TEACHING tip

The abstract thinking involved in Exploration 1 may be difficult for many students. Each point on the pre-image triangle can be represented by an ordered pair of the form (a, b). Then ordered pairs such as $(a + 1, b)$ move the triangle one unit to the right. The variables a and b may be easier for some student than x and y.

Exploration 1 *Translation*

You will need
Graph paper
Ruler
Protractor

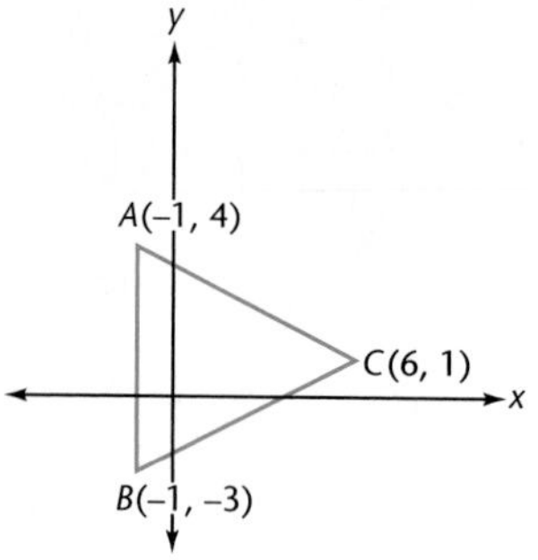

y
A(−1, 4)
C(6, 1)
x
B(−1, −3)

1. Draw triangle $\triangle ABC$ on graph paper.

2. Pick a number between −10 and +10 (but not zero). Then choose either x or y.

3. Depending on your choice, add the number you picked to either the x-or the y-coordinates of the points A, B, and C.

   Plot each of your new points and connect each pair with a segment. Does your new figure seem to be congruent to the original one? Measure the sides and angles of each triangle if you are not convinced.

4. The original figure has been translated to a new position. In what direction has it moved, and by how much? Draw a slide arrow for the translation. Compare your results with those of your classmates.

5. Formulate a conjecture about how to translate a figure horizontally or vertically on the coordinate plane. What would you do to each point in the figure?

6. Express your conjecture in the form of rules for moving a given point h units horizontally or v units vertically. Assume that h and v can be either positive or negative.

	Pre-image		Image
Horizontal movement	(x, y)	$\rightarrow$	(? , ?)
Vertical movement	(x, y)	$\rightarrow$	(? , ?) ❖

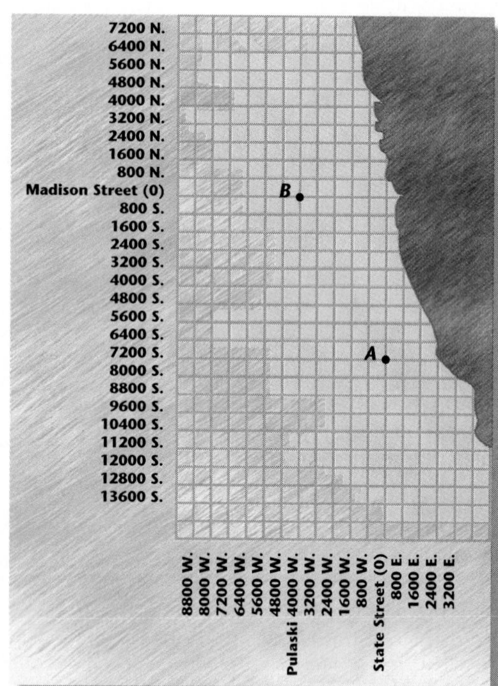

APPLICATION

On the map, each square is 8 blocks long on each side. Suppose you are driving from A at State and 72nd Street to B at Madison and Pulaski. The rule for the horizontal movement is $(x, y) \rightarrow (x - 40, y)$. The rule for the vertical movement is $(x, y) \rightarrow (x, y + 72)$. These rules can be combined to show both horizontal and vertical movement.

$(x, y) \rightarrow (x - 40, y + 72)$

What does the rule represent? ❖

Art Rigid transformations are used frequently in art and design, particularly for wallpaper, tile patterns, and fabrics. Have students collect examples and critique some of the patterns. When is repetition boring? When is variety confusing? Encourage students to use the language of transformations in their written work.

Exploration 2 *Reflection Over the x- or y-Axis*

You will need
Graph paper
Ruler
Protractor

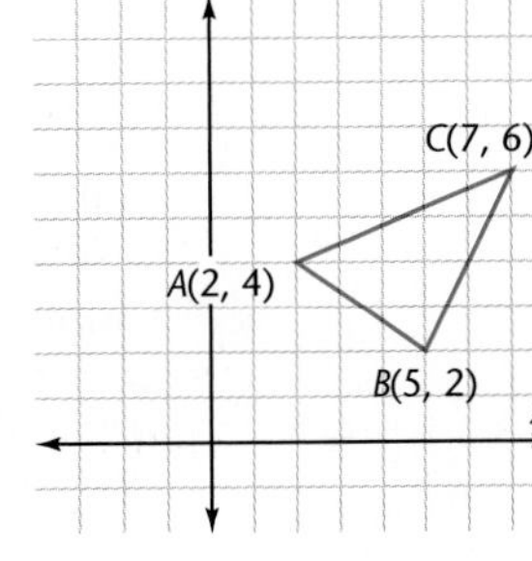

1. Draw the triangle on graph paper.
2. Multiply the x coordinate of each vertex by -1.
3. Plot each of the new points and draw a segment between each pair. Does the new triangle seem to be congruent to the original one? Measure each triangle.
4. Repeat Steps 2 and 3 using the y-coordinate instead.
5. The original triangle has been reflected in each case. What line is the axis of reflection, or "mirror", in each?
6. Formulate a conjecture about how to reflect a figure through the x- or y-axis. What would you do to each point in the figure?
7. Express your conjecture as a rule for reflecting a given point through the x- or y-axis.

	Pre-image		**Image**
Reflection through x-axis	(x, y)	$\rightarrow$	(? , ?)
Reflection through y-axis	(x, y)	$\rightarrow$	(? , ?) ❖

CRITICAL *Thinking*

Experiment with figures in other positions, such as the ones below. Do your conjectures seem to be true for all of them?

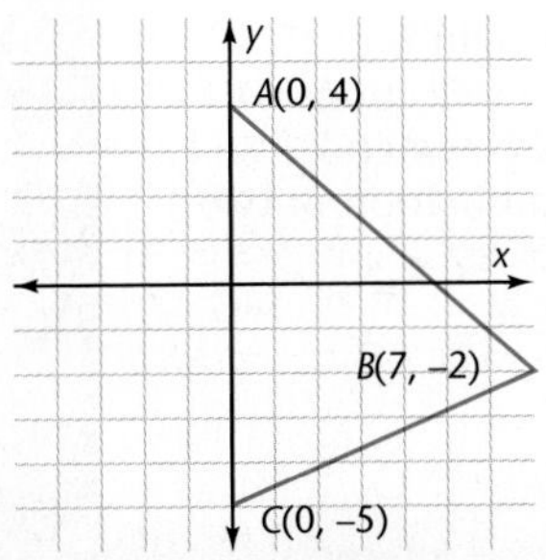

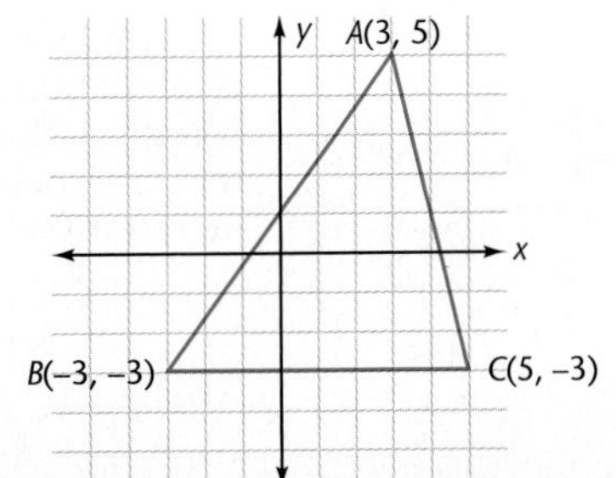

Use Transparency 11

Exploration 2 Notes

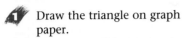

Have students do the activity with a triangle drawn in each of the four quadrants. If students are working in groups of four, each student can do one of the quadrants.

ongoing ASSESSMENT

6. **To reflect a figure over the x-axis, multiply the y-coordinate of each point in the figure by -1. To reflect a figure over the y-axis, multiply the x-coordinate of each point in the figure by -1.**

7. **Reflection over x-axis: $(x, y) \rightarrow (x, -y)$; reflection over y-axis: $(x, y) \rightarrow (-x, y)$**

CRITICAL *Thinking*

Yes. The students' methods should work for figures positioned anywhere on the coordinate plane.

TEACHING *tip*

You may also wish to have students explore the case in which just one vertex of the pre-image triangle lies on the line of reflection.

ENRICHMENT If a line can be drawn through a geometric figure so that the two halves are reflections of each other, then the line of reflection is also a *line of symmetry.* Challenge students to find figures that have more than one line of symmetry. (Examples include squares, equilateral triangles, and all the regular polygons.)

INCLUSION strategies **Using Cognitive Strategies** The abstract reasoning involved in associating ordered pairs of the form (x, y) with locations on a coordinate plane takes longer for some students to learn than others. For example, a common mistake is to graph (0, 3) instead of (3, 0.) Provide extra practice in plotting points and ask that students double-check all plotted points.

Exploration 3 Notes

In this exploration students discover a way to manipulate the numbers in ordered pairs to rotate a figure 180° about the origin. This is equivalent to reflecting a figure twice, once over each axis.

ongoing ASSESSMENT

180° rotation about the origin; $(x, y) \rightarrow (-x, -y)$

TEACHING tip

Technology Students will find geometry graphics software useful for exploring a wide variety of cases before making their conjectures.

CRITICAL Thinking

Yes, a rotation of 180° about the origin is the equivalent of a reflection over the x-axis followed by a reflection over the y-axis, or vice versa.

Exploration 3 *180° Rotation About the Origin*

You will need
Graph paper
Ruler
Protractor

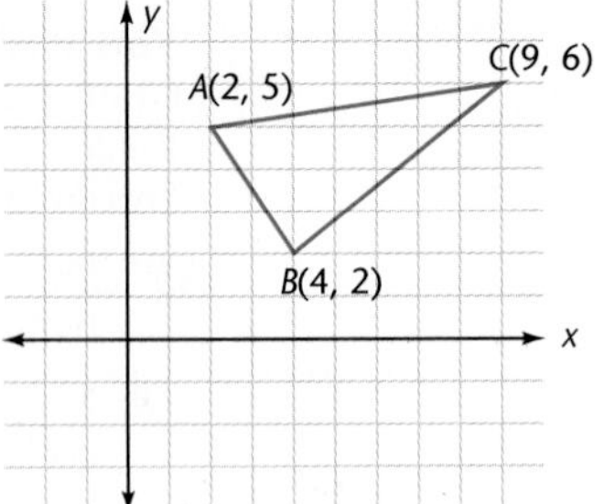

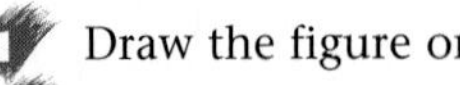

1. Draw the figure on graph paper.

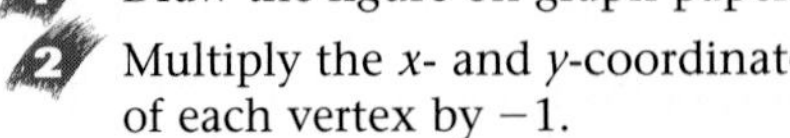

2. Multiply the x- and y-coordinate of each vertex by -1.
3. Plot each of your new ordered pairs as points and draw a segment between each one. Does the new figure seem to be congruent to the original one? Measure each triangle.
4. The original figure has been rotated. What is the turn center of the rotation? How much has the figure been rotated? Draw a turn arrow for the rotation.
5. Formulate a conjecture for an 180° rotation about the origin. Express your conjecture for rotating a given point in the original figure as a rule.

	Pre-image		**Image**
180° rotation about the origin	(x, y)	$\rightarrow$	(? , ?) ❖

CRITICAL Thinking

In the figure on the right, the upper left panel has been reflected first through the y-axis and then through the x-axis. Describe how a 180° rotation about the origin is a combination of these two reflections. Do you think this idea could be extended so that any rotation is a combination of two reflections?

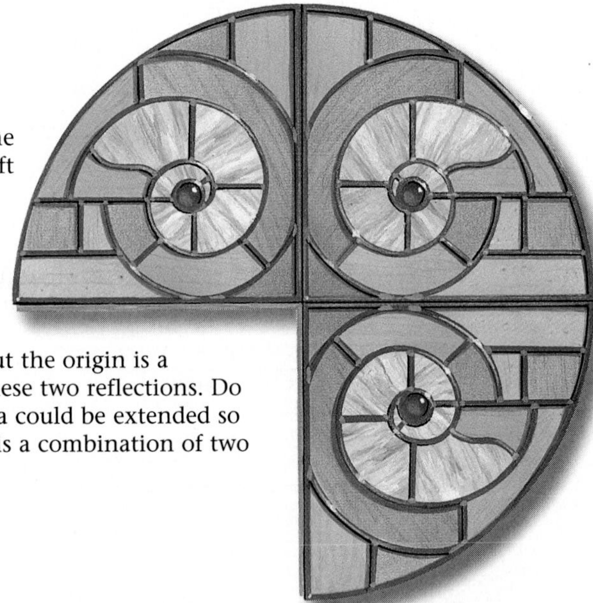

RETEACHING the lesson

Cooperative Learning Many students enjoy plotting points when the result is a picture of some recognizable figure. Have students work in pairs. Each one plots a capital letter or other simple figure. On a separate sheet of paper, they should record the sequence of ordered pairs needed to get the letter. Students should then trade instruction sheets and plot the points to discover the letter or figure their partner drew.

EXERCISES & PROBLEMS

Communicate

1. Explain how you would find the x- and y- coordinates of the given points on the coordinate plane.
2. Explain how to plot a point on a coordinate plane from its x- and y-coordinates.
3. Compare the graphs of the points (1, 8) and (8, 1). Would you say that the order of the points mattered? Why do you think the coordinates of a point are known as *ordered pairs?* Discuss.

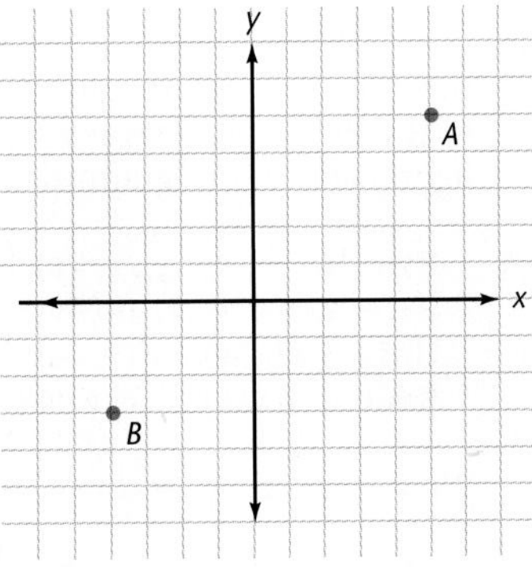

Practice & Apply

In Exercises 4–7, use the street map of Chicago to translate the shuttle bus as indicated. Write the rule you used in each case.

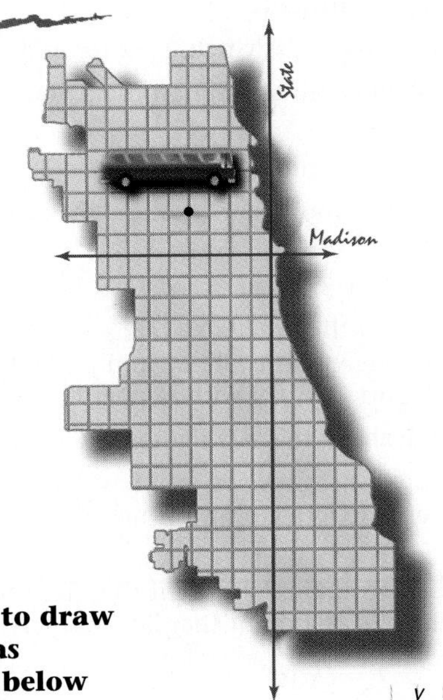

4. Translate the bus up 16 blocks.
5. Translate the bus down 32 blocks.
6. Translate the bus to the right 24 blocks.
7. Translate the bus to the left 8 blocks.
   $(x, y) \rightarrow (x - 8, y)$

In Exercises 8–12, use graph paper to draw the transformations of the figure as indicated. Write the rule you used below the transformed figure in each case.

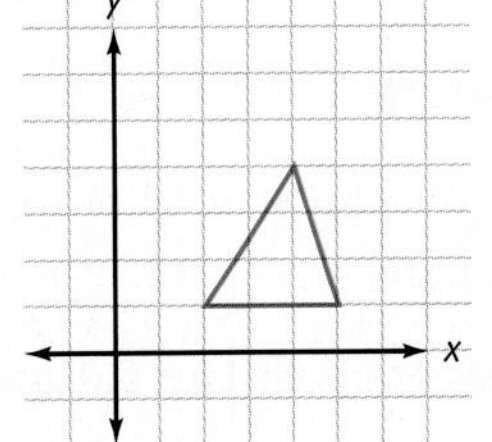

8. Reflect the figure through the x-axis.
9. Reflect the figure through the y-axis.
10. Rotate the figure 180° about the origin.
11. Reflect the figure first through the x-axis and then through the y-axis. What is the result?
12. Reflect the figure first through the y-axis and then through the x-axis. What is the result? How does your result compare with the result in the previous exercise? The result is the same as in Exercise 11.

4. $(x, y) \rightarrow (x, y + 16)$
5. $(x, y) \rightarrow (x, y - 32)$
6. $(x, y) \rightarrow (x, y + 24, y)$

ASSESS

Selected Answers

Odd-numbered Exercises 5–41

The answers to Communicate Exercises can be found in Additional Answers beginning on page 727.

Assignment Guide

Core 1–21, 31–44

Core Plus 1–3, 8–30, 43–47

Technology

Exercises 8–12 can be done with geometry graphics software on an x-y coordinate plane, or by using a graphing calculator.

Error Analysis

Before students begin Exercises 23–30, you may wish to have them use a graphics calculator to review graphing the lines $y = x$ and $y = -x$.

The answers to Exercises 8–11 can be found in Additional Answers beginning on page 727.

ALTERNATIVE ASSESSMENT

Performance Assessment

Students should draw any four-sided polygon in the first quadrant of a coordinate graph. Then they should find the image of the quadrilateral under each of these transformations:
$(x, y) \rightarrow (x + 3, y - 2)$,
$(x, y) \rightarrow (-x, y)$,
$(x, y) \rightarrow (x, -y)$, and
$(x, y) \rightarrow (-x, -y)$.
Have students label each transformation as a translation, a reflection, or a rotation.

13. translation 7 to the right
14. reflection through the y-axis
15. translation left 6 and up 7
16. translation left 8
20. translation left 7

The answers to Exercises 17, 19, 20, 23, 24, 26, and 27 can be found in Additional Answers beginning on page 727.

In Exercises 13–21, state the result of applying each rule to a figure on the standard coordinate plane.

13. $(x, y) \rightarrow (x + 7, y)$ **14.** $(x, y) \rightarrow (-x, y)$ **15.** $(x, y) \rightarrow (x - 6, y + 7)$
16. $(x, y) \rightarrow (x - 8, y)$ **17.** $(x, y) \rightarrow (x, y - 4)$ **18.** $(x, y) \rightarrow (x, y - 7)$ translation down 7
19. $(x, y) \rightarrow (-x, -y)$ **20.** $(x, y) \rightarrow (x - 7, y)$ **21.** $(x, y) \rightarrow (x, y + 2)$ translation up 2

22. Draw the figure on a piece of graph paper. Graph the transformation that results from applying the first rule to the figure. Then apply the second rule to the new figure, and so on, graphing each new figure.

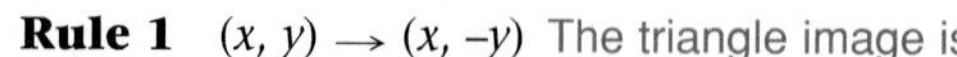

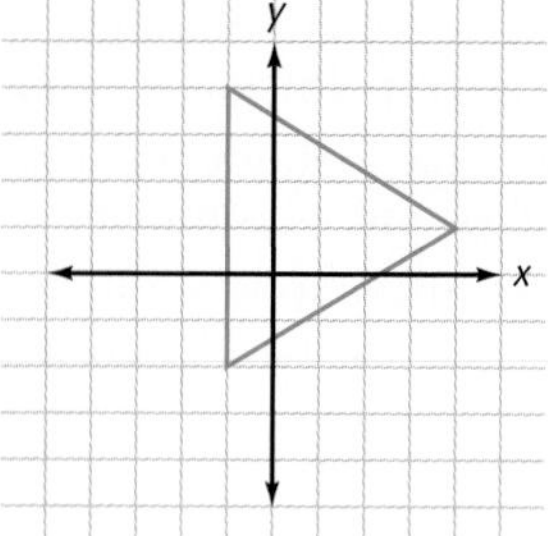

Rule 1 $(x, y) \rightarrow (x, -y)$ The triangle image is
Rule 2 $(x, y) \rightarrow (-x, y)$ where it started
Rule 3 $(x, y) \rightarrow (x, -y)$ after the fourth
Rule 4 $(x, y) \rightarrow (-x, y)$ transformation.

What is the final result of the series of transformations?

Algebra **Exercises 23–30 allow you to discover another rule for transforming an object.**

23. Fill in the table on the right, in which $y = x$ for each row.

y	x
?	?
?	?
?	?

24. Use the values in the table on the right to help you graph the line $y = x$.

25. What angle does the line form with the y-axis? 45°

26. Plot each of the following points on your graph. Connect the points with segments.
$A(2, 1)$ $B(5, 2)$ $C(6, 4)$

27. For each of the points in the previous exercise, reverse the order of the coordinates to get new points, A', B', and C'. Plot those points on your graph and connect them with segments.

28. What do you observe about the two figures you have drawn? They are congruent, and they are reflections of each other through the line $y = x$.

29. Write the rule for the transformation you have drawn. (**Hint:** Your rule will use just the letters x and y, no minus signs or numbers.) $(x, y) \rightarrow (y, x)$

30. What is true of the coordinates of a point that lies *below* the line $y = x$? of the coordinates of a point that lies *above* the line?
$y < x$ below the line, and $y > x$ above the line.

Look Back

Tell how you would find each of the following points of a triangle. [Lesson 1.3]

31. incenter **32.** centroid **33.** orthocenter **34.** circumcenter

Express each of the following as a compass heading. [Lesson 1.5]

35. W 270° **36.** NW 315° **37.** SW 225° **38.** SSW 202.5°

Describe each of the following transformations. [Lesson 1.6]

39. translation **40.** rotation **41.** reflection **42.** glide

31. Find the point where the angle bisectors meet.
32. Find the point where the medians meet.
33. Find the point where the altitudes meet.
34. Find the point where the side perpendicular bisectors meet.
39. A figure slides a given distance in a given direction.
40. A figure turns around a given point called the turn center.
41. A figure is transformed to its mirror image through a line.
42. A figure is reflected through a line as it is translated parallel to the line.

Practice Master

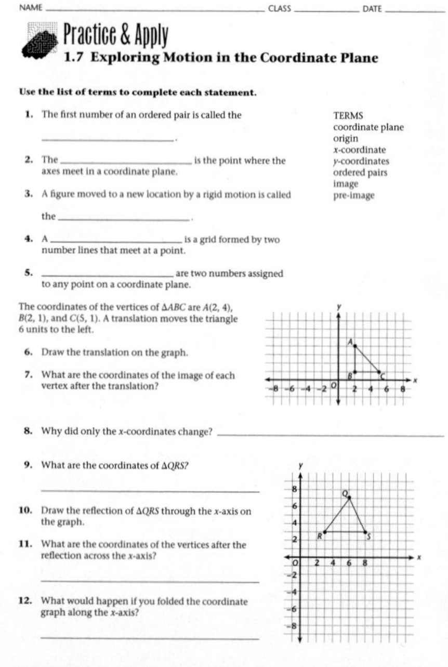

NAME ____________ CLASS ________ DATE ________

Practice & Apply
1.7 Exploring Motion in the Coordinate Plane

Use the list of terms to complete each statement.

TERMS: coordinate plane, origin, x-coordinate, y-coordinates, ordered pairs, image, pre-image

1. The first number of an ordered pair is called the ____________.
2. The ____________ is the point where the axes meet in a coordinate plane.
3. A figure moved to a new location by a rigid motion is called the ____________.
4. A ____________ is a grid formed by two number lines that meet at a point.
5. ____________ are two numbers assigned to any point on a coordinate plane.

The coordinates of the vertices of ΔABC are $A(2, 4)$, $B(2, 1)$, and $C(5, 1)$. A translation moves the triangle 6 units to the left.

6. Draw the translation on the graph.
7. What are the coordinates of the image of each vertex after the translation?
8. Why did only the x-coordinates change? ____________
9. What are the coordinates of ΔQRS? ____________
10. Draw the reflection of ΔQRS through the x-axis on the graph.
11. What are the coordinates of the vertices after the reflection across the x-axis? ____________
12. What would happen if you folded the coordinate graph along the x-axis? ____________

HRW Geometry — Practice & Apply 7

Look Beyond

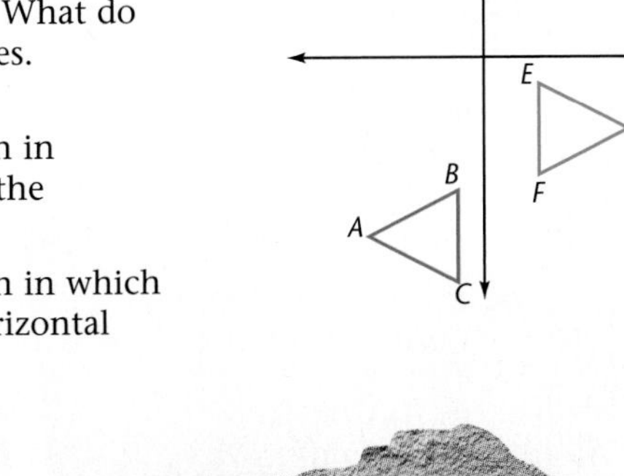

43. What two transformations would you apply to △ABC to get △DEF? Express these as a single rule.

44. Apply the *same rule* to the figure that results. What do you get? Repeat the process three or four times. Describe what happens.

45. Write a general rule for a glide transformation in which the axis of reflection is the y-axis and the vertical movement is v. $(x, y) \rightarrow (-x, y + v)$

46. Write a general rule for a glide transformation in which the axis of reflection is the x-axis and the horizontal movement is h. $(x, y) \rightarrow (x + h, -y)$

47. Cultural Connection: Africa An ear Egyptian drawing dating from 2650 B.(is a coordinate plane for the shape of ‹ rounded vault. The numbers, which aı the small vertical marks on the drawin give the height of the curve at interval of 1 cubit.

Make your own sketch of the curve using the dimensions on the drawing. Recall that a palm is equal to 4 fingers and a cubit is equal to 7 palms (or 28 fingers). Begin by filling in a table like the one below.

x-coordinate (in fingers)	y-coordinate (drawing inscription)	y-coordinate (in fingers)
0 cubit = 0 fingers	3 cubits, 3 palms, 2 fingers	98 fingers
1 cubit = ?	3 cubits, 2 palms, 3 fingers	?
2 cubits = ?	3 cubits	?
3 cubits = ?	2 cubits, 3 palms	?
4 cubits = ?	1 cubit, 3 palms, 1 finger	?
5 cubits = ?	?	?

Select an appropriate scale for a coordinate grid and plot the points from your table. Then draw a smooth curve through the points.

Look Beyond

Extension. After students have completed the drawing for Exercise 47, you can extend the activity by having them make a scale drawing of the vault using modern units of measure, either inches or centimeters.

47.

0	98
28	95
56	84
84	68
112	41
140	0

[Check student sketches]

Technology Master

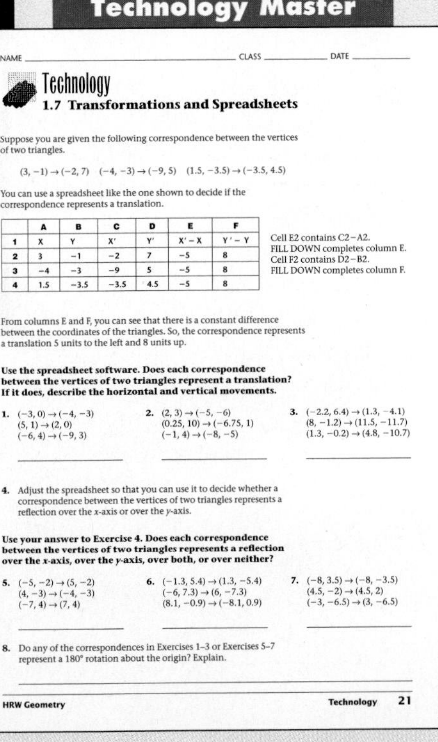

NAME ______________ CLASS ________ DATE ________

Technology

1.7 Transformations and Spreadsheets

Suppose you are given the following correspondence between the vertices of two triangles.

$(3, -1) \rightarrow (-2, 7)$ $(-4, -3) \rightarrow (-9, 5)$ $(1.5, -3.5) \rightarrow (-3.5, 4.5)$

You can use a spreadsheet like the one shown to decide if the correspondence represents a translation.

	A	B	C	D	E	F
1	X	Y	X'	Y'	X' − X	Y' − Y
2	3	−1	−2	7	−5	8
3	−4	−3	−9	5	−5	8
4	1.5	−3.5	−3.5	4.5	−5	8

Cell E2 contains C2−A2.
FILL DOWN completes column E.
Cell F2 contains D2−B2.
FILL DOWN completes column F.

From columns E and F, you can see that there is a constant difference between the coordinates of the triangles. So, the correspondence represents a translation 5 units to the left and 8 units up.

Use the spreadsheet software. Does each correspondence between the vertices of two triangles represent a translation? If it does, describe the horizontal and vertical movements.

1. $(-3, 0) \rightarrow (-4, -3)$
$(5, 1) \rightarrow (2, 0)$
$(-6, 4) \rightarrow (-9, 3)$
____

2. $(2, 3) \rightarrow (-5, -6)$
$(0.25, 10) \rightarrow (-6.75, 1)$
$(-1, 4) \rightarrow (-8, -5)$
____

3. $(-2.2, 6.4) \rightarrow (1.3, -4.1)$
$(8, -1.2) \rightarrow (11.5, -11.7)$
$(1.3, -0.2) \rightarrow (4.8, -10.7)$
____

4. Adjust the spreadsheet so that you can use it to decide whether a correspondence between the vertices of two triangles represents a reflection over the x-axis or over the y-axis.

Use your answer to Exercise 4. Does each correspondence between the vertices of two triangles represents a reflection over the x-axis, over the y-axis, over both, or over neither?

5. $(-5, -2) \rightarrow (5, -2)$
$(4, -3) \rightarrow (-4, -3)$
$(-7, 4) \rightarrow (7, 4)$
____

6. $(-1.3, 5.4) \rightarrow (1.3, -5.4)$
$(-6, 7.3) \rightarrow (6, -7.3)$
$(8.1, -0.9) \rightarrow (-8.1, 0.9)$
____

7. $(-8, 3.5) \rightarrow (-8, -3.5)$
$(4.5, -2) \rightarrow (4.5, 2)$
$(-3, -6.5) \rightarrow (3, -6.5)$
____

8. Do any of the correspondences in Exercises 1–3 or Exercises 5–7 represent a 180° rotation about the origin? Explain.

____

HRW Geometry — Technology 21

43. A reflection then a translation

44. The figure flips back and forth over the y-axis and translates up parallel to the y-axis.

Focus

Students will follow the steps to create an origami crane. The crane is made up of several geometric shapes and provides a physical example for some of the concepts discussed in Chapter 1.

Motivate

Origami is an ancient Japanese art in which figures are made from folded paper. The paper folding must be precise in order to create a pleasing figure. Books about origami are available at local libraries. You may wish to have some of these books available for students when they do the project.

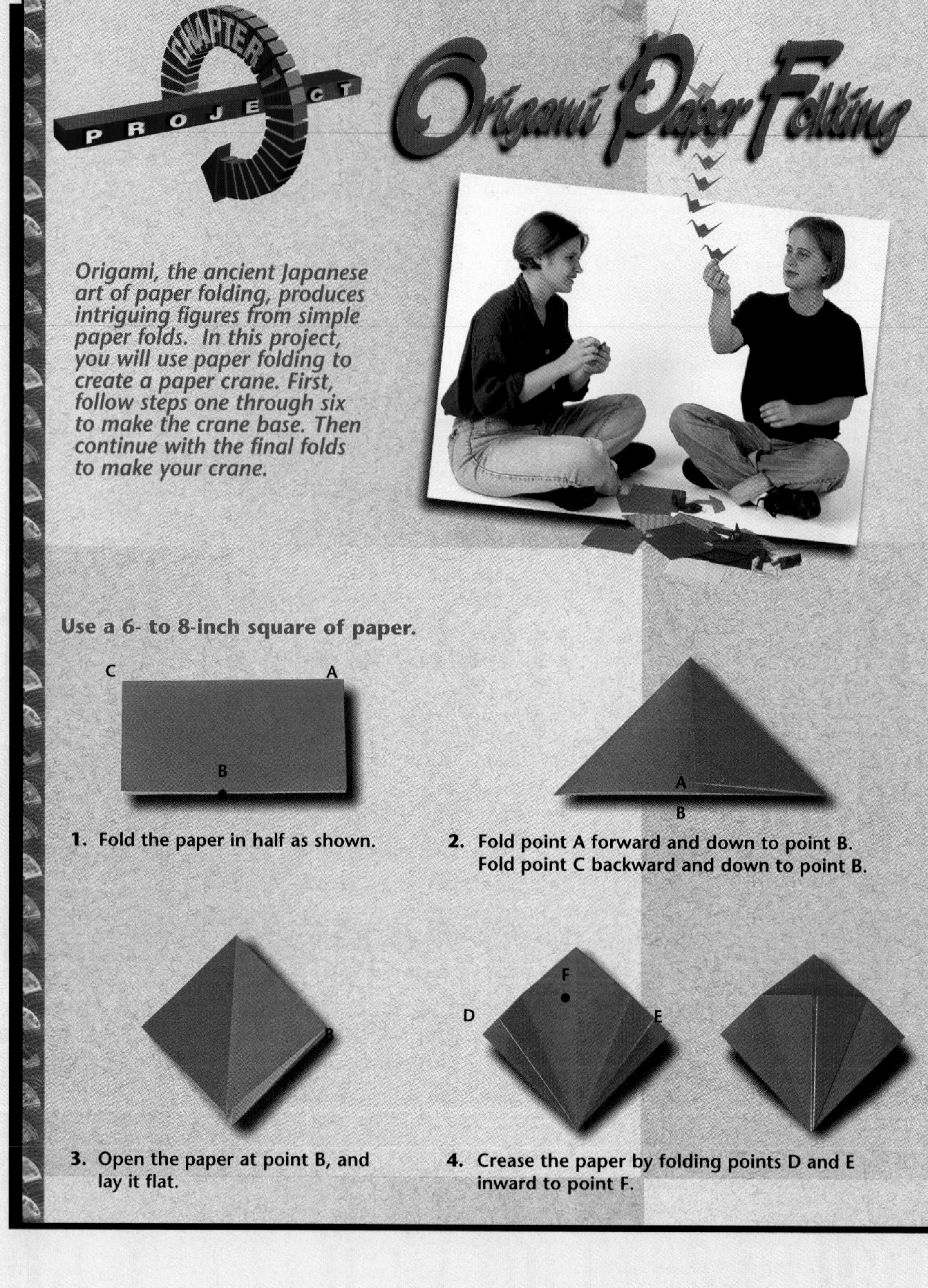

CHAPTER 1
PROJECT
Origami Paper Folding
Origami, the ancient Japanese art of paper folding, produces intriguing figures from simple paper folds. In this project, you will use paper folding to create a paper crane. First, follow steps one through six to make the crane base. Then continue with the final folds to make your crane.
Use a 6- to 8-inch square of paper.
C
A
B
1. Fold the paper in half as shown.
A
B
2. Fold point A forward and down to point B. Fold point C backward and down to point B.
B
3. Open the paper at point B, and lay it flat.
F
D
E
4. Crease the paper by folding points D and E inward to point F.

5. Open the figure and pull G upward, reversing the folds you made in Step 4.

6. Turn the paper over and repeat step 5.

You have now completed the crane base. Continue with the next steps to complete the crane.

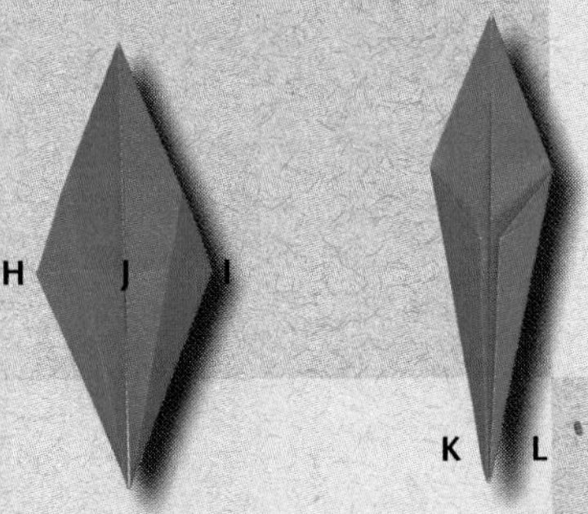

7. From the crane base, fold points *H* and *I* upward into point *J*. Turn the paper over and repeat.

8. Pull points *K* and *L* outward and reverse the folds. These will become the tail and neck of the crane.

9. Reverse-fold the neck to form the head and bend the wings down as shown. You can blow into the bottom of the figure to inflate its body and wings.

Your crane is finished.

10. Unfolding a paper crane will reveal a pattern of creases. Draw over and label geometric figures formed by the creases. Are any of the figures congruent with one another?

11. The crane base can be used to create many origami figures. Can you figure out how to fold a frog or fish?

12. Find an origami book at your library and expand your collection of origami figures.

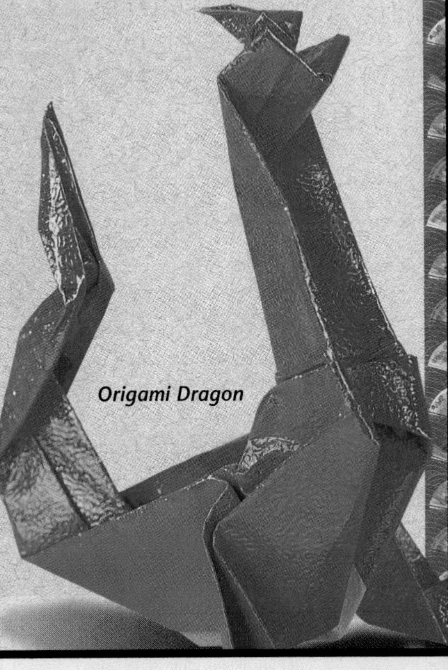

Origami Dragon

Cooperative Learning

Origami instructions are sometimes difficult to follow. Students working in groups will be able to consult with each other about their cranes. Provide each group with an origami book from the library and encourage them to try other shapes. Each group should create a menagerie of origami figures. Students can add these to their portfolios.

Discuss

Have students unfold some of their figures and study the lines created by the creases. Discuss the various geometric objects that appear. Introduce the concept of congruence by pointing out shapes that appear to be congruent. Ask students if there is something about the way they folded the paper that would assure that the shapes are congruent.

1. $\overline{KJ}$, $\overline{JL}$, $\overline{LK}$; $\angle JKL$, $\angle KLJ$, $\angle LJK$; Plane JKL
2. $\overline{MR}$, $\overline{MN}$, $\overline{MO}$, $\overline{MC}$, $\overline{MQ}$, $\overline{RM}$, $\angle RMN$, $\angle RMO$, RMC, $\angle RMQ$, $\angle NMO$, $\angle NMC$, $\angle NMQ$, $\angle OMC$, $\angle OMQ$, $\angle CMQ$
3. Points on an angle bisector are equidistant from the rays that form the angle.

Chapter 1 Review

Vocabulary

altitude	23	linear pair	40	pre-image	44
angle	11	median	23	ray	10
angle bisector	18	noncollinear	11	reflection	44
circumscribed circle	24	number line	29	right angle	17
collinear	11	parallel lines	17	rigid	44
congruent	31	perpendicular bisector	18	rotation	44
image	44	perpendicular lines	17	segment	10
inscribed circle	24	plane	9	translation	44
line	9	point	9	turn center	45

Key Skills and Exercises

Lesson 1.1

Key Skills

Identify and name parts of a figure.

In the figure, A and C are endpoints of the segment $\overline{AC}$, which is part of line $\overleftrightarrow{AC}$. $\overrightarrow{AB}$ is a ray; it starts at point A, passes through point B, and goes on forever. $\overrightarrow{AB}$ and $\overrightarrow{AC}$ form $\angle BAC$. Points A, B, and C lie on plane ABC.

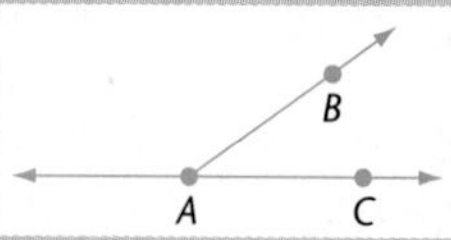

Exercises

1. Name the segments and angles in the triangle, and the plane that contains it.

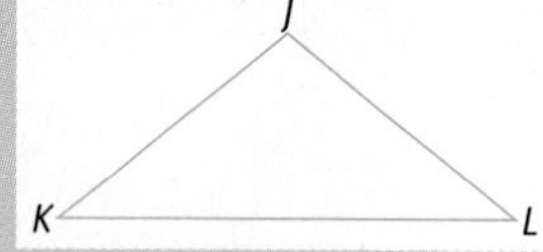

2. Name all the rays and angles in the figure.

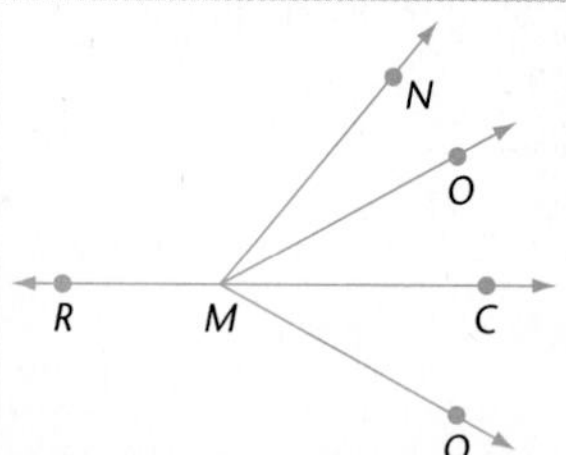

Lesson 1.2

Key Skills

Use paper folding to make conjectures about lines and angles.

Using folding paper, construct a perpendicular bisector of a segment $\overline{AB}$. Locate a point X on the perpendicular bisector. From measuring the distances XA and XB you might write the following conjecture: "Every point on the perpendicular bisector of a segment is equidistant from the endpoints of the segment."

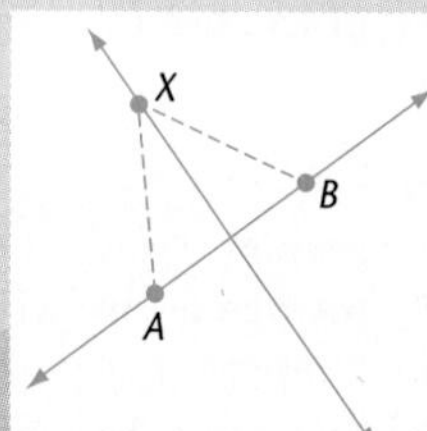

Exercise

3. Using folding paper, construct the bisector of an angle. Measure the perpendicular distances from a point on the bisector to the rays that form the angle. Write a conjecture about points on the bisector of an angle.

Lesson 1.3

➤ Key Skill

Indentify special parts of triangles and their intersection points.

Find the intersection point of the perpendicular bisectors of a triangle. Set your compass at the intersection and draw a circumscribed circle.

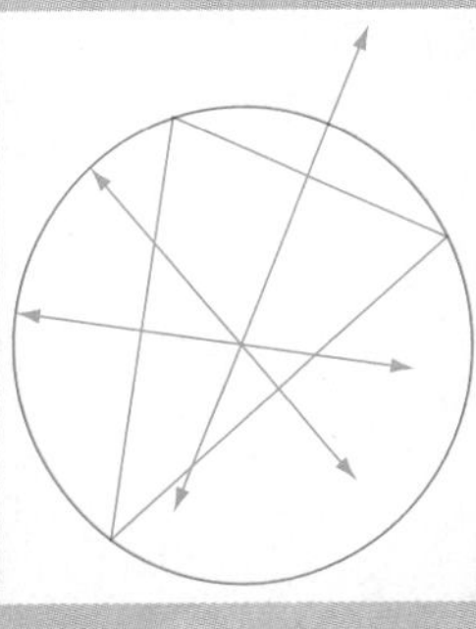

➤ Exercises

Use geometry technology to construct these figures.

4. A right triangle circumscribed by a circle.

5. An inscribed circle that touches each side of a triangle at the midpoint.

Lesson 1.4

➤ Key Skills

Determine if segments are congruent.

Are $\overline{CD}$, $\overline{DE}$, and $\overline{EF}$ congruent segments?

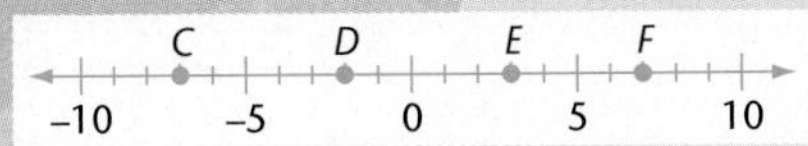

$mCD = |-7 - (-2)| = 5$

$mDE = |-2 - (-3)| = 5$

$mEF = |3 - 7| = 4$

$\overline{CD}$ and $\overline{DE}$ are congruent; $\overline{EF}$ is not congruent to the other segments.

Add measures of segments.

To add measures *AB* and *BX*, we first find the measure of each segment.

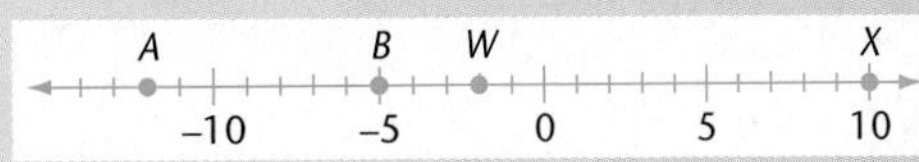

$mAB = |-12 - (-5)| = |-7| = 7$

$mBX = |-5 - 10| = |-15| = 15$

$mAB + mBX = 7 + 15 = 22$

➤ Exercises

6. Which segments, if any, are congruent?

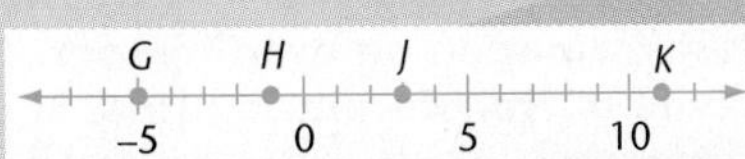

In Exercises 7–9, *A* is between points *R* and *P*. Sketch each figure and find the missing measure.

7. $RA = 25$ $AP = 13$ $RP =$ _?_ **8.** $RA =$ _?_ $AP = 39.5$ $RP = 200$

9. $RA = 121$ $AP =$ _?_ $RP = 350$

Lesson 1.5

➤ Key Skills

Find relationships between measures of angles.

In a parallelogram opposite angles are congruent; in other words, their measures are equal. In parallelogram *RSTU* the measure of $\angle URS$ and $\angle STU$ is twice the measure of $\angle TUR$ and $\angle RST$.

All the following relationships are true: $m\angle URS = m\angle STU = m\angle TUR + m\angle RST = 2m\angle TUR = 2m\angle RST$.

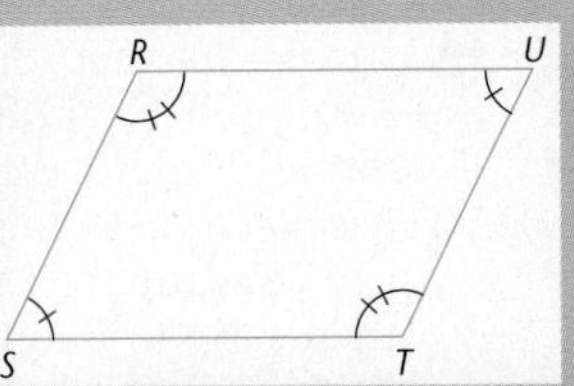

Chapter Review

6. $\overline{GH} \cong \overline{HJ}$; $\overline{GJ} \cong \overline{JK}$

7. $RP = 25 + 13 = 38$ units

8. $RA = 200 - 39.5 = 160.5$ units

9. $AP = 350 - 121 = 229$ units

4. $\overline{AB}$ is a diameter.

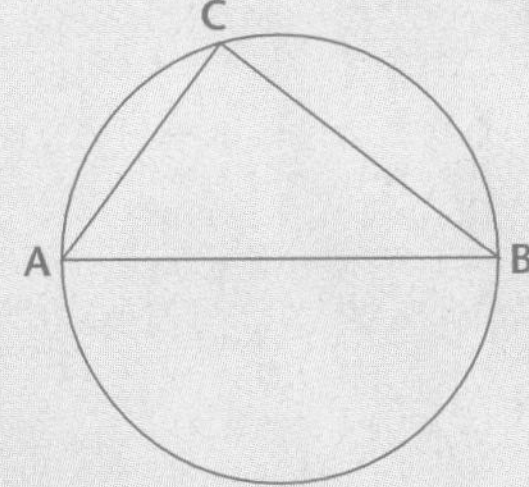

5. $\triangle ABC$ is equilateral.

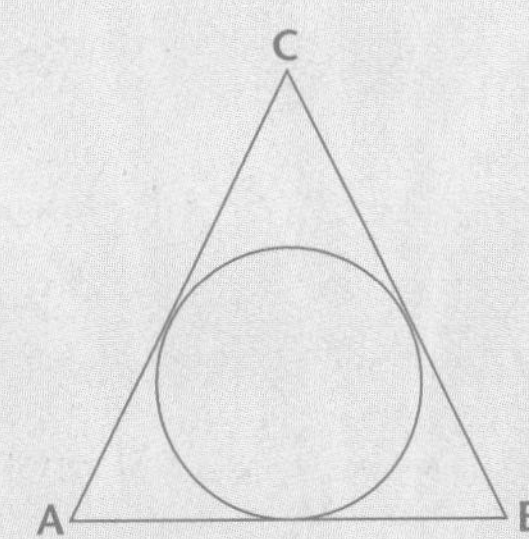

10. m∠EVB = 12° + 26° = 38°

11. m∠RVB = 126° – 63° = 63°

12. m∠EVR = 90° – 75° = 15°

13. In Exercise 11, m∠EVR = m∠RVB = 63°

14.

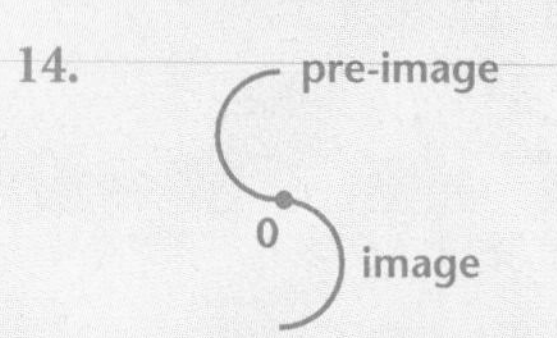

rotation of 180° about O

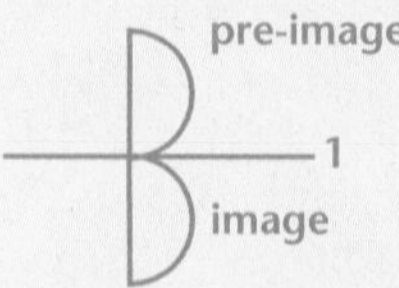

reflection in line 1

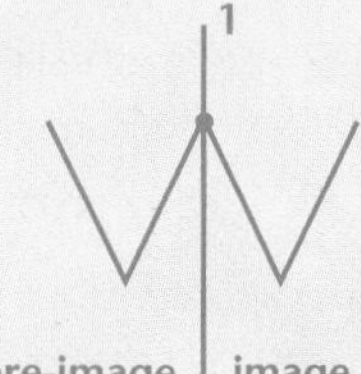

reflection in line 1

15.

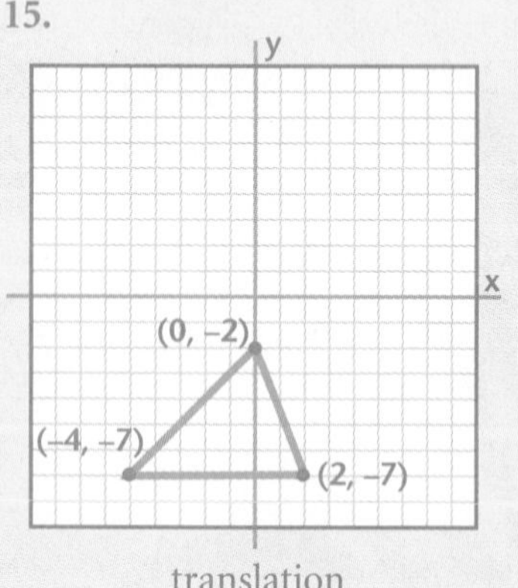

translation

➤ Exercises

In Exercises 10–12 find the missing measures.

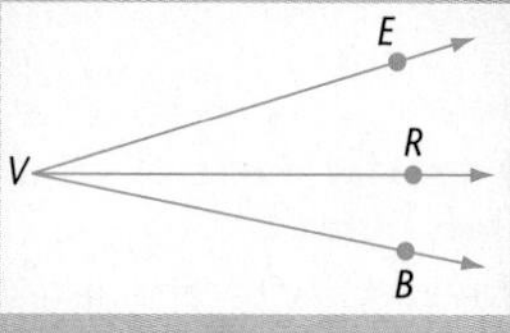

10. $m\angle EVR = 12°$ $m\angle RVB = 26°$ $m\angle EVB =$ __?__

11. $m\angle EVR = 63°$ $m\angle RVB =$ __?__ $m\angle EVB = 126°$

12. $m\angle EVR =$ __?__ $m\angle RVB = 75°$ $m\angle EVB = 90°$

13. For which exercises, if any, are $\angle EVR$ and $\angle RVB$ congruent in Exercises 9–11?

Lesson 1.6

➤ Key Skills

Identify transformations: translation, rotation, and reflection.

The figure below shows $\overline{AB}$ translated to $\overline{A'B'}$, reflected to $\overline{A''B''}$, and rotated to $\overline{AB'''}$.

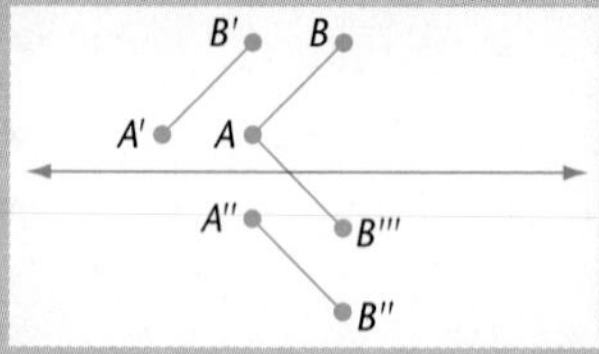

➤ Exercises

14. The letters *S*, *B*, and *W* can be thought of as transformations of the letters C, *P*, and *V*. Sketch these transformations and identify each as a reflection, translation, or rotation. Label the pre-image and the image.

Lesson 1.7

➤ Key Skills

Use the coordinate plane to transform geometric objects.

Using the rule $(x, y) \rightarrow (x + 3, y + 3)$, draw the transformed triangle. This represents a translation.

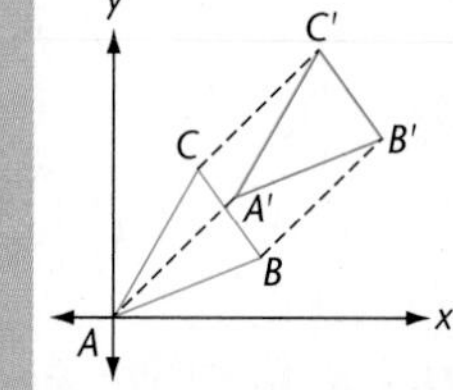

➤ Exercises

In Exercises 15–17 copy the figure and draw the transformation on graph paper. State what type of transformation it is.

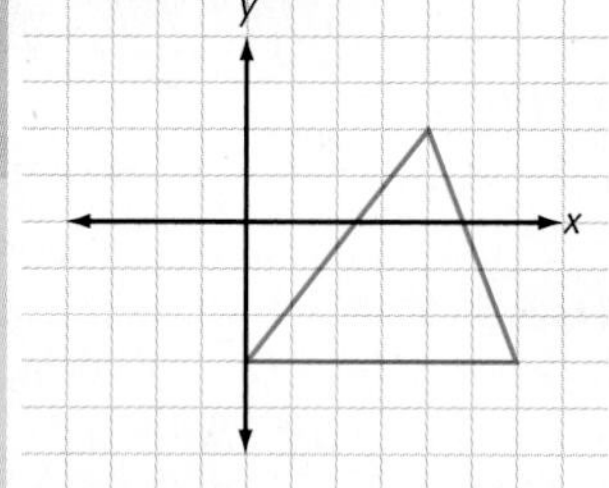

15. $(x, y) \rightarrow (x - 4, y - 4)$

16. $(x, y) \rightarrow (x, -y)$

17. $(x, y) \rightarrow (-x, -y)$

Applications

18. Describe how to find the center of gravity of a triangle. (Hint: Find the centroid.)

19. **Travel** Bastrop, Smithville, and LaGrange are towns on a highway that runs straight from Bastrop, through Smithville, to LaGrange. The distance from Bastrop to Smithville is 22 kilometers. The distance from Bastrop to LaGrange is 5 kilometers more than 3 times the distance from Bastrop to Smithville. Write an equation for the distances, and then solve to find the distance between Smithville and LaGrange and between Bastrop and LaGrange.

16.

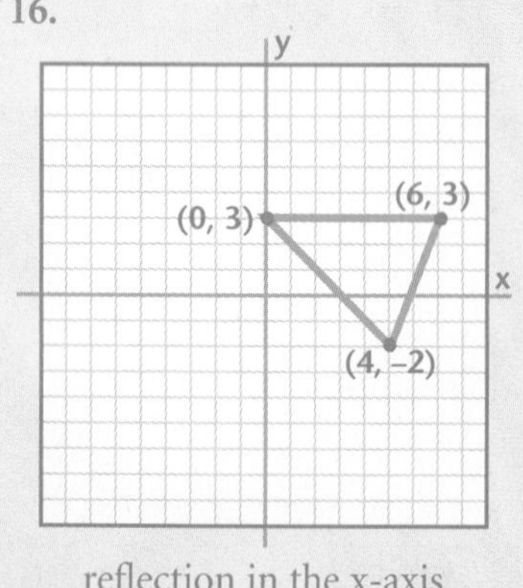

reflection in the x-axis

17.

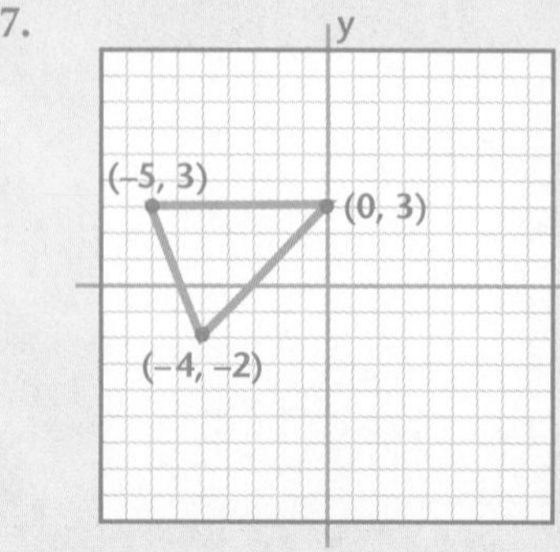

Rotation of 180° about (0, 0).

18. Draw the medians of the triangle. The point where the medians meet is the center of gravity.

19. BL = 5 + 3(BS) fi BL = 5 + 3(22) = 71 km; SL = 71 – 22 = 49 km

Chapter 1 Assessment

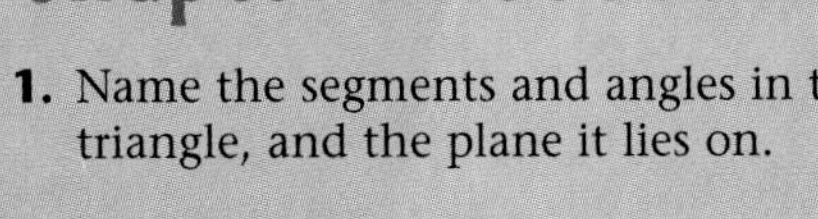

1. Name the segments and angles in the triangle, and the plane it lies on.

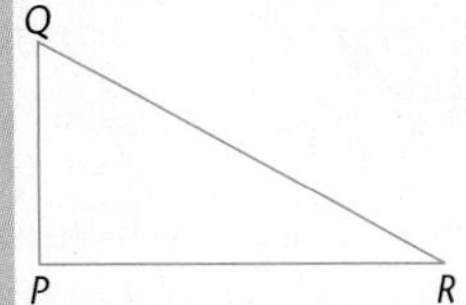

In the figure below, lines *l* and *m* are parallel.

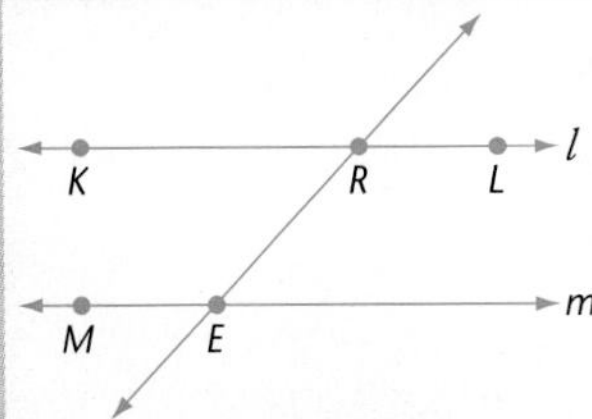

2. Write a conjecture about the relationship of $\angle LRE$ to $\angle MER$.
3. How would you test your conjecture?

4. Complete the conjecture: *The center of a circle inscribed in a triangle is the same point as the intersection of the triangle's* __?__.
   a. altitudes **b.** angle bisectors **c.** medians **d.** perpendicular bisectors

In Exercises 5–7, *N* is between points *A* and *D*. Sketch each figure and find the missing measure.

5. $AN = 51$ $ND = 6$ $AD =$ __?__
6. $AN =$ __?__ $ND = 14.7$ $AD = 24$
7. $AN = 111$ $ND =$ __?__ $AD = 180$
8. For which exercises, if any, are $\overline{AN}$ and $\overline{ND}$ congruent in Exercises 5–7?

In Exercises 9–11 find the missing measures.

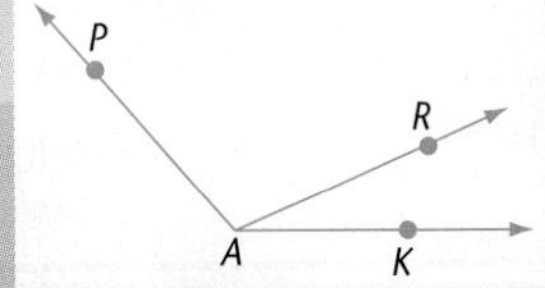

9. $m\angle PAR = 48$ $m\angle RAK = 52$ $m\angle PAK =$ __?__
10. $m\angle PAR = 89$ $m\angle RAK =$ __?__ $m\angle PAK = 126$
11. $m\angle PAR =$ __?__ $m\angle RAK = 45$ $m\angle PAK = 135$

12. A four-leaf clover can be thought of as transformations of a single leaf. Sketch the transformations and identify them as reflections, translations, or rotations. Label the pre-image and the image in each.

In Exercises 13–15 draw each transformation on graph paper and state what type of transformation it is. Use the pre-image shown.

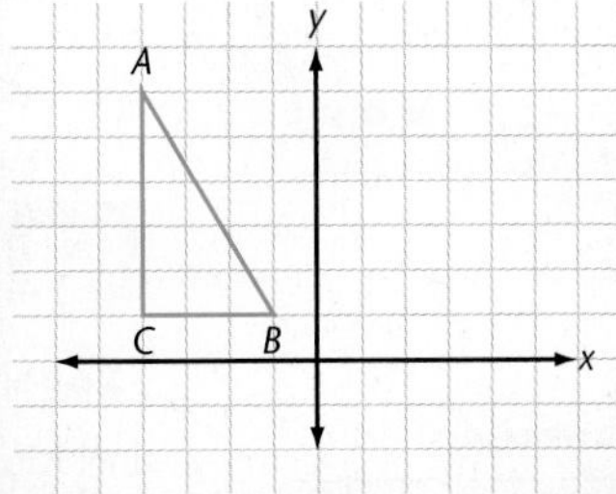

13. $(x, y) \rightarrow (x + 5, y + 5)$
14. $(x, y) \rightarrow (-x, -y)$
15. $(x, y) \rightarrow (-x, y)$

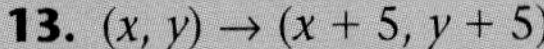

1. $\overline{QP}$, $\overline{PR}$, $\overline{RQ}$; $\angle QPR$, $\angle PRQ$, $\angle RQP$; Plane *PQR*
2. If line *l* is parallel to line m, then $m\angle LRE = m\angle MER$.
3. Make several different parallel lines cut by a third line by folding paper or using geometry graphics software. Then measure the angles and compare their measures.
4. b
5. $AD = 51 + 6 = 57$ units
6. $AN = 24 - 14.7 = 9.3$ units
7. $ND = 180 - 111 = 69$ units
8. none
9. $m\angle PAK = 48° + 52° = 100°$
10. $m\angle RAK = 126° - 89° = 37°$
11. $m\angle PAR = 135° - 45° = 90°$
12. reflections through perpendicular lines or rotations of 90°, 180° and 270°

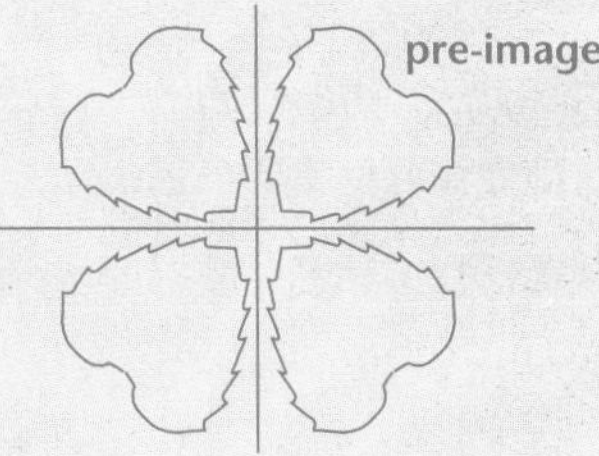

13. translation

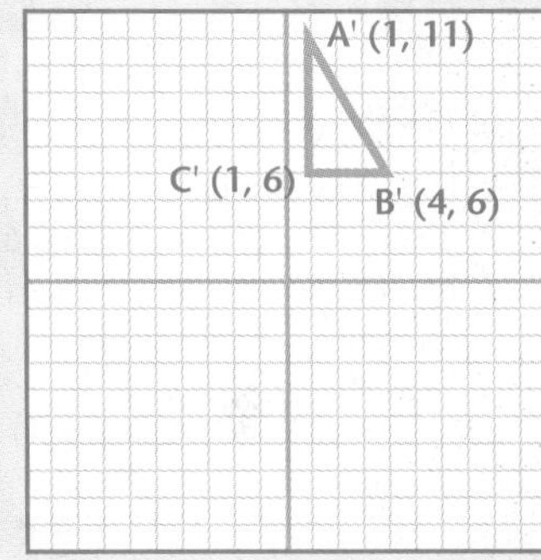

14. rotation of 180° about (0, 0)

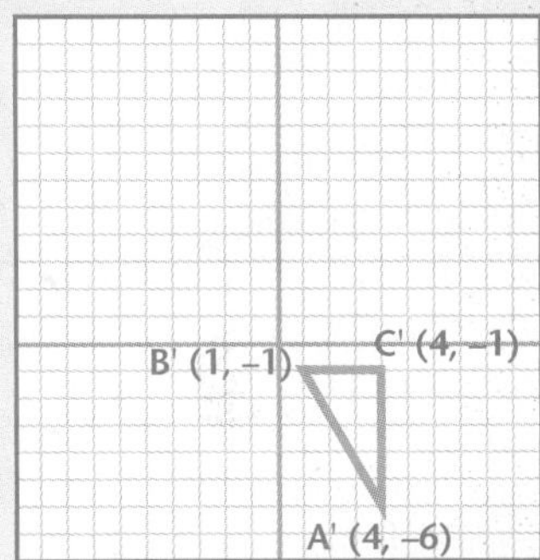

15. reflection in the y-axis

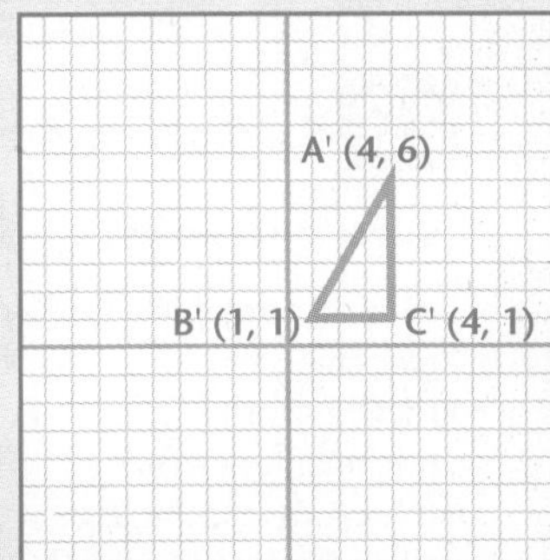

Reasoning in Geometry

Meeting Individual Needs

2.1 Exploring Informal Proofs

Core Resources	Core Plus Resources
Inclusion Strategies, p. 68 Reteaching the Lesson, p. 69 Practice Master 2.1 Enrichment Master 2.1 Technology Master 2.1 Lesson Activity Master 2.1	Practice Master 2.1 Enrichment, p. 68 Technology Master 2.1 Interdisciplinary Connection p. 67
[2 days]	[2 days]

2.2 Introduction to Logical Reasoning

Core Resources	Core Plus Resources
Inclusion Strategies, p. 74 Reteaching the Lesson, p. 75 Practice Master 2.2 Enrichment Master 2.2 Lesson Activity Master 2.2 Interdisciplinary Connection, p. 73	Practice Master 2.2 Enrichment, p. 74 Technology Master 2.2
[2 days]	[1 day]

2.3 Exploring Definitions

Core Resources	Core Plus Resources
Inclusion Strategies, p. 83 Reteaching the Lesson, p. 84 Practice Master 2.3 Enrichment, p. 83 Technology Master 2.3 Lesson Activity Master 2.3 Mid-Chapter Assessment Master	Practice Master 2.3 Enrichment Master 2.3 Technology Master 2.3 Mid-Chapter Assessment Master
[2 days]	[1 day]

2.4 Postulates

Core Resources	Core Plus Resources
Inclusion Strategies, p. 89 Reteaching the Lesson, p. 90 Practice Master 2.4 Enrichment Master 2.4 Lesson Activity Master 2.4	Practice Master 2.3 Enrichment, p. 89 Technology Master 2.4
[2 days]	[1 day]

2.5 Linking Steps in a Proof

Core Resources	Core Plus Resources
Inclusion Strategies, p. 94 Reteaching the Lesson, p. 95 Practice Master 2.5 Enrichment, p. 94 Lesson Activity Master 2.5 Interdisciplinary Connection, p. 93	Practice Master 2.5 Enrichment Master 2.5 Technology Master 2.5
[2 days]	[2 days]

2.6 Exploring Conjectures that Lead to Proof

Core Resources	Core Plus Resources
Inclusion Strategies, p. 101 Reteaching the Lesson, p. 102 Practice Master 2.6 Enrichment Master 2.6 Technology Master 2.6 Interdisciplinary Connection, p. 100	Practice Master 2.6 Enrichment, p. 101 Technology Master 2.6 Interdisciplinary Connection, p. 100
[2 days]	[2 days]

Chapter Summary

Core Resources	Core Plus Resources
Chapter 2 Project, pp. 106–107 Lab Activity Long Term Project Chapter Review, pp. 108–110 Chapter Assessment p. 111 Chapter Assessment, A/B Alternative Assessment Cumulative Assessment, pp. 112–113	Chapter 2 Project, pp. 106–107 Lab Activity Long Term Project Chapter Review, pp. 108–110 Chapter Assessment p. 111 Chapter Assessment, A/B Alternative Assessment Cumulative Assessment, pp. 112–113
[3 days]	**[2 days]**

Reading Strategies

Reading geometry, like reading any type of mathematics textbook, requires more concentration and more time than reading other types of material. Written mathematics is "dense" in that symbols often substitute for words, and key ideas are stated rather abruptly. A student who can comfortably read about 20 pages of a mystery story in an hour might well spend the same time reading just 2 pages of mathematics.

Use Lesson 2.2 to help students understand the importance of careful word choices. In particular, point out the connection between the "if-then" language used in conditionals and the inclusive "all." For example, these two statements have the same logical meaning: If a shape is a square, then it is also a rectangle. All squares are rectangles.

Visual Strategies

Visual discrimination skills are needed in Lesson 2.3. Here students are asked to look over a set of figures and identify common characteristics. Students who are not good visual thinkers may be helped by doing the activities in small groups. They can talk over each item in turn, pointing out and discussing key attributes.

Key postulates that are basic to ideas throughout the book are stated in Lesson 2.4 in verbal form. Have students make diagrams to illustrate each postulate. You may wish to have students save their drawings in a notebook or portfolio for future reference.

The drawings of the planes in Lesson 2.4 may be difficult for some students to understand. Modeling planes with pieces of cardboard can help students learn to "see" the three dimensions pictured in drawings of this type.

Hands-on Strategies

Paper folding may be used for the activities in Lesson 2.6 in which students create vertical angles and reflections of triangles across both parallel lines and intersecting lines. Paper that is reasonably transparent, such as tracing paper or wax paper, will be helpful when students work with the reflections.

This chapter includes several activities built around the game of Nim. Since this game is played with circular counters, you may wish to have a supply of such counters on hand.

Cooperative Learning

GROUP ACTIVITIES	
Strategies for Nim	Lesson 2.1, Exercise 29; Lesson 2.5, Exercise 35
Three Challenges	Lesson 2.1, Exploration
Fraction sequence	Lesson 2.1, Exercises 6–12
Writing logical statements	Lesson 2.2, Exercises 6–9
Too Tough for Computers	Eyewitness Math, pages 80–81
Writing definitions	Lesson 2.3, Explorations 1–2, Exercise 15
Using postulates	Lesson 2.4, Exercises 5–16
Overlapping angles	Lesson 2.5, Try This, p. 94
Conjectures	Lesson 2.6, Explorations 1, 2, and 3
Regions in a circle	Lesson 2.6, Exercises 33–34

You may wish to have students work in groups or with partners for some of the above activities. Additional suggestions for cooperative group activities are noted in the teacher's notes in each lesson.

Multicultural

The cultural references in this chapter include references to Europe and the Americas.

CULTURAL CONNECTIONS	
Africa: Euclid of Alexandria	Lesson 2.4 Opener
Europe: Russian-American Company	Lesson 2.5 Example 2

Portfolio Assessment

Below are portfolio activities for the chapter listed under seven activity domains which are appropriate for portfolio development.

1. **Investigation/Exploration** The exploration in Lesson 2.1 includes three informal proofs tied to interesting puzzles. The explorations in Lesson 2.3 challenge students to create their own definitions; those in Lesson 2.6 ask for student-created conjectures and informal proofs.
2. **Applications** Humor, Lesson 2.2, Exercise 13; Criminal Justice, Lesson 2.2, Exercise 19; Routing Problems, Eyewitness Math, pages 80–81; Photography, Lesson 2.4, Critical Thinking;
3. **Non-Routine Problems** Lesson 2.1, Exercise 28 (triangular number pattern); Lesson 2.6, Exercises 33–34 (circular regions pattern)
4. **Project** Caliban's Logic Puzzles: see pages 106–107. Students use trial and error and other strategies to solve two logic puzzles.
5. **Interdisciplinary Topics** Fine Arts, Lesson 2.2, Exercise 25; Chemistry, Lesson 2.3, Exercise 10; Algebra, Lesson 2.5, Example 1; Algebra, Lesson 2.6, Exercises 14–17;
6. **Writing** *Communicate* exercises offer excellent writing selections for the portfolio. Suggested selections include: Lesson 2.1, Exercises 3–4; Lesson 2.2, Exercises 4–5; Lesson 2.6, Exercise 4.
7. **Tools** In Chapter 2, students use geometry graphics software to explore conjectures about vertical angles and reflections.

Technology

Many of the activities in this book can be significantly enhanced by the use of geometry graphics software, but in every case the technology is optional. If computers are not available, the computer activities can, *and should,* be done by hands-on or "low-tech" methods, as they contain important instructional material. For more information on the use of technology, refer to the *HRW Technology Manual for High School Mathematics.*

Graphics Calculator

Using the ANS Function A number of the activities require, or are greatly facilitated by, the use of a graphics calculator. It is important for the students to have an easy command of this powerful mathematics tool.

Exercise 8 in Lesson 2.1 suggests the use of a scientific calculator to find sums of the series 1/2 + 1/4 + 1/8 + 1/16, ... This can be done by adding each of the terms for each new sum, but a much simpler method is to use the ANS function of the calculator, as indicated in the illustration.

When you want to use the last value that was calculated as the first part of a new entry, just enter the first operator—such as a plus or a minus—that would follow that value. The calculator will automatically enter the answer value for you, represented as **Ans.**

You can use the answer function anywhere in a calculator entry by pressing the answer key (which may require another keystroke, such as the 2nd key).

Computer Graphics Software

Lesson 2.6: Exploration 1 This exploration allows students to investigate the relationship between vertical angles and between angles that form linear pairs. Students dynamically investigate how the size of one angle affects the sizes of others as they change the angles between intersecting lines.

Lesson 2.6: Exploration 2 In this exploration, students reflect a triangle twice through two parallel lines and explore the relationships among the three triangles. Students change the shape of the pre-image and observe the effect the change has upon the other triangles. Students make conjectures about the relationships of the distances between the triangles and the lines of reflection.

Lesson 2.6: Exploration 3 Students reflect a triangle twice with respect to two intersecting lines and explore the relationships among the three triangles. Students change the shape of the pre-image and observe the effect the change has upon the image triangles. The technology measures angles of corresponding parts of the triangles, which allows the students to form conjectures about the reflections.

Hand-Held Computers

The geometry graphics capabilities of hand-held computers can be used in much the same way as the geometry software described above. Projection devices for classroom demonstration are available.

Reasoning in Geometry

About the Chapter

Background Information

This chapter introduces students to the ideas and concepts that provide a basis for understanding the deductive nature of geometry.

Chapter Resources

- Practice Masters
- Enrichment Masters
- Technology Masters
- Lesson Activity Masters
- Lab Activity Masters
- Long-Term Project Masters
- Assessment Masters
  - Chapter Assessments, A/B
  - Mid-Chapter Assessment
  - Alternative Assessment, A/B
- Teaching Transparencies
- Cumulative Assessment
- Spanish Resources

Chapter Objectives

- Define *informal proof.*
- Identify and construct informal proofs.
- Define and use *conditionals.*
- Define and use *logical chains.*
- Use Venn diagrams to determine definitions of objects.
- Use principles of logic to create definitions of objects.
- Define *theorem* and *postulate.*
- Given basic Euclidean postulates, determine the validity of elementary geometry statements.
- Identify and use the Addition Property of Equality, Reflexive Property of Equality, Symmetric Property of Equality, and Transitive Property of Equality.

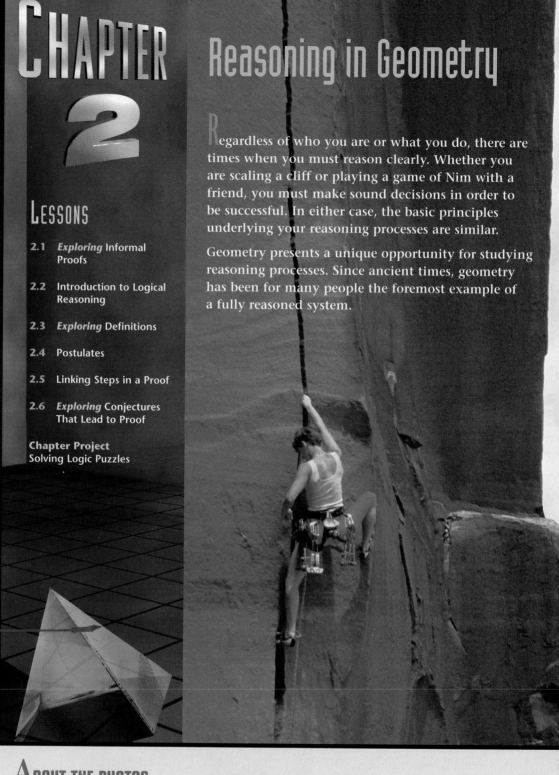

Chapter 2

Lessons

Reasoning in Geometry

Regardless of who you are or what you do, there are times when you must reason clearly. Whether you are scaling a cliff or playing a game of Nim with a friend, you must make sound decisions in order to be successful. In either case, the basic principles underlying your reasoning processes are similar.

Geometry presents a unique opportunity for studying reasoning processes. Since ancient times, geometry has been for many people the foremost example of a fully reasoned system.

About the Photos

Rock climbing is a physically demanding sport. A good rock climber is also constantly making decisions about where to make the next move. Besides physical strength, reasoning and logic are important in rock climbing.

PORTFOLIO ACTIVITY
The game of Nim is played by two persons, who take turns. Twelve counters are arranged in rows of 3, 4, and 5 as shown. On a player's turn, he or she must remove one or more counters from any one of the three rows. The object of the game is to take the last remaining counter.
Try to figure out a strategy that will ensure you of winning if you have the first turn. Present your strategy with an explanation of why it works.

- Link steps of geometric proof using properties and postulates.
- Define *formal proof.*
- Use conjectures to motivate the steps of a proof.

PORTFOLIO ACTIVITY

Students will need to play the game of Nim a number of times in order to determine a strategy that will ensure a victory on the first move. Each player should devise a strategy to win (and not discuss it with the other player), then try it out to see if it works. When both players think they have a winning strategy, they should write down an explanation for why it works.

Figuring out a strategy to win requires the use of inductive reasoning to make a conjecture (the strategy) about winning. The conjecture can be proved logically by a systematic examination of patterns. Inductive reasoning and conjectures are discussed in Lesson 2.6 of this chapter.

Additional Pupil's Edition portfolio activities can be found in the exercises for Lessons 2.1 and 2.5.

ABOUT THE CHAPTER PROJECT

The Chapter 2 Project, on pages 106–107, requires students to solve two logic puzzles providing them with an opportunity to practice logical thinking skills.

PREPARE

Objectives

- Define *informal proof.*
- Identify and construct informal proofs.

RESOURCES

• Practice Master	2.1
• Enrichment Master	2.1
• Technology Master	2.1
• Lesson Activity Master	2.1
• Quiz	2.1
• Spanish Resources	2.1

Assessing Prior Knowledge

1. The length of a square is 5 cm. What is the measure of its width? [**5 cm**]
2. What is the sum of the first 5 counting numbers? [**15**]
3. Solve the equation $2x + 1 = 5$. [$\boldsymbol{x = 2}$]

TEACH

This lesson builds upon students' prior experience with informal proofs to establish a foundation for understanding logical reasoning. Ask students to suggest an example of a statement they were asked to prove. Then ask how they proved it. Discuss with the class whether the argument used was logical or not.

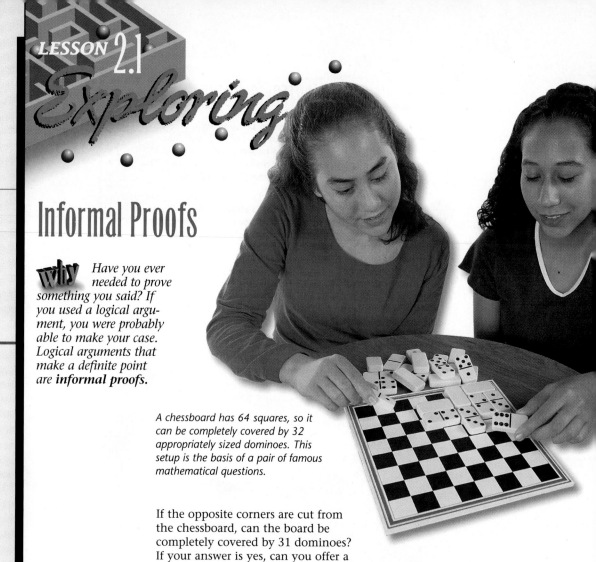

LESSON 2.1 Exploring

Informal Proofs

why *Have you ever needed to prove something you said? If you used a logical argument, you were probably able to make your case. Logical arguments that make a definite point are **informal proofs.***

A chessboard has 64 squares, so it can be completely covered by 32 appropriately sized dominoes. This setup is the basis of a pair of famous mathematical questions.

If the opposite corners are cut from the chessboard, can the board be completely covered by 31 dominoes? If your answer is yes, can you offer a method for showing that it is possible? If your answer is no, can you explain why it cannot be done? In either case you would be giving an informal proof of what you say.

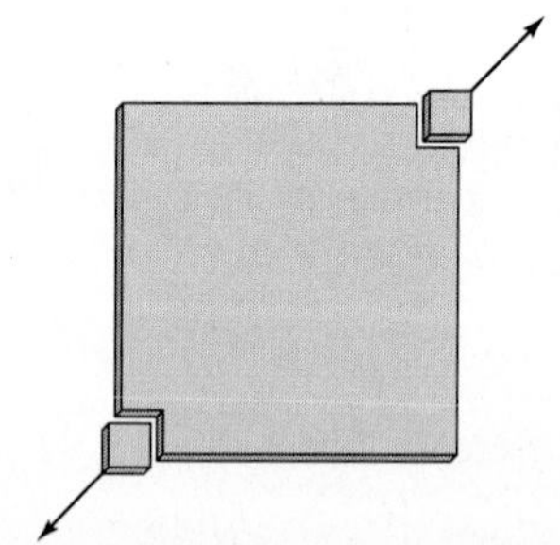

Here is an informal proof that the reduced chessboard cannot be covered by the dominoes:

"The squares that were cut off were of the *same color,* leaving more squares of one color than the other on the remaining board.

"Any arrangement you make with the dominoes must cover the same number of dark squares as light squares. This is because each domino will cover one dark square and one light square, no matter where you place it on the board. Therefore, it is not possible to cover the reduced board with the 31 dominoes."

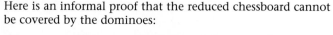

Using Visual Models Students can explore the use of logical arguments in the advertisements of products in newspapers and magazines. Bring some examples to class for discussion, or suggest that students find their own examples of logical arguments in product advertisements. Ask students if the advertisements prove their case.

Exploration Three Challenges

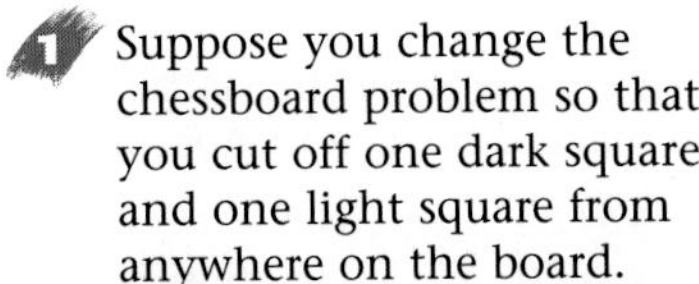

You will need
No special tools

The pathway through the maze suggests a proof.

1. Suppose you change the chessboard problem so that you cut off one dark square and one light square from anywhere on the board.

   Now the remaining board can be covered with dominoes. Use the diagram on the right and your own explanation to prove this is true.

2. How could you find the sum of the first n odd counting numbers without actually adding them up? Fill in a table like the one below and see if you can discover the answer.

n	First n odd numbers	Sum of the first n odd numbers
1	1	1
2	1, 3	4
3	1, 3, 5	9
4	1, 3, 5, 7	16
5	?	?
6	?	?
n	?	?

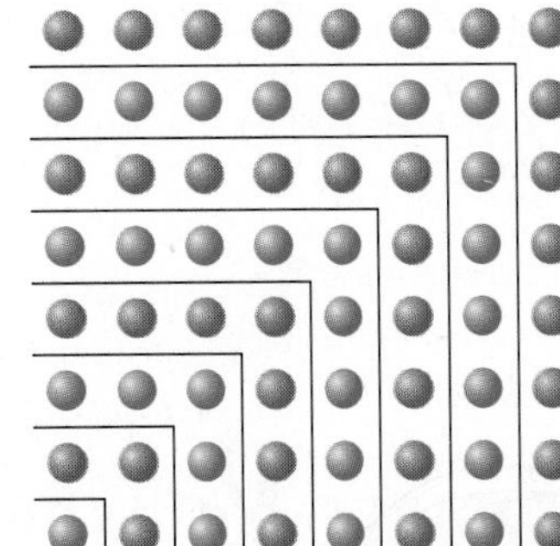

The diagram to the right of the table is a "proof without words" of the algebraic result you may have discovered. Explain in your own words how the diagram "proves" the result.

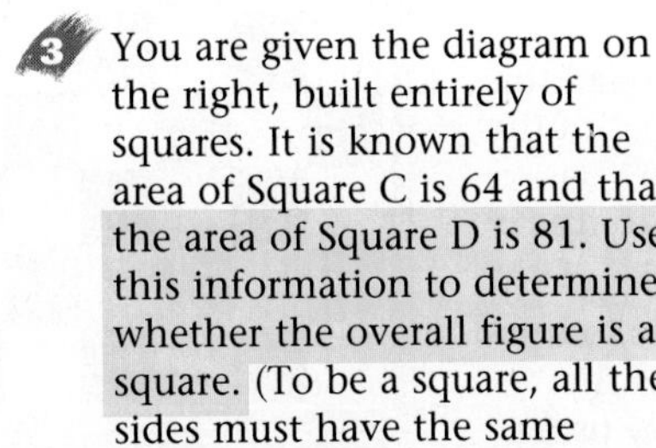

3. You are given the diagram on the right, built entirely of squares. It is known that the area of Square C is 64 and that the area of Square D is 81. Use this information to determine whether the overall figure is a square. (To be a square, all the sides must have the same length.) ❖

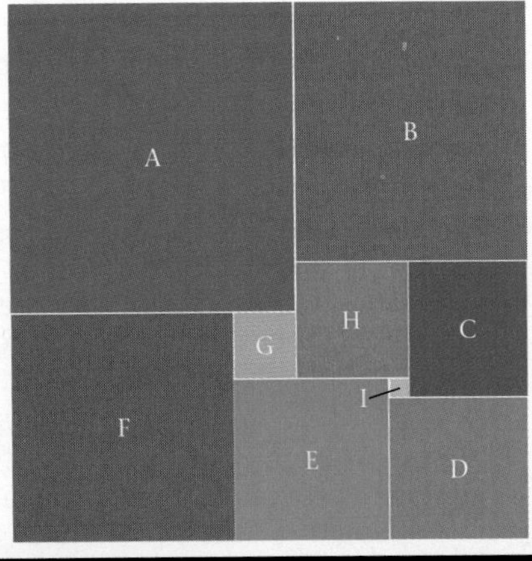

Exploration Notes

After students have discovered what they think is the answer to Step 2, suggest that they extend the values of n beyond 6 to check their answers.

Use Transparencies 12–14

ONGOING ASSESSMENT

1. **By laying dominoes end-to-end along the maze pattern, all 62 remaining squares can be covered. This can be done, no matter where the removed squares occur. Students should satisfy themselves that there is no problem with turning corners, etc.**
2. **Student tables should show that, for a sequence of consecutive integers starting with 1, the sum of the first n integers is n^2. The diagram shows that this is true for $n = 1$ to 8, and it should be clear that this would hold true for any value of n.**
3. **The figure is not a square. Students' arguments should consist of a series of linked statements, such as, "Since square C has sides = 8, and square D has sides = 9, square I has sides = 1," etc.**

interdisciplinary CONNECTION

Language Arts Authors of detective stories, such as Arthur Conan Doyle in his tales about Sherlock Holmes, use logical arguments to solve mysteries. Ask students if they can suggest any mystery stories they have read that use logical arguments to solve the mystery.

Bring a chessboard and dominoes to class so students can experiment with the challenge in Step 1.

What Is a Proof?

A proof is a fully convincing argument that a certain claim is true. But before you allow yourself to be convinced by a supposed proof, you should make sure that it is sound.

There are many different styles of proofs in mathematics. For example, the calculations below prove that for the given equation, $x = 4$.

$5x + 4 = 24$	Given
$5x = 20$	Subtraction Property of Equality
$x = 4$	Division Property of Equality

Formal Proof

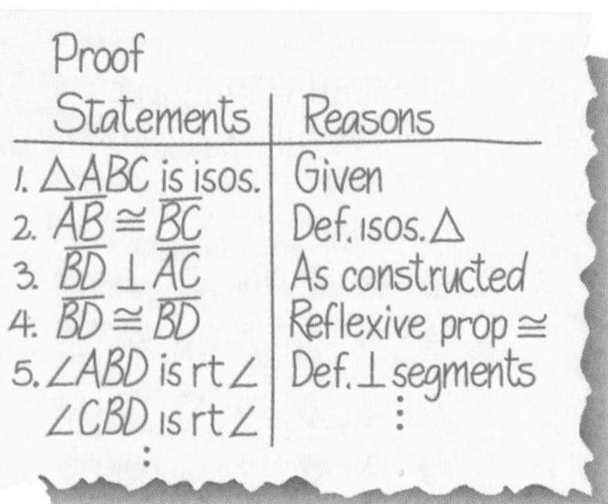

In geometry many proofs are formal in nature. Such proofs are carefully related to a *mathematical system* like the one you will be studying in this book. In such proofs, reasons or justifications are given for every statement that you make.

One kind of formal proof is known as a **two-column proof**. In proofs of this type, you write your statements in the left-hand column and your reasons in the right-hand column. For some proofs you may find this format to be the most convenient way to present your ideas in a simple, orderly fashion.

Communicate

1. In your own words, describe what a proof is.
2. How many different proofs of one claim might there be?
3. Explain why 31 dominoes could not cover a chessboard if you removed 2 dark squares.
4. Explain in your own words how you proved that the overall shape of the diagram in Step 2 of the exploration was or was not a square. Do you think it might be possible for someone to find a "hole" in your argument?
5. In Step 3 of the exploration, a geometric solution is given to an algebraic problem. Discuss the difference between algebra and geometry using this exploration to help you.

ENRICHMENT Ask students to list the first 10 even integers. Then ask them to prove informally that if n is any positive integer, then $2n$ is always an even integer.

English Language Development Students may need some special attention in learning new mathematical terms, particularly a term such as *argument* that is commonly used in everyday life. Discuss the everyday meaning of this word along with its mathematical meaning.

Math Connection Algebra

(p. 67) Students can find the sum of the first n odd counting numbers by observing the pattern in the chart. They can then use the diagrams to check their answers.

Cooperative Learning

Students can work together in groups of three or four to explore the three challenge activities. They can test their ideas with one another and arrive at a best answer. When each group has agreed upon the answer for each challenge, the solutions can be discussed with the entire class.

Math Connection Algebra

Have students compare the algebraic proof with the formal two-column proof.

ASSESS

Selected Answers

Odd–numbered Exercises 7–27

Assignment Guide

Core 1–4, 6–18, 24–27

Core Plus 1–12, 19–29

Technology

For Exercise 8, students first need to discover the pattern that can be used to generate the sums for terms five to eight. They can use their calculators to find.

Error Analysis

For Exercise 13, students might make simple arithmetic errors in finding the cubes of successive integers. Using a calculator to find the cubes can eliminate such errors.

Practice & Apply

Use the sequence $\frac{1}{2}, \frac{1}{4}, \frac{1}{8}, \frac{1}{16}, \ldots$ for Exercises 6–12.

6. **Algebra** If the pattern of the sequence is to double the denominator to find the next term, then what are the next four terms in the sequence? $\frac{1}{32}, \frac{1}{64}, \frac{1}{128}, \frac{1}{256}$

7. Are the numbers in the sequence getting larger or smaller? Explain your answer? Smaller, because 1 is successively divided by bigger numbers.

8. **Technology** Use your calculator to sum the first 8 terms of the sequence. Use a table like the one below to help you organize your sums.

Number of terms	Terms	Sum of terms
1	$\frac{1}{2}$	.5
2	$\frac{1}{2}+\frac{1}{4}$	.75
3	$\frac{1}{2}+\frac{1}{4}+\frac{1}{8}$	.875
4	$\frac{1}{2}+\frac{1}{4}+\frac{1}{8}+\frac{1}{16}$	.9375
5	_____	_____
6	_____	_____
7	_____	_____
8	_____	_____

9. What number do the sums seem to be getting closer and closer to? 1

10. You can demonstrate the infinite sum in Exercise 8 geometrically. Start by drawing a square with an area of 1 square unit. Label the lengths of the sides of the square.

11. Divide your square into the fractions of the sequence you identified in Exercise 6, starting with $\frac{1}{2}$. Use the diagram on the right as a guide.

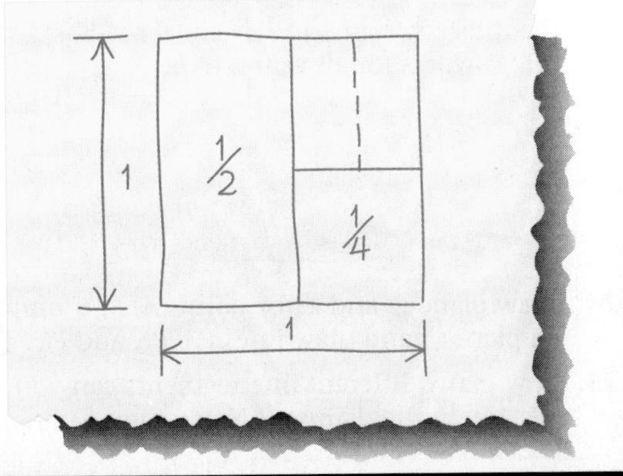

12. Explain how your illustration in Exercise 10 "proves" the sum of the infinite sequence. Is it a geometric or an algebraic proof?

Alternative Assessment

Performance Assessment

Students' understanding of logical argument and informal proof can be assessed by asking them to explain, in writing, why the sum of any even integer and odd integer is an odd integer.

[$2n + 2n + 1 = 4n + 1$. $4n$ contains a factor of 2 and therefore is even. An even number plus 1 is always odd.]

The answers to Exercises 8, 10, 11, and 12 can be found in Additional Answers beginning on page 727.

Using Patterns Have students continue the following sequence for five more numbers: 1, 3, 5, 7, 9, . . . Ask them to describe the rule they devised to continue the sequence. Then have the students come up with an informal proof for the following statement: If n is any positive integer, then $2n + 1$ is an odd integer.

18. The numbers that occur after the numbers in Column C are in Column A. Since 1000^3 is the number right after 999^3, and 999^3 is in Column C, 1000^3 will oc-cur in Column A.

Algebra Use the table for Exercises 13–18. The pattern of numbers is from the cubes of integers (i.e., $1^3 = 1$, $2^3 = 8$, $3^3 = 27$, $4^3 = 64$, etc.). The table continues infinitely.

A	B	C
1	8	27
64	125	216
•	•	•
•	•	•

13. In which column does the number 1 billion occur? One method for finding the appropriate column is to continue filling in the table until you reach 1 billion. What is the disadvantage of using this method? It takes a long time.

14. Begin to analyze the table. 1,000,000,000 is what number raised to the third power? (Use the following to help you if you are having trouble: $10^3 = 1000$, $100^3 =$ __?__) 1,000,000,000 = 1000^3

15. Continue to analyze the table. Notice that column A contains entries for the numbers 1^3, 4^3, 7^3, . . . What number entries occur in column B? in column C? 2^3, 5^3, 8^3, 11^3 . . .; 3^3, 6^3, 9^3, 12^3 . . .

16. What is true of every entry in column C? It is a multiple of 3 or it is divisible by 3.

17. In what column does the entry for the number 999 occur? C, because 999 is a multiple of 3.

18. How can you add to the above information to finish your proof that 1 billion occurs in column A? Write the final step in the proof.

Algebra Exercises 19–22 refer to the following conjecture: $1 + 2 + 3 + 4 + \ldots + n = \frac{n(n+1)}{2}$ where n can be any positive integer.

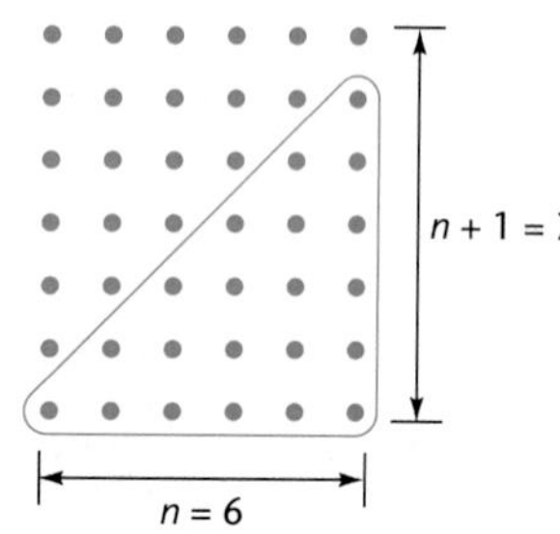

19. Check to verify that the conjecture holds for $n = 5$ and $n = 6$.

20. Draw a rectangle of dots that is 6 across and 7 down. How many dots are there on the whole rectangle? 42

21. If n represents the width of the rectangle and $n + 1$ represents the height, what is an expression for the number of dots in the whole rectangle? $n(n + 1)$

22. Divide the rectangle of dots into two equal parts by circling exactly half of them in the lower right-hand part of the rectangle, as shown. What is an expression, in terms of n, for the number of dots in the circled part of the rectangle? $\frac{n(n+1)}{2}$

23. How does the diagram suggest a proof for the claim that the sum of the integers from 1 to n is equal to $\frac{n(n+1)}{2}$? You should be sure that it works for all values of n.
It suggests relating the sum to the number of dots in the constructed triangle, and the pattern is the same for any size of the triangle, and hence n.

Look Back

24. Draw plane $\mathcal{Q}$ and three points A, B, C on plane $\mathcal{Q}$. Draw point P not on plane $\mathcal{Q}$ and draw lines $\overleftrightarrow{PA}$, $\overleftrightarrow{PB}$, and $\overleftrightarrow{PC}$. **[Lesson 1.1]**

25. How many different line segments can you identify in the diagram? Name them. **[Lesson 1.1]** 6; $\overline{GR}$, $\overline{GE}$, $\overline{GP}$, $\overline{RE}$, $\overline{RP}$, $\overline{EP}$

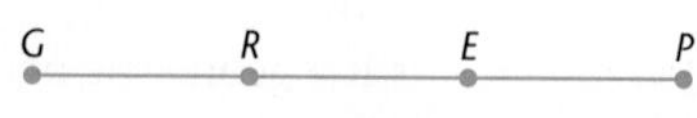

Practice Master

NAME ________ CLASS ______ DATE ______

Practice & Apply
2.1 Exploring Informal Proofs

Exactly one line segment can be drawn between these two points.

How many line segments can be drawn between 3, 4, 5, and 6 points? Draw them. Then record your data in the table. (Note: The points are noncollinear—that is, they are not on the same line.)

1. **2.** **3.** **4.**

Number of Points	Number of Line Segments
2	1
3	
4	
5	
6	
7	
8	
9	

5. What is the pattern? ________

6. If the pattern continues, how many line segments can be drawn between 9 noncollinear points? ________

Logical arguments that make a definite point are informal proofs.

W X Z Y

7. Design your own experiment to prove the conjecture: Opposite sides of a parallelogram are equal in measure. Test the conjecture by measuring the sides of several parallelograms. Show your work. Record your results in the table.

Trial	WX	ZY	WZ	XY

8. Were you able to prove the conjecture to be true? Explain.

HRW Geometry Practice & Apply 43

19. If $n = 5$, $\frac{n(n+1)}{2} = \frac{5 \cdot 6}{2} = 15$ and $1 + 2 + 3 + 4 + 5 = 15$.
If $n = 6$, $\frac{n(n+1)}{2} = \frac{6 \cdot 7}{2} = 21$ and $1 + 2 + 3 + 4 + 5 + 6 = 21$.

24.

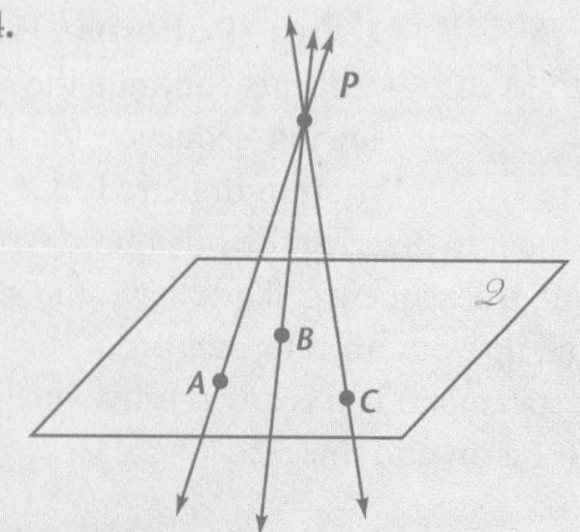

26. How many different angles can you identify in the diagram? **[Lesson 1.1]** 6

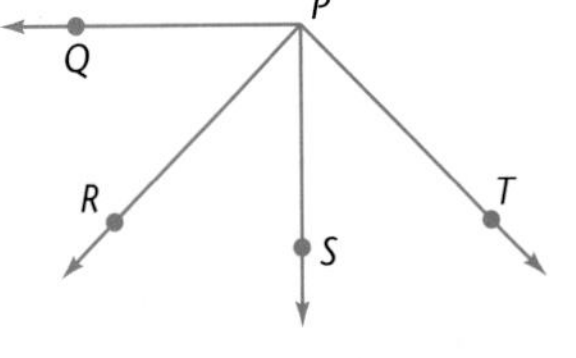

27. Use the diagram below to find the value of x and the measure of $\angle KIY$ and $\angle TIY$. **[Lesson 1.5]**

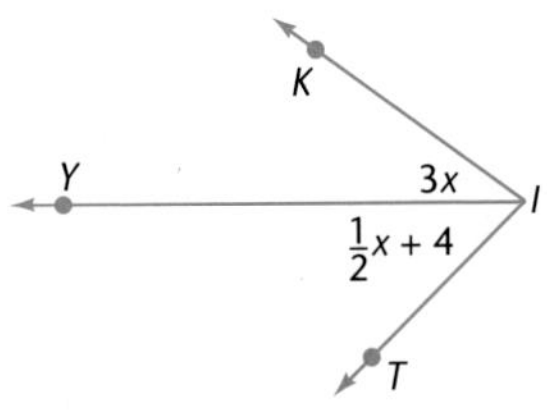

$\angle KIT = 80°$

$x = 21\frac{5}{7}$, $m\angle KIY = 65\frac{1}{7}°$, $m\angle TIY = 14\frac{6}{7}°$

Look Beyond

28. The positive integers are written in a triangular array as shown. In what row is the number 1000? 45th row

1
2 3
4 5 6
7 8 9 10
11 . . .

29.  **Portfolio Activity**

You can also play the game of Nim, on page 65, by changing the object of the game. In a second version of the game, the winner is the person who forces his or her opponent to take the last counter.

A clever strategist will discover that the game of Nim can be won by the player who succeeds in leaving just two rows with two counters in each row. This is true for either version of the game. Explain why this combination will always lead to a win.

The player on the right is removing one penny. Which player will win?

Look Beyond

Exercise 28 requires that students analyze a triangular array of positive integers to find the row in which the number 1000 appears. This analysis not only extends the work on logical arguments presented in this lesson, but also foreshadows future work on inductive reasoning and making conjectures that need to be proved. Exercise 29 introduces students to the game of Nim.

Technology Master

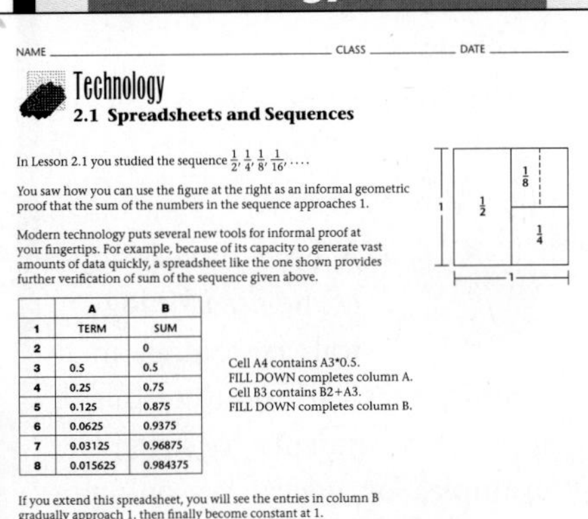

NAME ________ CLASS ________ DATE ________

Technology
2.1 Spreadsheets and Sequences

In Lesson 2.1 you studied the sequence $\frac{1}{2}, \frac{1}{4}, \frac{1}{8}, \frac{1}{16}, \ldots$.

You saw how you can use the figure at the right as an informal geometric proof that the sum of the numbers in the sequence approaches 1.

Modern technology puts several new tools for informal proof at your fingertips. For example, because of its capacity to generate vast amounts of data quickly, a spreadsheet like the one shown provides further verification of sum of the sequence given above.

	A	B
1	TERM	SUM
2		0
3	0.5	0.5
4	0.25	0.75
5	0.125	0.875
6	0.0625	0.9375
7	0.03125	0.96875
8	0.015625	0.984375

Cell A4 contains A3*0.5.
FILL DOWN completes column A.
Cell B3 contains B2+A3.
FILL DOWN completes column B.

If you extend this spreadsheet, you will see the entries in column B gradually approach 1, then finally become constant at 1.

Explain how to adjust the spreadsheet above to find the sum of the terms in each sequence. Then find each sum.

1. $\frac{1}{4}, \frac{1}{8}, \frac{1}{16}, \frac{1}{32}, \ldots$ ________
2. $\frac{1}{8}, \frac{1}{16}, \frac{1}{32}, \frac{1}{64}, \ldots$ ________
3. $\frac{1}{3}, \frac{1}{27}, \frac{1}{81}, \frac{1}{243}, \ldots$ ________
4. $\frac{1}{4}, \frac{1}{16}, \frac{1}{64}, \frac{1}{256}, \ldots$ ________
5. $\frac{1}{3}, \frac{1}{6}, \frac{1}{12}, \frac{1}{24}, \ldots$ ________
6. $\frac{1}{2}, \frac{1}{6}, \frac{1}{18}, \frac{1}{54}, \ldots$ ________

Make a sketch to provide a geometric demonstration of the sum of the sequence in the given exercise.

7. Exercise 1 ________
8. Exercise 2 ________
9. Exercise 3 ________

10. Create a sequence like those shown. What is the sum of the terms? Can you devise a geometric demonstration of the sum?

____

HRW Geometry — Technology 55

29. First Version – Winner takes last counter:

If your opponent takes one counter, you take one counter from the other row. Then your opponent must take exactly one of the remaining counters, leaving the last counter for you.

If your opponent takes both counters in one row, you take both counters in the other row.

Second Version – Winner forces opponent to take last counter:

If your opponent takes one counter, you take both counters from the other row, leaving the last counter for your opponent.

If your opponent takes both counters in one row, you take one counter, leaving the last counter for your opponent.

PREPARE

Objectives

- Define and use *conditionals.*
- Define and use *logical chains.*

- Practice Master 2.2
- Enrichment Master 2.2
- Technology Master 2.2
- Lesson Activity Master 2.2
- Quiz 2.2
- Spanish Resources 2.2

Assessing Prior Knowledge

Classify each of the following as true or false.

1. A statement is a sentence that is either true or false. **[True]**
2. If Tom lives in Boston, then he lives in Massachusetts. **[True]**
3. If Rose lives in Illinois, then she lives in Chicago. **[False]**
4. Venn diagrams can be used to show logical relationships. **[True]**

TEACH

A brief discussion of the *need* for logical reasoning in mathematics can motivate this topic. Ask students to suggest examples of logical reasoning from their previous courses in mathematics. Lead students to think about the fact that without logical reasoning, mathematics could not really exist.

LESSON 2.2 Introduction to Logical Reasoning

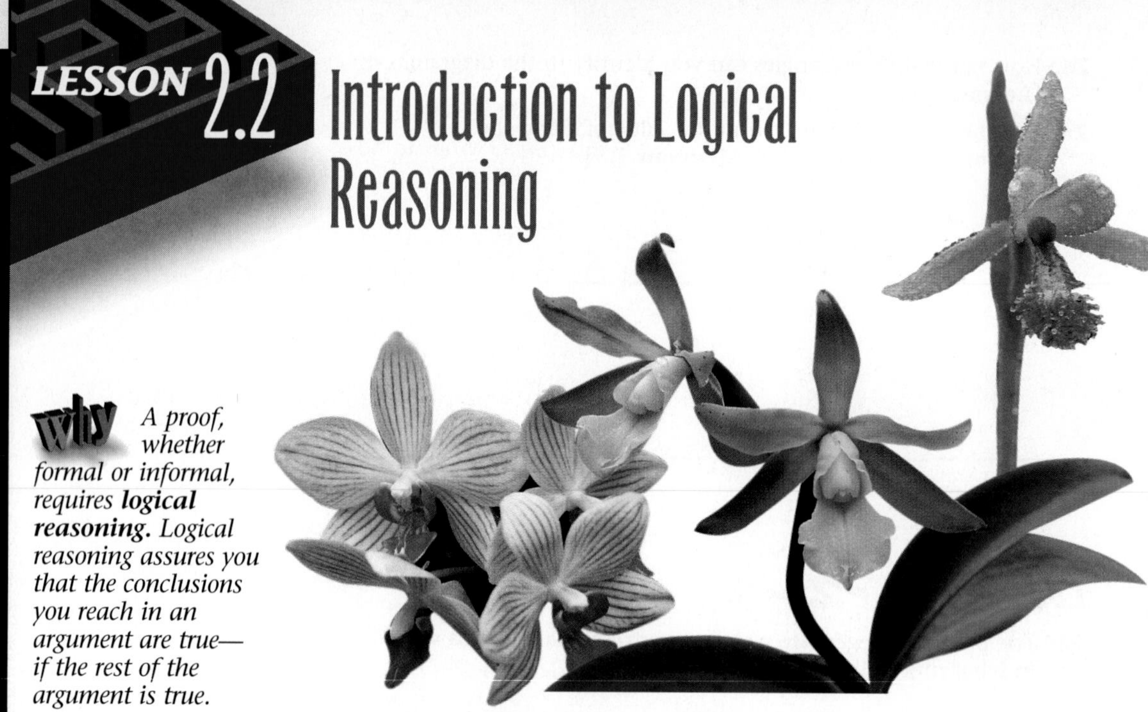

Why *A proof, whether formal or informal, requires* ***logical reasoning.*** *Logical reasoning assures you that the conclusions you reach in an argument are true—if the rest of the argument is true.*

Organisms can be classified according to their structure. For example, all of these flowers belong to the orchid family. Class organization, which also extends to man-made things, is the basis of logical reasoning.

Drawing Conclusions From Conditionals

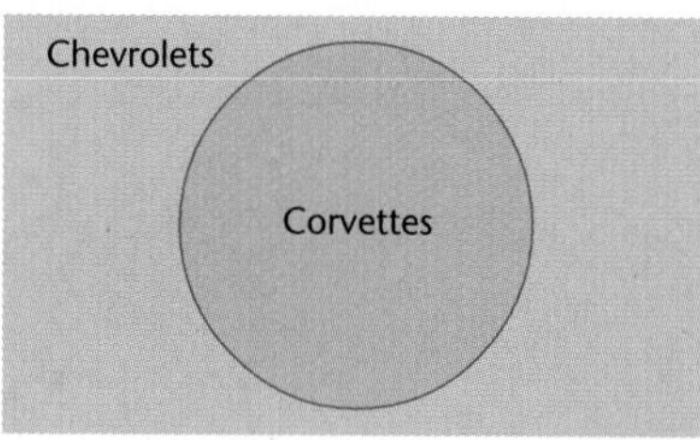

The force of logic comes from the way information is structured. For example, all Corvettes are Chevrolets, a fact which can be represented by a Venn diagram like the one on the left.

From the Venn diagram, it is easy to see that the following statement is true:

If a car is a Corvette, then it is a Chevrolet.

Statements like the one above are called **if-then**, or **conditional**, **statements**. In logical notation, conditional statements are usually written in the following form:

If *P* then *Q*
or $P \Rightarrow Q$ (read: "*P* implies *Q*")

In conditional statements the *P* part—following the word "if"—is the hypothesis. The *Q* part—following the word "then"—is the conclusion.

If a car is a Corvette, then it is a Chevrolet.
(hypothesis: *a car is a Corvette*; conclusion: *it is a Chevrolet*)

Cooperative Learning Working in small groups, students can write one statement per group that requires logical reasoning to prove. Groups can then exchange statements and try to prove the new statement they received. Statements and proofs can then be shared with the class. Acceptable statements can be mathematical or nonmathematical.

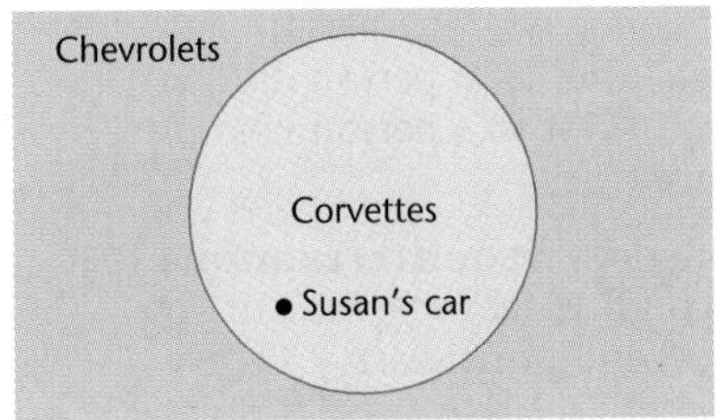

Now consider the following statement:

Susan's car is a Corvette.

By placing Susan's car into the Venn diagram, you can see that it is a Chevrolet. This result is a conclusion from a logical argument. Notice that the argument has three parts:

1. If a car is a Corvette, then it is a Chevrolet.
2. Susan's car is a Corvette.
3. Therefore, Susan's car is a Chevrolet.

The process of drawing conclusions using logical reasoning is known as **deductive reasoning** or **deduction**.

EXAMPLE 1

Use deduction to show that triangle ABC is an isosceles triangle. Illustrate your argument with a Venn diagram.

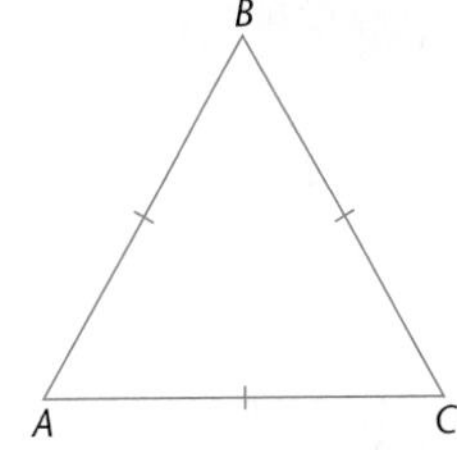

Solution ➤

An isosceles triangle has at least two congruent sides, and an equilateral triangle has three. From this information and the given drawing, you can write the following statements:

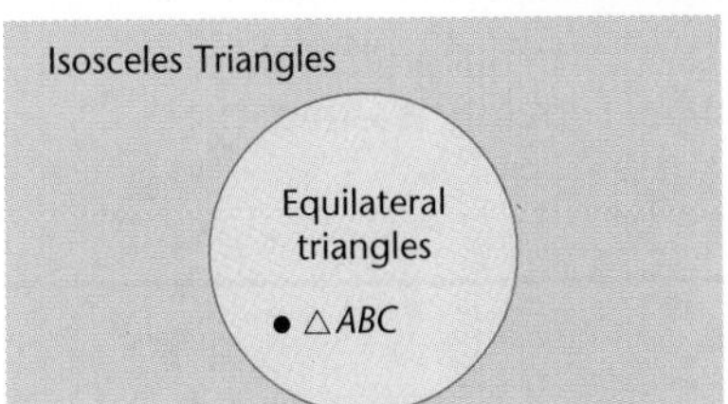

1. If a triangle is equilateral, then it is isosceles.
2. Triangle ABC is equilateral.

Then you can write a third statement that is the logical conclusion of Statements 1 and 2. (The Venn diagram illustrates the logic.)

3. Triangle ABC is isosceles. ❖

What Is a "True" Conditional?

In the above example, Statements 1 and 2 lead to Statement 3, a *logical conclusion*. A conditional (like Statement 1) that leads only to true conclusions is said to be a **true conditional**. Such conditionals accurately reflect the reality they are supposed to represent. So, if Statement 2 is true, you can be sure that Statement 3 is also true.

History Throughout history, people have been challenged to "Prove it!" Students can consider the following questions: What events proved the Earth was round and not flat? How can you prove who won the World Series in 1945? Is it possible to prove the exact date that Europeans first arrived on the shores of North America?

Cooperative Learning

Have students group themselves into categories chosen by the teacher. Draw Venn diagrams for these categories, and "act out" the relationships illustrated in the diagrams. Possible categories could be students who are multilingual, students who know how to water ski, and students who have birthdays in May.

Teaching tip

Be sure students understand that the tick marks on the triangle in Example 1 indicate congruence.

Alternate Example 1

Use logical reasoning to show that Tina lives in the state of California. Suppose you know that Tina lives in San Francisco. Illustrate your argument with a Venn diagram.

1. If a person lives in the city of San Francisco, then he or she lives in the state of California.
2. Tina lives in San Francisco.
3. Then Tina lives in California.

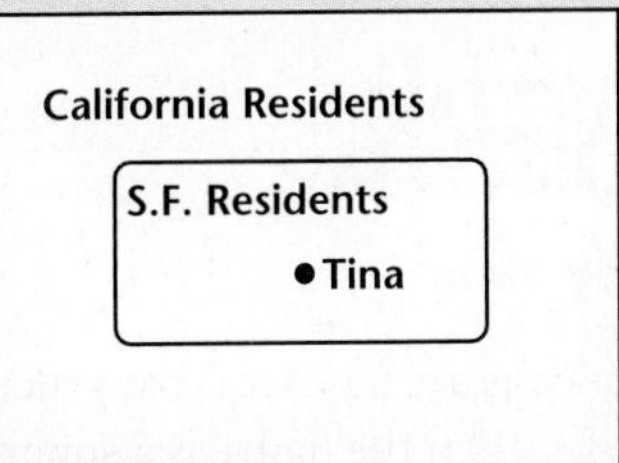

Alternate Example 2

Write a conditional with the conclusion *the person lives in California* and the hypothesis *a person lives in San Francisco.* Then write the converse of your statement.

[Statement: If a person lives in San Francisco, then the person lives in California.

Converse: If a person lives in California, then the person lives in San Francisco.]

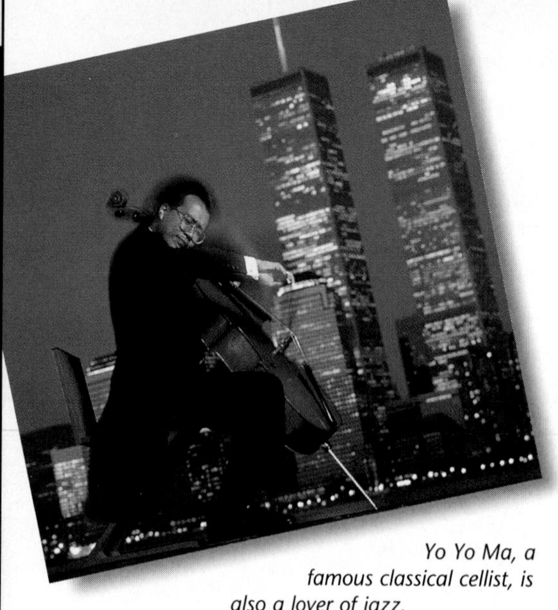

Yo Yo Ma, a famous classical cellist, is also a lover of jazz.

Suppose someone says, "If a person performs classical music, then that person dislikes jazz." Can you think of a person who proves that this is not true?

Yo Yo Ma provides a **counterexample** that proves this conditional to be false. If you represent the conditional with a Venn diagram, you will see that there is no place for Yo Yo Ma to go—either inside or outside of the diagram.

The diagram cannot be completed, because the conditional is false.

Reversing Conditionals

When you interchange the *if* and the *then* parts of a conditional statement, the new statement, which is also a conditional statement, is called the **converse** of the original statement.

Statement: If a car is a Corvette, then it is a Chevrolet.
Converse: If a car is a Chevrolet, then it is a Corvette.

The original statement is true. But what about its converse? If there is a Chevrolet that is not a Corvette—and there certainly is—then the converse is false.

EXAMPLE 2

Write a conditional with the conclusion *the triangle is isosceles* and the hypothesis *a triangle is equilateral.* Then write the converse of your statement.

Solution ➤

Statement: If a triangle is equilateral, then it is isosceles.
Converse: If a triangle is isosceles, then it is equilateral. ❖

Does your original statement seem to be true? Does the converse?

Try This Write a conditional with the hypothesis *the figure is a square* and the conclusion *a figure has four congruent sides.* Then write the converse. Does your original conditional seem to be true? Does its converse? If either seems to be false, find a counterexample.

ongoing ASSESSMENT

Yes, an equilateral triangle has at least two sides equal. No, a triangle with exactly two sides equal is isosceles but not equilateral.

ongoing ASSESSMENT

Try This

If a figure has four congruent sides, then the figure is a square. False

If a figure is a square, then the figure has four congruent sides. True

A rhombus can have four congruent sides and not be a square.

ENRICHMENT Some students may wish to consider the symbolic form of the logical reasoning presented in this lesson. Essentially, the reasoning is as follows: If a statement of the form *If P, then Q* is *assumed* to be true, and if *P* is known to be true, then *Q must* be true. Symbolically, this can be written in two ways:

If P, then Q	$P \Rightarrow Q$
and P	P
Therefore Q	$\therefore Q$

INCLUSION strategies **Guided Research** Students may wish to explore further the statement at the top of page 75 that "It does not matter whether the conditionals are actually true." By analyzing the logical arguments given in this lesson, students should be able to discern that it is the *form* or *structure* of the statements that is the key in applying logic to arrive at correct conclusions.

Logical Chains

Conditionals can be linked together. The result is a **logical chain**. In the following example, three different conditionals are linked together to form a chain. (It does not matter whether the conditionals are actually true.)

EXAMPLE 3

Given: **1.** If cats freak, then mice frolic.

2. If sirens shriek, then dogs howl.

3. If dogs howl, then cats freak.

Prove: Show that the statement below follows logically from the given statements.

If sirens shriek, then mice frolic.

Solution ➤

Identify the *if* clause of the statement you are trying to prove:

If sirens shriek . . .

Look for a statement that begins with "If sirens shriek":

If sirens shriek, then dogs howl.

Look for a statement that begins with "If dogs howl":

If dogs howl, then cats freak.

Look for a statement that begins with "If cats freak":

If cats freak, then mice frolic.

Finally, by linking the first *if* and the last *then* together, you can conclude:

If sirens shriek, then mice frolic.

Notice that there is a zigzag pattern in the resulting steps of the argument as you set it up:

If sirens shriek, then dogs howl.

If dogs howl, then cats freak.

If cats freak, then mice frolic. ❖

CRITICAL *Thinking*

Notice that we have not proven that mice are frolicking. What have we proven instead? What is lacking in the argument that would be necessary to prove that mice are actually frolicking?

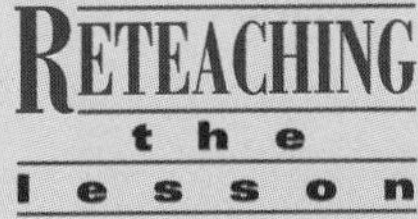

Using Cognitive Strategies The following statements can be used to help students think again about drawing conclusions from conditionals.

1. If you wish to travel to a country outside of the United States, then you need to apply for and be granted a passport.
2. You are now in France.

What conclusions can be drawn from these two statements?

Alternate Example 3

Given:

1. If there is a parade, then fireworks go off.
2. If it is July 4th, then flags are flying.
3. If flags are flying, then there is a parade.

Prove: If it is July 4th, then fireworks go off.

[**If it is July 4th, then flags are flying. If flags are flying, then there is a parade. If there is a parade, then fireworks go off. If it is July 4th, then fireworks go off.**]

CRITICAL *Thinking*

We have proven that *if* sirens shriek, then dogs howl. Students need to see that the conclusion is correct and is *independent* of the actual truth value of the statements involved. For the argument to prove something about the real world, we would have to know that sirens are actually shrieking (i.e., that the hypothesis of the first conditional is true). Also, the conditionals themselves must actually be true, or the conclusion will not be "forced" in the real world. (In fact, the conditionals in the example are not true.)

Some students will notice that dogs howling in the real world, or cats freaking, will also secure the result—in shortened versions of the argument.

CRITICAL *Thinking*

Since two conditionals can be linked to form a third conditional, the third conditional can be linked with a fourth conditional to form a fifth conditional, and so on.

Assess

Selected Answers

Odd-numbered Exercises 7–33

Assignment Guide

Core 1–13, 16–18, 26–38

Core Plus 1–5, 13–25, 29–38

Technology

Geometry graphics software can be used to manipulate the figures in Exercises 14–15 and 29–33.

Error Analysis

A common error students make in geometry is to assume that the converse of a true conditional is also true. The exercises and problems in this lesson should convince students that this is not the case.

6. If a person lives in Ohio, then the person lives in the United States.
7. Hypothesis: a person lives in Ohio
   Conclusion: the person lives in the United States

In proving the conditional, "If sirens shriek, then mice frolic," you have actually relied on an idea which has been assumed to be true without mentioning it. What is that idea?

IF-THEN TRANSITIVE PROPERTY

Suppose you are given:	You can conclude:
If A then B If B then C	If A then C

2.2.1

CRITICAL *Thinking*

Explain how you use the same property repeatedly when you link long chains of conditionals together.

Exercises & Problems

Communicate

1. Look at the satellite photo of the United States. Can you conclude that it rained all day in Miami. Why or why not?
2. Draw a Venn diagram that illustrates the statement given in the caption.
3. What is the converse of the given statement? Use your Venn diagram to illustrate whether or not the converse is true.
4. Explain how to write the converse of a given conditional statement.
5. Explain how to disprove a given conditional.

On March 21, it was cloudy all day in southern Florida.

Practice & Apply

For Exercises 6–9, refer to the following statement:

All people who live in Ohio live in the United States.

6. Rewrite the statement above as a conditional statement.
7. Identify the hypothesis and the conclusion of the statement.
8. Draw a Venn diagram that illustrates the statement.
9. Write the converse of the statement and construct its Venn diagram. If the converse is false, illustrate this with a counterexample.

8.

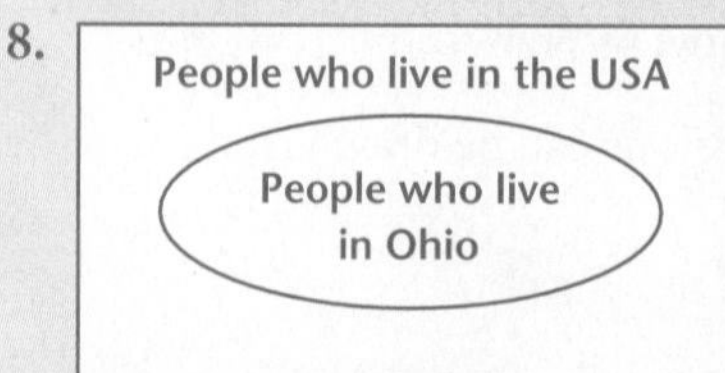

9. If a person lives in the United States, then the person lives in Ohio. False. A person who lives in Indiana is a counterexample.

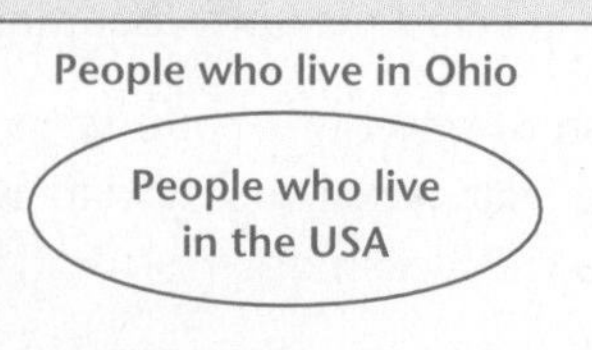

In Exercises 10–12, use the information given to draw a conclusion about the individual named. Then draw a Venn diagram to illustrate the solution.

10. If an animal is a mouse, then it is a rodent. "Mikey" is a mouse.

11. If someone is a human being, then he or she is mortal. Socrates is a human being.

12. If a person files his or her income tax early, then he or she will get an early refund. Susan filed her income tax at 7:00 A.M.

If the conclusion about Susan is not actually true, what might be the reason(s) for this?

13. Humor Can logic be used to do the impossible? Consider the following argument.

Is this man disappearing?

The independent farmer is disappearing.
That man is an independent farmer.
Therefore, that man is disappearing.

How could you write the above argument using a conditional statement? How would you criticize the argument?

For Exercises 14–15, write a conditional statement with the given hypothesis and conclusion, and then write the converse of the statement. Is the original statement true? Is the converse? If either is false, give a counterexample to show that it is false. (You may use the fact that the sum of the measures of the angles of a triangle is 180°.)

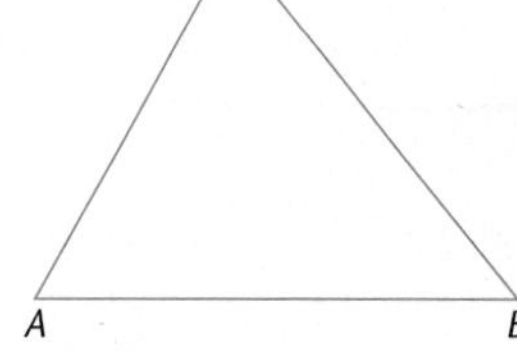

14. Hypothesis: "$m\angle A + m\angle B = 90°$"

Conclusion: "$m\angle C = 90°$"

15. Hypothesis: "$m\angle C = 90°$"

Conclusion: "triangle *ABC* is a right triangle"

In Exercises 16–18, arrange the three statements to form a logical claim. Then write the conditional statement that the argument as a whole proves.

16. If it is cold, then birds fly south.
If the days are short, then it is cold.
If it is winter, then the days are short.

17. If quompies pawn, then rhomples gleer.
If druskers leer, then homblers fawn.
If homblers fawn, then quompies pawn.

Do you think an argument has to make sense to be logical?

14. If $m\angle A + m\angle B = 90°$, then $m\angle C = 90°$. True
If $m\angle C = 90°$, then $m\angle A + m\angle B = 90°$. True

15. If $m\angle C = 90°$, then triangle *ABC* is a right triangle. True
If triangle *ABC* is a right triangle, then $m\angle C = 90°$. False; triangle *ABC* may have $m\angle A = 90°$.

16. If it is winter, then the days are short. If the days are short, then it is cold. If the days are cold, then birds fly south. If it is winter, then birds fly south.

17. If druskers leer, then homblers fawn. If homblers fawn, then quompies pawn. If quompies pawn, then rhomples gleer. If druskers leer, then rhomples gleer. No, as long as the reasoning chain is valid.

alternative ASSESSMENT

Performance Assessment

The ability to create a problem that satisfies certain criteria demonstrates a deep level of understanding of the ideas involved. To assess students' understanding of logical chains and the if-then transitive property, assign them the task of making up three statements that can be used to form a logical chain, as in Exercises 16–18.

10. "Mikey" is a rodent.

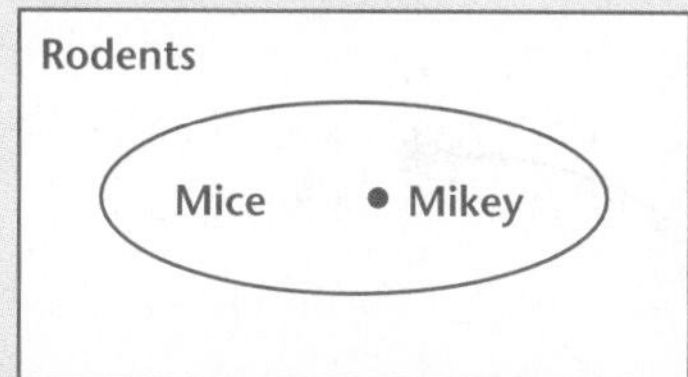

11. Socrates is mortal.

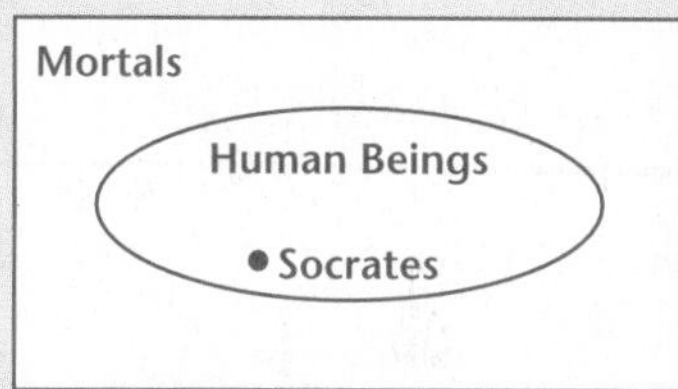

12. "Early" refers to a time of year rather than a time of day, and/or Susan might not be entitled to a refund.

13. If a person is an independent farmer, then the person is disappearing.

That man is an independent farmer. Therefore, that man is disappearing.

The statement does not mean that farmers are vanishing visually, only that many cannot continue as farmers.

18. If the police radar catches Tim speeding, then Tim gets a ticket.
If Tim drives a car, then Tim drives too fast.
If Tim drives too fast, then the police radar catches him speeding.

Should Tim's parents let him borrow the family car? Which statement(s) might Tim challenge to help him get the keys?

19. Criminal Justice A robber broke into an apartment. He left no fingerprints, but he did leave a mess behind, which included his personal comb. The police have a suspect in custody. You are the lawyer for the suspect, and upon meeting your client, you discover that he is completely bald. Using conditional statements, write a logical argument that you would present in court to prove the innocence of your client.

For Exercises 20–24, refer to the following statement:

If $m\angle A + m\angle B < 90°$, then the triangle is obtuse.

20. Identify the hypothesis. $m\angle A + m\angle B < 90°$

21. Identify the conclusion. The triangle is obtuse

22. Prove the statement.

23. Write the converse of the statement.

24. Prove or disprove the converse.

25. Fine Arts Choose one of the objects below. Suppose you were going to argue that the object is a work of art. Which of the following conditional statements would you use to lay the foundations for your argument? If you would like, supply your own conditional statement.

If an object displays form, beauty, and unusual perception, on the part of its creator, then the object is a work of art.

If an object displays creativity on the part of the creator, then the object is a work of art.

Sol Lewitt, *Two Open Modular Cubes/Half Off,* 1972

Jean Miro, *Tre Donne,* 1935

Write an outline of the argument you would use. Or if you prefer, write an argument that the object you choose is *not* a work of art.

18. If Tim drives a car, then Tim drives too fast. If Tim drives too fast, then the police radar catches him speeding. If the police radar catches Tim speeding, then Tim gets a ticket. If Tim drives a car, then Tim gets a ticket. Tim might say he will not drive too fast, or that it is unlikely police radar will catch him.

19. If a person has no hair, then that person will not carry a comb. If a person does not carry a comb, then the person cannot drop a comb. Therefore, if a person has no hair, then the person cannot drop a comb. The criminal dropped a comb, so the suspect is innocent.

22. $m\angle A + m\angle B + m\angle C$ must equal 180°.

If $m\angle A + m\angle B < 90°$ then $m\angle C > 90°$ to give a sum of 180°. If $m\angle C > 90°$, then $\triangle ABC$ is obtuse by definition.

23. If the triangle is obtuse, then $m\angle A + m\angle B < 90°$.

24. Let $\angle A$ be the obtuse angle. Then $m\angle A + m\angle B > 90°$. False

25. Answers will vary. If the student chooses the first statement, the student's argument should include conditional statements showing that the object displays form, beauty, and unusual perception on the part of the creator.

If the student chooses the second statement, the student's argument should include conditional statements showing that the object displays creativity on the part of the creator.

Practice Master

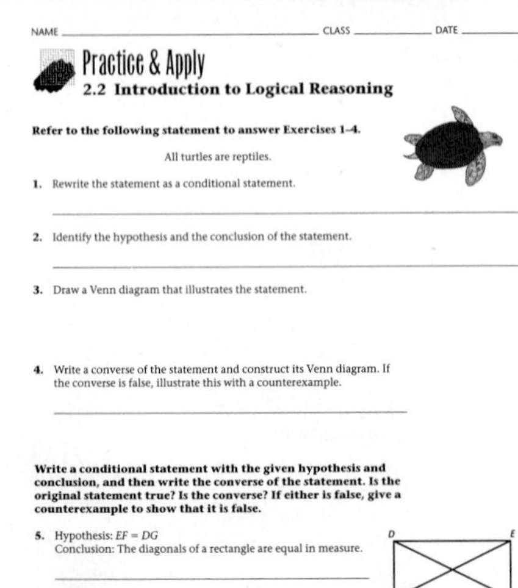

NAME ______________ CLASS ________ DATE ________

Practice & Apply

2.2 Introduction to Logical Reasoning

Refer to the following statement to answer Exercises 1–4.

All turtles are reptiles.

1. Rewrite the statement as a conditional statement.

2. Identify the hypothesis and the conclusion of the statement.

3. Draw a Venn diagram that illustrates the statement.

4. Write a converse of the statement and construct its Venn diagram. If the converse is false, illustrate this with a counterexample.

Write a conditional statement with the given hypothesis and conclusion, and then write the converse of the statement. Is the original statement true? Is the converse? If either is false, give a counterexample to show that it is false.

5. Hypothesis: $EF = DG$
Conclusion: The diagonals of a rectangle are equal in measure.

Arrange the three statements to form a logical chain. Then write the conditional statement that the argument as a whole proves.

6. If it is warm, then it is spring.
If flowers are blooming, it is warm.
If you see bees, then flowers are blooming.

44 Practice & Apply HRW Geometry

Look Back

26. A floor is modeled by what geometic figure? **[Lesson 1.1]** a plane

27. "Perpendicular" is defined to mean what? Using folding paper, construct a segment and its perpendicular bisector. **[Lesson 1.2]**

28. What is an angle bisector? Construct one using folding paper. **[Lesson 1.2]** An angle bisector divides an angle into equal angles. Fold one side of the angle onto the other. The crease is the angle bisector.

In a triangle, which of the following intersect in a single point? [Lesson 1.3]

29. altitudes single point

30. medians single point

31. perpendicular bisectors single point

32. angle bisectors single point

33. Reflect the drawing shown with respect to the given line. **[Lesson 1.6]**

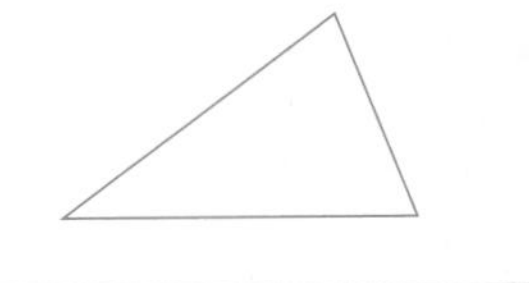

34. Name three types of transformations. **[Lesson 1.7]** translation, rotation, reflection

Look Beyond

Another kind of reasoning that is important in mathematics is called inductive reasoning. An inductive argument reaches conclusions based on the recognition of patterns. **Use inductive reasoning in Exercises 35–37 to find the next number in each sequence.**

35. 5, 8, 11, 14, ____ 17

36. 20, 27, 36, 47, 60, ____ 75

37. 3, 1, $\frac{1}{3}$, $\frac{1}{9}$, ____ $\frac{1}{27}$

38. The **Fibonacci sequence** has the pattern shown below. Numbers that occur in the sequence are known as Fibonacci numbers. What are the next five numbers in the sequence?

1, 1, 2, 3, 5, ____
8, 13, 21, 34, 55; add two previous terms

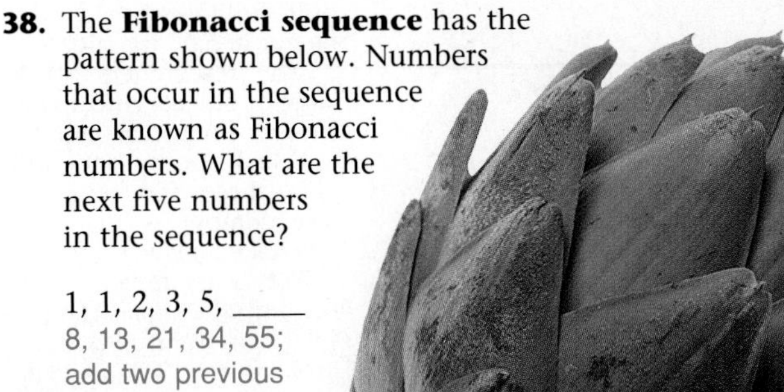

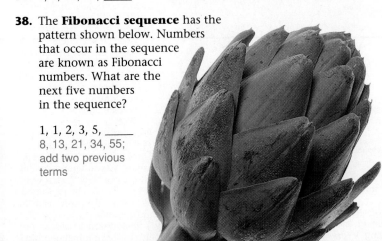

In artichokes and other plants, the number of spirals in each direction are Fibonacci numbers. This artichoke has 5 clockwise spirals and 8 counterclockwise spirals.

27. Two lines or segments are perpendicular if they form right angles when they meet. Fold one endpoint of the segment onto the other endpoint. The crease is the perpendicular bisector.

33.

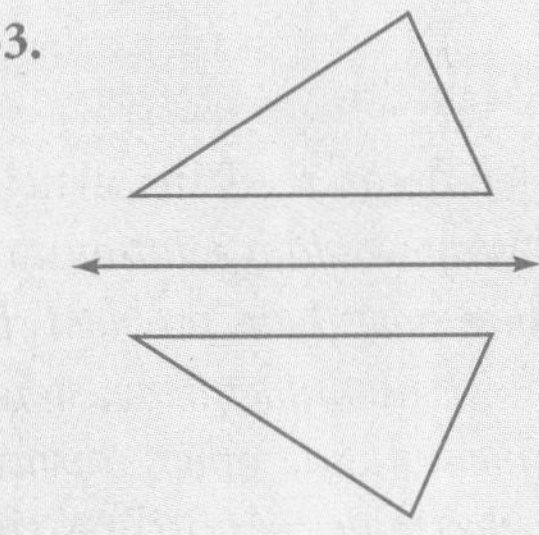

Look Beyond

Exercises 35–38 introduce students to inductive reasoning by using number patterns. In Lesson 2.6, students will learn that inductive reasoning is used to make conjectures, which may or may not be true.

Technology Master

NAME ________ CLASS ________ DATE ________

Technology
2.2 Creating and Testing Conditional Statements

A geometry class decided to explore this hypothesis.

If a triangle is equilateral, ...

Using geometry software, one group drew four equilateral triangles of different sizes, like the ones shown. Then they created this list of conclusions.

a. ... then the triangle has three equal sides.
b. ... then the triangle has three equal angles.
c. ... then the triangle has no obtuse angle.
d. ... then the sum of the angle measures is 180°.
e. ... then the triangle is not isosceles.
f. ... then the angle bisectors intersect at one point inside the triangle.
g. ... then the medians intersect at one point inside the triangle.
h. ... then the perpendicular bisectors of the sides intersect at one point inside the triangle.

Use geometry software to make a sketch of four equilateral triangles. Then refer to the list of conditional statements.

1. Test each conditional statement, using your sketch as necessary. Which statements are true? Which, if any, are false?
2. Write the converse of each conditional statement. Then test the converses. Which are true? Which are false?
3. Write two other true conditional statements to add to the list.
4. Write the converse of each statement that you wrote in Exercise 3. Is each converse true or false?
5. Begin with the hypothesis, "If a quadrilateral is a square, ..." Make a list of as many true conditional statements as you can.

56 Technology HRW Geometry

Focus

Students explore how the number of possible routes in a tour grows factorially with the number of cities.

Motivate

After students read the article, ask, *"Why would a solution to a traveling salesman problem be useful in manufacturing stereo equipment and other devices using electronic circuits?"* (A circuit board travels from point to point under a laser drill just as a salesman travels from town to town.)

Too Tough for Computers

Communication News

Math Problem, Long Baffling, Slowly Yields

By Gina Kolata, *New York Times*

A century-old math problem of notorious difficulty has started to crumble. Even though an exact solution still defies mathematicians, researchers can now obtain answers that are good enough for most practical applications.

The traveling salesman problem asks for the shortest tour around a group of cities. It sounds simple—just try a few tours out and see which one is shortest. But it turns out to be impossible to try all possible tours around even a small number of cities.

Companies typically struggle with traveling salesmen problems involving tours of tens of thousands or even hundreds of thousands of points.

For example, such problems arise in the fabrication of circuit boards, where lasers must drill tens to hundreds of thousands of holes in a board. What happens is that the boards move and the laser stays still as it drills the holes. Deciding what order to drill these holes is a traveling salesman problem.

Very large integrated circuits can involve more than a million laser-drilled holes, leading to a traveling salesman problem of more than a million "cities."

In the late 1970s, investigators were elated to solve 50-city problems, using clever methods that allow them to forgo enumerating every possible route to find the best one. By 1980, they got so good that they could solve a 318-city problem, an impressive feat but not good enough for many purposes.

Dr. David Johnson and Dr. Jon Bentley of AT&T Bell Laboratories are recognized by computer scientists as the world champions in solving problems by getting approximate solutions. They began by working on problems involving about 100,000 cities. By running a fast computer for two days, they can get an answer that is guaranteed to be either the best possible tour or less than 1 percent longer than the best possible one.

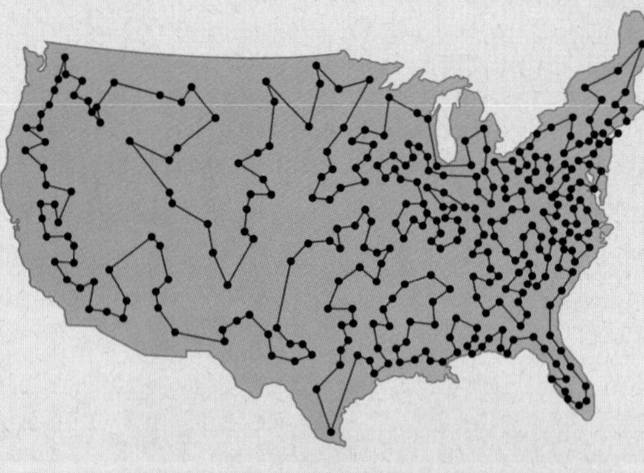

As of 1968, the computer solution for the shortest route connecting 532 cities with AT&T central offices looked like this. It was the biggest such problem calculated up to that time. Now, a route for 2,392 destinations has been computed, and mathematicians are working on a 3,038-city problem. (Source: New York University and Institute for Systems Analysis)

In most practical situations, an approximate solution is good enough, Dr. Johnson said. By just getting to within about 2 percent of the perfect solution of a problem involving drilling holes in a circuit board, the time to drill the holes can usually be cut in half, he said.

The researchers break large problems into many smaller ones that can be attacked one by one, and give these fragments to fast computers that can give exact answers.

For example, Dr. Bentley said, "If I ask you to solve a traveling salesman problem for 1,000 cities in the U.S., you would do it as a local problem. You might go from New York toward Trenton and then move to Philadelphia," he explained. Then the researchers would repeat this process from other hubs, like Chicago, and combine the results. "We end up calculating only a few dozen instances per point," Dr. Bentley said. "If you have a million cities, you might do only 30 million calculations."

1a. Let $Au \leftrightarrow$ Austin, $E \leftrightarrow$ El Paso, $Ab \leftrightarrow$ Abilene, $H \leftrightarrow$ Houston

There are two "shortest" paths:
$Au \Rightarrow H \Rightarrow E \Rightarrow Ab \Rightarrow Au$, 1550 miles
$Au \Rightarrow Ab \Rightarrow E \Rightarrow H \Rightarrow Au$, 1550 miles
Other routes are 1700, 1950, 1700, and 1950 miles.

b. $Au \Rightarrow H \Rightarrow Ab \Rightarrow E \Rightarrow Au$
$Au \Rightarrow H \Rightarrow E \Rightarrow Ab \Rightarrow Au$
$Au \Rightarrow Ab \Rightarrow H \Rightarrow E \Rightarrow Au$
$Au \Rightarrow Ab \Rightarrow E \Rightarrow H \Rightarrow Au$
$Au \Rightarrow E \Rightarrow Ab \Rightarrow H \Rightarrow Au$
$Au \Rightarrow E \Rightarrow H \Rightarrow Ab \Rightarrow Au$
All possible places for first stop, then second stop, etc., are listed systematically.

Mathematical Reasoning

The most obvious way of solving a "traveling-salesman" problem is to try all possible routes—which turns out to be very time consuming, even for the fastest computers. So there has been a search for methods of solving the problem that will reduce the number of steps. Finding shortcut methods involves conjecture and mathematical reasoning.

Cooperative Learning

Suppose that a computer can calculate a billion routes per second. How long do you think it would take the computer to calculate all possible routes for 50 cities? In Activities 1–3 you will find out.

1. A music group based in Austin plans to tour 3 cities, Abilene, Houston, and El Paso. The map at the right shows the distances in miles between cities.

a. What is the shortest possible route? (In these problems, a route means a path that starts and ends at one city and passes through every other city once and only once.)

b. List all the possible routes. How do you know you have listed them all?

Number of cities	Number of routes
3	?
4	?
5	?
6	?

2. Now see what happens as you change the number of cities.

a. Copy and complete the chart. Keep filling in rows until you identify a pattern that you are sure will continue.

b. Describe the pattern you have found.

c. If n is the number of cities, write a formula for R, the number of possible routes.

3. You can now use your formula to estimate the computer time required.

a. How many routes are possible for 50 cities?

b. Suppose a computer could calculate a billion route lengths every second. How many seconds would it take the computer to calculate the length of every route for 50 cities? How many years is that?

c. Suppose thousands of computers worked at the same time. Do you think they could make all the calculations in a reasonable time? Why or why not?

2a.

# of cities visited	# of routes
3	$3 \times 2 \times 1 = 6$
4	$4 \times 3 \times 2 \times 1 = 24$
5	$5 \times 4 \times 3 \times 2 \times 1 = 120$
6	$6 \times 5 \times 4 \times 3 \times 2 \times 1 = 720$

b. Starting with the number of cities to be visited, multiply by consecutive smaller integers through 1. Each number represents the number of places the band could choose to go at that stage of the tour.

c. $R = n(n-1)(n-2) \times \ldots \times 3 \times 2 \times 1 = n!$ Notice that n does not include the starting-point city.

Cooperative Learning

Have students work in pairs. Encourage students to take a systematic approach. By using a pattern in their lists, students can avoid duplicates and omissions. In Part 2a, have students start at the top with three cities. Encourage them to look for why a pattern develops as they add one city, then another, and so on. For added challenge, ask students to explain why the pattern will continue.

If students do not have a factorial function on their calculator, they can perform a chain calculation ($49 \cdot 48 \cdot 47 \cdot \ldots \cdot 1$) in a minute or two.

Discuss

Check the plausibility of the following statement: *If every electron in the universe were actually a computer that could calculate 1 billion routes a second, it would take all those computers more than a billion years to calculate all the routes for 200 cities.* You may want to give the students the number 10^{80} as an estimate of the number of electrons in the universe. Also, if their calculators do not handle numbers above 10^{100}, they will need to know that $200! \approx 7.89^{374}$. They can calculate the number of seconds in a 365.25-day year to be 31,557,600, or about $10^{7.5}$ ($\log_{10} 31{,}557{,}600 \approx 7.5$). Thus, in a billion years, the electron computers would have computed "only" $10^{105.5}$ routes—not even close to the required number!

PREPARE

Objectives

- Use Venn diagrams to determine definitions of objects.
- Use principles of logic to create definitions of objects.

- Practice Master 2.3
- Enrichment Master 2.3
- Technology Master 2.3
- Lesson Activity Master 2.3
- Quiz 2.3
- Spanish Resources 2.3

Assessing Prior Knowledge

1. Write as a conditional statement: A triangle is a figure with three sides. [**If a figure is a triangle, then it has three sides.**]
2. What is the converse of $A \Rightarrow B$? [$\boldsymbol{B \Rightarrow A}$]
3. If $A \Rightarrow B$ is a true statement, is $B \Rightarrow A$ a true statement? [**not necessarily**]

Use Transparency 15

TEACH

Many disagreements can usually be traced back to using different meanings for the same terms. Discuss with students why it is important in mathematics to know the definitions of the terms used.

why *Have you ever disagreed with someone, only to find out that you and the other person had different definitions of the same terms? In mathematics it is especially important to know the definitions of the terms you are using.*

Which of the figures in Card 3 are floppers? Simply by observing the differences between Card 1 and Card 2 figures, it is possible to write a definition of a flopper.

A flopper is a figure with one eye and two tails.

Using this definition, you can see that figures (d) and (e) are floppers.

Definitions have a special property when they are written as conditional statements. For example,

If a figure is a flopper, then it has one eye and two tails.

You can also write the converse of the statement by interchanging the hypothesis and conclusion:

If a figure has one eye and two tails, then it is a flopper.

Notice that both the original statement and its converse are true. This is the case with all definitions! The two true statements can be combined into a compact form joining the hypothesis and the conclusion with the phrase "if and only if," represented by $p \Leftrightarrow q$.

$$p \text{ if and only if } q, \quad \text{or} \quad p \Leftrightarrow q$$

Hands-On Strategy

Use an object that is readily available to all students in class, such as a pencil or a desk, and ask students how they would define the object. As students suggest various characteristics of the object, write them on the board. Have students agree on the least number of characteristics that can be used in a definition of the object.

By combining the statement and the converse, you create the following definition of a flopper, expressed in logical terms.

A figure is a flopper if and only if it has one eye and two tails.

Venn diagrams can be used to represent two parts of the definition.

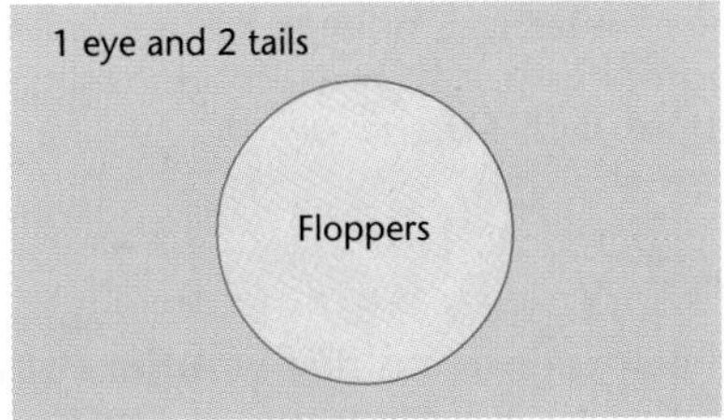

Statement

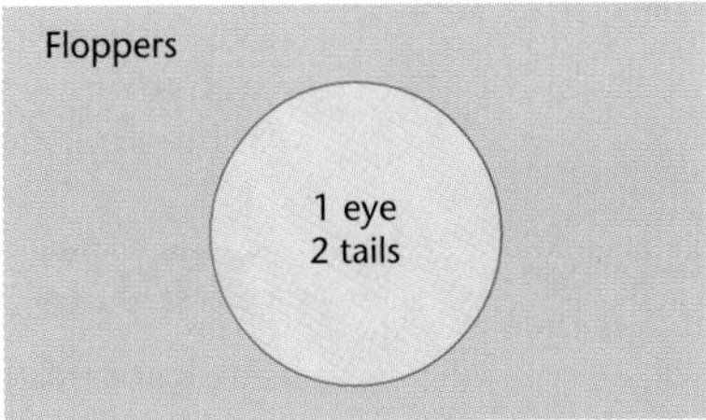

Converse

Can you imagine a Venn diagram that would represent the fact that both the statement and its converse are true? How would the "inner" and "outer" parts of the diagram be related? Discuss.

Exploration 1 Capturing the "Essence of a Thing"

You will need
No special tools

1. Look at the figure on the right. Make up your own name for the object. Then think: What must be true of a geometric figure in order for it to be a <u>(your name for the object)</u>?

2. According to your idea of a <u>(your name for the object)</u> which of the objects below would you consider to be one? There are no set rules for this. The conditions are up to you!

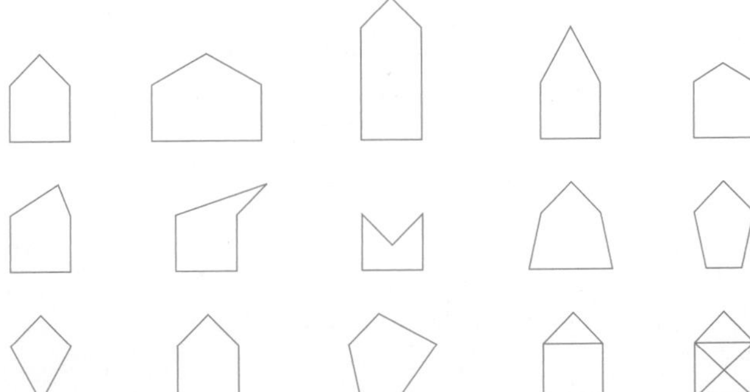

3. Write your own definition of a <u>(your name for the object)</u>. Base your definition on your answers to Step 1, or any other decisions you might want to make about the object. Test your conditions for the object to be sure they actually provide a true definition. ❖

ENRICHMENT Have students design and make games in which imaginary and geometric figures must be identified or drawn according to definitions. Students can add to the games throughout the year and play them as lesson reinforcement and review.

INCLUSION strategies **Using Cognitive Strategies** Students who are having difficulty with the abstract nature of the activities in this lesson can enhance their understanding of definitions by going back to Chapter 1 and making a list of all definitions of geometric terms introduced in that chapter.

TEACHING tip

Students need to understand that *p if and only if q* is a compact form of writing two conditionals: *If p then q* and *If q then p*. An abbreviated form of this is *p iff q*. Students can also read the symbolic expression $p \Leftrightarrow q$ as: *p implies q* and *q implies p*.

CRITICAL Thinking

The inner and outer parts would be the same and would contain both statements.

Exploration 1 Notes

These activities will help students understand that the essential nature of a geometric figure must be incorporated into its definition. They also learn that the name given to an object can be arbitrarily chosen. However, once named and defined, the definition must be able to distinguish that object from all others unambiguously.

ongoing ASSESSMENT

Sample answer: A <u>house-pentagon</u> is a five-sided polygon with a base made up of three sides of a rectangle and a "roof" made up of two sides of an isoceles triangle. Students should think of ways to make such proposed answers more precise, according to their conceptions of the class of objects in question.

Ongoing Assessment

Check student definitions; the better the definition, the more precisely the object is determined.

Teaching tip

Refer students to the third box and ask them to identify a pair of nonadjacent angles (∠3 and ∠5). Then refer them back to the other two boxes to identify other pairs of nonadjacent angles and to explain why the angles are not adjacent.

Use Transparency 16

Exploration 2 Notes

When discussing the definition of adjacent angles at the bottom of this page, you may wish to ask students how they would define the *interior* points of an angle.

Cooperative Learning

After completing the explorations, put students into groups and give them sets of attribute blocks. Tell them to divide all the attribute blocks into two categories, and have them write a definition for each category. Then have them group the blocks into three categories with three definitions, and so on. Ask students if it is easier to write definitions when there are fewer categories or more categories, and have them explain their reasoning.

Exploration 2 *Creating Your Own Objects*

You will need
No special tools

Create your own object; then give it a name and a definition. Your object need not be geometrical, or even mathematical, but it should be something you can draw. Test your definition to be sure it is a true one. Share your definitions with others. ❖

EXTENSION

An important geometry concept is that of **adjacent angles**. By examining the figures in the first two boxes, you should be able to form an idea about what adjacent angles are—and also what they are not. This information will enable you to write your own definition.

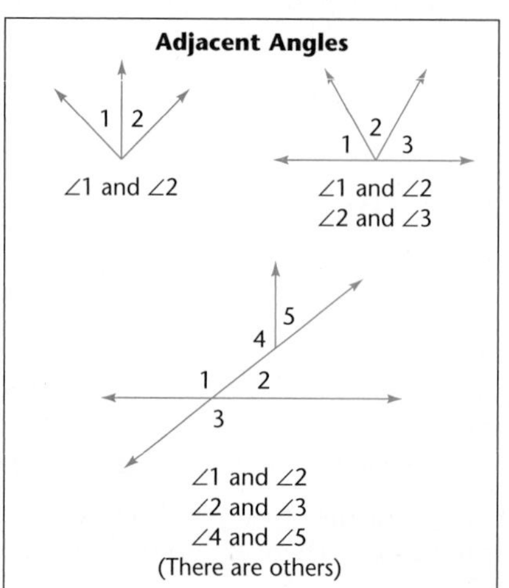

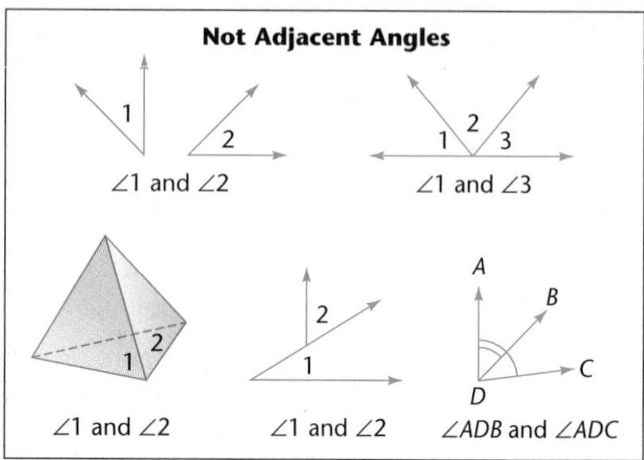

Notice that adjacent angles have a common vertex, a common ray, and do not overlap. So in the third box, angles 1 and 2 are adjacent angles, as are angles 3 and 4 and angles 4 and 5.

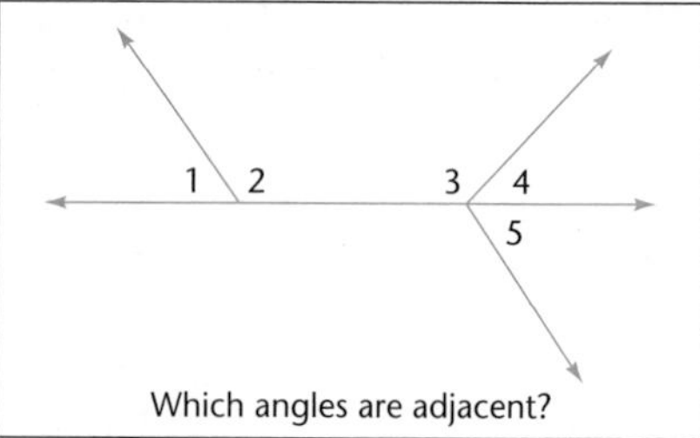

A difficulty in defining adjacent angles is to clearly express what is meant by *non-overlapping angles*. The following definition illustrates one way of doing this:

Adjacent angles are two angles in a plane that have their vertices and one ray in common, but no interior points in common.

Using Patterns The idea of a transformation can be used to reteach the lesson. For example, ask students how they would define a *rigid motion.* Have them write a definition of a rigid motion as a conditional statement, as the converse of the statement, and in "if and only if" form. Students can also take definitions from other subject areas, such as science or language arts, and write these as conditionals.

Communicate

1. Explain why a definition is a special form of if-then statement.
2. Why is it necessary to have undefined terms in geometry?
3. Explain why the following statement is not a definition:

   A tree is a plant with leaves.

4. These are blips. These are not blips

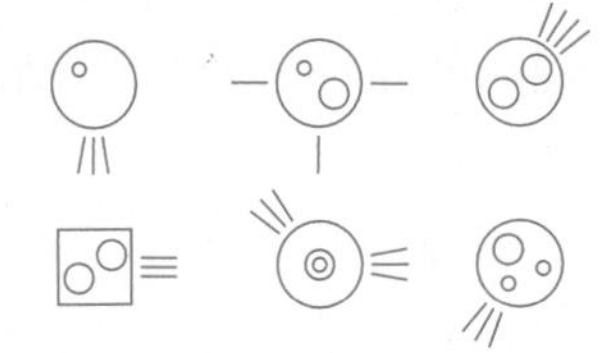

Find a definition for a blip and identify which of the following are blips.

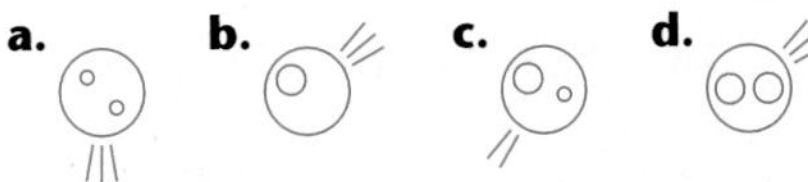

Practice & Apply

In Exercises 5–9, use the following steps to determine whether the given sentence is a definition. First write the sentence as a conditional statement. Then write the converse of the conditional. Finally, write an if-and-only-if statement. Does your final statement seem to be true? Discuss your answer. What is your conclusion? (Exercise 5 is partially worked.)

5. A teenager is a person over 12 years old.
   Conditional statement: If a person is a teenager, then . . .
   Converse: If a person is over 12, then . . .
   If-and-only-if: A person is a teenager if and only if . . .
6. A teenager is a person from 13 to 19 years old.
7. Zero is the integer between −1 and 1.
8. **Geology** Granite is a very hard crystalline rock.
9. **Music** A sitar is a lutelike instrument of India.
10. **Chemistry** Write a definition for propane as a conditional. Write the converse and then an if-and-only-if statement. Does your final statement seem to be true?

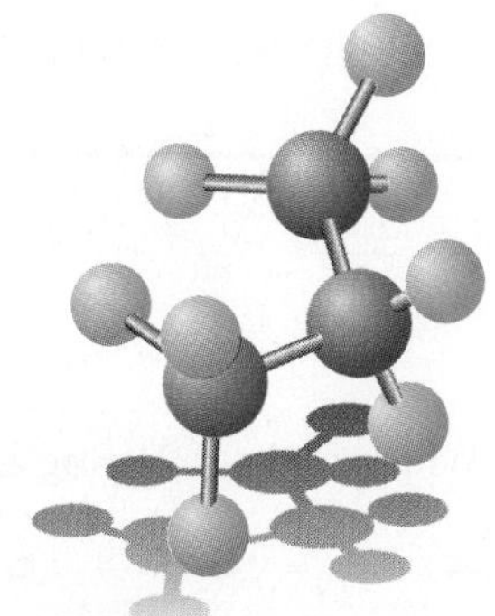

Propane is a colorless gas with the chemical formula C_3H_8.

5. . . . the person is over 12 years of age.
   . . . the person is a teenager.
   . . . the person is over 12 years of age.
   This is not a definition since the converse is false. A person who is 21 years old is not a teenager.

8. If a substance is granite, then the substance is a very hard crystalline rock. If a substance is very hard crystalline rock, then the substance is granite. A substance is granite if and only if it is very hard crystalline rock. This is not a definition since the converse is false. There are hard crystalline rocks other than granite—quartz and garnet, for example.

The answers to Exercises 6 and 7 can be found in Additional Answers beginning on page 727.

Assess

Selected Answers

Odd-numbered Exercises 5–23

Assignment Guide

Core 1–9, 11–14, 16–23

Core Plus 5–10, 13–15, 19–24

Technology

Geometry graphics tools can be used for Exercises 17–23.

Error Analysis

Some students may have difficulty writing the sentences in Exercises 5–10 as conditional statements. Have some students write their conditional statements on the board for each exercise. Seeing other students' conditionals will usually help all students to be successful.

9. If an instrument is a sitar, then it is a lutelike instrument of India. If an instrument is a lutelike instrument of India, then the instrument is a sitar. An instrument is a sitar if and only if it is a lutelike instrument of India. This is not a definition since the converse is false. There are other lutelike instruments of India–the vena, for example.

10. If a gas is propane, then it is colorless and has the chemical formula C_3H_8. If a gas is colorless and has the chemical formula C_3H_8, then it is propane. A gas is propane if and only if it is colorless and has the chemical formula C_3H_8. This is a definition, but one in terms of the label C_3H_8, which must also be defined.

alternative ASSESSMENT

Authentic Assessment

As a class activity, ask students to list the elements they think are important for a true definition to have. For example, it names the object being defined; it captures the "essence" of a thing; it can be expressed in "if and only if" form; it does not include unnecessary information. Then have them write several definitions as conditionals.

13. b, c

14. A figure is a regular polygon if and only if all its sides have equal length and all its angles have equal measure.

15. Student definitions should specify the object as uniquely as possible.

Practice Master

NAME ____ CLASS ____ DATE ____

Practice & Apply
2.3 Exploring Definitions

Write the given sentence in the forms requested. Conclude whether the if-and-only-if statement is true.

1. A high school sophomore is in the 10th grade.
   Conditional statement: ____
   Converse: ____
   If-and-only-if: ____
   Conclusion: ____
2. A diamond can cut glass.
   Conditional statement: ____
   Converse: ____
   If-and-only-if: ____
   Conclusion: ____
3. A 130° angle is an acute angle.
   Conditional statement: ____
   Converse: ____
   If-and-only-if: ____
   Conclusion: ____

These figures are hexagons. These figures are not hexagons.

4. Which of the following are hexagons? ____
   a. b. c. d.
5. Write a definition for a hexagon. ____

HRW Geometry Practice & Apply 45

11. The following are fliches:

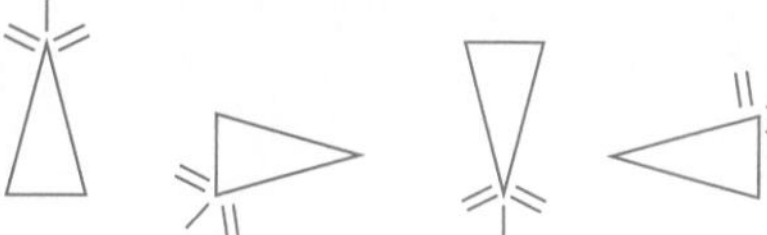

The following are not fliches:

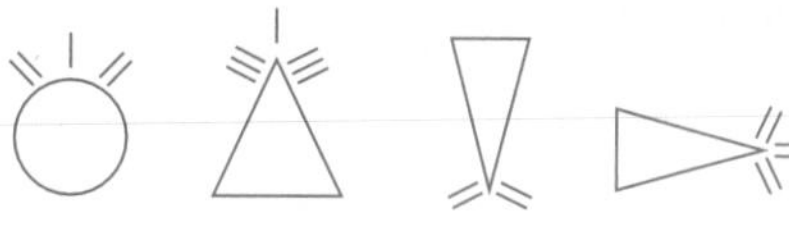

Which of the following are fliches? b, d

a. **b.** **c.** **d.**

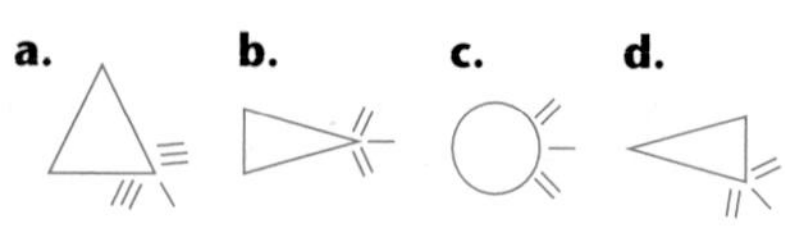

12. The following are zoobies:

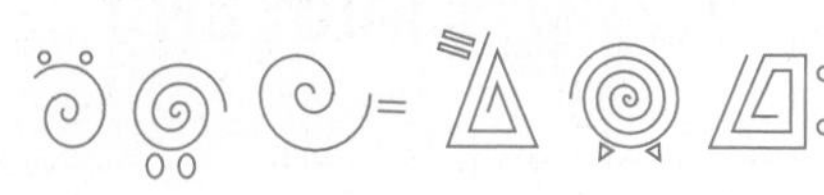

The following are not zoobies:

Which of the following are zoobies? a

a. **b.** **c.** **d.**

13. The following are parallelograms:

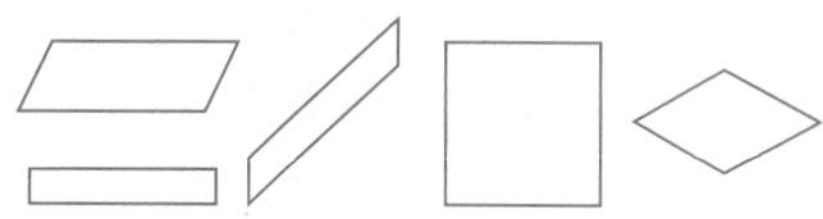

The following are not parallelograms:

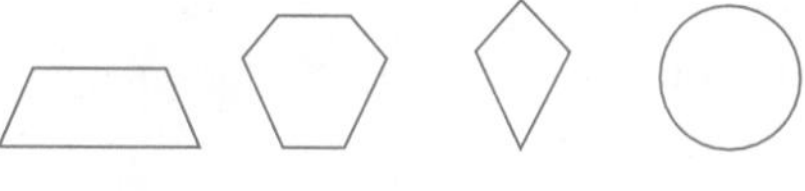

Which of the following are parallelograms?

a. **b.** **c.** **d.**

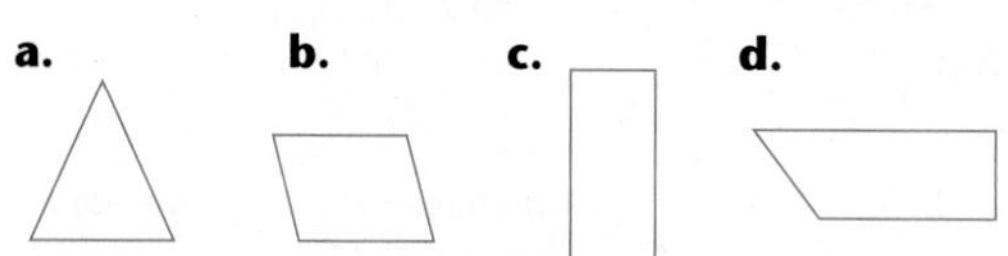

14. The following are regular polygons:

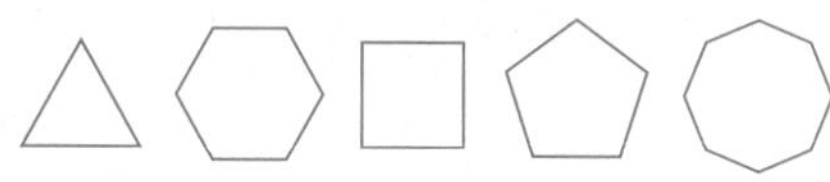

The following are not regular polygons:

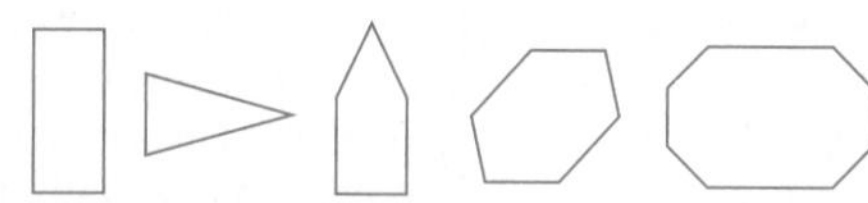

Write a definition for a regular polygon.

15. Make up problems that are similar to Exercises 11–14. Share them with your classmates.

17–18.

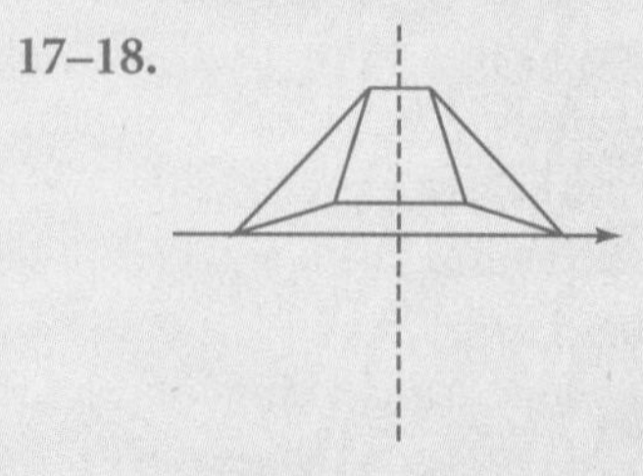

Look Back

16. Point A is the midpoint of $\overline{FB}$. Find x and $\overline{FA}$. **[Lesson 1.4]** $x = -3$; $FA = -2(-3) + 1 = 7$

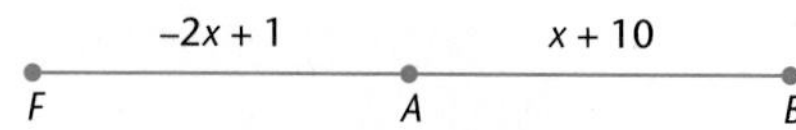

For exercises 17–20, refer to the diagram. [Lesson 1.6]

17. Reflect the triangle through the line of reflection drawn.

18. Draw line segments connecting corresponding parts.

19. What relationship appears to exist between the line segments you drew for Exercise 18? They are parallel.

20. What relationship appears to exist between the line segments you drew for Exercise 18 and the mirror of reflection? They are perpendicular.

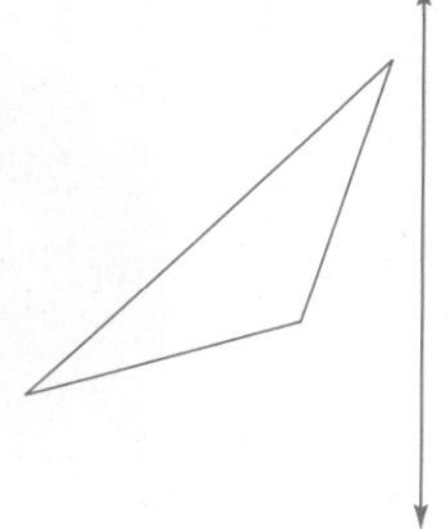

Refer to the diagram for Exercises 21–23. [Lesson 1.7]

21. On your own paper, slide the triangle five units to the right as shown by the arrow.

22. Draw line segments connecting corresponding parts of your drawing.

23. What relationship appears to exist between the segments you drew for Exercise 22? They are parallel.

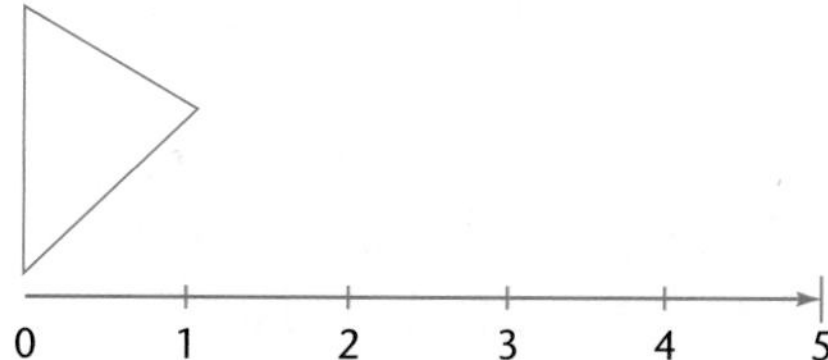

Look Beyond

24. Tonia, Julie, and Susanne are three exceptional students. Use the clues to answer the question below:

a. Two are exceptionally mathematical, two are exceptionally musical, two are exceptionally artistic, and two are exceptionally athletic.

b. Each has no more than three and no less than two exceptional characteristics.

c. If Tonia is exceptionally mathematical, then she is exceptionally athletic.

d. If Julie and Susanne are exceptionally musical, then they are exceptionally artistic.

e. If Tonia and Julie are exceptionally athletic, then they are exceptionally artistic.

Who is not exceptionally athletic? (Hint: Determine the students who are artistic first.)

?

Look Beyond

Nonroutine Problem. The Look Beyond exercise asks students to use logical reasoning to answer a question. This is the subject of Lesson 2.5, where students will be asked to link conditionals in a proof. (An additional hint to the puzzle is to link conditionals c and e first.)

Technology Master

NAME ______ CLASS ______ DATE ______

Technology
2.3 Geometry Software and Definitions

When you use geometry software, there may be two or more ways to create a given figure. The way the figure behaves depends upon the definitions associated with the features that you used to create it.

For example, Lee and Lu each used geometry software to make a sketch like the one shown. Their sketches appeared to be identical.

A C B

Using the *select* tool, Lee was able to grab point *C* and move it closer to point *A*, then closer to point *B*. But when Lu tried to do the same thing, point *C* seemed to be locked into position exactly halfway between point *A* and point *B*.

Refer to the situation described. Use geometry software.

1. Make a sketch that behaves like Lee's sketch. Describe how you made it. What definition does it depend on? Explain.

2. Make a sketch that behaves like Lu's sketch. Describe how you made it. What definition does it depend on? Explain.

Use geometry software. Find two different methods for drawing each figure. Describe each method. Explain how the resulting figures behave differently.

3. **4.** **5.**

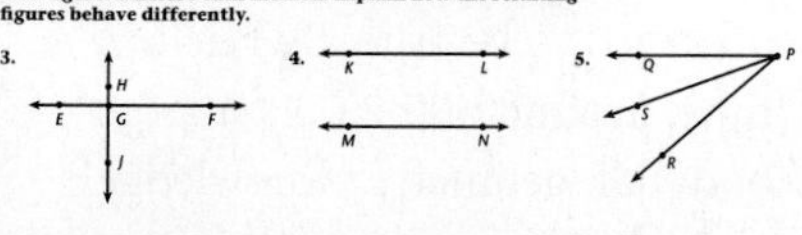

6. Using two different methods, draw a triangle *ABC* in which $\angle C$ is a right angle. Describe each method. Use the select tool to grab point *C* and move it. Describe how the figures behave differently.

HRW Geometry — Technology 57

21–22.

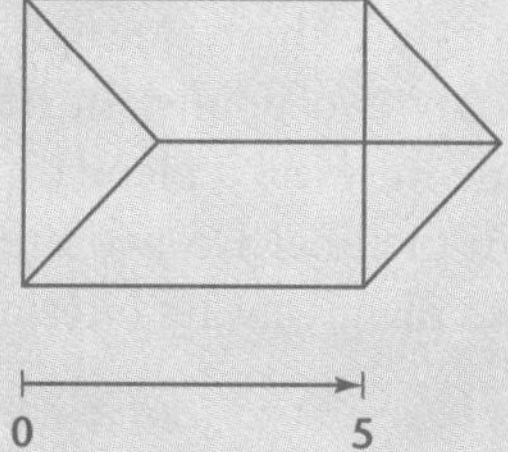

24. Julie is not exceptionally athletic.

	Math	Music	Art	Athletics
Tonia	no	yes	yes	yes
Julie	yes	yes	yes	no
Suzanne	yes	no	no	yes

PREPARE

Objectives

- Define *theorem* and *postulate.*
- Given basic Euclidean postulates, determine the validity of elementary geometry statements.

RESOURCES

- Practice Master 2.4
- Enrichment Master 2.4
- Technology Master 2.4
- Lesson Activity Master 2.4
- Quiz 2.4
- Spanish Resources 2.4

Assessing Prior Knowledge

Classify each statement as true or false.

1. A geometric point can be represented by a dot on a piece of paper. [**True**]
2. Geometric lines and planes have no thickness. [**True**]
3. A line segment has no endpoints. [**False**]

Use Transparency 17

TEACH

Many centuries ago, geometry was developed to solve practical problems in the real world. The first geometric ideas were based upon what people observed to be true. By Euclid's time, around 300 B.C.E., a large body of geometric knowledge existed, but no one had yet sorted out the basic postulates and then used them to prove known empirical facts as theorems.

LESSON 2.4 Postulates

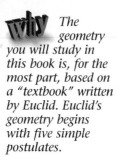

why *The geometry you will study in this book is, for the most part, based on a "textbook" written by Euclid. Euclid's geometry begins with five simple postulates.*

Viewed from a distance, roads may appear as straight lines. Which of Euclid's five postulates does this photo suggest?

The importance of Euclid's work lies not so much in what he discovered, as in the way he organized the knowledge of geometry that already existed in his time.

Euclid began with a few basic **postulates**, or statements that are accepted as true. Then, reasoning logically from his postulates, he was able to prove a number of **theorems**, or statements that are proven to be true. These theorems were then used to prove other theorems, until a vast body of knowledge was built up. *What do you think are the advantages of such a system?*

In Lesson 1.1 you discovered a number of "obvious truths" about points, lines, and planes. The list on the following pages states them formally as postulates. This list, while not the same as Euclid's list, should give you the flavor of the foundation of a deductive geometry system.

These postulates might seem as obvious to you as 2 + 2 = 4, but that's precisely what makes the geometry system built up from them so powerful.

POSTULATES	
Through any two points, there is exactly one line. *or* Two points determine exactly one line.	**2.4.1**
Through any three noncollinear points, there is exactly one plane. *or* Three noncollinear points determine exactly one plane.	**2.4.2**
If two points are on a plane, then the line containing them is also on the plane.	**2.4.3**
The intersection of two lines is exactly one point.	**2.4.4**
The intersection of two planes is exactly one line.	**2.4.5**

Using Discussion Students can develop a solid understanding of the postulates about points, lines, and planes by answering questions such as the following: How many lines are there through any one point? through any three points? How many planes are there through any one point? through any two points? In how many points does a line intersect a plane if the plane does not contain the line? Do two lines always intersect? Do two planes always intersect?

EXAMPLE 1

Answer each question and state the postulate that justifies your answer in each case.

A Name two points that determine line m.

B Is there a unique line through points C and D?

C Name three points that determine plane $\mathscr{R}$.

D Name the intersection of line l and line m.

E Does plane $\mathscr{R}$ contain $\overleftrightarrow{BE}$?

F What is the intersection of plane $\mathscr{Q}$ and plane $\mathscr{R}$?

G Do points B, C, and F determine a plane?

Solution ➤

A Points B and C. "Through any two points there is exactly one line." Points X and C or points X and B are also valid answers since they are also contained on the unique line m. Try to imagine another line besides m through any of these pairs of points if you are not convinced.

B Yes. "Two points determine exactly one line." Notice that the line does not have to be represented in the drawing for it to exist.

C Points E, B, and C. "Through any three noncollinear points, there is exactly one plane." Can you think of other valid answers?

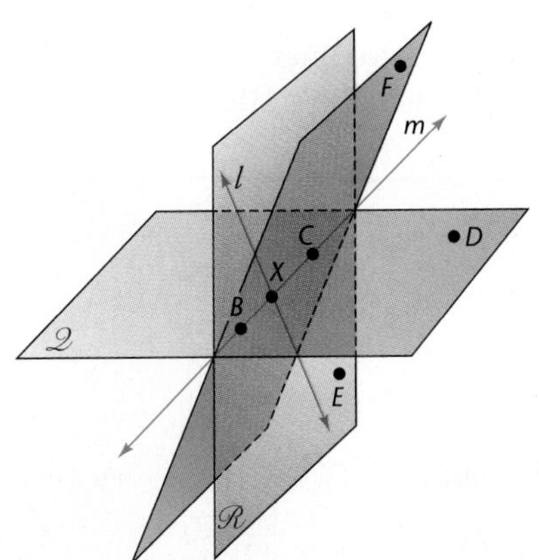

D Point X. "The intersection of two lines is exactly one point."

E Yes. "If two points are on a plane, then the line containing them is also on the plane." Which other postulate guarantees that there is a unique line through points B and E?

F $\overleftrightarrow{BC}$. "The intersection of two planes is exactly one line."

G Yes. "Three noncollinear points determine exactly one plane." The plane does not have to be represented in the drawing for it to exist mathematically. Here is how the drawing would look with the plane drawn. ❖

Try This If two lines intersect, must there be a plane that contains them?

Alternate Example 1

Answer each question and state the postulate number that justifies your answer in each case.

Use Transparency 18

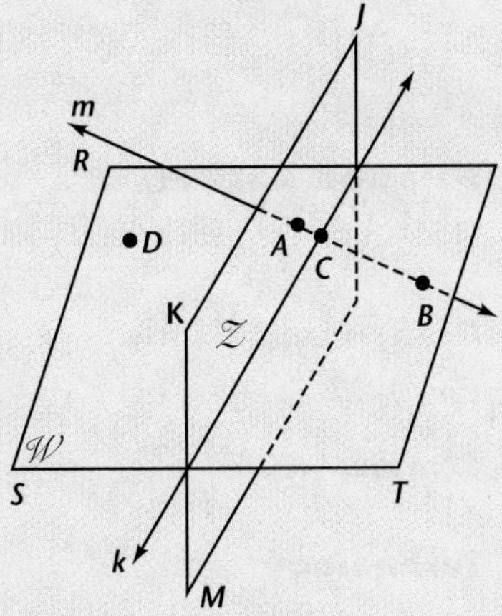

1. What line is determined by points A and B? **[line m, 2.4.1]**
2. What three points determine plane $\mathscr{Z}$? **[J, K, M, 2.4.2]**
3. Name the intersection of planes $\mathscr{W}$ and $\mathscr{Z}$. **[line k, 2.4.5]**
4. Name the intersection of lines k and m. **[point C, 2.4.4]**
5. Does plane $\mathscr{W}$ contain points A and B? **[no]**
6. Do points D, M, and B determine a plane? **[yes, 2.4.2]**
7. How many lines contain points J and M? **[one, 2.4.1]**

ENRICHMENT

Lines a and b intersect at point P to form angles 1, 2, 3, and 4. Suppose $m\angle 4 = 30°$. Use what you know about linear pairs of angles and logical reasoning to prove that $m\angle 1 = m\angle 3$.

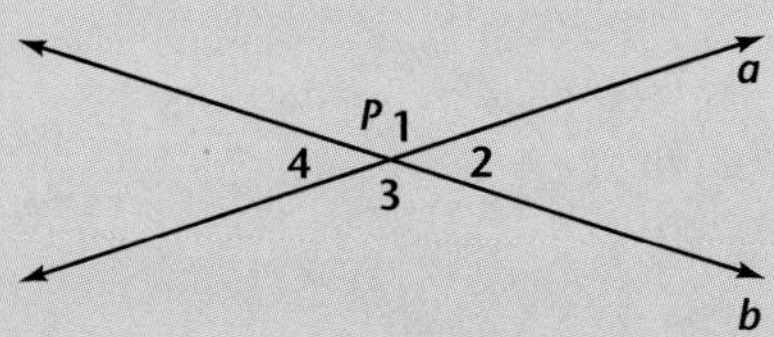

INCLUSION strategies

Using Models Many students have difficulty visualizing three-dimensional drawings. You can illustrate the drawings of intersecting planes on this page by using physical models of planes made of cardboard or other materials.

ongoing ASSESSMENT

Try This Yes. If two lines intersect, there must be three points: the point of intersection and another point to determine each line. Three noncollinear points determine exactly one plane.

CRITICAL *Thinking*

Since three points determine a plane, the legs of a tripod will lie in the plane of the surface on which they rest. A fourth point not in the same plane will cause an object, like a table, to wobble.

ASSESS

Selected Answers

Odd-numbered Exercises 5–23

Assignment Guide

Core 1–27

Core Plus 1–27

Technology

Students can use a computer graphics program to draw the pictures for Exercise 3.

Error Analysis

For Exercises 19 and 20, students may need help in drawing two or three intersecting planes correctly.

Practice Master

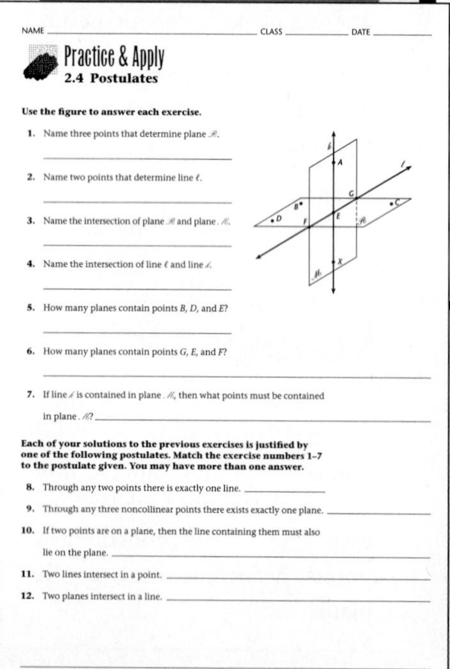

NAME ______ CLASS ______ DATE ______

Practice & Apply
2.4 Postulates

Use the figure to answer each exercise.

1. Name three points that determine plane $\mathscr{R}$. ______
2. Name two points that determine line ℓ. ______
3. Name the intersection of plane $\mathscr{R}$ and plane $\mathscr{M}$. ______
4. Name the intersection of line ℓ and line k. ______
5. How many planes contain points *B*, *D*, and *E*? ______
6. How many planes contain points *G*, *E*, and *F*? ______
7. If line k is contained in plane $\mathscr{M}$, then what points must be contained in plane $\mathscr{M}$? ______

Each of your solutions to the previous exercises is justified by one of the following postulates. Match the exercise numbers 1–7 to the postulate given. You may have more than one answer.

8. Through any two points there is exactly one line. ______
9. Through any three noncollinear points there exists exactly one plane. ______
10. If two points are on a plane, then the line containing them must also lie on the plane. ______
11. Two lines intersect in a point. ______
12. Two planes intersect in a line. ______

46 **Practice & Apply** **HRW Geometry**

Photographers use three-legged stands known as **tripods** to hold their cameras steady. You may have also noticed surveyors using tripods, because it is very important that their instruments not move when they measure angles or distances. Tripods are also used to provide stable bases for portable telescopes.

CRITICAL *Thinking* Why do you think tripods provide more stability than four-legged tables? Why is this particularly important when the surface on which they rest is rough or irregular?

EXERCISES & PROBLEMS

Communicate

1. Why is it necessary to use postulates in geometry?
2. Explain the difference between a postulate and a theorem.
3. Draw a picture to illustrate each postulate on page 88.
4. Describe a space in which it is possible to draw more than two lines through two different points. Hint: Consider the lines of longitude on a globe.

Practice & Apply

For Exercises 5–11, use the illustration to the right to answer each question.

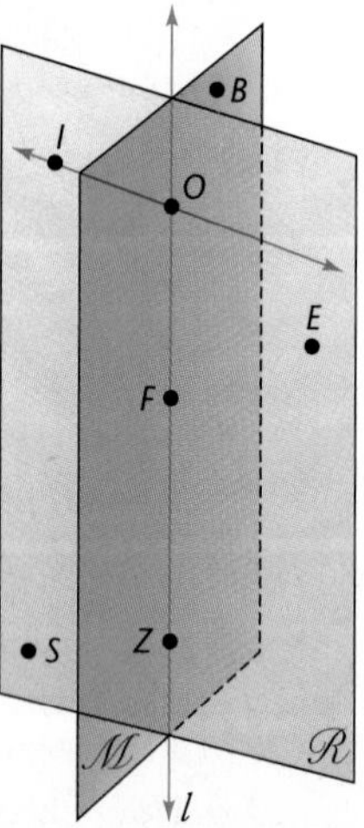

5. Name three points that determine plane $\mathscr{R}$. any 3 noncollinear points in plane $\mathscr{R}$
6. Name two points that determine line l. Z&F, Z&O, F&O
7. Name the intersection of plane $\mathscr{R}$ and plane $\mathscr{M}$. line l
8. Name the intersection of l and $\overline{IO}$. O
9. How many planes contain points *I*, *O*, and *E*? exactly 1
10. How many planes contain points *O*, *F*, and *Z*? an infinite number
11. If line l is contained in plane $\mathscr{M}$, then what points must be contained in plane $\mathscr{M}$? Z, F, O

RETEACHING the lesson

Using Analogies You can give the postulates stated on page 88 a physical interpretation to reteach their meanings. For example, the word *tree* can be used in place of *point*, *row* can be used in place of *line*, and *ground* can replace *plane*. Then postulate 2.4.1 can be reworded as *two trees are on exactly one row*. Have students reword the other four postulates using the words *tree*, *row*, and *ground*.

Each of your solutions to the previous exercises is justified by one of the postulates listed below. Match the exercise number (5–11) to the postulate given. You may have more than one answer.

12. Through any two points there is exactly one line. 6

13. Through any three noncollinear points there exists exactly one plane. 5, 9

14. If two points are on a plane, then the line containing them must also lie on the plane. 11

15. Two lines intersect in a point. 8

16. Two planes intersect in a line. 7

For each of the following statements, draw one diagram for which the statement is true and draw another diagram for a counterexample.

17. Two lines intersect in a point.

18. Three points determine a plane.

19. Two planes intersect in a line.

20. Three planes intersect in a line.

21. A line and a point determine a plane.

Look Back

22. Solve for x and find the values of the measures of the angles. **[Lesson 1.5]** $x = -15$, $m\angle ABD = 34°$, $m\angle DBC = 13°$

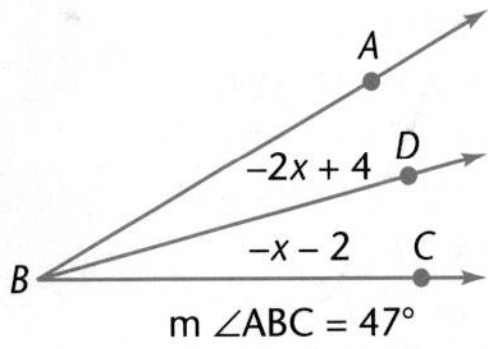

23. What do an angle bisector and the midpoint of a segment have in common? **[Lesson 1.5]**

24. Draw a Venn diagram to illustrate the following argument:

Every member of the Culver High School football team goes to Culver High School. Brady is a member of the Culver High School football team. Therefore, Brady goes to Culver High School. **[Lesson 2.2]**

Look Beyond

25. The following is a series of if-then statements. Construct a Venn diagram from these statements, and finish the concluding statement.

If you are taking geometry, then you are thinking hard.
If you are thinking hard, then you are developing good reasoning skills.
If you are developing good reasoning skills, then you will be able to succeed in many different careers.
If you are able to succeed in many different careers, then you will be able to choose your profession.
Therefore, if you are taking geometry, then . . .

26. Why do you think it is important to minimize the number of postulates in the Euclidean system of logic?

27. Any three noncollinear points determine a plane. The following box has 8 corners, each of which represents a point. Draw at least four other planes not shown in the diagram.

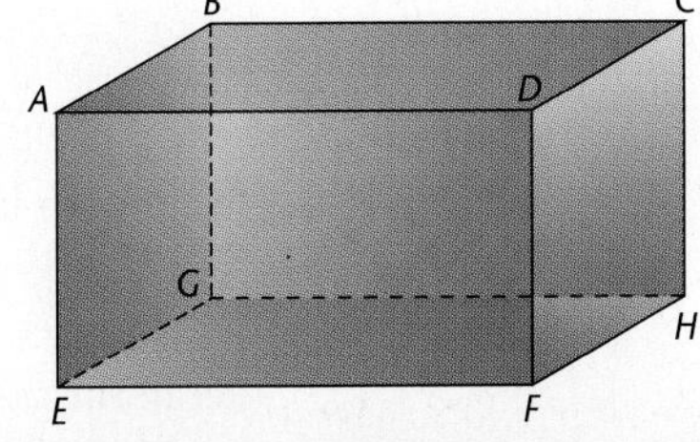

23. The midpoint of a segment is equidistant from the endpoints of the segment, and the angle bisector is the set of points equidistant from the sides of an angle.

26. Postulates are kept to a minimum in order to assume as little as possible without proof.

27. Student drawings may include: *ABHF, ADHG, CDEG, BCFE, BDFG, ACHE, BDE, ACG, BDH, ACF, AFG, DEH, CFG, BEH, HAC*

alternative ASSESSMENT

Performance Assessment

Call upon individual students to define the terms *postulate* and *theorem*. Other students can be asked to state one of the five postulates given on page 88 and illustrate it with a drawing on the board.

The answers to Exercises 17–21, 24, and 25 can be found in Additional Answers beginning on page 727.

Look Beyond

Exercises 25–27 look ahead to the next lesson on linking steps in a proof.

Technology Master

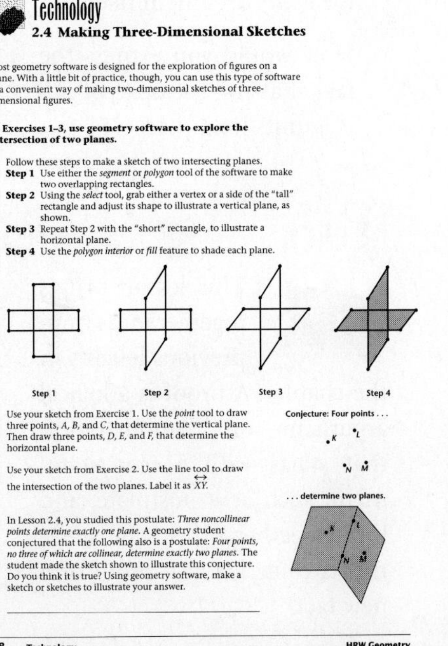
NAME ______ CLASS ______ DATE ______

Technology
2.4 Making Three-Dimensional Sketches

Most geometry software is designed for the exploration of figures on a plane. With a little bit of practice, though, you can use this type of software as a convenient way of making two-dimensional sketches of three-dimensional figures.

In Exercises 1–3, use geometry software to explore the intersection of two planes.

1. Follow these steps to make a sketch of two intersecting planes.
Step 1 Use either the *segment* or *polygon* tool of the software to make two overlapping rectangles.
Step 2 Using the *select* tool, grab either a vertex or a side of the "tall" rectangle and adjust its shape to illustrate a vertical plane, as shown.
Step 3 Repeat Step 2 with the "short" rectangle, to illustrate a horizontal plane.
Step 4 Use the *polygon interior* or *fill* feature to shade each plane.

Step 1 Step 2 Step 3 Step 4

2. Use your sketch from Exercise 1. Use the *point* tool to draw three points, *A, B,* and *C,* that determine the vertical plane. Then draw three points, *D, E,* and *F,* that determine the horizontal plane.

3. Use your sketch from Exercise 2. Use the line tool to draw the intersection of the two planes. Label it as $\overleftrightarrow{XY}$.

4. In Lesson 2.4, you studied this postulate: *Three noncollinear points determine exactly one plane.* A geometry student conjectured that the following also is a postulate: *Four points, no three of which are collinear, determine exactly two planes.* The student made the sketch shown to illustrate this conjecture. Do you think it is true? Using geometry software, make a sketch or sketches to illustrate your answer.

Conjecture: Four points . . .
K L N M
. . . determine two planes.

58 Technology HRW Geometry

PREPARE

Objectives

- Identify and use the Addition Property of Equality, Reflexive Property of Equality, Symmetric Property of Equality, and Transitive Property of Equality.
- Link steps of a geometric proof using properties and postulates.

RESOURCES

- Practice Master 2.5
- Enrichment Master 2.5
- Technology Master 2.5
- Lesson Activity Master 2.5
- Quiz 2.5
- Spanish Resources 2.5

Assessing Prior Knowledge

1. Solve the equation $x - 6 = 14$. [$x = 20$]
2. Does the set of real numbers include all rational and irrational numbers? [**yes**]
3. Does the symbol $\overline{AB}$ refer to the segment itself or to its measure? [**segment itself**]
4. How would you express the fact that the measure of $\overline{AB}$ is 5 units long? [$AB = 5$]

TEACH

This lesson brings together ideas from previous lessons in the chapter. A proof is a logical argument or demonstration that starts with a true statement, such as a postulate, and links together other true statements to arrive at a proof of a new fact.

LESSON 2.5 Linking Steps in a Proof

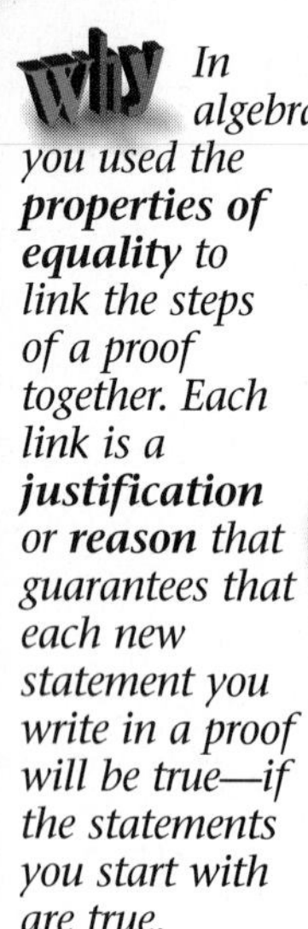

*In algebra, you used the **properties of equality** to link the steps of a proof together. Each link is a **justification** or **reason** that guarantees that each new statement you write in a proof will be true—if the statements you start with are true.*

Euclid's geometry starts with a list of twenty-three **definitions,** five **postulates**, and five other statements which Euclid called **Common Notions**. The Second Common Notion reads as follows:

IF EQUALS ARE ADDED TO EQUALS, THEN THE WHOLES ARE EQUAL.

What **algebraic property of equality** do you recognize in this statement?

The properties of equality are used to solve equations. For example, to solve $x - 3 = 5$, the Addition Property of Equality (Euclid's Second Common Notion) is used to add 3 to each side of the equation. That is,

$$x - 3 + 3 = 5 + 3$$
$$\text{or} \quad x = 8$$

In modern language, the Addition Property of Equality states:

For all real numbers a, b, and c,
if $a = b$ then $a + c = b + c$.

Recall the other properties of equality that you learned in algebra. State the properties that hold for subtraction, multiplication, and division.

Using Visual Models

Students can create line segment drawings of their own, such as the one at the top of page 93. Varying the numbers and working with many examples will help students develop an intuitive understanding of the proof given in Example 1 and of the Overlapping Segments Theorem.

The Algebraic Properties in Geometry

In the drawing, the lengths of the two outer segments are equal. What can you conclude about the lengths AC and BD of the two overlapping segments?

It's easy to see that

$$AC = 4 + 5 = 9$$
$$\text{and } BD = 5 + 4 = 9.$$

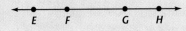

That is, $AC = BD$.

How does this conclusion illustrate the Addition Property of Equality? In the language of the Second Common Notion, what "equals" are added to what other "equals" in the two statements?

Linking Steps to Create a Proof

The following example illustrates how each of the steps in a detailed or "formal" proof can be linked, directly or indirectly, to the given information. The idea that is proved is the same as the one you have been working with. As you will see, one of the links or justifications in the proof is the Addition Property.

You are given that $AB = CD$. Prove that the lengths AC and BD are equal.

Solution ➤

The proof is given in two parts.

Part I

You are given that $AB = CD$. The Segment Addition Postulate can be used to write two equations about the lengths AC and BD.

$$AB + BC = AC \qquad BC + CD = BD$$

BC has been added to AB and CD. This gives a clue. Add the inner distance to each side of the given equation.

$AB = CD$	Given
$\underbrace{AB + BC}_{AC} = \underbrace{BC + CD}_{BD}$	Addition Property of Equality

There are several other properties of equality. One of them is called the Substitution Property. It states that if you have another name for a, such as b, you may replace a with b anywhere it appears in an expression.

Math Connection
Algebra

Algebra properties connect geometric figures, such as line segments, to the real numbers. Thus, all the properties of real numbers are now assumed and can be used to prove geometric statements.

Alternate Example 1

You are given that $EG = FH$. Prove that $EF = GH$.

$EG = FH$	**Given**
$EF + FG = FG + GH$	**Substitution Property**
$EF = GH$	**Subtraction Property**

TEACHING tip

This lesson introduces students to formal geometric proof. Some students may benefit from using a standard proof format.

interdisciplinary CONNECTION

Trade Occupations People working in trade occupations, such as carpenters, plumbers, and electricians, use the fact that "if equals are added to equals, then the wholes are equal" all the time. Students can probably suggest examples from their own experiences of when they also used this common notion.

ongoing ASSESSMENT

Try This

Statements	Reasons
$m\angle AOB = m\angle COD$	Given
$m\angle AOB + m\angle BOC = m\angle AOC$	Angle Addition Postulate
$m\angle COD + m\angle BOC = m\angle BOD$	Angle Addition Postulate
$m\angle AOB + m\angle BOC = m\angle COD + m\angle BOC$	Addition Property of Equality
$m\angle AOC = m\angle BOD$	Substitution Property of Equality

Overlapping Angles Theorem

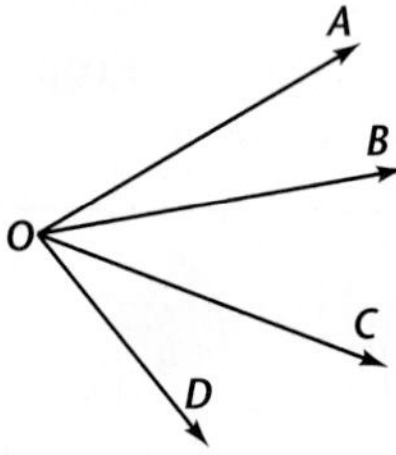

Given four rays with common endpoint arranged as shown, the following statements are true

- If $m\angle AOB = m\angle COD$, then $m\angle AOC = m\angle BOD$.
- If $m\angle AOC = m\angle BOD$, then

For all real numbers a and b,
if $a = b$, then either a or b may be replaced with the other in any expression.

Part II

Now you can substitute the distances AC and BD into the equation to get the conclusion you are looking for:

$AC = BD$ (Substitution Property) ❖

The results we have just proven, along with its converse, can be stated as a theorem. In your own future work you can use the theorem to justify a statement without going through the whole proof.

OVERLAPPING SEGMENTS THEOREM

Given a segment with points A, B, C, and D arranged as shown, the following statements are true:

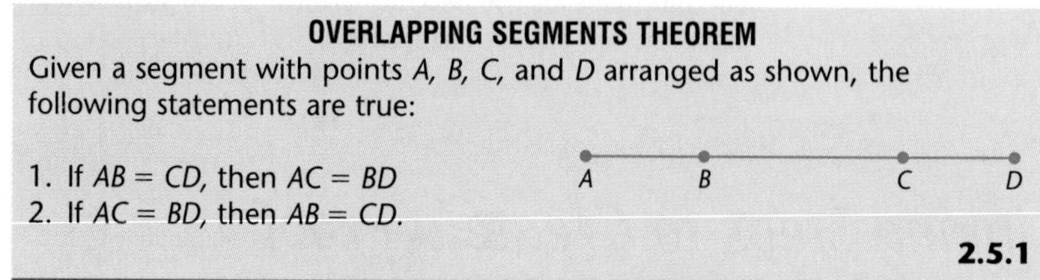

1. If $AB = CD$, then $AC = BD$
2. If $AC = BD$, then $AB = CD$.

2.5.1

Try This Use the pattern of the proof in the example to write the following proof on your own.

Given: $m\angle AOB = m\angle COD$

Prove: $m\angle AOC = m\angle BOD$

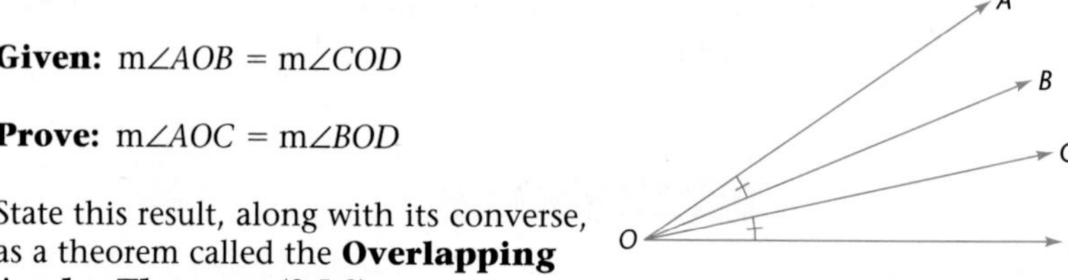

State this result, along with its converse, as a theorem called the **Overlapping Angles Theorem** (2.5.2).

The Equivalence Properties

In addition to the algebraic properties of equality given above, there are three very important properties known as the **Equivalence Properties of Equality**. They are so "obvious" that you probably don't even think of them when you use them.

EQUIVALENCE PROPERTIES OF EQUALITY

Reflexive Property of Equality	For all real numbers a, $a = a$.	**2.5.3**
Symmetric Property of Equality	For all real numbers a and b, if $a = b$, then $b = a$.	**2.5.4**
Transitive Property of Equality	For all real numbers a, b, and c, if $a = b$ and $b = c$, then $a = c$.	**2.5.5**

ENRICHMENT Students can explore the following relations to determine if they are equivalence relations.

1. is greater than or equal to
2. is less than
3. is approximately equal to
4. is the sister of

INCLUSION strategies **English Language Development** Some students may have difficulty reading and understanding mathematics. You can help these students by suggesting that they read the material in this lesson prior to discussing it in class. They can then write down their questions and note aspects of the lesson that are not clear. If these questions still exist after the lesson is taken up in class, you can meet individually with those students requiring further help.

Any relationship that satisfies the three properties of Reflexivity, Symmetry, and Transitivity is called an **equivalence relation**.

Try This Write three statements and call them the Equivalence Properties of Congruence. Instead of the real numbers a, b, and c, write *Figure A*, *Figure B*, and *Figure C*. And instead of "=," write "≅."

The drawings to the right illustrate the Equivalence Properties of Congruence. Tell which property each figure illustrates. Does congruence seem to be an equivalence relation?

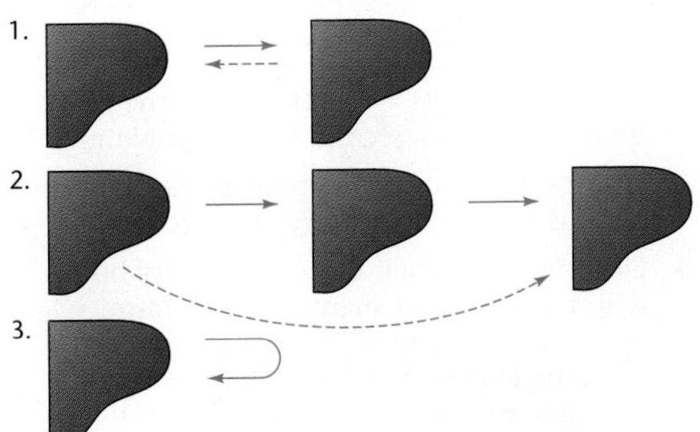

EXAMPLE 2

The first stamp pictured on the right measures 3 cm × 6 cm. The second stamp has the same measurements as the first. The third stamp is exactly the same as the second.

What can you conclude about the first and the third stamps from the given information? State your conclusions using geometry terms. What property discussed in this lesson justifies your conclusion?

Solution ➤

The first stamp is congruent to the third stamp. The justification for this conclusion (from the given information) is the transitive property of congruence. ❖

Russian stamps commemorating the Russian American Company

In Lesson 1.4 you used a compass to draw congruent segments on a line. What property of congruence guarantees that the segments you draw by the compass method are actually congruent? (Hint: Think of the segment that exists between the points of the compass even though it is not shown.)

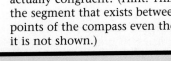

A B C

Cooperative Learning Arrange the class into small groups. Have each group prepare to "team teach" the lesson to the rest of the class. If there are several groups, have each group present just one part of the lesson. When the groups are finished, have them quiz their classmates over the lesson.

Ongoing Assessment

Try This

Equivalence Properties of Congruence:

I. Symmetric Property: If $A \cong B$, then $B \cong A$.

II. Transitive Property: If $A \cong B$ and $B \cong C$, then $A \cong C$.

III. Reflective Property: $A \cong A$

Yes, congruence seems to be an equivalence relation.

Alternate Example 2

The triangle pictured below has the measures shown. The second triangle has the same measurements as the first. The third triangle is exactly the same as the second.

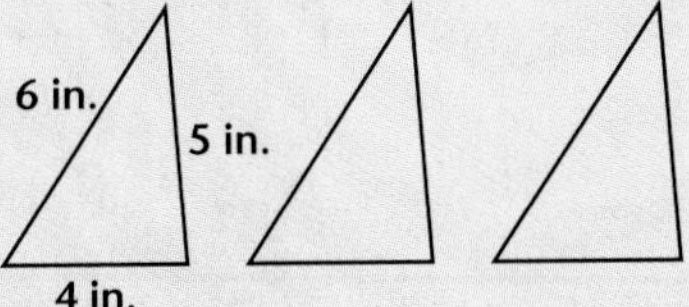

What can you conclude about the first and third triangles from the given information?

[The first triangle is congruent to the third triangle by the transitive property of congruence.]

Critical Thinking

Transitive Property. $\overline{AB}$ is congruent to the segment that exists between the points of the compass because the compass was used to draw the segment. Similarly, the segment between the compass is congruent to $\overline{BC}$. By the transitive property, $\overline{AB} \cong \overline{BC}$.

ASSESS

Selected Answers

Odd–numbered Exercises 5–33

Assignment Guide

Core 1–20, 27, 30–34

Core Plus 1–3, 15–22, 27–35

Technology

Students can use geometry graphics tools to create their own exercises similar to Exercises 7–12 and 15–20.

Error Analysis

In Exercises 27–29, students are asked to write their own proofs. Suggest that they model their proofs after those shown in Exercises 4–6.

Practice Master

NAME __________ CLASS __________ DATE __________

Practice & Apply
2.5 Linking Steps in a Proof

Identify the properties of equality that justify the conclusion.

1. $x + 9 = 16$ Given
   $x + 9 - 9 = 16 - 9$
   $x = 7$ __________

2. $\angle B \cong \angle C$ Given
   $\angle C \cong \angle D$
   $\angle B \cong \angle D$ __________

3. $\overline{AB} \cong \overline{CD}$ Given
   $\overline{CD} \cong \overline{AB}$ __________

4. $AB = CD$ Given
   $AB + BC = BC + CD$
   $AC = BD$ __________

Complete the proofs.

5. Given: $m\angle HGK = m\angle JGL$
   Prove: $m\angle 1 = m\angle 3$

Statements	Reasons
1. $m\angle HGK = m\angle JGL$	1. Given

6. Given: $\angle ABC \cong \angle EFG$
   $\angle 1 \cong \angle 3$
   Prove: $\angle 2 \cong \angle 4$

Statements	Reasons
1. $\angle ABC \cong \angle EFG$, $\angle 1 \cong \angle 3$	1. Given

HRW Geometry Practice & Apply 47

EXERCISES & PROBLEMS

Communicate

1. In the figure, $MO = NP$. How does the figure illustrate Euclid's Second Common Notion?
2. Discuss the following statement: Algebra can help us visualize geometry, and vice versa.
3. Fiona and Jada each have 22 mystery books which they do not share. The community library has 120 mystery books for everyone to share. Do Fiona and Jada have access to the same number of mystery books? What basic property enables you to answer this question without doing a calculation? Explain.

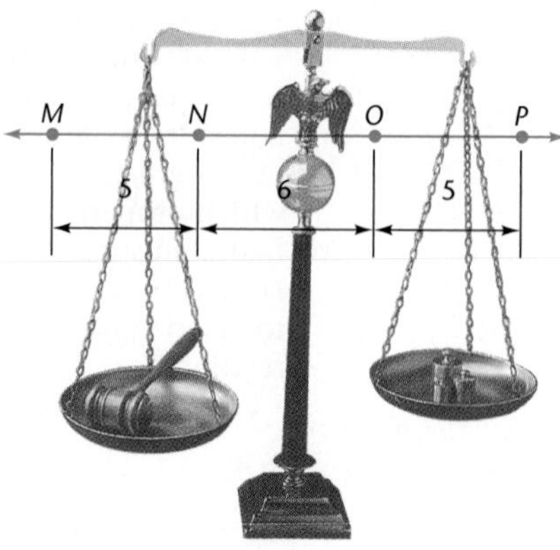

Explain how the balance scale illustrates Euclid's Common Notion that "if equals are added to equals, then the wholes are equal."

Practice & Apply

In Exercises 4–6, identify the properties of equality that justify the conclusion.

4. $x + 6 = 14$ Given
   $x + 6 - 6 = 14 - 6$ (Property?)
   $x = 8$ (Property?)
   Subtraction Property of Equality

5. $(a + b) = (c + d)$ Given
   $(c + d) = (e + f)$ Given
   $(a + b) = (e + f)$ (Property?)
   Transitive Property of Equality

6. $AB + CD = XY$ Given
   $CD + DE = XY$ Given
   $AB + CD = CD + DE$ (Property?)
   Substitution Property of Equality

In the diagram, $UT = AH$. Answer the questions in Exercises 7–12 to prove that $UA = HT$.

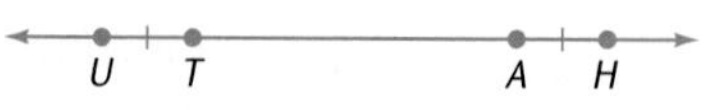

7. What two segments sum to UA? UT and TA
8. What property allows you to answer 7? Segment Addition Property
9. What two segments sum to HT? AH and TA
10. What property allows you to answer 9? Segment Addition Property
11. Write down the information that is given in the diagram about the lengths of segments UT and HA. Add TA to each side of your equation. What property allows you to add this length to both sides?
12. Use substitution and the results from Exercises 7 and 9 to get the conclusion you are looking for.
13. What theorem could you use to justify the result of Exercises 7–12 without doing the proof? Overlapping Segments Theorem

11. $UT = AH$ Given
    $UT + TA = AH + TA$ Add. Prop. of Equality

12. $UA = HT$ Sub. Prop. of Equality

14. Draw a diagram that illustrates a conclusion you could draw from Exercises 7–12 using the second part of the Overlapping Segments Theorem on page 94.

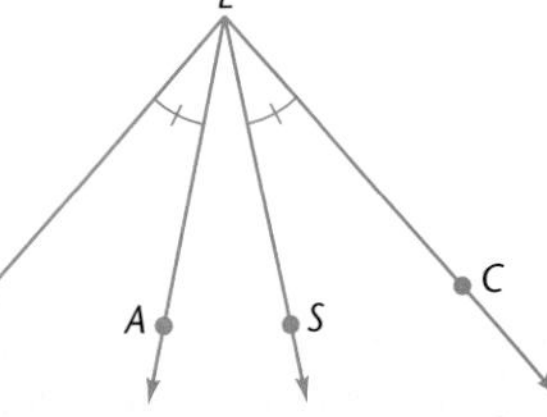

In the diagram, m∠*PLA* = m∠*SLC*. Answer the questions in Exercises 15–20 to prove that m∠*PLS* = m∠*CLA*.

15. What two angles sum to m∠*PLS*? ∠*PLA* and ∠*ALS*

16. What property allows you to answer Exercise 15?

17. What two angles sum to m∠*CLA*? m∠*SLC* and m∠*ALS*

18. What property allows you to answer Exercise 17?

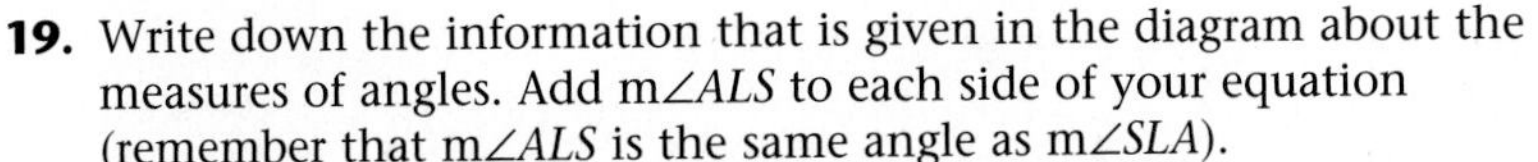

19. Write down the information that is given in the diagram about the measures of angles. Add m∠*ALS* to each side of your equation (remember that m∠*ALS* is the same angle as m∠*SLA*). Given m∠*PLA* = m∠*SLC* m∠*PLA* + m∠*ALS* = m∠*SLC* + m∠*ALS*

20. Now substitute angles to get the conclusion you are looking for (see your answers to questions 15 and 17). m∠*PLS* = m∠*CLA*

21. What theorem could you use to justify the result of Exercises 15–20 without doing the proof? Overlapping Angles Theorem

22. Draw a diagram that illustrates a conclusion you could draw from Exercises 15–20 using the second part of the Overlapping Angles Theorem.

Use the three triangles drawn below for Exercises 23–25.

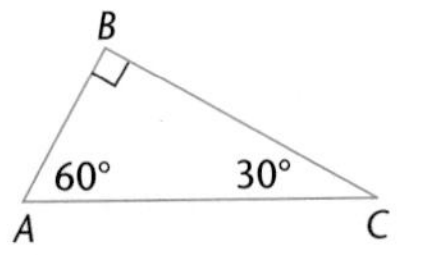

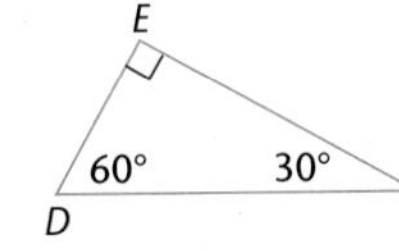

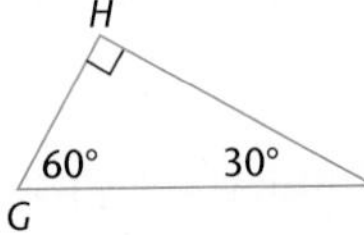

23. If *AB* = *DE* and *DE* = *GH*, what do you know about the relationship of *AB* and *GH*? *AB* = *GH*

24. What property did you use in 23?

25. Illustrate the same property using the same triangles but using angles of the triangles instead of sides.

26. JoAnn wears the same size hat as April. April's hat is the same size as Lara's. Will Lara's hat fit JoAnn? What property of equality supports your answer?

In Exercises 27–29, write your own proofs using any form you like.

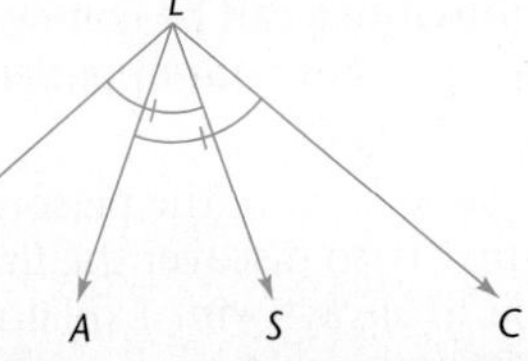

27. Given: m∠*PLS* = m∠*ALC*
Prove: m∠*PLA* = m∠*SLC*

14. If *AU* = *HT*, then *UT* = *AH*

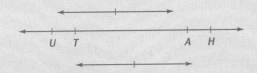

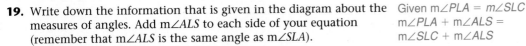

alternative ASSESSMENT

Performance Assessment

Suppose that *AC* = *BD*, *AD* = 18, and *BC* = 10. Find the length of $\overline{AB}$ and $\overline{CD}$. Have students write proofs to accompany their answers.

16. Angle Addition Postulate

18. Angle Addition Postulate

24. Transitive Property of Equality

The answers to Exercises 22, 25, 26, and 27 can be found in Additional Answers beginning on page 727.

Technology Master

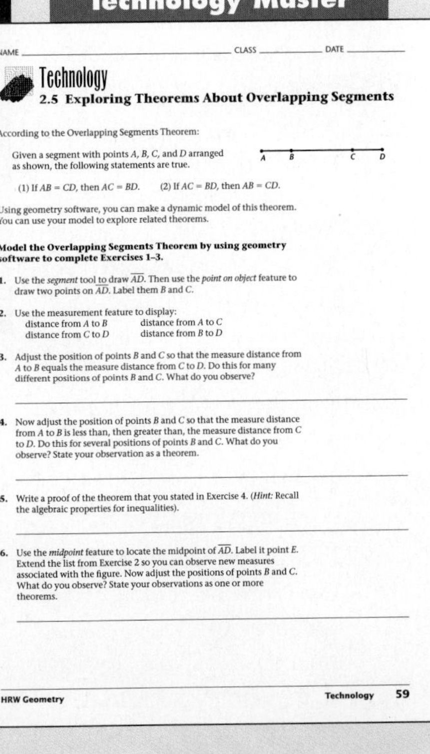

NAME ______ CLASS ______ DATE ______

Technology
2.5 Exploring Theorems About Overlapping Segments

According to the Overlapping Segments Theorem:

Given a segment with points *A*, *B*, *C*, and *D* arranged as shown, the following statements are true.

(1) If *AB* = *CD*, then *AC* = *BD*. (2) If *AC* = *BD*, then *AB* = *CD*.

Using geometry software, you can make a dynamic model of this theorem. You can use your model to explore related theorems.

Model the Overlapping Segments Theorem by using geometry software to complete Exercises 1–3.

1. Use the *segment* tool to draw $\overline{AD}$. Then use the *point on object* feature to draw two points on $\overline{AD}$. Label them *B* and *C*.
2. Use the measurement feature to display:
   distance from *A* to *B* — distance from *A* to *C*
   distance from *C* to *D* — distance from *B* to *D*
3. Adjust the position of points *B* and *C* so that the measure distance from *A* to *B* equals the measure distance from *C* to *D*. Do this for many different positions of points *B* and *C*. What do you observe?
4. Now adjust the position of points *B* and *C* so that the measure distance from *A* to *B* is less than, then greater than, the measure distance from *C* to *D*. Do this for several positions of points *B* and *C*. What do you observe? State your observation as a theorem.
5. Write a proof of the theorem that you stated in Exercise 4. (*Hint:* Recall the algebraic properties for inequalities).
6. Use the *midpoint* feature to locate the midpoint of $\overline{AD}$. Label it point *E*. Extend the list from Exercise 2 so you can observe new measures associated with the figure. Now adjust the positions of points *B* and *C*. What do you observe? State your observations as one or more theorems.

HRW Geometry — Technology 59

Look Beyond

The analysis of the game of Nim looks ahead to the next lesson on conjectures that lead to proof.

Version A

35. **3 in a row or 4 in a row** However many counters your opponent takes from a row, take the same number from the *other* row. Repeat until the game is over.

1, 2, and 3 counters in rows After your opponent has moved, you will be able to convert the combination that is left into two rows of 2 or two rows of 1. Either of these will win.

Version B

4 in a row If your opponent takes 1 from a row, you take 1 from the *other* row. This converts the pattern to two rows of 3. (See below.)

3 in a row If your opponent takes 1 from a row, you take 1 from the other row, leaving two rows of 2, which is winnable. (see p. 71).

If your opponent takes 2, you take 3. If your opponent takes 3, you take 2. Either of these situations forces your opponent to take the last counter.

1, 2, and 3 in rows After your opponent has moved, you may be able to convert the combination into two rows with 2 counters. (This will win.) If this is not possible, you will be able to convert the combination into one row of 1 or three rows of 1 will win.)

Winning Move (A or B)

Take 1 from the row with 3. You can convert what your opponent leaves into a combinations analyzed earlier, or an even quicker win.

28. Given: m∠CBD = m∠CDB
m∠ABD = 90°
m∠EDB = 90°

Prove: m∠ABC = m∠EDC

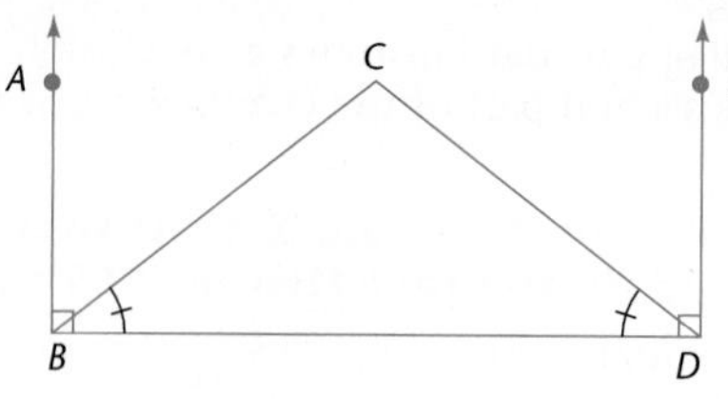

29. Given: m∠BAC + m∠ACB = 90°
m∠DCE + m∠DEC = 90°
m∠ACB = m∠DCE

Prove: m∠BAC = m∠DEC

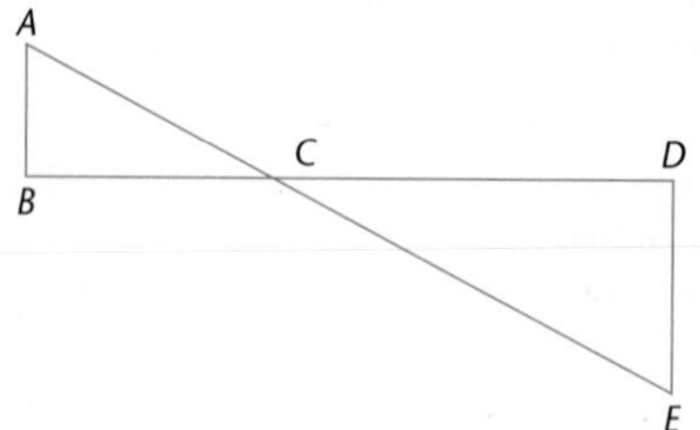

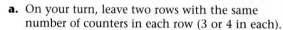

Look Back

30. Can three planes intersect in a point? If yes, draw an example. **[Lesson 1.1]**

31. Use the midpoint definition and the figure shown to find the value of x and the length of the segments. **[Lesson 1.4]**

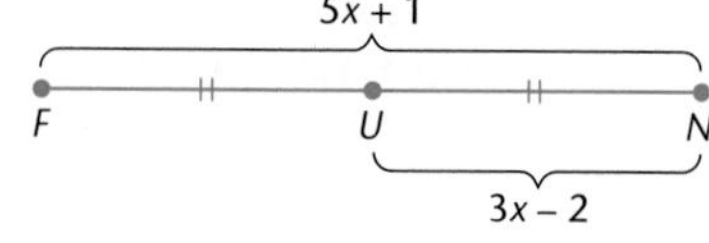

32. If you were to reflect a shape with respect to a line, would your original shape be congruent to the reflection? **[Lesson 1.6]** Yes

33. Write a true if-then statement whose converse is false. **[Lesson 2.2]**

34. In what special situation are the converse and the original statement both true? **[Lesson 2.2]** Definitions

Look Beyond

35. **Portfolio Activity** If you leave two rows with two counters each in a game of Nim, then you can win no matter what the other player does. The following strategies will allow you to achieve this winning combination—or an even quicker win, depending on what your opponent does.

a. On your turn, leave two rows with the same number of counters in each row (3 or 4 in each).

b. On your turn, leave three rows with 1, 2, and 3 counters, in any order.

Explain how these combinations can be converted to eventual wins, no matter what your opponent does, for either version of the game.

In either version of the game of Nim, the person who plays first can always win. Try to discover the first move that will allow you to always win. Explain.

These are winning combinations.

30. [Figure: planes with points C, D, A, B]

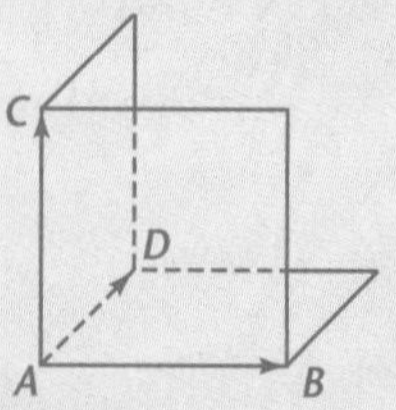

31. $x = 5$
$FN = 26$; $FU = UN = 13$

33. Example: If it is raining, then it is cloudy.
Converse: If it is cloudy, then it is raining.

The answers to Exercises 28, 29, and 35 can be found in Additional Answers beginning on page 727.

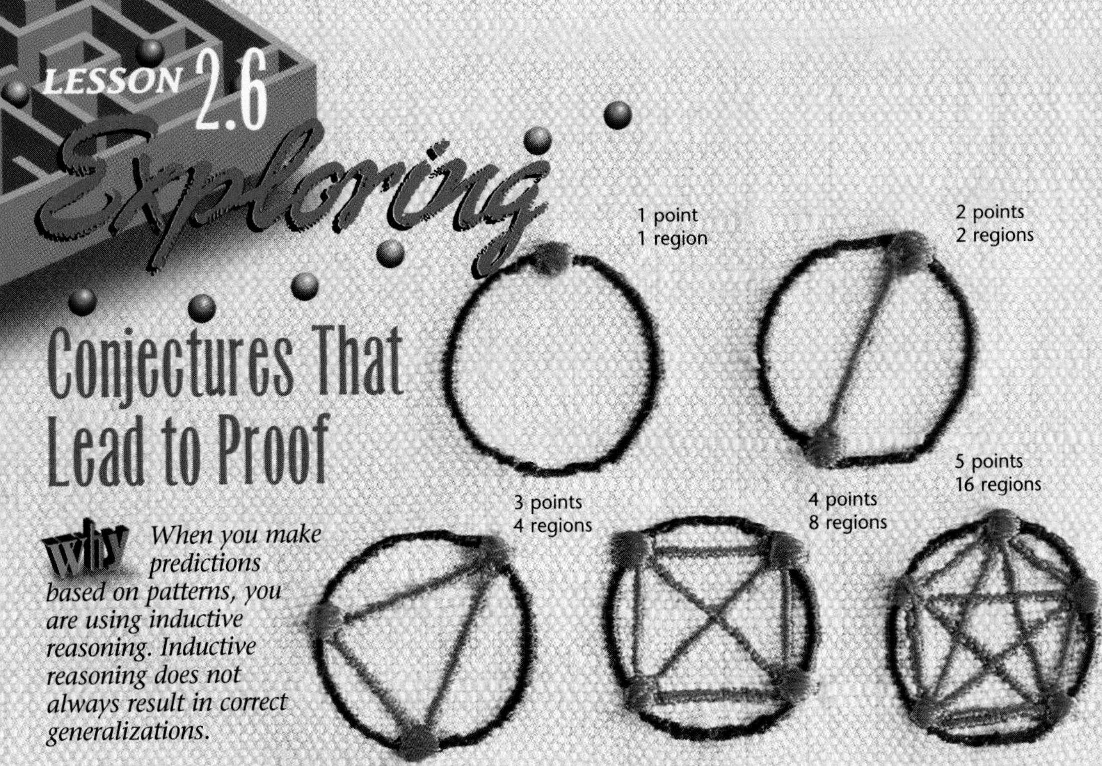

Conjectures That Lead to Proof

When you make predictions based on patterns, you are using inductive reasoning. Inductive reasoning does not always result in correct generalizations.

A Need for Proof

Most people who look at the pattern of circles shown above make the following generalization or conjecture:

The number of regions doubles each time a point is added. True or false?

This conjecture can be tested by drawing a picture. What do you discover? Is the conjecture true? We will address this question later.

For a conjecture to be considered true in mathematics, it must be proven true using deductive reasoning. That is, it must be shown to follow logically from statements that are already known or assumed to be true. In the following exploration you will fill in a simple proof of a conjecture.

Exploration 1 *The Vertical Angles Conjecture*

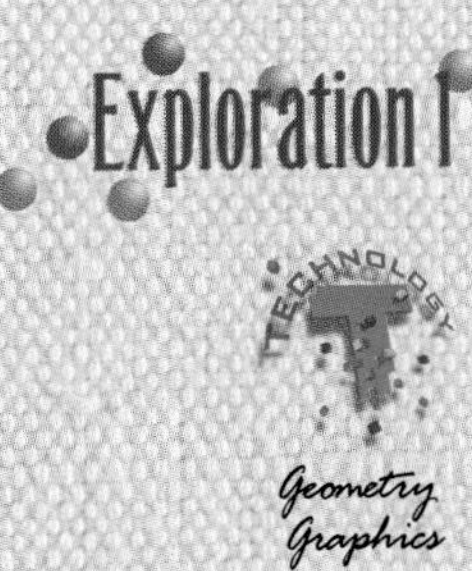

You will need
Geometry technology or
A ruler and protractor.

Vertical angles are the opposite angles formed wherever two lines intersect. You can think of a pair of scissors as forming vertical angles, with the blades making one of the angles and the handles the other.

Using Models Have students examine the sums of odd integers, make a conjecture about the sums, and prove their conjecture. (For example: The sum of two odd integers is even. Proof: If n is any integer, then $2n + 1$ is an odd integer; $2n + 1 + 2n + 1 = 4n + 2 = 2(2n + 1)$; the integer $2(2n + 1)$ has a factor of 2 and therefore is even.)

PREPARE

Objectives

- Define *formal proof.*
- Use conjectures to motivate the steps of a proof.

RESOURCES

- Practice Master 2.6
- Enrichment Master 2.6
- Technology Master 2.6
- Lesson Activity Master 2.6
- Quiz 2.6
- Spanish Resources 2.6

Assessing Prior Knowledge

Classify each statement as true or false.

1. A conjecture is a statement that you think is true but that may or may not be true. **[True]**
2. The sum of the measures of angles that form a linear pair is 90°. **[False]**
3. A reflection of a figure over a line changes the size of the figure but not its shape. **[False]**

TEACH

Since inductive reasoning is based on an examination of specific examples, it can never prove a generalization. Other methods that use deductive reasoning are needed to prove that a generalization is always true.

Exploration 1 Notes

Students develop an understanding of the congruence of vertical angles by measuring several pairs of vertical angles.

Ongoing ASSESSMENT

3. Vertical angles have equal measures.

Math Connection Algebra

In step 5, since $m\angle 1 + m\angle 3 = 180°$ and $m\angle 2 + m\angle 3 = 180°$, the substitution property of equality can be applied to state that $m\angle 1 + m\angle 3 = m\angle 2 + m\angle 3$. In step 6, the subtraction property of equality can be applied to conclude that $m\angle 1 = m\angle 2$.

Cooperative Learning

The three explorations in this lesson are ideal activities for small–group work. Using either geometry graphics software or a ruler and protractor with folding paper, students are led to make conjectures that they then try to prove. After the groups have completed their work, a discussion of the conjectures and the proofs would serve as an excellent summary of the content of this lesson.

1. Draw several pairs of vertical angles.
2. Measure each pair.
3. Make a conjecture about vertical angles.

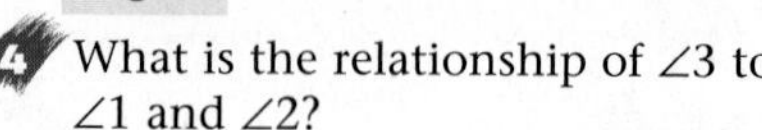

ALGEBRA Connection

4. What is the relationship of $\angle 3$ to $\angle 1$ and $\angle 2$?
5. Complete: $m\angle 1 + m\angle 3 = \underline{?}$
   $m\angle 2 + m\angle 3 = \underline{?}$
6. What property of equality lets you conclude that

   $m\angle 1 + m\angle 3 = m\angle 2 + m\angle 3$?
7. What property of equality lets you conclude that

   $m\angle 1 = m\angle 2$? ❖

Inductive and Deductive Reasoning

The steps in the exploration move from **inductive** to **deductive reasoning**. In Steps 1–3 you use inductive reasoning to make a conjecture. In Steps 4–7 you use deductive reasoning to complete an informal proof that all vertical angles are equal.

Throughout this book you will look for patterns to make generalizations and conjectures. As you gain experience in using the logic of deductive proofs, you will learn to write complete proofs of your own conjectures.

Here is an example of a formal proof of the theorem which states that all vertical angles have equal measure. A two-column format is used.

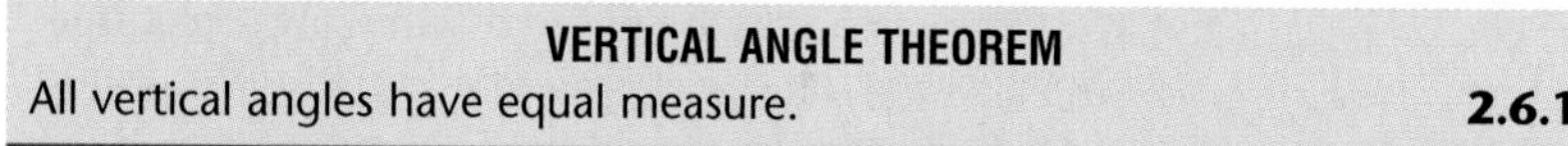

VERTICAL ANGLE THEOREM

All vertical angles have equal measure. **2.6.1**

Given: $\angle 1$ and $\angle 2$ are vertical angles.

Prove: $m\angle 1 = m\angle 2$

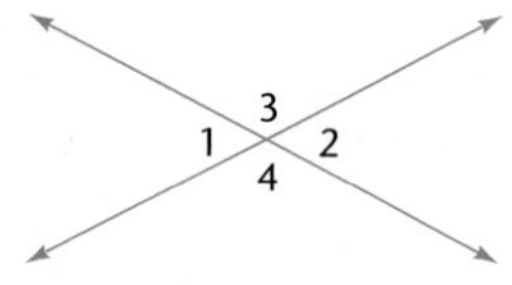

Proof

STATEMENTS	REASONS
1. $\angle 1$ and $\angle 2$ are vertical angles	Given
2. $m\angle 1 + m\angle 3 = 180$ $m\angle 2 + m\angle 3 = 180$	Angles that form linear pairs are supplementary
3. $m\angle 1 + m\angle 3 = m\angle 2 + m\angle 3$	Substitution Property
4. $m\angle 1 = m\angle 2$	Subtraction Property ❖

Law Lawyers use deductive reasoning when trying to prove that a person is innocent or guilty in a trial. They begin by assembling the known facts and then use these facts to reason logically to a conclusion.

Exploration 2 Reflections Through Parallel Lines

You will need
Geometry technology or
Folding paper and a ruler and protractor

Geometry Graphics

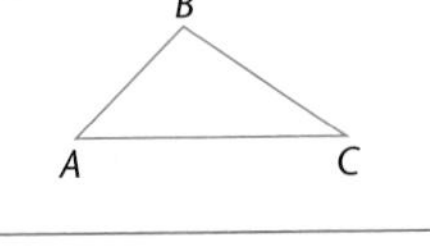

1. Draw 2 parallel lines, l_1 and l_2, and $\triangle ABC$ as shown in the figure.
2. Reflect $\triangle ABC$ through l_1. Call its image $\triangle A'B'C'$.
3. Reflect $\triangle A'B'C'$ through l_2. Call its image $\triangle A''B''C''$.
4. Study the relationship between the original triangle and the final triangle you drew. What single transformation would seem to produce the final image from the original image?
5. Measure the distance between each vertex in the original triangle and its image in the final triangle—that is, the distance AA'', BB'', and CC''. What do you discover?
6. Do all the points in the final triangle seem to have moved in the same direction? Explain your answer.
7. Measure the distance between l_1 and l_2. How does this distance compare with the distances you measured in Step 5.
8. Make a conjecture about the reflection of a figure through parallel lines. Include your results from Step 7 in your conjecture. **(Theorem 2.6.2)** ❖

What would you need to prove about the reflection of a point through two parallel lines to show that your conjecture is true?

An Informal Proof

The diagram suggests an informal proof of the result you discovered in Exploration 2. See if you can follow it.

1. Which distances are indicated as being equal? Why?
2. What is the distance between the two lines?
3. What does the expression $2D_1 + 2D_2$ or $2(D_1 + D_2)$ represent in the drawing? How does it compare with the distance between l_1 and l_2

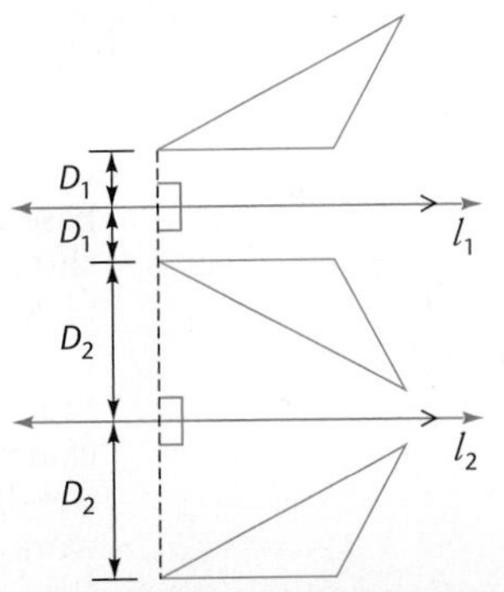

ENRICHMENT Some students may wish to try writing out the informal proof on this page as a formal proof, using statements and reasons. They can begin by first writing down the conjecture and then listing the given information.

INCLUSION strategies **Using Discussion** Some students find the concept of mathematical proof challenging, abstract, and unfriendly. Class discussions about every-day use of logic can help. Use a chair as an example. Ask students how they know that a chair won't collapse when sat on. Inductively, students know from previous experience with chairs. Deductively, students can consider construction and design features of the chair.

Exploration 2 Notes

Students construct a reflection of a triangle through a line and develop an understanding of the congruence of pre-image and image.

ongoing ASSESSMENT

4. **Translation**
8. **Reflection through two parallel lines is equivalent to a translation of twice the distance between the lines and perpendicular to the lines.**

CRITICAL *Thinking*

You would have to prove that the distance any figure moved when reflected through two parallel lines is always twice the distance between the parallel lines and that the figure is always translated perpendicular to the parallel lines.

CRITICAL *Thinking*

Yes: The direction of movement of every point is perpendicular to the mirror of reflection. Therefore, the direction lines are parallel. (Students should recall the paper-folding Exploration 1, page 17.)

Exploration 3 Notes

Students extend their understanding of how transformations can preserve congruence.

ongoing ASSESSMENT

4. **Rotation**
6. **Reflecting a figure through two intersecting lines is equivalent to rotating the figure about the point of intersection through twice the angle between the lines.**

ongoing ASSESSMENT

Try This

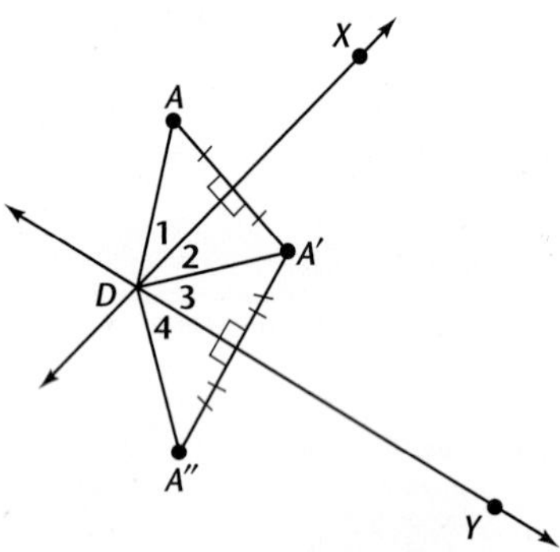

Need a theorem or postulate to justify

$m\angle 1 = m\angle 2, m\angle 3 = m\angle 4$

from the given information. Then,

$m\angle XDY = m\angle 2 + m\angle 3$

$m\angle ADA'' = 2m\angle 2 + 2m\angle 3$

$= 2(m\angle 2 + m\angle 3)$

[The theorem will be HL, a triangle congruence theorem.]

CRITICAL *Thinking* Do you think your result would hold true for any point you might choose on the figure? How could you prove that the direction of movement of each point is the same—that is, parallel?

Exploration 3 *Reflections Through Intersecting Lines*

You will need
Geometry technology or
Folding paper, ruler and protractor

1. Draw two intersecting lines l_1 and l_2, and $\triangle ABC$ as shown in the figure.
2. Reflect $\triangle ABC$ through l_1. Call its image $\triangle A'B'C'$.
3. Reflect $\triangle A'B'C'$ through l_2. Call its image $\triangle A''B''C''$.

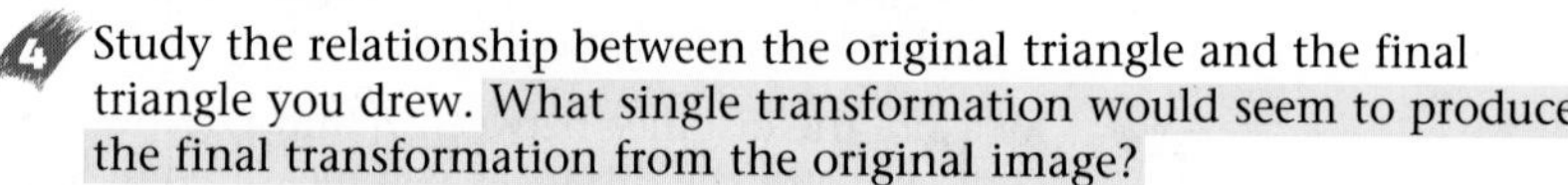

4. Study the relationship between the original triangle and the final triangle you drew. What single transformation would seem to produce the final transformation from the original image?
5. What seems to be special about the intersection point D of the lines l_1 and l_2 in the transformation you have created?
6. Make a conjecture about the reflection of a figure through two intersecting lines. Measure various angles with their vertices at point D, *such as* $\angle ADA'$ and $\angle A'DA''$. Include your results in your conjecture. **(Theorem 2.6.3)**❖

Try This Sketch a proof of your conjecture in the above exploration. It should be similar to the informal proof on the previous page. What is lacking from your present supply of theorems and postulates for a full proof?

The Amazing Thing About Proofs

Conjectures, like the prediction about the regions of the circle at the beginning of this lesson, may turn out to be false. This is because they are based on a limited number of observations or measurements. So, you should always ask whether the next case might prove your conjecture to be false.

Proofs are different. In the proof of the vertical angles conjecture, for example, it does not matter what the measures of the different angles on the drawing are. They could be *any size at all,* and the proof would still work. Therefore, a single proof covers *all possible cases.*

RETEACHING the lesson **Guided Research** Students can explore the intersection of two perpendicular lines by measuring all four angles at the point of intersection.Then ask them to make a conjecture and prove it. (If two lines are perpendicular, then all four angles at the point of intersection are right angles.)

Communicate

1. Discuss the advantages of deductive reasoning over inductive reasoning.
2. What is the value of inductive reasoning? Discuss, giving examples to support your position.
3. In the diagram, which pairs of angles are congruent? State a theorem that allows you to draw this conclusion without actually measuring the angles. Discuss.

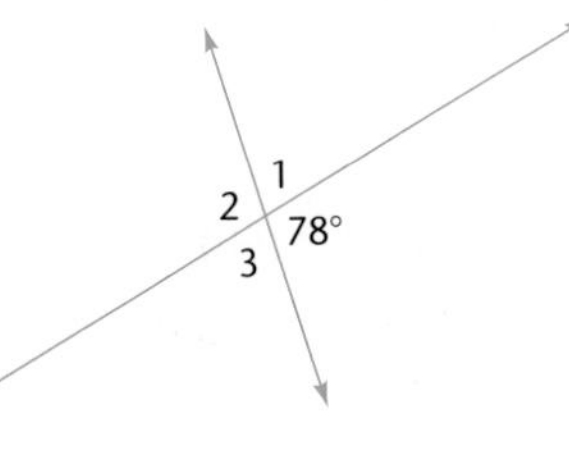

4. Explain how you could *translate* a given figure using only reflections.

Practice & Apply

5. Identify the pairs of congruent angles in the diagram to the right.

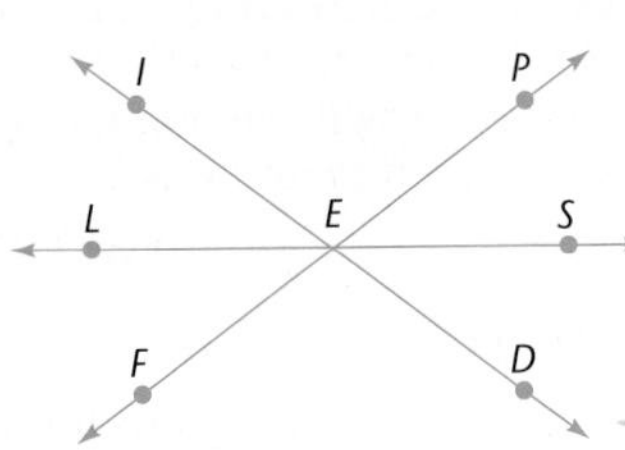

In the diagram to the right, which angle is congruent to the following:

6. ∠*LEI*
∠DES

7. ∠*SEF*
∠LEP

8. ∠*PED*
∠FEI

I
P
L
E
S
F
D

9. Find the measures of all the angles in the diagram.

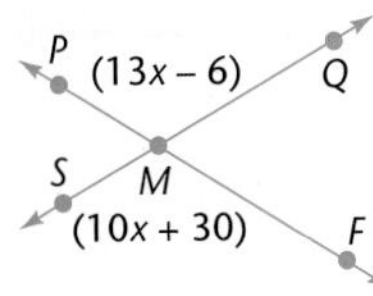

∠PMQ and ∠SMF measure 150°, ∠PMS and ∠QMF measure 30°

In Exercises 10–13, find the measure of ∠*ABC*.

10. m∠ABC = 160°

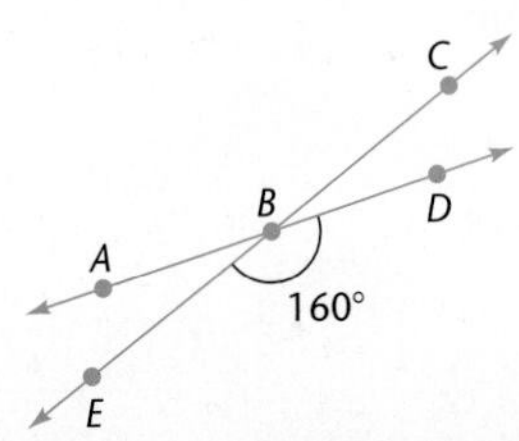

11. m∠ABC = 68°

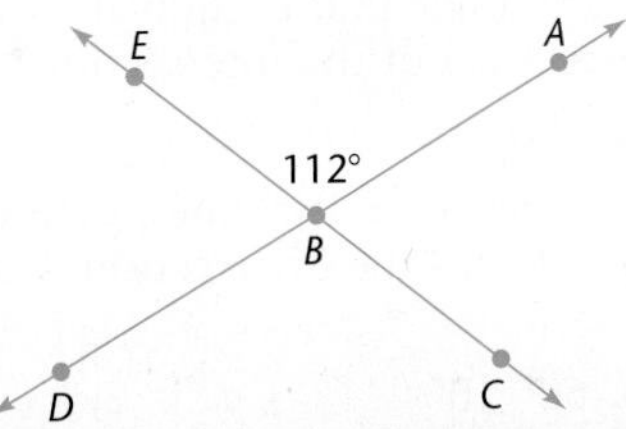

Assess

Selected Answers

Odd–numbered Exercises 5–31

Assignment Guide

Core 1–15, 18–20, 26–32

Core Plus 1–9, 16–34

Technology

For Exercises 21–23, students can use geometry graphics to create the appropriate images.

Error Analysis

In Exercises 14–17, students who get incorrect answers have most likely made simple algebraic errors in solving equations. Be sure to work through the solution of each exercise to review the algebra involved.

5. 1 and 3, 2 and 4, 5 and 7, 6 and 8

Alternative Assessment

Performance Assessment

Students should be able to write down their understanding of what the following terms mean: conjecture, inductive reasoning, deductive reasoning, informal proof, and formal proof.

12.

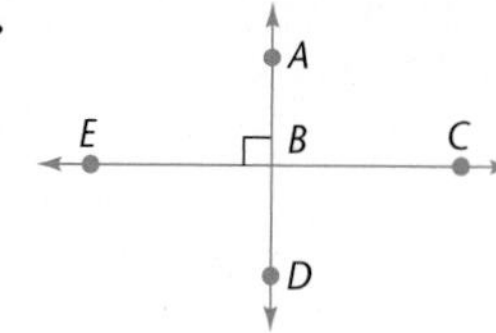

m∠ABC = 90°

13.

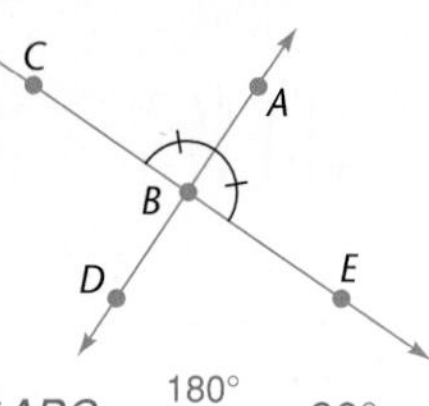

$m\angle ABC = \frac{180°}{2} = 90°$

Algebra **In Exercises 14–17, find the value of *x*.**

14.

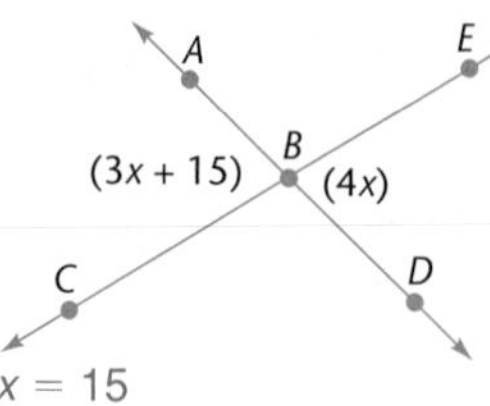

x = 15

15.

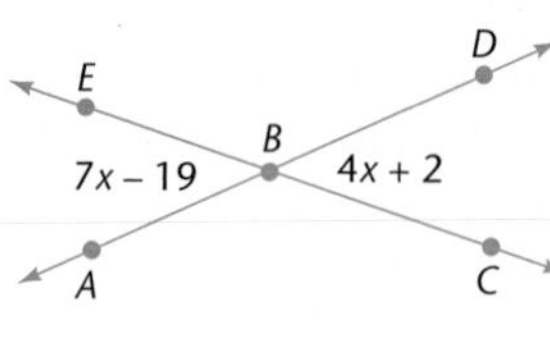

x = 7

16.

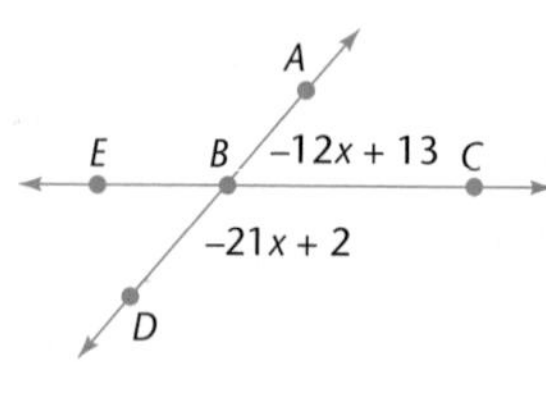

x = –5

17.

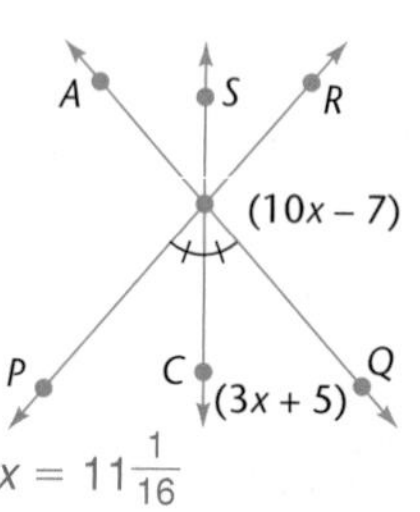

$x = 11\frac{1}{16}$

Use the results of the informal proof on page 101 for Exercises 18–20. If you reflect △*MNO* first through *m* and then through *n*, how far apart will the original figure and the final figure be if the distance between *m* and *n* is

18. 5 cm? 10 cm **19.** 10 cm? 20 cm **20.** *x* cm? 2*x*

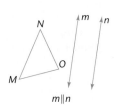

For Exercises 21–23, refer to the figure on the right.

21. If you wanted to use reflections to translate the figure 10 cm in the direction of the arrow, how far apart should your parallel lines be? 5cm

22. How would you determine the direction of the parallel lines in Exercise 21? They should be perpendicular to the direction of the arrow.

23. Is there more than one placememt of parallel lines that will give the translation described in Exercise 21? Use diagrams to illustrate your answer.

Use the results from Exploration 3 on page 102 for Exercises 24 and 25.

24. What angle of intersection of the two lines is needed to produce a double reflection of a triangle that is equivalent to a rotation of the given number of degrees about the intersection?

a. 190° 95° **b.** 90° 45° **c.** 2*x*° *x*°

25. For each angle of intersection of two lines, determine the degrees of an equivalent rotation about the intersection when a triangle is reflected twice.

a. 30° 60° **b.** 50° 100° **c.** *y*° 2*y*°

23. Yes. In addition to the more obvious case of two parallel mirrors on one side of the object, the case shown on the right is another possibility. In fact, an infinite number of arrangements are possible.

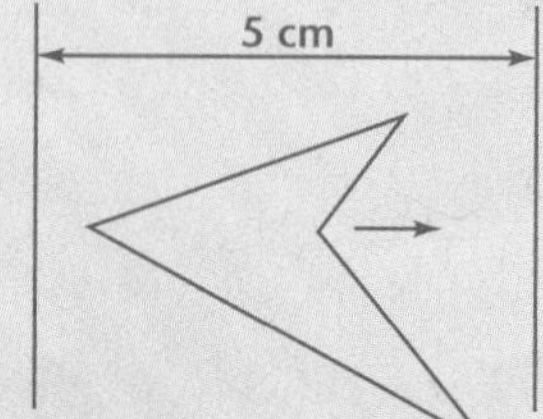

Practice Master

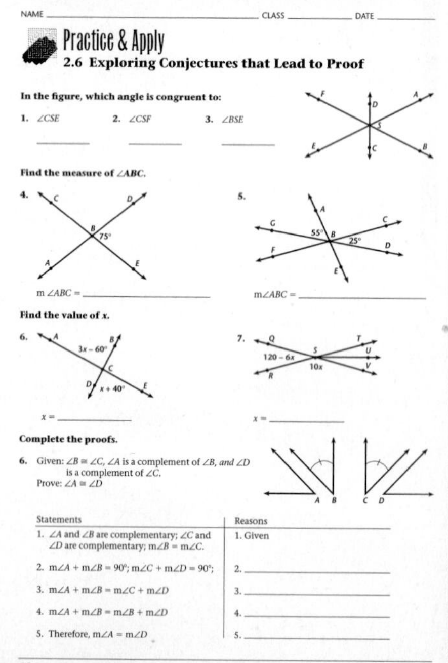

NAME ____________ CLASS ________ DATE ________

Practice & Apply

2.6 Exploring Conjectures that Lead to Proof

In the figure, which angle is congruent to:

1. ∠CSE **2.** ∠CSF **3.** ∠BSE

Find the measure of ∠ABC.

4. m ∠ABC = ____________

5. m∠ABC = ____________

Find the value of *x*.

6. x = ____________

7. x = ____________

Complete the proofs.

6. Given: ∠B ≅ ∠C, ∠A is a complement of ∠B, and ∠D is a complement of ∠C.
Prove: ∠A ≅ ∠D

Statements	Reasons
1. ∠A and ∠B are complementary; ∠C and ∠D are complementary; m∠B = m∠C.	1. Given
2. m∠A + m∠B = 90°; m∠C + m∠D = 90°;	2. ____________
3. m∠A + m∠B = m∠C + m∠D	3. ____________
4. m∠A + m∠B = m∠B + m∠D	4. ____________
5. Therefore, m∠A = m∠D	5. ____________

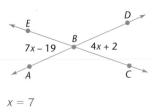

48 **Practice & Apply** **HRW Geometry**

The proof that follows is a proof of this theorem: *If two angles are supplements of congruent angles, then the two angles are congruent.* **Study the proof, then answer Exercises 26–28.**

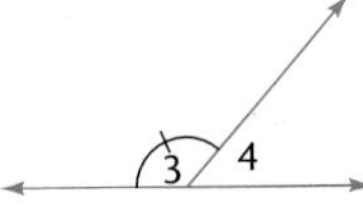

Given: m∠1 = m∠3, ∠1 and ∠2 are supplementary angles, ∠3 and ∠4 are supplementary angles.

Prove: m∠2 = m∠4

Proof

STATEMENTS	REASONS
1. m∠1 + m∠2 = 180°, m∠3 + m∠4 = 180°	Definition of Supplementary Angles
2. m∠1 + m∠2 = m∠3 + m∠4	__?__ Property of Equality
3. m∠1 = m∠3	__?__
4. m∠2 = m∠4	__?__ Property of Equality

26. Which property of equality is the justification for Step 2 of the proof? Transitive

27. What is the justification for Step 3 of the proof? Given

28. Which property of equality is the justification for Step 4 of the proof? Subtraction

Look Back

29. A segment, line, ray, or plane that intersects a segment at its midpoint is called a segment __?__ . **[Lesson 1.2]** bisector

In Exercises 30 and 31, refer to the following statements:

- If I am well rested in the afternoon, then I am in a good mood.
- If I sleep until 8:00 A.M., then I am well rested during the afternoon.
- If it is Saturday, then I sleep until 8:00 A.M.

30. Arrange the statements in a logical order. **[Lesson 2.2]**

31. Write the conditional statement that the three statements taken together prove. **[Lesson 2.2]**

32. What three groups of statements does Euclid's geometry begin with? **[Lesson 2.4]**

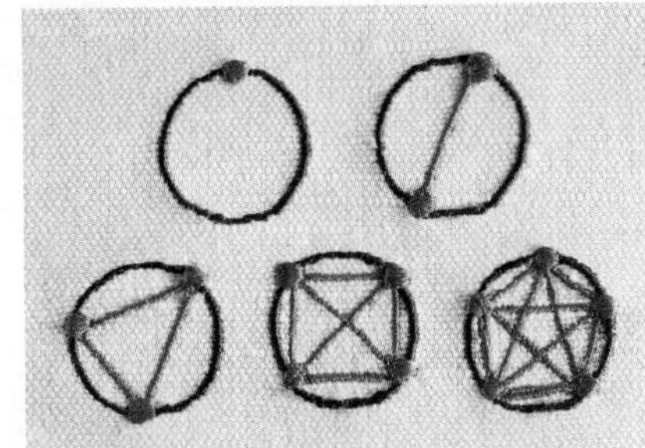

Look Beyond

33. The number of regions in the circle pattern on page 99 seem to double every time. Look for a possible deeper pattern in the diagram. Suppose the "real" pattern for the last row is 1, 2, 3, 4, . . . Use this pattern to fill in the numbers above the diagram. Do the numbers on the top row seem to match the actual regions in the circle figures? Test this by drawing circles with 6, 7, and 8 points. Place the points in an irregular way so that you get the maximum regions in each case.

34. If you found that the numbers in the diagram worked for a hundred, a thousand, or even more points on a circle, would you have proved that the pattern is correct? Explain why or why not.

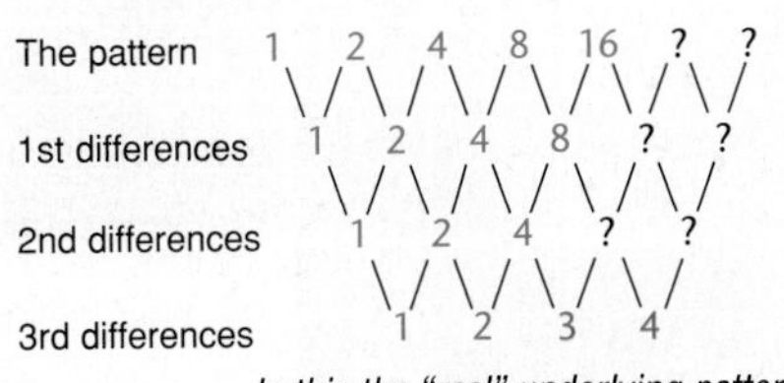

Is this the "real" underlying pattern?

Look Beyond

Extension. Exercises 33 and 34 reinforce the essential idea that inductive reasoning does not always result in correct generalizations because it is based on a limited number (which may be a thousand or more) of observations or measurements. In order to appreciate the need for using deductive reasoning to arrive at proofs of conjectures, students need to understand the nature of both inductive and deductive reasoning.

Technology Master

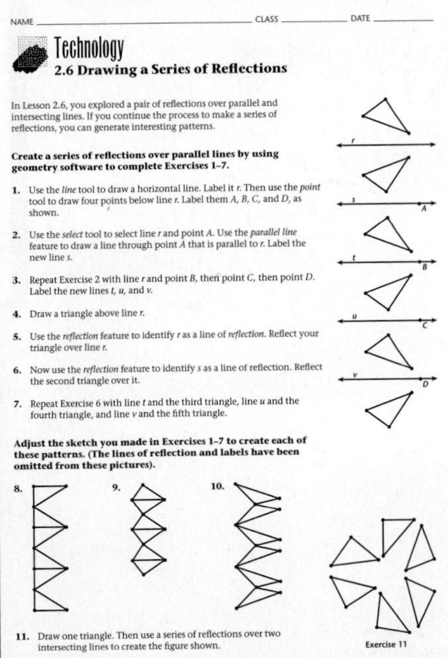

NAME ______________ CLASS ________ DATE ________

Technology

2.6 Drawing a Series of Reflections

In Lesson 2.6, you explored a pair of reflections over parallel and intersecting lines. If you continue the process to make a series of reflections, you can generate interesting patterns.

Create a series of reflections over parallel lines by using geometry software to complete Exercises 1–7.

1. Use the *line* tool to draw a horizontal line. Label it *r*. Then use the *point* tool to draw four points below line *r*. Label them *A*, *B*, *C*, and *D*, as shown.
2. Use the *select* tool to select line *r* and point *A*. Use the *parallel line* feature to draw a line through point *A* that is parallel to *r*. Label the new line *s*.
3. Repeat Exercise 2 with line *r* and point *B*, then point *C*, then point *D*. Label the new lines *t*, *u*, and *v*.
4. Draw a triangle above line *r*.
5. Use the *reflection* feature to identify *r* as a line of *reflection*. Reflect your triangle over line *r*.
6. Now use the *reflection* feature to identify *s* as a line of reflection. Reflect the second triangle over it.
7. Repeat Exercise 6 with line *t* and the third triangle, line *u* and the fourth triangle, and line *v* and the fifth triangle.

Adjust the sketch you made in Exercises 1–7 to create each of these patterns. (The lines of reflection and labels have been omitted from these pictures).

8. **9.** **10.**

11. Draw one triangle. Then use a series of reflections over two intersecting lines to create the figure shown.

Exercise 11

60 Technology HRW Geometry

30. If it is Saturday, then I sleep until 8:00 A.M. If I sleep until 8:00 A.M., then I am well rested during the afternoon. If I am well rested during the afternoon, then I am in a good mood.

31. If it is Saturday, then I am in a good mood.

32. undefined terms, postulates, definitions, and theorems

33.

The circle pattern	1 2 4 8 16 <u>31</u> <u>57</u> <u>99</u>
1st differences	1 2 4 8 <u>15</u> <u>26</u> <u>42</u>
2nd differences	1 2 4 <u>7</u> <u>11</u> <u>16</u>
3rd differences	1 2 3 4 5

Yes

34. No. The pattern must be proved for all possible numbers of circle points.

Focus

The objective of this project is to have students practice and sharpen their logical thinking skills.

Motivate

Students have learned a great deal in this chapter about logical thinking and how reasoning processes are used in geometry. They now know also that it is important to reason clearly and correctly in everyday situations. Logical reasoning skills, however, cannot be developed without studying what they are and without practicing them. Logic puzzles are an interesting and entertaining way to sharpen logic skills.

Solving Logic Puzzles

Geometry often involves using logical thinking to conclude what is possible or not for a given set of mathematical conditions. Logic puzzles provide a way to sharpen logic skills. Try the two logic puzzles by the British writer "Caliban."

Activity 1

Accomplishments

"My four granddaughters are all accomplished girls." Canon Chasuble was speaking with evident self-satisfaction. "Each of them, " he went on, "plays a different musical instrument and each speaks one European language as well as — if not better than — a native."

"What does Mary play?" asked someone.

"The cello."

"Who plays the violin?"

"D'you know," said Chasuble, "I've temporarily forgotten, alas! But I know it's the girl who speaks French."

The remainder of the facts which I elicited were of a negative character. I learned that the organist is not Valerie; that the girl who speaks German is not Lorna; and that Mary knows no Italian. Anthea doesn't play the violin; nor is she the girl who speaks Spanish. Valerie knows no French; Lorna doesn't play the harp; and the organist can't speak Italian.

What are Valerie's accomplishments?

Make a chart like the one below to determine the correct combinations. Mark an "x" for combinations which are not true and a "•" for combinations which are true. The first clue has been marked to help you get started.

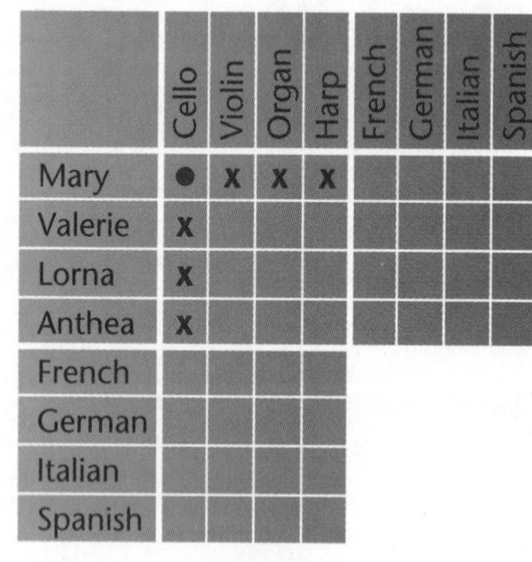

	Cello	Violin	Organ	Harp	French	German	Italian	Spanish
Mary	●	X	X	X				
Valerie	X							
Lorna	X							
Anthea	X							
French								
German								
Italian								
Spanish								

After you have filled in the "x's" and "•'s" from the initial clues, you will have to deduce the remaining relationships. For example, suppose you know that Mary plays the violin and the violinist doesn't speak German. Then you can conclude that Mary doesn't speak German. But this is only an example. Find out for yourself!

1. Valerie plays the harp and speaks Italian.

	Cello	Violin	Organ	Harp	French	German	Italian	Spanish
Mary	●	X	X	X	X	X	X	●
Valerie	X	X	X	●	X	X	●	X
Lorna	X	●	X	X	●	X	X	X
Anthea	X	X	●	X	X	●	X	X
French	X	●	X	X				
German	X	X	●	X				
Italian	X	X	X	●				
Spanish	●	X	X	X				

Activity 2

In this logic problem, each letter represents a digit.

Digits Are Symbols

"Digits are only symbols, " said Miss Piminy to her class. "We could use other symbols instead. Suppose, for example, that SE is the square of E. Then STET might also represent a perfect square."

"I don't see that, Miss Piminy," said Troublesome Tess, the *enfant terrible* of the class.

"Don't you, Tess? Then use your wits," said Miss Piminy. "Your comment," she added "gives me an idea." She wrote on the board:

```
  S E E
T E S S
-------
• • • •
```

"There's an addition sum, students. S, E, and T have the values mentioned already. What's the answer to my sum? — in my own notation of course."

How quickly can you produce the answer?

Write a step-by-step procedure which explains how to find the answer.

"Caliban" is the pen name of a famous British puzzle writer named Hubert Phillips. Many of his puzzles appeared regularly in British newspapers. A number of his puzzle books have been published in the United States.

Computer games often require logical thinking.

Two moves will solve this cube. What color will the bottom face be?

2. SE must be 25 or 36. The number cannot be 25 since there is no perfect square 2*x*5*x*. Therefore, SE is 36 and STET is 3969—the square of 63. The sum of SEE and TESS is the sum of 366 and 9633, or 9999. Therefore, the answer is TTTT.

Cooperative Learning

Students can work with a partner to do each logic puzzle. Have them first read each puzzle by themselves before trying to solve it together. For Activity 1, one student can make the chart and the other can fill it in. For Activity 2, both partners should try to write out the answers, discussing and comparing their results as they proceed. After writing out a step-by-step procedure, they should check it against their answer to make sure the procedure correctly describes what they have done.

Discuss

After students in each group complete both puzzles, have them write down and identify each step in which they used logical thinking to arrive at a conclusion. Then organize the groups of two into groups of four, and have students compare their answers and the conclusions arrived at through logical thinking.

1. The number of parts that result from any number of vertical lines that intersect the *S* is $3n + 1$ where n is the number of vertical lines.

2. If there are zero vertical lines, there is one part. Any vertical line that intersects the figure intersects the figure in three places. One vertical line divides the figure into four parts, the original part and three more. Any more vertical lines add three more parts—the three parts between the three points where that line hits the figure and the three points where the adjacent line hits the figure.

Chapter 2 Review

Vocabulary

Key Skills and Exercises

Lesson 2.1

➤ Key Skills

Give an informal proof for a conjecture.

In the figure the points on a circle have been connected with segments. The table shows the number of points and the number of segments connecting the points.

points	2 3 4 5
segments	1 3 6 10

Conjecture: *If there are n points on a circle, then the number of segments necessary to connect all the points is* $\frac{n}{2}(n-1)$. You can test this conjecture by sketching the next figure in the sequence, counting the number of segments, and comparing that to $\frac{6}{2}(6-1) = 15$.

➤ Exercises

The vertical line separates the S into four parts.

1. Make a conjecture about the number of parts that result from any number of vertical lines.
2. Write or draw an informal proof of your conjecture.

Lesson 2.2

➤ Key Skills

Draw a conclusion from a conditional.

What conclusion can you draw from these two statements?

If an animal is a cat, it has four legs.
Sandy is a cat.

Conclusion: *Sandy has four legs.*

State the converse of a conditional.

The converse of the previous conditional is *"If an animal has four legs, it is a cat."* A counterexample proves this converse is false: *Peek the dog has four legs.*

➤ Exercises

In Exercises 3–4 refer to the statements:

If a "star" doesn't flicker, then it's really a planet.
The Evening Star doesn't flicker.

3. Draw a conclusion about the Evening Star.

4. Write the converse of the conditional statement. Is it a true conditional?

Lesson 2.3

➤ Key Skills

Write and test a definition.

Is this a definition? *"A right angle is an angle whose measure is 90°."*

If the conditional form of a statement is true and its converse is also true, then the statement is a definition.

If an angle is a right angle, then its measure is 90°. ← **Statement is true**
If an angle measures 90°, then it is a right angle. ← **Converse is true**

Since both these conditionals are true, the original statement is a definition.

➤ Exercises

In Exercises 5–6 demonstrate whether the statement is a definition.

5. Parallel lines are lines that are always the same distance apart.

6. A goldenrod is a yellow wildflower.

Lesson 2.4

➤ Key Skills

Use postulates to justify a statement.

In the figure below, lines l and m intersect at point C. The postulate that justifies this statement is, "The intersection of two lines is exactly one point."

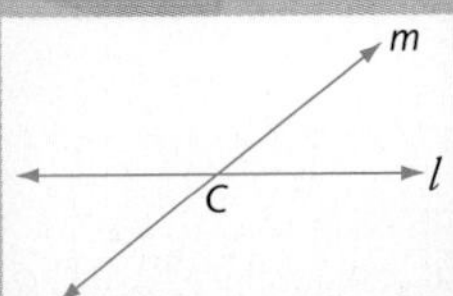

➤ Exercises

Use the figure to the right to answer each exercise. Then justify each answer with a postulate.

7. Name three points that determine plane $\mathscr{R}$.

8. Name the intersection of planes $\mathscr{R}$ and $\mathscr{M}$.

9. How many planes contain points A, C, and D?

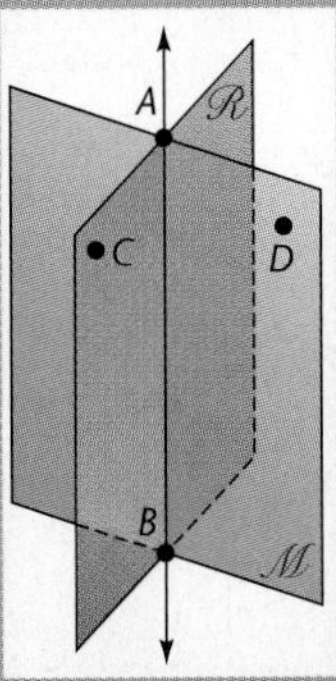

3. The Evening Star is a planet.
4. If a "star" is really a planet, then it doesn't flicker. True
5. If lines are parallel, then they are always the same distance apart. True
   If lines are always the same distance apart, then the lines are parallel. True
   Since both the conditional and its converse are true, the statement is a definition.

Chapter Review

6. If a wildflower is a Goldenrod, then it is yellow. True; If a wildflower is yellow, then it is a Goldenrod. False; Since the conditional statement and its converse are not both true, the statement is not a definition.
7. *ABC;* Three noncollinear points determine exactly one plane.
8. *AB;* The intersection of two planes is a line.
9. Exactly one. Three noncollinear points determine exactly one plane.

Lesson 2.5

➤ Key Skills

Use the properties of equality to justify statements.

If $\overline{AB} \cong \overline{CD}$, you know that the relationship between $\overline{AC}$ and $\overline{BD}$ is that they are congruent by the Addition Property of Equality.

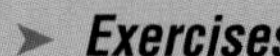

➤ Exercises

Given $\angle EZN \cong \angle PZO$.

10. Prove: $\angle EZO \cong \angle PZN$

Lesson 2.6

Key Skills

Use deductive reasoning to prove a conjecture.

In the figure m$\angle VOE > 90°$. Show that m$\angle VOT$ and m$\angle ROE$ are each less than 90°.

m$\angle VOT$ + m$\angle VOE$ = 180°, by the definition of supplementary angles. Because m$\angle VOE$ is greater than 90°, m$\angle VOT$ must be less than 90°, or the sum of the two angles would be greater than 180°. m$\angle VOT$ = m$\angle ROE$ because they are vertical angles, so m$\angle ROE$ must also be less than 90°.

➤ Exercises

Given that m$\angle SPR$ = m$\angle TRP$, and m$\angle QPS$ = m$\angle QRT$, prove that m$\angle QPR$ = m$\angle QRP$.

By __(11)__, m$\angle QPS$ + m$\angle QPR$ = m$\angle SPR$ and m$\angle QRT$ + m$\angle QRP$ = m$\angle TRP$. Since it is given that m$\angle SPR$ = m$\angle TRP$, then m$\angle QPS$ + m$\angle QPR$ = m$\angle QRT$+ m$\angle QRP$ by __(12)__. It is also given that m$\angle QPS$ = m$\angle QRT$. Therefore, m$\angle QPR$ = m$\angle QRP$ by __(13)__.

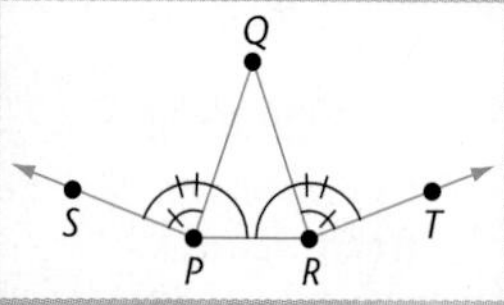

Applications

14. Computer Programming A programmer is trying to find the error in the logic of a program. In English the chain of logic is:

1. Is the first character on the screen a number or a letter?
2. If it is an E, then exit the program.
3. If it is any other letter, then go to the word-handling routine.

Find the logical gap and write a conditional statement that could bridge the gap.

15. Archaeology The symbols in the top row are from the ogham alphabet, found in archaeological digs in southern Ireland.

The symbols in the bottom row are not from the ogham alphabet.

Write a definition of ogham symbols.

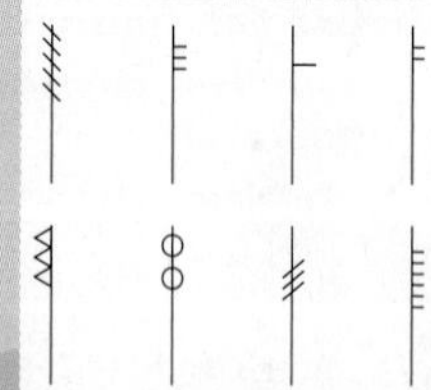

10. Statements
    1. $\angle EZN \cong \angle OZP$
    2. m$\angle EZN$ = m$\angle OZP$
    3. m$\angle NZO$ = m$\angle NZO$
    4. m$\angle EZN$ + $\angle NZO$ = m$\angle OZP$ + m$\angle NZO$
    5. m$\angle EZN$ + m$\angle NZO$ = m$\angle EZO$ and m$\angle OZP$ + m$\angle NZO$ = m$\angle PZN$
    6. m$\angle EZO$ = m$\angle PZN$
    7. $\angle EZO \cong \angle PZN$

    Reasons
    1. Given
    2. Definition of congruent angles
    3. Reflexive Property of Equality
    4. Addition Property of Equality
    5. Angle Addition Postulate
    6. Substitution Property of Equality
    7. Definition of congruent angle

11. the Angle Addition Postulate
12. the Transitive Property of Equality
13. the Subtraction Property of Equality
14. Sample answer: Insert the following after Conditional 1 or after Conditional 3: If it is a number, then go to the number-handling routine.
15. Sample answer: A symbol in the Ogham alphabet consists of one to five short lines perpendicular to a longer line (either crossing the line or stopping at the line) or crossing the line at a slant that is higher on the left side of the line.

Chapter 2 Assessment

In Exercises 1–2 refer to the sequence below.

$\frac{1}{2}, \frac{1}{4}, \frac{1}{8}, \frac{1}{16}, \ldots$

1. Make a conjecture about the sequence's relation to zero.
2. Write or draw an informal proof of your conjecture.

In Exercises 3–4 refer to the statements.

If an animal has six legs, it is an insect.
A flea has six legs.

3. Draw a conclusion about a flea.
4. Write the converse of the conditional statement. Is the converse a true conditional?

5. Demonstrate whether the statement is a definition:
A tuba is a brass musical instrument.

These objects are tori (singular, torus).

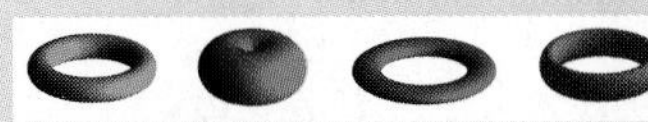

These objects are not tori.

6. Which of these objects are tori?
7. Write a definition of a torus.

In Exercises 8–9 use the figure to the right to answer each question. Then justify each answer with a postulate.

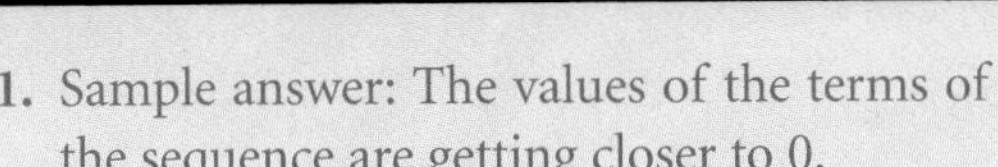

8. Name two points that determine line l.
9. Name the intersection of line l and $\overleftrightarrow{AD}$.

10. What property justifies this conclusion: $m\angle ABC = m\angle CBA$?
   a. Addition Property of Equality
   b. Reflexive Property of Equality
   c. Symmetric Property of Equality
   d. Transitive Property of Equality

1. Sample answer: The values of the terms of the sequence are getting closer to 0.
2. Sample answer: Dividing 1 by larger and larger numbers, results in smaller and smaller numbers, numbers that approach 0.
3. A flea is an insect.
4. If an animal is an insect, then it has six legs. True
5. If a musical instrument is a tuba, then it is brass. True
If a musical instrument is brass, then it is a tuba. False
Since the conditional statement and its inverse are not both true, the statement is not a definition.
6. The first and third figures are tori.
7. Sample answer: A torus is a solid that has a hole in it.
8. *B* and *C*; Two points determine exactly one line.
9. *C*; The intersection of two lines is a point.
10. b

COLLEGE ENTRANCE-EXAM PRACTICE

Multiple-Choice and Quantitative-Comparison Samples

The first half of the Cumulative Assessment contains two types of items found on standardized tests—multiple-choice questions and quantitative-comparison questions. Quantitative comparison items emphasize the concepts of equalities, inequalities, and estimation.

Free-Response Grid Samples

The second half of the Cumulative Assessment is a free-response section. A portion of this part of the Cumulative Assessment consists of student-produced response items commonly found on college entrance exams. These questions require the use of machine-scored answer grids. you may wish to have students practice answering these items in preparation for standardized tests.

Sample answer grid masters are available in the *Chapter Teaching Resources Booklets.*

Chapters 1–2 Cumulative Assessment

College Entrance Exam Practice

Quantitative Comparison. Exercises 1–4 consist of two quantities, one in Column A and one in Column B, which you are to compare as follows.

A. The quantity in Column A is greater.
B. The quantity in Column B is greater.
C. The two quantities are equal.
D. The relationship cannot be determined from the information given.

	Column A	Column B	Answers
1.	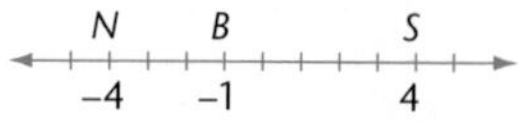 NB	BS	(A) (B) (C) (D) **[Lesson 1.4]**
2.	$\overline{WZ}$ is an angle bisector. 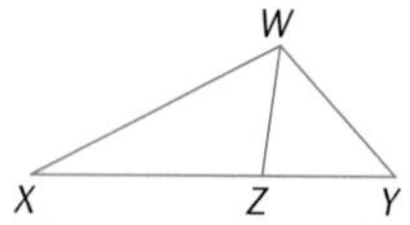 $m\angle XWZ$	$m\angle YWZ$	(A) (B) (C) (D) **[Lesson 1.5]**
3.	$\overline{EF}$ is congruent to $\overline{HG}$.  EH	FG	(A) (B) (C) (D) **[Lesson 2.5]**
4.	1, 4, n, . . . 9	n	(A) (B) (C) (D) **[Lesson 2.1]**

In Exercises 5–9 identify each as a translation, rotation, or reflection. [Lesson 1.7]

5. 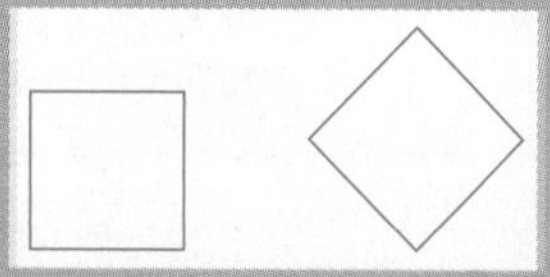

6. 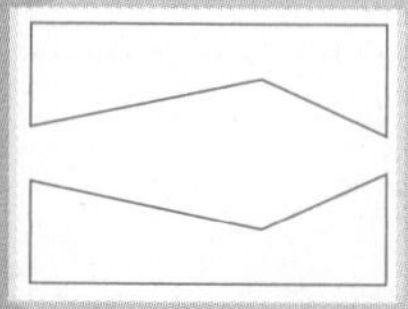

7. $(x, y) \rightarrow (-x, y)$

8. Which is a true definition? **[Lesson 2.3]**

a. The midpoint of a segment divides the segment into two equal parts.
b. An acute triangle has an angle less than 90°.
c. Two points determine exactly one line.
d. An equilateral triangle is an isosceles triangle.

1. b **2.** c **3.** c **4.** d

5. Rotation **6.** Reflection

7. Reflection in y-axis

8. a **9.** d

9. m∠1 = 35° and m∠2 = 35°. What property justifies this conclusion: m∠1 = m∠2? **[Lesson 2.5]**
   a. Addition Property of Equality
   b. Reflexive Property of Equality
   c. Subtraction Property of Equality
   d. Transitive Property of Equality

10. Write a conjecture about ∠*GYO* and ∠*RYE*. **[Lesson 1.2]**

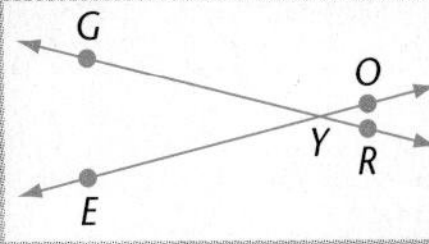

11. Construct an equilateral triangle and a circumscribed circle. **[Lesson 1.3]**

12. Copy the figure at the right and sketch the translation $(x + 4)$ of △*PQR*. **[Lesson 1.7]**

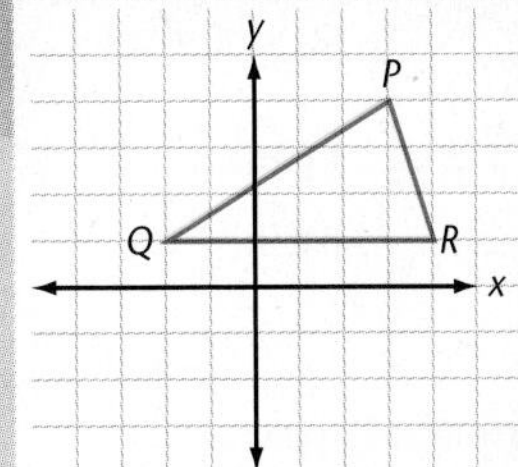

13. Write the converse of this statement: *If you have a piece of cake, then you can eat it.* Is the converse a true conditional? **[Lesson 2.2]**

14. Write a conclusion from these two statements: *If the maple leaves are falling, then it must be October. It is October.* **[Lesson 2.2]**

15. The following are splorts.

The following are not splorts.

Write a definition of a splort. **[Lesson 2.3]**

In Exercises 16–17 use the figure to answer each question. Then justify each answer with a postulate. [Lesson 1.1]

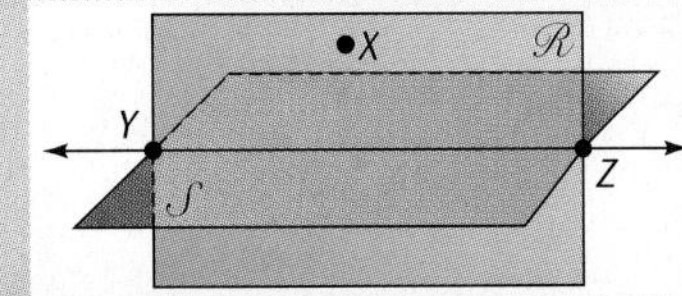

16. Name the intersection of planes ℛ and 𝒮.

17. How many planes contain points *X*, *Y*, and *Z*?

Free-Response Grid Exercises 18–20 may be answered using a free-response grid commonly used by standardized test services.

18. Hither, Thither, and Yon are towns on a highway that runs straight from Hither, through Thither, to Yon. The distance from Hither to Thither is 41 miles. The distance from Thither to Yon is 2 miles more than twice the distance from Hither to Thither. What is the distance between Thither and Yon? **[Lesson 1.4]**

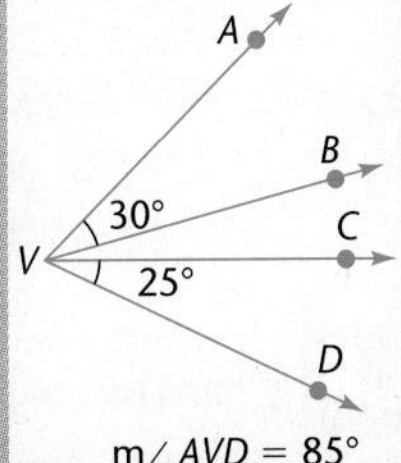

19. What is m∠*AVC*? **[Lesson 1.5]**

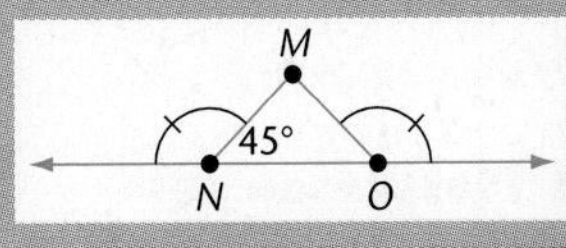

20. What is the measure of ∠*MON*? **[Lesson 1.5]**

Cumulative Assessment

13. If you can eat a piece of cake, then you can have it. False

14. The only conclusion that can be made is that it is October.

15. Sample answer: A splort is a figure and some of its reflections such that there are no points of intersection.

16. *YZ*; The intersection of two planes is a line.

17. Exactly one; Three noncollinear points determine exactly one plane.

18. *HT* = 41 mi
    TY = 2 + 2(*HT*) =
    2 + 2(41) = 84 mi

19. m∠*AVC* = 85° − 25° = 60°

20. m∠*MOW* = 45°

10. m∠*GYO* = m∠*RYE*

11.

12.

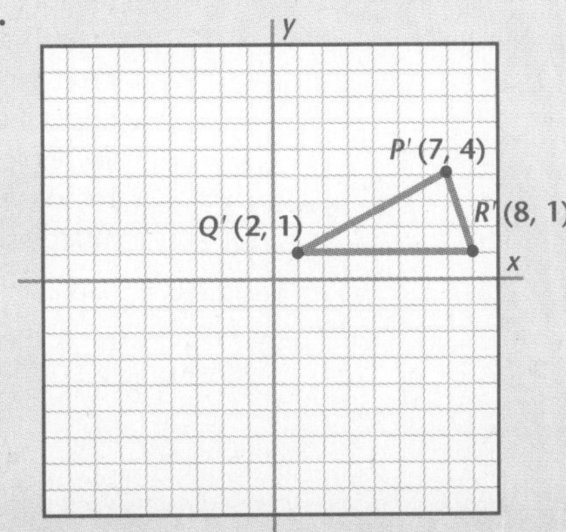

Parallels and Polygons

Meeting Individual Needs

3.1 Exploring Symmetry in Polygons

Core Resources	Core Plus Resources
Inclusion Strategies, p. 118 Reteaching the Lesson, p. 119 Practice Master 3.1 Enrichment, p. 118 Technology Master 3.1 Lesson Activity Master 3.1 Interdisciplinary Connection, p. 117	Practice Master 3.1 Enrichment Master 3.1 Technology Master 3.1 Interdisciplinary Connection, p. 117
[1 day]	[1 day]

3.2 Exploring Properties of Quadrilaterals

Core Resources	Core Plus Resources
Inclusion Strategies, p. 126 Reteaching the Lesson, p. 127 Practice Master 3.2 Enrichment, p. 126 Technology Master 3.2 Lesson Activity Master 3.2	Practice Master 3.2 Enrichment Master 3.2 Technology Master 3.2 Interdisciplinary Connection, p. 125
[3 days]	[2 days]

3.3 Parallel Lines and Transversals

Core Resources	Core Plus Resources
Inclusion Strategies, p. 134 Reteaching the Lesson, p. 135 Practice Master 3.3 Enrichment Master 3.3 Lesson Activity Master 3.3	Practice Master 3.3 Enrichment, p. 134 Technology Master 3.3
[2 days]	[2 days]

3.4 Proving Lines Parallel

Core Resources	Core Plus Resources
Inclusion Strategies, p. 139 Reteaching the Lesson, p. 140 Practice Master 3.4 Enrichment, p. 139 Technology Master 3.4 Lesson Activity Master 3.4 Mid-Chapter Assessment Master	Practice Master 3.4 Enrichment Master 3.4 Technology Master 3.4 Mid-Chapter Assessment Master
[2 days]	[2 days]

3.5 Exploring the Triangle Sum Theory

Core Resources	Core Plus Resources
Inclusion Strategies, p. 145 Reteaching the Lesson, p. 146 Practice Master 3.5 Enrichment, p. 145 Technology Master 3.5 Lesson Activity Master 3.5	Practice Master 3.5 Enrichment Master 3.5 Technology Master 3.5 Interdisciplinary Connection, p. 144
[2 days]	[1 day]

3.6 Exploring Angles in Polygons

Core Resources	Core Plus Resources
Inclusion Strategies, p. 151 Reteaching the Lesson, p. 152 Practice Master 3.6 Enrichment Master 3.6 Technology Master 3.6 Lesson Activity Master 3.6 Interdisciplinary Connection, p. 150	Practice Master 3.6 Enrichment, p. 151 Technology Master 3.6 Interdisciplinary Connection, p. 150
[2 days]	[2 days]

3.7 Exploring Midsegments of Triangles and Trapezoids

Core Resources	Core Plus Resources
Inclusion Strategies, p. 155 Reteaching the Lesson, p. 156 Practice Master 3.7 Enrichment Master 3.7 Lesson Activity Master 3.7	Practice Master 3.7 Enrichment p. 155 Technology Master 3.7
[1 day]	**[1 day]**

3.8 Analyzing Polygons Using Coordinates

Core Resources	Core Plus Resources
Inclusion Strategies, p. 162 Reteaching the Lesson, p. 163 Practice Master 3.8 Enrichment, p. 162 Technology Master 3.8 Lesson Activity Master 3.8 Interdisciplinary Connection, p. 161	Practice Master 3.8 Enrichment Master 3.8 Technology Master 3.8 Interdisciplinary Connection, p. 161
[2 days]	**[2 days]**

Chapter Summary

Core Resources	Core Plus Resources
Chapter 3 Project, pp. 166–167 Lab Activity Long-Term Project Chapter Review, pp. 168–170 Chapter Assessment, p. 171 Chapter Assessment, A/B Alternative Assessment	Chapter 3 Project, pp. 166–167 Lab Activity Long-Term Project Chapter Review, pp. 168–170 Chapter Assessment, p. 171 Chapter Assessment, A/B Alternative Assessment
[3 days]	**[2 days]**

Reading Strategies

Lesson 3.3 includes the use of proofs written in two-column form. The two-column form was introduced briefly in the last lesson of Chapter 2. Two-column proofs present special challenges to the reader. Make sure students understand that you do *not* read all the statements and then all the reasons. The proof should be read in "rows." Statement 1 is read, followed by its matching Reason; Statement 2 is read, followed by its Reason; and so forth. You may wish to have students draw horizontal lines across their own proofs so that these "rows" are more visually obvious.

Visual Strategies

Lessons 3.3 and 3.4 provide good examples of geometric figures used in association with exercises and problems. Have students preview and discuss the diagrams ahead of time, perhaps before they begin the chapter. They should point out how letters, numbers, arrowheads, lines, and dots are used in the drawings. Ask students to tell you which figures go with which exercises, and how they know. Students will also benefit from copying figures with paper and pencil or geometry graphics software.

Cooperative Learning

GROUP ACTIVITIES	
Symmetry in triangles and polygons	Lesson 3.1, Explorations 1 and 2
Properties of quadrilaterals	Lesson 3.2, Explorations 1, 2, 3, and 4
The Triangle Sum Theorem	Lesson 3.5, Exploration
Angle Sums in Polygons	Lesson 3.6, Explorations 1 and 2
String Figures	Chapter Project, pp. 166–167

You may wish to have students work in groups or with partners for some of the above activities. Additional suggestions for cooperative group activities are noted in the teacher's notes in each lesson.

Multicultural

The cultural references in this chapter include references to Europe and the Americas.

CULTURAL CONNECTIONS	
Europe: Tiling patterns in the Alhambra	Chapter Opener
Americas: American Indian rug	Lesson 3.1
Americas: Temple of the Giant Jaguar	Lesson 3.7

Portfolio Assessment

Below are portfolio activities for the chapter listed under seven activity domains which are appropriate for portfolio development.

1. **Investigation/Exploration** The explorations in Lesson 3.1 focus on symmetry in triangles and polygons. In Lesson 3.2, the explorations allow students to explore properties of special quadrilaterals. The exploration in Lesson 3.3 deals with two parallel lines cut by a transversal; the one in Lesson 3.5 with the Triangle Sum Theorem. In Lesson 3.6, the Explorations focus on angle sums in polygons. In Lesson 3.7, the topic is midsegments of triangles and trapezoids.
2. **Applications** Flood Control, Lesson 3.7, Exercises 23–26; Construction, Lesson 3.8, Exercise 17.
3. **Nonroutine Problems** in Lesson 3.3, Exercise 37; Lesson 3.6, Exercise 31; Lesson 3.8, Exercises 30.
4. **Project** String Figures: see pages 166–167. Students follow a specific set of steps to make a string figure called *Jacob's Ladder.*
5. **Interdisciplinary Topics** Students may choose from the following: Fine Art, Lesson 3.1, Exercise 3; Algebra, Lesson 3.1, Exercises 39–44; Geography, Lesson 3.1, Exercises 51–53; Algebra, Lesson 3.2, Exercises 25–26; Algebra, Lesson 3.3, Exercises 18–23; Algebra, Lesson 3.4, Exercises 19–26; Algebra, Lesson 3.6, Exercises 22–25; Algebra, Lesson 3.8, Exercises 18–19
6. **Writing** *Communicate* exercises offer writing selections for the portfolio. Lesson 3.2, Exercise 3; Lesson 3.6, Exercises 2 and 4; Lesson 3.8, Exercises 3 and 5
7. **Tools** In Chapter 3, students use geometry graphics software to make conjectures about properties of quadrilaterals, angles formed by a transversal and two parallel lines, and midsegments of triangles and trapezoids.

Technology

Many of the activities in this book can be significantly enhanced by the use of geometry graphics software, but in every case the technology is optional. If computers are not available, the computer activities can, *and should,* be done by hands-on or "low-tech" methods, as they contain important instructional material. For more information on the use of technology, refer to the *HRW Technology Manual for High School Mathematics.*

Tessellation Software

Exercise 35 in Lesson 3.2 and Exercise 30 in Lesson 3.8 invite students to create their own tessellations, following examples by M. C. Escher. Special geometry software, such as *TesselMania!*™, can greatly add to student enjoyment of these activities.

Students select a type of tessellation (translation, rotation, etc.) and use the software tools to modify basic polygons into new figures that will tessellate the plane. Through use of the paintbucket, the drawing tool, and selections of colors, textures, and stamps, students elaborate their figure outlines to create recognizable and entertaining figures—perhaps even works of art. A great portfolio idea.

Computer Graphics Software

Lesson 3.2 Exploration 1, Part 1 In this exploration, students investigate the relationships of segment lengths, angle measures and areas within parallelograms (including the diagonals). The geometry technology allows students to look at many different parallelograms quickly. Upon completion of this exploration, the students should have discovered the basic properties of parallelgrams.

Lesson 3.2: Exploration 2 In this exploration, students investigate the properties of rhombuses. The geometry technology quickly measures the sides, angles, and areas. Students are asked to find at least two new properties of rhombuses that are not generally true of parallelograms.

Lesson 3.2: Exploration 3 Students investigate the properties of rectangles by dragging a figure and making conjectures from the measurements shown. Students are asked to find two new properties of rectangles that are not generally true of parallelograms.

Lesson 3.2: Exploration 4 Students investigate how the properties of squares relate to the properties of parallelograms, rhombuses, and rectangles. Students are asked to make conjectures about squares using the measurements shown, and by dragging the figure.

Lesson 3.3: Exploration In this exploration, students look at the measures of the angles formed by a transversal intersecting parallel lines. For each special pair of angles the technology provides an update on the angle measures as the position of the transversal is changed. Students are asked to make conjectures about these special angle pairs.

Lesson 3.6: Exploration 2 In this exploration, students investigate exterior angle sums in convex polygons. The geometry technology allows the student to examine many different examples of exterior angle sums for triangles, quadrilaterals, and pentagons. In each case, the technology updates the angle measures as the figures change and provides evidence that the sum of the exterior angles of a convex polygon remains constant.

Lesson 3.7: Exploration 1 Students discover the properties of midsegments of a triangle. Students drag a triangle to change the displayed measurements, and then use these measurements to make a conjecture about the length of a triangle midsegment in relationship to its parallel side.

Lesson 3.7: Exploration 2 Students investigate the properties of the midsegment of a trapezoid. They are asked to drag the figure and use the displayed measurements to make a conjecture about the length of a trapezoid midsegment in relation to its bases.

Lesson 3.7: Exploration 3 In this exploration, students investigate the connection between the midsegment of a trapezoid and a midsegment of a triangle by dragging a vertex of a trapeqoid and changing the figure into a triangle (or a "trapezoid" with one base length equal to zero).

About the Chapter

Background Information

The content of this chapter introduces students to the concept of a polygon and explores in detail the properties of quadrilaterals. Parallel lines and transversals are studied, and their properties are then used to develop a number of important facts about the sums of interior and exterior angles of polygons, and about midsegments of triangles and trapezoids. The chapter concludes using coordinates to analyze polygons.

Chapter Resources

- Practice Masters
- Enrichment Masters
- Technology Masters
- Lesson Activity Masters
- Lab Activity Masters
- Long-Term Project Masters
- Assessment Masters
  - Chapter Assessments, A/B
  - Mid-Chapter Assessment
  - Alternative Assessments, A/B
- Teaching Transparencies
- Spanish Resources

Chapter Objectives

- Define and use *reflectional* and *rotational symmetry.*
- Define *polygon, regular polygon, center point, central angle,* and *axes of symmetry.*
- Define *quadrilateral, parallelogram, rhombus, rectangle, square,* and *trapezoid.*
- Identify and interrelate properties of quadrilaterals.

Chapter 3

Lessons

Parallels and Polygons

Rectangles, triangles and hexagons are examples of polygons. The crisscross interplay of parallel beams and polygons in the photograph of the Thorncrown Chapel suggests the theme of this chapter.

Thorncrown Chapel, Eureka Springs, Arkansas
Designed by E. Fay Jones

About the Photos

Polygon patterns have been used for centuries in art and architecture. The photos show patterns from buildings as diverse as the Alhambra in Spain and one designed by E. Fay Jones. Polygons also occur in the facets of the polished gemstones.

As you will learn, parallel lines and their properties provide the basis for classifying and exploring four-sided polygons known as quadrilaterals.
Patterns of polygons are often used for decorative purposes. The tiling pattern in the background is from the Alhambra, a famous Islamic fortress in Spain.
PORTFOLIO ACTIVITY
A repeating pattern that covers a surface is known as a tessellation. The word comes from the Latin word meaning "little square stones." Tessellations in the time of the Romans were mosaics made of small colored stones.
More recently, geometers have taken an interest in figures that cover a plane surface in ingenious ways, such as those displayed in the works of M. C. Escher. In this chapter you will examine some of the tessellations of Escher. You will also be invited to create tessellations of your own.

About the Chapter Project

In the Chapter 3 Project, on pages 166–167, students make string figures by following a carefully prescribed sequence of steps. Students can expand their repertoire of string figures by experimenting with different moves.

- Define *transversal, alternate interior angle, alternate exterior angle, consecutive angle,* and *corresponding angles.*
- Conjecture and prove theorems using postulates and properties of parallel lines and transversals.
- Identify and use the converse of the Corresponding Angle Theorem.
- Prove that lines are parallel using theorems and postulates.
- Identify and use the Parallel Postulate and Triangle Sum Theorem.
- Define *exterior* and *interior angles* of a polygon.
- Develop and use formulas for the sums of interior and exterior angles of a polygon.
- Define *midsegment of a triangle* and *midsegment of a trapezoid.*
- Develop and use formulas from the properties of midsegments.
- Develop and use the Equal Slope Theorem and the Slopes of Perpendicular Lines Theorem.
- Solve problems involving perpendicular and parallel lines in the coordinate plane using appropriate theorems.

Portfolio Activity

Discuss the meaning of the term *tessellation* with students. Ask students to look for patterns in magazines that they can bring to class as additional examples of tessellations. After students have examined some of the tessellations of Escher, they can create one or two of their own for inclusion in their portfolios. The portfolio activity is developed further in the exercises for Lessons 3.2 and 3.8.

PREPARE

Objectives

- Define and use *reflectional* and *rotational symmetry.*
- Define *polygon, regular polygon, center point, central angle,* and *axis of symmetry.*

- Practice Master 3.1
- Enrichment Master 3.1
- Technology Master 3.1
- Lesson Activity Master 3.1
- Quiz 3.1
- Spanish Resources 3.1

Use Transparency 19

Assessing Prior Knowledge

Refer to the diagram to answer the questions.

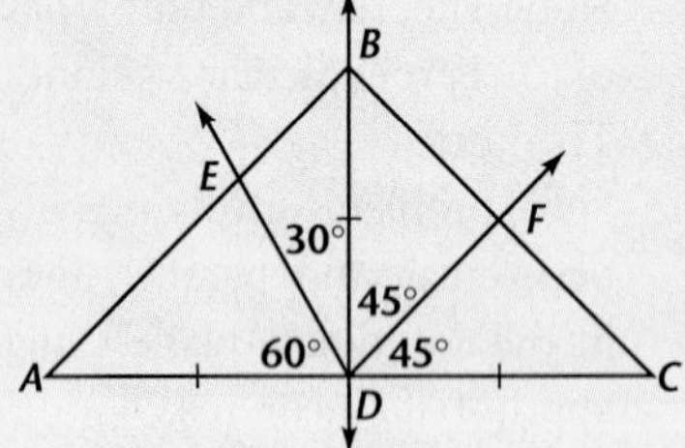

1. Name a ray that bisects AC. [**Any of these:** $\overrightarrow{DE}, \overrightarrow{DB}, \overrightarrow{DF}, \overrightarrow{BD}$]
2. Name the perpendicular bisector of $\overline{AC}$. [$\overleftrightarrow{BD}$]
3. Name the bisector of $\angle CDB$. [$\overrightarrow{DF}$]

TEACH

Students are intuitively familiar with the concept of symmetry from their experiences with symmetry in nature and real-world objects.

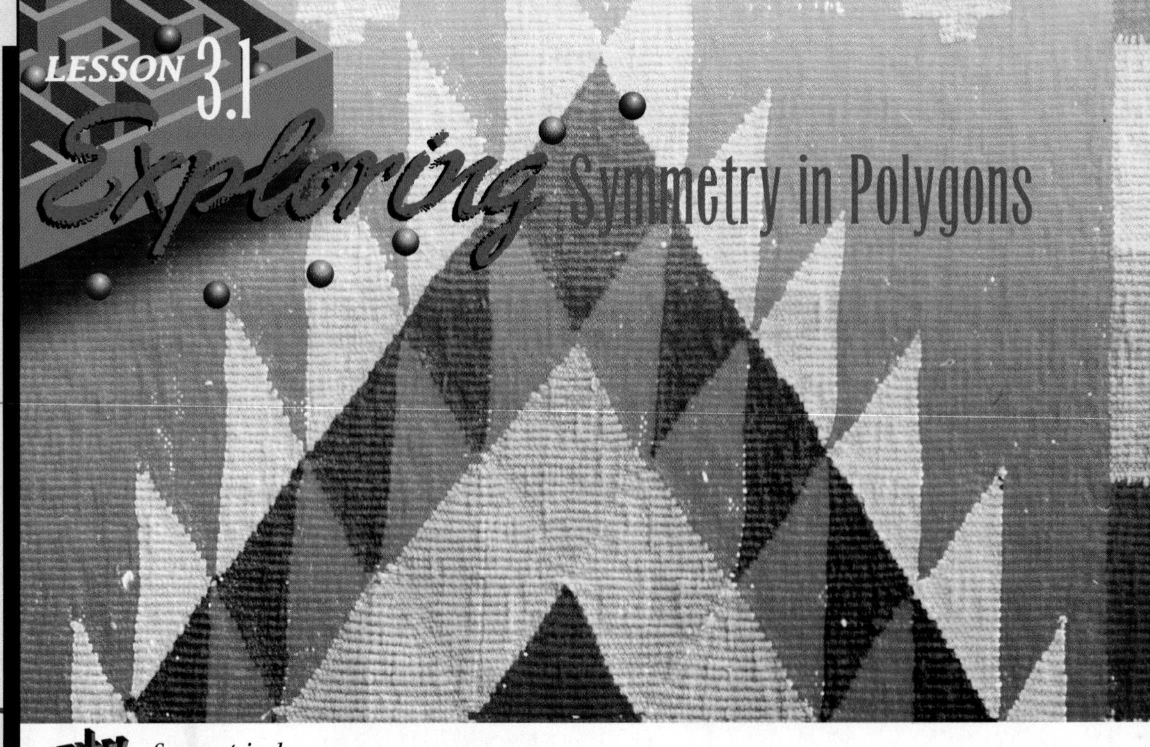

why *Symmetrical polygons give this American Indian blanket design an attractive appearance. They also have interesting mathematical properties.*

Defining Polygons

In Lesson 2.3 you learned to use examples and nonexamples to formulate definitions. Use the following figures to define a **polygon**.

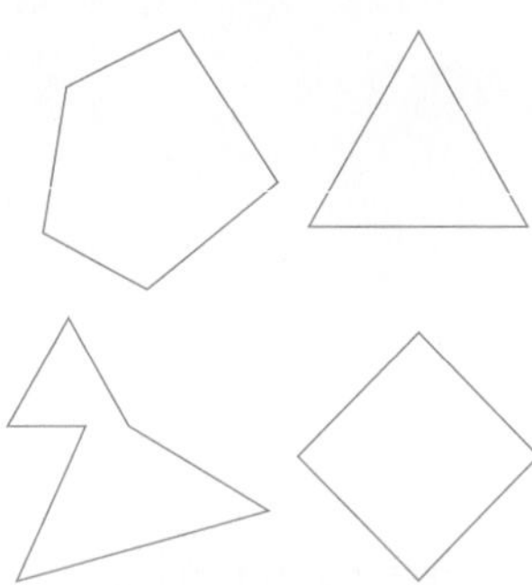

These are polygons.

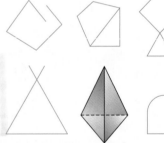

These are not polygons.

Compare your definition of a polygon with the definition below. Does your definition have more requirements than it needs?

POLYGON

A polygon is a closed-plane figure formed from three or more segments such that each segment intersects exactly two other segments, one at each endpoint. **3.1.1**

Technology First, engage students in a discussion of polygons and illustrate the concepts of reflectional and rotational symmetry using drawings of figures on the chalkboard. Then organize the class into small groups to investigate and explore the concepts of this lesson using geometry graphics tools.

A polygon is named according to the number of its sides. Familiarize yourself with the list below.

Polygons, With Number of Sides			
Triangle	3	Nonagon	9
Quadrilateral	4	Decagon	10
Pentagon	5	11-gon	11
Hexagon	6	Dodecagon	12
Heptagon	7	13-gon	13
Octagon	8	*n*-gon	*n*

Reflectional Symmetry

Is the cat's face symmetrical?

Hold your pencil in front of the photograph of the blanket on page 116 so that the part of the photo on one side of your pencil is the mirror image of the part on the other side. The line of your pencil is an **axis of symmetry**.

REFLECTIONAL SYMMETRY

A plane figure has **reflectional symmetry** if and only if its reflection image through a line coincides with the original figure. The line is called an **axis of symmetry**. **3.1.2**

Imagine giving each of the following figures a complete flip over the dotted lines shown. As you can see, the resulting figure will coincide exactly with the original figure. Therefore, each figure has reflectional symmetry, and each dotted line is an axis of symmetry.

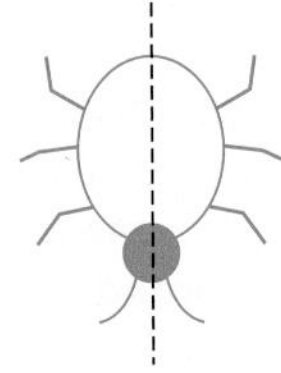

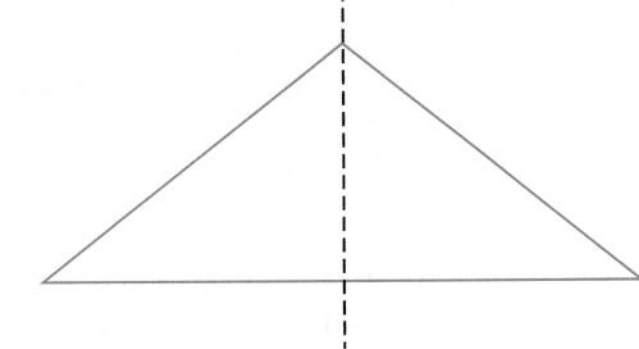

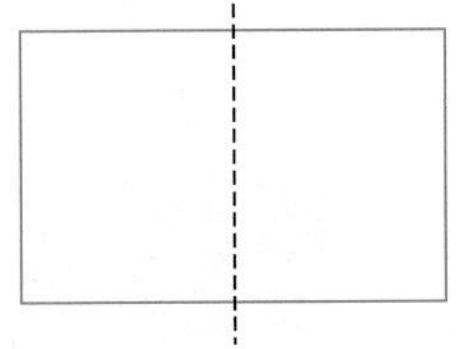

Exploration 1 Axes of Symmetry in Triangles

You will need
Folding paper

One way to name a triangle is by the number of its congruent sides:

Equilateral	3 congruent sides
Isosceles	at least 2 congruent sides
Scalene	0 congruent sides

Exploration 1 Notes

After you have discussed triangles, you may wish to extend this exploration to polygons with more than three sides. Ask students to make conjectures about the maximum number of lines of symmetry a given polygon can have. Is it possible for a polygon that has one or more axes of symmetry to have no congruent sides?

Use Transparency 20

interdisciplinary CONNECTION

Industrial Arts Nuts and bolts are used in a great variety of objects from kitchen tables to the Hubble Telescope. You may wish to discuss why the design of nuts and wrenches is frequently based on regular polygons. Discuss the kinds of regular polygons most often used. Ask why the number of sides is usually kept small.

ongoing ASSESSMENT

4. **In a scalene triangle, no angles are congruent. In an isosceles triangle, angles opposite congruent sides are congruent. In an equilateral triangle, all angles are congruent.**
5. **Each axis of symmetry is the perpendicular bisector of the side of the triangle it passes through.**
6. **Each axis of symmetry is the bisector of the vertex angle it passes through.**

You may find it helpful to ask students whether a triangular cake can always be easily divided into two equal pieces. If so, will the pieces necessarily be the same shape? Have students explain their thinking using diagrams at the chalkboard.

Equilateral

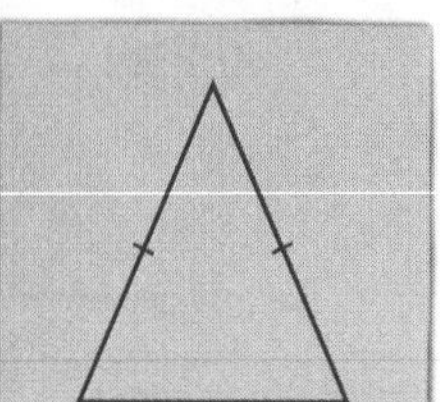

Isosceles

Scalene

As you do each of the following steps, recall geometry terms you studied in previous lessons, such as angles, bisectors, and altitudes of triangles. Make these terms part of your conjectures. *Write your conjectures in your notebook for future reference.*

1. Draw an example of each kind of triangle shown and find one or more axes of symmetry, where possible.
2. How many axes of symmetry does each triangle have?
3. Fold each figure through its axes of symmetry, if any. When you do, which angles coincide? Which segments coincide?
4. Make conjectures about angles that are congruent in the three different kinds of triangles.
5. How is each axis of symmetry related to the side of the triangle it passes through? Express your answers as conjectures.
6. How is each axis of symmetry related to the angle whose vertex it passes through? Express your answers as conjectures. ❖

Rotational Symmetry

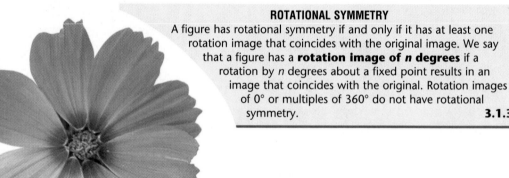

ROTATIONAL SYMMETRY

A figure has rotational symmetry if and only if it has at least one rotation image that coincides with the original image. We say that a figure has a **rotation image of *n* degrees** if a rotation by *n* degrees about a fixed point results in an image that coincides with the original. Rotation images of 0° or multiples of 360° do not have rotational symmetry. **3.1.3**

This flower has approximate rotational symmetry. How many times will it match its original image if it rotates one complete turn about its center?

ENRICHMENT Given a polygon of n sides, ask students how they could use it to draw a regular polygon having $2n$ sides.

INCLUSION strategies **Hands-On Strategies** Students can use cardboard cutouts of polygons to explore which are regular and which are not. They can do this by tracing around the polygons and changing the position of the cutout to see what changes of position give a perfect match.

Exploration 2 *Rotational Symmetry in Regular Polygons*

Central Angle

Center point

You will need
Tracing Paper

A **regular polygon** is a polygon in which all the sides and all the angles are congruent. A square, for example, is a regular polygon.

The **center** of a regular polygon is the point that is equidistant from all the vertices of the polygon.

A **central angle** of a polygon is an angle formed by two rays originating from the center and passing through adjacent vertices of the polygon.

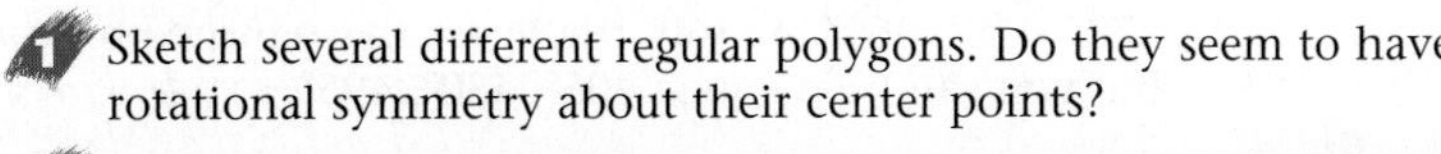

1. Sketch several different regular polygons. Do they seem to have rotational symmetry about their center points?

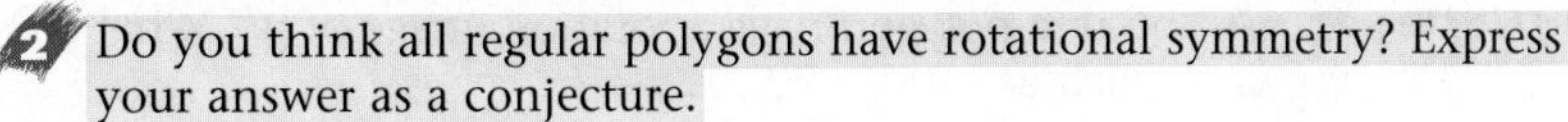

2. Do you think all regular polygons have rotational symmetry? Express your answer as a conjecture.
3. Draw or trace a regular pentagon such as the one in the illustration. Label the center P and the vertices Q, R, S, T, and U. Draw ray $\overrightarrow{PQ}$. Then copy the figure onto a sheet of tracing paper.
4. Use a pencil point to anchor one figure on top of the other at their centers so that the figures coincide.
5. Rotate the tracing paper counterclockwise until the trace of Q coincides with R. Continue rotating Q through each of the points S, T, U, and finally back to Q. How many moves did you make in all?
6. Notice that each of the turns rotates through one of the central angles of the pentagon. Do the central angles seem to be equal? Explain why or why not.

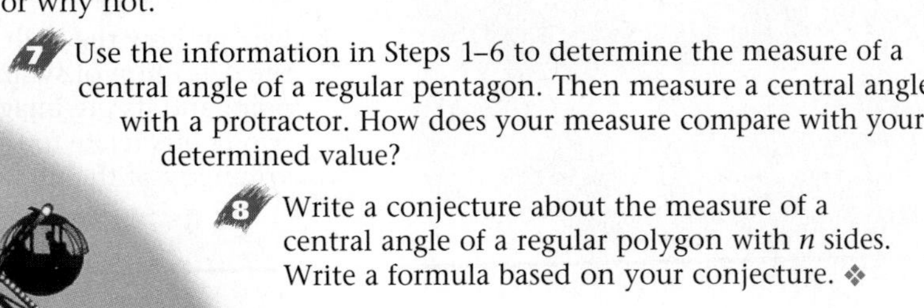

7. Use the information in Steps 1–6 to determine the measure of a central angle of a regular pentagon. Then measure a central angle with a protractor. How does your measure compare with your determined value?
8. Write a conjecture about the measure of a central angle of a regular polygon with n sides. Write a formula based on your conjecture. ❖

Q

R

U

P

S

T

Exploration 2 Notes

Have a class discussion about how to check whether a polygon is regular. Students may suggest tracings or the use of measuring tools such as rulers and protractors. Consider also how to check the diagram by using a compass alone. (For this, students may assume that an alleged center point is given.)

Ongoing Assessment

2. **All regular polygons have rotational symmetry.**

Cooperative Learning

Students can work together in groups of three or four to locate the center points of regular polygons. Students can explore various methods and then decide as a group which ones they think work best. Have groups present their methods to the class. Then discuss how to check the accuracy of the results.

Reteaching the lesson

Using Patterns You can review the main ideas of the lesson by using squares and rectangles for reflectional symmetry and a regular octagon for rotational symmetry. Discuss which conjectures from Explorations 1 and 2 apply to these figures.

CRITICAL *Thinking*

A negative angle of rotation must be a rotation in the clockwise direction. Every plane figure has a 0° rotation image, corresponding to no rotation.

Assess

Selected Answers

Odd-numbered Exercises 7–49

Assignment Guide

Core 1–38, 45–50

Core Plus 18, 22–44, 51–53

Technology

Students may use geometry graphics software for Exercises 19–21.

Error Analysis

In Exercise 14, students may make the mistake of sketching the image of the given part for only one 60° rotation. To correct, ask if the result of one rotation gives the whole figure or only a part of it.

APPLICATION

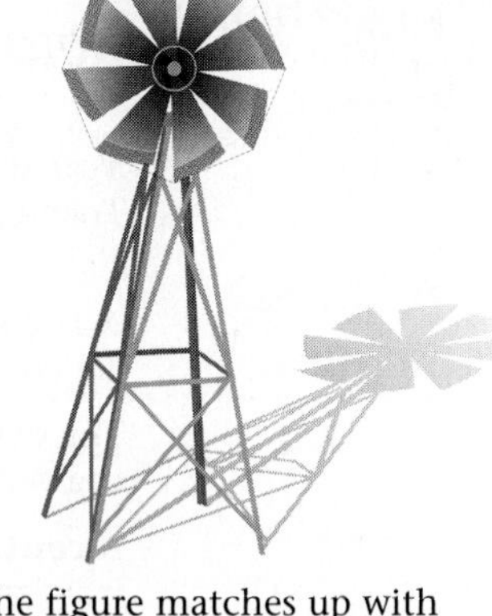

Engineering Jim is designing a special windmill for a wind-powered electrical generator. The windmill has eight blades, which will fit inside a regular octagon. What is the central angle of the blades? What is the rotational symmetry of the windmill?

Divide 360° by the number of sides of the octagon to find the central angle:

$$360° \div 8 = 45°$$

With each rotation through a central angle, the figure matches up with itself. So the figure has a rotational symmetry of 45°, 90°, 135°, 180°, 225°, 270°, 315°, 360°. At 360° the figure has returned to its original position. But you can keep going: 405°, 450°, 495°, . . . ❖

CRITICAL *Thinking* 

Can a polygon have a rotation image of 0°? What about negative degrees, such as −45°, −90°, . . .

Exercises & Problems

Communicate

1. What kind of symmetry can you find in the quilt on the left? Identify the axis of symmetry for some of the patterns.
2. Explain how the quilt illustrates the definition of symmetry. Identify a point and its pre-image. How do the two points relate to the axis of symmetry of the photograph?

3. Identify regions of the Escher woodcut at right that have rotational symmetry. Where are the centers of rotation?
4. Discuss the symmetry of a regular hexagon. How many axes of symmetry does it have? What are its rotational symmetries?
5. Can you draw a hexagon with just two axes of symmetry? Would it have rotational symmetry? Discuss.

6. 5 rotational; 72°, 144°, 216°, 288°, 360°

8. 2 reflectional; rotational, 180°

Practice & Apply

In Exercises 6–9, tell how many axes of symmetry, if any, each figure has. If the figure has rotational symmetry, make a list of its rotational symmetries.

6.

7.

1 reflectional

8.

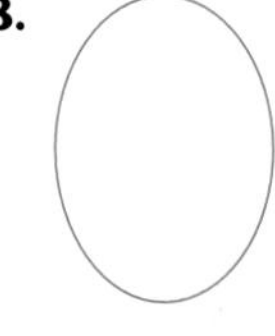

9.

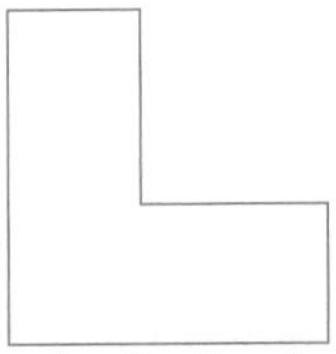

1 reflectional

10. How many axes of symmetry does a circle have? Explain your answer. Any line through the circle is an axis of symmetry.

Below are incomplete sketches of figures that have reflectional symmetry. **Copy and complete each figure.**

11.

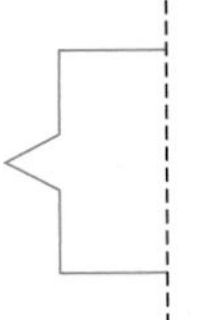

12.

13.

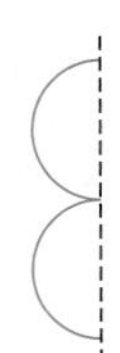

Below are incomplete sketches of figures that have rotational symmetry. **Copy and complete each figure.** The center and the amount of rotation are given.

14.

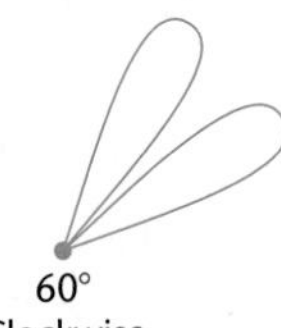

60°
Clockwise

15.

180°

16.

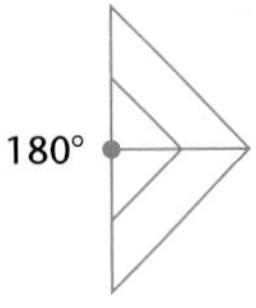

180°

17. Which completed figures in Exercises 11–13 also have rotational symmetry? Explain your answer. 11 and 13 have 180° rotational symmetry

18. Which completed figures in Exercises 14–16 also have reflectional symmetry? Explain your answer. 14, 15, and 16 have 3, 4, and 4 reflection axes

In Exercises 19–21, use a protractor and ruler. Draw $\overline{AB}$ with AB = 4 cm. (→ means "coincides with")

19. Draw an axis of symmetry for $\overline{AB}$ that passes through $\overline{AB}$ at a single point.

20. When you reflect $\overline{AB}$ over its axis of symmetry:

$A \rightarrow$ __?__ B $\quad B \rightarrow$ __?__ A $\quad \overline{AB} \rightarrow$ __?__ $\overline{BA}$

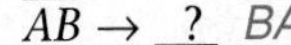

21. What is the relationship between segment $\overline{AB}$ and its axis of symmetry? It is the perpendicular bisector

alternative ASSESSMENT

Portfolio Assessment

Students should write a short paragraph to tell how they can use paper folding to check a figure for reflectional symmetry. They should also describe how to identify regular polygons and determine their rotational symmetries.

15.

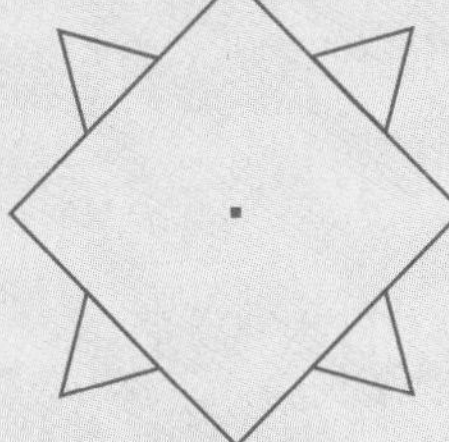

16.

19.

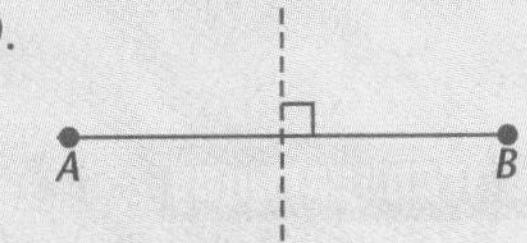

11.

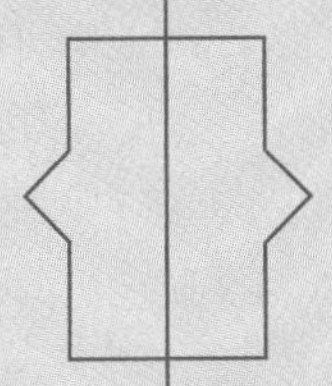

12.

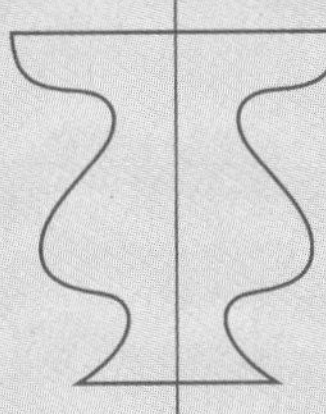

13.

14.

22. 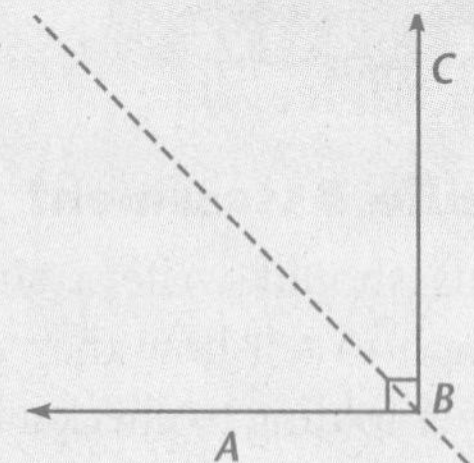

28. reflection through $\overline{AC}$; rotation of 180° around X

29. reflection through $\overline{AC}$; rotation of 180° clockwise around X

30. rotation of 180° around X

31. reflection through $\overline{BD}$

32. reflection through $\overline{AC}$

33. reflection through $\overline{AC}$; rotation of 180° clockwise around X

34. rotation of 180° around X

The answers to Exercises 39–44 can be found in Additional Answers beginning on page 727.

Use Transparency 22

Practice Master

NAME ______________ CLASS ________ DATE ________

Practice & Apply
3.1 Exploring Symmetry in Polygons

Use the terms in the list to complete each statement.

TERMS: reflection, central angle, polygon, axis of symmetry, regular polygon, center point

1. The ______________ of a regular polygon is equidistant from all the vertices of the polygon.
2. A plane figure formed by three or more segments such that each segment intersects exactly two other segments, one at each endpoint is called a ______________.
3. A ______________ of a polygon is formed by two rays originating from the center and passing through adjacent vertices of the polygons.
4. A plane figure that is both equiangular and equilateral is called a ______________.
5. A figure has an ______________ if there is a line such that its reflection maps the figure onto itself.

How many axes of symmetry, if any, does each figure have? If the figure has rotational symmetry, make a list of its rotational symmetries.

6. 7. 8.

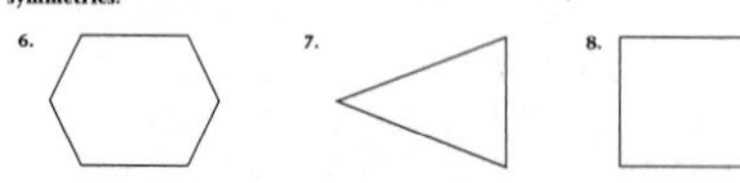

Each figure has reflectional symmetry. Complete each figure.

9. 10. 11.

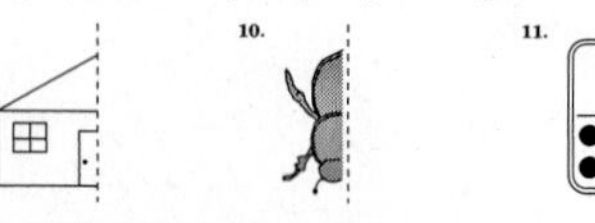

12. Describe all symmetries of a circle.

80 Practice & Apply — HRW Geometry

22. Draw $\angle ABC$ with $m\angle ABC = 90°$, and $AB = BC$. Then draw an axis of symmetry for $\angle ABC$.

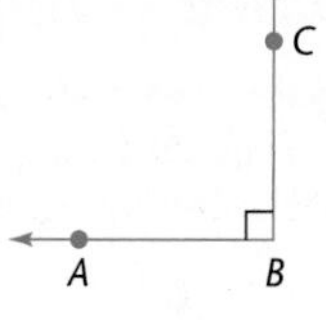

23. When you reflect $\angle ABC$ over your axis of symmetry,

a. $\overrightarrow{BA} \rightarrow$ _?_

b. $\overrightarrow{BC} \rightarrow$ _?_

c. $\angle ABC \rightarrow$ _?_ $\overline{BC}$, $\overline{BA}$, $\angle CBA$

24. What is the relationship between $\angle ABC$ and its axis of symmetry? It is the angle bisector.

Describe how to draw an axis of symmetry for the following.

25. an 80° angle Divide the angle into two 40° angles by drawing the bisector.

26. a 130° angle Divide the angle into two 65° angles by drawing the bisector.

27. In general, describe how to draw an axis of symmetry for any given angle. Draw the angle bisector.

ABCD is a rhombus; that is, its four sides are of equal measure. **Determine whether each of the following is a result of a rotation, a reflection, or both.**

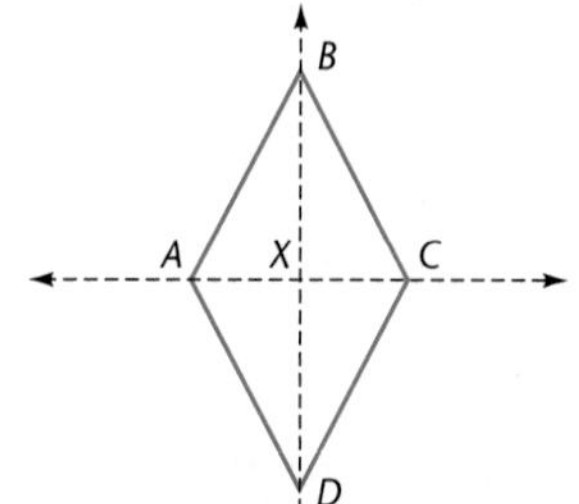

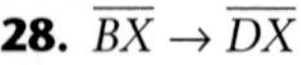

28. $\overline{BX} \rightarrow \overline{DX}$

29. $\overline{AD} \rightarrow \overline{AB}$

30. $\overline{AD} \rightarrow \overline{CB}$

31. $\angle BCX \rightarrow \angle BAX$

32. $\angle BCX \rightarrow \angle DCX$

33. $\angle BCX \rightarrow \angle DAX$

34. $\angle AXD \rightarrow \angle CXD$

Cultural Connection: Africa The designs below are taken from Egyptian bowls dating back to 3500 B.C.E. Describe the symmetries of each.

35.

36.

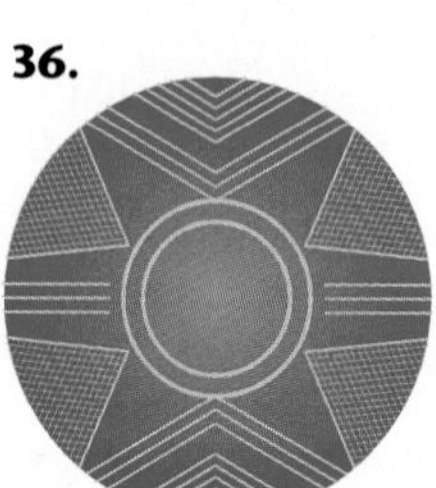

37.

38.

Algebra Use a graphics calculator or graph paper to draw each graph. Then write an equation for the axis of symmetry for each.

39. $y = (x - 1)^2 + 3$

40. $y = 2(x - 4)^2 + 3$

41. $y = -(x + 2)^2 + 3$

42. $y = -2(x + 5)^2 + 3$

43. $y = |x| + 3$

44. $y = |x + 3|$

35. 4 symmetry axes; rotational symmetries of 45°, 90°, 135°, 180°, 225°, 270°, 315°, 360°

36. 2 symmetry axes; rotational symmetries of 180°, 360°

37. no symmetry axes; rotational symmetries of 120°, 240°, 360°

38. 5 symmetry axes; rotational symmetries of 72°, 144°, 216°, 288°, 360°

Look Back

Use the diagram for Exercises 45–47. [Lesson 1.1]

45. Name the intersection of $\overleftrightarrow{AB}$ and $\overleftrightarrow{MN}$. *O*

46. Name three points that determine plane $\mathscr{T}$.

47. Name the intersection of planes $\mathscr{T}$ and $\mathscr{S}$. $\overleftrightarrow{AB}$

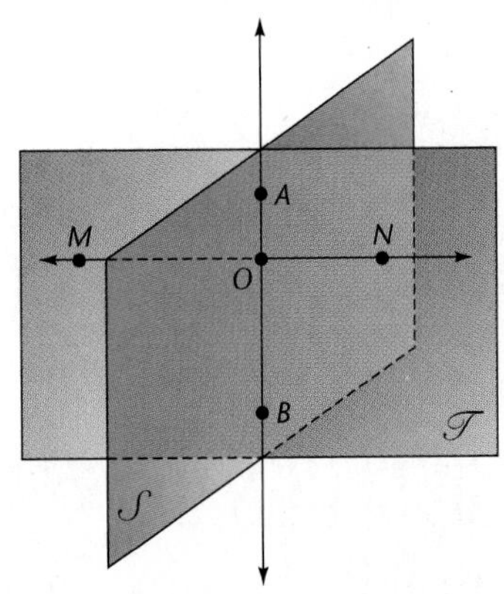

Find the missing measures in Exercises 48–50. [Lesson 1.5]

48. $m\angle ABD = 80°$, $m\angle ABC = 30°$, $m\angle CBD = ?$ 50°

49. $m\angle ABC = 25°$, $m\angle CBD = 60°$, $m\angle ABD = ?$ 85°

50. $m\angle ABC = m\angle CBD$, $m\angle ABD = 88°$, $m\angle ABC = ?$ 44°

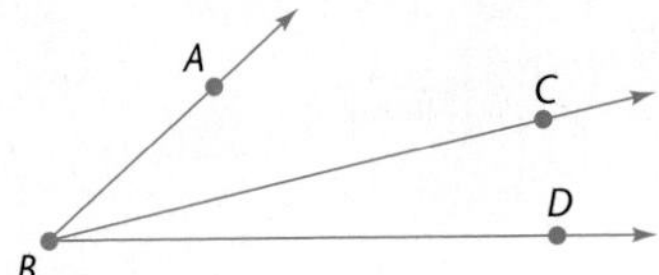

Look Beyond

51. Geography Name five countries whose flags have a horizontal axis of symmetry. Sketch each flag and identify the continent for each country.

52. Name five countries whose flags have a vertical axis of symmetry. Sketch each flag and identify the continent for each country.

53. Name five countries whose flags have rotation images of 180°. Sketch each flag and identify the continent for each country.

46. *AOM* or *BOM, AON, BON*

51. Sample answers:
Austria, Europe; Kenya, Africa; Indonesia, Asia; Bolivia, South America

52. Sample answers:
Canada, North America; Chad, Africa; Peru, South America; France, Europe

53. Sample answers:
Japan, Asia; Bangladesh, Asia; South Korea, Asia; Switzerland, Europe; United Kingdom, Europe

Look Beyond

Exercises 51–53 will enhance students' ability to recognize instances of reflectional and rotational symmetry in objects having the shapes of polygons. (Note: Though drawn in perspective, students should assume each flag is rectangular.)

Technology Master

NAME ________________ CLASS ________ DATE ________

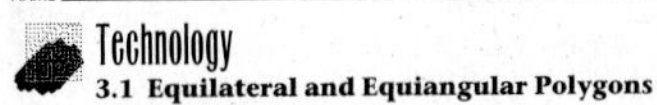

Technology
3.1 Equilateral and Equiangular Polygons

As you have learned, a regular polygon is a polygon in which all the sides and angles are congruent. Another way to say this is that a regular polygon is both *equilateral* and *equiangular*. Can a polygon be equilateral but not equiangular? equiangular but not equilateral? In this activity, you will explore these and several related questions.

In Exercises 1–3, use geometry software to explore equilateral and equiangular hexagons.

1. Use the *segment* tool or the *polygon* tool to draw hexagon *ABCDEF*. Use the *measurement* feature to display the measures of all the sides and angles. Then adjust the measures so that *ABCDEF* is regular, as shown. (Note: The software may have a *regular polygon* tool, but don't use it for this activity). Then make a copy of *ABCDEF*.

2. Using your original sketch, find new locations for points *B*, *C*, and *D* so that *ABCDEF* is equiangular but not equilateral. (Hint: Keep $\overline{AB}$ and $\overline{ED}$ perfectly horizontal). Sketch the result in the space at the right. Does your new figure have reflectional symmetry, rotational symmetry, or both? Explain.

3. Using your copy of *ABCDEF*, adjust it so that the hexagon is equilateral but not equiangular. (Hint: Start by moving $\overline{AB}$ and $\overline{ED}$ closer together). Sketch the result in the space at the right. Does the new figure have reflectional symmetry, rotational symmetry, or both? Explain.

Sketch each of the following figures, if possible.

4. an equilateral hexagon that has rotational symmetry, but not reflectional symmetry

5. an equiangular hexagon with neither reflectional nor rotational symmetry

6. an equilateral pentagon that is not equiangular

7. an equilateral quadrilateral that is not equiangular

8. an equilateral triangle that is not equiangular

9. Create a description of a polygon similar to the descriptions in Exercises 4–8. Trade descriptions with a partner. Each partner should then attempt to sketch the polygon described.

96 Technology HRW Geometry

PREPARE

Objectives

- Define *quadrilateral, parallelogram, rhombus, rectangle, square,* and *trapezoid.*
- Identify and interrelate properties of quadrilaterals.

RESOURCES

• Practice Master	3.2
• Enrichment Master	3.2
• Technology Master	3.2
• Lesson Activity Master	3.2
• Quiz	3.2
• Spanish Resources	3.2

Use Transparency 23

Assessing Prior Knowledge

Refer to the figure to answer the questions.

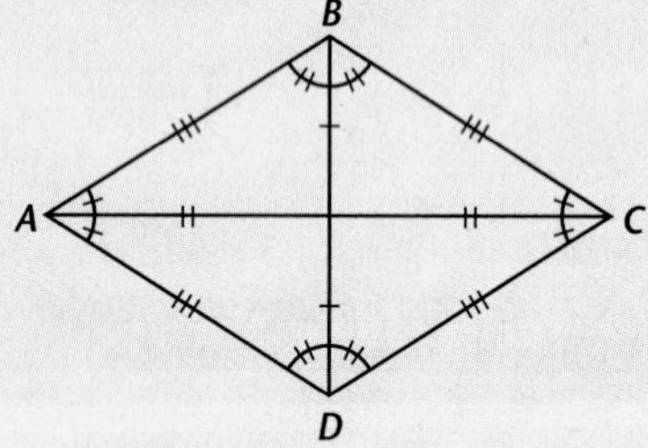

1. Name two angles that are bisected by AC. [**$\angle BAD$, $\angle BCD$**]
2. Name a segment that is bisected by BD. [**AC**]
3. Name four isosceles triangles. [**Triangles ABC, ABD, BCD, and ADC**]

TEACH

Point out that properties of quadrilaterals are used widely in determining property lines and in checking measurements for playing fields in sports.

LESSON 3.2 Exploring Properties of Quadrilaterals

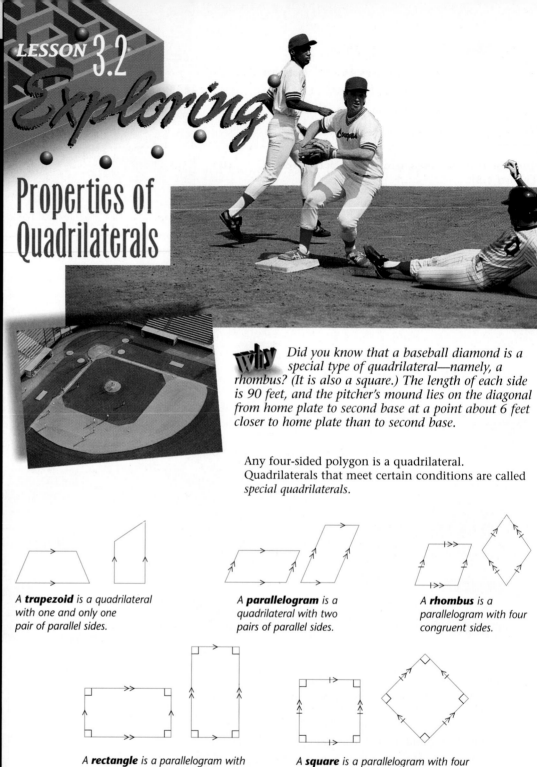

why *Did you know that a baseball diamond is a special type of quadrilateral—namely, a rhombus? (It is also a square.) The length of each side is 90 feet, and the pitcher's mound lies on the diagonal from home plate to second base at a point about 6 feet closer to home plate than to second base.*

Any four-sided polygon is a quadrilateral. Quadrilaterals that meet certain conditions are called *special quadrilaterals*.

*A **trapezoid** is a quadrilateral with one and only one pair of parallel sides.*

*A **parallelogram** is a quadrilateral with two pairs of parallel sides.*

*A **rhombus** is a parallelogram with four congruent sides.*

*A **rectangle** is a parallelogram with four right angles.*

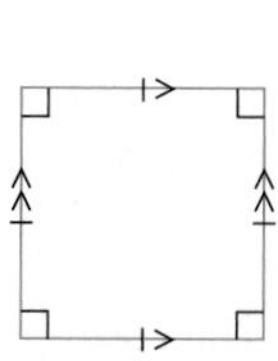

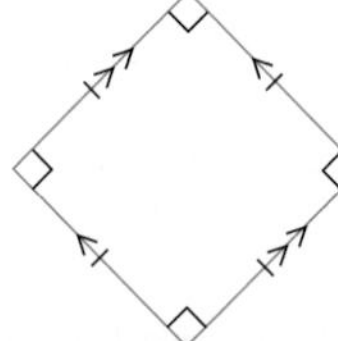

*A **square** is a parallelogram with four congruent sides and four right angles.*

Using Visual Models Students will enjoy doing the four explorations on a computer. However, they can also work through each exploration using graph paper, rulers, and protractors. For each exploration, suggest that they draw figures of different sizes to see if changing the size of a figure affects their conjectures.

Which of the quadrilaterals on the previous page seem to have rotational symmetry? Which seem to have reflectional symmetry? What can you learn about quadrilaterals from the symmetry properties? Keep this in mind as you do the explorations that follow.

The opposite sides of a parallelogram are parallel, by definition. However, a parallelogram has many more properties. For example, it appears that the *opposite sides of a parallelogram are congruent.*

In the following explorations you will make conjectures about other properties of parallelograms. Record your conjectures in your notebook for future reference.

Exploration 1 Parallelograms

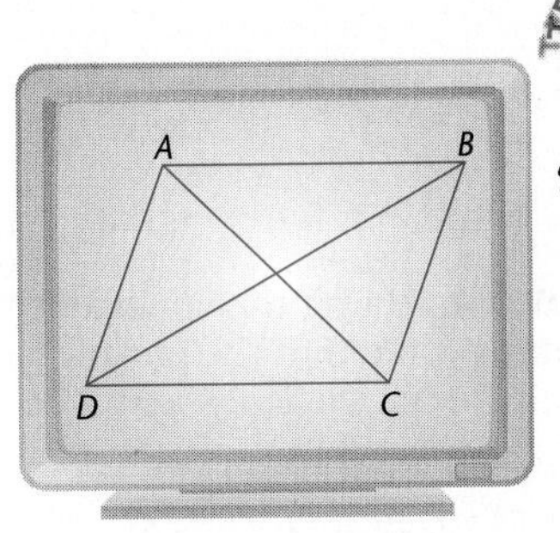

You will need
Geometry technology or
Ruler and protractor

Geometry Graphics

1. Draw a **parallelogram** that is not a rhombus, a rectangle, or a square. Measure the angles and the sides of the figure. What do you notice?

2. Draw diagonals connecting the opposite vertices. List any parts of your figure that appear to be congruent. You may want to measure these parts to verify that they are congruent. If you are using geometry technology, vary the shape of your figure by dragging one of the vertices.

3. What conjectures can you make about sides, angles, or triangles in the figure? (One was given just before this exploration. Try to find at least five conjectures in all.) ❖

Exploration 2 Rhombuses

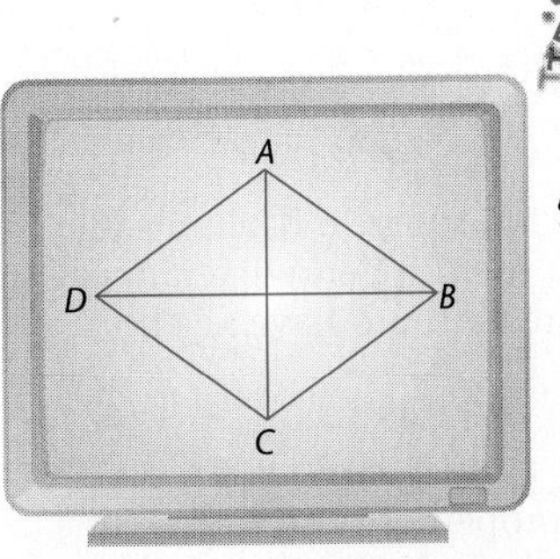

You will need
Geometry technology or
Ruler and protractor

Geometry Graphics

1. Draw a rhombus that is not a square. Draw diagonals connecting the opposite vertices. Make measurements as in the previous exploration.

2. Do the conjectures you made about parallelograms in Exploration 1 seem to be true of your rhombus? Discuss why they should or should not be true.

3. What new conjectures can you make about rhombuses that are not true of general parallelograms? (Try to find two new conjectures.) ❖

interdisciplinary CONNECTION **Industrial Design** Parallelograms, unlike triangles, are not rigid figures. Certain relationships between sides and angles do, however, stay the same, and this makes parallelograms effective for many special uses. For example, folding security gates for display windows in stores and doorways often have rhombuses in their design. Invite students to suggest other applications of flexible parallelograms.

CRITICAL Thinking

Parallelograms, rhombuses, rectangles, and squares appear to have rotational symmetry. Rhombuses, rectangles, and squares appear to have reflectional symmetry. Symmetry properties are related to congruent segments and angles in quadrilaterals.

Exploration 1 Notes

If sufficient time is not available for students to do the exploration on a computer, have them use graph paper, rulers, and protractors. In Step 1, discuss what the conditions imply about the side lengths and angle measures of the parallelogram.

Ongoing ASSESSMENT

3. **Sample Answers: Opposite sides and angles of a parallelogram are congruent. Parallelogram diagonals bisect each other and divide the parallelogram in half. Consecutive angles in a parallelogram are supplementary.**

Exploration 2 Notes

Ask which of the conditions from Step 1 of Exploration 1 have been changed for Step 1 of this exploration.

Ongoing ASSESSMENT

2. **Yes, a rhombus is a parallelogram, so it should have all properties of general parallelograms.**

3. **The diagonals of a rhombus are perpendicular and bisect its angles.**

Exploration 3 Notes

Contrast rhombuses with rectangles. Ask, "What is special about the sides of a parallelogram that is a rhombus?" Then ask, "What is special about the angles of a parallelogram that is a rectangle?"

ongoing ASSESSMENT

2. Yes, because a rectangle is a parallelogram.
3. The diagonals of a rectangle appear to be congruent.

Use Transparency 24

Exploration 4 Notes

Ask how the lengths of a pair of opposite sides of a rectangle that is not a square can be increased or decreased to change the rectangle into a square. Ask whether the rectangle remains a rectangle after the change.

ongoing ASSESSMENT

2. All previous conjectures are true for squares because a square is a parallelogram, a rhombus, and a rectangle.

CRITICAL Thinking

Special cases of a given class of figures have the properties that the most general kind of figures in the class have. The special cases thus inherit properties from the larger class.

Exploration 3 *Rectangles*

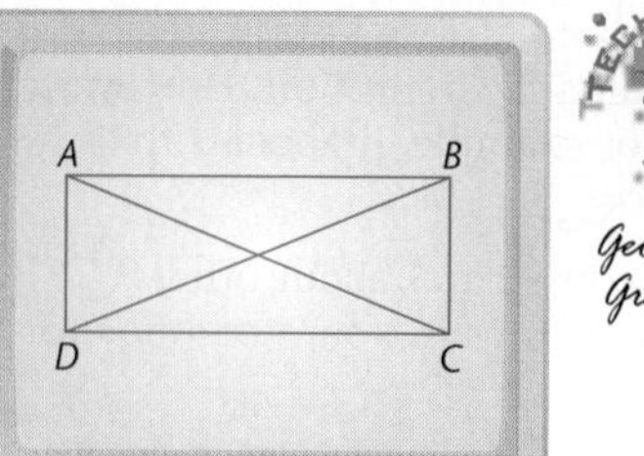

You will need
Geometry technology or
Ruler and protractor

Geometry Graphics

1. Draw a rectangle that is not a square. Draw diagonals connecting the opposite vertices. Make measurements as you did in the previous explorations.
2. Do the conjectures you made about parallelograms in Exploration 1 seem to be true for rectangles? Discuss why this should or should not be true.
3. What new conjecture can you make about rectangles that is not generally true of parallelograms? ❖

Exploration 4 *Squares*

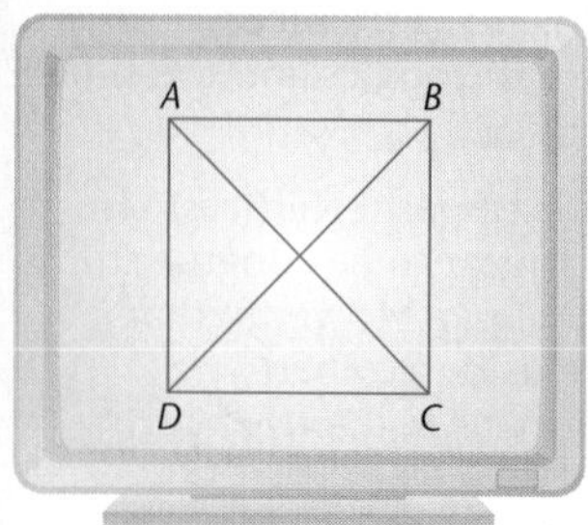

You will need
Geometry technology or
Ruler and protractor

Geometry Graphics

1. Draw a square with diagonals connecting the opposite vertices. Make measurements as you did in the previous explorations.
2. Which conjectures that you made in the previous explorations seem to be true of squares? Discuss. ❖

Summary

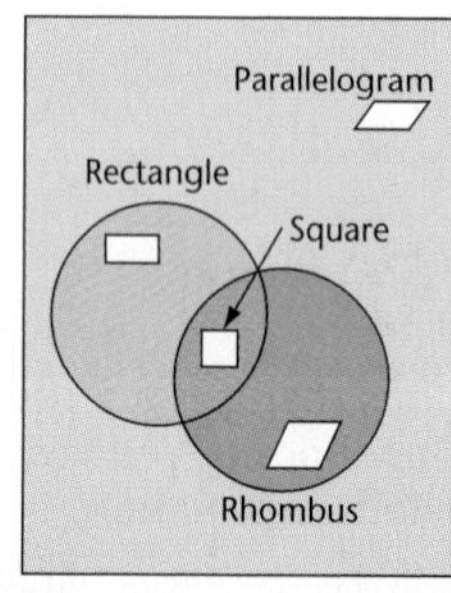

You may have noticed that there is a certain structure in the definitions and properties of quadrilaterals. This structure is revealed by the Venn diagram.

Is a square a rectangle? Is a square a rhombus? To answer these questions, use the Venn diagram of quadrilaterals. Do squares seem to have all the properties of rectangles? of rhombuses?

CRITICAL Thinking 

Certain regions of the Venn diagram "inherit" properties from the larger regions in which they are located. Explain what is meant by the "inheritance" of properties.

ENRICHMENT Have students write their conjectures for Explorations 1–4 in "if-then" form. Then have them write the converse of each conjecture. Have students investigate which of the converses seem to be true.

INCLUSION strategies **Hands-On Strategies** Students can use thin strips of tagboard (a strong cardboard) to make quadrilaterals of different sizes. You should precut the tagboard and punch holes in the ends of each strip. Students can put paper clips through the holes at the ends of four strips to create the sides of a quadrilateral. These nonrigid models can aid in the discussion of the Venn diagram on this page.

Communicate

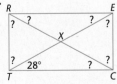

1. What type of special quadrilateral is the "least specialized"? Explain your answer.
2. Why is every rhombus also a quadrilateral and a parallelogram?
3. Explain why every square is also a quadrilateral, a rectangle, a rhombus, and a parallelogram.
4. How many examples of quadrilaterals can you find in your classroom? List them.

Practice & Apply

Decide whether each statement is true or false, and explain your answer. Use the Venn diagram given on page 126.

5. If a figure is a parallelogram, then it cannot be a rectangle.
6. If a figure is a rhombus, then it is a parallelogram.
7. If a figure is not a parallelogram, then it cannot be a square.
8. If a figure is a square, then it is a rhombus.
9. If a figure is a rectangle, then it cannot be a rhombus.
10. If a figure is a rhombus, then it cannot be a rectangle.

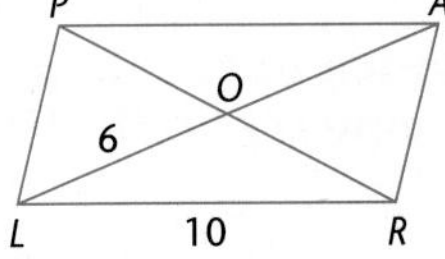

PARL is a parallelogram whose diagonals meet at point *O*.
In Exercises 11–16, find the indicated measures given the following: m∠*PLR* = 70°, m∠LAP = 20°, *LO* = 6, *LR* = 10. Give a justification for your answer using the parallelogram conjectures you made in this lesson.

11. m∠*PAR* 70°
12. m∠*LPA* 110°
13. *AO* 6
14. *PA* 10
15. m∠*ALR* 20°
16. m∠*PLO* 50°

***RECT* is a rectangle, *RHOM* is a rhombus, *SQUA* is a square. Copy each figure and fill in the missing angle measures.**

17.

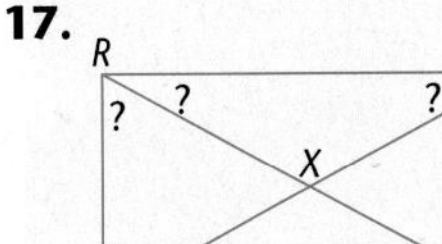

18.

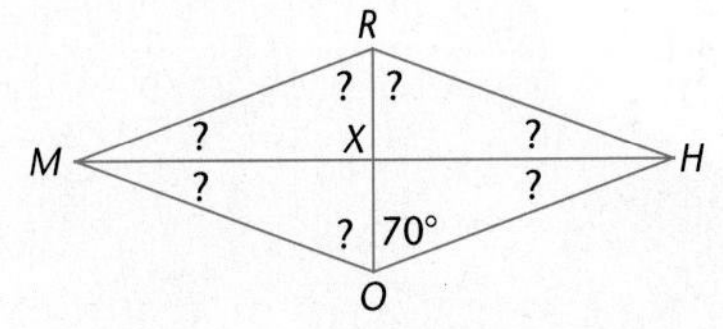

19.

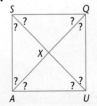

Cooperative Learning
Students can work in groups of three or four and begin by dividing the tasks for each exploration among themselves: drawing the figure, measuring segments and angles, and recording measurements.

Assess

Selected Answers
Odd-numbered Exercises 5–33

Assignment Guide
Core 1–16, 20–24, 27–35

Core Plus 1–26, 35

Technology
Students can use geometry graphics software to draw and investigate the figures for Exercises 17–24.

Error Analysis
Students may make errors in Exercises 5–10 as a result of confusing different types of special quadrilaterals. You can correct these errors by referring students to the diagrams and accompanying definitions on page 124.

9. False. A square is a rhombus that is also a rectangle.
10. False. A square is a rhombus that is also a rectangle.

The answers to Exercises 17–19 can be found in Additional Answers beginning on page 727.

Reteaching the Lesson

Using Visual Models Draw congruent acute angles on each of two acetate sheets, and use the overhead projector to show how these can be positioned to illustrate general parallelograms and rhombuses. Use a pair of right angles to discuss rectangles and squares.

5. False. A rectangle is a special parallelogram.
6. True. All rhombuses are parallelograms.
7. True. All squares are parallelograms.
8. True. All squares are rhombuses.

alternative ASSESSMENT

Performance Assessment

Check students' portfolios to evaluate the conjectures they have written for the explorations. Pay special attention to the care students have shown in their wording of the conjectures. Assumptions and conclusions should be easy to identify and understand. Diagrams should not show special cases that could cause students to assume more than is warranted by the stated assumptions.

27. If a device is a hypsometer, then it is used to measure atmospheric pressure and thus to determine height above sea level. If a device is used to measure atmospheric pressure and thus to determine height above sea level, then it is a hypsometer. It is not a definition because the converse is false. Other devices can be used.

Practice Master

NAME ______ CLASS ______ DATE ______

Practice & Apply

3.2 Exploring Properties of Quadrilaterals

Decide whether each statement is true or false. If false, explain your answer.

1. A rectangle is a quadrilateral with four right angles. ______
2. A trapezoid is a parallelogram. ______
3. A rhombus is a quadrilateral with two sides of equal length. ______
4. A square is a rectangle with four right angles and four sides of equal length. ______
5. A square is a type of rhombus. ______

***GRAM* is a parallelogram whose diagonals meet at point *O*. Find the indicated measures given the following:**

m∠*GMA* = 75°, m∠*MRG* = 35°, *MO* = 10, *MA* = 16.

6. m∠*GRA* ______
7. m∠*MGR* ______
8. *RO* ______
9. *GR* ______
10. m∠*RMA* ______
11. m∠*GMO* ______

***ABCD* is a rectangle, *HIJK* is a rhombus, and *QRST* is a square. Fill in the missing angle measures.**

12. 13. 14.

HRW Geometry — Practice & Apply 81

WXYZ is a rectangle. WZ = 10 cm; m∠QZY = 64°.

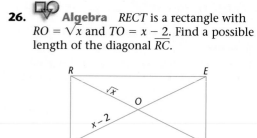

20. XY = ? 10
21. QY = ? 5
22. QZ = ? 5
23. What type of triangle is △*ZQY*? isosceles triangle
24. Name two different angles that must be congruent to ∠*QZY*. ∠*QYZ*, ∠*QXW*, or ∠*QWX*

25. **Algebra** *DIAM* is a rhombus with m∠*MIA* = x and m∠*ADI* = $2x$. Find the measure of each angle of *DIAM*.

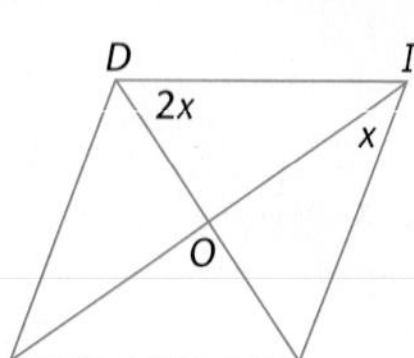

26. **Algebra** *RECT* is a rectangle with $RO = \sqrt{x}$ and $TO = x - 2$. Find a possible length of the diagonal $\overline{RC}$.

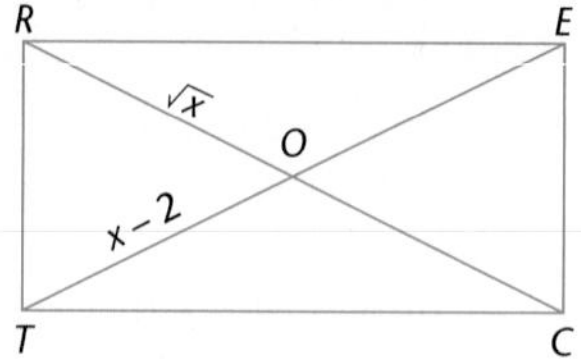

$x = 30°$; m∠*MDI* = 120°, m∠*MIA* = 30°, m∠*DIA* = m∠*ADI* = 60°

Look Back

Write each of the following statements as a conditional and then write the converse of the conditional. Is the original statement a definition? Explain your answer. [Lesson 2.3]

27. A hypsometer is a device used to measure atmospheric pressure and thus determine height above sea level.
28. An iris is a flower with sword-shaped leaves and three drooping sepals.
29. A whale is a mammal.

Which of the shapes in the objects below have rotational symmetry? reflectional symmetry? [Lesson 3.1]

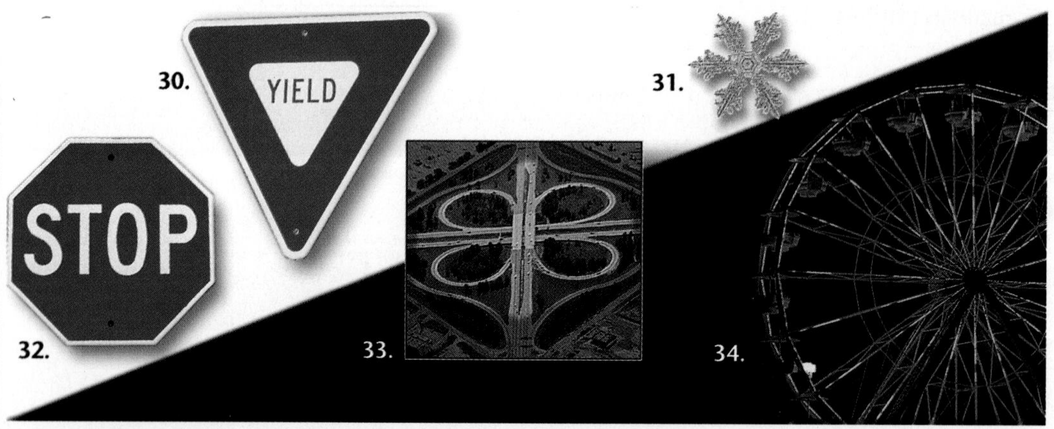

28. If a flower is an iris, then it has sword-shaped leaves and three drooping sepals. If a flower has sword-shaped leaves and three drooping sepals, then it is an iris. It is not a definition because the converse is false. Some flowers (for example, day lilies) that have sword-shaped leaves and three drooping sepals are not irises.

29. If an object is a whale, then it is a mammal. If an object is a mammal, then it is a whale. It is not a definition because the converse is false. Some mammals are not whales.

30. rotational and reflectional
31. rotational and reflectional
32. rotational and reflectional
33. rotational and reflectional
34. rotational and reflectional

Look Beyond

35. **Portfolio Activity** One of Escher's tessellation patterns is known as a **translation tessellation** because each of the repeating figures can be seen as a translation of other figures in the design. You can make your own translation tessellation by following the steps below. You may have to keep adjusting your curves until you get something recognizable!

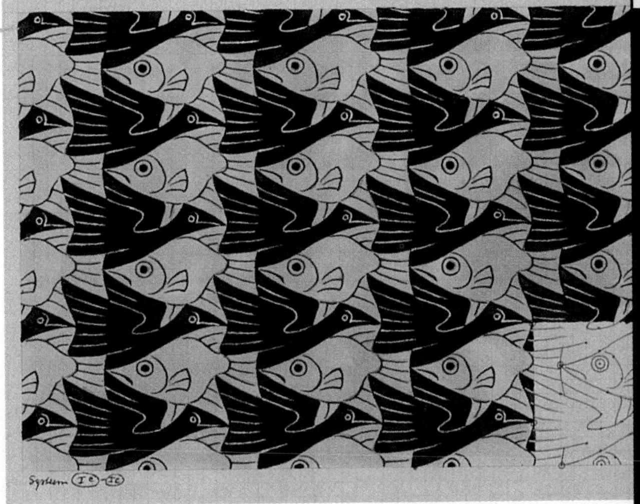

Draw your figures, as indicated below, on large-grid graph paper. The grid will help you copy the parts of your drawing. Or, if you like, you can use tracing paper. You can also do the steps using computer tessellation software.

a. Start with a square, rectangle, or other parallelogram. Replace one side of the parallelogram, such as $\overline{AB}$, by a broken line or curve.

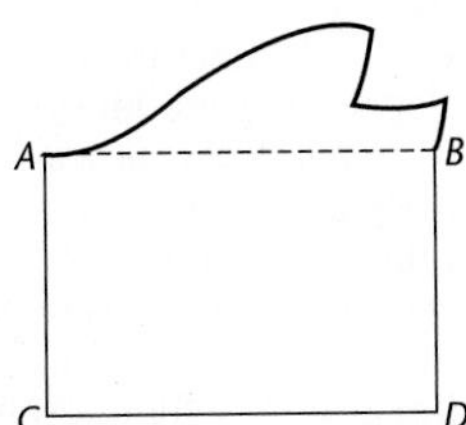

b. Translate the broken line or curve to the other side of the parallelogram. (If you are using a tracing of the curve, place it under your drawing and trace it again.)

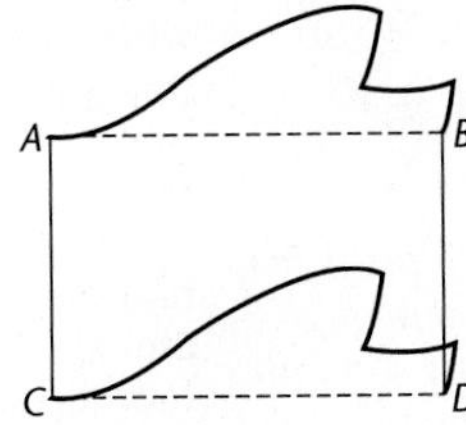

c. Repeat Steps A and B for the other two sides of your parallelogram.

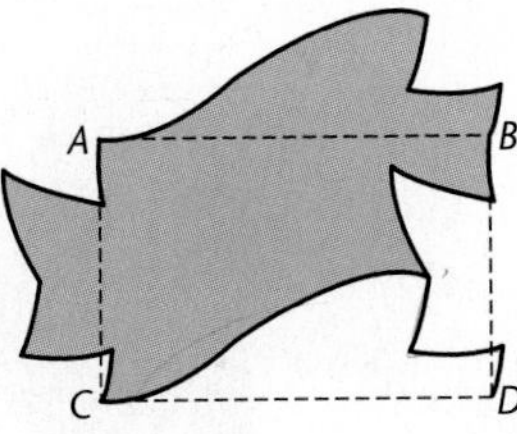

d. Your figure will now fit together with itself on all sides. Make repeated tracings of your figure in interlocking positions. You may want to add pictorial details.

Look Beyond

The activity in this exercise anticipates important later investigations of congruent figures and isometries (distance-preserving transformations).

35. Check students' drawings.

Use Transparency 25

Technology Master

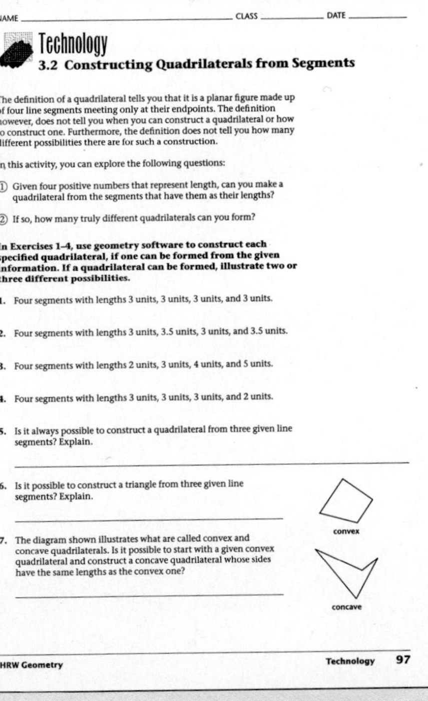

NAME ______ CLASS ______ DATE ______

Technology
3.2 Constructing Quadrilaterals from Segments

The definition of a quadrilateral tells you that it is a planar figure made up of four line segments meeting only at their endpoints. The definition however, does not tell you when you can construct a quadrilateral or how to construct one. Furthermore, the definition does not tell you how many different possibilities there are for such a construction.

In this activity, you can explore the following questions:

① Given four positive numbers that represent length, can you make a quadrilateral from the segments that have them as their lengths?

② If so, how many truly different quadrilaterals can you form?

In Exercises 1–4, use geometry software to construct each specified quadrilateral, if one can be formed from the given information. If a quadrilateral can be formed, illustrate two or three different possibilities.

1. Four segments with lengths 3 units, 3 units, 3 units, and 3 units.
2. Four segments with lengths 3 units, 3.5 units, 3 units, and 3.5 units.
3. Four segments with lengths 2 units, 3 units, 4 units, and 5 units.
4. Four segments with lengths 3 units, 3 units, 3 units, and 2 units.
5. Is it always possible to construct a quadrilateral from three given line segments? Explain.
6. Is it possible to construct a triangle from three given line segments? Explain.
7. The diagram shown illustrates what are called convex and concave quadrilaterals. Is it possible to start with a given convex quadrilateral and construct a concave quadrilateral whose sides have the same lengths as the convex one?

convex

concave

HRW Geometry — Technology 97

Focus

The design of a 25-foot-long metal egg highlights the difficulty in tessellating curved surfaces and illustrates how mathematical models can be adapted to fit economic and physical constraints.

Motivate

After students read both articles, ask questions like: What unexpected problems did Resch encounter in designing the egg? How did economic considerations affect Resch's mathematical model for the egg?

Egg Over Alberta

By Paul Hoffman

The town leaders of Vegreville, Alberta, contacted Ronald Dale Resch, a computer science professor, for a special project—to build a 31-foot Easter egg.

The problem Resch faced was that no one other than a chicken had ever built a chicken egg. With no formal training in mathematics, Resch relied on his ability to play with geometric abstractions in his mind, then with his hands or a computer, to turn those abstractions into physical reality.

Resch assumed that someone had developed the mathematics of an ideal chicken egg. He soon found, however, that there was no formula for an ideal chicken egg.

After four months of contemplation and simulation, Resch realized that he could tile the egg with 2,208 equilateral triangles of identical size and three pointed stars (equilateral but nonregular hexagons) that varied slightly in width, depending on their position on the egg.

For six weeks, Resch led a team of volunteers in assembling the egg. Residents of Vegreville were afraid it might blow down. Long after the egg was finished, Resch used a computer to analyze the egg's structural integrity and found that it was ten times stronger than it needed to be.

Many Ukrainians still practice the ancient tradition of painting pysanki, eggs decorated with colorful geometric patterns.

Computer generated drawing of egg

Cutting the egg tiles from aluminum

Partially assembled egg

1b. Identical polygons that tessellate the plane: squares, rectangles, equilateral triangles, and regular hexagons.

2a. Yes, travel in a straight line from one base of the cylinder to the other is possible. A cylinder is curved in only one direction.

2b. No, a sphere is curved in all directions.

2c. The paper must be bent in only one direction to form a cylinder, but in all directions to form a sphere.

Extension

On triangular dot paper make a flat tessellation somewhat like the curved one used for the Vegreville egg. The triangles and hexagons will not be equilateral.

Cooperative Learning

You can use an ordinary sheet of paper to help see why a giant egg was harder to design than other shapes.

1. a. Tessellate a sheet of paper with squares about 1 inch on a side. Then form the paper into a cylinder.
   b. Suppose you had to tile the curved surface of a cylindrical water tank. Describe the shape of the tiles you could use.
2. a. Can you travel in a straight line on the surface of a cylinder? Explain.
   b. Can you travel in a straight line on the surface of a sphere? Explain.
   c. Why is it easier to bend a flat sheet of paper into a cylinder than into a sphere?
3. a. How is a sphere curved differently from a cylinder?
   b. How is an ideal egg curved differently from a sphere?
   c. Why do you think a giant egg is harder to design than a giant sphere?

Cooperative Learning

Have students work in groups of four. The activities are sequenced to help students appreciate the difficulty of Resch's task. Students should examine tessellations in this order: plane surface (zero curvature), cylindrical surface, spherical surface, and egg surface (changing curvature).

Each group should proceed through all four steps. Each group can be assigned to report to the class on one of the steps.

Discuss

Discuss other polygons that can tessellate a plane, such as triangles, rectangles, and regular hexagons.

3a. A cylinder is straight between its bases and curved in the lateral direction. A sphere is curved in all directions.

3b. An egg curves differently in one direction than in another.

3c. Different curvatures in different directions of the giant egg made it difficult to tessellate.

PREPARE

Objectives

- Define *transversal, alternate interior angles, alternate exterior angles, same side interior angles,* and *corresponding angles.*
- Conjecture and prove theorems using postulates and properties of parallel lines and transversals.

RESOURCES

- Practice Master 3.3
- Enrichment Master 3.3
- Technology Master 3.3
- Lesson Activity Master 3.3
- Quiz 3.3
- Spanish Resources 3.3

Assessing Prior Knowledge

Use the diagram below for 1–2. The horizontal lines are parallel.

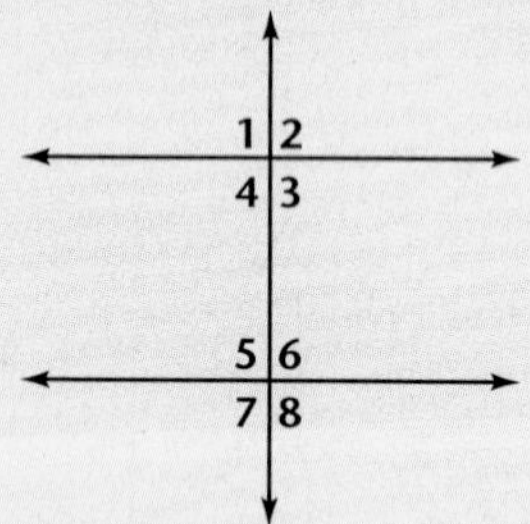

1. Is $\angle 3 \cong \angle 1$? If so, what postulate or theorem explains why? [**Yes, vertical $\angle$s are $\cong$.**]
2. If $\angle 6 \cong \angle 1$, what can you conclude about $\angle 1$ and $\angle 8$? [$\angle 8 \cong \angle 1$]

TEACH

Relationships between angles formed by lines and a transversal are used in construction and other applications to ensure that lines are parallel.

LESSON 3.3 Parallel Lines and Transversals

For practical as well as artistic reasons, parallel lines are often a prominent feature in architecture and engineering. The striking appearance of parallel cables, which help support the suspended portion of a bridge, inspired Joseph Stella's painting of the Brooklyn Bridge.

There's not much to go on.

Examine the parallel lines. What conjectures can you make? How many discoveries do you think you might make looking at just one pair of parallel lines?

The transversal changes the picture.

Draw a line that intersects the parallel lines. This line is called a **transversal**. Now there are many discoveries to make.

Transversals and Special Angles

TRANSVERSALS

A transversal is a line, ray, or segment that intersects two or more coplanar lines, rays, or segments, each at a different point. **3.3.1**

Notice that, according to the definition, the lines, rays, or segments that are cut by the transversal do not have to be parallel. This will be important in the next chapter.

Cooperative Learning Prepare four pieces of poster board by placing two parallel lines and a transversal on each. Divide the class into four groups and give one piece of poster board to each. Assign one of the four special pairs of angles to each group (alternate interior, alternate exterior, consecutive interior, corresponding). Have each group measure the angle pairs and develop a conjecture which completes the statement, "If two lines are cut by a transversal. then ____________." Groups should report their findings to the class.

Exploration Special Angle Relationships

You will need
Geometry technology or
Lined paper and protractor

Geometry Graphics

In each of the steps of this exploration, the terms interior and exterior will be used as shown.

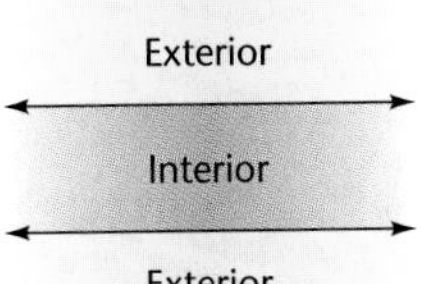

Start with two parallel lines. (If you are using lined paper, select two horizontal lines on the paper.) Then draw a third line that intersects both of them. Label each of the eight angles that are formed.

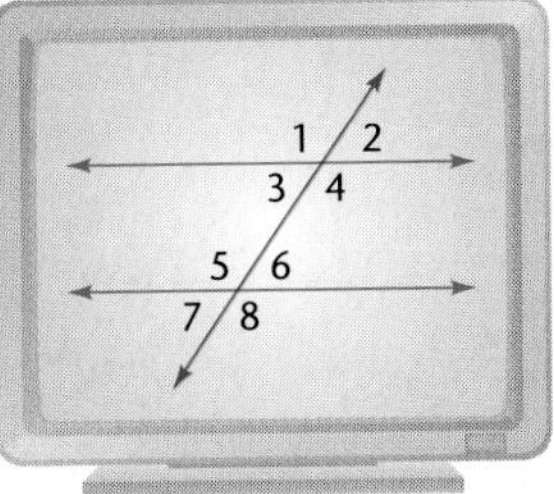

There are traditional names for certain special pairs of angles in the figure you drew. Measure each of the angles in the special pairs defined in the following steps:

1 Angles 3 and 6 are **alternate interior angles**, as are angles 4 and 5.

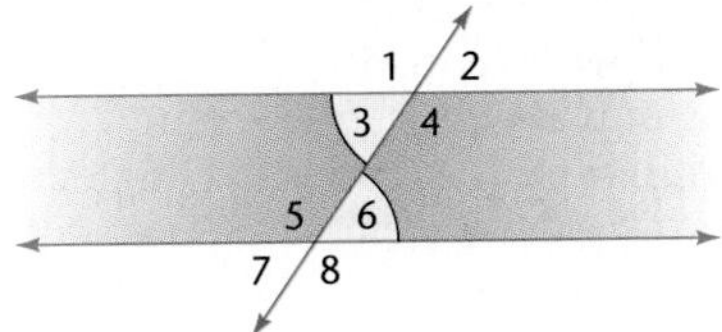

2 Angles 1 and 8 are **alternate exterior angles**. Name another pair of alternate exterior angles.

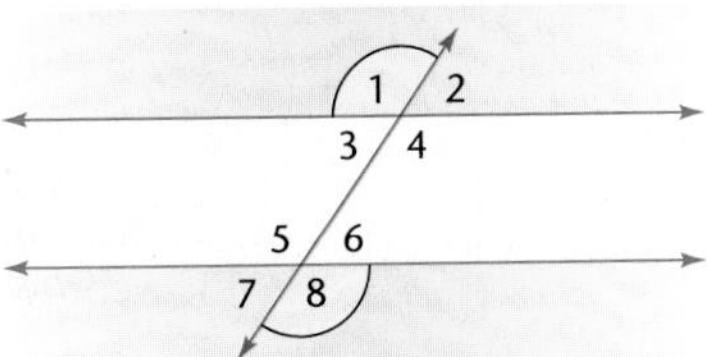

3 Angles 3 and 5 are **consecutive interior angles**. Name another pair of consecutive interior angles.

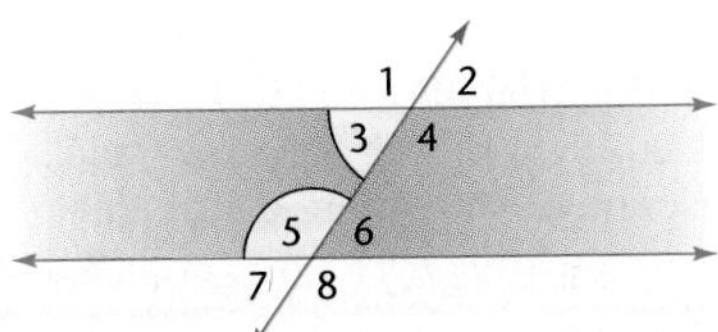

4 Angles 1 and 5 are **corresponding angles**. Name three other pairs of corresponding angles.

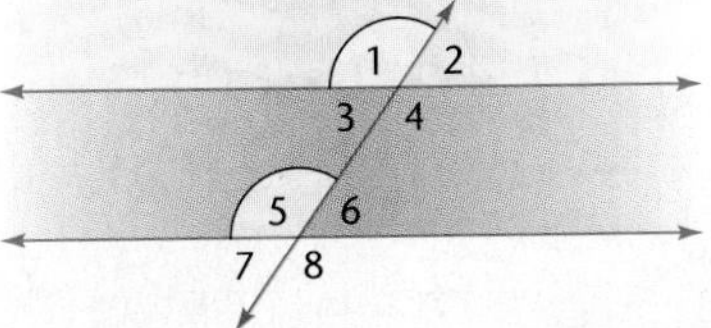

5 Make a conjecture for each special pair of angles. ❖

interdisciplinary CONNECTION **Physics** Knowledge of optical properties of materials and of angle relationships formed by parallel rays is essential in the design of precision optical instruments such as cameras, binoculars, and telescopes.

Use Transparency 26

Cooperative Learning

Have small groups of students investigate the angles formed by three or more parallel lines and a transversal. Ask the groups to list all the pairs of alternate interior angles, alternate exterior angles, same-side interior angles, and corresponding angles. Ask the groups to think about what conjectures they can make about the number of pairs in each category for *n* parallel lines cut by a single transversal. One member of each group can serve as its spokesperson and present the conjectures.

Exploration Notes

Ask students what terminology they might use for a pair of angles such as $\angle 2$ and $\angle 8$. Point out that angles such as $\angle 3$ and $\angle 5$ can be called consecutive interior angles as well as same-side interior angles.

ongoing ASSESSMENT

5. **If parallel lines are cut by a transversal, then alternate interior angles, alternate exterior angles, and corresponding angles are congruent. Also, consecutive interior angles are supplementary.**

Alternate Example

Indicate whether each is a pair of alternate interior, alternate exterior, same-side interior, or corresponding angles.

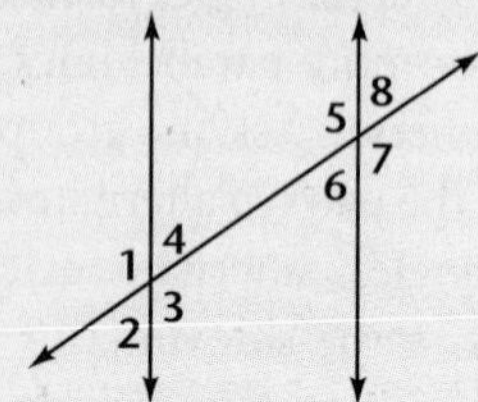

a. ∠1 and ∠7 **[alt. ext.]**

b. ∠6 and ∠2 **[corr.]**

c. ∠4 and ∠5 **[same-side int.]**

d. ∠3 and ∠5 **[alt. int.]**

Ongoing Assessment

Try This

Sample Answers:

∠8 and ∠4: corr. ang.

∠4 and ∠6: cons. int. ang.

∠6 and ∠3: alt. int. ang.

Ongoing Assessment

Yes. Since only a translation and no rotation is needed to match the corresponding angles, they must be congruent.

Ongoing Assessment

If parallel lines are cut by a transversal, then alternate interior angles are congruent.

If parallel lines are cut by a transversal, then alternate exterior angles are congruent.

If parallel lines are cut by a transversal, then consecutive interior angles are supplementary.

EXAMPLE

Indicate whether each is a pair of alternate interior, alternate exterior, consecutive interior, or corresponding angles:

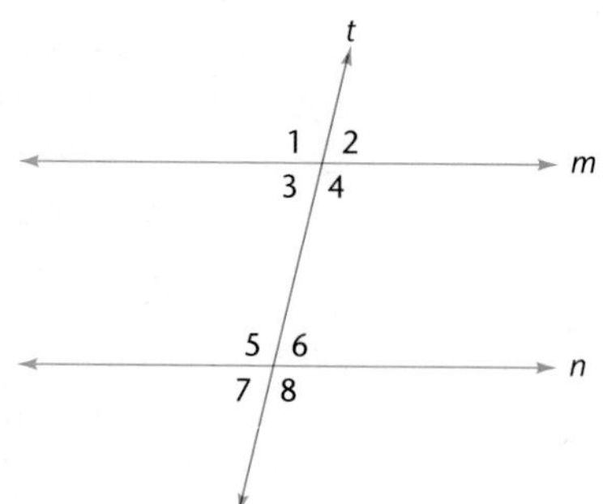

A ∠1 and ∠8

B ∠7 and ∠3

C ∠5 and ∠4

D ∠3 and ∠5

Solution ➤

A Alternate exterior angles

B Corresponding angles

C Alternate interior angles

D Consecutive interior angles ❖

Try This List three special pairs of angles not mentioned in the example.

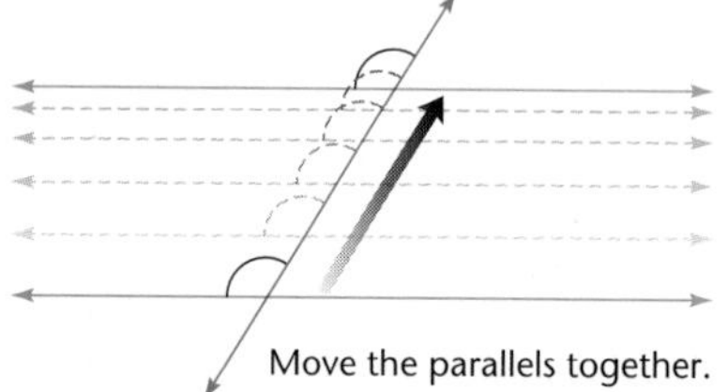

Move the parallels together.

In the exploration, one of the four conjectures you probably made is that corresponding angles are congruent.

Notice what happens if you slide (translate) one of the parallel lines closer to the other. Eventually, the highlighted corresponding angles will seem to come together as one. Do you think the corresponding angles will match exactly? Does this picture seem to support your conjecture? Explain.

One Postulate and Three Conjectures

We will not prove the Corresponding Angle Conjecture, but since it seems fully convincing, it will be a postulate in our system.

CORRESPONDING ANGLES POSTULATE

If two lines cut by a transversal are parallel, then corresponding angles are congruent.

3.3.2

There are three other conjectures that can be made in the exploration. Use the corresponding angles postulate as a model to state them formally.

ENRICHMENT Ask students to show that if any one of the four conjectures from the exploration is used as a postulate, then the other three can be proved as theorems.

INCLUSION strategies **Hands-On Strategies** Many students may find it easier to see the relationships between special angle pairs if they use tracings or cutout copies of the angles.

Once the corresponding angles conjecture has been accepted as a postulate, it can be used to prove theorems. In particular, you can use it to prove the other three conjectures as theorems.

ALTERNATE INTERIOR ANGLES THEOREM

If two lines cut by a transversal are parallel, then alternate interior angles are congruent. **3.3.3**

To prove the theorem, begin by drawing a diagram in which two parallel lines are cut by a transversal. Show that alternate interior angles, such as $\angle 1$ and $\angle 2$, are congruent.

Given: Line $l \parallel$ line m
Line p is a transversal.

Prove: $\angle 1 \cong \angle 2$

Plan: Study the diagram for ideas and plan your strategy.

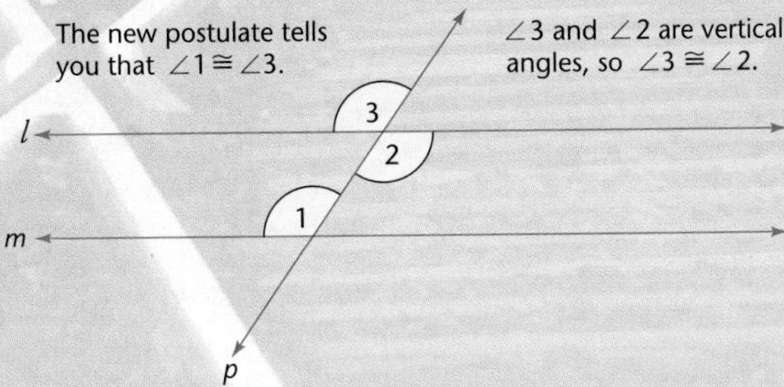

The new postulate tells you that ∠1 ≅ ∠3.
∠3 and ∠2 are vertical angles, so ∠3 ≅ ∠2.
3
2
1
l
m
p

Intersecting Railroad Tracks

Thus you have $\angle 1 \cong \angle 3$ and $\angle 3 \cong \angle 2$. So $\angle 1 \cong \angle 2$ (the desired result).

What postulate or property allows you to draw the final conclusion? (Recall the properties of congruence.)

Now you can write out your proof. In proofs of this nature you may find it convenient to use a two-column format.

Proof

Statements	Reasons
1. Line $l \parallel$ line m Line p is a transversal	Given
2. $\angle 1 \cong \angle 3$	If parallel lines are cut by a transversal, then corresponding angles are congruent.
3. $\angle 3 \cong \angle 2$	Vertical angles are congruent.
4. $\angle 1 \cong \angle 2$	Transitive Property of Congruence.

In writing proofs in geometry you may wish to develop abbreviations for use in stating reasons. For example, an abbreviation for the reason in Step 2 is

"$\parallel$s $\Rightarrow$ Corr. $\angle$s $\cong$."

If you do use abbreviations, be sure you are consistent.

TEACHING tip

Ask students if there is a pair of corresponding angles other than $\angle 1$ and $\angle 3$ that could be used to prove that $\angle 1 \cong \angle 2$.

ongoing ASSESSMENT

Transitive Property of Congruence

Using Discussion Point out that the Corresponding Angles Postulate states that for two parallel lines and a transversal, all four pairs of corresponding angles are congruent. Have a class discussion to show that assuming a single pair of corresponding angles is congruent will allow you to prove that the other three pairs of corresponding angles are congruent. Then show that if a single instance of each of the conjectures is true, all of the other three instances must also be true.

ASSESS

Selected Answers

Odd-numbered Exercises 5–35

Assignment Guide

Core 1–26, 33–37

Core Plus 1–4, 9–32, 37

Technology

Students may use geometry graphics software to complete Exercises 24-29.

Error Analysis

The diagonals in the diagram for Exercises 5–10 may cause some students to make errors in classifying the angle pairs.

Practice Master

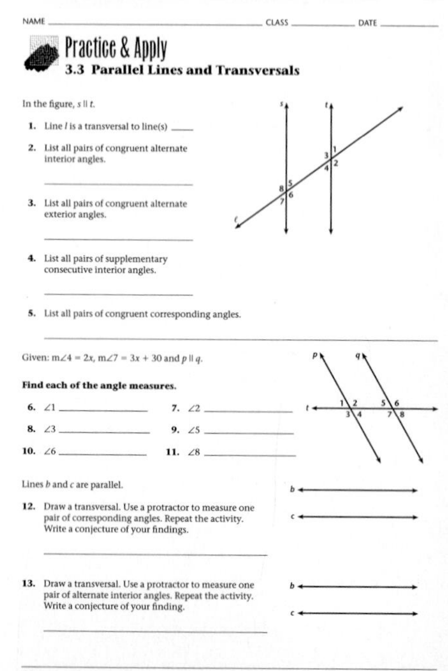

NAME ______ CLASS ______ DATE ______

Practice & Apply
3.3 Parallel Lines and Transversals

In the figure, $s \parallel t$.

1. Line l is a transversal to line(s) ____
2. List all pairs of congruent alternate interior angles.
3. List all pairs of congruent alternate exterior angles.
4. List all pairs of supplementary consecutive interior angles.
5. List all pairs of congruent corresponding angles.

Given: $m\angle 4 = 2x$, $m\angle 7 = 3x + 30$ and $p \parallel q$.

Find each of the angle measures.

6. $\angle 1$ ______ 7. $\angle 2$ ______
8. $\angle 3$ ______ 9. $\angle 5$ ______
10. $\angle 6$ ______ 11. $\angle 8$ ______

Lines b and c are parallel.

12. Draw a transversal. Use a protractor to measure one pair of corresponding angles. Repeat the activity. Write a conjecture of your findings.
13. Draw a transversal. Use a protractor to measure one pair of alternate interior angles. Repeat the activity. Write a conjecture of your finding.

82 Practice & Apply HRW Geometry

EXERCISES & PROBLEMS

Communicate

1. What is a transversal?
2. Why do you think parallel lines are not part of the definition of a transversal?
3. Did you make any conjectures in the exploration that were not discussed later in the lesson? If so, what were they? Can you justify them?
4. Identify each transversal in the figure below, and identify the lines to which each is a transversal.

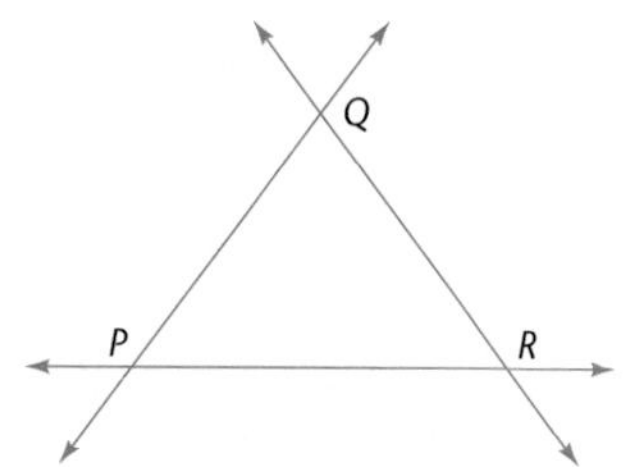

Can you identify transversals in the structure of the John Hancock building?

Practice & Apply

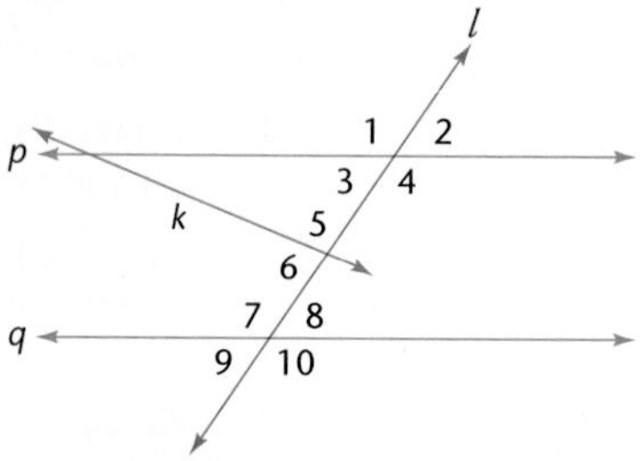

For Exercises 5–10, refer to the figure at right. In the figure, $p \parallel q$.

5. Line l is a transversal to lines __?__. p and q
6. Line k is a transversal to lines __?__. p, l, and q
7. List all pairs of congruent alternate interior angles.
8. List all pairs of congruent alternate exterior angles.
9. List all pairs of supplementary consecutive interior angles. $\angle 3$ & $\angle 7$; $\angle 4$ & $\angle 8$
10. List all pairs of congruent corresponding angles. $\angle 1$ & $\angle 7$; $\angle 2$ & $\angle 8$; $\angle 3$ & $\angle 9$; $\angle 4$ & $\angle 10$

11. **Carpentry** In the drawing of the partial garage frame, the ceiling joist $\overline{PQ}$ and soleplate $\overline{RS}$ are parallel. How is the "let-in" corner brace $\overline{PT}$ related to $\overline{PQ}$ and $\overline{RS}$? It is a transversal to them.
12. How are $\angle RTP$ and $\angle QPT$ related? They are alternate interior angles.
13. How is the corner brace $\overline{PT}$ related to the vertical studs it crosses? How are $\angle 1$ and $\angle 2$ related?

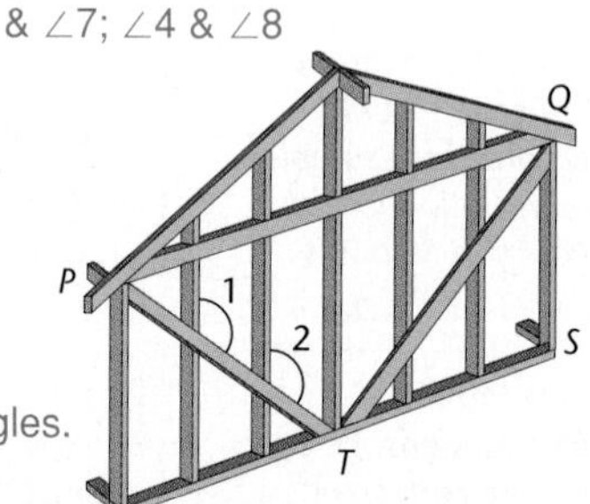

7. $\angle 3$ and $\angle 8$; $\angle 4$ and $\angle 7$

8. $\angle 1$ and $\angle 10$; $\angle 2$ and $\angle 9$

13. It is a transversal to them. They are corresponding angles.

Navigation A periscope is an optical instrument used on submarines to view above the surface of the water. A periscope works by placing two parallel mirrors facing each other.

14. Identify a transversal in the diagram. What does this line represent?

15. Are $\angle 1$ and $\angle 4$ alternate exterior angles? Why or why not?

16. Light rays reflect off the mirrors at right angles. How can you prove that $\angle 1 \cong \angle 4$?

17. If m $\angle 1 = 45°$, find the measures of angles 2, 3, and 4.

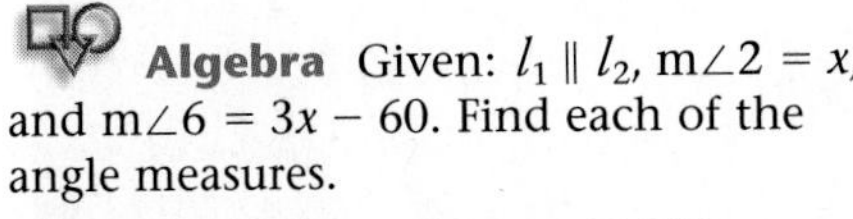

Algebra Given: $l_1 \parallel l_2$, m$\angle 2 = x$, and m$\angle 6 = 3x - 60$. Find each of the angle measures.

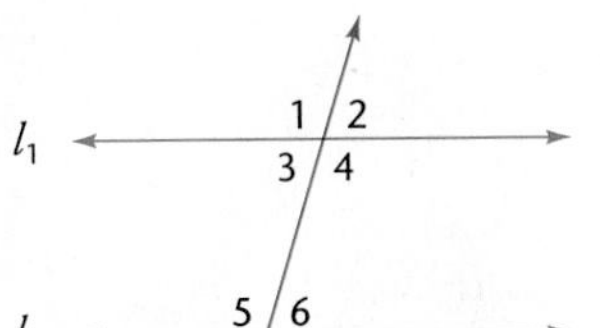

18. m$\angle 1$ 150°	**19.** m$\angle 3$ 30°
20. m$\angle 4$ 150°	**21.** m$\angle 5$ 150°
22. m$\angle 7$ 30°	**23.** m$\angle 8$ 150°

In Exercises 24–29, given $\triangle ABC$ with $\overline{DE} \parallel \overline{CB}$ and $\angle ADE \cong \angle AED$, find the indicated angle measures.

24. m$\angle ADE$ 50°	**25.** m$\angle AED$ 50°	**26.** m$\angle DEB$ 130°
27. m$\angle BDE$ 25°	**28.** m$\angle CDB$ 105°	**29.** m$\angle ABD$ 25°

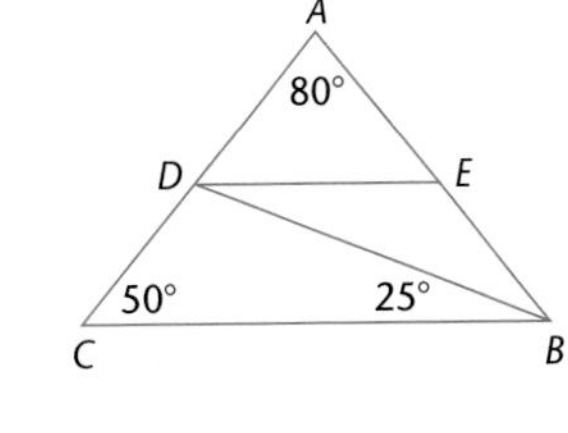

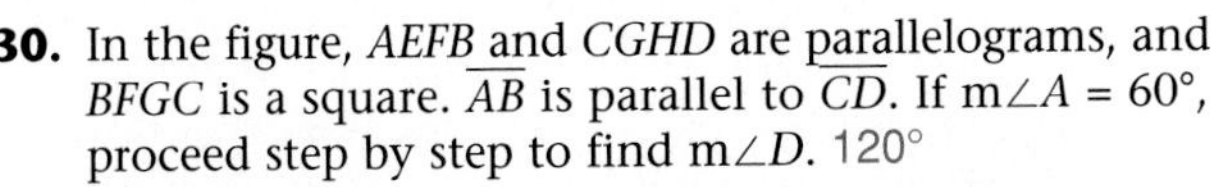

30. In the figure, *AEFB* and *CGHD* are parallelograms, and *BFGC* is a square. $\overline{AB}$ is parallel to $\overline{CD}$. If m$\angle A = 60°$, proceed step by step to find m$\angle D$. 120°

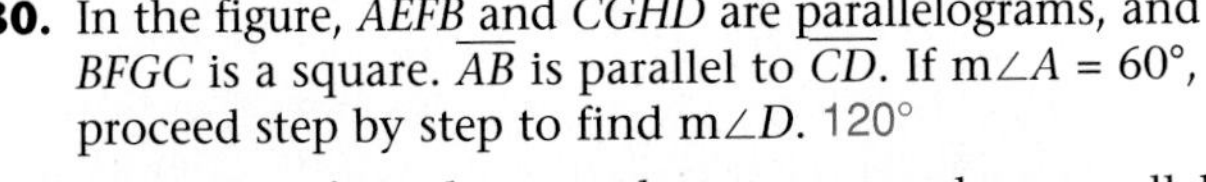

31. Prove that if two lines cut by a transversal are parallel, then alternate exterior angles are congruent.

32. Prove that if two lines cut by a transversal are parallel, then consecutive interior angles are supplementary.

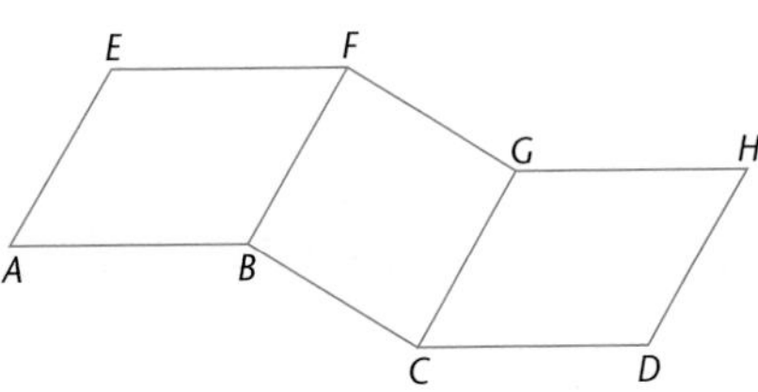

Look Back

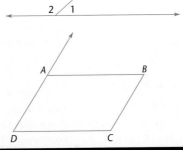

33. Lines that intersect at right angles are said to be ?. **[Lesson 1.2]** perpendicular

34. Describe what is meant when it is said that two lines are parallel. **[Lesson 1.2]** Parallel lines are coplanar and never intersect.

35. In the figure, the measure of $\angle 2$ is five times that of $\angle 1$. Find m$\angle 1$ and m$\angle 2$. **[Lesson 1.5]** m$\angle 1 = 30°$; m$\angle 2 = 150°$

36. Write this statement as a conditional: I have swimming practice on Tuesdays and Thursdays. **[Lesson 2.2]**
If it is Tuesday or Thursday, then I have swimming practice.

Look Beyond

Quadrilateral ABCD is a parallelogram.

37. Given: In quadrilateral *ABCD*, $\overline{CB} \parallel \overrightarrow{DA}$. Without drawing any additional lines, prove that the sum of the angles of the quadrilateral equals 360°.

14. It is the path light travels on between the mirrors of the periscope. It represents the line of sight in the periscope.

15. No. They are not formed from the same transversal.

16. m$\angle 1$ + m$\angle 2 = 90°$ and m$\angle 3$ + m$\angle 4 = 90°$. m$\angle 2$ = m$\angle 3$ because they are alternate interior angles for parallel lines. So, m$\angle 1$ = m$\angle 4$.

17. m$\angle 2$ = m$\angle 3$ = m$\angle 4 = 45°$

alternative ASSESSMENT

Performance Assessment

On the chalkboard, draw a pair of vertical parallel lines and a transversal. Label the angles. Ask students to sketch the figure and identify the transversal and to list and classify the pairs of special angles. Ask them to circle the pairs that are congruent according to the postulate that was introduced in the lesson. Ask them to draw a box around the pairs that are congruent according to the theorem. Finally, ask them to write *congruent* or *supplementary* beside each of the remaining pairs.

The answers to Exercises 31, 32, and 37 can be found in Additional Answers beginning on page 727.

Look Beyond

Exercise 37 anticipates results that will be developed for polygons in Lesson 3.6.

Technology Master

NAME ______ CLASS ______ DATE ______

Technology
3.3 Relating Angles Formed by a Transversal

In Lesson 3.3, you learned names for special pairs of angles seen in the diagram. For example, $\angle BAW$ and $\angle AWZ$ are alternate interior angles. You also learned special names for angles that are formed by two intersecting lines. For example, $\angle CAD$ and $\angle BAW$ are vertical angles.

Use geometry software for the following exercises.

1. Construct and label a diagram like the one shown. (In the diagram $\overleftrightarrow{BD} \parallel \overleftrightarrow{XZ}$). List the measures of the eight angles formed.

2. Suppose that $\overleftrightarrow{BD} \parallel \overleftrightarrow{XZ}$. Write a statement that gives a numerical relationship between an angle with vertex at *A* and an angle with vertex at *W*.

3. Tilt $\overleftrightarrow{CY}$ so that it is still a transversal but different from the one you sketched in Exercise 1. Do the numerical relationships from Exercise 2 remain true? Explain.

4. Suppose $\overleftrightarrow{LM}$ is added parallel to $\overleftrightarrow{BD}$ and $\overleftrightarrow{XZ}$ in the diagram. Does your conclusion from Exercise 2 apply to $\overleftrightarrow{LM}$? Explain.

5. Suppose that $\overleftrightarrow{BD}$ and $\overleftrightarrow{XZ}$ are cut by a transversal $\overleftrightarrow{CY}$ and the lines intersect the transversal at *A* and *W*. Suppose also that you select an angle with vertex at *A* at random and an angle with vertex at *W* at random and find that their measures are equal, can you conclude that $\overleftrightarrow{BD}$ and $\overleftrightarrow{XZ}$ are parallel? Explain your response with a diagram.

98 Technology HRW Geometry

PREPARE

Objectives

- Identify and use the converse of the Corresponding Angles Postulate.
- Prove that lines are parallel using theorems and postulates.

RESOURCES

- Practice Master 3.4
- Enrichment Master 3.4
- Technology Master 3.4
- Lesson Activity Master 3.4
- Quiz 3.4
- Spanish Resources 3.4

Assessing Prior Knowledge

1. Write the converse of this conditional: If Rosalia is 18 years old, then she can vote. [**If Rosalia can vote, then she is 18 years old.**]
2. If two parallel lines are cut by a transversal, how many pairs of corresponding angles are formed? [**4 pairs**]
3. Complete. If $\angle 1$ and $\angle 2$ are alternate angles formed by two parallel lines and a transversal, then $m\angle 1$ _?_ $m\angle 2$. [$\cong$]

TEACH

Discuss with students the fact that architects and contractors make extensive use of parallel lines in their plans and structures and require fail-safe ways of constructing them. Consequently, the geometry of parallel lines and knowing how to prove that lines are parallel is a significant aspect of the work that many people do.

LESSON 3.4 Proving Lines Parallel

Why *Suppose you needed to create a number of parallel lines, or you had to prove that certain lines were parallel. The converses of the transversal conjecture and theorems in Lesson 3.3 enable you to do these things.*

The parabolic mirrors of astronomical telescopes gather parallel light rays from distant stars and planets and direct them to a central point. In the photograph, technicians are inspecting the Hubble Space Telescope's mirror.

Forming the Converses

In Lesson 3.3 parallel lines were given in the hypotheses of the conjectures, and conclusions were made about certain special angles. What happens when special angles are given in the hypothesis, and conclusions are made about parallel lines? When a hypothesis and a conclusion are interchanged, the converse of the original statement is formed.

EXAMPLE 1

Write the converse of the Corresponding Angles Postulate.

Solution ➤

Identify the hypothesis and the conclusion of the Corresponding Angles Postulate. Then interchange the hypothesis and conclusion to form the converse.

Using Visual Models Draw a pair of intersecting lines on a transparency and number the angles. Call on students to come to the overhead projector and place a pencil on the transparency to show two parallel lines and a transversal. Discuss which angles were made equal to ensure that the lines are indeed parallel.

Statement: If two lines cut by a transversal are parallel, then corresponding angles are congruent.

Converse: If corresponding angles are congruent, then two lines cut by a transversal are parallel. ❖

Another way of stating the converse is as follows:

THE CONVERSE OF THE CORRESPONDING ANGLES THEOREM

If two lines are cut by a transversal in such a way that corresponding angles are congruent, then the two lines are parallel. **3.4.1**

Notice that the converse is labeled as a theorem. This is because you can prove it with the postulates and theorems you now know. However, the proof will be postponed because it involves a special form of reasoning.

Using the Converses

EXAMPLE 2

Suppose that m∠1 = 64° and m∠2 = 64° in the figure on the right. What can you conclude about lines *l* and *m*?

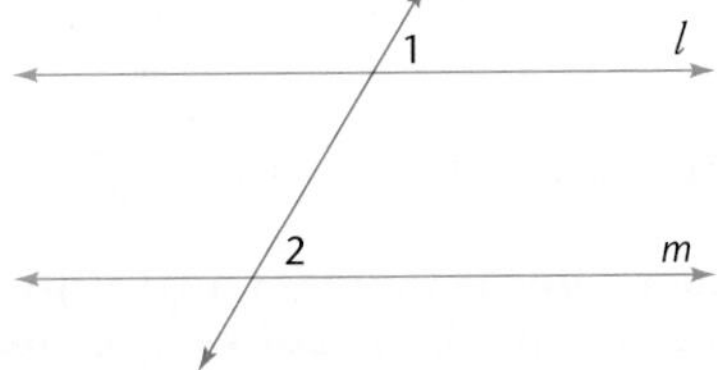

Solution ➤

∠1 and ∠2 are corresponding angles. The converse of the Corresponding Angles Postulate says that if such angles are congruent, then the lines are parallel. Conclusion: *l* ∥ *m*. ❖

EXAMPLE 3

Given a line *l* and a point *P* not on the line, draw a line through *P* parallel to *l*.

Solution ➤

Draw a line through point *P* and line *l*. Then measure ∠1.

Using a protractor, draw a line *m* through *P* so that the new angle corresponds to ∠1 and has the same measure as ∠1. By the converse of the Corresponding Angles Theorem, the new line *m* is parallel to *l*. ❖

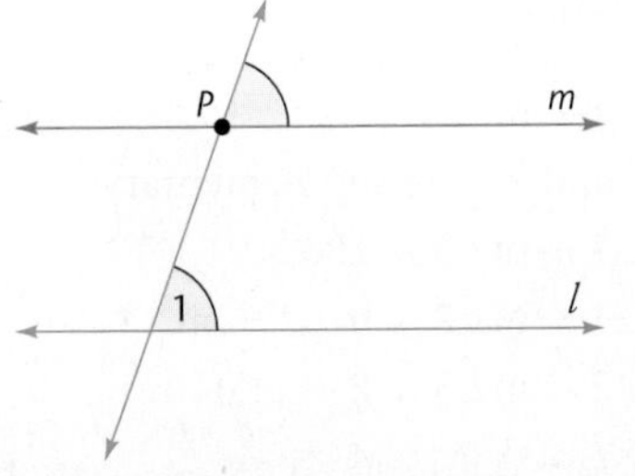

Alternate Example 1

Write the converse of this conditional: If two lines are cut by a transversal and are parallel, then alternate exterior angles are congruent. [**If two lines are cut by a transversal and alt. ext. ∠s are ≅, then the lines are ∥.**]

Cooperative Learning

Students can explore the converse of the Corresponding Angles Theorem in small groups. Provide each group with a copy of a diagram that shows two nonparallel lines and a transversal, with angles numbered. Ask students to tell which angle measures they would change to make the lines cut by the transversal parallel. Ask how changing the measures they specify would affect other measures. You may wish to call upon some groups to make a brief presentation to the class.

Alternate Example 2

Refer to the diagram for Example 2. If m∠2 = 90° and ∠1 and ∠2 are supplementary, what can you conclude? [**m∠1 = 90° and *l* ∥ *m*.**]

Alternate Example 3

Given an angle and a point on the angle not at the vertex, construct a line through the point parallel to a side of the angle. [**The construction procedure is similar to that in the solution for Example 3.**]

ENRICHMENT Students often think of parallel lines as having either a horizontal or vertical orientation in a plane. Expand their visual conceptualization of parallel lines by having them explore ways to draw parallel lines that are not horizontal or vertical on graph paper.

INCLUSION strategies **Hands-On Strategies** Provide students with cardboard cutouts of congruent angles. Have students position the cutouts along the edges of a ruler to obtain a model that shows parallel lines and a transversal.

Assess

Selected Answers

Odd-numbered Exercises 5–41

Assignment Guide

Core 1–18, 27–36, 38–46

Core Plus 1–10, 15–37, 43–48

Technology

Students may wish to use geometry graphics software to investigate the theorems presented in the exercises.

Error Analysis

In proofs, watch for students who make the mistake of citing a theorem or postulate as a reason when they should be using the converse.

The answers to Exercises 5–10 can be found in Additional Answers beginning on page 727.

Use Transparency 27

Practice Master

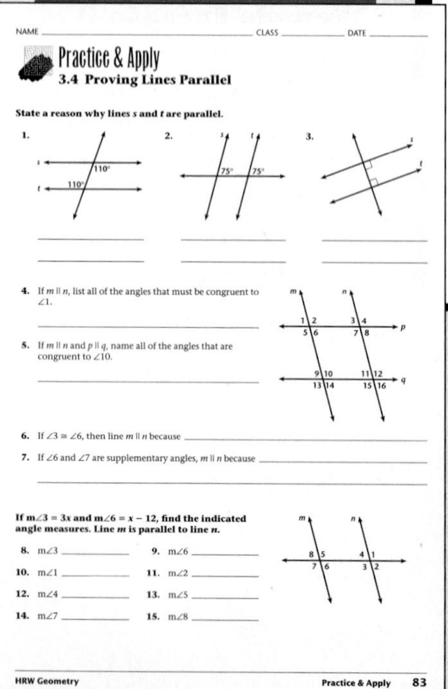

NAME ________ CLASS ________ DATE ________

Practice & Apply
3.4 Proving Lines Parallel

State a reason why lines *s* and *t* are parallel.

1. ________
2. ________
3. ________

4. If $m \parallel n$, list all of the angles that must be congruent to $\angle 1$. ________

5. If $m \parallel n$ and $p \parallel q$, name all of the angles that are congruent to $\angle 10$. ________

6. If $\angle 3 \cong \angle 6$, then line $m \parallel n$ because ________

7. If $\angle 6$ and $\angle 7$ are supplementary angles, $m \parallel n$ because ________

If $m\angle 3 = 3x$ and $m\angle 6 = x - 12$, find the indicated angle measures. Line *m* is parallel to line *n*.

8. $m\angle 3$ ________
9. $m\angle 6$ ________
10. $m\angle 1$ ________
11. $m\angle 2$ ________
12. $m\angle 4$ ________
13. $m\angle 5$ ________
14. $m\angle 7$ ________
15. $m\angle 8$ ________

HRW Geometry — Practice & Apply 83

Exercises & Problems

Communicate

In Exercises 1 and 2, state the converse of each theorem.

1. If two parallel lines are cut by a transversal, then corresponding angles are congruent.
2. If two parallel lines are cut by a transversal, then consecutive interior angles are supplementary.
3. What can you conclude about two lines if a transversal forms corresponding angles that are *not* congruent? Explain your answer.
4. Complete the statement: If two lines that are not parallel are cut by a transversal, then alternate interior angles are _?_.

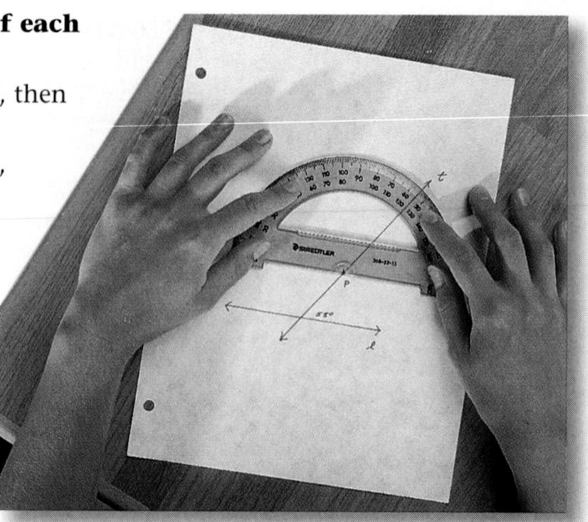

Practice & Apply

Copy the statements and supply the reasons in the proof of the theorem that follows. Refer to the diagram at right.

THEOREM

If two lines are cut by a transversal in such a way that consecutive interior angles are supplementary, then the two lines are parallel. **3.4.2**

Given: $\angle 1$ and $\angle 2$ are supplementary.

Prove: $l_1 \parallel l_2$

Proof:

STATEMENTS	REASONS
$m\angle 1 + m\angle 2 = 180°$.	**(5)**
$\angle 1$ and $\angle 3$ are supplementary.	**(6)**
$m\angle 1 + m\angle 3 = 180°$	**(7)**
$m\angle 1 + m\angle 2 = m\angle 1 + m\angle 3$	**(8)**
$m\angle 2 = m\angle 3$ ($\angle 2 \cong \angle 3$)	**(9)**
$l_1 \parallel l_2$	**(10)**

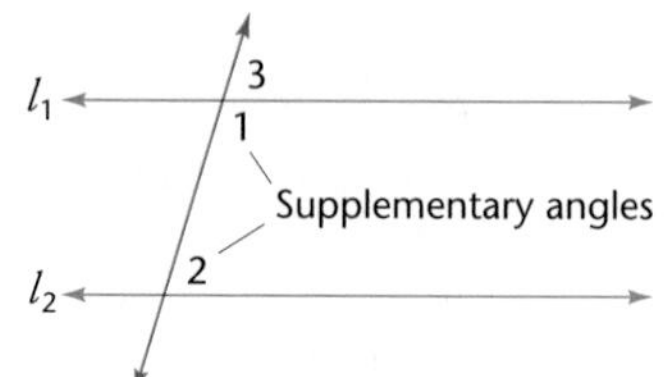

Reteaching the Lesson

Inviting Participation

Reteach the lesson by asking students questions that review the key concepts. (1) Ask, "How do you form the converse of statement?" (2) Draw two parallel lines cut by a transversal and ask students to identify the corresponding angles. (3) Ask students to state the converse of the Corresponding Angle Theorem. (4) Ask students to make up an example like Example 2.

Use the diagram for Exercises 11–14.

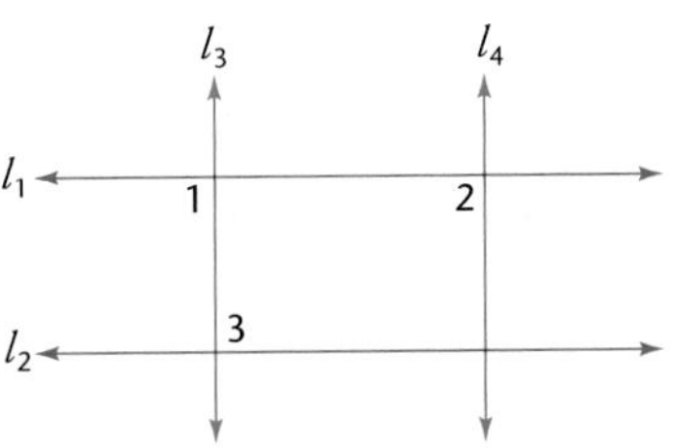

11. If m∠1 = m∠3, which pair of lines must be parallel? State a postulate or theorem to justify your conclusion.

12. If m∠1 = m∠2, which pair of lines must be parallel? State a postulate or theorem to justify your conclusion.

13. Redraw the given figure so that ∠1 ≅ ∠3 but ∠2 is not congruent to ∠1. Is $l_3 \parallel l_4$?

14. Redraw the given figure so that ∠1 ≅ ∠2 but ∠3 is not congruent to ∠1. Is $l_1 \parallel l_2$?

In Exercises 15–18, fill in the blanks. Use the figure at right.

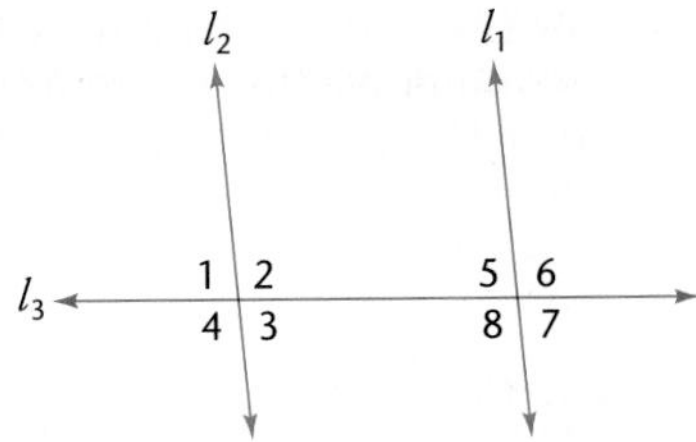

15. If ∠5 ≅ ∠3, then $l_1 \parallel l_2$ because _?_.

16. If ∠4 ≅ ∠8, then $l_1 \parallel l_2$ because _?_.

17. If ∠3 and ∠8 are supplementary, then $l_1 \parallel l_2$ because _?_.

18. If ∠5 ≅ ∠2 and $l_1 \parallel l_2$, then l_1 and l_2 are _?_ to l_3 because _?_.
Perpendicular; ∠2 and ∠5 are supplementary

Algebra If m∠4 = 2x and m∠5 = x + 6, find the indicated angle measures in Exercises 19–26. l_1 is parallel to l_2.

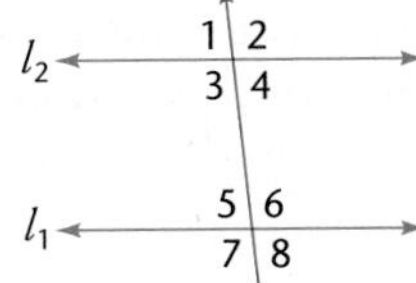

19. m∠4 12°
20. m∠5 12°
21. m∠1 12°
22. m∠2 168°
23. m∠3 168°
24. m∠6 168°
25. m∠8 12°
26. m∠7 168°

In Exercises 27–36 you will complete proofs of two theorems you will use in future chapters.

THEOREM

If two lines are perpendicular to the same line, the two lines are parallel to each other. **3.4.3**

Given: lines l, m, and p with $m \perp l$ and $p \perp l$

Prove: $p \parallel m$

Proof:

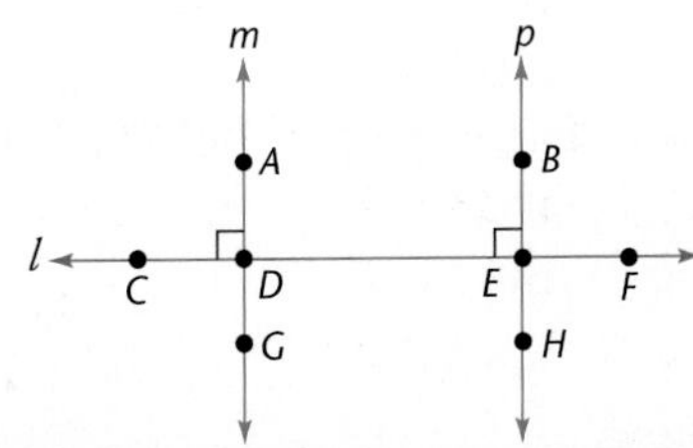

Line l is a __27__ of m and p, by definition. ∠ADC ≅ ∠BED because perpendicular lines form __28__. Therefore, __29__ is parallel to __30__ by the converse of the __31__ Angles Postulate.

11. l_1 and l_2; If two lines are cut by a transversal in such a way that alternate interior angles are congruent, then the two lines are parallel.

12. l_3 and l_4; If two lines are cut by a transversal in such a way that corresponding angles are congruent, then the two lines are parallel.

13.
14.

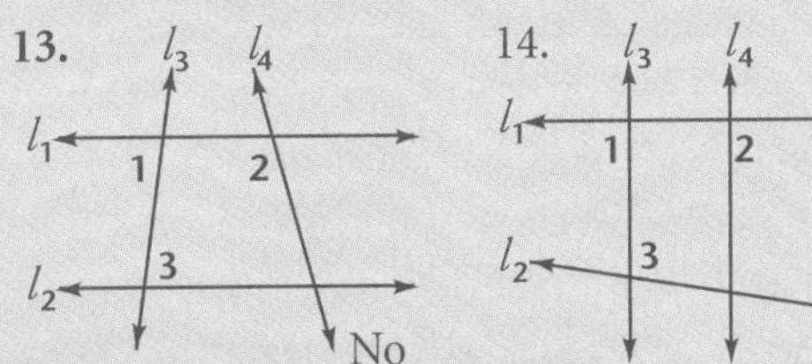

15. Congruent alternate interior angles imply parallel lines.

16. Congruent corresponding angles imply parallel lines.

alternative ASSESSMENT

Performance Assessment

Ask students to use three pencils to show what angles to make congruent or supplementary in order for the pencils to model a pair of parallel lines and a transversal.

17. Supplementary consecutive interior angles imply parallel lines.

27. transversal

28. right angles

29. m

30. p

31. Corresponding

Technology Master

NAME ______ CLASS ______ DATE ______

Technology
3.4 Opposite Sides of Regular Polygons

You can create the diagram shown in your geometry software to explore the relationships among the sides of certain regular polygons. One question you may ask is how your knowledge of alternate interior angles helps you discover relationships among their sides.

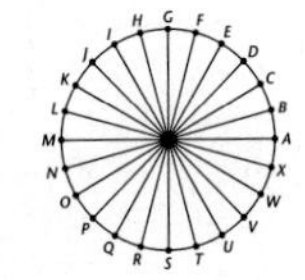

Use geometry software for the following exercises.

1. Construct and label a diagram like the one shown. In your sketch, make the angle between consecutive spokes of the wheel 15°.

2. Starting at point A, sketch line segments that join every fourth labeled point. What type of figure have you drawn?

3. List the measures of angles that allow you to conclude that opposite sides of the figure are parallel.

4. Starting at point A, sketch line segments that join every sixth labeled point. What type of figure have you drawn?

5. List the measures of angles that allow you to conclude that opposite sides of the figure are parallel.

6. Suppose that you have a regular polygon with n sides. Write a statement about the relationship between opposite sides of the polygon if n is even greater than 2. Test your hypothesis by starting at A and joining every third labeled point.

HRW Geometry Technology 99

32. ∠3

33. ∠3

34. ∠2

35. the Transitive Property of Congruence

36. the Corresponding Angle Theorem converse

38. If a figure is a rectangle, then it is a parallelogram.

39. Hypothesis: The figure is a rectangle. Conclusion: The figure is a parallelogram.

THEOREM

If two lines are parallel to the same line, then the two lines are parallel to each other. **3.4.4**

Given: lines l, m, and n, with transversal p; $l \parallel n$, and $m \parallel n$

Prove: $l \parallel m$

Proof:

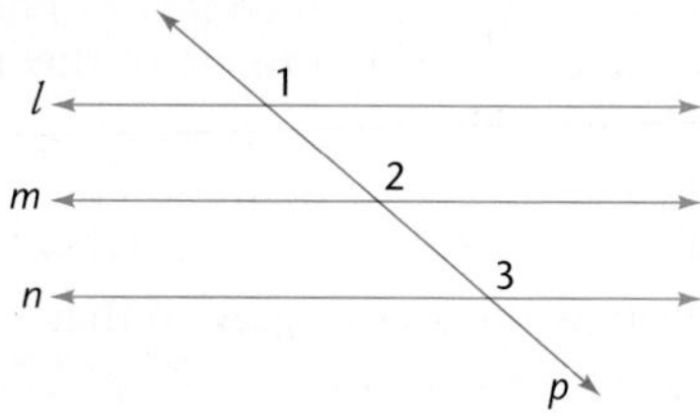

Since $l \parallel n$, then $\angle 1 \cong$ **32**. Since $m \parallel n$, then $\angle 2 \cong$ **33**. So $\angle 1 \cong$ **34** by **35**. Therefore, $l \parallel m$ by **36**.

37. Drafting A T square and a triangle can be used to draw parallel lines. While holding the T square steady, slide the triangle along the edge of the T square. How can you prove that the resulting lines l and m are parallel? By using the Corresponding Angle Theorem converse.

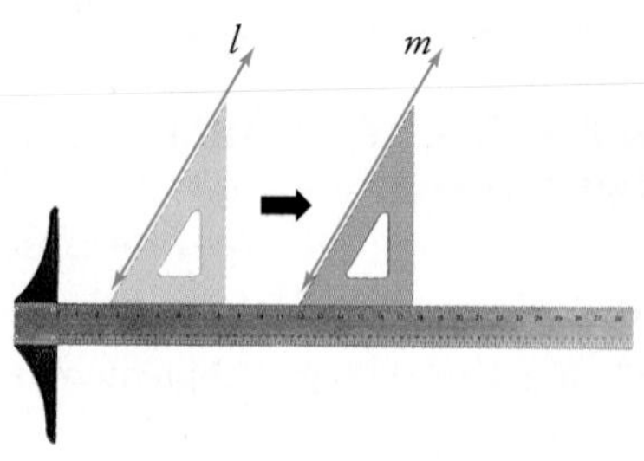

Look Back

For Exercises 38–42, refer to the statement: Every rectangle is a parallelogram. [Lesson 2.2]

38. Rewrite the statement above as a conditional statement.

39. Identify the hypothesis and conclusion of the statement.

40. Draw a Venn diagram that illustrates the statement.

41. Write the converse of the statement and construct its Venn diagram.

42. Is the converse statement in Exercise 41 false? If so, illustrate this with a counterexample.

Look Beyond

Exercises 43–48 lead students to results that can be derived by another approach once the Triangle Sum Theorem is proved in Section 3.5.

Look Beyond

Given: $l_1 \parallel l_2$; $\overrightarrow{PT}$ bisects $\angle RPQ$; $\overrightarrow{RT}$ bisects $\angle PRS$; $m\angle RPQ = 110°$. Find each of the following.

43. $m\angle 1$ 55° **44.** $m\angle 2$ 55° **45.** $m\angle PRS$ 70°

46. $m\angle 3$ 35° **47.** $m\angle 4$ 35° **48.** $m\angle 5$ 90°

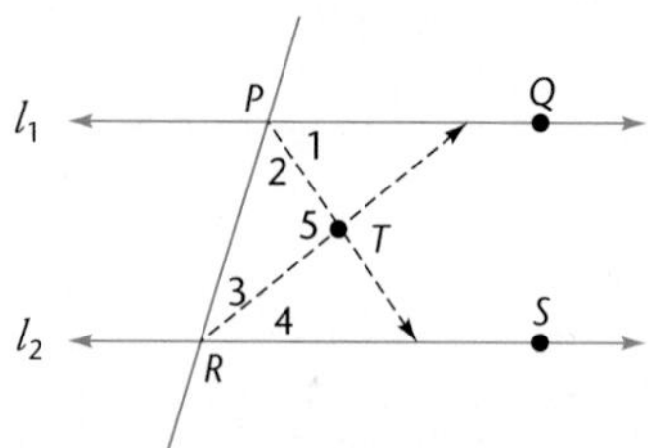

40.

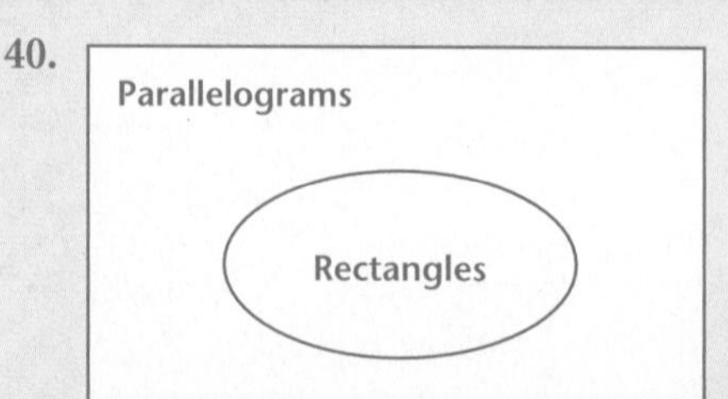

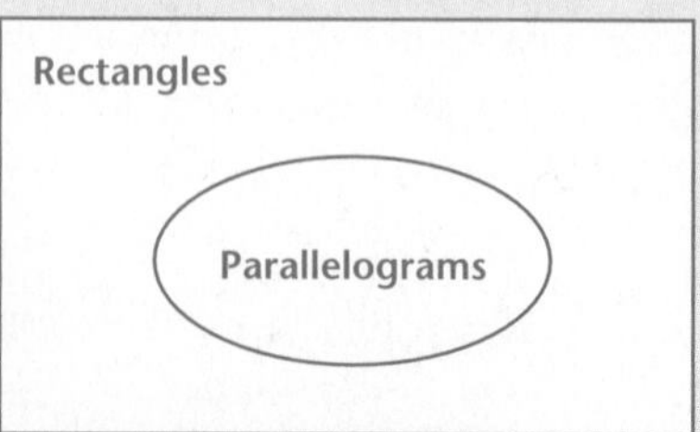

41. If a figure is a parallelogram, then it is a rectangle.

42. Converse is false. A rhombus is a parallelogram that is not a rectangle in general.

LESSON 3.5

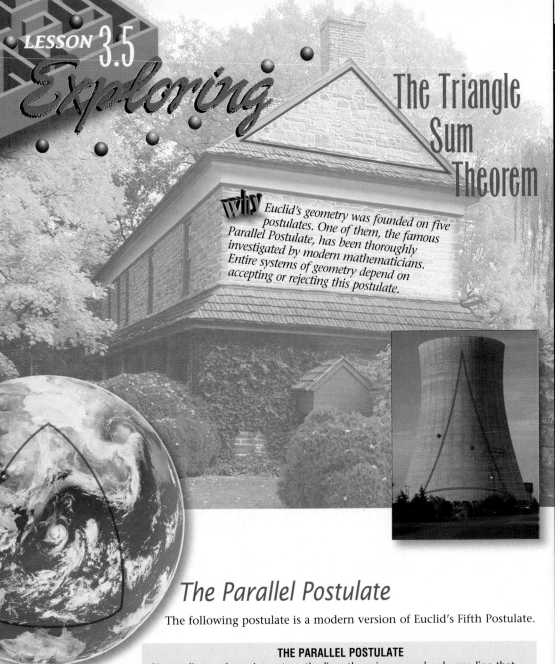

Exploring The Triangle Sum Theorem

Why *Euclid's geometry was founded on five postulates. One of them, the famous Parallel Postulate, has been thoroughly investigated by modern mathematicians. Entire systems of geometry depend on accepting or rejecting this postulate.*

The Parallel Postulate

The following postulate is a modern version of Euclid's Fifth Postulate.

> **THE PARALLEL POSTULATE**
> Given a line and a point not on the line, there is one and only one line that contains the given point and is parallel to the given line. **3.5.1**

In the previous lesson you constructed a line through a point parallel to a given line. You probably never questioned whether such a line actually existed, or whether there could be more than one such line. The assumption that there is exactly one such parallel line is the Parallel Postulate.

ALTERNATIVE teaching strategy

Technology Students can investigate the sum of the measures of the angles of a triangle by using geometry graphics tools. By drawing many triangles of different shapes and sizes and measuring their angles, they will discover the fact that the sum of the measures is always 180°. Students can then create a computer model of the drawing at the top of page 145 and use it to study the proof given.

PREPARE

Objectives

- Identify and use the Parallel Postulate and Triangle Sum Theorem.

RESOURCES

- Practice Master 3.5
- Enrichment Master 3.5
- Technology Master 3.5
- Lesson Activity Master 3.5
- Quiz 3.5
- Spanish Resources 3.5

Assessing Prior Knowledge

Refer to the following diagram for 1–3.

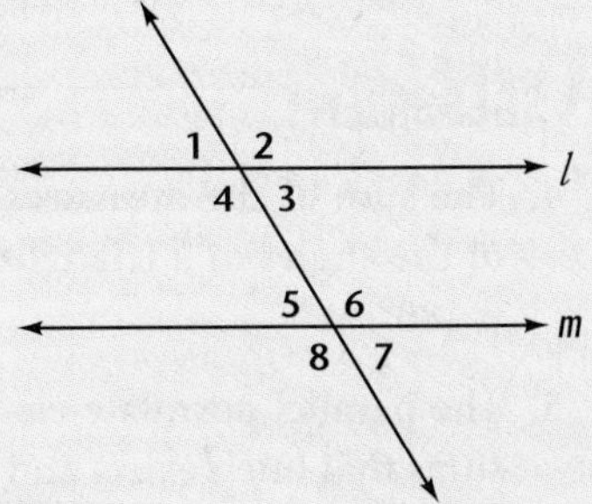

1. If $\angle 1$ and $\angle 5$ have different measures, are l and m parallel? [**no**]
2. If $m\angle 3 + m\angle 6 = 180°$, what can you conclude about lines l and m? [$l \parallel m$]

TEACH

Discuss with students the fact that Euclidean geometry is a useful model of the physical world on a "close-to-home," local scale. However, the Euclidean model needs to be modified when considering a geometry that can serve as a useful model of the universe.

Cooperative Learning

You can have members of cooperative groups perform the manipulative activity described in the exploration using triangles of several different shapes and sizes. Students can discuss what they have discovered and then compile their results for a presentation to the class.

Exploration Note

Before students tear off the corners of the triangle, you may wish to have them label the interiors of the angles with circled numbers. Otherwise, students who cut off the bottom angles with scissors may have trouble keeping track of which angles they are to place alongside angle 3.

ongoing ASSESSMENT

4. **The sum of the measures of the angles of a triangle is 180°.**
5. **The parallel postulate ensures that line *l* exists and is the only line through *P* that is parallel *AB*.**
7. **Yes, m∠3 + m∠4 + m∠5 is always 180°.**

CRITICAL Thinking

To prove the conjecture, the statement must be shown to be true for all possible triangles.

The Parallel Postulate is used to prove a theorem about the sum of the measures of the angles of a triangle. Before you look at this proof, try the following exploration.

Exploration *The Triangle Sum Theorem*

You will need
Scissors and paper

1. Cut out a triangle from the piece of paper.
2. Tear off two of the corners.
3. Position the two torn-off corners adjacent to the third angle as shown.
4. Make a conjecture about the sum of the measures of the angles of a triangle.

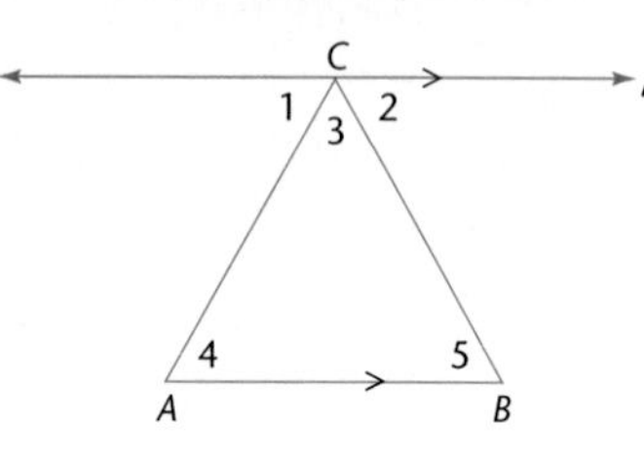

5. In the drawing on the right, line *l* has been drawn through a vertex of the triangle so that it is parallel to the opposite side. How does the Parallel Postulate relate to this figure?
6. Fill in the table for a figure like the one in the drawing. Use geometry theorems, not physical measurements.

m∠1	m∠2	m∠3	m∠4	m∠5	m∠3 + m∠4 + m∠5
40°	30°	?	?	?	?
20°	80°	?	?	?	?
30°	100°	?	?	?	?

7. Does the table support your conjecture? Explain. ❖

CRITICAL Thinking

Why do the activities of the exploration fall short of proving this conjecture?

One conjecture from the exploration is stated below as a theorem.

TRIANGLE SUM THEOREM
The sum of the measures of the angles of a triangle is 180°. **3.5.2**

interdisciplinary CONNECTION

Physics Non-Euclidean geometry plays an essential role in relativity theory. For some insights from a recent book for the general public, students may look up information on *geodesics* in *Black Holes and Time Warps: Einstein's Outrageous Legacy*, by Kip S. Thorne (W.W. Norton and Company, 1994).

Proving the Triangle Sum Theorem

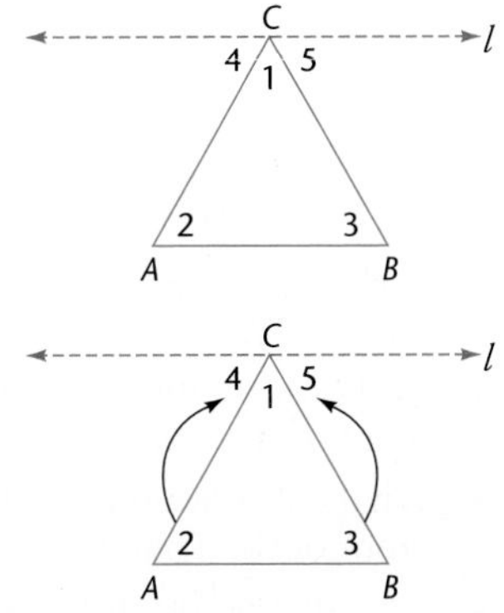

To prove the theorem, begin by drawing a line l through a vertex of $\triangle ABC$ so that it is parallel to the side opposite the vertex.

Given: $\triangle ABC$

Prove: $m\angle 1 + m\angle 2 + m\angle 3 = 180°$

Plan: Study the illustration, which is actually an informal proof of the theorem. You can use the ideas in it to write a more formal proof.

Proof:

STATEMENTS	REASONS
1. $m\angle 1 + m\angle 4 + m\angle 5 = 180°$	The angles fit together to form a straight line.
2. $l \parallel \overline{AB}$	As constructed (justification: the Parallel Postulate)
3. $\angle 2 \cong \angle 4$ $\angle 5 \cong \angle 3$	$\parallel$s $\Rightarrow$ alt int $\angle$s $\cong$
4. $m\angle 2 = m\angle 4$ $m\angle 3 = m\angle 5$	If $\angle$s are $\cong$, then their measures are =.
5. $m\angle 1 + m\angle 2 + m\angle 3 = 180°$	Substitution in Step 1.

Another Geometry

It is possible to create geometries in which the Parallel Postulate is not true. On the surface of a sphere, for example, where lines are defined differently from the way they are on flat surfaces, there are no parallel lines.

Geography On the surface of a sphere a "line" is defined as a **great circle**, which is a circle that lies on a plane that passes through the center of the sphere. The equator is a great circle on the surface of the earth. Lines of longitude, which run north and south, are also great circles. Notice that any two distinct "lines" (great circles) intersect in two points. Thus, there are no parallel lines on a sphere.

CRITICAL *Thinking*

Discuss the following statement: "On the surface of a sphere, the shortest path between two points is not a straight line." What is the shortest path?

TEACHING *tip*

After you have discussed the proof of the Triangle Sum Theorem, students can benefit from a discussion that compares and contrasts the proof with the exploration activity.

TEACHING *tip*

Flying a plane from New York City to London via Greenland is an application of a great circle route.

CRITICAL *Thinking*

The shortest path between two points on a sphere is along the great circle that contains the two points. (If the points are on opposite sides of the sphere, there will be an infinite number of great circles that contain the two points. But see Exercise 7, page 619.)

ENRICHMENT Students can investigate what happens if they draw a line through each vertex of a triangle parallel to the opposite side. Ask students to formulate conjectures about the sides and angles in the resulting figure.

INCLUSION strategies **Hands-On Strategies** Students may need to actually measure the angles listed in the table on page 144. Using both the geometry theorems and their measurements, students should be convinced that their answers are correct and that they do support their conjectures.

Assess

Selected Answers

Exercises 7, 9, 11, 13, 15, 17, 19, 23, 25, 27, 29, 31, 33, and 35

Assignment Guide

Core 1–3, 6–24, 29–35

Core Plus 1–17, 21–28, 36–38

Technology

Students can use geometry graphics software to study exterior angles of triangles in Exercises 21–24.

Error Analysis

Students may need a more detailed description of exterior angles for Exercise 22.

Practice Master

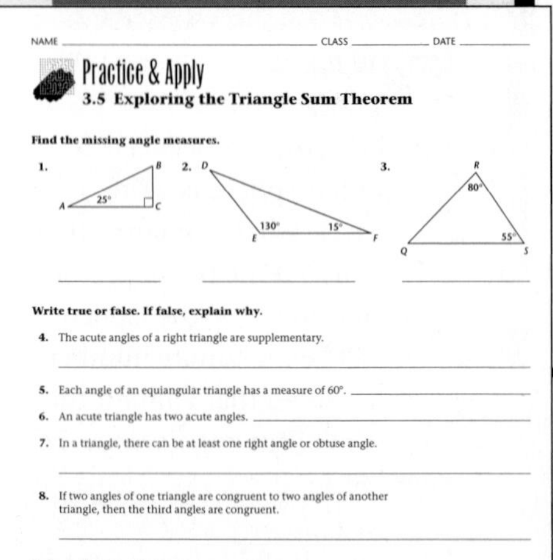

NAME ____________ CLASS ________ DATE ________

Practice & Apply

3.5 Exploring the Triangle Sum Theorem

Find the missing angle measures.

1. A 25° B C
2. D 130° E 15° F
3. R 80° Q 55° S

Write true or false. If false, explain why.

4. The acute angles of a right triangle are supplementary. ____________
5. Each angle of an equiangular triangle has a measure of 60°. ____________
6. An acute triangle has two acute angles. ____________
7. In a triangle, there can be at least one right angle or obtuse angle. ____________
8. If two angles of one triangle are congruent to two angles of another triangle, then the third angles are congruent. ____________

Refer to the triangle shown.

	m∠1	m∠2	m∠3	m∠4	m∠1 + m∠2
9.	____	35°	60°	____	____
10.	70°	____	____	____	120°
11.	58°	____	72°	____	____
12.	____	43°	55°	____	____
13.	45°	____	____	85°	____

R 2 Q 1 3 4 S

84 Practice & Apply HRW Geometry

Exercises & Problems

Communicate

1. Explain how the paper-tearing exercise is like the proof of the Triangle Sum Theorem.
2. Explain why the paper-tearing exercise is not a proof of the Triangle Sum Theorem.
3. What role does the Parallel Postulate play in the proof of the Triangle Sum Theorem?
4. Why do you think airline pilots like to follow a great-circle route on long flights?
5. One alternative to the Parallel Postulate is that there is no line through a given point parallel to a given line. What are some others? On what kind of surface do you think this might be true?

Practice & Apply

In Exercises 6 and 7, find the missing angle measures.

6. B 85° A C 45°

$m\angle A = ?$ 50°

7. N 90° M O 40°

$m\angle M = ?$ 50°

For Exercises 8 and 9, recall that an acute angle has a measure of less than 90° and an obtuse angle has a measure of greater than 90°.

8. Is it possible for a triangle to have more than one obtuse angle? Explain your reasoning.
9. If one of the angles of a triangle is a right angle, what must the other two angles be? Explain your reasoning.

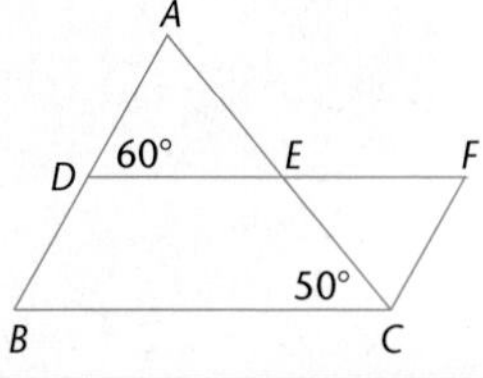

Given: $\overline{DE} \parallel \overline{BC}$; $\overline{CF} \parallel \overline{BD}$; $m\angle ADE = 60°$; $m\angle ACB = 50°$. Find the following angle measures.

10. $m\angle A$ 70° **11.** $m\angle B$ 60° **12.** $m\angle EDB$ 120° **13.** $m\angle AED$ 50°

14. $m\angle DEC$ 130° **15.** $m\angle FEC$ 50° **16.** $m\angle F$ 60° **17.** $m\angle ECF$ 70°

Reteaching the Lesson

Using Visual Models Redraw the diagram on page 145 that was used in the Proof of the Triangle Sum Theorem. This time, however, construct a line through vertex *B* parallel to segment $\overline{AC}$. Use this new diagram to prove the theorem. In the construction step, review the Parallel Postulate. Students should see that any vertex can be used to prove the theorem.

8. No, since an obtuse angle measures more than 90°, the sum of the measures of two or more obtuse angles is greater than 180°.
9. The sum of the other two angles is 90°, so each must be less than 90°, or acute.

For each of the triangles, find the indicated angle measures.

18.

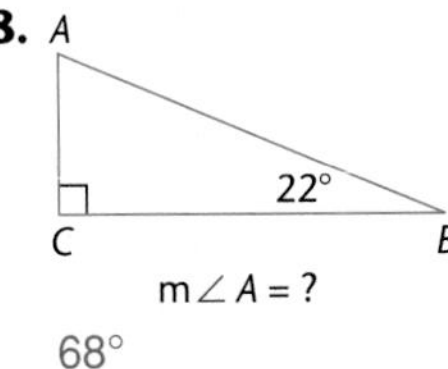

m∠A = ?

68°

19.

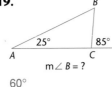

B
25°
85°
A
C

m∠B = ?

60°

20.

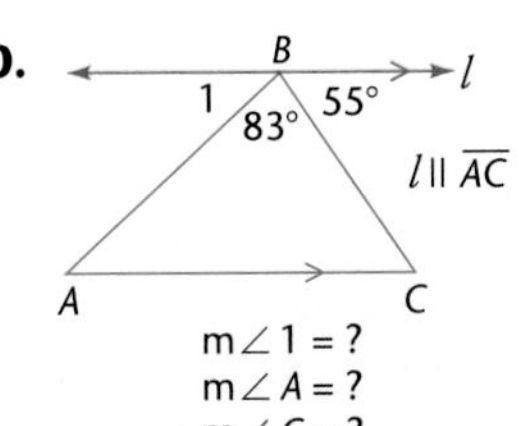

m∠1 = ?
m∠A = ?
m∠C = ?

m∠1 = 42°; m∠A = 42°; m∠C = 55°

In Exercise 21 you will discover and prove an important geometry theorem. **Begin by copying and filling in the following table, which refers to the triangle to the right.**

21.

m∠1	m∠2	m∠3	m∠4	m∠1 + m∠2
?	50°	70°	?	?
60°	?	?	120°	?
?	30°	?	100°	?
–	45°	65°	?	?
31°	–	75°	?	?
–	–	x	?	?
–	–	?	x	?

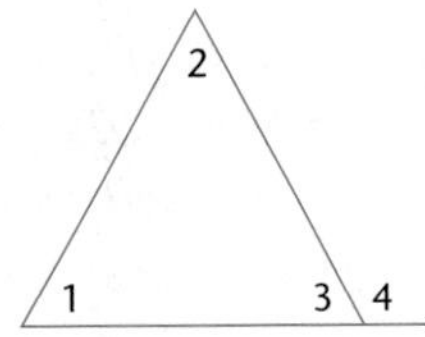

What do you notice about the values in the fourth and fifth columns of your table?

22. An angle like ∠4 is known as an **exterior angle** of a polygon. It is formed between one side of the polygon and the extended part of another side. How many exterior angles are possible at each vertex of a given polygon? Are their measures the same or different? Explain your answer.

23. In the triangle, angles ∠1 and ∠2 are known as the **remote interior angles** of ∠4, because they are the angles farthest from ∠4. State your observations from Exercise 21 as a theorem by completing the following statement.

EXTERIOR ANGLE THEOREM

The measure of an exterior angle of a triangle is equal to __?__. **3.5.3**

24. Explain how the last two lines of the table in Exercise 21 prove the theorem informally. Why do the first five lines fall short of proving the theorem?

21.

m∠1	m∠2	m∠3	m∠4	m∠1 + m∠2
60°	–	–	110°	110°
–	60°	60°	–	120°
70°	–	80°	–	100°
–	–	–	115°	115°
–	–	–	105°	105°
–	–	–	180 – x	180 – x
–	–	180 – x	–	x

alternative ASSESSMENT

Authentic Assessment

Ask cooperative learning groups to draw a diagram and devise exercises similar to Exercises 10–17. At least one exercise should require applying the Triangle Sum Theorem. Have groups trade and solve one another's exercises. Assess the construction of the exercises as well as the solutions.

The answers to Exercises 22, 23, and 24 can be found in Additional Answers beginning on page 727.

Technology Master

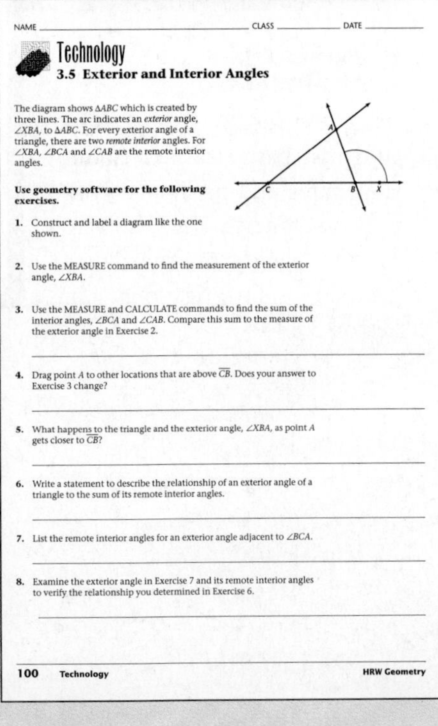
NAME ______ CLASS ______ DATE ______

Technology

3.5 Exterior and Interior Angles

The diagram shows △ABC which is created by three lines. The arc indicates an *exterior* angle, ∠XBA, to △ABC. For every exterior angle of a triangle, there are two *remote interior* angles. For ∠XBA, ∠BCA and ∠CAB are the remote interior angles.

Use geometry software for the following exercises.

1. Construct and label a diagram like the one shown.
2. Use the MEASURE command to find the measurement of the exterior angle, ∠XBA.
3. Use the MEASURE and CALCULATE commands to find the sum of the interior angles, ∠BCA and ∠CAB. Compare this sum to the measure of the exterior angle in Exercise 2.
4. Drag point A to other locations that are above $\overline{CB}$. Does your answer to Exercise 3 change?
5. What happens to the triangle and the exterior angle, ∠XBA, as point A gets closer to $\overline{CB}$?
6. Write a statement to describe the relationship of an exterior angle of a triangle to the sum of its remote interior angles.
7. List the remote interior angles for an exterior angle adjacent to ∠BCA.
8. Examine the exterior angle in Exercise 7 and its remote interior angles to verify the relationship you determined in Exercise 6.

100 Technology HRW Geometry

The proof below is an incomplete formal proof of the theorem you stated in Exercise 23. **Complete the proof by filling in the missing reasons.**

Given: $\triangle ABC$ with exterior angle $\angle BCD$ and remote interior angles $\angle A$ and $\angle B$

Prove: $m\angle BCD = m\angle A + m\angle B$

Proof:

STATEMENTS	REASONS
$\triangle ABC$ with exterior angle $\angle BCD$	Given
$m\angle BCD + m\angle BCA = 180°$	**(25)** $\angle BCD$ and $\angle BCA$ form a linear pair.
$m\angle A + m\angle B + m\angle BCA = 180°$	**(26)** Triangle Sum Theorem.
$m\angle BCD + m\angle BCA = m\angle A + m\angle B + m\angle BCA$	**(27)** Transitive Property of Equality
$m\angle BCD = m\angle A + m\angle B$	**(28)** Subtraction Property of Equality

Look Back

29. The set of points that two geometric figures have in common is called their _?_. **[Lesson 1.1]** intersection

30. _?_ points determine a line. **[Lesson 1.1]** two

31. Adjacent supplementary angles form a _?_ pair. **[Lesson 1.5]** linear

Give reasons for the steps in the proof. [Lesson 3.4]

Given: $l_1 \parallel l_2$; $\angle 2 \cong \angle 3$

Prove: $l_1 \parallel l_3$

Proof:

STATEMENTS	REASONS
$l_1 \parallel l_2$; $\angle 2 \cong \angle 3$	**(32)**
$\angle 1 \cong \angle 2$	**(33)**
$\angle 1 \cong \angle 3$	**(34)**
$l_1 \parallel l_3$	**(35)**

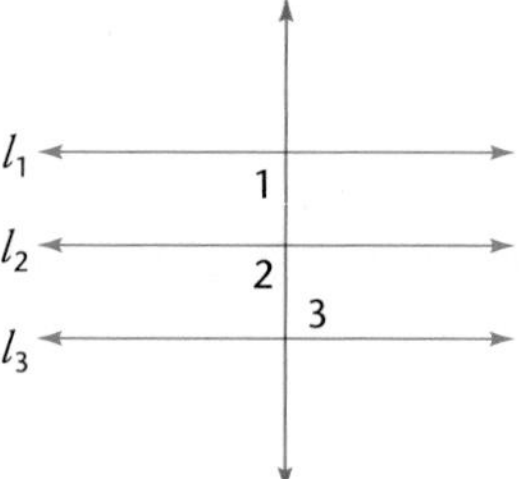

Look Beyond

A quadrilateral has vertices *P*(2, 6), *Q*(2, 10), *R*(7, 10), and *S*(7, 6).

36. What kind of quadrilateral is *PQRS*? Explain your reasoning.

37. What is the perimeter of *PQRS*? 18 units

38. What is the area of *PQRS*? 20 units^2

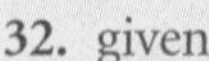

32. given

33. If parallel lines are cut by a transversal, then corresponding angles are congruent.

34. Transitive Property of Congruence

35. If corresponding angles are congruent, then the lines are parallel.

36. *PQRS* is a rectangle. The slopes of opposite sides are equal making the sides parallel. Opposite sides are also the same measure.

Look Beyond

Exercises 36–38 review what students have learned about slope, area and perimeter from previous courses. They also help set the stage for the coordinate geometry investigations in Lesson 3.8.

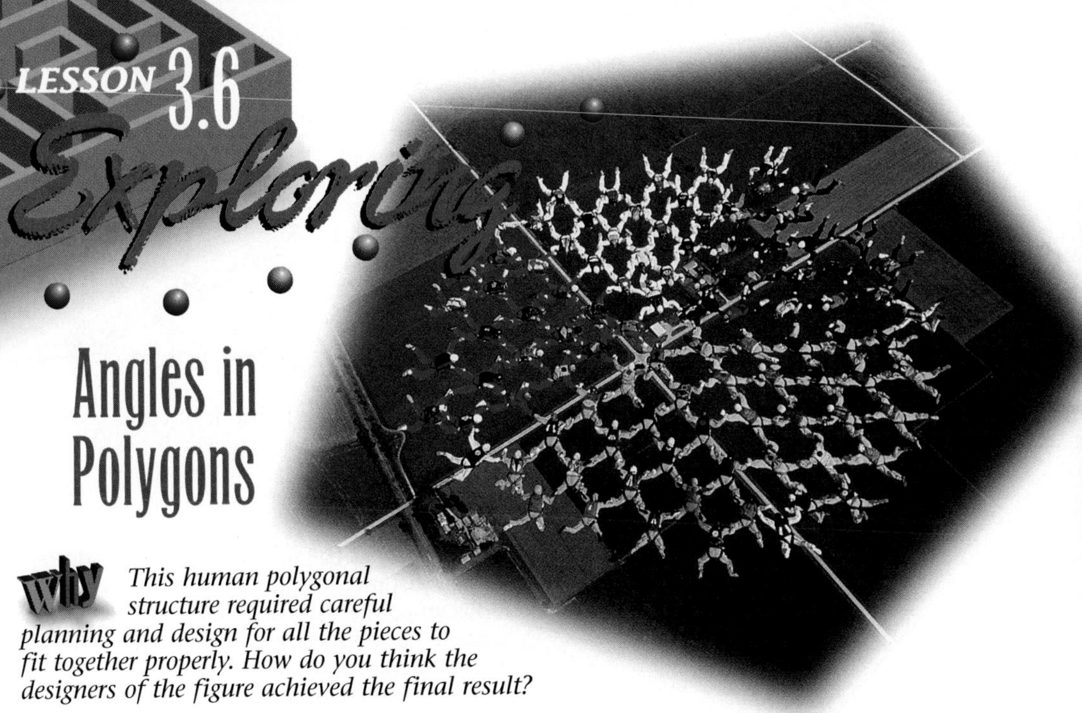

LESSON 3.6 Exploring

Angles in Polygons

Why *This human polygonal structure required careful planning and design for all the pieces to fit together properly. How do you think the designers of the figure achieved the final result?*

A **convex polygon** is one in which any line segment connecting two points of the polygon has no part outside the polygon. For a **concave polygon** this is not true. In this book the word "polygon" will mean a convex polygon, unless stated otherwise.

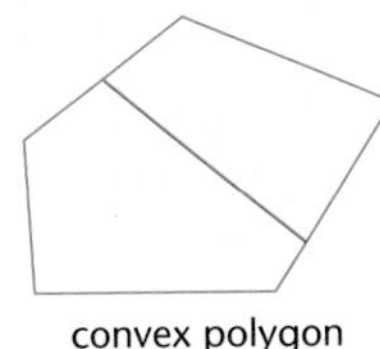
convex polygon

concave polygon

Exploration 1 Angle Sums in Polygons

You will need
Calculator (optional)

Pentagon $ABCDE$ has been divided into three triangular regions by drawing all possible diagonals from one vertex.

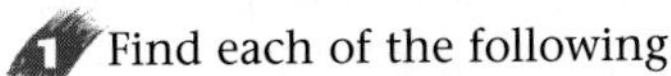

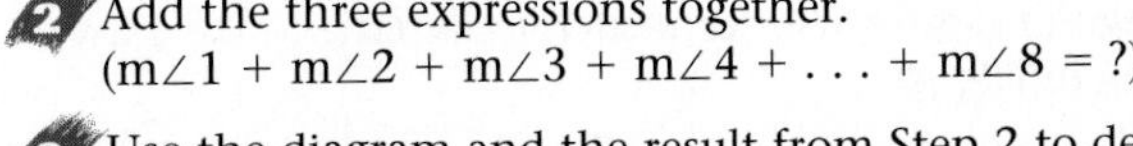

1. Find each of the following:
   $m\angle 1 + m\angle 2 + m\angle 3 = ?$
   $m\angle 4 + m\angle 5 + m\angle 9 = ?$
   $m\angle 6 + m\angle 7 + m\angle 8 = ?$

2. Add the three expressions together.
   $(m\angle 1 + m\angle 2 + m\angle 3 + m\angle 4 + \ldots + m\angle 8 = ?)$

3. Use the diagram and the result from Step 2 to determine the sum of the angles of a pentagon. $(m\angle EAB + m\angle B + m\angle BCD + m\angle CDE + m\angle E)$

ALTERNATIVE teaching strategy

Technology Geometry graphics tools can be used by students to conduct investigations and make conjectures about the sums of interior and exterior angles of polygons. Students can follow the development on pages 149–151 by actually drawing all of the polygons on the computer. Students can find the sum of the angle measures and the exterior angle sums and make conjectures based on the results.

PREPARE

Objectives

- Define *exterior* and *interior angles* of a polygon.
- Develop and use formulas for the sums of interior and exterior angles of a polygon.

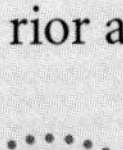

RESOURCES

• Practice Master	3.6
• Enrichment Master	3.6
• Technology Master	3.6
• Lesson Activity Master	3.6
• Quiz	3.6
• Spanish Resources	3.6

Assessing Prior Knowledge

1. If ΔABC has $m\angle A = 70°$, then $m\angle B + m\angle C = \underline{\ ?\ }$. [110°]
2. If ΔXYZ is a triangle all of whose angles are congruent, then $m\angle X = \underline{\ ?\ }$. [60°]

Use Transparency 28

TEACH

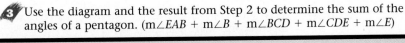

Point out to students that when polygons are used to create patterns, it is the angle relationships in the polygons that are important in determining the kinds of patterns that can be created. Polygonal patterns occur frequently in wallpaper, tiles, linoleum, and many other products for homes and business.

 Complete the table. Sketch each polygon and draw all possible diagonals from one vertex.

Polygon	Number of Sides	Number of Triangular Regions	Sum of Measures of Angles
Triangle	?	1	180°
Quadrilateral	?	?	?
Pentagon	?	3	540°
Hexagon	?	?	?
n-gon	?	?	?

5 Write a formula for the sum of the angles of a polygon in terms of the number of sides, n. ❖

EXTENSION

A **regular** polygon is one that has the measures of all its sides and angles equal. For instance, a square is a regular polygon. So is an equilateral triangle. In a square, each angle has a measure of 90°. In an equilateral triangle, each angle has a measure of 60°.

Fill in the chart below. Then write a formula for finding the measure of an angle of a regular n-gon.

Regular Polygon	Number of Sides	Sum of Measures of Angles	Measure of One Angle
Triangle	?	180°	?
Quadrilateral	?	?	90°
Pentagon	?	?	?
Hexagon	?	?	?
n-gon	?	?	?

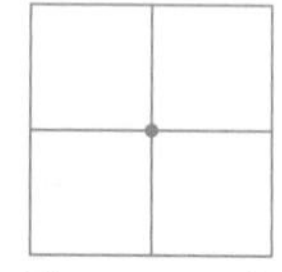

The squares fit together at a point.

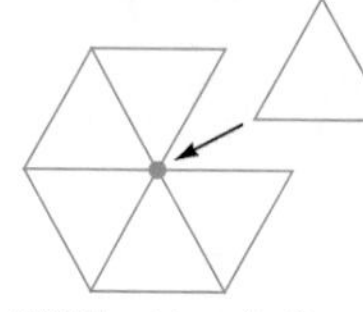

Will the triangles?

Some regular polygons fit together at a central point with no overlapping and no gaps. For example, notice how the squares fit together. Will the equilateral triangles fit together in the same way? Will regular pentagons? Will other n-gons?

For a regular n-gon to form a pattern like the one described above, the measure of one of its angles must be a factor of 360°. How many of the regular n-gons will form such a pattern? Make a list of them for future reference. ❖

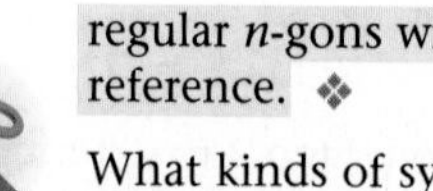

CRITICAL *Thinking*

What kinds of symmetries are present in patterns of polygons that fit together and share a single point?

Exploration 1 Notes

Some students may need a hint to fill in the table in Step 4. You may wish to suggest that students add a column headed "Sum of Measures of Angles of Each Triangle" before the last column.

Ongoing Assessment

5. **For a polygon of n sides, the sum of its interior angle measures is $180(n-2)$.**

Cooperative Learning

Challenge groups of students to discover a method of finding the sum of the angles of a polygon in terms of the number of sides. Provide hints, if necessary, to get the groups started. Your hints should lead students through Exploration 1. Encourage the groups to write a formula that summarizes their results.

Ongoing Assessment

$$\frac{180(n-2)}{n}$$

Ongoing Assessment

Triangles, squares, and hexagons are the only regular polygons that fit together.

CRITICAL *Thinking*

When regular polygons fit together, there are rotational symmetries about the shared points.

interdisciplinary CONNECTION

Design Polygons are the basic design element for many patterns that we see in wallpaper, fabrics, tiles, etc. Ask students to collect examples of polygonal patterns from magazines or other publications for their portfolios.

Exploration 2 Exterior Angle Sums in Polygons

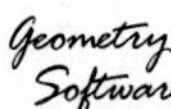

You will need
Geometry technology or
Protractor, scissors

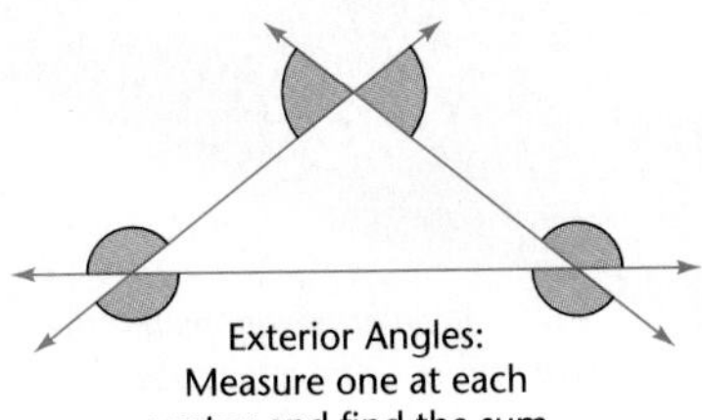
Exterior Angles: Measure one at each vertex and find the sum.

1. Draw a triangle and extend its sides to form exterior angles. Measure one exterior angle at each vertex. Find the sum of the three exterior angles that you measured. Record your results.
2. Cut out the exterior angles you measured and fit them together. Record your results.
3. Repeat Steps 1 and 2 for a quadrilateral.
4. Repeat Steps 1 and 2 for a pentagon.
5. Make a conjecture about the sum of the exterior angles of a polygon.

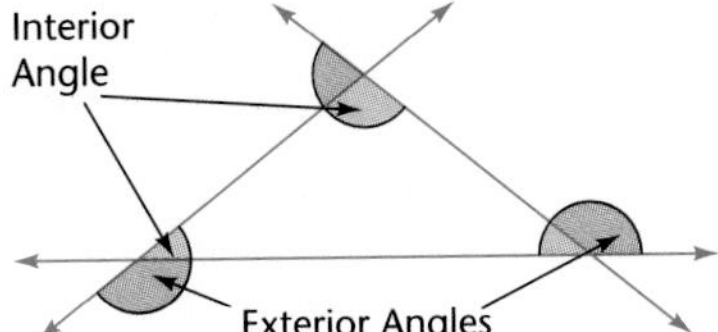

6. What is the sum of all the interior and exterior angles marked in the triangle?
7. What is the sum of all the interior and exterior angles marked in the quadrilateral?
8. Write a formula for the sum of the interior and exterior angles of an *n*-gon as in Steps 6 and 7.
9. Complete the table.

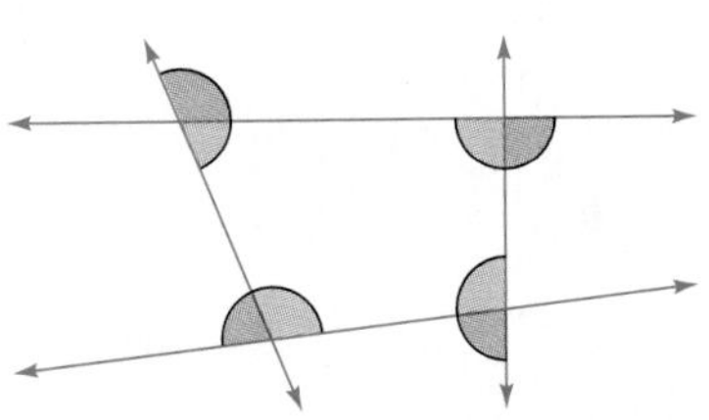

Polygon	No. Sides	Sum (Ext. + Int. Angles)	Sum Interior Angles	Sum Exterior
Triangle	3	540°	180°	360°
Quadrilateral	?	?	?	?
Pentagon	?	?	?	?
Hexagon	?	?	?	?
n-gon	?	?	?	?

10. Use the formula from Exploration 1 and from Step 8 above to write an expression for the sum of the exterior angles of an *n*-gon. Use algebra to simplify the expression. ❖

Use Transparency 29

Exploration 2 Note

When students are ready to fill in column three of the table for Step 9, check to be sure they understand what angles are to be included. They should use *all* interior angles but only one exterior angle at each vertex.

ongoing ASSESSMENT

5. The sum of the exterior angle measures of a polygon is 360°.

8. $180n$°

TEACHING tip

Be sure that students use only one exterior angle from each vertex when completing the chart in Step 9.

ENRICHMENT Students can investigate whether the angle-sum formula would work for concave polygons if we allowed interior angles to have measures greater than 180°. Ask what disadvantages such measures might have.

INCLUSION strategies **English Language Development** The geometry vocabulary presented in this lesson may be difficult for some students. Before teaching the lesson, prepare students by discussing the following words and providing an example of each: *convex polygon, concave polygon, regular polygon, interior polygon angles, exterior polygon angles, pentagon, hexagon, n-gon.* Students should sketch a picture and write the definition for each in their portfolios.

ASSESS

Selected Answers

Odd-numbered Exercises 5–29

Assignment Guide

Core 1–2, 5–13, 16–30

Core Plus 1–8, 12–21, 26–31

Technology

Students may wish to use geometry graphics software for Exercises 6–11.

Error Analysis

If you notice an unusual number of errors in Exercises 6–11, check whether students recall how single tics and double tics are used to indicate congruent angles in diagrams.

Practice Master

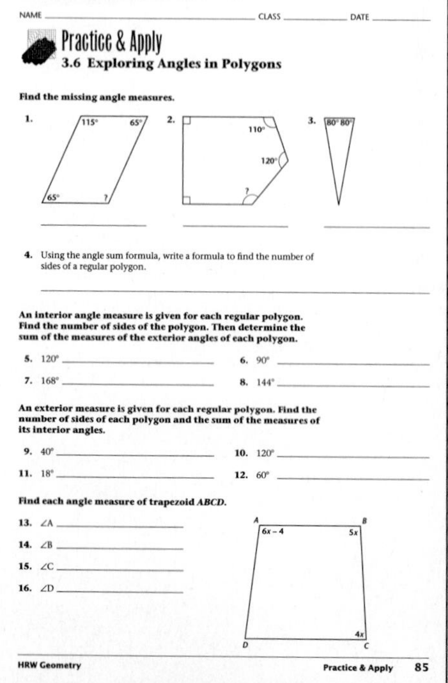

NAME ______ CLASS ______ DATE ______

Practice & Apply

3.6 Exploring Angles in Polygons

Find the missing angle measures.

1. 115° 65° 65° ? ______
2. 110° 120° ? ______
3. 80° 80° ______

4. Using the angle sum formula, write a formula to find the number of sides of a regular polygon.

______

An interior angle measure is given for each regular polygon. Find the number of sides of the polygon. Then determine the sum of the measures of the exterior angles of each polygon.

5. 120° ______
6. 90° ______
7. 168° ______
8. 144° ______

An exterior measure is given for each regular polygon. Find the number of sides of each polygon and the sum of the measures of its interior angles.

9. 40° ______
10. 120° ______
11. 18° ______
12. 60° ______

Find each angle measure of trapezoid *ABCD*.

13. ∠A ______
14. ∠B ______
15. ∠C ______
16. ∠D ______

A B 6x – 4 5x 4x D C

HRW Geometry — Practice & Apply 85

EXERCISES & PROBLEMS

Communicate

1. What is the formula that gives the sum of the interior angle measures of an n-gon?
2. Describe an exterior angle of a polygon.
3. Is it possible to draw a quadrilateral such that three of its angles each have a measure of 60°? Give a reason for your answer.
4. Explain why a pentagon can have 5 obtuse angles.

Regular hexagons are the most efficient shape for bees to use when constructing a hive.

Practice & Apply

5. In the figure, find the values of x, y, and z.

$x = 78°$
$y = 102°$
$z = 127°$

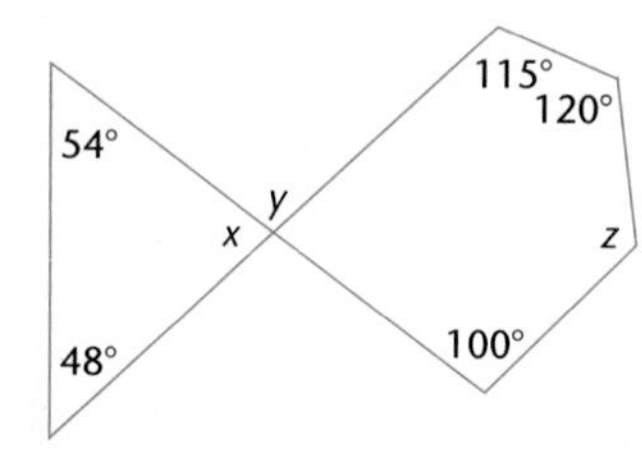

Find the missing angle measures in Exercises 6–11.

6. 90° 75° 85° ?

110°

7. 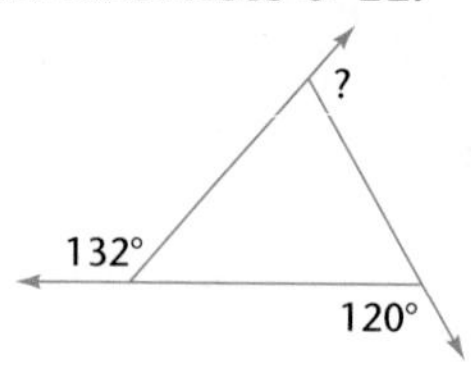

108°

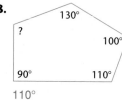

8. 130° ? 100° 90° 110°

110°

9. 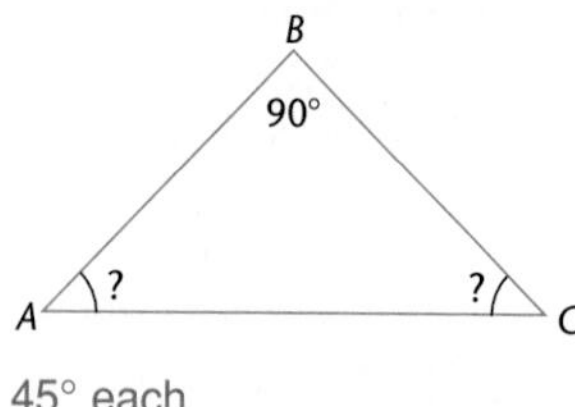

45° each

10. 120°

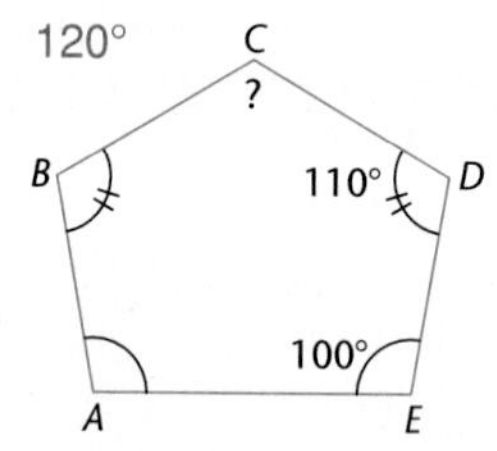

11. 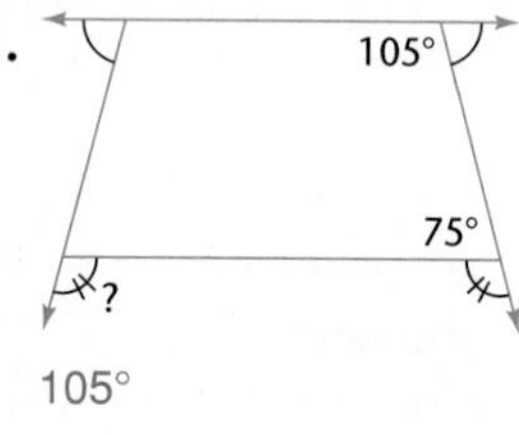

105°

12. Make a sketch and determine the measure of an interior angle for each of the following: a regular quadrilateral, a regular pentagon, and a regular hexagon. Quadrilateral, 90°; Pentagon, 108°; Hexagon, 120°
13. Determine the measure of an exterior angle for each of the figures in Exercise 12. Quadrilateral, 90°; Pentagon, 72°; Hexagon, 60°

Using Patterns You can connect an interior point of an n-gon to each vertex to form n triangles. The sum of their angles in $180n$. Subtract 360° for the angles that meet at the interior point. For the exterior angles, pick an exterior point P. In one direction around the polygon draw rays with endpoint P parallel to each side. Show how the resulting figure can be used in proving that the exterior angles have a sum of 360°.

In Exercises 14 and 15, use your results from Exercises 12 and 13.

14. Write a formula for finding the measure of one interior angle of a regular n-gon. $\frac{180(n-2)}{n}$

15. Write a formula for finding the measure of one exterior angle of a regular n-gon. $\frac{360}{n}$

For each regular polygon in Exercises 16–18, an interior angle measure is given. **Find the number of sides of the polygon. Then determine the sum of the measures of the exterior angles of each polygon.**

16. 135° 8; 360° **17.** 150° 12; 360° **18.** 165° 24; 360°

For each regular polygon in Exercises 19–21, an exterior angle measure is given. **Find the number of sides of each polygon and the sum of the measures of its interior angles.**

19. 60° 6; 720° **20.** 36° 10; 1440° **21.** 24° 15; 2340°

Algebra Find each angle measure of quadrilateral *QUAD*.

22. $\angle Q$ 36° **23.** $\angle U$ 72°

24. $\angle A$ 108° **25.** $\angle D$ 144°

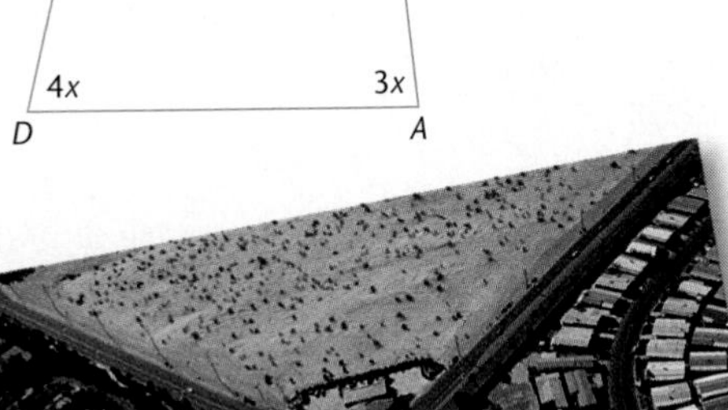

Look Back

26. Describe how the distance from a point to a line is determined. **[Lesson 1.4]**

27. In the photo, how many pairs of congruent angles are formed? What are these angle pairs called? **[Lesson 2.2]**

28. What is the meaning of the Transitive Property of Congruence? **[Lesson 2.5]**

29. Explain the meaning of the Parallel Postulate. **[Lesson 3.5]**

30. The measure of an exterior angle of a triangle is equal to the _?_ of the remote interior angles. **[Lesson 3.6]** sum

Look Beyond

31. Find the sum of the measures of the vertex angles of a 5-pointed star without using a protractor. Use the following hints.

- Recall the following theorem, which you proved in Lesson 3.5:

  The measure of an exterior angle of a triangle is equal to the sum of the measures of its remote interior angles.

- Using small triangles within the larger drawing, explain why two of the angles are labeled as 1 + 4 and 3 + 5.

5 1 4 3+5 1+4 3 + 5 2 1 + 4 3

---

26. Distance from a point to a line is measured along the perpendicular segment from the point to the line.

27. two pairs; vertical angles

28. If Figure A ≅ Figure B and Figure B ≅ Figure C, then Figure A ≅ Figure C.

29. For any line, there is one and only one line passing through a point not on the line that is parallel to the given line.

---

alternative ASSESSMENT

Authentic Assessment

Have students use a diagram to show how to find the sum of the interior angles and the sum of the exterior angles of an arbitrary polygon.

Look Beyond

Nonroutine problem. The result that students will discover in Exercise 31 is an excellent example of how deductive reasoning can lead to conclusions that are far from obvious at first glance.

31. The angle labeled "1 + 4" is an exterior angle for the triangle with remote interior angles labeled "1" and "4". The angle labeled "3 + 5" is an exterior angle for the triangle with remote interior angle labeled "3" and "5". The sum of the vertex angles of the star is 180°.

Use Transparency 30

Technology Master

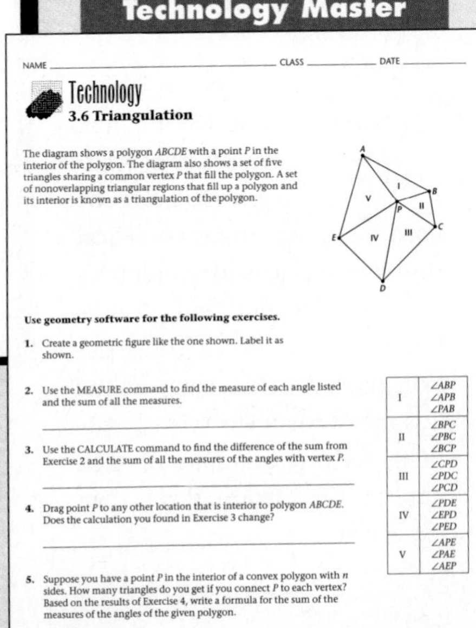

NAME ________ CLASS ________ DATE ________

Technology
3.6 Triangulation

The diagram shows a polygon *ABCDE* with a point *P* in the interior of the polygon. The diagram also shows a set of five triangles sharing a common vertex *P* that fill the polygon. A set of nonoverlapping triangular regions that fill up a polygon and its interior is known as a triangulation of the polygon.

Use geometry software for the following exercises.

1. Create a geometric figure like the one shown. Label it as shown.

2. Use the MEASURE command to find the measure of each angle listed and the sum of all the measures.

3. Use the CALCULATE command to find the difference of the sum from Exercise 2 and the sum of all the measures of the angles with vertex *P*.

4. Drag point *P* to any other location that is interior to polygon *ABCDE*. Does the calculation you found in Exercise 3 change?

5. Suppose you have a point *P* in the interior of a convex polygon with *n* sides. How many triangles do you get if you connect *P* to each vertex? Based on the results of Exercise 4, write a formula for the sum of the measures of the angles of the given polygon.

6. Show algebraically that your answer to Exercise 5 agrees with your answer to Question 5 of Exploration 1 of Lesson 3.6.

I	$\angle ABP$ $\angle APB$ $\angle PAB$
II	$\angle BPC$ $\angle PBC$ $\angle BCP$
III	$\angle CPD$ $\angle PDC$ $\angle PCD$
IV	$\angle PDE$ $\angle EPD$ $\angle PED$
V	$\angle APE$ $\angle PAE$ $\angle AEP$

HRW Geometry Technology 101

PREPARE

Objectives

- Define *midsegment of a triangle* and *midsegment of a trapezoid.*
- Develop and use formulas from the properties of triangle and trapezoid midsegments.

RESOURCES

- Practice Master 3.7
- Enrichment Master 3.7
- Technology Master 3.7
- Lesson Activity Master 3.7
- Quiz 3.7
- Spanish Resources 3.7

Assessing Prior Knowledge

1. If M is the midpoint of AB and $MA = 7$, then $AB =$ __?__ and $MB =$ __?__. [14; 7]
2. What is the average of 15 and 29? [22]
3. If the average of x and 20 is 16.25, what is x? [$x = 12.5$]

TEACH

Point out that properties of midsegments can be used to solve many practical problems in carpentry, architecture, model building, bridge building, and other construction-related fields. Later in the lesson you may wish to ask students for any examples of such applications from their own experience.

LESSON 3.7 Exploring Midsegments of Triangles and Trapezoids

Why *Can you think of a way to determine the width of the pyramid halfway up the stairs without actually measuring it at that point? The midsegment conjectures you make in this lesson will suggest a way.*

The Temple of the Giant Jaguar in Guatemala has four sloping trapezoidal sides. Along one side of the temple there is a stairway that rises 150 feet.

Exploration 1 Midsegments of Triangles

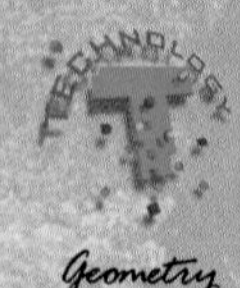

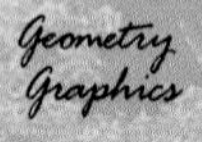

You will need
Geometry technology or
Ruler and protractor

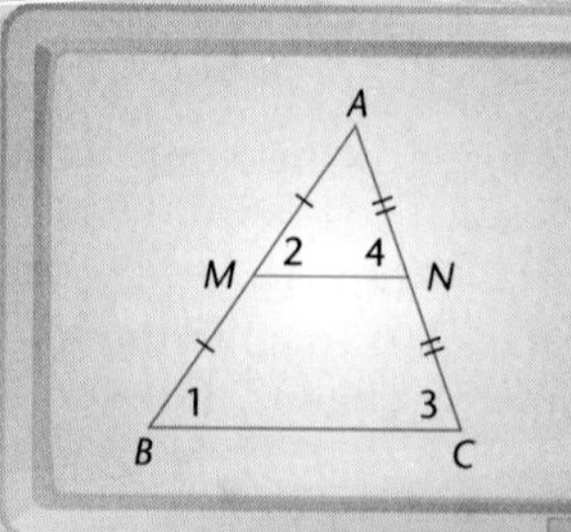

1. Draw $\triangle ABC$, which may be any shape and size. Find the midpoints of sides $\overline{AB}$ and $\overline{AC}$. Label them M and N as shown. Then draw the midsegment $\overline{MN}$.
2. Measure segments $\overline{MN}$ and $\overline{BC}$. What is the relationship between their lengths? Compare your results with those of your classmates.
3. Measure $\angle 1$ and $\angle 2$. What do you notice? Also measure $\angle 3$ and $\angle 4$. What do these measurements suggest about the relationship of $\overline{BC}$ and $\overline{MN}$? What postulate or theorem from an earlier lesson allows you to draw your conclusion?
4. Use your results from Steps 2 and 3 to write a conjecture about midsegments of triangles. ❖

Hands-On Strategies

You can have students use square grid paper for Explorations 1 and 2. You may wish to suggest that they make the bases of the figures horizontal and that they pick vertices at points where horizontal and vertical grid lines intersect.

APPLICATION

José is on his school swim team. During the summer he enjoys training on a small lake near his house. To evaluate his progress he needs to know the distance across the lake as shown. How can he use the conjecture from Exploration 1 to find this distance?

José can select a point A where he can measure segments $\overline{AX}$ and $\overline{AY}$. Then he can find the midpoints of $\overline{AX}$ and $\overline{AY}$ and measure the distance between them. Since this distance is one-half XY, José can double the number to find his answer. ❖

Exploration 2 *Midsegments of Trapezoids*

Geometry Graphics

You will need
Geometry technology or
Ruler and protractor

1. Draw trapezoid $ABCD$, which may be of any shape and size. Find the midpoints M and N of the nonparallel sides, and connect them. Draw the midsegment $\overline{MN}$.

A B M N D C 1 2 3 4 5 6

2. Measure the lengths of $\overline{AB}$, $\overline{DC}$, and $\overline{MN}$.
3. Find the average length of the bases. Compare your results with those of your classmates. What do you notice about the length of the midsegment?
4. Measure $\angle 1$ and $\angle 2$. What do you notice? Also measure $\angle 4$ and $\angle 5$. What do these measurements suggest about the relationship of $\overline{CD}$ and $\overline{MN}$?
5. Measure $\angle 2$ and $\angle 3$. What do you notice? Also measure $\angle 5$ and $\angle 6$. What do these measurements suggest about the relationship of $\overline{AB}$ and $\overline{MN}$?
6. Use your results from Steps 2–5 to write a conjecture about midsegments of trapezoids. ❖

Exploration 1 Notes

Students should be aware of the fact that their conjectures in Step 4 cannot be accepted as true statements until proved deductively.

Ongoing Assessment

4. **The midsegment of a triangle is half the length of, and parallel to, the third side of the triangle.**

Exploration 2 Notes

Here, as in Exploration 1, students should be aware that their response to Step 6 results in a conjecture that has yet to be proved as a theorem.

Ongoing Assessment

6. **The midsegment of a trapezoid is parallel to its bases, and its length is the average of the base lengths.**

ENRICHMENT Trapezoidal shapes are prevalent in many ancient pyramid ruins. As a research project, have students find various examples of trapezoids in ancient structures.

INCLUSION strategies **Using Discussion** Some students may need to be reminded about the definition of a trapezoid. Place a triangle, a trapezoid, and a parallelogram on the board. Ask students to discuss the differences between the figures and develop definitions of each for their portfolios. Next, introduce the concept of midsegment on the triangle and trapezoid.

Cooperative Learning

In Explorations 1 and 2, different members of each group can experiment with their own triangle or trapezoid. Encourage students to use a wide variety of shapes and sizes.

Exploration 3 Note

Students should draw each new trapezoid as the length of base AB is reduced. As the length of AB shrinks to 0, they will see clearly that the trapezoid starts to look more and more like a triangle.

Exploration 3 is an introduction to the mathematical concept of limit.

Use Transparency 31

APPLICATION

Archaeology The base of the Temple of the Giant Jaguar pyramid is a square 150 feet on a side. The top is a square 40 feet on a side. What is the width of the pyramid at a point midway between the base and the top?

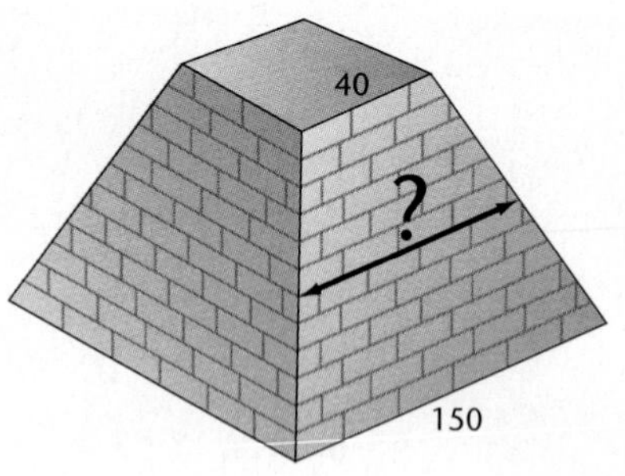

Solution ➤

Use the trapezoid midsegment conjecture:

$$\begin{aligned}\text{Length of midsegment} &= \tfrac{1}{2}(\text{base } 1 + \text{base } 2)\\ &= \tfrac{1}{2}(40 + 150)\\ &= \tfrac{1}{2}(190)\\ &= 95 \text{ ft}\end{aligned}$$ ❖

Exploration 3 *Making the Connection*

You will need

Geometry technology or
Graphics calculator
Ruler

1 Draw trapezoid $ABCD$, which may be of any size and shape, with midsegment $\overline{MN}$. Then fill in a table like the one below by gradually reducing the length of one of the bases. (Use your own measurements for AB and DC.)

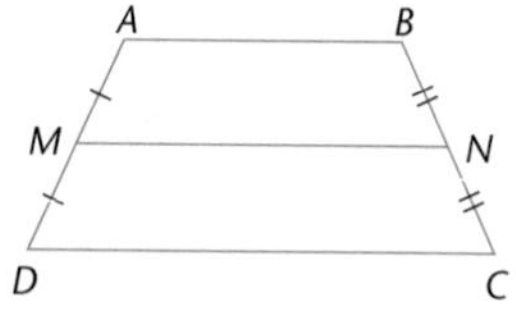

DC	AB	MN
6	5	?
6	4	?
6	3	?
6	2	?
6	1	?
6	.5	?
6	.1	?

You can use the table feature of a graphics calculator to find the values of MN as AB approaches zero. Enter the function as $y = (6 + x) \div 2$.

2 As AB shrinks to 0, what type of figure does the trapezoid become?

3 Write a formula for the length of the midsegment of a "trapezoid" with the measure of one base equal to 0. How does your formula relate to the Triangle Midsegment Conjecture?

Ongoing Assessment

2. a triangle
3. $\frac{1}{2}(\text{Base } 1 + 0) = \frac{1}{2}(\text{Base } 1)$; **The formula is the same as the one used for computing a triangle midsegment.**

Reteaching the Lesson

Hands-On Strategies

You may wish to do Exploration 1 as a paper-folding activity. To compare the length of the triangle midsegment with that of the third side, you can make appropriate marks along the edge of a 3 × 5 card. The conjecture for trapezoid midsegments can be derived from the conjecture for trapezoids by drawing a diagonal of the trapezoid.

EXERCISES & PROBLEMS

Communicate

1. How could you use the Trapezoid Midsegment Conjecture to find the length of the midsegment of a triangle?
2. Why is a triangle a "limiting case" of a trapezoid?
3. In the application on page 155, the triangle midsegment conjecture was used to find a distance that could not be measured directly. Describe a situation where the Trapezoid Midsegment Conjecture would be needed instead?
4. Do you think the midsegment conjectures would be useful in measuring a vertical distance, such as the height of a tree or a cliff? Explain your answer.
5. Suppose a student gave the following rule for finding the length of the midsegment of a trapezoid: "First you subtract the shorter base from the longer base. Then you take half the difference and add it to the shorter base." Would this method work? Discuss.

Practice & Apply

Find the indicated measures.

6. $AB = ?$ 40

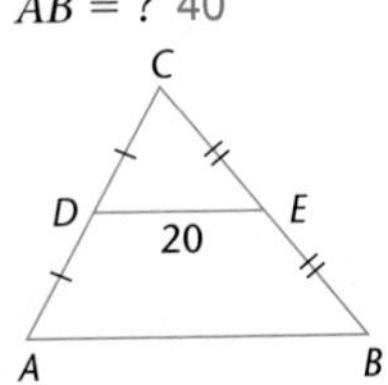

7. $DE = ?$ 25

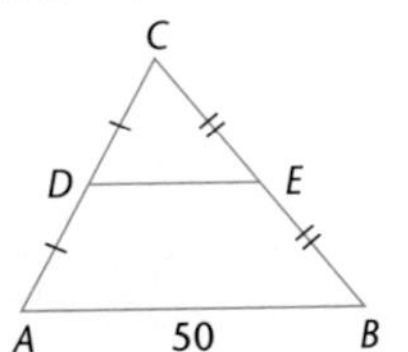

8. $EF = ?$ 45

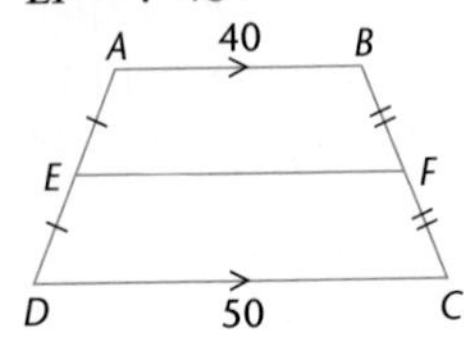

9. $AB = ?$ 80

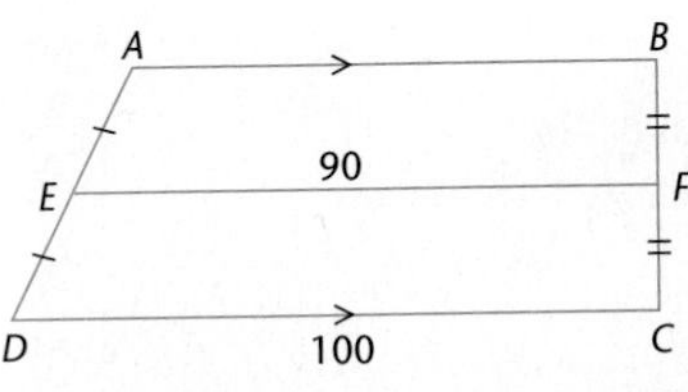

10. $HI = ?$ 30
$FG = ?$ 20
$DE = ?$ 10

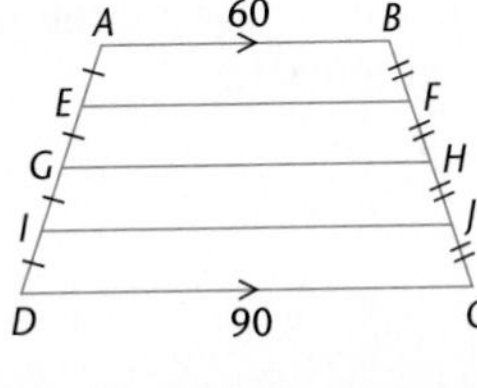

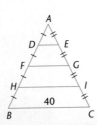

11. $EF = ?$ 67.5
$GH = ?$ 75
$IJ = ?$ 82.5

ASSESS

Selected Answers

Odd-numbered Exercises 7–31

ASSIGNMENT GUIDE

Core 1–4, 6–17, 22–32

Core Plus 1–26, 33–34

Technology

Exercises 18–21 can be studied with the aid of geometry graphics software.

Error Analysis

If students make errors in Exercises 6–11, they may be misapplying the midsegment conjectures. In Exercises 6 and 9, check to make sure they can write and solve appropriate equations. In Exercises 10 and 11, students may need a hint about which lengths to find first.

Alternative Assessment

Authentic Assessment

Ask students to write a short paragraph to explain how the midsegment of a trapezoid is related to the bases (the parallel sides) of the trapezoid. Ask for a second short paragraph that explains what happens if one base of the trapezoid stays the same while the other base shrinks to a point.

12. The lengths of the other segments can be found by multiplying 40 by $\frac{3}{4}$, $\frac{2}{4} = \frac{1}{2}$, and $\frac{1}{4}$.

13. The successive differences between the segment lengths are the same, 7.5.

16. $BC \parallel DE \parallel FG$ with $BC = \frac{1}{3}FG$ and $DE = \frac{2}{3}FG$.

The answers to Exercises 14, 15, 17 and 22 can be found in Additional Answers beginning on page 727.

Practice Master

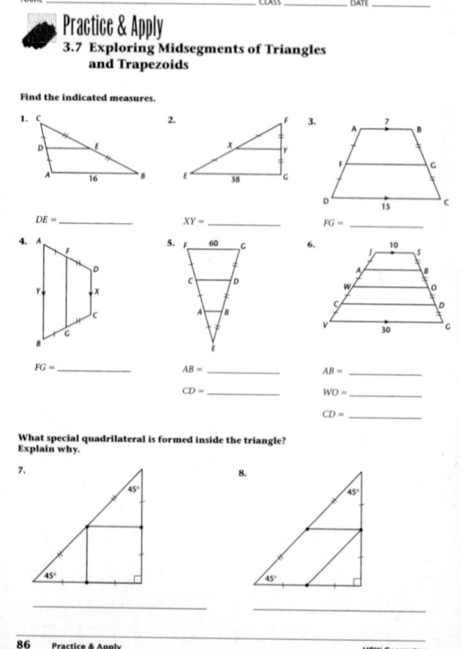
NAME ______ CLASS ______ DATE ______

Practice & Apply
3.7 Exploring Midsegments of Triangles and Trapezoids

Find the indicated measures.

1. DE = ______
2. XY = ______
3. FG = ______
4. FG = ______
5. AB = ______ CD = ______
6. AB = ______ WO = ______ CD = ______

What special quadrilateral is formed inside the triangle? Explain why.

7. ______
8. ______

86 Practice & Apply HRW Geometry

12. What pattern do you observe in the lengths of the segments in Exercise 10? Describe it.

13. What pattern do you observe in the lengths of the segments in Exercise 11? Describe it.

14. How would you find the lengths of the parallel segments of a triangle if its sides were divided into 2 congruent parts? 4 congruent parts? 16 congruent parts? *n* congruent parts? Is the rule the same for all?

15. Repeat Exercise 14 for trapezoids, instead of triangles.

16. The sides of $\triangle AFG$ are each divided into three congruent parts, with the points connected as shown. Write a conjecture about the connecting segments.

17. The figure at the right is formed by connecting the midpoints of all three sides. Find the perimeters of the inner and outer triangles. What is their relationship?

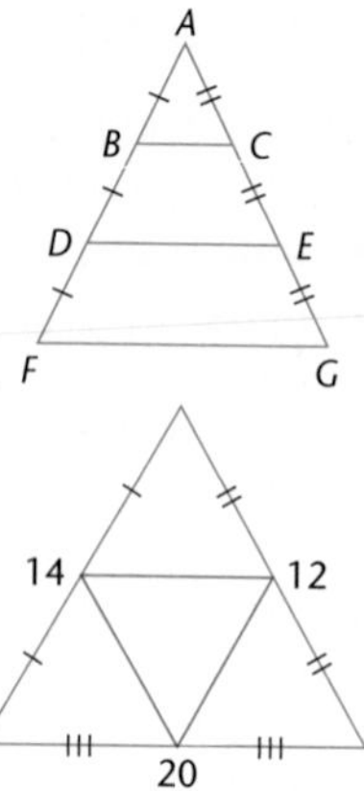

In Exercises 18–21 write an informal argument that explains why each special quadrilateral is formed.

18. Draw a scalene triangle. Connect the midpoints of two of the sides to the midpoint of the third side. What special quadrilateral is formed inside the triangle? Explain why.

19. Draw a right triangle that is not isosceles. Connect the midpoints of the two legs to the midpoint of the hypotenuse. What special quadrilateral is formed inside the triangle? Explain why.

20. Draw an isosceles triangle. Connect the midpoints of the two equal sides to the midpoint of the base. What special quadrilateral is formed inside the triangle? Explain why.

21. Draw an isosceles right triangle. Connect the midpoints of the legs to the midpoints of the hypotenuse. What special quadrilateral is formed inside the triangle? Explain why.

22. Structural Engineering Is $\overline{FC}$ the midsegment of trapezoid ABDE? Explain your answer. The overall trapezoidal shape of the tower gives it a broad, stable base. How are the trapezoids strengthened?

18. parallelogram, because the opposite sides are parallel

19. 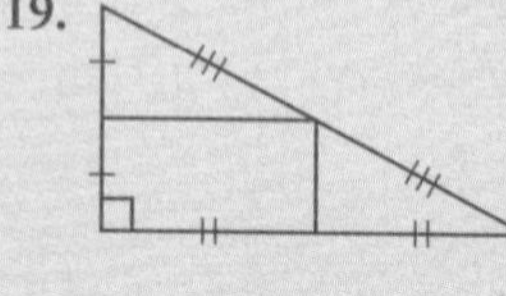
rectangle, because opposite sides are parallel and adjacent sides meet at right angles

20. 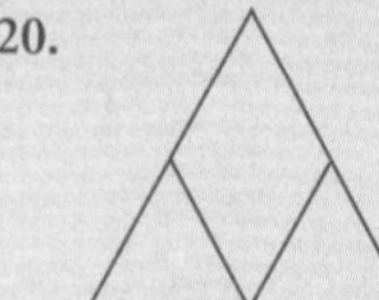
rhombus, because the opposite sides are parallel and all sides are congruent

21. 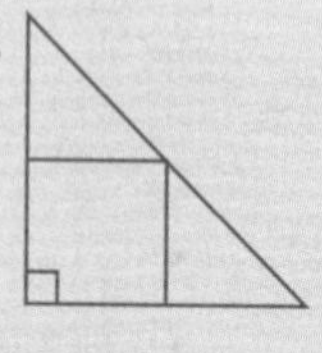
square, because opposite sides are parallel, all sides are congruent, and adjacent sides meet at right angles

Flood Control Exercises 23–26 refer to the cross section of a dam shown that has the shape of a trapezoid. Determine the following:

23. The length of the cross section halfway up 380 m

24. The length of the cross section three-quarters of the way up 418 m

25. If the floodgates should be opened when the distance of the waterline across the dam is 400 meters or more, should they be opened before the water reaches the three-quarter level? Yes, since 400 m < 418 m

26. Is it possible for the median of a trapezoid to be congruent to its bases? Explain your answer. No, if it were, the figure would be a parallelogram.

The cross section is 456 m long across the top and 304 m long across the bottom.

Look Back

Answer true or false for Exercises 27–29. [Lesson 3.2]

27. All rhombuses are squares. False

28. All rectangles are parallelograms. True

29. All squares are rectangles. True

In Exercises 30–32, find the missing angle measures. [Lesson 3.5]

30.

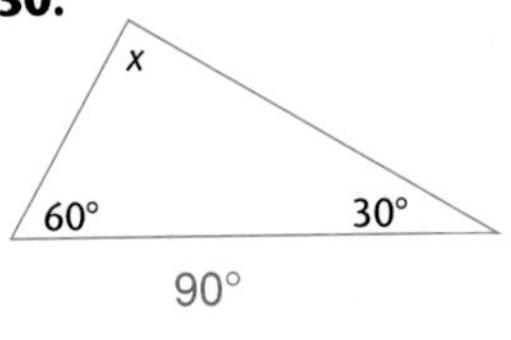

31.

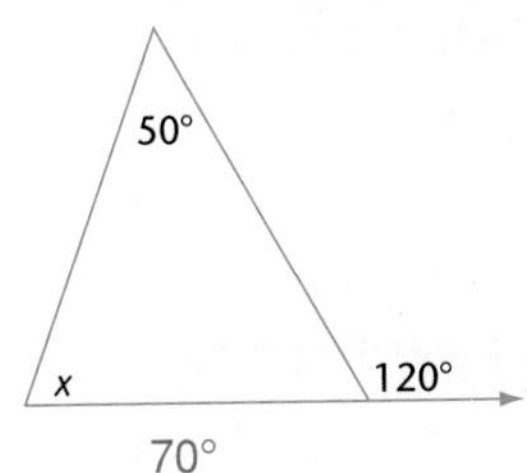

32.

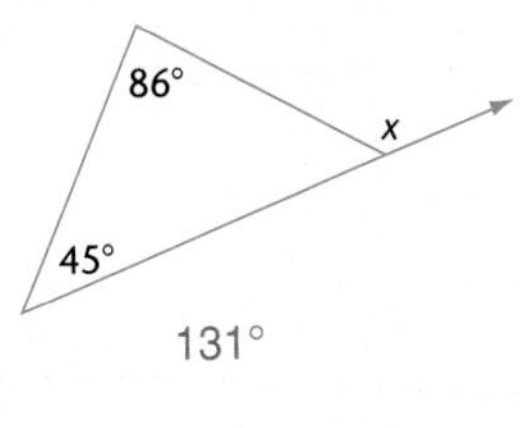

Look Beyond

Exercises 33 and 34 foreshadow concepts about congruent triangles that will be studied in detail in Chapter 4.

Look Beyond

33. Draw a triangle and connect the midpoints of the sides. Cut out each of the four triangles that are formed, and formulate a conjecture about them.

34. Use the Triangle Midsegment Conjecture as a basis for an informal argument that each triangle has sides of the same length. (Hint: It may be helpful to write the measures of the inner and outer triangle sides directly onto the figure you have drawn.)

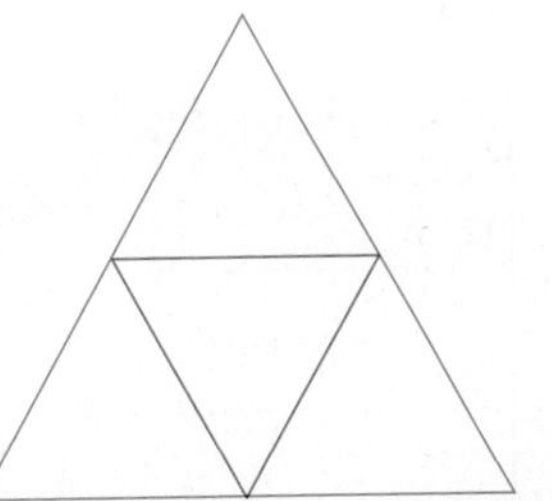

33. The four small triangles are congruent.

34. By the Midsegment Conjecture, each side of each of the small triangles is half the length of a corresponding side of the bigger triangle, so the corresponding sides of the small triangles are congruent.

Technology Master

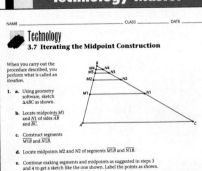

NAME ______________ CLASS ________ DATE ________

Technology

3.7 Iterating the Midpoint Construction

When you carry out the procedure described, you perform what is called an *iteration*.

1. a. Using geometry software, sketch $\triangle ABC$ as shown.

   b. Locate midpoints $M1$ and $N1$ of sides $\overline{AB}$ and $\overline{BC}$.

   c. Construct segments $\overline{M1B}$ and $\overline{N1B}$.

   d. Locate midpoints $M2$ and $N2$ of segments $\overline{M1B}$ and $\overline{N1B}$.

   e. Continue making segments and midpoints as suggested in steps 3 and 4 to get a sketch like the one shown. Label the points as shown.

2. Use the MEASURE command to list the lengths of $\overline{AC}$, $\overline{M1N1}$, $\overline{M2N2}$, $\overline{M3N3}$, and $\overline{M4N4}$.

____

3. Suppose that $\overline{MKNK}$ is the kth segment drawn according to the process described above. Write an equation that tells how long $\overline{MKNK}$ is in relation to $\overline{AC}$.

____

4. Viewing the diagram as a set of stacked trapezoids, locate and mark the midpoints of $\overline{AM1}$ and $\overline{CN1}$. Call the midpoints $\overline{R1}$ and $\overline{S1}$, respectively. Locate and mark the midpoints of $\overline{M1M2}$ and $\overline{N1N2}$, $\overline{M2M3}$ and $\overline{N2N3}$, and so on. Use the MEASURE command to list the lengths of $\overline{R1S1}$, $\overline{R2S2}$, and so on.

____

5. Suppose that $\overline{RKSK}$ is the kth segment drawn according to the process described in Exercise 4. Write an equation that tells how long $\overline{RKSK}$ is in relation to $\overline{AC}$.

____

102 **Technology** **HRW Geometry**

PREPARE

Objectives

- Develop and use the Equal Slope Theorem and the Slopes of Perpendicular Lines Theorem.
- Solve problems involving perpendicular and parallel lines in the coordinate plane using appropriate theorems.

RESOURCES

- Practice Master 3.8
- Enrichment Master 3.8
- Technology Master 3.8
- Lesson Activity Master 3.8
- Quiz 3.8
- Spanish Resources 3.8

Assessing Prior Knowledge

1. In the coordinate plane, to go from (2, 5) to (8, 16), you can go right _?_ units and up _?_ units. [6; 8]
2. Evaluate $\frac{d-c}{b-a}$ for $a = -3$, $b = 10$, $c = 0$, and $d = 39$. [3]
3. If you multiply any nonzero real number by its reciprocal, the product is _?_. [1]

TEACH

The concept of slope has applications in a wide variety of real-world situations. Examples are finding the steepness of stairs, the pitch of a roof, and the grade of a highway. You may wish to ask students to suggest examples of their own.

Alternate Example 1

Find the slope of the segment with endpoints at (–2, 5) and (4, 8). [1/2]

LESSON 3.8 Analyzing Polygons Using Coordinates

why *Mathematics also has a method of indicating steepness. This method, though not as colorful as the names of roller coasters, is precise.*

Amusement parks often have descriptive names for their roller coasters. Names such as "Shocker" or "Wild Thing" give riders an idea about the steepness of the falls they will encounter.

The **slope** of a line or surface tells you how it rises or falls. For example, a roller coaster hill with a slope of 4 is twice as steep as one with a slope of 2. If a slope of 1 describes a 45-degree incline, do you think a hill with a slope of 4 would be very steep?

Slope: A Measure of Steepness

Coordinate geometry is ideal for studying slope. You can define the slope of any line segment on a coordinate plane. The slope is the ratio of the vertical **rise** of the segment to the horizontal **run**.

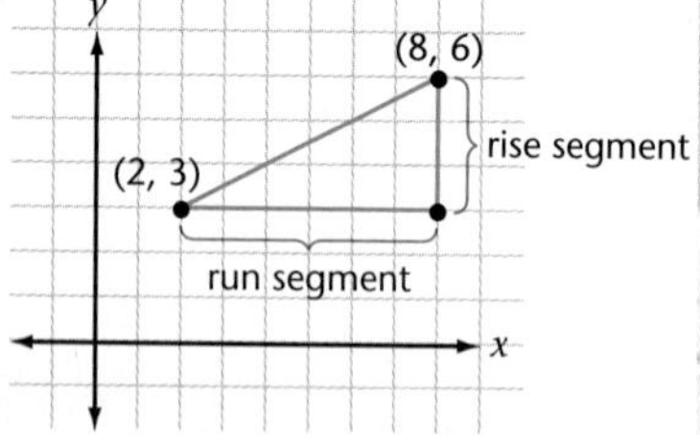

EXAMPLE 1

Find the slope of the segment with endpoints at (2, 3) and (8, 6).

Solution ➤

Draw a right triangle as shown. By counting squares, you can see that the rise is 3 and the run is 6. Thus, the slope is $\frac{3}{6} = \frac{1}{2}$, or 0.5. ❖

ALTERNATIVE teaching strategy

Hands-On Strategies You can use models of staircases made from same-size wooden or plastic cubes. Develop the concepts of rise and run, then move on to the more formal definition of slope. Rest strips of construction paper on two sets of side-by-side stairs to explore parallelism and perpendicularity.

How could you determine the slope of the figure by using just the coordinates—that is, without drawing a picture? The answer is found in the following definition.

SLOPE FORMULA

The slope of a segment $\overline{PQ}$ with endpoints $P(x_1, y_1)$ and $Q(x_2, y_2)$ is the following ratio:

$$\frac{y_2 - y_1}{x_2 - x_1}$$

3.8.1

CRITICAL *Thinking*

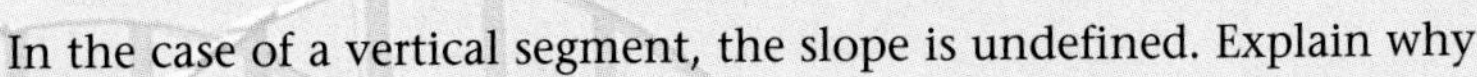

In the case of a vertical segment, the slope is undefined. Explain why.

EXAMPLE 2

Find the slope of segment $\overline{AB}$ with endpoints $A(-2, 3)$ and $B(-5, -3)$.

Solution ➤

$$\text{Slope} = \frac{y_2 - y_1}{x_2 - x_1} = \frac{-3 - 3}{-5 - (-2)} = \frac{-6}{-3} = 2$$ ❖

Parallel and Perpendicular Lines

ALGEBRA *Connection*

Recall from algebra that an equation in the form $y = mx + b$ has m as its slope. Create a system of equations consisting of two lines with the same slope and try to find the solution to the system. What happens? Such equations are called **inconsistent**.

In working with a quadrilateral in a coordinate plane, it is often useful to characterize its sides as parallel or perpendicular. The following four theorems will help you characterize the sides of a quadrilateral in a coordinate plane. The proofs involve using algebraic systems of equations and are not given here.

EQUAL SLOPE THEOREM

If two nonvertical lines are parallel, then they have the same slope. **3.8.2**

THE CONVERSE OF THE EQUAL SLOPE THEOREM

If two nonvertical lines have the same slope, then they are parallel. **3.8.3**

interdisciplinary CONNECTION **Industrial Arts** Many construction projects require implicit or explicit use of ideas related to rise, run, and slope. Students can look for examples in their school and in other familiar buildings. They can find more examples by consulting books on architecture.

Use Transparency **32**

Cooperative Learning

Students can work in small groups to explore the concept of slope by using diagrams. For example, have groups use staircase diagrams drawn on graph paper to illustrate the definitions and theorems in the lesson.

CRITICAL *Thinking*

The run is 0, and division by 0 is undefined.

Alternate Example 2

Find the slope of the segment MN with endpoints $M(-1, 6)$ and $N(-4, 0)$. [2]

TEACHING *tip*

Discuss whether it matters which point is used for (x_1, y_1) and which for (x_2, y_2) when the slope of a particular segment is found. Students will discover that the calculations give the same result for either choice. However, using (x_1, y_1) for the leftmost endpoint can help students to see whether the segment slants up, down, or is level as they go from left to right.

Math Connection Algebra

Have students investigate why two equations of the form $y = mx + b$ that have the same m-values but different b-values cannot have a common solution. They can use the substitution method to see that solving the equation to find a common solution would require dividing by zero, which is not defined as an operation.

TEACHING tip

Students will find it helpful to examine several pairs of segments drawn on graph paper when they investigate slopes of parallel and perpendicular lines. Simplify examples by using segments whose endpoints have integer coordinates.

Alternate Example 3

Plot triangle ABC with $A(2, 2)$, $B(5, 6)$, and $C(-3, 12)$. Draw the midsegment. What kind of triangle is ABC? Does the midsegment seem to be parallel to side AB? **[isosceles; yes]**

The slopes of perpendicular lines also have a special relationship to each other. This relationship is stated in the following two theorems.

SLOPES OF PERPENDICULAR LINES

If two nonvertical lines are perpendicular, then the product of their slopes is -1. **3.8.4**

THE CONVERSE OF SLOPES OF PERPENDICULAR LINES

If the product of the slopes of two nonvertical lines is -1, then the lines are perpendicular. **3.8.5**

The fact that the product of the slopes of perpendicular lines equals -1 reveals another relationship between the slopes. For the product of two numbers to equal -1, one number must be the negative reciprocal of the other. That is, if the slope of a line is $\frac{a}{b}$, then the slope of any line perpendicular to that line has slope $\frac{-b}{a}$. You can use this relationship to verify perpendicularity without multiplying slopes.

EXAMPLE 3

Plot quadrilateral $QUAD$ with $Q(1, 4)$, $U(7, 8)$, $A(9, 5)$, and $D(3, 1)$. What type of quadrilateral is $QUAD$?

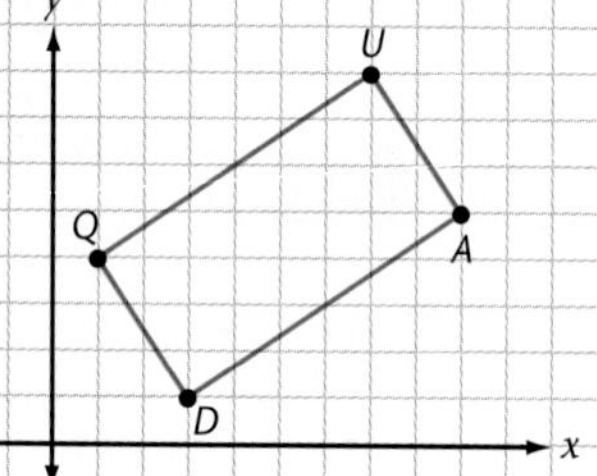

Based on the figure, it appears that $QUAD$ is a parallelogram, and perhaps a rectangle. You can test these conjectures by finding the slopes of each segment.

Slope of $\overline{QU} = \frac{8-4}{7-1} = \frac{4}{6} = \frac{2}{3}$ Slope of $\overline{DA} = \frac{5-1}{9-3} = \frac{4}{6} = \frac{2}{3}$

Slope of $\overline{DQ} = \frac{4-1}{1-3} = -\frac{3}{2}$ Slope of $\overline{UA} = \frac{5-8}{9-7} = -\frac{3}{2}$ ❖

Since the opposite sides of the quadrilateral have the same slope, the figure is a parallelogram.

Since the slopes of adjacent sides of the quadrilateral are negative reciprocals, the figure is a rectangle.

ENRICHMENT Have students use a graphics calculator to graph $y = 5000x - 9000$ using a graphing window with Xmin = −10, Xmax = 10, Ymin = −10, Ymax = 10. Why does the graph appear to be a vertical line? Is the graph really vertical? What is its slope?

INCLUSION strategies **Hands-On Strategies** Have students trace oblique segments and the related rise and run segments for right triangles drawn on graph paper. Use endpoints that have integer coordinates. Students can move the tracing to show that the rise and run are related when the hypotenuse for the tracing is parallel or perpendicular to the original hypotenuse.

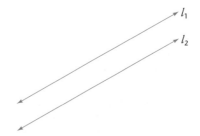

Communicate

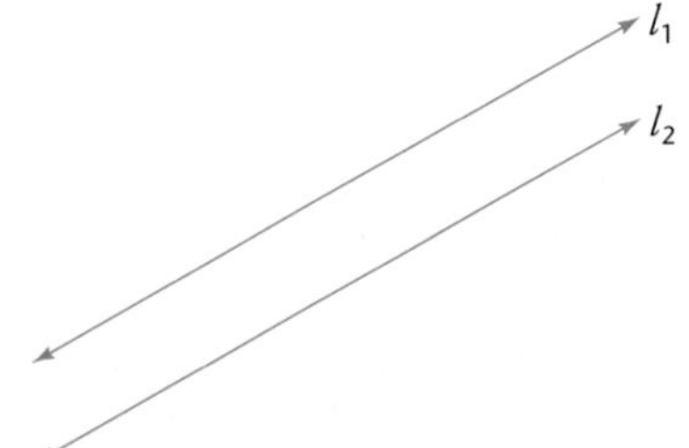

Lines l_1 and l_2 with $l_1 \parallel l_2$ as shown.

1. Suppose l_1 has a slope of 2. What is the slope of l_2? Explain why.

Given: Points $A(2, 5)$ and $B(8, 15)$

2. Draw $\overline{AB}$ on graph paper and show by a diagram how to find the slope by counting squares.

3. Explain how the definition of slope allows you to compute the slope of $\overline{AB}$ without using a diagram.

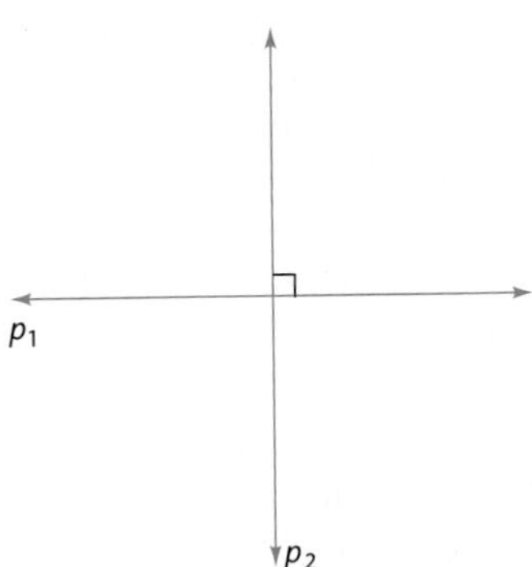

Given: Lines p_1 and p_2 with $p_1 \perp p_2$ as shown.

4. Suppose p_1 has slope m. What is the slope of p_2? Explain why.

5. Explain how you can use the slope to show that a quadrilateral is a rectangle.

Practice & Apply

In Exercises 6–8 use the points shown.

6. Find the slope of $\overline{SA}$. $\frac{-2}{3}$

7. Find the slope of $\overline{BT}$. $\frac{3}{2}$

8. Is $\angle 1$ a right angle? Explain your answer.
Yes. $\overline{SA} \perp \overline{BT}$ because the slopes of $\overline{SA}$ and $\overline{BT}$ are negative reciprocals.

For Exercises 9–12 plot the points on graph paper and connect them to form a polygon. Label the sides of the figure with their slopes. Then identify the type of polygon on the basis of the special properties it has and explain your reasoning. Write and then carry out a plan that proves that your identification is correct.

9. $C(0, -2)$, $R(5, -2)$, $A(5, 6)$, $B(0, 6)$

10. $S(0, 0)$, $A(2, 3)$, $I(5, 3)$, $N(7, 0)$

11. $R(0, 0)$, $A(-3, 4)$, $I(1, 7)$, $N(4, 3)$

12. $T(0, 6)$, $R(3, 9)$ $A(9, 3)$, $P(6, 0)$

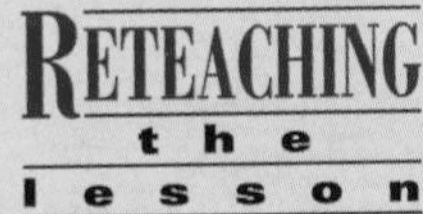

Using Visual Models Use the overhead projector to display a coordinate grid. Then use long pieces of spaghetti or long, very thin drinking straws to show lines in different positions. Compute slopes for several lines, including some that illustrate parallelism and others that illustrate perpendicularity.

ASSESS

Selected Answers

Odd-numbered Exercises 7–29

Assignment Guide

Core 1–16, 20–25, 30

Core Plus 1–19, 30

Technology

For Exercises 18 and 19, students can check their results by using a graphics calculator.

Error Analysis

In Exercises such as 6 and 7, students who make mistakes may have made a computational error or used subscripts inconsistently in applying the definition of slope. Have students check both possibilities to correct their work.

The answers to Exercises 9–12 can be found in Additional Answers beginning on page 727.

ALTERNATIVE ASSESSMENT

Performance Assessment

Have students use wooden or plastic cubes and strips of construction paper to model parallel or perpendicular pairs of lines.

13. One diagonal segment has a slope of 7. The other diagonal segment has a slope of $-\frac{1}{7}$.
Since $(7)(-\frac{1}{7}) = -1$, the diagonals are perpendicular.

The answers to Exercises 14–16 can be found in Additional Answers beginning on page 727.

Practice Master

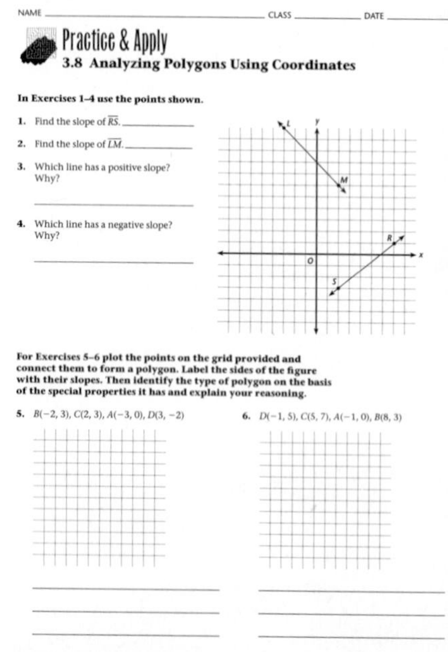
NAME ______ CLASS ______ DATE ______

Practice & Apply

3.8 Analyzing Polygons Using Coordinates

In Exercises 1–4 use the points shown.

1. Find the slope of $\overline{RS}$. ______
2. Find the slope of $\overline{LM}$. ______
3. Which line has a positive slope? Why?
______
4. Which line has a negative slope? Why?
______

For Exercises 5–6 plot the points on the grid provided and connect them to form a polygon. Label the sides of the figure with their slopes. Then identify the type of polygon on the basis of the special properties it has and explain your reasoning.

5. $B(-2, 3)$, $C(2, 3)$, $A(-3, 0)$, $D(3, -2)$
6. $D(-1, 5)$, $C(5, 7)$, $A(-1, 0)$, $B(8, 3)$

HRW Geometry Practice & Apply 87

$C(5, 2)$, and $D(2, -2)$. Use the coordinates to show that the diagonals of the square are perpendicular to each other. Explain your reasoning.

For Exercises 14–16 draw the indicated figure, using the given segment as one of its sides. Show the coordinates of the vertices of the figure and label the sides of the figure with their slopes.

14. Trapezoid *ABCD*. Endpoints of $\overline{AB} = A(3, 5), B(8, 5)$

15. Parallelogram *PLOG*. Endpoints of $\overline{PL} = P(3, 2), L(-1, 5)$

16. Rectangle *RCTG*. Endpoints of $\overline{RC} = R(1, 6), C(3, 2)$

17. Construction A house is 25.0 feet wide. The roof forms an isosceles triangle with the side walls. The height of the rectangular side wall is 10.5 feet, and the height of the peak of the roof is 23.0 feet. City building codes require that the pitch (slope) of a roof may be no more than 0.4. Use coordinates to determine whether the house is violating the building codes.

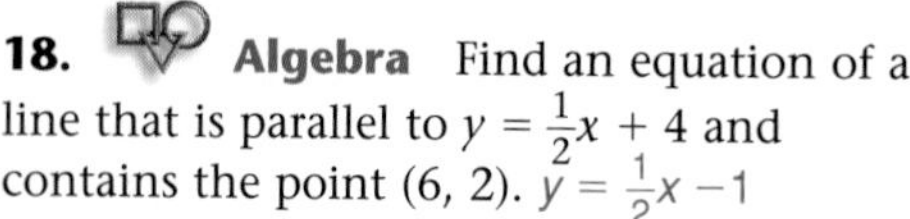

18. Algebra Find an equation of a line that is parallel to $y = \frac{1}{2}x + 4$ and contains the point (6, 2). $y = \frac{1}{2}x - 1$

19. Algebra Find an equation of a line that is perpendicular to $y = 3x - 1$ and contains the point (6, 5). $y = \frac{-1}{3}x + 7$

Look Back

Given $l_1 \parallel l_2$ and $\triangle ABC$ as shown, find the indicated angle measures. [Lesson 3.3]

20. $m\angle 1$ 125°

21. $m\angle 2$ 35°

22. $m\angle EBA$ 20°

23. $m\angle FBC$ 35°

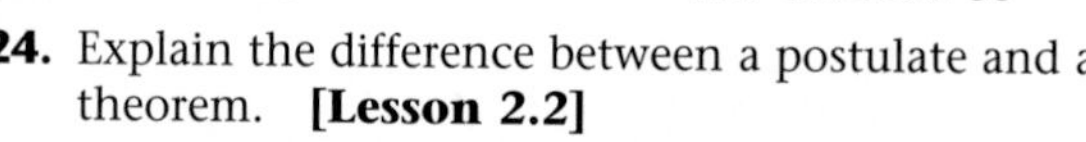

24. Explain the difference between a postulate and a theorem. **[Lesson 2.2]**

25. Upon what postulate does the proof of the Triangle Sum Theorem depend? **[Lesson 3.5]**

Given trapezoid TRAP shown, find the indicated angle measures. [Lesson 3.5]

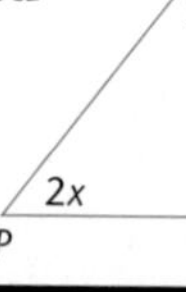

26. $m\angle R$ 90°

27. $m\angle T$ 108°

28. $m\angle P$ 72°

29. $m\angle A$ 90°

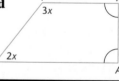

17. The pitch of the roof is 1.0, so the house is in violation of the building codes.

Look Beyond

30. **Portfolio Activity** Another tessellation pattern used by Escher is known as a **rotation tessellation**. Study the "reptile" tessellation and see how many different centers of rotation you can find.

You can make your own rotation tessellation by following the steps below. As in Lesson 3.2, use tracing paper, graph paper, or computer software—and keep trying until you get a figure you like.

a. Start with a regular hexagon. Replace one side of the hexagon, such as $\overline{AB}$, by a broken line or a curve. Rotate the curve about B so that A moves to point C. Draw the broken line or curve between B and C.

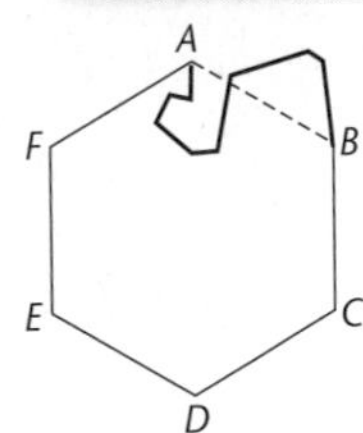

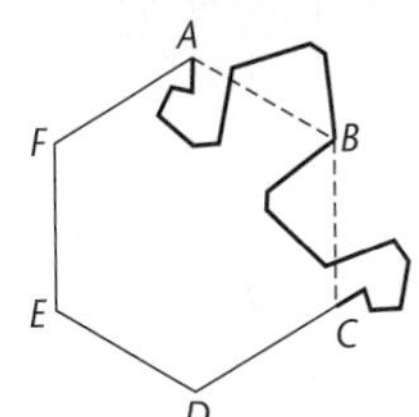

b. Draw a new curve at side $\overline{CD}$ and rotate it around point D to side $\overline{DE}$.

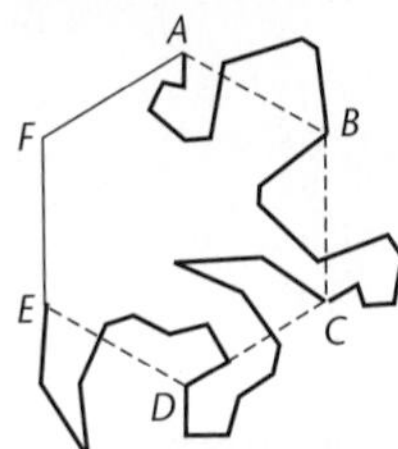

c. Draw a new curve at side $\overline{EF}$ and rotate it as before to side $\overline{FA}$.

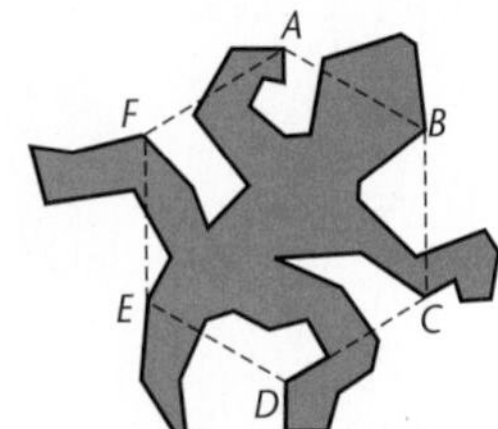

d. Your figure will now fit together with itself on all sides. Make repeated tracings of your figure in interlocking positions. Add pictorial details to make your figure more interesting.

24. A postulate is accepted without proof, whereas a theorem must be proved.

25. The Parallel Postulate. Given a line and a point not on the line, there is one and only one line that contains the given point and is parallel to the given line.

Look Beyond

Exercise 30 provides a useful review of basic understandings about rotations. It also previews ideas for the construction of transformations in Lesson 4.9.

30. Check students' drawings.

Use Transparency 33

Technology Master

NAME ______________________ CLASS __________ DATE __________

Technology
3.8 Making and Using Your Own Coordinate Plane

In Lesson 1.4, you learned how to make your own ruler by using a straightedge and compass. The key feature of a ruler is that the division marks are equally spaced apart. You can use the same notion of equal spacing to construct and use your own coordinate plane. In place of a compass, you can use a segment 1 unit long to lay out a set of horizontal and vertical lines.

1. **a.** In your geometry software, sketch a horizontal line about half way down the computer screen (verify slope = 0).
   b. Sketch a vertical line about half way across the computer screen (verify slope = undefined).
   c. Place a point at the intersection of the lines you drew in parts a and b. (Point represents intersection point of x- and y-axes.)
   d. Sketch a vertical line segment 0.5 in. long.
   e. Make copies of the line you drew in part a. Place them 0.5 in. apart, using your 0.5 in. line segment from part d as a spacer.
   f. Sketch a horizontal line segment 0.5 in. long.
   g. Make copies of the line you drew in part b. Place them 0.5 inches apart, using your 0.5 in. line segment from part f as a spacer.
   h. Finally, delete the 0.5 in. line segments used as spacers. Save the construction from parts a–g in a separate document with a name such as MY GRID.

0.5 in.

2. Explain how you know that each grid block is a square.

____

In your geometry software, (a) sketch line segment $\overline{AB}$, using the origin point identified, (b) use pencil and paper to find the slope of the line segment, and (c) use the SLOPE command to check your answer.

3. $A(0, 0)$ and $B(3, 2)$ ________
4. $A(-1, -2)$ and $B(3, 2)$ ________
5. $A(-1, 3)$ and $B(5, -2)$ ________

6. Sketch quadrilateral $ABCD$ with vertices $A(-2, 1)$, $B(3, 5)$, $C(4, 0)$ and $D(1, -2)$. Use the SLOPE command to find the slopes of $\overline{AB}$, $\overline{BC}$, $\overline{CD}$, and $\overline{DA}$. By moving point D, find the new coordinates of D that makes $ABCD$ a parallelogram.

____

HRW Geometry Technology 103

Focus

By following the steps, students can produces the elaborate string figure shown.

Motivate

String figures made with a loop of string and the hands have long been and intriguing game. Many different patterns and combinations are possible. Have students experiment with their own string figures.

String Figures

The following sequence and the resulting net have appeared in many parts of the world under different names: "Osage Diamonds" among the Osage Indians of North America, "The Calabash Net" in Africa, and "The Quebec Bridge" in Canada. In the United States it is commonly known as "Jacob's Ladder."

1. Start with a piece of string 4 to 5 feet in length. Tie the ends together. Loop the string around your thumb and little fingers as shown.

2. Use your right index finger to pick up the left palm string from below. In a similar way, use your left index finger to pick up the right palm string.

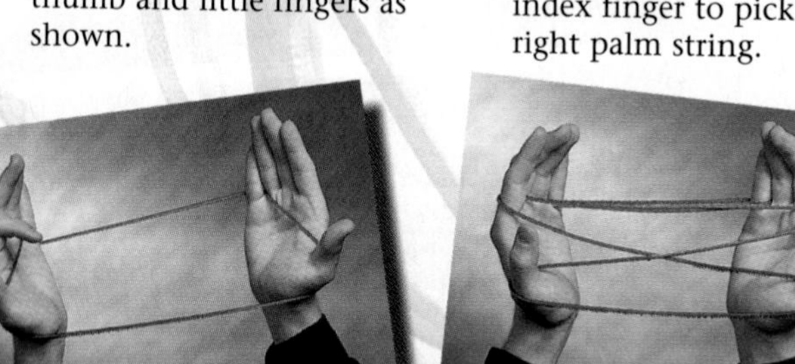

3. Let your thumbs drop their loop. Turn your hands so that the fingers face out. With your thumbs, reach under all the strings and pull the farthest string back toward you.

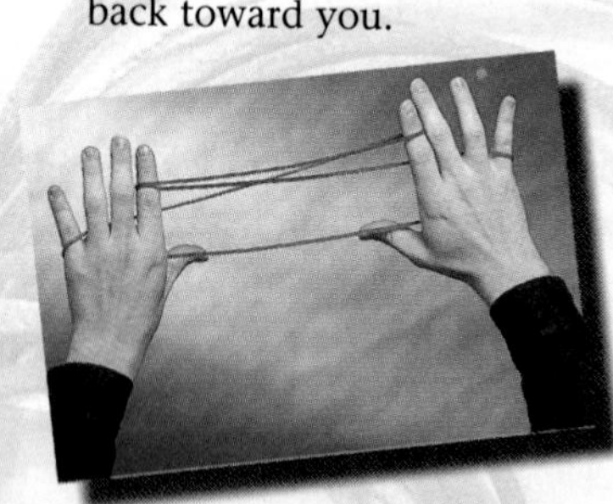

4. With your thumbs, go over the near index-finger string, reach under and get the far index-finger string, and return.

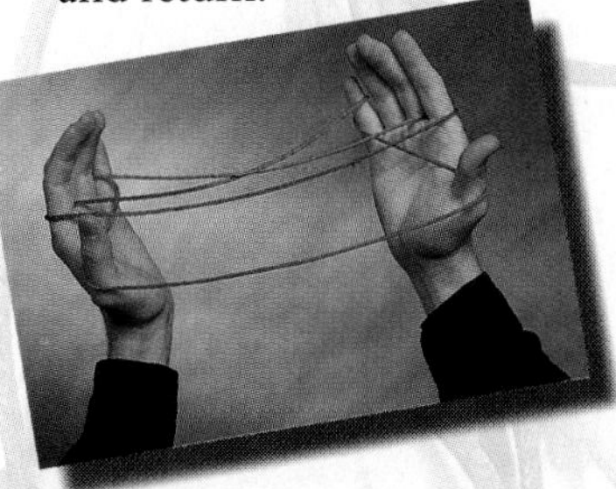

5. Drop the loops from your little fingers. Pass your little fingers over the index-finger string and get the thumb string closest to the little finger from below.

6. Drop the thumb loops. Pass your thumbs over the index-finger strings, get the near little-finger strings from below, and return.

7. Loosen the left index-finger loop with your right hand and place the loop over your thumb. Do the same with the right index-finger loop.

8. Each thumb now has two loops. Using your right hand, lift the lower loop of the left thumb up and over the thumb. Do the same with the lower loop of the right thumb.

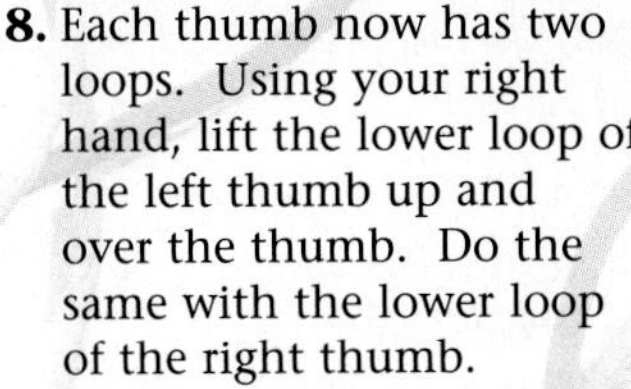

9. Bend your index fingers and insert the tips into the triangles that are near the thumbs.

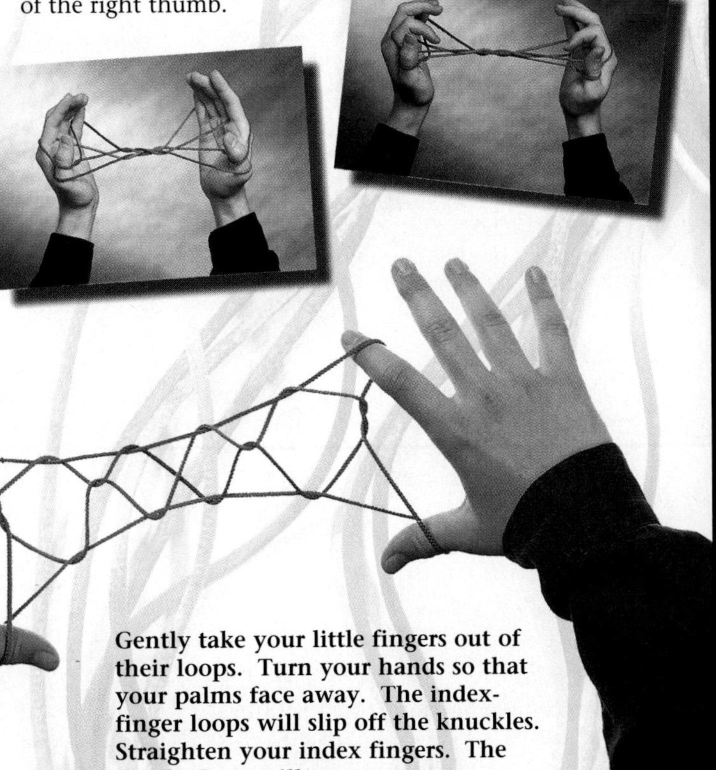

Gently take your little fingers out of their loops. Turn your hands so that your palms face away. The index-finger loops will slip off the knuckles. Straighten your index fingers. The finished net will appear.

Cooperative Learning

Students can work in small groups and create variations on the string figures. Provide students with books which describe other string figures. Students should sketch the final form of their string figures and identify the polygons. Students should also identify the lines of symmetry in their figures.

Discuss

Ask students to share their unique string figures with the rest of the class. Ask the students why some string figures are very elaborate and yet require relatively few steps. Also, ask why string figures have intrigued for so many centuries.

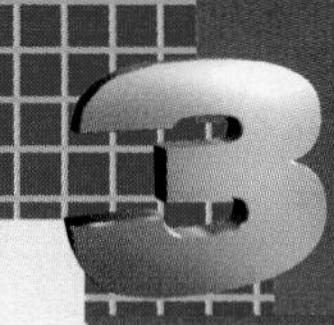

1. 2;

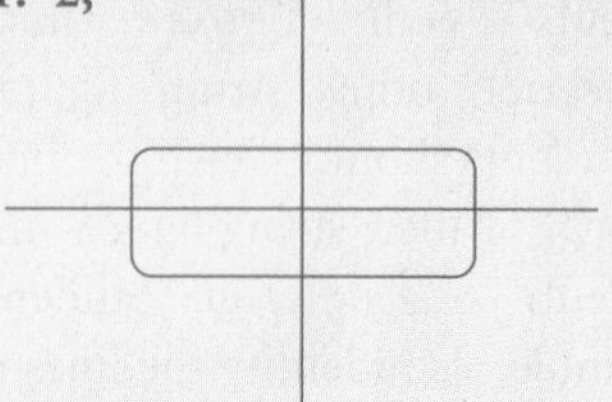

2.

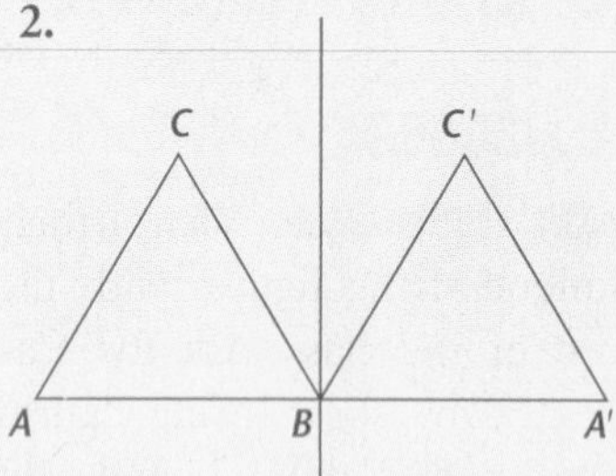

Chapter 3 Review

Vocabulary

alternate exterior angles	133	corresponding angles	133	rhombus	124
alternate interior angles	133	exterior angles	147	rise and run	160
axis of symmetry	117	parallelogram	124	rotational symmetry	118
central angle	119	polygon	116	slope	160
center point	119	rectangle	124	slope formula	161
concave polygon	149	reflectional symmetry	117	square	124
consecutive interior angles	133	regular polygon	119	transversal	132
convex polygon	149	remote interior angles	147	trapezoid	124

Key Skills and Exercises

Lesson 3.1

➤ Key Skills

Identify reflection and rotation of planar figures.

The dotted lines show the axes of symmetry in this triangle.

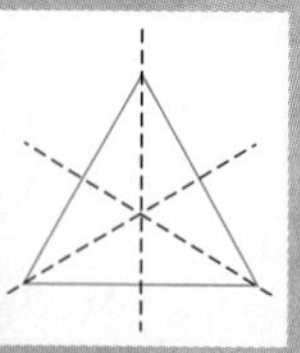

Note that the triangle has rotational symmetry of 120°; that is, rotating the image 120° about its center point makes it coincide with the original image.

Sketch rotation or reflection of a figure.

Sketch a reflection of the pre-image shown over a horizontal axis of symmetry. Use the pre-image to sketch a figure with 90° rotational symmetry.

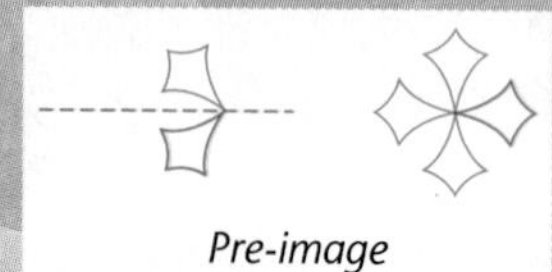
Pre-image

➤ Exercises

1. How many axes of symmetry does the figure below have? Copy the figure and draw all axes of symmetry.

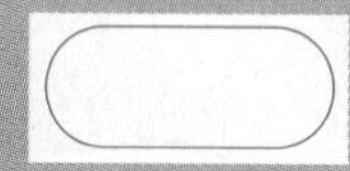

In Exercises 2–3 use the pre-image shown.

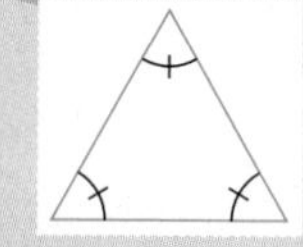

2. Sketch a reflection of the pre-image over a vertical axis of symmetry.
3. Use the pre-image to sketch a new figure that has rotational symmetry.

Lesson 3.2

➤ Key Skill

Conjecture properties of quadrilaterals.

If a problem gives a "quadrilateral," all you know is that it is a four-sided polygon. If a problem gives a "square," you have much more information to solve that problem. You know that the polygon has four sides of equal length, four vertex angles of 90°, and perpendicular diagonals of equal length.

3. **Sample answers:**

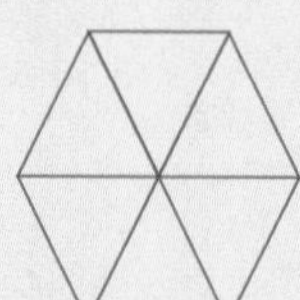

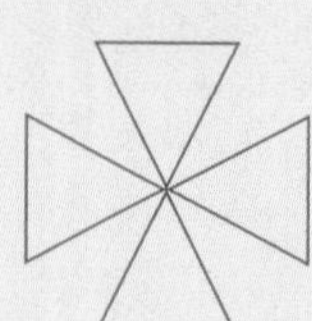

➤ Exercises

In Exercises 4–6 refer to figure *ABCD*.

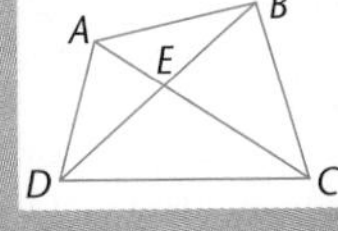

4. Given: *ABCD* is a rhombus. Find the measure of $\angle AEB$.

5. Given: *ABCD* is a rectangle. Find the measure of $\angle BCD$.

6. Given: *ABCD* is a rectangle. Name any sets of congruent line segments.

Lesson 3.3

➤ Key Skill

Identify angle relationships in parallel lines and transversals.

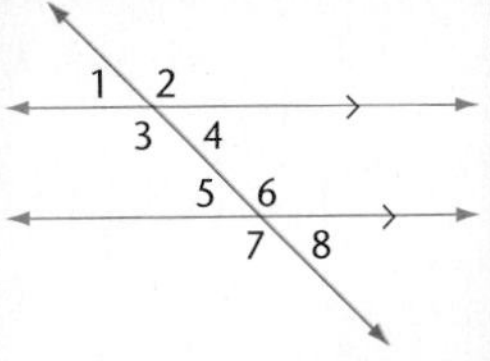

If you know one angle measure in the figure, you can determine all the others.

If $m\angle 4 = 40°$, then $m\angle 1$, $m\angle 5$, and $m\angle 8 = 40°$; $m\angle 2$, $m\angle 3$, $m\angle 6$, and $m\angle 7 = 140°$.

➤ Exercises

In Exercises 7–8 refer to the parallel lines and transversal above. If $m\angle 3 = x$ and $m\angle 4 = x - 60°$, find the angle measures.

7. $m\angle 5$ **8.** $m\angle 7$

Lesson 3.4

➤ Key Skill

Use converse theorems to show that lines are parallel.

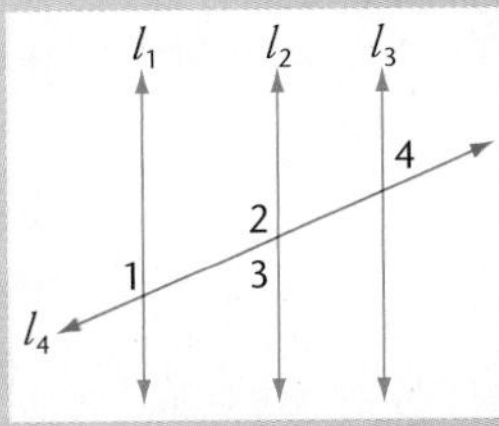

In the figure, if $m\angle 1 = m\angle 2$, then lines l_1 and l_2 are parallel by the converse of the Corresponding Angles Theorem. Similarly, if $m\angle 3 = m\angle 4$, then lines l_2 and l_3 are parallel.

➤ Exercise

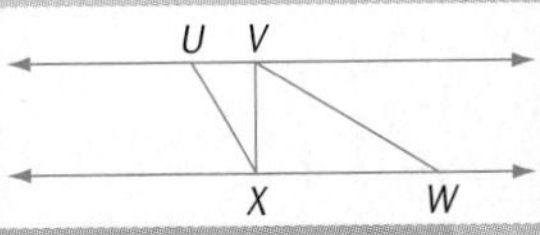

9. $\angle UVX$ and $\angle WXV$ are right angles. Is this enough information to show that $\overleftrightarrow{UV}$ and $\overleftrightarrow{XW}$ are parallel? Justify your answer.

Lesson 3.5

➤ Key Skills

Use triangle sum theorems to find angle measures.

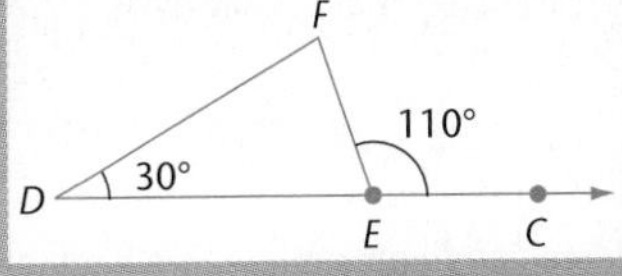

What is the measure of $\angle F$?

By the Exterior Angle Theorem $m\angle D + m\angle F = m\angle CEF$.

$30° + m\angle F = 110°$

$m\angle F = 110° - 30°$

$m\angle F = 80°$

➤ Exercises

In Exercises 10–12 refer to the figure on the right. $\overline{MP}$ is parallel to $\overline{LQ}$; $m\angle NOP = 105°$; $m\angle MNO = 50°$. Find the angle measure.

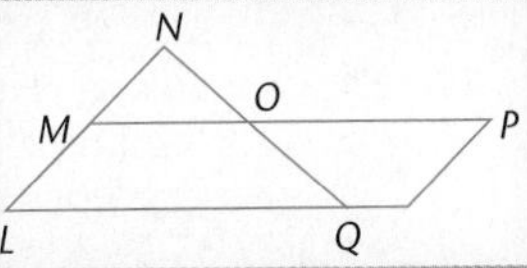

10. $m\angle NMO$ **11.** $m\angle MLQ$ **12.** $m\angle LQO$

Chapter Review

9. If alternate interior angles UVX and WXV are both 90°, then UV ∥ XW by the converse of the Alternate Interior Angles Theorem.

10. 55°

11. 55°

12. 75°

4. 90° 5. 90°

6. $\overline{AD} \cong \overline{BC}$; $\overline{AB} \cong \overline{CD}$; $\overline{AC} \cong \overline{BD}$; $\overline{AE} \cong \overline{EC} \cong \overline{DE} \cong \overline{EB}$

7. 60° 8. 120°

Chapter Review

13. 108°

14. 35°

15. 12

16. 6

17. 18

18. $\frac{3}{2}$

19. $\frac{-2}{3}$

Lesson 3.6

➤ Key Skills

Find interior angle measures and exterior angle measures using the angle sums for polygons.

From the formula you discovered in this lesson, the sum of the interior angles of a regular octagon is 1080°. Thus the missing interior angle measure is $1080 \div 8 = 135$.

You discovered in this lesson that the sum of the exterior angles of a polygon is always 360°. Because the given polygon is regular, simply divide 360° by the number of angles: $360° \div 8 = 45°$.

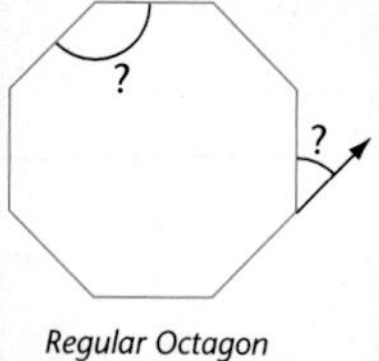

Regular Octagon

➤ Exercises

In Exercises 13–14 find the missing angle measure.

13.

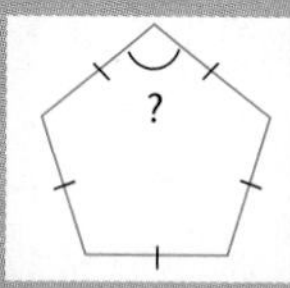

14.

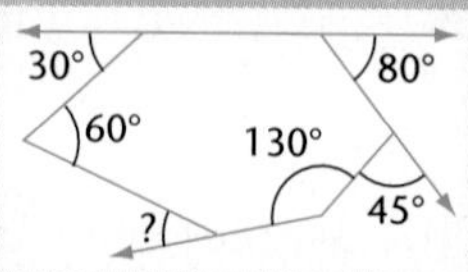

Lesson 3.7

➤ Key Skills

Use triangle and trapezoid midsegments to solve problems.

Find the measure of $\overline{FI}$.

$$FI = \frac{(\text{base}_1 + \text{base}_2)}{2} = \frac{(AB + DC)}{2} = \frac{(28 + 52)}{2} = 40$$

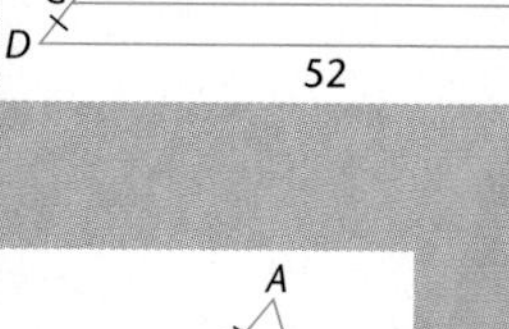

➤ Exercises

In Exercises 15–17 find the measures. Use △ABC.

15. EH **16.** DG **17.** FI

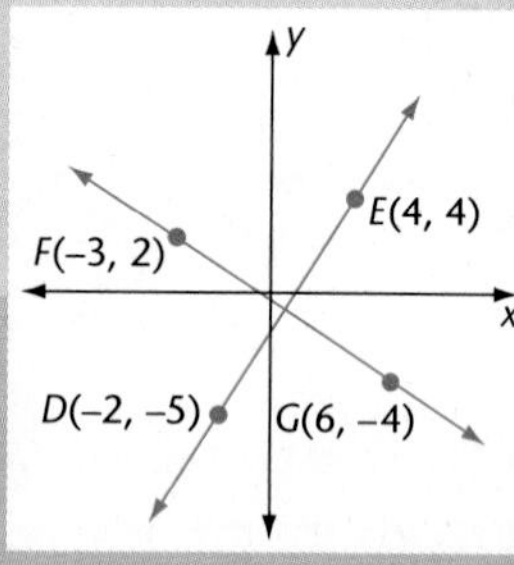

Lesson 3.8

➤ Key Skills

Find the slope of a line.

You need two points on a line to determine its slope. A line runs through the origin and point (4, −4). Use the formula $\frac{y_2 - y_1}{x_2 - x_1}$ to find the slope.

$\frac{0 - (-4)}{0 - 4} = \frac{4}{-4} = -1$. Thus, the line has a slope of −1.

➤ Exercises

In Exercises 18–19 refer to the graph.

18. Find the slope of $\overleftrightarrow{DE}$. **19.** Find the slope of $\overleftrightarrow{FG}$.

Applications

20. Paper Folding Use the properties of a square to fold and cut a square from an 8.5-by-11-inch sheet of paper. Use rotational symmetry to create a regular octagon by folding and cutting the square. What do the folds represent?

20. Use the properties that the diagonals of a square are perpendicular and equal in length to fold a square. Fold angle bisectors of the angles formed by the diagonals until the 8 lines of symmetry have been folded. Every other line of symmetry goes through two of the vertices of the regular octagon.

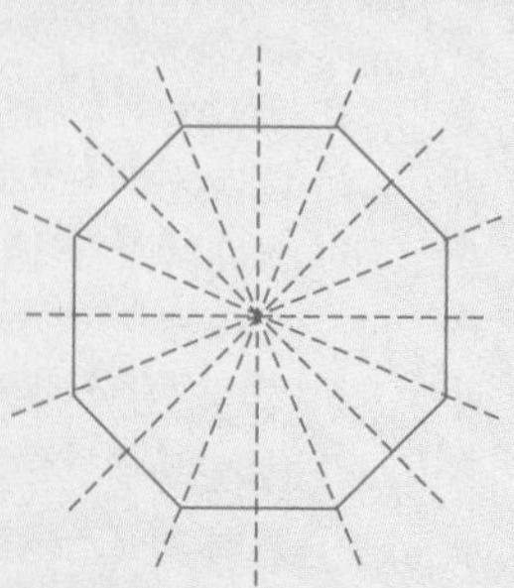

The folds are the 8 lines of symmetry.

Chapter 3 Assessment

In Exercises 1 and 2, refer to the figure at right.

1. How many axes of symmetry does the figure have?
2. If the figure has rotational symmetry, what is the smallest positive angle of rotation?

In Exercises 3 and 4, use the pre-image below.

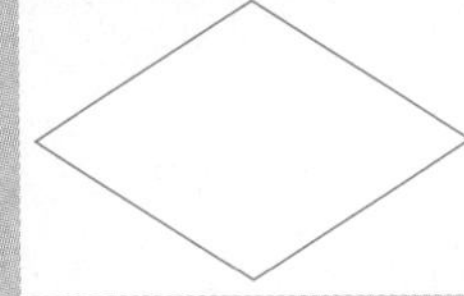

3. Sketch a reflection of the pre-image over a vertical line.
4. Sketch a figure that has rotational symmetry, using the pre-image.

5. In the trapezoid, what is the measure of $\angle BCD$?

a. 45° **b.** 135°
c. 60° **d.** 120°

6. $\triangle HIK$ and $\triangle JKI$ are equilateral triangles. Is this enough information to show that $\overleftrightarrow{HI}$ and $\overleftrightarrow{KJ}$ are parallel? Justify your answer.

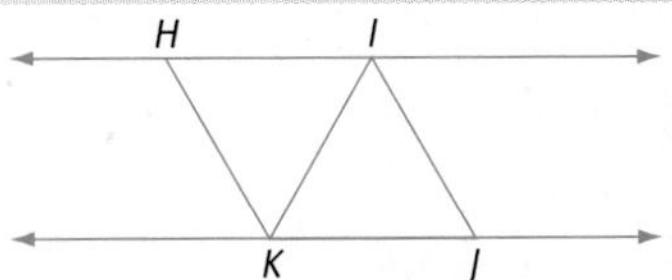

In Exercises 7–9, find the missing angle measures.

7.

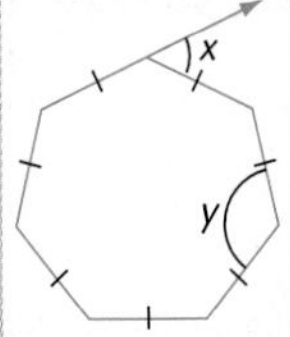

8.

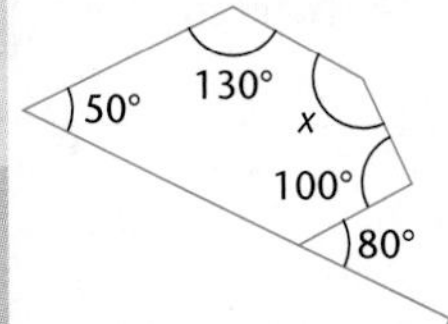

9.

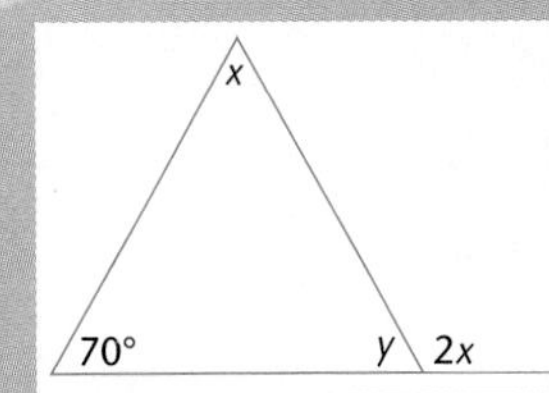

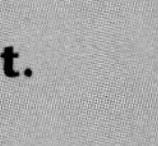

In Exercises 10 and 11, refer to the graph at right.

10. Is $\angle SPR$ a right angle?
11. Give the equation for a line that is parallel to $\overleftrightarrow{SP}$.

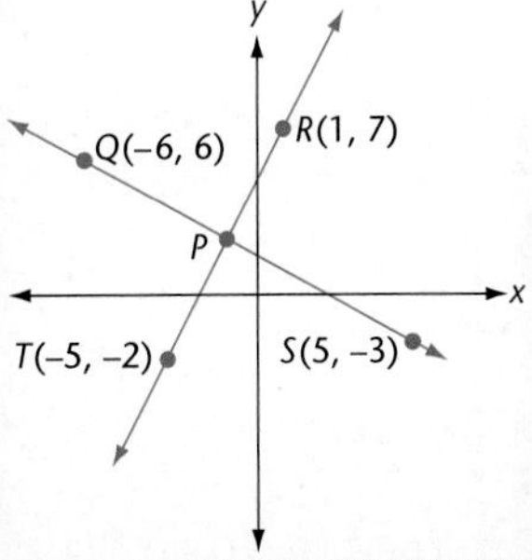

1. 5; One through each point of the star.
2. 72°
3.

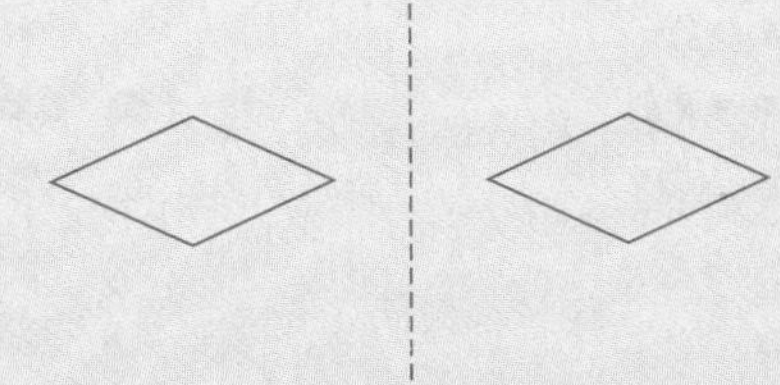

4. Sample answers;

5. d
6. Yes
Since $m\angle HIK = m\angle JKI = 60°$, HI ∥ KJ by the converse of the Alternate Interior Angles Theorem.
7. 51.4°
8. 160°
9. $x = 70°$, $2x = 140°$, $y = 40°$
10. Slope of $\overleftrightarrow{RT} = \frac{3}{2}$; Slope of $\overleftrightarrow{QS} = \frac{-9}{11}$; Since the slopes are not negative reciprocals, the lines are not perpendicular.
11. Any line parallel to $\overleftrightarrow{SP}$ has slope $\frac{-9}{11}$. Sample answer: $y = -\frac{9}{11}x + 2$.

CHAPTER 4

Congruent Triangles

Meeting Individual Needs

4.1 Polygon Congruence

Core Resources	Core Plus Resources
Inclusion Strategies, p. 175	Practice Master 4.1
Reteaching the Lesson, p. 176	Enrichment Master 4.1
Practice Master 4.1	Technology Master 4.1
Enrichment, p. 175	
Lesson Activity Master 4.1	
[2 days]	**[1 day]**

4.2 Exploring Triangle Congruence

Core Resources	Core Plus Resources
Inclusion Strategies, p. 182	Practice Master 4.2
Reteaching the Lesson, p. 183	Enrichment, p. 182
Practice Master 4.2	Technology Master 4.2
Enrichment Master 4.2	Interdisciplinary Connection, p. 181
Lesson Activity Master 4.2	
Interdisciplinary Connection, p. 181	
[2 days]	**[2 days]**

4.3 Analyzing Triangle Congruence

Core Resources	Core Plus Resources
Inclusion Strategies, p. 188	Practice Master 4.3
Reteaching the Lesson, p. 189	Enrichment, p. 188
Practice Master 4.3	Technology Master 4.3
Enrichment Master 4.3	
Lesson Activity Master 4.3	
Interdisciplinary Connection, p. 187	
[2 days]	**[1 day]**

4.4 Using Triangle Congruence

Core Resources	Core Plus Resources
Inclusion Strategies, p. 196	Practice Master 4.4
Reteaching the Lesson, p. 196	Enrichment, p. 188
Practice Master 4.4	Technology Master 4.4
Enrichment Master 4.4	Interdisciplinary Connection, p. 194
Technology Master 4.4	
Lesson Activity Master 4.4	
[2 days]	**[2 days]**

4.5 Proving Quadrilateral Properties

Core Resources	Core Plus Resources
Inclusion Strategies, p. 204	Practice Master 4.5
Reteaching the Lesson, p. 204	Enrichment p. 203
Practice Master 4.5	Technology Master 4.5
Enrichment Master 4.5	Interdisciplinary Connection, p. 202
Lesson Activity Master 4.5	Mid-Chapter Assessment Master
Mid-Chapter Assessment Master	
[2 days]	**[1 day]**

4.6 Exploring Conditions for Special Quadrilaterals

Core Resources	Core Plus Resources
Inclusion Strategies, p. 209	Practice Master 4.6
Reteaching the Lesson, p. 210	Enrichment Master 4.6
Practice Master 4.6	Interdisciplinary Connection, p. 208
Enrichment, p. 209	
Technology Master 4.6	
Lesson Activity Master 4.6	
[3 days]	**[2 days]**

4.7 Compass and Straightedge Constructions

Core Resources	Core Plus Resources
Inclusion Strategies, p. 216 Reteaching the Lesson, p. 216 Practice Master 4.7 Enrichment, p. 215 Lesson Activity Master 4.7 Interdisciplinary Connection, p. 214	Practice Master 4.7 Enrichment Master 4.7 Technology Master 4.7 Interdisciplinary Connection, p. 214
[2 days]	[2 days]

4.8 Exploring Congruence in the Coordinate Plane

Core Resources	Core Plus Resources
Inclusion Strategies, p. 222 Reteaching the Lesson, p. 223 Practice Master 4.8 Enrichment Master 4.8 Technology Master 4.8 Lesson Activity Master 4.8 Interdisciplinary Connection, p. 221	Practice Master 4.8 Enrichment, p. 222 Technology Master 4.8 Interdisciplinary Connection, p. 221
[2 days]	[1 day]

4.9 Exploring the Construction of Transformations

Core Resources	Core Plus Resources
Inclusion Strategies, p. 227 Reteaching the Lesson, p. 228 Practice Master 4.9 Enrichment, p. 227 Technology Master 4.9 Lesson Activity Master 4.9	Practice Master 4.9 Enrichment Master 4.9 Technology Master 4.9
[2 days]	[2 days]

Chapter Summary

Core Resources	Core Plus Resources
Chapter 4 Project, pp. 232–233 Lab Activity Long Term Project Chapter Review, pp. 234–238 Chapter Assessment p. 239 Chapter Assessment, A/B Alternative Assessment Cumulative Assessment, pp. 240–241	Chapter 4 Project, pp. 232–233 Lab Activity Long Term Project Chapter Review, pp. 234–238 Chapter Assessment p. 239 Chapter Assessment, A/B Alternative Assessment Cumulative Assessment, pp. 240–241
[3 days]	[2 days]

Hands-On Strategies

The compass and straightedge activities in Lessons 4.7 and 4.9 provide opportunities for students to create their own geometric figures. After students have mastered the basic constructions, have them go back through previous lessons in the chapter and duplicate the figures used with example proofs. If some students have difficulty learning to manipulate a compass and straightedge, paper folding can be used as an alternative hands-on method of creating particular triangles and quadrilaterals. If students have learned to fold paper to make perpendicular lines, parallel lines, congruent segments, and congruent angles, they can create many of the figures needed in this chapter. The exploration activities in Lesson 4.6 are particularly appropriate for paper folding.

Cooperative Learning

GROUP ACTIVITIES	
Developing triangle congruence postulates	Lesson 4.2, Explorations 1 and 2
Conditions for Special Quadrilaterals	Lesson 4.6, Exploration
Compass and straightedge constructions	Lesson 4.7, Examples 1–3, Exercises 6–21
Developing midpoint formulas	Lesson 4.8, Explorations 1 and 2
Constructing transformations	Lesson 4.9, Explorations 1 and 2 Exercises 5–33

You may wish to have students work in groups or with partners for some of the above activities. Additional suggestions for cooperative group activities are noted in the teacher's notes in each lesson.

Multicultural

The cultural references in this chapter include references to Europe and Africa.

CULTURAL CONNECTIONS	
Africa: Mozambican house construction	Lesson 4.6, Try This
Europe: Napoleon's construction	Lesson 4.7, Look Beyond

Portfolio Assessment

Below are portfolio activities for the chapter listed under seven activity domains which are appropriate for portfolio development.

1. **Investigation/Exploration** The explorations in Lessons 4.1 and 4.2 focus on congruence conditions for quadrilaterals and triangles. In Lesson 4.6, the exploration activities deal with conditions that determine whether or not a quadrilateral is a parallelogram, rectangle, or rhombus. The Lesson 4.8 explorations explore midpoint formulas; those in Lesson 4.9 deal with the construction of congruence transformations on the coordinate plane.

2. **Applications** Carpentry, Lesson 4.1, Exercise 16; Construction, Lesson 4.2, Exercise 12; Carpentry, Lesson 4.2, Exercise 13; Quilting, Lesson 4.2, Exercises 14–16; Surveying, Lesson 4.4, Exercises 25–26; Baseball, Lesson 4.5, Exercise 38; Interior Design, Lesson 4.7, Exercise 28.

3. **Nonroutine Problems** Lesson 4.4, Exercise 36 (alternate proof of the Isosceles Triangle Theorem); Lesson 4.5, Exercise 38 (baseball problem).

4. **Project** Flexagons: see pages 232–233. Students create folded paper polygons called *hexaflexagons* and *hexahexaflexagons.*

5. **Interdisciplinary Topics** Students may choose from the following: Coordinate Geometry, Lesson 4.1, Exercise 31; Algebra, Lesson 4.3, Exercises 28; Fine Art, Lesson 4.5, Exercises 13–19; Algebra, Lesson 4.5, Exercise 37; History, Lesson 4.7, Exercises 29–31

6. **Writing** *Communicate* exercises of the type where students are asked to state or explain postulates and theorems offer excellent writing selections for the portfolio. Suggested selections include: Lesson 4.2, Exercises 1–4; Lesson 4.7, Exercises 1–3; Lesson 4.9, Exercises 1–2.

7. **Tools** In Chapter 4, students use geometry graphics software to make conjectures about congruence properties, generate a variety of cases for theorems and postulates, and explore conditions for special quadrilaterals.

Technology

Many of the activities in this book can be significantly enhanced by the use of geometry graphics software, but in every case the technology is optional. If computers are not available, the computer activities can, *and should,* be done by hands-on or "low-tech" methods, as they contain important instructional material. For more information on the use of technology, refer to the *HRW Technology Manual for High School Mathematics.*

Spreadsheets

Spreadsheets have proven to be valuable far beyond their originally intended uses in accounting and business. Mathematicians and scientists are continually finding new and intriguing ways to put them to work. And part of the good news, from a teachers point of view, is that students find them fun! There should be little difficulty interesting students in learning and using this increasingly important tool.

Exercise 16 in Lesson 4.8 recommends the use of a spreadsheet to solve for midpoints of two given points. This would be an appropriate time for a classroom demonstration.

The details of the use of spreadsheet packages vary from one product to another, but their underlying principles are basically the same.

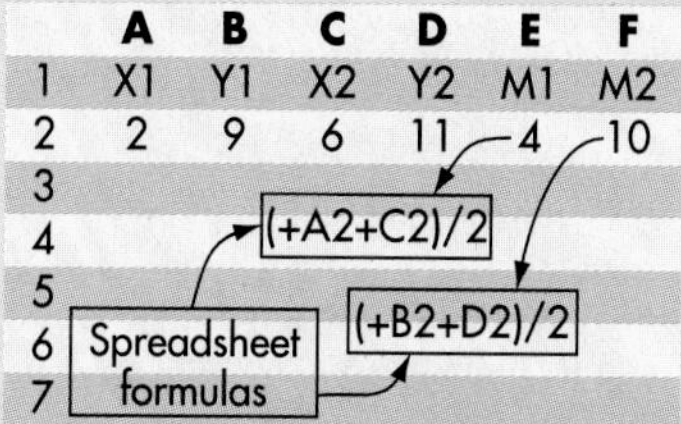

	A	B	C	D	E	F
1	X1	Y1	X2	Y2	M1	M2
2	2	9	6	11	4	10
3						
4						
5						
6						
7						

To use a spreadsheet, enter variable names in the cells in Row 1. In some cases the values for the variables will be supplied as data by the user. In other cases there is a formula, which is entered when the variable is created, that calculates the value of the variable from values in other cells. In the spreadsheet, notice that the variable names for the coordinates of the given points and the calculated midpoints are defined in the first row. The formulas that need to be entered for the calculated values are shown in boxes. The data and the values produced by the spreadsheet appear in the second row.

Students may want to experiment with ways of generating values for X1, Y1, X2, and Y2, perhaps by random-number algorithms, to fill a given number of rows of the spreadsheet.

Computer Graphics Software

Lesson 4.1: Exploration 1 Given a quadrilateral with specific side and angle measures, students drag another quadrilateral that matches the given figure. Students then complete the definition of polygon congruence.

Lesson 4.2: Exploration 1 Given a triangle with sides of a specific length, students create another triangle that has sides of the same length. In doing this, they discover that a triangle that has sides of lengths equal to the original triangle is a triangle of the same shape as the original triangle. This exploration demonstrates SSS congruence.

Lesson 4.2: Exploration 2, Part 1 Given a triangle with two sides of specific length and the included angle measure, students created another triangle that has these three parts the same measure. In doing this, they discover that the triangle that has these measures is a triangle of the same shape as the original triangle. This demonstrates SAS congruence.

Lesson 4.2: Exploration 2, Part 2 Given a triangle with two angles of specific measure and the included side length, students create another triangle that has these three parts the same measure. This demonstrates ASA congruence.

Hand-Held Computers

The geometry graphics capabilities of hand-held computers can be used in much the same way as the geometry software described above. Projection devices for classroom demonstration are available.

Exploring and Analyzing Congruent Triangles

About the Chapter

Background Information

This chapter introduces students to the triangle congruence postulates. First, students develop intuitive notions of triangle congruence through exploration. Later in the chapter, students apply the congruence postulates in formal proofs.

Chapter Resources

- Practice Masters
- Enrichment Masters
- Technology Masters
- Lesson Activity Masters
- Lab Activity Masters
- Long-Term Project Masters
- Assessment Masters
  Chapter Assessments, A/B
  Mid-Chapter Assessment
  Alternative Assessments, A/B
- Teaching Transparencies
- Cumulative Assessment
- Spanish Resources

Chapter Objectives

- Define *polygon congruence.*
- Solve problems using polygon congruence.
- Define triangle rigidity.
- Develop three triangle postulates—SSS, SAS, ASA—using construction.
- Identify and use triangle theorems: AAS, HL.
- Using counterexamples, prove that other combinations for triangle congruence are not valid.
- Use congruence of corresponding parts to prove congruence of triangles.
- Develop and use the Isosceles Triangle Theorem.

Chapter 4

Congruent Triangles

Lessons

All around you–in nature, art, and human technology–you find things that are the same shape and size. Such things are said to be **congruent.** In geometry,too, where you can construct models of things in the real world, you find the same idea.

When manufactured items are mass produced, like cars coming off the assembly line, it is essential that certain standard parts all have an exact shape and size, to a high degree of precision. If they do not, they may not fit together properly with other parts, or there may be mechanical failures.

About the Photos

Two objects are congruent if they have the same shape and size. The photos above show real-world examples of congruency. Remind the students of the importance of mass production in our society. Then discuss why identical parts are needed for mass production.

PORTFOLIO ACTIVITY
Use geometry technology to draw a triangle. Then, copy the triangle and use it to create your own geometric designs.
1. Make designs by overlapping the triangles. Do the overlapping triangles form other polygonal shapes?
2. Add color to your drawings. How does the color change the effect of your design?
Triangle congruence is basic to the study of more complex shapes.

- Prove quadrilateral conjectures using triangle congruence postulates and theorems.
- Develop conjectures for special quadrilaterals — parallelograms, rectangles, and rhombuses.
- Identify and use the Congruent Radii Theorem.
- Construct segment, angle, and triangle copies.
- Construct an angle bisector.
- Construct midpoints of segments by combining traditional construction methods.
- Rotate and reflect segments and polygons using a straightedge and compass.

PORTFOLIO ACTIVITY

In the Portfolio Activity, students use congruent triangles and other polygons to create geometric designs. The activity introduces students to the aesthetics of geometric congruence and provides students with experience using geometry graphics software.

Students can work together on the project to create designs for specific purposes such as a company logo or a school information poster.

Extend the activity by having students experiment with different color shadings using drawing software to create three-dimensional effects.

ABOUT THE CHAPTER PROJECT

The Chapter 4 Project, on pages 232–233, introduces students to the intriguing world of flexagons. Flexagons are made with folded paper and have the property of showing different faces, depending on how they are "flexed."

PREPARE

Objectives

- Define *polygon congruence.*
- Solve problems using polygon congruence.

RESOURCES

- Practice Master 4.1
- Enrichment Master 4.1
- Technology Master 4.1
- Lesson Activity Master 4.1
- Quiz 4.1
- Spanish Resources 4.1

Assessing Prior Knowledge

Display this figure on the overhead or chalkboard.

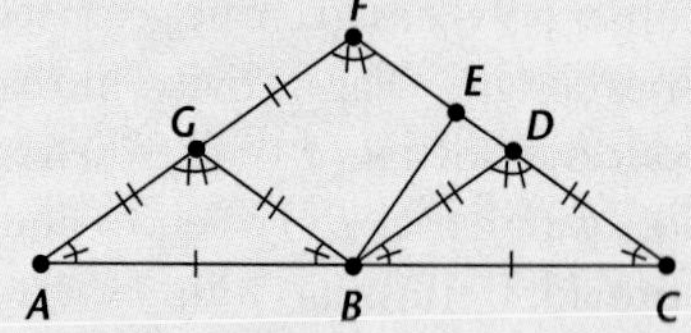

Have students identify marked segments congruent to $\overline{AB}$ and $\overline{AG}$; and, marked angles congruent to $\angle A$ and $\angle F$.

[$\overline{AB} \cong \overline{BC}$, $\overline{AG} \cong \overline{GB}$, $\overline{AG} \cong \overline{BD}$, $\overline{AG} \cong \overline{DC}$, $\angle A \cong \angle GBA$, $\angle A \cong \angle DBC$, $\angle A \cong \angle C$, $\angle F \cong \triangle AGB$, $\angle F \cong \angle BDC$]

TEACH

A definition of congruent polygons will be developed from the ideas students now have about congruent segments and angles.

ongoing ASSESSMENT

Two polygons are congruent if one can be moved on top of the other so that they match.

LESSON 4.1

Polygon Congruence

why *In earlier lessons you learned about congruent segments and angles. In this lesson you will develop a definition of congruent polygons.*

Andy Warhol, *One Hundred Cans, 1962*

Artists often use congruent shapes as an element of composition. Do you think this painting would be as interesting without the repeating pattern?

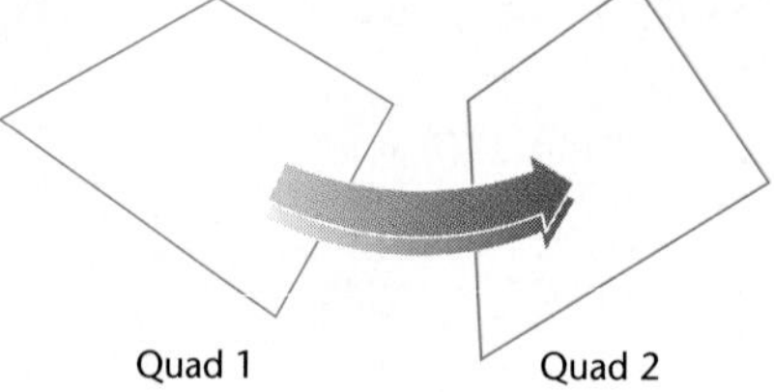

Quadrilaterals 1 and 2 are congruent. If you slide one on top of the other, you will see that you can make them match exactly. This suggests a definition of congruent polygons.

Can you say what it is?

In the exploration that follows, you will begin to develop a mathematical definition of congruent polygons.

Exploration *Polygon Congruence*

You will need

Geometry technology or
Ruler and protractor
Scissors

1 Construct quadrilateral *ABCD* using the following set of measures for the sides and angles.

$m\angle DAB = 104°$	$AB = 5.4$ cm
$m\angle ABC = 70°$	$BC = 7.0$ cm
$m\angle BCD = 72°$	$CD = 4.9$ cm
$m\angle CDA = 114°$	$DA = 3.7$ cm

ALTERNATIVE teaching strategy

Hands-On Strategies

Have students use straws cut to the specified lengths to build the quadrilateral in the Exploration. After the straws are strung together with thread or string, students should experiment by deforming the quadrilateral until they get the desired angles. Finished models can be glued to paper or tacked to cardboard.

Compare your quadrilateral with quadrilaterals made by other members of your class. Are they congruent? Do some of them have to be turned over before they will match?

If all the sides and angles of two polygons are congruent, will the polygons match? Would they match if one or more of the sides and angles failed to be congruent?

Complete the following preliminary definition of congruent polygons: "Two polygons are congruent if and only if . . ." ❖

Naming Polygons

When naming polygons, the rule is to go around the figure, either clockwise or counterclockwise, and list the vertices in order. It does not matter which vertex you pick to start with. What are some possible names for the hexagon on the right? How many possibilities are there?

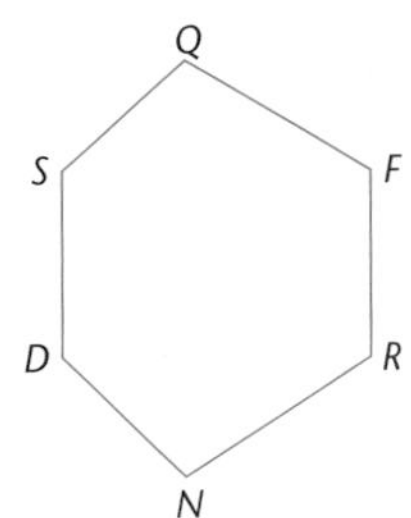

Corresponding Sides and Angles

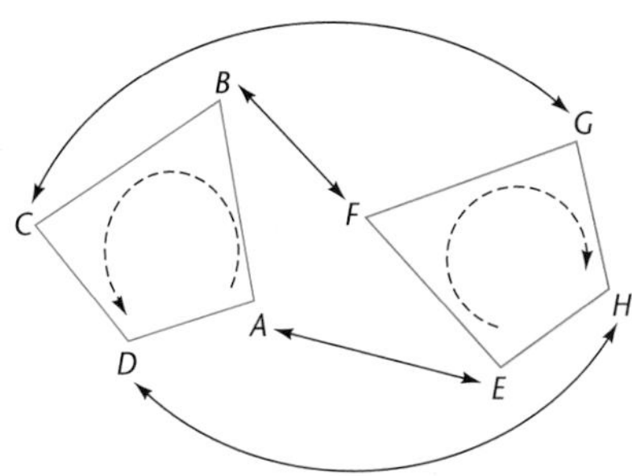

If two polygons have the same number of sides, it is possible to set up a correspondence between them by pairing off their parts. In quadrilaterals *ABCD* and *EFGH*, for example, you can pair off angles *A* and *E*, *B* and *F*, *C* and *G*, and *D* and *H*. Notice that you must go in order around each of the polygons.

The correspondence of the sides follows from the correspondence of the angles. In the present example, side $\overline{AB}$ corresponds to side $\overline{EF}$, and so on.

How many different ways are there of setting up a correspondence between the two quadrilaterals?

The polygons on the right are congruent. To show this fact mathematically, write a name for one of the polygons, followed by the congruence symbol. Then imagine the other polygon moved on top of the first one so that they match exactly. Finally, write the name of the second polygon to the right of the congruence symbol, with the angles listed in the order that they match:

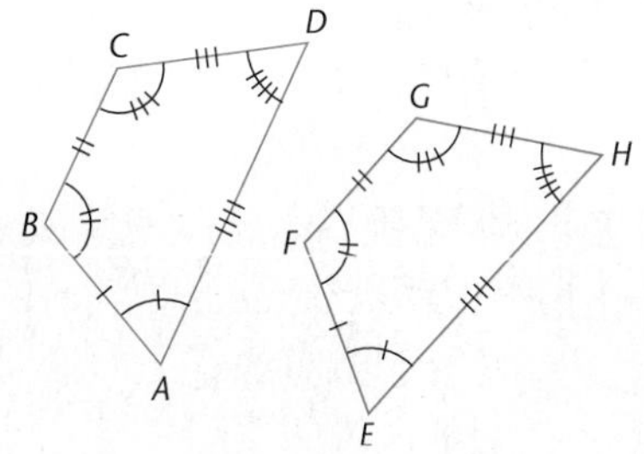

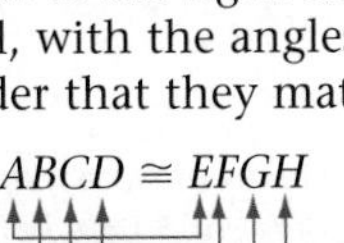

Exploration Notes

In this exploration, students will develop the logical criteria defining polygon congruence. It is important that all students place the vertices in the same order: *A*, *B*, *C*, and then *D*. Suggest that students use side $\overline{AB}$ for the base, with *A* at the left.

ongoing ASSESSMENT

3. .... if and only if all pairs of corresponding sides and angles are congruent.

TEACHING tip

Emphasize that the letters used to name polygons must be listed in either clockwise or counterclockwise order. The letters are not necessarily in *alphabetical* order.

ongoing ASSESSMENT

12 possibilities, 3 samples are: *QFRNDS, SQFRND, FRNDSQ*

CRITICAL Thinking

Each quadrilateral correspondence may be set up by naming the corresponding vertices either clockwise or counterclockwise, so there are 8 ways.

ENRICHMENT Have students cut a scalene triangle from cardboard and trace it to create many different pairs of congruent triangles. The pairs may share various sides, may be rotated or reflected with respect to each other, or may overlap in interesting ways. Students should try to create as wide a variety as possible. Each congruency is described with a statement such as $\triangle ABC \cong \triangle XYZ$.

INCLUSION strategies

Hands-On Strategies Have students work in pairs. They should cut any quadrilateral from cardboard and then trace it to make a congruent "twin." Then students should choose any eight letters. By assigning the letters to various vertices, they can write a variety of statements showing the congruency. Statements are checked by putting one quadrilateral on top of the other.

CRITICAL *Thinking*

BCDA ≅ *FGHE*
CBAD ≅ *GFEH*
ABCD ≅ *EFGH*
CDAB ≅ *GHEF*
DABC ≅ *HEFG*
ADCB ≅ *EHGF*
DCBA ≅ *HGFE*
BADC ≅ *FEHG*

Alternate Example

Given: line $j \perp$ line k; congruent segments and angles as marked in this figure

Prove: $\triangle ABC \cong \triangle EBF$

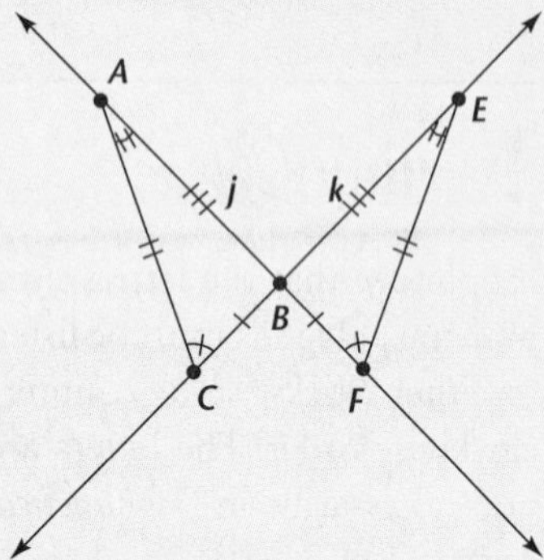

[Five pairs of congruent sides and angles can be listed from the figure: $\overline{AB} \cong \overline{EB}$, $\overline{BC} \cong \overline{BF}$, $\overline{AC} \cong \overline{EF}$, $\angle A \cong \angle E$, $\angle C \cong \angle F$. The four angles at point *B* are all right angles because the lines are perpendicular. Thus, $\angle ABC \cong \angle EBF$. This establishes the last pair of congruent angles.]

Cooperative Learning

Have students work in pairs or small groups. Each student should cut out a cardboard quadrilateral, trace two copies, and assign eight different letters to the vertices. Students should then exchange papers with a partner and write all the pairs of corresponding angles and sides.

There is more than one way to write a congruence statement for polygons *ABCD* and *EFGH*. Complete the congruence statements below, and then write all the other possibilities.

BCDA ≅ ?   *CBAD* ≅ ?

CONGRUENT POLYGONS

Two polygons are congruent if and only if there is a way of setting up a correspondence between their sides and angles, in order, so that

1. all pairs of corresponding angles are congruent, and
2. all pairs of corresponding sides are congruent. **4.1.1**

Proving that polygons, particularly triangles, are congruent is one of the most important things you will do in geometry.

Prove that $\triangle REX \cong \triangle FEX$ in the figure as marked.

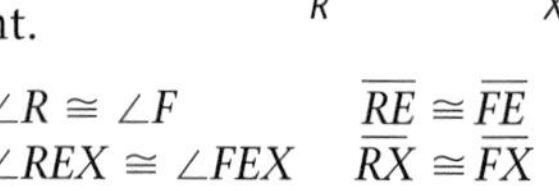

Solution ➤

List all the sides and angles that are given to be congruent.

$\angle R \cong \angle F$ $\overline{RE} \cong \overline{FE}$
$\angle REX \cong \angle FEX$ $\overline{RX} \cong \overline{FX}$
$\angle EXR \cong \angle EXF$

In all, there are six congruences that are required for triangles to be congruent—three pairs of angles and three pairs of sides. So one more pair of congruent sides is needed.

Notice that $\overline{EX}$ is shared by the two triangles. Use the Reflexive Property of Congruence to justify the statement that $\overline{EX} \cong \overline{EX}$. This gives the sixth congruence, so you can conclude that $\triangle REX \cong \triangle FEX$. ❖

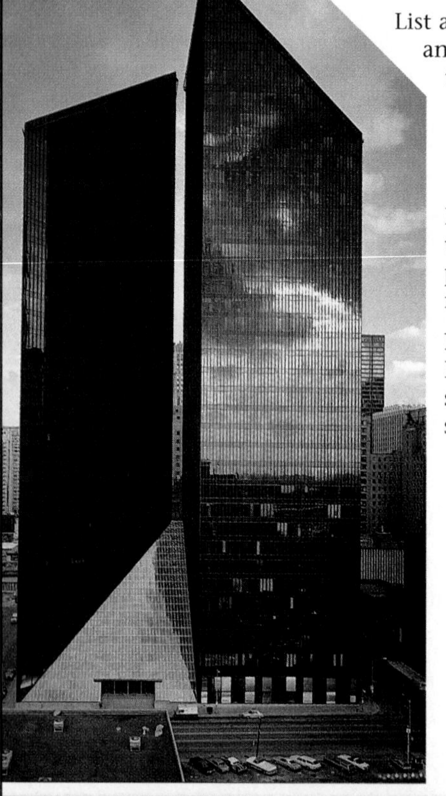

The Penzoil Building in Houston, Texas, features trapezoidal faces. Are the trapezoids congruent? How many congruent sides and angles do you see?

Using Symbols Give students statements such as $\triangle ABC \cong \triangle FIG$ or *BARN* ≅ *COWS*. Students should construct pairs of figures that illustrate the statements.

EXERCISES & PROBLEMS

Communicate

Explain whether the following pairs of figures appear to be congruent. Use Definition 4.1.1 to justify your answer.

1.

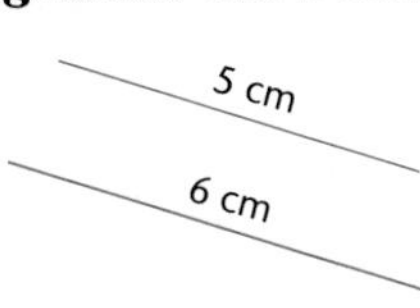

2.

3.

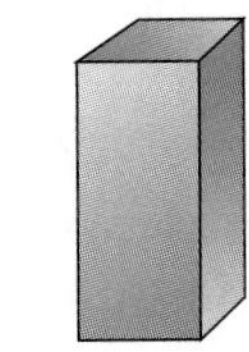

4.

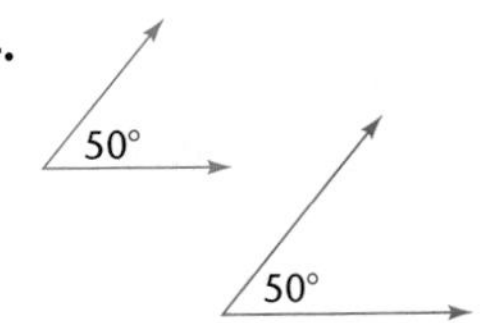

5.

6.

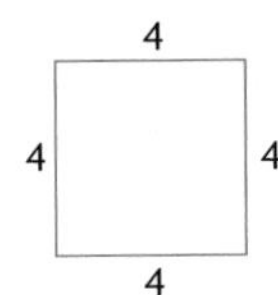

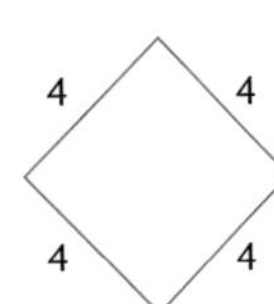

Practice & Apply

In Exercises 7–9, pentagon *UVWXY* ≅ pentagon *KMNTE*.

7. Draw a diagram and determine which angle is congruent to $\angle N$.

8. Describe how you would specify which angle is congruent to $\angle V$ without drawing a diagram.

9. Find a segment that is congruent to each of the segments below.

a. $\overline{WX} \cong$? $\overline{NT}$ **b.** $\overline{ET} \cong$? $\overline{YX}$ **c.** $\overline{UY} \cong$? $\overline{KE}$

10. Given: $\triangle REV \cong \triangle FOT$. Complete the following congruences.

a. $\angle FOT \cong$? $\angle REV$ **b.** $\angle EVR \cong$? $\angle OTF$ **c.** $\angle TFO \cong$? $\angle VRE$

Complete the congruences in Exercises 11–13.

11.

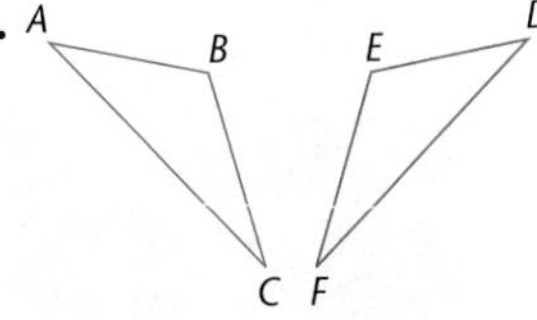

$\triangle ABC \cong \triangle DEF$

$\overline{AB} \cong \overline{DE}$ $\angle C \cong \angle F$

$\overline{AC} \cong$? ? $\cong \angle D$

$\overline{BC} \cong$? ? $\cong \angle E$ $\overline{AC} \cong \overline{DF}$; $\angle A \cong \angle D$

$\overline{BC} \cong \overline{EF}$; $\angle B \cong \angle E$

12.

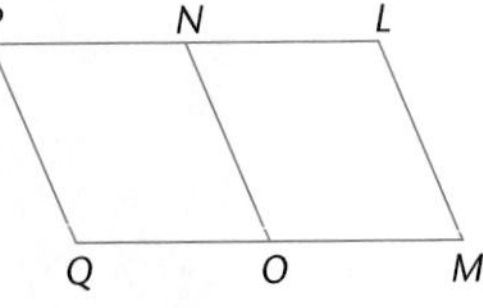

Quad *LMON* ≅ Quad *NOQP*

$\overline{NO} \cong \overline{LM}$ $\angle PNO \cong \angle NLM$

? $\cong \overline{MO}$ $\angle NOQ \cong$? $\overline{OQ} \cong \overline{MO}$; $\angle NOQ \cong \angle LMO$

? $\cong \overline{ON}$ $\angle OQP \cong$? $\overline{QP} \cong \overline{ON}$; $\angle OQP \cong \angle MON$

? $\cong \overline{NL}$ $\angle QPN \cong$? $\overline{PN} \cong \overline{NL}$; $\angle QPN \cong \angle ONL$

7. $\angle N \cong \angle W$

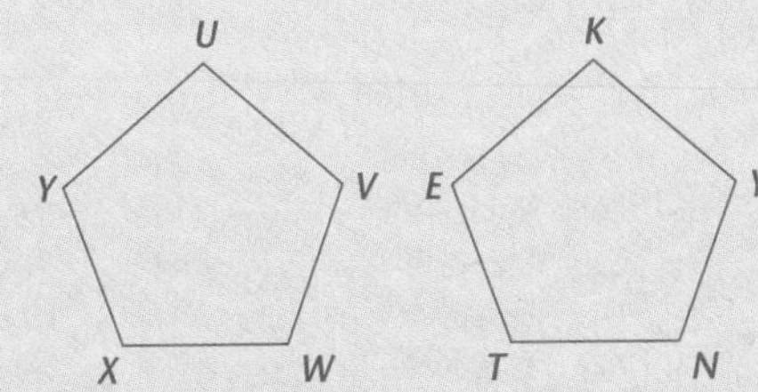

8. Find the letter that is in the same position in the name of the congruent pentagon.

11. $\overline{AC} \cong \overline{DF}$, $\angle A \cong \angle D$
$\overline{BC} \cong \overline{EF}$, $\angle B \cong \angle E$

ASSESS

Selected Answers

Odd-numbered Exercises 7–29

Assignment Guide

Core 1–13, 17–19, 24–30

Core Plus 1–10, 14–30, 31

Technology

Have students create figures for Exercises 11 and 13 with geometry graphics software. Students should use the reflection tool to make the triangles in Exercise 11 and the rotation tool to make the polygons in Exercise 13.

Error Analysis

For Exercise 15, students may not remember the name of the Transitive Property. Refer them to Lesson 2.5 for review.

alternative ASSESSMENT

Performance Assessment

Have students create convex hexagons with the vertices labeled *A, B, C, D, E,* and *F.* Then have them draw segments $\overline{AE}$ and $\overline{EC}$. Each student should trace his or her figure and use six new letters for the vertices. Students should write statements describing the congruent triangles and quadrilaterals in their figures. This activity can be done on a computer with geometry graphics or drawing software.

14. They are congruent.

15. Transitive Property of Congruence

The answers to Exercises 16, 17, and 20 can be found in Additional Answers beginning on page 727.

Practice Master

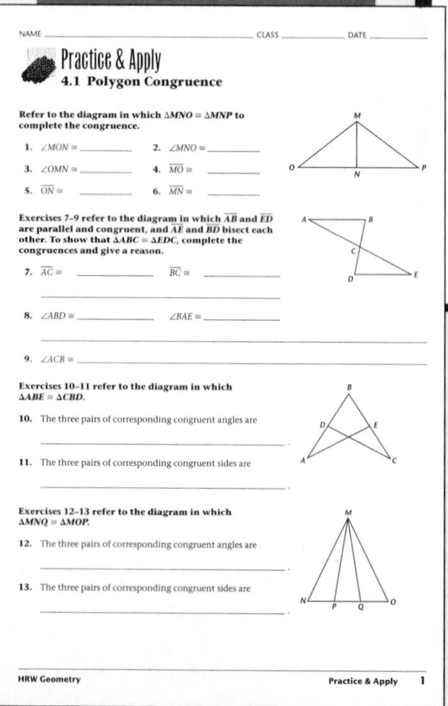
NAME ______ CLASS ______ DATE ______

Practice & Apply
4.1 Polygon Congruence

Refer to the diagram in which $\triangle MNO \cong \triangle MNP$ to complete the congruence.

1. $\angle MON \cong$ ______ 2. $\angle MNO \cong$ ______
3. $\angle OMN \cong$ ______ 4. $\overline{MO} \cong$ ______
5. $\overline{ON} \cong$ ______ 6. $\overline{MN} \cong$ ______

Exercises 7–9 refer to the diagram in which $\overline{AB}$ and $\overline{ED}$ are parallel and congruent, and $\overline{AE}$ and $\overline{BD}$ bisect each other. To show that $\triangle ABC \cong \triangle EDC$, complete the congruences and give a reason.

7. $\overline{AC} \cong$ ______ $\overline{BC} \cong$ ______
8. $\angle ABD \cong$ ______ $\angle BAE \cong$ ______
9. $\angle ACB \cong$ ______

Exercises 10–11 refer to the diagram in which $\triangle ABE \cong \triangle CBD$.

10. The three pairs of corresponding congruent angles are ______.
11. The three pairs of corresponding congruent sides are ______.

Exercises 12–13 refer to the diagram in which $\triangle MNQ \cong \triangle MOP$.

12. The three pairs of corresponding congruent angles are ______.
13. The three pairs of corresponding congruent sides are ______.

HRW Geometry Practice & Apply 1

13. Pentagon $LMNOP \cong$ Pentagon $QRSTU$

$\overline{LM} \cong \overline{QR}$	$\angle PLM \cong \angle UQR$
$\underline{?} \cong \overline{RS}$ $\overline{MN}$	$\angle LMN \cong \underline{?}$ $\angle QRS$
$\overline{NO} \cong \underline{?}$ $\overline{ST}$	$\underline{?} \cong \angle RST$ $\angle MNO$
$\underline{?} \cong \overline{TU}$ $\overline{OP}$	$\angle NOP \cong \underline{?}$ $\angle STU$
$\overline{PL} \cong \underline{?}$ $\overline{UQ}$	$\underline{?} \cong \angle TUQ$ $\angle OPL$

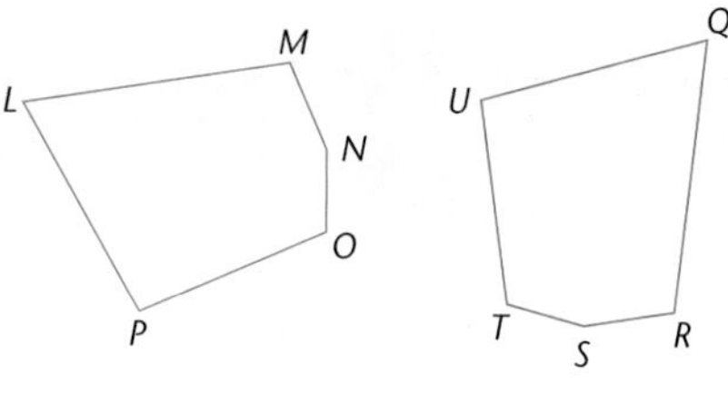

Given: $\triangle XYZ \cong \triangle FGH$ and $\triangle FGH \cong \triangle UMP$

14. What must be true about $\triangle XYZ$ and $\triangle UMP$?

15. Name a property that justifies your conclusion.

16. Carpentry A small shop makes house-shaped mailboxes. The faces opposite each other are rectangles, pentagons, or triangles of the same size and shape, as are the two slanted portions of the roof.

a. How many pairs of congruent polygons can you identify in this mailbox?

b. List at least 10 pairs of congruent angles.

c. List at least 10 pairs of congruent sides.

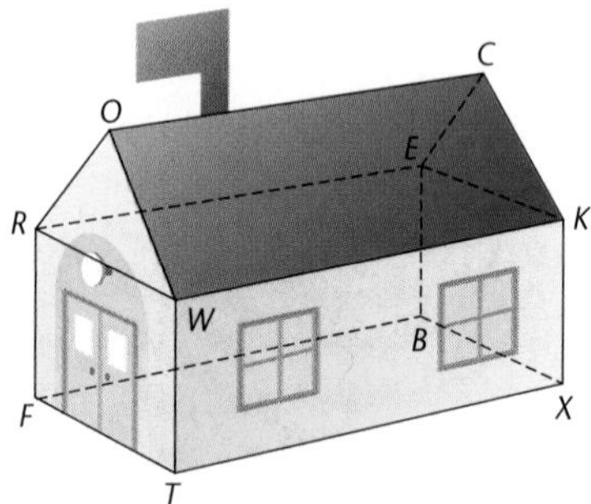

For Exercises 17–19, decide whether the polygons are congruent and write an explanation justifying your conclusion.

17.

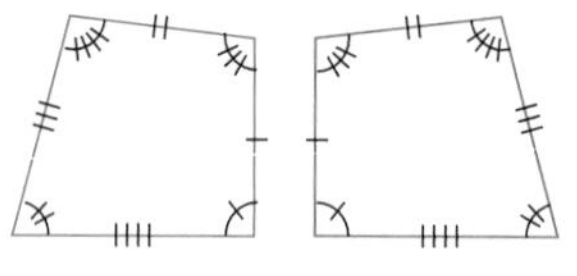

18.

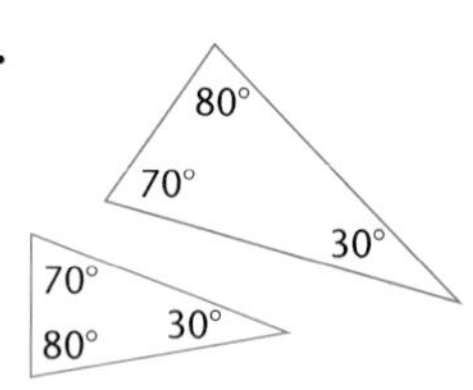

19.

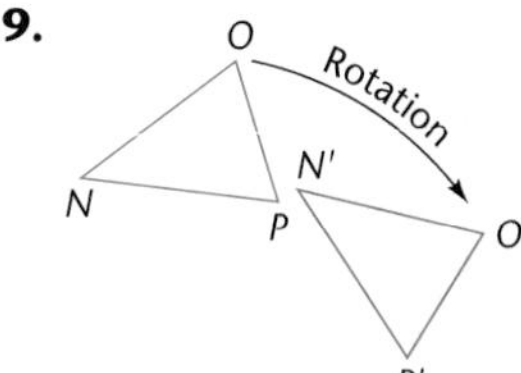

20. Quilting In the photograph of the quilt, how many different congruent shapes can you identify? How many different patterns of polygons did the quilter need to use to assemble the quilt? List them.

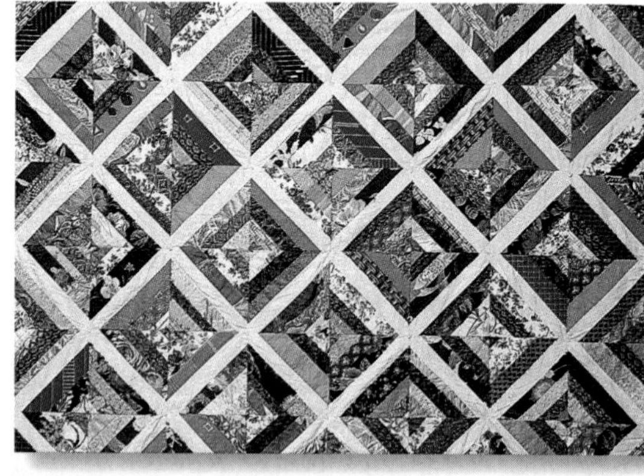

18. No. corresponding sides not congruent

19. Yes. Rotation is a rigid motion that preserves congruence.

Given: $\angle L \cong \angle P$, $\angle M \cong \angle O$, $\angle MRL \cong \angle OQP$, $\overline{LM} \cong \overline{PO}$, $\overline{MR} \cong \overline{OQ}$, $LQ = 5$ cm, $QR = 3$ cm, $RP = 5$ cm

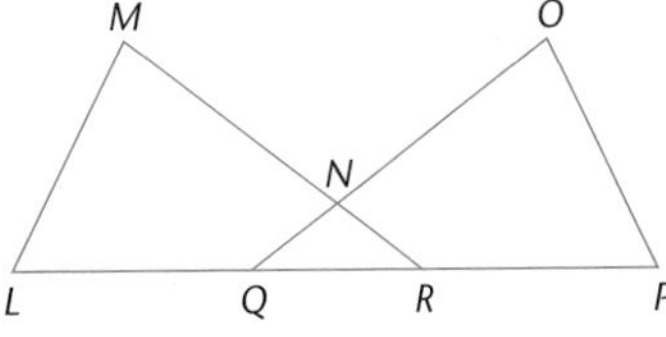

21. What is the length of $\overline{LR}$? 8 cm

22. What is the length of $\overline{QP}$? 8 cm

23. Is $\triangle LMR \cong \triangle POQ$? Why or why not?
Yes. All corresponding sides and angles are congruent.

Look Back

Given: *PARL* is a parallelogram. [Lessons 3.2, 3.3, 3.4]

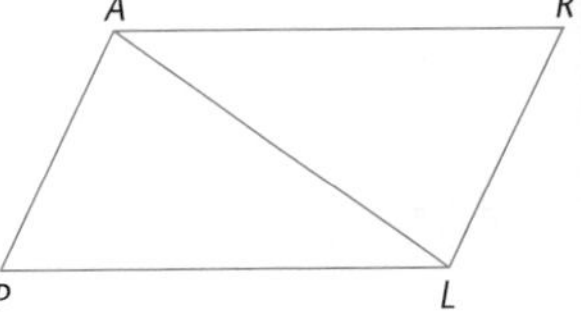

24. How do you know that $\overline{AR} \parallel \overline{PL}$ and $\overline{AP} \parallel \overline{RL}$?

25. How do you know that $\angle RAL \cong \angle PLA$ and $\angle RLA \cong \angle PAL$?

26. How do you know that $\overline{AL} \cong \overline{AL}$? Reflexive Property of Congruence

Determine whether each of the following is true or false. [Lessons 3.2, 3.3, 3.4]

27. A square is a rhombus. True

28. A rectangle is a parallelogram. True

29. The diagonals of a parallelogram bisect each other. True

30. Opposite angles of a rhombus are congruent. True

Look Beyond

31. **Coordinate Geometry** Graph the following points on a coordinate grid and connect the points in the order given.

Point $A(1, 2)$ Point $B(2, 4)$ Point $C(4, 4)$ Point $D(3, 1)$

Next, graph and connect the points resulting from the following translation on each point.

$(x, y) \rightarrow (x + 4, y - 2)$

Example: point $A(1, 2) \rightarrow (5, 0)$

Is the new figure congruent to the first figure? Why or why not?

Graph and connect the points resulting from the following transformation of the first figure.

$(x, y) \rightarrow (3x, -2y)$

Example: Point $A(1, 2) \rightarrow (3, -4)$

Is this new figure congruent to the first? Why or why not?

Look Beyond

Extension. In Exercise 31 students discover that translations preserve congruency, but that certain other transformations do not. You may also wish to have students experiment with ro-tated and reflected figures.

Technology Master

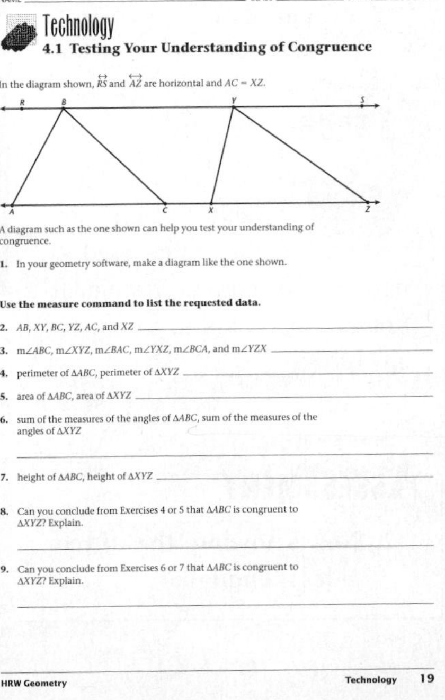

NAME ______ CLASS ______ DATE ______

Technology
4.1 Testing Your Understanding of Congruence

In the diagram shown, $\overleftrightarrow{RS}$ and $\overleftrightarrow{AZ}$ are horizontal and $AC = XZ$.

A diagram such as the one shown can help you test your understanding of congruence.

1. In your geometry software, make a diagram like the one shown.

Use the measure command to list the requested data.

2. AB, XY, BC, YZ, AC, and XZ ______
3. $m\angle ABC$, $m\angle XYZ$, $m\angle BAC$, $m\angle YXZ$, $m\angle BCA$, and $m\angle YZX$ ______
4. perimeter of ΔABC, perimeter of ΔXYZ ______
5. area of ΔABC, area of ΔXYZ ______
6. sum of the measures of the angles of ΔABC, sum of the measures of the angles of ΔXYZ ______
7. height of ΔABC, height of ΔXYZ ______
8. Can you conclude from Exercises 4 or 5 that ΔABC is congruent to ΔXYZ? Explain. ______
9. Can you conclude from Exercises 6 or 7 that ΔABC is congruent to ΔXYZ? Explain. ______

HRW Geometry Technology 19

---

24. definition of parallelogram

25. If parallel lines are cut by a transversal, then alternate interior angles are congruent.

31. $A(1, 2) \rightarrow A'(5, 0)$
$B(2, 4) \rightarrow B'(6, 2)$
$C(4, 4) \rightarrow C'(8, 2)$
$E(3, 1) \rightarrow E'(7, -1)$

PREPARE

Objectives

- Define *triangle rigidity.*
- Develop three triangle postulates—SSS, SAS, ASA—using construction.

- Practice Master 4.2
- Enrichment Master 4.2
- Technology Master 4.2
- Lesson Activity Master 4.2
- Quiz 4.2
- Spanish Resources 4.2

Assessing Prior Knowledge

Write all pairs of corresponding parts in the figures below.

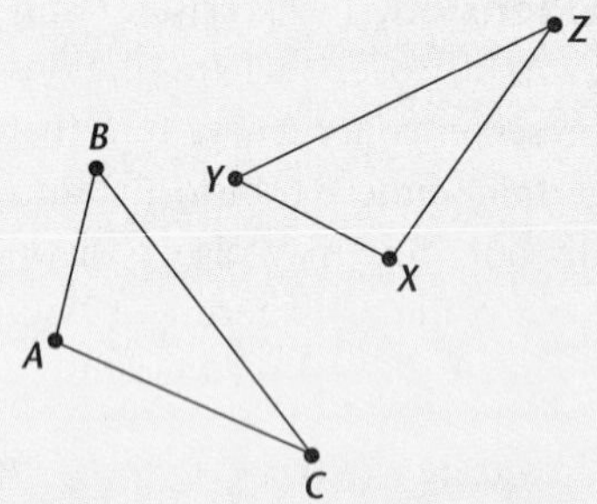

[$\overline{AB}$ and $\overline{XY}$, $\overline{BC}$ and $\overline{YZ}$, $\overline{AC}$ and $\overline{XZ}$, $\angle A$ and $\angle X$, $\angle B$ and $\angle Y$, $\angle C$ and $\angle Z$]

TEACH

Triangles are rigid figures and are thus often used as part of structures. Triangular braces can be found in almost any bridge.

ongoing ASSESSMENT

2. No. Knowing the three sides is enough.

Triangle Congruence

why *The structure of the Eiffel Tower includes thousands of triangles. Architects and engineers use triangular braces because triangles add rigidity. The rigidity of triangles has important mathematical consequences.*

Exploration 1 *Triangle Rigidity*

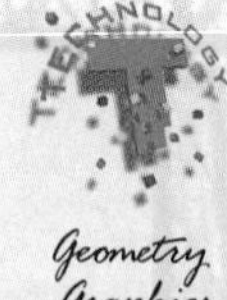

You will need
Geometry technology or
Modeling materials, such as geostrips, straws, or string
Ruler
Scissors

1. Using modeling materials or geometry technology, construct each triangle.

   $\triangle ABC$: $AB = 8$ units, $BC = 9$ units, $CA = 10$ units

   $\triangle XYZ$: $XY = 6$ units, $YZ = 8$ units, $ZX = 10$ units

2. Using the same measurements, try to draw more than one $\triangle ABC$ and $\triangle XYZ$ that look different from the first ones you drew. Compare the triangles you constructed in Step 1 with the new triangles. Are all triangles ABC congruent to each other? are triangles XYZ?

   Do you need to know the angle measures of a triangle to make a copy of it? Or is it enough to know the measures of the sides?

3. Given the lengths of the sides of a triangle, is there more than one shape that the triangle can have?

Using Visual Models

The explorations in this lesson may be done on grid paper. Students may wish to assign coordinates to the vertices of the figures. The work these students do should be saved and used for discussion during Lessons 4.8 and 4.9. These lessons explore congruence topics on the coordinate plane.

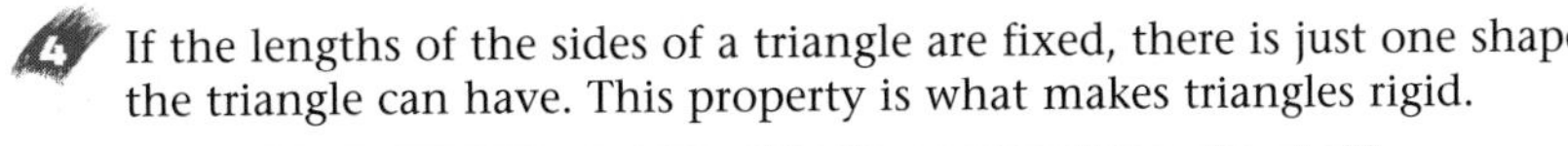

If the lengths of the sides of a triangle are fixed, there is just one shape the triangle can have. This property is what makes triangles rigid.

How do the triangles you drew show that triangles are rigid?

Write a rule, or **postulate**, that summarizes your results by completing the following statement.

If the __?__ of one triangle are congruent to the __?__ of another triangle, then the two triangles are congruent.

Give your postulate a descriptive name and record it in your notebook. ❖

CRITICAL *Thinking*

Are any polygons other than triangles rigid? If you repeated the exploration using quadrilaterals instead of triangles, do you think quadrilaterals with the same names would all be congruent?

Useful Geometry Tools

If you use the definition of congruent polygons on page 176 to show that two triangles are congruent, you must show that three pairs of sides and three pairs of angles are congruent. The postulate above provides a shortcut for showing that two triangles are congruent by showing that three pairs of sides are congruent.

There are still other shortcuts for proving triangle congruence. In the next exploration you will discover two more of them.

Exploration 2 Two More Congruence Postulates

Geometry Graphics

You will need

Geometry technology or
Ruler and protractor
Scissors

Part I

Draw and label each of the triangles.

$\triangle ABC$: $AB = 5$ units, $m\angle B = 45°$, $BC = 5$ units

$\triangle XYZ$: $XY = 5$ units, $m\angle Y = 30°$, $YZ = 7$ units

$\triangle MNO$: $MN = 8$ units, $m\angle N = 120°$, $NO = 6$ units

Cut out your triangles and compare them with triangles constructed by the other teams. If you are using a computer, draw additional triangles *ABC*, *XYZ*, or *MNO* using the above dimensions. Are all the triangles with the same names congruent?

Notice that in each of the triangles, the angles and the sides that you measured are arranged in a certain way. Describe that arrangement. (Hint: Ask yourself, what comes between what?)

Exploration 1 Notes

Students develop an understanding of SSS congruence in this exploration. Students using geometry graphics software can start by drawing segment $\overline{AB}$ eight units long. Then they should draw a circle centered at point *B* with a radius of nine units and a circle centered at point *A* with a radius of ten units. The intersection point of the circles is point *C*.

ongoing ASSESSMENT

4. A triangle constructed with given sides can have no other shape meaning that they are rigid.

CRITICAL *Thinking*

No. For example, many different quadrilaterals may be drawn having the same side lengths.

Exploration 2 Notes

In this exploration, students continue to develop an understanding of triangle congruence by exploring the SAS and ASA congruence postulates. The postulates will be formally defined in Lesson 4.3.

ongoing ASSESSMENT

2. The angle is between the two sides.

interdisciplinary CONNECTION

Architecture Have students collect photographs and illustrations of bridges for a bulletin board display. The way triangular elements are used for bracing will be more apparent than in other structures. And, the triangles found in the structures can be used to point out examples of congruency.

Cooperative Learning

Students doing the explorations with modeling materials or paper-and-pencil methods should work in pairs or small groups. In this way the work can be divided up and accomplished more quickly.

TEACHING tip

The descriptive names students give to the three postulates in this lesson will make a useful bulletin board display. As students study the next lesson, they can compare their names with the ones usually used in textbooks: Side-Side-Side (SSS), Side-Angle-Side (SAS), and Angle-Side-Angle (ASA).

3. Do you think that the above arrangements of measured sides and angles will determine the shape of a triangle? If you are not sure, try to find a set of similarly arranged measures that will not work. If you can draw two noncongruent triangles that have the same set of measures, you will have found a counterexample.

4. Write a postulate that summarizes your results by completing the following statement:

   If __?__ of one triangle are congruent to __?__ of another triangle, then the two triangles are congruent.

Give your postulate a descriptive name and record it in your notebook.

Part II

1. Construct and label each triangle.

   $\triangle ABC$: $m\angle A = 60°$, $AB = 6$ units, $m\angle B = 60°$

   $\triangle XYZ$: $m\angle X = 60°$, $XY = 6$ units, $m\angle Y = 45°$

   $\triangle MNO$: $m\angle M = 120°$, $MN = 6$ units, $m\angle N = 30°$

2. Answer the same three questions (Steps 2–4) for these measurements that you answered in Exploration 1. ❖

Using Your New Postulates

The triangle congruence postulates you discovered in the explorations can save steps in proofs. But much more important, they allow you to determine congruence from limited information.

EXTENSION

In each of the pairs below, the triangles are congruent. Tell which triangle congruence postulates allow you to conclude that they are congruent, based on the markings on the figures. ❖

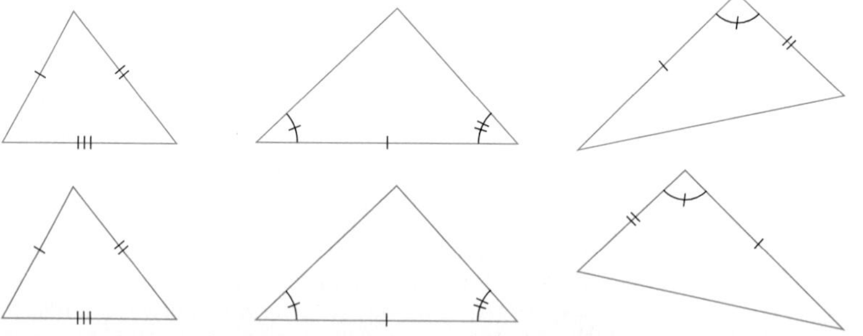

ENRICHMENT Have students check their reasoning about the following questions with geometry graphics software. Imagine $\triangle ABC$. Move point B toward point A along side $\overline{AB}$. 1. Which sides or angles in $\triangle ABC$ do not change? [**side $\overline{AC}$, $\angle A$**] 2. Which sides become shorter? [**side $\overline{AB}$, side $\overline{BC}$**] 3. Which sides become longer? [**none**] 4. Which angles decrease in size? [$\angle C$] 5. Which angles increase in size? [$\angle B$]

INCLUSION strategies

English Language Development Some students may have difficulty understanding what it means for a side to be "included between" two angles, or an angle to be "included between" two sides. Have students complete statements like the following: **1.** In triangle ABC, the angle included between sides $\overline{AB}$ and $\overline{BC}$ is ______________.

EXERCISES & PROBLEMS

Communicate

1. Explain triangle rigidity. What properties do triangles have that make them rigid?
2. Explain the congruence postulate you discovered in Exploration 1. Draw a diagram of two triangles that illustrates this postulate.
3. Explain the congruence postulate you discovered in Part I of Exploration 2. Draw a diagram of two triangles that illustrates this postulate.
4. Explain the congruence postulate you discovered in Part II of Exploration 2. Draw a diagram of two triangles that illustrates this postulate.
5. Use a counterexample to show that if corresponding angles of one triangle are congruent to corresponding angles of a second triangle, then the triangles are not necessarily congruent.

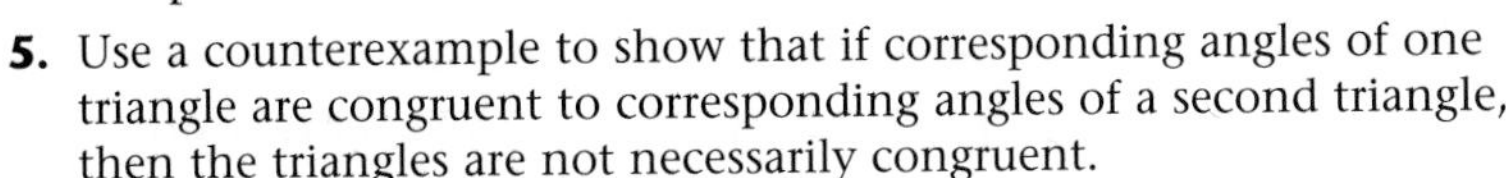

Practice & Apply

Decide whether each pair of triangles can be proven congruent using one of the three congruence postulates you discovered in the lesson. If so, write an appropriate congruence statement and name the postulate that it supports. If not, explain why.

6.

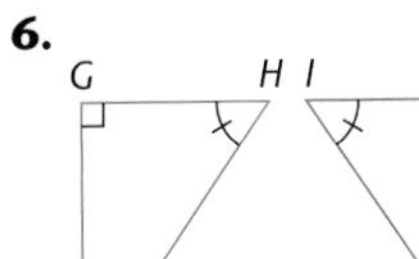

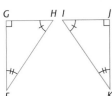

7.

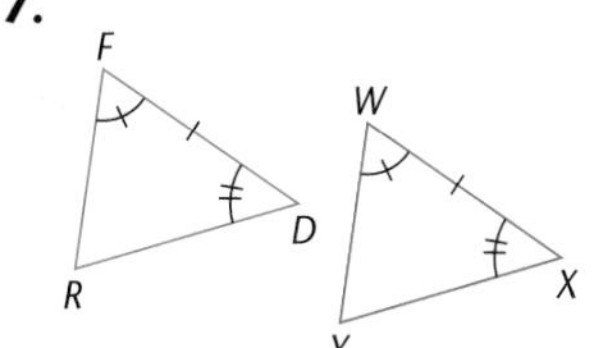

8.

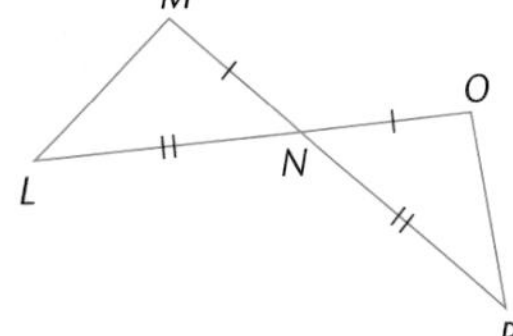

9.

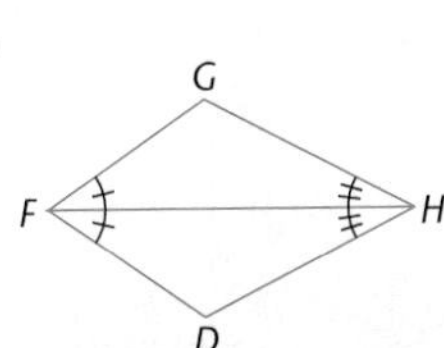

10.

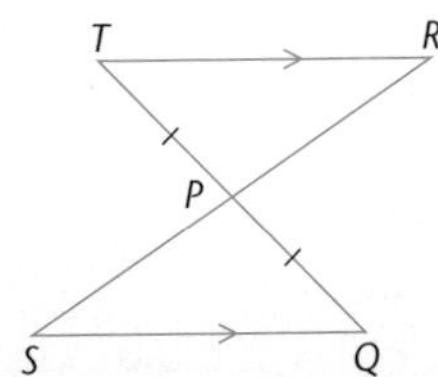

11.

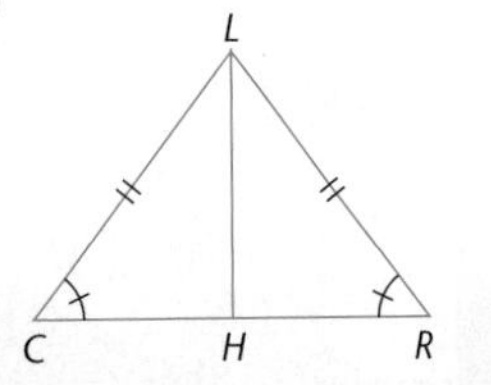

ASSESS

Selected Answers

Odd-numbered Exercises 7–29

Assignment Guide

Core 1–3, 6–20, 22–29

Core Plus 1–16, 21–31

Technology

Exercise 21 can be done with geometry graphics software. Have students start by drawing line $\overleftrightarrow{NT}$. Through a point P not on $\overleftrightarrow{NT}$, they should draw a line parallel to $\overleftrightarrow{NT}$. They should draw segment $\overline{PT}$ and find its midpoint O. Finally, they should draw a line through points N and O. This line meets the line through P in point H.

Error Analysis

For Exercises 14–16, the six sides of the hexagon are not marked as congruent. However, the quilter is using regular hexagons and therefore the sides are congruent.

6. No. Triangles can have the same angles, but sides with different lengths.
7. Yes; $\triangle DFR \cong \triangle XWY$; Angle-Side-Angle
8. Yes; $\triangle MLN \cong \triangle OPN$; Side-Angle-Side
9. Yes; $\triangle FDH \cong \triangle FGH$; Angle-Side-Angle
10. Yes; $\triangle TRP \cong \triangle QSP$; Angle-Side-Angle
11. No. The congruent angles are not between the two congruent pairs of corresponding sides.

RETEACHING the lesson

Cooperative Learning Have students work in pairs. (1) They should agree on three side lengths for a triangle. Then, independently, they each construct the triangle. They should use superposition, cutting if needed, to confirm that the triangles are congruent. Students should repeat the activity with (2) two sides and an included angle and (3) with two angles and an included side.

alternative ASSESSMENT

Performance Assessment

Allow students to choose any materials or methods to answer this question: Which of the following determines a unique triangle?

1. ΔPQR with side $PQ = 6$ units, side $QR = 8$ units [**no**]
2. ΔPQR with side $PQ = 6$ units, side $QR = 8$ units, side $RP = 12$ units [**yes**]
3. ΔPQR with side $PQ = 6$ units, $\angle Q = 50°$, side $QR = 8$ units [**yes**]

12. If the boards that form the triangles are all the same lengths, the triangles will be congruent by SSS.

13. The diagonal board creates two triangles, both of which are rigid.

Practice Master

NAME ______________ CLASS ________ DATE ________

Practice & Apply

4.2 Exploring Triangle Congruence

In Exercises 1–4, explain whether each pair of triangles can be proved congruent using one of the three congruence postulates. If so, write an appropriate congruence statement and name the postulate that supports it. If not, explain why.

1.

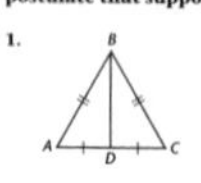

2.

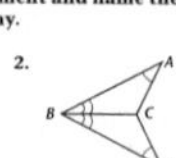

3.

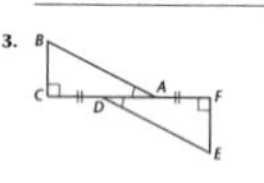

4.

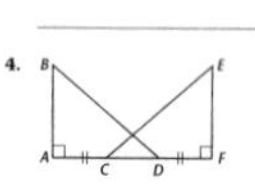

Exercises 5–7 refer to the diagram that shows ΔMNR and ΔPOQ in which $\overline{RN} \cong \overline{QO}$.

5. If $\overline{RN}$ and $\overline{QO}$ are each perpendicular to $\overline{MP}$, name the angles in the triangles that must be congruent. Why?

6. If $\overline{MO} \cong \overline{PN}$, which sides of the triangles must be congruent? Why?

7. What conclusion can you draw about the triangles? Write the postulate that supports your conclusion.

2 **Practice & Apply** **HRW Geometry**

12. Construction In the house shown in the photo, the triangles in the roof structure must be congruent. How can you be sure that the triangles are congruent without measuring any angles? Which triangle congruence postulate would you be using?

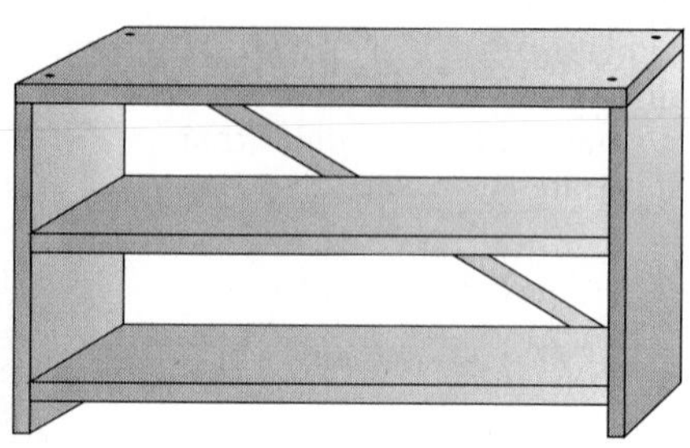

13. Carpentry A carpenter is building a rectangular bookshelf and finds that it wobbles from side to side. To stabilize the bookshelf, he nails a board that attaches the top left-hand corner to the bottom right-hand corner. Why will this diagonal board stabilize the bookshelf?

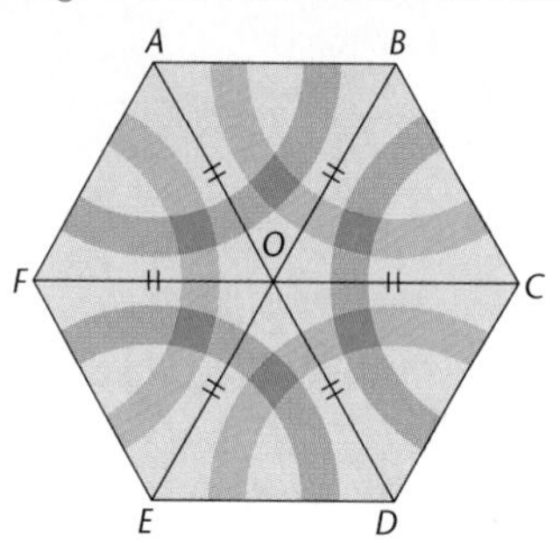

Quilting A quilter is making a quilt out of regular hexagons made from triangles as shown. Questions 14–16 refer to the quilt pattern.

14. Is $\triangle FOA \cong \triangle COD$? Why or why not?

15. Is $\triangle BOA \cong \triangle COD$? Why or why not?

16. Explain two ways to prove $\triangle FOE \cong \triangle COB$.

Recreation In Exercises 17–20, pairs of boat sails in the shape of right triangles are shown. Based on the congruences marked, are the sails necessarily congruent?

17. Yes

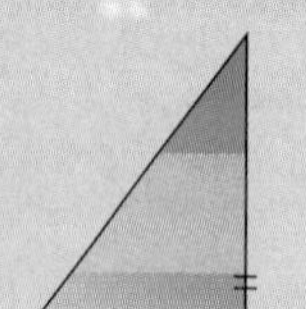

18. Yes

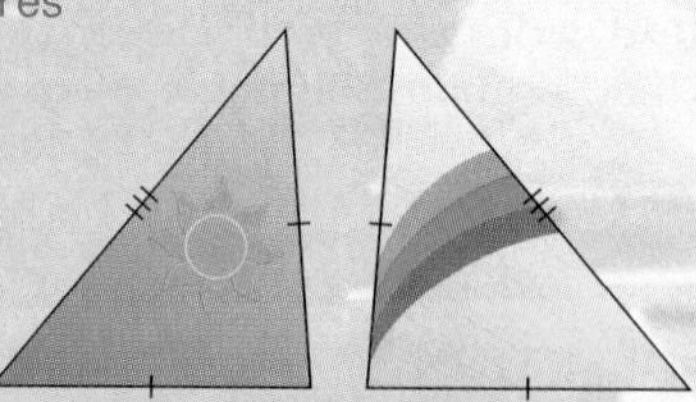

19. No

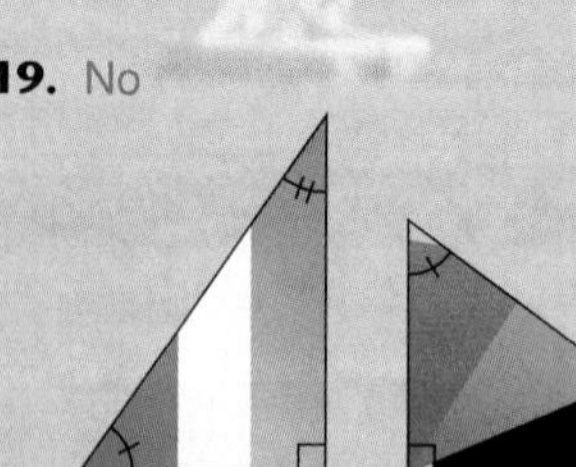

20. Yes

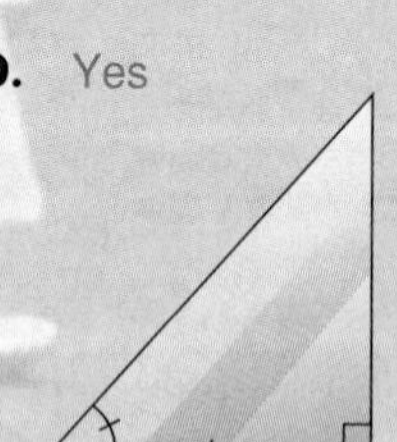

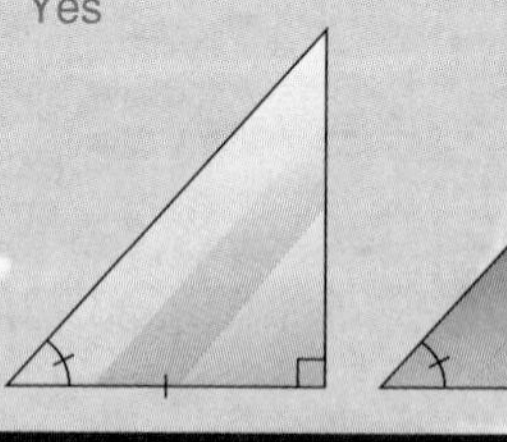

14. Yes. $\overline{AO} \cong \overline{DO}$ and $\overline{FO} \cong \overline{CO}$. $\overline{AF} \cong \overline{DC}$ because the sides of a regular hexagon are congruent. Also, $\angle AOF \cong \angle DOC$ because vertical angles are congruent. So $\triangle FOA \cong \triangle COD$ by Side-Side-Side.

15. Yes. $\overline{AO} \cong \overline{DO}$ and $\overline{BO} \cong \overline{CO}$. $\overline{AB} \cong \overline{DC}$ because the sides of a regular hexagon are congruent. So $\triangle BOA \cong \triangle COD$ by Side-Side-Side.

16. $\overline{EO} \cong \overline{BO}$ and $\overline{FO} \cong \overline{CO}$. $\overline{FE} \cong \overline{CB}$ because the sides of a regular hexagon are congruent. Also, $\angle FOE \cong \angle COB$ because vertical angles are congruent. Then $\triangle FOE \cong \triangle COB$ by either Side-Side-Side or Side-Angle-Side.

21. Use the diagram to help you answer the questions.

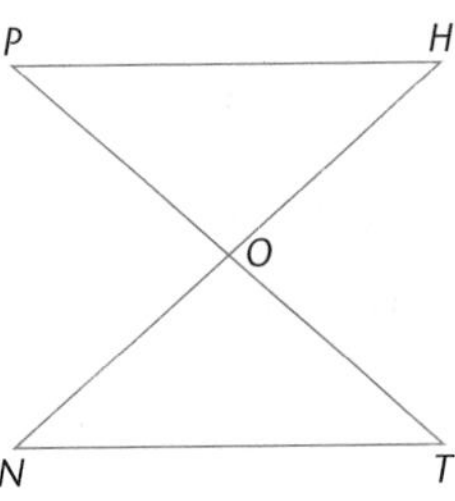

a. If O is the midpoint of $\overline{PT}$, then what two segments are congruent?

b. Because $\overline{PT}$ and $\overline{NH}$ intersect at point O, what angles must be congruent and why?

c. If $\overline{PH} \parallel \overline{NT}$, then $\angle T$ is congruent to what other angle and why?

d. If all of the above are true, what conclusion can you draw about these triangles?

Look Back

22. Points that lie on the same plane are called __?__. **[Lesson 1.1]** coplanar

23. If point B is between R and T, then what do you know about $RB + BT$ and why? **[Lesson 1.4]**

24. In the diagram, $\overrightarrow{AT}$ bisects $\angle RAM$, and $m\angle CAR = 122°$. Find the measures of $\angle RAM$, $\angle RAT$, and $\angle MAT$. **[Lesson 1.5]**

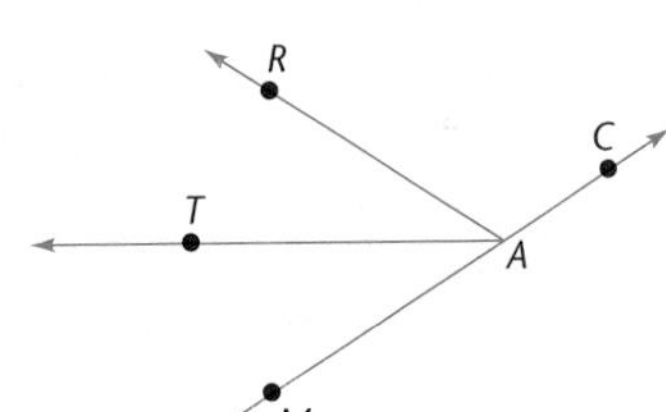

25. State two properties of the diagonals of a rhombus. **[Lesson 3.2]**

26. If the same-side interior angles of a transversal are congruent, then are the lines definitely parallel, possibly parallel, or never parallel? **[Lesson 3.4]** Possibly parallel

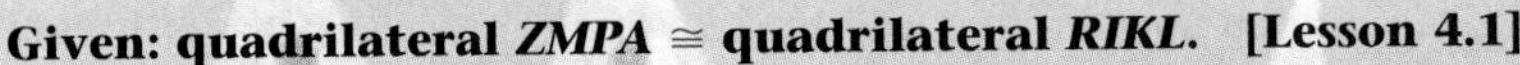

Given: quadrilateral *ZMPA* ≅ quadrilateral *RIKL*. [Lesson 4.1]

27. Identify an angle that is congruent to $\angle M$. $\angle I$

28. Name a segment that is congruent to $\overline{MP}$. $\overline{IK}$

29. Is it impossible, possible, or definite that $\angle R \cong \angle P$? Possible

Look Beyond

Study the triangles at the right.

30. Are the two triangles congruent? Why or why not?

31. What additional information would help you prove that the triangles are congruent?

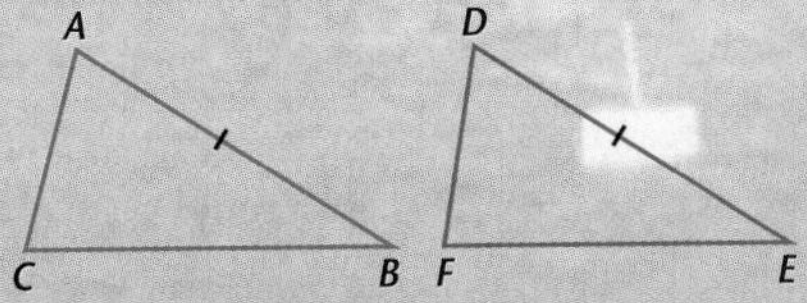

21. a. $\overline{PO} \cong \overline{TO}$

b. $\angle POH \cong \angle TON$ because vertical angles are congruent.

c. $\angle P$ because $\parallel$ s $\Rightarrow$ alternate interior $\angle$s $\cong$

d. $\triangle POH \cong \triangle TON$ by Angle-Side-Angle

23. $RB + BT = RT$ by the Segment Addition Postulate

24. $m\angle RAM = 58°$, $m\angle RAT = 29°$, $m\angle MAT = 29°$

25. The diagonals of a rhombus are perpendicular and bisect the rhombus angles as well as each other.

30. There is not enough information to conclude that the triangles are congruent.

31. $\overline{AC} \cong \overline{DF}$ and $\overline{CB} \cong \overline{FE}$; or $\overline{AC} \cong \overline{DF}$ and $\angle A \cong \angle D$; or $\angle A \cong \angle D$ and $\angle B \cong \angle E$

Look Beyond

Extension. Exercises 30 and 31 point out the danger in relying on appearances to determine congruence. Some students may believe that tracing the figures is a way to prove them congruent. Point out the inaccuracies in tracing. For example, if one side were just a tiny bit shorter than its corresponding sides, the triangles would not be congruent. Students should look for logical arguments to use in answering the question.

Technology Master

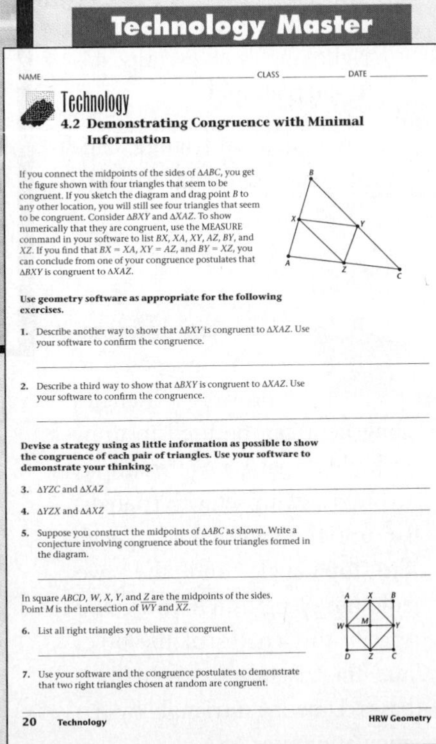

NAME ______________ CLASS ______ DATE ______

Technology

4.2 Demonstrating Congruence with Minimal Information

If you connect the midpoints of the sides of ΔABC, you get the figure shown with four triangles that seem to be congruent. If you sketch the diagram and drag point B to any other location, you will still see four triangles that seem to be congruent. Consider ΔBXY and ΔXAZ. To show numerically that they are congruent, use the MEASURE command in your software to list BX, XA, XY, AZ, BY, and XZ. If you find that $BX = XA$, $XY = AZ$, and $BY = XZ$, you can conclude from one of your congruence postulates that ΔBXY is congruent to ΔXAZ.

Use geometry software as appropriate for the following exercises.

1. Describe another way to show that ΔBXY is congruent to ΔXAZ. Use your software to confirm the congruence.

2. Describe a third way to show that ΔBXY is congruent to ΔXAZ. Use your software to confirm the congruence.

Devise a strategy using as little information as possible to show the congruence of each pair of triangles. Use your software to demonstrate your thinking.

3. ΔYZC and ΔXAZ ______

4. ΔYZX and ΔAXZ ______

5. Suppose you construct the midpoints of ΔABC as shown. Write a conjecture involving congruence about the four triangles formed in the diagram.

In square $ABCD$, W, X, Y, and Z are the midpoints of the sides. Point M is the intersection of $\overline{WY}$ and $\overline{XZ}$.

6. List all right triangles you believe are congruent.

7. Use your software and the congruence postulates to demonstrate that two right triangles chosen at random are congruent.

20 Technology HRW Geometry

Prepare

Objectives

- Identify and use triangle postulates: SSS, SAS, ASA, HL.
- Prove that other combinations for triangle congruence are not valid using counter examples.

Resources

- Practice Master 4.3
- Enrichment Master 4.3
- Technology Master 4.3
- Lesson Activity Master 4.3
- Quiz 4.3
- Spanish Resources 4.3

Assessing Prior Knowledge

Can each statement be disproved with a counterexample? If "yes," give the counterexample.

1. All rhombuses are square. **[yes; any rhombus without right angles]**
2. The diagonals of a square have different lengths **[no]**
3. No isosceles triangles have right angles. **[yes; any 45-45-90 triangle]**
4. An equilateral triangle cannot have right angles. **[no]**

Teach

The same information needed to prove two triangles congruent can be used to draw or build a triangle. For example, to build a swing set with triangular uprights, you could specify the top angle and the two adjoining lengths. Or you could specify the lengths of the poles and the distance between their bases. Drawing a triangle by triangulation uses ASA.

LESSON 4.3 Analyzing Triangle Congruence

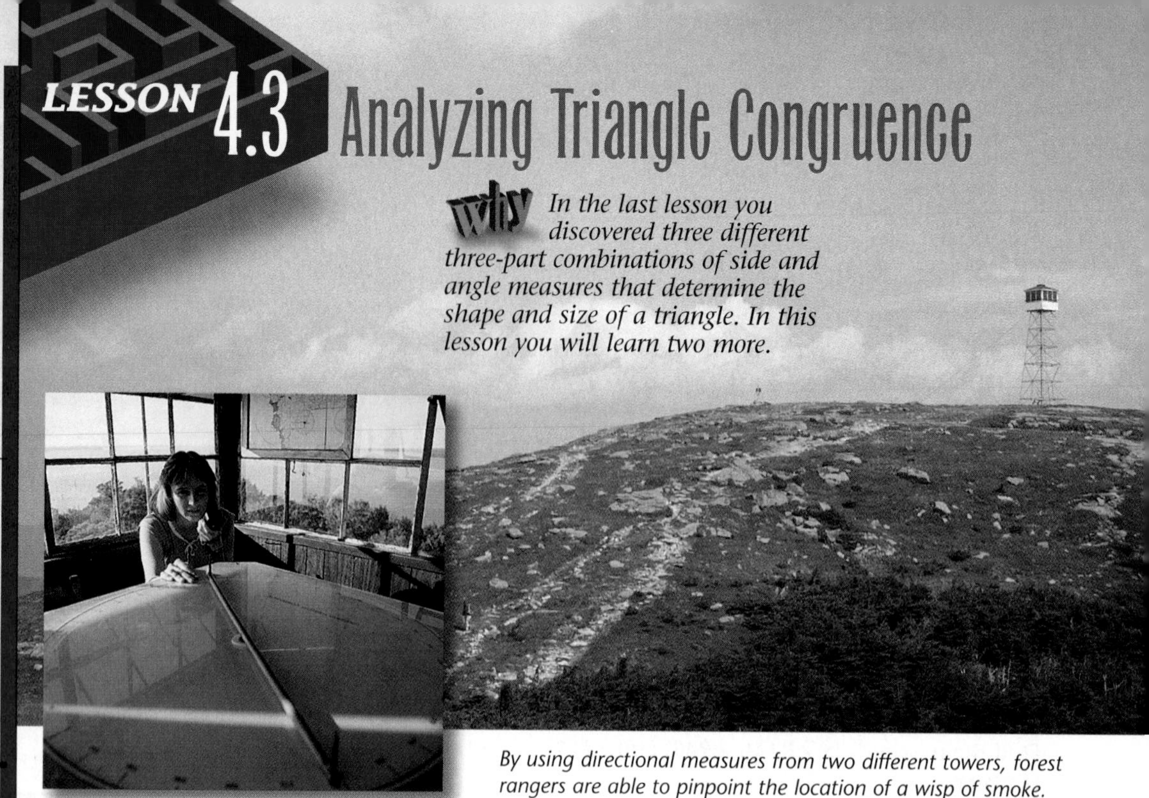

Why *In the last lesson you discovered three different three-part combinations of side and angle measures that determine the shape and size of a triangle. In this lesson you will learn two more.*

By using directional measures from two different towers, forest rangers are able to pinpoint the location of a wisp of smoke. The method is known as ***triangulation*** *because the given conditions determine the shape of a triangle.*

Three Congruence Postulates

The combinations of side and angle measures you learned in the previous lessons can be used to determine whether two triangles are congruent. In the explorations, you used these combinations to write your own congruence postulates. The traditional versions of the postulates are as follows:

SSS (SIDE-SIDE-SIDE) POSTULATE

If three sides in one triangle are congruent to three sides in another triangle, then the triangles are congruent. **4.3.1**

SAS (SIDE-ANGLE-SIDE) POSTULATE

If two sides and the angle between them in one triangle are congruent to two sides and the angle between them in another triangle, then the triangles are congruent. **4.3.2**

ASA (ANGLE-SIDE-ANGLE) POSTULATE

If two angles and the side between them in one triangle are congruent to two angles and the side between them in another triangle, then the triangles are congruent. **4.3.3**

Technology Many geometry graphics software packages come with activities for the three congruence postulates covered in this lesson. Students can use the files provided to explore the different cases.

APPLICATION

Navigation

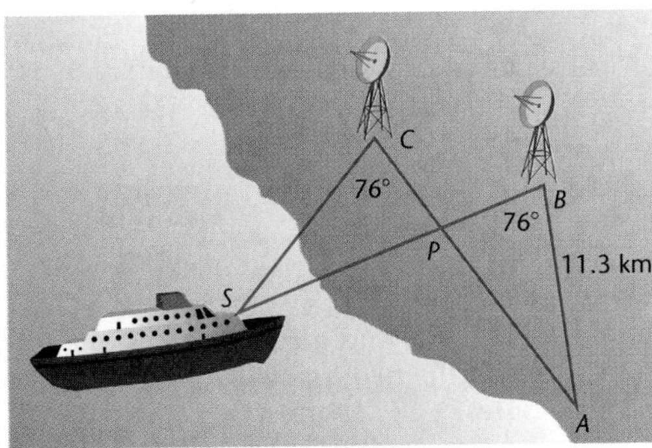

A ship at sea loses power and is stranded at S. The ship is located as shown, and point P is equidistant from stations B and C. The measurements shown are calculated from data obtained by radio and radar. How far is the ship from station C?

$\angle C \cong \angle B$ and $\overline{CP} \cong \overline{BP}$. $\angle CPS$ and $\angle BPA$ are congruent, because they are vertical angles. Therefore, by *ASA*, $\triangle PCS \cong \triangle PBA$. Since $\overline{BA}$ corresponds to $\overline{CS}$, the distance to the ship from the radar station is 11.3 km. ❖

Three Other Possibilities

You may have realized that the three-part combinations of sides and angles you studied in the last lesson are not the only combinations to consider. In fact, there are three others. Which of the following combinations do you think can establish triangle congruence?

1. **AAA combination** Three pairs of angles
2. **SSA combination** Two sides and an angle that is not between them (the angle is *opposite* one of the two sides)
3. **AAS combination** Two angles and a side that is not between them

You can quickly deal with the first two of the combinations by finding counterexamples. A simple proof will show that the third combination works.

EXAMPLE 1

Show that the AAA combination is not a valid test for triangle congruence.

Solution ➤

In order to get a better idea of what you need to disprove, state the combination in the form of a conjecture, as follows:

Conjecture: If three angles of one triangle are congruent to three angles in another triangle, then the triangles are congruent.

Counterexamples to this statement are easy to find. In the figures at right, there are three pairs of congruent angles (from one triangle to the next), but the two triangles are not congruent. Therefore, *the conjecture is false.* ❖

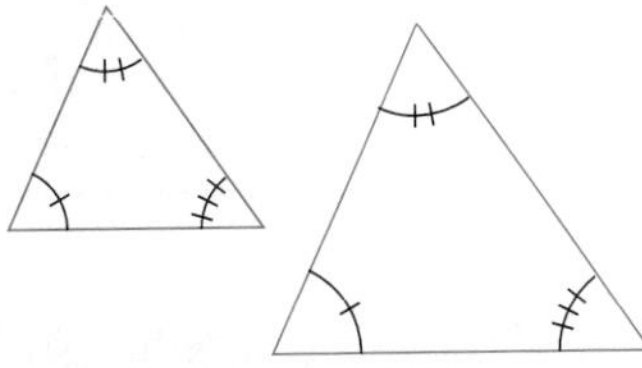

AAA Counterexample

ENRICHMENT It is not necessary to accept all of the conjectures SAS, SSS, and ASA as postulates. SSS and ASA can be proved from SAS. Challenge students to write the two proofs.

Using Discussion The concepts covered in this lesson can be approached using the question, "What three pieces of information are necessary and sufficient to enable you to draw or build a unique triangle?" Students will discover that the combinations SSS, ASA, AAS, and SAS work—AAA and SSA do not.

Use Transparency 34

TEACHING tip

The first three postulates in this lesson are the same as the ones students explored in the previous lesson. Have students compare the postulate names they wrote for Lesson 4.2 with those now given in the text.

Alternate Example 1

Have students draw these triangles and then tell which two disprove the AAA combination.

1. $\angle A = 40°$, $\angle B = 55°$, $\angle C = 85°$, $AC = 3.8$ cm
2. $\angle A = 40°$, $\angle B = 55°$, $\angle C = 85°$, $AB = 4.2$ cm
3. $\angle A = 40°$, $\angle B = 55°$, $\angle C = 85°$, $BC = 5.2$ cm

[Either triangles 1 and 3 or triangles 2 and 3 disprove AAA. Triangles 1 and 2 are congruent.]

TEACHING tip

Technology Counterexamples to the AAA conjecture in Example 1 can be generated using a dilation tool in geometry graphics software. Students can make tables listing the side and angle measures.

Alternate Example 2

Given: $\triangle PQR$ and $\triangle XYZ$ have congruent parts in the AAS pattern: $\angle P \cong \angle X$, $\angle Q \cong \angle Y$, $QR \cong YZ$

Prove: $\triangle PQR \cong \triangle XYZ$

[The givens tells us that $\angle P \cong \angle X$ and $\angle Q \cong \angle Y$. The third angles must also be congruent because the sum of the angles in each triangle is 180°. Thus, $\angle R \cong \angle Z$. Since $\overline{QR} \cong \overline{YZ}$ and $\angle Q \cong \angle Y$, the triangles are congruent by ASA.]

TEACHING *tip*

In Example 2, emphasize that it is not sufficient to find one pair of triangles where congruency is established using AAS. One example does not "prove" a theorem. Instead, one must use logical arguments. Compare Example 2 with Examples 1 and 3, in which one counterexample is enough to prove that the conjecture is false.

Use Transparency 35

CRITICAL *Thinking*

The sides of length 3 are not corresponding sides.

ongoing ASSESSMENT

The first three combinations were accepted without proof, while the AAS combination may be proved using ASA.

EXAMPLE 2

Show that the AAS combination is a valid test for triangle congruence.

Solution ➤

The AAS combination can be converted to the ASA combination as illustrated below.

Notice that the given sides are not between the given angles.

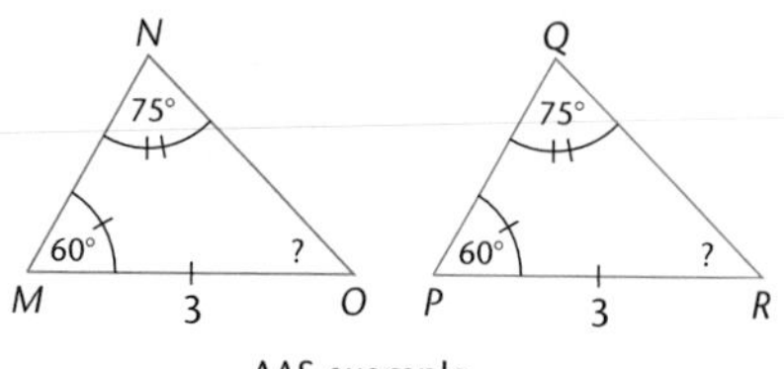

AAS example . . .

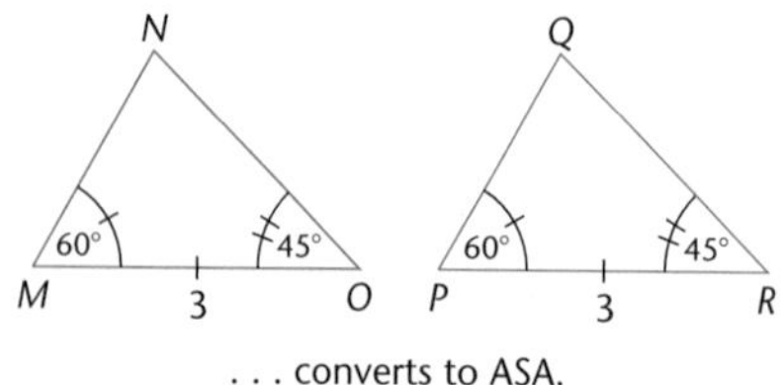

. . . converts to ASA.

The triangles on the right are an AAS combination. To convert this to an ASA combination, find the measures of the missing angles in the two triangles. Since the sum of the measures of the angles of a triangle is 180°:

$m\angle O = 180 - (60 + 75) = 45°$
$m\angle R = 180 - (60 + 75) = 45°$

The measures of the two missing angles are the same, so $\angle O \cong \angle R$. The given side of 3 units is between angles of 45° and 60° in each triangle. Therefore, the two triangles are congruent by the ASA postulate. ❖

CRITICAL *Thinking*

The two triangles below represent a "version" of AAS, but the triangles are not congruent. There's an important difference between the two arrangements. Can you say what it is?

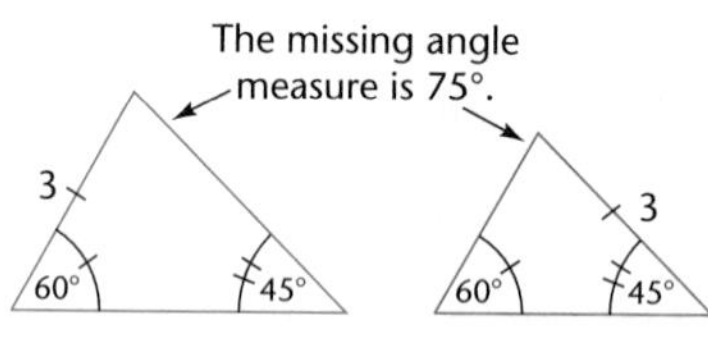

AAS Example ? ? ?

For AAS to be a valid test, the congruent parts must correspond.

AAS (ANGLE-ANGLE-SIDE) THEOREM

If two angles and a side that is not between them in one triangle are congruent to the corresponding two angles and the side not between them in another triangle, then the triangles are congruent. **4.3.4**

The first three combinations you studied were called postulates, but the AAS combination is called a theorem. Why?

RETEACHING the lesson

Using Models To help students remember that SSA doesn't work, have them construct this figure.

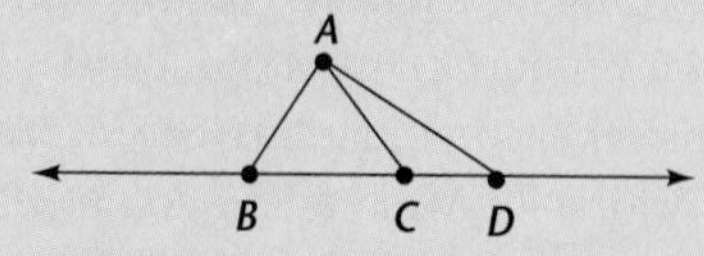

Students start with any isosceles triangle ABC. Then they draw segment $\overline{AD}$ so it meets the line through side $\overline{BC}$. Since $\overline{BA} \cong \overline{CA}$,and side $\overline{AD}$ and $\angle D$ are congruent to themselves, $\triangle BAD$ and $\triangle CAD$ satisfy SSA. The drawing and the phrase "bad cad" may help students remember to beware of SSA.

Try This Which pairs of triangles can be proven congruent by the AAS Theorem?

a.

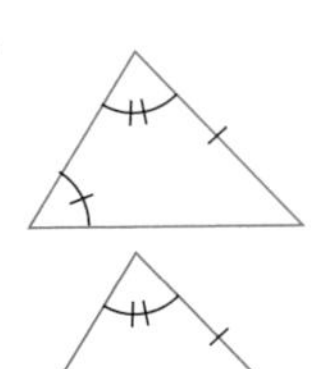

b.

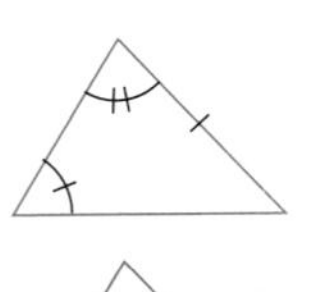

c.

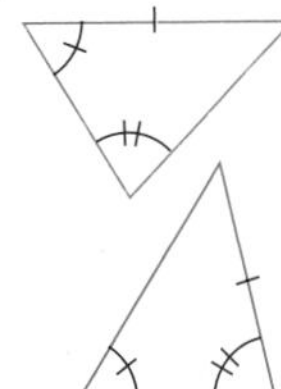

d.

EXAMPLE 3

Show that the SSA combination is not a valid test for triangle congruence.

Solution ➤

It is a bit more difficult to find a counterexample for this conjecture, but the figures on the right do the trick. The sides and angles are congruent as required, but the triangles are obviously not congruent. Therefore, the conjecture is false.

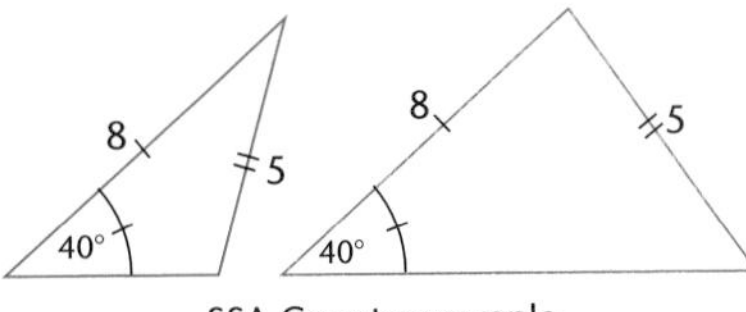

SSA Counterexample

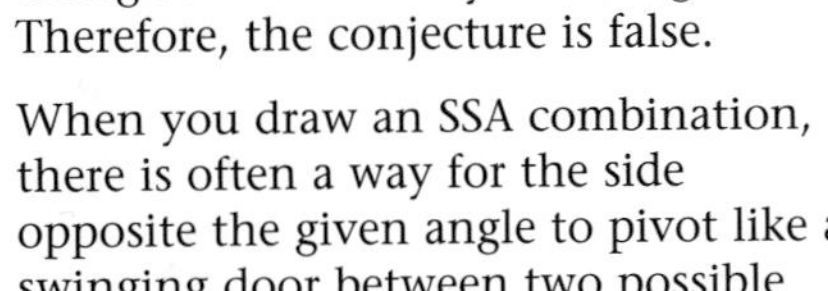

When you draw an SSA combination, there is often a way for the side opposite the given angle to pivot like a swinging door between two possible positions. ❖

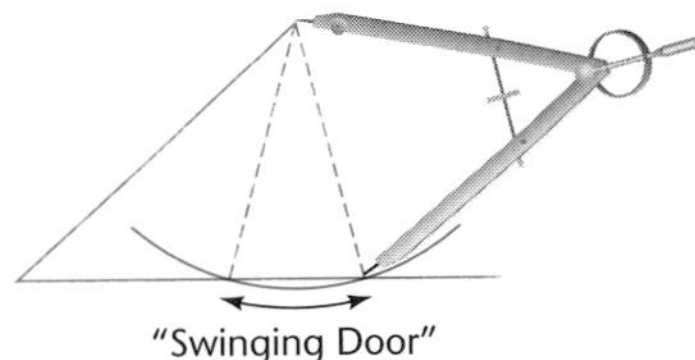

"Swinging Door"

A Special Case of SSA

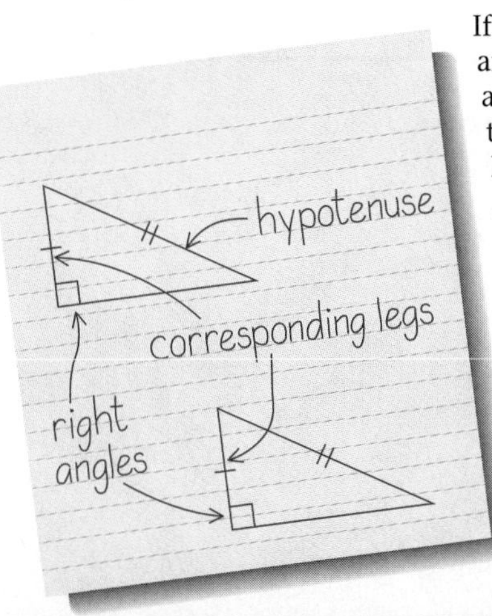

If the given angle in SSA is a right angle, the side opposite the angle cannot pivot in the "swinging-door" manner described above. So if the given angle is a right angle, SSA is a valid test for triangle congruence. In this case the test is called the Hypotenuse-Leg Theorem. A proof of this theorem will not be given here, but you will be able to prove it by adding an additional right triangle and using a property of isosceles triangles (see Lesson 4.4)

HL (HYPOTENUSE-LEG) THEOREM

If the hypotenuse and a leg of a right triangle are congruent to the hypotenuse and the corresponding leg in another right triangle, then the two triangles are congruent. **4.3.5**

ongoing ASSESSMENT

Try This

a, c, and d, not b.

TEACHING tip

The Try This exercises emphasize that the two angles and the side must be in the same order for AAS to apply.

Use Transparency 36

Cooperative Learning

Each group should work together to create a counterexample to the SSA conjecture. To do this, each student should draw a triangle with one 35° angle, one 7 centimeter side, and one 4 centimeter side. Students will find that two different triangles, one acute and one obtuse, satisfy the conditions.

Alternate Example 3

Have students draw these triangles, then tell which two disprove the SSA combination.

1. $\angle X = 40°$, $XZ = 9$ cm, $\angle Z = 30°$
2. $\angle X = 80°$, $XZ = 9$ cm, $\angle Z = 30°$
3. $\angle Y = 110°$, $\angle Z = 30°$, $ZX = 9$ cm

[Either triangles 1 and 2 or triangles 2 and 3 disprove SSA. Triangles 1 and 3 are congruent.]

ASSESS

Selected Answers

Odd-numbered Exercises 7–29

Assignment Guide

Core 1–3, 7–18, 24–31

Core Plus 1–6, 9–27

Technology

Students who can use geometry graphics software to duplicate segments and angles can be asked to replicate the figures in Exercises 7–12. Making the constructions will help reinforce the congruence postulates.

Error Analysis

Students may have difficulty following the directions in Exercise 16. Point out that point *K* is the midpoint of $\overline{TL}$.

Practice Master

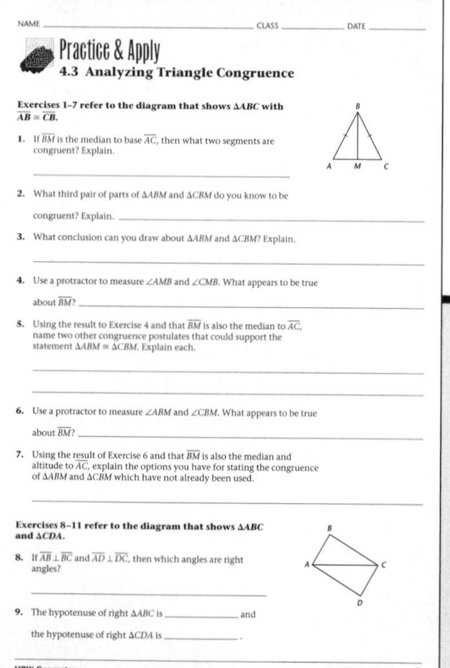

NAME ______________ CLASS ________ DATE ________

Practice & Apply

4.3 Analyzing Triangle Congruence

Exercises 1–7 refer to the diagram that shows $\triangle ABC$ with $\overline{AB} \cong \overline{CB}$.

1. If $\overline{BM}$ is the median to base $\overline{AC}$, then what two segments are congruent? Explain.
   ____
2. What third pair of parts of $\triangle ABM$ and $\triangle CBM$ do you know to be congruent? Explain. ______________
3. What conclusion can you draw about $\triangle ABM$ and $\triangle CBM$? Explain.
   ____
4. Use a protractor to measure $\angle AMB$ and $\angle CMB$. What appears to be true about $\overline{BM}$? ______________
5. Using the result to Exercise 4 and that $\overline{BM}$ is also the median to $\overline{AC}$, name two other congruence postulates that could support the statement $\triangle ABM \cong \triangle CBM$. Explain each.
   ____
   ____
6. Use a protractor to measure $\angle ABM$ and $\angle CBM$. What appears to be true about $\overline{BM}$? ______________
7. Using the result of Exercise 6 and that $\overline{BM}$ is also the median and altitude to $\overline{AC}$, explain the options you have for stating the congruence of $\triangle ABM$ and $\triangle CBM$ which have not already been used.
   ____

Exercises 8–11 refer to the diagram that shows $\triangle ABC$ and $\triangle CDA$.

8. If $\overline{AB} \perp \overline{BC}$ and $\overline{AD} \perp \overline{DC}$, then which angles are right angles?
   ____
9. The hypotenuse of right $\triangle ABC$ is ________ and the hypotenuse of right $\triangle CDA$ is ________.

HRW Geometry   Practice & Apply   3

EXERCISES & PROBLEMS

Communicate

1. Explain the SSS Postulate.
2. Explain the SAS Postulate.
3. Explain the ASA Postulate.
4. Explain why SSA does not show congruence.
5. **Triangulation** Towers A and B each sight a smoke plume in the directions shown in the diagram. What combination of sides and angles (AAS, SAS, etc.) are known in the triangle?

   If a triangle has the given measures, is there more than one shape the triangle can have? Explain how the information determines the location of the smoke.

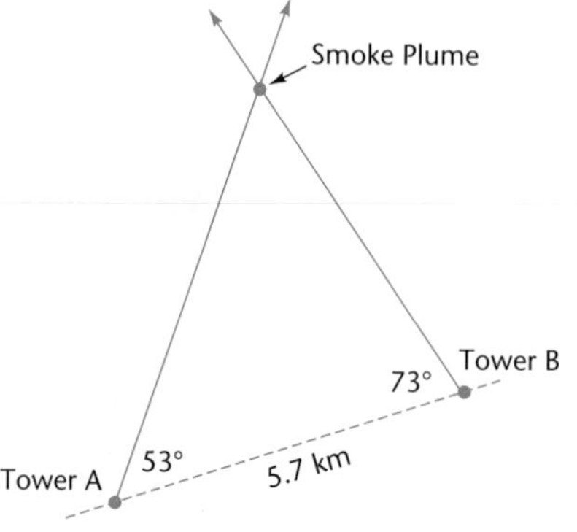

6. Which pairs of triangles are congruent and why?

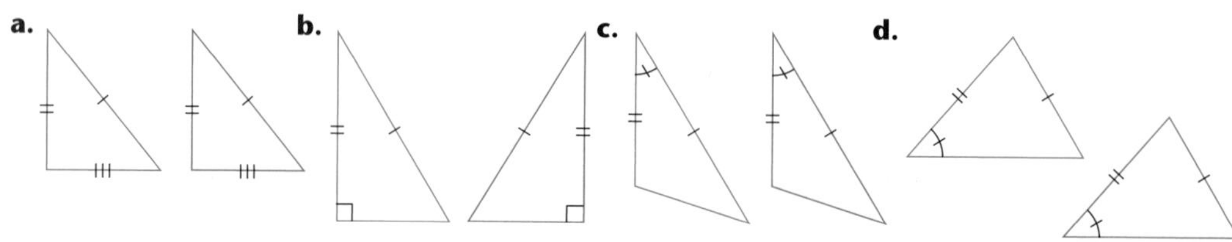

Practice & Apply

For Exercises 7–9, decide whether each pair of triangles are congruent. If so, write an appropriate congruence statement and name the theorem that supports it.

7.

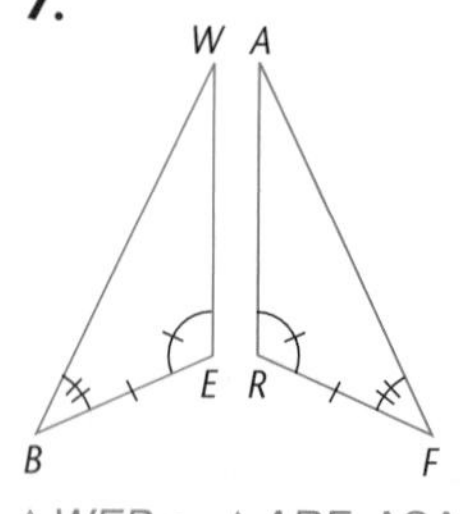

$\triangle WEB \cong \triangle ARF$; ASA

8.

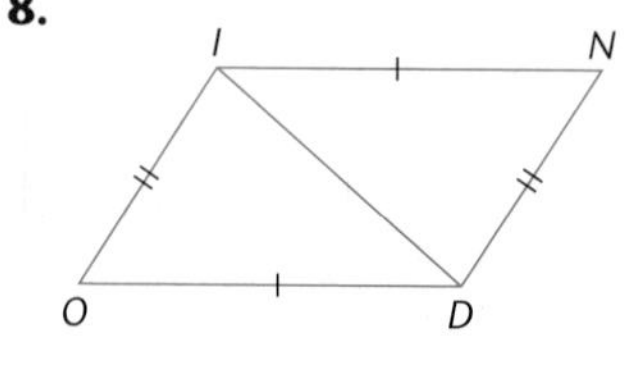

$\triangle IOD \cong \triangle DNI$; SSS

9.

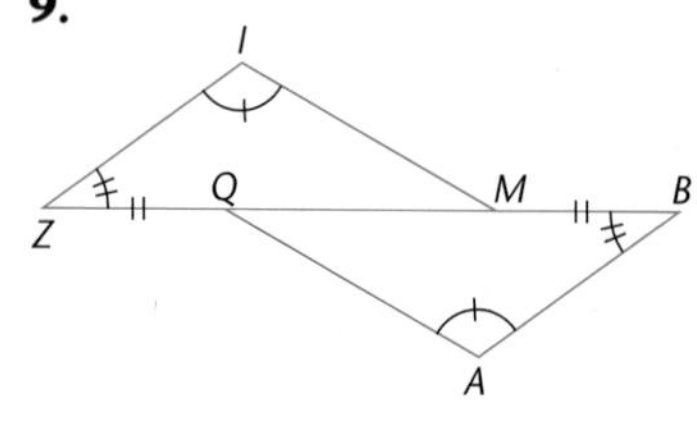

$\triangle MZI \cong \triangle QBA$; AAS

Copy the figures and label $\angle U$ and $\angle E$ as right angles for Exercises 10–12.

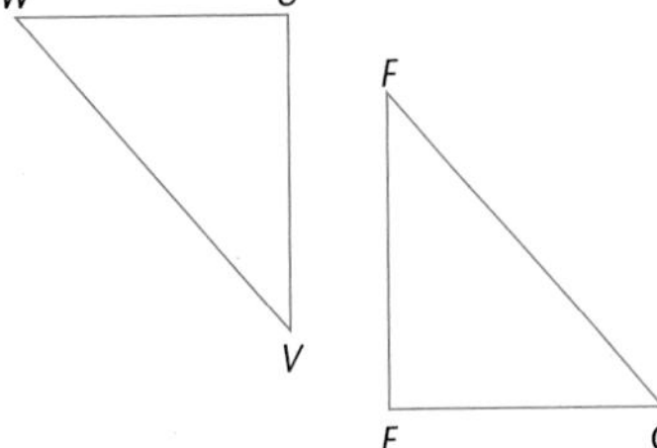

10. Identify the hypotenuse in each triangle. $\overline{VW}$; $\overline{FG}$

11. Identify the legs in each triangle. $\overline{UW}$ and $\overline{UV}$; $\overline{EF}$ and $\overline{EG}$

12. If $\overline{UV} \cong \overline{EF}$ and $\overline{UW} \cong \overline{EG}$, are the triangles congruent? Explain your answer. Yes, by SAS

F

E G

Copy the figures and label $\angle C$ and $\angle T$ as right angles for Exercises 13–15.

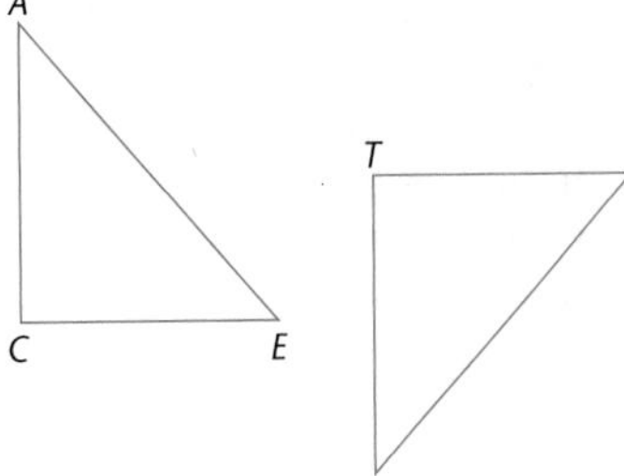

13. Identify the hypotenuse in each triangle. $\overline{AE}$; $\overline{RS}$

14. Identify the legs in each triangle. $\overline{CA}$ and $\overline{CE}$; $\overline{TR}$ and $\overline{TS}$

15. If $\angle A \cong \angle R$, are the triangles congruent? Explain your answer. No, at least one pair of congruent corresponding sides is needed.

T S

R

Triangle TAL is isosceles, with $\overline{AT} \cong \overline{AL}$. Copy the figure and label the congruent sides for Exercises 16–20.

A

T L

16. Draw $\overline{AK}$ with K on $\overline{TL}$ so that $\overline{AK} \perp \overline{TL}$. Label the right angles on your drawing.

17. You have now created two right triangles. Name the two right angles and identify the hypotenuse for each. $\angle TKA$, hypotenuse $\overline{TA}$; $\angle LKA$, hypotenuse $\overline{LA}$

18. What must be true about the two hypotenuses? Explain your reasoning. $\overline{TA} \cong \overline{LA}$ because they are the sides of an isoseles $\triangle$.

19. Why is $\overline{AK} \cong \overline{AK}$? Reflexive Property of Congruence

20. What do you know about $\triangle TAK$ and $\triangle LAK$? Why? $\triangle TAK \cong \triangle LAK$ by HL

Use the diagram below for Exercises 21 and 22.

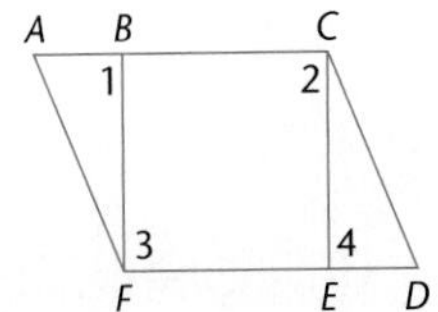

21. Given: $\angle A \cong \angle D$, $\overline{AF} \cong \overline{DC}$, $\overline{AC} \cong \overline{DF}$, $\angle BFA \cong \angle ECD$
Prove: $\triangle AFB \cong \triangle DCE$

22. Given: $\angle 1 \cong \angle 4$, $\overline{AF} \cong \overline{CD}$, $\angle A \cong \angle D$
Prove: $\triangle AFB \cong \triangle DCE$

16.

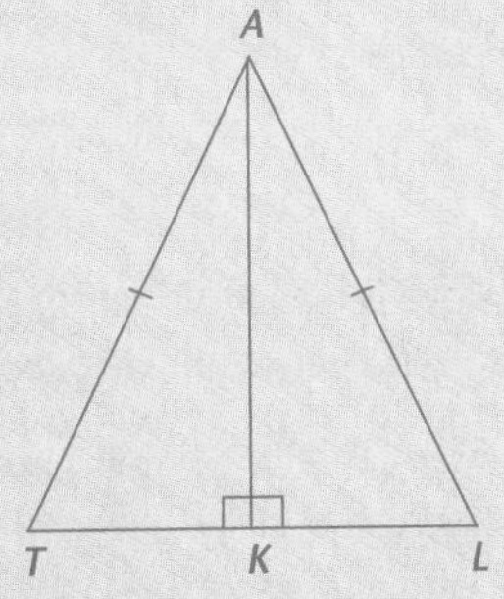

20. $\triangle TAK \cong \triangle LAK$ by HL.

alternative ASSESSMENT

Portfolio Assessment

Have students draw $\triangle XYZ$ congruent to any student-made $\triangle ABC$. Then, for each of the following postulates, they should identify three correspondences that would guarantee congruency.

1. SSS
2. ASA
3. AAS
4. SAS

21. $\angle A \cong \angle D$, $\overline{AF} \cong \overline{CD}$, and $\angle BFA \cong \angle ECD$ are given. Setting up the correspondence, then $\triangle AFB \cong \triangle DCE$ by ASA.

22. $\angle 1 \cong \angle 4$, $\overline{AF} \cong \overline{CD}$, and $\angle A \cong \angle D$ are given. Setting up the correspondence, then $\triangle AFB \cong \triangle DCE$ by AAS.

Technology Master

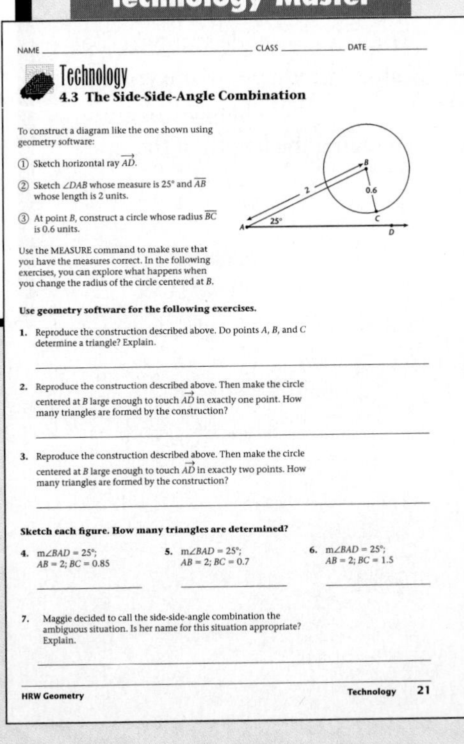

NAME ______________________ CLASS ____________ DATE ____________

Technology
4.3 The Side-Side-Angle Combination

To construct a diagram like the one shown using geometry software:

① Sketch horizontal ray $\overrightarrow{AD}$.

② Sketch $\angle DAB$ whose measure is 25° and $\overline{AB}$ whose length is 2 units.

③ At point B, construct a circle whose radius $\overline{BC}$ is 0.6 units.

Use the MEASURE command to make sure that you have the measures correct. In the following exercises, you can explore what happens when you change the radius of the circle centered at B.

Use geometry software for the following exercises.

1. Reproduce the construction described above. Do points A, B, and C determine a triangle? Explain.

2. Reproduce the construction described above. Then make the circle centered at B large enough to touch $\overrightarrow{AD}$ in exactly one point. How many triangles are formed by the construction?

3. Reproduce the construction described above. Then make the circle centered at B large enough to touch $\overrightarrow{AD}$ in exactly two points. How many triangles are formed by the construction?

Sketch each figure. How many triangles are determined?

4. $m\angle BAD = 25°$; $AB = 2$; $BC = 0.85$

5. $m\angle BAD = 25°$; $AB = 2$; $BC = 0.7$

6. $m\angle BAD = 25°$; $AB = 2$; $BC = 1.5$

7. Maggie decided to call the side-side-angle combination the ambiguous situation. Is her name for this situation appropriate? Explain.

HRW Geometry   Technology 21

23.

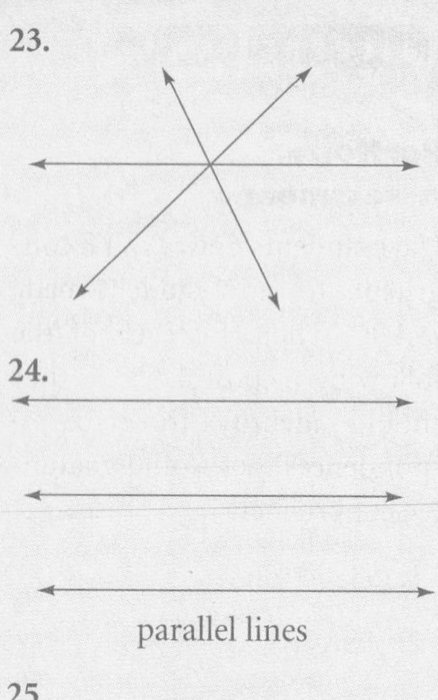

24.

parallel lines

25.

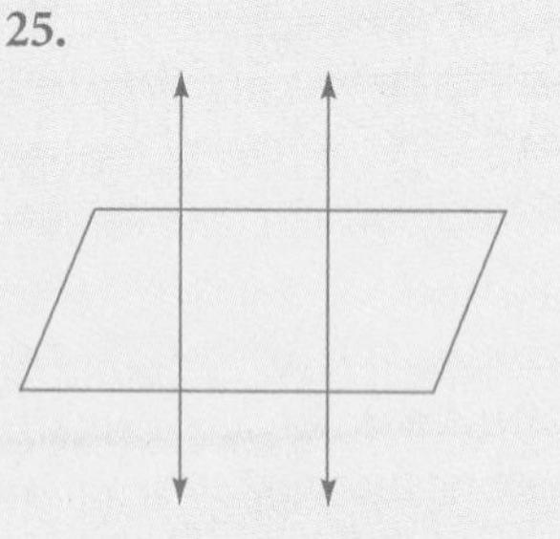

Look Beyond

Extension. The students should have no difficulty finding real-world examples of triangular bracing. In many instances students will find that 4-sided structural elements have one or two braces connecting opposite corners, as in the Forth Rail Bridge and the Eiffel Tower. Triangles provide support because the shape of the triangle can not be changed without changing the length of the sides.

Look Back

23. Draw three lines intersecting in a point. **[Lesson 1.1]**

24. Draw three lines that do not intersect each other. **[Lesson 1.1]**

25. Draw two lines that intersect a plane without intersecting each other. **[Lesson 1.1]**

26. To the right is a drawing of a transversal intersecting parallel lines. Label the lines as parallel and identify the congruent angles. **[Lesson 3.3]**

27. Identify two angles in the diagram that are supplementary. **[Lesson 3.3]**

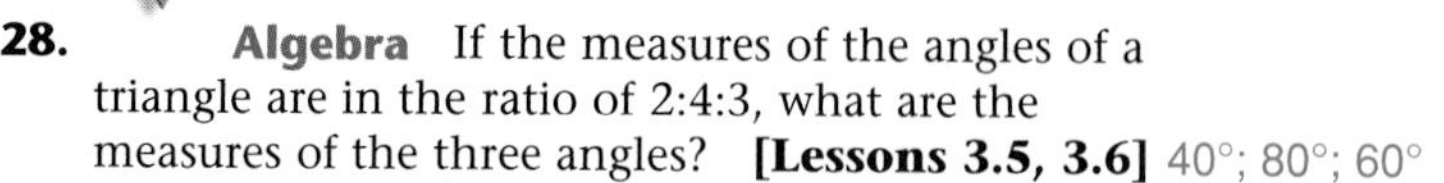

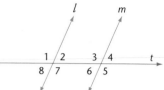

28. **Algebra** If the measures of the angles of a triangle are in the ratio of 2:4:3, what are the measures of the three angles? **[Lessons 3.5, 3.6]** 40°; 80°; 60°

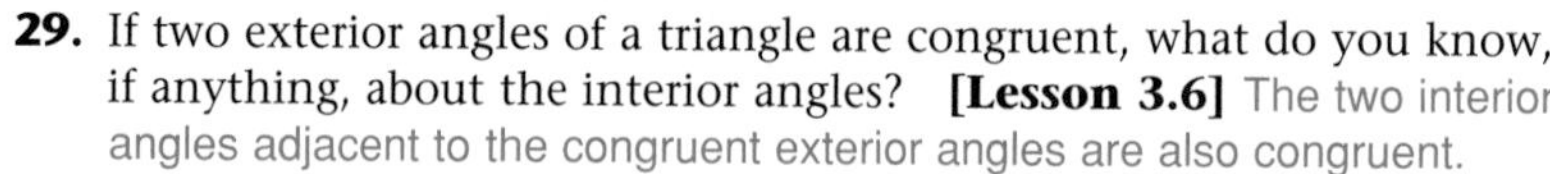

29. If two exterior angles of a triangle are congruent, what do you know, if anything, about the interior angles? **[Lesson 3.6]** The two interior angles adjacent to the congruent exterior angles are also congruent.

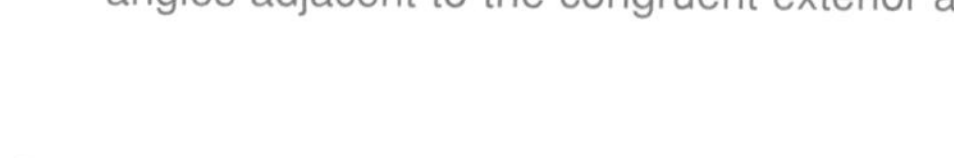

Look Beyond

30. **Construction** Triangles are often used in construction. Find a building, bridge, or house on which you can see the triangular supports, and draw an illustration of the supports. Identify the congruent triangles and explain why the triangles provide support.

The Forth Rail Bridge (1889) in Scotland is 5330 feet long and weighs 38,000 tons. Its strength rests in each of three cantilevered units that become a single truss. Can you see the triangles in the design?

26. $\angle 1 \cong \angle 3 \cong \angle 5 \cong \angle 7$; $\angle 2 \cong \angle 4 \cong \angle 6 \cong \angle 8$

27. Choose answers from: $\angle 1$ and $\angle 2$; $\angle 3$ and $\angle 4$; $\angle 5$ and $\angle 6$; $\angle 7$ and $\angle 8$; $\angle 1$ and $\angle 8$; $\angle 2$ and $\angle 7$; $\angle 3$ and $\angle 6$; $\angle 4$ and $\angle 5$; $\angle 1$ and $\angle 4$; $\angle 2$ and $\angle 3$; $\angle 5$ and $\angle 8$; $\angle 6$ and $\angle 7$.

30. Check student work.

LESSON 4.4 Using Triangle Congruence

why *The design elements of this famous woodcut by Dutch artist M.C. Escher are mathematically related. The size of one part determines the size of another. By using chains of mathematical reasoning, you can deduce things about one part of a figure from information about another part.*

In this course you will use chains of mathematical reasoning to discover or prove things about geometric figures. Your triangle congruence postulates and theorem will be some of your most powerful tools for doing this.

"CPCTC"

According to the definition of congruent triangles, *if two triangles are congruent, then their corresponding parts are congruent*. Therefore, if $\triangle ABC \cong \triangle DEF$, you can conclude that $\overline{AB} \cong \overline{DE}$ based on knowing that $\triangle ABC \cong \triangle DEF$. What other pairs of congruent sides and angles must be congruent?

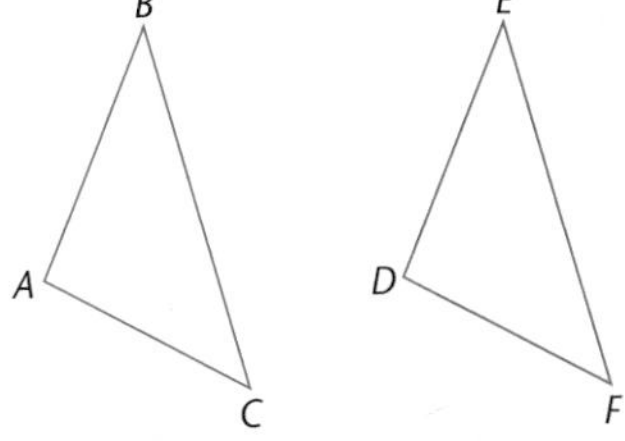

If $\triangle ABC \cong \triangle DEF$, then $\overline{AB} \cong \overline{DE}$, . . .

This idea is often stated in the following form: *Corresponding parts of congruent triangles are congruent,* abbreviated CPCTC. This statement is frequently used as a reason in proofs.

In each of the examples that follow, you will use a triangle congruence postulate or theorem to establish that two triangles are congruent. Then you will use CPCTC.

ALTERNATIVE teaching strategy

Hands-On Strategies Have students use tracing paper to trace the figures provided along with each example. Then they should carefully record the givens, marking pairs of corresponding congruent parts.

PREPARE

Objectives

- Use congruence of corresponding parts to prove congruence of triangles.
- Develop and use the Isosceles Triangle Theorem.

RESOURCES

- Practice Master 4.4
- Enrichment Master 4.4
- Technology Master 4.4
- Lesson Activity Master 4.4
- Quiz 4.4
- Spanish Resources 4.4

Assessing Prior Knowledge

1. In $\triangle AED$ and $\triangle BCD$, you are given that $\overline{AE} \cong \overline{BC}$, $\overline{AD} \cong \overline{BD}$, and $\overline{DE} \cong \overline{DC}$. Which triangles can you prove congruent and why? **[$\Delta AED \cong \Delta BCD$, SSS]**
2. In ΔAED and ΔBCD, $\overline{AE} \cong \overline{BC}$, $\overline{AD} \cong \overline{BD}$, and $\angle E \cong \angle C$. Why can't you prove that $\Delta AED \cong \Delta BCD$? **[correspondences shown are SSA]**

Use Transparency 37

TEACH

why Have students work together to discuss the Escher design given in the text. They should look for examples of congruent polygons. They may also be able to identify translations, reflections, and rotations.

Alternate Example 1

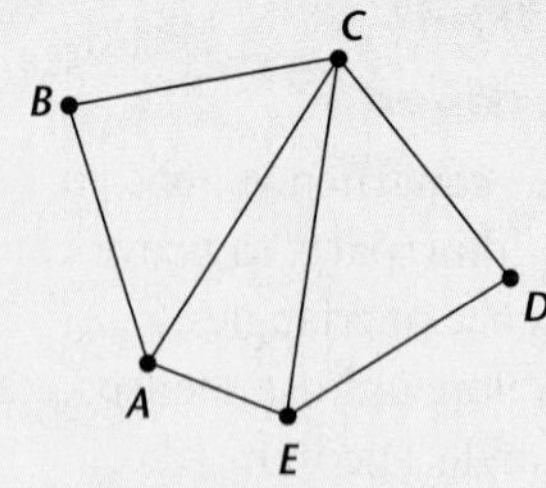

Given: $\overline{AB} \cong \overline{BC} \cong \overline{DC} \cong \overline{ED}$
$\angle B \cong \angle D$

Prove: $AC \cong EC$

[Statement
1. $\overline{AB} \cong \overline{ED}$
2. $\angle B \cong \angle D$
3. $\overline{BC} \cong \overline{DC}$
4. $\triangle ABC \cong \triangle EDC$
5. $\overline{AC} \cong \overline{EC}$

Reason
1. Given
2. Given
3. Given
4. SAS
5. CPCTC]

Alternate Example 2

Given: Two segments $\overline{MP}$ and $\overline{NQ}$ bisect each other in point O.

Prove: The segments joining the endpoints of the segments are congruent.

[Statement
1. $\overline{NO} \cong \overline{QO}$
2. $\angle NOM \cong \angle QOP$
3. $\overline{OM} \cong \overline{OP}$
4. $\triangle MNO \cong \triangle PQO$
5. $\overline{MN} \cong \overline{PQ}$

Reason
1. Def bisect
2. Vertical angles equal
3. Def bisect
4. SAS
5. CPCTC]

CRITICAL *Thinking*

If angles C and D were not known to be congruent, the given vertical angles would give SSA, which does not ensure triangle congruence.

EXAMPLE 1

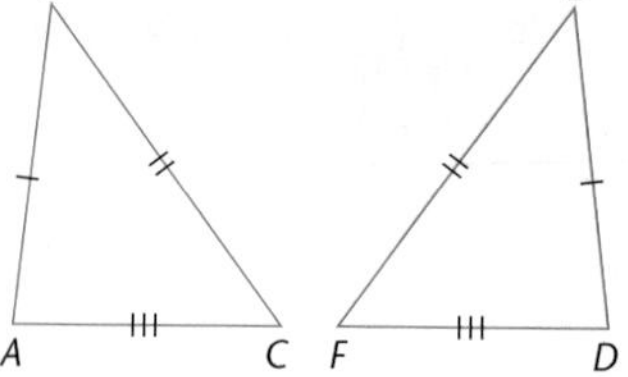

Given: $\overline{AB} \cong \overline{DE}$, $\overline{BC} \cong \overline{EF}$, $\overline{AC} \cong \overline{DF}$

Prove: $\angle A \cong \angle D$

Plan: Use a triangle congruence postulate to show that the two triangles are congruent. Then use CPCTC to establish $\angle A \cong \angle D$.

Solution ➤

Proof:

Statements	Reasons
1. $\overline{AB} \cong \overline{DE}$	Given
2. $\overline{BC} \cong \overline{EF}$	Given
3. $\overline{AC} \cong \overline{DF}$	Given
4. $\triangle ABC \cong \triangle DEF$	SSS
5. $\angle A \cong \angle D$	CPCTC ❖

In the next example, a simple congruence proof is a part of a larger proof.

EXAMPLE 2

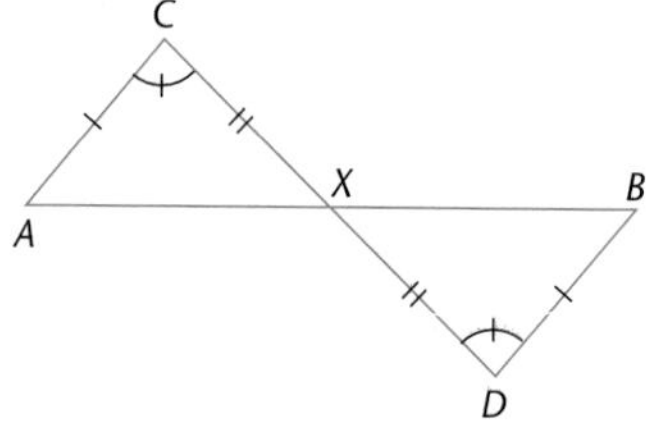

Given: $\overline{AC} \cong \overline{BD}$, $\overline{CX} \cong \overline{DX}$, $\angle C \cong \angle D$

Prove: X is the midpoint of $\overline{AB}$.

Plan: Use congruent triangles to show that $\overline{AX} \cong \overline{BX}$ (and so, $AX = BX$). Then use the definition of a midpoint.

Solution ➤

Proof:

Statements	Reasons
1. $\overline{AC} \cong \overline{BD}$	Given
2. $\overline{CX} \cong \overline{DX}$	Given
3. $\angle C \cong \angle D$	Given
4. $\triangle ACX \cong \triangle BDX$	SAS
5. $\overline{AX} \cong \overline{BX}$ $(AX = BX)$	CPCTC
6. X is the midpoint of $\overline{AB}$	Def. midpoint ❖

Why can't vertical angles $\angle AXC$ and $\angle BXD$ be used in this proof instead of $\angle C$ and $\angle D$?

Art Students who are interested in the Escher design that begins this lesson may enjoy trying to analyze some of Escher's work. Provide photocopies of Escher designs and tracing paper. Students try to find underlying triangular grid structures used to make the designs.

Overlapping Triangles

How many different triangles can you find in the figures?

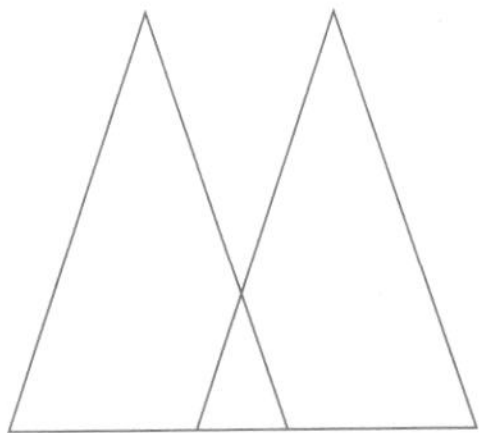
Figure 1

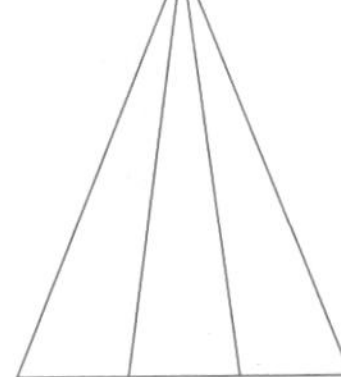
Figure 2

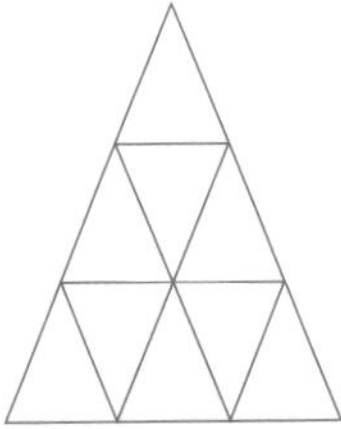
Figure 3

You will work with many diagrams such as Figure 2 in proving triangles congruent.

In the proof that follows, you will use the Overlapping Segment Theorem. Recall that for a figure like the one on the right you can conclude that $\overline{AC} \cong \overline{BD}$.

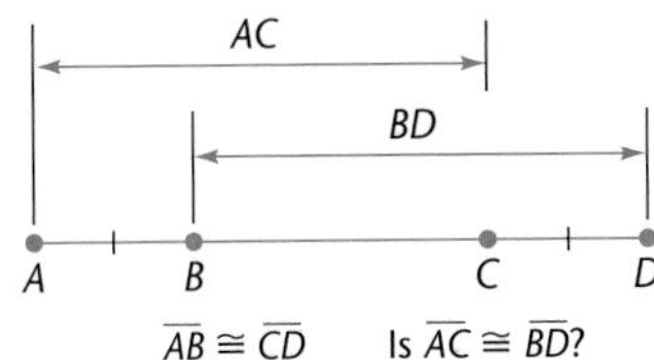

EXAMPLE 3

Given: $\overline{AB} \cong \overline{CD}$ $(AB = CD)$, $\overline{AE} \cong \overline{FD}$ $(AE = FD)$, $\angle A \cong \angle D$

Prove: $\overline{EC} \cong \overline{FB}$

Plan: Use the Overlapping Segments Theorem to show that $\overline{AC} \cong \overline{BD}$, and so $\triangle ACE \cong \triangle DBF$ by SAS. Then use CPCTC to show that $\overline{EC} \cong \overline{FB}$.

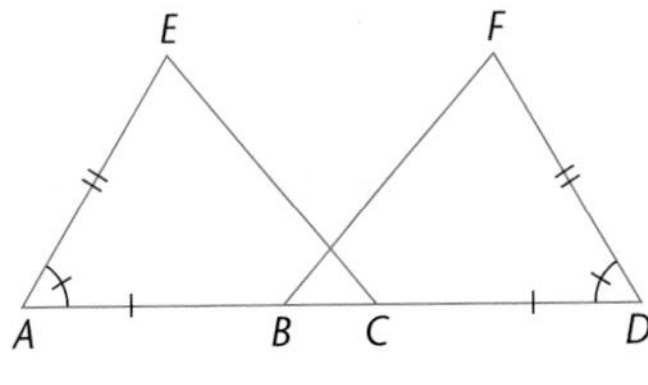

Solution ➤

Proof:

STATEMENTS	REASONS
1. $AB = CD$	Given
2. $AC = BD$ $(\overline{AC} \cong \overline{BD})$	Overlapping Segment Thm.
3. $\overline{AE} \cong \overline{FD}$	Given
4. $\angle A \cong \angle D$	Given
5. $\triangle ACE \cong \triangle DBF$	SAS
6. $\overline{EC} \cong \overline{FB}$	CPCTC ❖

Use Transparency 38

Alternate Example 3

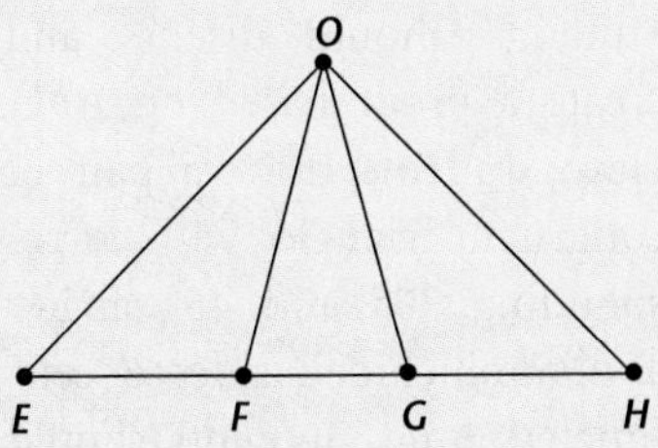

Given: $\angle EOF \cong \angle HOG$, $\angle OFE \cong \angle OGH$, $\overline{EG} \cong \overline{FH}$ Prove: ΔEOH is isosceles

[Statement
1. $\angle EOF \cong \angle HOG$
2. $\angle OFE \cong \angle OGH$
3. $\overline{EF} \cong \overline{HG}$
4. $\triangle EOF \cong \triangle HOG$
5. $\overline{EO} \cong \overline{HO}$ or $\angle E \cong \angle H$
6. $\triangle EOH$ is isosceles

Reason
1. Given
2. Given
3. Overlapping Segment Thm.
4. AAS
5. CPCTC
6. Def isosceles Δ]

Given: $\angle EAB \cong \angle CBA$, $\overline{EA} \cong \overline{CB}$, $\overline{DE} \cong \overline{DC}$, $\overline{BD} \cong \overline{AD}$

Prove: $\triangle BDE \cong \triangle ADC$

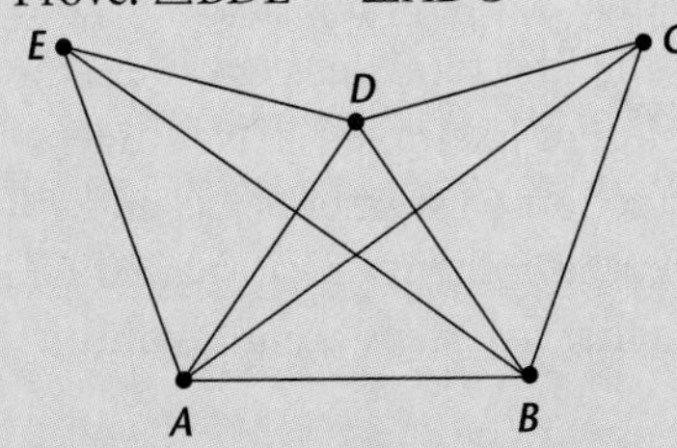

[Statement
1. $\overline{AB} \cong \overline{BA}$
2. $\angle EAB \cong \angle CBA$
3. $\overline{EA} \cong \overline{CB}$
4. $\triangle EAB \cong \triangle CBA$
5. $\overline{BE} \cong \overline{AC}$
6. $\overline{BD} \cong \overline{AD}$
7. $\overline{DE} \cong \overline{DC}$
8. $\triangle BDE \cong \triangle ADC$

Reason
1. Reflexive Prop
2. Given
3. Given
4. SAS
5. CPCTC
6. Given
7. Given
8. SSS]

Use Transparency 39

Cooperative Learning

Students should discuss and create a proof based on AAS. Have students draw a pair of congruent triangles with corresponding 30° and 45° angles and nonincluded sides 6 centimeters long. Students should choose another pair of corresponding sides and prove that they are congruent.

ongoing ASSESSMENT

Try This

Statements

1. $\overline{AC} \cong \overline{BC}$
2. $\overline{CD}$ is the bisector of $\angle ACB$
3. $\angle ACD \cong \angle BCD$
4. $\overline{CD} \cong \overline{CD}$
5. $\triangle ADC \cong \triangle BDC$
6. $\angle A \cong \angle B$

Reasons

1. Given
2. As constructed
3. Def angle bisector
4. Reflexive Prop of Congruence
5. SAS
6. CPCTC

TEACHING tip

Another example of a theorem and its converse: Every equilateral triangle is equiangular. Every equiangular triangle is equilateral. When a theorem and its converse are both true, they can be combined into a single theorem using *if and only if*. In this case: A triangle is equilateral if and only if it is equiangular.

The Isosceles Triangle Theorem

An **isosceles triangle** is a triangle with at least two congruent sides. The two congruent sides are known as the **legs** of the triangle, and the remaining side is known as the **base**. The angles at the base are known as the **base angles**, and the angle opposite the base is known as the **vertex angle**.

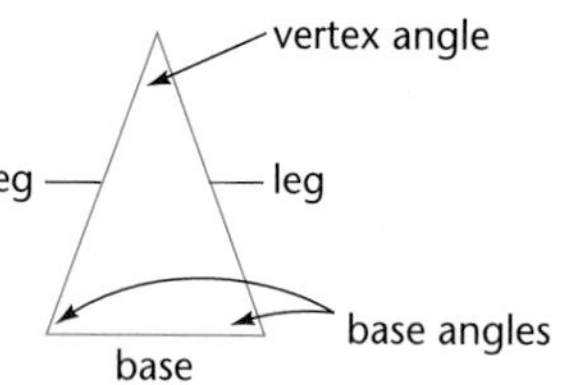

An **equilateral triangle** is a special type of triangle in which all three sides of the triangle are congruent. Is an equilateral triangle isosceles?

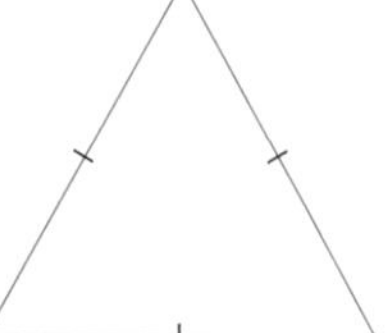

The following theorem, which is one of the great geometry classics, is one you may have already conjectured.

ISOSCELES TRIANGLE THEOREM

If two sides of a triangle are congruent, then the angles opposite those sides are congruent. **4.4.1**

Try This Prove the Isosceles Triangle Theorem using the plan provided below.

Given: $\overline{AC} \cong \overline{BC}$

Prove: $\angle A \cong \angle B$

Plan: Draw an angle bisector of the vertex angle and extend it to the base. Show that the two triangles that result are congruent by SAS. Then use CPCTC.

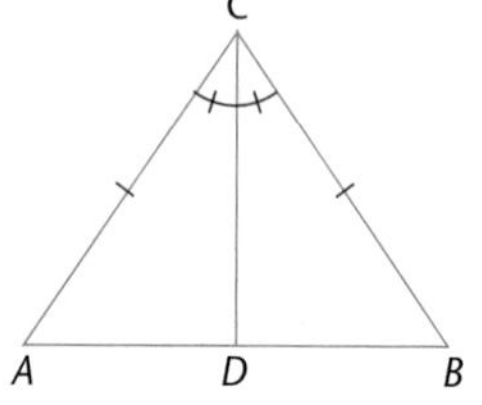

ISOSCELES TRIANGLE THEOREM: CONVERSE

If two angles of a triangle are congruent, then the sides opposite those angles are congruent. **4.4.2**

RETEACHING the lesson

Hands-On Strategies Overlapping figures such as those in Example 3 are particularly difficult for some students to read. Have students trace the figure and use two different colored pencils for the segments in each triangle. Students can also redraw the figures as two separate triangles.

INCLUSION strategies

Using Cognitive Strategies In Examples 2 and 3, other results can be proven from the given statements. For Example 2, have students prove that $\angle AXC$ and $\angle BXD$ are congruent angles. The first four steps are the same as in the example. For Example 3, have students mark the intersection of segments $\overline{EC}$ and $\overline{FB}$ as point O. Then have them prove that ΔBOC is isosceles. The first five steps are the same.

APPLICATION

Construction A resort owner wishes to place a gondola ride across a small canyon on her property. She needs to know the distance across the canyon in order to be well informed for construction bids. Study the diagram below. What is the distance across the canyon?

Since the 80° angle is an exterior angle, then $80° = 40° + m\angle x$, and so $m\angle x = 40°$ by the Exterior Angle Theorem. Since the base angles of the triangle are both 40°, then the sides opposite are congruent by the Converse of the Isosceles Triangle Theorem. Therefore, the distance across the canyon is 350 feet.

There are a number of corollaries to the Isosceles Triangle Theorem which you will also be able to prove.

COROLLARY

An equilateral triangle has three angles that measure 60°. **4.4.3**

COROLLARY

The bisector of the vertex angle of an isosceles triangle is the perpendicular bisector of the base. **4.4.4**

Communicate

1–3. Give the reason for each of the statements.

STATEMENTS	REASONS
$\overline{BA} \cong \overline{BC}$	Given
M is the midpoint of $\overline{AC}$	Given
$\overline{AM} \cong \overline{MC}$	**(1)**
$\overline{BM} \cong \overline{BM}$	**(2)**
$\triangle ABM \cong \triangle CBM$	**(3)**

4–5. Complete the statement of the theorem below, which the above proof establishes.

THEOREM

The bisector of the vertex angle of an isosceles triangle __(4)__ the triangle into __(5)__. **4.4.5**

ASSESS

Selected Answers

Odd-numbered Exercises 7–35

Assignment Guide

Core 1–12, 22, 25–36

Core Plus 1–5, 11–24, 30–37

Technology

Exercise 21 provides a good opportunity for using geometry graphics software. Students construct the figure and drag on line $\overleftrightarrow{AB}$, line $\overleftrightarrow{DE}$, or on any of the points *A*, *B*, *C*, or *E*. Students will see that *C* remains the midpoint of segment $\overline{BE}$ and that segments $\overline{AB}$ and $\overline{DE}$ remain congruent.

Error Analysis

For the proofs in Exercises 13–20, students will need to use properties of parallel lines cut by a transversal. If necessary, refer students to Lesson 3.3 for review.

alternative ASSESSMENT

Performance Assessment

Provide students with this figure and the following givens.

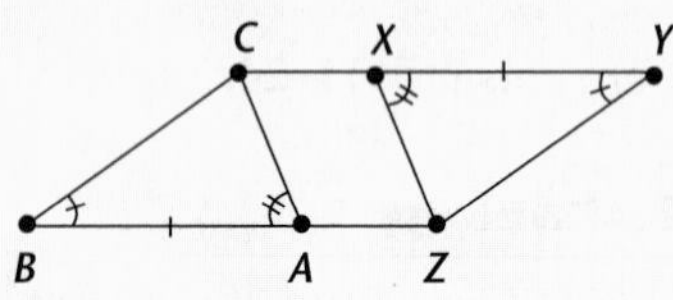

Given: $\angle B \cong \angle Y$, $\angle CAB \cong \angle ZXY$, $\overline{AB} \cong \overline{XY}$

[Statement
1. $\angle B \cong \angle Y$
2. $\overline{AB} \cong \overline{XY}$
3. $\angle CAB \cong \angle ZXY$
4. $\triangle CAB \cong \triangle ZXY$
5. $\overline{AC} \cong \overline{XZ}$

Reason
1. Given
2. Given
3. Given
4. ASA
5. CPCTC]

Practice Master

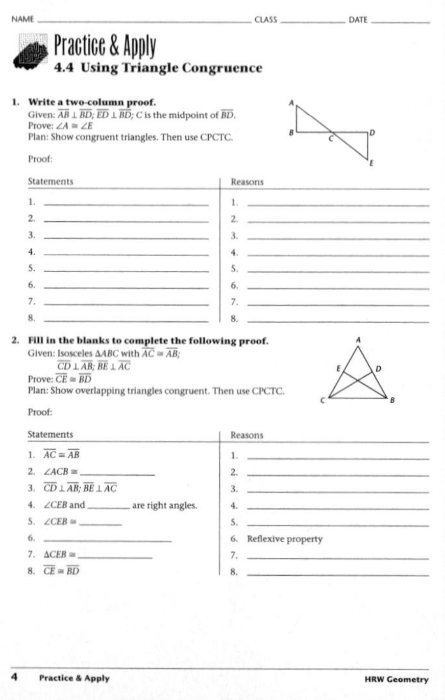

NAME ______________ CLASS ________ DATE ________

Practice & Apply
4.4 Using Triangle Congruence

1. **Write a two-column proof.**
Given: $\overline{AB} \perp \overline{BD}$; $\overline{ED} \perp \overline{BD}$; C is the midpoint of $\overline{BD}$.
Prove: $\angle A \cong \angle E$
Plan: Show congruent triangles. Then use CPCTC.

Proof:

Statements	Reasons
1. ________	1. ________
2. ________	2. ________
3. ________	3. ________
4. ________	4. ________
5. ________	5. ________
6. ________	6. ________
7. ________	7. ________
8. ________	8. ________

2. **Fill in the blanks to complete the following proof.**
Given: Isosceles $\triangle ABC$ with $\overline{AC} \cong \overline{AB}$;
$\overline{CD} \perp \overline{AB}$; $\overline{BE} \perp \overline{AC}$
Prove: $\overline{CE} \cong \overline{BD}$
Plan: Show overlapping triangles congruent. Then use CPCTC.

Proof:

Statements	Reasons
1. $\overline{AC} \cong \overline{AB}$	1. ________
2. $\angle ACB \cong$ ________	2. ________
3. $\overline{CD} \perp \overline{AB}$; $\overline{BE} \perp \overline{AC}$	3. ________
4. $\angle CEB$ and ________ are right angles.	4. ________
5. $\angle CEB \cong$ ________	5. ________
6. ________	6. Reflexive property
7. $\triangle CEB \cong$ ________	7. ________
8. $\overline{CE} \cong \overline{BD}$	8. ________

4 Practice & Apply HRW Geometry

Practice & Apply

6–7. Give the reasons for each of the statements.

STATEMENTS	REASONS
$\overline{AB} \cong \overline{BC}$	Given
$\angle ABX \cong \angle CBX$	Given
$\overline{BX} \cong \overline{BX}$	**(6)**
$\triangle ABX \cong \triangle CBX$	**(7)** SAS
$\overline{AX} \cong \overline{CX}$	**(8)** CPCTC

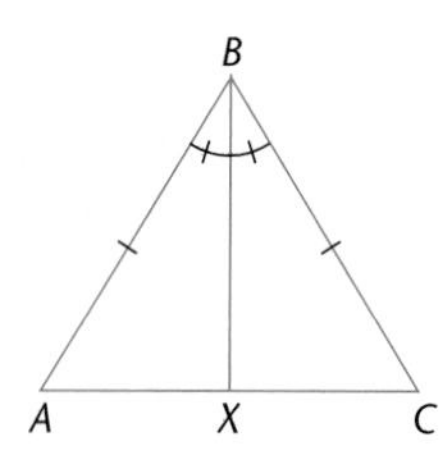

9–10. Complete the statement of the theorem below, which the above proof establishes.

THEOREM
The **(9)** of the vertex angle of an isosceles triangle **(10)**. **4.4.6**

bisector bisects the base

11. Draw the altitude to the base of an isosceles triangle and prove a new theorem about the relationship between the two triangles formed.

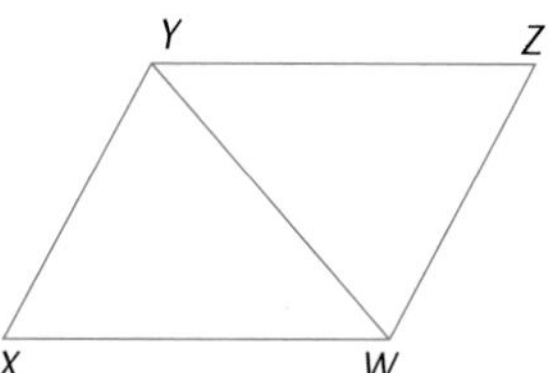

Use the diagram at the right for Exercises 12–17.

12. Given: $\overline{XY} \parallel \overline{ZW}$, $\overline{YZ} \parallel \overline{XW}$
Prove: $\overline{XY} \cong \overline{ZW}$, $\overline{YZ} \cong \overline{XW}$

13–14. Complete the statement of the theorem below, which your proof in Exercise 12 establishes.

THEOREM
The **(13)** of a parallelogram are **(14)**. **4.4.7**

opposite sides congruent

15. Given: $\overline{XY} \parallel \overline{ZW}$, $\overline{YZ} \parallel \overline{XW}$
Prove: $\angle XYZ \cong \angle XWZ$, $\angle YXW \cong \angle YZW$

16–17. Complete the statement of the theorem below, which your proof in Exercise 15 establishes.

THEOREM
The **(16)** of a parallelogram are **(17)**. **4.4.8**

opposite angles congruent

18. Prove Corollary 4.4.3 (page 197) using paragraph form.

19. Prove Corollary 4.4.4 (page 197) using paragraph form.

6. Reflexive Property of Congruence

11. Draw the altitude. The corresponding base angles are congruent and have measure of 90°. The corresponding sides are congruent by definition of an isosceles triangle. The triangles share an altitude leg, which implies the triangles are congruent by HL. Theorem: The two triangles formed by an altitude to the base of an isosceles triangle are congruent.

The answers to Exercises 12, 15, 18, and 19 can be found in Additional Answers beginning on page 727.

Use the diagram at the right for Exercises 20–22.

20. Given: $\overline{AB} \parallel \overline{DE}$, C is the midpoint of $\overline{BE}$

Prove: $\overline{AB} \cong \overline{DE}$

21. Given: $\triangle ABC \cong \triangle DEC$

Prove: $\overline{AB} \parallel \overline{DE}$

22. Given: $\angle A \cong \angle D$, $\overline{AB} \cong \overline{DE}$

Prove: C is the midpoint of $\overline{BE}$

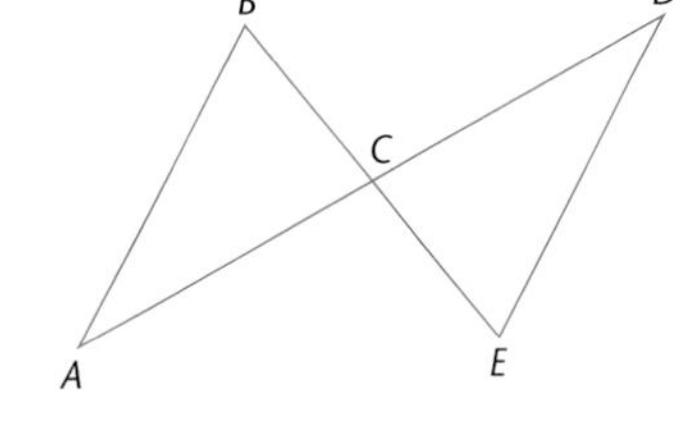

23. Given: $\angle A \cong \angle D$, $\overline{AB} \cong \overline{DE}$, $\overline{AF} \cong \overline{DC}$

Prove: $\angle B \cong \angle E$

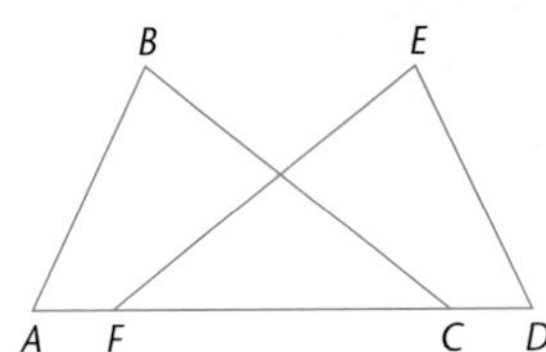

24. Given: $\overline{AB} \cong \overline{CB}$, $\overline{AD} \cong \overline{CE}$

Prove: $\triangle BDE$ is isosceles

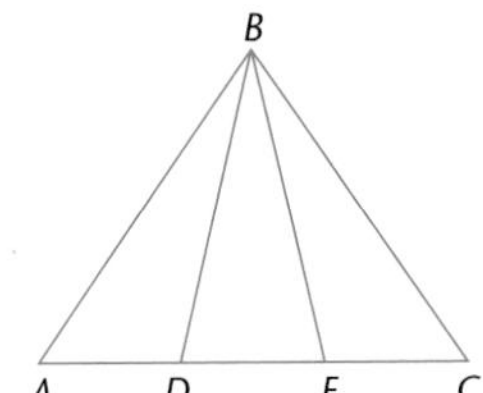

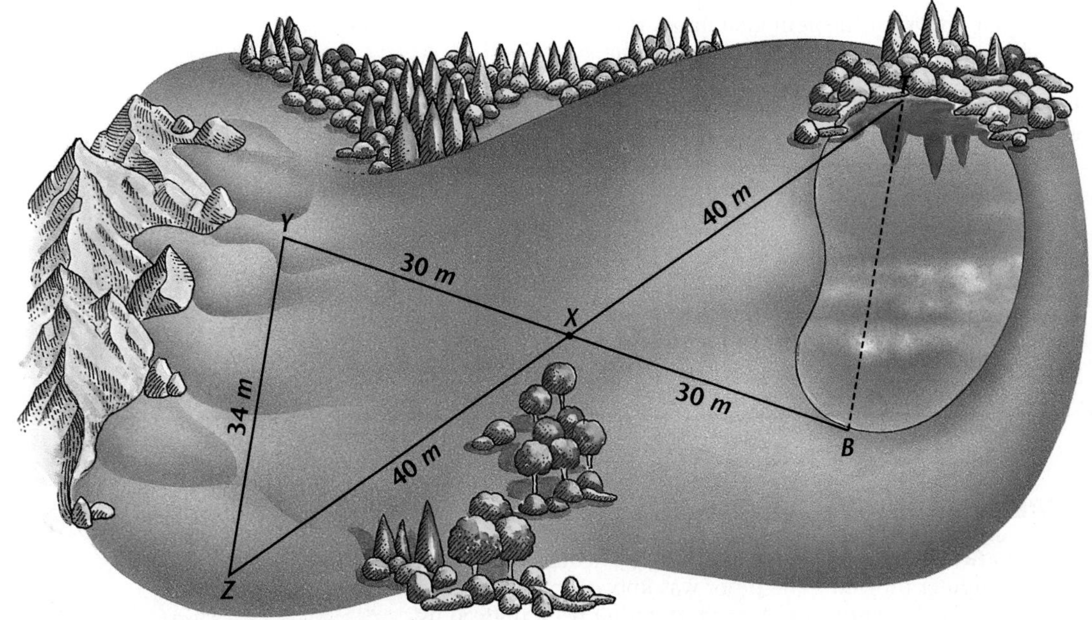

25. Surveying A surveyor needs to measure the distance across a pond from point T to point B. Describe how the measurements shown enable him to determine the distance.

26. What is the distance from T to B? 34 m

20. Statements
   1. $\overline{AB} \parallel \overline{DE}$, C is the midpoint of $\overline{BE}$
   2. $\overline{BC} \cong \overline{CE}$
   3. $\angle BAC \cong \angle EDC$, $\angle ABC \cong \angle DEC$
   4. $\triangle ABC \cong \triangle DEC$
   5. $\overline{AB} \cong \overline{DE}$

   Reasons
   1. Given
   2. Midpoint Definition
   3. $\parallel$ s $\Rightarrow$ Alt. int $\angle$s $\cong$
   4. AAS
   5. CPCTC

21. Statements
   1. $\triangle ABC \cong \triangle DEC$
   2. $\angle BAC \cong \angle EDC$
   3. $\overline{AB} \parallel \overline{DE}$

   Reasons
   1. Given
   2. CPCTC
   3. Alt. Int $\angle$s $\cong \Rightarrow \parallel$ s

25. $\triangle XYZ \cong \triangle XBT$ by SAS using the given sides and the vertical angles. Then $\overline{BT} \cong \overline{YZ}$ by CPCTC.

The answers to Exercises 22–24 can be found in Additional Answers beginning on page 727.

Technology Master

NAME ______ CLASS ______ DATE ______

Technology
4.4 Using Geometry Software to Check Yourself

When you sketch a geometric figure using only the numerical information given, you may obtain a figure that contains much hidden, or additional, information. To find additional information from the given information, you use geometric facts, such as theorems and definitions. Once you find information by deduction, you can use geometry software to see if you are right.

Use geometry software as called for in the following exercises.

1. Use geometry software to sketch the figure shown by using only the information given in the list. Use the MEASURE command to make sure you have faithfully reproduced the given information.

$AB = 4$
$DC = 4$
$FE = 4$
$AD = 3$
$BC = 3$
$FD = 3$
$EC = 3$
$m\angle DFE = 40°$
$m\angle ADC = 140°$
$m\angle BAD = 40°$

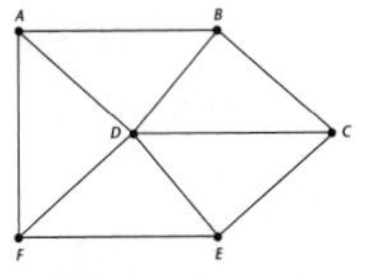

A B D C F E

(a) Use pencil and paper to determine the measures of each side and angle. (You may find it helpful to begin by writing the measures directly on the diagram.)
(b) Use the MEASURE command in your software to find each measure as a check on yourself.

2. $m\angle BCD$ ______ **3.** $m\angle ABC$ ______ **4.** $m\angle ADF$ ______

5. $m\angle DAF$ ______ **6.** $m\angle FEC$ ______ **7.** $m\angle AFE$ ______

8. From Exercise 7, what can you conclude about the relationship between $\overline{AF}$ and $\overline{FE}$? Use the software to confirm your conclusion.

______

9. From the information given and the additional information you found in Exercises 2–7, can you find AF? Explain.

______

22 **Technology** **HRW Geometry**

28. Yes. An equilateral octagon is not rigid.
Yes. An equiangular octagon that is not equilateral is shown.

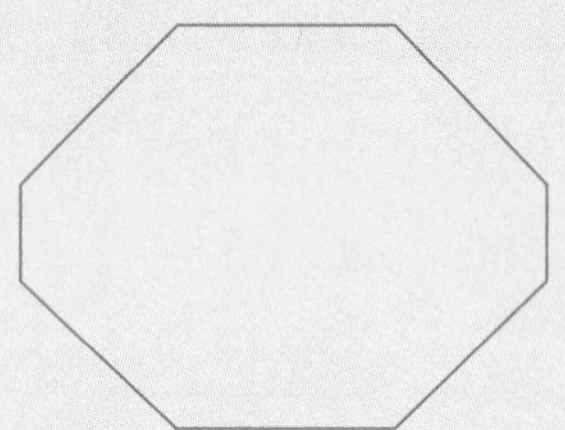

29. $\overline{AC} \cong \overline{DB}$; Overlapping Segment Theorem

31. $\angle ABE \cong \angle CBD$; Overlapping Angles Theorem

Look Beyond

Nonroutine problem. Euclid's proof of the Isosceles Triangle Theorem was so difficult for students in the Middle Ages that the proof acquired a nickname, *pons asinorum*, or "The Bridge of Asses" (also translated as "The Bridge of Fools).

27. Traffic Signs A yield sign is an equilateral triangle. What else do you know about it because it is equilateral? All angles are 60°

28. Traffic Signs A stop sign is a regular octagon; it is an eight-sided equilateral and equiangular polygon. Is it possible to have an equilateral octagon that is not equiangular? Is it possible to have an equiangular octagon that is not equilateral? Demonstrate and discuss your answers.

Look Back

Use the diagram for Exercises 29 and 30. [Lesson 2.5]

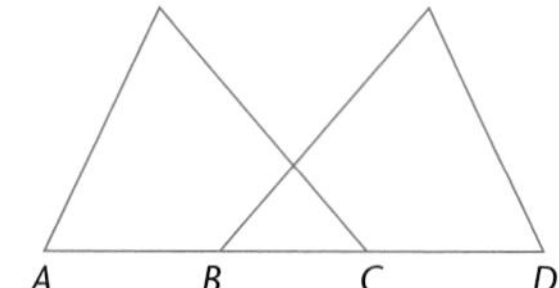

29. Given $\overline{AB} \cong \overline{DC}$, name another pair of congruent segments. What theorem guarantees this?

30. Given $\overline{AC} \cong \overline{DB}$, name another pair of congruent segments. What theorem guarantees this?
$\overline{AB} \cong \overline{DC}$; Overlapping Segments Theorem

Use the diagram for Exercises 31 and 32. [Lesson 2.5]

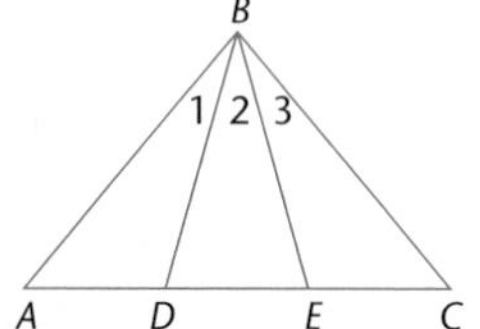

31. Given $\angle 1 \cong \angle 3$, name another pair of congruent angles. What postulate guarantees this?

32. Given $\angle ABE \cong \angle CBD$, name another pair of congruent angles. What postulate guarantees this?
$\angle 1 \cong \angle 3$; Overlapping Angles Theorem

Find the measure of each base angle of an isosceles triangle using the given vertex angle measures. [Lesson 3.5]

33. 24° 78°

34. 52° 64°

35. $x°$ $\frac{180 - x}{2}$

Look Beyond

36. Another proof of the Isosceles Triangle Theorem was "discovered" by a computer program. The proof was known to the Greek geometer Pappus of Alexandria (320 C.E.) but hadn't been noticed by mathematicians for centuries. It is much simpler than Euclid's proof (which is quite complicated) and even simpler than the one given in this chapter. See if you can rediscover it for yourself.

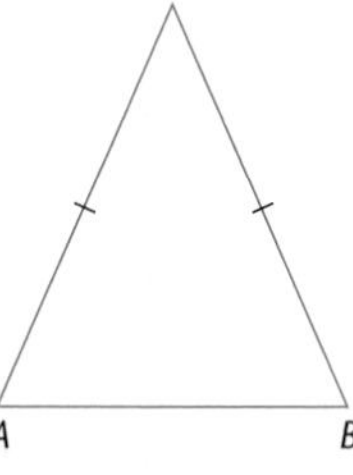

Given: $\overline{AC} \cong \overline{BC}$

Prove: $\angle A \cong \angle B$

Plan: Show that $\triangle ABC \cong \triangle BAC$

36. Statements
1. $\overline{AC} \cong \overline{BC}$
2. $\angle C \cong \angle C$ or $\overline{AB} \cong \overline{AB}$
3. $\triangle ABC \cong \triangle BAC$
4. $\angle A \cong \angle B$

Reasons
1. Given
2. Ref. Prop of Congruence
3. SAS or SSS
4. CPCTC

LESSON 4.5 Proving Quadrilateral Properties

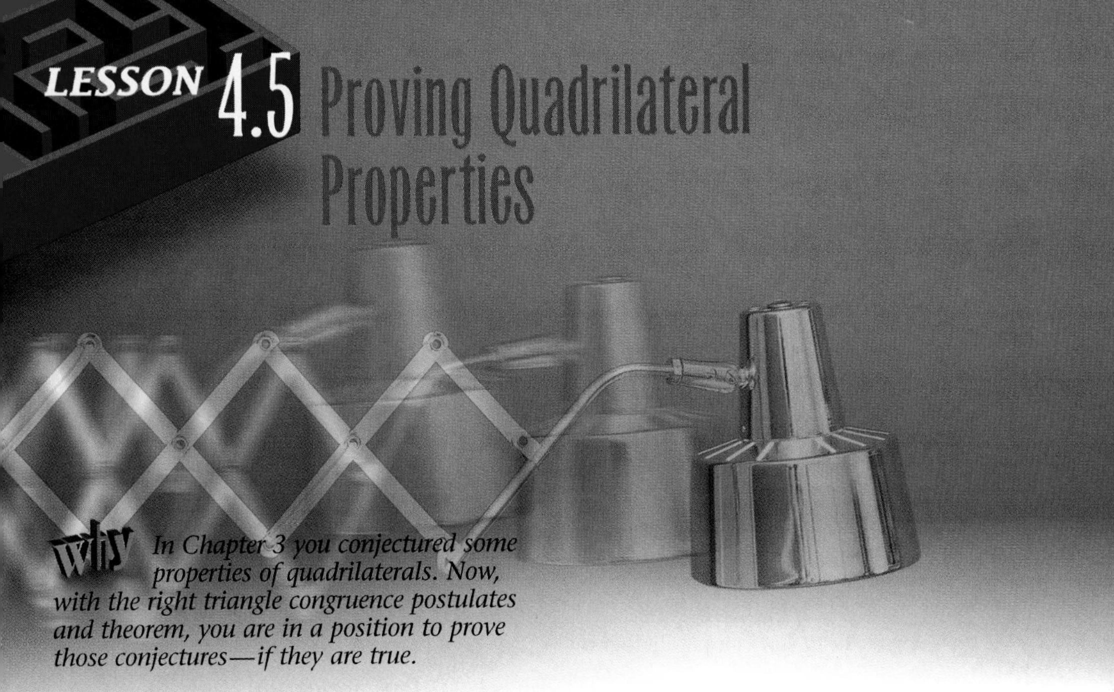

In Chapter 3 you conjectured some properties of quadrilaterals. Now, with the right triangle congruence postulates and theorem, you are in a position to prove those conjectures—if they are true.

A property of rhombuses, which are special parallelograms, explains why this lamp assembly stays perpendicular to the wall as it moves.

In Lesson 3.2 you may have conjectured that a parallelogram has 180° rotational symmetry. If this conjecture is true, you should be able to rotate the parallelogram through its center of rotation so that $\triangle MPL$ will swing around to match the original position of $\triangle LGM$. Therefore, the two triangles would seem to be congruent.

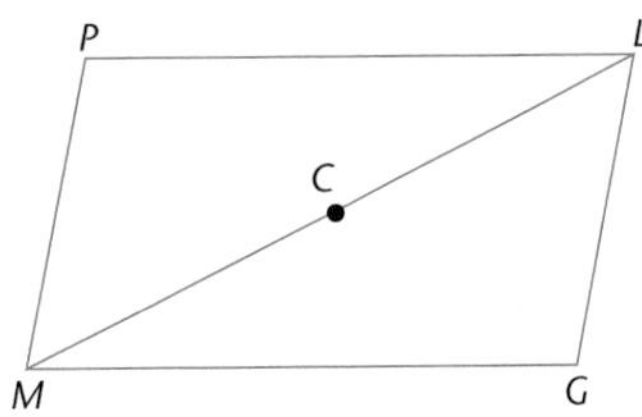

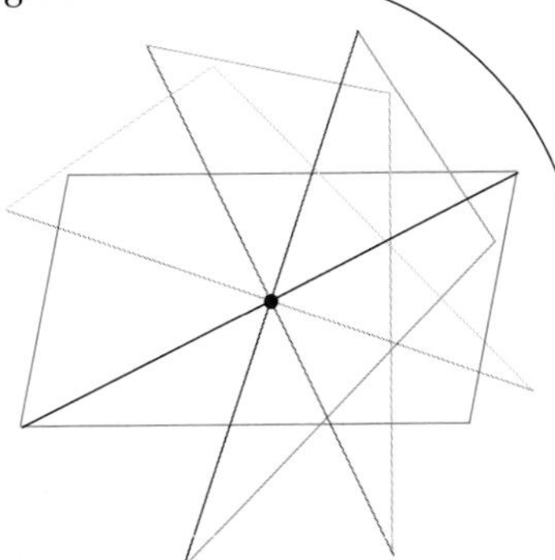

The above result can be stated as the conjecture shown below. It will be proved in Example 1. (Once proven, it becomes a theorem.)

A diagonal of a parallelogram divides the parallelogram into two congruent triangles. **(Theorem 4.5.1)**

Technology Have students use geometry graphics software to make the parallelograms for this lesson. Students should be sure to construct figures that retain their congruence properties when vertices or segments are dragged.

PREPARE

Objectives

- Prove quadrilateral conjectures using triangle congruence postulates and theorems.

RESOURCES

• Practice Master	4.5
• Enrichment Master	4.5
• Technology Master	4.5
• Lesson Activity Master	4.5
• Quiz	4.5
• Spanish Resources	4.5

Use Transparency 40

Assessing Prior Knowledge

How are the figures in each pair alike? How are they different?

1. a parallelogram and a square **[Alike: opposite sides parallel and congruent; Differ-ent: square has four right angles and four equal sides]**
2. a rhombus and a square **[Alike: four equal sides, opposite angles equal; Different: square has four right angles]**

TEACH

Drawing a diagonal in any quadrilateral creates two triangles. Because of this, students' experiences with triangle congruency will help them prove conjectures about properties of quadrilaterals.

TEACHING tip

When students are proving general properties as in Examples 1 and 2, have them set up their work by first stating the conjecture. For example:

Conjecture: A diagonal of a parallelogram divides the parallelogram into two congruent triangles.

Given: Any parallelogram *PLGM* with diagonal *LM*.

Prove: $\triangle LGM \cong \triangle MPL$.

Alternate Example 1

Does a diagonal divide each figure into two congruent triangles? If "yes," write a proof. If "no," give a counterexample.

1. a rectangle [**yes**]
2. a trapezoid [**no**]
3. a kite [**not necessarily**]
4. a square [**yes**]
5. a rhombus [**yes**]

[**Proofs are similar to those in Example 1.**]

CRITICAL Thinking

$\angle P \cong \angle G$ by CPCTC.

Alternate Example 2

Prove that the opposite angles of a parallelogram are congruent.

[**Given: Parallelogram *PLGM***

Prove: $\angle G \cong \angle P$ and $\angle M \cong \angle L$. Draw diagonal *LM* as shown in the text figure. Since the diagonal divides a parallelogram into two congruent triangles, $\triangle LGM \cong \triangle MPL$, and $\angle G \cong \angle P$ because CPCTC.

Now draw diagonal *PG*. $\triangle PGM \cong \triangle GPL$, and therefore $\angle M \cong \angle L$.]

EXAMPLE 1

Given: Parallelogram *PLGM* with diagonal $\overline{LM}$

Prove: $\triangle LGM \cong \triangle MPL$

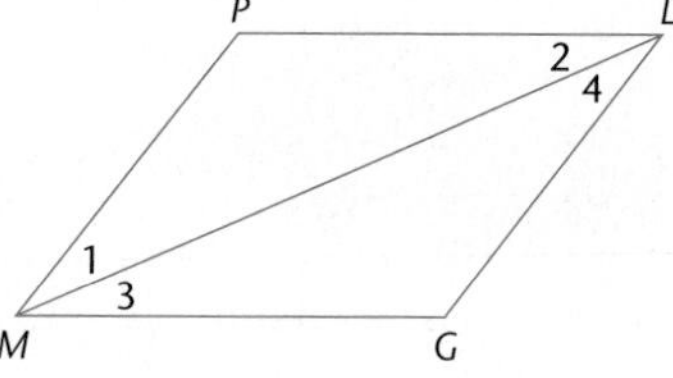

Solution ➤

Proof A (Paragraph form):

$\overline{PL}$ and $\overline{GM}$ are parallel, by definition of a parallelogram, so $\angle 3$ and $\angle 2$ are congruent because they are alternate interior angles. Similarly, $\angle 1$ and $\angle 4$ are congruent, because $\overline{PM}$ and $\overline{GL}$ are also parallel. Finally, diagonal $\overline{LM}$ is congruent to itself. Thus we have shown that two angles and the side between them are congruent in $\triangle LGM$ and $\triangle MPL$. We can conclude that the two triangles are congruent by ASA.

Proof B (Two-column form):

STATEMENTS	REASONS
1. *PLGM* is a parallelogram	Given
2. $\overline{PL} \parallel \overline{GM}$	Def. parallelogram
3. $\angle 3 \cong \angle 2$	$\parallel$s $\Rightarrow \cong$ alt. int. $\angle$s
4. $\overline{PM} \parallel \overline{GL}$	Def. parallelogram
5. $\angle 1 \cong \angle 4$	$\parallel$s $\Rightarrow \cong$ alt. int. $\angle$s
6. $\overline{LM} \cong \overline{LM}$	Reflexive prop. $\cong$
7. $\triangle LGM \cong \triangle MPL$	ASA ❖

CRITICAL Thinking

How could you use this result to show that $\angle P$ and $\angle G$ are congruent?

Another theorem, which you proved in Lesson 4.4, follows immediately from the theorem above. Explain.

The opposite sides of a parallelogram are congruent.

EXAMPLE 2

Construction John is building a rail for a stairway. Assume that the rail is parallel to the line of the steps and that the vertical bars are parallel to each other. How do you know that the vertical bars are congruent to each other?

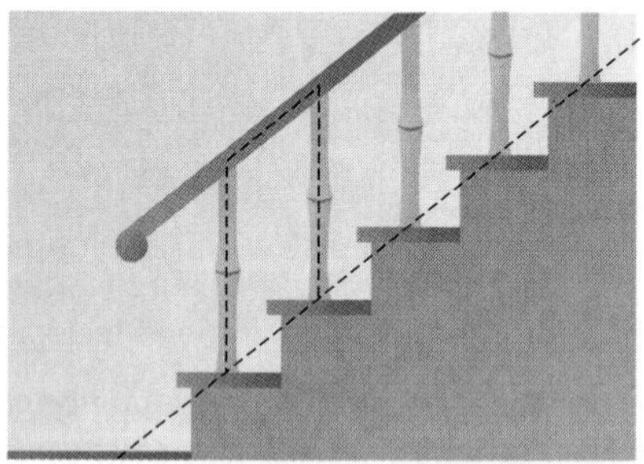

interdisciplinary CONNECTION **Physics** Parallelograms are used in physics to show vector addition. For example, the figure shows an object being pulled in the direction $\overrightarrow{AB}$ with a force of 50 pounds and in the direction $\overrightarrow{AD}$ with a force of 100 pounds.

B C
50
A 100 D

Have students look through their physical science textbooks for examples of these parallelogram diagrams.

Given: Parallelogram $PLGM$ with diagonal $\overline{LM}$

Prove: $\overline{PL} \cong \overline{GM}$, $\overline{PM} \cong \overline{GL}$

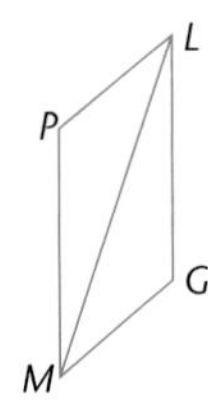

Solution ➤

Proof (Paragraph form):

Since a diagonal divides a parallelogram into congruent triangles, $\triangle LGM \cong \triangle MPL$. Therefore, $\overline{PL} \cong \overline{GM}$ and $\overline{PM} \cong \overline{GL}$ because they are corresponding parts of congruent triangles. ❖

Try This Write the Example 2 proof in two-column form.

Summary

A diagonal of a parallelogram divides the parallelogram into two congruent triangles. Opposite angles of a parallelogram are congruent. Opposite sides of a parallelogram are congruent.

EXERCISES & PROBLEMS

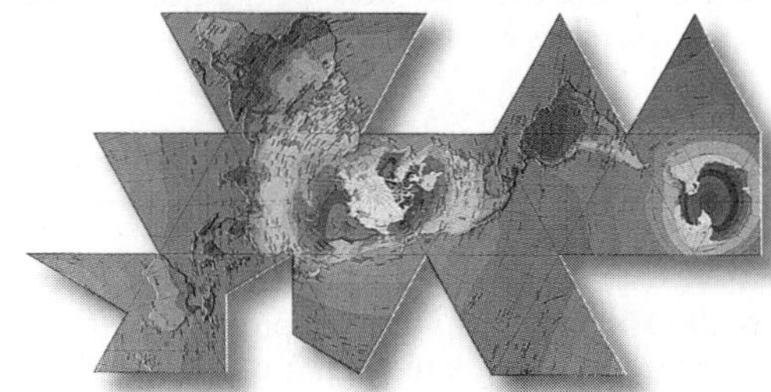

Communicate

1. Given parallelogram $PQRS$, explain how you know $\triangle SPQ \cong \triangle QRS$.
2. Explain why $\angle QRS \cong \angle SPQ$.
3. Explain whether each pair of triangles could fit together to form a parallelogram and why.

P Q S R

a.

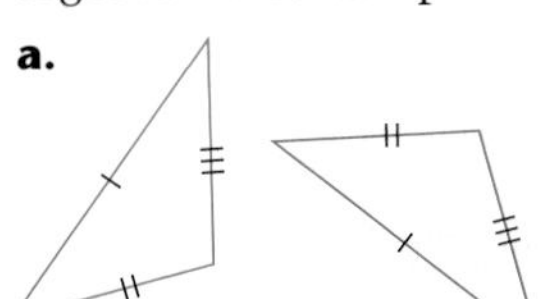

b.

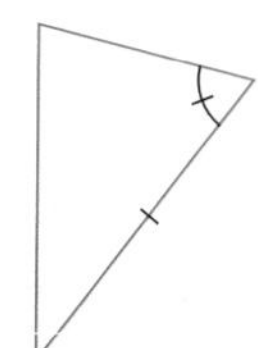

c.

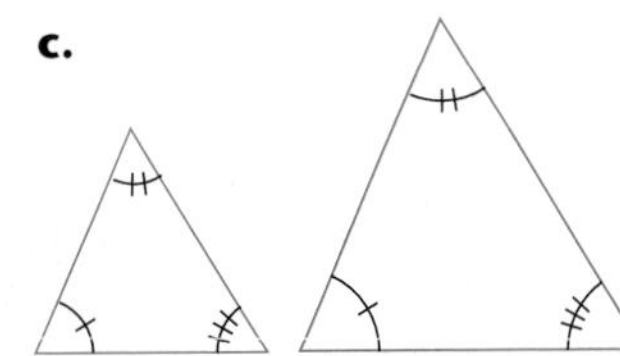

4. How many different pairs of congruent triangles are formed by the diagonals of a parallelogram?

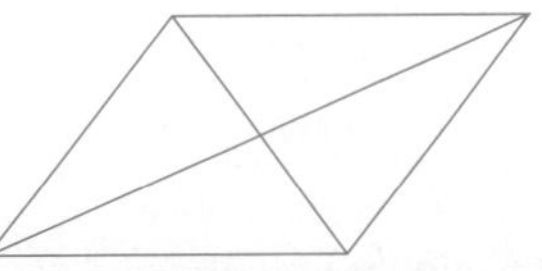

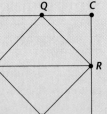

Cooperative Learning

Have students work in small groups to discuss the advantages and disadvantages of paragraph versus two-column proof. They should consider questions such as: Which type is easier to write? Which type is easier for someone else to read? Under what circumstances would you prefer one or the other?

ongoing ASSESSMENT

Try This

Statements

1. Parallelogram $PLGM$ with diagonal $\overline{LM}$
2. $\triangle LGM \cong \triangle MPL$
3. $\overline{PL} \cong \overline{GM}$; $\overline{PM} \cong \overline{GL}$

Reasons

1. Given
2. A diagonal of a parallelogram divides the parallelogram into two congruent triangles.
3. CPCTC

ASSESS

Selected Answers

Odd-numbered Exercises 5–37

Assignment Guide

Core 1–20, 32–37

ENRICHMENT

Given: $ABCD$ is a square. Points P, Q, R, and S are the midpoints of the sides.

Prove: $\triangle PQR \cong \triangle PSR$

B Q C
P R
A S D

[**Statements**

1. $\overline{BP} \cong \overline{PA} \cong \overline{AS} \cong \overline{SD} \cong \overline{DR} \cong \overline{RC} \cong \overline{CQ} \cong \overline{QB}$
2. $\angle A$, $\angle B$, $\angle C$, and $\angle D$ are all congruent
3. $\triangle PAS \cong \triangle SDR \cong \triangle RCQ \cong \triangle QBP$
4. $\overline{PS} \cong \overline{SR} \cong \overline{RQ} \cong \overline{QP}$
5. $\overline{PR} \cong \overline{PR}$
6. $\triangle PQR \cong \triangle PSR$

Reasons

1. Def midpoint
2. Def square
3. SAS
4. CPCTC

Technology

The figures for Exercises 5–40 can be constructed using geometry graphics software. Make sure students construct figures that retain their key properties when vertices or segments are dragged.

Error Analysis

In Exercises 5–11, students may think they are proving the same thing twice. Point out that the first proof deals with adjacent triangles and the second proof deals with opposite triangles.

The answers to Exercises 5–11 and 13–19 can be found in Additional Answers beginning on page 727.

Practice Master

NAME ______________ CLASS ______ DATE ______

Practice & Apply
4.5 Proving Quadrilateral Properties

1. **Write a two-column proof.**
Given: Parallelogram *ABCD* with points on diagonal $\overline{BD}$ so that $\overline{DE} \cong \overline{BF}$
Prove: $\overline{AE} \cong \overline{CF}$
Proof:

Statements	Reasons

2. **Write a paragraph proof.**
Given: Rhombus *ABCE* with $\overline{CE}$ extended to *F* and $\overline{AE}$ extended to *D* so that $\angle FAB \cong \angle DCB$
Prove: $\overline{FE} \cong \overline{DE}$
Proof:

HRW Geometry — Practice & Apply 5

Practice & Apply

5–10. Draw a picture to show the given conditions, then complete the proof.

Given: Parallelogram *ABCD*, with diagonals $\overline{AC}$ and $\overline{BD}$ intersecting at point *E*

Prove: Diagonal $\overline{AC}$ bisects diagonal $\overline{BD}$ at point *E*.

Proof:

$\overline{AB}$ and $\overline{DC}$ are parallel by **(5)**, so $\angle BDC$ and $\angle ABD$ are congruent **(6)** angles. Also, $\angle ACD$ and $\angle CAB$ are congruent alternate interior angles. $\overline{AB} \cong \overline{DC}$, because **(7)** sides of a parallelogram are **(8)**. $\triangle ABE \cong \triangle CDE$ by **(9)**. $\overline{BE} \cong \overline{DE}$, because **(10)**. Therefore, point *E* is the midpoint of $\overline{BD}$ by the definition of a midpoint, and so $\overline{AC}$ bisects $\overline{BD}$ at point *E*.

11. Write a paragraph proof or a two-column proof for the following: Given parallelogram *ABCD*, with diagonals $\overline{AC}$ and $\overline{BD}$ intersecting at point *E*, prove $\overline{BD}$ bisects $\overline{AC}$ at point *E*.

12. Complete the statement of the theorem below, which your two proofs in Exercises 5–11 establish.

THEOREM

The diagonals of a parallelogram **(12)** each other. **4.5.2**

bisect

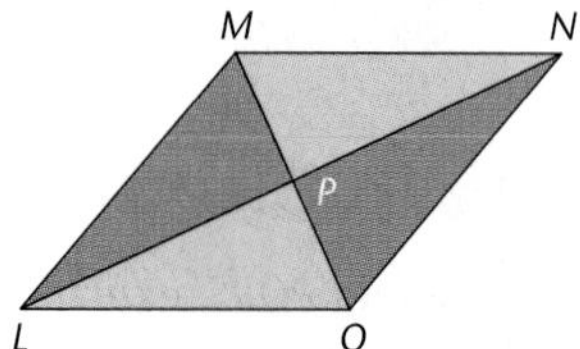

13–19. Fine Art An artist is preparing stained glass for a window using a repeating pattern of rhombuses with diagonals. The artist cuts several congruent triangles out of colored glass. Use the theorem you wrote in Exercise 12 for the following two proofs.

Given: Rhombus *LMNO* with diagonals $\overline{MO}$ and $\overline{LN}$ intersecting at point *P*

Prove: **A.** $\triangle MNP \cong \triangle ONP$, **B.** $\triangle MNP \cong \triangle OLP$

Part A:

Statements	Reasons
$\overline{MP} \cong \overline{PO}$	**(13)**
$\overline{PN} \cong \overline{PN}$	**(14)**
$\overline{MN} \cong \overline{ON}$	**(15)**
$\triangle MNP \cong \triangle ONP$	**(16)**

Part B:

Statements	Reasons
$\overline{MP} \cong \overline{OP}$; $\overline{NP} \cong \overline{LP}$	**(17)**
$\overline{MN} \cong \overline{LO}$	**(18)**
$\triangle MNP \cong \triangle OLP$	**(19)**

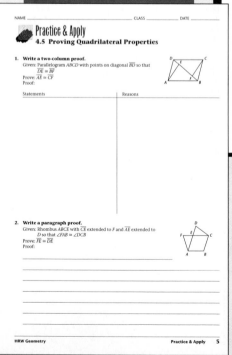

INCLUSION strategies **Inviting Participation** Some students may be more comfortable writing their two-column proofs using words, phrases, and sentences rather than symbols and abbreviations.

Cooperative Learning Have students draw the diagonals in an isosceles trapezoid and look for pairs of triangles that appear congruent. Have them work in small groups to write proofs of their conjectures.

20. What do you need to add to the proof in Exercises 13–19 to prove the theorem below?

> **THEOREM**
> The diagonals and sides of a rhombus form four congruent triangles. **4.5.3**

21–27. Use the theorem from Exercise 20 to help you prove the following.

Given: Rhombus *DEFG*, with diagonals $\overline{DF}$ and $\overline{GE}$ intersecting at *H*

Prove: The angles formed by the intersecting diagonals are right angles.

Proof:

$\triangle EHF \cong \triangle GHF$, because __(21)__. $\angle FHG \cong \angle FHE$, because __(22)__. Since $\angle FHG$ and $\angle FHE$ are a linear pair, the sum of the measures of the angles is __(23)__. Since $\angle FHG \cong \angle FHE$, their measures are equal and are therefore __(24)__ degree angles or __(25)__ angles. A similar relationship exists between $\triangle DHE$ and $\triangle DHG$. So $\angle DHE$ and $\angle DHG$ measure __(26)__ degrees also. Therefore, $\angle DHE$, $\angle EHF$, $\angle FHG$, and $\angle GHD$ are __(27)__ angles.

28. Complete the statement of the theorem below, which your proof in Exercises 21–27 establishes.

> **THEOREM**
> The diagonals of a rhombus are __(28)__ . **4.5.4**

perpendicular

29. Draw a picture to show the given conditions, then complete the proof. Hint: Since a rectangle is a parallelogram, you can use the theorems you proved about parallelograms in the Lesson 4.4 exercises.

Given: Rectangle *RSTU*, with diagonals $\overline{RT}$ and $\overline{US}$

Prove: $\overline{RT} \cong \overline{US}$.

30. Complete the statement of the theorem below, which your proof in Exercise 29 establishes.

> **THEOREM**
> The diagonals of a rectangle are __(30)__ . **4.5.5**

congruent

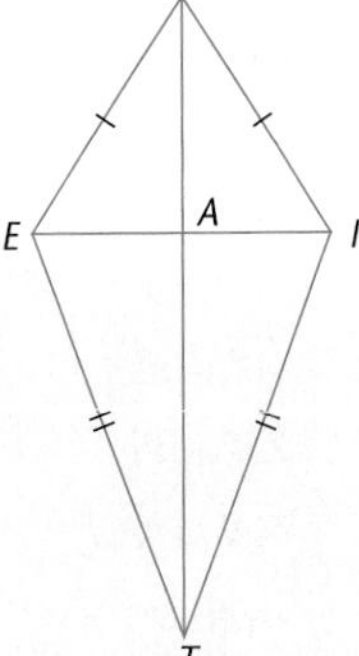

Kites A special quadrilateral called a kite has two pairs of congruent adjacent sides, and opposite sides are not congruent.

31. Using the given conditions and the diagram, write the required proof in paragraph or two-column form.

Given: Kite *KITE* with diagonals $\overline{KT}$ and $\overline{EI}$ intersecting at point *A*

Prove: $\overline{KT} \perp \overline{EI}$

21. they are triangles formed by sides and diagonals of a rhombus
22. CPCTC
23. 180°
24. 90
25. right
26. 90
27. right

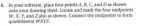

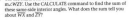

alternative ASSESSMENT

Portfolio Assessment

Have students keep the proof they have completed in their portfolios. Students should also write a theorem for each proof. Occasionally review the portfolios in class to show students their progress through the development of concepts of geometric proof.

The answers to Exercises 20, 29, and 31 can be found in Additional Answers beginning on page 727.

Technology Master

NAME ______ CLASS ______ DATE ______

Technology
4.5 Quadrilaterals Generated by Midpoints

The diagram shows what you get when you join the midpoints of the three sides of a triangle. The smaller triangles seem to be reduced copies of the original triangle.

Suppose that you construct a quadrilateral *ABCD*, locate the midpoints of the four sides, and connect the midpoints in order.

What happens if you carry out this construction on a quadrilateral?

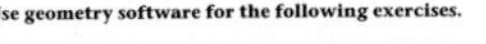

Use geometry software for the following exercises.

1. In your software, place four points *A*, *B*, *C*, and *D* as shown onto your drawing sheet. Locate and mark the four midpoints *W*, *X*, *Y*, and *Z* also as shown. Connect the midpoints to form quadrilateral *WXYZ*.

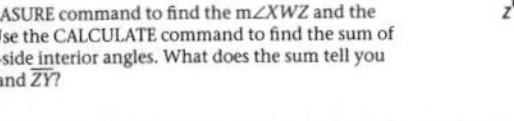

2. Use the MEASURE command to find the $m\angle XWZ$ and the $m\angle WZY$. Use the CALCULATE command to find the sum of these same-side interior angles. What does the sum tell you about $\overline{WX}$ and $\overline{ZY}$?
3. Use reasoning similar to that of Exercise 2 to draw a conclusion about $\overline{WZ}$ and $\overline{XY}$.
4. What do your responses to Exercises 2 and 3 tell you about quadrilateral *WXYZ*?
5. Drag any vertex of quadrilateral *ABCD* to form other quadrilaterals. Does your conclusion from Exercise 4 change? Explain.
6. For what quadrilateral *ABCD* will the sides of quadrilateral *WXYZ* be congruent?

HRW Geometry — Technology 23

32. Complete the statement of the theorem below, which your proof in Exercise 31 establishes.

THEOREM

The diagonals of a kite are __(32)__ . **4.5.6**

perpendicular

Look Back

33. Explain why if three angles of a triangle are congruent to three angles in another triangle, the triangles are not necessarily congruent. **[Lessons 4.2, 4.3]** The triangles are the same shape, but not necessarily the same size.

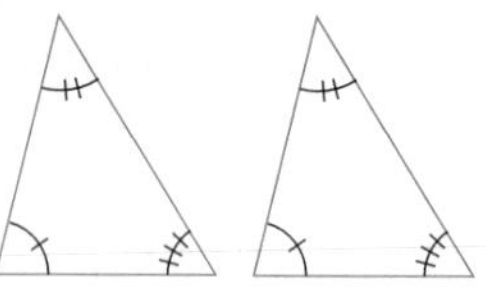

34. $\triangle ABC \cong \triangle DEF$ by which postulate? **[Lessons 4.2, 4.3]**

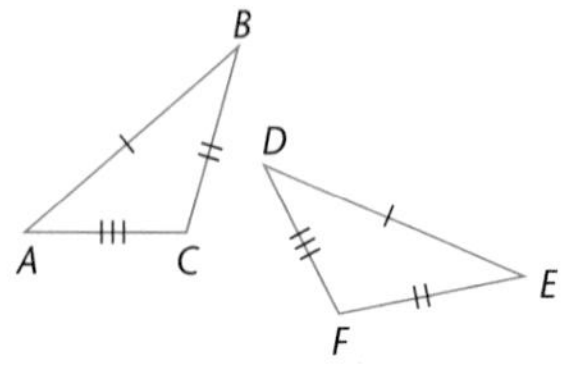

SSS

35. $\triangle GHI \cong \triangle JKL$ by which postulate? **[Lessons 4.2, 4.3]**

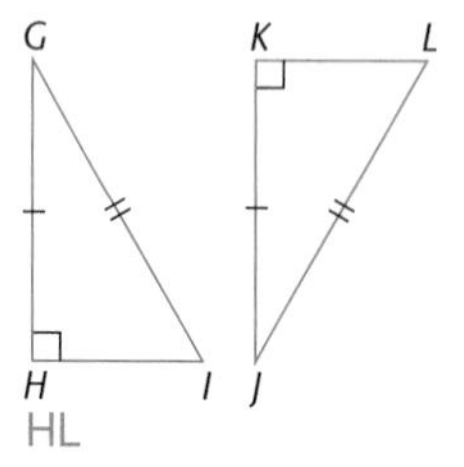

HL

36. $\triangle MNO \cong \triangle PQR$ by which postulate? **[Lessons 4.2, 4.3]**

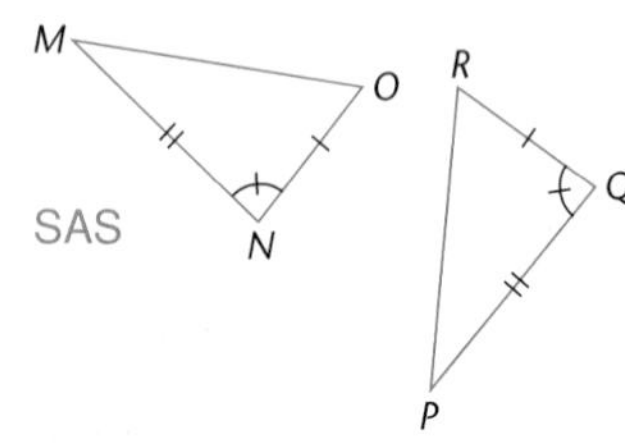

SAS

37. **Algebra** Given that $\triangle UVW \cong \triangle ABC$, find the value of x and y. **[Lesson 4.4]** $x = 20$, $y = 20$

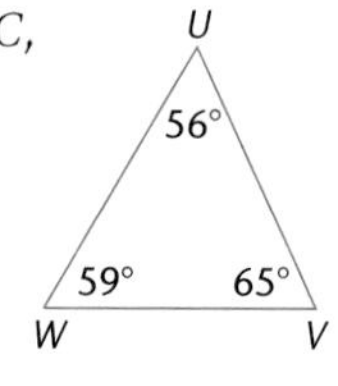

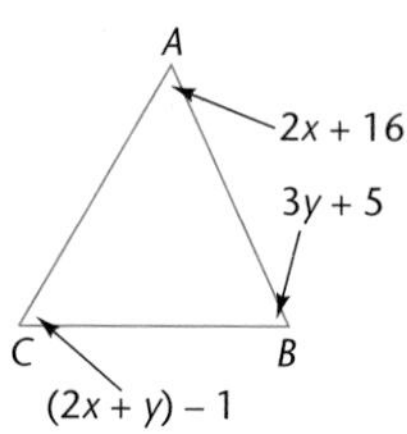

Look Beyond

38. **Baseball** A batter hits a ground ball at a 27° angle relative to the third-base line. The shortstop scoops up the ball and throws it to first base. The ball arrives at first base at an 18° angle relative to the first-to-second-base line. What angle did the shortstop turn to make the throw to first base? 45°

Hint: A baseball diamond is a square.

Look Beyond

In Exercise 38, make sure students know that a baseball "diamond" is actually a square. The vertices are home plate, first base, second base, and third base. The batter is standing at home plate.

LESSON 4.6 Exploring

Conditions for Special Quadrilaterals

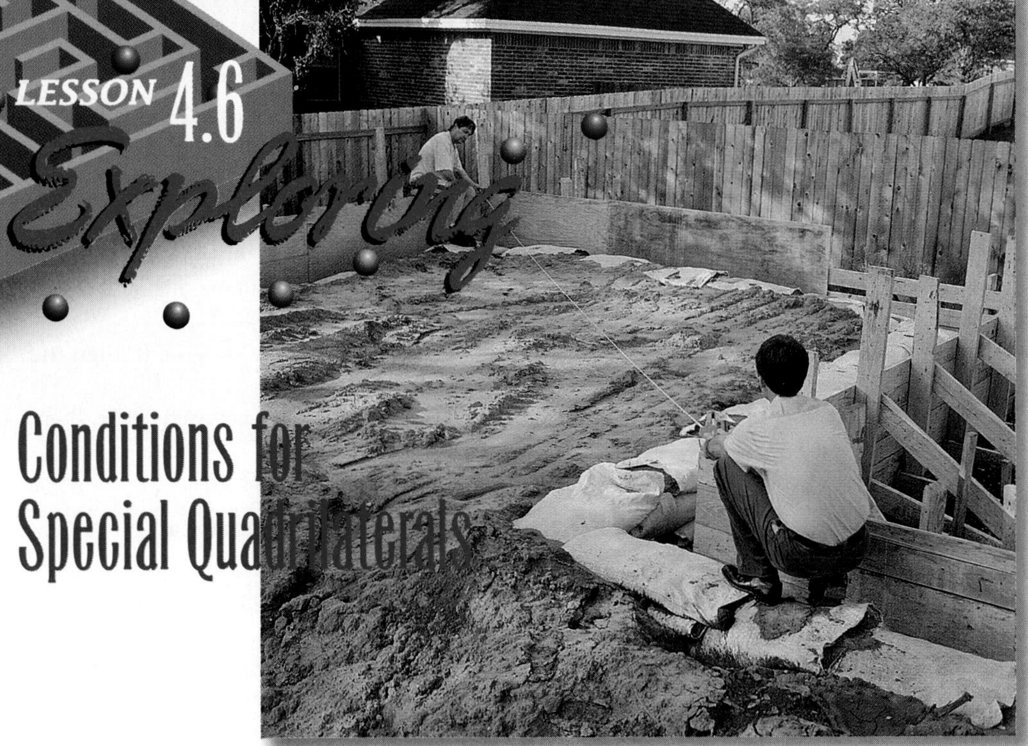

A rectangular foundation is essential to the construction of most modern buildings. Here workers are applying an important geometric principle to ensure that the foundation is a rectangle. In this lesson, you will discover what it is.

Why *How do you know if a figure is a parallelogram, rectangle, or rhombus? Checking the definition works only if the required special properties are given. In earlier lessons you were told what kind of special quadrilateral a given figure was at the beginning of a problem. In this lesson, the procedure is reversed. You will learn to answer the question, "What does it take to make . . . ?"*

Exploration *Parallelograms, Rectangles, and Rhombuses*

You will need

Toothpicks, popsicle sticks, or other materials to serve as sides for quadrilaterals or

Ruler and protractor

In the three parts that follow, decide whether each of the conjectures listed is true or false. If you believe a conjecture is false, prove that it is false by showing a counterexample. Make sketches of your results. If you believe a conjecture is true, consider how you would prove it.

ALTERNATIVE teaching strategy

Technology Geometry graphics software is a good tool for exploring the conjectures in this lesson. Make sure the figures students construct retain their essential properties when vertices or segments are dragged.

PREPARE

Objectives

- Develop conjectures for special quadrilaterals — parallelograms, rectangles, and rhombuses.

RESOURCES

- Practice Master 4.6
- Enrichment Master 4.6
- Technology Master 4.6
- Lesson Activity Master 4.6
- Quiz 4.6
- Spanish Resources 4.6

Assessing Prior Knowledge

Have students draw a quadrilateral *WXYZ* and list all pairs that satisfy each description.

1. opposite sides [$\overline{WX}$ and $\overline{ZY}$, $\overline{WZ}$ and $\overline{XY}$]
2. opposite angles [$\angle X$ and $\angle Z$, $\angle W$ and $\angle Y$]
3. adjacent sides [$\overline{WX}$ and $\overline{XY}$, $\overline{XY}$ and $\overline{YZ}$, $\overline{YZ}$ and $\overline{ZW}$, $\overline{ZW}$ and $\overline{WX}$]
4. consecutive angles [$\angle W$ and $\angle X$, $\angle X$ and $\angle Y$, $\angle Y$ and $\angle Z$, $\angle Z$ and $\angle W$]

TEACH

Why When building structures involving quadrilaterals, it is not necessary or convenient to measure four angles and four sides. The properties covered in this lesson can be applied to simplify and shorten projects such as the building foundation shown.

Exploration Notes

Have students read through all the conjectures and notice the types of segments or angles specified. They should see that key properties of a quadrilateral involve opposite sides or angles, adjacent sides or angles, and how the diagonals are related to each other.

Cooperative Learning

Have students work in small groups to create a chart summarizing the results of the exploration. For example, the chart might have four columns headed *Property, Parallelogram, Rectangle, Rhombus.* Under *Property*, students should write items such as "congruent diagonals." Then they should write *always, sometimes,* or *never* under each shape. When all groups are finished, lead a class discussion in which groups compare their various chart formats and content.

ongoing ASSESSMENT

Part I:	1. F, 2. T, 3. T, 4. F, 5. T
Part II:	1. F, 2. T, 3. F, 4. T 5. F, 6. F
Part III:	1. F, 2. T, 3. F, 4. T, 5. T

ongoing ASSESSMENT

(p. 209) If two pairs of opposite sides of a quadrilateral are congruent, then the quadrilateral is a parallelogram.

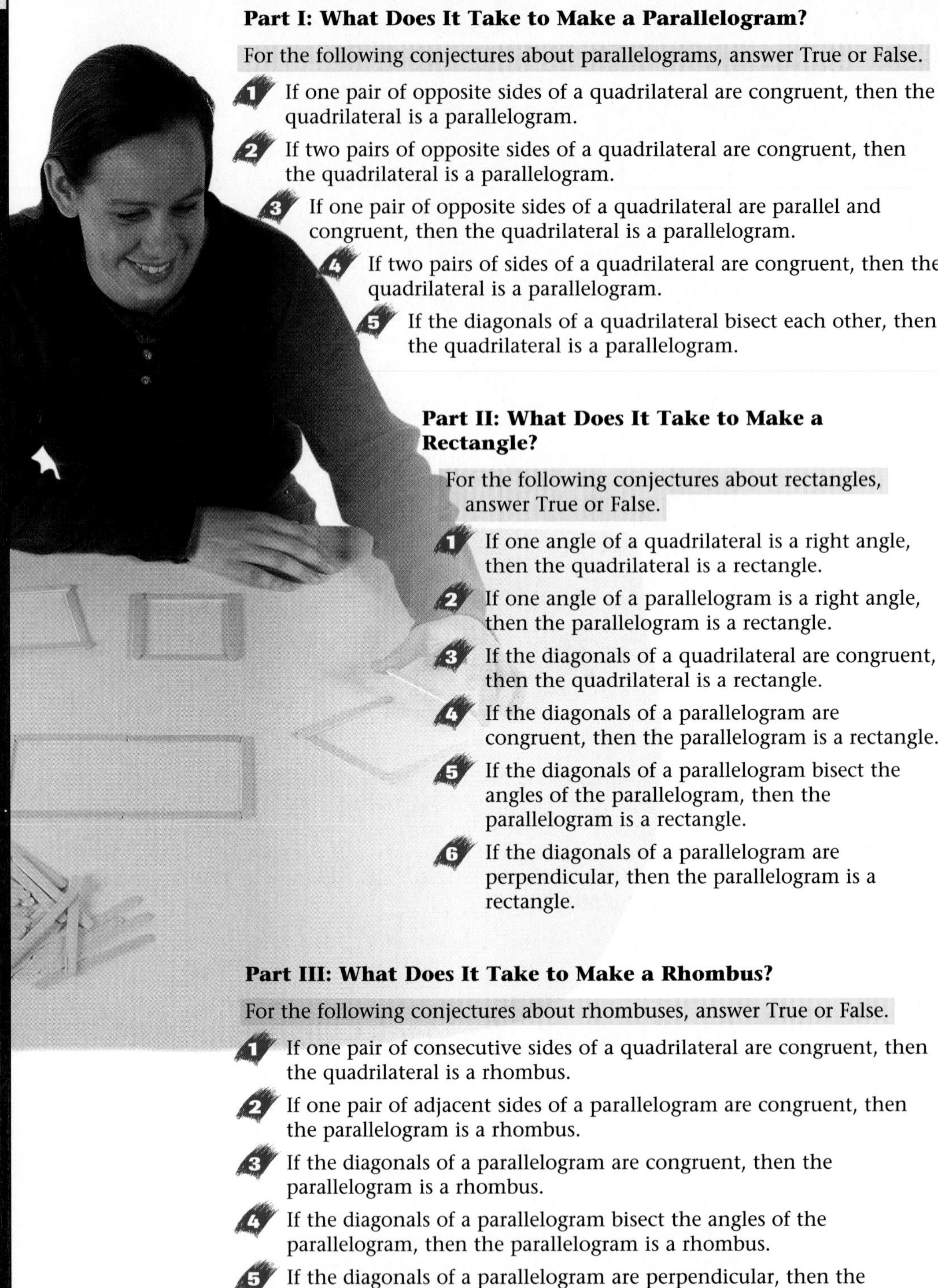

Part I: What Does It Take to Make a Parallelogram?

For the following conjectures about parallelograms, answer True or False.

1. If one pair of opposite sides of a quadrilateral are congruent, then the quadrilateral is a parallelogram.
2. If two pairs of opposite sides of a quadrilateral are congruent, then the quadrilateral is a parallelogram.
3. If one pair of opposite sides of a quadrilateral are parallel and congruent, then the quadrilateral is a parallelogram.
4. If two pairs of sides of a quadrilateral are congruent, then the quadrilateral is a parallelogram.
5. If the diagonals of a quadrilateral bisect each other, then the quadrilateral is a parallelogram.

Part II: What Does It Take to Make a Rectangle?

For the following conjectures about rectangles, answer True or False.

1. If one angle of a quadrilateral is a right angle, then the quadrilateral is a rectangle.
2. If one angle of a parallelogram is a right angle, then the parallelogram is a rectangle.
3. If the diagonals of a quadrilateral are congruent, then the quadrilateral is a rectangle.
4. If the diagonals of a parallelogram are congruent, then the parallelogram is a rectangle.
5. If the diagonals of a parallelogram bisect the angles of the parallelogram, then the parallelogram is a rectangle.
6. If the diagonals of a parallelogram are perpendicular, then the parallelogram is a rectangle.

Part III: What Does It Take to Make a Rhombus?

For the following conjectures about rhombuses, answer True or False.

1. If one pair of consecutive sides of a quadrilateral are congruent, then the quadrilateral is a rhombus.
2. If one pair of adjacent sides of a parallelogram are congruent, then the parallelogram is a rhombus.
3. If the diagonals of a parallelogram are congruent, then the parallelogram is a rhombus.
4. If the diagonals of a parallelogram bisect the angles of the parallelogram, then the parallelogram is a rhombus.
5. If the diagonals of a parallelogram are perpendicular, then the parallelogram is a rhombus. ❖

interdisciplinary CONNECTION

Geography Parallel dividers are used by navigators in drawing courses on maps. Angle measures are transferred from the compass to make specific course headings. Have a student research how this tool is used and give a presentation to the class.

EXTENSION

In Lessons 4.4 and 4.5 you studied a number of theorems about the properties of parallelograms and other special quadrilaterals. For example,

The opposite sides of a parallelogram are congruent. **(Theorem 4.4.7)**

In each theorem, the special quadrilateral is the "given," and the conclusion follows. An "if-then" version of the theorem may make this relationship clearer:

If a quadrilateral is a parallelogram, then its opposite sides are congruent.

As you may have noticed, one of the conjectures you tested in the exploration is the converse of this theorem. Which one is it? ❖

Try This Make a list of the exploration conjectures that you believe to be true. Write the converses of those conjectures. Which of these converses are theorems that you have already studied? State those theorems.

Cultural Connection: Africa Mozambican workers traditionally use long poles to establish the sides of their house foundations. They make a rectangular foundation in the following two steps:

1. The workers use two pairs of poles of the desired lengths for congruent opposite sides. This guarantees that the base of the house is a parallelogram—but not necessarily a rectangle.

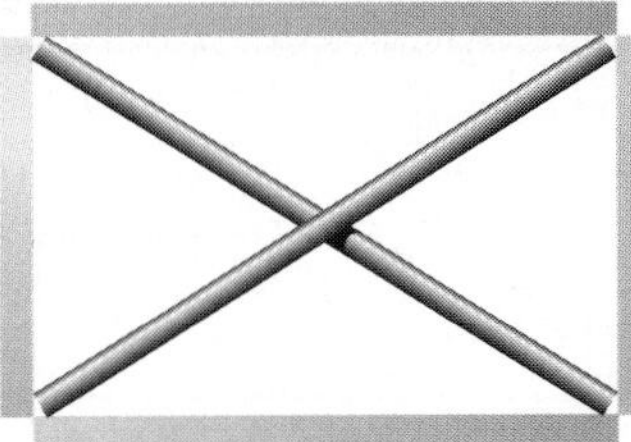

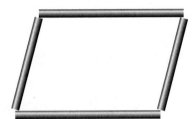

2. The workers measure the diagonals. If the diagonals are congruent, then the base of the house is a rectangle. If not, the workers reposition the poles as necessary to make the diagonals congruent. ❖

CRITICAL *Thinking*

The lamp in the picture remains parallel to the wall as it moves up and down. How does the first part of the Mozambican procedure explain the motion?

ONGOING ASSESSMENT

Try This

Parallelograms If two pairs of opposite sides of a quadrilateral are congruent, then the quadrilateral is a parallelogram. The converse has been proven. If one pair of opposite sides of a quadrilateral is parallel and congruent, then the quadrilateral is a parallelogram. The converse has been proven. If the diagonals of a quadrilateral bisect each other, then the quadrilateral is a parallelogram. The converse has been proven.

Rectangles If one angle of a parallelogram is a right angle, then the parallelogram is a rectangle. The converse is true by definition of a rectangle. If the diagonals of a parallelogram are congruent, then the parallelogram is a rectangle. The converse has been proven.

Rhombuses If one pair of adjacent sides of a parallelogram are congruent, then the parallelogram is a rhombus. The converse is true by the definition of a rhombus. If the diagonals of a parallelogram bisect the angles of the parallelogram, then the parallelogram is a rhombus. The converse is true but has not been proven at this point. If the diagonals of a parallelogram are perpendicular, then the parallelogram is a rhombus. The converse has been proven.

ENRICHMENT

Technology Have students use geometry graphics software to explore these questions for parallelograms, rhombuses, and rectangles.

1. Connect the midpoints of the sides. What figures result?
2. Construct the bisectors of the angles. What figures result?

INCLUSION strategies

Using Visual Models Have students use paper-and-pencil methods to make their drawings on grid paper. The grid lines will make it easier to construct parallels, perpendiculars, and congruent segments and angles.

Use Transparency 41

CRITICAL *Thinking*

Since opposite sides of the assembly are congruent, it is a parallelogram.

ASSESS

Selected Answers

Odd-numbered Exercises 3–31

Assignment Guide

Core 1–21, 31, 32

Core Plus 1–2, 9–37

Technology

Geometry graphics software may be used for Exercises 3–30. Have students construct each figure before completing the proof.

Error Analysis

Some students may find it easier to write paragraph proofs for the two conjectures in Exercises 3–17.

Practice Master

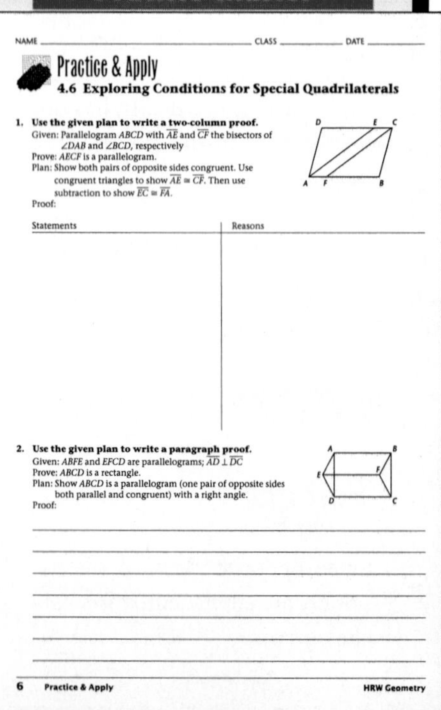
NAME ______ CLASS ______ DATE ______

Practice & Apply

4.6 Exploring Conditions for Special Quadrilaterals

1. **Use the given plan to write a two-column proof.**
Given: Parallelogram *ABCD* with $\overline{AE}$ and $\overline{CF}$ the bisectors of $\angle DAB$ and $\angle BCD$, respectively
Prove: *AECF* is a parallelogram.
Plan: Show both pairs of opposite sides congruent. Use congruent triangles to show $\overline{AE} \cong \overline{CF}$. Then use subtraction to show $\overline{EC} \cong \overline{FA}$.
Proof:

Statements	Reasons

2. **Use the given plan to write a paragraph proof.**
Given: *ABFE* and *EFCD* are parallelograms; $\overline{AD} \perp \overline{DC}$
Prove: *ABCD* is a rectangle.
Plan: Show *ABCD* is a parallelogram (one pair of opposite sides both parallel and congruent) with a right angle.
Proof:

6 Practice & Apply HRW Geometry

EXERCISES & PROBLEMS

Communicate

1. Summarize your investigations in this lesson by giving at least three answers to each question.

a. What does it take to make a parallelogram?

b. What does it take to make a rectangle?

c. What does it take to make a rhombus?

2. All of the figures below are parallelograms. Decide whether they must also be rectangles or rhombuses based on the given information (figures are not drawn to scale).

a.

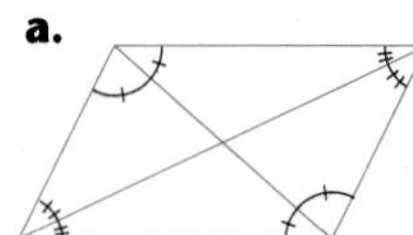

b.

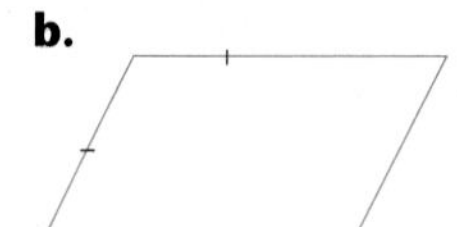

c.

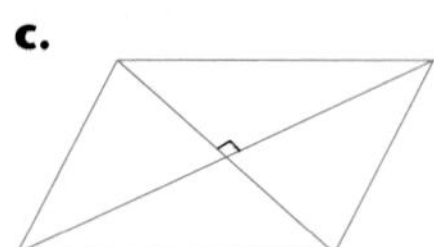

Practice & Apply

3–8. Complete the paragraph proof of the following theorem.

THEOREM

If two pairs of opposite sides of a quadrilateral are congruent, then the quadrilateral is a parallelogram. **4.6.1**

Given: $\overline{BA} \cong \overline{SE}$; $\overline{BE} \cong \overline{SA}$

Prove: *BASE* is a parallelogram

Plan: Use congruent sides to get $\triangle BAS \cong \triangle SEB$. This gives congruent corresponding angles, which can be used to get parallel lines, satisfying the definition of a parallelogram.

Proof:

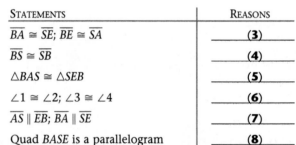

STATEMENTS	REASONS
$\overline{BA} \cong \overline{SE}$; $\overline{BE} \cong \overline{SA}$	**(3)**
$\overline{BS} \cong \overline{SB}$	**(4)**
$\triangle BAS \cong \triangle SEB$	**(5)**
$\angle 1 \cong \angle 2$; $\angle 3 \cong \angle 4$	**(6)**
$\overline{AS} \parallel \overline{EB}$; $\overline{BA} \parallel \overline{SE}$	**(7)**
Quad *BASE* is a parallelogram	**(8)**

RETEACHING the lesson

Cooperative Learning Have students work in small groups. Each group should be assigned one class of quadrilateral. Students should write down as many true statements as they can about their quadrilateral. Groups should then make a class presentation and decide on a concise list of true statements for each class of quadrilateral.

3. Given
4. Reflexive Property of $\cong$
5. SSS
6. CPCTC
7. If alt int $\angle$s $\cong \Rightarrow \parallel$ s
8. Parallelogram definition

9–17. Complete the proof of the following theorem.

THE "HOUSEBUILDER" RECTANGLE THEOREM

If the diagonals of a parallelogram are congruent, then the parallelogram is a rectangle. **4.6.2**

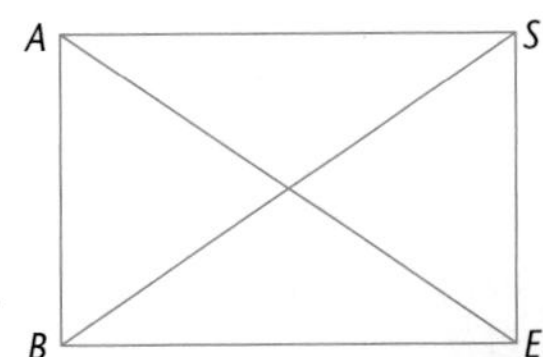

Given: *BASE* is a parallelogram; $\overline{BS} \cong \overline{EA}$

Prove: *BASE* is a rectangle.

Proof:

STATEMENTS	REASONS
BASE is a parallelogram; $\overline{BS} \cong \overline{EA}$	**(9)**
$\overline{BA} \cong \overline{ES}$	**(10)**
$\overline{BE} \cong \overline{EB}$	**(11)**
$\triangle ABE \cong \triangle SEB$	**(12)**
$\angle ABE \cong \angle SEB$	**(13)**
$\angle ABE$ and $\angle SEB$ are supplementary	**(14)**
$\angle ABE$ and $\angle SEB$ are right angles	**(15)**
$\angle ASE$ and $\angle SAB$ are right angles	**(16)**
BASE is a rectangle	**(17)**

In Exercises 18–21 prove each of the following theorems, using either a paragraph or two-column form. (Set up your own "Given" and "Prove" statements).

18.

THEOREM

If one pair of opposite sides of a quadrilateral are parallel and congruent, then the quadrilateral is a parallelogram. **4.6.3**

19.

THEOREM

If the diagonals of a quadrilateral bisect each other, then the quadrilateral is a parallelogram. **4.6.4**

20.

THEOREM

If one angle of a parallelogram is ight angle, then the parallelogram is a rectangle. **4.6.5**

21.

THEOREM

If one pair of adjacent sides of a parallelogram are congruent, then the quadrilateral is a rhombus. **4.6.6**

Alternative Assessment

Performance Assessment

For which of the three quadrilaterals—parallelogram, rectangle, rhombus—is each of the following properties true? Make drawings to justify your answers.

1. The diagonals bisect each other. [**parallelogram, rectangle, rhombus**]
2. The diagonals are congruent. [**rectangle**]
3. The diagonals are perpendicular. [**rhombus**]
4. The diagonals are congruent and perpendicular. [**none**]

The answers to Exercises 18–21 can be found in Additional Answers beginning on page 727.

Technology Master

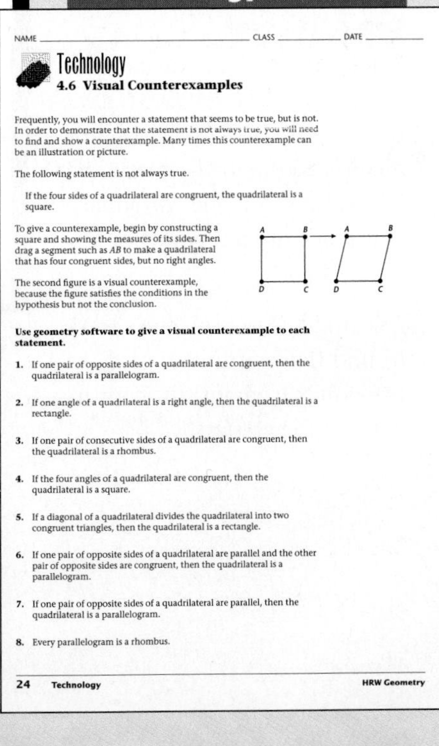

NAME ____________ CLASS ________ DATE ________

Technology

4.6 Visual Counterexamples

Frequently, you will encounter a statement that seems to be true, but is not. In order to demonstrate that the statement is not always true, you will need to find and show a counterexample. Many times this counterexample can be an illustration or picture.

The following statement is not always true.

If the four sides of a quadrilateral are congruent, the quadrilateral is a square.

To give a counterexample, begin by constructing a square and showing the measures of its sides. Then drag a segment such as *AB* to make a quadrilateral that has four congruent sides, but no right angles.

The second figure is a visual counterexample, because the figure satisfies the conditions in the hypothesis but not the conclusion.

Use geometry software to give a visual counterexample to each statement.

1. If one pair of opposite sides of a quadrilateral are congruent, then the quadrilateral is a parallelogram.
2. If one angle of a quadrilateral is a right angle, then the quadrilateral is a rectangle.
3. If one pair of consecutive sides of a quadrilateral are congruent, then the quadrilateral is a rhombus.
4. If the four angles of a quadrilateral are congruent, then the quadrilateral is a square.
5. If a diagonal of a quadrilateral divides the quadrilateral into two congruent triangles, then the quadrilateral is a rectangle.
6. If one pair of opposite sides of a quadrilateral are parallel and the other pair of opposite sides are congruent, then the quadrilateral is a parallelogram.
7. If one pair of opposite sides of a quadrilateral are parallel, then the quadrilateral is a parallelogram.
8. Every parallelogram is a rhombus.

24 **Technology** **HRW Geometry**

9. Given
10. Opposite sides of a parallelogram are congruent.
11. Reflexive Property of $\cong$
12. SSS
13. CPCTC
14. $\parallel$ s$\Rightarrow$ Consecutive interior $\angle$s supplementary
15. Congruent supplementary angles are right angles.
16. Opposite angles of a parallelogram are congruent.
17. Definition of a rectangle

The answer to Exercise 33 can be found in Additional Answers beginning on page 727.

22–30. Complete the proof of the following theorem.

THE TRIANGLE MIDSEGMENT THEOREM
If a segment joins the midpoints of two sides of a triangle, then it is parallel to the third side and its length is one-half the length of the third side. **4.6.7**

Given: $\triangle ABC$, $\overline{AD} \cong \overline{CD}$, $\overline{BE} \cong \overline{CE}$

Prove: $\overline{DE} \parallel \overline{AB}$ and $DE = \frac{1}{2}AB$

Add two segments to $\triangle ABC$ as shown, with $\overline{EF} \cong \overline{DE}$.

STATEMENTS	REASONS
$\triangle ABC$, $\overline{AD} \cong \overline{CD}$, $\overline{BE} \cong \overline{CE}$, $\overline{EF} \cong \overline{DE}$ E is the midpoint of $\overline{DF}$	Given
$\angle 2 \cong \angle 3$	**(22)** Vertical $\angle$s are $\cong$
$\triangle CED \cong \triangle EFB$	**(23)** SAS
$\angle 1 \cong \angle 4$	**(24)** CPCTC
$\overline{AC} \parallel \overline{FB}$	**(25)** If alt int $\angle$s are $\cong$, lines are $\parallel$
$\overline{CD} \cong \overline{BF}$	**(26)** CPCTC
$\overline{AD} \cong \overline{BF}$	Transitive Property of Congruence
$ABFD$ is a parallelogram	**(27)** If opposite sides of a quadrilateral are $\parallel$ and $\cong$, it is a parallelogram
$\overline{DF} \cong \overline{AB}$ $(DF = AB)$	**(28)** Opposite sides of a parallelogram are $\cong$
$DE = \frac{1}{2}DF$	Definition of midpoint
$DE = \frac{1}{2}AB$	**(29)** Substitution
$\overline{DE} \parallel \overline{AB}$	**(30)** Opposite sides of a parallelogram are $\cong$, segments contained in $\parallel$ segments are $\parallel$

Look Back

31. $\angle 1$ and $\angle 2$ form a linear pair. What special relationship exists between these two angles? **[Lesson 1.5]**

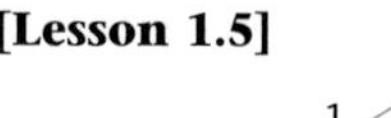

They are supplementary

32. Given: $\overline{BA} \cong \overline{BC}$; $\overline{BD} \cong \overline{BE}$
Prove: $\triangle ADC \cong \triangle CEA$
Hint: $\triangle ABC$ is isosceles.
[Lesson 4.4]

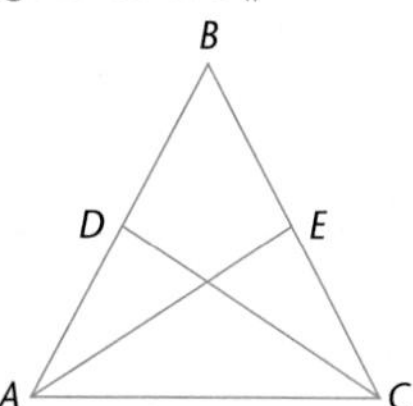

Use Transparency 42

Look Beyond

33. Plot the 4 points in a coordinate plane and draw quadrilateral $QUAD$ with diagonal $\overline{UD}$.
$Q(2, 4)$ $U(2, 9)$ $A(10, 9)$ $D(10, 4)$

34. What type of quadrilateral is $QUAD$? Rectangle

35. What is the perimeter of $QUAD$? 26 units

36. What is the area of $QUAD$? 40 units^2

37. What is the relationship between $\triangle QUD$ and $\triangle ADU$? They are congruent.

Look Beyond

Extension. After students have completed Exercises 33–37, have them draw the diagonals of quadrilateral $QUAD$, mark the intersection of the diagonals, and mark the midpoints of the sides.

Ask students to conjecture ways to find the coordinates of these five points from just the coordinates of the vertices.

32.
Statements	Reasons
1. $\overline{AB} \cong \overline{BC}$, $\overline{BD} \cong \overline{BE}$	1. Given
2. $\angle DAC \cong \angle ECA$	2. Isosceles Triangle Thm
3. $\overline{AC} \cong \overline{AC}$	3. Reflexive Property
4. $AB = AD + BD$, $BC = CE + BE$	4. Segment Addition Postulate
5. $AD + BD = CE + BE$	5. Substitution Property
6. $AD = CE$ $(\overline{AD} \cong \overline{CE})$	6. Subtraction Property
7. $\triangle ADC \cong \triangle CEA$	7. SAS

Compass and Straightedge Constructions

Why *Traditional drafting tools include an assortment of compasses. The use of a compass and straightedge involves the principles of triangle congruence. As you will learn, compasses are useful for much more than drawing circles.*

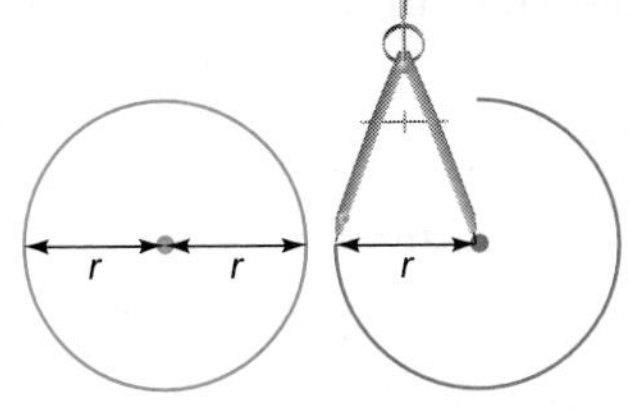

Using the same compass setting

Whenever you draw circles (or parts of circles) using the same compass settings, all the circles will have congruent radii. This is one of the keys to the classical constructions.

CONGRUENT RADII THEOREM

In the same circle, or in congruent circles, all radii are congruent. **4.7.1**

CRITICAL *Thinking*

Explain why Theorem 4.7.1 is true. (Use the fact that every point on a circle is the same distance from the center point.)

EXAMPLE 1

Given: $\overline{AB}$

Construct: A segment congruent to $\overline{AB}$

1. Using your straightedge, draw line l.
2. Select a point on l and label it O.
3. Set your compass equal to the distance AB. Use the point of your compass on O and draw an arc that intersects l. Label the intersection of the arc and the segment as point P.

   Conclusion: $\overline{OP} \cong \overline{AB}$ ❖

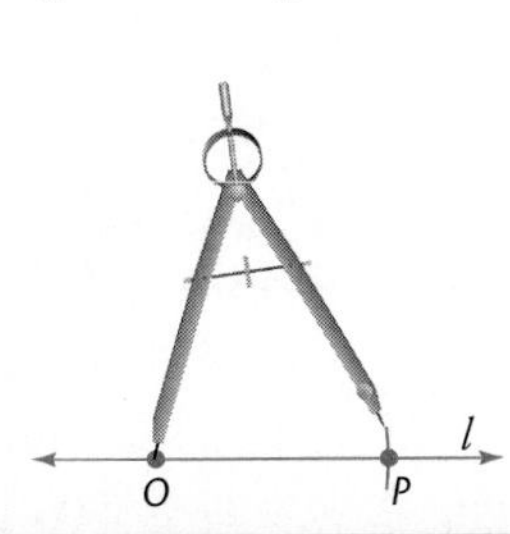

ALTERNATIVE teaching strategy

Technology If you plan on having students make extensive use of geometry graphics software, you may wish to do this entire lesson with a computer. The steps used for copying segments and angles may be different from compass-and-straightedge solutions. And, constructions such as bisecting an angle may be available as menu commands.

PREPARE

Objectives

- Identify and use the Congruent Radii Theorem.
- Construct segment, angle, and triangle copies.
- Construct an angle bisector.

RESOURCES

- Practice Master 4.7
- Enrichment Master 4.7
- Technology Master 4.7
- Lesson Activity Master 4.7
- Quiz 4.7
- Spanish Resources 4.7

Assessing Prior Knowledge

Use a compass and straightedge to construct the following:

1. a line segment
2. a circle using the line segment from #1 as the radius
3. a line segment and the perpendicular bisector of the line segment.

[Answers will vary.]

TEACH

Why A variety of tools are used to make accurate drawings. Ask students to name careers that require accurate drawings.

CRITICAL *Thinking*

Since every point on a circle is equidistant from the center of the circle, all radii have the same length. All radii in different congruent circles must be congruent to the radii in the original circle. $\overline{AB}$ and $\overline{OP}$ are radii of congruent circles, therefore they are congruent.

Alternate Example 1

Use paper folding to make a pair of congruent segments.

CRITICAL *Thinking*

Since each side in the constructed triangle is congruent to its corresponding side in the original triangle, SSS ensures congruence.

Alternate Example 2

Provide students with wax paper. Have them use paper folding to make a pair of congruent triangles.

[**Students use two sheets of wax paper. They fold any triangle on one sheet. Then they match the folds to duplicate the triangle on the second sheet.**]

CRITICAL *Thinking*

Side-Side-Side

TEACHING *tip*

After students have done Example 2 and its alternate, have them cut out the triangles to see which method was more precise. Answers will depend on the particular students. Paper folding can be almost as accurate as compass and straightedge.

How can you use the Congruent Radii Theorem to justify the conclusion of this construction? Where have you used this construction method before?

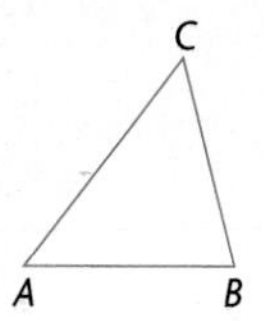

EXAMPLE 2

Given: $\triangle ABC$

Construct: A triangle congruent to $\triangle ABC$

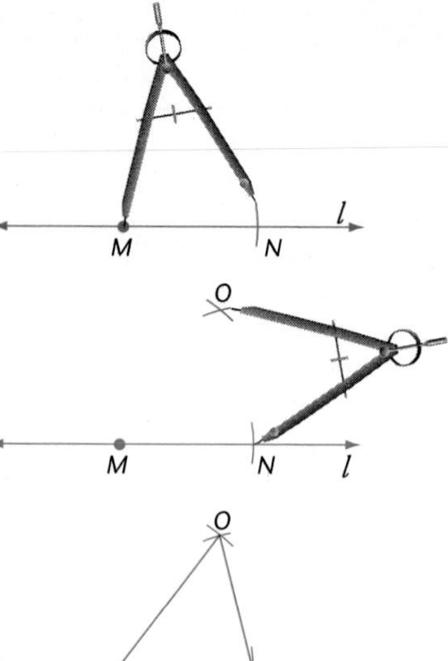

1. Using your straightedge, draw a line and label it l. Select a point on l and label it M.

   Set your compass equal to the distance AB. Place the point of your compass on M and draw an arc that intersects l. Label the intersection of the arc and the segment N.

2. Set your compass equal to the distance AC. Place the point of your compass on M and draw an arc in the area above.

   Set your compass equal to the distance BC. Place the point of your compass on N and draw an arc that intersects the arc you drew in Step 2. Label the intersection of the two arcs O.

3. Using your straightedge, draw a line segment that connects points M and O. Then draw a segment that connects points N and O.

   Conclusion: $\triangle MNO \cong \triangle ABC$ ❖

What postulate justifies the final conclusion of the construction?

EXAMPLE 3

Given: $\angle BAC$

Construct: An angle bisector of $\angle BAC$ using geometry software.

Geometry Graphics

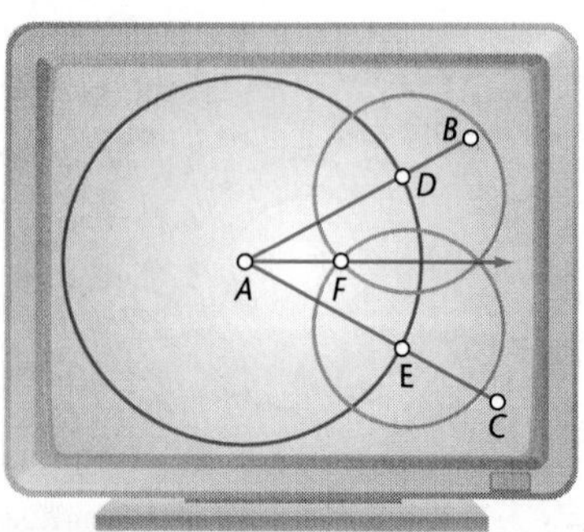

1. Construct $\angle BAC$.
2. Construct a circle with its center at A. Label the points of intersection between $\angle BAC$ and the circle as points D and E.

interdisciplinary CONNECTION **Architecture** Many Egyptian tombs had elaborate floor plans based on geometric figures. Have students research the floor plan for the tomb of Ramses IV. Students may enjoy analyzing and trying to construct the figure.

3. Using D and E as centers, construct two congruent circles. Use the measure menu to be sure that they are congruent. Make them large enough to intersect with each other. Label a point of intersection between them F.
4. Construct $\overline{AF}$. This is the angle bisector.
5. To check, measure $\angle BAF$ and $\angle CAF$. ❖

EXAMPLE 4

Given: The diagram on the right, which illustrates the compass and straightedge method of construction of the bisector of $\angle BAC$

Prove: $\angle BAF \cong \angle CAF$

Plan: Draw $\overline{BF}$ and $\overline{CF}$. Show $\overline{BF} \cong \overline{CF}$ since they are congruent radii, and $\triangle BAF \cong \triangle CAF$ by SSS. Then use CPCTC.

Proof:

STATEMENTS	REASONS
1. $\overline{AB} \cong \overline{AC}$	In the same or $\cong$ $\odot$s all radii are $\cong$ (Definition of a circle).
2. $\overline{CF} \cong \overline{BF}$	In the same or $\cong$ $\odot$s all radii are $\cong$ (Definition of a circle).
3. $\overline{AF} \cong \overline{AF}$	Reflexive Property of Congruence
4. $\triangle BAF \cong \triangle CAF$	SSS
5. $\angle BAF \cong \angle CAF$	CPCTC ❖

Communicate

1. Given $\overline{AB}$ as shown, tell how you would construct $\overline{XY}$ congruent to $\overline{AB}$.
2. State the theorem that justifies your construction in Exercise 1.
3. Explain how the Congruent Radii Theorem makes constructions with a straightedge and compass possible.
4. Explain how you would construct an angle bisector with a straightedge and compass.
5. Using part of the technique from Example 2 in the lesson, explain how you would copy an angle using only a straightedge and compass.

Alternate Example 3

Technology Use geometry graphics software to make a parallelogram. Construct the angle bisector of each angle. Drag on the vertices of the parallelogram to see what happens.

Cooperative Learning

The examples in this lesson are particularly appropriate for small group work if group members use different approaches and then compare the construction steps and precision of the final models. Possible methods include compass and straightedge; pencil, ruler, and grid paper; geometry graphics software; drawing software; and paper folding.

Alternate Example 4

Use paper folding to bisect an angle. Measure the angle and then use compass and straightedge to bisect it. Which method was more accurate?

ASSESS

Selected Answers

Odd-numbered Exercises 7–27

Assignment Guide

Core 1–10, 13, 17–21, 24–31

Core Plus 1–5, 8–31

ENRICHMENT The figure shows the beginnings of a regular pentagon. Have students carry out and complete the construction. Begin with $\overline{AB} \perp \overline{CD}$, and circle O through their intersection. Find the midpoint M of $\overline{OD}$, and then use $\overline{AM}$ as a radius to get point P. The radius $\overline{AP}$ gives P' and P''. $\overline{AP'}$ and $\overline{AP''}$ are the top two sides of the pentagon. The bottom two vertices are found by drawing circles with radii $\overline{AP}$ centered at P' and P''.

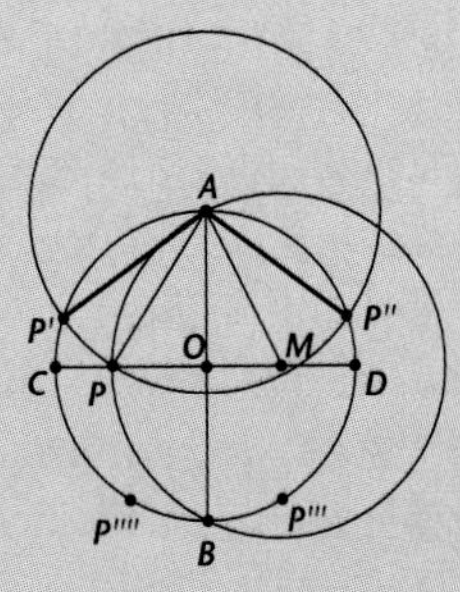

Technology

Students using geometry graphics software to construct perpendicular lines, perpendicular bisectors, and parallel lines should be told to follow the steps in the text rather than just using menu commands.

Error Analysis

For Exercises 22 and 23, students need the properties of parallel lines cut by a transversal. If students have forgotten these properties, refer them to Lessons 3.3, 3.4, and 3.5.

The answers to Exercises 6, 7, and 9 can be found in Additional Answers beginning on page 727.

Practice Master

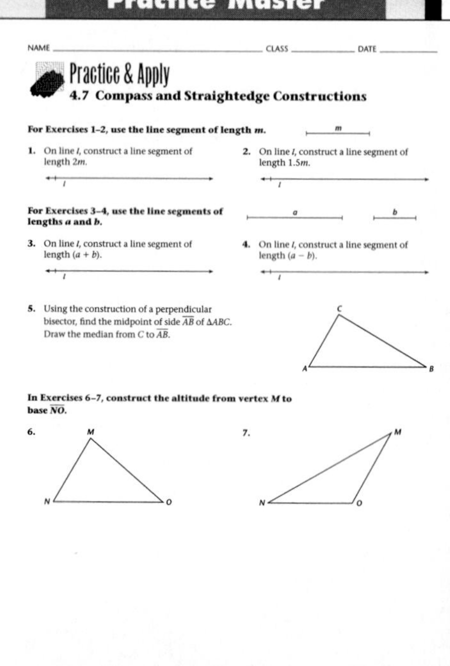

NAME ________ CLASS ________ DATE ________

Practice & Apply

4.7 Compass and Straightedge Constructions

For Exercises 1–2, use the line segment of length *m*.

1. On line *l*, construct a line segment of length $2m$.
2. On line *l*, construct a line segment of length $1.5m$.

For Exercises 3–4, use the line segments of lengths *a* and *b*.

3. On line *l*, construct a line segment of length $(a + b)$.
4. On line *l*, construct a line segment of length $(a - b)$.
5. Using the construction of a perpendicular bisector, find the midpoint of side $\overline{AB}$ of $\triangle ABC$. Draw the median from C to $\overline{AB}$.

In Exercises 6–7, construct the altitude from vertex *M* to base $\overline{NO}$.

6.
7.

HRW Geometry   Practice & Apply   7

Practice & Apply

Trace $\triangle GHJ$ onto your own paper and use it for Exercises 6–7.

6. Construct $\triangle DEF \cong \triangle GHJ$.
7. Construct the angle bisector of $\angle D$ in your constructed triangle.

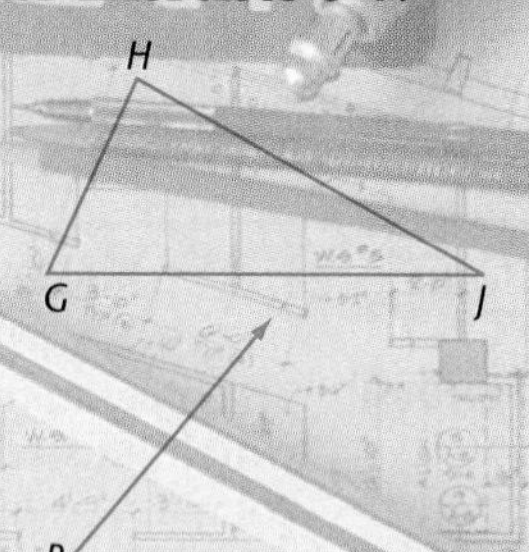

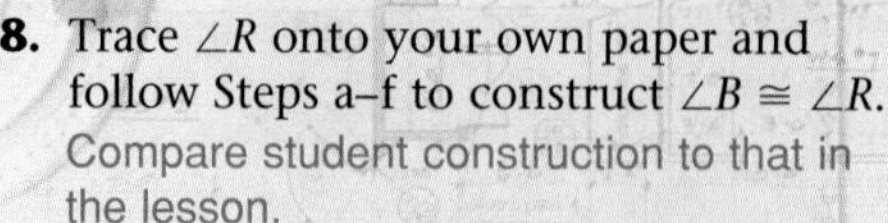

8. Trace $\angle R$ onto your own paper and follow Steps a–f to construct $\angle B \cong \angle R$.
Compare student construction to that in the lesson.

R

a. Using a straightedge, draw a ray with endpoint B.

b. Place your compass point on R in your original angle and draw an arc as shown. Label the intersection points Q and S.

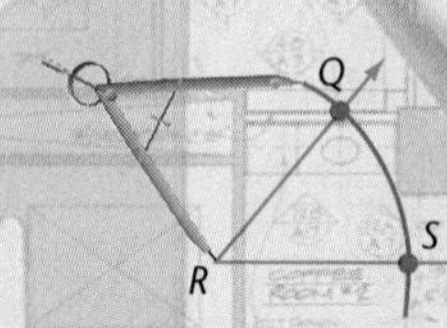

c. Use the same compass setting and place the point on B. Draw an arc crossing the ray and label the point of intersection C.

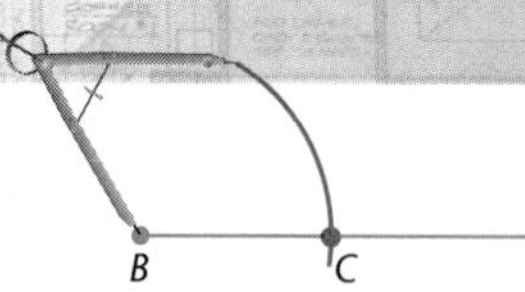

d. Set your compass equal to the distance QS in $\angle R$.

Q
S
R

e. Keep the same compass setting and place the point on C. Draw an arc crossing the first arc you made on ray $\overrightarrow{BC}$. Label the point of intersection A.

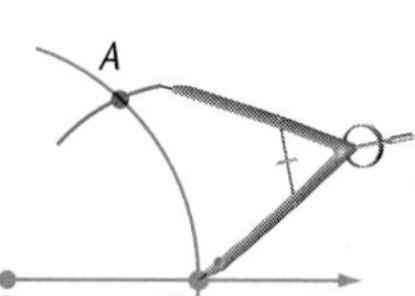

f. Draw ray $\overrightarrow{BA}$, forming $\angle B$.

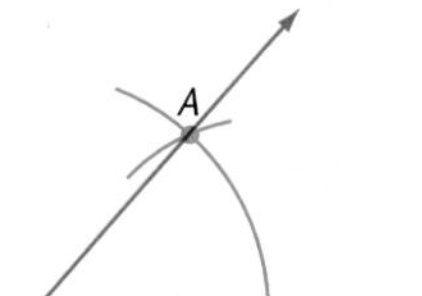

9. Refer to the construction in Exercise 8. Write a paragraph proving $\angle ABC \cong \angle QRS$. (Hint: Construct segments $\overline{QS} \cong \overline{AC}$.)

Hands-On Strategies A wide variety of compasses are now available, including some without sharp, pointed arms. Others are designed so that the pencil and pointed arm can be retracted for safety. Check a school supply catalog for alternatives to a traditional compass, particularly for students who lack manual dexterity.

Cooperative Learning Have students work in small groups to complete the following constructions. **1.** Copy a segment. **2.** Copy a triangle. **3.** Copy an angle. **4.** Bisect an angle. **5.** Construct the perpendicular bisector of a segment. **6.** Construct a pair of perpendicular lines. **7.** Construct a pair of parallel lines.

10. Trace $\overline{AC}$ onto your own paper and follow the steps to construct the perpendicular bisector of $\overline{AC}$. Check student constructions

a. Set your compass equal to a distance greater than half the length of $\overline{AC}$.

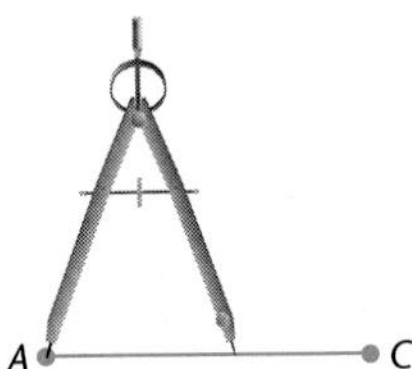

b. Place your compass point on A and draw an arc as shown.

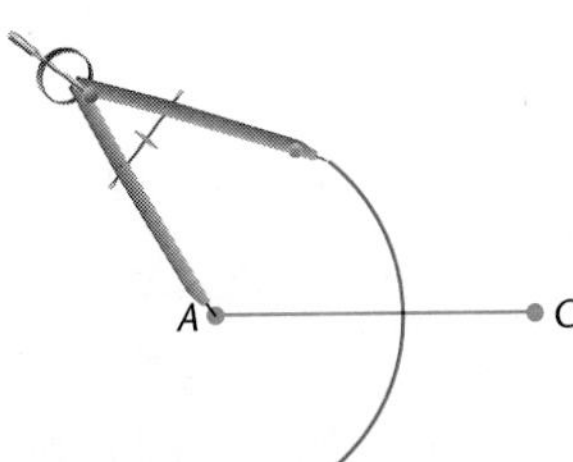

c. Using the same compass setting, place the compass point on C and draw a new arc. Label intersection points B and D.

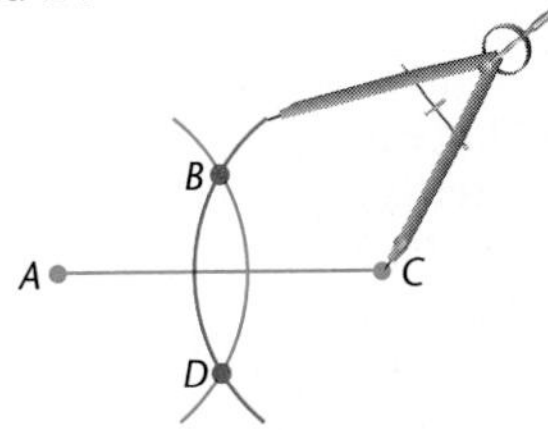

d. Use a straightedge to draw a line through points B and D. Label intersection E.

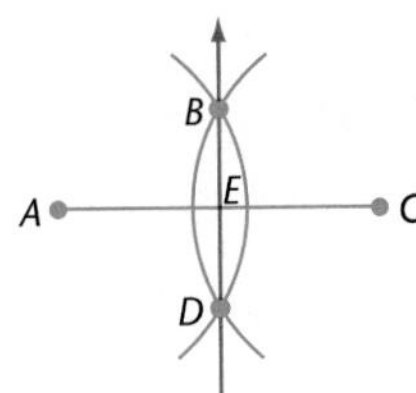

11–15. Using the construction in Exercise 10, prove that $\overline{BD}$ is the perpendicular bisector of $\overline{AC}$. Complete the paragraph proof.

$\overline{AB} \cong \overline{BC} \cong \overline{CD} \cong \overline{DA}$, because **(11)**. Therefore, quadrilateral $ABCD$ is a rhombus, by **(12)**. **(13)**, because the diagonals of a rhombus are perpendicular to each other. A rhombus is also a parallelogram, by **(14)**. Therefore, $\overline{BD}$ and $\overline{AC}$ bisect each other, because **(15)**.

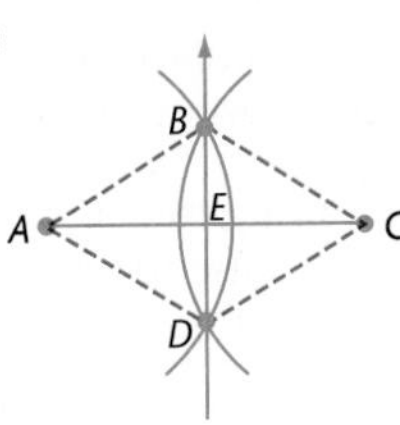

16. Draw a line l. Place point A above line l. Follow Steps a–d to construct a line perpendicular to l through A.

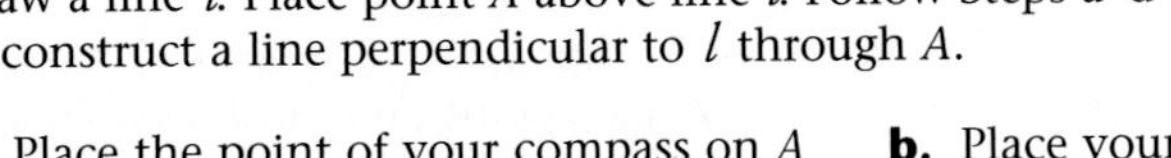

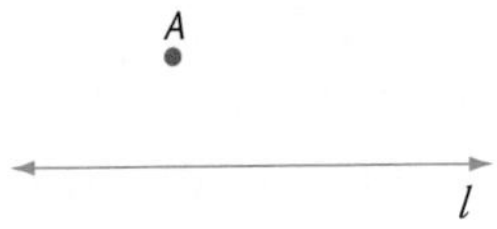

A

l

a. Place the point of your compass on A and draw an arc that intersects line l twice. Label the intersection points D and B. Check student constructions

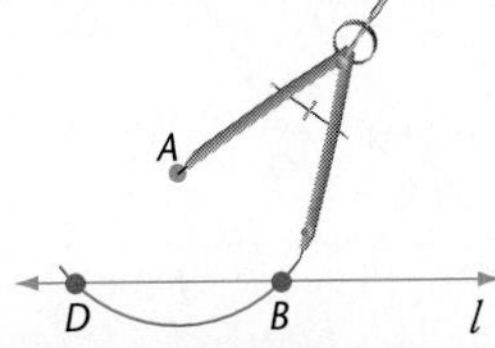

b. Place your compass point on D and draw an arc below line l. It is not necessary to keep the compass set at the same distance as Step a.

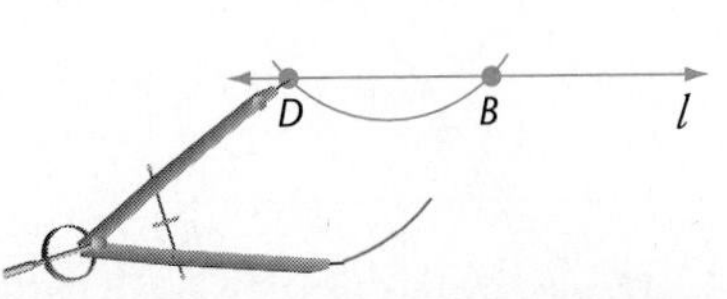

11. they are radii of congruent circles
12. definition of a rhombus
13. $\overline{AC} \perp \overline{BD}$
14. definition of a rhombus
15. the diagonals of a parallelogram bisect each other

alternative ASSESSMENT

Performance Assessment

1. Construct a right isosceles triangle using only a compass and an unmarked straightedge.
2. Write out the directions for this construction in paragraph form.
3. Justify or prove this construction.

Technology Master

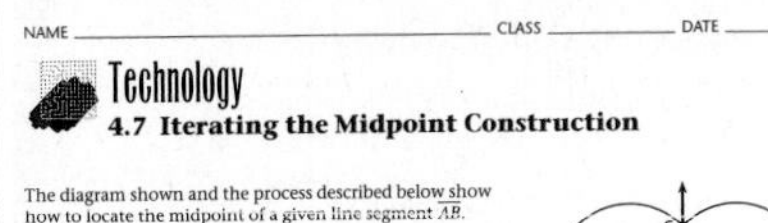

NAME ______ CLASS ______ DATE ______

Technology

4.7 Iterating the Midpoint Construction

The diagram shown and the process described below show how to locate the midpoint of a given line segment $\overline{AB}$.

① Construct circles with equal radii centered at A and B.

② Construct the line $\overleftrightarrow{CD}$ through the two points where the circles intersect.

③ Mark the intersection of $\overleftrightarrow{CD}$ and $\overline{AB}$ as M.

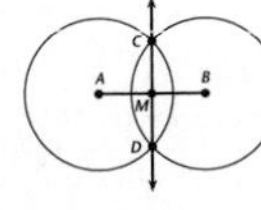

In order to make the construction work, you must choose the radius of the two circles to be more than half the length of $\overline{AB}$.

Use geometry software for the following exercises.

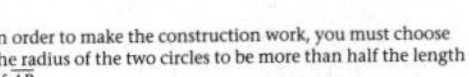

1. Sketch $\overline{AB}$ 4 in. long. At A and B, sketch two circles each having radius 1 in. and centers A and B. Does your sketch confirm the statement above about the length of the radii that must be used?
2. Sketch $\overline{AB}$ in your software. Make the segment horizontal and as long as your screen will allow. Complete the three-step construction and label the midpoint M_1.
3. Construct the midpoints of $\overline{AM_1}$ and $\overline{M_1B}$. Label them M_2 and M_3 respectively. List lengths AM_2, M_2M_1, M_1M_3, and M_3B.
4. If you carry out the construction process n times, how many midpoints will you have constructed? Write a formula in terms of n for the number of midpoints m constructed.
5. If you carry out the construction process n times, into how many small segments will you have divided $\overline{AB}$? How long will each segment be?

HRW Geometry Technology 25

22. Check student constructions.

c. Using the same compass setting as Step b, place the compass point at *B* and draw an arc which intersects the arc you drew in Step b. Label the point of intersection *C*.

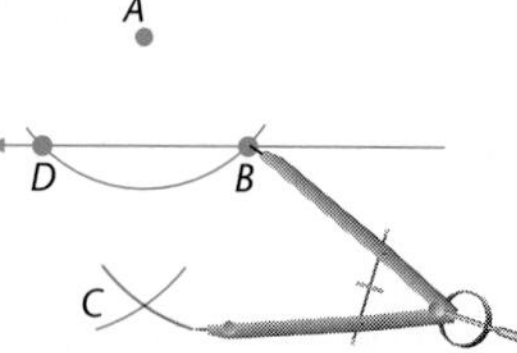

d. Using a straightedge, draw a line which passes through *A* and *C*. $\overline{AC}$ is perpendicular to *l*.

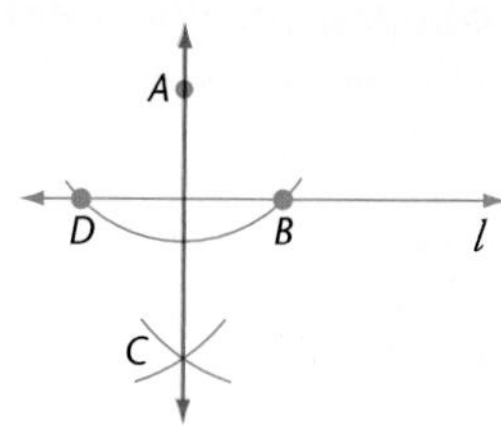

17–21. Kites You can create a kite by connecting the endpoints *A, B, C,* and *D* from the construction in Exercise 16. (Assume that a different compass setting was used to construct the arcs through *C* than was used to construct the arcs through *D* and *B*.) Prove that $\overline{AC} \perp \overline{DB}$.

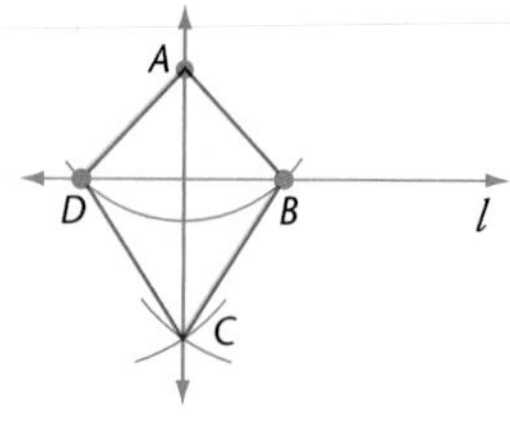

Proof:

STATEMENTS	REASONS
$\overline{AB} \cong$ __(17)__ and __(18)__ $\cong \overline{DC}$ $\overline{AD}$, $\overline{BC}$	**Congruent Radii Theorem**
$AB \neq DC$ and $AD \neq$ __(19)__ BC	Different compass setting used to construct each segment
Quadrilateral *ABCD* is a kite	__(20)__ Defn of kite
$\overline{AC} \perp \overline{DB}$	__(21)__ Diagonals of a kite are ⊥

22. Draw a line and place point *M* anywhere above *l*. Follow the steps to construct a line parallel to *l* through point *M*.

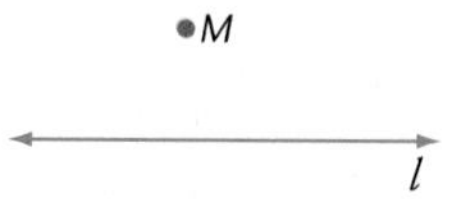

a. Using a straightedge construct a transversal line to *l* through *M*. Label the point of intersection *P*.

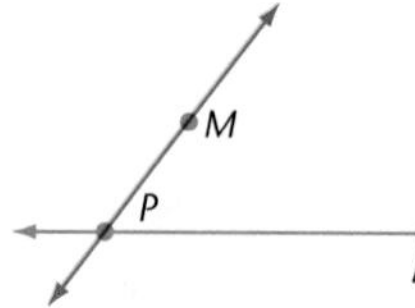

b. Place your compass point on *P* and draw an arc crossing both rays of ∠*P*. Label the intersection points *R* and *T*.

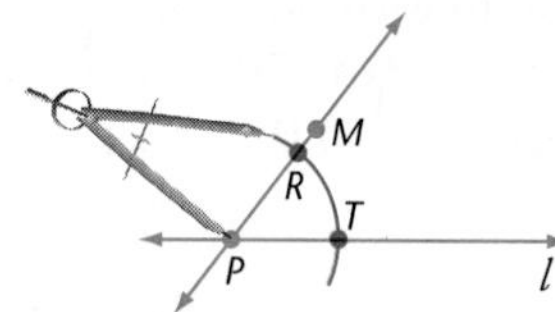

c. Using the same compass setting, place the compass point on *M* and draw an arc that crosses the transversal. Label the point of intersection *N*.

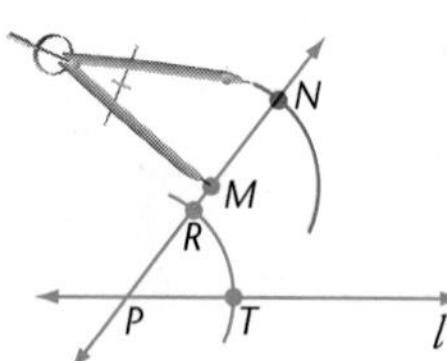

d. Set your compass to distance *RT*. Place the compass point on *N* and draw an arc that intersects the arc you drew in Step c. Label the point of intersection *O*.

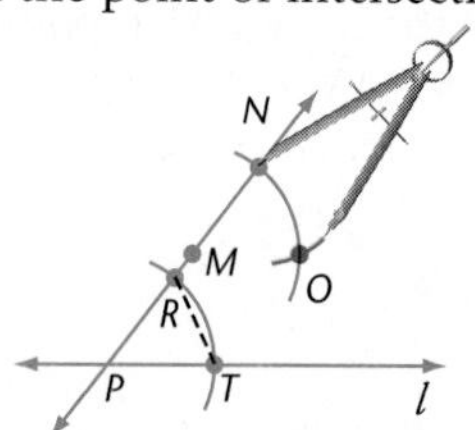

23. Statements

1. Line *l* and point M not on *l*.
2. $\overline{PR} \cong \overline{PT} \cong \overline{MN} \cong \overline{MO}$
3. $\overline{RT} \cong \overline{NO}$
4. $\triangle RPT \cong \triangle NMO$
5. $\angle RPT \cong \angle MNO$
6. line $u \parallel$ line l

Reasons

1. Given
2. Congruent Radii Theorem
3. Congruent Radii Theorem
4. SSS
5. CPCTC
6. Alt int ∠s ≅ ⇒ ∥ s

e. Using a straightedge draw the line which passes through points M and O. Label the line u. Line u is parallel to line l.

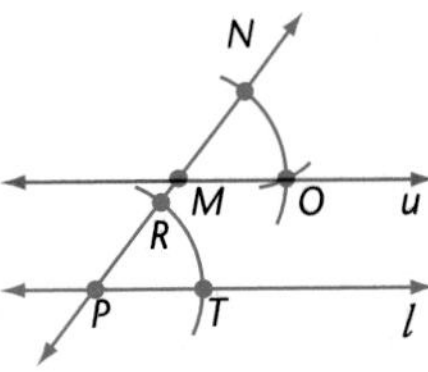

23. Write a two-column proof for the construction in Exercise 21. That is, prove that $u \parallel l$.

Look Back

Categorize each figure according to this list: parallelogram, quadrilateral, rectangle, square, rhombus, kite, or trapezoid. Use all appropriate terms with each. [Lessons 3.2, 3.3]

24. $\overline{AB} \parallel \overline{CD}$

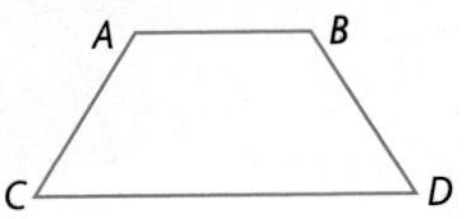

25.

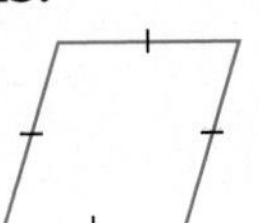

26.

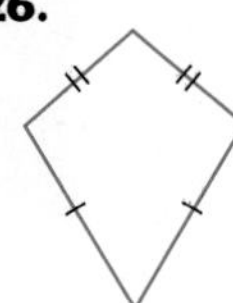

27. $\overline{AB} \parallel \overline{CD}$; $\overline{AC} \parallel \overline{BD}$

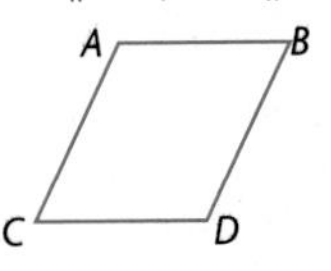

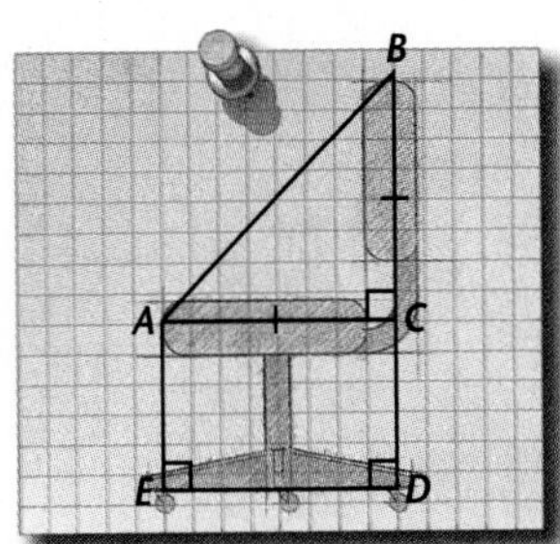

28. Interior Design Interior designers often use geometric figures to create interesting furniture designs. Identify at least 10 pairs of congruencies in the design of this chair. **[Lessons 4.1, 4.2, 4.3]**

The answers to Exercises 23 and 28–31 can be found in Additional Answers beginning on page 727.

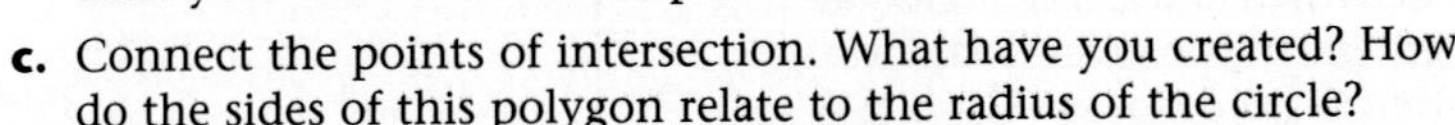

Look Beyond

History The French Revolutionary general Napoleon was an amateur mathematician. Napoleon found the following construction especially intriguing.

29. a. Draw a circle using your compass.

b. Without changing your compass setting start at the top of the circle and draw an arc that crosses the circle. Move your compass point to the point of intersection and draw another arc. Continue until your are back at the top of the circle.

c. Connect the points of intersection. What have you created? How do the sides of this polygon relate to the radius of the circle?

30. Using the results from Exercise 29, construct an equilateral triangle inscribed in a circle.

31. Explain how you could construct a square inscribed in a circle.

Look Beyond

Exercises 29–31 present some intriguing constructions. The six equally spaced points around a circle can be connected in two ways—to make a regular hexagon and to make a six-pointed star. Have students divide a circle into eight equal parts to get eight equally spaced points. They should then see how many different ways the points can be connected.

24. quadrilateral, trapezoid.

25. quadrilateral, parallelogram, rhombus.

26. quadrilateral, kite.

27. quadrilateral, parallelogram

PREPARE

Objectives

- Construct midpoints of segments by combining traditional construction methods with coordinate geometry.

- Practice Master **4.8**
- Enrichment Master **4.8**
- Technology Master **4.8**
- Lesson Activity Master **4.8**
- Quiz **4.8**
- Spanish Resources **4.8**

Assessing Prior Knowledge

1. Plot $H(4, 8)$, $T(8, 10)$, $P(10, 6)$, and $G(4, -2)$ in a coordinate plane. Draw the two diagonals. **[Check student graphs.]**
2. What is the name of quadrilateral *HTPG*? **[kite]**
3. What are the coordinates of point *O*, the intersection of the diagonals? **[$O(7, 7)$]**
4. Find all the pairs of congruent triangles. **[$\triangle HOT \cong \triangle POT$, $\triangle HOG \cong \triangle POG$, $\triangle HTG \cong \triangle PTG$]**

TEACH

Ask students for everyday examples in which a rectangular grid is a useful tool. One example is in sewing. Fabric stores carry large sheets of cardboard with 1-inch grids to help in cutting out material along the grain.

Congruence in the Coordinate Plane

Why *If you set up a coordinate plane with the same horizontal and vertical units, you can use techniques you learned in your study of algebra to analyze congruence.*

EXAMPLE

Given: $\triangle ABC$ with $A(2, 1)$, $B(5, 3)$, and $C(5, 1)$; $\triangle XYZ$ with $X(7, 2)$, $Y(10, 4)$, and $Z(10, 2)$

Prove: $\overline{AB} \cong \overline{XY}$

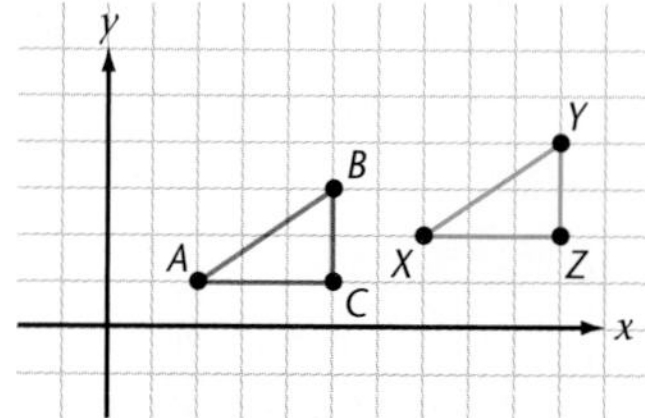

Solution ➤

The triangles each have one horizontal segment and one vertical segment. Therefore, the angles between these segments are right angles, because of the way standard coordinate axes are set up.

Find the measures of the horizontal segments by finding the difference in the *x*-coordinates. Then find the measures of the vertical segments by finding the absolute value of the difference in the *y*-coordinates.

Horizontal distances: $AB = |5 - 2| = 3$, $XY = |10 - 7| = 3$

Vertical distances: $BC = |1 - 3| = 2$, $YZ = |2 - 4| = 2$

Notice which sides and angles are congruent in the given triangles. Therefore, $\triangle ABC \cong \triangle XYZ$ by SAS and $\overline{AB} \cong \overline{XY}$ by CPCTC. ❖

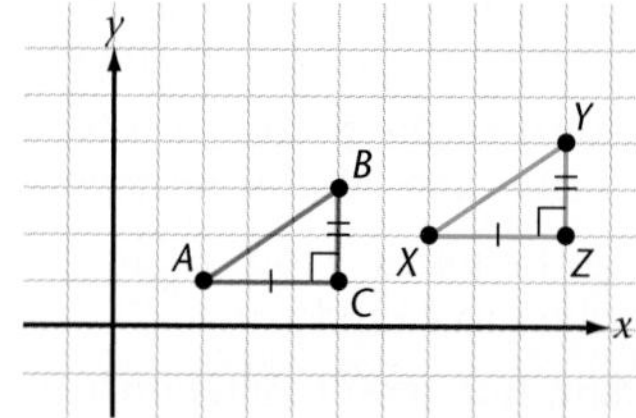

ALTERNATIVE teaching strategy

Hands-On Strategies A useful manipulative tool for this lesson and the next is a blank coordinate grid reproduced on an overhead transparency film. Students can use wipe-off pens to experiment with various figures. They can test for congruency by laying one grid on top of another.

Exploration 1 Midpoints of Horizontal Segments

You will need
Coordinate paper
Compass and straightedge or
Ruler

Plot each segment in a coordinate plane. Determine the coordinates of the midpoint of each segment. (You may use a compass and straightedge, a ruler, or paper folding.)

$\overline{AB}$, with $A(2, 3)$ and $B(10, 3)$ $\quad$ $\overline{XY}$, with $X(4, -2)$ and $Y(8, -2)$

$\overline{LN}$, with $L(-2, 5)$ and $N(10, 5)$ $\quad$ $\overline{CD}$, with $C(-10, 6)$ and $D(-2, 6)$

What rule can be applied to the x-coordinates of the endpoints of a horizontal segment to obtain the x-coordinates of the midpoint of the segment?

Apply your rule to find the coordinates of the midpoint of $\overline{KL}$, with $K(17, 432)$ and $L(95, 432)$. Find the average of the x-coordinates of K and L as follows: $(17 + 95) \div 2 = ?$ What do you notice about the result? ❖

CRITICAL *Thinking*

Write a formula for finding the midpoint of a horizontal segment with endpoints $A(x_1, y_1)$ and $B(x_2, y_2)$. Use the average of the x-coordinates. How could you find the formula for the midpoint of a vertical segment?

Exploration 2 Midpoints of General Segments

You will need
Coordinate paper
Ruler

Consider the following question. Streets in Springfield are laid out like a coordinate grid. Ron lives three blocks east and two blocks north of City Hall. Ron's friend Jaime lives 11 blocks east and eight blocks north of City Hall. They want to meet at the point midway between their locations. Where is this point located.

1. Locate City Hall at the origin of the coordinate plane and let the direction of the positive x-axis represent east and that of the positive y-axis represent north. Plot $R(3, 2)$ and $J(11, 8)$.

2. Draw a segment with endpoints that pass through points R and J. Then draw a horizontal line through R and a vertical line through J. What kind of geometric figure is formed?

The "origin" of the Chicago grid is at the intersections of Washington and State streets.

interdisciplinary CONNECTION **Computers** Although math texts almost always show the first quadrant in the upper right of a coordinate grid, the arrangement can be different if needed. Students who have worked with drawing or page-layout computer software will have examples in which the first quadrant is in the lower right; that is, the (0, 0) point is in the upper-left corner of the computer screen. It is important to always label each axis, show its direction with an arrowhead, and give at least one number on each axis to establish the scale.

Alternate Example

Given: $\triangle ABC$ with $A(5, 8)$, $B(9, 2)$, and $C(1, 2)$. Point M is the midpoint of CB.

Prove: $\triangle ABC$ is isosceles

[Draw $\overline{AM}$ connecting the vertex of the triangle with point M. $\overline{AM} \cong \overline{AM}$ because of the Reflexive Property. $\angle AMC \cong \angle AMB$ because they are right angles. $CM = 4$ units $= MB$. $\triangle ACM \cong \triangle ABM$ because of SAS. $\angle C \cong \angle B$ because CPCTC. Therefore, $\triangle ABC$ is isosceles.]

Exploration 1 Notes

Students develop a midpoint formula for horizontal segments in this exploration. Geometry graphics software that includes coordinate geometry features may be useful.

ongoing ASSESSMENT

2. Add the x-coordinates and divide by 2.

CRITICAL *Thinking*

Midpoint of horizontal segment $= \left(\frac{x_1 + x_2}{2}, y\right)$

Midpoint of vertical segment $= \left(x, \frac{y_1 + y_2}{2}\right)$

Exploration 2 Notes

Students develop a midpoint formula for general segments in this exploration.

Cooperative Learning

Have students work in pairs. They should use a number cube to generate three points. These will be the vertices of a triangle. (If the three points cannot be connected to make a triangle, students roll again.) Then students should draw the triangle on a 6-by-6 coordinate grid and use the midpoint formulas to find the midpoints of the three sides. By discussing which formula applies in each case, students will clarify their understanding of the various midpoint formulas.

Ongoing Assessment

8. Find the average of the *x*-coordinates and the average of the *y*-coordinates.

9. Midpoint of $\overline{AB} = \left(\frac{x_1 + x_2}{2}, \frac{y_1 + y_2}{2}\right)$.

Critical Thinking

$\triangle JVM \cong \triangle MHR$ by SAS, so $\overline{JM} \cong \overline{MR}$ by CPCTC.

3. Label the point where the horizontal and vertical lines intersect at *Q*. Find the coordinates of *Q* and label your drawing with them.

4. Use your knowledge of the midpoints of horizontal and vertical segments to find the coordinates of *V* (the midpoint of $\overline{JQ}$) and the coordinates of *H* (the midpoint of $\overline{RQ}$). Add these to your drawing.

5. Draw a vertical segment from *H* to intersect $\overline{RJ}$ and a horizontal segment from *V* to intersect $\overline{RJ}$. Label their intersection as point *M*. Does *M* lie on $\overline{RJ}$?

6. Measure the distances *RM* and *MJ*. What can you conclude about point *M*?

7. Find the coordinates of *M* and add them to your drawing. Compare the coordinates of *M* to the coordinates of *H* and *V*. What do you observe?

8. In your own words, write a rule for finding the midpoint of a segment that is not horizontal or vertical.

9. Express the rule you stated in words in Step 8 as an algebraic formula for finding the midpoint of a segment $\overline{AB}$ with endpoints $A(x_1, y_1)$ and $B(x_2, y_2)$. ❖

Critical Thinking

How can you use congruent triangles in your exploration drawing to prove that your midpoint rule is correct? (*MVQH* is a rectangle.)

Exercises & Problems

Communicate

1. Explain how to find the midpoint of a horizontal segment.
2. Explain how to find the midpoint of a vertical segment.
3. Explain the formula for finding the coordinates of the midpoint of a segment in general.
4. $\overline{AB}$ and $\overline{CB}$ are drawn on a coordinate grid with the following coordinates: $A(2, 3)$, $B(7, 3)$, $C(7, 8)$. What can you conjecture about $\overline{AB}$ and $\overline{CB}$?
5. Given: $\triangle CAT$ with $C(2, 3)$, $A(2, 7)$, and $T(4, 3)$; $\triangle DOG$ with $D(2, -3)$, $O(2, -7)$, and $G(4, -3)$. Explain how to prove $\triangle CAT \cong \triangle DOG$.

Enrichment Given a segment $\overline{AB}$ with endpoints $A(x_1, y_1)$ and $B(x_2, y_2)$, find algebraic formulas for three points that divide the segment in fourths.

[The midpoint formula gives the middle point. The other two points are

$x = \frac{3}{4}x_1 + \frac{1}{4}x_2, y = \frac{3}{4}y_1 + \frac{1}{4}y_2$

$x = \frac{1}{4}x_1 + \frac{3}{4}x_2, y = \frac{1}{4}y_1 + \frac{3}{4}y_2$]

Inclusion Strategies

Inviting Participation The notation used for the midpoint formulas will be difficult for some students. Allow them to express their formulas in words. For example: For a horizontal segment, average the *x*-coordinates. For a vertical segment, average the *y*-coordinates. For a slanted segment, average the *x*-coordinates and then average the *y*-coordinates.

Practice & Apply

Use a coordinate grid and determine whether the following pairs of segments are congruent.

6. $\overline{AB}$ and $\overline{CD}$, with $A(2, 6)$, $B(2, 10)$, $C(5, 1)$, and $D(5, 5)$ Yes; $AB = CD = 4$
7. $\overline{PQ}$ and $\overline{RS}$, with $P(4, 3)$, $Q(4, 0)$, $R(-2, 5)$, and $S(-2, 12)$ No; $PQ = 3$, $RS = 7$
8. $\overline{VE}$ and $\overline{HO}$, with $V(3, -1)$, $E(3, 5)$, $H(5, 4)$, and $O(11, 4)$ Yes; $VE = HO = 6$
9. $\overline{TO}$ and $\overline{OP}$, with $T(-3, 0)$, $O(0, 5)$, and $P(3, 0)$ Yes; $TO = OP$ by CPCTC

Use the midpoint formula from Exploration 2 to find the *x*- and *y*-coordinates of the midpoint for each segment.

10. $\overline{AB}$, with $A(1, 2)$ and $B(9, 2)$ $(5, 2)$
11. $\overline{ST}$, with $S(4, 5)$ and $T(4, -3)$ $(4, 1)$
12. $\overline{XY}$, with $X(4, 3)$ and $Y(9, 11)$ $\left(\frac{13}{2}, 7\right)$
13. $\overline{PQ}$, with $P(-2, 5)$ and $Q(-8, -7)$ $(-5, -1)$
14. Plot the points and draw the figures on the coordinate plane: $\triangle LCO$ and $\triangle RCO$ with $L(-5, 0)$, $C(0, 3)$, $O(0, 0)$, and $R(5, 0)$
15. Given the figures on the coordinate plane from Exercise 14, prove that $\triangle LCR$ is isosceles.
16. **Spreadsheets** Create a spreadsheet that will allow the user to specify the coordinates of the endpoints of a segment and that will compute the coordinates of the midpoint of the segment. Using the following column headings, insert the formulas for the *x*-coordinate and the *y*-coordinate of the midpoint in the appropriate cells of the spreadsheet.

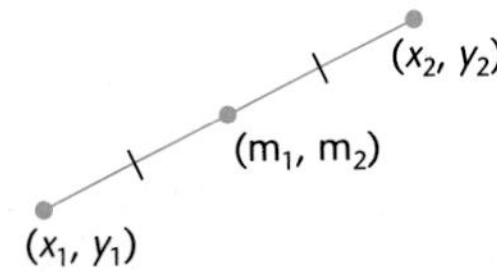

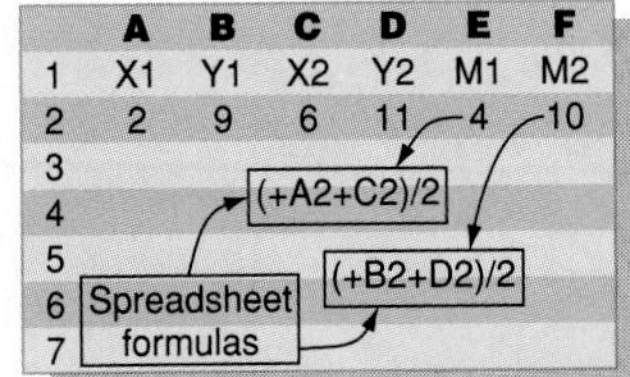

	A	B	C	D	E	F
1	X1	Y1	X2	Y2	M1	M2
2	2	9	6	11	4	10
3						
4						
5						
6						
7						

For Exercises 17–23, plot $\triangle PQR$, with $P(1, 2)$, $Q(3, 8)$, and $R(11, 4)$.

17. Let X be the midpoint of $\overline{PQ}$. Find its coordinates and plot it. $X(2, 5)$
18. Let Y be the midpoint of $\overline{QR}$. Find its coordinates and plot it. $Y(7, 6)$
19. Draw midsegment $\overline{XY}$. What type of figure is $PXYR$? Justify your answer.

ASSESS

Selected Answers

Odd-numbered Exercises 7–35

Assignment Guide

Core 1–16, 24–35

Core Plus 1–5, 8–11, 14–40

Technology

If geometry graphics software is used, make sure that it provides an *x-y* coordinate plane. Graphics calculators are needed for Exercises 24–27.

Error Analysis

For Exercises 24–27, students will likely find that their calculators ask for equations in the form "$y =$." If necessary, remind them of the general linear equation $y = Ax + B$, in which A is the slope of the equation and B is the *y*-intercept.

The answers to Exercises 15, 16, and 19 can be found in Additional Answers beginning on page 727.

Cooperative Learning Have students work in pairs. Independently, each should draw a triangle on a coordinate grid. Partners should tell each other just the coordinates of the vertices. Then each student should find the three side midpoints of the partner's triangle without making a drawing.

14.

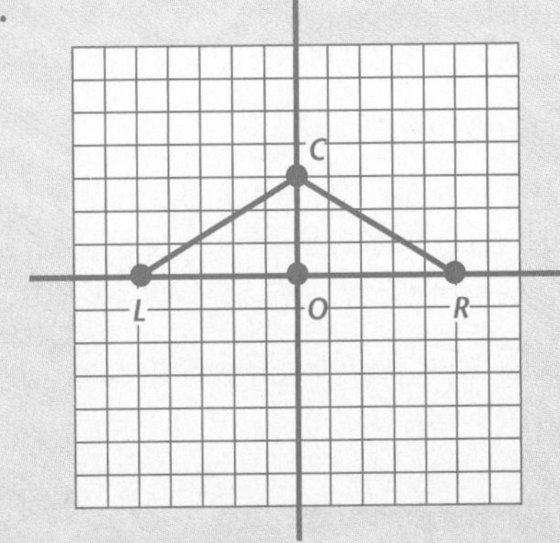

alternative ASSESSMENT

Performance Assessment

Have students describe the location of the point determined by each formula. The coordinates (x_m, y_m) are those of a point M on segment AB. The endpoints of $\overline{AB}$ are $A(x_1, y_1)$ and $B(x_2, y_2)$. Students should provide drawings along with their explanations.

1. $x_m = \frac{1}{2}(x_1 + x_2)$ **[the x-coordinate of the midpoint of a slanted segment]**
2. $y_m = \frac{1}{2}(y_1 + y_2)$ **[the y-coordinate of the midpoint of a slanted segment]**
3. $x_m = \frac{1}{2}(x_1 + x_2)$, $y_m = 0$ **[the coordinates of the midpoint of a horizontal segment]** $\frac{1}{2}$

The answers to Exercises 28–33 can be found in Additional Answers beginning on page 727.

Practice Master

NAME ______ CLASS ______ DATE ______

Practice & Apply
4.8 Exploring Congruence in the Coordinate Plane

In Exercises 1–2, plot the line segments on the grid and determine if the pairs of segments are congruent.

1. $\overline{RS}$ and $\overline{ST}$ with $R(2, 0)$, $S(5, 0)$ and $T(5, 2)$
   $\overline{RS} \cong \overline{ST}$? ______
2. $\overline{AB}$ and $\overline{CD}$ with $A(-2, 3)$, $B(-2, -1)$, $C(-2, 5)$, and $D(2, 5)$
   $\overline{AB} \cong \overline{CD}$? ______

In Exercises 3–4, use the midpoint formula to find the x- and y-coordinates of the midpoint for each segment.

3. $\overline{AB}$ with $A(2, -6)$, $B(2, -4)$ ______
4. $\overline{CD}$ with $C(3, -2)$, $D(1, -6)$ ______

Exercises 5–10 refer to the grid.

5. Draw $MNOP$ with $M(8, 5)$, $N(10, 0)$, $O(0, 0)$, and $P(3, 5)$.
6. What kind of figure is $MNOP$? ______
7. A is the midpoint of $\overline{OP}$, and B is the midpoint of $\overline{NM}$. Find the coordinates of A and B. ______
8. The length of $\overline{PM}$ is ______.
9. The length of $\overline{ON}$ is ______.
10. The length of $\overline{AB}$ is ______.
11. How is the length of $\overline{AB}$ related to the lengths of $\overline{PM}$ and $\overline{ON}$? ______
12. The line that joins the midpoints of the legs of a trapezoid is the median of the trapezoid. From your observations of $MNOP$, write a conclusion about the median of a trapezoid. ______

8 Practice & Apply HRW Geometry

20. Let Z be the midpoint of $\overline{PR}$. Find its coordinates and plot it. $Z(6, 3)$

21. Draw midsegment $\overline{YZ}$. What type of figure is $PXYZ$? Justify your answer.

22. Find two segments in the figure that are congruent to $\overline{PZ}$. $\overline{XY}$, $\overline{ZR}$

23. What is the relationship between the length of $\overline{PR}$ and the length of $\overline{XY}$ based on information obtained above? $PR = 2(XY)$

Technology Use a graphics calculator to create the following figures. Write the equations of the lines that you plot.

24. A parallelogram that is not a rhombus, with one pair of sides parallel to the x-axis

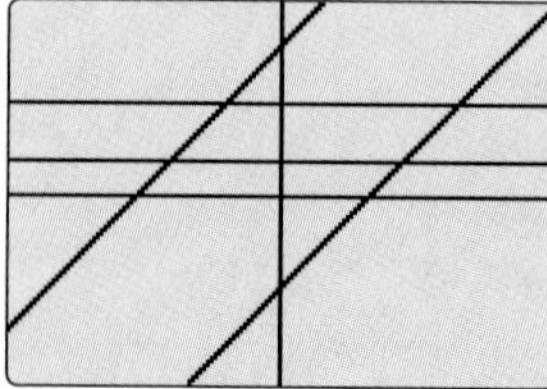
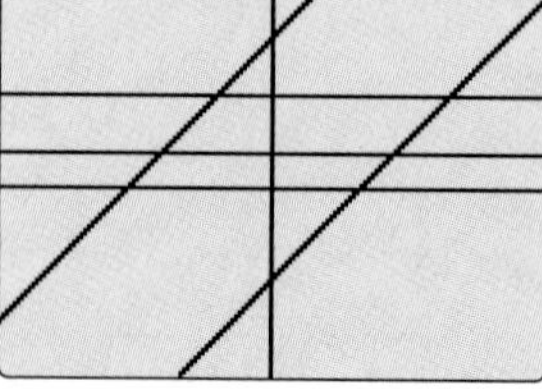

25. A parallelogram that is not a rhombus, with no sides parallel to the coordinate axes

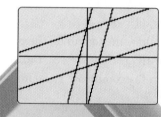

26. A rhombus that is not a square

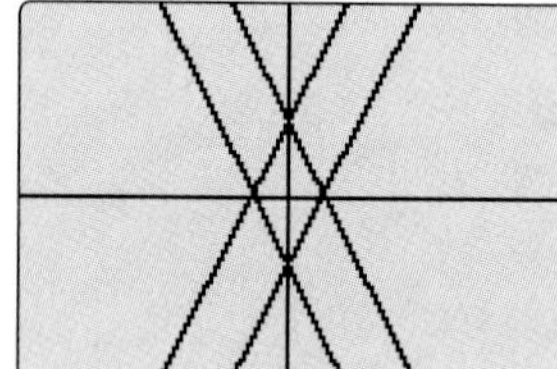

27. A square with no sides parallel to the coordinate axes

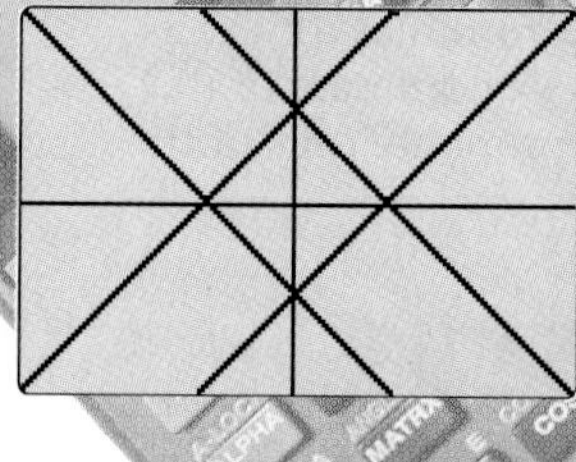

For Exercises 28–33, plot quadrilateral *QUAD* in a coordinate plane. Determine what type of quadrilateral *QUAD* is, and justify your answer.

28. $Q(-2, 1)$, $U(0, 4)$, $A(5, 4)$, $D(3, 1)$

29. $Q(1, 5)$, $U(3, 7)$, $A(3, 3)$, $D(1, 1)$

30. $Q(1, 3)$, $U(3, 6)$, $A(7, 7)$, $D(5, 4)$

31. $Q(-5, 1)$, $U(-2, 5)$, $A(2, 6)$, $D(3, 3)$

32. $Q(1, 6)$, $U(3, 10)$, $A(5, 6)$, $D(3, 2)$

33. $Q(-2, 1)$, $U(2, 5)$, $A(5, 2)$, $D(1, -2)$

21. Parallelogram. Both midsegments are parallel to a side of the triangle. Both pairs of opposite sides are parallel.

24. Graph two horizontal lines in the form $x = a$ and $x = b$ and two lines with the same slope in the form $y = mx + c$ and $y = mx + d$.

25. Graph two lines with the same slope in the form $y = m_1x + a$ and $y = m_1x + b$ and two lines with a different slope in the form $y = m_2x + c$ and $y = m_2x + d$.

26. Graph two lines with the same slope in the form $y = mx + a$ and $y = mx - a$ and two lines with a different slope in the form $y = -mx + a$ and $y = -mx - a$.

27. Graph two lines in the form $y = x + a$ and $y = x - a$ and two lines in the form $y = -x + a$ and $y = -x - a$.

Look Back

34. Write a two-column proof for the following:

Given: *ABCD* is an isosceles trapezoid with $\overline{BC} \parallel \overline{AD}$ and $\overline{AB} \cong \overline{DC}$; $\overline{BX} \perp \overline{AD}$; $\overline{CY} \perp \overline{AD}$

Prove: $\angle A \cong \angle D$

Hint: What must be true about $\overline{BX}$ and $\overline{CY}$?

[Lessons 4.5, 4.6]

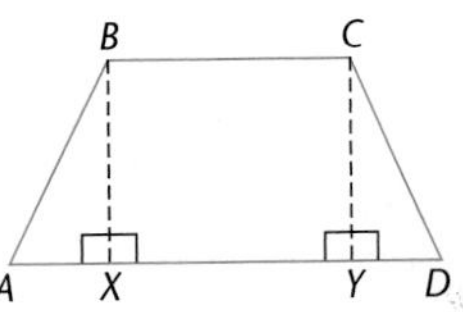

35. Write a two-column proof for the following:

Given: *PQRS* is a parallelogram; $\overline{QS}$, $\overline{PR}$, and $\overline{XY}$ intersect at *Z*

Prove: $\overline{XZ} \cong \overline{YZ}$

[Lessons 4.5, 4.6]

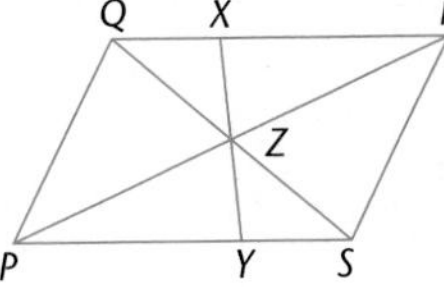

The answers to Exercises 34 and 35 can be found in Additional Answers beginning on page 727.

Look Beyond

Use the figure drawn in Exercise 36 to answer Exercises 37–40.

36. Draw an equilateral triangle 8 cm on a side. Then connect the midpoints of the sides of the triangle to form a second triangle inside the first. Next, draw a third triangle by connecting the midpoints of the sides of the second triangle. Continue with as many triangles as it is possible to draw.

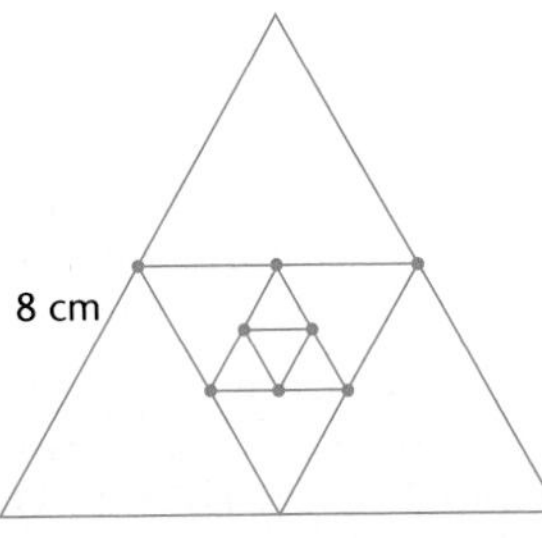

37. What type of triangles are the second and third triangles? Equilateral

38. What is the ratio of the perimeters of the first, second, and third triangles? $1 : \frac{1}{2} : \frac{1}{4}$

39. Find the sum of the perimeters of the second, third, fourth, and fifth triangles. You may wish to continue with additional triangles—sixth, seventh, eighth, etc.

40. How does the perimeter of the first triangle compare with the sum of the perimeters of the additional triangles? What happens to this answer as the perimeters of additional triangles are added to this sum?

Look Beyond

In Exercises 36–40, students may discover the pattern in the sequence of perimeters for each consecutive triangle.

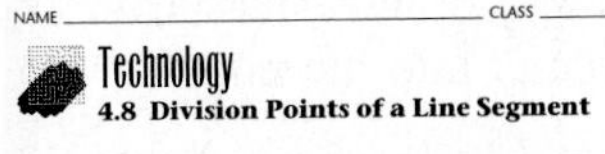

NAME ______ CLASS ______ DATE ______

Technology

4.8 Division Points of a Line Segment

Did you know that you can find the coordinates of the points *P*(*x*, *y*) on $\overline{AB}$, where *A* has coordinates *A*(−2, 1) and *B* has coordinates *B*(4, 5) by evaluating the formulas below for different values of *t*?

$$x = (-2)(1 - t) + (4)t \qquad y = (1)(1 - t) + (5)t$$

1. Find the coordinates of the midpoint of $\overline{AB}$ by using the midpoint formula given in Lesson 4.8.

2. Evaluate the two formulas above for $t = \frac{1}{2}$. Compare your answer with the one you found in Exercise 1.

In Exercises 3-6, use a spreadsheet.

3. Let *A* and *B* have coordinates *A*(−2, 1) and *B* (4, 5). Create a spreadsheet in which column *A* contains 0.00, 0.25, 0.50, 0.75, and 1.00 as values of *t*, column *B* contains the value of the formula for *x*, and column *C* contains the values of the formula for *y*.

4. Create a spreadsheet graph for the values in columns *B* and *C* of your spreadsheet in Exercise 3. From the graph, interpret the meaning of choosing *t* = 0.00, 0.25, 0.50, 0.75, and 1.00.

5. Let *A* and *B* have coordinates *A*(−2, 1) and *B*(4, 5). Modify your spreadsheet in Exercise 3 so that column *A* contains 0.0, 0.1, 0.2, 0.3,..., 0.8, 0.9, and 1.0 as values of *t*, column *B* contains the value of the formula for *x*, and column *C* contains the values of the formula for *y*.

6. Create a spreadsheet graph for the values in columns *B* and *C* of your spreadsheet in Exercise 5. From the graph interpret the meaning of choosing *t* = 0.0, 0.1, 0.2, 0.3,..., 0.8, 0.9, and 1.0.

26 **Technology** **HRW Geometry**

36. Check student drawings.

39. $12 + 6 + 3 + 1.5 = 22.5$;
$12 + 6 + 3 + 1.5 + 0.75 + 0.375 = 23.625$
$12 + 6 + 3 + 1.5 + 0.75 + 0.375 + 0.1875 = 23.8125$
$12 + 6 + 3 + 1.5 + 0.75 + 0.375 + 0.1875 + 0.09375 = 23.90625$

40. The sum of the perimeters of the additional triangles approaches the perimeter of the first triangle.

PREPARE

Objectives

- Translate segments and polygons using straightedge and compass.
- Rotate and reflect segments and polygons using straightedge and compass.

RESOURCES

- Practice Master 4.9
- Enrichment Master 4.9
- Technology Master 4.9
- Lesson Activity Master 4.9
- Quiz 4.9
- Spanish Resources 4.9

Assessing Prior Knowledge

Match each transformation with the information needed to perform it.

Transformation

1. rotation
2. dilation
3. translation
4. reflection

Needed Information

a. the distance moved, both horizontally and vertically
b. the center and the number of degrees
c. a "mirror" line
d. the center and a scale factor

[1-b, 2-d, 3-a, 4-c]

TEACH

Most students will know intuitively that rigid transformations preserve congruency. Discuss the difference between knowing something intuitively and proving something true. Intuition, although useful and important, can sometimes lead to incorrect results.

LESSON 4.9 Exploring

The Construction of Transformations

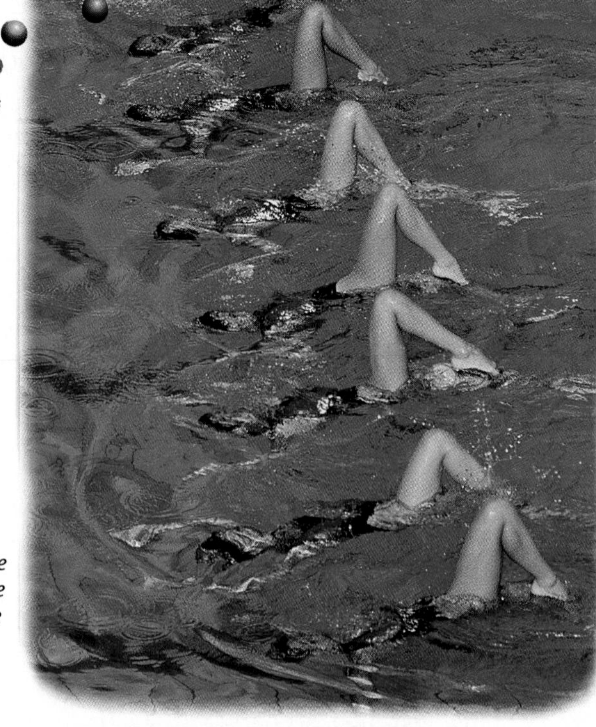

The ability to move in strict formation, without changing the size and shape of the pattern, is one of the requirements of a good synchronized swimming team.

Why *The transformations you have studied have all been rigid. That is, the size and shape of the objects that are transformed do not change. You can now prove that the transformations, as you have defined them, preserve congruence.*

In Lesson 1.6 you created your own definitions of three different transformations. A **translation**, for instance, is a transformation that moves every point of an object the same distance in the same direction. In the following exploration you will learn how to translate a segment by relocating and connecting the endpoints.

Exploration 1 *Translating a Segment*

You will need

Compass and straightedge
Ruler

Make your own drawing like the one on the right. The slide arrow shows the direction and distance of the translation you are to construct. (The distance is the overall length of the arrow from its "head" to its "tail.")

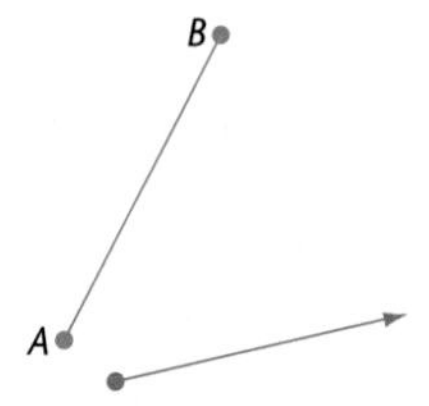

1. Construct a line l_1 through point A that is parallel to the slide arrow. Construct a line l_2 through point B that is parallel to the slide arrow. Are lines l_1 and l_2 parallel? What theorem justifies your answer?

2. Set your compass to the length s of the slide arrow. On the right hand side of $\overline{AB}$, construct points A' and B' that are a distance s from points A and B on lines l_1 and l_2, respectively.

ALTERNATIVE teaching strategy

Technology The explorations and exercises in this lesson are all appropriate for work with geometry graphics software. Although the software most likely will transform objects with menu commands, tell students to follow the steps as specified in each exploration. For example, in Exploration 1, students *construct* a translation of segment $\overline{AB}$; they should not use a menu command to translate the segment.

 Connect points A' and B'. Measure $\overline{A'B'}$. Is $\overline{A'B'}$ congruent to $\overline{AB}$?

4. Explain why the two segments must be congruent. Write a proof that they are. ❖

Exploration 1 shows you that the size and shape of a translated segment are exactly the same as in the pre-image. In Exploration 2, you will translate a polygon one line at a time, as in the previous method. Do you think each of the translated figures will have the same size and shape as each of the original figures?

Exploration 2 Congruence in Polygon Translations

You will need
No special tools

 If you translate each of the three segments that form triangle $\triangle ABC$ using the method of Exploration 1, would your new triangle be congruent to the original one? State the theorem or postulate that justifies your answer.

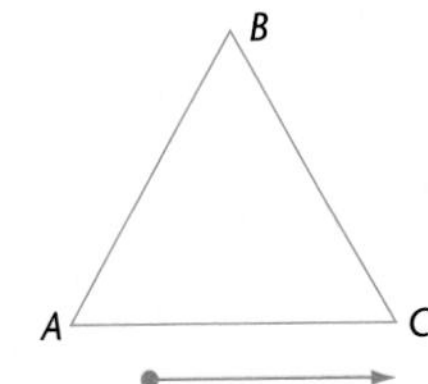

 If you translate each of the four segments that form quadrilateral $ABCD$ using the method of Exploration 1, would your new quadrilateral be congruent to the original one? State the theorem or postulate that justifies your answer. (Hint: Break the figure down into triangles by drawing a diagonal.)

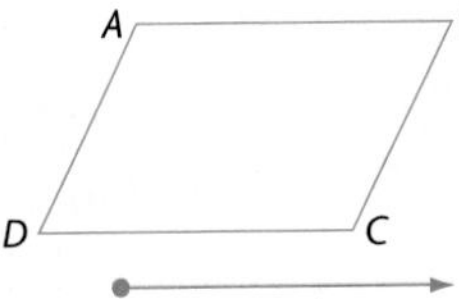

 Show how you can break down any polygon into a number of triangles by connecting the vertices with segments. Include both convex and concave polygons in your discussion. What can you conclude about the translation of any polygon using the method of Exploration 1?

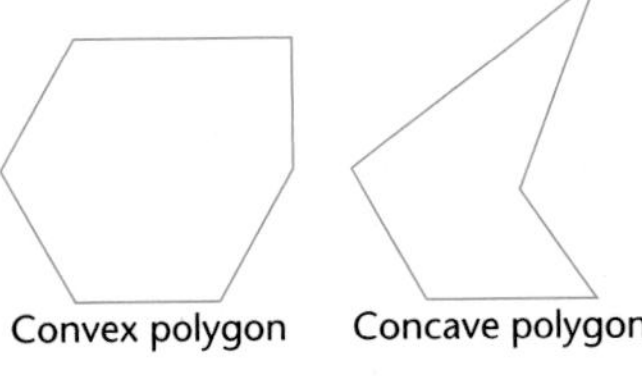
Convex polygon Concave polygon

4. What can you conclude about the translation of an "open" figure such as the one on the right? ❖

CRITICAL *Thinking*

Do you think your method of translating a figure can be applied to a curve? Hint: How can you use a number of segments to approximate a curve?

Exploration 1 Notes

Make sure students understand that the point of the exploration is to *prove* that a translation preserves congruency. The diagram is not constructed for itself; it is constructed for use in proving a conjecture.

ongoing ASSESSMENT

4. $\overline{AA'} \parallel \overline{BB'}$ because they are both parallel to the slide arrow. $\overline{AA'} \cong \overline{BB'}$ because they are radii of congruent circles. So $AA'BB'$ is a parallelogram because the opposite sides of one pair are both parallel and congruent. Therefore, $\overline{AB} \cong \overline{A'B'}$ because opposite sides of a parallelogram are congruent.

Exploration 2 Notes

For Step 3, students may need a reminder of the difference between convex and concave polygons. In a concave polygon, at least one of the interior angles is greater than 180°. A concave polygon "caves" inward.

ongoing ASSESSMENT

4. An open figure is congruent to its image under a translation.

ENRICHMENT **Technology** A double reflection has the same result as rotating a figure. Have students use geometry software to look for the relationship between the angle of rotation and the angle between the two reflecting lines. They will find that the first is double the second.

INCLUSION strategies **Using Cognitive Strategies** It is easy for students to get "lost" in the constructions and forget that the point is to prove a conjecture. Have students complete the following before beginning the proofs.

Conjecture: ____________________

Given: ____________________

Prove: ____________________

CRITICAL *Thinking*

Any curve is merely a large number of short segments joined together. By translating all the segments, a new congruent curve is formed.

CRITICAL *Thinking*

Because a figure is congruent to its image under a translation, the segment lengths and angle measures of the figure are preserved.

ongoing ASSESSMENT

Try This

Yes. $8.7750 + 9.6540 = 18.4290$, so $RS + TR = TS$. R is between T and S.

Cooperative Learning

Have students work together in small groups to construct the reflection of a triangle across the line $y = x$. After the construction is completed, students should discuss and write a paragraph proof of the construction.

Exploration 3 Notes

Translations can motivate some interesting mathematical deliberations about betweenness. Students can use the Segment Addition Postulate to show that X and X' are between their respective points.

ongoing ASSESSMENT

1. X' **appears to lie on** $\overline{A'B'}$ **and be between** A' **and** B'**.**

CRITICAL *Thinking*

It has been said that the rigid transformations, which are known as isometries, preserve Angles, Betweenness, Collinearity, and Distance ("ABCD"). How do Explorations 1 and 2 show that translations preserve angles and distances?

If point X is between points A and B in a pre-image of a figure, can you be sure that the image point X' will be between image points A' and B' in a translated image of the figure? Put another way, can you be sure that translations preserve betweenness?

To answer the question, you will need a mathematical way to determine whether one point is between two other points. The following postulate, which is the converse of a postulate you have already studied (1.4.2), gives a method.

> **CONVERSE OF THE SEGMENT ADDITION POSTULATE ("BETWEENNESS")**
> Given three points P, Q, and R, if $PQ + QR = PR$, then Q is between P and R. **4.9.1**

Try This For the points R, S, and T, $RS = 8.7750$, $TR = 9.6540$, and $TS = 18.4290$. Assuming the distances are exact, are the points collinear? If so, which point is between the other two?

Exploration 3 *Betweenness in Translations*

You will need
No special tools

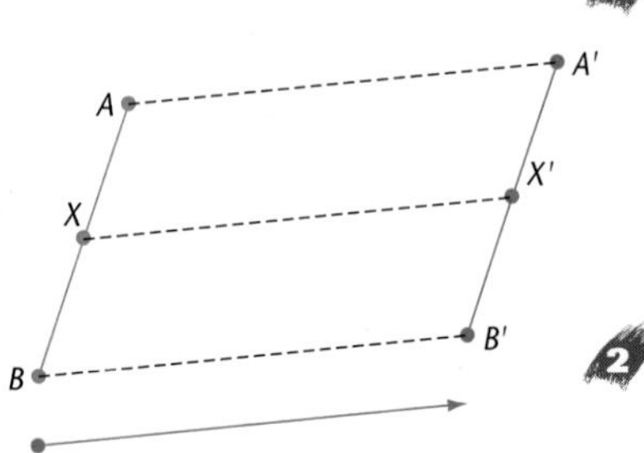

1. You are given a segment $\overline{AB}$ with a point X chosen anywhere between A and B. Translate each of the three marked points as indicated by the slide arrow. Then connect the points as in Exploration 1.

   Does X' lie on $\overline{A'B'}$? Is X' between A' and B'? You will answer these questions mathematically in the steps that follow.

2. What do you know about the distances AB and $A'B'$? About the distances AX and $A'X'$? about the distances BX and $B'X'$? Explain your reasoning.

3. From the Segment Addition Postulate you know that $AX + XB = AB$. What can you conclude about $A'X' + X'B'$? Explain your reasoning.

4. Does X' lie on $\overline{A'B'}$? (Is it collinear with A' and B'?) Is it between A' and B'? Explain your reasoning. ❖

You will consider the question of collinearity of points in translations in the exercise set.

RETEACHING the lesson

Hands-On Strategies Copy transparencies made from quarter-inch grid paper. Students should use grid paper and the transparent grids to translate, rotate, and/or reflect segments and/or triangles. Then students should write two-column proofs showing that congruency is preserved in each case.

Communicate

1. In your own words, state a theorem describing the congruence relationship in Exploration 1.
2. In your own words, state a theorem describing the congruence relationship in Exploration 2.
3. Explain the difference between copying a segment or triangle by construction and translating a segment or triangle by construction.
4. Explain some construction techniques that could be used to construct a rotation and a reflection.

Practice & Apply

They must all march the same distance along the given slide arrow.

5. During a marching-band show, a group of band members moves in a triangle shape across the field as shown. Players X, Y, and Z are the section leaders. What must the section leaders do to ensure that $\triangle XYZ \cong \triangle X'Y'Z'$?

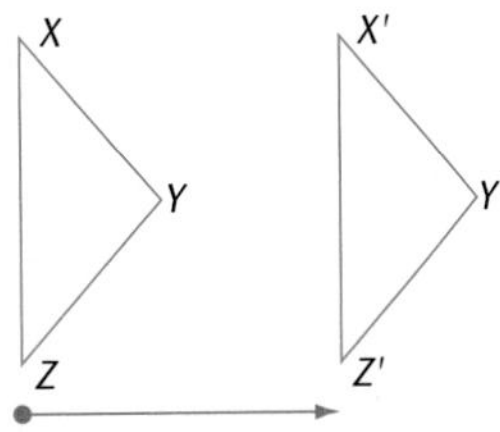

6. In Exercise 5, what must the other marchers who are not section leaders do to ensure that $\triangle XYZ \cong \triangle X'Y'Z'$?
They must stay between the section leaders by marching the same distance along the given slide arrow.

In $\triangle ABC$, B is *not* between A and C.

7. What seems to be true about $AB + BC$ compared with AC? $AB + BC > AC$
8. Suppose point B is relocated so that $AB + BC = AC$. What happens to $\triangle ABC$?
9. Use the results of Exercises 7 and 8 to complete the following statement, which is known as the Triangle Inequality Postulate.
greater than the length of the third side

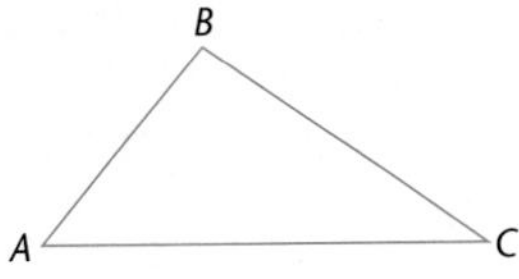

TRIANGLE INEQUALITY POSTULATE
The sum of the lengths of any two sides of a triangle is __?__. **4.9.2**

10. If one point is between two others, can you conclude that the three points are collinear? (Consider what happens to the triangle in Exercise 8.) If a transformation preserves betweenness, must it also preserve collinearity? Explain.

Assess

Selected Answers

Odd numbered Exercises 5–31

Assignment Guide

Core 1–3, 5–17, 28–34

Core Plus 1–34

Technology

Most of the exercises can be completed using geometry graphics software. You may wish to have a group of students work together to adapt the instructions for use with your particular geometry software.

Error Analysis

Students who have difficulty doing the constructions in a reasonable amount of time should be given completed figures. They should write the proof for each figure.

8. It collapses to a line.

10. Yes. If the sum of the lengths of two segments is greater than the length of a third segment, then the segments form a triangle. If the sum of the lengths of the two segments is equal to the length of the third segment, the triangle has collapsed to a line. A rigid transformation preserves betweenness, and also collinearity, but this is not necessarily true if the transformation does not preserve congruence.

alternative ASSESSMENT

Performance Assessment

Have students make a construction and write a proof for this conjecture: Rotating a triangle about a point outside the figure preserves congruency.

11. Check student drawings.
12. Check student drawings.
13. Check student drawings.
14. Check student drawings.

The answer to Exercise 18 can be found in Additional Answers beginning on page 727.

In a known rotation, *every point of a figure is moved the same angle about a fixed point known as the center of rotation.* **In Exercises 11–14 you will use this idea to rotate a segment 40° about a point. (You will need a compass and protractor.)**

11. You are given a segment $\overline{AB}$ and point P not on the segment. Draw or trace $\overline{AB}$ and P on your own paper.

B

A

P

12. Draw $\overline{PA}$. Set your compass at distance PA and draw an arc through A. Use a protractor to draw $\angle A'PA$ with a measure of 40°.

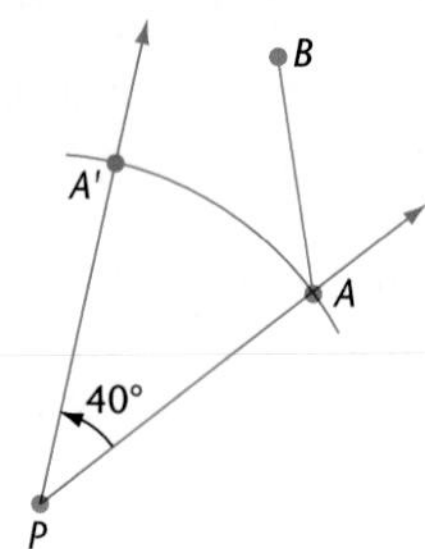

13. Draw $\overline{PB}$. Set your compass at distance PB and draw an arc through B. Use a protractor to draw $\angle BPB'$ with measure 40°.

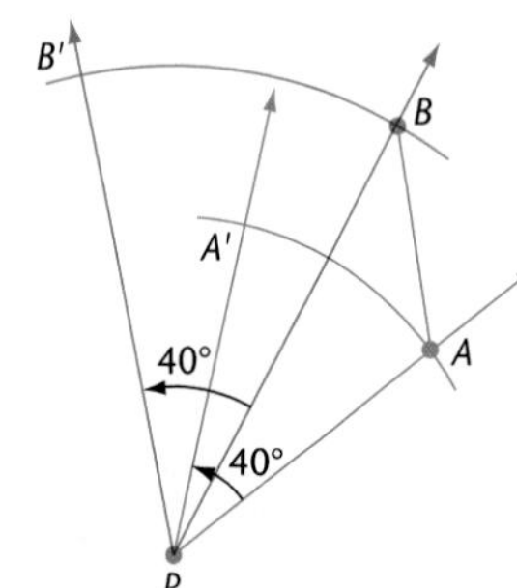

14. Draw $\overline{A'B'}$.

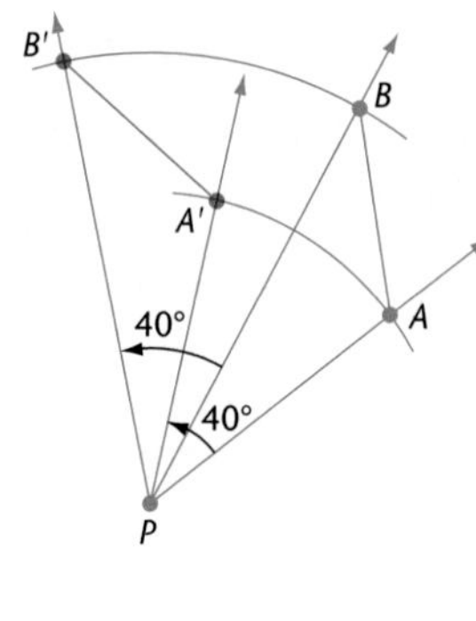

Answer the following questions (Exercises 15–17) about the above construction (Exercises 11–14).

15. What can you conclude about $\angle BPA$ and $\angle B'PA'$? (You will need to use the Overlapping Angle Theorem.) Explain.

16. What can you conclude about $\overline{AB}$ and $\overline{A'B'}$? Explain your reasoning.

17. How could rotations greater than 180° be drawn? (Hint: 360° − 290° = 70°)

In a reflection, *every segment connecting a point and its pre-image is bisected at right angles by a line known as the "mirror" of the transformation.* **In Exercises 18–27 you will use this idea in a proof.**

18. Draw or trace segment $\overline{AB}$, line l, and reflection $\overline{A'B'}$. Draw segments $\overline{AA'}$ and $\overline{BB'}$. Label the intersections with l as X and Y.

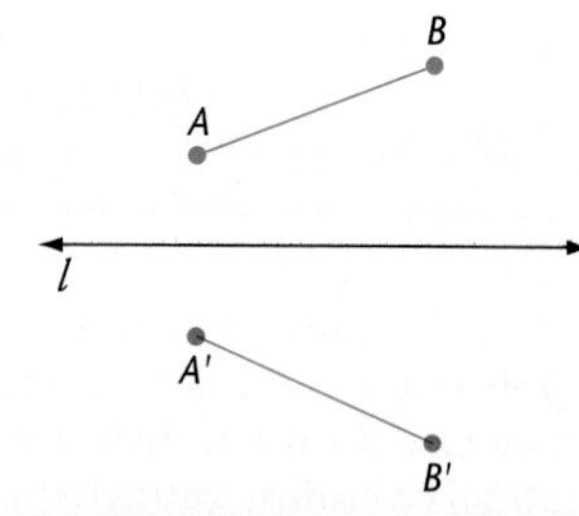

15. $\angle BPA \cong \angle B'PA'$; $\angle APB \cong \angle BPB' = 40°$ by construction. $\angle APB$ and $\angle BPB$ overlap at $\angle BPA'$. By the Overlapping Angle Theorem $\angle BPA \cong \angle B'PA'$.

16. $\overline{AB} \cong \overline{A'B'}$; $\overline{PA} \cong \overline{PA'}$ and $\overline{PB} \cong \overline{PB'}$ because they are radii of the same circle. $\angle BPA \cong \angle B'PA'$ from Exercise 15. So, $\triangle PAB \cong \triangle PA'B'$ by SAS and $\overline{AB} \cong \overline{A'B'}$ by CPCTC.

17. To construct a rotation of $n°$ counterclockwise where $n > 180°$, construct a rotation of $360° - n°$ clockwise.

Practice Master

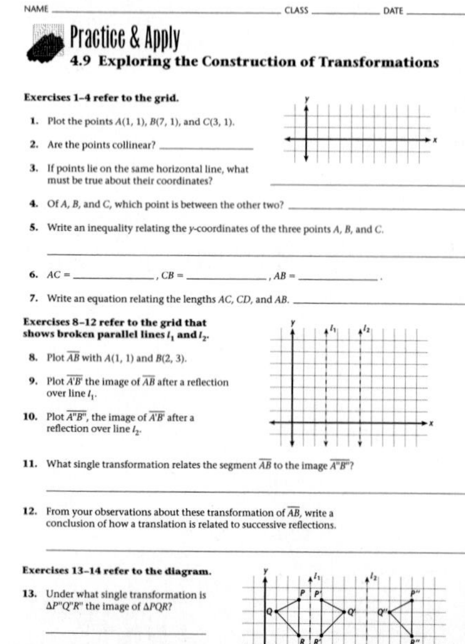

NAME ______________ CLASS ________ DATE ________

Practice & Apply

4.9 Exploring the Construction of Transformations

Exercises 1–4 refer to the grid.

1. Plot the points $A(1, 1)$, $B(7, 1)$, and $C(3, 1)$.
2. Are the points collinear? __________
3. If points lie on the same horizontal line, what must be true about their coordinates? __________
4. Of A, B, and C, which point is between the other two? __________
5. Write an inequality relating the y-coordinates of the three points A, B, and C. __________
6. $AC =$ ________, $CB =$ ________, $AB =$ ________.
7. Write an equation relating the lengths AC, CD, and AB. __________

Exercises 8–12 refer to the grid that shows broken parallel lines l_1 and l_2.

8. Plot $\overline{AB}$ with $A(1, 1)$ and $B(2, 3)$.
9. Plot $\overline{A'B'}$ the image of $\overline{AB}$ after a reflection over line l_1.
10. Plot $\overline{A''B''}$, the image of $\overline{A'B'}$ after a reflection over line l_2.
11. What single transformation relates the segment $\overline{AB}$ to the image $\overline{A''B''}$? __________
12. From your observations about these transformation of $\overline{AB}$, write a conclusion of how a translation is related to successive reflections. __________

Exercises 13–14 refer to the diagram.

13. Under what single transformation is $\Delta P''Q''R''$ the image of ΔPQR? __________
14. Under what composite transformation is $\Delta P''Q''R''$ the image of ΔPQR? __________

HRW Geometry — Practice & Apply 9

19. Label the right angles in your figure. Explain your reasoning.

20. Label the congruent segments in your figure. Explain your reasoning.

21. Label the parallel segments in your figure. Explain your reasoning.

22. Draw line m through A and parallel to l. Draw line n through A' and parallel to l. Label the points of intersection C and C' as shown. Check student drawings

23. What kind of quadrilaterals are $ACYX$ and $A'C'YX$? Explain your reasoning.

24. What kind of quadrilateral is $ACC'A'$? Explain your reasoning.

25. What can you conclude about $\overline{BC}$ and $\overline{B'C'}$? Explain your reasoning.

26. What can you conclude about $\triangle ABC$ and $\triangle A'B'C'$? Explain your reasoning.

27. What must be true about $\overline{AB}$ and $\overline{A'B'}$? Explain your reasoning.

28. Rotate a triangle about a point by 35°.

29. Use only a compass and a straightedge to rotate a triangle about a point. You may choose the amount of rotation. Remember to use the angle copying method from Lesson 4.7.

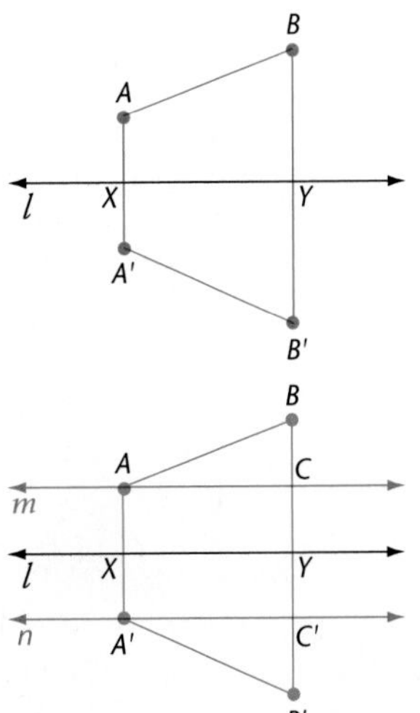

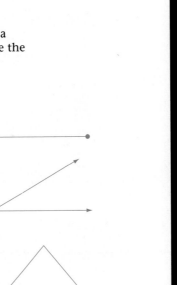

The answers to Exercises 19–21 and 25–32 can be found in Additional Answers beginning on page 727.

Look Back

30. Draw or trace the segment and construct the perpendicular bisector. **[Lesson 4.7]**

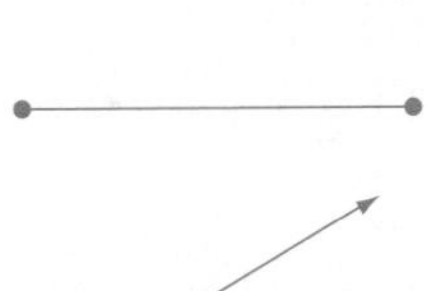

31. Draw or trace the angle and construct a copy of the angle. **[Lesson 4.7]**

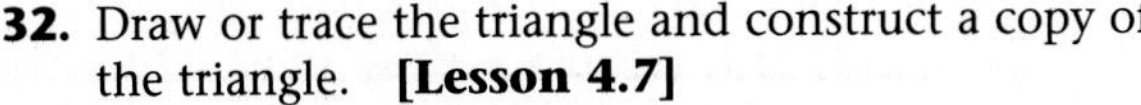

32. Draw or trace the triangle and construct a copy of the triangle. **[Lesson 4.7]**

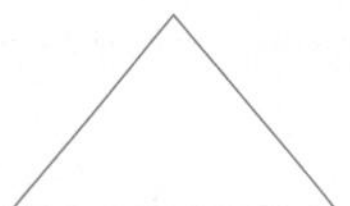

Look Beyond

Use the drawing of two rooms of a house for exercises 33 and 34.

33. Find the number of square yards of carpet you would need to carpet the two rooms. 45 yd² or 405 ft²

34. If carpet is $25.50 per square yard installed, how much will it cost to carpet the two rooms. $1,147.50

15 ft

15 ft

18 ft

9 ft

Look Beyond

This type of problem is fairly easy if the rooms are rectangles. For students who need more challenge, have them draw polygonal rooms on grid paper and compute the areas.

23. Parallelograms. Lines m and n are parallel to line l. $AA' \parallel BB'$ from Exercise 21. Since both pairs of opposite sides are parallel, $ACYX$ and $A'C'YX$ are parallelograms.

24. Parallelogram. Since lines m and n are both parallel to line l, line m is parallel to line n. $\overline{AA'} \parallel \overline{BB'}$ from Exercise 21. Since both pairs of opposite sides are parallel, $ACC'A'$ is a parallelogram.

Technology Master

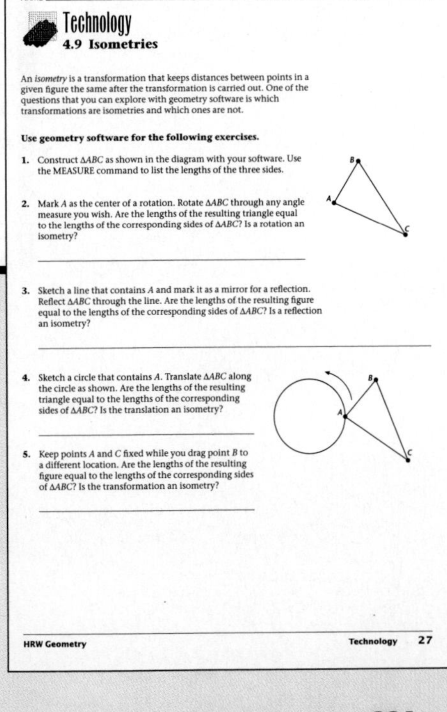

NAME ______ CLASS ______ DATE ______

Technology
4.9 Isometries

An *isometry* is a transformation that keeps distances between points in a given figure the same after the transformation is carried out. One of the questions that you can explore with geometry software is which transformations are isometries and which ones are not.

Use geometry software for the following exercises.

1. Construct ΔABC as shown in the diagram with your software. Use the MEASURE command to list the lengths of the three sides.

2. Mark A as the center of a rotation. Rotate ΔABC through any angle measure you wish. Are the lengths of the resulting triangle equal to the lengths of the corresponding sides of ΔABC? Is a rotation an isometry?

3. Sketch a line that contains A and mark it as a mirror for a reflection. Reflect ΔABC through the line. Are the lengths of the resulting figure equal to the lengths of the corresponding sides of ΔABC? Is a reflection an isometry?

4. Sketch a circle that contains A. Translate ΔABC along the circle as shown. Are the lengths of the resulting triangle equal to the lengths of the corresponding sides of ΔABC? Is the translation an isometry?

5. Keep points A and C fixed while you drag point B to a different location. Are the lengths of the resulting figure equal to the lengths of the corresponding sides of ΔABC? Is the transformation an isometry?

HRW Geometry — Technology 27

Focus

Flexagons are an interesting mathematical diversion. "Flexing" a flexagon reveals different faces with each flex. Analysis of the way the faces are revealed can be an instructive and absorbing experience.

Motivate

Trim the margins off of a few sheets of notebook paper to show students the type of strips of paper used to discover flexagons. Ask students if they have ever discovered something interesting while playing with discarded material.

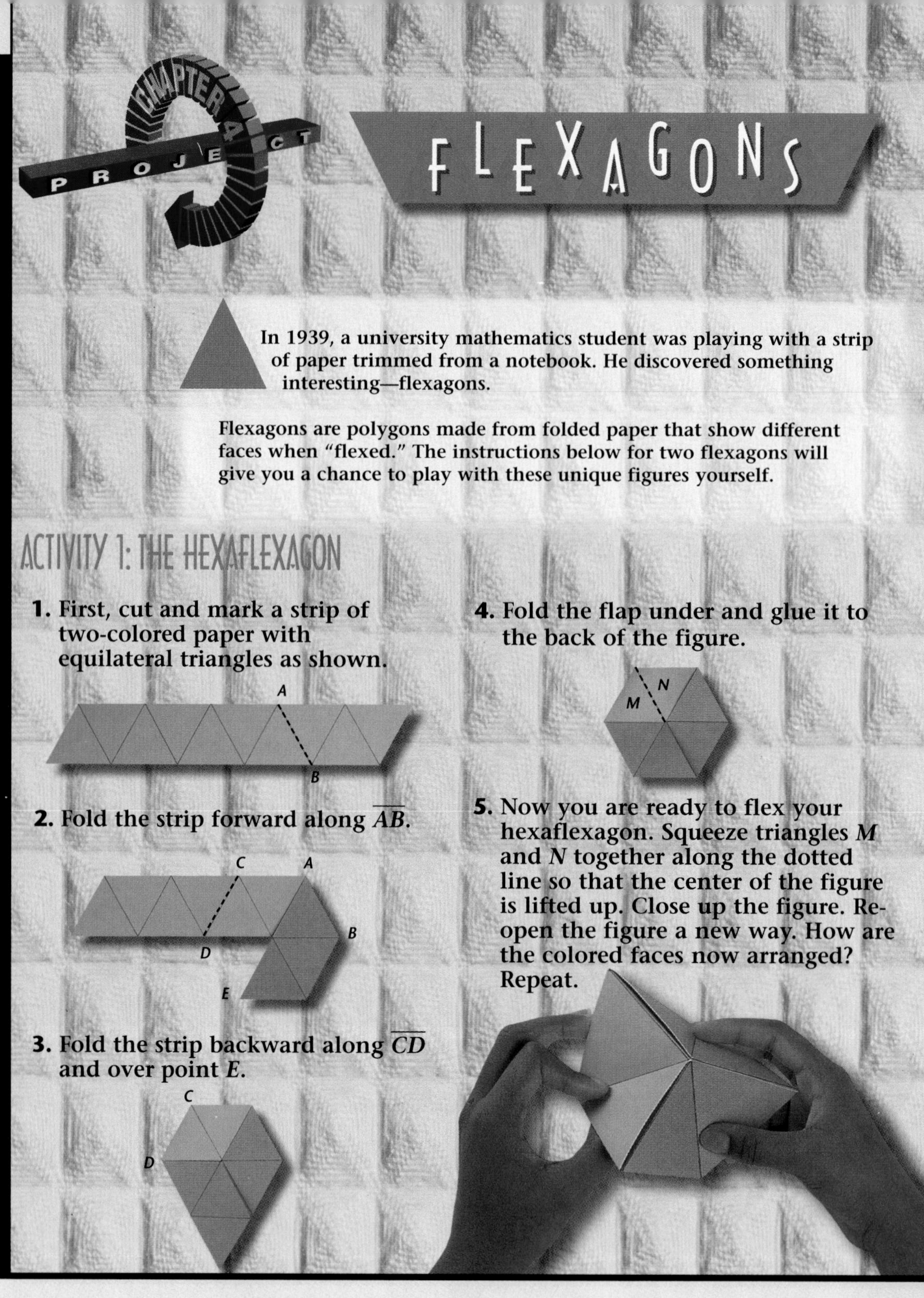

FLEXAGONS

In 1939, a university mathematics student was playing with a strip of paper trimmed from a notebook. He discovered something interesting—flexagons.

Flexagons are polygons made from folded paper that show different faces when "flexed." The instructions below for two flexagons will give you a chance to play with these unique figures yourself.

ACTIVITY 1: THE HEXAFLEXAGON

1. First, cut and mark a strip of two-colored paper with equilateral triangles as shown.

2. Fold the strip forward along $\overline{AB}$.

3. Fold the strip backward along $\overline{CD}$ and over point E.

4. Fold the flap under and glue it to the back of the figure.

5. Now you are ready to flex your hexaflexagon. Squeeze triangles M and N together along the dotted line so that the center of the figure is lifted up. Close up the figure. Reopen the figure a new way. How are the colored faces now arranged? Repeat.

ACTIVITY 2: THE HEXAHEXAFLEXAGON

Here is a flexagon that will show twice as many faces as the first. The hexahexaflexagon is a bit more complicated than the hexaflexagon, so follow the instructions carefully.

6. Start with a strip of two-colored paper that is cut and marked with 19 equilateral triangles on each side. Number each triangle on the top side 1, 2, and 3 as shown, skipping the last triangle.

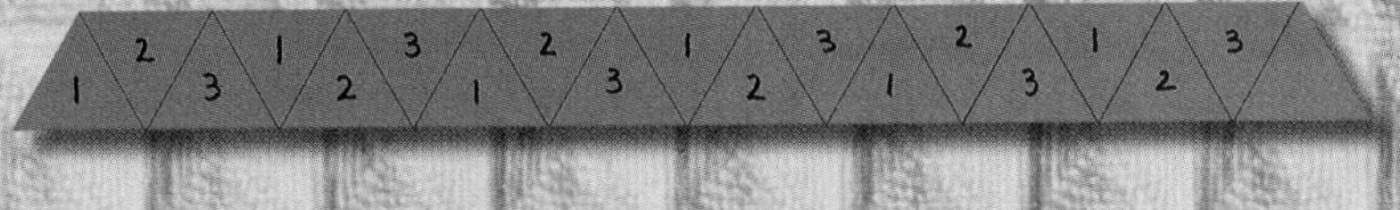

7. Without swapping the positions of the ends, turn the paper over. Number the triangles 4, 4, 5, 5, 6, 6 as shown, skipping the first triangle.

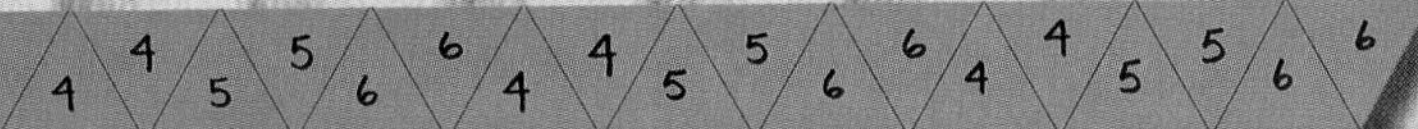

8. Fold the strip so that the 4s, 5s, 6s face each other as shown.

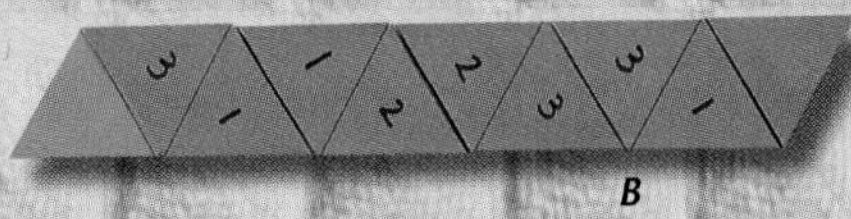

9. Now fold forward on $\overline{AB}$. Fold backward along $\overline{CD}$ and over point *E*.

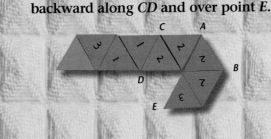

10. Glue the flap so that blank triangle meets blank triangle. The hexahexaflexagon flexes the same way as shown before.

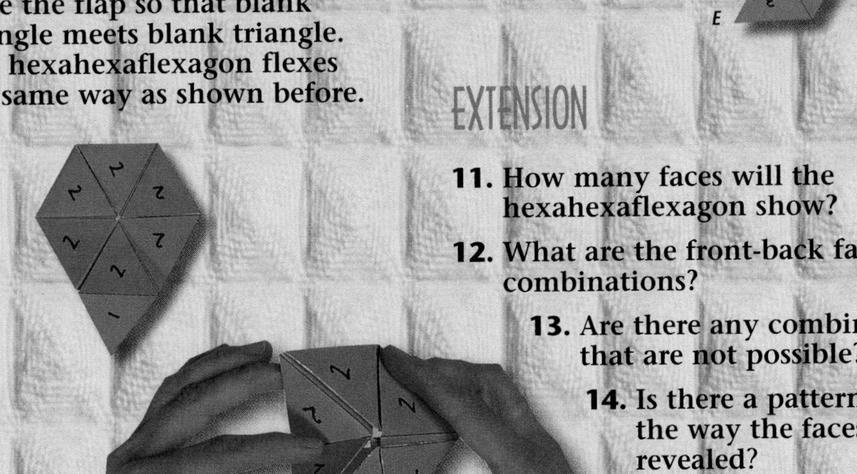

EXTENSION

11. How many faces will the hexahexaflexagon show?

12. What are the front-back face combinations?

13. Are there any combinations that are not possible?

14. Is there a pattern to the way the faces are revealed?

5. The hexaflexagon will show 18 different faces as it is flexed.

11. The hexahexaflexagon will show 36 different faces as it is flexed.

12. 1-2, 1-3, 1-5, 1-6, 2-3, 2-4, 2-5, 3-4, 3-6

13. 1-4, 2-6, 3-5

Cooperative Learning

Students doing the project activities with partners or in small groups can help each other interpret the directions. Gluing or taping the flaps may be easier if one person holds the model while the other applies the glue. However, each student should make a hexaflexagon and a hexahexaflexagon on his or her own.

DISCUSS

After all students have finished making their models, ask students to describe how the figures work. For the hexahexaflexagon, if five faces show the number 2, what number is on the other five faces? Which numbers are inside the figure? Talk about why it is important that all the faces be congruent triangles. [**The models would not flex smoothly and would not be as attractive if the faces varied in shape and size.**]

14. Student representations of the pattern will vary. One possible representation is shown below.

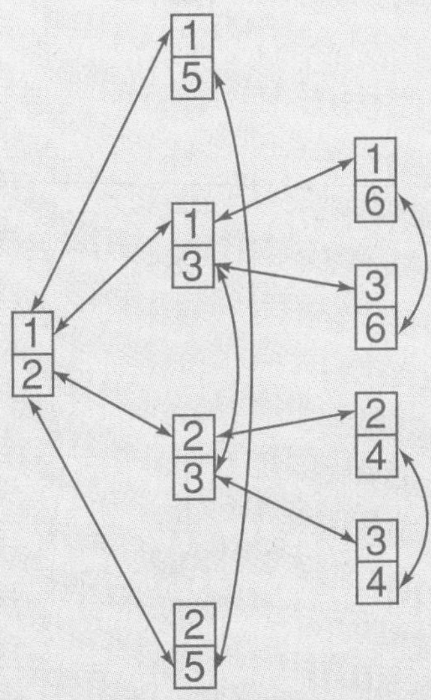

Chapter 4 Review

Vocabulary

base	196	CPCTC	193	rhombus	208
base angles	196	equilateral triangle	196	translation	226
betweenness	228	isosceles triangle	196	triangle rigidity	180
congruence	174	legs	196	triangulation	186
congruent polygons	176	parallelogram	208	vertex angle	196
corresponding sides, angles	175	rectangle	208		

Key Skills and Exercises

Lesson 4.1

➤ Key Skills

Name corresponding parts of congruent polygons.

Quadrilaterals *ABCD* and *QRST* are congruent. List all the corresponding sides and angles.

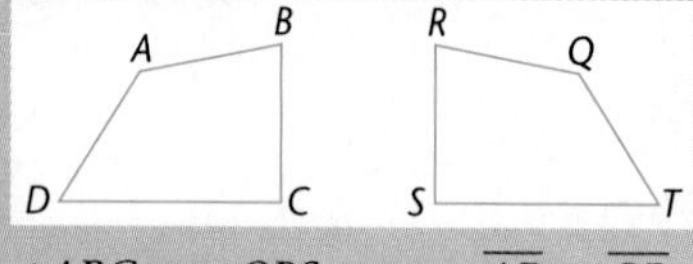

$\angle ABC \cong \angle QRS$ $\quad \overline{AB} \cong \overline{QR}$

$\angle BCD \cong \angle RST$ $\quad \overline{BC} \cong \overline{RS}$

$\angle CDA \cong \angle STQ$ $\quad \overline{CD} \cong \overline{ST}$

$\angle DAB \cong \angle TQR$ $\quad \overline{DA} \cong \overline{TQ}$

Show that polygons are congruent.

Prove that $\triangle EFG \cong \triangle GHE$.

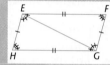

Five of the six necessary congruences are given:

$\angle HEG \cong \angle FGE$ $\quad \overline{EF} \cong \overline{GH}$

$\angle EGH \cong \angle GEF$ $\quad \overline{FG} \cong \overline{HE}$

$\angle GHE \cong \angle EFG$

By the Reflexive Property of Equality, $EG = EG$; therefore $\overline{EG} \cong \overline{EG}$. All six measures are congruent, so $\triangle EFG \cong \triangle GHE$.

➤ Exercises

1. $ABCDE \cong QRSTU$. Complete the congruences.

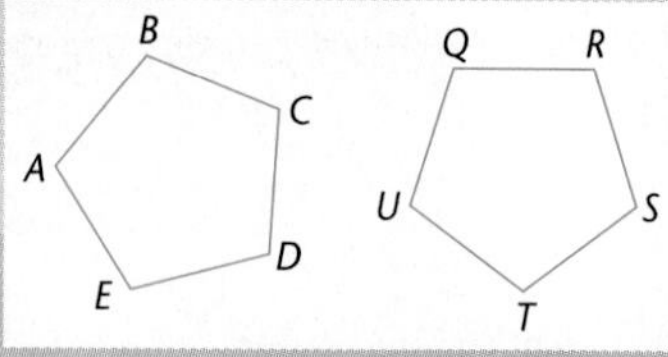

$\angle ABC \cong \angle QRS$ $\quad \overline{AB} \cong \overline{QR}$

$\angle BCD \cong$ ___?___ $\quad$ ___?___ $\cong \overline{RS}$

$\angle CDE \cong$ ___?___ $\quad$ ___?___ $\cong \overline{ST}$

$\angle DEA \cong$ ___?___ $\quad$ ___?___ $\cong \overline{TU}$

1. $\angle RST$ $\quad \overline{BC}$
 $\angle STU$ $\quad \overline{CD}$
 $\angle TUQ$ $\quad \overline{DE}$
 $\angle UQR$ $\quad \overline{EA}$

In Exercises 2–3 decide whether the triangles are congruent and justify your conclusion.

2. 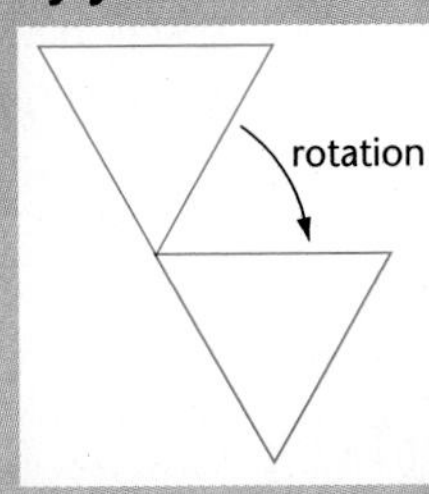

3. 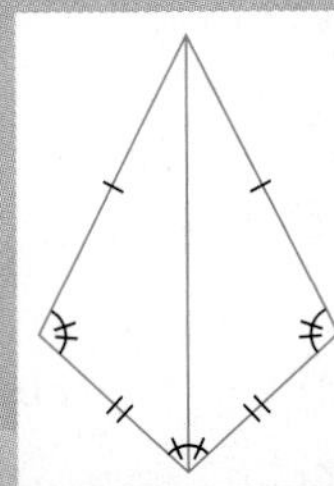

Lesson 4.2

➤ *Key Skills*

Apply SSS, SAS, and ASA congruence postulates.

With the information given you can prove three ways that $\triangle OPR \cong \triangle BNU$.

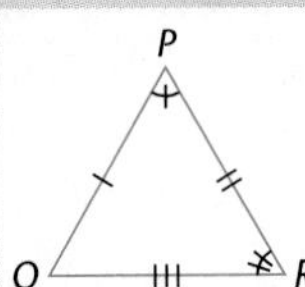

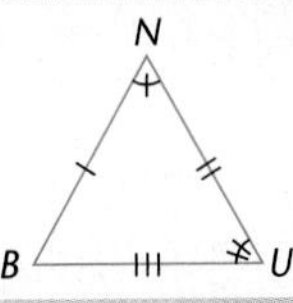

SSS: $\overline{BN} \cong \overline{OP}$, $\overline{NU} \cong \overline{PR}$, $\overline{UB} \cong \overline{RO}$
SAS: $\overline{OP} \cong \overline{BN}$, $\overline{PR} \cong \overline{NU}$, $\angle P \cong \angle N$ or
$\overline{OR} \cong \overline{BU}$, $\overline{RP} \cong \overline{UN}$, $\angle R \cong \angle U$,
ASA: $\angle P \cong \angle N$, $\angle R \cong \angle U$, $\overline{PR} \cong \overline{NU}$

➤ *Exercises*

In Exercises 4–6 decide whether the triangles are congruent and justify your answer.

4. 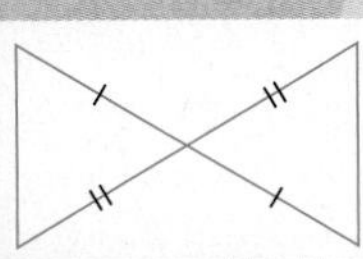

5. 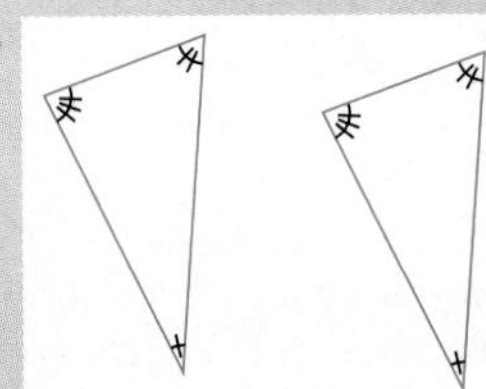

6. 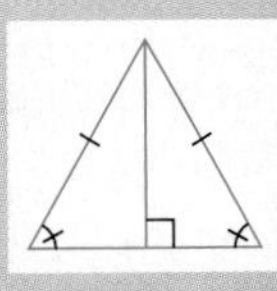

Lesson 4.3

➤ *Key Skills*

Apply AAS and HL congruence theorems.

With the information given you can prove two ways that $\triangle WXZ \cong \triangle YZX$.

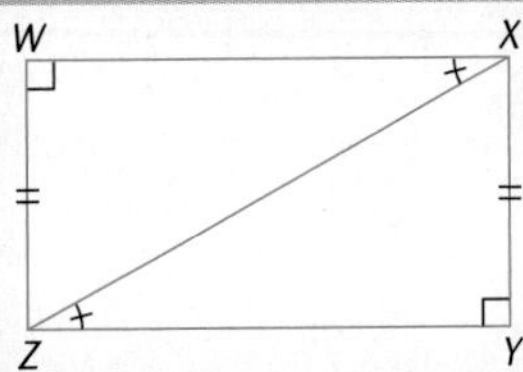

AAS: $\angle W \cong \angle Y$, $\angle WXZ \cong \angle YZX$, $ZX \cong XZ$
HL: $WZ \cong YX$, $ZX \cong XZ$

➤ *Exercises*

In Exercises 7–9 decide whether the triangles are congruent and justify your answer.

7. 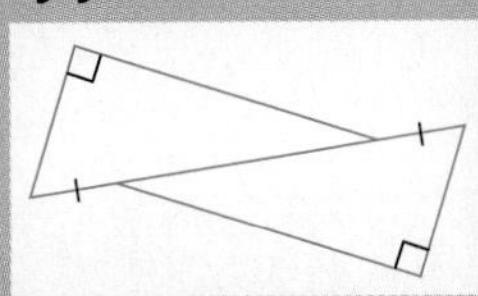

8. 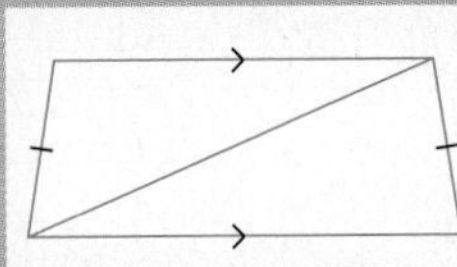

9. 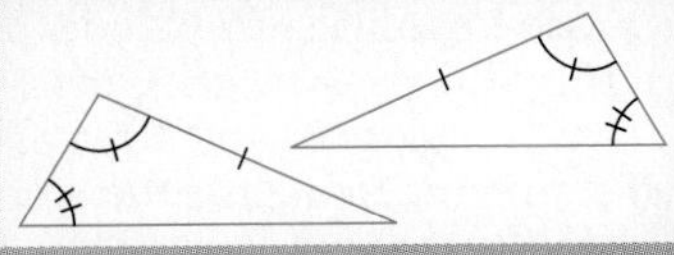

2. Yes. Rotation is a rigid motion preserving the measures of all angles and segments.
3. Yes. All corresponding sides and angles are congruent.
4. Congruent; SAS
5. Not enough information given to determine whether the triangles are congruent.
6. Congruent; HL
7. Not enough information given to determine whether the triangles are congruent.
8. Not enough information given to determine whether the triangles are congruent.
9. Congruent; AAS

Lesson 4.4

➤ Key Skill

Prove congruences in triangles.
In the figure, $\overline{AB} \cong \overline{DC}$ and $\overline{BD} \cong \overline{CA}$. Prove $\angle ABC \cong \angle DCB$.

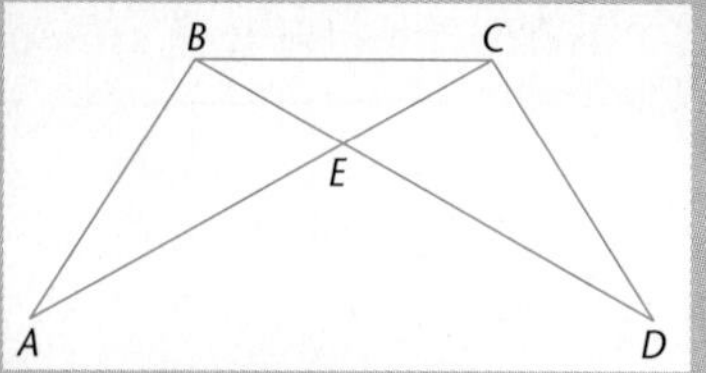

STATEMENTS	REASONS
1. $\overline{AB} \cong \overline{DC}$ and $\overline{BD} \cong \overline{CA}$	Given
2. $\overline{BC} \cong \overline{CB}$	Reflexive Property of Equality
3. $\triangle ABC \cong \triangle DCB$	SSS
4. $\angle ABC \cong \angle DCB$	CPCTC

➤ Exercises

10. Prove $\triangle ABD$ is isosceles.

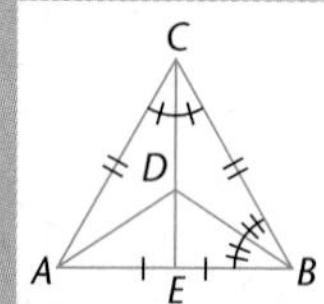

Lesson 4.5

➤ Key Skill

Prove properties of quadrilaterals.
FGHE is a rhombus. Prove that $\overline{ED} \cong \overline{GD}$.

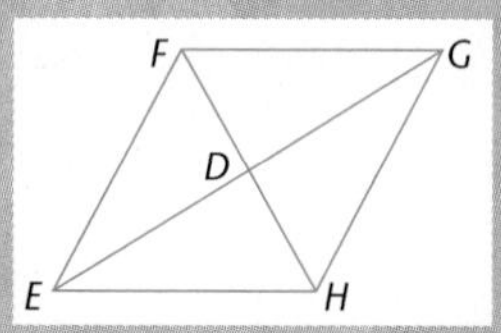

Since *FGHE* is a rhombus, $\overline{FG}$ is parallel and congruent to $\overline{HE}$. $\overline{FH}$, a transversal of $\overline{FG}$ and $\overline{HE}$, forms alternate interior angles that are congruent, so $\angle GFD \cong \angle EHD$. $\angle GDF \cong \angle EDH$ because they are vertical angles. By AAS $\triangle FGD \cong \triangle HED$. $\overline{ED} \cong \overline{GD}$ by CPCTC.

➤ Exercises

11. Prove that $\angle QPT \cong \angle SPT$.

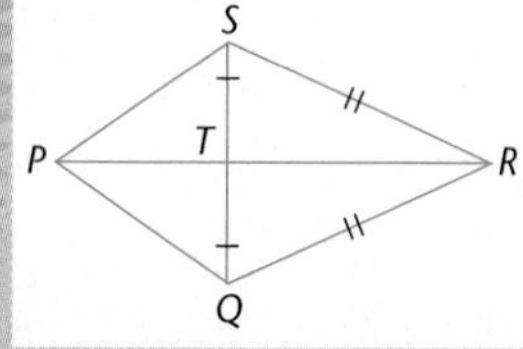

Lesson 4.6

➤ Key Skill

Determine whether a quadrilateral with given conditions is a special quadrilateral.

If two opposite sides of a quadrilateral are parallel and the other two sides are congruent, must the figure be a parallelogram? As the illustration shows, the answer is no, because the figure has the given conditions but is not a parallelogram.

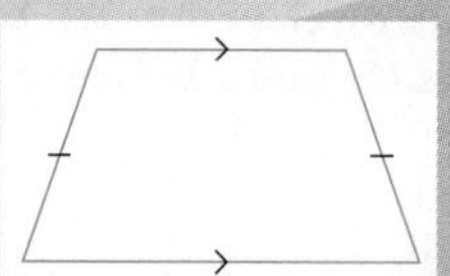

10. Encourage students to present an outline or paragraph proof such as the following.

Outline proof: First show that $\triangle CAD \cong \triangle CBD$ using SAS. Then, using linear pairs, show that $\angle ADE \cong \angle BDE$. Thus $\triangle ADE \cong \triangle BDE$ by AAS, and $\overline{AD} \cong \overline{BD}$ by CPCTC. $\triangle ABD$ is isoceles, by definition.

11. Encourage students to present an outline or paragraph proof such as the following:

Use the base $\angle$ property of isosceles $\triangle$'s to show that $\triangle QRT \cong \triangle SRT$ (SAS). Then use the linear pair property to show that $\angle STR$ and $\angle QTR$ are right $\angle$'s. Use the vertical $\angle$ property to show that $\angle STP$ and $\angle QTP$ are also right $\angle$'s. $\triangle STP \cong \triangle QTP$ (SAS or HL), and so $\angle QPT \cong \angle SPT$ by CPCTC.

Based only on the given conditions, tell what special quadrilateral each figure must be. (A figure may be of more than one type.)

12. 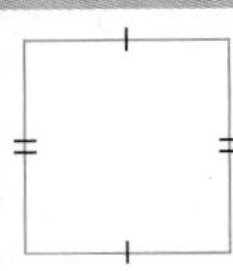

13. 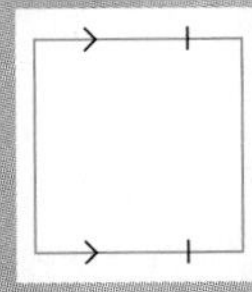

14. 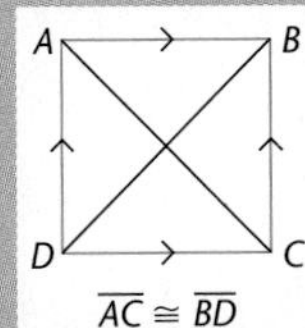

Lesson 4.7

➤ *Key Skills*

Construct copies of segments, angles, and triangles; construct a bisector of an angle.

Construct an angle twice the measure of $\angle A$.

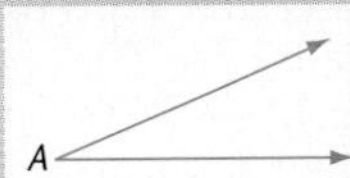

Using $\overrightarrow{BC}$ as a base, construct $\angle CBX \cong \angle A$. Using $\overrightarrow{BX}$ as a base, construct $\angle XBY \cong \angle A$.

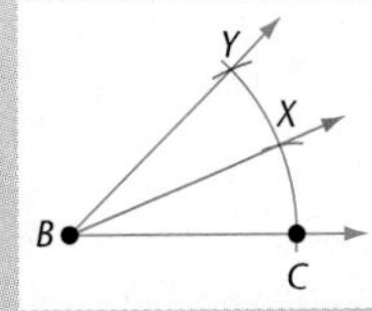

By the Addition Property of Equality $m\angle CBY = 2 \times m\angle A$.

➤ *Exercises*

15. Construct a line parallel to $\overleftrightarrow{AB}$ through a point C not on $\overleftrightarrow{AB}$.

Lesson 4.8

➤ *Key Skill*

Find the midpoints of segments on the coordinate plane.

EFGH is a square inscribed in square *ABCD*. Find the coordinates of each vertex of *EFGH*.

Find the midpoint of each side of *ABCD*.

midpoint of $\overline{AB} = E = (\frac{1}{2}[(-2) + 2], \frac{1}{2}[3 + 1])$

midpoint of $\overline{BC} = F = (\frac{1}{2}[2 + 4], \frac{1}{2}[3 + (-1)])$

midpoint of $\overline{CD} = G = (\frac{1}{2}[4 + 0], \frac{1}{2}[(-1) + (-3)])$

midpoint of $\overline{DA} = H = (\frac{1}{2}[0 + (-2)], \frac{1}{2}[(-3) + 1])$

The coordinates of vertex *E* are (0, 2), of *F* are (3, 1), of *G* are (2, –2), and of *H* are (–1, –1).

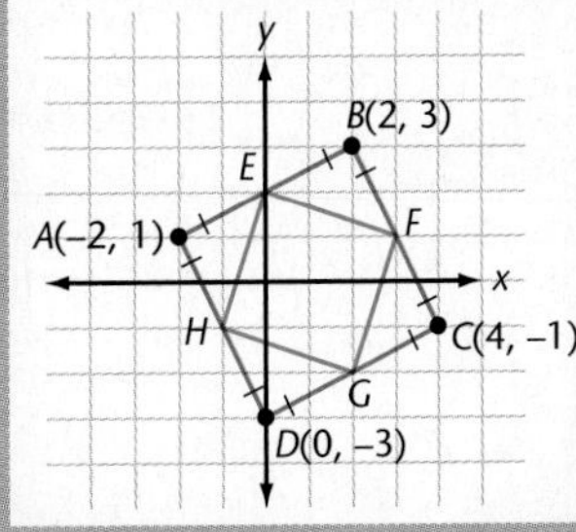

12. Parallelogram; Both pairs of opposite sides are congruent.

13. Parallelogram; one pair of sides is both parallel and congruent.

14. Parallelogram; Both pairs of opposite sides are parallel. Rectangle; the diagonals are congruent.

15. 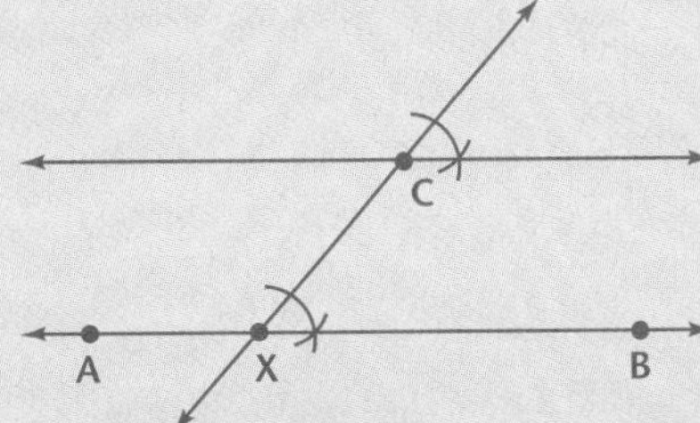

Students first draw a transversal through *C* that crosses $\overline{AB}$ at a point, say, *X*. Then they copy $\angle CXB$ as shown.

Chapter Review

16. Students can use the midpoint formula to show that S is midway between U and V. Hence, $US = SV$. Other distances needed to show congruence of sides of the triangles may be found using the distance formula.

17.

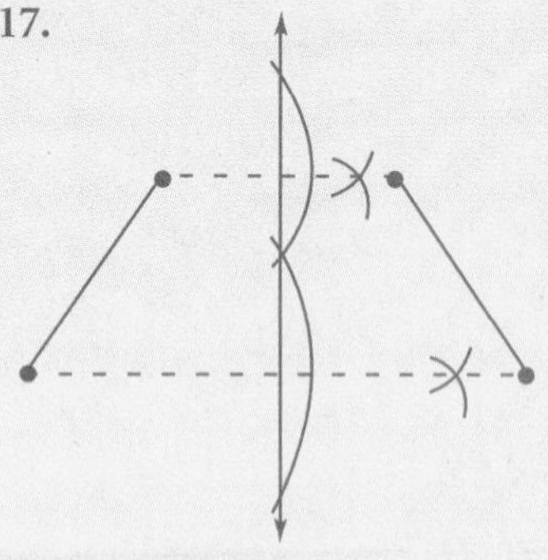

Student answers will vary according to the placement of the vertical line.

18. Measure $\overline{VY}$. $VY = WX$ because the triangles are congruent.

19. Check student drawings.

Exercises

16. In the figure, use the midpoint formula to show △UTS, △VWS, △TSZ, and △WSZ are congruent.

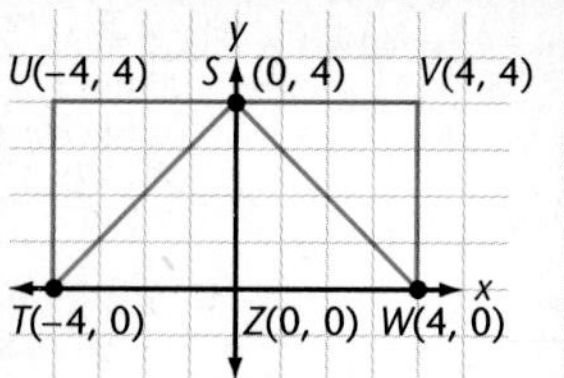

Lesson 4.9

Key Skills

Construct the rigid transformations.

Construct a rotation of $\triangle PQR$ through point P.

Construct arcs centered at P through points Q and R. Locate points Q' and R' so that $\angle QPQ' \cong \angle RPR'$.

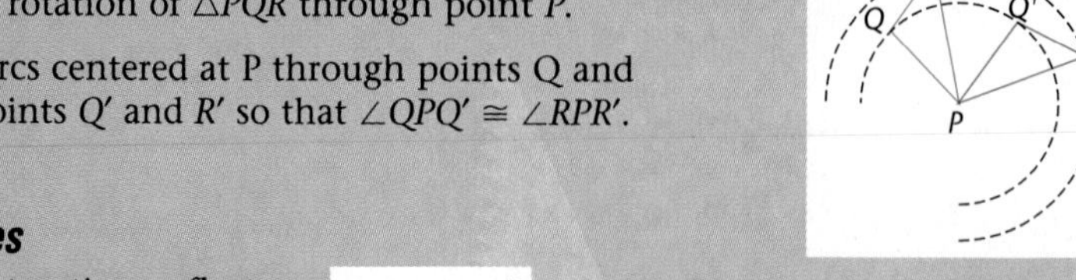

Exercises

17. By construction, reflect the segment through a vertical line.

Applications

18. Surveying For the figure, what measurement(s) would you need to make to determine WX, the distance across the lake?

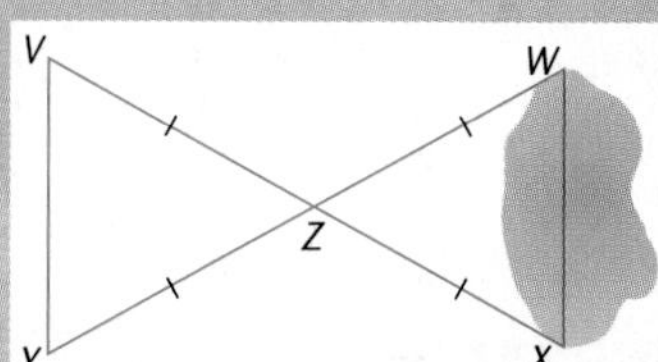

19. Architecture List five examples of congruent polygons in the building. Make your own sketch. Assume that the back and front are similar.

Arche de la Défense
Paris, France

20. Construct a congruent extension of this pattern.

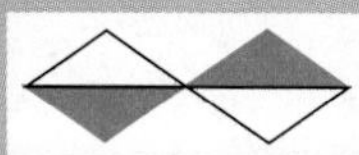

20. Check student drawings. Note that there are a number of artistically satisfying ways to extend the pattern—besides the obvious horizontal extension. Some students may realize that, mathematically speaking, there are infinitely many ways of extending the pattern (translations, rotations, and glides of any magnitude).

Chapter 4 Assessment

1. $NMRT \cong SLKB$. Complete the congruences.

$\angle RTN \cong \angle KBS$	$\overline{RT} \cong \overline{KB}$
$\angle TNM \cong$?	? $\cong \overline{BS}$
$\angle NMR \cong$?	? $\cong \overline{SL}$
$\angle MRT \cong$?	? $\cong \overline{LK}$

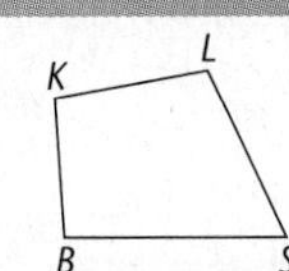
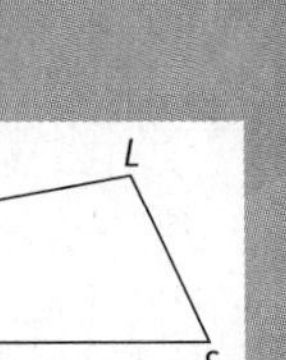

In Exercises 2–3 decide whether the polygons are congruent and justify your conclusion.

2. 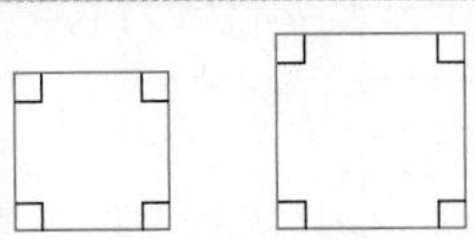

3. 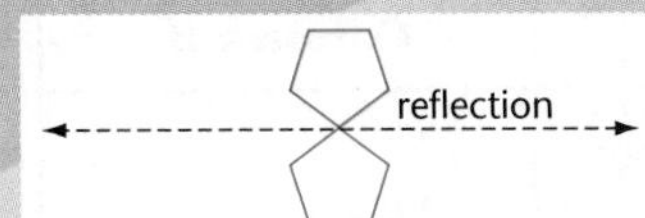

In Exercises 4–6 decide whether the triangles are congruent and justify your answer.

4. 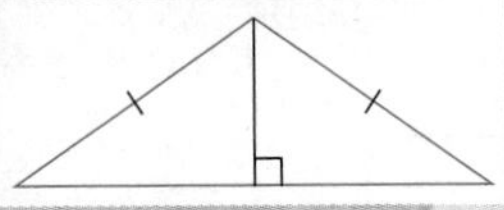

5. 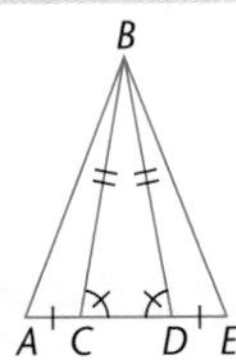

6. 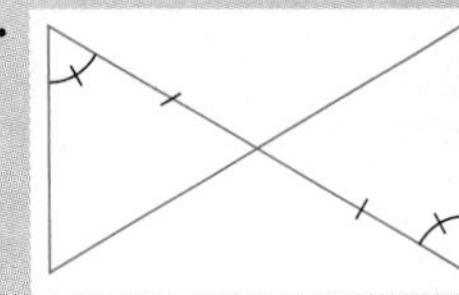

7. $\overline{AF}$, $\overline{BD}$, and $\overline{CE}$ are angle bisectors. Prove $\triangle ABC$ is equilateral.

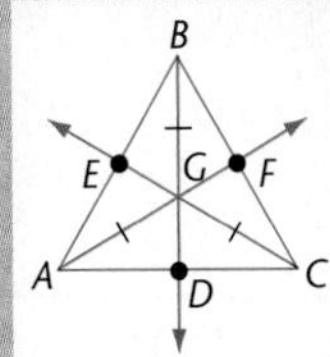

8. Which of the parallelogram(s) can be proven to be rectangles? rhombuses? squares?

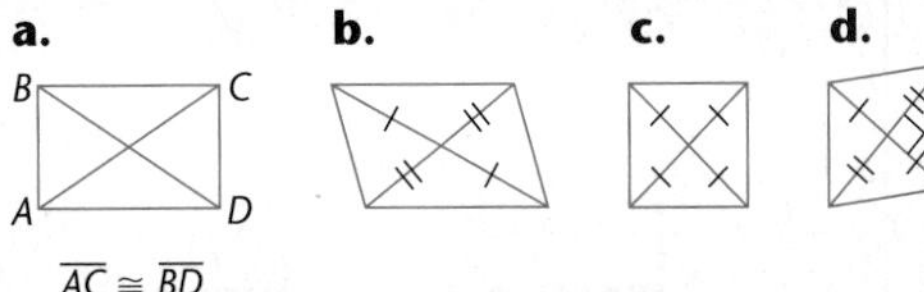

9. Are segments $\overline{FG}$ and $\overline{GH}$ congruent? Justify your answer.

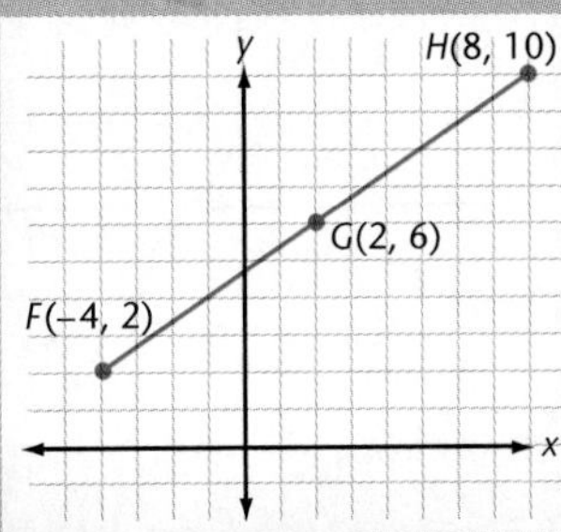

10. Construct a congruent triangle that shares point P.

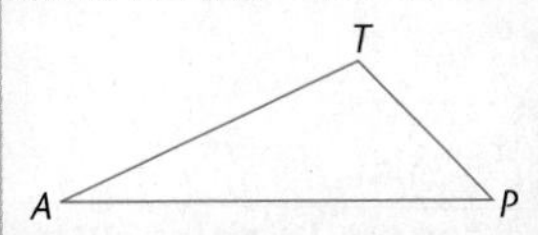

1. $\angle LSB$ $\overline{TN}$
   $\angle KLS$ $\overline{NM}$
   $\angle BKL$ $\overline{MR}$
2. They are not congruent. The corresponding sides are not congruent.
3. They are congruent. Reflection is a rigid motion that preserves all angle measures and segment lengths.
4. Congruent; AL
5. $\triangle ACB \cong \triangle EDB$; SAS
6. Congruent; ASA
8. a and c are rectangles; d is a rhombus; c is a square
9. The midpoint of $\overline{HF}$ is $\left(\frac{8+(-4)}{2}\right), \left(\frac{10+2}{2}\right) = (2,6)$ which is G. Since G is the midpoint of $\overline{HF}$, $\overline{FG} \cong \overline{GH}$.
10. Answers will vary. Sample answer:

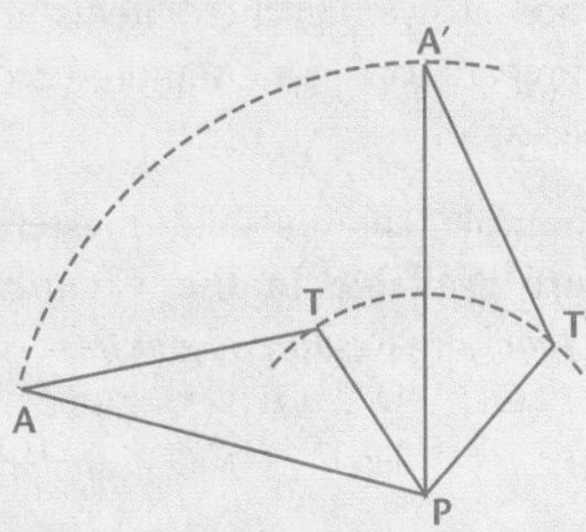

Students may use a protractor to copy angles.

COLLEGE ENTRANCE-EXAM PRACTICE

Multiple-Choice and Quantitative-Comparison Samples

The first half of the Cumulative Assessment contains two types of items found on standardized tests—multiple-choice questions and quantitative-comparison questions. Quantitative-comparison items emphasize the concepts of equality, inequality, and estimation.

Free-Response Grid Samples

The second half of the Cumulative Assessment is a free-response section. A portion of this part of the Cumulative Assessment consists of student-produced response items commonly found on college entrance exams. These questions require the use of machine-scored answer grids. You may wish to have students practice answering these items in preparation for standardized tests.

Sample answer grid masters are available in the *Chapter Teaching Resource Booklets.*

Chapters 1–4 Cumulative Assessment

College Entrance Exam Practice

Quantitative Comparison. Exercises 1–3 consist of two quantities, one in Column A and one in Column B, which you are to compare as follows.

A. The quantity in Column A is greater.
B. The quantity in Column B is greater.
C. The two quantities are equal.
D. The relationship cannot be determined from the information given.

	Column A	Column B	Answers
1.	Graph: $D(-7,7)$, $A(5,6)$, $E(-3,3)$, $B(1,-5)$ Slope of $\overline{DE}$	Slope of $\overline{AB}$	(A) (B) (C) (D) **[Lesson 3.8]**
2.	$\triangle MNO \cong \triangle PQR$ Triangle MNO: N 64°, M 60°, O 56° x	Triangle PQR: PR $(2x + y) - 1$, QR $2y + 14$, R $3x + 2$ y	(A) (B) (C) (D) **[Lesson 3.5]**
3.	Triangle XYZ $\angle XYZ$	Triangle TUV $\angle TUV$	(A) (B) (C) (D) **[Lesson 4.2, 4.3]**

4. Which set of points defines a line perpendicular to $\overline{MN}$?
a. (0, 7), (8, −4) **b.** (4, −7), (−4, 4) **[Lesson 3.8]**
c. (−7, 0), (4, 8) **d.** (7, −4), (−4, 4)

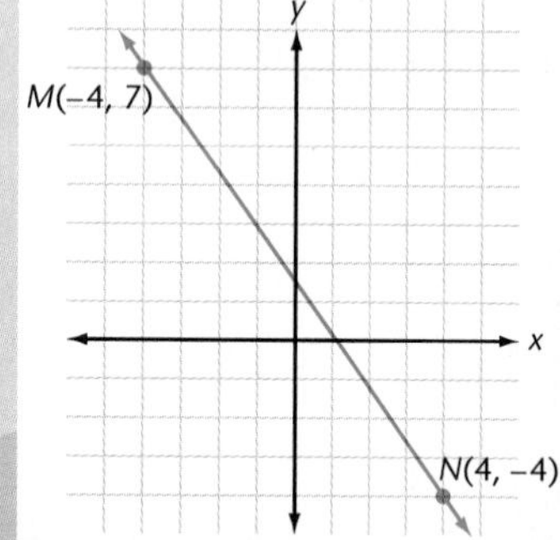

5. What is the measure of $\angle ACB$? **[Lesson 4.4]**

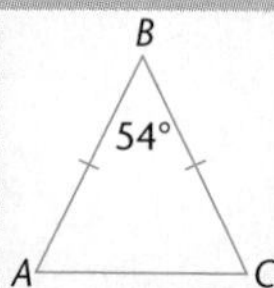

a. 42°
b. 126°
c. 63°
d. cannot be determined from the information given

6. The missing angle measure is **[Lesson 3.6]**
a. 120° **b.** 180°
c. 130° **d.** 100°

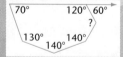

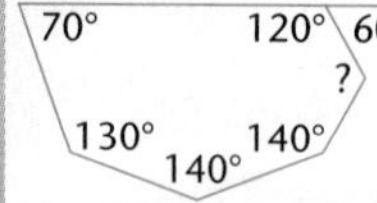

1. B 2. B 3. C 7. d
4. c 5. c 6. a

7. Supply the missing reason in the two-column proof. **[Lesson 4.2, 4.3]**

Given: $\angle NMP$ and $\angle OPM$ are right angles, and $\overline{NM} \cong \overline{OP}$. Prove that $\triangle MNP \cong \triangle POM$.

STATEMENTS	REASONS
$\overline{NM} \cong \overline{OP}$	Given
$m\angle NMP = m\angle OPM = 90°$	Given
$\angle NMP \cong \angle OPM$	If $\angle$s have = measure, they are $\cong$.
$\overline{MP} \cong \overline{PM}$	Reflexive Property of Congruence
$\triangle MNP \cong \triangle POM$	(7)

a. SSA **b.** SSS **c.** HL **d.** SAS

8. Which is not a feature of every rhombus? **[Lesson 4.6]**

a. parallel opposite sides **b.** equal diagonals
c. four equal sides **d.** 90° intersection of diagonals

9. Write the converse of this statement: If trees bear cones, then the trees are conifers. **[Lesson 2.2]**

10. Using the rule $(x, y) \rightarrow (x + 2, y - 2)$, transform the pre-image at the right. What type of transformation results? **[Lesson 1.7]**

11. Lines l and m are parallel. Find $m\angle 1$. **[Lesson 3.3]**

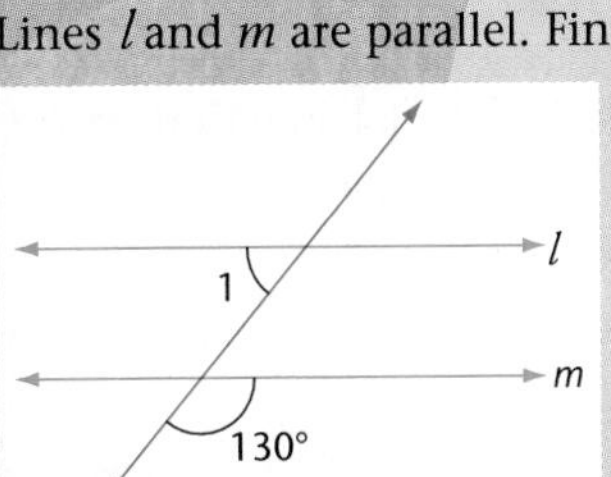

12. Are these triangles congruent? Justify your answer. **[Lesson 4.2, 4.3]**

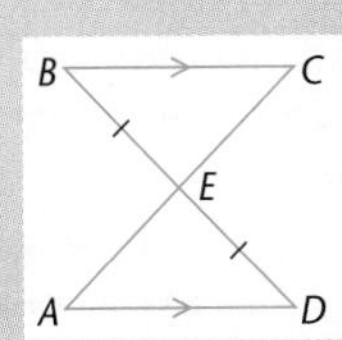

13. Find the midpoint of each side of $\triangle MSG$. **[Lesson 4.8]**

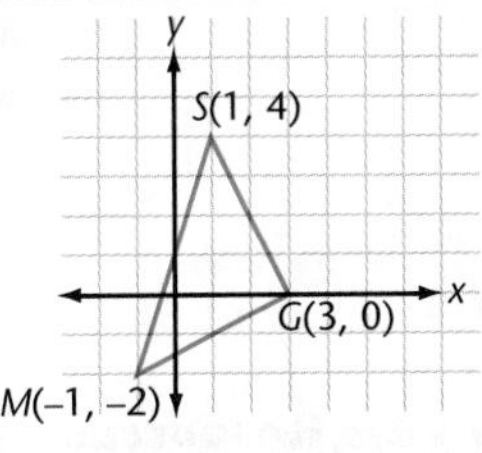

14. Construct a segment, a line, and a reflection of the segment over the line. **[Lesson 4.9]**

Free-Response Grid Questions 15–17 may be answered using a free-response grid commonly used by standardized test services.

15. *QUAD* is a rectangle. Find x. **[Lesson 4.5]**

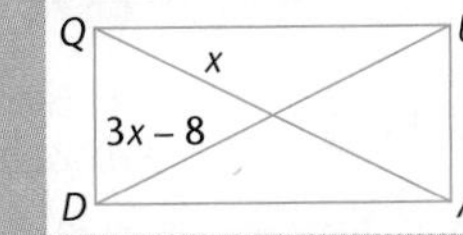

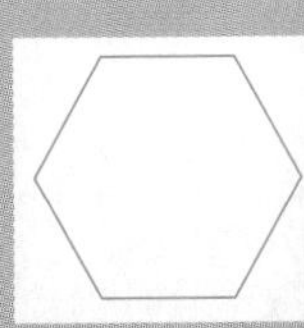

16. Find the interior angle measure of this regular polygon. **[Lesson 3.6]**

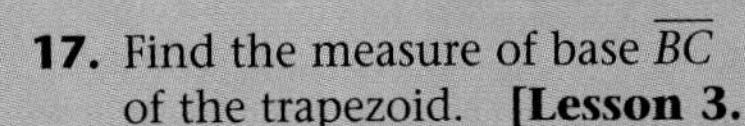

17. Find the measure of base $\overline{BC}$ of the trapezoid. **[Lesson 3.7]**

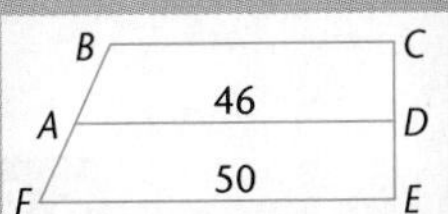

15. 4
16. 120°
17. 42

8. b
9. If the trees are conifers, then they bear cones.
10. A translation two units down and two units to the right.
11. 50°
12. Yes. Students may use two pairs of alternate interior angles and one pair of sides (AAS); or one pair of alternate interior angles, one pair of vertical angles, and one pair of sides (AAS or ASA).
13. (0, 1), (2, 2), (1, –1)
14. Student answers will vary. For a sample answer, see answer to Exercise 17, page 238.

CHAPTER 5

Perimeter and Area

Meeting Individual Needs

5.1 Exploring Perimeter and Area

Core Resources	Core Plus Resources
Inclusion Strategies, p. 246 Reteaching the Lesson, p. 247 Practice Master 5.1 Enrichment, p. 246 Technology Master 5.1 Lesson Activity Master 5.1	Practice Master 5.1 Enrichment Master 5.1 Technology Master 5.1 Interdisciplinary Connection, p. 245
[1 day]	[1 day]

5.2 Exploring Areas of Triangles, Parallelograms and Trapezoids

Core Resources	Core Plus Resources
Inclusion Strategies, p. 256 Reteaching the Lesson, p. 257 Practice Master 5.2 Enrichment Master 5.2 Technology Master 5.2 Lesson Activity Master 5.2 Interdisciplinary Connection, p. 255	Practice Master 5.2 Enrichment, p. 256 Technology Master 5.2 Interdisciplinary Connection, p. 255
[2 days]	[1 day]

5.3 Exploring Circumferences and Areas of Circles

Core Resources	Core Plus Resources
Inclusion Strategies, p. 263 Reteaching the Lesson, p. 264 Practice Master 5.3 Enrichment, p. 263 Lesson Activity Master 5.3 Interdisciplinary Connection, p. 262	Practice Master 5.3 Enrichment Master 5.3 Technology Master 5.3 Interdisciplinary Connection, p. 262
[1 day]	[1 day]

5.4 The "Pythagorean" Right-Triangle Theorem

Core Resources	Core Plus Resources
Inclusion Strategies, p. 269 Reteaching the Lesson, p. 270 Practice Master 5.4 Enrichment Master 5.4 Technology Master 5.4 Lesson Activity Master 5.1 Mid-Chapter Assessment Master	Practice Master 5.4 Enrichment, p. 269 Technology Master 5.4 Interdisciplinary Connection, p. 268 Mid-Chapter Assessment Master
[2 days]	[2 days]

5.5 Special Triangles, Areas of Regular Polygons

Core Resources	Core Plus Resources
Inclusion Strategies, p. 277 Reteaching the Lesson, p. 278 Practice Master 5.5 Enrichment, p. 277 Lesson Activity Master 5.5 Interdisciplinary Connection, p. 276	Practice Master 5.5 Enrichment Master 5.5 Technology Master 5.5 Interdisciplinary Connection, p. 268
[2 days]	[2 days]

5.6 The Distance Formula, Quadrature of a Circle

Core Resources	Core Plus Resources
Inclusion Strategies, p. 285 Reteaching the Lesson, p. 286 Practice Master 5.6 Enrichment, p. 285 Technology Master 5.6 Lesson Activity Master 5.6	Practice Master 5.6 Enrichment Master 5.6 Technology Master 5.6 Interdisciplinary Connection, p. 284
[2 days]	[1 day]

5.7 Geometric Probability

Core Resources	Core Plus Resources
Inclusion Strategies, p. 291 Reteaching the Lesson, p. 292 Practice Master 5.7 Enrichment Master 5.7 Lesson Activity Master 5.7 Interdisciplinary Connection, p. 290	Practice Master 5.7 Enrichment, p. 291 Technology Master 5.7
[1 day]	[1 day]

Chapter Summary

Core Resources	Core Plus Resources
Chapter 5 Project, pp. 296–297 Lab Activity Long-Term Project Chapter Review, pp. 298–300 Chapter Assessment, p. 301 Chapter Assessment, A/B Alternative Assessment	Chapter 5 Project, pp. 296–297 Lab Activity Long-Term Project Chapter Review, pp. 298–300 Chapter Assessment, p. 301 Chapter Assessment, A/B Alternative Assessment
[3 days]	[2 days]

Reading Strategies

Before starting this chapter, be sure students know the difference between perimeter and area. Because these topics are so often taught at the same time in elementary school, many students are often confused about which is which. Students will also need to understand that *perimeter* and *circumference* are terms for related ideas. Both describe the distance around a shape, the former for polygons and the latter for circles.

Lesson 5.4 includes an important part of mathematics that can be used to illustrate and discuss the special "tricks" used to read math. Have students find the statement of the "Pythagorean" Right-Triangle Theorem in this lesson. First have them read it silently, and then have them work in pairs to read the theorem aloud. Some students may need help reading or pronouncing the term *hypotenuse*. Then conduct a class discussion, dissecting the meaning of this sentence part by part. Phrases such as "for any right triangle," "square of the length," "sum of the squares," and "squares of the lengths of the legs" should each be discussed in turn. This activity not only will give students a better understanding of a key theorem, but also will help illustrate the procedures used to read any mathematical statement of this type.

Visual Strategies

This chapter includes activities dealing with maximum and/or minimum areas and perimeters. For fixed perimeter problems, you might have students use string of the given lengths. They then can experiment on grid paper to look for the maximum area enclosed by the string. For fixed area problems, centimeter squares or cubes can help convey the basic concept of the problems.

Cooperative Learning

GROUP ACTIVITIES	
The Circumference of a Circle	Lesson 5.3, Exploration 1
A "Monte Carlo" Method for Estimating π	Lesson 5.7, Exploration
Area of a Polygon	Chapter Project

You may wish to have students work in groups or with partners for some of the above activities. Additional suggestions for cooperative group activities are noted in the teacher's notes in each lesson.

Multicultural

The cultural references in this chapter include references to Asia, Africa, and Europe.

CULTURAL CONNECTIONS	
Asia: Ancient Chinese text	Eyewitness Math
Asia: Yin-Yang symbol	Lesson 5.3, Exercises 37
Europe: The Pythagoreans	Lesson 5.4
Asia: *Chiu Chang*	Lesson 5.4
Asia: Sulbastutras	Lesson 5.4, Exercise 21

Portfolio Assessment

Below are portfolio activities for the chapter listed under seven activity domains which are appropriate for portfolio development.

1. **Investigation/Exploration** The explorations in Lesson 5.1 allow students to investigate the relationship between perimeter and area. In Lessons 5.2 and 5.3, students explore and develop area formulas. The exploration in Lesson 5.4 deals with patterns found in Pythagorean Triples; that in Lesson 5.5 with relationships in 30-60-90 right triangles. The exploration in Lesson 5.6 describes a method for approximating the area of a quarter circle with rectangles, and the exploration in Lesson 5.7 shows students a way to compare theoretical and experimental probabilities.

2. **Applications** Construction, Lesson 5.1, Exercise 19; House Painting, Lesson 5.1, Exercise 20; Decorating, Lesson 5.2, Exercise 24; Farming, Lesson 5.2, Exercise 30; Irrigation, Lesson 5.3, Exercise 12; Automobile Engineering, Lesson 5.3, Exercises 13–17; Civil Engineering, Lesson 5.5, Exercises 19–20; Drafting, Lesson 5.5, Exercises 36–37; Environmental Protection, Lesson 5.6, Exercise 14; Skydiving, Lesson 5.7, Exercises 16–19.

3. **Nonroutine Problems** Exercises 31–33 (Heron's Formula); Lesson 5.5, Exercises 30–35 (Areas of Lunes); Lesson 5.7, Exercise 26 (Train Problem).

4. **Project** Area of a Polygon: see pages 296–297. Students look for patterns in data.

5. **Interdisciplinary Topics** Solar Energy, Lesson 5.1, Exercise 22; Maximum/Minimum, Lesson 5.2, Exercises 19–22; Algebra, Lesson 5.3, Exercises 5–7; Sports, Lesson 5.3, Exercise 25; Algebra, Lesson 5.4, Exercises 5–9, 13–20, 26, 34–36; Sports, Lesson 5.4, Exercise 12; Algebra, Lesson 5.6, Exercises 9–13; Fine Art, Lesson 5.7, Exercises 32–34.

6. **Writing** *Communicate* exercises: Lesson 5.1, Exercises 1, 2, 4; Lesson 5.4, Exercise 3; Lesson 5.5, Exercises 1–4; Lesson 5.6, Exercise 3.

7. **Tools** In Chapter 5, students use spreadsheets.

Technology

Maximum/Minimum problems are among the very best ways to expose students to the power and usefulness of mathematical functions—and technology definitely provides some of the best ways to investigate them. On a computer, a spreadsheet is an obvious choice, and the relatively large screen of a computer allows several students to observe while one or more students carry out and explain the procedures.

Graphics Calculators

Maximum/Minimum Graphics calculators, which are usually more readily available than computers, provide spreadsheet-like features. Graphics calculator tables, which are like simple spreadsheets, have all the features the students will need to do the spreadsheet explorations and exercises in this book. Furthermore, the relationships displayed in graphics calculator tables can be graphed as functions on the calculator screens.

Lesson 5.1 presents two Maximum/Minimum problems in exploration form. The table setup and appropriate y-values for Exploration 1 are shown below.

Students may notice that the equation for Y_2 is the product of two linear equations, one for the base and the other for the height of the rectangle: $b = x$, and $h = 12 - x$. That the resulting equation is quadratic suggests an important concept in the study of mathematical functions. The table for the values as entered is shown below. It can be scrolled down to find the desired results.

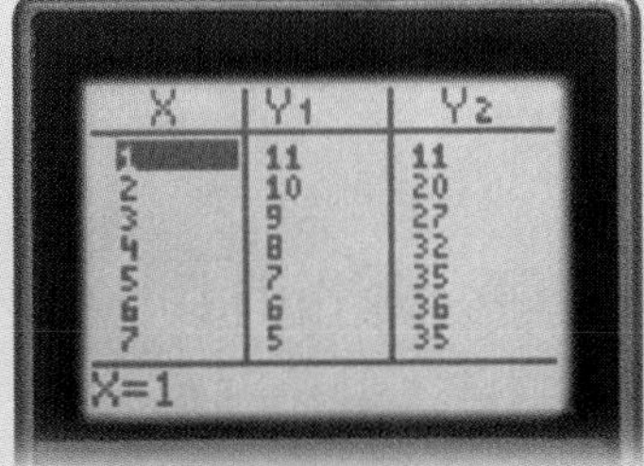

X	Y1	Y2
1	11	11
2	10	20
3	9	27
4	8	32
5	7	35
6	6	36
7	5	35

X=1

The present problem has an exact integer solution, but in other problems students might want to magnify a region of a table by resetting the **TblMin** and **ΔTbl** values in the Table Setup. An even more powerful method for finding precise solutions is to display the table as a graph. The window settings below are appropriate for this exploration.

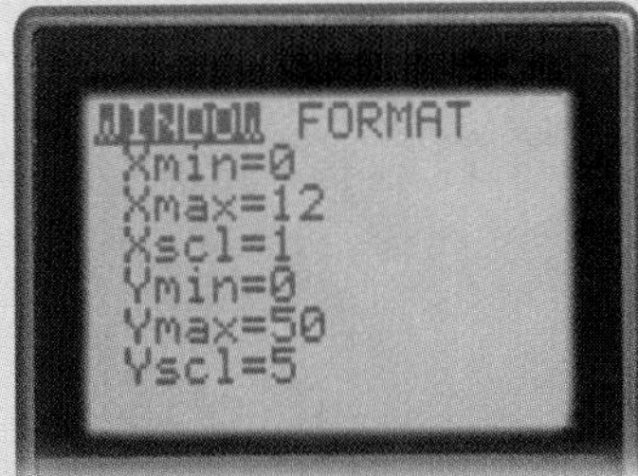

The graph produced for the table is shown below. Notice that there are two curves: a straight line representing the height of the rectangle, and a parabola representing its area. As the display indicates, the **maximum** function of the **CALC** feature of the calculator has been used to find precise values of x and y at the point where the parabola function is a maximum.

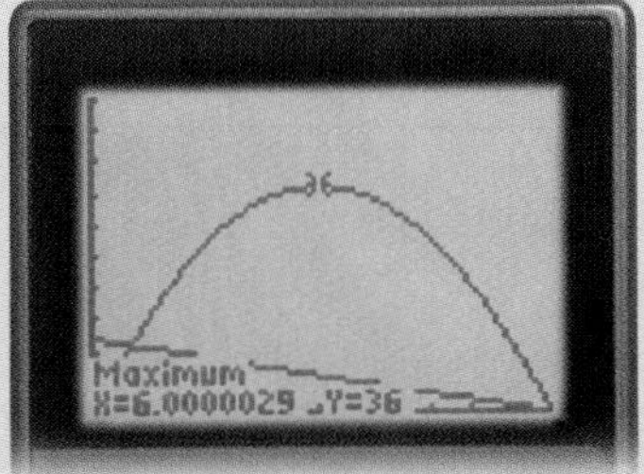

Integrated Software

f(g) Scholar™ is an integrated computer-based mathematics productivity tool that combines calculator, spreadsheet, and graphics capabilities. It provides a dynamic and interactive environment for explorations in mathematics. It is appropriate to use *f(g) Scholar*™ for any lesson needing a spreadsheet, calculator, graphics calculator, or any combination of the three.

About the Chapter

Background Information

This chapter builds on students' previous experiences to explore strategies for finding perimeter, circumference, and area. Students also learn key properties for right triangles, including proofs and applications of the "Pythagorean" Right-Triangle Theorem.

- Practice Masters
- Enrichment Masters
- Technology Masters
- Lesson Activity Masters
- Lab Activity Masters
- Long-Term Project Masters
- Assessment Masters
  - Chapter Assessments, A/B
  - Mid-Chapter Assessment
  - Alternative Assessments, A/B
- Teaching Transparencies
- Spanish Resources

Chapter Objectives

- Identify and use the Area of a Rectangle Postulate and Sum of Areas Postulate.
- Solve problems involving fixed perimeters and fixed areas.
- Identify formulas for the areas of triangles, parallelograms, and trapezoids.
- Solve problems using the formulas for the area of triangles, parallelograms, and trapezoids.
- Identify formulas for the circumference and area of a circle.

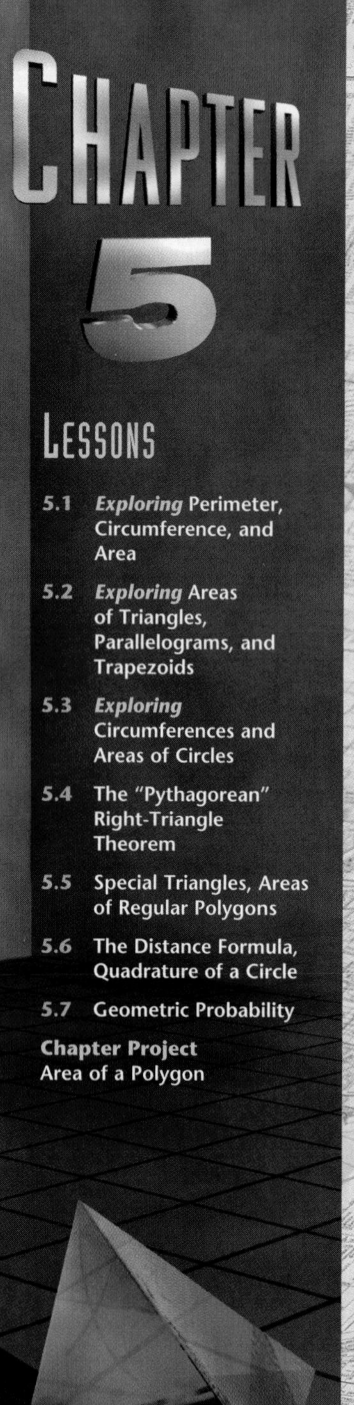

Chapter 5

Lessons

Perimeter and Area

The map of Houston reminds us of the many uses of measurement. How far is it from Hobby Airport to the center of the city? How long will it take to get from Jacinto City to Memorial Park? How many miles does one inch on the map represent?

Think of other questions about measurement that the map brings to mind, even if you do not know the answers. We could not get very far or achieve very much—in travel, construction, sports, or science—if it were not for measurement.

In this chapter you will use perimeter, such as the distance along Route 610 on the map; and area, such as the number of square miles inside Route 610. You will also learn about the postulates, theorems, and formulas that make it possible to take and use measurements.

Memorial Park

Satellite photo of Houston showing Loop 610.

Portfolio Activity

Lunes of Hippocrates

The figure shows a right isosceles triangle. Semicircles are constructed with centers at the midpoint of the hypotenuse and at the midpoint of each leg. The diameter of each semicircle has the length of a side of the triangle. The moonshaped crescents at the top of the figure are called **lunes**.

In this chapter you will learn how to find the area of the lunes—and the result may surprise you.

1 1 2 2 3

About the Photos

The photos should help remind students that perimeter and area are useful concepts for everyday life.

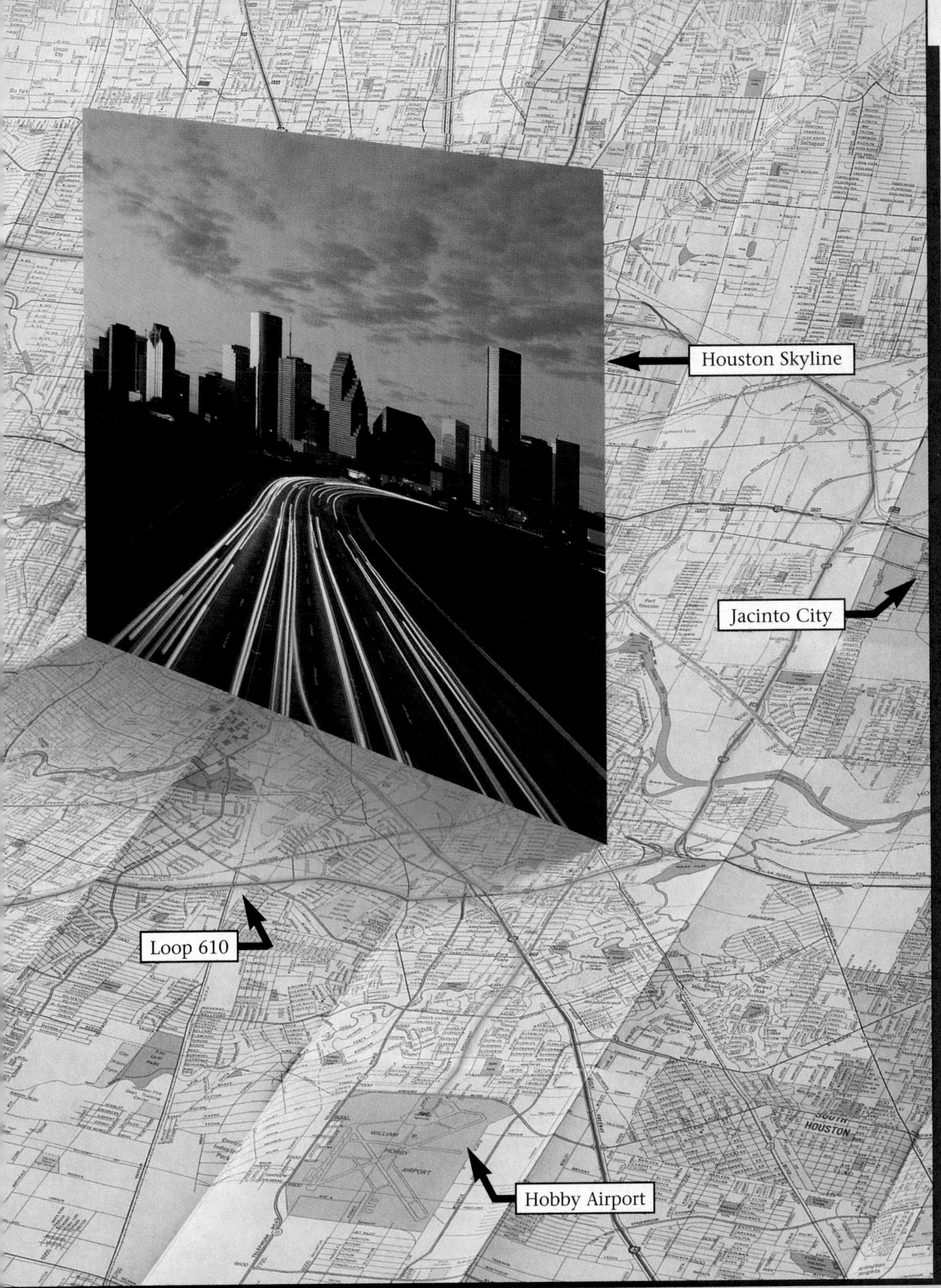

- Solve problems using the formulas for the circumference and area of a circle.
- Identify and apply the "Pythagorean" Right-Triangle Theorem and its converse.
- Solve problems using the "Pythagorean" Right-Triangle Theorem.
- Identify and use the 45-45-90 Right-Triangle Theorem and the 30-60-90 Right-Triangle Theorem.
- Identify and use the Area of a Regular Polygon Theorem.
- Develop and apply the Distance Formula and Mid-point Formula.
- Use the Distance Formula to develop techniques for estimating the area of a circle.
- Develop and apply the basic formula for geometric probability.

PORTFOLIO ACTIVITY

A *lune* is a figure made from two curves, each part of a circle. For centuries, mathematicians have worked to discover clever ways for finding areas of lunes. Have students use compass and straightedge, or geometry graphics software, to construct the figure given in the text. Have students conjecture ways of finding the area of the two upper lunes, or draw the figure on grid paper to do a rough estimate. Additional Pupil's Edition portfolio activities can be found in the exercises for Lessons 5.4, 5.5 and 5.7.

ABOUT THE CHAPTER PROJECT

The work in this chapter deals primarily with "tidy" two-dimensional shapes—circles, triangles, parallelograms, regular polygons, and so forth. In the Chapter 5 Project, on pages 296 – 297, students will collect data to derive a formula for polygons with irregular boundaries. Dot or graph paper is needed for the project.

PREPARE

Objectives

- Identify and use the Area of a Rectangle Postulate and Sum of Areas Postulate.
- Solve problems involving fixed perimeters and fixed areas.

RESOURCES

- Practice Master 5.1
- Enrichment Master 5.1
- Technology Master 5.1
- Lesson Activity Master 5.1
- Quiz 5.1
- Spanish Resources 5.1

Assessing Prior Knowledge

Ask each student to draw a rectangle with a length of 8 units and a width of 6 units.

1. What is the perimeter of this rectangle? [**28 units**]
2. What is the area of this rectangle? [**48 square units**]
3. Draw two other rectangles that have the same perimeter. [**Answers will vary.**]

TEACH

After students have read the text labeled Why, discuss what is meant by the "size" of each landholder's property. Point out that "size" can be interpreted to be length, width, perimeter or area.

CRITICAL Thinking

Redraw the figure with more sides. (Follow the shape of the curve as closely as possible.)

LESSON 5.1

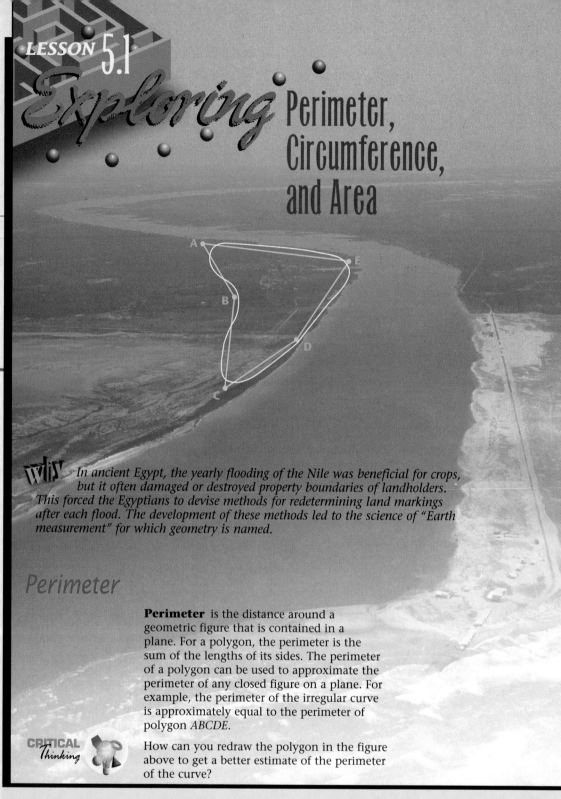

Exploring Perimeter, Circumference, and Area

A B C D E

why *In ancient Egypt, the yearly flooding of the Nile was beneficial for crops, but it often damaged or destroyed property boundaries of landholders. This forced the Egyptians to devise methods for redetermining land markings after each flood. The development of these methods led to the science of "Earth measurement" for which geometry is named.*

Perimeter

Perimeter is the distance around a geometric figure that is contained in a plane. For a polygon, the perimeter is the sum of the lengths of its sides. The perimeter of a polygon can be used to approximate the perimeter of any closed figure on a plane. For example, the perimeter of the irregular curve is approximately equal to the perimeter of polygon *ABCDE*.

CRITICAL Thinking How can you redraw the polygon in the figure above to get a better estimate of the perimeter of the curve?

Hands-On Strategies Using manipulative materials for the two explorations can provide a more kinesthetic approach. For Exploration 1, have students use 24 small paper clips to represent the 24 feet of fence. For Exploration 2, have students use 36 square tiles to solve a similar problem in which the area is 36 square feet. Using the manipulatives, students can actually move the objects around to maximize the area (Exploration 1) or minimize the perimeter (Exploration 2).

Area

The **area** of a plane figure is the number of nonoverlapping unit squares that will cover the interior of the figure. Using squares as a unit of measure makes it easy to find the area of rectangle *ABCD*.

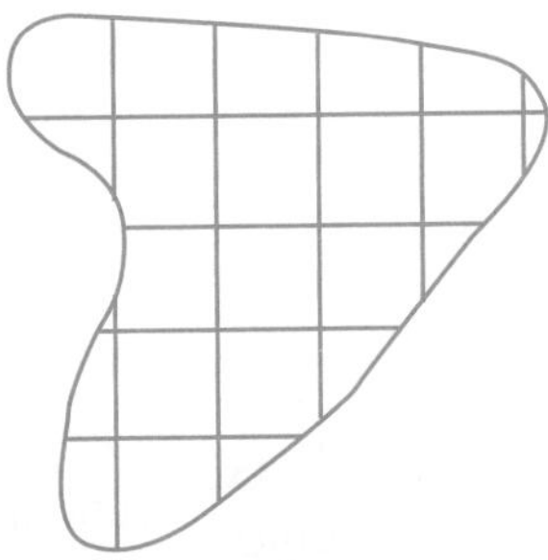

There are three unit squares in each of two rows. Thus, the area of rectangle *ABCD* is equal to 2×3, or 6 square units.

This leads us to the following Area Postulates.

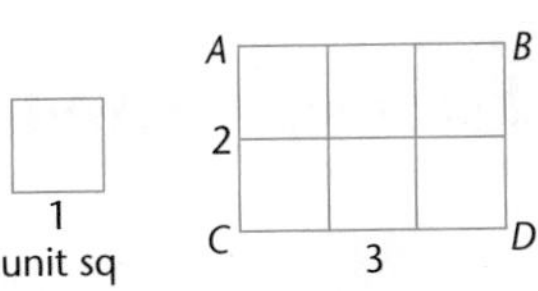

THE AREA OF A RECTANGLE

The area of a rectangle with base b and height h is $A = bh$. **5.1.1**

SUM OF AREAS

If a figure is composed of nonoverlapping regions *A* and *B*, then the area of the figure is the sum of the areas of regions *A* and *B*. (Regions are considered nonoverlapping if they share no common area, though they may share a common boundary.) **5.1.2**

EXAMPLE

Find the area of the following figures. Explain your method in each case.

A

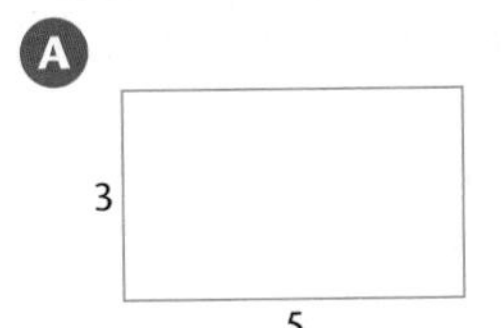

B

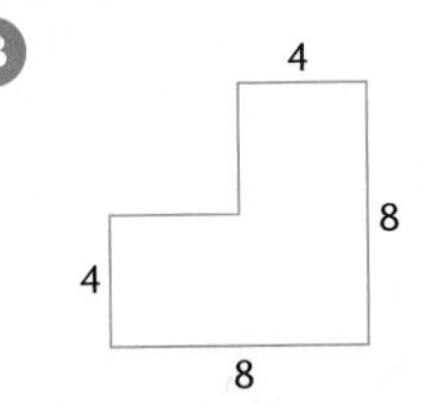

C

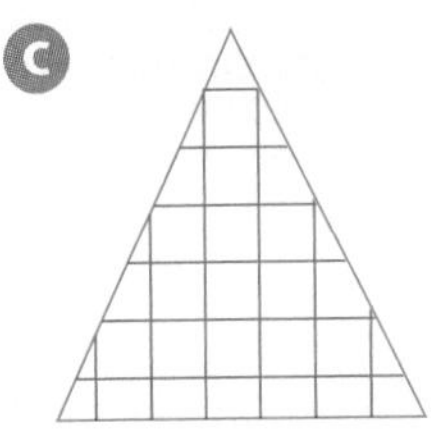

Solution ➤

A Multiply the base times the height. The area is 15 square units.

B Divide the figure into separate squares and/or rectangles and add the individual areas. (Three different ways of doing this are shown.) The area is 48 square units.

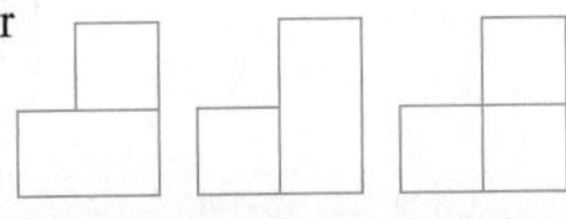

Geography When land is bought or sold, surveyors usually confirm the boundaries of the piece of property. Have students use library reference materials to find out why this is not always a simple process. They should list and discuss factors that make establishing boundary lines difficult.

Cooperative Learning

To reinforce the results of Explorations 1 and 2, have students work in small groups to check their conclusions for different initial conditions. For Exploration 1, they should start with a perimeter that is a multiple of 4. For Exploration 2, they should start with an area that is a perfect square. By working together and discussing other examples, students will clarify their understanding of the results of the explorations.

Teaching tip

For a rectangle or square, the distinction between the base and the height is arbitrary. Any of the four sides can be designated as the base. The height is then one of the sides perpendicular to this base.

Alternate Example

Provide students with graph paper.

1. Create a polygon like that in Example B. The sides must be either vertical or horizontal line segments.
2. Show two different ways to use the Sum of Areas Postulate to find the area of your polygon.

CRITICAL *Thinking*

In Example A, the Area of a Rectangle Postulate is used since the region is a rectangle. In Example B, the region is divided into non overlapping rectangles. The Area of a Rectangle Postulate is used to find the area of each rectangle. By the Sum of Areas Postulate, the area of the region is the sum of the areas of the non overlapping rectangles. In Example C, the region is divided into non overlapping squares and triangles. By the Sum of Areas Postulate, the area of the region is the sum of the areas of the non overlapping squares and triangles.

Exploration 1 Notes

Technology The sample spreadsheet shown in the text is just one way to set up this problem. If students seem confused by where the heading for the second column, $h = 12 - 2b$, comes from, suggest they use a spreadsheet with more columns. For example, the first four columns could be headed $b, 2b, 24 - 2b, (24 - 2b) \div 2$. The last column gives a value for h. This is then used to find the area.

Math Connection Algebra

Most students will have learned the perimeter formula $P = 2b + 2h$ in previous math courses. Ask students to explain how to use this formula to find the height if they are given the area and the base. This will remind them that formulas such as $P = 2b + 2h$ and $A = bh$ can be solved for any one of the three variables.

C. One way of estimating the result is to take half the number of squares that are partially inside the figure and add this to the number of squares that are entirely inside the figure. The area is approximately 21 square units. ❖

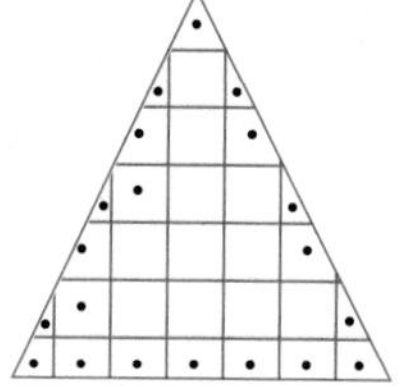

Place a dot in each partial square.

CRITICAL *Thinking* How are the two area postulates on page 245 used in the example?

Exploration 1 Fixed Perimeter, Maximum Area

You will need
Graph paper and
Graphics calculator or
Spreadsheet software

Gardening A gardener has material to make 24 ft of fencing for a miniature garden. What is the shape of a rectangle that will enclose the greatest area?

The formula for the perimeter of a rectangle is $P = 2b + 2h$.

MAXIMUM MINIMUM *Connection*

 **1** Trace three different rectangles that each have a perimeter of 24 units. What is the area of each one?

 **2** In the formula for the perimeter of a rectangle, substitute 24 for P and solve the equation for h.

 **3** Fill in a table like the one below. (Use a graphics calculator or a spreadsheet program if available.) What do you observe about the area values?

b	$h = 12 - b$	$A = bh$
1	11	11
2	?	?
3	?	?
•	•	•

ENRICHMENT The square has an area of 64 square units. The new figure has only 63 square units. What happened to the missing square? [**The two pieces do not fit together exactly. The diagonal overlaps slightly.**]

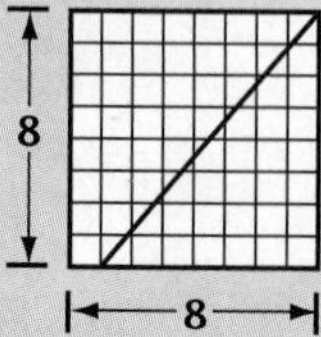

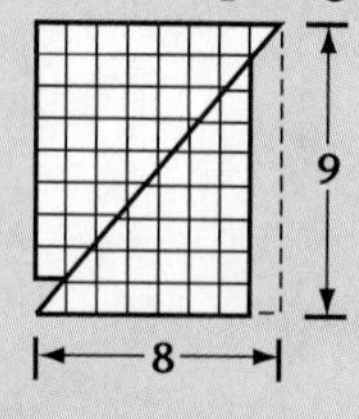

INCLUSION strategies **Using Models** Students who find the spreadsheets, formulas, and graphs in the explorations too abstract should use graph paper to draw a rectangle for each possible case.

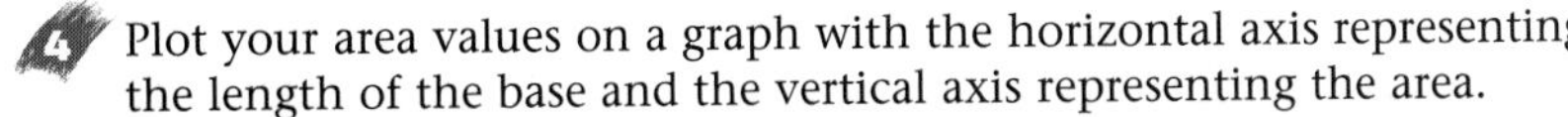
4. Plot your area values on a graph with the horizontal axis representing the length of the base and the vertical axis representing the area.

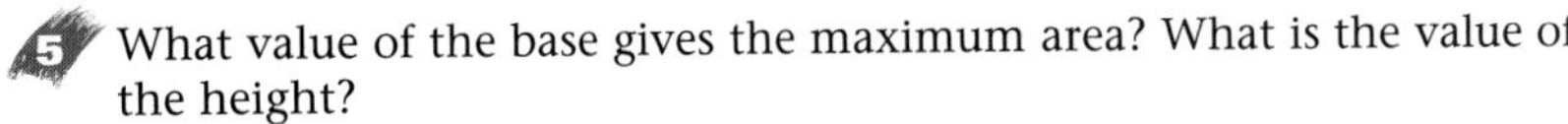
5. What value of the base gives the maximum area? What is the value of the height?

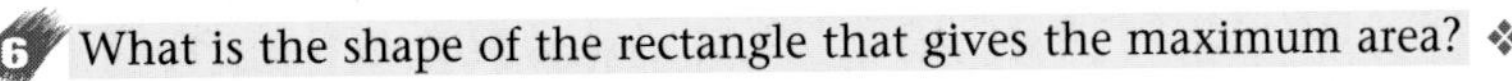
6. What is the shape of the rectangle that gives the maximum area? ❖

Do you think your result from Exploration 1 would hold true for a rectangle of any given perimeter? Explain.

Exploration 2 Fixed Area, Minimum Perimeter

You will need
Graph paper
Spreadsheet software or graphics calculator (optional)

A farmer wants to form a rectangular area of 3600 square ft with the minimum amount of fencing. What should the dimensions of the rectangle be?

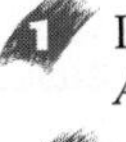
1. In the formula for the area of a rectangle ($A = bh$), substitute 3600 for A and solve for h.

2. Fill in a table like the one below. Values for b should range from 10 to 100. (Use a graphics calculator or spreadsheet software if available.)

b	$h = 3600 \div b$	$P = 2b + 2h$
10	360	380
20	?	?
30	?	?
•	•	•

3. Plot the perimeter values on a graph with the horizontal axis representing the length of the base and the vertical axis the perimeter.

4. After finding the value to the nearest ten, find the value to the nearest one. What value of the base gives the minimum perimeter? What is the value of the height?

5. What is the shape of the rectangle that gives the minimum perimeter for the given area? ❖

Do you think your result from Exploration 2 would hold true for a rectangle of any given area? Explain.

Using Models Have students use graph paper to draw as many rectangles with a perimeter of 32 units as they can. They should then number the rectangles in order of increasing area. Students should then draw as many rectangles with an area of 64 square units as they can. They should number these rectangles in order of decreasing perimeter.

ongoing ASSESSMENT

6. **A square**

CRITICAL Thinking

Yes. The maximum area will always be reached when the sides of the rectangle are equal. When the sides are not equal, more squares in one direction are lost than gained in the other direction. When the rectangle is changed from 6 by 6 to 5 by 7, you give up a row with 6 squares and only gain 1 square each in the other 5 rows.

Exploration 2 Notes

Technology With some spreadsheet utilities, students will need more than three columns. For example, the columns might progress in this order: b, $3600 \div b = h$, $2h$, $2b$, $2h + 2b$. Encourage students to set up the columns in any way that makes logical sense.

Math Connection
Algebra

Ask students to explain why $3600 = bh$ and $h = 3600 \div b$ are equivalent formulas.

ongoing ASSESSMENT

5. **A square. This is the same rectangle that has maximum area for a given perimeter.**

CRITICAL Thinking

Yes. The largest area that can be enclosed by a given perimeter has a square shape. Students can test different values of the area to see if this result seems to hold.

ASSESS

Selected Answers

Odd numbered Exercises 5–31

Assignment Guide

Core 1–15, 19–32

Core Plus 1–4, 11–35

Technology

Exercises 23 and 24 can be done using spreadsheet software. You may wish to have students work in pairs or small groups so that they can generate a number of possible cases.

Error Analysis

For Exercises 5–13, students may read the figure incorrectly and thus give an area or perimeter for the wrong polygon.

EXERCISES & PROBLEMS

Communicate

1. Explain how to estimate the perimeter of the swimming pool.

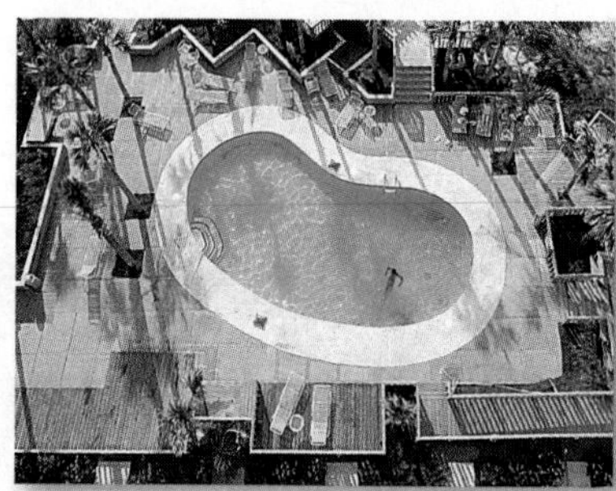

2. Explain how to estimate the top surface area of the airplane wing.

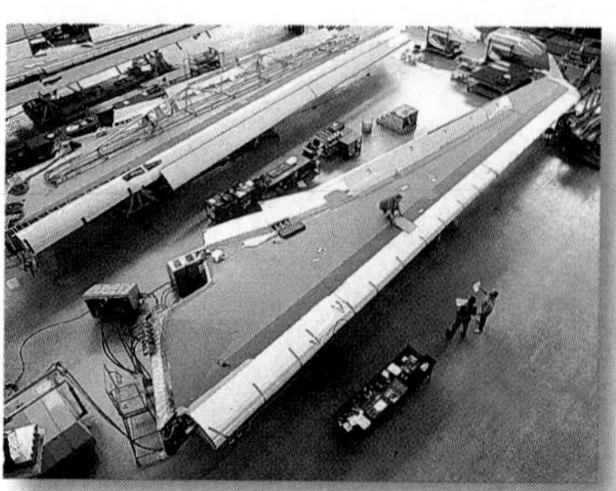

3. Explain why the definition of perimeter excludes overlapping pieces.

4. Many mathematics applications concern the area underneath a curve. Explain a way to estimate the area of the shaded part in the graph at right.

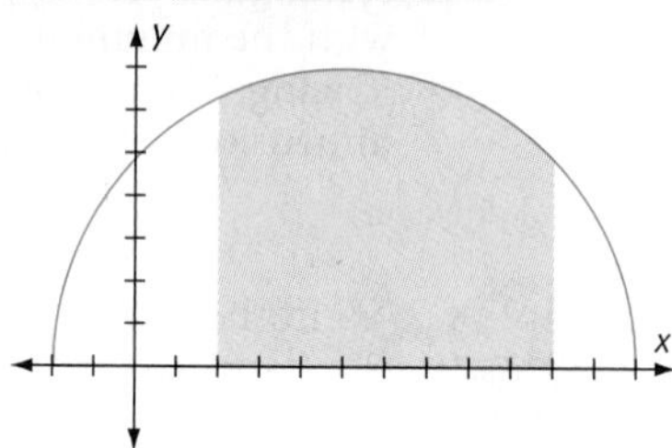

Practice & Apply

$AD = 15$ in. $AC = 13$ in.

$BD = 10$ in. $DL = 11$ in.

$EI = 3$ in. $CH = 4$ in.

Use the figure and measurements above to answer Exercises 5–15.

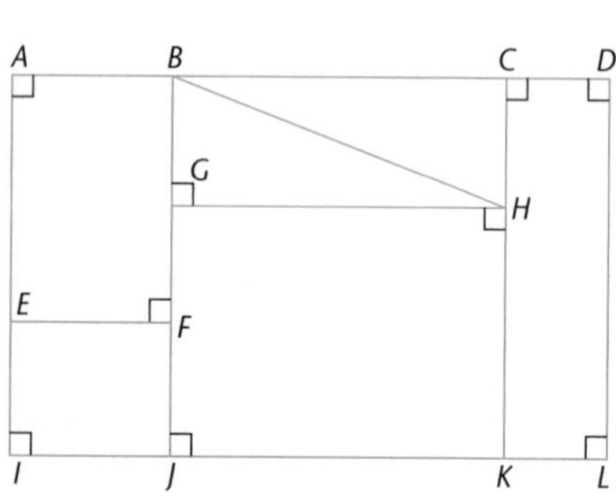

5. The perimeter of rectangle $ADLI$ = _?_. 52 in.

6. The area of rectangle $ADLI$ = _?_. 165 in.^{2}

7. The perimeter of rectangle $CDLK$ = _?_. 26 in.

8. The area of rectangle $CDLK$ = _?_. 22 in.^{2}

9. The perimeter of hexagon $GHCDLJ$ = _?_. 42 in.

10. The area of hexagon $GHCDLJ$ = _?_. 78 in.^{2}

11. The perimeter of rectangle $BCHG$ = _?_. 24 in.

12. The area of rectangle $BCHG$ = _?_. 32 in.^{2}

13. The area of $\triangle BHG$ = _?_ (use Exercise 12). 16 in.^{2}

Practice Master

NAME ______ CLASS ______ DATE ______

Practice & Apply

5.1 Exploring Perimeter, Circumference, and Area

Use the figure and the dimensions to answer Exercises 1–6.

$AD = 30$ in. $WD = AZ = FB = 24$ in. $WB = 12$ in.

1. The perimeter of rectangle $XAWF$ = ______.
2. The perimeter of rectangle $XAZE$ = ______.
3. The area of rectangle $XAWF$ = ______.
4. The area of ΔAWF = ______. (Use Exercise 3.)
5. The area of trapezoid $ADEF$ = ______. (Use Exercise 4.)
6. The area of hexagon $ABCDEF$ = ______. (Use Exercise 5.)

7. Suppose the perimeter of a rectangle is 108 cm, and the measure of the height is one-half the measure of the base. What is the area?

______

8. Two opposite sides of a 5-inch square were increased in measure to form a rectangle. If the area of the rectangle is 105 in.^{2}, by how much was one side of the square increased?

______

9. The Bixby Company wants to carpet the area of their lobby floor that is shown. Carpeting is sold by the square yard. How many square yards are there in the area shown? ($9 \text{ ft}^2 = 1 \text{ yd}^2$)

______

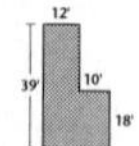

The number of square yards of carpeting that must be purchased depends on the width of the roll on which the carpet comes.

10. If the carpet chosen comes on 12-foot rolls, and lengths must be purchased in full square yards, how many square yards of carpet must Bixby buy?

______

HRW Geometry Practice & Apply 53

14. If points I and D were connected by a segment, what would the area of the resulting $\triangle ADI$ be? 82.5 in.^{2}

15. What theorems or postulates from previous chapters helped you to determine the perimeters and areas in Exercises 5–14?

16. **Algebra** Suppose the perimeter of a rectangle is 72 cm. The base measures three times the height. What are the lengths of the sides? What is the area? Base = 27 cm; height = 9 cm; area = 243 cm^2

17. **Algebra** Suppose the perimeter of a rectangle is $80x$. The base measures seven times the height. What are the lengths of the sides? What is the area? (Answers will be in terms of x.) base = $35x$; height = $5x$; area $175x^2$

18. **Algebra** Find the dimensions of a rectangle that has an area equal to its perimeter.

19. **Construction** A house has a roof with dimensions as shown. The roof will be built with plywood covered by shingles. Plywood comes in pieces 8 ft by 4 ft. How many pieces of plywood will be needed to cover the roof?

25 ft

42 ft

20. **House Painting** Brenda wants to paint her room which measures 14 ft by 16 ft with a height of 10 ft. She will give the walls and ceiling two coats of paint: one coat with a base paint that costs $10.00 per gallon and covers 500 sq ft, and the other coat with a finish paint that costs $20.00 per gallon and covers 250 sq ft. She has one window in her room that is 6 ft by 4 ft and a door (which will not be painted) that is 3 ft by 7 ft. How much will it cost to paint the room if taxes are 7.0%? Assume that paint is sold only in one-gallon cans.

21. **Landscaping** Leticia is resodding the lawn on the right. Given the dimensions, estimate the number of square feet of sod she will need.

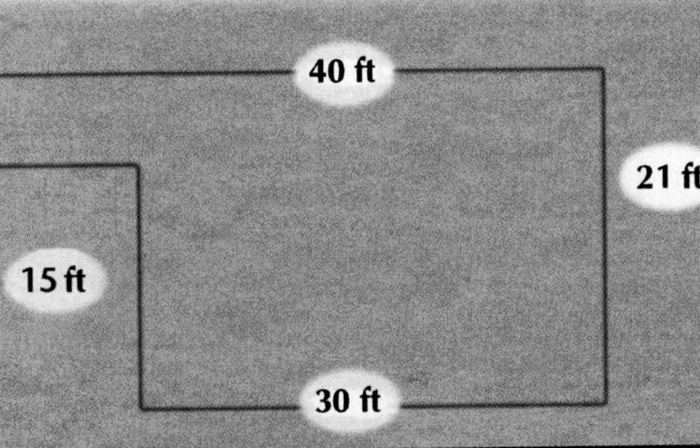

15. Segment Addition Postulate; Overlapping Segments Theorem

18. Any rectangle such that $b = \dfrac{2h}{h-2}$

Examples: $h = 4$ and $b = 4$; $h = 3$ and $b = 6$

19. 66

20. $107.00

21. 690

alternative ASSESSMENT

Performance Assessment

Have students work in pairs or small groups. Provide each team with several sheets of $8\frac{1}{2} \times 11$ graph paper. Students are to find a way to cut one sheet of paper to make a rectangle with the minimum perimeter for an area of 25 square inches. They should estimate the area and the perimeter, and describe in writing what procedures or strategies they used to solve the problem.

Technology Master

NAME ______________ CLASS ________ DATE ________

Technology

5.1 Studying Perimeter and Area on a Graphics Calculator

In many earlier courses in mathematics, you learned how to use formulas to find the perimeters and areas of familiar figures. For example, if a square has sides of length s, then perimeter P is given by $P = 4s$ and area A is given by $A = s^2$.

Have you ever asked yourself this question: *Is the area of a square always greater than its perimeter?* After all, when you look at a square, there appears to be more inside it than along its border. With a graphics calculator, you can study the question and others like it.

Use a graphics calculator in the exercises.

1. Graph $y = 4x$ and $y = x^2$, the functions that represent the perimeter and area of a square whose sides have length x, on the same calculator for positive values of x.

2. For what value(s) of x are the perimeter and the area equal?

3. For what value(s) of x is the perimeter greater than the area?

4. For what value(s) of x is perimeter less than area?

5. Suppose that a square has sides with length s. Use Exercises 1–4 as a model to analyze the relationship between perimeter and area if the lengths of the sides are doubled.

6. Suppose that a square has sides with length s. Use Exercises 1–4 as a model to analyze the relationship between perimeter and area if the lengths of the sides are multiplied by a positive number $n > 1$.

7. Suppose that a square has sides with length s. Use Exercises 1–4 as a model to analyze the relationship between perimeter and area if the lengths of the sides are divided by a positive number $n > 1$.

HRW Geometry — Technology 67

22. Mr. Venn needs 2 panels, each 10 feet by 4.5 feet or 20×2.25 feet.

22. Solar Energy Mr. Venn is converting some of his house's electric power to solar energy. He wishes to provide an additional 15,000 BTUs of heat for the house through 2 solar panels. He knows that he needs 6 square feet of solar panel per 1000 BTUs. Mr. Venn also wants the 2 panels to be equal in size, with each 10 feet in length. What are the dimensions of the panels he needs?

23. **Maximum/Minimum** You have 200 feet of fencing material to make a pen for your livestock. If you make a rectangular pen, what is the maximum area you can fence in? Extend the table below to determine the answer. Maximum area: 2500 ft^2

base	height	perimeter	area
1	99	200	99
2	98	200	196
5	95	200	475
20	80	200	1600
•	•	200	•

24. Agriculture To care properly for your livestock, you must provide a certain amount of area per animal. Suppose you need an area of 3600 sq ft. What is the minimum amount of fencing you need for a rectangular pen? Extend the table below to determine an answer. Minimum fencing: 240 ft

base	height	perimeter	area
1	3600	7202	3600
2	1800	3604	3600
5	720	1410	3600
20	360	740	3600
•	•	•	•

25. **Algebra** $2b + 2h = 100$ is the equation for a constant perimeter of 100. Solve for b or h and graph the equation. What type of function represents this relationship? What solution values for the function do not make sense for the perimeter example?

Look Back

26. Construct a Venn diagram to illustrate the relationships among parallelograms, rectangles, rhombuses, and squares. **[Lesson 3.2]**

27. Find the measure of an exterior angle of an equiangular triangle. **[Lesson 3.6]** 120°

28. If the sum of the measures of three angles of a quadrilateral equals 300°, find the measure of the fourth angle. **[Lesson 3.6]** 60°

29. Find the sum of the measures of the angles of a polygon with n sides. **[Lesson 3.6]** $(n-2)\ 180°$

30. Find the measure of an interior angle of a regular hexagon. **[Lesson 3.6]** 120°

31. Find the slope of the segment connecting points (2, 3) and (4, −1). **[Lesson 3.8]** −2

32. Find the midpoint of the segment connecting the points (−4, −6) and (6, 4). **[Lesson 4.8]** (1, −1)

Look Beyond

33. Which has a greater area, a square with a side of 4 in. or a circle with a diameter of 4 in.? Explain your answer.

34. The square at right has 1-inch sides. Explain how you could estimate the area of the circle inside the square.

35. Use the formula for the area of a circle ($A = \pi r^2$) to confirm your estimate.

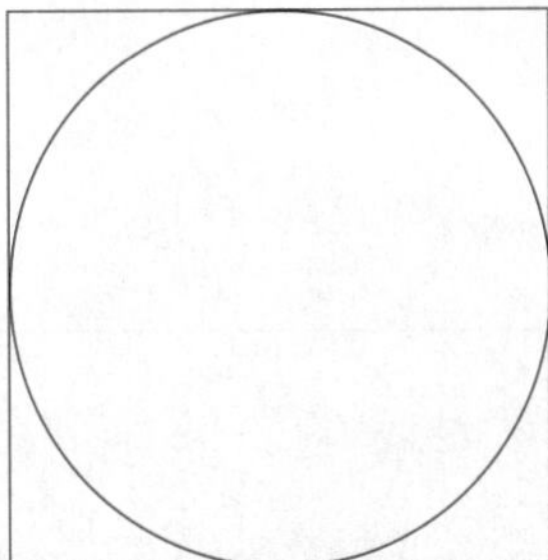

25. $b = 50 - h$ or $h = 50 - b$; Linear function ; b and h must both be positive and less than 50

26.

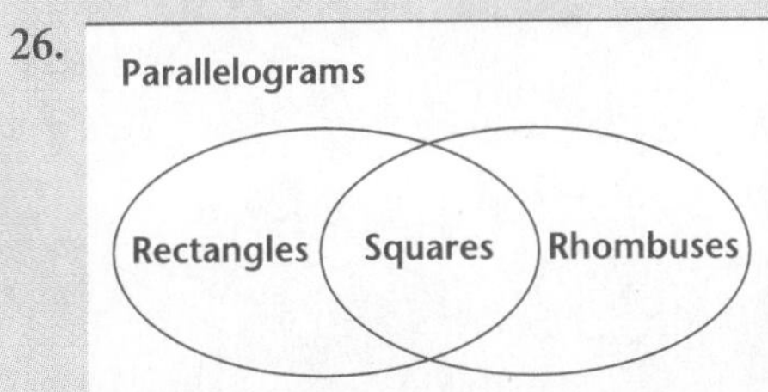

33. A square with side 4. A circle with diameter 4 will fit inside a square with side 4.

34.–35. Answers will vary, but should contain the idea of subdividing the square corner regions outside the circle, estimating their area, and subtracting from 1 to find the circle area. The better the corner estimate, the closer the area of the circle is to $\frac{\pi}{4}$.

Look Beyond

Exercises 33–35 foreshadow the circle activities students will do in Lesson 5.3.

Focus

A book of ancient Chinese mathematics ignored for centuries by Western civilization provides examples for exploring the solution of geometric problems by cutting up figures and rearranging the pieces.

Motivate

As you review the page, help students gain an appreciation for *The Nine Chapters on the Mathematical Art,* particularly the sophistication of the mathematics in relation to what else was going on in the world at that time. The book includes geometric and algebraic proofs credited to other mathematicians who didn't discover them until hundreds of years later.

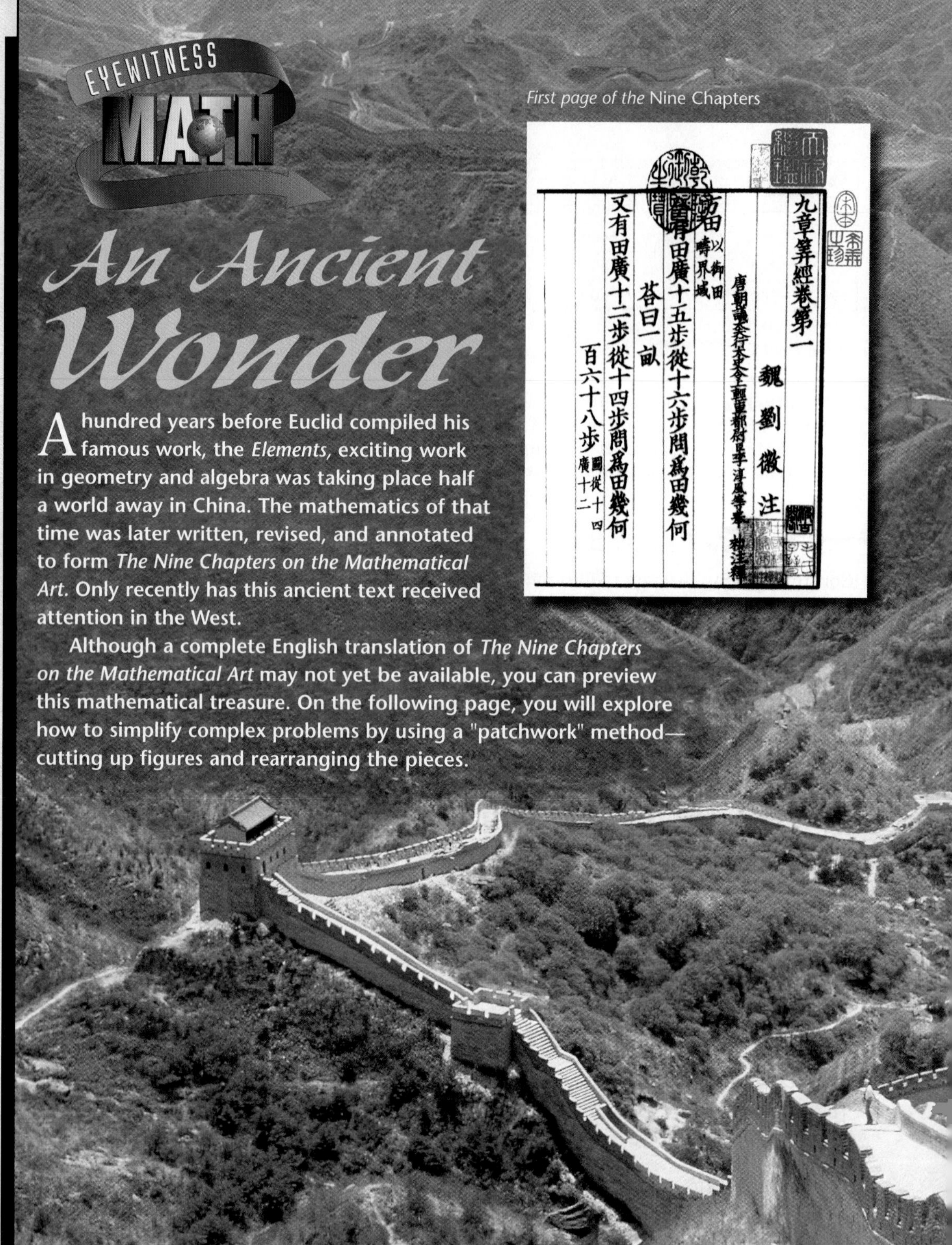

First page of the Nine Chapters

九章筭經卷第一

魏 劉 徽 注

唐朝議大夫行太史令上輕車都尉臣李淳風等奉 勅注釋

方田 以御田疇界域

有田廣十五步從十六步問爲田幾何

荅曰一畝

又有田廣十二步從十四步問爲田幾何

百六十八步 圖從十四廣十二

An Ancient Wonder

A hundred years before Euclid compiled his famous work, the *Elements,* exciting work in geometry and algebra was taking place half a world away in China. The mathematics of that time was later written, revised, and annotated to form *The Nine Chapters on the Mathematical Art.* Only recently has this ancient text received attention in the West.

Although a complete English translation of *The Nine Chapters on the Mathematical Art* may not yet be available, you can preview this mathematical treasure. On the following page, you will explore how to simplify complex problems by using a "patchwork" method—cutting up figures and rearranging the pieces.

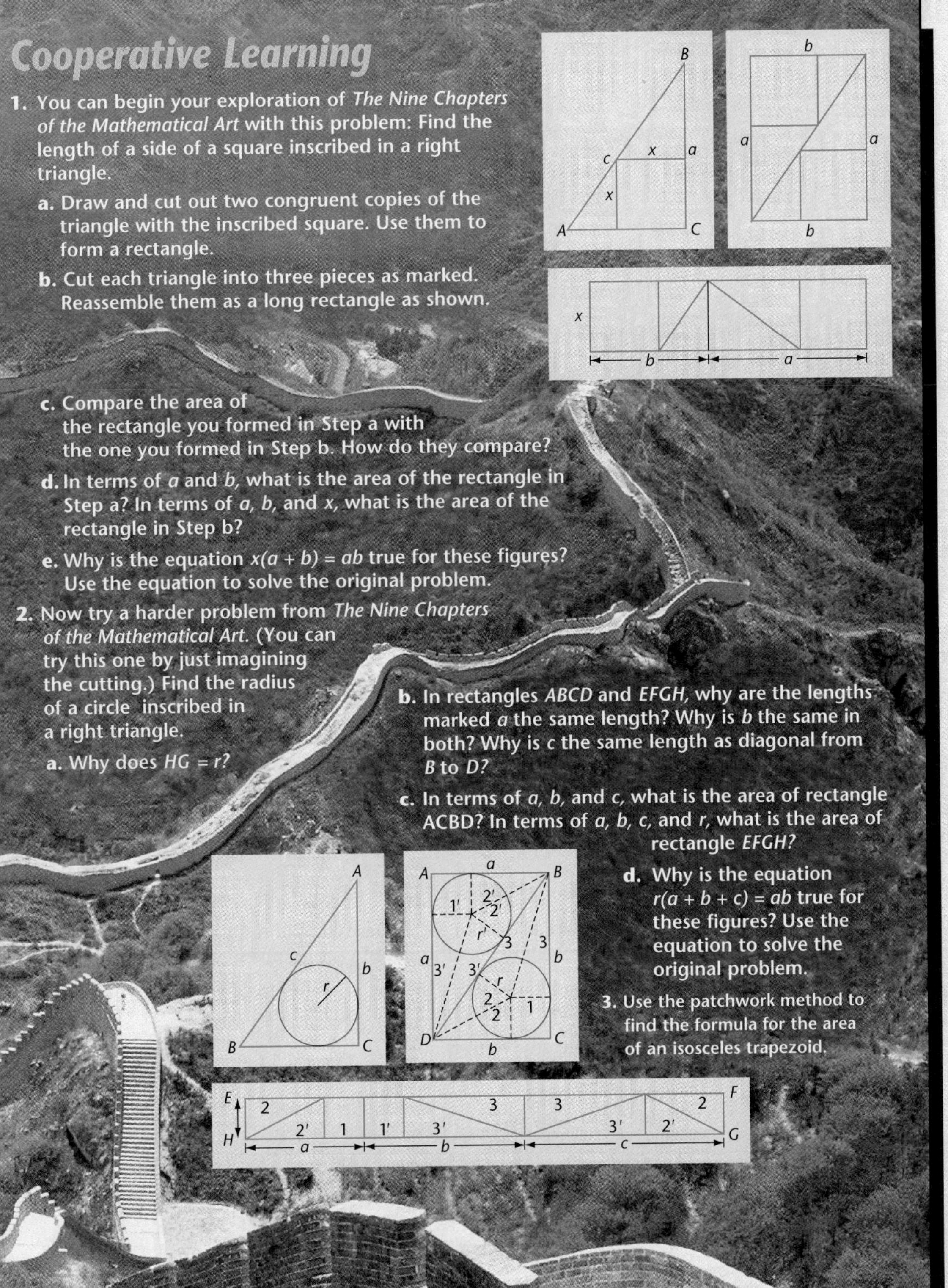

Cooperative Learning

1. You can begin your exploration of *The Nine Chapters of the Mathematical Art* with this problem: Find the length of a side of a square inscribed in a right triangle.
   a. Draw and cut out two congruent copies of the triangle with the inscribed square. Use them to form a rectangle.
   b. Cut each triangle into three pieces as marked. Reassemble them as a long rectangle as shown.
   c. Compare the area of the rectangle you formed in Step a with the one you formed in Step b. How do they compare?
   d. In terms of a and b, what is the area of the rectangle in Step a? In terms of a, b, and x, what is the area of the rectangle in Step b?
   e. Why is the equation $x(a + b) = ab$ true for these figures? Use the equation to solve the original problem.
2. Now try a harder problem from *The Nine Chapters of the Mathematical Art.* (You can try this one by just imagining the cutting.) Find the radius of a circle inscribed in a right triangle.
   a. Why does $HG = r$?
   b. In rectangles $ABCD$ and $EFGH$, why are the lengths marked a the same length? Why is b the same in both? Why is c the same length as diagonal from B to D?
   c. In terms of a, b, and c, what is the area of rectangle ACBD? In terms of a, b, c, and r, what is the area of rectangle $EFGH$?
   d. Why is the equation $r(a + b + c) = ab$ true for these figures? Use the equation to solve the original problem.
3. Use the patchwork method to find the formula for the area of an isosceles trapezoid.

The answers to Exercises 1–3 can be found in Additional Answers beginning on page 727.

Cooperative Learning

Have students work in pairs. Each group should be provided with paper, scissors, and rulers. Students will benefit most by actually cutting and rearranging the figures as directed. You may want to have the class proceed through one exercise at a time with a pause for discussion in between.

Discuss

Describe how to use the patchwork method to find the relationship between the sides of a right triangle.

PREPARE

Objectives

- Identify formulas for the areas of triangles, parallelograms, and trapezoids.
- Solve problems using the formulas for the area of triangles, parallelograms, and trapezoids.

RESOURCES

- Practice Master 5.2
- Enrichment Master 5.2
- Technology Master 5.2
- Lesson Activity Master 5.2
- Quiz 5.2
- Spanish Resources 5.2

Assessing Prior Knowledge

Define each figure by describing key properties of the sides and angles.

1. Right triangle **[3 sides, 1 right angle]**
2. Acute triangle **[3 sides, all angles < 90°]**
3. Parallelogram **[4 sides, opposite sides parallel and congruent]**

TEACH

By counting the number of squares in a grid-paper drawing of a figure, the area of the figure can be estimated. But there is often a quicker and more accurate method.

ongoing ASSESSMENT

5. $A = \frac{1}{2}bh$

LESSON 5.2

Exploring

Areas of Triangles, Parallelograms, and Trapezoids

Why *Designs drawn on grid paper, such as this knitting pattern, suggest methods of estimating the areas of geometric figures. But it is often more convenient to use exact formulas.*

Exploration 1 *Areas of Triangles*

You will need
Graph paper

Part I

1. Draw a rectangle on graph paper. Calculate its area.
2. Draw a diagonal of your rectangle. What kind of triangles are formed? Identify the heights or altitudes of each triangle.
3. What do you know about the two triangles from your study of special quadrilaterals? What is the area of each of the triangles?
4. If you are given a right triangle, can you form a rectangle by fitting it together with a congruent copy of itself? Illustrate your answer with examples.

5. Write a formula for the area of right triangle *ABC* in terms of its base, *b*, and its height (or altitude), *h*.

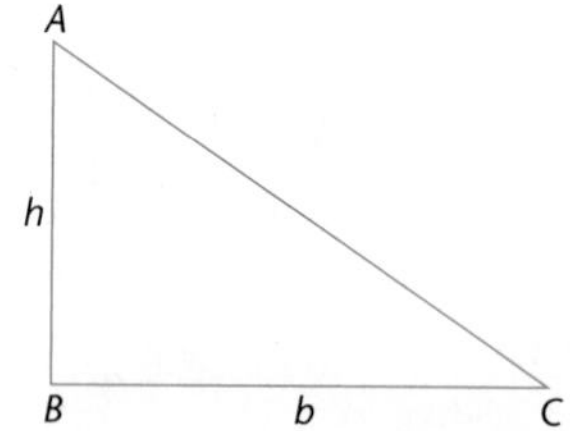

ALTERNATIVE teaching strategy

Technology Geometry graphics software can be used to explore the formulas in this lesson. For example, the figure shows the formula developed in Part 2 of Exploration 1. Students should construct two perpendicular lines, and then construct another line parallel to each of these. Four right angles are formed. Point *J* can move anywhere on the line containing points *G* and *H*. Students should drag point *J* and observe that the base, height, and area of $\triangle ABJ$ do not change.

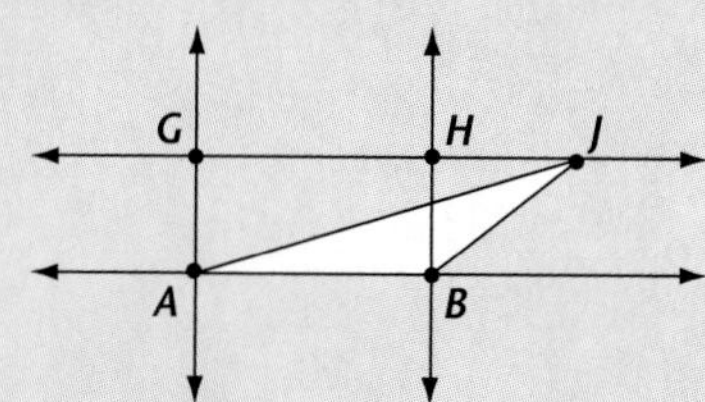

Part II

1. Make a copy of the drawing on graph paper.

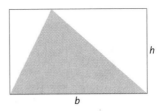

2. Draw an altitude of the triangle from the top vertex of the triangle to the base of the rectangle. Is the altitude parallel to the sides of the rectangle? What theorem justifies your answer?

3. The altitude divides the rectangle into two smaller rectangles. Each of these rectangles is divided into two congruent triangles. What theorem justifies this fact?

4. What is the relationship between the area of the shaded part of the large rectangle occupied by the original triangle and the area of the unshaded parts? Explain your answer.

5. Write a formula for the area, A, of a triangle in terms of its base, b, and its height (or altitude), h. ❖

APPLICATION

Theatre Arts You are building a triangular flat as a set for a school play. The flat needs to be covered with cloth. What is the area you need to cover?

$h = 22$ in.

$b = 30$ in.

$A = \frac{1}{2}bh \qquad A = \frac{1}{2}(30)(22) \qquad A = 330$ sq in. ❖

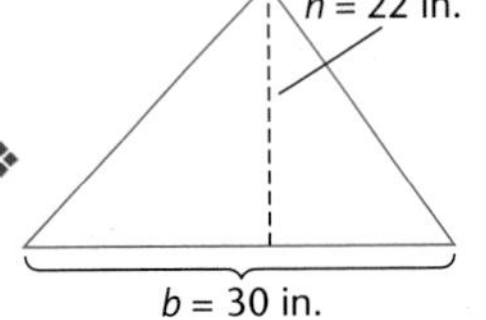

CRITICAL Thinking

Triangle $\triangle ABC$ is obtuse. How can you use the method of Exploration 1, Part 2, to show that the formula for the area of a triangle holds for an obtuse triangle?

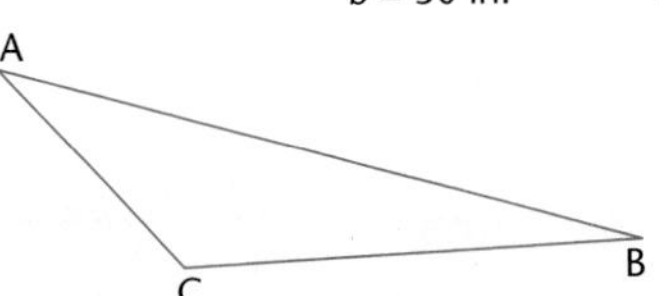

Exploration 2 *Areas of Parallelograms*

You will need

Graph paper

Scissors

1. Make a copy of the drawing on graph paper; b is the base of the parallelogram; h is the height of the parallelogram.

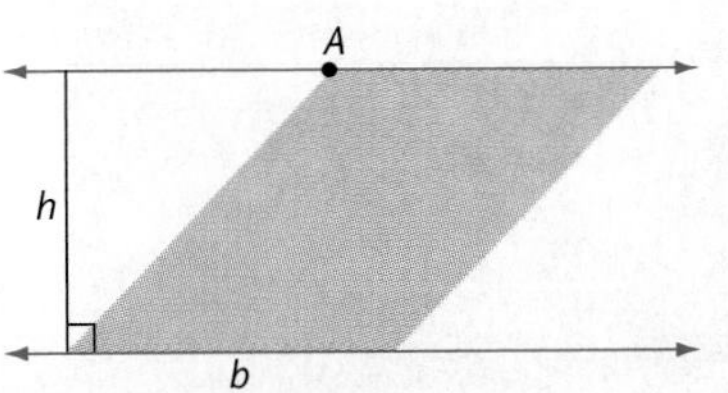

interdisciplinary CONNECTION **Social Studies** Triangles and parallelograms are frequently used in the design of flags. Have students use an atlas to find and copy interesting flag designs from a variety of countries. They can apply the area formula to estimate the amounts of colored cloth needed for flags of different sizes.

Math Connection Algebra

Stress that the base b and the height h must be perpendicular. If necessary, point out that the variable h in this situation does *not* stand for the hypotenuse of the right triangle.

Exploration 1 Notes

For Part 2 of the Exploration, emphasize that the height must be perpendicular to the base. Point out that any of the three sides of the triangle can be chosen as the base.

ongoing ASSESSMENT

5. $A = \frac{1}{2}bh$

Math Connection Algebra

Point out that only one side of the triangle, the base b, appears in the formula. This implies that the other two sides of the triangle can change in length without changing the area.

CRITICAL Thinking

Turn the triangle and draw the altitude from the obtuse angle. Then argue as before.

Exploration 2 Notes

Some students will think, logically enough, that the area of a parallelogram ought to equal the product of two adjoining sides. Have students use graph paper drawings or geometry graphics software to convince themselves that this strategy does not work.

Use Transparency 43

ONGOING ASSESSMENT

4. **$A = bh$. A parallelogram can be made into a rectangle by removing a right triangle from one side of the parallelogram and placing it on the other side.**

Math Connection
Algebra
Point out that the height h is the perpendicular distance between two opposite sides and thus can be drawn anywhere in the figure.

CRITICAL Thinking

The diagonal divides the parallelogram into two congruent triangles, the area of each is $A = \frac{1}{2}bh$.

Exploration 3 Notes

Technology If students use geometry graphics software for this activity, they should start by constructing a trapezoid. They will then reflect the trapezoid twice, once across a vertical line and once across a horizontal line. The final figure should then be fitted to the original figure to make a parallelogram.

ONGOING ASSESSMENT

3. **Two copies of a trapezoid can be put together to form a parallelogram with base = $b_1 + b_2$, height = h, and Area = $(b_1 + b_2)h$. Since there are two congruent trapezoids in the parallelogram, the area of one of the trapezoids will be $A = \frac{(b_1 + b_2)h}{2}$.**

2. Draw an altitude in the parallelogram from point A to its base. What figure is formed?

3. Cut out the parallelogram. Cut off the right triangle, move it to the other side, and fit it with the parallelogram. What figure is formed? What is the area, A, of the parallelogram in terms of b and h of the original figure?

4. Write a formula for the area of a parallelogram in terms of its base, b, and its height, h. Explain your answer in terms of the figure you formed in Step 3.

5. How do you know that the triangle will always fit, as in Step 3? To answer this question, first prove that $\triangle AEB \cong \triangle DFC$. Then prove that quadrilateral $ADFE$ is a rectangle. Write out your argument and save it in your portfolio. ❖

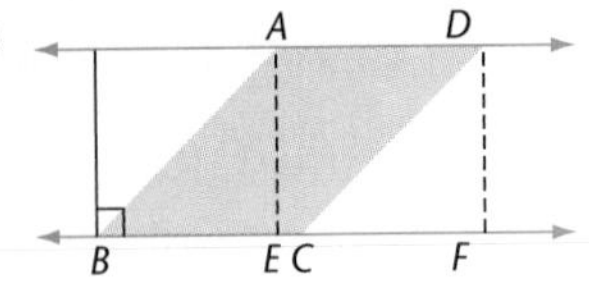

Use the fact that the area of a parallelogram is bh to find the formula for the area of a triangle.

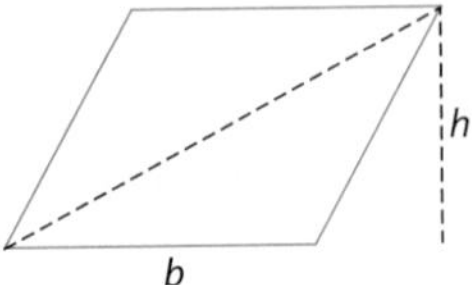

Exploration 3 Areas of Trapezoids

You will need
Graph paper

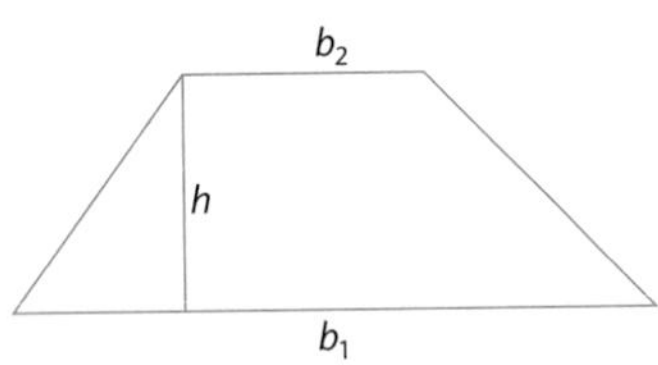

1. Make two copies of the trapezoid on graph paper: b_1 and b_2 are the bases of the trapezoid; h is the height of the trapezoid.

2. Find a way to fit the two trapezoids into a parallelogram.

3. Find several ways to explain why the area of a trapezoid is $A = \frac{(b_1 + b_2)h}{2}$. ❖

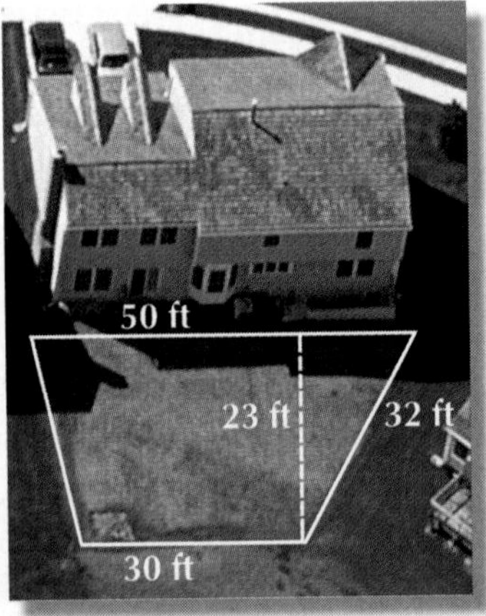

In order to apply the correct amount of fertilizer, you need to know the area of Mrs. Zapata's lawn. The dimensions of the trapezoidal lawn are as shown. What is the area of the lawn?

$$A = \frac{(b_1 + b_2)h}{2}$$
$$A = \frac{(30 + 50)(23)}{2}$$
$$A = \frac{80}{2}(23)$$
$$A = 40(23)$$
$$A = 920 \text{ sq ft}$$ ❖

ENRICHMENT Heron's Formula for the area of a triangle is introduced in Exercises 31–33. Have students use a spreadsheet to show that these formulas give the area of a trapezoid. The two parallel sides are a and b.

$$\text{Semiperimeter} = s = \frac{a+b+c+d}{2}$$

$$h = \frac{2}{a-b}\sqrt{(s(s-a)(s-b)(s-c)(s-d))}$$

$$\text{Area} = A = \frac{(a+b)h}{2}$$

INCLUSION strategies **Using Cognitive Strategies** Some students may attempt to simply memorize the area formulas without trying to really understand how the formulas are developed. Explain that they are likely to forget the formulas unless they also remember a diagram and the logical reasoning used to create the formula. For example, to remember the formula for the area of a triangle, they should think of a triangle drawn inside a rectangle, as shown in Part 2 of Exploration 1.

EXERCISES & PROBLEMS

Communicate

1. The area of a parallelogram is 14 sq in. What are the areas of the two triangles formed by the diagonal? Explain your answer.
2. Is it possible for two parallelograms to have the same area and not be congruent? Explain your answer. Is it possible for two triangles to have the same area and not be congruent? Explain your answer.
3. Do you need to know the lengths of the sides of a trapezoid to find its area? Explain your answer.

Practice & Apply

$\overline{AC} \parallel \overline{IL}$	$\overline{DH} \parallel \overline{IL}$
$\overline{AI} \parallel \overline{JB}$	$\overline{KC} \parallel \overline{JB}$
$BK = 15$ units	$KL = 14$ units
$IL = 32$ units	$DH = 25$ units
$FK = 7$ units	$EH = 17$ units
$JI = 8$ units	$DG = 18$ units

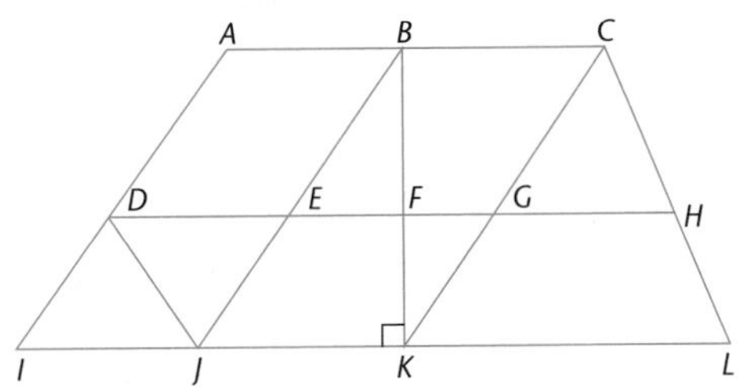

Use the diagram above to answer Exercises 4–15.

4. The area of $\triangle KCL$ = __?__. 105 units²
5. The area of $\triangle BJK$ = __?__. 75 units²
6. The area of $\triangle BCK$ = __?__. 75 units²
7. The area of $\triangle DIJ$ = __?__. 28 units²
8. The area of parallelogram $BCKJ$ = __?__.
9. The area of parallelogram $EGKJ$ = __?__.
10. The area of parallelogram $ABED$ = __?__.
11. The area of parallelogram $ACKI$ = __?__.
12. The area of trapezoid $EHLJ$ = __?__.
13. The area of trapezoid $BCHE$ = __?__.
14. The area of trapezoid $BCLJ$ = __?__.
15. The area of trapezoid $ACLI$ = __?__.
16. Which postulates and theorems from previous lessons helped you determine the areas in Exercises 4–15? Segment Addition Postulate; Overlapping Segments Theorem
17. **Algebra** Find the base measure of a triangle with height 10 cm and area 100 cm². 20 cm
18. **Algebra** Find the height of a parallelogram with a base 15 cm and area 123 cm². 8.2 cm

Cooperative Learning

Have students work in small groups to prove to themselves that any of the three sides of a triangle can be chosen as the base. They can use graph paper or geometry graphics software for this investigation.

ASSESS

Selected Answers

Odd-numbered Exercises 5–29

Assignment Guide

Core 1–18, 27–33

Core Plus 1–3, 17–33

Technology

Geometry graphics software may be used with Exercises 4–16. Have students replicate the figure. Then they can use the software to either find or verify answers.

Error Analysis

Students who are not able to do Exercises 25 and 26 working alone may be able to solve them working in small groups. The exchange of ideas and the brainstorming during discussion will help clarify the problems and suggest solution strategies.

The answers for Exercises 8–15 can be found in Additional Answers beginning on page 727.

Using Cognitive Strategies Most students will easily understand why the area of a right triangle is one-half the product of the base and the height. A different way to derive the area formula for any triangle is to think of cutting it into two right triangles.

If x and y are used for the two parts of the base b, then the total area is the sum of the areas of the two right triangles.

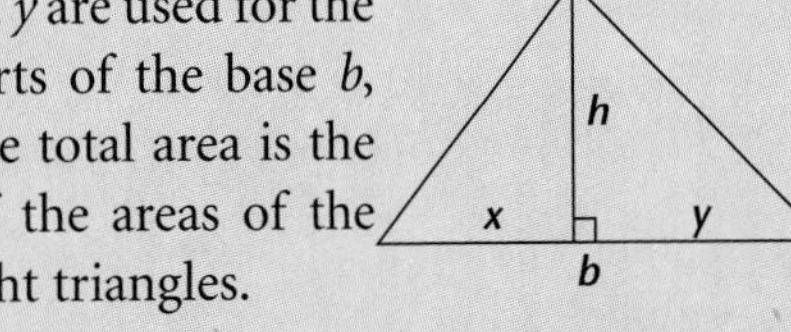

$$A = \frac{1}{2}xh + \frac{1}{2}yh = \frac{1}{2}h(x + y) = \frac{1}{2}hb$$

alternative ASSESSMENT

Authentic Assessment

Using materials of their choice, have students develop an area formula for a right isosceles triangle. They should assign variables to the various lengths, create figures to justify the formula, and show that the formula works in an application situation. The focus in this assessment is on the strategies and logic employed rather than on simply getting a valid formula.

19. Triangle a: $A = \frac{1}{2}(8)(6.93) = 27.72$ units^2. Triangle b. $A = \frac{1}{2}(6)(8) = 24$ units^2. Triangle a has the larger area.

Practice Master

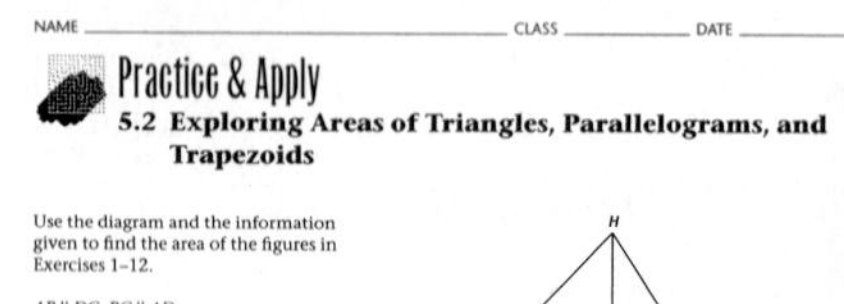

NAME ______________ CLASS ________ DATE ________

Practice & Apply

5.2 Exploring Areas of Triangles, Parallelograms, and Trapezoids

Use the diagram and the information given to find the area of the figures in Exercises 1–12.

$AB \parallel DC$, $BC \parallel AD$

$JD \parallel CE$, $HD \perp AE$

$BC = HD = 12$ units

$BG = HG = 7$ units

$DE = 8$ units

1. ΔGCD ________
2. ΔBCD ________
3. ΔABD ________
4. ΔAHD ________
5. ΔAHE ________
6. ΔHCD ________
7. parallelogram $ABCD$ ________
8. parallelogram $CDEF$ ________
9. parallelogram $JCED$ ________
10. trapezoid $ABCE$ ________
11. trapezoid $ABGD$ ________
12. trapezoid $GCED$ ________
13. Find the height of a triangle with base measure 12 cm and area 108 cm^2. ________
14. Find the base measure of a parallelogram with height 16 in. and area 192 cm^2. ________

This picture shows a roof consisting of two trapezoids and two triangles. The bases of the trapezoids measure 24 feet and 40 feet. The base measure of each triangle is 16 feet. The height of the trapezoids and the triangle is 8 feet.

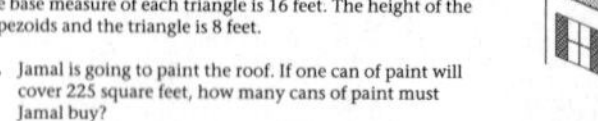

15. Jamal is going to paint the roof. If one can of paint will cover 225 square feet, how many cans of paint must Jamal buy?

____

54 Practice & Apply HRW Geometry

19. **Maximum/Minimum** Each of the following triangles has the same perimeter. Which has the larger area? Explain your answer.

a.

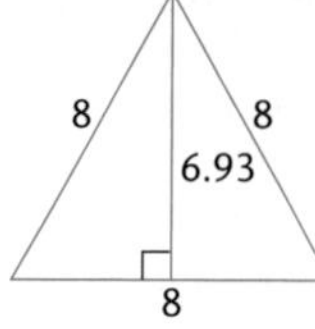

b.

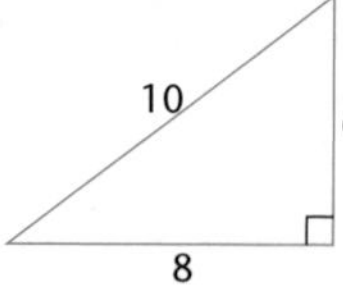

20. **Maximum/Minimum** Each of the following triangles has the same area. Which has the larger perimeter? Explain your answer.

a.

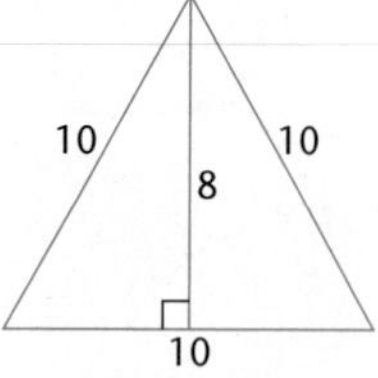

b.

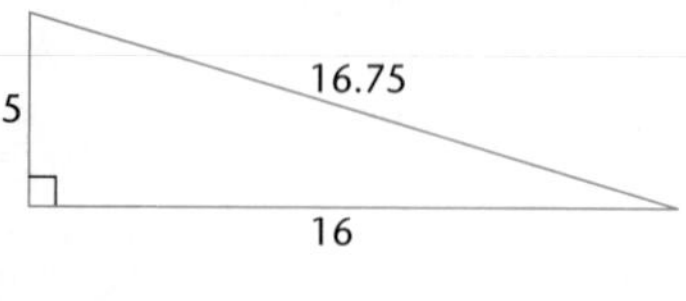

21. **Maximum/Minimum** Each of the following parallelograms has the same perimeter. Which has the larger area? Explain your answer.

a.

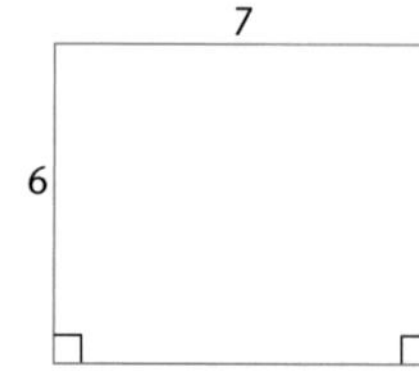

b.

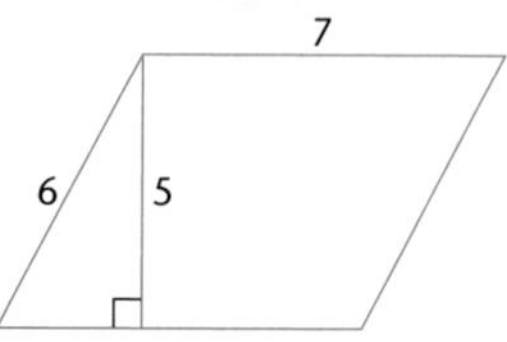

22. **Maximum/Minimum** Each of the following parallelograms has the same area. Which has the larger perimeter? Explain your answer.

a.

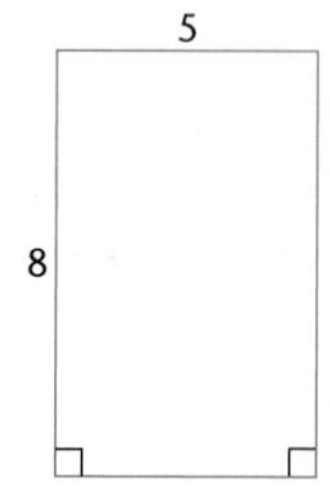

b.

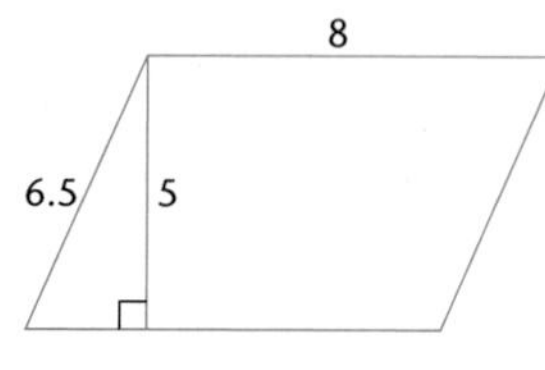

23. Find the area of $\triangle RST$.
1.5 units^2

R
1.5
2
S
2.5
T

20. Triangle a: $P = 10 + 10 + 10 = 30$ units Triangle b: $P = 5 + 16 + 16.75 = 37.75$ units Triangle b has the larger perimeter.

21. Parallelogram a: $A = (6)(7) = 42$ units^2. Parallelogram b. $A = (5)(7) = 35$ units^2. Parallelogram a has the larger area.

22. Parallelogram a: $P = 2(5) + 2(8) = 26$ units. Parallelogram b. $P = 2(8) + 2(6.5) = 29$ units. Parallelogram b has the larger perimeter.

24. Decorating Joe and his friends have to decorate a float for the homecoming parade. The float is 6 ft high and the length and width are 12 ft by 8 ft. Joe needs to wrap the sides of the float with chicken wire to form a frame for the decorating materials. How many square feet of chicken wire does he need? If chicken wire costs $2.00 per square yard, what is the cost of the chicken wire?

The answers to Exercises 27 and 28 can be found in Additional Answers beginning on page 727.

25. Algebra A **kite** is a quadrilateral in which two pairs of adjacent sides are congruent. The diagonals of a kite are perpendicular. Use the formula for the area of a triangle to prove that the area of a kite is one-half the product of the length of the diagonals.

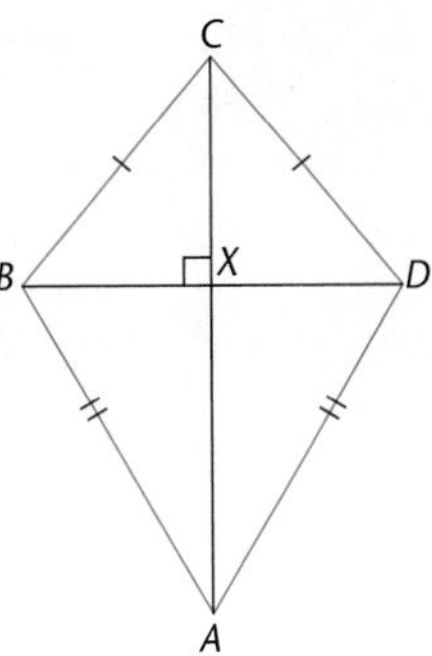

26. Algebra A line segment connecting the midpoints of the two nonparallel segments of a trapezoid is called the midsegment of a trapezoid. Develop a formula for the area of a trapezoid using the midsegment and the height of a trapezoid. (Hints: The midsegment is parallel to the bases. Use the triangles as shown in the diagram.)

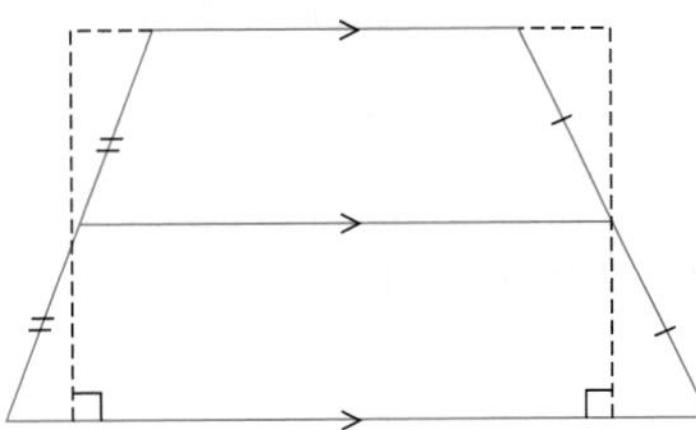

Look Back

27. Construct a Venn diagram to illustrate the relationship between scalene, isosceles, and equilateral triangles. **[Lesson 2.3]**

28. Given the congruences indicated in the diagram, prove $\triangle ABC \cong \triangle DEF$. **[Lesson 4.4]**

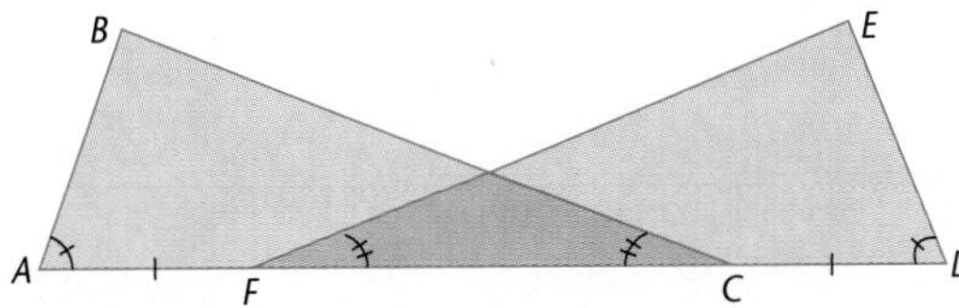

29. Algebra Find the area of a square whose side measures $x + y$. **[Lesson 5.1]**

24. Side wrap: 240 sq feet or $26\frac{2}{3}$ square yard
Cost: approximately $53.34

25. Area of $\triangle ACD = \frac{1}{2}(AC)(XD)$. Area of $\triangle ACB = \frac{1}{2}(AC)(XB)$. Area of $ABCD = \frac{1}{2}(AC)(XD) + \frac{1}{2}(AC)(XB) = \frac{1}{2}(AC)(XD + XB) = \frac{1}{2}(AC)(BD)$.

26. Let m = length of the midsegment: $m = \frac{b_1 + b_2}{2}$; Area of a trapezoid $= \frac{(b_1 + b_2)h}{2} = mh$

29. $x^2 + 2xy + y^2$

Technology Master

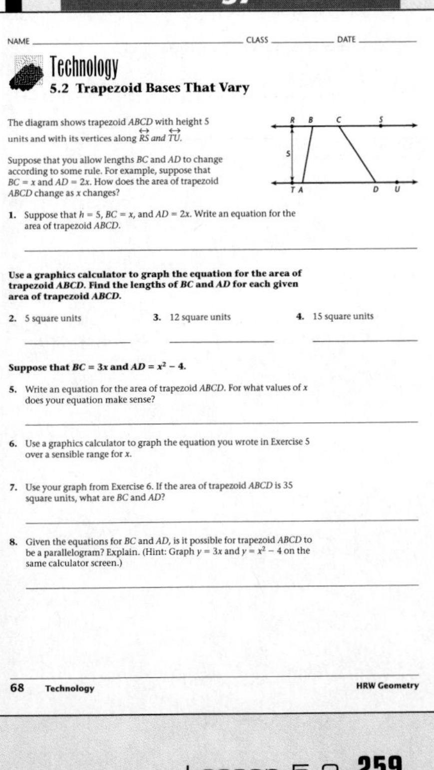

NAME ______ CLASS ______ DATE ______

Technology
5.2 Trapezoid Bases That Vary

The diagram shows trapezoid *ABCD* with height 5 units and with its vertices along $\overleftrightarrow{RS}$ and $\overleftrightarrow{TU}$.

Suppose that you allow lengths *BC* and *AD* to change according to some rule. For example, suppose that $BC = x$ and $AD = 2x$. How does the area of trapezoid *ABCD* change as x changes?

1. Suppose that $h = 5$, $BC = x$, and $AD = 2x$. Write an equation for the area of trapezoid *ABCD*.

Use a graphics calculator to graph the equation for the area of trapezoid *ABCD*. Find the lengths of *BC* and *AD* for each given area of trapezoid *ABCD*.

2. 5 square units
3. 12 square units
4. 15 square units

Suppose that $BC = 3x$ and $AD = x^2 - 4$.

5. Write an equation for the area of trapezoid *ABCD*. For what values of x does your equation make sense?

6. Use a graphics calculator to graph the equation you wrote in Exercise 5 over a sensible range for x.

7. Use your graph from Exercise 6. If the area of trapezoid *ABCD* is 35 square units, what are *BC* and *AD*?

8. Given the equations for *BC* and *AD*, is it possible for trapezoid *ABCD* to be a parallelogram? Explain. (Hint: Graph $y = 3x$ and $y = x^2 - 4$ on the same calculator screen.)

68 Technology HRW Geometry

30. 4,900,000; 112.4 acres

30. Farming In order to fertilize a field, a farmer needs to estimate its area. Estimate the area of the field indicated by the diagram. If an acre has 43,560 square feet, how many acres need fertilizer? **[Lesson 5.1]**

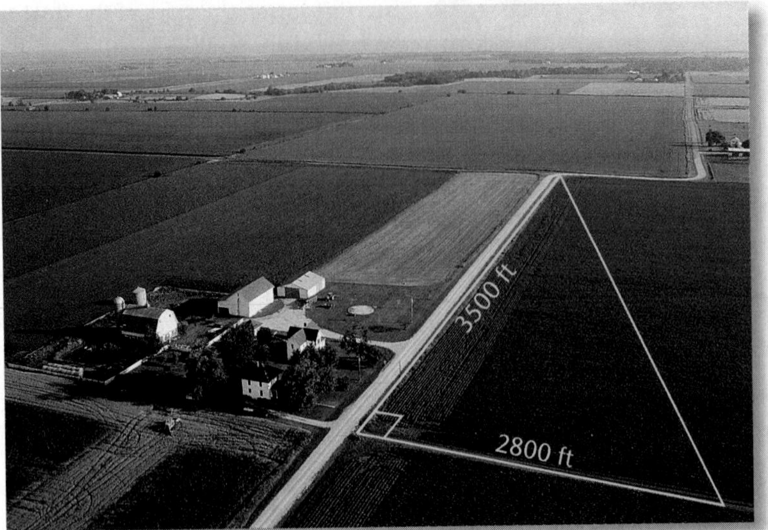

Look Beyond

In the past, Heron's Formula was used infrequently because of the tedious computation involved in its application. If students have programmable calculators, however, they can enter and store the formula. The advantage of using this formula is that it is not necessary to construct or compute the height.

Look Beyond

Cultural Connection: Africa There is a method for finding the area of a triangle using the lengths of the three sides. It is called Heron's Formula, named for a mathematician who lived in Alexandria in the first century. The formula is

$$A = \sqrt{s(s-a)(s-b)(s-c)}$$

where s is the semiperimeter—that is, half the perimeter—and a, b, and c are the lengths of the sides.

31. Find the semiperimeter of a triangle with sides 7 cm, 8 cm, and 9 cm. S = 12

32. Find the area of the triangle described in Exercise 31. 26.83 cm^2

33. Technology Set up a spreadsheet to find the area of a triangle when the lengths of the sides are given using Heron's Formula. Check student work.

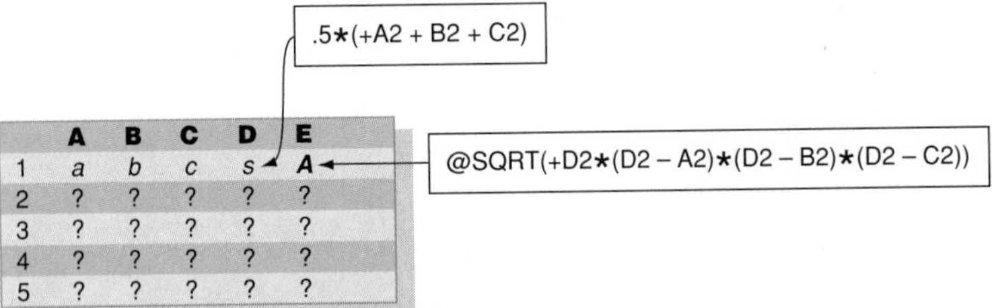

	A	B	C	D	E
1	a	b	c	s	A
2	?	?	?	?	?
3	?	?	?	?	?
4	?	?	?	?	?
5	?	?	?	?	?

LESSON 5.3

Exploring

Circumferences and Areas of Circles

Why *How do the areas of the larger craters of the moon compare with the area of your home town or state? A detailed photograph can be used to answer these questions. What else would you need to know?*

When you draw a circle with a compass, the distance from the point of the compass to the pencil or pen does not change. Therefore, every part of the circle you draw is the same distance from the **center of the circle**, the point where the compass is fixed. This leads us to the following definition.

CIRCLE

A **circle** is the figure that consists of all the points on a plane that are the same distance r from a given point known as the **center** of the circle. The distance r is the **radius** of the circle. The distance $d = 2r$ is known as the **diameter** of the circle. **5.3.1**

Technology Geometry graphics software can be used to create ways of approximating areas and circumferences of circles. For example, students might calculate the areas of an inner and outer square. They average these two areas and compare the result with that computed using π. Encourage students to use the software to create their own ways of approximating area and circumference.

PREPARE

Objectives

- Identify formulas for the circumference and area of a circle.
- Solve problems using the formulas for the circumference and area of a circle.

RESOURCES

- Practice Master 5.3
- Enrichment Master 5.3
- Technology Master 5.3
- Lesson Activity Master 5.3
- Quiz 5.3
- Spanish Resources 5.3

Assessing Prior Knowledge

Provide students with graph paper. Have each student draw a circle by using a compass or by tracing around a small, circular object.

1. Approximate the area of your circle in three different ways:
   a. Count square units.
   b. Square the radius and multiply by 3.
   c. Square the radius and multiply by 4.
2. Which of the three approximations is closest to the actual area? Justify your answer.

[Answers will vary.]

TEACH

From a knowledge of the diameter of the moon, students can estimate the diameter of the larger craters in the photograph. They will of course need the formula for the area of a circle.

Exploration 1 The Circumference of a Circle

You will need
String and ruler or
Tape measure
Various circular objects such as food cans, etc.

1. The distance around a circle is called its **circumference**. Measure the circumferences and diameters of several circular objects. Record the results in a table.

object	C	*d*	*ratio*: $\frac{C}{d}$
1. can	31.4	10	?
2. ?	?	?	?
3. ?	?	?	?

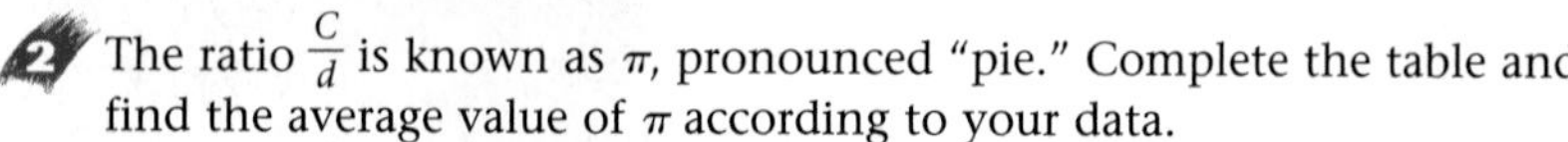

2. The ratio $\frac{C}{d}$ is known as π, pronounced "pie." Complete the table and find the average value of π according to your data.

3. Compare your result with the results of your classmates. How close is the result to 3.14, the approximate value of π? Use the π key on your calculator. What value does the calculator show?

ALGEBRA *Connection*

4. Write a formula for the circumference, C, of a circle. Begin by expressing π as a ratio. Then solve for C. Now write the formula in terms of the radius, r. ❖

Try This Use a calculator to find the circumferences of each of the following circles. Round your answers to two decimal places.

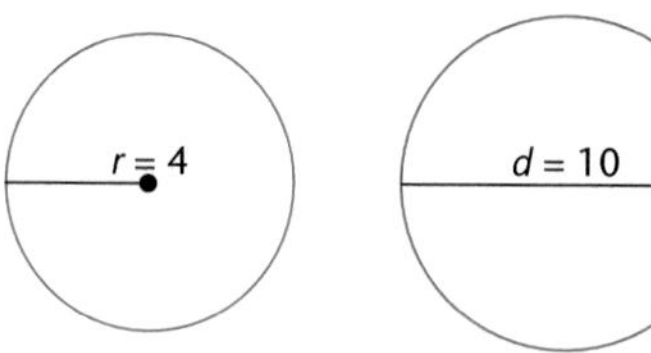

Exploration 2 The Area of a Circle

You will need
Ruler and compass
Protractor (optional)
Scissors

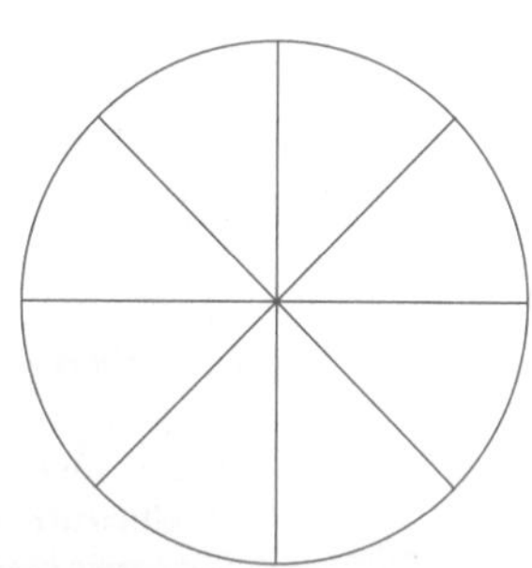

1. Draw a circle. Label its radius r. Using any method you like, such as paper folding, divide the circle into eight congruent pie-shaped parts, or **sectors**.

Cooperative Learning

Provide each group of students with one or two circular objects. Groups should exchange objects so that every group measures every object. Have students compare and discuss the measurements.

Exploration 1 Notes

The larger the number of measurements, the closer will be the approximation for π. After students have completed individual or group averages, compute a class average using all the collected data.

ongoing ASSESSMENT

4. $C = \pi d$; $C = 2\pi r$

Math Connection Algebra

For Step 4 in Exploration 1, students will need to convert the equation $\frac{C}{d} = \pi$ into $C = \pi d$. If necessary, remind them that the ratio $\frac{C}{d}$ can be written as the quotient $C \div d$. To solve $C \div d = \pi$ for C, students multiply both sides of this equation by d.

ongoing ASSESSMENT

Try This

a. $C = 2\pi(4) = 8\pi = 25.13$ units

b. $C = \pi(10) = 31.42$ units

Exploration 2 Notes

Students might imagine unrolling the eight segments of the circle, and then fitting the second four segments on top of the first four.

Use Transparency 44

interdisciplinary CONNECTION **History** Popular histories of mathematics almost always include stories of efforts to find approximations for the value of π. Have students use library materials to prepare a timeline showing the ever more precise values for π.

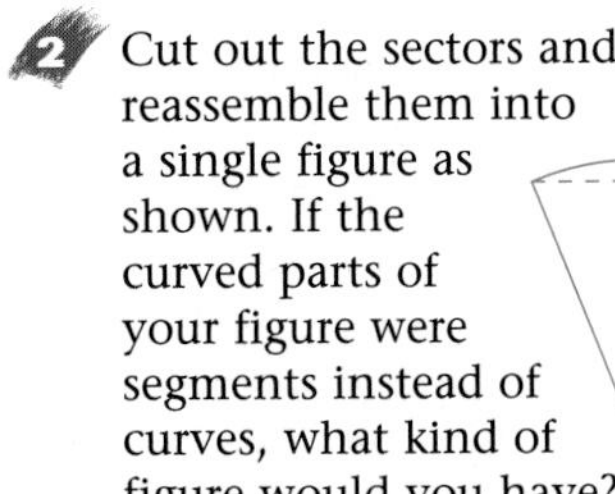

2. Cut out the sectors and reassemble them into a single figure as shown. If the curved parts of your figure were segments instead of curves, what kind of figure would you have?

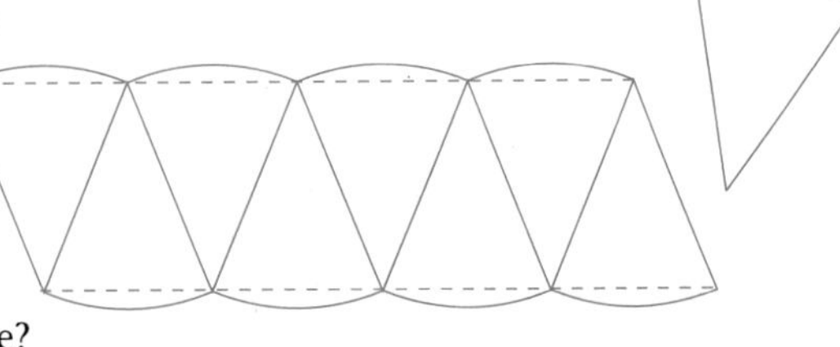

3. If you divided the circle into a larger number of congruent sectors, say 16, 32, or even more, would the curved parts seem more straight?

4. What geometric figures do your sectors in Step 3 resemble? The height of your assembled figure is approximately equal to r, the radius of the circle. What will happen to this approximation if the number of sectors increases infinitely?

5. The base of your assembled figure is equal to half the circumference of your original circle. Write an expression for the base of the figure in terms of π and the radius, r, of the original circle.

6. Write an expression for the area of the figure in terms of π and r. Should this be the formula for the area of a circle? Explain. ❖

CRITICAL *Thinking*

Why does the method you used become more realistic if you increase the number of sectors you cut?

Try This Use your calculator to find the areas of each of the following circles. Round your answers to two decimal places.

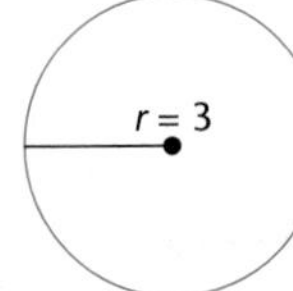

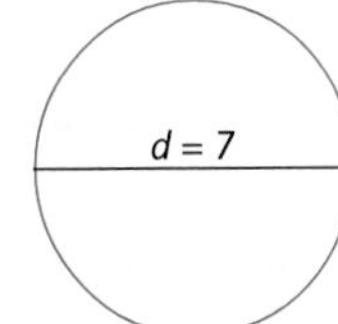

EXERCISES & PROBLEMS

Communicate

1. Suppose that you have 100 ft of fence to make a play area for your dog. Does a square yard or a circular yard provide the most area for your dog? Explain your answer.
2. When the cassette in the picture is rewinding, which point is moving faster? Explain your answer.

ONGOING ASSESSMENT

6. $A = \pi r^2$. Yes. As the number of sectors increases, the assembled sectors will more closely resemble a rectangle whose area is that of the circle.

CRITICAL *Thinking*

The curved parts become closer to straight lines; then straight lines more closely approximate the curved parts of the figure.

ONGOING ASSESSMENT

Try This

$r = 3$; 28.27 units^2

$d = 7$; 38.48 units^2

ASSESS

Selected Answers

Odd-numbered Exercises 5–29

Assignment Guide

Core 1–30

Core Plus 1–37

Technology

A calculator with a π key will be useful for the majority of these exercises. Geometry graphics software will be helpful in estimating or verifying answers for Exercises 23 and 24.

Error Analysis

Understanding and visualizing the problems in Exercises 13–17 may be difficult for many students. A bicycle wheel can be used to demonstrate what happens to various spots when the wheel is turned.

ENRICHMENT In the British Museum is a papyrus roll called the Rhind, or Ahmes Papyrus. Ahmes was the scribe who copied the work in about 1650 BCE. In Problem 50, Ahmes assumed that the area of a circular field with a diameter of 9 units is the same as the area of a square with a side of 8 units. Have students find the value of π implied by this assumption.

[Answer: $\pi(4.5)^2 = 8^2$; $\pi \approx 3.1605$]

INCLUSION strategies

English Language Development Have students write a short essay explaining the relationship among the radius, diameter, circumference, and area of a circle. They should first try to explain their ideas without using diagrams. They should then add drawings to clarify their work.

ALTERNATIVE ASSESSMENT

Portfolio Assessment

Students should explain in writing and with diagrams how to find each of the following:

1. The area of a quarter-circle if you know the radius **[Square the radius. Multiply by π. Divide by 4.]**
2. The distance around a quarter-circle if you know the radius **[Multiply the radius by π. Divide by 2.]**
3. The area of a circle if you know the circumference **[Divide the circumference by 2. Multiply by the radius.]**

Practice Master

NAME ______ CLASS ______ DATE ______

Practice & Apply
5.3 Exploring Circumferences and Areas of Circles

Use the π key on your calculator or use 3.14 for π. As needed, round answers to the nearest tenth.

1. If $r = 10$ cm, then the circumference = ______.
2. If $r = 10$ cm, then the area = ______.
3. If $d = 22$ cm, then the circumference = ______.
4. If $d = 22$ cm, then the area = ______.
5. If the area of a circle = 121π in.^{2}, then the radius = ______.
6. If the area of a circle = 121π in.^{2}, then the circumference = ______.
7. If the circumference = 42π cm, then the area = ______.
8. Phineas wants to circle the Earth in 80 days in his ship named *Phog*. If he plans to travel nonstop at a constant speed along the equator, at what speed should he travel? (Note: diameter of Earth ≈ 8000 miles)

For Exercises 9–10, find the area of the shaded region.

9.

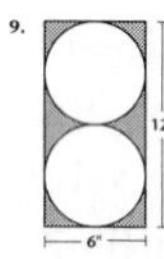

10.

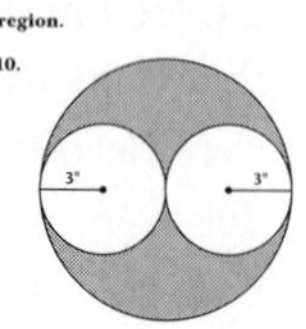

HRW Geometry — Practice & Apply 55

3. Sometimes π is approximated as $\frac{22}{7}$ and 3.14. Explain the differences between using one estimate or the other. Why is it necessary to use an estimate of π when calculating the circumference and area of a circle? What number does your calculator use for π?
4. Each of the figures below has an area of 9 sq units. Which figure has the smallest perimeter or circumference? Which figure has the largest perimeter or circumference? Explain your answers.

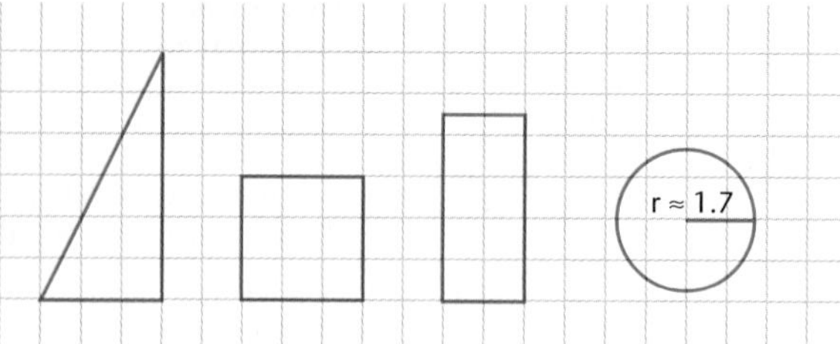

Practice & Apply

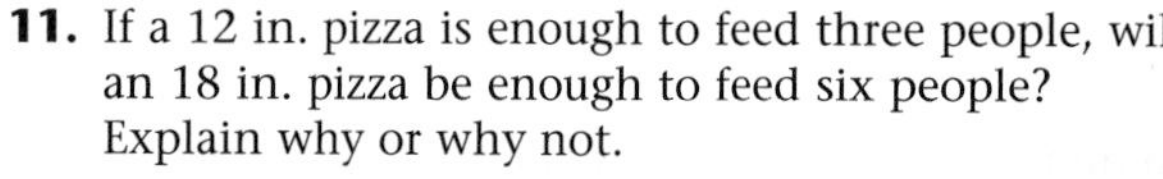

Algebra Find the circumference and area. Measurements are in inches. Use the π key on your calculator or use 3.14 for π. Round answers to the nearest tenth.

5. $r = 6$ $C = 37.7$ in.; $A = 113.1$ in.^{2}
6. $d = 4$ $C = 12.6$ in.; $A = 12.6$ in.^{2}
7. $d = 7$ $C = 22$ in.; $A = 38.5$ in.^{2}
8. If the area of a circle is 100π cm^2, find its radius and circumference. $r = 10$cm; $c = 62.83$cm
9. If the circumference of a circle is 50π m, find its area. 1963.5 m^2
10. Find the radius of a circle with area 221.7 m^2. 8.4m
11. If a 12 in. pizza is enough to feed three people, will an 18 in. pizza be enough to feed six people? Explain why or why not.
12. **Irrigation** In some parts of the world, farmers irrigate the land using a circle pattern. The picture shows sections of land irrigated in this way. If the area inside the square is one square mile, what is the area, in square feet, of the cultivated circle? 12,895,644 ft^2

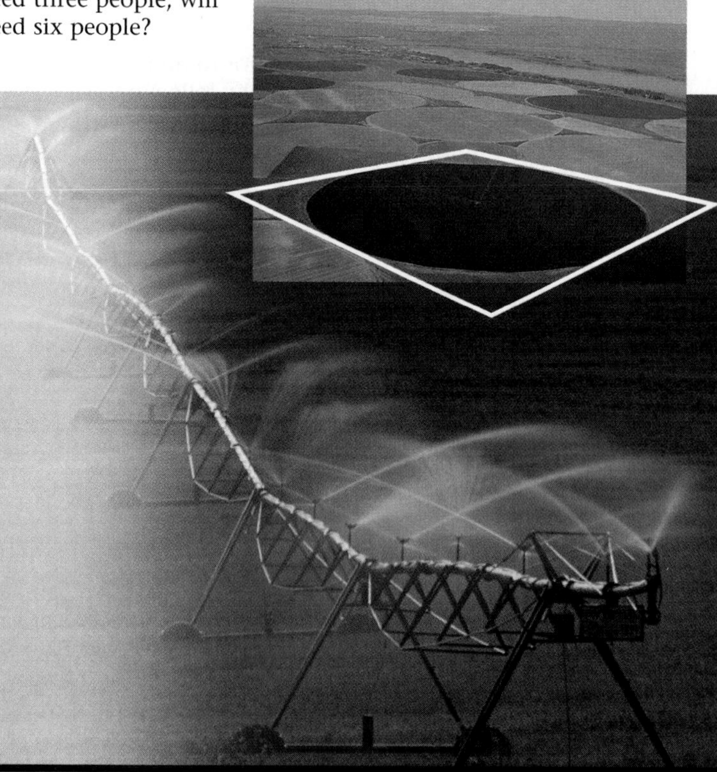

RETEACHING the lesson

Using Visual Models Provide students with graph paper. Each student should draw several circles on the paper and estimate each area by counting square units. Students should then compare their counts with areas computed using the formula.

11. Area of a pizza with $d = 12$ in.: 36p in^2
Area of a pizza with $d = 18$ in.: 81p in^2
The area of an 18 in pizza is more than twice the area of a 12 in pizza, so the 18 in pizza should feed six people.

Automobile Engineering Tires are tested for how well they stick to pavement by driving them around a circular track. Use the dimensions in the diagram below for Exercises 13–15.

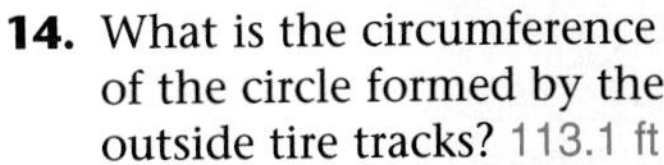

13. What is the circumference of the circle formed by the inside tire tracks? 78.5 ft

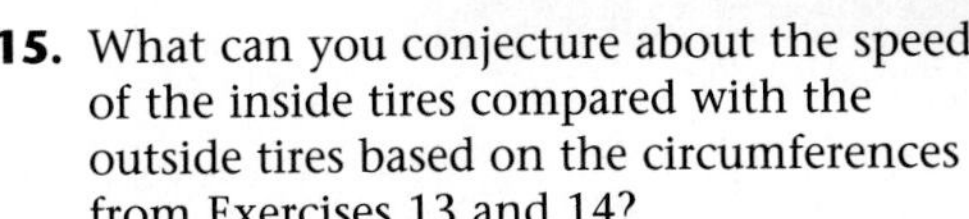

14. What is the circumference of the circle formed by the outside tire tracks? 113.1 ft

15. What can you conjecture about the speed of the inside tires compared with the outside tires based on the circumferences from Exercises 13 and 14?

Automobile Engineering Since the outside wheels of a car need to turn faster on a curve than the inside wheels, a car has a device called a "differential." The differential sends more power to the outside wheel, making it turn faster.

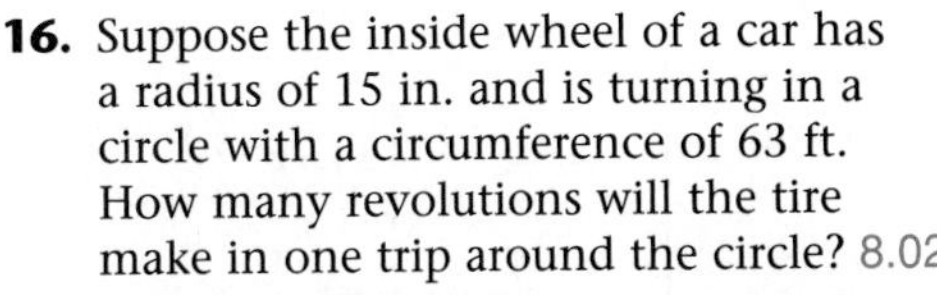

16. Suppose the inside wheel of a car has a radius of 15 in. and is turning in a circle with a circumference of 63 ft. How many revolutions will the tire make in one trip around the circle? 8.02

17. Suppose the outside wheel of the car in Exercise 16 has a radius of 15 in. and is turning in a circle with a circumference of 82 ft. How many revolutions will the tire make in one trip around the circle? 10.44

For Exercises 18–21, find the area of the shaded region.

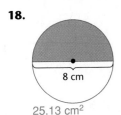

18. 8 cm — 25.13 cm^2

19. 8 in., 3 in. — 172.79 in.^{2}

20.

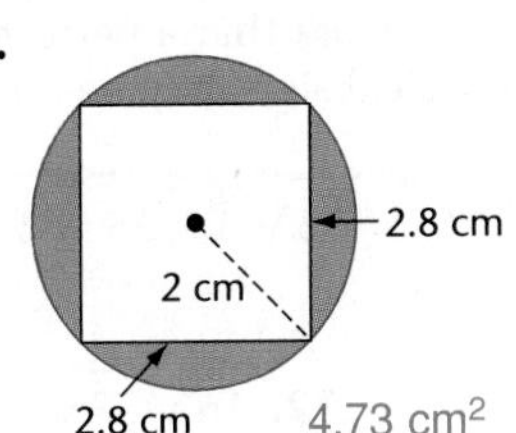

4.73 cm^2

21.

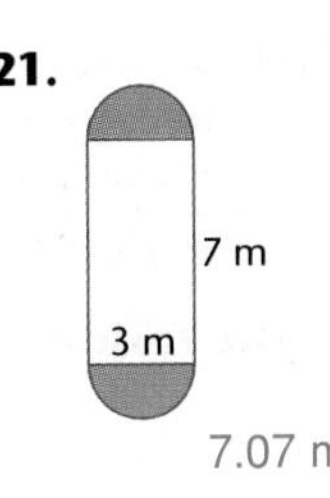

7.07 m^2

22. What happens to the area of a circle when the radius is doubled? The area is multiplied by 4.

23. **Maximum/Minimum** You have 50 m of fence to enclose a play area. Which would be a larger play area, a circular area or a square area? Explain your reasoning.

24. **Maximum/Minimum** Suppose a circle and a square have an area of 300 cm^2. Which has the greatest perimeter or circumference? Explain your reasoning.

15. The inside tires do not have to go as far, so they are not moving as fast.

23. Circular: $A = 199$ m^2
Square: $A = 156.25$ m^2
Circular area is greater.

24. Circle: $C = 61.4$ cm
Square: $P = 69.28$ cm
Square has the greater perimeter.

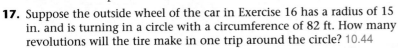

Technology Master

NAME ______ CLASS ______ DATE ______

Technology
5.3 Relating Circumferences and Areas of Circles

The diagram shows a set of *concentric circles*, circles having the same center. The innermost circle has radius 1. As you move outward, the radius of each circle increases by 1. The fourth circle out from the center, for example, has radius 4.

From the diagram, you can also see a set of annuli. An *annulus* is the region bounded by, or enclosed within, two concentric circles. For example, the space between the innermost and next circle out is an annulus.

To explore questions about the circumferences and areas of circles, you will find it helpful to use a spreadsheet. The needed formulas for a circle with radius r are shown.

Circumference $= 2\pi r$ Area $= \pi r^2$

1. Create a spreadsheet in which column A contains $r = 1, 2, 3, 4, 5$, column B contains the circumference of the circle with radius r, and column C contains the area of a circle with radius r.

2. Based on your spreadsheet from Exercise 1, what conjecture can you make about the ratio of the area of a circle with radius r and the circumference of a circle with radius r?

3. Extend your spreadsheet in Exercise 1 to include column D containing total circumference (sum of inner and outer circumferences) of each annulus and column E containing the area of each annulus in the diagram shown.

Find the total circumference and area of each annulus. (The first annulus is the innermost annulus.) Then find the ratio of the area of the annulus to the total circumference of it.

4. annulus 1 5. annulus 2 6. annulus 3 7. annulus 4

8. Based on your results from Exercises 4-7, make a conjecture about the ratio of the area of an annulus with inner radius r and outer radius $r + 1$ to the total circumference of the annulus.

HRW Geometry Technology 69

26. Statements

1. Parallelogram *ABCD* with diagonal $\overline{AC}$
2. $\overline{AB} \cong \overline{CD}$; $\overline{BC} \cong \overline{DA}$
3. $\angle ABC \cong \angle CDA$
4. $\overline{AC} \cong \overline{AC}$
5. $\triangle ABC \cong \triangle CDA$

Reasons

1. Given
2. Opposite sides of a parallelogram ≅.
3. Opposite angles in a parallelogram ≅.
4. Reflexive
5. SAS or SSS

Use Transparency 45

Look Beyond

Extension. The types of equations solved in Exercises 34–36 will be applied in Lesson 5.4 as students work with the "Pythagorean" Right-Triangle Theorem. The key to Exercise 37 is to realize that the figure is based on two smaller circles, each half the diameter of the larger circle.

25. Sports A basketball rim is 18 inches in diameter. What is the area encompassed by the basketball rim? 254.47 in.^{2}

Look Back

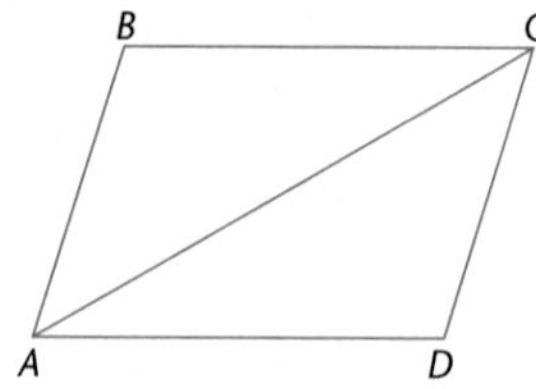

26. Given parallelogram *ABCD* with diagonal $\overline{AC}$, prove that $\triangle ABC \cong \triangle CDA$. **[Lesson 4.5]**

27. Find the area of a trapezoid with height 3 cm, and bases 6 cm and 5 cm. **[Lesson 5.2]**

Find the area of each figure below. [Lesson 5.2]

28.

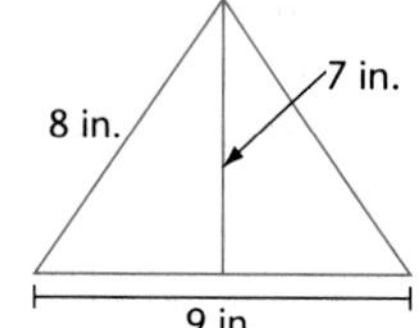

29.

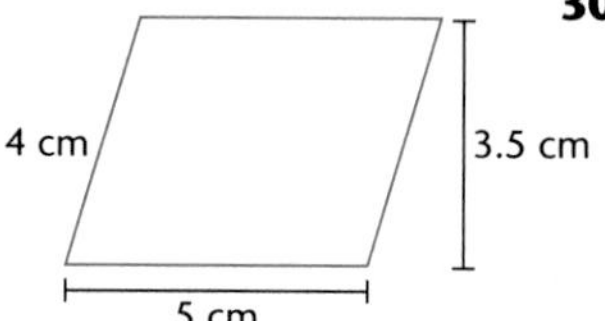

30.

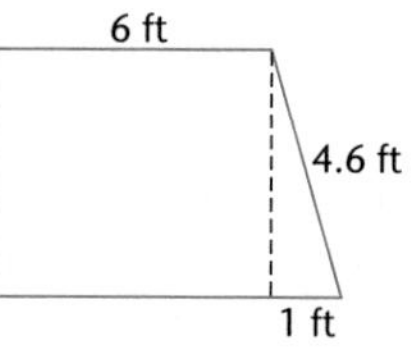

Look Beyond

 **Algebra** Recall from algebra that a perfect square that is a factor of a number under a radical sign may be removed by taking its square root. For example,

Simplify $2\sqrt{75}$ $\quad 2\sqrt{75} = 2\sqrt{3 \times 25} = 2 \times 5\sqrt{3} = 10\sqrt{3}$

Simplify the following.

31. $3\sqrt{75}$ $15\sqrt{3}$ **32.** $16\sqrt{32}$ $64\sqrt{2}$ **33.** $3\sqrt{500}$ $30\sqrt{5}$

Algebra Find the positive solution for x.

34. $x^2 + 16 = 25$ $x = 3$

35. $x^2 + 144 = 169$ $x = 5$

36. $x^2 + 12.25 = 13.69$ $x = 1.2$

37. Cultural Connection: Asia A yin-yang symbol consists of semicircles inside a circle, as shown. Which path from *A* to *B* is longer, the one through point *C* or the one through point *O*? Explain.

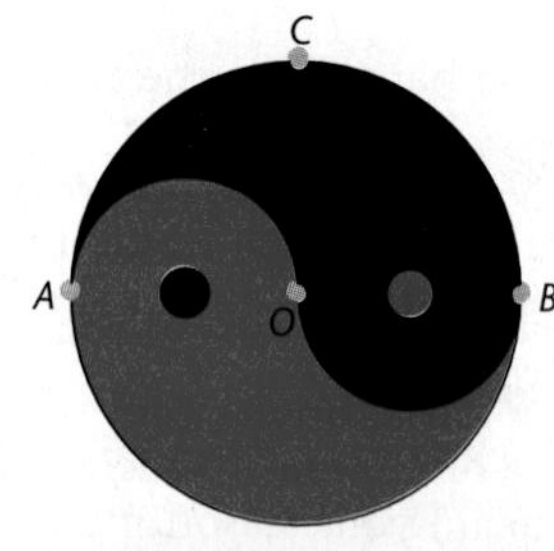

27. 16.5 units^2

28. 31.5 in^2

29. 17.5 cm^2

30. 29.2 ft^2

37. Let radius of smaller circle = r and radius of larger circle = $2r$. Distance through $O = 2\pi r$; Distance through $C = \dfrac{2\pi(2r)}{2} = 2\pi r$. The distances are the same.

LESSON 5.4 The "Pythagorean" Right-Triangle Theorem

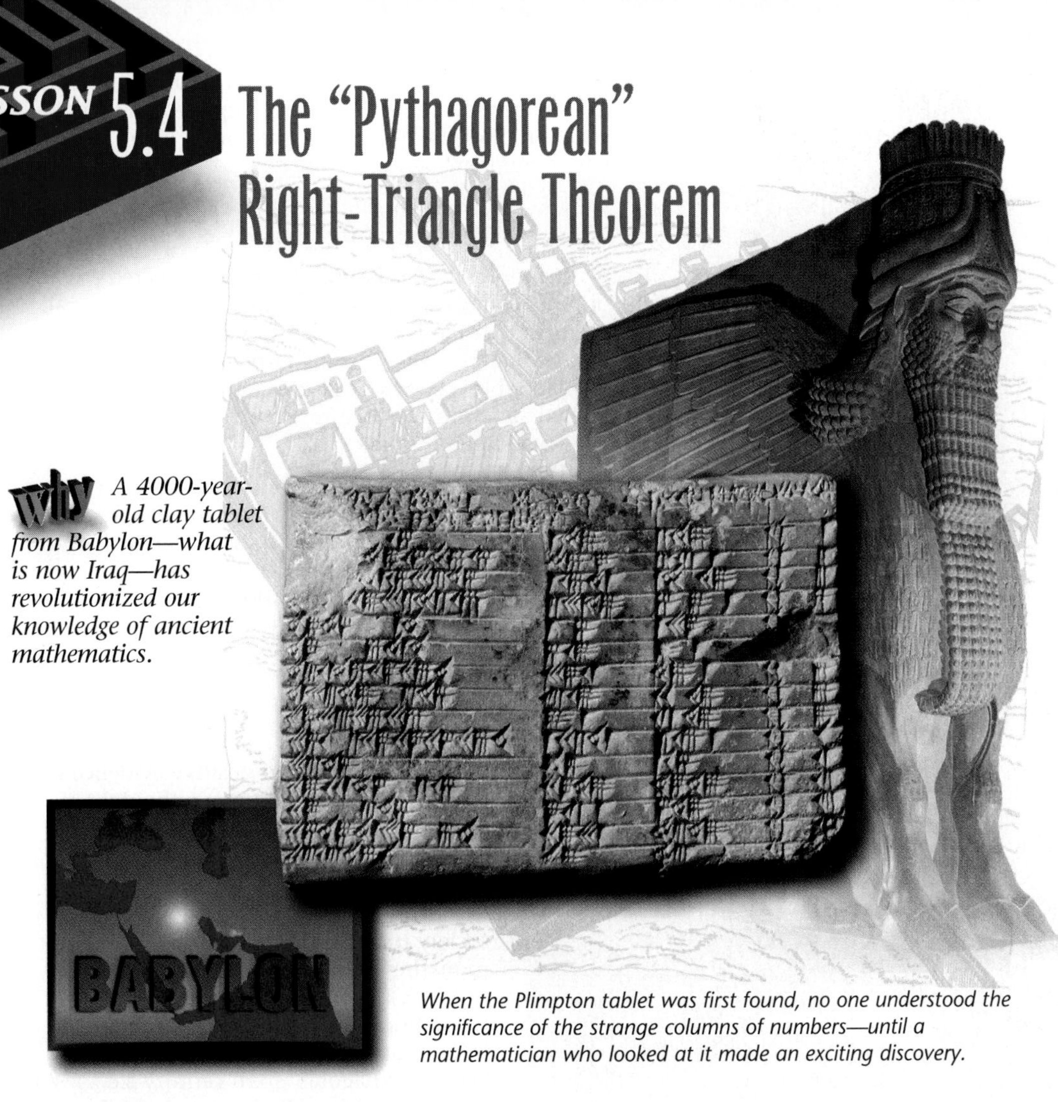

A 4000-year-old clay tablet from Babylon—what is now Iraq—has revolutionized our knowledge of ancient mathematics.

When the Plimpton tablet was first found, no one understood the significance of the strange columns of numbers—until a mathematician who looked at it made an exciting discovery.

Exploration *Solving the Puzzle*

You will need
A calculator

Cultural Connection: Asia The Babylonian tablet is a piece of a larger tablet. Part of it, including one column of numbers, has been broken off and lost. On the part that remains there are columns of numbers, including the two shown on the next page.

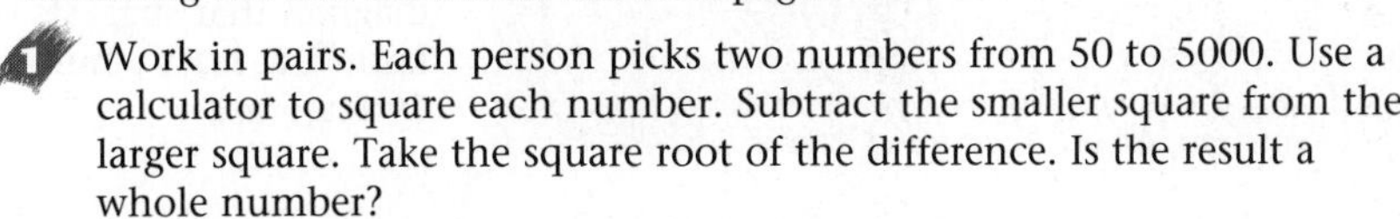

1. Work in pairs. Each person picks two numbers from 50 to 5000. Use a calculator to square each number. Subtract the smaller square from the larger square. Take the square root of the difference. Is the result a whole number?

2. Repeat Step 1 ten times. How often do you obtain a whole number?

PREPARE

Objectives

- Identify and apply the "Pythagorean" Right-Triangle Theorem and its converse.
- Solve problems using the "Pythagorean" Right-Triangle Theorem.

RESOURCES

- Practice Master 5.4
- Enrichment Master 5.4
- Technology Master 5.4
- Lesson Activity Master 5.4
- Quiz 5.4
- Spanish Resources 5.4

Assessing Prior Knowledge

Use a calculator to find each square root to two decimal places.

1. $\sqrt{72}$ [8.49]
2. $\sqrt{10}$ [3.16]
3. $\sqrt{120}$ [11.18]
4. $\sqrt{6^2 + 4^2}$ [7.21]
5. $\sqrt{14^2 - 5^2}$ [13.08]

TEACH

The now famous Plimpton tablet clearly reveals that the Babylonians knew the "Pythagorean" Right-Triangle over a thousand years before Pythagoras. Further study of the tablet (See Chapter 10 project) reveals that Babylonian mathematics was indeed very sophisticated.

ALTERNATIVE teaching strategy

Technology Geometry graphics software can be used to explore the "Pythagorean" Right-Triangle Theorem. Students should start with two perpendicular lines. By connecting various points on these lines they can generate many different right triangles. They should measure the three side lengths and confirm that the "Pythagorean" Right-Triangle Theorem is always true.

Exploration Notes

Exercises 19 and 20 give two different algebraic relationships for the three numbers in a Pythagorean triple. After students have completed the exploration, have them make conjectures about how the three numbers might be related.

ongoing ASSESSMENT

4. **Yes. It is unlikely that numbers with the property would be listed by chance.**

TEACHING tip

Students may be interested in researching recent efforts to prove "Fermat's Last Theorem." In this theorem, Fermat stated that there are no whole numbers x, y, and z such that $x^n + y^n = z^n$ for any n greater than 2. In particular, there are no whole numbers that satisfy $x^3 + y^3 = z^3$.

Cooperative Learning

Have students work in groups to draw the Chiu Chang diagram in varying sizes on graph paper. They should compute the areas of the five parts of each diagram, verifying that the "Pythagorean" Right-Triangle Theorem is true for a variety of right triangles. By discussing the work as a group, students will better internalize the way in which the formula relates to the diagram.

119	169
3367	4825
4601	6649
12709	18541
65	97
319	481
2291	3541
799	1249
481	769
4961	8161
45	75
1679	2929
161	289
1771	3229
56	106

Columns II and III of Plimpton 322

3. Square the numbers in each row of the tablet. Subtract the smaller square from the larger square and take the square root of the difference. How often is the result a whole number?

4. Do you think the Babylonians knew which whole numbers had this property? Explain. ❖

From your exploration you should have discovered that the original complete table was a list of what we now call "Pythagorean" triples—that is, sets of whole numbers a, b, and c such that $a^2 + b^2 = c^2$. But surprisingly, the Babylonian table was produced over a thousand years before Pythagoras, the person to whom the relationship has been attributed.

From these numbers, and from other evidence on the tablet, it is clear that the Babylonians knew that the sides of a right triangle had the property that we now know as the "Pythagorean" relationship.

Proving the Relationship

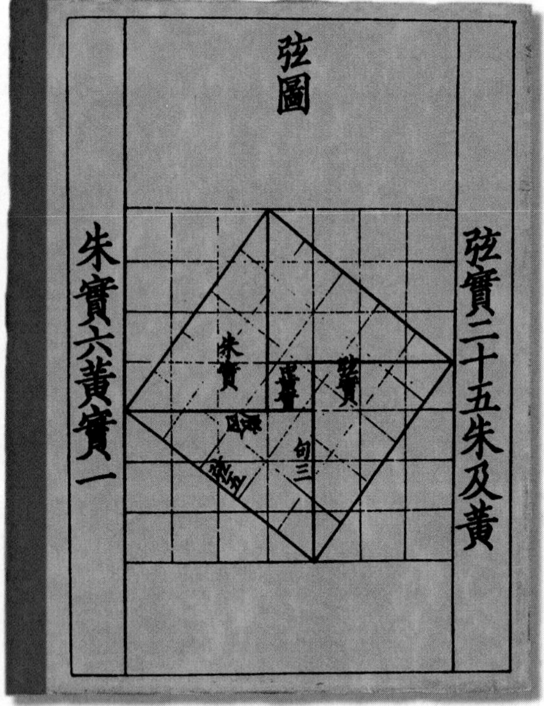

The area of the large tilted square is equal to the area of the 4 right triangles plus the area of the small square.

At present there is no positive evidence that the Babylonians could prove the right-triangle relationship.

Cultural Connection: Europe The Pythagoreans were members of a secret society in ancient Greece. They brought to mathematics a sense of reverence and mystery. In spite of their widespread reputation, there is very little factual information about the society and its leader, Pythagoras (sixth century B.C.E.). We do not know how Pythagoras actually proved the theorem that now bears his name, or whether his particular proof—and many are possible—was original with him.

Cultural Connection: Asia The earliest proof that we now know of is found in an early Chinese source, the *Chiu Chang*. No proof is actually given; however, there is a diagram that suggests that the proof was known at least a hundred years before Pythagoras.

interdisciplinary CONNECTION

Language Arts In science-fiction stories and movies, people looking for ways to prove to another life-form that humans are an intelligent species often use a diagram of the "Pythagorean" Right-Triangle Theorem. Lead a discussion in which students give reasons for or against using this theorem as a proof of intelligence. What other proofs might they use?

In the *Chiu Chang* diagram, four congruent right triangles have been assembled to form a large square with a smaller square in the center. The area of the larger square can be found by squaring the hypotenuse c of the right triangle. It can also be found by adding up the areas of the individual pieces of the figure. By setting these two equal to each other and simplifying, you obtain the famous result:

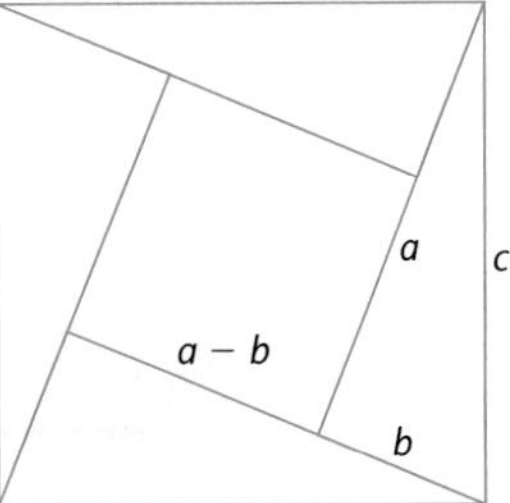

ALGEBRA
Connection

$$c^2 = 4(\tfrac{1}{2})ab + (a - b)^2$$
$$c^2 = 2ab + (a^2 - 2ab + b^2)$$
$$c^2 = 2ab + a^2 - 2ab + b^2$$
$$c^2 = a^2 + b^2$$

"PYTHAGOREAN" RIGHT-TRIANGLE THEOREM

For any right triangle, the square of the length of the hypotenuse is equal to the sum of the squares of the lengths of the legs.

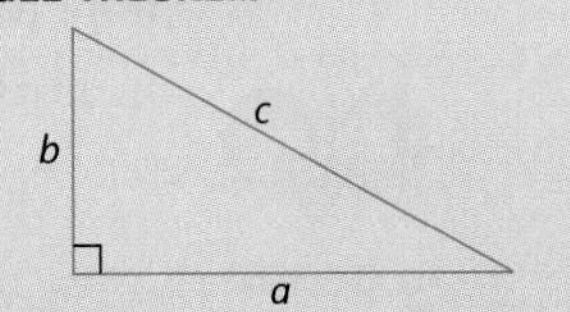

The **hypotenuse**, *c, is the side opposite the right angle.*

5.4.1

Does the shape of the right triangle matter in the proof above? How do you know that the large figure is in fact a square? What happens to the smaller square in the center if the right triangles are isosceles?

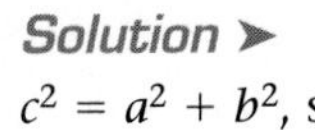

A plowed field forms a right triangle, with its hypotenuse along one road and one of its legs along another. If the legs have the lengths shown in the figure, find the length of the boundary along the roads.

Solution ➤

$c^2 = a^2 + b^2$, so

$c = \sqrt{a^2 + b^2} = \sqrt{1.7^2 + 3.4^2} = 3.8$ mi

$a + c = 1.7 + 3.8 = 5.5$ miles ❖

Math Connection Algebra

Check that students understand how the equation $c^2 = 4 \cdot \frac{1}{2}ab + (a - b)^2$ relates to the diagram. The area of one right triangle is $\frac{1}{2}ab$. The area of the square is $(a - b)^2$. Students may also need to be reminded of how to square an expression such as $(a - b)$. The procedure uses the distributive property: $(a - b)(a - b) = a(a - b) - b(a - b) = a^2 - ab - ba + b^2 = a^2 - 2ab + b^2$.

TEACHING tip

Technology Geometry graphics software may be used to explore the final question in the Critical Thinking question. Students will need to construct the figure in such a manner that they can enlarge and diminish the central square.

CRITICAL Thinking

No; all the sides are the same length. All the corner angles are made up of the two acute angles of the right triangles, so they must sum to 90°. ; The square disappears.

Alternate Example 1

A wooden bookcase 6 feet tall and 3 feet 6 inches wide needs to be braced with two diagonal cross braces. Find the length of each diagonal brace. [**$c = r(6^2 + 3.5^2) \approx 6.95$; about 7 ft**]

ENRICHMENT A text from the Old Babylonian period includes a diagram of a square and its diagonals. The number 30 is written along one side; the number 42.2535 appears along a diagonal. Explain the meanings of the numbers. [**42.2535 is an approximation of the length of the diagonal and shows that the text writer had some knowledge of the "Pythagorean" Right-Triangle Theorem. A calculator gives 42.4264 as a four-place value for the diagonal of a square with a side length of 30.**]

Hands-On Strategies The sides of the squares below can be arranged to form a right triangle and model a special case of the "Pythagorean" Right-Triangle Theorem. Students should look for ways to cut apart the two squares on the legs to show that their combined area equals the area of the square on the hypotenuse.

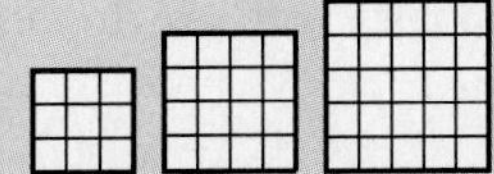

Emphasize that the relationship $a^2 + b^2 = c^2$ must hold *exactly* in order to prove that a triangle is right. If students are using rounded numbers (with or without a calculator) they may only be able to prove that a triangle is approximately right.

Math Connection Algebra

Given two real numbers, one of the following three situations must occur: they are equal, the first is less than the second, the first is greater than the second. This is called the Property of *Trichotomy*. A triangle must be right, obtuse, or acute. There is no fourth possibility.

Alternate Example 2

A triangle has sides 2.5 cm, 6 cm, and 6.5 cm. Is the triangle right, obtuse, or acute?

[$2.5^2 + 6^2 = 42.25$

$6.5^2 = 42.25$

$2.5^2 + 6^2 = 6.5^2$

The triangle is right.]

Alternate Example 3

A 10-foot long mirror leans against a wall. The top of the mirror is 8 feet 3 inches from the ground. How far is the base of the mirror from the wall? [$b = \sqrt{10^2 - 8.25^2} \approx 5.65$; about 5 ft 8 in.]

Using the Converse of the Theorem

The converse of the "Pythagorean" Right-Triangle Theorem is also true. It is useful in proving that two segments or lines are perpendicular.

THE "PYTHAGOREAN" RIGHT-TRIANGLE THEOREM: CONVERSE
If the square of the length of one side of a triangle equals the sum of the squares of the lengths of the other two sides, then the triangle is a right triangle. **5.4.2**

The figure at right shows the "Pythagorean" relationship. Two helpful inequalities can be derived from the relationship:

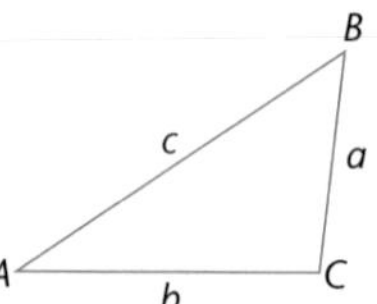

In any triangle, with c as its longest side,

If $c^2 = a^2 + b^2$, then $\triangle ABC$ is a right triangle.

If $c^2 > a^2 + b^2$, then $\triangle ABC$ is an obtuse triangle.

If $c^2 < a^2 + b^2$, then $\triangle ABC$ is an acute triangle.

EXAMPLE 2

A triangle has sides 7 inches, 8 inches, and 12 inches. Is the triangle right, obtuse, or acute?

Solution ➤

$12^2 \underline{\ ?\ } 7^2 + 8^2$

$144 > 113$. Therefore, the triangle is obtuse. ❖

EXAMPLE 3

If the bottom of a 15-foot ladder is 4 feet from the wall, how far up the wall does the ladder reach?

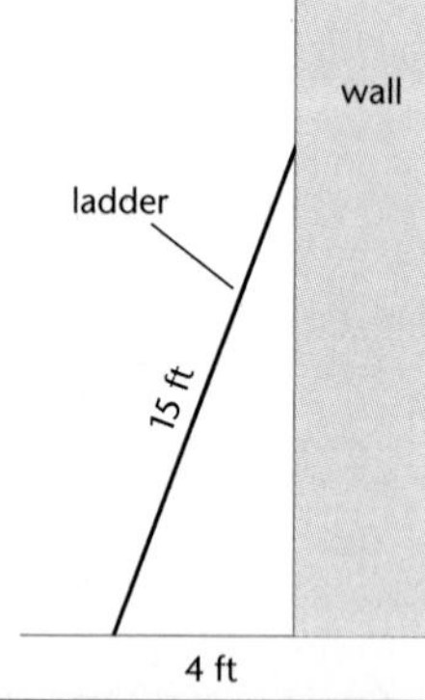

Solution ➤

$c^2 = a^2 + b^2$

$15^2 = 4^2 + b^2$

$225 = 16 + b^2$

$b^2 = 225 - 16$

$b = \sqrt{209} \approx 14.46$ ft ❖

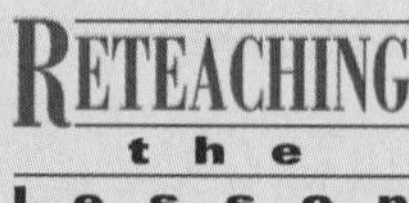

Cooperative Learning Have students work in pairs or small groups. They should use spreadsheet software to create their own list of Pythagorean triples. Either of the formulas given in Exercises 19 and 20 may be used to generate the data. Students should use graph paper or geometry graphics software to draw triangles that match the triples.

EXERCISES & PROBLEMS

Communicate

1. State the "Pythagorean" Right-Triangle Theorem in your own words.
2. Explain some of the practical uses of the "Pythagorean" Right-Triangle Theorem.
3. Explain how something called the "3-4-5 rule" would help carpenters square up corners.
4. What is the measure of the hypotenuse in the triangle at right?

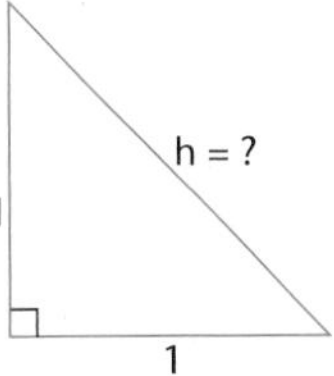

Practice & Apply

Algebra For Exercises 5–9, two lengths of sides of a right triangle are given. Find the missing length; a = leg 1, b = leg 2, and c = hypotenuse. Leave answers in radical form.

5. $a = 3$, $b = 4$, $c =$ __?__ 5
6. $a = 10$, $b = 15$, $c =$ __?__ $5\sqrt{13}$
7. $a = 46$, $b = 73$, $c =$ __?__ $\sqrt{7445}$
8. $a =$ __?__ , $b = 6$, $c = 8$ $2\sqrt{7}$
9. $a = 27$, $b =$ __?__ , $c = 53$ $4\sqrt{130}$

10. What is the length of a diagonal of a square whose sides measure 5 cm? 7.07 cm or $5\sqrt{2}$
11. The diagonal of a square measures 16 cm. Find its area. 128 cm^2

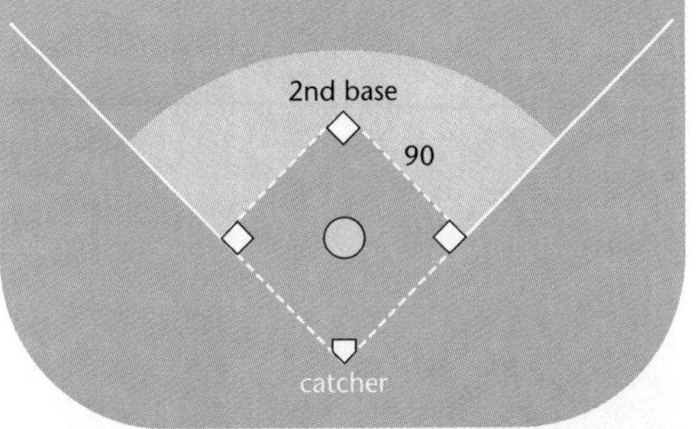

12. **Sports** A baseball diamond is a square with 90-foot sides. What is the approximate distance the catcher must throw from home to second base? 127.3 ft

Algebra Each of the following triples represents the sides of a triangle. Determine whether each triangle is right, obtuse, acute, or not a triangle. Give a reason for each answer.

13. 5, 9, 12
14. 13, 15, 17
15. 7, 24, 25
16. 7, 24, 26
17. 3, 4, 5
18. 25, 25, 30

13. $122 > 52 + 92$; Obtuse
14. $17^2 < 13^2 + 15^2$; Acute
15. $25^2 = 24^2 + 7^2$; Right
16. $26^2 > 24^2 + 7^2$; Obtuse
17. $3^2 + 4^2 = 5^2$; Right
18. $30^2 < 25^2 + 25^2$; Acute

ASSESS

Selected Answers

Odd-numbered Exercises 5–35

Assignment Guide

Core 1–20, 24–25, 30–36

Core Plus 1–4, 7–17, 19–38

Technology

Most of these exercises require a calculator with a square root key. Students who are fairly proficient with geometry graphics software can reproduce the more complicated diagrams used with Exercises 21 and 24–26.

Error Analysis

Students using calculators may inadvertently use the wrong order of procedures. For example, they may think they are entering $a^2 + b^2$, but instead may enter $(a^2 + b)^2$. Or, they may think they are computing $\sqrt{a^2 + b^2}$ but instead may calculate $\sqrt{a^2} + b^2$. You may wish to have students record intermediate steps in pencil to prevent these types of errors.

alternative ASSESSMENT

Performance Assessment

Have students create a diagram, on graph paper or using geometry graphics software, in which an isosceles right triangle is built on each of the three sides of a right triangle. Each side of the central right triangle should be one leg of an isosceles right triangle. Students should work in small groups to prove that the sum of the areas of the two smaller triangles is equal to the area of the triangle on the hypotenuse.

The answers to Exercises 19 and 20 can be found in Additional Answers beginning on page 727.

Practice Master

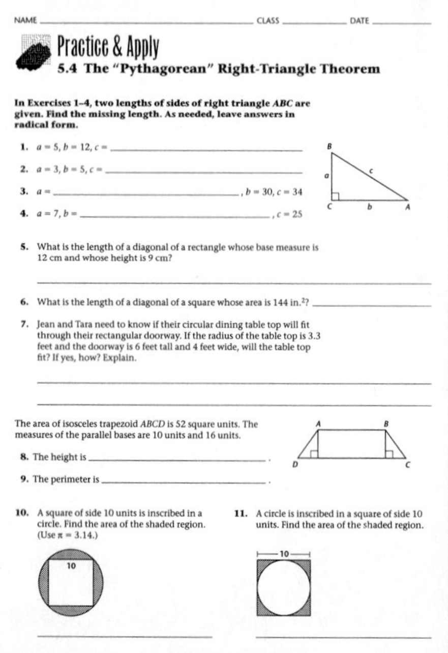
NAME ______ CLASS ______ DATE ______

Practice & Apply

5.4 The "Pythagorean" Right-Triangle Theorem

In Exercises 1–4, two lengths of sides of right triangle *ABC* are given. Find the missing length. As needed, leave answers in radical form.

1. $a = 5, b = 12, c =$ ______
2. $a = 3, b = 5, c =$ ______
3. $a =$ ______ $, b = 30, c = 34$
4. $a = 7, b =$ ______ $, c = 25$

5. What is the length of a diagonal of a rectangle whose base measure is 12 cm and whose height is 9 cm? ______

6. What is the length of a diagonal of a square whose area is 144 in.^{2}? ______

7. Jean and Tara need to know if their circular dining table top will fit through their rectangular doorway. If the radius of the table top is 3.3 feet and the doorway is 6 feet tall and 4 feet wide, will the table top fit? If yes, how? Explain. ______

The area of isosceles trapezoid *ABCD* is 52 square units. The measures of the parallel bases are 10 units and 16 units.

8. The height is ______.
9. The perimeter is ______.

10. A square of side 10 units is inscribed in a circle. Find the area of the shaded region. (Use $\pi = 3.14$.) ______

11. A circle is inscribed in a square of side 10 units. Find the area of the shaded region. ______

56 Practice & Apply HRW Geometry

Algebra Mathematicians have been fascinated with trying to generate "Pythagorean" triples. Two methods that generate sets of "Pythagorean" triples are shown below. Use each method to generate five sets of triples. Then use algebra to prove that the method will always work.

19. Method of Pythagoreans (m is an odd number greater than 1)

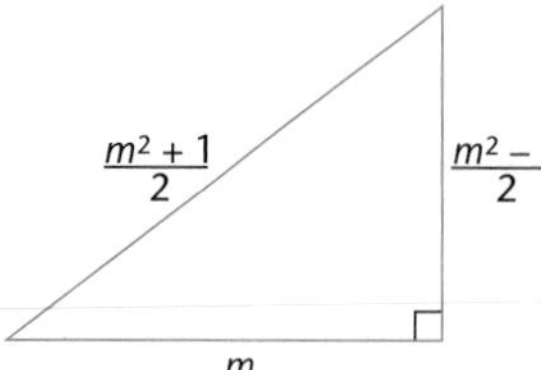

20. Method of Plato (m is any whole number)

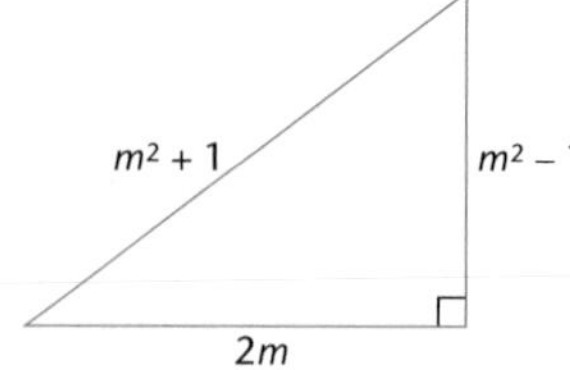

21. Cultural Connection: Asia The *Sulbastutras* books of India, written around 800 B.C.E., demonstrate a method for constructing a square whose area is equal to the sum of the areas of two given squares. If the two given squares are *ABCD* and *EFGH*, and if $\overline{AR}$ is constructed so that $AR = EH$, explain why *BRST* is the desired square.

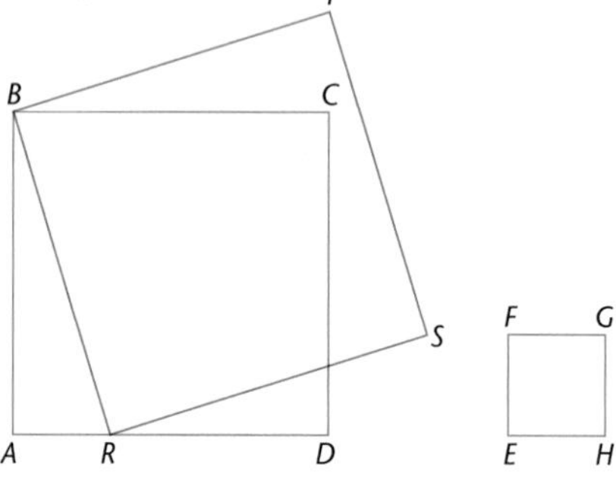

22. Use a method similar to the one in Exercise 21 to construct a square with double the area of a given square.

23. How high on the wall will the ladder reach? Assume that the base of the ladder is 8 feet off the ground. 43.6 ft

21. The area of the constructed square is, and is the area of *EFGH* and is the area of *ABCD*.

22. Use a diagonal of the square as the side of the bigger triangle.

The area of *BDEF* is twice the area of *ABCD*.

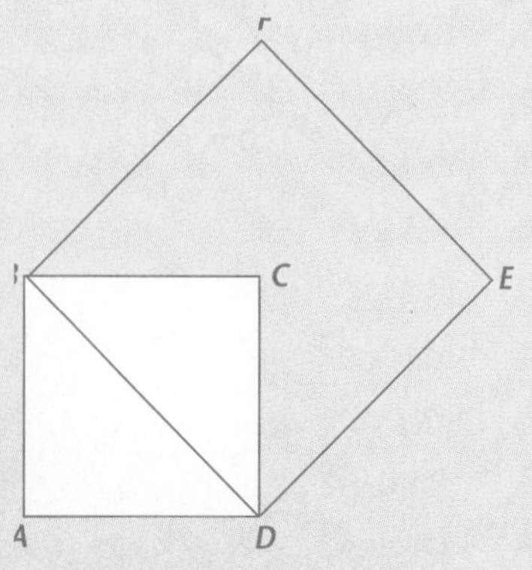

Mathematicians have provided many possible proofs of the "Pythagorean" Right-Triangle Theorem. Three are presented below.

24. History President Garfield, the twentieth president of the United States, devised a proof of the "Pythagorean" Right-Triangle Theorem using a trapezoid. Using the figure, find the area of the trapezoid in two different ways. Set your two expressions for the area equal to each other, and you will discover his proof.

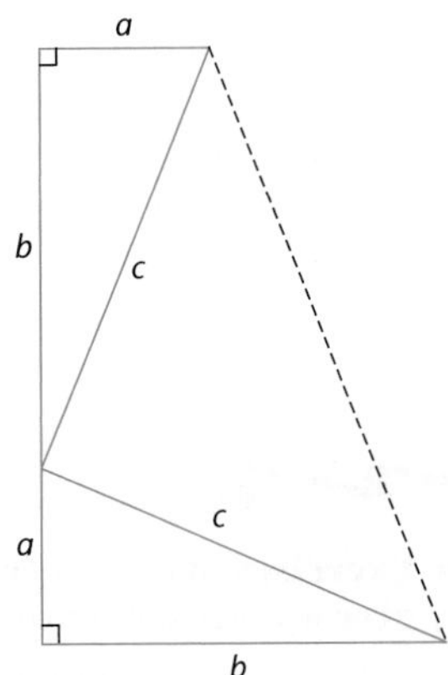

25. Portfolio Activity This is a proof of the "Pythagorean" Right-Triangle Theorem represented pictorially. Explain how the figures prove the theorem. You may wish to sketch the squares on a separate piece of paper and cut them into pieces to help explain the proof.

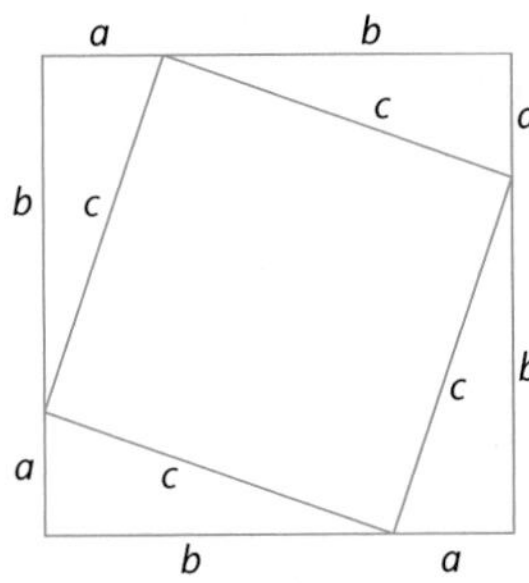

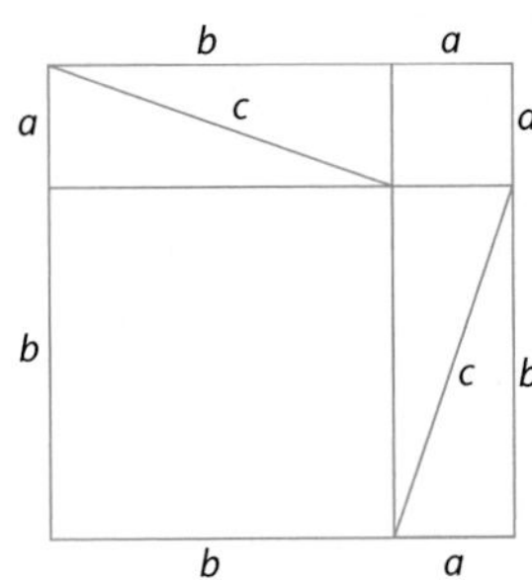

26. Algebra The diagram from the *Chiu Chang* suggest at least two different proofs of the "Pythagorean" Right-Triangle Theorem.

a. Find the equation for the area of square *EFGH* by subtracting the outer triangles of square *ABCD*:
$c^2 = (a + b)^2 - 2ab$. Why?

b. Find the equation for the area of square *EFGH* by adding the four inner right triangles to square *IJKL*.
$c^2 = (a - b)^2 + 2ab$. Why?

Simplify the equation in each part to obtain the desired results.

Why do you think the gridwork in the *Chiu Chang* was chosen with the small square in the center equal to one square unit?

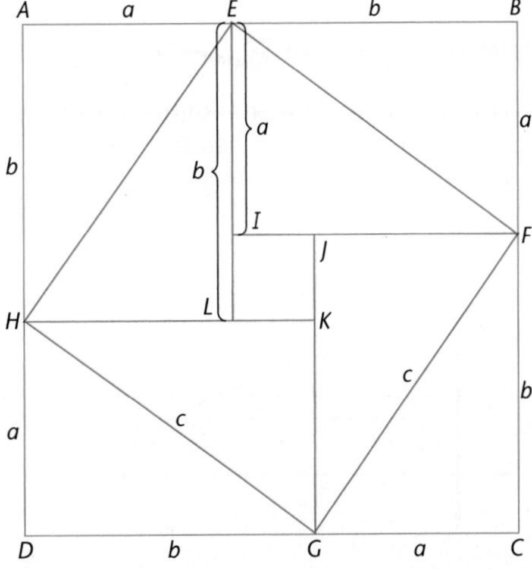

The answer to Exercise 26 can be found in Additional Answers beginning on page 727.

24. Using the formula for the area of a trapezoid: $A = \frac{(a+b)(a+b)}{2} = \frac{a^2 = 2ab = b^2}{2} = \frac{a^2}{2} + ab + \frac{b^2}{2}$. Finding the sum of the areas of the three right triangles in the trapezoid: $A = \frac{ab}{2} + \frac{ab}{2} + \frac{c^2}{2} = ab + \frac{c^2}{2}$. Setting the areas equal: $\frac{a^2}{2} + ab + \frac{b^2}{2} = ab + \frac{c^2}{2} \Rightarrow \frac{a^2}{2} + \frac{b^2}{2} = \frac{c^2}{2} \Rightarrow a^2 + b^2 = c^2$

25. The inner square has area $= c^2$. The area of the inner square is the area of the outer square minus the area of the 4 triangles. The area of a square with sides $a + b = (a+b)^2 = a^2 + 2ab + b^2$. Each triangle has area $\frac{1}{2}ab$, so the 4 triangles have total area of $2ab$. So the inner square has area $= a^2 + 2ab + b^2 - 2ab = a^2 + b^2 = c^2$.

Technology Master

NAME ______ CLASS ______ DATE ______

Technology
5.4 Pythagorean Triples

You know from your work with right triangles that there is a relationship between the lengths of the legs and the length of the hypotenuse, known as the Pythagorean Theorem.

$$a^2 + b^2 = c^2$$

Suppose that you look at the equation shown on its own terms, that is, without a connection to right triangles. Once you do that and focus on *a*, *b*, and *c* being integers, you can study the problem of finding Pythagorean triples. A *Pythagorean triple* is a set of integers $\{a, b, c\}$ that satisfy $a^2 + b^2 = c^2$.

Use paper and pencil to show that each triple is a Pythagorean triple.

1. $a = 3, b = 4, c = 5$
2. $a = 6, b = 8, c = 10$
3. $a = 5, b = 12, c = 13$

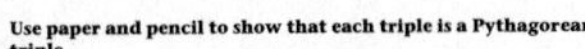

In Exercises 4–7, use a spreadsheet to analyze each statement. Use column A for the values of *k*, column B for the values of *a*, column C for the values of *b*, and column D for the values of *c*. Calculate the values of $a^2 + b^2$ in column E and c^2 in column F.

4. The set $\{6k, 8k, 10k\}$ is a Pythagorean triple for all positive integers *k*.
5. The set $\{5k, 12k, 13k\}$ is a Pythagorean triple for all positive integers *k*.
6. The set $\{3 + k, 4 + k, 5 + k\}$ is a Pythagorean triple if $k = 0$ and not a Pythagorean triple otherwise.
7. The set $\{3(2^k), 4(2^k), 5(2^k)\}$ is a Pythagorean triple for all positive integers *k*.
8. Suppose that $\{a, b, c\}$ is a Pythagorean triple. Give some examples to show that $\{3a + 4b, 4a - 3b, 5c\}$ is also a Pythagorean triple.
9. Use a spreadsheet to determine whether there is a right triangle that has one leg 7 units long. Justify your response.

70 Technology HRW Geometry

Find the area of the shaded region for each problem below. Round answers to the nearest tenth.

27.

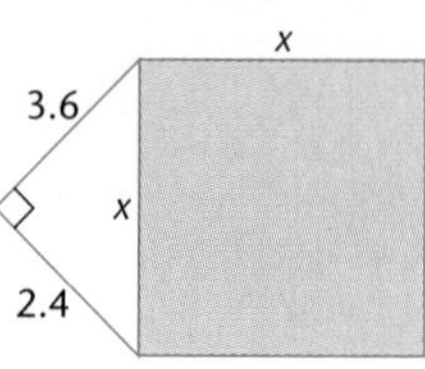

18.7 units²

28.

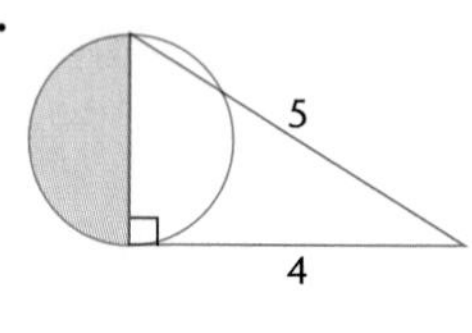

3.5 units²

29.

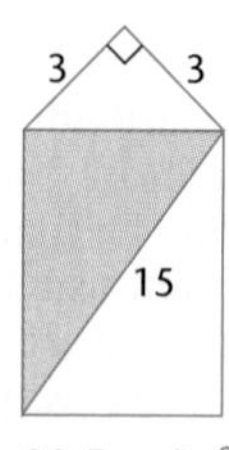

30.5 units²

Look Back

Answer Exercises 30–33 as true or false. If true, explain why; if false, give a counterexample. [Lesson 3.2]

30. Every rhombus is a rectangle.

31. Every rhombus is a parallelogram.

32. If a diagonal divides a quadrilateral into two congruent triangles, then the quadrilateral is a parallelogram.

33. The sum of the measures of the angles of a quadrilateral is 360°.

30. False. Consider a rhombus with angles 50° and 130°.

31. True. A rhombus is defined to be a parallelogram with four congruent sides.

32. False. Consider a kite.

33. True. The sum of the areas of the angles of a quadrilateral is (4 – 2)(180) = 360°.

Algebra Recall from algebra that a fraction with a radical in the denominator can be simplified by multiplying by a fraction equal to 1. For example,

Simplify $\frac{\sqrt{20}}{\sqrt{3}}$ $\qquad \frac{\sqrt{20}}{\sqrt{3}} = \frac{\sqrt{20} \times \sqrt{3}}{\sqrt{3} \times \sqrt{3}} = \frac{\sqrt{60}}{3} = \frac{2\sqrt{15}}{3}$

Simplify the following.

34. $\frac{\sqrt{30}}{\sqrt{5}}$ $\sqrt{6}$

35. $\frac{\sqrt{72}}{\sqrt{6}}$ $2\sqrt{3}$

36. $\frac{\sqrt{6}}{\sqrt{6}}$ 1

Look Beyond

Look Beyond

These exercises foreshadow the work with the Distance Formula in Lesson 5.6. Many students may find it easier to first work with a triangle that has its side lengths labeled.

37. Use the "Pythagorean" Right-Triangle Theorem to find the distance between points B and C. 9.85 units

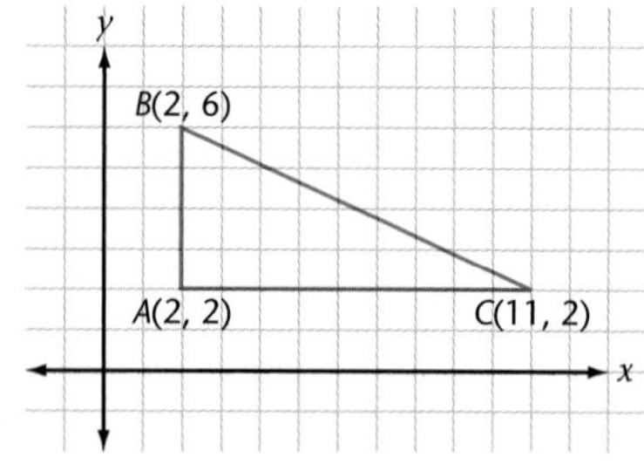

38. Based on the result from Exercise 37, devise a general formula for a distance between two points on a coordinate plane. Consider the distance between points K and I in Figure 2.
$d = \sqrt{(x_2 - x_1)^2 + (y_2 - y_1)^2}$

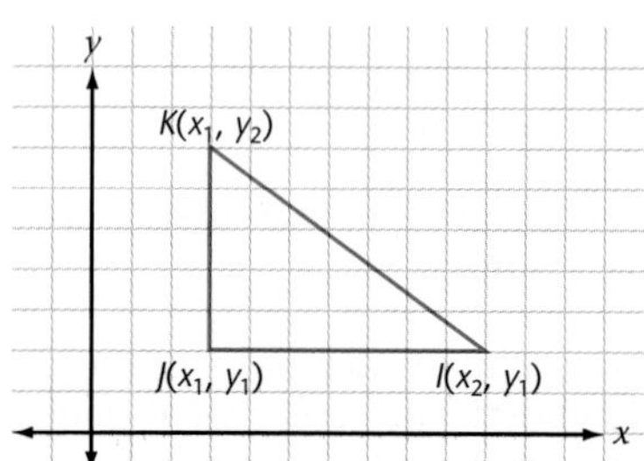

LESSON 5.5 Special Triangles, Areas of Regular Polygons

why *The traditional tools of mechanical drawing include a T square or parallel bar and two special triangles. One triangle has angles measuring 30°, 60°, and 90°. The other has angles measuring 45°, 45°, and 90°. The properties of these triangles make them especially useful in geometry as well as in drawing.*

45-45-90 Right Triangles

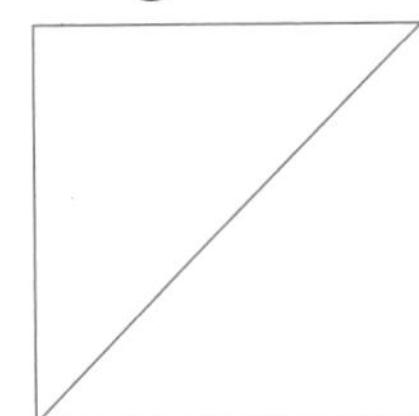

If you draw a diagonal of a square, two congruent isosceles triangles are formed. Since the diagonal is the hypotenuse of a right triangle, its length can be found by using the "Pythagorean" Right-Triangle Theorem.

EXAMPLE 1

Use the "Pythagorean" Right-Triangle Theorem to find the length of the hypotenuse of triangle $\triangle ABC$. What is the ratio of the hypotenuse to the side?

ALGEBRA Connection

Solution ➤

$h^2 = 10^2 + 10^2$

$h = \sqrt{200} = \sqrt{100} \times \sqrt{2} = 10\sqrt{2}$

The ratio of the hypotenuse to the side is $\frac{10\sqrt{2}}{10}$, or $\sqrt{2}$. ❖

CRITICAL Thinking

What is the length of the diagonal of a square with side s? What is the ratio of the diagonal to the side?

ALTERNATIVE teaching strategy

Using Technology Underlying the ideas in the lesson is the fact that all 45-45-90 triangles are similar, and that all 30-60-90 triangles are similar. Have students use geometry graphics software to create sets of similar triangles to be a visual reminder that the ratios are the same, no matter what the sizes of the triangles. Geometry graphics software may also be used to confirm the Area of a Regular Polygon Theorem for a wide variety of cases.

PREPARE

Objectives

- Identify and use the 45-45-90 Right-Triangle Theorem and the 30-60-90 Right-Triangle Theorem.
- Identify and use the Area of a Regular Polygon Theorem.

RESOURCES

- Practice Master 5.5
- Enrichment Master 5.5
- Technology Master 5.5
- Lesson Activity Master 5.5
- Quiz 5.5
- Spanish Resources 5.5

Assessing Prior Knowledge

Simplify each radical expression by writing the values for a and b.

1. $\sqrt{32} = \sqrt{a} \cdot \sqrt{2} = b\sqrt{2}$

   $[a = 16, b = 4]$

2. $\sqrt{54} = \sqrt{a} \cdot \sqrt{6} = b\sqrt{6}$

   $[a = 9, b = 3]$

3. $\sqrt{45} = \sqrt{a} \cdot \sqrt{5} = b\sqrt{5}$

   $[a = 9, b = 3]$

TEACH

Students will be able to find many examples of 45-45-90 and 30-60-90 triangles used in tessellation patterns. Have them collect pictures of patterns that show these triangles.

Alternate Example 1

A right isosceles triangle has a hypotenuse of 45 cm. Find the lengths of the other two sides of the triangle. $[\frac{\sqrt{2}}{1} = \frac{45}{x}, \sqrt{2}x = \frac{45}{x} = 45 \div \sqrt{2} \approx 31.8 \text{ cm}]$

Throughout this lesson you may wish to let students leave answers in the form $a\sqrt{b}$. If so, remind them that the simplest form of such expressions requires that (1) b have no factors that are perfect squares and (2) the denominator, if any, does not include a radical expression.

Math Connection
Algebra

(page 275) Point out that $\sqrt{2}$ is an exact number. In the same way that the ratio C/d in a circle is exactly π, the ratio of the hypotenuse to the side in an isosceles right triangle is exactly $\sqrt{2}$.

Alternate Example 2

A 30-60-90 triangle has a hypotenuse of 20 in. Find the lengths of the other two sides.

[shorter side = 20 ÷ 2 = 10 in., longer side = $\sqrt{20^2 - 10^2} = \sqrt{300}$ = $10\sqrt{3}$, ≈ 17.32 in.]

Math Connection
Algebra

Remind students of how to write $\sqrt{75}$ in simplest form. They should look for factors that are perfect squares. In this case, 75 = 25 · 3.

So, $\sqrt{75} = \sqrt{25 \cdot 3} = \sqrt{5^2 \cdot 3} = \sqrt{5} \cdot 3 = 5\sqrt{3}$.

Notice that the diagonal of the square forms a right triangle with two 45° base angles. This triangle is known as a 45-45-90 right triangle. Thus, the hypotenuse can be found by applying the "Pythagorean" Right-Triangle Theorem.

45-45-90 RIGHT-TRIANGLE THEOREM

In any 45-45-90 right triangle, the length of the hypotenuse is $\sqrt{2}$ times the length of a leg. **5.5.1**

30-60-90 Right Triangles

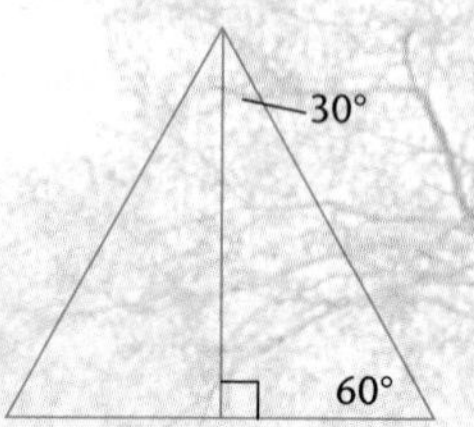

If you draw the altitude of an equilateral triangle, two congruent right triangles are formed. The acute angles of each right triangle measure 30° and 60°. These triangles are known as 30-60-90 right triangles. The length of the hypotenuse of each right triangle is two times the length of the shorter leg. (Why?)

EXAMPLE 2

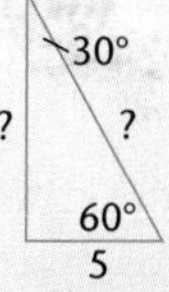

Find the missing lengths for the 30-60-90 right triangle to the right.

Solution ➤

In a 30-60-90 right triangle, the length of the hypotenuse is two times the length of the shorter leg. The length of the hypotenuse is 10.

Use the "Pythagorean" Right-Triangle Theorem to find the other leg.

$$5^2 + x^2 = 10^2$$
$$25 + x^2 = 100$$
$$x^2 = 100 - 25$$
$$x^2 = 75$$
$$x = \sqrt{75} = 5\sqrt{3} \approx 8.66$$ ❖

ENRICHMENT Interesting tessellations can be made using either 45-45-90 or 30-60-90 triangles. Have students work in small groups. They first design a black-and-white tessellation pattern. Then they can discuss how to use colors and patterns to enhance their designs.

Hands-On Strategies Some students may need to draw and measure examples of 45-45-90 triangles and 30-60-90 triangles to convince themselves that the ratios developed in the text always remain true.

Exploration 30-60-90 Right Triangles

You will need
No special tools

1. Use the "Pythagorean" Right-Triangle Theorem to fill in a table for 30-60-90 right triangles.

Shorter leg	Hypotenuse	Longer leg
1	2	?
2	?	?
3	?	?

2. Look for a pattern in the table regarding the longer leg and make a generalization.

3. If the length of the shorter leg is x:

   a. What is the length of the hypotenuse?

   b. What is the length of the longer leg?

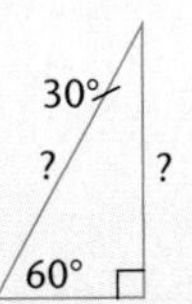

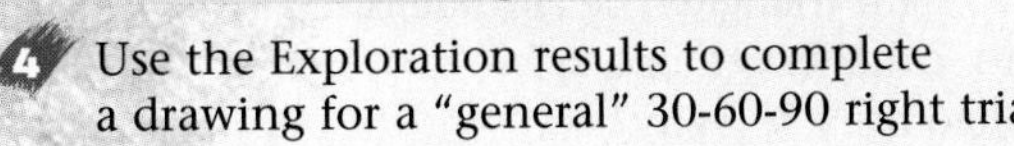

4. Use the Exploration results to complete a drawing for a "general" 30-60-90 right triangle. ❖

30-60-90 RIGHT-TRIANGLE THEOREM

In any 30-60-90 right triangle, the length of the hypotenuse is 2 times the length of the shorter leg, and the longer leg is $\sqrt{3}$ times the length of the shorter leg. **5.5.2**

APPLICATION

Measurement Jake is measuring the height of a tall tree his grandfather planted as a boy. Jake uses a special instrument to find a spot where a 30° angle is formed by the ground and a line to the top of the tree. How tall is the tree if Jake is 80 feet from the base of the tree?

The line of sight is the hypotenuse of a 30-60-90 right triangle. The longer leg of the triangle is 80 feet. The height of the tree represents the shorter leg. If x is the height of the tree:

$$x\sqrt{3} = 80$$

$$x = \frac{80}{\sqrt{3}} \approx 46.18 \text{ feet.}$$ ❖

Cooperative Learning Have students work in pairs. One student will construct a 45-45-90 triangle with legs that are 10 cm long. The other student will construct a 30-60-90 triangle with a 10-inch hypotenuse. Students should compute the lengths of the other sides and then measure to check both their drawings and their computation.

Exploration Notes

The data for the table should be computed using the "Pythagorean" Right-Triangle Theorem. However, you may also wish to have students draw a triangle for each row of the table. They can check their drawing precision by first making the two legs and then measuring the hypotenuse to see how close it comes to being twice the length of the shorter leg.

3. a. $2x$ b. $x\sqrt{3}$

ongoing ASSESSMENT

Try This

a. $x = \sqrt{3}$, b. $x = 2$, c. $x = 2\sqrt{3}$

Math Connection Algebra

Students should be careful when setting up the ratios in the Try This. Sketching a model triangle showing the ratios should help them set up the equations properly.

Cooperative Learning

Patterns involving sets of similar triangles can be created by successively drawing altitudes from the right angle in 45-45-90 and 30-60-90 triangles.

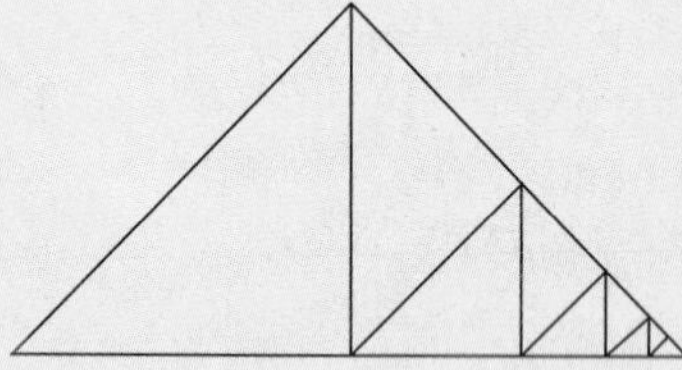

Have students work together to explore variations of these patterns. The patterns can be created for their aesthetic appeal alone. Or you can ask students to start with a large triangle of a specific size, for example, a 10-inch leg or hypotenuse. Students should then find the dimensions of all the line segments in the figure.

Alternate Example 3

Find the area of a regular hexagon with sides that are 10 inches long. [$150\sqrt{3}$ sq in.]

Note that the apothem of a regular polygon is also the radius of the inscribed circle.

Try This Find the missing measure x.

a.

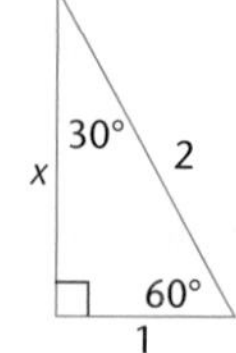

b.

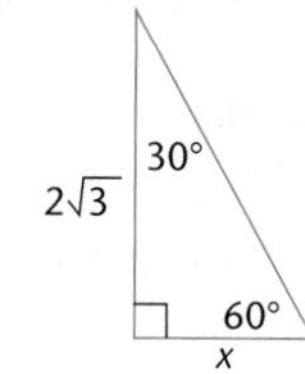

c.

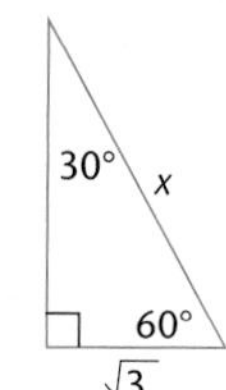

Areas of Regular Polygons

To find the area of a regular hexagon, divide the hexagon into six congruent nonoverlapping equilateral triangles. Find the area of one triangle and multiply by 6 for the area of the hexagon. Note that the altitude of one equilateral triangle is the longer leg of a 30-60-90 right triangle. The altitude is equal to $\frac{1}{2}$ the length of a side multiplied by $\sqrt{3}$. (Why?)

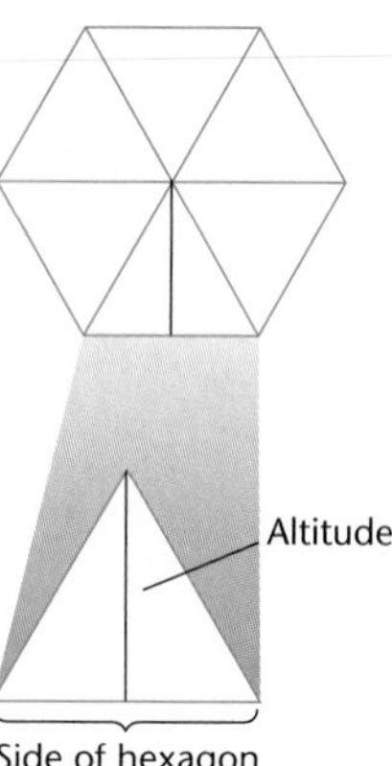

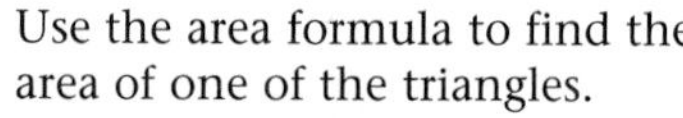

Find the area of a regular hexagon with sides of length 20 cm.

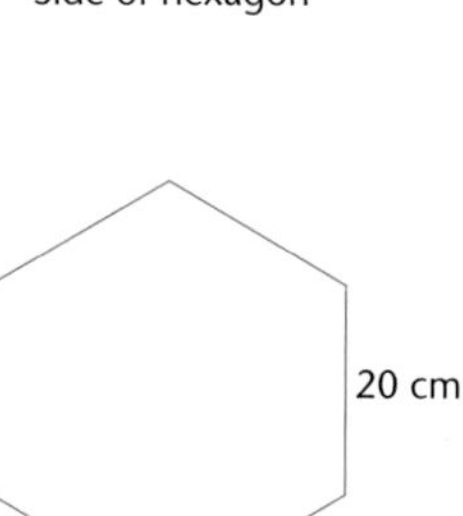

Solution ➤

Divide the hexagon into 6 equilateral triangles. Since an altitude of one of the triangles forms the longer leg of a 30-60-90 right triangle, and half of the side of the hexagon forms the shorter leg, the length of the altitude is $10\sqrt{3}$.

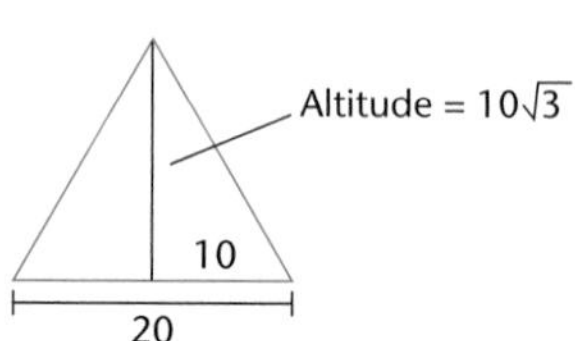

Use the area formula to find the area of one of the triangles.

$$A = \frac{1}{2}(20)(10\sqrt{3})$$
$$A = 100\sqrt{3}$$

Since the hexagon is composed of 6 congruent triangles, the area of the hexagon is

$$6(100\sqrt{3}) = 600\sqrt{3}.$$ ❖

The steps described in Example 3 can be applied to find the area of any regular polygon.

An n-sided polygon can be divided into n nonoverlapping congruent triangles. The altitude of each triangle is the segment from the center of the polygon to the midpoint of the side. In a regular polygon, this segment is called the apothem.

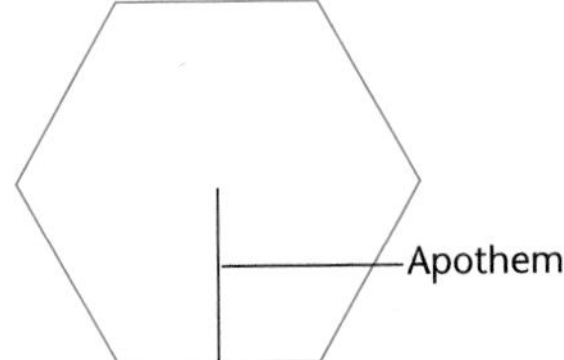

AREA OF A REGULAR POLYGON

The area of a regular polygon with apothem a and perimeter P is $A = \frac{1}{2}ap$, where a is the apothem and p is the perimeter. **5.5.3**

Try This Find the area of a regular polygon with a perimeter of 40 inches and an apothem of 5 inches.

Communicate

1. The leg of a 45-45-90 triangle is 6 centimeters long. Find the length of the hypotenuse and explain how you found it.
2. In a 30-60-90 triangle, the shorter leg has length 4 inches. Explain how to find the length of the longer leg and the length of the hypotenuse.
3. How do you find the length of the apothem of a regular polygon?
4. Explain how to find the area of a regular polygon?
5. What are the missing values in the chart for a 45-45-90 right triangle?

Leg 1	Leg 2	Hypotenuse
1	1	?
2	2	?
3	3	?
4	4	?
5	5	?

ongoing ASSESSMENT

Try This

$A = \frac{1}{2}(5)(40) = \mathbf{100\ in^2}$

ASSESS

Selected Answers

Odd-numbered Exercises 7–43

Assignment Guide

Core 1–44

Core Plus 1–5, 6–20 even, 21–46

Technology

Geometry graphics software may be used to construct the polygons in Exercises 23–29. It can also be used to explore the Portfolio Activity problem in Exercises 30–35.

Error Analysis

It is easy for students to incorrectly draw 30-60-90 triangles with one leg twice the length of the other leg. This triangle has acute angles of 63.43° and 26.56°, so it looks very much like a 30-60-90 triangle. Emphasize that the 30-60-90 triangle is one-half of an equilateral triangle. The altitude became the longer leg and thus has a length that is not a whole number.

alternative ASSESSMENT

Performance Assessment

1. What is the ratio of the apothem a of a regular hexagon to its side length s? $\left[\frac{a}{s} = \frac{\sqrt{3}}{2}\right]$
2. Use the results of Part 1 to find a formula for the area of a regular hexagon in terms of its side length s.
   $$\left[A = \frac{1}{2}ap = \frac{1}{2} \cdot \frac{\sqrt{3}s}{2} \cdot 6s = \frac{3s^2}{2}\sqrt{3}\right]$$

Practice Master

NAME ______ CLASS ______ DATE ______

Practice & Apply

5.5 Special Triangles, Areas of Regular Polygons

Exercises 1–3, refer to right $\triangle ABC$ shown. Using the given length, find the other two lengths.

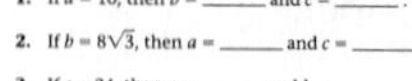
1. If $a = 10$, then $b =$ ______ and $c =$ ______.
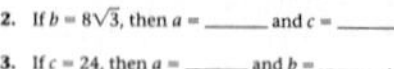
2. If $b = 8\sqrt{3}$, then $a =$ ______ and $c =$ ______.
3. If $c = 24$, then $a =$ ______ and $b =$ ______.

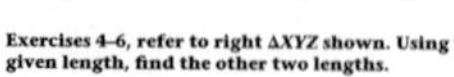
Exercises 4–6, refer to right $\triangle XYZ$ shown. Using the given length, find the other two lengths.

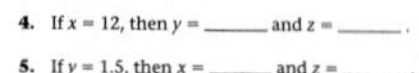
4. If $x = 12$, then $y =$ ______ and $z =$ ______.
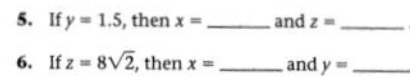
5. If $y = 1.5$, then $x =$ ______ and $z =$ ______.
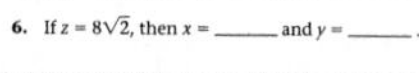
6. If $z = 8\sqrt{2}$, then $x =$ ______ and $y =$ ______.

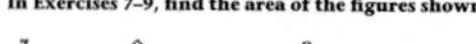
In Exercises 7–9, find the area of the figures shown.

7. 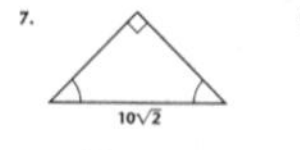

8. 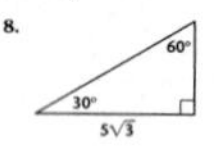

9. 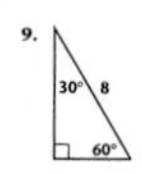

10. Find the height of an equilateral triangle whose side measures 18 cm.

11. Find the length of a side of an equilateral triangle whose height is $4\sqrt{3}$ cm.

HRW Geometry — Practice & Apply 57

Practice & Apply

Find the area of each figure.

6. 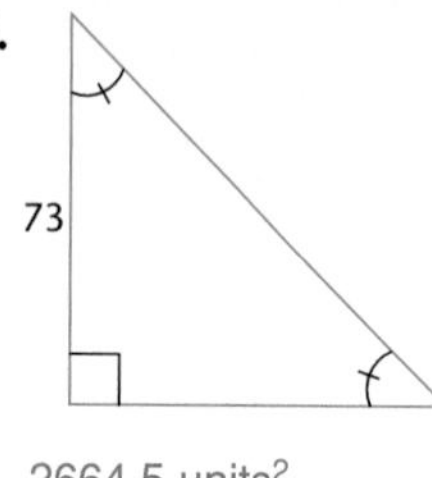

2664.5 units^2

7. 

25.25 units^2

8. 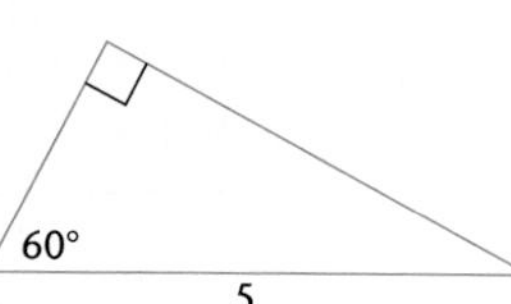

5.41 units^2

For the given length, find each of the remaining two lengths.

9. $x = 6$
10. $y = 6$
11. $z = 14$
12. $y = 4\sqrt{3}$

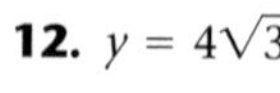

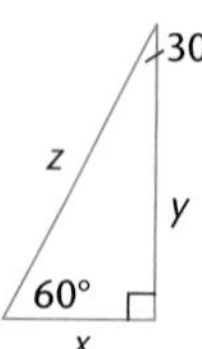

13. $p = 6$
14. $r = 6$
15. $q = 4\sqrt{2}$
16. $r = 10$

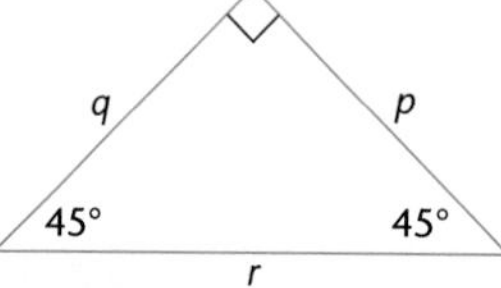

17. $k = 3.4$
18. $k = 6\sqrt{3}$
19. $g = 17$
20. $h = 2\sqrt{3}$

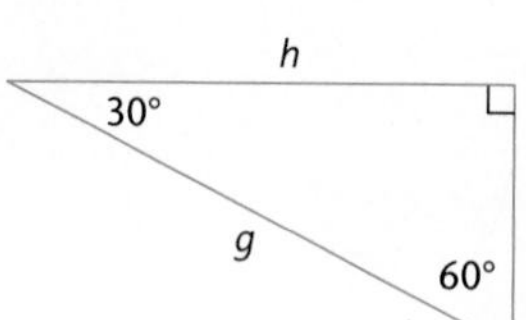

Civil Engineering An engineer is in charge of attaching guy wires to a tower.

21. One set of wires needs to extend at a 45° angle to the tower at a point on the ground 80 feet from the tower. What is the length of each wire? 113.14 ft
22. Another wire needs to make a 30° angle with the tower from a point on the ground, 80 feet from the tower. How high up the tower does the wire need to be? 138.56 ft

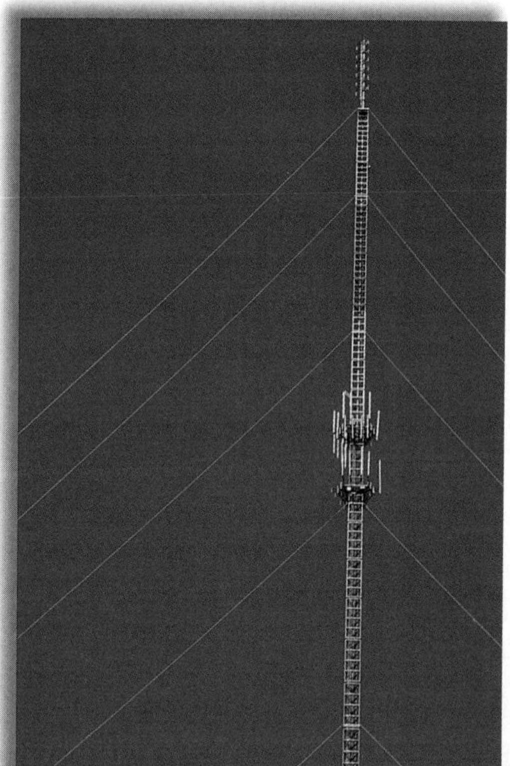

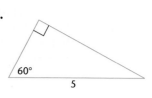

9. $y = 6\sqrt{3}$; $z = 12$
10. $x = 2\sqrt{3}$; $z = 4\sqrt{3}$
11. $x = 7$; $y = 7\sqrt{3}$
12. $x = 4$; $z = 8$
13. $q = 6$; $r = 6\sqrt{2}$
14. $p = q = 3\sqrt{2}$
15. $p = 4\sqrt{2}$; $r = 8$
16. $p = q = 5\sqrt{2}$
17. $h = 3.4\sqrt{3}$; $g = 6.8$
18. $h = 18$; $g = 12\sqrt{3}$
19. $k = 8.5$; $h = 8.5\sqrt{3}$
20. $k = 2$; $g = 4$

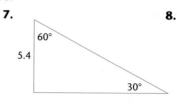

23. Find the area of a square if its perimeter is 16 cm and its apothem is 2 cm. 16 cm^2

24. Use a protractor and ruler to construct a regular hexagon with sides 2 cm. Measure the apothem with your ruler and find the area of a hexagon. $a = \sqrt{3}$ cm; $A = 10.39$ cm^2

25. Use a protractor and ruler to construct a hexagon with sides 6 cm. Measure the apothem with your ruler and find the area of the hexagon. $a = 5.2$ cm; $A = 93.6$ cm^2

26. Find the area of a regular hexagon whose sides measure 10 cm. 259.8 cm^2

27. Find the area and perimeter of an equilateral triangle whose sides measure 18 cm. $P = 54$ cm; $A = 140.3$ cm^2

28. Find the area and perimeter of an equilateral triangle whose altitude measures 6 cm. $P = 20.78$ cm; $A = 20.78$ cm^2

29. If the area of an equilateral triangle is 100 sq cm, find the length of a side. 15.2 cm^2

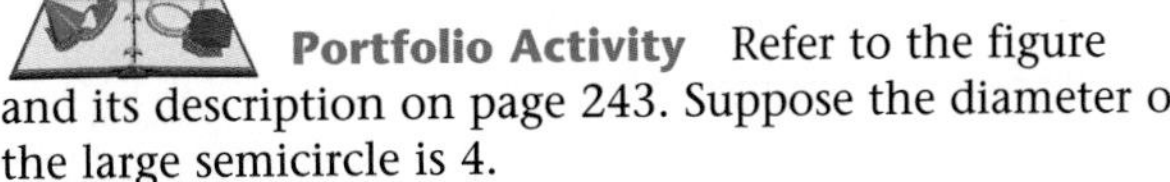

Portfolio Activity Refer to the figure and its description on page 243. Suppose the diameter of the large semicircle is 4.

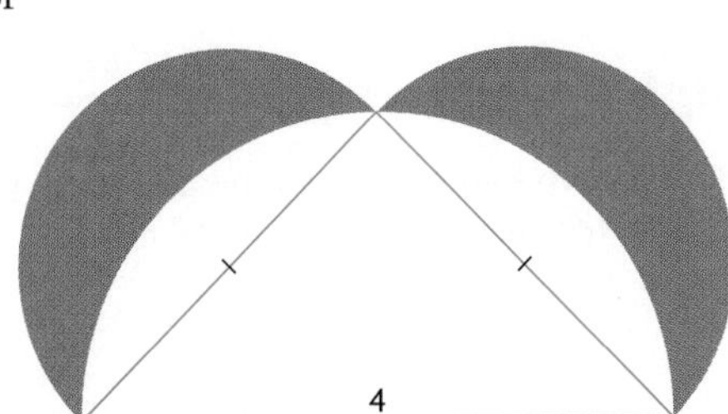

30. Find the area of the large right triangle. 4 units^2

31. Find the area of the small semicircles. π units each

32. Find the area of the large semicircle. 2π units^2

33. Combine the areas found to write an equation and find the area of the lunes.

34. What other area in the figure is the area of the lunes equal to?

35. What other figures can you construct based on the idea of semicircles or circles constructed on segments of polygons? Discover what you can about these figures. Use them to create patterns and designs.

Look Back

Drafting In drafting, a T square is used to help align drawings. The T square is held against the side of the drawing board and moved up and down. The angle of the straightedge stays perpendicular to the side of the drawing board. **[Lesson 3.4]**

36. An artist makes a series of horizontal lines by using the straightedge of the T square and moving it up 1 inch for each new line. What is the relationship between the lines? Why?

37. With the 30-60-90 triangle positioned as shown, the artist uses the hypotenuse to draw an angle of 60° with one of the horizontal lines. Then she slides the triangle along the T square and draws another angle the same way. How are the two angles related? Why?

33. Area of the lune = (Area of the isosceles right triangle) + 2(Area of the semicircle with diameter a leg of the triangle) – (Area of the semicircle with diameter the hypotenuse of the triangle); $A = 4 + 2\pi - 2\pi = 4$ units^2

34. The area of the right isosceles triangle

35. Lunes could be constructed on regular pentagons or hexagons, for example. Check student designs.

36. They are parallel. They are all perpendicular to the side of the table.

37. They are parallel. Corresponding angles are congruent.

Technology Master

NAME ____________ CLASS ________ DATE ________

Technology

5.5 Exploring the Existence of Regular Polygons

In Lesson 5.5, you saw that given a regular polygon, you can find its area A from information about its perimeter P and its apothem a. Since $P = ns$, where n is the number of sides of the polygon and s is the length of each side, you can write $A = \frac{1}{2}pa = \frac{1}{2}nsa$. In other words, if you have a regular polygon, you can find its area from n, s, and a.

Suppose you reverse the question: Given values of s and a, is there a regular polygon with n sides determined by s and a? To explore the question, you will need geometry software and a knowledge of reflections.

1. Construct a triangle like the one shown with $s = 4$ units and $a = 2$ units. Reflect the triangle over $\overline{OM}$. What do you notice?

In Exercises 2–7, use geometry software. For each value of s and a, begin by constructing a diagram such as the one shown. When you sketch the segment with length a, be sure to place the segment perpendicular to $\overline{LM}$ at its midpoint. Do the given values of a and s determine a regular polygon? Illustrate your answer.

2. $s = 4$ and $a = 2$

3. $s = 4$ and $a = 1$

4. $s = 4$ and $a = 3$

5. $s = 4$ and $a = 4$

6. Based on your results from Exercises 2-5, how can you tell whether given values of s and a will determine a regular polygon? If the values of s and a determine a regular polygon, how can you tell how many sides it has?

7. How can m$\angle LOM$ tell you if a regular polygon can be formed?

HRW Geometry Technology 71

38. Minimum area: $(1)(49) = 49\text{ m}^2$

Maximum area: $(25)(25) = 625\text{ m}^2$

40. ΔABC: $A = 6\text{ cm}^2$; $\Delta BCD = 11.25\text{ cm}^2$

38. Find the smallest possible area for a rectangle with a perimeter of 100 m and sides that are whole numbers. Then find the greatest possible area. State the dimensions of each rectangle. **[Lesson 5.1]**

39. The base of an isosceles triangle is 8 m and the legs are 6 m. Find the area and perimeter. **[Lesson 5.2]** $A = 17.89\text{ m}^2$; $P = 20\text{ m}$

40. Find the area of the triangles in the figure. **[Lesson 5.2]**

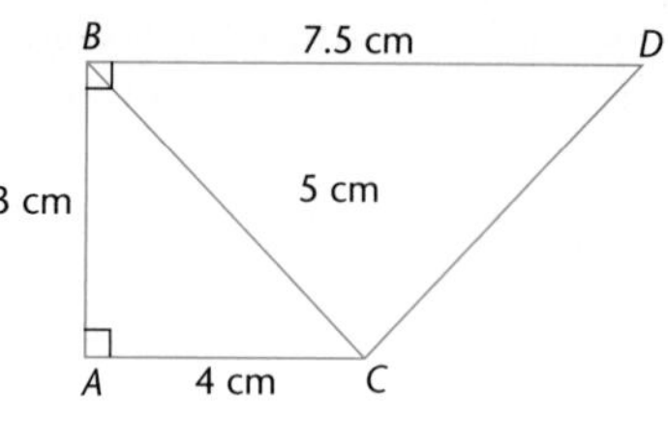

41. Find the height of a trapezoid with area 103.5 sq ft, with bases 17.5 ft and 5.5 ft. **[Lesson 5.2]** 9 ft

42. Find the area of an isosceles right triangle with leg $\sqrt{2}$ cm. **[Lesson 5.4]** 1 unit^2

43. Find the area of a right triangle with legs a and b. **[Lesson 5.4]** $\frac{1}{2}ab$

44. In $\triangle PQR$, find PR. **[Lesson 5.4]** 218.17 units

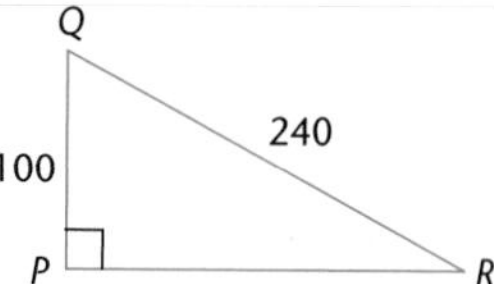

Look Beyond

Look Beyond

Extension. In earlier work students studied the relationship between perimeter and areas of rectangles. In Exercises 45 and 46, a 3-dimensional version of the 2-dimensional exercises is considered.

45. The two figures below have the same volume. Do they have the same surface area? Give the surface area for each figure.

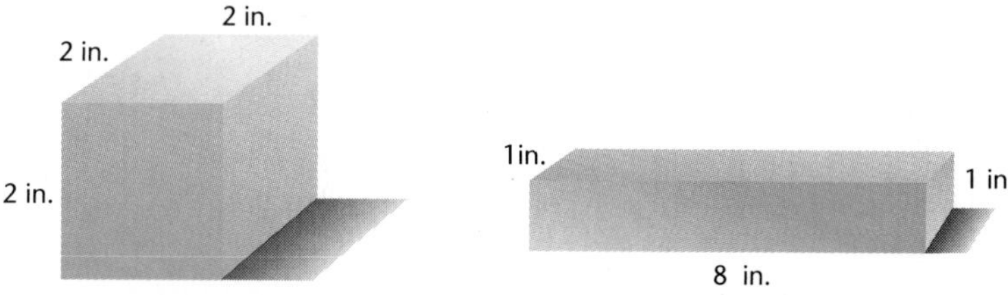

Cube: SA = 24 units^2; Rect Bar: SA = 34 units^2

46. Sketch another figure with the same volume as the figures in Exercise 45. Does your new figure have a surface area larger or smaller than those of the two figures?

46. The new solid has a surface area in between that of the figures shown.

LESSON 5.6 The Distance Formula, Quadrature of a Circle

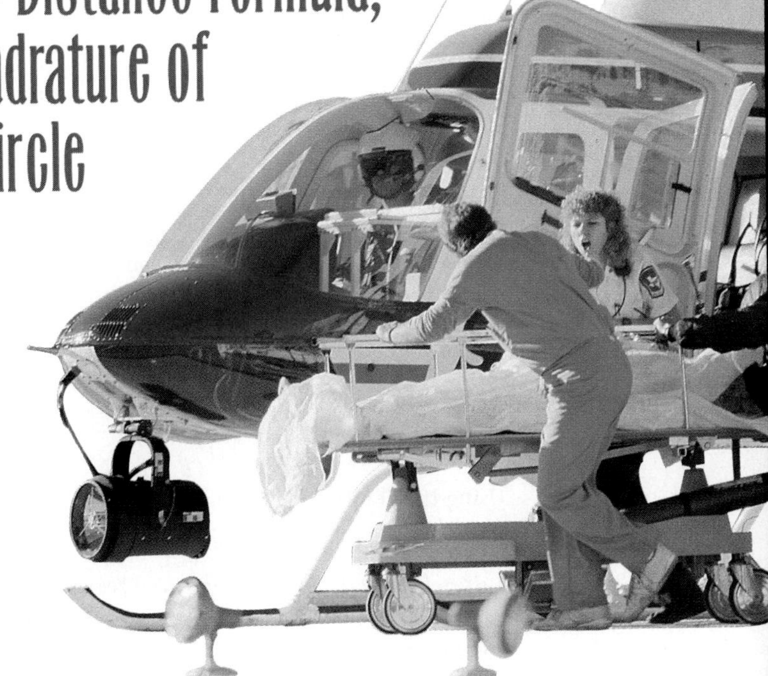

Why *To reach its destination, a helicopter will travel the shortest distance between two points. The ability to compute the distance between two points is important to know, especially in emergency situations. In some cases, an estimate based on measurement is sufficient. However, in other cases, great precision is necessary.*

The Distance Formula

The distance between two points on the same vertical or horizontal line can be found by taking the absolute value of the difference between the *x*- and *y*-coordinates. The vertical distance between points *A* and *B*, or length *AB*, is $|7 - 3| = |3 - 7| = 4$.

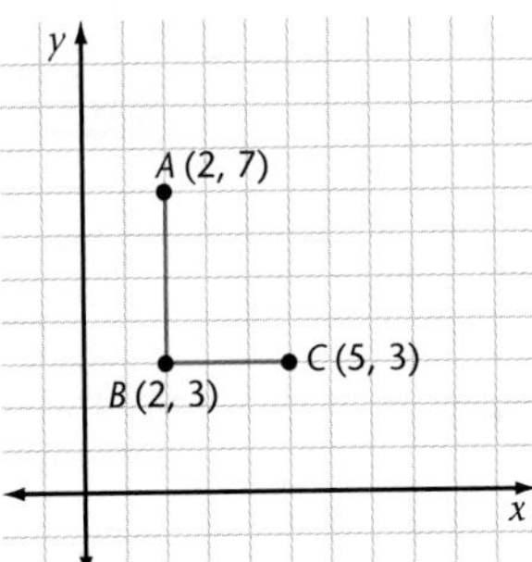

The horizontal distance between points *B* and *C*, or length *BC*, is $|2 - 5| = |5 - 2| = 3$.

Since $\overline{AC}$ forms the hypotenuse of right triangle $\triangle ABC$, its length can be found using the "Pythagorean" Right-Triangle Theorem.

$$(AC)^2 = 4^2 + 3^2$$

$$(AC)^2 = 25$$

$$AC = \sqrt{25} = 5$$

The "Pythagorean" Right-Triangle Theorem can be used to find the distance between any two points on the coordinate plane.

ALTERNATIVE teaching strategy

Technology Geometry graphics software that includes coordinate axes can be used to explore the ideas in this lesson. Students should start by replicating the figures in the text. Then they can drag points or line segments to see which relationships change and which remain constant.

PREPARE

Objectives

- Develop and apply the Distance Formula and Midpoint Formula.
- Use the Distance Formula to develop techniques for estimating the area of a circle.

RESOURCES

- Practice Master 5.6
- Enrichment Master 5.6
- Technology Master 5.6
- Lesson Activity Master 5.6
- Quiz 5.6
- Spanish Resources 5.6

Assessing Prior Knowledge

Use the two points $(x_1, y_1) = (2, 1)$ and $(x_2, y_2) = (6, 4)$. Find the value of each expression.

1. $|x_2 - x_1|$ [4]
2. $|x_2 - x_1|^2 + |y_2 - y_1|^2$ [25]
3. $\sqrt{(x_2 - x_1)^2 + (y_2 - y_1)^2}$ [5]

TEACH

On a plane, the distance between two points is defined as the straight-line distance between them. The students should compare another distance definition they studied in Lesson 1.2. In both cases the shortest path is chosen.

Before students read and discuss the proof of the Distance Formula, have them use graph paper and compute a few examples using triangles with just their side lengths labeled. This preliminary work may prevent confusion caused by the complicated-looking notation in the text.

Use Transparency 46

Math Connection Algebra

Remind students of the meaning and notation for absolute value: $|a| = a$ when a is greater than 0; $|a| = -a$ when a is less than 0. The absolute value of a nonzero quantity is always greater than 0.

Alternate Example 1

A waterfall in a park is located 6 miles east and 3 miles north of the Visitor's Center. There is a picnic site 1 mile east and 2 miles north of the Center. How far would you walk along a trail that goes directly from the waterfall to the picnic site? Include a drawing in your solution.

$[\sqrt{5^2 + 1^2} = \sqrt{26} \approx 5.1 \text{ mi}]$

DISTANCE FORMULA

The distance between two points (x_1, y_1) and (x_2, y_2) is

$$d = \sqrt{(x_2 - x_1)^2 + (y_2 - y_1)^2}$$ **5.6.1**

Proof:

Draw a right triangle with hypotenuse $\overline{AB}$, a right angle at C, and d equal to the distance between A and B.

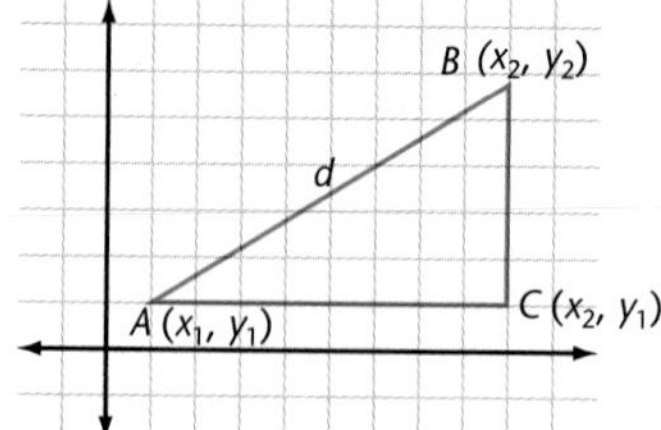

ALGEBRA Connection

The coordinates of C are x_2 and y_1. (why?)

The length AC is $|x_2 - x_1|$.

The length BC is $|y_2 - y_1|$.

Using the "Pythagorean" Right-Triangle Theorem,

$$d^2 = |x_2 - x_1|^2 + |y_2 - y_1|^2$$

$$d^2 = (x_2 - x_1)^2 + (y_2 - y_1)^2$$

You can drop the absolute value symbols because the quantities are being squared.

Solving for d,

$$d = \sqrt{(x_2 - x_1)^2 + (y_2 - y_1)^2}$$

Navigation A helicopter crew located 1 mile east and 3 miles north of Command Central must respond to an emergency located 7 miles east and 11 miles north of Command Central. How far must the helicopter travel to get to the emergency site?

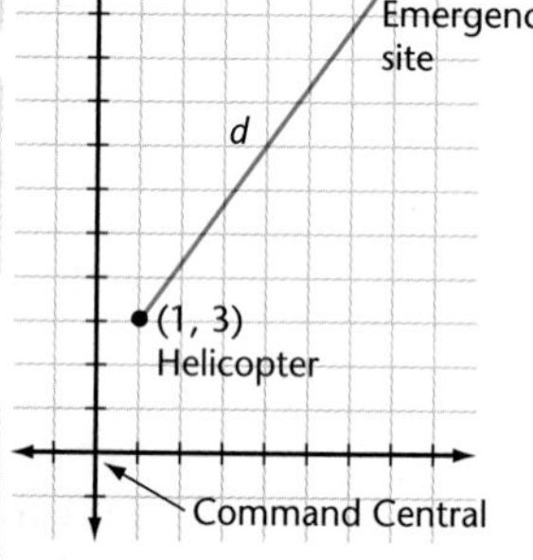

Solution

Let (0, 0) be the coordinates of Command Central. Then the helicopter crew coordinates are (1, 3). The emergency site coordinates are (7, 11). Using the distance formula,

$$d^2 = (7 - 1)^2 + (11 - 3)^2$$

$$d = \sqrt{6^2 + 8^2} = \sqrt{36 + 64} = \sqrt{100} = 10.$$

The helicopter must travel 10 miles to the emergency site. ❖

interdisciplinary CONNECTION

Physics Vector diagrams used to compute the results of combined forces apply the "Pythagorean" Right-Triangle Theorem and the Distance Formula. Have students find examples in physical science or physics textbooks of diagrams showing how winds or currents affect the speed and direction of airplanes or boats. These diagrams can be compared to those used in this geometry text.

Try This Find the distance between the points (−2, 4) and (3, −9).

Midpoint Formula

An Exploration in Lesson 4.8 led to the midpoint formula. Recall that the midpoint of two points on the same line can be found by averaging the values of the x- and y-coordinates.

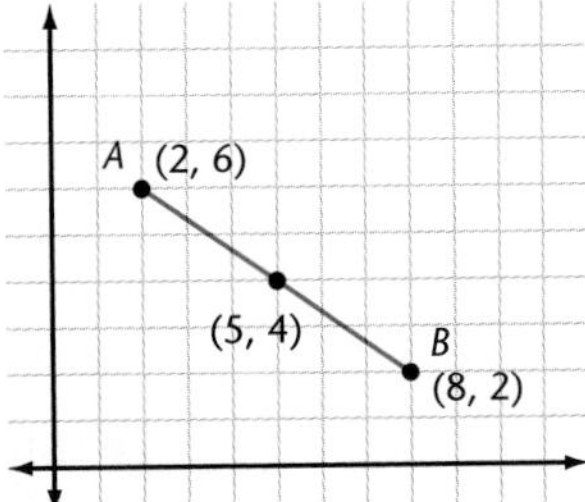

ALGEBRA *Connection*

$x_1 = 2, x_2 = 8 \quad x_m = \frac{8+2}{2} \quad x_m = 5$

$y_1 = 6, y_2 = 2 \quad y_m = \frac{6+2}{2} \quad y_m = 4$

The midpoint coordinates are (5, 4).

THE MIDPOINT FORMULA

For any two points (x_1, y_1) and (x_2, y_2) in the coordinate plane, the midpoint is given by

$$\left(\frac{x_1 + x_2}{2}, \frac{y_1 + y_2}{2}\right) \qquad \textbf{5.6.2}$$

Try This Find the midpoint of (8, 3) and (2, 6). Show that the lengths of the two segments formed by the midpoint are equal using the distance formula.

Methods of Quadrature

The area of an enclosed region can be approximated by breaking it down into a number of rectangles. This technique, called quadrature, is particularly important for finding the area under a curve.

Exploration: Estimating the Area of a Circle

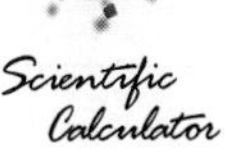

You will need
Graph paper
Scientific calculator
Spreadsheet software (optional)

Part I: Method A

1. Draw a quarter of a circle five units in diameter with its center at the origin (0, 0) of a coordinate plane. Add rectangles to completely cover the circle as shown.

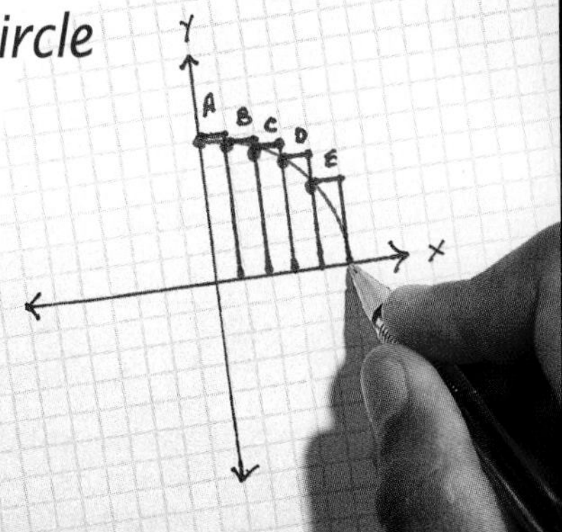

ENRICHMENT On a sphere, the shortest distance between two points lies on a great circle, a circle that divides the sphere in half. Have students work in pairs or groups to brainstorm ideas for estimating or computing distances on great-circle routes. Students can apply what they have learned about computing the circumference of a circle. They may also need to research information about latitude and longitude.

INCLUSION strategies **Using Symbols** The subscripted variables used in this lesson will be unfamiliar and therefore difficult for some students. Provide practice in solving simple linear equations, such as the following, to help students get comfortable with subscripted variables.

1. $4 + x_1 = 12$ [$x_1 = 8$]
2. $2y_2 - 1 = 9$ [$y_2 = 5$]
3. $3V_e = 18$ [$V_e = 6$]

ongoing ASSESSMENT

Try This

$d = \sqrt{(2-5)^2 + (6-4.5)^2}$
$= \sqrt{45} \approx 6.71$

Math Connection Algebra

Point out that absolute value signs are *not* used in the Midpoint Formula. Ask students why this is so. [The quantities in the Distance Formula must be positive numbers. In the Midpoint Formula, the quantities are coordinates and thus should be positive or negative, depending on where the segment lies in the coordinate plane.]

ongoing ASSESSMENT

Try This

$M = (5, 4.5)$

$d_1 = \sqrt{(8-5)^2 + (3-4.5)^2}$
$= \sqrt{3^2 + (-1.5)^2} = \sqrt{11.25}$

$d_2 = \sqrt{(2-5)^2 + (6-4.5)^2}$
$= \sqrt{(-3)^2 + (1.5)^2} = \sqrt{11.25}$

Exploration Notes

This activity, in which rectangles with a common width and varying heights are used to approximate the area under a curve, previews ideas basic to calculus. Although the method will work for any curved shape, a circle is used because students can apply the "Pythagorean" Right-Triangle Theorem to find the coordinates of points on a circle.

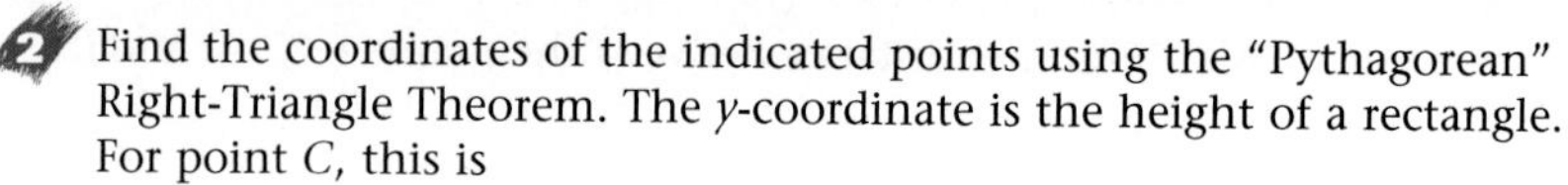

Cooperative Learning

Have students work in groups to apply the approximation method to a curve other than a circle.

ongoing ASSESSMENT

PART I

6. A = 25 π ≈ 78.5398
   E = 0.094

PART II

This method underestimates because parts of the circle are not included.

PART III

E = 0.033

TEACHING tip

Remind students that the purpose is not simply to find the area of a circle. The primary purpose is to learn a technique that can estimate areas of figures with irregular boundaries.

CRITICAL Thinking

Method A always overestimates, while Method B always underestimates. So the actual area of the circle will be between the two. The average gives an estimate halfway between the overestimate and the underestimate. If the actual area is closer to an estimate than to the average, the average will not give a better estimate.

2. Find the coordinates of the indicated points using the "Pythagorean" Right-Triangle Theorem. The y-coordinate is the height of a rectangle. For point C, this is

$$y = \sqrt{5^2 - 2^2} = \sqrt{21}$$

3. Find the area of each rectangle. (This is simplified by the fact that the base of each rectangle is 1.)

4. Find the sum of the areas of the rectangles by completing the sequence below.

$$\sqrt{25} + \sqrt{24} + \sqrt{21} + \ldots = ?$$

5. Multiply your sum by 4. The result is an estimate of the area of a circle with radius 5. Does it overestimate or underestimate the area? Explain why.

6. Calculate the true value of the area to four decimal places using $A = \pi r^2$. Find the relative error of your estimate using the formula

$$E = \frac{|V_e - V_t|}{V_t} \times 100$$

where V_e = estimated value, V_t = true value, and E = percent of error.

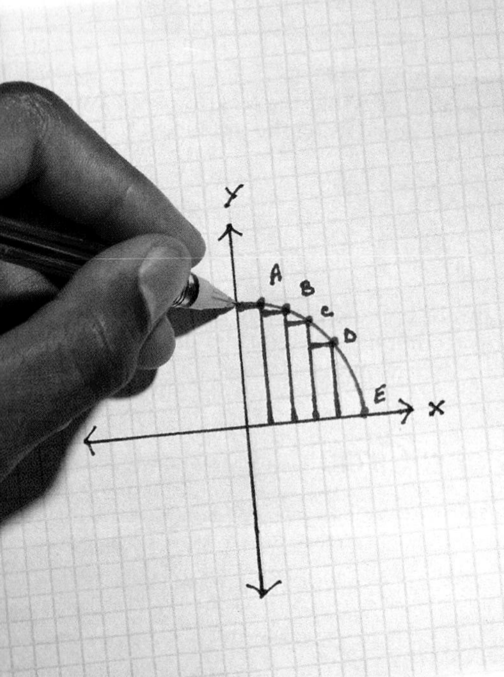

Part II: Method B

Repeat Part 1 using an arrangement of rectangles like the one shown at left. Does the new method overestimate or underestimate the area of the circle? Explain why.

Part III: Combining Methods

Average your results from Parts 1 and 2. What is the relative error of your new estimate? ❖

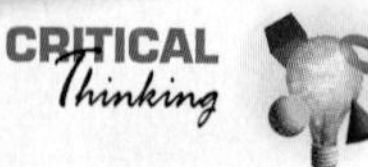

CRITICAL Thinking Do you think that the average of results from Methods A and B will always give more accurate estimates than either one by itself? Explain why or why not.

RETEACHING the lesson

Cooperative Learning Have students work in pairs. Provide each pair with graph paper. Each student should draw five line segments on a coordinate grid. The segments should not be horizontal or vertical. Students should then exchange papers. They should (1) find the segment lengths by counting squares and using the "Pythagorean" Right-Triangle Theorem and (2) find the segment lengths using endpoint coordinates and the Distance Formula. This will help students make the connection between the "Pythagorean" Right-Triangle Theorem and the Distance Formula.

EXERCISES & PROBLEMS

Communicate

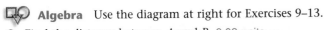

1. Explain why it is necessary to find the differences between x- or y-coordinates in the distance formula.
2. Explain how the distance formula is related to the "Pythagorean" Right-Triangle Theorem.
3. Explain two methods of estimating the shaded area under the curve.
4. Explain why the estimate for the area becomes more accurate as the rectangles are made smaller.

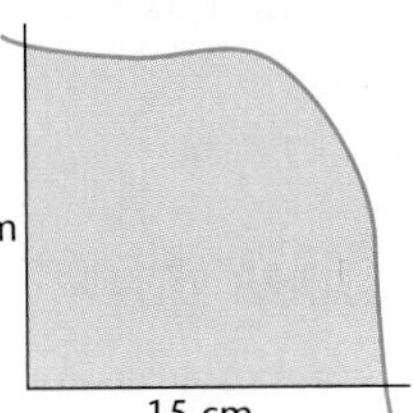

Practice & Apply

Find the distance between each pair of points.

5. (1, 4) and (3, 9) 5.39
6. (−3, −3) and (6, 12) 17.49
7. (−1, 4) and (−3, 9) 5.39
8. (−2, −3) and (−6, −12) 9.85

Algebra Use the diagram at right for Exercises 9–13.

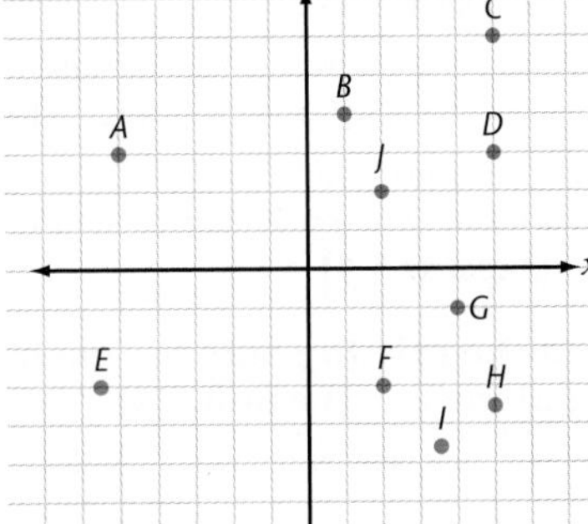

9. Find the distance between A and B. 6.08 units
10. Find the distance between B and I. 8.86 units
11. Is the triangle formed by B, C, and D isosceles? Why or why not?
12. Is the quadrilateral formed by F, G, H, and I a square? Why or why not?
13. Find the lengths of the sides of the triangle formed by E, J, and H.

14. **Environmental Protection** In order to minimize the potential for environmental damage, an oil pipeline is being rerouted so that less of the pipe will fall inside a national forest. What is the distance between point A and point B in the map on the right? 11.4 mi.

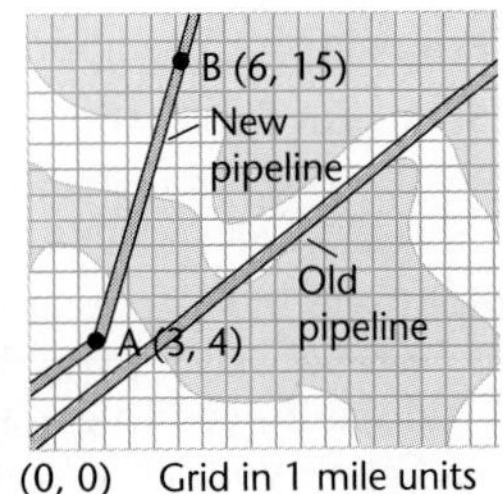

11. $BD = \sqrt{17}$, $BC = \sqrt{20}$, $CD = 3$. Since no two pairs of sides have the same length, the triangle is not isosceles.

12. No, the side lengths are not equal. $FG \neq IH$, for example.

13. $EJ = 9.01$ units ; $JH = 6.26$ units; $EH = 10.51$ units

ASSESS

Selected Answers

Odd-numbered Exercises 5–23

Assignment Guide

Core 1–10, 15–16, 19–25

Core Plus 1–4, 7–27

Technology

Students will need a calculator with a square-root key for most of these exercises. A graphics calculator will be helpful for Exercises 17 and 18. For Exercises 25–27, students might choose a combination of geometry graphics software, a hand-held graphics calculator, and graph paper.

Error Analysis

Some students may make computation errors in Exercises 5–13. If you suspect that this will happen, have them begin by copying the figure onto graph paper. Then they can count square units or measure to check their computation.

Practice Master

NAME ______ CLASS ______ DATE ______

Practice & Apply

5.6 The Distance Formula, Quadrature of a Circle

In Exercises 1–4, find the distance between each pair of points.

1. (1, 5) and (4, 1) ______
2. (−1, −3) and (−16, −11) ______
3. (−2, 7) and (10, 2) ______
4. (−2, −8) and (−8, −2) ______

In Exercises 5–7, use the diagram shown.

5. Is the triangle formed by A, F, and B isosceles? Why or why not? ______
6. What kind of quadrilateral is formed by A, B, C, and E? Explain. ______
7. What kind of quadrilateral is formed by A, B, D, and E? Explain. ______
8. Find the perimeter of this quadrilateral. ______

y x G F E D C A B

9. The length of diagonal $\overline{BE}$ is ______.
10. The length of diagonal $\overline{AD}$ is ______.
11. The midpoint of diagonal $\overline{BE}$ is ______.
12. The midpoint of diagonal $\overline{AD}$ is ______.
13. Use these results to draw a conclusion about the diagonals of this type of quadrilateral. Compare the characteristics of the diagonals of this type of quadrilateral to those of a parallelogram. ______

58 Practice & Apply HRW Geometry

Alternative ASSESSMENT

Portfolio Assessment

Have students write an explanation of the Distance Formula that includes how it works, why it works, and in what types of real-world situations it might be used. Students should organize their essay in logical order and include drawings to help communicate their ideas.

15. Place a 5–12–13 right triangle on the coordinate system. Show coordinates for each point and the distances for the two legs and the hypotenuse.

16. Select three points on the coordinate plane that will be the vertices of an isosceles triangle. Prove that the triangle is isosceles. Use the distance formula.

17. **Algebra** Use the rectangular method to estimate the area under $y = x^2 + 2$ and the x-axis when $0 \le x \le 2$.

18. **Algebra** Use the rectangular method to estimate the area under $y = -x^2 + 2$ and the x-axis when $0 \le x \le 1\frac{1}{2}$.

The answers to Exercises 15–19 can be found in Additional Answers beginning on page 727.

Look Back

19. Prove that $WXYZ$ is a rectangle using the coordinates $W(-1, 1)$; $X(7, -5)$; $Y(10, -1)$; $Z(2, 5)$. **[Lesson 3.8]**

20. Suppose you have 100 m of fence to enclose a play area. Which would be the larger play area, a square play area or a circular one? **[Lesson 5.1]**

21. Suppose a circle and a square both have an area of 225 sq cm. Which has the greater perimeter? **[Lesson 5.1]**

22. The sides of an equilateral triangle measure 8 cm. Find the area and perimeter. Find the area if the sides are s units in length. **[Lesson 5.4]**

Find the missing side for each right triangle. Side c is the hypotenuse. [Lesson 5.4]

23. $a = 4.5$, $b = 8$, $c =$ __?__ 9.18

24. $a =$ __?__, $b = 4$, $c = 5$ 3

Look Beyond

These exercises provide further work with area approximation using rectangles of common width. Students also use the formula for relative error given in Step 6 of the exploration.

Look Beyond

Technology Use appropriate technology to solve Exercises 25–27.

25. Repeat Exploration 1 using a larger number of rectangles. If you wish, you can use rectangles with a base of 1 and increase the diameter of the circle. Does your accuracy increase with larger numbers of rectangles? Yes

26. If you have access to a spreadsheet or other computer software, estimate using very large numbers of rectangles. How many rectangles do you need to get a 1 percent accuracy? 0.1 percent? .01 percent? Graph your accuracy for different numbers of rectangles. Check student work.

27. Is it always necessary to have extremely accurate answers to real-life problems? Why might you sometimes prefer a rough estimate to a more precise one?

Technology Master

NAME ____________ CLASS ________ DATE ________

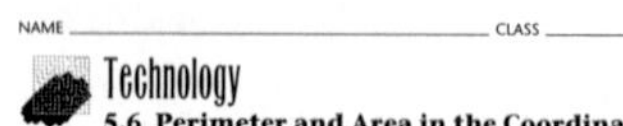

Technology

5.6 Perimeter and Area in the Coordinate Plane

The diagram shows a coordinate plane with ΔABC positioned in circle O, whose equation $x^2 + y^2 = 16$. In many problems, you will need to write y in terms of x. For the circle with equation $x^2 + y^2 = 16$, the equation you need for the upper half of the circle is $y = \sqrt{16 - x^2}$. Using the distance formula you can write expressions for AB and BC. These expressions are given here.

$$AB = \sqrt{25 + 6x} \qquad C = \sqrt{25 - 6x}$$

The equations for AB and BC in addition to the fact that $AC = 6$, give you the formula for the perimeter P of ΔABC.

$$P = 6 + \sqrt{25 + 6x} + \sqrt{25 - 6x}$$

Furthermore, the area A of ΔABC can be expressed by the following equation.

$$A = \frac{1}{2}\text{base} \times \text{height} = 0.5 \times 6 \times \sqrt{15 - x^2}$$

In Exercises 1 and 2, use a graphics calculator.

1. Graph the equation for values of x from -4 to 4.

   $P = 6 + \sqrt{25 + 6x} + \sqrt{25 - 6x}$

2. Graph the equation for values of x from -4 to 4.

   $A = 0.5 \times 6 \times \sqrt{16 - x^2}$

Using the calculator display from Exercises 1 and 2, find x such that ΔABC has the given perimeter. Then find the corresponding area.

3. $P = 16$ units ________ 4. $P = 15.5$ units ________ 5. $P = 15$ units ________

6. For what value(s) of x will the perimeter of ΔABC be a maximum? Find the area of the triangle(s) determined.

____

7. Use the distance formula to show that the perimeter of ΔABC is given by the equation $= 6 + \sqrt{25 + 6x} + \sqrt{25 - 6x}$.

____

72 Technology HRW Geometry

20. Area of square: 625 m^2
    Area of circle: 795.77 m^2
    The circular area is larger.

21. Perimeter of square: 60 cm
    Perimeter of circle: 53.17 cm
    Square has the greater perimeter.

22. $P = 24$ cm ; $A = 27.7$
    If a side is s, $A = \frac{s^2\sqrt{3}}{4}$

27. In many real-life problems the answers given by mathematical models are only estimates and an exact answer is unrealistic due to measurement uncertainties. For example, predicting the high temperature for tomorrow. In other cases an exact answer might be too difficult to find and not worth the trouble. For example, finding the exact distance between two cities.

LESSON 5.7

Geometric Probability

According to the fossil record, the Earth experienced a sudden, drastic change about 65 million years ago—resulting in the extinction of great dinosaurs like the Tyrannosaurus rex. *Many scientists believe that this upheaval was caused by an asteroid striking the Earth at the Yucatan Peninsula.*

The surface of the Earth consists of about 30 percent land and 70 percent water. Using these figures, and assuming that a comet or asteroid would be equally likely to strike anywhere on the Earth, what is the probability that such an object would have struck land instead of water?

Why Many scientists believe that an asteroid or comet caused the extinction of the dinosaurs. What is the likelihood that such an object would have struck land? You can use geometric probability to answer this question.

The Basic Formula

Mathematical intuition should tell you that there is a 30 percent chance, or probability, that the object would have struck land. **Probability** is the ratio that compares the number of "successful" outcomes with the total number of possible outcomes. For example, if there are six marbles in a bag and two are blue, the probability of picking a blue marble at random is $\frac{2}{6}$, or $\frac{1}{3}$.

The same idea applies to areas. Since the surface area of the Earth is 100 percent and about 30 percent of the Earth's surface is covered by land, the probability that the asteroid would have struck land is 30 percent, $\frac{30}{100}$, or 0.3.

Hands-On Strategies The probabilities computed in Example 1 and in many of the exercises can be converted into simulations in which students perform experiments, record data, and compute experimental probabilities. Comparing theoretical and experimental probabilities can help students better understand the difference between these two types of measures of chance.

PREPARE

Objectives

- Develop and apply the basic formula for geometric probability.

RESOURCES

• Practice Master	5.7
• Enrichment Master	5.7
• Technology Master	5.7
• Lesson Activity Master	5.7
• Quiz	5.7
• Spanish Resources	5.7

Assessing Prior Knowledge

Use graph paper for these problems:

1. Draw a rectangle in which the side lengths are whole numbers.
2. Draw a circle inside the rectangle with a whole number for a radius.
3. Find the ratio of the area of the circle to the area of the rectangle.

[Answers will vary.]

TEACH

Discuss with students how they would answer a question that begins, "What is the likelihood that . . ." They might use terms such as *very likely, somewhat likely, almost certain, almost impossible.* Point out that a number between 0 and 1 can be used to describe likelihood, and that this number is called a probability. Probabilities are written as fractions, as decimals, or as percentages.

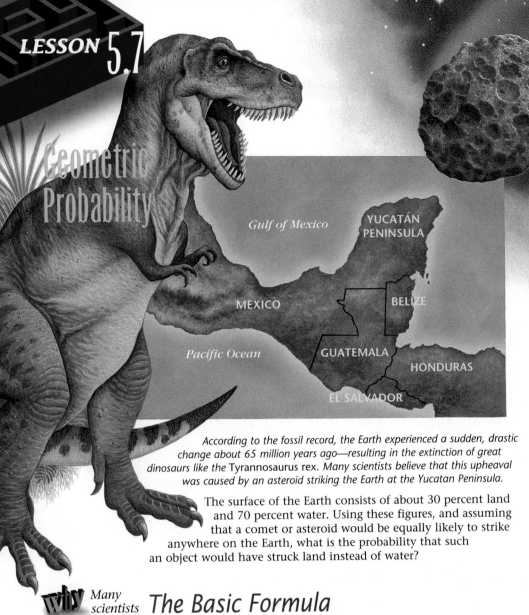

Math Connection Probability

If students need a reminder of probability concepts, conduct a simple demonstration in which a few counters or small objects are placed in a bag. Students predict the probability of drawing one object, and then compare predicted results with what actually happens.

Alternate Example 1

Draw three concentric circles on graph paper. Use 1, 3, and 5 units for the diameters; the middle circle is the shaded target area. What is the probability that a dart will land in the shaded area?

[shaded area = $\pi(3/2)^2 - \pi(1/2)^2 = \pi(9/4 - 1/4) = 2\pi$;
total area = $\pi(5/2)^2 = 25\pi/4$;
ratio of shaded to total = $2\pi \div 25\pi/4 = 8/25$]

ongoing ASSESSMENT

Try This

a. $\frac{8}{9}$ b. $\frac{5}{9}$ c. 0.84

TEACHING tip

Listen to students discuss Example 1 and the Try This problems to determine whether they understand that the probabilities are equal to the ratio of the shaded to unshaded areas.

Use Transparency 47

Geometric Probability

PROBABILITY *Connection*

Given a figure like the one at right, the probability that a point randomly selected from the figure will be in Area A is

$$\frac{\text{Area } A}{\text{Area } B}.$$

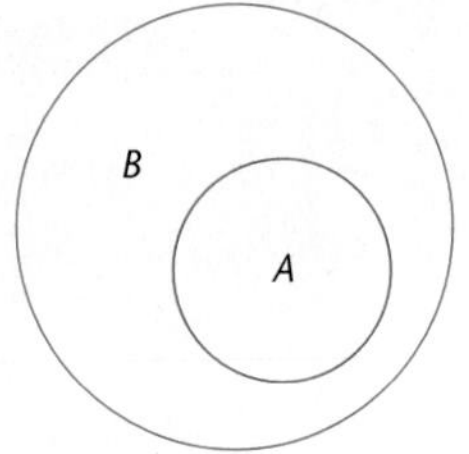

Area B is the total area of the figure.

EXAMPLE 1

If a dart lands in the shaded part of the board, a prize is given. What is the probability that a toss that hits the board at random will win?

Solution ➤

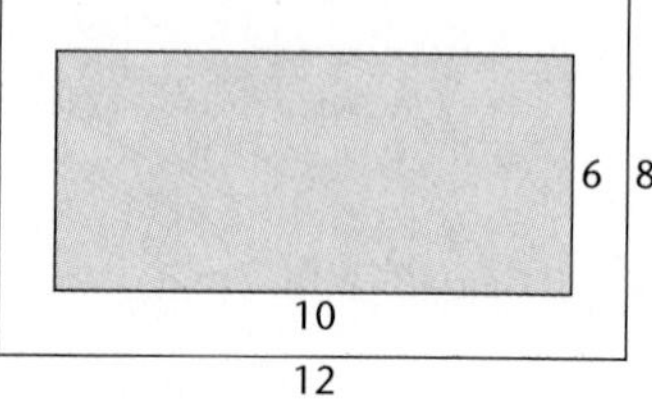

The area of the shaded part is 60.

The area of the entire region is 96.

The probability of landing on the shaded part is $\frac{60}{96}$, or $\frac{5}{8}$. ❖

Try This Find the probability that a dart will land in the shaded area.

a.

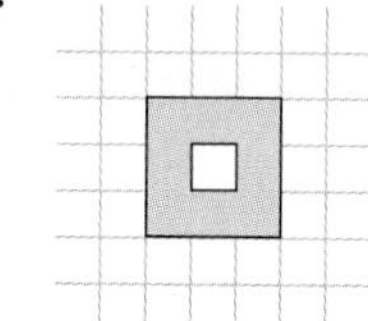

b.

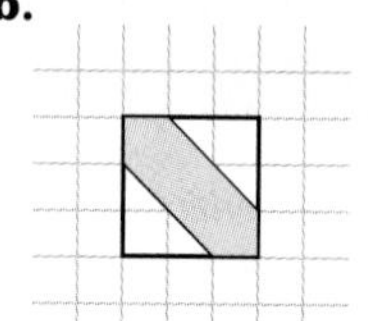

c.

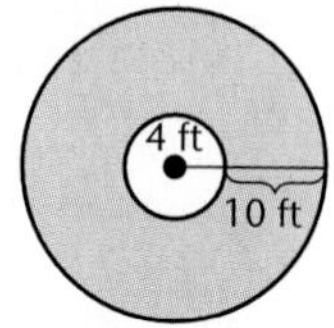

EXAMPLE 2

In a game, pennies are tossed onto a grid that has squares that are each the width of a penny. To win, a penny must not land on an intersection point of the lines that make up the grid. What is the probability that a random toss will lose?

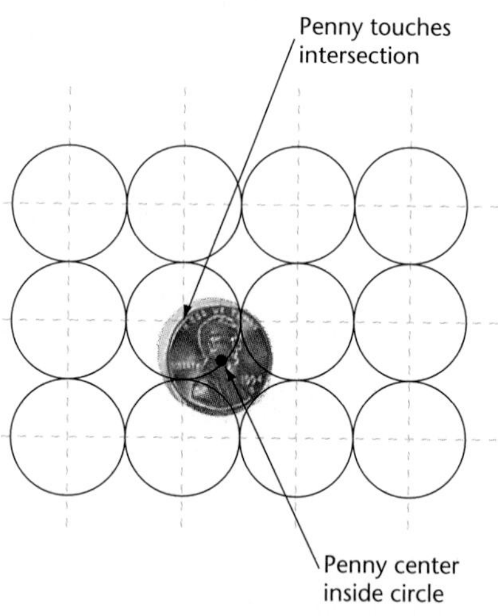

Solution ➤

Imagine that circles the size of a penny are drawn around each intersection point on a grid. Any time the center of a penny falls within one of the circles, the penny will touch an intersection point.

interdisciplinary CONNECTION **Earth Science** Probabilities in percentage form are often used in weather predictions. Have students use library materials to find out how these probabilities are computed and why, in many areas, weather predictions are often wrong.

The center of a penny that falls on the grid will land somewhere within one of the squares. What is the probability that it will be within the shaded region of the square?

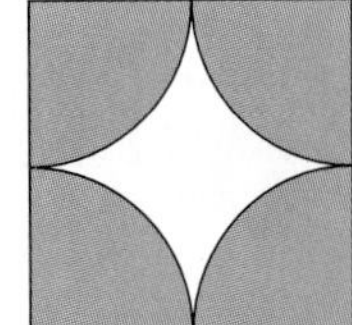

Let the radius of a penny equal 1 unit. Then the width of a square (and the diameter of a penny) will be 2. Thus,

Area of circle $A_c = \pi \times 1^2 = \pi$ sq units

Area of square $A_s = 2 \times 2 = 4$ sq units

The area of each quarter of a circle is $\frac{\pi}{4}$ sq units, so the area of the shaded region is π sq units. The probability P of the penny touching an intersection and of the player losing is

$$P = \frac{\text{Area of Shaded Region}}{\text{Area of Square}} = \frac{\pi}{4}. ❖$$

In the following exploration you will use the result of Example 2 to estimate the value of π.

Exploration A "Monte Carlo" Method for Estimating π

You will need
Grid paper with squares equal in width to the diameter of a penny
Pennies

1. Toss a coin onto the grid paper. Notice whether the coin touches an intersection point of two lines or not. Repeat 20 times. Record the number of tosses that fall on intersection points and the number of tosses that do not.

2. Share your results with the rest of the class. Tally the totals for the entire class. Calculate the ratio of tosses that land on intersections in your experiment to the total number of tosses, and call it E.

3. For a large number of coin tosses, you may assume that the value of E will be close to the probability P that was calculated in Example 2. Using this assumption, calculate an estimate for π:

   $E \approx \frac{\pi}{4}$ (assumed to be true for large numbers of tosses)

   So $\pi \approx E \times 4$

4. Compare your estimate with the known value of π. Determine the relative error of your estimate using the formula

$$E = \frac{|V_t - V_e|}{V_t},$$

where E = relative error, V_t = true value, and V_e = estimated value. ❖

Alternate Example 2

A coin equal in diameter to the side of each triangle is tossed onto a grid composed of equilateral triangles. What is the probability that the coin will touch an intersection point of the grid?

The probability equals the ratio of the three shaded areas to the area of the equilateral triangle.

$\left[\textbf{probability} = \frac{\pi}{2\sqrt{3}} = \frac{\pi\sqrt{3}}{6}\right]$

Exploration Notes

Any type of circular counters may be used instead of pennies. If you have counters 1 cm in diameter and centimeter grid paper, students will not need to draw the grid for the exploration.

Ongoing Assessment

4. The relative error should be 0.1 or lower.

Cooperative Learning

Have students work in small groups to apply the ideas in the Exploration to a variation of Buffon's Needle Problem. Each group will need a supply of toothpicks. They should prepare a large sheet of paper with parallel lines drawn such that the distance between adjacent lines is equal to the length of a toothpick. The probability of a dropped toothpick touching one of the lines is $P = \frac{2}{\pi}$. Students should compare experimental results with those predicted by the formula.

Enrichment

Have students work alone or with others to solve this problem: An infinite plane is covered with circles. Each circle touches four other circles, and the straight lines joining the centers of the circles divide the plane into congruent squares. A dart is thrown at random at the plane. What is the probability that the dart will land inside one of the circles? [**Students should realize that this problem is equivalent to the penny game problem in Example 2. The probability is $\pi/4$.**]

Inclusion Strategies

Hands-On Strategies It can be difficult for some students to understand why theoretical and experimental probabilities differ. Coin toss experiments can help convey the ideas involved. Students are likely to know that the chance of tossing heads with a fair coin is 50%. If they work in teams to toss coins a great number of times, they will see that as the number of tosses increases, the experimental results get closer and closer to 50%.

ASSESS

Selected Answers

Odd-numbered Exercises 5–31

Assignment Guide

Core 1–20, 27–31

Core Plus 1–8, 16–34

Technology

Geometry graphics software may be used to explore Exercises 16-19. Students who know how to use a random number generator can do Exercise 22 with a computer.

Error Analysis

It may confuse some students that there are no numbers on segment $\overline{AB}$ for Exercises 1–5. You may wish to have them draw a segment 8 cm long and label the points *A*, *B*, *C*, and *D*.

Practice Master

NAME ______ CLASS ______ DATE ______

Practice & Apply
5.7 Geometric Probability

In Exercises 1–6, a point *P* is selected at random from this segment.

0 3 6 9 12

1. The probability that $0 \le P \le 12$ is ______.
2. The probability that $P > 12$ is ______.
3. The probability that $0 \le P \le 6$ is ______.
4. The probability that $3 \le P \le 9$ is ______.
5. The probability that $3 \le P \le 6$ is ______.
6. The probability that $3 \le P \le 12$ is ______

In Exercises 7–9, convert the probability to a percent.

7. 0.20 ______ 8. $\frac{3}{4}$ ______ 9. 0.1666... ______

In Exercises 10–12, convert the percentage to a probability.

10. 20% ______ 11. $66\frac{2}{3}$% ______ 12. 100% ______

13. Can 150% represent a probability? Explain.

______

14. Two of the 4 points shown on the diagram are chosen at random. List the possibilities. Note that in this case, order is not important. That is, *XY* is the same selection as *YX*.

W X Y Z

______

______

15. What is the probability that both points chosen lie on $\overrightarrow{YW}$? Explain.

______

16. A point is chosen at random in rectangle *ABCD*. The probability that the chosen point is in rectangle *RQCS* is ______.

D S 4 C 3 Q R 2 A 9 B

HRW Geometry Practice & Apply 59

Communicate

Use the diagram of $\overline{AB}$ for Exercises 1–4. *M* is the midpoint of $\overline{AB}$ and *C* is the midpoint of $\overline{AM}$. For each exercise, a point on $\overline{AB}$ is picked at random.

1. What is the probability that the point is on $\overline{AM}$? Explain.
2. What is the probability that the point is on $\overline{AC}$? Explain.
3. What is the probability that the point is on $\overline{AB}$? Explain.
4. What is the probability that the point is not on $\overline{AC}$? Explain. (Hint: 1 minus the probability of the point being on $\overline{AC}$)

Practice & Apply

A point *Q* is selected at random from segment $\overline{AB}$ on the number line.

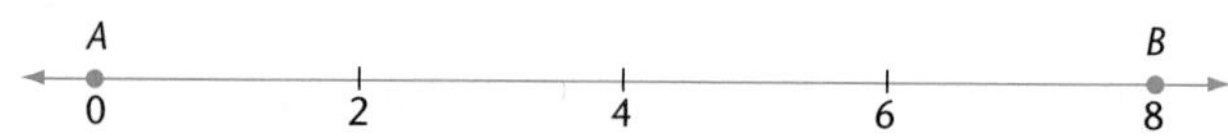

5. What is the probability that $0 \le Q \le 4$? $\frac{1}{2}$
6. What is the probability that $1 \le Q \le 4$? $\frac{3}{8}$
7. What is the probability that $5 \le Q \le 6.5$? $\frac{3}{16}$
8. What is the probability that $0 \le Q \le 8$? 1

Probability Probabilities are usually expressed as decimals or fractions ranging from 0 to 1. Probabilities can also be expressed as percents. For example, in Exercise 3, the probability is 1, or 100%, that the point will be on $\overline{AB}$. In Exercise 1, the probability is .5, $\frac{1}{2}$, or 50%, that the point is on $\overline{AM}$. Convert the following probabilities to percents.

9. 0.75 75% 10. $\frac{1}{4}$ 25% 11. $\frac{2}{3}$ $66\frac{2}{3}$%

Convert the following percentages to probabilities between 0 and 1.

12. 60% $\frac{3}{5}$ or 0.6 13. 50% $\frac{1}{2}$ or 0.5 14. $33\frac{1}{3}$% $\frac{1}{3}$ or $0.33\overline{3}$

15. **Meteorology** The weather forecaster predicts an 80% chance of rain. Express this as a probability between 0 and 1. $\frac{4}{5}$ or 0.8

Using Models Have students work in small groups to design target games on graph paper. They should compute the probabilities of an object landing on various regions and then perform experiments to see what actually happens.

Skydiving At the state fair, a sky diver jumps from an airplane and parachutes into a rectangular field as shown. Assume that he is equally likely to land anywhere in the field for Exercises 16–19.

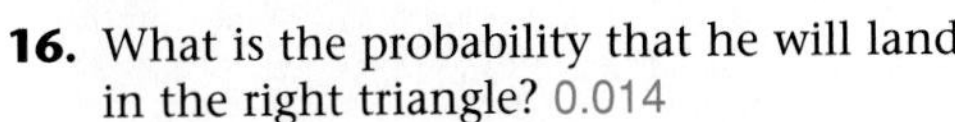

16. What is the probability that he will land in the right triangle? 0.014

17. What is the probability that he will land in the square? 0.089

18. What is the probability that he will land in the circle? 0.18

19. The sky diver will be successful if he lands in one of the shaded figures and not successful if he misses the figures. What is the probability that the sky diver will miss all the shaded areas? 0.717

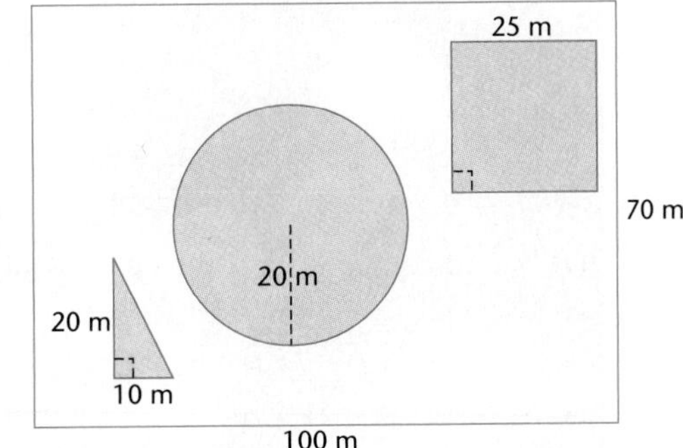

20. The probability of a success (P_s) and the probability of a failure or nonsuccess (P_f) is related by this formula: $P_f = 1 - P_s$. Find the probability that a penny will not land on a point of intersection in Example 2. 0.215

21. Suppose you randomly generate 20 ordered pairs with each coordinate being an integer from 0 to 6. For the graph on the right, what is the probability that an ordered pair lies inside the inner square? $\frac{2}{9}$

22. Conduct an experiment for Exercise 21 generating the ordered pairs using dice, a computer, or a calculator. What is the experimental result, and how does it compare with the theoretical probability you found in Exercise 21?

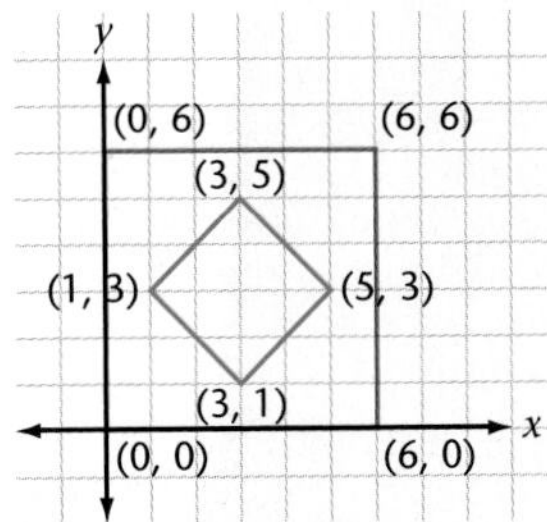

23. **Portfolio Activity** Design a dart-board in which the theoretical probability of a dart landing in a red circle is 50%.

24. **Portfolio Activity** Design a dart-board in which the theoretical probability of a dart landing in a red triangle is $33\frac{1}{3}$%.

25. If the square measures 4 inches per side, and the circle has a diameter of 3 inches, what is the theoretical probability that a dart will land inside the circle? In throwing darts, why might experimental result be higher than theoretical probability?

22. Answers will vary. The more ordered pairs generated, the closer the experimental probability to the theoretical probability.

23. Sample answer: Let the board be circular, with $r_{outer} = 0.707r_{inner}$

24. Sample answer: Let the board be square, with side s. Construct the base and height of the red triangle so that $A_{triangle} = \frac{100s^2}{3}$.

25. 0.44 Depending on the skill of the dart thrower, areas on the outer part of the board are less likely for the dart to land on.

alternative ASSESSMENT

Performance Assessment

Have students work in pairs or small groups to find the probability that a randomly tossed dart will land on the shaded area of this target. The diameter of each circle equals 1 unit.

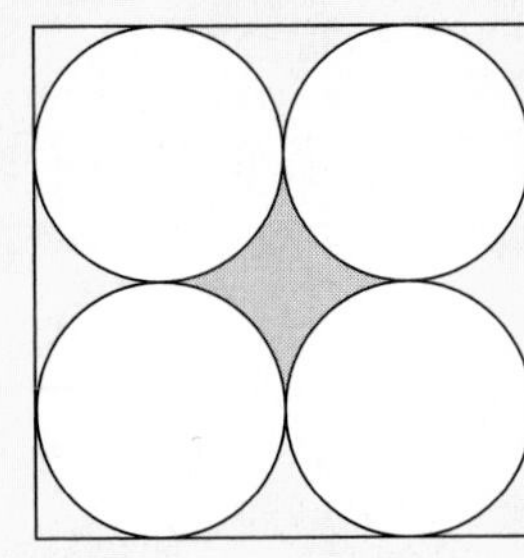

[Shaded area = one quarter of the square minus the area of one circle $= 1 - \frac{\pi}{4} = \frac{4-\pi}{4}$; probability $= \frac{4-\pi}{4} \div$ area of square $= \frac{4-\pi}{4} \div 4 = \frac{4-\pi}{16} \approx 0.054$]

Technology Master

NAME ______ CLASS ______ DATE ______

Technology
5.7 Maximizing Probabilities

To find the geometric probability that a point chosen at random from within the rectangle will be within the circle, you must find the ratio of the area of the circle to that of the rectangle.

Using geometry software, you can easily sketch the circle and rectangle, calculate the area of each figure, and compute the ratio of the areas.

The software enables you to do more, however. Suppose that you enlarge or shrink the circle inside the rectangle. What effect does this have on the probability a point randomly chosen will be inside the circle?

Sketch each diagram whose measurements are given. Then enlarge the circle to find the maximum probability a point chosen at random will be inside the circle.

1.
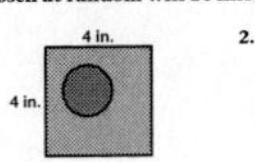

2.
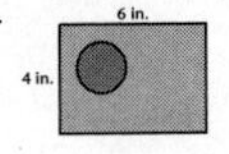

3.
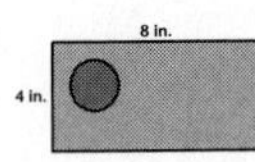

4. Suppose a rectangle has width 4 and length a, where $a \geq 4$. What happens to the maximum probability as a increases?

5. In your software, construct an equilateral triangle that is 5 inches on a side. Place a circle inside it. Calculate the areas of the circle and triangle. Find the maximum probability that a point chosen at random from within the triangle is also inside the circle.

6. Make a conjecture about the circle that maximizes the probability that a point chosen at random from inside a regular polygon will be inside the circle. Use your software to illustrate your conjecture in the case of a regular pentagon.

HRW Geometry Technology 73

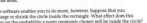

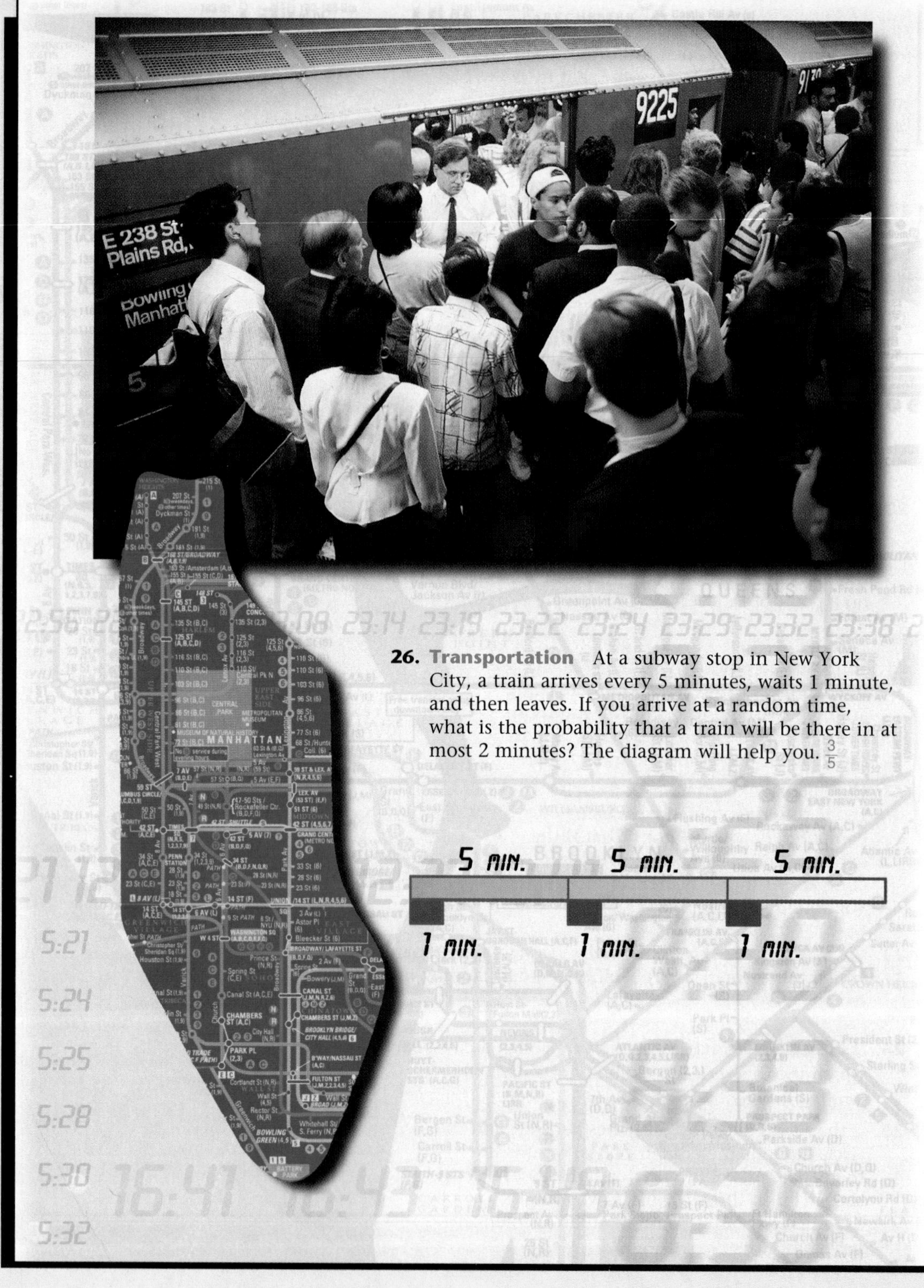

26. Transportation At a subway stop in New York City, a train arrives every 5 minutes, waits 1 minute, and then leaves. If you arrive at a random time, what is the probability that a train will be there in at most 2 minutes? The diagram will help you. $\frac{3}{5}$

Look Back

Find the area for each figure indicated below.

27. Triangle: Base = 4 in, Height = 7.5 in. **[Lesson 5.2]** 15 in.^{2}

28. Parallelogram: Base = 4 cm, Height = 7.5 cm **[Lesson 5.2]** 30 cm^2

29. Regular Polygon: Apothem = 3 ft, Perimeter = 18 ft **[Lesson 5.2]** 27 ft^2

30. Trapezoid: $\text{Base}_1 = 20$ yd, $\text{Base}_2 = 30$ yd, Height = 12.6 yd **[Lesson 5.2]** 315 yd^2

31. Circle: Radius = 16 mm (use $\frac{22}{7}$ for π) **[Lesson 5.3]** 804.57 units

Look Beyond

Fine Art Perspective is important in painting and drawing. Use the following picture to answer the questions below.

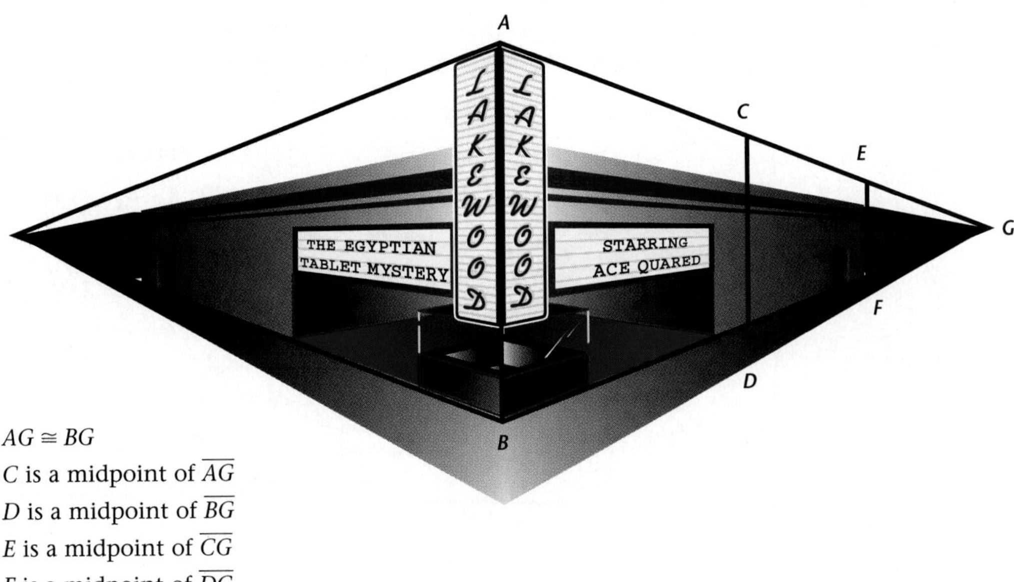

$AG \cong BG$
C is a midpoint of $\overline{AG}$
D is a midpoint of $\overline{BG}$
E is a midpoint of $\overline{CG}$
F is a midpoint of $\overline{DG}$

32. If $AB = 8$ in., $CD =$ __?__. 4

33. If $AB = 8$ in., $EF =$ __?__. 2

34. Explain your reasoning for Exercises 32 and 33.

Look Beyond

Exercises 32–34 encourages students to make conjectures about similar figures, which will be explored in Chapter 8. To work the exercises, students may intuitively use the idea that corresponding parts of similar figures are in proportion, or have the same ratio. They can also use measurements of a precise drawing.

32. $\triangle ABG$ seems to be twice the size of $\triangle CDG$, so it would seem that the base of $\triangle CDG$ is one-half the base of $\triangle ABG$. Similarly for $\triangle CDG$ and $\triangle EFG$.

Focus

In this activity, students generate data to derive an area formula sometimes called *Pick's Formula.* Students will find that the equation $A = \frac{Nb}{2} + Ni - 1$, or any equation equivalent to this, will relate the three variables in their data lists.

Motivate

Point out to students that they have learned the area formulas for circles, triangles, and a few other polygons. Ask them to describe how they would find the area of a polygon with an irregular boundary. [**Possible answer: Divide the polygon up into triangles, measure or compute an altitude for each triangle, find each triangle area, and add. Some students might suggest using geometry graphics software, which will find the area of any polygon.**]

After students have read the instructions, ask if they will end up with an area formula that will work for *any* polygon. The answer is "no." This formula only works if the vertices of the polygon are on the intersection points of a square grid.

Area of a Polygon

This project challenges you to find a formula for the areas of polygons drawn on dot paper—surprisingly, one formula will work for all! Work with a team of 3 or 4 classmates. You will probably find it helpful to divide up the work among your team's members.

Compute the areas for figures a–f in each of the three parts below. Then, on a separate sheet of paper, complete a table like the ones given in each part. For each figure, N_b is the number of points on the boundary of the figure, and N_i is the number of points in its interior. *A* is the area of the figure.

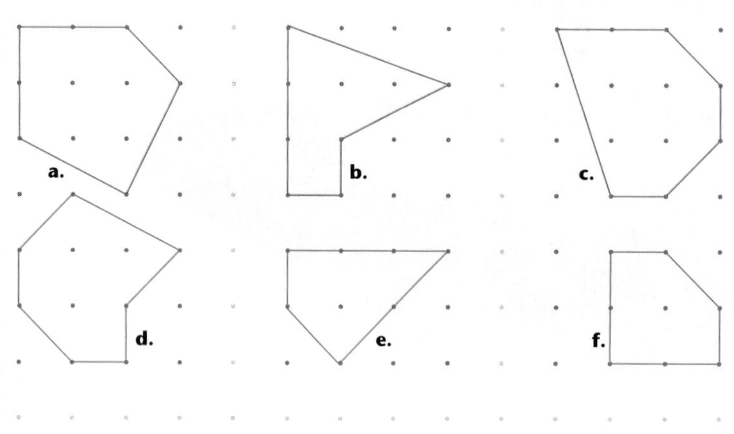

Part A

	N_b	N_i	A
a.	7	4	$6\frac{1}{2}$
b.	?	?	?
c.	?	?	?
d.	?	?	?
e.	?	?	?
f.	?	?	?

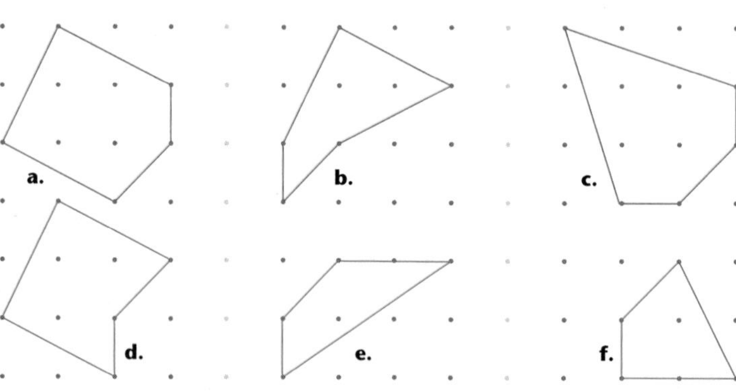

Part B

	N_b	N_i	A
a.	?	?	?
b.	?	?	?
c.	?	?	?
d.	?	?	?
e.	?	?	?
f.	?	?	?

Part A

	N_b	N_i	A
a.	7	4	6.5
b.	7	2	4.5
c.	7	4	6.5
d.	7	3	5.5
e.	7	1	3.5
f.	7	1	3.5

Part B

	N_b	N_i	A
a.	5	4	5.5
b.	5	2	3.5
c.	5	4	5.5
d.	5	3	4.5
e.	5	1	2.5
f.	5	1	2.5

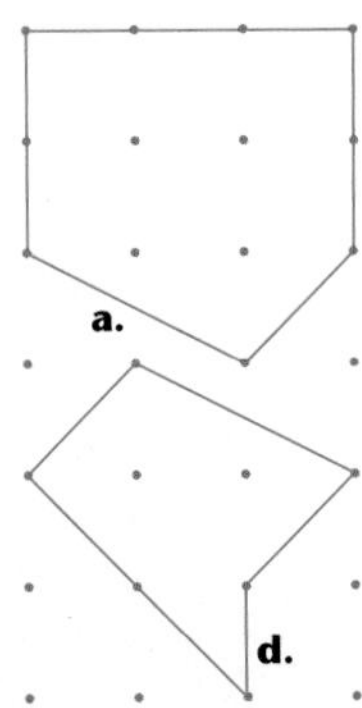

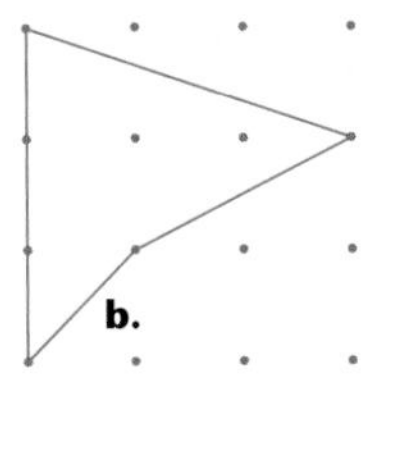

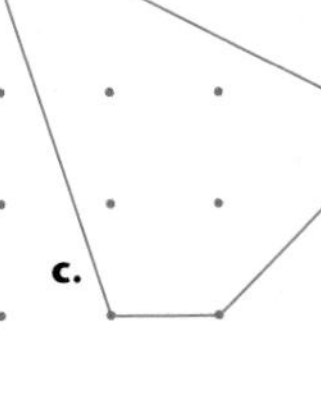

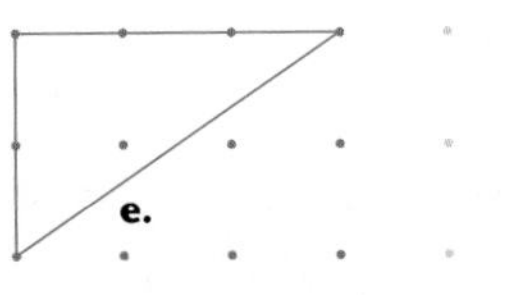

f.

Part C

	N_b	N_i	A
a.	?	?	?
b.	?	?	?
c.	?	?	?
d.	?	?	?
e.	?	?	?
f.	?	?	?

EXTENSION

To help you determine the formula, each part focuses on a fixed number of boundary points.

1. What is the pattern to calculate the area?
2. Write the formula.
3. Test your formula by creating new figures on dot paper.

The formula you have discovered was originally discovered by G. Pick in 1899.

Cooperative Learning

Students doing the Project with partners or in small groups can benefit from working each problem independently and then comparing results. It is easy to make an error in counting grid points or unit squares.

DISCUSS

After all students have finished Parts A, B, and C, lead a class discussion. Have students share the methods they used for finding areas. In some cases, for example, polygon c in Part B, it may be easier to start with 16 square units and subtract triangular regions.

Part C

	N_b	N_i	A
a.	6	4	6
b.	6	2	4
c.	6	4	6
d.	6	3	5
e.	6	1	3
f.	6	1	3

Extension

1. Take half the number of points on the boundary and add the number of points in the interior minus 1.
2. $A = \frac{1}{2} N_b + (N_i - 1)$
3. Check student work.

Chapter 5 Review

Vocabulary

Key Skills and Exercises

Lesson 5.1

➤ Key Skills

Find the perimeter of a polygon.

In the figure, find the perimeter of the hexagon.

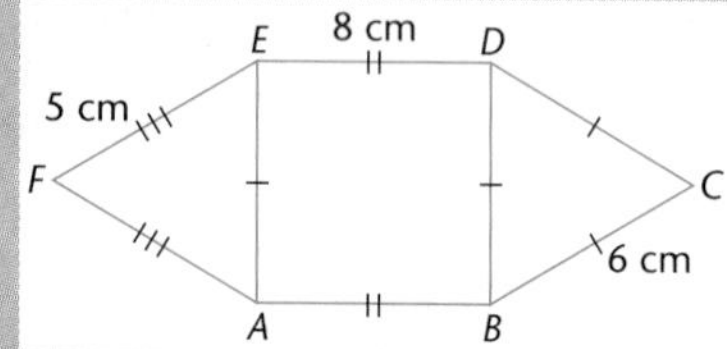

The perimeter equals the sum of all sides: $P = AB + BC + CD + DE + EF + FA$. We are given that $AB = DE = 8$ cm, $FA = EF = 5$ cm, and $CD = BC = 6$ cm. Thus, $P = 8 + 6 + 6 + 8 + 5 + 5 = 38$ cm.

Find the area of a rectangle.

In the figure above, find the area of rectangle *ABDE*.

The formula for the area of a rectangle is $A = bh$. Here $\overline{BD}$ is the height; $BD = BC = 6$ cm. $\overline{ED}$ is the base (8 cm). Thus, $A = 8 \times 6 = 48 \text{ cm}^2$.

➤ Exercises

Quadrilaterals ABFG and BCDE are rectangles.

1. Find the perimeter of *ACDEFG*.

2. Find the area of *ACDEFG*.

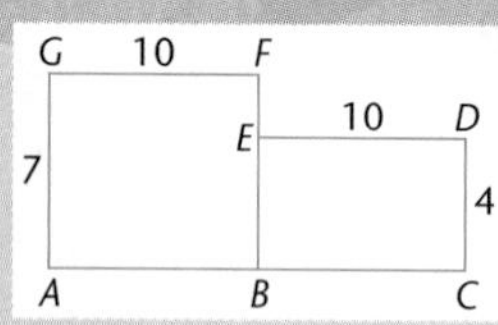

Lesson 5.2

➤ Key Skills

Solve problems using the formulas for the areas of triangles, parallelograms, and trapezoids.

This trapezoid has an area of 455 mm^2. What is its height?

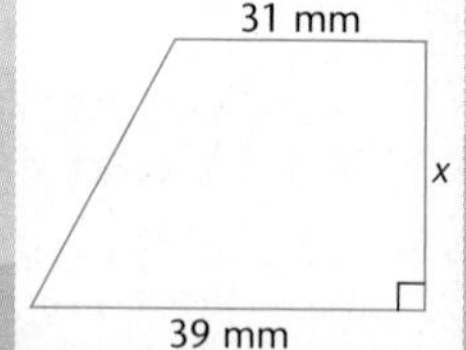

The formula for the area of a trapezoid is $A = \frac{(b_1 + b_2)h}{2}$.
Substitute the given dimensions and solve the equation.

$$455 = \frac{(31 + 39)x}{2}$$
$$910 = 70x$$
$$x = 13$$

The trapezoid's height is 13 mm.

1. 54 units

2. 110 units^2

Chapter Review

➤ Exercises

In Exercises 3–5, refer to the figure.

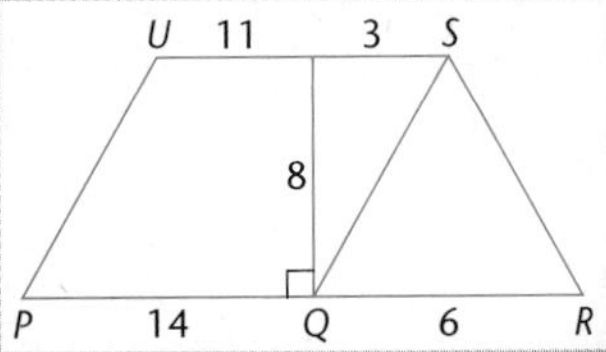

3. Find the area of parallelogram *PQSU*.
4. Find the area of trapezoid *PRSU*.
5. Find the area of $\triangle QRS$.

Lesson 5.3

➤ Key Skills

Solve problems using the formulas for the circumference and area of a circle.

A wheel makes 25 revolutions to go 4 meters. Find the radius of the wheel.

We know the circumference of the wheel is $\frac{4}{25}$ = .16 meters, or 16 centimeters. We use the formula for circumference, 2π times the radius, to find the radius.

$16 = 2\pi r \approx 2(3.14)r$

$r \approx 2.5$ centimeters.

➤ Exercises

6. Find the area of a circle whose diameter is 21 inches.
7. If the circumference of a circle is 3.14 yards, find its radius.

Lesson 5.4

➤ Key Skills

Solve problems using the "Pythagorean" relationship.

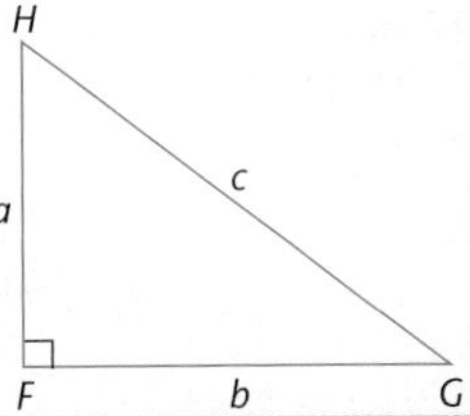

The "Pythagorean" relationship holds for any right triangle. The relationship of the length of the sides of $\triangle FGH$ is $a^2 + b^2 = c^2$.

➤ Exercises

In Exercises 8–9, find the missing measures. Round to the nearest tenth.

8. 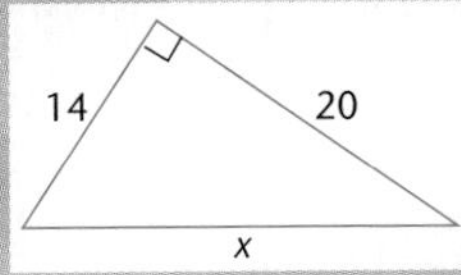

9. 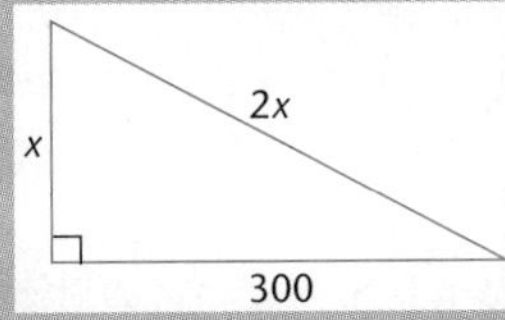

Lesson 5.5

➤ Key Skills

Use the properties of 45-45-90 and 30-60-90 right triangles to solve problems.

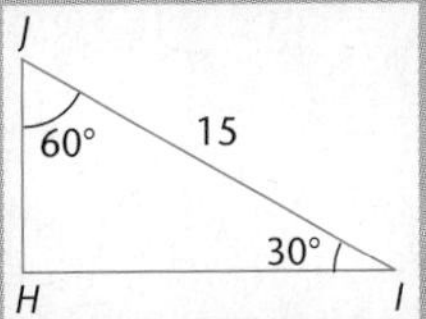

Find *HI* and *HJ*. Since two angles of $\triangle HIJ$ measure 30° and 60°, it is a 30-60-90 right triangle. In that type of triangle, the shorter leg is half the length of the hypotenuse, and the longer leg is $\sqrt{3}$ times the length of the shorter leg. In $\triangle HIJ$, $HJ = 7.5$, and $HI = 7.5\sqrt{3}$.

9. If alternate interior angles UVX and WXV are both 90°, then UV ∥ XW by the converse of the Alternate Interior Angles Theorem.
10. 55°
11. 55°
12. 75°

3. 112 units^2
4. 136 units^2
5. 24 units^2
6. ≈346.36 in^2
7. ≈0.5 yards
8. ≈24.4
9. $x \approx 173.2$, $2x \approx 346.4$

➤ Exercises

In Exercises 10–11, find the missing measures.

10.

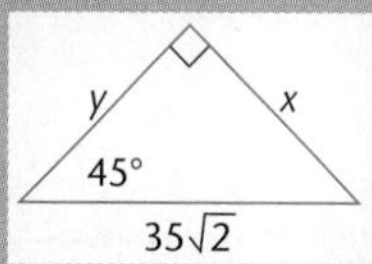

11.

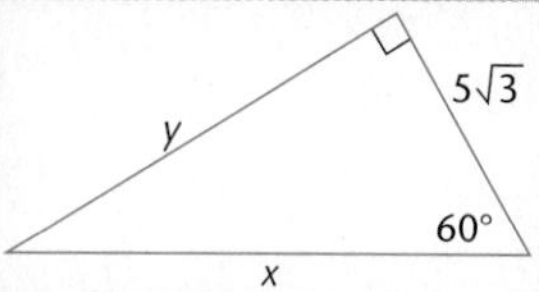

Lesson 5.6

➤ Key Skills

Determine the distance and the midpoint between two points on a coordinate plane.

Find the distance between points (6, 5) and (–6, 10).

You can use the distance formula, $d = \sqrt{(x_2 - x_1)^2 + (y_2 - y_1)^2}$.

$d = \sqrt{(-6 - 6)^2 + (10 - 5)^2} = \sqrt{(-12)^2 + (5)^2} = \sqrt{169} = 13$

The midpoint for two points is given by $\left(\frac{x_2 + x_1}{2}, \frac{y_2 + y_1}{2}\right)$. Thus, $x = \frac{-6 + 6}{2} = 0$, and $y = \frac{10 + 5}{2} = 7.5$. The midpoint is (0, 7.5).

➤ Exercises

In Exercises 12–13, refer to the figure.

12. Find the perimeter of $\triangle ABC$.

13. $\triangle DEF$ has its vertices at the midpoints of the sides of $\triangle ABC$. Give the coordinates of $\triangle DEF$'s vertices.

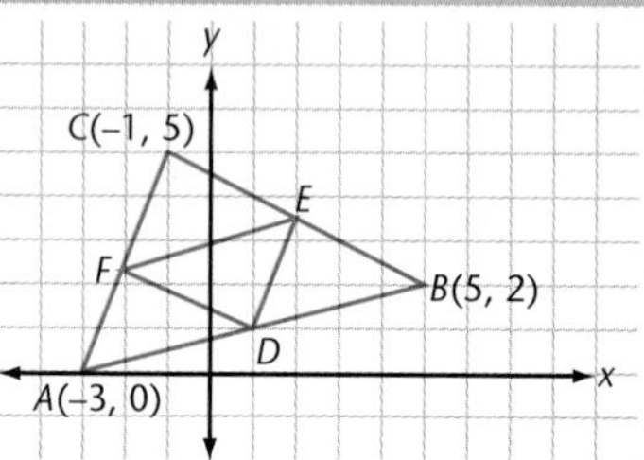

Lesson 5.7

➤ Key Skills

Find the probability of an event, using the ratio of two areas.

In the sketch, the circle represents a tree's crown, and the rectangle represents a reclining chair. If the tree drops an apple, what are the chances the chair will be hit? (Assume that the apple is equally likely to hit anywhere under the tree.)

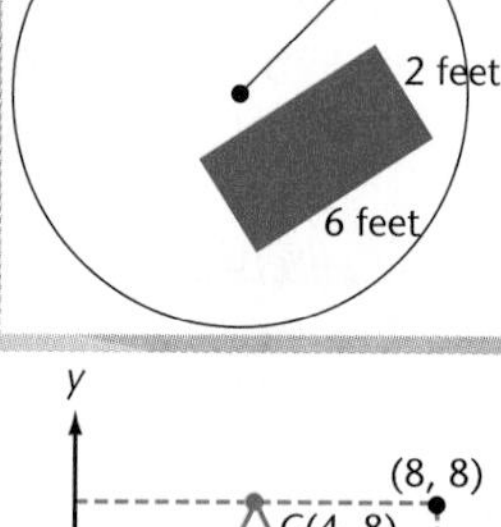

Find the ratio of the area of the rectangle to the area of the circle.

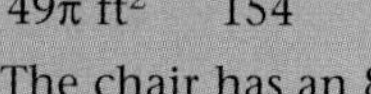

$\frac{12 \text{ ft}^2}{49\pi \text{ ft}^2} \approx \frac{12}{154} \approx .08$

The chair has an 8% chance of being hit by an apple.

➤ Exercises

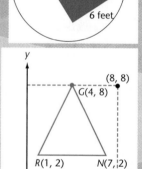

14. A random-number generator gives you ten pairs of integers, ranging from 0 to 8. What is the probability that $\triangle RNG$ includes one of the pairs?

y
(8, 8)
G(4, 8)
R(1, 2)
N(7, 2)
x
8

Applications

15. You have a 20-foot ladder to reach a roof that is 15 feet high. The top of the ladder must extend at least 3 feet above the roof's edge before you can safely step from the ladder onto the roof. To the nearest foot, what is the farthest from the house wall that you could set the bottom of the ladder?

10. $x = y = 35$

11. $x = 10\sqrt{3}; y = 15$

12. ≈ 20.34

13. $D(1, 1); E(2, 3.5); F(-2, 2.5)$

14. $\approx .28$ **15.** 8 ft

Chapter 5 Assessment

In Exercises 1–2, refer to the figure at the right.

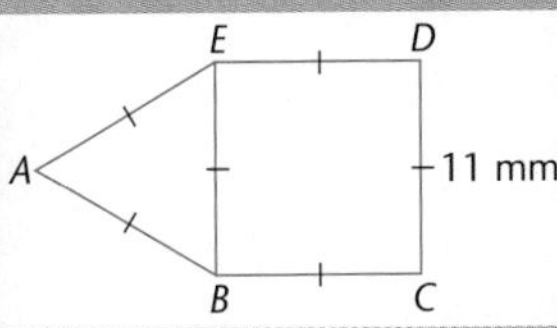

1. Find the perimeter of *ABCDE*.

2. Find the area of *ABCDE*.

3. Find the area of trapezoid *MNOP*.

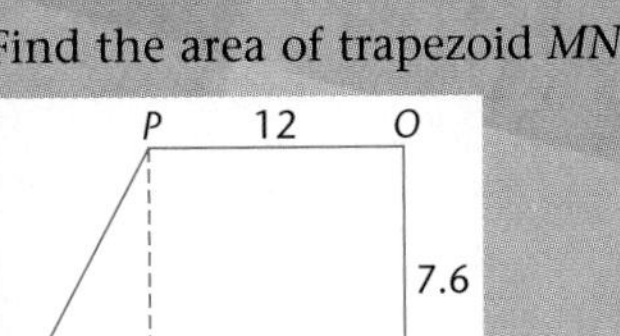

4. If the circumference of a circle is 30 millimeters, what is its area?

a. 706.9 mm^2 **b.** 289.4 mm^2 **c.** 72.3 mm^2 **d.** 94.2 mm^2

In Exercises 5–6, find the missing measures.

5.

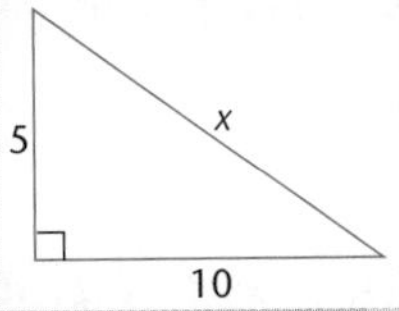

6.

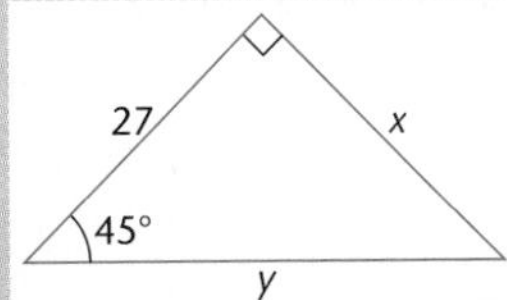

7. Find the area of a regular hexagon with a side of 6 inches and an apothem of $3\sqrt{3}$ inches.

In Exercises 8–9, refer to the figure at the right.

8. Find the perimeter of quadrilateral *ABCD*.

9. $\triangle EFG$ has its vertices at the midpoints of three sides of *ABCD*. Give the coordinates of $\triangle EFG$'s vertices.

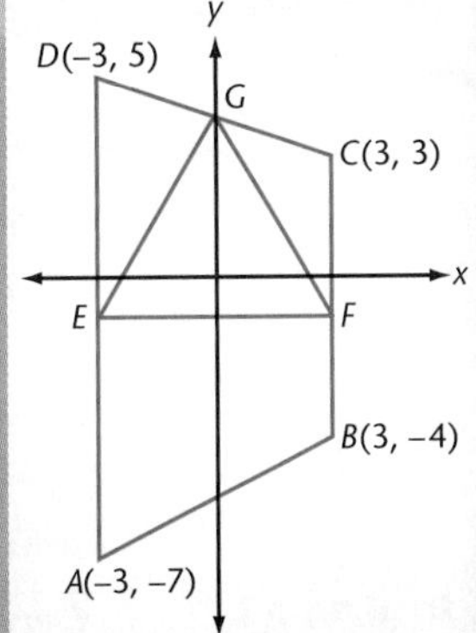

10. A point *P* is selected at random from the segment below. What is the probability that $2 \le P \le 8$?

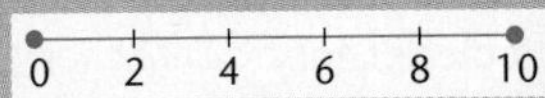

1. 55 mm
2. ≈173.39 mm^2
3. 117.8 units^2
4. c
5. ≈11.18
6. $x = 27$; $y = 27\sqrt{2}$
7. $x \approx 93.53$ in^2
8. ≈32.03
9. $E(-3, -1)$; $F(3, -0.5)$; $G(0, 4)$
10. $\frac{3}{5}$ or 0.6

CHAPTER 6 Shapes in Space

Meeting Individual Needs

6.1 Exploring Solid Shapes

Core Resources	Core Plus Resources
Inclusion Strategies, p. 306 Reteaching the Lesson, p. 307 Practice Master 6.1 Enrichment, p. 306 Technology Master 6.1 Lesson Activity Master 6.1 Interdisciplinary Connection, p. 305	Practice Master 6.1 Enrichment Master 6.1 Technology Master 6.1 Interdisciplinary Connection, p. 305
[1 day]	[1 day]

6.2 Exploring Spatial Relationships

Core Resources	Core Plus Resources
Inclusion Strategies, p. 313 Reteaching the Lesson, p. 314 Practice Master 6.2 Enrichment, p. 313 Technology Master 6.2 Lesson Activity Master 6.2 Interdisciplinary Connection, p. 312	Practice Master 6.2 Enrichment Master 6.2 Technology Master 6.2
[2 days]	[2 days]

6.3 An Introduction to Prisms

Core Resources	Core Plus Resources
Inclusion Strategies, p. 319 Reteaching the Lesson, p. 320 Practice Master 6.3 Enrichment, p. 319 Technology Master 6.3 Lesson Activity Master 6.3 Interdisciplinary Connection, p. 318 Mid-Chapter Assessment Master	Practice Master 6.3 Enrichment Master 6.3 Technology Master 6.3 Interdisciplinary Connection, p. 318 Mid-Chapter Assessment Master
[2 days]	[1 day]

6.4 Three-Dimensional Coordinates

Core Resources	Core Plus Resources
Inclusion Strategies, p. 327 Reteaching the Lesson, p. 328 Practice Master 6.4 Enrichment, p. 327 Lesson Activity Master 6.4	Practice Master 6.4 Enrichment Master 6.4 Technology Master 6.4
[2 days]	[2 days]

6.5 Lines and Planes in Space

Core Resources	Core Plus Resources
Inclusion Strategies, p. 333 Reteaching the Lesson, p. 334 Practice Master 6.5 Enrichment Master 6.5 Lesson Activity Master 6.5	Practice Master 6.5 Enrichment p. 333 Technology Master 6.5
[1 day]	[1 day]

6.6 Perspective Drawing

Core Resources	Core Plus Resources
Inclusion Strategies, p. 338 Reteaching the Lesson, p. 339 Practice Master 6.6 Enrichment, p. 338 Lesson Activity Master 6.6 Interdisciplinary Connection, p. 337	Practice Master 6.6 Enrichment Masters 6.6 Technology Master 6.6 Interdisciplinary Connection, p. 337
[2 days]	[2 days]

Chapter Summary

Core Resources	Core Plus Resources
Chapter 6 Project, pp. 344–345 Lab Activity Long Term Project Chapter Review, pp. 246–348 Chapter Assessment p. 349 Chapter Assessment, A/B Alternative Assessment Cumulative Assessment, pp. 350–351	Chapter 6 Project, pp. 344–345 Lab Activity Long Term Project Chapter Review, pp. 246–348 Chapter Assessment p. 349 Chapter Assessment, A/B Alternative Assessment Cumulative Assessment, pp. 350–351
[3 days]	**[2 days]**

Hands-On Strategies

Many students will have had almost no experience thinking about and working with three-dimensional shapes. It will be extremely helpful to have centimeter or inch cubes for Lesson 6.1 and a variety of prisms and other solids for the rest of the chapter. These solid models will also be useful in Chapter 7.

Students will also benefit from building their own models of solid shapes. The drawings of the different prisms in Lesson 6.3 can be used as a starting point, although students should be encouraged to build pyramids as well as prisms. Adequate models can be created by folding, cutting, and taping standard paper. More permanent models require thin cardboard.

Reading Strategies

Although the majority of the vocabulary used in this chapter is not new, it is used in a new setting. Students may clearly understand what it means for two lines to be parallel or perpendicular but parallel and perpendicular planes are completely different ideas. The chapter also uses vocabulary specific to the ideas involved with three-dimensional figures. Examples include *coplanar* and *dihedral angle*.

Visual Strategies

This chapter and Chapter 7 concentrate on properties of three-dimensional figures. In both chapters, it will be a visual challenge for students to learn to interpret the two-dimensional drawings used to show solid shapes. One type of activity that is helpful is for students to draw simple shapes from a variety of perspectives. For example, show a soup can or other simple cylinder and ask students to draw top, side, and other views.

Computer software programs are available that can be used to create and manipulate images of three-dimensional objects. Of particular interest is the capability of the software to rotate objects in space, so that the same object can be seen from different perspectives.

Cooperative Learning

GROUP ACTIVITIES	
Isometric drawings	Lesson 6.1, Explorations 1 and 2
Dihedral angles	Lesson 6.2, Exploration 3
Did the Camera Lie?	Eyewitness Math
Distances in the solar system	Lesson 6.4, Look Beyond
Perspective drawings	Lesson 6.6
Building an A-Shaped Level	Chapter Project

You may wish to have students work in groups or with partners for some of the above activities. Additional suggestions for cooperative group activities are noted in the teacher's notes in each lesson.

Multicultural

The cultural references in this chapter include references to Europe.

CULTURAL CONNECTIONS	
Europe: Euler's Formula	Lesson 6.3, Exercise 38
Europe: Perspective in art	Lesson 6.6

Portfolio Assessment

Below are portfolio activities for the chapter listed under seven activity domains which are appropriate for portfolio development.

1. **Investigation/Exploration** The explorations in Lesson 6.1 involve experiences in making two-dimensional drawings of solids built from unit cubes. In Lesson 6.2, students continue this work, focusing on segments, planes, angles, and parallel lines in space. The Lesson 6.3 exploration deals with the lateral faces of prisms.
2. **Applications** Navigation, Lesson 6.2, Exercises 16–19; Design, Lesson 6.6, Exercise 4; Carpentry, Chapter Project.
3. **Nonroutine Problems** Lesson 6.1, Exercise 29 (The Impossible Cube); Lesson 6.1, Exercises 38–41 (Orthographic Projections); Lesson 6.5, Exercises 24–25 (Surfaces of Revolution).
4. **Project** Building and Testing an A-Shaped Level: see pages 344–345. Students apply properties of perpendicular lines as they build and experiment with a level.
5. **Interdisciplinary Topics** Chemistry, Lesson 6.3, Application; Sports, Eyewitness Math; Astronomy, Lesson 6.4, Exercises 41–43; Algebra, Lesson 6.5, Exercises 5–18; Fine Arts, Lesson 6.6.
6. **Writing** *Communicate* exercises offer excellent writing selections for the portfolio. Suggested selections include: Lesson 6.1, Exercise 1; Lesson 6.2, Exercises 4–5; Lesson 6.4, Exercise 7; Lesson 6.5, Exercises 2–4; Lesson 6.6, Exercises 3–4.
7. **Tools** In Chapter 6, students use isometric grid paper to create drawings of three-dimensional objects. They can also use a graphics calculator, or any programmable calculator, for the computation involved in the distance formula.

Technology

What was once anxiety-provoking drudgery is increasingly becoming a relatively simple matter. The work of evaluating complex mathematical formulas can now be done quickly and precisely by computers and calculators. However, some mathematical formulas or procedures are used repeatedly in certain contexts, so that it is desirable to have a way of bypassing the routine parts of the calculations even when using a caclulator.

The three-dimensional distance formula, for example, (see pp. 328 and 330), is relatively simple to evaluate using a calculator. But for a person who has to use it frequently, a reduction in the number of steps required will result in considerable savings of time and energy. Such reductions can be achieved by programming a calculator or computer to do repetitive aspects of the calculation for the user.

Calculator Programs

The most straightforward calculator program for a distance formula will give a user a number of places to enter the coordinates of two different points. Then, after the coordinates have been entered, the program will calculate and display the distance between them. With this in mind, study the following calculator program for the three-dimensional distance formula.

```
PROGRAM:DIST3
:Disp "1st POINT COORDS"
:Prompt X
:X→A
:Prompt Y
:Y→B
:Prompt Z
:Z→C
:Disp "2nd POINT COORDS"
:Prompt X
:Prompt Y
:Prompt Z
:√((A-X)2+(B-Y)2+(C-Z)2)→D
:Disp "THE DISTANCE IS",D
:Stop
```

The first command (beginning with the first colon) causes the calculator to print "1ST POINT COORDS" on the display screen. The second command then gives a prompt for the variable *X*, which the user should input. At this point in the execution of the program the screen will be as shown on the left.

The user should now enter the x-coordinate of the first point. For Exercise 41 on page 330, for example, this might be .8189001, the *x*-coordinate of the Earth/Moon center's location in space.

Since the calculator has no provision for subscripted variables, the user's entry is stored into variable *A* (in the next step) so that the variable *X* can be used again for entering the coordinates of the second point.

In the third line from the end of the program, the value of the distance is calculated from the values stored in variables *A*, *B*, *C*, *X*, *Y*, and *Z* and is stored in the variable *D*. The next command causes the calculator to print the string "THE DISTANCE IS" followed by the value stored in *D*. The program then terminates. At this point, for the same exercise, the calculator screen would appear as shown.

Encourage students to write their own programs for the distance formula, which may differ in details from the one shown here. This experience will increase their enjoyment of the exercises and will introduce them to a valuable skill.

ABOUT THE CHAPTER

Background Information

This chapter introduces students to geometric figures in space. Students will extend ideas and concepts about plane figures to three-dimensional figures or spatial figures. Volume and surface area are introduced. A formal presentation of volume and surface area follows in Chapter 7.

CHAPTER RESOURCES

- Practice Masters
- Enrichment Masters
- Technology Masters
- Lesson Activity Masters
- Lab Activity Masters
- Long-Term Project Masters
- Assessment Masters
  Chapter Assessments, A/B
  Mid-Chapter Assessment
  Alternative Assessments, A/B
- Teaching Transparencies
- Spanish Resources

CHAPTER OBJECTIVES

- Use isometric grid paper to draw three-dimensional shapes built with cubes.
- Develop an understanding of orthographic projection.
- Develop a preliminary understanding of volume and surface area.
- Identify the unique relationships of points, lines, segments, planes, and angles in three-dimensional space.
- Define *dihedral angle.*
- Define *right* and *oblique prism.*

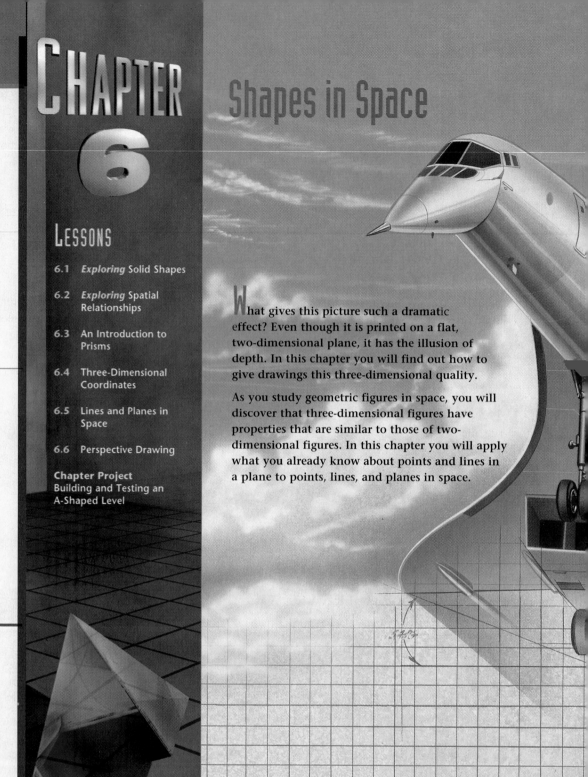

CHAPTER 6

Shapes in Space

LESSONS

What gives this picture such a dramatic effect? Even though it is printed on a flat, two-dimensional plane, it has the illusion of depth. In this chapter you will find out how to give drawings this three-dimensional quality.

As you study geometric figures in space, you will discover that three-dimensional figures have properties that are similar to those of two-dimensional figures. In this chapter you will apply what you already know about points and lines in a plane to points, lines, and planes in space.

ABOUT THE PHOTO

Geometry provides the aeronautical engineer with the representational tools necessary to draw and manipulate designs. The photo emphasizes how two-dimensional drawings can be an accurate representation of a three-dimensional object.

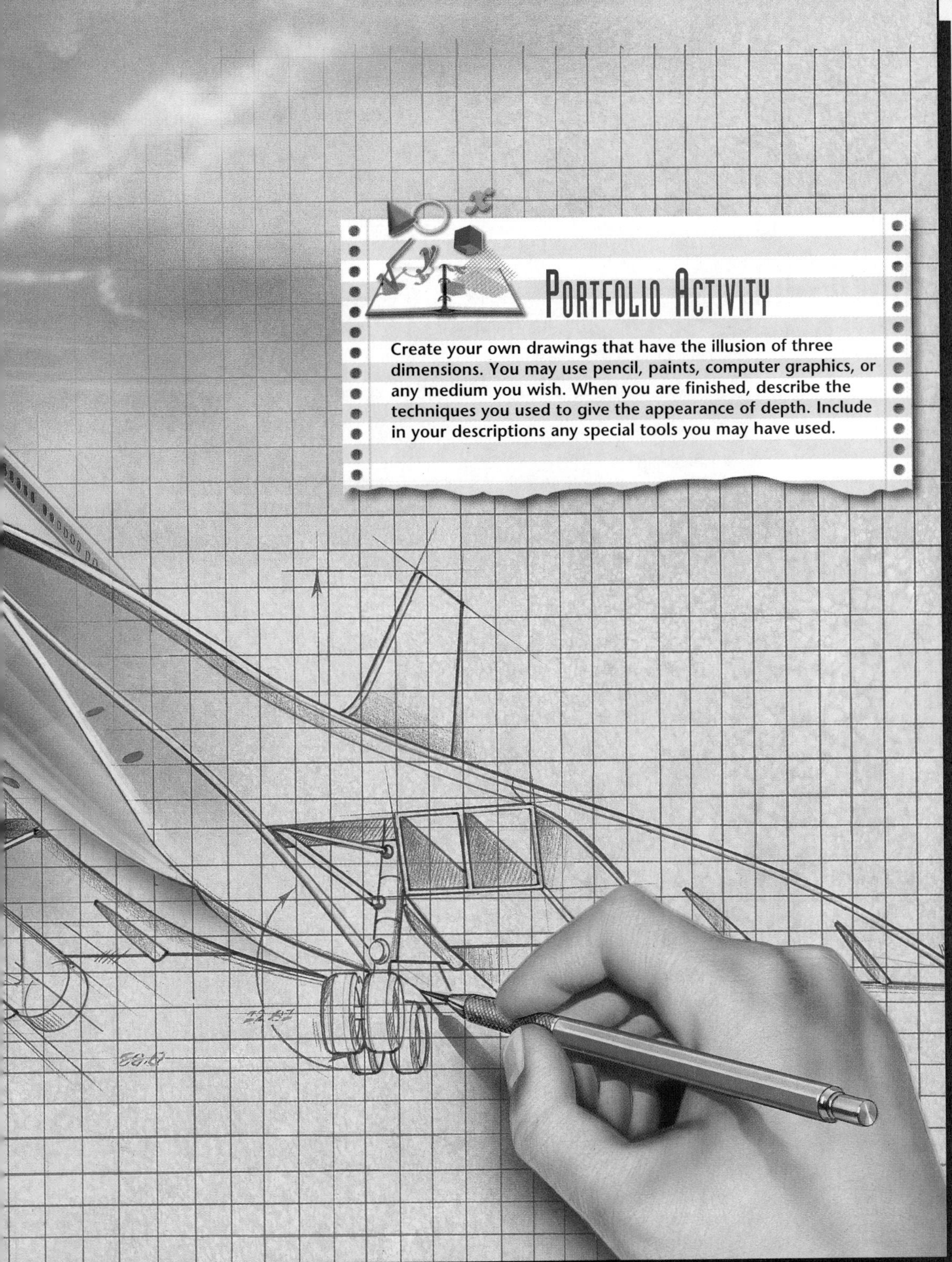

Portfolio Activity
Create your own drawings that have the illusion of three dimensions. You may use pencil, paints, computer graphics, or any medium you wish. When you are finished, describe the techniques you used to give the appearance of depth. Include in your descriptions any special tools you may have used.

About the Chapter Project

In the Chapter 6 Project, on pages 344–345, students use ordinary materials to build an A-shaped level that can be used in construction or surveying. Have students test the level on objects in the classroom and then use basic properties of geometry to discuss why the level works.

- Solve problems using the diagonal measure of a right prism.
- Identify the features of a three-dimensional coordinate system, including axes, octants, and coordinate notation.
- Solve problems using the distance formula in three dimensions.
- Define the *equation of a line* and the *equation of a plane* in space.
- Solve problems using the equations of lines and planes in space.
- Identify and define the elementary concepts of perspective drawing.
- Apply perspective drawing concepts by creating perspective drawings.

Portfolio Activity

Use this portfolio activity to provide an interdisciplinary connection to art and computer graphics.

Students should include preliminary sketches as well as final drawings in their portfolios and be prepared to discuss their techniques. Cooperative groups can be used to compare techniques and to provide assistance for students having difficulty drawing.

Extend the activity by having students draw the Platonic solids and other polyhedra. Compile the drawings into a class *Solid Shapes* book. Additional Pupil's Edition portfolio activities can be found in the exercises for Lessons 6.1 and 6.6. Additional Pupil's Edition portfolio activities can be found in the exercises for Lessons 6.1 and 6.6.

PREPARE

Objectives

- Use isometric grid paper to draw three-dimensional shapes built with cubes.
- Develop an understanding of orthographic projection.
- Develop a preliminary understanding of volume and surface area.

- Practice Master 6.1
- Enrichment Master 6.1
- Technology Master 6.1
- Lesson Activity Master 6.1
- Quiz 6.1
- Spanish Resources 6.1

Assessing Prior Knowledge

1. How many faces does a cube have? [6]
2. How many faces of a cube are ordinarily hidden from view? [3]
3. Find the area of a square with side lengths of 1 cm. [1 cm^2]

TEACH

Isometric grid paper drawings and other drawings and models used in this lesson will give students a concrete introduction to drawing solid figures and finding volumes and surface areas.

Use Transparency 48

LESSON 6.1 Exploring Solid Shapes

Why *Architects must be skilled in rendering their conceptions of solid shapes. There are many other careers that require this skill. Perhaps you have already applied some of these skills in some of your school courses.*

Mechanical Drawing

An **isometric drawing** is a type of three-dimensional drawing. In an isometric drawing of a cube, each of the nonvertical edges makes a 30° angle with a horizontal line on the plane of the picture.

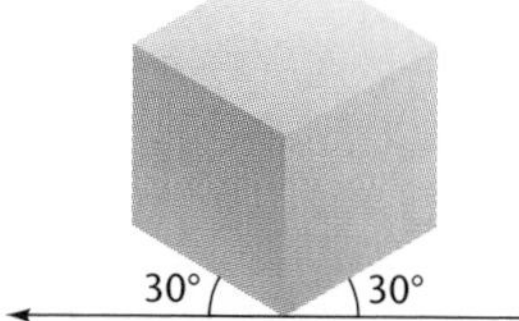

Drawing Cubes

Isometric grid paper can be used to draw a cube.

Ordinarily, not all six faces of a cube are visible. To make the drawing of the cube more realistic, the "hidden" segment can be erased. Three faces of the cube are visible: top, left, and right. The hidden parts of the cube can be represented with dashed lines, as though the faces in front were semitransparent.

A cube can be drawn so that it appears as if you are looking up at the bottom face.

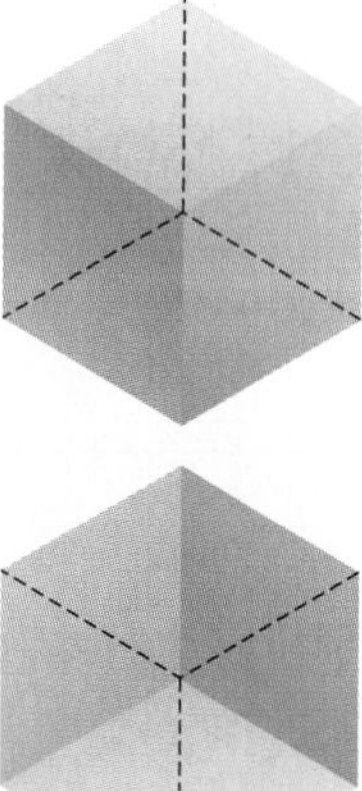

Hands-On Strategies

Students can explore different views of an object by taking apart various common boxes like cereal boxes or shoe boxes. Have them try to create a closed box by assembling it from nets cut out of grid paper. Calculate the surface area of the closed box by counting the number of squares of grid paper used to assemble it.

Exploration 1 Using Isometric Grid Paper

You will need
Isometric grid paper

1. Draw each of the solid figures on isometric grid paper. Then redraw the figure with the red cube or cubes removed.

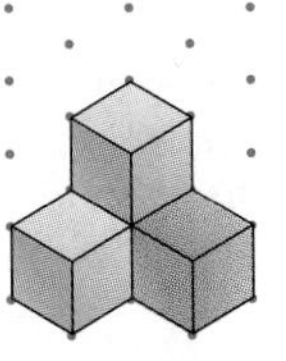

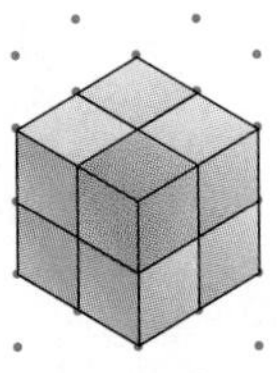

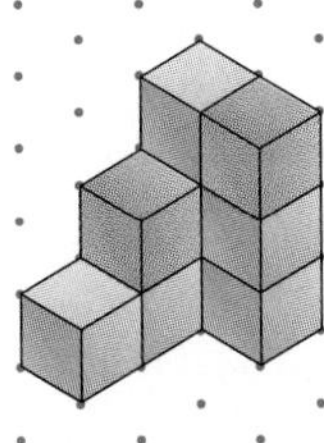

2. Draw each of the solid figures on isometric grid paper. Then add a cube to each red face and draw the new figure.

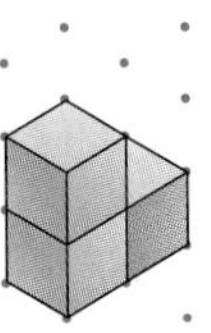

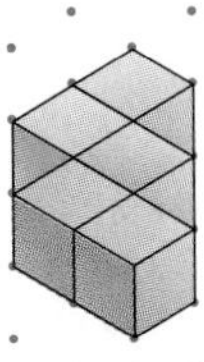

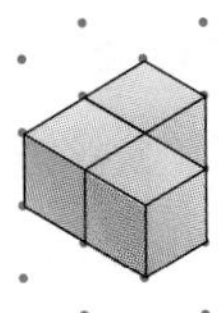

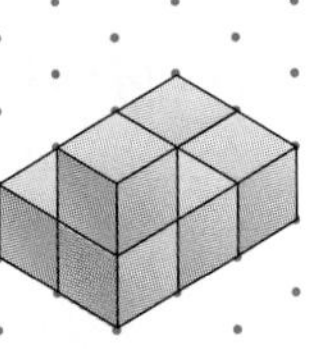

3. Were any of the drawings more difficult to make than others? What special problems did you encounter? Discuss. ❖

Exploration 2 Using Unit Cubes

You will need
Unit cubes
Isometric grid paper

1. Using unit cubes, build the three figures on the following page. Do the figures you built look like those in the pictures? Do the figures you built look like those built by your classmates? If not, why?

Exploration 1 Notes

This activity will give students practice drawing solids made out of cubes. Encourage students to show hidden lines, if necessary. Make the connection to surface area by pointing out which faces of the solids are exposed.

Use Transparency 49

Ongoing Assessment

3. **Drawings that require cubes to be drawn on top of each other are usually more difficult.**

Exploration 2 Notes

Students will draw their own shapes in this activity. Use student-designed shapes to introduce students to orthographic projection.

interdisciplinary CONNECTION **Industrial Arts** Some students in the class may have taken a course in mechanical drawing. Ask them to display some of their three-dimensional drawings and orthographic projections for the class and have them discuss various tools they use.

3. Yes. Parts of the figure might be hidden by the visible cubes, such as a cubic hole in the last figure.

Teaching tip

Using blocks with numbers or letters (like children's blocks) to create three-dimensional figures may help students distinguish different views of the figure.

Cooperative Learning

Have small groups of two or three students collaborate on how to draw different views of various objects in the classroom, including chairs, desks, file cabinets, etc. Small groups will provide the opportunity for students who are talented in drawing to help others.

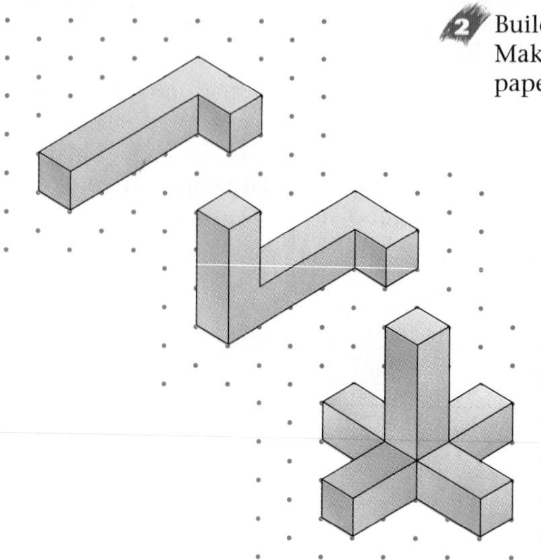

2 Build shapes of your own design from unit cubes. Make drawings of the shapes on isometric grid paper.

3 Do the representations in the illustrations leave any guesswork about the solid figures? Could there be parts of the figures that do not show in the drawings? Could there be any parts that do not show in the designs you created? Discuss. ❖

Orthographic Projections

Mechanical Drawing

An **orthographic projection** shows an object as it would appear if viewed from one of its sides. Such views are especially useful when an object's faces are at right angles to each other.

Typically, a figure may be drawn in six different orthographic projections or views: front, back, right, left, top, and bottom. For the figure at right, all of these views are shown. Dashed lines are used to show edges of the figure that cannot be seen in a particular view.

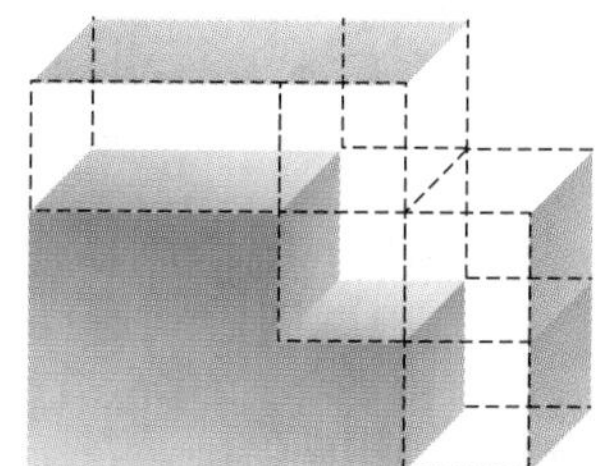

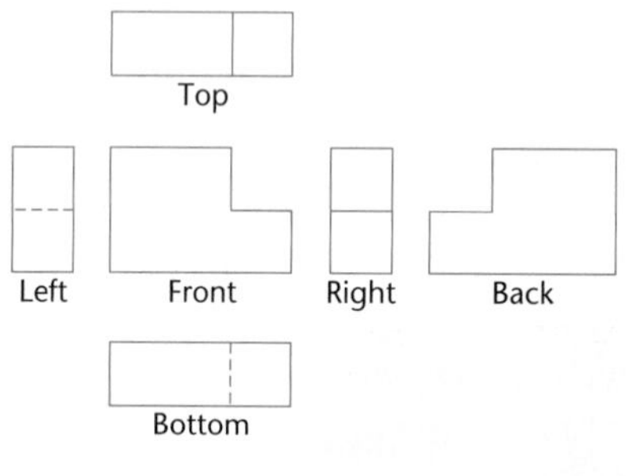

You may find it helpful to think of the different views of an object as an unfolded box with different views or projections on each face.

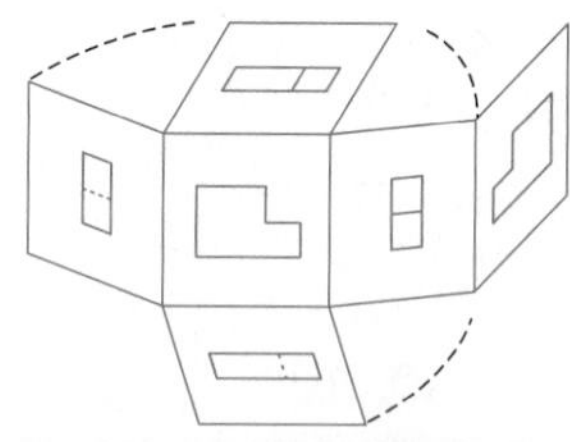

ENRICHMENT The figure below represents the top view of a solid made from cubes. The letters represent the corners of a piece of paper, and the numbers represent how high the cubes are stacked. Draw a view from corner *A*.

C		1	B
1	2	1	1
D		3	A

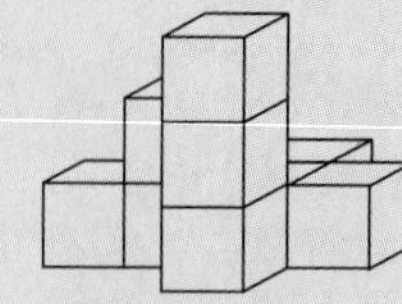

INCLUSION strategies

Hands-On Strategies

Some students may need extra practice drawing shapes made from unit cubes. The tactile experience of using actual cubes to make the figure they want to draw might help them to distinguish the views of the object.

Exploration 3 Volume and Surface Area

You will need
Graph paper
Unit cubes

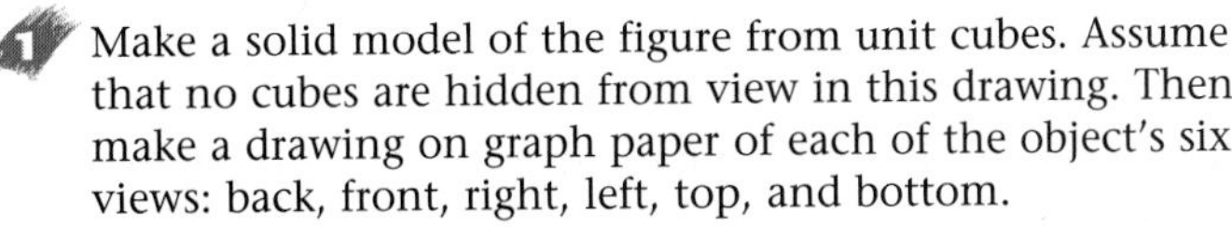

1. Make a solid model of the figure from unit cubes. Assume that no cubes are hidden from view in this drawing. Then make a drawing on graph paper of each of the object's six views: back, front, right, left, top, and bottom.

2. The volume of a solid figure is the number of unit cubes that it takes to completely fill it. How can you determine the volume of the figure you built from centimeter cubes?

3. The total area of the exposed surfaces of a figure is called its surface area. How can you use the six views of the object you built to determine its surface area? ❖

CRITICAL Thinking

In the six views you drew, each of the exposed cube faces of the object appeared in one—and only one—view. Do you think this would be true for any object you might build? Discuss, using drawings or models to illustrate your point.

EXERCISES & PROBLEMS

Communicate

1. Explain how a cube can be drawn so that three "hidden" faces are apparently visible.

Study the isometric drawing of a cube to answer Exercises 2–5.

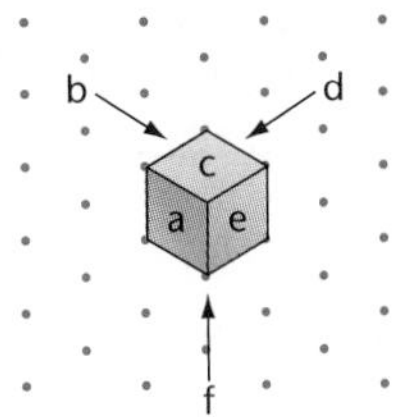

2. Which is the front-left face?
3. Which is the back-right face?
4. Which is the top face?
5. Which is the back-left face?

Use the isometric drawing of a rectangular solid to answer Exercises 6–9. Assume that the drawing shows all cubes.

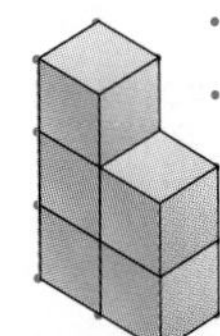

The magenta face of the solid is made up of two cube faces.

6. How many small cubes are used?
7. How many faces does the solid have?
8. How many cube faces are exposed in the solid?
9. Draw six orthographic views of the solid.

Exploration 3 Notes

In this activity, students draw an object from different views and relate their drawings to surface area and volume.

ongoing ASSESSMENT

2. Count the cubes used in constructing the figure.

3. Find the area of each of the six views and add them. This counts the area of each face once, and thus gives the total surface area.

CRITICAL Thinking

If an object has its faces at right angles, each exposed face will appear in exactly one view. If the sides don't meet at right angles, more than one side will appear in at least one of the six views. Thus, more views will be necessary if each exposed face is to appear in exactly one view.

ASSESS

Selected Answers

Exercises 17, 19, 21, 25, 29, 33, 35, and 37

Assignment Guide

Core 1–11, 14–22, 32–41

Core Plus 1–13, 16–41

Technology

Students can practice drawing shapes by using the copy and paste features of geometry graphics software.

RETEACHING the lesson

Have students draw the six different orthographic projections for the figure below, give the surface area, and find the volume.

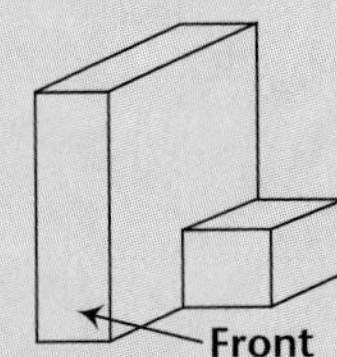

[Solution: surface area = 26 square units; volume = 7 cubes

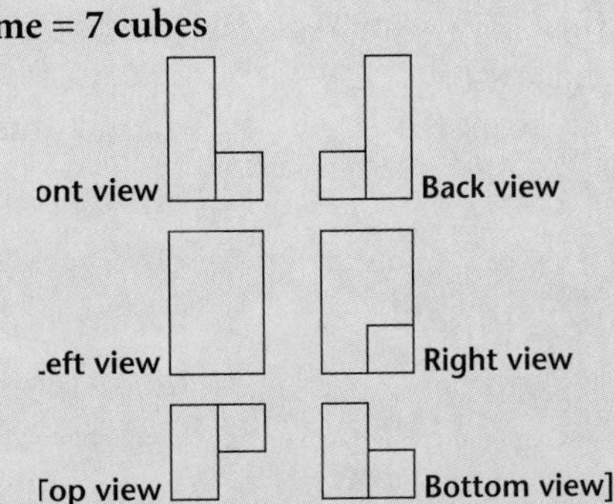

Error Analysis

Students may have difficulty making conjectures for Exercise 26. Encourage them to make a chart to keep track of the different numbers of cubes and the volume and surface area associated with them.

Practice & Apply

10. Use isometric grid paper to draw four different rectangular solids using eight cubes each.

11. How many faces does each solid have? How many cube faces are exposed?

12. Draw four more figures by removing exactly two cubes from each figure you drew in Exercise 10.

13. Repeat Exercise 11 for the four new solids you drew in Exercise 12.

14. **Portfolio Activity** Using isometric grid paper, design a building using 12 cubes.

15. Using isometric grid paper, draw six orthographic views of the building you designed in Exercise 14.

Study the isometric drawing to answer Exercises 16–20. Assume that no cubes are hidden from view in the drawing.

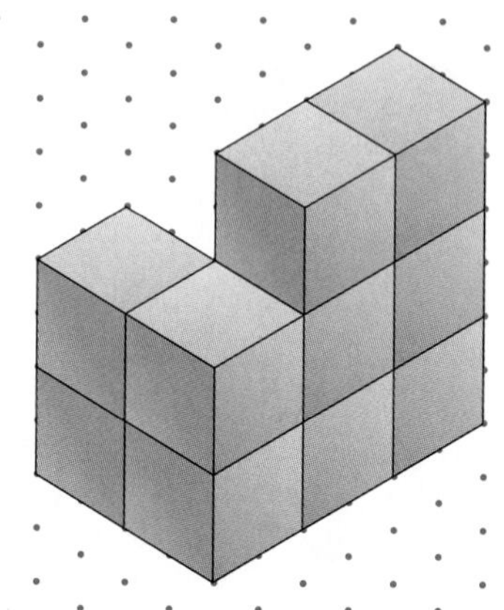

16. How many cubes are used in the drawing? 10

17. How many faces does the solid have? 10

18. Sketch the shape of the right-front face.

19. Does any other face of the solid have this shape? No

20. Are any two faces of the solid congruent? Explain your answer.

For Exercises 21–26, study the diagram and definitions.

The surface area of a solid built from unit cubes is the total area of all the exposed faces of the unit cubes.

The volume of a solid built from unit cubes is the total number of unit cubes.

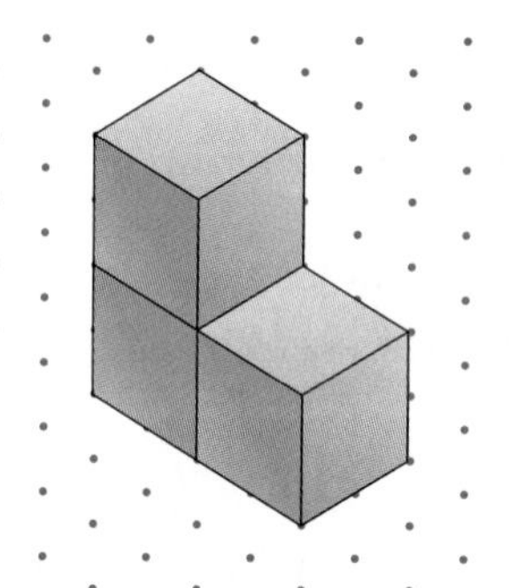

21. What is the surface area of the rectangular solid in the figure if a side of a cube is one unit? 14 units

22. What is the volume of the rectangular solid in the figure? 3 cubes

23. Use the drawing as a base. On isometric grid paper, add two cubes to the drawing.

24. What is the surface area of the solid you drew in Exercise 23?

25. What is the volume of the solid you drew in Exericse 23? 5 cubes

26. **Portfolio Activity** Experiment with surface area and volume using different numbers and arrangements of the cubes. Make conjectures about how surface area varies with volume.

The answers to Exercises 10–15, 18, 20, and 23–26 can be found in Additional Answers beginning on page 727.

Practice Master

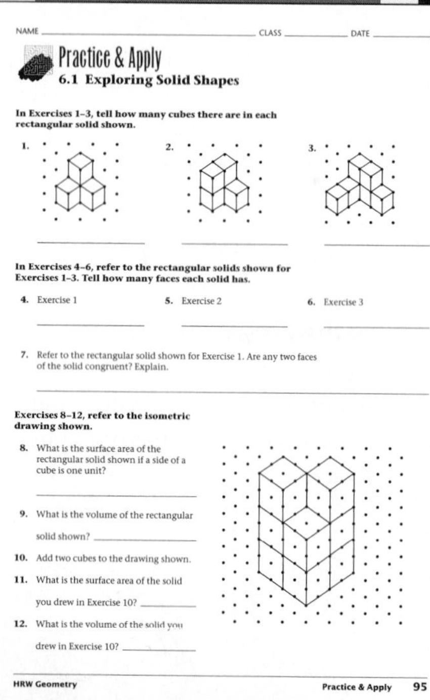
NAME ______ CLASS ______ DATE ______

Practice & Apply

6.1 Exploring Solid Shapes

In Exercises 1–3, tell how many cubes there are in each rectangular solid shown.

1. ______ 2. ______ 3. ______

In Exercises 4–6, refer to the rectangular solids shown for Exercises 1–3. Tell how many faces each solid has.

4. Exercise 1 ______ 5. Exercise 2 ______ 6. Exercise 3 ______

7. Refer to the rectangular solid shown for Exercise 1. Are any two faces of the solid congruent? Explain.

______

Exercises 8–12, refer to the isometric drawing shown.

8. What is the surface area of the rectangular solid shown if a side of a cube is one unit?

______

9. What is the volume of the rectangular solid shown? ______

10. Add two cubes to the drawing shown.

11. What is the surface area of the solid you drew in Exercise 10? ______

12. What is the volume of the solid you drew in Exercise 10? ______

HRW Geometry Practice & Apply 95

Recreation A wall tent is a good model of how planes intersect in space. Use the isometric drawing of a wall tent to answer Exercises 27 and 28.

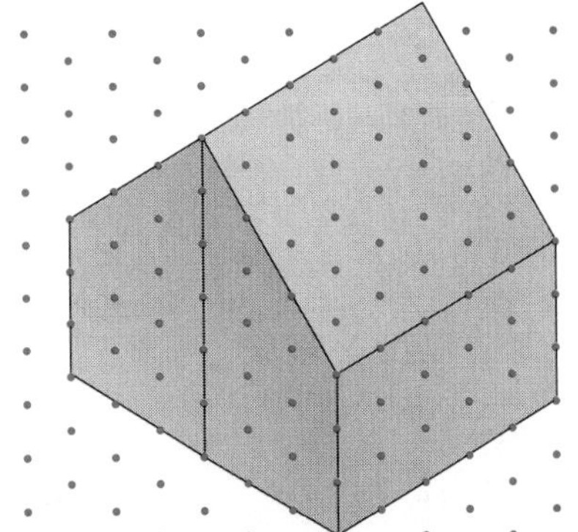

27. Sketch your own isometric drawing of a wall tent. Identify and label the various geometric shapes you find.

28. Think about the definitions, postulates, and theorems you have learned so far. State some conjectures about relationships that occur in the geometry of a wall tent.

29. When you look at the picture below you may see a cube that "pops out" toward you. But this would be spacially illogical with the rest of the picture. What does the picture "really" show?

Make orthographic projections of the building in the views indicated below.

30. Front view

31. Right side view

alternative ASSESSMENT

Authentic Assessment

To an architect, different views of a building are its elevations. Given the building below, have students draw its elevations.

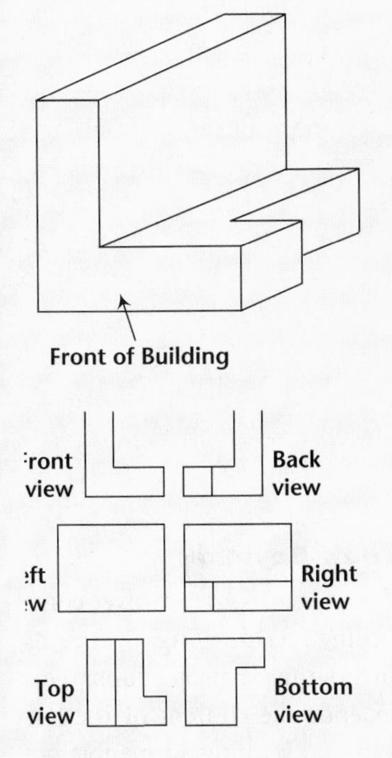

27. Check student drawings. A wall tent has rectanglular sides and bottoms, with trapezoidal front, back, and door flaps.

28. A wall tent has congruent front and back, roof sides, and lateral sides. The sides of the tent are perpendicular to the planes of the front and back of the tent. Also, the opening of the front of the tent bisects the angle at the top of the front.

29. Answers will vary.

The answers to Exercises 30 and 31 can be found in Additional Answers beginning on page 727.

Technology Master

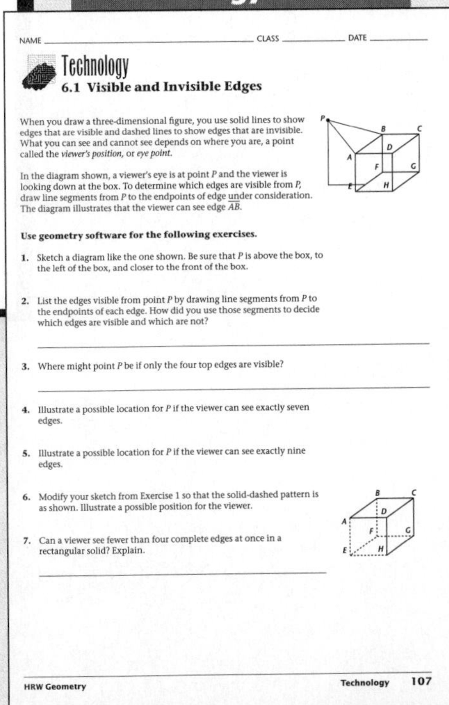

NAME ______ CLASS ______ DATE ______

Technology

6.1 Visible and Invisible Edges

When you draw a three-dimensional figure, you use solid lines to show edges that are visible and dashed lines to show edges that are invisible. What you can see and cannot see depends on where you are, a point called the *viewer's position*, or *eye point*.

In the diagram shown, a viewer's eye is at point P and the viewer is looking down at the box. To determine which edges are visible from P, draw line segments from P to the endpoints of edge under consideration. The diagram illustrates that the viewer can see edge $\overline{AB}$.

Use geometry software for the following exercises.

1. Sketch a diagram like the one shown. Be sure that P is above the box, to the left of the box, and closer to the front of the box.

2. List the edges visible from point P by drawing line segments from P to the endpoints of each edge. How did you use those segments to decide which edges are visible and which are not?

3. Where might point P be if only the four top edges are visible?

4. Illustrate a possible location for P if the viewer can see exactly seven edges.

5. Illustrate a possible location for P if the viewer can see exactly nine edges.

6. Modify your sketch from Exercise 1 so that the solid-dashed pattern is as shown. Illustrate a possible position for the viewer.

7. Can a viewer see fewer than four complete edges at once in a rectangular solid? Explain.

HRW Geometry Technology 107

Look Back

Find the area for each figure shown. [Lessons 5.2 and 5.3]

32.

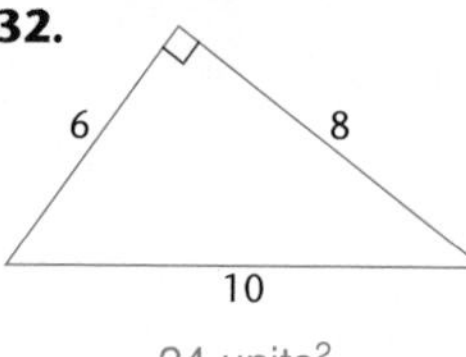

24 units^2

33.

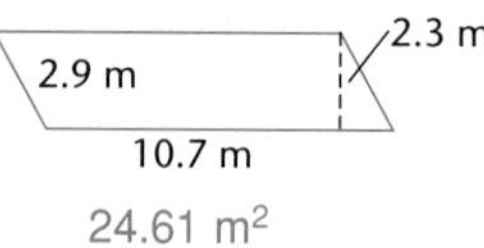

24.61 m^2

34.

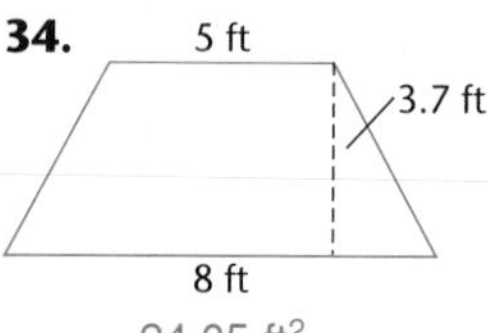

24.05 ft^2

35.

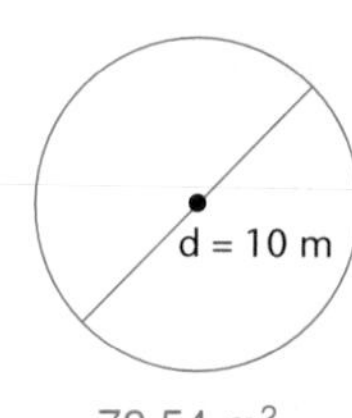

78.54 m^2

Find the length of the missing side of each right triangle below. [Lesson 5.4]

36.

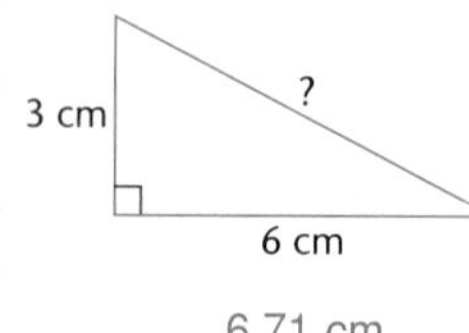

6.71 cm

37.

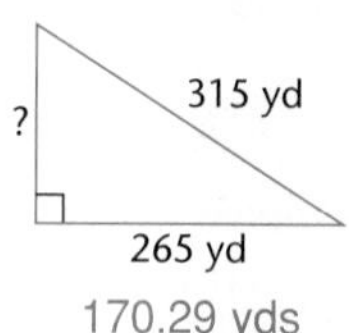

170.29 yds

Look Beyond

Exercises 38–41 involve composing a drawing from its orthographic projections. This extends the content of the lesson and requires more abstract thinking from students. Have students justify their drawings and explain how they know which faces are congruent.

Look Beyond

Use the orthographic projection at right for Exercises 38–41.

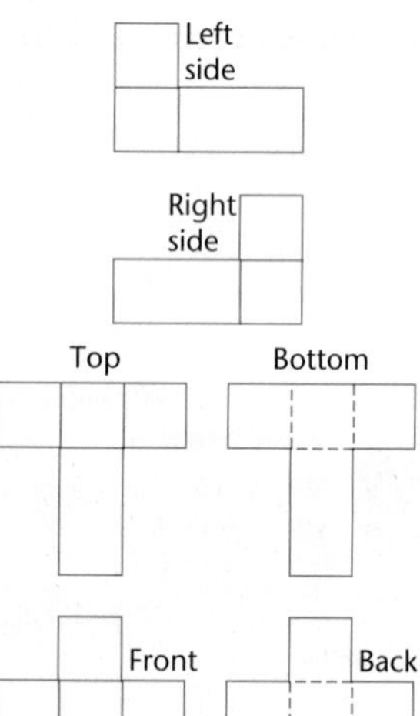

38. Use isometric grid paper to draw two different views of the rectangular solid.

39. Are any cubes hidden by your drawing?

40. How many faces does the solid have? 16

41. How many congruent faces does the solid have? Explain your answer.

38.

39. yes

41. The solid has one set of 11 congruent faces and one set of three. When counting congruent faces, you must disregard the individual blocks and view the object as a solid.

LESSON 6.2

Exploring Spatial Relationships

Why *An interplay of lines and planes is apparent in real-world objects such as this quartz crystal cluster. The relationships among lines and planes in three dimensions is essential to understanding the structure of matter.*

In the explorations that follow, you will discover and develop ideas about how points, lines, and planes relate to each other in three-dimensional space.

Exploration 1 *Parallel Lines and Planes in Space*

You will need
No special tools

Part I

1. Sketch a cube and label its vertices as shown. Identify the segments that form the vertical edges.

Do segments $\overline{AE}$ and $\overline{CG}$ seem to be coplanar? Do they seem to be parallel? (Do you think they would meet if they were extended infinitely?)

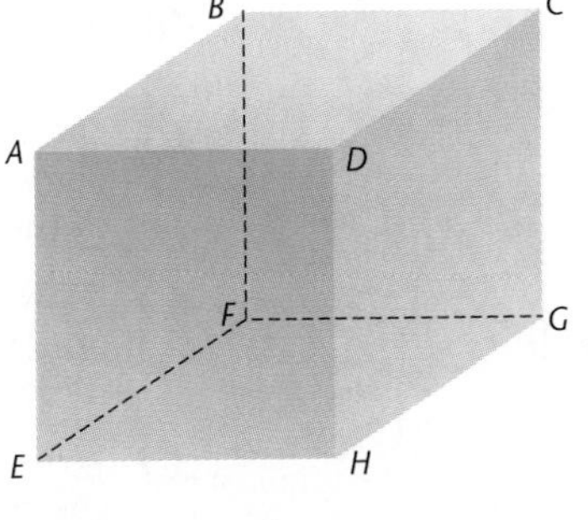

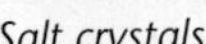

Salt crystals

ALTERNATIVE teaching strategy

Using Models Have students create model cubes out of index cards to help them visualize parallel planes and lines. Ask them to label the vertices and make a list of all pairs of parallel sides and parallel planes. Ask them to list all planes, including the nonvisible planes on the diagonal of the cube.

PREPARE

Objectives

- Identify the unique relationships of points, lines, segments, planes, and angles in three-dimensional space.
- Define *dihedral angle.*

RESOURCES

- Practice Master 6.2
- Enrichment Master 6.2
- Technology Master 6.2
- Lesson Activity Master 6.2
- Quiz 6.2
- Spanish Resources 6.2

Assessing Prior Knowledge

Classify each of the following as true or false.

1. Two lines which do not intersect are parallel. **[False]**
2. A cube has six faces. **[True]**
3. Parallel lines are coplanar. **[True]**

TEACH

Why Relating points, lines, and planes to each other in three-dimensional space will help students understand spatial relationships among solid figures.

Exploration 1 Notes

In this activity, students learn how to sketch a cube and label its vertices and faces. Emphasize that every set of three noncollinear points in the diagram determines a unique plane, even if the diagram does not show it.

Use Transparency 50

2. Which segments in the cube seem to be parallel? List them.

3. Are there segments in the cube that are not parallel and yet would never meet if they were extended infinitely in either direction? Explain how this can be. Such segments (or lines or rays) are said to be **skew**. Make a list of 4 pairs of skew segments in the cube.

Part II

1. How many faces does a cube have? Which faces seem to be parallel to each other?

2. Write your own definition of parallel planes by completing the sentence: "Two planes are parallel if and only if . . ." **(Def 6.2.1)**

3. How many sets of parallel planes are there in a cube? List the parallel planes in the figure you drew in Part I. ❖

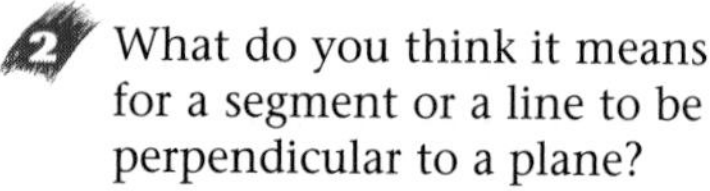

Exploration 2 Segments and Planes

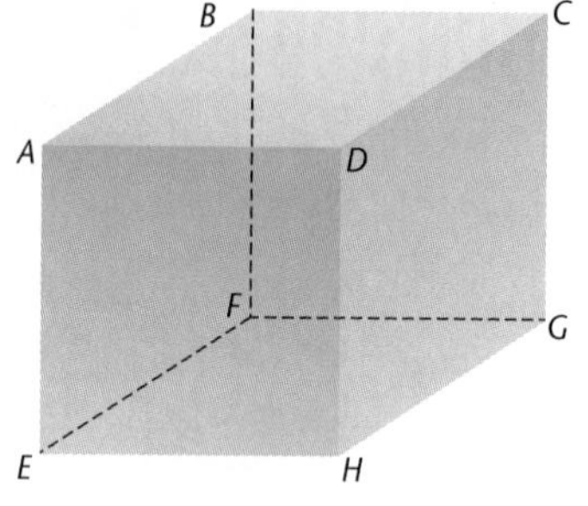

You will need
No special tools

Part I

1. Each of the segments in the cube is perpendicular to two different planes. Make a list of segments and the planes to which they are perpendicular.

2. What do you think it means for a segment or a line to be perpendicular to a plane?

Draw a line l on a piece of paper and label a point on it as P. Place your pencil on the point and hold it so that it is perpendicular to the paper. Is the pencil also perpendicular to line l?

3. Is it possible to tilt your pencil so that it is still perpendicular to line l but not to the paper? Make a sketch illustrating your answer.

4. Draw a new line m through point P. Place your pencil on P so that it is perpendicular to l and m at point P. What is the relation of the pencil to the plane of the paper?

5. Draw several other lines through point P. When your pencil is placed on P so that it is perpendicular to the paper, is it also perpendicular to these other lines?

6. Write your own definition of a line perpendicular to a plane by completing the following sentence: "A line is perpendicular to a plane if and only if. . ." **(Def 6.2.2)**

Ongoing Assessment

Part 1, Step 2. $\overline{AE} \parallel \overline{DH}$, $\overline{AE} \parallel \overline{CG}$, $\overline{AE} \parallel \overline{BF}$, $\overline{DH} \parallel \overline{CG}$, $\overline{DH} \parallel \overline{BF}$, $\overline{CG} \parallel \overline{BF}$, $\overline{AB} \parallel \overline{CD}$, $\overline{AB} \parallel \overline{GH}$, $\overline{AB} \parallel \overline{EF}$, $\overline{CD} \parallel \overline{GH}$, $\overline{CD} \parallel \overline{EF}$, $\overline{GH} \parallel \overline{EF}$, $\overline{AD} \parallel \overline{BC}$, $\overline{AD} \parallel \overline{FG}$, $\overline{AD} \parallel \overline{EH}$, $\overline{BC} \parallel \overline{FG}$, $\overline{BC} \parallel \overline{EH}$, $\overline{FG} \parallel \overline{EH}$

Part 2, Step 2. . . . they do not intersect.

Exploration 2 Notes

In this activity, students explore what it means for a line to be perpendicular to a plane. Students should discuss various ways to tilt a pencil so that it is perpendicular to a line. Have them clarify the terminology with each other and check each other's definition of a line perpendicular to a plane.

Teaching tip

You can help students understand the concepts in Exploration 2 by providing concrete models, including other noncubical rectangular prisms. Have students try to draw the models, label them, and make lists of all parallel segments and planes.

Ongoing Assessment

6. . . . the line (or segment or ray) is perpendicular to all lines in the plane that pass through the point where the line(or segment or ray) intersects the plane.

interdisciplinary CONNECTION

Architecture Parallel lines and planes in space are necessary and sometimes aesthetic components of architecture. Ask students to find interesting pictures that emphasize parallelism in architecture.

Part II

1. Each of the segments in the cube is parallel to two different planes. Make a list of segments and the planes to which they are parallel. What do you think it means for a segment or a line to be parallel to a plane?

2. Draw a line l on a piece of paper. Hold your pencil above l so that it is parallel to it. Does the pencil seem to be parallel to the plane of the paper?

3. Turn your pencil in a different direction so that it is still parallel to the paper but no longer parallel to the line. Do you think there is another line in the paper that is parallel to the pencil?

4. Write your own definition of a line (or segment or ray) parallel to a plane by completing the following sentence: "A line is parallel to a plane if and only if . . ." **(Def 6.2.3)** ❖

Exploration 3 *Measuring Angles Formed by Planes*

You will need
Stiff folding paper or an index card
Scissors

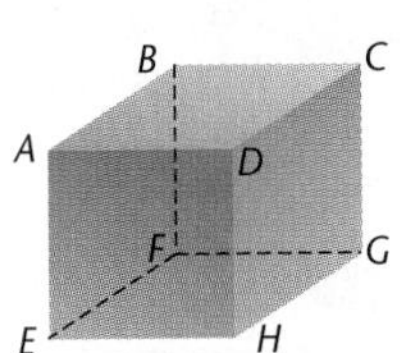

 Some of the faces of a cube form right angles (are perpendicular to each other). Each face of the cube is perpendicular to how many other faces of the cube?

 Draw a horizontal line l on a piece of paper. Mark and label points A, B, and C, with B between A and C. Make a crease through B so that line l folds onto itself. What is the relationship between l and the line of the crease?

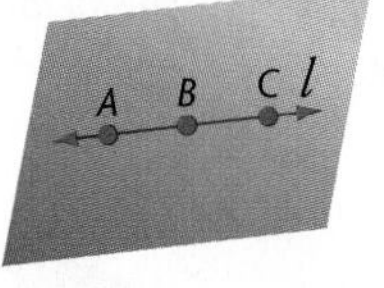

 Open the paper slightly. The angle formed by the sides of the paper is known as a **dihedral angle**. The measure of the dihedral angle is the measure of $\angle ABC$.

 Write your own definition of the measure of a dihedral angle by completing the following sentence: "The measure of a dihedral angle is the measure of the angle formed by . . ." **(Def 6.2.4)**

ONGOING ASSESSMENT

4. . . . it does not intersect the plane.

Or

. . . it is parallel to a line contained in the plane.

Exploration 3 Notes

This exploration will help students understand that dihedral angles are formed by intersecting planes.

ONGOING ASSESSMENT

4. . . . two rays that are perpendicular to the angle edge at the same point.

Cooperative Learning

Have small groups of students find different examples of dihedral angles in the classroom and outside of the classroom. Ask them to list the types of objects with dihedral angles equal to 90°, with dihedral angles less than 90°, and with dihedral angles over 90°.

ENRICHMENT Have students use the "net" below to predict the dihedral angles for the solid formed by the net. Have them check their answers by tracing and folding the net.

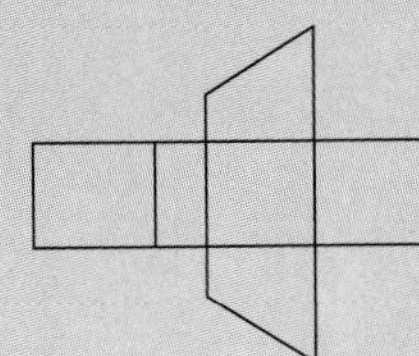

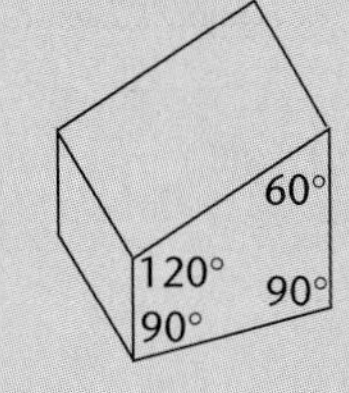

Using Visual Models You may help some students see relationships between segments and planes by allowing them to construct three-dimensional figures with geometry graphics software. Encourage them to use the measurement options of the software to verify the properties in this lesson.

CRITICAL Thinking

0°; 180°

ASSESS

Selected Answers

Odd-numbered Exercises 7–23

Assignment Guide

Core 1–9, 11–17, 20–27

Core Plus 1–15, 17–29

Technology

Geometry graphics software can be used to reinforce how perpendicular and parallel planes can be drawn to form cubes.

Error Analysis

Students may find Exercises 11–15 difficult to understand. Use the corner of a box to show students the relationships of lines and planes in space.

Use Transparency 51

8. *HLOK; KONJ; IMNJ; HIML*
9. $\overline{IJ}$; $\overline{HK}$; $\overline{MN}$; $\overline{LO}$ are perpendicular

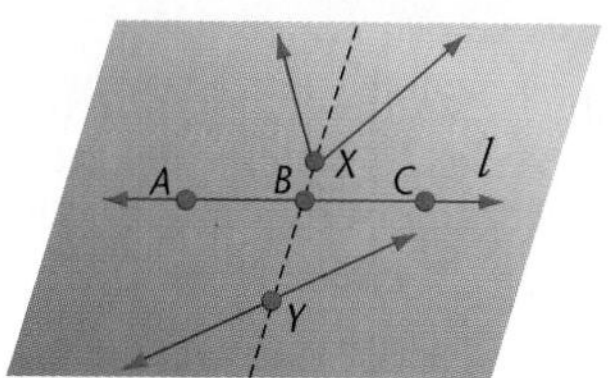

5 Open the paper and flatten it out. Add points *X* and *Y* to the line of the crease you made. Draw two rays from each point, with one ray on either side of the crease, as shown.

6 Fold the paper again on the same crease. Cut out pieces of paper that fit neatly into the angles formed by the different rays. Compare the shapes of the pieces of paper. Do the angles that are formed by the new rays have the same or different measures from $\angle ABC$? Discuss. ❖

Use pieces of paper to take the measure of the angles.

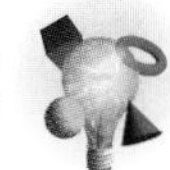

CRITICAL Thinking By measuring angles formed by different rays in Exploration 3, you can get a variety of results. What is the smallest angle you could measure? What is the largest?

EXERCISES & PROBLEMS

Communicate

Use the figure for Exercises 1–3.

1. Name three pairs of parallel segments, and explain why they are parallel.
2. Name two pairs of parallel planes, and explain why they are parallel.
3. Name three pairs of perpendicular planes, and explain why they are perpendicular.
4. If two lines are perpendicular to the same line, are the two lines parallel to each other? Why or why not?
5. If a line not on a plane is perpendicular to a line on a plane, is the line perpendicular to the plane also? Why or why not?

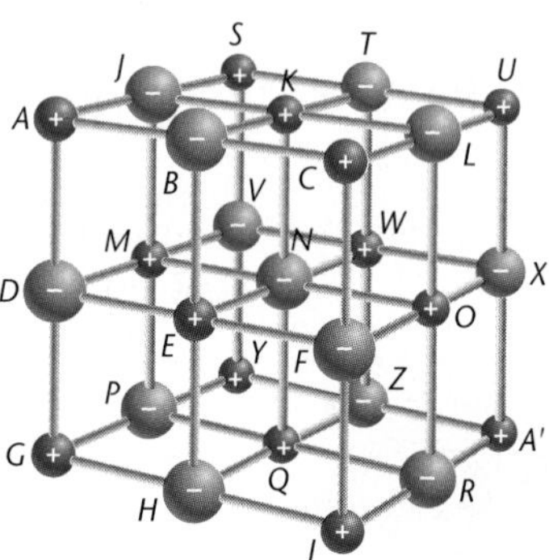

Sodium chloride (table salt) crystal lattice

Practice & Apply

Use the figure for Exercises 6–10.

6. Name all pairs of parallel planes. *HIJK* ∥ *LMNO*; *HIML* ∥ *KJNO*; *HLOK* ∥ *IMNJ*
7. Name two segments skew to $\overline{LM}$. $\overline{KO}$; $\overline{JN}$; $\overline{HK}$; $\overline{IJ}$ are skew
8. Name all planes perpendicular to plane *HIJ*.
9. Name two segments that are perpendicular to plane *JKO*.
10. Name two segments that are parallel to plane *HKO*. $\overline{MN}$; $\overline{IJ}$; $\overline{JN}$; $\overline{IM}$

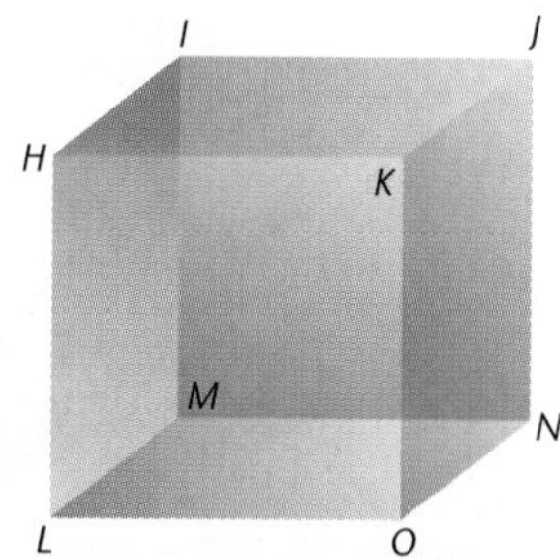

RETEACHING the lesson

Using Visual Models Have students name all the parallel segments and planes in the cube below.

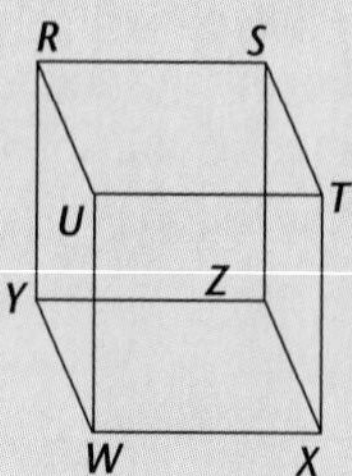

$\overline{RS}$ ∥ $\overline{UT}$ ∥ $\overline{WX}$ ∥ $\overline{YZ}$ (Note: these segments are not all coplanar); $\overline{RU}$ ∥ $\overline{ST}$ ∥ $\overline{ZX}$ ∥ $\overline{YW}$ (Note: these segments are not all coplanar); plane *RSTU* ∥ plane *YZXW*; plane *RUWY* ∥ plane *STXZ*; plane *RSZY* ∥ plane *UTXW*

Explain why the following statements are true or false for a figure in three-dimensional space.

11. In three-dimensional space, if two lines are parallel to a third line, then the lines are parallel.

12. If two planes are parallel to a third plane, then the planes are parallel.

13. If two planes are perpendicular to a third plane, then the planes are parallel.

14. If two planes are perpendicular to the same line, then the planes are parallel.

15. If two lines are perpendicular to the same plane, then the lines are parallel.

Navigation Ships navigate on an ocean surface, which can be modeled with a plane. Airplanes navigate in three-dimensional space. Exercises 16–19 concern navigation of ships and airplanes.

16. Two airplanes flying at the same altitude are flying in the same __?__. Plane

17. Airplane A is flying northwest at 10,000 ft, and airplane B is flying north at 5,000 ft. The lines formed by the flight paths are __?__. Skew lines

18. Ship A is going south at 20 knots, and ship B is going southeast at 15 knots. The lines formed by the ships' paths will eventually __?__. Intersect

19. An airplane takes off from an airport runway. The airplane's nose and wingtips define the "plane of flight." Describe the dihedral angles formed by the airplane's plane of flight and the runway. Is that angle acute, right, or obtuse? Explain your answer.

11. True. If the three lines are coplanar, then they will all be parallel. If the lines are not coplanar, then the first and second lines are contained in a plane parallel to the third line, and hence will never intersect.

12. True. Neither of the two planes intersect the third plane. If the two planes were not parallel, at least one would intersect the third plane. Thus, the planes must be parallel.

13. False. Consider the front wall and a side wall of your classroom. Both are perpendicular to the ceiling, but they are not parallel to each other.

14. True. Consider the roof and the floor of your classroom. If a line may be constructed perpendicular to both, then the planes of the roof and floor may never intersect.

alternative ASSESSMENT

Portfolio Assessment

The ability to create models that represent ideas presented in the lesson will reinforce many of its key ideas. Have students build paper models of cubes or prisms, label them, and summarize key ideas in the lesson in terms of their models.

15. True. There is a line in the plane that contains both the points where the two perpendicular lines intersect the plane. So both lines will be perpendicular to that line and are therefore parallel.

19. Acute. Normally, a plane will not take off at a very steep angle. The Harrier jump jet, which takes off straight up, is an exception.

Practice Master

NAME ______ CLASS ______ DATE ______

Practice & Apply
6.2 Exploring Spatial Relationships

1. Explain why a 3-legged stool will not wobble, even if the legs are of different lengths.

2. Explain why some 4-legged stools wobble.

In Exercises 3–5, name all the planes determined by the vertices of the figures shown.

3. 4. 5.

Exercises 6–12 refer to the figure.

6. Name a pair of parallel planes.

7. Is plane *ABF* parallel to plane *CBF*? Explain.

8. Is $\overline{DF}$ skew to $\overline{BF}$? Explain.

9. Is $\overline{DC}$ skew to $\overline{BF}$? Explain.

10. Name three segments skew to $\overline{BF}$.

11. Name all planes perpendicular to plane *ACD*.

12. Name two segments perpendicular to plane *BFD*.

96 Practice & Apply HRW Geometry

Use Exercise 20 to answer Exercises 21–23.

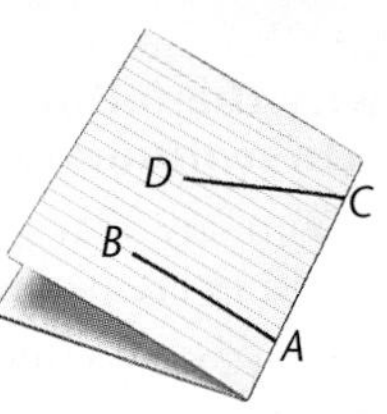

20. Use a couple of note cards to construct the following. Fold a note card in half and mark it with two segments $\overline{AB}$ and $\overline{CD}$ such that $\overline{AB}$ is perpendicular to the folded edge, and $\overline{CD}$ is not perpendicular. Cut halfway into each segment from the folded edge. Use two other pieces of note card to model intersecting planes as shown. Check student construction.

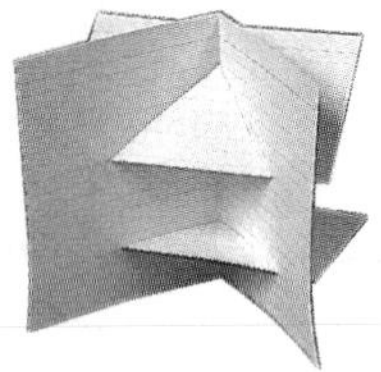

21. Which insert forms a plane which is not perpendicular to the edge?

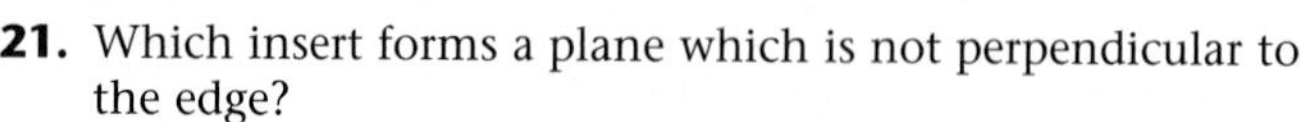

22. Which insert can be used to measure the dihedral angle formed by the two upright planes? The angle formed by the insert at $\overline{AB}$.

23. Which of the two angles formed by the inserts is larger and why? The dihedral angle formed by insert $\overline{AB}$ is larger than the angle formed by insert $\overline{CD}$.

Look Back

Use a compass and straightedge to construct the following. [Lesson 4.7]

24. Trace $\angle ABC$ on your own paper and construct a copy of the angle.

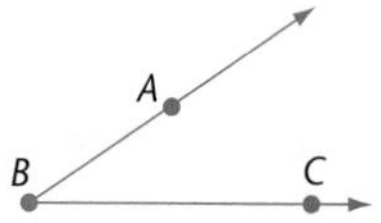

A
B
C

25. Trace segment $\overline{DE}$ on your own paper and construct the perpendicular bisector.

26. Trace segment $\overline{FG}$ and point H on your own paper and construct a line parallel to the segment through point H.

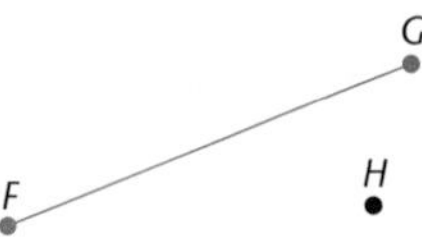

27. Trace triangle JKL on your own paper and construct a copy of the triangle.

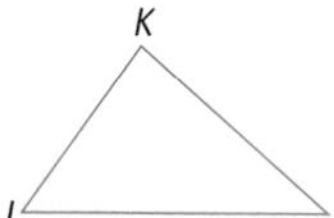

Look Beyond

28. Arrange the figure below into three triangles by moving only two toothpicks.

29. Arrange the figure below into three squares by moving only three toothpicks.

28.

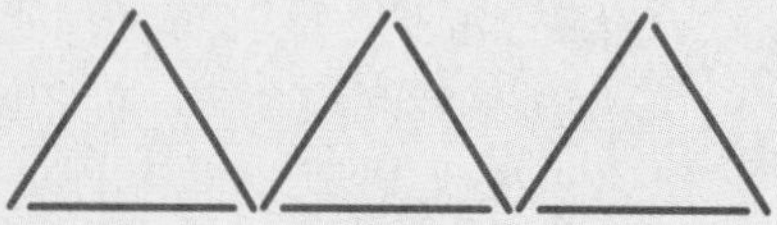

29.

---

21. insert in $\overline{CD}$

The answers to Exercises 24–27 can be found in the Additional Answers beginning on page 727.

Look Beyond

Exercises 28 and 29 give students experience with solving nonroutine problems.

Technology Master

NAME ______ CLASS ______ DATE ______

Technology
6.2 Combinations of Lines and Planes

You can represent a plane by a figure like one of those shown. The diagram shows planes that have a "horizontal" orientation, a "vertical" orientation, and an "oblique" orientation. Sketches like these are very useful if you want to sketch combinations of lines and planes.

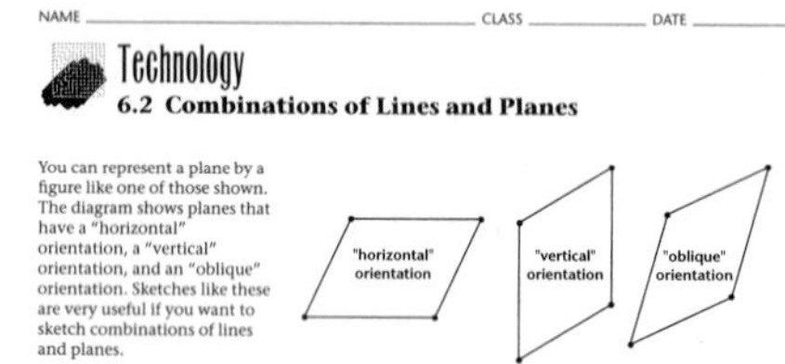

Use geometry software for the following exercises.

1. Sketch a plane $\mathscr{L}$ with a "horizontal" orientation. By adding the sketch of a second plane $\mathscr{M}$, illustrate all the ways the two planes can be related. Describe your illustrations in words.

2. Sketch a plane $\mathscr{L}$ with a "horizontal" orientation. By adding the sketches of two planes $\mathscr{M}$ and $\mathscr{N}$, illustrate the ways the three planes can be related. Describe your illustrations in words.

Illustrate each statement.

3. Through a point P not on plane $\mathscr{L}$, there are many lines parallel to $\mathscr{L}$.
4. Parallel planes $\mathscr{L}$ and $\mathscr{M}$ contain lines that are not parallel.

Give a counterexample to each statement by means of an illustration.

5. Points A, B, and C lying on plane $\mathscr{L}$ and point D not on plane $\mathscr{L}$ determine two planes.
6. Any line x on plane $\mathscr{L}$ perpendicular to plane $\mathscr{M}$ is perpendicular to exactly one line y on plane $\mathscr{M}$.

Illustrate each situation.

7. Sketch three lines x, y, and z that are perpendicular to each other. Sketch the planes that each pair determines.
8. You are standing on a ladder in one corner of a room and look at the opposite walls and the floor. Sketch the two walls and the floor.

108 Technology HRW Geometry

LESSON 6.3 An Introduction to Prisms

A beam of laser light is bent as it passes through the glass, triangular prisms in this student's experiment. Since laser light consists of just one color, it is not broken up into a spectrum.

why *Many familiar objects have the shape of three-dimensional geometric figures known as* **prisms.** *As you study prisms you will apply your knowledge of congruent segments, angles, and polygons.*

The figures below are prisms. A prism is named according to the shape of its base.

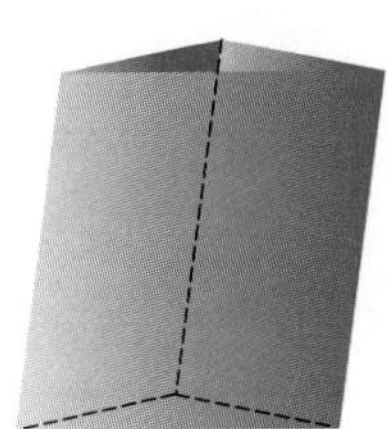

Triangular prism

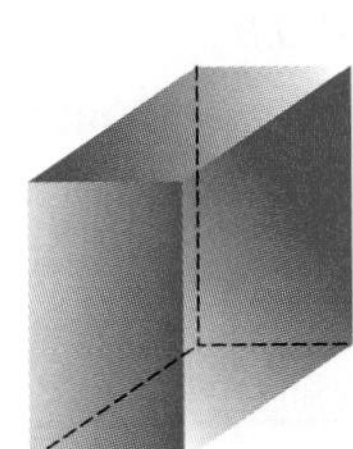

Rectangular prism

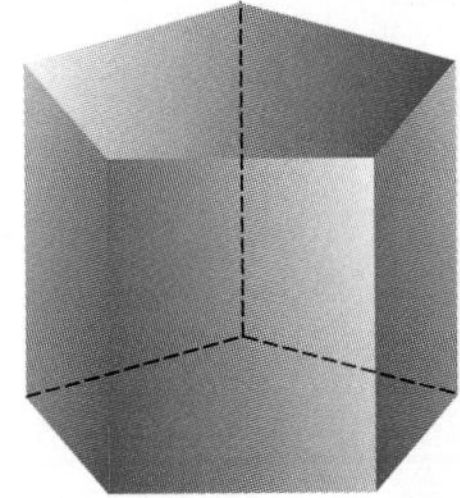

Pentagonal prism

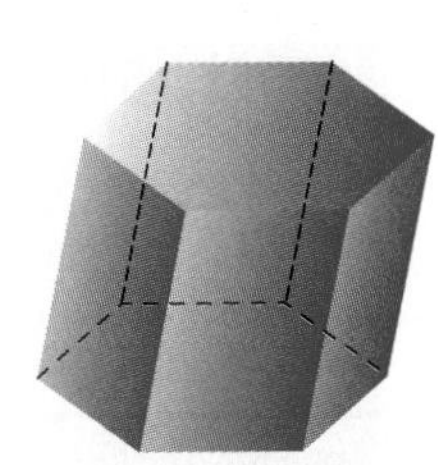

Hexagonal prism

A prism consists of a polygonal region, its translated image, and connecting line segments. The picture on the right suggests a way to think of constructing a prism.

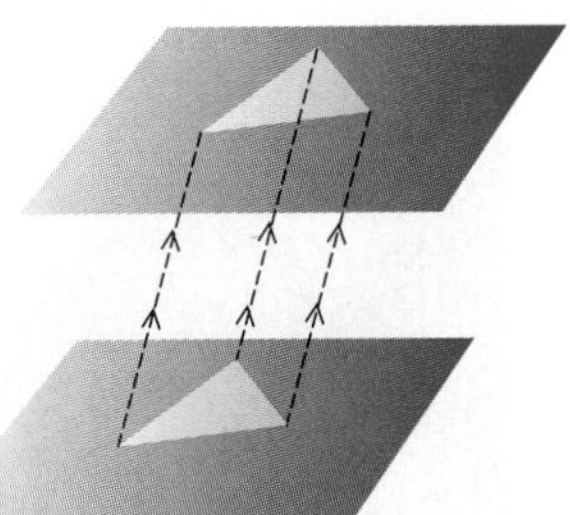

1. Start with a polygonal region in a plane.
2. Translate onto a parallel plane.
3. Connect the image and pre-image with segments.

PREPARE

Objectives

- Define *right* and *oblique prism.*
- Solve problems using the diagonal measure of a right prism.

RESOURCES

- Practice Master 6.3
- Enrichment Master 6.3
- Technology Master 6.3
- Lesson Activity Master 6.3
- Quiz 6.3
- Spanish Resources 6.3

Assessing Prior Knowledge

Name the planes in the figure below. Which planes are parallel?

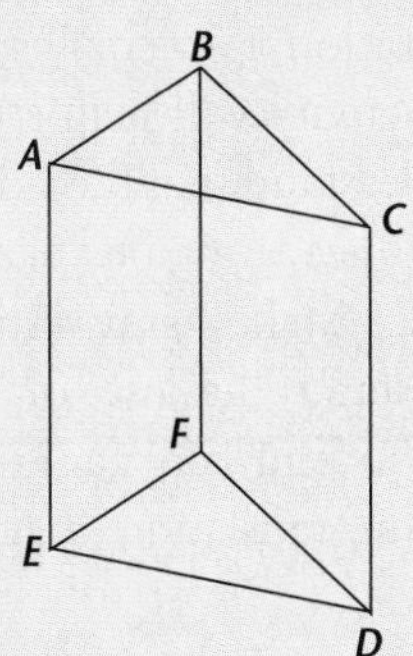

[planes: *ABC, EFD, BCDF, ACDE, ABFE*; planes *ABC* and *EFD* are parallel.]

TEACH

This lesson will introduce students to properties of prisms—three-dimensional figures that model many real-world objects.

Use Transparency 52

Hands-On Strategies

Have each student use thick paper to construct a right prism and a non-right prism. Include prisms with bases that are triangles, rectangles, pentagons, hexagons, or octagons. Have them label the vertices in their models, name the parallel planes, and name all other planes.

ongoing ASSESSMENT

If one pair of opposite sides of a quadrilateral are both congruent and parallel, then the quadrilateral is a parallelogram.

TEACHING tip

Use a deck of cards to illustrate an oblique rectangular prism. This will help students to make the connection later that the volume of a prism is equal to the area of its base times its height.

Exploration 1 Notes

Students will enjoy working on Exploration 1 in small groups. This activity will help students to understand that the sides of a prism need to be parallelograms. Make sure students understand the use of prime notation and can name which faces are translation images.

ongoing ASSESSMENT

Part 1, Step 3. a. rectangle, b. parallelogram, c. rectangle; For b, the deformation leaves the sides parallel but not at right angles to the adjacent sides.

Part 1, Step 4. All are still true.

Part 2, Step 3. The lateral faces are each contained in a single plane, and hence flat.

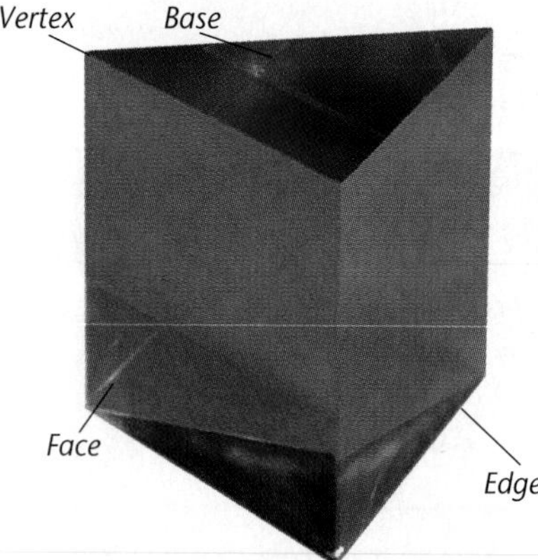

Each flat surface of a prism is called a **face**. Each line segment formed by the intersection of two faces is called an **edge**. Each point of intersection of the edges is called a **vertex**. The polygonal region and its translated image are each called a **base**. Each face of the prism that is not a base is called a **lateral face**. The triangular prism shown has two bases, three lateral faces (five faces total), and nine edges.

Each lateral face of a prism is a quadrilateral. The edges of the quadrilaterals that are part of the bases of the prism are congruent and parallel. Therefore, each face is a parallelogram. Explain why, referring to the description of the construction of a prism on the previous page.

Exploration *The Lateral Faces of Prisms*

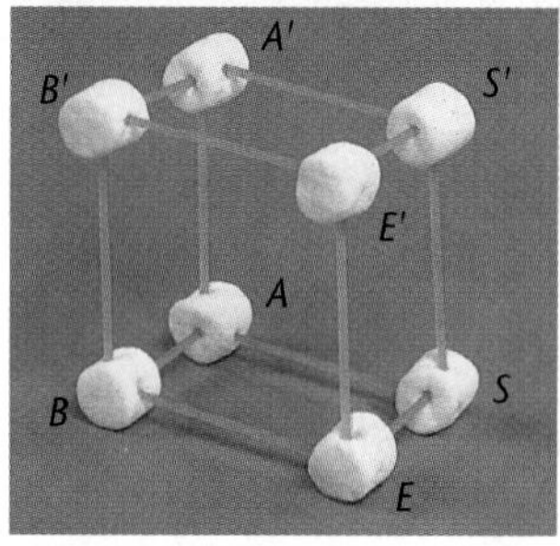

You will need
Thick, uncooked spaghetti
Miniature marshmallows

Part I

Use thick, uncooked spaghetti and miniature marshmallows to build a model of the rectangular prism shown.

1. What type of quadrilateral is each of the following? Explain your reasoning.

   a. *BASE* b. *B'E'EB* c. *E'S'SE*

2. List all pairs of congruent quadrilaterals.

   Rotate the lateral edges clockwise relative to *BASE* as shown.

3. What type of quadrilateral is each of the following? Explain your reasoning.

   a. *BASE* b. *B'E'EB* c. *E'S'SE*

4. Test your congruence statements from Step 2. Which ones are still true?

Part II

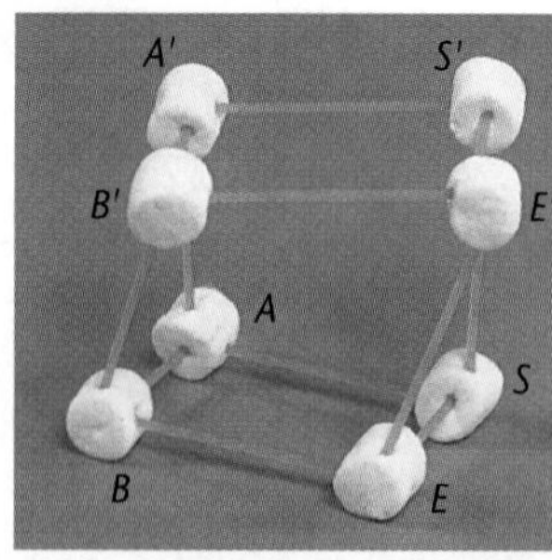

1. Return your prism model to its original upright position. Rotate, or twist, the upper base to the right or left as shown. Is your new figure still a prism? Explain why or why not.
2. Are the segments that form the "faces" of your twisted figure coplanar?
3. Why, do you think, are prisms defined so that one base is a translation of the other?

Chemistry Constructing prisms may remind students of lattice models of chemical compounds like sodium chloride (salt), which are known as *cubic isomorphs.* Ask students to use the library and their science books to find examples of prisms in science and nature.

Is it possible to manipulate your prism so that it remains a prism, but so that none of its lateral faces are rectangles? If your answer is "yes," explain why. ❖

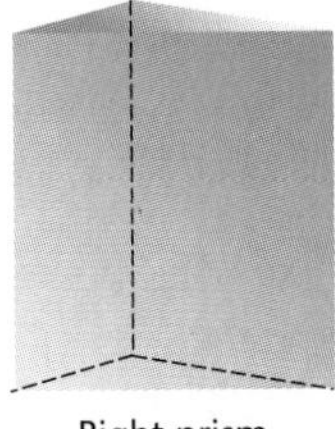

Right prism

Oblique prism

In Exploration 1, you encountered two types of lateral faces: rectangles and nonrectangular parallelograms. Prisms that appear to lean in one direction have at least one lateral face that is not a rectangle. Prisms that appear to be perfectly upright have lateral faces that are all rectangles. This gives us an additional classification for prisms:

A **right prism** is a prism in which all lateral faces are rectangles.

An **oblique prism** is a prism that has at least one nonrectangular parallelogram as a lateral face.

CRITICAL Thinking Why is an upright prism called a "right" prism?

Try This Use spaghetti and marshmallows to construct each of the prisms illustrated at the beginning of the lesson. Classify each as oblique or right.

The Diagonal of a Right Rectangular Prism

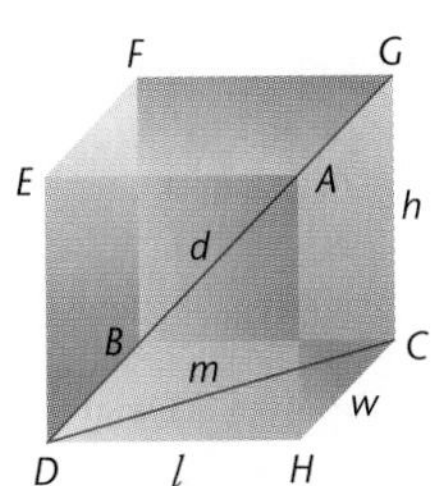

The length of the diagonal of a right rectangular prism with length l, width w, and height h is given by:

$$d = \sqrt{l^2 + w^2 + h^2}$$

Proof:

STATEMENTS	REASONS
$d^2 = m^2 + h^2$	"Pythagorean" Right-Triangle Theorem
$m^2 = l^2 + w^2$	"Pythagorean" Right-Triangle Theorem
$d^2 = l^2 + w^2 + h^2$	Substitution
$d = \sqrt{l^2 + w^2 + h^2}$	Take the square root of both sides

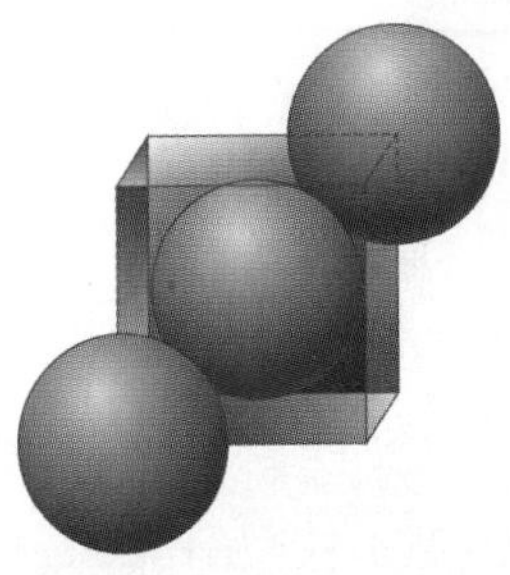

The diagram shows only the atoms along the diagonal. An iron crystal would have six additional atoms on the remaining corners of the cube.

Iron forms a cubic crystal with a unit-cell cube measuring 291 picometers (1×10^{-12} meters) on each edge. The unit cell is the smallest repeating unit of a crystal structure. The diagonal of an iron-crystal unit contains the diameters of two atoms, as shown. What is the radius of an iron atom?

First, find the length of the diagonal. Each edge of the unit-cell cube is 291 pm.

CRITICAL Thinking

The lateral edges are perpendicular to the planes containing the bases.

ONGOING ASSESSMENT

Try This

Check students' constructions.

TEACHING tip

It may be difficult for students to follow the proof of the formula for the diagonal of a prism. Encourage students to copy the diagram and use colored pencils to outline the segments and triangles.

Cooperative Learning

Have groups of two or three students collaborate on how to find the diagonal of a right rectangular prism. Have them make up examples and ask each other to explain how to find the diagonal. Small groups will give students a chance to verbalize how to solve a problem.

Use Transparency 53

ENRICHMENT Some students may be interested in learning about chemicals that have cubic crystals. Have them make a table of chemicals with cubic crystal structures. Include the name of the crystal, the length of an edge of the crystal, and a drawing of the crystal.

INCLUSION strategies

Cooperative Learning Have students design and play a game in which one student describes a prism in 10 words or less without naming the polygons used as faces or bases. Then a partner has to draw the prism, name it, and tell whether it is oblique or right. Students can use this game for lesson reinforcement and review.

ASSESS

Selected Answers

Exercises 9, 13, 15, 17, 19, 25, 27, 29, 31, 35 and 37

Assignment Guide

Core 1–27, 30–37

Core Plus 1–10, 14–38

Technology

Students may use geometry graphics software to draw various prisms. They may also use the software to experiment with other solids, including the Platonic solids.

Error Analysis

Students may find it difficult to generalize to n sides in Exercise 24. Have them extend the table to include more sides in the base of a prism if necessary.

$d = \sqrt{l^2 + w^2 + h^2}$

$d = \sqrt{291^2 + 291^2 + 291^2}$

$d = \sqrt{3(84681)}$

$d = 504$ pm

Since the diagonal contains two diameters or four radii, divide by 4.

$504 \div 4 \approx 126$ pm

The radius of an iron atom is approximately 126 picometers. ❖

EXERCISES & PROBLEMS

Communicate

Classify each prism and explain your classification.

1. 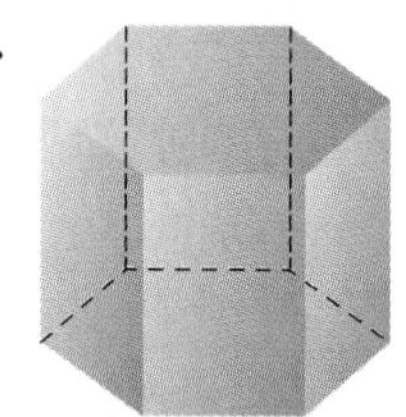

2. 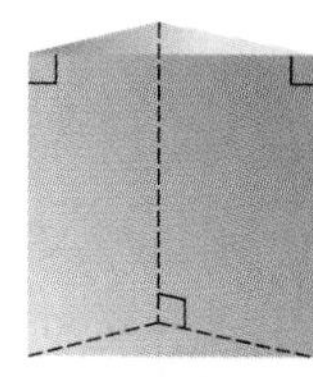

3. 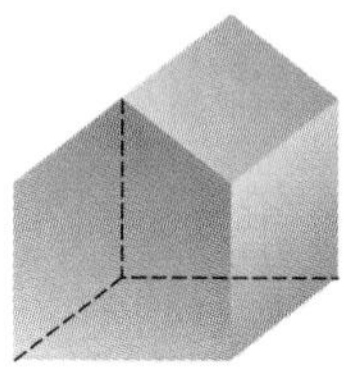

4. 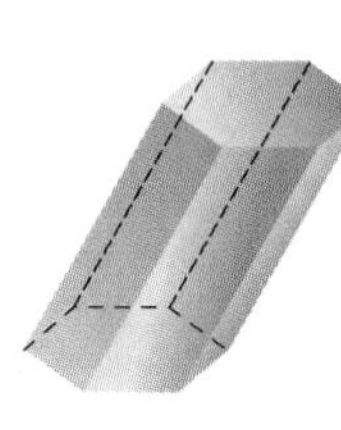

5. Name each face of the oblique prism, identifying the bases. What kind of polygon is each face?

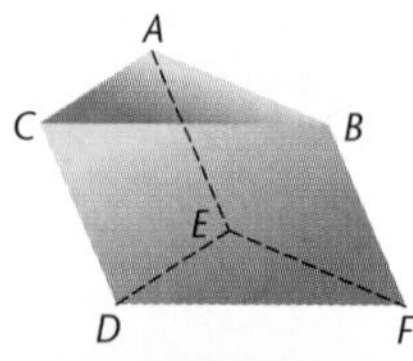

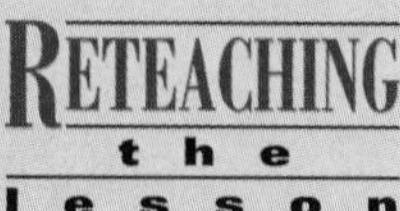

RETEACHING the lesson

Cooperative Learning

Have one student draw a picture of a right rectangular prism. Have another student add a measurement to the prism, such as the length of an edge, while naming the edge. Have students continue adding data until one student is able to name and find the length of a diagonal of the prism. Repeat the activity until all students have participated.

Practice & Apply

6. Draw a right triangular prism.
7. Draw an oblique pentagonal prism.

Use the right triangular prism for Exercises 8–13.

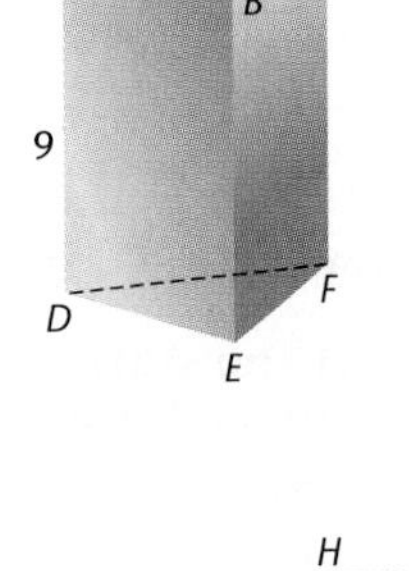

8. Name two faces that must be congruent.
$\triangle ABC \cong \triangle DEF$
9. Name all segments congruent to $\overline{BE}$.
$\overline{AD}$ and $\overline{CF}$
10. What type of quadrilateral is *ACFD*?
Rectangle
11. Name all segments congruent to $\overline{EF}$.
$\overline{BC}$; $\overline{DE}$; $\overline{AB}$
12. List all pairs of congruent lateral faces.
ABED and *BCFE*
13. List the angles whose measure must be 90°.
$\angle ADE$; $\angle ADF$; $\angle DAB$; $\angle DAC$; $\angle BED$; $\angle BEF$; $\angle EBC$; $\angle CFD$; $\angle CFE$; $\angle FCA$; $\angle FCB$; $\angle ABE$

Use the oblique rectangular prism for questions 14–19.

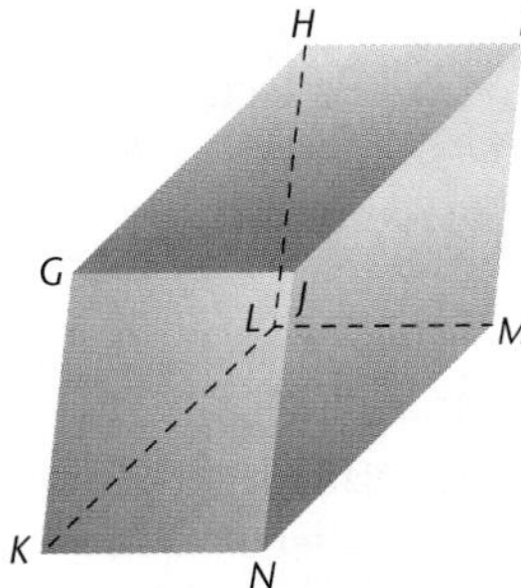

14. Name two rectangles that must be congruent.
KLMN $\cong$ *GHIJ*
15. Name all segments congruent to $\overline{GK}$.
$\overline{JN}$; $\overline{IM}$; $\overline{HL}$
16. What type of quadrilateral is *GJNK*?
Parallelogram
17. What type of quadrilateral is *JIMN*?
Parallelogram
18. List the lateral faces that are congruent to each other.
GKLH $\cong$ *JNMI*; *GKNJ* $\cong$ *HLMI*
19. List the angles whose measures are greater than 90°.
$\angle KNJ$; $\angle JGK$; $\angle LMI$; $\angle IHL$; $\angle NJI$; $\angle IMN$; $\angle KGH$; $\angle HLK$

For Exercises 20–24, fill in the table below for each prism indicated.

	Number of edges in the base	Number of faces	Number of vertices	Number of edges
20.	3	? 5	? 6	? 9
21.	4	? 6	? 8	? 12
22.	5	? 7	? 10	? 15
23.	6	? 8	? 12	? 18
24.	*n*	? $n + 2$	? $2n$	? $3n$

25. The base of a prism is a 20-sided polygon. How many faces, vertices, and edges does the prism have? 22 faces; 40 vertices; 60 edges

6. Check student drawings.
7. Check student drawings.

alternative ASSESSMENT

Portfolio Assessment

Ask students to draw an oblique and a right prism. Have them label the two prisms and measure all edges with a ruler. Ask them to write a description of how the prisms were constructed, including all characteristics.

Practice Master

NAME ________ CLASS ________ DATE ________

Practice & Apply
6.3 An Introduction to Prisms

1. In the space provided, draw an oblique triangular prism.

2. In the space provided, draw a regular right pentagonal prism.

For Exercises 3–9, use the regular right hexagonal prism.

3. Name the two bases. ________
4. Name all segments congruent to $\overline{BC}$. ________
5. In what manner are the two bases related? ________
6. List all the lateral faces. ________
7. Name all segments congruent to $\overline{CD}$. ________
8. What type of quadrilateral is *FEKJ*? ________
9. In what manner are the lateral faces related? ________

In Exercises 10–11, find the length of the diagonal of each right rectangular prism whose dimensions are given.

10. $l = 6, w = 9, h = 12$ ________
11. $l = 4.1, w = 7, h = 2.3$ ________

HRW Geometry — Practice & Apply 97

30. True for rectangles, false for general parallelograms.

31. False. Diagonals bisect the corner angles of a rhombus. The diagonals will only bisect the corner angles of a rectangle if the rectangle is a square.

32. False. Diagonals of a rhombus are perpendicular. The diagonals of a rectangle will only be perpendicular if the rectangle is a square.

Find the length of the diagonal of each right rectangular prism given below.

26. $l = 10$; $w = 5$; $h = 12$ 16.4

27. $l = 4.5$; $w = 2.3$; $h = 10$ 11.20

28.

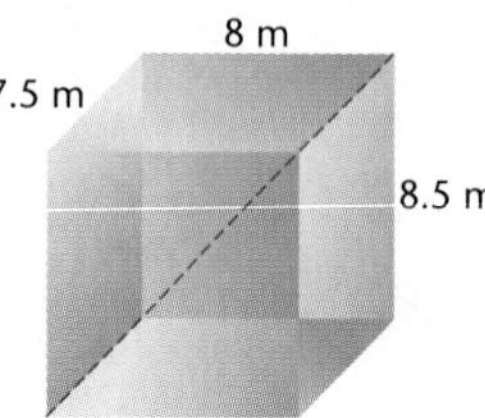

13.87 m

29.

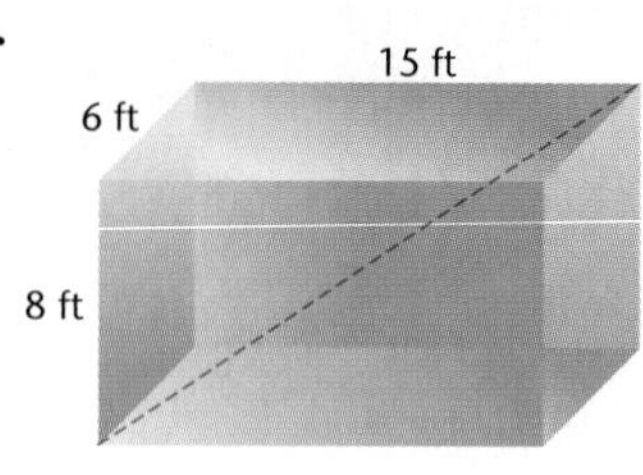

18.03 ft.

Look Back

Explain why each of the statements in Exercises 30–33 is true or false. [Lesson 4.6]

30. The diagonals of a rectangle that is not a square bisect each other.

31. The diagonals of a rectangle that is not a square bisect the angles.

32. The diagonals of a rectangle that is not a square are perpendicular to each other.

33. The diagonals of a rectangle that is not a square divide it into four congruent triangles.

34. A rectangular pizza is cut into four pieces along its diagonals. Which pieces are larger? Explain. **[Lesson 4.6]**
The same, all equal to $\frac{1}{4}$ *wl.*

Find the midpoint of a segment connecting each of the pairs of coordinates below. [Lesson 4.8]

35. (3, 4), (−3, −4)
(0, 0)

36. (4, −2), (−2, 3)
(1, 0.5)

37. (2, −2), (5, 6)
(3.5, 2)

Look Beyond

38. Cultural Connection: Europe The Swiss mathematician Leonard Euler (1707–1783) proved an interesting relationship between the parts of a prism: *F*, the number of faces; *V*, the number of vertices; and *E*, the number of edges. The relationship is called "Euler's Formula." Rediscover this formula by examining the relationship among the numbers in the last three columns of the table you created for Exercises 20–24. For a given prism, compare the number of faces and vertices against the number of edges. Write your formula in terms of *F*, *V*, and *E*. Vertices − Edges + Faces = 2

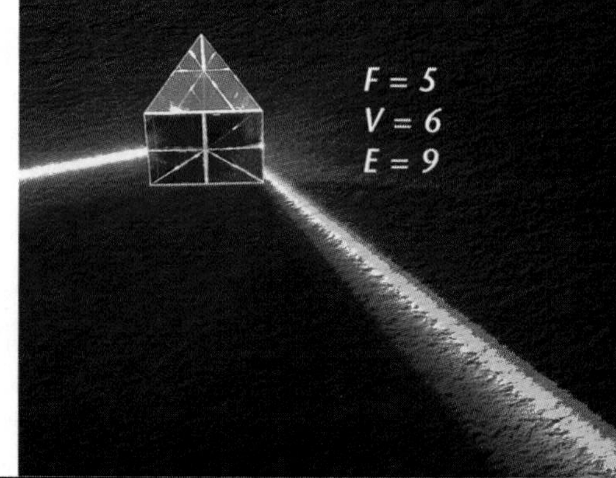

33. False. Diagonals of a rhombus divide the rhombus into four congruent triangles. The diagonals of a rectangle will only divide the rectangle into four congruent triangles if the rectangle is a square.

Look Beyond

Exercise 38 asks students to generalize the values in the table created for Exercises 20–24. Students who have a difficulty accepting Euler's formula should be encouraged to verify it with several values from the table.

Technology Master

NAME ______ CLASS ______ DATE ______

Technology
6.3 Prisms and Cross Sections

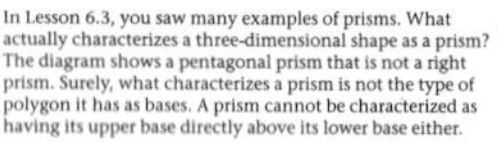

In Lesson 6.3, you saw many examples of prisms. What actually characterizes a three-dimensional shape as a prism? The diagram shows a pentagonal prism that is not a right prism. Surely, what characterizes a prism is not the type of polygon it has as bases. A prism cannot be characterized as having its upper base directly above its lower base either.

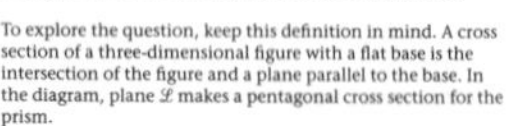

To explore the question, keep this definition in mind. A cross section of a three-dimensional figure with a flat base is the intersection of the figure and a plane parallel to the base. In the diagram, plane $\mathscr{L}$ makes a pentagonal cross section for the prism.

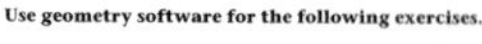

Use geometry software for the following exercises.

1. In your software, sketch a medium-sized parallelogram *ABCD* as the lower base of a three-dimensional figure. Sketch four congruent and parallel line segments that rise from parallelogram *ABCD* at its vertices.

2. Make multiple copies of parallelogram *ABCD*. Can you place those copies above parallelogram *ABCD* in such a way that their vertices coincide with points on the four line segments you drew in Exercise 1? Illustrate your answer.

3. Draw a conclusion about the three-dimensional figure generated by the instructions in Exercises 1 and 2.

4. Repeat Exercises 1–3 by starting with a pentagonal base.

5. Repeat Exercises 1–3 by starting with a triangular base.

In Exercises 6 and 7, is the three-dimensional figure shown a prism? Use an illustration to justify your response.

6.

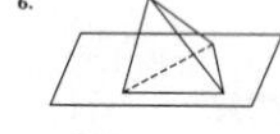

7.

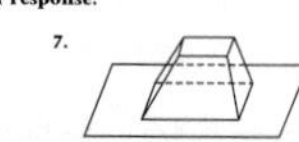

HRW Geometry Technology 109

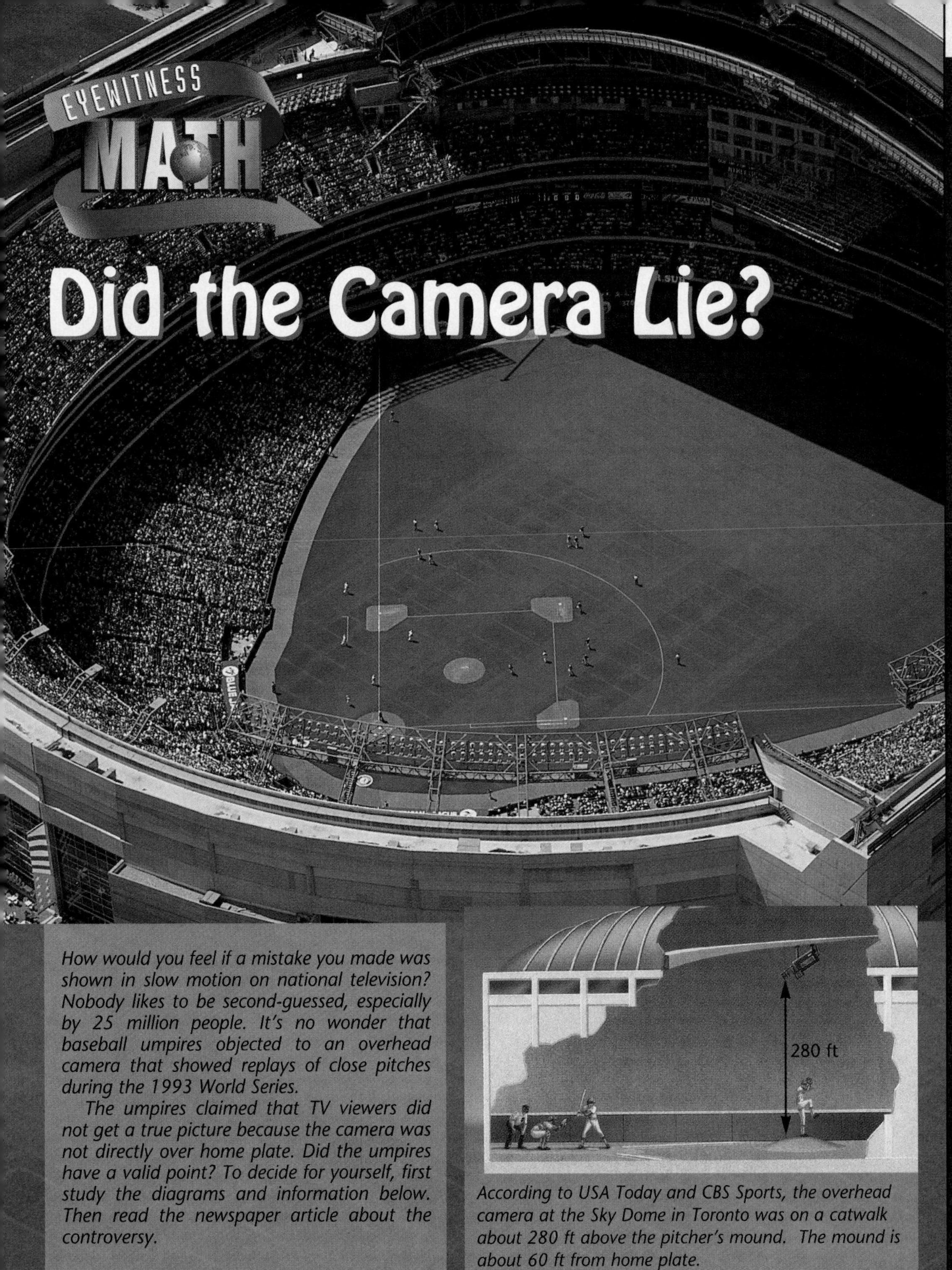

Did the Camera Lie?

How would you feel if a mistake you made was shown in slow motion on national television? Nobody likes to be second-guessed, especially by 25 million people. It's no wonder that baseball umpires objected to an overhead camera that showed replays of close pitches during the 1993 World Series.

The umpires claimed that TV viewers did not get a true picture because the camera was not directly over home plate. Did the umpires have a valid point? To decide for yourself, first study the diagrams and information below. Then read the newspaper article about the controversy.

According to USA Today and CBS Sports, the overhead camera at the Sky Dome in Toronto was on a catwalk about 280 ft above the pitcher's mound. The mound is about 60 ft from home plate.

Focus

Complaints made by umpires during the 1993 World Series sets the stage for exploring parallax (the apparent shift of position that occurs from different viewing angles). Students make a simple three-dimensional model, approximately to scale, to determine if the umpires were correct in claiming that overhead TV cameras gave a distorted view of pitches, and to determine if a news article used sound mathematical reasoning in dismissing the umpires claims.

Motivate

Have students experiment with parallax by focusing on objects around the classroom and then closing one eye. Some students will notice a shift in the object they are looking at depending on which eye is dominant. Ask students about sport activities which can be affected by parallax (throwing and catching a football, putting in golf).

Cooperative Learning

Materials: metersticks, yardsticks, or tape measures, tape, straws, toothpicks, pipe cleaners, coffee stirrers, or anything thin that can be cut and taped upright on the floor

Have students work in pairs or small groups. In the students' models, the tip of a pin, toothpick, or other object can represent the ball. The object should be cut to scale to represent 2 feet, the assumed height of the ball above home plate. Models need not be elaborate. The model need only be reasonable enough for students to see that the effect of parallax is negligible because the camera-to-ball distance is so much greater than the ball-to-plate distance.

Discuss

Ask the students the following questions.

1. What if the camera were directly over the center of home plate (instead of in front of or behind it)? Use similar triangles to show how much parallax there would be.
2. If you were the head of a sports broadcasting company, would you continue to use overhead cameras? Why or why not?

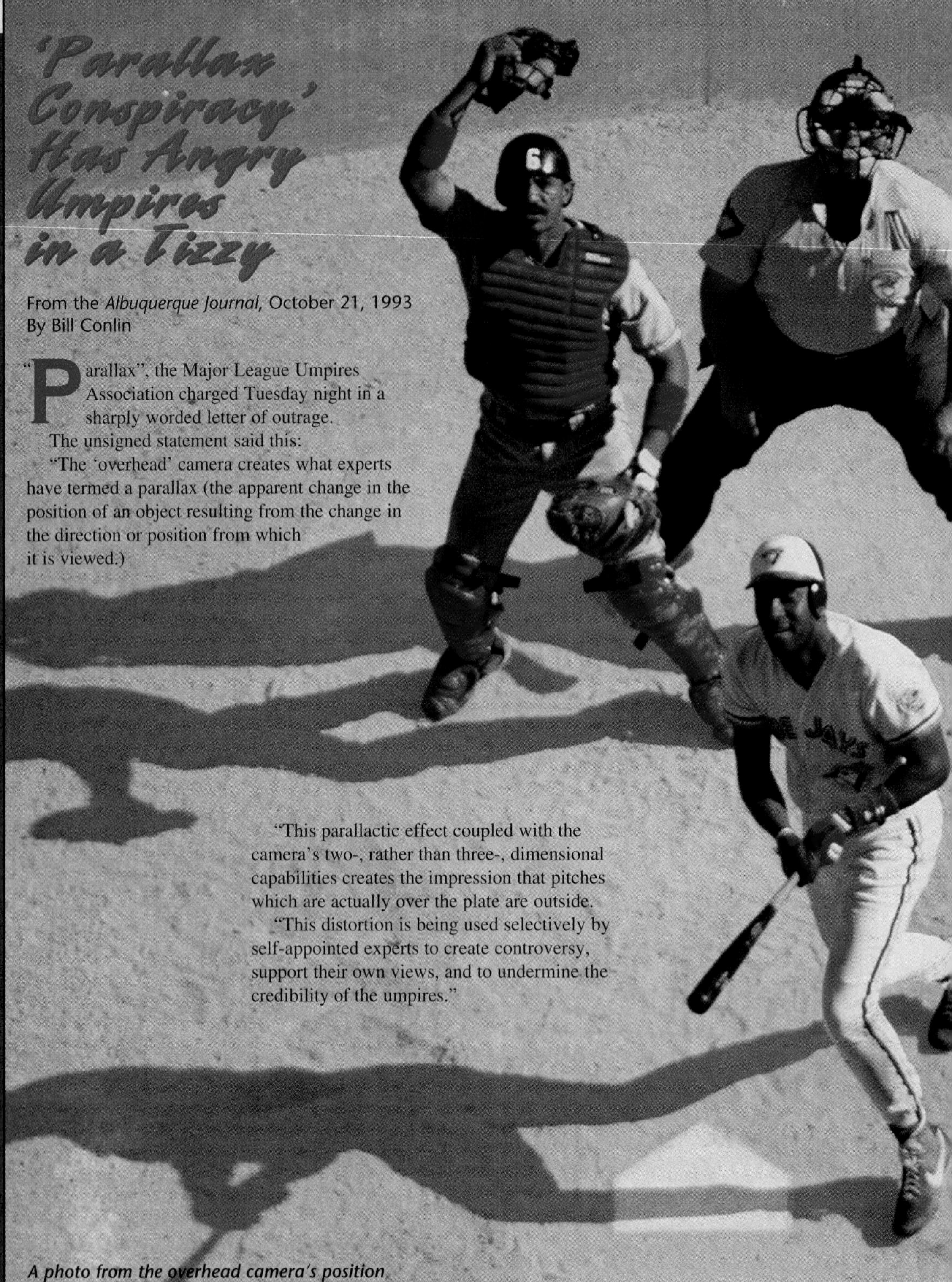

'Parallax Conspiracy' Has Angry Umpires in a Tizzy

From the *Albuquerque Journal*, October 21, 1993
By Bill Conlin

"Parallax", the Major League Umpires Association charged Tuesday night in a sharply worded letter of outrage.

The unsigned statement said this:

"The 'overhead' camera creates what experts have termed a parallax (the apparent change in the position of an object resulting from the change in the direction or position from which it is viewed.)

"This parallactic effect coupled with the camera's two-, rather than three-, dimensional capabilities creates the impression that pitches which are actually over the plate are outside.

"This distortion is being used selectively by self-appointed experts to create controversy, support their own views, and to undermine the credibility of the umpires."

A photo from the overhead camera's position

1. **a.** Answers will vary depending on height of eye above ground. For a student whose eyes are 5 feet high, the scale is about 5 feet: 280 feet, or 1:56.
   Answers for b–d will depend on the scale in (a). Sample answers are for a scale of 1:56.
   b. a little more than 1/4 inch (0.3125 or 5/16 in.)
   c. a little less than 1/2 inch (about 0.4 in.)
   d. a little more than 1 foot (about 26 in.)
2. No. The overhead camera cannot show the height of the ball over the plate relative to the batter. The umpire's view is able to encompass the total area of the strike zone. The overhead camera may be able to show whether a pitch is significantly outside, but only if the camera is placed directly above the plate.

What Do You Think?

The photos at the right illustrate parallax. You can apply this to the baseball situation by imagining that the far finger represents the edge of home plate and the closer finger represents the ball. The two different eye views represent two different views of the pitch.

The umpires argue that the camera's position causes it to view the ball and the plate differently. Complete the activity and determine for yourself whether the umpires have a case.

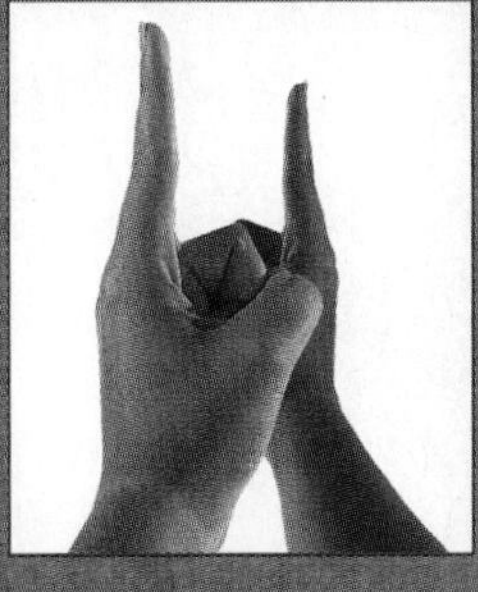

To see the parallax effect, line up your index fingers with your left eye. Then look at them with your right eye.

Cooperative Learning

1. Make a three-dimensional model to see what the camera sees. Answer these questions to help you set up the model.

a. Use your eye level when standing to represent the TV camera and the floor to represent the field. Based on your height, what is the scale of your model?

b. About how wide should home plate be in your model? What size is your model baseball (a standard baseball is 2.9 inches in diameter)?

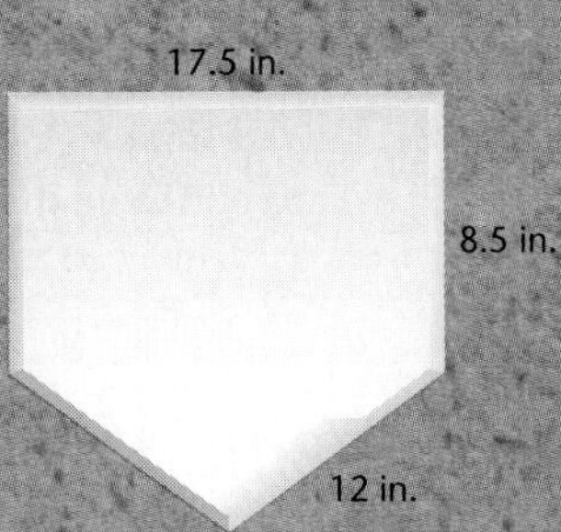

c. How high should the ball be placed to represent a pitch that is 2 ft above the plate?

d. On a baseball diamond, the picture's mound is 60 ft. 6 in. from home plate. Using your scale, how far should you stand from your model home plate to place your eye level above the pitcher's mound.

2. Comparing the view of the umpire with the view of the camera, is the camera in a better or worse position to judge strikes? Explain your answer.

3. The umpire has two eyes which bring a three dimensional capability to the observation of strikes. How does this affect the umpire's ability to judge strikes compared to the two dimensional imaging of the camera.

4. Suppose the camera shows a pitch to be 8 inches outside the strike zone and the umpire calls the pitch a strike. Do you think the parallax effect could cause this much difference? Explain your reasoning.

3. The three dimensional capability of two eyes adds a depth perception to the umpire's view. Depth perception is very important in determining the precise position of an object moving toward a person. For example, no baseball player bats with one eye closed.

4. The parallax effect is created by the overhead camera's position. However, the total effect is less than half an inch and therefore would not account for an eight inch discrepancy. In this case, it appears that the umpire made a bad call.

Prepare

Objectives

- Identify the features of a three-dimensional coordinate system, including axes, octants, and coordinate notation.
- Solve problems using the distance formula in three dimensions.

- Practice Master 6.4
- Enrichment Master 6.4
- Technology Master 6.4
- Lesson Activity Master 6.4
- Quiz 6.4
- Spanish Resources 6.4

Assessing Prior Knowledge

1. Find the midpoint of the segment joining points (3, −5) and (8, −2). [(5.5, −3.5)]
2. Name the axis containing point (−4, 0). [*x*-axis]
3. A square has vertices (5, 0), (−2, 0), (−2, 7). What is its fourth vertex? [(5, 7)]

Teach

Remind students that geometry was developed to solve practical problems. The three-dimensional coordinates studied in this lesson not only are practical but also are an extremely important tool for locating points in space.

CRITICAL Thinking

Yes, as long as rotating the *x*-axis toward the *y*-axis gives a thumb pointing in the *z*-direction.

Three-Dimensional Coordinates

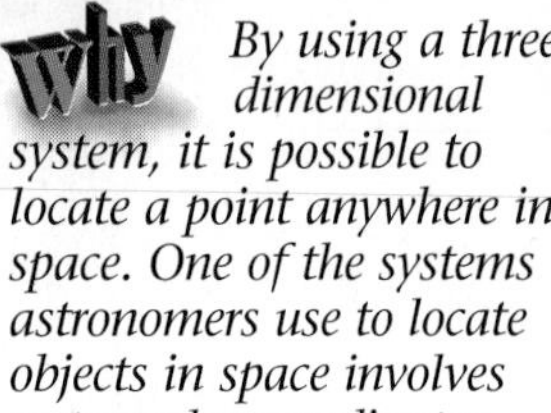

By using a three-dimensional system, it is possible to locate a point anywhere in space. One of the systems astronomers use to locate objects in space involves rectangular coordinates.

By using the *x*- and *y*-coordinate axes on a plane, you can give the location of a point anywhere on the plane. Since two numbers are required to do this, a plane is said to be **two-dimensional**. By adding a third coordinate axis at right angles, you can give the location of a point anywhere in space. Three numbers are now required, and so space is said to be **three-dimensional**.

The Arrangement of the Axes

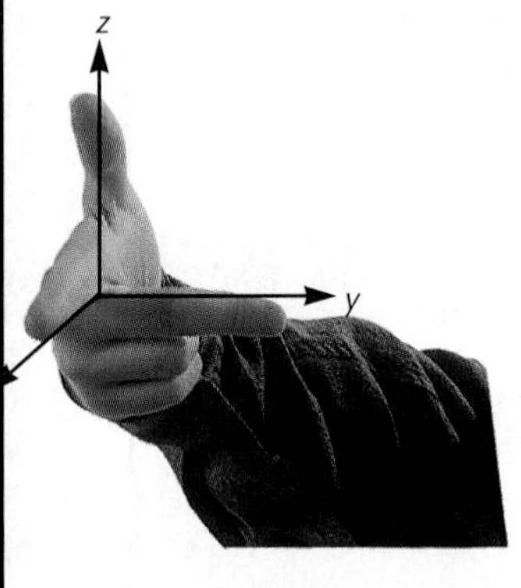

A number of arrangements of axes in space are possible. The one most commonly used by scientists and mathematicians is called the **right-handed system**. Let your right forefinger represent the positive direction of the *x*-axis.

Point with your forefinger and hold your thumb and middle finger so that they each make right angles with the line of your forefinger. Your middle finger will be pointing in the positive direction of the *y*-axis, and your thumb will be pointing in the positive direction of the *z*-axis.

Normally, a right-handed system of axes is pictured with the *y*-axis to the right on the page. Do you think there are other ways of drawing a right-handed system?

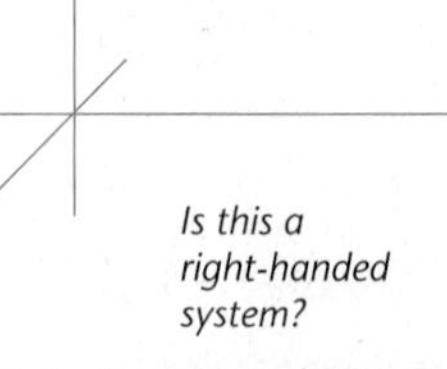

Is this a right-handed system?

Using Discussion Students can develop an understanding of how to locate points in space by considering questions such as the following:

1. How many coordinates must a point in space have? [3]
2. Which axis contains the point (0, −2, 0)? [*y*-axis]
3. Which plane contains the point (3, −2, 0)? [*xy*-plane]
4. Which axis is the intersection of planes *yz* and *xz*? [*z*-axis]
5. How many planes contain a line through (0, 0, 0) and (2, 3, 4)? [**infinitely many**]

EXAMPLE 1

Locate the point $P(1, 2, 3)$ in a three-dimensional coordinate system.

Solution ➤

1. Place your pencil at the origin. Count one unit in the x-direction. Make a mark or dot at the new position.
2. From the new position, count two units in the y-direction, drawing a light or dashed line to represent your path. Make a new mark or dot at the new position.
3. From the new position, count three units in the z-direction, drawing a light or dashed line to represent your path. Label your final position as point $P(1, 2, 3)$. ❖

The Arrangement of the Octants

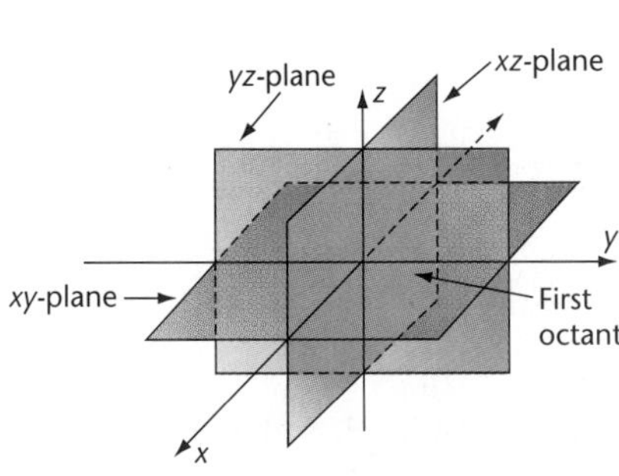

The x- and y-axes in the coordinate plane divide the plane into four quadrants. In a similar way, the x-, y-, and z-axes divide space into eight **octants**. The first octant is the one in which all three coordinates of the points in it are positive.

The octants can be referred to using the words *top, bottom, front, back, right,* and *left*. For example, the first octant is the top-front-right octant. Any point in this octant has three positive coordinates (+, +, +). What are the signs of the coordinates for points in the other seven octants?

Notice that there are three separate planes determined by the axes of a three-dimensional coordinate system.

EXAMPLE 2

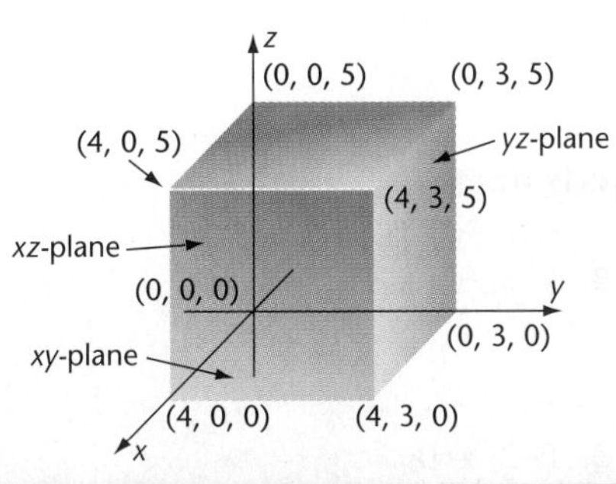

Draw a box in the first octant of a right-handed coordinate system. Place the box so that three of its faces are in the planes determined by the coordinate axes. Label the vertices with their coordinates. What is true of the points in the three axis planes?

Solution ➤

One possible solution is given in the diagram.
In the xy-plane, all the z-coordinates are 0.
In the xz-plane, all the y-coordinates are 0.
In the yz-plane, all the x-coordinates are 0. ❖

ENRICHMENT Air-traffic control towers need to keep accurate information on the whereabouts of all airplanes within their radar range. Suppose an airplane looks like it is located at point (8, 6) on a two-dimensional view screen with units in miles. However, the aircraft is really 2 miles above ground. Find the coordinates of the aircraft in three dimensions. [**Solution: (8, 6, 2)**]

Using Models Many students have difficulty visualizing three-dimensional coordinates. You can illustrate the eight octants of a three-dimensional system by using physical models of planes made of cardboard or other materials.

Alternate Example 1

Locate the point $P(2, 3, 6)$ in a three-dimensional coordinate system.

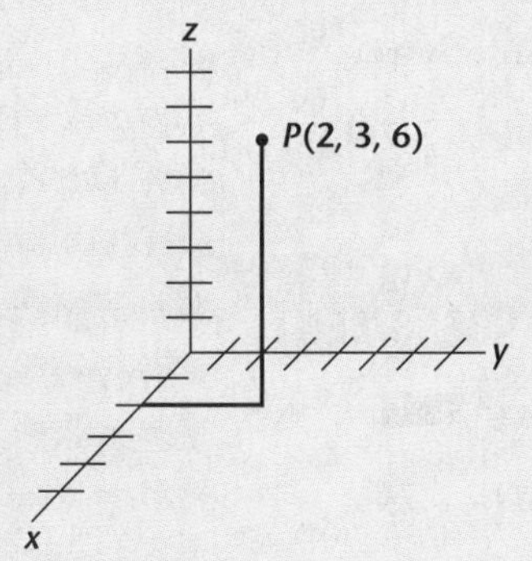

Cooperative Learning

Have groups of students quiz each other on the location of points in space, points on the axes, and points on the xy-plane, the yz-plane, and the xz-plane.

Use Transparencies 54–55

ongoing ASSESSMENT

Top–front–left: (+, –, +)
Top–back–right: (–, +, +)
Top–back–left: (–, –, +)
Bottom–front–right: (+, +, –)
Bottom–front–left: (+, –, –)
Bottom–back–right: (–, +, –)
Bottom–back–left: (–, –, –)

Alternate Example 2

Draw a box in the first octant of a right-handed coordinate system. Place the box so that three of its faces are in the planes determined by the coordinate axes. Label the vertices with their coordinates. Name the labeled coordinates contained in each of the following planes: xy-plane, xz-plane, yz-plane.

[**Answers will vary.**]

Alternate Example 3

Find the length of the segment with endpoints $P(2, -4, -2)$ and $Q(0, -3, 1)$.

[Solution

$\sqrt{14} \approx 3.74$]

Ongoing Assessment

Try This

$CD = 5.74$

Assess

Selected Answers

Odd-numbered Exercises 13–37

Assignment Guide

Core 1–6, 8–23, 28–40

Core Plus 1–21, 27–43

Technology

To find the distance between points in Exercises 21–24, have students store the distance formula in a function key of a graphing calculator, or $y1 = \sqrt{(D-A)^2 + (E-B)^2 + (F-C)^2}$. Enter A, B, C, D, E, F into the memory of the calculator (STORE key), and recall $y1$.

Error Analysis

For Exercises 29–34, students may not make the connection between finding the midpoint of a segment using algebra and finding the location of the midpoint in space.

The Distance Formula in Three Dimensions

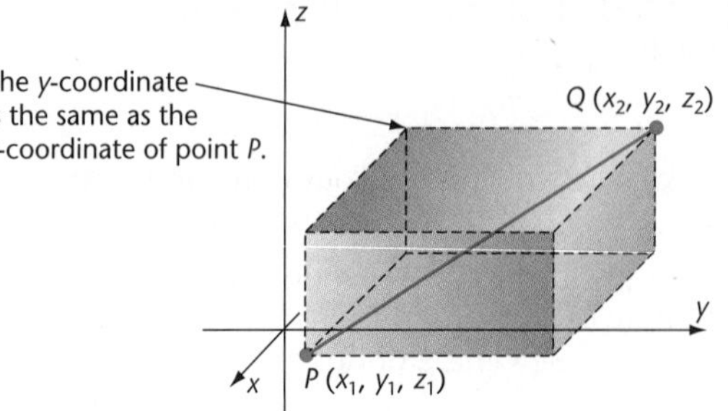

In a rectangular prism with length a, width b, and height c, the length of the diagonal can be found using the following formula.

$$l = \sqrt{a^2 + b^2 + c^2}$$

Any line in an x-y-z coordinate space can be viewed as a diagonal of a rectangular prism. The difference in the x-coordinates is the length of one side of the prism, and the differences in the y- and z-coordinates are the lengths of the other two sides.

Therefore, for points $P(x_1, y_1, z_1)$ and $Q(x_2, y_2, z_2)$, the distance PQ is

$$PQ = \sqrt{(x_2 - x_1)^2 + (y_2 - y_1)^2 + (z_2 - z_1)^2}.$$

EXAMPLE 3

ALGEBRA Connection

Find the length of the segment with endpoints $R(4, 6, -9)$ and $S(-3, 2, -6)$.

Solution ➤

$$RS = \sqrt{(-3-4)^2 + (2-6)^2 + (-6-(-9))^2}$$
$$= \sqrt{(-7)^2 + (-4)^2 + (3)^2}$$
$$= \sqrt{49 + 16 + 9} = \sqrt{74} \approx 8.6$$ ❖

Try This Find the length of the segment with endpoints $C(3, -4, -5)$ and $D(2, 0, -1)$.

Exercises & Problems

Communicate

In a three-dimensional coordinate system, on which axis does each point lie?

1. (0, 0, 7) **2.** (0, 3, 0) **3.** (−5, 0, 0)

In a three-dimensional coordinate system, on which coordinate plane does each point lie?

4. (2, 3, 0) **5.** (−3, 0, −2) **6.** (−2, 4, 0)

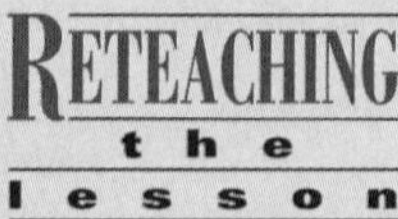

Reteaching the Lesson **Using Models** Have students give the coordinates of a right rectangular prism that has a vertex at the origin of a three-dimensional coordinate system and that has a height of 3, a width of 4, and a length of 5. Have them use those coordinates to give the length of a diagonal. [**Sample: (0, 0, 0), (0, 0, 3), (0, 4, 3), (0, 4, 0), (5, 4, 0), (5, 0, 0), (5, 0, 3); diagonal = $\sqrt{50}$**]

7. A rectangular prism has one vertex at the origin and an opposite vertex at (2,−3,−5). Three of its faces are parallel to the planes determined by the coordinate axes. Explain how to find the volume of the prism.

8. In the illustration, a white sphere is illuminated by two different light sources, one red and one blue. Where the red and blue lights mix, magenta (a pink color) is produced. What can you tell about the location of the light sources? Express your answer in terms of the coordinate axes.

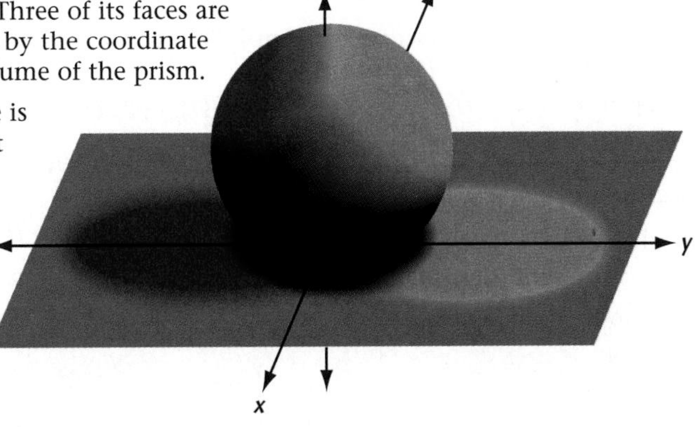

Practice & Apply

Locate each point on a three-dimensional coordinate axis.

9. (1, 2, 3) **10.** (1, −2, 3) **11.** (1, −2, −3) **12.** (−1, −2, −3)

Determine the octant that would contain each of the following points if *x*, *y*, and *z* are positive and not zero.

13. $(-x, y, z)$ Top-back-right
14. $(-x, -y, z)$ Top-back-left
15. $(x, y, -z)$ Bottom-front-right
16. $(-x, -y, -z)$ Bottom-back-left

Sketch a graph of each of the following.

17. Point A; (3, −2, 1)
18. Point B; (4, −6, 2)
19. Segment CD; $C(1, 4, -2)$, $D(2, -3, 0)$
20. Plane EFG; $E(4, -2, 3)$, $F(0, 1, 0)$, $G(-3, -1, 5)$

Find the distance between each pair of points.

21. (2, 1, 3), (5, −2, 7) 5.83
22. (1, 1, 1), (−1, −1, −1) 3.46
23. (7, −6, 5), (6, 4, 3) 10.25
24. (2, 0, 1), (−5, −6, −5) 11

The grid units of the coordinate system are the same as the measurement units of the box. Determine the coordinates of each point.

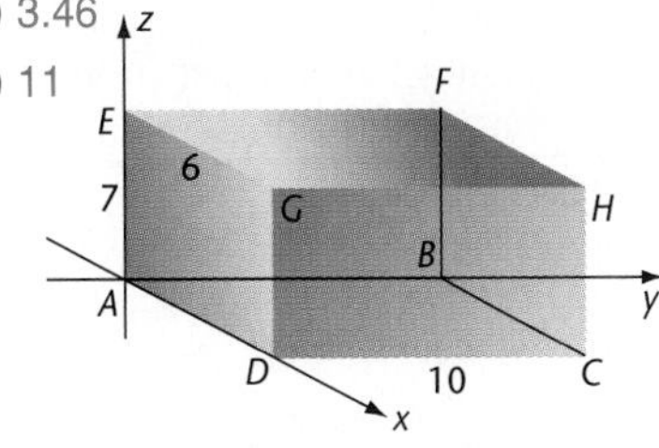

A is at the origin.

25. Point F $F(0, 10, 7)$
26. Point H $H(6, 10, 7)$
27. Point C $C(6, 10, 0)$
28. Point E $E(0, 0, 7)$

In a three-dimensional coordinate system, the midpoint of a segment is defined for points (x_1, y_1, z_1) and (x_2, y_2, z_2) as $\left(\frac{x_1 + x_2}{2}, \frac{y_1 + y_2}{2}, \frac{z_1 + z_2}{2}\right)$. Find the midpoint of each segment below.

29. $A(3, 2, 1)$, $B(1, 2, 3)$
30. $A(5, -2, 3)$, $B(6, -7, 4)$
31. $A(2, -1, 0)$, $B(0, 0, 1)$
32. $A(-1, -4, -5)$, $B(6, 1, 0)$
33. $A(2, 0, 2)$, $B(-1, 5, -3)$
34. $A(1, 1, 1)$, $B(-1, -1, -1)$

29. (2, 2, 2)

30. $\left(\frac{11}{2}, \frac{-9}{2}, \frac{7}{2}\right)$

31. $\left(1, \frac{-1}{2}, \frac{1}{2}\right)$

32. $\left(\frac{5}{2}, \frac{-3}{2}, \frac{-5}{2}\right)$

33. $\left(\frac{1}{2}, \frac{5}{2}, \frac{-1}{2}\right)$

34. (0, 0, 0)

alternative ASSESSMENT

Performance Assessment

Student presentations can be used to assess students' understanding of the objectives of this lesson. You may wish to call upon individual students to show where coordinates are located in a model of three-dimensional space. Other students can be asked to draw a right-handed coordinate system on the chalkboard, state the distance formula in three dimensions, and then give examples of how to find the distance between two points in space.

The answers to Exercises 9–12 and 17–20 can be found in Additional Answers beginning on page 727.

Practice Master

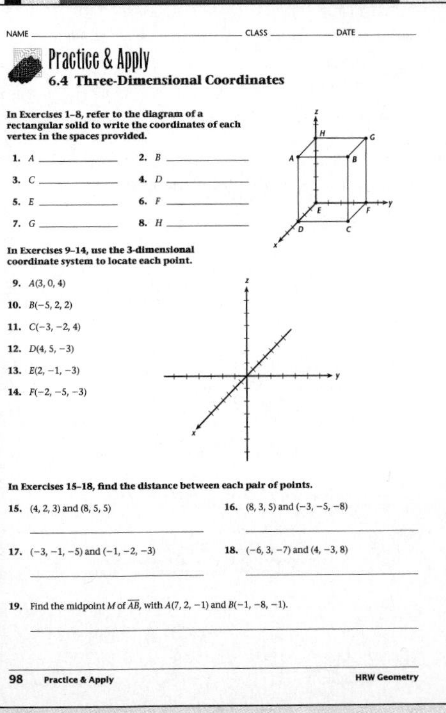
NAME ______ CLASS ______ DATE ______

Practice & Apply
6.4 Three-Dimensional Coordinates

In Exercises 1–8, refer to the diagram of a rectangular solid to write the coordinates of each vertex in the spaces provided.

1. A ______ 2. B ______
3. C ______ 4. D ______
5. E ______ 6. F ______
7. G ______ 8. H ______

In Exercises 9–14, use the 3-dimensional coordinate system to locate each point.

9. $A(3, 0, 4)$
10. $B(-5, 2, 2)$
11. $C(-3, -2, 4)$
12. $D(4, 5, -3)$
13. $E(2, -1, -3)$
14. $F(-2, -5, -3)$

In Exercises 15–18, find the distance between each pair of points.

15. (4, 2, 3) and (8, 5, 5) ______
16. (8, 3, 5) and (−3, −5, −8) ______
17. (−3, −1, −5) and (−1, −2, −3) ______
18. (−6, 3, −7) and (4, −3, 8) ______

19. Find the midpoint M of $\overline{AB}$, with $A(7, 2, -1)$ and $B(-1, -8, -1)$.
______

98 Practice & Apply HRW Geometry

Look Back

Find the distance between each point in a rectangular coordinate system. [Lesson 4.8]

35. (4, 7), (−3, 2) 8.60 **36.** (21, 37), (−2, 15) 31.83

Find the midpoint of the segment joining the two points. [Lesson 4.8]

37. $A(3, -4)$, $B(1, -6)$ (2, −5) **38.** $A(-2, 15)$, $B(11, -7)$ $(\frac{9}{2}, 4)$

39. Draw a right triangular prism. **[Lesson 6.3]** Check student drawings.

40. Draw an oblique pentagonal prism. **[Lesson 6.3]** Check student drawings.

Look Beyond

Exercises 41–43 give students examples of the distance formula in three dimensions, where each unit in space is an astronomical unit (au), the distance from the Earth to the sun. This unit was created so that the large numbers associated with outer space can be managed more easily, including the equations describing the orbits of the planets.

Look Beyond

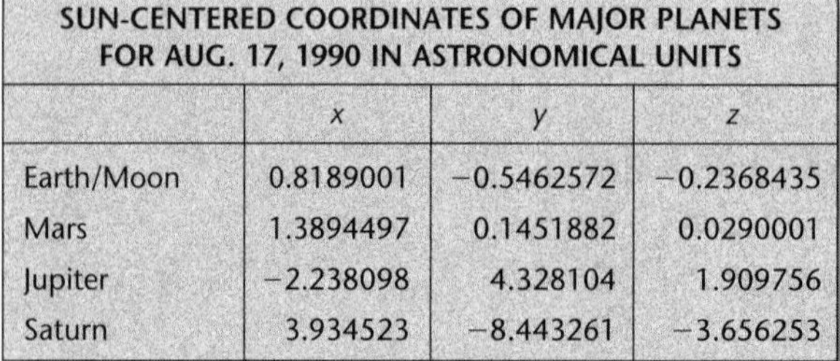

SUN-CENTERED COORDINATES OF MAJOR PLANETS FOR AUG. 17, 1990 IN ASTRONOMICAL UNITS			
	x	y	z
Earth/Moon	0.8189001	−0.5462572	−0.2368435
Mars	1.3894497	0.1451882	0.0290001
Jupiter	−2.238098	4.328104	1.909756
Saturn	3.934523	−8.443261	−3.656253

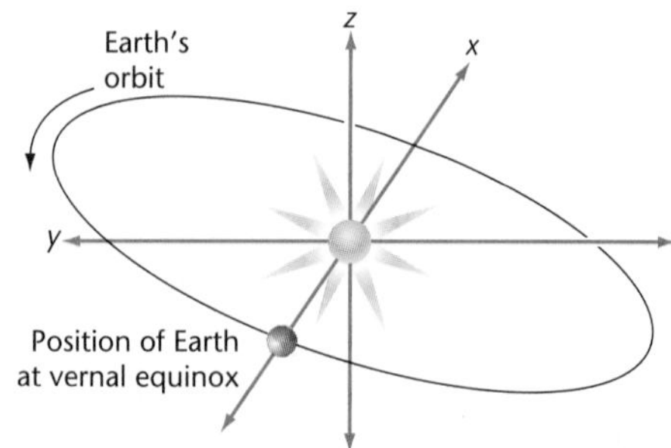

Astronomy The table gives three-dimensional, *x-y-z* coordinates for the planets using a coordinate system with the Sun at the origin. The *x*-axis is defined as lying in the direction of the Earth's position at the first moment of spring (the vernal equinox).

The *xy*-plane of the system is the equatorial plane, which corresponds to the plane of the equator of the Earth at the time of the equinox.

The unit of measure used in the table is the astronomical unit (au), which equals the mean distance of the Earth from the sun, about 92,900,000 mi, or 149,600,000 km.

The values given for the Earth/Moon are actually for a point called the center of mass of the two bodies, which is located between their centers.

41. Calculate the distance between the Earth/Moon center and Jupiter on August 17, 1990. Convert the astronomical units into miles.

42. Calculate the distance between Mars and Saturn on August 17, 1990. Convert the astronomical units to kilometers.

43. Calculate the distance between the Sun and Mars on August 17, 1990. Find a book that gives the mean distances of planets from the Sun. Does your calculation match? Why or why not?

41. 6.14 au = 570,000,000 mi

42. 9.87 au = 1,449,000,000 km

43. 1.41 au This may not be the mean distance because the orbit of Mars around the sun is elliptical.

Technology Master

NAME ______ CLASS ______ DATE ______

Technology
6.4 Polygonal Paths

A *polygonal path* in three-dimensional space is the set of points you get when you go from P_1 to P_2 to P_3 … to P_{n-1}, and to P_n along the line segments ___, ___, ___, …, ___.

If you enter the *x*-, *y*- and *z*-coordinates into columns A–C of a spreadsheet, you can use column D along with the distance formula for points in three-dimensional space to find the length of each piece of the path and the total length of the path. For reference, the distance *d* between (x_1, y_1, z_1) and (x_2, y_2, z_2) is stated below.

$$d = \sqrt{(x_2 - x_1 \quad + (y_2 - y_1 \quad (}$$

1. Create a spreadsheet based on the outline in which you can find the length of a polygonal path in three-dimensional space.

Use your spreadsheet to find the total length of each path.

2. $A(0, 0, 0)$, $B(1, 2, 3)$, $C(1, 3, -4)$ ______
3. $A(0, 0, 0)$, $B(2, -4, -1)$, $C(-3, 4, 2)$, $D(0, 0, 0)$ ______
4. $A(0, 0, 0)$, $B(-1, 4, 6)$, $C(2, 0, 0)$, $D(0, 3, 4)$, $E(0, 0, -2)$ ______
5. $A(5, -6, -4)$, $B(3, -5, -1)$, $C(1, -4, 2)$, $D(-1, -3, 5)$, $E(-3, -2, 8)$, $F(-5, -1, 11)$ ______
6. What can you say about the path formed by points *A*, *B*, *C*, and *D* if $AB = BC = CD = DA$ and the points lie in the same plane? ______
7. The points in Exercise 5 have *x*-coordinates that decrease. Compute *AF*. What can you say about points *A*, *B*, *C*, *D*, *E*, and *F*? ______
8. Find the total length of the path *ABCDEFGHI* if the points have coordinates $A(-4, 0, 0)$, $B(0, 0, 4)$, $C(4, 0, 0)$, $D(0, 0, -4)$, $E(-4, 0, 0)$, $F(0, 4, 0)$, $G(4, 0, 0)$, $H(0, -4, 0)$, and $I(-4, 0, 0)$. Describe the figure whose vertices are *A*, *B*, *C*, *D*, *E*, *F*, *G*, *H*, and *I*? ______

110 Technology HRW Geometry

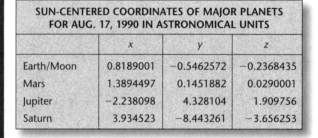

LESSON 6.5 Lines and Planes in Space

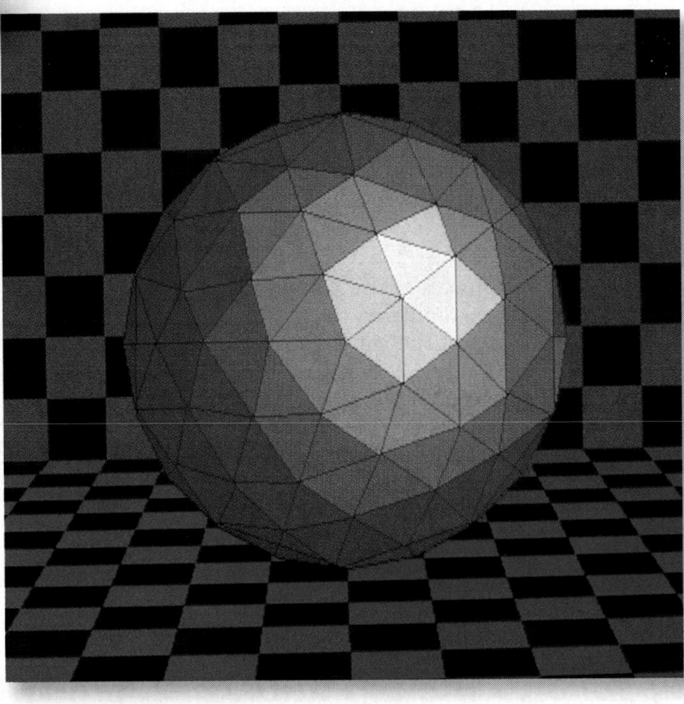

why *Planes in space may form three-dimensional shapes, as in this drawing. Each plane in the drawing can be described mathematically.*

The Equation of a Plane

Recall from algebra that the standard form of the equation of a line in a plane is

$$Ax + By = C.$$

ALGEBRA *Connection*

The equation of a plane in space resembles the equation of a line, with an extra variable for the added dimension of space:

$$Ax + By + Cz = D.$$

For example,

$$2x + 5y - 3z = 9$$

is the equation of a plane where $A = 2$, $B = 5$, $C = -3$, and $D = 9$.

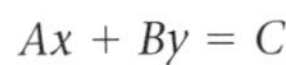

Using Intercepts in Graphing

In the coordinate plane, a line crosses the x- and y-axes at one or two points called **intercepts**. In coordinate space, a plane has one, two, or three intercepts at the x-, y-, and z-axes.

ALTERNATIVE teaching strategy

Using Visual Models Use clear plastic squares or thick acetate squares to create a model of the octants in a right-handed three-dimensional coordinate system. Cut slits in the squares to slide them into position. Label the axes. Using straws to represent lines, position the straws in various positions in the model. Have students identify intercepts of each (straw) line and give its equation. A sample equation would be $3x - 4y = 12$ in two dimensions and in three dimensions. Ask students to give the intercepts in both dimensions.

PREPARE

Objectives

- Define the *equation of a line* and the *equation of a plane* in space.
- Solve problems using the equations of lines and planes in space.

RESOURCES

- Practice Master 6.5
- Enrichment Master 6.5
- Technology Master 6.5
- Lesson Activity Master 6.5
- Quiz 6.5
- Spanish Resources 6.5

Assessing Prior Knowledge

1. Find the x- and y-intercepts of the line $2x + 3y = 6$. [**x-intercept = 3; y-intercept = 2**]
2. Is the point $(-2, 4)$ a point on the graph of $2x + 3y = 6$? [**no**]
3. Find the equation of a horizontal line through $(0, -3)$. [**$y = -3$**]

TEACH

This lesson brings together ideas from algebra to help students see the connections between graphing lines in two dimensions and graphing lines in three dimensions.

Math Connection Algebra

In algebra, students studied how to find and graph the equation of a line. It would be helpful to review those ideas at this time, including how to make a table of values that satisfy the equation of a line.

Alternate Example 1

Sketch the graph of the plane with equation $x + 4y + 2z = 8$.

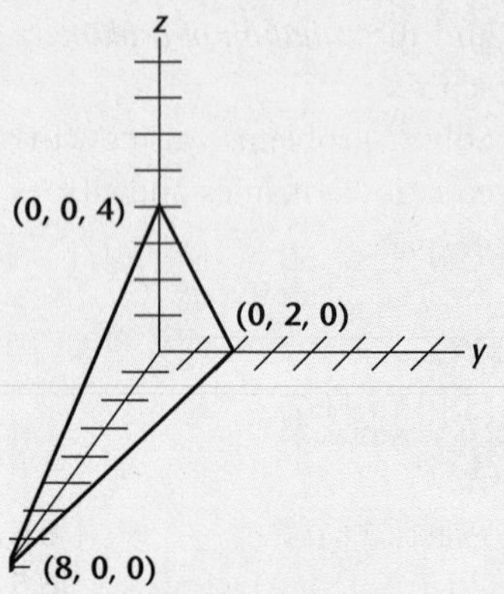

Math Connection Algebra

Have students discuss how to graph a line in the plane by first locating its intercepts. Ask them to discuss how intercepts in a plane may be related to intercepts in a three-dimensional system.

Math Connection Algebra

Students may find it difficult to understand that the same equation for a line in two dimensions gives a plane in three dimensions. Using the example $2x + 4y + 0z = 8$, make an extensive table of values to show that there are many values of z which can be used to give the same graph, the graph of a plane.

Cooperative Learning

The algebraic properties associated with three dimensions are so abstract that you may want to use groups to help students understand the concepts. Groups of three or four students can build models of lines and planes in space. Allow students to explain to each other how to locate points.

EXAMPLE 1

Sketch the graph of the plane with the equation $2x + y + 3z = 6$.

Solution ➤

ALGEBRA *Connection*

Find the intercepts where the plane crosses each of the axes. The x-intercept, for example, is the point where the plane crosses the x-axis. At this point, the y- and z-coordinates are 0. To find this point, set $y = 0$ and $z = 0$ and solve the equation for x:

$$2x + 0 + 3(0) = 6$$
$$2x = 6$$
$$x = 3$$

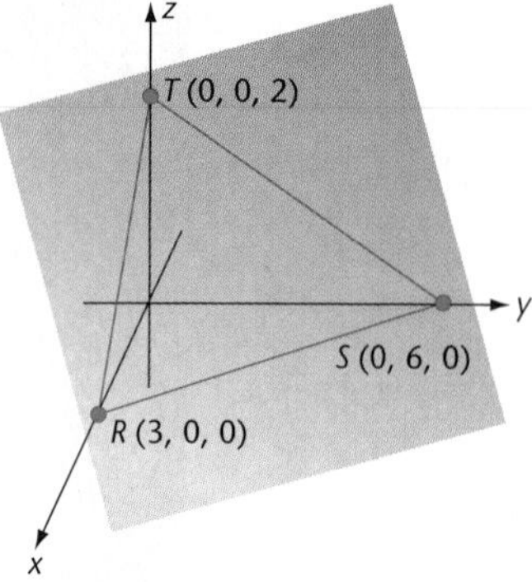

The coordinates of the x-intercept are (3, 0, 0). Similarly, the coordinates of the y-intercept are (0, 6, 0), and the coordinates of the z-intercept are (0, 0, 2). Plot these on the axes. These three noncollinear points determine the plane. To sketch the plane, connect the points with segments and add shading. ❖

What Happens to the Equation of a Line in Space?

ALGEBRA *Connection*

$2x + 4y = 8$ is the equation of a line in the plane. But in space it is a special case of the equation of a plane in which the coefficient of z is equal to 0.

Notice that this equation of a plane is unaffected by the values of z. Thus, for any values of x and y that satisfy the equation, any value of z will also work. So every point above and below a point on the line is in the graph of the plane.

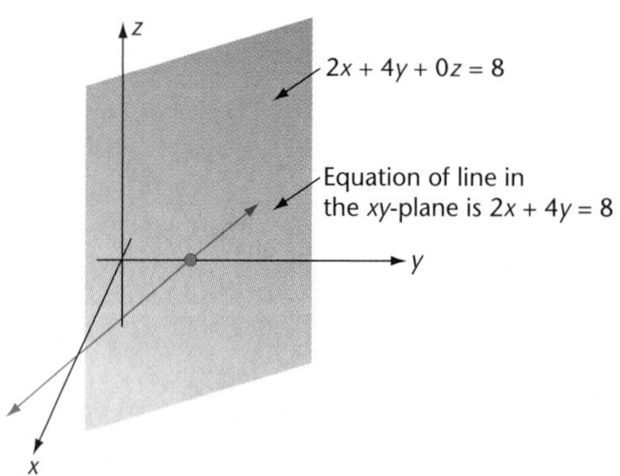

Art Computer artists frequently do their designing on coordinate grids, both in two dimensions and in three dimensions. Many special art effects can be reproduced by computers.

Lines in Space: A Step-by-Step Procedure

Imagine that you are able to plot one point of a graph every second according to a set of instructions. Your instructions give you one rule for the x-coordinates, another for the y-coordinates, and (if you are working in three dimensions) another for the z-coordinates.

Let $t = 1, 2, 3, \ldots$ represent the time (in seconds) at which you plot each point. Your rules for each coordinate will have the following form.

x = [an expression involving t]

y = [an expression involving t]

z = [an expression involving t]

(Once you get the idea, you can also think of 0 and negative values for t.)

EXAMPLE 2

Use the following rules to plot a line on a plane, where $t = \{1, 2, 3, \ldots\}$.

$$x = 2t$$

$$y = 3t + 1$$

Solution ➤

Fill in a table like the one below. Then plot the graph on an xy-coordinate plane. ❖

t	x	y
1	2	4
2	4	7

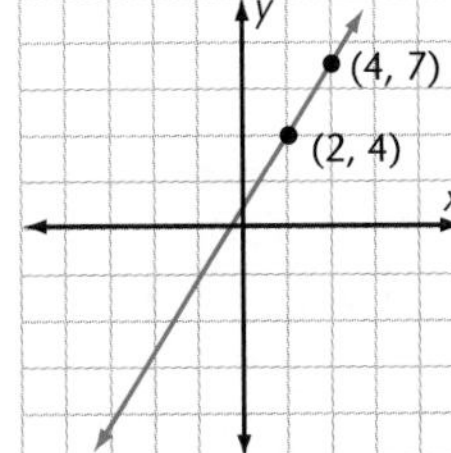

EXAMPLE 3

Use the following rules to plot a line in space.

$$x = 2t + 1$$

$$y = 3t - 5$$

$$z = 4t$$

Solution ➤

Fill in a table like the one below. Then plot the graph. ❖

t	x	y	z
1	3	−2	4
2	5	1	8

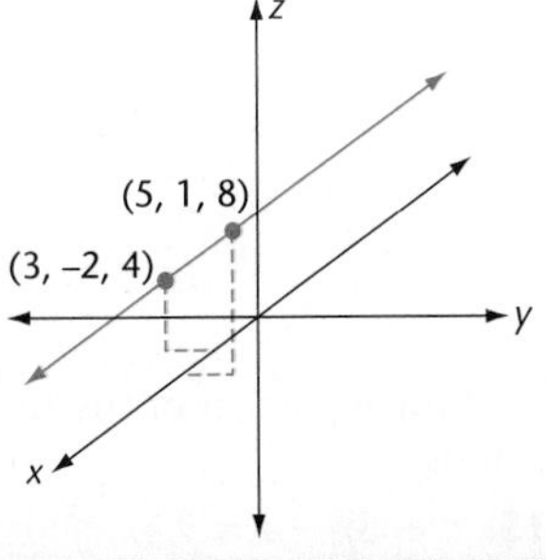

Math Connection Algebra

Using another variable, t, to help students plot a line or plane in space has connections to setting up parametric equations. These may be used to describe other kinds of graphs, especially graphs showing motion in space. You may want to use a graphing calculator in parametric mode to demonstrate graphing lines.

Alternate Example 2

Use the following rules to plot a line in a plane, where $t = \{1, 2, 3, \ldots\}$, $x = 3t$, and $y = 2t - 4$.

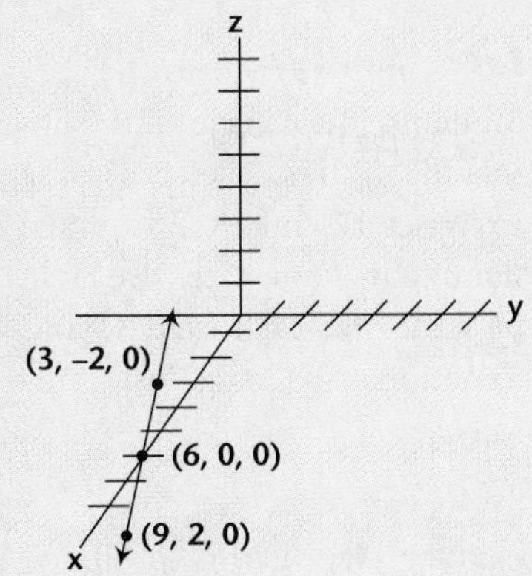

ENRICHMENT The radius of a sphere graphed in a three-dimensional coordinate system is $r = \sqrt{((x-i)^2+(y-j)^2+(z-k)^2)}$, where (i, j, k) is the center of the sphere. What is the radius of a sphere with the center at (3, −2, 4) and a point on the sphere at (93, −8, 4)? Give another point of the sphere. **[6; answers will vary.]**

INCLUSION strategies **English Language Development** Many students have difficulty reading and understanding mathematics on their own. Suggest that students read this lesson slowly and carefully, write notes, draw samples, and develop a list of questions they need answered. You can use cooperative groups to discuss the questions in class.

Assess

Selected Answers

Exercises 5, 9, 13, 15, 19, and 23

Assignment Guide

Core 1–22

Core Plus 1–18, 23, 24

Technology

Although the exercises and problems for this lesson do not require the use of computer technology, using software to create three-dimensional images may give students a better perspective of lines and planes in space.

Error Analysis

Students may have difficulty making the sketches for Exercises 17 and 18. Encourage them to make an extensive table of values for each exercise and to graph each equation carefully.

The answers to Exercises 5–14 can be found in Additonal Answers beginning on page 727.

Practice Master

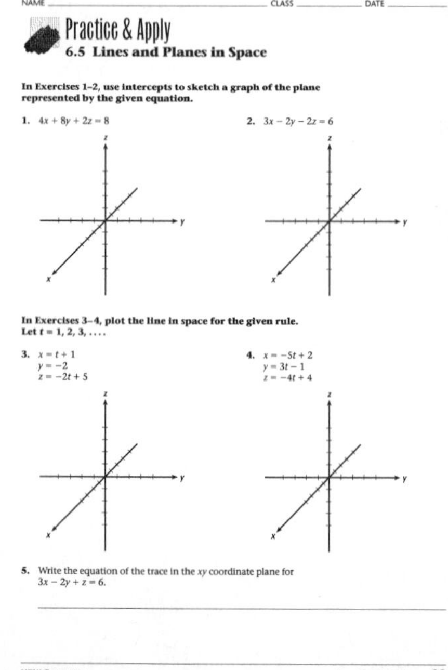

NAME ______ CLASS ______ DATE ______

Practice & Apply

6.5 Lines and Planes in Space

In Exercises 1–2, use intercepts to sketch a graph of the plane represented by the given equation.

1. $4x + 8y + 2z = 8$
2. $3x - 2y - 2z = 6$

In Exercises 3–4, plot the line in space for the given rule. Let $t = 1, 2, 3, \ldots$.

3. $x = t + 1$
   $y = -2$
   $z = -2t + 5$
4. $x = -5t + 2$
   $y = 3t - 1$
   $z = -4t + 4$

5. Write the equation of the trace in the xy coordinate plane for $3x - 2y + z = 6$.

HRW Geometry   Practice & Apply   99

Exercises & Problems

Communicate

1. State the standard form for an equation of a plane.
2. Describe the characteristics of a plane with an equation in which the coefficient of z is 0.
3. Describe the characteristics of a plane with an equation in which the coefficient of x is 0.
4. Describe the characteristics of a plane with an equation in which the coefficients of y and z are 0.

Practice & Apply

Algebra Use intercepts to sketch a graph of each of the planes represented by the equations below.

5. $3x + 2y + 7z = 4$ **6.** $2x - 4y + z = -2$ **7.** $x - 2y - 2z = -4$

8. $-3x + y = 4$ **9.** $x - 2y = 2$ **10.** $x = -4$

Algebra Plot the line in space for each of the following rules. Let $t = 1, 2, 3, \ldots$

11. $x = t$
$y = 2t$
$z = 3t$

12. $x = t + 1$
$y = 2t + 1$

13. $x = \frac{2t}{3}$
$y = t$
$z = 1 - t$

14. $x = 3t$
$z = 4t - 7$

Algebra The **trace** of a plane is the line of intersection of the plane with the xy-plane. To find the equation of the trace, remember that a point lies on the xy-plane if and only if the z-coordinate is 0. Set the z-coordinate equal to 0 for a given equation of a plane, and the resulting equation of a line is the trace of the plane. For example:

$2x + y + 3z = 6$	original equation
$2x + y + 3(0) = 6$	substitute 0 for z
$2x + y = 6$	equation of the trace

Find the equation of the trace for each equation of a plane below.

15. $x + 3y - z = 7$ ($x + 3y = 7$) **16.** $5x - 2y + z = 2$ ($5x - 2y = 2$)

Reteaching the lesson

Using Discussion Review the formula for the distance between two points in the plane, and then present the formula developed for similar concepts in three dimensions. Compare the formulas by discussing their similarities and differences.

 **Algebra** Sketch each of the following planes and indicate the trace.

17. $2x + 7y + 3z = 2$ **18.** $-4x - 2y + 2z = 1$

19. Navigation Suppose an airplane's initial climb out of takeoff is determined by the following parameters. Plot the line for the following rules:

$x = -t$

$y = t$

$z = 0.5t$

If the positive x-axis is to the north, in which direction did the airplane take off? Southwest

Look Back

20. Given: $\angle FEH \cong \angle FGH$; $\overline{EH} \cong \overline{FG}$
Prove: $\overline{EF} \cong \overline{HG}$ **[Lesson 3.3]**

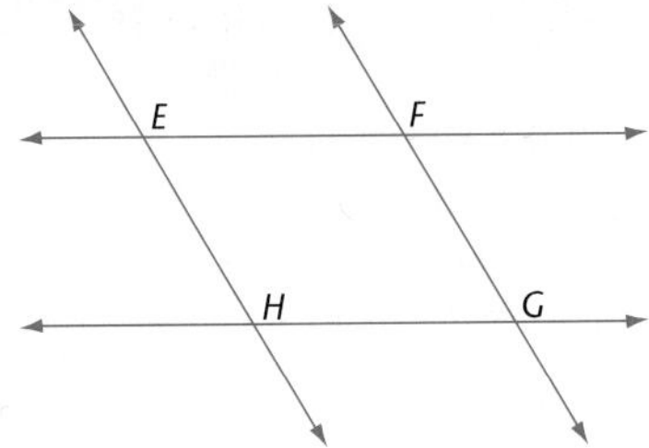

21. Given rectangle $ABCD$ with diagonals $\overline{AC}$ and $\overline{BD}$, prove $\triangle ADC \cong \triangle BCD$. **[Lesson 4.6]**

A B E D C

22. Given rectangle $ABCD$ with diagonals $\overline{AC}$ and $\overline{BD}$ intersecting at E, prove $\triangle AED \cong \triangle BEC$. **[Lesson 4.6]**

23. Given quadrilateral $DART$ with $\overline{DA} \cong \overline{TR}$ and $\overline{DT} \cong \overline{AR}$, prove $DART$ is a parallelogram. **[Lesson 4.6]**

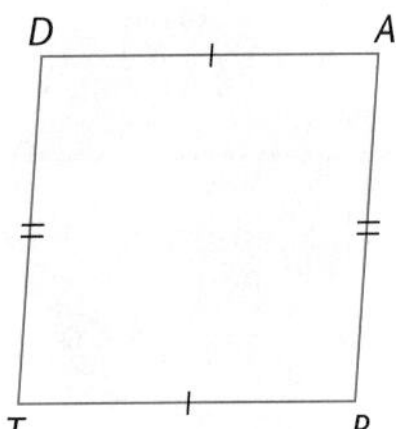

Look Beyond

24. Suppose a segment has endpoints with coordinates (4, 0, 0) and (4, 7, 0). What figure is formed by rotating the segment about the x-axis? Sketch the figure. A circular region with center (4, 0, 0) and radius 7.

25. What figure is formed by rotating the segment from Exercise 24 about the z-axis? Sketch the figure. A cylinder of radius 4 and height 7 centered on the z-axis.

alternative ASSESSMENT

Portfolio Assessment

Give students extra practice in drawing lines and planes in space by having them summarize this lesson through drawings. Have students label the drawings carefully and add notes where necessary.

The answers to Exercises 17, 18, and 20–23 can be found in Additional Answers beginning on page 727.

Look Beyond

In Exercises 24 and 25, students are asked to describe the figures formed by rotating segments in space. Suggest that they use physical models to help find the answers. You can extend the exercises by describing the figures formed by rotating the segment in other ways.

Technology Master

NAME ______ CLASS ______ DATE ______

Technology
6.5 Intersections of Lines and Planes in Space

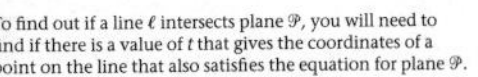

From Lesson 6.5, you know that a plane $\mathcal{P}$ is characterized by an equation of the form $Ax + By + Cz = D$ and that a line ℓ is characterized by three equations that give the x-, y-, and z-coordinates of the points on it: $x = mt + n$, $y = pt + q$, and $z = rt + s$.

To find out if a line ℓ intersects plane $\mathcal{P}$, you will need to find if there is a value of t that gives the coordinates of a point on the line that also satisfies the equation for plane $\mathcal{P}$.

How can you determine whether a given line intersects a given plane? To answer the question, complete the exercises.

1. Show that if $Ax + By + Cz = D$ and $x = mt + n$, $y = pt + q$, and $z = rt + s$, then $Amt + An + Bpt + Bq + Crt + Cs = D$. Use your result to derive the equation shown for t.

$$t = \frac{D - (An + Bq + Cs)}{Am + Bp + Cr}$$

Use a spreadsheet for the exercises.

2. Create a spreadsheet in which cells A1, B1, C1, and D1 contain the headings A, B, C, and D and cells A2, B2, C2, and D2 contain the values of A, B, C, and D. In similar fashion, use rows 3 and 4 for the headings and values of m, n, p, q, r, and s. In cell A5, enter the formula for t that you derived in Exercise 1.

3. Use the spreadsheet to see if the plane $3x + 5y + 6z = 12$ and the line $x = 2t - 3$, $y = t + 1$, $z = 2t + 3$ intersect.

4. Modify the spreadsheet from Exercise 2 to find the coordinates of the point where the line and plane in Exercise 3 intersect.

5. What are the possibilities for plane $\mathcal{P}$ and line ℓ if the entry in cell A5 is not defined?

6. Explain how to use a spreadsheet to distinguish the various possibilities you identified in Exercise 5.

HRW Geometry Technology 111

PREPARE

Objectives

- Identify and define the elementary concepts of perspective drawing.
- Apply perspective drawing concepts by creating perspective drawings.

- Practice Master 6.6
- Enrichment Master 6.6
- Technology Master 6.6
- Lesson Activity Master 6.6
- Quiz 6.6
- Spanish Resources 6.6

Assessing Prior Knowledge

1. Draw a rectangular prism. **[Answers will vary.]**
2. Name some ways to make a two-dimensional drawing look three-dimensional. **[Sample: Drawing parallelograms like the one below instead of rectangles for sides.]**

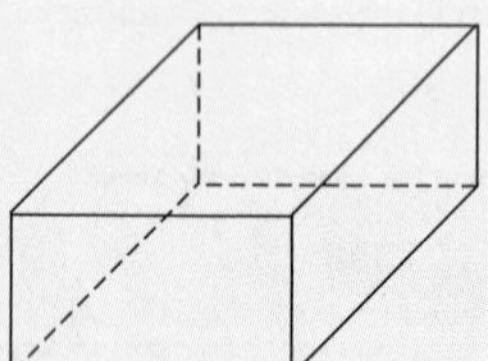

TEACH

Students will learn the basics of drawing in perspective. The theorems studied in this lesson will help them create illusions that are necessary for drawing in perspective and will reinforce ideas about parallelism and three-dimensional space.

LESSON 6.6 Perspective Drawing

European Renaissance artists of the fourteenth through sixteenth centuries rediscovered, from classical Greek and Roman art, how to create the illusion of depth in drawings and paintings. Compare the earlier work above with the Renaissance painting below it.

why *Objects that are far away appear smaller than they would if they were close to you. The study of perspective in drawing will show you the rules for making things (and parts of things) appear to be in the proper relationship to each other.*

Perspective Drawings: Windows to Reality

Modern perspective drawing methods, discovered by the Italian architect Brunelleschi (1377–1446), are based on the idea that a picture is like a window.

The artist creating the picture, or the person looking at the finished product, is thought of as looking through the picture to the reality it portrays. (The word perspective comes from the Latin words meaning "looking through.")

When someone looks at an object, there is a line of sight from every point

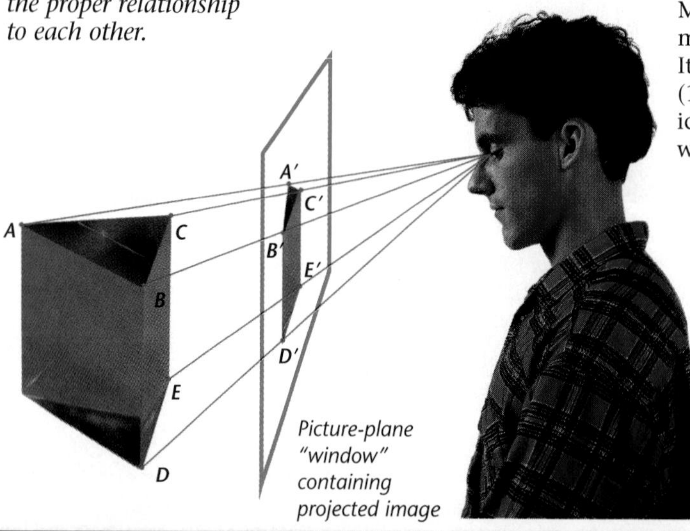

Picture-plane "window" containing projected image

Using Visual Models

To help demonstrate how a person looks at an object, use a flashlight to project the shadow of an object onto the wall. Draw imaginary lines from the center of the flashlight through the vertices of the object to the corresponding vertices of the shadow. Vary the position of the flashlight to show different shadows of the object.

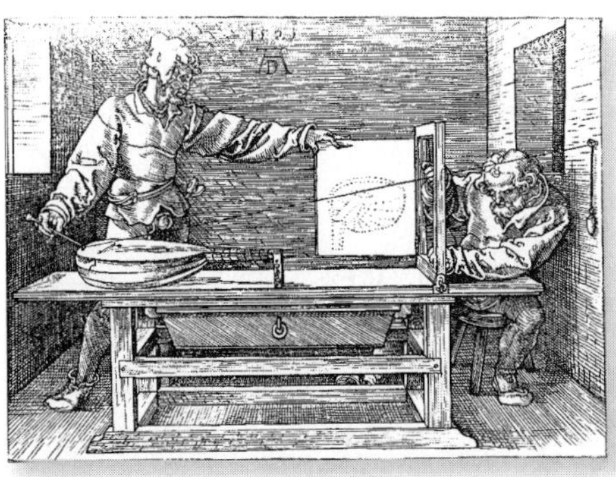

Albrecht Dürer

on the object to the eye. Imagine a plane, such as an empty canvas, intersecting the lines of sight. The points of intersection on the plane make up the image of the object, and the image is said to be **projected** onto the plane.

Albrecht Dürer (1471–1528), who visited Italy to learn the techniques of perspective drawing, produced a number of works that showed artists employing these techniques.

Parallel Lines and Vanishing Points

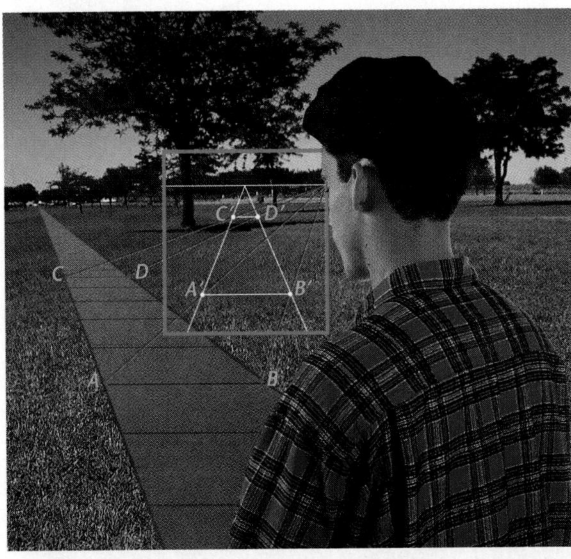

Segments $\overline{AB}$ and $\overline{CD}$ on the sidewalk are actually the same length. But when they are projected onto the student's picture plane, the image of one segment is longer than the image of the other.

Have you ever noticed how the rails of a railroad track or the sides of a highway seem to meet as they recede into the distance? The point at which parallel lines seem to meet, often on the horizon, is known in perspective drawing as a **vanishing point**.

The idea of a picture as a window provides a way to understand why parallel lines seem to meet as they vanish into the distance.

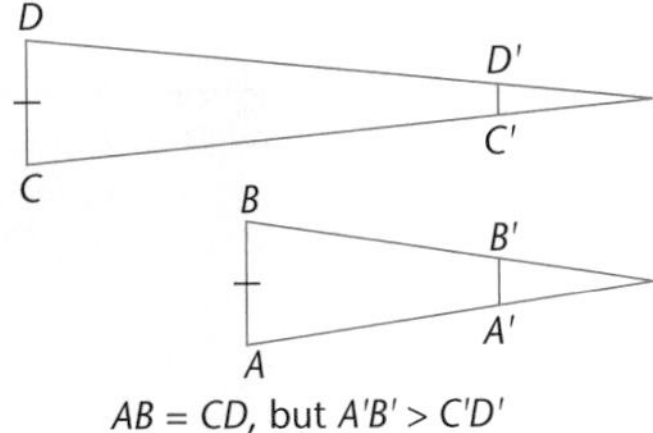

$AB = CD$, but $A'B' > C'D'$

Perspective Drawing Theorems

As Renaissance artists studied vanishing points in nature and in their perspective drawings, they developed a number of rules and procedures for making realistic drawings without using grids.

Cooperative Learning

Drawing figures in perspective is an ideal activity for small group work. Students can form a circle around an object and try to draw it using the perspective drawing theorems in this lesson. Each of the student drawings should show a different perspective of the object, especially from different distances.

Art Artists have used perspective in their work for hundreds of years. Have students bring in books showing the works of artists they like, and have them discuss how perspective is used by the artist.

The following principles, which are stated as theorems, are basic to an understanding of how perspective drawings are made. They can be informally demonstrated using sketches and your earlier geometry theorems.

THEOREM: SETS OF PARALLEL LINES

In a perspective drawing, all lines that are parallel to each other, but not to the picture plane, will seem to meet at the same point. **6.6.1**

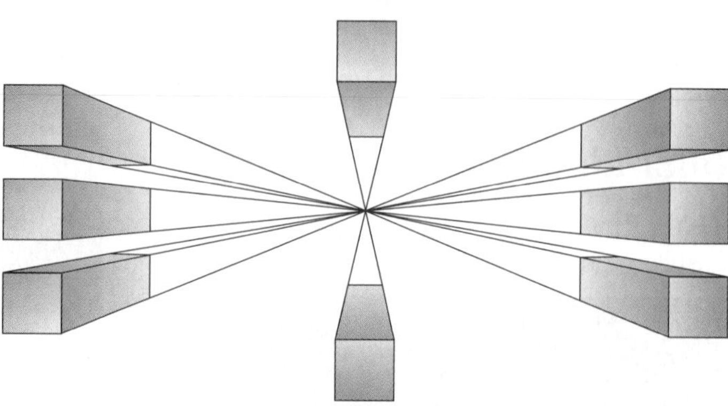

CRITICAL *Thinking* 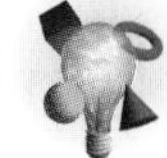

Do you think the point at which a set of parallel lines seem to meet would have to be somewhere in the drawing? Explain, using illustrations.

In the following theorem, the "ground" is assumed to be flat.

THEOREM: LINES PARALLEL TO THE GROUND

In a perspective drawing, a line on the plane of the ground will meet the horizon of the drawing if it is not parallel to the picture plane. Any line parallel to this line will meet at the same point on the horizon. **6.6.2**

CRITICAL *Thinking* 

Parallel lines that are parallel to the picture plane of a perspective drawing are usually represented with no vanishing point. In most cases this procedure causes no problems. Can you think of situations in which it would result in unrealistic drawings?

ENRICHMENT Computers make perspective drawings by labeling an object with three-dimensional coordinates that they then convert to two dimensions. Graphing two-dimensional coordinates in place of their three-dimensional counterparts is called a *projection.* Have students draw and label a rectangular prism in three dimensions. Then have them project its vertices onto the *xy*-plane and label them with prime notation. Redraw the new figure, the projection image.

INCLUSION strategies

English Language Development Students with limited English skills may have trouble understanding the theorems in this lesson. Demonstrating the theorems by using many illustrations may help them understand the concepts.

TEACHING *tip*

It may be very helpful to photocopy some illustrations and then project them onto an overhead screen. Use a ruler to show the lines of perspective.

CRITICAL *Thinking*

No. Depending on the position from which the artist was looking when making the drawing, the vanishing point might be outside the area framed by the drawing.

TEACHING *tip*

Review Lesson 6.1 by comparing how to draw a cube in perspective with how to draw a cube on isometric paper.

CRITICAL *Thinking*

In a proportionally tall (or wide) drawing in which an object extends very far vertically or horizontally, the method would not be realistic. The vertical edges of the side of a tall building, viewed from the ground, would in reality seem to converge toward a vanishing point.

Usually, such objects are drawn on normally proportioned surfaces with the picture plane tilted to represent the viewer looking up (or to the side).

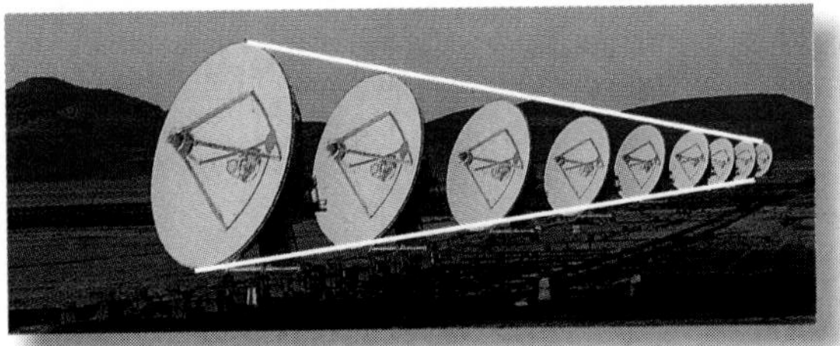

In perspective drawings, the principles of vanishing points apply even when there are no parallel lines actually shown in the drawings. In the row of telescope dishes, for example, there are imaginary lines through points at the tops and at the bottoms of the reflectors. These lines meet at the horizon. Explain why.

Ongoing Assessment

They meet at the horizon because the imaginary line containing the base of the radar dishes is on the ground.

Much of the subject matter in early perspective drawings was architectural. Buildings and houses are ideally suited to the development of the theories of perspective drawing because they usually contain many lines that are parallel to each other and to the ground. Eventually, perspective drawing began to influence architecture, as the Church of San Lorenzo in Florence, Italy, illustrates.

In his design for this church, Brunelleschi used the principles of perspective drawing. The vanishing point of the structural elements appears to be on the altar.

Critical Thinking

The two polar bears in the picture are actually the same size—measure them! And yet one appears much larger than the other. Explain why.

Critical Thinking

Because objects that are far away appear smaller than they would if they were close, our minds compensate for the distance and we think of and perceive the objects as larger. The bear in the photo appears to be far away, so we think of it as bigger even though it is the same size.

Reteaching the lesson

Using Cognitive Strategies You may want students to restate the theorems in the lesson while referring to some of their examples of drawings done in perspective. Have them identify the vanishing point of a drawing and draw the imaginary parallel lines that make up the drawing.

Assess

Selected Answers

Odd-numbered Exercises 17–25

Assignment Guide

Core 1–15, 19–25

Core Plus 1–27

Technology

For Exercise 14, students can use a drawing program to create the appropriate images. Tell them to locate the vanishing point first and then draw in guidelines to create the name. The guidelines can be removed later.

Error Analysis

For Exercise 15, you may want students to start with a less ambitious drawing and then work their way up to drawing the city.

Practice Master

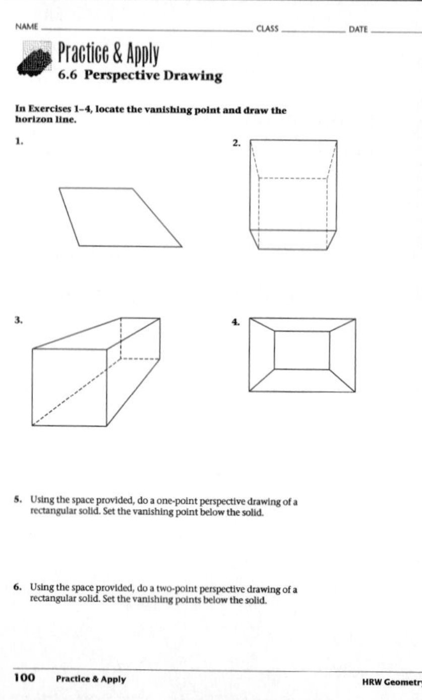

NAME ______ CLASS ______ DATE ______

Practice & Apply

6.6 Perspective Drawing

In Exercises 1–4, locate the vanishing point and draw the horizon line.

1.

2.

3.

4.

5. Using the space provided, do a one-point perspective drawing of a rectangular solid. Set the vanishing point below the solid.

6. Using the space provided, do a two-point perspective drawing of a rectangular solid. Set the vanishing points below the solid.

100 Practice & Apply HRW Geometry

Exercises & Problems

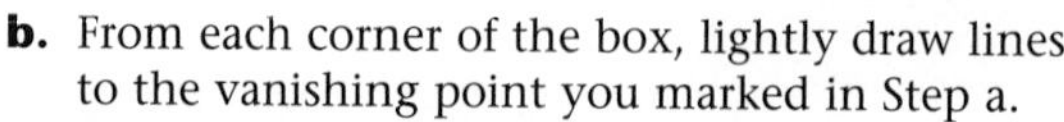

Communicate

Complete the phrase in Exercises 1–3.

1. In a perspective drawing, all lines that are parallel to each other but not parallel to the picture plane . . .
2. In a perspective drawing, all lines parallel to the ground but not parallel to the picture plane . . .
3. In a perspective drawing, lines not parallel to the ground or the picture plane . . .
4. Architects and interior designers sometimes use perspective techniques to create illusions of depth or height. Explain how these illusions work.
5. In perspective drawing, explain what is meant by "vanishing point."

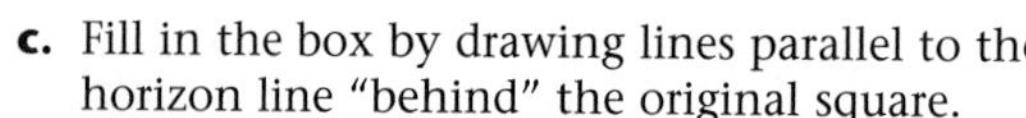

Practice & Apply

The exercises below take you through steps to produce various types of perspective drawing.

6. Follow the steps to produce a one-point perspective drawing of a cube. Check student drawings.
   a. Draw a square and a line above the square as the horizon line. Mark a vanishing point on the horizon line centered above the box.
   b. From each corner of the box, lightly draw lines to the vanishing point you marked in Step a.
   c. Fill in the box by drawing lines parallel to the horizon line "behind" the original square.
   d. You may erase the perspective lines, if you wish, to complete your drawing.

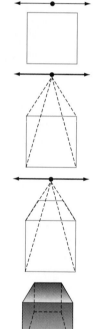

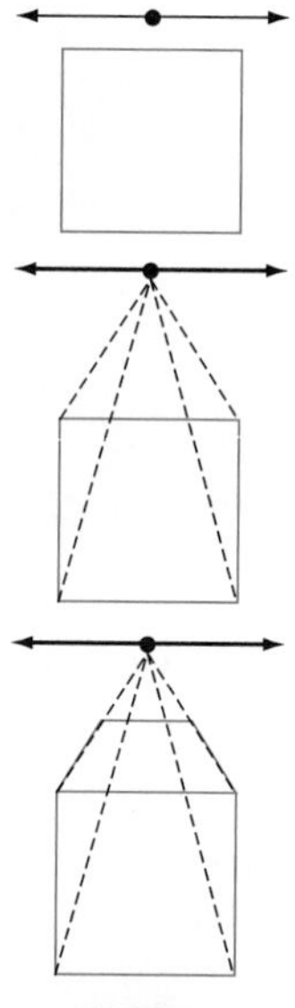

7. Follow the steps from Exercise 6, but draw the cube as if it were viewed from below. Set the horizon line and vanishing point below the square. Check student drawings.

8. Follow the steps from Exercise 6, but set the vanishing point to the right or left instead of the center. Check student drawings.

For Exercises 9 and 10, trace the figure on your own paper.

9. Locate the vanishing point.

10. Draw the horizon line.

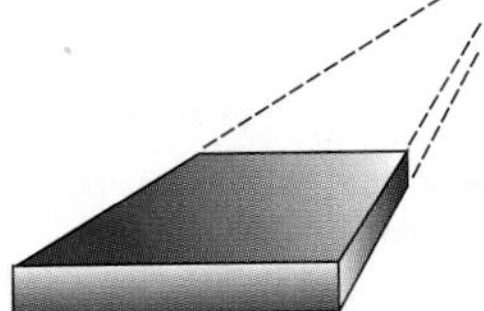

11. So far you have learned about one-point perspective. Perspective drawing can also be done with two-point perspective. Follow the steps below to complete a two-point perspective drawing. Check student drawings.

  a. Draw a vertical line. This is the front edge of your box. Draw a horizon line and two vanishing points on the line to the left and right of the box.

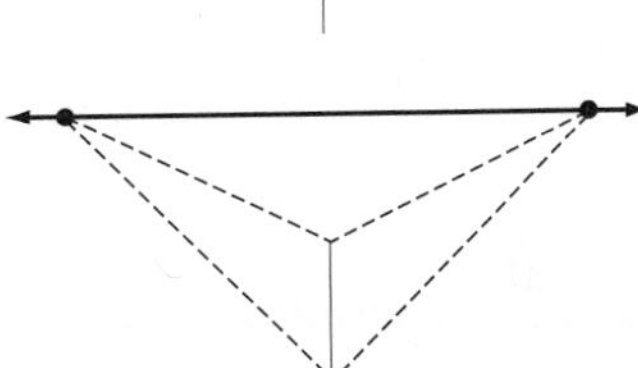

  b. Lightly draw lines back to the vanishing points from the vertical line as shown.

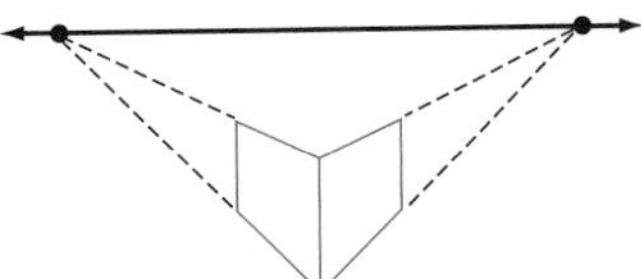

  c. Draw vertical segments to complete the sides of the box.

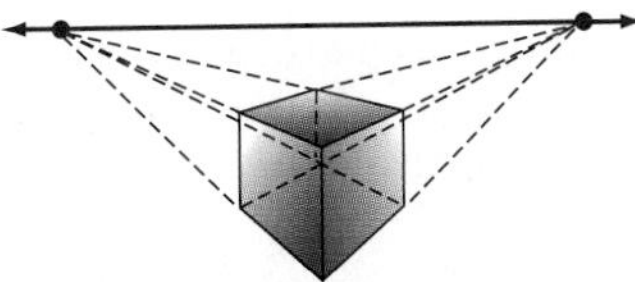

  d. Complete the other edges of the box as shown.

  e. You may erase the perspective lines to complete the drawing.

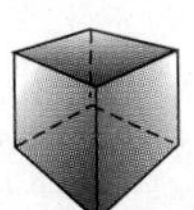

9–10.

Horizon line — Vanishing point

alternative ASSESSMENT

Authentic Assessment

Encourage students to experiment with the techniques presented in this lesson by making illustrations of their favorite things.

Technology Master

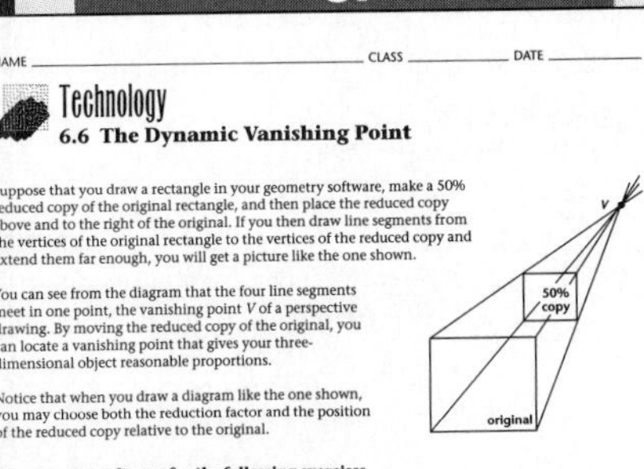
NAME ______ CLASS ______ DATE ______

Technology

6.6 The Dynamic Vanishing Point

Suppose that you draw a rectangle in your geometry software, make a 50% reduced copy of the original rectangle, and then place the reduced copy above and to the right of the original. If you then draw line segments from the vertices of the original rectangle to the vertices of the reduced copy and extend them far enough, you will get a picture like the one shown.

You can see from the diagram that the four line segments meet in one point, the vanishing point V of a perspective drawing. By moving the reduced copy of the original, you can locate a vanishing point that gives your three-dimensional object reasonable proportions.

Notice that when you draw a diagram like the one shown, you may choose both the reduction factor and the position of the reduced copy relative to the original.

Use geometry software for the following exercises.

1. Sketch a diagram like the one shown. (In some software programs, you make a "scaled" copy by using DILATE.)

2. What happens to the vanishing point if you move the reduced copy farther up and farther to the right? Farther up and farther to the left?

3. Describe the relationship between you, the original figure, and V if V is on the screen but to the left and below the original figure.

4. What happens to V when you make an enlargement of the original rectangle and then place it above and to the right of the original rectangle?

5. Repeat Exercises 1–4 using an equilateral triangle as the original.

6. Sketch rectangle $ABCD$. Make line segments from A, B, C, and D to a point V in the approximate center of the rectangle. Make a 50% copy of the rectangle centered about V.

7. Using the results of Exercise 6, drag V to different locations in and around the original rectangle. Use the MEASURE commands to compare the length of V to A with the length of V and the corresponding corner of the 50% copy rectangle.

112 Technology — HRW Geometry

For Exercises 12 and 13, trace the figure on your own paper.

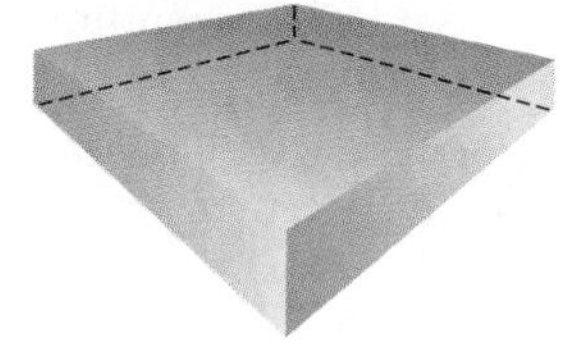

12. Locate the vanishing points.

13. Draw the horizon line.

14. **Portfolio Activity** You can use perspective techniques to create a block-letter drawing of your name. Follow the steps below. Check student drawings.

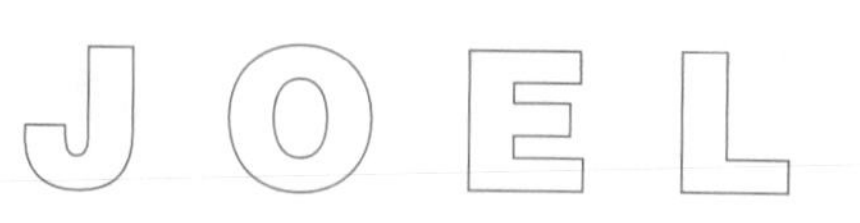

a. Draw "flat" block letters and a horizon line with a vanishing point.

b. Draw lines from all corners of the letters and appropriate curved edges as shown.

c. Fill in the edges.

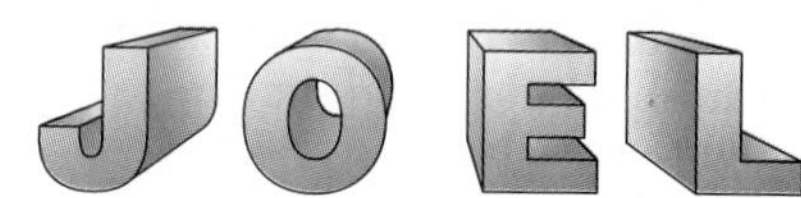

d. You may erase the perspective lines, if you wish, to complete your drawing.

15. **Portfolio Activity** Use a one-point or two-point perspective to draw a city. Start with boxes for buildings and add details. Check student drawings.

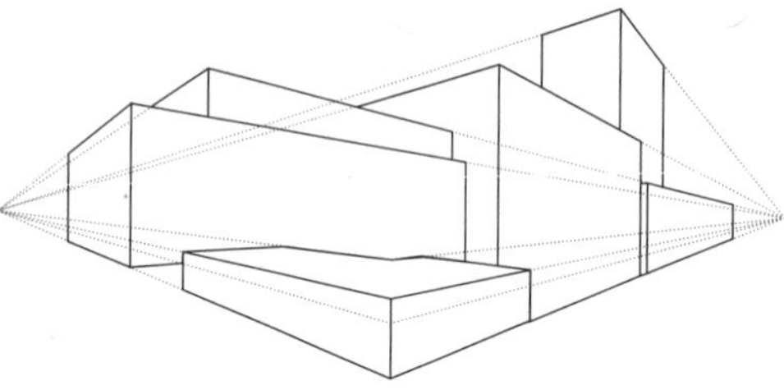

Fine Art Tiling was particularly intriguing to the artists who first completed perspective studies. The pictures (a–d) suggest a technique for creating a proper tile pattern. Study the pictures to answer Exercises 16–18.

a.

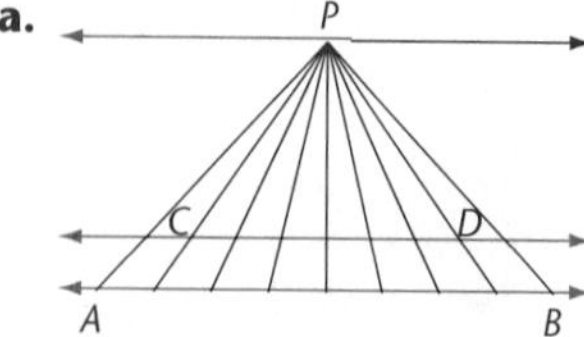

b.

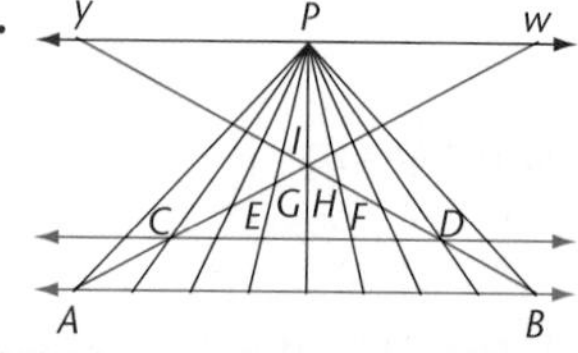

12-13.

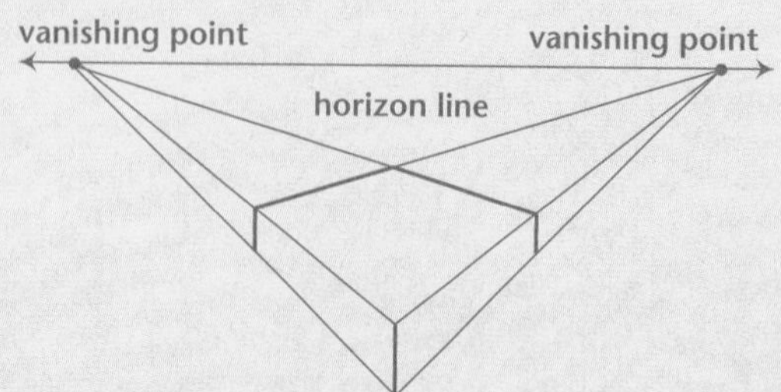

c.

d.

16. A method for finding the lines parallel to $\overleftrightarrow{AB}$ is suggested by the diagonals. Explain how the diagonals can be used to determine the parallel lines.

17. Explain how the tile pattern could be viewed from a corner using two-point perspective. How would the intersecting lines be determined?

18. Create your own one- or two-point perspective drawing of a square tile pattern.

Look Back

Locate and sketch each point on a three-dimensional coordinate system. [Lesson 6.4]

19. (5, −1, −2) **20.** (13, 0, 0) **21.** (−2, 0, 5)

Find the distance between the following points in a three-dimensional coordinate system. [Lesson 6.4]

22. (4, 3, 2), (−5, 2, −1) 9.54 **23.** (−1, 0, 1), (15, 6, −2) 17.35

Find the midpoint of the segments joining the following points in a three-dimensional coordinate system. [Lesson 6.4]

24. (5, 5, 5), (−3, −3, −3) (1, 1, 1) **25.** (0, 0, 0), (−1, 10, 9) $\left(\frac{-1}{2}, 5, \frac{9}{2}\right)$

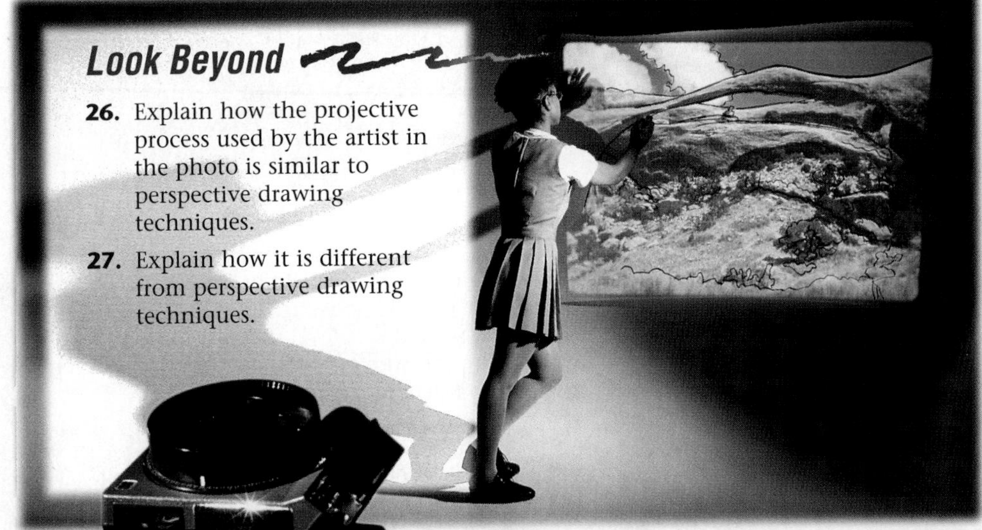

Look Beyond

26. Explain how the projective process used by the artist in the photo is similar to perspective drawing techniques.

27. Explain how it is different from perspective drawing techniques.

The answers to Exercises 19–21 can be found in Additional Answers beginning on page 727.

26. Both use "lines of sight." In perspective drawing, the line of sight from the eye to points on the object is used to project the object onto the plane of the picture. When using a projector, the light from the projector bulb is used to project the picture onto the plane of the wall or screen. The bulb could be thought of as the vanishing point or the eye.

27. In perspective drawing, a three-dimensional object in space is projected onto a two-dimensional plane. Objects are projected onto the plane of the picture at the intersection of the line of sight. When a projector is used, a two-dimensional picture is projected onto another two-dimensional picture. In perspective drawing, the picture is usually smaller than the object being drawn. The image is not the same shape but appears to have the same shape. When a projector is used, the image is the same shape but larger.

Look Beyond

Exercises 26 and 27 look beyond to projective geometry. Have students discuss their answers in small groups.

16. Segments radiating from point P divide $\overleftrightarrow{AB}$ into a number of congruent segments. $\overleftrightarrow{CD}$ is drawn parallel to $\overleftrightarrow{AB}$. Diagonals $\overline{AW}$ and $\overline{BY}$ are drawn through points C and D, respectively. The intersections of the diagonals with the segments from point P determine the vertical placement of the parallel lines.

17. Simplest method: Make a perspective drawing of a square with horizontal and vertical diagonals. Divide the horizontal diagonal equally. Pass lines from the vanishing points through the division points.

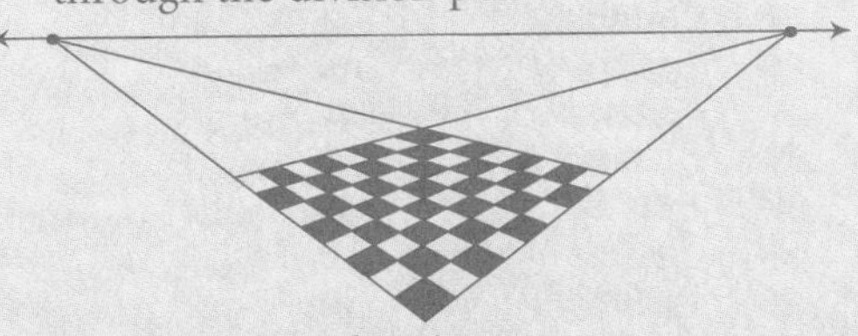

18. Check student drawings.

Focus

Students build and test an A-shaped level with properties of geometry.

Motivate

In this project, students will review the idea of a line being perpendicular to a plane by building and testing the properties of an A-shaped level. The level can then be used in the classroom to determine which desks are level.

Building and Testing an A-shaped Level

Many carpentry tools use basic properties of geometry. One example is the early Egyptian A-shaped level which has been used for thousands of years for construction and surveying. In this project you will build and test the various properties of an A-shaped level.

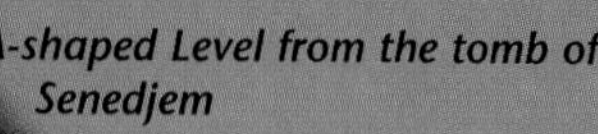

A-shaped Level from the tomb of Senedjem

Activity 1

Part of the A-shaped level is made with a string and a weight forming what is called a plumb line. Test the behavior of a plumb line by attaching a washer or similar weight to a string as shown. Suspend your plumb line above a surface that you believe is level, such as the floor of your classroom. Gradually lower the plumb line until the weight makes contact with the floor. Imagine that the weight is touching the floor at point F, the intersection of two segments as shown below. Then answer the questions that follow.

1. $m\angle AFP = ?$
2. $m\angle AFQ = ?$
3. Hold a pencil at arm's length and suspend the plumb line directly above the pencil. If you hold the pencil so that it is level (parallel to the floor), what is the measure of the angles formed between the plumb line and the pencil?
4. Formulate a conjecture:

   A segment is level if and only if it is __?__ to a plumb line.

1. 90°
2. 90°
3. 90°
4. perpendicular

Activity 2

An A-shaped level is easy to make from three popsicle sticks, glue, a thumbtack, string, and a small weight. Build and test an A-shaped level by following the steps below.

1. Glue two popsicle sticks together to form the legs of the level.
2. Glue the crossbar to the legs forming an isosceles triangle above the crossbar and two congruent segments below it.
3. Carefully mark the midpoint M on the crossbar.
4. Put a thumbtack at the vertex of your level and tie your plumb line to the thumbtack.
5. Place your A-shaped level along several different segments that you believe are level. Observe and record the behavior of your plumb line $\overline{AW}$. What is the relationship between $\overline{AW}$ and M?
6. Place your A-shaped level along several different segments that you believe are not level. Observe and record the behavior of $\overline{AW}$. What is the relationship between $\overline{AW}$ and M?
7. Use your results and the picture above formulate a conjecture:

   $\overline{DE}$ is level if and only if __?__ (a statement about $\overline{AW}$ and M).

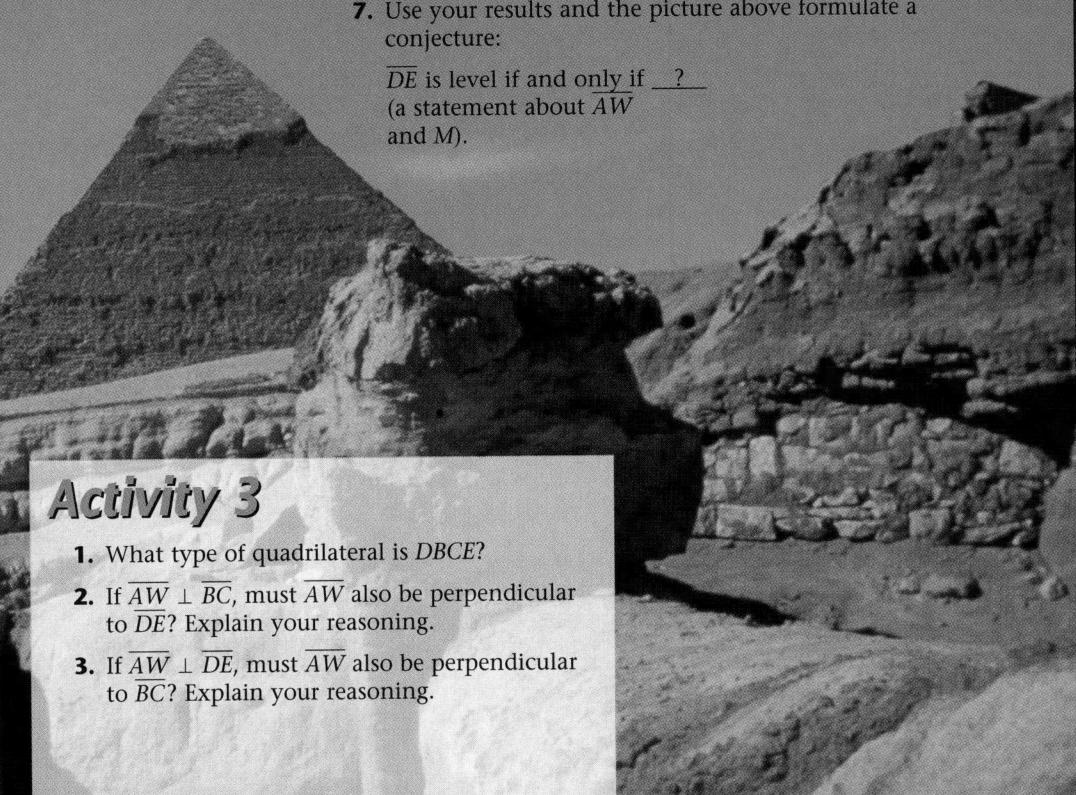

Activity 3

1. What type of quadrilateral is $DBCE$?
2. If $\overline{AW} \perp \overline{BC}$, must $\overline{AW}$ also be perpendicular to $\overline{DE}$? Explain your reasoning.
3. If $\overline{AW} \perp \overline{DE}$, must $\overline{AW}$ also be perpendicular to $\overline{BC}$? Explain your reasoning.

5. M is on $\overline{AW}$.

6. M is not on $\overline{AW}$.

7. M is on $\overline{AW}$.

Extension

1. Trapezoid with $\overline{BC} \parallel \overline{DE}$.
2. Yes. Since $\overline{BC} \parallel \overline{DE}$, if $\overline{AW} \perp \overline{BC}$, then $\overline{AW} \perp \overline{DE}$ also.
3. Yes. Since $\overline{BC} \parallel \overline{DE}$, if $\overline{AW} \perp \overline{DE}$, then $\overline{AW} \perp \overline{BC}$ also.

Cooperative Learning

Cooperative learning groups are very appropriate for this activity because it is "hands-on." In Activity 1, students build a plumb line out of string with a weight attached. Have each student in the group do a different part of the activity, but they should decide on the conjecture as a group. For Activity 2, students can help each other build the A-shaped level and answer the questions. They should formulate a conjecture as a group.

Discuss

What is the advantage of using the A-shaped level compared with other levels (like commercially available ones)?

What are the disadvantages?

How accurate is this level?

How does geometry help the reliability of this particular level?

Chapter 6 Review

Vocabulary

Key Skills and Exercises

Lesson 6.1

➤ Key Skills

Create and interpret isometric drawings.
Assuming that no cubes are hidden in this drawing, what is the surface area of the solid?

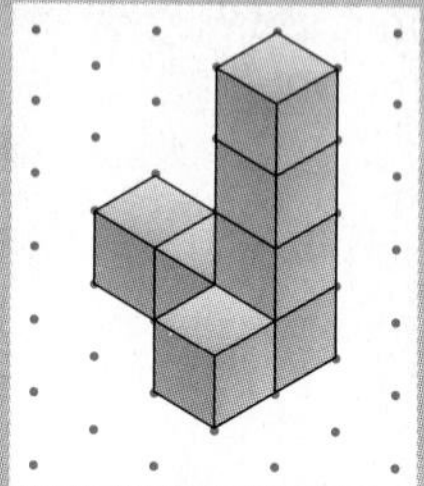

A cube standing alone would have six faces showing, each face with a surface area of 1 cm^2. The cubes in this drawing show different numbers of faces: three cubes show five faces, three cubes show four faces, and one cube shows three faces. Adding up the exposed faces at 1 cm^2 each gives a total of 30 cm^2.

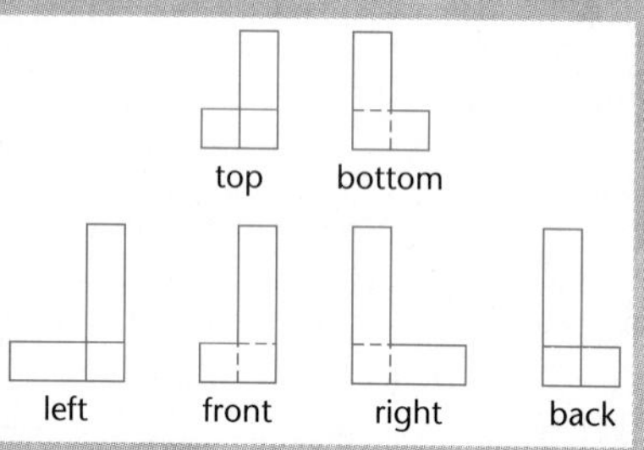

Orthographic projections that show each face of the object exactly once demonstrate that 30 cube faces make up the total surface area.

➤ Exercises

In Exercises 1–2, refer to the isometric drawing. Assume that no cubes are hidden in the drawing.

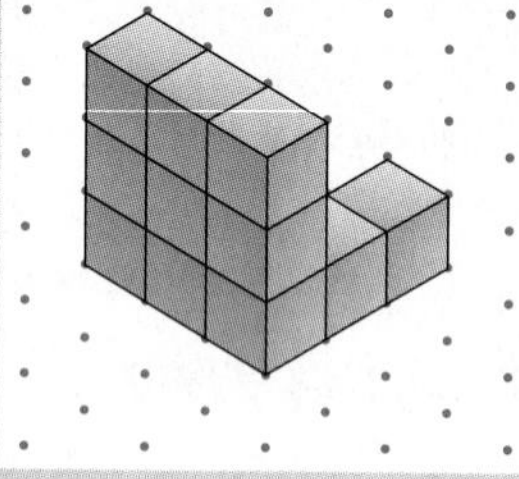

1. What is the volume (in cubes) of the solid?
2. Draw orthographic projections of the faces of the object.

Lesson 6.2

➤ Key Skills

Identify spatial relationships among lines and planes in space.
In the sketch of a cube, name one set of parallel planes, perpendicular planes, skew segments, and segments perpendicular to a plane.

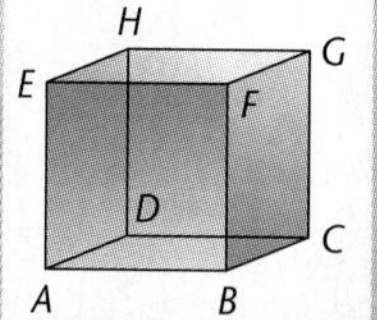

Three points name a plane, so you can arbitrarily choose three points from the four planes available. Planes *ADE* and *BCF* are parallel. Both *ADE* and *BCF* are perpendicular to *ABF*, *DCG*, *EFG*, and *ABC*. $\overline{EH}$ and $\overline{BF}$ are a set of skew segments. Perpendicular to *ADE* and *BCF* are $\overline{AB}$, $\overline{HG}$, $\overline{DC}$, and $\overline{EF}$.

1. 11 cubes
2. 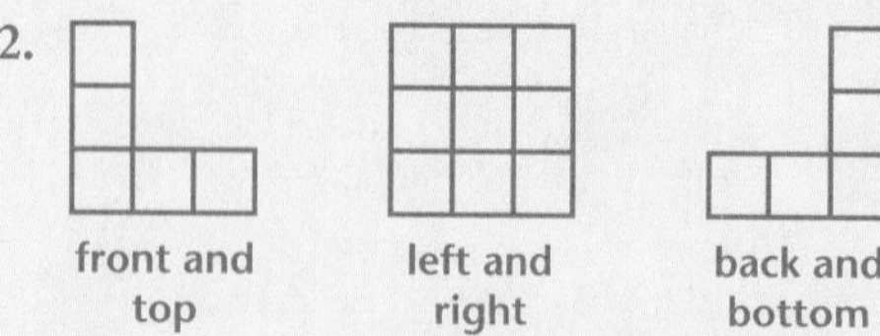

➤ Exercises

In Exercises 3–5, refer to the figure.

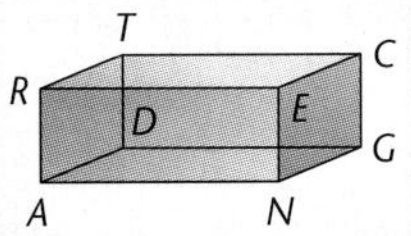

3. Name all the pairs of parallel planes.
4. Name a pair of skew segments.
5. Name all the pairs of perpendicular planes.

Lesson 6.3

➤ Key Skills

Recognize congruences in a prism.

For the oblique prism at the right, name the segments that are congruent to $\overline{AD}$ and the lateral faces that are congruent to each other.

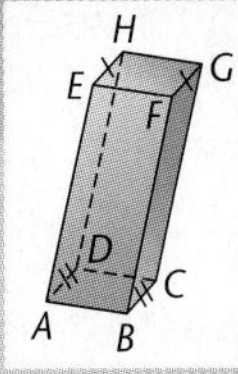

You're given that $\overline{EH} \cong \overline{FG}$ and $\overline{AD} \cong \overline{BC}$. Because the bases of a prism are congruent, $\overline{EH} \cong \overline{AD}$; thus $\overline{EH} \cong \overline{FG} \cong \overline{AD} \cong \overline{BC}$. Lateral faces *ADHE* and *BCGF* are congruent because the base edges, lateral edges, and angles are congruent.

Find the length of a diagonal of a right rectangular prism.

A right rectangular prism has a length of 20, a height of 24, and a width of 10. What is the length of its diagonal?

The length of the diagonal of a right rectangular prism is $\sqrt{l^2 + w^2 + h^2}$. In this case, $d = \sqrt{20^2 + 10^2 + 24^2} = \sqrt{1076} \approx 32.8$.

➤ Exercises

In Exercises 6–7, refer to the figure.

6. Name the segments congruent to $\overline{RC}$.
7. Name the lateral faces that are congruent.
8. A right rectangular prism has a length of 12, a width of 2, and a height of 12. Find the length of the prism's diagonal.

Lesson 6.4

➤ Key Skills

Describe and sketch lines and planes on a three-dimensional coordinate system.

The figure below represents a cube, drawn in a three-dimensional coordinate system. If $\overline{OP}$ connects the midpoints of $\overline{RQ}$ and $\overline{TS}$, find its end points and its length.

The midpoint formula for a segment is

$$\left(\frac{x_1 + x_2}{2}, \frac{y_1 + y_2}{2}, \frac{z_1 + z_2}{2}\right).$$

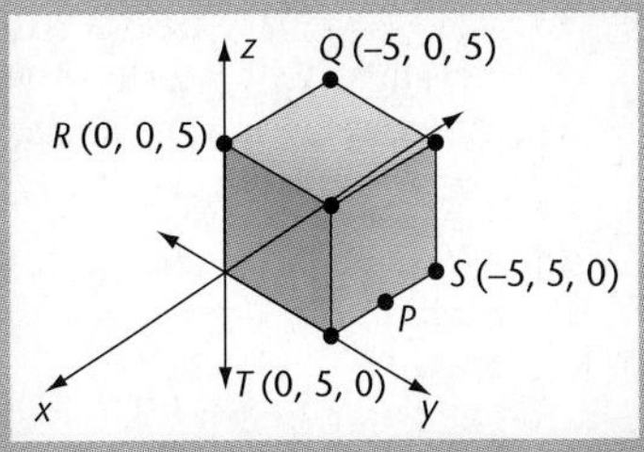

Using this formula for $\overline{RQ}$ and $\overline{TS}$, we get (−2.5, 0, 5) and (−2.5, 5, 0) as the endpoints of segment $\overline{OP}$. Using the distance formula,

$$d = \sqrt{(x_2 - x_1)^2 + (y_2 - y_1)^2 + (z_2 - z_1)^2},$$

$$OP = \sqrt{(-2.5 - (-2.5))^2 + (0 - 5)^2 + (5 - 0)^2}$$

$$= \sqrt{50} = 5\sqrt{2}$$

6. $\overline{CO}$; $\overline{TK}$; $\overline{TE}$
7. *RCTK* and *OCTE*
8. ≈ 17.09

3. *RTCE* ∥ *ADGN*; *RTDA* ∥ *ECGN*; *RENA* ∥ *TCGD*
4. Answers will vary. Sample answers: $\overline{RA}$ and $\overline{NG}$; $\overline{RA}$ and $\overline{TC}$; $\overline{RA}$ and $\overline{DG}$; $\overline{RA}$ and $\overline{CE}$
5. *ANGD* ⊥ *RANE*; *ANGD* ⊥ *ENGC*; *ANGD* ⊥ *TCGD*; *ANGD* ⊥ *TDAR*; *RECT* ⊥ *RANE*; *RECT* ⊥ *ENGC*; *RECT* ⊥ *TCGD*; *RECT* ⊥ *TDAR*; *ENGC* ⊥ *RENA*; *ENGC* ⊥ *TCGD*; *TDAR* ⊥ *RENA*; *TDAR* ⊥ *TCGD*

Chapter Review

9.

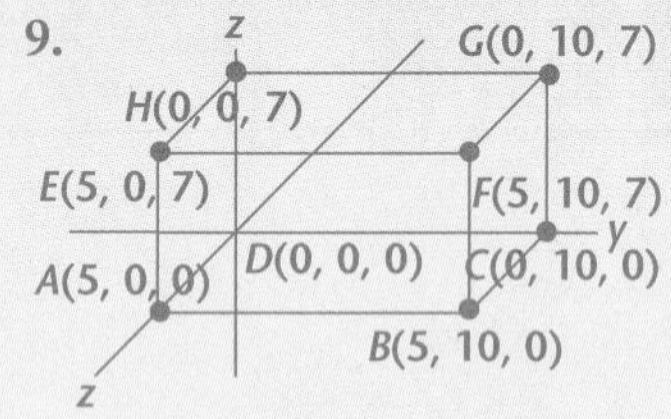

10. ≈ 13.19

11.

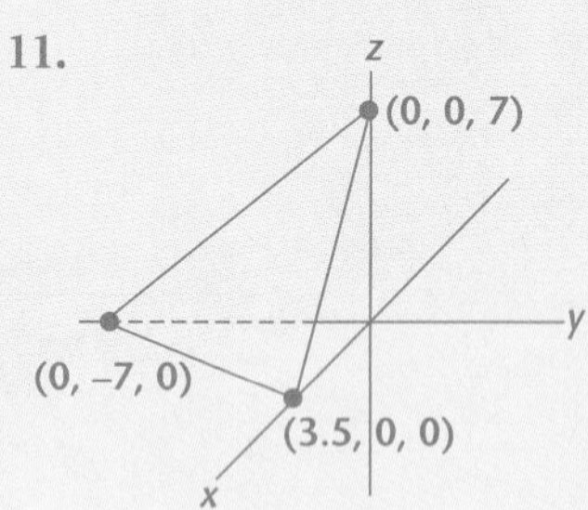

➤ Exercises

In Exercises 9–10, refer to the figure.

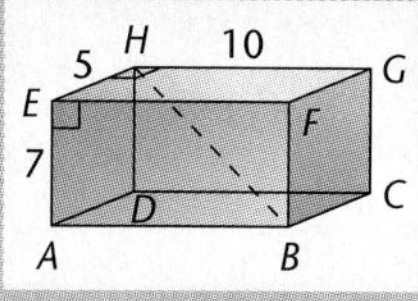

9. Sketch the box on a coordinate system. Give the coordinates for each vertex.

10. Find the length of $\overline{BH}$.

Lesson 6.5

➤ Key Skills

Plot planes on a three-dimensional coordinate system.

Sketch the graph of the plane with the equation $x + 2y - 3z = 18$.

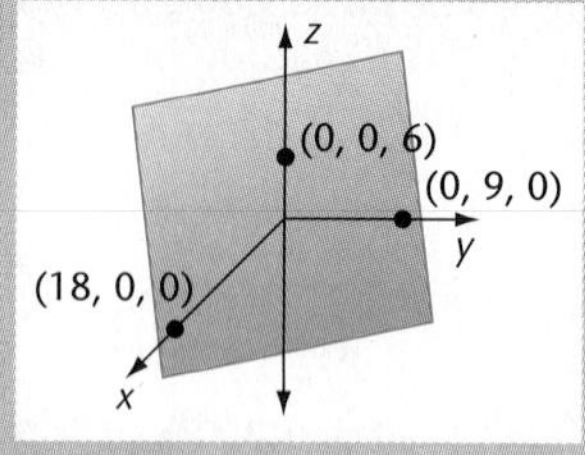

You need three noncollinear points to graph a plane. You can use the three points where the plane intercepts the three axes. When y and z are 0, $x + 2(0) + 3(0) = 18$. Thus $x = 18$. So the coordinates for the x-intercept are (18, 0, 0). By a similar process, you find the y-intercept is (0, 9, 0) and the z-intercept is (0, 0, 6). Now you can plot the points and sketch the graph of the plane.

➤ Exercises

11. Sketch the graph of the plane with the equation $2x - y + z = 7$.

Lesson 6.6

➤ Key Skills

Use vanishing points to make perspective drawings.

The drawings below illustrate two ways to make the letter T look three-dimensional. Each uses only one vanishing point: in the left-hand drawing the point is above and to the right of the T, in the right-hand drawing beside the T to the left.

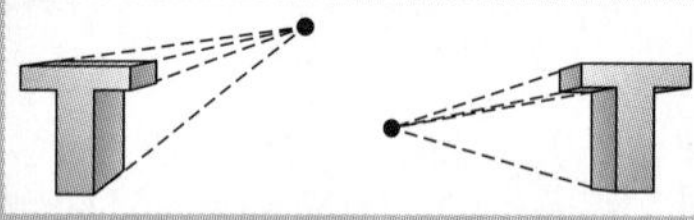

➤ Exercises

12. Make a perspective drawing of a cube. Show the horizon, the vanishing point(s), and the perspective lines you use.

Applications

13. **Geology** Topaz is a mineral that forms orthorhombic crystals. This type of crystal has noncongruent edges that meet at right angles. Use isometric grid paper to sketch an orthorhombic crystal.

14. Here are two views of a table. Make a perspective drawing of it.

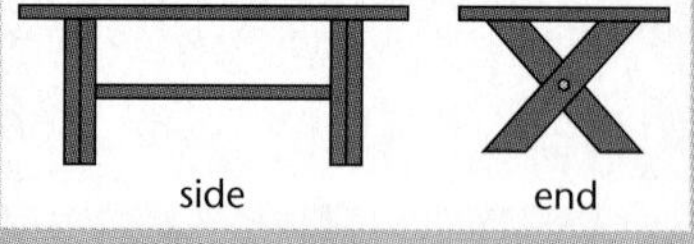

12. Sample answer. (Students could also make a 2-point perspective drawing.)

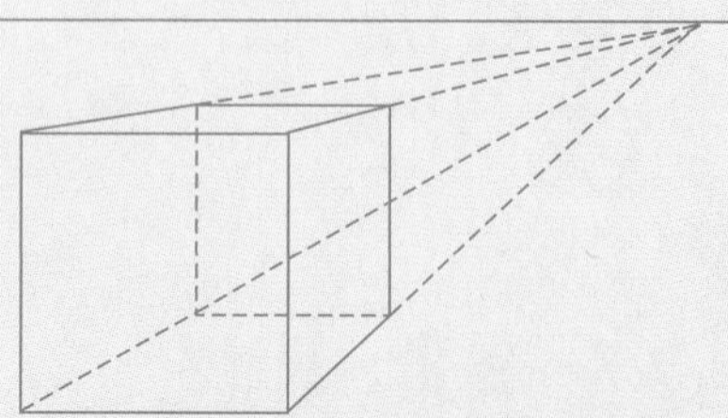

13. Check student drawings.

14. Check student drawings.

Chapter 6 Assessment

In Exercises 1–2, refer to the isometric drawing. Assume that no cubes are hidden in the drawing.

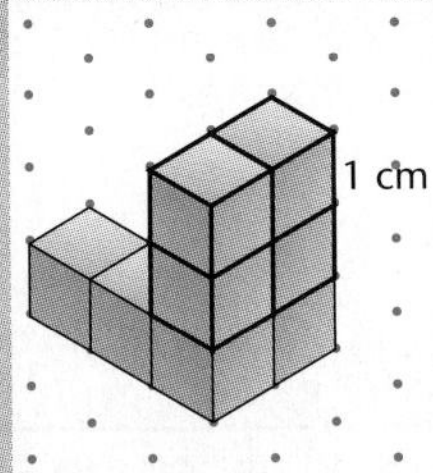

1. What is the surface area of the solid?
2. Draw orthographic projections of the faces of the object.

In Exercises 3–5, refer to the figure below.

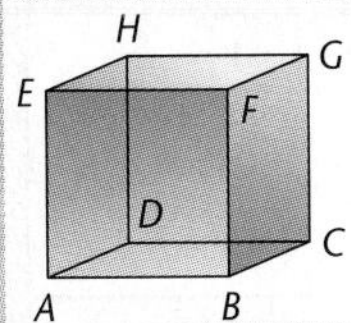

3. Name a pair of parallel planes.
4. Name a pair of skew segments.
5. Name a segment perpendicular to plane *ABC*.

In Exercises 6–8, refer to the right rectangular prism below.

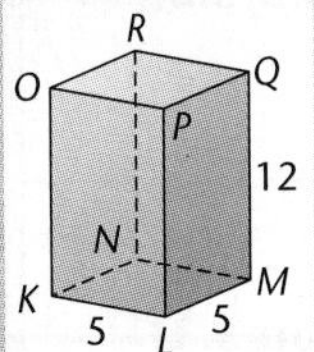

6. Name the segments congruent to $\overline{RQ}$.
7. Name the lateral faces that are congruent.
8. Find the length of the prism's diagonal.
9. Sketch the box below on a coordinate system. Give the coordinates for each vertex.

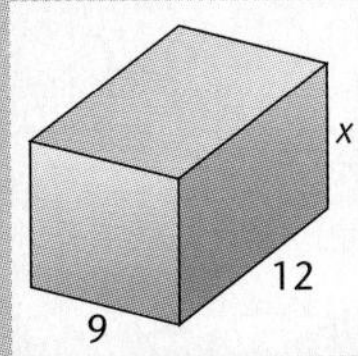

10. Sketch the graph of the plane with the equation $3x - y + 4z = 24$.

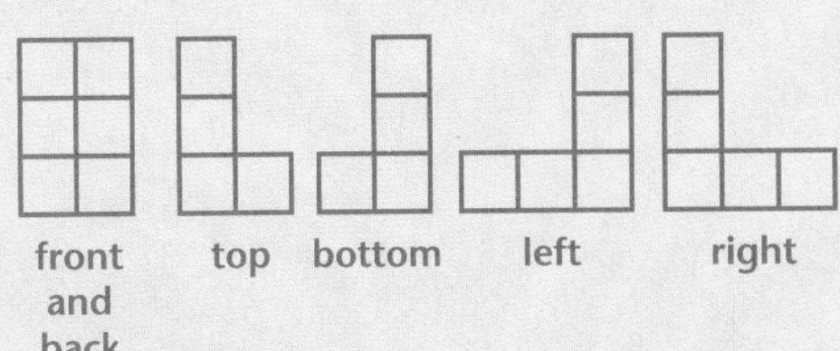

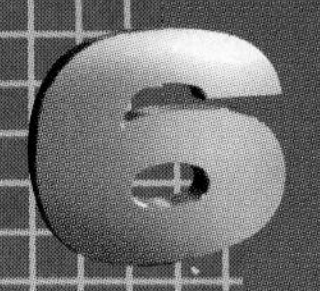

3. *ABFE* ∥ *DCGH*; *ABCD* ∥ *EFGH*; *ADHE* ∥ *BCGF*
4. Answers will vary. Sample answers: $\overline{EA}$ and $\overline{BC}$; $\overline{EA}$ and $\overline{GF}$; $\overline{EA}$ and $\overline{HG}$; $\overline{EA}$ and $\overline{DC}$
5. $\overline{EA}$; $\overline{FB}$; $\overline{GC}$; $\overline{HD}$
6. $\overline{QP}$; $\overline{OP}$; $\overline{RO}$; $\overline{NM}$; $\overline{ML}$; $\overline{LK}$; $\overline{KN}$
7. *OPLK* ≅ *QPLM* ≅ *RQM* ≅ *ORNK*
8. ≈ 13.93
9. 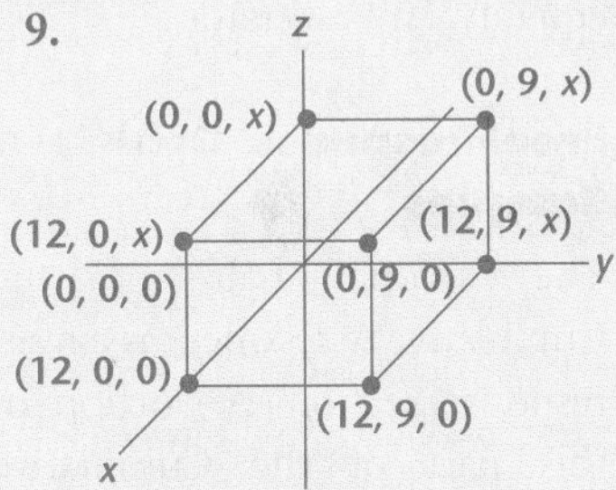

10. 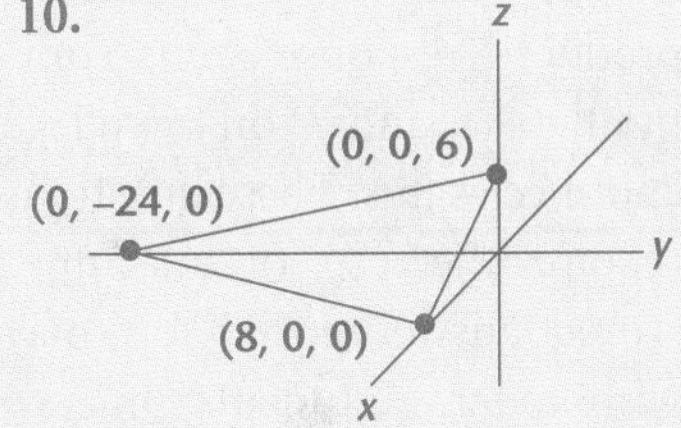

COLLEGE ENTRANCE-EXAM PRACTICE

Multiple-Choice and Quantitative-Comparison Samples

The first half of the Cumulative Assessment contains two types of items found on standardized tests — multiple-choice questions and quantitative-comparison questions. Quantitative-comparison items emphasize the concepts of equality, inequality, and estimation.

Free-Response Grid Samples

The second half of the Cumulative Assessment is a free-response section. A portion of this part of the Cumulative Assessment consists of student-produced response items commonly found on college entrance exams. These questions require the use of machine-scored answer grids. You may wish to have students practice answering these items in preparation for standardized tests.

Sample answer grid masters are available in the *Chapter Teaching Resources Booklets.*

Chapters 1–6 Cumulative Assessment

College Entrance Exam Practice

Quantitative Comparison Questions 1–4 consist of two quantities, one in Column A and one in Column B, which you are to compare as follows:

A. The quantity in Column A is greater.
B. The quantity in Column B is greater.
C. The two quantities are equal.
D. The relationship cannot be determined from the information given.

	Column A	Column B	Answers
1.	Right triangle ABC, right angle at B, $BC = 2$, angle C = 60° length of $\overline{AB}$	Right triangle XYZ, right angle at Y, $YZ = 2$, angle Z = 45° length of $\overline{XZ}$	(A) (B) (C) (D) **[Lesson 5.5]**
2.	distance between (9, 9, 9) and (0, 0, 0)	distance between (15, 0, 19) and (6, 9, 10)	(A) (B) (C) (D) **[Lesson 7.7]**
3.	Trapezoid EFGH, $FG = 14$, height 6 area of *EFGH*	Parallelogram JKLM, $KL = 10$, height 6 area of *JKLM*	(A) (B) (C) (D) **[Lesson 5.2]**
4.	Circle with radius 3 area of the circle	Square with half-diagonal 3 area of the square	(A) (B) (C) (D) **[Lesson 5.3]**

5. What is the measure of an interior angle of a regular nonagon? **[Lesson 3.6]**
a. 40° **b.** 100° **c.** 140° **d.** 160°

6. The area of a circle is 154 square inches. What is its circumference? **[Lesson 5.3]**
a. 22 inches **b.** 44 inches
c. 51 inches **d.** 77 inches

7. What are the coordinates of the midpoint of a segment whose endpoints are (9, 12) and (−12, −9)? **[Lesson 5.7]**
a. (−1.5, 1.5) **b.** (4.5, 6) **c.** (−6, −4.5) **d.** (10.5,

1. A **2.** C **3.** D **4.** A
5. c **6.** b **7.** a

8. What are the coordinates of the midpoint of a segment whose endpoints are (2, 10, 0) and (11, 10, −6)? **[Lesson 6.4]**
   a. (6.5, 0, −3) **b.** (4.5, 10, 3)
   c. (6.5, 10, −3) **d.** (5.5, 5, 3)

9. In the oblique rectangular prism below, what is the measure of $\angle x$? **[Lesson 3.5]**
   a. 60°
   b. 80°
   c. 100°
   d. 120°

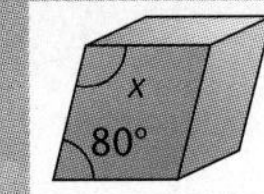

10. What is the surface area of the stack of cubes at the right? Assume that no cubes are hidden. **[Lesson 6.1]**
   a. 38 cm^2 **b.** 45 cm^2
   c. 54 cm^2 **d.** 66 cm^2

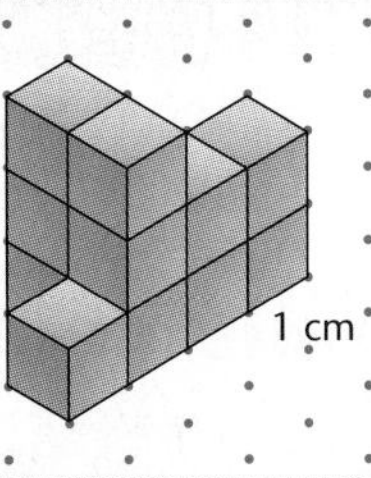

11. "Skew lines are not parallel and never intersect." Is this statement a definition? Explain. **[Lesson 2.3]**

12. A circle and a square both have an area of 441 square inches. What are their perimeter and circumference? **[Lesson 5.3]**

13. Use the slope formula to show that quadrilateral *ABCD* is a rectangle. **[Lesson 3.8]**

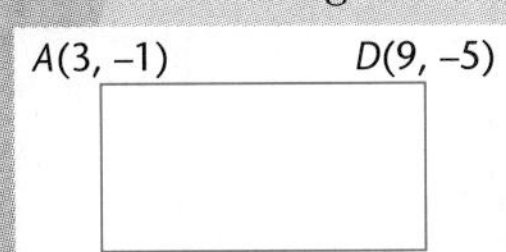

14. Find the area of the regular hexagon. **[Lesson 5.5]**

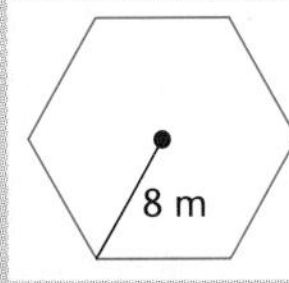

15. Draw a rhombus and construct a reflection of it over a horizontal line. **[Lesson 4.9]**

Free-Response Grid Questions 16–19 may be answered using a free-response grid commonly used by standardized test services.

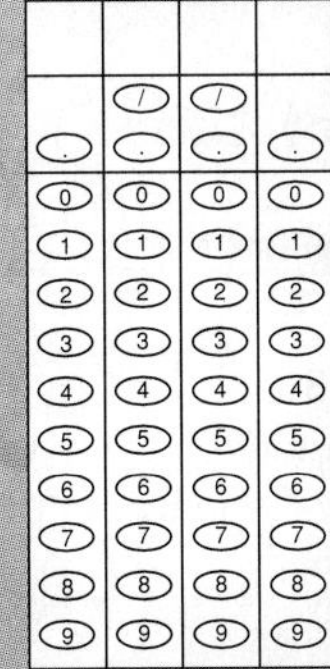

16. Find the area of triangle *PEG*. **[Lesson 5.2]**

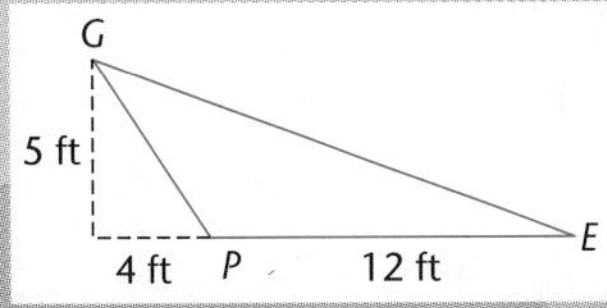

17. Find the total area of the three rectangles. **[Lesson 5.1]**

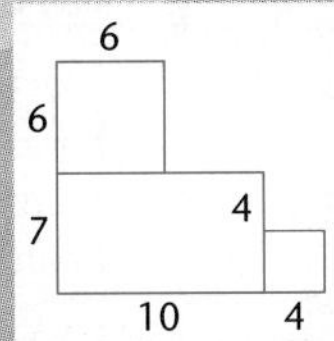

18. Find the missing angle measure. **[Lesson 3.6]**

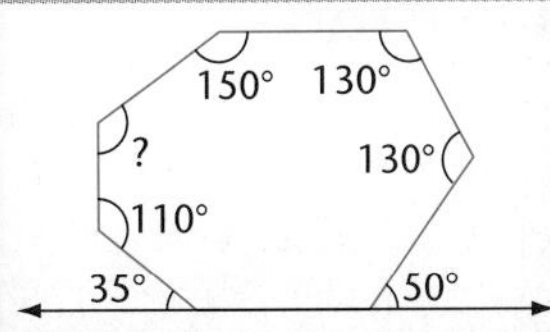

19. A point *P* is selected at random from the segment below. What is the probability that $2 \le P \le 2.5$? **[Lesson 5.7]**

Cumulative Assessment

14. ≈ 166.3 m^2

15.

16. 30 ft^2

17. 118 units^2

18. 105°

19. $\frac{5}{6} = \frac{1}{12}$

8. c 9. b 10. a

11. If two lines are skew lines, then they are not parallel and never intersect. If two lines are not parallel and never intersect, then they are skew lines. Since both the statement and its converse are true, the statement is a definition.

12. Square: $s = \sqrt{441} = 21$; $p = (21)(4) = 84$ in
Circle $r = \sqrt{\frac{441}{\pi}} \approx 11.85$; $C = 2(11.85)(\pi) \approx 74.46$ in

13. slope of $\overline{AD} = \frac{-2}{3}$; slope of $\overline{BC} = \frac{-2}{3}$; slope of $\overline{AB} = \frac{3}{2}$; slope of $\overline{DC} = \frac{3}{2}$; Since the slopes of $\overline{AD}$ and $\overline{BC}$ and the slopes of $\overline{AB}$ and $\overline{DC}$ are equal, the sides are parallel and *ABCD* is a parallelogram. Since the slopes of $\overline{AD}$ and $\overline{AB}$ are opposite reciprocals, $\overline{AD} \perp \overline{AB}$ and *ABCD* is a rectangle.

CHAPTER 7

Surface Area and Volume

Meeting Individual Needs

7.1 Exploring Surface Area and Volume

Core Resources	Core Plus Resources
Inclusion Strategies, p. 356 Reteaching the Lesson, p. 357 Practice Master 7.1 Enrichment Master 7.1 Technology Master 7.1 Lesson Activity Master 7.1 Interdisciplinary Connection, p. 355	Practice Master 7.1 Enrichment, p. 356 Technology Master 7.1
[1 day]	[1 day]

7.2 Surface Area and Volume of Prisms

Core Resources	Core Plus Resources
Inclusion Strategies, p. 361 Reteaching the Lesson, p. 362 Practice Master 7.2 Enrichment, p. 361 Lesson Activity Master 7.2 Interdisciplinary Connection, p. 360	Practice Master 7.2 Enrichment Master 7.2 Technology Master 7.2
[2 days]	[1 day]

7.3 Surface Area and Volume of Pyramids

Core Resources	Core Plus Resources
Inclusion Strategies, p. 368 Reteaching the Lesson, p. 369 Practice Master 7.3 Enrichment, p. 368 Lesson Activity Master 7.3	Practice Master 7.3 Enrichment Master 7.3 Technology Master 7.3 Interdisciplinary Connection, p. 367
[2 days]	[1 day]

7.4 Surface Area and Volume of Cylinders

Core Resources	Core Plus Resources
Inclusion Strategies, p. 375 Reteaching the Lesson, p. 376 Practice Master 7.4 Enrichment Master 7.4 Lesson Activity Master 7.4 Mid-Chapter Assessment Master	Practice Master 7.4 Enrichment, p. 375 Technology Master 7.4 Interdisciplinary Connection, p. 374 Mid-Chapter Assessment Master
[2 days]	[1 day]

7.5 Surface Area and Volume of Cones

Core Resources	Core Plus Resources
Inclusion Strategies, p. 381 Reteaching the Lesson, p. 382 Practice Master 7.5 Enrichment Master 7.5 Technology Master 7.5 Lesson Activity Master 7.5	Practice Master 7.5 Enrichment, p. 381 Technology Master 7.5 Interdisciplinary Connection, p. 380
[2 days]	[1 day]

7.6 Surface Area and Volume of Spheres

Core Resources	Core Plus Resources
Inclusion Strategies, p. 389 Reteaching the Lesson, p. 390 Practice Master 7.6 Enrichment Master 7.6 Lesson Activity Master 7.6	Practice Master 7.6 Enrichment, p. 389 Technology Master 7.6
[2 days]	[2 days]

7.7 Exploring Three-Dimensional Symmetry

Core Resources	Core Plus Resources
Inclusion Strategies, p. 398 Reteaching the Lesson, p. 399 Practice Master 7.7 Enrichment, p. 398 Technology Master 7.7 Lesson Activity Master 7.7	Practice Master 7.7 Enrichment Master 7.7 Technology Master 7.7
[2 days]	**[2 days]**

Chapter Summary

Core Resources	Core Plus Resources
Chapter 7 Project, pp. 404–405 Lab Activity Long-Term Project Chapter Review, pp. 406–408 Chapter Assessment, p. 409 Chapter Assessment, A/B Alternative Assessment	Chapter 7 Project, pp. 404–405 Lab Activity Long-Term Project Chapter Review, pp. 406–408 Chapter Assessment, p. 409 Chapter Assessment, A/B Alternative Assessment
[3 days]	**[2 days]**

Hands-On Strategies

Some of the activities in this chapter ask students to build solid models and then estimate volume using centimeter cubes or a material such as dry cereal. The extra effort required to set up activities of this sort will pay off as students gain a better understanding of how the volume and surface area formulas are derived.

Reading Strategies

This chapter, like Chapter 6, employs a variety of vocabulary specific to the study of three-dimensional shapes. Examples include *altitude, base, face, vertex, lateral surface, surface area,* and *volume.* Have students add any new terms to vocabulary lists in their notebooks. And encourage students to use the glossary whenever they are not clear on the meaning of a term.

Various adjectives are used to describe specific types of prisms and pyramids. The terms *right* and *oblique* deserve special attention. These are introduced in Lesson 7.2 in connection with prisms and in Lesson 7.3 for pyramids. Adjectives such as *triangular, pentagonal,* and *hexagonal* are also used in describing these solids. They tell us the shape of the base. To understand expressions such as *oblique triangular prism*, matching exercises can be used. Have students work in teams to create drawings and matching descriptions on pairs of index cards. The cards are then used for practice. A game format, such as that used in Concentration, can be employed.

Cooperative Learning

GROUP ACTIVITIES	
Surface area and volume	Lesson 7.1, Explorations 1 and 2
Volume of a pyramid	Lesson 7.3, Exploration
Surface area of a cone	Lesson 7.7, Explorations 1 and 2
Treasure of King Tut's Tomb	Eyewitness Math
Identifying three-dimensional symmetry	Lesson 7.7, Explorations 2 and 4,
Polyhedra	Chapter 7 Project

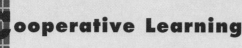

You may wish to have students work in groups or with partners for some of the above activities. Additional suggestions for cooperative group activities are noted in the teacher's notes in each lesson.

Multicultural

The cultural references in this chapter include references to Africa.

CULTURAL CONNECTIONS	
Africa: Great Pyramids at Gisa	Portfolio Activity
Africa: Pyramid of Khufu	Lesson 7.3, Application
Africa: Treasure of King Tut's Tomb	Eyewitness Math

Portfolio Assessment

Portfolio activities for the chapter listed below under seven activity domains.

1. **Investigation/Exploration** In Lesson 7.1, the explorations deal with relationships between surface area and volume. In the explorations for Lessons 7.3 and 7.4, students investigate the volumes of pyramids and cylinders. Surface areas of cones are the focus of the explorations in Lesson 7.5; rotations and reflections in three-dimensions are explored in Lesson 7.7.
2. **Applications** Cooking, Lesson 7.1, Exercise 17; Recreation, Lesson 7.2, Exercise 13; Product Packaging, Lesson 7.2, Exercise 29; Construction, Lesson 7.3, Exercise 21; Consumer Awareness, Lesson 7.4, Exercise 23; Maximum/Minimum, Lesson 7.4, Exercise 25; manufacturing, Lesson 7.5, Exercises 19–21; Hot Air Ballooning, Lesson 7.6, Examples 1 and 2; Sports, Lesson 7.6, Exercises 10–13; Food, Lesson 7.6, Exercises 14–16.
3. **Nonroutine Problems** Lesson 7.4, Exercises 32 (Marbles Problem); Lesson 7.6, Exercises 26–29 (Cube Inscribed in Sphere); Lesson 7.7, Exercises 32–33 (Spinning Fan).
4. **Project** Polyhedra: see pages 404–405. Students make models of the five regular polyhedra.
5. **Interdisciplinary Topics** Biology, Lesson 7.1, Exercise 11; Physiology, Lesson 7.1, Exercise 12; Botany, Lesson 7.1, Exercises 13–14; Architecture, Lesson 7.2, Exercise 30; Coordinate Geometry, Lesson 7.2, Exercises 38–39; Geology, Lesson 7.5, Example 2.
6. **Writing** *Communicate* exercises offer excellent writing selections for the portfolio. Lesson 7.2, Exercises 1–4; Lesson 7.3, Exercises 1–2; Lesson 7.5, Exercises 1–2; Lesson 7.6, Exercises 1–2.
7. **Tools** In Chapter 7, students can use graphics calculators to organize data in spreadsheets.

Technology

Graphics calculators, which are usually more readily available than computers, provide spreadsheet-like features. Calculator tables, which are like simple spreadsheets, have all the features the students will need to do the spreadsheet explorations and exercises in this book. Furthermore, the relationships displayed in graphics calculator tables can be graphed as functions on the calculator screens.

Solving a "Calculus" Problem

Lesson 7.1: Exploration 2

In this activity, students find the shape of a box of maximum volume made from a sheet of paper with dimensions 11 x 14. The box is made by removing sqares from the corners of the sheet and folding up the resulting flaps.

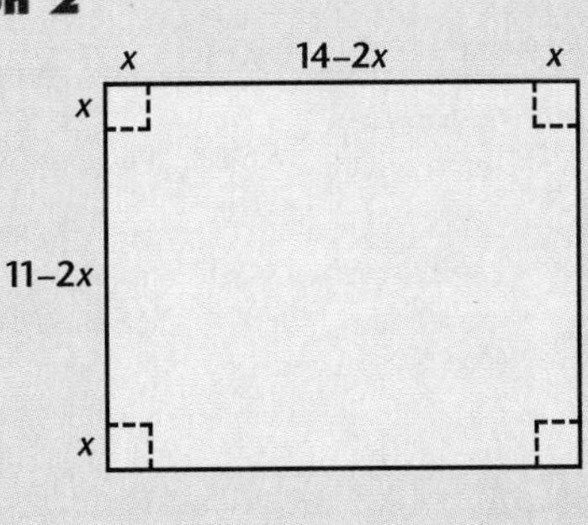

The students are asked to fill in the following table for the dimensions and volume of the paper box.

Side of Square x	Length l	Width w	Height h	Volume lwh
1	12	9	1	108
2	?	?	?	?
3	?	?	?	?
x	?	?	?	?

The expressions for the values to be filled in are:

Length l	$14 - 2x$
Width w	$11 - 2x$
Height h	x
Volume lwh	$(14 - 2x)(11 - 2x)x$
or	$4x^3 - 50x + 154$

These values are entered into the calculator as functions Y_1, Y_2, Y_3, and Y_4.

The resulting table, using increments of 1 and a minimum x value of 0, is

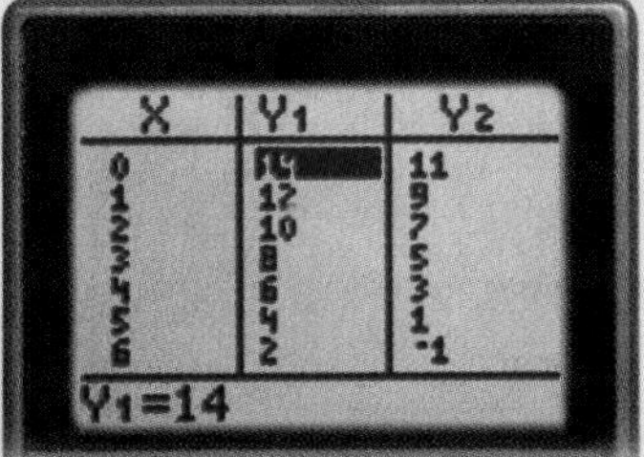

To see the values of Y_3 and Y_4, students should use the right-arrow key.

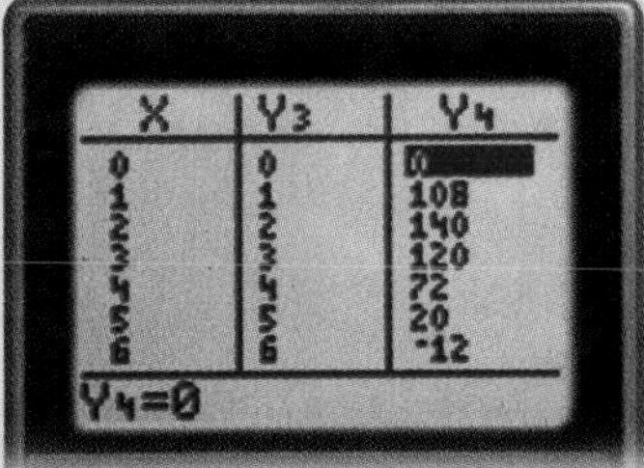

Students will notice that there is a maximum value of 140 for Y_4, the volume function, near $x = 2$. As they scroll farther down the table, they will see that the volume value becomes negative (thus physically meaningless) at $x = 6$ and then increases indefinitely for increasing values of x. (Remind the students of the general shape of the graphs of cubic functions.)

Students should understand that the table has physical meaning for just those values of x for which l, w, and h are all positive. This occurs for values of x between 0 and 5, approximately. These considerations suggest the window range settings for the graph.

The resulting graph, with the x-coordinate of the maximum value shown (the CALC feature of the calculator has been used), is shown below. The answer is accurate to five decimal places, rounded.

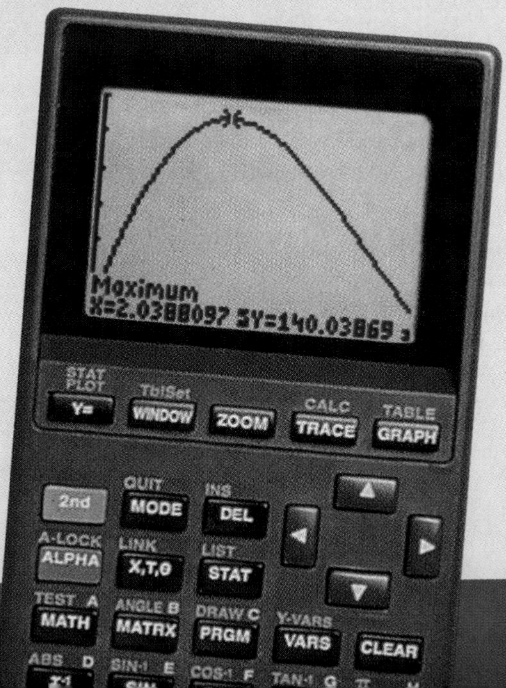

About the Chapter

Background Information

In this chapter students will calculate the surface area and volume of prisms, pyramids, cylinders, cones, and spheres. Students also explore how to maximize the volume of a prism while minimizing its surface area, and three-dimensional symmetry.

- Practice Masters
- Enrichment Masters
- Technology Masters
- Lesson Activity Masters
- Lab Activity Masters
- Long-Term Project Masters
- Assessment Masters
  Chapter Assessments, A/B
  Mid-Chapter Assessment
  Alternative Assessments, A/B
- Teaching Transparencies
- Spanish Resources

Chapter Objectives

- Explore ratios of surface area to volume.
- Develop an understanding of the concept of maximizing volume while minimizing surface area.
- Define and use the formula for finding the surface area of a right rectangular prism.
- Define and use the formula for finding the volume of a right rectangular prism.
- Use Cavalieri's Principle to develop the general formula for finding the volume of an oblique rectangular prism.

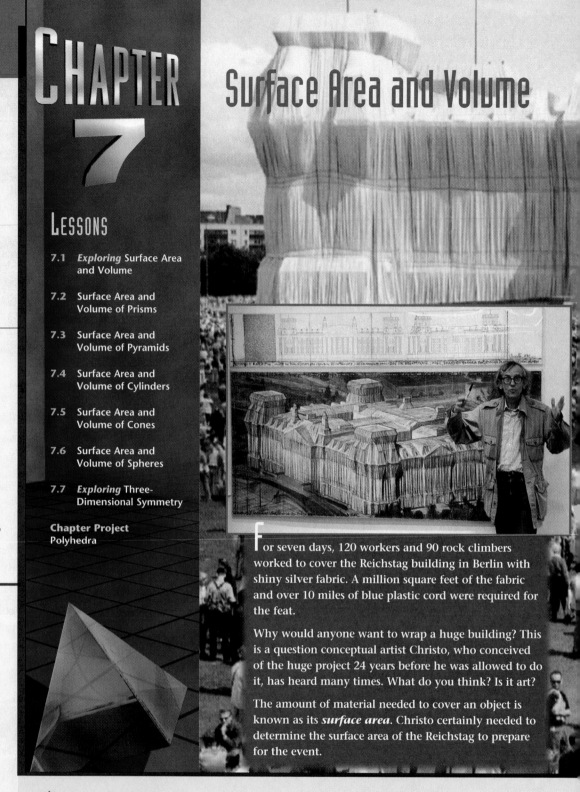

Chapter 7

Surface Area and Volume

Lessons

For seven days, 120 workers and 90 rock climbers worked to cover the Reichstag building in Berlin with shiny silver fabric. A million square feet of the fabric and over 10 miles of blue plastic cord were required for the feat.

Why would anyone want to wrap a huge building? This is a question conceptual artist Christo, who conceived of the huge project 24 years before he was allowed to do it, has heard many times. What do you think? Is it art?

The amount of material needed to cover an object is known as its ***surface area***. Christo certainly needed to determine the surface area of the Reichstag to prepare for the event.

About the Photos

The wrapping of the Reichstag building reveals features of the surface of the building which might otherwise go unnoticed. Ask students to imagine a one-million square foot square of fabric.

The base of the Great Pyramid of Giza is about 756 feet on each side. The pyramid itself is about 481 feet. If you wanted to wrap it in cloth, how much cloth would it take? How does this compare with the amount of cloth Christo used to wrap the Reichstag?

The Pyramids of Gisa. The Great Pyramid is the one farthest away.

- Define and use the formula for finding the surface area of a pyramid.
- Define and use the formula for finding the volume of a pyramid.
- Define and use the formula for finding the surface area of a cylinder.
- Define and use the formula for finding the volume of a cylinder.
- Define and use the formula for finding the surface area of a cone.
- Define and use the formula for finding the volume of a cone.
- Define and use the formula for finding the surface area of a sphere.
- Define and use the formula for finding the volume of a sphere.
- Define various transformations in three-dimensional space.
- Solve problems using transformations in three-dimensional space.

PORTFOLIO ACTIVITY

This portfolio activity provides an interdisciplinary connection to archeology, construction, and history.Extend the activity by having students build a model pyramid with the same proportions as the Great Pyramid. The ratio of the perimeter of the base to twice the height is π, and the ratio of the slant height to one-half the length of the base approximates the golden ratio, 1.618. Additional portfolio activities relative to surface area and volume are found in the exercises in Lessons 7.2, 7.4, and 7.5.

ABOUT THE CHAPTER PROJECT

In the Chapter 7 Project, on pages 404–405, students construct models of the five Platonic solids from net patterns that they trace or design. Students can also create models of polyhedra from straws.

PREPARE

Objectives

- Explore ratios of surface area to volume.
- Develop an understanding of the concept of maximizing volume while minimizing surface area.

- Practice Master 7.1
- Enrichment Master 7.1
- Technology Master 7.1
- Lesson Activity Master 7.1
- Quiz 7.1
- Spanish Resources 7.1

Assessing Prior Knowledge

1. Find the surface area and volume of a cube with sides of 4 units in length.

   [96 units^2 ; 64 units^3]

2. Find the surface area and volume of a rectangular prism with a length of 5 inches, a width of 4 inches, and a height of 2 inches.

   [76 units^2; 40 units^3]

3. Suppose there are 18 girls and 12 boys in geometry class. What is the ratio of boys to girls? [2:3]

TEACH

Maximizing volume while minimizing surface area is an important concept for students to understand because of all the applications in science.

Math Connection Algebra

Students should remember how to evaluate a formula. You may wish to review the steps.

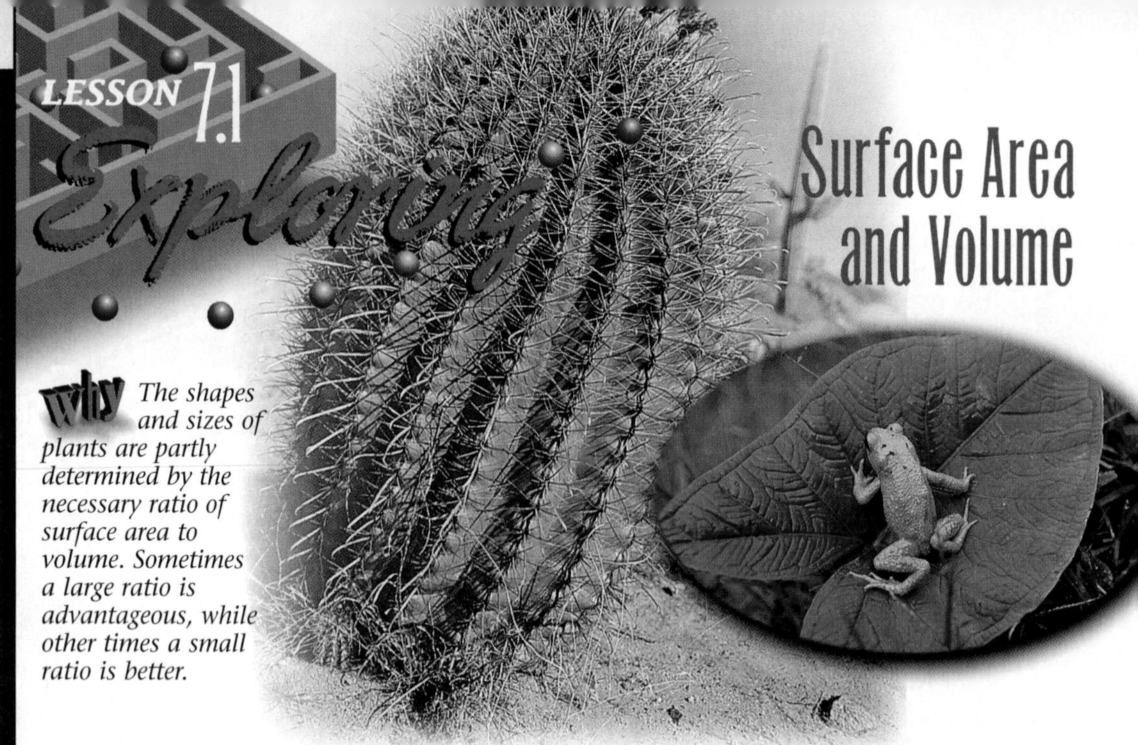

Why *The shapes and sizes of plants are partly determined by the necessary ratio of surface area to volume. Sometimes a large ratio is advantageous, while other times a small ratio is better.*

Desert plants must conserve water but have plentiful light. Tropical plants have plentiful water but must often compete for light. How does climate affect the shapes of desert and tropical plants?

Exploration I Ratio of Surface Area to Volume

You will need

Isometric grid paper or centimeter cubes

Spreadsheet software or graphics calculator (optional)

Technology Spreadsheet

Part I

 Draw or build rectangular prisms with the shape $n \times 1 \times 1$ and fill in a table like the one below. A spreadsheet may be used to complete the data table.

 If $n = 100$, what is the ratio of surface area to volume? What happens to the ratio of surface area to volume as the length of n increases? Is there a number that the ratio approaches? ❖

Length n	Surface Area $2lw + 2lh + 2wh$	Volume lwh	$\frac{SA}{Vol}$
1	6	1	?
2	?	?	?
3	?	?	?
n	?	?	?

Cooperative Learning Students can explore the ratio of surface area to volume by working in small groups. Have each student build and measure one or more rectangular prisms and then compile the data into one large table for the group. Have each group discuss their conjectures about the ratio of surface area to volume until they agree. Share the conjectures with the entire class.

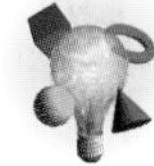

On sunny days a snake may lie in the sun to absorb heat. At night, a snake may coil up tightly to retain its body heat. Explain why this strategy works.

Part II

Draw or build rectangular prisms with the shape $n \times n \times n$ and fill in a table like the one below.

MAXIMUM MINIMUM *Connection*

Length n	Surface Area $2lw + 2lh + 2wh$	Volume lwh	$\frac{SA}{Vol}$
1	6	1	?
2	?	?	?
3	?	?	?
n	?	?	?

1 If $n = 100$, what is the ratio of surface area to volume? As the value for n gets larger, what happens to the ratio of surface area to volume?

2 What conclusions can you draw about the surface area–volume ratios of smaller cubes compared with the ratios of larger cubes? ❖

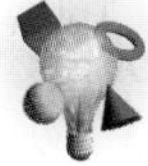

In cold weather, why might a very large animal be able to maintain its body heat more easily than a smaller one? Why do you think small, warmblooded animals, such as mice and birds, need to have higher metabolisms than larger ones, such as cows or elephants?

Exploration 2 *Maximizing Volume*

You will need

Graphics calculator or graph paper

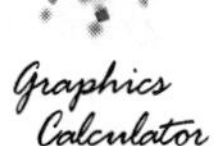

Graphics Calculator

1 Make an 11- × 14-inch piece of paper into an open box by cutting squares out of the corners and folding up the sides. How large should the squares be to maximize the volume of the box? Fill in a table like the one shown to help determine the answer.

Side of Square x	Length l	Width w	Height h	Volume lwh
1	12	9	1	108
2	?	?	?	?
3	?	?	?	?
x	?	?	?	?

Industrial Technology The design of a package may reflect environmental and health factors as well as the economics of maximizing volume. Have students discuss what a packaging engineer may need to know to do his/her job.

Exploration 1 Notes

In this activity, students explore the ratio of surface area to volume for a rectangular prism.

ongoing ASSESSMENT

Part 1, Step 2. $\frac{402}{100}$; The ratio of surface area to volume decreases as n increases. The ratio approaches 4.

CRITICAL *Thinking*

When the snake stretches out, it maximizes its surface area and its surface area to volume ratio. There is more area through which to absorb energy. When the snake coils up, it minimizes the surface area through which heat can escape.

ongoing ASSESSMENT

Part 2, Step 2. As the length of a side increases, the ratio of surface area to volume decreases.

CRITICAL *Thinking*

Because larger animals have a smaller ratio of surface area to volume. Larger animals thus have relatively little surface area through which to lose heat compared to their body weight. Small animals have a higher surface area to volume ratio. So, smaller animals need to generate more heat because they lose heat faster.

Exploration 2 Notes

In this activity, students explore how to maximize the volume of an open box by tracing the volume function on a graphing calculator.

Math Connection Maximum/Minimum

Maximum/minimum problems are common in real-world math applications. Calculus deals extensively with maximum/minimum concepts.

3. a. $x = 2$ b. $V = 140$

CRITICAL *Thinking*

The largest possible ratio of volume to surface area ratio possible minimizes cost for packaging materials.

Cooperative Learning

Students may work on Exploration 2 in small groups. Working together may help students who have difficulty in finding a complete viewing window for a graph.

Assess

Selected Answers

Odd-numbered Exercises 7–25

Assignment Guide

Core 1–14, 19–25

Core Plus 1–268

Technology

Students can use the trace features of a graphing calculator to help them explain their answers to Exercises 15 and 16. Encourage students to use the table features to explore the change in y-values of a graph as x-values increase rapidly.

Error Analysis

Make sure that students have the correct functions to graph for Exercises 15 and 16.

MAXIMUM MINIMUM *Connection*

2. Use a graphics calculator to graph the Side-of-Square data on the x-axis and Volume data on the y-axis. Hint: Note that if a square with side length x is removed, the length of the box is $14 - 2x$.

3. Trace along the graph and determine the following:

a. x-value that produces the largest volume;

b. largest volume. ❖

CRITICAL *Thinking*

Why are food processing companies concerned with maximizing volume for a given surface area for product packaging?

APPLICATION

Product Packaging

A cereal company is choosing between two box designs with dimensions as shown

The company would like to keep the amount of cardboard used for the box to a minimum because it is more cost effective and better for the environment. The company must choose between two box designs. Which box provides less surface area for the same volume?

Both boxes provide a volume of 160 cubic inches. The surface area of box A is

$2(2.5)(8) + 2(2.5)(8) + 2(8)(8) = 208$ cubic inches.

The surface area of box B is

$2(2)(8) + 2(2)(10) + 2(8)(10) = 232$ cubic inches.

Box A has less surface area. ❖

Exercises & Problems

Communicate

For each situation, determine whether maximum volume or maximum surface area is appropriate. Explain your reasoning.

1. building a storage bin with a limited amount of lumber
2. designing soup cans to hold 15 oz of soup
3. building solar energy panels

ENRICHMENT Have students explore the relationship between the area and volume of similar figures. Have cooperative groups to build pairs of similar rectangular prisms. Have them find the scale factor k between corresponding sides of the prisms and answer the following questions: What is the ratio of the surface areas for each similar pair? [$k : k^2$] What is the ratio of the volumes for each similar pair? [$k : k^3$]

Hands-On Strategies Have students use small cubes to model the surface area and volume data for Exploration 1. Students can glue together sugar cubes to create their own models for their portfolios. Students should also model Exploration 2 using several sheets of 11- × 14-inch paper. You may wish to have the students create a poster to display their work on these explorations.

4. designing the "heat" tiles on the space shuttle

5. designing tires for 4-wheel-drive vehicles

Practice & Apply

Use the charts from Explorations 1 and 2 for Exercises 6–10.

6. If a rectangular prism has the dimensions $3 \times 1 \times 1$, what is the ratio of surface area to volume? $\frac{14}{3}$

7. If a rectangular prism has the dimensions $3 \times 3 \times 3$, what is the ratio of surface area to volume? $\frac{2}{1}$ or 2

8. Find the dimensions of a cube with a volume of 1000 cubic centimeters. What is the ratio of surface area to volume? $s = 10; \frac{\text{surface area}}{\text{volume}} = .6$

9. Find the ratio of surface area to volume for a cube with a volume of 64 cubic inches. $s = 4; \frac{\text{surface area}}{\text{volume}} = 1.5$

10. What is the surface area of the cube in Exercise 9? 96 units^2

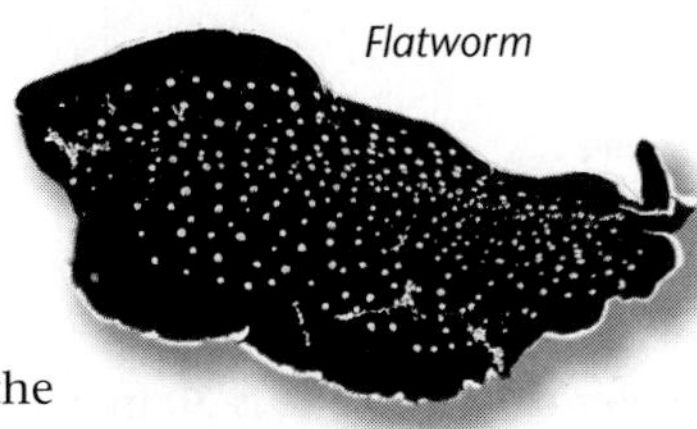
Flatworm

11. **Biology** Flatworms absorb oxygen through their skin. Explain why a flatworm's shape helps it to live.

12. **Physiology** Human lungs are subdivided into thousands of little air sacs. How does this help the absorption of oxygen when we breathe?

13. **Botany** The short, squat form of cactus (barrel cactus) has a unique function. Explain how the shape helps a cactus thrive in the desert.

14. **Botany** Why do tall trees with large, broad leaves not usually grow in the desert?

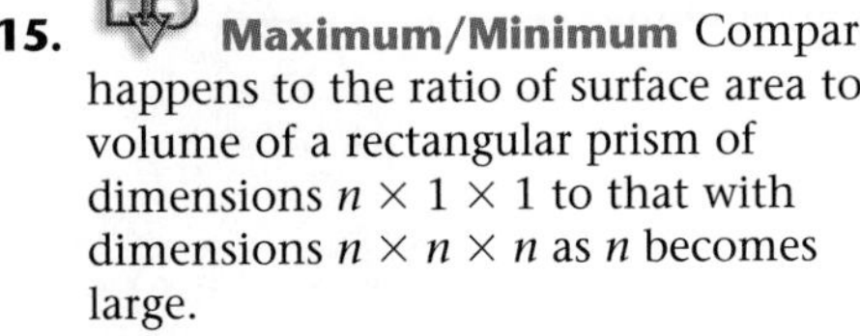

15. **Maximum/Minimum** Compare what happens to the ratio of surface area to volume of a rectangular prism of dimensions $n \times 1 \times 1$ to that with dimensions $n \times n \times n$ as n becomes large.

16. **Maximum\Minimum** Using the result of Exercise 15, make a conjecture about what type of rectangular prism has the smallest surface area for a given volume.

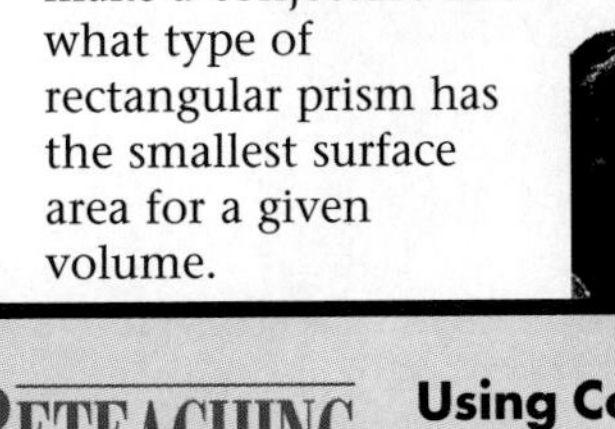

RETEACHING the lesson

Using Cognitive Strategies Students can review the explorations by working problems "backwards." Give students a surface area for a rectangular prism and ask them to find the volume or vice versa. For example, have students find the maximum volume for a rectangular prism with surface area 384 in^2. [512 in^3]

alternative ASSESSMENT

Portfolio Assessment

As a writing assignment, ask students to describe their work in Explorations 1 and 2. Ask them to state the principles they learned and to include examples that illustrate the principles.

The answers to Exercises 11–16 can be found in Additional Answers beginning on page 727.

Practice Master

NAME ______ CLASS ______ DATE ______

Practice & Apply
7.1 Exploring Surface Area and Volume

Use the charts from Explorations 1 and 2 for Exercises 1–5.

1. If a rectangular prism has the dimensions $4 \times 1 \times 1$, what is the ratio of surface area to volume? ______
2. If a rectangular prism has the dimensions $4 \times 4 \times 4$, what is the ratio of surface area to volume? ______
3. Find the dimensions of a cube with a volume of 8000 cubic centimeters. ______
4. Find the surface area of a cube with a volume of 216 cubic inches. ______
5. What is the ratio of surface area to volume for the cube in Exercise 4? ______
6. The wood shop class assignment was to make a paperweight in the shape of a rectangular prism, and then to decorate the prism with 1-inch square tiles. Tanya and Robyn's finished paperweights are shown. Which girl used more tiles? Explain. ______

4" 4" 4" TANYA — 4" 5" 3" ROBYN

You can graph a rectangular prism by following these steps.

7. Plot the points $A(1, 1)$, $B(4, 2)$, $C(3, 3)$, $D(0, 2)$.
8. Translate each point A, B, C, D, 4 units up on the graph to the image points A', B', C', D'.
9. Connect points A, B, C, and D.
10. Connect points A', B', C', and D'.
11. Draw $\overline{AA'}$, $\overline{BB'}$, $\overline{CC'}$, and $\overline{DD'}$.
12. Find the dimensions of the rectangular prism. ______
13. Find the surface area of the rectangular prism. ______
14. Find the volume of the rectangular prism. ______

HRW Geometry — Practice & Apply 1

17. **Cooking** A roast is finished cooking when the inside of the meat reaches a certain temperature. Explain why a large roast takes longer to cook than a small roast.

18. **Physiology** Considering surface area and volume, why does thoroughly chewing your food help make digestion easier?

19. **Aeronautical Engineering** Thrust is the force caused by air escaping from the back of a jet engine. Thrust propels the jet forward. Currently, two different engine designs are used on commercial aircraft. One engine pushes a relatively small volume of air very quickly to produce thrust. Another engine pushes a much larger volume of air more slowly to produce thrust. Which engine do you think is more fuel efficient? Why? Which engine is quieter? Why?

Look Back

20. Show that AAA cannot be used to prove triangles congruent. **[Lesson 4.3]**

21. If the perimeter of a rectangle is 20 m, what are five possible areas if the sides measure an integer number of meters? **[Lesson 5.2]**

22. If two triangles have the same area, are they necessarily congruent? If not, use a counterexample to show why not. **[Lesson 5.2]**

23. The circumference of a circle is 10π. Find the area. **[Lesson 5.3]** 25π units^2

24. The legs of an isosceles right triangle each measure 6 cm. Find the measure of the hypotenuse. **[Lessons 5.4, 5.5]** 8.49 cm

25. Find the area of a trapezoid in which one base measures 10 in., the other base measures 7 in., and the height measures 8 in. **[Lesson 5.2]** 68 cm^2

Look Beyond

26. What information would you need in order to compute the surface area of a triangular pyramid?

17. The ratio of surface area to volume is smaller in a larger roast, so there is relatively little surface area through which heat can enter. There is also more meat to heat in a larger roast.

18. Digestive juices work on the surface of food. Chewing food more completely increases the surface area of the chewed food. The larger the surface area, the faster digestion can take place.

Look Beyond

Exercise 26 introduces surface area of a pyramid, a solid they will study later in the chapter. Have students draw pyramids, and ask them to describe the faces.

Technology Master

NAME ______ CLASS ______ DATE ______

Technology

7.1 Surface Area by Counting Visible Faces

When you "fly around" a cube, all six of its faces will be visible to you. If, however, you place two or three identical cubes in a row, you will see 10 and 14 visible faces, respectively. There is a pattern to the number of visible faces in a row of n identical cubes. If you now arrange m rows of n identical cubes in each row, you will have a rectangular array whose number of visible faces is also patterned.

Cubes in a row: 1, 2, 3

6, 10, 14

Visible Faces During Fly Around

Row 1, Row 2, Row 3

1. In terms of n, write a formula for the number of visible faces of n identical cubes aligned in a row.

______

2. Suppose you make an array consisting of m rows of n identical cubes in each row. In terms of m and n, write a formula for the number of visible faces in the rectangular array.

______

3. Suppose a cube has edges that are 1 unit long. Write a statement that relates the surface area S of the prism made of m rows of n unit cubes in each row.

______

4. Create a spreadsheet in which row 1 contains $n = 1, 2, 3, \ldots$, column A contains $m = 1, 2, 3, \ldots$, and each cell contains the value of the formula you wrote in Exercise 2.

5. Use the spreadsheet to find the surface area of 120 unit cubes arranged in 5 rows each row having 24 cubes.

______

6. From the spreadsheet, is it possible to arrange a set of unit cubes to make a rectangular prism 1 cube deep with a surface area of 1148 square units?

______

7. Use the spreadsheet to show that it is *not possible* to arrange a set of unit cubes in a rectangular array 1 cube deep in such a way that the surface area will be 236 square units.

______

HRW Geometry — Technology 15

19. The engine that pushes a larger amount of air slowly is quieter and more fuel efficient. It is quieter because the speed of the air coming out of the engine is not as great. It is more efficient because it delivers more power per weight of the engine.

20. Consider two equilateral triangles, one with side of length 5 and the other with side of length 10. All the angles of both triangles will measure 60°, but the triangles are not congruent because the sides aren't congruent.

21. $1 \times 9 = 9$ m^2; $2 \times 8 = 16$ m^2; $3 \times 7 = 21$ m^2; $4 \times 6 = 24$ m^2; $5 \times 5 = 25$ m^2

22. No. Sample example: Consider a triangle with base 10 and height 10 and another triangle with base 20 and height 5. They both have area of 50 units^2, but they are not congruent.

26. The dimensions of each face.

LESSON 7.2 Surface Area and Volume of Prisms

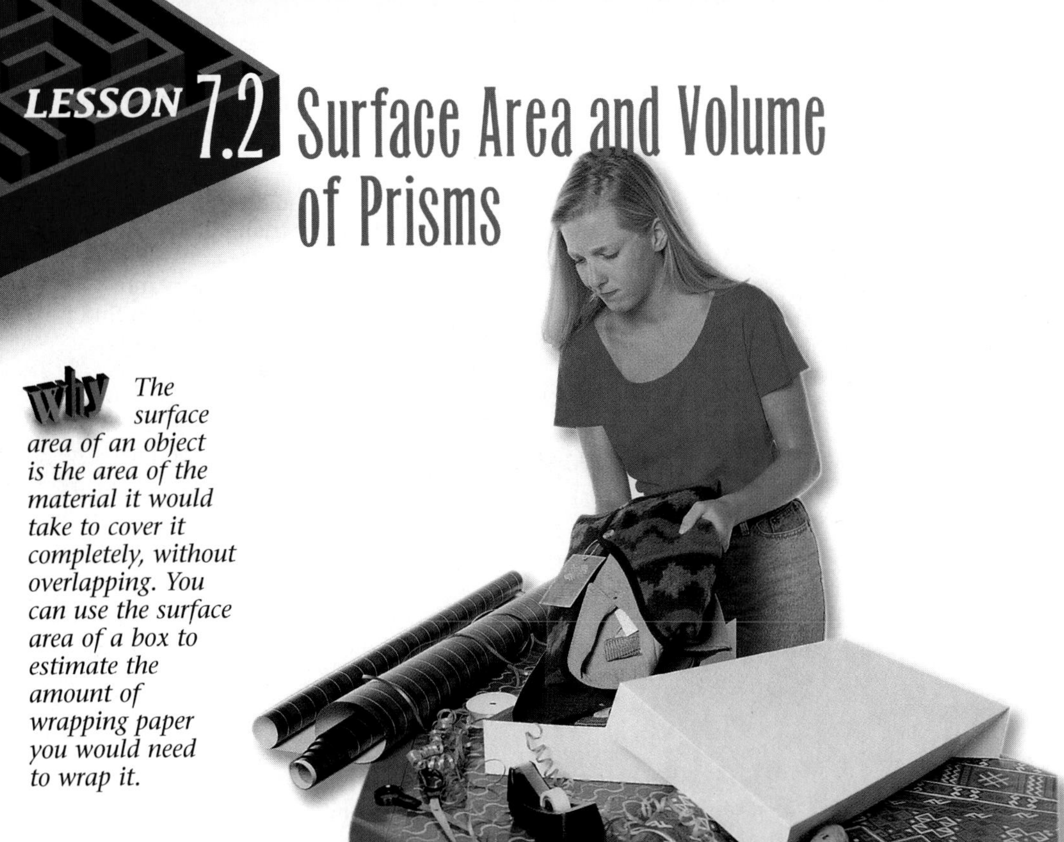

The surface area of an object is the area of the material it would take to cover it completely, without overlapping. You can use the surface area of a box to estimate the amount of wrapping paper you would need to wrap it.

You may recall from Lesson 6.3 that a **prism** is a polyhedron with two bases that form polygons that are parallel and congruent. The lateral faces are parallelograms formed by the segments connecting the corresponding vertices of the bases. A **right prism** has lateral edges that are perpendicular to the planes of the bases.

An **altitude** of a prism is any segment perpendicular to the planes containing the bases with endpoints in these planes. The **length of an altitude** is the height of a prism.

The **lateral area** of a prism is found by adding all the areas of the lateral faces. The **surface area** of a prism is found by adding the lateral area to the areas of the two bases of a prism. Since the bases of a prism are congruent, the base area is twice the area of one base.

ALTERNATIVE teaching strategy

Hands-On Strategies Have students use grid paper to create their own net patterns for prisms. Ask them to construct the prisms from the nets and measure each face to calculate the surface area and volume.

PREPARE

Objectives

- Define and use the formula for finding the surface area of a right rectangular prism.
- Define and use the formula for finding the volume of a right rectangular prism.
- Use Cavalieri's Principle to develop the general formula for finding the volume of an oblique rectangular prism.

RESOURCES

• Practice Master	7.2
• Enrichment Master	7.2
• Technology Master	7.2
• Lesson Activity Master	7.2
• Quiz	7.2
• Spanish Resources	7.2

Use Transparency 56

Assessing Prior Knowledge

1. Find the area of a triangle with height 6 and base 4.

   [12 units^2]

2. Find the distance between points (0, 3, –2) and (4, –1, 3) in a three-dimensional coordinate system.

   [$\sqrt{57}$]

3. If $s = 8$, find s^2 and s^3.

   [64, 256].

TEACH

This lesson helps students extend their ideas about surface area and volume by introducing the idea that the volume of a prism is equal to the area of its base times its height.

To analyze the surface area of certain solid figures, you can think of them as **nets**—flat figures that can be folded to enclose a particular solid figure. By calculating the area of the net, you can determine the surface area of a solid.

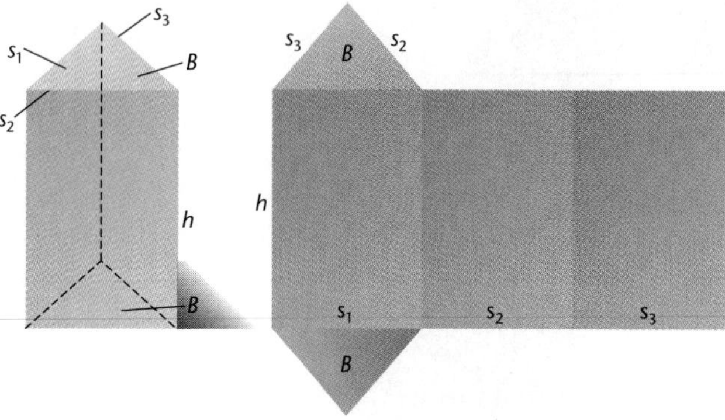

To find a formula for the surface area of a right triangular prism with height h and the area of a base B, think of cutting the prism and flattening it to form a net. To find the lateral area L, find the area of the rectangular faces.

$$L = h(s_1 + s_2 + s_3)$$

But $(s_1 + s_2 + s_3)$ is the perimeter p of a base, so

$$L = hp.$$

To find the total surface area S of the prism, add $2B$ to the lateral area.

$$S \text{ (surface area)} = L \text{ (lateral area)} + 2B \text{ (area of the bases)} = hp + 2B$$

LATERAL AREA OF A RIGHT PRISM

The lateral area L of a right prism with height h and perimeter of a base p is

$$L = hp \qquad \textbf{7.2.1}$$

SURFACE AREA OF A RIGHT PRISM

The surface area S of a right prism with lateral area L and the area of a base B is

$$S = L + 2B \qquad \textbf{7.2.2}$$

CRITICAL *Thinking*

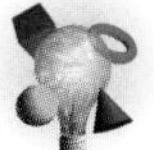

Explain why some of the letters in the formulas above are lowercase and other letters are uppercase.

Volumes of Right Prisms

To find the volume V of a right rectangular prism, use the formula

$$V = lwh,$$

where l is the length, w is the width, and h is the height of the prism.

TEACHING *tip*

Emphasize that a prism is named by the shape of its base. The faces that are not bases are called lateral faces and are always parallelograms. Point out that the bases of a prism are always congruent.

TEACHING *tip*

You may want to use models of right prisms with different bases to emphasize the applicability of the surface area formula. Have students draw prisms on the chalkboard, measure them, and find their surface area and volume.

CRITICAL *Thinking*

Lower case letters are used for length measures. Capital letters are used for areas.

interdisciplinary CONNECTION **Manufacturing** Cardboard boxes (rectangular prisms) often are manufactured as nets with extra flaps attached to help keep the boxes closed when assembled. Have students create examples of nets by taking boxes apart. Find the surface area of each net and compare it to finding the surface area of a rectangular prism with the same dimensions.

EXAMPLE 1

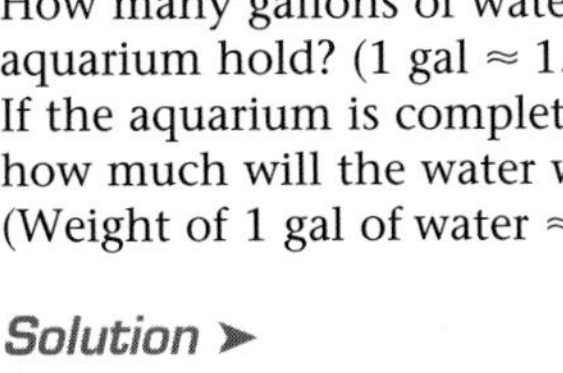

Marine Biology

Find the volume of an aquarium that is 110 ft by 50 ft by 7 ft in size. How many gallons of water will the aquarium hold? (1 gal $\approx$ 1.6 cu ft.) If the aquarium is completely filled, how much will the water weigh? (Weight of 1 gal of water $\approx$ 8.33 lb)

Solution ➤

The volume of the aquarium is

$$V = lwh = 110 \times 50 \times 7 = 38{,}500 \text{ cu ft.}$$

To find the volume in gallons, divide by 1.6:

$$V \text{ (in gallons)} = \frac{38{,}500}{1.6} \approx 24{,}062.5 \text{ gal}$$

To find the weight, multiply by 8.33:

$$\text{Weight} = 24{,}062.5 \times 8.33 \approx 200{,}441 \text{ lb}$$ ❖

In the formula for the volume of a right rectangular prism, $V = lwh$, notice that lw is the area of a base. Setting $B = lw$, the formula becomes $V = Bh$. This formula is true for all prisms.

CRITICAL *Thinking*

Write a volume formula for a right prism with bases that are regular polygons.

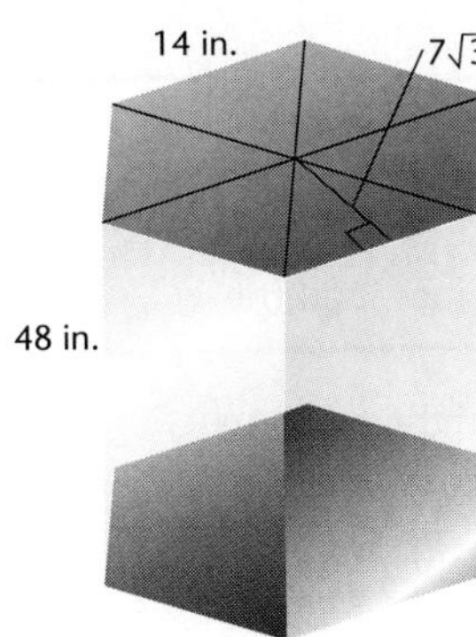

EXAMPLE 2

An aquarium is in the shape of a right regular hexagonal prism. Each side of its base is 14 in. If the aquarium is 48 in. tall, find the volume.

Solution ➤

Find the area of a base of the aquarium. Notice that each base can be divided into six equilateral triangles. The altitude of each triangle is $7\sqrt{3}$ in., which is also the measure of the apothem of the hexagon. The perimeter of the hexagon is 84 in.

The area of the base is

$$\frac{1}{2}ap = \frac{1}{2} \times 7\sqrt{3} \times 84 = 509.22 \text{ sq in.}$$

To find the volume of the aquarium, multiply this result by the height h:

$$V = Bh = 509.22 \times 48 = 24{,}442.56 \text{ cu in.}$$ ❖

Alternate Example 1

Find the volume of a cereal box that is 19 cm by 7.6 cm by 30.5 cm. If a cubic centimeter of cereal weighs 0.1 g, how much weight will the box hold?

[4404.2 cm^3; 440.42 g]

CRITICAL *Thinking*

$A = \frac{1}{2}aph$, where a is the apothem of a base.

Alternate Example 2

A cookie jar comes packaged in a hexagonal box that is a right regular hexagonal prism. Each side of the base is 8.5 in. If the box is 9 in. tall, find the volume of the box.

[1689.4 in^3]

ENRICHMENT Have students give counterexamples to each of the following statements by sketching figures that show the statement to be false. (1) A solid with opposite faces parallel is a prism. (2) A solid with one pair of opposite faces congruent is a prism. (3) A solid with length 12, width 6, and height 8 is a right prism. (4) If the surface area of two prisms is the same, the volume is the same.

INCLUSION strategies **Using Models** You may help some students understand the formula for the surface area of a right prism by providing them with examples of nets of right prisms. Have them find the area of the net, cut it out, form a right prism, and find the surface area again. Compare it to the original area.

TEACHING tip

You may need to spend extra time explaining "cross-sectional area" and reviewing the various formulas for the area of a polygon. Emphasize the general nature of the volume formula for a prism, $V = Bh$, where B is the area of the polygon that is the base of the prism.

Cooperative Learning

Use cooperative groups to help students understand Cavalieri's Principle. Provide the groups with stacks of index cards. First, have the groups stack the cards straight to represent a right rectangular prism. Next, have the groups lean and/or rotate the cards to create different stacks. Have the groups sketch each new solid they create. Then ask students whether they think that the volume of their new solids is the same as that of the original prism. Have each group explain their reasoning.

Oblique Prisms

In an oblique prism the lateral edges are not perpendicular to the planes of the bases. For such prisms there is no simple general formula for the surface area. However, the formula for the volume is the same as for a right prism. To understand why this is so, consider the following idea.

Stack a set of index cards in the shape of a right rectangular prism. If you push the stack over into an oblique prism, the volume of the figure does not change because the number of cards does not change.

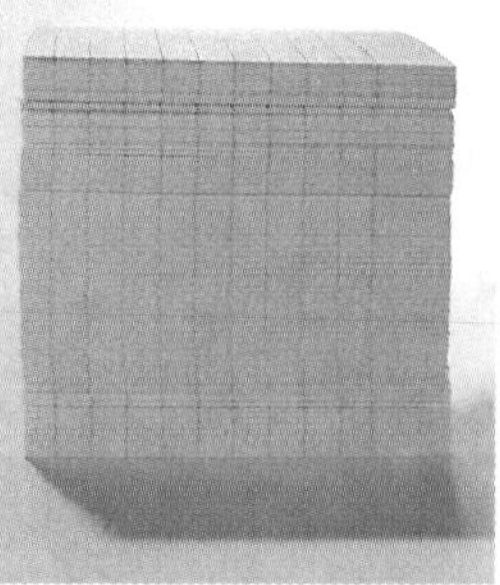

In the picture, both stacks have the same number of cards, and so each prism is the same height. Also, since every card has the same size and shape, they all have the same area. Any card in either stack represents a cross section of each prism.

This idea illustrates an important geometry concept.

CAVALIERI'S PRINCIPLE

If two solids have equal heights, and if the cross sections formed by every plane parallel to the bases of both solids have equal areas, then the two solids have the same volume. **7.2.3**

The solids pictured above will have equal volumes if all the cross-sectional areas are equal.

Since an oblique prism can be compared to a right prism that has been pushed over like the deck of cards, the base area, the height, and the volume do not change. Therefore, the formula for the volume of an oblique prism is the same as the formula for the volume of a right prism.

VOLUME OF A PRISM

The volume V of a prism with height h and the area of a base B is

$$V = Bh$$ **7.2.4**

Using Summary Summarize the lesson with the following important facts about prisms:

(1) The bases are congruent and parallel polygons. (2) The lateral faces are parallelograms. (3) The edges of right prisms are perpendicular to the bases. Edges of oblique prisms are not perpendicular to the bases. (4) The lateral faces of right prisms are rectangles. (5) The height of a right prism is the same as the length of a lateral edge. (6) The lateral area of a prism is the height times the perimeter of the base, or $L = hp$. (7) The surface area of a prism is the lateral area plus twice the area of the base, or $S = L + 2B$. (8) The volume of a prism is the area of the base times the height, or $V = Bh$.

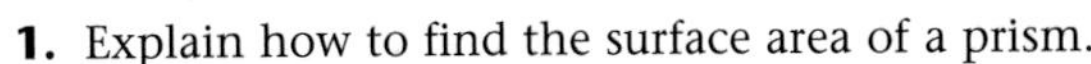

Communicate

1. Explain how to find the surface area of a prism.
2. Explain the formula for the volume of a right prism.
3. Explain Cavalieri's principle and how it can be used to find the volume of an oblique prism.
4. Can Cavalieri's principle be used for two prisms of the same height if the base of one is a triangle and the base of the other is a hexagon? Why or why not?
5. What does doubling the height of a prism do to the volume if the base is unchanged?

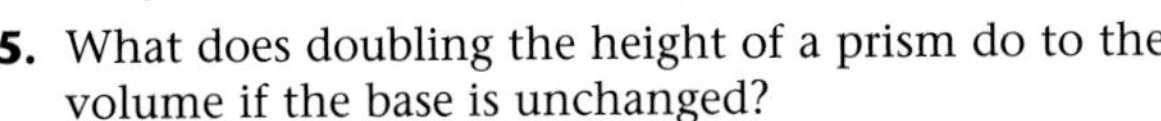

Practice & Apply

Draw the net for each of these figures. Assume them to be right (not oblique).

6. Cube (square prism)
7. Equilateral triangular prism
8. Regular hexagonal prism
9. Rectangular prism
10. Find the surface area of a rectangular prism measuring 18 cm by 15 cm by 14 cm. 1464 cm^2
11. Find the height of a rectangular prism with a surface area of 286 sq in. and base measuring 7 in. by 9 in. 5 in.
12. Find the surface area of a right regular hexagonal prism whose altitude is 20 cm. The apothem of the hexagonal base is $4\sqrt{3}$. 1292.55 cm^2
13. **Recreation** Suppose a tent in the shape of an isosceles triangular prism is resting on a flat surface. Find the surface area of the tent, including the floor.

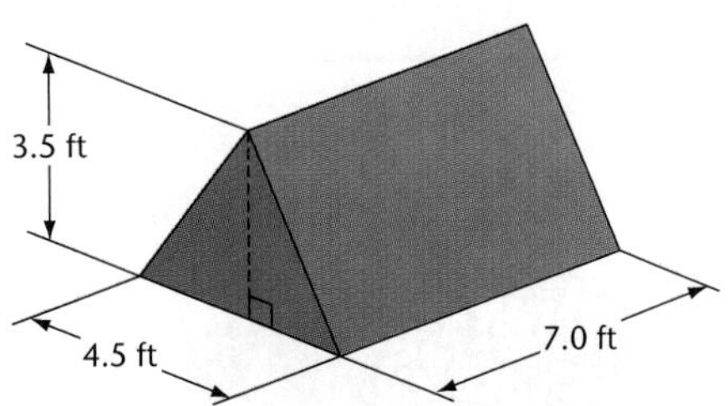

14. Find the surface area of a right regular hexagonal prism. The sides of the regular hexagon are 18 cm and its altitude is 25 cm. 4383.55 cm^2

Find the volume of each prism.

15. $B = 7 \text{ cm}^2$, $h = 5$ cm 35 cm^3
16. $B = 9 \text{ m}^2$, $h = 6$ m 54 m^3
17. $B = 17 \text{ cm}^2$, $h = 23$ cm 391 cm^3
18. $B = 32 \text{ ft}^2$, $h = 17$ ft 544 ft^3

6. Sample answer:

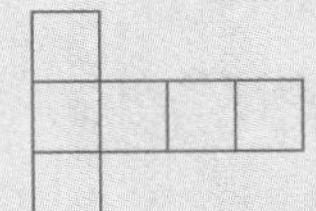

8. Sample answer:

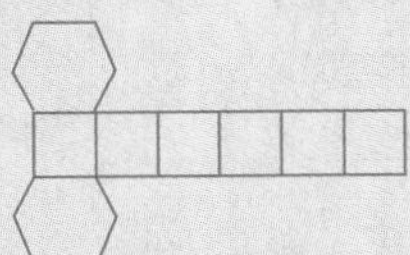

7. Sample answer:

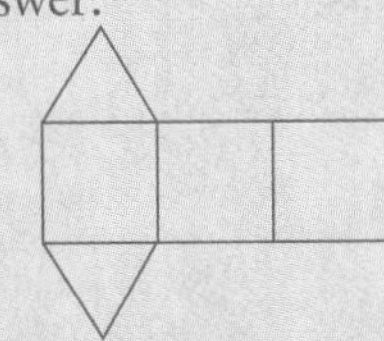

9. Sample answer:

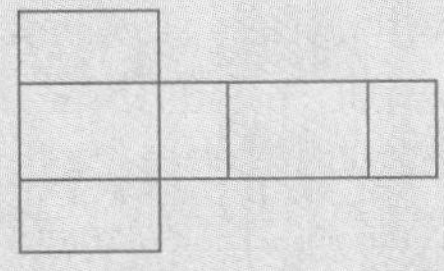

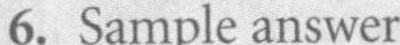

ASSESS

Selected Answers

Odd-numbered Exercises 15–37

Assignment Guide

Core 1–21, 24–28, 31–37

Core Plus 1–16, 19–30, 37–39

Technology

Students can use a scientific calculator for most of the exercises. Establish rules for rounding. The provided answers are rounded to the nearest hundredth.

Error Analysis

Students will be alternating between finding volumes and surface areas. Encourage students to be very careful with their calculations and double-check all answers.

13. 105.5 sq. ft.

alternative ASSESSMENT

Portfolio Assessment

The ability to relate drawings to ideas presented in the lesson will reinforce many of its key ideas. Have students draw various prisms, label them, and summarize the lesson's key ideas with their drawings.

Find the volume of each prism.

19.

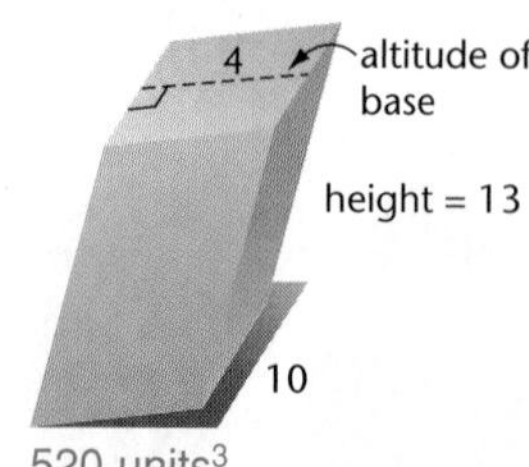

520 units^3

20.

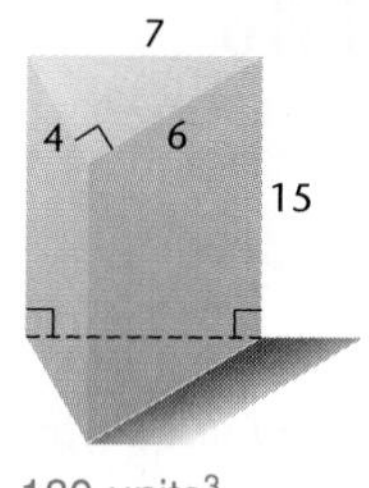

180 units^3

21. If a cube has a volume of 343 cu yds, what is its surface area? 294 yd^2

22. Find the volume of a right trapezoidal prism. The parallel sides of the trapezoids measure 6 m and 8 m and the trapezoid's altitude is 7 m. The altitude of the prism is 18 m. 882 m^3

23. Find the volume of a right hexagonal prism. The length of the sides of the hexagonal bases are 8 cm and the altitude of the prism is 12 cm. 1995.32 cm^3

24. Find the volume of a right triangular prism whose base is an isosceles right triangle with hypotenuse 10 cm. The altitude of the prism is 23 cm. 575 cm^3

25. Find the surface area and volume of a right triangular prism whose base is an isosceles triangle with sides 4 in., 4 in., and 6 in. The height of the prism is 13 in. SA = 197.87 in.^{2}; V = 103.18 in.^{3}

26. **Portfolio Activity** Using 100 sq in. of cardboard, a pair of scissors, and masking tape, make a box without a top. What are the measures of its sides? What is the volume?

27. If the edges of a cube are doubled, what happens to the surface area? to the volume? If the edges are tripled, what happens to the surface area? to the volume?

28. When does a cube have the same numerical value for its surface area and volume? When $s = 6$ units

29. **Product Packaging** Find the surface area and volume of each box of breakfast food.

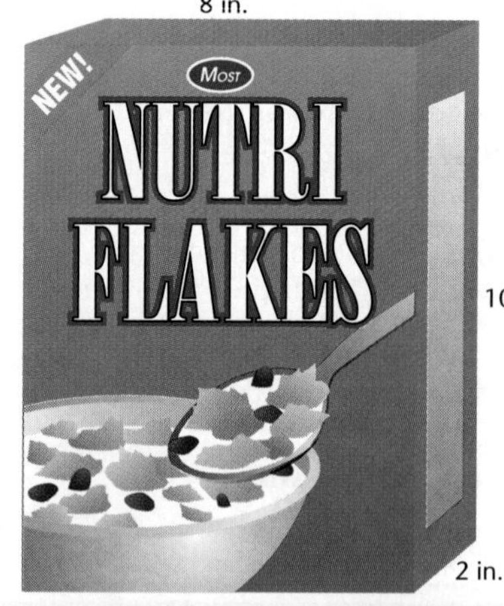

26. Answers will vary, but the dimensions of the box must satisfy

$2lh + 2wh + lw = 100.$

27. If doubled: The surface area is multiplied by $2^2 = 4$ and the volume is multiplied by $2^3 = 8$. If tripled: The surface area is multiplied by $3^2 = 9$ and the volume is multiplied by $3^3 = 27$.

29. Power flakes: SA = 158 in^2 V = 120 in^3

Nutri Flakes: SA = 232 in^2 V = 160 in^3

Practice Master

NAME ______ CLASS ______ DATE ______

Practice & Apply

7.2 Surface Area and Volume of Prisms

In Exercises 1–2, draw the net below each figure given.

1.

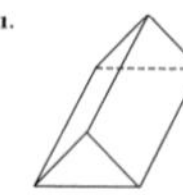

2.

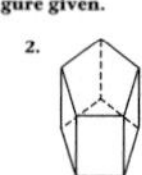

In Exercises 3–4, find the surface area of the given figure.

3. A right equilateral triangular prism with altitude 18 cm. The altitude of the triangular base is $6\sqrt{3}$ cm.

4. A right regular pentagonal prism with altitude 24 cm. The apothem of the pentagonal base is 8.3 cm and a side is 12 cm.

In Exercises 5–7, find the volume of the given prism.

5. $B = 8$ cm^2, $h = 7$ cm

6. 11" 6.4" 14"

7. A rectangular prism measuring 8.2 cm by 6 cm by 10.7 cm.

8. Find the volume of a right isosceles trapezoidal prism. The shorter base of the trapezoid measures 12 cm, a base angle measures 60°, and the altitude of the trapezoid is $6\sqrt{3}$ cm. The altitude of the prism is 16 cm.

2 Practice & Apply HRW Geometry

30. Architecture Estimate how much glass was used to cover the outside walls of this building. The distance from one floor to the next is 12 feet and the base of the building is a square which measures 48 feet on each side. 23,040 ft^2

Look Back

Simplify each radical expression. Leave answers in simplified radical form. [Lesson 5.4]

31. $\sqrt{20}$ $2\sqrt{5}$

32. $(\sqrt{18})(\sqrt{2})$ 6

33. $(5\sqrt{7})^2$ 175

34. $(2\sqrt{2})(3\sqrt{8})$ 24

In a 30-60-90 triangle, for Exercises 35 and 36, the hypotenuse measures 10 in. [Lesson 5.5]

35. Find the measure of the shorter leg. 5 in.

36. Find the measure of the longer leg. 8.66 in.

37. Find the apothem of a regular hexagon with each side measuring 18 in. 15.59 in.

Look Beyond

38. **Coordinate Geometry** Point (x, y) is in a plane while point (x, y, z) is in space. Find the volume of the prism shown. 120 units^3

39. **Coordinate Geometry** Find the length of the diagonal joining (4, 0, 0) and (0, 5, 6). 8.77 units

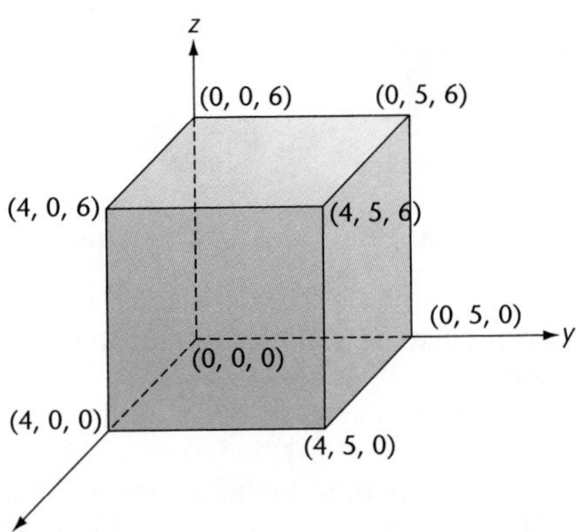

Look Beyond

Extension. Exercises 38 and 39 should remind students of the three-dimensional coordinates they studied in Chapter 6. You may want students to do several more examples and then summarize how to find the volume of a prism whose vertices are coordinates.

Technology Master

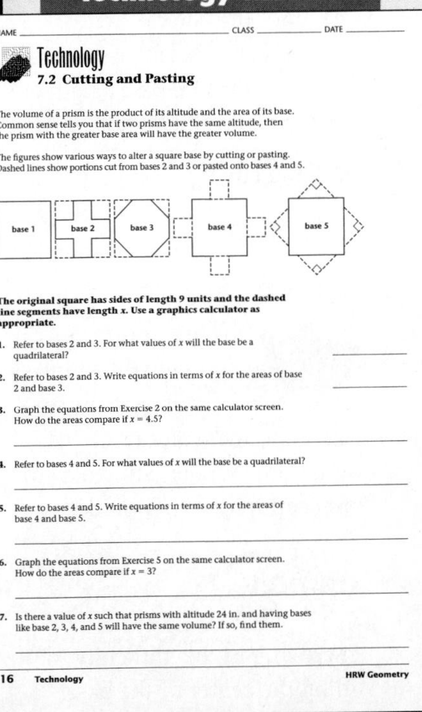
NAME ______ CLASS ______ DATE ______

Technology

7.2 Cutting and Pasting

The volume of a prism is the product of its altitude and the area of its base. Common sense tells you that if two prisms have the same altitude, then the prism with the greater base area will have the greater volume.

The figures show various ways to alter a square base by cutting or pasting. Dashed lines show portions cut from bases 2 and 3 or pasted onto bases 4 and 5.

The original square has sides of length 9 units and the dashed line segments have length x. Use a graphics calculator as appropriate.

1. Refer to bases 2 and 3. For what values of x will the base be a quadrilateral? ______
2. Refer to bases 2 and 3. Write equations in terms of x for the areas of base 2 and base 3. ______
3. Graph the equations from Exercise 2 on the same calculator screen. How do the areas compare if $x = 4.5$? ______
4. Refer to bases 4 and 5. For what values of x will the base be a quadrilateral? ______
5. Refer to bases 4 and 5. Write equations in terms of x for the areas of base 4 and base 5. ______
6. Graph the equations from Exercise 5 on the same calculator screen. How do the areas compare if $x = 3$? ______
7. Is there a value of x such that prisms with altitude 24 in. and having bases like base 2, 3, 4, and 5 will have the same volume? If so, find them. ______

16 Technology HRW Geometry

PREPARE

Objectives

- Define and use the formula for finding the surface area of a pyramid.
- Define and use the formula for finding the volume of a pyramid.

- Practice Master 7.3
- Enrichment Master 7.3
- Technology Master 7.3
- Lesson Activity Master 7.3
- Quiz 7.3
- Spanish Resources 7.3

Use Transparency 57

Assessing Prior Knowledge

1. Find the area of a triangle with height 6 and base 3.

   [9 units^2]

2. Find the height of an isosceles triangle with sides 8, 8, and 6. [$\sqrt{55}$]

3. Find the lateral area of a right triangular prism with an equilateral base length of 5 cm and a height of 10 cm.

   [150 cm^2]

TEACH

This lesson gives students formulas for finding the surface area and volume of pyramids. Encourage students to give examples of pyramids.

CRITICAL *Thinking*

There might be difficulty defining the center of the base.

LESSON 7.3 Surface Area and Volume of Pyramids

Why *Across cultures, the pyramid has an enduring allure. From ancient Egyptian tombs to the Trans-America Tower in San Francisco, the pyramid shape has strength and simple beauty.*

A **pyramid** consists of a **base**, which is a polygon, and a number of **lateral faces**, which are triangles. The lateral faces meet at a point called the **vertex** of the pyramid. The **altitude** of a pyramid is the segment from the vertex perpendicular to the plane of the base. The height of a pyramid is the length of its altitude.

A **regular** pyramid is a pyramid whose base is a regular polygon. Regular pyramids may be **right** or **oblique**. In a right pyramid the altitude intersects the base at its center, which is the point equidistant from the vertices. In an oblique pyramid the altitude intersects the plane of the base at some point other than the center.

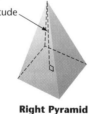

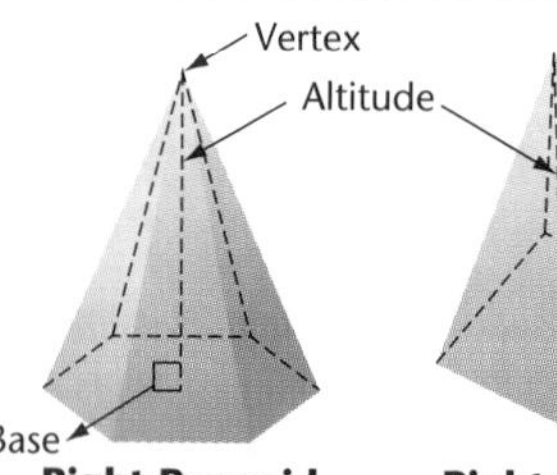

Right Pyramid **Right Pyramid**

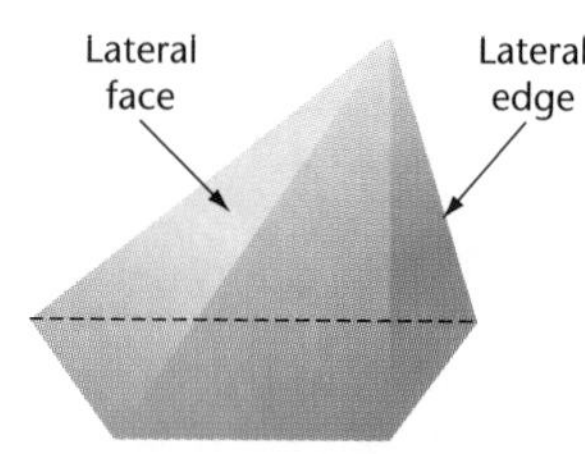

Oblique Pyramid

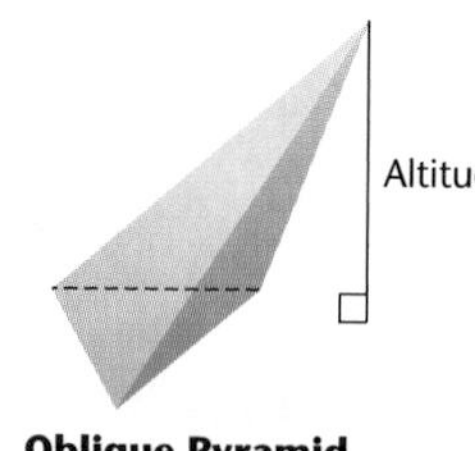

Oblique Pyramid

In a right pyramid, all the faces are congruent triangles. In an oblique pyramid, the faces are not congruent.

CRITICAL *Thinking*

What difficulty is there in defining a right pyramid whose base is not a regular polygon?

Using Models Use modeling clay to help students visualize how prisms can be cut into pyramids. Form the clay into a right prism. Slice it with a cut parallel to a base to form another prism. Slice it along a diagonal to form a pyramid. Include prisms with bases that are triangles, rectangles, pentagons, hexagons, or octagons. Have students identify the faces, edges, and base of each pyramid.

The Surface Area of a Pyramid

A pyramid can be "unfolded" and analyzed as a net as shown. The slant height l of a pyramid is the length of the altitudes of the lateral faces (the triangles). The figure shows a right regular pyramid with a square base. To find the surface area of the pyramid, add the lateral area (the area of the faces) to the area of the base.

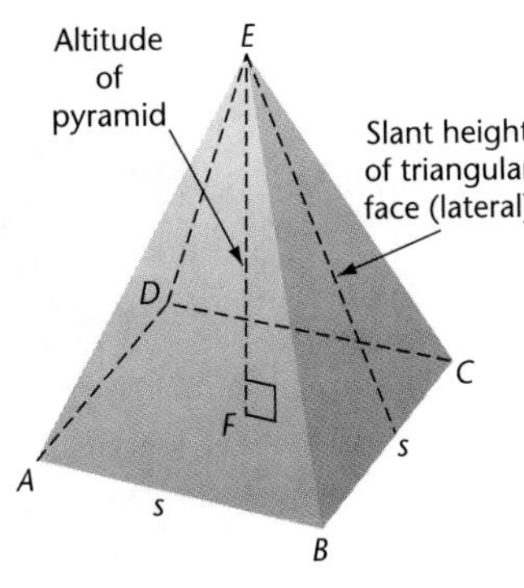

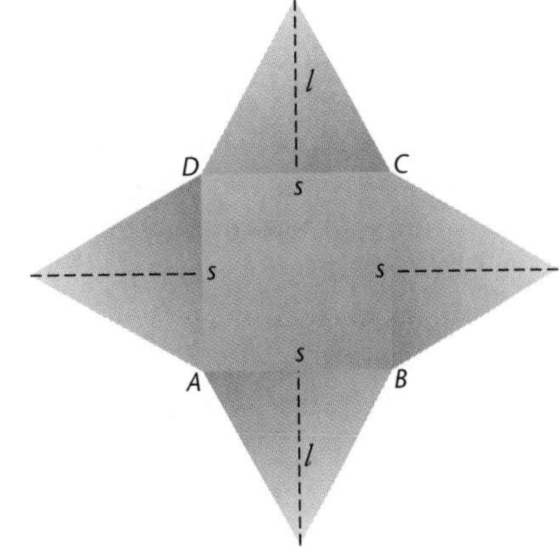

Surface Area = Lateral Area + Area of the Base

$$S = L + B$$

$$S = 4(\tfrac{1}{2}sl) + s^2 = \tfrac{1}{2}l(4s) + s^2$$

Since $4s$ is the perimeter p of the base,

$$S = \tfrac{1}{2}lp + s^2.$$

For any right regular pyramid, the lateral area is $\frac{1}{2}lp$. The total surface area is found by adding the base area.

LATERAL AREA OF A RIGHT REGULAR PYRAMID

The lateral area L of a right regular pyramid with slant height l and perimeter p of a base is

$$L = \tfrac{1}{2}lp$$

7.3.1

Note: For pyramids other than right regular pyramids, the areas of the triangular faces must be calculated individually and then added together.

SURFACE AREA OF A PYRAMID

The surface area S of a pyramid with lateral area L and area of a base B is

$$S = L + B$$

7.3.2

TEACHING tip

Emphasize the difference between the height of the pyramid and the slant height of a triangular face. The height of the pyramid is used to find its volume, and the slant height is used to find the lateral. The formula for the lateral area of a pyramid applies to right pyramids because the slant height is the same for each triangular face.

Use Transparency 58

Math Connection Algebra

It would be helpful to review the Commutative Property of Multiplication and the Substitution Principle at this time. They are used to simplify the expression for lateral area of a pyramid. Point out that formulas are created to help shorten steps in problem solving.

Chemistry Tetrahedra, pyramids with all triangular faces, are found in carbon-based molecules like methane and carbon tetrachloride. Ask students to use straws to construct models of molecules based on pyramids that they find in chemistry books. Ask students if they can think of a reason for this shape. [The arrangement makes the atoms attached to the carbon atom be as far apart as possible, where each atom is the same distance from the carbon center.]

Alternate Example 1

The highway department of a city stores its salt for use on icy roads in a pyramid-shaped structure. If the base is a regular 20-gon 14 ft on a side, and the slant height is 27 ft, find the lateral area of the pyramid. If it costs $5.95 per square foot to recover the sides of the pyramid with tile, how much does it cost to cover the sides?

[LA = 3780 ft^2 ; $22,491]

Exploration Notes

Students may not believe the outcome of the exploration with only one prism. Use as many different models of prisms as necessary to make them believe the conjecture. Extend the conjecture to include nonright prisms.

Cooperative Learning

Use the Exploration as a cooperative group activity. Provide each group with the required materials. Have each group post their conjectures for Step 3 on the chalkboard or on sheets of butcher paper. Conclude by having a class discussion based on the various results.

ONGOING ASSESSMENT

3. $V = \frac{1}{3}Bh$

EXAMPLE

Construction The roof of a gazebo is a right square pyramid. The square base of the pyramid is 12 ft on each side and the slant height is 16 ft. Find the area of the roof. If roofing material costs $3.50 per sq ft, how much will it cost to cover the roof.

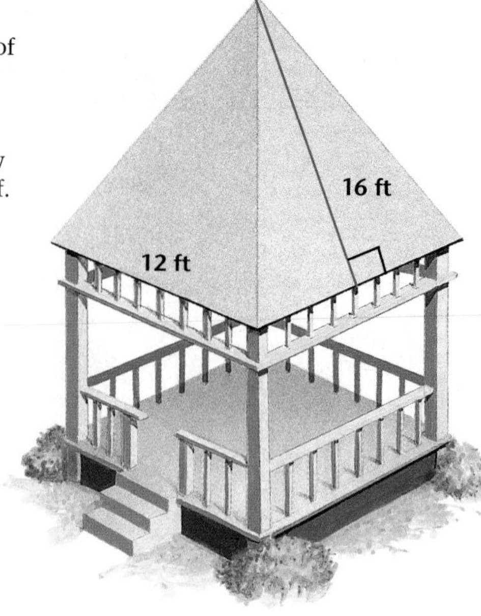

Solution ➤

The area of the roof is the lateral area L of the pyramid.

$$L = \frac{1}{2}lp$$
$$= \frac{1}{2}16(4 \times 12)$$
$$= 384 \text{ sq ft}$$

384 sq ft × $3.50 = $1,344.00 ❖

The Volume of a Pyramid

To find the volume of a pyramid, compare the pyramid with a prism that has the same base and height as the pyramid.

Exploration *Pyramids and Prisms*

You will need

Stiff construction paper
Scissors, tape, ruler
Dry cereal or packing material

1. Construct a right square prism and a right square pyramid with the same base and the same height. Draw each of the nets on construction paper and tape the tabs in place. Seal the necessary edges with tape.
2. Fill the pyramid with dry cereal or packing material. Pour the contents into the right prism. Repeat as many times as necessary to fill the prism completely. How many times did you have to fill the pyramid in order to fill the prism?
3. Make a conjecture about the relationship of the volume of a pyramid to the volume of a prism with the same base and height. Express your conjecture as a formula for the volume of a pyramid. ❖

ENRICHMENT Have students use straws to design and construct various polyhedra, like an octahedron from pyramids. Have them describe how to find the area and volume.

INCLUSION strategies

Hands-On Strategies Some students may be interested in exploring the various polyhedra by taking cross sections of a prism, including ones that form pyramids. Have them use modeling clay and a knife to create these models. Ask them to diagram their results and describe the polygon that is the cross section.

The relationship can be studied mathematically by dividing a prism into three pyramids that have the same volume. Although the pyramids are not all congruent to one another, it can be shown that, in pairs, they have equal volumes.

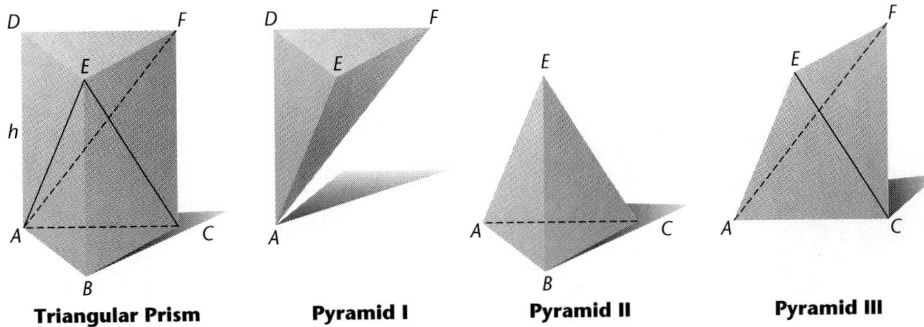

Triangular Prism **Pyramid I** **Pyramid II** **Pyramid III**

(Each pyramid is named by its vertex, followed by the polygon that forms its base.)

Consider the two following ways of pairing the pyramids.

A. I Pyramid *ADEF* has base *DEF*, altitude *h*.

II Pyramid *EABC* has base *ABC*, altitude *h*.

Since the bases are congruent and the altitudes are equal, these two pyramids have equal volumes.

B. I Pyramid *EABC* has base *EBC*.

II Pyramid *AEFC* has base *EFC*.

These two pyramids have congruent bases, since diagonal $\overline{EC}$ divides rectangle *BEFC* into two congruent triangles. The vertices of the pyramids are the same, as are the altitudes. Therefore, the two pyramids also have equal volumes.

By transitivity, all three pyramids have the same volume, and each must be one-third of the original prism. This suggests the following formula:

VOLUME OF A PYRAMID

The volume V of a pyramid with area of its base B and altitude h is

$$V = \frac{1}{3}Bh$$

7.3.3

Does the proof above work for all pyramids? If not, what is needed to make it a completely general proof?

RETEACHING the lesson

Cooperative Learning Have cooperative groups construct models of pyramids from straws and use the models to summarize the lesson for each other. Ask them to create sample questions to be used as a quiz for the whole class. Collate the questions and use them for review.

Students should be reminded that, by Cavalieri's principle, two pyramids with congruent bases and equal heights will have equal volume. Also point out to them that in Part A of the proof, face *ABC* is identified as the base of Pyramid II (*EABC*), but in Part B face *EBC* is the base of the same pyramid.

CRITICAL Thinking

Triangular Pyramids. Notice that the height *h* of the prism can be as desired, and the bases can be any triangular shape. Thus the base and height of Pyramid II, whose base is the same as a base of the prism, can be a triangular pyramid with any desired triangular base and any desired height *h*. (Point out that the vertex E can be moved without changing the height and volume.)

All other pyramids. A prism with a base of *n* sides can be cut by planes into $(n-2)$ triangular prisms. In the case of a pentagonal prism, for example, the intersection of the cutting planes with the lower base might be as shown below.

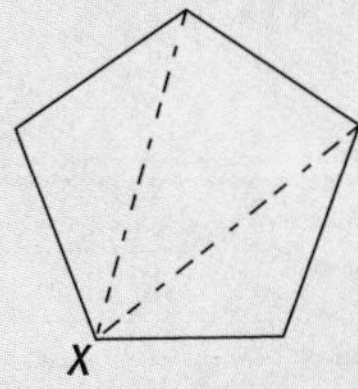

Each of the triangular prisms can then be sectioned so that the vertices of all "Pyramid II's" meet at vertex of the upper face of the prism corresponding to *X*. These pyramids form a single pyramid with height *h* and the given pentagonal base.

Assess

Selected Answers

Odd-numbered Exercises 9–25

Assignment Guide

Core 1–13, 16–20, 23–26

Core Plus 1–27

Technology

Students should use scientific calculators to perform the calculations for the exercises. Be sure to establish rules for rounding. Some students may wish to create calculator programs for finding surface area and volume.

Error Analysis

Students may confuse the formulas for surface area and volume. Encourage students to double-check all answers and be sure the answer they are providing is what was asked for in the exercise.

Practice Master

NAME ____________ CLASS ________ DATE ________

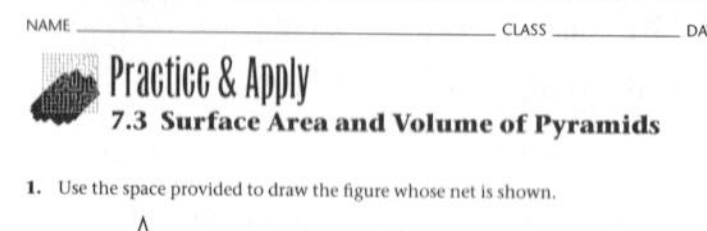

Practice & Apply

7.3 Surface Area and Volume of Pyramids

1. Use the space provided to draw the figure whose net is shown.

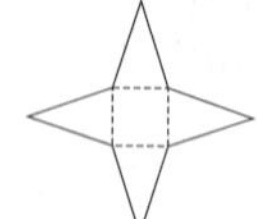

In Exercises 2–5, find the surface area or the volume for the given pyramids. As needed, round answers to the nearest tenth.

2. A right triangular pyramid: the side of the equilateral triangular base is 8 cm, the altitude of the pyramid is 12 cm, and its slant height is $4\sqrt{10}$ cm. Find its surface area.

3. A right rectangular pyramid: the rectangular base is 9 cm by 12 cm, and the altitude of the pyramid is 12 cm. Find its volume.

4. A right hexagonal pyramid: the apothem of the regular hexagonal base is 9 cm, the altitude of the pyramid is 12 cm, and its slant height is 15 cm. Find its surface area.

5. A right square pyramid: the side measures 10 cm, the altitude is 12 cm, and its slant height is 13 cm. Find its volume.

6. A pyramid has a right triangle as its base. The legs are 10 cm and 20 cm, and the volume of the pyramid is 600 cm^3. Find the altitude of the pyramid.

HRW Geometry — Practice & Apply 3

APPLICATION

Cultural Connection: Africa What is the weight of the pyramid of Khufu?

The pyramid of Khufu is constructed of stones with weight of approximately 167 pounds per cubic foot. The dimensions of the pyramid are as follows:

Base side = 775.75 ft

Height = 481 ft

What is the volume of the pyramid of Khufu in cubic feet? What is the weight of the pyramid of Khufu in pounds? in tons? (1 ton = 2000 lb)

The volume of the pyramid is

$$V = \frac{1}{3}Bh = \frac{1}{3}(775.75)^2(481) \approx 96{,}486{,}686 \text{ cubic feet.}$$

The weight in pounds is

$$96{,}486{,}686 \cdot 167 \approx 16{,}113{,}000{,}000 \text{ lb, which is } 8{,}056{,}638 \text{ tons.}$$ ❖

Exercises & Problems

Communicate

1. Explain how to find the total surface area of a right pyramid.
2. Explain how to find the volume of a pyramid.
3. In a pyramid, which is longer in measure, its altitude (height) or its slant height? Why?
4. Explain the relationship between the volume of a pyramid and the volume of a rectangular prism.
5. Where does the altitude of a regular right pyramid intersect the base?

Practice & Apply

Draw the net for each pyramid below.

6. a square pyramid
7. an equilateral triangular pyramid
8. a regular hexagonal pyramid
9. a regular pentagonal pyramid

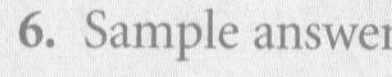

6. Sample answer:

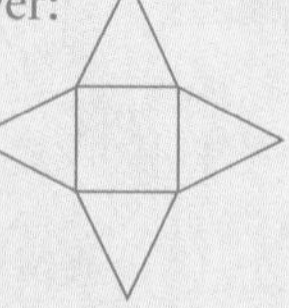

7. Sample answer:

8. Sample answer:

9. Sample answer:

Find the total surface area of each right pyramid.

10.

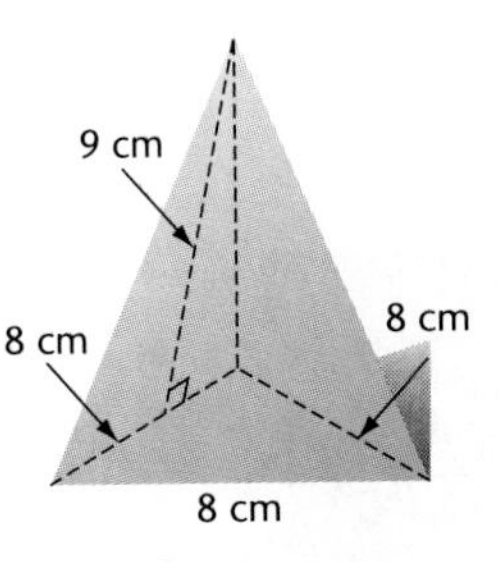

140 cm²

11.

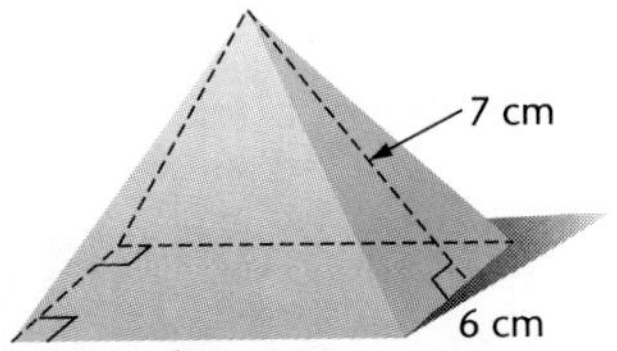

120 cm²

12.

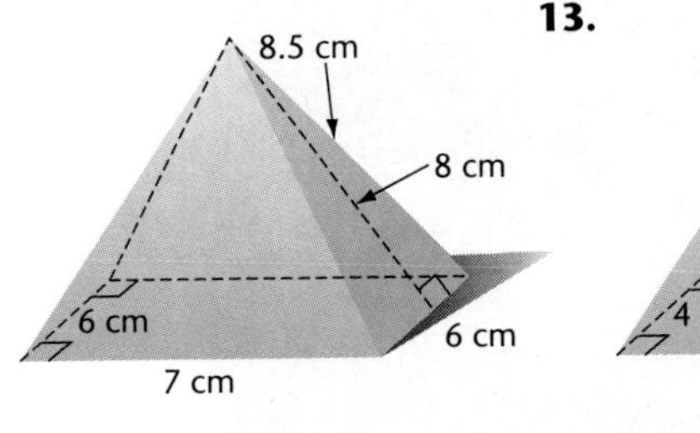

144.32 cm²

13.

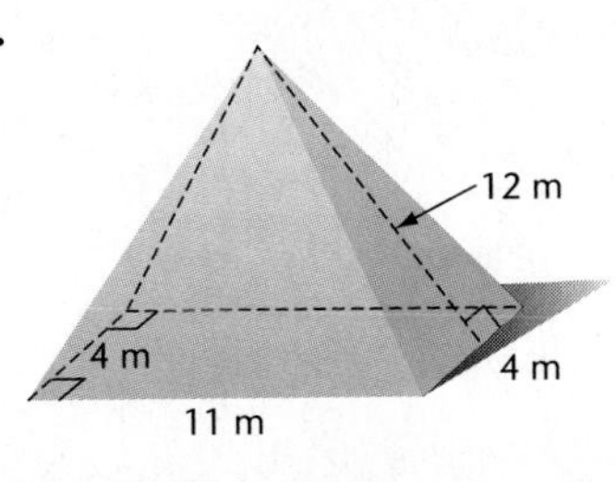

211.4 m²

For Exercises 14–15, find the volume. Draw the figure and label the parts.

14. a right rectangular pyramid with base 5 m by 7 m and height 11 m 128.33 m³

15. a right triangular pyramid whose base dimensions are 5, 12, and 13 and whose altitude is 10 100 units³

Find the volume for each right pyramid.

16.

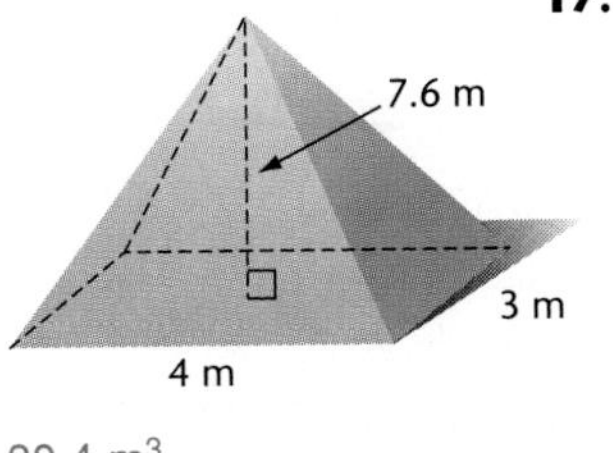

30.4 m³

17.

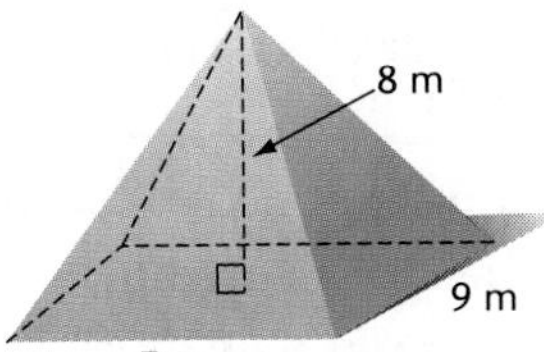

168 m³

18.

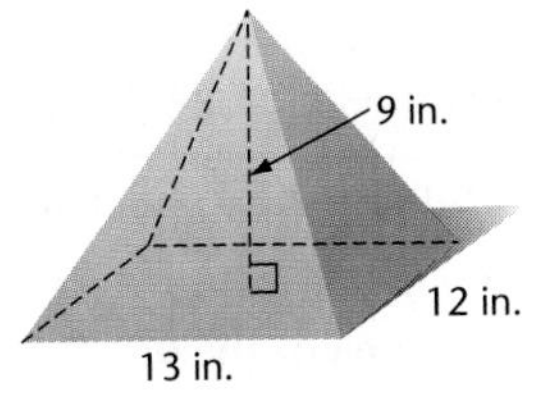

468 in.³

19. Find the surface area and volume of a right square pyramid. The base measures 18 cm on each side. The slant height measures 15 cm.
SA = 864 cm²; V = 1296 cm³

20. Find the surface area and volume of a right square pyramid. The base measures 1 ft on each side. The slant height measures 10 in. SA = 384 in.²; V = 384 in.³

alternative ASSESSMENT

Portfolio Assessment

Ask students to draw various pyramids with the same height and with rectangles for bases. Include several rectangles with the same area. Have students label the pyramids, find the volumes, and describe which pyramids have the same volume.

Technology Master

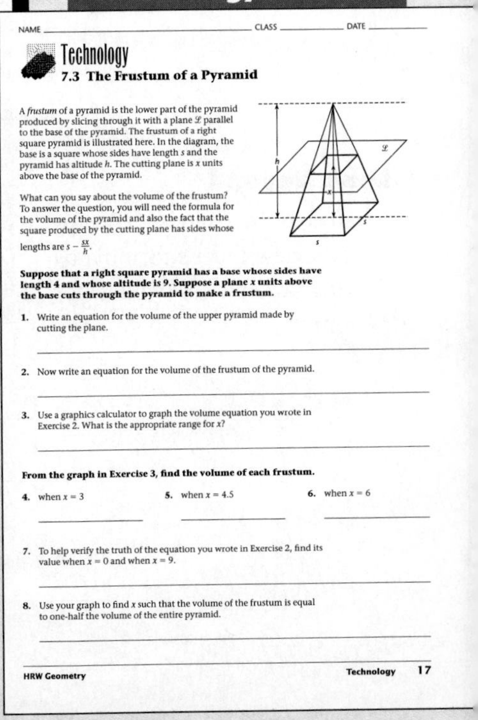

NAME ______ CLASS ______ DATE ______

Technology
7.3 The Frustum of a Pyramid

A *frustum* of a pyramid is the lower part of the pyramid produced by slicing through it with a plane $\mathscr{P}$ parallel to the base of the pyramid. The frustum of a right square pyramid is illustrated here. In the diagram, the base is a square whose sides have length s and the pyramid has altitude h. The cutting plane is x units above the base of the pyramid.

What can you say about the volume of the frustum? To answer the question, you will need the formula for the volume of the pyramid and also the fact that the square produced by the cutting plane has sides whose lengths are $s - \frac{sx}{h}$.

Suppose that a right square pyramid has a base whose sides have length 4 and whose altitude is 9. Suppose a plane x units above the base cuts through the pyramid to make a frustum.

1. Write an equation for the volume of the upper pyramid made by cutting the plane.
2. Now write an equation for the volume of the frustum of the pyramid.
3. Use a graphics calculator to graph the volume equation you wrote in Exercise 2. What is the appropriate range for x?

From the graph in Exercise 3, find the volume of each frustum.

4. when $x = 3$
5. when $x = 4.5$
6. when $x = 6$

7. To help verify the truth of the equation you wrote in Exercise 2, find its value when $x = 0$ and when $x = 9$.
8. Use your graph to find x such that the volume of the frustum is equal to one-half the volume of the entire pyramid.

HRW Geometry — Technology 17

21. Construction An architect is designing a building to resemble the entrance of the Louvre in France (shown below). If the height of the structure is 15 meters, and the area of the base is 225 square meters, what will be the volume of the building? 1125 m^3

22. A right pyramid has an isosceles right triangle as its base. The legs are 20 cm and the volume of the pyramid is 3000 cubic cm. Find the altitude of the pyramid. 45 cm

Look Back

23. Angles Y and Z are complementary. Angle Y measures 51°. What is the measure of angle Z? **[Lesson 1.5]** 39°

A quadrilateral has coordinates (0, 0), (5, 0), (2, 6), and (7, 6).

24. Find the perimeter of the quadrilateral. **[Lesson 5.6]** 22.65 units

25. Find the area of the quadrilateral. **[Lesson 5.6]** 30 $units^2$

26. Give the coordinates of a quadrilateral that has the same area as the one above, but a different perimeter. **[Lesson 5.6]**
Sample answer: (0, 0), (5, 0), (0, 6), (5, 6)

Look Beyond

Extension. In Exercise 27, students review Euler's formula by making a table of values for pyramids.

Look Beyond

27. Consider pyramids with bases of 3, 4, 5, 6, 7, 8, and n sides. Count the vertices, faces, and edges. Make a table. Does the generalization for n satisfy Euler's formula?

27.

Base Sides	Vertices	Faces	Edges
3	4	4	6
4	5	5	8
5	6	6	10
6	7	7	12
7	8	8	14
8	9	9	16
n	n+1	n+1	2n

$V + F - E = 2$; $(n + 1) + (n + 1) - 2n = 2$; Yes

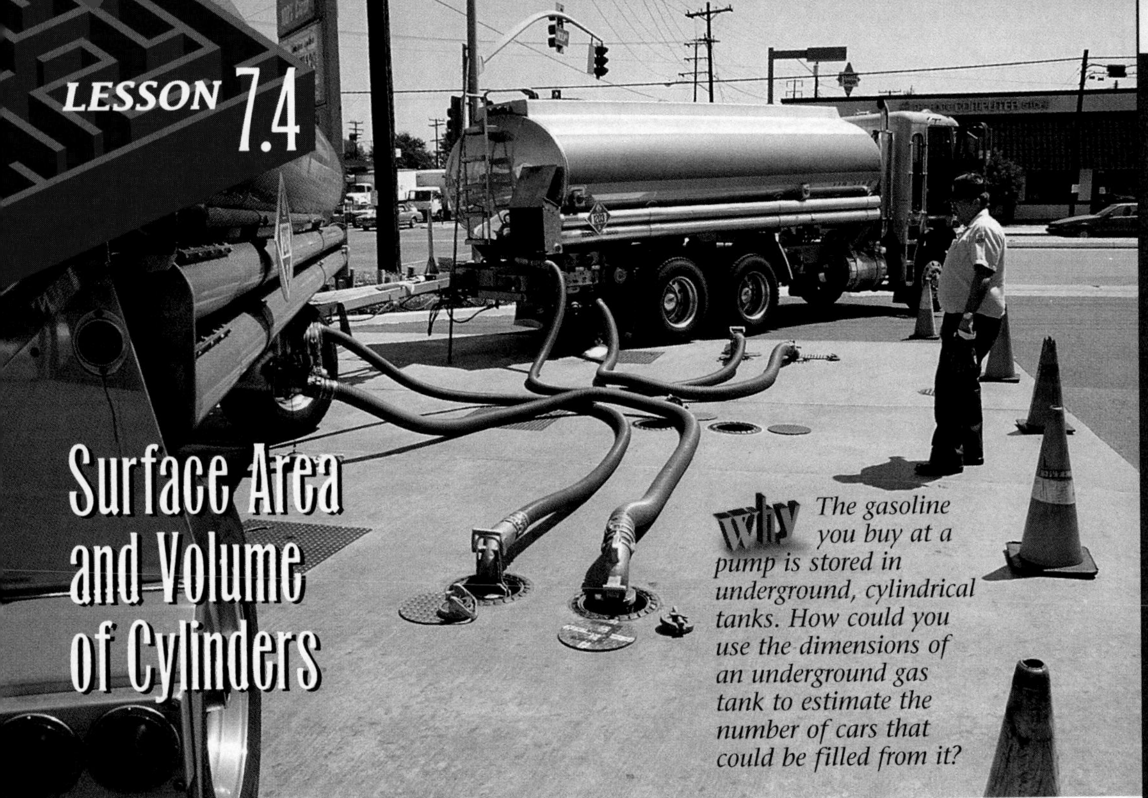

LESSON 7.4

Surface Area and Volume of Cylinders

Why *The gasoline you buy at a pump is stored in underground, cylindrical tanks. How could you use the dimensions of an underground gas tank to estimate the number of cars that could be filled from it?*

A cylinder is a three-dimensional figure with two congruent circular **bases** in parallel planes, connected by a curved **lateral surface**.

The **altitude** of a cylinder is a segment that is perpendicular to the planes of the bases, from the plane of one base to the plane of the other. The length of the altitude is the height of the cylinder.

The **axis** of a cylinder is the segment joining the centers of the two bases.

Axis
Base
Altitude
Lateral surface

If the axis of a cylinder is perpendicular to the bases, then the cylinder is a **right cylinder**. If not, it is an **oblique cylinder**.

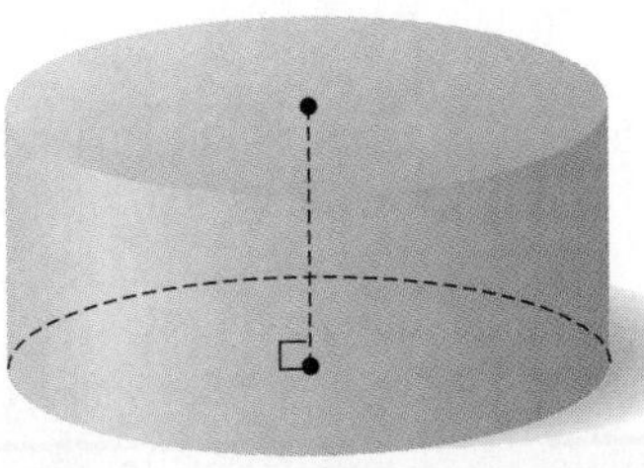

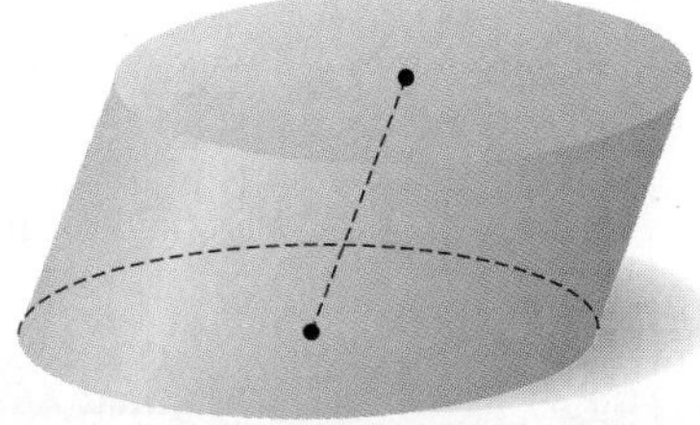

ALTERNATIVE teaching strategy

Using Models Students can develop an understanding of lateral and surface area of a cylinder by cutting out models that can be assembled into cylinders. Have them cut out a rectangle and two circles for the bases whose circumference equals a side of the rectangle. Then tape together the three pieces. For an oblique cylinder, use a nonright parallelogram and circles whose circumference equals a side of the cylinder. The lateral area equals the rectangle (or parallelogram) and the surface area equals the sum of the three pieces. Ask them to think about whether the choice of side of the rectangle or parallelogram may affect the volume.

PREPARE

Objectives

- Define and use the formula for finding the surface area of a cylinder.
- Define and use the formula for finding the volume of a cylinder.

RESOURCES

- Practice Master 7.4
- Enrichment Master 7.4
- Technology Master 7.4
- Lesson Activity Master 7.4
- Quiz 7.4
- Spanish Resources 7.4

Use Transparency 59

Assessing Prior Knowledge

1. Find the circumference and area of a circle with radius 6. **[37.7 units, 113.1 units^2]**
2. Find the surface area and volume of a right rectangular prism whose base is a square with side length 6 and whose height is 10. **[312 units^2; 360 units^3]**

TEACH

Finding the surface area and volume of a cylinder is a natural extension of finding the surface area and volume of a prism. Help students make the connection between polygons and circles by describing circles as polygons with an infinite number of sides.

TEACHING tip

Emphasize the similarity in the lateral area formulas for a prism and cylinder. Since the circumference of a circle is its perimeter, the formula is the perimeter of the base times height.

Math Connection Algebra

Verifying the formula for the lateral and surface area of a cylinder shows a connection between algebra and geometry. Since the net for a right cylinder includes a rectangle, one pair of sides of the rectangle has to equal the circumference of the circular base. The other pair of sides is the height of the cylinder.

Cooperative Learning

Use the Exploration as a cooperative group activity. Have students post their theorems on the board or on butcher paper. Have each group explain the reasoning behind their theorems.

Cylinders and Prisms

As the number of sides of a regular polygon increases, the figure becomes more and more like a circle.

Similarly, as the number of sides of a regular polygonal prism increases, the figure becomes more and more like a cylinder.

The illustrations suggest the formulas for the surface areas and volumes of prisms and cylinders are similar.

The Surface Area of a Cylinder

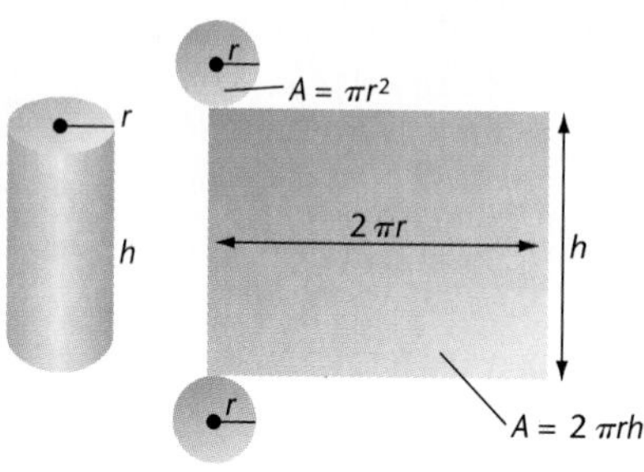

The surface area of a right cylinder is found by a method similar to that used for a right prism. The net for a right cylinder includes the two circular bases and the flattened lateral surface, which becomes a rectangle. The length of this rectangle is the circumference of the base. The width of the rectangle is the height of the cylinder.

The lateral area of the cylinder equals $2\pi rh$, and the area of each base is πr^2. Thus,

$$S \text{ (surface area)} = L \text{ (lateral area)} + 2B \text{ (area of bases)} = 2\pi rh + 2\pi r^2$$

LATERAL AREA OF A RIGHT CYLINDER

The lateral area L of a right cylinder with radius r and height h is

$$L = 2\pi rh$$

7.4.1

SURFACE AREA OF A RIGHT CYLINDER

The surface area S of a right cylinder with radius r, height h, area of a base B, and lateral area L is

$$S = L + 2B \quad \text{or} \quad S = 2\pi rh + 2\pi r^2$$

7.4.2

Science Scientists use test tubes, beakers, and graduated cylinders to find the volume of liquids. Borrow some graduated cylinders from the science department and have students measure them and calculate their surface area and volume. Determine how well the measurements compare to ones determined by the markings on the cylinders. Have students use the cylinders to do a surface-area-and-volume experiment similar to Exploration 1 in Lesson 7.1.

CRITICAL *Thinking*

Simplify the formula $S = 2\pi rh + 2\pi r^2$ by factoring. Is the new formula easier or harder to compute? Is it easier or harder to remember?

APPLICATION

Find the surface area of a right circular cylinder with a radius of 4 cm and a height of 7 cm.

Solution ➤

The surface area is

$S = 2\pi rh + 2\pi r^2$, or

$S = 2\pi(4)(7) + 2\pi(4)(4) \approx 276.46 \text{ cm}^3$ ❖

The Volume of a Cylinder

In the following exploration, recall the method you used to find the area of a circle.

Exploration Analyzing the Volume of a Cylinder

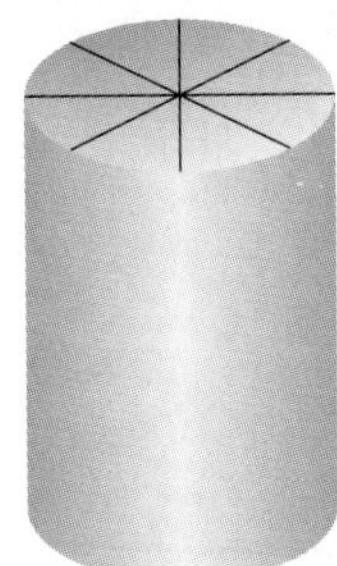

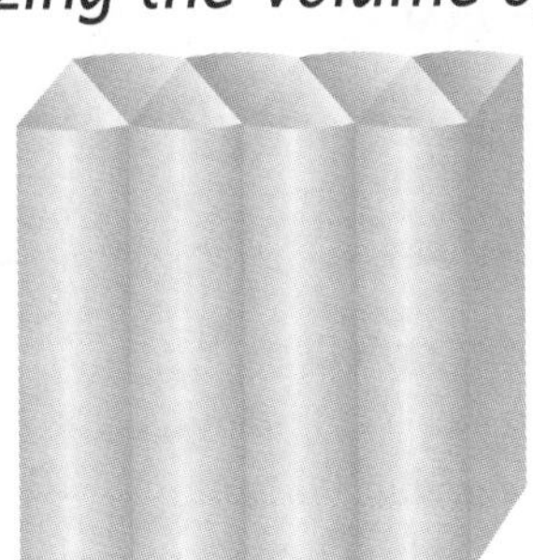

You will need

No special tools

The formula for the area of a circle was found by cutting a large number of sections and fitting them together to approach a rectangular shape. The same idea can be used with a cylinder. Use the figure to derive the formula for the volume of a cylinder.

1. What geometric solid does the cylinder approximate when the sections are organized as they are above?
2. Using the length, width, and height of the figure above, write a formula for the volume of a cylinder.
3. Summarize your exploration by writing a theorem for the volume of a right cylinder. ❖

CRITICAL *Thinking*

Is the formula for the volume of an oblique cylinder the same as or different from the formula for a right cylinder? What principle justifies your answer? Illustrate your answer with sketches.

ENRICHMENT Cylindrical food cans are often delivered to a grocery store packed tightly in boxes that are rectangular prisms. Ask students to find the ratio of the volume of six cans to the volume of the box containing them.

[π: 4]

INCLUSION strategies

English Language Development Some students may have difficulty understanding all the terminology in this lesson. Have them draw and label sample cylinders with each of the following parts: bases, radius, altitude, axis, and lateral surface. Include oblique cylinders. Extend the exercise by having them draw and label a net for each cylinder.

CRITICAL *Thinking*

$SA = 2\pi r(h + r)$; This is easier to compute, and some students may find it easier to memorize. But for the student who understands the unfactored form, that form should be easier to recall because of its obvious logic.

Exploration Notes

Students may discuss the exploration in groups. Students should section paper circles representing the bases of cylinders until they are convinced that the area of the base of a cylinder approaches the area of a parallelogram. Have them check their algebraic derivation for the volume formula for a right cylinder.

Ongoing ASSESSMENT

3. **$V = \pi r^2 h$, where r is the radius and h is the height of the cylinder.**

CRITICAL *Thinking*

The formula for volume will be the same for right and oblique cylinders. This follows from Cavalieri's Principle.

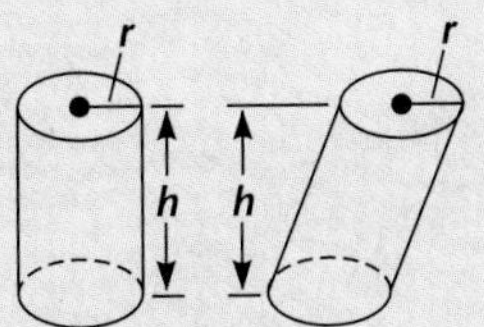

Alternate Example 2

Find the volume of a right cylindrical water tank whose radius is 12 ft and whose altitude is 30 ft. How many gallons of water will the tank hold?

[$13{,}571.7 \text{ ft}^3$; $\approx 101{,}524$ gal]

ASSESS

Selected Answers

Odd-numbered Exercises 9–31

Assignment Guide

Core 1–16, 19–23, 27–32

Core Plus 1–12, 17–32

Technology

Some students may want to use the following graphing calculator program to find the surface area and volume of a cylinder.

```
:Input "BASE RADIUS", R
:Input "HEIGHT", H
:2(3.1416)RH + 2(3.1416)R²→ S
:3.1416R²H → V
:Disp "SURFACE AREA", S
:Disp "VOLUME", V
```

Error Analysis

For Exercise 25, you may want to review the algebra necessary to work backwards from the volume of a cylinder to the dimensions used to find the volume.

6.

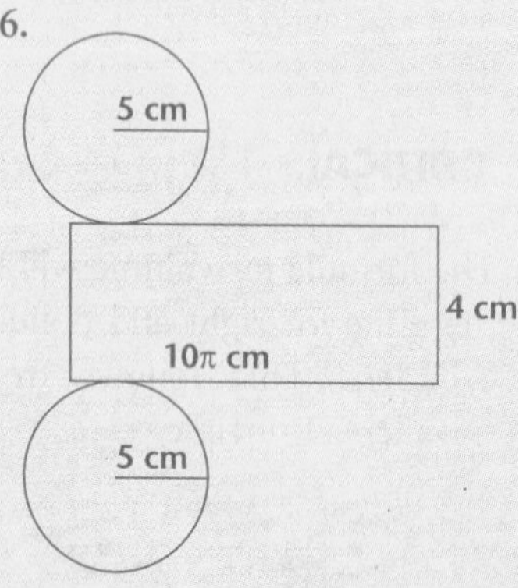

VOLUME OF A CYLINDER

The volume V of a cylinder with radius r, height h, and area of a base B, is

$V = Bh$ or $V = \pi r^2 h$

7.4.3

EXAMPLE

Gasoline Storage Find the volume of a cylindrical underground gasoline tank with a radius of 6 ft and an altitude of 25 ft. How many gallons of gasoline will the tank hold? (1 gal = 0.13368 cu ft)

Solution ➤

$$V = \pi r^2 h = \pi(6^2)(25) = 900\pi \approx 2827.4 \text{ cu ft}$$

To find the amount in gallons, divide by 0.13368.

$$V \approx \frac{2827.4}{0.13368} \approx 21{,}151 \text{ gal}$$ ❖

EXERCISES & PROBLEMS

Communicate

1. State the formula for the volume of a cylinder.
2. Explain how to find the surface area of a cylinder.
3. Explain how a cylinder is like a rectangular prism.
4. What does doubling the radius of a cylinder do to the volume?
5. What does doubling the height of a cylinder do to the surface area?

Practice & Apply

Draw a net for each of the cylinders shown. Label the known parts.

6.

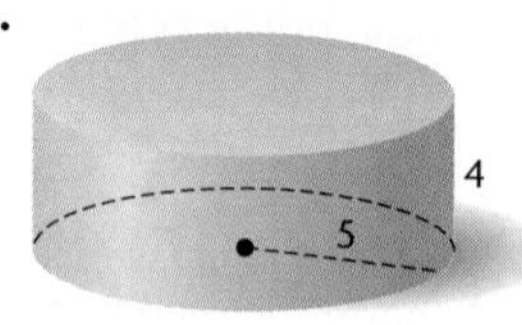

7.

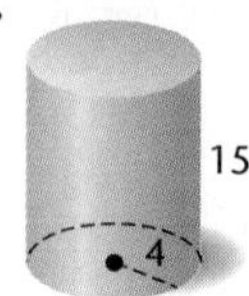

RETEACHING the lesson Bring example cylinders to class and have the students work in groups to measure the cylinders' radii and heights. Ask each group to calculate the lateral area, surface area, and volume for each of their cylinders.

7.

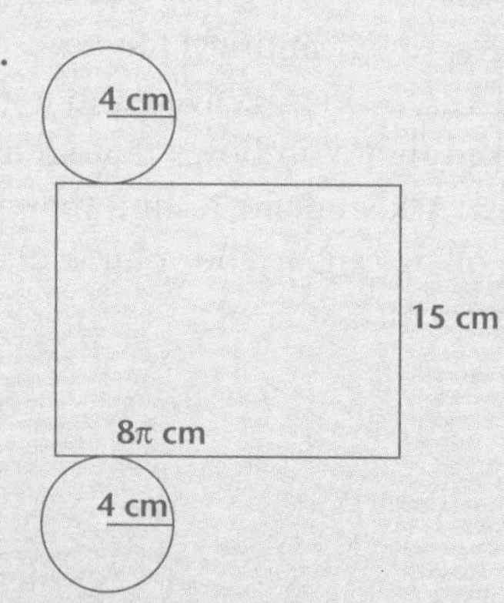

8. Find the surface area for the figure in Exercise 6. 282.74 units^2
9. Find the surface area for the figure in Exercise 7. 477.52 units^2
10. Find the volume of the figure in Exercise 6. 314.16 units^3
11. Find the volume of the figure in Exercise 7. 753.98 units^3
12. The surface area of a cylinder is 200 sq cm. The diameter and altitude are equal. Find the radius. 3.26 cm
13. Find the surface area of a right cylinder whose diameter is 10 cm and whose altitude is 20 cm. 785.4 cm^2
14. Find the surface area of a right cylinder whose radius is 10 cm and whose height is 30 cm. If the dimensions of the cylinder are doubled, what happens to the surface area? 2513.27 cm^2; SA is multiplied by 4.
15. Find the volume of a cylinder whose diameter is 5 m and whose height is 8 m. 157.08 m^3
16. If the dimensions of the cylinder in Exercise 15 are doubled, what happens to the volume? The volume is multiplied by 8.
17. If the volume of a cylinder is 360π cubic mm and its altitude is 10 mm, find the circumference of the base. 37.7 mm
18. Georgia wants to wrap a large Christmas present in a cylindrical package. She bought 1 roll of wrapping paper which contains 30 square feet of paper. If the diameter of the cylinder is 2 feet and the height is 4 feet, will she have enough wrapping paper? SA = 31.4 ft^2; No, Georgia would need over 30 ft^2.
19. Which cylinder has the larger lateral area? Which has the larger surface area? Explain your answers.

   cylinder x: radius = 3, altitude = 6   cylinder y: radius = 6, altitude = 3

20. A cylindrical glass is full of water and is to be poured into a rectangular pan. The base of the pan is 15 in. by 10 in. and the height of the pan is 3 in. The height of the cylinder is 15 in. What must the radius of the cylinder be so that the water fills the pan but does not spill? 3.09 in.

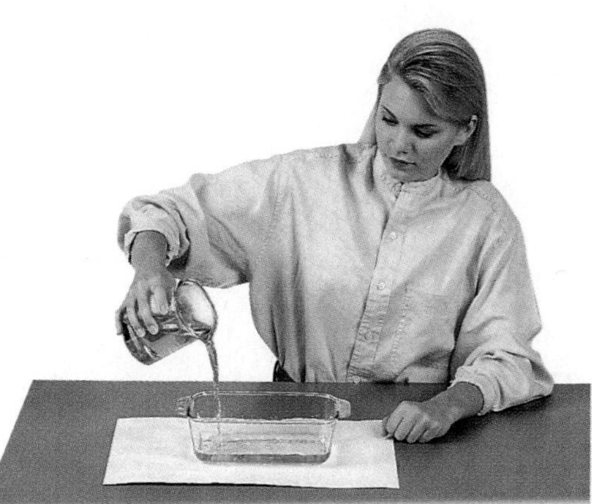

21. A straw is 25 cm long and 4 mm in diameter. How much liquid can the straw hold? 3141.59 mm^3
22. **Portfolio Activity** Find a cylindrical can of vegetables or fruit in your home. Take the measurements to find the surface area (sq in.) and volume. Check student work.
23. **Consumer Awareness** Many grocery shoppers assume that a taller can or a wider can holds more. Why is this not always true?
24. Find the volume of a semicircular cylinder formed by cutting a cylinder in half on its diameter. The diameter is 8 ft and the height of the cylinder is 10 ft. 251.33 ft^3

19. Cylinder x: LA = 113.10; SA = 165.65
Cylinder y: LA = 113.10; SA = 226.20
The lateral areas are the same. Cylinder y has greater surface area.

23. The volume of a cylinder depends on both the radius and the height of the cylinder. Both dimensions must be taken into consideration when finding volume.

alternative ASSESSMENT

Performance Assessment

Student presentations can help you assess students' understanding of the objectives of this lesson. You may wish to call upon individual students to describe the key dimensions of a cylinder. Other students can be asked to draw nets for the cylinders, calculate the lateral area, surface area, or volume. You may want one student to describe the similarities and differences between a prism and a cylinder.

Practice Master

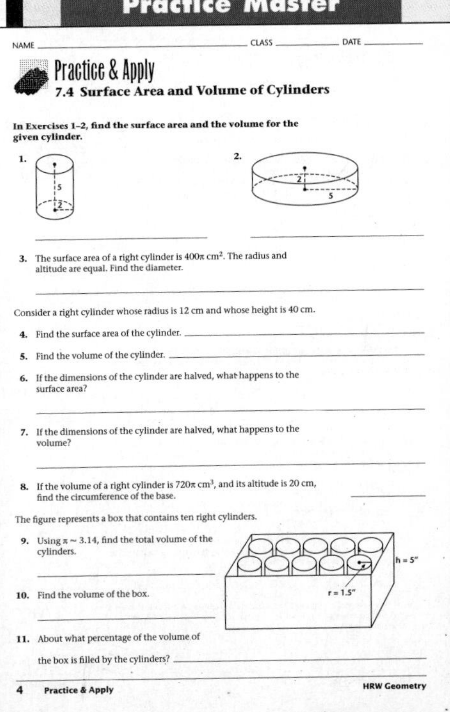

NAME ________ CLASS ________ DATE ________

Practice & Apply

7.4 Surface Area and Volume of Cylinders

In Exercises 1–2, find the surface area and the volume for the given cylinder.

1.

2.

3. The surface area of a right cylinder is 400π cm^2. The radius and altitude are equal. Find the diameter.

Consider a right cylinder whose radius is 12 cm and whose height is 40 cm.

4. Find the surface area of the cylinder. ________
5. Find the volume of the cylinder. ________
6. If the dimensions of the cylinder are halved, what happens to the surface area?
7. If the dimensions of the cylinder are halved, what happens to the volume?
8. If the volume of a right cylinder is 720π cm^3, and its altitude is 20 cm, find the circumference of the base. ________

The figure represents a box that contains ten right cylinders.

9. Using $\pi \sim 3.14$, find the total volume of the cylinders.
10. Find the volume of the box.
11. About what percentage of the volume of the box is filled by the cylinders? ________

4 Practice & Apply   HRW Geometry

25. **Maximum/Minimum** A city zoo needs a cylindrical aquarium that holds 100,000 cubic feet of water. In order to save money, the city wants to use the least amount of materials possible to build the aquarium. What is the smallest surface area the aquarium can have and still hold 100,000 cubic feet of water?
About 9500 ft^3

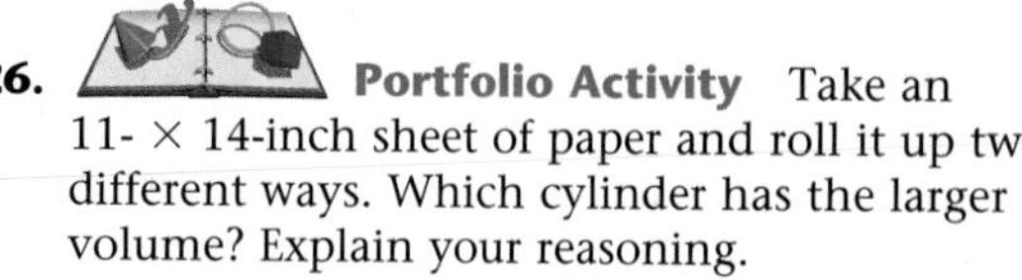

26. **Portfolio Activity** Take an 11- × 14-inch sheet of paper and roll it up two different ways. Which cylinder has the larger volume? Explain your reasoning.

Look Back

Find the surface area and volume for each figure. [Lessons 7.2, 7.3]

27.

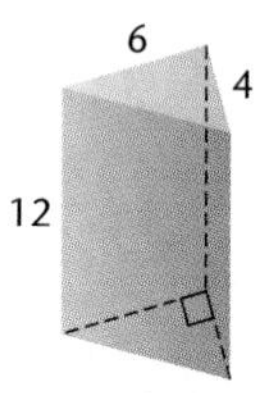

SA = 223.79 units^2; V = 144 units^3

28.

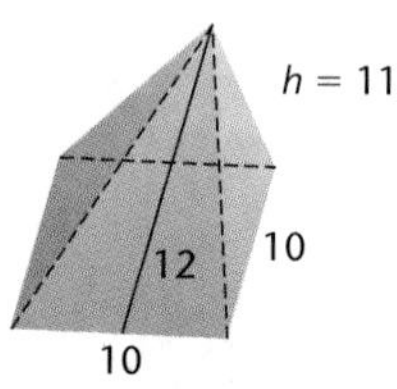

SA = 340 units^2; V = 383.33 units^3

29.

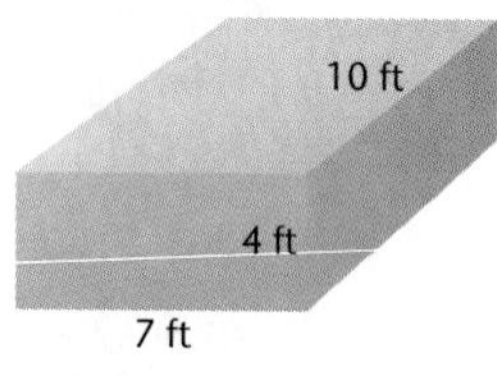

SA = 276 ft^2; V = 280 ft^3

30.

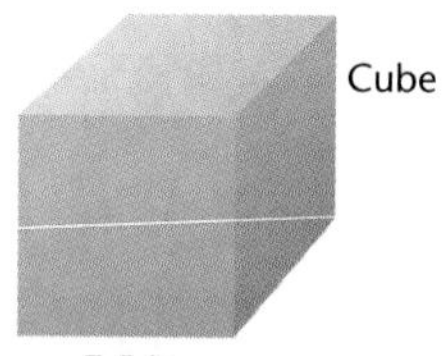

SA = 337.5 in.^{2}; V = 421.88 in.^{3}

31. State Cavalieri's principle and use a diagram to support it. **[Lesson 7.2]**

Look Beyond

32. In a guessing contest at a school fair, students are asked to guess the number of marbles in a jar. The inner dimensions of the jar are 12.4 cm in diameter and 28.6 cm tall. The diameters of the marbles average 1.55 cm. List as many different ways as you can for estimating the number of marbles in the jar. State your own estimate of the number. Your teacher has the actual number.

Look Beyond

Nonroutine problem. Exercise 32 challenges students to estimate the number of marbles in a glass cylinder. Using the dimensions given and the appropriate formulas, students may be surprised at how close their calculations are to the actual number of marbles.

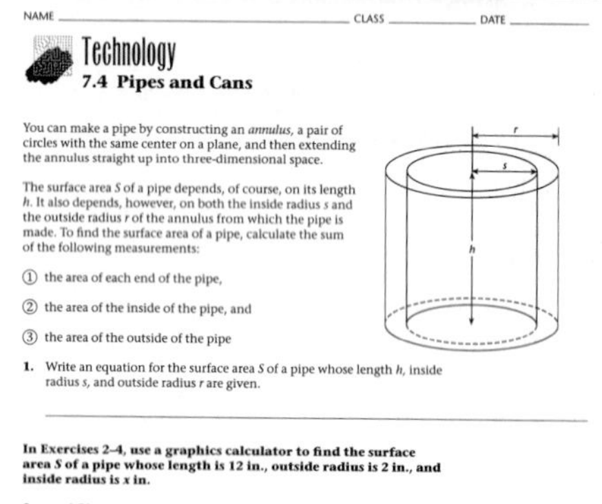

NAME ______ CLASS ______ DATE ______

Technology
7.4 Pipes and Cans

You can make a pipe by constructing an *annulus*, a pair of circles with the same center on a plane, and then extending the annulus straight up into three-dimensional space.

The surface area S of a pipe depends, of course, on its length h. It also depends, however, on both the inside radius s and the outside radius r of the annulus from which the pipe is made. To find the surface area of a pipe, calculate the sum of the following measurements:

① the area of each end of the pipe,

② the area of the inside of the pipe, and

③ the area of the outside of the pipe

1. Write an equation for the surface area S of a pipe whose length h, inside radius s, and outside radius r are given.

In Exercises 2–4, use a graphics calculator to find the surface area S of a pipe whose length is 12 in., outside radius is 2 in., and inside radius is x in.

2. $x = 1.75$ 3. $x = 1.5$ 4. $x = 1.25$

In Exercises 5–7, use a graphics calculator to find the inside radius x of a pipe whose length is 12 in., outside radius is 2 in., and surface area S is given.

5. $S = 280$ in.^{2} 6. $S = 270$ in.^{2} 7. $S = 250$ in.^{2}

8. An *open can* is a pipe whose bottom is plugged by a thin cylinder whose height is h' and whose radius is s, the inside radius of the pipe. Write an equation for the surface area of the can.

9. Use a graphics calculator to find the inside radius of a can whose height is 6 in., outside radius is 3.5 in., bottom is 0.1 in. thick and whose surface area is 335 in.^{2}

18 Technology HRW Geometry

26. Volume when base has circumference 11 in $= \pi\left(\frac{11}{2p}\right)^2(14) = 134.80$ in^3
Volume when base has circumference 14 in $= \pi\left(\frac{14}{2p}\right)^2(11) = 171.57$ in^3
The volume is larger when the base has diameter 14 in.

31. If two solids have equal heights, and if the cross-sections formed by every plane parallel to the bases of both solids have equal areas, then the two solids have the same volume. Check student drawings.

32. The actual count is 1159 marbles. Accept all reasonable estimates.

LESSON 7.5 Surface Area and Volume of Cones

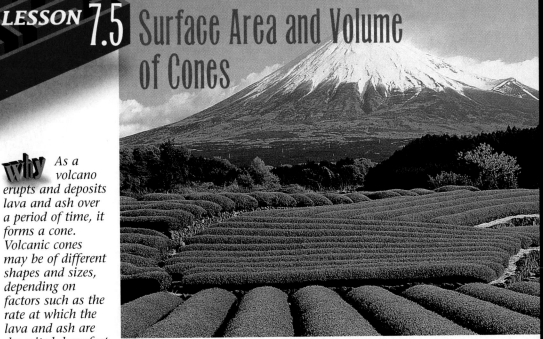

why *As a volcano erupts and deposits lava and ash over a period of time, it forms a cone. Volcanic cones may be of different shapes and sizes, depending on factors such as the rate at which the lava and ash are deposited, how fast the lava cools, etc. However, all of them can be modeled by geometric cones. By using the properties of geometric cones, geologists can study the physical structure of volcanoes.*

A circular **cone** is an object that consists of a circular **base** and a curved **lateral** surface which extends from the base to a single point called the **vertex**.

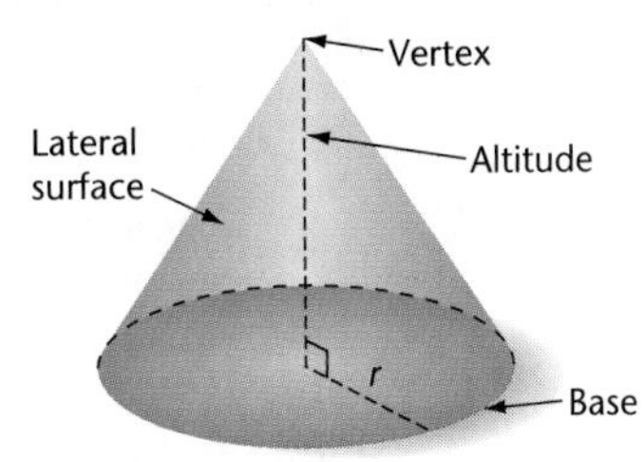

The **altitude** of a cone is the segment from the vertex perpendicular to the plane of the base. The height of the cone is the length of the altitude.

If the altitude of a cone intersects the base of the cone at its center point, the cone is a **right cone**. If not, it is an **oblique cone**.

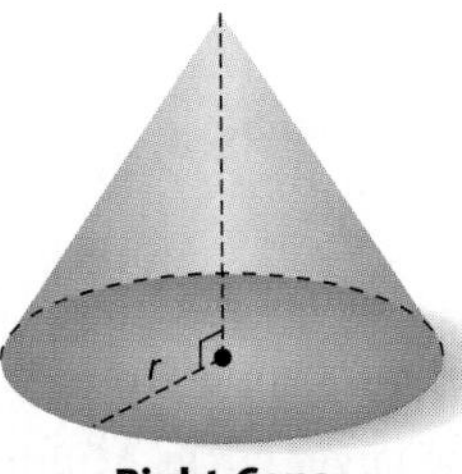

Right Cone

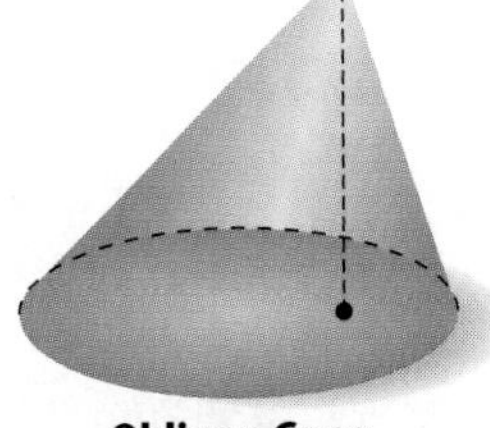
Oblique Cone

Just as a cylinder resembles a prism, a cone resembles a pyramid. As the number of sides of the base of a pyramid increases, the figure becomes more and more like a cone.

Hands-On Strategies
Some students will benefit from building cones from nets. Use nets of right circular cones to emphasize that the radius of a sector that forms a cone equals its slant height. The same net can be used to show that the area of the sector equals the lateral area of the cone. Ask students to find a net for an oblique cone and give its dimensions.

PREPARE

Objectives

- Define and use the formula for finding the surface area of a cone.
- Define and use the formula for finding the volume of a cone.

RESOURCES

• Practice Master	7.5
• Enrichment Master	7.5
• Technology Master	7.5
• Lesson Activity Master	7.5
• Quiz	7.5
• Spanish Resources	7.5

Use Transparency 60

Assessing Prior Knowledge

Classify each of the following as true or false.

1. The area of a circle with radius 8 is 64π. **[True]**
2. The ratio of the volume of a pyramid to a prism is 1:3. **[False]**
3. A prism may be divided into three pyramids that have the same volume. **[True]**
4. The slant height a pyramid is used to find its volume. **[False]**

TEACH

This lesson brings together ideas about surface area and volume to help students see the connection between pyramids, cylinders, and cones. Motivate students by having them list real-world examples of cones.

Exploration 1 Notes

In Exploration 1 students find the surface area of a right cone. Point out that the slant height is the radius of the larger circle used to make the net for the cone. Make sure students understand that the slant height is used to find the surface area while the height of the cone is used to find the volume.

Cooperative Learning

Finding the surface area of a cone in Exploration 1 is abstract enough that you may want to divide the class into cooperative groups to help students understand the concepts. Groups of three or four students can build models of cones. Allow students to explain to each other how to compose the cones from nets. Extend the ideas to include oblique cones.

ongoing ASSESSMENT

5. $B = \pi(7^2) = 49\pi$; $105\pi + 49\pi = 154\pi$; The surface area of the cone.

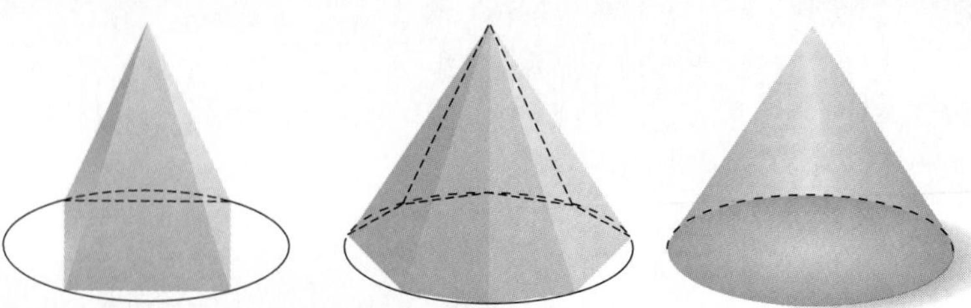

The illustrations suggest that the formulas for the surface areas and volumes of prisms and cylinders are similar.

Exploration 1 *The Surface Area of a Right Cone*

You will need
No special tools

The surface area of a right cone is found by a method similar to the one used for a right pyramid. The net for a cone includes the circular base and the flattened lateral surface, which becomes a portion of a circle known as a sector. The radius l of the sector is the slant height of the cone.

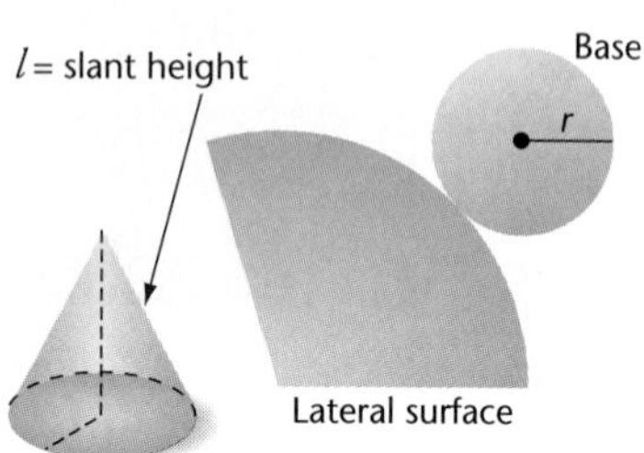

The surface area of a cone can be found by adding the area of the lateral surface and the area of the base.

1. The lateral surface of a cone is part of a larger circle. The arc labeled c matches the circumference of the base of a cone in a three-dimensional figure. So c is equal to the circumference of the base. Find c using r, the radius of the base.

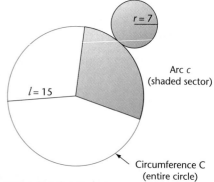

2. The radius of the larger circle is l. Find C, the circumference of the larger circle.

$r = 7$

$l = 15$

Arc c (shaded sector)

Circumference C (entire circle)

3. Divide c by C. This number tells you what fractional part the lateral surface occupies in the larger circle.

4. Find the area of the larger circle. Multiply this number by the fraction from Step 3. The result is the lateral surface of the cone. This is the lateral area L.

5. Find B, the area of the base of the cone. Add this to the area of the lateral surface L (Step 4). What does your answer represent? ❖

Physical Science Open pits formed during mining operations often have the shape of an inverted cone. Ask students to draw diagrams of why this may be so. Have them observe that sand passing through an hour-glass forms a cone much like the piles of salt, sand, coal, or asphalt piled up by construction trucks.

EXAMPLE 1

Find the surface area of a cone with the measures shown.

$l = 15$ m
$r = 7$ m

Solution ➤

Step 1 $c = 2\pi r = 14\pi$

Step 2 $C = 2\pi l = 30\pi$

Step 3 $\frac{c}{C} = \frac{14\pi}{30\pi} = \frac{7}{15}$

Step 4 area of larger circle

$\pi l^2 = 225\pi$

$L = \frac{7}{15} \cdot 225\pi = 105\pi$

Step 5 $B = \pi r^2 = 49\pi$

$B + L = 49\pi + 105\pi = 154\pi \text{ m}^2 \approx 483.8 \text{ m}^2$ ❖

Exploration 2 The Surface Area Formula for a Cone

You will need
No special tools

1. Follow the steps in the example above and on the previous page to find the surface area of a general cone. That is, use the variables r, h, and l instead of numerical measures. Your answer will be a formula for the surface area of a cone.

2. Use algebra to simplify your formula. Show your steps in an organized way and save your work in your notebook.

3. Compare your results with the formula given below. Are they the same? Explain. ❖

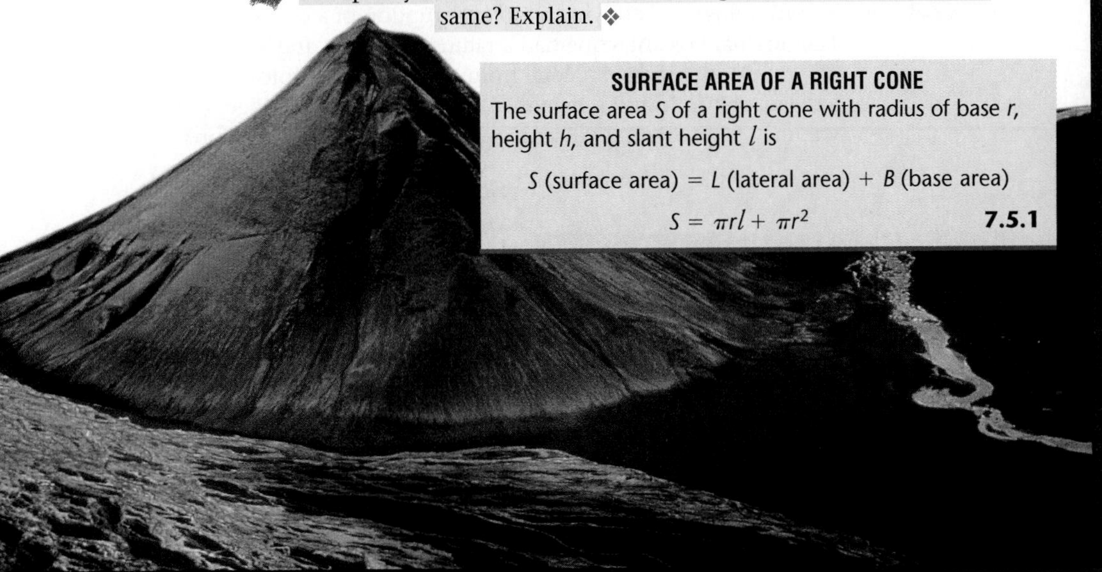

SURFACE AREA OF A RIGHT CONE

The surface area S of a right cone with radius of base r, height h, and slant height l is

S (surface area) = L (lateral area) + B (base area)

$$S = \pi r l + \pi r^2 \quad \textbf{7.5.1}$$

Alternate Example 1

Find the surface area of a cone with radius 6 m and slant height 10 m.

[$96\pi \text{ m}^2 \approx 301.59 \text{ m}^2$]

Math Connection Algebra

The solution in Example 1 makes the connection between the geometry of Exploration 1 and the algebra used to calculate the surface area of a cone.

Exploration 2 Notes

Emphasize the connections between the steps in the exploration and the use of variables in the formula for the surface area of a cone. Draw examples of cones on the chalkboard and use the formula to find their surface areas.

Ongoing ASSESSMENT

The results are the same. Solving the problem with variables in place of all specific values gives the answer full generality. The process is a derivation (in effect, a proof)of the formula.

ENRICHMENT Find the converse of the following true statement and prove or disprove it: If the altitudes of a cone and cylinder are equal and the diameters of their bases are equal, then the volume of the cone is equal to one-third of the volume of the cylinder.

INCLUSION strategies **English Language Development** Many students have difficulty reading and understanding mathematics on their own. Suggest that students read this lesson slowly and carefully, compare it to the lesson on pyramids, and write a list of questions they need answered. You can use cooperative groups to discuss the questions in class.

The Volume of a Cone

In Lesson 7.3 you found that a pyramid had one-third the area of a prism with the same base and height. Now imagine performing a similar experiment with a cone and a cylinder. The result of the experiment should be the same—you can try it—because cones and cylinders are like many-sided pyramids and prisms.

This leads us to the following formula for the volume of a cone.

> **VOLUME OF A CONE**
> The volume V of a cone with radius r and height h is
> V (volume) $= \frac{1}{3} B$ (base area) $\times h$ (height): $V = \frac{1}{3}\pi r^2 h$ **7.5.2**

Is the volume of an oblique cone the same as the formula for a right cone? What principle justifies your answer? Illustrate your answer with sketches.

EXAMPLE 2

Geology A vulcanologist is studying a violent eruption of a cone-shaped volcano. The original volcano cone had a radius of 5 miles and a height of 2 miles. The eruption removed a cone from the top of the volcano. This cone had a radius of 1 mile and a height of $\frac{1}{2}$ mile. What percentage of the total volume of the original volcano was destroyed by the eruption?

Solution ➤

The volume of the original volcano is

$V = \frac{1}{3}\pi r^2 h = \frac{1}{3}\pi(5^2)(2) \approx 52.4$ cubic miles.

The volume of the destroyed cone is

$V = \frac{1}{3}\pi r^2 h = \frac{1}{3}\pi(1^2)(0.5) \approx 0.52$ cubic miles.

The percentage of the original volcano is

$\left(\frac{0.52}{52.4}\right)(100) \approx 1\%$. ❖

CRITICAL *Thinking*

The formulas for the volumes of right cones and oblique cones are the same. This follows from Cavalieri's Principle.

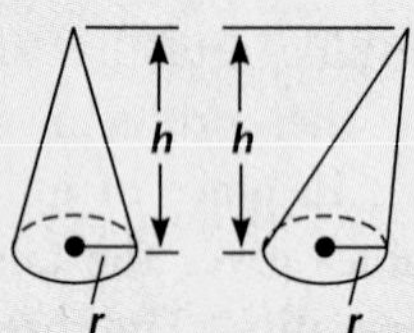

Alternate Example 2

A sugar cone for ice cream has a radius of 3 cm and a height of 15 cm. A cake cone looks like a sugar cone with the tip removed. If the volume of a cake cone is 24π cm^3, how much larger is the sugar cone than the cake cone? Express your answer as a percent.

[187.5%]

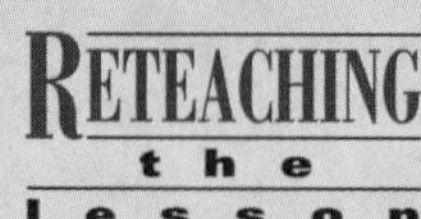

Hands-On Strategies Have students work in cooperative groups and repeat the exploration from Lesson 7.3 (p. 368) using cylinders and cones instead of prisms and pyramids. Have the groups express their results as a conjecture for the volume of a cylinder. Have the students make a list of similarities between prism, cylinders, pyramids, and cones.

EXERCISES & PROBLEMS

Communicate

1. Explain how to find the surface area of a cone.
2. Explain how to find the volume of a cone.
3. In a cone, which is longer in measure, the altitude or the slant height? Explain your reasoning.
4. What happens to the lateral area of a cone if the radius is doubled? if the slant height is doubled? if both are doubled?
5. What happens to the volume of a cone if the radius is doubled? if the slant height is doubled? if both are doubled?

Practice & Apply

Find the surface area for each right cone.

6.

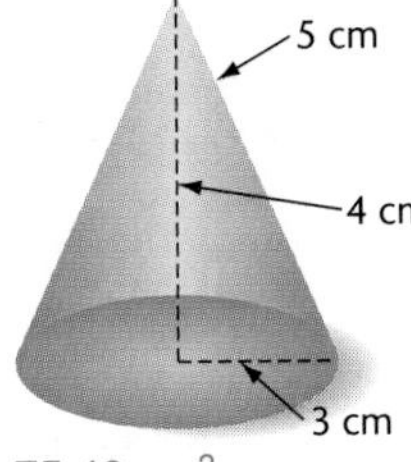

75.40 cm²

7.

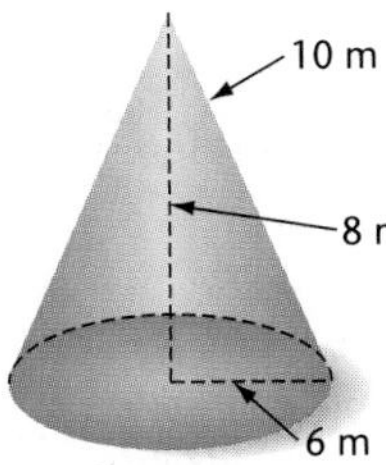

301.59 m²

8.

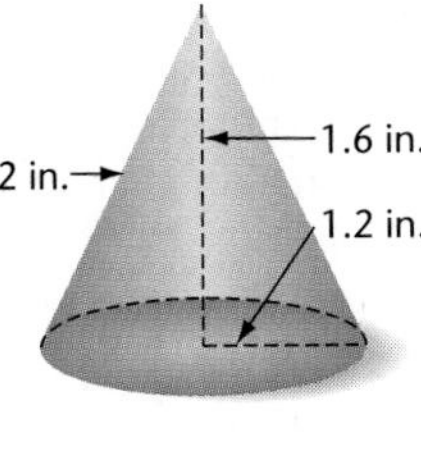

12.06 in.²

9. A right cone has a radius of 5 in. and a surface area of 180 sq in. What is the slant height of the cone? 6.46 in.

Find the volume of each right cone.

10.

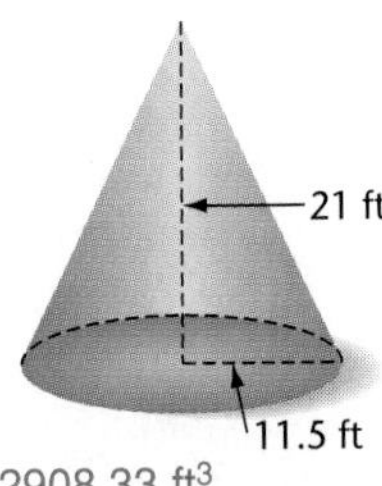

2908.33 ft³

11.

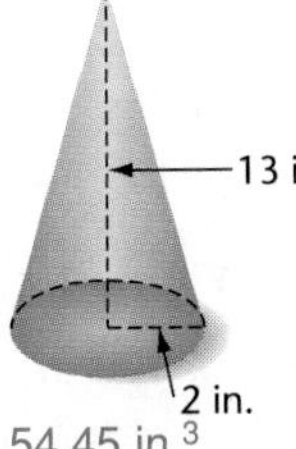

54.45 in.³

12.

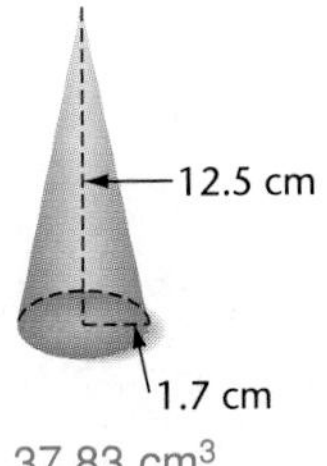

37.83 cm³

13. Find the surface area of a right cone whose base area is 25π sq cm and whose altitude is 13 cm. 297.35 cm²
14. An oblique cone has a volume of 1000 cm³ and a height of 10 cm. What is the radius of the cone? 9.77 cm

ASSESS

Selected Answers

Odd-numbered Exercises 7–35

Assignment Guide

Core 1–15, 24–33, 35–40

Core Plus 1–23, 25–40

Technology

For Exercises 32 and 33, after graphing the volume function on a graphing calculator, use the built-in option that finds the maximum value of a function in order to find the largest volume.

Error Analysis

Students may have difficulty doing Exercises 19–21. You may want students to build models of the cone-shaped paper cups from the suggested patterns.

alternative ASSESSMENT

Portfolio Assessment

Give students extra practice in finding the area and volume of cones by exploring real-world examples of cones. Have students determine the measurements of the cones and find their areas and volumes. Students should keep notes about their research in their portfolios.

15. **Portfolio Activity** Draw a circle with a radius of 5 cm and cut it out with scissors. At the center of the circle, draw a right angle and cut it out. Use tape and make a cone. What is the area of the sector that was cut out? What is the surface area of the cone? A = 19.63 cm^2; SA = 24.53 cm^2

16. A cone and a cylinder have bases with the same area and equal altitudes; why doesn't Cavalieri's principle hold?

17. If the triangular region *ABC* is rotated about *AB*, what solid figure is formed? Find its surface area. Cone; SA = 3522.35 units^2

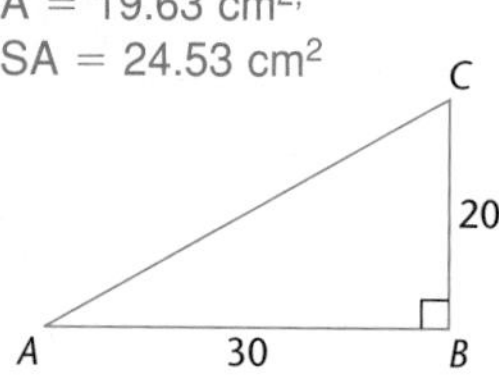

18. A right triangle has sides 10, 24, and 26. Rotate the triangle about each leg and find the volume of the figures formed.

Manufacturing Cone-shaped paper cups can be manufactured from patterns in the shape of sectors of circles. The figures below show two patterns. The first is cut with a straight angle. The second is cut with an angle measuring 120°. Use the information provided for Exercises 19–21.

19. If each sector has a radius of 6 cm, what will be the area of each sector? (Hint: What part of the circle is used by the sector?)

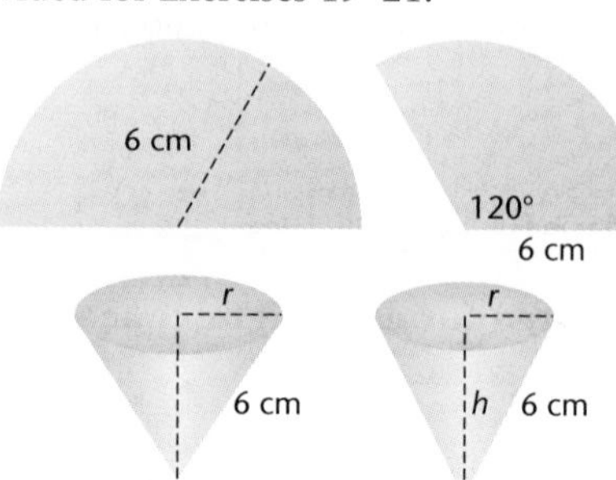

Algebra To find the volume of a cone-shaped cup formed from the first sector, we must first consider that the length of the arc of the sector becomes the circumference of the base of the cone.

$$\text{Arc length} = \frac{1}{2} \times 2\pi \times 6 = 6\pi = \text{circumference of cone base.}$$

Using the circumference formula again, we find the radius of the base.

$$\text{If } C = 2\pi r = 6\pi, \text{ then } r = 3.$$

Using the Pythagorean Theorem, we can find the height of the cone:

$$h = \sqrt{36-9} = \sqrt{27} = 3\sqrt{3}.$$

We can now find the volume of the cone.

$$V = \frac{1}{3}\pi r^2 h = \frac{1}{3}\pi \times 3^2 \times 3\sqrt{3}$$

$$= \pi \times 9 \times \sqrt{3}$$

$$= 48.97 \text{ cm}^3$$

20. Find the volume of the cup formed by the sector measuring 120°. 23.71 cm^3

21. If the radius of the larger sector increases by 50%, by how much does the volume of the cup formed from it increase? 134.3 cm^3

16. Their cross sections do not have equal areas. The cross sections of the cylinder are circles with the same radius, while the cross sections of the cone are circles with decreasing radii as the vertex is approached.

18. About short leg: V = 2513.27 units^3
About long leg: V = 6031.86 units^3

19. Straight angle: A = 56.55 cm^2
120° angle: A = 37.70 cm^2

Practice Master

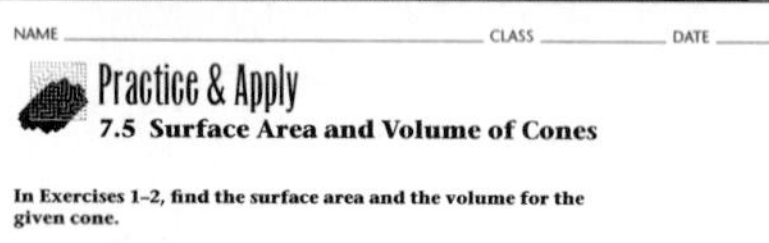

NAME ______ CLASS ______ DATE ______

Practice & Apply
7.5 Surface Area and Volume of Cones

In Exercises 1–2, find the surface area and the volume for the given cone.

1.

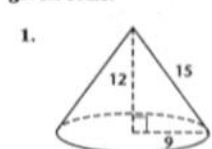

2.

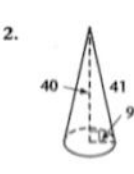

3. The surface area of a right cone is 300π cm². The diameter of the base and the slant height of the cone are equal. Find the radius. ______

Consider a right cone with radius 12 cm, slant height 37 cm, and altitude 35 cm.

4. Find the surface area of the cone. ______
5. Find the volume of the cone. ______
6. If the dimensions of the cone are doubled, what happens to the surface area? ______
7. If the dimensions of the cone are doubled, what happens to the volume? ______
8. If the volume of a cone is 360π cm³, and its altitude is 30 cm, find the circumference of the base. ______

The measures of the sides of right ΔABC are 7 mm, 24 mm, and 25 mm.

9. If ΔABC is rotated about its shorter leg, what is the volume of the figure formed? ______
10. If ΔABC is rotated about its shorter leg, what is the surface area of the figure formed? ______
11. If ΔABC is rotated about its longer leg, what is the volume of the figure formed? ______
12. If ΔABC is rotated about its longer leg, what is the surface area of the figure formed? ______

HRW Geometry Practice & Apply 5

22. Small Business The owner of an ice cream parlor is choosing bowls for her new store. Of the two bowls shown, which has the greater volume? Express the difference in the volumes in terms of π. How do the amounts of ice cream heaped above the tops of the cups compare?

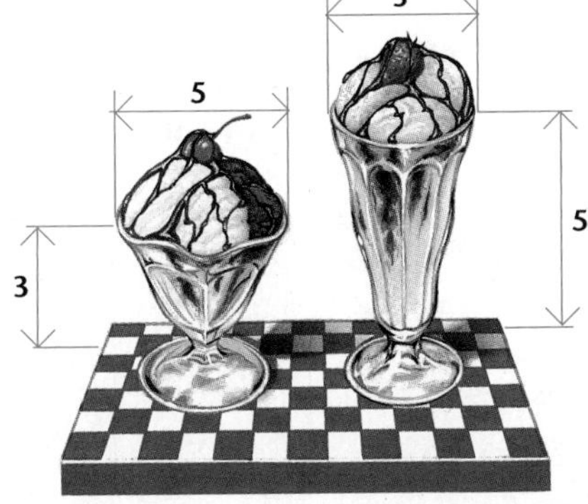

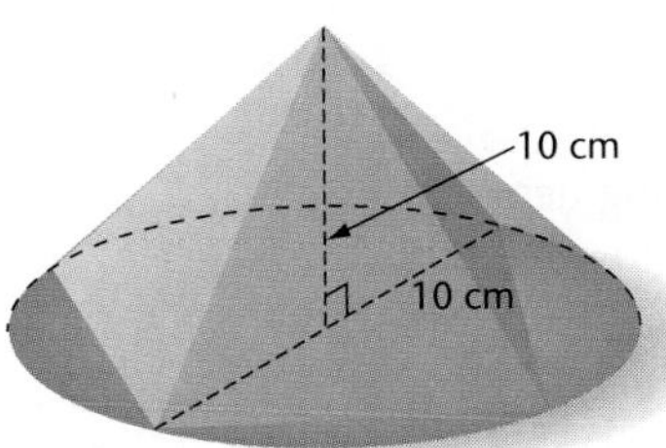

23. A right square pyramid is inscribed in a right cone of the same height. A radius of the base and the altitude of the cone are each 10 cm long. What is the ratio of the volume of the pyramid to the volume of the cone? $\frac{2}{\pi}$

24. Find the surface area and volume of a cone with a radius of 6 cm and an altitude of 9 cm. SA = 316.99 cm^2; V = 339.29 cm^3

Maximum/Minimum A cone has a slant height of 10. What should the radius and height be to maximize its volume? Fill in the chart for Exercises 25–31.

Radius	Height	Volume
1	$\sqrt{99}$	$\frac{1}{3}\pi \times 1^2 \times \sqrt{99} \approx 10.4$
2	$\sqrt{96}$	**25.** ___?___
3	**26.** ___?___	**27.** ___?___
4	**28.** ___?___	**29.** ___?___
x	**30.** ___?___	**31.** ___?___

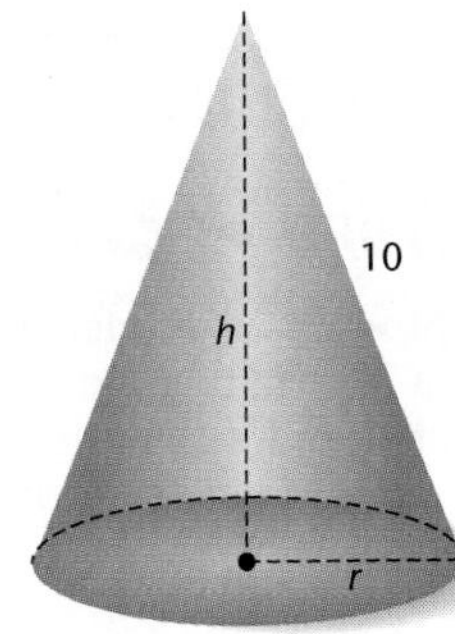

Technology Use a graphics calculator or software and the information from Exercises 25–31 to graph the radius on the x-axis and the volume on the y-axis. Set the viewing window so that Ymax is at least 410. Trace the graph and find the following.

32. the x-value that produces the largest volume 8.2 units

33. the largest volume 403.1 units^2

34. Recreation At an amusement park, an oblique conical tower is to be built. The vertex is to be directly above the edge of the circular base. The dimensions are as shown. What is the volume of the cone? 6702.06 ft^3

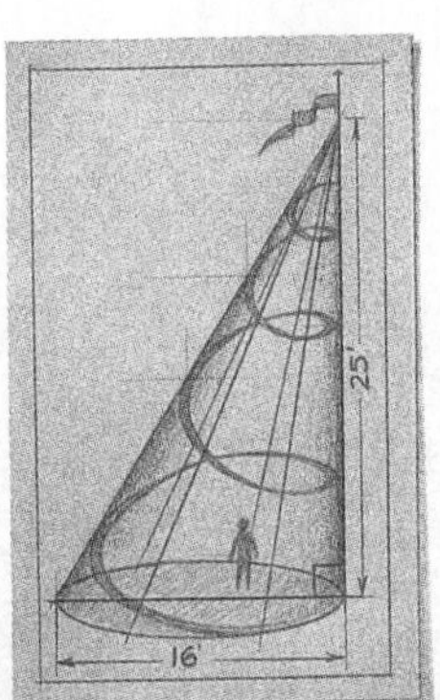

22. The shorter bowl has the greater volume.; volume of short bowl = 625π units2, volume of tall bowl = 3.75π units2, difference = 2.5π; The amount of ice cream heaped over the top of the short bowl is greater.

25. 41.04 26. $\sqrt{91}$ 27. 89.91

28. $\sqrt{84}$ 29. 153.56 30. $\sqrt{100 - x^2}$

31. $V = \frac{1}{3}\pi(x^2)(\sqrt{100 - x^2})$

Technology Master

NAME ______________ CLASS ________ DATE ________

Technology
7.5 The Uniformly Expanding Cone

The diagram shows a right circular cone with radius r_0 and altitude is h_0 getting larger with time. If both radius and height increase at the same rate, you have a uniformly expanding cone. Suppose that $r = 2.5t$ and $h = 2.5t$, where t is time in minutes. In terms of r and h, the formulas for surface area S and volume V are shown.

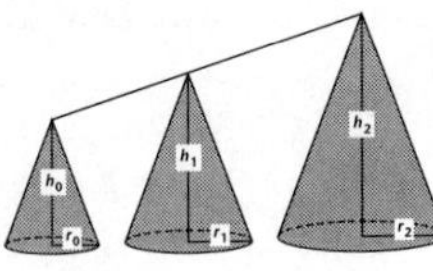

$S = \pi r\sqrt{r^2 + h^2} + \pi r^2 \quad V = \frac{1}{3}\pi r^2 h$

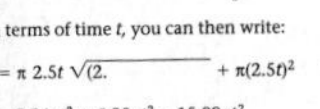

In terms of time t, you can then write:

$S = \pi\, 2.5t\, \sqrt{(2.} \quad + \pi(2.5t)^2 \qquad V = \frac{1}{3}\pi(2.5t)^2(2.5t)$

$= 8.84\pi t^2 + 6.25\pi t^2 = 15.09\pi t^2 \qquad = 5.21\pi t^3$

Use a graphics calculator for the following exercises.

1. Graph $S = 15.09\pi t^2$ and $V = 5.21\pi t^3$ on the same calculator screen.

2. For what value(s) of t will the surface area and the volume be numerically equal? ____________

3. Analyze graphically the relationship between the surface area and the volume over time t.

____

4. Use your graphs to determine whether there is a time t at which volume will be 1000 cubic units while the surface area will be 950 square units.

____

5. Graph the equation determined by the ratio of S to V. For what value(s) of t will the surface area be twice the volume? Find the corresponding radius and altitude.

____

6. Graph the equation determined by $V - S$. For what value(s) of t will the volume be 50 more than the surface area? Find the corresponding radius and altitude.

____

HRW Geometry Technology 19

Look Back

Use the given coordinates to find the coordinates of point *Q*. [Lesson 3.8]

35. (2*a*, 0)

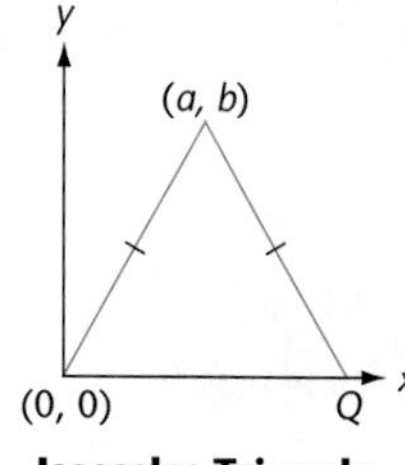

36. ($c + d$, 2)

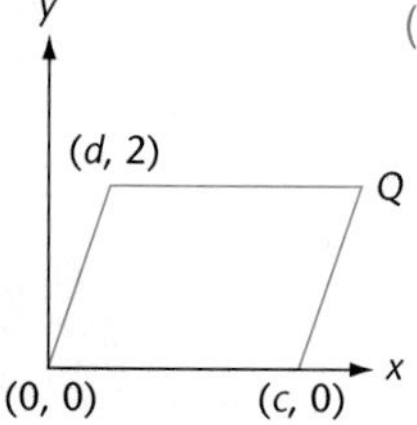

37. Prove that a diagonal divides a parallelogram into two congruent triangles. **[Lesson 4.6]**

38. Find the distance between (2, 4) and (–5, 8). **[Lesson 5.6]** 8.06

39. **Maximum/Minimum** A manufacturing company makes cardboard boxes of varying sizes by cutting out square corners from rectangular sheets of cardboard that are 12 in. by 18 in. The cardboard is then folded to form a box without a lid. If a customer wants a box with the greatest possible volume, what dimensions should be used? Use your graphics calculator. **[Lesson 7.1]** $V = x(18 - 2x)(12 - 2x)$ Max. volume = 228 in.^{3} when $x \approx 2.35$ in.

Look Beyond

Exercise 40 foreshadows the study of similar 3-dimensional figures. Students should have no difficulty finding the answers intuitively.

Look Beyond

40. **Scale Models** A model train engine is $\frac{1}{36}$ the size of a real train engine. If the model's width is 4 inches and its length is 16 inches, what are the width and length of the real engine? w = 12 ft; *l* = 48 ft

37. Given: Parallelogram ABCD with diagonal $\overline{AC}$

Prove: $\triangle ADC \cong \triangle CBA$

Statements: 1. Parallelogram ABCD with diagonal $\overline{AC}$, 2. $\overline{AD} \cong \overline{CB}$; $\overline{DC} \cong \overline{BA}$, 3. $\overline{AC} \cong \overline{AC}$, 4. $\angle ADC \cong \angle CBA$, 5. $\overline{DC} \parallel \overline{BA}$; $\overline{AD} \parallel \overline{CB}$, 6. $\angle DAC \cong \angle BCA$; $\angle DCA \cong \angle BAC$, 7. $\triangle ADC \cong \triangle CBA$;

Reasons: 1. Given, 2. Opposite sides of a parallelogram are ≅, 3. Reflexive, 4. Opposite angles in a parallelogram are ≅, 5. Parallelogram definition, 6. If lines ||, then alt. int. ∠s ≅, 7. SSS or SAS or ASA or AAS.

LESSON 7.6 Surface Area and Volume of Spheres

To make a hot-air balloon, individual pieces of cloth are sewn together. How much cloth do you think is needed? The "envelope" of an inflated balloon is not a perfect sphere, but you can use the information about spheres to get good approximate answers to this and other questions about spherical objects.

The Volume of a Sphere

To find the formula for the volume of a sphere, we first show that a sphere has the same volume as a special cylinder with a double cone cut out of it. Then, by using the formulas for cones and cylinders, we derive the formula for the volume of each figure. We begin with a numerical calculation before generalizing.

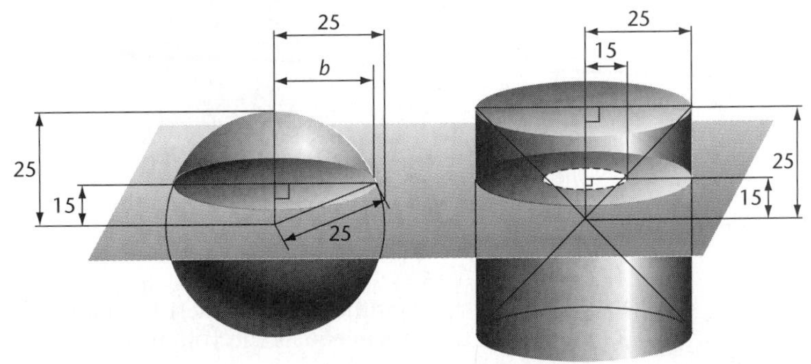

The sphere has a radius of 25 units. The cylinder has a radius of 25 units and a height of 50 units, and a double cone is cut out of it. A plane cuts through both figures 15 units above the center of each. We can show that the area of the shaded circular disk in the sphere is equal to the area of the shaded **annulus** (the ring-shaped figure) in the cylinder.

PREPARE

Objectives

- Define and use the formula for finding the surface area of a sphere.
- Define and use the formula for finding the volume of a sphere.

RESOURCES

- Practice Master 7.6
- Enrichment Master 7.6
- Technology Master 7.6
- Lesson Activity Master 7.6
- Quiz 7.6
- Chapter Resources 7.6

Use Transparency 61

Assessing Prior Knowledge

1. Find the area of a circle with radius 8 cm.

   [64π cm$^2 \approx 201.1$ cm^2]

2. Find the volume of a cylinder with radius 6 in. and height 8 in. **[88π in$^3 \approx 276.5$ in^3]**

3. Find the volume of a right cone with radius 4 cm and height 6 cm.

   [2π cm$^3 \approx 6.3$ in^3]

TEACH

Spherical objects are so common that students should understand their geometry thoroughly. Emphasize how common spheres are by having each student in the room give an example of a sphere.

ALTERNATIVE teaching strategy

Using Visual Models Bring in various examples of spheres to class. Have students measure the spheres and calculate their volumes. Calculate the radii by working backwards from the circumferences.

Math Connection Algebra

You may need to review the "Pythagorean" Right-Triangle Theorem before deriving the formula for the volume of a sphere.

Review the general nature of Cavalieri's Principle before discussing the formula for the volume of a sphere. To provide students with a usable concrete image, make a clay model of a cylinder with a double cone removed and cut it into cross-sections. The sphere can then be formed and measured from the remainder of the clay cylinder.

The large triangle formed in the cylinder by the side of the cone, the base radius, and half the height is isosceles with two legs 25 units in length. Since the plane is parallel to the bases, congruent corresponding angles are formed with the side of the cone, so the small triangle is similar to the large triangle, and is also isosceles with legs 15 units in length. Thus the radius of the inner circle of the annulus is 15 units.

Area of the circle in the sphere

$$b^2 + 15^2 = 25^2$$
$$b^2 + 225 = 625$$
$$b^2 = 400$$
$$b = 20 \text{ units}$$
$$\text{Area} = \pi 20^2 = 400\pi \text{ sq units}$$

Area of the annulus

area of large circle – area of small circle

$$= \pi 25^2 - \pi 15^2$$
$$= \pi(25^2 - 15^2)$$
$$= \pi(625 - 225)$$
$$= \pi(400)$$
$$\text{Area} = 400\pi \text{ sq units}$$

Now generalize the procedure to get the proof for a sphere and a cylinder of radius r.

Area of the circle

$$b^2 + y^2 = r^2$$
$$b^2 = r^2 - y^2$$
$$b = \sqrt{r^2 + y^2}$$
$$\text{Area} = \pi\left(\sqrt{r^2 - y^2}\right)^2 = \pi(r^2 - y^2)$$

Area of the annulus

$$\text{Area} = \pi r^2 - \pi y^2$$
$$\text{Area} = \pi(r^2 - y^2)$$

We now know that the result is true for all planes parallel to the bases of each figure. The result is true for all values of y.

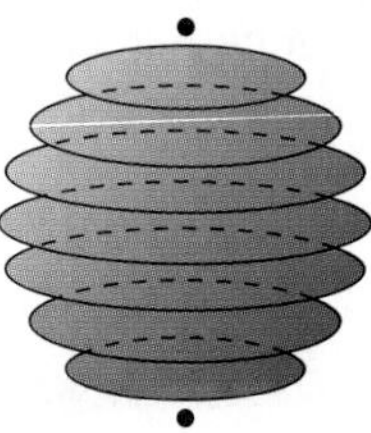

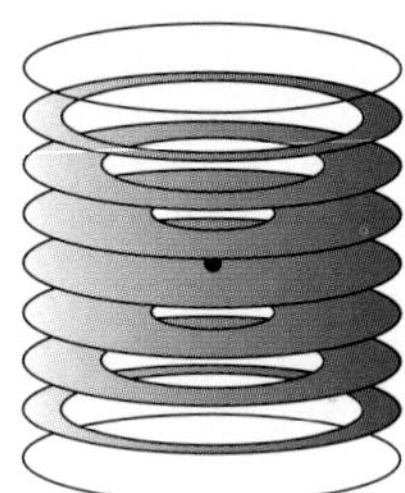

The corresponding cross sections have equal areas.

Therefore, the conditions of Cavalieri's Principle are satisfied. And so the volume of the sphere equals the volume of the cylinder with the double cone removed. That is,

$$V\text{ (Sphere)} = V\text{ (Cylinder)} - V\text{ (Cones)}$$
$$V\text{ (Sphere)} = \pi r^2(2r) - 2(\tfrac{1}{3}\pi r^2)(r)$$
$$= 2\pi r^3 - \tfrac{2}{3}\pi r^3$$
$$= \tfrac{4}{3}\pi r^3$$

interdisciplinary CONNECTION

Art Sculptors start with a ball of clay when they make clay bowls or figurines. Have students discuss different ways to estimate the volume of a hemispherical bowl that is made out of a sphere of clay that is 6 in. in diameter.

VOLUME OF A SPHERE

The volume V of a sphere with radius r is

$$V = \frac{4}{3}\pi r^3 \qquad \textbf{7.6.1}$$

CRITICAL Thinking When the cutting plane in the proof of the volume formula cuts through the center of the sphere, the distance y is 0. What happens to the annulus? How does this affect the calculations? Explain.

EXAMPLE 1

Hot Air Ballooning If the envelope of a hot-air balloon has a radius of 27 ft when fully inflated, approximately how many cubic ft of gas can it hold?

Solution ➤

$$\begin{aligned} V &= \frac{4}{3}\pi r^3 \\ &= \frac{4}{3}\pi 27^3 \\ &= \frac{4}{3}(19{,}683)\pi \\ &= 26{,}244\pi \text{ cubic ft} \approx 82{,}448 \text{ cubic ft} \end{aligned}$$ ❖

The Surface Area of a Sphere

Cartography In previous lessons you analyzed the surface areas of three-dimensional figures by unfolding them to form a net on a flat surface. But, as map makers well know, a sphere will not unfold smoothly onto a flat surface. The most common map of the world uses the Mercator projection of the Earth's surface onto a flat plane. On this map, the landmasses near the North and South Poles, such as Greenland, appear much larger than they should.

CRITICAL Thinking

The annulus becomes a circle. This has no effect on the calculations because when $y = 0$, the plane intersects the sphere and the cylinder in congruent circles.

Alternate Example 1

The Pantheon has a hemispherical dome 142 ft wide. It was copied in Rome around 118 AD. What is its volume?

[~749,607 ft³]

ENRICHMENT From physics, students learn that a soap bubble has tension at its surface that holds in its volume of air. If it does not burst, the bubble changes from a sphere to a hemisphere when it lands on a table. Have students determine how much larger the diameter of the hemisphere is.

[$\sqrt[3]{2}$ times as much]

INCLUSION strategies **Guided Research** Some students may be interested in the chemistry connection between the volume of a metal sphere and its density. The density of an object is the ratio of its mass to its volume, or Mass = Density × Volume. Have students find the mass of some common metal balls, given their density and volume.

Teaching tip

Cut up a piece of paper into triangles and glue them onto a large ball. Verify the area formula for a sphere by comparing the area of the wrapping paper to 4π times the square of the radius of the ball.

Cooperative Learning

Divide the students into five cooperative groups and assign one of the Communicate problems to each. Have each group present their answers to the whole class.

Doppler radar

A formula for the surface area can nevertheless be derived using some clever techniques.

Imagine the surface of a sphere as a large number of congruent regular polygons, as in the geodesic dome in the photo. The smaller these polygons are, the more closely they will approximate the surface area of the sphere.

Consider each polygon to be the base of a pyramid with its vertex at the center of the sphere. The volumes of these pyramids taken together will approximate the volume of the sphere. The height of each pyramid is the radius of the sphere. Therefore, the volume of each pyramid is $\frac{1}{3}Br$, where B is the area of the base of the pyramid and r is the radius of the sphere. So,

$$\text{Volume of sphere} \approx \tfrac{1}{3}B_1r + \tfrac{1}{3}B_2r + \ldots \tfrac{1}{3}B_nr = \tfrac{1}{3}r(B_1 + B_2 + \ldots B_n).$$

If the total area of the bases of the pyramids is assumed to equal the surface area S of the sphere, then

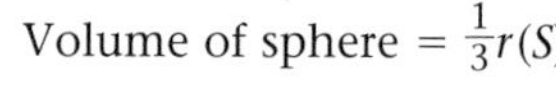

Solving for the surface area S,

$$S = \frac{3V}{r},$$

where V is the volume of the sphere.

You now have a formula for the surface area of a sphere in terms of its volume. Substituting the formula for V from above,

$$S = 3\frac{V}{r}$$

$$= 3\,\frac{\left(\frac{4}{3}\pi r^3\right)}{r} \quad \text{(Substitution)}$$

$$= 4\pi r^2$$

The surface area S of a sphere with radius r is

$$S = 4\pi r^2$$

7.6.2

Inviting Participation Have one student draw a sphere on the chalkboard. Have another student give the measure of its radius or diameter. Ask a third student to calculate its surface area and a fourth student to calculate its volume. Repeat the process until all students have participated.

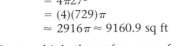

EXAMPLE 2

Hot Air Ballooning

The envelope of a hot-air balloon is 54 ft in diameter when inflated. The cost of the fabric used in making the envelope is $1.31 per sq ft. What is the approximate total cost of the fabric for the balloon envelope?

Solution ➤

Approximate the surface area of the inflated balloon envelope. Since the diameter is 54 ft, the radius will be 27 ft.

$$\begin{aligned} S &= 4\pi r^2 \\ &= 4\pi 27^2 \\ &= (4)(729)\pi \\ &\approx 2916\pi \approx 9160.9 \text{ sq ft} \end{aligned}$$

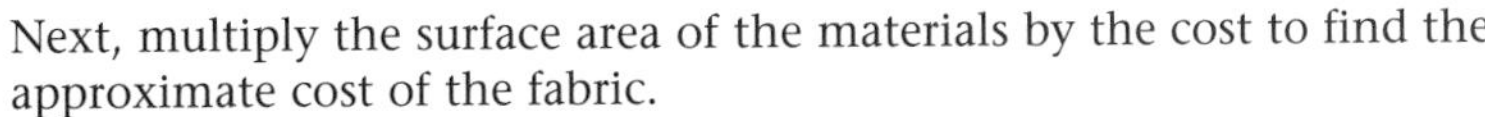

Next, multiply the surface area of the materials by the cost to find the approximate cost of the fabric.

9160.9 sq ft · \$1.31 ≈ \$12,000 ❖

EXERCISES & PROBLEMS

Communicate

1. Explain how to find the surface area of a sphere.
2. Explain how to find the volume of a sphere.
3. What happens to the surface area of a sphere when the radius is doubled?
4. What happens to the volume of a sphere when the radius is doubled?
5. Explain how the "Pythagorean" Right-Triangle Theorem is used to help derive the formula for the volume of a sphere.

Alternate Example 2

A geodesic dome is in the shape of a hemisphere and has a diameter of 50 ft. If the cost of constructing the dome is \$2.42 per square foot, how much does it cost to construct?

[~\$79,194]

ASSESS

Selected Answers

Odd-numbered Exercises 7–25

Assignment Guide

Core 1–4, 6–16, 20–25

Core Plus 1–13, 17–29

Technology

Students will benefit from the use of a graphing or scientific calculator for most of the exercises.

Error Analysis

In Exercise 16 be sure students understand that some of the surface area of both figures is removed to determine the total surface area.

Practice & Apply

Find the volume and surface area of each sphere. All measurements are in centimeters.

6. $r = 8$ **7.** $r = 7$ **8.** $d = 10$ **9.** $r = 9$

10. Sports A basketball has a radius of approximately 4.75 in. when filled. How much material is needed to make one (round to the nearest tenth)? How much air will it hold? If the basketball is stored in a box that is a cube whose edges are 9.5 in., what percent of the box is not filled by the basketball?

11. Sports A can of tennis balls has 3 balls in it. The approximate height of the stacked balls is 8 in. and the approximate radius of each ball is 1.5 in. How much space do the tennis balls occupy (to the nearest tenth)? 42.4 $in.^3$

12. Sports Find the surface area of the baseball. 19.6 $in.^2$

13. Sports Find the volume of the softball. 22.4 $in.^3$

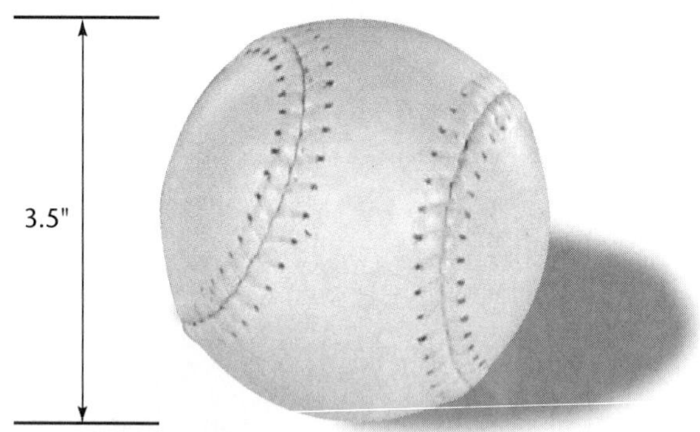

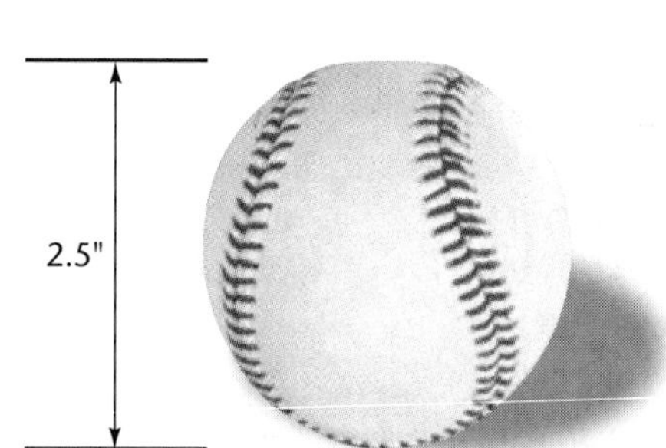

Food Hosea buys an ice cream cone. The ice cream is a sphere with a radius of 1.25 in. The cone has a height of 8 in. and a diameter of 2.5 in.

14. What is the volume of the ice cream? 8.18 $in.^3$

15. What is the volume of the cone? 13.09 $in.^3$

16. If the ice cream sits on top of the cone to form a hemisphere, what is the total surface area of the ice cream cone? 41.6 $in.^2$

alternative ASSESSMENT

Portfolio Assessment

Ask students to summarize the lesson by writing what criteria must be met for a solid to be a sphere. Ask them to describe how to derive the formula for the surface area by breaking it up into smaller and smaller pyramids.

6. SA = 804.25 cm^2
   V = 2144.66 cm^3

7. SA = 615.75 cm^2
   V = 1436.76 cm^3

8. SA = 314.16 cm^2
   V = 523.60 cm^3

9. SA = 1017.85 cm^2
   V = 3053.54 cm^3

10. SA = 283.5 in^2 is the amount of material needed.
    V = 448.92 in^3 is the volume of air.
    47.6% of the box is not filled by the ball.

Practice Master

NAME ______ CLASS ______ DATE ______

Practice & Apply

7.6 Surface Area and Volume of Spheres

In Exercises 1–3, find the surface area and volume of the sphere.

1. $r = 6$ cm $S =$ ______ $V =$ ______

2. $d = 18$ in. $S =$ ______ $V =$ ______

3. $r = 2.5$ ft $S =$ ______ $V =$ ______

Consider a sphere with radius 16 cm.

4. Find the surface area of the sphere. ______

5. Find the volume of the sphere. ______

6. If the radius of the sphere is doubled, what happens to the surface area? ______

7. If the radius of the sphere is doubled, what happens to the volume? ______

8. If the radius of the sphere is halved, what happens to the surface area? ______

9. If the radius of the sphere is halved, what happens to the volume? ______

The figure represents a sphere that is inscribed in a cube (the sphere touches each face of the cube).

10. If each edge of the cube is 12 cm in length, find the volume of the cube. ______

11. If each edge of the cube is 12 cm in length, find the volume of the sphere. ______

12. About what percentage of the volume of the cube does the sphere fill? ______

13. A lead sphere of radius 8" is melted and cast into a right cone 8" high. Find the diameter of the base of the cone. ______

6 **Practice & Apply** **HRW Geometry**

17. Suppose that a cube and a sphere both have a volume of 1000 cubic units. Compare the surface areas.

18. Suppose a cube and a sphere both have the same surface area. Compare their volumes. Let the surface area equal 864 sq units.

19. What is the volume of the largest ball that would fit into a cubical box with edges 12 in.? 904.78 in.^{3}

17. Cube: 600 units^2
Sphere: 483.1 units^2
The cube has a larger surface area.

18. Cube: 1728 units^3
Sphere: 2386.4 units^3
The sphere has a larger volume

Look Back

Find the surface area of each figure. [Lessons 7.2, 7.4, 7.5]

20. right rectangular prism: $l = 3$ in., $w = 10$ in., $h = 5$ in. 190 in.^{2}

21. right cone: $r = 15$ cm, slant height $= 45$ cm 2827.43 cm^2

22. right cylinder: $r = 9$ m, $h = 10$ m 1074.42 m^2

Find the volume of each figure. [Lessons 7.3, 7.4, 7.5]

23. oblique rectangular pyramid: $l = 15$ ft, $w = 7$ ft, $h = 12$ ft 420 ft^3

24. right cone: $r = 5$ in., $h = 10$ in. 261.80 in.^{3}

25. oblique cylinder: $r = 7.5$ m, $h = 20$ m 3534.29 m^3

Look Beyond

A cube is inscribed inside a sphere with a diameter of 2 units.

26. What is the volume of the cube? 1.54 units^3

27. What is the ratio of the volume of the sphere to the volume of the cube? 2.72

28. What is the surface area of the cube? 8 units^2

29. What is the ratio of the surface area of the sphere to the surface area of the cube? 1.57

Look Beyond

Exercises 26–29 have students compare the volumes and surface areas of a cube inscribed in a sphere with the sphere.

Technology Master

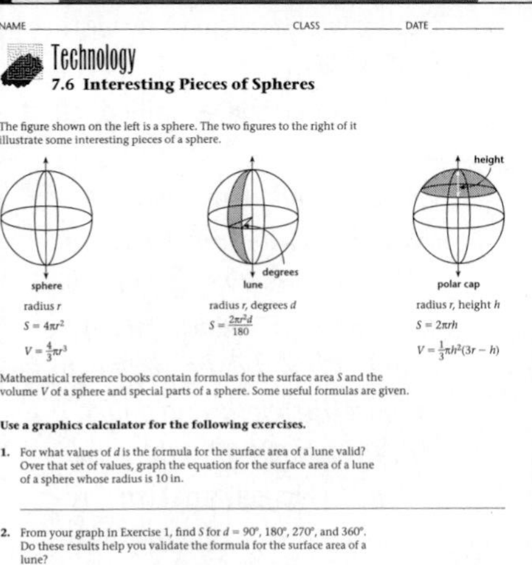

NAME ______ CLASS ______ DATE ______

Technology
7.6 Interesting Pieces of Spheres

The figure shown on the left is a sphere. The two figures to the right of it illustrate some interesting pieces of a sphere.

sphere	lune	polar cap
radius r	radius r, degrees d	radius r, height h
$S = 4\pi r^2$	$S = \frac{2\pi r^2 d}{180}$	$S = 2\pi rh$
$V = \frac{4}{3}\pi r^3$		$V = \frac{1}{3}\pi h^2(3r - h)$

Mathematical reference books contain formulas for the surface area S and the volume V of a sphere and special parts of a sphere. Some useful formulas are given.

Use a graphics calculator for the following exercises.

1. For what values of d is the formula for the surface area of a lune valid? Over that set of values, graph the equation for the surface area of a lune of a sphere whose radius is 10 in.

______

2. From your graph in Exercise 1, find S for $d = 90°$, $180°$, $270°$, and $360°$. Do these results help you validate the formula for the surface area of a lune?

______

3. For what values of h is the formula for the surface area of a polar cap valid? Graph the equation for the surface area of a polar cap of a sphere whose radius is 10 in. over the appropriate set of values.

______

4. From your graph in Exercise 3, find S for $h = 0$, 5, and 10. Do these results help you validate the formula for the surface area of a polar cap?

______

5. On a separate piece of paper outline a strategy for using a graphics calculator to find the surface area of that portion of a hemisphere below a polar cap given various heights of a polar cap on a sphere of radius 10 in.

20 Technology HRW Geometry

Focus

The actual reconstruction of a robe found in the tomb of Tutankhamun provides a case study of geometry and number in archaeology.

Motivate

After students read the news excerpt, discuss why the discovery of Tutankhamun's tomb was so significant. Note the following:

- Despite some early plundering, the tomb and its contents were left virtually intact, especially compared to the ravaged state of other discovered tombs of ancient Egyptian rulers.
- The mummy of Tutankhamun was still there
- Many archaeologists believed that the search for the tomb of Tutankhamun would be fruitless.

To help students understand the background of the reconstruction. Ask:

- *What would happen if Carter had picked up the robe when he first opened the box?* (The cloth would have crumbled, the beads would have spilled all over, etc.)
- *Carter could have poured a solidifying substance over the robe to preserve it in the condition in which it was found. Why did he decide to sacrifice the cloth and make a reconstruction of the robe instead?* (He felt that the reconstruction showing the overall size and design of the robe would tell more about the ancient Egyptian culture than preserved clumps of cloth and disorganized beads and sequins.)

TREASURE of King Tut's Tomb

GEM-STUDDED RELICS AMAZE EXPLORERS

SPECIAL CABLE TO THE NEW YORK TIMES

LONDON, Nov. 30, 1922 — The Cairo correspondent of *The London Times,* in a dispatch to his paper, describes how Lord Carnarvon and Howard Carter unearthed below the tomb of Rameses VI, near Luxor, two rooms containing the funeral paraphernalia of King Tutankhamun, who reigned about 1350 B.C.

A sealed outer door was carefully opened, then a way was cleared down some sixteen steps and along a passage of about twenty-five feet, A door to the chambers was found to be sealed as the outer door had been and as the

Funerary mask of gold, lapis, cloisonné, and quartz

The excitement of unearthing the tomb was followed by three years of hard work. Unpacking a box of priceless relics took three painstaking weeks. Items stored for thousands of years are easily damaged forever if not handled with great care.

A crumpled robe in Box 21 presented a dilemma for Howard Carter. Should he preserve the garment as is and lose the chance to learn its full design? Or should he handle it and sacrifice the cloth in order to examine the complete robe?

Excerpt from Howard Carter's notes.

...by sacrificing the cloth, picking it carefully away piece by piece, we could recover, as a rule, the whole scheme of decoration. Later, in the museum, it will be possible to make a new garment of the exact size, to which the original ornamentation—bead-work, gold sequins, or whatever it may be—can be applied. Restorations of this kind will be far more useful, and have much greater archaeological value, than a few irregularly shaped pieces of preserved cloth and a collection of loose beads and sequins.

1. a. 1.096.

b. 54.8 (Carter folded over the garment to determine a "width")

c. 0.9162 d. 1.096 e. 83 to 85 cm

f. 84 cm g. 100 cm

Cooperative Learning

The robe in Box 21, which had a cylindrical shape, was covered by a pattern of glass beads and gold sequins, with a lower band of pendant strings. Howard Carter recorded how he figured out the robe's size. For the following activity, some of the data have been removed from his notes.

1. Figure out the missing numbers in Howard Carter's notes and show how you found them.
2. Based on its size, do you think the robe was an adult's garment or a child's? Why?

The bead pattern on the ceremonial robe, enlarged

We know the distance between the pendant strings of the lower band to have averaged 8 mm. There are 137 of these pendants. Therefore the circumference of the lower part of the garment must have been (a) m, which would make one width about (b) cm.

Now there are 3054 gold sequins. Three sequins in the pattern require 9 square centimeters. Therefore, total area of network = (c) square meters. As circumference was (d), this would make the height work out at about (e) cm.

Thus size of garment would work out at about (f) X (g).

Incised cartouche of King Tutankhamun, gold, and gems

Gold collar of Vulture Goddess, Nekhbet

Cooperative Learning

Have students work in pairs or small groups on the activity. Students may find it helpful to use calculators. In his notes, Carter's use of the terms *circumference* and *width* might be confusing. Students should use a diagram to see that the "width" of the garment is half the "circumference." Remind students that one square meter is 10,000 square centimeters.

Discuss

Have the class make up their own "reconstruction" problem about an ancient garment, piece of jewelry, or other relic. Be sure the students include enough information about the contents and pattern of the discovered artifact so the size or appearance of the original article can be determined.

2. Answers may vary, though the length of about 30 inches strongly suggests a child's garment. Note—According to Howard Carter, the pattern of the decoration seemed to be a design for a female—perhaps the boy king was dressed as a girl on some occasions. In another box in the tomb, Carter found two much larger men's garments, which he called shirts, sizes 137 cm × 85 cm and 138 cm × 103 cm.

PREPARE

Objectives

- Define various transformations in three-dimensional space.
- Solve problems using transformations in three-dimensional space.

RESOURCES

• Practice Master	7.7
• Enrichment Master	7.7
• Technology Master	7.7
• Lesson Activity Master	7.7
• Quiz	7.7
• Spanish Resources	7.7

Assessing Prior Knowledge

1. Find the reflection image of (–2, 4) in the *y*-axis.

   [(2,4)]

2. Find the reflection image of (3, –4) in the *x*-axis.

   [(3, 4)]

3. Point *P*(0, –2, 4) is located in which plane of a three-dimensional coordinate system?

   [*yz*-plane]

TEACH

This lesson extends the ideas of symmetry to include reflectional and rotational symmetry in space. Students need to be able to describe a three-dimensional object as well as find its area and volume. This lesson also presents many examples of objects that have symmetry.

Three-Dimensional Symmetry

why *So far, the definitions of symmetry have been limited to a plane. But three-dimensional figures, like the tiger in the photo, may also have symmetry. You are ready to extend your ideas of symmetry to solid or three-dimensional figures.*

Tiger, Tiger, burning bright,
In the forests of the night;
What immortal hand or eye,
Could frame thy fearful symmetry?
—William Blake

A three-dimensional figure may be reflected through a plane, just as a two-dimensional figure may be reflected through a line. What happens to each point in the pre-image of a figure as a result of a reflection through a plane? In Exploration 1 you will investigate this question.

Exploration 1 Reflections in Coordinate Space

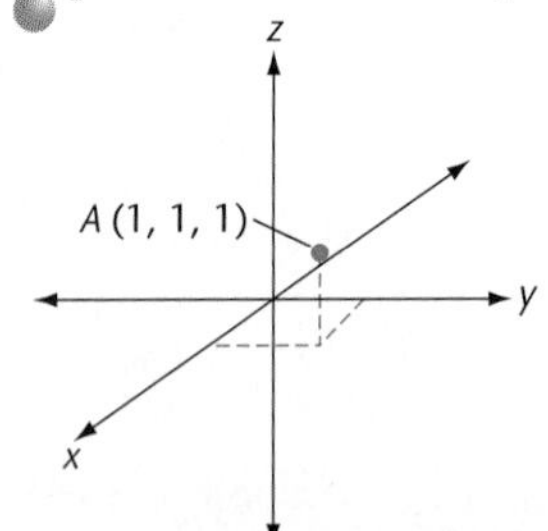

You will need
No special tools

Part I

1. Graph the point $A(1, 1, 1)$ in a three-dimensional coordinate space. Use dashed lines to make the location of the point in space evident.

2. Multiply the *x*-coordinate of point *A* by -1 and graph the resulting point. Label the new point A'.

ALTERNATIVE teaching strategy

Using Visual Models Orient a rectangular box in the first octant of a three-dimensional coordinate system with one vertex at (0, 0, 0) and one face of the box in the *xy*-plane. Measure the box and determine the coordinates of its vertices. Find the reflection images of each of its vertices in the *xy*-plane, the *xz*-plane, and the *yz*-plane. Rotate the box so that it is in a different octant with one vertex still at (0, 0, 0) and one face in a plane. Repeat the reflections. Are the image points the same? **[No]**

3. The point A' is the reflection image of point A through a plane. Name this plane. If you connect points A and A' to form a segment $\overline{AA'}$, what is the relationship of this segment to the reflection plane?

4. Write your own definition for the reflection of a point through a plane in space.

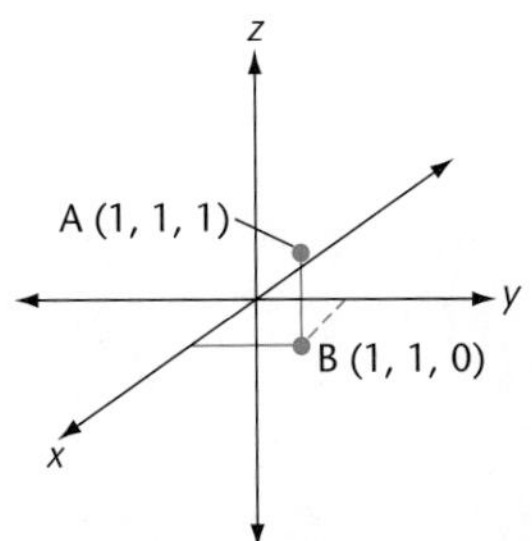

5. Experiment with other reflections of point A. What happens if you multiply the y-coordinates by -1? the z-coordinates?

6. Now study the reflection of an entire segment, such as $\overline{AB}$, through a plane. (For example, multiply the x-coordinates of points A and B by -1 and connect the endpoints of the resulting points.) Experiment with the reflections of other figures, such as cubes.

7. Write your own definition of the reflection of a figure in space through a plane.

Part II

1. Fill in the table below. Use the terms *front, back, left, right, top,* and *bottom* to describe the octant of a point.

		Octant of image	Coordinates of image
$A(2, 3, 4)$	Reflection image through the xy-plane	front-right-bottom	$(2, 3, -4)$
	Reflection image through the xz-plane	?	?
	Reflection image through the yz-plane	?	?
$B(-4, 5, 6)$	Reflection image through the xy-plane	?	?
	Reflection image through the xz-plane	?	?
	Reflection image through the yz-plane	?	?

2. Generalize your findings, given $P(x, y, z)$.

a. What are the coordinates of the reflection image of P through the xy-plane?

b. What are the coordinates of the reflection image of P through the xz-plane.

c. What are the coordinates of the reflection image of P through the yz-plane. ❖

ONGOING ASSESSMENT

7. ***A reflection of a figure through a plane in space is a transformation that produces an image of the figure with the following property:***

Each point of the figure and its image are on opposite sides of the reflection plane and are equidistant from it. Or,

The reflection plane is the perpendicular bisector of all the line segments connecting points of the figure to their images.

Note: If the figure touches the reflection plane, then the image and the preimage of a point of the figure that falls in the plane are the same point.

Exploration 1 Notes

Students may need more practice finding reflections in coordinate space. Have them repeat the exploration with as many points as necessary to generalize the coordinates of a reflection image in space. It may also help to repeat the exploration with a segment in space instead of a single point.

Cooperative Learning

Divide the students into cooperative groups to complete the explorations in this lesson. Have every group do all four explorations. One group for each exploration should be chosen to present their results to the class.

interdisciplinary CONNECTION

Art Computer artists work with the symmetry of figures using a coordinate grid. Have students use computer drawing programs to practice drawing solid figures .

Exploration 2 Reflectional Symmetry in Space

You will need
No special tools

1. The three-dimensional figures below have reflectional symmetry. Explain why. Where is the reflection "mirror" in each case?
2. Write your own definition of reflectional symmetry in space. Use your earlier definition of reflectional symmetry in the plane as a model. ❖

Exploration 3 Rotations in Coordinate Space

z
A(−1, 1, 1)
B(−1, 1, 0)
y
x

You will need
No special tools

1. Graph segment $\overline{AB}$ in a three-dimensional coordinate space. Use dashed lines to make the location of the segment evident.
2. Multiply the *x*- and *y*-coordinates of points *A* and *B* by −1 and graph the resulting points *A′* and *B′*. Connect the new points to form segment $\overline{A'B'}$.
3. Segment $\overline{A'B'}$ is a reflection image of segment $\overline{AB}$ about the *z*-axis. What is the reflection "mirror" in each case?
4. Through what point on the *z*-axis has endpoint *A* been rotated? endpoint *B*?

Exploration 2 Notes

This activity will help students understand that some three-dimensional figures also have reflectional symmetry. They are asked to find the line of symmetry for real objects and to write their own definition of reflectional symmetry in space.

ongoing ASSESSMENT

2. ***A figure in space has reflectional symmetry if and only if one side of the figure is the image of the other side when reflected through a plane drawn through the figure.*** **If a plane can be drawn through a figure so that every point in the figure has an image point on the other side of the plane such that the plane is the perpendicular bisector of the line segments connecting the points and their images, the figure has reflectional symmetry.**

TEACHING tip

It may be very helpful to photocopy some illustrations and then have students draw in lines of symmetry.

Exploration 3 Notes

This activity will help students extend their ideas about symmetry in space by having them rotate coordinates in space. This generates a three-dimensional figure. Have students compare their definitions in step 6 with those of their classmates.

ENRICHMENT Have students do a collage of figures with symmetry. Ask them to select a specific theme and use magazines to find examples of the theme. Possible themes include: nature items with reflectional symmetry, nature items with rotational symmetry, faces, flowers with rotational symmetry, buildings with reflection symmetry, food with rotational symmetry.

INCLUSION strategies

Hands-On Strategies Some students may have trouble finding image points under reflectional or rotational symmetry. Have students work with model prisms and experiment with the effects of a rotation on a model three-dimensional coordinate system made with drinking straws.

5 Imagine the segment rotating about the z-axis, as suggested by the picture. Does it seem to you that each of the rotation images of point A is in the same plane? of point B? What is the relationship between these planes and the z-axis?

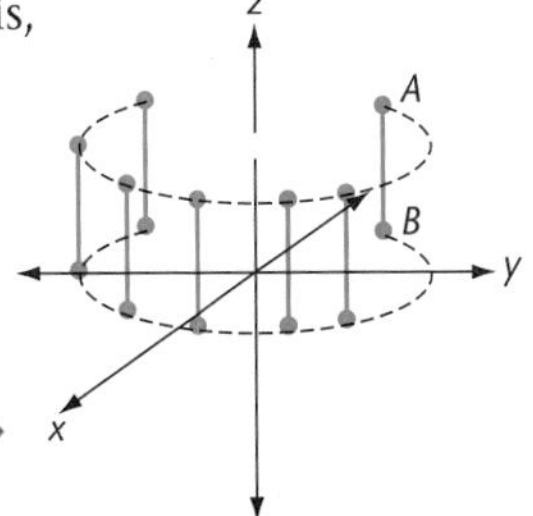

6 Write your own definition of the rotation of a figure about an axis in space. ❖

Exploration 4 Rotational Symmetry in Space

You will need
No special tools

1 These three-dimensional figures each have rotational symmetry. Explain why. Where is the axis of rotation in each case?

2 Write your own definition of rotational symmetry in space. Use your earlier definition of reflectional symmetry in the plane as a model. ❖

Revolutions in Coordinate Space

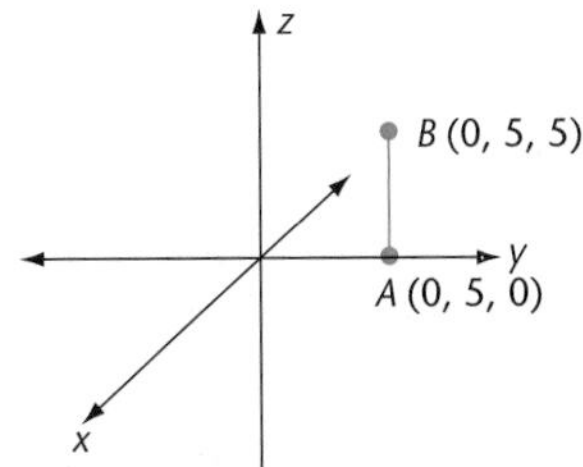

If you rotate a plane figure about an axis in space, the result is called a **solid of revolution**.

EXAMPLE

You are given the segment $\overline{AB}$ from $A(0, 5, 0)$ to $B(0, 5, 10)$. Sketch, describe, and give the dimensions of the figure that results if

A $\overline{AB}$ is rotated about the z-axis.

B $\overline{AB}$ is rotated about the y-axis.

RETEACHING the lesson

Using Discussion You may want to have students restate the definitions they wrote in this lesson while referring to some examples of their drawings. Have students identify the preimage and image points.

ONGOING ASSESSMENT

6. *The rotation of a figure about an axis in space is a transformation that produces an image of the figure with the following property:*

Each point in the figure moves a given amount around the axis of rotation, in a circle that is contained in a plane perpendicular to the axis. (The center of the circle is on the axis of rotation, and the angle of rotation is the same for all points.)

Note: If the figure touches the axis of rotation, then the image and preimage of a point of the figure that falls on the axis are the same point.

Exploration 4 Notes

In this exploration, students explain why figures have three-dimensional symmetry.

ONGOING ASSESSMENT

2. *A figure in space has rotational symmetry if and only if one side of the figure is the image of the other side when rotated around a line drawn through the figure.* If a line can be drawn through a figure so that every point in the figure has an image point in the figure that is a rotation of a given amount on a circle perpendicular to the line and centered on the line, the figure has rotational symmetry and the line is the axis of rotation.

ongoing ASSESSMENT

Try This

sphere; cone; "doughnut" or torus

Use Transparency 62

ASSESS

Selected Answers

Exercises 9, 13, 21, 23, 25, 27, and 29

Assignment Guide

Core 1–14, 18–31

Core Plus 1–33

Technology

Have students use the copy and paste features of a geometry graphics program to experiment with reflection images and rotation images for Exercises 20 and 23.

Error Analysis

Students need to be sure that they are creating the correct solid when performing the rotations in Exercises 20–23. Encourage students to double-check each answer.

A

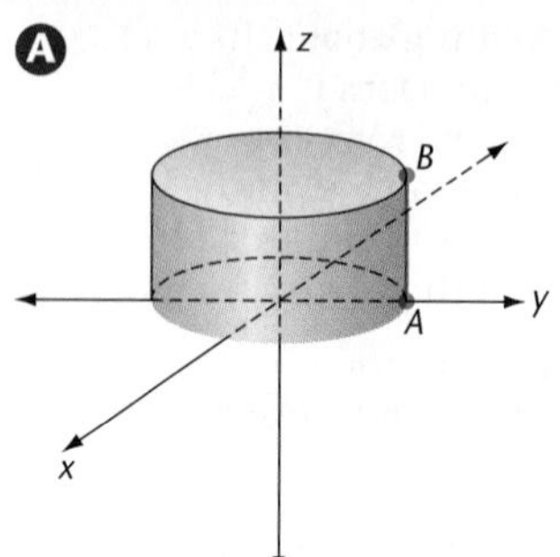

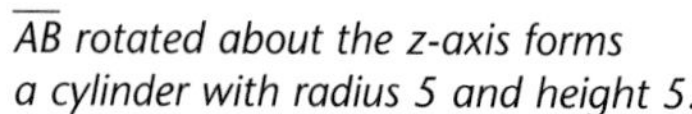

$\overline{AB}$ rotated about the z-axis forms a cylinder with radius 5 and height 5.

B

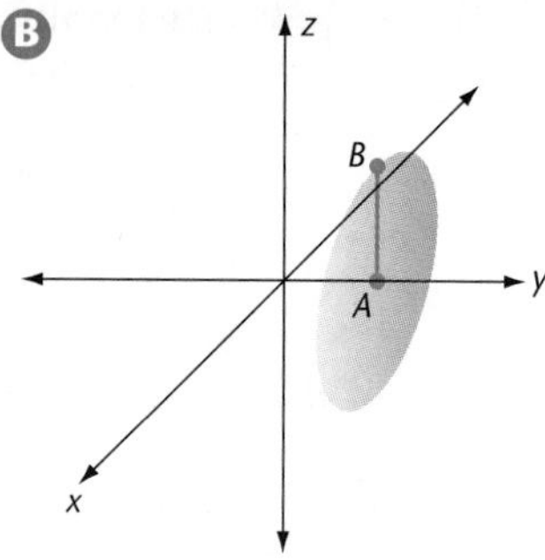

$\overline{AB}$ rotated about the y-axis forms a circle with radius 5, and its interior.

Try This Describe or sketch the solid of revolution that would be formed by rotating each of the plane figures about the dotted line.

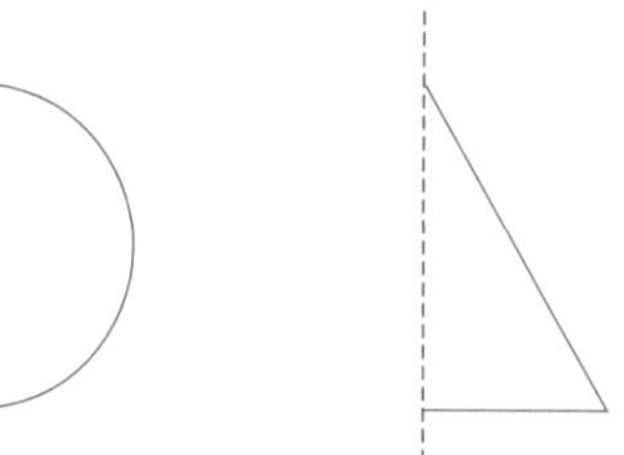

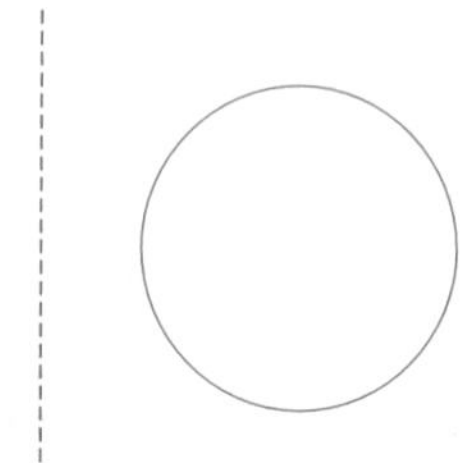

EXERCISES & PROBLEMS

Communicate

1. Describe the similarities and differences between three-dimensional symmetry and two-dimensional symmetry.

2. What geometric object is formed by rotating rectangle $ABCD$ about $\overline{AD}$?

3. What geometric figure is formed by rotating $\triangle EFG$ about $\overline{EG}$?

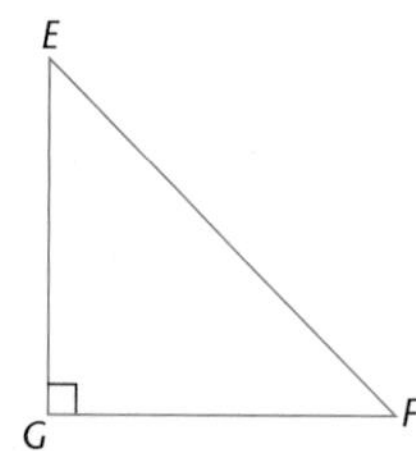

4. Describe the effect of multiplying each of the coordinates of the endpoints of a segment by -3.

5. Name some objects in your classroom that have three-dimensional rotational symmetry.

Practice & Apply

Draw three-dimensional coordinate systems and graph the segments with the given endpoints. Reflect each segment by multiplying each *y*-coordinate by -1.

6. $(4, -2, 3); (-2, -3, 2)$
7. $(-5, 2, 1); (1, 1, 1)$
8. $(1, -2, -3); (-1, 5, 2)$

What is the octant and coordinate of the reflection image if each of the points below is reflected over the *xy*-plane?

9. $(6, 5, 8)$
10. $(-2, 3, -1)$
11. $(1, 1, 1)$

What is the octant and coordinate of the reflection image if each point below is reflected over the *xz*-plane?

12. $(6, -2, 8)$
13. $(-4, -1, -1)$
14. $(1, 0, 1)$

15. Identify five physical objects that have rotational symmetry in space.

A potter is making pots according to certain patterns. For each half below, sketch what the complete pot will look like. (Rotate about the dotted line.)

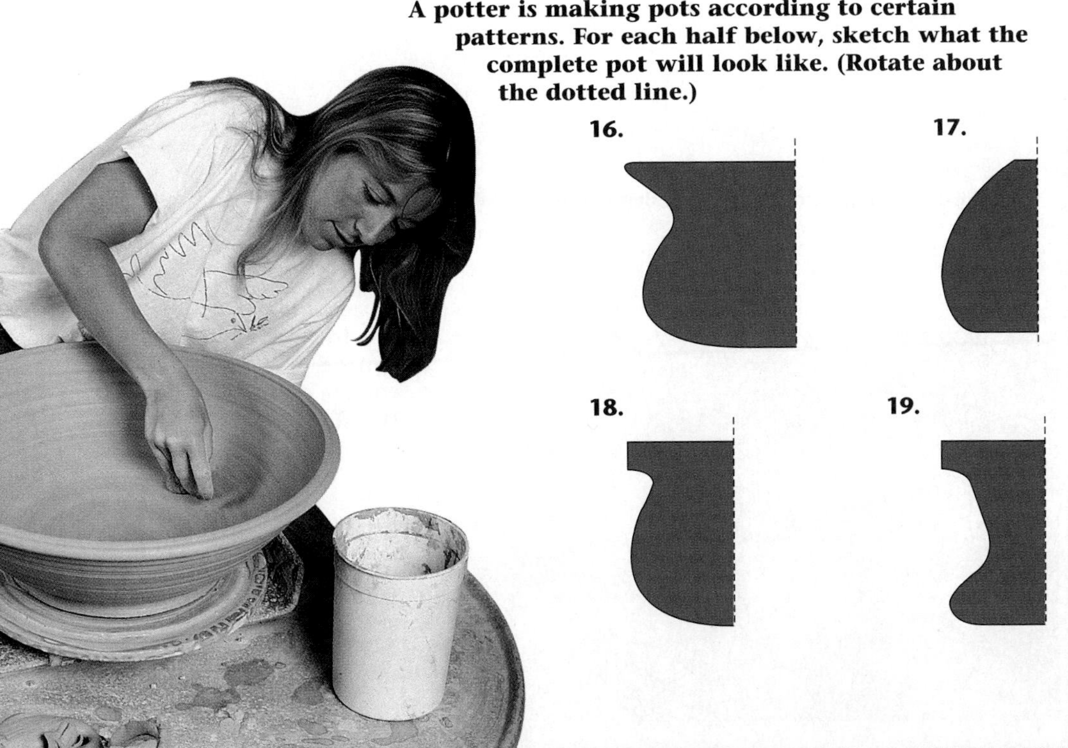

alternative ASSESSMENT

Authentic Assessment

Encourage students to find the symmetries in real objects. Have them list objects from their experience and describe their symmetries.

The answers to Exercises 6–8 and 16–19 can be found in Additional Answers beginning on page 727.

9. front-right-bottom; (6, 5, –8)
10. back-right-top; (–2, 3, 1)
11. front-right-bottom; (1, 1, –1)
12. front-right-top; (6, 2, 8)
13. back-right-bottom; (–4, 1, –1)
14. (1, 0 ,1) In the *xz* plane.
15. Sample answers: tires, steering wheels, cans, bottles, plates, glasses, bats, balls. Also, pizza and cheeseburgers under certain circumstances.

20. A circle with center (5, 0, 0) and radius 10 plus its interior.

21. 314.16 units^2

24. A cone with radius 4, height 4, and slant height $4\sqrt{2}$.

For Exercises 20–23, segment $\overline{AB}$ has coordinates $A(5, 0, 10)$ and $B(5, 0, 0)$.

20. What figure is formed by rotating $\overline{AB}$ about the x-axis?

21. What is the area of the figure formed by rotating $\overline{AB}$ about the x-axis? 314.16 units^2

22. What figure is formed by rotating $\overline{AB}$ about the z-axis? Cylinder with $r = 5$, h = 10.

23. What is the volume of the figure formed by rotating $\overline{AB}$ about the z-axis? 785.40 units^3

For Exercises 24 and 25, $\overline{CD}$ has coordinates $C(4, 0, 0)$ and $D(0, 0, 4)$.

24. What figure is formed by rotating $\overline{CD}$ about the z-axis?

25. What is the volume of the figure formed by rotating $\overline{CD}$ about the z-axis? 67.02 units^3

26. **Maximum/Minimum** The area of a right triangle for a given perimeter is maximized when the triangle is a 45-45-90 triangle. Suppose you rotate a 45-45-90 triangle about a leg to create a cone. Does the cone have the maximum volume? Why or why not? (See Exercises 25–33 in Lesson 7.5, p. 385.)

Look Back

Find the volume and surface area of each prism. [Lesson 7.2]

27.

7
11
15

Right-triangular prism

28.

6
12

Regular hexagonal prism

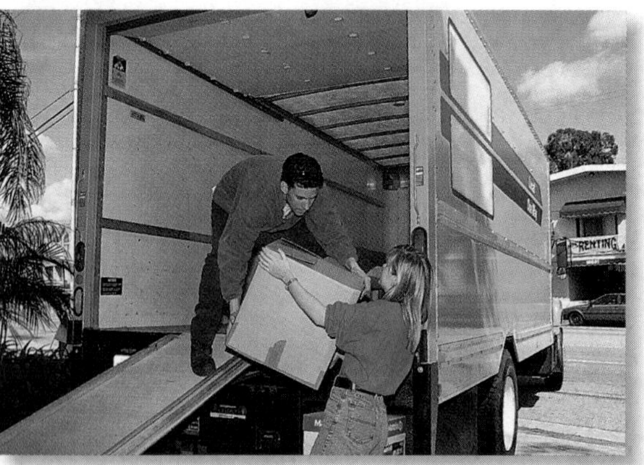

29. The truck's storage space measures 8 m × 5 m × 4 m. What is the volume of the storage space? **[Lesson 7.2]** 160 m^3

30. Find the surface area and volume of a square pyramid with area of base $B = 36$ cm^2, height $h = 5$ cm, and slant height $l = 8$ cm. **[Lesson 7.3]** SA = 132 cm^2; V = 60 cm^3

26. No, for example, if the slant height is held constant at 4, the volume function is $y = \frac{1}{3}\pi(x^2)(\sqrt{16 - x^2})$, where x is the cone height. This has a maximum at about $x = 3.27$, not at $\frac{4}{\sqrt{2}} = 2.83$.

27. SA = 456.63 units^2
V = 445.48 units^3

28. SA = 619.06 units^2
V = 1122.36 units^3

Practice Master

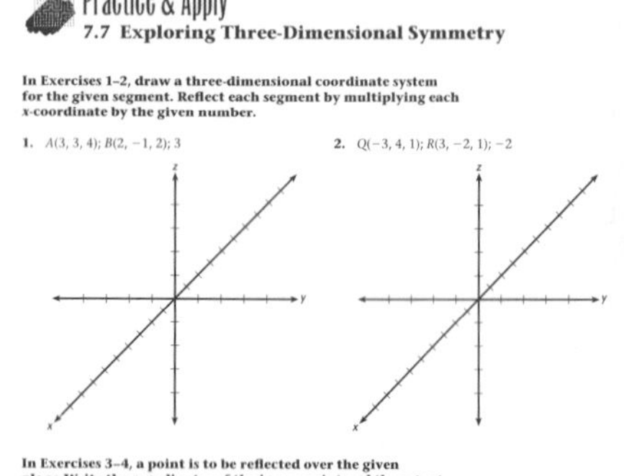

NAME ______________ CLASS ________ DATE ________

Practice & Apply

7.7 Exploring Three-Dimensional Symmetry

In Exercises 1–2, draw a three-dimensional coordinate system for the given segment. Reflect each segment by multiplying each x-coordinate by the given number.

1. $A(3, 3, 4)$; $B(2, -1, 2)$; 3

2. $Q(-3, 4, 1)$; $R(3, -2, 1)$; −2

In Exercises 3–4, a point is to be reflected over the given plane. Write the coordinates of the image point and the octant in which it will lie.

		Octant of image	Coordinates of image
3. (4, 3, −2)	over xy-plane	______	______
4. (4, 3, −2)	over xz-plane	______	______
5. (−2, 1, −3)	over yz-plane	______	______
6. (−2, 1, −3)	over xy-plane	______	______

In Exercises 5–6, describe the solid of revolution that would be formed by rotating the plane figure about the dotted line.

7. b, a

8. b, c, a

HRW Geometry — Practice & Apply 7

31. Noticing the barbecue smokers next to the potting soil, Ms. Solis decides to make a planter of her old smoker at home. She notices that the center of a smoker comes to her waist, or half her height, which is 5′-0″. Using this information to estimate the dimensions, determine how much soil it will take to fill the smoker to the centerline of the barrel from which it is made. **[Lesson 7.4]**

Look Beyond

A spinning fan creates the illusion of being a solid. A strobe light, which emits a flash at certain time intervals, can "freeze" the action.

32. Suppose a 4-bladed fan is turning at 12 revolutions per second. How often will the strobe need to flash to make the fan appear to stop?

33. If a strobe flashes at 36 flashes per second, at what speeds can a 4-bladed fan be going and still appear to be stopped?

31. Answers will vary. A reasonable estimate is approximately 5 cubic feet.

32. 3, 6, 12, 24, 48 . . . times per second

33. 9, 18, 36 . . . revolutions per second

Look Beyond

Extension. Exercises 32 and 33 extend symmetry ideas by including motion as a factor in making an object appear to have symmetry. Have students think of other exam-ples and discuss them in small groups.

NAME ______________________ CLASS ____________ DATE ____________

Technology

7.7 Building Symmetric Three-Dimensional Figures

The diagram shown is a three-dimensional coordinate system you can build with your geometry software. Using the segment and midpoint tools, you can make segments that are reflections of one another in a plane.

In the diagram, for example, $\overline{OP}$ and $\overline{OP'}$ are reflections of each other in the origin O in the xy plane. Using the rotate tool, you can create segments that appear to be perpendicular. For example, $\overline{OP''}$ in the xz plane appears to be perpendicular to both $\overline{OP}$ and $\overline{OP'}$ in the xy plane. Then using these and other tools, you can build your own three-dimensional symmetric figures.

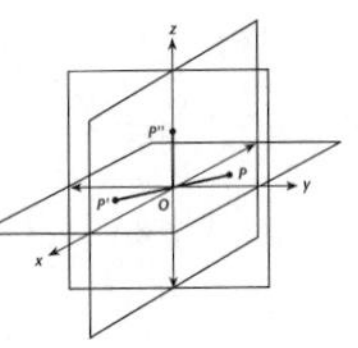

Use geometry software in the following exercises.

1. In your software, sketch three planes perpendicular in pairs. Insert the three lines where the planes meet in pairs, and label the lines x, y, and z as shown.

2. Using the results of Exercise 1, sketch $\overline{OP}$, $\overline{OP'}$ and $\overline{OP''}$ as suggested in the diagram. Join P, P', and P''. Describe the figure you made.

____

3. Using the results of Exercises 1 and 2, construct $\overline{AB}$ on the xy plane perpendicular and congruent to $\overline{PP'}$ with midpoint O. Join P, P', and P'', A and B to form a three-dimensional solid. Describe the figure you have made.

____

4. Using the results of Exercises 1–3, add in $\overline{OP'''}$ congruent to $\overline{OP''}$ but in the opposite direction to $\overline{OP''}$. Describe the figure you have formed.

____

5. Using as few segments and software commands as you can, construct a cube that is symmetric about the origin O.

____

HRW Geometry Technology 21

Focus

Build a polyhedron from a net pattern.

Motivate

In this project, students will use net patterns to help them build paper models of polyhedra. Have the class make nets out of different colors of paper and hang the resulting polyhedra from clothes hangers with strings of different lengths to form a mobile. Encourage students to include their own patterns.

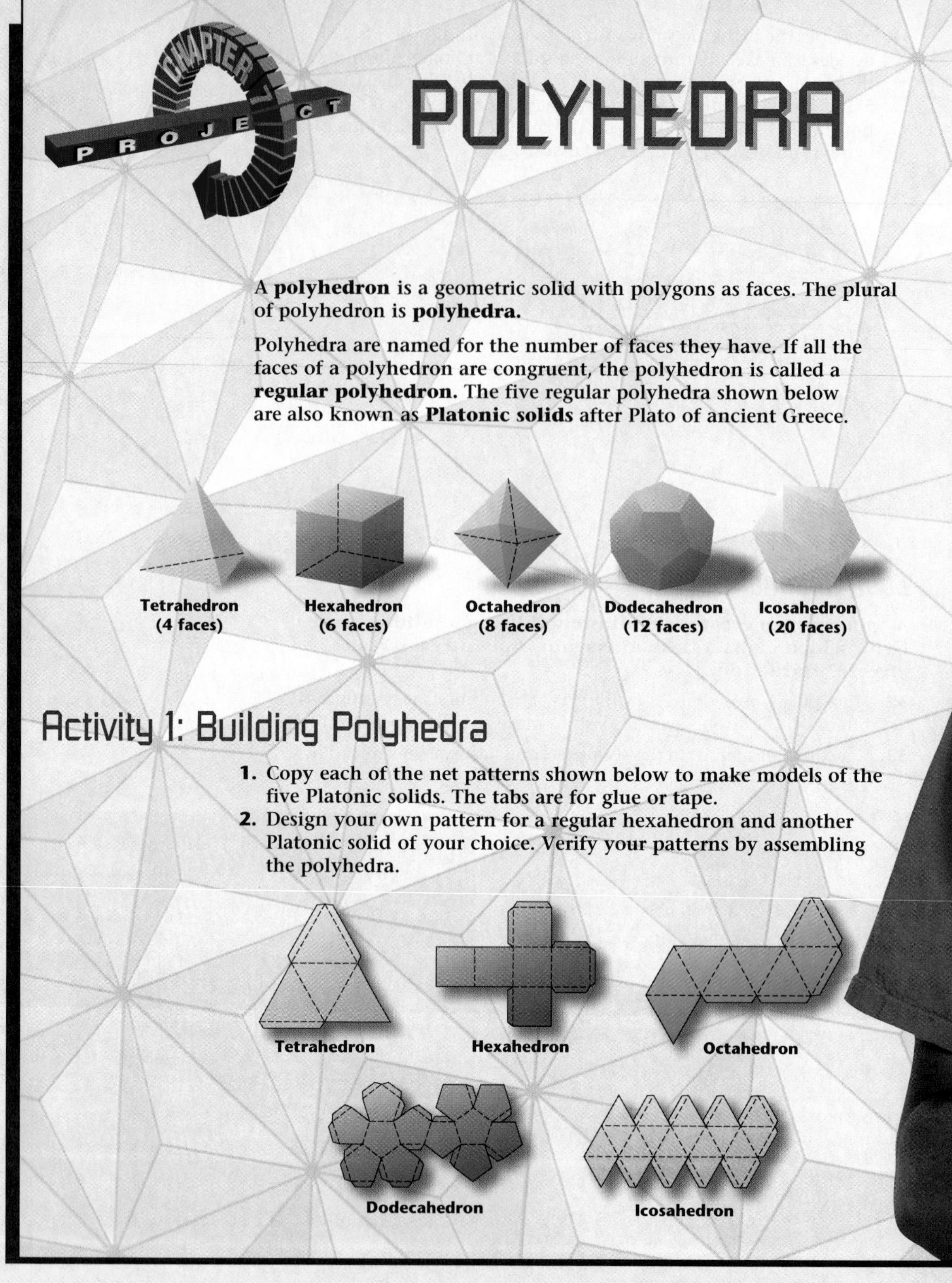

POLYHEDRA

A **polyhedron** is a geometric solid with polygons as faces. The plural of polyhedron is **polyhedra.**

Polyhedra are named for the number of faces they have. If all the faces of a polyhedron are congruent, the polyhedron is called a **regular polyhedron.** The five regular polyhedra shown below are also known as **Platonic solids** after Plato of ancient Greece.

Tetrahedron (4 faces) | Hexahedron (6 faces) | Octahedron (8 faces) | Dodecahedron (12 faces) | Icosahedron (20 faces)

Activity 1: Building Polyhedra

1. Copy each of the net patterns shown below to make models of the five Platonic solids. The tabs are for glue or tape.
2. Design your own pattern for a regular hexahedron and another Platonic solid of your choice. Verify your patterns by assembling the polyhedra.

Tetrahedron | Hexahedron | Octahedron | Dodecahedron | Icosahedron

Activity 2: "Pop-up" Dodecahedron

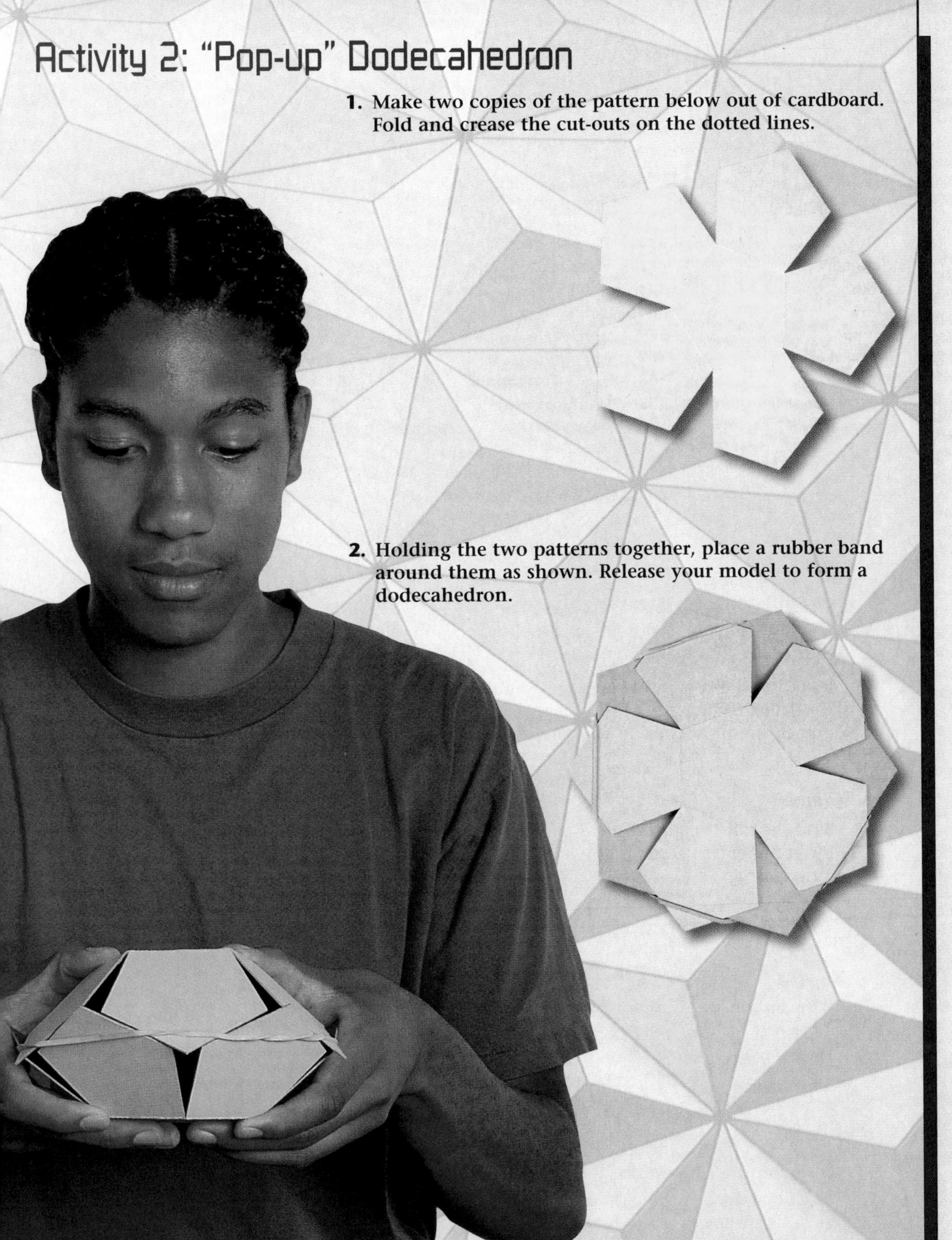

1. Make two copies of the pattern below out of cardboard. Fold and crease the cut-outs on the dotted lines.

2. Holding the two patterns together, place a rubber band around them as shown. Release your model to form a dodecahedron.

Cooperative Learning

Cooperative learning groups are very appropriate for this hands-on activity. Have students construct the polyhedra and offer to each other ideas about original patterns.

Discuss

How does symmetry help when you are designing a net for a solid? What are some difficulties encountered in designing a net? Can there be more than one net design for a polyhedron?

1. Cube: V = 64 units^3 ; SA = 96 units^2 ;
$\frac{SA}{V} = \frac{96}{64} = \frac{3}{2} = 1.5$
Prism: V = 84 units^3 ; SA = 122 units^2 ;
$\frac{SA}{V} = \frac{122}{84} \approx 1.45$
The ratio is larger for the cube.

Chapter 7 Review

Vocabulary

Key Skills and Exercises

Lesson 7.1

➤ ***Key Skills***

Solve problems using the ratio of surface area to volume.

A cube has a volume of 27,000 cubic millimeters. What is the ratio of its surface area to its volume?

The length of the cube's side must be 30 mm (30 × 30 × 30 = 27,000). The area of a face of the cube is 900 mm^2. Multiply that by 6 to get the surface area, 5,400 mm^2. Thus the ratio is 5,400 ÷ 27,000, or 0.2.

➤ ***Exercise***

1. Find the ratio of surface area to volume for the rectangular prism and the cube. Which ratio is larger?

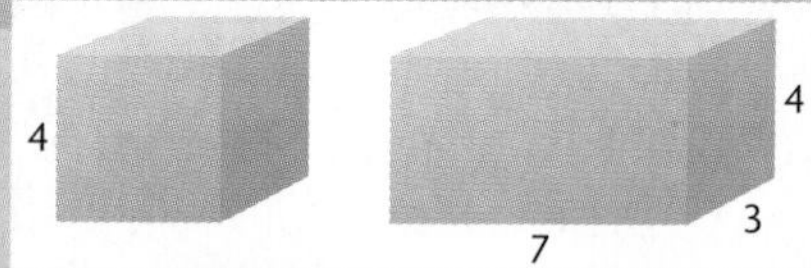

Lesson 7.2

➤ ***Key Skills***

Find the surface area of a right prism. $S = L + 2B$ **or** $ph + 2B$

The surface area of the prism is the lateral area plus the area of the bases. Multiply the height, 20 cm, by the perimeter of the base, 12 cm, to get lateral area, 240 cm^2. The bases happen to be right triangles; the base area is $\frac{1}{2}$ (3 × 4), or 6 cm^2. Total surface area is 240 + 6 + 6 = 252 cm^2.

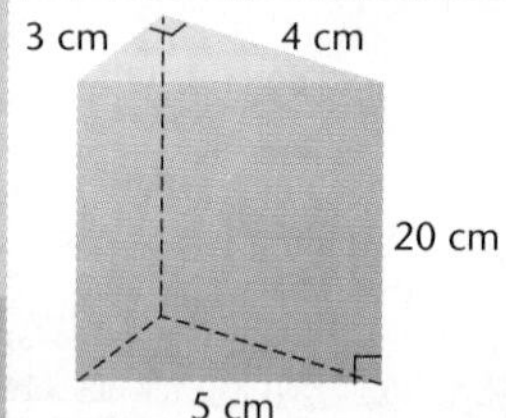

Find the volume of a prism. $V = Bh$

The volume of the prism equals the area of the base times the height of the prism. The triangular base has an area of 6 cm^2, and the height is 20 cm. Therefore, the volume of the prism is 6 × 20 = 120 cm^3.

➤ Exercises

In Exercises 2–3, refer to the right rectangular prism.

2. Find the surface area of the prism. **3.** Find the volume of the prism.

9 in.

23 in.

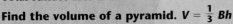

11 in.

Lesson 7.3

➤ Key Skills

Find the surface area of a right pyramid. $S = L + B$ **or** $\frac{1}{2}pl + B$

The surface area of the pyramid is the lateral area plus the area of the base. Multiply the slant height, 13 ft, by the perimeter of the base, 40 ft, and by $\frac{1}{2}$ to get lateral area, 260 ft^2. The base is a square; the base area is thus 100 ft^2. Total surface area is $260 + 100 = 360$ ft^2.

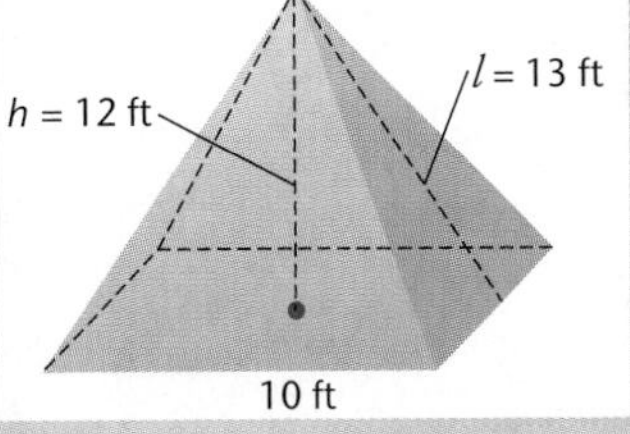

Find the volume of a pyramid. $V = \frac{1}{3}Bh$

The volume of the pyramid above equals the area of the base times the height times $\frac{1}{3}$, or $100 \times 12 \times \frac{1}{3} = 400$ ft^3.

➤ Exercises

In Exercises 4–5, refer to the right regular pyramid.

4. Find the surface area of the pyramid.

5. Find the volume of the pyramid.

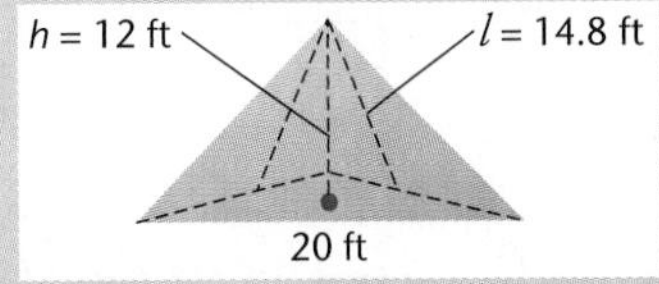

Lesson 7.4

➤ Key Skills

Find the surface area of a right cylinder. $S = L + 2B$**, or** $2\pi rh + 2\pi r^2$

The surface area of the cylinder above is the lateral area plus the area of the bases. Multiply the height, 35 m, by the circumference of the base, 44 m, to get lateral area, 1,540 m^2. The base area is πr^2, or approximately 154 m^2. Total surface area is $1,540 + 154 + 154 \approx 1,848$ m^2.

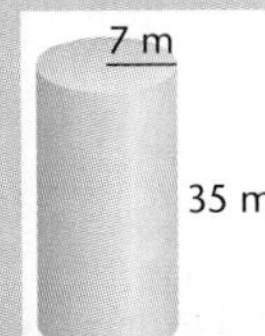

Find the volume of a cylinder. $V = Bh$**, or** $V = \pi r^2 h$

The volume of the cylinder above equals the area of the base times the height of the cylinder, or $154 \times 35 \approx 5,390$ m^3.

➤ Exercises

In Exercises 6–7, refer to the right cylinder.

6. Find the surface area of the cylinder.

7. Find the volume of the cylinder.

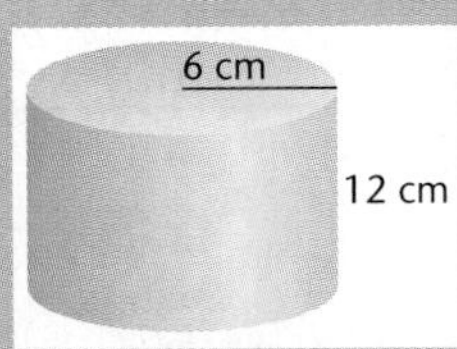

Lesson 7.5

➤ Key Skills

Find the surface area of a right cone.
$S = L + B$**, or** $S = \pi rl + \pi r^2$

The surface area of the cone is the lateral area plus the area of the base. Multiply the slant height, 5 in., by the radius of the base, 3 in., by π, ~3.14, to get lateral area, 47.1 in.^{2}. The base area is approximately 28.3 in.^{2}. Total surface area is $47.1 + 28.3 \approx 75.4$ in.^{2}.

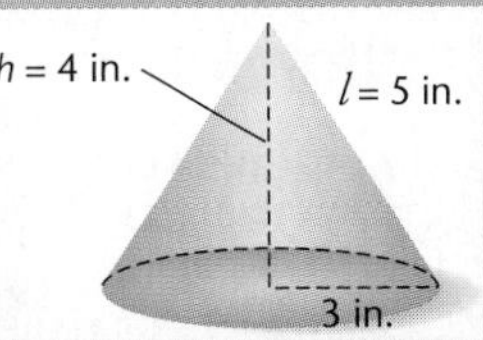

Find the volume of a cone. $V = \frac{1}{3}Bh$**, or** $V = \frac{1}{3}\pi r^2 h$

The volume of the cone above equals the area of the base times the altitude of the cone times $\frac{1}{3}$, or approximately 37.7 in.^{3}.

2. $SA = 2(99) + 23(40) = 1118$ in^2

3. $V = (9)(11)(23) = 2277$ in^3

4. 617.2 ft^2

5. 692.8 ft^3

6. $SA = (2\pi)(6^2) + (2\pi)(6)(12) = 216\pi \approx 678.58$ cm^2

7. $V = (\pi)(6^2)(12) = 432\pi = 1357.17$ cm^3

8. SA = $\pi(x)(xr(2)\ ) + \pi x^2 = px^2\sqrt{2} + \pi x^2$

9. $V = \frac{1}{3}\pi(x^2)(x) = \frac{1}{3}\pi x^3$

10. ~ 201.06 nanometers^2

11. ~ 65.45 mm^3

12.

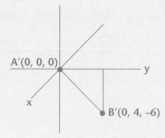

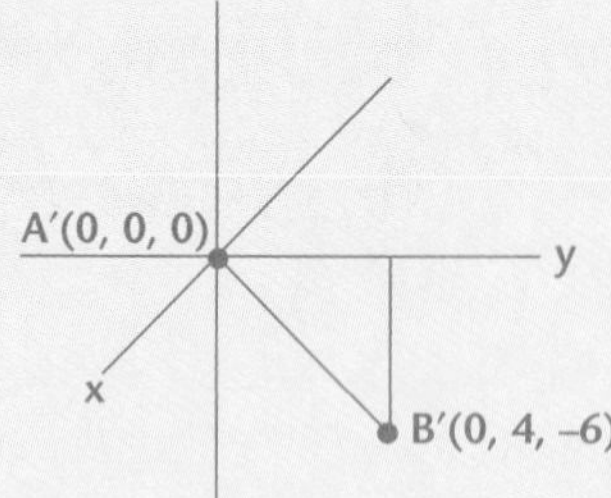

3. $V = \frac{1}{3}(6^2\pi)(4) = 48\pi \approx 150.8$

14. The largest tank can hold: $(25^2\pi)(25) \approx 49087.4$ ft^3 which is $(49087.4)(7.48) \approx 367{,}173.8$ gallons. So the company will need 2 tanks to hold 500,000 gallons.

15. ~155.03 m^2

➤ *Exercises*

In Exercises 8–9, refer to the right cone.

8. Find the surface area of the cone.

9. Find the volume of the cone.

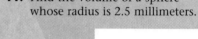

Lesson 7.6

➤ *Key Skills*

Find the surface area of a sphere. $S = 4\pi r^2$

For a sphere with radius 21 feet, the surface area is

$4 \times \frac{22}{7} \times (21)^2 = 5{,}544$ square feet.

Find the volume of a sphere. $V = \frac{4}{3}\pi r^3$

For a sphere with radius 21 feet, the volume is approximately

$\frac{4}{3} \times \frac{22}{7} \times (21)^3 = 38{,}808$ cubic feet.

➤ *Exercises*

10. Find the surface area of a sphere whose radius is 4 nanometers.

11. Find the volume of a sphere whose radius is 2.5 millimeters.

Lesson 7.7

➤ *Key Skills*

Reflect a line in a three-dimensional coordinate system.

In the coordinate system at the right, a segment with endpoints at (5, 5, –2) and (5, 5, 2) is reflected through the *xz*-plane.

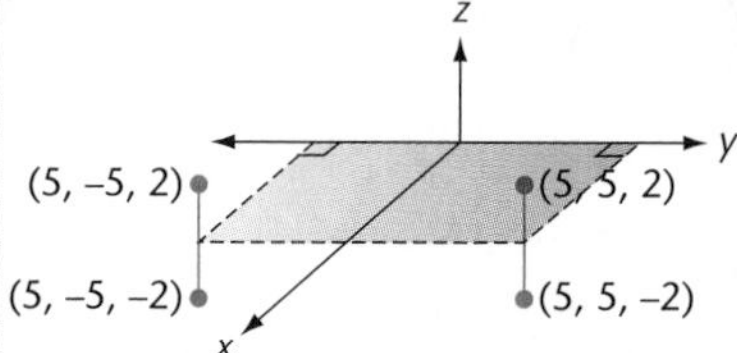

Sketch a solid of revolution, given a line.

If you rotate the segment above about the *z*-axis, you produce the lateral surface of a cylinder of height 4 and radius 5 shown at the right.

➤ *Exercises*

In Exercises 12–13, refer to the segment at the right.

12. Draw a reflection of segment $\overline{AB}$ through the *xy*-plane. Give the coordinates of the reflection's end points.

13. Find the volume of the figure created by rotating the segment around the *y*-axis.

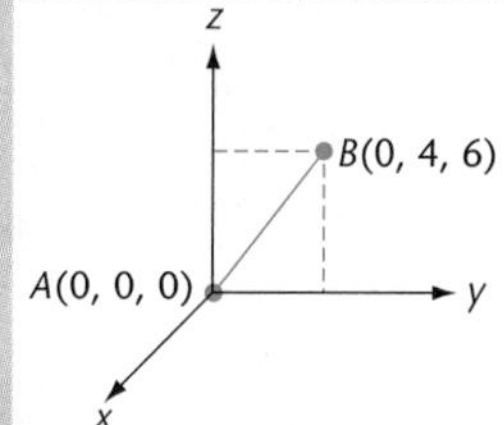

Applications

14. Manufacturing A processing plant needs storage tanks to hold at least half a million gallons of waste water. For a combination of visual and engineering reasons, the interior of the tanks must be no higher than 25 feet and no wider than 50 feet. How many cylindrical tanks must be built to hold the waste water? (1 cubic foot = 7.48 gallons)

15. Find the surface area of the truncated cone at the right.

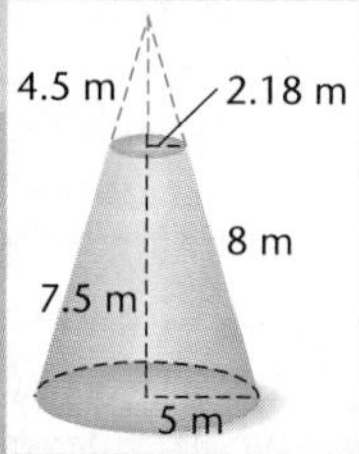

Chapter 7 Assessment

1. Find the ratio of surface area to volume for a sphere with a radius of 120 feet.

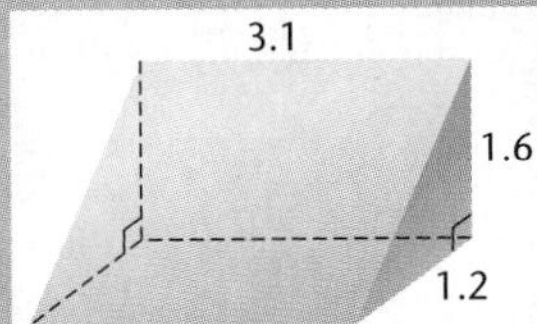

In Exercises 2–3, refer to the right rectangular prism.

2. Find the surface area of the prism.
3. Find the volume of the prism.

In Exercises 4–5, refer to the pyramid below.

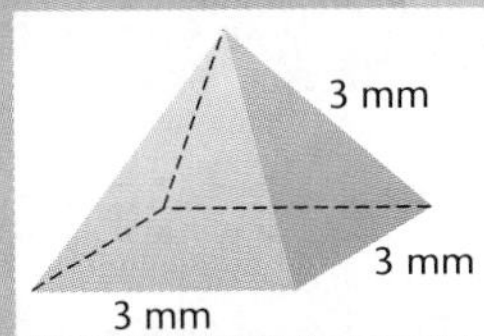

4. Find the surface area of the pyramid.
5. Find the volume of the pyramid.

In Exercises 6–7, refer to the right cylinder.

6. Find the surface area of the cylinder.
7. Find the volume of the cylinder.

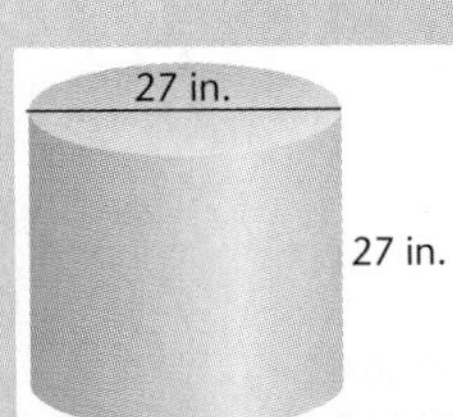

In Exercises 8–9, refer to the right cone below.

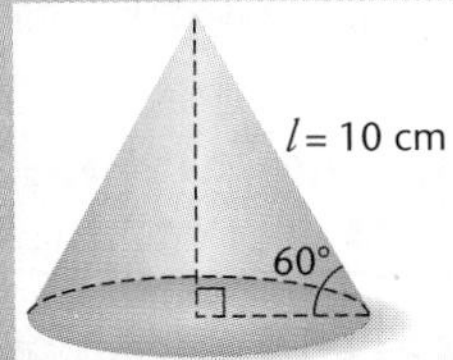

8. Find the surface area of the cone.
9. Find the volume of the cone.

10. Give the coordinates for the end points of the reflection through the *yz*-plane of the segment to the right.

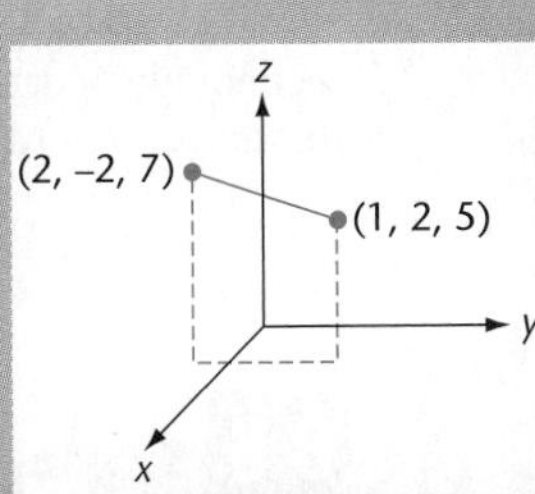

10.

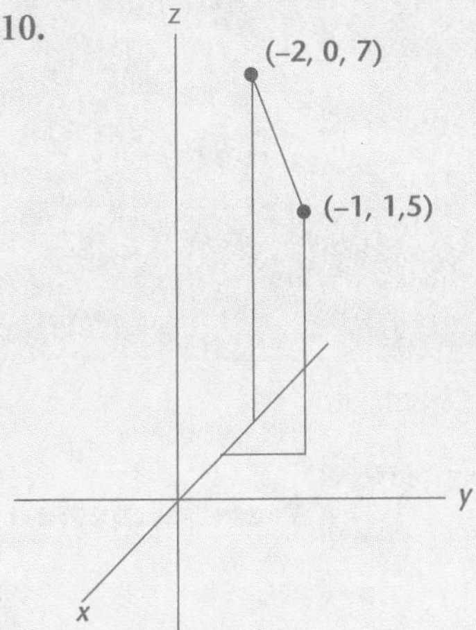

1. ~ .025
2. 16.8 units^2
3. 2.976 units^3
4. ~24.59 mm^2
5. ~ 6.36 mm^3
6. ~ 3435.33 in^2
7. ~ 15458.99 in^3
8. ~ 235.62 cm^2
9. ~ 226.72 cm^3

Similar Shapes

Meeting Individual Needs

8.1 Transformations and Scale Factors

Core Resources

Inclusion Strategies, p. 414
Reteaching the Lesson, p. 415
Practice Master 8.1
Enrichment Master 8.1
Lesson Activity Master 8.1
Interdisciplinary Connection, p. 413

[2 days]

Core Plus Resources

Practice Master 8.1
Enrichment, p. 414
Technology Master 8.1
Interdisciplinary Connection, p. 413

[2 days]

8.2 Exploring Similar Polygons

Core Resources

Inclusion Strategies, p. 421
Reteaching the Lesson, p. 422
Practice Master 8.2
Enrichment, p. 421
Technology Master 8.2
Lesson Activity Master 8.1

[2 days]

Core Plus Resources

Practice Master 8.2
Enrichment Master 8.2
Technology Master 8.2

[1 day]

8.3 Exploring Triangle Similarity Postulates

Core Resources

Inclusion Strategies, p. 431
Reteaching the Lesson, p. 432
Practice Master 8.3
Enrichment Master 8.3
Lesson Activity Master 8.1
Interdisciplinary Connection, p. 430
Mid-Chapter Assessment Master

[2 days]

Core Plus Resources

Practice Master 8.3
Enrichment, p. 431
Technology Master 8.3
Interdisciplinary Connection, p. 430
Mid-Chapter Assessment Master

[2 days]

8.4 The Side-Splitting Theorem

Core Resources

Inclusion Strategies, p. 438
Reteaching the Lesson, p. 439
Practice Master 8.4
Enrichment, p. 438
Technology Master 8.4
Lesson Activity Master 8.1
Interdisciplinary Connection, p. 437

[2 days]

Core Plus Resources

Practice Master 8.4
Enrichment Master 8.4
Technology Master 8.4
Interdisciplinary Connection, p. 437

[1 day]

8.5 Indirect Measurement, Additional Similarity Theorems

Core Resources

Inclusion Strategies, p. 445
Reteaching the Lesson, p. 446
Practice Master 8.5
Enrichment Master 8.5
Technology Master 8.5
Lesson Activity Master 8.1

[2 days]

Core Plus Resources

Practice Master 8.5
Enrichment, p. 445
Technology Master 8.5
Interdisciplinary Connection, p. 444

[2 days]

8.6 Exploring Area and Volume Ratios

Core Resources

Inclusion Strategies, p. 454
Reteaching the Lesson, p. 455
Practice Master 8.6
Enrichment, p. 454
Lesson Activity Master 8.1
Interdisciplinary Connection, p. 453

[2 days]

Core Plus Resources

Practice Master 8.6
Enrichment Master 8.6
Technology Master 8.6
Interdisciplinary Connection, p. 453

[1 day]

Chapter Summary

Core Resources	Core Plus Resources
Chapter 8 Project, pp. 460–461 Lab Activity Long Term Project Chapter Review, pp. 462–464 Chapter Assessment p. 465 Chapter Assessment, A/B Alternative Assessment Cumulative Assessment, pp. 466–467	Chapter 8 Project, pp. 460–461 Lab Activity Long Term Project Chapter Review, pp. 462–464 Chapter Assessment p. 465 Chapter Assessment, A/B Alternative Assessment Cumulative Assessment, pp. 466–467
[3 days]	**[2 days]**

Reading Strategies

This chapter includes a good example of a mathematical term that has a different meaning in everyday life. The key term *similar* is often used in common speech to mean that two or more things share a characteristic. For example, consider a five-pointed star. If a student is asked to draw a figure similar to this star, he or she might well draw a six-pointed star. But in mathematics, the word *similar* has a very specific meaning connected with the ratios of corresponding sides—these ratios must be equal. And in similar figures, corresponding angles must be equal.

Have students work in small groups to brainstorm math vocabulary terms that have different meanings in everyday life. They can use the glossary to get ideas for their lists. Amusing sentences, puns, or drawings can be created by using words in the wrong context. One example is a drawing showing the "legs" of a triangle.

Visual Strategies

As students work on the activities in this chapter, they will begin to learn to estimate whether two polygons look similar. Although a visual guess should always be verified with appropriate ratios, postulates, or theorems, it is still helpful for students to be able to visually identify possible cases of similarity.

An overhead projector can be used to demonstrate sets of similar figures. Use a variety of rectangles and triangles on a transparency. By moving the projector closer to or farther from the screen, figures will change in size but not in shape. You might have volunteers trace sets of similar polygons during this type of demonstration.

Show students that similar rectangles and triangles can be "nested" in a specific way. That is, if a set of similar rectangles is stacked up with one vertex coinciding, the diagonals of all the rectangles line up. In a nested set of similar triangles, the sides opposite the coinciding vertices are all parallel. This idea may help students learn to recognize similar shapes more easily. Also, students can copy figures in the text on tracing paper so that both polygons can be viewed in the same orientation.

Cooperative Learning

GROUP ACTIVITIES	
Defining Similar Polygons	Lesson 8.2, Exploration
Indirect measurement	Lesson 8.2, Look Beyond
Postulates for similar triangles	Lesson 8.3, Explorations 1–3
Area and volume ratios	Lesson 8.6, Explorations 1–3
Indirect Measurement and Scale Models	Chapter 8 Project

You may wish to have students work in groups or with partners for some of the above activities. Additional suggestions for cooperative group activities are noted in the teacher's notes in each lesson.

Multicultural

The cultural references in this chapter include references to Africa.

CULTURAL CONNECTIONS	
Africa: Egyptian geometry problems	Lesson 8.4, Example 2
Africa: Eratosthenes' Estimate	Lesson 8.5, Example 2
Africa: Kush Pyramids	Lesson 8.5, Exercise 21

Portfolio Assessment

Below are portfolio activities for the chapter listed under seven activity domains which are appropriate for portfolio development.

1. **Investigation/Exploration** In the explorations for Lesson 8.1, students investigate properties of transformations. The Lesson 8.2 exploration applies ideas of proportionality to similar polygons, then postulates for similar triangles are developed in the explorations for Lesson 8.3. In the Lesson 8.6 explorations, students explore ratios of areas and volumes.

2. **Applications** Optics, Lesson 8.1, Exercises 15–17; Wildlife Management, Lesson 8.2, Exercises 24–26; Advertising, Lesson 8.2, Exercise 31; Surveying, Lesson 8.3, Exercises 17–19; Packaging, Lesson 8.6, Exercise 40; Sports, Lesson 8.6, Exercise 42.

3. **Nonroutine Problems** Lesson 8.2, Exercises 41–44 (Width of the River); Lesson 8.6, Exercise 52 (Nested Squares).

4. **Project** Indirect Measurement and Scale Models: see pages 460–461.

5. **Interdisciplinary Topics** Earth Science, Lesson 8.1, Exercises 37–38; Algebra, Lesson 8.2, Exercises 13–23, 38; Marine Biology, Lesson 8.2, Exercise 27; Fine Art, Lesson 8.2, Exercises 29–30; Algebra, Lesson 8.3, Exercises 22–25; Algebra, Lesson 8.5, Exercises 5–6; Astronomy, Lesson 8.5, Exercises 13–17.

6. **Writing** *Communicate* exercises offer excellent writing selections for the portfolio. Suggested selections include: Lesson 8.3, Exercise 4; Lesson 8.4, Exercises 3–5.

7. **Tools** In Chapter 8, students can use graphics calculators or spreadsheet software to organize data and look for relationships. Geometry graphics software will be helpful as students investigate properties of similar triangles and other polygons.

Technology

Many of the activities in this book can be significantly enhanced by the use of geometry graphics software, but in every case the technology is optional. If computers are not available, the computer activities can, *and should,* be done by hands-on or "low-tech" methods, as they contain important instructional material. For more information on the use of technology, refer to the *HRW Technology Manual for High School Mathematics.*

When instructing students in the use of the software, you may find that the best approach is to give students a few well-chosen hints about the different tools and let them discover their use on their own. Today's computer-savvy students can be quite impressive in their ability to discover the uses of computer software! Once the students are familiar with the geometry software, they can design and create their own computer sketches for the explorations.

A second approach is for you, the teacher, to demonstrate the creation of the exploration sketches "from scratch" using a projection device. The students will find this exciting to watch, and they will be eager to learn to use the software on their own.

In the interest of convenience or time-saving, you may want to use the *Geometry Investigations Software* available from HRW. Each of the files on the disks contains a "sketch," custom-designed for an *HRW Geometry* Exploration. The sketches are to be used with *Geometer's Sketchpad*™ (Key Curriculm Press) or *Cabri Geometry II*™ (Texas Instruments).

Computer Graphics Software

Lesson 8.1: Exploration 1 Students examine the properties of a dilation by dragging a dilated segment on a coordinate plane. They discover the relationship of the coordinates of the image to the coordinates of the preimage and then the scale factor.

Lesson 8.1: Exploration 2 Students make conjectures about the distance of a pre-image from the center of dilation, the distance of the dilated image from the center of dilation, and the scale factor of the image. Students observe changes in measurements as they drag the vertices of the pre-image.

Lesson 8.3: Exploration 1 Given the measures of the angles, students investigate two similar triangles. After they find the lengths of the sides they can complete the wording of a similarity postulate (AA Similarity).

Lesson 8.3: Exploration 2 Given the lengths of the sides, students investigate two similar triangles. Once they have determined the ratio of the sides and the angle measures, they are asked to complete the wording of a similarity postulate (SSS similarity).

Lesson 8.3: Exploration 3 Given the lengths of two sides and the measures of the included angles, students investigate two similar triangles. After they determine the measures and the ratios of the remaining parts of the triangles they can complete the wording of a similarity postulate (SAS similarity).

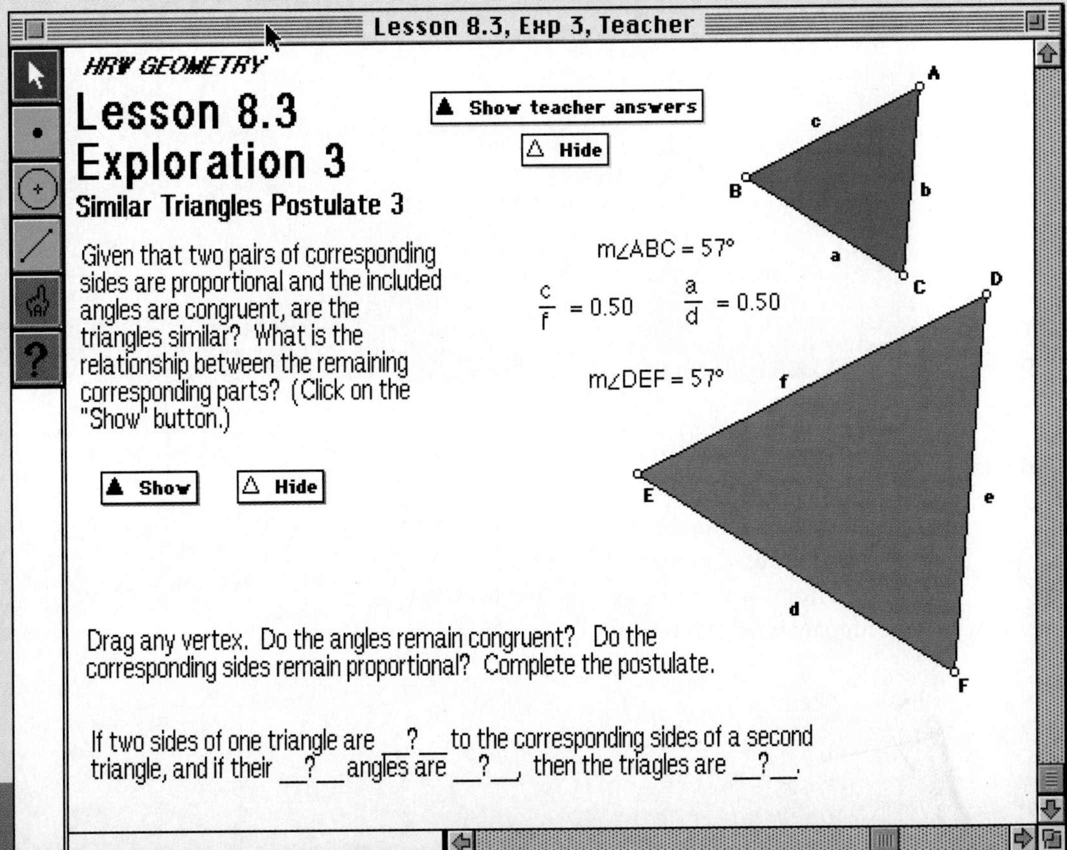

About the Chapter

Background Information

In this chapter, students explore similar figures through transformations in the plane, construction, and formal proof. Students use similarity to measure distances indirectly and explore the area and volume of similar figures.

Chapter Resources

- Practice Masters
- Enrichment Masters
- Technology Masters
- Lesson Activity Masters
- Lab Activity Masters
- Long-Term Project Masters
- Assessment Masters
  - Chapter Assessments, A/B
  - Mid-Chapter Assessment
  - Alternative Assessments, A/B
- Teaching Transparencies
- Cumulative Assessment
- Spanish Resources

Chapter Objectives

- Construct a translation of a segment and a point using a scale factor.
- Construct a dilation of a closed plane figure.
- Define *polygon similarity.*
- Use properties of proportion and scale factor with similar polygons.
- Develop three triangle similarity postulates—AA, SSS, SAS.
- Formally define the three triangle similarity postulates —AA, SSS, SAS.
- Develop and use the Side-Splitting Theorem.

Chapter 8

Lessons

Similar Shapes

What geometric idea is illustrated by the different sizes on these pages? The car and the model are not congruent because they do not have the same size. However, they do have the same shapes. Such figures are known as **similar figures.** You will find many opportunities to study similar figures both in nature and in art.

Artists and designers often enlarge or reduce figures without changing their shapes. For example, a muralist may cover an entire wall with a painting made from a small photograph. You will learn the mathematics of such procedures in this chapter.

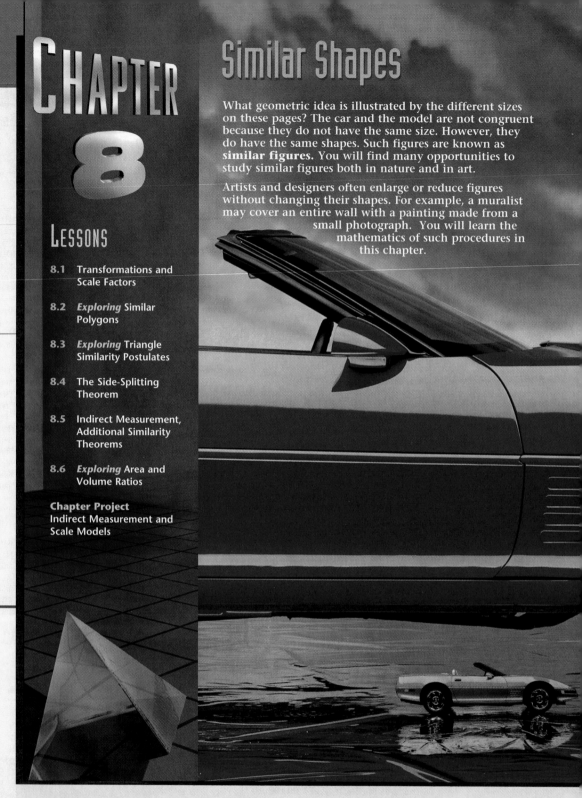

About the Photo

Similarity is demonstrated in the photo of the full-sized car and the model. Properties of similarity are used in many diverse human endeavors from scale-drawings in architectures to miniature models used for film making.

- Use triangle similarity to measure distances indirectly.
- Develop and use the Proportional Altitudes and Proportional Medians Theorems.
- Develop and use ratios for areas of similar figures.
- Develop and use ratios for volumes of similar solids.
- Explore relationships between cross-sectional area, weight, and height.

PORTFOLIO ACTIVITY

This portfolio activity has an interdisciplinary connection to art and photography. Students may practice the enlargement or reduction process by using newspaper comics. First draw a grid on each cartoon panel. Carefully copy the contents of each grid to another larger or smaller grid. After the method seems clear, students can then enlarge or reduce a more complicated picture.

An additional Pupil's Edition portfolio activity can be found in the exercise for Lesson 8.2.

ABOUT THE CHAPTER PROJECT

In the Chapter 8 Project, on pages 460–461, students use a variety of different indirect methods to find the dimensions of a building or of another structure that would otherwise be difficult to measure. They also build a scale model of a structure, explain how to find its dimensions, and explain how they constructed the model.

PREPARE

Objectives

- Construct a translation of a segment and a point using a scale factor.
- Construct a dilation of a closed plane figure.

RESOURCES

- Practice Master 8.1
- Enrichment Master 8.1
- Technology Master 8.1
- Lesson Activity Master 8.1
- Quiz 8.1
- Spanish Resources 8.1

Assessing Prior Knowledge

1. Find the distance between points (–2, 1) and (4, 3). [$2\sqrt{10}$]
2. What is the slope of the line $y = -3x + 2$? [**–3**]
3. The points (–2, 4), (3, 4), and (3, –1) are vertices of a square. Find the fourth vertex. [**(–2, –1)**]
4. Find the equation of the line containing (5, –1), (6, –2). [$\mathbf{y = -x + 4}$]
5. True or False: Points (5, –1), (6, 2) and (–1, 3) are collinear. [**False**]

TEACH

Students need to understand the properties of transformations because they are the analytical basis for constructing similar figures and drawing scale models.

LESSON 8.1

Exploring Transformations and Scale Factors

why *The principle of the camera obscura was discovered by the Iraqi scientist Ibn al Haitham (965–1039 C.E.), probably from thinking about inverted images of objects projected through small holes in tents. Transformations can be used to explain why the images are inverted.*

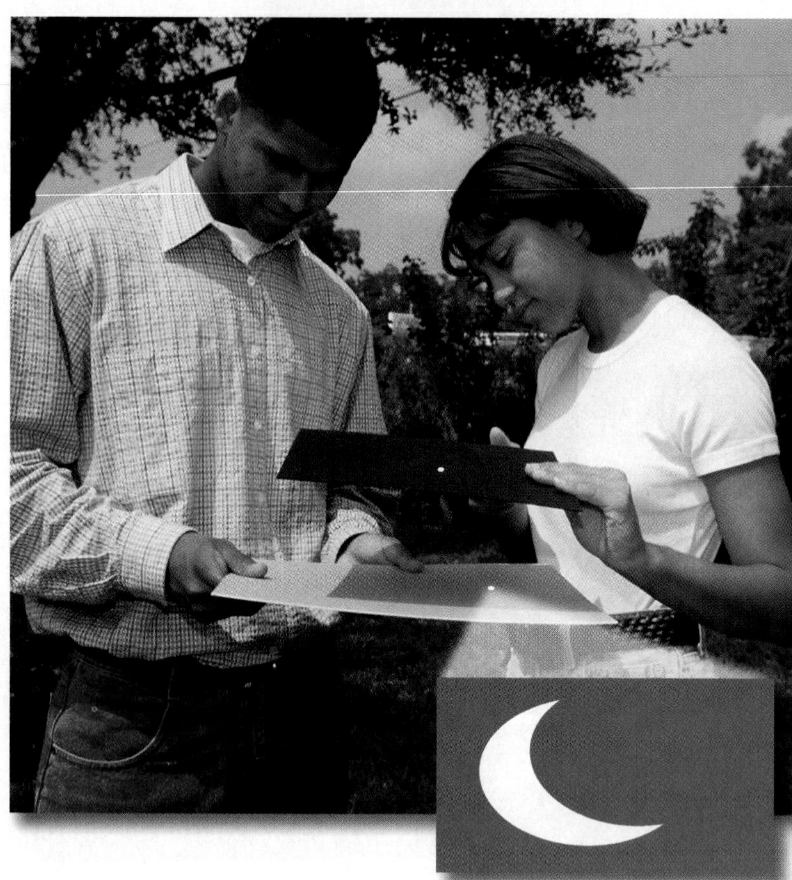

The camera obscura, which was the forerunner of the modern camera, and the pinhole used by these students to observe an eclipse of the sun operate on the same principles.

Scale Factors

What happens to a point when you multiply both coordinates of a point by the same real number?

Pre-image		**Image**
$A(2, 3)$	$\rightarrow$	$A'(2 \times 4, 3 \times 4) = A'(8, 12)$
	Multiply by 4	

The result is a *transformation* of the original point. The number that multiplied the coordinates is called the **scale factor**.

ALTERNATIVE teaching strategy

Technology Students can explore dilations and other transformations with geometry graphics software, which can also be used to measure the lengths of image segments and to explore the collinearity of transformed points. Ask students to verify their conjectures with computer-drawn images.

Exploration 1 *Transforming a Point Using a Scale Factor*

You will need
Geometry technology or
Graph paper
Ruler and calculator (optional)

In this exploration, you will observe what happens when you transform a segment using different scale factors.

Part I

1. Draw segment $\overline{AB}$ on a coordinate plane, with endpoints $A(2, 4)$ and $B(6, 1)$. Multiply the coordinates of A and B by a scale factor of 2.

2. Plot your transformed points A' and B' on your graph. Connect them to form segment $\overline{A'B'}$.

3. Use scale factors of 0.5 and -1 to create additional new segments $\overline{A''B''}$ and $\overline{A'''B'''}$. Construct these segments on your graph and label each one with its scale factor.

Part II

1. Look at just the point A and its images A', A'', and A'''. What is the simplest geometric figure that contains all of these points and the *origin O as well*? Add this figure to your graph.

2. Use the Distance Formula or a ruler to find the distances OA and OA'. Find the ratio $\frac{OA'}{OA}$. How does this ratio compare with the scale factor that gives the point A'?

Repeat for all the transformations of point A. Then copy and complete the table.

A	Scale factor	A'	OA	OA', etc.	$\frac{OA'}{OA}$, etc.
(2, 4)	−1	?	?	?	?
(2, 4)	0.5	?	?	?	?
(2, 4)	2	?	?	?	?
(2, 4)	n	?	?	?	?

3. Write a conjecture about the effect of a scale factor transformation on a point. Include all your findings from Steps 1 and 2.

4. What will happen if you transform a point by a scale factor of n?

Pre-image		**Image**
$P(x, y)$	$\rightarrow$	$P'(x \times n, y \times n) = P'(xn, yn)$
	Multiply by n	

Exploration 1 Notes

In this exploration, students will transform points and segments using different scale factors. This activity will help students understand that the ratio of the image distance to the preimage distance (from the center of the transformation) is equal to the scale factor.

Teaching tip

Prime notation used to label image points confuses some students. Point out that it is used to make the correspondence between preimage points and image points more obvious.

Ongoing Assessment

3. **The image of a point under a scale factor transformation on n is |n| times as far from (0, 0).**

Photography Photographers use dilations when they reduce or enlarge photographs. Have students research the techniques and types of equipment used by photographers to do enlargements. Have them calculate the area of a 3 in. by 5 in. photograph and compare it with the area of an enlargement done with a scale factor of 2. [**The enlargement will have 4 times the area, or 60 sq in.**]

CRITICAL Thinking

If the distance of the original point is d, then the distance of the transformed point is nd, if n is the scale factor. The ratio is $\frac{nd}{d} = n.$

TEACHING tip

Have students verify their conjectures about scale factor transformations by redoing Exploration 1 with various centers of dilation using fractional and negative scale factors.

ongoing ASSESSMENT

3. **The image of a segment under a scale factor transformation of n is $|n|$ times as long as the pre-image. The image has the same slope as the pre-image and is parallel to the pre-image.**

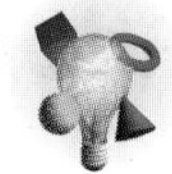

CRITICAL Thinking What is the distance of the original point from the origin? of the transformed point from the origin? What is the ratio of the second distance to the first distance?

Part III

1. Consider each segment in the table below. Find the length of each of the segments. How do the lengths of the transformed segments compare to the length of the original segment $\overline{AB}$? Copy and complete the table.

A	B	Scale factor	A', etc.	B', etc.	Length AB	Length $A'B'$	$\frac{A'B'}{AB}$, etc.
(2, 4)	(6, 1)	−1	?	?	?	?	?
(2, 4)	(6, 1)	0.5	?	?	?	?	?
(2, 4)	(6, 1)	2	?	?	?	?	?
(2, 4)	(6, 1)	n	?	?	?	?	?

2. Find the slope of each of the segments on your graph.
3. Write a conjecture about the effect of a scale factor transformation on a segment. Include all of your findings from Steps 1 and 2.
4. What is the effect of a scale factor transformation on a segment if the scale factor is n? Let $P(x_1, y_1)$ and $Q(x_2, y_2)$ be endpoints of the segment. What is the length of the original segment? of the transformed segment? What is the slope of each? How do these compare? ❖

Dilations

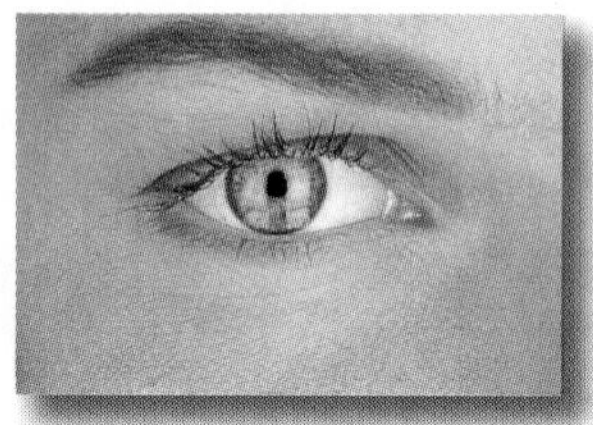

Muscles in the iris dilate the pupil to let in more light and contract to let in less light.

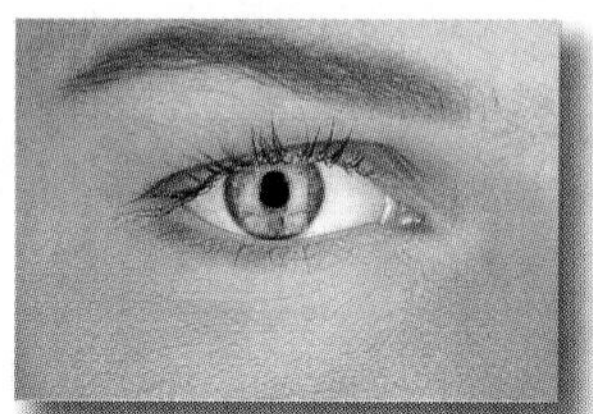

The transformations you have been exploring in this lesson are known as **dilations**. The name applies to all transformations of this kind. However, when transformed figures are reduced in size, they are often called **contractions**.

Every dilation has a point known as the **center of dilation**. In Exploration 1, that point was the origin.

CRITICAL Thinking If you want to use a center of dilation that is not at the origin, how can you adapt the procedure in Exploration 1 to your purposes? You will need to do some other transformations in the process.

CRITICAL Thinking

Translate the center of the dilation so that it is the new origin (0, 0).

ENRICHMENT A pantograph is an instrument that is used to reproduce maps and other drawings. Have students research how it works and write a report. Extend the assignment by having students build a pantograph.

INCLUSION strategies **Using Cognitive Strategies** Help students make generalizations by having them keep track of scale factor versus preimage distance versus image distance for each transformation they do. Encourage them to make conjectures about the properties of dilations as they do each exploration and verify each conjecture as they go along.

Exploration 2 Constructing a Dilation

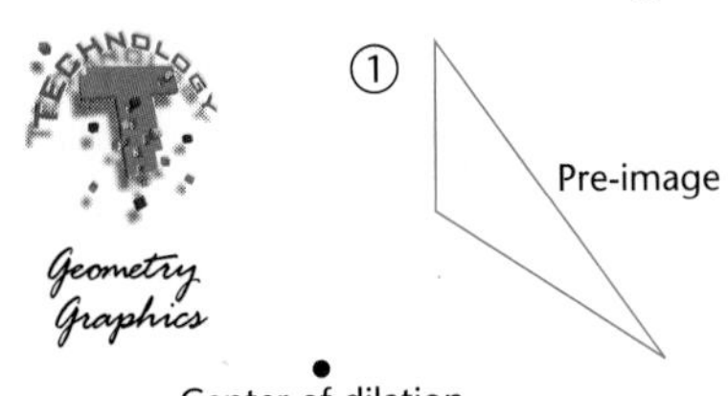

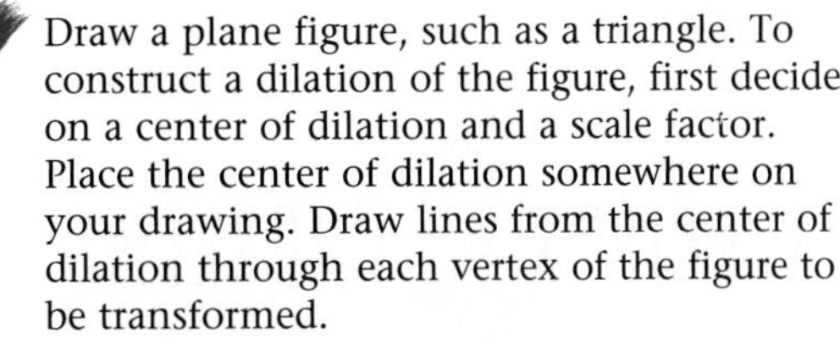

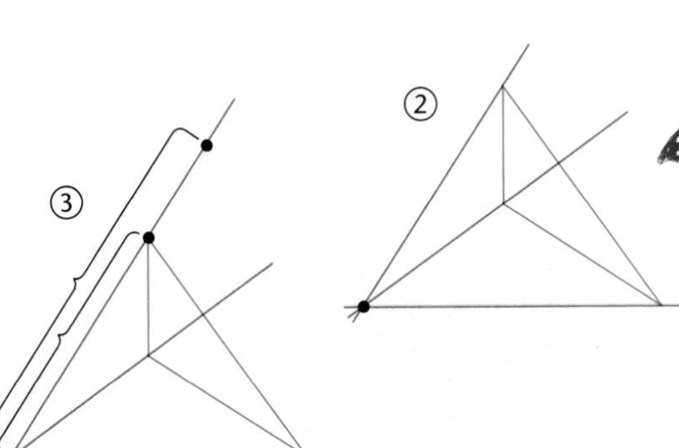

You will need
Geometry technology or
Straightedge and ruler

1. Draw a plane figure, such as a triangle. To construct a dilation of the figure, first decide on a center of dilation and a scale factor. Place the center of dilation somewhere on your drawing. Draw lines from the center of dilation through each vertex of the figure to be transformed.

2. Measure the distance from the center of dilation to a vertex. Multiply this distance by the scale factor.

   Using the new distance you just determined, plot the image point on the line containing the center and the vertex. (Measure the distance from the center of dilation. If the number is negative, measure *away* from the pre-image point.)

3. Repeat Step 2 for each of the remaining pre-image points.

4. Connect the image points to form the transformed image of the original figure. ❖

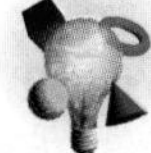

How could you construct an approximate dilation of a figure that was composed of curves rather than segments?

Exploration 3 The Position of Transformed Points

You will need
No special tools

In Exploration 1, you may have noticed that the pre-images and images of a point you transformed using different scale factors were collinear, and that the line passed through the origin (0, 0).

You can prove both of these results using the equation of a line,

$$y = mx + b,$$

where m is the slope of the line and b is the y-intercept.

Recall from algebra that when you are given points (x_1, y_1) and (x_2, y_2) on a line, the equation of the line can be written as

$$y = \left(\frac{y_2 - y_1}{x_2 - x_1}\right) x + b.$$

Inviting Participation
Have a student draw a regular polygon on an overhead or chalkboard grid and choose a center of dilation and a scale factor. Have the rest of the students perform the transformation and compare their results. Have another student draw another polygon and repeat the activity. Continue until a variety of polygons, scale factors, and centers of dilation are used.

Use Transparency 63

Exploration 2 Notes
Students construct a dilation in this activity. To help students generalize the properties of a dilation, extend the exploration to include different preimage shapes.

ongoing ASSESSMENT

4. Check student work.

CRITICAL Thinking
Dilate several points on each curve and connect the images with curves.

TEACHING tip
Point out that the center and scale factor of a dilation are needed if coordinates are not given. Make sure students measure the image distance from the center rather than from the preimage point, and if the center of a dilation is one of the preimage points, then the image distance is 0.

Exploration 3 Notes
In this activity, students learn that the center of a dilation is collinear with the preimage points and image points.

Math Connection
Algebra
In algebra, students studied how to find the slope of a line and the equation of a line in slope-intercept form. It may be helpful to review those ideas at this time.

ongoing ASSESSMENT

1. $y = \frac{4}{3}x$

(6, 8) is on the line because it works in the equation of the line: $8 = \frac{4}{3}(6)$.

(9, 12) is on the line because it works in the equation of the line: $12 = \frac{4}{3}(9)$.

TEACHING tip

Summarize dilations by pointing out that if k is the scale factor with $|k| > 1$, the dilation will be an expansion. If $0 < |k| < 1$, then the dilation will be a contraction.

Cooperative Learning

Students may work on Exercises 1 to 4 in small groups. Have them discuss their answers and reach a consensus.

ASSESS

Selected Answers

Exercises 5, 9, 15, 17, 21, 23, 25, 27, 29, and 37

Assignment Guide

Core 1–7, 23–30, 35–40

Core Plus 1–4, 11–26, 31–43

Technology

Students can use geometry graphics software to dilate the figures in Exercises 31–34. Encourage students to experiment with other scale factors.

Error Analysis

Students may have difficulty with Exercises 23–26. You may want to review how to find the equation of a line from two points.

1. Find an equation of the line that contains the origin (0, 0) and the point (3, 4). Is the point (6, 8) on the line? Is the point (9, 12) on the line? Explain your reasoning.

2. Generalize as follows: Find the equation of the line that contains the points (0, 0) and (a, b). Is the point $(2a, 2b)$ on the line? $(3a, 3b)$? (na, nb)? ❖

EXERCISES & PROBLEMS

Communicate

1. What is a scale factor? How does it affect the points in a transformation?
2. What is a dilation? How is a dilation different from other transformations you have studied?
3. What is the result of a dilation when the scale factor is one? when the scale factor is zero? when the scale factor is negative? when the scale factor is a fraction?
4. In the photo, what would the scale factor tell you about the relationship of the pre-image to the dilated images? Assume the center bowl is the pre-image.

Practice & Apply

In Exercises 5–7, the dashed-line pre-image has been transformed to form the solid-line image. What is the scale factor for each dilation?

5.

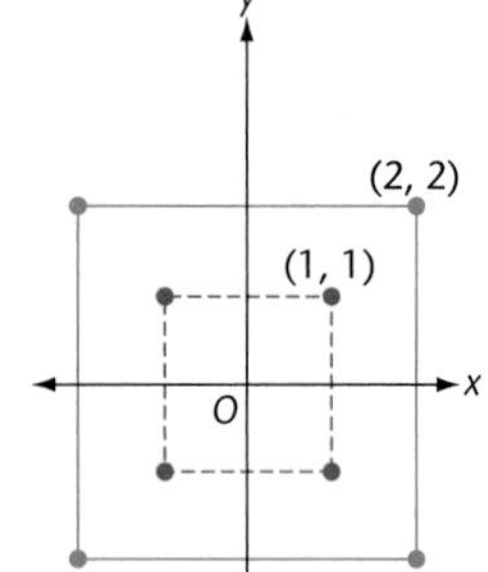

6.

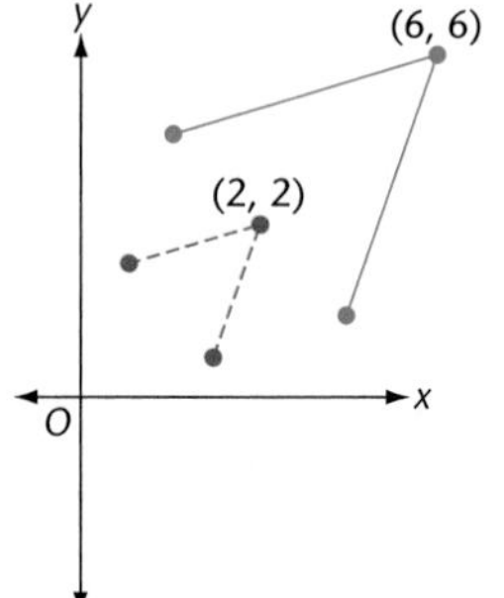

7.

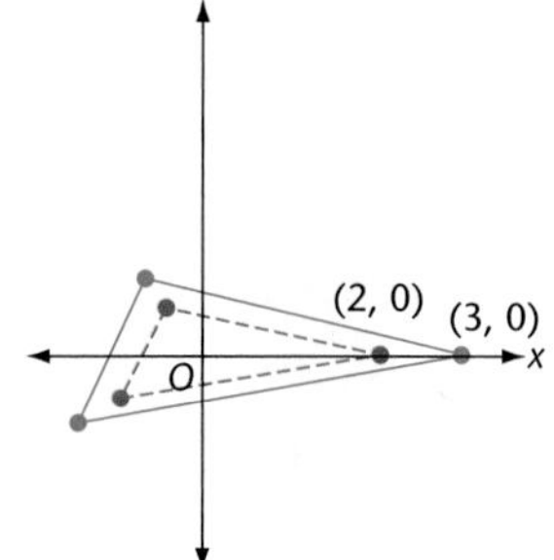

Transform the points on graph paper using the given scale factors.

8. (3, 5) scale factor 3 (9, 15) **9.** (−2, 6) scale factor $\frac{1}{3}$ $(-\frac{2}{3}, 2)$ **10.** (4, −12) scale factor $-\frac{1}{3}$ $(-\frac{4}{3}, 4)$

5. 2

6. 3

7. 1.5.

Trace the figures below onto your own paper. Draw a dilation of each figure. Choose a point for the center of dilation and use the scale factors indicated for each dilation.

11. scale factor 3

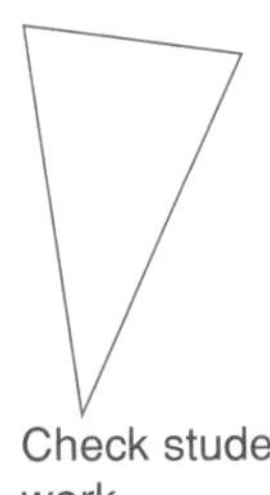

Check student work

12. scale factor 4

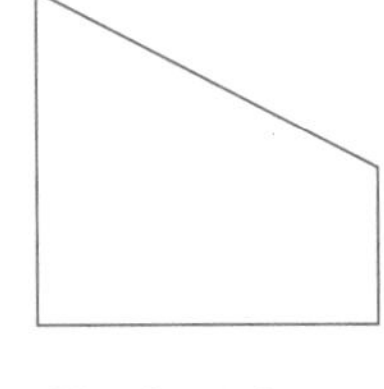

Check student work

13. scale factor −2

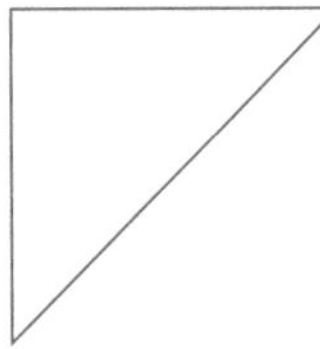

Check student work

14. scale factor −3

Check student work

Optics In the drawing of a camera obscura, the projected image is an example of a dilation.

15. What part of the camera acts as the center of dilation? the pinhole

16. Is the scale factor positive or negative? Explain your answer.

17. Explain why the projected image is inverted.

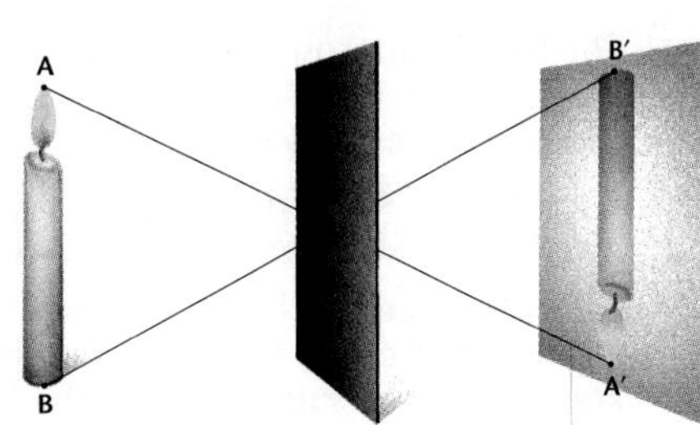

Draw the image of △*ABC* under the given mapping.

18. $(x, y) \rightarrow (2x, 2y)$

19. $(x, y) \rightarrow (3x, 3y)$

20. $(x, y) \rightarrow \left(1\frac{1}{2}x, 1\frac{1}{2}y\right)$

21. $(x, y) \rightarrow \left(\frac{3}{4}x, \frac{3}{4}y\right)$

22. What scale factor will determine the given dilation image of △*ABC* if the sides of the image are three times as long as the sides of the original figure? 3

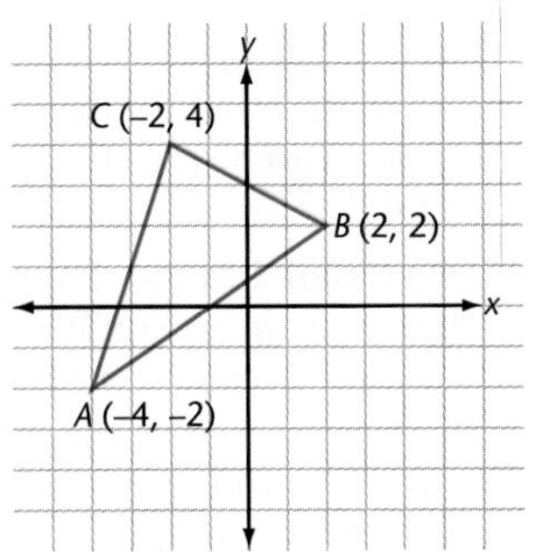

Algebra Find an equation of the line that contains the following points.

23. (0, 0), (2, 5) $y = \frac{5}{2}x$

24. (3, 7), (1, 6) $y = \frac{1}{2}x + \frac{11}{2}$

25. (−8, 2), (4, 5) $y = \frac{1}{4}x + 4$

26. $(a, b), (4a, 4b)$ $y = \frac{b}{a}x$

The following points were transformed from point *A* using different scale factors. Show that these points are collinear.

27. $A(5, 9)$ $A'(10, 18)$ $A''(25, 45)$

28. $B(6, 10)$ $B'(3, 5)$ $B''(24, 40)$

29. $C\left(\frac{2}{3}, \frac{1}{5}\right)$ $C'\left(\frac{8}{3}, \frac{4}{5}\right)$ $C''\left(2, \frac{3}{5}\right)$

30. $D(-3, 7)$ $D'(-6, 14)$ $D''\left(\frac{-3}{4}, \frac{7}{4}\right)$

27. Line containing $A(5, 9)$ and $A'(10, 18)$ has equation $y = \frac{9}{5}x$. $A''(25, 45)$ is on the line because $45 = \frac{9}{5}(25)$.

28. Line containing $B(6, 10)$ and $B'(3, 5)$ has equation $y = \frac{5}{3}x$. $B''(24, 40)$ is on the line because $40 = \frac{5}{3}(24)$.

29. Line containing $C(\frac{2}{3}, \frac{1}{5})$ and $C'(\frac{8}{3}, \frac{4}{5})$ has equation $y = \frac{3}{10}x$. $C''(2, \frac{3}{5})$ is on the line because $\frac{3}{5} = \frac{3}{10}(2)$.

30. Line containing $D(-3, 7)$ and $D'(-6, 14)$ has equation $y = -\frac{7}{3}x$. $D''(\frac{-3}{4}, \frac{7}{4})$ is on the line because $\frac{7}{4} = \frac{-7}{3}\left(\frac{-3}{4}\right)$.

alternative ASSESSMENT

Portfolio Assessment

Ask students to list the steps in constructing a dilation. Students should include examples that illustrate the steps and a list of the properties of a dilation.

The answers to Exercises 18–21 can be found in Additional Answers beginning on page 727.

16. The scale factor is negative because the image is on the opposite side of the center of dilation as the pre-image.

17. Light travels in straight lines. The light rays containing the points of the pre-image and the image cross at the pinhole of the camera, making the image inverted.

Practice Master

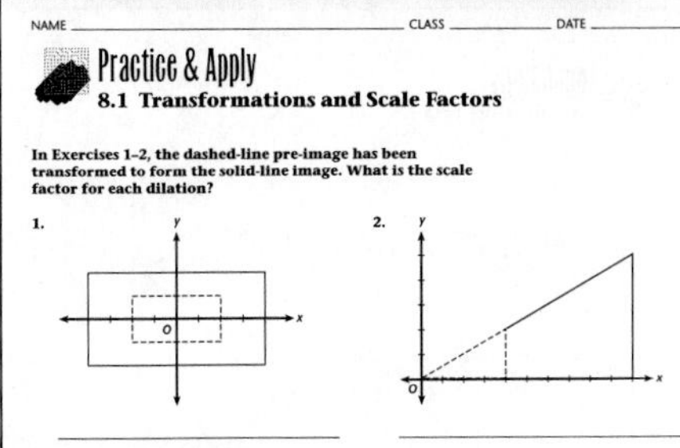

NAME ______ CLASS ______ DATE ______

Practice & Apply

8.1 Transformations and Scale Factors

In Exercises 1–2, the dashed-line pre-image has been transformed to form the solid-line image. What is the scale factor for each dilation?

1.

2.

In Exercises 3–4, graph the given point and use the given scale factor to transform the point.

3. $A(2, 4)$ scale factor 1.5

4. $P(-3, 6)$ scale factor $\frac{1}{3}$

In Exercises 5–6, find an equation of the line that contains the given pair of points.

5. (0, −6), (2, 0) ______

6. (1, 4), (−3, 2) ______

7. Point *P* has been transformed using different scale factors. Are the points P, P', P'' collinear? Explain. $P(4, 6), P'(2, 3), P''(10, 15)$

HRW Geometry Practice & Apply 43

Draw a dilation of each figure using the given scale factor.

31. scale factor 3

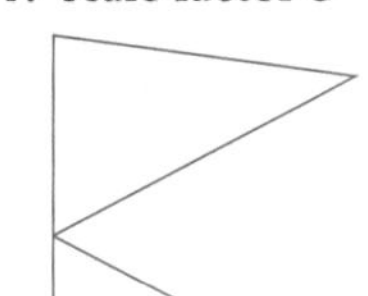

Check student work

32. scale factor $\frac{3}{4}$

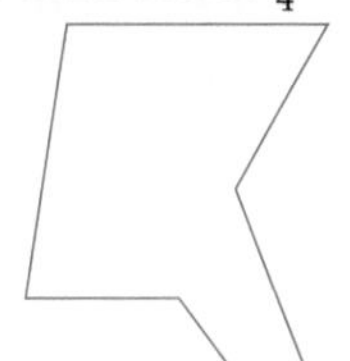

Check student work

33. scale factor 2

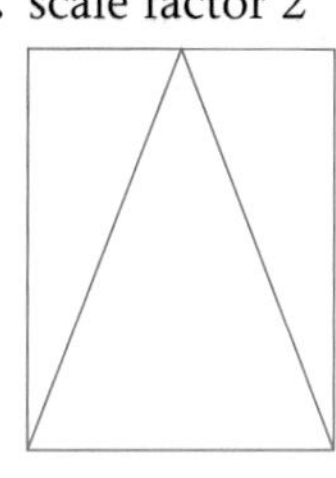

Check student work

34. scale factor $\frac{1}{2}$

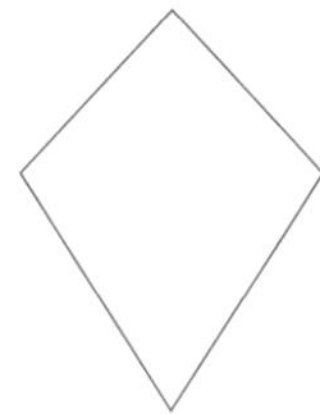

Check student work

Look Back

35. The base of an isosceles triangle is 6 meters and the legs are each 8 meters. Find the area and perimeter of the triangle. **[Lessons 5.1, 5.2]** P = 22m; A = 22.25 m^2

36. A leg of a 45-45-90 triangle is 7 cm long. What is the length of the hypotenuse? **[Lesson 5.4]** $7\sqrt{2}$ or 9.9 cm

37. Earth Science If the circumference of a great circle of the Earth is about 40,000 km, what is the surface area of the Earth? **[Lesson 7.6]** ~509,295,817 km^2

38. Earth Science The atmosphere of the Earth has an altitude of about 550 km. Use Exercise 37 to find the volume of the Earth and its atmosphere. **[Lesson 7.6]** ~1.3857×10^{12} km^3

The Earth's atmosphere, as photographed by a Russian cosmonaut.

39. A steel gas tank has the shape of a sphere. A radius of the inner surface of the tank is 2 feet long. The tank itself is made of $\frac{1}{4}$-inch-thick steel. Find the difference between the outside area and inside area of the tank. **[Lesson 7.6]** ~151.58 in^2

40. If a gallon of paint will cover 400 square feet, how many gallons of paint will be needed to paint both the inside and the outside of the tank described in Exercise 39? **[Lesson 7.6]** 0.26 gallon

Look Beyond

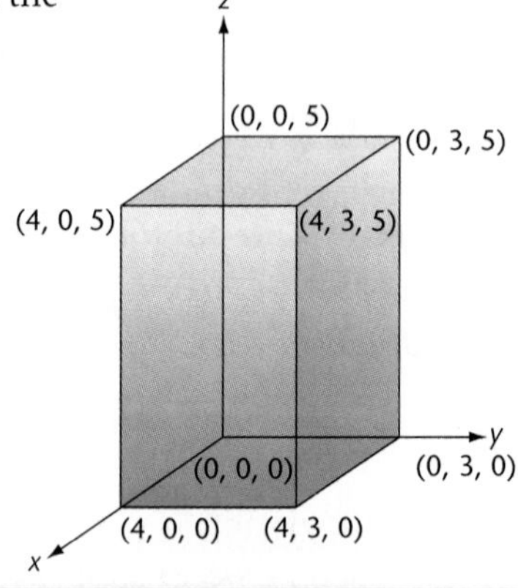

41. Draw the dilation for the box using a scale factor of 2. Write the coordinates for the vertices.

42. How do the lengths of the sides of the pre-image compare to the lengths of the sides of the image? The sides are twice as long.

43. Find the volumes of both boxes. How do they compare?

Look Beyond

Exercises 41–43 extend dilations in the plane to dilations in space. In Exercise 43, students are asked to compare the volumes of the preimage and image shapes Have them make a conjecture about the volumes before calculating them.

Use Transparency 64

Technology Master

NAME ______________ CLASS ________ DATE ________

Technology

8.1 Dilation Images

The diagram shows a point P on a coordinate plane. Its coordinates are $P(3, 2)$. To find the image of P under a dilation with the origin as the center of the dilation and scale factor r, you multiply each coordinate of P by r. If, for example, $r = 2$, then the image of $P(3, 2)$ is $P'(2 \times 3, 2 \times 2) = P'(6, 4)$. To study the effect of different choices for r, you can use a spreadsheet and an examination of the resulting table.

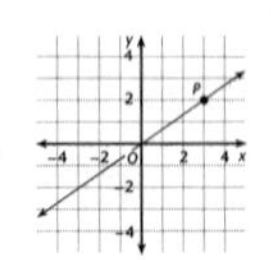

1. Create a spreadsheet in which column A contains $r = 3.0, 2.5, 2.0, \ldots, -2.0, -2.5, -3.0$, cells B1 and C1 contain 3 and 2, respectively, and cells D1 and E1 contain A1*B1 and A1*C1, respectively. Then fill the first thirteen rows.

2. From the spreadsheet in Exercise 1, describe the position of the image of $P(3, 2)$ relative to $P(3, 2)$ and the origin when $r = 2.5$, $r = 0.5$, $r = -0.5$, and $r = -2.5$.

____

____

Given the coordinates of P, use your spreadsheet from Exercise 1 to describe the position of the image of $P(x, y)$ relative to $P(x, y)$ and the origin when $r = 2.5$, $r = 0.5$, $r = -0.5$, and $r = -2.5$.

3. $P(x, y) = P(-4, 1)$ ______________________________

____

4. $P(x, y) = P(-3, -5)$ ______________________________

____

5. Based on your results from Exercises 2–4, describe the position of the image of $P(x, y)$ if you apply a dilation with the origin as the center of rotation and with a scale factor of r.

____

____

HRW Geometry — Technology 55

41.

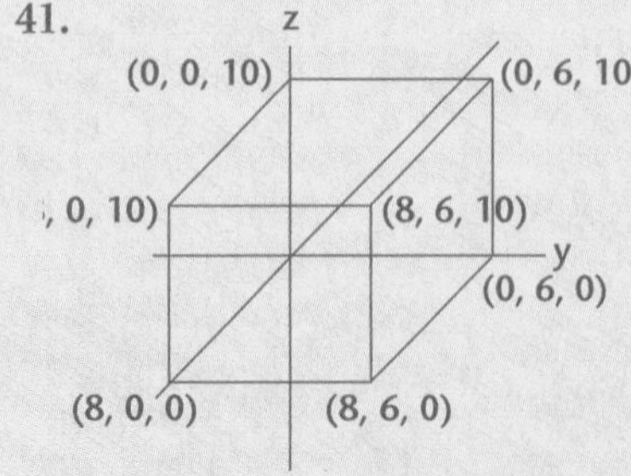

43. The volume of the image is 8 times the volume of the pre-image.

LESSON 8.2 *Exploring* Similar Polygons

why *Graphic designers use the geometric-concept of similarity to enlarge and reduce figures in flyers, posters, and newsletters.*

Figures that have the same shape, but not necessarily the same size, are called **similar figures**. In the exploration that follows, you will develop a definition for **similar polygons**.

Exploration *Defining Similar Polygons*

You will need
Ruler and protractor

1. $\triangle ABC$ and $\triangle DEF$ seem to have the same shape. In mathematical terms, they appear to be similar. (Notation: $\triangle ABC \sim \triangle DEF$.) Measure the sides and angles in the two figures.

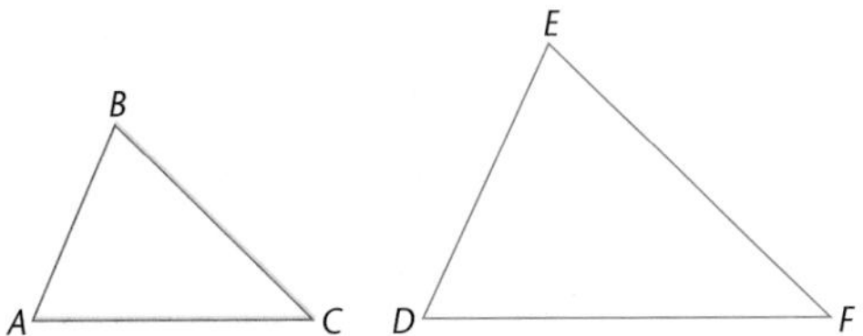

2. Are any of the angles congruent? If so, state which ones.

Using Visual Models Use a photocopier to enlarge and reduce a cartoon or drawing. Transfer the copies to transparency acetate and superimpose the similar figures on the overhead projector. Measure and compare the sides and angles of the similar pairs.

PREPARE

Objectives

- Define *polygon similarity.*
- Use properties of proportion and scale factor with similar polygons.

RESOURCES

- Practice Master 8.2
- Enrichment Master 8.2
- Technology Master 8.2
- Lesson Activity Master 8.2
- Quiz 8.2
- Spanish Resources 8.2

Assessing Prior Knowledge

1. Given $\Delta ABC \cong \Delta DEF$, name the congruent angles and sides. [$\angle ABC \cong \angle DEF$; $\angle ACB \cong \angle DFE$; $\angle BAC \cong \angle EDF$; $AB \cong DE$; $AC \cong DF$; $BC \cong EF$]
2. The dilation image of a segment with length 4 has length 6. What is the scale factor? [1.5]
3. Solve the equation $6x = 42$. [7]

TEACH

Creating similar polygons is important for students because of the application to photography, maps, scale drawings, and scale models. The lesson helps students make connections to algebra and can be generalized to include three-dimensional figures.

Exploration Notes

In this exploration, students compare polygons and develop a definition for *similar polygons.* Make sure students understand that corresponding angles must be congruent and corresponding sides must be proportional for two figures to be similar. Make sure students understand that congruence is a special case of similarity.

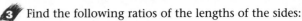

Have several students write their definitions of *similar polygons* on the chalkboard. Have the class compare them and agree on a definition. Discuss the importance of including "if and only if" in the definition.

ongoing ASSESSMENT

6. **Two polygons are similar if and only if their corresponding angles are congruent and their corresponding sides are proportional.**

CRITICAL Thinking

Student definitions should include the ideas of corresponding sides being proportional and corresponding angles being congruent for two polygons to be similar.

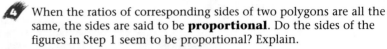

3. Find the following ratios of the lengths of the sides:

$$\frac{AB}{DE} \qquad \frac{BC}{EF} \qquad \frac{AC}{DC}$$

What do you notice about the ratios?

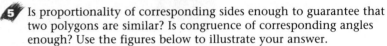

4. When the ratios of corresponding sides of two polygons are all the same, the sides are said to be **proportional**. Do the sides of the figures in Step 1 seem to be proportional? Explain.

5. Is proportionality of corresponding sides enough to guarantee that two polygons are similar? Is congruence of corresponding angles enough? Use the figures below to illustrate your answer.

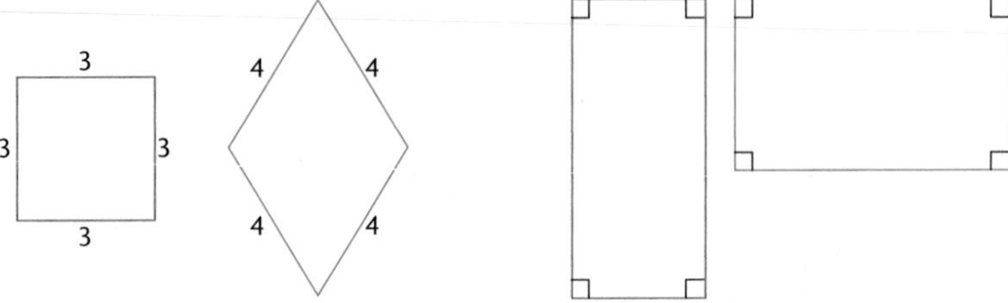

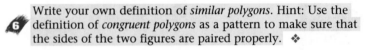

6. Write your own definition of *similar polygons*. Hint: Use the definition of *congruent polygons* as a pattern to make sure that the sides of the two figures are paired properly. ❖

CRITICAL Thinking

Can triangles be used in Step 4 of the exploration, instead of quadrilaterals, to illustrate the argument? Explain why or why not.

A formal definition of *similar polygons* is stated below. How does your definition from the exploration compare with it?

> **SIMILAR POLYGONS**
>
> Two polygons are similar if, and only if, there is a way of setting up a correspondence between their vertices so that
>
> **1.** the corresponding angles are congruent, and
> **2.** the corresponding sides are proportional. **8.2.1**

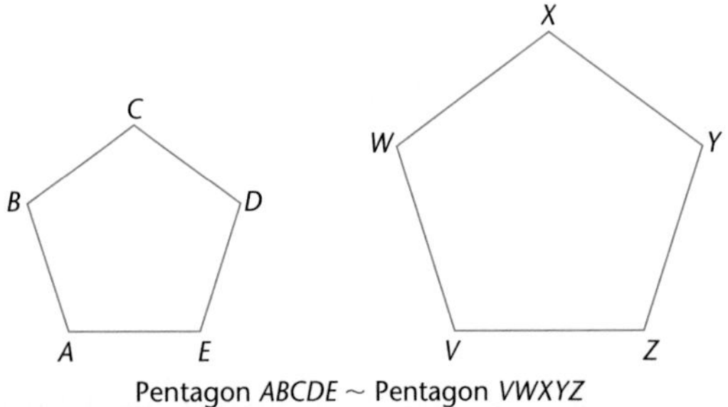

Pentagon *ABCDE* ~ Pentagon *VWXYZ*

interdisciplinary CONNECTION

Biology Microscopes are used to study the characteristics of plant and animal cells that cannot be seen with the naked eye. Borrow several microscopes and some prepared slides from the science department and have students draw what they see. Find the scale factor between the specimens and the drawings.

Proportions and Scale Factors

Recall from Lesson 8.1 that transforming a figure by a scale factor (dilating the figure) causes the length of its sides to be multiplied by that scale factor. What scale factor transforms $\triangle ABC$ to $\triangle XYZ$? $\triangle XYZ$ to $\triangle ABC$?

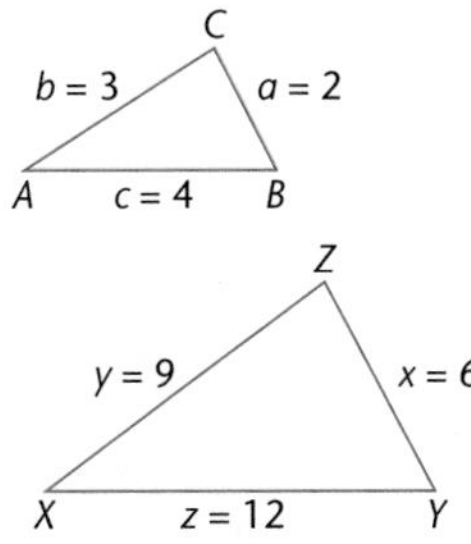

The sides of $\triangle ABC$ and $\triangle XYZ$ are in proportion. That is,

$$\frac{x}{a} = \frac{6}{2} = 3,$$

$$\frac{y}{b} = \frac{9}{3} = 3,$$

$$\frac{z}{c} = \frac{12}{4} = 3,$$

and so, $\frac{x}{a} = \frac{y}{b} = \frac{z}{c}$,

which is the required condition for proportionality of sides. Notice that each of the ratios (the fractions in the above example) reduces to 3, which is the scale factor of the transformations.

It is often useful to think of similar figures in terms of dilations. When you do, you may find it convenient to place the sides of the pre-image figure in the denominators of the fractions and the sides of the image figure in the numerators as in the example above. Then, the value of each fraction is the scale factor. (But, as you will see later, either way of arranging numerators and denominators results in a true statement.)

Similiar polygons appear in many structures.

Try This Find the scale factor that would transform $\triangle ABC$ into $\triangle DEF$.

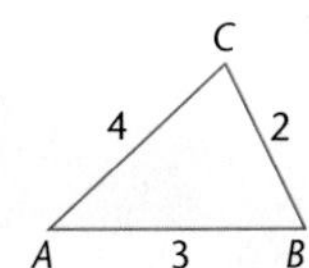

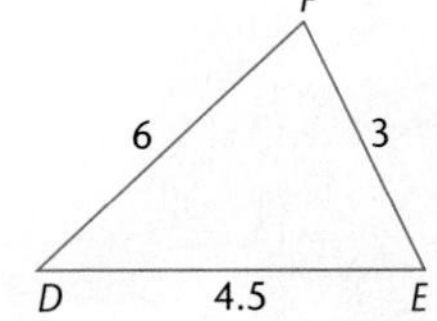

CRITICAL *Thinking* If you are given that the sides of two figures are proportional, is it necessary to find all the ratios of the sides to determine the scale factor? Explain why or why not.

TEACHING *tip*

Emphasize that $\frac{x}{a} = \frac{y}{b} = \frac{z}{c}$. is a short way to write $\frac{x}{a} = \frac{y}{b}, \frac{y}{b} = \frac{z}{c}$, and $\frac{x}{a} = \frac{z}{c}$.

ongoing ASSESSMENT

Try This

$\frac{a}{d} = \frac{b}{e} = \frac{c}{f} = \frac{2}{3}$**; so the scale factor is** $\frac{3}{2}$**.**

CRITICAL *Thinking*

No, since the ratios of all the corresponding sides are the same, it is only necessary to find one ratio.

ENRICHMENT Have students draw a simple floor plan of a house using a scale of 1 inch to 16 feet. Ask them to list the proportions relating to the measurements on the floor plan to the actual dimensions of the house.

INCLUSION strategies

English Language Development Some students may need to see many examples of the properties of proportions. This will help them relate the operations involved to the specific property. Emphasize that properties are used to simplify their work with similar figures. Point out that in everyday language, the word *proportion* is sometimes used to mean a single fraction. For example, "the proportion of students getting A's is $\frac{1}{5}$".

Math Connection Algebra

In algebra, students studied how to solve an equation of the form $\frac{a}{b} = \frac{c}{d}$ by cross-multiplication. It would be helpful to review those ideas at this time, including how to solve literal equations of the form $ad = bc$ and $d(a + b) = b(c + d)$.

Point out that in Method B of the application, the scale factor is found first. Finding the scale factor is an important skill for estimating sizes.

The Properties of Proportions

When working with similar figures it is often helpful to know the following properties of proportions.

PROPERTIES OF PROPORTION

Cross-Multiplication Property

If $\frac{a}{b} = \frac{c}{d}$ and $b, d \neq 0$, then $ad = bc$. **8.2.2**

Reciprocal Property

If $\frac{a}{b} = \frac{c}{d}$ and $a, b, c, d \neq 0$, then $\frac{b}{a} = \frac{d}{c}$. **8.2.3**

Exchange Property

If $\frac{a}{b} = \frac{c}{d}$ and $a, b, c, d, \neq 0$, then $\frac{a}{c} = \frac{b}{d}$. **8.2.4**

"Add-One" Property

If $\frac{a}{b} = \frac{c}{d}$ and $b, d \neq 0$, then $\frac{a+b}{b} = \frac{c+d}{d}$. **8.2.5**

APPLICATION

Some students are sizing photos for the yearbook. An original slide measures 1.25 inches wide by 0.75 inches long. In the yearbook, the photo needs to fill a space that is 5 inches wide. How would the students determine the length of the sized photo?

To find the length, they would set up a proportion and solve for the missing length. $\frac{x}{5} = \frac{0.75}{1.25}$

Method A

By observing the denominators, determine that the scale factor is 4. Multiply the numerator and denominator by this factor.

$$\frac{x}{5} = \frac{0.75 \times 4}{1.25 \times 4} = \frac{3}{5} \quad \text{So, } x = 3.$$

Method B

$$\frac{x}{5} = \frac{0.75}{1.25}$$

$$1.25x = 3.75$$

$$x = 3 \text{ in. long}$$

Using Visual Models Have each student draw and label a polygon and measure each side and angle. Ask them to construct a similar polygon with scale factor 3 and another polygon with scale factor $\frac{1}{3}$. Verify the properties of proportions with the corresponding sides of the similar figures.

Using the Reciprocal Property

You are given the following proportionality statement for two triangles.

$$\frac{a}{x} = \frac{b}{y} = \frac{c}{z}$$

Is the following statement also true?

$$\frac{x}{a} = \frac{y}{b} = \frac{z}{c}$$

To answer the question, consider the first two members of the original statement.

$$\frac{a}{x} = \frac{b}{y}$$

By the Reciprocal Property, you get the following proportion.

$$\frac{x}{a} = \frac{y}{b}$$

Apply the Reciprocal Property to the second and third members of the original statement to obtain the following.

$$\frac{y}{b} = \frac{z}{c}$$

Then you can conclude that $\frac{x}{a} = \frac{y}{b} = \frac{z}{c}$. (Why?)

Using the Exchange Property

You are given the following proportionality statement for two triangles.

$$\frac{a}{x} = \frac{b}{y} = \frac{c}{z}$$

By using the Exchange Property on two members of the statement at a time, you can show that the following statements are also true.

$$\frac{a}{b} = \frac{x}{y} \qquad \frac{b}{c} = \frac{y}{z} \qquad \frac{a}{c} = \frac{x}{z}$$

The three statements just proven show another way of thinking about similarity. That is, the parts of one figure are in the same ratios within the figure as the corresponding parts within the other figure.

Math Connection Algebra

The content of Using the Reciprocal Property shows a connection between algebraic and geometric proofs that you may want to emphasize. Reinforce some of the ideas from Chapter 2 by pointing out how logical reasoning can be used to justify statements.

Cooperative Learning

Have groups of five students to review the properties of proportions. Have one student write a proportion and each of the other four students restate the proportion in terms of the properties. Working in groups provides an informal but thorough way to review material.

ASSESS

Selected Answers

Exercises 9, 11, 13, 15, 17, 19, 21, 23, 33, 35, and 41

Assignment Guide

Core 1–19, 20–34 even, 35–40

Core Plus 1–7, 8–20 even, 21–44

Technology

Use geometry graphics software to demonstrate how to enlarge or reduce figures to scale as is done in Exercises 29 and 30.

Error Analysis

Exercises 20–23 may be difficult to understand without the use of concrete examples. Encourage students to restate the questions with numbers in place of the variables.

APPLICATION

Architecture Amber and Adrienne are making a scale model of a building with a rectangular foundation. If they want the longer sides of the model to be 24 inches, what should the length of the shorter sides be?

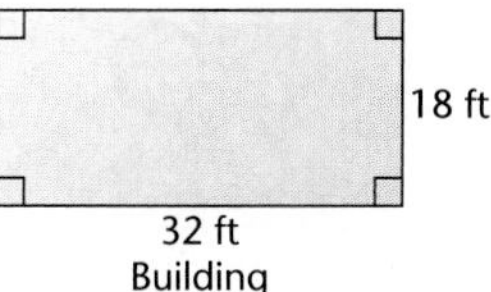

Since the model and the building must have the same shape, the figures are similar.

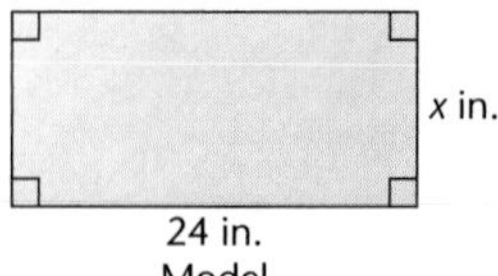

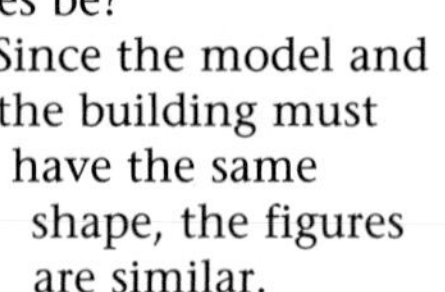

$$\frac{x}{18} = \frac{24}{32}$$

$$\frac{x}{18} = \frac{3}{4}$$

$$4x = 54$$

$$x = 13.5 \text{ in.}$$ ❖

EXERCISES & PROBLEMS

Communicate

1. If $\triangle RTW \sim \triangle IOU$, name all pairs of proportional sides.
2. If quadrilateral $KMPQ \sim$ quadrilateral $RTAW$, show the ratios between the sides.
3. In $\triangle FRG$ and $\triangle NBY$, $\frac{FR}{BY} = \frac{RG}{YN} = \frac{FG}{BN}$. State the similarity showing the correct correspondence.
4. In $\triangle ZXC$ and $\triangle VML$, $\frac{CX}{ML} = \frac{XZ}{LV} = \frac{ZC}{VM}$. State the similarity showing the correct correspondence.

5. If $\triangle WER \cong \triangle POI$, is $\triangle WER \sim \triangle POI$? Explain.

6. Use the diagram at right to determine whether quadrilateral $GHJK \sim$ quadrilateral $KLMN$. Explain your conclusion.

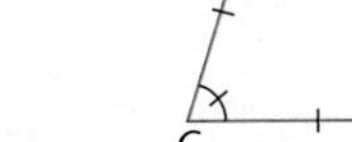

7. Explain why the "Add-One" Property has that name.

Practice & Apply

8. Write the ratio $\frac{3}{4} = \frac{15}{20}$ using the Cross-Multiplication Property. $(3)(20) = (4)(15)$

9. Write the ratio $\frac{5}{2} = \frac{10}{4}$ using the Reciprocal Property. $\frac{2}{5} = \frac{4}{10}$

10. Write the ratio $\frac{2}{5} = \frac{4}{10}$ using the Exchange Property. $\frac{2}{4} = \frac{5}{10}$

11. Write the ratio $\frac{6}{9} = \frac{2}{3}$ using the "Add-One" Property. $\frac{6+9}{9} = \frac{2+3}{3}$ or $\frac{15}{9} = \frac{5}{3}$

12. If $\frac{x}{4} = \frac{y}{8}$, find $\frac{x}{y}$. $\frac{x}{y} = \frac{4}{8}$ using the Exchange Property

Algebra Solve for x.

13. $\frac{6x}{24} = \frac{27}{9}$ $x = 12$

14. $\frac{4.8}{x} = \frac{6}{8.4}$ $x = 6.72$

15. $\frac{\frac{2}{5}}{8} = \frac{\frac{7}{10}}{x}$ $x = 14$

16. $\frac{6}{x} = \frac{x}{150}$ $x = \pm 30$

17. $\frac{3}{x-4} = \frac{7}{x+5}$ $x = 10\frac{3}{4}$

18. $\frac{5-2x}{12} = \frac{3x+1}{4}$ $x = \frac{2}{11}$

19. Rectangle $TGHF \sim$ rectangle $NBKJ$. Use proportions to find GH. $GH = 15$

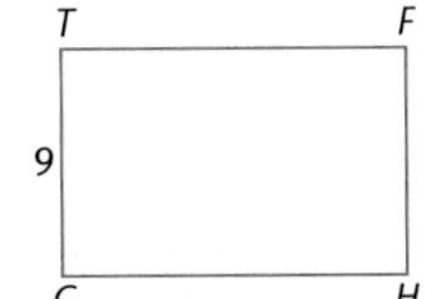

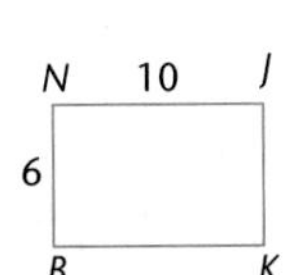

Algebra Tell whether the proportion is true for all values of the variables. If the proportion is false, give a numerical counterexample.

20. If $\frac{x}{y} = \frac{r}{s}$, then $\frac{x+c}{y} = \frac{r+c}{s}$.

21. If $\frac{x}{y} = \frac{r}{s}$, then $\frac{x+a}{y+a} = \frac{r+a}{s+a}$.

22. If $\frac{x}{y} = \frac{r}{s}$, then $\frac{x}{x+y} = \frac{r}{r+s}$.

23. If $\frac{x}{y} = \frac{r}{s} = \frac{m}{n}$, then $\frac{x}{y} = \frac{x+r+m}{y+s+n}$.

20. False ; $\frac{2}{3} = \frac{4}{6}$ but $\frac{2+5}{3} \neq \frac{4+5}{6}$

21. False ; $\frac{2}{3} = \frac{4}{6}$ but $\frac{2+5}{3+5} \neq \frac{4+5}{6+5}$

22. True ; If $\frac{x}{y} = \frac{r}{s} \Rightarrow xs = yr$; $\frac{x}{x+y} = \frac{r}{r+s} \Rightarrow$ $xr + xs = xr + yr \Rightarrow xs = yr$

23. True : $xs = yr$, $xn = ym \Rightarrow xy + xs + xn =$ $xy + yr + ym \Rightarrow \frac{x}{y} = \frac{x+r+m}{y+s+m}$

Alternative Assessment

Portfolio Assessment

The ability to describe how to find the scale factor between similar figures will reinforce many of the key ideas of the lesson. Have students find the scale factor between various pairs of similar shapes and summarize the key ideas in writing.

The answers to Exercises 24 and 25 can be found in Additional Answers beginning on page 727.

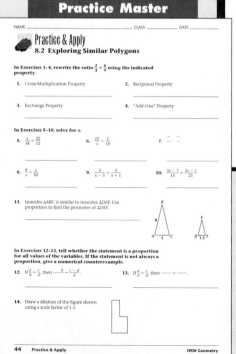

The Aransas National Wildlife Refuge, on the Texas coast, contains important nesting sites for endangered whooping cranes. The delicate salt-marsh environment must be carefully maintained in order to help this species, and other species of wildlife, survive.

Wildlife Management Use the map for Exercises 24–26.

24. Use the scale of miles and the map above to estimate the number of square miles contained in the wildlife refuge. If there are 640 acres in 1 square mile, how many acres does the refuge contain? Describe your method for estimating the area of the refuge.

25. Animal overpopulation within the refuge must be monitored in order to prevent diseases and environmental damage. A deer census reveals that there are an average of 3 deer on every 20 acres of the refuge. Give an estimate for the number of deer that live in the wildlife refuge.

26. Most of the endangered whooping cranes spend their summers in Canada and their winters in or near the Aransas National Wildlife Refuge. If there are 123 whooping cranes alive in the wild and 98 of them are spotted inside the wildlife refuge, what percentage of the whooping crane population is inside the refuge? 79.6%

27. **Marine Biology** In a marine environment, 1 pound of sea water consists of 6.5% minerals and organic material. How many pounds of pure water are there in 57 pounds of sea water? 53.3 lbs

Practice Master

NAME ______ CLASS ______ DATE ______

Practice & Apply
8.2 Exploring Similar Polygons

In Exercises 1–4, rewrite the ratio $\frac{2}{3} = \frac{6}{9}$ using the indicated property.

1. Cross-Multiplication Property
2. Reciprocal Property
3. Exchange Property
4. "Add-One" Property

In Exercises 5–10, solve for x.

5. $\frac{x}{18} = \frac{22}{12}$
6. $\frac{22}{x} = \frac{2}{18}$
7.
8. $\frac{8}{x} = \frac{x}{50}$
9. $\frac{3}{x-3} = \frac{5}{x+1}$
10. $\frac{3x-7}{15} = \frac{2x-1}{21}$

11. Isosceles $\triangle ABC$ is similar to isosceles $\triangle DEF$. Use proportion to find the perimeter of $\triangle DEF$.

B, 8, A, 3, C; E, D, 1.5, F

In Exercises 12–13, tell whether the statement is a proportion for all values of the variables. If the statement is not always a proportion, give a numerical counterexample.

12. If $\frac{a}{b} = \frac{c}{d}$, then $\frac{\quad b}{} = \frac{c-d}{d}$.
13. If $\frac{a}{b} = \frac{c}{d}$, then —— = ——.

14. Draw a dilation of the figure shown using a scale factor of 1.5.

44 **Practice & Apply** **HRW Geometry**

28. Landscaping Using the landscape design plans and the given scale, determine how many feet apart the trees must be planted. Express your answer in meters and feet (1m = 3.28 ft). Scale: 1 cm = 4.5 m.

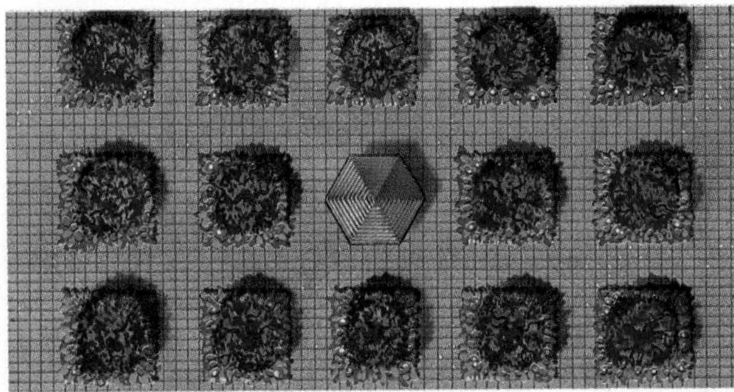

1.7 cm

29. Fine Art Brenda is attempting to paint an accurate reproduction of the *Mona Lisa* from a print that hangs in her art class. The dimensions of the print are 16 in. by 24 in., but Brenda's canvas is 15 in. by 20 in. What modifications must she make to her canvas to ensure that her painting is proportional to the print?

30. Fine Art On the classroom print of the *Mona Lisa,* the length of the woman's face from the hairline to the chin is 6 in. The width of the face is 4 in. What width and length must Brenda use for the dimensions of the woman's face in her reproduction?
width = $3\frac{1}{3}$ in.; length = 5 in.

31. Advertising An athletic shoe manufacturer wants to put the company logo on a billboard. An artist has an 8-inch-by-12-inch picture of the logo and must reproduce it onto the billboard using a scale factor of 25. What are the dimensions of the billboard logo?
200 in. x 300 in., or 16 ft 8 in. x 25 ft

32. Model-Building An artist builds a scale model of a clipper ship. The lifeboats on the model ship are 1 inch long, but on the real clipper ship they were about 24 feet long. If the length of the model is 10 inches, what was the length of the real ship? 240 ft

33. What is the scale factor used in Exercise 32 for the model ship? $\frac{1}{288}$

34. The perimeter of a rectangle is 60 cm. The ratio of the base to the altitude of the rectangle is 5 to 7. Find the lengths of the base and altitude. $17\frac{1}{2}$

28. 7.65 m or 25.09 ft

29. Brenda will have to make a similar, smaller drawing using a scale factor of $\frac{20}{24} = \frac{5}{6}$. Her painting could have the dimensions $13\frac{1}{3}$ in. by 20 in.

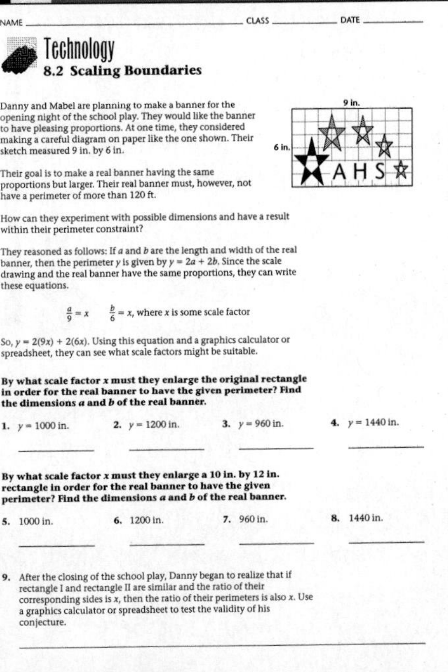

Technology Master

NAME ______ CLASS ______ DATE ______

Technology
8.2 Scaling Boundaries

Danny and Mabel are planning to make a banner for the opening night of the school play. They would like the banner to have pleasing proportions. At one time, they considered making a careful diagram on paper like the one shown. Their sketch measured 9 in. by 6 in.

Their goal is to make a real banner having the same proportions but larger. Their real banner must, however, not have a perimeter of more than 120 ft.

How can they experiment with possible dimensions and have a result within their perimeter constraint?

They reasoned as follows: If a and b are the length and width of the real banner, then the perimeter y is given by $y = 2a + 2b$. Since the scale drawing and the real banner have the same proportions, they can write these equations.

$$\frac{a}{9} = x \qquad \frac{b}{6} = x, \text{ where } x \text{ is some scale factor}$$

So, $y = 2(9x) + 2(6x)$. Using this equation and a graphics calculator or spreadsheet, they can see what scale factors might be suitable.

By what scale factor x must they enlarge the original rectangle in order for the real banner to have the given perimeter? Find the dimensions a and b of the real banner.

1. $y = 1000$ in. ______
2. $y = 1200$ in. ______
3. $y = 960$ in. ______
4. $y = 1440$ in. ______

By what scale factor x must they enlarge a 10 in. by 12 in. rectangle in order for the real banner to have the given perimeter? Find the dimensions a and b of the real banner.

5. 1000 in. ______
6. 1200 in. ______
7. 960 in. ______
8. 1440 in. ______

9. After the closing of the school play, Danny began to realize that if rectangle I and rectangle II are similar and the ratio of their corresponding sides is x, then the ratio of their perimeters is also x. Use a graphics calculator or spreadsheet to test the validity of his conjecture.

______

56 Technology HRW Geometry

35. **Portfolio Activity** The ratio $\frac{1+\sqrt{5}}{2}$ is called the Golden Ratio because rectangles with length and width in the ratio $\frac{l}{w} = \frac{1+\sqrt{5}}{2}$ are thought to be pleasing to the eye. Make a collage of rectangles whose lengths and widths use this ratio.
Check student work.

Look Back

36. **Algebra** The angles of a triangle have degree measures $x + 5$, $5x + 12$, and $2x + 3$. Find the measures of the angles. **[Lesson 3.6]** $x = 20$; 25°, 112°, 43°

37. Find the measure of the base angles of an isosceles triangle if its vertex angle is 92°. **[Lesson 4.4]** 44° each

38. A right triangle has legs of length 5 cm and 7 cm. Find the length of its hypotenuse. **[Lesson 5.4]** $\sqrt{74} \approx 8.6$ cm

39. A right triangle has a hypotenuse of length 7 cm and a leg of length 5 cm. Find the length of the other leg. **[Lesson 5.4]** $2\sqrt{6} \approx 4.9$ cm

40. **Probability** A blindfolded person throws a dart at the target at right. Assuming that it is equally likely that the dart will land anywhere on the rectangle, find the probability of the following: **[Lesson 5.7]**

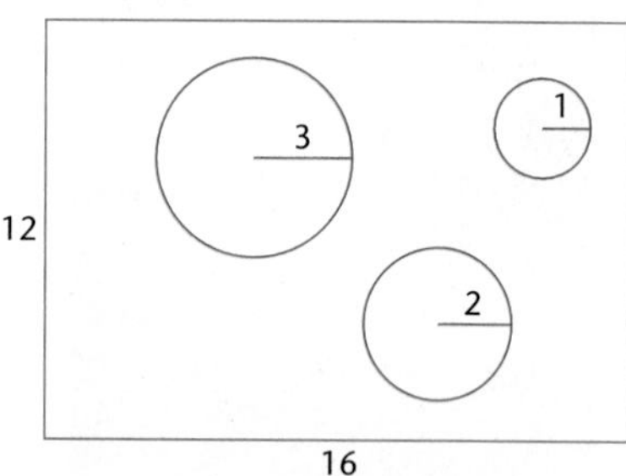

a. The person will hit a circle. 0.23 or 23%

b. The person will hit the biggest circle. 0.15 or 15%

Look Beyond

Indirect Measurement Anthony wants to find the width of the river. First he pulls the visor of his cap down over his eye until a spot R, on the opposite shore, is in his line of vision. Without changing the position of his cap, he turns and sights along the visor to another spot, A, on his side of the river.

41. Explain how Anthony can now find the width of the river.

42. Which segments in the figure will have the same length as $\overline{NR}$? $\overline{NA}$

43. Which postulate can be used to prove that $\triangle SNR \cong \triangle SNA$? Explain your answer.

44. What are some possible problems with using this method of indirect measurement? How accurate do you think this method is?

Look Beyond

Exercises 41–44 introduces students to indirect measurement which will be studied later in the chapter.

41. Anthony can find the distance from where he is standing to point A which is equal to the distance across the river.

43. ASA : The distance from the ground to Anthony's eyes is the same, Anthony's body forms the same angle with the ground in both triangles and the angle that Anthony sights along using his cap is the same in both triangles.

44. Anthony might not keep his head at exactly the same angle to his body and he might not keep his body at the same angle to the ground when turning. The ground might not be perfectly flat.

LESSON 8.3

Exploring Triangle Similarity Postulates

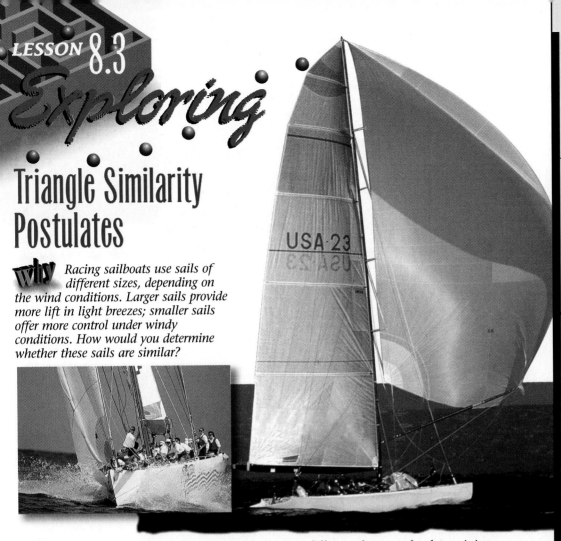

Why *Racing sailboats use sails of different sizes, depending on the wind conditions. Larger sails provide more lift in light breezes; smaller sails offer more control under windy conditions. How would you determine whether these sails are similar?*

The explorations below suggest different shortcuts for determining whether triangles are similar. The shortcuts can be formalized into triangle similarity postulates. To provide a basis for the explorations, recall the definition of similar polygons from the previous lesson (Definition 8.2.1).

Exploration 1 *Similar Triangles Postulate 1*

You will need

Geometry technology or

Ruler, protractor, and compass

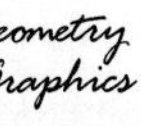

1. Construct $\triangle ABC$ with $m\angle A = 45°$ and $m\angle B = 65°$. Measure the sides and remaining angle.
2. Construct $\triangle DEF$ with $m\angle D = 45°$, $m\angle E = 65°$, and sides greater than the sides of $\triangle ABC$. Measure the sides and remaining angle.

ALTERNATIVE teaching strategy

Hands-On Strategies Some students may benefit from the tactile-kinesthetic experience of doing hands-on measurements. Have those students use a ruler and a protractor to measure the sides and angles of pairs of similar triangles. Make a chart showing the measurements and the ratio of the corresponding sides. Include triangles that are acute, obtuse, and right. Have students do as many examples as necessary to formulate a postulate for similarity between two triangles.

PREPARE

Objective

- Develop three triangle similarity postulates-AA, SSS, SAS.

RESOURCES

- Practice Master 8.3
- Enrichment Master 8.3
- Technology Master 8.3
- Lesson Activity Master 8.3
- Quiz 8.3
- Spanish Resources 8.3

Assessing Prior Knowledge

1. List the triangle congruence postulates.

   [SSS, SAS, ASA]

2. What is the ratio of corresponding sides for two congruent triangles?

   [1:1]

3. What is the ratio of corresponding angles for two congruent triangles?

   [1:1]

TEACH

It is important for students to learn how to determine whether two figures are similar. Developing the postulates themselves through exploration should help students understand the relationships between corresponding angles and between corresponding sides of similar figures.

Exploration 1 Notes

It may be helpful to compare the postulate developed in this exploration with the ASA congruence postulate for triangles. Emphasize the similarities and differences between them.

Cooperative Learning

Have students discuss in small groups the postulates they formulate from the explorations. Have each group share their postulates with the class and agree to the wording of the postulates as a class. Collaborating in groups helps students develop their verbal skills.

ongoing ASSESSMENT

5. Two triangles are similar if two pairs of corresponding angles are congruent.

ongoing ASSESSMENT

Try This

The triangles in 2 are similar because two pairs of corresponding angles are congruent.

Exploration 2 Notes

In this exploration, students formulate the SSS similarity postulate. This activity will help students understand that when all pairs of corresponding sides are in the same proportion, two triangles are similar. Ask students to compare this postulate with the SSS congruence postulate.

 3. Use the measures to calculate the ratios on the chart below.

	$\triangle DEF$	$\triangle ABC$	Ratio of the sides $\frac{\triangle DEF}{\triangle ABC}$
Measures of corresponding sides	DE = ? EF = ? DF = ?	AB = ? BC = ? AC = ?	$\frac{DE}{AB}$ = ? $\frac{EF}{BC}$ = ? $\frac{DF}{AC}$ = ?
Measures of corresponding angles	$m\angle D = 45°$ $m\angle E = 65°$ $m\angle F$ = ?	$m\angle A = 45°$ $m\angle B = 65°$ $m\angle C$ = ?	

4. What is the relationship between corresponding angles?

5. What is the relationship between corresponding sides?

6. Are the triangles similar according to the definition of *similar polygons*?

7. State a postulate for similarity between two triangles based on the result of the exploration. ❖

Try This Tell whether each pair of triangles is similar and why.

1. 45°, 105°   47°, 105°

2.

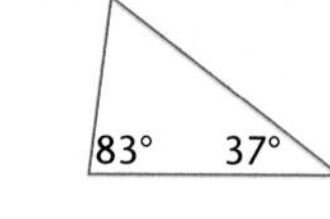

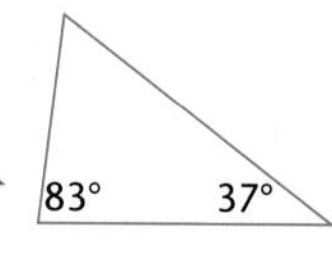

Exploration 2 *Similar Triangles Postulate 2*

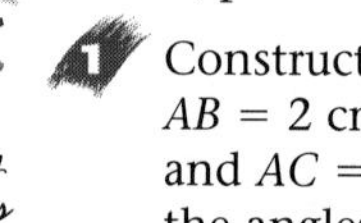

You will need
Geometry technology or
Ruler, protractor, and compass

 1. Construct $\triangle ABC$ with $AB = 2$ cm, $BC = 3$ cm, and $AC = 4$ cm. Measure the angles.

2. Construct $\triangle DEF$ with $DE = 6$ cm, $EF = 9$ cm, and $DF = 12$ cm. Measure the angles.

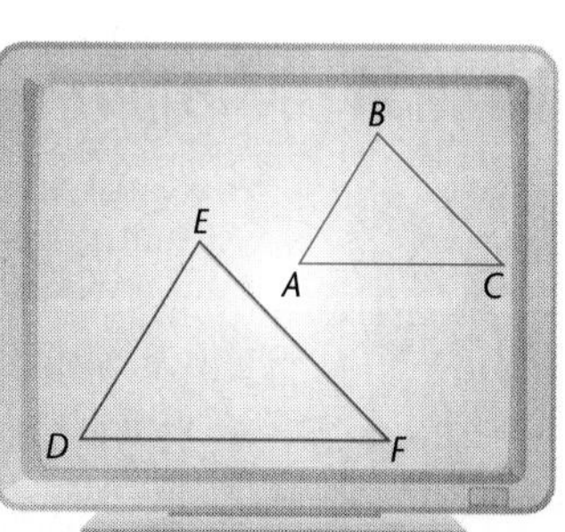

interdisciplinary CONNECTION **Arts** Much graphic design can be done with computers. Ask students to find examples of computer-drawn graphic designs in newspapers and magazines. Ask them to describe how designs and photographs are enlarged or reduced to fit a space on a newspaper page.

3 Use the measures to calculate the ratios on the chart below. For example, $\overline{DE}$ corresponds to $\overline{AB}$, so the ratio is $\frac{6}{2}$ or $\frac{3}{1}$.

	$\triangle DEF$	$\triangle ABC$	Ratio of the sides $\frac{\triangle DEF}{\triangle ABC}$
Measures of corresponding sides	$DE = 6$ $EF = 9$ $DF = 12$	$AB = 2$ $BC = 3$ $AC = 4$	$\frac{3}{1}$ $\frac{3}{1}$?
Measures of corresponding angles	m∠D = ? m∠E = ? m∠F = ?	m∠A = ? m∠B = ? m∠C = ?	

4 What is the relationship between corresponding angles?

5 What is the relationship between corresponding sides?

6 Are the triangles similar according to the definition of *similar polygons*?

7 State a postulate for similarity between two triangles based on the result of the exploration. ❖

Try This Tell whether each pair of triangles is similar and why.

1.

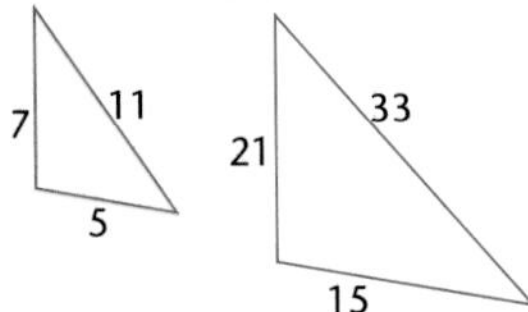

2.

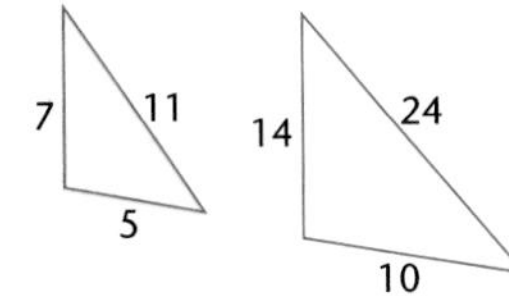

Exploration 3 *Similar Triangles Postulate 3*

You will need
Geometry technology or
Ruler and protractor

1 Construct $\triangle ABC$ with $AB = 3$ cm, m∠B = 90°, and $BC = 4$ cm. Measure the sides and the angles.

2 Construct $\triangle DEF$ with $DE = 6$ cm, m∠E = 90°, and $EF = 8$ cm. Measure the sides and the angles.

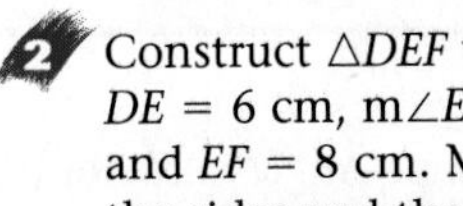

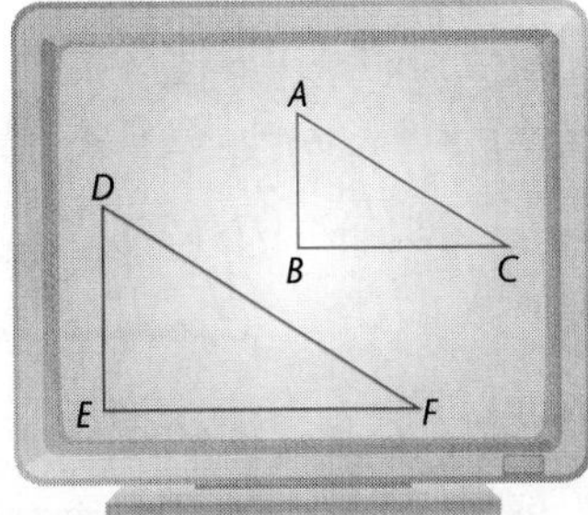

ONGOING ASSESSMENT

7. Two triangles are similar if all three pairs of corresponding sides are in the same proportion to each other.

ONGOING ASSESSMENT

Try This

The triangles in 1 are similar because all three pairs of corresponding sides are in the same proportion to each other. The larger triangle sides are three times bigger.

Exploration 3 Notes

In this exploration, students discover the SAS similarity postulate. Use geometry graphics software to draw two random triangles. Use its measurement options to determine whether the triangles are similar. If so, identify which similarity postulate was used.

TEACHING tip

Summarize the lesson by comparing each similar triangle postulate to its corresponding congruent triangles postulate. Have students verify the similarity postulates by drawing pairs of similar figures on the chalkboard and relating them to the postulates.

ENRICHMENT Many books on fractal geometry show the Sierpinski Triangle, a triangle containing an infinite number of self-similar triangles. Have students use the library to find books on fractals. Use a computer to generate a color drawing of the Sierpinski Triangle.

INCLUSION strategies **Inviting Participation** Some students may be more comfortable using a more analytical approach to learning the triangle similarity postulates. Have these students to look ahead to the postulate summary on page 432. Ask them to use technology or a ruler and protractor to verify the postulates and to give examples.

ongoing ASSESSMENT

7. Two triangles are similar if two pairs of corresponding sides are in the same proportion and the included angles are congruent.

ongoing ASSESSMENT

Try This

The triangles in 1 are similar because two pairs of corresponding sides are in the same proportion and the included angles are congruent.

CRITICAL Thinking

Two corresponding angles are enough for similarity, so a side isn't necessary.

3 Use the measures to calculate the ratios on the chart below. For example, $\overline{DE}$ corresponds with $\overline{AB}$, so the ratio is $\frac{8}{4}$ or $\frac{2}{1}$.

	△*DEF*	△*ABC*	Ratio of the sides $\frac{\triangle DEF}{\triangle ABC}$
Measures of corresponding sides	$DE = 6$ $EF = 8$ $DF =$?	$AB = 3$ $BC = 4$ $AC =$?	$\frac{2}{1}$ $\frac{2}{1}$?
Measures of corresponding angles	$m\angle D =$? $m\angle E =$? $m\angle F =$?	$m\angle A =$? $m\angle B =$? $m\angle C =$?	

4 What is the relationship between corresponding angles?

5 What is the relationship between corresponding sides?

6 Are the triangles similar according to the definition of *similar polygons*?

7 State a postulate for similarity between two triangles based on the result of the exploration. ❖

Try This Tell whether each pair of triangles is similar and why.

1.

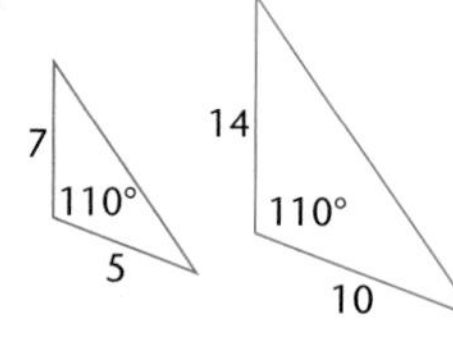

2.

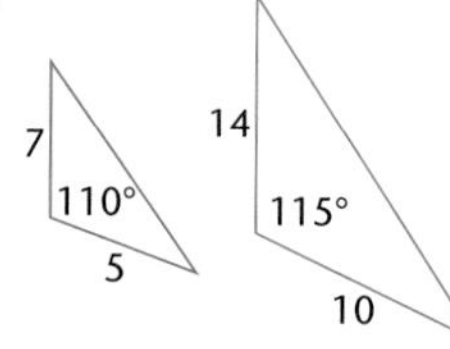

The three similarity postulates you explored above could be called the Angle-Angle, Side-Side-Side, and Side-Angle-Side postulates.

Summary

Two triangles are similar if

1. two pairs of corresponding angles are congruent, or
2. all three pairs of corresponding sides are in the same proportion, or
3. two pairs of corresponding sides are in the same proportion and the included angles are congruent.

CRITICAL Thinking

Why are Angle-Side-Angle and Angle-Angle-Side not included in a list of useful triangle similarity postulates?

RETEACHING the lesson

Cooperative Learning Divide the class into small groups. Have each group construct similar triangles from straws and use them to summarize the lesson to each other. Ask them to create sample questions to be used as a review for the whole class.

EXERCISES & PROBLEMS

Communicate

1. Explain the similarity postulate that you discovered in Exploration 1. What would you call this postulate?
2. Explain the similarity postulate that you discovered in Exploration 2. What would you call this postulate?
3. Explain the similarity postulate that you discovered in Exploration 3. What would you call this postulate?
4. Use a counterexample to show that if corresponding angles of one quadrilateral are congruent to corresponding angles of a second quadrilateral, then the quadrilaterals are not necessarily similar.

Practice & Apply

Can each pair of triangles be proven similar? Explain why or why not.

5. Yes; AA-Similarity

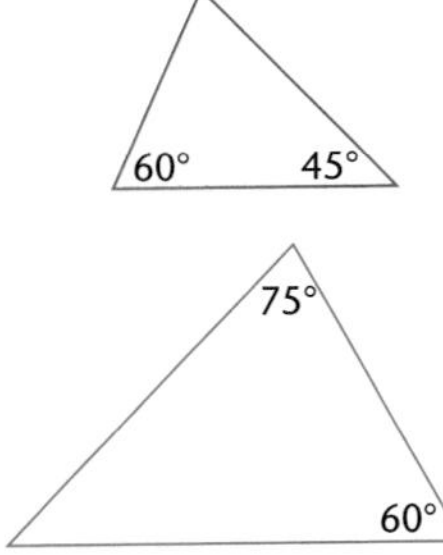

6. Yes; SAS-Similarity

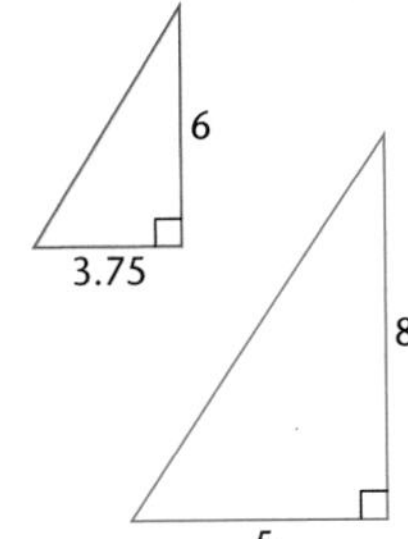

7. Yes; SSS-Similarity

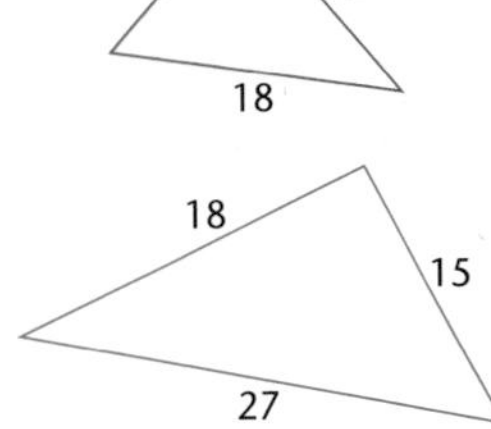

8. $\triangle ABC \sim \triangle DEF$. The perimeter of isosceles triangle ABC is 99 and the perimeter of isosceles triangle DEF is 48. If the length of the base of $\triangle ABC$ is 33, what is the length of the base of $\triangle DEF$? 16 units
9. The ratio of the corresponding sides of two triangles is equal to the ratio of the perimeters. If the perimeters of two triangles are 16 cm and 24 cm, what is the ratio of the perimeters? What is the ratio of the sides? $\frac{2}{3}, \frac{2}{3}$

ASSESS

Selected Answers

Exercises 5, 7, 9, 11, 13, 15, 21, 23, and 25

Assignment Guide

Core 1–9, 12–17, 20–25

Core Plus 1–4, 6–18 even, 20–26

Technology

Students may use geometry graphics software to draw the similar triangles in Exercises 10 and 11. They may also use the software to help them answer other questions in these exercises.

Error Analysis

Students may have difficulty explaining their answers for Exercises 5–7. Ask them to mark all angles that are congruent and then check to see that the corresponding sides are proportional.

alternative ASSESSMENT

Portfolio Assessment

As a writing assignment, ask students to draw pairs of similar triangles. Have them label the congruent angles and give the scale factor between the corresponding sides. Ask them to summarize the triangle similarity postulates in their drawings.

10. Yes; The ratio between all pairs of corresponding sides is 2. SSS-Similarity

Use $\triangle ABC$ to complete Exercises 10 and 11.

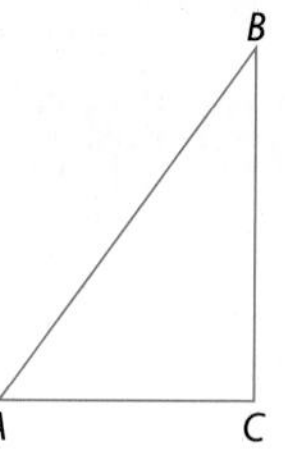

10. Draw a triangle in which each side is twice as long as each side of $\triangle ABC$. Is this triangle similar to $\triangle ABC$? Why or why not?

11. Draw a triangle in which each side is 3 cm longer than each side of $\triangle ABC$. Is this triangle similar to $\triangle ABC$? Why or why not?

In Exercises 12 and 13, indicate which figures are similar and explain why.

12.

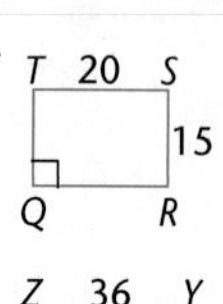

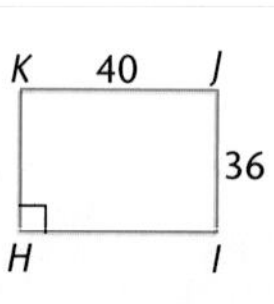

Z 36 Y, 27, W X

13.

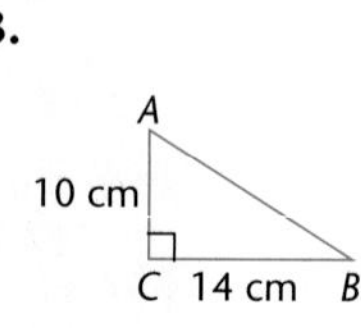

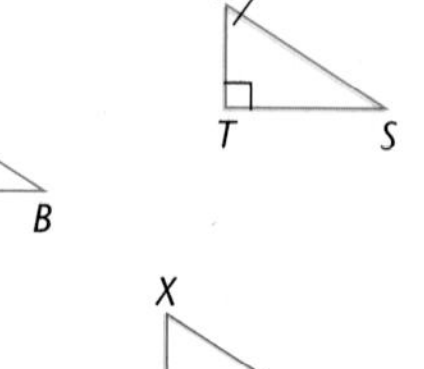

X, 12 cm, 32°, Z, 16 cm, Y

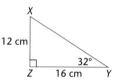

$\triangle RST \sim \triangle XYZ$; AA-Similarity

14. Can each set of triangles be proven similar? Why or why not?

a.

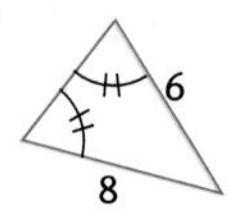

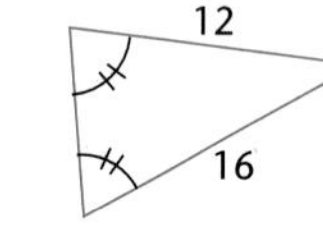

Yes; AA-Similarity

b.

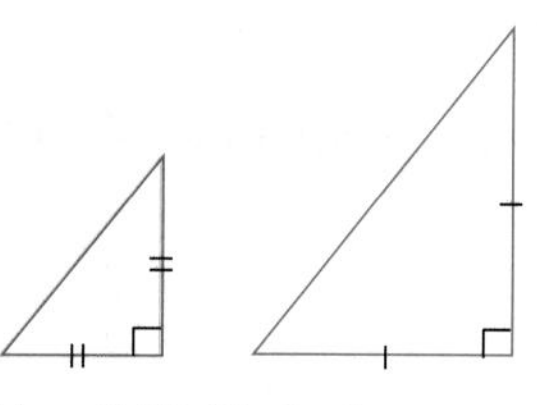

Yes; SAS-Similarity

15. $\triangle ABC \sim \triangle XYZ$. Find the lengths of sides $\overline{XZ}$ and $\overline{YZ}$.

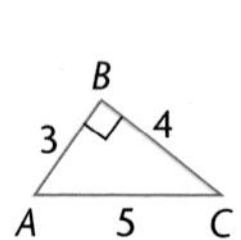

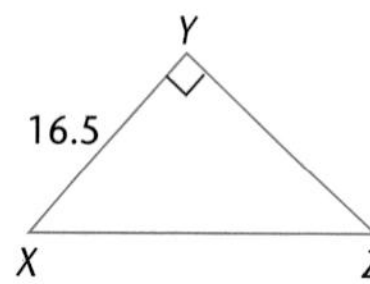

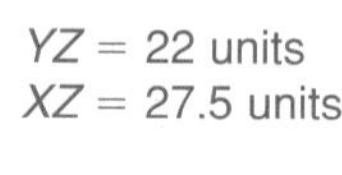

$YZ = 22$ units
$XZ = 27.5$ units

16. What is the ratio of the perimeters of the two similar triangles? What is the ratio of the corresponding sides? $\frac{1}{3}$; $\frac{1}{3}$

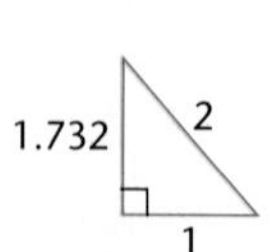

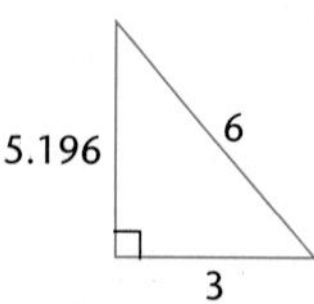

11. No; The ratios between pairs of corresponding sides will not necessarily be equal.

12. $TSRQ$ is similar to $ZYXW$. Corresponding angles are congruent (all 90°) and corresponding sides are in the proportion $\frac{5}{9}$.

Practice Master

NAME ______ CLASS ______ DATE ______

Practice & Apply

8.3 Exploring Triangle Similarity Postulates

1. Can $\triangle ABC$ be proven similar to $\triangle DEF$? If yes, state the postulate. If no, write a condition that will make the proof possible.

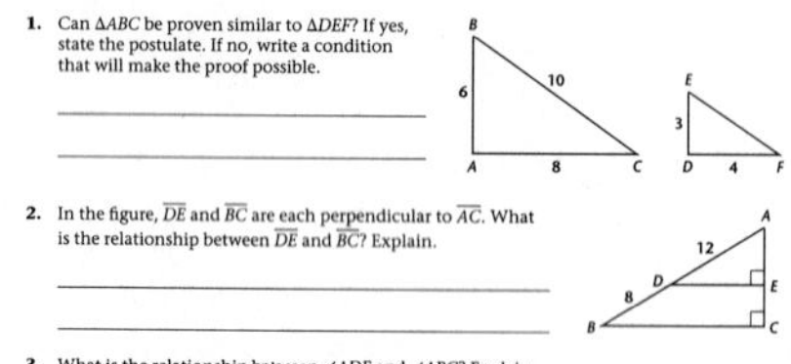

2. In the figure, $\overline{DE}$ and $\overline{BC}$ are each perpendicular to $\overline{AC}$. What is the relationship between $\overline{DE}$ and $\overline{BC}$? Explain.

3. What is the relationship between $\angle ADE$ and $\angle ABC$? Explain.

4. Name a pair of similar triangles and state the reason for similarity.

5. Use a proportion to find BC.

6. Two corresponding sides of two similar triangles are 6 cm and 14 cm. If the perimeter of the first triangle is 21 cm, find the perimeter of the second triangle.

7. The sides of a triangle measure 6″, 8″, and 10″. Find the length of the smallest side of a similar triangle whose perimeter is 16″.

8. The lengths of the sides of a triangle are 5′, 6′ and 7′. If the perimeter of a similar triangle is 36′, find the length of its shortest side.

9. The sides of a triangle are 8′, 20′, and 24′. The longest side of a similar triangle is 12′. Find the perimeter of the smaller triangle.

10. Corresponding sides of two similar triangles are in the ratio 2:3. What is the ratio of the areas of the triangles?

HRW Geometry — Practice & Apply 45

Surveying Surveyors sometimes use similar triangles to measure inaccessible distances. A surveyor could find distance *AB* by setting up similar triangles $\triangle ABC$ and $\triangle EDC$. Assuming that all lengths may be directly measured except *AB*, the surveyor can use these measures to set up a proportion and solve for *AB*.

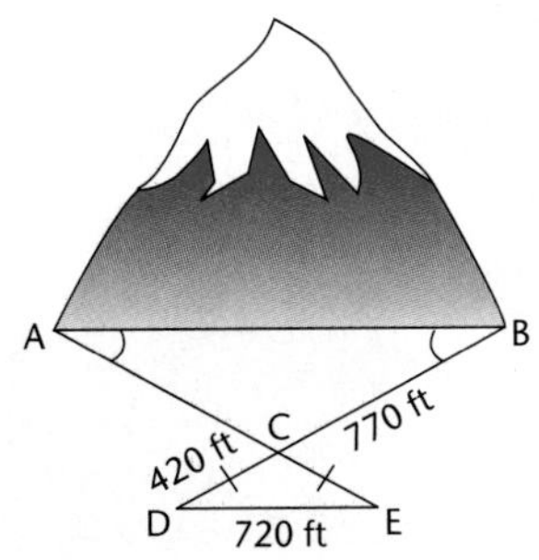

17. How does the surveyor make $\triangle ABC$ similar to $\triangle EDC$?

18. Set up a proportion and solve for *AB*.

19. Find the areas of the two triangles in Exercise 17. Compare the ratio of the areas to the ratio of the perimeters. What do you notice?

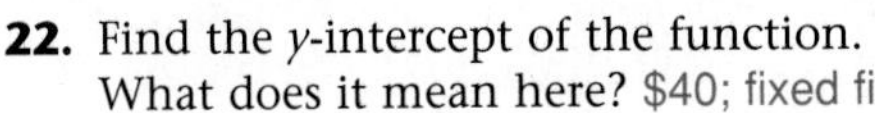

Look Back

20. In the diagram lines *j* and *k* are parallel; $\overline{AB} \cong \overline{BC}$. Find *v*, *x*, *y*, and *z*. **[Lesson 3.3]**

21. Find the distance between the points (−6, 2) and (3, 8). **[Lesson 5.6]** $\sqrt{117} \approx 10.82$

Algebra In Exercises 22–25, the fine for speeding in Marshall County is found using the function $f(n) = 40 + 15n$. The variable *n* is the number of miles per hour over the speed limit and $f(n)$ is the fine for going *n* miles per hour over the limit.

22. Find the *y*-intercept of the function. What does it mean here? $40; fixed fine

23. Find the slope of the function. What does it mean here?

24. How many miles per hour over the limit will result in a $100 fine?

25. Find the equation of the line that contains the points (−6, 2) and (3, 8). $y = \frac{2}{3}x + 6$

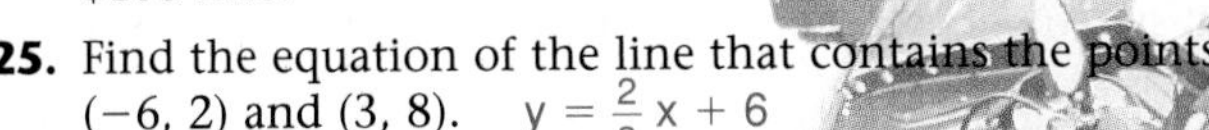

Look Beyond

26. Copy the figures onto a piece of paper. In each picture, draw lines connecting *A* to *A′*, *B* to *B′*, and *C* to *C′*. Extend the lines until they meet. Label the point where they meet *O*. Compare the lengths of segments $\overline{OA}$ and $\overline{OA'}$. Compare the lengths of segments $\overline{AB}$ and $\overline{A'B'}$. What do you notice? Compare other lengths. What did you find?
The ratios are the same.

Use Transparency 65

Look Beyond

Exercise 26 asks students to review dilations in terms of similar triangles. Ask students to use the triangle similarity postulates to help them list some properties of dilations.

Technology Master

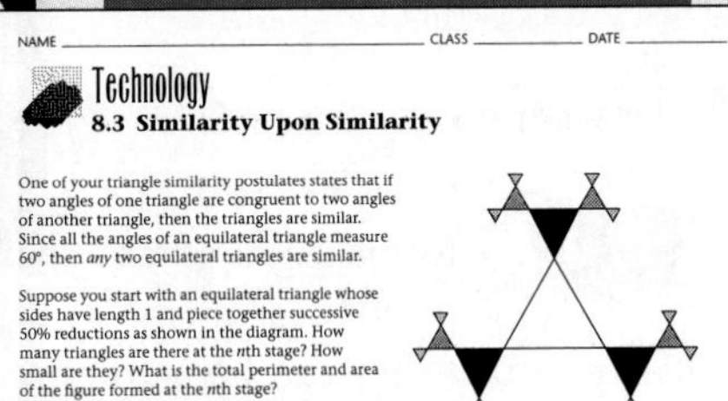

NAME ______ CLASS ______ DATE ______

Technology
8.3 Similarity Upon Similarity

One of your triangle similarity postulates states that if two angles of one triangle are congruent to two angles of another triangle, then the triangles are similar. Since all the angles of an equilateral triangle measure 60°, then *any* two equilateral triangles are similar.

Suppose you start with an equilateral triangle whose sides have length 1 and piece together successive 50% reductions as shown in the diagram. How many triangles are there at the *n*th stage? How small are they? What is the total perimeter and area of the figure formed at the *n*th stage?

Using your knowledge of similarity and a spreadsheet, you will be able to answer these questions and more.

Write formulas for each expression in terms of *n*.

1. the number of triangles added at the *n*th stage of the process ______
2. the length of a side of a triangle added at the *n*th stage of the process ______
3. the perimeter of a triangle added at the *n*th stage of the process ______
4. Create a spreadsheet in which column A contains the stage number $n = 1, 2, 3, \ldots$, column B contains the number of triangles added at the *n*th stage, column C contains the length of each side of a triangle added at the *n*th stage, and column D contains the perimeter of each triangle added at the *n*th stage. What do you notice in your spreadsheet? ______
5. In your spreadsheet from Exercise 4, add columns E and F to compute the total of the perimeters of all triangles added at the *n*th stage and in all *n* stages of the process. What do you notice? ______
6. The area *A* of an equilateral triangle whose sides have length *s* is $A = 0.433s^2$. Add columns G and H to compute the total area of all triangles added at the *n*th stage and the total of all areas of all triangles added in all *n* stages of the process. What do you notice? ______

HRW Geometry — Technology 57

17. By making $\angle BDE \cong \angle AED$. Then $\angle CED \cong \angle ABD$ and $\angle CDE \cong \angle EAB$. Then $\triangle EAB \sim \triangle EAB$ by AA similarity.

18. $AB = 1320$ ft

19. $A_{CDE} = 155{,}757.6 \text{ ft}^2$
$A_{CAB} = 523{,}010.4 \text{ ft}^2$
$\frac{P_{CDE}}{P_{CAB}} \approx 0.55$; $\frac{A_{CDE}}{P_{CAB}} \approx 0.30$
The area ratio appears to be the square of the perimeter ratio.

20. $v = 60°$; $x = 135°$; $y = 45°$; $z = 60°$

23. $15; Fine for every mile over the speed limit.

24. 4 miles over the speed limit

PREPARE

Objectives

- Formally define the three triangle similarity postulates —AA, SSS, SAS.
- Develop and use the Side-Splitting Theorem.

RESOURCES

• Practice Master	8.4
• Enrichment Master	8.4
• Technology Master	8.4
• Lesson Activity Master	8.4
• Quiz	8.4
• Spanish Resources	8.4

Assessing Prior Knowledge

State three ways triangles are similar.

[If two pairs of corresponding angles are congruent; if all three pairs of corresponding sides are in the same proportion to each other; if two pairs of corresponding sides are in the same proportion and the included angles are congruent.]

TEACH

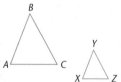

Students need to formalize their postulates for similarity so that they can be used in proofs. This includes the Side-Splitting Theorem and its corollary.

LESSON 8.4 The Side-Splitting Theorem

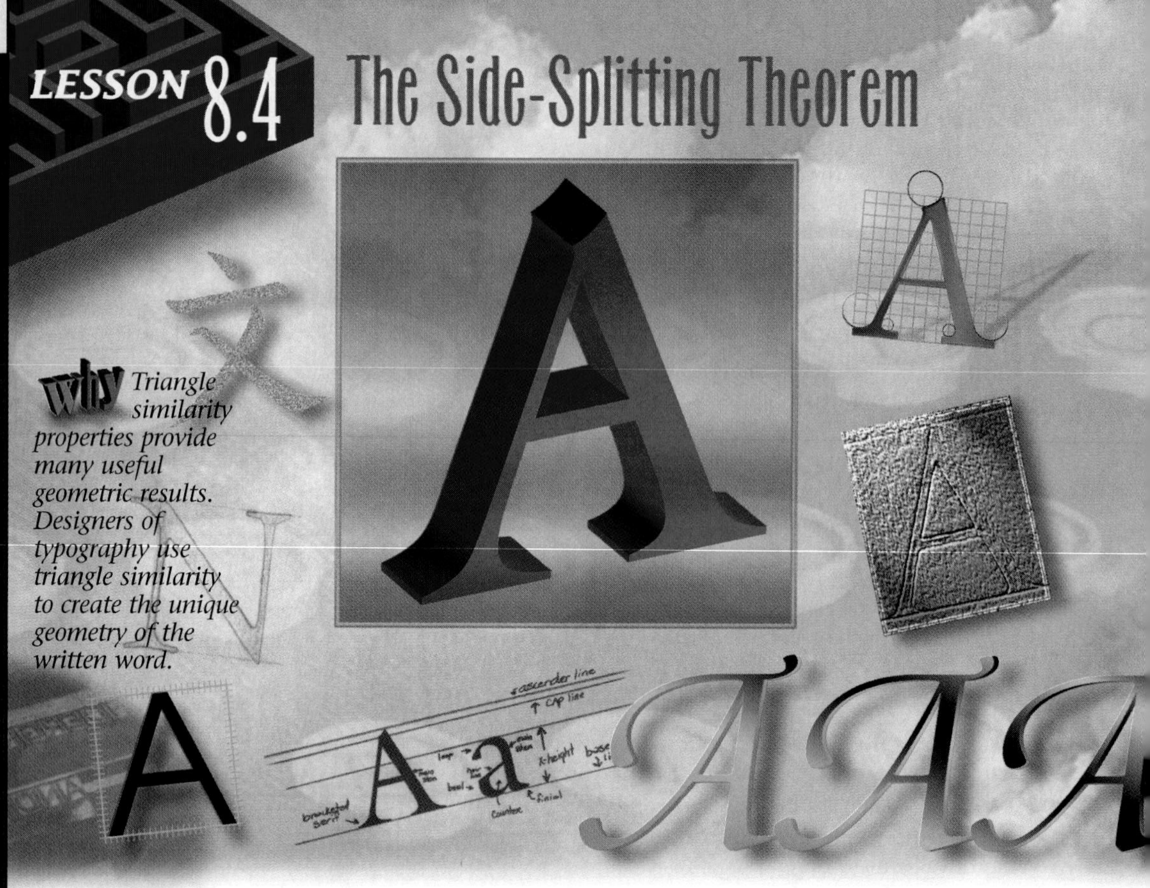

why *Triangle similarity properties provide many useful geometric results. Designers of typography use triangle similarity to create the unique geometry of the written word.*

In the last lesson you investigated three different combinations of sides and angles that allow you to conclude triangle similarity from limited information. Formal statements of each combination are given below as postulates.

AA (ANGLE-ANGLE) SIMILARITY POSTULATE

If two angles of one triangle are equal in measure to two angles of another triangle, then the triangles are similar. **8.4.1**

In $\triangle ABC$ and $\triangle XYZ$, if $m\angle A = m\angle X$ and $m\angle B = m\angle Y$, then $\triangle ABC \sim \triangle XYZ$.

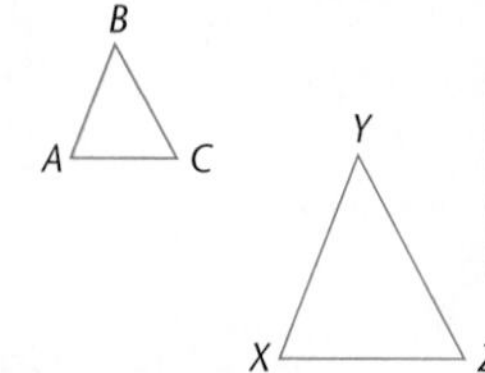

SSS (SIDE-SIDE-SIDE) SIMILARITY POSTULATE

If the measures of pairs of corresponding sides of two triangles are proportional, then the two triangles are similar. **8.4.2**

In $\triangle ABC$ and $\triangle XYZ$, if $\frac{AB}{XY} = \frac{AC}{XZ} = \frac{BC}{YZ}$, then $\triangle ABC \sim \triangle XYZ$.

Hands-On Strategies

Students can develop an understanding of the Side-Splitting Theorem by cutting out paper models of triangles. They should cut each triangle along a parallel to one side of the triangle. Then students should superimpose the angles to show that they are congruent. They should measure the corresponding sides to show that they are proportional.

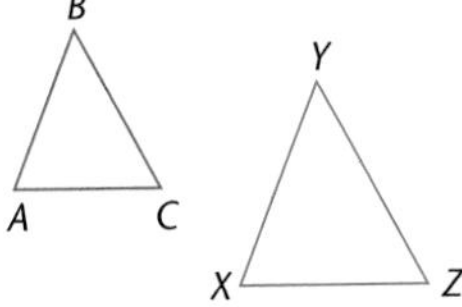
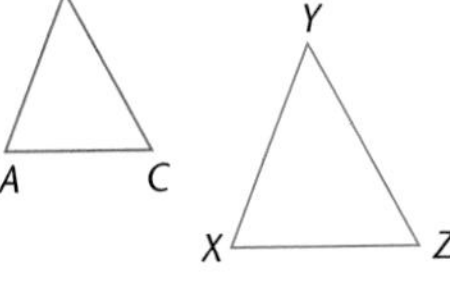

SAS (SIDE-ANGLE-SIDE) SIMILARITY POSTULATE

If the measures of two pairs of corresponding sides of two triangles are proportional and the measures of the included angles are equal, then the triangles are similar. **8.4.3**

In $\triangle ABC$ and $\triangle XYZ$, if $\frac{AB}{XY} = \frac{AC}{XZ}$ and $m\angle A = m\angle X$, then $\triangle ABC \sim \triangle XYZ$.

Can the proportion above be written $\frac{XY}{AB} = \frac{AC}{XZ}$? Explain.

In Lesson 3.7 you learned a theorem about the midsegment of a triangle. Are there proportional relationships between segments created by drawing other segments parallel to the side of a triangle?

SIDE-SPLITTING THEOREM

A line parallel to one side of a triangle divides the other two sides proportionally. **8.4.4**

Given: $\overline{EC} \parallel \overline{AB}$

Prove: $\frac{AE}{ED} = \frac{BC}{CD}$

Proof:

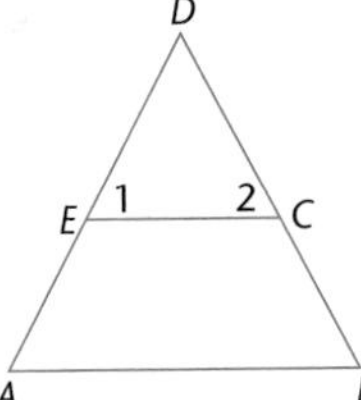

STATEMENTS	REASONS
1. $\overline{EC} \parallel \overline{AB}$	Given
2. $m\angle A = m\angle 1$ and $m\angle B = m\angle 2$	If ∥ segments are cut by a transversal, corresponding angles are congruent.
3. $\triangle ADB \sim \triangle EDC$	AA Similarity Postulate
4. $\frac{AD}{ED} = \frac{BD}{CD}$	Definition of Similar Triangles
5. $AD = AE + ED$, $BD = BC + CD$	Segment Addition Postulate
6. $\frac{AE + ED}{ED} = \frac{BC + CD}{CD}$	Substitution
7. $\frac{AE}{ED} = \frac{BC}{CD}$	"Add-One" Property of Proportions

TEACHING tip

It may be helpful to compare the similarity postulates, word for word, with the congruence postulates. Give examples, and point out differences and similarities.

CRITICAL Thinking

Yes; Reciprocal Property of Proportions

Cooperative Learning

Have student work in pairs. Each student should draw and label an example to illustrate the Side-Splitting Theorem. Ask them to measure the angles and sides and verify each step of the proof of the theorem in terms of their example. Have students switch papers and check each other's work.

Use Transparency 66

Architecture Triangles are often used to brace roofs and other structures because they are rigid. The triangles themselves may be braced along a parallel to one side. Have students find examples of triangular bracing in architecture books.

CRITICAL *Thinking*

$\frac{DE}{DA} = \frac{DC}{DB}; \frac{AE}{DA} = \frac{BC}{DB};$

$\frac{DE}{DA} = \frac{EC}{AB}; \frac{DC}{DB} = \frac{EC}{AB}$

Alternate Example 1

Solve for x using Theorem 8.4.4.

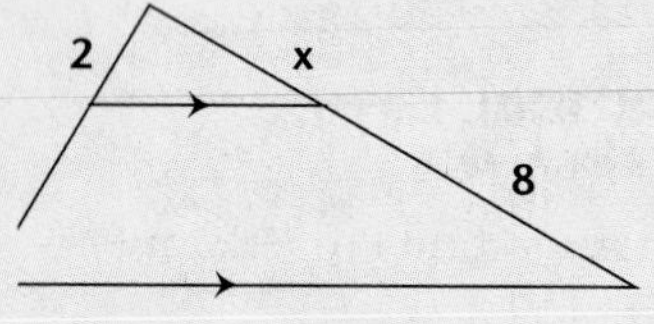

[3.2]

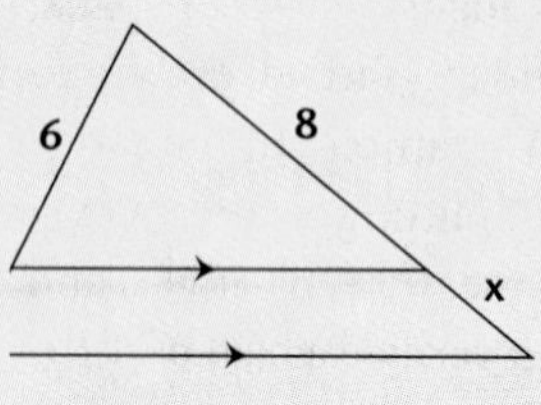

$[5\frac{1}{3}]$

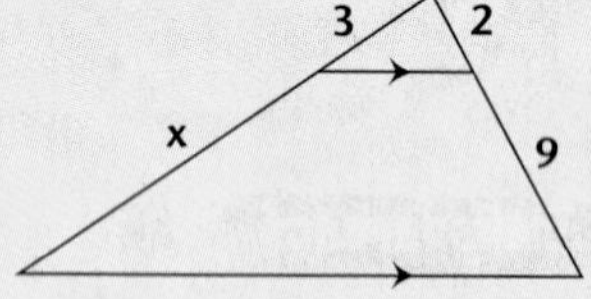

[3.5]

CRITICAL *Thinking*

What other proportions can you find in $\triangle ADB$ using the Side-Splitting Theorem?

The diagram at right suggests an easier way to remember the proportions that arise from the Side-Splitting Theorem. Using the diagram, the following proportions result:

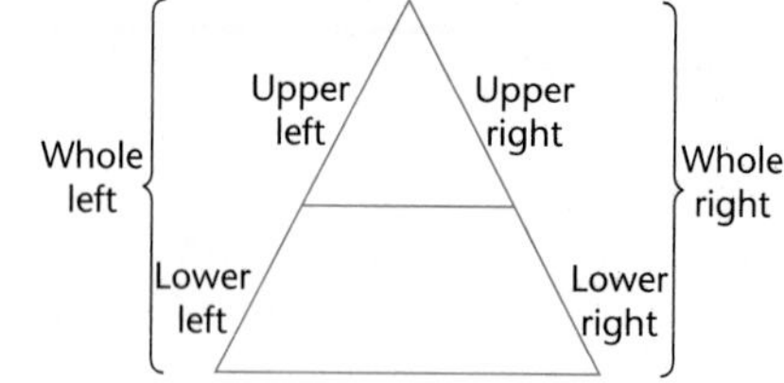

$$\frac{\text{Upper left}}{\text{Whole left}} = \frac{\text{Upper right}}{\text{Whole right}}$$

$$\frac{\text{Upper left}}{\text{Lower left}} = \frac{\text{Upper right}}{\text{Lower right}}$$

$$\frac{\text{Lower left}}{\text{Whole left}} = \frac{\text{Lower right}}{\text{Whole right}}$$

EXAMPLE 1

Solve for x using Theorem 8.4.4.

A

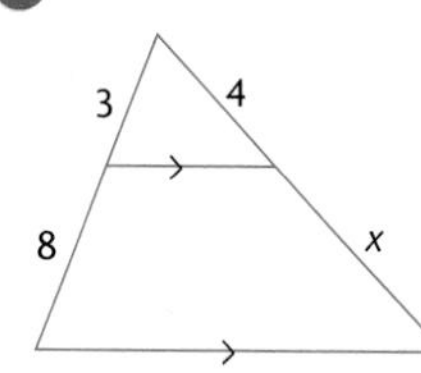

B

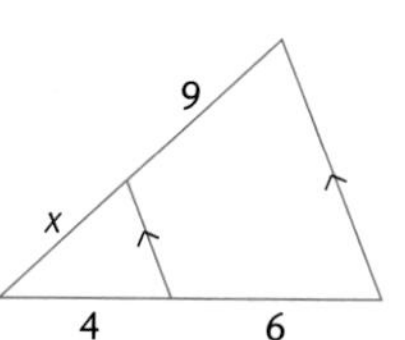

C

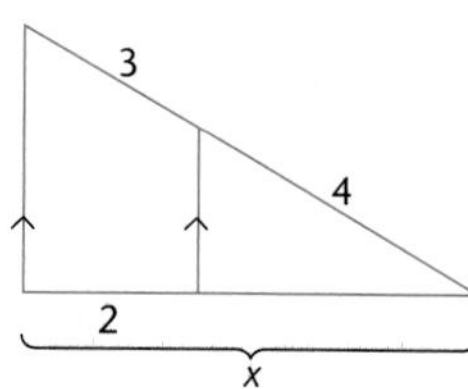

Solution

A $\frac{3}{8} = \frac{4}{x}$

$3x = 32$

$x = 10\frac{2}{3}$

B $\frac{4}{6} = \frac{x}{9}$

$36 = 6x$

$x = 6$

C $\frac{x}{2} = \frac{7}{3}$

$3x = 14$

$x = 4\frac{2}{3}$ ❖

A corollary is a theorem that is easily proven from another theorem. A corollary that follows directly from the Side-Splitting Theorem is given on the next page.

ENRICHMENT The Eiffel Tower offers many examples of the Side-Splitting Theorem and triangular bracing. Ask students to write a report about the Eiffel Tower.

INCLUSION strategies **English Language Development** Some students may have difficulty understanding the terminology in this lesson. Have them list all concepts that are confusing and then verify the postulates and theorems in terms of those concepts.

TWO TRANSVERSAL PROPORTIONALITY COROLLARY
Two or more parallel lines divide two transversals proportionally. **8.4.5**

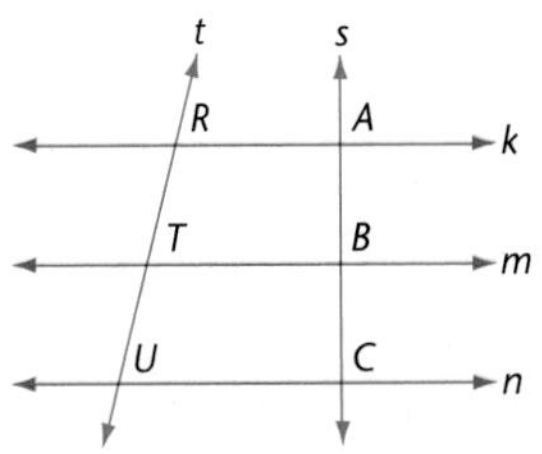

If $k \parallel m \parallel n$, with transversals s and t as shown, then $\frac{AB}{BC} = \frac{RT}{TU}$.

Use the words *upper, lower,* and *whole* to describe the proportions in Corollary 8.4.5.

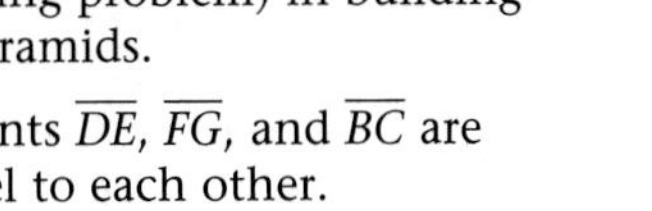

Cultural Connection: Africa
Students in ancient Egypt studied geometry to solve practical problems (such as the following problem) in building the pyramids.

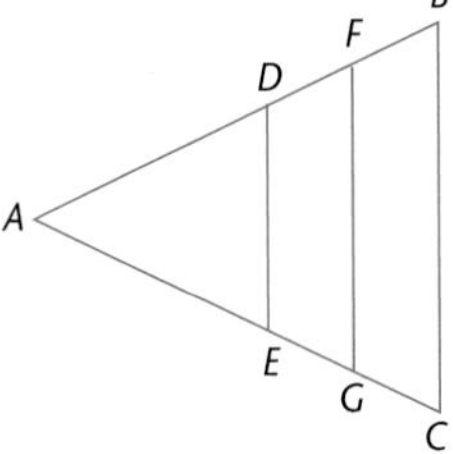

Segments $\overline{DE}$, $\overline{FG}$, and $\overline{BC}$ are parallel to each other.

$AD = AE = 7$ cubits

$DF = FB = 3\frac{1}{2}$ cubits

$DE = 2\frac{1}{4}$ cubits

Solution ➤

$\frac{AD}{DF} = \frac{AE}{EG}$ $\quad \frac{7}{3.5} = \frac{7}{EG}$ $\quad EG = 3.5$

$\frac{DF}{FB} = \frac{EG}{GC}$ $\quad \frac{3.5}{3.5} = \frac{3.5}{GC}$ $\quad GC = 3.5$

$\frac{AD}{AF} = \frac{DE}{GF}$ $\quad \frac{7}{10.5} = \frac{2.5}{GF}$ $\quad GF = 3.75$

$\frac{AD}{AB} = \frac{DE}{BC}$ $\quad \frac{7}{14} = \frac{2.5}{BC}$ $\quad BC = 5$ ❖

Cooperative Learning Use cooperative groups to review the terminology used in this lesson, including *parallel, transversal,* and *proportional.* Ask each group to create examples of the Side-Splitting Theorem and its corollary, and use them as review for the entire class. Having students work together may help students make connections between the algebra and geometry used in this lesson.

CRITICAL *Thinking*

$\frac{\text{upper left}}{\text{lower left}} = \frac{\text{upper right}}{\text{lower right}}$;
$\frac{\text{upper left}}{\text{whole left}} = \frac{\text{upper right}}{\text{whole right}}$;
$\frac{\text{lower left}}{\text{whole left}} = \frac{\text{lower right}}{\text{whole right}}$

Alternate Example 2

A street was laid out according to the plan shown below with lot boundaries parallel to each other. Find the remaining lengths.

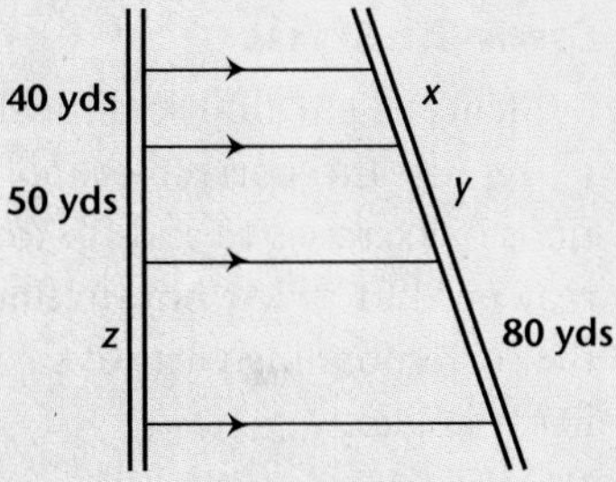

[$x = 51.2$ yds; $y = 64$ yds, $z = 62.5$ yds]

Assess

Selected Answers

Exercises 7, 9, 11, 13, 15, 19, 21, and 23

Assignment Guide

Core 1–10, 12–24

Core Plus 1–27

Technology

A scientific calculator will be helpful for many of the exercises.

Error Analysis

Students might find it difficult to set up the correct proportions in Exercises 13 and 14. You may need to review how to find the corresponding parts of similar triangles.

Practice Master

NAME ______ CLASS ______ DATE ______

Practice & Apply
8.4 The Side-Splitting Theorem

In Exercises 1–2, state which triangles are similar.

1. $\overline{AB} \perp \overline{BC}$, $\overline{BD} \perp \overline{AC}$
2. $\angle EDA \cong \angle DAC$
3. $\overline{DE} \parallel \overline{AB}$. Find x.
4. $\overline{DE} \parallel \overline{AB}$. Find x.
5. $\overline{AB} \parallel \overline{CD} \parallel \overline{EF}$. Find x.
5. $\overline{AB} \parallel \overline{CD} \parallel \overline{EF}$. Find x.
7. In $\triangle ABC$, D is the midpoint of $\overline{AB}$ and $\overline{DE} \parallel \overline{BC}$. Use a proportion to find EC, and draw a conclusion about this result.
8. If $BC = 14$, find DE, and draw a conclusion about this result.

46 Practice & Apply HRW Geometry

Exercises & Problems

Communicate

1. Why isn't there an AAA Postulate for similarity?
2. What is the relationship between the ratios of the sides and the ratios of the perimeters of two similar triangles? Do you think the same relationship will occur for similar quadrilaterals? Explain.
3. Are all isosceles triangles similar? Explain or give a counterexample.
4. Are all equilateral triangles similar? Explain or give a counterexample.
5. Are all right triangles similar? Explain or give a counterexample.

Practice & Apply

State which triangles are similar in Exercises 6–8. Which similarity postulate supports your answer? Determine the scale factor where possible.

6.

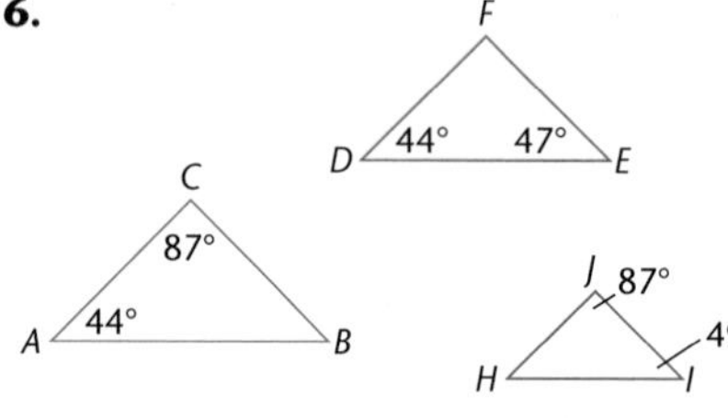

7. $\overline{AB} \parallel \overline{DE}$

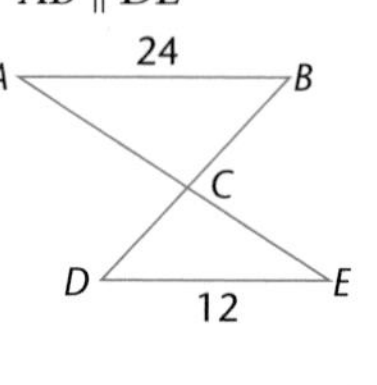

8.

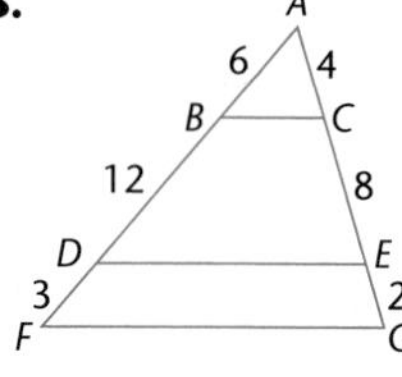

9. $\overline{TS} \parallel \overline{QR}$. Find x. $x = 9.6$

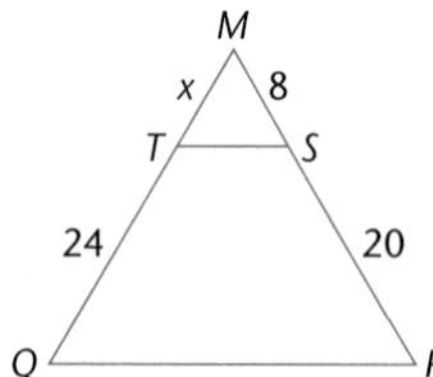

10. $\overline{DC} \parallel \overline{EB}$. Find x and y. $x = 9$; $y = 15$

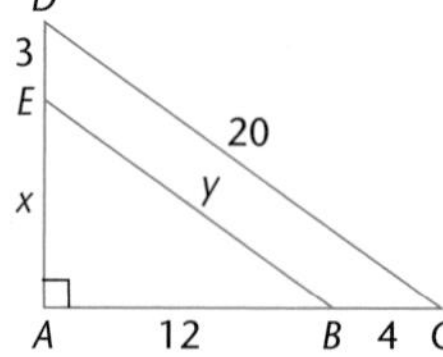

11. $\triangle ABC \sim \triangle AED$. Find x. $x = 2.4$

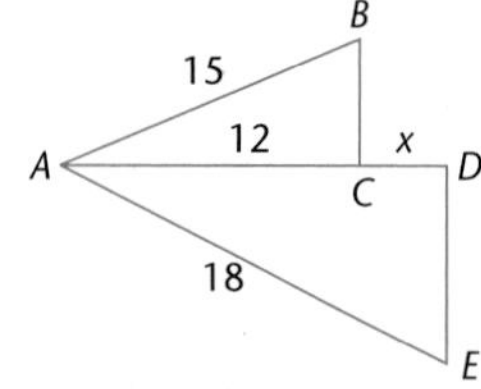

12. $QRST$ is a parallelogram. Find x and y. $x = 16$; $y = 15$

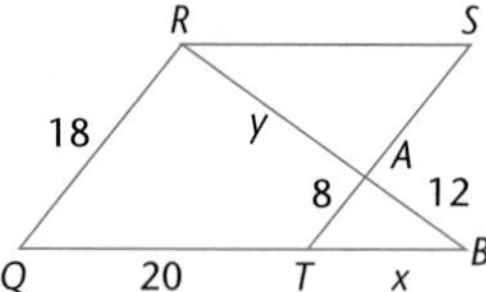

6. $\Delta ABC \sim \Delta HIJ$; AA-Similarity
7. $\Delta ABC \sim \Delta EDC$; AA-Similarity; Scale factor $= 2$ or $\frac{1}{2}$
8. $\Delta ABC \sim \Delta ADE$; SAS-Similarity; Scale factor $= 3$ or $\frac{1}{3}$; $\Delta ABC \sim \Delta AFG$; SAS-Similarity; Scale factor $= \frac{7}{2}$ or $\frac{2}{7}$; $\Delta ADE \sim \Delta AFG$; SAS-Similarity; Scale factor $= \frac{7}{6}$ or $\frac{6}{7}$

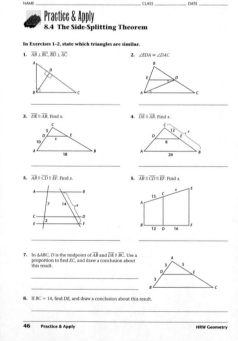

Indirect Measurement You can estimate the height of an object by placing a mirror on level ground so that you can see the top of the object in the mirror. By the laws of optics, light bounces off the mirror at an angle equal to the angle at which it strikes it, so $\angle HRE \sim \angle TRB$.

13. By which postulate is $\triangle TBR \sim \triangle HER$?
AA – Similarity

14. A person whose eyes are 5 feet above the ground places a mirror so that the dinosaur's nose is visible in it. The mirror is 12 feet from the point directly below the dinosaur's nose and 3 feet from the person. How high is the dinosaur's nose above the ground?
20ft

13. The AA Similarity Postulate

14. 20 ft

Use Transparency 67

alternative ASSESSMENT

Performance Assessment

Student presentations can be used to assess students' understanding of this lesson. Call on individual students to describe how the triangle similarity postulates are used. Other students can be asked to give examples of the Side-Splitting Theorem. Have them use visual aids if possible.

Technology Master

NAME ______________ CLASS ________ DATE ________

Technology

8.4 Extensions of the Side-Splitting Theorem

The Side-Splitting Theorem presented in Lesson 8.4 is illustrated in the diagram. The line parallel to $\overline{AC}$ divides the other sides proportionately so that

$$\frac{AA'}{A'B} = \frac{CC}{CB}$$

Does this theorem extend to polygons other than triangles? To explore the validity of any extension of the theorem, geometry software will come in handy.

Use geometry software in the following exercises. In Exercises 1 and 2, refer to the diagram at the left. In Exercises 3 and 4, refer to the diagram at the right.

1. Sketch a rectangle and a parallel line through it as shown. Mark points where the line intersects the sides of the rectangle. Find the ratios of the lengths of the corresponding sides of rectangle $ABCD$ and $A'BCD'$. Does the Side-Splitting Theorem still hold?

   ____

2. Experiment with the construction and calculations you made in Exercise 1 by raising or lowering $\overleftrightarrow{A'D'}$ to other positions. Do you get similarity of rectangles for any position of $\overleftrightarrow{A'D'}$?

   ____

3. Sketch a trapezoid and a line through it as shown. Mark points where the line intersects the sides of the trapezoid. Find the ratios of the lengths of the corresponding sides of trapezoid $ABCD$ and $A'BCD'$.

   Does the Side-Splitting Theorem still hold? ______________

4. Experiment with the construction and calculations you made in Exercise 1 by raising or lowering $\overleftrightarrow{A'D'}$ to other positions. Do you get similarity of trapezoids for any position of $\overleftrightarrow{A'D'}$?

   ____

58 Technology HRW Geometry

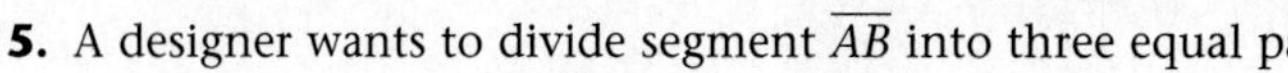

15. A designer wants to divide segment $\overline{AB}$ into three equal parts.

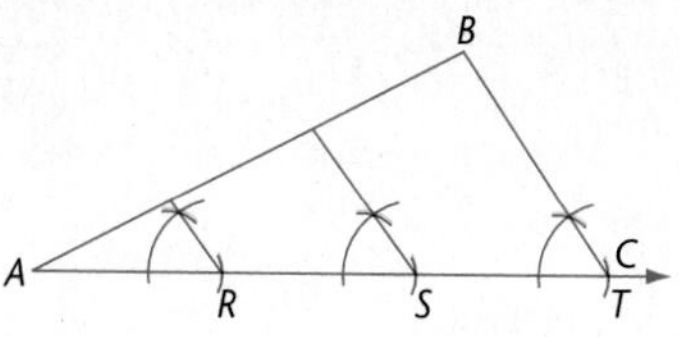

a. First he draws ray $\overrightarrow{AC}$ and marks three equal segments on the ray using his compass.

b. He then makes triangle $\triangle ABT$.

c. Finally, the designer constructs parallels through R and S. Explain why the construction works.

16. A rectangular strip of cloth $6\frac{1}{4}$ inches long must be divided into 7 congruent parts, and $6\frac{1}{4} \div 7$ involves a fraction that does not appear on a ruler. How can the Side-Splitting Theorem be applied to divide the cloth.

Look Back

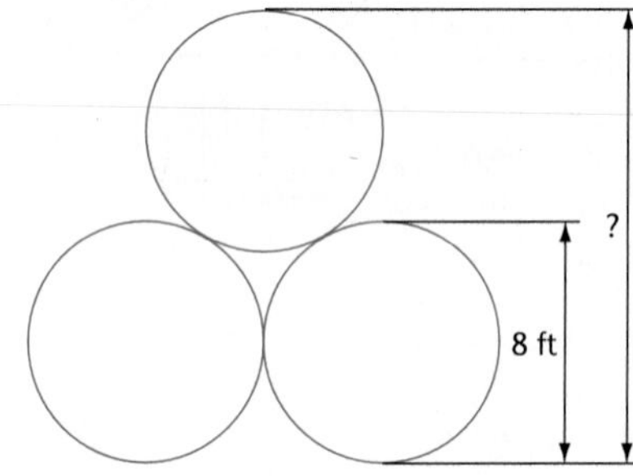

17. Find the height of the stack of pipes if the diameter of the outside of each pipe is 8 feet. (Hint: Assume that segments connecting the centers of the circles form an equilateral triangle whose sides measure 8 feet.) **[Lesson 5.4]** $4\sqrt{3} + 8$ ft

Use $\triangle ABC$, with $A(3, 5)$, $B(6, 1)$ and $C(-2, -7)$ for Exercises 18–24. [Lesson 5.6]

18. Graph $\triangle ABC$.

19. Find the perimeter of $\triangle ABC$. 29.31

20. Find the midpoints of each side. Label them D, E, and F.

21. Find the perimeter of $\triangle DEF$. 14.66

22. Draw the median of $\triangle ABC$ from A, and find its length using the Distance Formula. 8.06

23. Draw the median of $\triangle ABC$ from B, and find its length using the Distance Formula. 5.85

24. Draw the median of $\triangle ABC$ from C, and find its length using the Distance Formula. 11.93

Look Beyond

Use $\triangle XYZ$ and $\triangle XJK$ for Exercises 25–27.

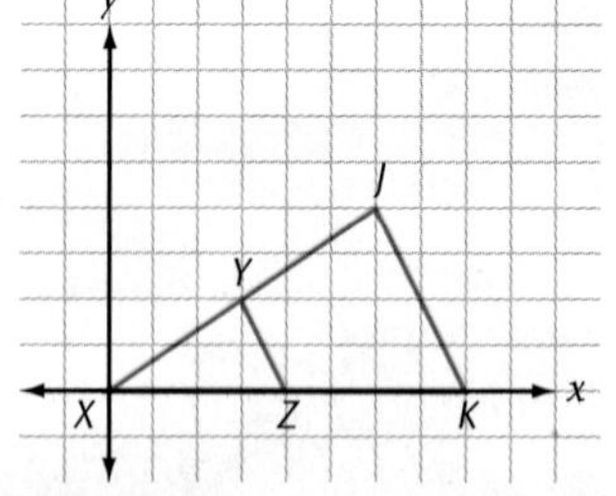

25. Find the ratios of the corresponding sides of the two triangles. Are they equal? 1:2; yes

26. Find the midpoints of sides $\overline{XZ}$ and $\overline{XK}$. Find the lengths of the medians of $\triangle XYZ$ and $\triangle XJK$. 2.24 and 4.48

27. Find the ratio of the corresponding medians in Exercise 26. 1:2

15. By the Side-Splitting Theorem, the segments cut by $\overline{AC}$ and $\overline{AB}$ are proportional. By construction, the scale factor is 1. Therefore, $\overline{AB}$ is cut into congruent segments.

16. Place the ruler at an angle with one side of each piece of cloth. Mark off 7 congruent marks on the ruler, such as every inch. Connect the 7 inch mark to the corner of the cloth. Create parallels to this segment at every inch mark. This will divide the side of the cloth into 7 congruent pieces.

Use Transparency 68

Look Beyond

Exercises 25–27 apply the triangle similarity postulates to triangles drawn in the coordinate plane. Students use coordinate geometry to learn that corresponding medians will have the same ratio as the corresponding sides.

18.

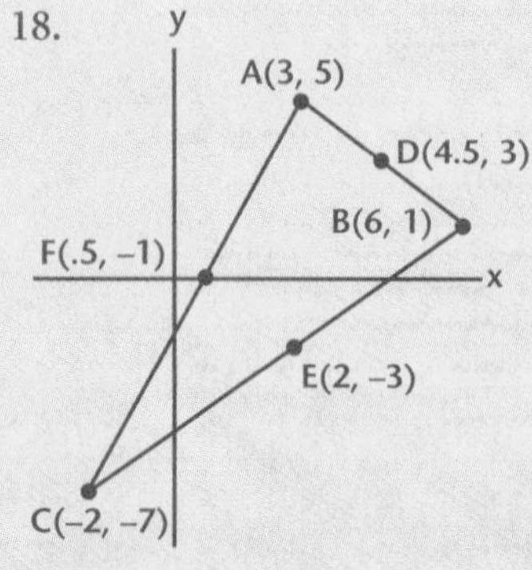

LESSON 8.5 Indirect Measurement, Additional Similarity Theorems

Why *If you needed to measure the breadth of a lake or the height of a mountain, you certainly could not do it directly with a tape measure. Instead, you would need to use a method that could approximate the mountain's height using indirect measurement.*

Using Similar Triangles to Measure Distances

Civil Engineering Inaccessible distances can often be measured by using similar triangles. For example, a military engineer must build a temporary bridge across a river. To do this he must find the distance across the river.

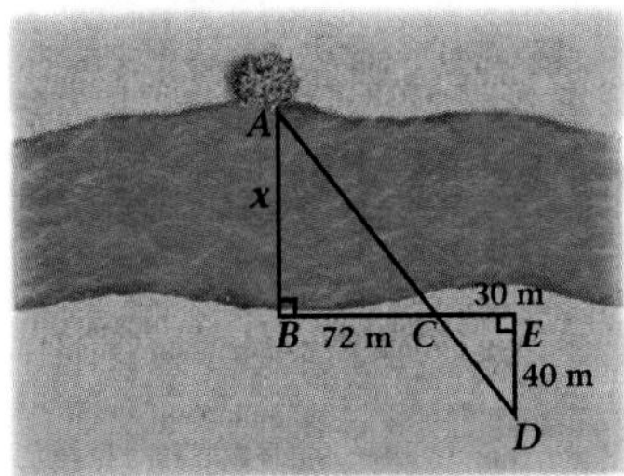

Sighting a tree across the river, the engineer sets up right triangles along the bank of the river.

Method 1 Using the right angles and their vertical angles, the engineer knows that $\triangle ABC \sim \triangle DEC$ because of AA similarity.

Therefore, $\frac{AB}{DE} = \frac{BC}{EC}$.

Then $\frac{x}{40} = \frac{72}{30}$.

$$30x = 2880$$
$$x = 96 \text{ m}$$

Method 2 Since $\angle ABE \cong \angle ACD$ and $\angle A \cong \angle A$, we know that $\triangle ABE \sim \triangle ACD$ because of AA similarity.

So, $\frac{x}{x + 32} = \frac{60}{80}$.

$$80x = 60x + 1920$$
$$20x = 1920$$
$$x = 96 \text{ m}$$

CRITICAL Thinking

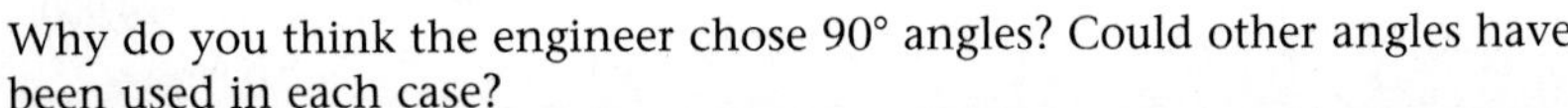

Why do you think the engineer chose 90° angles? Could other angles have been used in each case?

ALTERNATIVE teaching strategy

Cooperative Learning Divide the class into cooperative groups and have each group devise an experiment to estimate the size of something in or near the school building that would otherwise be difficult to measure. Examples include the height of the classroom, the height of a closet, the height of a stairwell, or the height of a flagpole. Ask each group to describe their experiment and their results to the class.

PREPARE

Objectives

- Use triangle similarity to measure distances indirectly.
- Develop and use the Proportional Altitudes and Proportional Medians Theorems.

RESOURCES

• Practice Master	8.5
• Enrichment Master	8.5
• Technology Master	8.5
• Lesson Activity Master	8.5
• Quiz	8.5
• Spanish Resources	8.5

Assessing Prior Knowledge

1. Solve $\frac{3}{x} = \frac{15}{20}$. [4]
2. $\triangle ABC \sim \triangle DEF$. Find the ratio of their sides. $\left[\frac{AB}{DE} = \frac{AB}{DF} = \frac{BC}{EF}\right]$
3. $\triangle ABC$ has sides 5, 5, and 6. Find the altitude from the vertex angle. [4]

Use Transparency 69

TEACH

Over the centuries similarity has been important to surveryors and map makers. Students need to learn how to use similarity in order to measure distances that would not otherwise be measurable.

CRITICAL Thinking

It is easy to make 90° angles. Yes, but other angles would have to be measured.

Use Transparency 70

Alternate Example 1

A flagpole casts a shadow of 128 in. at the same time a student standing next to the flagpole casts a shadow of 24 in. If the student is 5 ft 3 in. tall, how tall is the flagpole?

[28 ft]

Math Connection Algebra

Example 1 includes solving a proportion. You may want to review the Cross-Multiplication Property at this time.

Alternate Example 2

About 2500 years ago, the Greek mathematician Thales measured the height of the Great Pyramid by using shadows and similar triangles. He measured the shadow of a pole of known length and the shadow of the pyramid. Using similar triangles he determined the height of the pyramid. If the shadow of the pyramid is 65 m and the shadow of a 2 m pole is 0.9 m, how tall is the pyramid?

[144.4 m]

Math Connection Algebra

In algebra, students use absolute value to find the distance between two numbers. It may be necessary to review absolute value in order to calculate relative error.

EXAMPLE 1

Recreation Kim wants to know whether the basketball hoop in the new playground is the correct height, which is 10 feet. She notices the shadow of the pole that holds the hoop and wonders whether the shadow could be used to check the height of the hoop. Kim is 5 feet 10 inches (70 inches) tall.

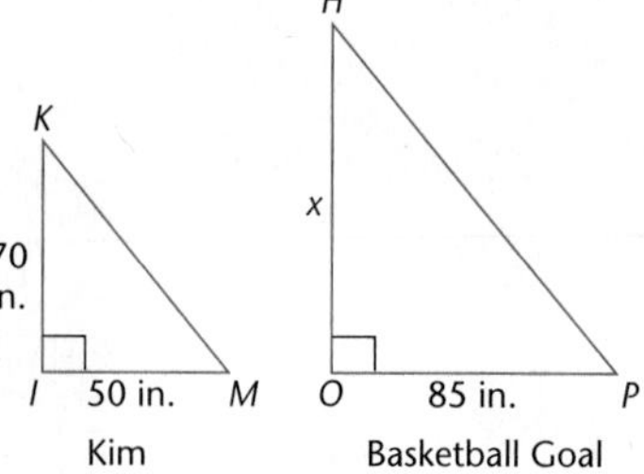

Solution ➤

Kim measures her shadow and the shadow of the pole. Notice that $\angle I \cong \angle O$, and since the sun's rays travel parallel to each other, $\angle M \cong \angle P$. Thus, $\triangle KIM \sim \triangle HOP$ because of AA similarity.

$$\frac{x}{70} = \frac{85}{50}$$

$$50x = 5950$$

$x = 119$ in. $= 9$ ft 11 in. The hoop is a little low. ❖

EXAMPLE 2

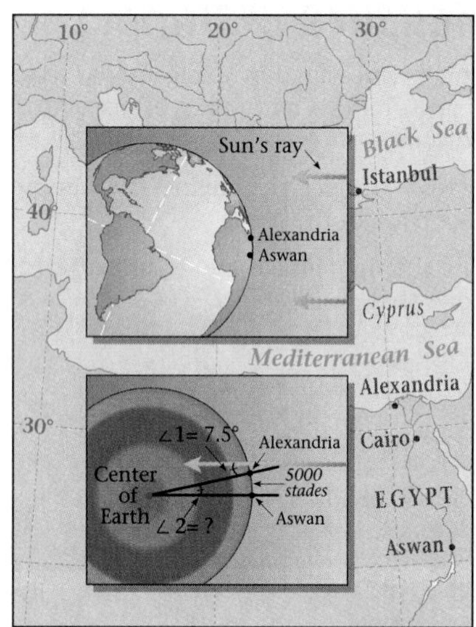

Cultural Connection: Africa About 200 B.C.E., Eratosthenes, a Greek astronomer who lived in North Africa, estimated the circumference of the Earth using indirect measurement. He knew that a vertical rod in the city of Aswan would not cast a shadow at noon at the time of the summer solstice. In Alexandria, which is 5000 stades (575 miles) from Aswan, a vertical rod would cast a shadow at the same time. He measured the angle the Sun's rays formed with the vertical rod as 7.5°.

A Explain why $\angle 1 \cong \angle 2$. Remember that the Sun's rays are considered to be parallel.

B Find what Eratosthenes estimated the Earth's circumference to be.

C Find the relative error of Eratosthenes' measurement. (The circumference of the Earth is approximately 24,900 miles.)

Solution ➤

ALGEBRA
Connection

A $\angle 1 \cong \angle 2$ because they are alternate interior angles.

B $\frac{7.5}{360} = \frac{575}{c}$ $\quad c = 27{,}600$ mi

C $E_R = \frac{|24{,}900 - 27{,}600|}{24{,}900} \times 100\%$

$\approx 10.8\%$ ❖

interdisciplinary CONNECTION **Surveying** Surveyors use special instruments to measure distances that cannot be measured directly. Have students research the type of tools used by surveyors, including modern laser instruments and standard transits. Ask them to explain how similar triangles are used in the measurements.

Additional Similarity Theorems

The sides of similar triangles are proportional. But, as you will see, other parts of similar triangles are also in proportion.

PROPORTIONAL ALTITUDES THEOREM

If two triangles are similar, then their corresponding altitudes have the same ratio as the corresponding sides. **8.5.1**

For example, if $\triangle ABC \sim \triangle XYZ$, $\overline{AD} \perp \overline{BC}$, and $XM \perp YZ$ then $\frac{AD}{XM} = \frac{AB}{XY}$.

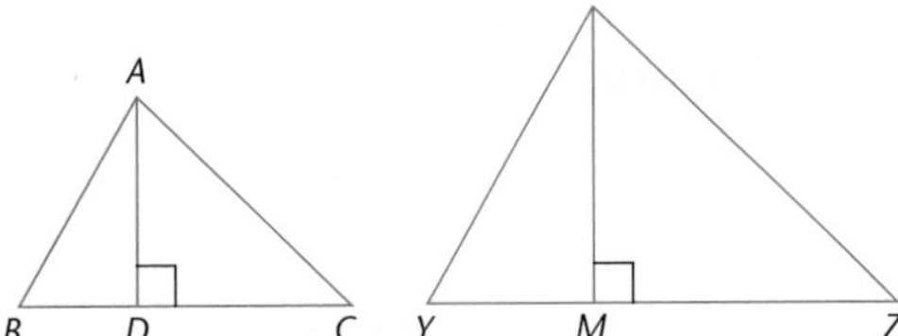

If an altitude of a triangle is not a side of the triangle, then it may be either inside or outside of the triangle. Therefore, a proof of the Proportional Altitudes Theorem must cover both possibilities. The proof below considers only the case of an altitude inside a triangle.

Given: $\triangle ABC \sim \triangle XYZ$; $\overline{AD}$ is an altitude of $\triangle ABC$; $\overline{XM}$ is an altitude of $\triangle XYZ$.

Prove: $\frac{AD}{XM} = \frac{AB}{XY}$

Proof:

Case 1: The altitude is inside the triangle. $\angle B \cong \angle Y$, by definition of similar triangles. $\angle BDA$ and $\angle YMX$ are right angles, by definition of an altitude of a triangle, and so $\angle BDA \cong \angle YMX$. Then $\triangle BDA \sim \triangle YMX$ by AA similarity, and so $\frac{AD}{XM} = \frac{AB}{XY}$.

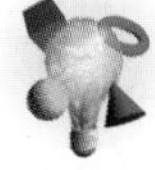

Given: $\triangle ABC \sim \triangle XYZ$; $\overline{AD}$ is an altitude of $\triangle ABC$; $\overline{XM}$ is an altitude of $\triangle XYZ$. Write a plan to prove Case 2 of the Proportional Altitudes Theorem, in which the altitude is outside the triangle.

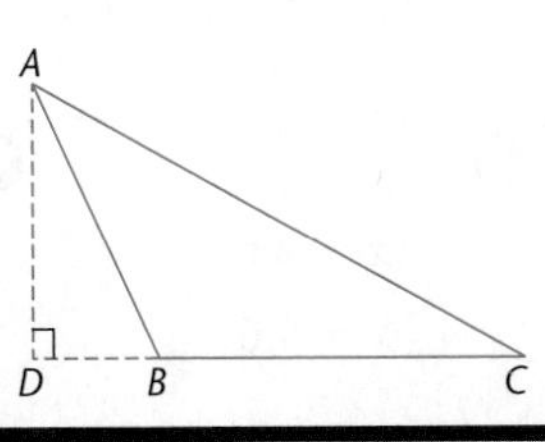

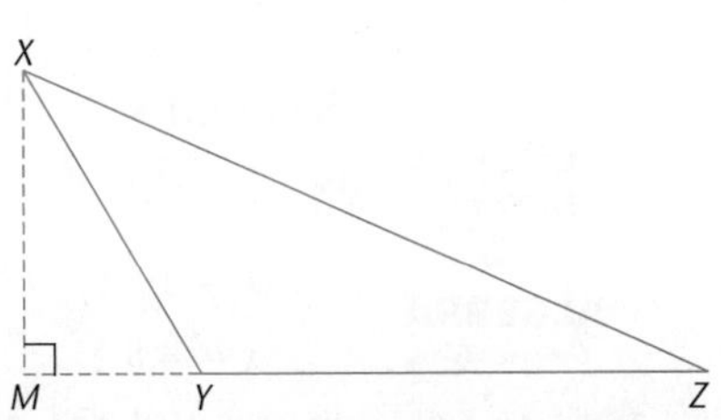

Cooperative Learning

Use small groups both to review altitudes and medians in a triangle and to discuss the proof of the Proportional Altitudes Theorem. Emphasize the necessity of proving the theorem by cases, since altitudes can be outside a triangle. Group work will allow students to explain and refine their steps in the proof.

CRITICAL Thinking

First show that $\angle DBA \cong \angle MYX$ using supplementary angles. Then use AA similarity to show that $\triangle ADB \sim \triangle XMY$, from which it follows that $\frac{AD}{XM} = \frac{AB}{XY}$.

ENRICHMENT Mount Everest and other mountains are measured indirectly by using surveying instruments at various points along their faces. Research how Mount Everest was measured to within 26 feet of its actual height over 140 years ago and present a report to the class.

INCLUSION strategies **English Language Development** Many students have difficulty reading and understanding mathematics textbooks. Suggest that students read this lesson slowly and carefully and write a list of questions that they need answered. You can use cooperative groups to discuss the questions in class.

Alternate Example 3

Solve for *x*.

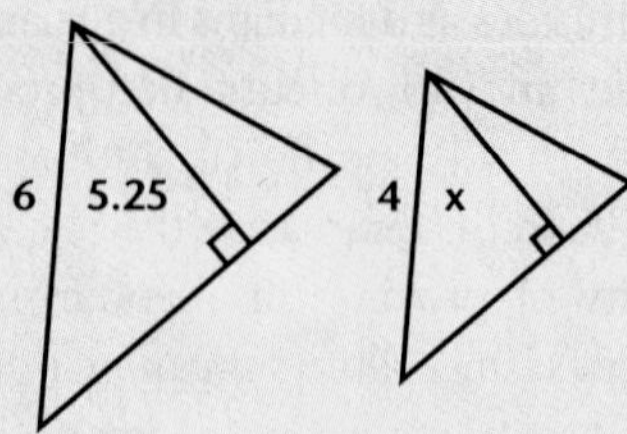

[Solution 3.5]

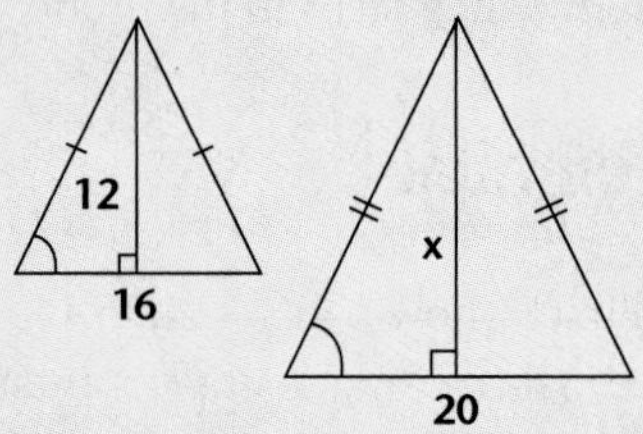

[Solution 15]

Math Connection Algebra

Remind students that although different geometric segments are used in the different examples, the algebra is the same.

Alternate Example 4

Solve for *x*.

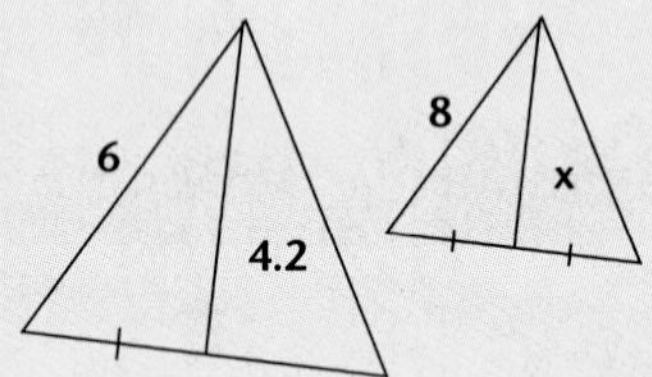

[Solution 5.6]

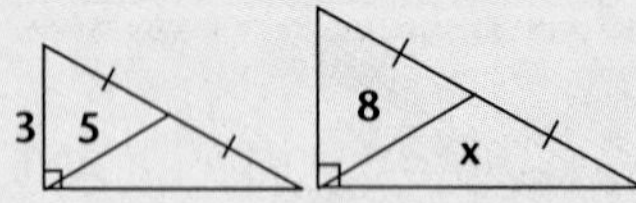

$\left[\text{Solution } 13\frac{1}{3}\right]$

EXAMPLE 3

Solve for *x*.

A

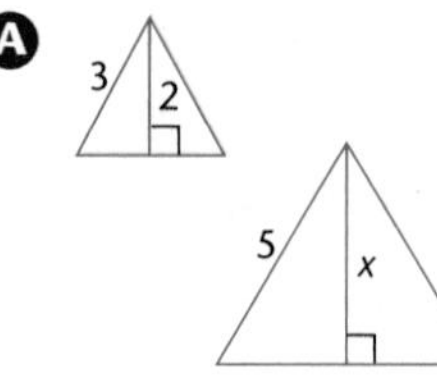

B

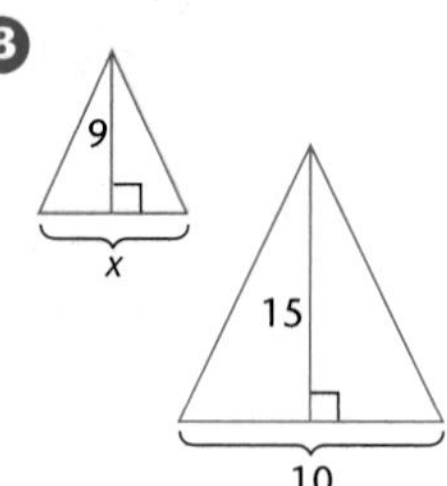

C

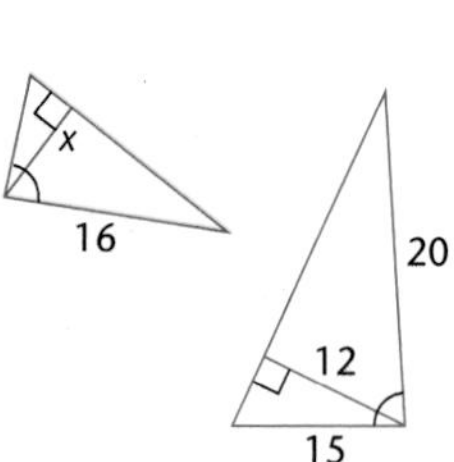

ALGEBRA Connection

Solution ➤

A $\frac{2}{3} = \frac{x}{5}$
$10 = 3x$
$x = 3\frac{1}{3}$

B $\frac{9}{15} = \frac{x}{10}$
$90 = 15x$
$x = 6$

C $\frac{x}{12} = \frac{16}{20}$
$20x = 192$
$x = 9.6$ ❖

PROPORTIONAL MEDIANS THEOREM

If two triangles are similar, then their corresponding medians have the same ratio as the corresponding sides. **8.5.2**

For example, if $\triangle ABC \sim \triangle XYZ$, D is the midpoint of $\overline{BC}$, and M is the midpoint of $\overline{YZ}$ then $\frac{AD}{XM} = \frac{AB}{XY}$.

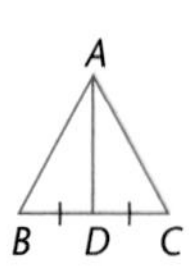

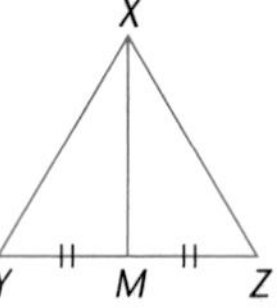

You will be asked to complete a proof of this theorem in the exercise set.

EXAMPLE 4

Solve for *x*.

A

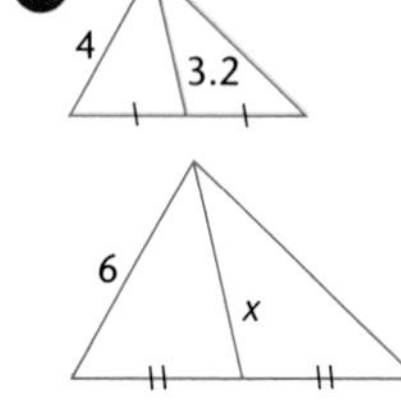

B

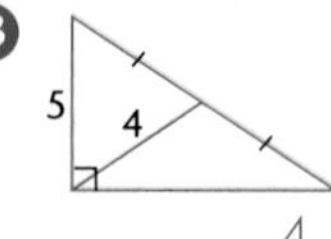

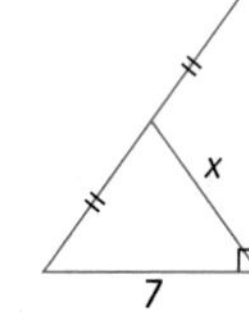

C

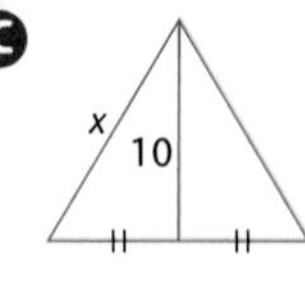

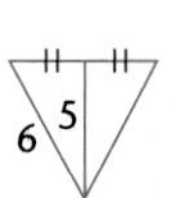

ALGEBRA Connection

Solution ➤

A $\frac{4}{6} = \frac{3.2}{x}$
$4x = 19.2$
$x = 4.8$

B $\frac{5}{7} = \frac{4}{x}$
$5x = 28$
$x = 5.6$

C $\frac{6}{x} = \frac{5}{10}$
$60 = 5x$
$x = 12$ ❖

RETEACHING the lesson Have students list all ratios for the similar triangles below that equal the scale factor between the triangles. Include the ratios for altitudes *BE* and *SV* and medians *BD* and *SU*.

$\left[\frac{AB}{AS} = \frac{BC}{ST} = \frac{AC}{AT} = \frac{BE}{SV} = \frac{BD}{SV}\right]$

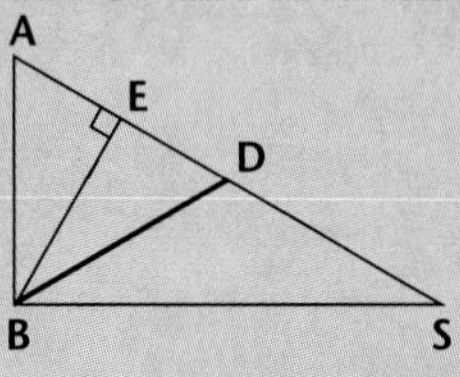

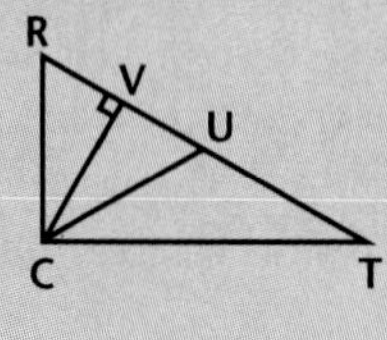

EXERCISES & PROBLEMS

Communicate

1. Explain why it is convenient to use similarity for measuring the height of a mountain.
2. If $\triangle ABC \sim \triangle DEF$, will the medians of the two triangles be proportional? Why?
3. If $\triangle ABC \sim \triangle DEF$, will the altitudes of the two triangles be proportional? Why? If the triangles were isosceles, how would that affect your answer? If the triangles were equilateral, how would that affect your answer?
4. Explain how Eratosthenes found the approximate circumference of the Earth. What postulate or theorem did he apply?

Practice & Apply

Algebra Solve for x in Exercises 5 and 6.

5. $x = 72.2$

6. $x = 245$ m

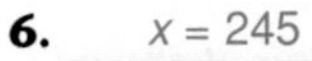

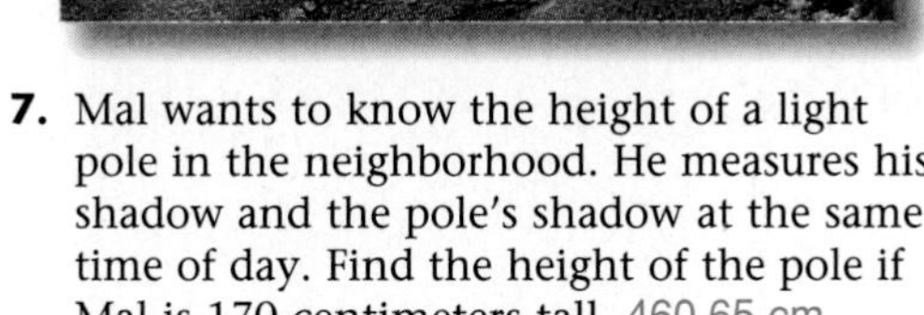

7. Mal wants to know the height of a light pole in the neighborhood. He measures his shadow and the pole's shadow at the same time of day. Find the height of the pole if Mal is 170 centimeters tall. 460.65 cm

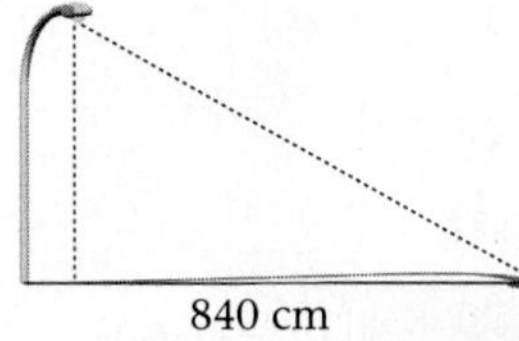

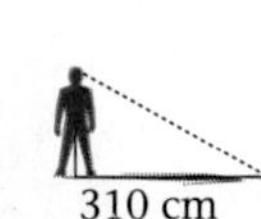

ASSESS

Selected Anwers

Exercises 9, 11, 15, 17, 23, 25, 27, 31, 35, 39, 43, 47, 51, and 57

Assignment Guide

Core 1–4, 5–21 odd, 22–32, 54–59

Core Plus 1–4, 6–20 even, 21–62

Technology

You may want students to compose proofs with geometry graphics software. Have students prove the Proportional Angle Bisectors Theorem in Exercises 33–44.

Error Analysis

Students may have difficulty doing the proofs in Exercises 22–53. Have students go over their proofs in small groups. Students may be able to help each other understand the logic used in the proofs.

alternative ASSESSMENT

Authentic Assessment

Give students extra practice in measuring an object indirectly by asking them to use the techniques shown in this lesson. Have them measure something of their choice in or near their home. Ask them to draw a schematic diagram of their work and to show all calculations.

11. 704 m

12. 531.43 m

Use Transparency 71

Practice Master

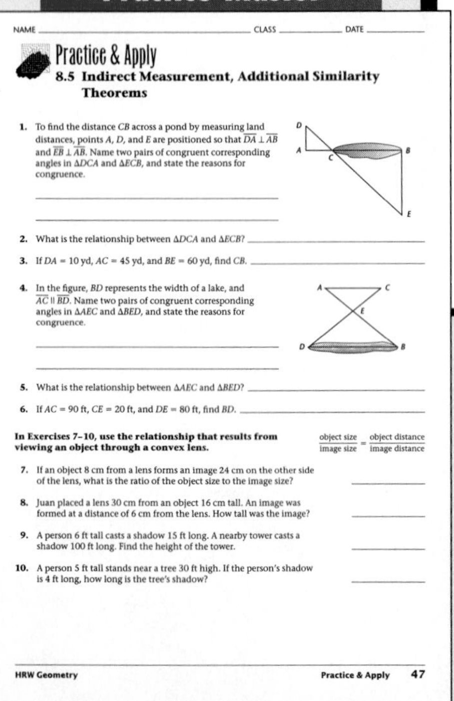

NAME ______ CLASS ______ DATE ______

Practice & Apply
8.5 Indirect Measurement, Additional Similarity Theorems

1. To find the distance *CB* across a pond by measuring land distances, points *A*, *D*, and *E* are positioned so that $\overline{DA} \perp \overline{AB}$ and $\overline{EB} \perp \overline{AB}$. Name two pairs of congruent corresponding angles in $\triangle DCA$ and $\triangle ECB$, and state the reasons for congruence.

______

______

2. What is the relationship between $\triangle DCA$ and $\triangle ECB$? ______
3. If $DA = 10$ yd, $AC = 45$ yd, and $BE = 60$ yd, find *CB*. ______
4. In the figure, *BD* represents the width of a lake, and $\overline{AC} \parallel \overline{BD}$. Name two pairs of congruent corresponding angles in $\triangle AEC$ and $\triangle BED$, and state the reasons for congruence.

______

______

5. What is the relationship between $\triangle AEC$ and $\triangle BED$? ______
6. If $AC = 90$ ft, $CE = 20$ ft, and $DE = 80$ ft, find *BD*. ______

In Exercises 7–10, use the relationship that results from viewing an object through a convex lens. $\frac{\text{object size}}{\text{image size}} = \frac{\text{object distance}}{\text{image distance}}$

7. If an object 8 cm from a lens forms an image 24 cm on the other side of the lens, what is the ratio of the object size to the image size? ______
8. Juan placed a lens 30 cm from an object 16 cm tall. An image was formed at a distance of 6 cm from the lens. How tall was the image? ______
9. A person 6 ft tall casts a shadow 15 ft long. A nearby tower casts a shadow 100 ft long. Find the height of the tower. ______
10. A person 5 ft tall stands near a tree 30 ft high. If the person's shadow is 4 ft long, how long is the tree's shadow? ______

HRW Geometry Practice & Apply 47

In Exercises 8–10, $\triangle ADE \sim \triangle RVW$, $AC = CD$, and $RT = TV$.

8. Find *RW*. 12.5
9. Find *EB*. 14.4
10. Find *WV*. 17.5

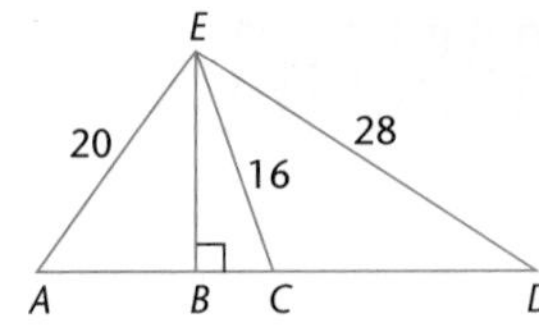

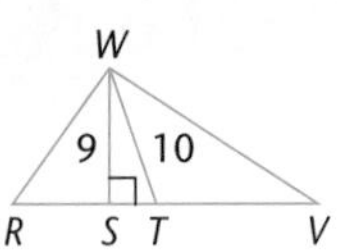

In Exercises 11 and 12, find the distance across the lake. Justify your solution.

11. $\overline{AB}$ is parallel to $\overline{DE}$.

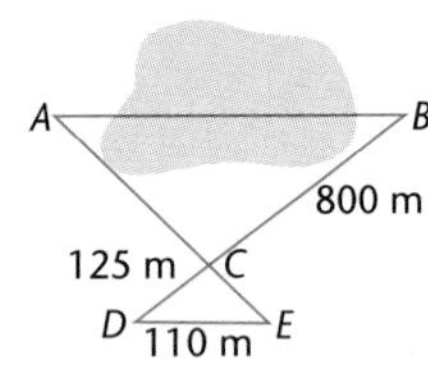

12. $\overline{EB}$ is parallel to $\overline{DC}$.

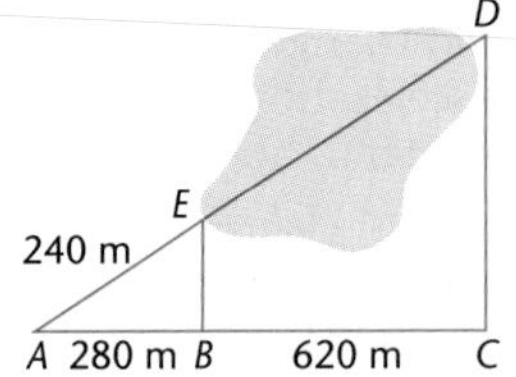

Astronomy Astronomers usually specialize in one of the many branches of their science, such as instruments and techniques. This includes devices used in the formation of images.

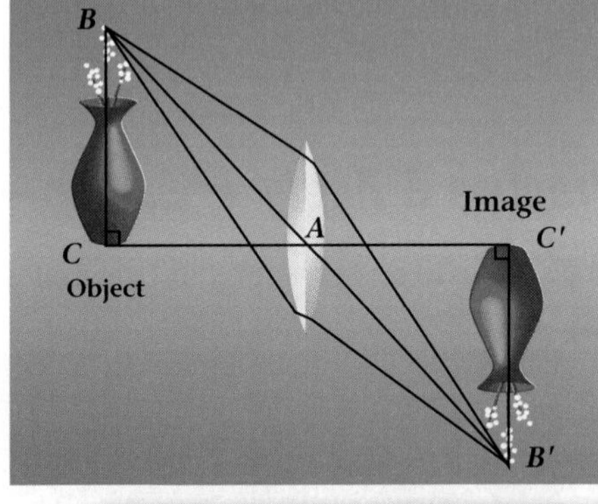

The figures at the right show that a convex lens (a lens that is thicker at the middle than at the edges) can bring parallel light rays together to form an image on the opposite side of the lens from the object. Rays that strike the edge of a lens bend by a certain amount. Rays that pass through the center of a lens are not bent. Thus, two similar triangles $\triangle ABC$ and $\triangle AB'C'$ are formed, and

$$\frac{BC}{B'C'} = \frac{AC}{AC'} \text{ or}$$

$$\frac{\text{Object size}}{\text{Image size}} = \frac{\text{Object distance}}{\text{Image distance}}$$

13. If an object 6 centimeters from a lens forms an image 15 centimeters on the other side of the lens, what is the ratio of the object size to the image size? $\frac{6}{15} = \frac{2}{5}$
14. Rosa placed a lens 25 centimeters from an object 10 centimeters tall. An image was formed at a distance of 5 centimeters from the lens. How tall was the image? 2 cm
15. In the given figures, prove that $\triangle ABC \sim \triangle AB'C'$.
16. How could you arrange an object, a lens, and an image so that the image is 20 times taller than the object to which it corresponds?
17. How would you arrange an object, a lens, and an image, so that the object and the image are the same size?

15. $m\angle BCA = m\angle B'C'A = 90°$; $m\angle BAC = m\angle B'AC'$ because vertical angles have equal measures; then $\triangle ABC \sim \triangle AB'C'$ by AA-Similarity.

16. The distance from the lens to the image should be twenty times the distance from the lens to the object.

17. The lens should be half the distance from the object to the image, then the object distance is the same as the image distance.

Surveying A county wants to connect two towns on opposite sides of a forest with a road. Surveyors have laid out the map shown. The road can be built through the forest or around the forest through Point F. The line joining Town A to Town B is parallel to $\overline{CD}$.

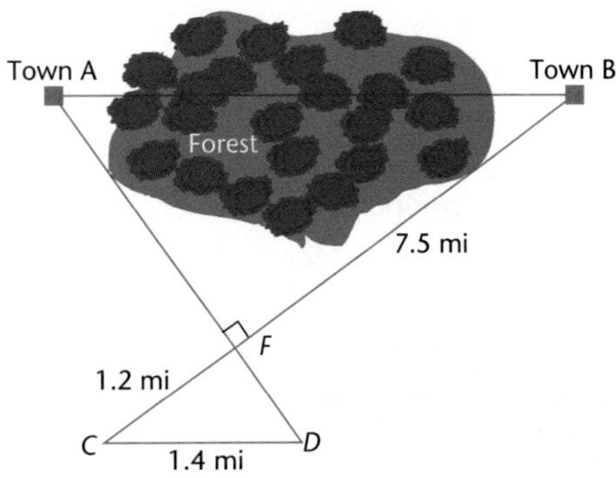

18. Find the distance between the towns through the forest. 8.75 mi

19. Find the distance from Town A to Town B through Point F. 12.01 mi

20. If it costs \$4.5 million per mile to build the road around the forest and \$6.2 million per mile to build the road through the forest, which road would be cheaper to build? the road around the forest

21. Cultural Connection: Africa These pyramids were built over 2200 years ago in the African Kingdom of Kush, in what is now the Sudan. Over time they have lost their tops. Describe a method that uses similar triangles to find how high they were when they were built.

According to the Proportional Medians Theorem, If two triangles are similar, then the corresponding medians are proportional to the corresponding sides.

Complete the proof of the Proportional Medians Theorem on the following page.

Given: $\triangle ABC$ with median $\overline{AD}$, $\triangle XYZ$ with median $\overline{XM}$, $\triangle ABC \sim \triangle XYZ$

Prove: $\frac{AD}{XM} = \frac{BA}{YX} = \frac{BD}{YM} = \frac{BC}{YZ}$

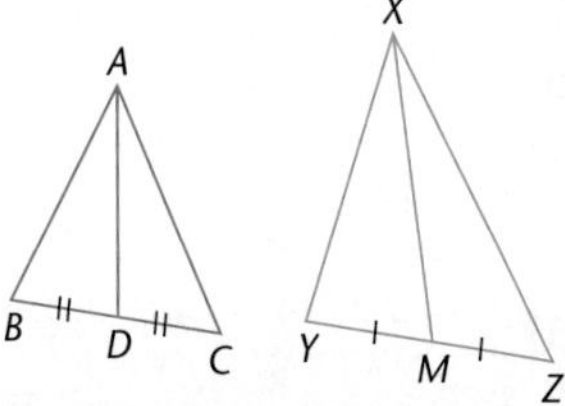

21. Assuming that the top, base and the present height (h_{pres}) can be measured, one can make a scale drawing of the pyramid. By extending the slant side lines to the original vertex on the scale drawing, the eroded height (h_{eroded}) can be determined. Then $h_{orig} = h_{eroded} + h_{pres}$.

Technology Master

NAME ______________ CLASS ________ DATE ________

Technology
8.5 Artistic Designs from Altitudes

The diagram shows an isosceles right triangle. In it, the altitude from point C, where the right angle is located, has been sketched. From the point where the altitude meets the hypotenuse, two altitudes have been added to the sketch. If you wish, you can continue constructing altitudes in the four small triangles that are determined.

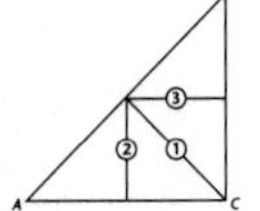

According to the theorem in Lesson 8.5, similar triangles have corresponding altitudes with the same ratios as corresponding sides. The figure that results has a pleasing appearance. By using lines of reflection, you can flip the figure to make an even more artistic design from altitudes.

Geometry software makes the construction of the design enjoyable. If you have the capability to print your design, you can add color to the printout.

Use geometry software in the following exercises. For each triangle, create a design like the one above by constructing altitudes.

1. Start with an equilateral triangle. As the first step, construct the altitude from one vertex.

2. Start with an isosceles nonright triangle. As the first step, construct the altitude from one vertex.

Use reflections in a line to extend each design. Print and color your final design.

3. Exercise 1

4. Exercise 2

Refer to the diagram shown.

5. Sketch an obtuse scalene triangle like the one shown. (The dashed lines will help you get a sketch that looks something like the one shown.)

6. Sketch the altitude from C. Sketch altitudes from the point where it intersects $\overline{AB}$. Finally, use reflections in a line to extend your design. Print and color your final design.

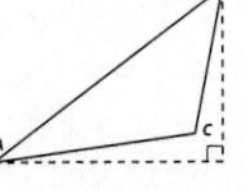

HRW Geometry Technology 59

22. $\triangle ABC$ with median $\overline{AD}$, $\triangle XYZ$ with median $\overline{XM}$, $\triangle ABC \sim \triangle XYZ$

25. $BD + DC = BC$; $YM + MZ = YZ$

36. Angle Addition Postulate

37. Substitution and addition

Proof:

STATEMENTS	REASONS
(22)	Given
D is the midpoint of $\overline{BC}$. M is the midpoint of $\overline{YZ}$.	**(23)** Median Definition
$BD = DC$, $YM = MZ$	**(24)** Midpoint Definition
(25)	Segment Addition
$BC = 2(BD)$, $YZ = 2(YM)$	**(26)** Midpoint Theorem
$\frac{BC}{YZ} = \frac{2(BD)}{2(YM)} = \frac{BD}{YM}$	**(27)** Substitution
$\frac{BC}{YZ} = \frac{BA}{YX}$, $\angle B \cong \angle Y$	**(28)** Property of Similar Polygons
$\frac{BD}{YM} = \frac{BA}{YX}$	**(29)** Substitution
(30) $\triangle ABD \sim \triangle XYM$	SAS Similarity Postulate
(31) $\frac{AD}{XM} = \frac{BA}{YX}$	Property of Similar Polygons
$\frac{AD}{XM} = \frac{BA}{YX} = \frac{BD}{YM} = \frac{BC}{YZ}$	**(32)** Substitution and Property of Similar Polygons

Complete the proof of the following theorem:

THE PROPORTIONAL ANGLE BISECTORS THEOREM

If two triangles are similar then the corresponding angle bisectors are proportional to the corresponding sides. **8.5.3**

Given: $\triangle ABC \sim \triangle XYZ$, $\overline{AD}$ bisects $\angle BAC$, and $\overline{XM}$ bisects $\angle YXZ$

Prove: $\frac{AD}{XM} = \frac{AC}{XZ} = \frac{AB}{XY} = \frac{BC}{YZ}$

Proof:

STATEMENTS	REASONS
(33)	Given
$\angle C \cong \angle Z$, $\angle BAC \cong \angle YXZ$, $\frac{AC}{XZ} = \frac{BC}{YZ} = \frac{BA}{YX}$	**(34)** Property of Similar Polygons
$\angle BAD \cong \angle DAC$, $\angle YXM \cong \angle MXZ$	**(35)** Angle Bisector Definition
$m\angle BAC = m\angle BAD + m\angle DAC$, $m\angle YXZ = m\angle MXZ + m\angle YXM$	**(36)**
$m\angle BAC = 2(m\angle DAC)$, $m\angle YXZ = 2(m\angle MXZ)$	**(37)**
(38) $2(m\angle DAC) = 2(m\angle MXZ)$	Substitution
(39) $m\angle DAC = m\angle MXZ$	**(40)** Division Property of Equality
(41) $\triangle CAD \sim \triangle ZXM$	AA Similarity Postulate
(42) $\frac{AD}{XM} = \frac{AC}{XZ}$	Property of Similar Polygons
(43) $\frac{AD}{XM} = \frac{AC}{XZ} = \frac{AB}{XY} = \frac{BC}{YZ}$	**(44)** Substitution and Property of Similar Polygons

THE PROPORTIONAL SEGMENTS THEOREM
The angle bisector of a triangle divides the opposite side into segments proportional to the other two sides of the triangle. **8.5.4**

Given: $\triangle ABC$, $\overline{AD}$ bisects $\angle CAB$

Prove: $\frac{DC}{DB} = \frac{AC}{AB}$

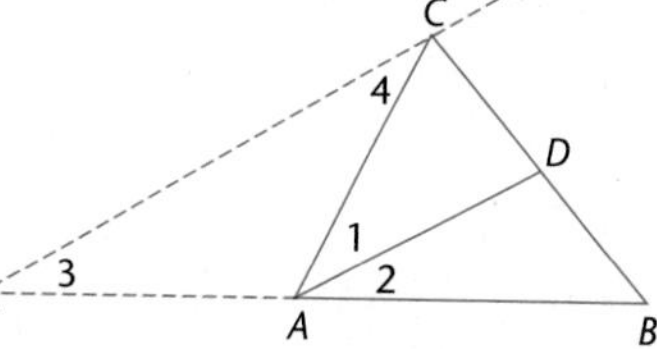

Proof: To prove this theorem, first construct $\overleftrightarrow{CE}$ parallel to $\overline{AD}$, and extend $\overrightarrow{BA}$ to intersect with $\overleftrightarrow{CE}$ at point E.

STATEMENTS	REASONS
(45)	Given
Draw $\overleftrightarrow{CE}$ parallel to $\overline{AD}$.	Parallel Postulate
Extend $\overrightarrow{BA}$ to intersect $\overleftrightarrow{CE}$ at point E.	2 lines intersect in 1 point.
In $\triangle CBE$, $\frac{DC}{DB} = \frac{AE}{AB}$.	**(46)** Side Splitting Theorem
(47) $\angle 3 \cong \angle 2$	If ∥ lines are intersected by a transversal, corresponding angles are congruent.
(48) $\angle 4 \cong \angle 1$	If ∥ lines are intersected by a transversal, alternate interior angles are congruent.
$\angle 1 \cong \angle 2$	**(49)** Angle Bisector Definition
$\angle 3 \cong \angle 4$	**(50)** Substitution
$\overline{AE} \cong \overline{AC}$	**(51)** Isosceles Triangle Theorem
(52) $\frac{DC}{DB} = \frac{AC}{AB}$	**(53)** Substitution

Look Back

54. Find the slope of the line containing (4, 5) and (6, 1). **[Lesson 3.8]** −2

55. Find the equation of the line containing (0, 2) and (−3, 0). **[Lesson 3.8]** $y = \frac{2}{3}x + 2$

For Exercises 56–59, write Sometimes, Always, or Never. [Lesson 4.6]

56. A rhombus is a square. Always

57. Consecutive angles of a trapezoid are supplementary. Sometimes

58. Consecutive angles of a rectangle are complementary. Never

59. The diagonals of a kite bisect each other. Sometimes

Look Beyond

60. Are the two squares similar? Explain.

61. What is the ratio of their sides? their perimeters? their areas?

62. What is the relationship between the ratio of the perimeters and the ratio of the areas?

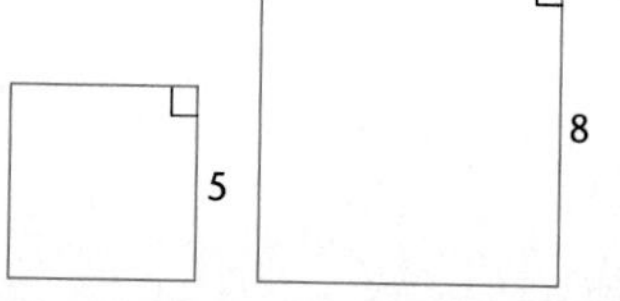

45. $\triangle ABC$, $\overline{AD}$ bisects $\angle CAB$

60. Yes; All corresponding angles are 90° and therefore congruent.

Corresponding sides are in proportion because the sides of squares are equal.

61. $\frac{5}{8}, \frac{5}{8}, \frac{25}{64}$

62. The ratio of their areas is the ratio of the perimeters squared.

Look Beyond

Exercises 60–62 foreshadow the next lesson by asking students to think about the ratio of the areas of similar figures. Extend the exercises by asking them to prove that the perimeters have the same ratio as the sides.

PREPARE

Objectives

- Develop and use ratios for areas of similar figures.
- Develop and use ratios for volumes of similar solids.
- Explore relationships between cross-sectional area, weight, and height.

- Practice Master 8.6
- Enrichment Master 8.6
- Technology Master 8.6
- Lesson Activity Master 8.6
- Quiz 8.6
- Spanish Resources 8.6

Assessing Prior Knowledge

1. Find the area of a circle with radius 4 cm. [**16π cm^2**]
2. Find the volume of a rectangular prism with sides 4 cm and 6 cm and 7 cm.

   [**168 cm^3**]
3. Find the volume of a right cylinder with radius 5 cm and height 6 cm. [**150π cm^3**]

TEACH

In this lesson, students compare the areas or volumes of similar figures. Students must know this if they are to understand the effect that a dimension change has on area or volume. Emphasize that the area goes up by the square and the volume goes up by the third power of the dimension change.

Area and Volume Ratios

A town's water-treatment facility uses a spherical container with a radius of 4 meters. The town council wants to install a new container with double the volume. They can find the dimensions of the new container using ratios.

Exploration 1 *Ratios of Areas of Similar Figures*

You will need
Calculator

For each pair of similar figures below, the ratio of a pair of corresponding segments is given. Find the ratio of the areas for each.

 Two squares

$$\frac{\text{Side of square A}}{\text{Side of square B}} = \frac{3}{1}$$

$$\frac{\text{Area of square A}}{\text{Area of square B}} = \frac{?}{?}$$

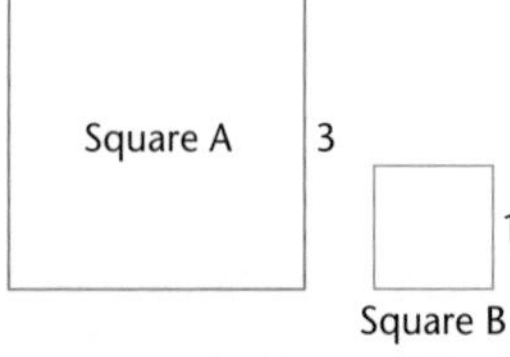

 Two triangles

$$\frac{\text{Side of triangle A}}{\text{Side of triangle B}} = \frac{2}{1}$$

$$\frac{\text{Area of triangle A}}{\text{Area of triangle B}} = \frac{?}{?}$$

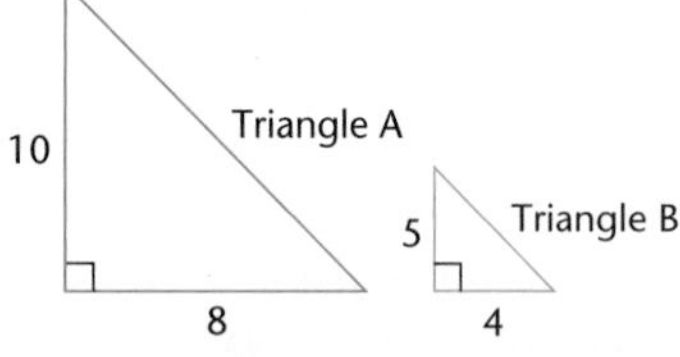

Hands-On Strategies Have students use grid paper to explore the ratio of the areas of similar figures. Ask them to cut two similar figures from grid paper and superimpose them. Then the students should count the squares that make up their area, and find the ratio of their areas. They can likewise use cubes to explore the ratio of the volumes of two similar figures. Create two similar shapes with cubes, count the cubes that make up their respective volumes, and find the ratio of their volumes.

 Two rectangles

$$\frac{\text{Side of rectangle A}}{\text{Side of rectangle B}} = \frac{3}{2}$$

$$\frac{\text{Area of rectangle A}}{\text{Area of rectangle B}} = \frac{?}{?}$$

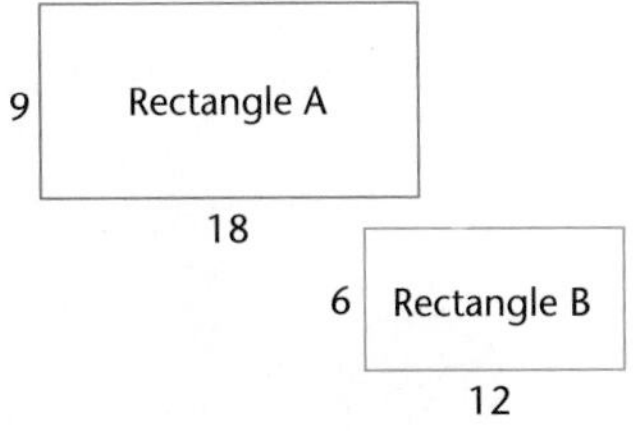

4 Two circles

$$\frac{\text{Radius of circle A}}{\text{Radius of circle B}} = \frac{4}{3}$$

$$\frac{\text{Area of circle A}}{\text{Area of circle B}} = \frac{?}{?}$$

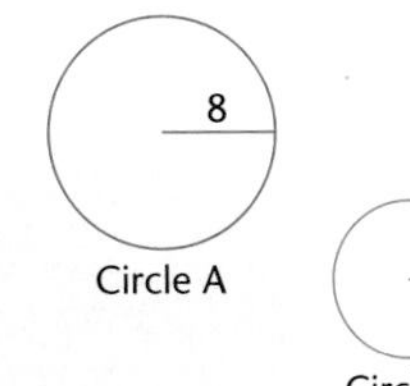

6
Circle B

 Use the results above to complete the following conjecture: Two similar figures with corresponding sides in the ratio $\frac{a}{b}$ have areas in the ratio __?__. ❖

Exploration 2 Ratios of Volume

You will need
Calculator

Solids are similar if they are the same shape and all the corresponding dimensions are proportional. For example, two rectangular prisms are similar if the dimensions for each corresponding face and base are proportional. For each pair of similar solids below, the ratio of a pair of corresponding segments is given. Find the ratio of the volumes for each.

 Two cubes

$$\frac{\text{Side of cube A}}{\text{Side of cube B}} = \frac{3}{1}$$

$$\frac{\text{Volume of cube A}}{\text{Volume of cube B}} = \frac{?}{?}$$

 Two rectangular prisms

$$\frac{\text{Side of prism A}}{\text{Side of prism B}} = \frac{3}{2}$$

$$\frac{\text{Volume of prism A}}{\text{Volume of prism B}} = \frac{?}{?}$$

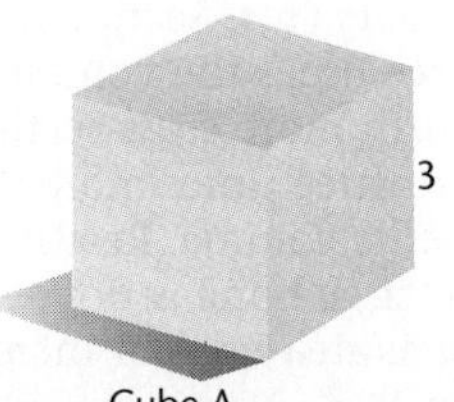

Cube A

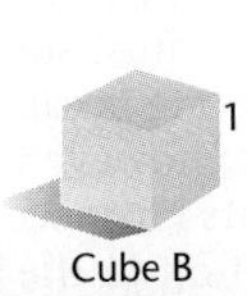

Cube B

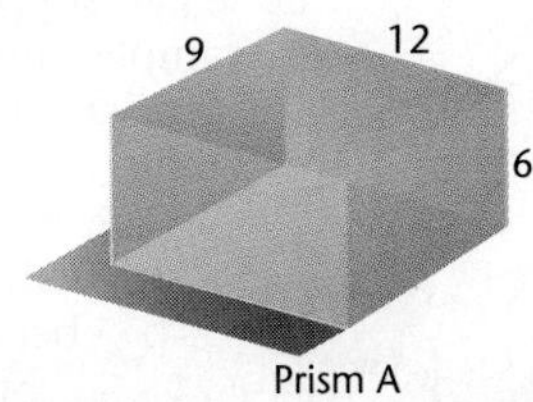

Prism A

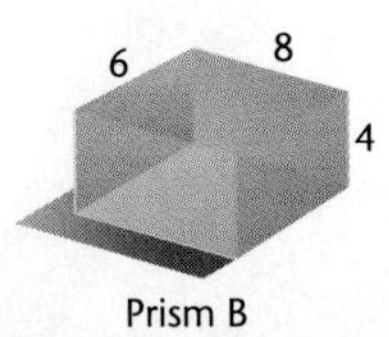

Prism B

interdisciplinary CONNECTION **Art** Sculptors often use clay or other heavy materials when they make figurines. Have students discuss how the volume, and therefore the weight, of a figure changes as the dimensions change. Ask them to estimate the weight of a statue that is five times taller than a similar statue that weighs 2 pounds.

Exploration 1 Notes

This activity will help students understand that similar figures with scale factor $\frac{a}{b}$. have areas with ratio $\frac{a^2}{b^2}$. Use similar figures cut from transparent graph paper to create visual models of similar figures for the overhead projector. Students can count the squares in each figure to verify the ratio of the areas.

ongoing ASSESSMENT

5. $\left(\frac{a}{b}\right)^2 = \frac{a^2}{b^2}$

Use Transparency 72

Exploration 2 Notes

This activity will help students understand that similar figures with scale factor $\frac{a}{b}$. have volumes with ratio $\frac{a^3}{b^3}$.

TEACHING tip

Try to include oblique figures in the discussion about similar figures. Emphasize that the shape of similar figures doesn't matter. As long as two figures are similar, their area increases by the square of the dimension change and their volume increases by the cube.

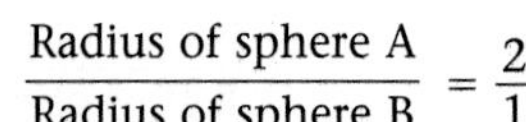

3 Two spheres

$$\frac{\text{Radius of sphere A}}{\text{Radius of sphere B}} = \frac{2}{1}$$

$$\frac{\text{Volume of sphere A}}{\text{Volume of sphere B}} = \frac{?}{?}$$

4 Two cylinders

$$\frac{\text{Radius of cylinder A}}{\text{Radius of cylinder B}} = \frac{2}{1}$$

$$\frac{\text{Volume of cylinder A}}{\text{Volume of cylinder B}} = \frac{?}{?}$$

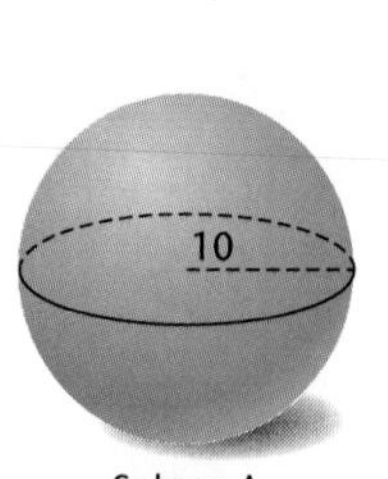

Sphere A

Sphere B

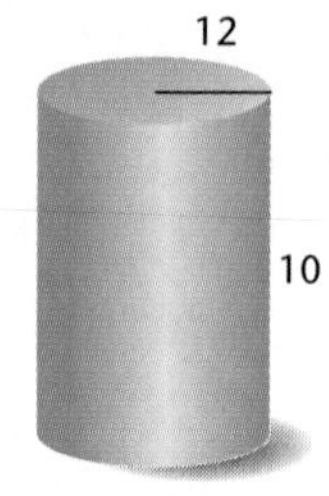

Cylinder A

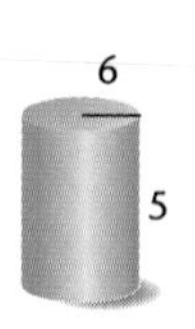

Cylinder B

5 Use the results above to complete the following conjecture: Two similar solids with corresponding sides in the ratio $\frac{a}{b}$ have volumes in the ratio __?__. ❖

CRITICAL Thinking Are all cubes similar? Why or why not? Are all spheres similar? Why or why not? Are all cylinders similar? Why or why not?

Try This

1. The corresponding sides of two similar triangles are in the ratio $\frac{5}{9}$. What is the ratio of their areas?
2. The corresponding areas of two similar rectangular prisms are in the ratio $\frac{9}{4}$. What is the ratio of their sides?
3. One sphere has a radius of 5 meters. Another sphere has a radius of 15 meters. What is the ratio of their volumes?

Cross-Sectional Areas, Weight, and Height

Many folk tales feature accounts of human giants. However, some mathematical investigation will show that there is a limit to just how tall a human or an animal can be. The amount of weight that a bone can support is proportional to its cross-sectional area. For example, the larger bone has three times the volume of the smaller bone, but only one-fourth more cross-sectional area. The larger bone must support three times more weight than the smaller bone because of its volume. Relative to the small bone, however, the large bone is not as strong because its cross-sectional area has not increased in its proportion to the size.

ongoing ASSESSMENT

5. $\left(\frac{a}{b}\right)^3 = \frac{a^3}{b^3}$

CRITICAL Thinking

All cubes are similar because they have the same shape and, since all sides of a cube have the same length, all corresponding sides will have the same ratio. All spheres are similar because they have the same shape and, since lengths on a sphere are determined by radius, all corresponding lengths in any two spheres will have the same ratio as the ratio between the radii of the spheres. All cylinders are not similar because the ratio of the radii of the bases of two cylinders could be different than the ratio of their heights.

ongoing ASSESSMENT

Try This

1. $\left(\frac{5}{9}\right)^2 = \frac{5^2}{9^2} = \frac{25}{81}$
2. $\sqrt{\frac{9}{4}} = \frac{3}{2}$
3. $\left(\frac{5}{15}\right)^3 = \frac{1^3}{3^3} = \frac{1}{9}$

ENRICHMENT Robert Wadlow was the tallest man who ever lived (8 ft 11.1 in.). Have students research his life and relate his height and weight statistics to the content of this lesson. Ask students to determine whether a model pyramid in the classroom is similar to the Great Pyramid. Its dimensions are 756 ft on a side and 481 ft tall.

INCLUSION strategies **Hands-On Strategies** Some students may be interested in the effect that increasing size dimensions has on cross-sectional area and volume. Have them create similar solid shapes from clay. and compare the cross-sectional areas of the figures and the weights of the figures. Students should also compare the sizes and volumes of two drinking cups from a fast-food restaurant. Does the increase in size dimensions match the increase in price?

APPLICATION

A young elephant that is 4 ft tall at the shoulder has leg bones with a cross-sectional radius of 4 cm. How much thicker would the leg bones need to be to support an elephant 8 ft tall at the shoulder? By what scale factor does the original radius need to be multiplied, to provide an adequate cross-sectional area?

The two-dimensional size of the elephant increases by a factor of 2. However, the volume of the elephant increases by a factor of 8. Thus, the larger elephant would need a leg bone with a cross-sectional area 8 times larger than the original elephant.

So, $8\pi r^2$ gives the new cross-sectional area.

ALGEBRA Connection

Let R represent the radius of the larger elephant. Then $\pi R^2 = 8\pi r^2$, so $R^2 = 8r^2$, or $R = \sqrt{8}r$. Thus the radius of the larger bone is $\sqrt{8}$ or about 2.8 times the radius of the smaller bone. ❖

Exploration 3 Increasing Height and Volume

Have you read stories of giants hundreds of feet tall? Do you think they could be realistic? The elephant example shows the effect that increasing size (linear) dimensions has on volume and on the dimensions of the structures that must support the increased bulk. Think about the question of giants as you do this exploration.

1 Complete the chart to determine the necessary scale factors for the cross-sectional radii of leg bones of giant horses.

Height scale factor	Volume scale factor	Cross-sectional radiu scale factor
2	8	$\sqrt{8}$
3	27	$\sqrt{27}$
4	64	?
5	?	?
100	?	?

Math Connection Algebra

You may want to review the algebra needed to find the radius of a circle if the area is known. Point out that the radius is the scale factor used for the cross-sectional area of the leg bone in the application.

Exploration 3 Notes

Use the chart the students create in this exploration to generalize the ratios of the cross-sectional area and volume of similar solids. If the size dimensions of similar solids have ratio $1:n$, then their cross-sectional areas have ratio $1:n^2$ and their volumes have ratio $1:n^3$. Have students verify the measurements in the chart with these ratios.

Cooperative Learning

Use cooperative groups to verify that the weight of an object increases by the third power of the dimension change. Have each group plan an experiment with weight and size, carry out the experiment, and then report the results to the class. Blocks or other standardized shapes are convenient to use. Encourage them to keep a chart of data as shown in Exploration 3. Working in groups helps students develop their social skills as well as their communication skills.

RETEACHING the lesson

Using Models Ask students to do the following activity: (1) Draw a pair of similar rectangular prisms. (2) Measure the length, width, and height of each. (3) Find the surface area of each. (4) Predict the ratio of the surface areas and then verify the prediction. (5) Find the volume of each. (6) Predict the ratio of the volumes and then verify the prediction.

ONGOING ASSESSMENT

3. **The legs would be proportionally thicker because their thickness must increase by more than the increase in height.**

2. If a normal-sized horse requires a leg bone with a cross-sectional radius of 2.5 centimeters for support, what must the cross-sectional radius of a leg bone be for a horse 20 times taller? 50 times taller? 100 times taller?

3. What effect does increasing the size of the horse have on the relative proportions of the legs? Would the legs be proportionally thicker or thinner than those of a normal-sized horse? ❖

ASSESS

Selected Answers

Odd-numbered Exercises 7–47

Assignment Guide

Core 1–7, 11–20, 27–33, 41–50

Core Plus 1–10, 18–43, 45–51

Technology

Use geometry graphics software to create similar solids in Exercises 21–30. Use the measurement features of the software to find their volumes.

Error Analysis

Students may have difficulty answering Exercises 24–30 because they have to find the scale factor between the figures. You may want to remind them that if the ratio of the areas is known, they can take its square root to find the scale factor. If the ratio of the volumes is known, they can take its cube root to find the scale factor.

EXERCISES & PROBLEMS

Communicate

1. What is the relationship between the ratio of the sides of two cubes and the ratio of their volumes?

2. Why are all spheres and cubes similar? Is this true for other three-dimensional figures? If so, name or describe them.

3. If you know the ratio of the areas of two similar prisms, explain how you can find the ratio of the volumes.

4. What happens to the volume of a cylinder if the radius is doubled but the height stays the same?

5. How does the cross-sectional area of a bone relate to an animal's weight and height?

Practice & Apply

6. A 12 ounce box of noodles is 15 cm wide, 20 cm high, and 4 cm deep. If the noodle company would like to sell a 24 ounce box, would it have to double all the dimensions of the 12 ounce box? Explain.

7. A new pizzeria sells an 8-inch diameter pizza for $4 and a 16-inch diameter pizza for $6. Which pizza is the better deal? Explain.

The ratio between the sides of two similar pyramids is $\frac{7}{5}$. Find the ratio between the following.

8. the perimeters of their bases $\frac{7}{5}$

9. the areas of their bases $\frac{49}{25}$

10. their volumes $\frac{343}{125}$

6. No. If all the dimensions were doubled, the volume of the box would be multiplied by 8 and not 2.

7. The 16-in pizza. If the radius of a pizza is doubled, the area of the pizza is multiplied by 4. The 16-in pizza is approximately 4 times as big as the 8-in pizza, but it costs $6, which is not 4 times as much as the $4 cost of the 8-in pizza.

Two spheres have radii 5 cm and 7 cm. Find the ratio between the following.

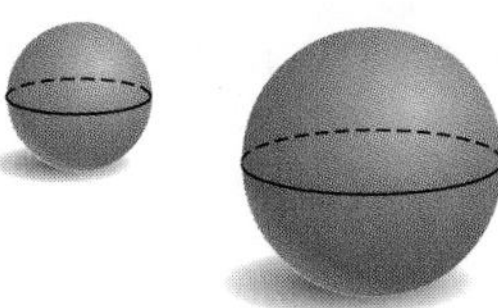

11. the circumferences of their great circles $\frac{5}{7}$
12. their volumes $\frac{125}{343}$

The ratio between the areas of the bases of two similar cones is $\frac{16}{25}$. Find the ratio between the following.

13. their heights $\frac{4}{5}$
14. the circumferences of their bases $\frac{4}{5}$
15. their volumes $\frac{64}{125}$

In the figure, $\overline{EC} \parallel \overline{AB}$.

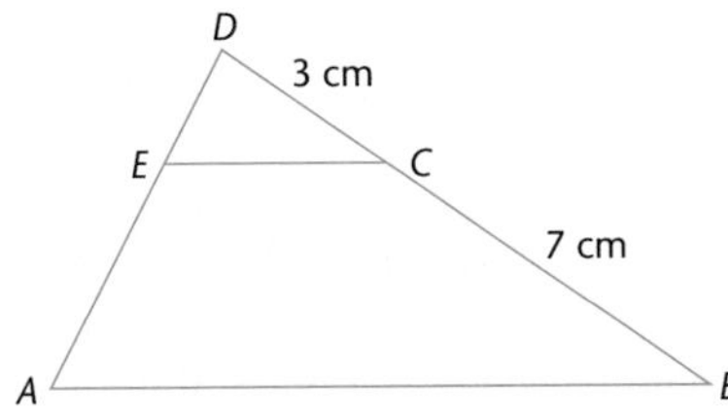

16. Find the ratio between the perimeters of $\triangle ABD$ and $\triangle ECD$. $\frac{10}{3}$
17. Find the ratio between the areas of $\triangle ABD$ and $\triangle ECD$. $\frac{100}{9}$

***ABCDE* ~ *QRSTU*. *ABCDE* has perimeter 24 meters and area 50 square meters.**

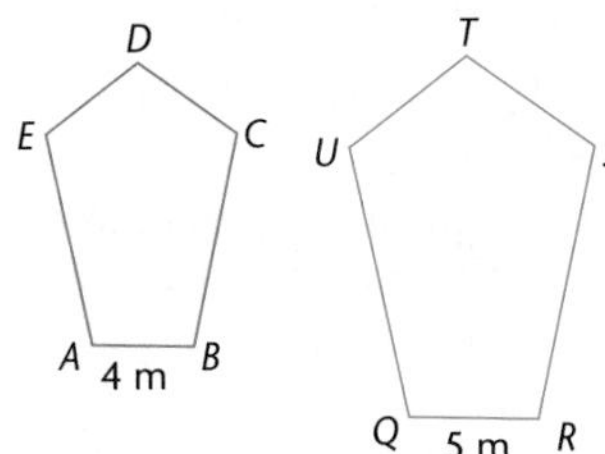

18. Find the perimeter of *QRSTU*. 30 m
19. Find the area of *QRSTU*. 78.125 m^2
20. Two similar cylinders have bases with areas 16 cm^2 and 49 cm^2. If the larger cylinder has height 21 cm, find the height of the smaller cylinder. $x = 12$ cm

If the ratio between the heights of two similar cones is $\frac{7}{9}$, find the ratio between the following.

21. the circumferences of their bases $\frac{7}{9}$
22. their volumes $\frac{343}{729}$
23. the areas of their bases $\frac{49}{81}$

The ratio between the surface areas of two spheres is 144:169. Find the ratio between the following.

24. their radii $\frac{12}{13}$
25. their volumes $\frac{1728}{2197}$
26. the circumferences of their great circles $\frac{12}{13}$

If the ratio between the volumes of two similar prisms is $\frac{64}{125}$, find the ratio between the following.

27. their surface areas $\frac{16}{25}$
28. the perimeters of their bases $\frac{4}{5}$
29. the diagonals of their bases
30. the areas of their bases $\frac{16}{25}$

alternative ASSESSMENT

Portfolio Assessment

Ask students to summarize the lesson in writing by describing what effect increasing a dimension has on the cross-sectional area of a solid and on the volume of the solid. Ask them to describe how to find the scale factor if the sizes of the similar solids are known. Encourage them to include sample problems.

29. $\frac{4}{5}$

39. the diameters are in ratio $\frac{201}{50}$, or about 4, and the volumes are in ratio $\left(\frac{201}{50}\right)^3$, or about 64.

40. If people continue to use the same length of toothpaste, volume of toothpaste sold would increase by $\left(\frac{3}{2}\right)^2 = \frac{9}{4}$. People would use over twice as much toothpaste.

Practice Master

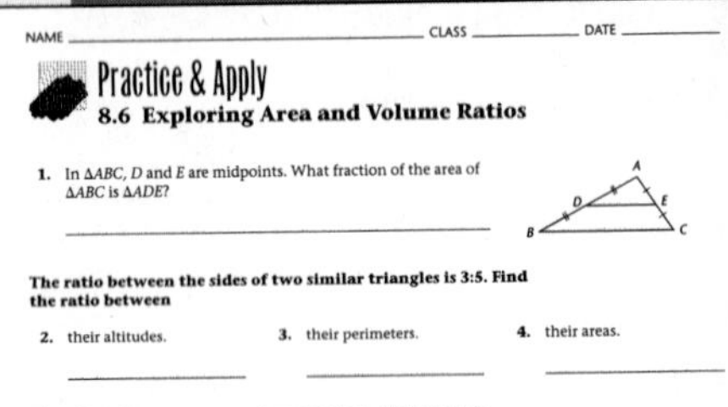

NAME ______ CLASS ______ DATE ______

Practice & Apply
8.6 Exploring Area and Volume Ratios

1. In ΔABC, D and E are midpoints. What fraction of the area of ΔABC is ΔADE? ______

The ratio between the sides of two similar triangles is 3:5. Find the ratio between

2. their altitudes. ______ 3. their perimeters. ______ 4. their areas. ______

The sides of two squares are 4 cm and 9 cm. Find the ratio between

5. their diagonals. ______ 6. their perimeters. ______ 7. their areas. ______

Two spheres have radii 6 cm and 8 cm. Find the ratio between

8. the circumferences of their great circles. ______

9. their surface areas. ______ 10. their volumes. ______

Consider a cube C of edge 5 cm.

11. If the edge of cube C is multiplied by 2, what is the ratio of the volume of the new cube Q to the volume of cube C? ______

12. If the edge of cube C is multiplied by 3, what is the ratio of the volume of the new cube R to the volume of cube C? ______

13. If the edge of cube C is multiplied by 4, what is the ratio of the volume of the new cube T to the volume of cube C? ______

14. Use these results to write a general rule for the ratio of the volumes of a cube C of edge e and a cube K of edge ke. ______

15. Write a general rule for the ratio of the volumes of a rectangular solid S of dimensions l, w, h, and a rectangular solid K of dimensions kl, kw, kh. ______

48 Practice & Apply HRW Geometry

31. $ABCDEF \sim PQRSTU$. $ABCDEF$ has perimeter 42 meters and area 96 square meters. Find the perimeter and area of $PQRSTU$. P = 28 m; A = $42\frac{2}{3}$ m^2

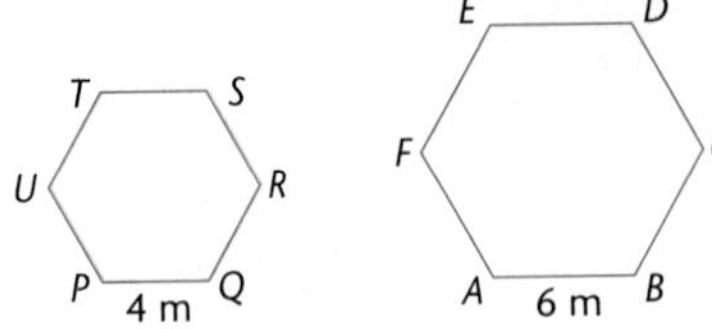

Two similar cones have surface areas of 225 cm^2 and 441 cm^2.

32. If the height of the larger cone is 12 cm, find the height of the smaller cone. $8\frac{4}{7}$ cm

33. If the volume of the smaller cone is 250 cm^3, find the volume of the larger cone. 686 cm^3

34. A leg bone of a horse has a cross-sectional area of 19.6 cm^2. What is the diameter of the leg bone? 5 cm

35. How much longer must the diameter of the leg bone in Exercise 34 be to support a horse twice as tall? 2.8 times as long.

36. A 100-foot cylindrical tower has a cross-sectional radius of 26 feet. What would be the radius of a 350-foot tower if it were proportional to the 100-foot tower? 91 ft

Cone *EDC* is formed by cutting off the bottom of cone *EAB* parallel to the base.

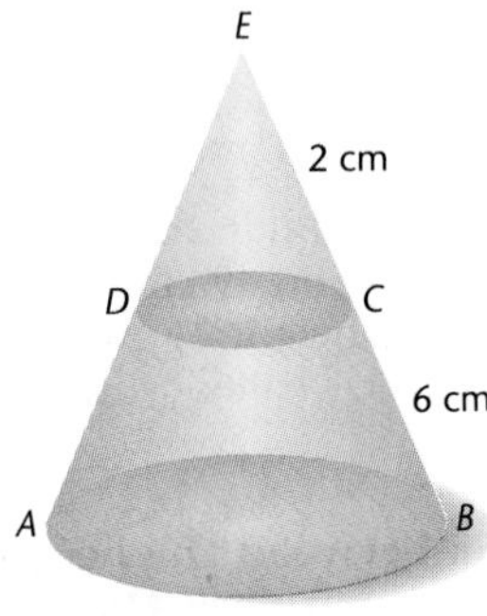

37. If cone EAB has volume 288 cm^3, find the volume of cone EDC. 4.5 cm^3

38. If the base of cone EDC has area 3.6 cm^2, find the area of the base of cone EAB. 57.6 cm^2

39. If the Earth and the Moon were perfect spheres, their equators would be approximately 40,200 km and 10,000 km respectively. Compare the diameters and the volumes of the Earth and the Moon.

40. **Packaging** A toothpaste company packages its toothpaste in a tube with a circular opening with radius 2 mm. The company increases the radius to 3 mm. Predict what will happen to the amount of toothpaste used if the same number of people continue to use the toothpaste. Explain your reasoning.

41. The area of the parking lot at Jerome's Restaurant is 400 m^2. Jerome buys some land and is able to make the parking lot 1.5 times as wide and 1.75 times as long. Find the area of the expanded lot. 1050 m^2

42. **Sports** The diameter of a standard basketball is approximately 9.5 inches. A company that makes standard-sized basketballs contracts to make a basketball with diameter 5 inches for a restaurant promotion. If the materials to make a standard-sized ball cost the company \$1.40, how much will it cost the company to make the promotional basketball if it uses the same materials? 39¢

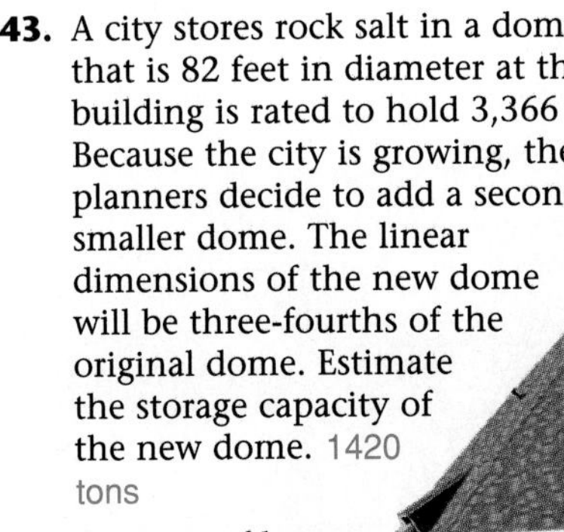
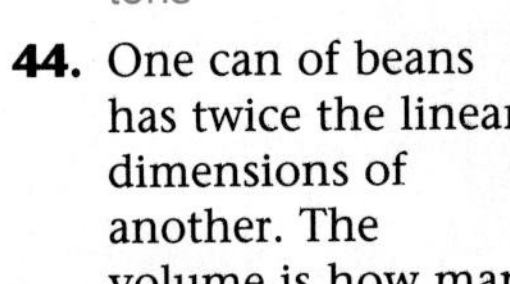

43. A city stores rock salt in a dome-shaped building that is 82 feet in diameter at the base. The building is rated to hold 3,366 tons of salt. Because the city is growing, the city planners decide to add a second, smaller dome. The linear dimensions of the new dome will be three-fourths of the original dome. Estimate the storage capacity of the new dome. 1420 tons

44. One can of beans has twice the linear dimensions of another. The volume is how many times greater? 8 times greater

Look Back

45. Each interior angle of a regular polygon measures 165°. How many sides does the polygon have? **[Lesson 3.6]** 24 sides

46. The exterior angles of a regular polygon each measure 40°. How many sides does the polygon have? **[Lesson 3.6]** 9 sides

47. Which of these lines are parallel? **[Lesson 3.8]** (b) and (c)

a. $5x + 4y = 18$ **b.** $-10x + 8y = 21$ **c.** $30x - 24y = 45$

48. Write the equation of the line parallel to $5x + 4y = 18$ and through the point (8, 3). **[Lesson 3.8]** $y = \frac{-5}{4}x + 13$

49. **Algebra** Graph: $y = x^2 - 6x - 4$ **[Lesson 3.8]**

50.  **Algebra** Solve for x: $8x^2 + 5x - 8 = 5x^2 - 2x + 4$

Look Beyond

Algebra ***ABCD* is a square with side length 1 and perimeter 4. The squares inside *ABCD* are formed by connecting the midpoints of the bigger square.**

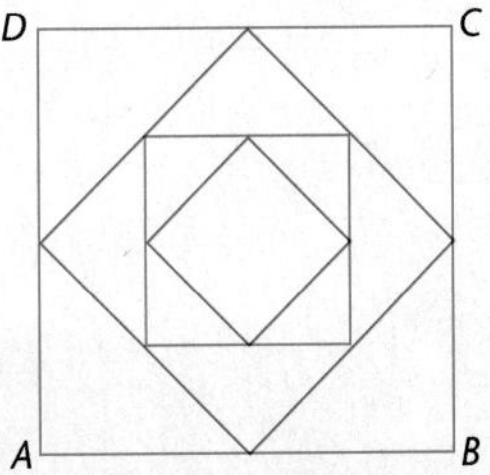

51. What happens to the perimeter of each successive square? Using a calculator, add the perimeters of the first eight squares together. Keep track of the sum after each number is added. Do you think the sum of all the squares generated this way will ever reach 14? Explain.

Look Beyond

Exercise 51 introduces the concept of the convergence of a geometric series. Encourage students to try to find the limit of the sum. They can be given the hint that the result from Lesson 2.1 Exercises 8–12, page 69, is the key.

Technology Master

NAME ______ CLASS ______ DATE ______

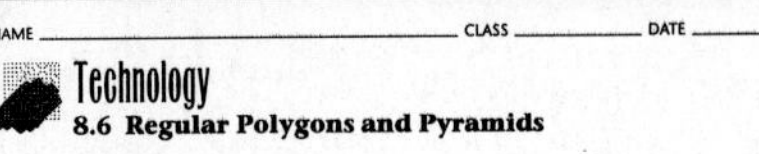

Technology

8.6 Regular Polygons and Pyramids

As you know, a regular polygon with n sides is a polygon with n sides all equal in length and n angles all equal in measure. The definition itself suggests that all regular polygons with a given number of sides are similar. In particular, all equilateral triangles are similar and all squares are similar.

To explore how the areas of regular polygons are related to the lengths of their sides, it would be helpful to have formulas for the areas of the different types of regular polygons. Some of those formulas are listed in the table.

Polygon with Length of Side (s)	Number of Sides	Area (Square Units)
Pentagon	5	$1.72048s^2$
Hexagon	6	$2.59808s^2$
Octagon	8	$4.82843s^2$
Nonagon	9	$6.18182s^2$
Decagon	10	$7.69421s^2$
Dodecagon	12	$11.19615s^2$

1. Suppose you wish to study pentagons. Create a spreadsheet in which column A contains $s = 0.5, 1.0, 1.5, \ldots, 3.0, 3.5$, and 4.0 and column B contains the area formula for each entry in column A. In cell C2, enter the formula for the square of the ratio of two consecutive s values. In cell D2, enter the ratio of the areas corresponding to consecutive choices of s. What can you conclude?

______

2. From the table, choose a regular polygon other than a pentagon. Modify your spreadsheet to answer the question in Exercise 1.

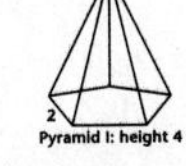

3. Refer to the pyramids shown. Write an equation for y_1, the ratio of the volume of pyramid II to that of pyramid I, an equation for y_2, the cube of the ratio of the height of pyramid II to that of pyramid I, and an equation for y_3, the cube of the ratio of the length of a side of the base of pyramid II to that of pyramid I.

______

4. Graph the three equations you wrote in Exercise 3. For what value(s) of h will the three graphs intersect.

______

5. Write a statement that relates pyramids I and II, their volumes, heights, and lengths of the sides of their bases.

______

______

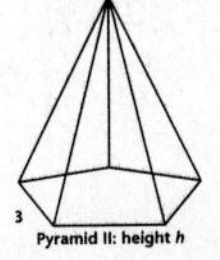

60 **Technology** **HRW Geometry**

49. $y = x^2 - 6x - 4$

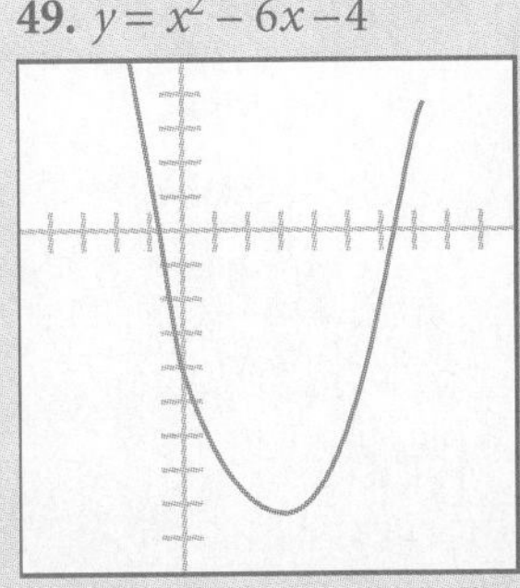

50. $x = \dfrac{-7 \pm \sqrt{193}}{6}$

51. The sum of the first 8 terms is about 13.05. The sum of an infinite number of terms is

$$4[1 + \tfrac{1}{2}\sqrt{2} + \tfrac{1}{2} + \tfrac{1}{4}\sqrt{2} + \tfrac{1}{4} + \tfrac{1}{8}\sqrt{2} + \tfrac{1}{8} + \ldots..] =$$

$$4[1 + \sqrt{2}(\tfrac{1}{2} + \tfrac{1}{4} + \tfrac{1}{8} + \ldots..) + (\tfrac{1}{2} + \tfrac{1}{4} + \tfrac{1}{8} + \ldots..)] =$$

$4[1 + \sqrt{2}(1) + 1] \approx 13.65685$.....Thus the sum of the perimeters will never reach 14.

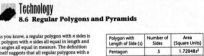

Focus

Measure a large structure by indirect methods and build a scale model of a structure.

Motivate

In this project, students will measure a large structure indirectly and build a scale model of a structure. Ask students to discuss, display, or show pictures of scale models that they already own. Display the student-created models from this project in the classroom. Encourage theater students to discuss how building scale models is important in creating sets and in reading plans for sets.

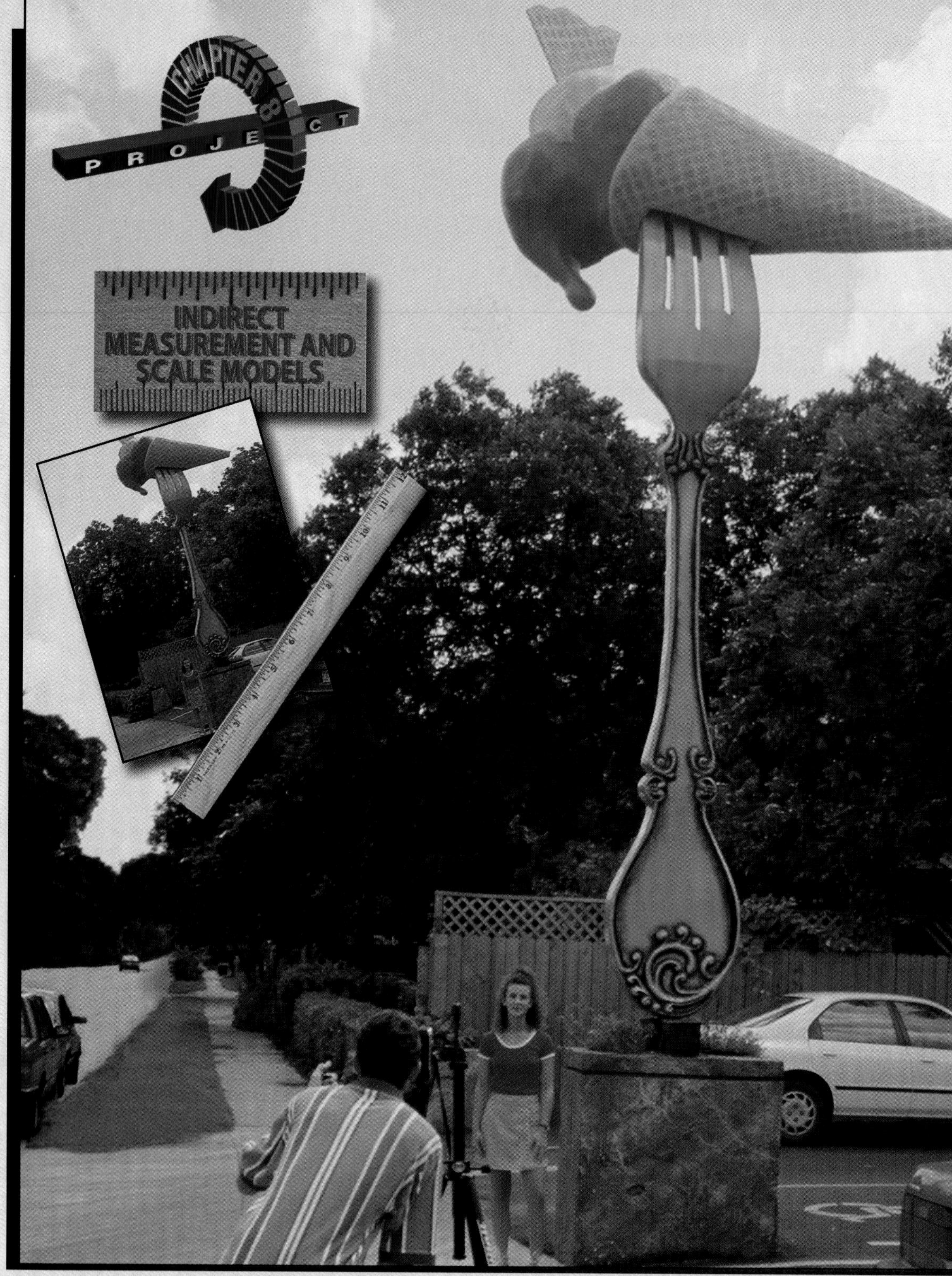

Activity 1: Measuring Buildings

1. Use several different methods, including photographs, shadows, and mirrors, to indirectly find the dimensions of buildings or other structures at your school or in your neighborhood.
2. Draw a diagram of the building with the direct measurements labeled. Include the photograph if you used a photograph.
3. Explain how you made the measurements.
4. Explain how you calculated the dimensions you did not measure directly.
5. If available, find the known dimensions of the structures you measured. Explain how you can account for any differences between what you calculated and the known dimensions.

Activity 2: Scale Models

1. Obtain information on the dimensions of a structure near your school or in your neighborhood. Or, use one of the structures from Activity 1.
2. Build a scale model of the building using cardboard or some other suitable modeling material.
3. Explain the process you used to determine the dimensions of your model and how you constructed the model.

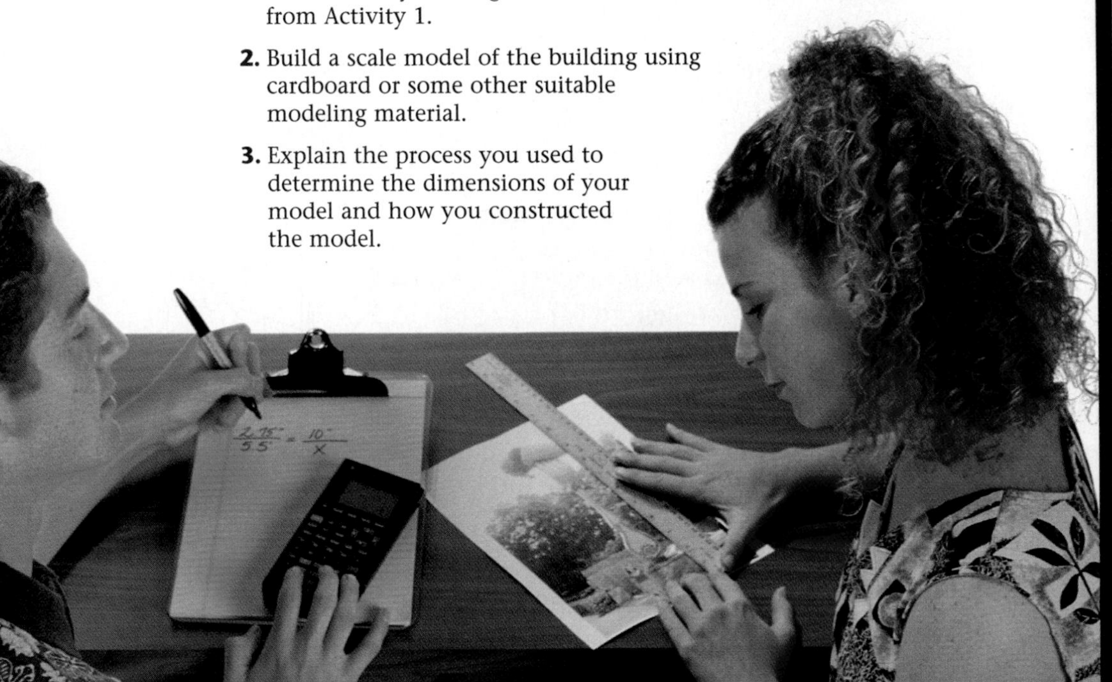

Activities 1 and 2. Check student work. Encourage students to work up presentations to share in class or for their portfolios.

Cooperative Learning

Cooperative learning groups are recommended for this activity because it is hands-on. Have students help each other measure the structures and build the scale models. Ask groups to keep a record of the process they used to find the dimensions of the model and to be prepared to describe the process to the class. The group should also include a description of how each student contributed to the project.

Discuss

What types of hobbies depend on building scale models?

Do scale-model airplanes fly just like their real counterparts?

What are problems associated with greatly enlarging a figure?

1.

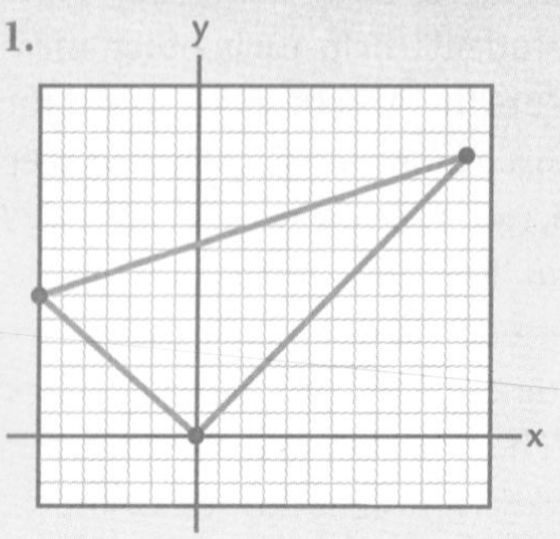

2. 2

Chapter 8 Review

Vocabulary

Key Skills and Exercises

Lesson 8.1

➤ Key Skills

Given a scale factor, draw a dilation.

Dilate the figure by a scale factor of 0.5.

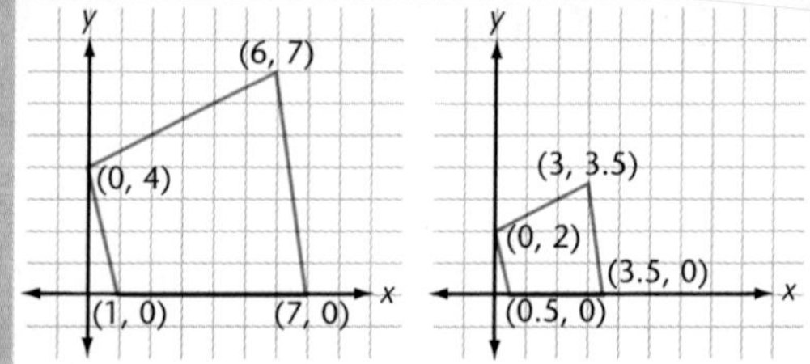

Multiply the coordinates of the vertices of the figure by 0.5, and sketch the dilation. In this case it is a contraction.

➤ Exercises

In Exercises 1–2, refer to the figure at the right.

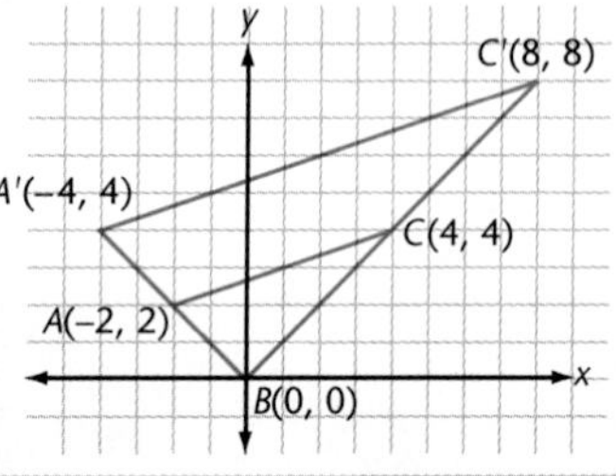

1. Dilate $\triangle ABC$ by a scale factor of 3 and sketch the result, giving coordinates of the vertices.
2. What is the scale factor for $\triangle A'BC'$?

Lesson 8.2

➤ Key Skills

Determine if polygons are similar.

In the pair of triangles, $EA = \frac{2}{3}$ AC, DA $= \frac{2}{3}$ BA, $m\angle D = m\angle B$, and $\angle DAE$ and $\angle BAE$ are right angles. Are the triangles similar?

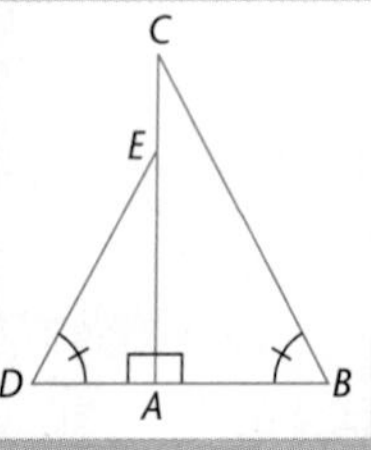

You are given that two angles are congruent, so by the Triangle Sum Theorem the third angles are congruent. Two sets of corresponding sides are proportional with a ratio of 2:3. Use the Pythagorean Theorem to show that the hypotenuses are in the same proportion, thus showing that corresponding sides are all proportional.

Use the properties of proportions to solve equations.

Solve for x.

$$\frac{11}{x+3} = \frac{3}{x-5}$$

By the Cross-Multiplication Property,

$$11(x - 5) = 3(x + 3)$$
$$11x - 55 = 3x + 9$$
$$8x = 64$$
$$x = 8$$

a b regular pentagon regular pentagon c

➤ Exercises

3. Which of the pairs of polygons above are similar? Justify your answer.

In Exercises 4–5, solve for *x*.

4. $\frac{26}{x} = \frac{2x}{13}$ **5.** $\frac{3 + 5x}{7} = \frac{3x - 4}{5}$

Lesson 8.3

➤ Key Skills

Show that triangles are similar.

Which triangles can be shown to be similar?

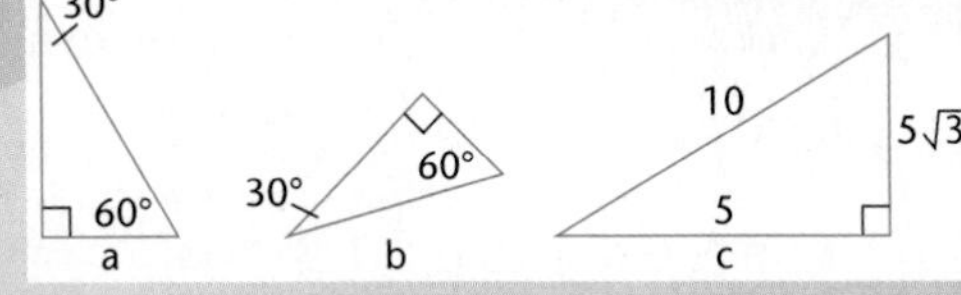

Using the Angle-Angle Similarity Postulate, $\triangle A$ and $\triangle B$ are similar because they have two equal angles. What about $\triangle C$? The proportions of the lengths of the sides of the triangle tell us that *a triangle with one leg equal to half the longest side and one leg equal to $\sqrt{3}$ times half the longest side is a 30-60-90 right triangle.* Thus $\triangle C$ is similar to $\triangle A$ and $\triangle B$.

➤ Exercises

In Exercises 6–7, are the triangles similar? Justify your answer.

6.

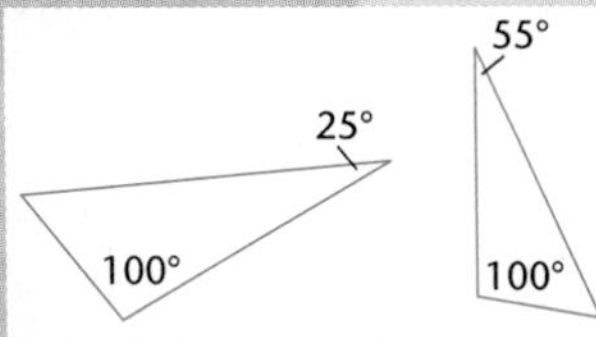

7.

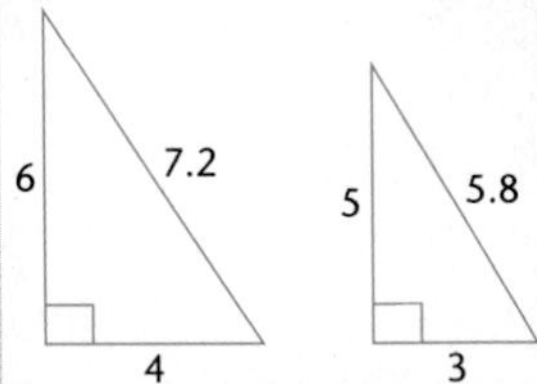

Lesson 8.4

➤ Key Skills

Show that triangles are similar using the similarity postulates.

In the triangle below, $\overline{DE}$ is parallel to $\overline{AB}$. Show that $\triangle DEC$ is similar to $\triangle ABC$.

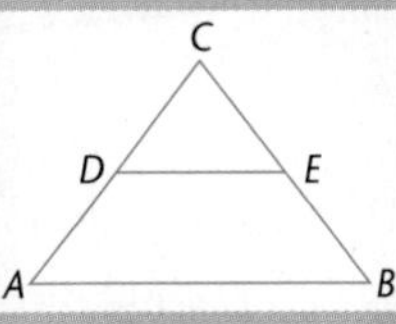

By the Side-Splitting Theorem, we know that sides $\overline{AC}$ and $\overline{BC}$ are divided proportionally. Thus, $\overline{AC}$ is proportional to $\overline{DC}$, and $\overline{BC}$ is proportional to $\overline{EC}$. $\angle C$ is equal to itself, so the SAS Similarity postulate applies, and $\triangle DEC$ is similar to $\triangle ABC$.

➤ Exercises

In Exercises 8–9, which triangles are similar? Justify your answer.

8.

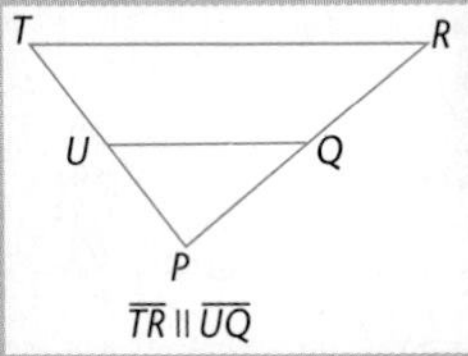

9.

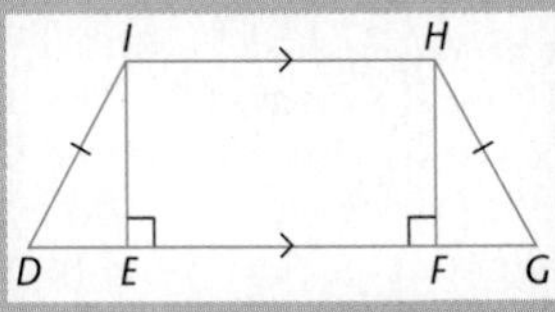

3. a. Not similar. Corresponding sides not in proportion.

   b. Similar. Both are 45°–45°–90° triangles, so all corresponding angles are congruent and corresponding sides are in proportion.

   c. Similar. All sides and all angles are congruent in both, so all angles are 120° and all corresponding sides are in proportion.

4. $x = \pm 13$ 5. $x = -10.75$

6. Triangles are similar. Both triangles have angles measuring 100°, 25° and 55°, so they are similar by AA Postulate.

7. Triangles are not similar. The ratios between corresponding sides are not equal.

8. $\triangle PUQ \sim \triangle PTR$; $\angle P \cong \angle P$, $\angle PUQ \cong -PTR$ because they are corresponding angles and $\overline{TR} \parallel \overline{UQ}$, $\triangle PUQ \sim \triangle PTR$ by AA Postulate

9. $\triangle DEI \sim \triangle GFH$; AA Postulate

10. $h = 8$ m

11. $8:1$ or $\frac{8}{1}$;

$2:1$ or $\frac{2}{1}$

12. $T = 27$ m; AA Postulate

Lesson 8.5

➤ Key Skills

Use similar triangles to measure distance.

A right rectangular pyramid has a base length of 40 feet. If the pyramid casts a shadow 60 feet long and a yardstick casts a shadow 6 feet long, what is the height of the pyramid?

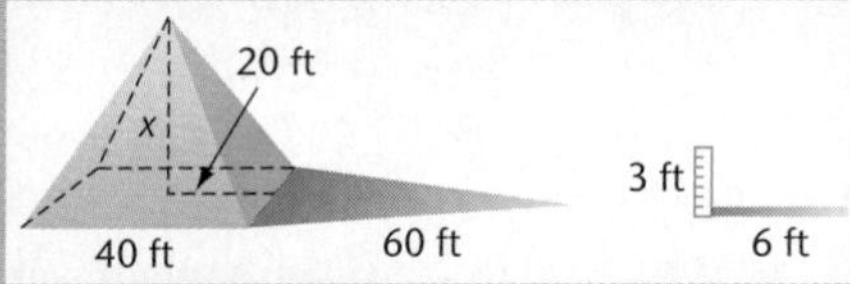

We use similar triangles to find the proportions of the two heights. The horizontal side of the small triangle is six feet; of the larger, 60 feet + 20 feet, or 80 feet. Set up the proportion:

$$\frac{6}{80} = \frac{3}{x}$$
$$6x = 240$$
$$x = 40$$

Thus the height of the pyramid is 40 feet.

➤ Exercises

10. You are 1.6 meters tall and cast a shadow of 3.5 meters. If a house simultaneously casts a shadow of 17.5 meters, how tall is the house?

Lesson 8.6

➤ Key Skills

Given similar figures, find ratios between their volume, area, etc.

The two cylinders to the right are similar. What is the ratio of their lateral areas?

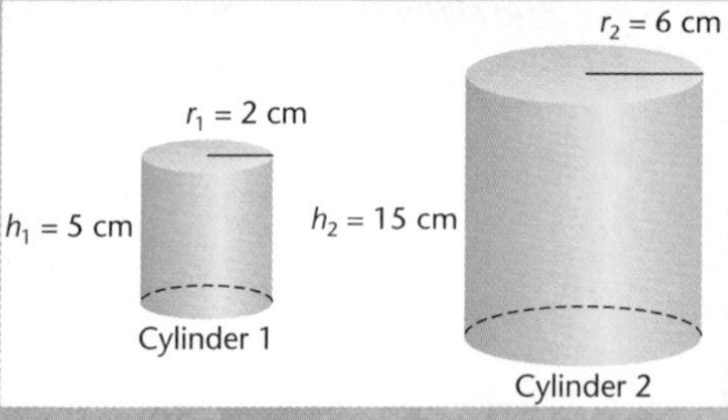

Method 1: Calculate the lateral areas L_1 and L_2. $L_1 \approx 62.5 \text{ cm}^2$, and $L_2 \approx 565.5 \text{ cm}^2$. The ratio of L_1 to L_2 is $\frac{62.5}{565.5}$, which simplifies to approximately $\frac{1}{9}$.

Method 2: Notice that $r_2 = 3r_1$ and $h_2 = 3h_1$. Thus,

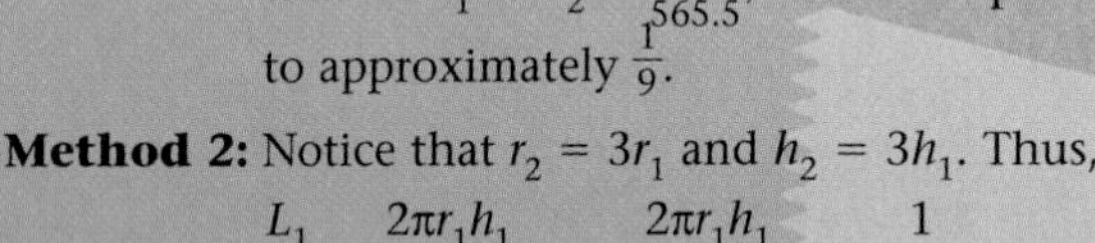

$$\frac{L_1}{L_2} = \frac{2\pi r_1 h_1}{2\pi r_2 h_2} = \frac{2\pi r_1 h_1}{2\pi(3r_1)(3h_1)} = \frac{1}{9}$$

➤ Exercises

11. The two prisms shown are similar. What is the ratio of their volumes? of their heights?

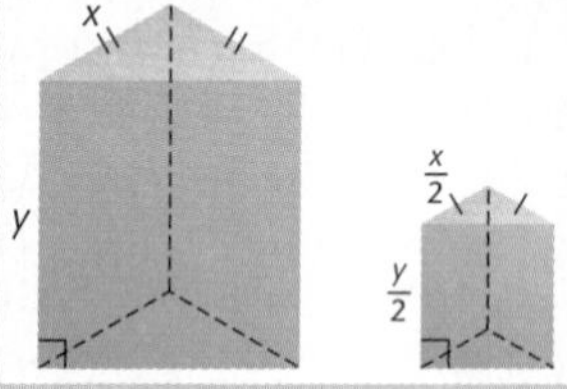

Applications

12. Suppose you put a mirror on the ground 15 meters from the base of a tree. If you stand 1 meter from the mirror, you can see the top of the tree. If your eyes are 1.8 meters above the ground, how tall is the tree? Which theorems or postulates prove that this method of indirect measurement works?

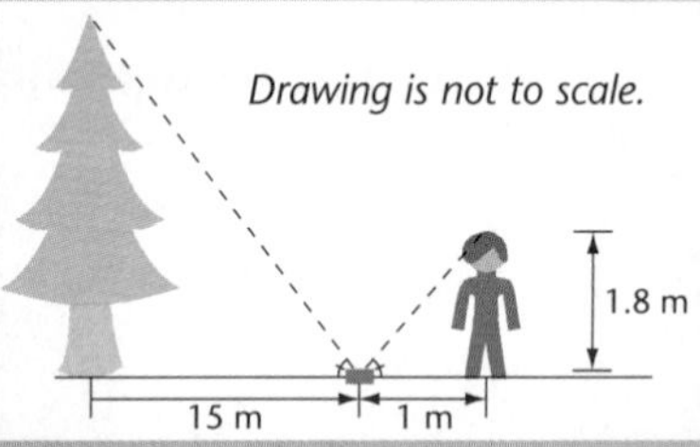

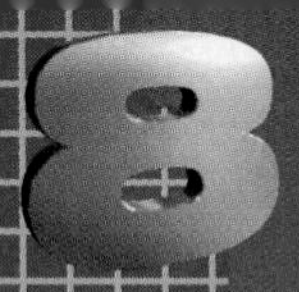

Chapter 8 Assessment

In Exercises 1–2, refer to the figure.

1. What is the scale factor for $F'G'H'I'$?
2. Dilate *FGHI* by a scale factor of 1.5 and sketch the result, giving coordinates of the vertices.

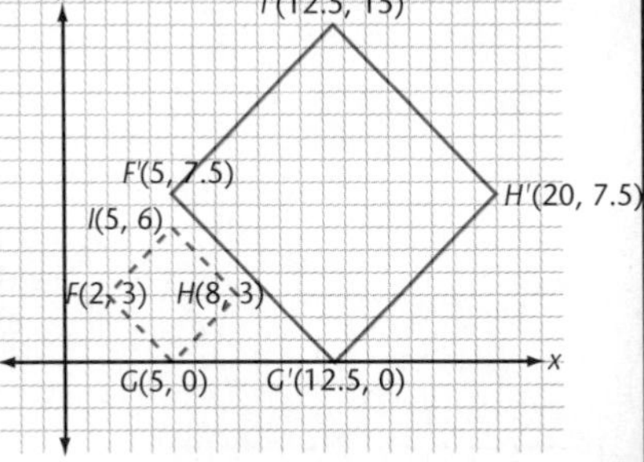

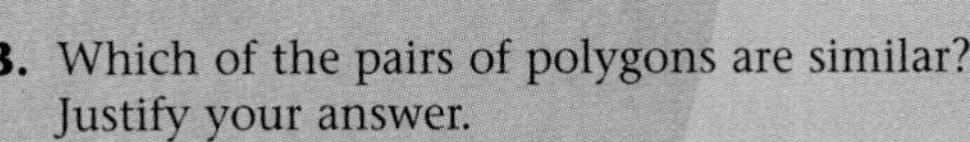

3. Which of the pairs of polygons are similar? Justify your answer.

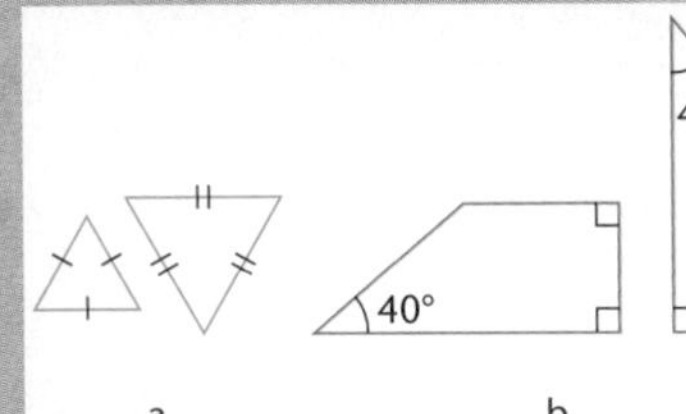

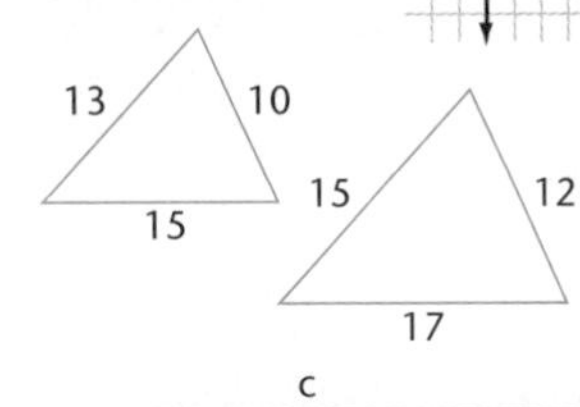

In Exercises 4–5, solve for *x*.

4. $\frac{x}{18} = \frac{21}{4}$

5. $\frac{10}{2 - 3x} = \frac{6}{x + 6}$

In Exercises 6–7, state which triangles are similar. Justify your answer.

6.

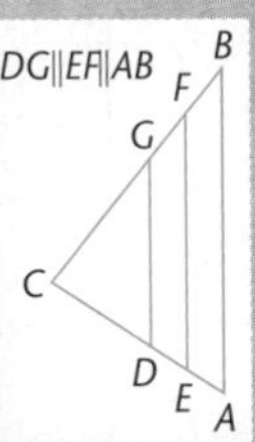

7.

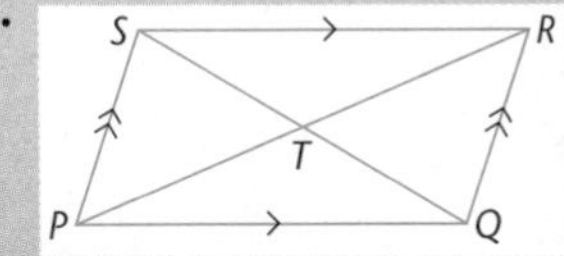

8. A surveyor made the measurements shown below. $\overline{ZY}$ is parallel to $\overline{WX}$. Find the distance across the base of the hill.

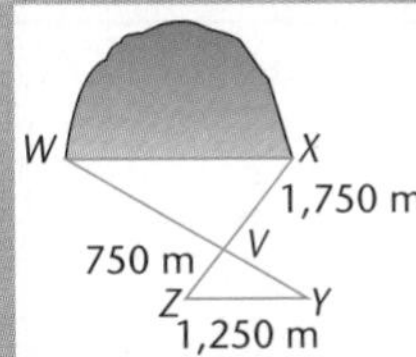

9. The two right cones below are similar. What is the ratio of their lateral areas?

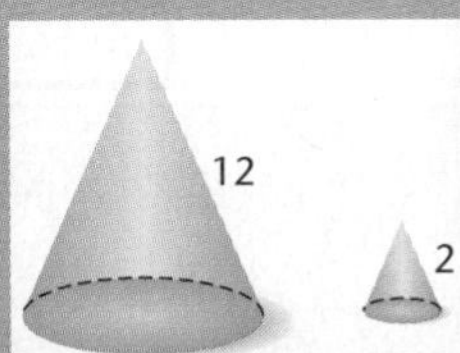

10. The surface areas of two spheres have the ratio 169:900. If the smaller sphere has a volume of 4,599.05, what is the volume of the larger?

4. $x = 94.5$
5. $x = \frac{-12}{7}$
6. $\triangle CGD \sim \triangle CFE \sim \triangle CBA$; $\angle C \cong \angle C \cong \angle C$; $\angle CGD \cong \angle CFE \cong \angle CBA$ because they are corresponding angles and $\overline{DG} \parallel \overline{EF} \parallel \overline{AB}$, so $\triangle CGD \sim \triangle CFE \sim \triangle CBA$ by AA Postulate
7. $\triangle STR \sim \triangle QTP$, AA Postulate or SAS Similarity or SSS Similarity; $\triangle STP \sim \triangle QTR$, AA Postulate or SAS Similarity or SSS Similarity; $\triangle SRQ \sim \triangle QPS$, AA Postulate or SAS Similarity or SSS Similarity; $\triangle SRP \sim \triangle QPR$, AA Postulate or SAS Similarity or SSS Similarity
8. $WX = 2916\frac{2}{3}$ m
9. $\left(\frac{6}{1}\right)^2 = \frac{36}{1}$
10. $V = 56519.96$ units^3

1. 2.5
2.

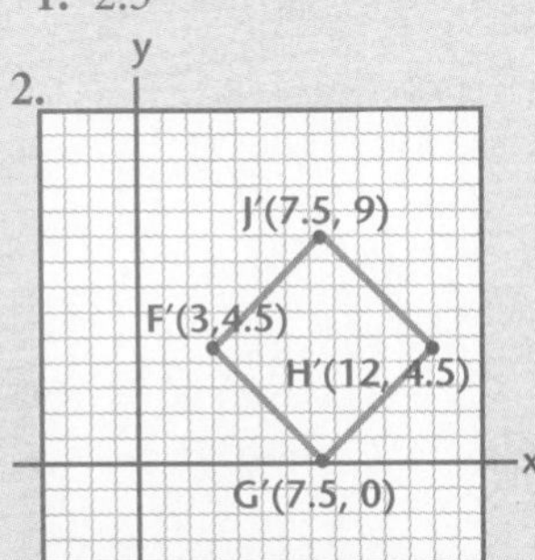

3. a. Triangles are similar. Ratios between corresponding sides are equal and all angles measure 60°, so corresponding angles are congruent.

   b. Trapezoids not similar or can't determine. Corresponding angles are congruent, but can't determine if the ratios between corresponding sides are equal.

   c. Triangles not similar. Ratios between corresponding sides are not equal.

College Entrance-Exam Practice

Multiple-Choice and Quantitative-Comparison Samples

The first half of the Cumulative Assessment contains two types of items found on standardized tests—multiple-choice questions and quantitative-comparison questions. Quantitative comparison items emphasize the concepts of equalities, inequalities, and estimation.

Free-Response Grid Samples

The second half of the Cumulative Assessment is a free-response section. A portion of this part of the Cumulative Assessment consists of student-produced response items commonly found on college entrance exams. These questions require the use of machine-scored answer grids. you may wish to have students practice answering these items in preparation for standardized tests.

Sample answer grid masters are available in the *Chapter Teaching Resources Booklets.*

Chapters 1–8 Cumulative Assessment

College Entrance Exam Practice

Quantitative Comparison Exercises 1–3 consist of two quantities, one in Column A and one in Column B, which you are to compare as follows:

A. The quantity in Column A is greater.
B. The quantity in Column B is greater.
C. The two quantities are equal.
D. The relationship cannot be determined from the information given.

	Column A	Column B	Answers
1.	Assume there are no hidden cubes. Exterior surface area of A	Exterior surface area of B	(A) (B) (C) (D) **[Lesson 6.1]**
2.	Circle A with radius 5. Ratio of A's area to B's area	Circle B with radius 6. Ratio of B's area to A's area	(A) (B) (C) (D) **[Lesson 5.3]**
3.	$\overline{EF}$ is congruent to $\overline{CA}$. Triangle ABC: AC = 4, CB = 5. $m\angle A$	Triangle DEF: DF = 2, FE = 4, DE = $2\sqrt{3}$. $m\angle D$	(A) (B) (C) (D) **[Lessons 5.4, 5.5]**

4. Which are consecutive interior angles? **[Lesson 3.3]**
a. 1 and 5 **b.** 2 and 3
c. 3 and 6 **d.** 4 and 6

5. Which angle is not an angle of rotational symmetry of a regular octagon? **[Lesson 3.6]**
a. 135° **b.** 45° **c.** 60° **d.** 90°

1. B 2. B 3. D
4. d 5. c

6. Two similar pyramids have slant heights in the ratio of 4:7. The ratio of their surface areas is **[Lesson 7.1]**
   a. 4:7 **b.** 8:14 **c.** 16:28 **d.** 16:49

7. Which property of proportion justifies this conditional? **[Lesson 8.2]**
   If $\frac{x}{2y} = \frac{a}{b}$, then $\frac{(x + 2y)}{2y} = \frac{(a + b)}{b}$.
   a. "Add-One" Property **b.** Cross-Multiplication Property
   c. Exchange Property **d.** Reciprocal Property

8. Is this a true conditional? "If a quadrilateral is equilateral, then it is a square." If false, provide a counterexample. **[Lesson 2.2]**

9. Mon, Tues, and Wed are towns on a road that runs straight from Mon to Wed. The distance from Mon to Tues is 20 kilometers. The distance from Mon to Wed is 5 kilometers less than three times the distance from Tues to Wed. What is the distance between Tues and Wed? **[Lesson 1.4]**

10. In the figure at the right, name all the pairs of triangles that can be shown to be congruent. **[Lessons 4.2, 4.3]**

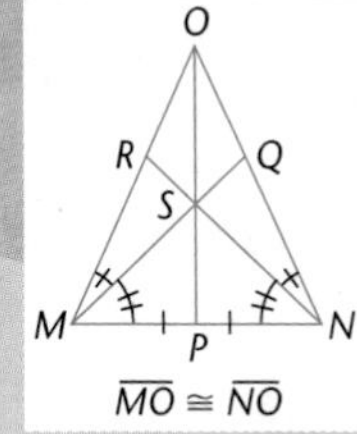

11. Describe the intersection of plane *FGH* and plane *PQR*. **[Lesson 1.2]**

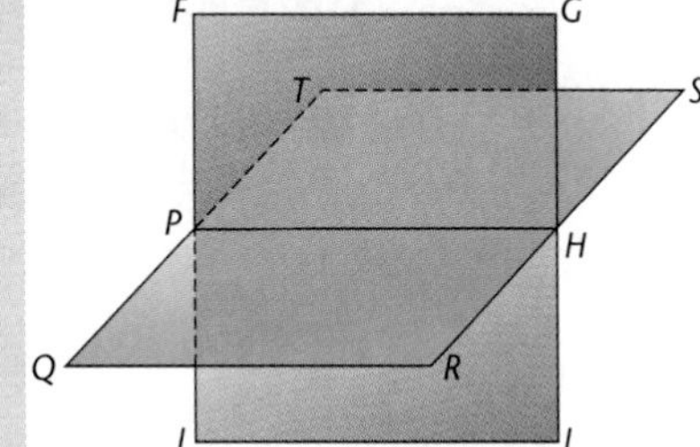

12. Find the volume of the sphere below. **[Lesson 7.6]**

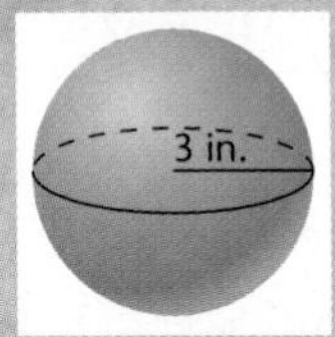

13. An equilateral triangle with a side length of 5 units is inscribed in a circle with a radius of 3 units. If a point is picked at random anywhere inside the circle, what is the probability that the point will not be inside the triangle? **[Lesson 5.7]**

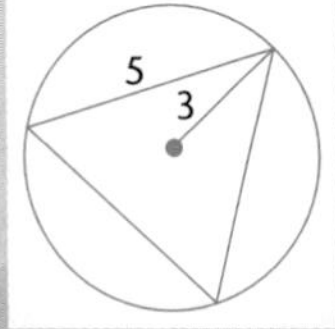

Free-Response Grid Exercises 14–17 may be answered using a free-response grid commonly used by standardized test services.

14. Find the slope of a line that passes through coordinates (2, 6) and (9, 12). **[Lesson 3.8]**

15. $AC = 12$ m, and $DB = 10$ m. What is the area of the rhombus? **[Lessons 5.2, 5.4]**

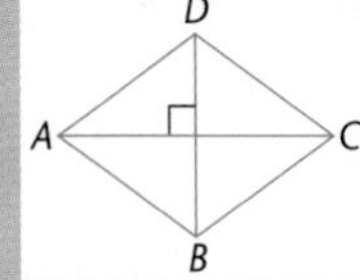

16. What is the area of the regular octagon? **[Lesson 5.5]**

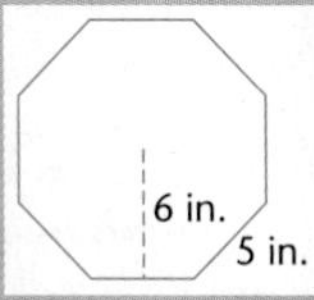

17. The dimensions of a right rectangular prism are length = 14 mm, width = 3 mm, and height = 11 mm. Find the length of the prism's diagonal. **[Lesson 6.3]**

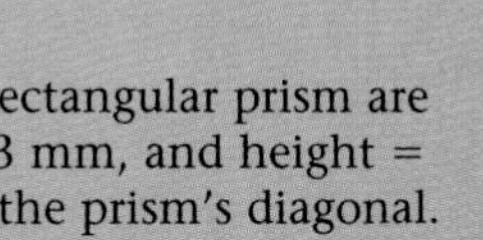

Cumulative Assessment

15. 60 m^2

16. 120 in^2

17. ≈ 18.06 mm

6. d 7. a

8. False. A rhombus with acute angles of 50° is an equilateral quadrilateral that is not a square.

9. Let x be the distance from Tues to Wed. $20 + x = 3x - 5 \Rightarrow 2x = 25 \Rightarrow x = 12.5$ km

10. $\triangle OMP \cong \triangle ONP$; $\triangle SNP \cong \triangle SMP$; $\triangle QSN \cong \triangle RSM$; $\triangle QMN \cong \triangle RNM$; $\triangle OSR \cong \triangle OSQ$; $\triangle OSM \cong \triangle OSN$; $\triangle OQM \cong \triangle ORN$

11. $\overleftrightarrow{PH}$

12. $V \approx 113.1$ in^3

13. ≈ 0.38

14. $m = \frac{6}{7}$

Circles

Meeting Individual Needs

9.1 Exploring Chords and Arcs

Core Resources	Core Plus Resources
Inclusion Strategies, p. 472 Reteaching the Lesson, p. 473 Practice Master 9.1 Enrichment, p. 472 Technology Master 9.1 Lesson Activity Master 9.1 Interdisciplinary Connection, p. 471	Practice Master 9.1 Enrichment Master 9.1 Technology Master 9.1 Interdisciplinary Connection, p. 471
[2 days]	[1 day]

9.2 Exploring Tangents to Circles

Core Resources	Core Plus Resources
Inclusion Strategies, p. 480 Reteaching the Lesson, p. 481 Practice Master 9.2 Enrichment, p. 480 Technology Master 9.2 Lesson Activity Master 9.2	Practice Master 9.2 Enrichment Master 9.2 Technology Master 9.2
[2 days]	[2 days]

9.3 Exploring Inscribed Angles and Arcs

Core Resources	Core Plus Resources
Inclusion Strategies, p. 486 Reteaching the Lesson, p. 487 Practice Master 9.3 Enrichment, p. 480 Technology Master 9.3 Lesson Activity Master 9.3 Mid-Chapter Assessment Master	Practice Master 9.3 Enrichment, p. 486 Technology Master 9.3 Interdisciplinary Connection, p. 485 Mid-Chapter Assessment Master
[2 days]	[2 days]

9.4 Exploring Circles and Special Angles

Core Resources	Core Plus Resources
Inclusion Strategies, p. 493 Reteaching the Lesson, p. 494 Practice Master 9.4 Enrichment, p. 493 Technology Master 9.4 Lesson Activity Master 9.4 Interdisciplinary Connection, p. 492	Practice Master 9.4 Enrichment Master 9.4 Technology Master 9.4 Interdisciplinary Connection, p. 485
[3 days]	[2 days]

9.5 Exploring Circles and Special Segments

Core Resources	Core Plus Resources
Inclusion Strategies, p. 504 Reteaching the Lesson, p. 505 Practice Master 9.5 Enrichment Master 9.5 Lesson Activity Master 9.5 Interdisciplinary Connection, p. 503	Practice Master 9.5 Enrichment, p. 504 Technology Master 9.5 Interdisciplinary Connection, p. 503
[2 days]	[2 days]

9.6 Circles in the Coordinate Plane

Core Resources	Core Plus Resources
Inclusion Strategies, p. 511 Reteaching the Lesson, p. 512 Practice Master 9.6 Enrichment, p. 511 Lesson Activity Master 9.6	Practice Master 9.6 Enrichment Master 9.6 Technology Master 9.6 Interdisciplinary Connection, p. 510
[2 days]	[1 day]

Chapter Summary

Core Resources	Core Plus Resources
Chapter 9 Project, pp. 516–517 Lab Activity Long-Term Project Chapter Review, pp. 518–520 Chapter Assessment, p. 521 Chapter Assessment, A/B Alternative Assessment	Chapter 9 Project, pp. 516–517 Lab Activity Long-Term Project Chapter Review, pp. 518–520 Chapter Assessment, p. 521 Chapter Assessment, A/B Alternative Assessment
[3 days]	[2 days]

Reading Strategies

This chapter introduces vocabulary used in the study of circles and the relationships among circles, lines, and angles. Examples include *chord, arc, central angle, inscribed angle, exterior angle, secant,* and *tangent.*

Have students try to define each key term in their own words. They should write down their definitions and then compare them with those given in the text or in the glossary. Make sure figures are included with each student-made definition. Student-made definitions will often be imprecise or vague. By working in small groups and discussing how to improve their definitions, students will not only learn the meanings of key terms, they will also improve their ability to communicate mathematical ideas.

Hands-On Strategies

The activities in this chapter call for either a compass and straightedge or geometry graphics software as students investigate relationships among circles, arcs, chords, tangents, secants, and related figures. Although in some cases a rough sketch may be sufficient, encourage students to make careful drawings to improve their geometric drawing ability.

Careful measurement of angles plays a key role in this chapter. Encourage students to make very large drawings of figures in which they need to measure angles. This will help give more accurate measurements. If necessary, review how to use the edge of a piece of paper to extend the side of an angle before measuring.

Cooperative Learning

GROUP ACTIVITIES	
Chords and Arcs Theorem	Lesson 9.1, Exploration 3
Radii, chords, and tangents	Lesson 9.2, Explorations 1 and 2
Intercepted Arcs	Lesson 9.3, Explorations 1 and 2
Circles and special angles	Lesson 9.4, Explorations 1–3
Point of Disaster	Eyewitness Math
Circles and special segments	Lesson 9.5, Explorations 1–3

You may wish to have students work in groups or with partners for some of the above activities. Additional suggestions for cooperative group activities are noted in the teacher's notes in each lesson.

Multicultural

The cultural references in this chapter include references to Asia, and the Americas.

CULTURAL CONNECTIONS	
Asia: Yin-Yang symbol	Portfolio Activity
Americas: Yucatan impact crater	Lesson 9.5

Portfolio Assessment

Below are portfolio activities for the chapter listed under seven activity domains which are appropriate for portfolio development.

1. **Investigation/Exploration** The explorations in Lesson 9.1 focus on chords, arcs, and designed created with circles; those in Lesson 9.2 on secants, tangents, and their properties. In the explorations for Lesson 9.3, students investigate the relationship between the measure of an inscribed angle and its intercepted arc. Other angles are the focus of the explorations in Lesson 9.4. Segment relationships are covered in Lesson 9.5.

2. **Applications** Civil Engineering, Lesson 9.1, Exercise 21; Communication, Lesson 9.2, Exercise 8; Space Flight, Lesson 9.2, Exercise 11; Stained Glass, Lesson 9.3, Exercise 23; Communications, Lesson 9.4, Exercise 23; Machining, Lesson 9.5, Exercise 22.

3. **Nonroutine Problems** Lesson 9.5, Exercise 36 (visual proof); Chapter 9 Project, Activity 2 (marching band problem).

4. **Project** The Olympic Symbol: see pages 516–517.

5. **Interdisciplinary Topics** Statistics, Lesson 9.1, Exercise 22; Algebra, Lesson 9.1, Exercises 24–27, 33–34; Algebra, Lesson 9.2, Exercises 12–16; Algebra, Lesson 9.3, Exercises 7–22; Algebra, Lesson 9.4, Exercises 5–11; Algebra, Lesson 9.5, Exercises 6–21; Algebra, Lesson 9.6, Exercises 31–34.

6. **Writing** *Communicate* exercises offer excellent writing selections for the portfolio. Lesson 9.1, Exercises 1–5; Lesson 9.2, Exercise 1; Lesson 9.3, Exercise 6.

7. **Tools** Students can use geometry graphics software to make and test conjectures. In Lessons 9.3 and 9.4, spreadsheet software may be used to organize data and test for relationships. A graphics calculator is used in Lesson 9.6.

Technology

Many of the activities in this book can be significantly enhanced by the use of geometry graphics software, but in every case the technology is optional. If computers are not available, the computer activities can, *and should,* be done by hands-on or "low-tech" methods, as they contain important instructional material. For more information on the use of technology, refer to the *HRW Technology Manual for High School Mathematics.*

When instructing students in the use of the software, you may find that the best approach is to give students a few well-chosen hints about the different tools and let them discover their use on their own. Today's computer-savvy students can be quite impressive in their ability to discover the uses of computer software! Once the students are familiar with the geometry software, they can design and create their own computer sketches for the explorations.

A second approach is for you, the teacher, to demonstrate the creation of the exploration sketches "from scratch" using a projection device. The students will find this exciting to watch, and they will be eager to learn to use the software on their own.

In the interest of convenience or time-saving, you may want to use the *Geometry Investigations Software* available from HRW. Each of the files on the disks contains a "sketch," custom-designed for an *HRW Geometry* Exploration. The sketches are to be used with *Geometer's Sketchpad*™ (Key Curriculm Press) or *Cabri Geometry II*™ (Texas Instruments).

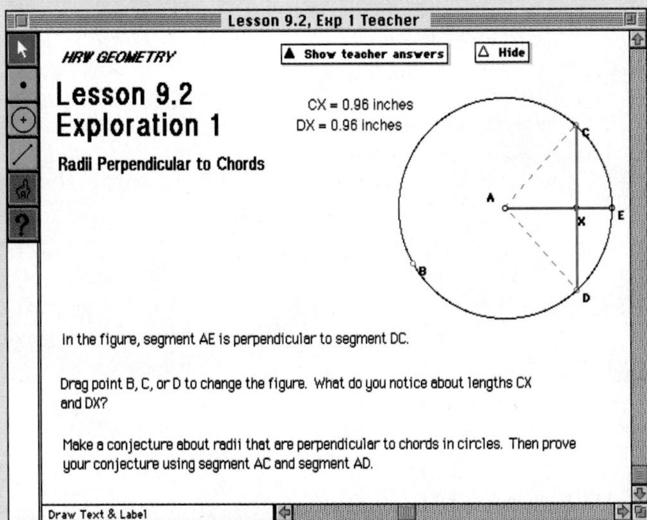

Computer Graphics Software

Lesson 9.1 Exploration 2 In this exploration, students investigate a "flower" formed from a hexagon construction. The geometry technology allows students to investigate many such flowers formed from the construction very quickly. Students can drag one vertex of the flower and change its size.

Leson 9.1: Exploration 3 Students explore the relationship between congruent chords of circles and their arcs. Students can look at many different cases very quickly by dragging points on the figure.

Lesson 9.2: Exploration 1 By dragging a figure and observing the changes in measurements, students can make a conjecture about radii that are perpendicular to chords.

Lesson 9.2: Exploration 2 Students are instructed to drag a secant line until it becomes a tangent. The geometry technology measures the angle formed by the secant line and a radius. Students can see the angle increasing to a right angle as the secant line becomes a tangent.

Lesson 9.3: Exploration 1 In this exploration, students examine the relationship between the measure of an inscribed angle and its intercepted arc. The geometry technology allows them to look at many different arcs and measurements, leading them to a conjecture. They can use the calculator tool to verify this conjecture.

Lesson 9.5: Exploration 1 Students investigate the tangent segments to a circle from a common external point. The geometry technology allows them to quickly make measurements of a variety of tangent segments. It also allows them to discover and explain a special case—as they drag the point around a circle, the measurements will disappear under certain conditions.

Lesson 9.5: Exploration 2 This is a two-part exploration in which students investigate secant and tangent lines intersecting outside a circle. Part 1 deals with secant lines. The geometry technology allows students to make measurements and change the size of the circle and the length of the secant lines while also changing the measurements. The calculator tool calculates the cross products and adjust them as the students alter the sizes of the figures.

Lesson 9.5: Exploration 2 As a continuation of Part 1, the second part of the exploration has students find the cross product when a tangent line and a secant intersect outside a circle. Students drag one intersection point of a secant until it becomes a tangent.

Lesson 9.5: Exploration 3 In this exploration, students investigate chords which intersect inside a circle. The calculator tool calculates cross products and adjusts them while students drag the endpoints of the chords around the circle.

About the Chapter

Background Information

This chapter focuses on the relationships among parts of circles and between circles and various angles, segments, and arcs. Students develop key theorems and rules through exploration activities, and then use those rules in problem-solving situations.

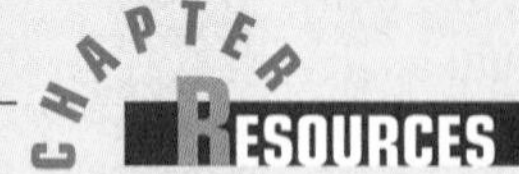

- Practice Masters
- Enrichment Masters
- Technology Masters
- Lesson Activity Masters
- Lab Activity Masters
- Long-Term Project Masters
- Assessment Masters
  Chapter Assessments, A/B
  Mid-Chapter Assessment
  Alternative Assessments, A/B
- Teaching Transparencies
- Spanish Resources

Chapter Objectives

- Define parts of a circle—*center, radius, chord, diameter, major* and *minor arcs,* and *arc.*
- Develop a theorem concerning congruent chords and arcs.
- Define *secant, tangent,* and *point of tangency.*
- Develop and use the Tangent Theorem.
- Define *inscribed angle* and *intercepted arc.*
- Develop and use the Inscribed Angle Theorem and Corollary.
- Explore special angles relative to the circle including secant-tangent, secant-secant, and chord-chord.

Chapter 9

Lessons

Circles

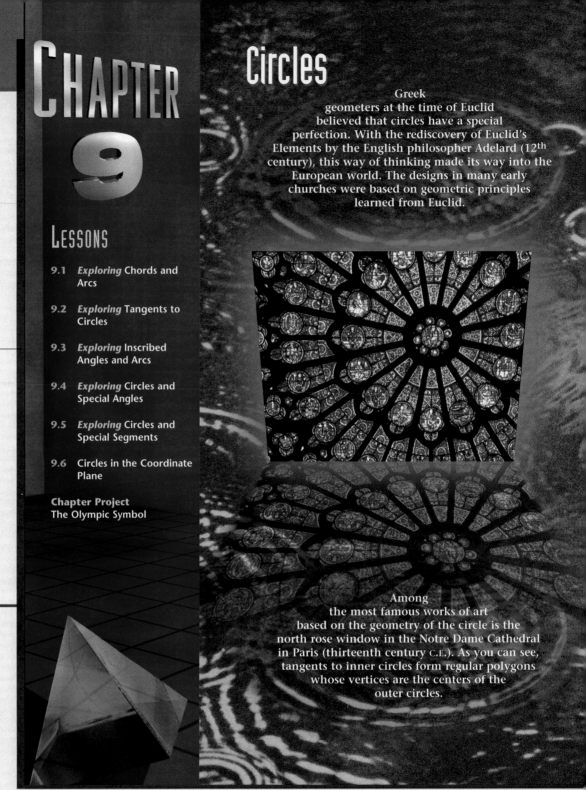

Greek geometers at the time of Euclid believed that circles have a special perfection. With the rediscovery of Euclid's Elements by the English philosopher Adelard (12th century), this way of thinking made its way into the European world. The designs in many early churches were based on geometric principles learned from Euclid.

Among the most famous works of art based on the geometry of the circle is the north rose window in the Notre Dame Cathedral in Paris (thirteenth century C.E.). As you can see, tangents to inner circles form regular polygons whose vertices are the centers of the outer circles.

About the Photos

Circles have fascinated people for centuries. Many cultures of the world attach mystical significance to circles. The photos show works of humans in which the circle holds special meaning.

Portfolio Activity

Cultural Connection: Asia According to the Chinese philosophy of Taoism, all reality consists of an interplay of opposite forces yin and yang. The diagram, which symbolizes the tao (meaning "the way") represents the interplay of the two principles geometrically. The figure has a number of striking geometric features that you will discover in this chapter.

To this day, circles figure prominently in many logos and symbols. Find examples of these to analyze and include in your portfolio.

In contrast with the rest of Pueblo architecture, which is usually rectangular, the circular shape of these Kivas in the ruins at Chaco Canyon suggests their special purpose. The walls of Kivas were often decorated with ceremonial murals.

- Develop and use rules for special angles including secant-tangent, secant-secant, and chord-chord.
- Explore special segments relative to the circle including secant-secant, secant-tangent, and chord-chord.
- Develop and use rules for special segments including secant-tangent, secant-secant, and chord-chord.
- Develop and use the equation of a circle.
- Adjust the parameters of a circle equation to move the center on a coordinate plane.

Portfolio Activity

Have students discuss the yin and yang diagram, both in terms of what it means and how it's constructed. Then have students work in pairs or small groups to construct the diagram using compasses or geometry graphics software. If students need a hint, have them start by drawing two congruent circles that touch at just one point.

Additional portfolio activities in the Pupil Edition can be found in the exercises for Lessons 9.1 and 9.6.

About the Chapter Project

The activities in the Chapter 9 Project, on pages 516–517, explore a symbol based on congruent circles—the Olympic Games symbol.

LESSON 9.1 Exploring Chords and Arcs

Mathematical relationships of circles and segments make many interesting compass-and-straight edge constructions possible.

Defining the Parts of a Circle

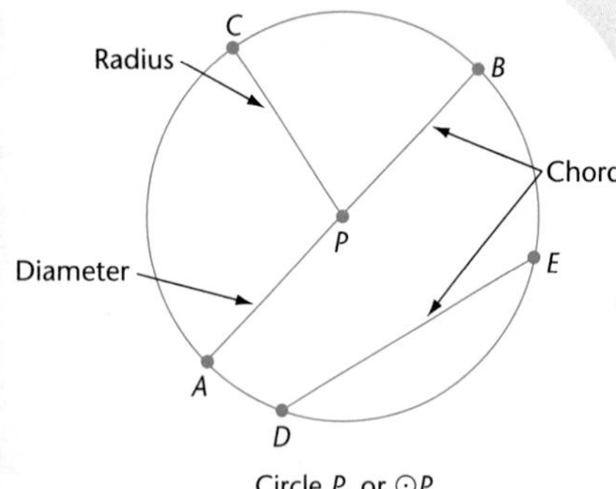

Circle P, or $\odot P$

CIRCLE

A **circle** consists of the points in a plane that are equidistant from a given point known as the **center** of the circle. A **radius** (plural, radii) is a segment from the center of a circle to a point on the circle. A **chord** is a segment whose endpoints lie on a circle. A **diameter** is a chord that passes through the center of a circle. **9.1.1**

When you use circles in constructions, you make use of the fact that congruent circles have congruent radii. This fact follows from the definition of a circle.

A circle is named using the symbol $\odot$ and the center of the circle.

PREPARE

Objectives

- Define parts of a circle—*center, radius, chord, diameter, major* and *minor arcs,* and *arc.*
- Develop a theorem concerning congruent chords and arcs.

RESOURCES

- Practice Master 9.1
- Enrichment Master 9.1
- Technology Master 9.1
- Lesson Activity Master 9.1
- Quiz 9.1
- Spanish Resources 9.1

Assessing Prior Knowledge

Define the following terms.

1. circle
   [A set of points in a plane equidistant from a point]
2. radius
   [A line segment connecting the center of a circle with any point on the circle]
3. diameter
   [A line segment connecting two points on a circle and passing through the center]

Use Transparency 73

TEACH

After students have examined the circle flowers in the text, point out that the original overlapping circles are often just the beginning. Colors can be added, parts erased, etc.

ALTERNATIVE teaching strategy

Technology Geometry graphics software may be used for the constructions in this lesson and for measuring central angles in degrees. Some software programs will give arc measures in linear units; some will not.

Exploration 1 Constructing a Regular Hexagon

You will need
Compass and straightedge

1. Draw a circle with a compass. Label the center *P*. Choose a point on the circle and label it *A*. (Figure a.)

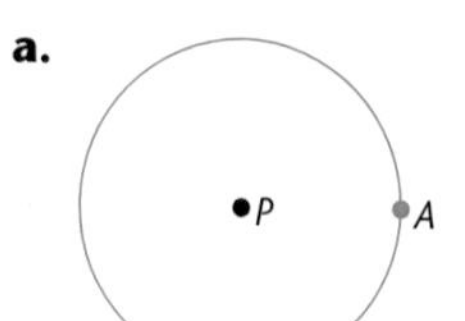

2. Without changing your radius setting, set the point of your compass on *A*. Draw an arc that intersects the circle at a new point. Label the new point *B*. (Figure b.)

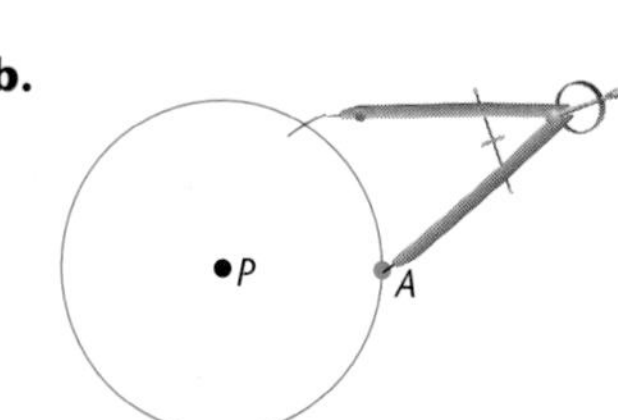

3. Draw chord $\overline{AB}$. Draw radii $\overline{PA}$ and $\overline{PB}$. (Figure c.)

4. What kind of triangle is $\triangle ABP$? What are the measures of its angles? Explain your reasoning.

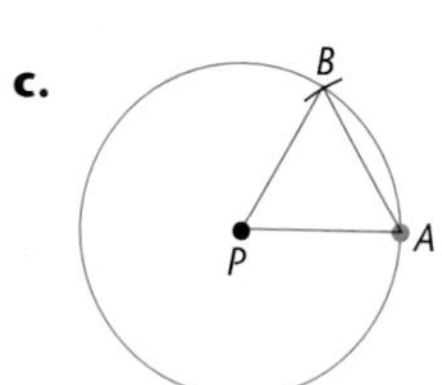

5. Without changing your radius setting, set the point of your compass on *B*. Draw an arc that intersects the circle at a new point. Label the new point *C*. (Figure d.)

6. Draw chord $\overline{BC}$ and the new radius $\overline{PC}$.

7. Continue drawing new points, chords, and radii until you have completed a figure like the one shown. (Figure e.)

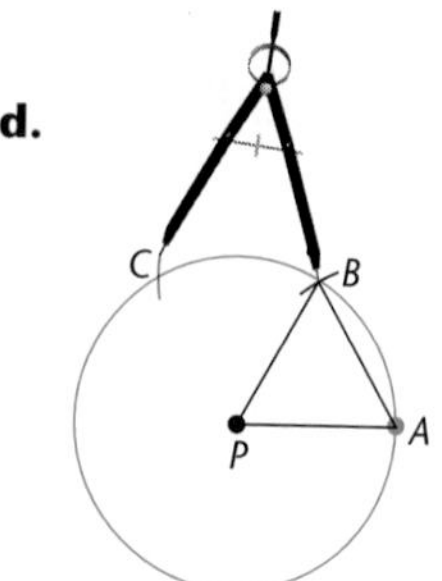

8. Is the polygon *ABCDEF* a regular hexagon? Explain your answer.

9. An angle such as $\angle APB$ is known as a **central angle** of a circle (and also of a regular polygon). Are the central angles of the circle (and of *ABCDEF*) congruent? Does the sum of their measures equal 360°? Explain your reasoning. ❖

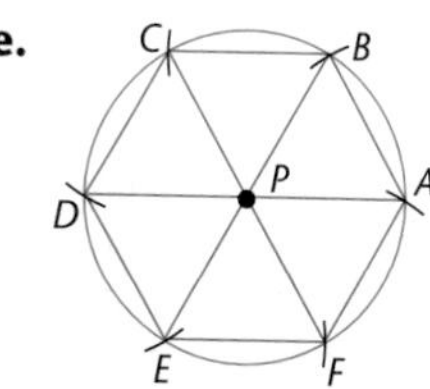

If two central angles of a circle (such as $\angle APB$ and $\angle CPD$) are congruent, what can you conclude about their chords? Does the size of the central angles matter? State your answer as a theorem and explain your reasoning.

Exploration 1 Notes

Exploration 1 provides students with an introduction to the study of circles. Students can use geometry graphics software to complete this exploration.

ongoing ASSESSMENT

4. **Equilateral triangle. All angles measure 60° because all sides are congruent.**

8. **Yes. All the sides are the same length as the radius of the circle. All the interior angles measure 120°, twice the measure of an angle in an equilateral triangle.**

TEACHING tip

Before students answer the Critical Thinking questions, make sure that they understand the meanings of—and distinctions among—these terms:

1. central angle
2. measure of a central angle
3. an arc formed by a central angle
4. a chord formed by a central angle

CRITICAL Thinking

If two central angles of a circle are congruent, then their chords are congruent. If the two central angles of a circle are congruent, then they intercept congruent arcs of the circle, whose lengths are the same fraction of the circle. Thus, the segments that connect the points on the circle where the angles intercept the circle will also be congruent.

interdisciplinary CONNECTION

Art Designs based on congruent circles, either overlapping or separate, are found in textile and rug patterns from many different cultures. Have students collect samples of such designs for discussion. Of particular interest may be the portions of the original circles that are hidden or erased to make the final designs.

Use Transparency 74

Exploration 2 Notes

After students have finished the flower construction, have them experiment to make flowers with other than 6 petals.

ongoing ASSESSMENT

5. **All six of the outer circles meet at the center of the original circle. A regular hexagon would be formed if the points where the petals of the flower meet the circle were connected. All the petals of the flower are congruent.**

TEACHING tip

Before students read and discuss the definitions following Exploration 2, make sure that all students understand what a *circle* is. A circle is merely the line around the boundary. Some students mistakenly think that the term *circle* also means the region inside the boundary. You may wish to introduce the term *sector* at this time to help clarify *arc*. An arc is part of a circle; a sector is part of a circular region.

CRITICAL Thinking

You would not need a third point if you were interested in the degree measure or length of the semicircle because both semicircles have the same degree measure and length. A third point would be necessary to designate a specific semicircle.

Exploration 2 Constructing a "Circle Flower"

You will need
Geometry technology or
Compass

1. Draw a circle with a compass or geometry technology. Choose a point on the circle and label it A.
2. Using A as the center point, draw another circle congruent to the first.
3. Pick one of the points where circle A intersects the original circle. Label this point B. Draw circle B congruent to circle A.
4. Repeat Step 3, going around the circle in order, until you have completed the flower.
5. Discuss the mathematical features of the flower. For example, do all six of the outer circles meet at a single point? How is the flower related to the hexagon in Exploration 1? ❖

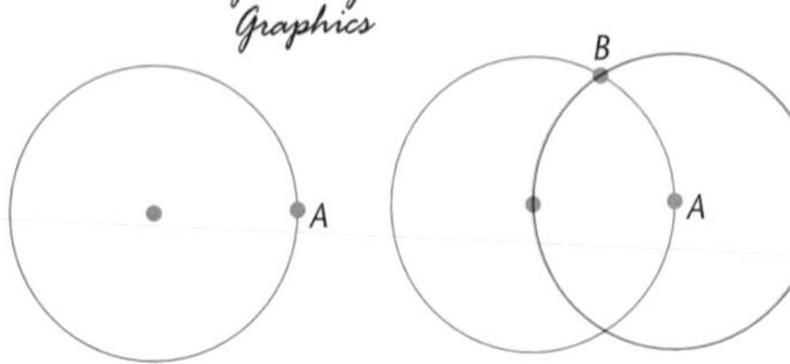

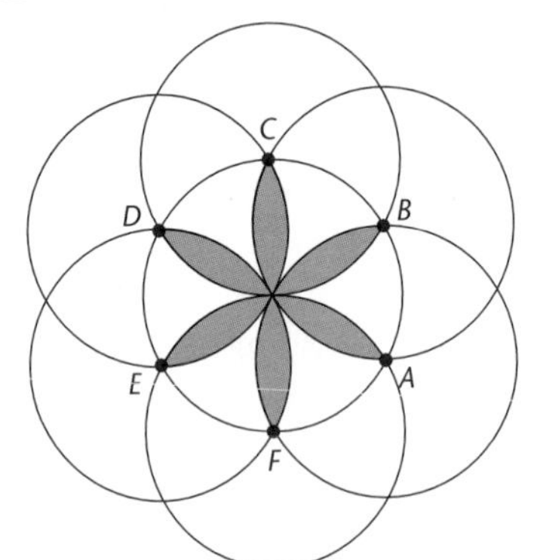

Major and Minor Arcs

An **arc** is a part of a circle. An arc is named using its endpoints and an arc symbol.

Notice that there are two different arcs that have their endpoints at Q and R. One arc (known as the **minor arc**) is less than a half-circle. The other arc (known as the **major arc**) is greater than a half-circle. When an arc is designated by only two letters, it is assumed to be a minor arc.

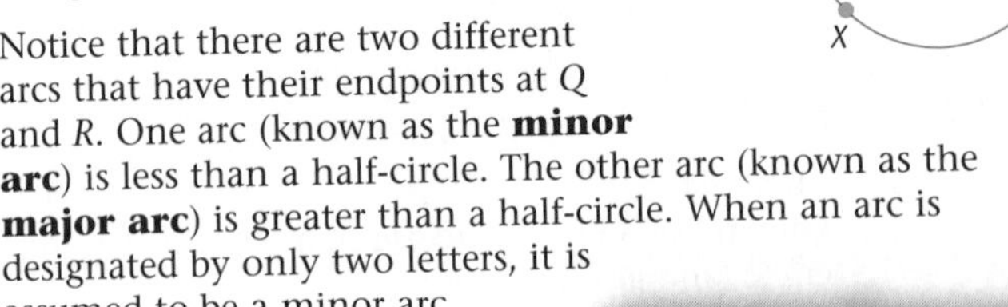

To designate a major arc, use the two endpoints and a third point that lies on the arc. Write the third point between the endpoints.

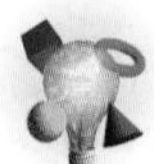

CRITICAL Thinking

A **semicircle** is an arc that is exactly half of a circle. Would you need a third point in your notation of a semicircle? Explain your reasoning.

ENRICHMENT Challenge students to construct the pinwheel design shown below to the left. Students should explain how the pinwheel construction relates to the circle flower construction in Exploration 2. (The illustration to the right shows the relationship.)

INCLUSION strategies **Hands-On Strategies** Some students may be confused by a figure such as the one for the flower in Exploration 2. Have them cut seven congruent circles from transparent paper and use the circles to make the design. You can also have students use four pencils of different colors: black for the central circle; red, blue, and yellow for pairs of opposite circles.

Measurement of Arcs: Degrees

Arcs may be measured in two different ways: by degree measure or by linear units.

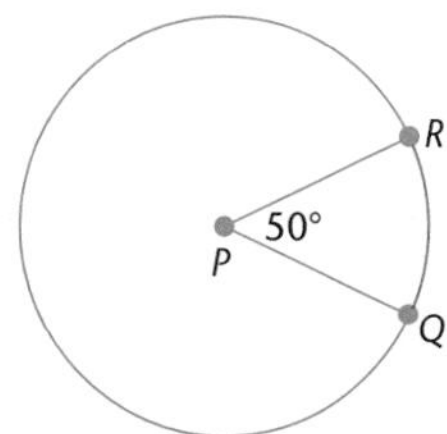

To find the degree measure of arc $\widehat{RQ}$, draw a central angle whose rays pass through the endpoints of the arc and whose vertex is at the center of the circle. Measure the angle.

Imagine rotating an endpoint of the arc, such as Q in the illustration, about the center of the circle until it coincides with the other endpoint R. The amount of rotation, which is the measure of the central angle, is the degree measure of the arc:

$$m\widehat{RQ} = 50°.$$

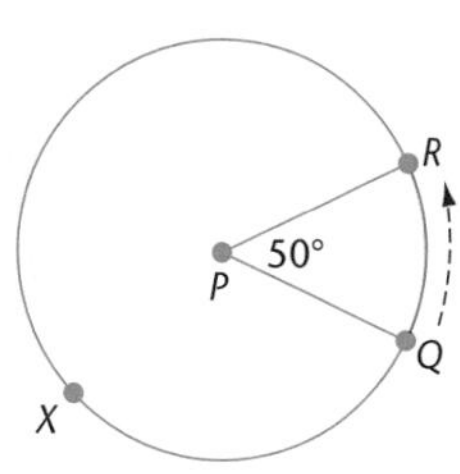

A complete rotation is 360°, so the measure of the major arc is

$$m\widehat{RXQ} = 360° - m\widehat{RQ}$$
$$= 360° - 50° = 310°.$$

Measurement of Arcs: Length

Notice that arc $\widehat{XY}$ is part of the circumference of circle P. Since the measure of the central angle is 90° (one-fourth of a complete rotation), the measure of the arc is one-fourth the circumference C of the circle ($C = 2\pi r$):

X
6
90°
P
Y

$$\text{Length of } XY = \frac{90°}{360°}(2)(\pi)(6)$$
$$= \frac{1}{4}(2)(\pi)(6) = 3\pi \text{ units.}$$

In general,

$$\text{arc length} = \frac{A}{360°} \times \text{circumference},$$

where A is the measure of the central angle.

Is it possible for two arcs to have the same degree measures but different lengths? Explain why or why not.

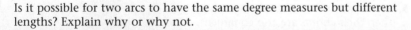

Using Review Have students list all the mathematical terms used or introduced in this lesson. Then, for each term, they should do the following:

1. Write a definition of the term.
2. Make several drawings illustrating the term.
3. Write a sentence using the term's geometry meaning.

Cooperative Learning

The explorations in this lesson are particularly appropriate for cooperative learning groups. Tell students to find more than one way to achieve each construction. The regular hexagon, in particular, can be created in many different ways.

Math Connection Algebra

An alternative way to develop the arc length formula is to use ratio and proportion:

$$\frac{\text{arc length}}{\text{circumference of circle}} = \frac{\text{degrees in central } \angle}{360°}$$

$$\frac{\text{arc length}}{2\pi r} = \frac{A}{360°}$$

$$\text{arc length} = \frac{A}{360°} \times 2\pi r$$

Tell students that arc lengths are often expressed using the symbol π. For example, a quarter circle with a radius of 3 cm has an arc length of $\frac{9\pi}{4}$ cm.

CRITICAL Thinking

Yes. Consider arcs in circles with different radii but with the same central angles. Arcs with the same degree measure are longer in the larger circle.

Exploration 3 Notes

To create a slightly more challenging version of this exploration have the drawing show the two chords intersecting. If students are doing the activity in small groups, you might have half the team use the drawing in the text; while the other half uses a drawing in which the chords intersect.

ASSESS

Selected Answers

Exercises 7, 9, 15, 17, 19, 21, 25, 27, 29, 31, 33, 35, and 37

Assignment Guide

Core 1–14, 16–32 even, 38

Core Plus 1–5, 6–14 even, 15–32, 34, 36–40

Technology

Geometry graphics software will be helpful for the figures in Exercises 13 and 14.

Error Analysis

For Exercises 24–28, have students make a careful drawing for each problem. This will help them visualize the relationships.

Exploration 3 *Chords and Arcs Theorem*

You will need

Compass and straightedge

1. In the figure, chords $\overline{AB}$ and $\overline{CD}$ are congruent. Do you think that arcs $\overset{\frown}{AB}$ and $\overset{\frown}{CD}$ have equal measures? To find the answer, copy the figure and construct central angles $\angle APB$ and $\angle CPD$.

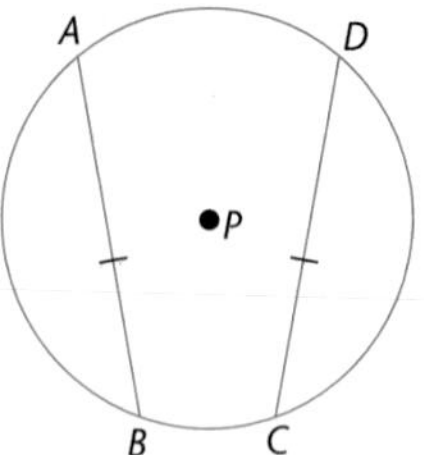

2. Prove that triangles $\triangle APB$ and $\triangle CPD$ are congruent. What can you conclude about central angles $\angle APB$ and $\angle CPD$? about arcs $\overset{\frown}{AB}$ and $\overset{\frown}{CD}$?

3. In a circle, if two chords are congruent, what may you conclude about their arcs? State your conclusion as a theorem. ❖

EXERCISES & PROBLEMS

Communicate

In Exercises 1–5, classify each statement as true or false and explain your reasoning.

1. Every diameter of a circle is also a chord of the circle.
2. Every radius of a circle is also a chord of the circle.
3. Every chord of the circle contains exactly two points of the circle.
4. If two chords of a circle are congruent, then their arcs are congruent.
5. If two arcs of a circle are congruent, then their chords are also congruent.

Practice & Apply

Use the figure for Exercises 6–12.

6. Name the center of the circle. *P*

7. Name a radius of the circle. $\overline{PA}$; $\overline{PB}$; $\overline{PC}$ are three

8. Name a chord of the circle. $\overline{AC}$; $\overline{ED}$

9. Name a diameter of the circle. $\overline{AC}$

10. Name a central angle of the circle. ∠*APB*; ∠*BPC*; ∠*APC*

11. Name a minor arc and a major arc of the circle.

12. Describe the conditions necessary for two arcs to be congruent.

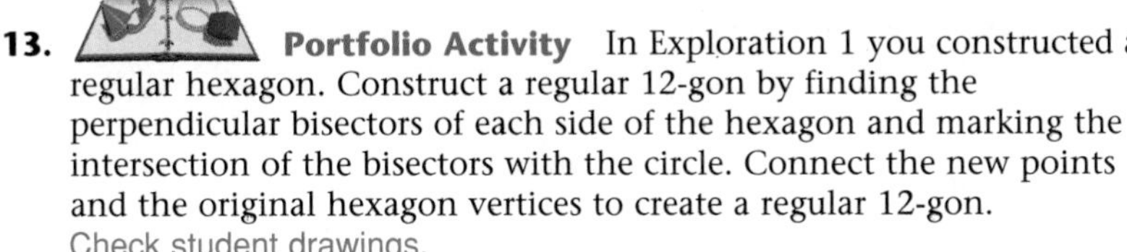

13. **Portfolio Activity** In Exploration 1 you constructed a regular hexagon. Construct a regular 12-gon by finding the perpendicular bisectors of each side of the hexagon and marking the intersection of the bisectors with the circle. Connect the new points and the original hexagon vertices to create a regular 12-gon. Check student drawings.

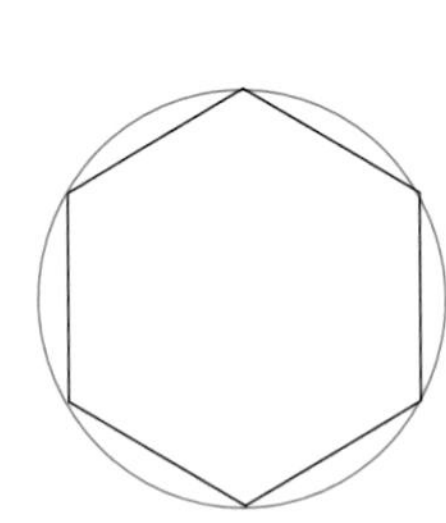

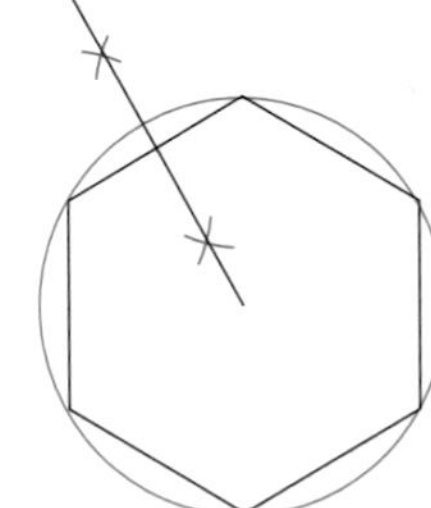

14. **Portfolio Activity** Use the 12-gon from Exercise 13 to construct a 12-petal circle flower. Check student drawings.

Find the measure of each arc using the central angle measures given on circle *Q*.

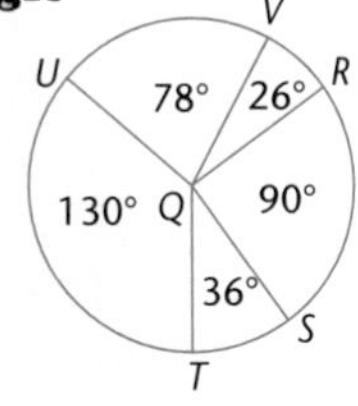

15. $\overset{\frown}{RT}$ 126° **16.** $\overset{\frown}{UR}$ 104° **17.** $\overset{\frown}{VS}$ 116°

18. $\overset{\frown}{VTR}$ 334° **19.** $\overset{\frown}{US}$ 166° **20.** $\overset{\frown}{SUV}$ 244°

21. **Civil Engineering** An engineer needs to calculate the length of a circular section of road. Use the measurements given to determine the length. 90 ft

alternative ASSESSMENT

Performance Assessment

Students should create at least two different ways to construct a regular hexagon. They may use any tools they wish—compass, straightedge, ruler, protractor, paper folding, or geometry graphics software. For each construction, students should describe the steps in paragraph or outline form. Written work should include terms for parts of circles such as center, radius, chord, arc, diameter, and central angle.

11. Sample answers: $\overset{\frown}{AB}$, $\overset{\frown}{BC}$, $\overset{\frown}{CD}$, $\overset{\frown}{DE}$, $\overset{\frown}{BE}$, $\overset{\frown}{CE}$, $\overset{\frown}{AEB}$, $\overset{\frown}{BCA}$, $\overset{\frown}{DAB}$.

12. Their central angles are congruent and their intercepted chords are congruent.

22. Freshmen: 104.5°

Sophomores: 87.1°

Juniors: 92.9°

Seniors: 75.5°

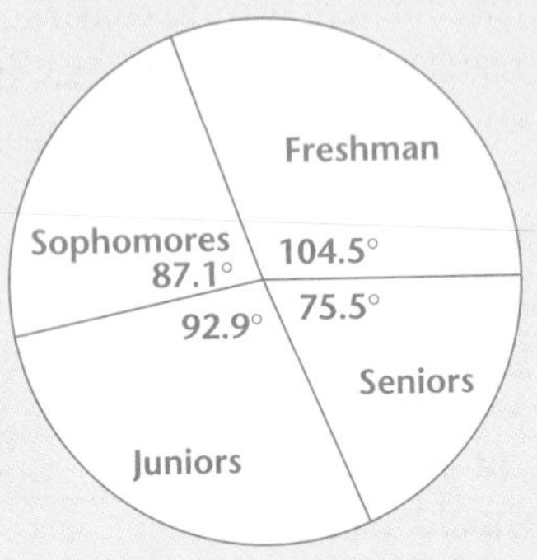

The answer to Exercise 23 can be found in Additional Answers beginning on page 727.

22. **Statistics** At Smith High School, 450 students are freshmen, 375 students are sophomores, 400 students are juniors, and 325 students are seniors. Create a pie chart showing the distribution of the students. First, find the percentage for each class of the total student body. Multiply each percentage by 360°, and then find the measure of the central angle.

23. In a circle, if two arcs are congruent, what can you conclude about their chords? State your conclusion as a theorem and prove it.

Algebra For each arc length measure and radius given below, find the measure of the central angle.

24. 14π; $r = 70$ 36°

25. 20π; $r = 100$ 36°

26. 3π; $r = 15$ 36°

27. 5π; $r = 25$ 36°

28. In $\odot P$, if $m\angle APB = 43°$ and $AB = 5$ cm, find XY. 5 cm

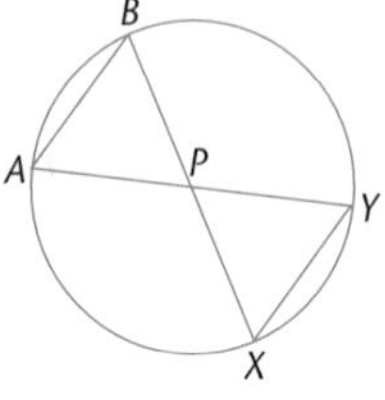

Use circle X, with $\angle MXN \cong \angle NXO$, for Exercises 29–31.

29. Explain why $\overline{MN} \cong \overline{NO}$.

30. Explain why $\overline{MX} \cong \overline{NX}$ and $\overline{NX} \cong \overline{OX}$.

31. Explain why $\triangle MNX \cong \triangle NOX$.

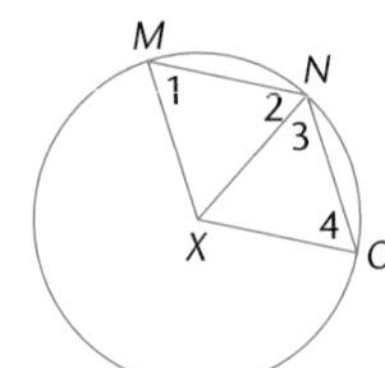

32. **Portfolio Activity** Look in magazines for examples of circles used as symbols or product logos to add to your portfolio. Write a short description that identifies a unique feature of each symbol. Check student work.

29. If central angles of a circle are congruent, their chords are congruent.

30. Definition of a circle. All radii of a circle are congruent.

31. Their central angles and arcs are congruent.

Practice Master

NAME ______ CLASS ______ DATE ______

Practice & Apply

9.1 Exploring Chords and Arcs

Exercises 1–3 refer to the diagram of circle *O*.

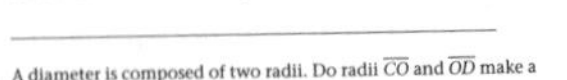

1. Is $\overline{OM}$ a radius of circle *O*? Explain.

2. A diameter is composed of two radii. Do radii $\overline{CO}$ and $\overline{OD}$ make a diameter of circle *O*? Explain.

3. Using 3 letters, name a major arc of circle *O*. ______

4. Use the construction for a regular hexagon to construct an equilateral triangle that will be inscribed in circle *P*.

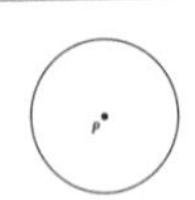

Exercises 5–11 refer to circle *O*, where $\overset{\frown}{AQ} \cong \overset{\frown}{QB}$.

5. Explain why $\overline{AQ} \cong \overline{QB}$. ______

6. Explain why $\angle AOQ \cong \angle BOQ$.

7. Explain why $\overline{OA} \cong \overline{OB}$. ______

8. Are the three conditions established in Exercises 5–7 sufficient to conclude that $\Delta AOQ \cong \Delta BOQ$? Explain.

9. Explain why the third pair of sides of $\angle AOQ$ and $\angle BOQ$ are congruent.

10. Explain why now you can say that $\Delta AOQ \cong \Delta BOQ$.

11. Explain why $\angle QAO \cong \angle QBO$.

80 Practice & Apply HRW Geometry

Look Back

33. What effect does doubling the radius of a sphere have on the volume? **[Lessons 7.1, 7.6]** The volume is multiplied by 8.

Algebra Find the volume and surface area for each solid given below. **[Lessons 7.2, 7.6]**

34.

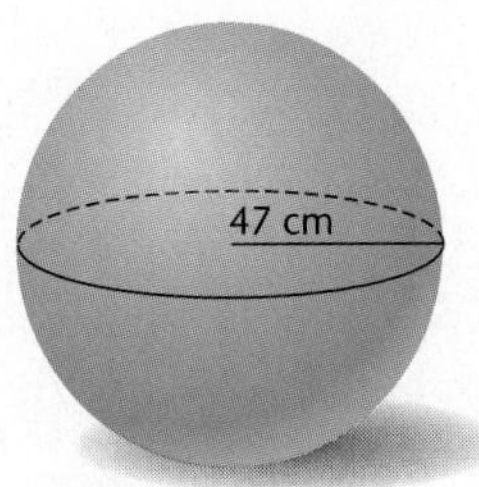

SA = 27,759.11 cm^2 V = 434,892.77 cm^3

35.

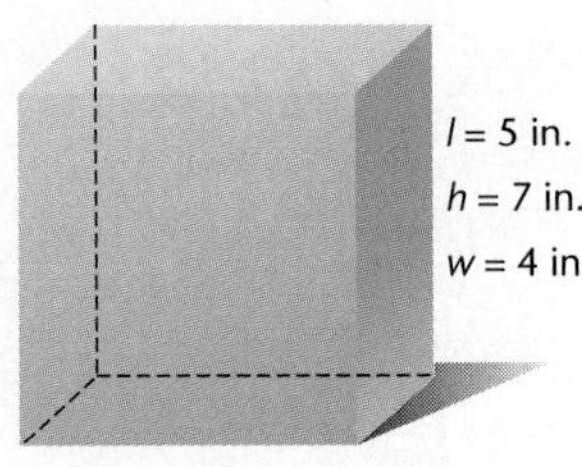

SA = 166 in.^{2}; V = 140 in.^{3}

Are the triangles similar? Why or why not? [Lessons 8.3, 8.4]

36.

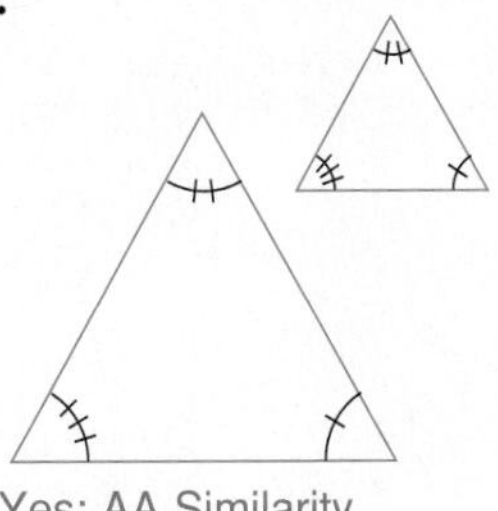

Yes; AA Similarity

37.

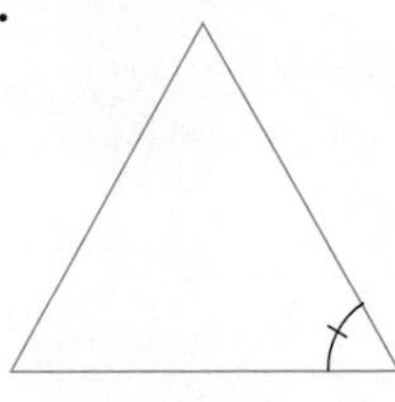

No; not enough information

38.

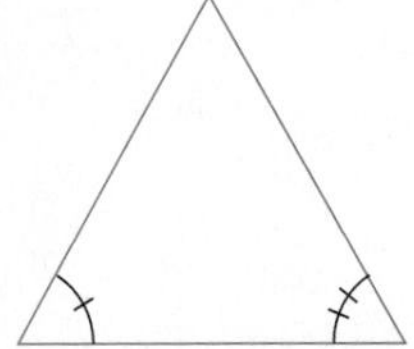

Yes; AA Similarity

Look Beyond

39. Imagine a line and a circle in the same plane. Describe three ways that the line and circle can intersect. The line and the circle can intersect in 0 points, 1 point, or in 2 points.

40. A line that touches a circle in only one point is called a tangent. The perpendicular to the tangent at the point of intersection passes through which point? The center of the circle.

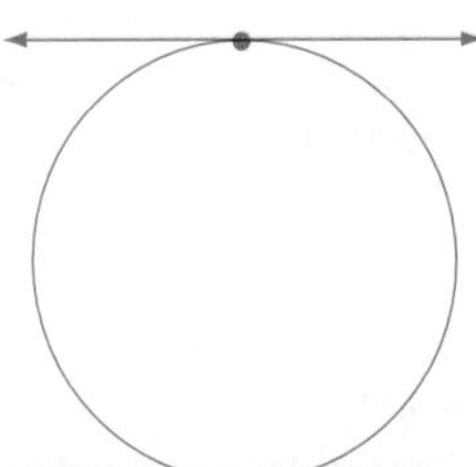

Look Beyond

Exercises 39 and 40 provide students with an introduction to tangents of circles. If a tangent to a circle is drawn correctly, it will be perpendicular to the radius at the point of tangency.

Technology Master

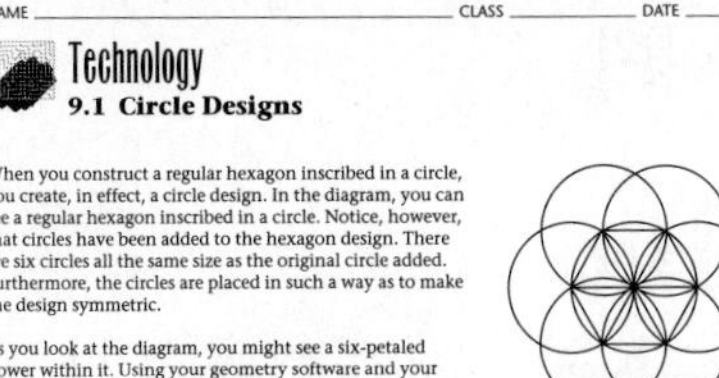

NAME ______________ CLASS ________ DATE ________

Technology
9.1 Circle Designs

When you construct a regular hexagon inscribed in a circle, you create, in effect, a circle design. In the diagram, you can see a regular hexagon inscribed in a circle. Notice, however, that circles have been added to the hexagon design. There are six circles all the same size as the original circle added. Furthermore, the circles are placed in such a way as to make the design symmetric.

As you look at the diagram, you might see a six-petaled flower within it. Using your geometry software and your creative talents, you too can produce circle designs.

Use geometry software in the following exercises.

1. Use the steps outlined in Lesson 9.1 to create a regular hexagon inscribed in a circle. Use MEASURE to simulate the use of a compass to assure that the six line segments are equal to the radius of your circle.
2. To your diagram from Exercise 1, add six circles whose radii are one-half that of your original circle to form a symmetric design. Place these circles on your diagram in such a way that they all pass through the center of the original circle.
3. To your diagram from Exercise 1, add six circles whose radii are one-half that of your original circle to form a symmetric design. Place these circles on your diagram in such a way that they are outside the original circle and touch it in just one point.
4. Inscribe a square in a circle, like the one shown. (The dashed line segments will help you locate the vertices of the square.)

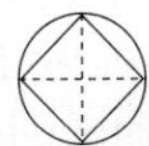

In Exercises 5 and 6, use the sketch you made in Exercise 4.

5. Create a four-petaled flower in your design. The petals should all be the same size.
6. Create an eight-petaled flower. Make your petals alternate between large and small.

92 Technology HRW Geometry

PREPARE

Objectives

- Define *secant, tangent,* and *point of tangency.*
- Develop and use the Tangent Theorem.

- Practice Master 9.2
- Enrichment Master 9.2
- Technology Master 9.2
- Lesson Activity Master 9.2
- Quiz 9.2
- Spanish Resources 9.2

Assessing Prior Knowledge

How many points are described in each case stated below?

1. the intersection of a circle and one of its chords

 [2]

2. the intersection of a circle and one of its radii

 [1]

3. the intersection of a circle and one of its central angles

 [2]

TEACH

You may wish to review or introduce the definition of *inertia.* The inertia of a body is its tendency to preserve its state of rest or uniform motion in a straight line. The direction of motion is a straight line unless other forces act upon it. So, an object moving in a circle at the end of a string will start to move in a straight line if the string breaks. This line of motion is a tangent to the circle.

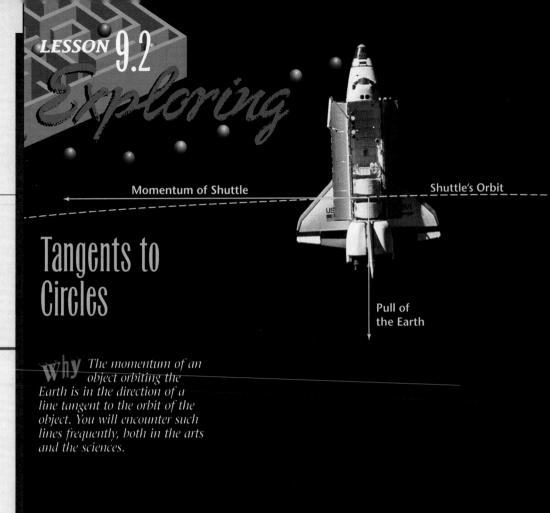

LESSON 9.2 *Exploring*

Momentum of Shuttle

Shuttle's Orbit

Pull of the Earth

Tangents to Circles

why *The momentum of an object orbiting the Earth is in the direction of a line tangent to the orbit of the object. You will encounter such lines frequently, both in the arts and the sciences.*

Secants and Tangents

A line can intersect a circle in three ways.

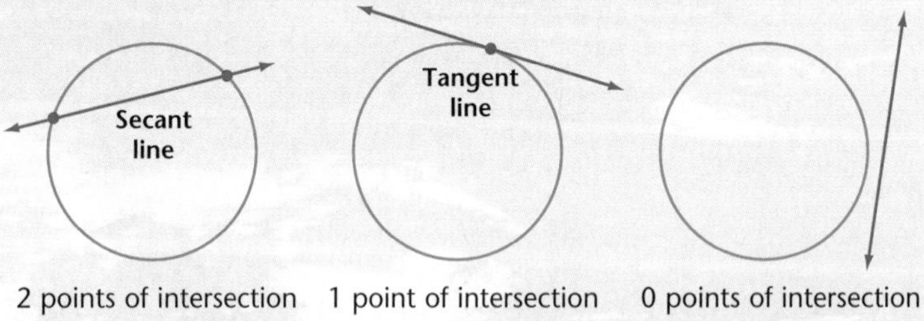

Secant line — 2 points of intersection

Tangent line — 1 point of intersection

0 points of intersection

Hands-On Strategies

Circular geoboards can be used for the explorations and some of the exercises in this lesson. The circular geoboards can be used for modeling problems and measuring angles.

SECANTS AND TANGENTS

A **secant** to a circle is a line that intersects a circle at two points. A **tangent** to a circle is a line that intersects a circle at just one point, which is known as the **point of tangency.** **9.2.1**

The word *tangent* comes from the Latin word meaning "to touch." The word *secant* comes from the Latin word meaning "to cut." Why are these words appropriate names for the lines they describe?

Exploration 1 Radii Perpendicular to Chords

Technology

Geometry Graphics

You will need

Geometry technology or
Ruler, compass, and protractor

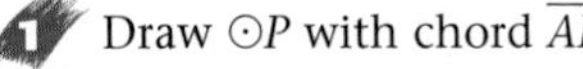

1. Draw $\odot P$ with chord $\overline{AB}$.

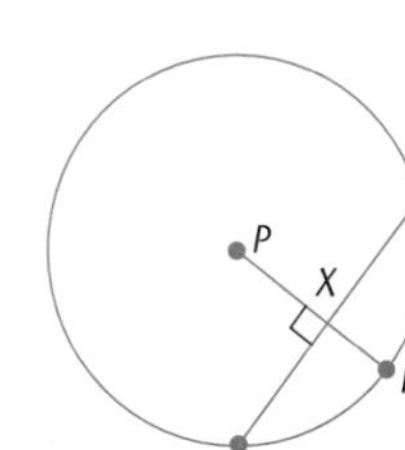

2. Construct a radius that is perpendicular to the chord. Label the point of intersection X.

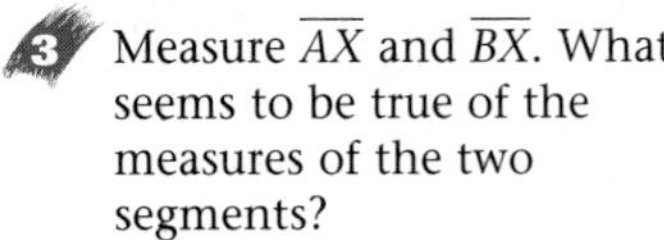

3. Measure $\overline{AX}$ and $\overline{BX}$. What seems to be true of the measures of the two segments?

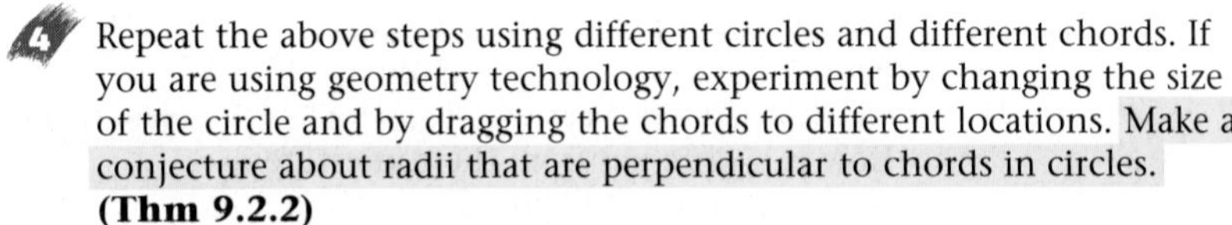

4. Repeat the above steps using different circles and different chords. If you are using geometry technology, experiment by changing the size of the circle and by dragging the chords to different locations. Make a conjecture about radii that are perpendicular to chords in circles. **(Thm 9.2.2)**

5. Draw segments $\overline{PA}$ and $\overline{PB}$ in one of your circles. Using this diagram, write out a paragraph proof of your conjecture. ❖

EXTENSION

ALGEBRA *Connection*

$\odot P$ has a radius of 5 inches and PX is 3 inches. $\overline{PR}$ is perpendicular to $\overline{AB}$ at point X. Find AB.

By the "Pythagorean" Right-Triangle Theorem,

$$(AX)^2 + 3^2 = 5^2$$
$$(AX)^2 = 5^2 - 3^2$$
$$(AX)^2 = 16$$
$$AX = 4.$$

By the conjecture you proved in Exploration 1, $\overline{PR}$ bisects $\overline{AB}$, so $BX = 4$. Therefore, $AB = AX + BX = 4 + 4 = 8$. ❖

interdisciplinary CONNECTION **Physics** The inertial tendency of an object to travel in a straight line is demonstrated by a stone in a sling. The sling is swung in a circular motion and then released. The stone travels in a line tangent to the point of release.

ongoing ASSESSMENT

A secant to a circle cuts through the circle. A tangent touches the circle once.

Use Transparency 75

Exploration 1 Notes

Students discover that a radius perpendicular to a chord bisects that chord. After they have completed the exploration, discuss whether or not conjectures remain true if the chord is a diameter.

ongoing ASSESSMENT

4. In a circle, if a radius is perpendicular to a chord of the circle, then the radius bisects the chord.

5. Because $\overline{PX} \perp \overline{AB}$, $\angle AXP \cong \angle BXP$. $\overline{PA} \cong \overline{PB}$ because the radii of a circle are congruent. $\overline{PX} \cong \overline{PX}$ by the Reflexive Property. $\triangle APX \cong \triangle BPX$ by HL.

Math Connection Algebra

Help students understand the "work backwards" reasoning used in the solution to this problem: To find the length AB, you could first find AX. To find AX, you can draw the right triangle PAX. To find the hypotenuse of $\triangle PAX$, you use the fact that the hypotenuse equals the radius of circle P.

Exploration 2 Notes

In this exploration, students discover that a tangent to a circle is perpendicular to the radius at the point of tangency.

Cooperative Learning

Students should work in small groups so that several different cases for each exploration can be constructed, analyzed, and discussed.

ongoing ASSESSMENT

4. **If a radius is drawn to the point of tangency, then the radius is perpendicular to that line.**

CRITICAL Thinking

A and *B* get closer together, while remaining equidistant from *X*. When *X* reaches *R*, the points *A*, *B*, and *R* coincide and the line is tangent to the circle. Conjecture: If a line is perpendicular to a radius of a circle at its endpoint on the circle, then the line is tangent to the circle. This is the converse of the previous conjecture.

TEACHING tip

Technology Have students use geometry graphics software to create the figure from the proof of the Tangent Theorem. The figure should be constructed so that students can drag on point *Q* to see how the length of segment *OQ* changes. The lengths *OP* and *OQ* can be measured using a "length" command or tool.

Exploration 2 Radii and Tangents

You will need
Geometry technology or
Ruler, compass, and protractor

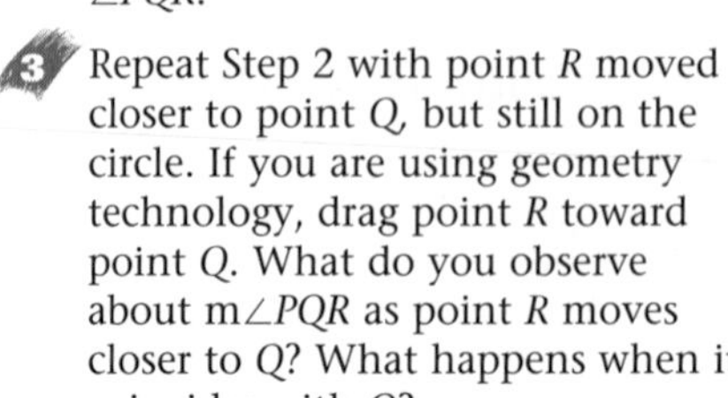

1. Draw $\odot P$ with radius $\overline{PQ}$.
2. Locate a point *R* on the circle and draw a line $\overleftrightarrow{QR}$. Measure $\angle PQR$.
3. Repeat Step 2 with point *R* moved closer to point *Q*, but still on the circle. If you are using geometry technology, drag point *R* toward point *Q*. What do you observe about $m\angle PQR$ as point *R* moves closer to *Q*? What happens when it coincides with *Q*?

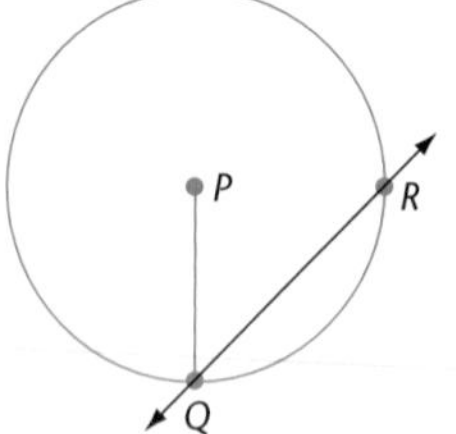

4. Make a conjecture about the relationship between a tangent to a circle and a radius drawn to the point of tangency. ❖

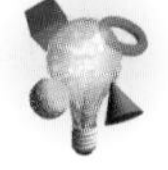

In the diagram, which is like the one from Exploration 1, imagine moving point *X* toward point *R*. What happens to intersection points *A* and *B* as *X* gets closer and closer to *R*? What happens when *X* touches *R*? What conjecture does this suggest to you? How is it different from the conjecture you made in Exploration 2?

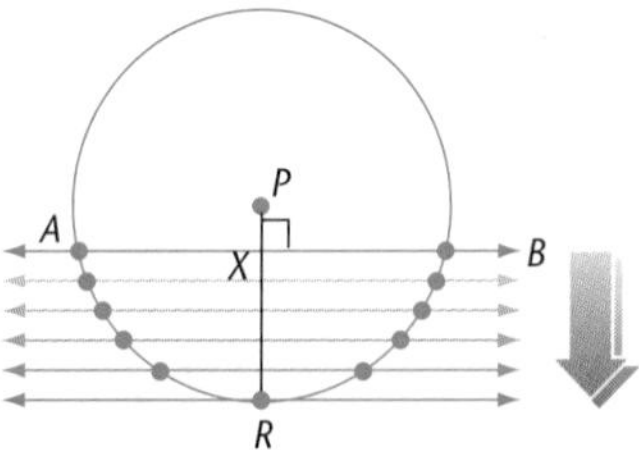

An Unusual Proof

The following proof is related to the conjecture you made in Exploration 2. It uses the fact that the hypotenuse is the longest side of a triangle. Its method is unlike the proofs you have studied so far in this book.

Given: Point *P* is on $\odot O$, and $\overline{OP}$ is perpendicular to $\overline{AB}$.

Prove: $\overline{AB}$ is tangent to $\odot O$ at *P*.

Proof: Choose any point on $\overline{AB}$ other than *P* and label it *Q*. Draw right triangle *OPQ*. Since $\overline{OQ}$ is the hypotenuse of the triangle, it is longer than $\overline{PO}$, which is a radius of the circle. Therefore, the point *Q* does not lie on the circle. This is true for all points on $\overline{AB}$ except point *P*, so $\overline{AB}$ touches the circle just at the one point. By definition, $\overline{AB}$ is tangent to $\odot O$ at *P*.

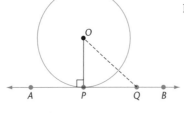

The result proven here can be stated as a theorem.

ENRICHMENT Line designs, sometimes called *string art* because they can be made from string, use the idea of the tangent to a circle or other curve. A set of tangent lines can be drawn so that they suggest the shape of a curved.

INCLUSION strategies **English Language Development** Some students may need extra help with the new vocabulary introduced in this chapter. Draw a circle on the chalkboard and have students demonstrate their understanding of various terms by referring to or adding to the drawing.

TANGENT THEOREM
If a line is perpendicular to a radius of a circle at its endpoint, then the line is tangent to the circle. **9.2.3**

The converse of the theorem, which is also true, will be proven in Look Beyond, Exercises 22–25.

APPLICATION

An artist wants to draw the largest circle that will fit into a square.

1. She connects the midpoints of each of the sides of the square by segments parallel to the sides.
2. She measures the distance from the intersection P of the new segments to the sides of the square. Using this distance as the setting of her compass, she draws a circle with its center at the point P. ❖

CRITICAL Thinking How can you prove that the artist's method gives the desired result? How can you show that no part of the circle lies outside of the square?

EXERCISES & PROBLEMS

Communicate

1. Explain the three ways in which a line and a circle intersect in a plane.
2. Explain how a secant intersects a circle.
3. Explain how a tangent intersects a circle.
4. Explain what "point of tangency" is.

Practice & Apply

Use ⊙R, with $\overline{RY} \perp \overline{XZ}$ at W, for Exercises 5–7.

5. $\overline{XW} \cong$ __?__. $\overline{WZ}$
6. If $RY = 7$ and $RW = 2$, what is the length of $\overline{XW}$? of $\overline{WZ}$? $XW = WZ = 6.71$ units
7. If $RY = 3$ and $RW = 2$, what is the length of $\overline{XW}$? of $\overline{WZ}$? $XW = WZ = 2.24$ units

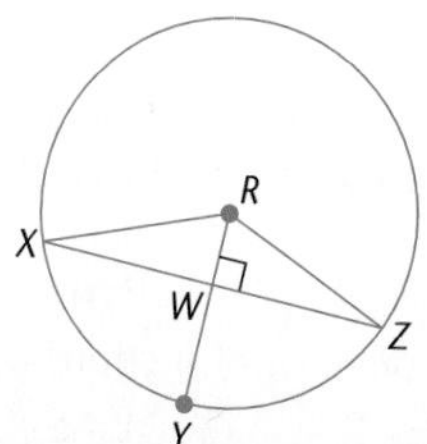

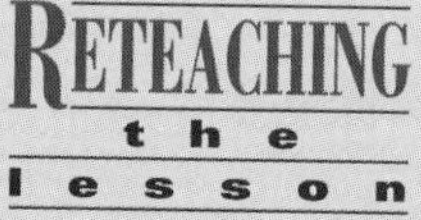

Cooperative Learning Have students work in pairs. Have them copy the statement of the Tangent Theorem as given in their text and then create five different ways of stating the theorem. For example, they might begin, "If a line passing through the endpoint of a radius . . . " or "A radius of a circle is perpendicular to a line . . . " Students must be prepared to defend, orally or in writing, the fact that their new statements have exactly the same meaning as the statement given in the text.

CRITICAL Thinking

The segment from the point of intersection is perpendicular to a side of the square. The perpendicular distance is the shortest distance from a point to a line. Any other length would be the hypotenuse of a right triangle with the perpendicular as one leg. The hypotenuse of a right triangle is always longer than a leg of the triangle. So, any other point on the square is farther from the point of intersection and thus outside the circle.

ASSESS

Selected Answers

Exercises 5, 7, 11, 13, 15, 19, and 21

Assignment Guide

Core 1–11, 18–21

Core Plus 1–25

Technology

Geometry graphics software can be used for the proof in Exercise 17. For example, students using *Cabri* should construct a circle and any chord on that circle. Then they construct the perpendicular bisector of the chord. They should select the center of the circle and the perpendicular bisector and use the *Member* tool from the *Check Property* toolbox to see whether the center of the circle is on the perpendicular bisector.

Error Analysis

Be sure students know that they need to convert between miles and feet in Exercise 8.

alternative ASSESSMENT

Performance Assessment

Have students work together to create a clover-like figure.

1. Draw any two perpendicular segments. Mark their point of intersection O.
2. Construct four circles so each is (a) centered at an endpoint of one segment and (b) tangent to the other segment at point O.
3. Describe how your figure would change if (a) point O bisects both segments, (b) point O coincides with the endpoint of one of the segments.

Use Transparency 76

Practice Master

NAME ______________ CLASS ________ DATE ________

Practice & Apply
9.2 Exploring Tangents to Circles

1. In the diagram of circle O, diameter $\overline{AB} \perp$ chord $\overline{CD}$ at E. If $CD = 8$ and $OE = 3$, find the length of the radius.

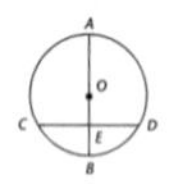

2. In the diagram of circle P, the radius is 10, and the distance from center P to chord $\overline{RS}$ is 6. Find RS.

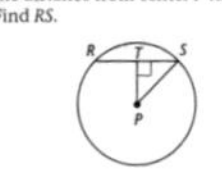

In Exercises 3–6, refer to the diagram in which $\overrightarrow{PA}$ and $\overrightarrow{PB}$ are tangent to circle O. Radii $\overline{OA}$ and $\overline{OB}$ are drawn to the points of tangency, and $\overline{OP}$ is drawn.

3. Explain why $\overline{OA} \perp \overrightarrow{PA}$ and $\overline{OB} \perp \overrightarrow{PB}$.
4. Explain why $\Delta AOP \cong \Delta BOP$.
5. Explain why $\overline{PA} \cong \overline{PB}$.
6. State the results of Exercises 3–5 as a theorem.

Use the theorem you wrote in Exercise 6 to answer Exercises 7–8.

7. From A, $\overline{AC}$ and $\overline{AB}$ are tangent to the circle and from E, $\overline{EC}$ and $\overline{ED}$ are tangents. Find AE.

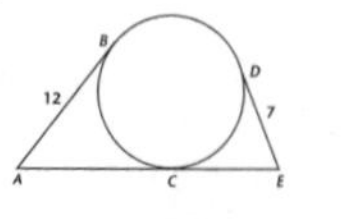

8. At A, B, and C, the circle is tangent to each side of the triangle drawn around it. Find the perimeter of the triangle.

HRW Geometry — Practice & Apply 81

8. **Communications** A radio station installs a VHF radio tower that stands 1500 feet tall. What is the maximum effective signal range for the tower? The diagram below suggests a way to utilize tangents to solve the problem. Use the "Pythagorean" Right-Triangle Theorem to find d. 47.67 mi

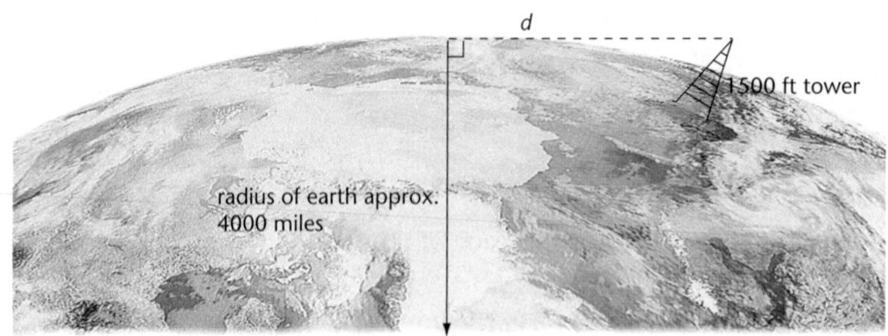

9. How many tangent lines can a circle have? Explain your reasoning.
10. How many lines are tangent to a circle at a given point on the circle? Explain your reasoning.
11. **Space Flight** The space shuttle orbits at 155 miles above the Earth. If the diameter of the Earth is approximately 8000 miles, how far is it from the shuttle to the horizon? 1124.29 mi

Algebra Use the diagram to find the indicated lengths. Line r is tangent to $\odot B$ at T. $BT = 2$, $BS = 1$, and $WT = 5$.

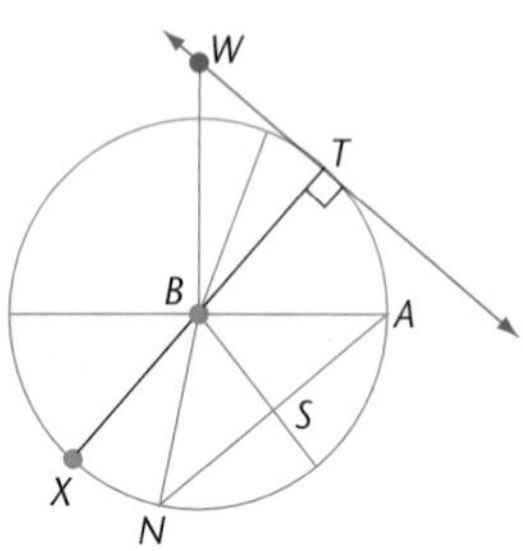

12. $BA = ?$ 2
13. $SA = ?$ 1.73
14. $SN = ?$ 1.73
15. $BW = ?$ 5.39
16. $XT = ?$ 3.46
17. In Exploration 1 you proved a conjecture about radii that are perpendicular to a chord. Prove the following converse.

THEOREM
The perpendicular bisector of a chord passes through the center of the circle. **9.2.4**

9. A circle could have an infinite number of tangent lines, one at each of the infinite number of points on the circle.

10. Exactly one. Any other line would not be perpendicular to the radius drawn to the point of tangency and would go through the circle.

The answer to Exercise 17 can be found in Additional Answers beginning on page 727.

Look Back

18. **Algebra** A triangle has a perimeter of 24 cm and an area of 24 sq cm. What is the perimeter and area of a larger similar triangle if the scale factor is $\frac{2}{1}$? **[Lessons 8.1, 8.6]** A = 96 cm^2; P = 48 cm

19. **Algebra** A rectangle has a perimeter of 22 ft and an area of 22 sq ft. What is the perimeter and area of a larger similar rectangle if the scale factor is $\frac{8}{3}$? **[Lessons 8.1, 8.6]** P = $58\frac{2}{3}$ ft; A = $156\frac{4}{9}$ ft^2

20. **Algebra** A rectangular prism has dimensions $l = 12$ in., $w = 8$ in., and $h = 15$ in. What is the volume of a larger similar rectangular prism if the scale factor is $\frac{5}{3}$? **[Lessons 8.1, 8.6]** V = $6666\frac{2}{3}$ in.^{3}

21. **Algebra** A cylinder has a radius of 4 ft and a height of 27 ft. What is the surface area of a larger similar cylinder if the scale factor is $\frac{4}{1}$? **[Lessons 8.1, 8.6]** SA = 12,465.84 ft^2

Look Beyond

Answer the questions below to prove the following theorem.

> **THE CONVERSE OF THE TANGENT THEOREM**
> If a line is tangent to a circle, then it is perpendicular to a radius of the circle at the point of tangency. **9.2.5**

22. Suppose that the theorem is *false.* That is, suppose that m is tangent to circle O at point A, but m is *not* perpendicular to $\overline{OA}$. If this is true, then there *is* some segment, different from $\overline{OA}$, that is perpendicular to m. Call that segment $\overline{OB}$. Then $\triangle OBA$ is a right triangle. (What is the hypotenuse of $\triangle OBA$?)

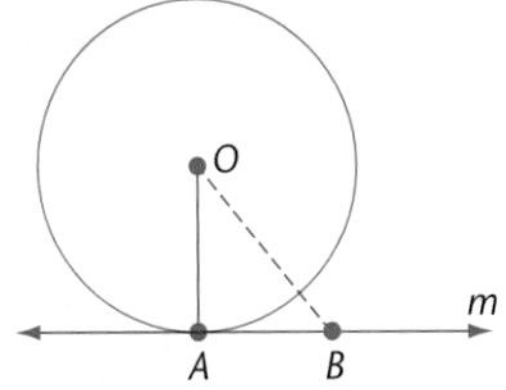

Which segment would be longer, $\overline{OA}$ or $\overline{OB}$?

23. Point B must be in the exterior of the circle, since m is a tangent line. What does this imply about the relative lengths of $\overline{OA}$ and $\overline{OB}$? Explain your answer.

24. Compare your answers to Exercises 22 and 23. What do you observe?

25. When an assumption leads to a contradiction, then the assumption must be rejected. This is the basis for a type of proof known as an **indirect proof**. Explain how the argument above leads to the desired conclusion.

Look Beyond

Exercises 22–25 introduce students to the concept of indirect proof. Have students work in pairs or small groups to complete the exercises.

Technology Master

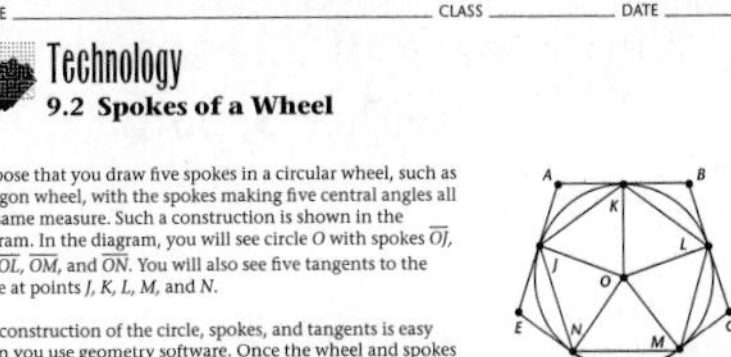
NAME ________ CLASS ________ DATE ________

Technology
9.2 Spokes of a Wheel

Suppose that you draw five spokes in a circular wheel, such as a wagon wheel, with the spokes making five central angles all the same measure. Such a construction is shown in the diagram. In the diagram, you will see circle O with spokes $\overline{OJ}$, $\overline{OK}$, $\overline{OL}$, $\overline{OM}$, and $\overline{ON}$. You will also see five tangents to the circle at points J, K, L, M, and N.

The construction of the circle, spokes, and tangents is easy when you use geometry software. Once the wheel and spokes are drawn, you can begin to explore polygons that fit snugly inside and outside the circle.

Use geometry software in the following exercises.

1. Create a circle and place three points on it. Sketch three spokes. MEASURE the angles formed. Drag the points around the circle until the measures of the three central angles are equal.

2. Construct three tangents to the circle at the endpoints of the spokes.(Your software may have a tool called PERPENDICULAR AT.)

3. Sketch segments that join the three points on the circle. Describe the polygon formed. Mark the three points where the tangents intersect. What polygon do they form?

____

4. Use the RATIO tool to find the ratios of the corresponding sides of the two polygons you identified in Exercise 3. How are the polygons related?

____

5. Repeat Exercises 1–4 by starting with a circle having five equally spaced spokes. Do your conclusions still hold true?

____

6. The diagrams you drew in Exercises 1–5 suggest a sequence of geometric figures: polygon, circle, polygon, ... Do you think the process can be continued indefinitely?

____

7. Use your sketch in Exercise 5 to show the next two steps in the sequence.

____

HRW Geometry | Technology 93

22. OA is the hypotenuse of $\triangle OBA$, so $OA > OB$

23. OA is a radius, so any point in the exterior of the circle is farther from the center. Therefore $OA < OB$.

24. The answer to Ex 22 is opposite to that of Ex 23.

25. The argument leads to $OA > OB$ and $OA < OB$, which is a contradiction. Therefore the *only* perpendicular segment to the tangent line is OA.

PREPARE

Objectives

- Define *inscribed angle* and *intercepted arc.*
- Develop and use the Inscribed Angle Theorem and Corollary.

RESOURCES

- Practice Master 9.3
- Enrichment Master 9.3
- Technology Master 9.3
- Lesson Activity Master 9.3
- Quiz 9.3
- Spanish Resources 9.3

Assessing Prior Knowledge

A circle and an angle are drawn in the same plane. The vertex of the angle is on the circle.

1. Find and sketch all the possible ways the two figures can be arranged. [**Possible arrangements: Neither, both, or one ray of the angle passes through the circle.**]
2. For each arrangement, give the number of intersection points. [**Neither, 3; both, 2; one; 1**]

TEACH

The construction described in the text is explained in the application found on page 488. Ask students to brainstorm various ways of finding the center of a circle.

LESSON 9.3

Exploring Inscribed Angles and Arcs

Why *You can find the center of a circular object, such as the one shown here, by using a carpenter's square. The idea involves inscribed angles.*

A carpenter's square can be used to find the center of a circle.

Inscribed Angles

An **inscribed angle** is a special type of angle whose vertex lies on the circle and whose rays each extend from the vertex and intersect the circle again. The arc from one of these intersection points to the other is called the intercepted arc of the angle. (The vertex of the angle does not lie on the intercepted arc.)

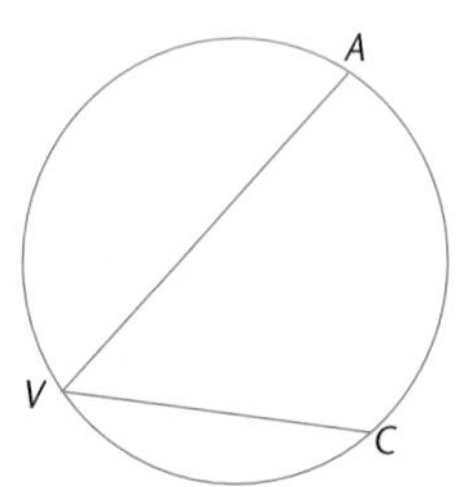

$\angle AVC$ is inscribed in $\overset{\frown}{AVC}$ and is said to ***intercept*** *(cut off) $\overset{\frown}{AC}$. $\overset{\frown}{AC}$ is the* ***intercepted arc*** *of $\angle AVC$.*

ALTERNATIVE teaching strategy

Hands-On Strategies

Many students will benefit from constructing the figures given in this lesson and creating different cases for each exploration. Students can use a compass and straightedge, geometry graphics software, or paper folding.

A special relationship exists between the measure of an inscribed angle and the degree measure of its intercepted arc.

Exploration 1 Intercepted Arcs

You will need
Geometry technology or
Ruler, compass, and protractor

Geometry Graphics

1. Draw three different figures in which inscribed angle AVC intercepts an arc of the circle. Include one major arc, one minor arc, and one semicircle.

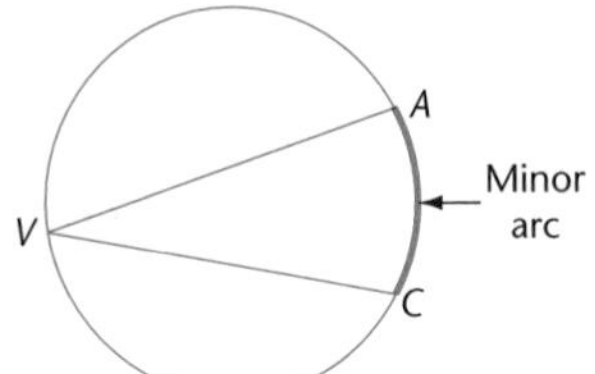

2. Measure the inscribed angle and the intercepted arc in each figure. (You will need to add central angles to determine the measures of the arcs.)

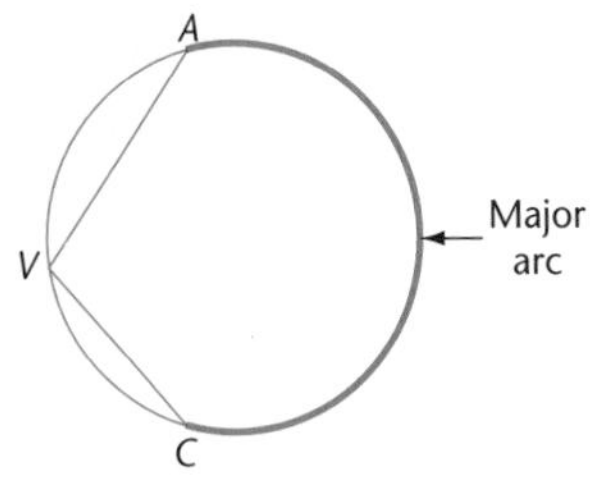

3. Compare the measure of the inscribed angle and its intercepted arc in each case.

4. If $m\widehat{AC} = 148°$, what is $m\angle AVC$? If $m\angle AVC = 53°$, what is $m\widehat{AC}$?

5. Make a conjecture about the relationship between the measure of an inscribed angle and its intercepted arc. ❖

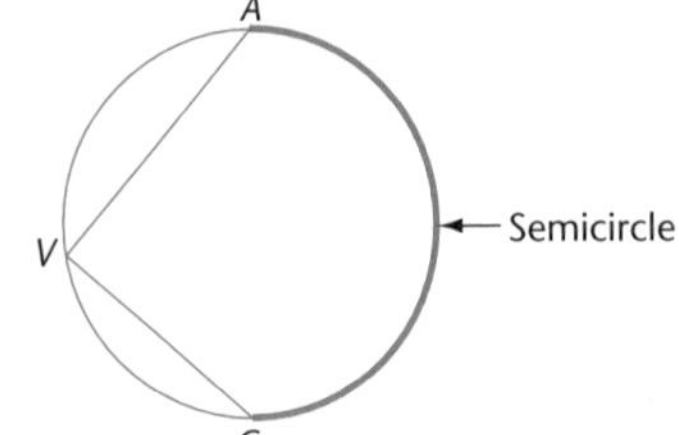

Exploration 2 Proving the Relationship

You will need
No special tools

Part I

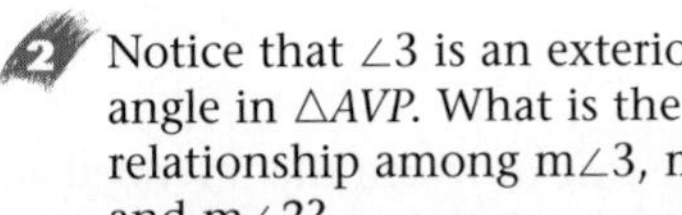

1. In the figure, one ray of the inscribed angle contains the center of the circle. What is the relationship between $m\angle 1$ and $m\angle 2$?

2. Notice that $\angle 3$ is an exterior angle in $\triangle AVP$. What is the relationship among $m\angle 3$, $m\angle 1$, and $m\angle 2$?

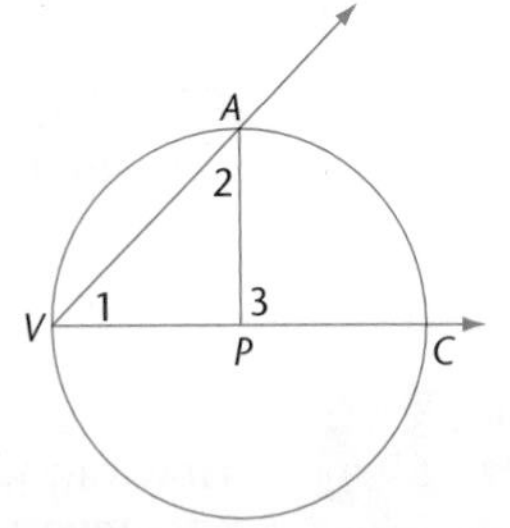

interdisciplinary CONNECTION **Physics** Ask students to suppose that the inside of a cylinder has a reflective surface. A beam of light starts on the rim of a circular cross-section of the cylinder and moves towards any other point on the circle. When the beam hits the circle, it is reflected so that the angle of incidence equals the angle of reflection. Have students make a sketch to model the beam of light. Students should discover that the second chord on the path of the beam must cut off an arc equal in length to the first arc.

TEACHING tip

Contrast the definition of *inscribed angle* with that of *central angle* introduced in Lesson 9.1. The first has a vertex on the circle; the second has a vertex at the center of the circle. Emphasize that students need to remember which type of angle is which.

Exploration 1 Notes

Students discover that the measure of an inscribed angle is equal to one-half the measure of its intercepted arc. Before they begin the activity, ask them to explain these terms: major arc, minor arc, semicircle, and intercepted arc. If necessary, review the meanings of these important terms.

ongoing ASSESSMENT

5. The measure of an inscribed angle is half the measure of its intercepted arc.

Use Transparency 77

Exploration 2 Notes

Students prove the relationship they discovered in Exploration 1. The figure can be created using geometry graphics software. Students can drag on points A or C to change the measures of $\angle AVC$.

Cooperative Learning

Have students work on the explorations in groups of three. Students should display their group work on large sheets of paper and share their conjectures with the whole class.

ongoing ASSESSMENT

4. $m\angle 1 = \frac{1}{2} m\overset{\frown}{AC}$

 Fill in a table like the one below. For each entry in the last row of the table, give a reason.

m∠1	m∠2	m∠3	$m\overset{\frown}{AC}$
20°	?	?	?
30°	?	?	?
40°	?	?	?
x°	?	?	?

 What does your table show about the relationship between m∠1 and $m\overset{\frown}{AC}$?

Part II

 In the figure at right, the center of the circle is in the interior of the inscribed angle. What is the relationship between m∠1 and $m\overset{\frown}{AX}$? between m∠4 and $m\overset{\frown}{CX}$?

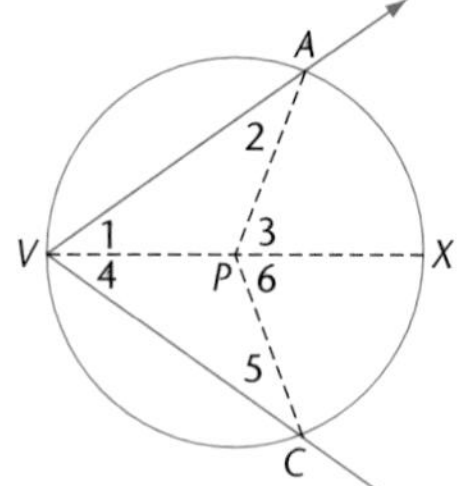

 Fill in a table like the one below. For each entry in the last row of the table, give a reason.

m∠1	$m\overset{\frown}{AX}$	m∠4	$m\overset{\frown}{CX}$	m∠AVC	$m\overset{\frown}{AXC}$
20°	?	20°	?	?	?
30°	?	20°	?	?	?
40°	?	50°	?	?	?
x°	?	y°	?	?	?

 What does your table show about the relationship between m∠AVC and $m\overset{\frown}{AC}$? ❖

ongoing ASSESSMENT

3. $m\angle AVC = \frac{1}{2} m\overset{\frown}{AC}$

CRITICAL Thinking

How can you prove the case in which the inscribed angle does not encompass the center of the circle?

CRITICAL Thinking

Draw a ray from the vertex of the inscribed angle through the center of the circle and subtract angle and arc measures.

ENRICHMENT Followup the Interdisciplinary Connection from the Teacher's Edition page 485 by asking the following questions. What types of paths can the beam follow if it is going to return to its starting point? What type of starting condition is necessary so that the beam eventually returns to its starting point? Remind students that each time the beam of light hits the rim of a circle, the angle of incidence equals the angle of reflection.

INCLUSION strategies **Hands-On Strategies** Many of the figures in this lesson can be created on circular geoboards. The hands-on experience involved in creating the figures with rubber bands may help some students understand the relationships between various angles and the circle.

The results of the explorations are stated in the following theorem.

INSCRIBED ANGLE THEOREM
An angle inscribed in an arc has a measure equal to one-half the measure of the intercepted arc. **9.3.1**

EXTENSION

Find the measure of $\angle A$.

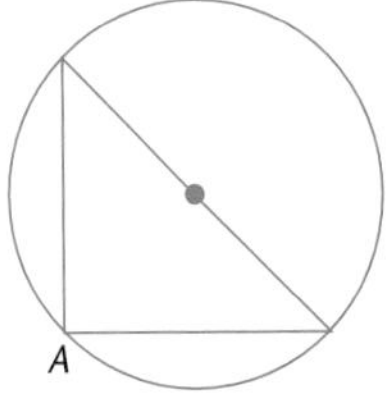

$\angle A$ is inscribed in a half-circle; therefore, the arc it intercepts is also a half-circle. The measure of the intercepted arc is thus 180° (why?), and so the measure of $\angle A$ is $\frac{1}{2} \times 180°$, or 90°. ❖

The extension above illustrates the following corollary.

INSCRIBED ANGLE COROLLARY
An angle inscribed in a half-circle is a right angle. **9.3.2**

EXTENSION

Find the measure of $\angle B$.

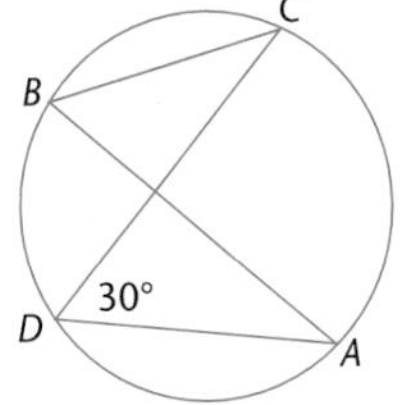

The measure of $\overset{\frown}{AC}$ is twice the measure of $\angle D$, or $2 \times 30° = 60°$. The measure of $\angle B$ is one-half the measure of $\overset{\frown}{AC}$, or $\frac{1}{2} \times 60° = 30°$. ❖

The extension above illustrates an important principle. Notice that $\angle D$ and $\angle B$ intercept the same arc. As you saw, they have the same angle measure. This leads to the following theorem.

THEOREM
If two inscribed angles intercept the same arc, then they have the same measure. **9.3.3**

TEACHING tip

The corollary and the theorem illustrated in the first two extensions can be developed through experiment, conjecture, and logical reasoning. If time permits, you may wish to have students work in small groups to develop these ideas on their own.

TEACHING tip

Many students will have difficulty interpreting the figure for the theorem in the second extension. Have them sketch the figure twice. In the first drawing they should shade the interior of $\angle CBA$; in the second drawing, the interior of $\angle CDA$. This may help them see how both angles intercept the same arc, arc $\overset{\frown}{CA}$.

Cooperative Learning Have students work in pairs to review the lesson. Each student should draw a figure such as the ones used in the explorations. For example, the figure could show an inscribed angle and the related central angle. Students then exchange papers. Each should measure just one angle, and then use the lesson concepts to determine the measures of all the other angles.

EXTENSION

Given $\odot P$ with diameter $\overline{AB}$, $m\widehat{CB} = 110°$, and $m\widehat{BD} = 130°$. Find the measures of $\angle 1$, $\angle 2$, $\angle 3$, $\angle 4$, $\angle APC$, $\angle ADB$, and $\angle CAD$.

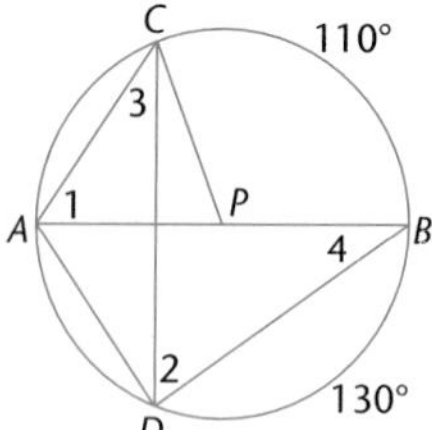

Copy the figure and add information as you work through the solution.

Use Transparency 78

Begin by filling in as many arc measures as you can. Since $\overline{AB}$ is a diameter, $\widehat{ACB}$ and $\widehat{ADB}$ each measure 180°. Therefore, $m\widehat{AC} = 70°$ and $m\widehat{AD} = 50°$.

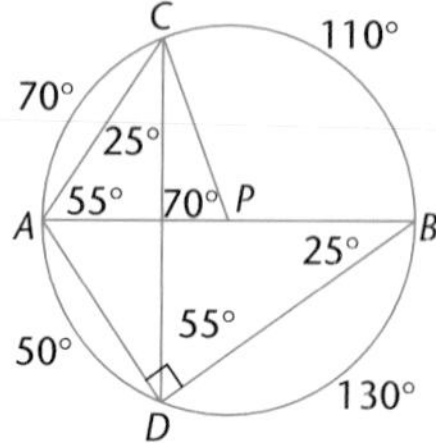

Angles 1 and 2 intercept $\widehat{CB}$. Thus $m\angle 1 = m\angle 2 = \frac{1}{2}(110°) = 55°$.

Angles 3 and 4 intercept $\widehat{AD}$. Thus $m\angle 3 = m\angle 4 = \frac{1}{2}(50°) = 25°$.

$\angle APC$ is a central angle. Thus $m\angle APC = 70°$.

$\angle ADB$ is inscribed in a semicircle. Thus $m\angle ADB = 90°$.

$\angle CAD$ intercepts $\widehat{CBD}$. Thus $m\angle CAD = \frac{1}{2}(110° + 130°) = 120°$. ❖

APPLICATION

Carpentry A student needs to find the center of a small table top. How can she do this using a carpenter's square? The student inscribes two right angles, $\angle ABC$ and $\angle DEF$, inside the circle. She then adds the hypotenuses of the triangles by connecting points A and C and points D and F. The intersection of $\overline{AC}$ and $\overline{DF}$ is the center of the table. ❖

CRITICAL Thinking

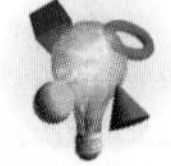

How can the student check the accuracy of her procedure by carrying the process a step further? Discuss.

CRITICAL Thinking

Draw a ray from the vertex of the inscribed angle through the center of the circle and subtract angle and arc measures.

EXERCISES & PROBLEMS

Communicate

Refer to circle *O* in the photo for Exercises 1–5.

1. Explain how to find m∠1.
2. Explain how to find $m\overset{\frown}{AC}$.
3. Name an inscribed angle on circle *O*.
4. Explain how to find m∠2.
5. Explain how to find $m\overset{\frown}{BD}$.
6. Explain why two angles inscribed in the same arc have the same measure.

Practice & Apply

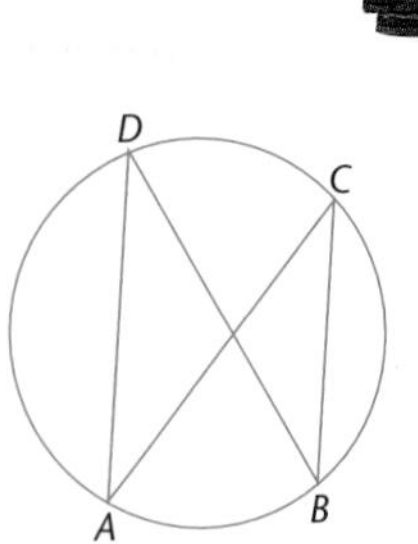

Algebra For Exercises 7–10, refer to the circle at right.

7.	$m\overset{\frown}{AB} = 68°$	m∠C = _?_ 34°	m∠D = _?_ 34°
8.	m∠D = 30°	$m\overset{\frown}{AB}$ = _?_ 60°	m∠C = _?_ 30°
9.	$m\overset{\frown}{CD} = 87°$	m∠B = _?_ 43.5°	m∠A = _?_ 43.5°
10.	m∠B = a°	$m\overset{\frown}{DC}$ = _?_ (2a)°	m∠A = _?_ a°

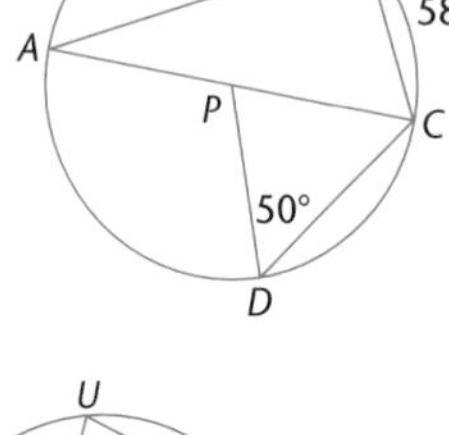

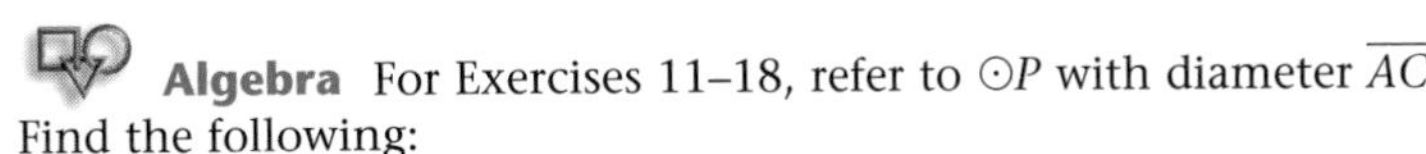

Algebra For Exercises 11–18, refer to ⊙P with diameter $\overline{AC}$. Find the following:

11. m∠A 29° **12.** m∠B 90° **13.** m∠BCA 61°

14. $m\overset{\frown}{AB}$ 122° **15.** m∠PCD 50° **16.** m∠CPD 80°

17. $m\overset{\frown}{DC}$ 80° **18.** $m\overset{\frown}{AD}$ 100°

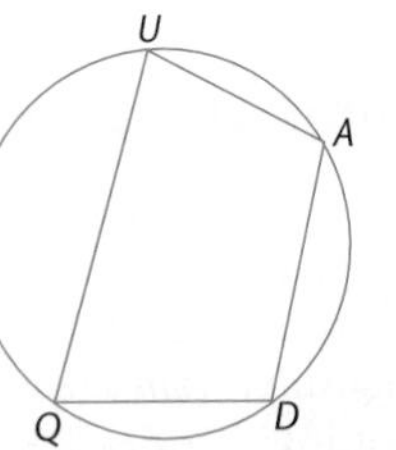

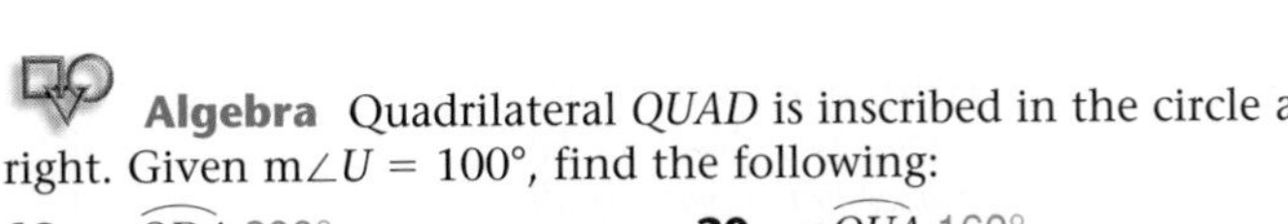

Algebra Quadrilateral *QUAD* is inscribed in the circle at right. Given m∠U = 100°, find the following:

19. $m\overset{\frown}{QDA}$ 200° **20.** $m\overset{\frown}{QUA}$ 160°

21. m∠D 80° **22.** m∠Q + m∠A 180°

ASSESS

Selected Answers

Odd-numbered Exercises 7–25

Assignment Guide

Core 1–22, 24–26

Core Plus 1–27

Technology

Geometry graphics software can be used to create the figures for Exercises 7-22. Students can use the software to check angle measures they find through computation or logical reasoning.

Error Analysis

If students are using geometry graphics software for Exercises 7–22, remind them that they must determine the measures based on the information given in the problem and then confirm their answers with the software.

Practice Master

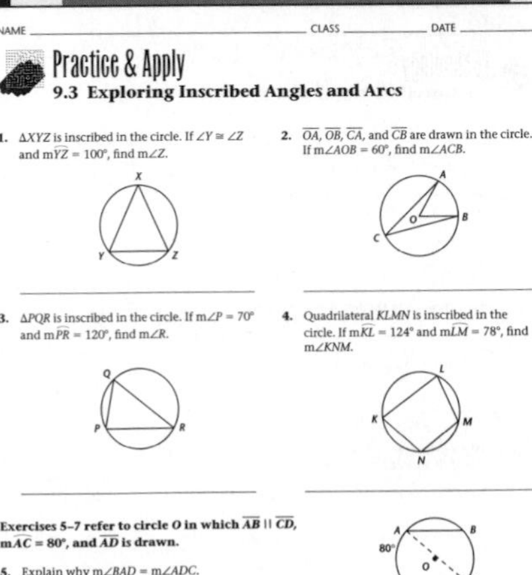

NAME ______ CLASS ______ DATE ______

Practice & Apply

9.3 Exploring Inscribed Angles and Arcs

1. ΔXYZ is inscribed in the circle. If ∠Y ≅ ∠Z and $m\overset{\frown}{YZ}$ = 100°, find m∠Z.

2. $\overline{OA}$, $\overline{OB}$, $\overline{CA}$, and $\overline{CB}$ are drawn in the circle. If m∠AOB = 60°, find m∠ACB.

3. ΔPQR is inscribed in the circle. If m∠P = 70° and $m\overset{\frown}{PR}$ = 120°, find m∠R.

4. Quadrilateral KLMN is inscribed in the circle. If $m\overset{\frown}{KL}$ = 124° and $m\overset{\frown}{LM}$ = 78°, find m∠KNM.

Exercises 5–7 refer to circle O in which $\overline{AB} \parallel \overline{CD}$, $m\overset{\frown}{AC}$ = 80°, and $\overline{AD}$ is drawn.

5. Explain why m∠BAD = m∠ADC.

6. Explain why $m\overset{\frown}{BD}$ = 80°.

7. State the results of Exercises 5–6 as a theorem.

Use the theorem you wrote in Exercise 7 to answer Exercise 8.

8. In the circle $\overline{AB} \parallel \overline{CD}$. If $m\overset{\frown}{AB}$ = 100° and $m\overset{\frown}{AC}$ = 100°, find $m\overset{\frown}{CD}$.

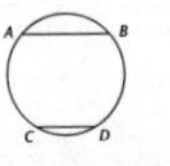

82 Practice & Apply HRW Geometry

Performance Assessment

Ask students to construct a circle, its center, and its diameter. Then have them mark one point on each semicircle. Students should connect the five points (center, endpoints of diameter, two points on semicircles) in whatever way they wish to create a diagram that interests them. Students should then use what they have learned about angle relationships to find all angle measures in the figure. They should measure as few angles as possible. Students' work should include a description of how all angle measures were found.

Look Beyond

Exercise 27 is a type of logical argument which will be studied extensively in Chapter 12. Make sure students understand that the question is not whether the object is a circle. The question is whether the argument is valid.

Technology Master

NAME ______________ CLASS ________ DATE ________

Technology
9.3 Inscribed Angles and Triangles

If you place four points on a circle and connect them as shown in the diagram, you will have a figure from which you can collect a wealth of information. Because geometry software allows you to measure angles and lengths, you will find geometry software invaluable in the exercises that follow.

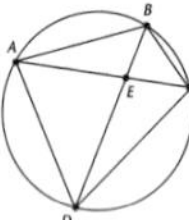

Use geometry software in the following exercises.

1. Create a diagram something like the one shown. Use MEASURE to find $m\angle ABD$, $m\angle ACD$, $m\angle BAC$, and $m\angle BDC$. Also measure the arcs they intercept. Do your measurements confirm a theorem you learned in Lesson 9.3? What theorem?

2. Use MEASURE to find the ratios $AB : DC$, $AE : DE$, and $BE : CE$. What can you conclude about $\triangle ABE$ and $\triangle DCE$? Explain.

3. Can you draw the same conclusion about $\triangle AED$ and $\triangle BEC$ as the conclusion you drew in Exercise 2? Explain.

4. Based on your results from Exercises 2 and 3, what can you say about triangles formed by four points placed on any circle?

5. Move points A, B, C, and D along the circle until all four triangles are congruent. What figure is formed by A, B, C, and D?

6. Move points A, B, C, and D along the circle until $m\angle BEA = 90°$. What can you say about the four arcs determined by A, B, C, and D? What figure is formed by A, B, C, and D?

94 Technology HRW Geometry

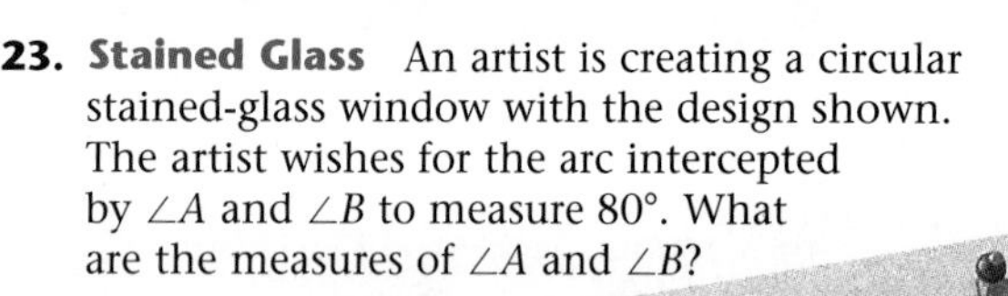

23. Stained Glass An artist is creating a circular stained-glass window with the design shown. The artist wishes for the arc intercepted by $\angle A$ and $\angle B$ to measure 80°. What are the measures of $\angle A$ and $\angle B$?

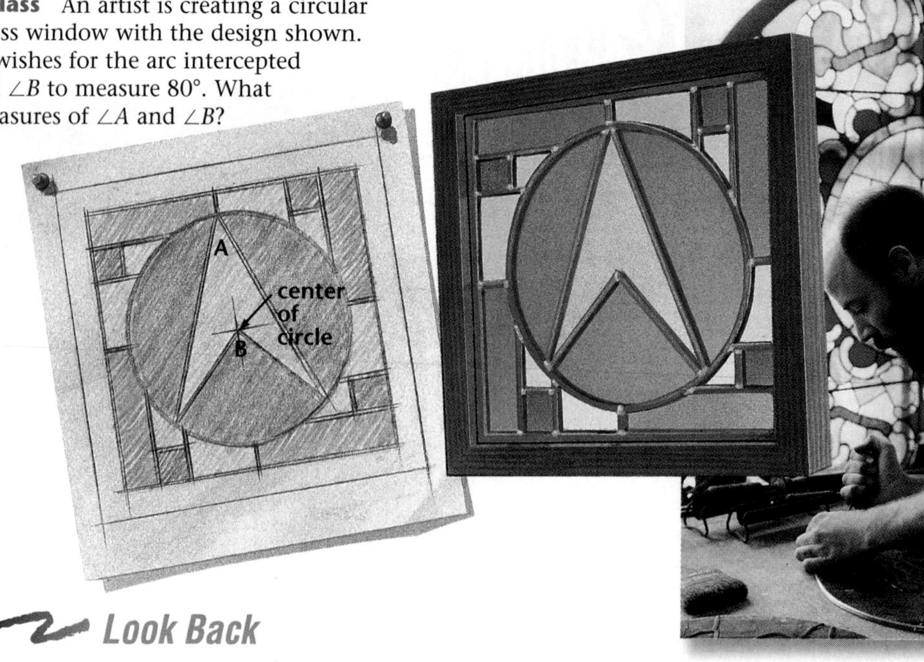

Look Back

24. Given: 1. If the bees are happy, then it pours.
2. If flowers grow, then the bees are happy.
3. If it rains, then flowers grow.

Prove: "If it rains, then it pours" by arranging the three given statements according to the If-Then Transitive Property. **[Lesson 2.2]**

25. Given: $ABCD$ is a rhombus with $AC = 6$ and $BD = 8$. Find the perimeter of $ABCD$. **[Lesson 5.4]** 20 units

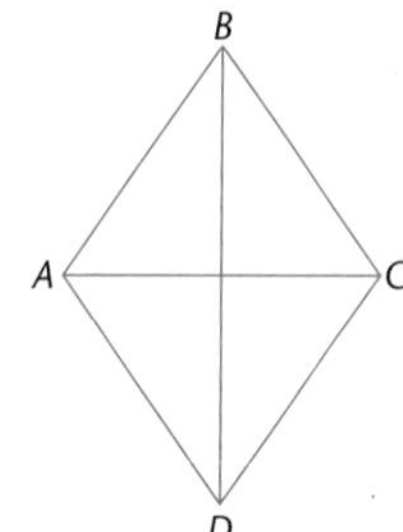

26. Given: In $\odot P$, $AB = 12$; $PX = 4$. Find the length of the radius $\overline{PQ}$. **[Lesson 5.4]** 7.21 units

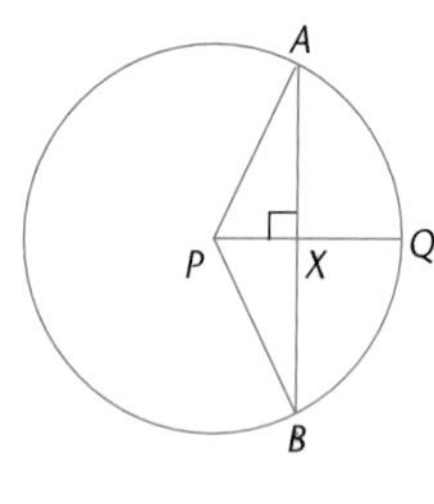

Look Beyond

27. Consider the following argument: *If an object consists of a set of points equidistant from a given point, then the object is a circle. The object at the right is not a circle. Therefore, the object at the right does not consist of all points equidistant from a given point.* Is the argument valid? Explain your reasoning.

23. $m\angle A = 40°$; $m\angle B = 80°$

24. If it rains, then flowers grow.
If flowers grow, then the bees are happy.
If the bees are happy, then it pours.
Therefore, if it rains, then it pours.

27. Yes. The argument is the contrapositive of the given definition. If the object did consist of points equidistant from a given point, then it would be a circle.

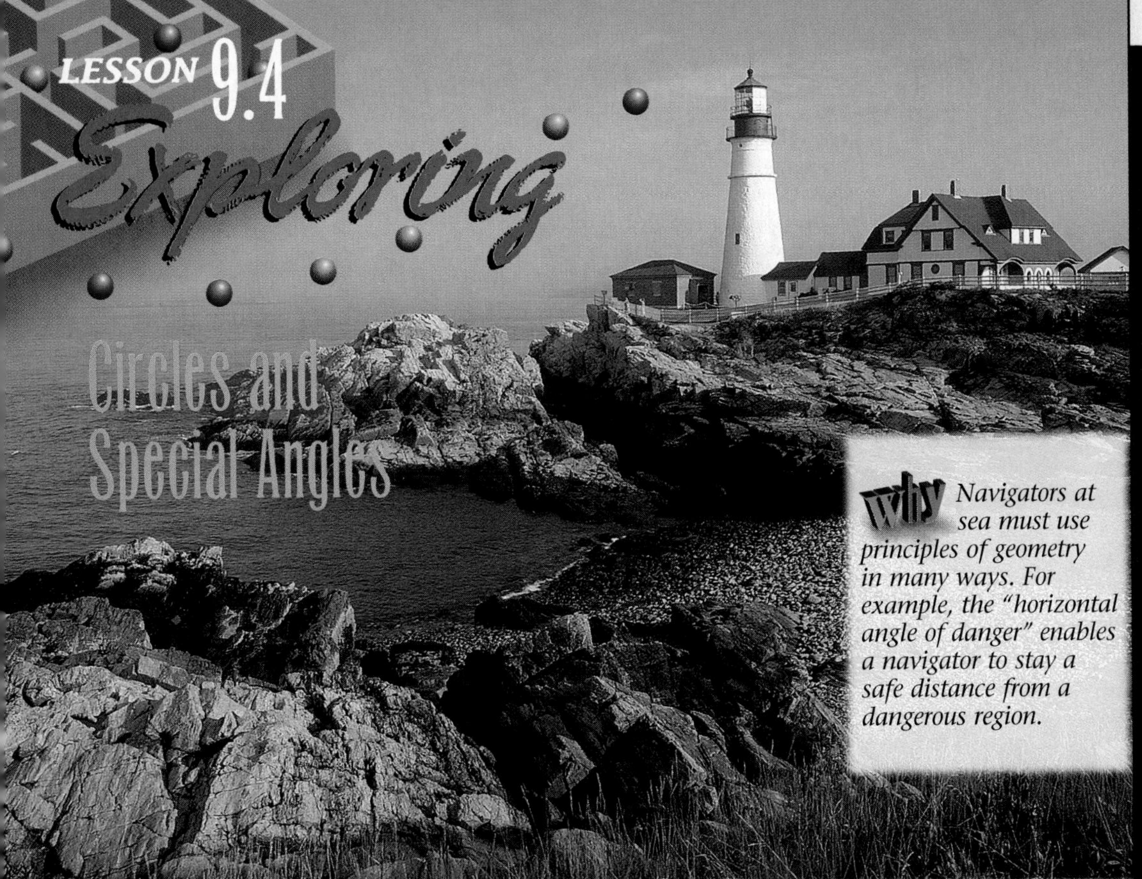

Why Navigators at sea must use principles of geometry in many ways. For example, the "horizontal angle of danger" enables a navigator to stay a safe distance from a dangerous region.

Lighthouses are landmarks for navigators at night, and they often serve as warnings of dangerous shoals or reefs.

Angles or intersecting lines that touch a circle in two or more places can be studied systematically. There are three cases to consider, according to the placement of the vertex of the angles.

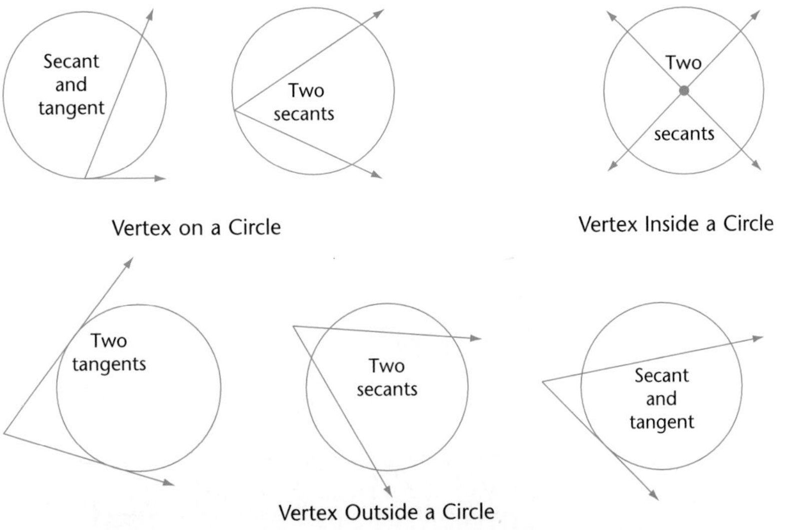

Technology Students can use geometry graphics software for the three explorations and many of the exercises. You may need to model for students how to find the measure of an arc in degrees. For example, in *The Geometer's Sketchpad,* students must select the two endpoints of the arc and the circle that contains the arc. Then they use **Arc Angle** from the **Measure** menu to find the number of degrees in the arc.

PREPARE

Objectives

- Explore special angles relative to the circle including secant-tangent, secant-secant, and chord-chord.
- Develop and use rules for special angles including secant-tangent, secant-secant, and chord-chord.

RESOURCES

- Practice Master 9.4
- Enrichment Master 9.4
- Technology Master 9.4
- Lesson Activity Master 9.4
- Quiz 9.4
- Spanish Resources 9.4

Assessing Prior Knowledge

A circle and an angle are drawn in the same plane. Find all the possible ways in which there are 2 intersection points.

[intersections are vertex and point on one ray, circle inside angle with intersection points at tangents to circle, one ray intersects circle in two points other than the vertex]

Use Transparency 91

TEACH

The application on page 494 focuses on the concept of *horizontal angle of danger.* Ask students to brainstorm about other navigation situations in which angle measures might be important.

Exploration 1 Vertex on a Circle, Secant-Tangent Angles

You will need
No special tools

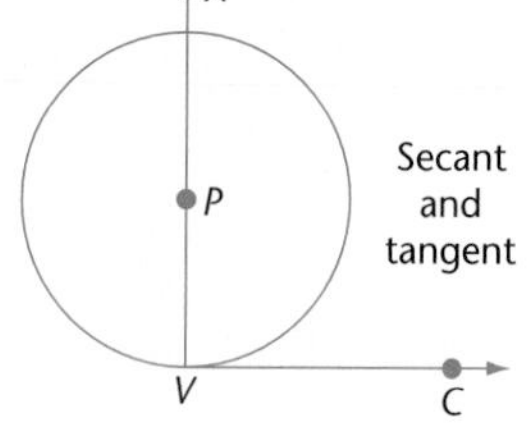

In this exploration you will examine three special cases of secant-tangent angles.

Case 1: The secant-tangent angle is a right angle. The secant line contains the center of the circle.

1. $m\angle AVC = ?$ $\quad m\overarc{AV} = ?$

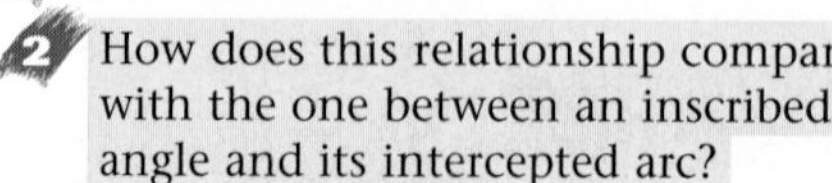

2. How does this relationship compare with the one between an inscribed angle and its intercepted arc?

Case 2: The secant-tangent angle is acute.

1. Copy and complete the following table.

$m\overarc{AV}$	$m\angle 1$	$m\angle 2$	$m\angle PVC$	$m\angle AVC$
120°	120°	30°	?	60°
100°	?	?	?	?
80°	?	?	?	?
60°	?	?	?	?
x°	?	?	?	?

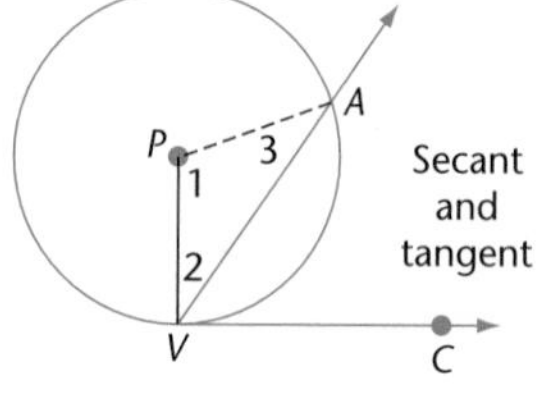

2. Complete the following statement:

The measure of an acute secant-tangent angle with its vertex on a circle is ___?___ its intercepted arc.

Case 3: The secant-tangent angle is obtuse.

1. Copy and complete the following table.

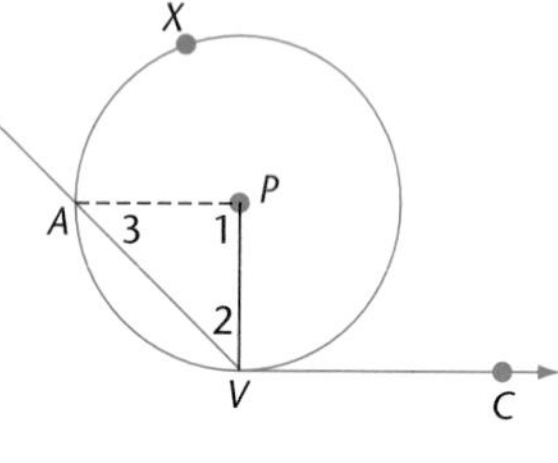

$m\overarc{AXV}$	$m\angle 1$	$m\angle 2$	$m\angle PVC$	$m\angle AVC$
200°	160°	10°	?	100°
220°	?	?	?	?
240°	?	?	?	?
260°	?	?	?	?
x°	?	?	?	?

TEACHING tip

Technology A spreadsheet can be used to record, organize, and analyze the data students generate in the three explorations.

Use Transparency 80

ongoing ASSESSMENT

2. **The relationship is the same as between an inscribed angle and its intercepted arc:**
$$m\angle AVC = \frac{1}{2} m\overarc{AV}$$

Exploration 1 Notes

In this exploration, students discover that the measure of a secant-tangent angle with its vertex on a circle is one-half the measure of the intercepted arc. Before students begin, explain that the angle in this activity is called a secant-tangent angle because one ray is a secant and one ray is a tangent. Ask students which is which to confirm they know the meanings of the terms *secant* and *tangent*. This secant-tangent situation is one possibility for the "vertex on the circle" arrangement.

ongoing ASSESSMENT

2. **one-half**

interdisciplinary CONNECTION

Language Arts Have students collect examples of poetry or song lyrics that mention circles. Examples, perhaps accompanied by student-made illustrations or cartoons, will make an interesting bulletin board display.

Complete the following statement:

The measure of an obtuse secant-tangent angle with its vertex on a circle is _?_ its intercepted arc.

How does this relationship compare with that of an inscribed angle and its intercepted arc?

Complete the theorem below. ❖

THEOREM

If a tangent and a secant (or a chord) intersect on a circle, at the point of tangency, then the measure of the angle formed is _?_ the measure of its intercepted arc. **9.4.1**

Exploration 2 *Lines Intersect Inside a Circle*

You will need
No special tools

1. In the given figure, note that $\angle AVC$ is an exterior angle of $\triangle ADV$. What is the relationship between the measure of $\angle AVC$ and the measures of $\angle 1$ and $\angle 2$?

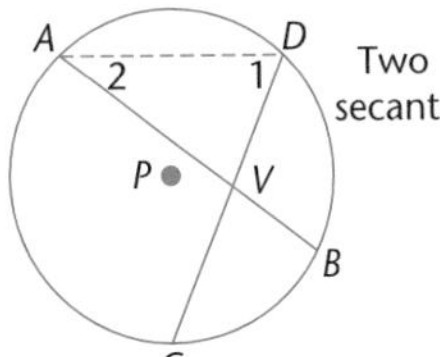

2. Copy and complete the following table.

m$\overset{\frown}{AC}$	m$\overset{\frown}{BD}$	m∠1	m∠2	m∠AVC	m∠DVB
160°	40°	80°	20°	100°	100°
180°	70°	?	?	?	?
x_1°	x_2°	?	?	?	?

3. Complete the theorem below. ❖

THEOREM

The measure of an angle formed by two secants or chords intersecting in the interior of a circle is _?_ of the _?_ of the measures of the arcs intercepted by the angle and its vertical angle. **9.4.2**

Exploration 3 *Vertex of an Angle Is Outside a Circle*

You will need
No special tools

1. In the given figure, $\angle 1$ is an exterior angle of $\triangle BVC$. What is the relationship between the measure of $\angle 1$ and the measures of $\angle 2$ and $\angle AVC$?

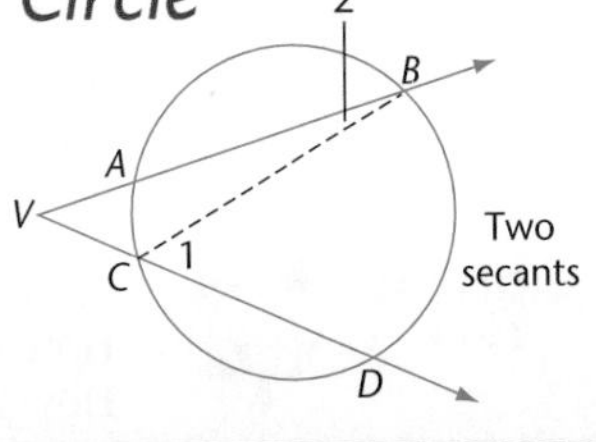

Cooperative Learning

Any or all of the three explorations lend themselves to cooperative learning groups. Students can divide up the work needed to complete the tables. And, by discussing the relationships in the figures, students will better understand the key ideas being developed.

Ongoing Assessment

2. one-half

Exploration 2 Notes

This activity focuses on the chord-chord case. Students discover that the angle formed by two chords intersecting within a circle equals one-half the sum of the measures of the intercepted arcs.

Ongoing Assessment

2. one-half, sum

Exploration 3 Notes

The angle in this activity has its vertex outside the circle and is formed from two secants. Students discover that the angle equals one-half the difference of the measures of the intercepted arcs. Two other vertex-outside-the-circle cases are covered in the exercises: secant-tangent and tangent-tangent.

Enrichment

Have students construct a circle that passes through three noncollinear points. [**Possible method: Choose any points *A*, *B*, and *C* so that the points are not on the same line. Draw segments *AB* and *BC*. Construct the perpendicular bisectors of the segments. The bisectors meet in a point *O*. This point is the center of the required circle.**]

Inclusion strategies

Using Visual Models Students may understand the concepts but become confused by the *names* of the various lines and angles. Suggest that they use color to help identify special parts of the figures; for example, red for tangent and blue for secant.

 Copy and complete the following table.

m$\overset{\frown}{BD}$	m$\overset{\frown}{AC}$	m∠1	m∠2	m∠AVC
200°	40°	100°	20°	80°
250°	60°	?	?	?
100°	50°	?	?	?
80°	20°	?	?	?
x_1°	x_2°	?	?	?

 Complete the theorem below. ❖

THEOREM

The measure of an angle formed by two secants intersecting in the exterior of a circle is ___?___ of the ___?___ of the measures of the intercepted arcs. **9.4.3**

You will explore the two remaining configurations in the exercises.

APPLICATION

Navigation The illustration shows a ship at point *F* and two lighthouses at points *A* and *B*. A circle has been drawn that encloses a region of dangerous rocks. The captain wants to avoid the danger by staying outside the circle.

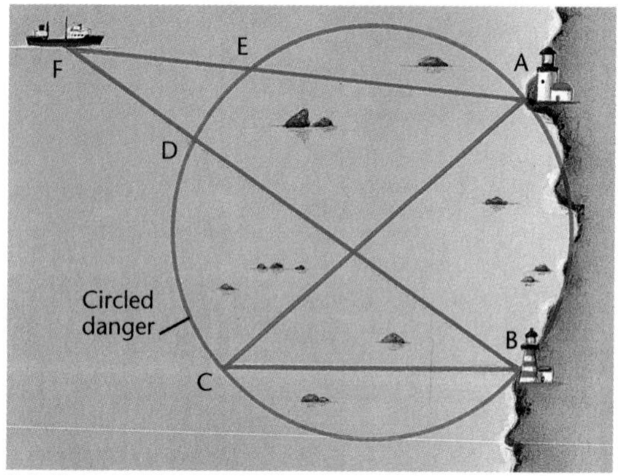

The ship's navigator measures ∠*F*, the angle between his lines of sight to the two lighthouses. He knows that if m∠*F* is less than m∠*C*, then the ship is outside the circle of danger. ∠*C* is known as the "horizontal angle of danger." Why does this method work?

Notice that ∠*C* is an inscribed angle and that it intercepts arc $\overset{\frown}{AGB}$. So if the ship is on the circle, then ∠*F* ≅ ∠*C*. It should be obvious that if m∠*F* < m∠*C*, the ship is outside the circle. ❖

CRITICAL Thinking

How can you use theorems from this lesson to prove this "obvious" fact? How can you prove that if ∠*F* > ∠*C*, then the ship is inside the circle?

ongoing ASSESSMENT

The measure of an angle formed by two secants intersecting in the exterior of a circle is half of the difference of the measures of the intercepted arcs.

CRITICAL Thinking

Case 1

For a point C on the circle, m∠C = $\frac{1}{2}$m$\overset{\frown}{AB}$. For a point *F* outside the circle, m∠*F* = $\frac{1}{2}$(m$\overset{\frown}{AB}$ − m$\overset{\frown}{DE}$). Since m*DE* > 0, m∠*F* < m∠*C*.

Case 2

For a point *G* inside the circle, m∠*G* = $\frac{1}{2}$(m$\overset{\frown}{AB}$ + m$\overset{\frown}{HI}$) where $\overset{\frown}{HI}$ is the arc intercepted by the vertical angle of ∠*F*. Since m$\overset{\frown}{HI}$ > 0, m∠*F* < m∠*G*.

TEACHING tip

Technology Students can use geometry graphics software to copy the figure in the extension. The software can then be used to confirm the results found through computation, logical reasoning, and the application of angle-circle relationships.

RETEACHING the lesson

Cooperative Learning Use the drawings of the possible arrangements shown at the beginning of the lesson. Explain that students are looking for relationships between various angles and their intercepted arcs. Have students work in pairs or small groups to copy the figures. Then they should describe in writing at least one relationship for each drawing.

The following extension illustrates how to use the angle-arc relationships explored in this lesson.

EXTENSION

Given: $\overline{TU}$ is tangent to $\odot P$ at T.

$m\widehat{QR} = 90°$;

$m\widehat{RT} = 150°$;

$m\widehat{QS} = 50°$.

Find the following:

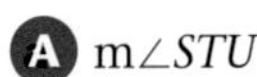
A $m\angle STU$ 
B $m\angle 1$ **C** $m\angle 2$

Make a sketch of the figure. As you obtain new information, add it to the figure.

A Since the vertex of $\angle STU$ is on $\odot P$, and is formed by a tangent and a secant line, $m\angle STU = \frac{1}{2} m\widehat{ST}$. To find $m\widehat{ST}$, note that $50° + 90° + 150° + m\widehat{ST} = 360°$. Thus, $m\widehat{ST} = 70°$ and $m\angle STU = \frac{1}{2}(70°) = 35°$.

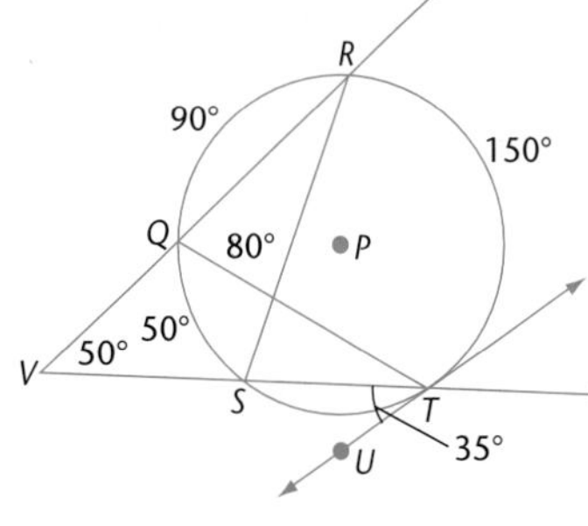

B Since the vertex of $\angle 1$ is in the interior of $\odot P$ and is formed by two intersecting chords, $m\angle 1 = \frac{1}{2}(m\widehat{QR} + m\widehat{ST}) = \frac{1}{2}(90° + 70°) = 80°$.

C Since the vertex of $\angle 2$ is in the exterior of the circle and is formed by two secants, $m\angle 2 = \frac{1}{2}(m\widehat{RT} - m\widehat{QS}) = \frac{1}{2}(150° - 50°) = 50°$. ❖

EXERCISES & PROBLEMS

Communicate

1. The measure of an acute secant-tangent angle with its vertex on a circle is __?__.
2. The measure of an obtuse secant-tangent angle with its vertex on a circle is __?__.
3. If two secants (or chords) intersect in the interior of a circle, the measure of any angle formed is __?__.
4. If two secants intersect in the exterior of a circle, the measure of the angle formed is __?__.

Math Connection
Algebra

Students are accustomed to using a single letter for a variable in an equation, often *x* or *n*. Here, the variables are written as "m∠*STU*" or "m*QR*."

ASSESS

Selected Answers

Exercises 5, 7, 9, 11, 19, 25, 27, 29, and 31

Assignment Guide

Core 1–30, 33–37

Core Plus 1–4, 12–23, 31–43

Technology

The two angle-circle cases explored in Exercises 12–22 may be investigated using geometry graphics software. A spreadsheet can be used to organize and analyze the data for the activities in Exercises 12–22.

Error Analysis

If you do not assign Exercises 13 and 22, you may want to discuss the exercises in class so that students can see the theorems.

Practice & Apply

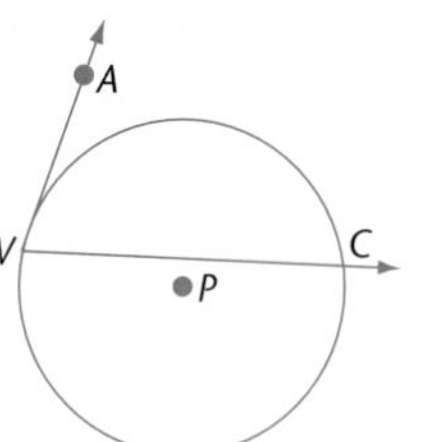

Algebra In the figure, $\overline{VA}$ is tangent to $\odot P$ at V.

5. If $m\widehat{VC} = 150°$, find $m\angle AVC$. 75°

6. If $m\angle AVC = 80°$, find $\widehat{VC}$. 160°

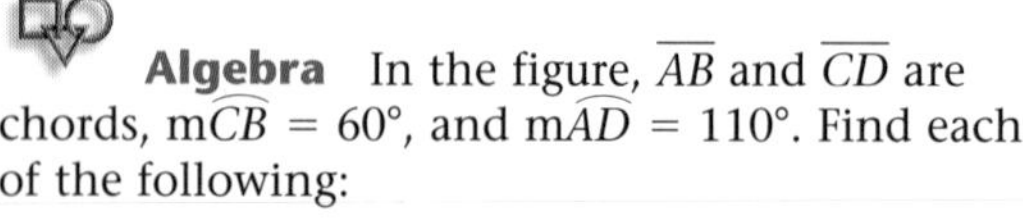

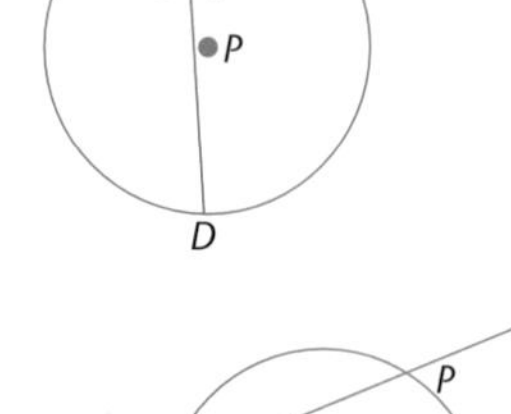

Algebra In the figure, $\overline{AB}$ and $\overline{CD}$ are chords, $m\widehat{CB} = 60°$, and $m\widehat{AD} = 110°$. Find each of the following:

7. $m\angle 1$ 85°

8. $m\angle 2$ 95°

9. $m\angle 3$ 85°

10. $m\angle 4$ 95°

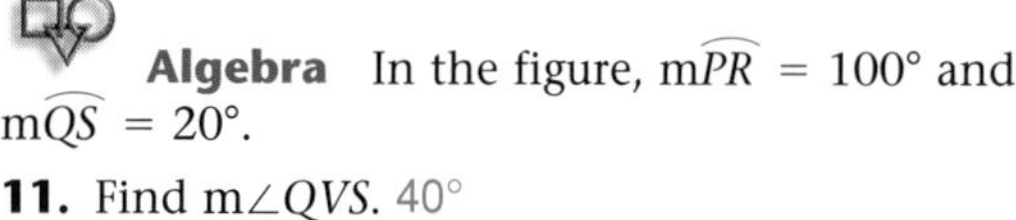

Algebra In the figure, $m\widehat{PR} = 100°$ and $m\widehat{QS} = 20°$.

11. Find $m\angle QVS$. 40°

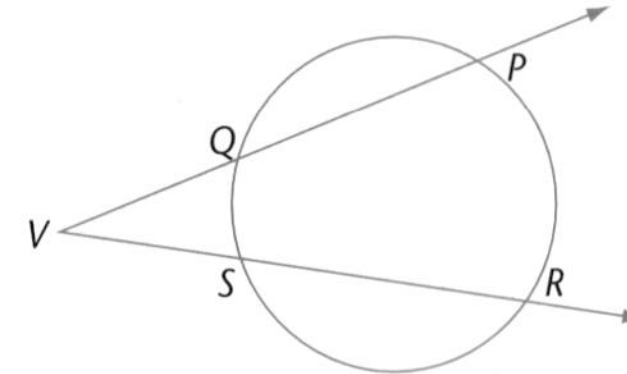

Exercises 12–22 explore two additional angle-circle cases.

Case 1: A secant and a tangent intersect with a vertex that is outside the circle. Use $\odot P$ for Exercises 12–15. $\overrightarrow{VC}$ is tangent to $\odot P$ at C.

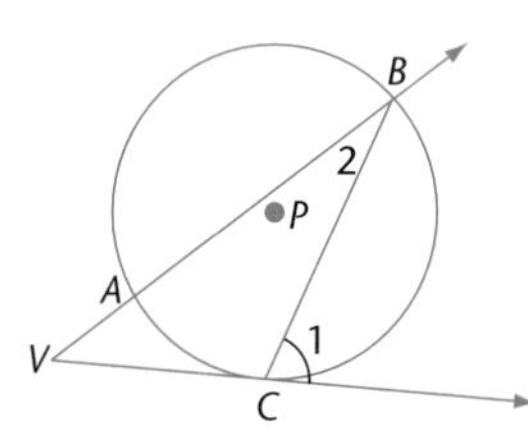

12. What is the relationship between $m\angle 1$ and the measures of $\angle 2$ and $\angle AVC$? $m\angle 1 = m\angle 2 + m\angle AVC$

13. Copy and complete the table.

$m\widehat{BC}$	$m\widehat{AC}$	$m\angle 1$	$m\angle 2$	$m\angle AVC$
250°	60°	(a)	(b)	(c)
200°	40°	(d)	(e)	(f)
130°	40°	(g)	(h)	(i)
70°	30°	(j)	(k)	(l)
$x_1°$	$x_2°$	(m)	(n)	(o)

14. Write an equation that describes $m\angle AVC$ in terms of $m\widehat{BC}$ and $m\widehat{AC}$. $m\angle AVC = \frac{m\widehat{BC} - m\widehat{AC}}{2}$

15. Complete the following statement.
half of the difference between the measures of the two intercepted arcs

THEOREM

The measure of a secant-tangent angle with its vertex outside the circle is ___?___. **9.4.4**

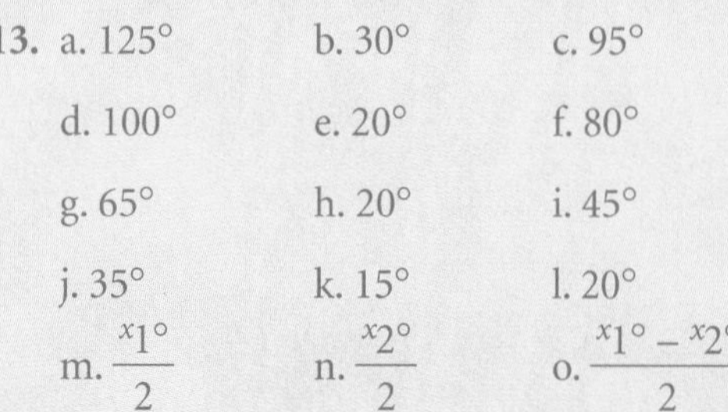

13. a. 125° b. 30° c. 95°
d. 100° e. 20° f. 80°
g. 65° h. 20° i. 45°
j. 35° k. 15° l. 20°
m. $\frac{x_1°}{2}$ n. $\frac{x_2°}{2}$ o. $\frac{x_1° - x_2°}{2}$

Practice Master

NAME ____________ CLASS ________ DATE ________

Practice & Apply
9.4 Exploring Circles and Special Angles

In Exercises 1–3, find the measure of $\angle x$.

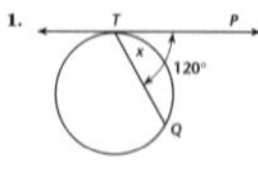
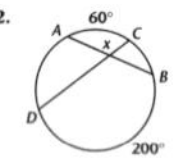
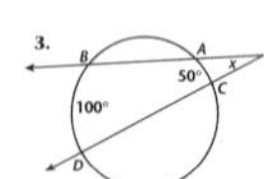

In Exercises 4–6, find the measure of arc y.

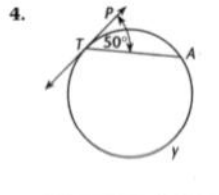
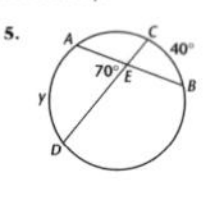
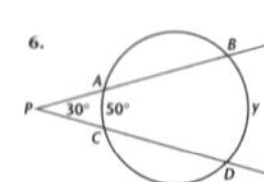

Exercises 7–11 refer to the diagram. $\overline{PA}$ and $\overline{PD}$ are secants to the circle. Chords $\overline{AC}$ and $\overline{BD}$ intersect at E. $\overline{BA} \cong \overline{CD}$, $m\widehat{BC} = 40°$, and $m\angle ABD = 60°$.

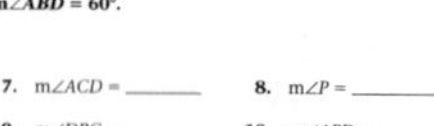

7. $m\angle ACD$ = ______ 8. $m\angle P$ = ______
9. $m\angle DBC$ = ______ 10. $m\angle AED$ = ______ 11. $m\angle PCB$ = ______

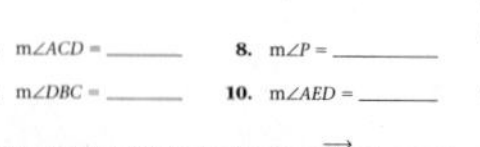

Exercises 12–16 refer to the diagram. $\overrightarrow{PD}$ is tangent to circle O, B is the midpoint of $\widehat{AC}$, $\overline{PB}$ is a secant, and $m\widehat{BA} : m\widehat{AD} : m\widehat{DC} = 2 : 3 : 5$.

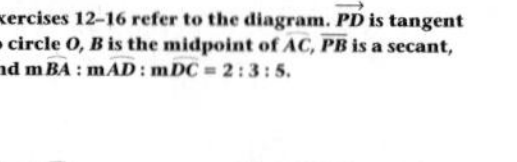

12. $m\widehat{BC}$ = ______ 13. $m\angle ADC$ = ______
14. $m\angle AEB$ = ______ 15. $m\angle ADP$ = ______ 16. $m\angle P$ = ______

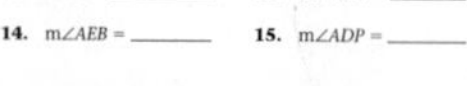
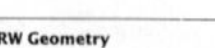

HRW Geometry Practice & Apply 83

Case 2: Two tangents intersect with a vertex that is outside the circle.

Use $\odot M$ for Exercises 16–22. $\overrightarrow{VA}$ and $\overrightarrow{VC}$ are tangent to $\odot M$ at A and C, respectively.

16. What is the relationship between $m\angle 1$ and the measures of $\angle 2$ and $\angle AVC$? $m\angle 1 = m\angle 2 + m\angle AVC$

17. The measure of $\angle 1$ is half the measure of its intercepted arc. Name the arc. $\overparen{AXC}$

18. The measure of $\angle 2$ is half the measure of its intercepted arc. Name the arc. $\overparen{AC}$

19. Suppose $m\overparen{AXC} = 260°$; find $m\overparen{AC}$. 100°

20. Copy and complete the table.

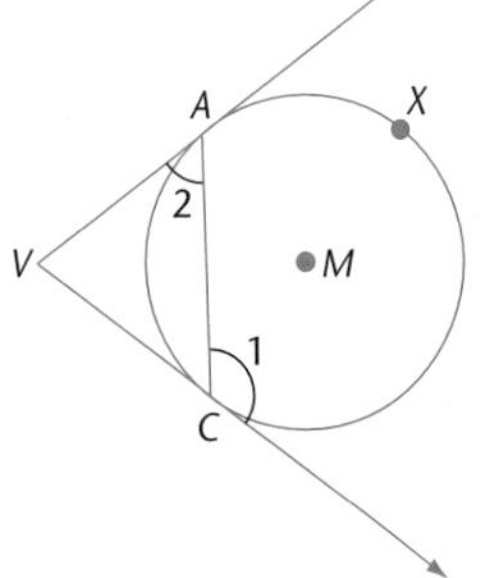

$m\overparen{AXC}$	$m\overparen{AC}$	$m\angle 1$	$m\angle 2$	$m\angle AVC$
220°	(a)	(b)	(c)	(d)
300°	(e)	(f)	(g)	(h)
100°	(i)	(j)	(k)	(l)
80°	(m)	(n)	(o)	(p)
$x°$	(q)	(r)	(s)	(t)

$$m\angle AVC = \frac{m\overparen{AXC} - m\overparen{AC}}{2} = m\overparen{AXC} - 180°$$

21. Write an equation that describes $m\angle AVC$ in terms of $m\overparen{AXC}$ and $m\overparen{AC}$.

22. Complete the following statement.

half of the difference between the measures of the two intercepted arcs, or the measure of the larger arc minus 180°

THEOREM

If two tangents intersect in the exterior of a circle, the measure of the angle formed is __?__. **9.4.5**

23. Communications The maximum distance a radio signal can reach directly is the length of the segment tangent to the earth's curve. If the tangent radio signals from a tower form angles with the tower to the horizon of 89.5°, what is the measure of the intercepted arc of the Earth? Use your answer and the information in the photo to estimate the area of coverage of the signal.

Note: Drawing is not to scale.

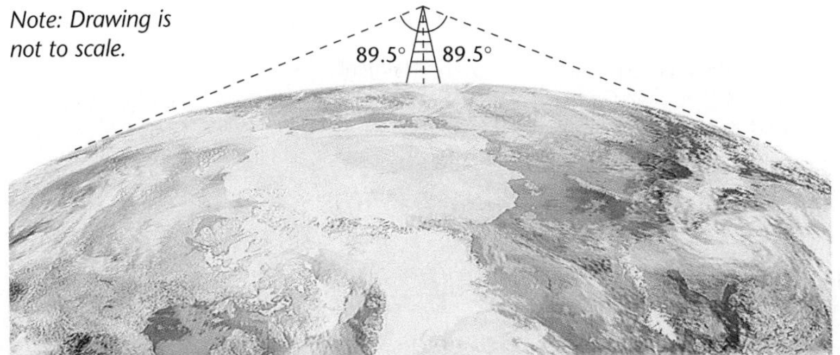

20. a. 140° b. 110° c. 70° d. 40°
e. 60° f. 150° g. 30° h. 120°
i. 260° j. 130° k. 50° l. 80°
m. 280° n. 140° o. 40° p. 100°
q. $360°-x°$ r. $\frac{x°}{2}$ s. $\frac{360°-x°}{2}$ t. $x°-180°$

Alternative Assessment

Portfolio Assessment

Students should explain in writing what is meant by the title of this lesson, Exploring Circles and Special Angles. Their written work should include several types of "special" angles, explain why the angles are "special," and point out any significant relationships among angles, arcs, chords, and so forth. Included in the students' work should be accurate constructions with appropriate degree measures noted.

23. 1°; 3,491 sq mi

Technology Master

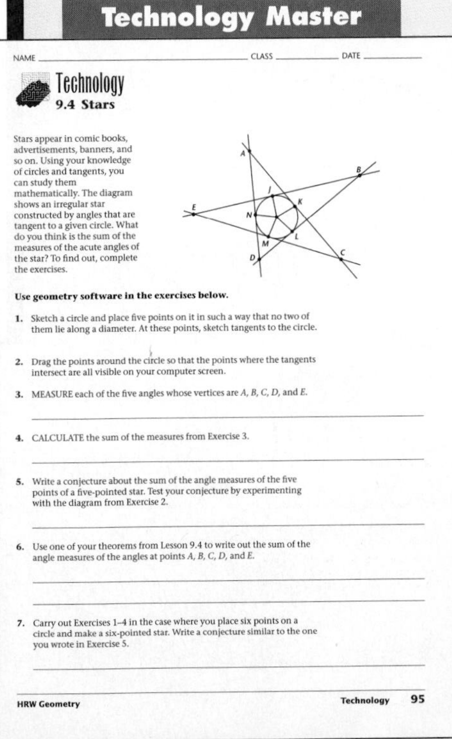

NAME ________ CLASS ________ DATE ________

Technology
9.4 Stars

Stars appear in comic books, advertisements, banners, and so on. Using your knowledge of circles and tangents, you can study them mathematically. The diagram shows an irregular star constructed by angles that are tangent to a given circle. What do you think is the sum of the measures of the acute angles of the star? To find out, complete the exercises.

Use geometry software in the exercises below.

1. Sketch a circle and place five points on it in such a way that no two of them lie along a diameter. At these points, sketch tangents to the circle.
2. Drag the points around the circle so that the points where the tangents intersect are all visible on your computer screen.
3. MEASURE each of the five angles whose vertices are A, B, C, D, and E.
4. CALCULATE the sum of the measures from Exercise 3.
5. Write a conjecture about the sum of the angle measures of the five points of a five-pointed star. Test your conjecture by experimenting with the diagram from Exercise 2.
6. Use one of your theorems from Lesson 9.4 to write out the sum of the angle measures of the angles at points A, B, C, D, and E.
7. Carry out Exercises 1–4 in the case where you place six points on a circle and make a six-pointed star. Write a conjecture similar to the one you wrote in Exercise 5.

HRW Geometry Technology 95

24. Inscribed angle; $m\angle AVC = \frac{1}{2} m\overset{\frown}{AC}$

25. Acute secant-tangent angle, vertex on circle; $m\angle AVC = \frac{1}{2} m\overset{\frown}{AC}$

26. Obtuse secant-tangent angle, vertex on circle; $m\angle AVC = \frac{1}{2} m\overset{\frown}{AC}$

Exercises 12–22 completed the investigations of angle-arc relationships in circles. Exercises 24–30 summarize the angles studied.

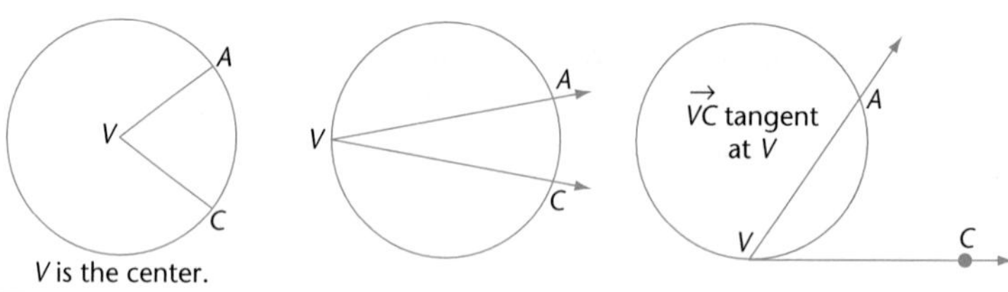

Description of angle	Central angle	**24.** ________	**25.** ________
formula: m∠AVC = ?	$m\angle AVC = m\overset{\frown}{AC}$	________	________

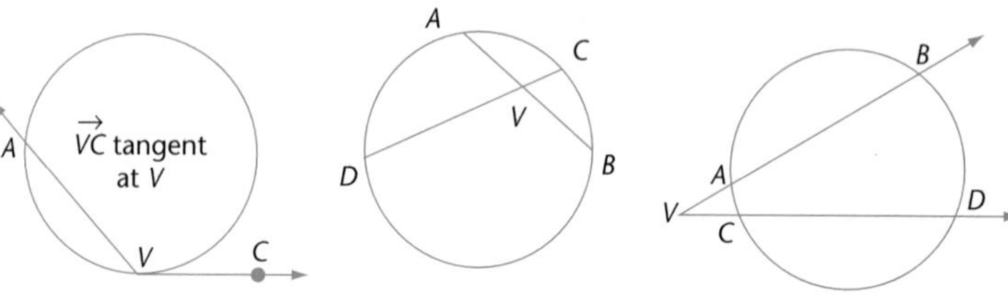

Description of angle	**26.** ________	**27.** ________	**28.** ________
formula: m∠AVC = ?	________	________	________

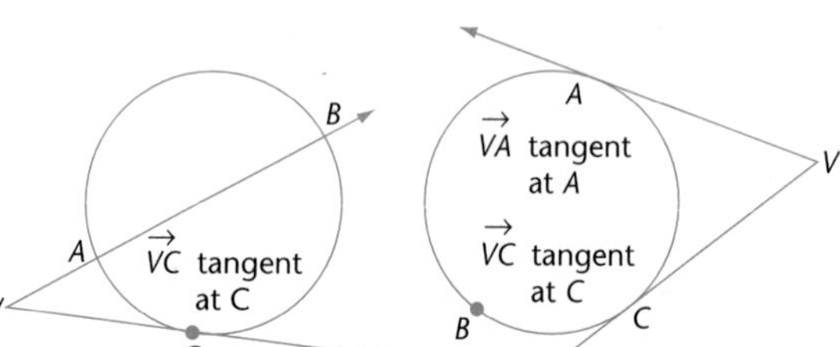

Description of angle	**29.** ________	**30.** ________
formula: m∠AVC = ?	________	________

27. Chord-chord angle, vertex inside circle; $m\angle AVC = \frac{1}{2}(m\overset{\frown}{AC} + m\overset{\frown}{BD})$

28. Secant-secant angle, vertex outside circle; $m\angle AVC = \frac{1}{2}(m\overset{\frown}{BD} - m\overset{\frown}{AC})$

29. Secant-tangent angle, vertex outside circle; $m\angle AVC = \frac{1}{2}(m\overset{\frown}{BC} - m\overset{\frown}{AC})$

30. Tangent-tangent angle; $m\angle AVC = \frac{1}{2}(m\overset{\frown}{ABC} - m\overset{\frown}{AC}) = m\overset{\frown}{ABC} - 180°$

Use $\odot O$ for Exercises 31 and 32. $\overleftrightarrow{AF}$ is tangent at A, m$\overset{\frown}{CD}$ = 105° and m$\overset{\frown}{BC}$ = 47°. $\triangle BDC \cong \triangle CAB$; m$\overset{\frown}{AB}$ = m$\overset{\frown}{DC}$.

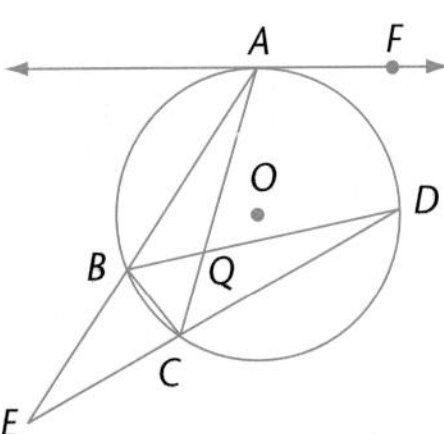

31. Find m$\angle CQD$. 105°

32. Find m$\angle BQC$. 75°

Look Back

$ABCD$ is a parallelogram. [Lessons 5.2, 5.5]

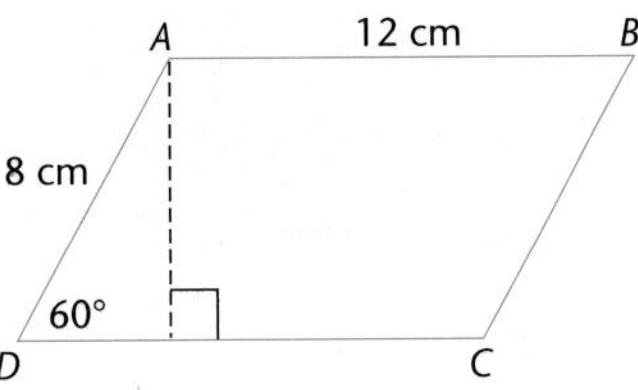

33. Find the perimeter. 40 cm

34. Find the area. 83.14 cm^2

A photocopier was used to reduce $\triangle ABC$ by a factor of 0.75.

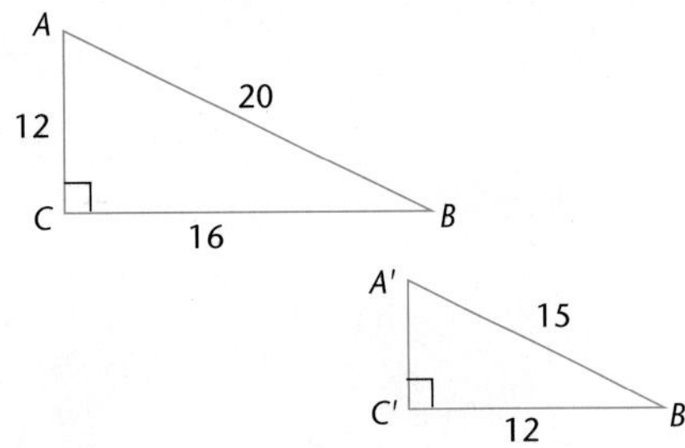

35. Are triangles ABC and $A'B'C'$ similar? Explain your reasoning. **[Lesson 8.2]** Yes; SSS-Similarity

36. Find the value of the following ratios: $\frac{A'B'}{AB}$, $\frac{A'C'}{AC}$, and $\frac{C'B'}{CB}$. **[Lesson 8.6]** All = $\frac{3}{4}$

37. Find $A'C'$. **[Lesson 8.6]** $A'C' = 9$

Look Beyond

A **conic section** is the intersection of a right double cone and a cutting plane. A circle, for example, can be described as a conic section. Other geometric figures result when the angle at which the plane cuts the double cone is adjusted.

Make a sketch showing how each of the following geometric figures can be produced as a conic section.

38. ellipse

39. parabola

40. point

41. two intersecting lines

42. line

43. hyperbola

Use Transparency 81

Look Beyond

In Exercises 38–43 students investigate and sketch the intersection of a double cone and a plane. When the plane is perpendicular to the axis of the cone, the result is a circle. You may need to review the shapes of an ellipse, a parabola, and a hyperbola before students do the activities.

The answers to Exercises 38–43 can be found in Additional Answers beginning on page 727.

Focus

The 1994 earthquake that struck Los Angeles serves as a case study for developing an equation that relates seismic wave data to the distance from the epicenter. Students use the geometry of circles to plot the location of the epicenter.

Motivate

Discuss the LA earthquake. Students may remember reading or hearing about it. Those in the area will of course remember experiencing it. Discuss the following points about seismic waves with students:

- they are vibrations of the earth.
- the two types used in this lesson are called s waves and p waves.
- s waves are transverse (the vibrations are perpendicular to the path of the wave) like ocean waves.
- p waves are longitudinal (the vibrations are along the path of the wave) like sound waves.
- seismographs are instruments that record seismic waves, the display produced is called a seismogram.

Severe Earthquake Hits Los Angeles

Collapsed Freeways Cripple City

Imagine that you are a geologist in the Los Angeles area. You have records from three different stations (A, B, and C) of waves produced by the earthquake. From these seismograms you must pinpoint the epicenter, the place on the earth's surface directly above the origin of the earthquake.

To find the epicenter, you will use two types of seismic waves, **S** waves and **P** waves. Because **S** waves and **P** waves travel at different speeds, you can use the difference in their travel times to determine how far they have gone. You are now ready to begin.

Distance in km (d)	Travel time in secs for s wave (t_s)	Travel time in secs for p wave (t_p)	Time difference btwn s and p waves (D_t)
10	2.86	1.67	1.19
20	5.71	3.33	2.38
30	8.57	5.00	3.57
40	11.43	6.67	4.76

D_t gets larger, because if you imagine an s and p wave starting at the same time and same place, the p wave will keep increasing its lead as the two waves keep traveling

b. $t_s = \dfrac{d}{vs}$ $\qquad t_p = \dfrac{d}{vp}$

$$Dt = ts - tp = \left(\frac{d}{vs}\right) - \left(\frac{d}{vp}\right)$$

c. $d = D_t \cdot (v_s)(v_p) \backslash (v_p - v_s)$

COOPERATIVE LEARNING

WAVE SPEED VALUES

Speed of **P** wave
v_p = 6 km/s

Speed of **S** wave
v_s = 3.5 km/s

1. First, determine how the time between the arrival of the **S** waves and **P** waves depends on how far you are from the origin of the earthquake.

   a. Copy and complete the table. Use the wave speed values shown. What happens to $\mathbf{D_t}$ as **d** increases? Why?

Distance in kilometers	Travel time in seconds for S wave (t_s)	Travel time in seconds for P wave (t_p)	Time difference between S and P waves (D_t)
10.0	?	?	?
20.0	?	?	?
30.0	?	?	?
40.0	?	?	?

STATION DATA
D_t (seconds)

Station A	4.4
Station B	5.5
Station C	6.3

   b. Write an equation for $\mathbf{D_t}$ in terms of v_s, v_p, and **d.** (**Hint:** First write equations for $\mathbf{t_s}$ and $\mathbf{t_p}$.)

   c. Solve your equation for **d**.

2. Use your equation from Part (c) to find the distances in kilometers from the earthquake origin.

   a. Distance from station A ($\mathbf{d_A}$).

   b. Distance from station B ($\mathbf{d_B}$).

   c. Distance from station C ($\mathbf{d_C}$).

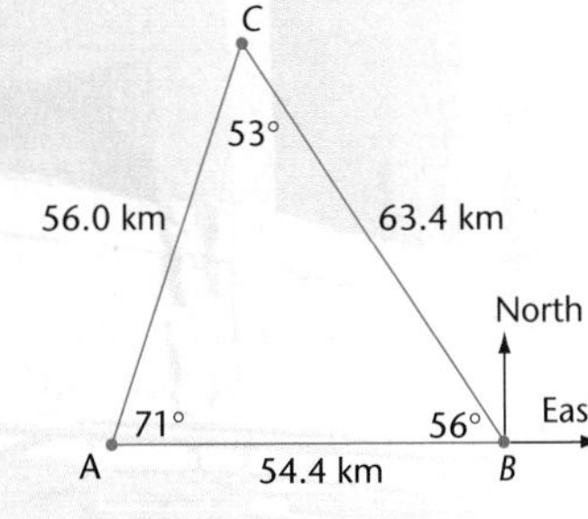

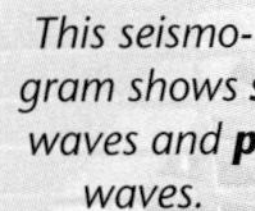

*This seismogram shows **s** waves and **p** waves.*

3. Now you are ready to locate the epicenter. Copy the map of stations A, B, and C.

   Draw three circles to scale:

   Radius = $\mathbf{d_A}$, center at A
   Radius = $\mathbf{d_B}$, center at B
   Radius = $\mathbf{d_C}$, center at C.

   The epicenter is in the region where the three circles overlap.

   For more precision, draw chords connecting the intersections of each pair of circles. The epicenter is where the three chords intersect. How far and in what direction from station B is the epicenter?

   The seismograph readings used to locate the epicenter of the 1994 Los Angeles earthquake were actually from stations all southeast of the epicenter. How would that make finding the location of the epicenter more difficult?

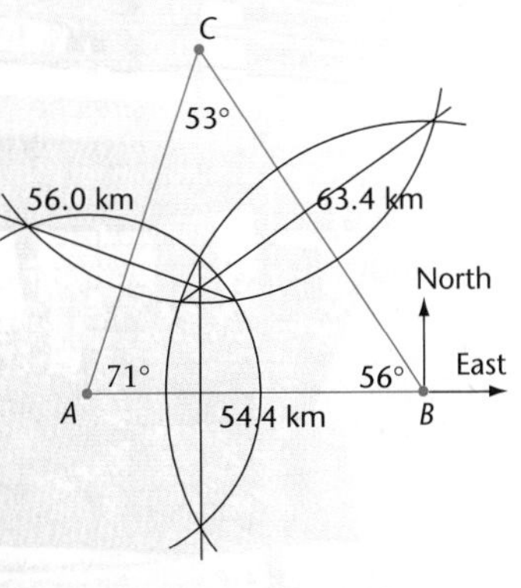

Cooperative Learning

Have students work in pairs or small groups using a compass, a ruler, and a protractor for the constructions. You may need to discuss the algebraic notation in the activities. Point out that the notation is not difficult but students need to be careful to use the appropriate values when substituting into the formulas. Data for the students' calculations are provided in the boxes on the page. Note that the times provided are for three recording stations that were actually used to locate the epicenter of the quake. To simplify the activity, the locations of stations A and C have been changed so that the three stations surround the epicenter thus making the drawings of circles and chords much clearer.

Discuss

What are the advantages of having a global network of seismograph stations? What is the minimum number of stations needed to locate the epicenter of a quake? Would more stations make locating quake epicenters more precise? Why or why not?

2. a. $d_A \approx 37.0$ km

   b. $d_B \approx 46.2$ km

   c. $d_C \approx 52.9$ km

3. The epicenter is between 37 and 38 km from station B.in a direction 16° North of West from station B.

   The circles would be closer together making it harder to distinguish their points of intersection.

PREPARE

Objectives

- Explore special segments relative to the circle including secant-secant, secant-tangent, and chord-chord.
- Develop and use rules for special segments including secant-tangent, secant-secant, and chord-chord.

RESOURCES

- Practice Master 9.5
- Enrichment Master 9.5
- Technology Master 9.5
- Lesson Activity Master 9.5
- Quiz 9.5
- Spanish Resources 9.5

Assessing Prior Knowledge

Use compass and straightedge or geometry graphics software.

1. Construct circle P and two secant lines that pass through the circle. The lines intersect at a point X that is outside of the circle.

   [check student constructions]

2. Start from point X and move along one secant line towards the circle. Label the first intersection point C and the second intersection point A.

   [check student constructions]

TEACH

why Ask for volunteers to describe ways of finding the center of any arc. If necessary, remind students that this point is the center of the circle that contains the arc.

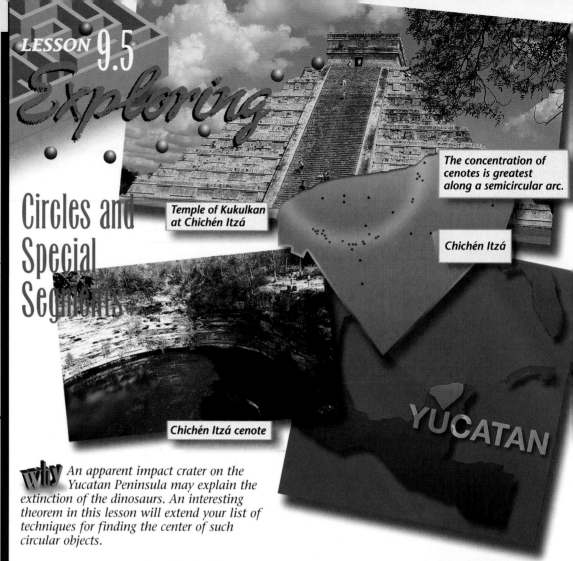

LESSON 9.5 *Exploring*

Circles and Special Segments

Temple of Kukulkan at Chichén Itzá

The concentration of cenotes is greatest along a semicircular arc.

Chichén Itzá

YUCATAN

Chichén Itzá cenote

why *An apparent impact crater on the Yucatan Peninsula may explain the extinction of the dinosaurs. An interesting theorem in this lesson will extend your list of techniques for finding the center of such circular objects.*

A semicircular arc of natural wells (cenotes) *such as this one at Chichén Itzá, stretches across the Yucatan Peninsula. Some scientists believe that the* cenote *ring is evidence of an impact crater from an asteroid or comet.*

In previous lessons you investigated special angles and arcs formed by secants and tangents to circles. Do you think segments formed by secants and tangents might also have special properties? The following terms will be used in this lesson.

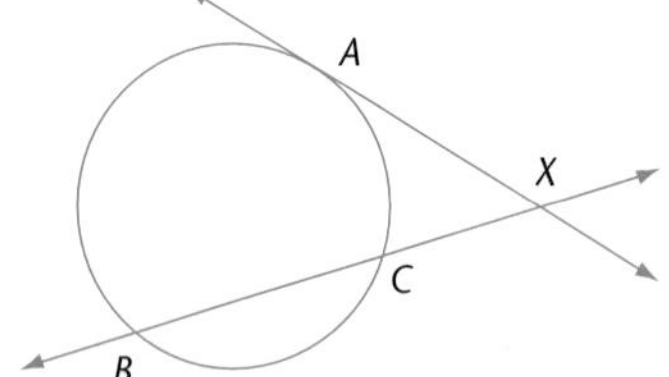

If $\overline{XA}$ is a tangent line and $\overline{XB}$ is a secant line, then

$\overline{XA}$ is a **tangent segment**,

$\overline{XB}$ is a **secant segment**,

$\overline{XC}$ is an **external secant segment**,

and $\overline{BC}$ is a **chord**.

ALTERNATIVE teaching strategy

Hands-On Strategies

The ideas developed in the three explorations can be investigated using paper folding. Wax paper works particularly well. Students who find geometry software intimidating and compass-straightedge work cumbersome may enjoy using paper-folding strategies as an alternative.

Exploration 1 *Lines Formed by Tangents*

You will need

Geometry technology or
Ruler and compass

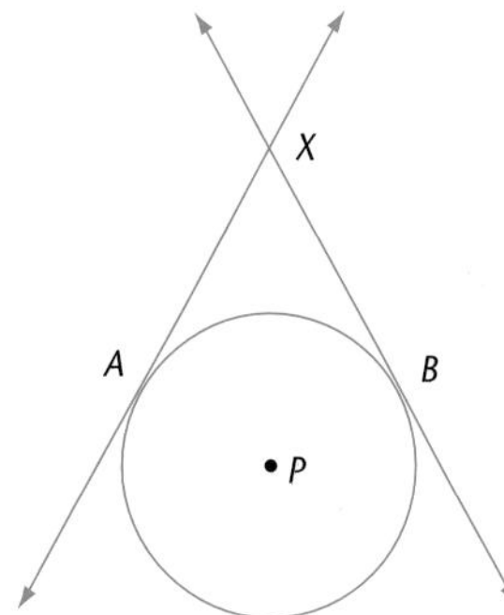

1. You are given $\odot P$ with tangent lines $\overleftrightarrow{XA}$ and $\overleftrightarrow{XB}$. Construct the figure.
2. Measure the lengths of $\overleftrightarrow{XA}$ and $\overleftrightarrow{XB}$.
3. Write a conjecture about the lengths of two segments that are tangent to a circle from the same external point.
4. Add segments $\overline{AP}$, $\overline{BP}$, and $\overline{XP}$ to your figure. Use the resulting figure to prove your conjecture. ❖

Exploration 2 *Lines Formed by Secants*

You will need

Geometry technology or
Ruler and compass

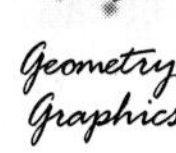

Part I

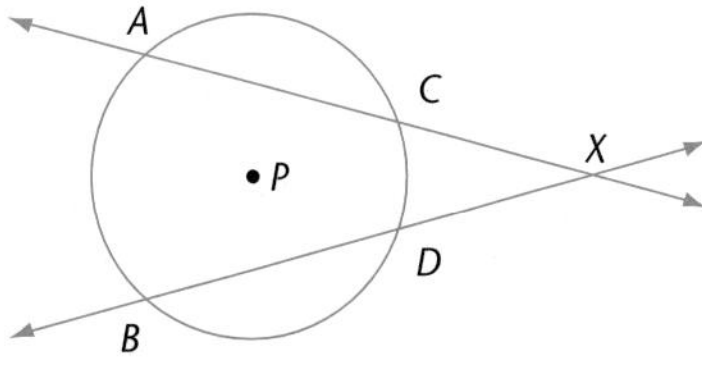

1. You are given a circle with secant lines $\overleftrightarrow{XA}$ and $\overleftrightarrow{XB}$. Construct the figure.
2. Construct segments $\overline{AD}$ and $\overline{BC}$ that intersect at a point O. Name the two triangles that are formed inside the circle.
3. Name two angles of the triangles that intercept the same arc of the circle. What can you conclude about these angles?
4. What other angles of the triangle can you show to be congruent? What can you conclude about the triangles inside the circle?
5. Name the two large triangles in your figure that each have a vertex at X. What can you conclude about them? Complete the similarity statement:

$$\Delta AXD \sim \underline{\ ?\ }$$

6. Complete the proportion by relating two sides of one triangle to two sides of the other triangle:

$$\frac{AX}{?} = \frac{XD}{?}$$

Exploration 1 Notes

Students discover that the two lengths AX and BX in the figure are equal. Students using geometry graphics software may believe they can "prove" this conjecture by measuring segments $\overline{AX}$ and $\overline{BX}$, then showing the lengths are equal no matter how they drag on parts of the construction. These students may say that Step 4 in the exploration is unnecessary. This is a very convincing demonstration, but it is not the same as a logical geometry proof, which must be based on known facts about angles, arcs, segments, triangles, and other parts of the figures.

Ongoing Assessment

3. The lengths of the two segments that are tangent to a circle from the same external point are equal.

Exploration 2 Notes

In both Explorations 2 and 3, students are looking for four segments so that they can write a relationship involving products of segments. That is, if the segments are a, b, c, and d, students will end up with an equation like $ab = cd$ or $ac = bd$.

Earth Science Craters such as the one shown at the beginning of this lesson are formed by waves spreading out from a center point. Have students share what they have learned about wave motions in science courses.

TEACHING tip

For Explorations 2 and 3, the use of colored pencils can help make relationships in the figures more obvious. Students can, for example, say that the product of the two red segments must always equal the product of the two blue segments.

ongoing ASSESSMENT

7. ... the product of the lengths of the other secant segment and its external segment.

Math Connection Algebra

For the last step in Part 2, make sure students understand how to interpret the expression XB^2 which means "the square of the length of segment $\overline{XB}$." It does not mean "X times the square of B."

ongoing ASSESSMENT

2. ... the length of the tangent segment squared.

Exploration 3 Notes

The proportion in Step 5 of the exploration may be more evident to students if the figure shows two very obvious scalene similar triangles. Students using geometry software can drag on point X to change the shapes of the triangles.

7. Cross-multiply and state your result.

Based on your result, complete the following theorem.

THEOREM

If two secants intersect outside of a circle, then the product of the lengths of one secant segment and its external segment equals __?__ .
(Whole × Outside = Whole × Outside.) **9.5.1**

8. Measure segments $\overline{XA}$, $\overline{XC}$, $\overline{XB}$, and $\overline{XD}$. Calculate the products $XA \cdot XC$ and $XB \cdot XD$. Do your results agree with the theorem?

Part II

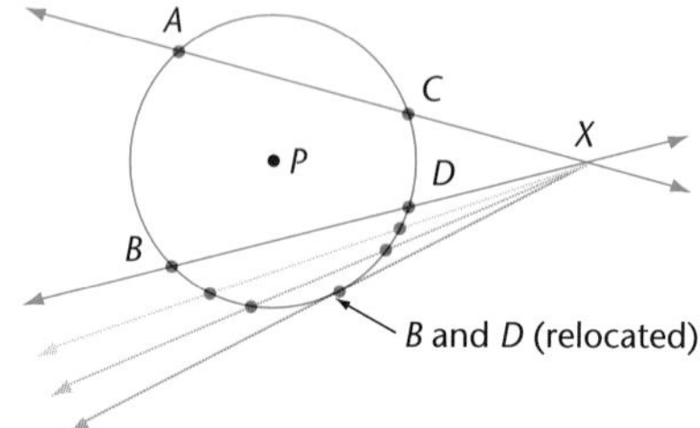

1. Imagine moving point B along the circle so that $\overleftrightarrow{XB}$ becomes a tangent line. You can think of points B and D as coinciding at the point of tangency. What is the relationship between $\overline{XB}$ and $\overline{XD}$?

2. Substitute $\overleftrightarrow{XB}$ for $\overleftrightarrow{XD}$ in your result from Part 1, Step 6. Complete the following theorem.

THEOREM

If a tangent and a secant intersect outside of a circle, then the product of the length of the secant segment and its external segment equals __?__ .
(Whole × Outside = Whole Squared) **9.5.2**

3. Construct the figure with point B moved. Measure segments $\overline{XA}$, $\overline{XC}$, and $\overline{XB}$. Calculate the product $XA \cdot XC$ and XB^2. Do your results agree with the theorem? ❖

Exploration 3 Intersection of Chords Inside a Circle

You will need
Geometry technology or
Ruler and compass

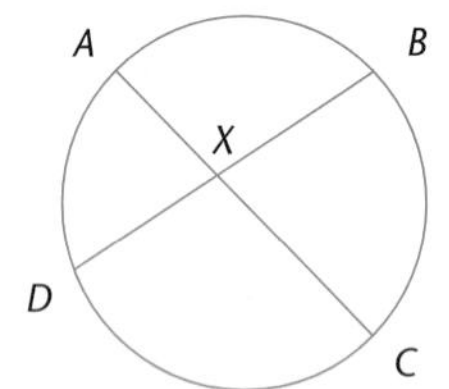

1. You are given a figure with chords $\overline{AC}$ and $\overline{DB}$ intersecting at point X. Construct the figure.

2. Draw segments $\overline{AD}$ and $\overline{BC}$. Name the two triangles that are formed.

ENRICHMENT Students are to construct a tangent to a circle from a point outside the circle. The resulting figure will look like the diagram for Exploration 1. **Possible method: Given circle O and point A outside the circle, draw segment $\overline{AO}$. Bisect $\overline{AO}$ to get its midpoint M. Draw circle M with radius $\overline{MO}$. This circle meets circle O in points B and C. Lines $\overleftrightarrow{AB}$ and $\overleftrightarrow{AC}$ are both tangent to circle O.**

INCLUSION strategies **English Language Development** It is particularly important that students master the vocabulary given on the first page of the lesson. If the vocabulary continues to present difficulties, have students use different colors to highlight and describe particular segments.

3. Name two angles of the triangles that intercept the same arc of the circle. What can you conclude about these angles?

4. What other angle of the triangles can you show to be congruent? What can you conclude about the triangles?

5. Complete the proportion by relating two sides of one triangle to two sides of the other triangle.

$$\frac{DX}{XA} = \frac{?}{?}$$

Cross-multiply and state your result.

6. Based on your result, complete the following theorem.

> **THEOREM**
>
> If two chords intersect inside a circle, then the product of the lengths of the segments of one chord equals ___?___ . **9.5.3**

7. Measure segments $\overline{BX}$, $\overline{DX}$, $\overline{AX}$, and $\overline{CX}$. Calculate the product $BX \cdot DX$ and $AX \cdot CX$. Do your results agree with the theorem? ❖

APPLICATION

Geology On this map of the *cenote* ring, the distance from point A on the coast to point B near Chichén Itzá measures 3.8 cm. You can use the theorem you have just learned to find the center and diameter of the *cenote* ring.

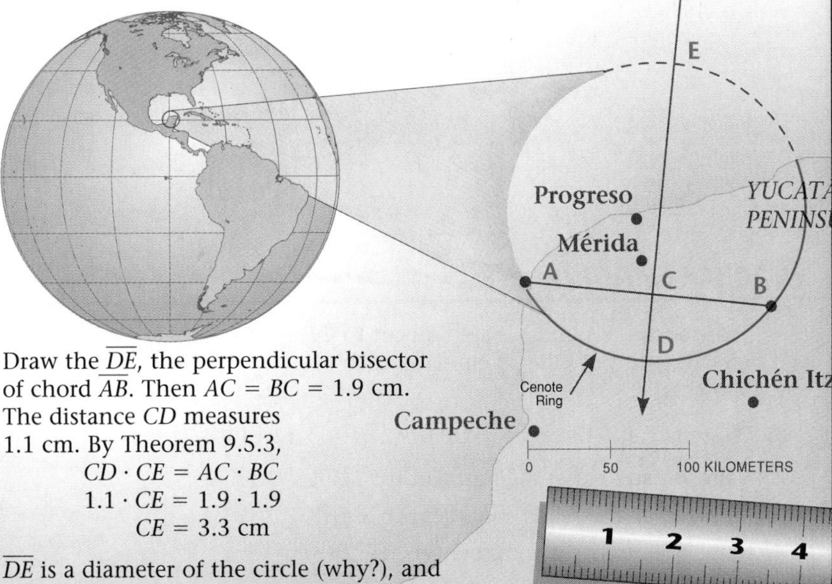

Draw the $\overline{DE}$, the perpendicular bisector of chord $\overline{AB}$. Then $AC = BC = 1.9$ cm. The distance CD measures 1.1 cm. By Theorem 9.5.3,

$$CD \cdot CE = AC \cdot BC$$
$$1.1 \cdot CE = 1.9 \cdot 1.9$$
$$CE = 3.3 \text{ cm}$$

$\overline{DE}$ is a diameter of the circle (why?), and

$$DE = 1.1 + 3.3 = 4.4 \text{ cm}$$

Using the map scale, the actual diameter is about 170 km.
Using a radius of 2.2 cm for the *cenote* ring, you can locate the center point on $\overline{DE}$. This point is 0.7 cm east of Progreso on the map. Using the map scale, the actual distance is a little over 15 km east of Progreso.

Cooperative Learning

Most students will benefit if Explorations 2 and 3 are done in small groups. One student might be responsible for reading the directions for each step, while two students carry out the instructions independently of each other, and the fourth student checks and compares results. Students can change roles and repeat the activity. Exploration 2, in particular, might well be done two or three times.

ONGOING ASSESSMENT

6. . . . the product of the lengths of the other segment.

Technology The key theorem developed in Exploration 3 can be demonstrated using geometry graphics software. You can present this demonstration yourself. Or, you can assign the construction to small groups of students. Construct two rectangles with side lengths a, b, c, and d, corresponding to the four chord-segments. Then create two rectangles, one with sides a and b, and the other with sides c and d. Confirm the cross-product relationship by showing that the areas of the rectangles are equal.

Assess

Selected Answer

Exercises 7, 9, 13, 15, 19, 21, 27, 29, and 31

Assignment Guide

Core 1–19, 23–31

Core Plus 1–6, 13–16, 18–32

Technology

For Exercises 8–21, students can use geometry graphics software to construct the figures. The software can then be used to confirm results found through application of angle-circle relationships and computation.

Error Analysis

For Exercises 8–21, students may benefit from sketching figures and color coding segments with highlighter pens.

Practice Master

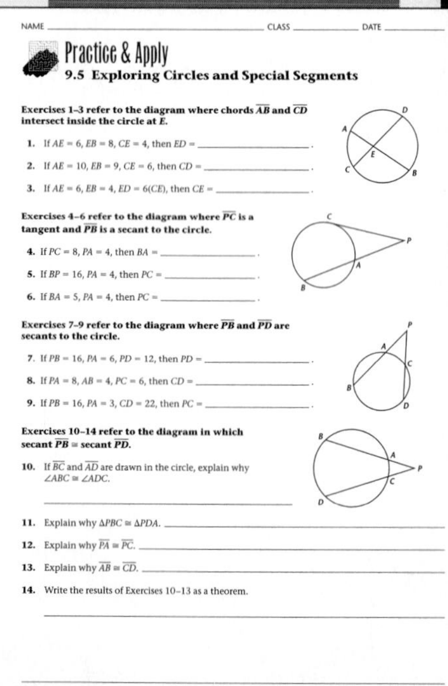

NAME ______ CLASS ______ DATE ______

Practice & Apply

9.5 Exploring Circles and Special Segments

Exercises 1–3 refer to the diagram where chords $\overline{AB}$ and $\overline{CD}$ intersect inside the circle at E.

1. If $AE = 6$, $EB = 8$, $CE = 4$, then $ED =$ ______.
2. If $AE = 10$, $EB = 9$, $CE = 6$, then $CD =$ ______.
3. If $AE = 6$, $EB = 4$, $ED = 6(CE)$, then $CE =$ ______.

Exercises 4–6 refer to the diagram where $\overline{PC}$ is a tangent and $\overline{PB}$ is a secant to the circle.

4. If $PC = 8$, $PA = 4$, then $BA =$ ______.
5. If $BP = 16$, $PA = 4$, then $PC =$ ______.
6. If $BA = 5$, $PA = 4$, then $PC =$ ______.

Exercises 7–9 refer to the diagram where $\overline{PB}$ and $\overline{PD}$ are secants to the circle.

7. If $PB = 16$, $PA = 6$, $PD = 12$, then $PD =$ ______.
8. If $PA = 8$, $AB = 4$, $PC = 6$, then $CD =$ ______.
9. If $PB = 16$, $PA = 3$, $CD = 22$, then $PC =$ ______.

Exercises 10–14 refer to the diagram in which secant $\overline{PB} \cong$ secant $\overline{PD}$.

10. If $\overline{BC}$ and $\overline{AD}$ are drawn in the circle, explain why $\angle ABC \cong \angle ADC$. ______
11. Explain why $\Delta PBC \cong \Delta PDA$. ______
12. Explain why $\overline{PA} \cong \overline{PC}$. ______
13. Explain why $\overline{AB} \cong \overline{CD}$. ______
14. Write the results of Exercises 10–13 as a theorem. ______

84 Practice & Apply HRW Geometry

Communicate

$\overleftrightarrow{AB}$ is tangent at point C. Identify each of the following in $\odot R$.

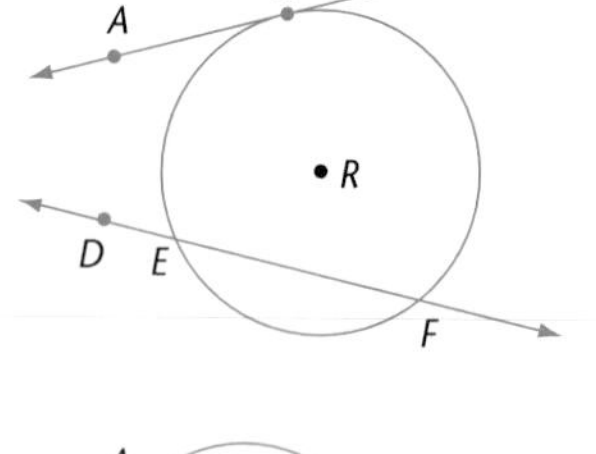

1. a tangent segment
2. a secant segment
3. an external secant segment

Identify each of the following in $\odot S$.

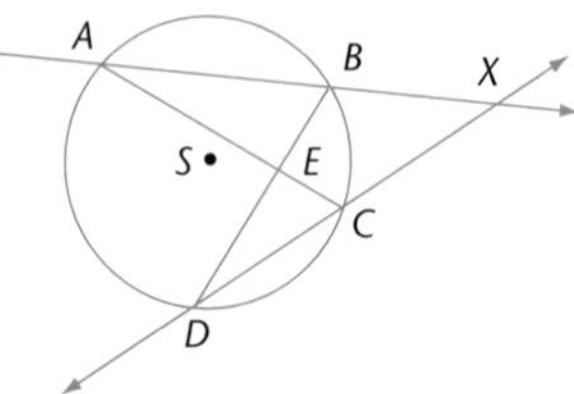

4. a pair of similar triangles
5. 2 pairs of congruent angles

Algebra Find x in each of the following.

6.

7.

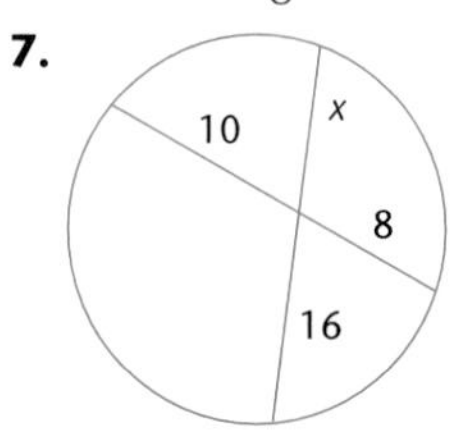

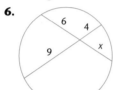

Practice & Apply

Algebra $\overrightarrow{VA}$ and $\overrightarrow{VB}$ are tangent to $\odot P$. $VA = 6$ cm; radius of $\odot P = 3$ cm. Find each of the following.

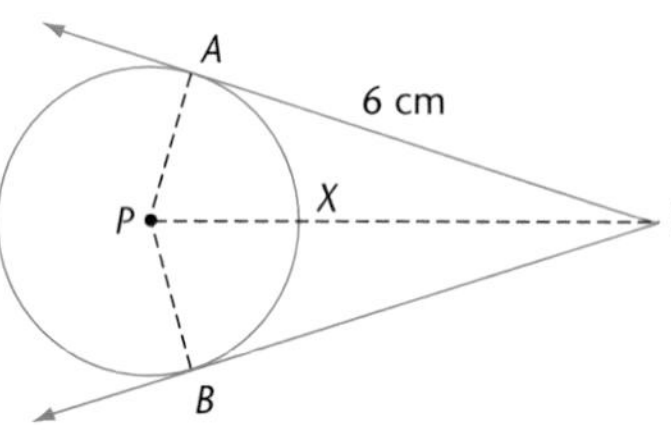

8. VB 6 cm **9.** AP 3 cm **10.** BP 3 cm **11.** PV 6.71 cm

12. Name an angle that is congruent to $\angle AVP$. $\angle BVP$

13. Name an angle that is congruent to $\angle APV$. $\angle BPV$

14. Name an arc that is congruent to $\overset{\frown}{AX}$. $\overset{\frown}{BX}$

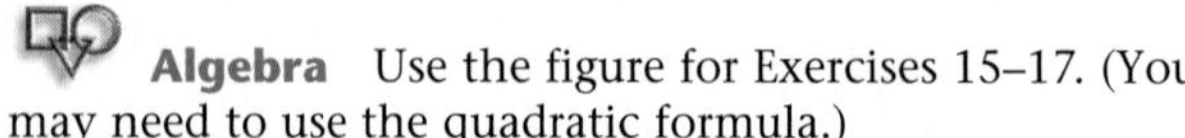

Algebra Use the figure for Exercises 15–17. (You may need to use the quadratic formula.)

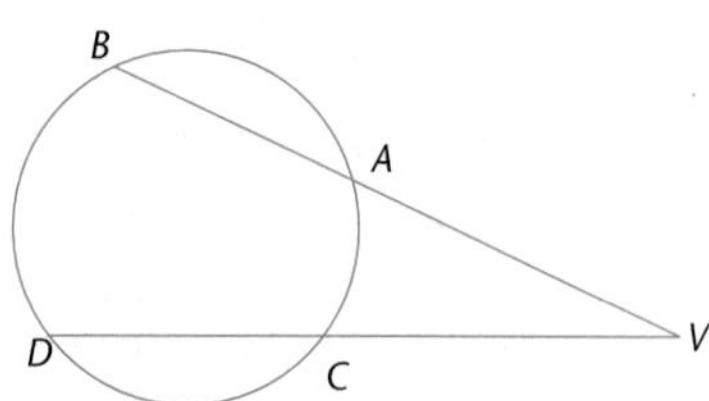

15. Given: $VA = 4$; $VB = 10$; $VC = 5$. Find CD. 3

16. Given: $VB = x$; $VA = 6$; $VD = x + 3$; $VC = 5$. Find x. 15

17. Given: $VB = x$; $VA = x - 16$; $VD = 8$; $VC = 5$. Find x. 18.2

Algebra Use the figure for Exercises 18 and 19.

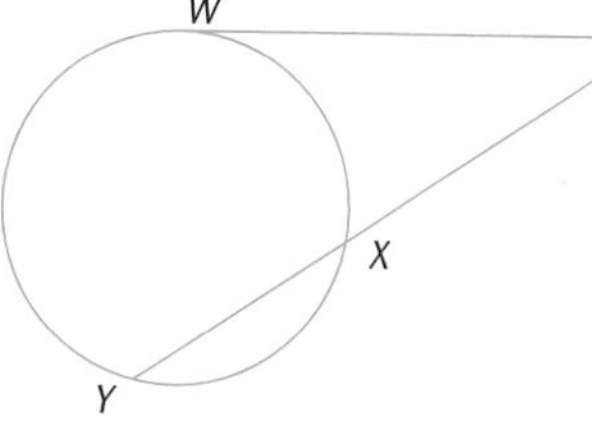

18. Given: $VY = 16$; $VX = 4$.
Find VW. 8 units

19. Given: $VW = 10$; $VX = 8$.
Find VY. 12.5 units

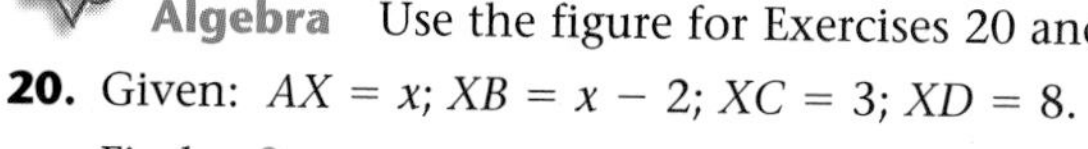

Algebra Use the figure for Exercises 20 and 21.

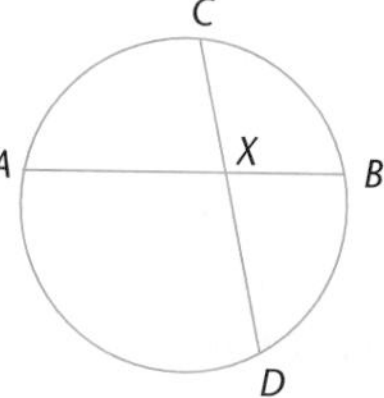

20. Given: $AX = x$; $XB = x - 2$; $XC = 3$; $XD = 8$.
Find x. 6

21. Given: $AB = 10$; $CX = 2$; $CD = 12$.
Find AX. 7.24 or 2.76

22. Machining Jeff is restoring a clock and needs a new gear drive to replace the broken one shown. To machine a new gear, he must determine the diameter of the original. In the picture, F is the midpoint of $\overline{BD}$. Use the product of chord segments to find the original diameter. Diameter = 12.5 cm

$BD = 10$ cm
$EF = 2.5$ cm

alternative ASSESSMENT

Authentic Assessment

Have students design a belt-and-pulley system. The students' mechanical creations will include tangents to various circles. Have students identify in writing the circle-angle relationships in their designs, using terms and concepts learned in this chapter.

Technology Master

NAME ______ CLASS ______ DATE ______

Technology
9.5 Kites

A kite is a simple geometric figure that you have seen before. An illustration of kite *KLMN* inscribed in a circle, whose radius is 6, is shown.

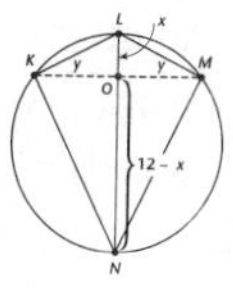

Using one of the theorems you learned about intersecting chords and a spreadsheet, you can successfully study length, perimeter, and area of a kite as x changes.

1. Use one of the theorems of Lesson 9.5 to write an equation for y in terms of x.

Use the Pythagorean theorem to write an expression in terms of x for each length.

2. KL 3. LM 4. MN 5. NK

6. Use the results of Exercises 2–5 to write expressions for the perimeter and area of kite *KLMN*.

7. Create a spreadsheet in which column A contains $x = 0.0, 0.5, 1.0, \ldots, 11.0, 11.5,$ and 12.0, column B contains your expression for perimeter and column C contains your expression for area.

Use your spreadsheet to answer each question.

8. Describe how the perimeter changes as x increases.

9. Describe how the area changes as x increases.

10. Describe the kite inscribed in a circle of radius 6 whose perimeter and area is a maximum.

96 Technology HRW Geometry

23.

24.

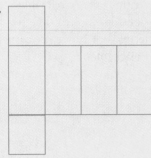

25.

Look Beyond

In Exercise 32, make sure students understand that the circle is centered at the bottom left vertex of the right triangle. Point out the two radii labeled c to show this. The diameter of the circle and the chord formed by extending leg b are the two intersecting chords. Using chord-segment products, students should write $b^2 = (c + a)(c - a)$, which simplifies to $b^2 = c^2 - a^2$.

Look Back

Draw a net for each solid. [Lesson 6.1]

23.

24.

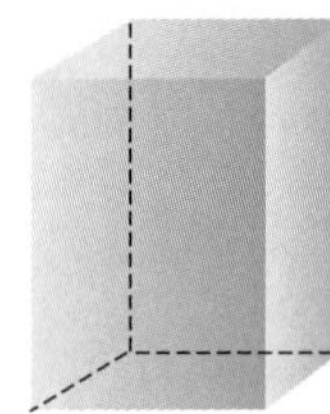

25.

Given: m$\overset{\frown}{AD}$ = 140°, m$\overset{\frown}{BC}$ = 40° and m$\overset{\frown}{CD}$ = 120°. Find the following: [Lesson 9.4]

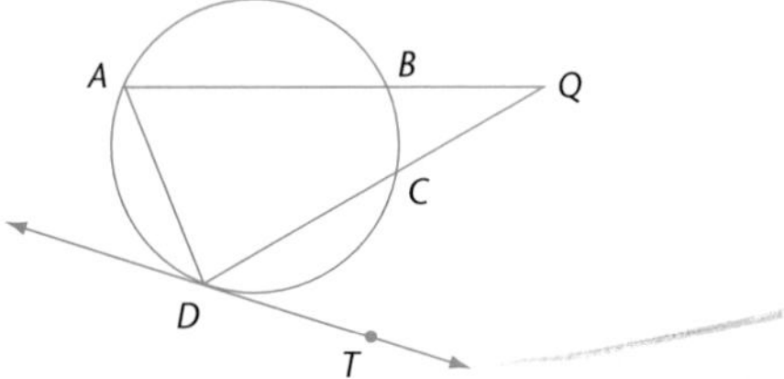

26. m∠AQD 50°

27. m∠QDT 60°

28. m∠QAD 80°

29. m∠ADQ 50°

30. Are there any perpendicular segments in the figure? Explain. No. There are no 90° angles.

31. Are there any parallel segments in the figure? Explain.

Look Beyond

32. **Algebra** The drawing below suggests a visual proof of the "Pythagorean" Right-Triangle Theorem. Use chord-segment products to explain how the proof works.

$(c + a)(c - a) = (b)(b) \Rightarrow c^2 - a^2 = b^2 \Rightarrow c^2 = a^2 + b^2$

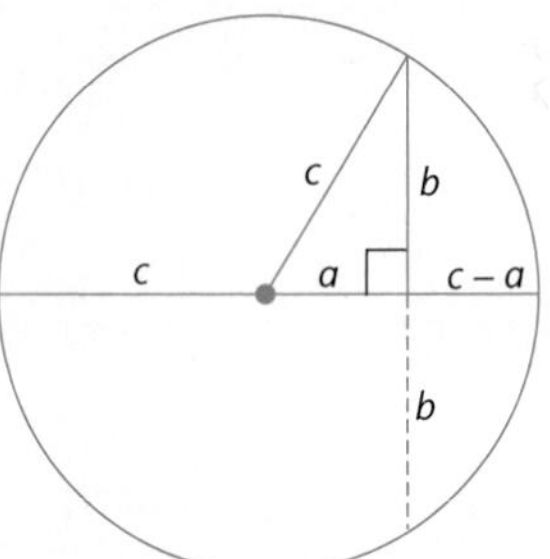

31. No. No angle congruencies exist which would indicate parallel lines.

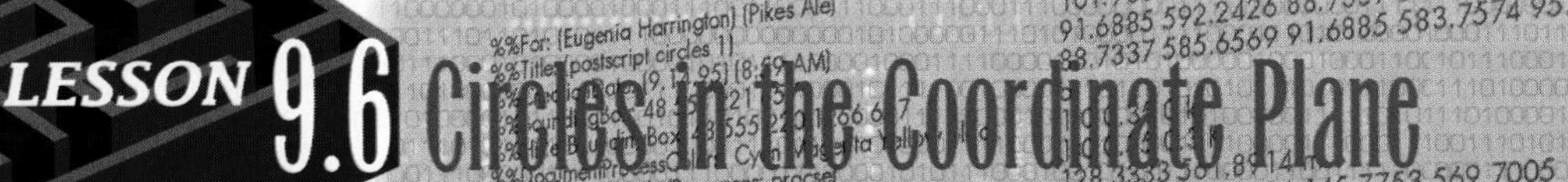

LESSON 9.6 Circles in the Coordinate Plane

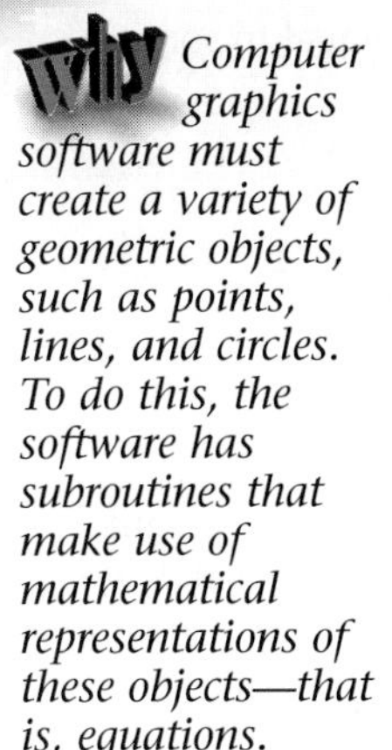

Computer graphics software must create a variety of geometric objects, such as points, lines, and circles. To do this, the software has subroutines that make use of mathematical representations of these objects—that is, equations.

Graphing a Circle From an Equation

ALGEBRA *Connection*

In your work in algebra, you may have investigated graphs of equations such as $y = 2x - 3$ (a line), $y = x^2 - 3$ (a parabola), and $y = 3 \cdot 2^x$ (an exponential curve). In this lesson you will investigate equations in which both x and y are squared.

EXAMPLE 1

Given: $x^2 + y^2 = 25$

Graphics Calculator

Sketch and describe the graph by finding ordered pairs that satisfy the equation. Use a graphics calculator to verify your sketch.

Solution ➤

When sketching the graph of an equation, it is often helpful to investigate its intercepts, which is where the graph crosses the x- and y-axes. To find the x-intercept(s), find the value(s) of x when $y = 0$.

$$x^2 + 0^2 = 25$$
$$x^2 = 25$$
$$x = \pm 5$$

Thus the graph has two x-intercepts, (5, 0) and (−5, 0).

Technology Geometry graphics software that includes a coordinate system can help students investigate and understand the equations for circles in the coordinate plane. By dragging on the center and on the circle, students can see how the equation changes with the size and position of the circle.

PREPARE

Objectives

- Develop and use the equation of a circle.
- Adjust the parameters of a circle equation to move the center on a coordinate plane.

RESOURCES

• Practice Master	9.6
• Enrichment Master	9.6
• Technology Master	9.6
• Lesson Activity Master	9.6
• Quiz	9.6
• Spanish Resources	9.6

Assessing Prior Knowledge

In Exercises 1–3, solve for x or y.

1. $x^2 + 4^2 = 5^2$ $[x = \pm 3]$
2. $x^2 + 12^2 = 13^2$ $[x = \pm 5]$
3. $3^2 + y^2 = 5^2$ $[y = \pm 4]$
4. Solve $x^2 + y^2 = 4^2$ for y. $[y = \pm\sqrt{16 - x^2}]$

TEACH

Investigating circle relationships when the circle is graphed on a coordinate plane helps students relate geometric ideas to those they have learned in algebra.

Math Connection Algebra

You may need to review solving an equation such as $x^2 + 4^2 = 5^2$. Remind them that they get both a positive and a negative value when taking the square root.

Alternate Example 1

Graph the circle with:

1. x-intercepts $(3, 0)$ and $(-3, 0)$.
2. y-intercepts $(0, 4.5)$ and $(0, -4.5)$.
3. its center at the origin and a radius of 6 units.

Technology Students using the TI-82 can define the viewing window and its format using the values and settings below. These choices will work well for most of the activities in the lesson.

WINDOW	FORMAT
Xmin = −18.8	
Xmax = 18.8	
Xscl = 2	
Ymin = −12.4	
Ymax = 12.4	
Yscl = 2	

WINDOW	**FORMAT**
RectGC	PolarGC
CoordOn	CoordOff
GridOff	**GridOn**
AxesOn	AxesOff
LabelOff	LabelOn

The radius would be 7 units, 9 units, $\sqrt{51} \approx 7.14$ units, respectively.

To find the y-intercept(s), find the value(s) of y when $x = 0$.

$$0^2 + y^2 = 25$$
$$y^2 = 25$$
$$y = \pm 5$$

Thus the graph has two y-intercepts, $(0, 5)$ and $(0, -5)$.

Make a table. Next, set x equal to some other value, such as 3.

$$3^2 + y^2 = 25$$
$$y^2 = 16$$
$$y = \pm 4$$

Thus there are two points with an x-value of 3: $(3, 4)$ and $(3, -4)$.

Similarly, by choosing other convenient values for x, a table like the one below is obtained.

x	y	**Points on Graph**
3	±4	(3, 4), (3, −4)
−3	±4	(−3, 4), (−3, −4)
4	±3	(4, 3), (4, −3)
−4	±3	(−4, 3), (−4, −3)

Add these new points to the graph. A circle with radius 5 and its center at the origin (0, 0) begins to appear. Sketch in the curve. ❖

How does the graph change if 25 is changed to 49? to 81? to 51?

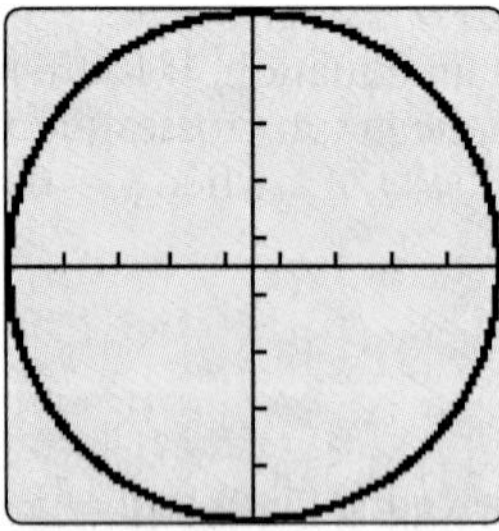

You can also graph the curve using a graphics calculator or a computer with graphics software. You can use the trace function of the computer software or calculator to find the coordinates of individual points on the graph. Note: Some graphing technology requires that you write your equation in the form $y =$ __?__. If your technology requires this, you will need to solve your equation for y.

Social Studies The latitude and longitude lines shown on a globe or map form a type of three-dimensional coordinate system. Have students research how the grid works. Suppose a circle with a radius of 69 miles is centered at the (0, 0) point—on the equator directly south of Greenwich, England. What places would this circle include? What latitude and longitude lines would it intersect or meet? Have students experiment with circles centered at other places on the Earth.

Deriving the Equation of a Circle

In a circle, all the points are a certain distance r from a fixed point. In the simplest case, that fixed point is the origin, as shown.

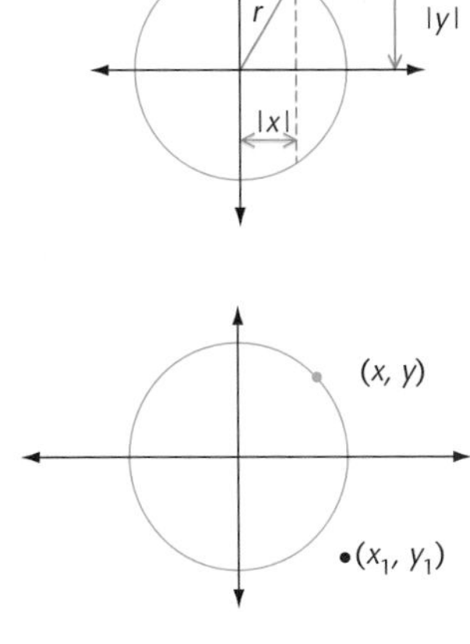

For any point (x, y) on the circle that is not on the x- or y-axis, you can draw a right triangle whose legs have the values $|x|$ and $|y|$. The length of the hypotenuse is the distance r of the point from the origin. So for any such point,

$$x^2 + y^2 = r^2. \quad \text{(Equation 1)}$$

By substituting $y = 0$ or $x = 0$ into the equation, you can see that it is also true for points (x, y) that are on the x- or y-axis.

If point (x_1, y_1) is not on the circle, its distance from the origin is some value not equal to r (why?), so the equation will not be true. That is, for the same value of r,

$$x_1^2 + y_1^2 \neq r^2.$$

Explain why this is true.

Notice that Equation 1 satisfies the following two conditions:

1. It is true of all points (x, y) that are on the circle.

2. It is not true of any point (x, y) that is not on the circle.

Thus Equation 1 is the equation of the circle.

Moving the Center of the Circle

To find the general equation of a circle centered at a point (h, k) that is not at the origin, study the diagram at right. For such a circle,

$$(x - h)^2 + (y - k)^2 = r^2. \quad \text{(Equation 2)}$$

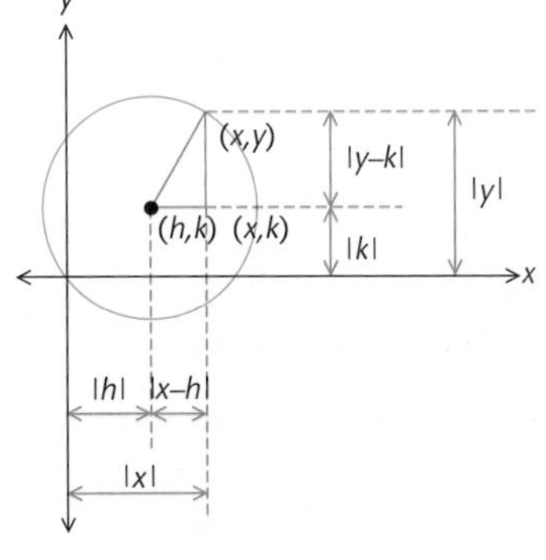

It should also be clear that the equation is not true for a point that does not lie on the equation of the circle.

How can you show that the relationship given in the diagram holds for points (h, k) in quadrants II, III, and IV?

Use Transparency 82

Cooperative Learning

Before students read the text on this page, have them work in small groups to graph a variety of circles. Each graph should be labeled with its equation; for example, $x^2 + y^2 = r^2$. After students have finished, ask them to suggest a general equation for the graph of any circle with center at the origin.

TEACHING tip

Introduce the derivation of the circle formulas by first showing just the right triangles. Ask students how they can use the "Pythagorean" Right-Triangle Theorem to relate the sides of the triangles.

ongoing ASSESSMENT

$x^2 + y^2$ is the square of the distance of the point (x_1, y_1) from the origin. Since (x_1, y_1) is not on the circle, $x^2 + y^2 \neq r$.

Use Transparency 83

CRITICAL Thinking

Draw the circle in each quadrant and check to see that the relationships hold.

ENRICHMENT Students can investigate equations of the form $(x - h)^2 - (y - k)^2 = r^2$. (They will get hyperbolas for the graphs.) Students describe in writing what graphs result and how the values of h, k, and r change the graphs.

Hands-On Strategies

Have students make a large-square coordinate system using four pieces of standard-size paper. Students then cut circles from transparent paper. The diameters should be expressed in whole numbers. Students can move their "see-through" circles around on the large grid to center them at various spots. Find the radius, center coordinates, and equations for each placement.

Alternate Example 2

Find the center and the radius of the circle with the equation $(x+4)^2 + y^2 = 20$.

$\left[\text{Center: } (-4, 0); \text{radius: } +\sqrt{20}\right]$

Math Connection Algebra

Some students may need help understanding why they need to rewrite the expression $(y + 3)$ as $(y - (-3))$. Point out that the general circle equation has minus signs before the values of h and k. So, to find the coordinates of the center of a circle with equation $(x + a)^2 + (y + b)^2 = r^2$, the equation must be written as $(x - (-a))^2 = (y - (-b))^2 = r^2$.

Ongoing Assessment

Try This

a. Center = (–3,3) ; r = 7 units

b. Center = (4,5) ; r = $\sqrt{30}$ units

c. Center = (3,–3) ; r = $\sqrt{50}$ units

d. Center = (–2, 5) ; r = $\sqrt{50}$ units

Assess

Selected Answers

Exercises 13, 15, 17, 19, 21, 27, 29, 33, 37, 41, and 43

Assignment Guide

Core 1–20, 25–30, 37–43, 46–51

Core Plus 1–11, 17–36, 42–55

Technology

A graphics calculator is suggested for Exercises 31–34, although the problems may also be done with pencil and graph paper. It may be helpful to students to do some problems in both ways. Exercises 36–44 suggest that students use geometry graphics software.

EXAMPLE 2

ALGEBRA *Connection*

Given: $(x - 7)^2 + (y + 3)^2 = 36$

Find the center of the circle and its radius.

Solution ➤

Comparing the given equation with the general equation of the circle, you find the following correspondences:

The general equation		*The given equation*
$(x - h)^2$	$\leftrightarrow$	$(x - 7)^2$
$(y - k)^2$	$\leftrightarrow$	$(y + 3)^2$ or $(y - (-3))^2$
r^2	$\leftrightarrow$	36

From this you can conclude that

$$h = 7,\ k = -3, \text{ and } r = 6.$$

That is, the radius of the circle is 6 units, and the center is at $(7, -3)$. ❖

Try This For each of the equations, find the radius and the center of the circle represented. Graph the equation, and compare the graph with your values for the radius and the center of the circle.

a. $(x + 3)^2 + (y - 3)^2 = 49$

b. $(y - 4)^2 + (x - 5)^2 = 30$

c. $(x - 3)^2 + (y + 3)^2 + 1 = 50$

d. $(x + 2)^2 + (y - 5)^2 = 50$

Exercises & Problems

Communicate

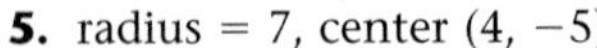

Find the x- and y-intercepts for each equation.

1. $x^2 + y^2 = 100$
2. $x^2 + y^2 = 64$
3. $x^2 + y^2 = 50$
4. State the general form for the equation of a circle.

State the formula for each circle with the center and radius given.

5. radius = 7, center $(4, -5)$
6. radius = 2.5, center $(0, 0)$
7. radius = 10, center $(-1, -7)$

Reteaching the lesson

Technology Using a graphics calculator, have students enter and graph the following equations:

$Y_1 = \sqrt{(36 - (X - 7)^2)} - 3$

$Y_2 = -\sqrt{(36 - (X - 7)^2)} - 3$

The equations will give the graph of the circle in Example 2. Once the equations are entered, students should experiment by changing the 7 and the 3. They should change both numerical values and the signs in front of the numbers. Students will get circles with a radius of 6, but with different centers. To change the radius, students should enter different values for 36.

In Exercises 8–11, which equations are circles? How can you tell?

8. $x^2 = 10 - y^2$ **9.** $x + y^2 = 100$

10. $x^3 + y^3 = 64$ **11.** $x^2 - y^2 = 9$

Practice & Apply

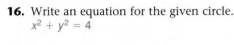

 **Algebra** Find the center and radius of each circle.

12. $x^2 + y^2 = 100$
center = (0, 0); $r = 10$

13. $x^2 + y^2 = 101$
center = (0, 0); $r = \sqrt{101}$

Write an equation of a circle with the given characteristics.

14. center = (0, 0); radius = 6
$x^2 + y^2 = 36$

15. center = (0, 0); radius = $\sqrt{13}$
$x^2 + y^2 = 13$

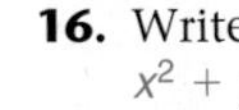

16. Write an equation for the given circle.
$x^2 + y^2 = 4$

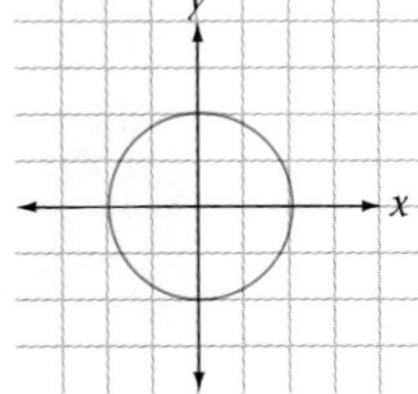

 **Algebra** Find the center and radius of each circle.

17. $(x - 6)^2 + y^2 = 9$ **18.** $x^2 + (y + 3)^2 = 4$

19. $(x + 5)^2 + (y - 2)^2 = 16$ **20.** $(x + 1)^2 + (y + 3)^2 = 19$

21. $x^2 + y^2 = 36$ **22.** $y^2 + (x + 3)^2 = 49$

Algebra Write an equation of a circle with the given characteristics.

23. center = (2, 3); radius = 4 **24.** center = (−1, 5); radius = 7

25. center = (0, 6); radius = 5 **26.** center = (4, −3); radius = $\sqrt{7}$

 **Algebra** Write an equation for the given circle.

27.

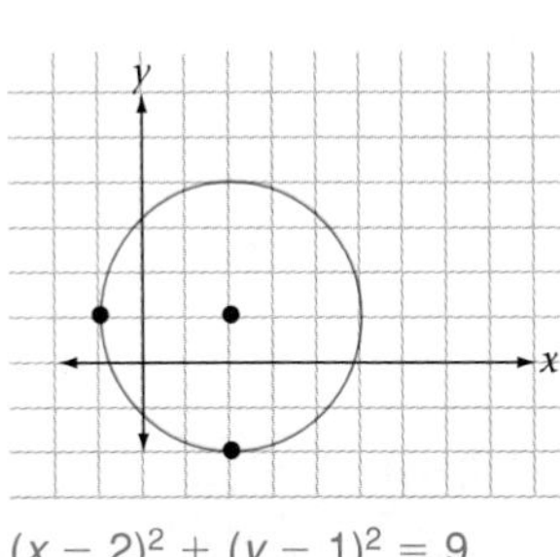

$(x - 2)^2 + (y - 1)^2 = 9$

28.

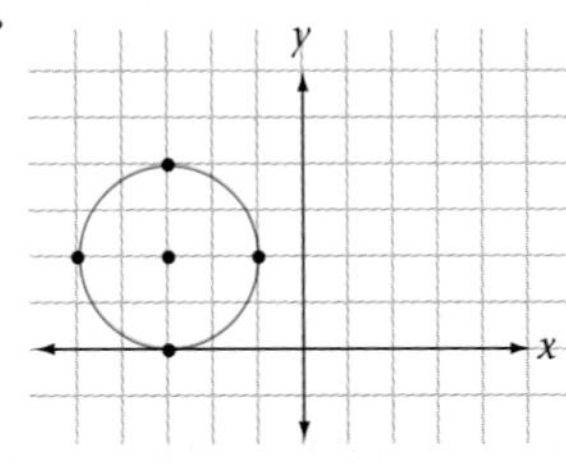

$(x + 3)^2 + (y - 2)^2 = 4$

Algebra Use a graphics calculator to find an equation of a circle with the given characteristics, or sketch a graph.

29. center = (5, 3); tangent to the x-axis $(x - 5)^2 + (y - 3)^2 = 9$

30. center = (5, 3); tangent to the y-axis $(x - 5)^2 + (y - 3)^2 = 25$

Error Analysis

Students using graphing calculators need to be sure that the window is set to a square grid. Otherwise, their circles will appear to be ellipses.

25. $x^2 + (y - 6)^2 = 25$

26. $(x - 4)^2 + (y + 3)^2 = 7$

17. Center = (6, 0); $r = 3$ units

18. Center = (0, −3); $r = 2$ units

19. Center = (−5, 2); $r = 4$ units

20. Center = (−1, −3); $r = \sqrt{19} = 4.36$ units

21. Center = (0, 0); $r = 6$ units

22. Center = (−3, 0); $r = 7$ units

23. $(x - 2)^2 + (y - 3)^2 = 16$

24. $(x + 1)^2 + (y - 5)^2 = 49$

Alternative Assessment

Performance Assessment

Provide students with grid paper and have them do the following:

1. Draw a number of different circles on one sheet of paper.
2. No two circles can have the same radius.
3. Circles should be tangent to other circles, but not intersect.

The students' written work should include equations for all the circles.

The answer for Exercise 35 can be found in Additional Answers beginning on page 727.

Algebra Use a graphics calculator or graph paper to find the following:

31. Given: $x^2 + y^2 = 9$
Find the x-intercepts. (3, 0), (−3, 0)
Find the y-intercepts. (0, 3), (0, −3)

32. Given: $(x - 2)^2 + y^2 = 9$
Find the x-intercepts. (5, 0), (−1, 0)
Find the y-intercepts. $(0, \sqrt{5})$, $(0, -\sqrt{5})$

33. Given: $x^2 + (y - 4)^2 = 25$
Find the x-intercepts. (3, 0), (−3, 0)
Find the y-intercepts. (0, 9), (0, −1)

34. Given: $(x - 3)^2 + (y - 4)^2 = 25$
Find the x-intercepts. (0, 0), (6, 0)
Find the y-intercepts. (0, 0), (0, 8)

35. Structural Design John is designing a wheelchair to use in wheelchair basketball. He wants the push rim to be 6.75 inches less in diameter than the wheel rim. The wheel diameter is 24.5 inches. Using the wheel hub as the origin and inches as the measurement unit, draw a graph for the push rim and wheel rim. Write an equation for each.

Wheelchair basketball is a highly competitive, organized sport. The players' custom-made chairs cost up to $2000.

You may use geometry technology for Exercises 36–44.

36. Sketch a graph of $(x - 3)^2 + (y - 5)^2 = 4$.
Reflect the graph over the x-axis and sketch the image circle. Write an equation for the image circle.

37. Sketch a graph of $(x - 4)^2 + (y - 2)^2 = 1$.
Reflect the graph over the y-axis and sketch the graph of the image circle. Write an equation for the image circle.

38. Sketch a graph of $(x - 2)^2 + y^2 = 9$.
Translate the graph 6 units to the right and sketch a graph of the image circle. Write an equation for the image circle.

39. Sketch a graph of $(x - 6)^2 + (y - 4)^2 = 9$.
Translate the graph 2 units to the left and then 1 unit down. Sketch a graph of the image circle and write its equation.

40. Sketch a graph of $(x - 5)^2 + (y - 4)^2 = 9$.
Rotate the graph 180° about the origin. Sketch a graph of the image circle and write its equation.

41. Find an equation of a circle with center (2, 3) and containing the point (8, 3). $(x - 2)^2 + (y - 3)^2 = 36$

36. $(x - 3)^2 + (y + 5)^2 = 4$; center at (3, –5) with a radius of 2

37. $(x + 4)^2 + (y - 2)^2 = 9$; center at (–4, 2) with a radius of 3

38. $(x - 8)^2 + y^2 = 9$; center at (8, 0) with a radius of 3

9. $(x - 4)^2 + (y - 3)^2 = 9$; center at (4, 3) with a radius of 3

40. $(x + 5)^2 + (y + 4)^2 = 9$; center at (–5, –4) with a radius of 3

Practice Master

NAME ________________ CLASS ________ DATE ________

Practice & Apply
9.6 Circles in the Coordinate Plane

In Exercises 1–4, find the center and radius of each circle.

1. $x^2 + y^2 = 64$
center ________
radius ________

2. $x^2 + y^2 = 65$
center ________
radius ________

3. $(x - 3)^2 + y^2 = 81$
center ________
radius ________

4. $(x + 5)^2 + (y - 2)^2 = 16$
center ________
radius ________

In Exercises 5–6, write an equation of a circle with the given characteristics.

5. center (0, 0); radius = 5
____

6. center (2, −3); radius = 7
____

In Exercises 7–8, find the x-intercepts and y-intercepts of the circle whose equation is given.

7. $x^2 + y^2 = 25$
x-intercepts ________
y-intercepts ________

8. $(x - 1)^2 + y^2 = 4$
x-intercepts ________
y-intercepts ________

In Exercises 9–11, use the grid to do your sketches. Label your circles A, A', A''.

9. Sketch the graph of circle A whose equation is $(x - 3)^2 + (y - 2)^2 = 4$.

10. Reflect the graph of circle A over the y-axis and sketch its image A'. Write the equation of circle A'. ________

11. Reflect the graph of circle A over the x-axis and sketch its image A''. Write the equation of circle A''. ________

y
x

HRW Geometry — Practice & Apply 85

42. Find an equation of a circle with center (2, 3) and containing the point (8, 11). $(x - 2)^2 + (y - 3)^2 = 100$

43. Find an equation of the tangent to the circle $x^2 + y^2 = 100$ at the point (−6, 8). $y = \frac{3}{4}x + 12.5$

44. Show that $x^2 + 6x + y^2 - 4y + 12 = 0$ is a circle.
By completing the square, the equation becomes $(x + 3)^2 + (y - 2)^2 = 1$

45. **Portfolio Activity** Draw a design on a coordinate grid consisting of circles and lines. Then list the equations of the geometric figures in the picture on another sheet of paper. Trade lists with a classmate and reproduce the picture by graphing the equations. Check student designs.

Look Back

The wheel of a bicycle has a radius of 14 in. [Lessons 5.3, 9.3]

46. If the wheel makes 3 complete revolutions, how far will the bicycle travel? 263.89 in.

47. If the wheel rotates through 45°, how far will the bicycle travel? 11 in.

48. Find *DE*. **[Lesson 9.5]**
9 units

49. Find *DE*. **[Lesson 9.5]**
$6\frac{6}{7}$ units

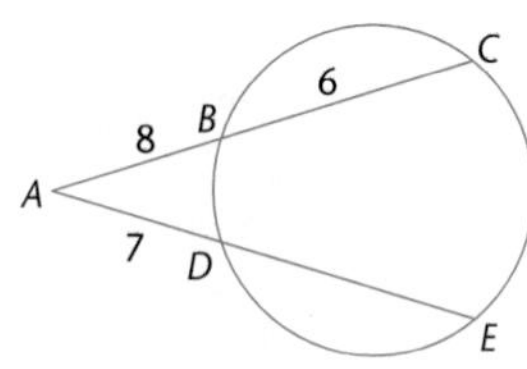

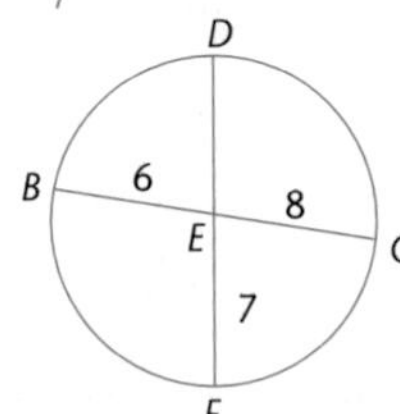

Look Beyond

Unit circle is the name given to the circle with center at the origin and radius 1.

50. Write an equation for the unit circle. $x^2 + y^2 = 1$

51. What are the coordinates of *A*, *B*, *C*, and *D*?
$A(1, 0)$; $B(0, 1)$; $C(-1, 0)$; $D(0, -1)$

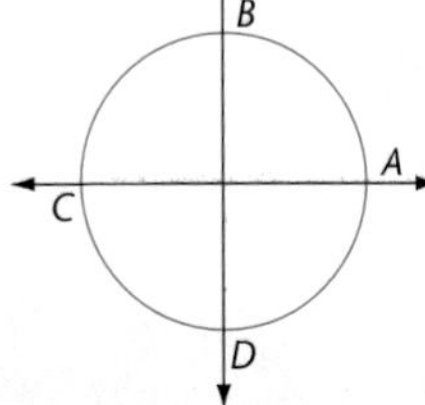

52. What is the circumference of the unit circle?
6.28 units

Use the unit circle to find the following.

53. $m\overset{\frown}{AB}$ 90°
length of $\overset{\frown}{AB}$ 1.57 units

54. $m\overset{\frown}{ABC}$ 180°
length of $\overset{\frown}{ABC}$ 3.14 units

55. $m\overset{\frown}{ABD}$ 270°
length of $\overset{\frown}{ABD}$ 4.71 units

Look Beyond

Exercises 50–55 introduce the unit circle which will play an important role in the development of the trigonometric ideas introduced in the next chapter.

Technology Master

NAME ______________ CLASS ________ DATE ________

Technology
9.6 Another Look at Chords

There are many ways to study chords in circles. A very fruitful approach is to use coordinates. A circle is shown whose center is the origin and whose radius is 5. Its equation is $x^2 + y^2 = 25$. Also shown is point *P* with coordinates $P(-5, 0)$. Notice that point *P* is on the circle, since $(-5)^2 + 0^2 = 25$. To study chords in circles, you will need an equation for the length *d* of $\overline{PQ}$, where Q is any other point on the circle. An equation for *d* in terms of *x* is derived below.

Q (x, y)
P (−5, 0)
O

$d = \sqrt{(x$

$= \sqrt{(x \quad 5)^2 + (25 - x^2)}$

$= \sqrt{50 + 10x}$

If you let *x* take on different values from −5 to 5, you can see how the value of *d* changes as Q moves around the circle. When you use a graphics calculator to graph the equation, you will need to think of *d* as Y1.

Use a graphics calculator in the following exercises.

1. Graph $d = \sqrt{50 + 10x}$ for values of *x* between −5 and 5.

Use the graph to find the length of $\overline{PQ}$ for each value of *x*.

2. $x = -2$ ______
3. $x = 0$ ______
4. $x = 2$ ______

5. For what value(s) of *x* will *PQ* be equal to the radius of the circle?

6. For what value(s) of *x* will *PQ* be equal to 10? What does this tell you about the relationship between chords and diameters of a given circle? number of diameters of the circle that contains $P(-5, 0)$?

7. Use algebra to show that $(x + 5)^2 + (25 - x^2) = 50 + 10x$.

HRW Geometry — Technology 97

Focus

In these activities, students analyze, construct, and apply the five-circle figure used as the official symbol for the Olympic Games. The second activity also involves making a scale drawing.

Motivate

Lead a class discussion about the Olympic Games. (You may want to give students research assignments before this discussion.) Ask students whether they recognize the symbol used on flags and other materials connected with the games. If no one volunteers, sketch the five-circle symbol and ask students to speculate as to why it was chosen and what it means.

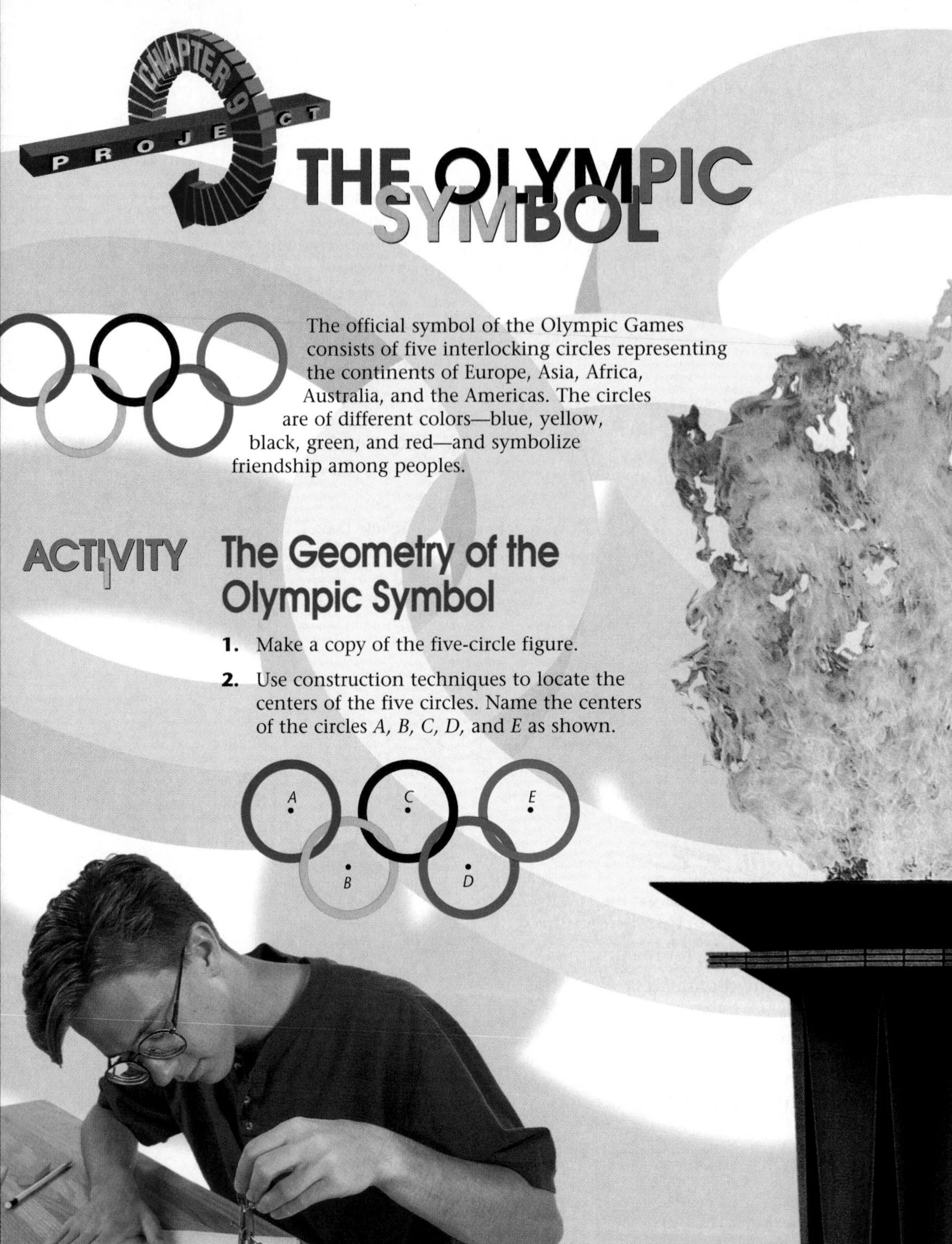

The Olympic Symbol

The official symbol of the Olympic Games consists of five interlocking circles representing the continents of Europe, Asia, Africa, Australia, and the Americas. The circles are of different colors—blue, yellow, black, green, and red—and symbolize friendship among peoples.

Activity 1 The Geometry of the Olympic Symbol

1. Make a copy of the five-circle figure.
2. Use construction techniques to locate the centers of the five circles. Name the centers of the circles *A, B, C, D,* and *E* as shown.

3. The perpendicular bisector of $\overline{AE}$ through *C* is a line of symmetry.
4. Draw the line through the points where the circle with center A intersects the circle with center B.
5. a. $\angle ABC = 60°$ b. $\angle CBD = 60°$
   c. $\angle CDE = 60°$ d. $\angle BAC = 60°$
   e. $\angle BCA = 60°$ f. 60°
6. a. $\frac{1}{1}$ b. $\frac{1}{1}$
7. If all the centers of the circle are connected, the result is three equilateral triangles that together form an isosceles trapezoid. *B* is on the perpendicular bisector of $\overline{AB}$ and *D* is on the perpendicular bisector of $\overline{CE}$. The circle with center *C* is the reflection image of the circle with center *A* over the perpendicular bisector of $\overline{AB}$. The circle with center *E* is the reflection image of the circle with center *C* over the perpendicular bisector of $\overline{CE}$.

3. Describe any symmetries that you see in the figure.

4. $\odot B$ can be viewed as a reflection image of $\odot A$. Draw the reflection line.

5. Measure the following angles.
   a. $m\angle ABC$ b. $m\angle CBD$
   c. $m\angle CDE$ d. $m\angle BAC$
   e. $m\angle BCA$
   f. the angle with vertex at A and containing the points of intersection of $\odot A$ and $\odot B$

6. Find the following ratios.
   a. $\frac{AB}{BC}$ b. $\frac{AC}{AB}$

7. Describe the interesting features of the figure. Look for congruent angles, segments, and triangles. Look for transformations and special polygons.

8. Suppose that North America and South America each want their own circle. Add a circle to your figure and describe the method that you used.

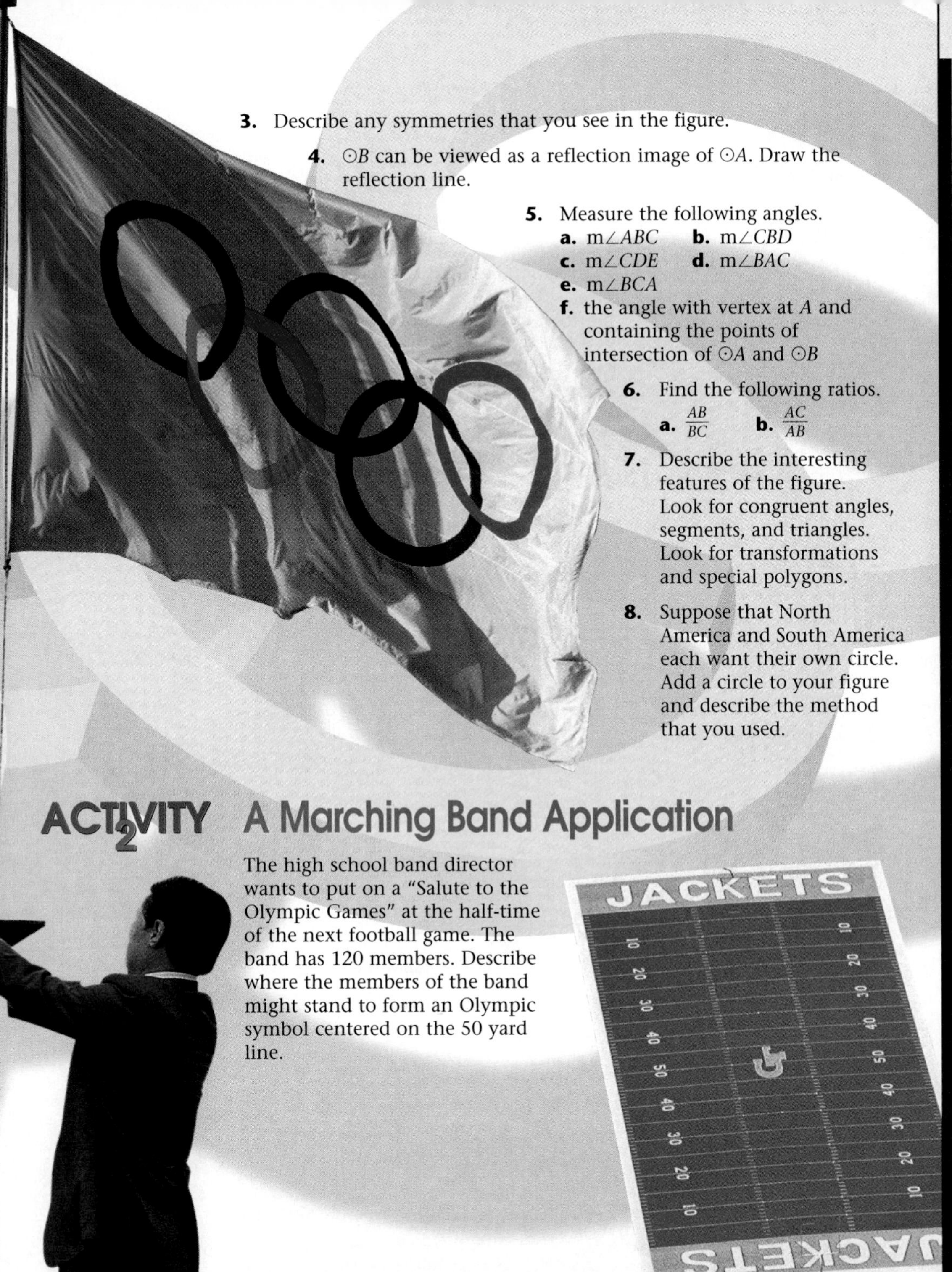

ACTIVITY 2 A Marching Band Application

The high school band director wants to put on a "Salute to the Olympic Games" at the half-time of the next football game. The band has 120 members. Describe where the members of the band might stand to form an Olympic symbol centered on the 50 yard line.

Activity 2
Each circle will use 24 band members. The radius of each circle should be about 15 yards. Three circles should be centered on one hash mark. The center of the middle circle should be at the 50 yard line and the centers of the other two circles should be at each 25 yard line. The other two circles should be centered on the second hash mark. The centers of the circles should be between the 37 and the 38 yard lines.

Cooperative Learning

Have students work in pairs or small groups for the activities. Students in each group can use different construction methods such as compass and straightedge, paper folding, or geometry graphics software.

Discuss

After Activity 1 is completed, have students share their results and approaches. They should comment on which methods seemed most useful for individual problems, and why. For the class discussion following Activity 2, have students explain what scale factor they used and why they chose that factor. The instruction that the symbol be "centered on the 50 yard line" may have caused difficulty for some groups of students. Include discussion on how students found the center of the symbol.

Chapter 9 Review

Vocabulary

arc	472	external secant segment	502	radius	470
center	470	inscribed angle	484	secant	479
central angle	471	intercepted arc	484	secant segment	502
chord	470	major arc	472	semicircle	472
circle	470	minor arc	472	tangent	479
diameter	470	point of tangency	479	tangent segment	502

Key Skills and Exercises

Lesson 9.1

Key Skills

Identify parts of a circle.

In circle M, $\overline{AB}$ and $\overline{CD}$ are chords. $\overline{CD}$ is also a diameter. $\overline{MC}$, $\overline{ME}$, and $\overline{MD}$ are radii. Central angle $\angle CME$ intercepts minor arc $\widehat{CE}$.

Find arc and central-angle measures.

In circle N, what is the measure of $\angle ONP$? What is the length of $\widehat{OP}$? The total measure of the central angles of a circle is 360°. For circle N,

$$90° + 60° + 90° + m\angle ONP = 360° \quad m\angle ONP = 360° - 240° = 120°$$

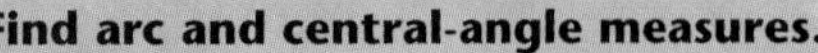

The length of an arc is its angle divided by 360° times the circle's circumference.

$$\widehat{OP} = \frac{120°}{360°} \times 2\pi(21 \text{ cm}) = 44 \text{ cm}$$

Exercises

In Exercises 1–2, refer to circle *P*.

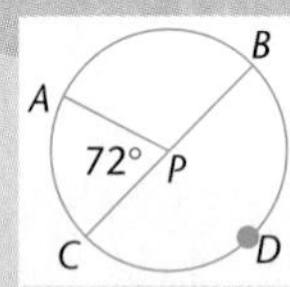

1. Name a chord, a radius, a central angle, and a major arc.
2. If the circumference is 75 inches, how long is arc $\widehat{AC}$?

Lesson 9.2

Key Skills

Use secants and tangents and their perpendicular radii to solve problems.

In the figure, what length is FH?

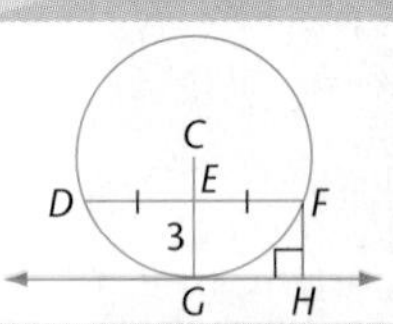

$\overline{CG}$ is a radius of circle C and bisects secant $\overline{DF}$, so it is perpendicular to $\overline{DF}$. $\overline{CG}$ intersects tangent $\overleftrightarrow{GH}$ at G, so it is perpendicular to $\overleftrightarrow{GH}$. Thus $\overline{DF}$ and $\overleftrightarrow{GH}$ are parallel, by the Corresponding Angles Theorem. $EFHG$ is a rectangle, and $EG = 3$. Therefore, $FH = 3$.

1. $\overline{BC}$; $\overline{PA}$, $\overline{PB}$, $\overline{PC}$; $\angle APB$, $\angle APC$; $\widehat{ACB}$, $\widehat{ABC}$
2. Length of $\widehat{AC} = \frac{72°}{360°}(75) = 15$ in

➤ Exercises

In Exercises 3–4, refer to the figure. Lines l and m are tangent to circle O.

3. Find y. Explain your answer.

4. Find x. Explain your answer.

Lesson 9.3

➤ Key Skills

Find the measure of an inscribed angle and its intercepted arc.

In the circle, find $m\angle C$ and $m\widehat{AD}$.

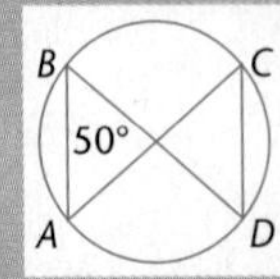

Two angles that intercept the same arc have the same measure. Since $\angle B$ and $\angle C$ both intercept arc $\widehat{AD}$, and $m\angle B$ is 50°, then $m\angle C$ is 50°. By the Inscribed Angle Theorem, $m\widehat{AD}$ is twice $m\angle B$ or $m\angle C$, or 100°.

➤ Exercises

In Exercises 5–7, refer to the circle.

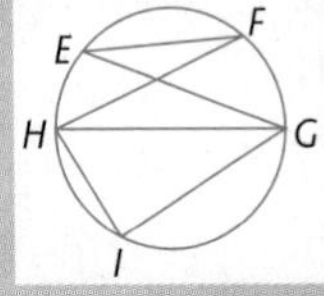

5. $m\widehat{HE} = 40°$ $m\angle F =$ __?__ **6.** $m\angle FHG = 25°$ $m\widehat{FG} =$ __?__

7. $\overline{HG}$ is a diameter. $m\angle I =$ __?__

Lesson 9.4

➤ Key Skills

Solve problems using the angle-arc relationships between secants and tangents.

Find the measure of arc x in the figure.

The angle between two secants is half the difference between the two intercepted arcs. Thus,

$$\frac{(80° - x)}{2} = 20°$$

$$x = 40°$$

➤ Exercises

8. Find the measure of minor arc $\widehat{YZ}$.

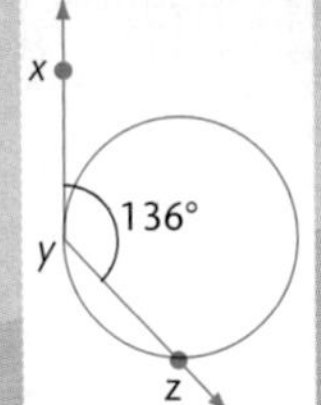

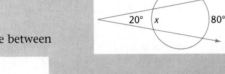

Lesson 9.5

➤ Key Skills

Solve problems using the angle-arc relationships among secant segments, tangent segments, and chords.

$\overrightarrow{AB}$ and $\overrightarrow{AC}$ are tangent segments to circle P. Show that $\triangle PAB \cong \triangle PAC$.

$\overline{BP} \cong \overline{CP}$ because all radii of a circle are congruent. $\overline{PA} \cong \overline{PA}$ by the Reflexive Property of Equality. Since $\overrightarrow{AB}$ and $\overrightarrow{AC}$ are tangent segments, $\angle PBA$ and $\angle PCA$ are both right angles. By the Hypotenuse-Leg Postulate, $\triangle PAB \cong \triangle PAC$.

➤ Exercises

9. $\overline{PM} \cong \overline{PQ}$. Rays $\overrightarrow{QS}$ and $\overrightarrow{QP}$ are tangent to circle L. Find LM.

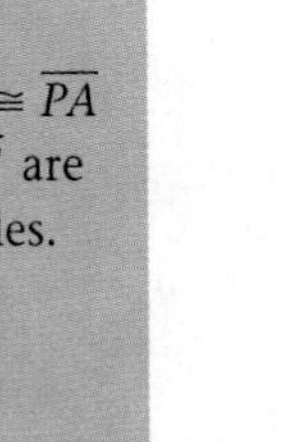

Chapter Review

5. 20°

6. 50°

7. 90°

8. 88°

9. ≈ 16.14

3. $y = 12\sqrt{2}$; The tangents are perpendicular to the radii at the points of tangency. Since the tangents are also perpendicular to each other, each tangent is parallel to the radius it doesn't intersect. Since the figure is a parallelogram, $\angle O$ must be a right angle. The triangle in the circle is a 45°–45°–90° triangle since the radii of a circle are congruent.

4. $x = 12$; The figure formed by the radii and the tangents is a square because it is a parallelogram with right angles and two consecutive sides congruent. All the sides of the square have length 12.

10. $(x-6)^2+(y+2)^2=93$

11. Center $= (0, 7)$; $r = 5$

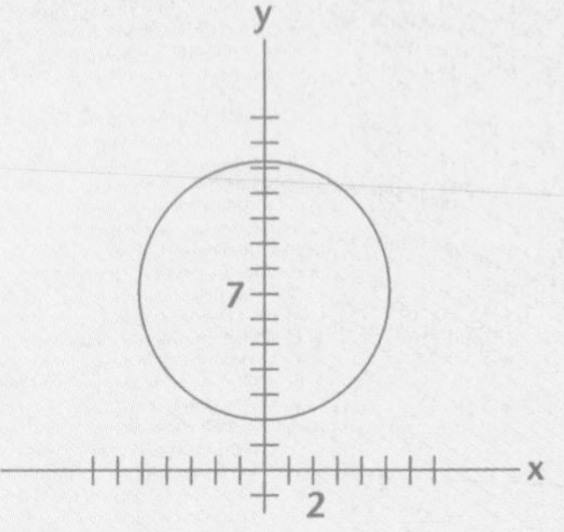

12.

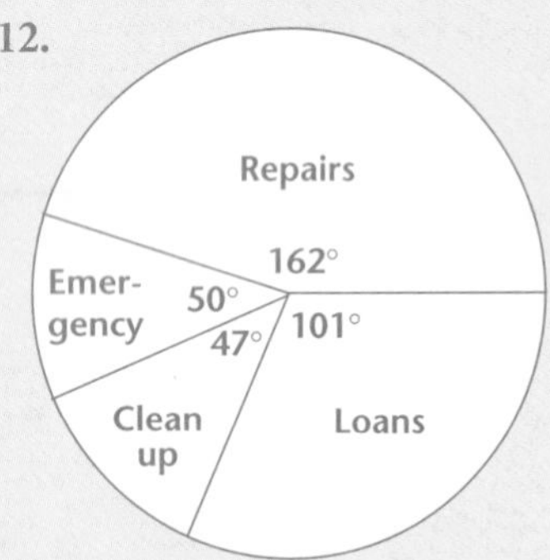

13. 11.35 miles

14. ≈ 1068.14 in ≈ 89 ft.

Lesson 9.6

➤ Key Skills

Write an equation for a circle.

Write an equation for a circle with its center at (5, 0) and a radius of 6.

Use the general equation for a circle, $(x - h)^2 + (y - k)^2 = r^2$. Here $h = 5$, $k = 0$, and $r = 6$, so we have $(x - 5)^2 + y^2 = 36$.

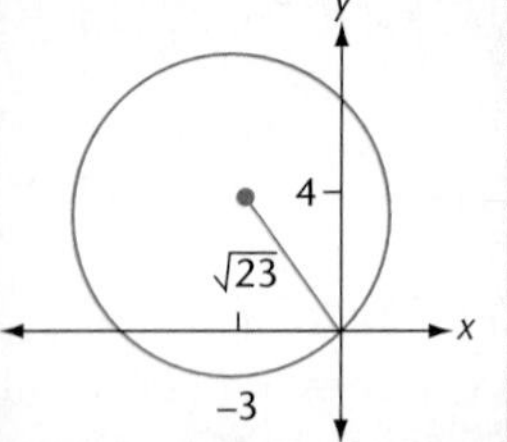

Sketch a circle from its equation.

Sketch a circle that has the equation $(x + 3)^2 + (y - 4)^2 = 23$.

The circle's center is (–3, 4); its radius is $\sqrt{23}$.

➤ Exercises

10. Write an equation for a circle with its center at (6, –2) and a radius of $\sqrt{93}$.

11. Sketch a circle that has the equation $x^2 + (y - 7)^2 = 25$.

Applications

12. The city of New Rockford received $1.5 million in federal disaster relief after Hurricane Alex. It was paid out so that 28% went to low-interest loans for individuals; 45% went to the city to repair a breakwater and dredge the harbor entrance; 14% paid for emergency shelter, water, and food; and 13% went toward cleaning up the wreckage of the storm. Sketch a pie chart of the data, noting the central angles of each portion.

13. **Navigation** Two kayakers are rowing toward a lighthouse. If the light is mounted 85 feet above sea level, how far away are the kayakers when they first sight the light? Assume calm seas and excellent visibility. (1 mile = 5,280 feet)

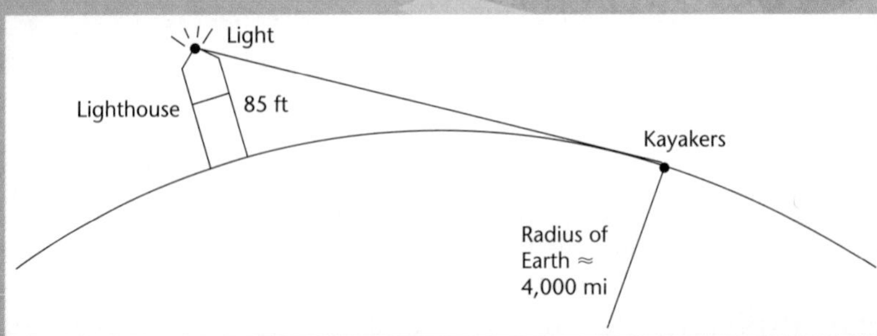

14. **Surveying** A simple way to get a rough measure of property boundaries is to use a measuring wheel. Attach an 8-inch-diameter wheel to a stick, put a flag on the edge of the wheel, and count the number of rotations it makes as you walk along the property line. If the flag goes around 42.5 times, how long is the line?

Chapter 9 Assessment

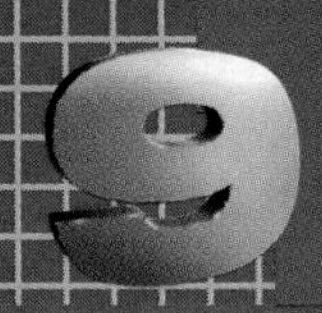

1. If arc $\widehat{AC}$ is 12 mm, what is the circumference of circle R?

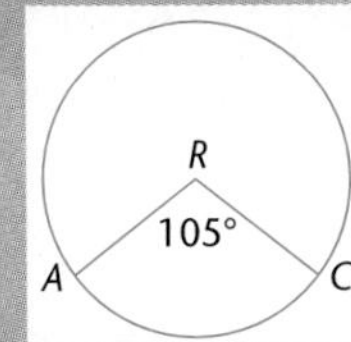

In Exercises 2–3, refer to the figure at the right.

2. Find $\angle WYZ$.
3. Find x.

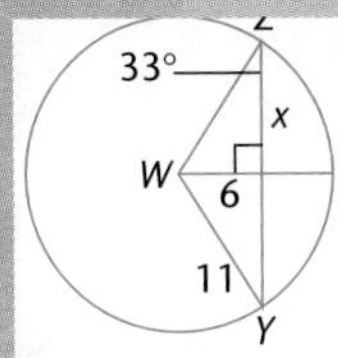

In Exercises 4–6, refer to circle Q below.

4. $m\widehat{AB} = 132°$ $m\angle AQB =$ __?__
5. $m\angle A = 37°$ $m\widehat{DC} =$ __?__
6. $\overline{AC}$ is a diameter, and $\overline{QB}$ is a radius; $m\angle QCB =$ __?__

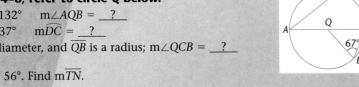

7. $\angle TAN = 56°$. Find $m\overline{TN}$.

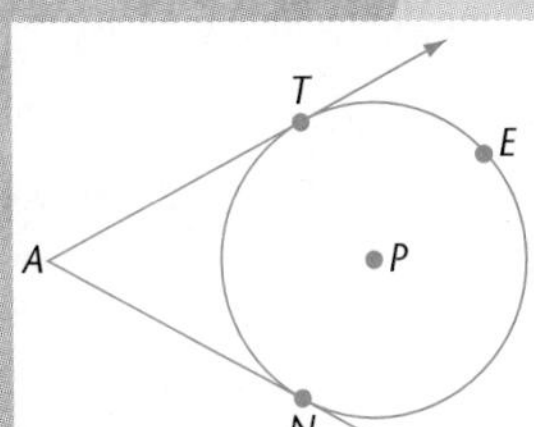

8. Find x.

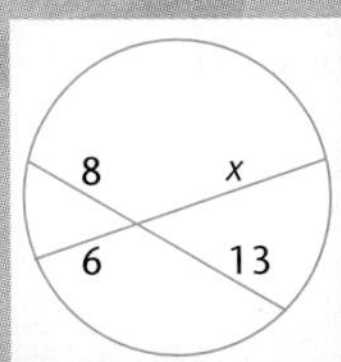

9. Write an equation for a circle with its center at (–5, 0) and a radius of 27.
10. Sketch a circle that has the equation $(x - 20)^2 + (y - 10)^2 = 121$.

10. Center = (20, 10); $r = 11$

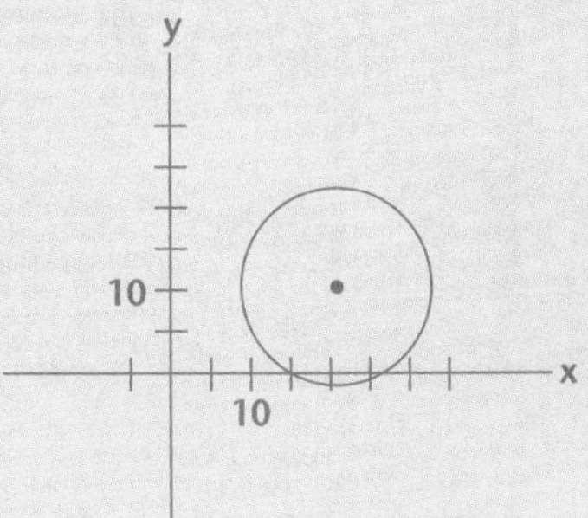

1. 41.14 mm
2. 33°
3. ≈ 9.22
4. 132°
5. 74°
6. 67°
7. 124°
8. $17\frac{1}{3}$
9. $(x+5)^2 + y^2 = 729$

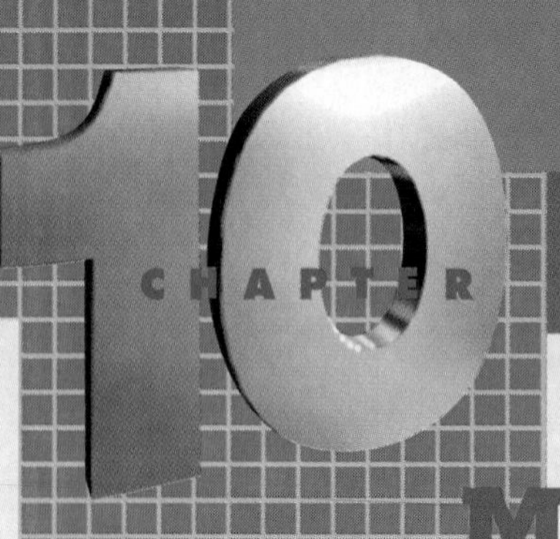

Trigonometry

Meeting Individual Needs

10.1 Exploring Tangent Ratios

Core Resources	Core Plus Resources
Inclusion Strategies, p. 526 Reteaching the Lesson, p. 527 Practice Master 10.1 Enrichment, p. 526 Technology Master 10.1 Lesson Activity Master 10.1	Practice Master 10.1 Enrichment Master 10.1 Technology Master 10.1 Interdisciplinary Connection, p. 525
[2 days]	**[1 day]**

10.2 Exploring Sines and Cosines

Core Resources	Core Plus Resources
Inclusion Strategies, p. 535 Reteaching the Lesson, p. 536 Practice Master 10.2 Enrichment Master 10.2 Technology Master 10.2 Lesson Activity Master 10.2	Practice Master 10.2 Enrichment, p. 535 Technology Master 10.2 Interdisciplinary Connection, p. 534
[2 days]	**[2 days]**

10.3 Exploring the Unit Circle

Core Resources	Core Plus Resources
Inclusion Strategies, p. 543 Reteaching the Lesson, p. 544 Practice Master 10.3 Enrichment Master 10.3 Lesson Activity Master 10.3 Interdisciplinary Connection, p. 542	Practice Master 10.3 Enrichment, p. 543 Technology Master 10.3
[3 days]	**[2 days]**

10.4 Exploring Rotations with Trigonometry

Core Resources	Core Plus Resources
Inclusion Strategies, p. 550 Reteaching the Lesson, p. 551 Practice Master 10.4 Enrichment, p. 550 Technology Master 10.4 Lesson Activity Master 10.4 Mid-Chapter Assessment Master	Practice Master 10.4 Enrichment Master 10.4 Technology Master 10.4 Interdisciplinary Connection, p. 549 Mid-Chapter Assessment Master
[2 days]	**[2 days]**

10.5 The Law of Sines

Core Resources	Core Plus Resources
Inclusion Strategies, p. 557 Reteaching the Lesson, p. 558 Practice Master 10.5 Enrichment, p. 557 Lesson Activity Master 10.5 Interdisciplinary Connection, p. 556	Practice Master 10.5 Enrichment Master 10.5 Technology Master 10.5 Interdisciplinary Connection, p. 556
[2 days]	**[1 day]**

10.6 The Law of Cosines

Core Resources	Core Plus Resources
Inclusion Strategies, p. 566 Reteaching the Lesson, p. 567 Practice Master 10.6 Enrichment Master 10.6 Technology Master 10.6 Lesson Activity Master 10.6	Practice Master 10.6 Enrichment, p. 566 Technology Master 10.6
[2 days]	**[1 day]**

10.7 Vectors in Geometry

Core Resources	Core Plus Resources
Inclusion Strategies, p. 573 Reteaching the Lesson, p. 574 Practice Master 10.7 Enrichment Master 10.7 Technology Master 10.7 Lesson Activity Master 10.7 Interdisciplinary Connection, p. 571	Practice Master 10.7 Enrichment, p. 572 Technology Master 10.7
[2 days]	[1 day]

Chapter Summary

Core Resources	Core Plus Resources
Chapter 10 Project, pp. 576–577 Lab Activity Long-Term Project Chapter Review, pp. 578–580 Chapter Assessment, p. 581 Chapter Assessment, A/B Alternative Assessment Cumulative Assessment, pp. 582–583	Chapter 10 Project, pp. 576–577 Lab Activity Long-Term Project Chapter Review, pp. 578–580 Chapter Assessment, p. 581 Chapter Assessment, A/B Alternative Assessment Cumulative Assessment, pp. 582–583
[2 days]	[2 days]

Hands-On Strategies

This chapter includes activities in which students construct right triangles to derive and investigate trigonometric ratios. Compass-and-straightedge drawings may be made on grid paper to ensure that right angles are accurate. Encourage students to make fairly large figures—this will give them better accuracy when measuring angles. Students may also use geometry graphics software to create needed figures. In fact, students will benefit from doing some activities twice—once with graph paper and once with the software. Results and procedures can then be compared using two different approaches.

Lesson 10.4 deals with rotations. Geometry graphics software may be used to create the figures. Or, if students are using paper-and-pencil methods, tracing paper can be helpful to actually demonstrate the motion involved in a rotation.

Cooperative Learning

GROUP ACTIVITIES	
The tangent ratio	Lesson 10.1, Exploration 1
Sines and Cosines	Lesson 10.2, Exploration 1
Redefining the Trigonometric Ratios	Lesson 10.3, Exploration 2
Rotating a Point	Lesson 10.4, Exploration 1
Law of Sines	Lesson 10.5, Exploration

You may wish to have students work in groups or with partners for some of the above activities. Additional suggestions for cooperative group activities are noted in the teacher's notes in each lesson.

Multicultural

The cultural references in this chapter include references to Asia, Africa, and Europe.

CULTURAL CONNECTIONS	
Africa: Ahmes Papyrus	Lesson 10.1, Application
Europe: Stonehenge	Lesson 10.2, Look Beyond
Asia: Plimpton 322 Revisited	Chapter 10 Project

Portfolio Assessment

Below are portfolio activities for the chapter listed under seven activity domains which are appropriate for portfolio development.

1. **Investigation/Exploration** The explorations in Lesson 10.1 focus on the tangent ratio; those in Lessons 10.2 and 10.3 deal with the tangent as well as other trigonometric ratios. Rotations using trigonometry and matrices are the focus of the explorations in Lesson 10.4. The Law of Sines is explored in Lesson 10.5.

2. **Applications** Surveying, Lesson 10.1, Exploration 3; Engineering, Lesson 10.1, Exercises 40–41; Recreation, Lesson 10.2, Application; Astronomy, Lesson 10.3, Exercises 35–37; Navigation, Lesson 10.5, Application; Surveying, Lesson 10.5, Exercises 14–16; Architecture, Lesson 10.5, Exercise 17; Wildlife Management, Lesson 10.5, Exercise 18; Sports, Lesson 10.7, Exercise 20; Navigation, Lesson 10.7, Exercise 22.

3. **Nonroutine Problems** Lesson 10.3, Look Beyond (moons of Jupiter); Lesson 10.6, Exercise 26 (network problem); Chapter 10 Project (Babylonian tablet).

4. **Project** Plimpton 322 Revisited: see pages 576–577.

5. **Interdisciplinary Topics** Algebra, Lesson 10.1, Explorations 2 and 3; Archeology, Lesson 10.2, Look Beyond; Algebra, Lesson 10.5, Example 1; Physics, Eyewitness Math; Algebra, Lesson 10.6, Examples 1 and 2; Physics, Lesson 10.7.

6. **Writing** *Communicate* exercises offer excellent writing selections for the portfolio. Suggested selections include: Lesson 10.1, Exercise 2; Lesson 10.2, Exercises 3–4; Lesson 10.5, Exercise 2; Lesson 10.6, Exercise 4.

7. **Tools** A scientific calculator is needed in Chapter 10. In Lessons 10.1 and 10.5, students use geometry software to explore the tangent ratio and the Law of Sines. In Lessons 10.3 and 10.4, this software is used for rotations.

Technology

Many of the activities in this book can be significantly enhanced by the use of geometry graphics software, but in every case the technology is optional. If computers are not available, the computer activities can, *and should,* be done by hands-on or "low-tech" methods, as they contain important instructional material. For more information on the use of technology, refer to the *HRW Technology Manual for High School Mathematics.*

When instructing students in the use of the software, you may find that the best approach is to give students a few well-chosen hints about the different tools and let them discover their use on their own. Today's computer-savvy students can be quite impressive in their ability to discover the uses of computer software! Once the students are familiar with the geometry software, they can design and create their own computer sketches for the explorations.

A second approach is for you, the teacher, to demonstrate the creation of the exploration sketches "from scratch" using a projection device. The students will find this exciting to watch, and they will be eager to learn to use the software on their own.

In the interest of convenience or time-saving, you may want to use the *Geometry Investigations Software* available from HRW. Each of the files on the disks contains a "sketch," custom-designed for an *HRW Geometry* Exploration. The sketches are to be used with *Geometer's Sketchpad*™ (Key Curriculm Press) or *Cabri Geometry II*™ (Texas Instruments).

Computer Graphics Software

Lesson 10.1: Exploration 1 In this exploration, students drag a leg of a right triangle while the technology calculates the lengths of the opposite and adjacent sides. The ratio of these sides is also displayed. Students make a conjecture about the ratio of the legs in a right triangle.

Lesson 10.1: Exploration 2 Students change the shape of a right triangle while the technology calculates and displays the tangent of one of the angles for measures between 0 and 90 degrees. The students record specific tangent ratios using the table feature of the software. They use the information in this table to sketch a graph of the tangent function.

Lesson 10.1: Exploration 3 A model of an application of the tangent ratio is given. Students estimate a distance using the tangent, then they display the solution on the screen. Students can change the measurements and complete the activity as many times as they need to master the technique.

Lesson 10.2: Exploration 1 Students change the shape of a right triangle while the technology calculates and displays the sine and cosine of one of the angles for measures between 0 and 90 degrees. The students record specific sine and cosine ratios using the table feature of the software. They use the information in this table to sketch a graph of the sine and cosine functions.

Lesson 10.2: Exploration 2 Students investigate relationships among the sine, cosine, and tangent by dragging a vertex of a right triangle while the technology calculates and displays the values of these functions. Students are then able to make conjectures that include the basic tangent identity and the "Pythagorean" identity.

Lesson 10.5: Exploration Students drag any vertex of a triangle while the technology displays the measurements of the parts of the triangle and the ratio of the angle to the side opposite. From this displayed information, students are led to conjecture the Law of Sines.

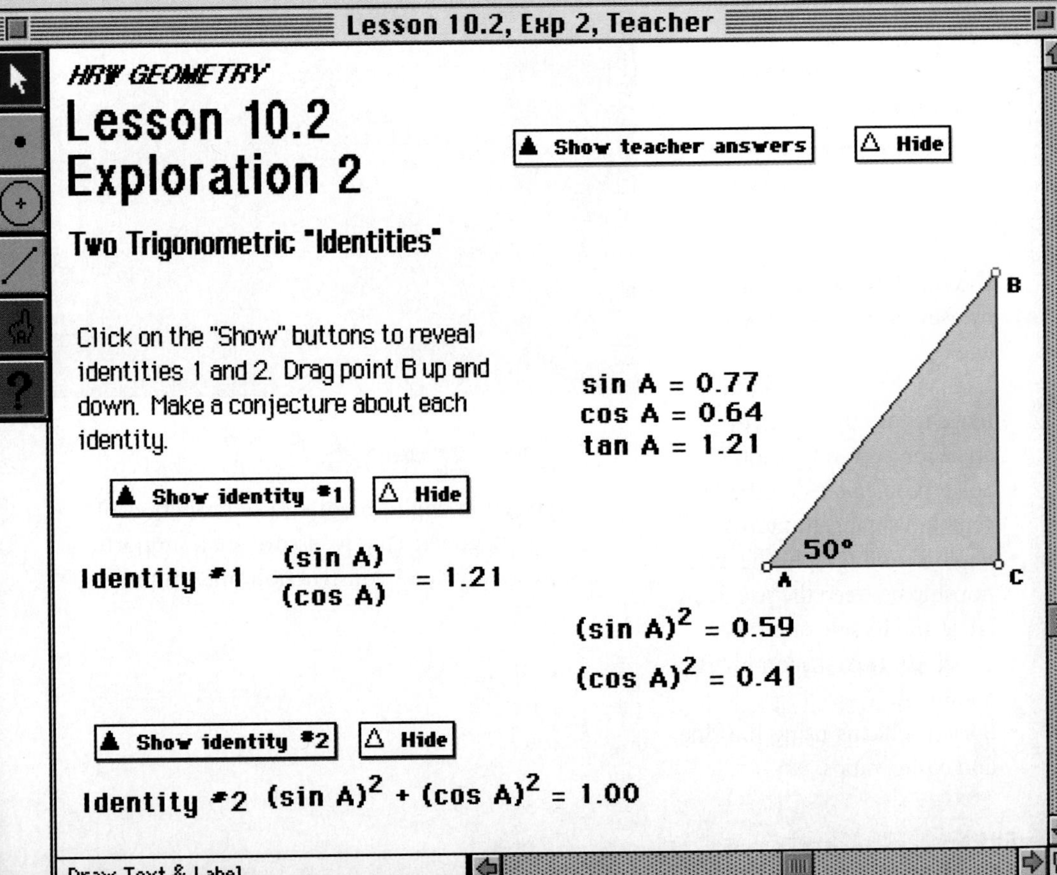

About the Chapter

Background Information

In this chapter students develop the tangent, sine, and cosine ratios from both right triangles and unit circles. They explore some trigonometric identities, apply trigonometric rotation equations, and use the Law of Sines and the Law of Cosines to solve problems. The chapter concludes with an introduction to vectors and vector addition.

- Practice Masters
- Enrichment Masters
- Technology Masters
- Lesson Activity Masters
- Lab Activity Masters
- Long-Term Project Masters
- Assessment Masters
  - Chapter Assessments, A/B
  - Mid-Chapter Assessment
  - Alternative Assessments, A/B
- Teaching Transparencies
- Cumulative Assessment
- Spanish Resources

Chapter Objectives

- Develop the tangent ratio by measuring sides of right triangles.
- Use a graph or chart to find the tangent of an angle or the angle for a given tangent.
- Solve problems using the tangent and cotangent ratios.
- Explore and graph the relationship between the size of an angle and its sine or cosine.
- Develop two trigonometric identities.
- Solve problems using the sine and cosine ratios.

Chapter 10

Lessons

Trigonometry

Have you ever wondered how highway engineers are able to be sure that one section of a freeway or overpass will correctly match up with another section that is under construction a considerable distance away? Accurate measurements and calculations are necessary to ensure success. In this kind of work, trigonometry is an indispensable tool.

Trigonometry, like much of geometry, depends upon triangles. The simple study of relationships between the parts of right triangles quickly leads ultimately to more sophisticated calculation techniques that are well within your grasp at this time in your mathematical studies.

As you progress in your study of trigonometry, you will be introduced to vectors, which are among the most important of all of mathematical concepts for scientists and engineers.

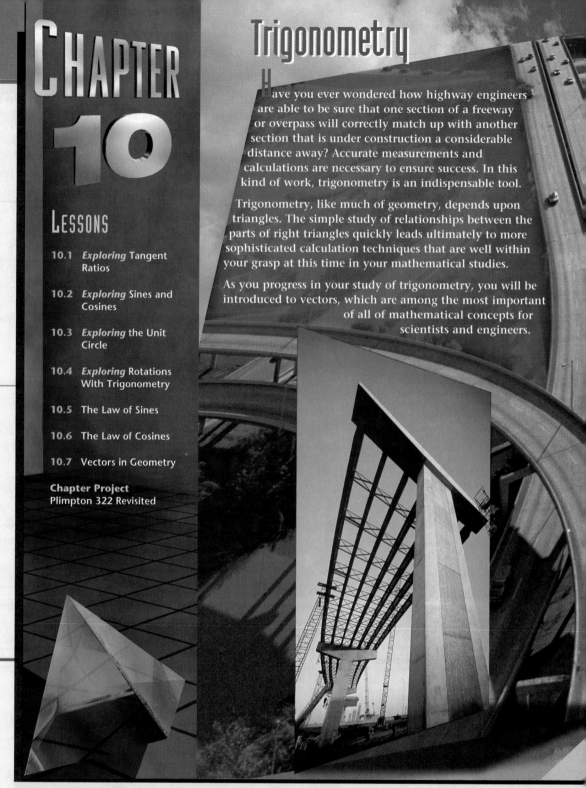

About the Photos

Construction engineers often use properties of triangle trigonometry to assure that supports and beams in road construction line up properly.

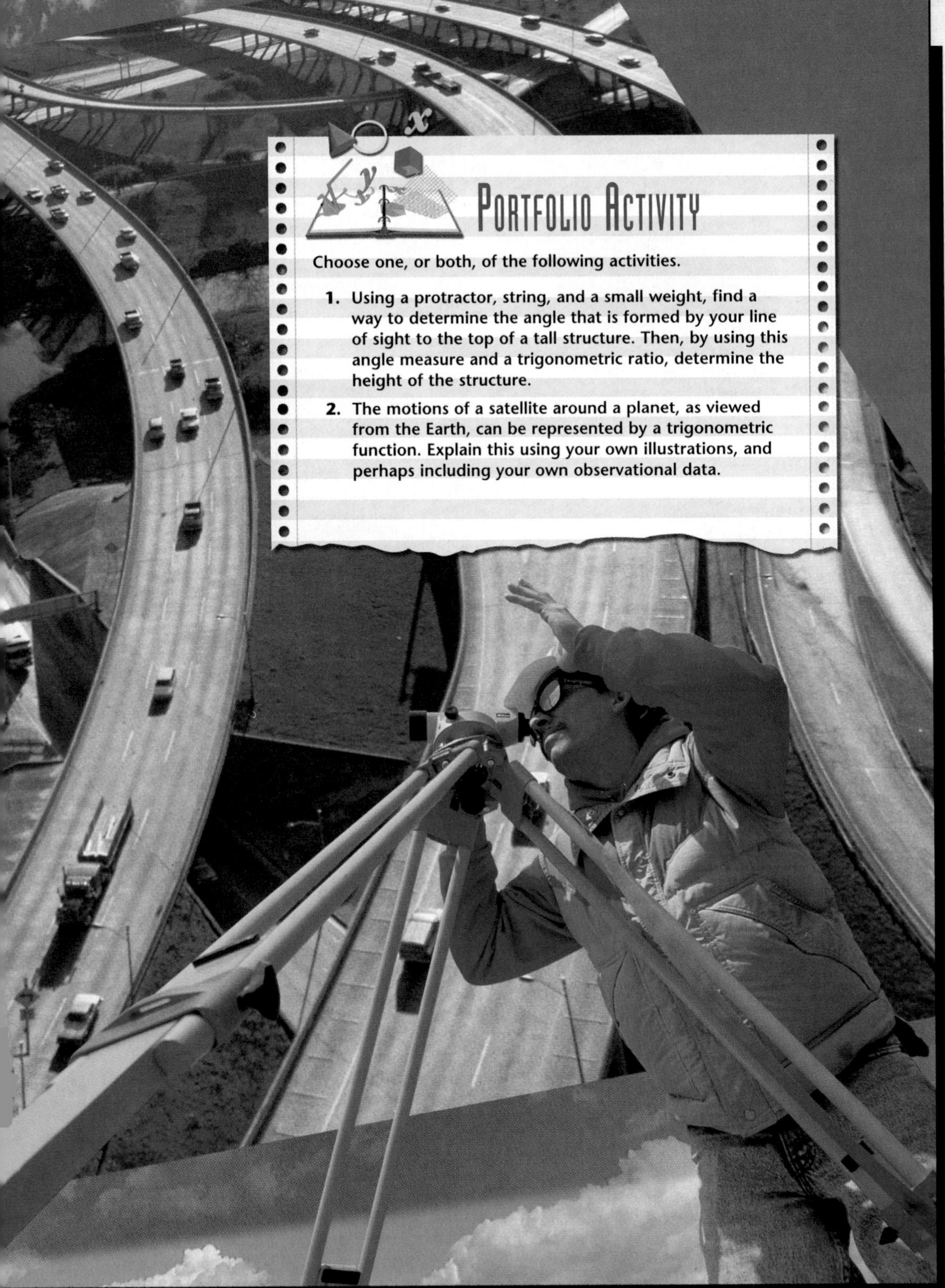

PORTFOLIO ACTIVITY
Choose one, or both, of the following activities.
1. Using a protractor, string, and a small weight, find a way to determine the angle that is formed by your line of sight to the top of a tall structure. Then, by using this angle measure and a trigonometric ratio, determine the height of the structure.
2. The motions of a satellite around a planet, as viewed from the Earth, can be represented by a trigonometric function. Explain this using your own illustrations, and perhaps including your own observational data.

- Use a rotating ray on a coordinate plane to define angles measuring more than 180° or less than 0°.
- Define *sine, cosine, and* tangent *for angles of any size.*
- Extend the graphs of the sine, cosine, and tangent functions.
- Use two transformation equations to rotate points.
- Use a rotation matrix to rotate points or polygons.
- Develop the Law of Sines.
- Use the Law of Sines to find the measure of an angle or a side in a triangle.
- Use the Law of Cosines to find the measure of an angle or a side in a triangle.
- Use the Law of Sines and the Law of Cosines to solve problems.
- Define *vector.*
- Add two vectors.
- Use vectors and vector addition to solve problems.

PORTFOLIO ACTIVITY

The Portfolio Activities set the stage for this chapter by focusing students' attention on indirect measurement and surveying, situations which often involve application of trigonometric ratios and functions. You may wish to assign library research to augment these activities. Students can investigate various surveying techniques and instruments, and present their results in oral reports. Additional portfolio activities can be found in the exercises for Lessons 10.2 and 10.3.

ABOUT THE CHAPTER PROJECT

In the Chapter 10 Project, on pages 576–577, students return to the Plinton 322 tablet to discover an intriguing fact about how the list of Pythagorean triples is organized.

PREPARE

Objectives

- Develop the tangent ratio by measuring sides of right triangles.
- Use a graph or chart to find the tangent of an angle or the angle for a given tangent.
- Solve problems using the cotangent and tangent ratios.

RESOURCES

- Practice Master 10.1
- Enrichment Master 10.1
- Technology Master 10.1
- Lesson Activity Master 10.1
- Quiz 10.1
- Spanish Resources 10.1

Assessing Prior Knowledge

Construct any right triangle. Label the right angle C. Use A and B for the other two vertices. Name each of the following.

1. the hypotenuse [$\overline{AB}$]
2. the side opposite $\angle A$ [$\overline{BC}$]
3. the side adjacent to $\angle A$ [$\overline{AC}$]
4. the side opposite $\angle B$ [$\overline{AC}$]
5. the side adjacent to $\angle B$ [$\overline{BC}$]

TEACH

Ideas studied in trigonometry relate the sides and angles of a single right triangle. Since right triangles appear so frequently in surveying problems and other applications, trigonometric concepts have extremely wide applications in real-life problem-solving situations.

LESSON 10.1 Exploring Tangent Ratios

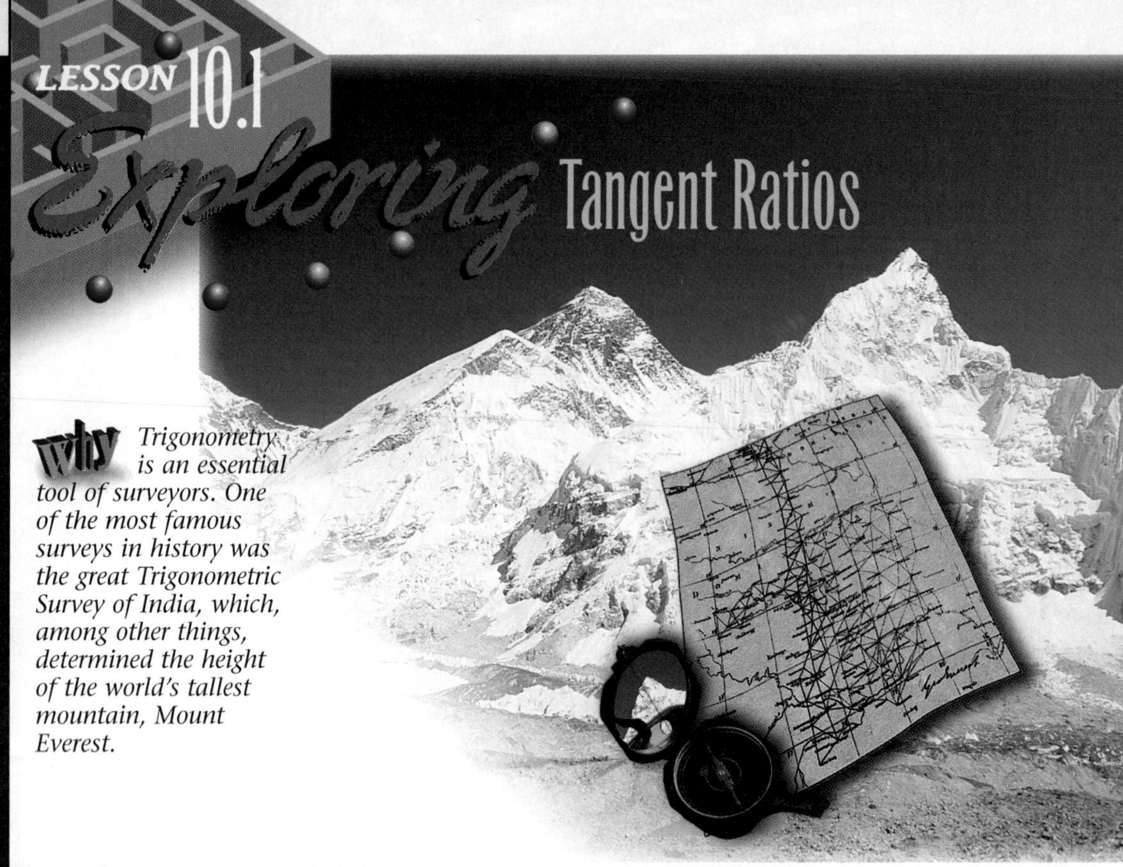

Why *Trigonometry is an essential tool of surveyors. One of the most famous surveys in history was the great Trigonometric Survey of India, which, among other things, determined the height of the world's tallest mountain, Mount Everest.*

Exploration 1 A Familiar Ratio

You will need
Geometry technology or
Ruler and protractor
Calculator

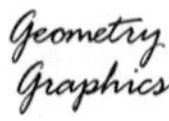

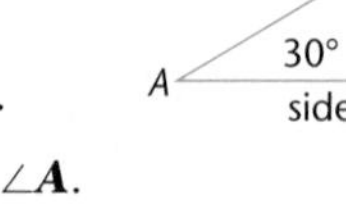

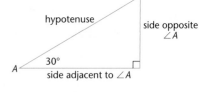

1. Draw a right triangle with a 30° angle. Label this angle A.
2. Measure the **side opposite $\angle A$**. Record your result.
3. One of the sides of $\angle A$ is the hypotenuse of the right triangle. The other side of $\angle A$ is called the **adjacent side**. Measure the adjacent side. Record your result.
4. Divide the measure of the side opposite $\angle A$ by the side adjacent to $\angle A$. Record your result.
5. Repeat Steps 2–4 for other right triangles with 30° angles. Or compare your result in Step 4 with other class members' results. What do you notice? Make a conjecture about the ratio of the side opposite an acute angle of a right triangle to the side adjacent to the angle. ❖

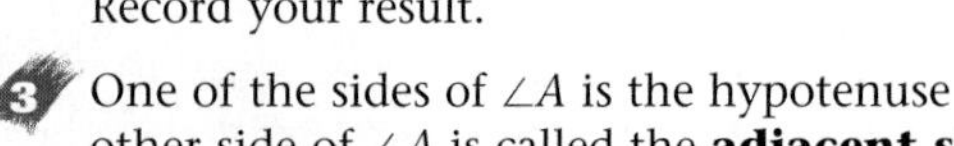

Technology The first two explorations are appropriate for use with spreadsheet software. Students should discuss how to set up the columns. Then class data can be compiled and, for the second exploration, graphed.

In an earlier lesson you divided the "rise" of a segment by its "run" to determine its slope. How is the ratio you calculated in the exploration related to the concept of slope? How is it different?

The Trigonometric Ratios

In the right triangles $\triangle ABC$ and $\triangle MNO$, $\angle A$ is congruent to $\angle M$. Therefore, $\triangle ABC$ is similar to $\triangle MNO$. (Explain.)

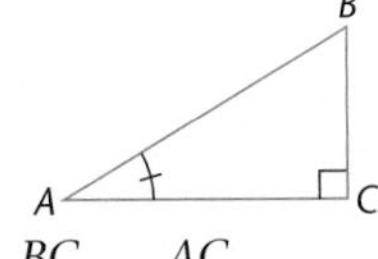

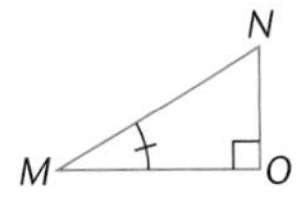

Because the triangles are similar, $\frac{BC}{NO} = \frac{AC}{MO}$.

Then, by the Exchange Property of Proportions, $\frac{BC}{AC} = \frac{NO}{MO}$.

In each ratio, the side opposite the angle is the numerator and the side adjacent to the angle is the denominator. The result shows us that if you construct a right triangle with an angle of a given measure, *the ratio of the side opposite the angle to the side adjacent to the angle will always have the same value.* This ratio is the **tangent ratio** of the angle.

The tangent is one of six different ratios that can be formed by the sides of a right triangle. These six ratios form the basis for the study of **trigonometry**. The ideas of trigonometry date back to ancient times. The word itself comes from the Greek word meaning "triangle measurement."

List the six possible side ratios for $\angle A$ in terms of "opposite," "adjacent," and "hypotenuse." Identify the tangent ratio.

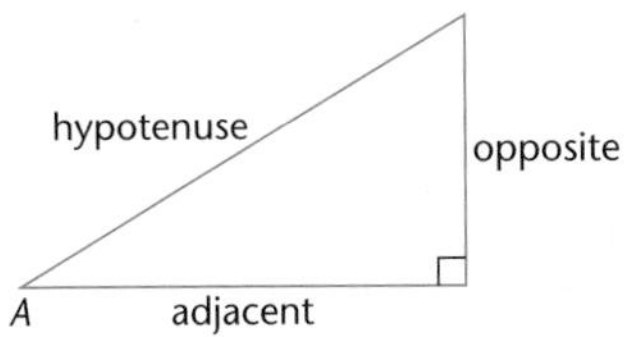

Exploration 2 Graphing the Tangent Ratio

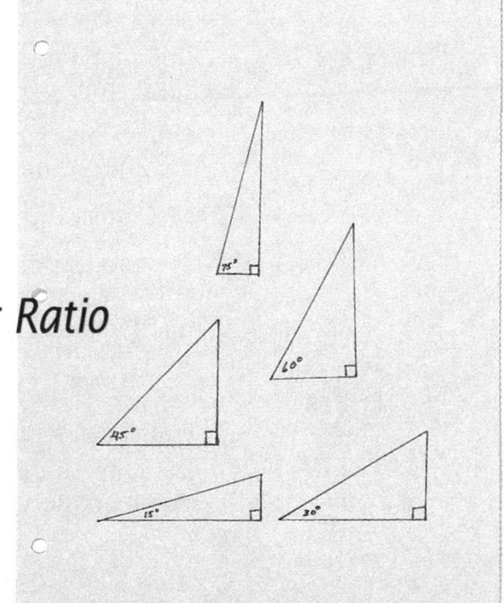

You will need

Geometry technology or
Ruler and protractor
Graph paper

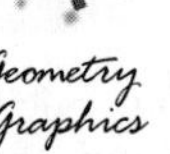

1 Draw five different right triangles, as shown, with m$\angle A$ ranging from 15° to 75° at 15° intervals.

Language Arts In "The Musgrave Ritual," a Sherlock Holmes story by Arthur Conan Doyle, Holmes uses similar triangles to find the height of a tree. Have students rewrite the episode, using the tangent ratio to find the tree's height.

Exploration 1 Notes

Students discover that the tangent ratio depends only on the size of an acute angle. After they have finished the exploration, discuss why their result will be true for *all* right triangles with one acute 30° angle. Students should mention the properties of similar triangles.

ongoing ASSESSMENT

5. The ratio of the side opposite to the acute angle of a right triangle to the side adjacent to the acute angle is constant for an acute angle of 30° (and perhaps for other given measures).

CRITICAL Thinking

If the right triangle in the exploration were placed with $\angle A$ at the origin and the side adjacent to $\angle A$ along the x-axis, then the tangent ratio found for $\angle A$ would be the slope of the line containing the hypotenuse. Both acute angles of a right triangle have tangent ratios. If a line is rotated, its slope changes, while a rotation of a right triangle will not change the tangent ratios of its acute angles.

Use Transparency 84

CRITICAL Thinking

$\frac{opp}{adj}$; $\frac{adj}{opp}$; $\frac{opp}{hyp}$; $\frac{hyp}{opp}$; $\frac{adj}{hyp}$; $\frac{hyp}{adj}$; Tangent ratio $= \frac{opp}{adj}$

Exploration 2 Notes

If students use a graphics calculator for this activity, have them plot points for the five ordered pairs and try using a best-fit line to get the smooth curve. They may find that paper and pencil methods work better than the calculator.

ongoing ASSESSMENT

5. The graph is increasing at an increasing rate. While the difference between the measures of the angles used are the same, the differences between the tangent ratios for the angles are increasing.

Math Connection Algebra

In Part 5 of Exploration 2, students may not be sure what is meant by a graph increasing "at a steady rate." A distance-time graph showing the motion of a moving car can help convey the idea—the steepness of the graph represents the speed of the car.

Exploration 3 Notes

Have students read through the entire activity before they begin. Explain that they are to find the width of the canyon and that it is too wide to be measured directly.

ongoing ASSESSMENT

4. $\tan \angle ZXY = \dfrac{YZ}{XY} \Rightarrow$

$(XY)(\tan \angle ZXY) = YZ \Rightarrow$

$XY = \dfrac{YZ}{\tan \angle ZXY}$

2. In each of the triangles that you formed in Step 1, measure the following:

   ∠A,
   the side opposite ∠A, and
   the side adjacent to ∠A.

3. For each angle, compute its tangent by dividing the measure of the side opposite ∠A by the measure of the side adjacent to ∠A. Record your results.

ALGEBRA *Connection*

4. Plot the ordered pairs (angle measure, tangent) for the different angles you drew. Draw a smooth curve through the points.

5. Does your graph seem to consistently increase or decrease? If so, does it do so at a steady rate? Describe its behavior. ❖

Exploration 3 Using the Tangent Ratio

Calculator

Surveying

You will need

A ruler and protractor
A calculator
Your graph from Exploration 2

In this exploration you will simulate or model the work of a surveying crew as they measure the distance across a canyon. To make your answers realistic, you can let 1 centimeter = 10 meters.

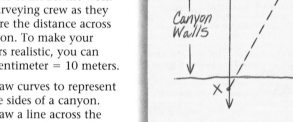

1. Draw curves to represent the sides of a canyon. Draw a line across the canyon as shown. This line represents a line of sight from point X to point Y.

2. 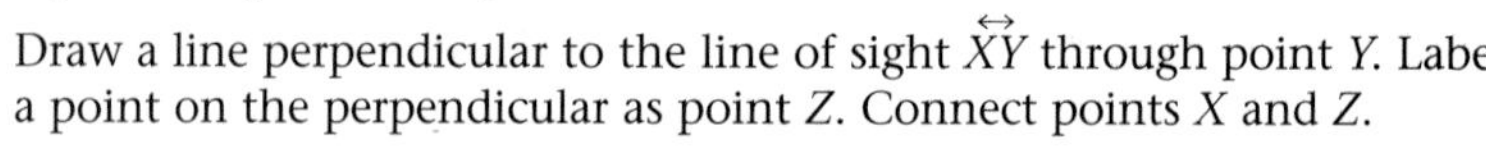
Draw a line perpendicular to the line of sight $\overleftrightarrow{XY}$ through point Y. Label a point on the perpendicular as point Z. Connect points X and Z.

3. Measure $\overline{YZ}$ and $\angle ZXY$.

4. The tangent ratio of $\angle ZXY$ is the measure of the side opposite the angle ($\overline{YZ}$) divided by the measure of the side adjacent to the angle ($\overline{XY}$). In mathematical notation,

$$\tan \angle ZXY = \frac{YZ}{XY}.$$

ALGEBRA *Connection*

Use your graph from Exploration 2 to find an approximate value for the tangent of $\angle ZXY$. Substitute this value and the measure of $\overline{YZ}$ into your equation. Solve for XY.

5. Measure $\overline{XY}$. Compare your measured value with your calculated value from Step 4. How accurate is your answer? What do you think are the main sources of error in your work? ❖

ENRICHMENT Have students work in pairs or small groups to construct a scale model of the pyramid illustrated in the Application that occurs just before the Exercises & Problems. Students may choose any convenient scale factor, for example, 1 cm = 18 cubits. Determining the altitudes for the triangles that form the lateral sides will be a challenging and interesting problem.

English Language Development Students can easily become confused by the word *tangent.* (The same term was used previously for a line that touches a circle in one point.) In this lesson, a tangent is both a ratio of two side lengths and a single number, the decimal equivalent of that ratio. These students may need a review of the idea of ratio, specifically that a ratio can be converted to a fraction or a decimal.

Try This Find the indicated measure in each figure by using the given equation.

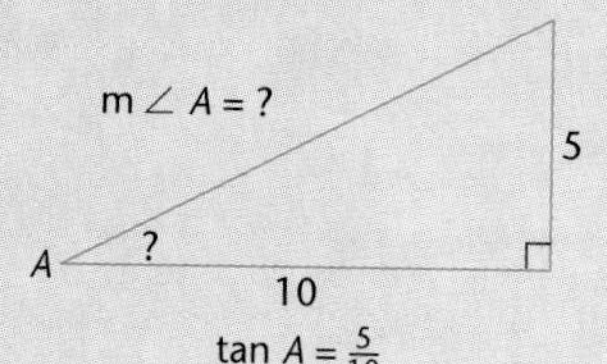

$\tan A = \frac{5}{10}$

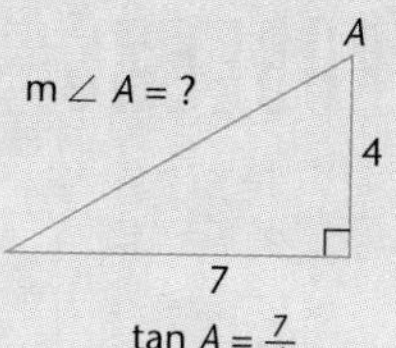

$\tan A = \frac{7}{4}$

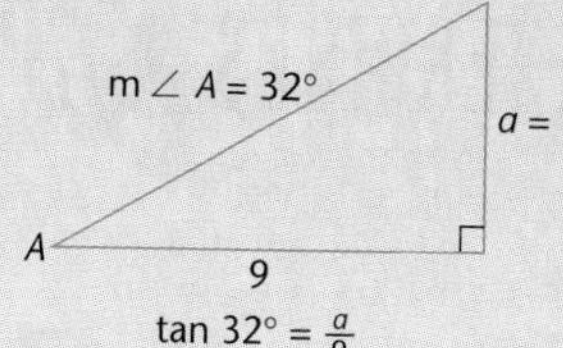

$\tan 32° = \frac{a}{9}$

APPLICATION

Cultural Connection: Africa Trigonometry has been used for more than 4,000 years. A certain trigonometric ratio was used as early as 2650 B.C.E. in ancient Egypt.

To build smooth-sided, square pyramids, the ancient Egyptians knew that the ratio of the "rise" and the "run" of the sides must be kept constant.

In modern times, the tangent ratio is generally used for such purposes—that is, the rise divided by the run. The pyramid builders used the reciprocal of the tangent, which is known today as the **cotangent**. The ancient Egyptians called this value the **seked** of the sloping surface. A typical problem from an ancient papyrus reads as follows:

> IF A PYRAMID IS 250 CUBITS HIGH AND THE SIDE OF ITS BASE IS 360 CUBITS LONG, WHAT IS ITS SEKED?
>
> —PROBLEM 56 OF THE AHMES PAPYRUS

The pyramid is assumed to be a right square pyramid. One-half the length of a side of the base, or 180 cubits, is the run of the shaded triangle. The rise is 250 cubits.

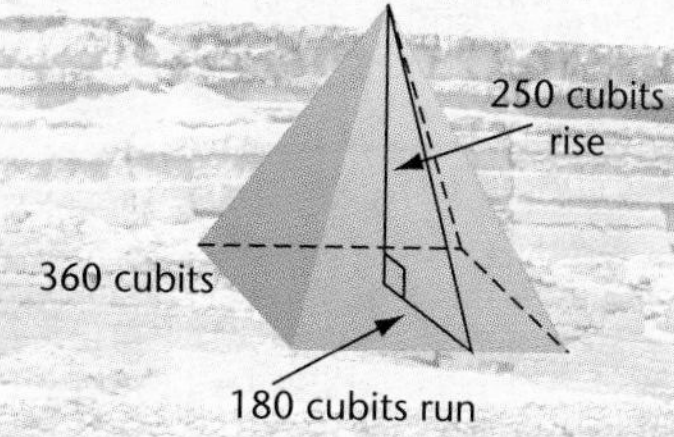

Thus, seked of pyramid $= \frac{\text{run}}{\text{rise}} = \frac{180}{250} = \frac{18}{25}$.

Unlike our trigonometric ratios, which have no units, the Egyptian ratio was written as though it were a length measure, cubits. A fractional part of a cubit was converted to palms. (There are seven palms to a cubit.) Thus the papyrus reads,

Seked is $5\frac{1}{25}$ palms. ❖

Cooperative Learning

Technology Have students work in groups and use geometry graphics software to add more data points to the graphs made in Exploration 2. They can change the acute angle by 5°, or even 1°, increments to better see the curved shape of the resulting graph.

Math Connection

Algebra

Algebra is necessary to determine the various values of the angles and sides in trigonometry applications.

Ongoing Assessment

Try This

m∠A = 26.57°

m∠A = 60.26°

a = 5.62

Teaching tip

The application introduces the cotangent ratio as the reciprocal of the tangent. The cotangent equals the adjacent side length divided by the opposite side length.

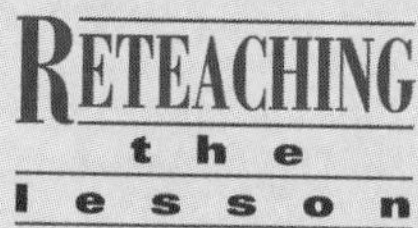

Cooperative Learning Have students work in small groups to draw a variety of right triangles on grid paper. The legs of the triangles should be whole numbers so that the tangent ratios are easy to compute. For each triangle, students should measure the appropriate acute angle to the nearest degree. They then should compare their results to those given using a scientific calculator.

Assess

Selected Answers

Odd-numbered Exercises 7–49

Assignment Guide

Core 1–16, 19–36, 41–50

Core Plus 1–6, 12–18, 30–40, 44–52

Technology

Most scientific calculators can be used to find both the tangent for a given angle and the angle for a given tangent.

Error Analysis

Students should sketch the triangles for Exercises 35–39 to help identify the correct placement of values in the equation.

Practice Master

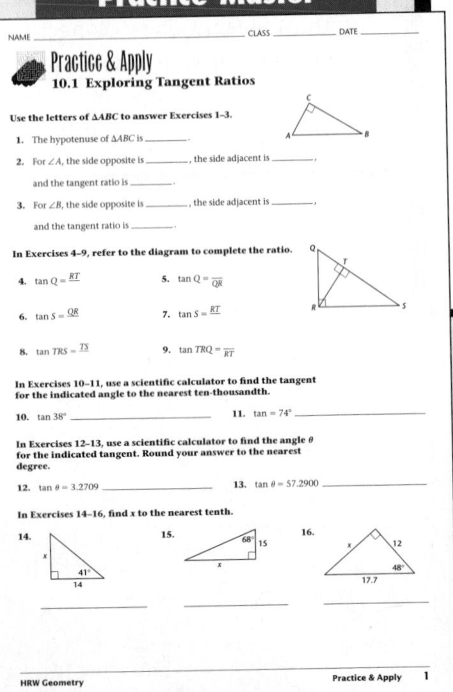
NAME ________ CLASS ________ DATE ________

Practice & Apply

10.1 Exploring Tangent Ratios

Use the letters of ΔABC to answer Exercises 1–3.

1. The hypotenuse of ΔABC is ________.
2. For $\angle A$, the side opposite is ________, the side adjacent is ________, and the tangent ratio is ________.
3. For $\angle B$, the side opposite is ________, the side adjacent is ________, and the tangent ratio is ________.

In Exercises 4–9, refer to the diagram to complete the ratio.

4. $\tan Q = \frac{RT}{\quad}$
5. $\tan Q = \frac{\quad}{QR}$
6. $\tan S = \frac{QR}{\quad}$
7. $\tan S = \frac{RT}{\quad}$
8. $\tan TRS = \frac{TS}{\quad}$
9. $\tan TRQ = \frac{\quad}{RT}$

In Exercises 10–11, use a scientific calculator to find the tangent for the indicated angle to the nearest ten-thousandth.

10. tan 38° ________
11. tan = 74° ________

In Exercises 12–13, use a scientific calculator to find the angle θ for the indicated tangent. Round your answer to the nearest degree.

12. $\tan \theta = 3.2709$ ________
13. $\tan \theta = 57.2900$ ________

In Exercises 14–16, find x to the nearest tenth.

14. 15. 16.

HRW Geometry — Practice & Apply 1

Exercises & Problems

Communicate

1. What is the difference between a tangent ratio and the ratio that the ancient Egyptians called a seked?
2. Does the tangent ratio increase or decrease as an angle gets larger? Explain your answer.

Which sides are opposite and adjacent to $\angle A$ in each right triangle below. Explain your answers. Then find the tangent ratio for $\angle A$.

3.

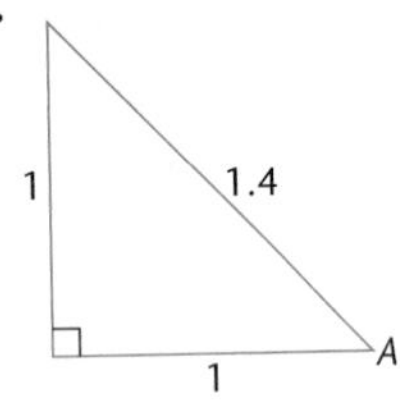

4.

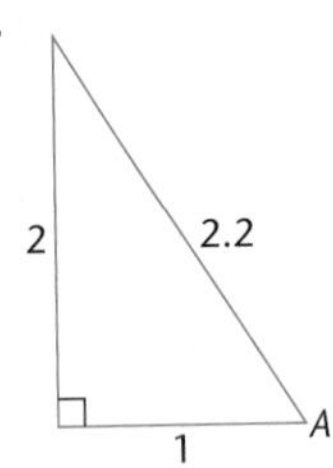

5.

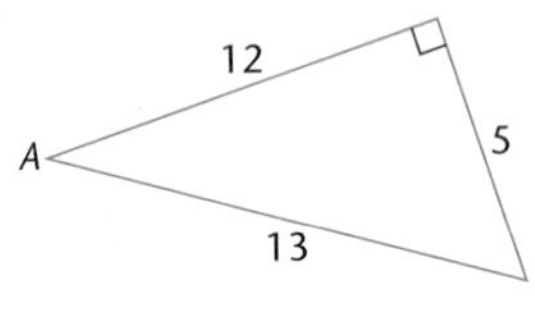

Practice & Apply

Find the tangent ratio for $\angle A$ in each right triangle below.

6.

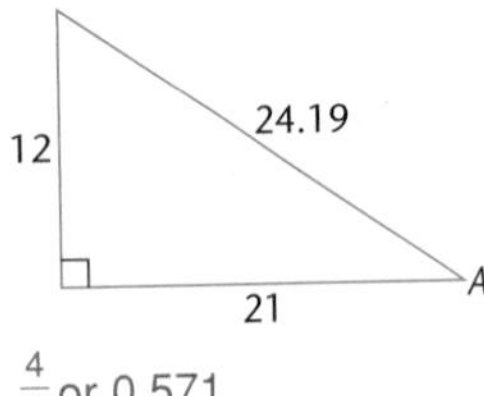

$\frac{4}{7}$ or 0.571

7.

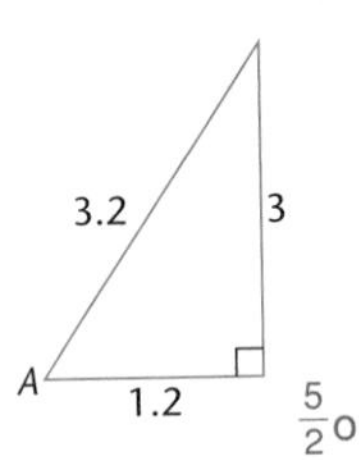

$\frac{5}{2}$ or 2.5

8.

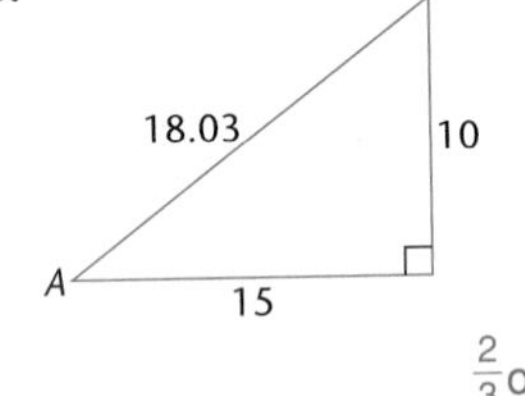

$\frac{2}{3}$ or 0.667

9.

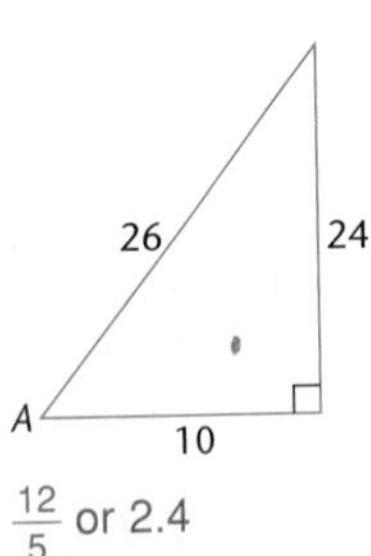

$\frac{12}{5}$ or 2.4

10.

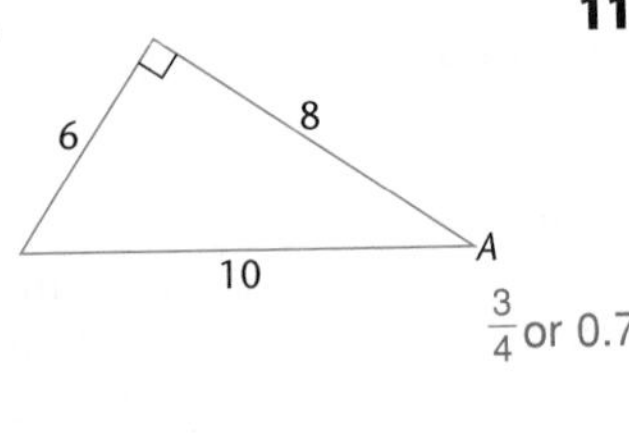

$\frac{3}{4}$ or 0.75

11.

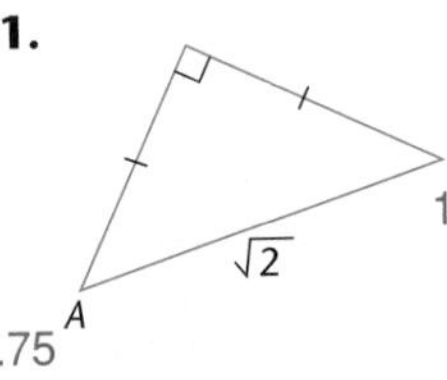

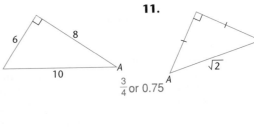

Find the tangent ratio for $\angle C$ in each right triangle below.

12.

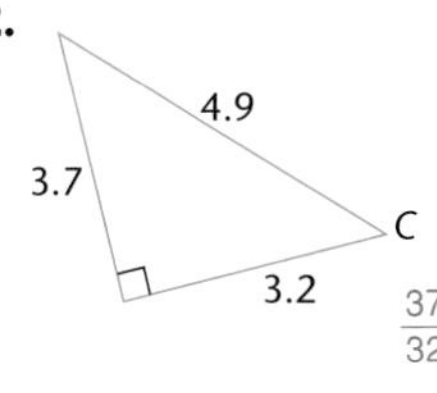

$\frac{37}{32}$

13.

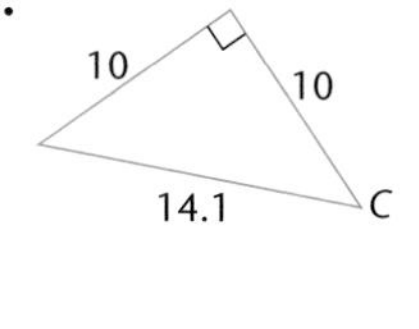

1

14.

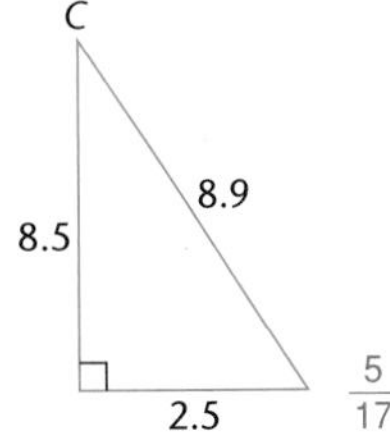

$\frac{5}{17}$

15.

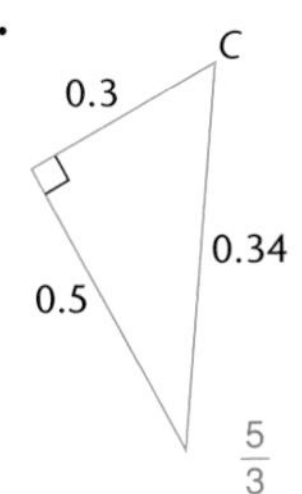

$\frac{5}{3}$

16.

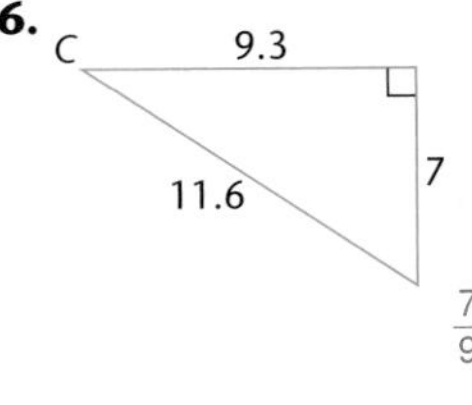

$\frac{70}{93}$

17.

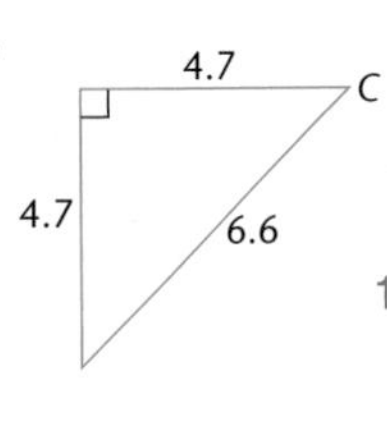

1

Use the graph from Exploration 2 to estimate the angle associated with each tangent ratio.

18. 0.5 About 26°–27° **19.** 1.5 About 56°–57° **20.** 0.85 About 40°–41°

Technology Use a calculator with scientific functions to find the tangent ratio or tangent for each angle indicated.

21. 67° 2.3559 **22.** 25° 0.4663 **23.** 87° 19.0811 **24.** 19° 0.3443

25. 47° 1.0724 **26.** 53° 1.3270 **27.** 21° 0.3839 **28.** 75° 3.7321

Technology Use a calculator with scientific functions to find the angle θ for each tangent. Use the inverse tangent function ($\tan^{-1}$). Round to the nearest degree.

29. $\tan \theta = .4663$ 25°

30. $\tan \theta = .6009$ 31°

31. $\tan \theta = 2.3559$ 67°

32. $\tan \theta = 1$ 45°

ALTERNATIVE ASSESSMENT

Performance Assessment

Students should construct kite *ABCD* and draw its diagonals with intersection point *O*. Students should then find the tangent of each of the eight acute angles in the figure. Applying what they have learned previously about the properties of kites will make the project go faster, as some of the angles will be congruent.

Technology Master

NAME ______ CLASS ______ DATE ______

Technology
10.1 Pitch

The word pitch has many meanings. In carpentry, pitch refers to the ratio of the rise of a roof to the length it spans. The house pictured has a roof that has two different pitches. The portion of the roof at the left of the diagram has pitch $\frac{i}{run} = \frac{B}{AC}$ and the portion at the right has pitch $\frac{i}{run} = \frac{BC}{CD}$. You might notice that these ratios are the tangents of the roof angles at *A* and at *D*.

B, rise, run, run, A, C, D, F, E

In a geometry software program, you can experiment with different positions for *B* and the different roof pitches that result.

Use geometry software in the following exercises.

1. Sketch and label a figure like the one shown. (In your sketch, use the PERPENDICULAR command to construct the line through *B* and *C*.)
2. MEASURE $m\angle BAC$, *BC*, and *AC*. CALCULATE the ratio $BC : AC$. ______
3. MEASURE $m\angle BDC$, *BC*, and *DC*. CALCULATE the ratio $BC : DC$. ______
4. Find the two pitches of the roof given that $m\angle BAC = 30°$ and $m\angle BDC = 45°$. ______
5. A homeowner wants a roof that has a pitch of 0.6 in ΔBAC and a pitch of 0.4 in ΔBDC. What are the $m\angle BAC$ and $m\angle BDC$? ______
6. Is it possible to construct a roof with $m\angle BAC = 30°$, $m\angle BDC = 45°$ and $BC = 10$? Explain. ______

HRW Geometry Technology 15

33. As the angles get larger, the tangent values get larger. As the angles get smaller, the tangent values get smaller.

34. For angles larger than 90° but less than 180°, as the angle measure increases the absolute value of the tangent ratio decreases. However, it is negative, so it is increasing in this interval.

Use a table of trigonometric values or a scientific calculator to answer Exercises 33 and 34.

33. What happens to the tangent values as the angles get larger?

34. What happens to the pattern in Exercise 33 for angles larger than 90°?

Technology For Exercises 35–39, write an equation and solve for x. Use a scientific calculator to find the tangent ratio. An example is given below.

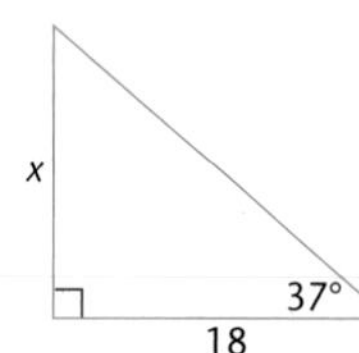

$$\tan 37° = \frac{x}{18}$$

$$0.7536 = \frac{x}{18}$$

$$x = 18(0.7536)$$

$$x = 13.56$$

35.

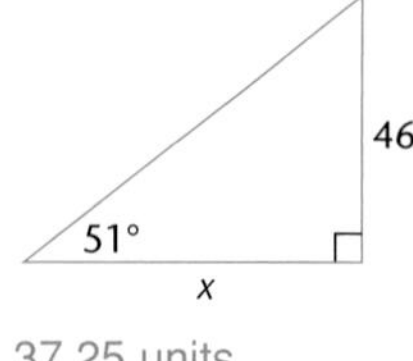

37.25 units

36.

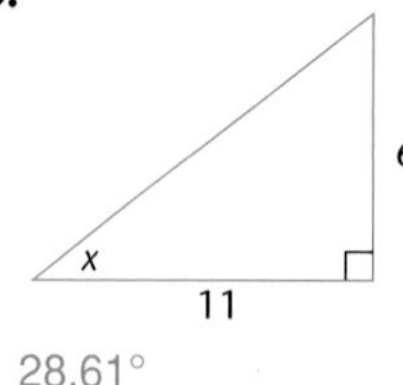

28.61°

37.

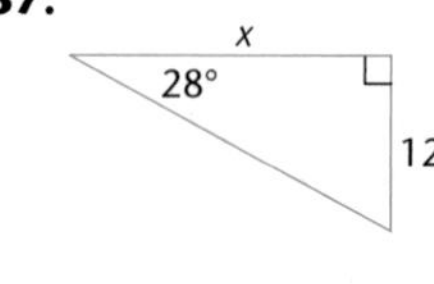

22.57 units

38.

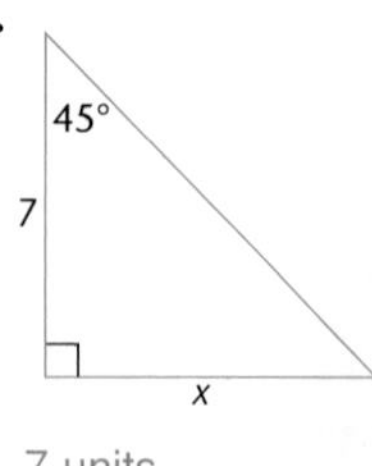

7 units

39.

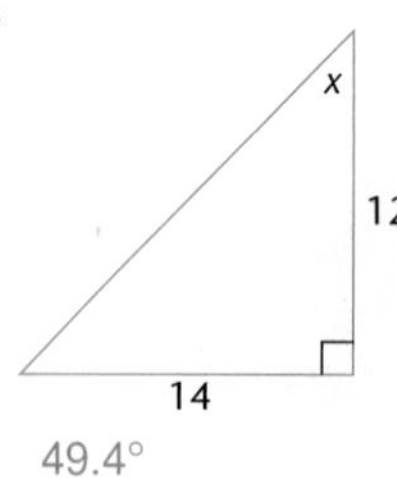

49.4°

40. Engineering In order to construct a bridge across a lake, an engineer wishes to determine distance AB. A surveyor found that $AC = 530$ meters and $m\angle C = 43°$. If $\angle A$ is a right angle, find AB, the distance across the lake. 494.2 m

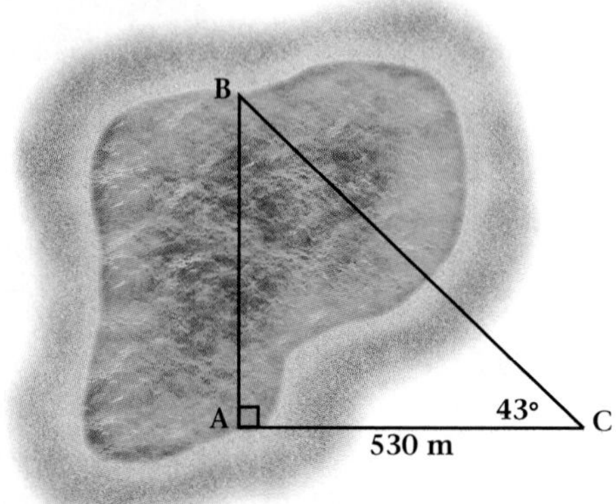

Technology Use a scientific calculator for Exercises 40–43. Express the answer to the nearest tenth or nearest degree.

41. **Engineering** The steepness, or grade, of a highway or railroad is expressed as a percent. A railway that rises 18 feet for 100 feet of horizontal run has a grade of 18 percent, as in the photograph of the cog railway at Pike's Peak, Colorado. What is θ, the angle of inclination of the railway in the photograph? 10.2°

The maximum grade of the railway at Pike's Peak is 25 percent. What angle is this? 14°

42. Determine the height of the flagpole shown in the figure at the right. 18.0 m

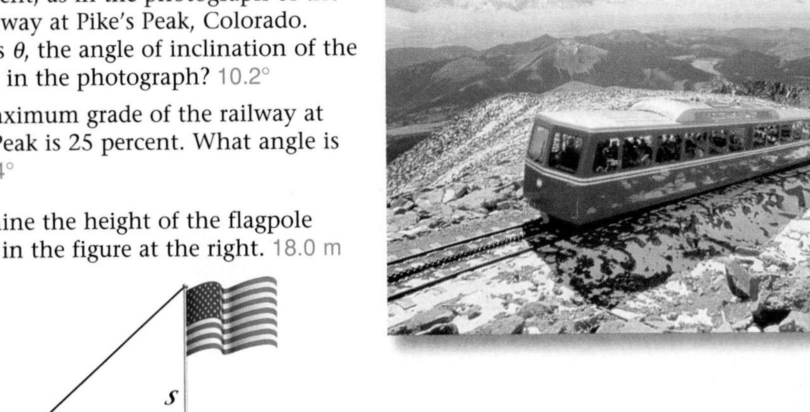

43. **Surveying** Points P and Q are on the north and south rims, respectively, of Glen Canyon, with Q directly south of P. Point R is located 300 feet from P and $\overline{QP} \perp \overline{PR}$. The measure of $\angle PRQ$ is 75°. Find PQ, the width of the canyon.

43. 1119.62 ft

Use Transparency 85

Look Back

Find the volume and surface area of each.

44. right prism **[Lesson 7.2]** **45.** cylinder **[Lesson 7.4]**

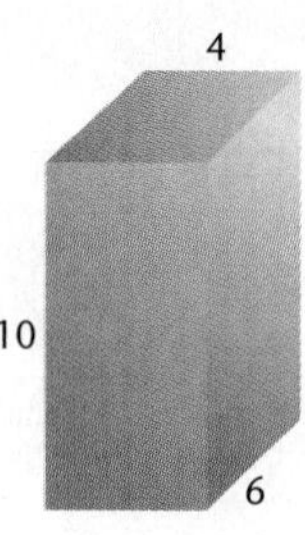

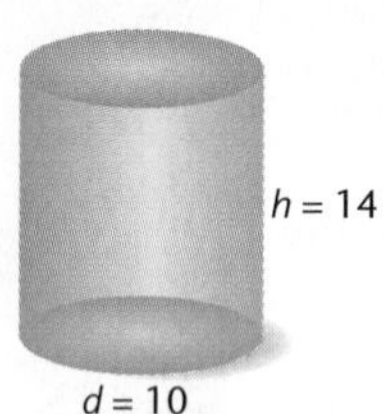

V = 240 units^3
SA = 248 units^2

V = 1099.56 units^3
SA = 596.90 units^2

46. cone **[Lesson 7.5]** **47.** sphere **[Lesson 7.6]**

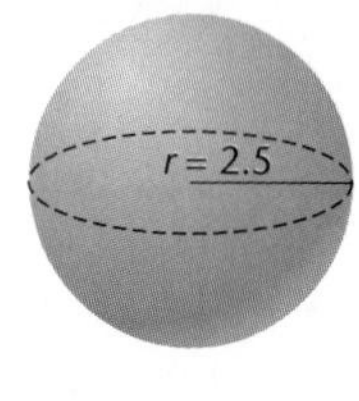

V = 728.64 units^3
SA = 502.09 units^2

V = 65.45 units^3
SA = 78.54 units^2

Use ⊙*P*, with $\overline{MN} \perp \overline{RP}$ at *Q*, for Exercises 48–50. [Lessons 9.4, 9.5]

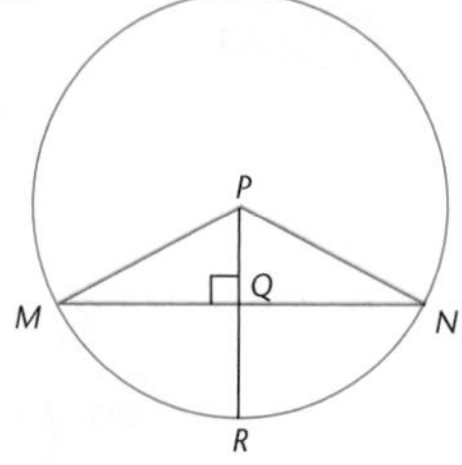

48. $\overline{MQ} \cong$ __?__ $\overline{NQ}$

49. If $PR = 8$ and $PQ = 3$, what is the length of $\overline{MQ}$? of $\overline{QN}$? $MQ = 7.42$ units $= QN$

50. If $PR = 12$ and $PQ = 4$, what is the length of $\overline{MQ}$? of $\overline{QN}$? $MQ = 11.31$ units $= QN$

Look Beyond

51. Does tan 20° + tan 60° = tan 80°? Why or why not?

52. If a pyramid is 300 cubits high and the side of its base is 420 cubits long, what is its seked? (Hint: Use the cotangent as on page 527.)
$\frac{7}{10}$ cubits

Look Beyond

Extension. Exercise 51 demonstrates that tangent measure for an angle is like an "alternative protractor"—the angle addition postulate would not hold for such a "protractor." Exercise 52 is appropriate for small group work because of the difficulty of understanding and converting the unfamiliar units of measurement.

51. No. The counterexample proves that the tangent of a sum is not the sum of the tangents.

Sines and Cosines

A paraskier is able to fly quite high above the water. If you know the length of the rope and can estimate its angle, you can use trigonometry to estimate the skier's height above the water.

Trigonometric Ratios

In the previous lesson you learned two different ratios for an angle in a right triangle.

$$\text{tangent of } \angle A\text{, or } \tan A = \frac{\text{length of side opposite } \angle A}{\text{length of side adjacent to } \angle A}$$

$$\text{cotangent of } \angle A\text{, or } \cot A = \frac{\text{length of side adjacent to } \angle A}{\text{length of side opposite } \angle A}$$

Now consider the following two additional ratios:

$$\textbf{sine of } \angle \boldsymbol{A}\text{, or } \sin A = \frac{\text{length of side opposite } \angle A}{\text{length of hypotenuse}}$$

$$\textbf{cosine of } \angle \boldsymbol{A}\text{, or } \cos A = \frac{\text{length of side adjacent to } \angle A}{\text{length of hypotenuse}}$$

These ratios can also be written $\sin \theta$, $\cos \theta$, etc., where θ (theta) is the measure of the angle.

The cotangent is the reciprocal of the tangent. Is the cosine the reciprocal of the sine? Explain.

Technology The explorations may be done using geometry graphics software such as *The Geometer's Sketchpad.* Students should construct any right triangle and then measure one acute angle and all three sides. Then they use *Calculate* from the *Measure* menu to compute the sine, cosine, and other needed relationships. By dragging on the vertices of the triangle to change the size of the acute angle, students can make and test conjectures.

PREPARE

Objectives

- Explore and graph the relationship between the size of an angle and its sine or cosine.
- Develop two trigonometric identities.
- Solve problems using the sine and cosine ratios.

• Practice Master	10.2
• Enrichment Master	10.2
• Technology Master	10.2
• Lesson Activity Master	10.2
• Quiz	10.2
• Spanish Resources	10.2

Assessing Prior Knowledge

One acute angle in a right triangle is increasing by 5° increments. What happens to the other acute angle? Include sketches with your answer. **The other acute angle decreases by 5° increments.**

TEACH

In Lesson 10.1, students learned to use the tangent of an angle to find missing parts of a right triangle. Here, they will see how the sine and cosine of an angle can be used in similar situations.

CRITICAL *Thinking*

No. The reciprocal of the sine is $\frac{\textbf{hyp}}{\textbf{opp}}$, not $\frac{\textbf{adj}}{\textbf{hyp}}$.

TEACHING tip

Students should have a scientific or graphics calculator for most of this lesson. Have students make sure their calculators are set for degrees rather than radians. If scientific calculators are not available, students can use tables of trigonometric functions as an alternative.

Exploration 1 Notes

In this exploration, students investigate the relationship between the size of an angle and its sine or cosine. You may wish to have students use graphics calculators or a computer graphic program for the graphs in Step 5 of Exploration 1. Ask students to predict the shapes of the curves before they make the graphs. Students should be able to tell from their data that they will not get straight lines for the graphs; that is, the functions are not linear.

ongoing ASSESSMENT

3. As an angle θ increases from 0° to 90°, the value of sin θ increases, while the value of cos θ decreases.

Exploration 1 Sines and Cosines

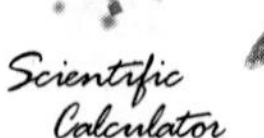

You will need
Scientific calculator
Graph paper (small grid)

1. As $\angle A$ increases in the illustration below, what happens to the value of sin θ? Does it increase or decrease? Write a conjecture about the sine of 0° and of 90°.

$$\sin \theta = \frac{opp.}{hyp.}$$

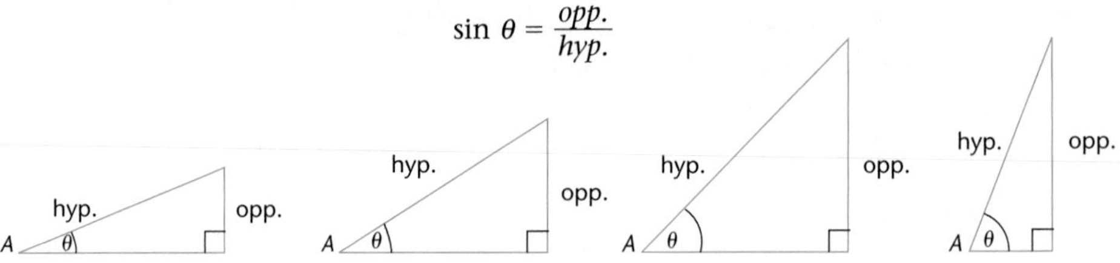

2. As $\angle A$ increases in the illustration below, what happens to the value cos θ? Does it increase or decrease? Write a conjecture about the cosine of 0° and 90°.

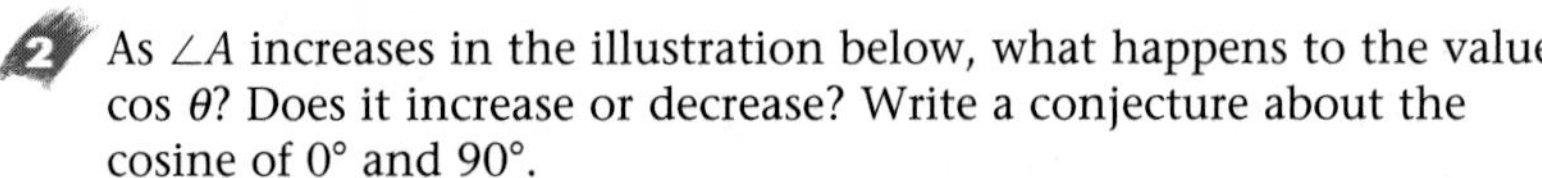

$$\cos \theta = \frac{adj.}{hyp.}$$

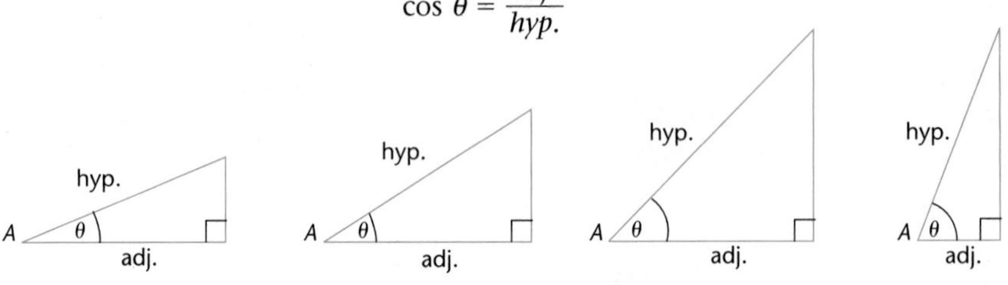

3. Write a statement that contrasts the behavior of the sine and cosine ratios as an angle increases from 0° to 90°.

4. Use a scientific calculator to find the sine and cosine of each angle indicated. Fill in a table like the one below. Three figures to the right of the decimal points is sufficient for this purpose. (Be sure your calculator is set to the "degree" mode.)

θ	sin θ	cos θ
0°	?	?
10°	?	?
20°	?	?
.	.	.
90°	?	?

5. Plot the values of the sine ratios for the angles in the table (angle measure, sine). Draw a smooth curve through the points. Repeat for the cosine ratios. Compare your graphs with your statement in Step 3. ❖

interdisciplinary CONNECTION

Physics Snell's Law describes how a ray of light is refracted as it passes from air or a vacuum into any denser material such as water, oil, or glass. One expression of this principle is $\frac{\sin \angle 1}{\sin \angle 2} = \frac{n_2}{n_1}$. The right-hand ratio is the two indices of refraction.

normal
light ray
1
air
water
2

Have students use science textbooks or library resources to research Snell's Law. Their written reports should include diagrams.

APPLICATION

Recreation

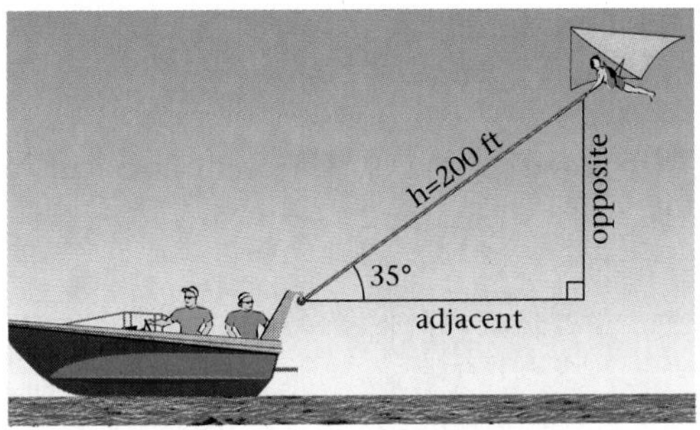

A paraskier is towed behind a boat with a 200-foot rope. The spotter in the boat estimates the angle of the rope to be 35° above the horizontal. Estimate the skier's height above the water.

Label the hypotenuse and the opposite and adjacent sides for the given angle. The hypotenuse is known, and the side you want to find is the opposite side. Use the sine ratio.

$$\sin 35° = \frac{o}{h}$$

Solve for a and substitute the values for the sine ratio and the length of the rope.

$$o = h(\sin 35°)$$
$$= 200(0.5736) \approx 115 \text{ ft}$$ ❖

APPLICATION

Astronomy

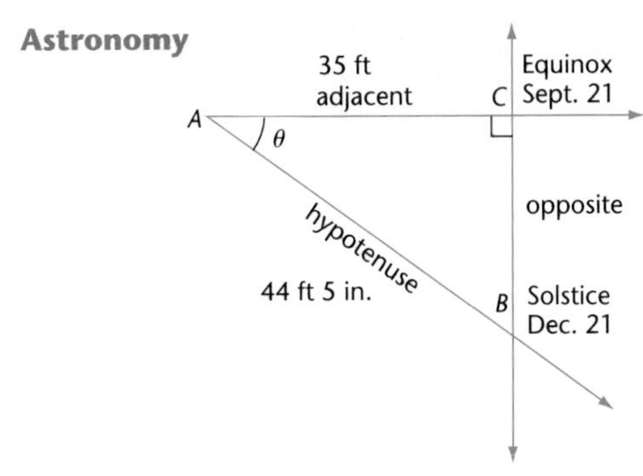

The members of an astronomy club keep records of their observations of the position of the sunrise as viewed from a fixed point A in a parking lot. They set up lines of sight from point A in the directions of sunrise at the autumnal equinox and the winter solstice. Then they set up a line perpendicular to the solstice line as shown. Finally, they measure the distances AB and AC. What is the measure of angle θ?

$$\cos \theta = \frac{AC}{AB} = \frac{35 \text{ ft } 0 \text{ in.}}{44 \text{ ft } 5 \text{ in.}}$$

$$= \frac{35.0}{44.417} \approx 0.788$$

To find the angle whose cosine is 0.788, you can use a trigonometry table. To find the angle using a scientific calculator, use the $\cos^{-1}$ key (the inverse cosine key).

.788 ≈ 38° ❖

Use Transparency 86

Cooperative Learning

Students will benefit from doing both of the applications in small groups. Hand out the statements of the problems, without solutions, to each group. Students should work together to analyze the problems, interpret the figures, and devise possible solutions. When all groups are finished, lead a class discussion in which students report on their experiences. Then they can compare their solution strategies with those used in the text.

TEACHING tip

Read the first application aloud to help students develop their visual thinking skills. Students should listen carefully and try to draw the figure for the problem. Then they should compare their figures with the one given in the text.

ENRICHMENT Have students prove the following: The ratio of two sides of any triangle is equal to the ratio of the sines of the angles opposite those sides. Sketches will help students determine the answer.

[Solution: Since $\sin A = \frac{h}{b}$ and $\sin B = \frac{h}{a}$, $h = b \sin A = a \sin B$, and $\frac{a}{b} = \frac{\sin A}{\sin B}$.]

INCLUSION strategies **Using Cognitive Strategies** Many students have difficulty remembering which side lengths to use for the sine, cosine, and tangent ratios. Here is one memory device: Some Old Hen Caught Another Hen Taking One Away. The initial letters of the words show the ratios sine = opposite/hypotenuse, cosine = adjacent/hypotenuse, and tangent = opposite/adjacent. Students may enjoy making up memory devices of their own.

Teaching tip

Explain that an *identity* is an equation that is true for all possible values of the variables. For example, $(a+b)(a-b)=a^2-b^2$ is true for all real numbers a and b. Identities are useful in simplifying expressions and in converting them to more convenient forms.

Exploration 2 Notes

Students create data tables to develop two trigonometric identities. In Part 1, if students try to use 90° for the angle, they will get $\frac{1}{0}$ for the sine-cosine ratio and an error message for the tangent on a calculator. Ask students why 90° is not an acceptable value to use in this investigation.

Ongoing Assessment

2. $\tan\theta = \dfrac{\sin\theta}{\cos\theta}$

Math Connection
Algebra

In Step 4 of Part 2 of the Exploration, make sure students are careful with the parentheses in the expressions.

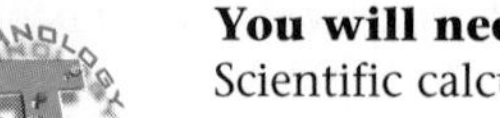

Exploration 2 Two Trigonometric "Identities"

You will need
Scientific calculator

Part I

Scientific Calculator

1 Using the SIN, COS, and TAN keys on a scientific calculator, copy and complete the table.

θ	$\sin\theta$	$\cos\theta$	$\tan\theta$	$\frac{\sin\theta}{\cos\theta}$
20°	?	?	?	?
40°	?	?	?	?
60°	?	?	?	?

2 What do you notice about the values in the tangent column? Make a conjecture about a relationship between the tangent of an angle and the sine and cosine of the angle.

3 In the equation below, what happens when you simplify the second expression?

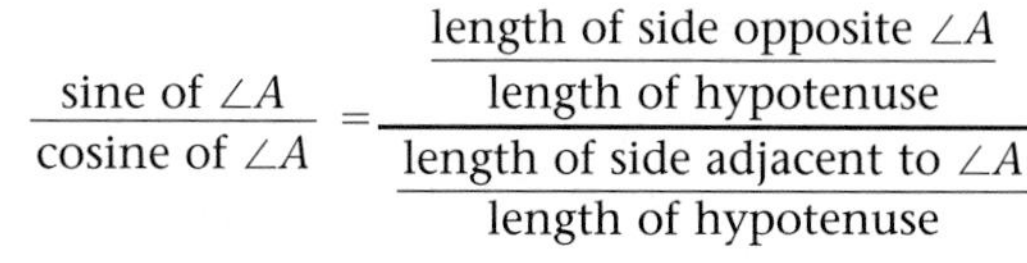

$$\frac{\text{sine of } \angle A}{\text{cosine of } \angle A} = \frac{\dfrac{\text{length of side opposite } \angle A}{\text{length of hypotenuse}}}{\dfrac{\text{length of side adjacent to } \angle A}{\text{length of hypotenuse}}}$$

Does the equation prove your conjecture? Discuss.

Part II

1 Using the SIN and COS keys on a scientific calculator, fill in a table like the one below.

θ	$\sin\theta$	$\cos\theta$	$(\sin\theta)^2 + (\cos\theta)^2$
20°	?	?	?
40°	?	?	?
60°	?	?	?

Cooperative Learning Students should work in pairs. They should draw four similar right triangles of different sizes. The triangles should not be isosceles; that is, they should not have two 45° acute angles. Have students use subscripted variables for the vertices: C_1, C_2, C_3, C_4 for the four right angles; A_1, A_2, A_3, A_4 for one set of congruent acute angles; and B_1, B_2, B_3, B_4 for the other set of congruent acute angles.

Students should measure side lengths in tenths of centimeters. Then they should write sine, cosine, and tangent ratios for all the acute angles. Using the four triangles of different sizes will help emphasize that the trigonometric ratios are related only to the size of the acute angle. They are independent of the side lengths.

2 If a graphics calculator is available, graph the following equation.

$$y = (\sin x)^2 + (\cos x)^2$$

3 What do you notice about the values of y? Make a conjecture about the value of $(\sin x)^2 + (\cos x)^2$ for any given angle x.

ALGEBRA Connection

4 Give a reason for each step in the proof below.

$$(\text{sine of } \angle A)^2 + (\text{cosine of } \angle A)^2$$
$$= \left(\frac{a}{c}\right)^2 + \left(\frac{b}{c}\right)^2$$
$$= \frac{(a^2 + b^2)}{c^2}$$

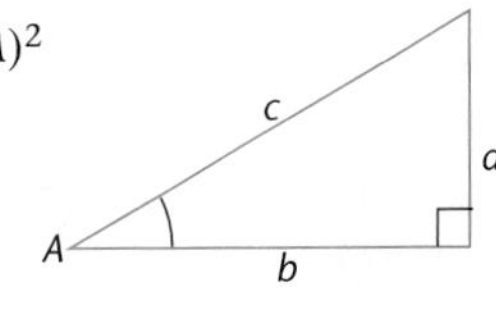

Compare the numerator and the denominator of the resulting fraction. What do you observe? How does this prove the conjecture? ❖

ONGOING ASSESSMENT

3. The value of y is always 1; $(\sin \theta)^2 + (\cos \theta)^2 = 1$

ASSESS

Selected Answers

Odd-numbered Exercises 7–49

Assignment Guide

Core 1–31, 39–49

Core Plus 1–15, 31–51

Technology

A scientific calculator is needed for most of the exercises. If these calculators are not available, students can use a chart of the trigonometric functions. Geometry graphics software can be used for the cofunction investigations in Exercises 34–37.

Error Analysis

The notation used in Exercises 25–29 may confuse some students. Explain that an expression such as *$\sin^{-1} 0.3$* means the angle with a sine equal to 0.3. Since expressions of this type look so different from the notation $\angle ABC$, many students will need extra practice learning to understand the notation.

EXERCISES & PROBLEMS

Communicate

Complete each statement.

1. How is the sine of an angle found in a right triangle?
2. How is the cosine of an angle found in a right triangle?
3. Does the sine ratio increase or decrease as the angle gets larger? Explain your answer.
4. Does the cosine ratio increase or decrease as the angle gets larger? Explain your answer.

Complete each trigonometric identity.

5. Describe two ways of finding the tangent of an angle in a right triangle.
6. Why does $(\sin A)^2 + (\cos A)^2$ equal 1?

B

A C

alternative ASSESSMENT

Authentic Assessment

Students should work in small groups and they apply what they have learned about sines, cosines, and tangents to find the height of a building or tall tree. As part of the assessment, students will need to invent some instrument for measuring the angle of sight to the top of the structure. You may have all groups use the same structure, or you may choose to let each group choose their own tree or building for the project. Written reports should include descriptions of strategies used, including any methods that did not work and why.

Practice & Apply

Find the following for the triangle on the right.

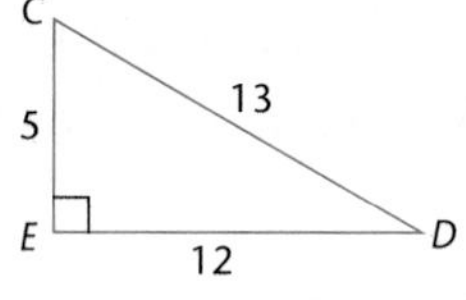

7. sin C $\frac{12}{13}$ **8.** cos D $\frac{12}{13}$ **9.** sin D $\frac{5}{13}$

10. cos C $\frac{5}{13}$ **11.** tan C $\frac{12}{5}$ **12.** tan D $\frac{5}{12}$

13. $m\angle C$ (use a scientific calculator) 67.38°

14. $m\angle D$ 22.62°

Find the following for the triangle on the right.

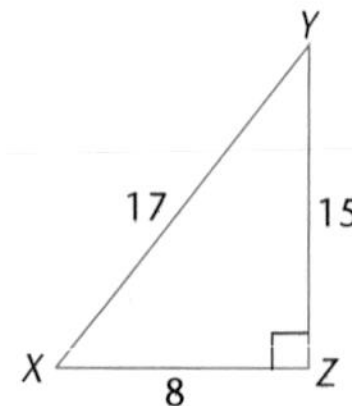

15. $\frac{15}{17}$ = cos ___?___ $\angle Y$

16. $\frac{8}{17}$ = sin ___?___ $\angle Y$

17. $\frac{8}{17}$ = cos ___?___ $\angle X$

18. $\frac{15}{17}$ = sin ___?___ $\angle X$

19. $\frac{8}{15}$ = tan ___?___ $\angle Y$

Use a scientific calculator to find the following.

20. sin 35° **21.** cos 72° **22.** tan 37°

23. sin 57° **24.** cos 52°

Use a scientific calculator to find the angles for each of the following. Round to the nearest degree.

25. $\sin^{-1} 0.3$ 17° **26.** $\sin^{-1} 0.8775$ 61° **27.** $\cos^{-1} 0.56$ 56°

28. $\cos^{-1} 0.125$ 83° **29.** $\tan^{-1} 2.432$ 68°

30. Recreation A straight water slide is 25 m long and starts at the top of a tower that is 21 m high. Find the angle between the tower and the slide. 32.9°

Use a calculator to answer Exercises 31–33.

31. Choose any angle measure between 0° and 90°. Find the sine of your angle and then apply the $\sin^{-1}$ key to the sine value. Repeat this procedure for four more angles. Look for a pattern and write a conjecture based on the pattern.

32. Repeat Exercise 31 but replace the sine with tangent and replace $\sin^{-1}$ with $\tan^{-1}$. What do you notice?

33. Repeat Exercise 31 for cosine and $\cos^{-1}$. What do you notice?

Practice Master

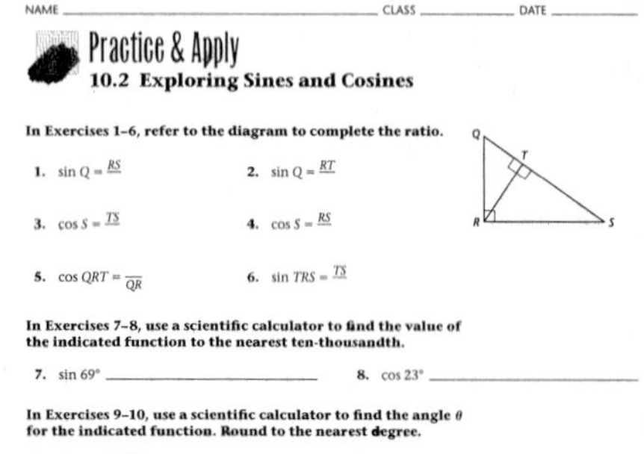
NAME ______ CLASS ______ DATE ______

Practice & Apply

10.2 Exploring Sines and Cosines

In Exercises 1–6, refer to the diagram to complete the ratio.

1. sin Q = $\frac{RS}{}$
2. sin Q = $\frac{RT}{}$
3. cos S = $\frac{TS}{}$
4. cos S = $\frac{RS}{}$
5. cos QRT = $\frac{}{QR}$
6. sin TRS = $\frac{TS}{}$

In Exercises 7–8, use a scientific calculator to find the value of the indicated function to the nearest ten-thousandth.

7. sin 69° ______
8. cos 23° ______

In Exercises 9–10, use a scientific calculator to find the angle θ for the indicated function. Round to the nearest degree.

9. $\sin \theta = 0.9925$ ______
10. $\cos \theta = 0.3584$ ______

In Exercises 11–13 find x and y to the nearest tenth.

11.

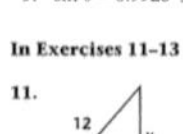

12.

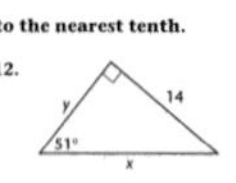

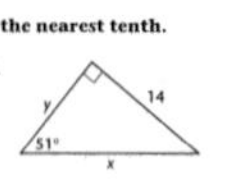

13.

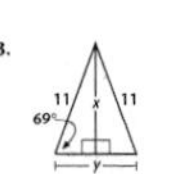

14. In parallelogram $ABCD$, altitude $\overline{DH}$ is drawn to base $\overline{AB}$. $m\angle A = 47°$, $AD = 16$ ft, and $AB = 43$ ft. Use the space provided to draw a diagram.

15. To the nearest tenth of a foot, DH = ______.

16. To the nearest square foot, the area of $ABCD$ is ______.

17. An 18-ft ladder leans against a house that stands on level ground. The foot of the ladder is 10 ft from the house. Use the space provided to draw a diagram.

18. To the nearest degree, the angle that the foot of the ladder makes with the ground is ______.

2 **Practice & Apply** **HRW Geometry**

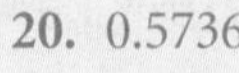

20. 0.5736

21. 0.3090

22. 0 .7536

23. 0.8387

24. 0.6157

31. $\sin^{-1}(\sin \theta) = \theta$. If you find the sine of an angle and the inverse sine of the result, you return to the original angle.

32. $\tan^{-1}(\tan \theta) = \theta$. If you find the tangent of an angle and the inverse tangent of the result, you return to the original angle.

33. $\cos^{-1}(\cos \theta) = \theta$. If you find the cosine of an angle and the inverse cosine of the result, you return to the original angle.

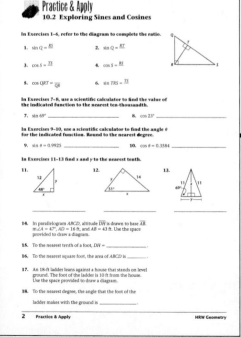

Exercises 34–37 lead through the development of the cofunction identities for sine and cosine. Use a scientific calculator.

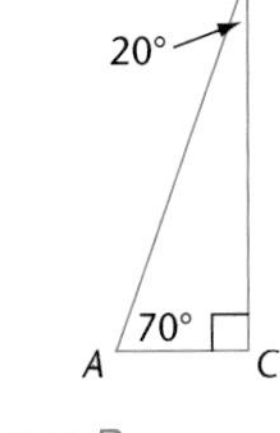

34. In $\triangle ABC$, what is the sine of $\angle A$? What is the cosine of $\angle B$? What do you notice about these two values?

35. Repeat the steps in Exercise 34 for another set of complimentary angles. (Remember that the acute angles of a right triangle sum to 90°.) What do you notice about the relationship between sin A and cos B? sin A = cos B

36. Complete the conjecture below.

In a right triangle, the __?__ of one acute angle is __?__ to the __?__ of the other acute angle. sine; equal to; cosine

37. Since the acute angles A and B of a right triangle sum to 90°, $\sin A = \cos B$ can also be written $\sin A = \cos (90° - m\angle A)$. Complete the following cofunction identities. (The first one is done for you.)

$\sin A = \cos (90° - m\angle A)$

$\cos B =$ __?__ sin (90° − m∠B)

$\sin B =$ __?__ cos (90° − m∠B)

$\cos A =$ __?__ sin (90° − m∠A)

38. Building Codes For easy access, a sidewalk ramp is supposed to have an angle of 15° or less to the ground. A ramp to a landing 1.5 m high is 5 m long. Is the angle 15° or less? Explain your reasoning.

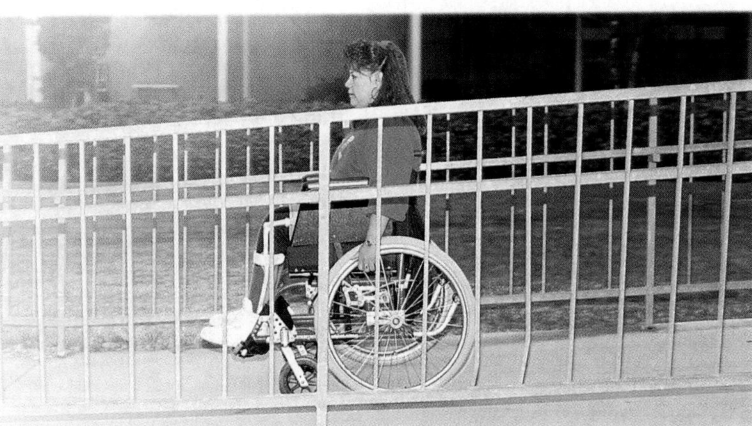

The angle is 17.5°, which is over 15°.

39. Use trigonometry to find the area of the triangle.

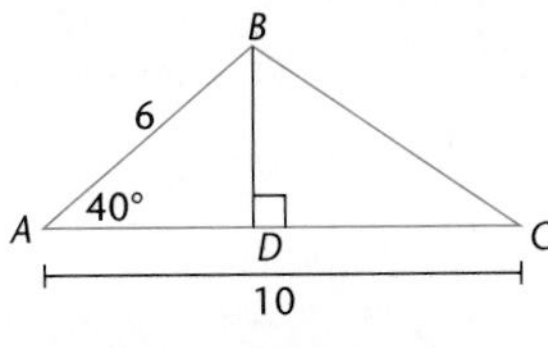

Area = 19.28 units^2

40. Use trigonometry to find the area of the parallelogram.

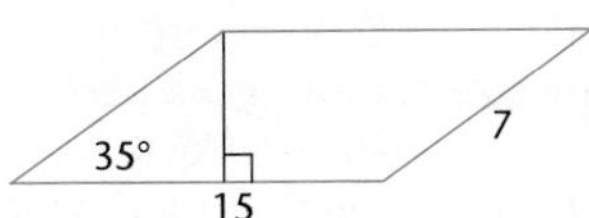

Area = 60.23 units^2

34. sin 70°= .9397; cos 20° = .9397; They are the same.

Use Transparency 87

Technology Master

NAME ______________ CLASS ________ DATE ________

Technology
10.2 Analyzing Sin θ and Cos θ

You can easily construct a diagram like the one shown with your geometry software. Then using your diagram, you can study how sin θ and cos θ change as the diagram is dynamically changed.

To construct the diagram, follow these steps.

① Construct a horizontal line $\overleftrightarrow{AB}$. Place P on it.

② Construct $\overleftrightarrow{XY}$ perpendicular to $\overleftrightarrow{AB}$ at P. Place point L on $\overleftrightarrow{XY}$.

③ Construct $\overleftrightarrow{RS}$ parallel to $\overleftrightarrow{AB}$ through L.

④ Construct $\overline{AL}$.

Use geometry software in the following exercises.

1. Follow the steps above to sketch a diagram like the one shown.

2. MEASURE m∠LAP, LP, AP, and AL. CALCULATE the ratios that give sin A and cos A.

3. Move L along $\overleftrightarrow{XY}$. What can you say about sin A as m∠LAP changes? What can you say about sin 0° and sin 90°?

4. Move P along $\overleftrightarrow{AB}$. What can you say about cos A as m∠LAP changes? What can you say about cos 0° and cos 90°?

5. Can you move points L and P in such a way that sin A = cos A? If so, what is m∠LAP?

6. Can you move point L in such a way that sin A = 0.5? If so, what type of triangle will you have?

7. By moving points L and P in your diagram, do you think that sin A and cos A have maximum values? If so, what are they?

16 Technology HRW Geometry

Look Back

For the given length, find each of the remaining two lengths. [Lesson 5.5]

41. $x = 7$

42. $z = 13$

43. $y = 14$

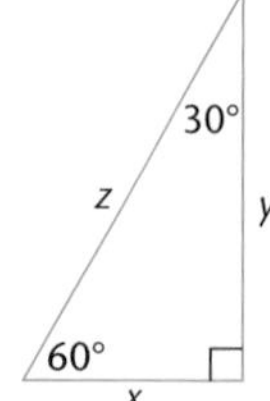

44. $q = 3$

45. $p = 1$

46. $r = 16$

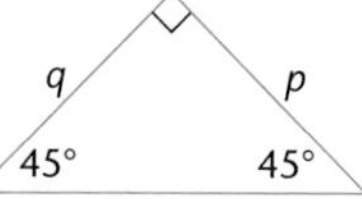

For each arc length and radius given below, find the measure of the central angle. [Lesson 9.1]

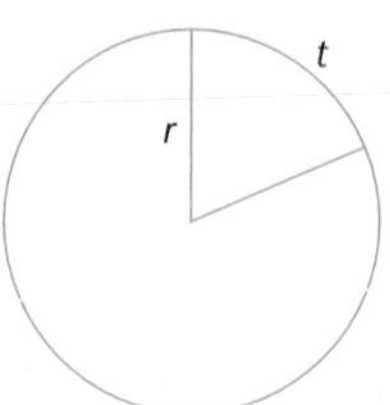

47. $t = 12\pi;\ r = 20$ 108°

48. $t = 10\pi;\ r = 100$ 18°

49. $t = 3\pi;\ r = 25$ 21.6°

Look Beyond

Exercises 50 and 51 extend the lesson concepts with a study of sunrises from an astronomical point of view. Students may need to be told that a person observing the sun at sunrise would be located at the dividing line between light and shadow (vertical in the drawing).

Look Beyond

Portfolio Activity The placement of certain stones at **Stonehenge** suggests that they were markers for the rising and setting positions of the sun and the moon. Most important, the "Heel Stone" marks the position of the rising sun on the day of the summer solstice as seen from the center of the stone circle. At the latitude of Stonehenge in southern England, this angle is about 50° east of true north.

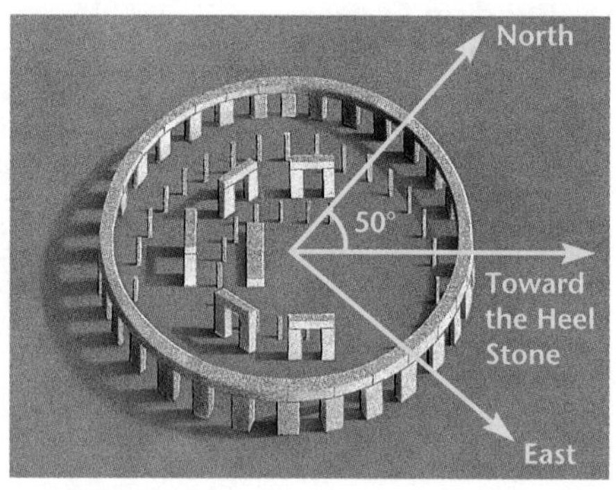

Use the diagram to answer Exercises 50 and 51. A globe may also be helpful.

On the day of the summer solstice, the Earth's North Pole is inclined toward the sun at an angle of 23.5° with the vertical.

50. What is the direction of the summer solstice sunrise for (a) a person on the equator and (b) a person at the Arctic Circle (66.5° N)? (North is the direction toward the North Pole along a line of longitude.)

51. Describe the motion of the sun in the sky on the day of the summer solstice for a location on the Arctic Circle.

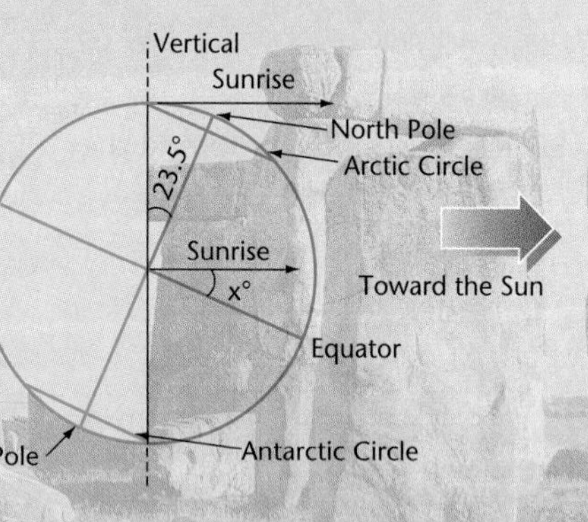

41. $Y = 7\sqrt{3}$; $Z = 14$

42. $X = 6.5$; $Y = 6.5\sqrt{3}$

43. $X = \dfrac{14\sqrt{3}}{3}$; $Z = \dfrac{28\sqrt{3}}{3}$

44. $p = 3$; $r = 3\sqrt{2}$

45. $\theta = 1$; $r = \sqrt{2}$

46. $= 8\sqrt{2}$; $p = 8\sqrt{2}$

50. a. 23.5°North of East

b. 0°East of North or due North

51. The sun travels in a circle in the sky. The sun doesn't rise or set completely. At the end of the day the sun returns to the point where it started.

The curve shown on the oscilloscope is a graph of the sound wave generated by the synthesizer. Such a curve is known as a sine curve.

Are Trigonometric Ratios Limited to Acute Angles?

So far, trigonometric ratios have been defined in terms of acute angles of right triangles. Such definitions seem to limit trigonometric ratios to angles greater than 0° and less than 90°. But you have already seen that a calculator gives values for the sines and cosines of 0° and 90°.

Exploration 1 Trigonometric Ratios on a Calculator

You will need
Scientific calculator

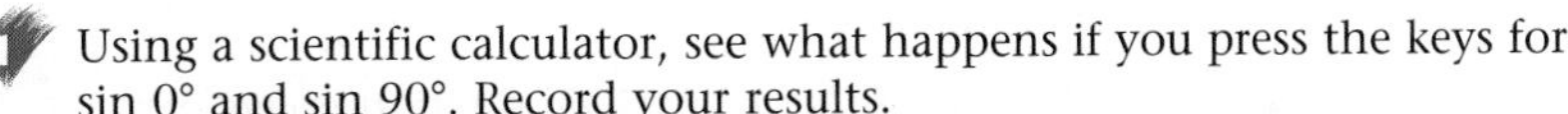

1. Using a scientific calculator, see what happens if you press the keys for sin 0° and sin 90°. Record your results.

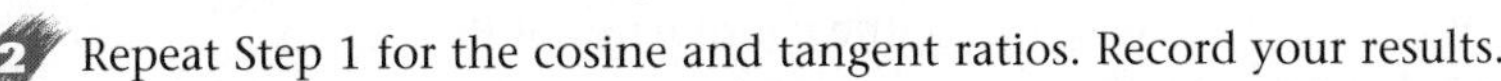

2. Repeat Step 1 for the cosine and tangent ratios. Record your results.

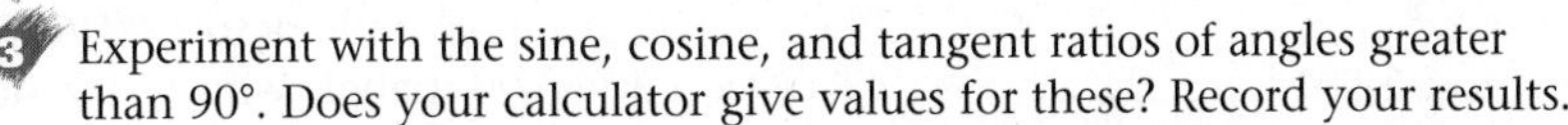

3. Experiment with the sine, cosine, and tangent ratios of angles greater than 90°. Does your calculator give values for these? Record your results.
4. Experiment with the sine, cosine, and tangent ratios of angles with negative measures. Does your calculator give values for these? Record your results.

Technology Geometry graphics software that includes a coordinate system may be used for activities with the unit circle. If students are using *Cabri Geometry II*, they choose the *Show Axes* and the *Define Grid* tools from the *Draw* toolbox to set up a coordinate system. They should construct a circle of any size centered at the origin, and a segment from the origin to a point on the circle.
Students measure the angle between this segment and a segment on the *x*-axis and then use *Calculate* from the *Measure* toolbox to find the sine and cosine of this angle. By moving the segment around on the circle, students can see the signs of the sine and cosine change in the different quadrants.

PREPARE

Objectives

- Use a rotating ray on a coordinate plane to define angles measuring more than 180° or less than 0°.
- Define *sine, cosine,* and *tangent* for angles of any size.
- Extend the graphs of the sine, cosine, and tangent functions.

RESOURCES

• Practice Master	10.3
• Enrichment Master	10.3
• Technology Master	10.3
• Lesson Activity Master	10.3
• Quiz	10.3
• Spanish Resources	10.3

Assessing Prior Knowledge

Use a protractor to draw an angle with each measure.

1. 90°
2. 180°
3. 270°
4. 360°
5. 0°

[Check students' drawings.]

TEACH

The text and figure illustrate one example of a *periodic* function. There are many physical phenomena that are periodic, recurring or repeating at regular intervals. Ask students to give examples of things that happen at repeating intervals. The motions of yo-yos or pendulums are two such examples.

Exploration 1 Notes

Students use a scientific calculator to investigate values of the trigonometric ratios for angles greater than 90°. By experimenting with various angles, they may begin to discover that the sine, cosine, and tangent functions are periodic functions. The sine and cosine have periods of 360°; the tangent has a period of 180°.

A function $f(x)$ is periodic if $f(x+k) = f(x)$ for every value of k. The constant k is called the period of the function $f(x)$.

ongoing ASSESSMENT

5. The sine of an angle and its supplement are equal. Both the cosine and tangent of an angle and its supplement are opposites.

Use Transparency 88

CRITICAL Thinking

Rotation in the clockwise direction should be negative, the opposite of rotation in the counterclockwise direction.

 Find the sine, cosine, and tangent ratios of 65° and 115°. Repeat for angles of 70° and 110°. Record your results. Do you observe a pattern? If so, describe it. ❖

Extending the Trigonometric Ratios

Begin on a coordinate plane and focus on the ray consisting of the origin and the positive x-axis. Imagine this ray rotating counterclockwise about the origin. As it rotates, it sweeps through a certain number of degrees, say $\theta°$. Although measures of angles are restricted between 0° and 180°, the ray is free to rotate any number of degrees. The illustration shows that a rotation can, in fact, be greater than 360°.

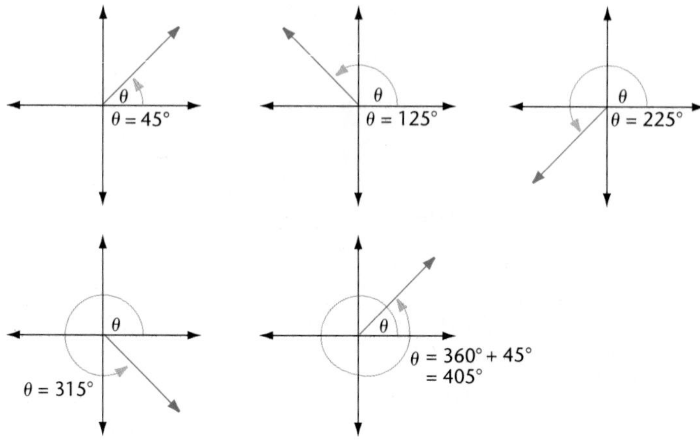

Explain how a rotation can have a negative measure.

EXTENSION

A **unit circle** is a circle with its center at the origin and radius 1. It consists of all the rotation images of $P(1, 0)$ about the origin. Find the coordinates of P'.

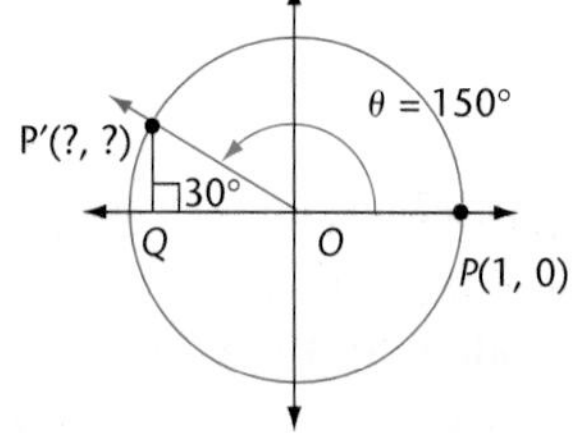

P' is on the 150° rotation image of $P(1, 0)$ about the origin. Find the coordinates of P'.

Since $\angle P'OQ$ is supplementary to $\angle P'OP$, its measure is 30°. Thus, $\triangle P'OQ$ is a 30-60-90 triangle with its hypotenuse equal to 1. $P'Q = \frac{1}{2}$ and $QO = \frac{\sqrt{3}}{2}$. The x-coordinate of P' is $-\frac{\sqrt{3}}{2} \approx -0.866$. The y-coordinate of P' is $\frac{1}{2} = 0.5$. Thus, the coordinates of P′ are $(-0.866, 0.5)$. ❖

interdisciplinary CONNECTION

Music When a string or wire with fixed endpoints is plucked, a standing wave is created. For the simplest vibration of the string, the graph of the vertical position of the center point of the string vs. time is a sine curve. However, the string may also vibrate in different ways at the same time. This type of vibration is related to the sounds made by many musical instruments. Students may enjoy carrying out simple experiments with stretched string or wire to see what kinds of waves they can create.

Exploration 2 *Redefining the Trigonometric Ratios*

You will need
Scientific calculator
Graph paper

Scientific Calculator

1 P' is the 30° rotation image of $P(1, 0)$ about the origin. Use the rules from the 30-60-90 Right-Triangle theorem to find the coordinates of P'.

2 P' is the 210° rotation image of $P(1, 0)$ about the origin. Find the coordinates of P'.

3 P' is the 330° rotation image of $P(1, 0)$ about the origin. Find the coordinates of P'.

4 Use the results of Example 1, Exploration 2, and a scientific calculator to complete the table.

θ	x-coordinate of image point	y-coordinate of image point	cos θ	sin θ
30°	?	?	?	?
150°	?	?	?	?
210°	?	?	?	?
330°	?	?	?	?

5 Write definitions for sin θ and cos θ based on the information in the table. Complete the figure to illustrate your definition.

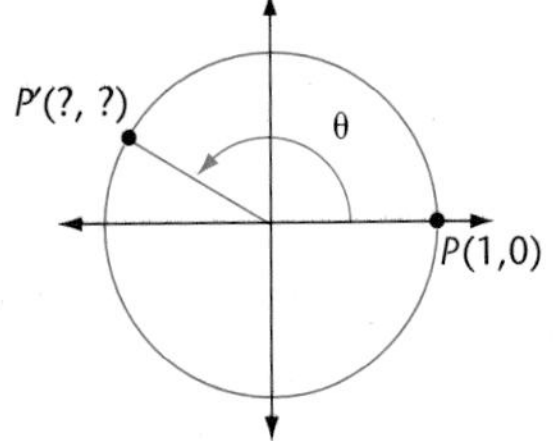

6 Do your values for sin 30° and cos 30° agree with those obtained using the right-triangle definitions? Explain. ❖

Before giving an official definition for sin θ and cos θ, we need to adopt a convention: a counterclockwise rotation has positive measure and a clockwise rotation has negative measure.

Let θ be the measure of a rotation of $P(1 ,0)$ about the origin. Sin θ is the y-coordinate of the image point. Cos θ is the x-coordinate of the image point.

The definition for tan θ obtained in Lesson 10.2 can be generalized to give the following definition:

Let θ be the measure of a rotation of $P(1, 0)$ about the origin.

$$\tan \theta = \frac{\sin \theta}{\cos \theta}$$

Use Transparency 89

Exploration 2 Notes

Before students begin the investigation, direct their attention to the diagrams on the previous page. Point out how the ray is "sweeping out" an angle. The angle is always measured in the counterclockwise direction, starting at the x-axis. In Exploration 2, students investigate values for the sines and cosines of these "swept out" angles.

ongoing ASSESSMENT

6. Yes. The image of (1, 0) under a rotation of 30° is in the first quadrant, so the values of sin θ and cos θ are positive and can be found using a triangle.

ENRICHMENT Explain that angles can be measured in radians rather than in degrees. Have students work together to find out what a radian is. Then they redraw the graphs from Exploration 3 using radians rather than degrees on the horizontal axis. Students write to discuss the advantages and disadvantages of radians versus degrees.

INCLUSION strategies **English Language Development** The fact that *sign* and *sine* are pronounced the same can cause confusion during this lesson. One way to avoid problems is to not use the term *sign* in spoken discussion. Ask students instead to talk about whether the sine of an angle formed by a rotated ray is positive or negative.

Teaching tip

Technology A graphics calculator will be very helpful for Exploration 3, particularly in Step 8.

Exploration 3 Notes

Display a figure like the unit circle diagram for Exploration 2. Point out that point P is moving through the four quadrants of the x-y coordinate system. The point starts in quadrant I, then moves through quadrants II, III, and IV. If necessary, remind students of the Roman numeral notation for the four quadrants. As students complete the exploration, they will discover patterns that involve the two angles with measures θ and $180° - \theta$.

Teaching tip

Point out how the coordinate system used in Step 2 of Exploration 3 is different from the x-y coordinate system used before. Here, the horizontal axis is labeled with the Greek letter θ. The vertical axis shows the sine, cosine, or tangent of θ. However, the two systems are describing the same motion. As the point P rotates around the circle on the x-y system, the angle θ increases in size and its sine, cosine, or tangent changes.

Ongoing Assessment

8. **No, $\cos 60° = 0.5$ and $\cos 120° = -0.5$. Modify as $\cos \theta = -\cos (180° - \theta)$.**

Exploration 3 Graphing the Trigonometric Ratios

Technology — Scientific Calculator

You will need
Scientific calculator
Graph paper

1. Complete the table; θ is the measure of a rotation of $P(1, 0)$ about the origin.

Quadrant of image point	sign of $\sin\theta$	sign of $\cos\theta$	sign of $\tan\theta$
I	+	?	?
II	+	?	?
III	–	?	?
IV	?	?	?

2. Extend the graphs of each of the trigonometric functions below to $\theta = 360°$ by plotting points at intervals of 30°.

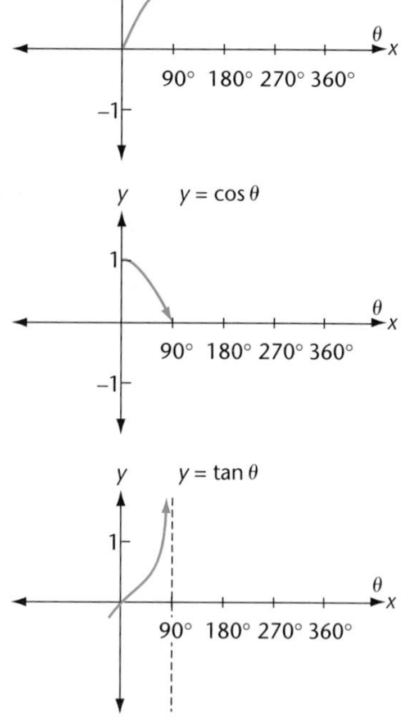

3. Use your graphs to determine the quadrants in which the sine, cosine, and tangent are positive and those in which they are negative. Does this information agree with the table that you completed in Part 1?
4. In Exploration 1 you found that $\sin 65° = \sin 115°$ and that $\sin 70° = \sin 110°$. Locate these values on your graph of the sine curve. List at least three other pairs of angle measures that share the same sine ratios.
5. Use the cosine curve. List at least three pairs of angle measures that have the same cosine ratios.
6. Use the tangent curve. List at least three pairs of angle measures that have the same tangent ratios.
7. Consider the following identity:

   $\sin \theta = \sin (180° - \theta)$, for all values of θ.

   Do the results from Step 1 agree with this identity? Explain.
8. Does $\cos \theta = \cos (180 - \theta)$ for all values of θ in the table above? If yes, explain why. If no, provide a counterexample and modify the equation to make a true statement. ❖

Reteaching the Lesson

Using Visual Models It is essential that students understand how the two coordinate systems used in Explorations 2 and 3 are related. To demonstrate the relationship, draw the x-y system with the unit circle on the left of the chalkboard. On the right, draw a θ-sine θ coordinate system. Get a student volunteer to help you. As you demonstrate the rotating radius vector with a ruler, the student traces out the corresponding sine curve. Repeat with the cosine and tangent.

APPLICATION

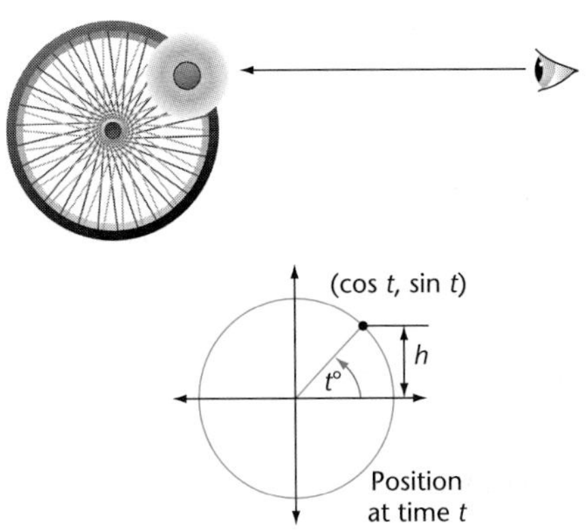

A wheel with a 1-foot radius is turning slowly at a constant velocity of one degree per second and has a light mounted on its rim. A distant observer watching the wheel from the side sees the light moving up and down in a vertical line. Write an equation for the vertical position of the light starting from the horizontal position shown at time $t = 0$. What will be the vertical position of the light after 1 minute? After 5 minutes? After 24 hours?

Imagine a coordinate system with the origin at the center of the wheel as shown. After t seconds have elapsed, the value of θ is $t°$. The coordinates of the light are $(\cos t, \sin t)$. Notice that $\sin t$, the y coordinate, is the vertical position of the light. Thus, at time t the vertical position h of the light is given by the equation,

$$h = \sin t.$$

At $t = 1$ minute, or 60 seconds, $h = \sin 60° \approx 0.867$ units.

At $t = 5$ minutes, or 300 seconds, $h = \sin 300° \approx -0.867$ units.

At $t = 24$ hours, or 86,400 seconds, $h = \sin 86{,}400° = 0$ units. ❖

Try This Write the equation for the vertical position of the light if the wheel is turning at the rate of one complete rotation per second. Write its equation if the diameter of the wheel is 2 units. (Assume t is in seconds.)

Exercises & Problems

Communicate

1. How is the unit circle used to extend the trigonometric ratios beyond right triangles?
2. Explain how to find sin 90°, cos 90°, and tan 90°.
3. Explain how to find sin 180°, cos 180°, and tan 180°.

Explain how to find the sign of the following:

4. cosine in quadrant III
5. sine in quadrant II
6. tangent in quadrant III
7. Explain how a trigonometric function can represent the vertical motion of point X on the oil-rig pump.

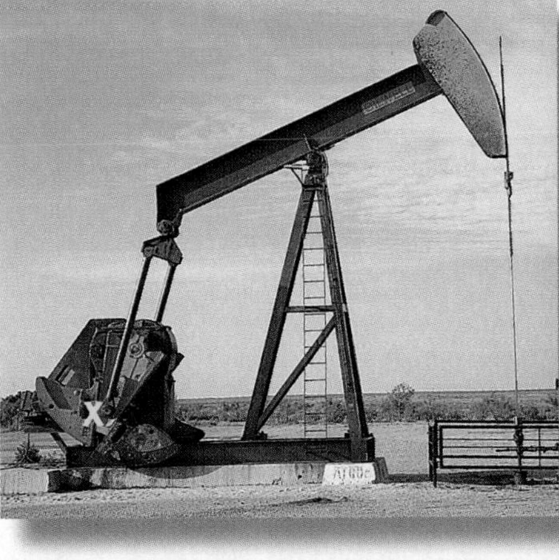

Cooperative Learning

The application following Exploration 3 is directly related to the ideas investigated in Explorations 2 and 3. Have students work in small groups to study, discuss, and solve the problem. Ask that they create a model of some sort to demonstrate the movement of the light as the wheel turns. Creating rotating wheels of their own will help students understand the lesson.

Ongoing Assessment

Try This

$h = \sin(360t)$; $h = 2\sin(360t)$

Assess

Selected Answers

Odd-numbered Exercises 9–33

Assignment Guide

Core 1–16, 19–25, 27–34

Core Plus 1–10, 15–21, 26–37

Technology

A scientific calculator is needed for most of the exercises and problems. If scientific calculators are not available, students will need to use a chart of the trigonometric functions.

Error Analysis

For Exercises 16 and 17, students can use the $\sin^{-1}$ and the $\cos^{-1}$ functions on a scientific calculator to find the first values of θ. If they then get stuck, suggest that they draw a unit-circle graph to help them find the other values for θ.

Alternative Assessment

Portfolio Assessment

Students write to compare the graphs of $y = \sin \theta$ and $y = \cos \theta$. How are the graphs alike? How are they different? Students also include some logical reasons for the similarities and differences. Sketches of graphs should be included with the written work.

The answers to Exercises 8–11 can be found in Additional Answers beginning on page 727.

Use Transparency 90

Practice & Apply

Use graph paper and a protractor or geometry software to sketch a unit circle and a ray with the given angle θ. **Estimate the sine, cosine, and tangent of θ to the nearest hundredth. Include the sketch with your answers.**

8. $\theta = 45°$ **9.** $\theta = 110°$ **10.** $\theta = 175°$ **11.** $\theta = 450°$

For Exercises 12–15, use a scientific calculator to find the following values.

12. sin 110° .9397 **13.** cos 157° −.9205 **14.** tan 230° 1.1918 **15.** sin 480° .8660

16. If $\sin \theta = .4756$, what are all the possible values of θ between 0° and 360° rounded to the nearest degree? 28°, 152°

17. If $\cos \theta = -.7500$, what are all the possible values of θ between 0° and 360° rounded to the nearest degree? 139°, 221°

Find the coordinates of P', the rotation image of $P(1, 0)$, for the following angles of rotation.

18. $\theta = 145°$ **19.** $\theta = 45°$ **20.** $\theta = 60°$ **21.** $\theta = 72°$

22. $\theta = 470°$ **23.** $\theta = 250°$ **24.** $\theta = 840°$ **25.** $\theta = 760°$

26. Astronomy An astronomer observes that a satellite moving around a planet moves one degree per hour. Assuming that the radius of the satellite's orbit is one unit, what is the horizontal position, in coordinate form, of the satellite after 2 days? after 5 days?

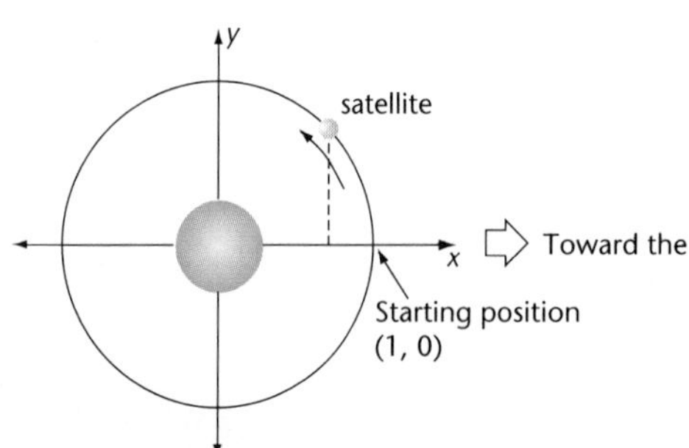

View of planet from above.

Practice Master

NAME ______ CLASS ______ DATE ______

Practice & Apply

10.3 Exploring the Unit Circle

In Exercises 1–2, on the grid with its unit circle, use a protractor to sketch a unit vector with the angle θ. From the graph, estimate the values of the sine, cosine, and tangent of θ.

1. $\theta = 60°$ 2. $\theta = 150°$

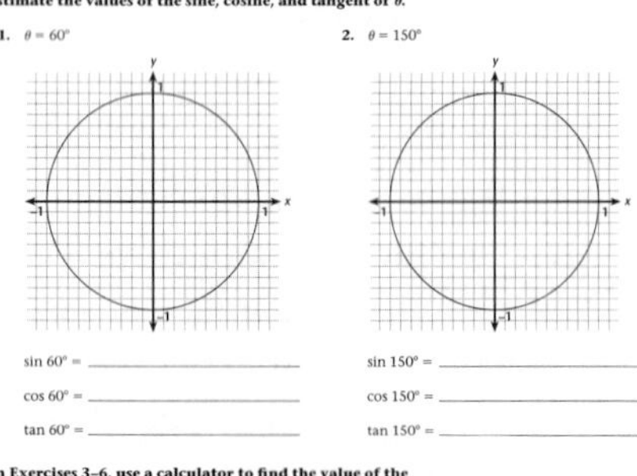

sin 60° = ______ sin 150° = ______

cos 60° = ______ cos 150° = ______

tan 60° = ______ tan 150° = ______

In Exercises 3–6, use a calculator to find the value of the indicated function to the nearest ten-thousandth.

3. sin 140° = ______ 4. cos 168° = ______

5. tan 250° = ______ 6. sin 320° = ______

For the given function value in Exercises 7–10, find all possible values of θ between 0° and 360° to the nearest degree.

7. $\sin \theta = 0.4695$ ______ 8. $\cos \theta = 0.8572$ ______

9. $\tan \theta = 1.0355$ ______ 10. $\tan \theta = 9.5144$ ______

In Exercises 11–16, find the coordinates of P', the rotation image of $P(1, 0)$, for each given angle θ of rotation to the nearest tenth.

11. $\theta = 30°$ ______ 12. $\theta = 120°$ ______ 13. $\theta = 210°$ ______

14. $\theta = 300°$ ______ 15. $\theta = 420°$ ______ 16. $\theta = 540°$ ______

HRW Geometry Practice & Apply 3

18. $P'(-0.8192, 0.5756)$
19. $P'(0.7071, 0.7071)$
20. $P'(0.5, -0.8660)$
21. $P'(0.3090, 0.9511)$
22. $P'(-0.3420, 0.9397)$
23. $P'(-0.3420, 0.9397)$
24. $P'(-0.5, 0.8660)$
25. $P'(0.7660, 0.6428)$
26. After 2 days: (0.6691, 0.7431); After 5 days: (−0.5, 0.866)

Look Back

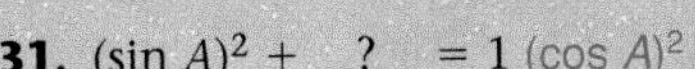

Use $\triangle ABC$ for Exercises 27–34. [Lessons 10.1, 10.2]

27. $\sin A = \frac{?}{c}$ a

28. $\cos B = \frac{?}{c}$ a

29. $\tan A = \frac{a}{?}$ b

30. $\cos A = \frac{?}{c}$ b

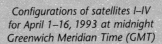

B, c, a, A, b, C

31. $(\sin A)^2 +$ __?__ $= 1$ $(\cos A)^2$

32. $\frac{\sin A}{\cos A} =$ __?__ $\tan A$

33. $\sin A = \cos$ (__?__) $(90° - m\angle A)$

34. $\cos B = \sin$ (__?__) $(90° - m\angle B)$

Look Beyond

The graph shows the positions of the four Galilean Moons of Jupiter from the point of view of the Earth at midnight on April 1–16, 1993. The parallel lines in the center of the graph represent the visible width of Jupiter.

Configurations of satellites I–IV for April 1–16, 1993 at midnight Greenwich Meridian Time (GMT)

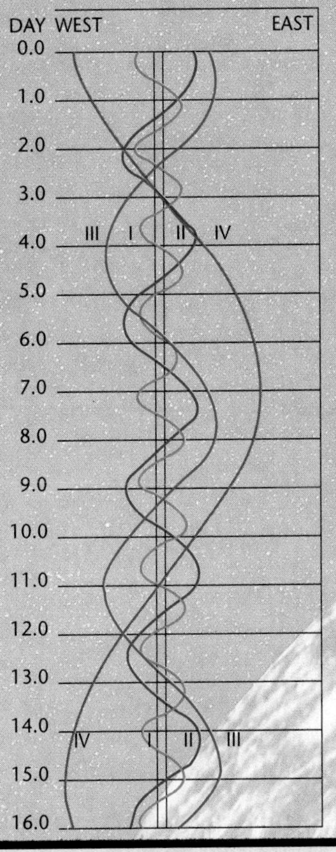

35. **Portfolio Activity** Galileo was able to observe Jupiter's four largest satellites. Viewed from the Earth, they appear to move back and forth in an approximate straight line through the center of the planet. Today these four satellites are known as the "Galilean Moons" of Jupiter. Their names are

I. Io
II. Europa
III. Ganymede
IV. Callisto

Sketch the positions of the planet Jupiter and its four Galilean Moons as they would appear to a person with a small telescope or a pair of binoculars on (a) April 4, (b) April 8, and (c) April 12. Use dots for the moons and a circle for Jupiter.

36. Use the graph to estimate the orbital periods for each of the four Galilean Moons of Jupiter.

37. What kind of curves do the satellite lines on the graph appear to be? Explain why they have these shapes. (Note: The orbits of the four satellites are nearly circular.)

Look Beyond

Exercises 35–37 use the motions of the moons of Jupiter to extend the concept of sine curves.

Technology Master

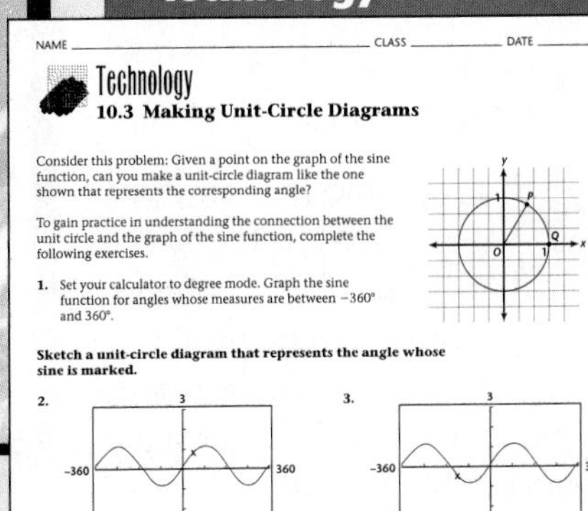

NAME ________ CLASS ________ DATE ________

Technology
10.3 Making Unit-Circle Diagrams

Consider this problem: Given a point on the graph of the sine function, can you make a unit-circle diagram like the one shown that represents the corresponding angle?

To gain practice in understanding the connection between the unit circle and the graph of the sine function, complete the following exercises.

1. Set your calculator to degree mode. Graph the sine function for angles whose measures are between $-360°$ and $360°$.

Sketch a unit-circle diagram that represents the angle whose sine is marked.

2. 3.

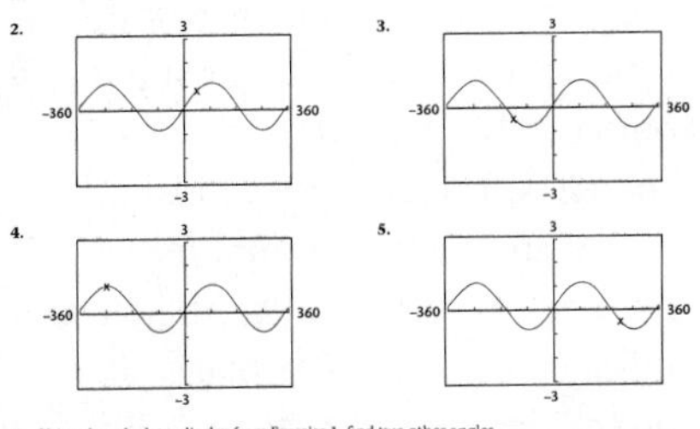

4. 5.

6. Using the calculator display from Exercise 1, find two other angles between $-360°$ and $360°$ that have the same sine as the angle in Exercise 2. Sketch these angles in unit-circle diagrams.

HRW Geometry — Technology 17

35. a. III ○ II IV

b. ○ II I III IV

c. IV III II I ○

In (a) Moon I is behind Jupiter.

36. I, Io: 1.75 days ; II, Europa: 3.5 days ; III, Ganymede: 7 days ; IV, Callisto: 16 days

37. The curves are sine curves, because they are repeatedly traveling around Jupiter in a circular motion that can be modeled by a point rotating around the unit circle.

PREPARE

Objectives

- Use two transformation equations to rotate points.
- Use a rotation matrix to rotate points or polygons.

- Practice Master 10.4
- Enrichment Master 10.4
- Technology Master 10.4
- Lesson Activity Master 10.4
- Quiz 10.4
- Spanish Resources 10.4

Assessing Prior Knowledge

Use compass and straightedge for these problems.

1. Draw an x-y coordinate system. Draw $\triangle DEF$ in the first quadrant with vertices on grid intersection points.

[Check students' drawings.]

2. Rotate $\triangle DEF$ 60° counterclockwise about the origin.

[Check students' drawings.]

TEACH

Students who have used geometry graphics software or standard computer-graphics programs will be familiar with tools or commands for rotating objects. Discuss the need to choose both a center for a rotation and to specify the amount of the rotation in degrees. A positive number such as 30° rotates an object in one direction; –30° rotates it in the opposite direction.

LESSON 10.4 Exploring Rotations With Trigonometry

Why *Computers use geometric transformations to show moving objects. In this lesson you will add rotations to the transformations you can do on a coordinate plane.*

Astronauts prepare for the Hubble Space Telescope repair mission by studying "virtual-reality" simulations of the planned event.

In earlier lessons you studied translations, reflections, and dilations in a coordinate plane or space. You also studied a number of special cases of rotations. Now, with trigonometry as a tool, you are ready to rotate a point in a coordinate system by *any desired amount*.

Transformation Equations

ALGEBRA Connection

The two equations below are known as transformation equations.

$$x' = x \cos \theta - y \sin \theta$$

$$y' = x \sin \theta + y \cos \theta$$

To use the equations, you need to know the coordinates x and y of the point you want to rotate and also the measure of the amount of rotation θ about the origin. The coordinates x' and y' are the coordinates of the rotated image point.

Technology Geometry graphics software can be used to explore rotations of points and figures on a coordinate plane. Although students do not need to use trigonometry to manipulate the computer software, an understanding of *how* the computer performs the rotations will be useful to students in other types of problem-solving situations.

Exploration 1 *Rotating a Point*

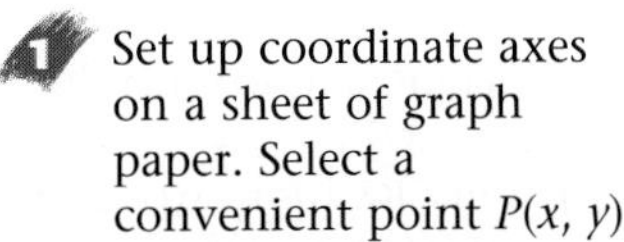

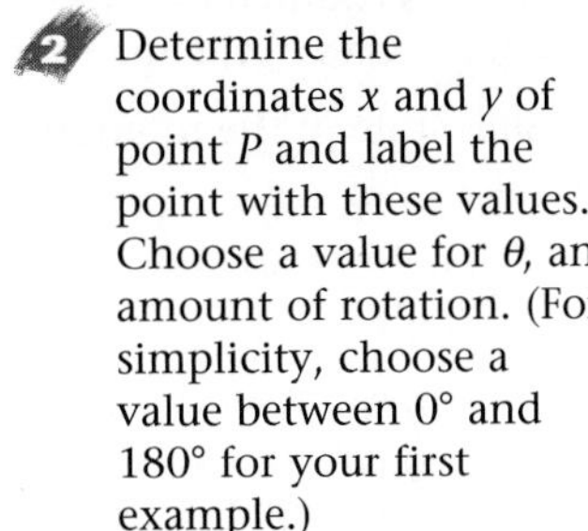

You will need
Scientific calculator
Graph paper (small grid)
Ruler and protractor

Scientific Calculator

Part I

1. Set up coordinate axes on a sheet of graph paper. Select a convenient point $P(x, y)$.

2. Determine the coordinates x and y of point P and label the point with these values. Choose a value for θ, an amount of rotation. (For simplicity, choose a value between 0° and 180° for your first example.)

3. Substitute your values for x, y, and θ into the transformation equations on the previous page. Use a calculator to determine x' and y', the coordinates of the image point P'. Plot point P′ on your graph and measure the rotation angle. Discuss your results.

Part II

1. Find the values of the sines and cosines in the table below.

θ	0°	90°	180°	270°	360°
Sine	?	?	?	?	?
Cosine	?	?	?	?	?

2. For each value of θ, substitute the pair of sine and cosine values into the transformation equations. Use the resulting equations to state a rule in words for rotations of 0°, 90°, 180°, 270°, and 360°.

3. Experiment with negative values for θ, such as −90°, and values greater than 360°, such as 540°. Describe your results. ❖

TEACHING tip

Remind students of other transformation equations they may have used. For example, $x' = x - h$ and $y' = y - k$ can be used to translate a point on a coordinate plane.

Math Connection
Algebra

Have students identify the five x variables used in the two rotation equations: x, y, θ, x' and y'.

Exploration 1 Notes

Students use two transformation equations to rotate a point on a coordinate system. This activity should be done first using graph paper and a scientific calculator.

ongoing ASSESSMENT

2.1 A rotation of 0° or 360° has no effect on the coordinates of the figure being rotated. Under a rotation of 90°, the *x*-coordinate of the image is the opposite of the *y*-coordinate of the pre-image and the *y*-coordinate of the image is equal to the *x*-coordinate of the pre-image. Under a rotation of 180°, the *x*-coordinate of the image is the opposite of the *x*-coordinate of the pre-image and the *y*-coordinate of the image is the opposite of the *y*-coordinate of the pre-image. Under a rotation of 270°, the *x*-coordinate of the image is equal to the *y*-coordinate of the pre-image and the *y*-coordinate of the image is the opposite of the *x*-coordinate of the pre-image.

interdisciplinary CONNECTION

Physics Remind students that motions of bodies in the solar system are often described by ellipses. One application of rotation equations is to simplify the form of equations for ellipses (and other conic sections). For example, the equation $34x^2 - 24xy + 41y^2 = 200$ has an ellipse for its graph. By rotating the figure through an angle θ, where $\tan 2\theta = \frac{24}{7}$, the equation simplifies to $x'^2 + 2y'^2 = 8$. Students can confirm these results with a graphics calculator.

Cooperative Learning

Technology After students have solved the application problem using the methods given in the text, have them work in pairs or small groups to use geometry graphics software to solve the same problem in a different way. Students should talk about how the approaches differ, and compare paper-and-pencil versus computer techniques.

Teaching tip

If students have had no previous work with matrices, you can omit the two sections on Rotation Matrices and Rotating Polygons. Exploration 2 can then be modified so that students investigate the effect of two rotations without using matrix techniques.

CRITICAL Thinking

0° and 360°: $\begin{bmatrix} 1 & 0 \\ 0 & 1 \end{bmatrix}$

90°: $\begin{bmatrix} 0 & -1 \\ 1 & 0 \end{bmatrix}$

180°: $\begin{bmatrix} -1 & 0 \\ 0 & -1 \end{bmatrix}$

270°: $\begin{bmatrix} 0 & 1 \\ -1 & 0 \end{bmatrix}$

Identity matrix: $\begin{bmatrix} 1 & 0 \\ 0 & 1 \end{bmatrix}$

APPLICATION

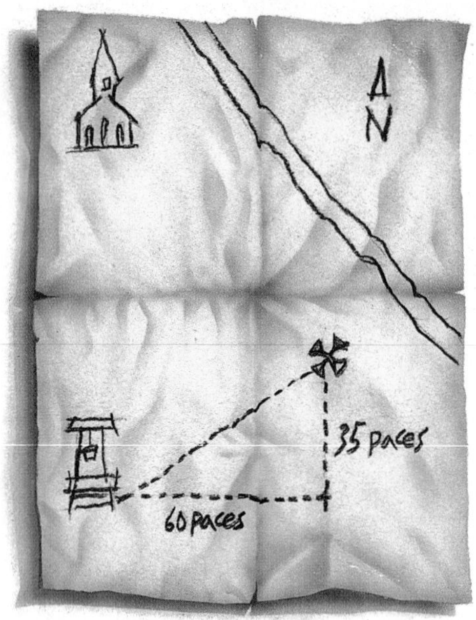

According to an old family story just uncovered, some jewels are buried at a point 60 paces east and 35 paces north of the old water well. Before visiting the site, the family draws a coordinate map of the property with true north as the direction of the positive y-axis.

If magnetic north for the location of the site were used instead of true north, point J (the jewelry location) would need to be rotated $5\frac{1}{2}$ degrees counterclockwise. What would the coordinates of the new point J' be?

Substitute in the rotation equations:

$$\begin{aligned} x' &= 60 \cos 5\tfrac{1}{2}° - 35 \sin 5\tfrac{1}{2}° \\ &\approx 60(0.9954) - 35(0.0958) \\ &\approx 56 \text{ paces} \end{aligned}$$

$$\begin{aligned} y' &= 60 \sin 5\tfrac{1}{2}° + 35 \cos 5\tfrac{1}{2}° \\ &\approx 60(0.0958) + 35(0.9954) \\ &\approx 41 \text{ paces} \end{aligned}$$

The coordinates of J' are (56, 41). ❖

Rotation Matrices

Recall from algebra that a matrix is an effective tool for visually storing data. Matrices can be added, subtracted, and multiplied. Matrices also provide a convenient way to represent the rotation of a point in a coordinate system.

$$\begin{bmatrix} \cos\theta & -\sin\theta \\ \sin\theta & \cos\theta \end{bmatrix}$$

2 × 2 rotation matrix

To rotate a point (x, y) about the origin by an amount θ, place the coordinates of the point in a 2 × 1 column matrix. Then multiply the column matrix by the 2 × 2 **rotation matrix** as shown below.

$$\begin{bmatrix} \cos\theta & -\sin\theta \\ \sin\theta & \cos\theta \end{bmatrix} \times \begin{bmatrix} x \\ y \end{bmatrix} = \begin{bmatrix} x\cos\theta - y\sin\theta \\ x\sin\theta + y\cos\theta \end{bmatrix}$$

2 × 2 rotation matrix *2 × 1 matrix* *2 × 1 matrix*

Notice that the expressions in the product matrix agree with the expressions in the rotation equations at the beginning of this lesson.

CRITICAL Thinking 

Give the rotation matrices for θ = 0°, 90°, 180°, 270°, and 360° by filling in the sine and cosine values. Compare and contrast the matrices. Do you see a 2 × 2 **identity matrix** among them? Discuss.

ENRICHMENT The activities and problems in this lesson dealt with rotating objects about the origin. Challenge students to find a way to rotate a triangle about a point other than the origin. Students should work in pairs or groups on the project. Their solutions should be algebraic in nature; that is, they should not use technology such as geometry graphics software.

INCLUSION strategies **Using Models** Some students may understand the concepts better if they think in terms of rotating the entire coordinate system, not just the point or polygon. Have students draw an x-y coordinate system and plot any $\triangle ABC$. Then they trace the axes and triangle. By rotating the tracing paper about the origin, they can see what is happening to the axes and the polygons.

Rotating Polygons

The convenience of rotation matrices becomes especially apparent when more than one point is rotated at a time. In the matrix multiplication below, the 2 × 2 matrix rotation matrix moves the three vertices of $\triangle ABC$ 60° counterclockwise about the origin.

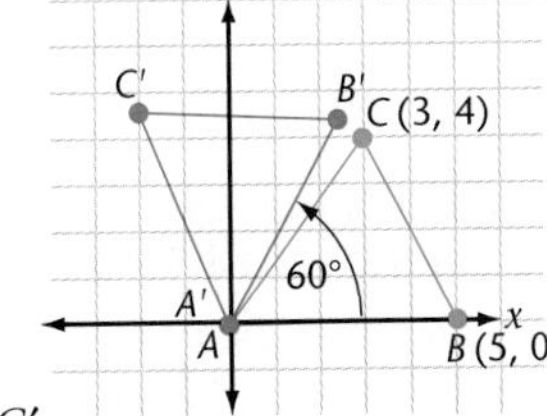

Create a 60° rotation matrix.

$$\begin{bmatrix} \cos 60° & -\sin 60° \\ \sin 60° & \cos 60° \end{bmatrix} = \begin{bmatrix} .5 & -.866 \\ .866 & .5 \end{bmatrix}$$

Multiply this matrix times the triangle matrix.

$$\begin{bmatrix} .5 & -.866 \\ .866 & .5 \end{bmatrix} \begin{matrix} \begin{matrix} A & B & C \end{matrix} \\ \begin{bmatrix} 0 & 5 & 3 \\ 0 & 0 & 4 \end{bmatrix} \end{matrix} = \begin{matrix} \begin{matrix} A' & B' & C' \end{matrix} \\ \begin{bmatrix} 0 & 2.5 & -1.964 \\ 0 & 4.33 & 4.598 \end{bmatrix} \end{matrix}$$

Try This Rotate $\triangle ABC$ 90°, 180°, and 270° about the origin.

Exploration 2 Combining Rotations

Calculator

You will need
Calculator
Graph paper
Ruler and protractor

If you rotate point P first 30° counterclockwise and then 40° counterclockwise about the origin, what will the final result be? In this exploration you will experiment with combined rotations using matrices.

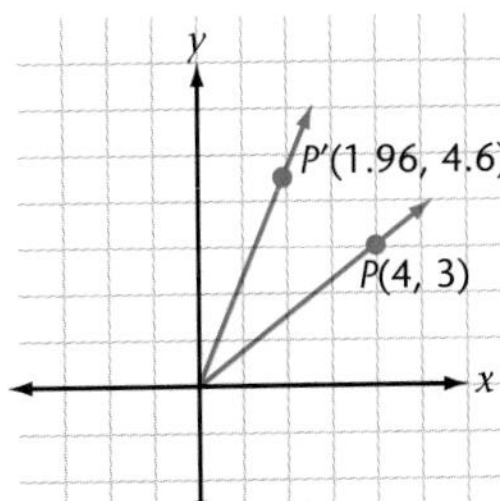

1. Pick a point in the first quadrant of the coordinate plane. Rotate the point 30° counterclockwise using the rotation matrix. Record your results. Plot your new point on the coordinate plane and check the angle of rotation with a protractor.

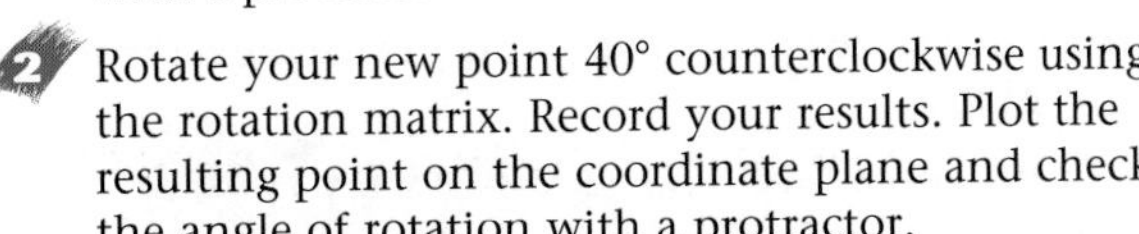

2. Rotate your new point 40° counterclockwise using the rotation matrix. Record your results. Plot the resulting point on the coordinate plane and check the angle of rotation with a protractor.

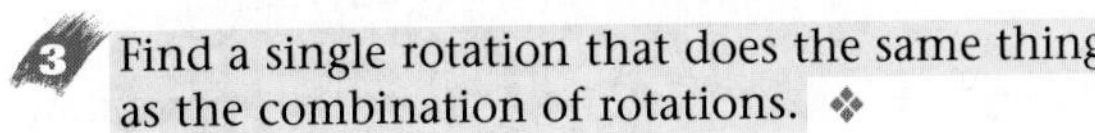

3. Find a single rotation that does the same thing as the combination of rotations. ❖

CRITICAL Thinking Does the order of the rotations in Steps 1 and 2 make a difference? How can you demonstrate your answer? When would the order of rotations make a difference?

Teaching tip

Have students write a matrix for the triangle. The ordered pairs for the three vertices are $A(0, 0)$, $B(5, 0)$, and $C(3, 4)$. Each vertex pair becomes a column in the triangle matrix.

Ongoing Assessment

Try This

$$\begin{bmatrix} 0 & 0 & -4 \\ 0 & 5 & 3 \end{bmatrix}$$

$$\begin{bmatrix} 0 & -5 & -3 \\ 0 & 0 & -4 \end{bmatrix}$$

$$\begin{bmatrix} 0 & 0 & 4 \\ 0 & -5 & -3 \end{bmatrix}$$

Exploration 2 Notes

Students investigate the effect of combined rotations on the coordinates of a point. This activity can be done without using matrices. Students can use paper-and-pencil techniques or geometry graphics software.

Ongoing Assessment

A single rotation of 70° is equal to a 30° rotation followed by a 40° rotation. Rotations in a plane can be added.

Reteaching the lesson

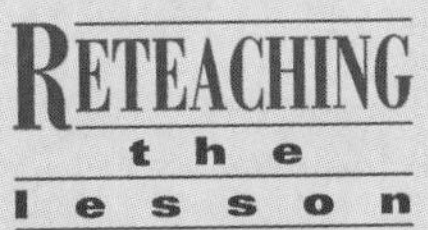

Technology Have students use graphics calculators to explore the rotation equations from an algebraic starting point. They should begin by entering the two rotation equations given at the beginning of the lesson. Then they choose a measure for θ and create a table of values for three to five ordered pairs. By graphing the ordered pairs both before and after the rotation, students will be able to see the effect of the rotation equations. The ordered pairs can be connected to form polygons if students wish.

CRITICAL Thinking

No. Rotate the figure 40° first, then rotate the image 30°. The order of rotation would never make a difference. (Note that matrix multiplication is not commutative in general, but *rotation* matrices commute.)

Assess

Selected Answers

Odd-numbered Exercises 5–29 and Exercises 39, 43, and 47

Assignment Guide

Core 1–9, 13–25, 38–47

Core Plus 1–6, 10–48

Technology

Students will need scientific calculators for most of the exercises.

Error Analysis

For Exercises 4-27, students who are having trouble mastering the algebraic techniques can do a few problems in each group using grid paper and protractors. Plotting points and measuring angles may help make the ideas more concrete.

The answers to Exercises 24–27 can be found in Additional Answers beginning on page 727.

Practice Master

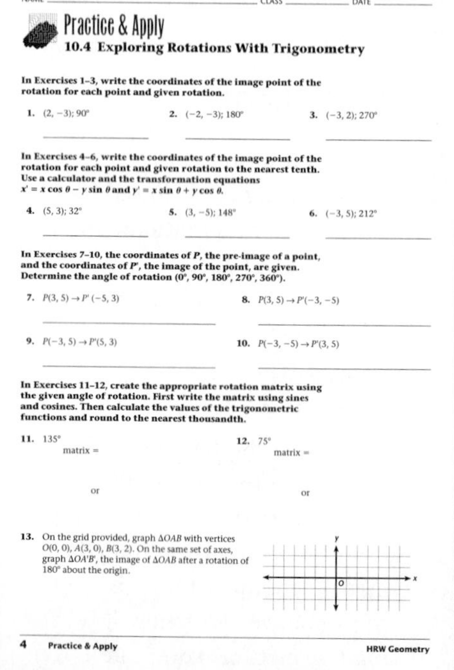

NAME ______ CLASS ______ DATE ______

Practice & Apply

10.4 Exploring Rotations With Trigonometry

In Exercises 1–3, write the coordinates of the image point of the rotation for each point and given rotation.

1. (2, −3); 90° ______
2. (−2, −3); 180° ______
3. (−3, 2); 270° ______

In Exercises 4–6, write the coordinates of the image point of the rotation for each point and given rotation to the nearest tenth. Use a calculator and the transformation equations $x' = x\cos\theta - y\sin\theta$ and $y' = x\sin\theta + y\cos\theta$.

4. (5, 3); 32° ______
5. (3, −5); 148° ______
6. (−3, 5); 212° ______

In Exercises 7–10, the coordinates of P, the pre-image of a point, and the coordinates of P', the image of the point, are given. Determine the angle of rotation (0°, 90°, 180°, 270°, 360°).

7. $P(3, 5) \rightarrow P'(-5, 3)$ ______
8. $P(3, 5) \rightarrow P'(-3, -5)$ ______
9. $P(-3, 5) \rightarrow P'(5, 3)$ ______
10. $P(-3, -5) \rightarrow P'(3, 5)$ ______

In Exercises 11–12, create the appropriate rotation matrix using the given angle of rotation. First write the matrix using sines and cosines. Then calculate the values of the trigonometric functions and round to the nearest thousandth.

11. 135°
    matrix =

    or

12. 75°
    matrix =

    or

13. On the grid provided, graph ΔOAB with vertices $O(0, 0)$, $A(3, 0)$, $B(3, 2)$. On the same set of axes, graph $\Delta OA'B'$, the image of ΔOAB after a rotation of 180° about the origin.

4 Practice & Apply HRW Geometry

Communicate

1. What is the identity matrix for 2 × 2 matrices? What happens when you multiply it by another matrix? Explain why.
2. Give an angle of rotation that produces a 2 × 2 identity matrix when substituted into a rotation matrix.
3. Give examples of two rotations, one clockwise and one counterclockwise, that produce the same result.

Practice & Apply

Find the image point for each pre-image and rotation angle given.

4. (−3, 2); 270° (2, 3)
5. (4, −5); 90° (5, 4)
6. (5, 10); 180° (−5, −10)
7. (−1, −4); 360 (−1, −4)
8. (1, −4); 270° (−4, −1).
9. (3, 0); 90° (0, 3)

Technology For each point and given rotation, name the image point of the rotation. Use a calculator and the transformation equations from Exploration 1.

10. (4, 2); 27°
11. (5, −10); 160°
12. (−1, −1); 15°
13. (15, 10); 230°
14. (−5, −6); 295°
15. (−2, 3); 85°

Given the pre-image and image below, determine the angle of rotation (0°, 90°, 180°, 270°, or 360°).

	Pre-image	Image
16.	(4, 2) 90°	(−2, 4)
17.	(3, 7) 180°	(−3, −7)
18.	(−10, −9) 270°	(−9, 10)
19.	(3, −2) 0° or 360°	(3, −2)

Find the image point for each pre-image point and rotation angle given.

20. (−10, 2); 270° (2, 10)
21. (−3, 5); 65° (−5.7994, −0.6058)
22. (2, −7); 150° (1.7679, 7.0622)
23. (2, 1); 150° (−2.2321, 0.134)

Technology For Exercises 24–27, create the appropriate rotation matrix. First write the matrix using sines and cosines. Then calculate the values and round them to the nearest thousandth.

24. 90°
25. 270°
26. 45°
27. 60°

10. (2.656, 3.598)
11. (−1.278, 11.107)
12. (−0.7071, −1.225)
13. (−1.981, −17.919)
14. (−7.551, 1.996)
15. (−3.163, −1.731)

Technology For Exercises 28–35, the use of a graphics calculator is recommended to perform the matrix multiplication.

28. Rotate $\triangle ABC$ with $A(2, 0)$, $B(4, 4)$, and $C(5, 2)$ 45° about the origin. Call the image triangle $\triangle A'B'C'$ and plot both triangles on the same set of axes.

29. Rotate $\triangle A'B'C'$ 45° about the origin. Call the image triangle $A''B''C''$ and plot it along with $\triangle ABC$ and $\triangle A'B'C'$.

30. Describe the transformation $\triangle ABC \rightarrow \triangle A''B''C''$ as a single rotation. Write the rotation matrix for this rotation. Check your answers by multiplying this rotation matrix times the point matrix for $\triangle ABC$.

31. Let $[R_{60}]$ denote a 60° rotation matrix. Fill in a table like the one below. Plot the points in the last column on graph paper along with the point $P(4, 0)$.

60° 60° 60° 60° 60° 60°
P(4, 0)

n	$[R_{60}]^n$	$[R_{60}]^n \begin{bmatrix} 4 \\ 0 \end{bmatrix}$
1	?	?
2	?	?
3	?	?
4	?	?
5	?	?
6	?	?

32. Perform the following matrix multiplication. Work from right to left as you multiply.

$$[R_{60}][R_{60}][R_{60}][R_{60}][R_{60}][R_{60}] \begin{bmatrix} 4 \\ 0 \end{bmatrix}$$

33. Repeat Exercise 32, but this time work from left to right as you multiply. Compare the results of Exercises 32 and 33 with $[R_{60}]^6 \begin{bmatrix} 4 \\ 0 \end{bmatrix}$, the last entry in your table in Exercise 31.

34. One vertex of a regular pentagon centered at the origin is (4, 0). Describe a procedure for finding the coordinates of the other four vertices. Carry out your procedure and plot all five vertices.

35. If A and B are 2 × 2 rotation matrices, does AB = BA? For example, is it true that $[R_{30}]\,[R_{60}] = [R_{60}]\,[R_{30}]$? Is the same thing true for 2 × 2 matrices other than rotation matrices? Illustrate your answer with an example.

ALTERNATIVE ASSESSMENT

Performance Assessment

Students should work in pairs to graph a square on a coordinate system and choose any angle θ between 0° and 90°. Students should then rotate the square about the origin through both θ and $-\theta$. They should predict the positions of vertex coordinates before they begin, then compare results to predictions.

The answers to Exercises 28–33 and 35 can be found in Additional Answers beginning on page 727.

34. Multiply $\begin{bmatrix} 4 \\ 0 \end{bmatrix}$ repeatedly by the rotation matrix for a rotation of 72° until all the vertices are found.

$\begin{bmatrix} 1.24 \\ 3.80 \end{bmatrix}; \begin{bmatrix} -3.24 \\ 2.35 \end{bmatrix}; \begin{bmatrix} -3.24 \\ -2.35 \end{bmatrix}; \begin{bmatrix} 1.24 \\ -3.80 \end{bmatrix}$

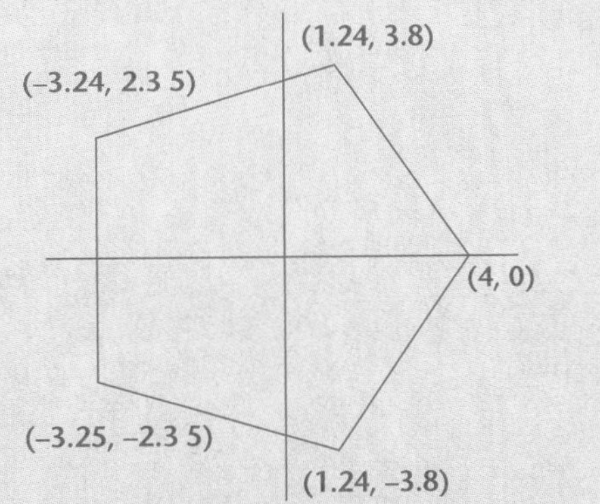

Technology Master

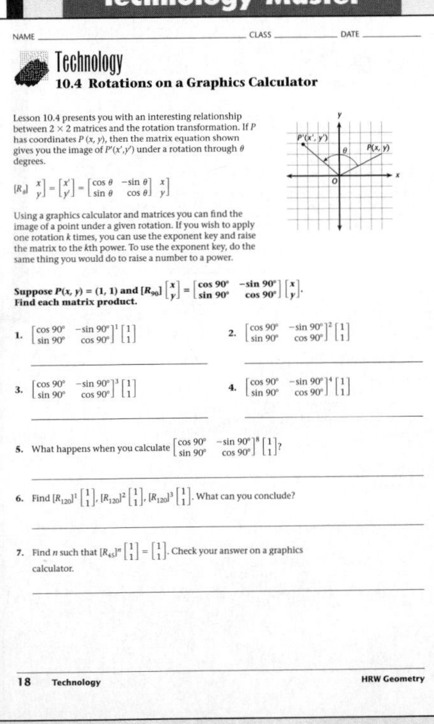
NAME ______ CLASS ______ DATE ______

Technology
10.4 Rotations on a Graphics Calculator

Lesson 10.4 presents you with an interesting relationship between 2 × 2 matrices and the rotation transformation. If P has coordinates $P(x, y)$, then the matrix equation shown gives you the image of $P'(x', y')$ under a rotation through θ degrees.

$[R_\theta]\begin{bmatrix} x \\ y \end{bmatrix} = \begin{bmatrix} x' \\ y' \end{bmatrix} = \begin{bmatrix} \cos\theta & -\sin\theta \\ \sin\theta & \cos\theta \end{bmatrix}\begin{bmatrix} x \\ y \end{bmatrix}$

Using a graphics calculator and matrices you can find the image of a point under a given rotation. If you wish to apply one rotation k times, you can use the exponent key and raise the matrix to the kth power. To use the exponent key, do the same thing you would do to raise a number to a power.

Suppose $P(x, y) = (1, 1)$ and $[R_{90}]\begin{bmatrix} x \\ y \end{bmatrix} = \begin{bmatrix} \cos 90° & -\sin 90° \\ \sin 90° & \cos 90° \end{bmatrix}\begin{bmatrix} x \\ y \end{bmatrix}$. Find each matrix product.

1. $\begin{bmatrix} \cos 90° & -\sin 90° \\ \sin 90° & \cos 90° \end{bmatrix}^1 \begin{bmatrix} 1 \\ 1 \end{bmatrix}$

2. $\begin{bmatrix} \cos 90° & -\sin 90° \\ \sin 90° & \cos 90° \end{bmatrix}^2 \begin{bmatrix} 1 \\ 1 \end{bmatrix}$

3. $\begin{bmatrix} \cos 90° & -\sin 90° \\ \sin 90° & \cos 90° \end{bmatrix}^3 \begin{bmatrix} 1 \\ 1 \end{bmatrix}$

4. $\begin{bmatrix} \cos 90° & -\sin 90° \\ \sin 90° & \cos 90° \end{bmatrix}^4 \begin{bmatrix} 1 \\ 1 \end{bmatrix}$

5. What happens when you calculate $\begin{bmatrix} \cos 90° & -\sin 90° \\ \sin 90° & \cos 90° \end{bmatrix}^k \begin{bmatrix} 1 \\ 1 \end{bmatrix}$?

6. Find $[R_{120}]^1\begin{bmatrix} 1 \\ 1 \end{bmatrix}$, $[R_{120}]^2\begin{bmatrix} 1 \\ 1 \end{bmatrix}$, $[R_{120}]^3\begin{bmatrix} 1 \\ 1 \end{bmatrix}$. What can you conclude?

7. Find n such that $[R_{45}]^n\begin{bmatrix} 1 \\ 1 \end{bmatrix} = \begin{bmatrix} 1 \\ 1 \end{bmatrix}$. Check your answer on a graphics calculator.

18 Technology HRW Geometry

Translations Recall that to translate the point $P(2, 1)$ 5 units to the right and 3 units down, you can add 5 units to the x-coordinate and -3 units to the y-coordinate:

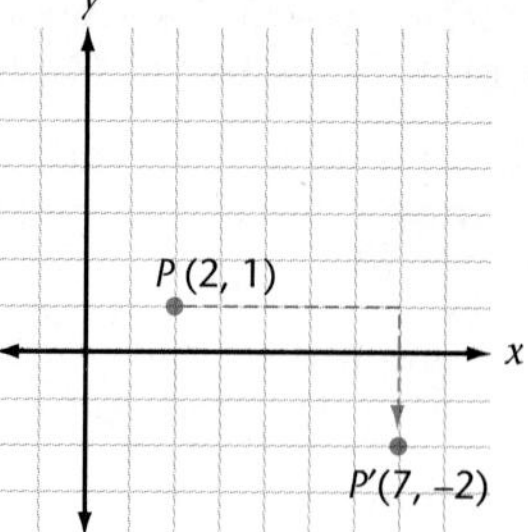

$$P' = (2 + 5, 1 + -3) = (7, -2)$$

Addition of matrices can be used to perform this translation.

36. Fill in the missing entries.

$\begin{bmatrix} ? \\ ? \end{bmatrix} + \begin{bmatrix} 2 \\ 1 \end{bmatrix} = \begin{bmatrix} 7 \\ -2 \end{bmatrix}$ $\begin{bmatrix} 5 \\ -3 \end{bmatrix}$

37. Translate $\triangle PQR$, with $P(2, 1)$, $Q(4, 8)$, and $R(8, 5)$, five units to the right and three units down. Fill in the missing entries.

$$\begin{bmatrix} ? & ? & ? \\ ? & ? & ? \end{bmatrix} + \overset{P\ Q\ R}{\begin{bmatrix} 2 & 4 & 8 \\ 1 & 8 & 5 \end{bmatrix}} = \overset{P'Q'R'}{\begin{bmatrix} ? & ? & ? \\ ? & ? & ? \end{bmatrix}}$$ $\begin{bmatrix} 5 & 5 & 5 \\ -3 & -3 & -3 \end{bmatrix}$; $\begin{bmatrix} 7 & 9 & 13 \\ -2 & 5 & 2 \end{bmatrix}$

Look Back

Find the angle θ between 0° and 90° for each trigonometric ratio given. Round to the nearest degree. [Lesson 10.1]

38. $\sin \theta = 0.767$ 50° **39.** $\cos \theta = 0.3565$ 69° **40.** $\tan \theta = 2.676$ 70° **41.** $\sin \theta = 0.3565$ 21°

Find the sine, cosine, and tangent for each angle below. Round to the nearest hundredth. [Lesson 10.1]

42. 37° 0.60; 0.80; 0.75 **43.** 187° –0.12; –0.99; 0.12 **44.** 90° 1; 0; undefined

45. 250° –0.94; –0.34; –2.75 **46.** 400° 0.64; 0.77; 084 **47.** 1800° 0; 1; 0

Look Beyond

Exercise 48 expands the concepts presented in the lesson and anticipates the Law of Sines and Law of Cosines.

Look Beyond

48. A triangle has a right angle C, two other angles A and B, and $\sin A = \cos A$. What are the measures of $\angle A$ and $\angle B$? Classified by its side lengths, what type of triangle is the triangle? Expressed in terms of the hypotenuse c, what are the lengths of sides a and b?

45; isosceles triangle; $\sqrt{\frac{c^2}{2}}$

LESSON 10.5

The Law of Sines

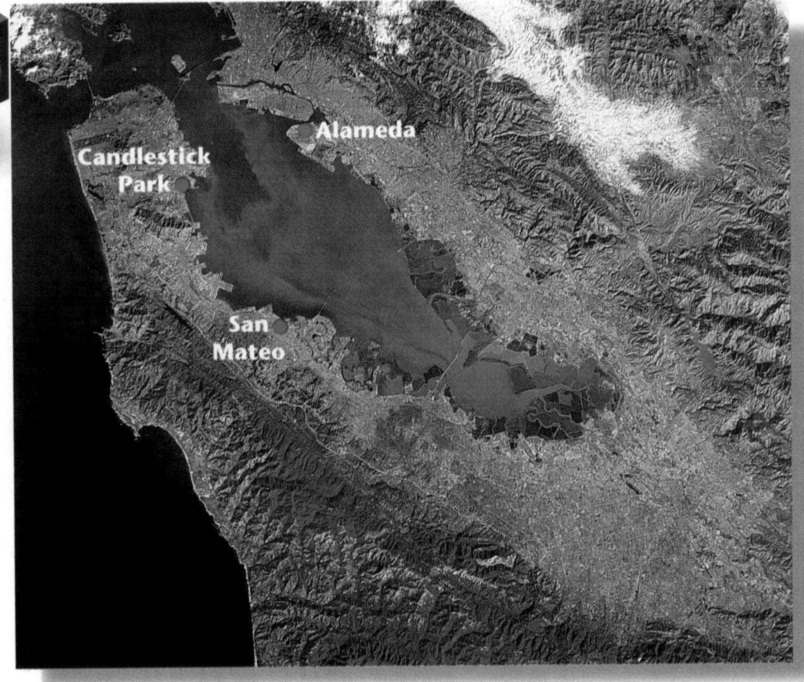

Satellite photo of the San Francisco Bay area

Why *Although the triangle formed by the three points on the photo is not a right triangle, its measures can be found using trigonometry.*

Exploration Law of Sines

You will need
Geometry technology or
Centimeter ruler, protractor, and scientific calculator

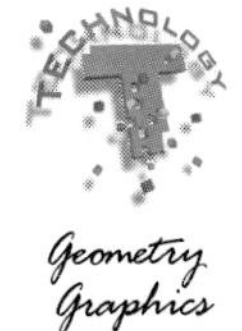
Geometry Graphics

1. Construct an acute triangle, a right triangle, and an obtuse triangle. Measure the sides and angles of each triangle.

2. Use the measurements from Step 1 to complete a table like the one below.

	m∠A	m∠B	m∠C	a	b	c	$\frac{\sin A}{a}$	$\frac{\sin B}{b}$	$\frac{\sin C}{c}$
Acute Triangle	?	?	?	?	?	?	?	?	?
Right Triangle	?	?	?	?	?	?	?	?	?
Obtuse Triangle	?	?	?	?	?	?	?	?	?

3. Write a conjecture using the data in the last three columns of the table. ❖

PREPARE

Objectives

- Develop the Law of Sines.
- Use the Law of Sines to find the measure of an angle or a side in a triangle.

RESOURCES

- Practice Master 10.5
- Enrichment Master 10.5
- Technology Master 10.5
- Lesson Activity Master 10.5
- Quiz 10.5
- Spanish Resources 10.5

Assessing Prior Knowledge

1. Draw any non-right triangle. Label the vertices A, B, and C. Use a, b, and c to label the sides, with side a opposite ∠A, side b opposite ∠B, and side c opposite ∠C. **[Check students' drawings.]**

2. List all combinations for two angles and one side. **[∠A, ∠B, a; ∠A, ∠B, b; ∠A, ∠B, c; ∠A, ∠C, a; ∠A, ∠C, b; ∠A, ∠C, c; ∠B, ∠C, a; ∠B, ∠C, b; ∠B, ∠C, c]**

Ongoing ASSESSMENT

3. $\frac{\sin A}{a} = \frac{\sin B}{b} = \frac{\sin C}{c}$

Technology The examples and application in this lesson, as well as many of the exercises, can be explored or confirmed using geometry graphics software. If the software includes trigonometric functions in the menu or toolbox, computations can be checked directly. If the trigonometric functions are not available, students can construct triangles with the given measures and confirm results by measuring sides or angles.

TEACH

If students know at least one side and any two other measures, they can find the missing three measures using the Law of Sines or the Law of Cosines. In this lesson, students develop the Law of Sines.

Exploration Notes

Students experiment with various triangles to discover the Law of Sines. Point out that Step 1 asks for a right, an acute, and an obtuse triangle. Ask students why all three types are needed.

Point out that the Law of Sines is actually three separate "laws" or formulas. Have students take the ratios in pairs to write out the three equations.

Alternate Example 1

a. Have students construct $\triangle ABC$ with $\angle A = 30°$, $\angle C = 112°$, and side $b = 360$ ft.

b. Ask how this situation differs from Example 1 in the text. [**Triangle is obtuse. The given side is included between the two $\angle$s.**]

c. Students solve the triangle; that is, they find the three missing measures. [**Students first find $\angle B$. $180° - (112° + 30°) = 38°$. Use $\angle A$, $\angle B$, and b to find a. $a \approx 292.4$ ft. Use $\angle B$, $\angle C$, and b to find c. $c \approx 542.2$ ft.**]

Your conjecture should be equivalent to the following theorem.

THE LAW OF SINES

For any triangle ABC $\frac{\sin A}{a} = \frac{\sin B}{b} = \frac{\sin C}{c}$. **10.5.1**

Proof: Acute case

Each angle of $\triangle ABC$ is an acute angle. The measures of the altitudes from C and B are h_1 and h_2, respectively.

In right triangles $\triangle ACD_1$ and $\triangle BCD_1$,

$$\sin A = \frac{h_1}{b} \text{ and } \sin B = \frac{h_1}{a}.$$

Thus, $b \sin A = h_1$ and $a \sin B = h_1$.

Since h_1 can be written in two different ways, $b \sin A = a \sin B$.

Dividing both sides of the equation by ab gives

$$\frac{\sin A}{a} = \frac{\sin B}{b}.$$

Similarly, in right triangles $\triangle ABD_2$ and $\triangle CBD_2$,

$$\sin A = \frac{h_2}{c} \text{ and } \sin C = \frac{h_2}{a}.$$

Thus, $\frac{\sin A}{a} = \frac{\sin C}{c}$.

Therefore, $\frac{\sin A}{a} = \frac{\sin B}{b} = \frac{\sin C}{c}$.

The proof for right triangles and for obtuse triangles is left for you to try in the exercises.

The Law of Sines can be used to find the measures of sides and angles in any triangle, as long as you know certain combinations of measures in the triangle.

EXAMPLE 1

Find b and c.

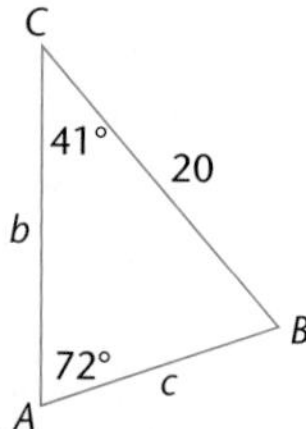

Solution

Set up a proportion based on the Law of Sines in which 3 of the 4 terms are known. Look for a known angle whose opposite side is also known, in this case $\angle A$ and a.

Physics Trigonometry applications in astronomy often involve an extremely long "skinny" triangle. The base might be a segment between two points on the surface of the Earth. But the opposite vertex is at some very distant point, and the angle at this vertex is extremely small. The two long sides of the triangle are almost parallel. Have students sketch this situation and discuss the types of application problems that might occur. If students are interested, suggest they do research to find out how the "parallax" method is used in situations of this type.

ALGEBRA *Connection*

Since m∠C is known, we have

$$\frac{\sin A}{a} = \frac{\sin C}{c}, \text{ or } \frac{\sin 72°}{20} = \frac{\sin 41°}{c}.$$

Cross-multiplication gives

$$c = \frac{20 \sin 41°}{\sin 72°} \approx 13.8.$$

To find b, we must first find m ∠B: 180 – (72 + 41) = 67°. Again using the known angle-with-side pair,

$$\frac{\sin A}{a} = \frac{\sin B}{b}, \text{ or } \frac{\sin 72°}{20} = \frac{\sin 67°}{b}.$$

Cross-multiplication gives

$$b \sin 72° = 20 (\sin 67°).$$

$$b = \frac{20 \sin 67°}{\sin 72°} \approx 19.36. ❖$$

The Law of Sines can also be used to find angles.

EXAMPLE 2

Find m∠Q.

Solution ➤

$$\frac{\sin 82°}{34} = \frac{\sin x}{27}$$

$$\sin x = \frac{27 \sin 82°}{34}$$

$$\sin x \approx 0.78639$$

$$x = \sin^{-1}(0.78639)$$

$$x \approx 52°$$

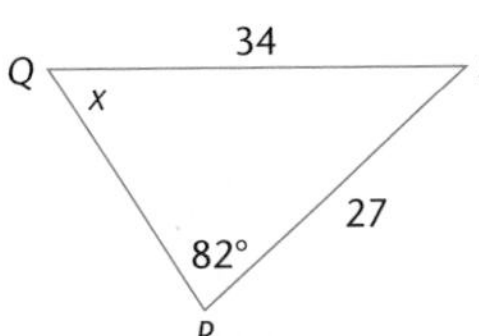

Caution: In solving for an angle using the Law of Sines, you will need to know in advance whether the angle is acute or obtuse. This is because an acute and an obtuse angle may have the same sine ratio. For example, sin 50° = sin 130° = 0.7660. However, your calculator will give only one value for the angle whose sine is 0.7660. (Which one does it give?)

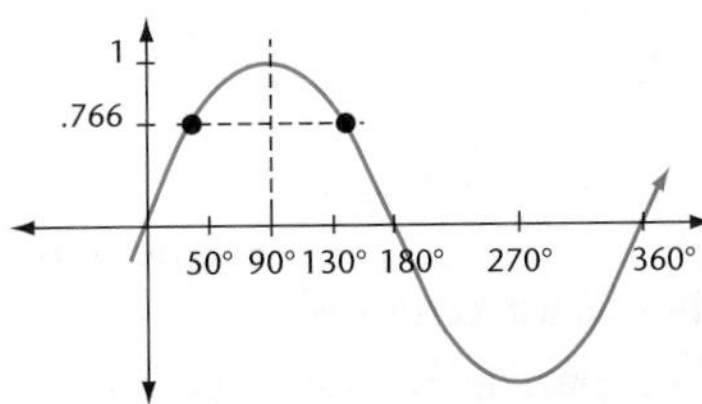

In general, the Law of Sines can be used in any triangle if you know the measures of

1. two angles and a side of a triangle or
2. two sides and an angle that is opposite one of the sides of a triangle. ❖

Math Connection Algebra

For Example 1, remind students how to use cross-multiplication to convert a proportion into an equation. For example, $\frac{a}{b} = \frac{c}{d}$ can be converted into $ad = bc$.

Teaching tip

Remind students of their work with the triangle congruent postulates, emphasizing that the SSA situation does not guarantee congruence. This type of ambiguous case can occur when using the Law of Sines. If two sides and an angle not included between them are given, there can be two possible triangles. Alternate Example 2 illustrates this case.

Use Transparency 91

Alternate Example 2

a. Have students construct △ABC with ∠A = 30°, side $a = 15$ in., and side $b = 20$ in.

b. Ask how this situation differs from Example 1 in the text. **[There are two possible triangles that meet these conditions.]**

c. Use the Law of Sines to find the measure of ∠B. **[41.81° or 138.19°]**

Enrichment

Alternate Example 2 shows one ambiguous case of the Law of Sines. In another type of ambiguous case, there is no triangle at all. Have students try to find the measure of ∠B for △ABC with ∠C = 55°, $c = 9$ cm, and $b = 25$ cm. **[Students will get sin B ≈ 2.2754, which is impossible. The sine of an angle cannot be greater than 1. Trying to construct the triangle will also show that the dimensions will not work.]**

Inclusion strategies

Technology Students will most likely be using graphics or scientific calculators for the exercises and problems in this lesson. They may find that it is easier to apply the Law of Sines if they convert it to forms like the following:

$a = \sin A \cdot b \div \sin B$	$\sin A = a \cdot \sin B \div b$
$b = \sin B \cdot c \div \sin C$	$\sin B = b \cdot \sin C \div c$
$c = \sin C \cdot a \div \sin A$	$\sin C = c \cdot \sin A \div a$

Ongoing Assessment

The calculator gives the value for the angle between 0° and 90° (in this case, 50°).

Cooperative Learning

The exploration in this lesson is particularly appropriate for small-group work because of the large number of measurements that need to be made. Have students read through the entire activity before they begin, discussing any steps that are not clear. They then agree on how to divide up the work required.

Math Connection Algebra

Students may need help interpreting the phrase "Solving the ratios in pairs." Use the simpler statement $x = y = z$ to illustrate what is meant. The statement can be converted into three separate equations by taking the variables "in pairs." The equations are $x = y$, $y = z$, and $z = x$.

APPLICATION

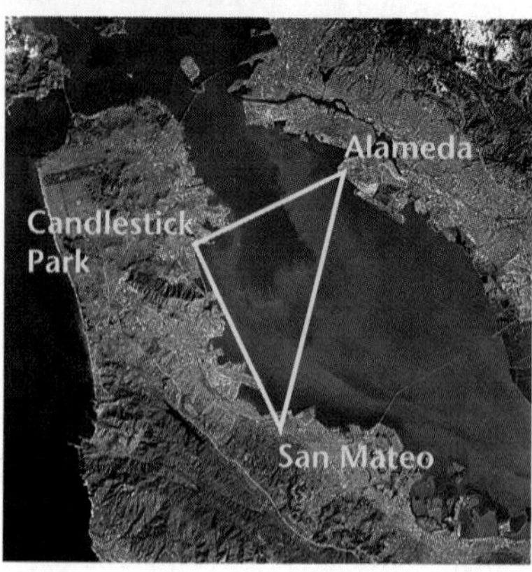

A boater regularly travels across San Francisco Bay from San Mateo to Alameda, a distance of approximately 12.8 miles. If the boater decides to travel first to Candlestick Park, what is the distance to Alameda?

The angle formed at Candlestick Park is 95°, at Alameda 53°. Setting up the proportion:

$$\frac{\sin 95°}{12.8} = \frac{\sin 53°}{b} = \frac{\sin 32°}{c}$$

Solve the ratios in pairs:

$$\frac{\sin 95°}{12.8} = \frac{\sin 53°}{b}$$
$$b\sin 95° = 12.8 \sin 53°$$
$$b = \frac{12.8 \sin 53°}{\sin 95°} = \frac{12.8(0.7986)}{0.9962} \approx 10.26$$

$$\frac{\sin 95°}{12.8} = \frac{\sin 32°}{c}$$
$$c \sin 95° = 12.8 \sin 32°$$
$$c = \frac{12.8 \sin 32°}{\sin 95°} = \frac{12.8(0.5299)}{0.9962} \approx 6.81$$

Total distance = 17.1 miles

The total boating distance from San Mateo to Alameda via Candlestick Park is approximately 17.1 miles. ❖

Exercises & Problems

Communicate

1. State the Law of Sines. How is it used?
2. In solving for an angle using the Law of Sines, why must you know in advance if the angle is acute or obtuse?
3. Give two examples of triangle measures that can be used to apply the Law of Sines.
4. The length of the side and the angle opposite it must be known to apply the Law of Sines. Why is this true?
5. Draw and label a triangle for which the Law of Sines cannot be applied. Use your diagram to give an explanation.

Reteaching the Lesson

Cooperative Learning Students should work in small groups to verify the Law of Sines for a specific triangle. Students should agree on three angle measures and one side length. Each constructs the triangle and measures the other two sides. Then the sines of the three angles can be found with a calculator or a table. Students should compute the three ratios to verify the Law of Sines.

Practice & Apply 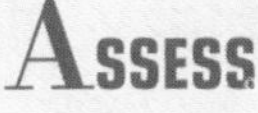

For Exercises 6–13, sketch triangle *ABC* roughly to scale using the given triangle measures. Then find the measure of the indicated triangle part.

6. Find side c given $m\angle B = 24°$, $m\angle A = 56°$, and $b = 1.22$ cm. 2.954 cm

7. Find side b given $m\angle B = 73°$, $m\angle C = 85°$, and $a = 3.14$ cm. 8.02 cm

8. Find side a given $m\angle A = 35°$, $m\angle B = 44°$, and $c = 24$ mm. 14.02 mm

9. Find side b given $m\angle C = 72°$, $m\angle A = 53°$, and $c = 2.34$ cm. 2.02 units

10. Find $\angle A$ given $a = 3.12$, $b = 3.28$, and $m\angle B = 29°$. 27.46°

11. Find side c in Exercise 10. 5.64 cm

12. Find the measure of $\angle C$ given $c = 2.13$ cm, $b = 7.36$ cm, and $m\angle B = 67°$. 15.45°

13. Find side a in Exercise 12. 7.93 cm

Surveying You are driving toward a mountain. At a rest stop, you use a clinometer (a pocket-size device that can measure angles of elevation) to measure the angle of inclination from your eye to the top of the mountain. The angle measures 11.5°. You then drive toward the mountain on level ground for 5 miles. You stop and measure the angle of inclination to be 27.5°.

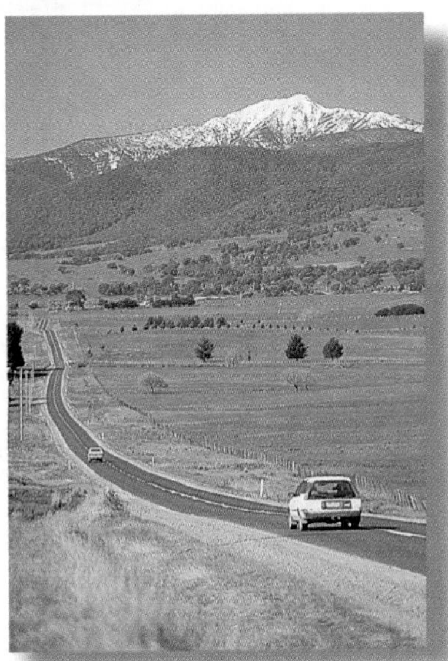

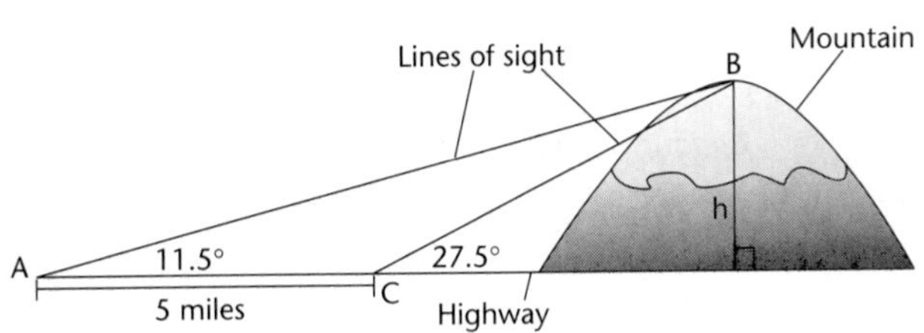

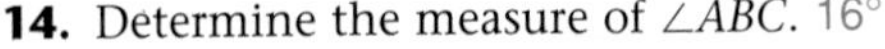

14. Determine the measure of $\angle ABC$. 16°

15. Find the distance BC. 3.62 miles

16. Using the results from Exercises 14 and 15, what is the height, h, of the mountain. 1.67 miles

17. Architecture A real-estate developer wants to build an office tower on a triangular lot between Oak Street and Third Avenue. The lot measures 42 m along Oak Street and 39 m along Third Avenue. Oak Street and Third Avenue meet at a 75° angle and the third side of the lot is 48.8 m long. The architects want to know the other angles formed by the sides of the lot before they start drawing the blueprints. Find the other angles.

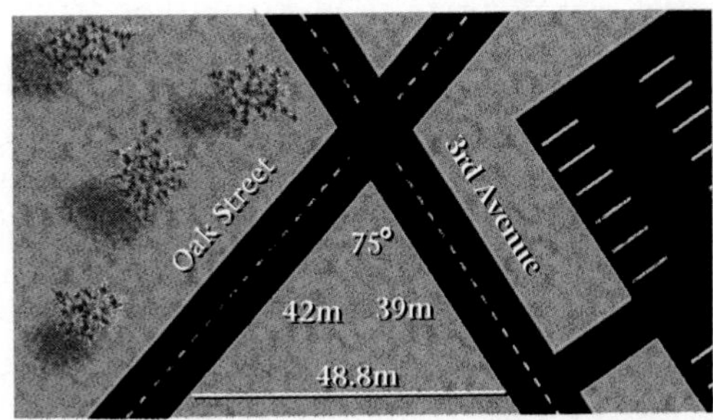

17. Let A be the angle opposite Oak Street and angle B be the angle opposite 3rd Ave.

$m\angle A = 56.24°$; $m\angle B = 48.76°$

ASSESS

Selected Answers

Odd-numbered Exercises 7–41

Assignment Guide

Core 1–4, 6–16, 34–42

Core Plus 1–5, 10–44

Technology

Geometry graphics software can be used to check results obtained from the Law of Sines. Students construct triangles with the given measures. Then they measure the needed angle or side to check their computational results.

Use Transparency 92

Error Analysis

It is extremely easy to make computational errors when using the Law of Sines. One way to check answers is to use geometry graphics software, as suggested above. Another technique is to have students work in pairs or groups. Each student solves a problem independently. Then results are compared and discussed.

alternative ASSESSMENT

Performance Assessment

Provide students with this figure.

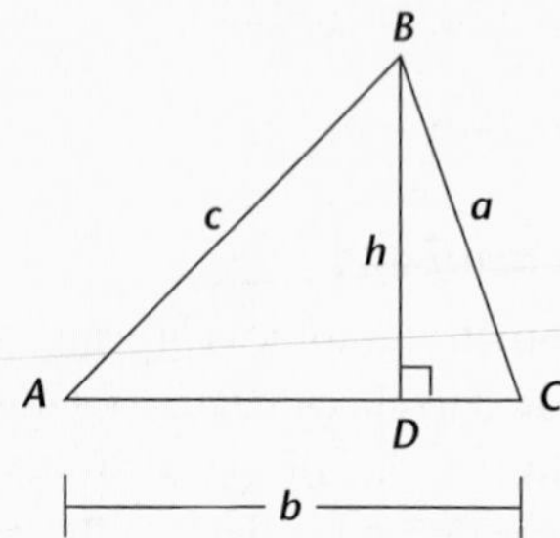

Students should use the figure to derive the Law of Sines using any strategy they wish. Written work should include an explanation of why they approached the project the way they did. **[Possible method: In $\triangle ADB$, $\sin A = \frac{h}{c}$. In $\triangle BDC$, $\sin C = \frac{h}{a}$. Solve both equations for h. Result: $h = c \sin A = a \sin C$. Divide both sides by $\sin A \sin C$. The result is $\frac{a}{\sin A} = \frac{c}{\sin C}$.]**

Practice Master

NAME ______ CLASS ______ DATE ______

Practice & Apply
10.5 The Law of Sines

Exercises 1–3 refer to $\triangle ABC$.

1. If $m\angle A = 41°$, $m\angle B = 58°$, and $a = 14$ cm, then the length of side b = ______.
2. If $m\angle A = 72°$, $m\angle B = 40°$, and $c = 15$ in., then the length of side a = ______.
3. If $m\angle A = 35°$, $a = 8$ cm, and $b = 12$ cm, then the measure of $\angle B$ = ______.
4. In $\triangle XYZ$, $m\angle Y = 35°$, $m\angle Z = 41°$, $YZ = 23$ cm. The length of the shortest side is ______.
5. Two angles of a triangle are 25° and 58°. The longest side of the triangle is 39 cm. The shortest side is ______.

Exercises 6–8 refer to $\triangle ABC$.

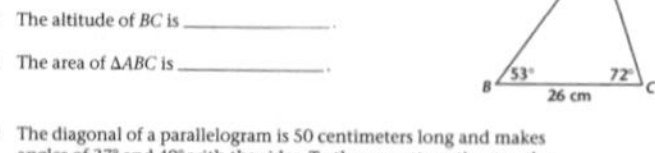

6. The length of AB is ______.
7. The altitude of BC is ______.
8. The area of $\triangle ABC$ is ______.
9. The diagonal of a parallelogram is 50 centimeters long and makes angles of 37° and 49° with the sides. To the nearest centimeter, the length of the shorter side of the parallelogram is ______.

Exercises 10–12 refer to the diagram in which a surveyor on the ground has taken two readings of the angle of elevation to the top of a tower.

10. Use the Law of Sines in $\triangle ABD$ to find the length of $\overline{AD}$ to the nearest tenth of a foot. ______
11. Use the result of Exercise 10 to find the height of the tower, $\overline{AC}$, to the nearest foot. ______
12. Use the result of Exercise 11 to find the distance CB. ______

HRW Geometry   Practice & Apply 5

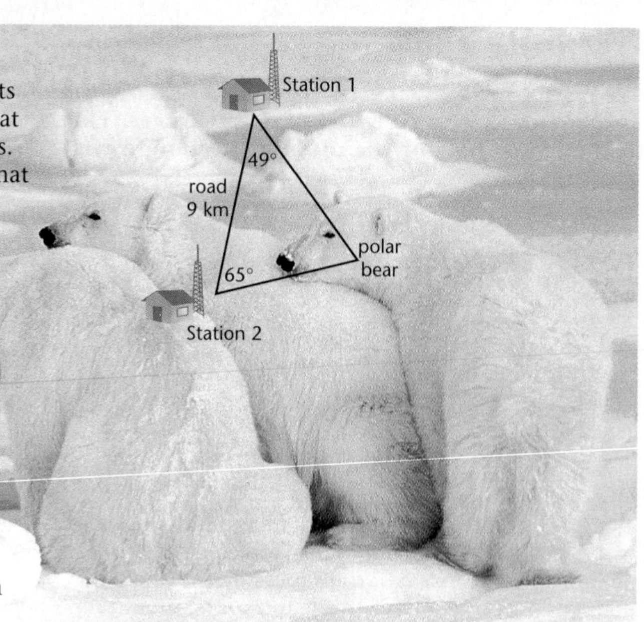

18. Wildlife Management Scientists are tracking several polar bears that have been fitted with radio collars. The scientists have two stations that are connected by a straight road that is 9 km long. The scientists find that the line between Station 1 and a polar bear forms a 49° angle with the road, while the line between Station 2 and the polar bear forms a 65° angle with the road.

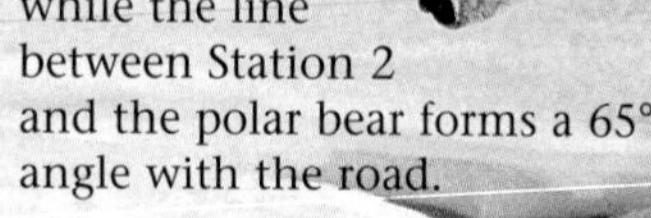

a. How far is the polar bear from Station 2.

b. Find the shortest distance from the polar bear to the road.

Complete the proof for the obtuse case of the Law of Sines in Exercises 19–33.

Given: In $\triangle ABC$, $\angle ABC$ is obtuse. The measures of the altitudes from C and B are h_1 and h_2, respectively.

Prove: $\frac{\sin A}{a} = \frac{\sin B}{b} = \frac{\sin C}{c}$

Proof:

STATEMENTS	REASONS
In $\triangle BCD_1$ $\sin B = \frac{h_1}{a}$ and in $\triangle ACD_1$ $\sin A = \frac{h_1}{b}$	Definition of sine
$h_1 =$ **(19)** and $h_1 =$ **(20)**	**(21)**
(22)	Substitution property of equality
$\frac{\sin B}{b} = \frac{\sin A}{a}$	**(23)**
In $\triangle CD_2B$ $\sin C =$ **(24)** and in $\triangle BD_2A$ $\sin A =$ **(25)**	**(26)**
$h_2 =$ **(27)** and $h_2 =$ **(28)**	**(29)**
$c \sin A = a \sin C$	**(30)**
(31)	Division property of equality
(32)	**(33)**

18. a. 7.44 km b. 6.74 km
19. $a \sin B$
20. $b \sin A$
21. Multiplication Property of Equality
22. $a \sin B = b \sin A$
23. Division (by ab) Property of Equality
24. $\frac{h_2}{a}$
25. $\frac{h_2}{c}$
26. Definition of sine
27. $a \sin C$
28. $c \sin A$
29. Multipliplication Property of Equality
30. Substitution Property of Equality
31. $\frac{\sin A}{a} = \frac{\sin C}{c}$
32. $\frac{\sin A}{a} = \frac{\sin B}{b} = \frac{\sin C}{c}$
33. Transitive Property of Equality

Use the diagram to prove the Law of Sines for right triangles for Exercises 34–36.

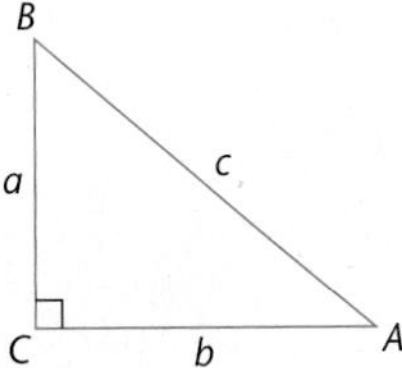

34. Write the sine of angles A, B, and C in terms of a, b, and c.

35. Form the three ratios and simplify $\frac{\sin A}{a}$, $\frac{\sin B}{b}$, and $\frac{\sin C}{c}$.

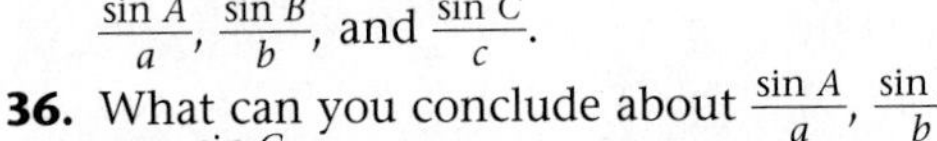

36. What can you conclude about $\frac{\sin A}{a}$, $\frac{\sin B}{b}$, and $\frac{\sin C}{c}$?

$\frac{\sin A}{a} = \frac{\sin B}{b} = \frac{\sin C}{c}$

Look Back

37. Find the surface area and volume of a sphere with a radius of 15 m. What would happen to the volume and surface area of the sphere if the radius were tripled? **[Lesson 7.6]**

38. Find x. **[Lesson 8.3]**

51, 45, 24, x

11.3 units

39. $\overline{BC}$ is parallel to $\overline{DE}$. Find x. **[Lesson 8.4]**

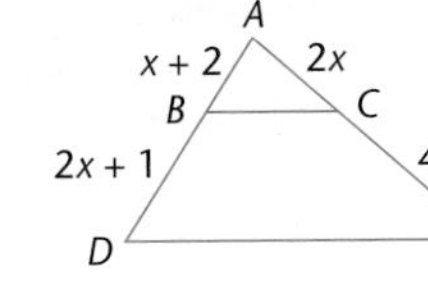

$x = 4$

Find these without using a calculator. [Lesson 10.2]

40. $\sin^{-1}\left(\frac{\sqrt{3}}{2}\right)$

60°

41. $\tan^{-1}\sqrt{3}$

60°

42. $\cos^{-1}\left(\frac{\sqrt{2}}{2}\right)$

45°

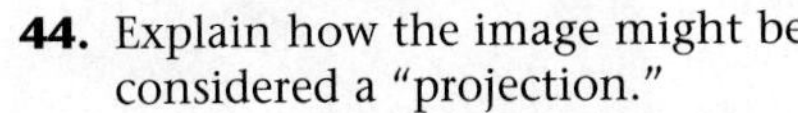

Look Beyond

43. The image of the rabbit is distorted. It can be seen in proper perspective if you hold the book so that your line of sight forms a certain angle to the page. Determine the approximate angle for your line of sight in order to see the image properly.

44. Explain how the image might be considered a "projection."

34. $\sin A = \frac{a}{c}$; $\sin B = \frac{b}{c}$

35. $\frac{\sin A}{a} = \frac{1}{c}$; $\frac{\sin B}{b} = \frac{1}{c}$; $\frac{\sin C}{c} = \frac{1}{c}$

37. SA= 2827.4 m^2;

V = 14,137.2 m^3;

If the radius were tripled, the surface area would be multiplied by $3^2 = 9$ and the volume would be multiplied by $3^3 = 27$.

43. about 30° to 40°

44. The reader's eye is the center of projection and the lines of sight are the projection rays, so all the properties of a projection will hold. Collinear points are mapped to collinear points, straight lines are mapped to straight lines, intersecting lines are mapped to intersecting lines and parallel lines are mapped to parallel lines.

Use Transparency 93

Look Beyond

Exercises 43 and 44 feature an art special effect known as an anamorphosis. The exercises foreshadow projective geometry.

Technology Master

NAME ______ CLASS ______ DATE ______

Technology
10.5 Wedges

Any isosceles triangle can be a model of a wedge. If you specify the maximum distance across a wedge, you can use the Law of Sines to experiment with different sizes of wedges based on the measure θ of the vertex angle, $\angle APB$ in the diagram shown. From the Law of Sines, you can write this equation.

$$\frac{x}{\sin(90° - 0.5\theta)} = \frac{6}{\sin\theta}$$

$$x = \frac{6\sin(90° - 0.5\theta)}{\sin\theta}$$

A graphics calculator will be helpful to you in dealing with the following exercises. Be sure your calculator is set to degree mode.

Use a graphics calculator in the following exercises.

1. For what values of θ does the equation for x above make sense? Graph the equation for x over that range.

Use the graph to find each value of x or θ given each value of θ or x.

2. $\theta = 30°$ 3. $\theta = 45°$ 4. $x = 5$ 5. $x = 12.2$

6. Write an equation for the perimeter P of ΔAPB in terms of θ. Graph your equation.

7. Verify graphically that $P = 18$ when $\theta = 60°$.

8. Verify graphically that there is no ΔAPB with $AB = 6$ and $P = 9$.

9. Use your knowledge of triangles to show that in ΔAPB above

$x = \frac{6\sin(90° - 0.5\ )}{\sin}$

HRW Geometry Technology 19

Focus

The NASA mission to repair a flawed mirror in the Hubble Telescope serves as a context for exploring very small angles and the geometry of reflection.

Motivate

Have students read the news excerpts to get an overview of the Hubble Space Telescope rescue mission.

Emphasize these points in discussion:

- Our atmosphere, which makes stars appear to twinkle, also makes telescopic pictures blurry. A space-based telescope provides clearer pictures because it avoids the interference produced by the earth's atmosphere.
- After Hubble was launched, scientists discovered that its main mirror was flawed — it was too flat at the edges by about $\frac{1}{50}$th of the thickness of a sheet of paper.
- NASA took advantage of a planned servicing mission to install new optics in the telescope to correct for the flaws.

Astronauts Snare Hubble Telescope for Vital Repairs

By John Noble Wilford
Special to the *New York Times*

HOUSTON, Dec. 4—Arriving for a house call 357 miles above Earth, the astronauts of the space shuttle Endeavor today reached out with a mechanical grappling arm and easily snared the Hubble space telescope and prepared to treat the crippled spacecraft in five days of the most complex orbital repairs yet attempted.

The Hubble telescope was deployed in orbit by another shuttle in April, 1990. The fanfare of early expectations faded in the following weeks as astronomers were dismayed to see the blurry pictures. Tests traced the problem to an improperly ground mirror.

The flawed curvature meant that the mirror could not focus incoming light to a precise point. As a result, observations were not as clear as desired and the effective range was limited to about 4 billion light years; astronomers had been promised views out to 10 to 14 billion light years.

The Big Fix

By Ron Cowen
Science News, Nov. 6, 1993

Astronomers had long awaited the ultrasharp images that the Hubble Space Telescope seemed poised to offer. But the fuzzy pictures that the craft began radioing back to Earth dashed their hopes. The telescopes' primary mirror was hopelessly flawed. Instead of focusing most of the light from faint stars and galaxies into a tight circle with a radius of 0.1 arc second, the mirror spread the concentration of incoming rays. Scientists later learned that the flaw stemmed from a spherical defect introduced when a contractor ground the mirror's surface to its final shape.

With that bad news came one ray of hope: The flaw was so precise that additional mirrors could be introduced into the telescope's light path might correct for the error. Researchers suggested that if the removed one of Hubble's four instruments–phone-booth-size devices that lie parallel to Hubble's optical axis–they could replace it with a similarly shaped compartment mounted with corrective mirrors. These mirrors would intercept the fuzzily focused light bounced from the flawed primary mirror before it had a chance to enter any of the deflectors.

1. Yes. Sample explanation:
2 millionths of a meter = 2×10^{-6} m = 2×10^{-4} cm ≈ $\frac{(2 \times 10^{-4})}{2.54 \text{ in.}} \approx 7.9 \times 10^{-5}$, or 0.000079 in.

2. Sample explanation:

The description states that the pointing system will not stray by more than 0.007 arcsec, which is $\frac{0.07}{(60 \cdot 60)}$ degrees, or about 1.9×10^{-6} degrees. If the system is aimed at the center of the dime, it can stray no more than 0.9 cm, or it will no longer be aimed at the dime.
To find that angle (a), find the measure of the angle whose tangent is equal to $\frac{0.9 \text{ cm}}{560}$ km.
560 km = 5.6×10^5 m = 5.6×10^7 cm
$\tan a = \frac{0.9}{(5.6 \times 10^7)} \approx 1.6 \times 10^{-8}$
$a = \tan^{-1} 1.6 \times 10^{-8} \approx 9.2 \times 10^{-7}$ degrees
So, the system according to NASA's figures, is just about precise enough to stay aimed at the dime.

1. According to NASA, the main mirror in the Hubble telescope was ground too flat by about 2 microns, or 2 millionths of a meter. But *Newsweek* Magazine reported the amount as 0.000039 inch. Is there a discrepancy between these figures? Explain.

2. In the NASA description below, is the last sentence consistent with the data in the description on the telescope's stability? Explain your reasoning. Use the data on the right.

1 arcminute = 1/60°
1 arcsecond = 1/3600°

Los Angeles to San Francisco is about 560 km

> *A unique Pointing Control Subsystem aligns the spacecraft to point to and remain locked on any target. The pointing subsystem is designed to hold the Telescope with 0.007 arcsec stability for up to 24 hours while the Telescope continues to to orbit the Earth at 17,000 mph. If the Hubble Telescope were in Los Angeles, it could hold a beam of light on a dime in San Francisco, without the beam straying from the coin's diameter.*
>
> —*from NASA Media Reference Guide*

3. To get an idea of how the shape of a mirror affects the way it works, examine the diagram below. A vertical ray of light $\overrightarrow{FG}$ reflects off a tilted mirror $\overline{AC}$. If $\overleftrightarrow{DE} \perp \overline{AC}$, $\overleftrightarrow{FG} \parallel \overleftrightarrow{BC}$, and $m\angle BAC = a$, find $m\angle FGD$ and $m\angle DGH$. (Hint: Extend $\overleftrightarrow{FG}$.)

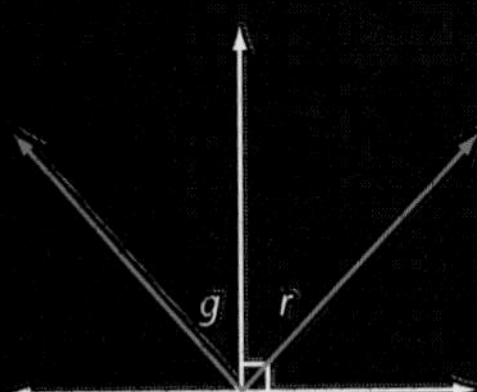

Light reflects at the same angle as its approach. In the picture, $\angle g \cong \angle r$.

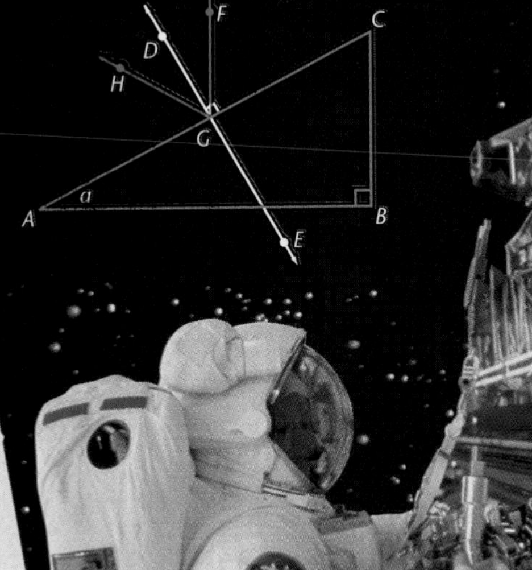

Shuttle's Crew Ends Repairs to Telescope

By John Noble Wilford
Albuquerque Journal, Dec. 1993

HOUSTON, Dec, 8—Wrapping up the most critical repairs to the Hubble Space Telescope with surprising ease and alacrity, the space-walking shuttle astronauts have now given astronomers everything they asked for: two new sets of corrective mirrors designed to overcome the telescope's flawed vision and give clear sight to its two cameras and ... instruments.

Cooperative Learning

Have students work in pairs or small groups. Have students imagine aiming at the center of a circular target. By how much of an angle can they move their aim and still be aimed somewhere on the target? Review the units arcminute and arcsecond. Students will need a calculator with an inverse tangent (arctan) function to find angles with very small tangents. Encourage students to measure or estimate the diameter of a dime (about 1.8 cm).

Discuss

Discuss the diagram so students understand that when a light ray reflects off a surface, the incoming angle is the same size as the outgoing angle (as measured from a line perpendicular to the surface).

3. Sample:

Extend $\overline{FG}$ and label as shown.

$a + m = 90$

$n + m = 90$

So, $a = m$

$m\angle FGD = a$ ($\angle FGD$ and $\angle EGI$ are vertical angles)

$m\angle DGH = a$ (reflecting light)

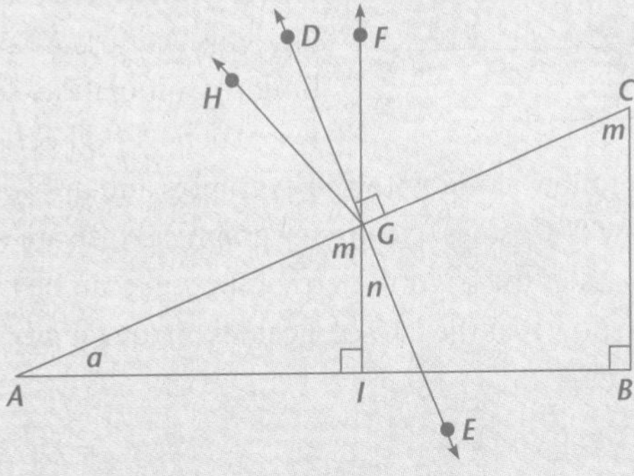

PREPARE

Objectives

- Use the Law of Cosines to find the measure of an angle or a side in a triangle.
- Use the Law of Sines and the Law of Cosines to solve problems.

RESOURCES

- Practice Master 10.6
- Enrichment Master 10.6
- Technology Master 10.6
- Lesson Activity Master 10.6
- Quiz 10.6
- Spanish Resources 10.6

Assessing Prior Knowledge

1. Draw any non-right triangle. Label the vertices A, B, and C. Use a, b, and c to label the sides, with side a opposite $\angle A$, side b opposite $\angle B$, and side c opposite $\angle C$. **[Check students' drawings.]**
2. List all possible combinations for two sides and the angle included between the sides. **[a, b, $\angle C$; a, c, $\angle B$; b, c, $\angle A$]**

TEACH

There are problem-solving situations in which the Law of Sines cannot be used because you are not given the appropriate information. Have students try to solve the problem using the Law of Sines. They will find that none of the three proportions can be used to find the needed facts.

LESSON 10.6 The Law of Cosines

Why *Will a parking attendant at Station 1 be able to talk directly to a parking attendant at Station 3? The Law of Cosines will enable you to determine the answer.*

At a football stadium there are three parking attendants directing traffic into the parking areas. When one area is full, the attendant at one station notifies the attendants at the other two stations. The parking attendants are equipped with walkie-talkies that have a range of 1,320 feet.

In the triangle formed by the three parking attendants, two sides and the included angle are known (SAS). But the Law of Sines will not work in this situation because any proportion in the Law of Sines would have two unknowns.

The Law of Cosines provides a way to find measures of sides and angles of triangles when two sides and an included angle or all three sides of a triangle are known. (A proof of the acute case of Law of Cosines is included at the end of the lesson.)

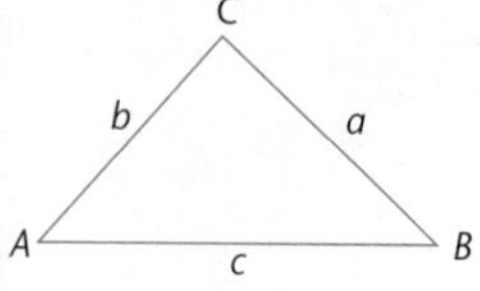

THE LAW OF COSINES

For any triangle ABC,

$$a^2 = b^2 + c^2 - 2bc \cos A$$
$$b^2 = a^2 + c^2 - 2ac \cos B$$
$$c^2 = a^2 + b^2 - 2ab \cos C \qquad \textbf{10.6.1}$$

The important thing to notice about the Law of Cosines is that the length of the side used on the left side of the equation is opposite the angle used in the equation.

Using Technology Students can use algebraic manipulation software such as *f(g) Scholar* to follow along with the examples and apply the Law of Cosines. Geometry graphics software that includes the trigonometric functions can be used to show that the Law of Cosines is true for any triangle.

EXAMPLE 1

Find b.

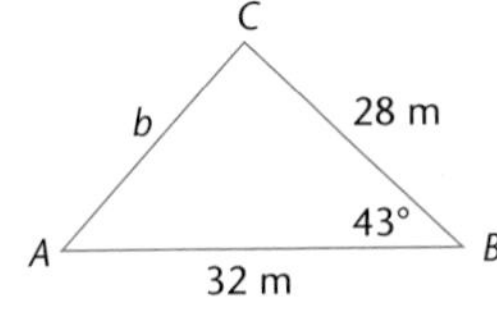

Solution ➤

Using the Law of Cosines,

$$b^2 = a^2 + c^2 - 2ac \cos B$$
$$b^2 = 28^2 + 32^2 - 2(28)(32) \cos 43°$$
$$b^2 \approx 1808 - 1310.586$$
$$b^2 \approx 497.414$$
$$b \approx 22.303$$

So b is 22.303 m long. ❖

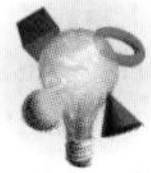

Would the Law of Sines have worked in Example 1? Explain your reasoning.

EXAMPLE 2

Find m∠D, m∠E, and m∠F.

Solution ➤

Part I

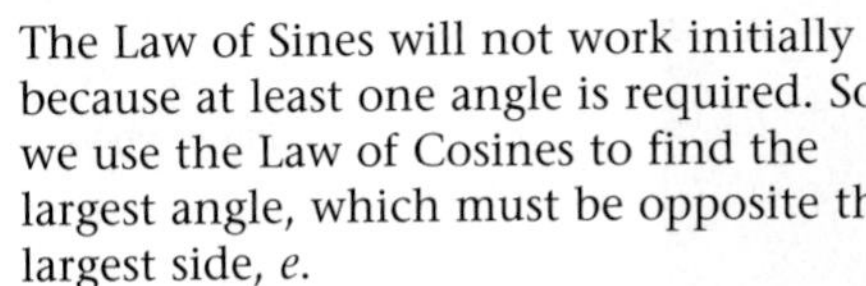

The Law of Sines will not work initially because at least one angle is required. So we use the Law of Cosines to find the largest angle, which must be opposite the largest side, e.

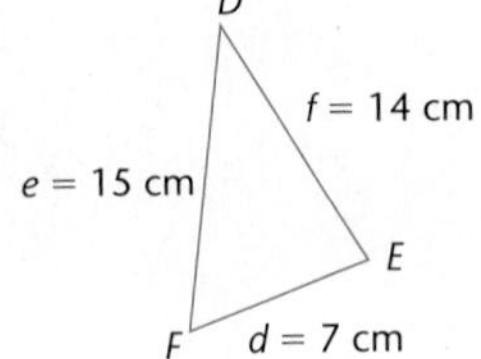

Thus, we will use $e^2 = d^2 + f^2 - 2df \cos E$.

$$15^2 = 7^2 + 14^2 - 2(7)(14) \cos E$$
$$225 = 49 + 196 - 196 \cos E$$
$$-20 = -196 \cos E$$
$$E = \cos^{-1}(0.10204) \approx 84°$$

Part II

Since we now know m∠E, we can use the Law of Sines to find m∠F.

$$\frac{\sin F}{14} = \frac{\sin 84°}{15}$$
$$\sin F = \frac{14 \sin 84°}{15}$$
$$\sin F \approx 0.9282$$
$$F = \sin^{-1}(0.9282) \approx 68°$$

Part III

$$m\angle D = 180 - (m\angle E + m\angle F) \approx 180 - (84 + 68) \approx 28°$$ ❖

interdisciplinary CONNECTION **Physics** When designing highways, railroads, or buildings, architects sometimes use formulas involving the trigonometric functions to calculate various stresses on foundations and other supporting structures. Invite an architect or engineer to visit the class the and show examples of formulas involving the sine, cosine, and tangent functions.

TEACHING tip

Point out that the Law of Cosines is actually three separate equations.

Alternate Example 1

a. Construct $\triangle XYZ$ with $\angle X = 50°$, $XY = 8$ cm, and $ZX = 14$ cm.

b. Relabel the sides so that you can use the Law of Cosines to find side $\overline{YZ}$. **[Possible labels: change $\overline{YZ}$ to x, $\overline{XY}$ to z, $\overline{XZ}$ to y. Then use the Law of Cosines in the form $x^2 = y^2 + z^2 - 2yz \cos X$.]**

c. Find $\overline{YZ}$. **[10.77 cm]**

Math Connection Algebra

In Example 1, emphasize the step in which students take the square root of both sides of the equation. Students often forget this step and in-advertently get b^2 rather than b.

CRITICAL Thinking

No. Any proportion in the Law of Sines would have contained two variables. For example, $\frac{\text{Sin } A}{28} = \frac{\text{Sin } 43°}{6}$.

Math Connection Algebra

In Example 2, you may wish to have students first solve the equation for cos E. This may prevent possible computational errors.

Alternate Example 2

a. Construct $\triangle PQR$ with $PQ = 1.25$ in., $QR = 2.75$ in., and $RP = 3$ in.

b. Change the variables in the Law of Cosines to show how you can find $\angle P$. **[Write PQ as r, QR as p, and RP as q. Then use the Law of Cosines in the form $\cos P = (q^2 + r^2 - p^2) \div 2qr$.]**

c. Find $\angle P$. **[66.4°]**

Cooperative Learning

After students have finished Example 2, have them work in groups on another application of the Law of Cosines. Using grid paper, students draw any triangle with vertices on the grid intersection points. First, they measure all sides and angles as carefully as they can. Then they use the Distance Formula to compute the sides and the Law of Cosines to compute the angles. Computed and measured quantities are compared.

TEACHING *tip*

Ask students what happens if the angle used in the Law of Cosines is 90°. They should see that the cosine 90° is 0 and that the formula reduces to the "Pythagorean" Right-Triangle Theorem. This makes sense because the triangle is a right triangle.

Summary

In general, the Law of Cosines is used in triangles when you know the measures of

1. two sides and the included angle of the triangle (SAS) or
2. all three sides of the triangle (SSS).

APPLICATION

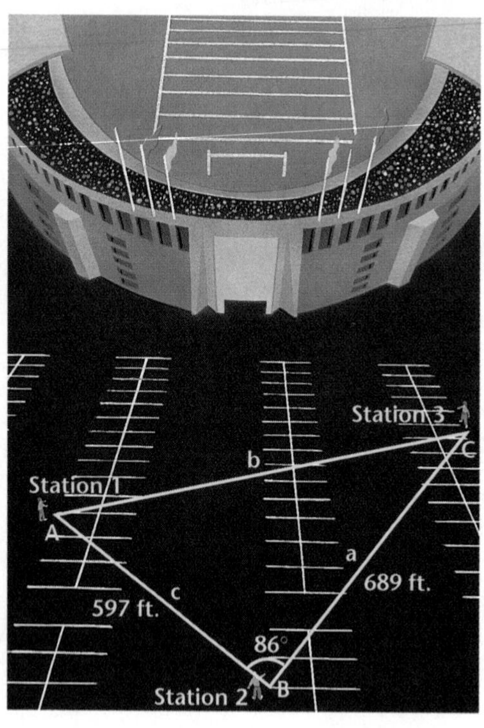

Let's take another look at the question posed at the beginning of the lesson. The parking attendant at Station 1 is 597 feet away from the attendant at Station 2. The parking attendant at Station 2 is 689 feet away from the attendant at Station 3. The angle between these segments is 86°, as shown. The parking attendants communicate with walkie-talkies that have a range of 1,320 feet. Will the parking attendant at Station 1 be able to talk directly to the parking attendant at Station 3?

Label the figure and then apply the Law of Cosines:

$$b^2 = a^2 + c^2 - 2ac \cos B$$
$$b^2 = 689^2 + 597^2 - 2(689)(597) \cos 86°$$
$$b^2 = 474{,}721 + 356{,}409 - 822{,}666 \cos 86°$$
$$b^2 = 831{,}130 - 822{,}666 \cos 86°$$
$$b^2 \approx 831{,}130 - 57{,}386.28$$
$$b^2 \approx 773{,}743.72$$
$$b \approx 880 \text{ ft}$$

Yes, the parking attendant will be able to communicate directly with the walkie-talkies. ❖

ENRICHMENT Students should use the figure below to prove the Law of Cosines using coordinates.

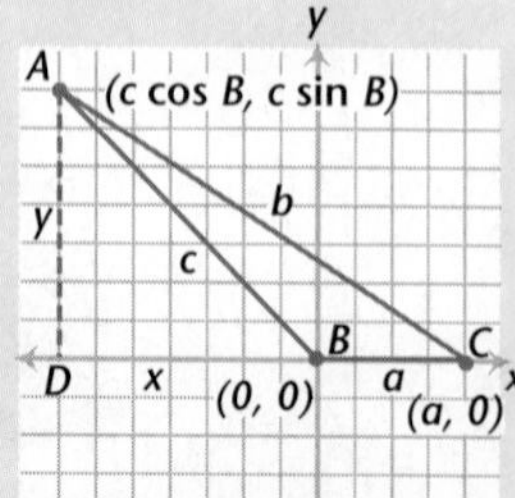

INCLUSION strategies **Using Models** One difficulty for students in using the Law of Cosines is keeping track of the sides and angles. Have students work in pairs and draw 10 different triangles. A different triple of capital letters should be used for each set of vertices. Sides are labeled with the appropriate lowercase letters. With a highlighter or colored pencil, students mark one side in each triangle. Then they write out the Law of Cosines with the variables needed to solve for that side.

Proof of the Law of Cosines: The Acute Case

To prove the Law of Cosines, first consider whether the triangle in the proof is acute or obtuse. The following proof is of the acute case. A proof of the obtuse case appears in the exercises.

Each angle of $\triangle ABC$ is an acute angle, and the measure of the altitude from C is h. Segment $\overline{AB}$ is divided into two segments, with length c_1 and c_2 as indicated in the figure. From the definition of cosine, you know that

$$\cos A = \frac{c_1}{b}.$$

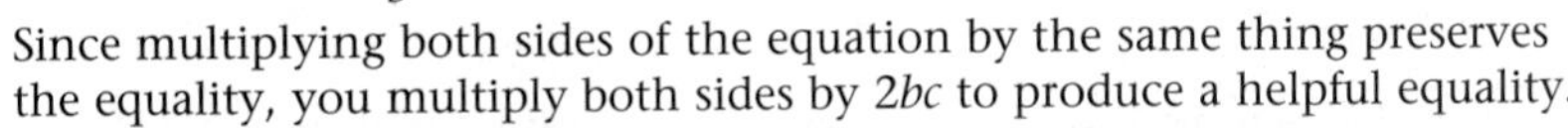

Since multiplying both sides of the equation by the same thing preserves the equality, you multiply both sides by $2bc$ to produce a helpful equality.

$$(2bc)\cos A = \frac{(2bc)c_1}{b}$$

$$2bc\cos A = 2cc_1$$

From the "Pythagorean" Right-Triangle Theorem,

$$a^2 = h^2 + c_2^2.$$

Since $c_2 = c - c_1$, you can substitute

$$a^2 = h^2 + (c - c_1)^2$$

$$a^2 = h^2 + (c_1^2 + c^2 - 2cc_1)$$

$$a^2 = (h^2 + c_1^2) + c^2 - 2cc_1.$$

Since $b^2 = h^2 + c_1^2$, you can substitute

$$a^2 = b^2 + c^2 - 2cc_1.$$

From above you know that $2cc_1 = 2bc\cos A$. Substituting produces

$$a^2 = b^2 + c^2 - 2bc\cos A,$$

which is the Law of Cosines.

Communicate

Refer to the general triangle shown in Exercises 1–4.

1. Summarize the Law of Cosines.
2. Explain why the Law of Sines will not work to solve a triangle when two sides and an included angle are known (SAS). Will it work when two angles and the side between them are known (ASA)?
3. What parts of a triangle need to be known in order for the Law of Cosines to be applied?
4. Explain how right-triangle trigonometry and the "Pythagorean" Right-Triangle Theorem are used in the proof of the Law of Cosines for acute triangles.

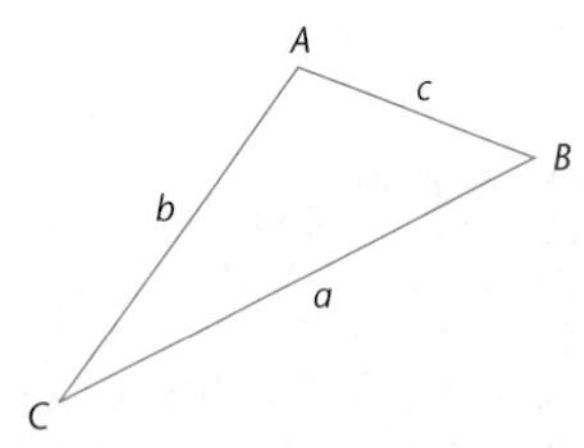

Math Connection Algebra

Have students rewrite the Proof of the Law of Cosines using a triangle with different letters for the vertices and sides. Watch to see if the subscripted variables are confusing students. If so, suggest that they use variables without subscripts.

ASSESS

Selected Answers

Exercises 5, 9, 13, 15, 17, 18, 19, 23, and 25

Assignment Guide

Core 1–3, 5–16, 20–25

Core Plus 1–26

Technology

As in the previous lesson, geometry graphics software can be used to check computed results. Students construct triangles with the given measures. Then they measure the needed sides or angles to check their computation.

Error Analysis

When using calculators, many students will make errors if they try to solve equations such as those in Examples 1 and 2 in just one step. Have them write out solutions in several steps. Then if they do not get the correct answer, they will be better able to find out where they entered a wrong number or operation.

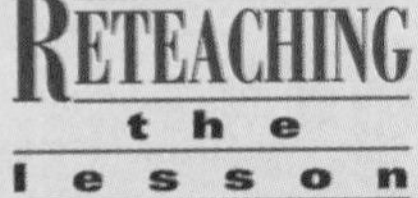

Cooperative Learning Have students work in pairs to find all three angles in a triangle with sides measuring 10, 16, and 20 units. They should use the Law of Cosines to find the largest angle first. This angle is opposite the 20-unit side and has a cosine of –0.1375, so it is obtuse. Make sure students know how to interpret the negative number they get for the cosine. A second angle can be found using the Law of Sines, the third can be found by subtracting from 180°. [**The three angles measure 97.9°, 52.4°, and 29.7°.**]

Alternative Assessment

Portfolio Assessment

Students should make a chart showing the four possible cases covered by the Law of Sines and the Law of Cosines. The cases may be abbreviated with the initials SAA, SSA, SSS, and SAS. For each case, students describe the given parts, draw a figure, and write the appropriate equation or equations used to solve the triangle.

Practice & Apply

Algebra For Exercises 5–9, sketch $\triangle ABC$ roughly to scale using the information given. Then find the indicated side or angle measure.

5. Find side b given $a = 12$, $c = 17$, and $B = 33°$. 9.53 units

6. Find the measures of angles $\angle C$ and $\angle A$ in Exercise 5. $m\angle A = 43.3°$; $m\angle C = 103.7°$

7. Find side c given $a = 2.2$, $b = 4.3$, and $m\angle C = 52°$. 3.42 units

8. Find the measures of angles $\angle A$ and $\angle B$ in Exercise 7. $m\angle A = 30.5°$; $m\angle B = 97.5°$

9. Find side a given $b = 68.2$, $c = 23.6$, and $m\angle A = 87°$. 70.99 units

10. Mark and Stephen walk into the woods along lines that form a 72° angle. If Mark walks at 2.8 miles per hour and Stephen walks at 4.2 miles per hour, how far apart will they be after 3 hours? 12.80 miles

11. On a standard baseball field, the bases form a square with sides 90 ft long. The pitcher's mound is 60.5 ft from home plate on a diagonal to second base. Find the distance from the pitcher's mound to first base. 63.72 ft

12. On a standard softball field, the bases form a square with sides 60 ft long, and the pitcher stands 40 ft from home plate. Find the distance from the pitcher to first base. 42.5 ft

13. The distance from Greenfield to Brownsville is 37 km, and the distance from Greenfield to Red River is 25 km. The angle between the road going from Greenfield to Brownsville and the road going from Greenfield to Red River is 42°. The state decides to build a road directly from Brownsville to Red River.

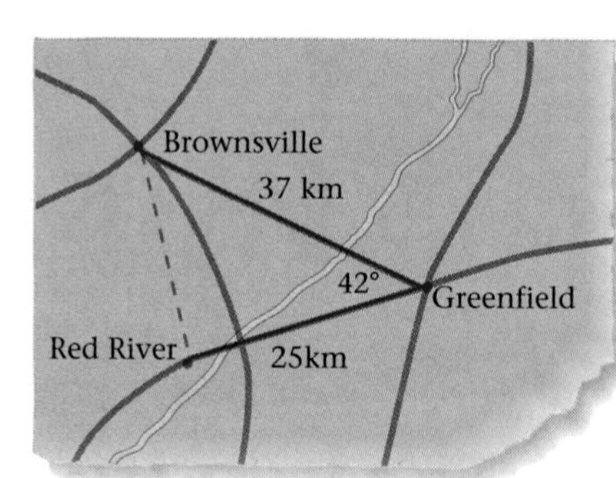

a. How long will the road be if it is straight? 24.88 km

b. At what angle to the road from Greenfield to Brownsville should the new road be built? 42.25°

14. Given parallelogram $ABCD$ as shown, use the Law of Cosines to find the measures of the interior angles.

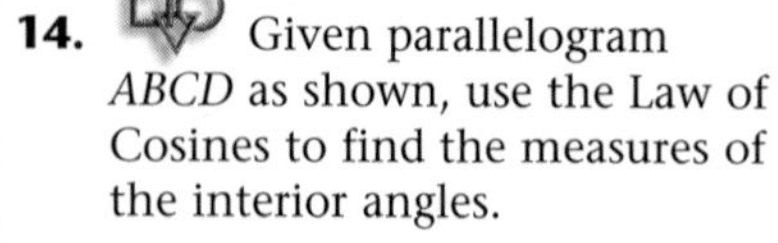

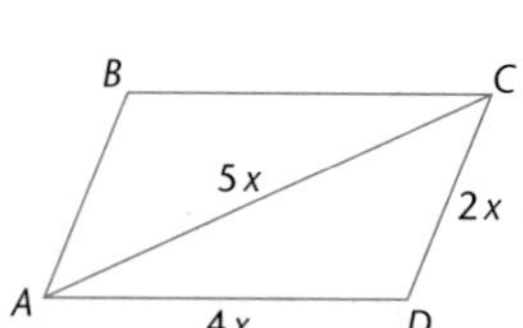

14. Two interior angles measure 108.21°and the other two measure 180° – 108.21° = 71.79°.

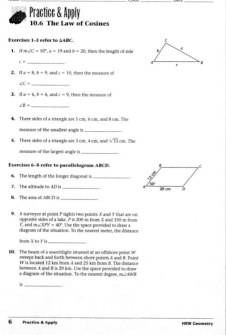

Practice Master

NAME ______ CLASS ______ DATE ______

Practice & Apply
10.6 The Law of Cosines

Exercises 1–3 refer to ΔABC.

1. If $m\angle C = 50°$, $a = 19$ and $b = 20$, then the length of side $c =$ ______.
2. If $a = 8$, $b = 9$, and $c = 10$, then the measure of $\angle C =$ ______.
3. If $a = 6$, $b = 6$, and $c = 9$, then the measure of $\angle B =$ ______.
4. Three sides of a triangle are 5 cm, 6 cm, and 8 cm. The measure of the smallest angle is ______.
5. Three sides of a triangle are 3 cm, 4 cm, and $\sqrt{33}$ cm. The measure of the largest angle is ______.

Exercises 6–8 refer to parallelogram $ABCD$.

6. The length of the longer diagonal is ______.
7. The altitude to AD is ______.
8. The area of $ABCD$ is ______.
9. A surveyor at point P sights two points X and Y that are on opposite sides of a lake. P is 200 m from X and 350 m from Y, and $m\angle XPY = 40°$. Use the space provided to draw a diagram of the situation. To the nearest meter, the distance from X to Y is ______.
10. The beam of a searchlight situated at an offshore point W sweeps back and forth between shore points A and B. Point W is located 12 km from A and 25 km from B. The distance between A and B is 29 km. Use the space provided to draw a diagram of the situation. To the nearest degree, $m\angle AWB$ is ______.

6 Practice & Apply HRW Geometry

15. How does this diagram differ from the diagram shown in the proof of the Law of Cosines given in the lesson?

16. Explain why the proof given in the lesson does not work for this case.

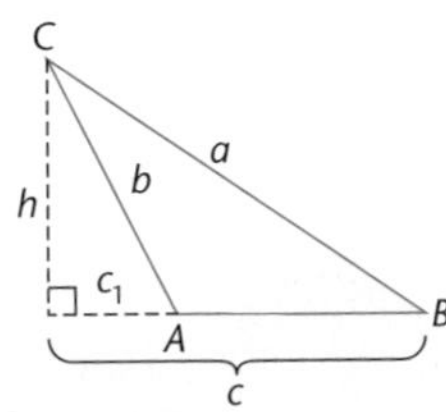

In Exercises 17–19, you will prove the obtuse case of the Law of Cosines. Use the diagram in Exercises 15 and 16.

17. Use the "Pythagorean" Right-Triangle Theorem to show that $a^2 = h^2 + (c + c_1)^2$. Then use algebra to show that $a^2 = h^2 + c^2 - 2cc_1 + c_1^2$.

18. By right-triangle trigonometry, show that $h = b \sin A$ and $c_1 = b \cos A$.

19. By substituting the results of Exercise 18 into the result of Exercise 17, prove that $a^2 = b^2 + c^2 - 2bc \cos A$ for the obtuse triangle ABC. (Hint: Use the identity $(\sin A)^2 + (\cos A)^2 = 1$.)

Look Back

For Exercises 20–22, draw the image of $\triangle ABC$ under the given mapping. [Lesson 8.1]

20. $(x, y) \rightarrow (3x, 3y)$

21. $(x, y) \rightarrow (x - 2, y - 2)$

22. $(x, y) \rightarrow \left(\frac{3x}{4}, \frac{3y}{4}\right)$

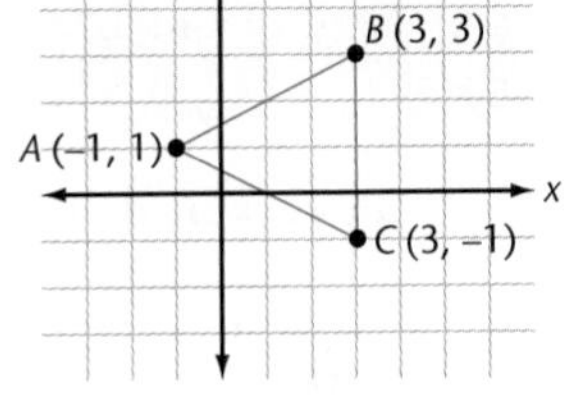

23. The perimeter of a rectangle is 80 cm. The ratio of the length to the width of the rectangle is 1 to 5. Find the length and width. **[Lesson 8.2]**
length = $6\frac{2}{3}$ cm; width = $33\frac{1}{3}$ cm

Determine if the triangles in each pair are similar. Explain why or why not. [Lesson 8.3]

24.

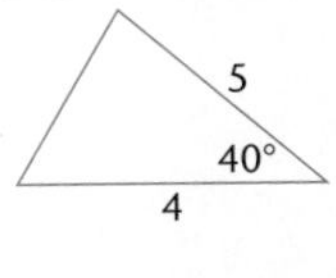

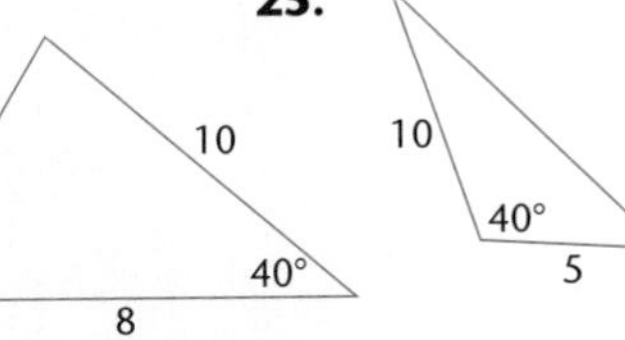

25.

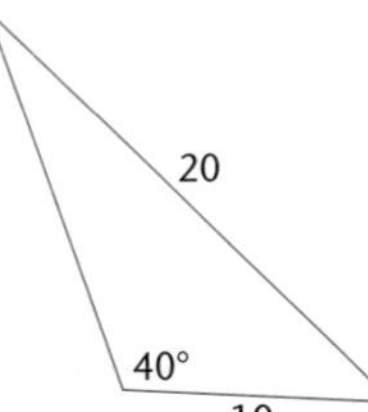

20
40°
10

Look Beyond

26. Can you draw this figure without lifting your pencil or retracing? Why or why not?

15. The triangle is obtuse.

16. Because $\angle CAB$ is not a right angle, $\cos \angle CAB \neq \frac{c_1}{b}$.

24. Yes, SAS-Similarity

25. No, the congruent angles are not between the corresponding sides in both triangles.

26. Sample answer:

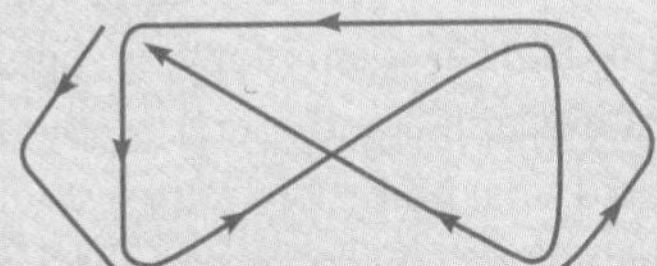

The answers to Exercises 17–22 can be found in Additional Answers beginning on page 727.

Look Beyond

This exercise foreshadows the investigation of networks in Lesson 11.6. It can be answered by showing a path that meets the question's requirements. Proving that a traversal is impossible for a given network is more difficult, and will be treated in the network lesson.

Technology Master

NAME ______ CLASS ______ DATE ______

Technology

10.6 Subdividing a Side of a Triangle

The diagram shows ΔABC whose sides have length $AB = 8.53$ units and $AC = 10.17$ units, and whose angle at A measures 30°. The diagram also shows a dashed line segment from B to $\overline{AC}$. The point of intersection is labeled Z. If you let $AZ = 10.17z$, where $0 \leq z \leq 1$, then you can study ΔABZ and find BZ by using the Law of Cosines and a graphics calculator or a spreadsheet.

According to the Law of Cosines, you can write the equation shown.

$BZ = \sqrt{(AB)^2 + (A}$

$= \sqrt{8.53^2 + (1}$

Use a spreadsheet in the following exercises.

1. Create a spreadsheet in which column A contains $z = 0.0, 0.1. 0.2, \ldots, 0.8, 0.9$, and 1.0 and column B contains the corresponding values of BZ.

From your spreadsheet, find each length BZ for the given values of z.

2. $z = 0.0$ ______ 3. $z = 0.3$ ______ 4. $z = 0.5$ ______ 5. $z = 1.0$ ______

6. What information does your answer in Exercise 4 give you about ΔABC?

7. Modify your spreadsheet from Exercise 1 to find the length of $\overline{BZ}$ when z is such that Z divides $\overline{AC}$ into quarters. Find the corresponding values of BZ.

8. Describe what happens if you let $z = 2$.

20 Technology HRW Geometry

PREPARE

Objectives

- Define *vector.*
- Add two vectors.
- Use vectors and vector addition to solve problems.

- Practice Master 10.7
- Enrichment Master 10.7
- Technology Master 10.7
- Lesson Activity Master 10.7
- Quiz 10.7
- Spanish Resources 10.7

Assessing Prior Knowledge

1. Draw an x-y coordinate system on grid paper. Choose any number a and draw a segment from (0, 0) to (a, 0). **[Check student drawings.]**
2. What is the length of this segment? **[Students use the "Pythagorean" Right-Triangle Theorem or the Distance Formula.]**

TEACH

Vectors are frequently found in problem-solving situations, particularly in the physical sciences and engineering. Any quantity that has both a magnitude and a direction can be represented by a vector. Examples include speed and force. Have students collect vector situations from science textbooks for discussion and display.

LESSON 10.7 Vectors in Geometry

A goalie attempting to kick the ball to a forward downfield must take the wind into account as she takes aim. A stadium flag gives her a visual indication of the wind's ***direction*** *and its* ***magnitude****, or strength. Because the wind has these two properties, it can be represented as a vector.*

A **vector** is a mathematical "object" that has both **magnitude** (a certain numerical measure) and **direction**. Arrows are used to represent vectors, because arrows have both magnitude and direction. The length of a vector arrow represents the magnitude of the vector.

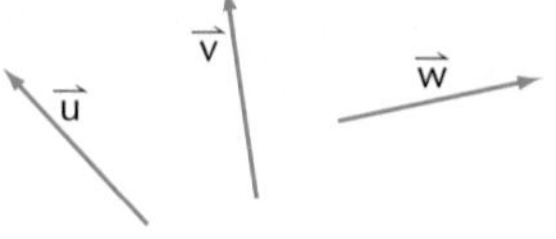

$\vec{u}$, $\vec{v}$, and $\vec{w}$ are vectors

Physics Anything that has both magnitude and direction can be modeled by a vector. In the pictorial examples on this and the following pages, vectors are used to represent a velocity, a force, and a displacement (a relocation). Describe the magnitude and direction of the vector in each.

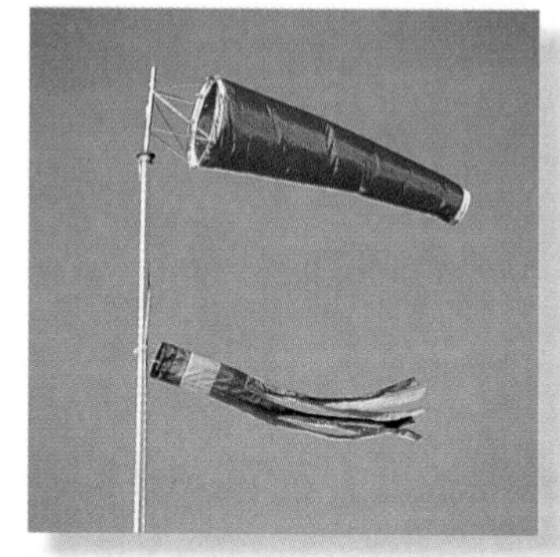

*(**Velocity**) The wind sock indicates the strength of the wind.*

ALTERNATIVE teaching strategy

Using Technology Geometry graphics software can be used to construct vectors and find vector sums. Have students copy the figures used in the text, examples, and exercises. By dragging on initial and final points of vectors, students can see what happens to the magnitude and direction as the vectors change.

(Force) *A tugboat applies a force of 2000 kg on a barge in the direction the barge is pointed.*

(Displacement) *The car speeds down the quarter-mile track.*

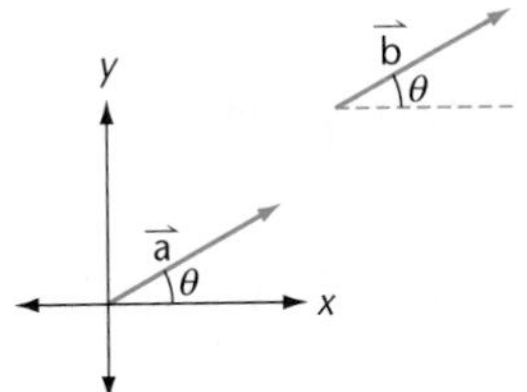

To describe the direction of a vector, you need a direction to use as a reference. On a coordinate grid this is usually the positive *x*-axis direction. (What are the reference directions in the pictorial examples above?)

Vectors with the same direction, such as vectors $\vec{a}$ and $\vec{b}$ in the illustration, are said to be **parallel**. Vectors at right angles are perpendicular.

Vector Sums

In many situations it is appropriate to combine vectors to get a new vector called the **resultant**. The process of combining vectors represented in the examples is called **vector addition**. The resultant is called the **vector sum**. For example,

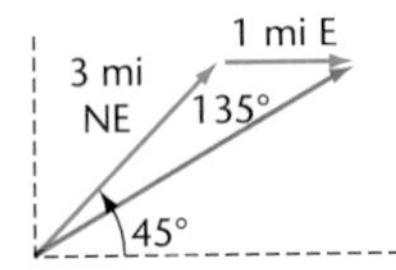

A hiker travels 3 miles northeast, then 1 mile east. The resultant vector (red) is the total distance traveled in a specific direction from the starting point.

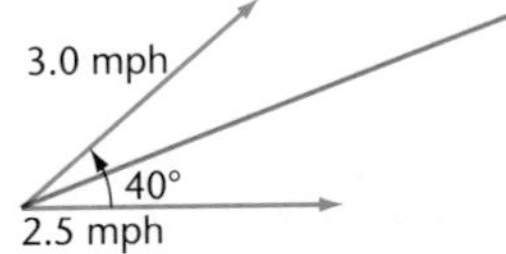

The vector of a current in a lake is 2.5 mph east. The vector of a swimmer is 3.0 mph (relative to the water) at an angle of 40° with the direction of the current. The actual velocity of the swimmer in the direction shown is the resultant vector.

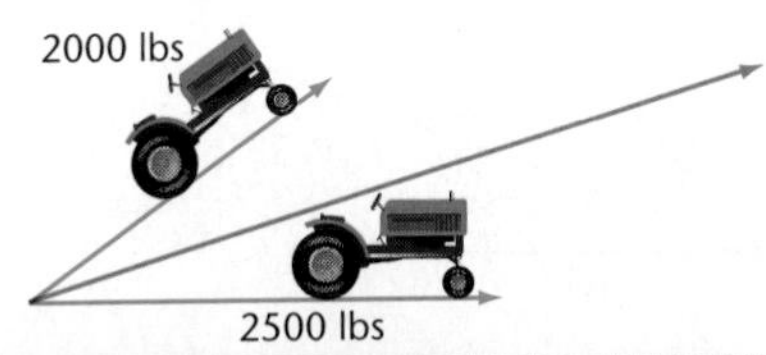

Two tractors pull on a tree stump with the forces and directions shown. The combined force exerted by the tractors in the direction shown is the resultant vector.

Ongoing Assessment

(p. 570)

Wind sock: magnitude = the velocity of the wind, direction = direction opposite the way the wind sock is pointed

Tugboat: magnitude = 2000 kg, direction = 0° with the direction of the barge

Car: magnitude = the velocity of the car, direction = direction of the track

(p. 571)

For the wind sock the reference direction is the direction that the end is pointed.

For the tugboat, the reference direction is the direction the barge is pointed.

For the car, the reference direction is the direction of the track.

Interdisciplinary Connection

Geography The north-south-east-west symbol used on maps is called a *compass rose.* Have students find examples of this symbol on road maps or maps in social studies books. Working in pairs, students should choose two places on a map for an imaginary trip. Then they should make a sketch that includes a vector to describe their trip. The direction of the vector can be described in terms such as "about 30° north of east."

Cooperative Learning

Students have already encountered vectors in their work on translations. Have students work in pairs. They each draw a triangle on an *x-y* coordinate system and a translation vector. They then translate the triangles using the information represented by the vector. Students write to explain how this use of vectors is similar to, or different from, the examples given in the lesson.

ongoing ASSESSMENT

Try This

The vector sum is 0. The sum has magnitude of the absolute value of the difference of the magnitudes of the forces and direction the direction of the greater force.

CRITICAL Thinking

Drawing in the opposite side of a parallelogram is the same as placing the second vector at the end of the first vector because the opposite sides of a parallelogram are congruent and parallel. The opposite side has the same magnitude and direction as the second vector.

TEACHING tip

Before students study the examples, ask them to suggest ways of adding vectors that do *not* involve constructing a diagram. Possible ideas might include using the "Pythagorean" Right-Triangle Theorem if the two vectors make right angles, and using trigonometry for vector pairs that include angles other than 90°.

The Head-to-Tail Method of Vector Addition

Physics In the hiker example, each vector represents a displacement—that is, a relocation of the hiker from a given point. In such cases, the **head-to-tail method** is a natural way to do vector addition.

To use the head-to-tail method of vector addition, place the *tail* of one vector at the *head* of the other. The vector sum is a new **displacement vector** from the tail of the first vector to the head of the second vector.

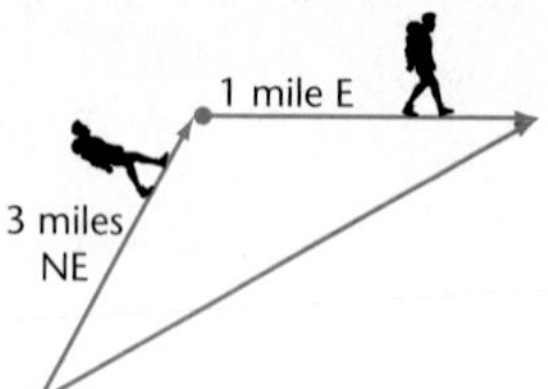

The Parallelogram Method of Vector Addition

Physics In many cases the **parallelogram method** of vector addition is a natural way to think of combining vectors. This is especially true when two forces act on the same point.

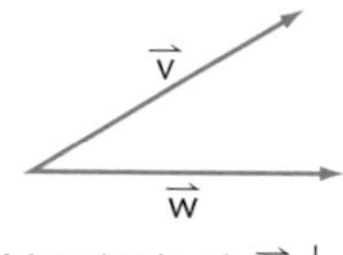

Magnitude of $\vec{v} = |v|$
Magnitude of $\vec{w} = |w|$

A lowercase letter with a vector arrow over it may be used to indicate a vector. The same symbol between double bars represents the magnitude of the vector.

To find the sum of vectors $\vec{a}$ and $\vec{b}$, complete a parallelogram by adding two segments to the figure. The vector sum is a vector along the diagonal of the parallelogram starting at the common point.

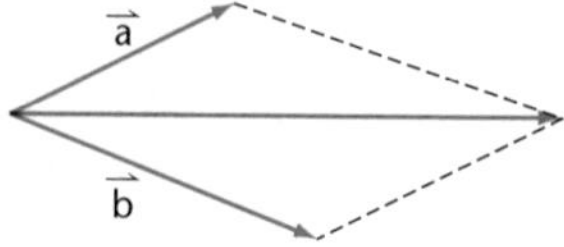

Try This What is the vector sum if two equal forces are applied to a point in opposite directions? What happens to the sum if two such forces are unequal?

CRITICAL Thinking

The parallelogram method and the head-to-tail method are equivalent since they produce the same resultant. How does the diagram show that the methods are equivalent?

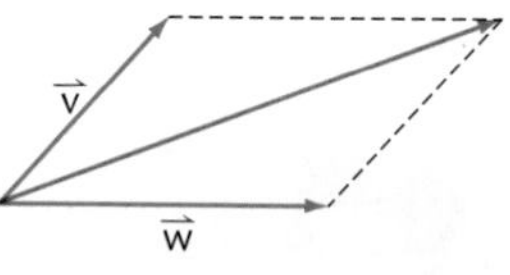

ENRICHMENT How steep does a plane have to be before an object resting on it will slide? In the diagram, *w* is the weight of an object, and *w* sin *β* is the force acting on the object in a direction parallel to the surface of the plane. If *w* sin *β* is greater than **f**, the frictional force that tends to hold keep the object from sliding, then the object will move. A related idea is the **angle of repose** for a material. This is the greatest base angle a conical heap of the material can have.

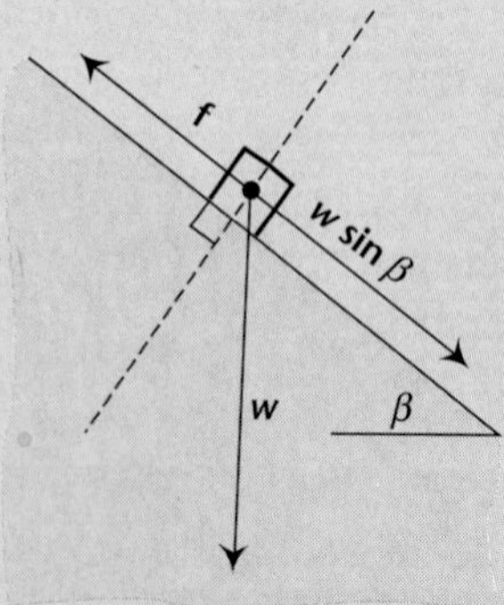

EXAMPLE 1

A swimmer swims perpendicular to a 3-mph current in a lake. Her speed in still water is 2.5 mph. Find the actual speed and actual direction of the swimmer.

Solution ➤

The parallelogram method for adding the two vectors gives a rectangle in this case. Thus, you can solve for x using the "Pythagorean" Right-Triangle Theorem.

$$x^2 = 3^2 + 2.5^2$$

$$x^2 = 15.25$$

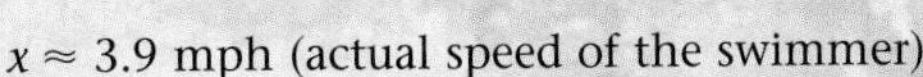

$$x \approx 3.9 \text{ mph (actual speed of the swimmer)}$$

There is more than one way to find θ. For example,

$$\tan \theta = \frac{3}{2.5} = 1.2$$

$\theta = \tan^{-1}(1.2) \approx 50°$. The swimmer actually swims at an angle 50° to the current. ❖

EXAMPLE 2

The swimmer in Example 1 changes direction so that she is swimming against the current at an angle of 40° with the direction of the current. What is her actual speed and direction?

Solution ➤

One simple and often quite effective way to solve such problems is to draw an accurate vector diagram and measure the resultant vector.

If you wish to find x mathematically, use the law of cosines.

$$x^2 = 3^2 + 2.5^2 - 2(3)(2.5)\cos 40°$$

$$x^2 \approx 3.759$$

$$x \approx 1.94 \text{ mph}$$

Use the Law of Sines to find θ.

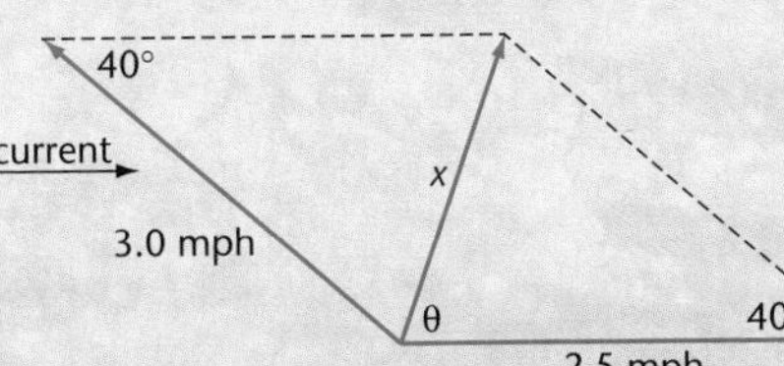

$$\frac{\sin \theta}{3.0} = \frac{\sin 40°}{1.94}$$

$$\sin \theta = \frac{3.0 \sin 40°}{1.94} \approx 0.9940$$

$\theta \approx \sin^{-1}(0.9940) \approx 84°$ with the current. ❖

INCLUSION strategies

Using Models Students draw a north-south-east-west symbol and a vector that begins at the intersection of the two lines. Have them use two different angles to describe the direction of the vector, such as 30° north of east or 60° east of north. This activity will help to illustrate how the reference angle for a vector can vary in its placement.

Alternate Example 1

The current in a lake is moving directly east. A boat heads directly north, but is pushed off course by the current so that it is moving on a line 40° north of east at an actual speed of 2.5 mph. How fast is the current moving? [**2.5 cos 40° ≈ 1.9 mph**]

Math Connection Algebra

Have students point out the three segments used for the "Pythagorean" Right-Triangle Theorem. You may wish to have them label the right-hand dotted segment as 3.0 to make the triangle more obvious.

Alternate Example 2

Two ropes are attached to a heavy object. Two people pull on the ropes exerting forces of 15 and 22 newtons. The angle between the ropes is 100°. Find the magnitude of the force exerted on the object. [**A parallelogram diagram shows the acute angles of the parallelogram to be 80°. Substituting in the Law of Cosines gives $15^2 + 22^2 - 2(15)(22)\cos 80° =$ 24.4 newtons.**]

Math Connection Algebra

Remind students that they are solving for x and not x^2. So they need to take the square root of both sides as the last step in the solution.

ASSESS

Selected Answers

Odd-numbered Exercises 15, 17, 19, 21, 23, 25, and 27

Assignment Guide

Core 1–4, 6–28

Core Plus 1–31

Technology

Exercises 14-19 are particularly appropriate for use with geometry graphics software. Check that your software includes commands for constructing and adding vectors.

Error Analysis

For Exercises 6-13, students will need tracing or grid paper to copy the vectors exactly.

The answers to Exercises 6–16 and 19 can be found in Additional Answers beginning on page 727.

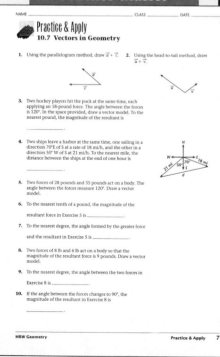

Practice Master

NAME ______________ CLASS ________ DATE ________

Practice & Apply

10.7 Vectors in Geometry

1. Using the parallelogram method, draw $\vec{a} + \vec{c}$.
2. Using the head-to-tail method, draw $\vec{u} + \vec{v}$.

$\vec{a}$ $\vec{c}$ $\vec{u}$ $\vec{v}$

3. Two hockey players hit the puck at the same time, each applying an 18-pound force. The angle between the forces is 120°. In the space provided, draw a vector model. To the nearest pound, the magnitude of the resultant is ________.
4. Two ships leave a harbor at the same time, one sailing in a direction 70°E of S at a rate of 18 mi/h, and the other in a direction 50° W of S at 21 mi/h. To the nearest mile, the distance between the ships at the end of one hour is ________.

N W E S 21 mi 18 mi 50° 70° d

5. Two forces of 28 pounds and 35 pounds act on a body. The angle between the forces measure 120°. Draw a vector model.
6. To the nearest tenth of a pound, the magnitude of the resultant force in Exercise 5 is ________.
7. To the nearest degree, the angle formed by the greater force and the resultant in Exercise 5 is ________.
8. Two forces of 8 lb and 6 lb act on a body so that the magnitude of the resultant force is 9 pounds. Draw a vector model.
9. To the nearest degree, the angle between the two forces in Exercise 8 is ________.
10. If the angle between the forces changes to 90°, the magnitude of the resultant in Exercise 8 is ________.

HRW Geometry Practice & Apply 7

EXERCISES & PROBLEMS

Communicate

1. What is the "magnitude" of a vector. Explain.
2. What is the "direction" of a vector. Explain.

Describe the magnitude and direction of the vector or vectors in each of the following.

3. an airplane flying northwest at 175 knots
4. a boat going 15 knots upstream against a current of 3 knots
5. two equal tug-of-war teams pulling on opposite ends of a rope

Practice & Apply

Copy these vectors and draw $\vec{a} + \vec{b}$ using the parallelogram method. You may have to translate one of the vectors.

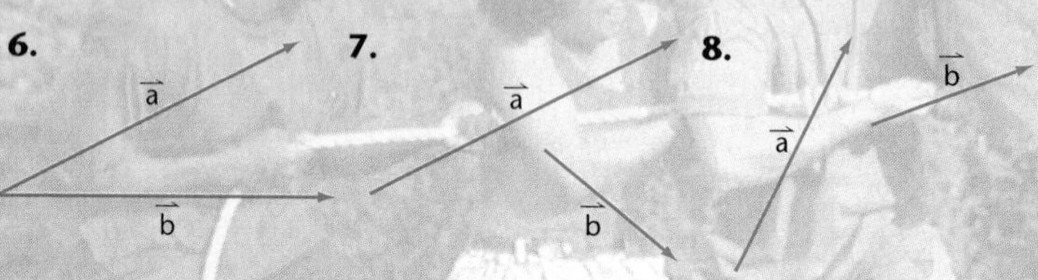

Copy these vectors and draw $\vec{a} + \vec{b}$ using the head-to-tail method. You may need to translate one of the vectors.

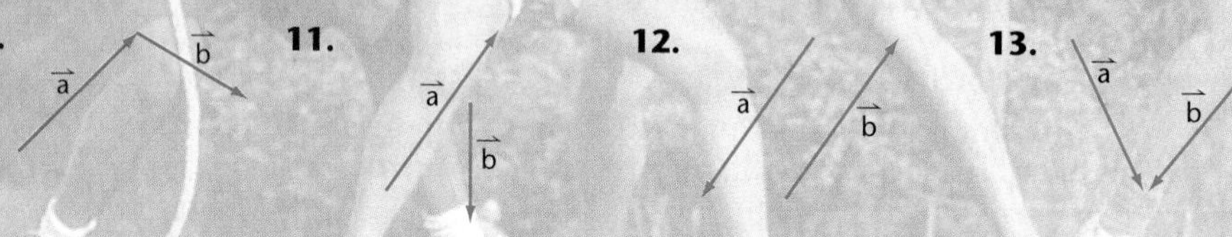

Use the diagram below for Exercises 14–18.

$\vec{a}$ 18 $\vec{b}$ 12 60° 25°

14. Draw a parallelogram using $\vec{a}$ and $\vec{b}$.
15. Find the angles of the parallelogram.
16. Draw vector $\vec{c}$, the resultant of $\vec{a}$ and $\vec{b}$.
17. Find the magnitude of $\vec{c}$. 20.74 units
18. Find the angle that $\vec{c}$ makes with $\vec{b}$. 59.83
19. On a coordinate grid, vector $\vec{a}$ has an initial point of (3, 14) and a terminal point (where the arrow goes) of (8, 6). Vector $\vec{b}$ has an initial point of (−2, 3) and a terminal point of (3, −5). Are the magnitude of $\vec{a}$ and $\vec{b}$ equal? Explain your reasoning.

Using Models To help students understand why the study of vectors relates to trigonometry, explain that it is often convenient to resolve a vector into horizontal and vertical components. For example, the vector F represents the force exerted as the person pulls on the wagon. To resolve vector F into the horizontal and vertical vectors F_x and F_y, students use these relationships:

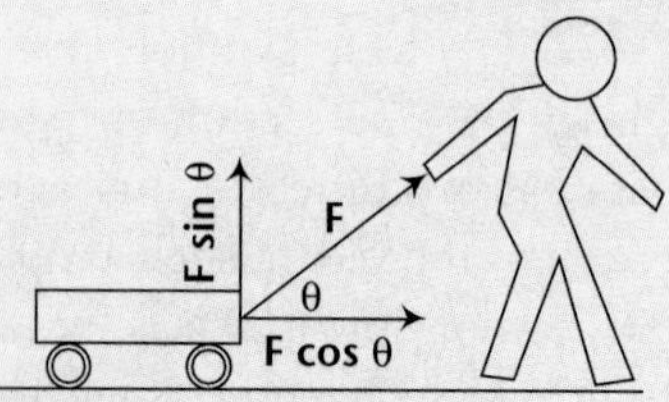

Have students work in pairs to draw five different vectors and resolve them into horizontal and vertical components.

For Exercises 20–22, draw a vector model and solve.

20. In football, receivers run pass routes that can be thought of as vectors. Ahmed's pass route calls for him to start behind the line of scrimmage and run on a line parallel to the line of scrimmage at 3.5 meters per second for 4 seconds. When the ball is snapped, Ahmed must spin around sprint downfield on a line that forms a 60° angle with his original line at 8.2 meters per second for 1.5 seconds. How far from his original position will Ahmed be?

21. Dan is a Navy SEAL investigating a boat sunk at the mouth of a river. To reach the wreck, Dan must swim against the current of 2.7 miles per hour. Suppose Dan dives and starts swimming at 4.1 miles per hour (still water speed) at a 15° angle with the water's surface. Find the speed Dan will swim as a result of the current.

22. A fishing boat leaves port and sails 5.6 miles per hour for 3.5 hours. The boat then turns at a 57° angle and sails at 4.9 miles per hour for 4.25 hours.

  a. If the boat sails for port at 6 miles per hour, how long will it take the boat to reach port?

  b. At what angle will the boat have to turn to head directly back to port?

Look Back

Use the Law of Sines and $\triangle ABC$ for Exercises 23–25. [Lesson 10.5]

23. Find side c given $m\angle B = 27°$, $m\angle A = 40°$, and $b = 142$. 287.92 units

24. Find side b given $m\angle B = 79°$, $m\angle C = 81°$, and $a = 3.14$. 9.01 units

25. Find side c given $a = 7.4$, $b = 9.5$, and $m\angle B = 79°$. 7.53 units

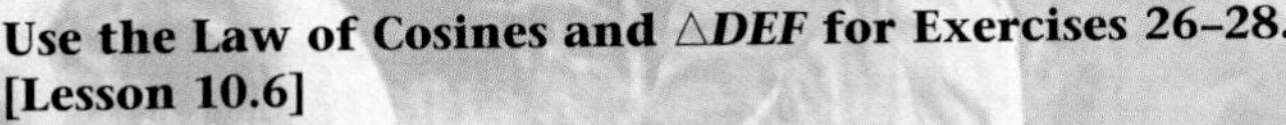

Use the Law of Cosines and $\triangle DEF$ for Exercises 26–28. [Lesson 10.6]

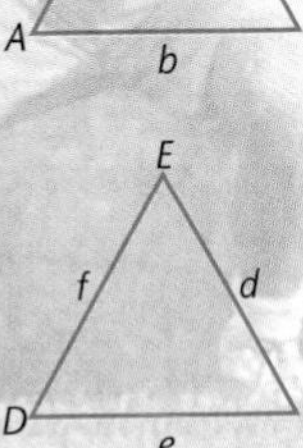

26. Find side d given $e = 13$, $f = 19$, and $m\angle D = 40°$. 12.31 units

27. Find side f given $d = 3.2$, $e = 4.5$, and $m\angle F = 55°$. 3.74 units

28. Find the measures of angles $\angle D$ and $\angle E$ in Exercise 27.
$m\angle D = 44.5°$ $m\angle E = 80.5°$

Look Beyond

Describe the conditions necessary to make each statement true.

29. It is a sunny day, and my car has a flat tire. Both parts must be true.

30. It is a sunny day, or my car has a flat tire. At least one of the parts must be true.

31. Today is Tuesday and the lawn needs mowing, or cats have kittens and dogs have fleas.

ALTERNATIVE ASSESSMENT

Performance Assessment

Students should work in small groups. On grid paper, have them draw a "lake" and a selection of vectors. Each vector represents a heading in still water, either for a swimmer or for a boat. Students should then imagine a river entering the lake and creating a current. Have them show this current with a vector and then find the actual speed and heading for each swimmer and boat.

Look Beyond

Exercises 29–31 look ahead to lessons about logic in Chapter 12. You may want to have students use truth tables to determine the answers.

Technology Master

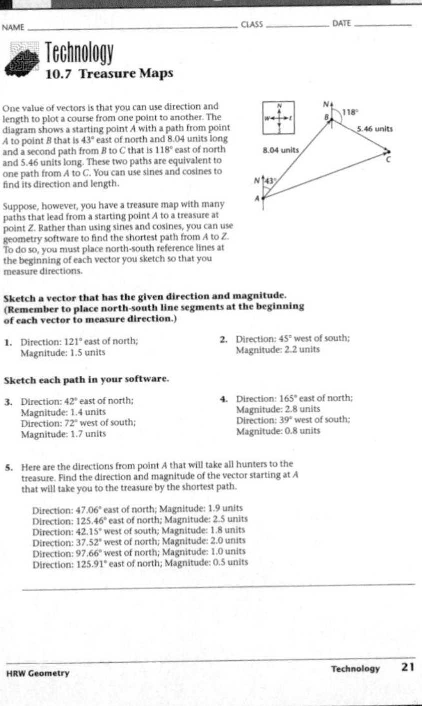

NAME ____________ CLASS ________ DATE ________

Technology
10.7 Treasure Maps

One value of vectors is that you can use direction and length to plot a course from one point to another. The diagram shows a starting point A with a path from point A to point B that is 43° east of north and 8.04 units long and a second path from B to C that is 118° east of north and 5.46 units long. These two paths are equivalent to one path from A to C. You can use sines and cosines to find its direction and length.

Suppose, however, you have a treasure map with many paths that lead from a starting point A to a treasure at point Z. Rather than using sines and cosines, you can use geometry software to find the shortest path from A to Z. To do so, you must place north-south reference lines at the beginning of each vector you sketch so that you measure directions.

Sketch a vector that has the given direction and magnitude. (Remember to place north-south line segments at the beginning of each vector to measure direction.)

1. Direction: 121° east of north; Magnitude: 1.5 units
2. Direction: 45° west of south; Magnitude: 2.2 units

Sketch each path in your software.

3. Direction: 42° east of north; Magnitude: 1.4 units; Direction: 72° west of south; Magnitude: 1.7 units
4. Direction: 165° east of north; Magnitude: 2.8 units; Direction: 39° west of south; Magnitude: 0.8 units
5. Here are the directions from point A that will take all hunters to the treasure. Find the direction and magnitude of the vector starting at A that will take you to the treasure by the shortest path.

Direction: 47.06° east of north; Magnitude: 1.9 units
Direction: 125.46° east of north; Magnitude: 2.5 units
Direction: 42.15° west of south; Magnitude: 1.8 units
Direction: 37.52° west of north; Magnitude: 2.0 units
Direction: 97.66° west of north; Magnitude: 1.0 units
Direction: 125.91° east of north; Magnitude: 0.5 units

HRW Geometry — Technology 21

20. 13.23 m

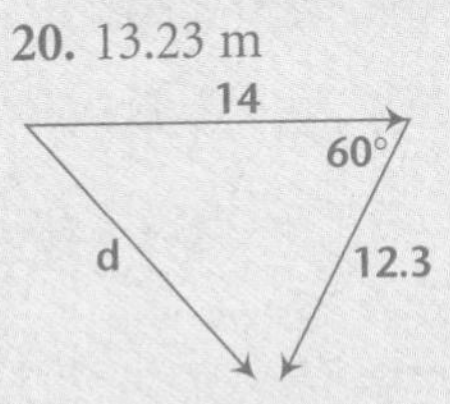

21. 1.65 mph

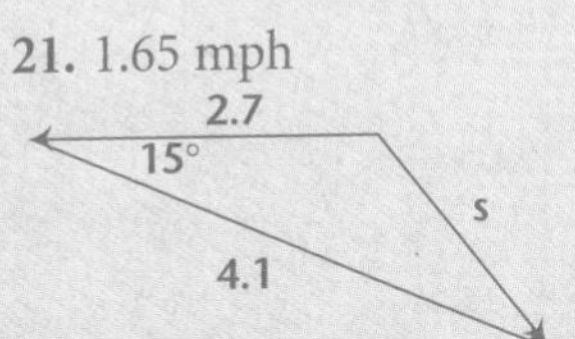

22. a. 3.22 hours b. 58.3°

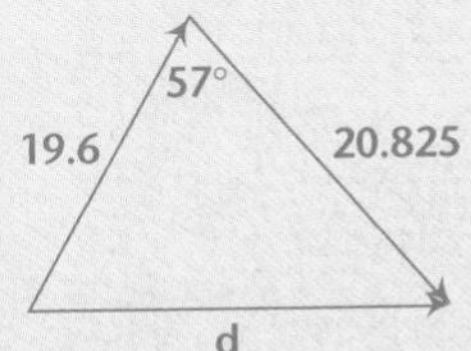

31. At least one of the following must be true: Both today is Tuesday and my lawn needs mowing must be true *or* both cats have kittens and dogs have fleas must be true.

Focus

The clay tablet from ancient Babylon known as Plimpton 322 is the focus of study in this project. The numbers on the tablet were identified in Chapter 5 as being Pythagorean triples. A close study of the arrangement of the numbers reveals an organizing scheme for the tablet.

Motivate

Students should use a scientific calculator for the calculations on the tablet. As the pattern emerges, ask students make conjectures about the pattern.

The values of θ, in degrees, are 44.8, 44.3, 43.8, 43.3, 42.1, 41.5, 40.3, 39.8, 38.7, 37.4, 36.9, 35.0, 33.9, 33.3, 31.9. The average spacing is about 1 degree, which suggests the Babylonian division of the circle into 360 parts.

PLIMPTON 322 REVISITED

The arrangement of the numbers in the tablet known as Plimpton 322 (see Lesson 5.4) may seem to be quite haphazard. Large numbers are mixed with smaller ones. But there is an underlying method to the arrangement.

Cuneiform Tablet, ca. 1900-1600 B.C

"Pythagorean" Triples

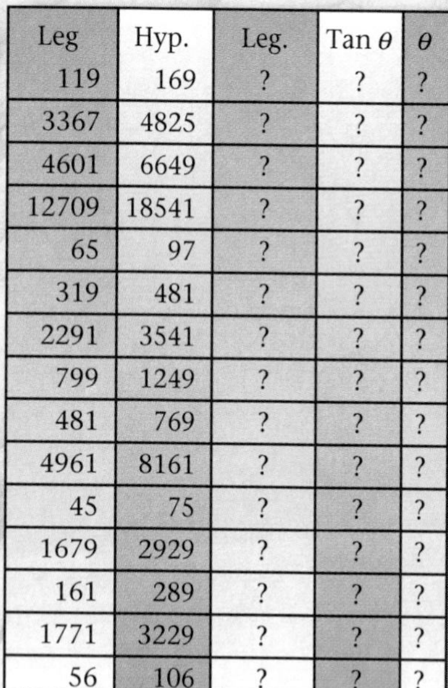

Leg	Hyp.	Leg.	Tan θ	θ
119	169	?	?	?
3367	4825	?	?	?
4601	6649	?	?	?
12709	18541	?	?	?
65	97	?	?	?
319	481	?	?	?
2291	3541	?	?	?
799	1249	?	?	?
481	769	?	?	?
4961	8161	?	?	?
45	75	?	?	?
1679	2929	?	?	?
161	289	?	?	?
1771	3229	?	?	?
56	106	?	?	?

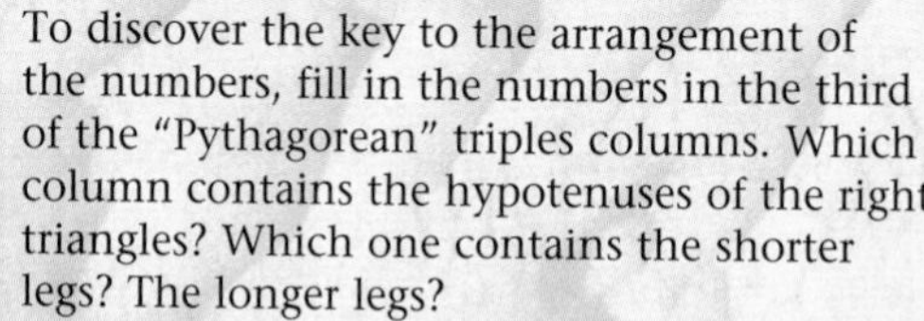

To discover the key to the arrangement of the numbers, fill in the numbers in the third of the "Pythagorean" triples columns. Which column contains the hypotenuses of the right triangles? Which one contains the shorter legs? The longer legs?

Divide the length of the shorter leg by the length of the longer leg in each of the triangles and place these numbers in the column with the heading Tan θ. Then find the values of θ and add them to your table.

What do you notice about the arrangement of the values of the angles. What does the average difference between the sizes of the angles seem to be?

The Babylonian number system seems to have originally been a base 10 system like our own, as the **cuneiform** ("wedge-shaped") numbers below illustrate.

1, 2, 3, 4, 5, 6, 7, 8, 9, 10, 11, 12, 13, 14, 15, 16, 17, 18, 19, 20, 30, 40, 50, 60, 70, 80, 90, 100, 200, 1000, 2000

The numbers on Plimpton 322 are in base 60. The first three entries in Column II are:

1, 59 56, 7 1, 16, 41

To understand the base 60 number system, compare a base 10 number with a base 60 number.

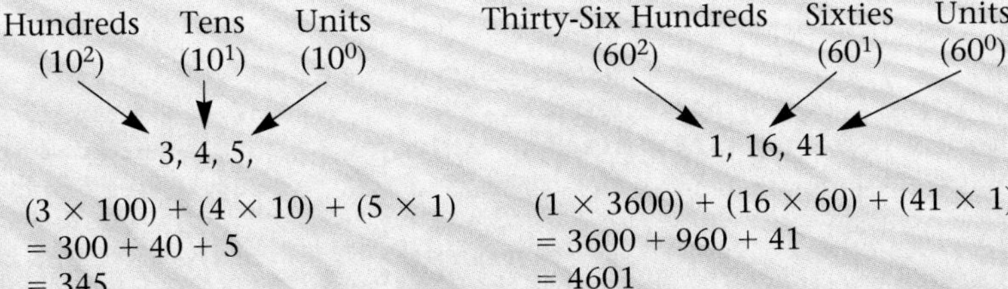

$(3 \times 100) + (4 \times 10) + (5 \times 1)$
$= 300 + 40 + 5$
$= 345$

$(1 \times 3600) + (16 \times 60) + (41 \times 1)$
$= 3600 + 960 + 41$
$= 4601$

The transcription of the cuneiform numbers on Plimpton 322 is shown. (Some of the damaged numbers have been supplied by researchers.) Compare the numbers in the transcription with the table on the previous page. Four of the entries in columns II and III are errors. Can you find them?

I	II	III	IV
1,59,15	1,59	2,49	ki-1
1,56,56,58,14,50,6,15	56,7	3,12,1	ki-2
1,55,7,41,15,33,45	1,16,41	1,50,49	ki-3
1,5,3,10,29,32,52,16	3,31,49	5,9,1	ki-4
1,48,54,1,40	1,5	1,37	ki-5
1,47,6,41,40	5,19	8,1	ki-6
1,43,11,56,28,26,40	38,11	59,1	ki-7
1,41,33,59,3,45	13,19	20,49	ki-8
1,38,33,36,36	9,1	12,49	ki-9
1,35,10,2,28,27,24,26,40	1,22,41	2,16,1	ki-10
1,33,45	45	1,15	ki-11
1,29,21,54,2,15	27,59	48,49	ki-12
1,27,3,45	7,12,1	4,49	ki-13
1,25,48,51,35,6,40	29,31	53,49	ki-14
1,23,13,46,40	56	53	ki-15

The numbers in the first column are fractional values. Notice for example, the number 1,38,33,36,36 in line 9. The decimal value of this number can be found as shown.

$$1 + \frac{38}{60} + \frac{33}{3600} + \frac{36}{216{,}000} + \frac{36}{12{,}960{,}000} = 1.6266944444\ldots$$

Compare this value with the following result:

$$\left(\frac{769}{600}\right)^2 = 1.64266944444\ldots$$

What does the first column in the Babylonian table seem to represent? Make your own conversions like the one above to test your idea. Do the numbers exactly match your conversions? Or are there errors in the table? If so, which entries are in error?

Cooperative Learning

Students working in groups can divide tasks. One student can complete the calculations while another records the results. Students can help each other when converting base 60 numbers into base 10. One student can act as a reporter for the group to share results

Discuss

Ask students to speculate about why the arrangement of numbers was useful for the Babylonians.

The errors are Row 2, column III; Row 9, column II; Row 13, column II; Row 15, column III.

The values in column I, in modern terms, are the squares of the secants of θ, that is, $\left(\frac{\text{hyp}}{\text{adj}}\right)^2$. Errors are found in Rows 1, 4, 8, and 13. All the rest of the entries agree with calculator results in all 12 digits displayed. If students have access to other mathematical software they may want to carry out the calculations further—to 1000 digits or even more. A decimal conversion of the Row 10, column 1 entry, for example, reveals a decimal value with a cycle of 729 digits that *exactly* agrees with the decimal value of the squared fraciton. (Students might well be impressed by the compactness of the Babylonian base-60 system for representing high-precision values.)

1. $\frac{12}{5} \approx 2.4$

2. $\frac{5}{13} \approx 0.385$

3. $\frac{5}{13} \approx 0.385$

4. $\angle F \approx 22.6°$

Chapter 10 Review

Vocabulary

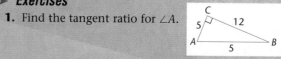

cosine	533	magnitude	570	tangent ratio	525
cotangent	527	parallelogram method	572	unit circle	542
direction	570	resultant	571	vector	570
displacement vector	572	rotation matrix	550	vector addition	571
head-to-tail method	572	sine	533	vector sum	571

Key Skills and Exercises

Lesson 10.1

➤ Key Skills

Find the tangent ratio of a triangle.

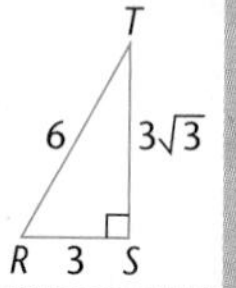

In $\triangle RST$ the tangent ratio of $\angle T$ is the length of the opposite side, RS, divided by the length of the adjacent side, TS: $\frac{3}{(3\sqrt{3})} = \frac{1}{\sqrt{3}} \approx 0.577$.

➤ Exercises

1. Find the tangent ratio for $\angle A$.

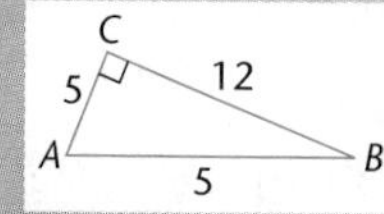

Lesson 10.2

➤ Key Skills

Find the sine and cosine of a triangle.

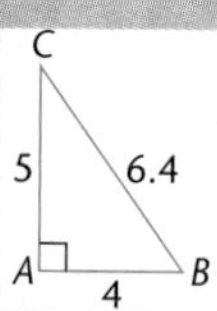

In $\triangle ABC$ the sine of $\angle B$ is the length of the opposite side, CA, divided by the length of the hypotenuse, CB: $\frac{5}{6.4} \approx 0.781$. The cosine of $\angle B$ is the length of the adjacent side, BA, divided by the length of the hypotenuse, CB: $\frac{4}{6.4} \approx 0.625$.

Find the measure of an angle.

In $\triangle ABC$ above, find $m\angle B$ from either the sine, 0.781, or the cosine, 0.625: $m\angle B \approx 51.3°$.

➤ Exercises

In Exercises 2–4, refer to the figure at the right.

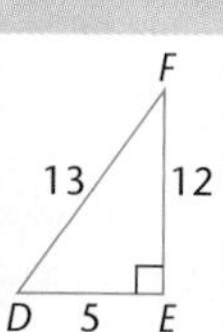

2. Find $\cos \angle D$.
3. Find $\sin \angle F$.
4. Find $m\angle F$.

Lesson 10.3

➤ Key Skills

P′ is on the 60° rotation image of P(1, 0) about the origin. Find the coordinates of P′.

The coordinates of P′ are (0.5, 0.866).

➤ Exercises

In Exercises 5–7, find the ratio.

5. P′ is the 420° rotation image of P(1, 0) about the origin. Find the coordinates of P′.
6. Estimate the ratio cos 390° to the nearest hundredth.
7. Estimate the ratio sin 235° to the nearest hundredth.

Lesson 10.4

➤ Key Skills

Name the image point of a rotation.

Find the vertices of the image of line $\overleftrightarrow{FG}$ rotated 25° clockwise about the origin.

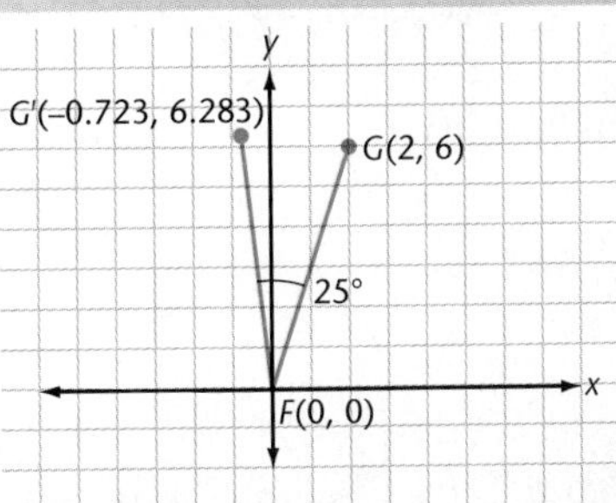

Create a 25° rotation matrix.

$$\begin{bmatrix} \cos 25° & -\sin 25° \\ \sin 25° & \cos 25° \end{bmatrix} \approx \begin{bmatrix} .906 & -.423 \\ .423 & .906 \end{bmatrix}$$

Multiply this matrix by the line matrix.

$$\begin{bmatrix} .906 & -.423 \\ .423 & .906 \end{bmatrix} \times \begin{bmatrix} 0 & 2 \\ 0 & 6 \end{bmatrix} = \begin{bmatrix} 0 & -0.723 \\ 0 & 6.283 \end{bmatrix}$$

Point F is at (0, 0), and point G' is at (−0.723, 6.283).

➤ Exercises

In Exercises 8–9, name the image point of the rotation.

8. Point (5, 8) rotated 270°
9. Point (−2, 3) rotated 34°

Lesson 10.5

➤ Key Skills

Apply the Law of Sines to find measures in a triangle.

$$\frac{\sin A}{a} = \frac{\sin B}{b} = \frac{\sin C}{c}$$

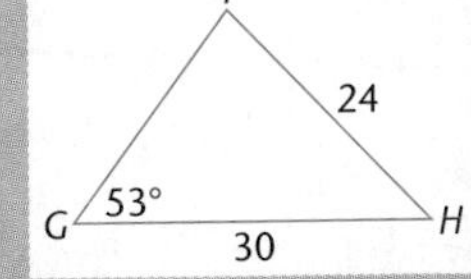

Find m∠I.

Set up a proportion with the sine of an angle and the length of the side opposite: $\frac{\sin I}{30} = \frac{\sin 53°}{24}$.

$$\sin I = 30\left(\frac{\sin 53°}{24}\right) = .998$$

We find m∠$I \approx 86.37°$.

➤ Exercises

In Exercises 10–11, refer to the figure.

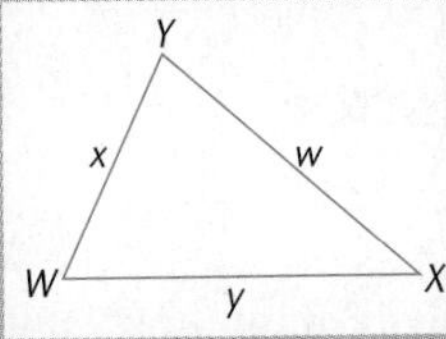

10. Find side y when ∠X = 64°, ∠W = 41°, and w = 100.
11. Find m∠X when y = 412, w = 533, and m∠W = 39°.

5. ≈ (0.5, 0.866)
6. ≈ 0.87
7. ≈ −0.82
8. (8, −5)
9. (−3.336, 1.369)
10. ~ 147.23
11. ~ 111.9°

12. ≈ 22.38

13. ≈ 63.25°

14.

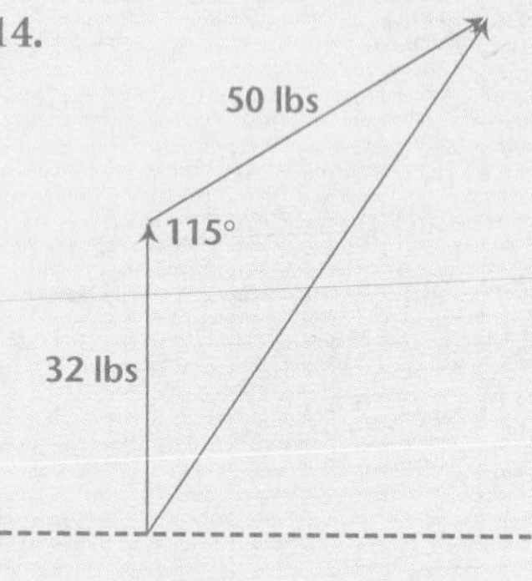

15.

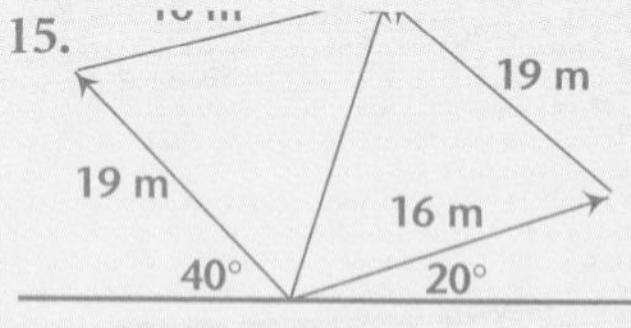

16. ≈ 110.72 miles

Lesson 10.6

➤ Key Skills

Apply the Law of Cosines to find measures in a triangle.

$c^2 = a^2 + b^2 - 2ab\cos C$

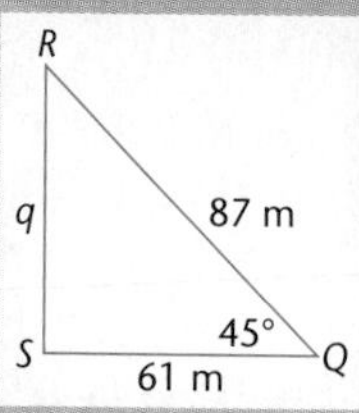

Find the length of side q.

We know the length of two sides and the measure of the included angle, so we can use the Law of Cosines:

$q^2 = r^2 + s^2 - 2rs(\cos \angle Q)$

$q^2 = 61^2 + 87^2 - 2(61)(87)\cos 45°$

$q^2 \approx 3{,}784.8 \text{ m}^2$

$q \approx 61.5 \text{ m}$

➤ Exercises

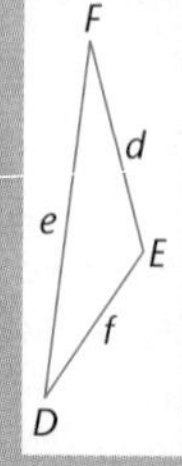

In Exercises 12–13, refer to the figure.

12. Find side f when $d = 23$, $e = 41$, and $\angle F = 25°$.

13. Find $m\angle D$ when $e = 321$, $f = 233$, and $d = 300$.

Lesson 10.7

➤ Key Skills

Add vectors using two methods.

A swimmer is moving 30° west of north at 2 miles per hour (relative to the water) through a current that flows north at 1.3 miles per hour. Find the actual speed and direction of the swimmer.

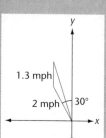

You can use vectors to represent current and swimming velocity and add the vectors to get actual speed and direction. Using the head-to-tail method, draw an accurate vector diagram and measure the resultant.

Using the parallelogram method, sketch a parallelogram and use the Law of Cosines to find the magnitude and the Law of Sines to find the direction.

$x^2 = 2^2 + 1.3^2 - 2(2)(1.3)\cos 150°$; $x \approx 3.2$ miles

$\dfrac{\sin y}{1.3} = \dfrac{\sin 150°}{3.2}$; $y \approx 11.7°$

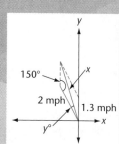

➤ Exercises

In Exercises 14–15, add the vectors below, drawing your solution.

14. Use the head-to-tail method.

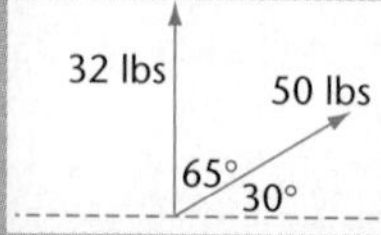

15. Use the parallelogram method.

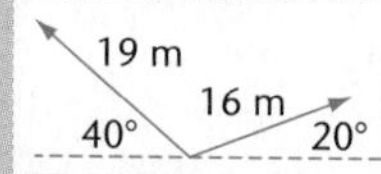

Application

16. Navigation Two planes set off from an airport, one traveling 45° northeast at 100 mph, the other at 30° southeast at 115 mph. After 40 minutes, how far apart will the planes be?

Chapter 10 Assessment

1. Which ratio does $\frac{5}{12}$ represent?

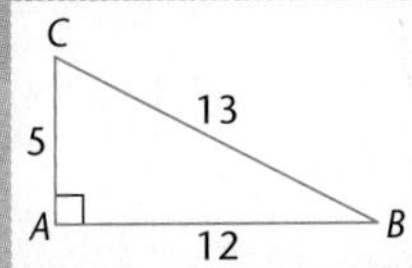

a. $\tan B$ **b.** $\cot B$
c. $\cos C$ **d.** $\sin A$

2. Find $m\angle R$.

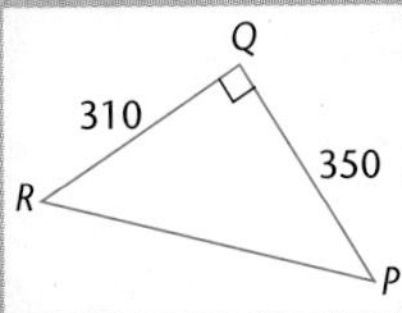

3. Find the cosine of 140°.

In Exercises 4–5, name the image point of the rotation.

4. Point (7, −9) rotated 450°

5. Point (−8, 3) rotated 73°

In Exercises 6–9, refer to the figure below.

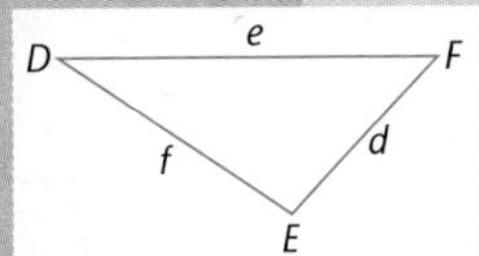

6. Find side e when $\angle E = 86°$, $\angle D = 28°$, and $d = 42$.

7. Find $m\angle D$ when $e = 211$, $f = 263$, and $m\angle F = 75°$.

8. Find side f when $d = 1.7$, $e = 3.1$, and $\angle F = 125°$.

9. Find $m\angle F$ when $e = 440$, $f = 240$, and $d = 340$.

10. Add the vectors below, drawing your solution.

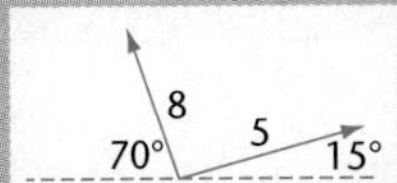

12.

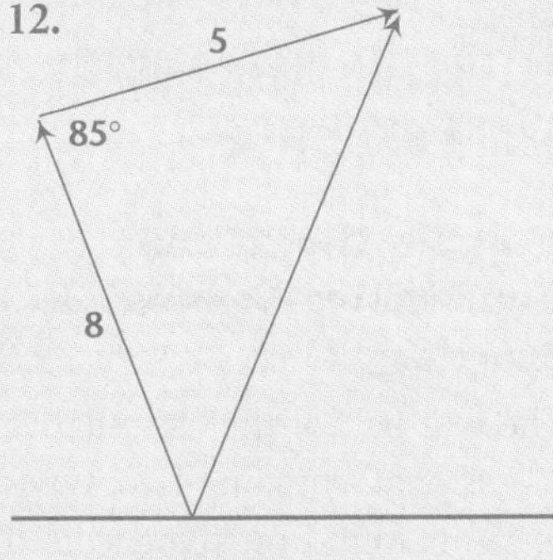

or

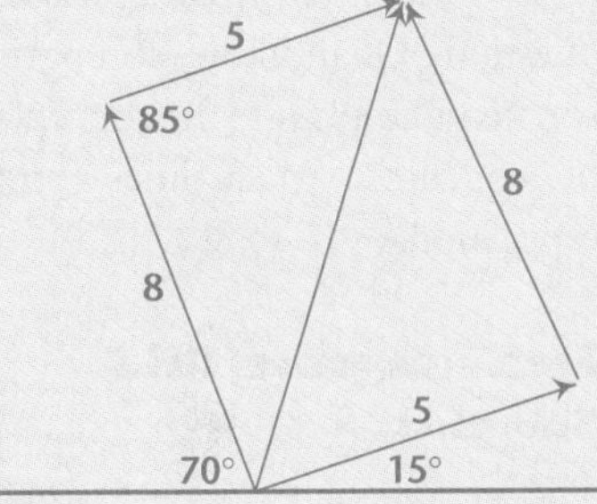

1. a
2. ~ 48.46°
3. –0.766
4. (9, 7)
5. (–5.21, –6.77)
6. ~ 89.24
7. ~ 54.2°
8. ~ 4.3
9. ~ 32.76°

College Entrance-Exam Practice

Multiple-Choice and Quantitative-Comparison Samples

The first half of the Cumulative Assessment contains two types of items found on standardized tests—multiple-choice questions and quantitative-comparison questions. Quantitative-comparison items emphasize the concepts of equalities, inequalities, and estimation.

Free-Response Grid Samples

The second half of the Cumulative Assessment is a free-response section. A portion of this part of the Cumulative Assessment consists of student-produced response items commonly found on college entrance exams. These questions require the use of machine-scored answer grids. you may wish to have students practice answering these items in preparation for standardized tests.

Sample answer grid masters are available in the *Chapter Teaching Resources Booklets.*

Chapters 1–10 Cumulative Assessment

College Entrance Exam Practice

Quantitative Comparison Exercises 6–9 consist of two quantities, one in Column A and one in Column B, which you are to compare as follows:

A. The quantity in Column A is greater.
B. The quantity in Column B is greater.
C. The two quantities are equal.
D. The relationship cannot be determined from the information given.

	Column A	Column B	Answers
1.	$\triangle DEF$ is similar to $\triangle GHI$. 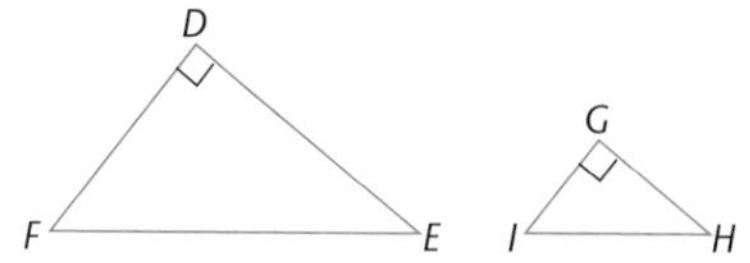 tan $\angle E$	tan $\angle H$	(A) (B) (C) (D) **[Lesson 10.1]**
2.	Angle $\theta > 45°$ cos θ	sin θ	(A) (B) (C) (D) **[Lesson 10.2]**
3.	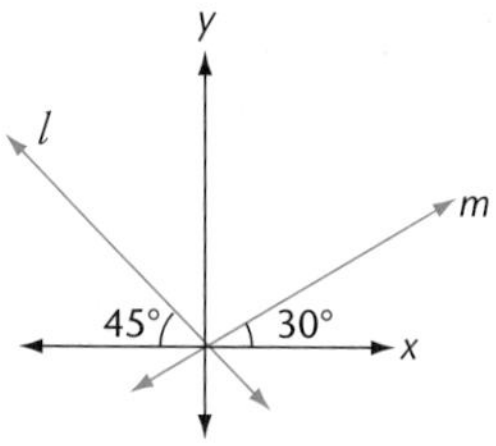 Slope of l	Slope of m	(A) (B) (C) (D) **[Lesson 3.8]**
4.	$\overline{CB} \cong \overline{DB}$ 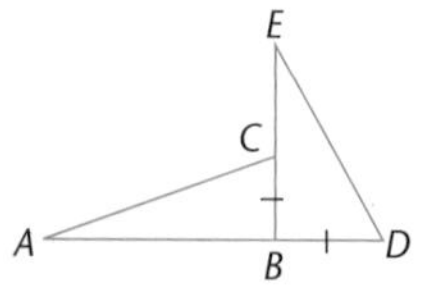 Area of $\triangle ACB$	Area of $\triangle BDE$	(A) (B) (C) (D) **[Lesson 5.2]**

5. Choose the most complete, accurate description of the two polygons. **[Lesson 4.6]**

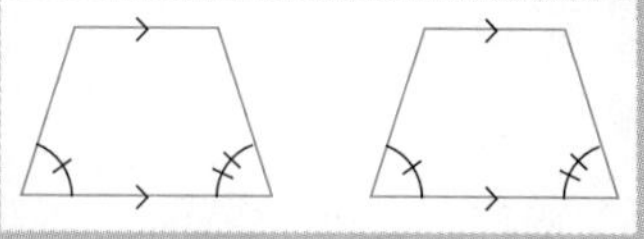

a. quadrilaterals
b. trapezoids
c. similar trapezoids
d. congruent trapezoids

1. C **2.** B **3.** B
4. D **5.** b

6. A surveyor has taken the measures shown below. What method can he use to find x? **[Lessons 10.5, 10.6]**

a. Cross-Multiplication Property
b. Exchange Property
c. Law of Sines
d. Law of Cosines

7. Which ratio does $\frac{20}{13}$ represent? **[Lessons 10.1, 10.2]**

a. tan B
b. cot B
c. cos C
d. sin A

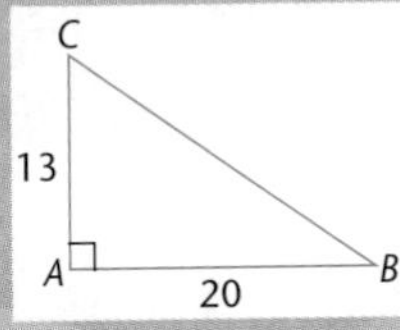

8. In circle O, which angle or arc measures 60°? **[Lesson 9.3]**

a. $\angle ABD$
b. $\angle BDC$
c. $\overset{\frown}{AD}$
d. $\overset{\frown}{BC}$

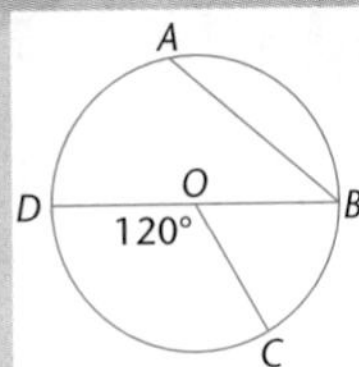

9. Find the volume of the oblique pyramid. **[Lesson 7.3]**

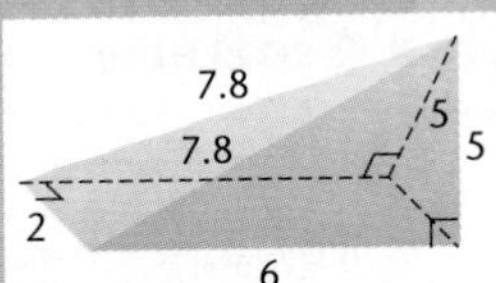

10. The ratio between the volumes of two spheres is 27:1. If the smaller sphere has a radius of 15 inches, what is the radius of the larger sphere? **[Lessons 7.1, 7.6]**

In Exercises 11–12, refer to the figure at the right.

11. Construct a rotation of the segment below. Rotate the segment 30° counterclockwise about its endpoint (0, 0). **[Lesson 4.9]**

12. Give the coordinates of the endpoints of the rotated segment. **[Lesson 10.4]**

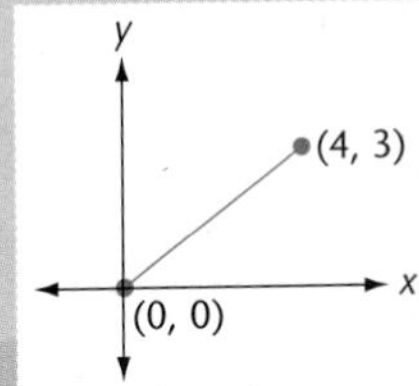

13. Write a paragraph proof that $\triangle PYW$ and $\triangle PYX$ in circle P are congruent. **[Lessons 4.2, 4.3]**

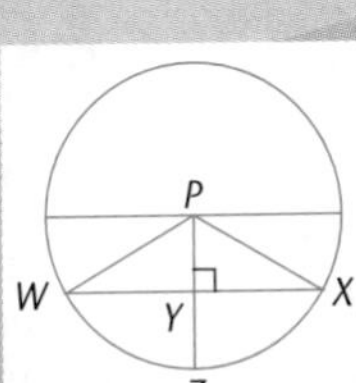

Free-Response Grid Exercises 14–17 may be answered using a free-response grid commonly used by standardized test services.

14. Find the volume of a cylinder with a radius of 3 cm and a height of 10 cm. **[Lesson 7.4]**

15. What is the cosine of 240°? **[Lesson 10.2]**

16. Find the area of the parallelogram. **[Lesson 5.2]**

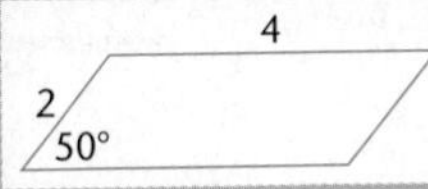

17. What is the area of a regular pentagon whose sides are 1 inch long? **[Lesson 5.5]**

Cumulative Assessment

6. c 7. b 8. d

9. ~ 19.6 units^3

10. $r = 45$ in

11.

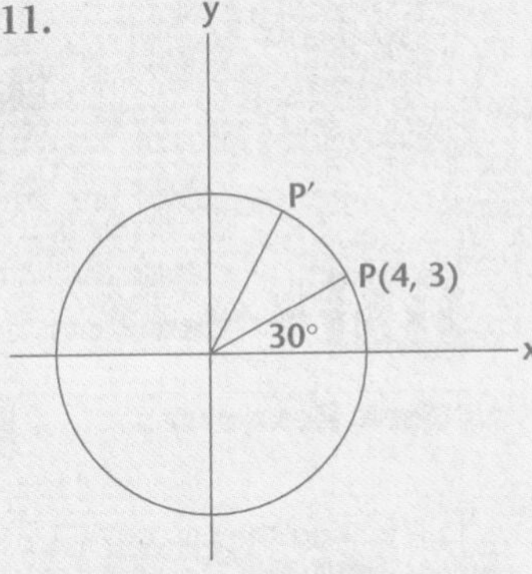

Use Transparency 95

12. P′(1.96, 4.60)

13. Sample answer:

$PW = PX$ because radii of a circle have equal length. $\overline{PZ}$ is given to be perpendicular to $\overline{WX}$, so $m\angle PYW = m\angle PYX = 90°$. By the Reflexive Property, $PY = PY$. Thus, $\triangle PYW \cong \triangle PYX$ by HL.

14. ~ 282.74 cm^3

15. –.5

16. ~ 6.13 units^2

17. ~ 1.72 in^2

CHAPTER 11

Taxicabs, Fractals, and More

Meeting Individual Needs

11.1 Golden Connections

Core Resources	Core Plus Resources
Inclusion Strategies, p. 588 Reteaching the Lesson, p. 589 Practice Master 11.1 Enrichment, p. 588 Lesson Activity Master 11.1 Interdisciplinary Connection, p. 587	Practice Master 11.1 Enrichment Master 11.1 Technology Master 11.1 Interdisciplinary Connection, p. 587
[2 days]	[2 days]

11.2 Taxicab Geometry

Core Resources	Core Plus Resources
Inclusion Strategies, p. 596 Reteaching the Lesson, p. 597 Practice Master 11.2 Enrichment Master 11.2 Lesson Activity Master 11.1 Interdisciplinary Connection, p. 595	Practice Master 11.2 Enrichment, p. 596 Technology Master 11.2 Interdisciplinary Connection, p. 595
[2 days]	[2 days]

11.3 Networks

Core Resources	Core Plus Resources
Inclusion Strategies, p. 602 Reteaching the Lesson, p. 603 Practice Master 11.3 Enrichment, p. 602 Technology Master 11.3 Lesson Activity Master 11.1 Interdisciplinary Connection, p. 601	Practice Master 11.3 Enrichment Master 11.3 Technology Master 11.3 Interdisciplinary Connection, p. 601
[2 days]	[2 days]

11.4 Topology: Twisted Geometry

Core Resources	Core Plus Resources
Inclusion Strategies, p. 609 Reteaching the Lesson, p. 610 Practice Master 11.4 Enrichment Master 11.4 Lesson Activity Master 11.1 Mid-Chapter Assessment Master	Practice Master 11.4 Enrichment, p. 609 Technology Master 11.4 Mid-Chapter Assessment Master
[2 days]	[2 days]

11.5 Euclid Unparalleled

Core Resources	Core Plus Resources
Inclusion Strategies, p. 616 Reteaching the Lesson, p. 617 Practice Master 11.5 Enrichment, p. 616 Technology Master 11.5 Lesson Activity Master 11.1	Practice Master 11.5 Enrichment Master 11.5 Technology Master 11.5
[3 days]	[2 days]

11.6 Exploring Projective Geometry

Core Resources	Core Plus Resources
Inclusion Strategies, p. 623 Reteaching the Lesson, p. 624 Practice Master 11.6 Enrichment Master 11.6 Lesson Activity Master 11.1 Interdisciplinary Connection, p. 622	Practice Master 11.6 Enrichment, p. 623 Technology Master 11.6 Interdisciplinary Connection, p. 622
[2 days]	[2 days]

11.7 Fractal Geometry

Core Resources	Core Plus Resources
Inclusion Strategies, p. 631 Reteaching the Lesson, p. 632 Practice Master 11.7 Enrichment, p. 631 Technology Master 11.7 Lesson Activity Master 11.1 Interdisciplinary Connection, p. 630	Practice Master 11.7 Enrichment Master 11.7 Technology Master 11.7 Interdisciplinary Connection, p. 630
[3 days]	[2 days]

Chapter Summary

Core Resources	Core Plus Resources
Chapter 11 Project, pp. 636–637 Lab Activity Long-Term Project Chapter Review, pp. 638–640 Chapter Assessment, p. 641 Chapter Assessment, A/B Alternative Assessment	Chapter 11 Project, pp. 636–637 Lab Activity Long-Term Project Chapter Review, pp. 638–640 Chapter Assessment, p. 641 Chapter Assessment, A/B Alternative Assessment
[3 days]	[2 days]

Hands-On Strategies

The golden ratio activities in Lesson 11.1 call for either a compass and straightedge or geometry graphics software. In Lesson 11.2, students will use grid paper for explorations in taxicab geometry. In some cases, a geoboard can be used as an alternative to the grid paper.

In Lesson 11.4, students will need paper, scissors, and tape to create Moebius strips and Klein bottles. Adding machine tape works well for making Moebius strips.

Visual Strategies

In Lesson 11.1, students investigate properties of the Golden Rectangle. This rectangle is thought to be especially attractive from an aesthetic point of view. Prepare an overhead transparency with a collection of different rectangles. Include just one that has the golden ratio; others should have a variety of length-width ratios. Hold a class vote as to which rectangle has the more attractive proportions.

Lesson 11.2 may be the first experience students have in seeing drawings of networks. Help them to understand that the distances between vertices do not matter in a network. Whether or not the network is traversable is unrelated to the specific lengths and angle measures.

Lessons 11.4 and 11.5 include illustrations of some difficult three-dimensional figures. Examples include a torus, a tetrahedron distorted into a sphere, a Klein bottle, orthogonal arcs on a sphere, and the "saddle" used to discuss hyperbolic geometry. You may wish to have students study and discuss these figures ahead of time.

The fractal illustrations in Lesson 11.7 are some of the most intriguing drawings in mathematics because they include the idea of an infinite number of iterations. Students will need to imagine an established procedure repeated over and over again. Be sure students create enough iterations to see how the pattern develops.

Cooperative Learning

GROUP ACTIVITIES	
Star of Pythagoras, Moebius Strip, Koch Snowflake	Portfolio Activity
Golden rectangle	Lesson 11.1, Exploration 1
Fibbonacci Sequence	Lesson 11.1, Look Beyond
Taxicab geometry	Lesson 11.2, Explorations 1 and 2
Networks	Lesson 11.3, Exploration
The Moebius Strip	Lesson 11.4, Exercises 12–19
The Klein Bottle	Lesson 11.4, Exercises 24–26
Pascal and Sierpenski	Lesson 11.7, Exploration 2

You may wish to have students work in groups or with partners for some of the above activities. Additional suggestions for cooperative group activities are noted in the teacher's notes in each lesson.

Multicultural

The cultural references in this chapter include references to Europe, Africa, and Asia.

CULTURAL CONNECTIONS	
Europe: Parthenon	Lesson 11.1
Europe: Königsberg Bridge	Lesson 11.3
Asia: Kolam	Chapter 11.7, Exercises 13–16

Portfolio Assessment

Below are portfolio activities for the chapter listed under seven activity domains which are appropriate for portfolio development.

1. **Investigation/Exploration** The explorations in Lesson 11.1 deal with various aspects of the golden ratio; those in Lesson 11.2 focus on taxicab geometry. Students explore networks in Lesson 11.3, investigate affine transformations and two theorems from projective geometry in Lesson 11.6, and create fractal designs in Lesson 11.7.

2. **Applications** Transportation, Lesson 11.2, Exploration 1; Law Enforcement, Lesson 11.3, Exercises 17–19; Materials Handling, Lesson 11.4, Exercise 15; Communication, Lesson 11.7; Candy Making, Lesson 11.7, Look Beyond.

3. **Nonroutine Problems** Lesson 11.3, Look Beyond (balloon problem); Lesson 11.6, Look Beyond (Nine Coin Puzzle).

4. **Project** Two Random Process Games: see pages 636–637.

5. **Interdisciplinary Topics** Algebra, Lesson 11.1; History, Lesson 11.7; Algebra, Lesson 11.7, Explorations 1 and 2.

6. **Writing** *Communicate* exercises offer excellent writing selections for the portfolio. Suggested selections include: Lesson 11.1, Exercise 2; Lesson 11.2, Exercise 4; Lesson 11.3, Exercise 5; Lesson 11.7, Exercise 2.

7. **Tools** In Chapter 11, students can use geometry graphics software for the constructions in Lessons 11.1 and 11.6.

Technology

The transformational geometry theorems in Exploration 2, Lesson 11.6, are particularly well-suited for geometry software. The instructions for the hands-on execution of of the exploration, as given in the text, are also appropriate for a geometry software activity.

Geometry Software

A Theorem of Pappus Begin by drawing two rays from a single point. (Select the ray tool by clicking and holding the fourth tool-bar button and then moving to the right to the ray symbol and releasing.)

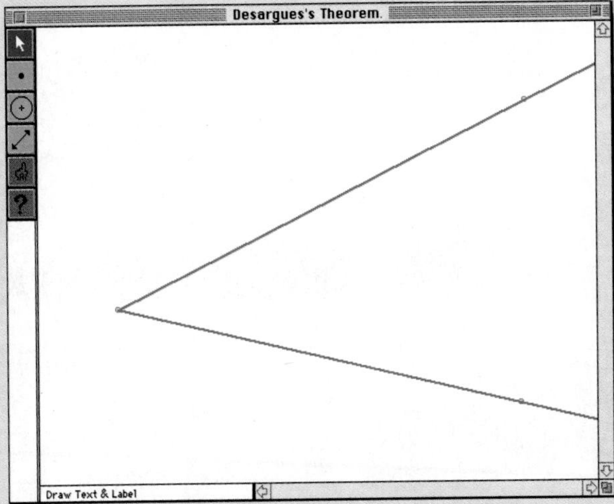

Notice that there is a point on each ray. (By dragging these points students can rotate the rays about their common endpoint.) Drag these points toward the edge of the screen to avoid confusion.

Use the pointer tool to place three points on each ray. Label the points using the text ("hand") tool. Then, if desired, change the labels to the ones suggested in the text. (Double-click on the labels and edit them.)

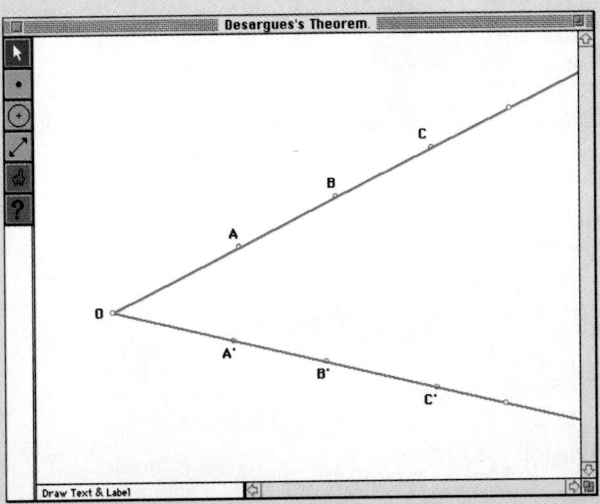

Use the segment tool to connect the points by segments, as shown. Use the point tool to place points on the indicated intersections. Label the points *X*, *Y*, and *Z*.

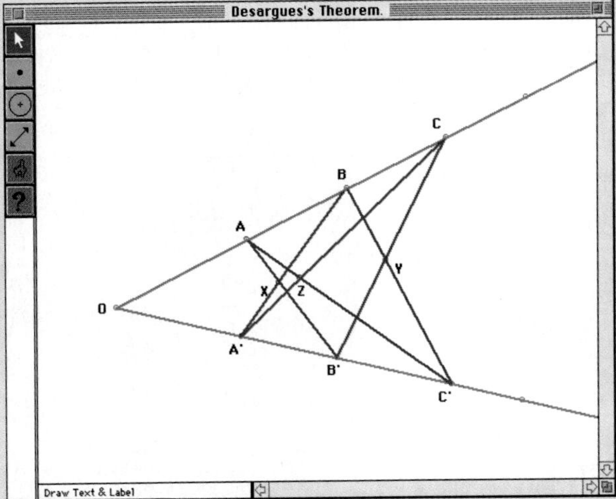

Use the line tool to draw a line through points X and Y. (Click on X; then, still holding the mouse button down, move the cross hairs to point Y and release.)

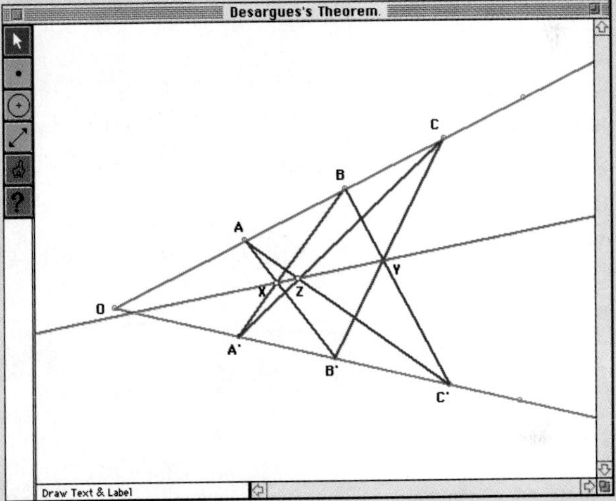

Students will notice that the line just drawn passes through point Z. They should drag the configuration into various shapes to demonstrate that the result is always true.

A Theorem of Desargues Desargues's Theorem, though more complex than Pappus's Theorem, is handled the same way. The required operations are the same. The only difficulty students may experience will be in adjusting the figure so that it will fit onto the screen.

About the Chapter

Background Information

In this chapter, students explore different topics in mathematics, including the golden ratio, taxicab geometry, networks, topology, two non-Euclidean geometries, projective geometry, and fractals.

- Practice Masters
- Enrichment Masters
- Technology Masters
- Lesson Activity Masters
- Lab Activity Masters
- Long-Term Project Masters
- Assessment Masters
  - Chapter Assessments, A/B
  - Mid-Chapter Assessment
  - Alternative Assessments, A/B
- Teaching Transparencies
- Spanish Resources

Chapter Objectives

- Discover the relationship known as the golden ratio.
- Solve problems using the golden ratio.
- Develop a non-Euclidean geometry based on taxi movements on a street grid and known as taxicab geometry.
- Solve problems within a taxicab geometry system.
- Develop the concept of Euler-type networks.
- Solve problems involving network traverses and Euler circuits.
- Explore and develop general notions for Riemanian, Lobachevskian, and Poincaréan geometries.

Chapter 11

Lessons

Taxicabs, Fractals, and More

Can you imagine a geometry in which a coffee cup is equivalent to a donut? Or one in which there is more than one "shortest path" between two points? There are many such strange ideas in this chapter.

By questioning commonly held assumptions, or by taking unusual imaginative leaps, mathematicians create entirely new branches of mathematics that often prove to be rewarding fields of study.

Before branching out to more recent discoveries in mathematics, you will first study an idea that goes back to classical times—the "golden ratio." This is an idea so powerful that it appears again and again in mathematics and nature.

About the Photos

Chapter 11 covers many interesting additional geometry topics. The photo collage shows visuals representing such topics as fractals, the golden ratio, and taxicab geometry. You may wish to discuss the photo with your students before proceeding with the chapter. Ask the students to speculate about the meaning of the various images.

- Develop informal proofs and solve problems using concepts of non-Euclidean geometries.
- Develop the concepts of affine transfromation and geometric projection.
- Solve problems and make conjectures using the Theorem of Pappus and the Theorem of Desargues.
- Discover the basic properties of fractals including self-similarity and iterative processes.
- Build fractal designs using iterative steps.

Portfolio Activity

Portfolio activities are keyed to three different lessons in this chapter. The topics covered are the Fibonacci sequence in Lesson 11.1, Moebius strips in Lesson 11.4, and fractals in lesson 11.7.

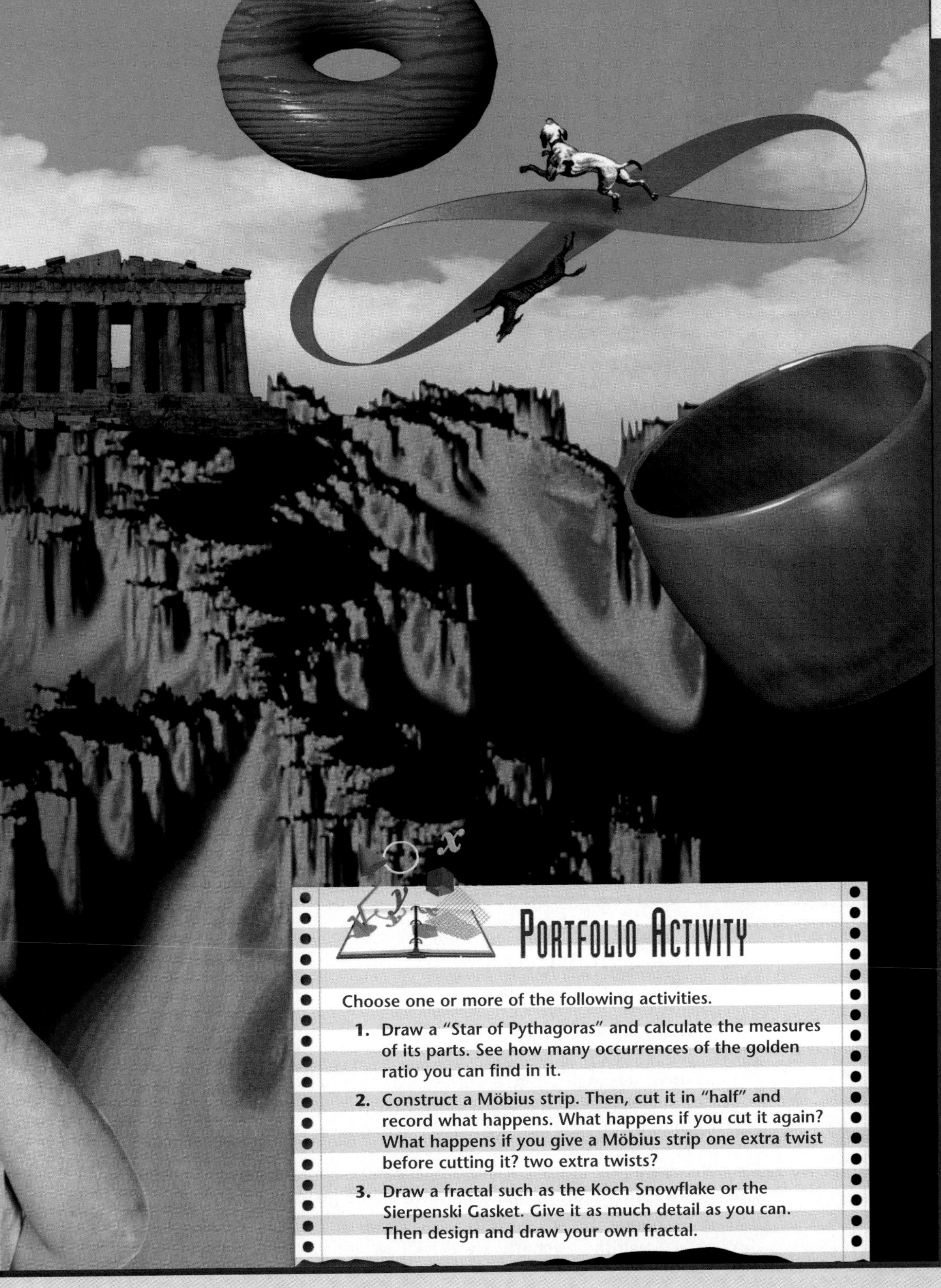

Portfolio Activity

Choose one or more of the following activities.

1. Draw a "Star of Pythagoras" and calculate the measures of its parts. See how many occurrences of the golden ratio you can find in it.
2. Construct a Möbius strip. Then, cut it in "half" and record what happens. What happens if you cut it again? What happens if you give a Möbius strip one extra twist before cutting it? two extra twists?
3. Draw a fractal such as the Koch Snowflake or the Sierpenski Gasket. Give it as much detail as you can. Then design and draw your own fractal.

About the Chapter Project

The Chapter 11 Project, on pages 636–637, stresses the iterative nature of fractals. Geometry graphics software or a graphics calculator will be helpful for completing the project.

PREPARE

Objectives

- Discover the relationship known as the golden ratio.
- Solve problems using the golden ratio.

RESOURCES

- Practice Master 11.1
- Enrichment Master 11.1
- Technology Master 11.1
- Lesson Activity Master 11.1
- Quiz 11.1
- Spanish Resources 11.1

Assessing Prior Knowledge

1. Solve the proportion $\frac{6}{x} = \frac{12}{8}$. [4]
2. Use the quadratic formula to solve the equation $x^2 - 2x - 4 = 0$. $[1 \pm \sqrt{5}]$
3. Use a calculator to find $\frac{1 + \sqrt{5}}{2}$. [≈**1.618033989**]
4. A right triangle has legs 1 and 6. Find the length of the hypotenuse. $[\sqrt{37}]$

TEACH

Studying the golden ratio should help students appreciate how seemingly unrelated objects may have a mathematical connection. The connection to the golden ratio is especially important because of its use in art and architecture.

LESSON 11.1 Golden Connections

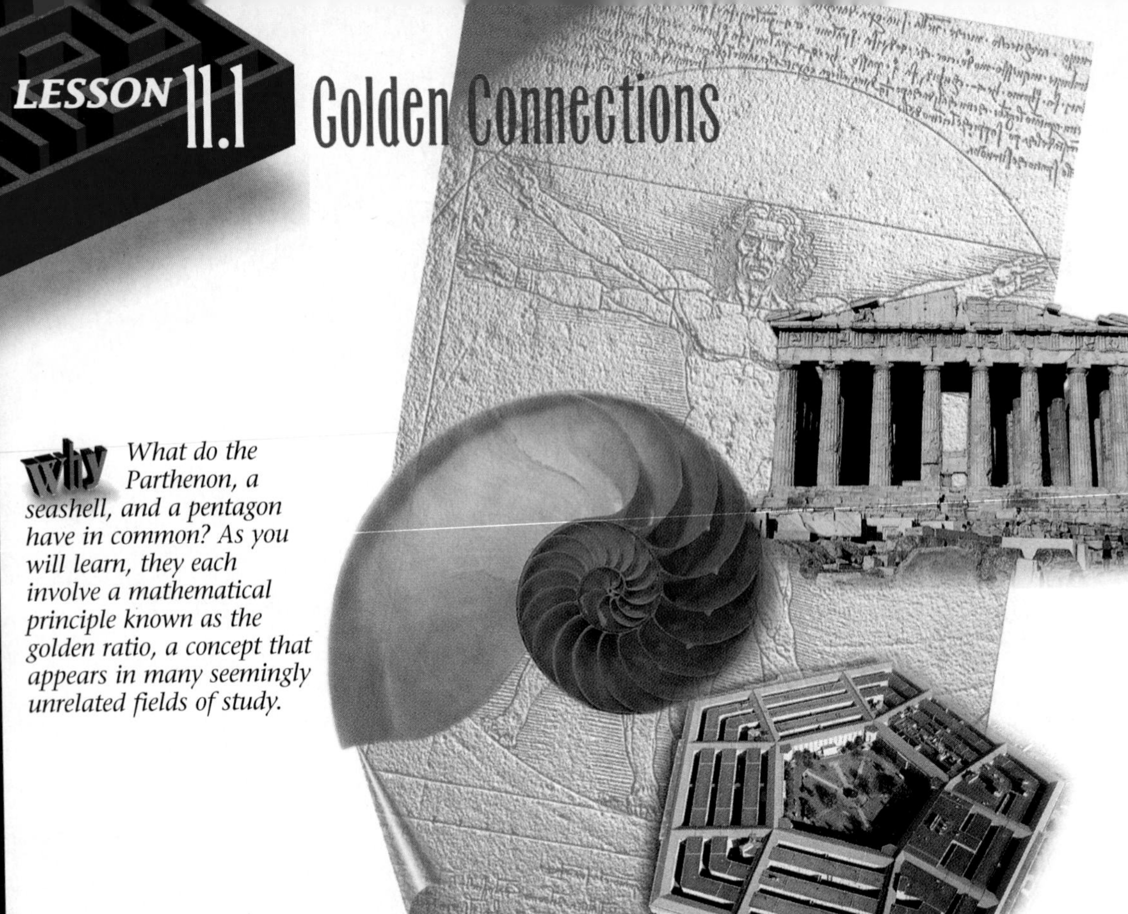

Why *What do the Parthenon, a seashell, and a pentagon have in common? As you will learn, they each involve a mathematical principle known as the golden ratio, a concept that appears in many seemingly unrelated fields of study.*

Golden Rectangles

The golden rectangle, which is considered to have pleasing proportions, has been used by artists and architects for centuries.

To test whether a rectangle is golden, you can construct a square inside it as shown. A rectangle is considered golden if the large and small rectangles in the resulting diagram are similar.

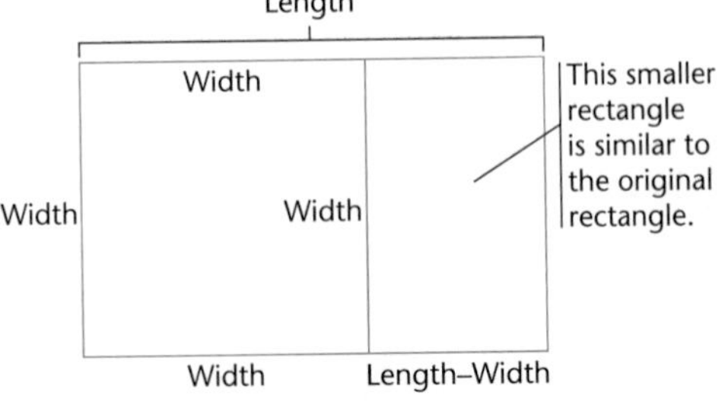

The similarity of the two rectangles can be expressed by the proportion below. The resulting value is known as the **golden ratio**.

$$\frac{\text{length}}{\text{width}} = \frac{\text{width}}{\text{length} - \text{width}}$$

Using Discussion Begin by reviewing the quadratic formula. Use Exploration 1 as an outline for a class presentation. An overhead graphics calculator will help students see the pattern in the ratio. Conclude the discussion by computing the golden ratio as on page 588. Students should then be able to complete Explorations 2 and 3 on their own.

Exploration 1 *The Dimensions of a Golden Rectangle*

You will need
Ruler
Graphics calculator

Graphics Calculator

1. Draw a vertical line segment 16 cm long. This segment will be the width of the golden rectangle you will draw.
2. For your rectangle to be "golden," it must satisfy the proportion on the previous page. Using a table like the one below, experiment with different values for l until the third and fourth columns match closely. Your final value for l should be accurate to the nearest tenth of a centimeter.

l	w	$\frac{l}{w}$	$\frac{w}{l-w}$
32	16	2	1
?	16	?	?
?	16	?	?
?	16	?	?
?	16	?	?

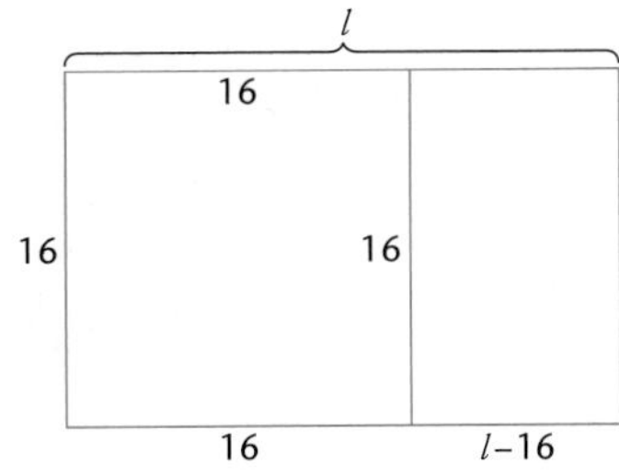

3. Draw the rest of the golden rectangle with the correct dimension l.
4. Complete the following statement:
The ratio $\frac{\text{length}}{\text{width}}$ in a golden rectangle is _?_. ❖

Exploration 2 *Seashells*

You will need
Geometry technology or
Ruler
Compass
Your golden rectangle from Exploration 1

Geometry Graphics

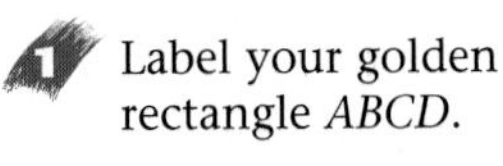

1. Label your golden rectangle *ABCD*.

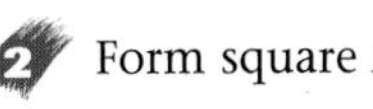

2. Form square *EBCF*.

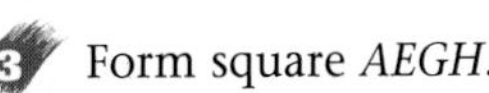

3. Form square *AEGH*.

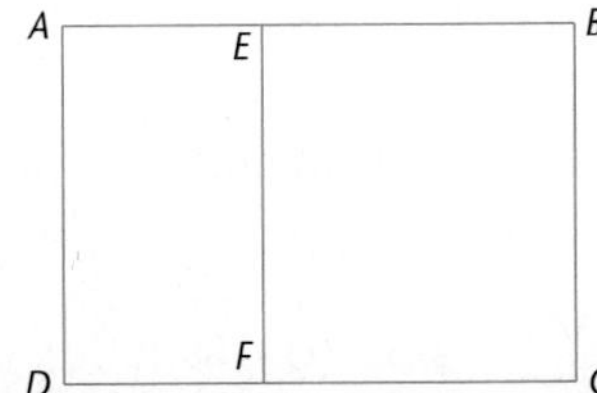

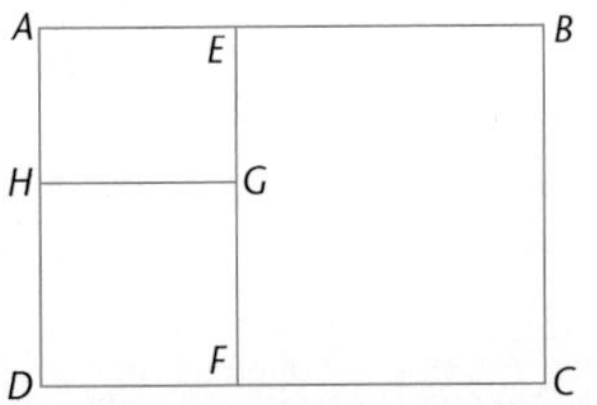

interdisciplinary CONNECTION **Architecture** Have students find examples of the golden ratio in architecture, especially skyscrapers. Ask interested students to make a simple model of a building using the golden ratio.

TEACHING tip

Motivate students by having them find the ratio of their height to their navel height, a ratio that may be golden. Point out that the golden ratio depends on finding the ratio between lengths of line segments.

Exploration 1 Notes

In this exploration, students will construct a golden rectangle from a vertical segment given as the width of the rectangle. The activity emphasizes that a rectangle is golden if the ratio of its length to its width is a certain proportion.

ongoing ASSESSMENT

4. **Approximately 1.62**

Exploration 2 Notes

Students will use a compass to break up the golden rectangle from Exploration 1 into similar nested golden rectangles. The activity will help students see how nesting golden rectangles creates a spiral model like those found in nature.

Math Connection
Algebra

To compute the golden ratio algebraically, students need to use the quadratic formula to find the length of a segment. Review the quadratic formula at this time, including how to simplify radicals.

ongoing ASSESSMENT

4. **The rectangles appear golden. Sample data: $\frac{25.9}{16} \approx 1.62$, $\frac{16}{9.9} \approx 1.62$, $\frac{9.9}{6.1} \approx 1.62$, $\frac{6.1}{3.8} \approx 1.61$, $\frac{3.8}{2.3} \approx 1.65$**

TEACHING tip

Point out that the length of a golden rectangle with width 1 is $\frac{1+\sqrt{5}}{2}$, the golden ratio. The algebraic derivation of the length on this page validates the geometric construction of the ratio. Emphasize that the properties of proportions guarantee that as the width increases in a golden rectangle, so does the length, by the same scale factor.

4. Continue this process until you have five nested rectangles. Do the rectangles seem to be golden? Measure the length and width of each. What is the ratio $\frac{l}{w}$ in each case?

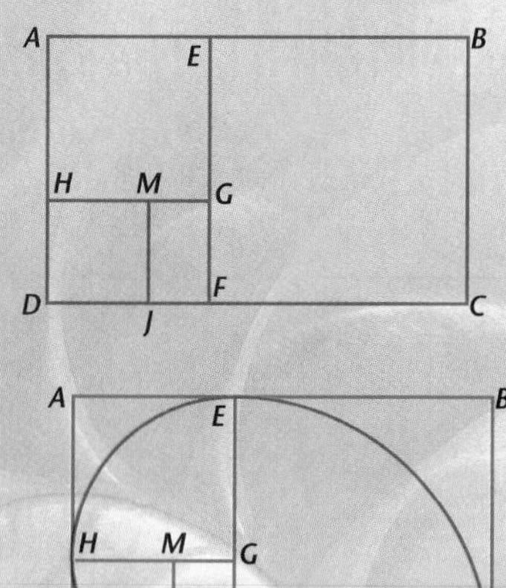

5. Use a compass to make quarter-circles in each square. The resulting equiangular or logarithmic spiral models the growth pattern of seashells such as the chambered nautilus. ❖

Computing the Golden Ratio

The proportions of the golden ratio can be solved algebraically to find the numerical value for $\frac{l}{w}$. First recognize that in a golden rectangle, the proportion must be true regardless of the value of w. For example, if $w = 1$, we can find a value for l that satisfies the proportion.

$$\frac{l}{1} = \frac{1}{(l-1)}$$

$$l(l-1) = 1$$

$$l^2 - l - 1 = 0$$

Notice that the resulting equation is a quadratic equation of the form

$Ax^2 + Bx + C = 0$, where $A = 1$, $B = -1$, and $C = -1$.

Substitute these values into the quadratic formula,

$$\frac{-B \pm \sqrt{B^2 - 4AC}}{2A}.$$

The result is

$$l = \frac{-(-1) \pm \sqrt{(-1)^2 - 4(1)(-1)}}{2(1)} = \frac{1 \pm \sqrt{5}}{2} \approx 1.618$$

(for length, use the positive value only).

Since $w = 1$,

$$\frac{l}{w} = \frac{l}{1} = \frac{1.618}{1} = 1.618.$$

This ratio is called the golden ratio. It is often represented by the Greek letter ϕ (phi):

$$\phi = \frac{1+\sqrt{5}}{2} = 1.618033989\ldots$$

ENRICHMENT Construct a pyramid from straws so that the pyramid has the same proportions as the Great Pyramid. Find the ratio of the slant height to one-half the base length. It should be very close to the golden ratio.

English Language Development Encourage students to make a list of all new vocabulary words. Explain how each of the words on their lists applies to the golden ratio in rectangles. Make sure students understand the steps in the construction of a golden rectangle using a compass and straightedge.

Do you think you would get the same result for $\frac{l}{w}$ if a value for w other than 1 is used? Try other values for w. What do you discover?

Exploration 3 Constructing a Golden Rectangle

You will need
Geometry technology or
Compass
Straightedge

Geometry Graphics

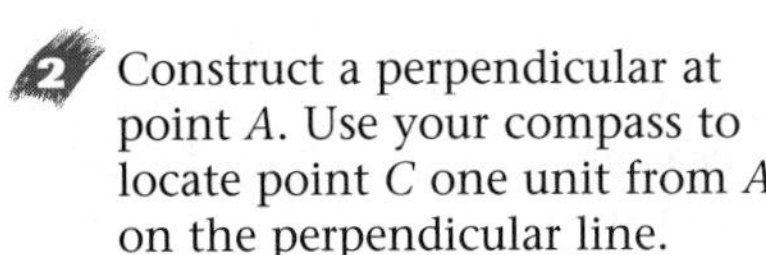

Draw a vertical segment and label points $\overline{AB}$ as shown. Find the midpoint M. Label the distance AB 2 units.

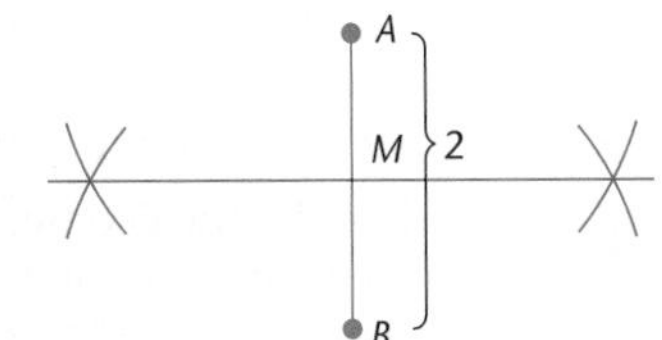

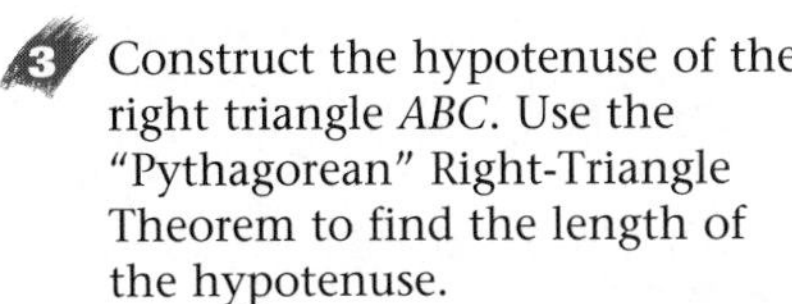

Construct a perpendicular at point A. Use your compass to locate point C one unit from A on the perpendicular line.

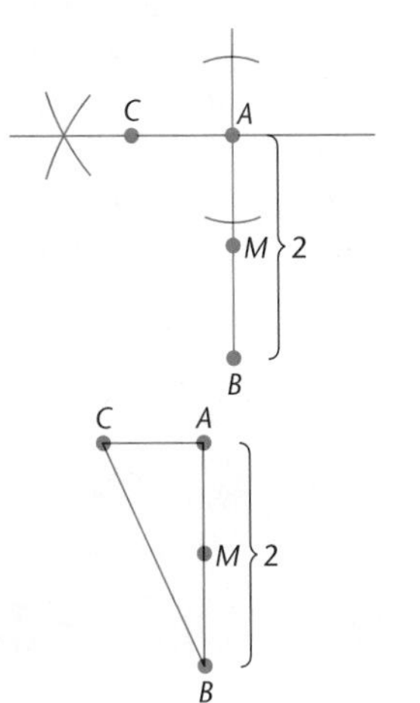

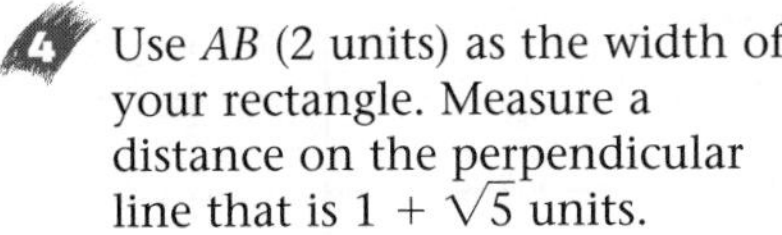

Construct the hypotenuse of the right triangle ABC. Use the "Pythagorean" Right-Triangle Theorem to find the length of the hypotenuse.

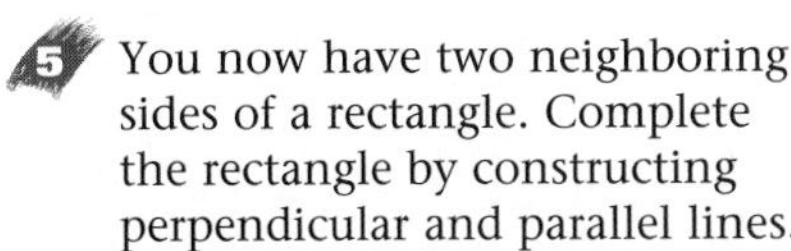

Use AB (2 units) as the width of your rectangle. Measure a distance on the perpendicular line that is $1 + \sqrt{5}$ units.

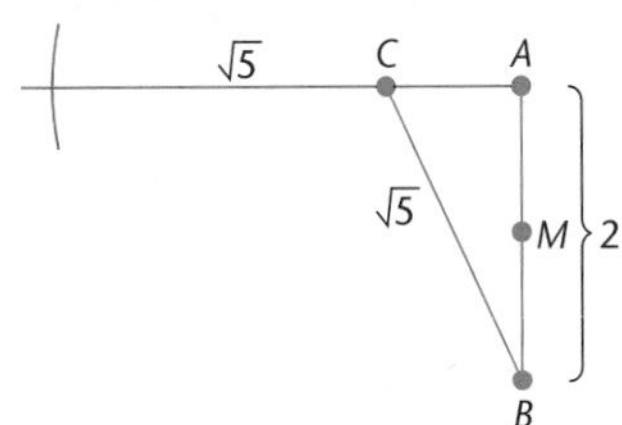

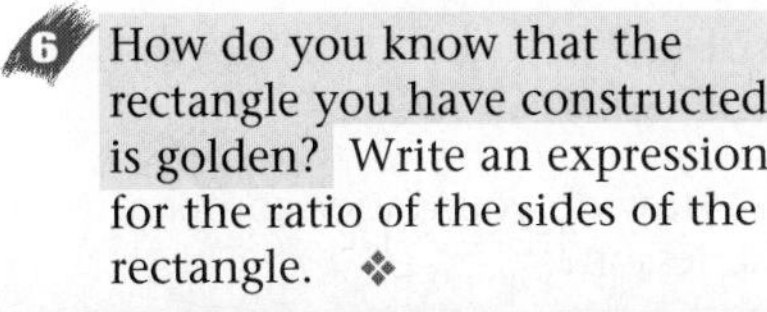

You now have two neighboring sides of a rectangle. Complete the rectangle by constructing perpendicular and parallel lines.

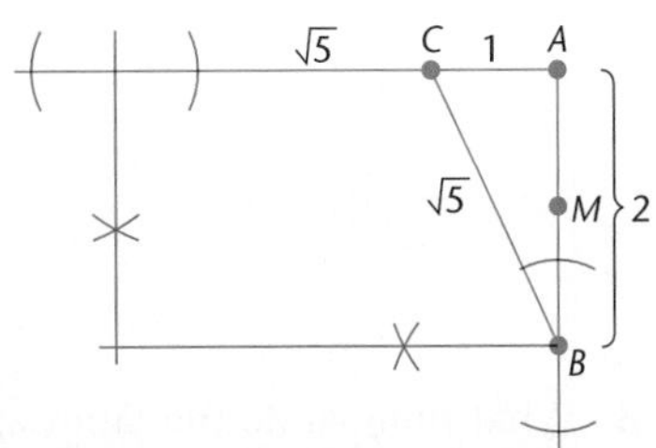

How do you know that the rectangle you have constructed is golden? Write an expression for the ratio of the sides of the rectangle. ❖

Hands-On Strategies

Have each student draw a vertical line of any length on a sheet of paper. Use the line to construct a golden rectangle with a compass and straightedge or by using a ruler. Ask them to measure the rectangle precisely and explain in writing how they know they constructed a golden rectangle.

CRITICAL *Thinking*

The ratio $\frac{l}{w}$ is the same for any value of w.

Use Transparency 95

Exploration 3 Notes

In this activiy, students construct a golden rectangle using a compass and straightedge. Point out the use of the "Pythagorean" Right-Triangle Theorem to construct a segment whose length is irrational. Have students discuss how to construct a golden rectangle with a different width.

ongoing ASSESSMENT

6. The ratio between the sides of the rectangle is $\frac{l}{w} = \frac{1+\sqrt{5}}{2}$, which is the golden ratio.

Cooperative Learning

Have groups of three students review the lesson. Have each student describe one of the explorations in detail to the other members of the groups.

Assess

Selected Answers

Odd-numbered Exercises 11–21

Assignment Guide

Core 1–16, 20–28

Core Plus 1–5, 9–30

Technology

Students can use geometry graphics software to construct a golden rectangle for Exercise 23. Use the measurement options to verify the golden ratio in the constructed rectangle.

Error Analysis

Students may have difficulty with Exercises 13–16. Use small discussion groups to review how the Law of Sines is used to calculate the sides of a triangle.

Practice Master

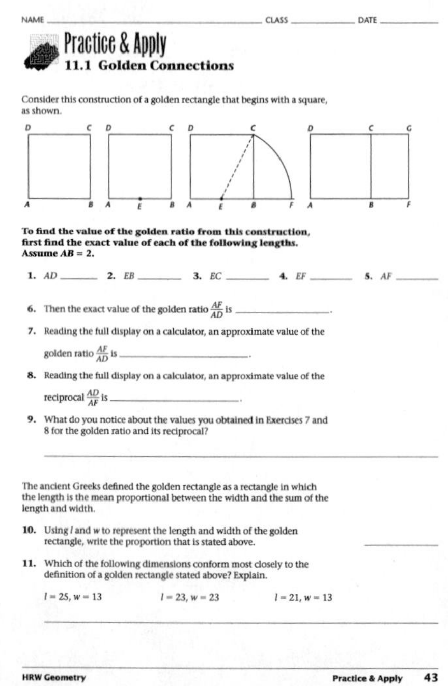

NAME ______ CLASS ______ DATE ______

Practice & Apply

11.1 Golden Connections

Consider this construction of a golden rectangle that begins with a square, as shown.

To find the value of the golden ratio from this construction, first find the exact value of each of the following lengths. Assume $AB = 2$.

1. AD ______ 2. EB ______ 3. EC ______ 4. EF ______ 5. AF ______

6. Then the exact value of the golden ratio $\frac{AF}{AD}$ is ______.
7. Reading the full display on a calculator, an approximate value of the golden ratio $\frac{AF}{AD}$ is ______.
8. Reading the full display on a calculator, an approximate value of the reciprocal $\frac{AD}{AF}$ is ______.
9. What do you notice about the values you obtained in Exercises 7 and 8 for the golden ratio and its reciprocal?

The ancient Greeks defined the golden rectangle as a rectangle in which the length is the mean proportional between the width and the sum of the length and width.

10. Using l and w to represent the length and width of the golden rectangle, write the proportion that is stated above. ______
11. Which of the following dimensions conform most closely to the definition of a golden rectangle stated above? Explain.

$l = 25, w = 13$ $l = 23, w = 23$ $l = 21, w = 13$

HRW Geometry Practice & Apply 43

Exercises & Problems

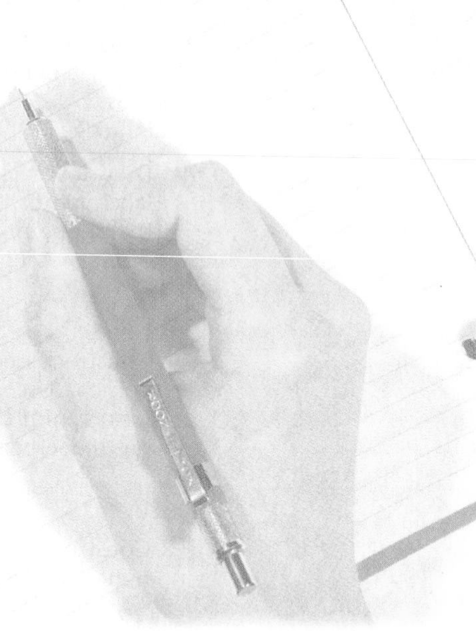

Communicate

1. Explain how to construct a golden rectangle. How is the "Pythagorean" Right Triangle Theorem important to this construction?
2. Describe places in art, architecture, and nature where the golden ratio can be seen.
3. Describe the geometric relationship between the sides of a golden rectangle. How does this relationship relate to what you know about similar figures?
4. How is the golden rectangle self-replicating?
5. The golden rectangle is supposed to be pleasing to the eye. Do you agree or disagree? Explain your answer.

Practice & Apply

Use the five nested golden rectangles you constructed in Exploration 2 for Exercises 6–8.

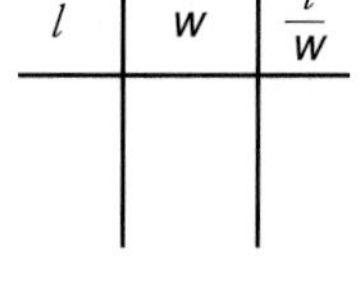

6. Using a ruler, measure the dimensions of each of the five rectangles. Put your measurements in a chart like the one shown.
7. Compute the ratio $\frac{l}{w}$ for each of the five rectangles. Student results should be close to 1.62.

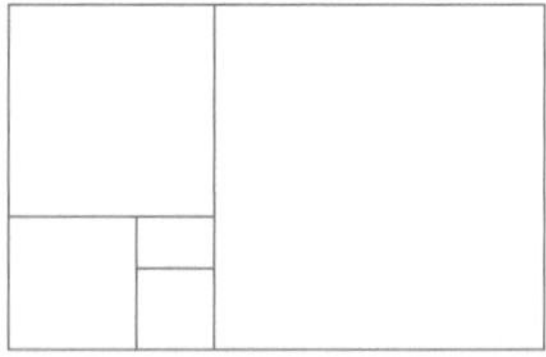

8. What number do the ratios approach? How do your ratios compare with the golden ratio computed in the lesson?

6. (answers in centimeters)

l	w	$\frac{l}{w}$
3.24	2	1.62
2	1.24	1.62
1.24	.76	1.62
.76	.47	1.62
.47	.29	1.62

8. The number approaches the golden ratio.

In Exercises 9–12, you will construct a regular pentagon and determine its relationship to the golden ratio.

9. Construct a regular pentagon using geometry technology or a compass and straightedge. Check student drawings.

a. Draw circle A and diameter $\overline{BC}$.

b. Construct the perpendicular bisector of $\overline{AB}$ in order to mark point M, the midpoint of $\overline{AB}$.

c. Label N, a point where the perpendicular bisector intersects the circle.

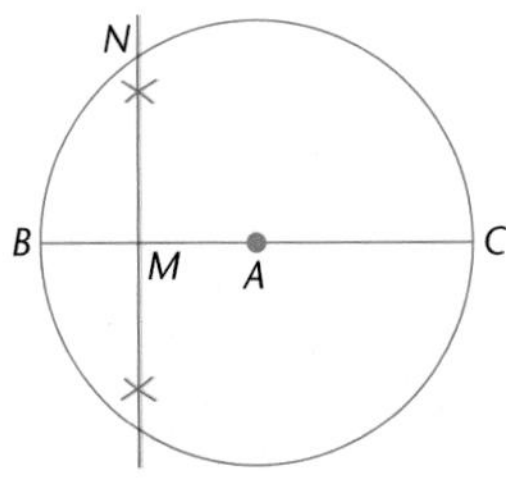

d. Construct the perpendicular bisector of $\overline{BC}$ and label D, a point where it intersects the circle.

e. Starting at M, mark off distance MD on $\overline{BC}$. Label the point O.

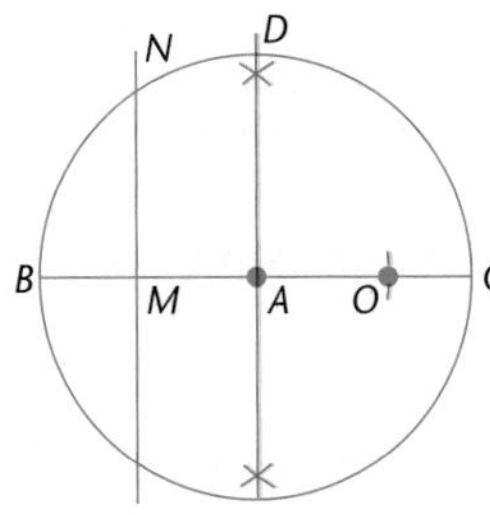

f. DO is the length of the pentagon's side. Starting at D, mark off length DO around the circle. Draw segments connecting the arcs.

g. The result is regular pentagon $DEFGH$.

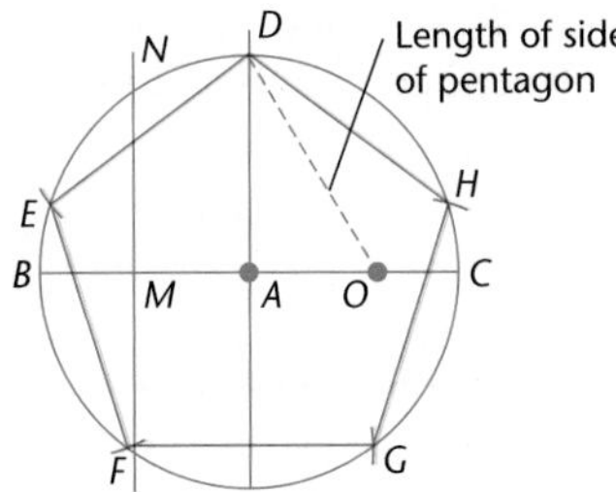

10. Draw $\overline{EH}$. Measure a and b to the nearest tenth of a centimeter.

11. Compute $\frac{a}{b}$ to the nearest tenth. 1.6

12. Write a conjecture about the result in Exercise 11.

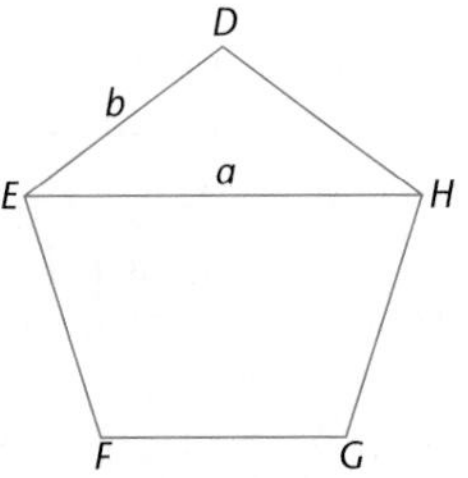

10. Answers will vary.

12. The ratio between the lengths of a diagonal and a side of a regular pentagon is the golden ratio.

alternative ASSESSMENT

Portfolio Assessment

Ask students to list the steps in the construction of a golden rectangle using a compass and straightedge. Students should include examples that illustrate the steps.

Technology Master

NAME ______ CLASS ______ DATE ______

Technology
11.1 Searching for the Golden Ratio

In Lesson 11.1, you learned that if a golden rectangle has length x and width y, then $\frac{x}{y} = \frac{y}{x-y}$. If $y = 1$, then $\frac{x}{1} = \frac{1}{x-1}$ and $x^2 - x - 1 = 0$. Using the quadratic formula, you found that $x = \frac{1 \pm \sqrt{5}}{2}$. As you will see, there are other ways to find x.

1. Justify each step. If $x^2 - x - 1 = 0$, then $x^2 = x + 1$. If $x^2 = x + 1$, then $x = \sqrt{x + 1}$.

______

2. Create a spreadsheet in which cell A1 contains 0.5, cell B1 contains SQRT(A1+1), and cell A2 contains B1. Then FILL DOWN the first several rows of columns A and B. What number do the values in column B tend to approach?

______

3. Justify each step. If $x^2 - x - 1 = 0$, then $x(x - 1) - 1 = 0$. If $x(x - 1) - 1 = 0$, then $x(x - 1) = 1$. If $x(x - 1) = 1$, then $x = \frac{1}{x-1}$.

______

4. Create a spreadsheet in which cell A1 contains 0.5, cell B1 contains 1/(A1−1), and cell A2 contains B1. Then FILL DOWN the first several rows of columns A and B. What number do the values in column B tend to approach?

______

5. Create a spreadsheet in which column A contains 1, 2, 3, ... and column B contains $x_n = \frac{1}{\sqrt{5}}\left[\left(\frac{1+\sqrt{5}}{2}\right)^n - \left(\frac{1-\sqrt{5}}{2}\right)^n\right]$.

Enter (1/SQRT(5))*(((1+SQRT(5))/2)^A1−((1−SQRT(5))/2)^A1) in cell B1. Then FILL DOWN. What familiar numbers from Lesson 11.1 does column B give you?

______

HRW Geometry Technology 57

21. $\frac{610}{377} = 1.618037135$;
$\frac{377}{233} = 1.618025751$;
$\frac{233}{144} = 1.618055556$;
$\frac{144}{89} = 1.617977528$;
$\frac{89}{55} = 1.618181818$;
$\frac{55}{34} = 1.617647059$;
$\frac{34}{21} = 1.619047619$;
$\frac{21}{13} = 1.615384615$;
$\frac{13}{8} = 1.625$; $\frac{8}{5} = 1.6$;
$\frac{5}{3} = 1.666666667$;
$\frac{3}{2} = 1.5$; $\frac{2}{1} = 2$; $\frac{1}{1} = 1$
After the first two terms the number approaches the golden ratio.

Use Transparency 96

22. $\frac{514229}{317811} = 1.618033989$;
The ratio between a term of the Fibonacci sequence and the previous term in the sequence approaches the golden ratio.

Confirm the conjecture you made in Exercise 12 using trigonometry. Exercises 13–16 lead you through the steps.

13. Use the pentagon you constructed in Exercise 9. Find the measure of $\angle D$ by recalling the formula from Lesson 3.6 for the interior angles of a regular polygon. 108°

14. Since the sides of a regular polygon are congruent, $\triangle EDH$ is isosceles. Use this fact and the measure of $\angle D$ to compute $m\angle DEH$ and $m\angle DHE$. $m\angle DEH = m\angle DHE = 36°$

15. Use the Law of Sines and properties of proportions to calculate $\frac{a}{b}$ to as many places as your calculator will display. $\frac{a}{b} \approx 1.618033989$

16. How does the ratio from Exercise 15 compare with the one you computed in the lesson? They are the same.

Star of Pythagoras Extending the sides of a regular pentagon creates a figure known as the Star of Pythagoras. The five small triangles in the star are isosceles and congruent.

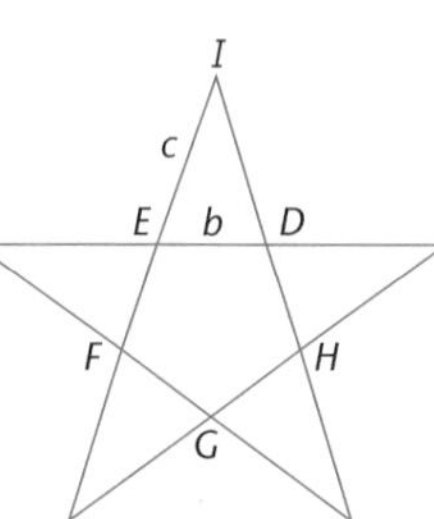

17. Use a protractor and find the measures of the angles of $\triangle DEI$. $m\angle DEI = m\angle EDI = 72°$; $m\angle EID = 36°$

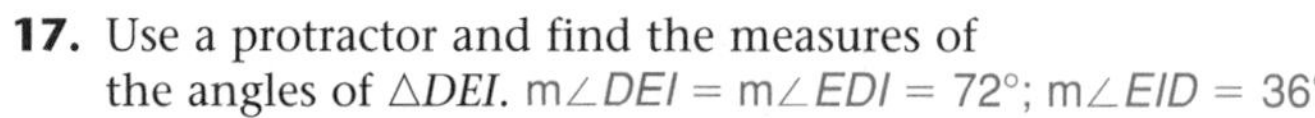

18. Use the Law of Sines to compute $\frac{c}{b}$. ≈ 1.618

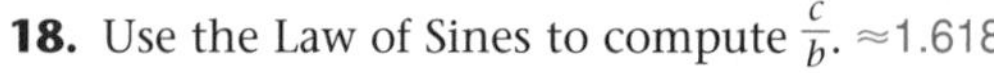

19. Complete the conjecture for a triangle from a star of Pythagoras.

The ratio of the __?__ to the base is equal to the __?__.
congruent sides; Golden Ratio

Fibonacci Sequence The Fibonacci Sequence is a series of numbers that appear in many places in nature. Each term is the sum of the two terms preceding it. The following is a Fibonacci sequence.

1 1 2 3 5 8 13 21 . . .

20. Find the next seven terms of the sequence shown above. 34, 55, 89, 144, 233, 377, 610

21. Divide each number in the sequence by the number that immediately precedes it. Record your results to as many decimal places as your calculator will display. What do you notice?

22. The 28th term in the Fibonacci sequence is 317,811, and the 29th term is 514,229. Find the ratio of these numbers and write a conjecture.

23. In the lesson, you constructed a golden rectangle from the ratio $\frac{(1 + \sqrt{5})}{2}$. Here is an alternative construction.
Check student drawings.

a. Begin by constructing square $ABCD$. Construct the midpoint of $\overline{AB}$. Call the midpoint E.

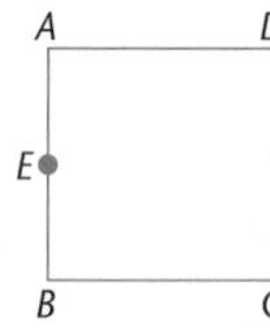

b. Extend the segment on $\overleftrightarrow{AB}$ as shown. Using E as the center of a circle and $\overline{EC}$ as the radius, construct an arc that extends the shorter distance from point C to $\overleftrightarrow{AB}$. Call the point of intersection of the arc with $\overleftrightarrow{AB}$ point F.

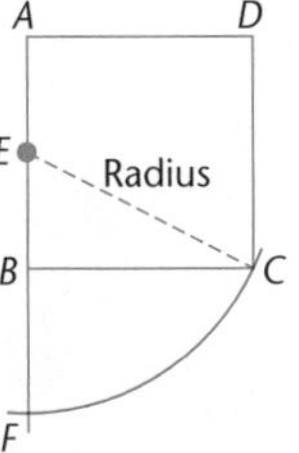

c. Construct the perpendicular through F. Extend the perpendicular and extend the segment on $\overleftrightarrow{DC}$ until they intersect. Call the intersection G.

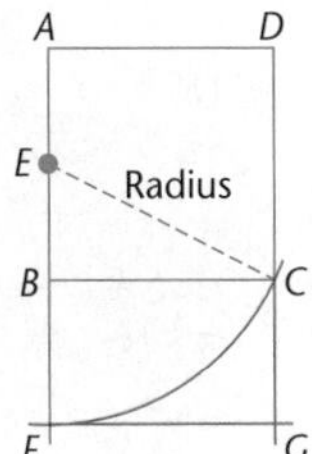

24. Explain why $AFGD$ is a golden rectangle.

Look Back

Technology Use geometry technology or graph paper and a protractor to sketch a unit circle and a radius with the given angle θ. Estimate the sine, cosine, and tangent of θ to the nearest hundredth. Include the sketch with your answers. **[Lesson 10.3]**

25. $\theta = 17°$ **26.** $\theta = 387°$ **27.** $\theta = 65°$ **28.** $\theta = 122°$

Answers for Exercises 25–28 can be found in Additional Answers beginning on page 727.

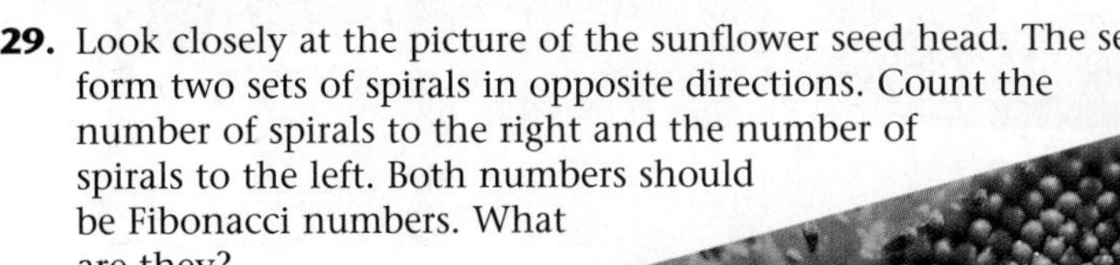

Use Transparency 97

Look Beyond

29. Look closely at the picture of the sunflower seed head. The seeds form two sets of spirals in opposite directions. Count the number of spirals to the right and the number of spirals to the left. Both numbers should be Fibonacci numbers. What are they?

30. **Portfolio Activity** Certain flowers, leaf and stem patterns, and fruits such as pineapples all have mathematical structures related to the Fibonacci sequence. Write a short research report about the Fibonacci sequence in nature.
Check student portfolio.

Look Beyond

Exercises 29 and 30 explore the Fibonacci sequence and the golden ratio in nature.

24. Since E is the midpoint of $\overline{AB}$ and $ABCD$ is a square, $BC = 2(EB)$. By the "Pythagorean" Right-Triangle Theorem,
$EC = \sqrt{EB^2 + BC^2} = \sqrt{EB^2 + (2EB)^2} = \sqrt{5(EB)^2} = EB\sqrt{5}$. Since $EF = EC$,
$\frac{AF}{AD} = \frac{AE + EF}{AD} = \frac{EB + EB\sqrt{5}}{2(EB)} = \frac{1+\sqrt{5}}{2} =$ Golden Ratio.

29. 34 and 55 ; These numbers are in the Fibonacci sequence.

PREPARE

Objectives

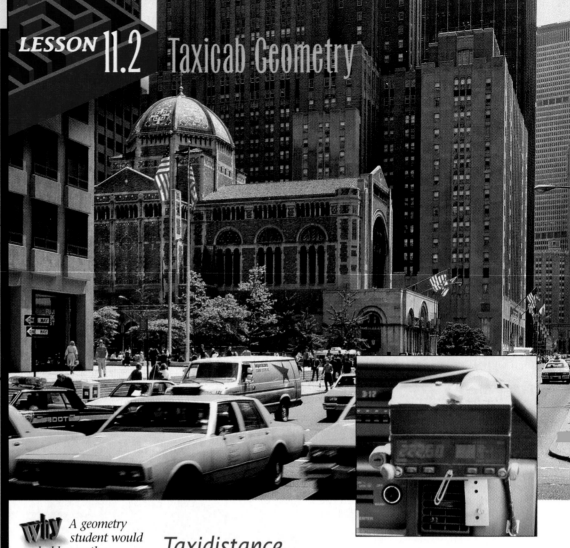

- Develop a non-Euclidean geometry based on taxi movements on a street grid and known as taxicab geometry.
- Solve problems within a taxicab geometry system.

RESOURCES

• Practice Master	11.2
• Enrichment Master	11.2
• Technology Master	11.2
• Lesson Activity Master	11.2
• Quiz	11.2
• Spanish Resources	11.2

Assessing Prior Knowledge

1. Points (–4, 4, (1, 4), (1, –1), (–4, –1) are vertices of a quadrilateral. Name the type of quadrilateral.

   [square]

2. Find the shortest distance between points (1, 1) and (5, 7). **[$2\sqrt{13}$]**

3. Find the distance between the points (1, 1) and (5, 7) along grid lines of graph paper.

   [10 units]

TEACH

Exploring taxicab geometry is important not only because of its application to real-life paths but also because it is an introduction to a non-Euclidean geometry. The lesson helps students make the connection that geometric systems can exist which are logical and consistent as well as applicable to the real world.

LESSON 11.2 Taxicab Geometry

why *A geometry student would probably use the "Pythagorean" Right-Triangle Theorem to find the distance between two points on a city map. However, a taxicab driver might have a very different idea. As you will see, it is possible to develop a logically consistent geometry around a taxicab driver definition of the distance between two points.*

Taxidistance

In "taxicab" geometry, points are located on a special kind of map or coordinate grid. The horizontal and vertical lines of the grid represent streets. But unlike points on a traditional coordinate plane, points on a taxicab grid can be only at intersections of two "streets." So the coordinates are always integers.

In taxicab geometry, the distance between two points, known as the **taxidistance**, is the smallest number of grid units, or "blocks," a taxi must travel to go from one point to the other. On the map shown, the taxidistance between the two points is 5.

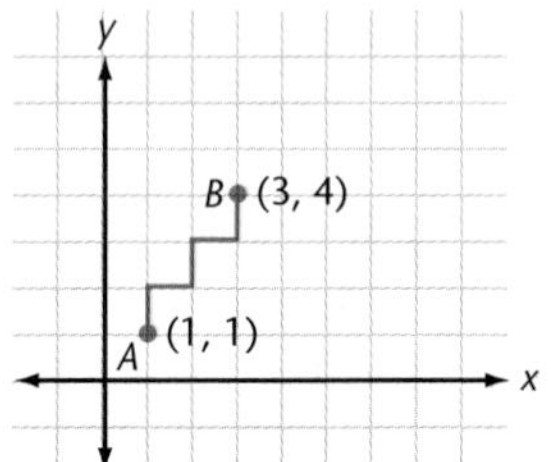

How many other ways can you travel five "blocks" from point A to point B?

Visual Models Photocopy a portion of a real local map onto a transparent grid. Demonstrate taxicab "pathways" that are familiar to students on the overhead projector. Measure and compare the shortest distance between two known points on the map with the taxicab distance.

Exploration 1 *Exploring Taxidistances*

Transportation

You will need
Graph paper (large grid)

Part I: The taxidistance from Central Dispatch

Assume that all taxis leave for their destinations from a central terminal at point $O(0, 0)$.

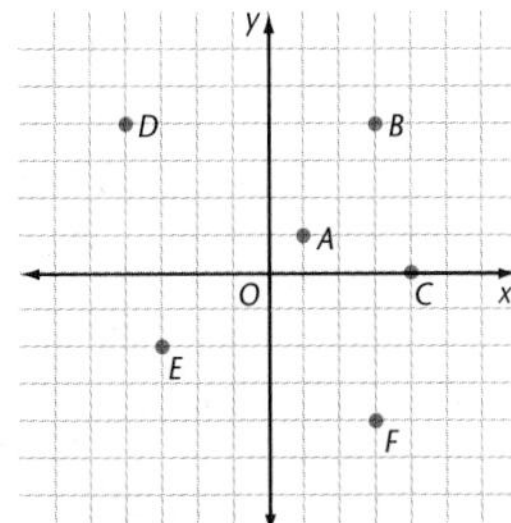

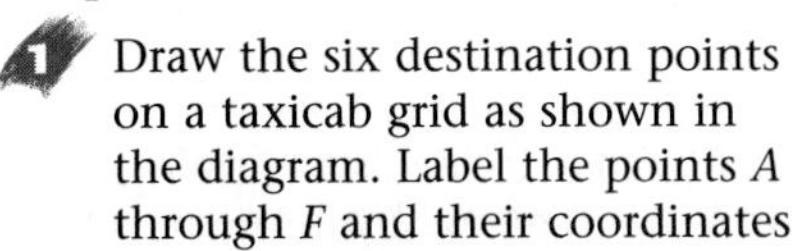

1. Draw the six destination points on a taxicab grid as shown in the diagram. Label the points A through F and their coordinates.

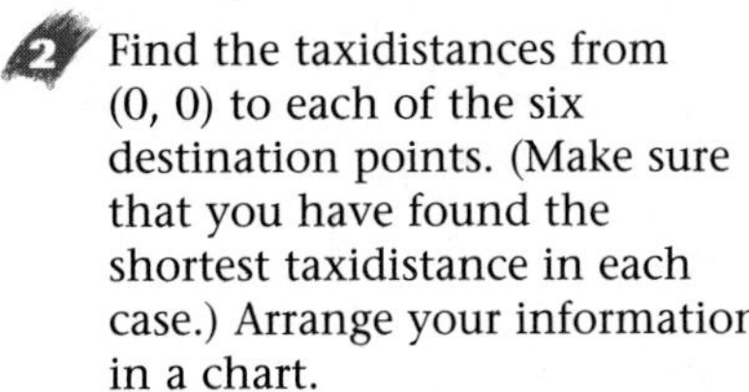

2. Find the taxidistances from (0, 0) to each of the six destination points. (Make sure that you have found the shortest taxidistance in each case.) Arrange your information in a chart.

Destination point coordinates	Taxi distance from O
A (?, ?)	
B	
C	
D	
E	
F	

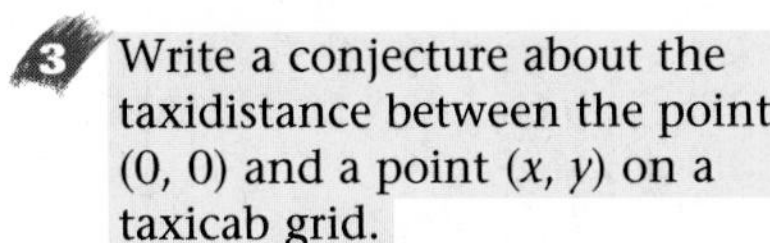

3. Write a conjecture about the taxidistance between the point (0, 0) and a point (x, y) on a taxicab grid.

Part II: The taxidistance between any two points

1. Using the diagram from Part 1, find the taxidistance between at least six pairs of points in addition to the pair illustrated. Use different combinations of points in the four quadrants to make sure all cases are covered. Arrange your information in a chart.

x_1, y_1	x_2, y_2	x_1	x_2	y_1	y_2	Taxi distance
A (1, 1)	B (3, 4)	1	3	1	4	5

2. Write a conjecture about the taxidistance between a point (x_1, y_1) and a point (x_2, y_2) on a taxicab grid. ❖

Two Points Determine . . . ?

For two points that are a given taxidistance apart, how many minimum-distance pathways are there between them?

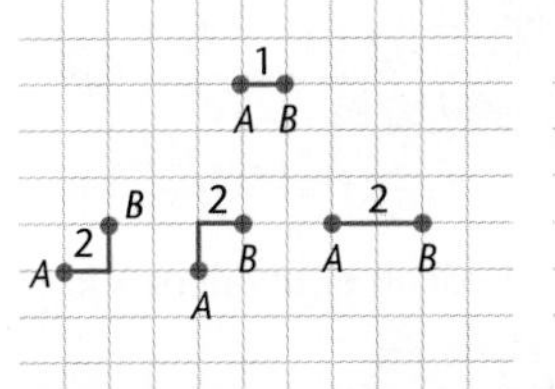

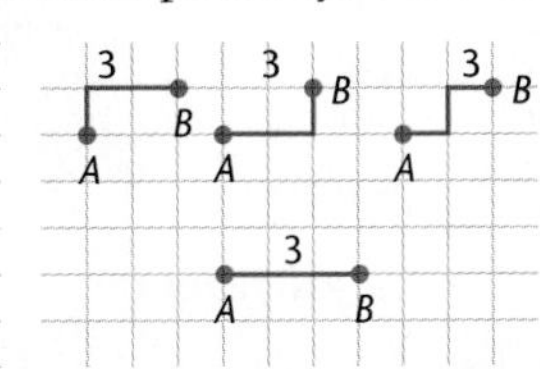

If the taxidistance between the points is just one unit, there is just one minimum pathway.

If the taxidistance between the points is two units, there are three minimum pathways.

If the taxidistance between the points is three units, there are four minimum pathways.

interdisciplinary CONNECTION **Geography** Cartography is the science of map and chart making. Have students study a road map that includes the shortest "road distance" between two points. Although the map is not a square grid, it represents taxicab distance nonetheless. Have students write a report about how a road map has its own logically consistent taxicab geometry.

TEACHING tip

Stress the importance of using integers for the coordinates in a taxicab geometry system. Ask students to find instances when negative integers, as well as positive integers, would be applicable.

Use Transparency 98

Exploration 1 Notes

In this activity, students measure distances between points in a coordinate system as taxicab distances. This activity will help students understand that there is a minimum taxidistance between two points of a grid.

ongoing ASSESSMENT

3. **The taxidistance between (0,0) and (x, y) = $|x| + |y|$.**

ongoing ASSESSMENT

2. **Taxidistance between (x_1, y_1) and (x_2, y_2) = $|x_2 - x_1| + |y_2 - y_1|$.**

Cooperative Learning

Divide students into groups of four to explore the number of minimum pathways that are possible as the taxidistance between two points increases. Have each student draw a small grid that corresponds to a map and make sketches corresponding to the minimum pathways between two points of the map.

Ongoing Assessment

Try This:

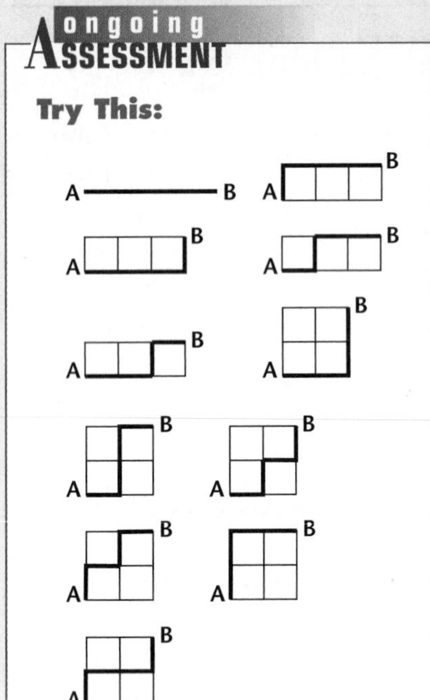

Try This This problem quickly gets complicated as the taxidistance between the points increases. Make sketches to show all of the minimum paths between points separated by a distance of 4. There are 11!

Exploration 2 *Taxicab Circles*

You will need
Grid paper

In Euclidean geometry, a circle consists of points that are a given distance from a given point. What happens if this definition is applied to points in taxicab geometry?

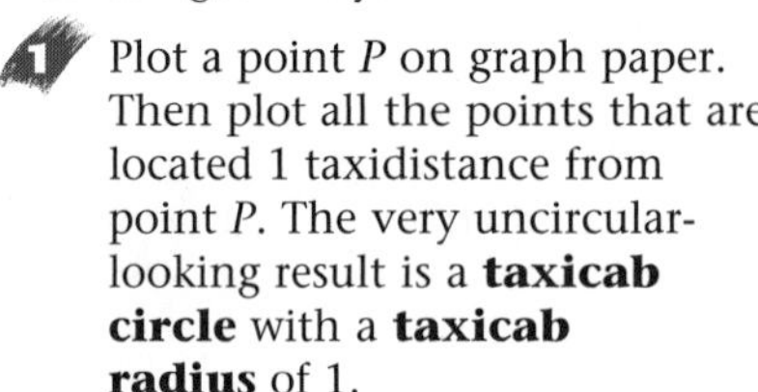

1. Plot a point P on graph paper. Then plot all the points that are located 1 taxidistance from point P. The very uncircular-looking result is a **taxicab circle** with a **taxicab radius** of 1.

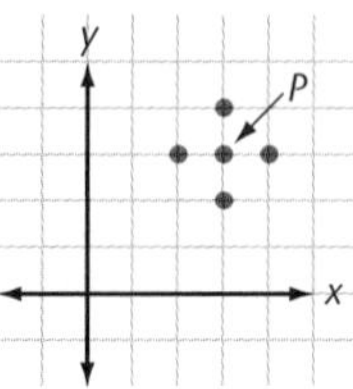

A taxicab circle with radius 1

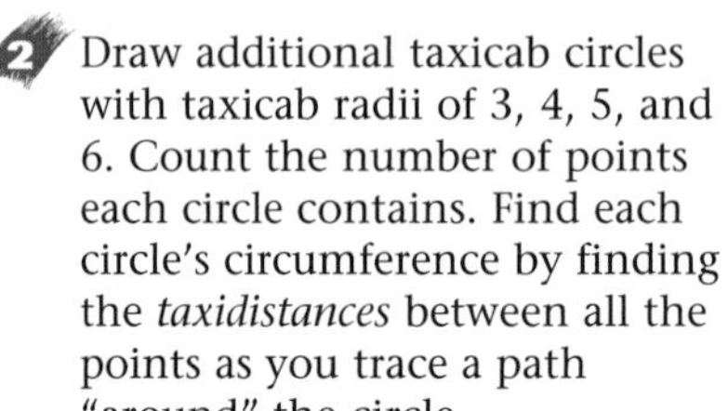

2. Draw additional taxicab circles with taxicab radii of 3, 4, 5, and 6. Count the number of points each circle contains. Find each circle's circumference by finding the *taxidistances* between all the points as you trace a path "around" the circle.

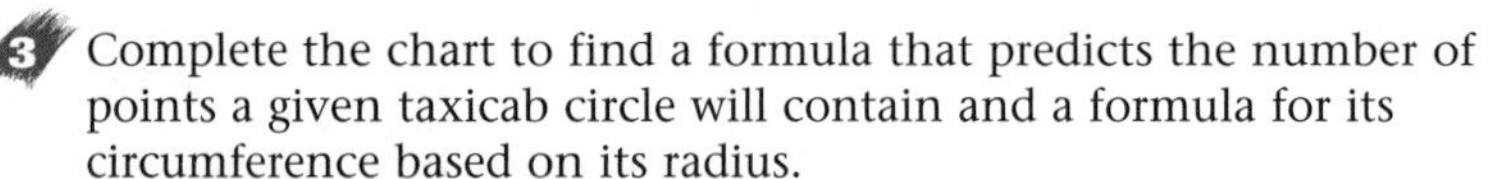

3. Complete the chart to find a formula that predicts the number of points a given taxicab circle will contain and a formula for its circumference based on its radius.

Radius	Number of points in circle	Circumference
1	5	8
2	?	?
3	?	?
4	?	?
5	?	?
r	?	?

4. Use the information in the chart to determine a taxicab equivalent for π. (Hint: $\pi = \frac{\text{circumference}}{\text{diameter}}$) ❖

Exploration 2 Notes

In this activity, students learn how to draw a taxicab circle. This activity will help students understand that a circle, and by extension each polygon, can be redefined logically within another system of geometry.

Ongoing Assessment

4. $\pi = \frac{C}{d} = \frac{8r}{2r} = 4$

ENRICHMENT Extend Exploration 2 by having students develop or verify the area formulas for taxicab triangles, quadrilaterals (including special ones), and hexagons.

INCLUSION strategies **Using Models** Students may benefit from developing their own taxicab maps. Have them use graph paper to develop a map of the area between the school building and their home. Ask them to use the map to show the pathway corresponding to the minimum distance between their home and school.

EXERCISES & PROBLEMS

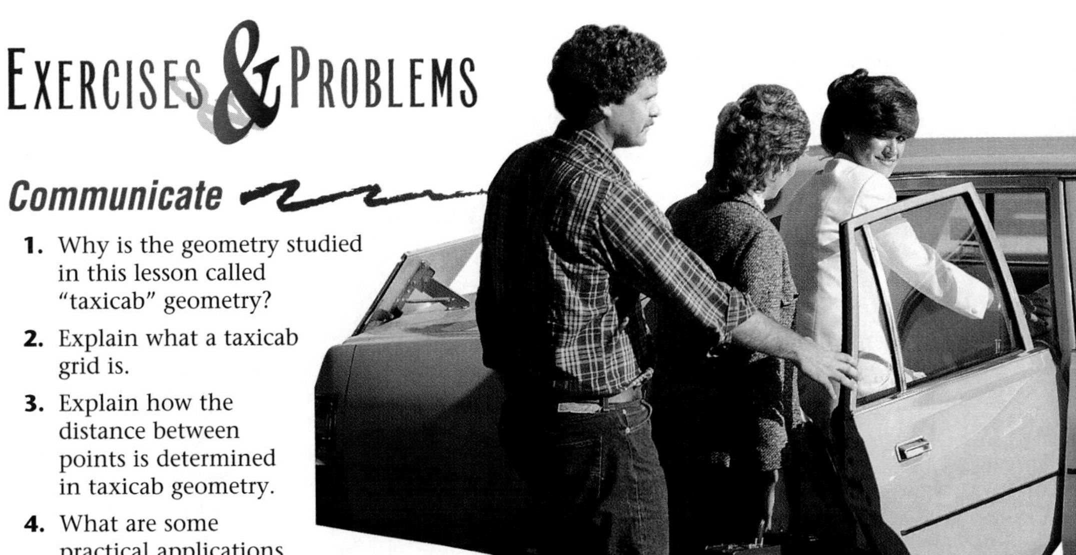

How does taxicab geometry affect the final bill for the ride?

Communicate

1. Why is the geometry studied in this lesson called "taxicab" geometry?
2. Explain what a taxicab grid is.
3. Explain how the distance between points is determined in taxicab geometry.
4. What are some practical applications of taxicab geometry?

Practice & Apply

5. Draw a grid that could be used in taxicab geometry and identify the points A(5, −3) and B(−2, 4). What is the distance between these points? How would you find the distance without counting movement around the squares?
6. Without plotting, find the taxidistance between the points (−11, 4) and (−3, 9). 13
7. Find the taxidistance between (1, 7) and (−2, −5). 15
8. Draw a grid that could be used in taxicab geometry. On that grid, find a "circle" with its center at (0, 0) and a radius of 3 units.
9. Identify two points on a taxicab grid that have a taxidistance of 4. Verify, using the formula, that the taxidistance is 4.
10. Without plotting, find the taxidistance between the points (−129, 43) and (152, 236). 474
11. Draw a grid that could be used in taxicab geometry. On the grid, find the "circle" with its center at (5, −2) and a radius of 4 units.

Though π is not used in taxicab geometry, there is a number that serves a similar purpose in finding the circumference of a taxicab circle. **Exercises 12–21 explore taxicab circles and help you to determine what that number is.**

12. Why is π not used in taxicab geometry?
13. Plot any point P on a taxicab grid. Now find all the points that are located a taxidistance of 1 from this point. Check student drawings.
14. Using different colored pencils, draw taxicab circles with radii of 2, 3, 4, 5, and 6. Check student drawings.

RETEACHING the lesson

Using Visual Models Have each student draw and label a taxicab grid. Using the origin as the center point, locate all points that are a taxidistance of 3 units from the center. Describe the resulting shape. [**a taxicab circle of radius 3**]

ASSESS

Selected Answers

Odd-numbered Exercises 5–27

Assignment Guide

Core 1–11, 14–16, 22–29

Core Plus 1–9, 12–23, 26–29

Technology

Geometry graphics software is appropriate for most of the exercises. The coordinate geometry mode is necessary for completing taxicab problems. Be sure that students use the taxicab geometry properties for finding distances.

Error Analysis

Making conjectures in Exercises 16–21 may be difficult for students. Encourage students to discuss their answers in small groups.

The answers to Exercises 5, 8, and 11 can be found in Additional Answers beginning on page 727.

9. Sample answer:

$A(2,5)$ and $B(3,8)$;

$AB = |3-2| + |8-5| = 4$

12. $\pi = \frac{C}{d}$ of a circle. However, a circle is defined differently in taxicab geometry, so π is not used.

alternative ASSESSMENT

Portfolio Assessment

The ability to describe how to find the taxidistances between any two points on a taxicab grid will demonstrate the student's understanding of taxicab geometry. Have students summarize key ideas and include diagrams as examples.

The answers to Exercises 22 and 23 can be found in Additional Answers beginning on page 727.

Practice Master

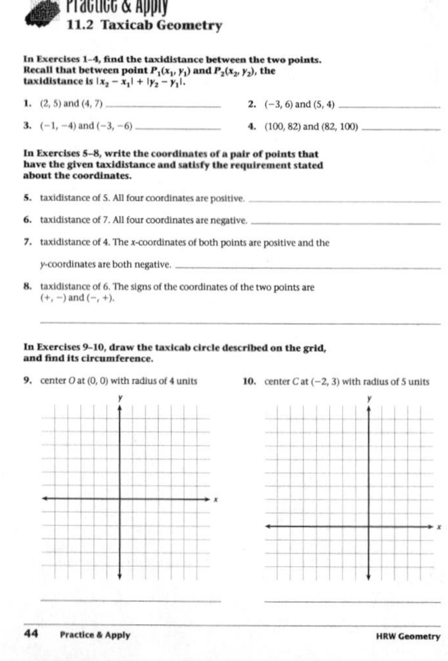
NAME ______ CLASS ______ DATE ______

Practice & Apply

11.2 Taxicab Geometry

In Exercises 1–4, find the taxidistance between the two points. Recall that between point $P_1(x_1, y_1)$ and $P_2(x_2, y_2)$, the taxidistance is $|x_2 - x_1| + |y_2 - y_1|$.

1. (2, 5) and (4, 7) ______
2. (−3, 6) and (5, 4) ______
3. (−1, −4) and (−3, −6) ______
4. (100, 82) and (82, 100) ______

In Exercises 5–8, write the coordinates of a pair of points that have the given taxidistance and satisfy the requirement stated about the coordinates.

5. taxidistance of 5. All four coordinates are positive. ______
6. taxidistance of 7. All four coordinates are negative. ______
7. taxidistance of 4. The x-coordinates of both points are positive and the y-coordinates are both negative. ______
8. taxidistance of 6. The signs of the coordinates of the two points are (+, −) and (−, +). ______

In Exercises 9–10, draw the taxicab circle described on the grid, and find its circumference.

9. center O at (0, 0) with radius of 4 units
10. center C at (−2, 3) with radius of 5 units

44 Practice & Apply HRW Geometry

15. Using the following table as a way to organize your results, count the number of points each circle contains. Then find the circumference of the taxicab circle by finding the taxidistances between all the points as you trace a path "around" the circle. The first two are done for you.

Radius	Number of points	Circumference
1	4	8
2	8	16
3	?	?
4	?	?
5	?	?
6	?	?
r	?	?

16. Making observations about the pattern in your table, predict the number of points and the circumference for a taxicab circle of radius r.

17. The number π is a ratio of what two parts of a circle? $\pi = \frac{\text{circumference}}{\text{diameter}}$

18. How could you define the diameter of a taxicab circle? Diameter = $2r$

19. In Euclidean geometry, the relationship between the radius and the diameter of a circle is $2r = d$. Is this also true in a taxicab circle? Why or why not?

20. In terms of d (diameter), what is the circumference of a taxicab circle? $c = 4d$

21. What is the taxicab geometry equivalent of π? $\frac{c}{d} = 4$ is the taxicab equivalent of π.

22. In Chapter 4, you learned that a point is equidistant from a segment's endpoints if and only if it lies on the segment's perpendicular bisector. You can use this theorem to discover what a taxicab perpendicular bisector might look like by doing the following steps.

a. On a taxicab grid, plot the points $A(0, 0)$ and $B(4, 2)$. Locate all the points that are a taxidistance of 2 from point A and from point B.

b. On the same diagram, locate all the points that are a taxidistance of 3 from both A and B.

c. Locate points that are a taxidistance of 4 from both A and B.

d. Continue locating points that are the same taxidistance from both A and B until you have constructed the "perpendicular."

23. How is the perpendicular you constructed in Exercise 22 similar to a perpendicular in Euclidean geometry? How is it different?

15.

Radius	# of Points	Circ
3	12	24
4	16	32
5	20	40
6	24	48
r	$4r$	$8r$

16. If radius = r, number of points = $4r$ and circumference = $8r$.

19. Yes. The distance from a point on the circle to another point on the circle on a line through the center of the circle is twice the distance from a point on the circle to the center of the circle.

Look Back

Find y for Exercises 24 and 25. [Lesson 9.5]

24.

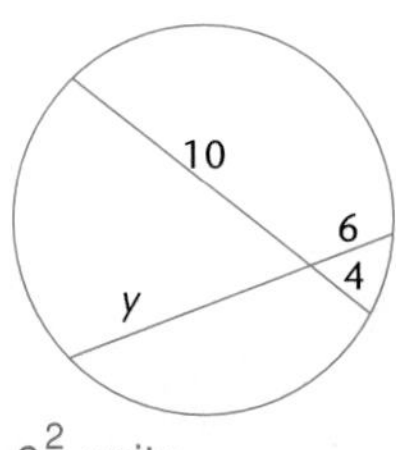

$6\frac{2}{3}$ units

25.

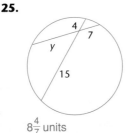

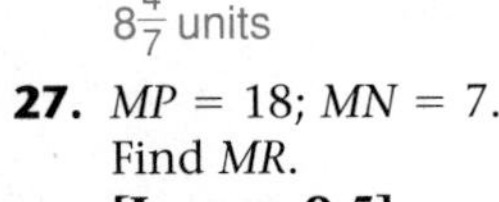

$8\frac{4}{7}$ units

26. $BC = 5$; $AC = 22$; $CD = 6$. Find DE. **[Lesson 9.5]**

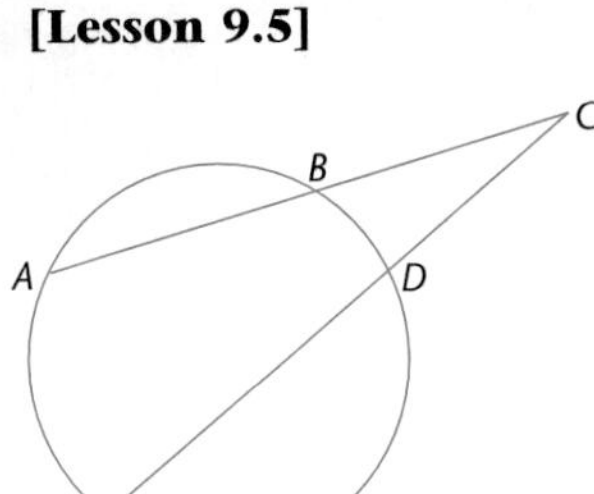

$12\frac{1}{3}$ units

27. $MP = 18$; $MN = 7$. Find MR. **[Lesson 9.5]**

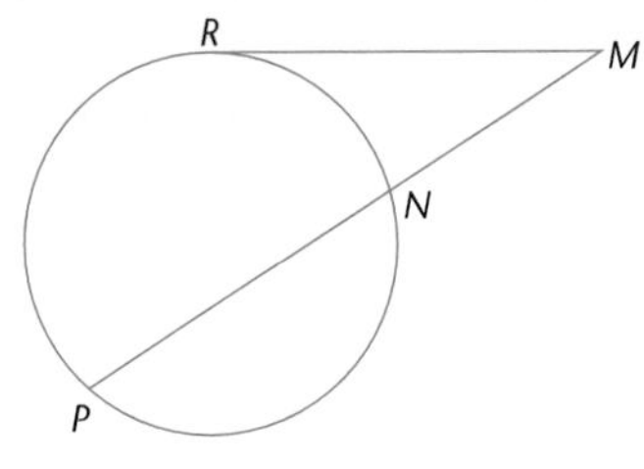

11.22 units

Look Beyond

Consider the following statements:

If today is February 30, then dogs can drive taxicabs.

Today is February 30.

Therefore, dogs can drive taxicabs.

28. Does this argument follow a proper order of reasoning? Explain your answer.

29. Can you determine if the argument is true or false? Explain your answer.

28. Yes. If a conditional statement is true, then the conclusion follows if the hypothesis is true.

29. The argument is false, because there is no February 30.

Look Beyond

Exercises 28 and 29 will help prepare students for Chapter 12. The statements are an example of modus ponens, the principle commonly used to make deductions in geometry proofs.

Technology Master

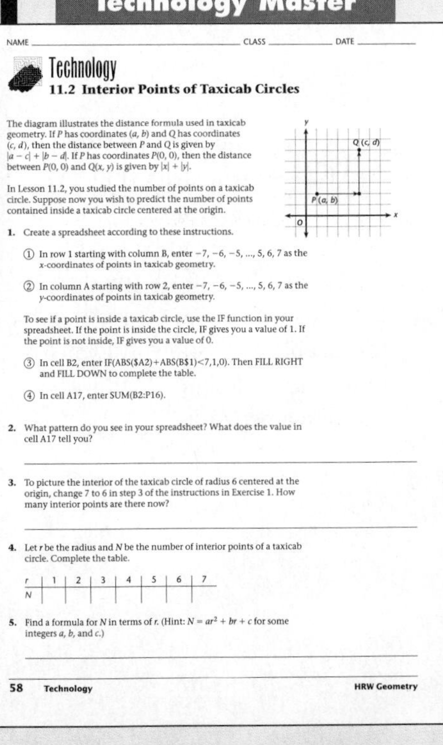

NAME ______________ CLASS ________ DATE ________

Technology

11.2 Interior Points of Taxicab Circles

The diagram illustrates the distance formula used in taxicab geometry. If P has coordinates (a, b) and Q has coordinates (c, d), then the distance between P and Q is given by $|a - c| + |b - d|$. If P has coordinates $P(0, 0)$, then the distance between $P(0, 0)$ and $Q(x, y)$ is given by $|x| + |y|$.

In Lesson 11.2, you studied the number of points on a taxicab circle. Suppose now you wish to predict the number of points contained inside a taxicab circle centered at the origin.

1. Create a spreadsheet according to these instructions.

   ① In row 1 starting with column B, enter −7, −6, −5, ..., 5, 6, 7 as the x-coordinates of points in taxicab geometry.

   ② In column A starting with row 2, enter −7, −6, −5, ..., 5, 6, 7 as the y-coordinates of points in taxicab geometry.

   To see if a point is inside a taxicab circle, use the IF function in your spreadsheet. If the point is inside the circle, IF gives you a value of 1. If the point is not inside, IF gives you a value of 0.

   ③ In cell B2, enter IF(ABS($A2)+ABS(B$1)<7,1,0). Then FILL RIGHT and FILL DOWN to complete the table.

   ④ In cell A17, enter SUM(B2:P16).

2. What pattern do you see in your spreadsheet? What does the value in cell A17 tell you?

____

3. To picture the interior of the taxicab circle of radius 6 centered at the origin, change 7 to 6 in step 3 of the instructions in Exercise 1. How many interior points are there now?

____

4. Let r be the radius and N be the number of interior points of a taxicab circle. Complete the table.

r	1	2	3	4	5	6	7
N							

5. Find a formula for N in terms of r. (Hint: $N = ar^2 + br + c$ for some integers a, b, and c.)

____

58 Technology HRW Geometry

PREPARE

Objectives

- Develop the concept of Euler-type networks.
- Solve problems involving network traverses and Euler circuits.

RESOURCES

- Practice Master 11.3
- Enrichment Master 11.3
- Technology Master 11.3
- Lesson Activity Master 11.3
- Quiz 11.3
- Spanish Resources 11.3

Assessing Prior Knowledge

1. How is the term network commonly used? [**Answers will vary.**]
2. How is the number of vertices of a polygon related to the number of sides?

   [**They are equal.**]

Use Transparency 99

TEACH

why Studying networks will help students recognize how points and lines can be used to create mathematical models. Learning whether a network is traversable will show students how to design pathways through a network.

LESSON 11.3 Networks

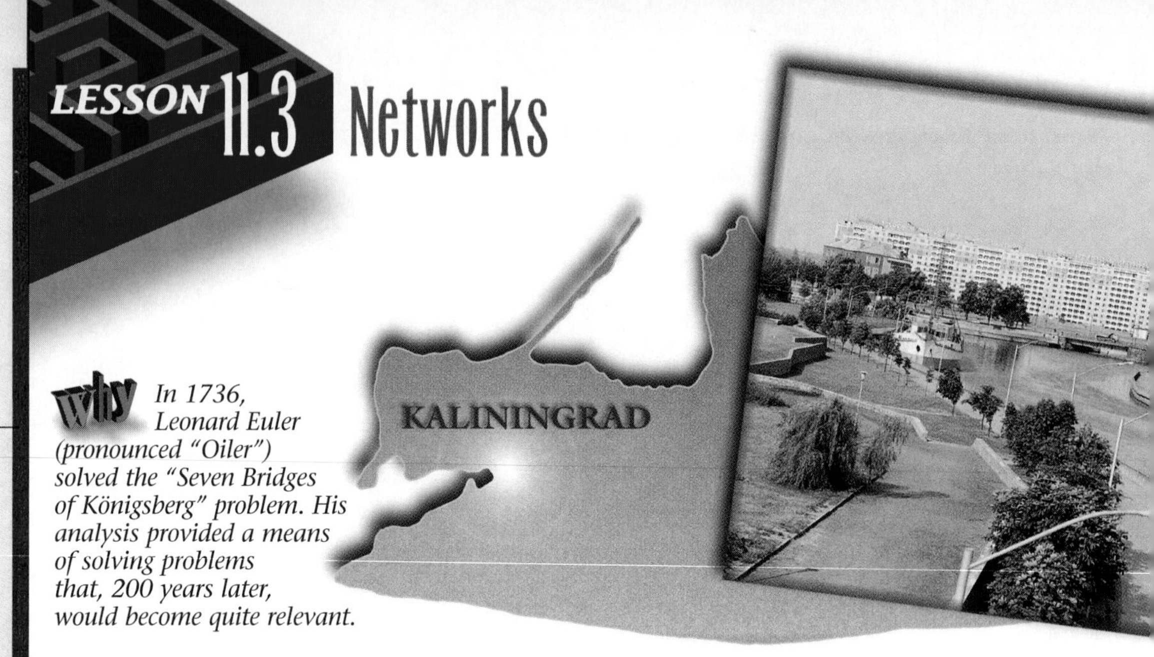

why In 1736, Leonard Euler (pronounced "Oiler") solved the "Seven Bridges of Königsberg" problem. His analysis provided a means of solving problems that, 200 years later, would become quite relevant.

The Bridge Problem

On Sunday afternoons in the 1700s, people in the old city of Königsberg (now Kaliningrad, Russia) developed an interesting mathematical pastime. As people took their Sunday strolls, they wondered if it was possible to cross each of the seven bridges of the city once and only once. The diagram below represents the city. It shows the two islands in the river that runs through the city, as well as the seven connecting bridges.

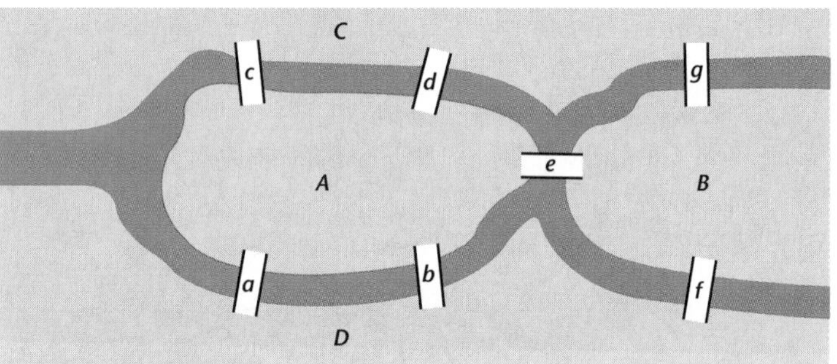

Euler represented the problem by a diagram called a **network**. The points represent the land masses, and the line segments and arcs represent the bridges. The bridge problem becomes that of traversing the network—that is, of going over every path once and only once—without lifting your pencil from the paper. The points can be crossed as many times as you like.

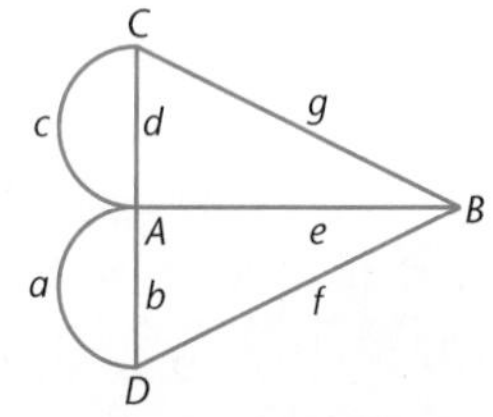

The shapes and sizes of the land masses are irrelevant.

ALTERNATIVE teaching strategy

Technology For extra practice, use software that creates networks in order to demonstrate some of the concepts in this lesson. Encourage students to experiment with different paths as they try to traverse a network on the computer.

Odd and Even Vertices

You may have tried to traverse the "Seven Bridges" network and decided that it was impossible. The people of Königsberg certainly thought so, but no one could prove it with logical argument until Euler presented his proof.

As Euler discovered, the traversability of a network can be determined by classifying its points, or vertices, as **even** or **odd**. Odd vertices have an odd number of paths (segments or arcs) going to them. Even vertices have an even number of paths going to them.

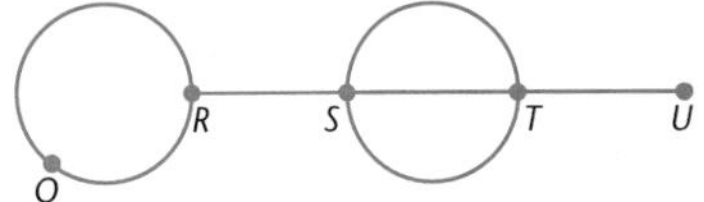

In the diagram, *U* has one path going to it, so it is an odd vertex. *R* is also an odd vertex, with three segments going to it. *Q*, *S*, and *T* are even vertices.

The network is traversable. One possible route begins at *U* and traces a path represented by the letter sequence *UTSTSRQR*. What other paths are possible?

Exploration Networks

You will need
No special tools

Which of the networks seem to be untraversable? Make a table like the one at the bottom of the page and see if you can discover Euler's proof.

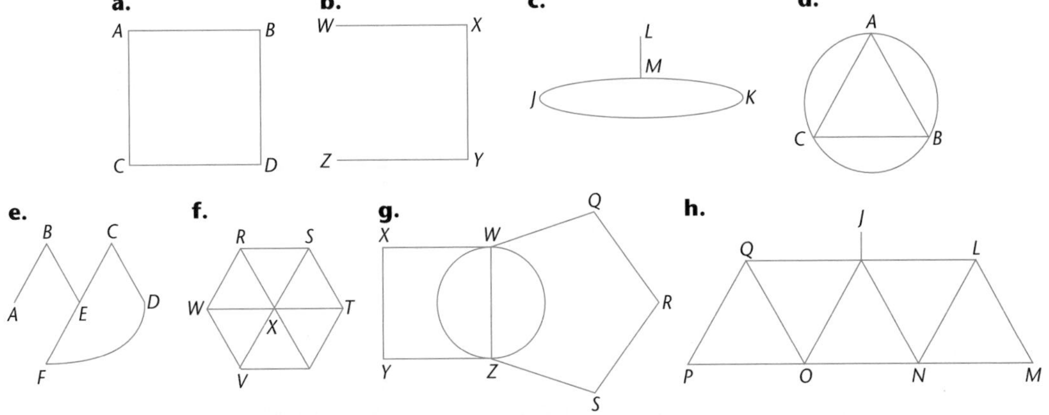

Number of vertices	Number of odd vertices	Number of even vertices	Can the network be traversed?
a. ?	?	?	?
b. ?	?	?	?
c. ?	?	?	?
•	•	•	•

Use Transparency 100

Teaching tip

Make sure students understand how the network at the bottom of the previous page represents the "Seven Bridges" problem. Point out that in a network, a large body of land may be treated as a single point.

Cooperative Learning

Use cooperative groups to explore the idea of vertices being even or odd. Have students in the group draw various networks that represent even or odd vertices. Group members should share and discuss the networks.

Exploration Notes

In this exploration, students create a chart of even and odd vertices to help them discover the content of Euler's proof. This activity will help students understand that if all vertices in a network are even, the network is traversable.

interdisciplinary CONNECTION

Social Studies One of the jobs of a city planner is to make sure roads are designed so that many parts of the city are connected. Ask students to find examples of cities that are well planned and cities that are not well planned. Ask them to describe some of the difficulties that city planners often face when planning cities.

Alternate Example 1

Make a network from the house plan shown below, indicating all possible routes between the rooms.

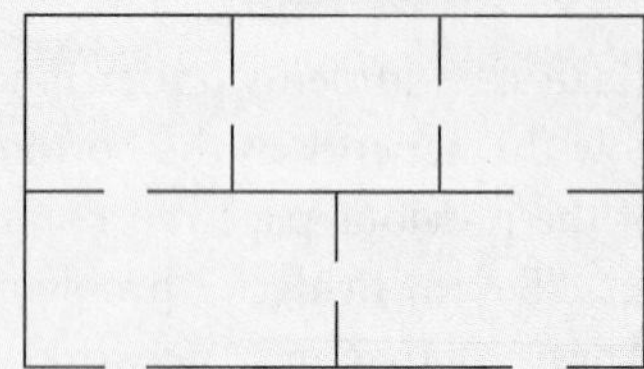

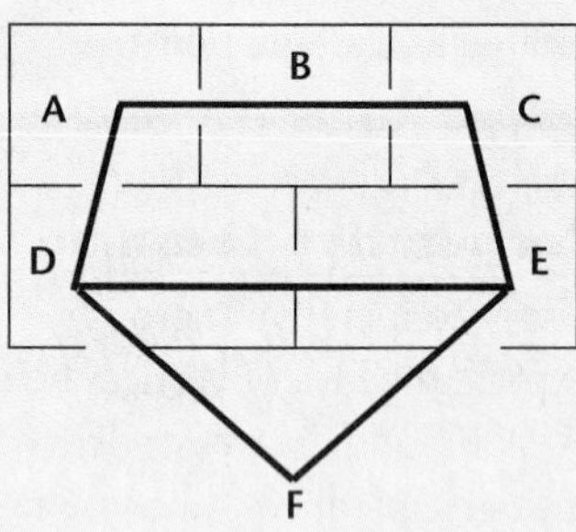

CRITICAL Thinking

There is just one "space" outside the house, so it is convenient and simpler to represent it as a single point.

EXAMPLE

Make networks from the plans of the two houses, indicating all possible routes between the rooms.

A **B**

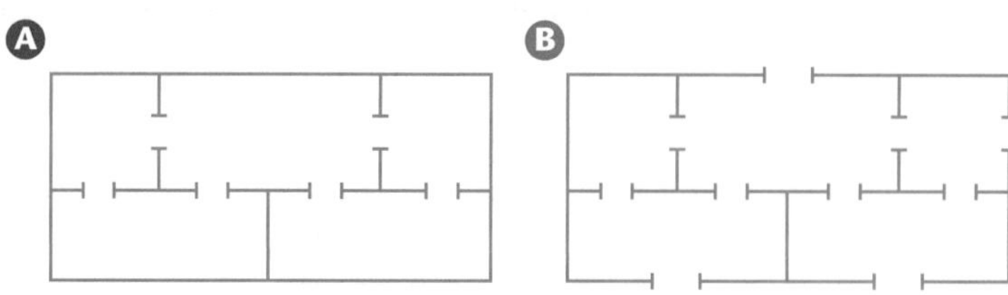

Solution ➤

A Draw and label five points to represent each of the rooms. Connect the points to indicate doorways from one room to the next. Notice that there is no connection between *D* and *E* in the network because there is no doorway between the rooms.

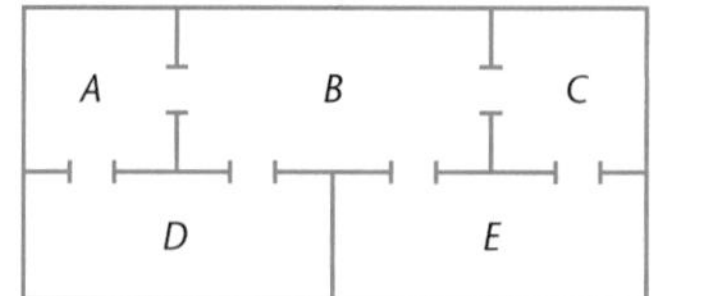

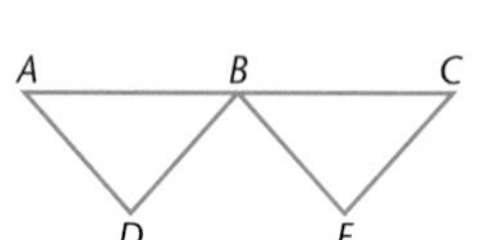

B Two possibilities for representing the space outside the house are shown below, along with their networks. ❖

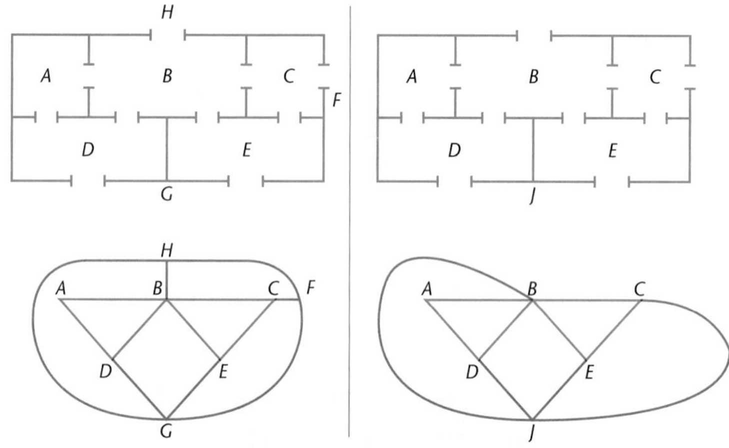

CRITICAL Thinking

Which of the two possibilities for representing the space outside the house do you prefer? Which do you think is the better abstraction of the problem? Give reasons for your answer. Does the "shape" of the space outside the house matter? Is there more than one "space" outside the house?

ENRICHMENT An *Euler path* is a path drawn so that it starts and ends at two different vertices and no edge is passed more than once. Draw an example of an Euler path.

INCLUSION strategies **Using Cognitive Strategies** Some students may be more comfortable using a more analytical approach to analyzing a network. (1) Identify the vertices or nodes of the network. (2) Count the number of even vertices and the number of odd vertices. (3) The number of odd vertices of a network must be even, namely 0 or 2, for the network to be traversable. (4) A network with more than 2 odd vertices cannot be traversed. (5) If the network has 2 vertices start at one odd vertex and end up at the other. (6) If the network has 0 odd vertices, start anywhere.

Starting and Stopping

The following problem may help you understand Euler's proof.

While hiking one snowy day, you startle a rabbit. It scampers out of the woods and runs from bush to bush, hiding under each one. It covers all the paths between the bushes without going over the same path more than once. Then it stops. Which bush is the rabbit hiding under?

After a few tries you should see that the rabbit always ends up under bush *E*, an odd vertex. The only other odd vertex is the woods, where the rabbit's journey begins.

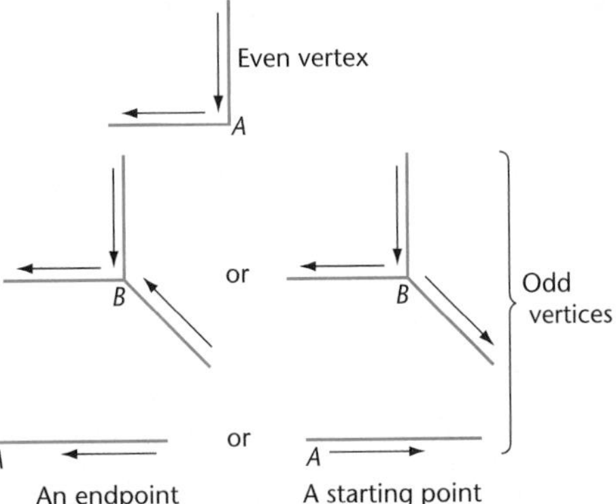

Consider the difference between an odd and even vertex when traversing a network: *An even vertex always allows a path into the point and a different path out of it.*

Even vertex
A

If you are attempting to traverse a network, one of the paths of an odd vertex will eventually lead into a vertex with no unused path leading out.

This means that an odd vertex can only be the beginning or ending point of a route in a network. Otherwise a retracing will occur, or a part of the network will remain untraversed.

B or B
Odd vertices
A or A
An endpoint
A starting point

This insight about odd vertices provided Euler with a convincing explanation that solved the problem of whether a given network can be traversed. From the data in your exploration, it appears that networks with only even vertices can always be traversed. This is in fact true, since there will always be an unused path out of an even vertex if it is needed to complete a route.

Networks with odd vertices can be traversed, but such points must either begin or end the route. Otherwise a retracing will occur, or a part of the network will remain untraversed. When a network contains exactly two odd vertices, as in the rabbit problem, one odd vertex must begin the journey and the other must end it.

Networks with more than two odd vertices are impossible to traverse. Do you think it is possible for a network to have just one odd vertex?

Further Developments From Euler's Result

Euler's analysis of the "Seven Bridges" problem led to the development of two new areas of mathematics: topology and graph theory. You will learn about topology in the next lesson.

Teaching tip

Represent the hallways of the school with a network and ask students to decide whether it is traversable using Euler's analysis.

Ongoing Assessment

No. If all the other vertices were even, then one path out of the odd vertex would be unconnected.

Reteaching the lesson

Cooperative Learning Arrange the class in small groups. Have members of each group draw a network and use it to summarize the lesson to each other. Ask each group to draw its network on the chalkboard and have the class decide whether it is traversable.

Assess

Selected Answers

Exercises 7, 9, 11, 12, 14, 23, 25, and 27

Assignment Guide

Core 1–8, 12–15, 17–27

Core Plus 1–28

Technology

Students may use computer-aided design programs to draw floor plans of houses for Exercises 12–15 or for other floor plans.

Error Analysis

Suggest that students who are having difficulty with Exercises 17–19 draw simpler diagrams (with fewer dots) to represent the parking meters. These simpler diagrams may help them see the patterns.

Practice Master

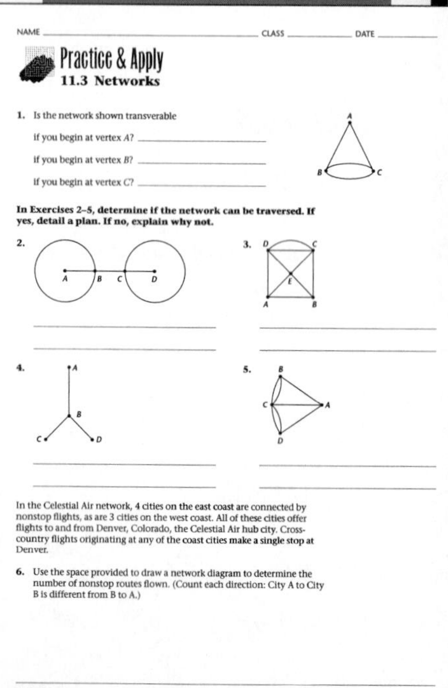

NAME ______ CLASS ______ DATE ______

Practice & Apply
11.3 Networks

1. Is the network shown transversable
   if you begin at vertex *A*? ______
   if you begin at vertex *B*? ______
   if you begin at vertex *C*? ______

In Exercises 2–5, determine if the network can be traversed. If yes, detail a plan. If no, explain why not.

2. 3. 4. 5.

In the Celestial Air network, 4 cities on the east coast are connected by nonstop flights, as are 3 cities on the west coast. All of these cities offer flights to and from Denver, Colorado, the Celestial Air hub city. Cross-country flights originating at any of the coast cities make a single stop at Denver.

6. Use the space provided to draw a network diagram to determine the number of nonstop routes flown. (Count each direction: City A to City B is different from B to A.)

HRW Geometry — Practice & Apply 45

Exercises & Problems

Communicate

1. What was the inspiration for Euler's method of analyzing networks?
2. Describe when a network can be traversed.
3. If the soccer ball traverses the network, which player must be the starting point? the ending point? Explain why.
4. Can the seven bridges of Königsberg be traversed? Explain your reasoning.
5. Describe some current applications of network theory.

Practice & Apply

6. Draw a network illustrating all possible routes between rooms in Figure 1.
7. Can the network in Figure 1 be traversed? Explain your reasoning. Yes. There are 2 odd vertices.
8. Draw a network illustrating all possible routes between rooms in Figure 2. Notice that outside doors have been added. Can this network be traversed? Explain your reasoning.

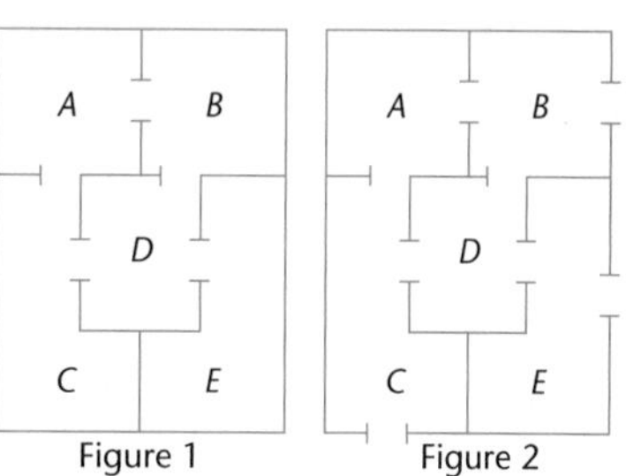

Figure 1 Figure 2

Below is a drawing adapted from Euler's original paper. He used this as an example to help illustrate the concept behind the bridges problem. Use this illustration for Exercises 9–11.

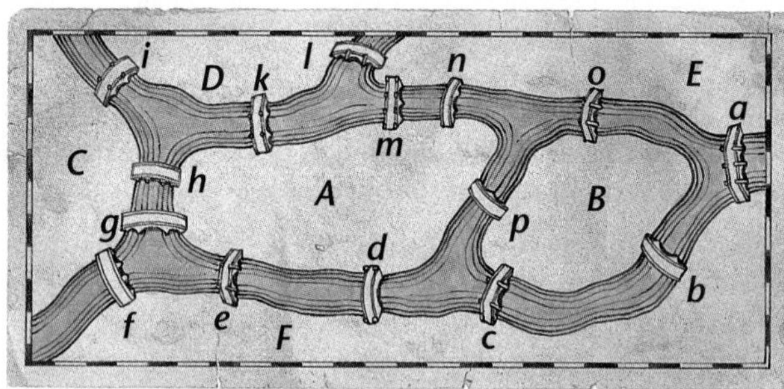

9. Draw Euler's diagram as a simplified network.
10. Determine if this network can be traversed. Explain your reasoning.
11. If you were to traverse this network, where would you begin and where would you end?

6. A B D C E

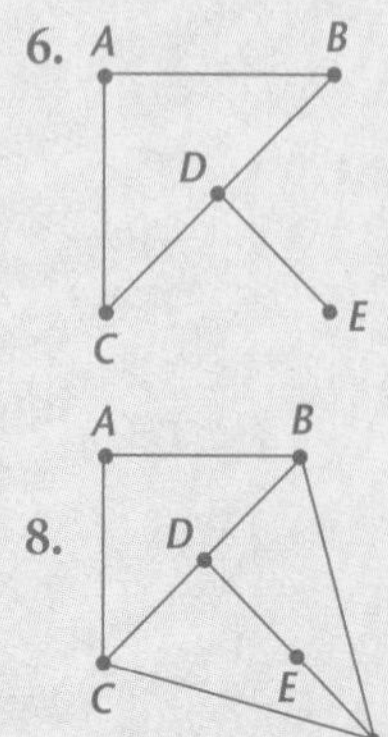

8. No. There are more than two odd vertices.

9.

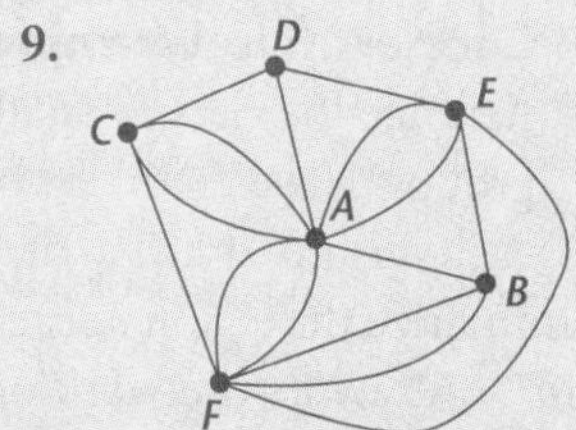

10. Yes. There are exactly two odd vertices—*D* and *E*.
11. Start at *D* and end at *E*. Or, start at *E* and end at *D*.

12. The following is a floor plan of a house. Draw a network that represents the circulation in the house.

13. Is it possible to walk through the house in Exercise 12 and pass through each door exactly once? Must you start or end at a particular place, or can you start anywhere? Explain your reasoning.

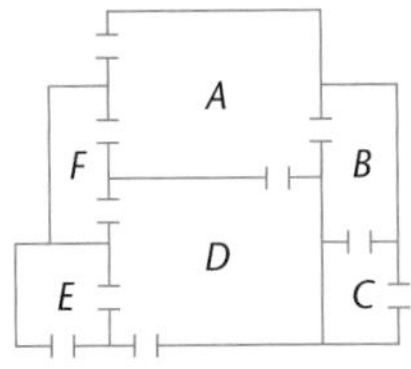

14. The following is a floor plan of a house. Draw a network that represents the circulation in the house.

15. Is it possible to walk through the house in Exercise 14 and pass through each door exactly once? Must you start or end at a particular place, or can you start anywhere? Explain your reasoning.

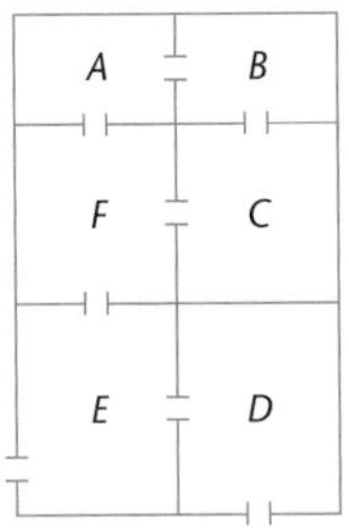

16. An **Euler circuit** is a network that has only even vertices. What is the benefit of an Euler circuit over a network that can simply be traversed?

Law Enforcement A member of the police force is collecting money from the parking meters on the streets diagramed below. The dots represent the meters. If meters occur on both sides of the street, the police officer must go down the street twice. Use this diagram to complete Exercises 17–19.

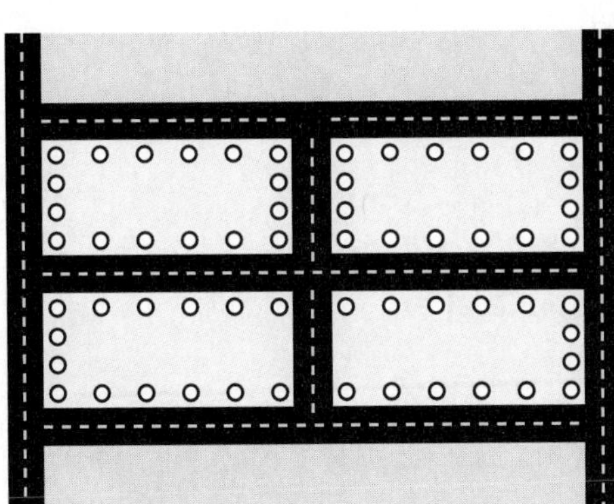

17. Draw a network that represents the parking meter problem.

18. Is this network traversable? Is it an Euler circuit? Explain your reasoning.

19. Where should the officer park in order to start and end at the same point, and to take the most efficient route possible? Explain your reasoning.

12.

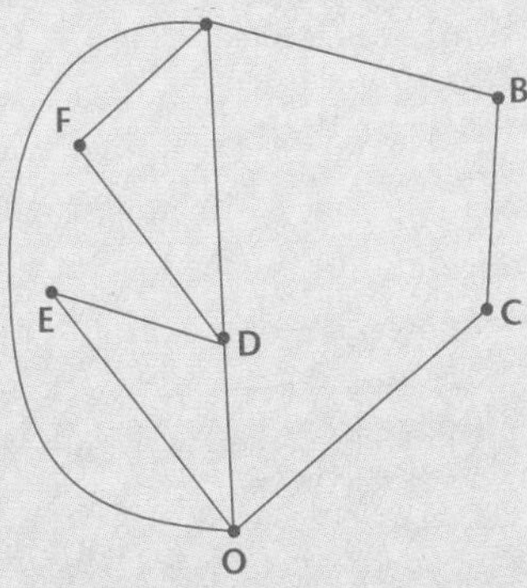

13. Yes. All the rooms and the outside are even vertices – each has an even number of doors. Since every vertex is even, you can start anywhere. There will always be a door to leave through for every door you come in through.

alternative ASSESSMENT

Authentic Assessment

Ask students to draw a network of local bus routes, newspaper routes, communication channels, or other real-life examples of networks. Ask them to determine whether the networks are traversable.

15. Yes, there are two odd vertices, *E* and *F*. Since two of the vertices are odd, you must start at one end or the other.

16. An Euler Circuit can be traversed starting anywhere in the circuit. If there are odd vertices, you must start and end at the odd vertices to traverse the network.

The answers to Exercises 14 and 17–19 can be found in Additional Answers beginning on page 727.

Technology Master

NAME ____________ CLASS ________ DATE ________

Technology
11.3 Communication Matrices

The diagram shows a group of five people in a network who, in some way, can communicate with one another. For example, persons *A* and *B* can communicate back and forth directly. Person *D* can communicate directly with person *C*, but person *C* cannot communicate directly with person *D*.

Matrix *M* represents the lines of communication. An entry of 1 indicates direct communication from one member to another. An entry of 0 indicates that there is no direct line of communication from the member to another.

1. Enter matrix *M* into your graphics calculator.

$$M = \begin{array}{c} \\ A \\ B \\ C \\ D \\ E \end{array} \begin{array}{c} \begin{array}{ccccc} A & B & C & D & E \end{array} \\ \begin{bmatrix} 0 & 1 & 1 & 0 & 1 \\ 1 & 0 & 1 & 0 & 0 \\ 1 & 0 & 0 & 0 & 1 \\ 1 & 0 & 1 & 0 & 1 \\ 1 & 0 & 0 & 0 & 0 \end{bmatrix} \end{array}$$

Use matrix *M* and the exponent key to find each matrix power. Record your entries in the space given.

2. $M^2 =$ [ ]

3. $M^3 =$ [ ]

4. How many communication paths are there from member *B* to member *E* via one intermediate group member? Compare this number to the entry in row 2 column 5 of M^2.

____

5. How many communication paths are there from member *C* to member *A* via two intermediate group members? (You may enter and exit *A* along the way if you wish.) Compare this number to the entry in row 3 column 1 of M^3.

____

6. Experiment with other lines of communication that require one intermediary. What do you think M^2 tells you about the network?

____

7. Experiment with other lines of communication that require two intermediaries. What do you think M^3 tells you about the network?

____

HRW Geometry Technology 59

You may be familiar with a game where the challenge is to draw a figure in one stroke without lifting the pencil from the page or retracing steps. Use the figures for Exercises 20–22.

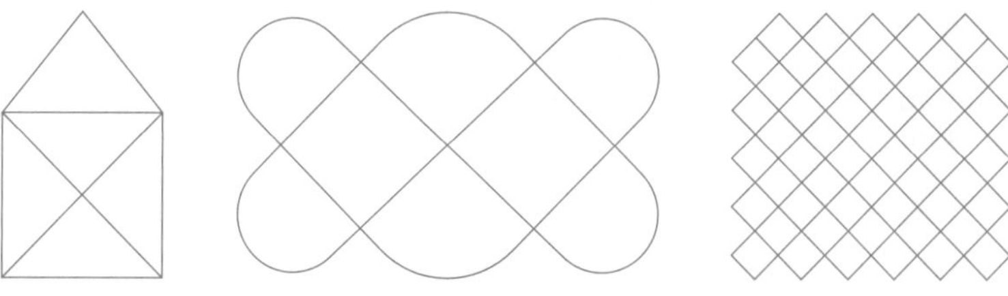

20. Which, if any, are traversable? How can you tell?

21. Which, if any, are Euler circuits? How can you tell?

22. If any of the above are traversable but are not Euler circuits, where must you start and end? Why?

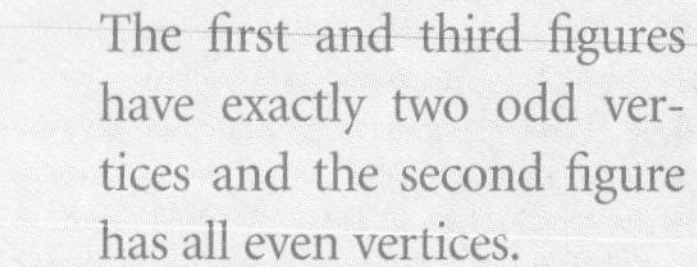

20. All three are traversable. The first and third figures have exactly two odd vertices and the second figure has all even vertices.

21. The second figure is an Euler Circuit because all its vertices are even.

22. You must start at one of the odd vertices and end at the other odd vertex. The odd vertices don't have a way out for every way in. When you start you only need a way out and when you end you only need a way in, so an odd vertex will work.

The answers to Exercises 23 and 24 can be found in Additional Answers beginning on page 727.

Look Back

Copy and reflect each figure with respect to the given line. [Lesson 1.6]

23.

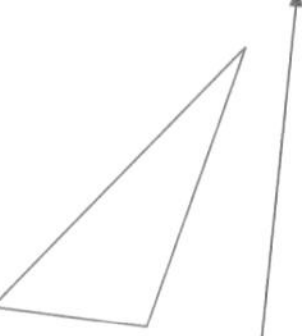

24.

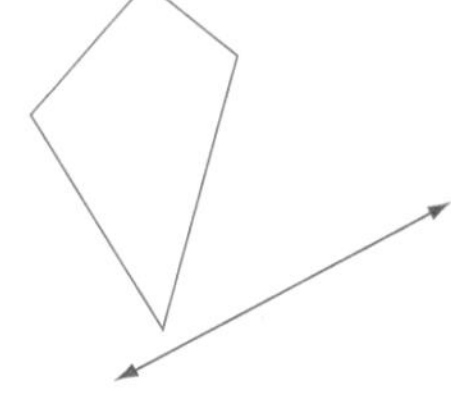

Write each statement below as a conditional. [Lesson 2.2]

25. All people who live in California live in the United States.

26. The measures of the angles of a triangle sum to 180°.

27. A square is a parallelogram with four congruent sides and four congruent angles.

Look Beyond

28. Suppose you draw a triangle on a non-inflated balloon and then add air to the balloon. Will the angles of the triangle still sum to 180°? Why or why not?

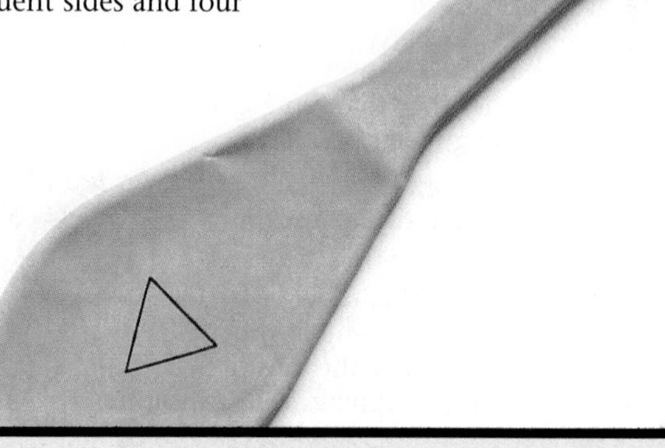

Look Beyond

Exercise 28 anticipates the study of Riemanian geometry in Lesson 11.5.

25. If a person lives in California, then the person lives in the USA.

26. If a figure is a triangle, then the sum of the measures of its angles is 180°.

27. If a parallelogram has four congruent sides and four congruent angles, then it is a square.

28. No. The sum of the measures of the angles of a triangle will be 180° as long as the triangle is on a flat surface – a plane. When the balloon is inflated, the angles will get bigger and their sum will be greater than 180°.

LESSON 11.4 Topology: Twisted Geometry

Are the faces of the panda and the tiger topologically equivalent?

why *Euler's idea of a network, which showed only connections between parts, was extended to geometric shapes in the branch of mathematics known as* ***topology***—*with some startling results. This field of study has proven to be useful in such fields as chemistry, biology, and cosmology.*

Knots, Pretzels, Molecules, and DNA

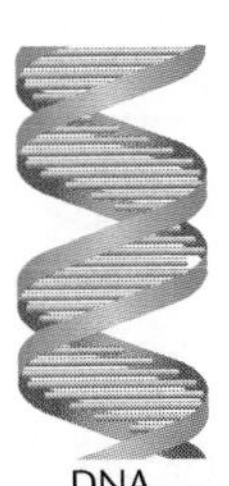

DNA

Möbius molecule

Genetics The shape at right is known as a **Möbius strip**, a topological classic with intriguing properties. (You will explore Möbius strips in the exercises.) The mathematical analysis of such objects has proven useful in other fields, such as chemistry and biology.

David Walba and his co-workers at the University of Colorado have synthesized a molecule in the shape of a Möbius strip. Structures like this help them understand why certain drugs with the same molecular structure can have vastly different effects.

Researchers have also used a branch of topology called **knot theory** to study the structures of DNA. The strand of DNA has the same structure as a **trefoil knot**.

ALTERNATIVE teaching strategy

Hands-On Strategies Help students understand topological equivalence by having them compare real-life objects.

PREPARE

Objectives

- Explore and develop general notions about geometric topology including knots, Moebius Strips, and toruses.
- Use Jordan's Theorem and other properties to solve problems in topology.

RESOURCES

- Practice Master 1.4
- Enrichment Master 1.4
- Technology Master 1.4
- Lesson Activity Master 1.4
- Quiz 1.4
- Spanish Resources 1.4

Assessing Prior Knowledge

1. Imagine a line drawn from the North Pole to a point on the equator. Is the line straight? Why or why not? Is it the shortest distance? Explain. [**Sample: The line is "straight" because it is the shortest distance.**]
2. Imagine a balloon with a printed message on it. Draw how the message would look if the balloon were deflated. [**Answers will vary.**]

TEACH

why Topology is a good introduction to non-Euclidean geometry because students may already be familiar with some of the applications to chemistry, biology, and graphic design.

TEACHING *tip*

Carefully explain the circumstances under which two shapes are topologically equivalent. Clay can be used to demonstrate the classic example that a doughnut is topologically equivalent to a coffee cup.

Cooperative Learning

Have each student in a group draw and label two figures that have invariant properties. Ask them to draw a standard polygon and then distort it. List all invariant properties. Also create solid shapes with invariant properties. Present unusual examples to the class. Working in groups gives students time to explore abstract principles.

CRITICAL *Thinking*

Yes. The definitions could be adapted to three dimensions. A simple closed figure could enclose a distinct space without crossing itself in a self-intersection. Other definitions could be adapted to three dimensions in the same way.

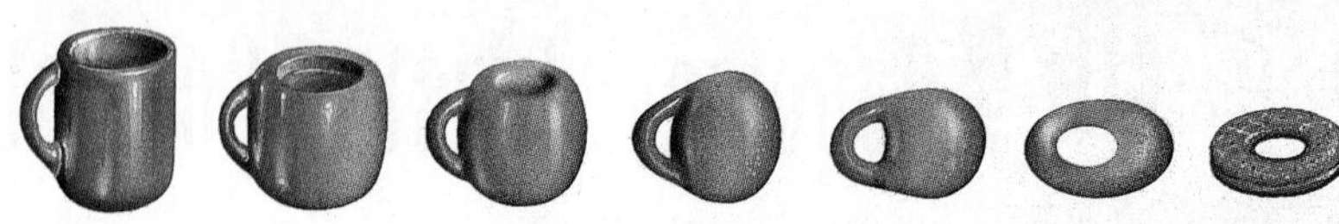

Imagine that the coffee cup is made of extremely elastic material so that it can be easily stretched or squashed. As you see, it can be deformed into a doughnut shape.

Do you think it is possible to unlink the rings of the "pretzel" without tearing or cutting in any way?

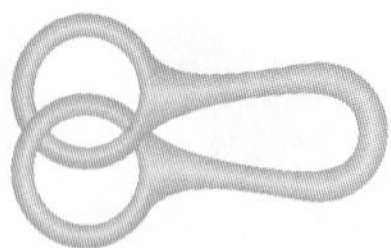

Topological Equivalence

In topology, two shapes are equivalent if one of them can be stretched, shrunk, or otherwise distorted into the other without cutting or tearing. In a plane, all of the shapes below are topologically equivalent.

Such shapes are called **simple closed curves**. They enclose a distinct region, and their segments or curves do not cross themselves.

Curves or shapes whose lines cross over each other form other categories of shapes.

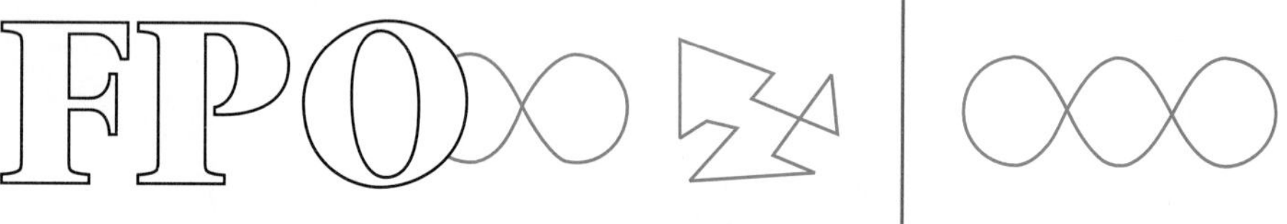

The two shapes on the left are topologically equivalent to each other but not to the shape on the right, since the latter contains two intersections instead of one.

CRITICAL *Thinking*

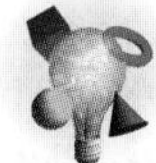

Do you think that there is a categorization, involving "loops" and intersections, that occurs for 3-dimensional objects similar to the 2-dimensional objects above? Explain.

interdisciplinary CONNECTION **Chemistry** A new branch of chemistry called chemical topology studies the configuration of chemical molecules. Have students create a model molecule from straws and marshmallows and use it to explain some topological properties.

Jordan's Theorem

A fundamental theorem in topology was first stated by the French mathematician Camille Jordan (1838–1922) in the nineteenth century.

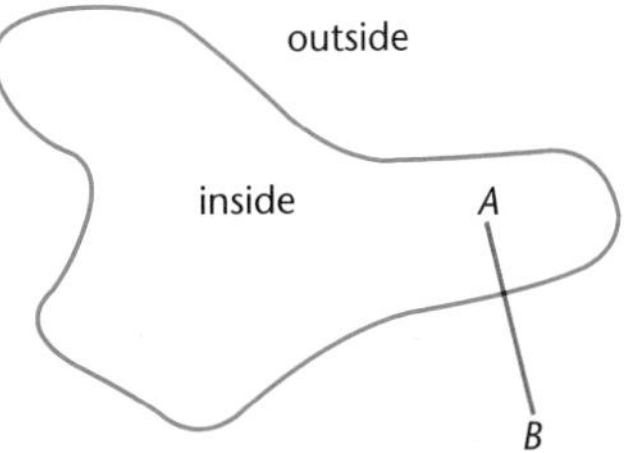

JORDAN'S THEOREM

Every simple closed curve divides the plane into two distinct regions. To connect a point inside the region with a point outside, you must "cut" the curve. **11.4.1**

The theorem, which is very difficult to prove mathematically, seems obvious for the curve shown above. But for some curves it is not so simple. Are points *P* and *Q* inside or outside this "simple" closed curve?

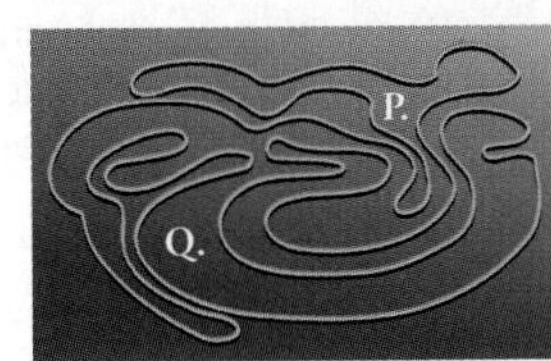

Jordan's Theorem is also true for surfaces other than planes. For example, the equator divides the Earth into two regions. To go from one region to the other you must cross or "cut" the equator.

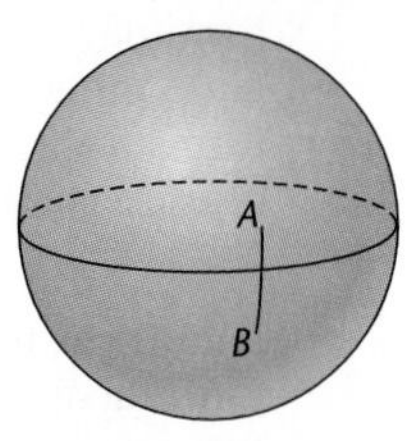

On the surface of a donut shape, however, Jordan's Theorem does not hold. Use the illustration to explain why.

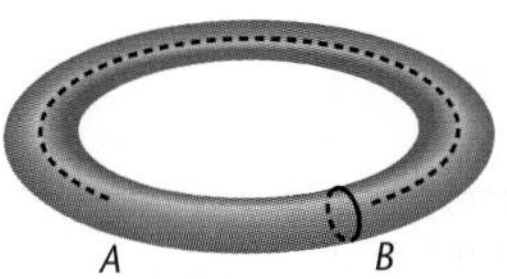

Topologists call a doughnut-shaped surface a **torus**. Because Jordan's Theorem is true for a sphere but not for a torus, topologists can conclude that the two shapes are not topologically equivalent without trying to see if they can "deform" one into the other.

Invariants

Properties that stay the same no matter how a figure is deformed are called **invariants**. If a pentagon is distorted into a curve, the distances between the vertices may change, but the order of the points around the circle stays the same. The order of the points is an invariant.

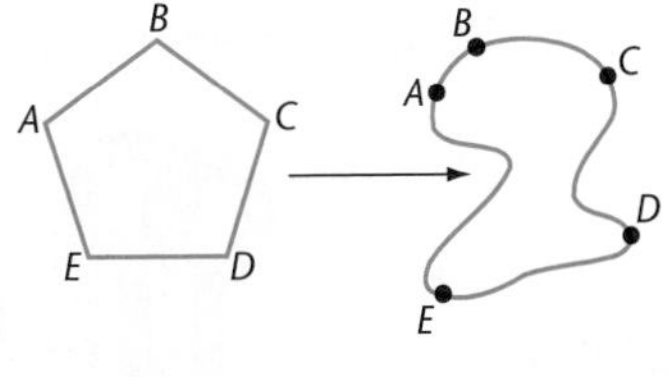

TEACHING *tip*

Using string models may help students distinguish which properties of plane figures are invariant.

ongoing ASSESSMENT

The circle around the torus does not divide the surface of the torus into two distinct regions because it is possible to connect *A* to *B* without cutting the circle by going around the torus.

ENRICHMENT Knot theory is a branch of three-dimensional topology. Have students find out the basic definition of a knot and examples of when two knots are topologically equivalent.

English Language Development Some students may have difficulty understanding the terminology in this lesson. Have them list new terms and concepts that are confusing to them. Ask them to draw an example of how each term is applied in this lesson.

TEACHING tip

Point out that Euler's formula applies to networks as the students learned them in Lesson 11.3. Ask students to explain how the examples on this page are also networks.

One of the most important invariants in topology comes from Euler's formula for solids.

> **EULER'S FORMULA**
> For any polyhedron with vertices V, edges E, and faces F, $V - E + F = 2$. **11.4.2**

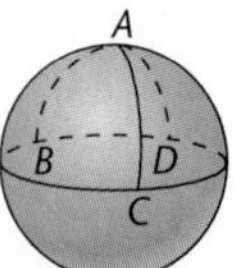

Imagine distorting the pyramid into a sphere. You can see that the sphere is separated into four regions corresponding to the four faces of the pyramid. (The face, BCD, is the entire lower hemisphere.)

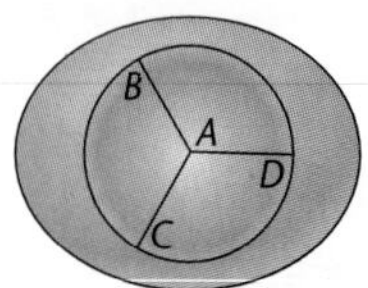

Next imagine deflating the sphere and flattening it out like a pancake. You can verify that Euler's formula works for the resulting topologically equivalent shape. Notice that the bottom face BCD is the region surrounding the circle in the flattened-out shape.

If you draw a number of vertex points on a plane or a sphere and connect them into a single figure, Euler's formula will always yield the number 2. (Connecting a single vertex to itself forms a simple closed curve.) The area outside the figure is considered a face.

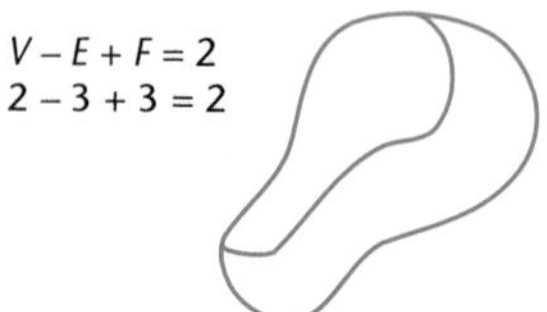

The region outside the shape is counted as the third face.

However, it is possible to draw figures on a torus for which a different formula applies:

$$V - E + F = 0$$

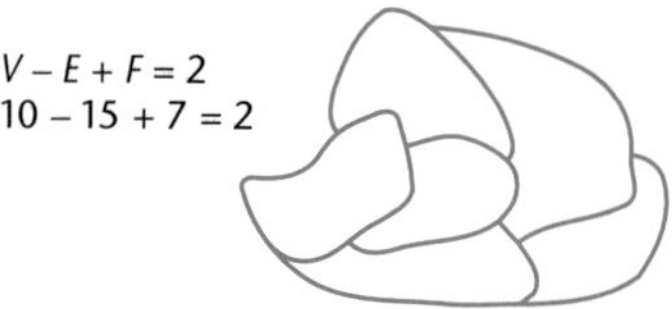

Hence, the Euler number of a torus is said to be 0. Topologists use the number resulting from Euler's formula as an invariant, allowing them to classify different types of surfaces mathematically. This is important, because some shapes that seem to be topologically equivalent in fact are not equivalent.

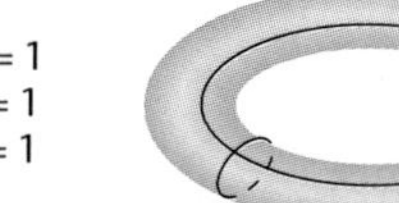

Finally, after everything you have learned about topology, it may not surprise you to learn that the "pretzel" shown at the beginning of this lesson can indeed be unlinked.

Cooperative Learning Use cooperative groups to review the terminology relating to topology used in this lesson, including equivalence, invariant, simple closed curve, and the Euler number. Ask each group to create examples of Jordan's Theorem and use them as a review for the entire class. Having students work together may help them make connections between geometry and topology.

EXERCISES & PROBLEMS

Communicate

1. Explain what is meant by two shapes being topologically equivalent.

Decide if each of the following pairs is topologically equivalent. Explain your reasoning.

2. 

3. 

4. What is an invariant?

5. Explain Jordan's Theorem.

Practice & Apply

Decide if the following are topologically equivalent. Explain your reasoning.

6.  

7.  

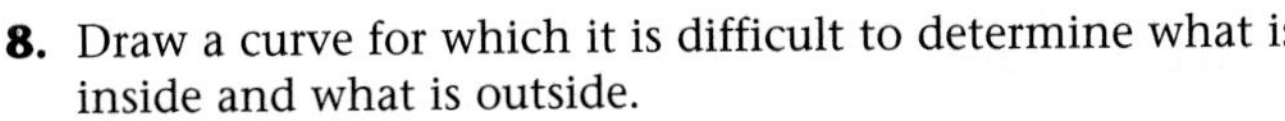

8. Draw a curve for which it is difficult to determine what is inside and what is outside.

9. Imagine distorting a cube into a sphere the same way that the pyramid in the lesson was distorted. Draw a picture of the cube distorting into a sphere.

10. Use your illustration from Exercise 9 to draw the vertices and edges flattened onto a plane (see p. 610).

11. Count the vertices, edges, and faces of your figure in Exercise 10 (remember to count the "outside" area as a face). Does your calculation verify Euler's Theorem?

Portfolio Activity Exercises 12–19 explore the Möbius strip. You will need paper, tape, scissors, and a pencil.

12. Construct a Möbius strip by following these instructions:

a. Cut a piece of paper about 3 cm wide and 28 cm long (the length of your notebook paper will work well).

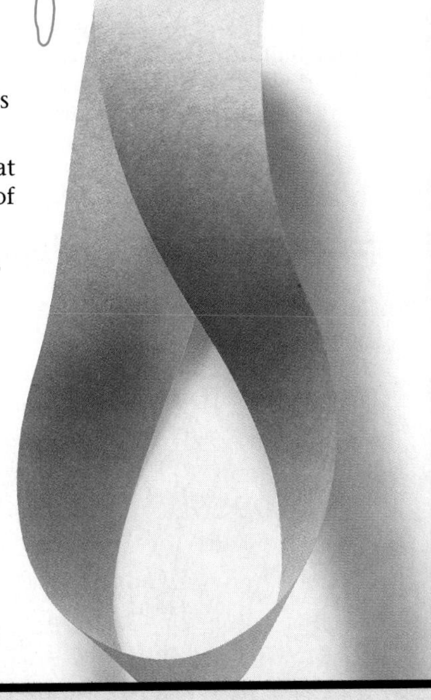

6. No. The curve on the left is not connected while the figure on the right is made up of closed curves. The figure on the right would have to be cut to make it equivalent to the figure on the left.

7. Yes. They are both simple closed curves.

11. $V = 8$, $E = 12$, $F = 6$; $8 - 12 + 7 = 2$; Yes

12. Check student constructions.

ASSESS

Selected Answers

Odd-numbered Exercises 7–23

Assignment Guide

Core 1–15, 20–23

Core Plus 1–26

Technology

For Exercises 6 and 7, use geometry graphics software to draw figures that are topologically equivalent. Various graphics options can be used to distort shapes.

Error Analysis

Exercises 16–19 may be done in cooperative groups. Have students create the strips together and discuss how to put them back together. While the strips may look different, have students explain how their results are the same.

The answers to Exercises 8–10 can be found in Additional Answers beginning on page 727.

alternative ASSESSMENT

Performance Assessment

Oral presentations can be used to assess students' understanding of this lesson. Call upon individual students to describe how they can tell whether two objects are topologically equivalent. Ask students to give examples of shapes that are equivalent. Have them use visual aids if possible.

13. The path traverses what would have been both sides of the original strip of paper, but the path never goes over the edge of the strip. The path finishes where it began. The strip has only one side.

14. The path traverses what would have been both sides of the original strip of paper, but the path never goes over the edge of the strip. The path always stays near the edge of the strip.

Practice Master

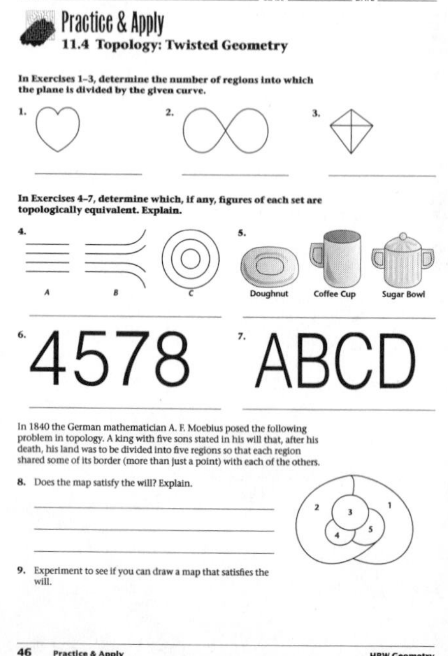
NAME ______ CLASS ______ DATE ______

Practice & Apply
11.4 Topology: Twisted Geometry

In Exercises 1–3, determine the number of regions into which the plane is divided by the given curve.

1. ______ 2. ______ 3. ______

In Exercises 4–7, determine which, if any, figures of each set are topologically equivalent. Explain.

4. A B C ______
5. Doughnut Coffee Cup Sugar Bowl ______
6. 4578 ______
7. ABCD ______

In 1840 the German mathematician A. F. Moebius posed the following problem in topology. A king with five sons stated in his will that, after his death, his land was to be divided into five regions so that each region shared some of its border (more than just a point) with each of the others.

8. Does the map satisfy the will? Explain.

9. Experiment to see if you can draw a map that satisfies the will.

46 Practice & Apply HRW Geometry

b. Connect the ends of the piece of paper together with tape so that your loop has a half-twist in it. Be sure to tape the side completely so that it will stay together.

13. Starting at the tape, use your pencil to draw the path of an imaginary bug that crawls in the center of the strip. Describe your results.

14. Let your bug crawl on the strip starting at the tape again, but this time draw the bug's path close to an edge of the strip. Describe your observations.

15. **Materials Handling** What do you think would be the advantage of conveyor belts shaped as Möbius strips?

16. Now you are going to construct a double Möbius strip. Take two rectangles of paper, 3 cm by 28 cm, and lay them on top of each other. Put a half-twist on both strips and tape the ends together to produce two nested Möbius strips. Check student constructions.

17. Using your nested Möbius strips, put your finger in between the two and run it all the way around the strip to feel the separation. Pay attention to the "ceiling" and the "floor" as if you were a little person walking. What do you observe?

18. Pull the strips apart. What is the result? As a challenge, try to nest the strips back together again.

19. Take a long strip of paper (about 3 cm wide and 50 cm long) and put three half-turns into the strip. Connect the ends with tape. Cut the strip down the middle, lay your shape flat on the table, and draw a picture of the result. This result is a **trefoil knot**.

Look Back

20. Identify the pairs of congruent angles in the diagram below, given $l_1 \parallel l_2$. **[Lesson 3.3]**

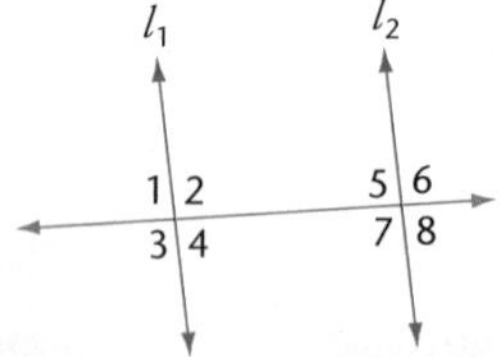

21. Find the measures of $\angle ABC$ and $\angle DBE$. **[Lesson 2.6]**
Both angles measure 60°

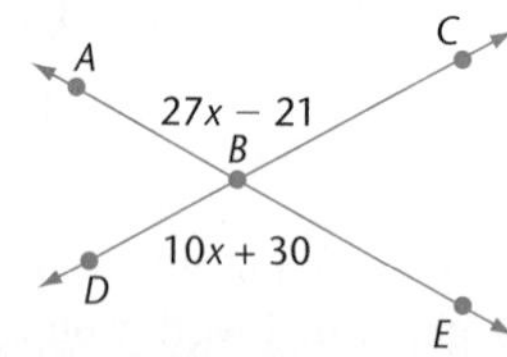

15. Sample answer: Only one side would have to be cleaned and maintained. Workers would not have to go "under" the conveyor belt to clean or repair it. The belt would wear evenly—one side would not wear more quickly than the other.

17. The strip that is at the top of your finger when you start is at the bottom of your finger when you finish going around the nested strips.

18. The result is a longer twisted strip that has two loops.

19. The result is a twisted strip of paper that has three loops.

20. $\angle 1 \cong \angle 4$; $\angle 2 \cong \angle 3$; $\angle 5 \cong \angle 8$; $\angle 6 \cong \angle 7$, $\angle 1 \cong \angle 5$; $\angle 2 \cong \angle 6$; $\angle 3 \cong \angle 7$; $\angle 4 \cong \angle 8$, $\angle 2 \cong \angle 7$; $\angle 4 \cong \angle 5$, $\angle 1 \cong \angle 8$; $\angle 3 \cong \angle 6$

Use the Law of Cosines and $\triangle ABC$ for Exercises 22 and 23. [Lesson 10.6]

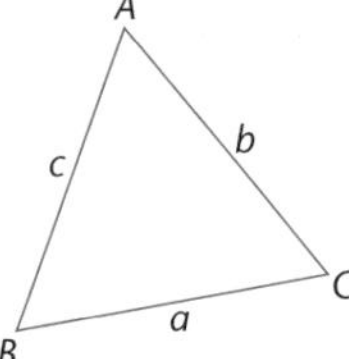

22. Find side a given $b = 25$, $c = 20$, and $A = 55°$. 21.25

23. Find side b given $a = 104$, $c = 47$, and $B = 92°$. 115.61 units

Use Transparency 101

Look Beyond

Klein Bottle The shape called a Klein bottle is an example of a closed surface (like a sphere) that has only one side (like a Möbius strip). An illustration of a Klein bottle drawn in a traditional way is shown. To understand the Klein bottle takes some imagination. Imagine a tube that is stretched at the bottom and pulled up and twisted into itself. It is not actually possible to construct a "real" Klein bottle in three dimensional space—or even to draw one—but the following paper model will give you a sample of some of its properties.

Look Beyond

Exercises 24–26 give students the opportunity to explore the famous Klein bottle. Students learn that a solid closed surface may only have one side like a Moebius strip.

24. Paper model of a Klein bottle: Check student constructions.

a. Take a rectangular piece of paper 4 in. by 11 in. and fold it lengthwise down the center. Tape the entire length together to produce a flattened tube.

b. Holding the flattened tube vertically, cut a horizontal slot the width of the paper, about a quarter of the way down the paper, through only the side of the tube nearest you.

c. Fold the paper up and insert the bottom edge of the tube through the slot. Align both ends of the tube together and tape the two pairs of edges together to produce one "hole" from above. This is your model of a Klein bottle. Study the picture and compare the drawing to your construction.

25. When you think that you have an adequate understanding of your bottle and how it has the same interior as exterior, lay the paper flat and cut up the middle. Without cutting or tearing your paper, unfold the two halves of your bottle as much as you can. You should discover two familiar shapes. What are they? Two Möbius strips are created.

26. It is possible to cut the paper model into one Möbius strip. Can you discover how it is done?

26. Cut around the self-intersection.

Technology Master

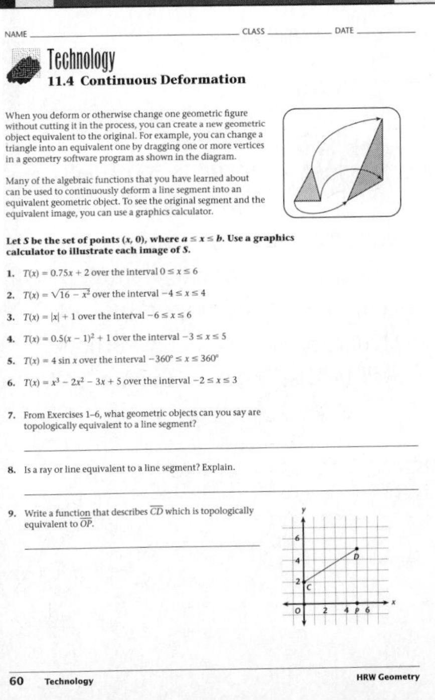

NAME ______ CLASS ______ DATE ______

Technology

11.4 Continuous Deformation

When you deform or otherwise change one geometric figure without cutting it in the process, you can create a new geometric object equivalent to the original. For example, you can change a triangle into an equivalent one by dragging one or more vertices in a geometry software program as shown in the diagram.

Many of the algebraic functions that you have learned about can be used to continuously deform a line segment into an equivalent geometric object. To see the original segment and the equivalent image, you can use a graphics calculator.

Let S be the set of points $(x, 0)$, where $a \le x \le b$. Use a graphics calculator to illustrate each image of S.

1. $T(x) = 0.75x + 2$ over the interval $0 \le x \le 6$
2. $T(x) = \sqrt{16 - x^2}$ over the interval $-4 \le x \le 4$
3. $T(x) = |x| + 1$ over the interval $-6 \le x \le 6$
4. $T(x) = 0.5(x - 1)^2 + 1$ over the interval $-3 \le x \le 5$
5. $T(x) = 4 \sin x$ over the interval $-360° \le x \le 360°$
6. $T(x) = x^3 - 2x^2 - 3x + 5$ over the interval $-2 \le x \le 3$
7. From Exercises 1–6, what geometric objects can you say are topologically equivalent to a line segment?
8. Is a ray or line equivalent to a line segment? Explain.
9. Write a function that describes $\overline{CD}$ which is topologically equivalent to $\overline{OP}$.

60 Technology HRW Geometry

PREPARE

Objectives

- Explore and develop general notions for Reimanian, Lobachevskian, and Poincaréan geometries.
- Develop informal proofs and solve problems using concepts of non-Euclidean geometries.

RESOURCES

- Practice Master 11.5
- Enrichment Master 11.5
- Technology Master 11.5
- Lesson Activity Master 11.5
- Quiz 11.5
- Spanish Resources 11.5

Assessing Prior Knowledge

1. Find the approximate measure of the angles of a triangle drawn on the surface of a sphere. Is the measure higher than 180° or lower? [**higher**]
2. Can you distort a triangle so that the sum of its angles is less than 180°? [**yes**]

TEACH

Learning that denying a postulate can lead to an entirely different branch of geometry is crucial to an understanding of the importance of structure in geometry. Proofs will be discussed in Chapter 12.

LESSON 11.5 Euclid Unparalleled

why *Many revolutionary concepts in math and science begin by questioning traditional assumptions.* Circle Limit X, *a woodcut by M. C. Escher, is an artist's conception of a non-Euclidean geometry. Such geometries reject Euclid's assumption about parallel lines.*

Imagine that you live on this two-dimensional surface. As you move outward from the center, everything gets smaller. Would you ever reach the edge of your "universe"?

The Non-Euclidean Geometries

Euclid's geometry rested upon five assumptions or postulates:

1. A straight line may be drawn between any two points.
2. Any terminated straight line may be extended indefinitely.
3. A circle may be drawn with any given point as center and any given radius.
4. All right angles are equal.
5. If two straight lines lying in a plane are met by another line, and if the sum of the internal angles on one side is less than two right angles, then the straight lines will meet.

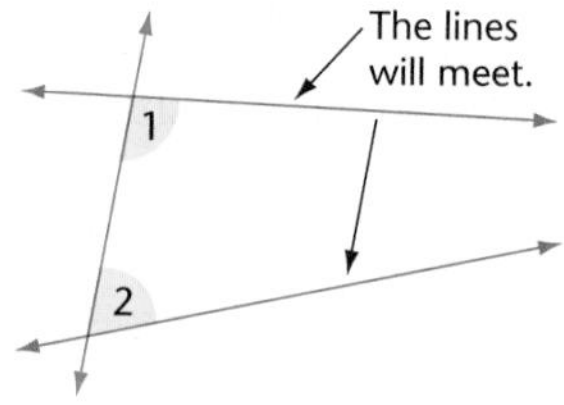

$m\angle 1 + m\angle 2 < 180°$

Cooperative Learning Have cooperative groups each devise an experiment to measure the angles of a triangle drawn on a globe. Have groups compare their results and make a statement about a "spherical triangle." Ask each group to describe their experiment and their results to the class.

Because the Fifth Postulate seemed less obvious than the four that came before it, many mathematicians wished to prove it in terms of the earlier four. Although they never succeeded in doing this, several statements were found to be equivalent to the Fifth Postulate:

1. If a straight line intersects one of two parallels, it will intersect the other.
2. Straight lines parallel to the same straight line are parallel to each other.
3. Two straight lines that intersect one another cannot be parallel to the same line.
4. If given, in a plane, a line l and a point P not on l, then through P there exists one and only one line parallel to l.

These statements are equivalent to the Fifth Postulate because any one of them (with the help of the first four postulates) can be used to prove the Fifth. Statement 4 is the version most mathematicians now think of as "Euclid's Fifth Postulate" or the "Parallel Postulate."

$m\angle 1 + m\angle 2 + m\angle 3 = 180°$?

For hundreds of years mathematicians tried to prove the Parallel Postulate, until finally it was shown that the task is impossible. Many even wondered whether the Parallel Postulate is actually true.

If the Parallel Postulate is not true, then the theorems that depend on it must be questioned. One such theorem is the Triangle Sum Theorem. (To prove the theorems, you must construct a line through one vertex parallel to the opposite base.)

The great mathematician Karl Friedrich Gauss (1777–1855) went so far as to measure the angles of a triangle formed by points on three different mountaintops to see if their measures added up to 180°. (Within the limits of the accuracy of his measurements, they did.)

Some mathematicians adopted a different attitude. They found that they could develop entirely new systems of geometry *without using the Parallel Postulate*. Systems in which the Parallel Postulate does not hold are called **non-Euclidean geometries**. To their surprise, mathematicians found no major flaws in these new geometries.

Have students list theorems and postulates that may change if the Parallel Postulate is not true.

Physics Outer space has often been described as non-Euclidean because a hyperbolic geometry model may fit it. Have students research the paths that may be followed by a probe which travels through the solar system.

ongoing ASSESSMENT

Their paths will get closer together. On a plane, their paths would stay the same distance apart.

Use Transparency 102

Spherical Geometry

Geometry on the Earth's surface is an example of a type of non-Euclidean geometry known as **Riemannian geometry**, after its discoverer, G. F. B. Riemann (1826–1866).

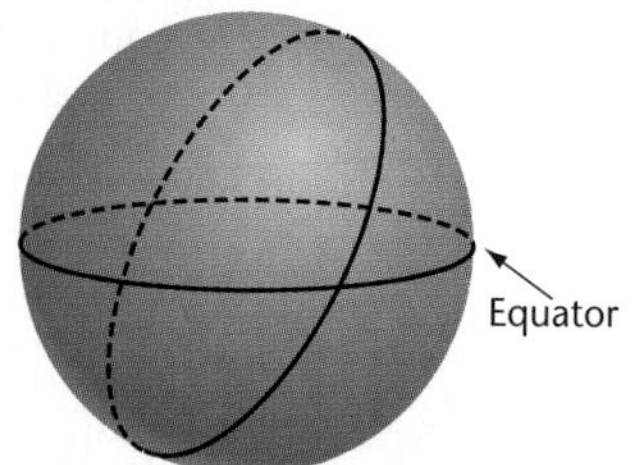

All great circles intersect.

In spherical geometry, a line is defined as a great circle of the sphere—that is, a circle that divides the sphere into two equal halves. In this geometry, as with any Riemannian geometry, there are *no parallel lines* at all, since all great circles intersect.

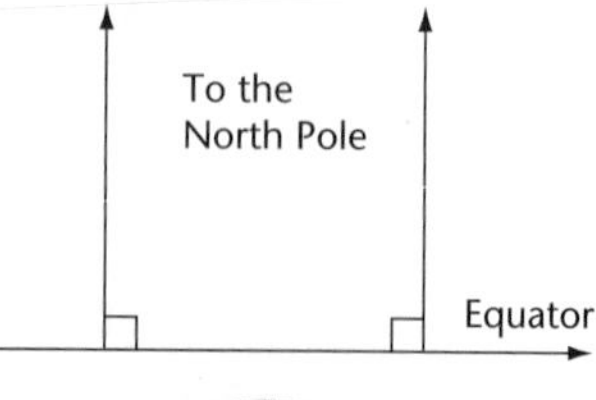

Imagine two superhuman runners who start running at the equator. Their paths form right angles to the equator. What will happen to their paths as they approach the North Pole? How does this differ from the result you would expect on a plane?

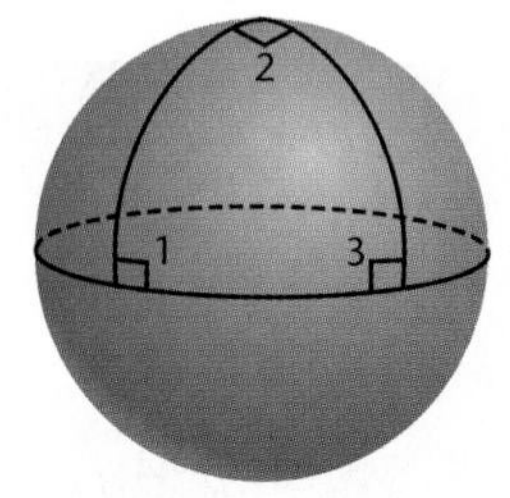

Theorems that depend on the Parallel Postulate for their proof may actually be false in spherical geometry. On a sphere, for example, the sum of the measures of the angles of a triangle is greater than 180°, as in the spherical triangle at right.

$m\angle 1 + m\angle 2 + m\angle 3 = 270°$

CRITICAL Thinking

Do the first four postulates of Euclid seem to be true on the surface of a sphere? If they are, should the theorems that follow them all be true on a sphere?

CRITICAL Thinking

Yes. Theorems that depend on only the first four postulates would also be true on a sphere.

Cooperative Learning

Begin class by introducing the idea of denying the Parallel Postulate by asking students to consider lines on a sphere. Then, divide the class into small groups and assign one of the three types of geometries to each. Have the groups read the text and develop a class presentation about their particular geometry.

Hyperbolic Geometry on a "Saddle"

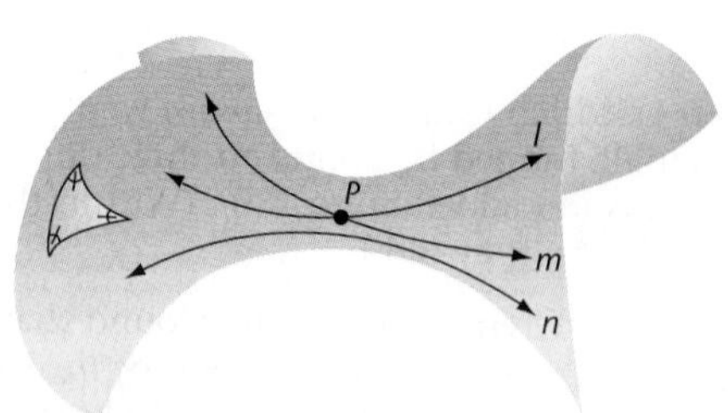

Unlike the surface of a sphere, which curves *outward*, the surface of a "saddle" curves *inward*. On such a surface, there is more than one line through a point that is parallel to a given line. In fact, there are infinitely many.

The geometry of a saddle is an example of **hyperbolic**, or **Lobachevskian, geometry**. Lobachevski (1773–1856) was one of two mathematicians who discovered this type of geometry independently. Janos Bolyai (1802–1860) was the other.

ENRICHMENT M. C. Escher used hyperbolic models in some of his graphic work. Have students find examples of Escher graphics.

INCLUSION strategies

English Language Development Many students have difficulty interpreting mathematics from textbooks. Suggest that students read this lesson slowly and carefully, and redraw the diagrams if necessary. Students should explain each diagram in terms of the model it represents.

In hyperbolic geometry, a line is defined as the shortest path between two points. In the illustration, lines *l* and *m* pass through the point *P*, and both are parallel to line *n* (because they will never intersect line *n*).

Once again, the Triangle Sum Theorem does not hold. The sum of the angles of a triangle on a saddle is less than 180°.

Poincaré's Model of Hyperbolic Geometry

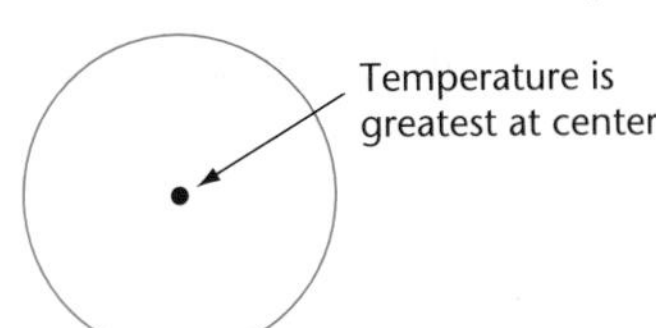

The "universe" in the Escher woodcut on the first page of this lesson is actually a representation of a three-dimensional universe imagined by the French mathematician Henri Poincaré (1854–1912). Poincaré, whose many interests included physics and thermodynamics, imagined physical reasons for the geometric properties of his imagined universe.

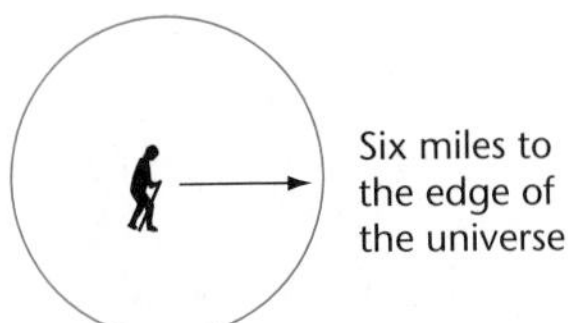

The temperature of Poincaré's universe is greatest at the center, and it falls away to absolute zero at the edges. According to Poincaré's rules for his system, no one would be aware of temperature changes.

In Poincaré's universe, the sizes of objects are directly proportional to their temperatures. An object would grow smaller and smaller as it moved away from the center.

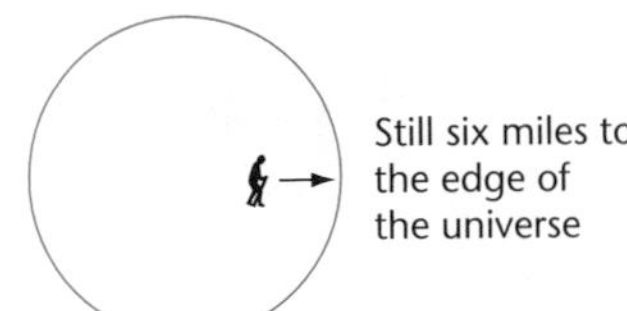

Since everything, including rulers, would be shrinking in size, there would be no way of detecting the change. A table that came up to your waist when you were at the center of the circle would still come up to your waist if you moved away from the center. In fact, since distance measures keep shrinking, you would never get any closer to the edge of the universe, no matter how long you traveled.

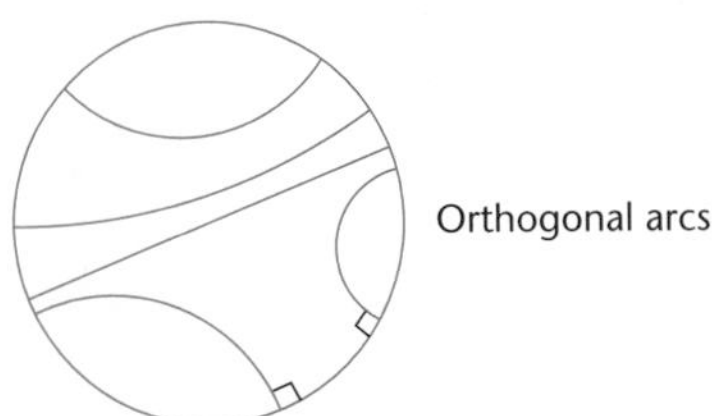

The arcs inside the diagram represent the curving paths that rays of light are supposed to travel in Poincaré's imagined universe. These are the "lines" of the system. Light rays that stay close to the circumference of the circle have a greater curvature than those that pass close to the center, which are relatively straight. Another feature of the lines in Poincaré's system is that they are **orthogonal**, or "perpendicular," to the outer circle.

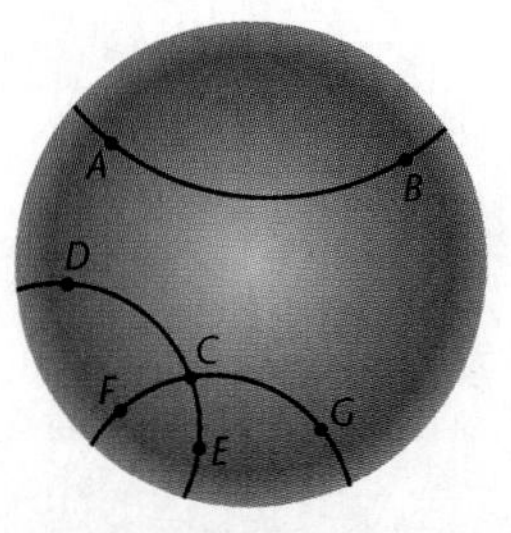

In Poincaré's system, a "line segment" is the orthogonal arc connecting two points. This is the shortest distance between the two points. If you wanted to walk from *A* to *B*, the curved line would be shorter than the straight line. This is because your steps would get larger as you approached the center of the circle. You would cover more distance with a single step if you moved toward the center instead of walking directly "to the point."

Teaching tip

Make sure students understand the sense in which both Euclidean and Poincare "lines" go on forever. The distance to the edge of the "universe" along a Poincare line is infinite because the measure keeps shrinking as you approach (but do not reach) the edge.

Reteaching the lesson

Using Discussion Have students restate the Parallel Postulate. Then change the words "exactly one parallel" to "no parallels." Describe the geometry that results. Change the same words to "more than one parallel" and describe the geometry that results.

ongoing ASSESSMENT

Parallel lines are lines that do not intersect, no matter how far they are extended. Lines $\overleftrightarrow{DE}$ and $\overleftrightarrow{FG}$ are already extended as far as they can be in Poincare's system, and yet they do not intersect line $\overleftrightarrow{AB}$. Therefore, they are both parallel to line $\overleftrightarrow{AB}$.

ASSESS

Selected Answers

Odd-numbered Exercises 7–29

Assignment Guide

Core 1–13, 19–29

Core Plus 1–32

Technology

You may want students to draw spheres with three-dimensional geometry software. Have them use the spheres for explorations in Exercises 6–10.

Error Analysis

Students may have difficulty doing Exercises 14–18. Use these exercises to help explain the Poincaré model. Explain that orthogonal arcs are perpendiculars in the model.

Notice also on the previous page that there is more than one line through a point drawn parallel to a given line. For example, lines $\overleftrightarrow{DE}$ and $\overleftrightarrow{FG}$ pass through the point *C*, and both are parallel to line $\overleftrightarrow{AB}$ (Why?) Thus the geometry of Poincaré's system is an example of a hyperbolic, or Lobachevskian, geometry.

Finally, notice that the sum of the measures of the angles of a triangle in Poincaré's system is less than 180°, just as it is on the surface of a saddle.

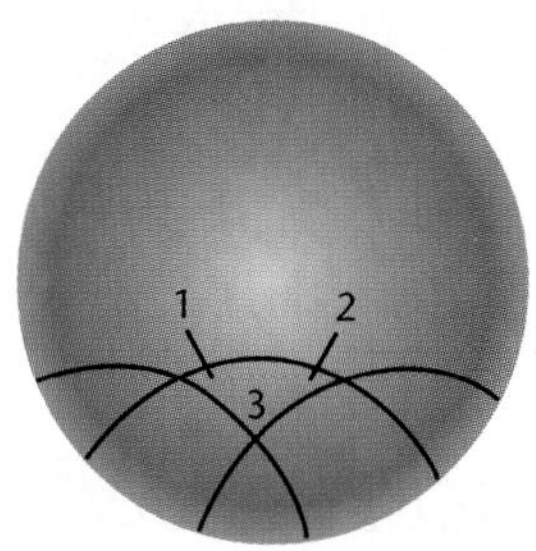

$m\angle 1 + m\angle 2 + m\angle 3 < 180°$

Years Later—Applications

In Einstein's General Theory of Relativity, space is not Euclidean. In fact, owing to the influence of gravity or—equivalently—acceleration, space is curved. The non-Euclidean geometries discovered many years earlier proved both an inspiration and a useful tool to Einstein in formulating fundamental laws of the universe.

Albert Einstein, 1879–1955

EXERCISES & PROBLEMS

Communicate

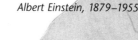

1. Which postulate of Euclid is the one that spawned all the debate? State the postulate in your own words.
2. Euclid's parallel postulate states one parallel. How many parallels exist in sphere geometry? How many exist in Poincaré's system of geometry?
3. Euclid stated in the parallel postulate that there was one parallel. Why do you think this statement went uncontested for nearly 2,000 years?
4. What are the two options to one parallel that have yielded the two non-Euclidean geometries?
5. In sphere geometry, how does a line have infinite length?

Practice & Apply

6. If you were to cut an orange in half around the "equator," and then cut each half twice at right angles on the poles, you would divide the orange peel into 8 "triangles." If you were to sum the angles of each of these triangles, what would you get? Why does this seem odd?

7. In sphere geometry, a pair of points opposite each other, such as the north and south poles of the Earth, are defined to be a *single point*. Is it true, then, that any two "lines" on a sphere intersect in just a single point? Is it true that there is just one line between any *two* points on a sphere? Illustrate your answers.

8. Draw a two-sided polygon in sphere geometry. Represent it on a sphere. What must be true about the vertices of this shape? How do you know?

9. In sphere geometry, how many degrees can a 2-gon have?

10. Draw a 4-gon on a sphere. Include one diagonal in your illustration.

11. How many degrees does your quadrilateral for Exercise 10 have? How do you know? How might you prove it?

12. Draw an illustration of a triangle in Poincaré's model.

13. In Poincaré's model of geometry, how does a line have infinite length?

14. On a plane, construct circle A. This circle will represent the "plane" of Poincaré. Pick any point P on circle A and construct (a) the radius of circle A and (b) the tangent to circle A at point P. Call the tangent line l.

15. "Lines" in Poincaré's model are represented by arcs of orthogonal circles. Circles are orthogonal if the radii at the points of intersection are perpendicular. If a radius of one circle is perpendicular to a radius of another, what do you know about the lines that contain the radii of the circles? The lines that contain the radii are perpendicular.

16. In your illustration for Exercise 14, the center of a set of circles that are orthogonal to circle A lies on l and intersects point P. How many circles fit this characteristic? Drawing lightly, construct a circle whose center lies relatively near P and one whose center lies relatively far away from P. Darken the arcs of the orthogonal circles that fall inside circle A. These arcs represent the lines of Poincaré's model.

17. How many points could you have chosen to initially draw the tangent? How many points could you have chosen on the tangent to be the center of the orthogonal circle? How many "lines" are possible in Poincaré's plane?

18. Draw a line l and point Q not on the line in Poincaré's model. How many lines can you draw through point Q that do not intersect line l? How is this illustration relevant to Euclid's Parallel Postulate?

6. $90° + 90° + 90° = 270°$; The sum is not 180°, as it would be for a plane triangle.

9. Each of the angles of a 2-gon could measure nearly 180°, so the sum of the measures of the angles of a 2-gon < 360°.

11. The sum of the measures of the angles of a quadrilateral is greater than 360°. The quadrilateral can be divided into two triangles with a diagonal. The sum of the measures of the angles of each triangle is greater than 180°, so the sum of the measures of the angles of the quadrilateral is greater than 360°.

13. Since distance measure keeps shrinking as you come closer to the edge of the circle, you would never come to the edge of the circle when traveling along a line.

17. An infinite number.; An infinite number.; An infinite number.

alternative ASSESSMENT

Performance Assessment

Have students describe verbally how non-Euclidean geometries came about. Have them discuss the Parallel Postulate and how changing a few words leads to other geometries. Ask them to draw diagrams as visual aids.

The answers to Exercises 7, 8, 10, 12, 14, 16, and 18 can be found in Additional Answers beginning on page 727.

Practice Master

NAME ______________ CLASS ________ DATE ________

Practice & Apply
11.5 Euclid Unparalleled

In Exercises 1–2, refer to the figure that shows line *l* on the surface of a sphere.

1. Are perpendicular lines possible in spherical geometry? ________
2. Are parallel lines possible in spherical geometry? ________

In Exercises 3–4, illustrate your answer with a drawing. When two lines *l* and *m* intersect, how many points do they have in common?

3. in Euclidean geometry
4. in spherical geometry

In Exercises 5–6, illustrate your answer with a drawing.

If lines m, l, and n lie in the same plane and m and n are each perpendicular to line l, what is the relationship between m and n?

5. in Euclidean geometry
6. in spherical geometry

In Exercises 7–8, refer to the figure that shows spherical triangle *ABC*.

7. What is the sum of the measures of the angles of a triangle in Euclidean Geometry? ________
8. What is the sum of the measures of the angles of a triangle in Spherical Geometry? ________

HRW Geometry — Practice & Apply 47

The answers to Exercises 19, 23, 24, 25, and 32 can be found in Additional Answers beginning on page 727.

31. Assume that the triangles are not congruent.

19. Draw three lines in Poincaré's model that intersect to form a triangle. Draw one of the midsegments of that triangle. Use this drawing to answer Exercises 20–22.

20. What is the sum of the angles in the triangle you drew in Exercise 19? How does this compare with the sum of the angles in Euclidean and in sphere geometry?

21. What appears to be true about the length of this midsegment?

22. Does the midsegment appear to be parallel to one of the bases of the triangle? Why or why not?

23. Draw a quadrilateral on a Poincaré model.

24. How many degrees does your quadrilateral for Exercise 23 have? How do you know? How might you prove it?

25. Draw a right triangle in Poincaré's model. What do you know about the acute angles of a right triangle in this geometry?

Look Back

26. Draw a floor plan for your house or apartment. Draw a network that represents the circulation in the house. Remember that the rooms are points and the doors represent the segments. **[Lesson 11.3]** Check student work.

27. Is your network from Exercise 26 traversable? Why or why not? **[Lesson 11.3]** Check student work.

28. Is a soup bowl topologically equivalent to a coffee cup? to a tennis ball? In each case explain your reasoning. **[Lesson 11.4]**

29. Describe three objects that are topologically equivalent to a coffee cup. **[Lesson 11.4]** Sample answers: a doughnut, a life preserver, a key.

Look Beyond

Look Beyond

Exercises 30–32 foreshadow the next chapter by asking students to do an indirect proof. They prove that similar triangles are also congruent in Lobachevski's geometry.

Hyperbolic Geometry In Lobachevski's geometry, similar triangles are also congruent. To prove this, you need to attack it indirectly—i.e., temporarily assume the opposite of what you would like to prove and then seek a contradiction.

Given: $\triangle ABC \sim \triangle DEF$

Prove: $\triangle ABC \cong \triangle DEF$

30. What do you know about the corresponding angles? The corresponding angles are congruent.

31. What do you want to temporarily assume?

32. Place $\triangle ABC$ onto $\triangle DEF$ such that $\angle B$ lies on top of $\angle E$.

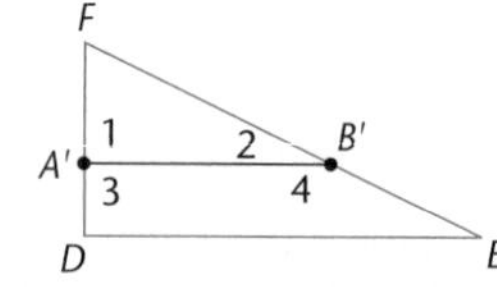

Explore the relationship of the angles and lines until you reach a contradiction about quadrilateral $A'B'ED$.

What do you now know?

20. The sum of the measures of the triangle is less than 180°. The sum of the measures of a triangle in Euclidean geometry is 180°. The sum of the measures of a triangle in Sphere geometry is greater than 180°.

21. The length of the midsegment appears to be close to half the length of the third side of the triangle.

22. Yes. The midsegment appears to be a part of a "line" that will not intersect the third side of the triangle.

28. No, a soup bowl cannot be distorted to have the handle of the coffee cup. No, the soup bowl distorted into a spherical shape would have a "hole" in it.

Technology Master

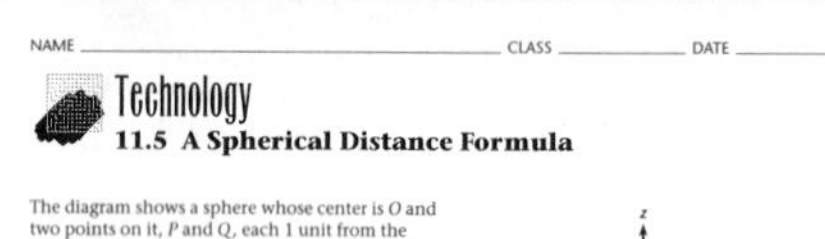

NAME ________ CLASS ________ DATE ________

Technology

11.5 A Spherical Distance Formula

The diagram shows a sphere whose center is O and two points on it, P and Q, each 1 unit from the origin. The sphere becomes a model of spherical geometry whose extent is 1 unit.

For simplicity, assume the xz-plane is oriented vertically. Suppose that P has coordinates $P(x_1, y_1, z_1)$ and Q has coordinates $Q(x_2, y_2, z_2)$. Using the Distance Formula you learned, you can find PQ, the Euclidean distance between P and Q.

Then by applying the Law of Cosines to ΔOPQ, you can write a formula for θ, the angle between $\overline{OP}$ and $\overline{OQ}$.

$$\cos \theta = 1 - 0.5(PQ)^2$$

The length of $\overset{\frown}{PQ}$ is the value of θ.

1. Create a spreadsheet in which cells A1, B1, and C1 contain the x-, y-, and z-coordinates of point P and cells A2, B2, and C2 contain the x-, y-, and z-coordinates of point Q. In cell A3 enter the distance formula that will find the Euclidean distance between points P and Q. In cell B3, enter the formula ACOS(1−0.5*A3^2) to find the spherical distance between points P and Q. (ACOS() gives you the value of the angle whose cosine is known.)

Use your spreadsheet from Exercise 1 to find the Euclidean and spherical length of each "line segment."

2. $P(0.500, 0.400, 0.768)$ and $Q(0.300, 0.700, 0.648)$ ________
3. $P(0.500, 0.400, 0.768)$ and $Q(0.600, 0.600, 0.529)$ ________
4. $P(0.400, -0.600, 0.693)$ and $Q(-0.100, 0.400, 0.693)$ ________
5. $P(-0.500, -0.300, 0.812)$ and $Q(0.300, 0.700, 0.648)$ ________

6. What can you say about the relationship between the Euclidean distance between P and Q and the spherical distance between P and Q?

____

HRW Geometry Technology 61

LESSON 11.6 Exploring

Projective Geometry

why *In spite of the deformities of a Mercator projection, shapes are still recognizable. This is because certain things are invariant when figures are transformed by projections. The study of such invariants is the subject matter of projective geometry.*

The three rigid transformations—reflections, rotations, and translations—preserve the sizes and shapes of objects. Dilations, on the other hand, preserve shape but not size. But there are types of transformations that shrink, expand, or stretch an object in different directions, so that neither shape nor size is preserved.

Affine Transformations

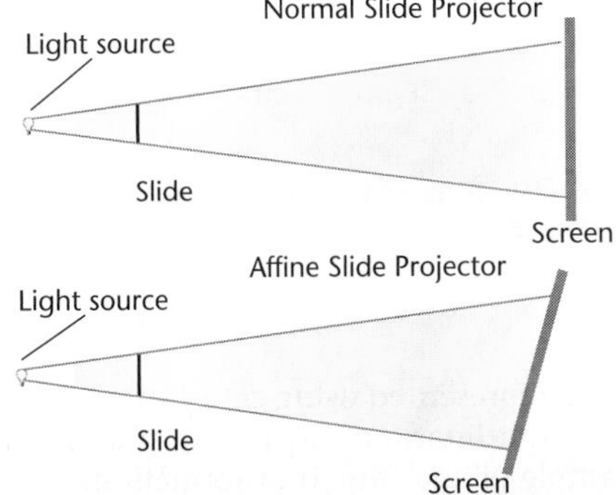

Consider a normal slide projector. The projected image on the screen is geometrically similar to the image on the slide. The original image has been dilated.

Now consider what happens to the image if you tilt the slide or the screen so that the planes of the slide and of the screen are no longer parallel. The image on the screen will be distorted. The resulting transformations are examples of **affine transformations**.

Technology Use geometry graphics software to transform plane figures. Ask students to draw a figure on the coordinate grid and use the affine transformation tool to transform it. While in the dialog box, have them define the transformation of (x, y) to (x', y').

PREPARE

Objectives

- Develop the concepts of affine transformation and geometric projection.
- Solve problems and make conjectures using the Theorem of Pappus and the Theorem of Desargues.

RESOURCES

- Practice Master 11.6
- Enrichment Master 11.6
- Technology Master 11.6
- Lesson Activity Master 11.6
- Quiz 11.6
- Spanish Resources 11.6

Assessing Prior Knowledge

1. List the transformations that preserve the size and shape of an object. **[reflections, rotations, and translations]**
2. Which is preserved by dilations, size or shape? **[shape]**
3. List the transformations that preserve parallelism. **[translations, dilations]**

TEACH

Studying projective geometry helps students understand how to draw in perspective and how to represent objects as projections (like the Earth projected onto a flat piece of paper). Point out that projective geometry studies the properties of an object that are invariant.

Exploration 1 Notes

In this exploration, students construct an affine transformation both by compass and on the coordinate plane. Have students do the same exploration with a variety of shapes.

Use Transparency 103

TEACHING tip

When discussing affine transformations, include a discussion about which properties of a figure are preserved. Use Exploration 1 to emphasize that the property of being a line or being a circle is invariant under an affine transformation.

ongoing ASSESSMENT

6. The parallelogram figure is an affine transformation of the square figure. Notice that:
1. All collinear points in the square figure are also collinear in the parallelogram figure.
2–4. Similarly for straight lines, intersecting lines, and parallel lines.

AFFINE TRANSFORMATION

An affine transformation maps all pre-image points P in a plane to image points P' so that

1. collinear points project to collinear points.
2. straight lines project to straight lines.
3. intersecting lines project to intersecting lines.
4. parallel lines project to parallel lines.

11.6.1

Exploration 1 *Affine Transformations*

You will need
Graph paper (large grid)
Compass

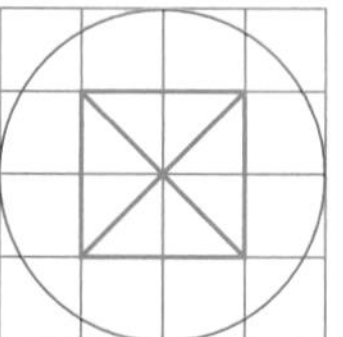

Part I

1. Mark off a 4 × 4 square grid and a 4 × 4 grid of parallelograms as shown. (Use more than one unit square to make the grid large enough to work with.)

2. Inscribe a circle in the square grid. Draw a square with its diagonals inside the circle.

3. Mark the points of intersection of the circle with the grid. Mark the vertices of the square.

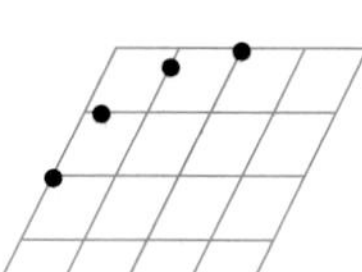

4. Mark the corresponding points of intersection in a parallelogram grid like the one shown. The first few are marked in the diagram.

5. To see how the figure would look under an affine transformation, connect the points of the "square" by segments and draw a smooth curve through the points of the "circle."

6. How does your resulting figure illustrate each of the four points in the definition of an affine transformation?

Part II

Affine transformations can also be represented using coordinates. For example, multiplying the x and y coordinates of the points of a figure by two different numbers is an example of an affine transformation.

interdisciplinary CONNECTION **Art** Artists need to have a thorough understanding of projective geometry in order to draw objects in perspective. Computer graphic artists have to use a coordinate grid to draw objects in perspective. Have students discuss how principles in this chapter can be translated directly to an artist's work.

Pre-image point	Multiply x by 2, y by 3	Image point
(x, y)	$\rightarrow$	$(2x, 3y)$

1. Draw a square and its diagonals in the first quadrant of a coordinate plane.

2. Multiply the x-coordinates of the vertices of your figure by 2 and the y-coordinates of the vertices by 3. Plot the resulting points and connect them with segments. Describe the resulting figure.

3. Multiply the x-coordinates of the vertices of your figure by 3 and the y-coordinates of the vertices by 2. Plot the resulting points and connect them with segments. Describe the resulting figure.

4. How do your resulting figures illustrate each of the four points in the definition of an affine transformation?

5. How do these affine transformations differ from the one in Part I? How are they alike? ❖

Projections and Projective Geometry

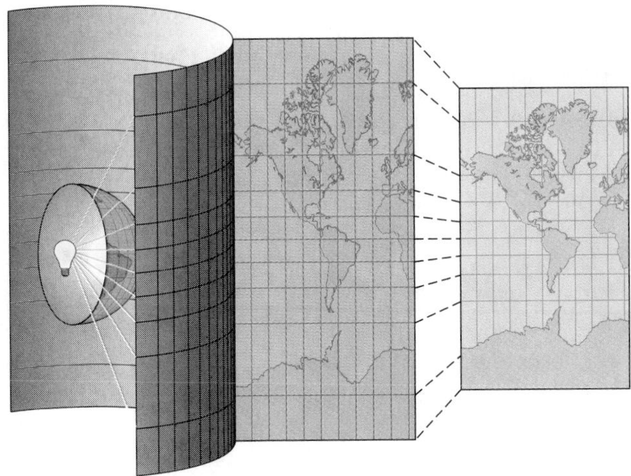

Geography students have long pointed out that Greenland is really much smaller than it appears to be on Mercator maps. Deformities like these result because it is impossible to flatten out the Earth's spherical surface into a rectangle without some distortion of sizes, shapes, distances, and directions.

The diagram shows a projection between two lines in the same plane. The points on line l are projected onto line m from a center of the projection called point O. Rays are drawn from the center of the projection. These lines intersect l at points A, B, and C. The rays are extended, and their intersections with line m determine the locations of the projected points A', B', and C'.

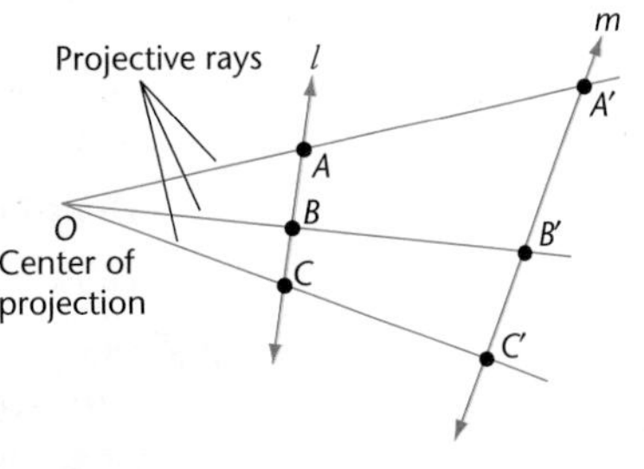

ONGOING ASSESSMENT

4. 1. All collinear points in the square figure are also collinear in the rectangular figure.
2–4. Similarly for straight lines, intersecting lines, and parallel lines.

Cooperative Learning

Have cooperative groups discuss how to do an affine transformation on the coordinate plane. Have each student in the group draw a figure in the plane and transform it by a rule decided on by the group (for example, $(x, y) \longrightarrow (3x, 4y)$). Ask each student to describe how to apply each part of the definition of *affine transformation* on page 622. Working in groups may help students understand the definitions in this lesson.

ENRICHMENT Draw a right-handed three-dimensional coordinate system (see Chapter 6). Measure the angle $a°$ made by the x-axis with the y-axis. Then, the projection of the point (x, y, z) in the three-dimensional coordinate system is (x', y') with $x' = x(-\cos a°) + y$ and $y' = x(-\sin a°) + z$. Ask students to use a calculator to find the projection image of (2, 3, 4). **[Sample: If a = 135°, (2, 3, 4) —> (4.41, 2.59).]**

INCLUSION strategies

Hands-On Strategies

Ask students to draw several geometric shapes in the first quadrant and project them onto the x-axis. Which shapes project as a line segment? Conversely, suppose the projection image of several shapes onto the x-axis is a line segment. What may some of the pre-image shapes look like?

Artists used the concept of projection to give their works depth and realism. Meanwhile, mathematicians used projections to develop an entire system of geometry.

Summary

THE MAIN FEATURES OF PROJECTIVE GEOMETRY

1. It is the study of the properties of figures that do not change under a projection.
2. There is no concept of size, measurement, or congruence.
3. Its theorems state facts about such things as the positions of points and the intersections of lines.
4. An unmarked straightedge is the only tool allowed for drawing figures.

Exploration 2 Two Projective Geometry Theorems

Geometry Graphics

You will need
Unlined paper
Straightedge or
Geometry technology

Part I: A Theorem of Pappus

Cultural Connection: Africa Pappus was a mathematician who lived in Alexandria in the third century B.C.E. His work was very important in the development of projective geometry many centuries later. In the steps that follow, you will discover one of his theorems.

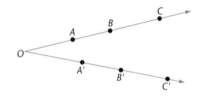

1. Orient the paper horizontally. Mark a point O toward the left edge of the paper.
2. Draw two rays from point O.
3. From point O, mark A, B, and C (in that order) on one ray.
4. In the same manner, mark A', B', and C' on the other ray.
5. Draw $\overline{AB'}$ and $\overline{A'B}$. Label their intersection X.

   Draw $\overline{BC'}$ and $\overline{B'C}$. Label their intersection Y.

   Draw $\overline{CA'}$ and $\overline{C'A}$. Label their intersection Z.
6. What appears to be true about X, Y, and Z? Compare your results with your classmates' results. (If you are using geometry technology, drag the rays in various ways to see if your result still holds.) State your results as a conjecture. **(Thm 11.6.2)**

Exploration 2 Notes

This exploration will help students understand two of the basic theorems of projective geometry. Encourage students to compare their results with those of their classmates. Repeat the exploration with a different set of points.

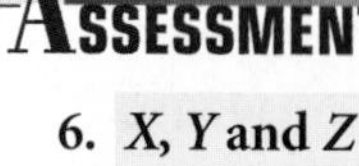

6. X, Y and Z are collinear. If A, B, and C are three distinct points on one line and A', B', and C' are three distinct points on a second line, then the intersections of $\overline{AB'}$ and $\overline{BA'}$, $\overline{AC'}$ and $\overline{CA'}$, and $\overline{BC'}$ and $\overline{CB'}$ are collinear.

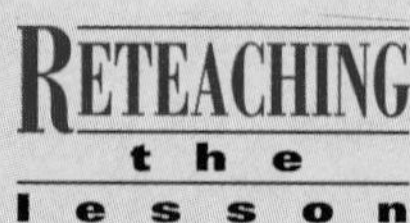

Using Visual Models

Summarize the lesson by having students draw a geometric shape on the coordinate plane. Label the vertices as preimage points. Ask them to find the image of the shape under an affine transformation $(x, y) \longrightarrow (2x, 4y)$. (They should use relatively small numbers for the coordinates of the points of the triangle so the image will fit on the page.)

Part II: A Theorem of Desargues

Girard Desargues (1593–1662) was a French mathematician whose ideas are among the most basic in projective geometry. Carefully follow each step to discover one of his most important theorems.

1. Draw $\triangle ABC$ approximately in the center of the paper. This can be any type of triangle, but it should be small so that the resulting construction will fit on your paper.
2. Mark a point outside $\triangle ABC$ and label it O. This is the center of projection.
3. Mark a random point A' on ray $\overrightarrow{OA}$. (For clarity of construction, it should be farther from point O than from point A.) A' is the projection of A.
4. Mark B', the projection of B, in a random spot on ray $\overrightarrow{OB}$.
5. Extend $\overleftrightarrow{AB}$ and $\overleftrightarrow{A'B'}$ until they intersect, and label the point of intersection X. The drawing shows the construction completed through this step.
6. Mark C', the projection of C, in a random spot on ray $\overrightarrow{OC}$.
7. Extend $\overleftrightarrow{AC}$ and $\overleftrightarrow{A'C'}$ until they intersect, and label the point of intersection Y.
8. Extend $\overleftrightarrow{BC}$ and $\overleftrightarrow{B'C'}$ until they intersect, and label the point of intersection Z. (In some cases the extended segments will intersect at points off the paper. If this happens, reposition B' and C' so that points X, Y, and Z all fall on the paper.)
9. What appears to be true about points X, Y, and Z? Compare your results with your classmates' results. (If you are using geometry technology, drag the rays and points in various ways to see if your result still holds.) State your result as a conjecture about two triangles when one is the projection of the other. **(Thm 11.6.3)** ❖

TEACHING tip

Students may need to be encouraged to persevere with the instructions in Part II. They may need to adjust their choices for the original triangle ABC or the randomly chosen points on the rays so that all the intersection points will fit on the page. They should compare their results with those of their classmates.

ongoing ASSESSMENT

9. X, Y, and Z are collinear. If one triangle is a projection of another triangle, then the lines containing the corresponding sides of the two triangles meet at points on a line.

Assess

Selected Answers

Exercises 5, 9, 11, 13, 16, 19, 21, and 23

Assignment Guide

Core 1–10, 14-23

Core Plus 3-17, 21, 23, 24

Technology

You may want students to use geometry graphics software to do Exercises 14–17. Have students label points as they draw them to avoid errors.

Error Analysis

Students may have difficulty with Exercise 17 because they have to find the original center of a projection.

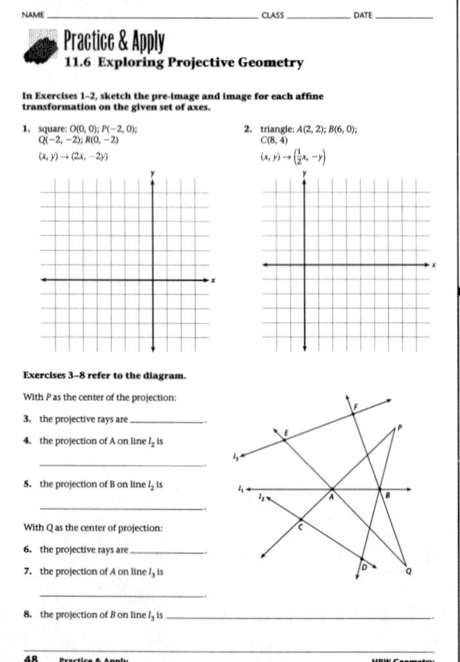
Practice Master

NAME ______ CLASS ______ DATE ______

Practice & Apply
11.6 Exploring Projective Geometry

In Exercises 1–2, sketch the pre-image and image for each affine transformation on the given set of axes.

1. square: $O(0, 0)$; $P(-2, 0)$; $Q(-2, -2)$; $R(0, -2)$
$(x, y) \to (2x, -2y)$

2. triangle: $A(2, 2)$; $B(6, 0)$; $C(8, 4)$
$(x, y) \to \left(\frac{1}{2}x, -y\right)$

Exercises 3–8 refer to the diagram.

With P as the center of the projection:

3. the projective rays are ______.
4. the projection of A on line l_2 is ______.
5. the projection of B on line l_2 is ______.

With Q as the center of projection:

6. the projective rays are ______.
7. the projection of A on line l_3 is ______.
8. the projection of B on line l_3 is ______.

48 Practice & Apply HRW Geometry

Exercises & Problems

Communicate

1. Describe what an affine transformation is.
2. Which of the following mappings are affine transformations? Explain your reasoning.
   a. $(x, y) \to (2x, 4y)$
   b. $(x, y) \to (y, x)$
   c. $(x, y) \to (-4x, -4y)$
   d. $(x, y) \to \left(\frac{x}{5}, \frac{y}{3}\right)$
3. Describe the four main features of projective geometry.
4. State the theorems of Pappus and Desargues.

Practice & Apply

Sketch the pre-image and image for each affine transformation below.

5. square (0, 0), (4, 0), (4, 4), (0, 4)
   $(x, y) \to (3x, -2y)$
6. rectangle (0, 0), (5, 0), (5, 8), (0, 8)
   $(x, y) \to (2x, 0.5y)$
7. triangle (4, 7), (−1, −1), (0, 8)
   $(x, y) \to \left(\frac{5}{2}x, 5y\right)$

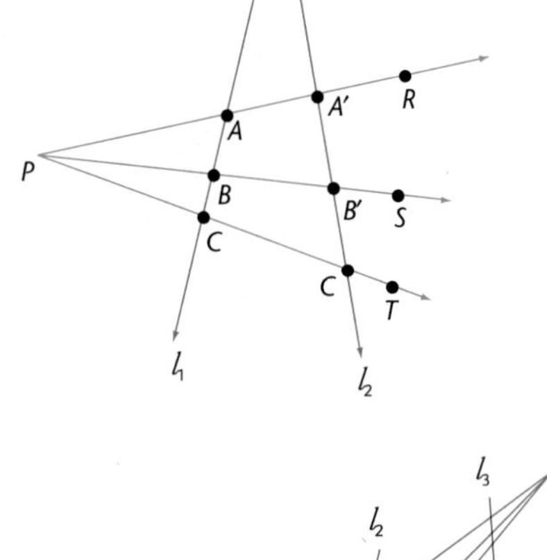

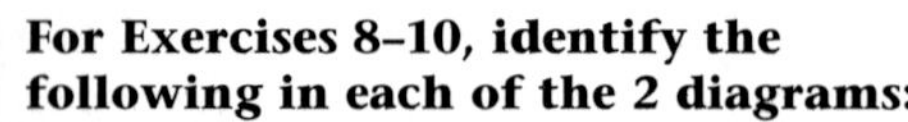

For Exercises 8–10, identify the following in each of the 2 diagrams:

a. the center of projection
b. the projective rays

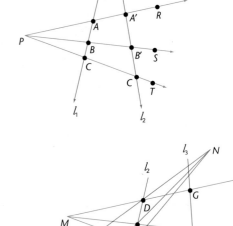

8. Projection of points on l_1 onto l_2. Use the figure above right. *P*; $\overrightarrow{PR}$, $\overrightarrow{PS}$, $\overrightarrow{PT}$
9. Projection of points on l_2 onto l_1. Use the figure at right. *N*; $\overrightarrow{NJ}$, $\overrightarrow{NK}$, $\overrightarrow{NL}$
10. Projection of points on l_2 onto l_3. *M*; $\overrightarrow{MG}$, $\overrightarrow{MH}$, $\overrightarrow{MI}$

The answers to Exercises 5–7 can be found in Additional Answers beginning on page 727.

11. If A is the center of projection, then the projection of B on $\overline{FK}$ is __?__, the C projection of I on $\overline{FK}$ is __?__, and the H projection of J on $\overline{FK}$ is __?__. K

12. If L is the center of projection, then the projection of H on $\overline{MA}$ is __?__ and the B projection of J on $\overline{MA}$ is __?__. A

13. If F is the center of projection, then the projection of __?__ on $\overline{AG}$ is I, and the B projection of __?__ on $\overline{AG}$ is G. D or E

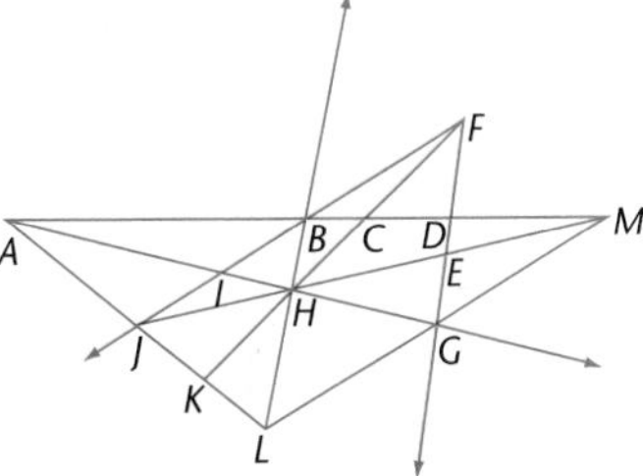

Copy the diagram and draw each projection.

14. Using center of projection O, project points A, B, C, and D onto line l_2. Label the projected points A', B', C', and D'.

15. Using center of projection P, project points A', B', C', and D' onto line l_3. Label the projected points A'', B'', C'', and D''.

16. Draw another line l and label points A, B, and C on it. Draw lines l_2 and l_3, and label a center of projection O.
   a. Sketch the projection of points A, B, and C on l_2. Label the projected points A', B', and C'.
   b. Sketch the projection of points A', B', and C' on l_3. Label the projected points A'', B'', and C''.

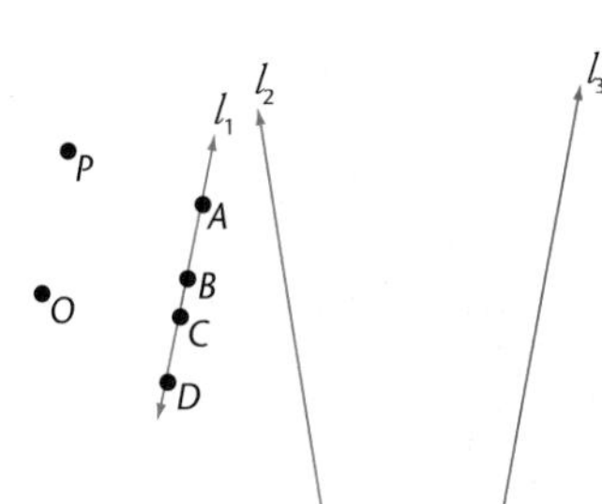

17. **Converse of the Desargues Theorem** This construction begins where the activity on the Desargues Theorem ends. Working backward, you are able to locate the original center of projection.
   a. Draw l_1 and label X and Y at arbitrary locations on it.
   b. Draw ray $\overrightarrow{XB}$ and mark a random point A on it.
   c. Draw ray $\overrightarrow{XB'}$ and mark a random point A' on it.
   d. Draw ray $\overrightarrow{YA}$ and mark a random point C on it (see diagram).
   e. Draw ray $\overrightarrow{CB}$. This will determine the location of point Z on l_1. Label point Z on l_1.
   f. Draw ray $\overrightarrow{YA'}$ and ray $\overrightarrow{ZB'}$. Label their point of intersection C'.
   g. What can you conclude about C and C'. (Hint: compare $\triangle ABC$ to $\triangle A'B'C$.)

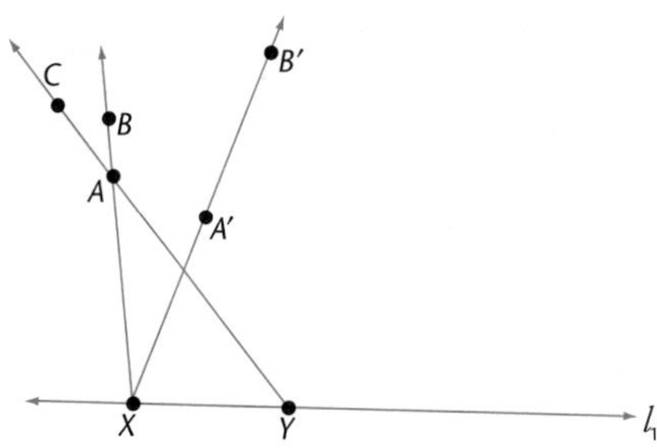

17. a-f Check student constructions.

g. CB and $C'B'$ are corresponding sides of triangles ABC and $A'B'C'$. The lines containing the sides of these triangles intersect in a line. $A'B'C'$ is the projection ABC.

alternative ASSESSMENT

Portfolio Assessment

Ask students to summarize the lesson in writing by listing and describing the properties of an affine transformation and of projective geometry. Encourage them to include sample problems.

The answers to Exercises 14–16 can be found in Additional Answers beginning on page 727.

Technology Master

NAME ____________ CLASS ________ DATE ________

Technology
11.6 Affine Transformations and Matrices

The diagram shows a new coordinate system that you can make by rotating the existing x- and y-axes through θ degrees counterclockwise. As you know, you can represent the rotation transformation by means of a matrix.

There are other coordinate systems you can make from an xy system. In an affine coordinate system, collinear points will remain collinear and parallel lines will remain parallel.

Use a graphics calculator as appropriate.

1. If $M((x, y)) = \begin{bmatrix} -3 \\ -3 \end{bmatrix} \quad \begin{bmatrix} 2 & 2 \\ 0 & 2 \end{bmatrix} \begin{bmatrix} x \\ y \end{bmatrix}$ use a graphics calculator and matrix multiplication and addition to find the image of $O(0, 0)$, $P(1, 0)$, and $Q(0, 1)$. Plot P and Q and the images O', P', and Q' on the coordinate plane provided. Then sketch $\overleftrightarrow{O'P'}$ and $\overleftrightarrow{O'Q'}$. (These lines are the axes in a new coordinate system.)

2. On the xy-coordinate plane in Exercise 1, plot $S(0, 4)$, $T(3, 4)$, and $U(3, 0)$. Sketch rectangle $OSTU$. Find the images S', T', and U' of S, T, and U under the transformation in Exercise 1. Plot S', T', and U' on the original xy-coordinate plane in Exercise 1.

3. Does the transformation preserve parallelism? That is, are $\overline{O'S'}$ and $\overline{U'T'}$ parallel? Explain.

4. Does the transformation preserve right angles? That is, is $m\angle S'O'U'$ a right angle? Explain.

5. Does the transformation preserve distance? That is, are $O'S'$ and $U'T'$ equal? Explain.

6. Does the transformation preserve collinearity? Explain.

62 Technology HRW Geometry

18. The golden ratio is the ratio of length to width in a golden rectangle, which is exactly $\frac{1+\sqrt{5}}{2}$.

Look Beyond

Nonroutine problem. Exercise 24 is a puzzle that can be proved using Pappus's Theorem. Encourage students to use models to solve the problem.

Look Back

18. Describe what is meant by the "Golden Ratio." **[Lesson 11.1]**

Find the "taxidistance" for each pair of points. [Lesson 11.2]

19. (4, 3), (2, 1) 4 **20.** (−3, 2), (1, 1) 5 **21.** (1, 3), (5, 5) 6

22. Can the figure be traversed? Why or why not? **[Lesson 11.3]** Yes. All vertices are even.

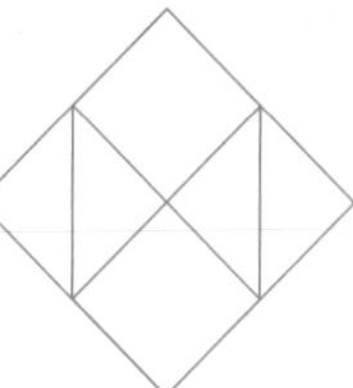

23. In spherical geometry, how many lines are parallel to another line through a point not on the line? **[Lesson 11.5]** None

Look Beyond

24. **The Nine Coin Puzzle** The nine coins are arranged in eight rows of three. Can you rearrange them into ten rows of three? Hint: You can use Pappus's Theorem to solve this puzzle.

24.

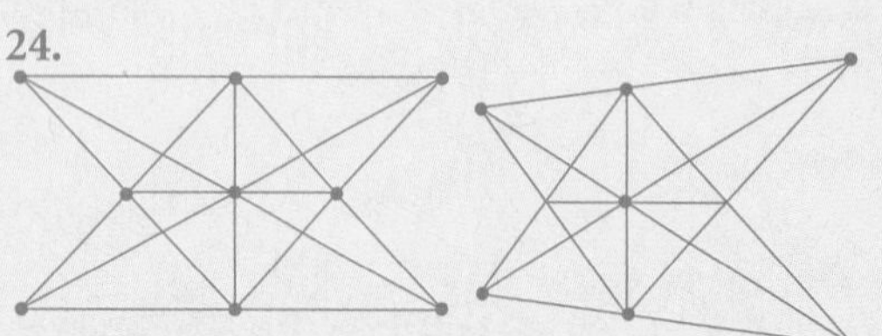

Fractal Geometry

why *A **fractal**, such as this computer-generated "fern," is a self-similar structure. Notice how each subdivision of the leaves of the fern has basically the same structure as the leaves themselves—all the way down to the curving tips. This self-similarity enables a computer programmer to write a program for drawing such structures in just a few lines of code.*

Self-Similarity in Fractals

A fractal, like the **Menger Sponge**, is a geometric object that exhibits some degree of **self-similarity**. This means that the object always looks the same, whether seen in an extremely close view, from middle distance, or from far away. If you were to cut off a small, cube-shaped portion of the Menger Sponge and examine it, you would find it to be a miniature copy of the entire sponge.

In mathematically created fractals like the Menger Sponge, this process can be theoretically continued forever, and the self-similarity will always be evident.

Fractals are created by doing a simple procedure over and over. The Menger Sponge, for example, is created by starting with a certain shape (a cube) and changing it according to a certain rule (removing a part of the cube). This same rule is applied to the newly changed shape. The process is then continued. This repetitive application of the same rule is called **iteration**.

Technology Have students use fractal programs to create images. Ask them to use the computer to explore the Mandelbrot Set and to create the Sierpinski Triangle.

PREPARE

Objectives

- Discover the basic properties of fractals including self-similarity and iterative processes.
- Build fractal designs using iterative steps.

RESOURCES

• Practice Master	11.7
• Enrichment Master	11.7
• Technology Master	11.7
• Lesson Activity Master	11.7
• Quiz	11.7
• Spanish Resources	11.7

Assessing Prior Knowledge

Describe the mathematical qualities of plants such as ferns and broccoli.

[Answers will vary. Students with previous experience with fractals will be likely to mention the fractal qualities.]

TEACH

The study of fractals gives students an exciting introduction to the world of mathematics enhanced by computers. Point out that while fractal images closely resemble nature, they are created purely by a computer and not by a camera.

TEACHING tip

Point out that the self-similarity seen in a coastline can also be seen in tree-like structures in parts of the human body (like the veins in a kidney). The same tree-like structures can be seen on a much larger scale in a satellite photograph of land taken from space. Emphasize that a fractal includes self-similar shapes which repeat themselves over and over again with different scale factors.

How Long is a Coastline?

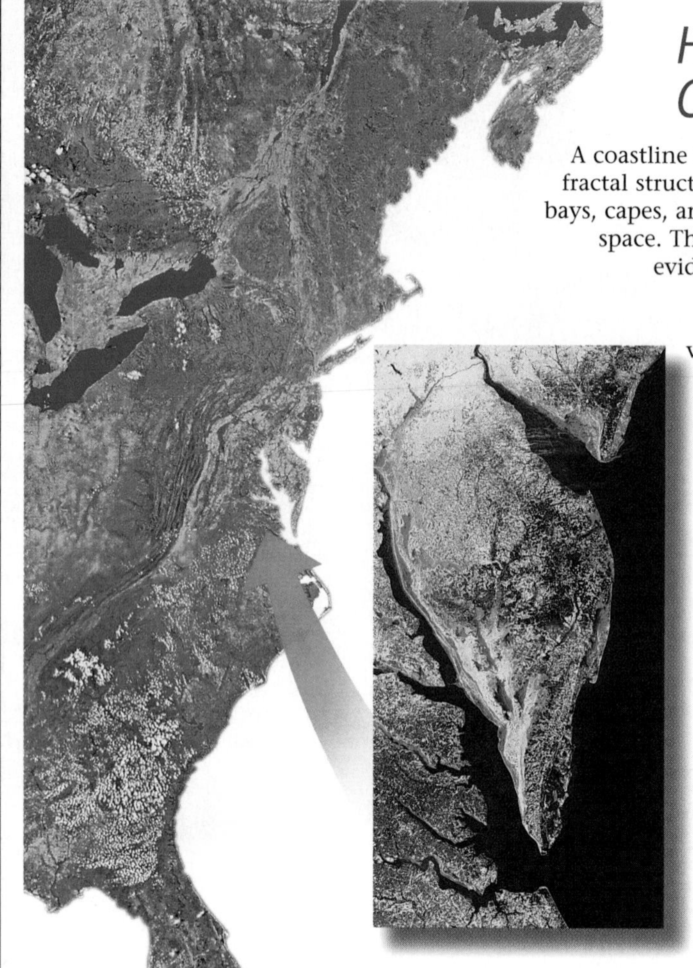

A coastline is a good example of a self-similar fractal structure in nature. Jagged irregularities, bays, capes, and inlets can be seen from outer space. The same basic structures are evident over a smaller section of a coastline viewed from a plane, or even a smaller section viewed during a walk or a drive.

How long is a coastline? Unlike ordinary geometric segments or curves, which can often be readily measured, the "length" of a fractal coastline depends on how closely you move in to measure it.

History During the American Revolution, the British Royal Navy attempted to blockade the American coastline. Although fairly successful, the blockade was not able to prevent shipping in and out of many of the harbors. What does a close examination of the coastline reveal about its "length" to explain the difficulty faced by the British to completely blockade it?

In a theoretical fractal coastline, where the depth of self-similarity is endless, the length is considered to be infinite.

Many objects found in nature also exhibit self-similarity, although to a lesser degree than in mathematical fractals. Notice, for instance, the self-similarity of the broccoli.

interdisciplinary CONNECTION **Art** Graphic artists often use fractal images in the creation of their artwork. The actual computer commands to do the drawings may be quite simple, but they are done repetitively, perhaps millions of times (which is why computers are needed). Have students find examples of fractal images in art.

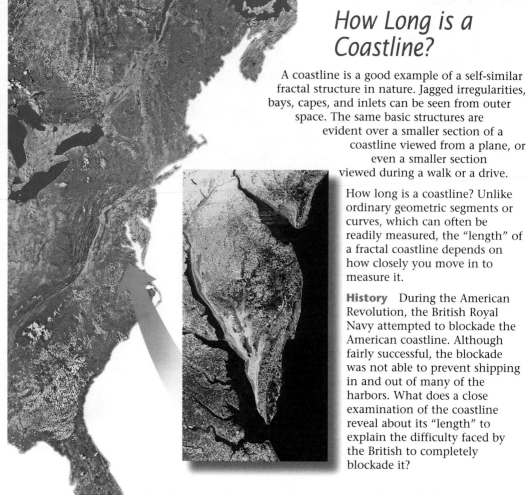

Exploration 1 Creating a "Cantor Dust"

One of the simplest fractals was discovered by Georg Cantor (1845–1918) years before fractals were defined and studied. As you will see, Cantor's fractal is a one-dimensional version of the Menger Sponge.

You will need
A ruler

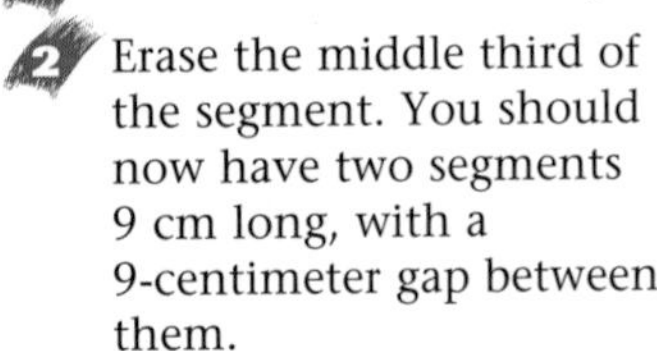

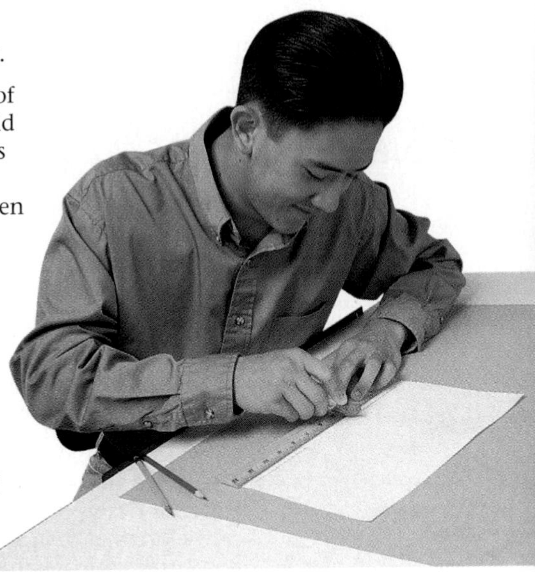

1. Draw a line 27 cm long.
2. Erase the middle third of the segment. You should now have two segments 9 cm long, with a 9-centimeter gap between them.
3. Erase the middle third of each of the two 9-centimeter segments. Now you will have four segments, each 3 cm long.
4. Continue erasing the middle third of each of these segments until you are left with a scattering of pointlike segments.

27 cm
9 cm — 9 cm — Iteration 1
3 cm — 3 cm — 3 cm — 3 cm — Iteration 2
Iteration 3
Iteration 4
Iteration 5

ALGEBRA Connection

5. Calculate the number of segments and their combined lengths after each iteration.

Iteration	0	1	2	3	4	5	*n*
Number of segments	?	?	?	?	?	?	?
Combined length	?	?	?	?	?	?	?

6. As the number of iterations increases, what happens to the number of segments? What happens to the combined lengths of the segments? Do you think there is a limit greater than zero of the combined lengths of the segments as the process continues? ❖

Use Transparency 104

Exploration 1 Notes

In this exploration, students draw a simple fractal. Use this activity to introduce the concept of iteration. You may want to use a spreadsheet to extend the chart.

Math Connection Algebra

Review from algebra the idea of a sequence of terms and compare it to a sequence of iterations. Point out that a sequence of terms or iterations can extend indefinitely.

ONGOING ASSESSMENT

6. The number of segments doubles with each iteration. The combined lengths of the segments is reduced by $\frac{1}{3}$ with each iteration. No, the limit of the combined lengths of the segments approaches 0 as the number of iterations increases.

ENRICHMENT Have students make a collage of fractal images that they find in magazines or that they create with a computer. Extend the activity by having individual groups use a theme for the fractals they choose. For example, one group could find images based on coastlines; another could find images based on fern-like structures. Display the collages in the classroom.

INCLUSION strategies

Using Visual Models
Use a computer and fractal software to demonstrate how a fractal image is created slowly, by drawing self-similar images. Vary the inputs to create different images.

Exploration 2 Notes

This exploration will give students some hands-on experience at creating a fractal-like image. Encourage students to compare their results with those of their classmates.

Math Connection Algebra

In algebra, students studied how to expand a binomial with Pascal's Triangle. It would be helpful to review those ideas at this time, including how Pascal's Triangle is generated. Point out that the triangle can be extended indefinitely.

Use Transparency 105

Cooperative Learning

Use cooperative groups to discuss the different types of fractal images students have seen in this lesson. Have each student in the group describe how he or she knows that an object or image is self-similar. Working in groups may help students understand the importance of self-similarity in a fractal image.

Use Transparency 106

Communication Technology

Communications When an electric current transmits data over a wire, a certain amount of "noise" occurs, which can cause errors in the transmission. The noise seems to occur in random bursts, with "clean" spaces in between. Benoit Mandelbrot, the discoverer of fractal geometry, showed that the noise patterns closely matched the pattern of a Cantor dust. This mathematical representation allowed the development of new strategies for reducing transmission noise to a minimum.

The Sierpinski Gasket

The Sierpinski Gasket, like the Menger Sponge and the Cantor dust, is created by applying the same rule to an initial shape. The rules are as follows:

1. Start with an isosceles or an equilateral triangle. Find the midpoints of each of the sides.
2. Connect the midpoints of the sides to form four congruent isosceles triangles in the interior of the original triangle. This constitutes one iteration.
3. Each new iteration is performed on the three outer triangles while the triangle in the middle is left alone.

Mathematical ideas often turn out to have surprising connections to seemingly unrelated fields. The following exploration shows a connection between a Sierpinski Gasket and Pascal's Triangle.

Exploration 2 Pascal and Sierpinski

You will need
Graph paper

Recall that Pascal's Triangle is a triangular number array beginning with 1 at the top. Each number in each row is the sum of the two numbers directly above it. The pattern is repeated endlessly.

1
1 1
1 2 1
1 3 3 1
1 4 6 4 1
1 5 10 10 5 1

Using Discussion Review the lesson by showing students examples of fractal images on the overhead projector. Have students trace out self-similar images on the fractals. Ask them to think of analogous objects in nature.

1. Build a Pascal's Triangle with at least 24 rows in a triangle of squares as indicated in the diagram. The more rows you can complete, the more impressive the result will be.

2. Darken each box occupied by an odd number, and leave each box occupied by an even number blank. Describe your results. ❖

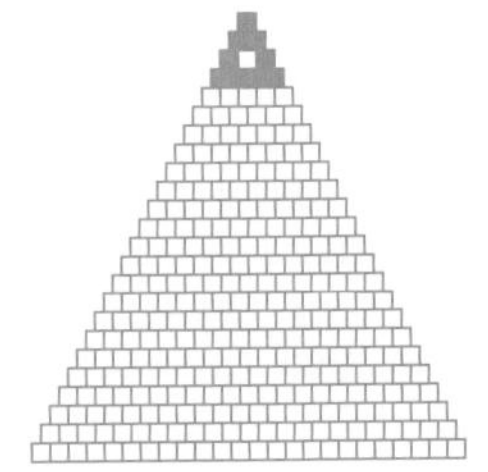

Ongoing Assessment

2. **The result resembles a Sierpinski Gasket.**

Exercises & Problems

Communicate

1. What is self-similarity? If it helps you to describe an object to explain self-similarity, do so.
2. Name an object from nature that is self-similar, and describe how it is self-similar.
3. Why do you think a set of segments called Cantor dust is called that?
4. How many iterations does any fractal have?

Practice & Apply

5. Mimi, Arnold, and Roberto measure the rocky ocean frontage of a hotel using paces. Arnold's stride is 3 ft long and he counts 24 paces. Roberto's stride is $2\frac{1}{2}$ ft long and he counts 36 paces. Mimi's stride is 2 ft long and she counts 47 paces. How long is the coastline to each? Explain your answer.
6. Explain how a fractal coastline could be considered to have an infinite length.

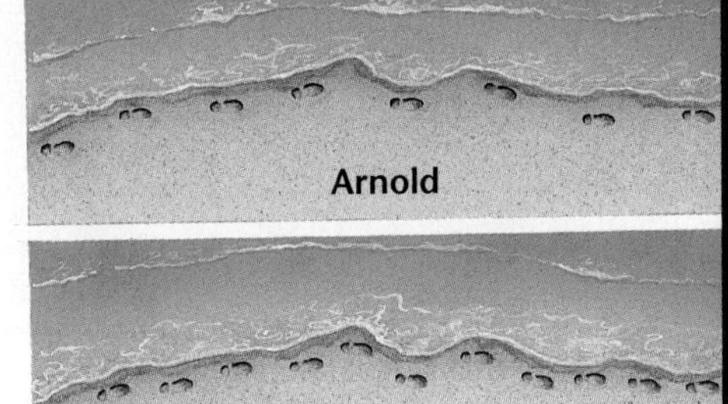

Two different measurements of a "coastline."

Portfolio Activity Exercises 7–11 ask you to construct a fractal called the Koch Snowflake.

7. Construct the first iteration of the Koch Snowflake using the directions below. Check student constructions.
   a. Construct an equilateral triangle with 18-cm sides.
   b. Divide each side of the triangle into three 6-cm segments.
   c. Using the middle 6-cm segment on each side of your original triangle, construct an equilateral triangle outside the triangle with the middle segment as the base. Erase the middle segment. You should now have a six-pointed star. This completes the first iteration.

5. Distance for Arnold = (3)(24) = 72 ft.

   Distance for Mimi = (2)(47) = 94 ft.

   Distance for Roberto = (2.5)(36) = 90 ft.

   The distance is greater as the stride length gets shorter because a person with smaller strides will be able to go along more of the irregularities in the coast than a person with longer strides.

6. Since the depth of self-similarity is endless, it would be impossible to measure around all the "edges" of the coastline. There would always be another similar shape to measure around.

Assess

Selected Answers

Exercises 6, 10, 15, 17, 19, and 21

Assignment Guide

Core 1–24

Core Plus 1–24

Technology

You may want students to use fractal software to draw the Koch Snowflake in Exercise 8. Have them do many iterations. Students may also use the script features of geometry graphics programs to create fractal images.

Error Analysis

Students may have difficulty with Exercises 13–16 because of the length of time it takes to create the "monster curve." Use these exercises to emphasize the importance of computers in drawing fractal images.

Use Transparency 107

alternative ASSESSMENT

Authentic Assessment

Ask students to find examples of fractal images in nature and include them in their portfolio. Encourage them to include a description of the starting shape of the analogous fractal on the computer (the shape before any iterations were done).

8. Continue the construction of the Koch Snowflake. Repeat the steps by dividing each segment into three congruent segments, constructing new smaller equilateral triangles and erasing the bases. Do at least three iterations. Check student constructions.

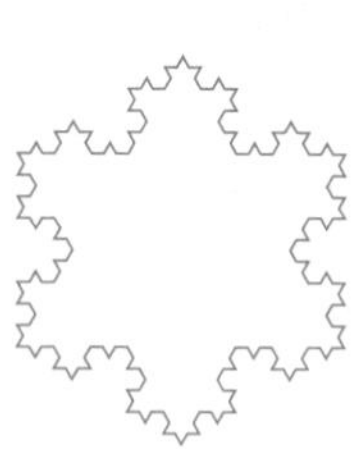

9. Find the perimeter of the Koch Snowflake for the first two iterations. Remember the snowflake starts with a perimeter of $18 \times 3 = 54$.

10. Is the perimeter increasing or decreasing? If it is increasing, is it increasing by greater leaps each iteration, or by smaller leaps—i.e., is it increasing at an increasing or decreasing rate? What does this tell you about the perimeter?

11. Look at your snowflake and consider the changes it would make if you completed many more iterations. What do you think happens to the area of the snowflake as the iterations increase?

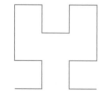

Cultural Connection: Asia People of India have been using fractals as art for many centuries. They are taught to draw a *Kolam* very quickly. A Kolam has an algorithm very similar to a curve known as the Hilbert Curve, which you will construct in Exercises 12–15.

12. Start with a square with no bottom side. Divide the top side into three congruent segments, and construct a square inward from the middle segment. Erase the middle segment of the top of the original square. Check student constructions.

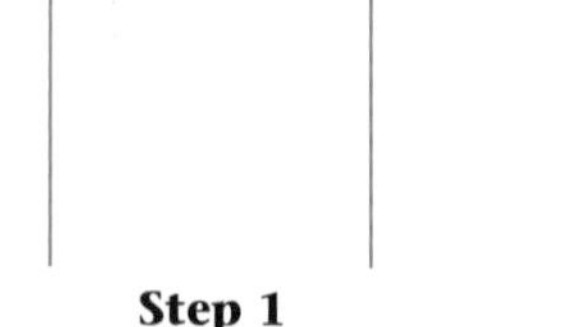

Step 1 **Step 2**

13. Divide the sides into three segments. Construct an inward square using the lowest segment on each of the sides. Erase the outside segment from each of these squares. This is the first iteration.

Practice Master

NAME ______ CLASS ______ DATE ______

Practice & Apply
11.7 Fractal Geometry

Here are the first two levels of the construction of a fractal.

Level 0: Begin with a line segment.

Level 1: Divide the segment into thirds. Replace the middle third with two segments whose lengths are each one-third of the original segment.

1. In the space provided, construct the next level of the fractal by replacing each segment of the figure with a one-third replica of Level 1.

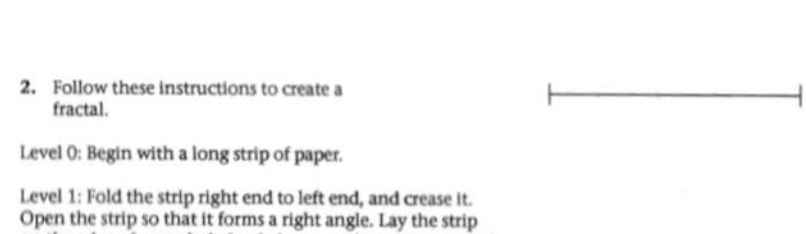

2. Follow these instructions to create a fractal.

Level 0: Begin with a long strip of paper.

Level 1: Fold the strip right end to left end, and crease it. Open the strip so that it forms a right angle. Lay the strip on the edge of your desk. Look down on the open strip and sketch this view.

Level 2: Start over. Refold the strip as in Level 1. Now fold the right end to the left end a second time to make a second crease. Open the strip to form right angles. Lay the strip on your desk, look down on it, and sketch this view.

Level 3: Start over again. Refold the strips as in Level 1 and Level 2. Add Level 3 by folding the right end to the left end a third time. Open the strip to form right angles. Lay the strip on your desk, look down on it, and sketch this view.

Level 4: Start over. Refold Level 1, then Level 2, then Level 3. Add Level 4 by folding again. Open, stand, sketch.

HRW Geometry — Practice & Apply 49

9. Each iteration results in 4 times as many sides as the previous iteration and each side is $\frac{1}{3}$ the length of the sides from the previous iteration. So the perimeter after each iteration is $\frac{4}{3}$ times the perimeter after the previous iteration. First iteration: $(54)\left(\frac{4}{3}\right) = 72$ cm; Second iteration: $(72)\left(\frac{4}{3}\right) = 96$ cm

10. The perimeter is increasing at an increasing rate with each iteration. The perimeter will approach infinity with continued iterations.

11. The area of the snowflake will approach the area of some circle as the iterations increase. Thus the area of the Koch Snowflake is finite but the perimeter is infinite.

13. Check student constructions.

14. Repeat this process on the two squares now extending upward at the top of where your original square was, and on the two squares that turn inward at the bottom of your original square. Repeat this process three times.

15. This is known as a space-filling curve, or "monster curve." If you were to continue this process indefinitely, what would the result look like? The entire plane would appear to be filled.

Look Back

Find the distance between each pair of points. [Lesson 5.6]

16. (4, −2), (2, −1) 2.24 units

17. (5, −10), (−2, 3) 14.76 units

18. (15, 2), (−6, 5) 21.21 units

Find the distance between each point in a three-dimensional coordinate system. [Lesson 6.4]

19. (4, 3, 2), (2, −3, 5) 7 units

20. (18, 1, 0), (0, −1, 5) 18.79 units

21. (5, 1, −5), (2, −12, 0) 14.25 units

22. Find the measure of the diagonal of a cube with its corner at (0, 0, 0), lying in the top-front-right octant, and with coordinates as shown. **[Lesson 6.4]** 13.86 units

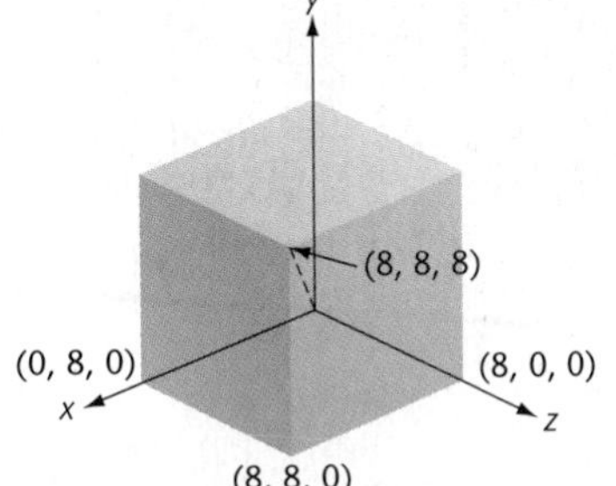

Look Beyond

Candy Making Dissolve $2\frac{1}{2}$ cups of sugar into a cup of boiling water. Hang a string into the sugar water and leave it untouched for at least a day. Sugar crystals will form on the string, resulting in "rock candy."

23. How is the resultant "rock candy" a fractal?

24. What else do you see in nature that is the same kind of process that results in a fractal?

Look Beyond

Extension. Exercises 23 and 24 give students a visual way to compare the crystallization of sugar to a fractal.

Technology Master

NAME ______ CLASS ______ DATE ______

Technology

11.7 The Sierpinski Triangle Revisited

The four diagrams shown illustrate the first four stages in the construction of the Sierpinski triangle based on an initial triangle that is equilateral. Notice that the triangle develops holes in it as the construction process continues. The higher the stage number, the more holes and the more boundary lines around the holes.

To explore the Sierpinski triangle in detail, a spreadsheet will be very helpful. Before you can use a spreadsheet, you must use your pattern finding skills to represent important variables.

Stage 1 Stage 2 Stage 3 Stage 4

An equilateral triangle initially has sides 1 unit long. Let n = stage number, s = the number of shaded triangles added in stage n, L = the length of one leg of each shaded triangle added in stage n, and P = perimeter of all the triangles added in stage n. Write a formula involving n for the specified variable.

1. s ______ 2. L ______ 3. P ______

4. Create a spreadsheet in which column A contains the values of n, column B contains the values of s, column C contains the values of L, and column D contains the values of P. What can you conclude about P when n becomes large?

5. The area A of an equilateral triangle whose sides are L units long is given by $A = 0.43301\ L^2$. Modify your spreadsheet from Exercise 4 to find the total area of all shaded triangles at stage n. What can you say about the total area of the shaded triangles when n becomes large?

6. Modify your spreadsheet from Exercise 5 to find the total area of all unshaded triangles at stage n. What can you say about the total area of the unshaded triangles when n becomes large?

HRW Geometry Technology 63

14. Check student constructions.

23. The large crystals are formed by combining smaller crystals of the same shape. The large crystals can be broken down into smaller crystals that have the same shape. The crystals are self-similar.

24. Sample answers: The growth of coral reefs. The formation of gems and other stones.

Focus

Two random process games are the focus of this chapter project. The Chaos Game will produce the Sierpinski Triangle. The Forest Fire Simulation shows the pattern of the spread of a fire.

Motivate

Review fractal concepts from Lesson 11.7. Remind students that fractal are a result of thousands to millions of iterations of a simple formula or algorithm. Students may realize immediately that they need the power of a computer or calculator to be able to complete the games. Use the TI-82 program given below as a model for a computer or graphics calculator program which simulates the Chaos Game.

```
PROGRAM:SIERPINS
:FnOff :ClrDraw
:PlotsOff
:AxesOff
:0→Xmin : 1→Xmax
:0→Ymin : 1→Ymax
:rand→X : rand→Y
:For (K, 1, 3000)
:rand→N
:If N≤1/3
:Then
:.5X→X
:.5Y→Y
:End
:If 1/3 <N and N≤2/3
:Then
:.5(.5+X)→X
:.5(1+Y)→Y
:End
:If 2/3 <N
:Then
:.5(1+X)→X
:.5Y→Y
:End
:Pt-On(X, Y)
:End
:StorePic Pic6
```

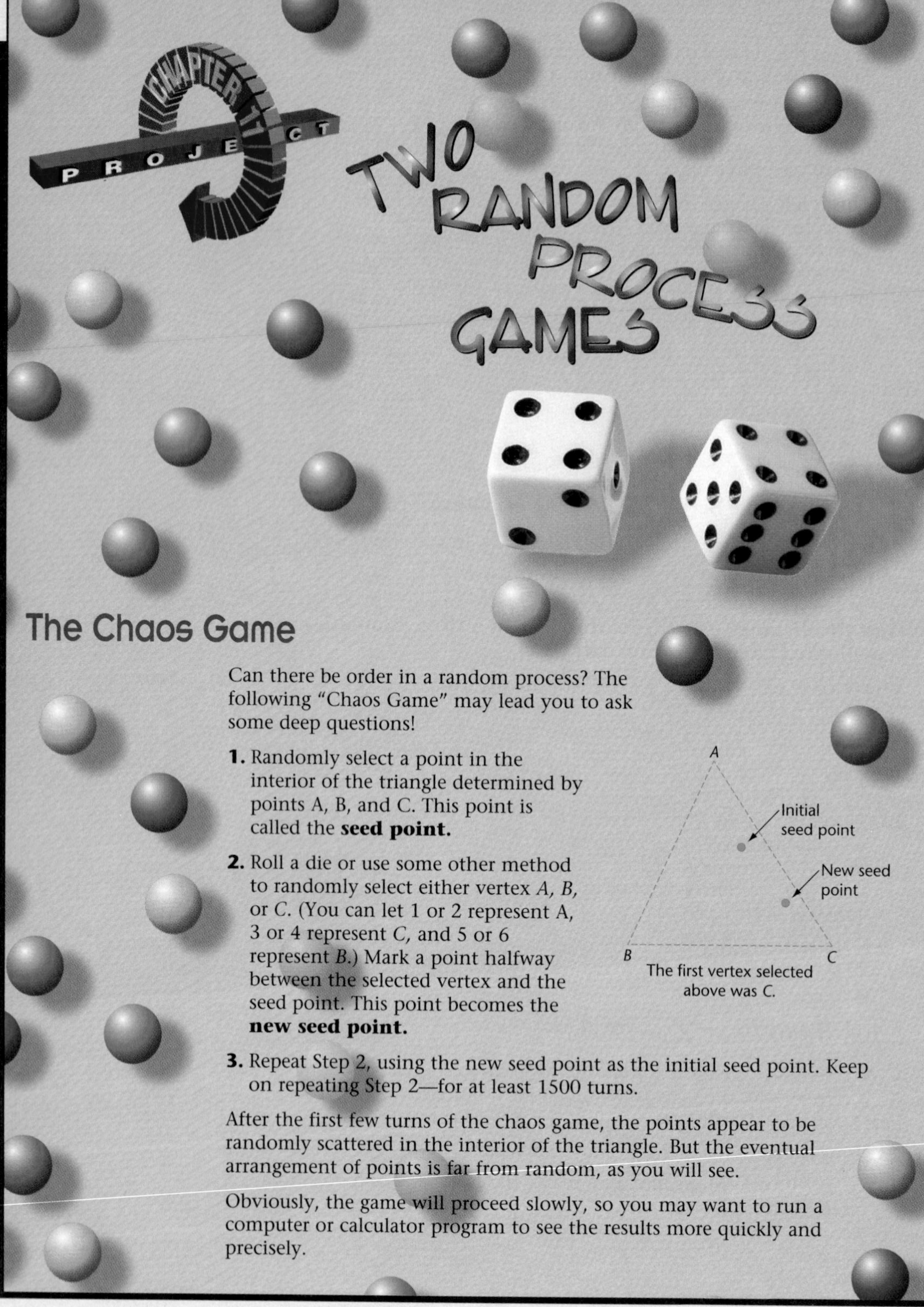

The Chaos Game

Can there be order in a random process? The following "Chaos Game" may lead you to ask some deep questions!

1. Randomly select a point in the interior of the triangle determined by points A, B, and C. This point is called the **seed point.**
2. Roll a die or use some other method to randomly select either vertex *A, B,* or *C.* (You can let 1 or 2 represent A, 3 or 4 represent *C,* and 5 or 6 represent *B.*) Mark a point halfway between the selected vertex and the seed point. This point becomes the **new seed point.**
3. Repeat Step 2, using the new seed point as the initial seed point. Keep on repeating Step 2—for at least 1500 turns.

After the first few turns of the chaos game, the points appear to be randomly scattered in the interior of the triangle. But the eventual arrangement of points is far from random, as you will see.

Obviously, the game will proceed slowly, so you may want to run a computer or calculator program to see the results more quickly and precisely.

The Chaos Game

Students should discover that the points plotted in the triangle generate an image of a Sierpinski gasket.

A Forest Fire Simulation

A forest fire is less likely to spread through a forest where trees are sparse than where they are densely concentrated. But what is the exact density and pattern of trees that guarantees the spread of the fire?

In this simplified model, you will try to discover the critical density of trees required for a forest fire to spread. Follow the steps below. (It will be necessary to pool your data with the rest of the class to obtain valid results.)

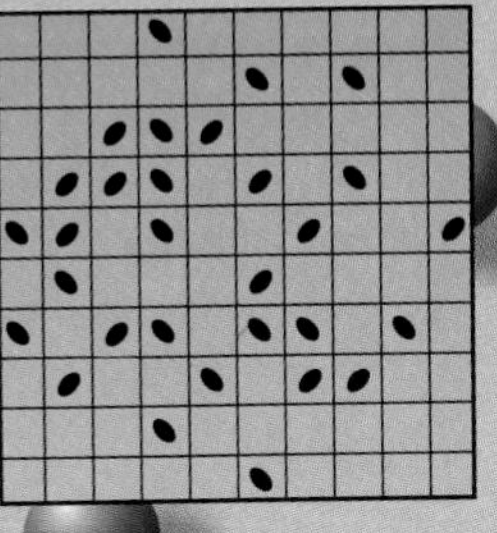

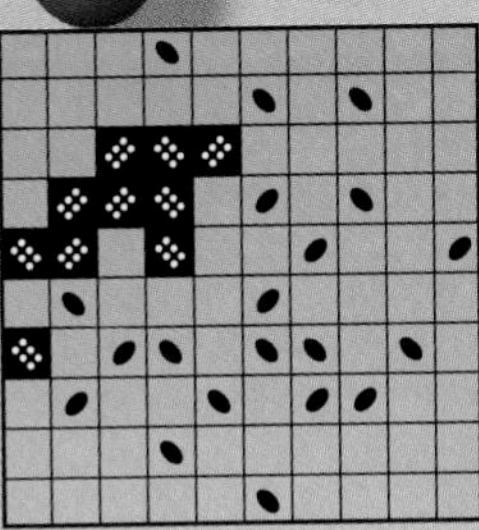

1. Your forest will be a 10×10 grid, which you will randomly plant with a certain percentage of trees. To have a forest with 30% trees, you will need to fill 30 of the 100 grid units with trees. One way to do this randomly is to number small squares of paper from 1 to 100 and mix them up. Pick a number for each square on the grid. If the number picked is 1–30, place a "tree" in that grid unit.
2. To simulate a fire's spread, start the "fire" with all the trees on the left side of the grid. The fire can spread to a neighboring tree directly above, below, to the right, or to the left of a burning tree—not diagonally.
3. See if the fire spreads from the left side of the grid to the right side. In the diagram, the fire dies out before it reaches the right side of the grid.
4. Each person or group in the class should simulate fires with 10%, 20%, 30%, 40%, 50%, 60%, 70%, 80%, 90% tree concentrations (0% and 100% tree concentrations are obvious).
5. Collect and tabulate the data of the class. Does there appear to be a critical percentage that assures the spread of the fire across the grid?

Instead of the numbered squares suggested in Step 1 you can use a calculator program. The following program will work on a graphics calculator. First enter the following code: **int (100*rand)+1.** Then press enter. Each time you press enter you will get a different random number between 1 and 100.

Cooperative Learning

Students should work in pairs. Assign one game to each group. Students should divide responsibilites. One student can perform the caclultions and the other can record the results. You may want to compare results from a group doing the activities by hand with those from a group doing the activities with a computer simulation.

Discuss

Fractals have only been recently discovered in mathematics. This is due partially to the fact that fractals require massive amounts of computing (i.e., iterations) before a clear pattern emerges. Ask students to discuss how increased computing capacity has changed mathematics.

Forest Fire Simulation

Collect class data. Discuss various ways of presenting and interpreting the data. For example, students might experimentally conjecture the probabilities of forest fire spread across the grid for each of the concentrations tested.

Chapter 11 Review

Vocabulary

Key Skills and Exercises

Lesson 11.1

➤ Key Skills

Define and construct golden rectangles.

Is the quadrilateral a golden rectangle?

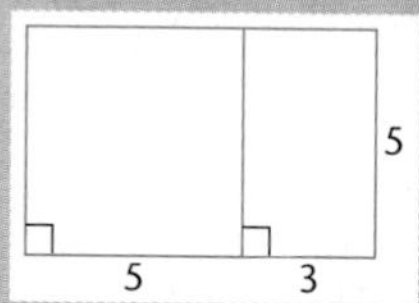

A rectangle is golden if it is constructed from a square and a similar rectangle. Set up a proportion to see if the rectangle is golden.

$$\frac{8}{5} \stackrel{?}{=} \frac{5}{3}$$

$$1.6 \neq 1.67$$

Or you can compare the ratio of length to width with the golden ratio, $\frac{1 \pm \sqrt{5}}{2}$.

$$\frac{8}{5} \stackrel{?}{=} \frac{(1 + \sqrt{5})}{2}$$

$$1.6 \neq 1.62$$

Both methods give an inequality, so the rectangle is not golden.

➤ Exercises

1. The figure at the right is a sand dollar. In what ways does a sand dollar's shape reflect the golden ratio?

Lesson 11.2

➤ Key Skills

Using taxicab geometry, find the distance between two points.

Find the taxidistance between points (9, 5) and (3, −4).

To find taxidistance, you calculate the differences between the *x* values and between the *y* values and add the absolute value of the differences.

$|9 - 3| = 6 \qquad |5 - (-4)| = 9$

$6 + 9 = 15$. The taxidistance is 15.

1. The golden ratio occurs in regular pentagons and a regular pentagon is formed in the sand dollar.

Create taxicab circles.

The taxicab circle has a radius of 2; that is, the points shown are all the possible ways of going two units from the center of the circle. The circumference is 16.

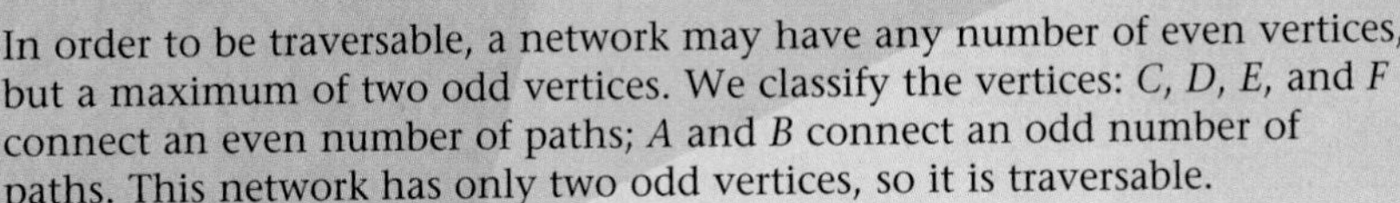

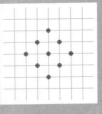

➤ *Exercises*

2. Draw a taxicab circle with a radius of 3. Find the circumference of the circle.

Lesson 11.3

➤ *Key Skills*

Determine whether a network is traversable.

Can you traverse the network at the right?

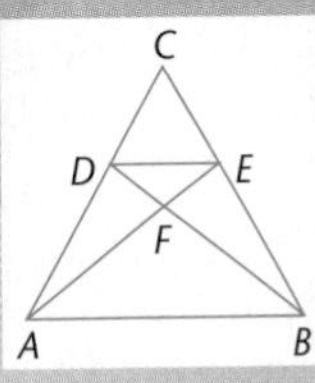

In order to be traversable, a network may have any number of even vertices, but a maximum of two odd vertices. We classify the vertices: *C*, *D*, *E*, and *F* connect an even number of paths; *A* and *B* connect an odd number of paths. This network has only two odd vertices, so it is traversable.

➤ *Exercises*

3. Which of these networks can be traversed?

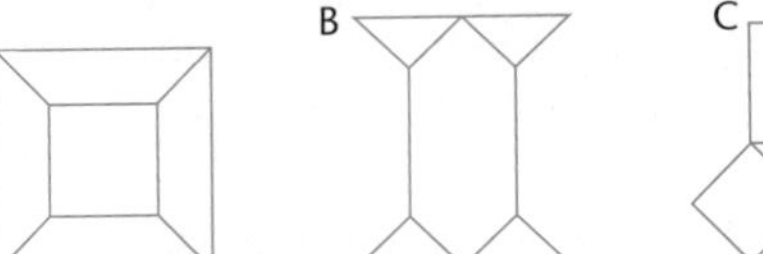

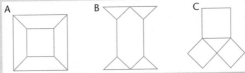

Lesson 11.4

➤ *Key Skills*

Determine the topological equivalence of two figures.

Are the two closed loops topologically equivalent?

To be equivalent, the shapes must be transformable, one to the other, without tearing or cutting them. The left-hand shape can be pulled into a circle, but it will have no knot in it. The right-hand shape cannot lose its knot without being cut. So the two shapes are not equivalent.

➤ *Exercises*

4. Which figures are topologically equivalent? Explain your answer.

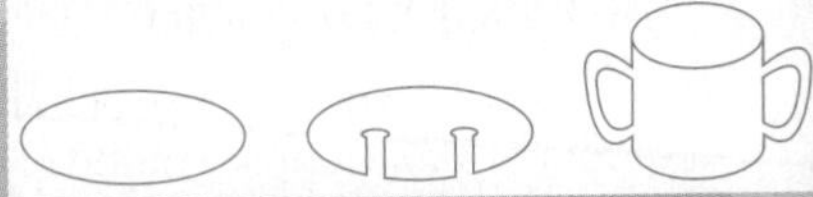

5. Draw a topological equivalent to a water hose.

Lesson 11.5

➤ *Key Skills*

Find the consequences of omitting the Parallel Postulate from a geometry.

In Riemannian geometry, which is one example of a non-Euclidean geometry, a line is a great circle on a sphere. Since all great circles intersect, there are no parallel lines in this geometry. Thus any theorem whose proof depends on the Parallel Postulate may be false in Riemannian geometry.

➤ *Exercises*

6. In Reimannian geometry, the sum of the angles in a triangle is greater than what number?

Chapter Review

5. Check student's drawings.

6. Sum must be greater than 180°.

2. C = 24 units

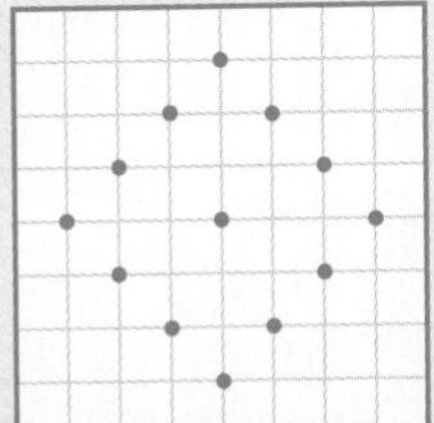

3. C

4. The first two are equivalent because the first can be stretched to make the second. The third is not equivalent because it has two "holes" that the other two don't have. The first two cannot be stretched to make the holes.

7.

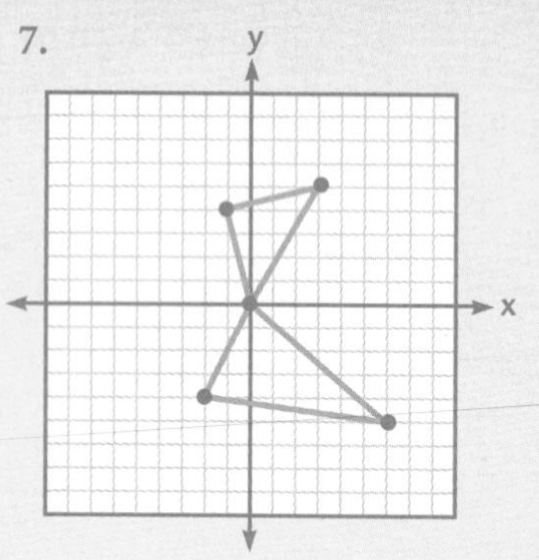

8.

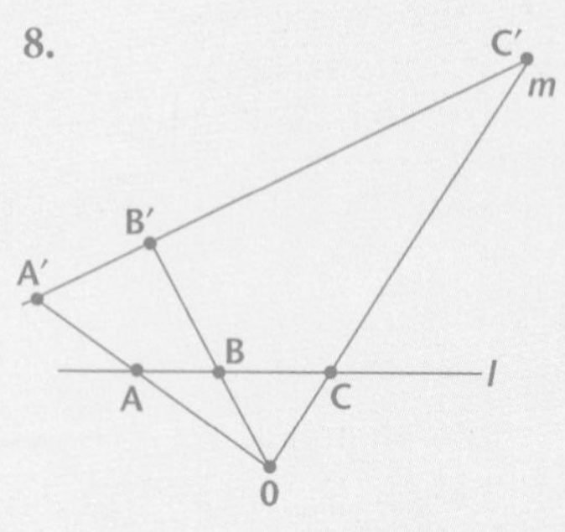

Lesson 11.6

➤ Key Skills

Make an affine transformation of an image.

Sketch the pre-image and image for the affine transformation $(x, y) \rightarrow (3x, 4y)$ of rectangle (0, 0), (0, 6), (3, 6), (3, 0).

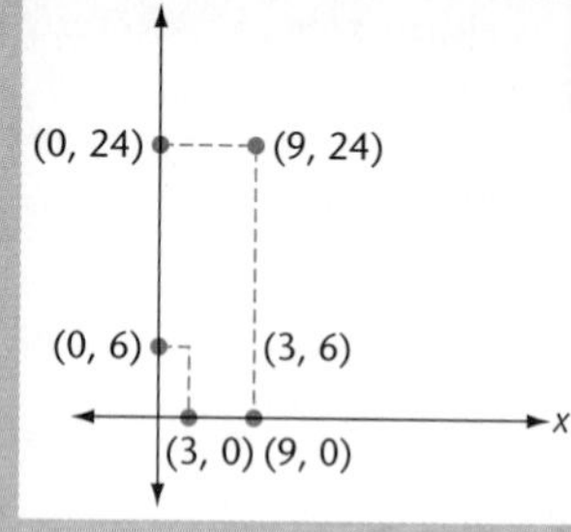

Plot the original points and draw the rectangle. Multiply the x-coordinates by 3 and the y-coordinates by 4, and plot the resulting points for the image.

Sketch the projection of points.

Using center of projection O, project points A and B onto line m.

Use a straightedge to draw rays from the center of projection through A and B on line l to line m. The new points A' and B' lie at the intersection of the rays and line m.

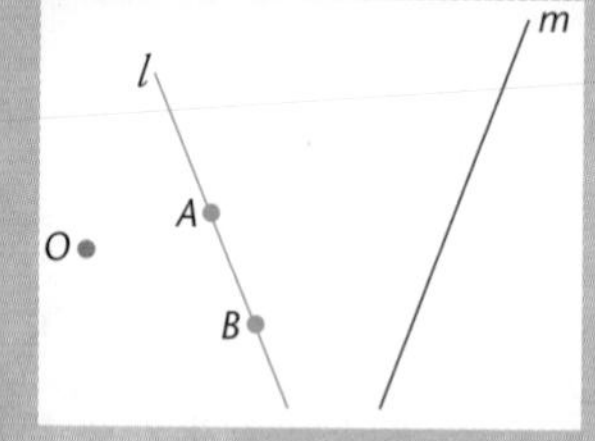

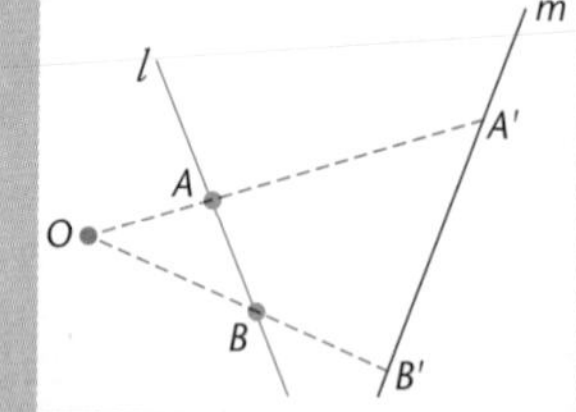

➤ Exercises

7. Sketch the pre-image and image for the affine transformation $(x, y) \rightarrow (2x, -y)$ of triangle (0, 0), (3, 5), (−1, 4).

8. Using center of projection O, project points A, B, and C onto line m.

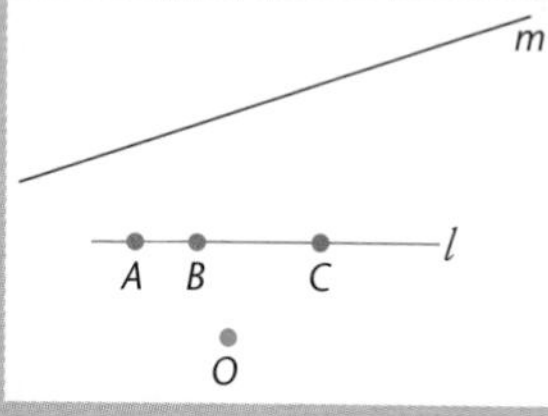

Lesson 11.7

➤ Key Skills

Use fractal concepts to solve problems.

The pilot of a small airplane going over the Fractal Mountains reports that it is 2.5 miles from point A to point B. A hiker traveling from point A to point B reports the distance as 3.5 miles. Explain the difference in reports.

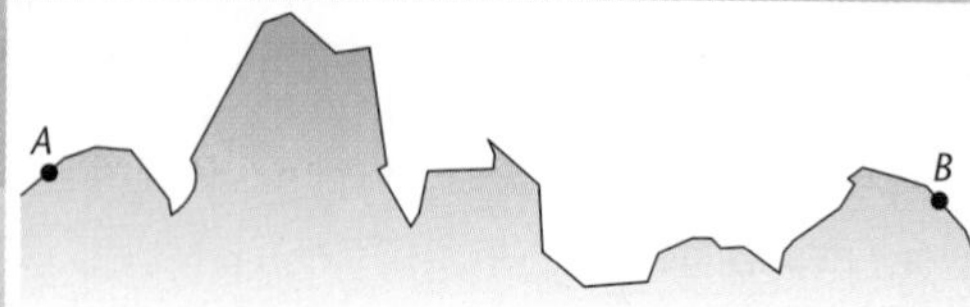

The reports depend on how closely the measurements follow the contours of mountains. The hiker follows the contours very closely. The airplane does not.

➤ Exercises

9. You and I would like to know how long the coast of Oregon is. You find a satellite photo of the coast and find it is 490 kilometers long. I use a series of aerial photos of the coast and find it is 560 kilometers long. Explain the difference in our results.

Application

10. Make a perspective drawing of the checkered floor.

9. A coastline is jagged and irregular. However, when viewed from far away it looks smoother than when viewed from close up. When using the satellite photo which is taken from farther away, fewer of the irregularities or contours in the coastline are measured than when using the aerial photo.

10. Check student's drawings.

Chapter 11 Assessment

1. Which quadrilaterals are golden rectangles, given that *ACEH* is a golden rectangle?

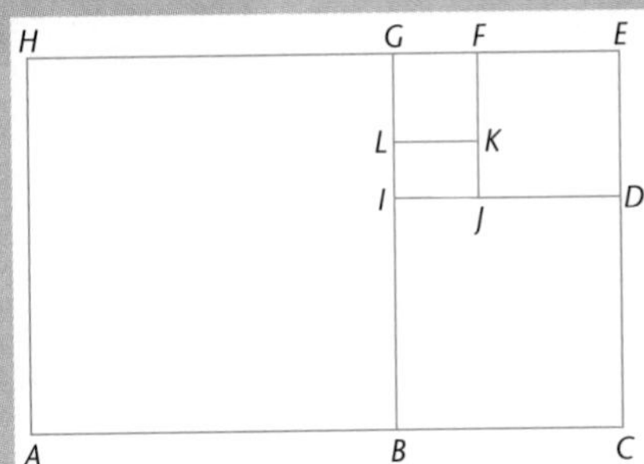

2. Find the taxidistance between points (7, −3) and (−1, −6).

3. Draw a taxicab circle with a radius of 4.

4. Which of these networks are traversable?

A 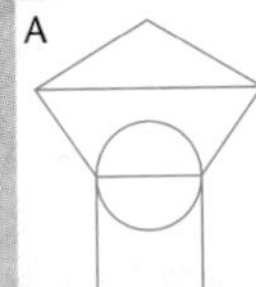 B 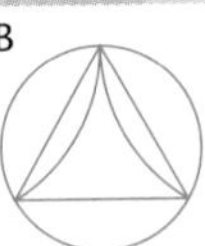 C 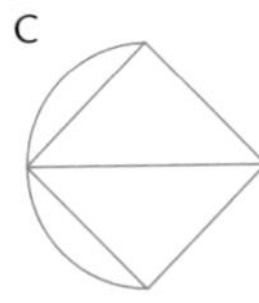

5. Which figures at the right are topologically equivalent? Explain your answer.

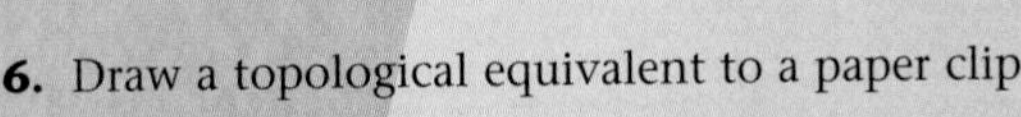

6. Draw a topological equivalent to a paper clip.

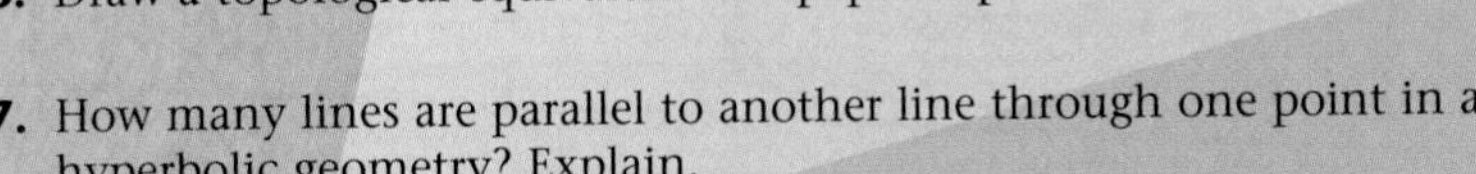

7. How many lines are parallel to another line through one point in a hyperbolic geometry? Explain.

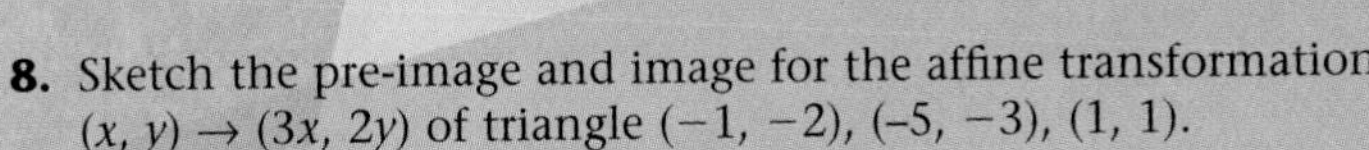

8. Sketch the pre-image and image for the affine transformation $(x, y) \rightarrow (3x, 2y)$ of triangle (−1, −2), (−5, −3), (1, 1).

9. Using center of projection *O*, project points *A*, *B*, and *C* onto line *m*.

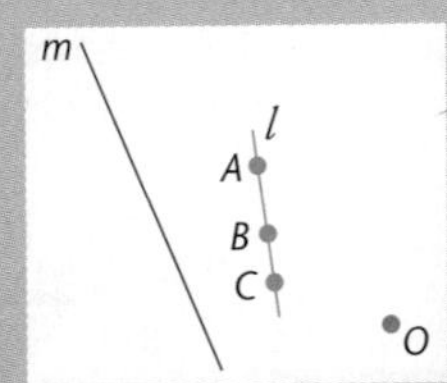

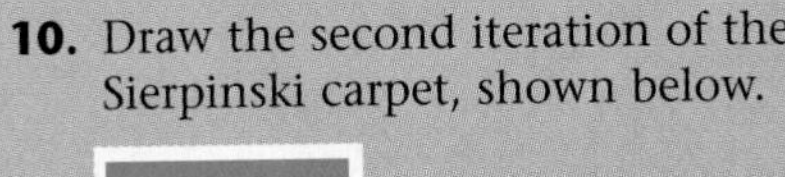

10. Draw the second iteration of the Sierpinski carpet, shown below.

1. *ACEH; GBCE; GIDE; GIJF; LIJK*

2. $x = |7 - -1| + |-3 - -6| = 11$

3. 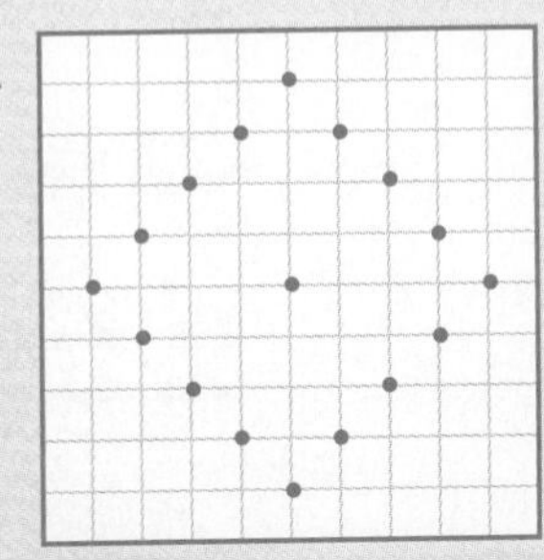

4. B

5. All three are equivalent. Each can be stretched or transformed into the others.

6. Check student's answers.

7. An infinite number. Because of the curvature in the space used in hyperbolic geometry, through a point not on a line an infinite number of lines can be drawn that do not intersect the line.

8. 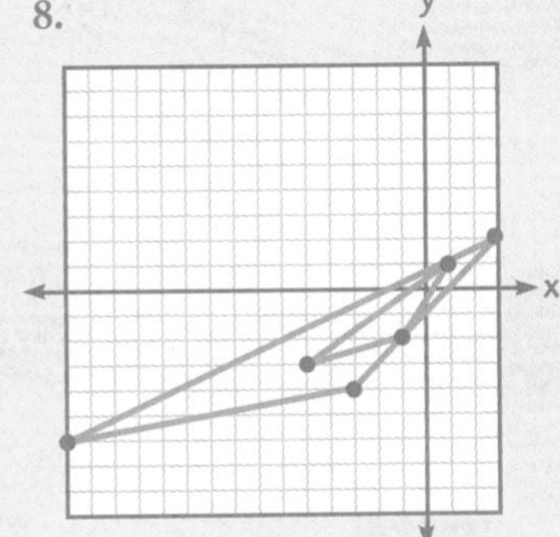

9. 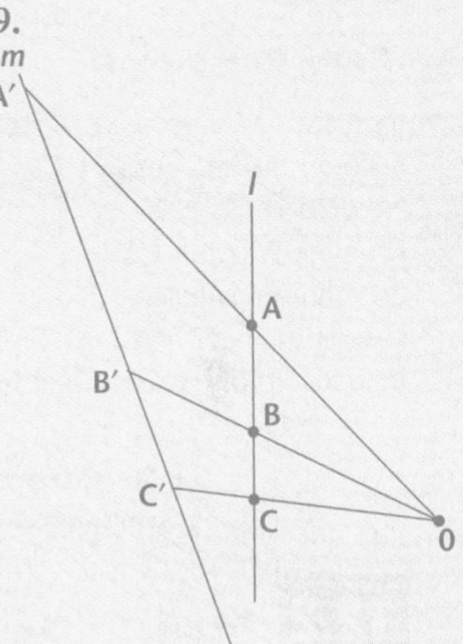

10. 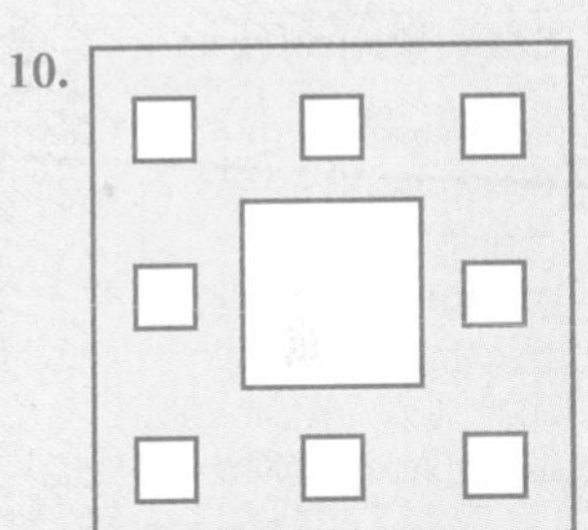

A Closer Look at Proof and Logic

Meeting Individual Needs

12.1 Truth and Validity in Logical Arguments

Core Resources

Inclusion Strategies, p. 646
Reteaching the Lesson, p. 647
Practice Master 12.1
Enrichment Master 12.1
Lesson Activity Master 12.1
Interdisciplinary Connection, p. 645

[2 days]

Core Plus Resources

Practice Master 12.1
Enrichment, p. 646
Technology Master 12.1
Interdisciplinary Connection, p. 645

[1 day]

12.2 "And," "Or," and "Not" in Logic

Core Resources

Inclusion Strategies, p. 652
Reteaching the Lesson, p. 653
Practice Master 12.2
Enrichment, p. 652
Technology Master 12.2
Lesson Activity Master 12.2

[2 days]

Core Plus Resources

Practice Master 12.2
Enrichment Master 12.2
Technology Master 12.2

[1 day]

12.3 A Closer Look at If-Then Statements

Core Resources

Inclusion Strategies, p. 660
Reteaching the Lesson, p. 661
Practice Master 12.3
Enrichment Master 12.3
Lesson Activity Master 12.3
Mid-Chapter Assessment Master

[2 days]

Core Plus Resources

Practice Master 12.3
Enrichment, p. 660
Technology Master 12.3
Interdisciplinary Connection, p. 659
Mid-Chapter Assessment Master

[2 days]

12.4 Indirect Proof

Core Resources

Inclusion Strategies, p. 667
Reteaching the Lesson, p. 668
Practice Master 12.4
Enrichment, p. 667
Technology Master 12.4
Lesson Activity Master 12.4

[2 days]

Core Plus Resources

Practice Master 12.4
Enrichment, p. 667
Technology Master 12.4

[2 days]

12.5 Computer Logic

Core Resources

Inclusion Strategies, p. 673
Reteaching the Lesson, p. 674
Practice Master 12.5
Enrichment Master 12.5
Technology Master 12.5
Lesson Activity Master 12.5
Interdisciplinary Connection, p. 672

[2 days]

Core Plus Resources

Practice Master 12.5
Enrichment, p. 673
Technology Master 12.5
Interdisciplinary Connection, p. 672

[2 days]

12.6 Exploring Proofs Using Coordinate Geometry

Core Resources

Inclusion Strategies, p. 681
Reteaching the Lesson, p. 682
Practice Master 12.6
Enrichment Master 12.6
Technology Master 12.6
Lesson Activity Master 12.6
Interdisciplinary Connection, p. 680

[2 days]

Core Plus Resources

Practice Master 12.6
Enrichment, p. 673
Technology Master 12.6
Interdisciplinary Connection, p. 672

[2 days]

Chapter Summary

Core Resources	Core Plus Resources
Chapter 12 Project, pp. 685–686	Chapter 12 Project, pp. 685–686
Lab Activity	Lab Activity
Long-Term Project	Long-Term Project
Chapter Review, pp. 688–690	Chapter Review, pp. 688–690
Chapter Assessment, p. 691	Chapter Assessment, p. 691
Chapter Assessment, A/B	Chapter Assessment, A/B
Alternative Assessment	Alternative Assessment
Cumulative Assessment, pp. 692–693	Cumulative Assessment, pp. 692–693
[3 days]	[2 days]

Reading Strategies

The terms *argument* and *conclusion* as used in this chapter have meanings different from those used in everyday conversation. Remind students how the term *similar,* used extensively in Chapter 8, has a specific meaning in mathematics that is different from its meaning in ordinary conversation.

It is important that students clearly understand the distinction between the exclusive and the inclusive "or." Explained in Lesson 12.2, this distinction is used in many parts of mathematics, not just in the study of logic. Have students work in small groups to write several examples of each type of "or." Groups can exchange writing samples and sort the statements into the two appropriate categories, one for the exclusive "or," and one for the inclusive "or."

Visual Strategies

To introduce or accompany this chapter, collect a supply of books or magazines on logic puzzles. Included with the puzzles themselves will be various types of charts used in solving the puzzles. The charts help a solver use trial and error and the process of elimination to arrive at solutions.

Although logic puzzles of this type are not directly covered in the chapter, they use the same type of reasoning as many of the activities and exercises. And in working some of the puzzles, students will see the benefit of using chart-type drawings to clarify their thinking.

The truth tables introduced in Lesson 12.2 are one type of visual representation often used in the study of logic. Make sure students have enough practice with these tables to understand how they show the various possibilities for multipart statements. A key idea in a truth table is to list *all* possible variations so that each can be considered in turn.

Another visual representation sometimes used to test statements in logic is the Venn diagram. Students who are visual thinkers may find these diagrams useful and interesting. Examples can be found in introductory logic textbooks.

Cooperative Learning

GROUP ACTIVITIES	
The Negation of a Conjunction	Lesson 12.2, Exploration
The Ends of Time	Eyewitness Math
Truth or lie?	Lesson 12.3, Look Beyond
On-Off Tables	Lesson 12.5, Exploration
Proofs in coordinate geometry	Lesson 12.6, Explorations 1–3

You may wish to have students work in groups or with partners for some of the above activities. Additional suggestions for cooperative group activities are noted in the teacher's notes in each lesson.

Multicultural

The cultural references in this chapter include references to the Americas and Europe.

CULTURAL CONNECTIONS	
Americas: Mayan Calendar	Eyewitness Math
Europe: The Pythagoreans	Chapter 12 Project

Portfolio Assessment

Below are portfolio activities for the chapter listed under seven activity domains which are appropriate for portfolio development.

1. **Investigation/Exploration** The exploration in Lesson 12.2 focuses on the negation of a conjunction; that in Lesson 12.5 uses on-off tables to simulate the operation of a computer. In Lesson 12.6, students use coordinate geometry to investigate and prove three theorems.
2. **Applications** Recommended to be included are any of the following: Home Design, Lesson 12.4, Look Beyond; Environmental Science, Lesson 12.5, Exercises 39–40.
3. **Nonroutine Problems** Interesting and nontraditional activities are included in Lesson 12.2, Look Beyond (truth or lie?); Lesson 12.3, Exercise 29 (Who is telling the truth?)
4. **Project** Two Famous Theorems: see pages 685–686. Students study two famous number theory proofs.
5. **Interdisciplinary Topics** Students may choose from the following: Archaeology, Eyewitness Math.
6. **Writing** *Communicate* exercises of the type where students are asked to explain why an argument is valid or a statement is true offer excellent writing selections for the portfolio. Suggested selections include: Lesson 12.1, Exercises 1–5; Lesson 12.2, Exercises 3–6; Lesson 12.4, Exercise 5.
7. **Tools** In Lesson 12.6, students can use geometry graphics software to create the figures for the first two proofs. The software may also be used for the third proof and some of the exercises.

Technology

By studying simple computer or calculator programs, students can gain insight into the ways logical statements are implemented in "smart" machines. Students who have never programmed before will find that programs written in the traditional BASIC language are easy to understand and write.

If computers are available, encourage your students to write programs of their own. They can also write programs for graphics or other programmable calculators, which, though limited in comparison to computers, can be quite sophisticated.

Computers

Flow Charts and Logical Reasoning A computer program is a list of instructions to be performed by a computer. Before writing a program, a programmer must have a fairly definite idea of the procedure or algorithm to be followed. A flow chart is often used in designing and illustrating such procedures.

The flow chart below illustrates a procedure that can be followed by a computer. The purpose of the procedure is to determine whether a triangle with any three given sides is equilateral, isosceles but not equilateral, or scalene.

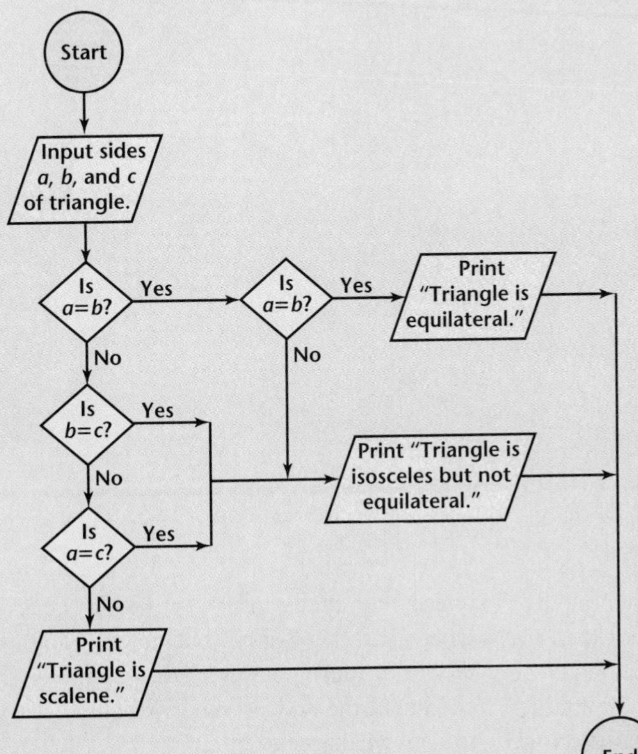

Each of the diamond-shaped boxes indicate *decisions.* At these points in the flow of a program, a *condition* is checked to determine what command is excecuted next. If the condition is true, one path is taken. If it is false, the program control procedes in another direction. Students should realize that decision boxes are like conditionals in logic.

The parallelogram boxes indicate input or output to the program. The first parallelogram box after the start circle is the point at which the user inputs the data for the program to act upon.

Students should discuss the flow chart to be sure they understand its logic. They should be able to answer questions like the following: In going from the start of the program to the output message for a scalene triangle, what decision boxes do you go through? What must the answer be to each of the questions in order to follow this path? (Yes.)

A BASIC Program The computer code below is a BASIC program to perform the procedure specified by the flow chart. Students should be able to understand it even if they don't know the BASIC programming language.

```
100   INPUT A, B, C
110   IF A = B THEN GOTO 160
120   IF B = C THEN GOTO 200
130   IF A = C THEN GOTO 200
140   PRINT "TRIANGLE IS SCALENE."
150   GOTO 210
160   IF B = C THEN GOTO 180
170   GOTO 200
180   PRINT "TRIANGLE IS EQUILATERAL."
190   GOTO 210
200   PRINT "TRIANGLE IS ISOSCELES BUT
      NOT EQUILATERAL."
210   END
```

Program statements 110, 120, 130, and 160 are IF-THEN statements, which correspond to the decision boxes in the flow chart. If the if-then statement is true, then the command after THEN is executed. If it is false, the program flow goes to the next statement (the one below).

In order to understand the flow of the program better, students will find it helpful to label the different boxes in the flow diagram with the numbers of the corresponding statements in the program. (There is no program statement corresponding to the Start circle.)

Ask the students to discuss what would happen if they entered impossible combinations of sides such as 1, 1, 10 or 0, 0, 0. (The program would respond with "isosceles" in the first case and "equilateral" in the second.) Discuss what might be done to the program to remedy the situation. Suggest that students write their own code to solve the problem.

About the Chapter

Background Information

Students explore the validity of arguments and the different rules of formal logic, including if-then statements. They also do indirect proofs, explore computer logic, and do proofs using coordinate geometry.

Chapter Resources

- Practice Masters
- Enrichment Masters
- Technology Masters
- Lesson Activity Masters
- Lab Activity Masters
- Long-Term Project Masters
- Assessment Masters
  Chapter Assessments, A/B
  Mid-Chapter Assessment
  Alternative Assessments, A/B
- Teaching Transparencies
- Cumulative Assessment
- Spanish Resources

Chapter Objectives

- Define *valid argument* and *invalid argument*.
- Develop and use the Law of Indirect Reasoning.
- Define *conjunction*, *disjunction*, and *negation*.
- Solve logic problems using conjunction, disjunction, and negation.
- Define *conditional*, *converse*, *inverse*, and *contrapositive*.
- Use if-then statements and forms of valid argument for problems involving logical reasoning.
- Develop the concept of indirect proof (*reductio ad absurdum*, proof by contradiction).

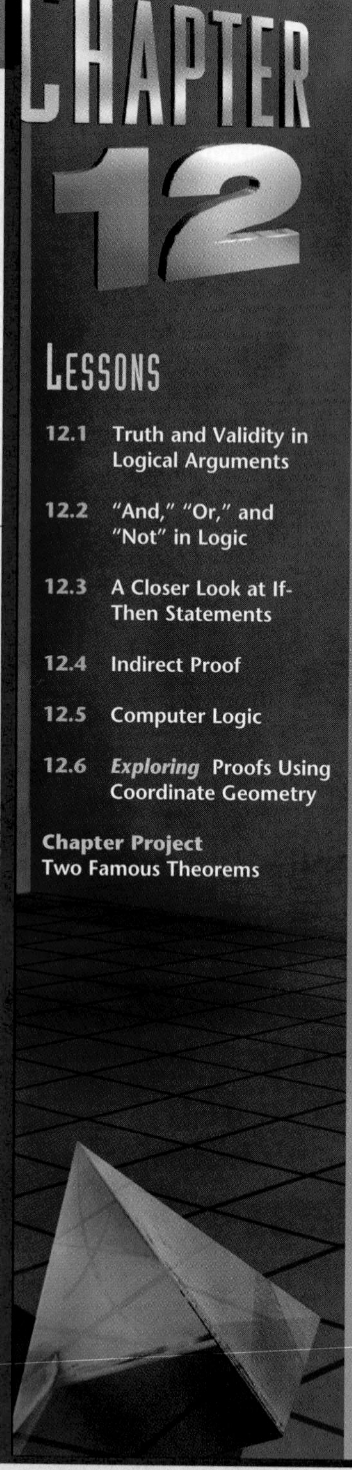

Chapter 12

Lessons

A Closer Look at Proof and Logic

How can a machine be intelligent? Some people believe that computers cannot actually think—and never will be able to. However, machines can certainly follow the rules of logic and make decisions based on given conditions. The decisions made by computer chess players and other "smart" machines are made possible by formal logic.

Formal logic reduces logical procedures to their essential elements, which can, in turn, be implemented by computers or other machines.

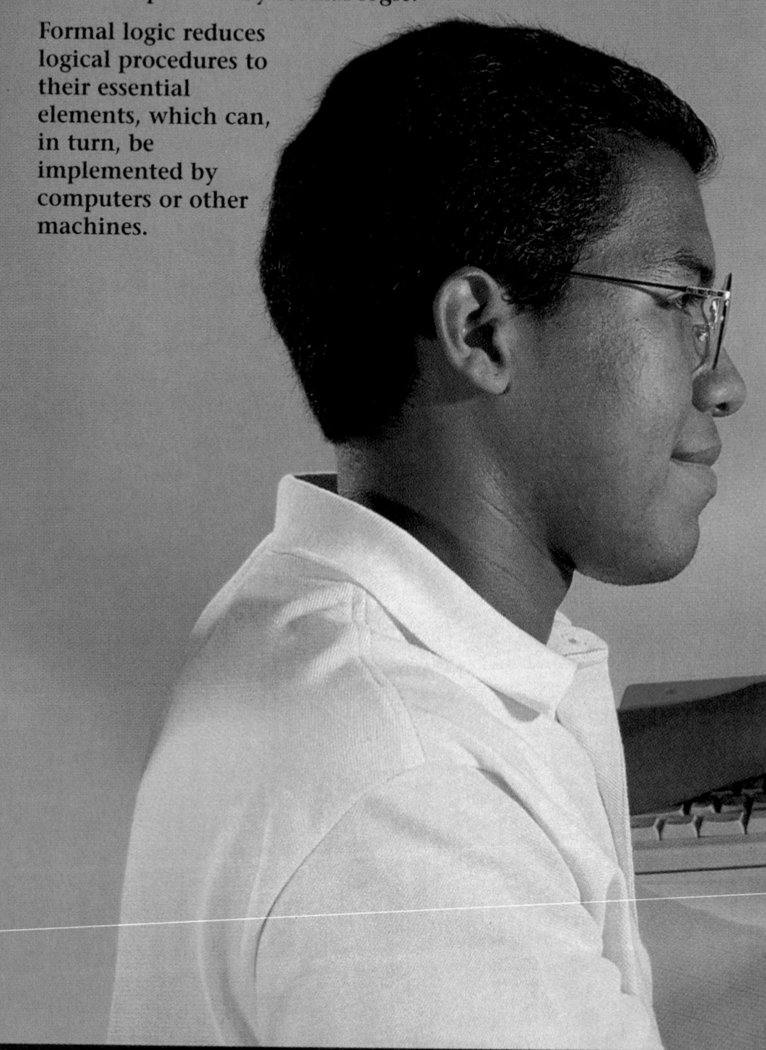

About the Photo

The rules of chess can be formalized logically. For example, if a player's King is in check then the player can, in addition to the options listed in the photo, (a) interpose a piece between the attacker and the King or (b) resign. In the game on the screen, the student can capture the attacking Queen with his Bishop (but not his King). On the next move, however, the computer will checkmate the student with its Knight. (Then the student will have *no* options.)

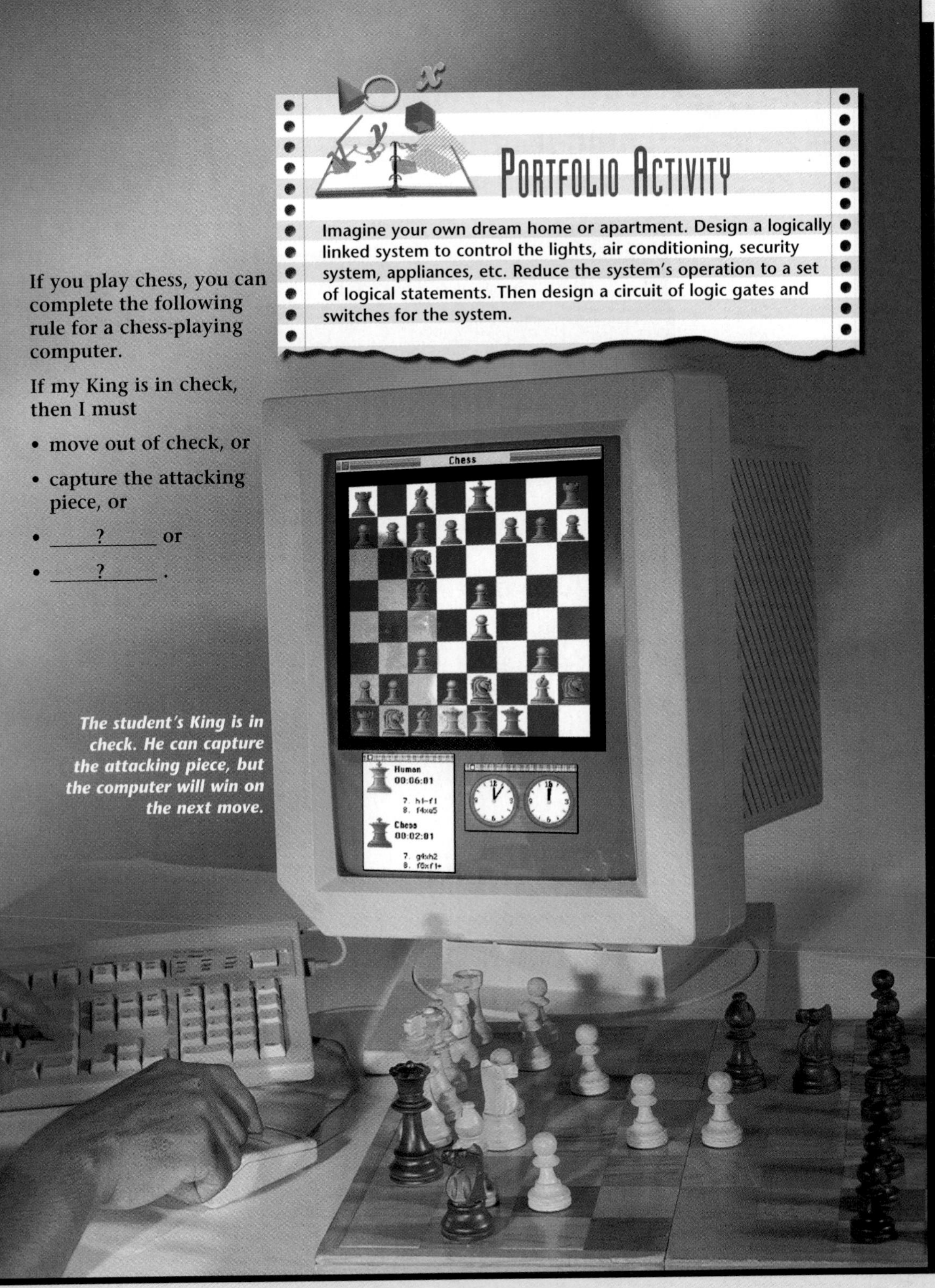

The student's King is in check. He can capture the attacking piece, but the computer will win on the next move.

- Use indirect proof with problems involving logical reasoning.
- Explore on-off tables, logic gates, and computer logic networks.
- Solve problems using computer logic.
- Develop coordinate proofs for the Triangle Midsegment Theorem, the Diagonals of a Parallelogram, and the reflection of a point about the line $y = x$.
- Use the concepts of coordinate proof to solve problems on the coordinate plane.

PORTFOLIO ACTIVITY

In this portfolio activity, students design a network to control electricity usage in a home. They will then represent the network as a system of logical statements that can be translated into a computer model. The portfolio activity is further developed in the Look Beyond exercise for Lesson 12.4. Extend the activity by showing the plans for a commercially available network.

ABOUT THE CHAPTER PROJECT

Students explore two famous proofs from number theory in the Chapter 12 Project on pages 685–687.

PREPARE

Objectives

- Define *valid argument* and *invalid argument.*
- Develop and use the Law of Indirect Reasoning.

RESOURCES

- Practice Master 12.1
- Enrichment Master 12.1
- Technology Master 12.1
- Lesson Activity Master 12.1
- Quiz 12.1
- Spanish Resources 12.1

Assessing Prior Knowledge

1. Convert the following statement to if-then form: A frog is an amphibian. [**Sample: If an animal is a frog, then it is an amphibian.**]
2. Write the converse of the following statement: If Jamie scores 95% on the test, she will get an A in the class.

   [**If Jamie gets an A in the class, then she scores 95% on the test.**]

TEACH

Studying the rules of formal logic should help students recognize the importance of rules in determining whether arguments are valid or invalid. Valid arguments are a precursor to stating or writing convincing arguments in real applications like law.

LESSON 12.1 Truth and Validity in Logical Arguments

WHY Newspaper reporters, politicians, lawyers, and even baseball managers may use logic to convince others of their views. Recognizing valid arguments as well as invalid ones will help you to think clearly in confusing situations.

Arguments: Valid and Invalid

In logic, an **argument** consists of a sequence of statements. The final statement of the argument is called the **conclusion**, and the statements that come before it are known as **premises**. The following is an example of a logical argument:

If an animal is an amphibian, then it is a vertebrate. } ← premises
Frogs are amphibians.
Therefore, frogs are vertebrates. ← conclusion

In this argument, the conclusion is said to *follow logically* from the premises. The premises force the conclusion. An argument of this kind is known as a valid argument, and the conclusion of such an argument is said to be a valid conclusion.

A valid argument makes the following "guarantee": *If the premises are all true, then the conclusion is true as well.* In the valid argument above, both of the premises are true. Therefore, the conclusion must be true.

Using Visual Models

Students may be able to visualize the hypothesis and conclusion of an argument with a Venn diagram. To do so, it may be helpful to rephrase the if-then sentence as about the members of the class. For example, "If a car is a Corvette, then it is a Chevrolet" can be rephrased as "All Corvettes are Chevrolets." The rephrased version may be easier for the students to diagram.

Now consider a different argument:

Some vertebrates are warmblooded. Frogs are vertebrates.	← premises
Therefore, frogs are warmblooded.	← conclusion

This *argument is invalid.* Both of the premises are true, but the conclusion is false. *This can never happen in a valid argument.*

Suppose that you did not know that a frog is a coldblooded animal. Would you have questioned the conclusion of the second argument anyway? Is there something basically "wrong" with the argument? If so, try to describe what it is.

Try This Write your own examples of valid and invalid arguments.

The Form of an Argument

Logicians can tell whether an argument is valid or invalid without knowing anything about the validity of its premises or of its conclusions. They are able to do this by analyzing its form. The valid argument on the previous page, for example, has the following form:

If p then q. p	← premises
Therefore, q.	← conclusion

This argument form is sometimes referred to by its medieval Latin name, ***modus ponens***—the "proposing mode." Any argument that has this form is valid, regardless of the statements that are substituted for p and q. The following nonsense argument is valid:

If flivvers twiddle then bokes malk. Flivvers twiddle.	← premises
Therefore, bokes malk.	← conclusion

This argument's form (*modus ponens*) guarantees that if the first two statements should somehow turn out to be true, then the third statement (the conclusion) will be true as well.

False Premises

If the premises of an argument are false, then there can be no "guarantee" that the conclusion is true, even though the argument might have a valid form. The following *modus ponens* argument is valid, but its conclusion is false.

If an animal is an amphibian, then it can fly. A frog is an amphibian.	← premises
Therefore, a frog can fly.	← conclusion

interdisciplinary CONNECTION

Law Lawyers are trained to recognize and use logical reasoning. This includes using information that is known and using the deductive process to reach a logical conclusion. Ask students to write a short argument that a lawyer might use in a court of law.

TEACHING tip

Emphasize the difference between a valid argument and an invalid argument. A valid argument guarantees that if the premises are all true, then the conclusion is also true. An argument in which the premises are true but the conclusion is false is invalid. Point out that the form of an argument rather than the content of the argument determines its validity.

CRITICAL Thinking

The first premise only holds for some vertebrates. In order to draw conclusions, the premises must always be true, not just in some cases. If the premises are only true in some cases, the conclusion is only true in some cases.

ongoing ASSESSMENT

Try This Check student arguments.

Cooperative Learning

Have groups of students practice composing arguments that use the *modus ponens* or *modus tollens* pattern. Have group members critique each other's arguments.

CRITICAL *Thinking*

Yes. Sample Examples:

Premises: If an animal can swim, then it's a fish. A tuna can swim.

Conclusion: A tuna is a fish.

(The first premise—the conditional—is false, but the conclusion is true. Stress that the argument is still valid.)

Premises: If an animal is a fish, then it can swim. A whale is a fish.

Conclusion: A whale can swim.

(In this case, the second premise is false—a whale is actually a mammal. The conclusion is true, and the argument is valid.)

TEACHING *tip*

Review many examples of denying the antecedent and *modus tollens* in detail so that students will understand the difference between them.

ongoing ASSESSMENT

Try This

Sample answers:

If an animal is a fish, then it can swim. A whale is not a fish. Therefore, a whale can not swim. (True premises, false conclusion.)

If an animal is a fish then it can swim. A robin is not a fish. Therefore, a robin can not swim. (True premises, true conclusion.)

Stress that both arguments are invalid, regardless of the truth or falsity of the conclusions.

Notice that *the conclusion, though false, is a valid conclusion* because the argument has a valid form. But since one of the premises is false, the conclusion is not guaranteed to be true. Remember, a valid argument guarantees that a conclusion is true *if the premises are true*. There is no guarantee if one or more of the premises are false.

The last argument has a false premise and a false conclusion. Do you think it is possible for a valid argument to have a false premise and a true conclusion? If so, give an example. If not, explain why.

The Law of Indirect Reasoning

Consider the following argument.

If a shirt is a De Morgan, then it has a blackbird logo.
This shirt does not have a blackbird logo. } ← premises
Therefore, this shirt is not a De Morgan. ← conclusion

Does this argument seem valid to you? If you knew the premises to be true, would you be convinced that the conclusion was true? The argument has the following form:

If p then q.
Not q. } ← premises
Therefore, not p. ← conclusion

This is, in fact, a valid argument form. It is sometimes referred to by its medieval Latin name, ***modus tollens***—the "removing mode." In more recent times it has come to be known as the Law of Indirect Reasoning.

You should be careful not to confuse the *modus tollens* form with the following form:

If p then q.
Not p. } ← premises
Therefore, not q. ← conclusion

This is the form of a common logical mistake, or **fallacy**, known as "denying the antecedent." The conclusion does not follow logically from the premises even *if it is true*. Be sure you understand the difference between this form and the *modus tollens* form.

Try This Give two examples of an argument that has the "denying the antecedent" form. In your first example, make the premises true and the conclusion false. In your second example, make the premises true and the conclusion true. Explain why the form is a logical fallacy, or mistake. (Does it make a guarantee about the conclusion?)

ENRICHMENT Research Lewis Carroll's use of logical arguments in *Symbolic Logic*. Explain how the use of nonsensical statements supports the importance of form in a logical argument.

INCLUSION strategies **Using Symbols** Encourage students to use symbols to represent premises and conclusions. This will emphasize the importance of form in determining when an argument is valid. Make sure students understand how the form of an argument relates to the validity of the conclusion.

EXERCISES & PROBLEMS

Communicate

For Exercises 1–4, determine if the argument is valid. Explain why or why not.

1. If today is Wednesday, then the cafeteria is serving beef stew.
   Today is Wednesday.
   Therefore, the cafeteria is serving beef stew.
2. If pigs fly, then today is February 30.
   Today is February 30.
   Therefore, pigs fly.
3. If Jon is a man, then Jon is mortal.
   Jon is not mortal.
   Therefore, Jon is not a man.
4. If $y > x$, then $a > b$.
   $y > x$.
   Therefore, $a > b$.
5. Is it possible for a valid argument to have a false conclusion? Explain your reasoning.

Practice & Apply

In Exercises 6–9, write a valid conclusion from the given premises.

6. If a person belongs to Party c, then the person is a conservative.
   Maria is a member of Party c.
7. If the team wins on Saturday, the team will be in the playoffs.
   The team did not make the playoffs.
8. If Evan beats Rob, then Evan plays Mario.
   If Evan plays Mario, then Mario will win the tournament.
   Evan beats Rob.
9. If a then b.
   Given: a conclusion: b

ASSESS

Selected Answers

Odd-numbered Exercises 7–19

Assignment Guide

Core 1–13, 15–20

Core Plus 1–14, 18–23

Technology

Students can use geometry graphics software to construct a quadrilateral for Exercise 13. Use the measurement options to help verify the conclusions.

Error Analysis

Students may have difficulty with Exercise 14 because of its use of symbols to represent premises. Encourage them to replace the symbols with statements in English if necessary.

Inviting Participation Play the following game as a review. Have one student write a conditional sentence as a premise on the chalkboard. Have another student add another premise to the argument. Ask a third student to continue the game by writing a conclusion from the two premises. Ask a fourth student to state whether the argument has the form *modus ponens*, *modus tollens*, or neither. Have a fifth student argue the validity of the argument. Ask a sixth student to start the game again with another premise. Continue in this fashion until all students have participated.

alternative ASSESSMENT

Portfolio Assessment

As a writing assignment ask students to write a logical argument using the forms *modus ponens* or *modus tollens.* Ask them to state each premise and conclusion clearly.

The If-Then Transitive Property is a form of argument used in proofs throughout the text. It was introduced in Lesson 2.2.

If-Then Transitive Property

If p then q.
If q then r.
Therefore, if p then r.

For Exercises 10 and 11, list all valid conclusions that can be drawn from the given premises.

10. If x then y.
If y then k.
Given: not k.

If x then k.
Not y.
Not x.

11. If n then m.
If q then r.
If m then q.
Given: n.

If n then r.
If n then q.
If m then r.
r; q; m

12. Given premises:
If you study, then you will succeed.
Eleanor studied.
Tamara did not study.
José succeeded.
Mary did not succeed.
Which of the following conclusions are valid? Explain your reasoning.

a. Eleanor will succeed.
b. Eleanor will not succeed.
c. Tamara will succeed.
d. Tamara will not succeed.
e. José studied.
f. José will not study.
g. Mary studied.
h. Mary did not study.

13. Consider the following argument:
If a quadrilateral is a parallelogram, then its diagonals are congruent.
PQRS is a parallelogram.
Therefore, the diagonals of *PQRS* are congruent.

a. Is the conclusion valid? Explain your reasoning.
b. Is the first premise true or false?
c. If *PQRS* is actually a square, is the conclusion true or false? Explain your reasoning.
d. If *PQRS* is a parallelogram that is not a rectangle or a square, is the conclusion false? Explain your reasoning.

Practice Master

NAME ________________ CLASS ________ DATE ________

Practice & Apply
12.1 Truth and Validity in Logical Arguments

In Exercises 1–6, write a valid conclusion from the given premises.

1. If it is a weekend, then José is not at work. It is a weekend.
2. If it is a weekday, then José is at work. José is not at work.
3. If José is not at work, then he is with Anna. José is not with Anna.
4. If José is at work, then he is not with Anna. José is with Anna.
5. If it is a weekday, then José is at work. If José is at work, then he is wearing a tie. It is a weekday.
6. If it is a weekend, then José is not at work. If José is not at work, then he is wearing jeans. It is a weekend.
7. Given these premises:

If you exercise, then you are energized.
Ian exercised.
Jon did not exercise.
Jean was energized.
Johan was not energized.

Which of the following conclusions are valid? Explain.

a. Ian was energized. **b.** Ian was not energized.
c. Jon was energized. **d.** Jon was not energized.
e. Jean had exercised. **f.** Jean had not exercised.
g. Johan had exercised. **h.** Johan had not exercised.

HRW Geometry — Practice & Apply 85

12. a and h. Part a is valid because it follows from the first and second premises using the form *modus ponens.*

Part h is valid because it follows from the first and fourth premises using the form *modus tollens.*

13. a. The conclusion is valid because the argument is in the form *modus ponens.*

b. False

c. The conclusion is true. The diagonals of a square are congruent.

d. The conclusion is false. The diagonals of a parallelogram are not always congruent.

14. A valid argument has premises a, b, c, and d, and conclusion r.

a. Does validity guarantee that all four premises and r are true? Explain your reasoning.

b. If the four premises are true, must r be true? Explain your reasoning.

c. If the argument is valid, under what circumstances might r be false?

Look Back

Determine if each pair of triangles is congruent. Explain why or why not. [Lessons 4.2, 4.3]

15.

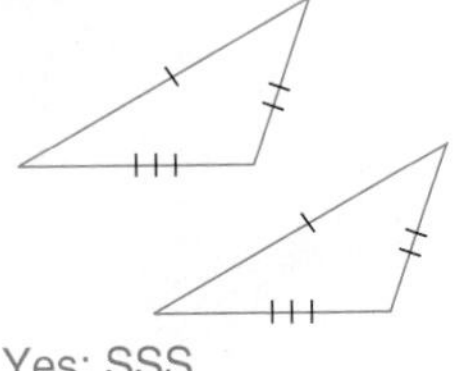

Yes; SSS

16.

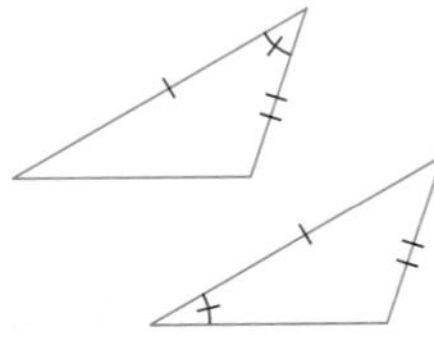

17.

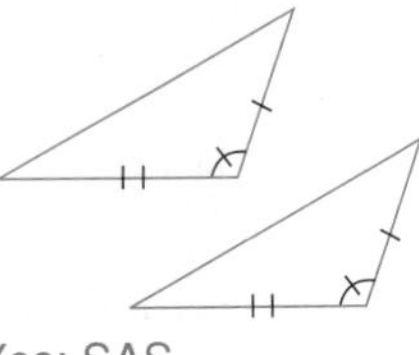

Yes; SAS

Use the circle at right for Exercises 18–20. [Lessons 9.3, 9.4]

$m\angle 1 = 20°$; $m\angle 2 = 35°$; $m\overset{\frown}{SR} = 80°$

Find each of the following.

18. $m\overset{\frown}{QR}$ 40°

19. $m\overset{\frown}{PS}$ 110°

20. $m\overset{\frown}{PQ}$ 130°

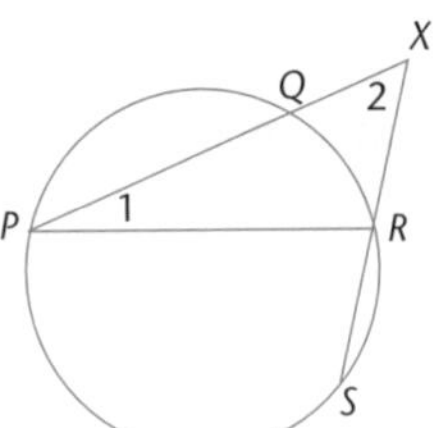

Look Beyond

The **contrapositive** of an if-then statement is written by negating the two parts of the statement and by switching the *if* and *then* parts.

If-then statement:	If Joe goes home, then Joe will watch TV.
Contrapositive:	If Joe does not watch TV, then Joe did not go home.

Write the contrapositive for each statement below.

21. If I make an A on the geometry test, then my father will give me $5.

22. If Bill is from France, then he can speak French.

23. If the piano has been moved, then the piano will be badly out of tune.

Use Transparency 108

Look Beyond

Exercises 21–23 ask students to write the contrapositive of a statement, something they will also formally do in Lesson 12.3. Emphasize that the content of each sentence does not affect how to form the contrapositive.

Technology Master

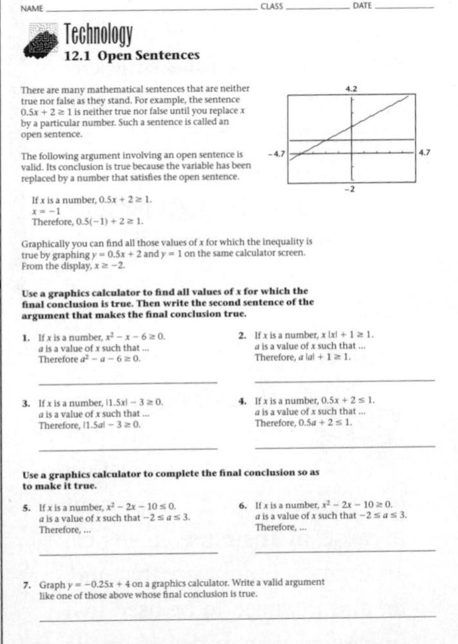

NAME ____________ CLASS ________ DATE ________

Technology

12.1 Open Sentences

There are many mathematical sentences that are neither true nor false as they stand. For example, the sentence $0.5x + 2 \geq 1$ is neither true nor false until you replace x by a particular number. Such a sentence is called an open sentence.

The following argument involving an open sentence is valid. Its conclusion is true because the variable has been replaced by a number that satisfies the open sentence.

If x is a number, $0.5x + 2 \geq 1$.
$x = -1$
Therefore, $0.5(-1) + 2 \geq 1$.

4.2, −4.7, 4.7, −2

Graphically you can find all those values of x for which the inequality is true by graphing $y = 0.5x + 2$ and $y = 1$ on the same calculator screen. From the display, $x \geq -2$.

Use a graphics calculator to find all values of x for which the final conclusion is true. Then write the second sentence of the argument that makes the final conclusion true.

1. If x is a number, $x^2 - x - 6 \geq 0$.
a is a value of x such that ...
Therefore $a^2 - a - 6 \geq 0$.

2. If x is a number, $x|x| + 1 \geq 1$.
a is a value of x such that ...
Therefore, $a|a| + 1 \geq 1$.

3. If x is a number, $|1.5x| - 3 \geq 0$.
a is a value of x such that ...
Therefore, $|1.5a| - 3 \geq 0$.

4. If x is a number, $0.5x + 2 \leq 1$.
a is a value of x such that ...
Therefore, $0.5a + 2 \leq 1$.

Use a graphics calculator to complete the final conclusion so as to make it true.

5. If x is a number, $x^2 - 2x - 10 \leq 0$.
a is a value of x such that $-2 \leq a \leq 3$.
Therefore, ...

6. If x is a number, $x^2 - 2x - 10 \geq 0$.
a is a value of x such that $-2 \leq a \leq 3$.
Therefore, ...

7. Graph $y = -0.25x + 4$ on a graphics calculator. Write a valid argument like one of those above whose final conclusion is true.

HRW Geometry — Technology 97

14. a. No. Validity only guarantees that the conclusion is true if all the premises are true.

b. Yes. In a valid argument, true premises guarantee a true conclusion.

c. r might be false if one or more of the premises is false.

16. No, the angle in the bottom triangle is not between the two sides.

21. If my father does not give me $5, then I did not make an A on the Geometry test.

22. If Bill cannot speak French, then he is not from France.

23. If the piano is not badly out of tune, then the piano has not been moved.

PREPARE

Objectives

- Define *conjunction, disjunction,* and *negation.*
- Solve logic problems using conjunction, disjunction, and negation.

RESOURCES

• Practice Master	12.2
• Enrichment Master	12.2
• Technology Master	12.2
• Lesson Activity Master	12.2
• Quiz	12.2
• Spanish Resources	12.2

Assessing Prior Knowledge

1. Rewrite the two sentences below as one sentence connected by the word *and.* It is raining today. School is in session. [**It is raining today, and school is in session.**]
2. Rewrite the two sentences in Exercise 1 as one sentence connected by the word *or.* [**It is raining today, or school is in session.**]
3. Add the word *not* in order to rewrite the following sentence with the opposite meaning. Albert got permission to see a movie. [**Albert did not get permission to see a movie.**]

LESSON 12.2 "And," "Or," and "Not" in Logic

Why *The words "and," "or," and "not" are used frequently in everyday conversation. These words have precise meaning in logical arguments.*

In logic, a **statement** is a sentence that is either true or false. For example, "Belinda orders pepperoni on her pizza" is a statement, because it is either true or false. A **compound statement** in logic is formed when two statements are connected by **and** or by **or**.

Conjunctions

A compound statement that uses the word **and** is called a **conjunction**.

Sentence p:	*Today is Tuesday.*
Sentence q:	*Tonight is the first varsity basketball game.*
Conjunction p **AND** q:	*Today is Tuesday **and** tonight is the first varsity basketball game.*

A conjunction is true when both of its statements are true. If one or both of its statements are false, the conjunction is false. The four possibilities for a conjunction can be illustrated with a **truth table**.

All possible combinations of truth values for the two statements that form the conjunction are placed in the first two columns. The last column indicates the truth values for the conjunction. In the first line, for example, both of the statements that make up the conjunction are true. In this case, the conjunction is true.

p	q	p and q
T	T	T
T	F	F
F	T	F
F	F	F

Cooperative Learning Have groups of students form the conjunction and disjunction of statements they know from mathematics. Ask them to determine as a group whether the compound sentences they form are true or false. Have them compare their sentences with those of other groups.

TEACH

Exploring the conjunction, disjunction, and negation of statements is important because of their use in logical arguments. Emphasize that *and, or,* and *not* need to be used in a precise way in order to make arguments valid.

EXAMPLE 1

Determine if the following conjunctions are true:

A George Washington was the first president of the United States, and John Adams was the second.

B The sum of the measures of the angles of a triangle is 200°, and blue is a color.

Solution ➤

A The conjunction is true because both of its statements are true.

B The conjunction is false because one of its statements is false. ❖

Disjunctions

Two statements may also be combined into a single statement by the word **or**. In logic, **OR** statements are known as **disjunctions**. When used in everyday language, **OR** often means "one or the other, but not both." For example, if a waitress says,

"You may have soup **or** salad with your pizza,"

she means that you may choose just one. This kind of *or* is known as the **exclusive OR**.

However, in mathematics and logic, **OR** means "one *or* the other, *or* both." This kind of *or* is known as the **inclusive OR**. If someone asks how John spends his Saturday afternoons, the answer might be,

"He goes swimming **or** bowling."

The sentence can be written in logical form as p **OR** q, where p and q are identified as shown:

$$\underbrace{\text{He goes swimming}}_{p} \textbf{ or } \underbrace{\text{[he goes] bowling.}}_{q}$$

The statement is false only if John does neither one. Notice that in the truth table, only the fourth combination gives a value of false for the conjunction. If he does either or both it is true, as the values for the first three combinations show.

p	q	p OR q
T	T	T
T	F	T
F	T	T
F	F	F

TEACHING tip

Stress that a conjunction of two sentences is true only when both statements are true. Encourage students to learn how to apply the truth table on this page to test truth or falsity of conjunctions.

Alternate Example 1

Determine whether the following conjunctions are true.

a. Alaska was the 49th state to be admitted to the United States, and Hawaii was the 50th.

b. The hypotenuse of a right triangle is its longest side, and the vertex angle of an isosceles triangle is the largest angle.

[a. The conjunction is true because both of its statements are true.

b. The conjunction is false because one of its statements is false.]

TEACHING tip

Stress that the disjunction of two sentences is false only when both parts are false. Otherwise the disjunction is true. Encourage students to apply the truth table on this page to test the truth or falsity of a disjunction.

Business Checking accounts at a bank can be set up so that one or the other of two persons on an account has to endorse a check before it is valid. Have students explain how this is like a disjunction. Ask them to explain how to set up the checking account so that it is like a conjunction.

Alternate Example 2

Determine whether each of the following disjunctions is true.

a. John F. Kennedy was the 39th or 40th president of the United States.

b. An obtuse triangle can be a right triangle, or a triangle contains at least two acute angles.

[a. The disjunction is false because both of its statements are false.

b. The disjunction is true because one of its statements is true.]

CRITICAL *Thinking*

~(p AND q) ; ~(p OR q)

Exploration Notes

In this exploration, students learn how to find the negation of a conjunction and of a disjunction. Point out that the negation of a conjunction of two statements is logically equivalent to the disjunction of the negation of each statement. Make sure students understand that symbols are used for statements so that truth can be determined easily by form rather than by content.

Use Transparency 109

Cooperative Learning

Use groups of four to explore the truth or falsity of conjunctions and disjunctions. Emphasize the usefulness of truth tables by having students use them to determine the truth or falsity of the statements in Examples 1 and 2. Check tables with group members.

EXAMPLE 2

Determine whether each of the following disjunctions is true:

A Teddy Roosevelt liked to go hunting or horseback riding.

B Dogs can fly or $5 - 3 = 2$.

Solution ➤

A The disjunction is true because both statements are true. (Teddy Roosevelt liked to go hunting. He also liked to go horseback riding).

B The disjunction is true because one of the statements is true. ❖

Negation

It is raining outside. *It is* **not** *raining outside.*

One of the statements above is the **negation** of the other. If p is a statement, then **not** p is its negation. The negation of p is written $\sim p$.

It is raining outside. *It is* **not** *raining outside.*

p $\sim p$

When a statement p is true, its negation $\sim p$ is false.

When a statement p is false, its negation $\sim p$ is true.

CRITICAL *Thinking* 

How do you write the negation of a conjunction, **p AND q**, or of a disjunction, **p OR q**? Hint: You will need to use parentheses.

Exploration *The Negation of a Conjunction*

You will need

No special tools

1. Copy and complete the truth table for the negation of a conjunction. Remember that the fourth column represents a negation of the third column.

p	q	p AND q	~(p AND q)
T	T	T	F
T	F	?	?
F	T	?	?
F	F	?	?

ENRICHMENT Have students find other examples of De Morgan's Laws. Ask them to translate the symbols into example sentences.

INCLUSION strategies **Using Models** Students may benefit from giving concrete examples of conjunctions, disjunctions, and negations. Have them apply truth tables to their examples to determine their truth or falsity.

2. Copy and complete the truth table for a disjunction formed by two negations. The first row is completed for you. Remember that the values for **~*p*** and **~*q*** are used to determine the truth tables for **~*p* OR ~*q***.

p	*q*	~*p*	~*q*	~*p* OR ~*q*
T	T	F	F	F
T	F	?	?	?
F	T	?	?	?
F	F	?	?	?

3. Compare the last column from Step 1 with the last column from Step 2. Explain what you observe.

4. When two logic statements have the same truth values they are said to be **logically equivalent**. Complete the following statement, which is one of **De Morgan's Laws**.

~(*p* AND *q*) is logically equivalent to __?__. ❖

In the two truth tables in the exploration, why is it important that the combinations of T and F be listed in exactly the same way in the first two columns of each? If you list T and F values for the three statements ***p***, ***q***, and ***r***, how many different combinations of T and F will there be? What is a good order for these combinations?

EXERCISES & PROBLEMS

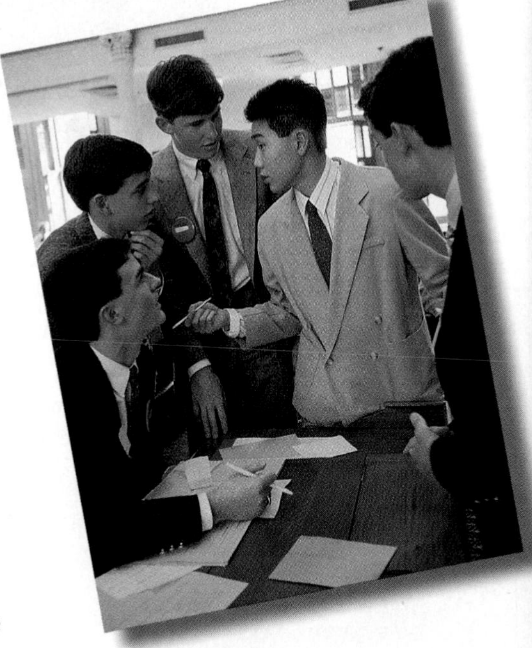

Team strategy meeting before a debate.

Communicate

1. Explain the conditions necessary for a conjunction to be true.
2. Explain the conditions necessary for a disjunction to be true.

Indicate whether the statement is true or false. Explain your reasoning.

3. $4 + 5 = 9$ and $4 \cdot 5 = 9$.
4. All triangles have three angles or all triangles have four angles.
5. Three noncollinear points determine a unique plane, and a segment has two endpoints.
6. All squares are hexagons or all triangles are squares.

RETEACHING the lesson

Using Review Have each student write the conjunction and disjunction of the following statements. *p*: $\triangle ABC$ is a right triangle. *q*: $\triangle ABC$ has two complementary angles.

[Conjunction: $\triangle ABC$ is a right triangle, and $\triangle ABC$ has two complementary angles. Disjunction: $\triangle ABC$ is a right triangle, or $\triangle ABC$ has two complementary angles.]

ongoing ASSESSMENT

3. The truth values are the same for ~(p AND q) and for ~p OR ~q.

CRITICAL Thinking

The last columns of each can be compared easily. If the first two columns were not in the same order, comparing the last column would be much more difficult. $2^3 = 8$ combinations.

p	q	r
T	T	T
T	T	F
T	F	T
T	F	F
F	T	T
F	T	F
F	F	T
F	F	F

ASSESS

Selected Answers

Odd-numbered Exercises 9–37

Assignment Guide

Core 1–17, 19–27, 31–37

Core Plus 5–14, 18–30, 34–38

Technology

Use the TEST LOGIC menu of a graphics calculator to help create a truth table for Exercises 18 and 28–30. First store values for *p*, *q*, and *r* in memory. Use parentheses and the logic menu to find the truth value of individual expressions.

Error Analysis

For Exercises 28–30, you may need to stress the importance of the order of operations when constructing the tables. Compound expressions in parentheses need to be done first.

alternative ASSESSMENT

Portfolio Assessment

Have each student list all truth tables they have learned thus far. Ask them to summarize each truth table by stating when the truth table is true and when the truth table is false.

9. A carrot is a vegetable and Florida is a state. True

10. A ray has only one endpoint and kangaroos can fly. False

11. The sum of the measures of the angles of a triangle is 180° and two points determine a line. True

Practice Master

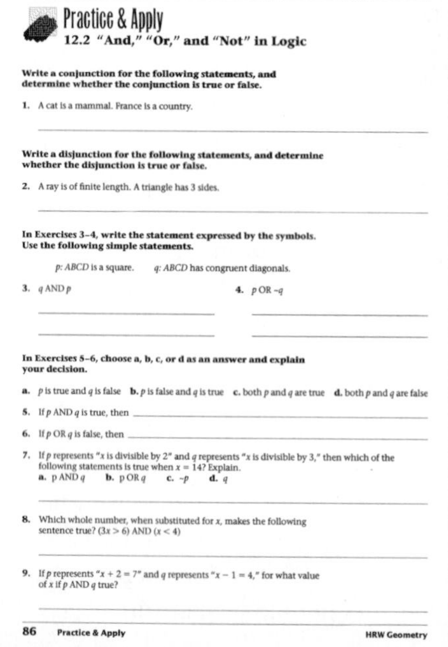

NAME ______________ CLASS ________ DATE ________

Practice & Apply

12.2 "And," "Or," and "Not" in Logic

Write a conjunction for the following statements, and determine whether the conjunction is true or false.

1. A cat is a mammal. France is a country.

____

Write a disjunction for the following statements, and determine whether the disjunction is true or false.

2. A ray is of finite length. A triangle has 3 sides.

____

In Exercises 3–4, write the statement expressed by the symbols. Use the following simple statements.

p: *ABCD* is a square. *q*: *ABCD* has congruent diagonals.

3. *q* AND *p*

4. *p* OR ~*q*

In Exercises 5–6, choose a, b, c, or d as an answer and explain your decision.

a. *p* is true and *q* is false **b.** *p* is false and *q* is true **c.** both *p* and *q* are true **d.** both *p* and *q* are false

5. If *p* AND *q* is true, then ______________

6. If *p* OR *q* is false, then ______________

7. If *p* represents "*x* is divisible by 2" and *q* represents "*x* is divisible by 3," then which of the following statements is true when $x = 14$? Explain.
a. p AND *q* **b.** p OR *q* **c.** ~*p* **d.** *q*

____

8. Which whole number, when substituted for *x*, makes the following sentence true? $(3x > 6)$ AND $(x < 4)$

____

9. If *p* represents "$x + 2 = 7$" and *q* represents "$x - 1 = 4$," for what value of *x* if *p* AND *q* true?

____

86 Practice & Apply HRW Geometry

State the negation of each statement.

7. The state is Alaska.
8. The weather is not rainy.

Practice & Apply

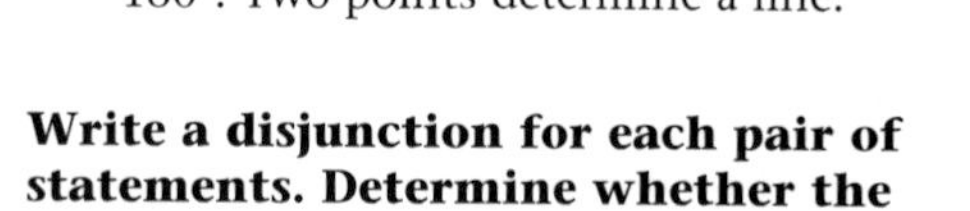

Write a conjunction for each pair of statements. Determine whether the conjunction is true or false.

9. A carrot is a vegetable.
Florida is a state.
10. A ray has only one endpoint. Kangaroos can fly.
11. The sum of the measures of the angles of a triangle is 180°. Two points determine a line.

Write a disjunction for each pair of statements. Determine whether the disjunction is true or false.

12. An orange is a fruit.
Cows have kittens.
13. Triangles are circles.
Squares are parallelograms.
14. Points equidistant from a given point form a circle. The sides of an equilateral triangle are congruent.

Write the negation of each statement.

15. The figure is a rectangle. The figure is not a rectangle.
16. My client is not guilty. My client is guilty.
17. Rain makes the road slippery. Rain does not make the road slippery.

18. **a.** Copy and complete the truth table for **~(~*p*)**.

p	~*p*	~(~*p*)
T	?	?
F	?	?

b. What is the statement **~(~*p*)** logically equivalent to? Explain your reasoning.

12. An orange is a fruit or cows have kittens. True

13. Triangles are circles or squares are parallelograms. True

14. All points equidistant from a given point form a circle or the sides of an equilateral triangle are congruent. True

18. a.

p	~p	~(~p)
T	F	T
F	T	F

b. ~(~p) is logically equivalent to p because they both have the same truth values.

For Exercises 19–26, write the statement expressed by the symbols. Use the following simple statements.

p**:** $\triangle ABC$ is isosceles.

q**:** $\triangle ABC$ has two equal angles.

r**:** $\angle 1$ and $\angle 2$ are adjacent.

s**:** $\angle 1$ and $\angle 2$ are acute angles.

19. ~*p* **20.** *q* OR *p* **21.** *p* AND *q* **22.** ~*q*

23. ~*s* **24.** *r* OR *s* **25.** *s* AND ~*r* **26.** *q* OR ~*s*

27. Explain all the logical possibilities that make the following statement true:

This weekend we will go camping and hiking, or it will rain and we will cancel the trip.

28. Construct a truth table for **(*p* AND *q*) OR (*r* AND *s*)**. When is **(*p* AND *q*) OR (*r* AND *s*)** false?

29. Construct a truth table for **(*p* OR *q*) OR *r***. When is **(*p* OR *q*) OR *r*** false?

30. Construct a truth table for **(*p* AND *q*) AND *r***. When is **(*p* AND *q*) AND *r*** true?

Look Back

Is each pair of triangles similar? Explain why or why not. [Lessons 8.3, 8.4]

31.

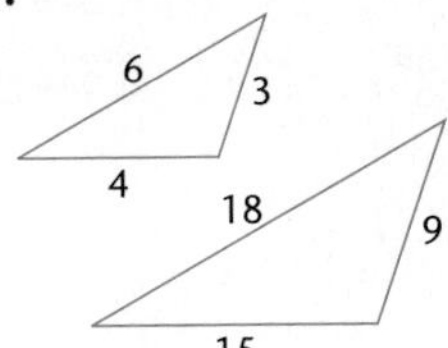

32.

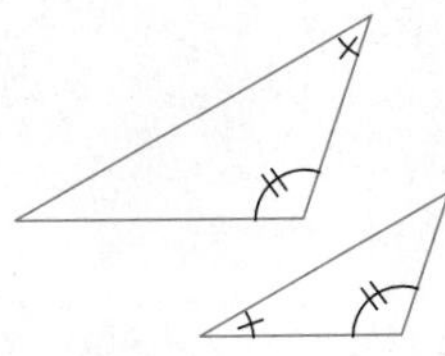

Yes; AA-similarity.

33.

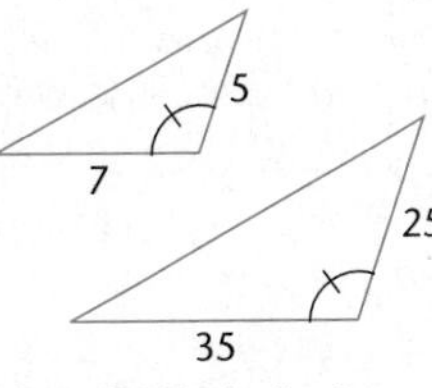

Yes; SAS-similarity.

Find the following for the triangle below. [Lessons 10.1, 10.2]

34. sin *A* $\frac{4}{5}$

35. cos *B* $\frac{4}{5}$

36. tan *B* $\frac{3}{4}$

37. cos *A* $\frac{3}{5}$

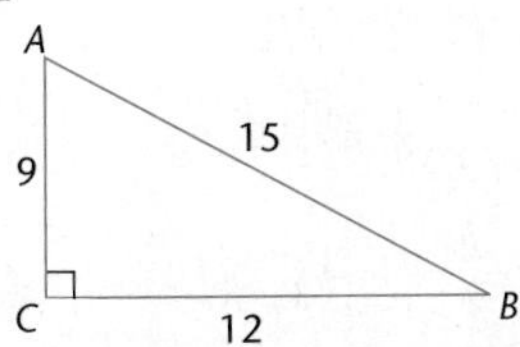

Look Beyond

38. Nonroutine Can you determine if the following sentence is true or false? Explain your reasoning.

I never tell the truth.

---

19. $\triangle$ABC in not isosceles. or $\triangle$ABC is scalene.

20. $\triangle$ABC has two equal angles or $\triangle$ABC is isosceles.

21. $\triangle$ABC is isosceles and $\triangle$ABC has two equal angles.

22. $\triangle$ABC has does not have two equal angles.

23. $\angle 1$ and $\angle 2$ are not acute angles.

or

$\angle 1$ is not an acute angle or $\angle 2$ is not an acute angle.

24. $\angle 1$ and $\angle 2$ are adjacent or $\angle 1$ and $\angle 2$ are acute angles.

25. $\angle 1$ and $\angle 2$ are acute angles and $\angle 1$ and $\angle 2$ are not adjacent angles.

26. $\triangle$ABC has two equal angles or $\angle 1$ and $\angle 2$ are not acute angles.

or

$\triangle$ABC has two equal angles or $\angle 1$ is not an acute angle or $\angle 2$ is not an acute angle.

27. This weekend we will go hunting and fishing.

This weekend it will rain and we will cancel the trip.

The answers to Exercises 28–30 and 38 can be found in Additional Answers beginning on page 727.

Look Beyond

Exercise 38 is a nonroutine problem in reasoning by negation. Extend the exercise by having students give similar examples using negation.

Technology Master

NAME ______________ CLASS ________ DATE ________

Technology
12.2 Combining Connectives

The spreadsheet shows the truth tables for *p* AND *q* and *p* OR *q*. To create the spreadsheet, enter the T/F combinations for *p* and *q* in columns A and B. Represent T by 1 and F by 0. Enter the formulas shown in cells C2 and D2, respectively. Then FILL DOWN.

Cell C2 IF(AND(A2=1,B2=1),1,0)

Cell D2 IF(OR(A2=1,B2=1),1,0)

	A	B	C	D
1	P	Q	P AND Q	P OR Q
2	1	1	1	1
3	1	0	0	1
4	0	1	0	1
5	0	0	0	0

To explore the logical equivalence (≡) of more complicated expressions such as *p* AND (*q* OR *r*) and (*p* AND *q*) OR (*p* AND *r*), you will need to make IF formulas using both the connectives AND and OR.

Show that each pair of expressions are logically equivalent.

1. Create a spreadsheet in which *p* = 1, 1, 1, 1, 0, 0, 0, 0, in column A, *q* = 1, 1, 0, 0, 1, 1, 0, 0, in column B and *r* = 1, 0, 1, 0, 1, 0, 1, 0, in column C. Consider *p* AND (*q* OR *r*) and (*p* AND *q*) OR (*p* AND *r*). In cell D2, enter IF(AND(A2=1,OR(B2=1,C2=1)),1,0). In cell E2, enter IF(OR(AND(A2=1,B2=1),AND(A2=1,C2=1)),1,0).

2. Consider *p* OR (*q* AND *r*) and (*p* OR *q*) AND (*p* OR *r*). Make the following changes in your spreadsheet from Exercise 1. In cell D2, enter IF(OR(A2=1,AND(B2=1,C2=1)),1,0). In cell E2, enter IF(AND(OR(A2=1,B2=1),OR(A2=1,C2=1)),1,0).

3. Consider *p* AND (*q* AND *r*) and (*p* AND *q*) AND *r*.

4. Consider *p* OR (*q* OR *r*) and (*p* OR *q*) OR *r*.

5. Modify your spreadsheet in Exercise 3 to consider −(*p* AND *q* AND *r*) and −*p* OR −*q* OR −*r*. (Use NOT().)

6. Modify your spreadsheet in Exercise 4 to consider −(*p* OR *q* OR *r*) and −*p* AND −*q* AND −*r*. (Use NOT().)

7. If you think of AND in terms of × and OR in terms of +, what property from algebra does the logical equivalence in Exercise 1 look like?

____

98 Technology HRW Geometry

Focus

A news article about the use of computers in translating Mayan symbols leads into a study of the Mayan calendar. Students informally explore modular arithmetic and greatest common factors to understand a time-keeping system based on two years of different lengths.

Motivate

Before students read the article, discuss some background about the Mayan civilization:

- The Mayan civilization reached its peak around 800 AD.
- In what is now Mexico, Guatemala, and Honduras, the Maya built great centers of religion and politics.
- The Maya had advanced systems of agriculture and astronomy.
- Hundreds of years before Spanish explorers reached the Americas, for reasons still unclear, the Maya abandoned their cities and monuments.

After students read the article, discuss how a double calendar would allow you to identify many more dates without specifying the year.

EYEWITNESS MATH
THE ENDS OF TIME
Message of the Maya in Modern Times
By Greg Stec
Special to the Christian Science Monitor
Discovering the titles of Mayan royalty, the names of their gods and their food, the dates of important events, and all the other things great and small that make up an advanced society has taken more than a hundred years of digging and probing. At last, though, a computer program has been developed to calculate Mayan dates.
The Mayans worshiped time and numbers. The current thinking holds that by dating the elite's birth, accomplishments, and death with unimpeachable accuracy, the person's position and rank would be permanently established.
Computer programs wade through a sea of Mayan dates, saving investigators effort that can be used to examine other translations.
Imagine having two calendars on your wall, one of them 260 days long and the other 365 days long. Each day would have two dates, usually different. Your friend's birthday, for example, might be August 4 and April 29.
The Mayan calendar was similar in that it had two different years simultaneously. To get a better idea of how it worked, you can use diagrams of wheels and gears. To keep track of dates, you'll use letters for months instead of the actual Mayan names.

1. **a.** 8H; Sample: In our calendar we go through every day of a month before going on to the next month. The Maya system would be like having February 5th followed by March 6th.

   b. 11K; 7T; 1G

   c. Yes. **d.** No. See **e**, below.

   e. Sample Answer: The numbers of sections on the two wheels must have no common factors except 1.

every date occurs	every date does not occur
13, 20	4, 6
2, 3	3, 3
3, 5	2, 4
4, 7	3, 6
5, 6	6, 9

Cooperative Learning

1. The Mayan 260-day year, called the Sacred Round, can be represented by the two smaller wheels in the diagram. Each day, the wheels move one notch to the next section.

   a. The date shown in the diagram is 7G. What will the date be 1 day later? How is that different from our calendar?

   b. What will the date be 4 days after 7G? 13 days after 7G? 20 days after 7G?

   c. If the wheels keep turning, will every one of the 260 possible dates occur? How do you know?

   d. Suppose the outer wheel had only 6 letters and the inner wheel had only 4 numbers. Would all 24 pairs of letters and numbers occur? How do you know?

   e. For every date to occur, what must be true about the number of sections on the wheels? Hint: Experiment with small numbers. Make a table of your results. Look for common factors.

Every date occurs	Every date does not occur
13, 20	4, 6

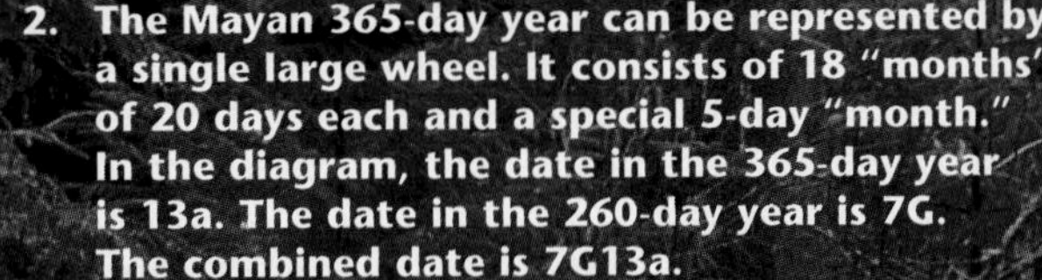

2. The Mayan 365-day year can be represented by a single large wheel. It consists of 18 "months" of 20 days each and a special 5-day "month." In the diagram, the date in the 365-day year is 13a. The date in the 260-day year is 7G. The combined date is 7G13a.

   a. The first day of the 365-day year is 0a. The 21st day is 0b. What is the 365th day of the year?

   b. What will the date be 1 day after 7G13a? 20 days after 7G13a? 260 days after 7G13a? 365 days after 7G13a?

   c. Does every possible date occur? Explain? Hint: Use your answer to question 1d.

The logic may be simple, but the Mayan round calendar was not. It employed two intermeshing years, one 260 days long, and the other 365 days long...

Super Challenge

3. How many times will the large wheel turn before the date 7G13a occurs again? How many days is that?

2. a. 4S **b.** 8H14a; 1G13b; 7G13n; 8L13a

c. No. 260 and 365 have a common factor (5)

Super Challenge 52 complete turns; 18,720 days

Cooperative Learning

Since the activities are sequential, have each cooperative group do all three activities.

Part 1a Discuss the fact that, with each new day, both wheels advance in the directions shown by the arrows.

Part 1c As an aid, students might write the dates in an array:

A	B	C	D	E	F	G	H	I
1	2	3	4	5	6	7	8	9

and so on.

Part 1e This is a difficult notion for most students to discover without guidance. Help them focus on common factors and encourage them to try out at least 5 or 6 pairs of numbers.

Part 2 Discuss the structure of the large wheel. The first 18 months (a-r) have 20 days each. The last month has only 5. In this system (unlike the 260-day Mayan year and more like our own) you go through all the days in a month before proceeding to the next month. Students can think of the two small wheels as a single wheel with 260 dates.

Part 2a Encourage students to look for shortcuts. Ask: *What is a shortcut for finding the date that comes 31 days after January 19th.*

Super Challenge Suggest students work with very small numbers and look for a pattern that can then be applied to the pair 260 and 365.

Discuss

Ask students what they think was the benefit of the Mayan calendar. Have students research another Mayan time-keeping system called the *Long Count,* a base-20 system for counting the number of days since the last "creation."

PREPARE

Objectives

- Define *conditional, converse, inverse,* and *contrapositive.*
- Use if-then statements and forms of valid argument for problems involving logical reasoning.

RESOURCES

• Practice Master	12.3
• Enrichment Master	12.3
• Technology Master	12.3
• Lesson Activity Master	12.3
• Quiz	12.3
• Spanish Resources	12.3

Assessing Prior Knowledge

1. Write the following statement as a conditional.

   Warm-blooded animals are mammals.

   [If an animal is warm-blooded, then it is a mammal.]

2. Write the converse of the conditional in question 1.

   [If an animal is a mammal, then it is warm-blooded.]

TEACH

Studying the converse, inverse, and contrapositive of a conditional statement is important because of their use in an argument. Students need to be aware of when a conditional sentence is true so that it can be used as part of the reasoning process.

LESSON 12.3 A Closer Look at If-Then Statements

why *In courtrooms, as in everyday life, if-then statements are a very important part of our language. They are also used in mathematical reasoning. The statement, "If a rhombus has four right angles, then it is a square" is just one of many examples of if-then statements to be found in this book.*

Lawyers use logic to convince a jury: "If the defendant committed a crime, then he must have been at 4th Avenue and Crescent between 10 P.M. and 11 P.M."

Conditionals

Various forms of if-then statements have been used throughout this book. You may recall from Lesson 2.2 that if-then statements are called conditionals.

Suppose your neighbor makes the following promise:

If $\underbrace{\text{you mow his lawn}}_{p}$, then $\underbrace{\text{he will give you \$10}}_{q}$.

Four possible situations can occur:

1. *You mow the lawn and your neighbor gives you $10.*

   Since your neighbor kept his promise, we agree that the conditional statement is true.

2. *You mow the lawn. Your neighbor does not give you $10.*

   The promise is broken. Therefore, we agree that the conditional statement is false.

ALTERNATIVE teaching strategy

Using Technology Use the following program to demonstrate how if-then logic may be used in a graphing-calculator program. This program uses the converse of the "Pythagorean" Right-Triangle Theorem to determine whether a triangle is a right triangle.

```
PROGRAM:PYTHAG
:Input "INPUT SIDE A", A
:Input "INPUT SIDE B", B
:Input "INPUT SIDE C", C
:If A^2 + B^2 = C^2
:Then : Disp "RIGHT"
:Else : Disp "NOT RIGHT"
:End
```

3. *You do not mow the lawn. Your neighbor gives you $10.*

The promise is not broken, so we agree that the conditional is true.

4. *You do not mow the lawn. Your neighbor does not give you $10.*

The promise is not broken, so we agree that the conditional is true.

You can think of a conditional as a promise. In logic, if the "promise" is broken, the conditional is said to be false. Otherwise, it is said to be true.

The truth table below summarizes the truth values for the conditional. Recall that the logical notation for "if p then q" is $p \Rightarrow q$ (read: "**p implies q**").

p	q	$p \Rightarrow q$
T	T	T
T	F	F
F	T	T
F	F	T

The first two columns list all possible combinations of T and F for the two statements p and q. Notice that the only time $p \Rightarrow q$ is false is when the promise of $10 for mowing the lawn is broken.

The Converse

Recall from Lesson 2.2 that the converse of a conditional results from interchanging its "if" and "then" parts. Consider the following conditional:

If Tamika lives in Montana, then she lives in the United States.

(Is this statement true?)

The converse of the conditional is:

If Tamika lives in the United States, then she lives in Montana.

(Is this statement true?)

The truth table below summarizes the truth values for the conditional and its converse. When studying the table, notice that q and p are reversed in the last column.

p	q	$p \Rightarrow q$	$q \Rightarrow p$
T	T	T	T
T	F	F	T
F	T	T	F
F	F	T	T

TEACHING tip

Write the truth tables for a conditional and its converse side by side. Use the fact that the truth values are not the same in order to emphasize that it is an error to assume the converse is true when the conditional is true.

ongoing ASSESSMENT

The statement is true.

The converse is false. Tamika might live in Florida, for example.

Computer Programming One of the most powerful aspects of a computer program is its ability to branch to a different part of the program based on if-then logic. Ask students who are interested in computer programming to explain how branching logic works.

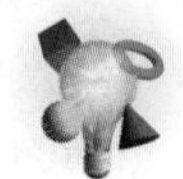

Two statements are logically equivalent if they have the same truth values. Compare the truth values for the conditional and the converse. Are they logically equivalent? Explain your reasoning.

The Inverse

The **inverse** of a conditional is formed by negating both the hypothesis and the conclusion. The inverse of the original conditional is as follows:

If Tamika does not live in Montana, then she does not live in the United States. (Is this statement true?)

The truth table below summarizes the truth values for the inverse. Notice that extra rows are required for the negations of ***p*** and ***q***.

p	q	$\sim p$	$\sim q$	$\sim p \Rightarrow \sim q$
T	T	F	F	T
T	F	F	T	T
F	T	T	F	F
F	F	T	T	T

CRITICAL *Thinking*

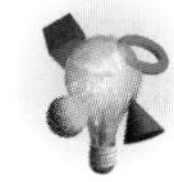

Are a conditional and its inverse logically equivalent? Explain your reasoning.

The Contrapositive

The **contrapositive** of a conditional is formed by interchanging the "if" and "then" parts and negating each part. The contrapositive of the original conditional is as follows:

If Tamika does not live in the United States, then she does not live in Montana. (Is this statement true?)

The truth table below summarizes the truth values for the contrapositive of a conditional. When studying the table, notice that ***q*** and ***p*** are reversed and negated in the last column.

p	q	$\sim q$	$\sim p$	$\sim q \Rightarrow \sim p$
T	T	F	F	T
T	F	T	F	F
F	T	F	T	T
F	F	T	T	T

CRITICAL *Thinking*

A conditional and its converse are not logically equivalent because their truth values are not the same.

ongoing ASSESSMENT

The inverse is false. Tamika might live in Indiana, for example.

CRITICAL *Thinking*

A conditional and its inverse are not logically equivalent because their truth values are not the same.

ongoing ASSESSMENT

The statement is true.

Cooperative Learning

Divide the class into groups of three students, so that they can practice forming the converse, inverse, and contrapositive of conditional sentences. Have each member of the group write three conditionals and pass one of them to each of the other members of the group. Ask one person to write the converse of each statement, one person to write the inverse, and one person to write the contrapositive. The group members should share the answers.

ENRICHMENT Write a truth table for the following statement:

(p and q) or (p and r) implies (p or q or r). If the statement is always true, then it is called a tautology. Is this statement a tautology?

[Yes, it is a tautology.]

Using Symbols Some students may be more comfortable using symbols and truth tables to find the truth value for statements and arguments. Encourage them to represent the hypothesis and the conclusion of each statement with variables and then use the variables in truth tables.

Notice that the final columns for the truth tables of the original conditional and those of its contrapositive are the same. Recall from Lesson 12.2 that this means the two statements are logically equivalent. This is a very useful piece of information. Whenever a conditional is true, its contrapositive must also be true. Thus, the contrapositive of every postulate and theorem that can be written in if-then form must also be true!

In addition to a conditional and its contrapositive, what other two forms of a conditional are logically equivalent?

Summary

An affine if-then statement has three related forms.

If-then statement:	If p then q.	$p \Rightarrow q$
Converse:	If q then p.	$q \Rightarrow p$
Inverse:	If ~p then ~q.	$\sim p \Rightarrow \sim q$
Contrapositive:	If ~q then ~p.	$\sim q \Rightarrow \sim p$

Try This Write the converse, inverse, and contrapositive of each conditional below. State whether each new statement is true or false.

a. If a triangle is equilateral, then it is an isosceles triangle.

b. If a quadrilateral is a rhombus, then it is a square.

EXERCISES & PROBLEMS

Communicate

For Exercises 1–4, state the converse, inverse, and contrapositive for each conditional.

1. If today is February 30, then the moon is made of green cheese.
2. If all the sides of a triangle are congruent, then the triangle is equilateral.
3. If I do not go to the market, then I will not buy cereal.
4. If the car starts, then I will not be late to school.
5. Describe the circumstances that would make a statement "If a then b" false.

What is green cheese, anyway?

RETEACHING the lesson

Using Review Write the following conditional on the chalkboard. If *ABCD* is a square, then *ABCD* is a parallelogram. Have each student write the converse, inverse, and contrapositive of the conditional. Ask them to tell the truth value of each.

[Converse: If *ABCD* is a parallelogram, then *ABCD* is a square. False. Inverse: If *ABCD* is not a square, then *ABCD* is not a parallelogram. False. Contrapositive: If *ABCD* is not a parallelogram, then *ABCD* is not a square. True.]

CRITICAL Thinking

The converse and the inverse of a conditional are logically equivalent.

ongoing ASSESSMENT

Try This

a. **Converse: If a triangle is isosceles, then it is equilateral. False**

Inverse: If a triangle is not equilateral, then it is not isosceles. False

Contrapositive: If a triangle is not isosceles, then it is not equilateral. True

b. **Converse: If a quadrilateral is a square, then it is a rhombus. True**

Inverse: If a quadrilateral is not a rhombus, then it is not a square. True

Contrapositive: If a quadrilateral is not a square, then it is not a rhombus. False

ASSESS

Selected Answers

Exercises 7, 11, 17, 19, 21, and 25

Assignment Guide

Core 1–13, 16–21, 24–28

Core Plus 1–15, 19–23, 26–29

Technology

Students may use geometry graphics programs to draw shapes for the conditionals in Exercises 6 and 8.

Error Analysis

Suggest that students use concrete examples of conditionals.

Practice & Apply

For each conditional in Exercises 6–10, write the converse, inverse, and contrapositive. Decide whether each conditional, converse, inverse, and contrapositive is true or false and explain your reasoning.

6. If a figure is a square, then it is a rectangle.

7. If $a = b$, then $a^2 = b^2$.

8. If three angles of a triangle are congruent to three angles of another triangle, then the triangles are congruent.

9. If p and q are even numbers, then $p + q$ is an even number.

10. If water freezes, then its temperature is less than or equal to 32°F.

11. Given: *If p then q.* Write the contrapositive of this statement. Then write the contrapositive of the contrapositive. What may you conclude about the contrapositive of the contrapositive of an if-then statement?

12. Suppose the following statement is true:

If the snow exceeds 6 inches, then school will be canceled.

Which of the following statements must also be true? Explain your reasoning.

a. If the snow does not exceed 6 inches, then school will not be canceled.

b. If school is not canceled, then the snow does not exceed 6 inches.

c. If school is canceled, then the snow exceeds 6 inches.

13. State the "Pythagorean" Right-Triangle Theorem, its converse, inverse, and contrapositive. Determine whether each is true or false and explain your reasoning.

14. Choose a postulate or theorem from Chapter 3 that is written in if-then form. Write its converse, inverse, and contrapositive and decide whether these statements are true or false.

15. Choose a postulate or theorem from Chapter 9 that is written in if-then form. Write its converse, inverse, and contrapositive and decide whether these statements are true or false.

Some statements that are not written in if-then form can be rewritten in if-then form. For example,

Every rectangle is a parallelogram

can be rewritten as follows:

If a figure is a rectangle, then it is a parallelogram.

For Exercises 16–19, write each statement in if-then form.

16. All seniors must report to the auditorium.

17. A point on the perpendicular bisector of a segment is equidistant from the endpoints of the segment.

18. Call me if you expect to be late.

alternative ASSESSMENT

Authentic Assessment

Ask students to find examples of conditional, converse, inverse, or contrapositive statements in newspapers or magazine articles. Ask them to determine whether the statements are true or false in the context of the article.

The answers to Exercises 8–18 can be found in Additional Answers beginning on page 727.

Practice Master

NAME ______________ CLASS ________ DATE ________

Practice & Apply
12.3 A Closer Look at If-Then Statements

For the conditional given, write the converse, inverse, and contrapositive. Determine if each of the four statements is true or false, and explain your reasoning.

1. Conditional: If Aja lives in Texas, she lives in the United States.

   ____

   Converse: ______________

   Inverse: ______________

   Contrapositive: ______________

In Exercises 2–4, write the statement in if-then form.

2. Cindy does not eat meat if she is a vegetarian.

   ____

3. 3 is an odd number.

   ____

4. I will eat when I am hungry.

   ____

5. Which of the following is the inverse of $\sim p \rightarrow \sim q$? ________

   a. $p \rightarrow \sim q$ **b.** $q \rightarrow \sim p$ **c.** $p \rightarrow q$ **d.** $\sim q \rightarrow \sim p$

6. Which of the following is the contrapositive of $p \rightarrow q$? ________

   a. $q \rightarrow p$ **b.** $\sim p \rightarrow q$ **c.** $\sim q \rightarrow \sim p$ **d.** $\sim p \rightarrow \sim q$

In Exercises 7–8, draw a conclusion from the given premises, if possible. Explain your reasoning.

7. If I do not sleep well, then I am tired. I am not tired.

   ____

8. If I sleep well, then I feel refreshed. I did not sleep well.

   ____

HRW Geometry Practice & Apply 87

6. The conditional is true. All squares are rectangles. Converse: If a figure is a rectangle, then it is a square. False, all rectangles are not squares. Inverse: If a figure is not square, then it is not a rectangle. False, a rectangle with sides 6 cm and 8 cm is not a square. Contrapositive: If a figure is not a rectangle, then it is not a square. True, all squares are rectangles. The statement is true because the original statement is true.

7. Conditional is true by the Multiplication Property of Equality. Converse: If $a^2 = b^2$, then $a = b$. False. $(-2)^2 = (2)^2$. Inverse: If $a \neq b$, then $a^2 \neq b^2$. False. $(-2)^2 = (2)^2$. Contrapositive: If $a^2 \neq b^2$, then $a \neq b$. True. If the conditional is true, the contrapositive is also true.

19. Doing mathematics homework every night will improve your grade in mathematics.

20. Look for three if-then statements (or statements that can be written in if-then form) in a newspaper, in a magazine, or on TV. Write the statement, its converse, inverse, and its contrapositive. Determine if each is true or false and explain your reasoning.

21. Consider the following statement:

You will make the honor roll only if you get at least a B in mathematics.

Which of the following statements appear to convey the same meaning as the original statement? Explain your reasoning.

a. If you made the honor roll, then you must have gotten at least a B in mathematics.

b. If you get at least a B in mathematics, then you will make the honor roll.

c. If you do not make the honor roll then you did not get at least a B in mathematics.

d. If you do not get at least a B in mathematics then you will not make the honor roll.

22. The statement ***p* if and only if *q***, written ***p* iff *q***, is equivalent to two statements:

p if ***q*** and ***p*** *only if* ***q***.

Suppose the statement ***p* iff *q*** is true. Which of the following must be true? Explain your reasoning.

a. If ***p*** then ***q***.

b. If ***q*** then ***p***.

c. If ~***p*** then ~***q***.

d. If ~***q*** then ~***p***.

23. Statements using "if and only if" are sometimes called **biconditionals**. Based on the results from Exercise 22, what can you conclude about many of the theorems in this book. Explain your reasoning and provide examples from the text.

766·TCF
IF YOU CAN READ THIS
YOU'RE TOO CLOSE

23. Many of the theorems can be written in biconditional form. Some examples follow. A quadrilateral is a parallelogram if and only if its diagonals bisect each other. A triangle is a right triangle with hypotenuse c and legs a and b if and only if $a^2 + b^2 = c^2$.

Technology Master

NAME ______ CLASS ______ DATE ______

Technology
12.3 Logical Equivalents to If-Then Statements

In Lesson 12.2, you learned about sentences involving AND, OR, and NOT. In Lesson 12.3, you learned the truth table for the conditional $p \Rightarrow q$. As you can see in the table, $p \Rightarrow q$ is logically equivalent to NOT p OR q. Thus, you can explore conditional sentences by studying compound sentences using spreadsheets. See Technology 12.3 for information about IF formulas.

	A	B	C	D
1	P	Q	P ⇒ Q	NOT P OR Q
2	1	1	1	1
3	1	0	0	0
4	0	1	1	1
5	0	0	1	1

Use a spreadsheet in the following exercises.

1. Create a spreadsheet in which $p = 1, 1, 0, 0$ in column A and $q = 1, 0, 1, 0$ in column B. Fill column C as shown. Consider the converse of $p \Rightarrow q$, $q \Rightarrow p$. Write a logical expression you think is logically equivalent to $q \Rightarrow p$. Check your answer by entering your expression in cell D2, and then filling down.

Write a logical expression you think is logically equivalent to the given conditional. Check your answer by entering your expression in cell D2, and then filling down.

2. the inverse of $p \Rightarrow q$; NOT $p \Rightarrow$ NOT q ______
3. the contrapositive of $p \Rightarrow q$; NOT $q \Rightarrow$ NOT p ______
4. Which of the conditionals in Exercises 1–3 are logically equivalent to $p \Rightarrow q$? Explain. ______
5. Create a spreadsheet in which $p = 1, 1, 1, 1, 0, 0, 0, 0$ in column A, $q = 1, 1, 0, 0, 1, 1, 0, 0$ in column B and $r = 1, 0, 1, 0, 1, 0, 1, 0$ in column C.

Enter the truth values for the given conditional in column D of your spreadsheet from Exercise 5. Then write an expression logically equivalent to the given conditional. Enter your expression in cell E2, and FILL DOWN. Are you correct?

6. p AND $q \Rightarrow r$ ______
7. p OR $q \Rightarrow r$ ______

HRW Geometry — Technology 99

19. If you do your mathematics homework every night, then your mathematics grade will improve.

20. Check answers depending on whether the converse of statements are true.

21. b is the same form as the given statement. ; c is the contrapositive of the given statement.

22. All are true.

Statement *a* is true because *p* is true only if *q* is true. So, if *p* is true, then *q* must also be true.

Statement *b* is true because *p* if *q* is the same as if *q*, then *p*.

Statement *c* is the contrapositive of *a* and also true if *a* is true.

Statement *d* is the contrapositive of *b* and also true if *b* is true.

Look Back

Use the triangle at right and the Law of Sines to determine the missing measure. [Lesson 10.5]

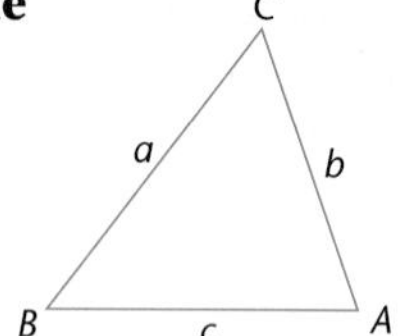

24. Find side c given $m\angle B = 37°$, $m\angle A = 50°$, and $b = 100$. 165.94 units

25. Find side b given $m\angle C = 65°$, $m\angle A = 47°$, and $c = 3.45$. 3.53 units

Copy the vectors below and draw the resultant using the head-to-tail method. You may have to translate one of the vectors. [Lesson 10.7]

26.

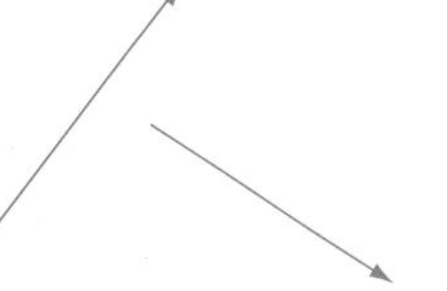

27.

28.

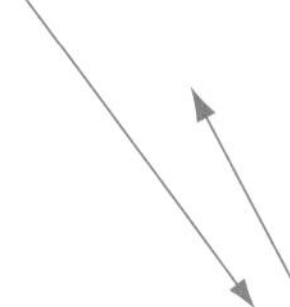

Look Beyond

The Look Beyond exercise asks students to analyze a situation in terms of the truth or falsity of statements people make. Have students discuss their answers to this question.

26.

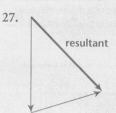

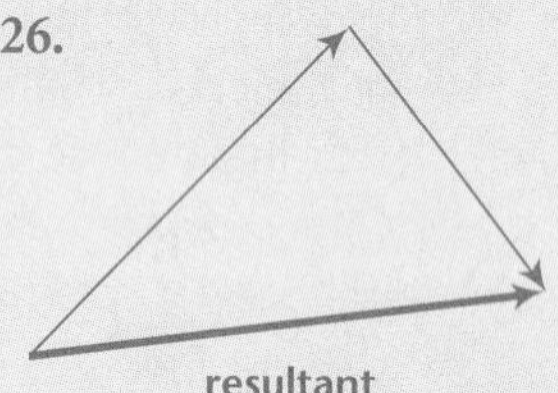

27.

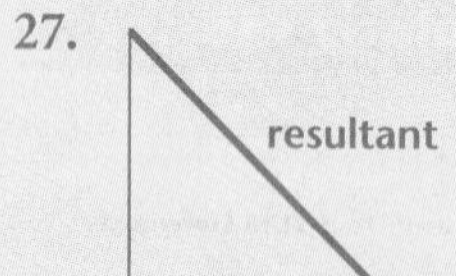

28.

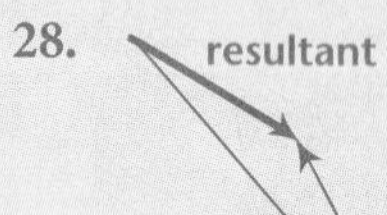

Look Beyond

29. You are on an island and are trying to determine whether you should go east or west in order to get back to the boat dock. Two different groups of people live on the island. One group always tells the truth. The other group always lies. The groups dress differently, but you have not been able to determine which is which. You approach two strangers who are dressed differently to ask directions to the dock. What one question can you ask that will provide you with the correct direction?

29. Ask one of the people what the other would say if he or she were asked the correct direction. Then take the opposite of the answer. If you asked the truthful person, the person would give the incorrect answer because the person would truthfully report what the liar would say. If you asked the liar, the liar would give the incorrect answer.

LESSON 12.4

Indirect Proof

why *Lewis Carroll, the author of* Alice in Wonderland *and* Through the Looking-Glass, *was a logician who was fond of absurdity as a form of entertainment. But does absurdity have any real place in logic or mathematics? In this lesson you will see that it can, in fact, be quite relevant.*

'The time has come,' the Walrus said,
'To talk of many things:
Of shoes—and ships—and sealing wax—
Of cabbages—and kings—
And why the sea is boiling hot
And whether pigs have wings.'

— Lewis Carroll

Have you ever heard an expression like,

"If you are twenty-one, then pigs have wings!"

The speaker, perhaps without realizing it, is inviting you to use the Law of Indirect Reasoning, which you studied in Lesson 12.1. Since it is certainly not true that pigs have wings, the statement in question (in the "if" part) must be false.

If you are twenty-one, then pigs have wings.	(If ***p*** then ***q***.)
Pigs do not have wings.	(Not ***q***.)
Therefore, you are not twenty-one.	(Therefore, not ***p***.)

Indirect Proofs

A closely related form of argument is known by its Latin name ***reductio ad absurdum***—literally, "reduction to absurdity." In this type of argument, an assumption is shown to lead to an absurd or impossible conclusion. In this case, the assumption must be rejected.

ALTERNATIVE teaching strategy

Using Visual Models Help students understand the method of proof by contradiction by having them draw a diagram like a "flow proof." At the top of the paper show the statements *p* and *not p* for a given proof. From each of these statements, write conclusions that follow deductively. Follow the flow of each branch until one branch leads to a contradiction. Reject that branch of the argument and accept the other branch.

PREPARE

Objectives

- Develop the concept of indirect proof (*reductio ad absurdum,* proof by contradiction).
- Use indirect proof with problems involving logical reasoning.

RESOURCES

• Practice Master	12.4
• Enrichment Master	12.4
• Technology Master	12.4
• Lesson Activity Master	12.4
• Quiz	12.4
• Spanish Resources	12.4

Assessing Prior Knowledge

1. Write the contrapositive of the following statement. If the defendant was in his car at 10:00 P.M., then he was in an automobile accident.

   [If the defendant was not in an automobile accident, then he was not in his car at 10:00 P.M.]

2. If a triangle is a right triangle, then is it possible for it also to be an obtuse triangle? Explain. **[Sample: No, it is not possible because if the sum of the angles is 180°, then it cannot contain both an angle of 90° and an angle greater than 90°.]**

TEACH

Indirect proof is a very common form of argument and is used when normal deduction is difficult. Proof by contradiction is especially useful in proving that things *are not* true.

TEACHING tip

Carefully explain circumstances under which the Law of Indirect Reasoning applies to an argument. Have students write examples of the law on the chalkboard and discuss them in class.

Cooperative Learning

Have groups of students write several direct and indirect proofs separately on note cards. Ask them not to label the kind of proof done. Shuffle the cards and pass them out to other groups. Ask them to classify each proof as direct or indirect and explain to each other the reasoning behind their classifications. Working in groups gives students extra help in understanding difficult concepts.

Impossible spatial configurations

In formal logic and mathematics, certain proofs use a *reductio ad absurdum* strategy, but with an important twist. In such proofs, you assume the *opposite*, or in logical terms, the *negation*, of the statement that you want to prove. If this assumption leads to an impossible result, then you can conclude that the assumption was false. (Then you know that the original statement was true.) Such proofs are known as **proofs by contradiction**.

Proofs by Contradiction

What is meant by an "absurd" or "impossible" result? In logic, a **contradiction** is such a result. A contradiction has the following form:

p **and** $\sim p$.

That is, a contradiction asserts that a statement and its negation are both true. The following compound statement is a contradiction:

A horse is a vegetarian, and a horse is not a vegetarian.

In formal logic and in mathematics, any assumption that leads to a contradiction must be rejected. Contradictions turn out to be very useful in indirect proofs.

PROOF BY CONTRADICTION

To prove s, assume $\sim s$. Then the following argument form is valid:

If $\sim s$ then (t and $\sim t$).
Therefore, s. **12.4.1**

Corresponding Angles Revisited

The following proof uses a contradiction to prove the converse of the Corresponding Angles Postulate. The converse, which is itself a theorem, states:

If two lines are cut by a transversal in such a way that corresponding angles are congruent, then the two lines are parallel. **(Thm 3.4.1)**

In the proof, the "if" part of the theorem is the given.

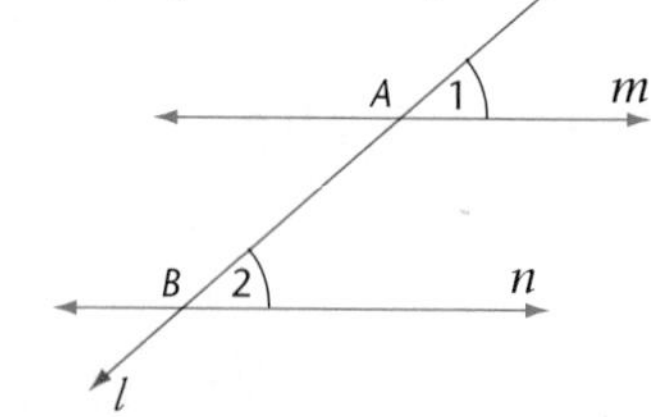

Given: Line l is a transversal that passes through lines m and n. $\angle 1 \cong \angle 2$

Prove: $m \parallel n$

Law Enforcement Detectives often use indirect reasoning to eliminate suspects or to implicate suspects in a crime. Have students research the types of courses taken by people studying to become a detective.

Proof:

Assume that m is *not* parallel to n.

Since, by assumption, m is not parallel to n, the two lines will meet at some point C, as shown in the redrawn figure.

Since $\angle 1$ is an exterior angle of $\triangle ABC$,

$$m\angle 1 = m\angle 2 + m\angle 3.$$

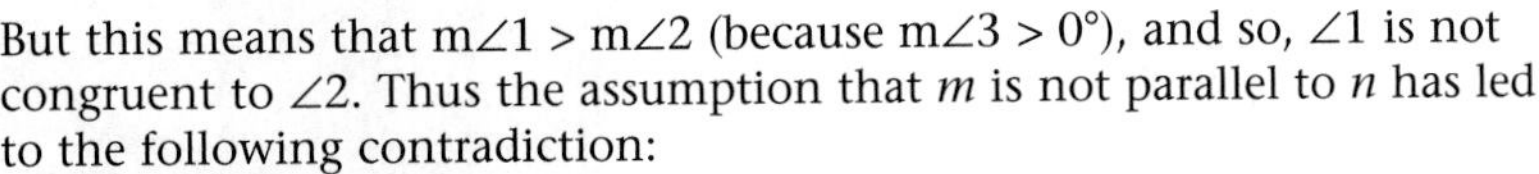

But this means that $m\angle 1 > m\angle 2$ (because $m\angle 3 > 0°$), and so, $\angle 1$ is not congruent to $\angle 2$. Thus the assumption that m is not parallel to n has led to the following contradiction:

$$\angle 1 \cong \angle 2 \text{ and not } (\angle 1 \cong \angle 2).$$

The assumption must therefore be false. Thus, the conclusion is

$$m \parallel n.$$

Alibis and Indirect Proof

Arguments using the Law of Indirect Reasoning are more common than you might think. In a court of law, for example, a lawyer might want to show that a claim on the part of the prosecution contradicts the evidence—the "given." Arguments like the following are quite common:

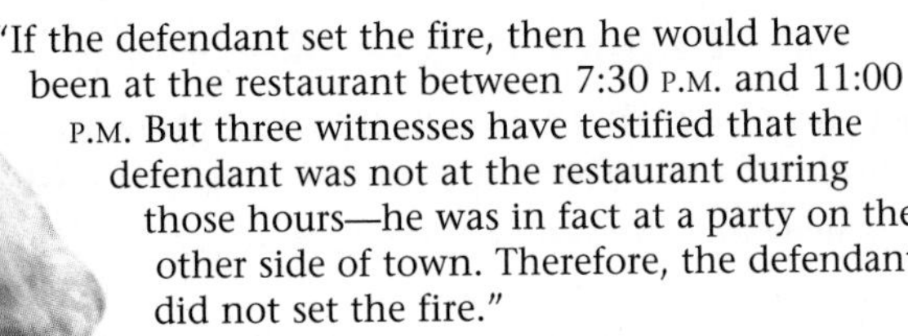

"If the defendant set the fire, then he would have been at the restaurant between 7:30 P.M. and 11:00 P.M. But three witnesses have testified that the defendant was not at the restaurant during those hours—he was in fact at a party on the other side of town. Therefore, the defendant did not set the fire."

The form of the argument is as follows:

1. If the defendant set the fire, then he was at the restaurant between 7:30 P.M. and 11:00 P.M. **(If p then q.)**
2. The defendant was not at the restaurant between 7:30 P.M. and 11:00 P.M. **($\sim q$.)**
3. Therefore, the defendant did not set the fire. **(Therefore, $\sim p$.)**

CRITICAL *Thinking*

Show how the argument above could be made into a proof by contradiction. Let the first two statements be the given. Then assume the opposite of what you want to prove. What contradiction emerges?

TEACHING *tip*

Emphasize that statements made in paragraph proofs must also be justified.

CRITICAL *Thinking*

Given: The person who set the fire was at the restaurant between 7:30 and 11:00 P.M. The defendant was not at the restaurant between 7:30 and 11:00 P.M.

Prove: The defendant did not set the fire.

Assume that the defendant set the fire. Then the defendant was at the restaurant between 7:30 and 11:00 P.M. But this contradicts the fact that the defendant was not at the restaurant between 7:30 and 11:00 P.M. The assumption must therefore be false. So the defendant did not set the fire.

ENRICHMENT Have students simulate a defense lawyer's job by writing a short argument using both deduction and the method of indirect proof. Have students share their arguments with the class.

INCLUSION strategies **Using Cognitive Strategies** Some students may have difficulty understanding the methods of proof in this lesson. Have students approach the procedures one step at a time by asking them to write each step of an indirect proof on a separate note card. Ask them to explain on the card how each step was determined.

Assess

Selected Answers

Exercises 9, 11, 13, 17, 21, 23, 25, 27, 31, 35, and 39

Assignment Guide

Core 1–28, 31, 34–40

Core Plus 1–33, 41

Technology

For Exercises 26–28, use geometry proof programs. Various options can be used to test different possible conclusions.

Error Analysis

The proofs in Exercises 26–28 may be difficult for students to write. Ask pairs of students to check each other's work.

The answers to Exercises 6–10 can be found in Additional Answers beginning on page 727.

Practice Master

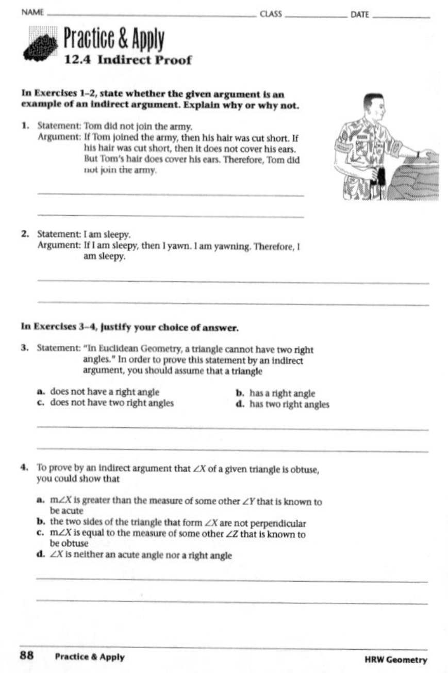
NAME ______ CLASS ______ DATE ______

Practice & Apply
12.4 Indirect Proof

In Exercises 1–2, state whether the given argument is an example of an indirect argument. Explain why or why not.

1. Statement: Tom did not join the army.
   Argument: If Tom joined the army, then his hair was cut short. If his hair was cut short, then it does not cover his ears. But Tom's hair does cover his ears. Therefore, Tom did not join the army.

2. Statement: I am sleepy.
   Argument: If I am sleepy, then I yawn. I am yawning. Therefore, I am sleepy.

In Exercises 3–4, justify your choice of answer.

3. Statement: "In Euclidean Geometry, a triangle cannot have two right angles." In order to prove this statement by an indirect argument, you should assume that a triangle
   a. does not have a right angle
   b. has a right angle
   c. does not have two right angles
   d. has two right angles

4. To prove by an indirect argument that $\angle X$ of a given triangle is obtuse, you could show that
   a. $m\angle X$ is greater than the measure of some other $\angle Y$ that is known to be acute
   b. the two sides of the triangle that form $\angle X$ are not perpendicular
   c. $m\angle X$ is equal to the measure of some other $\angle Z$ that is known to be obtuse
   d. $\angle X$ is neither an acute angle nor a right angle

88 Practice & Apply HRW Geometry

Exercises & Problems

The artist René Magritte (1898–1967), like other Surrealists, loved to create images that seem to defy logic and common sense. The words in this famous painting (1929) mean, "This is not a pipe." Do you feel a sense of contradiction as you look at the words and the picture?

Communicate

1. What is a contradiction?
2. Give two real contradictory statements. Explain why they are contradictory.
3. Give a mathematical example of a contradiction.
4. Describe the form of argument known as *reducto ad absurdum*.
5. Summarize the steps for writing an indirect proof.

Practice & Apply

For Exercises 6–10, state whether the given argument is an example of an indirect argument. Explain why or why not.

6. Statement: It is raining.
Argument: If it were not raining, there would be no puddles on the ground. But I see puddles on the ground. Therefore, it is raining.

7. Statement: You are ill.
Argument: If you were not ill, then you would eat a large dinner. But you did not eat a large dinner. Therefore, you must be ill.

8. Statement: The sun is shining.
Argument: I see my shadow. If I see my shadow then the sun must be shining. Therefore the sun is shining.

9. Statement: I am in New York.
Argument: If I am not in the United States, then I am not in New York. Therefore, I am in New York.

10. Statement: My client is innocent.
Argument: If my client were guilty then he would look guilty. Since my client does not look guilty, then he must be innocent.

Reteaching the Lesson

Using Review Summarize how to write an indirect proof with the following statements: (1) Assume that the negation of what you are trying to prove is true. (2) Use deduction to show that the assumption in Step 1 leads to a contradictory statement (usually of a given statement). (3) Conclude that the original assumption in Step 1 must be false and therefore its negation (what you are trying to prove) must be true.

Fill in the blanks to prove that if ∠1 is not congruent to ∠2, then ∠1 and ∠2 are not vertical angles.

Given: __(11)__ $\angle 1 \not\cong \angle 2$

Prove: __(12)__ ∠1 and ∠2 are not vertical angles.

Assume that ∠1 and ∠2 are __(13)__. Then ∠1 __(14)__ ∠2. But the given states that ∠1 __(15)__ ∠2, which is a __(16)__. Therefore, the assumption that ∠1 and ∠2 are vertical angles is false, and ∠1 and ∠2 are __(17)__.

Fill in the blanks to prove that if m∠1 ≠ m∠2, then line *l* is not perpendicular to line *m*.

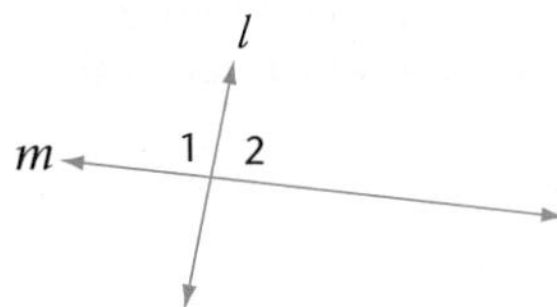

Given: __(18)__ $m\angle 1 \neq m\angle 2$

Prove: __(19)__ *l* not perpendicular to *m*

Assume that __(20)__. Then $\angle 1 \cong \angle 2$ because __(21)__. If $\angle 1 \cong \angle 2$, then m∠1 __(22)__ m∠2 because __(23)__. But the given states that __(24)__, which is a contradiction. Therefore, __(25)__.

Write an indirect proof for Exercises 26–28.

26. Given: $\overline{BD}$ bisects ∠ABC; $\overline{BD}$ is not a median.
Prove: $\overline{AB}$ is not congruent to $\overline{BC}$.

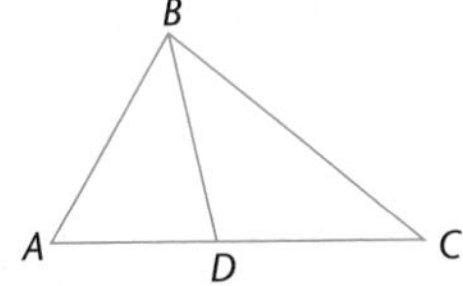

27. Given: $\overline{CT}$ is not congruent to $\overline{BK}$.
Prove: $\overline{BC}$ and $\overline{KT}$ do not bisect each other.

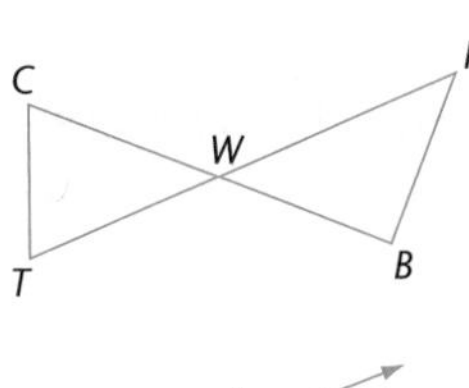

28. Given: ∠1 is not congruent to ∠2.
Prove: $m\angle 1 \neq 90°$

Construct an indirect argument in support of the given statement from the evidence provided.

29. Statement: My client is innocent.
Evidence: My client was miles away from the scene of the crime at the time it happened.

30. Statement: This is not Elm Street.
Evidence: Elm Street has a brick house on the corner. All the houses on this street are made from wood.

31. Statement: The temperature must be 32°F or higher.
Evidence: The water on the sidewalk is not frozen.

13. vertical angles
14. ≅
15. ≅
16. contradiction
17. not vertical angles.
20. $l \perp m$
21. perpendicular lines form congruent adjacent angles.
22. =

Alternative Assessment

Performance Assessment

Oral presentations can be used to assess students' understanding of indirect proof. Call upon individual students to describe how to prove a statement indirectly. Ask other students to offer alternative suggestions.

23. congruent angles have equal measures.
24. $\angle 1 \neq \angle 2$
25. *l* not perpendicular to *m*.

The answers to Exercises 26–31 can be found in Additional Answers beginning on page 727.

Technology Master

NAME ______ CLASS ______ DATE ______

Technology
12.4 Consistent and Inconsistent Data Sets

The compound statement *p* AND (NOT *p*) is always false as you can see from the truth table. You can say that the set of sentences {*p*, −*p*} is an inconsistent set of data.

p	*not p*	*p and (not p)*
1	0	0
1	0	0
0	1	0
0	1	0

In many geometry problems, you may encounter sets of measurements that are inconsistent or self contradictory. When you attempt to construct a desired figure with the specified measurements, you will find the construction to be impossible. In such a case, you may need to modify some of the data to make the data set consistent.

In Exercises 1–4, you are given a set of measurements that may or may not be consistent. Use geometry software to attempt a construction of each figure. If the data is consistent, change the least amount of data to make the data set consistent.

1. Given three points *A*, *B*, and *C* in the plane: *AB* = 1.206 units, *BC* = 1.414 units, *AC* = 1.708 units, m∠*BAC* = 54.85°, and m∠*ABC* = 60.95°.
2. Given three points *X*, *Y*, and *Z* in the plane: *XY* = 1.800 units, *YZ* = 2.177 units, *XZ* = 1.806 units, m∠*YXZ* = 38.11°, m∠XYZ = 71.18°, and m∠YZX = 70.71°.
3. In quadrilateral *RSTU*: *SR* = 1 unit, *RU* = 1.806 units and $\overline{SR} \perp \overline{RU}$, *ST* = 1.264 units and $\overline{SR} \perp \overline{ST}$, m∠STU = 118.44°, and *TU* = 0.637 units
4. In quadrilateral *KLMN*: *KL* = 0.665 units, *LM* = 1.019 units, *MN* = 1.182 units, and *NK* = 1.587 units. In addition, m∠*LKN* = 91.23°, m∠*KLM* = 107.70°, m∠*LMN* = 103.70°, and m∠*MNK* = 57.37°.

In Exercises 5–7, you are given a pair of linear equations. If they have a solution, the set is called consistent. If they do not, they are called inconsistent. Use a graphics calculator to classify each data set. How can you tell by looking at each pair which classification is appropriate?

5. $\begin{cases} y = 1.2x - 1 \\ y = 1.2x + 1 \end{cases}$ ______
6. $\begin{cases} y = 1.2x + 1 \\ y = -0.5x + 2 \end{cases}$ ______
7. $\begin{cases} y = -0.5x + 1 \\ y = -0.5x + 2 \end{cases}$ ______

100 Technology HRW Geometry

32. Assume that the intersection of a line and a plane not containing it contains more than 1 point. If the intersection contains more than one point, then the intersection contains at least two points. If a plane contains two points of a line, then the plane contains the whole line. But that contradicts the given statement that plane does not contain the line. The assumption must therefore be false and the intersection of a line and a plane not containing it contains exactly one point.

The answers to Exercises 33, 37, and 38 can be found in Additional Answers beginning on page 727.

32. Statement: If a line intersects a plane not containing it, then the intersection contains exactly one point.

Evidence: If a plane contains two points of a line, then the plane contains the whole line.

33. Statement: If two lines are cut by a transversal so that alternate interior angles are congruent, then the lines are parallel.

Evidence: Use the given "strange" figure and postulates and theorems found earlier in this textbook.

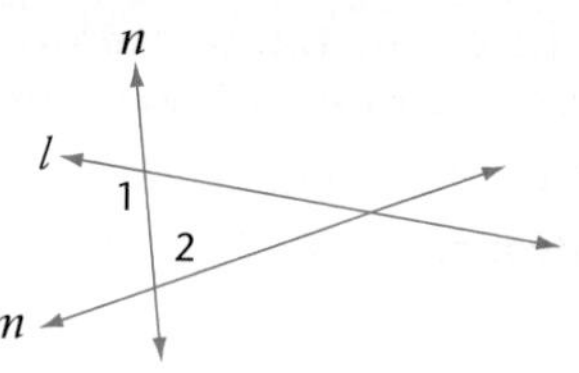

Look Back

Compute the taxidistance between two points on a taxicab grid. [Lesson 11.2]

34. (0, 0) and (3, 3) 6

35. (1, 2) and (3, 4) 4

36. (1, 1) and (3, 4) 5

Use grid paper for the following exercises. Draw a taxicab circle on a grid with the indicated taxicab radius. [Lesson 11.2]

37. taxicab radius = 4

38. taxicab radius = 7

39. List two items that are topologically equivalent to a thumbtack. **[Lesson 11.4]**

40. List two items that are topologically equivalent to a drinking straw. **[Lesson 11.4]**

Look Beyond

In Exercise 41 students write a series of if-then statements to control the air conditioning and security system of a house.

Look Beyond

41.  **Portfolio Activity** You are designing a system for your dream home that will do the following: turn on the air conditioner if the outside temperature is greater than 85°, turn the lights on in the living room from 2 A.M. to 6 A.M., turn the lights on in the entryway when the door opens, and arm the security system when the front door is locked. Write a series of if-then statements logically describing the system's operation.

39. Sample answers: cube, pencil

40. Sample answers: a large section of pipe, a cylinder

41. If the outside temperature is greater than 85°, then turn on the air conditioning. If the time is between 2:00 a.m. and 6:00 a.m., then turn on the living room lights. If the door opens, then turn on the entryway lights. If the front door is locked, then arm the security system.

LESSON 12.5 Computer Logic

A single computer chip in this "mother board" contains hundreds of thousands of logic gates like the one shown here, greatly magnified.

Why *Logic provides the foundation for the decision-making and arithmetic processes of many "smart" electronic devices. The fundamental units of logical circuits are logic gates, which function like logical operators such as AND, OR, and NOT.*

A computer is an example of a device that uses electric impulses and the **binary number system** to operate. **Binary** means "having two parts," and the binary number system is based on two numbers, 1 and 0. Think of a computer as a series of electrical switches that exist in one of two states, *on* or *off*. When a switch is *on*, it can be represented by 1. When a switch is *off*, it can be represented by 0.

Exploration *On-Off Tables*

You will need
No special tools

The tables in this exploration are a simulation of how computers work. The individual columns of each table represent electrical switches that work together. Let 1 = ON and 0 = OFF. Work in pairs and complete each table.

PREPARE

Objectives

- Explore on-off tables, logic gates, and computer logic networks.
- Solve problems using computer logic.

RESOURCES

• Practice Master	12.5
• Enrichment Master	12.5
• Technology Master	12.5
• Lesson Activity Master	12.5
• Quiz	12.5
• Spanish Resources	12.5

Assessing Prior Knowledge

1. Give examples of electronic devices with on-off switches. **[Sample: VCR, CD-player]**

2. Write a truth table for p and q.

[p	q	p and q
T	T	T
T	F	F
F	T	F
F	F	F]

3. Write a truth table for $\sim p$.

[p	$\sim p$
T	F
F	T]

TEACH

This lesson will help students make connections between computer logic, networks, and logical arguments.

ALTERNATIVE teaching strategy

Cooperative Learning Have each group write an example of a simple on-off network like a tape recorder. Encourage them to write a description of the options available through the network. Have groups compare their answers and make a statement about computer networks. Ask each group to describe their network to the class.

Exploration Notes

In this exploration students create tables that simulate the on-off nature of computer gates. This activity will help students understand that computer logic can be represented by the binary condition of *off* (0 or false) or *on* (1 or true).

ongoing ASSESSMENT

4. The combination of the values in the first two columns were considered. Both of the buttons must be pushed for the tape recorder to record in Step 2, while only one of the brakes must be pushed to stop the car in Step 3.

1. The power button on a TV remote control will turn the TV on if it is off, and vice-versa. Determine whether the TV will be on or off after pressing the power button by filling in the second column of the table.

TV	TV after pressing POWER button
1	?
0	?

2. To record on a video recorder, press both PLAY and RECORD. Determine whether the video recorder will record or not by filling in the last column of the table.

PLAY button	RECORD button	Video Recorder
1	1	?
1	0	?
0	1	?
0	0	?

3. The Student Driver car at Dover High School has been equipped with two brake pedals so that either the student driver or the instructor can stop the car. Complete the last column of the table, determining when the brake is on or off.

Student pedal	Instructor pedal	Brake
1	1	?
1	0	?
0	1	?
0	0	?

4. What did you consider as you filled in the last column of each table? Explain why Steps 2 and 3 have different answers even though the tables are set up in the same way.

5. How can the situation described in Step 3 be changed to get the same answers as in Step 2? Why is this new situation impractical in real life? ❖

CRITICAL Thinking

How do the tables in the exploration compare with the truth tables you have constructed before?

CRITICAL Thinking

The table in Step 2 is identical to the truth table for a conjunction. The table in Step 3 is identical to the truth table for a disjunction.

interdisciplinary CONNECTION

Electronics Designing electronic circuitry requires a good working knowledge of logic gates and on-off switches. Have students studying electronics display and describe some of their network designs for the class.

Logic Gates

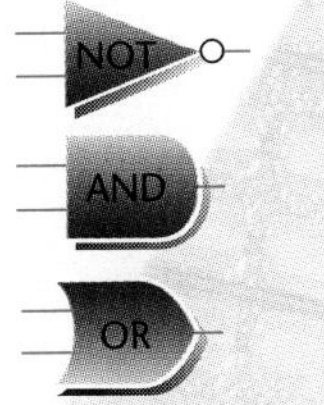

Each table in the exploration corresponds to a particular type of electronic circuitry called a **logic gate**. Logic gates are the building blocks for all "smart" electronic devices. Each logic gate has a special symbol representing **NOT**, **AND**, or **OR**.

In the diagram, p represents the electric input pulse. The **input-output table**, which is like a truth table, records what happens to the input as it passes through the logic gate. If input p has a value of 1, when it passes through a **NOT** gate it will have the opposite value, 0. This input-output table corresponds to the table in Step 1 of the exploration.

NOT Logic Gate

Input-Output Table	
Input	Output
p	NOT p
1	0
0	1

An **AND** logic gate needs two input pulses, which are represented in the diagram by p and q. Notice in the table that in order to get an output value of 1, both pulses must work together. Both p and q must have an input value of 1 in order for the output to be 1. This input-output table corresponds to the table in Step 2 of the exploration.

AND Logic Gate

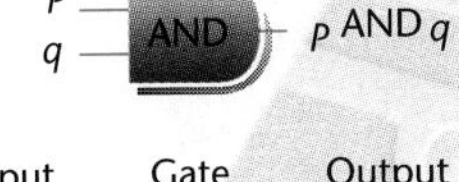

Input-Output Table		
Input		Output
p	q	p AND q
1	1	1
1	0	0
0	1	0
0	0	0

An **OR** logic gate works differently from an AND gate because the input pulses do not have to work together. In order to get an output value of 1, *either p or q* has to have an input value of 1. Notice that to have an output value of 0, *neither p nor q* can have a value of 1. The input-output table on the following page corresponds to the table in Step 3 of the exploration.

Cooperative Learning

Use small groups to compare input-output tables with truth tables and with the tables in Exploration 1. Have students list their similarities and differences. Group work may give students individual help.

ENRICHMENT Ask students to research the work of George Boole and Boolean algebra. Have them describe how Boolean algebra is used to solve problems in a computer system.

Using Cognitive Strategies Some students may have difficulty representing electronic circuitry as logic gates. Suggest that they read this lesson very carefully, take notes, and explain each diagram in terms of the model it represents. You can use cooperative groups to discuss the diagrams.

Using chips and blocks and other models may be helpful in designing a logic network. Ask students to describe how to translate the network into a physical model.

Alternate Example 1

Construct an input-output table for the following network.

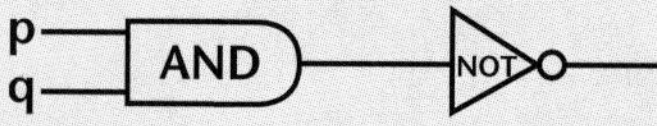

p	q	p and q	not (p and q)
1	1	1	0
1	0	0	1
0	1	0	1
0	0	0	1

OR Logic Gate

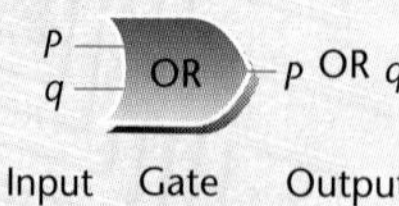

Input-Output Table		
Input		Output
p	q	p OR q
1	1	1
1	0	1
0	1	1
0	0	0

Networks

Logic gates can be combined to form networks. To determine how a network operates, you can use input-output tables.

EXAMPLE 1

Construct an input-output table for the following network.

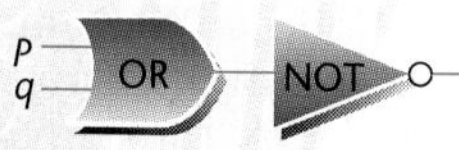

Solution ➤

Read from left to right. The first gate is p or q. Use parentheses to capture the output from this gate: (p OR q). Then perform the NOT operation: NOT (p OR q). To construct the input-output table, all possible input combinations must be considered. The numbers are filled in as you would fill in a truth table, where 1 is *true* and 0 is *false*. ❖

p	q	p OR q	NOT (p OR q)
1	1	1	0
1	0	1	0
0	1	1	0
0	0	0	1

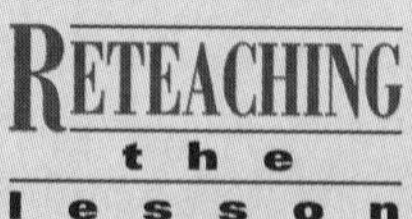

Using Discussion Have each student give a simple logical expression using NOT, OR, and AND. Ask them to create a network for the logical expression and describe to the class how to use the network.

EXAMPLE 2

Create a logical expression that corresponds to the following network.

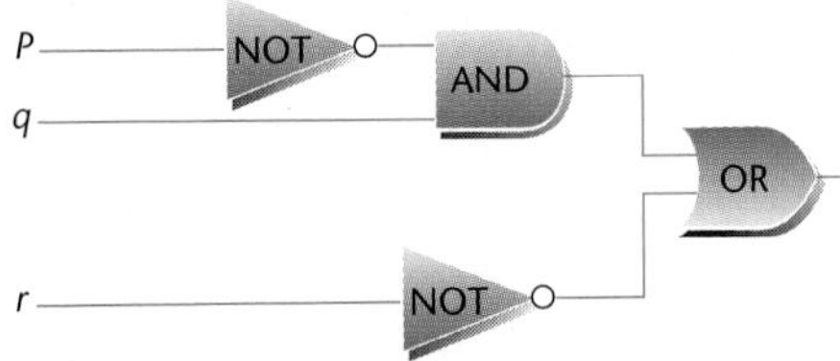

Solution ➤

Read from left to right, using parentheses to capture output.

NOT appears first: (NOT p)

The AND gate gives ((NOT p) AND q)

The bottom branch gives (NOT r)

The OR gate takes the output from the three preceding steps:

((NOT p) AND q) OR (NOT r) ❖

How many rows and how many columns will an input-output table require for the network in Example 2? Explain your reasoning.

EXERCISES & PROBLEMS

Communicate

Draw the symbol that corresponds to each of the following.

1. NOT **2.** AND **3.** OR

Use the logic gates to answer each question.

4. If $p = 1$ and $q = 0$, what will the output be?

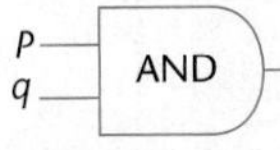

5. If $p = 1$ and $q = 0$, what will the output be?

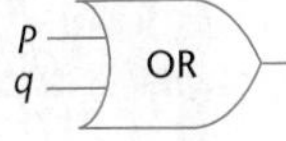

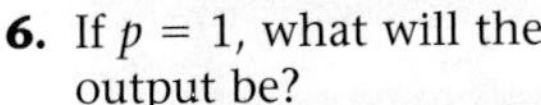

6. If $p = 1$, what will the output be?

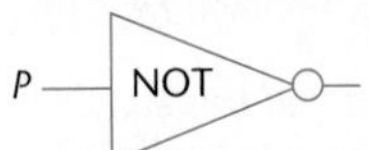

7. If $p = 1$, what will the output be?

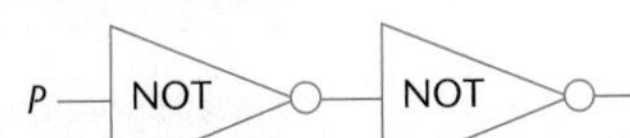

Use Transparency 110

Alternate Example 2

Create a logical expression that corresponds to the following network.

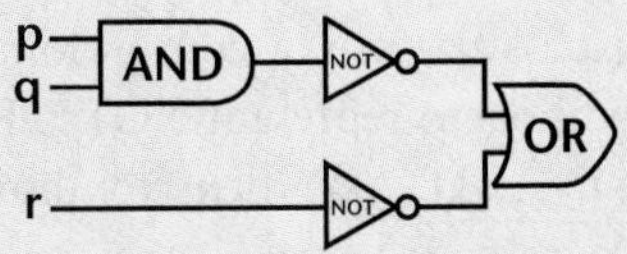

[Solution: (NOT (p and q)) OR (NOT r)]

CRITICAL Thinking

$[(2)(2)](2) = 8$. Each variable has 2 possible values, so there are 4 ways to combine the values for the first two variables and 8 ways to combine those 4 with the 2 possible values for the third variable.

ASSESS

Selected Answers

Exercises 9, 11, 15, 19, 23, 25, 27, 31, and 35

Assignment Guide

Core 1–13, 16–23, 29–38

Core Plus 1–28, 33, 37–40

Technology

Students can use the TEST LOGIC menu of a graphing calculator to give truth values for compound sentences in Exercises 16–23. Store the values of p, q, and r as 0 or 1 and use the menu to calculate the truth value of the compound sentence.

Error Analysis

Students may have difficulty doing Exercises 16–23 if they do not pay attention to the parentheses. Emphasize the use of parentheses in determining order of operations.

Performance Assessment

Have students describe verbally how to create a network for a logical expression. Have them discuss the use of conventional symbols to represent NOT, OR, and AND. Ask them to draw diagrams as visual aids.

Practice Master

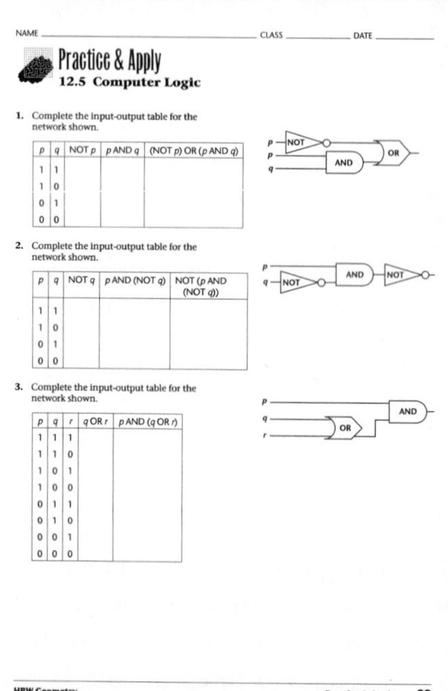

NAME ______ CLASS ______ DATE ______

Practice & Apply
12.5 Computer Logic

1. Complete the input-output table for the network shown.

p	q	NOT p	p AND q	(NOT p) OR (p AND q)
1	1			
1	0			
0	1			
0	0			

2. Complete the input-output table for the network shown.

p	q	NOT q	p AND (NOT q)	NOT (p AND (NOT q))
1	1			
1	0			
0	1			
0	0			

3. Complete the input-output table for the network shown.

p	q	r	q OR r	p AND (q OR r)
1	1	1		
1	1	0		
1	0	1		
1	0	0		
0	1	1		
0	1	0		
0	0	1		
0	0	0		

HRW Geometry   Practice & Apply 89

Practice & Apply

For Exercises 8–11, complete the given input-output table for the following network.

	p	q	p AND q	NOT (p AND q)
8.	1	1	? 1	? 0
9.	1	0	? 0	? 1
10.	0	1	? 0	? 1
11.	0	0	? 0	? 1

p, q → AND → NOT

12. Complete an input-output table for the network.

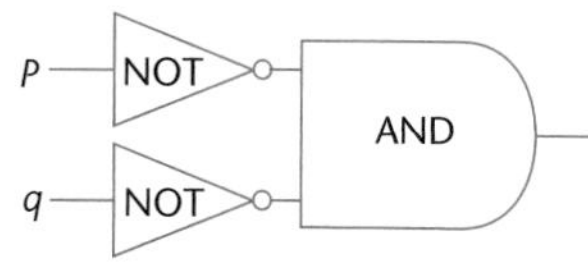

p	q	NOT p	NOT q	(NOT p) AND (NOT q)
?	?	?	?	?
?	?	?	?	?
?	?	?	?	?
?	?	?	?	?

13. Complete an input-output table for the following network.

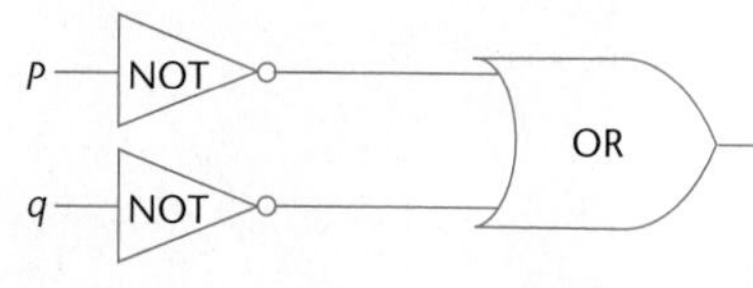

14. Two of the logical expressions shown are functionally equivalent. That is, when given the same input they will produce the same output. Identify the two functionally equivalent logical expressions and complete the equivalence statement.

NOT (p AND q)

(NOT p) OR (NOT q)

p AND (NOT q)

? ≡ ?

15. Decide which two logical expressions are functionally equivalent and write an equivalence statement.

p OR (NOT q)

NOT (p OR q)

(NOT p) AND (NOT q)

Charles Babbage (1791–1871) invented this early computer, which he called the Difference Engine.

12.

p	q	NOT p	NOT q	(NOT p) AND (NOT q)
1	1	0	0	0
1	0	0	1	0
0	1	1	0	0
0	0	1	1	1

13.

p	q	NOT p	NOT q	(NOT p) OR (NOT q)
1	1	0	0	0
1	0	0	1	1
0	1	1	0	1
0	0	1	1	1

14. NOT(p AND q) = (NOT p) OR (NOT q)

15. NOT(p OR q) = (NOT p) AND (NOT q)

For Exercises 16–23, complete an input-output table for the following network.

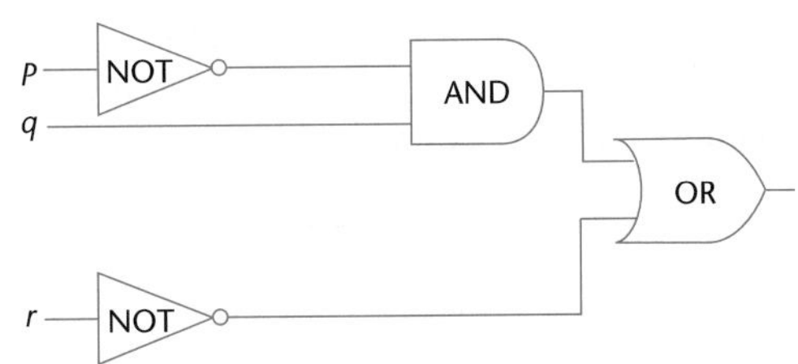

	p	q	r	NOT p	(NOT p) AND q	NOT r	((NOT p) AND q) OR (NOT r)
16.	1	1	1	? 0	? 0	? 0	? 0
17.	1	1	0	? 0	? 0	? 1	? 1
18.	1	0	1	? 0	? 0	? 0	? 0
19.	1	0	0	? 0	? 0	? 1	? 1
20.	0	1	1	? 1	? 1	? 0	? 1
21.	0	1	0	? 1	? 1	? 1	? 1
22.	0	0	1	? 1	? 0	? 0	? 0
23.	0	0	0	? 1	? 0	? 1	? 1

Construct a network diagram for the following expressions:

24. *p* AND (NOT *q*) **25.** NOT (*p* OR *q*)

Construct a logical expression and an input-output table for the following networks.

26.

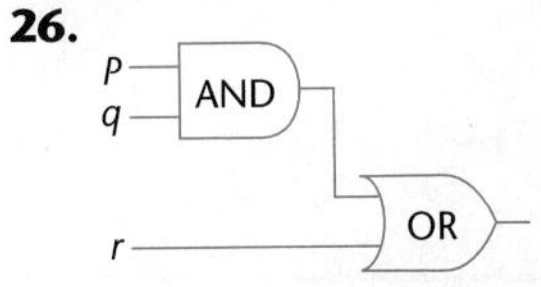

27.

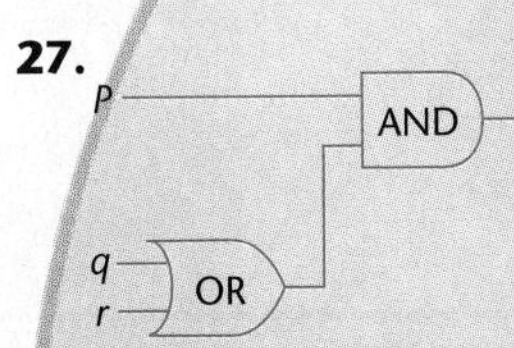

28. Are the two networks in Exercises 26 and 27 functionally equivalent? Explain your reasoning.

In Exercises 29–32, decide whether the situation is best described with a NOT, AND, or OR gate. Create a logical expression, a network diagram, and an input-output table to describe the given situation.

29. To run a particular dishwasher, the door must be locked and the power button must be on.

30. The living-room lights can be turned on by either a switch in the living room or a switch in the hallway.

31. A CD is placed in a CD player. To operate the player, you must first hit the POWER button, and then you must hit the PLAY button to start the music.

The answers to Exercises 26, 27, 29, 30, and 31 can be found in Additional Answers beginning on page 727.

24.

25.

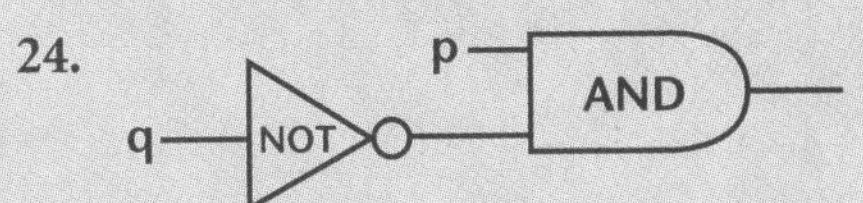

28. No. When given the same input, they do not always produce the same output.

Technology Master

NAME ______ CLASS ______ DATE ______

Technology
12.5 Exploring Logic Gates on the Computer

The three basic logic gates are shown. When you arrange a collection of them into a logical system, you make more sophisticated circuits, such as those in the following exercises.

For each logical system, you can associate a logical expression, whose truth table you can construct in a spreadsheet. In the spreadsheet, you represent ON by 1 and OFF by 0. The IF function enables you to determine the ON/OFF status for a given set of ON/OFF inputs.

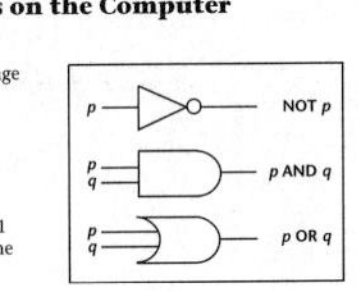

Use a spreadsheet in the following exercises.

1. Create a spreadsheet in which column A contains $p = 1, 1, 0, 0$ and column B contains $q = 1, 0, 1, 0$. In cells C2, D2, and E2, enter expressions that represent activity at gates 1, 2, and 3. Then FILL DOWN to determine the ON/OFF status of the system.

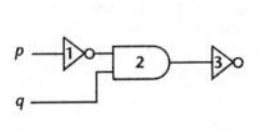

Use a modification of your spreadsheet in Exercise 1 to determine the activity at each gate and the ON/OFF status of each system.

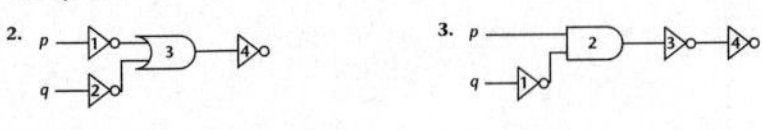

4. Sketch a simpler logical system that has the same ON/OFF status as the system in Exercise 2.

5. Sketch a simpler logical system that has the same ON/OFF status as the system in Exercise 3.

6. Create a spreadsheet in which column A contains $p = 1, 1, 1, 1, 0, 0, 0, 0$, column B contains $q = 1, 1, 0, 0, 1, 1, 0, 0$, and column C contains $r = 1, 0, 1, 0, 1, 0, 1, 0$. In cells D2, E2, F2, and G2, enter expressions that represent the activity at gates 1, 2, 3, and 4. Then FILL DOWN to determine the ON/OFF status of the system.

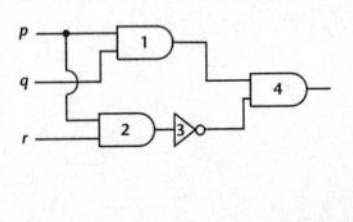

HRW Geometry Technology 101

32. Using a certain word processing program, if the BOLD icon is selected, the print style will change from regular to bold or vice versa.

33.  **Portfolio Activity** Design a "machine" that must use logic in order to operate. Draw a picture and write a description of your machine. Then draw a network diagram of the machine's circuitry and complete an input-output table for the network. Answers will vary.

Look Back

In Exercises 34 and 35, tell whether the proportion is true for all values of the variables. If the proportion is false, give a numerical counterexample. [Lesson 8.2]

34. If $\frac{x}{y} = \frac{r}{s}$, then $\frac{(x + c)}{y} = \frac{(r + c)}{s}$. False; For example: $\frac{1}{2} = \frac{2}{4}$ but $\frac{1+2}{2} \neq \frac{2+2}{4}$.

35. If $\frac{x}{y} = \frac{r}{s}$, then $\frac{x}{y} = \frac{(x + r + m)}{(y + s + m)}$. False; For example: $\frac{1}{2} = \frac{2}{4}$ but $\frac{1}{2} \neq \frac{1+2+3}{2+4+3}$.

36. The ratio of the corresponding sides of two similar triangles is equal to the ratio of the perimeters. If the perimeters of two similar triangles are 18 cm and 25 cm, what is the ratio of the perimeters? What is the ratio of the sides? **[Lesson 8.6]** $\frac{18}{25}$, $\frac{18}{25}$

Look Beyond

Look Beyond exercises 39 and 40 ask students to calculate probabilities for geometric figures using area.

The answers to Exercises 32, 39 and 40 can be found in Additional Answers beginning on page 727.

Look Beyond

37. A 50 ft pipe carries water from a well to a house. The pipe has sprung a leak. What is the probability that the leak will be within 5 ft of the house? $\frac{1}{10}$

38. A speck of dust lands on the diagram. What is the probability that the speck of dust will land inside the triangle? $\frac{1}{6}$

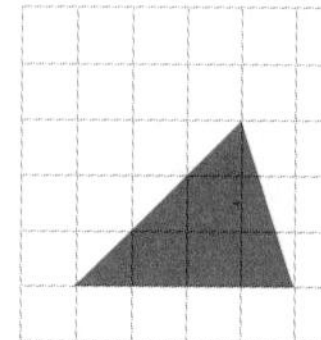

Environmental Science An aerial photograph shows workers cleaning up an oil spill.

39. How can you use objects in the photo to estimate the area of the spill? (Hint: First estimate the size of the squares in the yellow grid.)

40. Estimate the area of the spill.

Meteorologists use latitude and longitude readings to track hurricanes on a map of the Earth. Near the Equator, these coordinates are very similar to x-y coordinates.

Why *Properties of geometric figures can be represented in a coordinate geometry drawing. For this reason, coordinate geometry can be used to prove geometry theorems.*

To prove a theorem using coordinate geometry, begin by drawing a figure in a coordinate plane. If the figure is a polygon, it is usually convenient to place one vertex at the origin and at least one side along an axis.

Exploration 1 *Triangle Midsegment Theorem Revisited*

You will need
Graph paper or
Geometry technology

Geometry Graphics

1 Use the coordinates of the three triangles given in the table to prove the following: *The segment joining the midpoints of two sides of a triangle is half the length of the third side.* Draw the first two triangles in a coordinate plane. Use a different xy-axis for each triangle.

Vertices of triangle	Coordinates of M	Coordinates of S	Slope of $\overline{MS}$	Slope of $\overline{AC}$	Length of $\overline{MS}$	Length of $\overline{AC}$
1. $A(0, 0)$ $B(2, 6)$ $C(8, 0)$	?	?	?	?	?	?
2. $A(0, 0)$ $B(6, -8)$ $C(10, 0)$	?	?	?	?	?	?
3. $A(0, 0)$ $B(2p, 2q)$ $C(2r, 0)$	?	?	?	?	?	?

ALTERNATIVE teaching strategy

Technology Ask students to use geometry graphics software to draw a figure on a coordinate grid and use the measurement options to verify their proof statements.

PREPARE

Objectives

- Develop coordinate proofs for the Triangle Midsegment Theorem, the Diagonals of a Parallelogram, and the reflection of a point about the line $y = x$.
- Use the concepts of coordinate proof to solve problems on the coordinate plane.

RESOURCES

- Practice Master 12.6
- Enrichment Master 12.6
- Technology Master 12.6
- Lesson Activity Master 12.6
- Quiz 12.6
- Spanish Resources 12.6

Assessing Prior Knowledge

1. Find the slope of the line containing (–2, 4) and (–1, 3). [–1]
2. Find the length of the segment with endpoints (4, 7) and (–2, –1). [10]
3. What is the image of (–2, 4) under a reflection about the line $y = x$? [(4, –2)]

TEACH

Why Studying coordinate proofs helps students describe geometric shapes analytically and demonstrates the strong connection between algebra and geometry.

Exploration 1 Notes

In this exploration, students prove the Triangle Midsegment Theorem by using coordinates. Students learn how to use coordinates so that calculations are more convenient.

ongoing ASSESSMENT

4. $\overline{MS} \parallel \overline{AC}$ and $MS = \frac{1}{2}AC$.

CRITICAL *Thinking*

The coordinate proof appears easier. The slopes and lengths of the segments are found directly. The earlier proof required proving that lines were parallel triangles were similar first.

Exploration 2 Notes

In this exploration, students verify that the diagonals of a parallelogram bisect each other. Have students do the same exploration with a variety of parallelograms.

TEACHING *tip*

When labeling polygons on the coordinate plane, use convenient coordinates if possible. For example, if a segment is to be bisected, use $2a$ and $2b$ as its endpoints. Then its midpoint uses coordinates a and b. Other strategies include using the first quadrant for figures and the axes for vertices.

Cooperative Learning

Divide students into cooperative groups and assign one exploration to each. Have the groups work through the steps of their exploration and present their results to the whole class.

2 Label the vertices of each triangle, indicating the coordinates. For each triangle, find the midpoints of $\overline{AB}$ and $\overline{BC}$, labeling these points M and S, respectively. Draw the midsegment $\overline{MS}$ for each triangle and record the coordinates of the midpoint in a table like the one on the previous page.

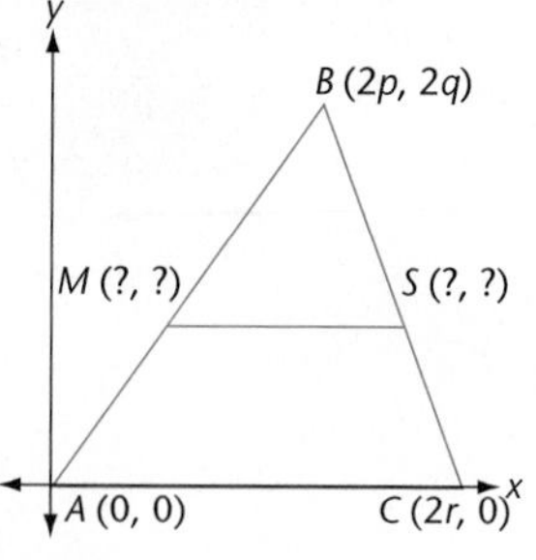

3 Copy and complete the table for the first two triangles. Then use the figure shown to state the general case of the Triangle Midsegment Theorem in row three of the table. The general case, along with the completed table, is the proof of the theorem.

4 Based on the information in the last row of your table, what can you conclude about the relationship between $\overline{MS}$ and $\overline{AC}$? ❖

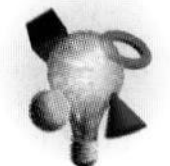

CRITICAL *Thinking* Compare this proof of the Triangle Midsegment Theorem with the one that appears in Lesson 3.7. Which proof appears easier? Explain why?

Exploration 2 *The Diagonals of a Parallelogram*

You will need
Graph paper or
Geometry technology

1 Use the coordinates given in the table to prove the following: *The diagonals of a parallelogram bisect each other.* Draw the first two parallelograms in a coordinate plane. Use a different xy-axis for each parallelogram. Three vertices are given. Find the fourth vertex to complete each figure.

3 Vertices of parallelogram	Fourth vertex	Midpoint of $\overline{BD}$	Midpoint $\overline{AC}$
1. $A(0, 0)$ $B(2, 6)$ $D(10, 0)$	$C(?, ?)$	?	?
2. $A(0, 0)$ $B(4, -8)$ $D(10, 0)$	$C(?, ?)$	?	?
3. $A(0, 0)$ $B(2p, 2q)$ $D(2r, 0)$	$C(?, ?)$	?	?

2 For each parallelogram, draw diagonals $\overline{BD}$ and $\overline{AC}$, and determine their midpoints.

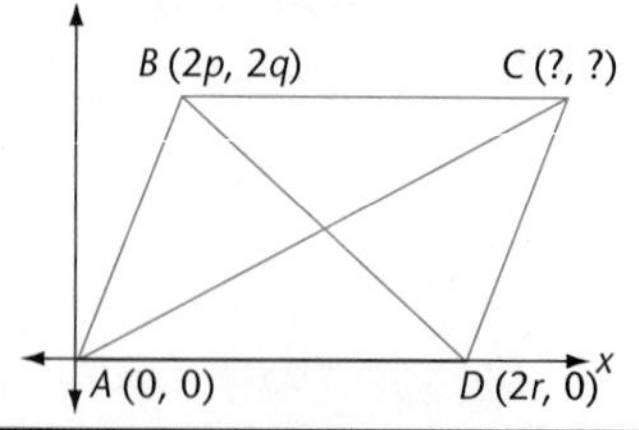

interdisciplinary CONNECTION **Art** Computer artists need a thorough understanding of coordinates in order to draw figures on the computer screen. Have students discuss how principles in this lesson can be applied directly to an artist's work.

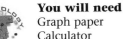

3. Copy and complete the table for the first two parallelograms. Then use the figure shown to state a general case for the theorem in row three of the table. The general case, along with the completed table, is the proof of the theorem.

4. Based on the information in the last row of your table, what may you conclude about the diagonals of a parallelogram? ❖

How can you prove this theorem without using coordinate geometry? Which kind of proof appears easier? Explain why?

Exploration 3 Reflection About the Line $y = x$

You will need
Graph paper
Calculator

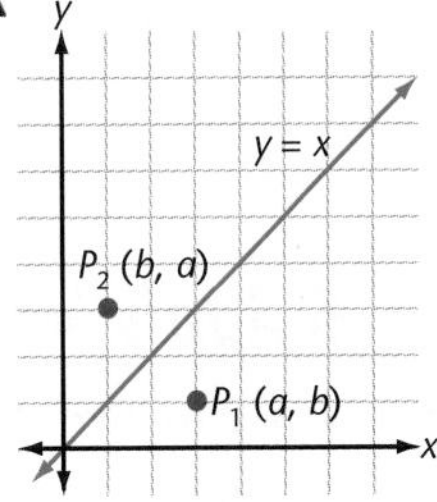

You may recall that the effect of reversing the coordinates of a point was to reflect the point through the line $y = x$. You can use coordinate geometry to prove this result.

1. If you know the x-coordinate of a point on the line $y = x$, what can you conclude about the y-coordinate? (Filling in a table like the one on the right will reveal the pattern.)

x	y
1	?
2	?
0	?
–1	?
P	?

2. Find the midpoint between the point $P_1(a, b)$ and $P_2(b, a)$ using the midpoint formula. Does this point lie on the line $y = x$? Explain your reasoning.

3. Pick two points on the line $y = x$ and use them to find the slope of the line $y = x$. Record your result.

4. Find the slope of the line that passes through the points $P_1(a, b)$ and $P_2(b, a)$. Record your result. Hint: You can rewrite $b - a$ as $-(a - b)$.

5. Compare your results in Steps 3 and 4. What can you conclude about the relationship between the two lines?

6. Explain how your results prove that the effect of reversing the coordinates of a point is to reflect the point through the line $y = x$. (Recall the definition of a reflection.) ❖

ONGOING ASSESSMENT

4. The diagonals of a parallelogram bisect each other.

CRITICAL Thinking

By showing that certain triangles are congruent and then using CPCTC to show the diagonals bisect each other. The coordinate proof appears easier. The midpoints are found directly without having to first prove triangles congruent.

Exploration 3 Notes

In this exploration students verify that the reflection image of the point (x, y) in the line $y = x$ is the point (y, x). This activity will help students to understand the algebraic connection of reflections to transformations. Make sure students understand the use of subscript notation, which is used to identify points.

ONGOING ASSESSMENT

6. The line $y = x$ is the perpendicular bisector of the line through (a, b) and (b, a). Therefore, (a, b) and (b, a) are equidistant from the line and on opposite sides of the line. By definition of reflection through a line, (a, b) and (b, a) are the images of each other when reflected through $y = x$.

ENRICHMENT Suppose a circle centered at $O(0, 0)$ with radius 1 is drawn on the coordinate plane. Draw segment OA to any point $A(x, y)$ on the circle in the first quadrant. Find the sine, cosine, and tangent of the angle B that OA makes with the x-axis. [**sine $B = y$, cosine $B = x$, tangent $B = 1$**]

Using Cognitive Strategies Students may have difficulty distingusihing between synthetic (or traditional) geometric proof and coordinate proof. Create a chart listing different features. For example, the synthetic approach usually refers to angle or side congruence, the coordinate approach may refer to equivalence of measure of sides or angles.

Exercises & Problems

Communicate

1. In proofs using coordinate geometry, why is it best to place one vertex of the figure at the origin and one side along the x-axis?
2. How could you use coordinate geometry to prove that opposite sides of a parallelogram are equal in length?
3. How could you use coordinate geometry to prove that the diagonals of a rhombus are perpendicular?
4. Explain how you would prove that the diagonals of a rhombus are perpendicular without using coordinate geometry.

Practice & Apply

Use graph paper or geometry technology for Exercises 5–9.

5. $ABCD$ is a rectangle.
   a. Given: $A(0, 0)$, $B(0, 3)$, $D(7, 0)$
   Find the coordinates of C. $C(7, 3)$
   b. Given: $A(0, 0)$, $B(0, p)$, $D(q, 0)$
   Find the coordinates of C. $C(q, p)$

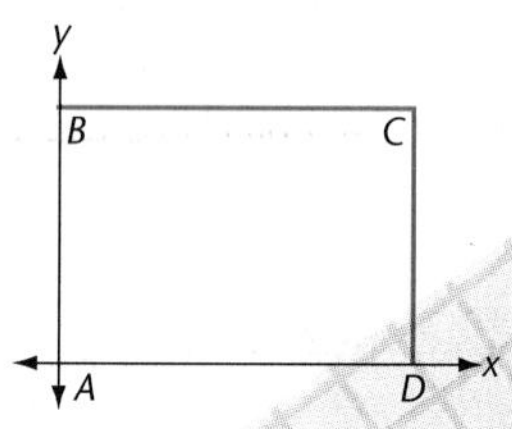

6. ABC is an isosceles triangle with $\overline{AB} \cong \overline{BC}$.
   a. Given: $A(0, 0)$, $B(4, 3)$
   Find the coordinates of C. $C(8, 0)$
   b. Given: $A(0, 0)$, $B(p, q)$
   Find the coordinates of C. $C(2p, 0)$

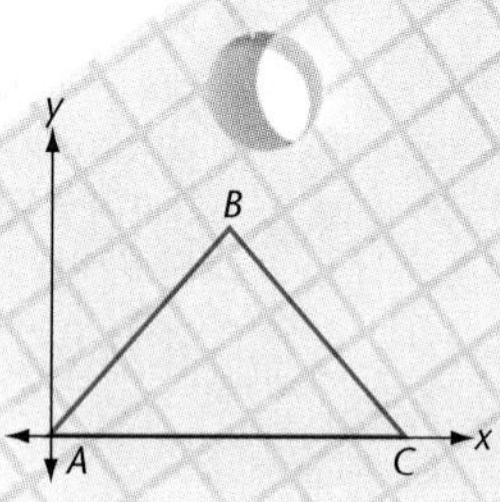

7. $ABCD$ is a parallelogram.
   a. Given: $A(0, 0)$, $B(3, 5)$, $D(7, 0)$
   Find the coordinates of C. $C(10, 5)$
   b. Given: $A(0, 0)$, $B(p, q)$, $D(r, 0)$
   Find the coordinates of C. $C(p + r, q)$

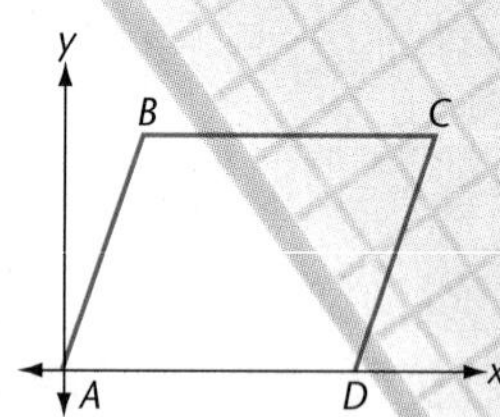

Assess

Selected Answers

Exercises 5, 7, 11, 13, 15, 17, 19, 25, 27, and 29

Assignment Guide

Core 1–16, 19–21, 25–29

Core Plus 1–5, 8–24, 30–31

Technology

You may want students to use geometry graphics software to do Exercises 19–21. Have students use the measurement options to verify the statements.

Error Analysis

Students may have difficulty with Exercise 18 because the figure does not have sides parallel to the axes.

Practice Master

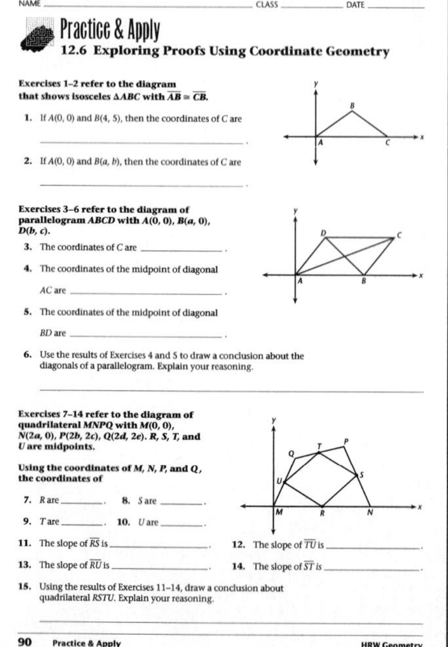

NAME ______________ CLASS ________ DATE ________

Practice & Apply

12.6 Exploring Proofs Using Coordinate Geometry

Exercises 1–2 refer to the diagram that shows isosceles $\triangle ABC$ with $\overline{AB} \cong \overline{CB}$.

1. If $A(0, 0)$ and $B(4, 5)$, then the coordinates of C are ______________.
2. If $A(0, 0)$ and $B(a, b)$, then the coordinates of C are ______________.

Exercises 3–6 refer to the diagram of parallelogram $ABCD$ with $A(0, 0)$, $B(a, 0)$, $D(b, c)$.

3. The coordinates of C are ______________.
4. The coordinates of the midpoint of diagonal AC are ______________.
5. The coordinates of the midpoint of diagonal BD are ______________.
6. Use the results of Exercises 4 and 5 to draw a conclusion about the diagonals of a parallelogram. Explain your reasoning.

____

Exercises 7–14 refer to the diagram of quadrilateral $MNPQ$ with $M(0, 0)$, $N(2a, 0)$, $P(2b, 2c)$, $Q(2d, 2e)$. R, S, T, and U are midpoints.

Using the coordinates of M, N, P, and Q, the coordinates of

7. R are ________. 8. S are ________.
9. T are ________. 10. U are ________.
11. The slope of $\overline{RS}$ is ______________. 12. The slope of $\overline{TU}$ is ______________.
13. The slope of $\overline{RU}$ is ______________. 14. The slope of $\overline{ST}$ is ______________.
15. Using the results of Exercises 11–14, draw a conclusion about quadrilateral $RSTU$. Explain your reasoning.

____

90 Practice & Apply HRW Geometry

Reteaching the Lesson

Using Visual Models Summarize the lesson by having students draw a polygon on the coordinate plane with one side on the x-axis and one vertex at (0, 0). Label the vertices with convenient coordinates like $(2a, 2b)$. Ask them to list all the statements they can prove about their polygon. Have them describe how they would use coordinates to prove each statement.

8. *ABCD* is a rhombus.

a. Given: $A(-2, 0)$, $B(0, 5)$ $C(2, 0)$, $D(0, -5)$

Find the coordinates of C and D.

b. Given: $A(-p, 0)$, $B(0, r)$

Find the coordinates of C and D. $C(p, 0)$; $D(0, -r)$

9. *ABCD* is a square.

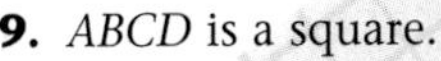

a. Given: $A(-3, 0)$ $B(0, 3)$; $C(3, 0)$; $D(0, -3)$

Find the coordinates of B, C, and D.

b. Given: $A(-p, 0)$

Find the coordinates of B, C, and D.
$B(0, p)$; $C(p, 0)$; $D(0, -p)$

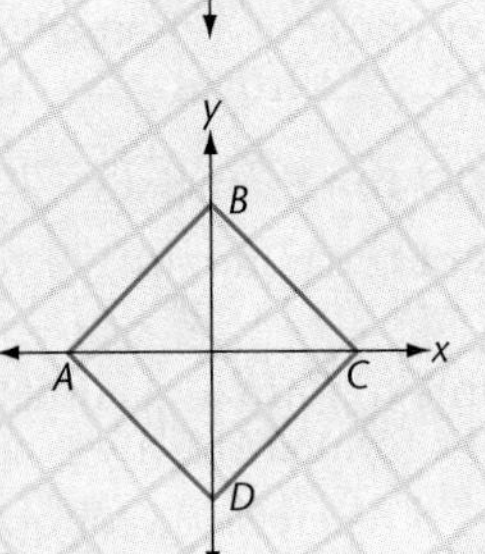

For Exercises 10–13, *ABCD* is a trapezoid with coordinates $A(0, 0)$, $B(2, 6)$, $C(8, ?)$, and $D(12, ?)$.

10. Find the missing coordinates for C and D. $C(8, 6)$; $D(12, 0)$

11. Find the coordinates of M and S, the midpoints of $\overline{AB}$ and $\overline{CD}$. $M = (1, 3)$; $S = (10, 3)$

12. Find the lengths of $\overline{AD}$, $\overline{BC}$, and the midsegment $\overline{MS}$.

13. How is the length of a midsegment of a trapezoid related to the length of the two bases of the trapezoid?

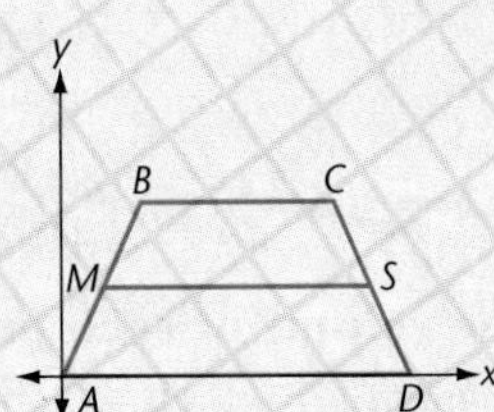

For Exercises 14–17, *ABCD* is a trapezoid with coordinates $A(0, 0)$, $B(2p, 2q)$, $C(2r, ?)$, and $D(2s, ?)$.

14. Find the missing coordinates for C and D in terms of p, q, r, and s.

15. Find the coordinates of M and S, the midpoints of $\overline{AB}$ and $\overline{CD}$, in terms of p, q, r, and s.

16. Prove that the length of the midsegment of a trapezoid is equal to the average of the lengths of the two bases.

17. Prove that the midsegment of a trapezoid is parallel to the bases.

18. What type of quadrilateral is given in the figure below? Justify your answer.

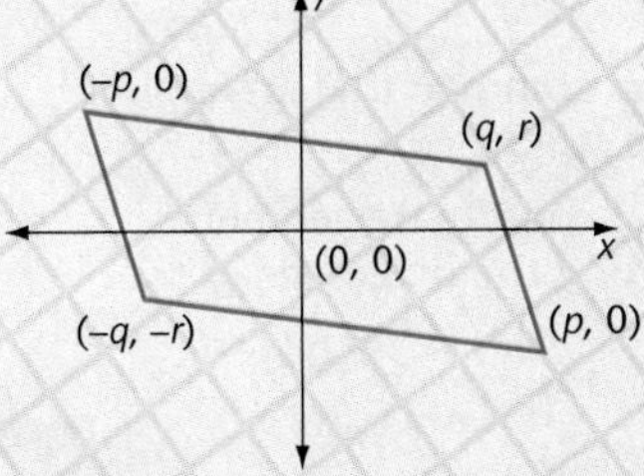

12. $AD = 12$; $BC = 6$; $MS = 9$

13. $MS = \dfrac{AD + BC}{2}$. The length of the median of a trapezoid is equal to the average of the lengths of its bases.

14. $C(2r, 2q)$; $D(2s, 0)$

15. $M = (p, q)$; $S = (r + s, q)$

18. Parallelogram. The slopes of opposite sides are equal. Therefore, opposite sides are parallel and the figure is a parallelogram by definition.

alternative ASSESSMENT

Portfolio Assessment

Ask students to summarize the lesson in writing by listing and describing how to do a coordinate proof. Ask them to include sample problems with complete solutions.

The answers to Exercises 16 and 17 can be found in Additional Answers beginning on page 727.

Technology Master

NAME ____________ CLASS ________ DATE ________

Technology
12.6 Quadrilateral Data Sheets

The diagram shows a quadrilateral whose coordinates are given. To draw conclusions logically about the quadrilateral, you can create a quadrilateral data sheet. In the sheet, you can enter formulas for measures such as length, slope, and angle. You can also enter formulas for the coordinates of midpoints, and so on. By reading the entries in the data sheet, you can then tell which segments are congruent, parallel, perpendicular, have the same midpoint, and so on. A model spreadsheet is shown. You may wish to enter a variety of formulas before you work your first exercise.

	A	B	C	D	E	F	G
1		X COORD	Y COORD		SEGMENT	LENGTH	SLOPE
2	P				PQ		
3	Q				QR		
4	R				RS		
5	S				SP		
6					PR		
7	ANGLE MEASURES				QS		
8	SPQ	PQR	QRS	RSP			
9							
10							
11		MIDPOINTS			OTHER INFORMATION		
12		X COORD	Y COORD				
13	PQ						
14	QR						
15	RS						
16	SP						

Use your spreadsheet to draw conclusions about *PQRS*.

1. $P(1, 2)$, $Q(3, 6)$, $R(9, 7)$, $S(7, 3)$ ____________
2. $P(-1, 1)$, $Q(1, 7)$, $R(3, 6)$, $S(3, 1)$ ____________
3. $P(1, 1)$, $Q(2, 6)$, $R(7, 7)$, $S(6, 2)$ ____________
4. $P(2, 0)$, $Q(0, 2)$, $R(5, 7)$, $S(7, 5)$ ____________
5. Although none of the quadrilaterals in Exercises 1–4 have exactly the same classification, there is something true about the quadrilaterals formed by joining the midpoints of the sides in order. What type of quadrilateral is formed by each set of midpoints?

____

102 Technology HRW Geometry

22. mdpnt of $\overline{AB} = (1, 3)$

mdpnt of $\overline{BC} = (4, 5)$

mdpnt of $\overline{CD} = (7, 2)$

mdpnt of $\overline{DA} = (4, 0)$

23. Parallelogram. The slopes of opposite sides are equal. Therefore, the opposite sides are parallel and the figure is a parallelogram by definition.

The answers to Exercises 21, 24, and 30 can be found in Additional Answers beginning on page 727.

Use $\triangle ABC$ for Exercises 19–21.

19. Find the coordinates of the midpoint of the hypotenuse. $M = (3, 2)$

20. Show that the midpoint of the hypotenuse is equidistant from the three vertices. $AM = AB = CM = \sqrt{13}$

21. Use a general case of $\triangle ABC$ to prove that the midpoint of the hypotenuse of a right triangle is equidistant from the three vertices.

y
B(0, 4)
A(0, 0)
C(6, 0)
x

Use quadrilateral *ABCD* for Exercises 22–24.

22. Find the coordinates of the midpoints of the four sides.

23. Connect the midpoints to form a quadrilateral. What type of quadrilateral is formed? Justify your answer.

24. Generalize the results of Exercise 23 for quadrilateral *ABCD* with $A(0, 0)$, $B(2p, 2q)$, $C(2r, 2s)$, and $D(2t, 0)$.

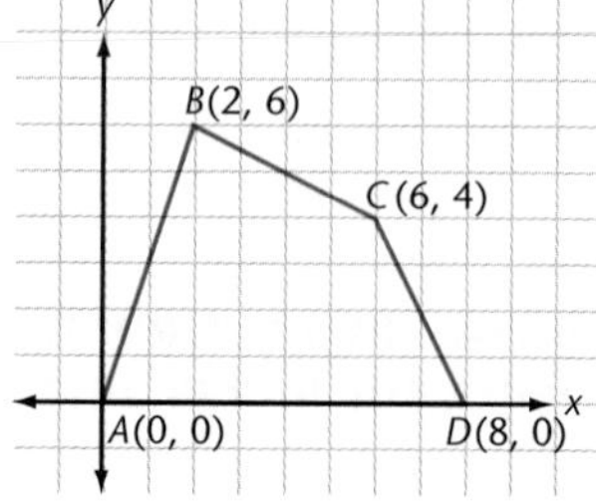

Look Back

Find the volume of each prism. [Lesson 7.2]

25. $B = 6 \text{ cm}^2$, $h = 5$ cm 30 cm³ **26.** $B = 12 \text{ cm}^2$, $h = 17$ cm 204 cm³

27. Find the volume of a semicircular cylinder formed by cutting a cylinder in half on its diameter. The diameter is 5 in. and the height of the cylinder is 14 in. **[Lesson 7.4]** 137.44 in.³

Find the surface area of each right cone. [Lesson 7.5]

28.

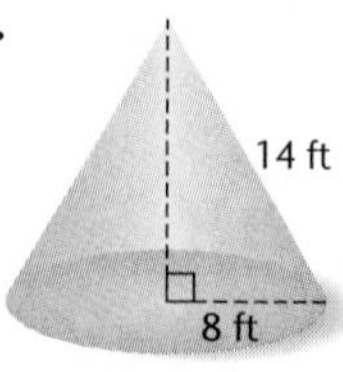

552.92 ft²

29.

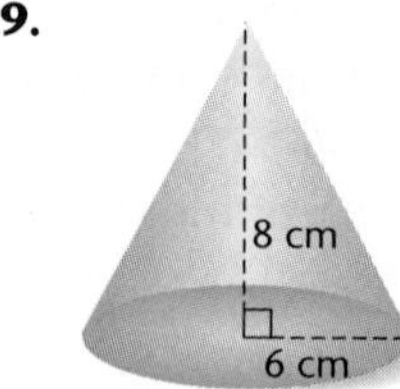

301.59 cm²

Look Beyond

30. Use coordinate geometry to prove that an angle inscribed in a semicircle is 90°.

31. Using three-dimensional coordinate geometry, show a relationship among the diagonals of the cube.

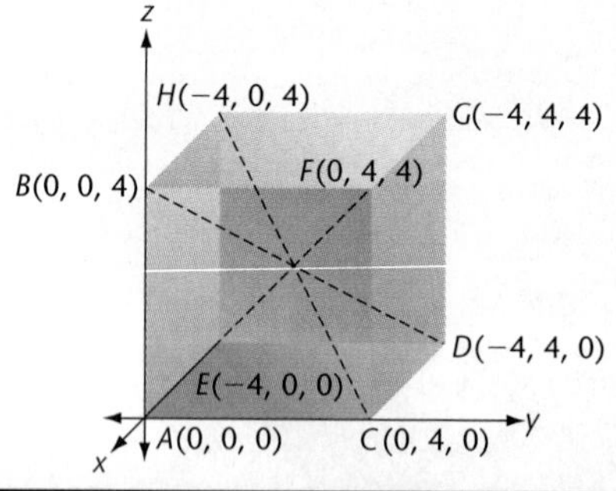

Look Beyond

Exercises 30 and 31 extend the ideas of coordinate proof to include three-dimensional coordinates. Have students compare their proofs in groups.

31. $AG = BD = CH = \sqrt{48}$; The diagonals of a cube are congruent; mdpnt of $\overline{AG} = (2, 2, 2)$; mdpnt of $\overline{BD} = (2, 2, 2)$; mdpnt of $\overline{CH} = (2, 2, 2)$. The diagonals of a cube all bisect each other.

Two Famous Theorems

Two of the most famous theorems in mathematics come to us from classical times. Both of them involve indirect proof, which you studied in Lesson 4 of this chapter. Both of them also involve the following theorem, which is known as the "Fundamental Theorem of Number Theory." It is not difficult to prove, but here it will be assumed.

Every number has a unique prime factorization.

The Infinity of the Primes

Euclid's proof that there are infinitely many primes has captured the imagination of people over the ages. To help you understand the proof, which is actually very brief, a numerical example is given first.

The proof will assume that there is only a finite number of primes. Suppose, for example, that there are only 5 primes, so that 11 is the largest prime number. Then the list of all the prime numbers would read:

2, 3, 5, 7, 11.

Now form a new number m by multiplying the prime numbers together and adding 1.

$$m = (2)(3)(5)(7)(11) + 1 = 2311$$

The new number m must be composite, according to the assumption, which states that 11 is the largest prime number. Therefore, m must be factorable into a combination of prime numbers, which by the assumption range from 2 to 11. But none of the numbers in the list of primes will divide the number evenly, because you will always get a remainder of 1.

Focus

Students explore two famous number theory proofs.

Motivate

Review prime factorization with your students. Remind them that a number is either prime or has a unique prime factorization. Review what rational and irrational means in mathematics. Ask students what they think is meant by the word "incommensurability."

Cooperative Learning

Each of the activities can be individually assigned to a cooperative group. Have groups work on their activity as a long-term project and present their results to class.

Discuss

Students have probably taken the number system for granted. Lead a discussion contrasting mathematics as an invented discipline with mathematics as discovered fact.

The Infinity of the Primes

1. The number $(p_1)(p_2)(p_3)\ldots(p_n)$ is exactly divisible by any one of the primes in the list. If 1 is added to the number, then 1 will be left over in the division.
2. If $(p_1)(p_2)(p_3)\ldots(p_n) - 1$ is divided by any one of the primes in the list, say p_x, then there will be a remainder of $p_x - 1$.
3. The **Goldbach Conjecture** states that any even number greater than 4 is the sum of two *odd* primes (i.e., not including 2). Thus 12 = 7 + 5; 16 = 7 + 9; 32 = 13 + 19; etc. The so-called "weak form" of the conjecture states that any odd number greater than 7 is the sum of three odd primes. Thus 11 = 5 + 3 + 3, etc.

   Twin primes are primes that differ by exactly 2. Examples are 11 and 13, 17 and 19, etc. The **Twin Primes Conjecture** states that there are infinitely many such pairs of primes.

For example, $5\overline{)2311}$ = 462 r. 1

But if 2311 is not divisible by one of the primes in the list, then it must be divisible by some other prime greater than 11, which contradicts the assumption that 11 is the largest prime.

Once you understand the example, you should be able to follow the generalization of it, which is given below.

Prove: There are infinitely many primes.
Proof: Assume that there are only finitely many primes, say, n of them. Then the list of the primes would be

$$p_1, p_2, p_3, \ldots p_n.$$

Form a new number $m = (p_1)(p_2)(p_3) \cdots (p_n) + 1$.
By assumption, m must be composite. But m is not divisible by any of the numbers in the list of primes, so there must be a larger prime than p, which contradicts the assumption. Thus there are infinitely many primes. ❖

The "Incommensurability" of the Square Root of 2

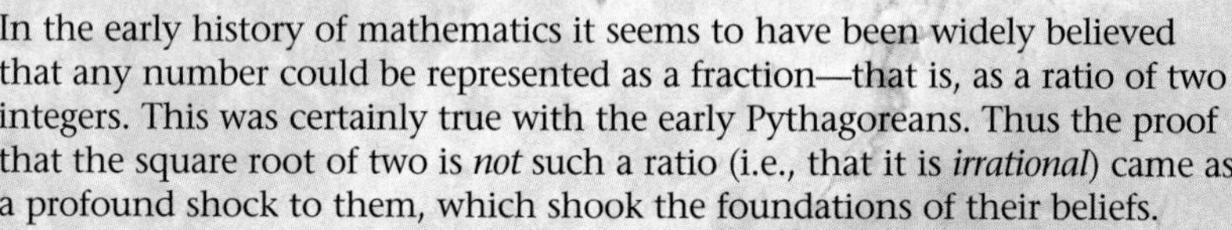

In the early history of mathematics it seems to have been widely believed that any number could be represented as a fraction—that is, as a ratio of two integers. This was certainly true with the early Pythagoreans. Thus the proof that the square root of two is *not* such a ratio (i.e., that it is *irrational*) came as a profound shock to them, which shook the foundations of their beliefs.

Before studying the theorem, you will need to know three simple results from number theory. *The following three theorems use the prime factorization theorem.*

Prove: The square of an even number is even.
Proof: If a number is even, then it has 2 as one of its prime factors. Otherwise, it would not be divisible by 2. (This is because 2 is the only even prime number.) When the number is squared, then, the number 2 will appear at least twice in the prime factorization of the result. Therefore, the result is divisible by 2.

Prove: The square of an even number is divisible by 4.
Proof: The proof is left as an exercise for the reader.

Prove: The square of an odd number is odd.
Proof: The proof is left as an exercise for the reader.

You are now ready to tackle the "incommensurability" theorem.

Prove: The square root of 2 cannot be written as a ratio of two integers.
Proof: Assume that the square root of 2 can be represented as the ratio of two numbers—that is, as a fraction. If this fraction is not in lowest terms then there is a fraction in lowest terms, to which it can be reduced. Thus the assumption can be written in the following way, which is

Chapter 12 Peoject 686

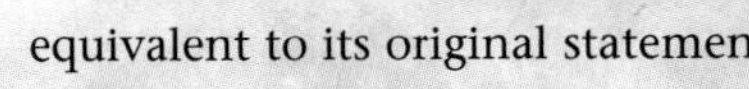
equivalent to its original statement:

> *Assume that the square root of 2 can be represented as a fraction in lowest terms.*

Let p be the numerator and q the denominator of the fraction which is assumed to exist. That is,

$$\frac{p}{q} = \sqrt{2} \quad (p \text{ and } q \text{ have no common factors.})$$

and so, $$\left(\frac{p}{q}\right)^2 = (\sqrt{2})^2 = 2$$

and $$p^2 = 2q^2.$$

The last equation implies that p^2 is an even number (why?). Therefore, p^2 is the square of an even number. This implies that p^2 is divisible by 4.

But if p^2 is divisible by 4, then q^2 must be an even number (why?). Therefore, q is even, as well.

But if p and q are both even, then they must have a common factor of 2, which contradicts the original assumption. Therefore, the assumption must be false, and so the theorem has been proven. ❖

Activities

The Infinity of the Primes

1. Repeat the numerical example in the proof using different numbers of primes. Explain why you always get a remainder of 1 when you divide m by one (or more) of the primes in your list.
2. If you subtract 1 instead of adding 1 to obtain the number m in the proof, how would the proof be affected?
3. Do some research on the number theory and find some of the unproven 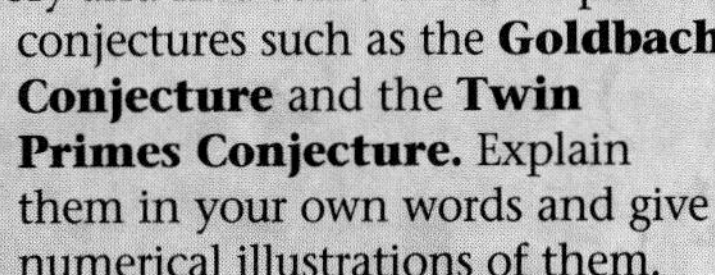conjectures such as the **Goldbach Conjecture** and the **Twin Primes Conjecture.** Explain them in your own words and give numerical illustrations of them.

The "Incommensurability" of the Square Root of 2

1. Prove that the square of an even number is divisible by 4.
2. Prove that the square of an odd number is odd.
3. In the incommensurability proof, there are two points at which the reader is asked to explain why a result follows. Explain why in your own words.

The "Incommensurability" of the Square Root of 2

1. If a number is even, then it has 2 as one of its prime factors. When the number is squared, 2 will appear at least twice in the prime factorization of the result. Since (2)(2) = 4, the result is divisible by 4.
2. If a number is odd, then 2 is not one if its prime factors. When the number is squared, 2 will not appear in the prime factorization of the result. Therefore, the result is not divisible by 2 and thus is odd.
3. p^2 is even, because 2 is at least one of its prime factors (as the equation indicates).

   Since p^2 is divisible by 4, 4 is one of its factors (not prime, of course). Thus $p^2 = 4j$ for some integer j. So,

   $$p^2 = 2q^2$$
   $$4j = 2q^2 \quad \text{(Subst.)}$$
   $$2j = q^2 \quad \text{(Division)}$$

   Therefore q^2 is even (because it has 2 as one of its factors).

Chapter 12 Review

Vocabulary

Key Skills and Exercises

Lesson 12.1

➤ Key Skills

Write a valid conclusion based on premises.

We have two premises:

If the Memorial Day parade passes the house, then the house is on Main Street.

The house is not on Main Street.

A valid conclusion would be: *Therefore, the Memorial Day parade does not pass the house.*

This argument is in the form: *If p then q; not q; therefore, not p.*

Spot false arguments.

What is wrong with the following argument?

If the Memorial Day parade passes the house, then the house is on Main Street.

The Memorial Day parade does not pass the house.

Therefore, the house is not on Main Street.

This is a false argument known as denying the antecedent. This fallacy has the form: *If p then q; not p; therefore, not q.*

➤ Exercises

1. Write a valid conclusion for the following premises:

If groms are plamous, then they are rute.
Merts are plamous groms.

2. Is the following argument valid or invalid? Explain.

Some butterflies migrate.
The monarch is a butterfly.
Therefore, the monarch migrates.

1. Merts are rute.
2. The argument is invalid. The premises are true but the conclusion is false.

Lesson 12.2

➤ Key Skills

Write a conjunction, disjunction, and negation.

You have two statements: (*p*) *We eat muffins*, and (*q*) *We have breakfast at the diner*. The conjunction of these two statements is *p* AND *q*:

We eat muffins, and we have breakfast at the diner.

The truth table for this conjunction is shown above right.

p	*q*	*p* AND *q*
T	T	T
T	F	F
F	T	F
F	F	F

For statements *p* and *q* above, the disjunction is *p* OR *q*:

We eat muffins or we have breakfast at the diner.

The truth table for this disjunction is shown below right.

p	*q*	*p* OR *q*
T	T	T
T	F	T
F	T	T
F	F	F

The negations of the statements above are (~*p*) *We do not eat muffins*, and (~*q*) *We do not have breakfast at the diner.*

➤ Exercises

For Exercises 3-5, create two statements and follow the directions.

3. Write a conjunction.
4. Write a disjunction.
5. Write a negation of your conjunction in Exercise 3. Give the truth table for the negation.

Lesson 12.3

➤ Key Skills

Write the converse, inverse, and contrapositive of a statement.

Statement: If Socks is purring, then he is happy.
Converse: If Socks is happy, then he is purring.

Statement: If Socks is purring, then he is happy.
Inverse: If Socks is not purring, then he is not happy.

Statement: If Socks is purring, then he is happy.
Contrapositive: If Socks is not happy, then he is not purring.

➤ Exercises

In Exercises 6–8, use the statement "If a quadrilateral is a trapezoid, then it has two parallel sides." Is each new statement true or false.

6. Write the converse.
7. Write the inverse.
8. Write the contrapositive.

Lesson 12.4

➤ Key Skills

Use indirect reasoning in a proof.

Prove indirectly that $\overline{SQ}$ and $\overline{PT}$ do not bisect each other inside $\triangle RPQ$.

Assume that the segments intersect each other. They are diagonals of *PQTS*, and if diagonals of a quadrilateral bisect each other, then the quadrilateral is a parallelogram. If *PQTS* is a parallelogram, $\overline{PS}$ is parallel to $\overline{QT}$. Thus $\overline{PQ}$ is a transversal, and $\angle RPQ$ and $\angle RQP$ are supplementary angles. That means $m\angle RPQ + m\angle RQP = 180°$. However, $\triangle RPQ$ has three nonzero angles that sum to 180°. This contradicts $m\angle RPQ + m\angle RQP = 180°$, indirectly proving that the segments do not bisect each other.

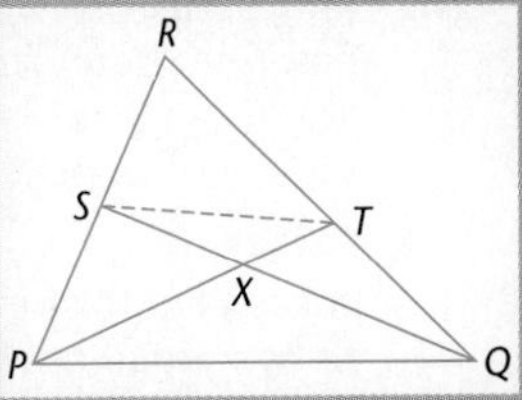

Chapter Review

3. Check student's answers. Statements should be joined by and.

4. Check student's answers. Statements should be joined by or.

5. Negation should be in the form ~p OR ~q.

p	q	~p	~q	~p OR ~q
T	T	F	F	F
T	F	F	T	T
F	T	T	F	T
F	F	T	T	T

~(p AND q)

p	q	p AND q	~(p AND q)
T	T	T	F
T	F	F	T
F	T	F	T
F	F	F	T

6. If a quadrilateral has two parallel sides, then it is a trapezoid. True.

7. If a quadrilateral is not a trapezoid, then it doesn't have two parallel sides. True.

8. If a quadrilateral doesn't have two parallel sides, then it is not a trapezoid. True.

➤ Exercises

9. Prove by indirect reasoning that the bisector of an angle of a scalene triangle is not perpendicular to the opposite side.

Lesson 12.5

➤ Key Skills

Construct input-output tables.

Create an input-output table for this network.

p	q	(NOT p) AND q
1	1	0
1	0	0
0	1	1
0	0	0

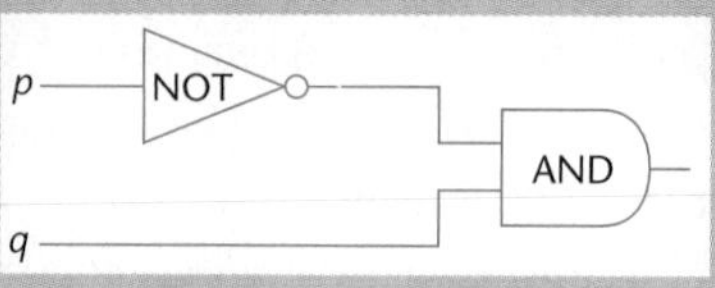

Write a logical expression for a network.

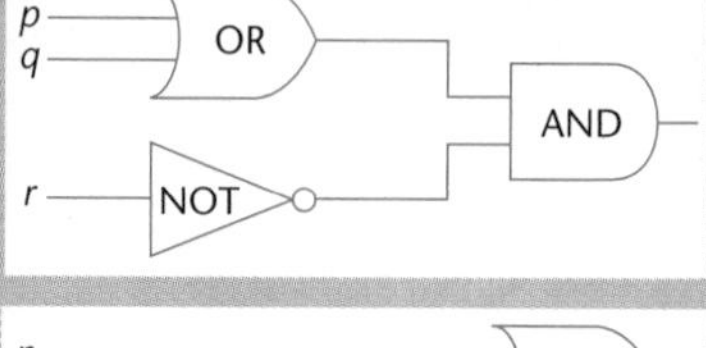

We can create a logical expression that represents the network above. Read from left to right, top to bottom. The left-hand terms, top to bottom, are (p OR q), (NOT r). The right-hand term combines these: OR q) AND (NOT r).

➤ Exercises

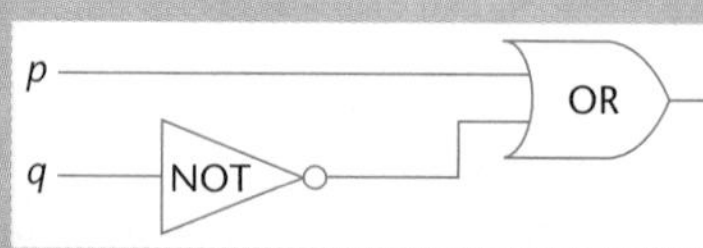

10. Write a logical expression for the network.

11. Construct an input-output table for the network.

Lesson 12.6

➤ Key Skills

Use coordinate geometry in proofs.

Prove that the angle inscribed in the semicircle measures 90°.

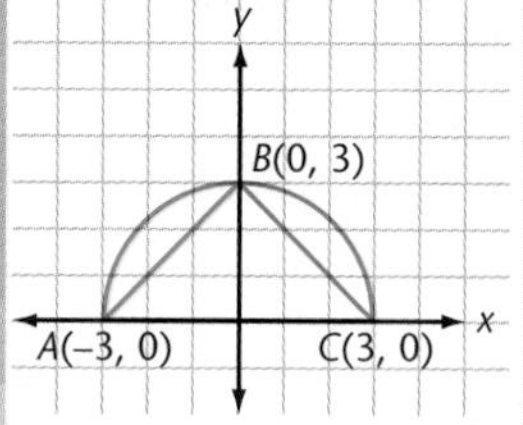

Use the coordinates of the endpoints to find the slopes of $\overline{AB}$ and $\overline{BC}$. If the slopes of $\overline{AB}$ and $\overline{BC}$ are negative reciprocals, the lines are perpendicular and $\angle ABC$ is a right angle.

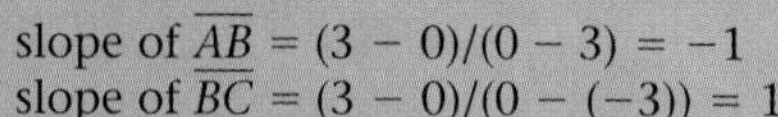

slope of $\overline{AB} = (3 - 0)/(0 - 3) = -1$
slope of $\overline{BC} = (3 - 0)/(0 - (-3)) = 1$

The slopes are negative reciprocals, and so inscribed $\angle ABC$ is a right angle.

➤ Exercises

12. Using coordinate geometry, prove that if the diagonals of a parallelogram are equal, then the figure is a rectangle.

Applications

13. Mario, his sister, his daughter, and his son are sprinters. Two facts describe these people: (1) The worst sprinter's twin and the best sprinter are of opposite sex. (2) The worst sprinter and the best sprinter are the same age. Which one of the four is the best sprinter?

14. **Entertainment** The cable TV signal goes into my VCR, which is connected to the TV set. I want to watch cable channel 24. Create a network diagram and a logical expression to describe what I must do before I can select and watch channel 24.

---

9. Given: $\triangle ABC$ is scalene; $\overline{AD}$ bisects $\angle CAB$

Prove: $\overline{AD}$ is not perpendicular to $\overline{BC}$

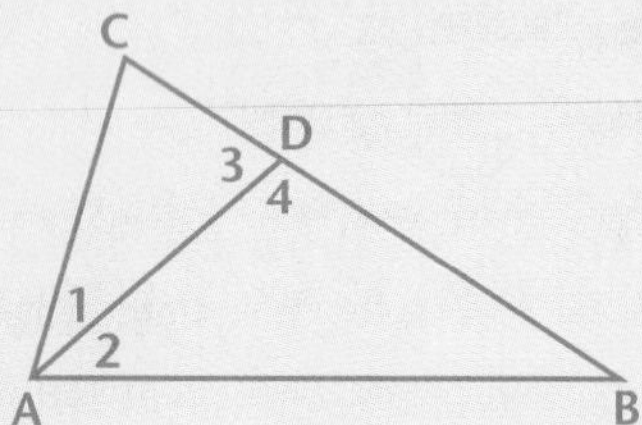

Statements: 1. $\triangle ABC$ is scalene; $\overline{AD}$ bisects $\angle CAB$, 2. Suppose $\overline{AD} \perp \overline{BC}$, 3. $\angle 1 \cong \angle 2$, 4. $\angle 3 \cong \angle 4$, 5. $\overline{AD} \cong \overline{AD}$, 6. $\triangle ADC \cong \triangle ADB$, 7. $\overline{AC} \cong \overline{AB}$

Reasons: 1. Given, 2. assuming the opposite, 3. Angle bisector definition, 4. Perpendicular segments form right angles, 5. Reflexive, 6. ASA, 7. CPCTC. The last step contradicts the given fact that $\triangle ABC$ is scalene. Therefore, $\overline{AD}$ is not perpendicular to $\overline{BC}$.

10. p OR (NOT q)

11.

p	q	NOT q	p OR (NOT q)
1	1	0	1
1	0	1	1
0	1	0	0
0	0	1	1

12.

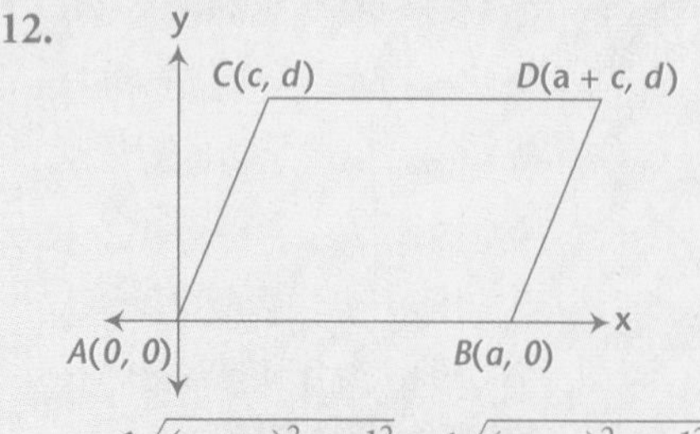

$$\sqrt{(a+c)^2 + d^2} = \sqrt{(c-a)^2 + d^2}$$

$$(a+c)^2 + d^2 = (c-a)^2 + d^2$$

$$a^2 + 2ac + c^2 = c^2 - 2ac + a^2$$

$$4ac = 0$$

Since $a > 0$, $c = 0$

Therefore $m\angle A = 90°$ because C is on the y-axis. And so, $\square ABCD$ is a rectangle.

13. Mario's sister is the best sprinter.

14. Both the TV AND VCR must be on

Chapter 12 Assessment

1. Write a valid conclusion for these premises:
   If it was raining when I arrived, my umbrella is wet.
   My umbrella is dry.

2. Write a disjunction for these statements:
   Farah makes a shot.
   Richard blocks the shot.

3. Write a negation of your disjunction from Exercise 2. Give the truth table for the negation.

In Exercises 4–6, use the statement "If an angle is inscribed in a semicircle, then it is a right angle."

4. Write the converse. Is the new statement true or false?
5. Write the inverse. Is the new statement true or false?
6. Write the contrapositive. Is the new statement true or false?

7. Prove by indirect reasoning that a triangle can have at most one obtuse angle.

In Exercises 8–9, refer to the network below.

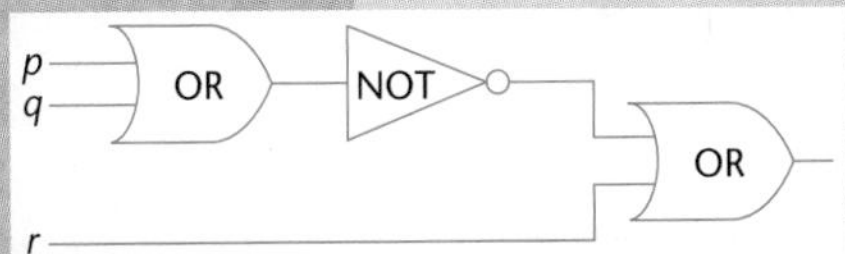

8. Write a logical expression for the network.
9. Construct an input-output table for the network.

10. Prove that the triangle below is isosceles.

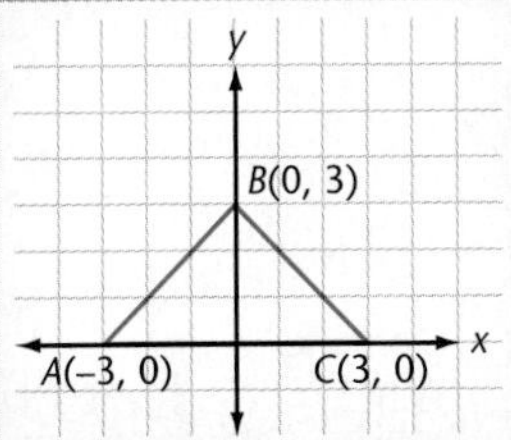

1. It was not raining when I arrived.

2. Farah makes a shot or Richard blocks the shot.

3. Farah misses a shot and Richard doesn't block the shot.

p	q	p OR q	~(p OR q)
T	T	T	F
T	F	T	F
F	T	T	F
F	F	F	T

p	q	~p	~q	~p AND ~q
T	T	F	F	F
T	F	F	T	F
F	T	T	F	F
F	F	T	T	T

4. If an angle is a right angle, then it can be inscribed in a semicircle. True.

5. If an angle can't be inscribed in a semicircle, then it is not a right angle. True.

6. If an angle is not a right angle, then it can't be inscribed in a semicircle. True.

7. Suppose a triangle has two obtuse angles. Since an obtuse angle measures more than 90°, the sum of the measures of two obtuse angles must be greater than 180°. Thus the sum of the measures of the angles of the triangle is greater than 180°. This contradicts the theorem that the sum of the measures of a triangle is 180° (in Euclidean Geometry). Therefore, a triangle can't have two obtuse angles. A similar argument applies for three obtuse angles. So, a triangle can have at most one obtuse angle.

8. NOT(p OR q) OR r

9. (see page 692 for answer)

10. $AB = \sqrt{(0 - -3)^2 + (3 - 0)^2} = \sqrt{18}$

    $CB = \sqrt{(0 - 3)^2 + (3 - 0)^2} = \sqrt{18}$

    Since $AB = CB$, $\triangle ABC$ is isosceles by definition.

College Entrance-Exam Practice

Multiple-Choice and Quantitative-Comparison Samples

The first half of the Cumulative Assessment contains two types of items found on standardized tests—multiple-choice questions and quantitative-comparison questions. Quantitative-comparison items emphasize the concepts of equality, inequality, and estimation.

Free-Response Grid Samples

The second half of the Cumulative Assessment is a free-response section. A portion of this part of the Cumulative Assessment consists of student-produced response items commonly found on college entrance exams. These questions require the use of machine-scored answer grids. You may wish to have students practice answering these items in preparation for standardized tests.

Sample answer grid masters are available in the *Chapter Teaching Resource Booklets.*

Chapters 1–12 Cumulative Assessment

College Entrance Exam Practice

Quantitative Comparison Exercises 1–3 consist of two quantities, one in Column A and one in Column B, which you are to compare as follows:

A. The quantity in Column A is greater.
B. The quantity in Column B is greater.
C. The two quantities are equal.
D. The relationship cannot be determined from the information given.

<table>
<tr><th></th><th>Column A</th><th>Column B</th><th>Answers</th></tr>
<tr><td>1.</td><td colspan="2">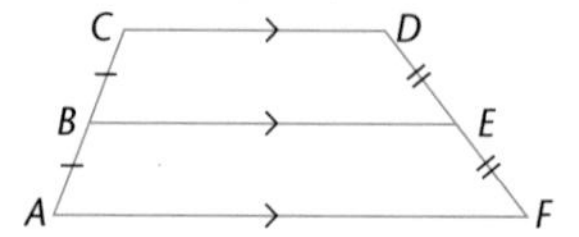

BE | AF/2</td><td>(A) (B) (C) (D)
[Lesson 3.7]</td></tr>
<tr><td>2.</td><td colspan="2">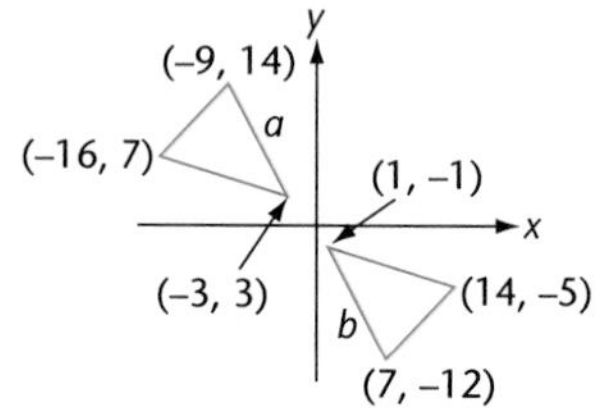

Side A | Side B</td><td>(A) (B) (C) (D)
[Lesson 1.7]</td></tr>
<tr><td>3.</td><td colspan="2">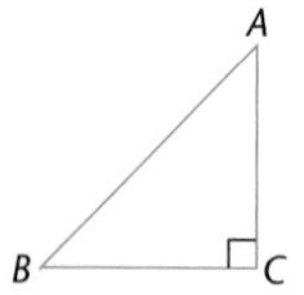

AB + BC | AC + BC</td><td>(A) (B) (C) (D)
[Lesson 5.4]</td></tr>
</table>

4. Which is a definition? **[Lesson 2.3]**
a. A monarch is an orange butterfly.
b. A line segment is the shortest path between two points.
c. A pile of loose rubble is a hazard.
d. A rhombus is a parallelogram.

5. Line $\overleftrightarrow{AF}$ contains points $A(13, 15)$ and $F(9, 20)$. Which points define a line parallel to $\overleftrightarrow{AF}$? **[Lesson 3.8]**
a. (0, 0), (25, 16) **b.** (15, 13), (20, 9)
c. (4, −6), (8, −1) **d.** (4, 3), (12, −13)

1. A 2. C 3. A
4. b 5. c

9. (page 691)

p	q	r	p OR q	~(p OR q)	~(p OR q) OR r
1	1	1	1	0	1
1	1	0	1	0	0
1	0	1	1	0	1
1	0	0	1	0	0
0	1	1	1	0	1
0	1	0	1	0	0
0	0	1	0	1	1
0	0	0	0	1	1

6. Which value is largest in $\triangle ABC$? **[Lessons 5.5, 10.2]**
a. $\cos A$ **b.** $\sin B$ **c.** $\cos B$ **d.** $\sin C$

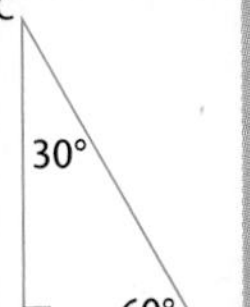
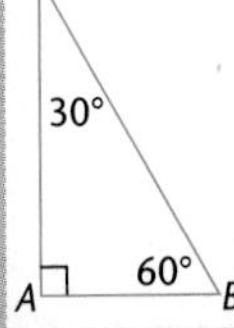

7. Identify the correct expression for length EC. **[Lesson 9.5]**

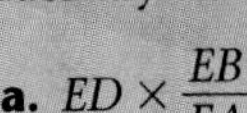

a. $ED \times \dfrac{EB}{EA}$ **b.** $EB \times \dfrac{EA}{ED}$

c. $EB \times \dfrac{CD}{AB}$ **d.** $EB \times \dfrac{AB}{CD}$

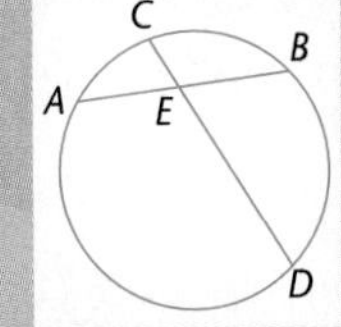

8. Which statement is true in a hyperbolic geometry? **[Lesson 11.5]**
a. No lines are parallel.
b. Two-sided polygons exist.
c. The sum of the angles of a triangle is less than 180°.
d. The shortest distance between two points is a great circle.

In Exercises 9–10, refer to the triangle at right. [Lesson 5.6]
9. Make a conjecture about $\overline{DE}$.
10. Use coordinate geometry to prove your conjecture from Exercise 9.

In Exercises 11–13, use these statements: [Lesson 2.2]
a. If the fog has lifted, the boat can leave the harbor.
b. The boat is leaving the harbor.
11. Write a valid conclusion.
12. Write a contrapositive of statement (a).
13. Project points A, B, and C onto line n. **[Lesson 11.6]**
14. Write an equation for a circle with radius 6 and center at (11, −13). **[Lesson 9.6]**

Free-Response Grid Exercises 15–17 may be answered using a free-response grid commonly used by standardized test services.

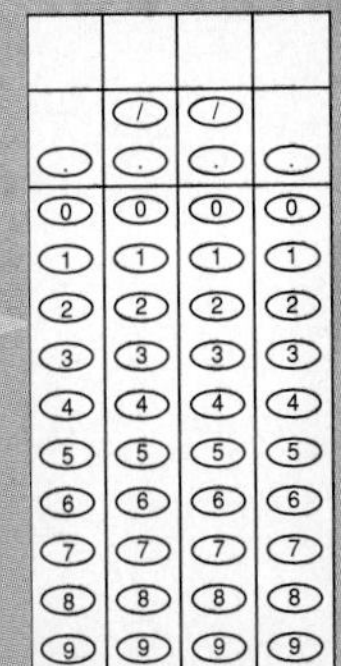

15. I parachute into a taxicab circle of radius 6 that includes the dangerously busy intersection of Fifth and Broadway. If I have an equal chance of landing at any intersection within or on the circle, what are the chances that I'll land at Fifth and Broadway? **[Lessons 5.7, 11.2]**
16. What is the length of segment $\overline{AB}$? **[Lessons 10.5, 10.6]**
17. How long is $\overline{FG}$? **[Lesson 5.6]**

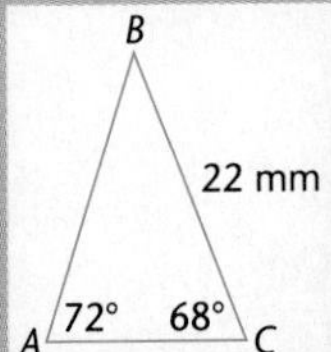

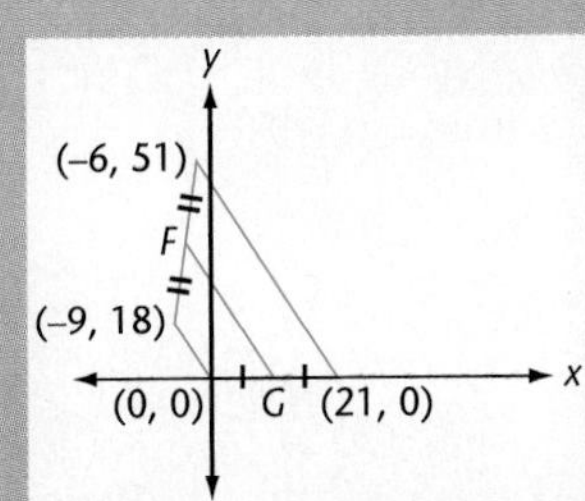

14. $(x-11)^2 + (y+13)^2 = 36$
15. $6.1^2 = (x-11)^2 + (y+13)^2$
16. $\dfrac{1}{85}$
17. AB ≈ 21.45 mm
18. FG ≈ 38.91

6. b **7.** b **8.** c
9. $\overline{DE} \parallel \overline{AB}$; $DE = \dfrac{1}{2}AB$
10. slope of $\overline{AB} = 0$; slope of $\overline{DE} = 0$
Since their slopes are equal, $\overline{DE} \parallel \overline{AB}$.
$AB = 9$; $DE = 4.5$: $DE = \dfrac{1}{2}AB$
11. The fog has not lifted.
12. If the boat can not leave the harbor, then the fog has not lifted.
13.

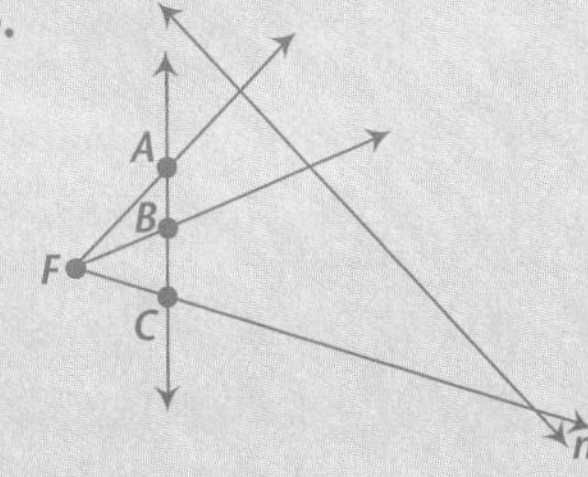

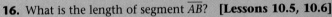

Info Bank

Tables

Table of Squares, Cubes, Square and Cube Roots

No.	Squares	Cubes	Square Roots	Cube Roots	No.	Squares	Cubes	Square Roots	Cube Roots
1	1	1	1.000	1.000	51	2,601	132,651	7.141	3.708
2	4	8	1.414	1.260	52	2,704	140,608	7.211	3.733
3	9	27	1.732	1.442	53	2,809	148,877	7.280	3.756
4	16	64	2.000	1.587	54	2,916	157,464	7.348	3.780
5	25	125	2.236	1.710	55	3,025	166,375	7.416	3.803
6	36	216	2.449	1.817	56	3,136	175,616	7.483	3.826
7	49	343	2.646	1.913	57	3,249	185,193	7.550	3.849
8	64	512	2.828	2.000	58	3,364	195,112	7.616	3.871
9	81	729	3.000	2.080	59	3,481	205,379	7.681	3.893
10	100	1,000	3.162	2.154	60	3,600	216,000	7.746	3.915
11	121	1,331	3.317	2.224	61	3,721	226,981	7.810	3.936
12	144	1,728	3.464	2.289	62	3,844	238,328	7.874	3.958
13	169	2,197	3.606	2.351	63	3,969	250,047	7.937	3.979
14	196	2,744	3.742	2.410	64	4,096	262,144	8.000	4.000
15	225	3,375	3.873	2.466	65	4,225	274,625	8.062	4.021
16	256	4,096	4.000	2.520	66	4,356	287,496	8.124	4.041
17	289	4,913	4.125	2.571	67	4,489	300,763	8.185	4.062
18	324	5,832	4.243	2.621	68	4,624	314,432	8.246	4.082
19	361	6,859	4.359	2.668	69	4,761	328,509	8.307	4.102
20	400	8,000	4.472	2.714	70	4,900	343,000	8.367	4.121
21	441	9,261	4.583	2.759	71	5,041	357,911	8.426	4.141
22	484	10,648	4.690	2.802	72	5,184	373,248	8.485	4.160
23	529	12,167	4.796	2.844	73	5,329	389,017	8.544	4.179
24	576	13,824	4.899	2.884	74	5,476	405,224	8.602	4.198
25	625	15,625	5.000	2.924	75	5,625	421,875	8.660	4.217
26	676	17,576	5.099	2.962	76	5,776	483,976	8.718	4.236
27	729	19,683	5.196	3.000	77	5,929	456,533	8.775	4.254
28	784	21,952	5.292	3.037	78	6,084	474,552	8.832	4.273
29	841	24,389	5.385	3.072	79	6,241	493,039	8.888	4.291
30	900	27,000	5.477	3.107	80	6,400	512,000	8.944	4.309
31	961	29,791	5.568	3.141	81	6,561	531,441	9.000	4.327
32	1,024	32,768	5.657	3.175	82	6,724	551,368	9.055	4.344
33	1,089	35,937	5.745	3.208	83	6,889	571,787	9.110	4.362
34	1,156	39,304	5.831	3.240	84	7,056	592,704	9.165	4.380
35	1,225	42,875	5.916	3.271	85	7,225	614,125	9.220	4.397
36	1,296	46,656	6.000	3.302	86	7,396	636,056	9.274	4.414
37	1,369	50,653	6.083	3.332	87	7,569	658,503	9.327	4.431
38	1,444	54,872	6.164	3.362	88	7,744	681,472	9.381	4.448
39	1,521	59,319	6.245	3.391	89	7,921	704,969	9.434	4.465
40	1,600	64,000	6.325	3.420	90	8,100	729,000	9.487	4.481
41	1,681	68,921	6.403	3.448	91	8,281	753,571	9.539	4.498
42	1,764	74,088	6.481	3.476	92	8,464	778,688	9.592	4.514
43	1,849	79,507	6.557	3.503	93	8,649	804,357	9.644	4.531
44	1,936	85,184	6.633	3.350	94	8,836	830,584	9.695	4.547
44	2,025	91,125	6.708	3.557	95	9,025	857,375	9.747	4.563
46	2,116	97,336	6.782	3.583	96	9,216	884,736	9.798	4.579
47	2,209	103,823	6.856	3.609	97	9,409	912,673	9.849	4.595
48	2,304	110,592	6.928	3.634	98	9,604	941,192	9.899	4.610
49	2,401	117,649	7.000	3.659	99	9,801	970,299	9.950	4.626
50	2,500	125,000	7.071	3.684	100	10,000	1,000,000	10.000	4.642

Table of Trigonometric Ratios

Angle	sin	cos	tan
0°	0.0000	1.0000	0.0000
1°	0.0175	0.9998	0.0175
2°	0.0349	0.9994	0.0349
3°	0.0523	0.9986	0.0524
4°	0.0698	0.9976	0.0699
5°	0.0872	0.9962	0.0875
6°	0.1045	0.9945	0.1051
7°	0.1219	0.9925	0.1228
8°	0.1392	0.9903	0.1405
9°	0.1564	0.9877	0.1584
10°	0.1736	0.9848	0.1763
11°	0.1903	0.9816	0.1944
12°	0.2079	0.9781	0.2126
13°	0.2250	0.9744	0.2309
14°	0.2419	0.9703	0.2493
15°	0.2588	0.9659	0.2679
16°	0.2756	0.9613	0.2867
17°	0.2924	0.9563	0.3057
18°	0.3090	0.9511	0.3249
19°	0.3526	0.9455	0.3443
20°	0.3420	0.9397	0.3640
21°	0.3584	0.9336	0.3839
22°	0.3746	0.9272	0.4040
23°	0.3907	0.9205	0.4245
24°	0.4067	0.9135	0.4452
25°	0.3420	0.9063	0.4663
26°	0.4384	0.8988	0.4877
27°	0.4540	0.8910	0.5095
28°	0.4695	0.8829	0.5317
29°	0.4848	0.8746	0.5543
30°	0.5000	0.8660	0.5774
31°	0.5150	0.8572	0.6009
32°	0.5299	0.8480	0.6249
33°	0.5446	0.8387	0.6494
34°	0.5592	0.8290	0.6745
35°	0.5736	0.8192	0.7002
36°	0.5878	0.8090	0.7265
37°	0.6018	0.0786	0.7536
38°	0.6157	0.7880	0.7813
39°	0.6293	0.7771	0.8098
40°	0.6428	0.7660	0.8391
41°	0.6561	0.7547	0.8693
42°	0.6691	0.7431	0.9004
43°	0.6820	0.7314	0.9325
44°	0.6947	0.7193	0.9657
45°	0.7071	0.7171	1.0000
45°	0.7071	0.7071	1.0000
46°	0.7193	0.0647	1.0355
47°	0.7314	0.6820	1.0724
48°	0.7431	0.6691	1.1106
49°	0.7547	0.6561	1.1504
50°	0.7660	0.6428	1.1918
51°	0.7771	0.6293	1.2349
52°	0.7880	0.6157	1.2799
53°	0.7986	0.6018	1.3270
54°	0.8090	0.5878	1.3764
55°	0.8192	0.5736	1.4281
56°	0.8290	0.5592	1.4826
57°	0.8387	0.5446	1.5399
58°	0.8480	0.5299	1.6003
59°	0.8572	0.5150	1.6643
60°	0.8660	0.5000	1.7321
61°	0.8746	0.4848	1.8040
62°	0.8829	0.4695	1.8807
63°	0.8910	0.4540	1.9626
64°	0.8988	0.4384	2.0503
65°	0.9063	0.4226	2.1445
66°	0.9135	0.4067	2.2460
67°	0.9205	0.3907	2.3559
68°	0.9272	0.3746	2.4751
69°	0.9336	0.3584	2.6051
70°	0.9397	0.3420	2.7475
71°	0.9455	0.3256	2.9042
72°	0.9511	0.3090	3.0777
73°	0.9563	0.2924	3.2709
74°	0.9613	0.2756	3.4874
75°	0.9659	0.2588	3.7321
76°	0.9703	0.2419	4.0108
77°	0.9744	0.2250	4.3315
78°	0.9781	0.2079	4.7046
79°	0.9816	0.1908	5.1446
80°	0.9848	0.1736	5.6713
81°	0.9877	0.1564	6.3138
82°	0.9903	0.1392	7.1154
83°	0.9925	0.1219	8.1443
84°	0.9945	0.1045	9.5144
85°	0.9962	0.0872	11.4301
86°	0.9976	0.0698	14.3007
87°	0.9986	0.0523	19.0811
88°	0.9994	0.0349	28.6363
89°	0.9998	0.0175	57.2900
90°	1.0000	0.0000	∞

POSTULATES, THEOREMS, AND DEFINITIONS

Def 1.4.1 **Measure of Segment *AB*** Let A and B be points on a number line with coordinates a and b. Then the measure of segment AB, which is called the length, is $|a - b|$ or $|b - a|$. The measure of segment AB is written as mAB or simply AB. (30)

Post 1.4.2 **Segment Addition Postulate** On segment PQ, if R is between points P and Q, then $PR + PQ = PQ$. (32)

Def 1.5.1 **Measure of Angle** Suppose the vertex V of $\angle AVB$ is placed in the center point of a half-circle with coordinates from 0 to 180. Let a and b be the coordinates of the points where VA and VB cross the half-circle. Then the measure of $\angle AVB$, written as $m\angle AVB$ is $|a - b|$ or $|b - a|$. (38)

Post 1.5.2 **Angle Addition Postulate** If point S is in the interior of $\angle PQR$, then $m\angle PQS + m\angle SQR = m\angle PQR$. (39)

Def 1.5.3 **Special Angle Sums** If the sum of the measures of two angles is 90°, then the angles are **complementary**. If the sum of the measures of two angles is 180°, then the angles are **supplementary**. (39)

Post 2.2.1 **If-Then Transitive Property** Suppose you are given: If A then B and if B then C. You can conclude: If A then C. (76)

Post 2.4.1 **Postulate** Through any two points there is exactly one line, *or* two points determine exactly one line. (88)

Post 2.4.2 **Postulate** Through any three noncollinear points, there is exactly one plane, *or* three noncollinear points determine exactly one plane. (88)

Post 2.4.3 **Postulate** If two points are on a plane, then the line containing them is also on the plane. (88)

Post 2.4.4 **Postulate** The intersection of two lines is exactly one point. (88)

Post 2.4.5 **Postulate** The intersection of two planes is exactly one line. (88)

Thm 2.5.1 **Overlapping Segments Theorem** Given a segment with points A, B, C, and D arranged as shown, the following statements are true:

1. If $AB = CD$, then $AC = BD$
2. If $AC = BD$, then $AB = CD$

A B C D

(94)

Thm 2.5.2 **Overlapping Angles Theorem** Given four rays with common endpoint arranged as shown, the following statements are true:

1. If $m\angle AOB = m\angle COD$, then $m\angle AOC = m\angle BOD$
2. If $m\angle AOC = m\angle BOD$, then $m\angle AOB = m\angle COD$

A B C D O

(94)

Post 2.5.3 **Reflexive Property of Equality** For all real numbers a, $a = a$. (94)

Post 2.5.4 **Symmetric Property of Equality** For all real numbers a and b, if $a = b$, then $b = a$. (94)

Post 2.5.5 **Transitive Property of Equality** For all real numbers a, b, and c, if $a = b$ and $b = c$, then $a = c$. (94)

Thm 2.6.1 **Vertical Angles Theorem** All vertical angles have equal measure. (100)

Thm 2.6.2 **Reflection Through Parallel Lines** Reflection twice through a pair of parallel lines is equivalent to a translation by twice the distance between the lines in a direction perpendicular to the lines. (101)

Thm 2.6.3 **Reflection through Intersecting Lines** Reflection twice through a pair of intersecting lines is equivalent to a rotation about the intersection point. The rotation angle measure is double that of the angle formed by the lines. (102)

Def. 3.1.1 **Polygon** A polygon is a closed plane figure formed from three or more segments such that each segment intersects exactly two other segments, one at each endpoint. (116)

Def. 3.1.2 **Reflectional Symmetry** A plane figure has reflectional symmetry if and only if its reflection image through a line coincides with the original figure. The line is called an **axis of symmetry**. (117)

Def. 3.1.3 **Rotational Symmetry** A figure has rotational symmetry if and only if it has at least one rotation image that coincides with the original image. We say that a figure has a **rotation image of *n* degrees** if a rotation by n degrees about a fixed point results in an image that coincides with the original. Rotation images of 0° or multiples of 360° do not have rotational symmetry. (118)

Def. 3.3.1 **Transversals** A transversal is a line, ray, or segment that intersects two or more coplanar lines, rays, or segments, each at a different point. (132)

Post 3.3.2 **Corresponding Angles Postulate** If two lines cut by a transversal are parallel, then corresponding angles are congruent. (134)

Thm 3.3.3 **Alternate Interior Angles Theorem** If two lines cut by a transversal are parallel, then alternate interior angles are congruent. (135)

Thm 3.4.1 **Converse of the Corresponding Angles Theorem** If two lines are cut by a transversal in such a way that corresponding angles are congruent, then the two lines are parallel. (139)

Thm 3.4.2 **Theorem** If two lines are cut by a transversal in such a way that consecutive interior angles are supplementary, then the two lines are parallel. (140)

Thm 3.4.3 **Theorem** If two lines are perpendicular to the same line, the two lines are parallel to each other. (141)

Thm 3.4.4 **Theorem** If two lines are parallel to the same line, then the two lines are parallel to each other. (142)

Post 3.5.1 **The Parallel Postulate** Given a line and a point not on the line, there is one and only one line that contains the given point and is parallel to the given line. (143)

Thm 3.5.2 **Triangle Sum Theorem** The sum of the measures of the angles of a triangle is 180°. (144)

Thm 3.5.3 **Exterior Angle Theorem** The measure of an exterior angle of a triangle is equal to the sum of the remote interior angles. (147)

Thm 3.8.1 **Slope Formula** The slope of segment *PQ*, with endpoints $P(x_1, y_1)$ and $Q(x_2, y_2)$ is the following ratio:

$$\text{Slope} = \frac{y_2 - y_1}{x_2 - x_1}. \quad (161)$$

Thm 3.8.2 **Equal Slope Theorem** If two nonvertical lines are parallel, then they have the same slope. (161)

Thm 3.8.3 **Converse of the Equal Slope Theorem** If two nonvertical lines have the same slope, then they are parallel. (161)

Thm 3.8.4 **Slopes of Perpendicular Lines** If two nonvertical lines are perpendicular, then the product of their slopes is −1. (162)

Thm 3.8.5 **Converse of Slopes of Perpendicular Lines** If the product of the slopes of two nonvertical lines is −1, then the lines are perpendicular. (162)

Def 4.1.1 **Congruent Polygons** Two polygons are congruent if and only if there is a way of setting up a correspondence between their sides and angles, in order, so that

1. all pairs of corresponding angles are congruent, and
2. all pairs of corresponding sides are congruent. (176)

Post 4.3.1 **SSS(Side-Side-Side) Postulate** If three sides in one triangle are congruent to three sides in another triangle, then the triangles are congruent. (186)

Post 4.3.2 **SAS(Side-Angle-Side) Postulate** If two sides and the angle between them in one triangle are congruent to two sides and the angle between them in another triangle, then the triangles are congruent. (186)

Post 4.3.3 **ASA(Angle-Side-Angle) Postulate** If two angles and a side between them in one triangle are congruent to two angles and the side between them in another triangle, then the triangles are congruent. (186)

Thm 4.3.4 **AAS(Angle-Angle-Side) Theorem** If two angles and a side that is not between them in one triangle are congruent to the corresponding two angles and the side not between them in another triangle, then the triangles are congruent. (188)

Thm 4.3.5 **HL (Hypoteneuse Leg) Theorem** If the hypotenuse and a leg of a right triangle are congruent to the hypotenuse and the corresponding leg in another right triangle, then the two triangles are congruent. (189)

Thm 4.4.1 **Isosceles Triangle Theorem** If two sides of a triangle are congruent, then the angles opposite those sides are congruent. (196)

Thm 4.4.2 **Isosceles Triangle Theorem: Converse** If two angles of a triangle are congruent, then the sides opposite those angles are congruent. (196)

Cor 4.4.3 **Corollary** An equilateral triangle has three angles with measure 60°. (197)

Cor 4.4.4 **Corollary** The bisector of the vertex angle of an isosceles triangle is the perpendicular bisector of the base. (197)

Thm 4.4.5 **Theorem** The bisector of the vertex angle of an isosceles triangle divides the triangle into two congruent triangles. (197)

Thm 4.4.6 **Theorem** The bisector of the vertex angle of an isosceles triangle bisects the base. (198)

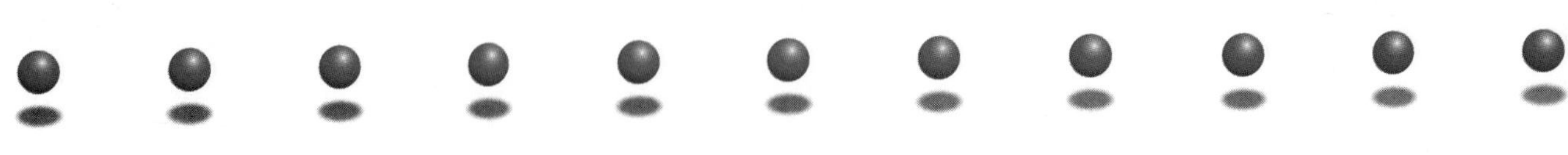

Thm 4.4.7 **Theorem** The opposite sides of a parallelogram are congruent. (198)

Thm 4.4.8 **Theorem** The opposite angles of a parallelogram are congruent. (198)

Thm 4.5.1 **Theorem** A diagonal of a parallelogram divides the parallelogram into two congruent triangles. (201)

Thm 4.5.2 **Theorem** The diagonals of a parallelogram bisect each other (204)

Thm 4.5.3 **Theorem** The diagonals and sides of a rhombus form four congruent triangles. (205)

Thm 4.5.4 **Theorem** The diagonals of a rhombus are perpendicular. (205)

Thm 4.5.5 **Theorem** The diagonals of a rectangle are congruent. (205)

Thm 4.5.6 **Theorem** The diagonals of a kite are perpendicular. (206)

Thm 4.6.1 **Theorem** If two pairs of opposite sides of a quadrilateral are congruent, then the quadriateral is a parallelogram. (210)

Thm 4.6.2 **The "Housebuilder" Rectangle Theorem** If the diagonals of a parallelogram are congruent, then the parallelogram is a rectangle. (211)

Thm 4.6.3 **Theorem** If one pair of opposite sides of a quadrilateral are parallel and congruent, then the quadrilateral is a parallelogram. (211)

Thm 4.6.4 **Theorem** If the diagonals of a quadrilateral bisect each other, then the quadrilateral is a parallelogram. (211)

Thm 4.6.5 **Theorem** If one angle of a parallelogram is a right angle, then the parallelogram is a rectangle. (211)

Thm 4.6.6 **Theorem** If one pair of adjacent sides of a parallelogram are congruent, then the quadrilateral is a rhombus. (211)

Thm 4.6.7 **Triangle Midsegment Theorem** If a segment joins the midponits of two sides of a triangle, then it is parallel to the third side and its length is one-half the length of the third side. (212)

Thm 4.7.1 **Congruent Radii Theorem** In the same circle, or in congruent circles, all radii are congruent. (213)

Post 4.9.1 **Converse of the Segment Addition Postulate ("Betweenness")** Given three points *P*, *Q*, and *R*, if $PQ + QR = PR$ then *Q* is between *P* and *R*. (228)

Post 4.9.2 **Triangle Inequality Postulate** The sum of the lengths of any two sides of a triangle is larger than the length of the other side. (229)

Post 5.1.1 **The Area of a Rectangle** The **area of a rectangle** with base *b* and height *h* is $A = bh$. (245)

Post 5.1.2 **Sum of Areas** If a figure is composed of nonoverlapping regions *A* and *B*, the area of the figure is the sum of the areas of regions *A* and *B*. (245)

Def 5.3.1 **Circle** A circle is the figure that consists of all the points on a plane that are the same distance *r* from a given point known as the center of the circle. The distance *r* is the radius of the circle and $d = 2r$ is the diameter. (261)

Thm 5.4.1 **"Pythagorean" Right-Triangle Theorem** For any right triangle, the square of the length of the hypotenuse is equal to the sum of the squares of the lengths of the legs. (269)

Thm 5.4.2 **Converse of the "Pythagorean" Right-Triangle Theorem** If the square of the length of one side of a triangle equals the sum of the squares of the lengths of the other two sides, then the triangle is a right triangle. (270)

Thm 5.5.1 **45-45-90 Right-Triangle Theorem** In any 45-45-90 right triangle, the length of the hypotenuse is $\sqrt{2}$ times the length of a leg. (276)

Thm 5.5.2 **30-60-90 Right Triangle Theorem** In any 30-60-90 right triangle, the length of the hypotenuse is two times the length of the shorter leg, and the longer leg is $\sqrt{3}$ times the length of the shorter leg. (277)

Thm 5.5.3 **Area of a Regular Polygon** The area of a regular polygon with apothem a and perimeter p is $A = \frac{1}{2}ap$. (279)

Thm 5.6.1 **Distance Formula** The distance between two points (x_1, y_1) and (x_2, y_2) is

$$d = \sqrt{(x_2 - x_1)^2 + (y_2 - y_1)^2}. \quad (284)$$

Thm 5.6.2 **The Midpoint Formula** For any two points (x_1, y_1) and (x_2, y_2) in the coordinate plane, the midpoint is given by $\left(\frac{x_1 + x_2}{2}, \frac{y_1 + y_2}{2}\right)$. (285)

Def 6.2.1 **Definition** Two planes are parallel if and only if they do not intersect, no matter how far they are extended. (312)

Def 6.2.2 **Definition** A line is perpendicular to a plane if and only if it is perpendicular to every line in the plane that intersects it. (312)

Def 6.2.3 **Definition** A line is parallel to a plane if and only if it is parallel to a line in the plane. (313)

Def 6.2.4 **Definition** The measure of a dihedral angle is the measure of the angle formed by a line in one of the two planes, perpendicular to the intersection of the planes, and a line in the other plane which intersects the first line and is also pependicular to the intersection of the planes. (313)

Thm 6.6.1 **Sets of Parallel Lines** In a perspective drawing, all lines that are parallel to each other, but not to the picture plane, will seem to meet at the same point. (338)

Thm 6.6.2 **Lines Parallel to the Ground** In a perspective drawing, a line on the plane of the ground will meet the horizon of the drawing, if it is not parallel to the picture plane. Any line parallel to this line will meet at the same point on the horizon. (338)

Thm 7.2.1 **Lateral Area of a Right Prism** The lateral area L of a right prism with height h and perimeter of a base p is

$$L = hp \quad (360)$$

Thm 7.2.2 **Surface Area of a Right Prism** The surface area S of a right prism with lateral area L and area of a base B is

$$S = L + 2B \quad (360)$$

Thm 7.2.3 **Cavalieri's Principle** If two solids have equal heights, and if the cross-sections formed by every plane parallel to the bases of both solids have equal areas, then the two solids have the same volume. (362)

Thm 7.2.4 **Volume of a Prism** The volume V of a prism with height h and the area of a base B is

$$V = Bh \quad (362)$$

Thm 7.3.1 **Lateral Area of a Right Regular Pyramid** The lateral area L of a right regular pyramid with slant height l and primeter p of a base is

$$L = \frac{1}{2}lp. \quad (367)$$

Thm 7.3.2 **Surface Area of a Pyramid** The surface area S of a pyramid with lateral area L and area of base B is

$$S = L + B \quad (367)$$

Thm 7.3.3 **Volume of a Pyramid** The volume V of a pyramid with area of its base B and altitude h is

$$V = \frac{1}{3}Bh \quad (369)$$

Thm 7.4.1 **Lateral Area of a Right Cylinder** The lateral area L of a right cylinder with radius r and height h is

$$L = 2\pi rh. \quad (374)$$

Thm 7.4.2 **Surface Area of a Right Cylinder** The surface area S of a right cylinder with radius r, height h, area of base B, and lateral area L is

$$S = L + 2B \quad \text{or} \quad S = 2\pi rh + 2\pi r^2. \quad (374)$$

Thm 7.4.3 **Volume of a Cylinder** The volume V of a cylinder with radius r, height h, and area of a base B, is

$$V = Bh \quad \text{or} \quad V = \pi r^2 h. \quad (376)$$

Thm 7.5.1 **Surface Area of a Right Cone** The surface area S of a right cone with radius of base r, height h, and slant height l is

$$S = L + B \quad \text{or} \quad S = \pi rl + \pi r^2. \quad (381)$$

Thm 7.5.2 **Volume of a Cone** The volume V of a cone with radius r and height h is

$$V = \frac{1}{3}Bh \quad \text{or} \quad V = \frac{1}{3}\pi r^2 h. \quad (382)$$

Thm 7.6.1 **Volume of a Sphere** The volume V of a sphere with radius r is

$$V = \frac{4}{3}\pi r^3. \quad (389)$$

Thm 7.6.2 **Surface Area of a Sphere** The surface area S of a sphere with radius r is

$$S = 4\pi r^2. \quad (390)$$

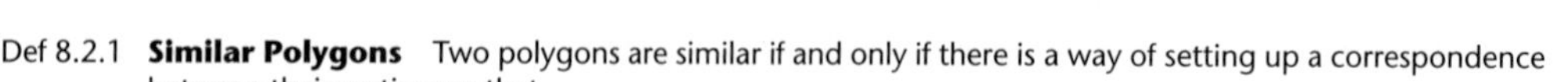

Def 8.2.1 **Similar Polygons** Two polygons are similar if and only if there is a way of setting up a correspondence between their vertices so that

1. the corresponding angles are congruent
   and
2. the corresponding sides are proportional. (420)

Thm 8.2.2 **Cross-Multiplication Property** If $\frac{a}{b} = \frac{c}{d}$ and $b, d \neq 0$, then $ad = bc$. (422)

Thm 8.2.3 **Reciprocal Property** If $\frac{a}{b} = \frac{c}{d}$ and $a, b, c, d \neq 0$, then $\frac{b}{a} = \frac{d}{c}$. (422)

Thm 8.2.4 **Exchange Property** If $\frac{a}{b} = \frac{c}{d}$ and $a, b, c, d \neq 0$, then $\frac{a}{c} = \frac{b}{d}$. (422)

Thm 8.2.5 **"Add-One" Property** If $\frac{a}{b} = \frac{c}{d}$ and $b, d \neq 0$, then $\frac{a+b}{b} = \frac{c+d}{d}$. (422)

Post 8.4.1 **AA (Angle-Angle) Similarity Postulate** If two angles of one triangle are equal in measure to two angles of another triangle, then the triangles are similar. (436)

Post 8.4.2 **SSS (Side-Side-Side) Similarity Postulate** If the measures of pairs of corresponding sides of two triangles are proportional, then the two triangles are similar. (436)

Post 8.4.3 **SAS (Side-Angle-Side) Similarity Postulate** If the measures of two pairs of corresponding sides of two triangles are proportional and the measures of the included angles are equal, then the triangles are similar. (437)

Thm 8.4.4 **Side-Splitting Theorem** A line parallel to one side of a triangle divides the other two sides proportionally. (437)

Cor 8.4.5 **Two Transversal Proportionality Corollary** Two or more parallel lines divide two transversals proportionally. (439)

Thm 8.5.1 **Proportional Altitudes Theorem** If two triangles are similar, then their corresponding altitudes have the same ratio as the corresponding sides. (445)

Thm 8.5.2 **Proportional Medians Theorem** If two triangles are similar, then their corresponding medians have the same ratio as the corresponding sides. (446)

Thm 8.5.3 **Proportional Angle Bisectors Theorem** If two triangles are similar then the corresponding angle bisectors are proportional to the corresponding sides. (450)

Thm 8.5.4 **Proportional Segments Theorem** The angle bisector of a triangle divides the opposite side into segments proportional to the other two sides of the triangle. (451)

Def 9.1.1 A **circle** consists of the points in a plane that are equidistant from a given point known as the **center** of the circle. A **radius** is a segment from the center of the circle to a point on the circle. A **chord** is a segment whose endpoints lie on the circle. A **diameter** is a chord that passes through the center of the circle. (470)

Thm 9.1.2 **Theorem** If two chords are congruent, the arcs they cut are congruent. (474)

Def 9.2.1 A **secant** to a circle is a line that intersects a circle at two points. A **tangent** to a circle is a line that intersects a circle at just one point, known as the **point of tangency.** (479)

Thm 9.2.2 **Theorem** A radius perpendicular to a chord bisects the chord. (479)

Thm 9.2.3 **Tangent Theorem** If a line is perpendicular to a radius of a circle at its endpoint, then the line is tangent to the circle. (481)

Thm 9.2.4 **Theorem** The perpendicular bisector of a chord passes through the center of the circle. (482)

Thm 9.2.5 **The Converse of the Tangent Theorem** If a line is tangent to a circle, then it is perpendicular to a radius of a circle at a point of tangency. (483)

Thm 9.3.1 **Inscribed Angle Theorem** An angle inscribed in an arc has a measure equal to one-half the measure of the intercepted arc. (487)

Cor 9.3.2 **Inscribed Angle Corollary** An angle inscribed in a half-circle is a right angle. (487)

Thm 9.3.3 **Theorem** If two inscribed angles intercept the same arc, then they have the same measure. (487)

Thm 9.4.1 **Theorem** If a tangent and a secant (or a chord) intersect on a circle at the point of tangency, then the measure of the angle formed is one-half the measure of its intercepted arc. (493)

Thm 9.4.2 **Theorem** The measure of an angle formed by two secants or chords intersecting in the *interior* of a circle is one-half of the *sum* of the measures of the arcs intercepted by the angle and its vertical angle. (493)

Thm 9.4.3 **Theorem** The measure of an angle formed by two secants or chords intersecting in the *exterior* of a circle is one-half of the *difference* of the measures of the intercepted arcs. (494)

Thm 9.4.4 **Theorem** The measure of a secant-tangent angle with its vertex *outside* the circle is one-half the *difference* of the measures of the intercepted arcs. (496)

Thm 9.4.5 **Theorem** If two tangents intersect in the *exterior* of a circle, the measure of the angles formed is one-half the *difference* of the measures of the intercepted arcs. (497)

Thm 9.5.1 **Theorem** If two secants intersect outside of a circle, then the product of the lengths of one secant segment and its external segment equals the product of the lengths of the other secant segment and its external secant segment. (504)

Thm 9.5.2 **Theorem** If a tangent and a secant intersect outside of a circle, then the product of the length of the secant segment and its external segment equals the square of the length of the external secant segment. (504)

Thm 9.5.3 **Theorem** If two chords intersect inside a circle, then the product of the lengths of the segments of one chord equals the product of the divided lengths of the other chord. (505)

Thm 10.5.1 **The Law of Sines** For any triangle *ABC*: $\frac{\sin A}{a} = \frac{\sin B}{b} = \frac{\sin C}{c}$. (556)

Thm 10.6.1 **The Law of Cosines** For any triangle *ABC*:

$$a^2 = b^2 + c^2 - 2bc\cos A$$
$$b^2 = a^2 + c^2 - 2ac\cos B$$
$$c^2 = a^2 + b^2 - 2ab\cos C. \quad (564)$$

Thm 11.4.1 **Jordan's Theorem** Every simple closed curve divides the plane into two distinct regions. (609)

Thm 11.4.2 **Euler's Formula** For any polyhedron with vertices *V*, edges *E*, and faces *F*,

$$V - E + F = 2. \quad (610)$$

Thm 11.6.1 **Affine Transformation** An affine transformation maps all pre-image points *P* in a plane to image points *P'* so that

1. collinear points project to collinear points.
2. straight lines project to straight lines.
3. intersecting lines project to intersecting lines.
4. parallel lines project to parallel lines. (622)

Thm 11.6.2 **Theorem of Pappus** If A,B, and C are three distinct points on one line and A′,B′, and C′ are three distinct points on a second line, then the intersections of $\overline{AB'}$ and $\overline{BA'}$, $\overline{AC'}$ and $\overline{CA'}$, and $\overline{BC'}$ and $\overline{CB'}$ are collinear. (624)

Thm 11.6.3 **Theorem of Desargues** If one triangle is a projection of another triangle, then the intersections of the lines containing the corresponding sides of the two triangles are collinear. (625)

Def 12.4.1 **Proof by Contradiction** To prove *s*, assume ~*s*. Then the following argument form is valid:

If ~*s* then (*t* and ~*t*)

Therefore, *s*. (666)

acute triangle A triangle with three acute angles. (555)

adjacent angles Two angles in a plane that share a common vertex and a common side but have no interior points in common. (84)

affine transformation A tranformation in which all pre-image points are mapped to image points so that collinear points, straight lines, intersecting lines, and parallel lines remain as such. (621)

alternate exterior angles Two nonadjacent exterior angles which lie on opposite sides of a transversal. (133)

alternate interior angles Two nonadjacent interior angles which lie on opposite sides of a transversal. (133)

altitude of a cone A segment from the vertex perpendicular to the plane of the base. (379)

altitude of a cylinder A segment joining the two base planes and perpendicular to both. (373)

altitude of a prism A segment joining the two base planes and perpendicular to both. (359)

altitude of a pyramid A segment from the vertex perpendicular to the plane of the base. (366)

altitude of a triangle A line segment from a vertex drawn perpendicular to the line containing the opposite side. (23)

angle A figure formed by two rays that have the same endpoint. (11)

angle bisector A ray that divides an angle into two congruent angles. (18)

annulus The region between two circles which have the same center but different radii. (387)

arc A part of a circle. (472)

area The number of nonoverlapping unit squares that will cover the interior of a figure. (245)

argument A sequence of statements. (644)

axis of cylinder The segment joining the centers of the two bases. (373)

axis of symmetry The line that divides a figure into two symmetrical halves. (117)

Base of isosceles triangle The side opposite the vertex angle. (196)

base angles of isosceles triangle The angles opposite the legs. (196)

bases of a prism Two congruent polygonal faces that lie in parallel planes. (318)

base of a cone The circular face of the cone. (379)

bases of a cylinder The two congruent circular faces that lie in parallel planes. (373)

base of a pyramid The polygonal face that is opposite the vertex. (366)

betweenness Given three points, A, B, and C, if $AB + BC = AC$, then B is between A and C. (228)

binary number system A number system based on the digits 0 and 1. (671)

Cavalieri's Principle If two solids have equal heights, and if the cross-sections formed by every plane parallel to the bases of both solids have equal areas, then the two solids have the same volume. (362)

center of a circle The point inside the circle that is equidistant from all the points on the circle. (261)

center of mass See centroid. (25)

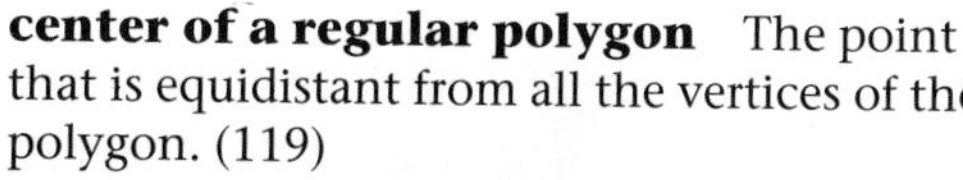

center of a regular polygon The point that is equidistant from all the vertices of the polygon. (119)

central angle of a circle An angle formed by two rays orginating from the center and passing through two points on the circle. (471)

central angle of a polygon An angle formed by two rays orginating from the center and passing through adjacent vertices of the polygon. (119)

centroid The point where the three medians of a triangle intersect. (25)

chord A segment whose endpoints lie on a circle. (470)

circle The set of points in the plane that are equidistant from a given point known as the center of the circle. (261)

circumcenter The point where the three perpendicular bisectors of the sides of a triangle intersect; it is equidistant from the three vertices of the triangle and is the center of the circumscribed circle. (25)

circumference The distance around a circle. (262)

circumscribed circle A circle is circumscribed about a polygon if each vertex of the polygon lies on the circle. (24)

collinear Points that lie on the same line. (11)

complementary angles Two angles whose measures have a sum of 90°. (39)

compound statement A statement formed when two statements are connected by "and" or by "or". (650)

concave polygon A polygon that is not convex. (149)

conclusion The phrase in a conditional statement following the word "then". (72)

conclusion The final statement of an argument. (644)

conditional statement A statement that can be written in the form "If p, then q," where p is called the hypothesis, and q is called the conclusion. (72)

cone An object that consists of a circular base and a curved lateral surface which extends from the base to a single point called the vertex. (379)

congruence The relationship between figures having the same shape and same size. (31)

congruent polygons Two polygons are congruent if their vertices can be matched such that corresponding angles are congruent, and corresponding sides are congruent. (176)

conic section The plane curves that can be formed by the intersection of a plane with a right circular cone; they include the circle, ellipse, parabola, and hyperbola. (499)

conjecture An "educated guess" based on observation. (18)

conjunction A compound statement that uses the word "and". (650)

consecutive interior angles Two interior angles which lie on the same side of a transversal. (133)

contraction A dilation where the figure that is transformed is reduced in size. (414)

contradiction A contradiction asserts that a statement and its negation are both true. (666)

contrapositive The statement formed by interchanging the hypothesis and conclusion of a conditional statement and negating both parts. (660)

converse The statement formed by interchanging the hypothesis and conclusion of a conditional statement. (74)

convex polygon A polygon in which any line segment connecting two points of the polygon has no part outside the polygon. (149)

coordinate plane The plane of the x- and y-axes. (51)

corresponding angles Two nonadjacent angles, one interior and one exterior, that lie on the same side of the transversal. (133)

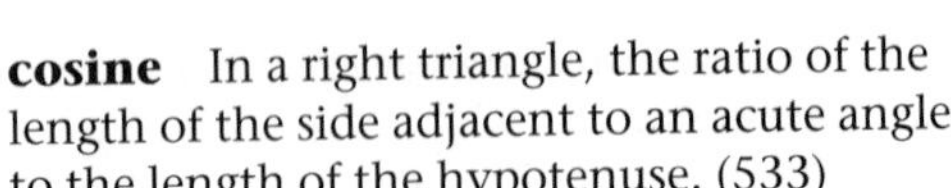

cosine In a right triangle, the ratio of the length of the side adjacent to an acute angle to the length of the hypotenuse. (533)

cotangent In a right triangle, the ratio of the length of the side adjacent to an acute angle to the length of the opposite side. (527)

counterexample An example which proves that a conditional statement is false in that the hypothesis is true but the conclusion is false. (74)

cube A prism with six square faces. (304)

deductive reasoning The process of drawing conclusions by using logical reasoning. (73)

diagonal A segment that joins two nonadjacent vertices of a polygon. (201)

diameter A chord that passes through the center of a circle. (261)

dihedral angle An angle formed by the intersection of planes. (313)

dilation A dilation with center C and scale factor k is a tranformation that maps every point P to a point P' determined as follows

(1) if P is point C, then $P = P'$,
(2) otherwise, P' lies such that $CP' = k \bullet CP$, where $k > 0$ and $k \neq 1$. (414)

direction of vector The component of a vector, usually indicated by an arrow, that indicates orientation. (570)

disjunction A compound statement that uses the word "or". (651)

Distance Formula The distance between two points in the plane containing (x_1, y_1) and (x_2, y_2) is

$d = \sqrt{(x_2 - x_1)^2 + (y_2 - y_1)^2}$. (284)

edge A segment formed by the intersection of two faces of a polyhedron. (318)

equation of a circle A circle with its center at (h, k) and a radius of length r has an equation

$$(x - h)^2 + (y - k)^2 = r^2. \text{ (511)}$$

equilateral triangle A triangle in which all three sides are congruent. (196)

Equivalence Properties of Equality The reflexive, symmetric, and transitive properties of equality. (94)

equivalence relation Any relation that satisfies the three properties of reflexivity, symmetry, and transitivity. (95)

even vertices The vertices of a network that have an even number of paths going to them. (601)

exclusive *or* Indicating either one or the other, but not both. (651)

exterior angle of a polygon An angle formed between one side of polygon and the extension of an adjacent side. (147)

external secant segment The portion of a secant segment that lies outside the circle. (502)

face of a prism Each flat surface of a prism. (318)

fractal A structure which is self-similar in that each subdivision has the same structure as the whole. (629)

glide A combination of a translation and a reflection. (49)

Golden Ratio See Golden Rectangle. (586)

Golden Rectangle The rectangle in which the length l and the width w satisfy the proportion $\frac{l}{w} = \frac{w}{l - w}$. The ratio lw is called the golden ratio.

great circle The intersection of a sphere with a plane that passes through the center of the sphere. (145)

head-to-tail method To find the sum of two vectors place the tail of one vector at the head of the other; the vector drawn from the tail of the first to the head of the second represents the vector sum. (572)

heptagon A polygon with seven sides. (117)

hexagon A polygon with six sides. (117)

hyperbolic geometry The geometry of a surface that curves inward like a "saddle". (616)

hypotenuse The side opposite the right angle in a right triangle. (189)

hypothesis The phrase in a conditional statement following the word "if". (72)

identity matrix A square marix in which all the entries on the main diagonal (from upper left to lower right) are one, and all other entries are zero. (550)

image See transformation. (44)

incenter The point where the three angle bisectors of a triangle intersect; it is equidistant from the three sides of the triangle and is the center of the inscribed circle. (25)

inclusive *or* Indicating either one or the other, or both. (651)

indirect proof Proving a conjecture by showing that the opposite of the conjecture is impossible. (665)

inductive reasoning Forming conjectures on the basis of an observed pattern. (99)

input-output table A table which records what happens to the input as it passes through a logic gate. (673)

inscribed angle An angle whose vertex is on the circle and whose sides contain chords of the circle. (484)

inscribed circle A circle is inscribed in a polygon if each side of the polygon is tangent to the circle. (24)

intercepts The points where a line in the coordinate plane crosses the x- and y- axes. (331)

intercepted arc The arc that lies in the interior of an angle inscribed in a circle. (484)

invariant Properties that stay the same regardless of how a figure is deformed. (609)

inverse The statement formed by interchanging the hypothesis and conclusion of a conditional statement. (660)

isometric drawing A drawing on graph paper that has three rather than two sets of parallel lines; therefore, three-dimensional objects can be represented. (304)

isosceles triangle A triangle with at least two congruent sides. (196)

kite A quadrilateral in which two pairs of adjacent sides are congruent. (259)

knot theory A branch of topology which investigates a curve formed by looping and interlacing a piece of string and then joining the ends together. (607)

lateral area The sum of the areas of the lateral faces. (359)

lateral faces The faces of a prism or pyramid that are not bases. (318)

lateral surface The curved surface of a cylinder or cone. (373)

legs of a right triangle The sides adjacent to the right angle. (189)

legs of an isosceles triangle The congruent sides of an isosceles triangle. (196)

line An undefined term in geometry, a line is understood to be straight, contain an infinite number of points, extend infinitely in two directions, and have no thickness. (10)

linear pair of angles Two adjacent angles whose noncommon sides are opposite rays. (40)

logic gate An electronic circuit that represents "not," "and," or "or". (673)

logical chain Linking several conditionals together. (75)

logical reasoning Linking true conditionals together to form a valid conclusion. (72)

logically equivalent Two logic statements that have the same truth values. (653)

magnitude of vector The length of the vector arrow. (570)

major arc The major arc consists of points A and B and all points of the circle in the exterior of central angle *AOB*. (472)

median A segment from a vertex to the midpoint of the opposite side in a triangle. (23)

midpoint of a segment The point that divides the segment into two congruent segments. (221)

midsegment of a trapezoid The segment that connects the midpoints of the nonparallel sides. (155)

midsegment of a triangle A segment that connects the midpoints of two sides. (155)

minor arc The minor arc $\widehat{AB}$ of central angle *AOB* consists of all the points on the circle that lie in the interior of the central angle. (472)

Modus Ponens In logic, a valid argument of the following form

If p, then q.
p, therefore q.

Modus Tollens In logic, a valid argument of the following form

If p, then q.
Not q, therefore not p.

Möbius strip The one-sided surface formed by taking a long rectangular strip of paper and pasting its two ends together after giving it half a twist. (607)

Monte Carlo Method A simulation technique used to obtain a probability in an experiment which has n number of equally likely outcomes. (291)

negation If p is a statement then not p is its negation. (652)

nets Flat figures that can be folded to enclose a particular solid figure. (359)

network A collection of points called vertices some of which may be connected by edges. (600)

non-Euclidean geometries A system of geometry in which the Parallel Postulate does not hold. (615)

noncollinear Three or more points not all of which lie on the same line. (11)

number line A line whose points correspond with the set of real numbers. (29)

oblique cone A cone that is not a right cone. (379)

oblique cylinder A cylinder that is not a right cylinder. (373)

oblique prism A prism that is not a right prism. (319)

oblique pyramid A pyramid that is not a right pyramid. (366)

obtuse triangle A triangle which has one obtuse angle. (555)

octagon A polygon with eight sides. (117)

octant One of the eight spaces into which the whole of space is divided by the x-, y-, and z-axes. (327)

odd vertices The vertices of a network that have an odd number of paths going to them. (601)

orthocenter The point where the three lines containing the altitudes of a triangle intersect. (25)

orthogonal Perpendicular to. (617)

orthographic projections A parallel projection with all rays perpendicular to the plane of projection. (306)

parallel lines Two coplanar lines that do not intersect. (17)

parallel planes Two planes that do not intersect. (312)

parallelogram A quadrilateral in which opposite sides are parallel. (124)

pentagon A polygon with five sides. (117)

perimeter The distance around a geometric figure that is contained in a plane. (244)

perpendicular bisector A line that is perpendicular to a segment at its midpoint. (18)

perpendicular lines Two lines that intersect to form a right angle. (16)

plane An undefined term in geometry, a plane is understood to be a flat surface that extends infinitely in all directions. (11)

Platonic Solids The five regular polyhedra which are the tetrahedron, the hexahedron. (cube), the octahedron, the dodecahedron, and the icosahedron. (404)

point An undefined term in geometry, a point can be thought of as a dot that represents a location on a plane or in space. (10)

point of tangency The one point at which a tangent intersects a circle. (478)

polygon A closed plane figure formed by three or more segments such that each segment intersects exactly two other segments, one at each endpoint. (116)

polyhedron A geometric solid with polygons as faces. (404)

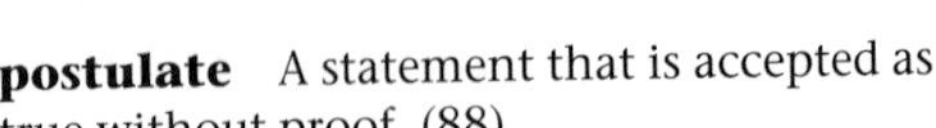

postulate A statement that is accepted as true without proof. (88)

pre-image See transformation. (44)

premise The statements in an argument that come before the conclusion. (644)

prism A polyhedron that has two parallel faces called bases; all other faces, called lateral faces, are parallelograms formed by joining corresponding vertices of the bases. (317)

probability A ratio that compares the number of successful outcomes with the total number of possible outcomes. (289)

projective geometry The study of the properties of figures that do not change under projection. (624)

proof An organized series of statement used to form a convincing argument that a given statement is true. (66)

proof by contradiction See indirect proof. (666)

proportional sides Sides are proportional if all the ratios of corresponding sides of two polygons are the same. (420)

pyramid A solid figure which consists of one base and lateral faces which are triangles. (366)

Pythagorean triple A set of three positive integers, a, b, and c such that $a^2 + b^2 = c^2$. (268)

quadrilateral A polygon with four sides. (124)

radius A segment which connects the center of a circle with a point on the circle. (261)

ratio A comparison of two numbers by division. (354)

ray Consists of an initial point A and all points on that lie on the same side of A as B does. (10)

rectangle A parallelogram with four right angles. (124)

reflection A transformation in which a line of reflection acts as a mirror reflecting points to their images. (46)

reflectional symmetry A plane figure has reflectional symmetry if its reflection image through a line coincides with the original figure. (117)

Reflexive Property of Equality For all real numbers a, $a = a$. (94)

regular polygon A polygon which is both equilateral and equiangular. (119)

regular pyramid A pyramid whose base is a regular polygon. (366)

remote interior angle An interior angle of a triangle that is not adjacent to a given exterior angle. (147)

resultant vector The vector that represents the sum of two given vectors. (571)

rhombus A parallelogram with four congruent sides. (124)

Riemannian geometry A geometry in which there are no parallel lines, such as on the surface of a sphere. (616)

right angle An angle with a measure of 90°. (17)

right cone A cone in which the altitude intersects the base at its center. (379)

right cylinder A cylinder whose axis is perpendicular to the bases. (373)

right prism A prism in which all the lateral faces are rectangles. (319)

right pyramid A pyramid in which the altitude intersects the base at its center. (366)

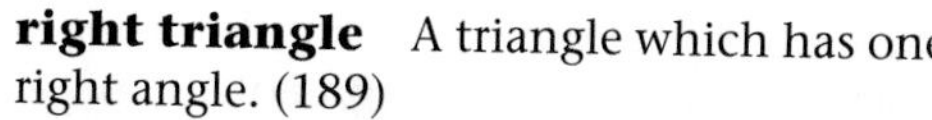

right triangle A triangle which has one right angle. (189)

right-handed system The most common arrangement of the three axes in a space coordinate system, as illustrated in the figure. (326)

rigid transformation A transformation that does not change the size or shape of a figure. (44)

rise and run The vertical and horizontal distances between two points in the plane. (160)

rotation The transformation that moves a geometric figure about a point known as the turn center. (45)

rotation matrix A matrix used to rotate a figure around a given point (x, y) through an angle. (550)

rotational symmetry A figure has rotational symmetry if it has at least one rotation image that coincides with the original image. (118)

scale factor In a transformation the number used to multiply each coordinate of the pre-image to obtain the coordinates of the image. (412)

secant to a circle A line that intersects a circle at two points. (478)

secant segment A portion of a secant with two endpoints. (502)

sector of a circle A region of a circle bounded by two radii and their intercepted arc. (262)

segment A portion of a line with two endpoints. (10)

segment bisector A line that divides a segment into two congruent segments. (18)

self-similarity A structure where each subdivision has the same structure as the whole. (629)

semicircle The arc of a circle that is cut off by a diameter. (243)

similar figures Two figures that have the same shape; if they are polygons, the corresponding angles are congruent and the corresponding sides are proportional. (419)

simple closed curve A figure that encloses a distinct region of the plane and whose segments or curves do not cross themselves. (608)

sine In a right triangle, the ratio of the length of the side opposite an acute angle to the length of the hypotenuse. (533)

skew lines Lines that are not coplanar. (312)

slide arrow An arrow that shows the direction and motion of points in a translation. (45)

slope The ratio of the vertical rise to the horizontal run of a segment. (160)

slope formula The slope of a segment whose endpoints have coordinates $(x_1, y_1)$and (x_2, y_2) is the ratio $\frac{y_2 - y_1}{x_2 - x_1}$. (161)

solid of revolution The object formed by rotating a plane figure about an axis in space. (399)

sphere The set of points in space that are equidistant from a given point known as the center of the sphere. (387)

square A parallelogram with four congruent sides and four right angles. (124)

supplementary angles Two angles whose measures have a sum of 180°. (39)

surface area of a prism The sum of the areas of all the faces of a prism. (359)

Symmetric Property of Equality For all real numbers a and b, if $a = b$, then $b = a$. (94)

tangent to a circle A line that intersects a circle at a single point. (478)

tangent ratio In a right triangle, the ratio of the length of the side opposite an acute angle to the length of the adjacent side. (525)

tangent segment A portion of a tangent with two endpoints. (502)

taxicab circle The set of points in the plane that are at a given taxidistance from a given point known as the center of the circle. (596)

taxicab radius The distance between the center of a taxicab circle and any point on the taxicab circle. (596)

taxidistance In taxicab geometry the smallest number of grid units that must be traveled to go from one point to another point. (594)

theorem A statement that can be proved true. (88)

topology The geometry that deals with the properties of figures that are unchanged by distortion. (607)

torus A donut-shaped surface in topology. (609)

transformation The movement of a figure in the plane from its original position, the *preimage*, by translation, rotation, or reflection to a new position, the *image*. (44)

Transitive Property of Equality For all real numbers a, b, and c, if $a = b$ and $b = c$, then $a = c$. (94)

translation A transformation that glides all points in the plane the same distance in the same direction. (44)

transversal A line, ray, or segment that intersects two or more coplanar lines, rays, or segments, each at a different point. (132)

trapezoid A quadrilateral with only one pair of parallel sides. (124)

trefoil knot A knot which has the shape of a plane figure made of congruent arcs of a circle arranged on an equilateral triangle so that the figure is symmetrical about the center of the triangle and the ends of the arcs are on the triangle. (607)

triangle A polygon with three sides. (23)

triangle rigidity The property that if the sides of a triangle are fixed, there is only one shape the triangle can have. (180)

triangulation The process of using known measurements in a triangle to find other measurements indirectly rather than by direct measurement. (186)

trigonometry The geometry of the triangle. (525)

true conditional A conditional statement that leads only to a true conclusion. (73)

truth table A table used to list all possible combinations of truth values for a given statement or compound statement. (650)

turn center The point about which a figure is rotated. (45)

two-column proof A proof in which the statements are written in the left-hand column and the reasons in the right-hand column. (68)

unit circle A circle with its center at the origin of the coordinate plane and with a radius of 1. (542)

vanishing point In a perspective drawing the point at which the parallel lines seem to meet. (337)

vector A mathematical quantity that has both direction and magnitude (or length). (570)

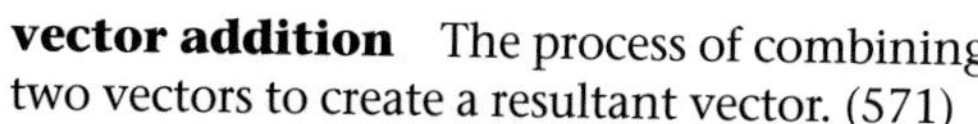

vector addition The process of combining two vectors to create a resultant vector. (571)

vector sum The resultant vector created by vector addition. (571)

vertex angle of an isosceles triangle The angle opposite the base of the triangle. (196)

vertex of a cone The single point opposite the base of the cone. (379)

vertex of a prism A point where three or more edges meet. (318)

vertex of a pyramid The point where the lateral triangular faces meet. (366)

vertical angles The opposite angles formed when two lines intersect. (99)

volume The number of nonoverlapping unit cubes that will fill the interior of a solid figure. (360)

INDEX

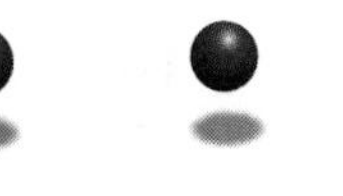       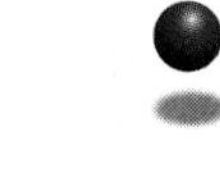

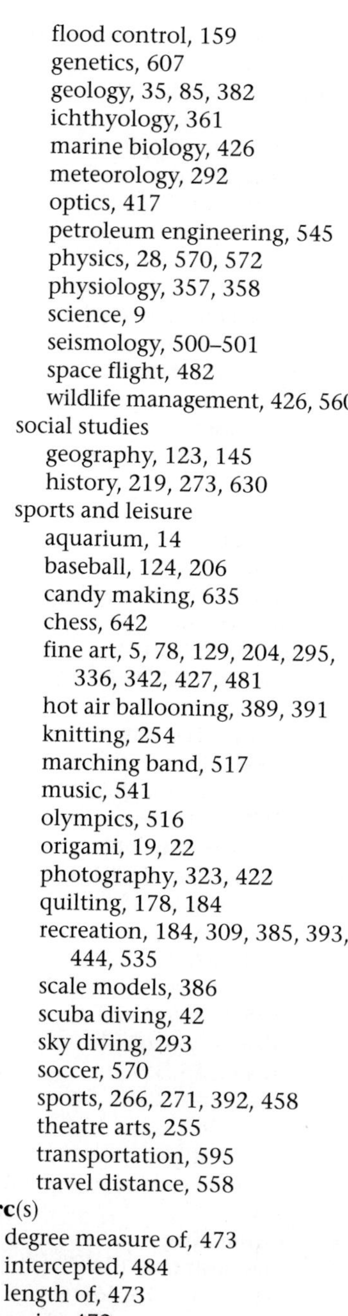

B

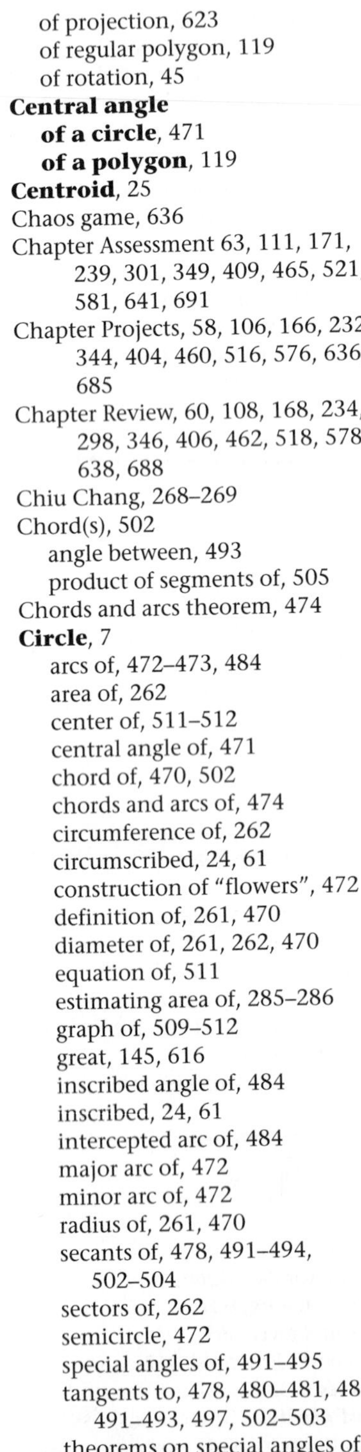

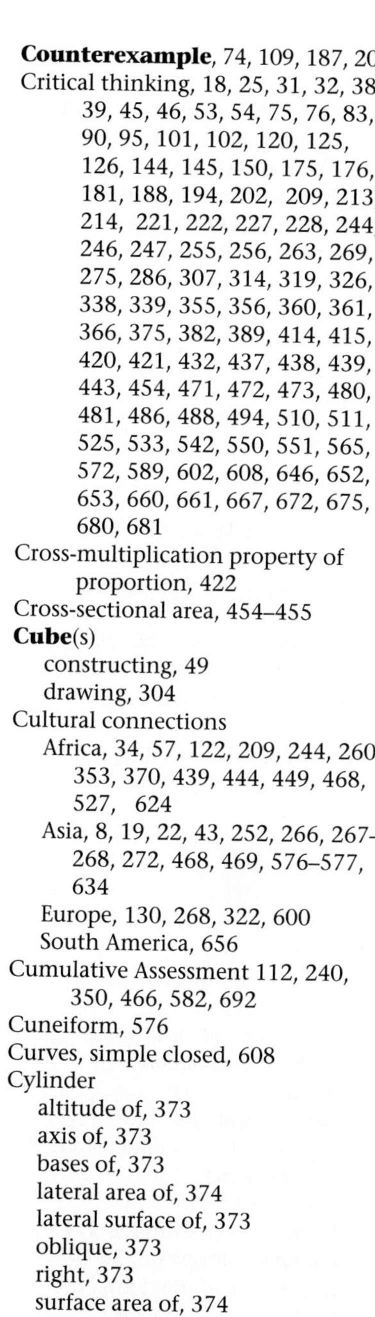

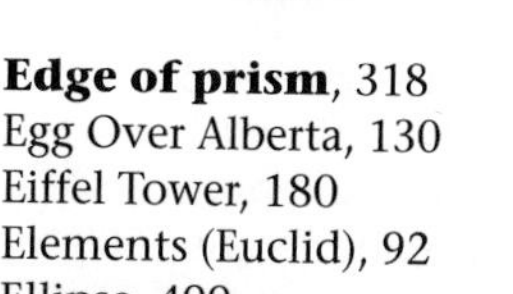

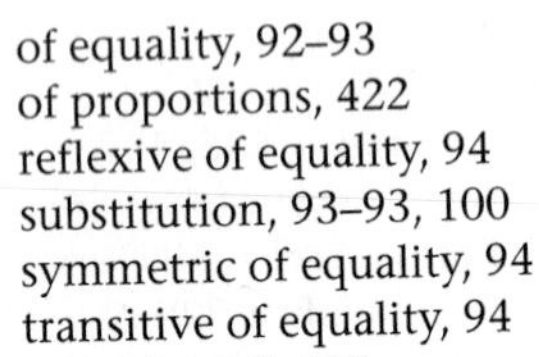

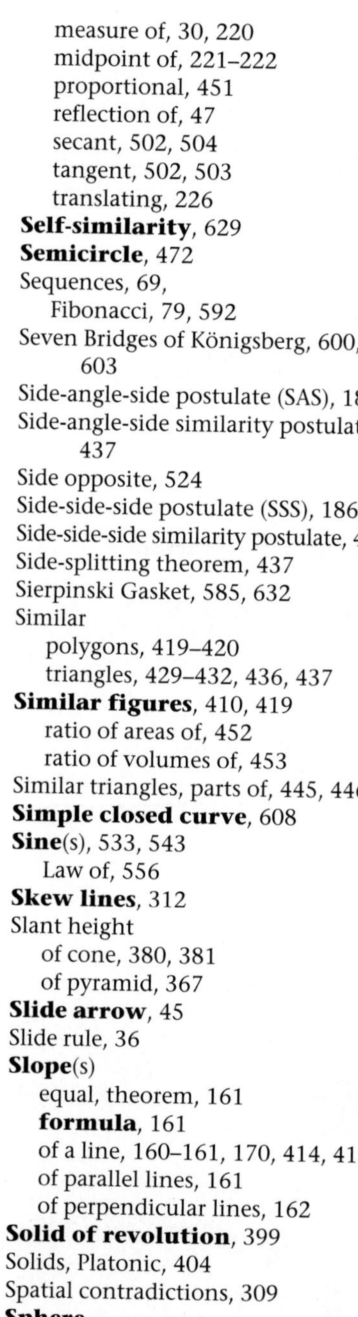

CREDITS

PHOTOS

Abbreviations used: (t) top, (c) center, (b) bottom, (l) left, (r) right, (bckgd) background, (bdr) border.

FRONT COVER: (l), Julie Vanderchmitt/AllSport, France; (c), Homer Smith © 1994 Art Matrix; (r), David Ladd Nelson. **TABLE OF CONTENTS:** Page vi(tr), Dennis Fagan/HRW Photo; vi(cr), Michael Thompson/COMSTOCK; vi(cr), Denis Valentine/The Stock Market; vi(cr), COMSTOCK; vi(bl), Dennis Fagan/HRW Photo; vi(br), Coco McCoy/Rainbow; vii(tr), Gary A. Conner/ PhotoEdit; vii(bl), Tomas Pantin/HRW Photo; vii(br), Alex Bartell/Science Photo Library/PhotoResearchers, Inc.; viii(tr), George Whiteley/Photo Researchers Inc.; viii(cl), Guy Sauvage/Photo Researchers, Inc.; ix(tl), Bob Daemmrich/The Image Works; ix(cr), Tony Freeman/PhotoEdit; ix(br), Bob Daemmrich/Stock, Boston; x(tl), Dennis Fagan/HRW Photo; x(cr), R. Sydney/ The Image Works; x(br), Okoniewski/The Image Works; xi(tl), Tomas Pantin/HRW Photo; xi(cr), xi(bl), Dennis Fagan/HRW Photo; xi(br), H. L. Romberg. **CHAPTER ONE:** Page: 2-3(t), COMSTOCK; 2(c), COMSTOCK; 2-3(b), Scott Van Osdol/HRW Photo; 4(t) (c), Georgia Tblisi/Sovfoto/Eastphoto; 4-8(bckgd), Scott Van Osdol/HRW Photo; 5(tl), Lisa R. Glass/Still Life Stock; 5(cr), Giraudon/Art Resource; 5(bl), Judy Canty/Stock, Boston; 6(tl), Planet Art Collection; 7(c), Dennis Fagan/HRW Photo; 7(br), Russell Dian/Monkmeyer Press Photo; 8(cl), Macduff Everton/The Image Works; 8(cr), Addison Geary/Stock, Boston; 9(bckgd), (tr), (bl), Fermilab Visual Media Services; 10-11(bckgd), Magrath Photography/Science/Photo Library/Photo Researchers Inc.; 12(bl), Myrleen Ferguson Cate/PhotoEdit; 13(tr), Lowell Georgia/Photo Researchers, Inc.; 14(tr), Thomas J. Magno/Still Life Stock, Inc.; 15(br), 16(t), (br), 17(all), Dennis Fagan/HRW Photo; 18(tl), Mug Shots/The Stock Market; 18(bl), Dennis Fagan/HRW Photo; 19(cl), (bl), (br), Dennis Fagan/HRW Photo; 20(br), Michael Newman/PhotoEdit; 21(cr), 22(cr), (br), Dennis Fagan/HRW Photo; 23(tr), Dassault-Breguet/Science Photo Library/Photo Researchers, Inc.; 24(br), Rogers Fund, 1918/The Metropolitan Museum of Art; 25(br), Scott Van Osdol/HRW Photo; 28(tr), Dennis Fagan/HRW Photo; 28(br), 29(tr), Dennis Fagan/HRW Photo; 29 D.P. Wilson/Science Source/Photo Researchers, Inc.; 30(br), COMSTOCK; 33(tr), Scott Van Osdol/HRW Photo; 34(br), Courtesy of Mathematics in the Time of the Pharaohs, by Gillings; 35(br), Courtesy of USGS; 36(cr); 36(br), Dennis Fagan/HRW Photo; 37(tr), Joe Towers/The Stock Market; 37(tl), Ron Sherman/Stock, Boston; 37(br), (bckgd), Dennis Fagan/HRW Photo; 40(br), Wellzenbach/The Stock Market; 42(bckgd), Mark M. Lawrence/The Stock Market; 43(br), Babylonian Collection (YBC 7289)/ Yale University; 44(bl), David Ulmer/Stock, Boston; 44(br), Lisa R. Glass/Still Life Stock, Inc.; 44(tr), F. Gohier/Photo Researchers, Inc.; 44(bl), Thomas Zimmerman/FPG International; 45(br), Alex Bartell/Science Photo Library/ PhotoResearchers, Inc.; 46(tr), 48(cl), (cr), Dennis Fagan/HRW Photo; 48(bckgd), Stephen Kaufman/Tony Stone Images; 49(cr), Paul Steel/The Stock Market; 49(br), Rod Planck/Photo Researchers, Inc.; 50(tr), Courtesy of Quarto Publishing, England; 50(br), Frank Siteman/Stock, Boston; 51(t), (bckgd), Vee Sawyer/MertzStock; 57(cr), Building in Egypt, by Dieter Arnold; 58(all), 59(all), Dennis Fagan/HRW Photo. **CHAPTER TWO:** Page 64-65(Bckgd), Chris Noble/Tony Stone Images; 65(br), (tl), 66(tr), 67(tr), 68(br), 69(tr), 71(br), Dennis Fagan/HRW Photo; 72(tl), Michael Thompson/COMSTOCK; 72(t), (c), Denis Valentine/The Stock Market; 72(tr), COMSTOCK; 73(tl), Gerald Gusthall/FPG International; 74(tl), Courtesy of Karen Kuehn/Sony Music; 76(br), NASA/The Image Works; 77(tl), Andy Sacks/Tony Stone Images; 78(bl), Tate Gallery, London/Art Resource; 78(br), Art Resources, N.Y.; 79(b), Joseph Sachs/Still Life Stock, Inc.; 80-81(bckgd), George Diebold/The Stock Market; 82(tr), Dennis Fagan/HRW Photo; 87(br), Michael Newman/PhotoEdit; 88(tr), Llewellyn/Uniphoto Picture Agency; 90(tr), Coco McCoy/Rainbow; 92(tr),Vee Sawyer/MertzStock; 95(cr), Ernest Manewal/FPG International; 96(tr), Randy Duchaine/The Stock Market; 97(br), Michael Newman/PhotoEdit; 98(br), Dennis Fagan/HRW Photo; 99(bckgd), (br), 105(cr), Dennis Fagan/HRW Photo; 107(br), Courtesy of Broderbund Software; 107(br), Dennis Fagan/HRW Photo; 107(bl), Scott Van Osdol/HRW Photo; 106-107(b), (t), Sam Dudgeon/HRW Photo. **CHAPTER THREE:** Page 114-115(bckgd), Photo Edit; 114(t), Greg Hursley; 115(tl), COMSTOCK; 115(br), Murray Alcosser/The Image Bank; 116(t), Coco McCoy/Rainbow; 118(tr), Dennis Fagan/HRW Photo; 118(bl), IPA/The Image Works; 119(bl), Color Box 1992/FPG International; 120(cl), Steve Elmore/The Stock Market; 120(br), © 1995 M.C. Escher/Cordon Art–Baarn–Holland. All rights reserved; 123(b), (bckgd), Vee Sawyer/ MertzStock; 124(t), Jerry Wachter/Photo Researchers, Inc.; 124(bl), © Jim Richardson; 127(tr), Dennis Fagan/HRW Photo; 128(bl), Michael Newman/ PhotoEdit; 128(tl), Michael Newman/PhotoEdit; 128(b), (c), VanBucher/Photo Researchers, Inc.; 128(tr), Scott Camazine/Photo Researchers, Inc.; 128(br), Chromosohn/The Stock Market; 129(tr), © 1995 M.C. Escher/Cordon Art–Baarn–Holland. All rights reserved; 130-131(bckgd), Annette Del Zoppo; 130(tl), (bl), (b) Annette Del Zoppo; 131(br), Annette Del Zoppo; 130-131(c), Robin White/Fotolex Associates; 132(bckgd), Robert Rathe/Scock, Boston; 132(tr), Joseph Stella, Bridge, 1936. Oil on canvas; 50-1/8 x 30-1/8 (127.3 x 76.5 cm). San Francisco Museum of Modern Art, WPA /Federal Arts Project allocation to the San Francisco Museum of Art 3760.43; 134(bl), Robert Frerck/Tony Stone Images; 135(t), (bckgd), Gabe Palmer/Mug Shots/The Stock Market; 136(tr), Peter Pearson/Tony Stone Images; 138(tr), Courtesy of NASA; 140(tr), 142(tr), Dennis Fagan/HRW Photo; 143(bl), Tony Stone Images; 143(t), (bckgd), Blair Seitz/Photo Researchers, Inc.; 143(br), Crandall/The Image Works; 144(tr), Dennis Fagan/HRW Photo; 145(bl), Tony Craddock/Science Photo Library/Photo Researchers, Inc.; 146(tr), Dennis Fagan/HRW Photo; 149(tr), Tom Sanders/The Stock Market; 152(tr), Brownie Harris/The Stock Market; 153(br), © Jim Richardson; 154(bckgd), Larry Nelson/Stock, Boston; 155(tr), Bob Daemmrich/Stock, Boston; 155(tl), Harvey Lloyd/The Stock Market; 157(cr), HRW Photo/Tomas Pantin; 158(bckgd), Lester Lefkowitz/Tony Stone Images; 159(tr), Chris Rogers/Rainbow; 160-361(bckgd), Joseph Nettis/Stock, Boston; 160(tr), Hank Morgan/Rainbow; 162(bckgd), EuniceHarris/Photo Researchers, Inc.; 164(tr), Steve Weber/Stock, Boston; 165(br), Tomas Pantin/HRW Photo; 165 TessellMania!; 165(tr), © 1995 M.C. Escher/Cordon Art–Baarn–Holland. All rights reserved; 166(tr), (bl), Dennis Fagan/HRW Photo; 166(b), (c), (br), 167(all), Dennis Fagan/HRW Photo. **CHAPTER FOUR:** Page 172(cr), Mark Joseph/Tony Stone Images; 172-173(t), McConnell, McNamara & Company; 173(c), Ron Kimball; 172-173(b), Art Wolfe/Tony Stone Images; 174(t), Andy Warhol, 100 Cans. 1962; Synthetic polymer paint and silkscreen, ink on canvas. © The Andy Warhol Foundation. Collection of Albright-Knox Art Gallery. Gift of Semour Knox; 176(bl), Richard Payne; 178(br), Gary A. Conner/PhotoEdit; 179(br), Dennis Fagan/HRW Photo; 180(t), G. Zimbel/Monkmeyer; 180-181(bckgd), Lee Snider/The Image Works; 183(tr), Tomas Pantin/HRW Photo; 184(tl), M. Antman/The Image Works; 184-185(bckgd), Cary Wolinsky/Stock, Boston; 186(bl), Sullivan/ Texastock; 186(tr), Daniel Brody/Stock, Boston; 192(b), Landform Slides; 193(tr), (bl), © 1995 M.C. Escher/Cordon Art–Baarn–Holland. All rights reserved; 197(br), Tomas Pantin/HRW Photo; 200(tl), (tr), Michael Newman/ PhotoEdit; 201 HRW Photo; 203(cr), Dennis Fagan/HRW Photo; 206(b), Dennis O'Clair/Tony Stone Images; 207(tr), Scott Van Osdol/HRW Photo; 208(tl), Dennis Fagan/HRW Photo; 210(br), 211(tr), Dennis Fagan/HRW Photo; 213(tr), Scott Van Osdol/HRW Photo; 216(t) (bckgd), Michael Newman/ PhotoEdit; 220(tr), Tomas Pantin/HRW Photo; 222(tl), 224(br), Dennis Fagan/HRW Photo; 226(tr), Jean Marc Barey/Photo Researchers, Inc.; 232-233(bckgd), Phototone; 232-233(all) Scott Van Osdol/HRW Photo; 238(bl), Luis Villota/ The Stock Market. **CHAPTER FIVE:** Page 242-243(bckgd), Scott Van Osdol/HRW Photo; 242(c), NASA; 243(c), Richard Stockton; 244(bckgd), Kazuyoshi Nomachi/Photo Researchers, Inc.; 245(tr), Kazuyoshi Nomachi/ Photo Researchers, Inc.; 246(cr), J. Barry O'Rorke/The Stock Market; 247(cr), Frank Siteman/Stock, Boston; 248(tl), Steve Elmore/The Stock Market; 248(tr), Michael Rosenfeld/Tony Stone Images; 249(b), Carroll Seghers/Photo Researchers, Inc.; 249 John Running/Stock, Boston; 250(t), Conklin/ Monkmeyer Press Photo; 252-253(bckgd), Boden/Ledingham/MASTERFILE; 254(tr), Jonathan Nourok/PhotoEdit; 254(tl), (br), Vee Sawyer/Mertzstock; 256(bl), Robert Rathe/Stock, Boston; 257(tr), Tomas Pantin/HRW Photo; 259(tl), Gary A. Connor/PhotoEdit; 260(tr), Gary Irving/Tony Stone Images; 261(tr), NASA; 263(bl), Dennis Fagan/HRW Photo; 264(tr), Gary Braasoh; 264(b), © Jim Richardson; 265(tr), Tony Stone Images; 266(tr), Bob Daemmrich/The Image Works; 267(c) , Columbia University Rare Book and Manuscript Library. Gift of George A. Plimpton; 267(bckgd), Vee Sawyer/Mertzstock; 267(r), Art Resource; 268(bl), Vee Sawyer/Mertzstock; 269(br), G. Gardner/The Image Works; 271(cr), Dennis Fagan/HRW Photo; 272(br), Frank Siteman/Monkmeyer Press Photo; 275(tr), Tomas Pantin/HRW Photo; 276-277(bckgd), Bill Binzen/Rainbow; 279(br), Tomas Pantin/HRW Photo; 280(br), Vee Sawyer/Mertzstock; 281(br), Dennis Fagan/HRW Photo; 283(tr), L. Kolvoord/The Image Works; 284(b), (bckgd), SIU/Photo Researchers, Inc.; 285(br), 286(bl), Dennis Fagan/HRW Photo; 287(b), (c), Edward Pieratt/Stock, Boston; 291(bl), Dennis Fagan/HRW Photo; 293(tr), Guy Sauvage/Photo Researchers, Inc.; 294(t), David M. Grossman/Photo Researchers, Inc.; 294(bckgd), (bl),Vee Sawyer/Mertzstock; 295(c), H. L. Romberg; 297(b), Scott Van Osdol/HRW Photo. **CHAPTER SIX:** Page 304(tr), K. Kai/FujiFotos/The Image Works; 305(bl), Dennis Fagan/HRW Photo; 309(tl), H. L. Romberg; 309(bl), Scott Van Osdol/HRW Photo; 311(tr), George Whiteley/Photo Researchers Inc.; 311(bl), Charles M. Falco/Photo Researchers, Inc.; 312(cr), 313(tr), Dennis Fagan/HRW Photo; 315(b), Bair/Monkmeyer Press Photo; 316(tr), Sam Dudgeon/HRW Photo; 316(cr), (bl), (br), Dennis Fagan/HRW Photo; 317(tr), © Jim Richardson; 318(tl), Sam Dudgeon/HRW Photo; 318(bl), 320(br), 322(cr), Dennis Fagan/HRW Photo; 322(br), David Parker/Science Photo Library/Photo Researchers, Inc.; 323(bckgd), 324(bckgd), © Chuck Solomon; 325(tl), Dennis Fagan/HRW Photo; 326(tr), Frank Rossotto/The Stock Market; 326(bl), Dennis Fagan/HRW Photo; 330(bckgd), David A. Hardy/Science Photo Library/Photo Researchers, Inc.; 331(tr), Click Chicago/Tony Stone Images; 336(tr), "Grandes Chroniques de France", Castres, Musee Goya/Giraudon/Art Resource; 336(c), "Veduci di citta ideale"/Art Resource; 336(bc), Dennis Fagan/HRW Photo; 336(bl), Sam Dudgeon/HRW Photo; 337(tl), Fotomas Index; 337(cr), Donovan Reese/Tony Stone Images; 337(bl), Dennis Fagan/HRW Photo; 337(bl), Cindy Verheyden/ HRW Photo; 338(br), H. L. Romberg; 339(tl), Stan Osolinski/Tony Stone Images; 339(cl), Brunelleschi, "Interno verso l'altare S. Lorenzo"/Art Resource; 339(b), Kathy Bushue/Tony Stone Images; 343(b), Scott Van Osdol/HRW Photo; 344(tl), Building in Egypt, Pharaonic Stone Masonry ; 344(bl), Michelle Bridwell/Frontera Fotos; 344-345(bckgd), Doug Armand/Tony Stone Images; 345(tr), Dennis Fagan/HRW Photo. **CHAPTER SEVEN:** Page 352-353(bckgd), Fabrizio Bensch REUTERS/Bettman; 352(c), Lutz Schmidt REUTERS/Bettman; 353(bl), Kurgan-Lisnet/Liaison International; 354(tl), Phillip Bailey/The Stock Market; 354(tr), Paul Conklin/Monkmeyer Press Photo; 355(tr), Jany Sauvanet/Photo Researchers, Inc.; 355(cr), W. K. Almond/Stock, Boston; 357(tr), Courtesy of NASA; 357(cr), Fred McConnaughey/Photo Researchers, Inc.; 357(br), G. W. Ott/The Stock Market; 358(tr), Tom Tracy/The Stock Market; 358(br), Sam Dudgeon/HRW Photo; 359(tr), Scott Van Osdol/HRW Photo; 361(tr), Mark Stouffer/Animals Animals; 362(tr), Scott Van Osdol/HRW Photo; 363(tr), Tomas Pantin/HRW Photo; 363(br), B. Mahoney/The Image Works; 365(tr), Mark Antman/The Image Works; 366(t), 368(bl), 369(cl), Dennis Fagan/HRW Photo; 370(tl), Will and Deni McIntyre/Photo Researchers, Inc.; 372(t), Spencer Jones/FPG International; 373(t), Tony Freeman/PhotoEdit; 377(br), Tomas Pantin/HRW Photo; 378(tr), David Wells/The Image Works; 378(br), Dennis Fagan/HRW Photo; 379(tr), Tony Stone Images; 381(b), Bilderberg/The Stock Market; 382(bl), Courtesy of US GS; 386(br), Peter Menzel/Stock, Boston; 386(br), Dean Abramson/Stock, Boston; 387(tr), Stephen Agricola/Stock, Boston; 389(cl), Trask/Stock Imagery, Inc; 389(bl), Imtek Imagineering;/MASTERFILE; 390(tl), Matthew McVay/Tony Stone Images; 391(tr), Bob Daemmrich/The Image Works; 391(br), Tomas Pantin/HRW Photo; 392(c), (tr), Dennis Fagan/HRW Photo; 394(bckgd), Scott Van Osdol/HRW Photo; 394, 395 BPL; 396(tr), Frank Rossotto/The Stock Market; 396(bl), Dennis Fagan/HRW Photo; 398(bcl), Fay Torresyap/Stock, Boston; 398(collage),Vee Sawyer/MertzStock; 399(t), Biophoto Associates/Science Source/Photo; Researchers, Inc.; 399(tr), (tl), (bcr), Patrice Ceisel/Stock, Boston; 399(bcl), (tc), (tcl), (bckgd), Vee Sawyer/MertzStock; 401(bl), Bob Daemmrich/Stock, Boston; 402(bl), Amy Etra/PhotoEdit; 403(tl), Pun Nio/HRW Photo; 403(bl), (br), Dennis Fagan/HRW Photo; 404-405(bckgd), Richard Pasley/Stock, Boston; 405(tr), (br), (bl), Scott Van Osdol/HRW Photo. **CHAPTER EIGHT:** Page 410-411(bckgd), Ron Kimball; 412(tr), Michelle Bridwell/HRW Photo; 413(tl), Tomas Pantin/HRW Photo; 414(br), Oscar Burriel/Latin Stock/Science Photo; 414(bl), Oscar Burriel/Latin Stock/Science Photo; 416(tr), Felicia Martinez/PhotoEdit; 418(cr), World Perspectives; 419(tr), Scott Van Osdol/HRW Photo; 419(bl), Dennis Fagan/HRW Photo; 421(cl), Ed Michaels/Photo Researchers, Inc.; 422(bl), Tony Freeman/PhotoEdit; 424(tl), Donald Dietz/Stock, Boston; 426 Courtesy of Texas Highway Department; 426 Courtesy of Texas Parks and Wildlife; 429(tr), Stephen Dunn/ALLSPORT USA; 429(tl), Al Bello/ALLSPORT USA; 433(tr), Tomas Pantin/HRW Photo; 435(br), Michael Newman/PhotoEdit; 436(t), Vee Sawyer/MertzStock; 441(bckgd), Ken Reid/FPG International; 441(bl), Tomas Pantin/HRW Photo; 443(tr), Michael Collier/Stock, Boston; 447(cl), Peter Miller/Photo Researchers, Inc.; 447(cr), Dick Dietrich/FPG International; 448(br), Daemmrich/UNIPHOTO Picture Agency; 449(cr), Robert Caputo/ Stock, Boston; 452(tr), Tim Barnwell/Stock, Boston; 454(bl), Dennis Fagan/HRW Photo; 455(tl), Kevin Schafer/Tony Stone Images; 456(cr), Dennis Fagan/HRW Photo; 459(tr), Vic Bider/PhotoEdit; 460(bckgd), (cl), 461(b), Scott Van

Osdol/HRW Photo. **CHAPTER NINE:** Page 468-469(bckgd), COMSTOCK; 468(c), 469(c), COMSTOCK; 470(tr), Scott Van Osdol/HRW Photo; 471(cl), Dennis Fagan/HRW Photo; 472(br), LeDuc/Monkmeyer Press Photo; 474(br), Scott Van Osdol/HRW Photo; 475(bl), LeDuc/Monkmeyer Press Photo; 476(br), Dennis Fagan/HRW Photo; 478(bckgd), Courtesy of NASA; 482(t), (b), Tony Stone Images; 484(t), 488(br), Dennis Fagan/HRW Photo; 489(tr), Tomas Pantin/HRW Photo; 490(tr), Okoniewski/The Image Works; 491(t), John Henley/The Stock Market; 497(b), Tony Stone Images; 500(tl), Steven Starr/Stock, Boston; 500-501(bckgd), Spencer Grant/Monkmeyer Press Photo; 501(bl), Krafft/Explorer/Photo Researchers inc.; 502(t), Clark James Mishler/ Westlight; 502(bl), Peter Menzel/Stock, Boston; 507(bckgd), Peter Saloutos/ Tony Stone Images; 507(t), Peter Saloutos/Tony Stone Images; 507(br), Randy Duchaine/The Stock Market; 507(br), Vee Sawyer/MertzStock; 509(b), Tomas Pantin/HRW Photo; 510(tl), (cl), 512(br), Dennis Fagan/HRW Photo; 514(tr), Bob Daemmrich/Stock, Boston; 516(bl), Dennis Fagan/HRW Photo; 516(cr), Bob Daemmrich/Stock, Boston; 517(tl), Mark Green/Tony Stone Images; 517(br), Alan Schein/The Stock Market. **CHAPTER TEN:** Page 522-523 COMSTOCK; 523(br), Eric Crossan; 522(b) left, Zigy Kaluzny/Tony Stone Images; 522-523(bckgd), Pete Saloutos/The Stock Market; 524(tr), (bckgd), Paul Steel/The Stock Market; 524(bl), Vee Sawyer/MertzStock; 525(br), Dennis Fagan/HRW Photo; 527(bckgd), Carolyn Brown/Photo Researchers, Inc.; 529(br), Jeff Greenberg/PhotoEdit; 531(tr), Martin Frick/M&PPRY; 531(b), Michael Collier/Stock, Boston; 533(t), Ken Straiton/The Stock Market; 536(bl), Dennis Fagan/HRW Photo; 537(br), R. Sydney/The Image Works; 538(r), Alain Evrard/Photo Researchers, Inc.; 539(cr), Amy Etra/PhotoEdit; 540(bckgd), Geoff Dore/Tony Stone Images; 541(tl), Eastcott & Momatiuk/The Image Works; 541(br), Mark Antman/The Image Works; 543(bckgd), Vandystadt/Photo Researchers, Inc.; 545(br), Bob Daemmrich/The Image Works; 546(br), Dennis Fagan/HRW Photo; 547(bckgd), NASA/Photo Researchers, Inc.; 548(tr), NASA/Photo Researchers, Inc.; 549(cr), Dennis Fagan/HRW Photo; 555(tr), 558(tl), Earth Satellite Corporation/Science Photo Library/Photo Researchers, Inc.; 559(cr), Bill Bachman/Photo Researchers, Inc.; 560(t) (bckgd), Kathy Bushue/Tony Stone Images; 561(br), HRW Photo; 562(c), NASA/Science Photo Library/Photo Researchers, Inc; 563(br), NASA/Frank Rossotto/ The Stock Market; 562-563(bckgd), Digital Art; 564(c), Scott Van Osdol/HRW Photo; 564(bckgd), Robert Rathe/Stock, Boston; 564(bcdgd), (b), Vee Sawyer/MertzStock; 570(t), David Young Wolff/PhotoEdit; 570(br), Tony Freeman/PhotoEdit; 571(tl), Tom Tracy/The Stock Market; 571(tr), Ed Pritchard/Tony Stone Images; 573(bckgd), Bob Abraham/The Stock Market; 574-575(bckgd), George Disario/The Stock Market; 576-577(bckgd), COMSTOCK; 576(tr), Columbia University Rare Book and A. Plimpton. **CHAPTER ELEVEN:** Page 584-585(bckgd) Gregory Sams/Science Photo Library/Photo Researchers,Inc; 584(bl), Lisa R. Glass/Still Life Stock; 584(cl), Michelle Bridwell 584(br),Scott Van Osdol/HRW Photo; 585(tl), Judy Canty/Stock, Boston; 586(cl), Lisa R. Glass/Still Life Stock; 586(cr), Judy Canty/Stock, Boston; 586(bl),Photri Inc.; 586(bckgd),Planet Art Collection; 588(bckgd), Jim Erickson/The Stock Market; 590(bckgd), Scott Van Osdol/HRW Photo; 593(br), Rue/MonkMeyer Press Photo; 594(t) Ed Pritchard/The Stock Market; 594(br), Robert Brenner/PhotoEdit; 596(tr), Robert Brenner/PhotoEdit; 597(tr), Chad Ellers/Tony Stone Worldwide; 598 (tl), Susan McCartney/Photo Researchers, Inc.; 599(b), Robert Estall/Tony Stone Worldwide; 599(bl), Mike Mazzaschi/ Stock, Boston; 600(tr) Sovfoto/Eastfoto; 604(tr), Scott Van Osdol/HRW Photo; 605(br), Spencer Grant/Stock, Boston; 606(br), Dennis Fagan/HRW Photo; 607(tl), Tony Stone Worldwide; 607(tr),Michael Burgess/The Stock Market; 611(br), Dennis Fagan/HRW Photo 612(tr), Tomas Pantin/HRW Photo; 614(t) © 1995 M.C. Escher/Cordon Art–Baarn–Holland. All rights reserved.; 615(cl), Wayne Scherr/Photo Researchers, Inc.; 618(br), The Image Works; 619(tr), Dennis Fagan/HRW Photo; 621(tr), Don Hay; 621(tr), Scott Van Osdol/HRW Photo; 625(br), 628(b) Dennis Fagan/HRW Photo; 629(bckgd), Scott Van Osdol/HRW Photo; 629(bl),Fractal Geometry of Nature by Mandelbrot/ Freeman; 630(tl), Worldsat International, Inc./Science Photo Library/Photo Researchers, Inc.; 630(c), Courtesy of NASA; 630(b), Sam Dudgeon/HRW Photo; 631(tr), Tomas Pantin/HRW Photo; 632(bckgd), Stephen Johnson/Tony Stone Worldwide; 634 Przemyslaw Prusinkiewicz; 636(br), Seattle Times/Gamma-Liaison International; 637(tr), Sam Dudgeon/HRW Photo. **CHAPTER TWELVE:** Page 642-643(bckgd), Scott Van Osdol/HRW Photo; 644(tr), Joe Arcure; 645(bl), Zefa-Bach/The Stock Market; 647(tl), John Lund/Tony Stone Worldwide; 647(br), Scott Van Osdol/HRW Photo; 648(tr), Bob Daemmrich/The Image Works; 650(t), Tony Arruza/Tony Stone Worldwide; 651(bckgd), Ronnie Kaufman/The Stock Market; 652(bckgd), The Kobal Collection; 653(br), Bob Daemmrich/Tony Stone Worldwide; 654(cr), Gary Lefever/Grant Heilman; 656(bckgd), Graham French/MASTERFILE ; 658(t), Shooting Star International Photo Agency; 661(br), Vee Sawyer/ MertzStock; 663(br), Frank Siteman/Stock, Boston; 664(b), Esbin-Anderson/The Image Works; 665(t), Vee Sawyer/MertzStock; 666(tl), H.L. Romberg; 667(b), Paramount Pictures/The Kobal Collection; 668(tr), Magritte,*"Ceci n'est pas une pipe"*, 1929. Oil on canvas, 60x81cm. © ARS, NY. Los Angeles County Museum of Art, Los Angeles, CA/Giraudon/Art Resource; 668-669(bckgd), Eric Neurath/Stock, Boston; 671(tr), Courtesy of Motorola; 672(t), Tony Freeman/PhotoEdit; 672(c), Spencer Jones/FPG International; 672(b),David Woods/The Stock Market; 673-674(bckgd),SB Photography/Tony Stone Worldwide; 675(br), Tomas Pantin/HRW Photo; 676(br), David Parker/ Science Museum/Science Photo Library/Photo Researchers Inc.; 677(br), David Bassett/Tony Stone Worldwide; 678(br), J. B. Diderich/Contact Press Images/The Stock Market; 679(t), Photri/The Stock Market; 681(bl), 682-683(bckgd)Dennis Fagan/HRW Photo; 685(br), 687(bl), Scott Van Osdol/HRW Photo; 685-687 Letraset Phototone.

ILLUSTRATIONS/DESIGN

Abbreviations used: (t) top, (c) center, (b) bottom, (l) left, (r) right, (bckgd) background, (bdr) border.

Brooks, Janet pages 252-253, 394-395
Cericola, Anthony pages 443, 447, 448 (c), 530, 531, 535, 559, 609, 613
Claunch, David/Liason pages 172-173, 242-243, 468-469, 562-563, 656-657
Design Island 584-585
Davis, Will pages 385 (t), 417, 550, 608, 656-657
Effler, Jim pages 315, 323, 427, 540, 566
Farrell, Russell pages 289
Fischer, David pages X, 54, 219, 368, 385, 395, 417, 490, 494, 604, 633
Kell, Leslie pages 500-501, 516-517, 576-577, 636-637, 685-687
Lee, Kanokwalee pages 232-233, 296-297, 404-405-460-461
Maryland Carto Graphics pages 426, 444, 505, 568, 623
Nigro, Lisa pages costume 419
Obershan, Mike pages 2-8, 46, 58-59,106-107, 130-131, 302-303, 323-325, 329 (t), 330 (c)(r), 336 (b), 337 (b)(l), 344-345, 410-411, 468-469, 642-643
Pembroke, Richard pages 428
Randazzo, Tony pages 302-303
Sawyer, Vee pages 35, 51, 52, 55, 267, 507, 524, 600, 650 Collages: 6, 37, 44, 92, 123, 128, 143, 244, 245, 249, 254, 267, 294, 354, 398, 399, 434, 436, 500, 502, 507, 509, 524, 564, 586, 647, 650, 661, 665
Scrofani, Joe pages 75, 199, 670
Sullivan, Alicia pages 57, 64-65, 80-81, 114-115, 166-167, 172-173, 252-253, 468-469, 522-523, 656-657
Szetela, Chris page 603
Tenniel, Sir John pages 665

PERMISSIONS

Grateful acknowledgment is made to the following sources for permission to reprint copyrighted material.

Discover Syndication, a division of Disney Magazine Publishing, Inc.: Excerpt by Paul Hoffmann and five photographs by Annette Del Zoppo from "Egg Over Alberta" from *Discover,* May 1988, pp. 37, 38, 39, and 42. Copyright © 1988 by Discover Magazine and Annette Del Zoppo.

Dover Publications, Inc.: "6 Accomplishments" and "31 Digits Are Symbols" from *My Best Puzzles in Logic and Reasoning* by Hubert Phillips. Copyright © 1961 by Dover Publications, Inc.

Griffith Institute, Ashmolean Museum: From Howard Carter's Notebook and sketch of "Band A."

Harcourt Brace & Company: From "Law of Sines: Case 1" on page 165 from *Trigonometry,* Revised Edition by Arthur F. Coxford. Copyright © 1981, 1987 by Harcourt Brace & Company.

Holt, Rinehart and Winston, Inc.: From "Sun-Centered Coordinates of Major Planets for Aug. 17, 1990 in Astronomical Units" p. 690 from *HRW Algebra with Trigonometry.* Copyright © 1986, 1992 by Holt, Rinehart and Winston, Inc. Exercises 11–13 with drawing, p. 104; from "Critical Thinking Questions," pp. 110 and 159; drawing from p. 110, and graph from p. 159 from *HRW Geometry,* Annotated Teacher's Edition. Copyright © 1991, 1986 by Holt, Rinehart and Winston, Inc. "Brainteaser," on p. 622, from *Holt Algebra with Trigonometry.* Copyright © 1986, 1992 by Holt, Rinehart and Winston, Inc.

Kentucky Department of Education: "Kentucky Mathemathics Porfolio Holistic Scoring Guide" (Retitled: "Portfolio Holistic Scoring Guide") from *Kentucky Department of Education,* 1994–1995. Copyright © 1994 by Kentucky Department of Education.

Knight-Ridder Tribune News Service: From "'Parallax Conspiracy' Has Angry Umpires in a Tizzy" by Bill Conlin, as it appears in the *Albuquerque Journal,* October 21, 1993. Copyright © 1993 by Tribune Media Services.

NASA and Goddard Space Flight Center: From section 5.1.4 "Pointing Control Subsystem (PCS)" from *Hubble Space Telescope Media Reference Guide: 1st Serving Mission.*

National Council of Teachers of Mathematics: Adapted from "Activities: Spatial Visualization" by Glenda Lappan, Elizabeth A. Phillips, and Mary Jean Winter from *Mathematics Teacher,* vol. 77, no. 8, November 1984. Copyright © 1984 by the National Council of Teachers of Mathematics.

The New York Times Company: From "Gem Studded Relics in Egyptian Tomb Amaze Explorers" from *The New York Times,* December 1, 1922. Copyright © 1922 by The New York Times Company. From "Math Problem, Long Baffling, Slowly Yields" by Gina Kolata (with map "The Efficient Traveling Salesman") from *The New York Times,* March 12, 1991. Copyright © 1991 by The New York Times Company. From "Main Telescope Repairs are Completed" by John Noble Wilford from *The New York Times,* December 9, 1993. Copyright © 1993 by The New York Times Company. Headlines "Severe Earthquake Hits Los Angeles," "Collapsed Freeways Cripple City," "Scientists Say Unknown Fault Deep Within Earth Probably Caused Tremor" and map "Damage: First the Quake, Then Aftershocks and Fires" from *The New York Times,* January 18, 1994. Copyright © 1994 by The New York Times Company. From "Astronauts Snare Hubble Telescope for Vital Repairs" by John Noble Wilford from *The New York Times,* December 5, 1993. Copyright © 1993 by The New York Times Company.

Oxford University Press, London: "Fig. 2.5" on p. 36; "(h) prism" on p. 44; illustrations on p. 47; and "Fig. 3.7" on p. 71 from *Chinese Mathematics: A Concise History* by Lǐ Yǎn and Dù Shírán, translated by John N. Crossley and Anthony W.-C. Lun. Translation copyright © 1987 by John N. Crossley and Anthony W.-C. Lun.

Science News, The Weekly Newsmagazine of Science: From "The Big Fix" by Ron Cowen from *Science News®,* vol. 144, no. 19, November 6, 1993, pp. 296-297. Copyright © 1993 by Science Service, Inc.

Greg Stec: From "Message of the Maya in Modern Translation" by Greg Stec (Retitled "Message of the Maya in Modern Times") from *The Christian Science Monitor,* June 22, 1989. Copyright © 1989 by Greg Stec.

United States Olympic Committee: "Olympic Rings" by The U.S. Olympic Committee.

Routledge, a division of International Thomson Publishing Services: From *Mysticism and Logic* by Bernard Russell. Copyright 1917 by The Bertrand Russell Peace Foundation.

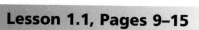

Additional Answers

Lesson 1.1, Pages 9–15

Exploration

1. *A,B,C,D,E,F,G,H,* point; The intersection of two lines is a point.
3. No limit because lines have no thickness and may be oriented in any direction.
2. $\overleftrightarrow{AB}$, $\overleftrightarrow{BC}$, $\overleftrightarrow{CD}$, $\overleftrightarrow{DA}$, $\overleftrightarrow{AE}$, $\overleftrightarrow{BF}$, $\overleftrightarrow{CG}$, $\overleftrightarrow{DH}$, $\overleftrightarrow{EF}$, $\overleftrightarrow{FG}$, $\overleftrightarrow{GH}$, $\overleftrightarrow{HE}$, lines;
   The intersection of two planes is a line.
3. 1 line passes through both points. No
   Through any two points there is exactly 1 line.
4. 1 plane. No
   Through any three noncolinear points there is exactly 1 plane.
5. Sample answers. Plane *ABC*, $\overleftrightarrow{AB}$; plane *DCG*, $\overleftrightarrow{GH}$. Yes
   If two points are in a plane, then the line containing them lies in the plane.

Communicate

1. Geometric figures are mental constructions, and are not physical. For example, planes and lines have no thickness and points have no size at all.
2. Depending on the particular drawing, one name might be more clear than another.
3. First count all the segments that have A as the left endpoint. Then count all segments that have *B* as the left endpoint. Finally, count the segments that have *C* as the left endpoint. Yes, you could start by counting the segments that have *D* as the right endpoint, for example.
4. First count all the angles that have $\overrightarrow{VA}$ as one side. Then count the angles that have $\overrightarrow{VB}$ as one side not counted in the first step. Finally count the angles that have $\overrightarrow{VC}$ as one side, excluding angles counted in the first two steps. Yes, you could start by counting the angles that have $\overrightarrow{VD}$ as one side, for example.
5. Rays must be named with 2 points in diagrams like this. The first point in the name of a ray gives its starting point and the second point gives its direction. Switching the letters in the name of a ray changes both its starting point and its direction.
6. Depending on the particular drawing, one way of naming an angle may be better than another. Also, confusion may be avoided when more than one angle has the same vertex.

Lesson 1.2, Pages 16–22

Exploration 1

3. ...perpendicular lines. The lines meet at right angles.
6. They are parallel.

Exploration 2

2. The distances from the endpoints of the segment to the perpendicular bisector are equal.
3. Line n bisects $\angle BAC$
7. A point on the angle bisector is equidistant to the angle sides.

Communicate

1. All angle pairs formed match up, because each angle is a right angle. Perpendicular lines form 4 right angles when they intersect.
2. 8 right angles are formed. If 2 lines in a plane are perpendicular to the same line, then they are parallel.
3. The two parts of $\overline{AB}$ must have equal length because they fold onto each other and match exactly.
4. Measure the distances from the line points to the endpoints of the segment. If the distances are equal, then the line is the perpendicular bisector. Measure the distances from the line points to the sides of the angle. If the distances are equal, then the line is the angle bisector. Measure from at least two points, since two lengths must be compared in each case.

Lesson 1.3, Pages 23–28

Exploration 1

5. The angle bisectors of a triangle meet at one point.
   The perpendicular bisectors of the triangle sides meet at one point.
   The medians of a triangle meet at one point.
   The altitudes of a triangle meet at one point.

Exploration 2

A circle that circumscribes the triangle can be drawn if its center is the point where the triangle perpendicular bisectors meet.
A circle inscribed in the triangle can be drawn if its center is the point where the triangle angle bisectors meet.

Communicate

1. No. If the circle touches all three sides of the triangle, the center must be inside the circle.
2. Yes. In an obtuse triangle, the circumcenter is outside the triangle.
3. No. All the medians are inside the triangle, so the intersection of the medians will also be inside the triangle.
4. Yes. An obtuse triangle has altitudes that are completely outside the triangle, so the point of intersection must also be outside the triangle.

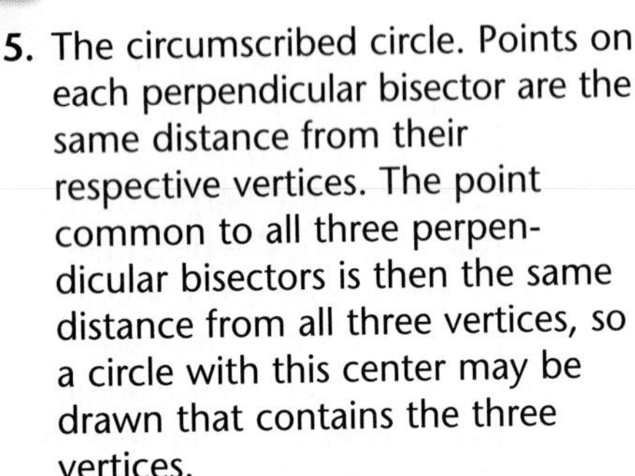

5. The circumscribed circle. Points on each perpendicular bisector are the same distance from their respective vertices. The point common to all three perpendicular bisectors is then the same distance from all three vertices, so a circle with this center may be drawn that contains the three vertices.

Lesson 1.4, Pages 29–36

Communicate

1. When a segment was folded onto itself, the two parts of the segment were congruent because they matched exactly, and the segment was bisected.
2. Congruent segments always have equal measure only if they are measured with a ruler with equal divisions.
3. Common answers: meters, centimeters, millimeters, kilometers, feet, inches, yards, miles.
4. Very large distances or very small distances may be easier to measure in larger or smaller units.
5. $\overline{MN} + \overline{OP} = 30$ cm does not make sense, since segments may not be added to produce a number. $MN + OP = 30$ cm and $m\overline{MN} + m\overline{OP} = 30$ cm both make sense, since numbers are added to produce numbers.
6. In order to measure segments that are not an integral number of units long to increase precision.

Lesson 1.5, Pages 37–43

Exploration

1. Sample Table:

m∠1	m∠2	m∠1+m∠2
60°	120°	180°
50°	130°	180°
80°	100°	180°
$x°$	$y°$	180°

2. The sum of the angle measures of a linear pair is 180°
3. Yes. The sum of the angle measures must be the angle measure of a half circle, or 180°.
4. 180° The two protractor angles are linear pairs.

Communicate

1. Sample Answers: Highway names, street names, maps, landmarks.
2. The complement is $90° - 43° = 47°$, while the supplement is $180° - 43° = 137°$.
3. Both use uniform units, and measure the absolute value of the difference between coordinates. The measure of an angle is based on a circle, while the measure of a segment is based on a line.
4. $m\angle A$ and $m\angle 2$ are numbers, while $\angle X$ and $\angle Y$ are geometric figures, so only $m\angle A + m\angle 2 = 190°$ makes sense.
5. They are both right angles. If they are congruent and supplementary, their measure must be $\frac{180}{2} = 90°$.
6. When the line is folded onto itself, the crease and the line then form two linear pairs above and below the crease. Since the angle pairs are congruent and form a linear pair, they are 90° angles.

Lesson 1.6, 44–50

Exploration 1

4. l is the perpendicular bisector of $\overline{PP'}$.

   Since the line of reflection is the perpendicular bisector of the pre-image and image segments, the image point is the mirror image of the pre-image point across the line of reflection.

Exploration 2

2. A segment and its reflection image are congruent.
4. A triangle and its reflection image are congruent.

Communicate

1. Translation, since the canoe "slides" down the river.
2. Rotation, because the canoe is "rolling" or rotating.
3. Rotation, because the center of the Earth is turning around the sun.
4. Translation, because they are "gliding" or sliding through the air.
5. The point is translated along the road and rotated around the center of the wheel simultaneously.
6. The one on the right, since it shows "CAT" and its mirror image.
7. It is not rigid, since the image is not congruent to the pre-image.

Practice and Apply

16.

17.

MOM MOM | MOM

MOM | MOM MOM / WOW

21. Translate $ABCD$ one side unit up, one unit down, and one unit to the left. Then slide the square twice to the right by one unit.
22. Reflect square $ABCD$ through $\overline{CB}$, $\overline{CD}$, $\overline{DA}$, and $\overline{BA}$. Then reflect the square at $BEFA$ through $\overline{EF}$.

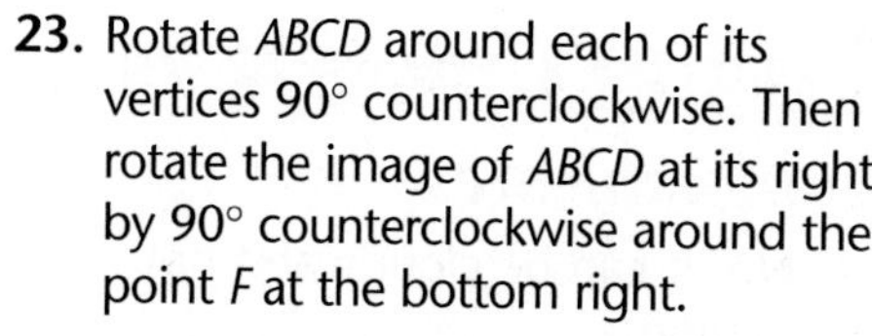

23. Rotate *ABCD* around each of its vertices 90° counterclockwise. Then rotate the image of *ABCD* at its right by 90° counterclockwise around the point *F* at the bottom right.

Lesson 1.7, Pages 51–57

Exploration 1

3. The new figure is congruent to the original one.
4. It moves the chosen number of units up, down, right, or left, depending on the coordinate picked and its sign.
5. To translate a figure horizontally, add the same number to the *x*-coordinate of each point in the figure. To translate a figure vertically, add the same number to the *y*-coordinate of each point in the figure.
6.

Horizontal movement
$(x,y) \rightarrow (x + h,y)$

Vertical movement
$(x,y) \rightarrow (x,y + v)$

Exploration 2

3. The new triangle appears congruent to the original one.
4. The new triangle appears congruent to the original one.
5. *y*-axis in step 3, *x*-axis in step 4
6. To reflect a figure through the *x*-axis, multiply the *y*-coordinate of each point in the figure by -1. To reflect a figure through the *y*-axis, multiply the *x*-coordinate of each point in the figure by -1.
7. Reflection over *x*-axis
$(x,y) \rightarrow (x,-y)$
Reflection over *y*-axis
$(x,y) \rightarrow (-x,y)$

Exploration 3

3. Yes
4. The turn center is the origin. The rotation seems to be 180°
5. 180° rotation about the origin
$(x,y) \rightarrow (-x,-y)$
A rotation of 180° about the origin is equivalent to a reflection through the *x*-axis followed by a reflection through the *y*-axis or vice versa. Yes, but the reflection lines are not always perpendicular to each other.

Communicate

1. To find the *x*-coordinate, start at the origin and find the *x*-coordinate of a vertical line containing the point. To find the *y*-coordinate, start at the origin and find the *y*-coordinate of a horizontal line containing the point.
2. Start at the point (0, 0). The *x*-coordinate tells you how many spaces to move horizontally left or right. The *y*-coordinate tells you how many spaces to move vertically up or down. The combination of movements will reach the point.
3. Since the horizontal coordinate always comes first, the graph shows that changing the order of the numbers changes the position of the point by exchanging "horizontal" and "vertical".

Practice and Apply

8. $(x,y) \rightarrow (x,-y)$

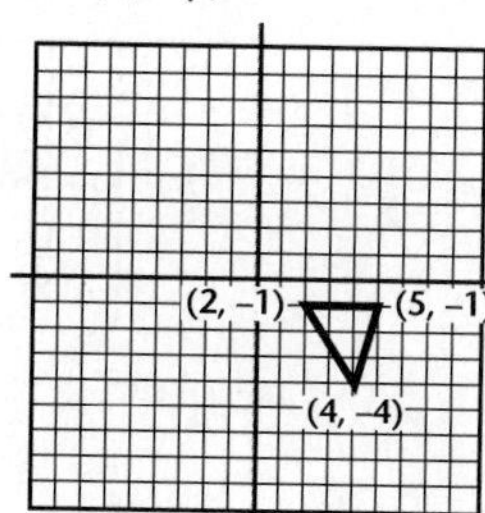

9. $(x,y) \rightarrow (-x,y)$

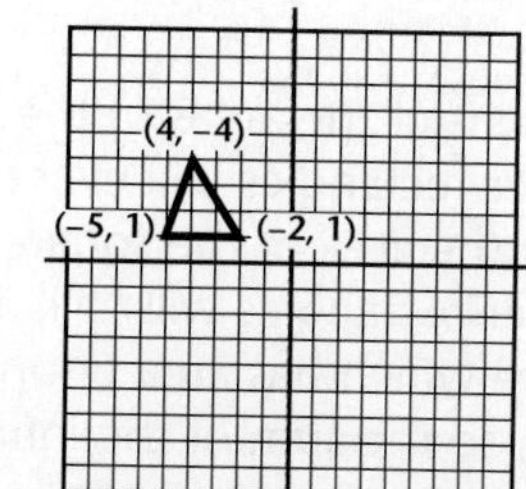

10. $(x,y) \rightarrow (-x,-y)$

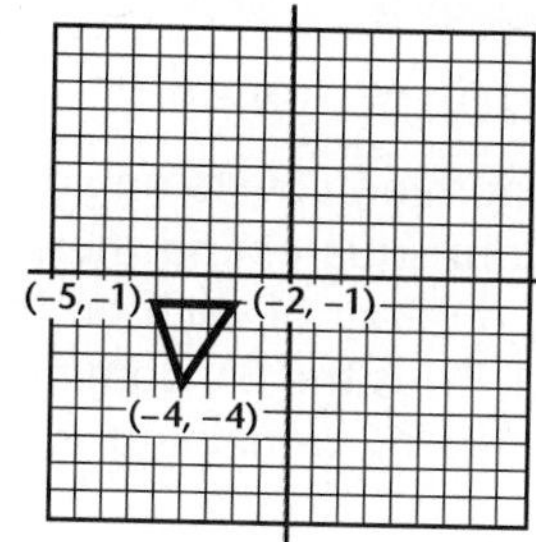

11. A 180° rotation about the origin.

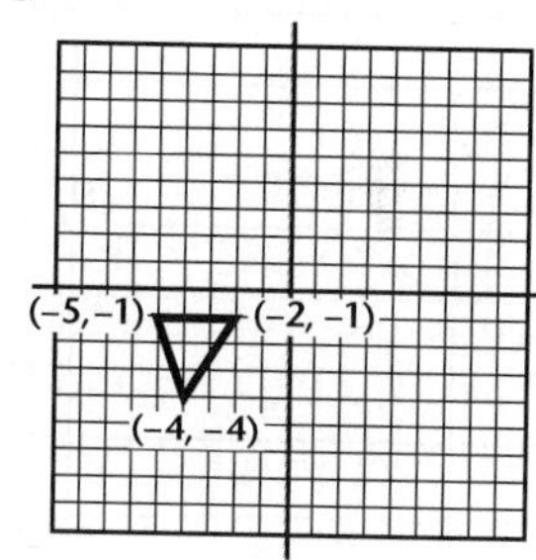

17. translation down 4
19. rotation of 180° about the origin
20. translation left 7
23. Sample table

x	y
0	0
2	2
5	5

24.

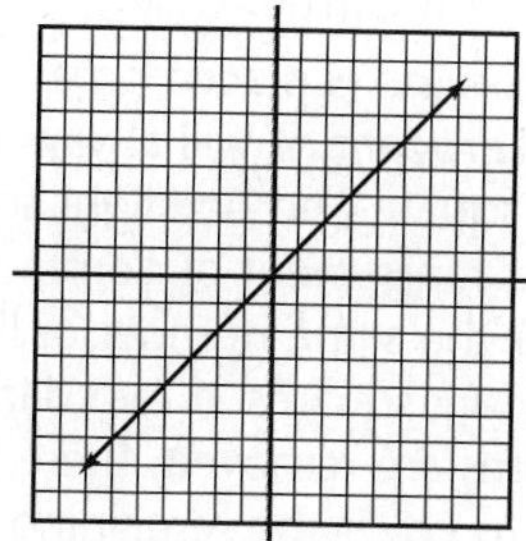

26,27.

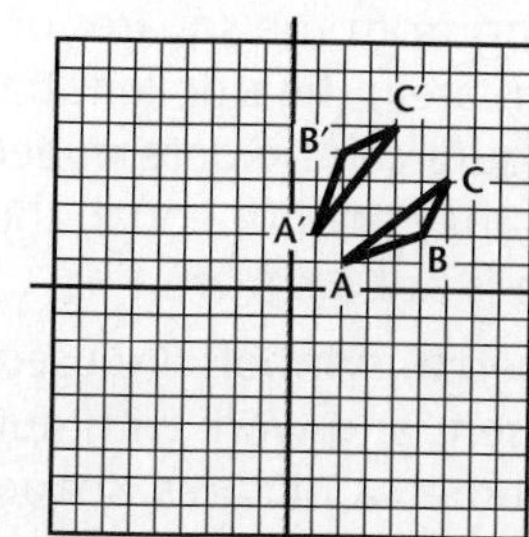

Lesson 2.1, Pages 66–71

Exploration

1. Each domino covers one dark and one light square, and there are an equal number of each on the new board. Thus the 31 dominoes can cover the 62 remaining squares.
2. The diagram shows that the sum of the first n odd numbers is the same as the number of balls arranged in a square with n balls on a side. Student tables should show that the sum of the first n numbers $= n^2$

3.

Square	Length of Side	Reason
C	8	$\sqrt{64}$
D	9	$\sqrt{81}$
E	10	Length of $D + 1$
H	7	Length of $C - 1$
B	15	Length of C + Length of H
G	4	Length of E − (Length of $H - 1$)
F	14	Length of G + Length of E
A	18	Length of F + Length of G

Length of $A + B = 18 + 15 = 33$
Length of $A + F = 18 + 14 = 32$
The sides are not the same length, so the figure is not a square.

Communicate

1. A proof is a sound, fully convincing argument that a claim is true.
2. None if it is false, but possibly many if it is true.
3. If a domino is placed on a checkerboard, it will always cover 1 dark square and one light square. Any arrangement of dominoes will cover the same number of light and dark squares. If two dark squares are removed, there is no way to cover the remaining 30 dark and 32 light squares.
4. Starting from the squares of known area, the side lengths of the figure can be determined. Since they are not equal, the figure is not a square.
5. In algebra, symbols are used to represent and work with numbers. In geometry, pictures, shapes, or diagrams are often used.

Practice and Apply

8.

Number of terms	sum
5	0.96875
6	0.984375
7	0.9921875
8	0.99609375

10–11.

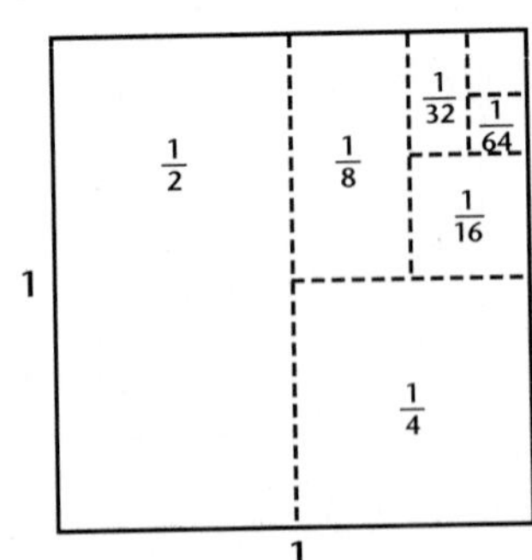

12. The sum of the areas of the smaller rectangles will eventually equal the area of the square. The smaller rectangles will "fill" the square which has area of one unit^2, so the sum of the infinite sequence is 1.

Lesson 2.2, Pages 72–79

Communicate

1. No. It can be cloudy and not raining.
2.

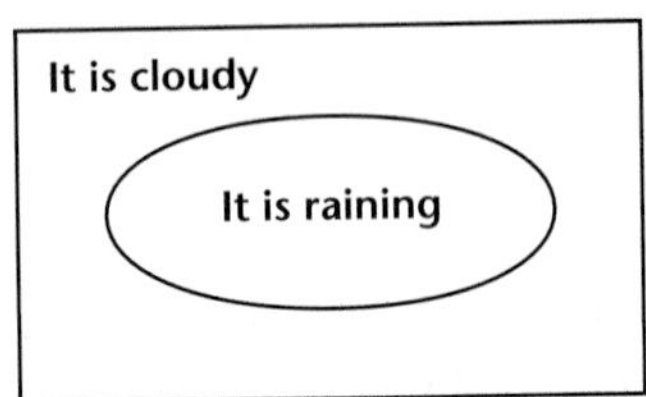

3. If it is raining, then it is cloudy. True.
4. Interchange the *if* and *then* parts of the statement.
5. Find a counterexample.

Lesson 2.3, Pages 82–87

Exploration 1

In 1–3, answers will vary, a sample is given.

1. Right angle pentagon, it has 5 sides and 2 right angles.
2. 6
3. A right angle pentagon has 5 straight sides and 2 right angles.

Exploration 2

Check student definitions; the better the definitions, the more precisely the object is determined.

Communicate

1. In a definition, both the original statement and its converse are true.
2. Everything can't be defined in terms of something simpler. Undefined terms give a starting point from which other terms may be defined.
3. The converse is not always true. Some plants with leaves are not trees.
4. A figure is a blip if and only if it is a circle containing two smaller

circles and has a "tail" consisting of 3 dashes from near some point on the circle. a and d are blips

Practice and Apply

6. If a person is a teenager, then the person is from 13 to 19 years of age.
   If a person is from 13 to 19 years of age, then the person is a teenager.
   A person is a teenager if and only if the person is from 13 to 19 years of age.
   A definition since both statement and converse are true
7. If a number is 0, then the number is an integer between −1 and 1.
   If a number is an integer between −1 and 1, then the number is 0.
   A number is 0 if and only if the number is an integer between −1 and 1.

Lesson 2.4, Pages 88–91

Communicate

1. It is impossible to prove everything from something more basic, a starting point must be found. Postulates give fundamental building blocks for use in proving more complex theorems.
2. Postulates are accepted as true, while theorems are proven to be true.
3.

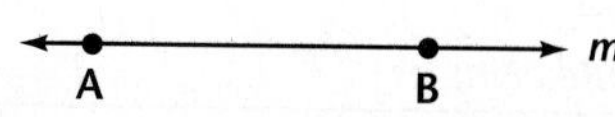

Points A and B determine line m.

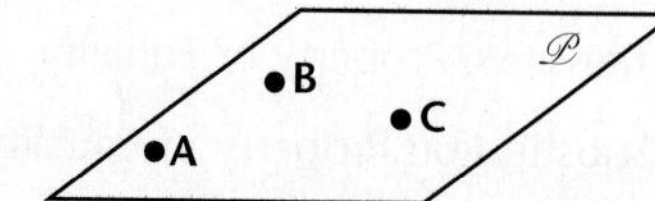

Noncollinear points $A, B,$ and C determine plane P.

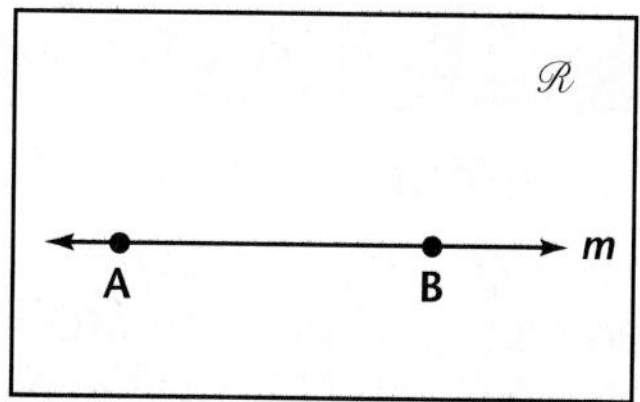

If plane R contains points A and B, then plane R contains line m.

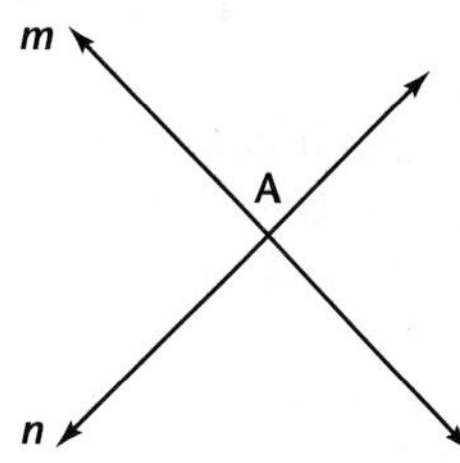

The intersection of line m and line n is point A.

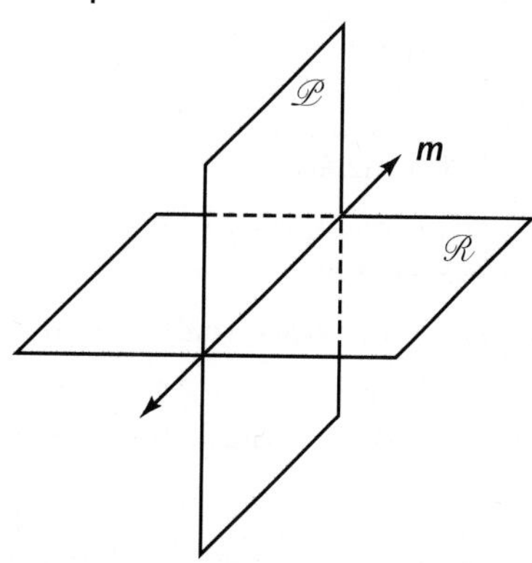

The intersection of plane P and plane R is line m.

4. This possible on a curved surface like that of the Earth The "lines" of longitude all go through both the North Pole and the South Pole.

Practice and Apply

17.

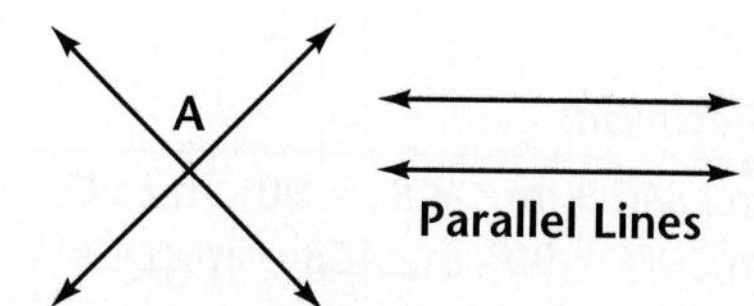

18.

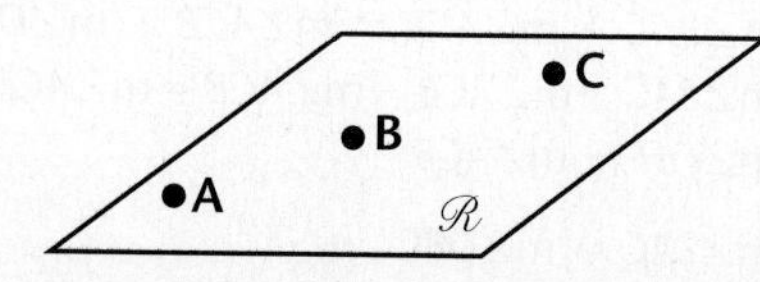

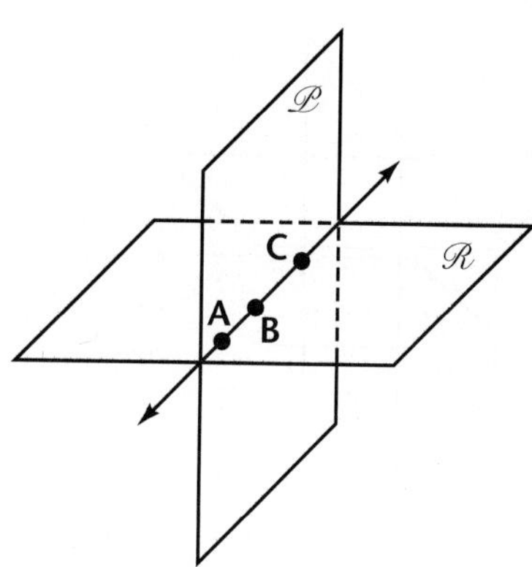

Collinear points $A, B,$ and C are contained in more than one plane.

19.

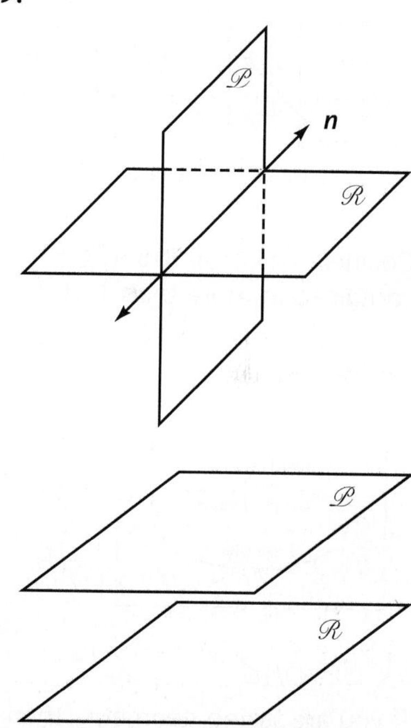

Parallel planes

20.

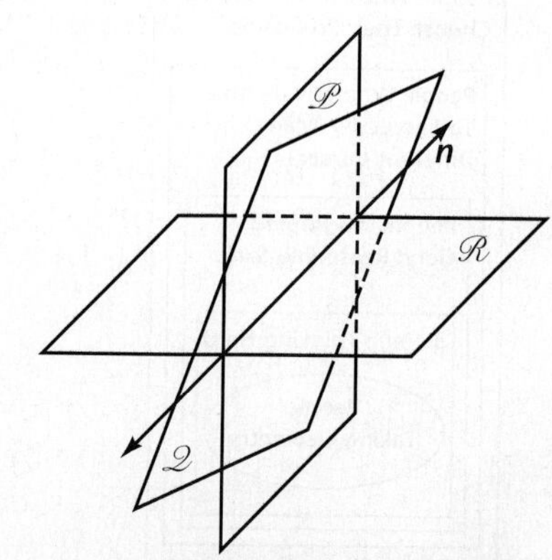

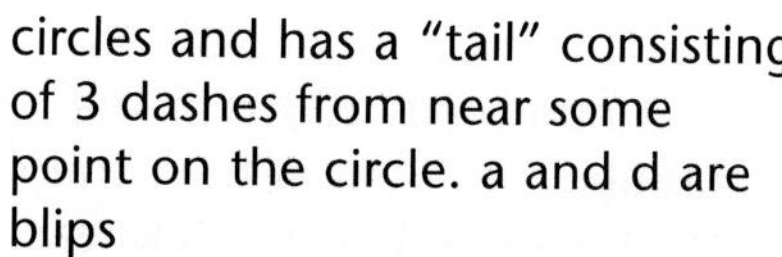

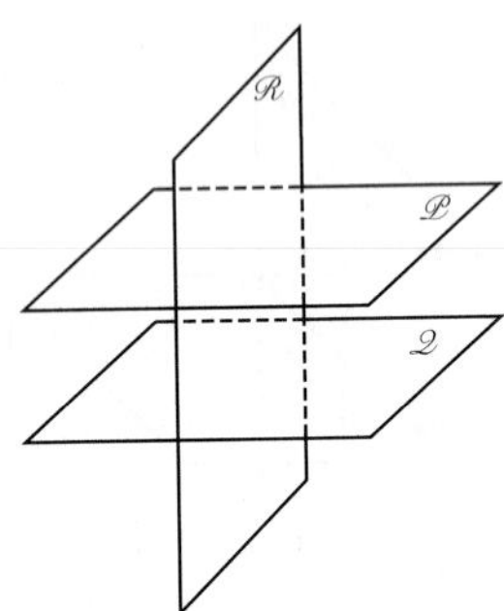

21.

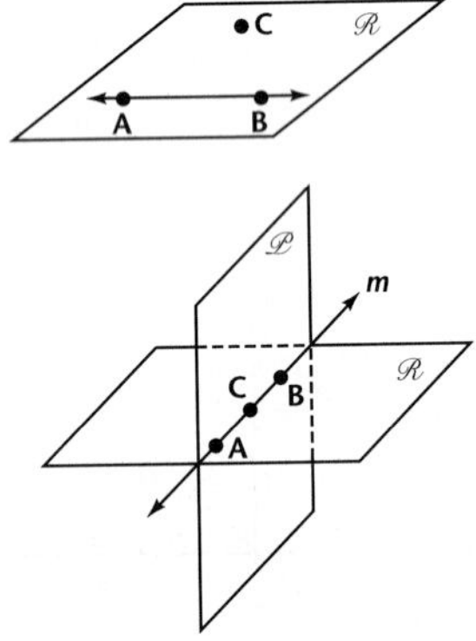

Collinear points *A*, *B*, and *C* are contained in more than 1 plane.

Look Back

24.

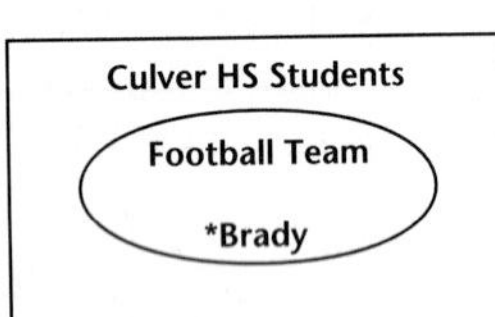

Look Beyond

25. If you are taking geometry, then you will be able to choose your profession.

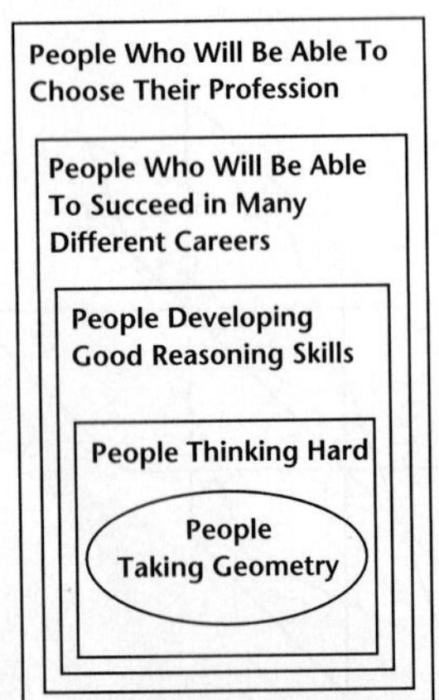

Lesson 2.5, Pages 92–98

Communicate

1. Since $MN = OP$ and $NO = NO$, then $MO = NP$ because $MO = MN + NO$ and $NP = OP + NO$. If equals are added to equals, then the wholes are equal.
2. Algebra can make relationships between parts of a figure easier to see because it can be used in calculations based on the relationships. Geometry can make algebraic relationships easier to understand by giving them a concrete picture.
3. Since Fiona and Jada have an equal number of books, when that number is added to the number of books in the library, the total number of books each has access to is the same. The Addition Property of Equality guarantees this.

Practice and Apply

22. If $m\angle PLS = m\angle CLA$, then $m\angle PLA = m\angle SLC$.

25. If $m\angle BAC = m\angle EDF$ and $m\angle EDF = m\angle HGI$, then $m\angle BAC = m\angle HGI$.

26. Joe wears the same size helmet as Lara by the Transitive Property of Congruence.

27.

Statements	*Reasons*
1. $m\angle PLS = m\angle ALC$	1. Given
2. $m\angle PLA + m\angle ALS = m\angle PLS$, $m\angle SLC + m\angle ALS = m\angle ALC$	2. Angle Addition Postulate
3. $m\angle PLA + m\angle ALS = m\angle SLC + m\angle ALS$	3. Substitution Property of Equality
4. $m\angle PLA + m\angle ALS - m\angle ALS = m\angle SLC + m\angle ALS - m\angle ALS$, $m\angle PLA = m\angle SLC$	4. Subtraction Property of Equality

28.

Statements	*Reasons*
1. $m\angle CBD = m\angle CDB$, $m\angle ABD = 90°$, $m\angle EDB = 90°$	1. Given
2. $m\angle ABD = m\angle EDB$	2. Transitive Property of Equality
3. $m\angle ABC + m\angle CBD = m\angle ABD$, $m\angle EDC + m\angle CDB = m\angle EDB$	3. Angle Addition Postulate
4. $m\angle ABC + m\angle CBD = m\angle EDC + m\angle CDB$	4. Substitution Property of Equality
5. $m\angle ABC + m\angle CDB - m\angle CBD = m\angle EDC + m\angle CBD \angle m\angle CBD$, $m\angle ABC = m\angle EDC$	5. Subtraction Property of Equality

29.

Statements	*Reasons*
1. $m\angle BAC + m\angle ACB ,= 90°$, $m\angle DCE + m\angle DEC = 90°$, $m\angle ACB = m\angle DCE$	1. Given
2. $m\angle BAC + m\angle ACB = m\angle DCE + m\angle DEC$	2. Transitive Property of Equality
3. $m\angle BAC + m\angle ACB = m\angle ACB + m\angle DEC$, $m\angle BAC + m\angle ACB - m\angle ACB = m\angle ACB + m\angle DEC - m\angle ACB$	3. Substitution Property of Equality
4. $m\angle BAC = m\angle DEC$	4. Subtraction Property of Equality

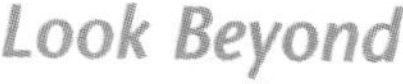

Look Beyond

35a. Take from the other row the same number of counters as your opponent takes. If this leaves 2 counters in each row, then follow the strategy given before. If this leaves 1 counter is each row, your opponent must take the counter in 1 row, leaving the last one for you.

b. Three rows with 1,2, and 3 counters

If your opponent takes all the counters in the row with 3 counters, take the second counter from the row with 2 counters. This leaves 1 counter in each row, your opponent must take the counter in 1 row, leaving the last one for you.

If your opponent takes both counters in the row with 2 counters, take 2 counters from the row with 3 counters. This leaves 1 counter in each row.

If your opponent takes the counter in the row by itself, take 1 counter from the row with 3 counters. This leaves 2 counters in each row.

If your opponent takes 1 counter from the row with 3 counters, take the counter in the row by itself. This leaves 2 counters in 2 rows.

If your opponent takes 1 counter from the row with 2 counters, take all 3 counters from the row with 3 counters. This leaves 1 counter in 2 rows.

If your opponent takes 2 counters from the row with 3 counters, take both counters from the row with 2 counters. This leaves 1 counter in 2 rows. Three rows with 2,3, and 4 counters

Whatever your opponent does, take the number of counters to reduce the situation to one of the ones discussed above.

Take all the counters in the middle row.

Lesson 2.6, Pages 99–105

Exploration 1

3. Vertical angles have equal measures.

4. They are linear pairs and supplementary.

5. $m\angle 1 + m\angle 3 = \underline{180^\circ}$

$m\angle 2 + m\angle 3 = \underline{180}$

6. Transitive Property of Equality.

7. Subtraction Property of Equality.

Exploration 2

4. Translation down

5. The distances are the same.

6. Yes. They all moved perpendicular to the parallel lines.

7. It is half the distance.

8. Reflection over two parallel lines is the same as a translation of twice the distance between the lines in a direction perpendicular to the lines.

Exploration 3

4. Rotation

5. It is the center of rotation.

6. Reflecting a figure over two intersecting lines is the same as rotating the figure about the point of intersection through twice the angle between the lines.

Communicate

1. Inductive reasoning is based on a limited number of examples. It does not establish a result for all possible examples.
Deductive reasoning, however establishes a result for all possible examples, even if there are infinitely many of them.

2. Inductive reasoning allows for discovery of conjectures which may afterwards turn out to be provable as theorems by deduction.

3. $\angle 1$ and $\angle 3$ are congruent, as are $\angle 2$ and the angle labeled with measure 78°. The vertical angles theorem allows us to make this conclusion from the figure. Also, we can conclude from the figure that $m\angle 2 = 78$ and that $m\angle 1 = m\angle 3 = 82^\circ$.

4. Place two parallel lines in the neighborhood of the figure. Reflect the figure twice, first through one of the parallel lines, and then through the other.

Lesson 3.1, Pages 116–123

Exploration 1

2. Scalene triangle: 0
Isosceles triangle: 1
Equilateral triangle: 3

3. The angles opposite congruent segments and the congruent segments coincide.

4. In a scalene triangle, none. In an isosceles triangle, angles opposite congruent sides are congruent.

In an equilateral triangle, all angles are congruent.

5. Each axis of symmetry is the perpendicular bisector of the side of the triangle it passes through.

6. Each axis of symmetry is the bisector of the vertex angle it passes through.

Exploration 2

1. Yes

2. All regular polygons have rotational symmetry.

5. 5

6. Yes. The lengths of the sides are equal, so a full revolution can be equally divided by the central angle measure.

7. 72° close

8. Measure of the central angle of a regular n-gon $= \frac{360^\circ}{n}$.

Communicate

1. Rotational and Reflectional

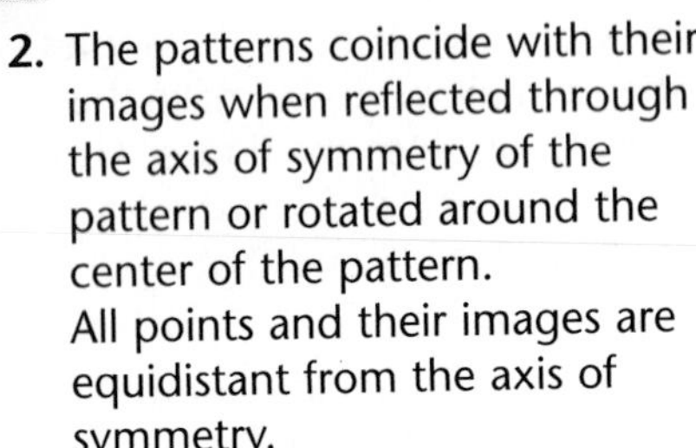

2. The patterns coincide with their images when reflected through the axis of symmetry of the pattern or rotated around the center of the pattern.
All points and their images are equidistant from the axis of symmetry.
3. Answers will vary.
4. The hexagon has 6 distinct axes of symmetry; 60°, 120°, 180°, 240°, 300°, 360°. Other possible angles give repeated axes.
5. Yes
The figures shown have rotational symmetry of 180°, 360° around the point where the axes of symmetry intersect.

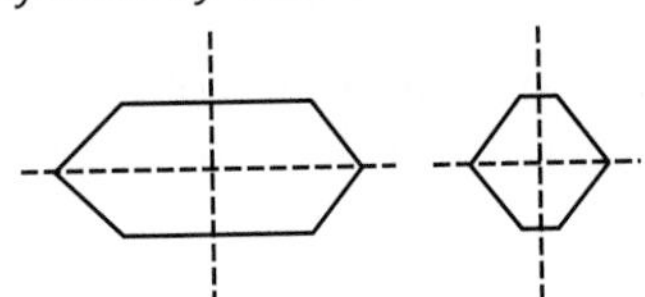

Practice and Apply

39. $x = 1$

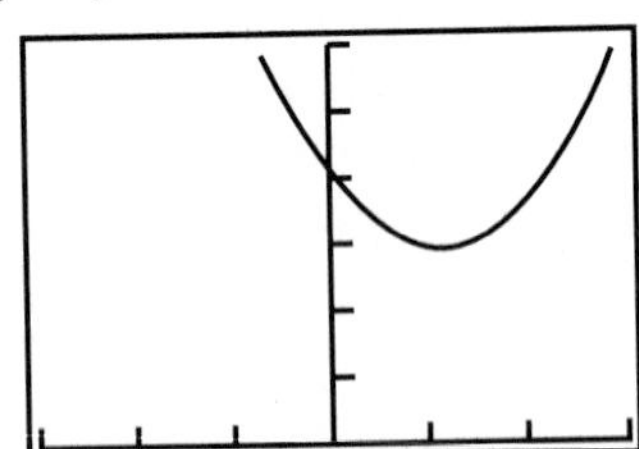

40. $x = 4$

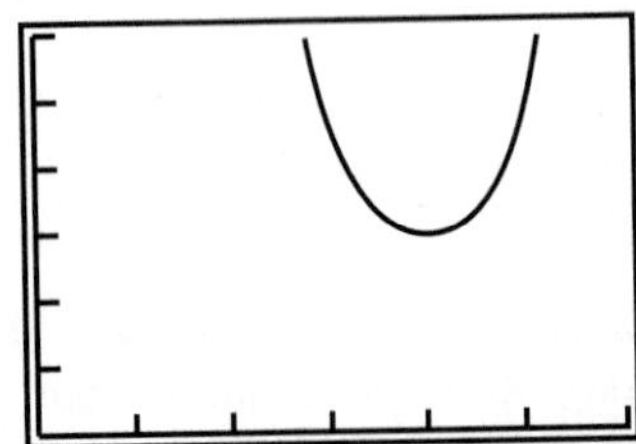

41. $x = -2$

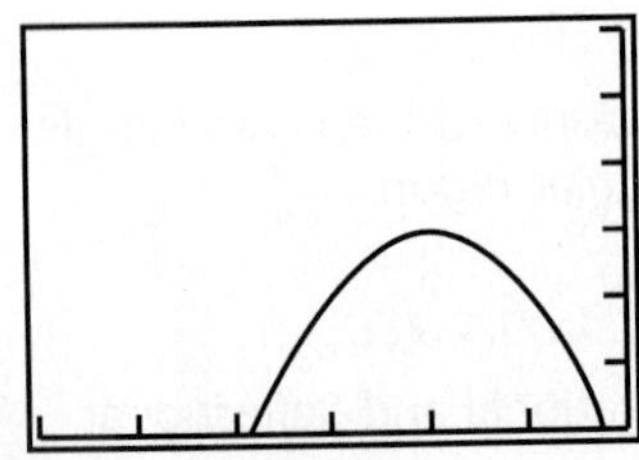

42. $x = -5$

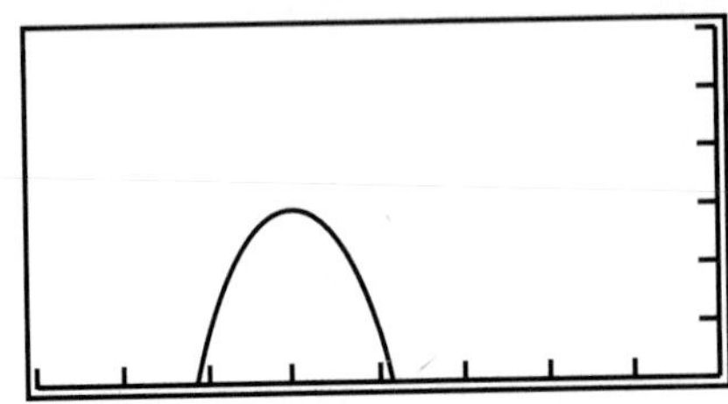

43. $x = 0$

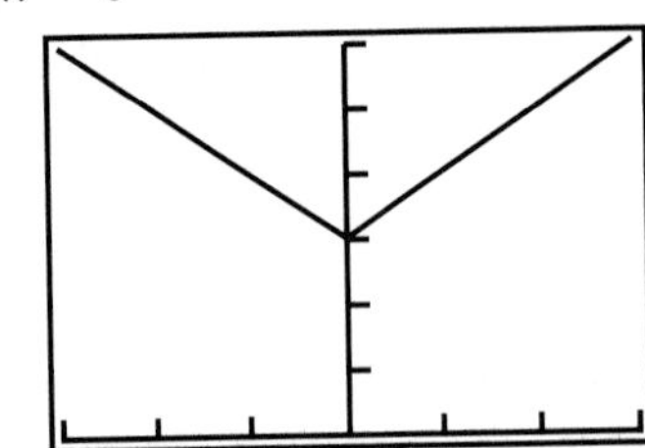

44. $x = -3$

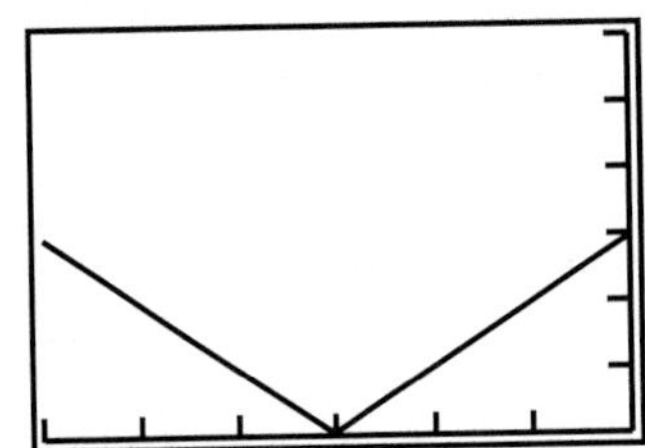

Lesson 3.2, Pages 124–129

Exploration 1

3. Sample Answers: Opposite sides and angles of a parallelogram are congruent. Parallelogram diagonals bisect each other and divide the parallelogram in half. Consecutive angles in a parallelogram are supplementary.

Exploration 2

2. Yes, a rhombus is a parallelogram, so it should have all properties of general parallelograms.
3. The diagonals of a rhombus are perpendicular and bisect its angles.

Exploration 3

2. Yes, because a rectangle is a parallelogram.
3. The diagonals of a rectangle appear to be congruent.

Exploration 4

2. All previous conjectures are true for squares since a square is a parallelogram, a rhombus, and a rectangle.

Communicate

1. A trapezoid is the least specialized since only one condition, that of having exactly one pair of parallel sides, specifies a trapezoid. All the other special quadrilaterals require more than one condition.
2. A rhombus is a kind of parallelogram, which is a special quadrilateral.
3. A square is parallelogram with 4 congruent sides and 4 right angles. A square is a quadrilateral because a parallelogram is a quadrilateral. A square is a rhombus because it has 4 equal sides. A square is a rectangle because it has 4 right angles.
4. Sample answers: floors, walls, ceiling, windows, desktops, posters

Practice and Apply

17.

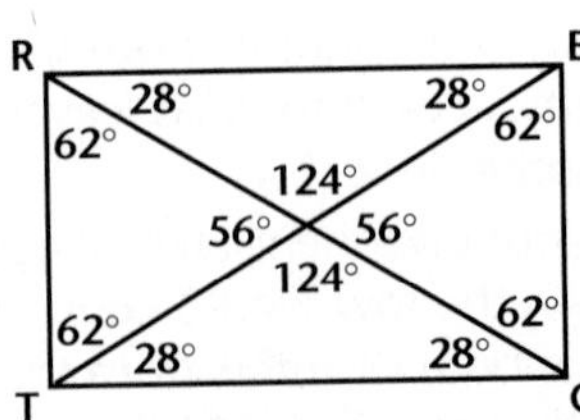

18.

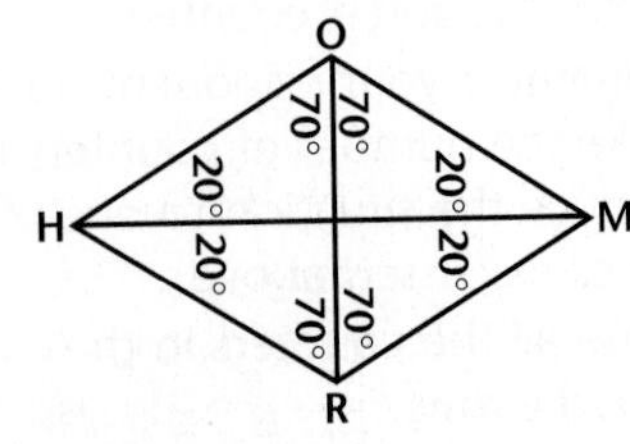

19.

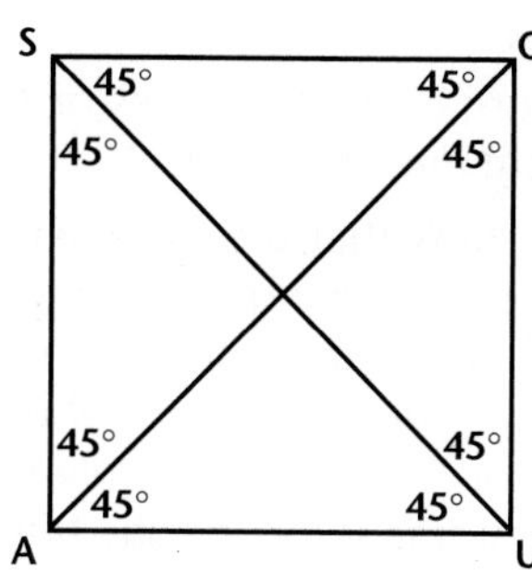

Lesson 3.3, Pages 132–137

Exploration

5. If parallel lines are cut by a transversal, then alternate interior angles, alternate exterior angles, and corresponding angles are congruent. Also, consecutive interior angles are supplementary.

Communicate

1. A transversal is a line, ray, or segment that intersects two or more coplanar lines, rays, or segments, each at a different point.
2. A transversal can cut any two lines at different points, regardless of whether or not the lines are parallel.
3. If parallel lines are cut by a transversal, then alternate exterior angles are congruent. This can be justified by noting that one exterior angle forms a vertical angle with a corresponding angle to the other exterior angle.
If parallel lines are cut by a transversal, then consecutive interior angles are supplementary. This can be justified by noting that one interior angle is supplementary to a corresponding angle to the other interior angle.
4. *PQ* is the transversal for *RQ* and *RP*. *QR* is the transversal for *PQ* and *PR*. *PR* is the transversal for *QP* and *QR*.

Practice and Apply

31. Given: Line $l \parallel$ line m

Line p is a transversal.

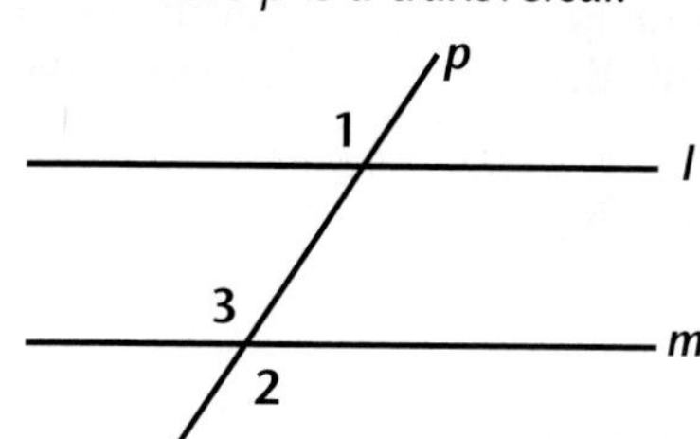

Prove: $\angle 1 \cong \angle 2$

Statements	*Reasons*
1. Line $l \parallel$ line m, Line p is a transversal	1. Given
2. $\angle 1 \cong \angle 3$	2. $\parallel$s⇒Corr.$\angle$s $\cong$
3. $\angle 3 \cong \angle 2$	3. Vertical $\angle$s $\cong$
4. $\angle 1 \cong \angle 2$	4. Transitive Prop of Congruence

32. Given: Line $l \parallel$ line m

Line p is a transversal.

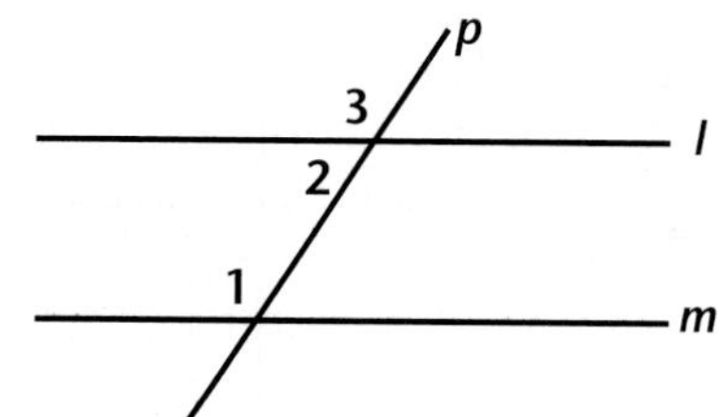

Prove: $m\angle 2 + m\angle 1 = 180°$

Statements	*Reasons*
1. Line $l \parallel$ line m, Line p is a transversal	1. Given
2. $m\angle 2 + m\angle 3 = 180°$	2. $\angle 2$ and $\angle 3$ are a linear pair
3. $m\angle 1 = m\angle 3$	3. $\parallel$s⇒Corr.$\angle$s =
4. $m\angle 2 + m\angle 1 = 180°$	4. Substitution Prop of Equality

Look Beyond

37.

Statements	*Reasons*
1. Quadrilateral *ABCD*, $\overline{BC} \parallel \overline{AD}$	1. Given
2. $m\angle D + m\angle C = 180°$, $m\angle B + m\angle DAB = 180°$	2. $\parallel$s⇒consec int $\angle$s supplementary
3. $m\angle D + m\angle C + m\angle B + m\angle DAB = 360°$	3. Addition

Lesson 3.4, Pages 138–142

Communicate

1. If two lines are cut by a transversal in such a way that corresponding angles are congruent, then the two lines are parallel.
2. If two lines are cut by a transversal in such a way that consecutive interior angles are supplementary, then the two lines are parallel.

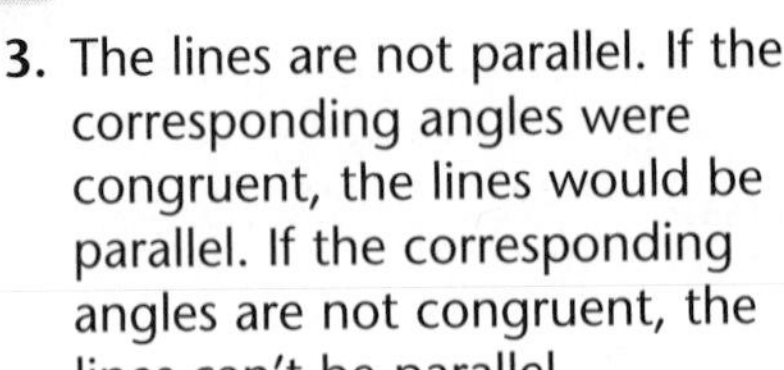

3. The lines are not parallel. If the corresponding angles were congruent, the lines would be parallel. If the corresponding angles are not congruent, the lines can't be parallel.
4. corresponding angles are not congruent

5. An alternative may assert that there are two or more lines through a given point parallel to a given line. This could take place on a saddle surface.

Practice and Apply

22. Two exterior angles are possible at each vertex. Their measures are the same because both are supplementary to the vertex angle.
23. the sum of the measures of its remote interior angles
24. They show that the statement is true for any triangle by using the variable x, which could be any positive number. The first five lines fall short because they only prove the statement for specific triangles.

Lesson 3.5, Pages 143–148

Exploration

4. The sum of the measures of the angles of a triangle is 180°.
5. The parallel postulate insures that line l exists and is the only line through P that is parallel to $\overline{AB}$.
6.

m∠3	m∠4	m∠5	m∠3 + m∠4 + m∠5
110°	40°	30°	180°
80°	20°	100°	180°
50°	30°	100°	180°

7. Yes, m∠3 + m∠4 + m∠5 is always 180°.

Communicate

1. It showed that the three angles of a triangle fit together to form a straight edge so that the angle measures have a sum of 180°. The proof used a constructed line parallel to one side of the triangle and depended on the fact that the sum of the measures of the angles forming that parallel line is 180°.
2. The paper-tearing exercise shows that the property holds only for the particular triangles used, not for all triangles.
3. If there were not exactly one line parallel to a triangle side through its opposite vertex, the sum of the angle measures would vary. The sum would be different for different parallel lines.
4. Because great-circle routes are the shortest between two points on a sphere.

Lesson 3.6, Pages 149–153

Exploration 1

1. Each sum is 180°.
2. 540°
3. The sum of the pentagon angle measures is 540°.
4.

Polygon	Number of Sides	Number of Triangular Regions	Sum of Angle Measures
Triangle	3	1	180°
Quadrilateral	4	2	360°
Pentagon	5	3	540°
Hexagon	6	4	720°
Septagon	7	5	900°
Octagon	8	6	080°
n-gon	n	$n-2$	$180(n-2)°$

5. For a polygon of n sides, the sum of its interior angle measures is $180(n-2)°$.

Exploration 2

5. The sum of the exterior angle measures of a polygon is 360°.
6. 540°
7. 720°
8. $180n°$
9.

Polygon	No. Sides	Sum(Ext+Int Angles)	Sum Interior	Sum Exterior
Triangle	3	540°	180°	360°
Quadrilateral	4	720°	360°	360°
Pentagon	5	900°	540°	360°
Hexagon	6	1080°	720°	360°
n-gon	n	180n°	180(n−2)°	360°

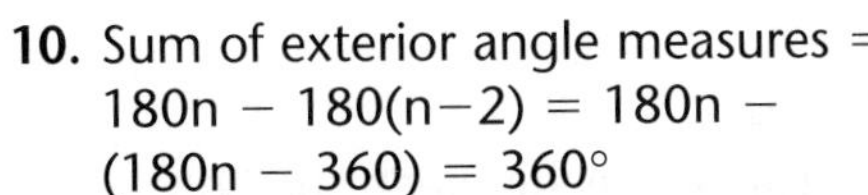

10. Sum of exterior angle measures = 180n − 180(n−2) = 180n − (180n − 360) = 360°

Communicate

1. 180(n−2)°
2. An exterior angle is formed by a side of a polygon and an extension of its adjacent side.
3. No. The sum of the angle measures of a quadrilateral is 360°. So the sum of three 60° angle measures is 180°, which means the 4th angle would have to measure 180°.
4. The sum of 5 obtuse angles would be greater than 450°. The sum of the angle measures of a pentagon is 540°, so that 5 obtuse angles are possible in a pentagon.

Lesson 3.7, Pages 154–159

Exploration 1

2. $MN = \frac{1}{2}BC$
3. m∠1 = m∠2; m∠3 = m∠4; $\overline{MN} \parallel \overline{BC}$ by the corresponding angle converse.
4. The midsegment of a triangle is half the length of and parallel to its third side.

Exploration 2

3. The length of the midsegment is the average of the base lengths.
4. m∠1 = m∠2; m∠4 = m∠5; $\overline{MN} \parallel \overline{CD}$
5. m∠2 + m∠3 = 180°; m∠5 + m∠6 = 180°; $\overline{MN} \parallel \overline{AB}$
6. The midsegment of a trapezoid is parallel to its bases and its length is the average of the base lengths.

Exploration 3

1.

MN
5.5
5
4.5
4
3.5
3.25
3.05

2. A triangle
3. $\frac{1}{2}$ (base 1 + 0) = $\frac{1}{2}$ base 1; The formula is the same as the one used for computing a triangle midsegment.

Communicate

1. Substitute 0 for the length of one of the bases.
2. As the length of one base of a trapezoid approaches 0, the trapezoid approaches a triangle.
3. See Application on page 156 for a sample answer.
4. Yes. The same principle applies if distance is measured vertically.
5. Yes. Assume base 1 is the longer base.

   $\frac{1}{2}$(base 1 − base 2) + base 2 = $\frac{1}{2}$ base 1 − $\frac{1}{2}$ base 2 + base 2 = $\frac{1}{2}$ base 1 + $\frac{1}{2}$ base 2 = $\frac{1}{2}$(base 1 + base 2)

Practice and Apply

14. Let the base have length b. Then the length of each successive parallel segment decreases by $\frac{b}{n}$.
15. Let the bases have lengths b_1 and b_2. Then the length of each successive parallel segment decreases by $\frac{b_1-b_2}{n}$, if b_1 is the longer base.
17. Perimeter of the outer triangle is 46. Perimeter of the inner triangle is 23. Perimeter of the inner triangle is half the perimeter of the outer triangle
22. No. Because (70 + 25) 4 2 = 47. 5 not 42. The trapezoids are strengthened by adding cross-pieces to make triangles which are rigid.

Lesson 3.8, Pages 160–165

Communicate

1. 2; Parallel lines have the same slope.
2.

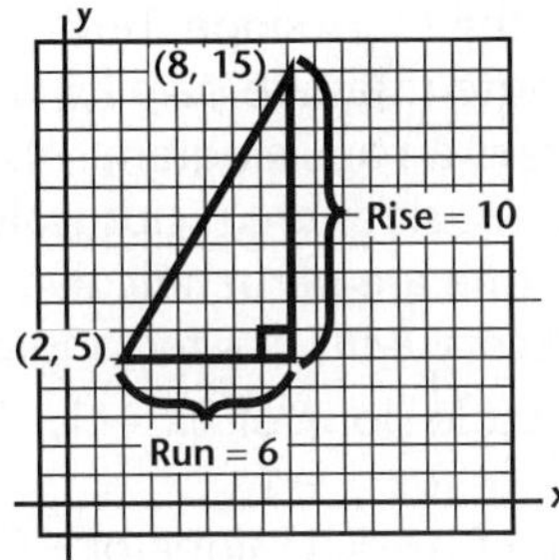

slope = $\frac{10}{6} = \frac{5}{3}$

3. $y_2 - y_1 = 15 - 5 = 10$ gives the vertical rise.
   $x_2 - x_1 = 8 - 2 = 6$ gives the horizontal run.
   The slope is found by dividing the rise by the run.
4. $\frac{-1}{m}$; The slopes of perpendicular lines are negative reciprocals of each other.
5. Find the slopes of the adjacent sides of the quadrilateral. If the slopes are negative reciprocals, then the sides are perpendicular, meaning that the quadrilateral is a rectangle. If the slopes are 0 and undefined, then the lines are horizontal and vertical and therefore perpendicular.

Practice and Apply

9. Rectangle slope of $\overline{CR}$ = 0; slope of $\overline{RA}$ is undefined slope of $\overline{AB}$ = 0; slope of $\overline{BC}$ is undefined
   Vertical lines have undefined slopes and horizontal lines have slopes = 0. Because vertical lines

and horizontal lines are perpendicular, *CRAB* is a rectangle.

10. Trapezoid slope of $\overline{SA} = \frac{3}{2}$;
slope of $\overline{AI} = 0$
slope of $\overline{IN} = \frac{-3}{2}$; slope of $\overline{NS} = 0$
Since one pair of opposite sides have the same slope and one pair doesn't, only one pair of opposite sides is parallel and the figure is a trapezoid.
11. Rectangle slope of $\overline{RA} = \frac{-4}{3}$;
slope of $\overline{AI} = \frac{3}{4}$
slope of $\overline{IN} = \frac{-4}{3}$; slope of $\overline{NR} = \frac{3}{4}$
Since both pairs of opposite sides have the same slope, both pairs of opposite sides are parallel and the figure is a parallelogram. Also, adjacent sides are reciprocals and therefore are perpendicular. Therefore *RAIN* is a rectangle.
12. Rectangle slope of $\overline{TR} = 1$;
slope of $\overline{RA} = -1$
slope of $\overline{AP} = 1$; slope of $\overline{PT} = -1$
Since the slopes of adjacent sides are negative reciprocals, the sides are perpendicular and the figure is a rectangle.

Lesson 4.1, Pages 174–179

Exploration

2. Yes. Probably yes.
3. Yes. No.
4. … if and only if all pairs of corresponding sides and angles are congruent.

Communicate

1. No, a 5 cm segment cannot match a 6 cm segment.
2. No, corresponding sides not congruent.
3. No, corresponding sides not congruent.
4. Yes, the angles can be matched exactly.
5. Yes, the segments can be matched exactly.
6. Yes, corresponding sides and angles are congruent.

Practice and Apply

16a. 1 pair of triangles, 3 pairs of rectangles, 1 pair of pentagons

b. $\angle ROW \cong \angle ECK$; $\angle FRO \cong \angle BEC$; $\angle TFR \cong \angle XBE$; $\angle WTF \cong \angle KXB$; $\angle OWT \cong \angle CKX$; $\angle ORW \cong \angle CEK$; $\angle OWR \cong \angle CKE$; $\angle FRW \cong \angle BEK$; $\angle RWT \cong \angle EKX$; $\angle WTF \cong \angle KXB$; $\angle TFR \cong \angle XBA$; $\angle FRE \cong \angle TWK$; $\angle REB \cong \angle WKX$; $\angle EBF \cong \angle KXT$; $\angle BFR \cong \angle XTW$; $\angle OCE \cong \angle OCK$; $\angle CER \cong \angle CKW$; $\angle ERO \cong \angle KWO$; $\angle ROC \cong \angle WOC$

c. $\overline{FR} \cong \overline{BE}$; $\overline{RO} \cong \overline{EC}$; $\overline{OW} \cong \overline{CK}$; $\overline{WT} \cong \overline{KX}$; $\overline{TF} \cong \overline{XB}$; $\overline{RW} \cong \overline{EK}$; $\overline{TX} \cong \overline{FB}$; $\overline{WK} \cong \overline{\ E}$; $\overline{FR} \cong \overline{TW}$; $\overline{RW} \cong \overline{FT}$; $\overline{WT} \cong \overline{KX}$; $\overline{WK} \cong \overline{TX}$; $\overline{RE} \cong \overline{FB}$; $\overline{FR} \cong \overline{BE}$; $\overline{EK} \cong \overline{BX}$; $\overline{BE} \cong \overline{XK}$

17. Yes; Corresponding sides are congruent and corresponding angles are congruent
20. Answers will vary.

Lesson 4.2, Pages 180–185

Exploration 1

2. Yes. Yes.
No, knowing the three sides is enough.
3. No
4. A triangle constructed with given sides can have no other shape, meaning that they are rigid.
5. If the sides of one triangle are congruent to the sides of another triangle, then the two triangles are congruent.

Exploration 2

Part I

2. Yes. The angle is between the two sides.
3. Yes
4. If two sides and the angle between them of one triangle are congruent to two sides and the angle between them of another triangle, then the two triangles are congruent.

Part II

1. Yes
2. The side is between the two angles.
3. Yes
4. If two angles and the side between them of one triangle are congruent to two angles and the side between them of another triangle, then the two triangles are congruent.

Communicate

1. If the lengths of the sides of a triangle are fixed, there is just one shape it can have, meaning that it is rigid.
2. Given the lengths of the sides of a triangle, you can make exactly one triangle. Check student drawings.
3. Given the lengths of two sides and the measure of the angle between them, you can make exactly one triangle. Check student drawings.
4. Given the measures of two angles and the length of the side between them, you can make exactly one triangle. Check student drawings.
5. In equiangular triangles, all the angle measures are 60°, but many equiangular triangles of different size may be made.

Lesson 4.3, Pages 186–192

Communicate

1. To prove two triangles congruent, it is enough to know that all corresponding sides are congruent.
2. To prove two triangles congruent, it is enough to know that two pairs of corresponding sides and the angles between them are congruent.
3. To prove two triangles congruent, it is enough to know that two pairs of corresponding angles and the sides between them are congruent.
4. Given two sides and an angle

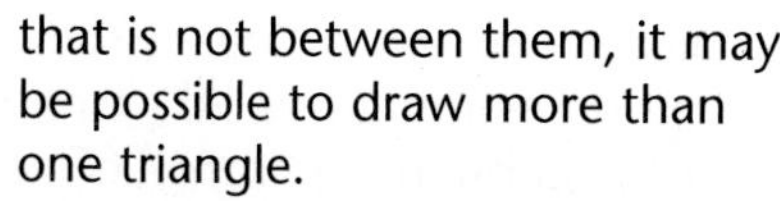

that is not between them, it may be possible to draw more than one triangle.

5. The known parts of the triangle are in the ASA configuration. A triangle with the given combination of measures can have just one shape. The triangle can be drawn on a map of the forest, and the smoke will be at the third vertex of the triangle (the one that is not at a station).

6a. Congruent by SSS

b. Congruent by HL

c. Congruent by SAS

d. Not congruent

Lesson 4.4, Pages 193–200

Communicate

1. Midpoint Definition

2. Reflexive Property of Congruence

3. SSS

4,5. The bisector drawn from the vertex angle of an isosceles triangle (4)divides the triangle into (5)two congruent triangles.

Practice and Apply

12.

Statements	*Reasons*
1. $\overline{XY} \parallel \overline{ZW}$, $\overline{YZ} \parallel \overline{XW}$	1. Given
2. $\overline{YW} \cong \overline{YW}$	2. Reflexive Property of Congruence
3. $\angle ZYW \cong \angle XWY$, $\angle ZWY \cong \angle XYW$	3. $\parallel$s $\Rightarrow$ Alt. int $\angle$s
4. $\triangle XWY \cong \triangle ZYW$	4. ASA
5. $\overline{XY} \cong \overline{ZW}$, $\overline{YZ} \cong \overline{XW}$	5. CPCTC

15.

Statements	*Reasons*
1. $\overline{XY} \parallel \overline{ZW}$, $\overline{YZ} \parallel \overline{XW}$	1. Given
2. $\overline{YW} \cong \overline{YW}$	2. Reflexive Property of Congruence
3. $\angle ZYW \cong \angle XWY$	3. $\parallel$s $\Rightarrow$ Alt. int $\angle$s $\cong$
4. $\overline{YZ} \cong \overline{XW}$	4. Opp sides of parl are $\cong$
5. $\triangle XWY \cong \triangle ZYW$	5. SAS
6. $\angle X \cong \angle Z$	6. CPCTC

By drawing diagonal $\overline{XZ}$ and repeating the argument, $\angle XYZ \cong \angle XWZ$. This completes the proof.

18. For any two sides of an equilateral triangle, the angles opposite those sides are congruent, since it is isosceles. For the other side and one of the two sides, the angles opposite them are congruent. From the transitive property of congruence, the three angles are congruent. Thus they must each measure 60°, since the sum of the measures of a triangle is 180°.

19. The vertex angle bisector divides the isosceles triangle into two congruent triangles. The two angles formed by the base and the bisector are congruent by CPCTC. They are also a linear pair, which means their measures are $\frac{180}{2} = 90°$. Thus the bisector is perpendicular to the base

22.

Statements	*Reasons*
1. $\angle A \cong \angle D$, $\overline{AB} \cong \overline{DE}$	1. Given
2. $\angle ACB \cong \angle DCE$	2. Vertical $\angle$s $\cong$
3. $\triangle ABC \cong \triangle DEC$	3. AAS
4. $\overline{BC} \cong \overline{EC}$	4. CPCTC
5. C is the midpoint of $\overline{BE}$	5. Midpoint definition

23.

Statements	*Reasons*
1. $\angle A \cong \angle D$, $\overline{AB} \cong \overline{DE}$, $\overline{AF} \cong \overline{DC}$	1. Given
2. $\overline{AC} \cong \overline{DF}$	2. Overlapping Segment Thm
3. $\triangle ABC \cong \triangle DEF$	3. SAS
4. $\angle B \cong \angle E$	4. CPCTC

24.

Statements	*Reasons*
1. $\overline{AB} \cong \overline{CB}$, $\overline{AD} \cong \overline{CE}$	1. Given
2. $\angle A \cong \angle C$	2. Isosceles Triangle Thm
3. $\triangle ABD \cong \triangle CBE$	3. SAS
4. $\overline{BD} \cong \overline{BE}$	4. CPCTC
5. $\triangle BDE$ is isosceles	5. Isosceles Triangle Def

Lesson 4.5, Pages 201–206

Communicate

1. The diagonal $\overline{SQ}$ of parallelogram *PQRS* divides it into the two congruent triangles $\triangle SPQ$ and $\triangle QRS$.

2. $\angle QRS$ and $\angle SPQ$ are corresponding angles of congruent triangles $\triangle SPQ$ and $\triangle QRS$ and are therefore congruent by CPCTC.

3a. Yes. The triangles are congruent by SSS, so they can be put together to form parallelogram.

b. No. There is not enough information to prove the

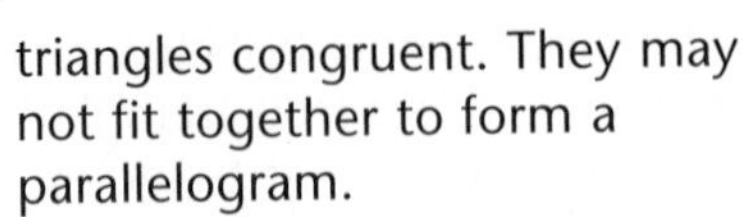

triangles congruent. They may not fit together to form a parallelogram.

c. No. At least one congruent side pair is needed to prove two triangles congruent. They may not fit together to form a parallelogram.

4. Two pairs, the upper and lower pair, and the left and right pair.

Practice and Apply

5. the definition of a parallelogram
6. alternate interior
7. opposite
8. congruent
9. ASA
10. CPCTC
11. $\overline{AB} \parallel \overline{DC}$ by definition of parallelogram, so $\angle BDC$ and $\angle ABD$ are congruent alternate interior angles. Also, $\angle ACD$ and $\angle CAB$ are congruent alternate interior angles. $\overline{AB} \cong \overline{DC}$ because opposite sides of a parallelogram are congruent. $\triangle ABE \cong \triangle CDE$ by ASA, and $\overline{AE} \cong \overline{CE}$ since CPCTC. Therefore, point E is the midpoint of $\overline{AC}$ by definition of midpoint and $\overline{BD}$ bisects $\overline{AC}$ at point E.

    Or

Statements	Reasons
1. Parallelogram $ABCD$ with diagonals, $\overline{AC}$ and $\overline{BD}$ intersecting at E	1. Given
2. $\overline{AB} \parallel \overline{DC}$	2. Parallelogram Definition
3. $\angle BDC$ and $\angle ABD$, $\angle ACD$ and $\angle CAB$	3. $\parallel$s $\Rightarrow$ Alt int $\angle$s $\cong$
4. $\overline{AB} \cong \overline{DC}$	4. Opposite sides of a Parallelogram are $\cong$
5. $\triangle ABE \cong \triangle CDE$	5. ASA
6. $\overline{AE} \cong \overline{CE}$	6. CPCTC
7. E is the midpoint of $\overline{AC}$	7. Midpoint definition
8. $\overline{BD}$ bisects $\overline{AC}$ at point E	8. Bisector definition

13. The diagonals of a parallelogram bisect each other.
14. Reflexive Prop of $\cong$
15. Definition of rhombus.
16. SSS
17. Diagonals of a parallelogram bisect each other.
18. Definition of rhombus or opposite sides of a parallelogram are congruent.
19. SSS
20. First prove that triangle MLP is congruent to triangle OLP by a proof similar to Part A. Then use the Transitive Property of Congruence to establish the congruence of all four triangles.
29. $\overline{RU} \cong \overline{ST}$ and $\overline{RS} \cong \overline{UT}$ because opposite sides of a parallelogram are congruent. By the definition of a rectangle, $\angle URS$ and $\angle STU$ are congruent since they are both right angles. Therefore $\triangle URS \cong \triangle STU$ by SAS, so $\overline{RT} \cong \overline{SU}$ because CPCTC.
31. $\overline{KI} \cong \overline{KE}$ and $\overline{TE} \cong \overline{TI}$ is given. Therefore $\angle KEA \cong \angle KIA$ and $\angle TEA \cong \angle TIA$ because base angles of an isosceles triangle are congruent. Thus $m\angle E = m\angle KEA + m\angle TEA$ ***and*** $m\angle KIA + m\angle TIA = m\angle I$. Therefore, $m\angle E = m\angle I$ by angle addition.
    Then $\triangle KET \cong \triangle KIT$ by SAS, so that $\angle EKA \cong \angle IKA$ because CPCTC. Next, $\triangle KEA \cong \triangle KIA$ by ASA, and $\angle KAE \cong \angle KAI$ by CPCTC. But the last two angles form a linear pair, which means they must each measure 90°. Finally, $\overline{KT} \perp \overline{EI}$ by the definition of perpendicular lines.

Lesson 4.6, Pages 207–212

Exploration Part 1

1. False
2. True
3. True
4. False
5. True

Exploration Part 2

1. False
2. True
3. False
4. True
5. False
6. False

Exploration Part 3

1. False
2. True
3. False
4. True
5. True

Communicate

1a. A quadrilateral with any one following conditions satisfied:
both pairs of opposite sides parallel.
both pairs of opposite sides congruent.
both pairs of opposite angles congruent.
one pair of opposite sides both parallel and congruent.
diagonals that bisect each other.

b. A parallelogram with one of the following conditions satisfied:
one right angle.
congruent diagonals.

c. A parallelogram with one of the following conditions satisfied:
one pair of adjacent sides congruent.
diagonals that are perpendicular.
diagonals that bisect the angles of the parallelogram.

2a. Rhombus, the diagonals bisect the angles, not a rectangle.

b. Rhombus, one pair of adjacent sides is congruent, may be a rectangle.

c. Rhombus, the diagonals are perpendicular, may be a rectangle.

Practice and Apply

18. Given: Quadrilateral $ABCD$ with diagonal $\overline{AC}$, $\overline{AB} \parallel \overline{DC}$, $\overline{AB} \cong \overline{DC}$

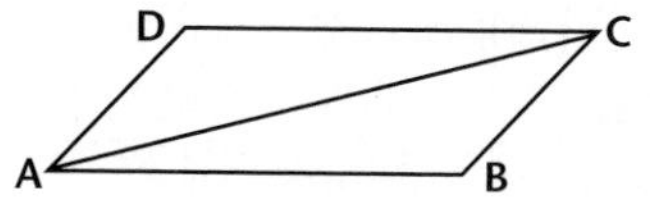

Prove: *ABCD is a parallelogram*

Statements	Reasons
1. Quadrilateral $ABCD$ with diagonal $\overline{AC}$, $\overline{AB} \parallel \overline{DC}$, $\overline{AB} \cong \overline{DC}$	1. Given
2. $\overline{AC} \cong \overline{AC}$	2. Reflexive Prop of $\cong$
3. $\angle BAC \cong \angle DCA$	3. $\parallel$ s $\Rightarrow$ Alt int $\angle$s $\cong$
4. $\triangle ABC \cong \triangle CDA$	4. SAS
5. $\angle BCA \cong \angle DAC$	5. CPCTC
6. $\overline{DA} \parallel \overline{CB}$	6. Alt int $\angle$s $\cong \Rightarrow \parallel$ s
7. $ABCD$ is a parallelogram	7. Parallelogram definition

19. Given: Quadrilateral $ABCD$ with diagonals $\overline{AC}$ and $\overline{BD}$ meeting at E, $\overline{DE} \cong \overline{BE}$, $\overline{AE} \cong \overline{CE}$

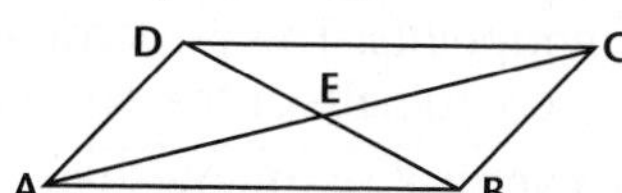

Prove: *ABCD* is a parallelogram
Quadrilateral $ABCD$ with diagonals $\overline{AC}$ and $\overline{BD}$ meeting at E is given. It is also given that the diagonals bisect each other, so $\overline{DE} \cong \overline{BE}$ and $\overline{AE} \cong \overline{CE}$. $\angle DEC \cong \angle AEB$ and $\angle DEA \cong \angle BEC$ because they are vertical angles. Thus $\triangle DEC \cong \triangle BEA$ and $\triangle DEA \cong \triangle BEC$ by SAS. $\angle DCA \cong \angle BAC$ and $\angle DAC \cong \angle BCA$ by CPCTC. Because alternate interior angles are congruent, $\overline{AB} \parallel \overline{DC}$ and $\overline{AD} \parallel \overline{CB}$. Therefore $ABCD$ is a parallelogram by definition.

20. Given: Parallelogram $ABCD$ with right angle $\angle A$

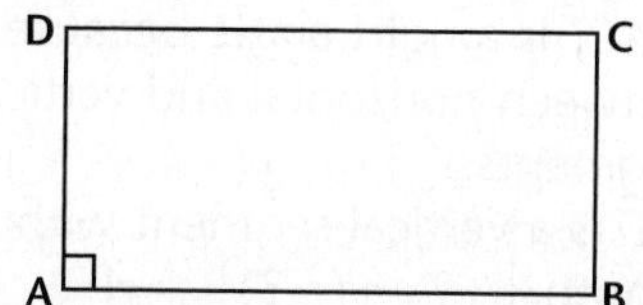

Prove: *ABCD* is a rectangle

It is given that $ABCD$ is a parallelogram with right angle $\angle A$. $\overline{AB} \parallel \overline{DC}$ and $\overline{AD} \parallel \overline{BC}$ by definition of parallelogram. Therefore, $\angle A$ and $\angle B$ and $\angle A$ and $\angle D$ are supplementary because they are same-side interior angles. Thus $\angle B$ and $\angle D$ are right angles, and $\angle C$ is also a right angle because opposite angles in a parallelogram are congruent. Since all the angles in $ABCD$ are right angles, it is a rectangle by definition.

21. Given: Parallelogram $ABCD$, $\overline{AB} \cong \overline{BC}$

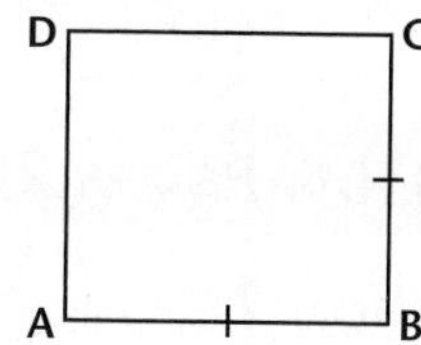

Prove: *ABCD* is a rhombus

Statements	Reasons
1. Parallelogram $ABCD$, $AB \cong BC$	1. Given
2. $\overline{AB} \cong \overline{CD}$, $\overline{BC} \cong \overline{DA}$	2. Opposite sides of a parallelogram are $\cong$
3. $\overline{AB} \cong \overline{BC} \cong \overline{CD} \cong \overline{DA}$	3. Transitive Prop of $\cong$
4. $ABCD$ is a rhombus	4. Rhombus definition

Look Beyond

33.

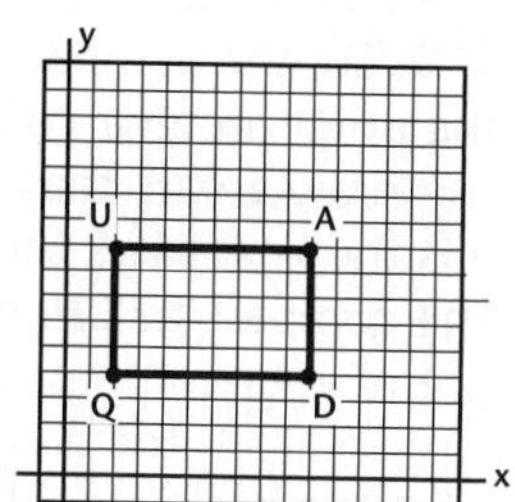

Lesson 4.7, Pages 213–219

Communicate

1. Draw a line and mark a point on the line, say X. Set the compass to AB. Put the compass point on X and draw an arc that intersects the line. Label the point of intersection Y, then $\overline{AB} \cong \overline{XY}$.
2. In the same circle, or in congruent circles, all radii are congruent.
3. Because the radii in a circle or in congruent circles are congruent, it is possible to use a compass to make duplicate congruent segments.
4. Construct an angle, say $\angle ABC$. Draw a circle with center B and label the points where the circle intersects the sides of the angle D and E. Using D and E as centers, construct two congruent circles that intersect. Label the point of intersection that is in the interior of $\angle ABC$ as F. Draw $\overline{BF}$, it is the angle bisector of $\angle ABC$.
5. Construct $\angle ABC$, the angle to be copied. Draw a line l and mark a point Y on the line to be the vertex of the copy. Draw a circle with center B and a congruent circle with center Y. Mark the points where the circle intersects $\angle ABC$ as D and E. Mark the point where the congruent circle intersects line l as Z. Set the compass to DE. Make a circle with center Z and radius DE on the copy. Mark the point where the circle with center Y intersects the circle with center Z as X. Draw $\overline{YX}$, then $\angle ABC \cong \angle XYZ$.

Practice and Apply

6. Draw a line and mark a point on the line D. Set the compass to GJ. Place the point on your compass on D and draw an arc that intersects the line. Mark the point F. Set the compass to GH. Place the point of your compass on D and draw an arc in the area above. Set the compass to JH. Place the point of the compass on F and draw an arc the intersects the arc drawn using D. Label the point of intersection E. Draw $\triangle DEF$, then $\triangle DEF \cong \triangle GHJ$.

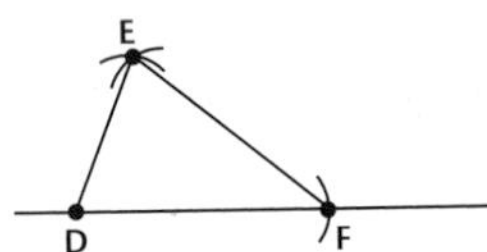

7. Draw a circle with center D. Mark the points where the circle intersects the sides of the angle as Y and X, then draw congruent circles with centers X and Y. Mark the point Z in the interior of the angle where the circles meet. Draw $\overline{DZ}$, the bisector of $\angle EDF$.

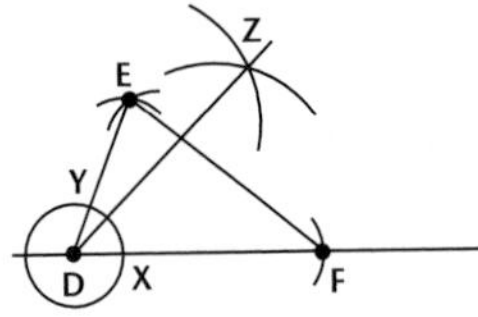

9. $\overline{RS} \cong \overline{RQ} \cong \overline{BC} \cong \overline{BA}$ and $\overline{QS} \cong \overline{AC}$ because they are radii of congruent circles. $\triangle QRS \cong \triangle ABC$ by SSS. So $\angle ABC \cong \angle QRS$ by CPCTC.

23.

Statements	Reasons
1. Line l and point M not on l.	1. Given
2. $\overline{PR} \cong \overline{PT} \cong \overline{MN} \cong \overline{MO}$	2. Congruent Radii Theorem
3. $\overline{RT} \cong \overline{NO}$	3. Congruent Radii Theorem
4. $\triangle RPT \cong \triangle NMO$	4. SSS
5. $\angle RPT \cong \angle MNO$	5. CPCTC
6. line $u \parallel$ line l	6. Corresponding $\angle$s $\cong \Rightarrow \parallel$ s

Look Back

28. $AC \cong ED$, $AC \cong BC$, $AE \cong CD$, $ED \cong BC$, $\angle D \cong \angle ACD$, $\angle D \cong \angle E$, $\angle E \cong \angle ACD$, $\angle D \cong \angle EAC$, $\angle ACD \cong \angle ACB$, $\angle B \cong \angle BAC$

Look Beyond

29c. Regular hexagon; They are congruent.

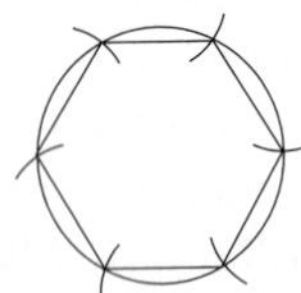

30. Connect alternate points of intersection.

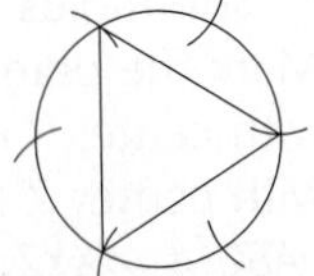

31. Draw perpendicular diameters. Connect the points where the diameters intersect the circle.

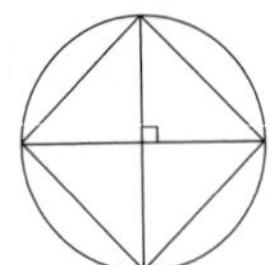

Lesson 4.8, Pages 220–225

Exploration 1

1. Midpoint of $\overline{AB}$ = (6,3); Midpoint of $\overline{XY}$ = (6,−2) Midpoint of $\overline{LN}$ = (4,5); Midpoint of $\overline{CD}$ = (−6,6)
2. Add the x-coordinates and divide by 2.
3. Midpoint of $\overline{KL}$ = (56,432) The midpoint of a horizontal segment is the average of the x-coordinates.

Exploration 2

2. Right Triangle
3. $Q(11,2)$
4. $V(11,5)$; $H(7,2)$
5. Yes
6. M is the midpoint of $\overline{RJ}$.
7. $M(7,5)$ The x-coordinate of M is the same as the x-coordinate of H and the y-coordinate of M is the same as the y-coordinate of V.
8. Find the average of the x-coordinates and the average of the y-coordinates.
9. Midpoint of $\overline{AB} = \left(\frac{x_1 + x_2}{2}, \frac{y_1 + y_2}{2}\right)$

Communicate

1. For a horizontal segment, the midpoint x-coordinate is the average of the x-coordinates of the endpoints. The y-coordinate is the y-coordinate of the endpoints.
2. For a vertical segment, the y-coordinate is the average of the y-coordinates of the endpoints. The x-coordinate is the x-coordinate of the endpoints.
3. The x-coordinate is the average of the x-coordinates of the endpoints. The y-coordinate is the average of the y-coordinates of the endpoints.
4. They are congruent, both have length 5.
5. $\overline{CA}$ is a vertical segment with length $|7 - 3| = 4$.
   $\overline{CT}$ is a horizontal segment with length $|4 - 2| = 2$.
   $\angle ACT$ is a right angle because it is between horizontal and vertical segments.
   $\overline{DO}$ is a vertical segment with length $|-7 - (-3)| = 4$.
   $\overline{DG}$ is a horizontal segment with length $|4 - 2| = 2$.
   $\angle ODG$ is a right angle because it is between horizontal and vertical segments.
   $\triangle CAT \cong \triangle DOG$ by SAS

Practice and Apply

15.

Statements	Reasons
1. Points $L(-5,0)$, $C(0,3)$, $O(0,0)$, $R(5,0)$	1. Given
2. $LO = OR = 5$	2. Definition of length of horizontal segments
3. $OC = OC$	3. Reflexive Prop of Equality
4. $m\angle COL = m\angle COR = 90°$	4. Rt $\angle$s are between vertical and horizontal segments
5. $\triangle COL \cong \triangle COR$	5. SAS
6. $CL = CR(\overline{CL} \cong \overline{CR})$	6. CPCTC
7. $\triangle LCR$ is isosceles	7. Isosceles $\triangle$ definition

16. Sample:

X1	Y1	X2	Y2	MPX	MPY
2	9	6	11	4	10
X1	Y1	X2	Y2	(X1+X2)/2	(Y1+Y2)/2

19. Trapezoid; The midsegment is parallel to the third side but half the length. So one pair of opposite sides is parallel, but not the other.

28. The graph is a parallelogram. The slope of $\overline{QU} = \frac{3}{2}$; slope of $\overline{UA} = 0$; slope of $\overline{AD} = \frac{3}{2}$; slope of $\overline{DQ} = 0$. Both pairs of opposite sides have the same slope and are therefore parallel.

29. The graph is a parallelogram. The slope of $\overline{QU} = 1$; slope of $\overline{UA}$ is undefined; slope of $\overline{AD} = 1$; slope of $\overline{DQ}$ is undefined. Both pairs of opposite sides have the same slope and are therefore parallel.

30. The graph is a parallelogram. The slope of $\overline{QU} = \frac{3}{2}$; slope of $\overline{UA} = \frac{1}{4}$; slope of $\overline{AD} = \frac{3}{2}$; slope of $\overline{DQ} = \frac{1}{4}$. Both pairs of opposite sides have the same slope and are therefore parallel.

31. The graph is a trapezoid. The slope of $\overline{QU} = \frac{4}{3}$; slope of $\overline{UA} = \frac{1}{4}$; slope of $\overline{AD} = -3$; slope of $\overline{DQ} = \frac{1}{4}$. Only one pair of opposite sides have the same slope and are therefore parallel.

32. The graph is a rhombus. The slope of $\overline{QU} = 2$; slope of $\overline{UA} = -2$; slope of $\overline{AD} = 2$; slope of $\overline{DQ} = -2$. Also, $QU = UA = AD = DQ = 2\sqrt{5}$ Both pairs of opposite sides have the same slope and are therefore parallel, and all the sides have the same length.

33. The graph is a rectangle. The slope of $\overline{QU} = 1$; slope of $\overline{UA} = -1$; slope of $\overline{AD} = 1$; slope of $\overline{DQ} = -1$. Slopes of adjacent sides are negative reciprocals and therefore perpendicular.

Look Back

34.

Statements	Reasons
1. $ABCD$ is an isosceles trapezoid, $\overline{BC} \parallel \overline{AD}$, $\overline{AB} \cong \overline{DC}$, $\overline{BX} \perp \overline{AD}$, $\overline{CY} \perp \overline{AD}$	1. Given
2. $\angle BXA$ and $\angle CYD$ are right $\angle$s	2. Perpendicular definition
3. $\overline{BX} \parallel \overline{CY}$	3. Both $\perp$ to $\overline{AD}$
4. $XYCB$ is a parallelogram	4. Parallelogram definition
5. $\overline{BX} \cong \overline{CY}$	5. Opposite sides of a parallelogram are $\cong$
6. $\triangle ABX \cong \triangle DCY$	6. HL
7. $\angle A \cong \angle D$	7. CPCTC

35.

Statements	Reasons
1. Parallelogram $PQRS$, $\overline{OS}$, $\overline{PR}$, $\overline{XY}$ intersect at Z	1. Given
2. $\overline{RZ} \cong \overline{PZ}$	2. Diagonals of a parallelogram bisect each other
3. $\overline{QR} \parallel \overline{SP}$	3. Parallelogram definition
4. $\angle RXY \cong \angle PYX$	4. $\parallel$ s $\Rightarrow$ Alt int $\angle$s $\cong$
5. $\angle XZR \cong \angle PZY$	5. Vertical $\angle$s $\cong$
6. $\triangle XZR \cong \triangle YZP$	6. AAS
7. $\overline{XZ} \cong \overline{YZ}$	7. CPCTC

Lesson 4.9, Pages 226–231

Exploration 1

1. Yes. If two lines are both parallel to a third line, then they are parallel to each other.
3. Yes.
4. $\overline{AA'} \parallel \overline{BB'}$ because they are both parallel to the slide arrow. $\overline{AA'} \cong \overline{BB'}$ because they are radii of congruent circles. So, $AA'B'B$ is a parallelogram because one pair of opposite sides is both parallel and congruent. Therefore, $\overline{AB} \cong \overline{A'B'}$ because opposite sides of a parallelogram are congruent.

Exploration 2

1. Yes, SSS
2. Yes, by applying SSS twice.
3. A polygon is congruent to its image under a translation, since any polygon may be divided into non-overlapping triangles. Each triangle remains congruent under the translation by SSS.

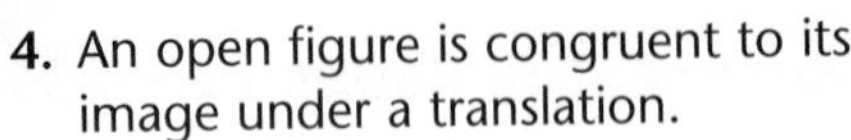

4. An open figure is congruent to its image under a translation.

Exploration 3

1. Yes
2. $AB = A'B'$; $AX = A'X'$; $BX = B'X'$; A segment is congruent to its image under a translation.
3. $A'X' + X'B' = A'B'$; using the Substitution Property of Equality
4. Yes; Yes; Because of the Segment Addition Postulate Converse

Communicate

1. A line segment is congruent to its image under a translation.
2. A triangle or any polygon is congruent to its image under a translation.
3. In a construction the segment lengths are used to make congruent segments using the properties of circles. In a translation the segment endpoints are moved to make congruent segments using the properties of parallelograms.
4. To construct a rotation, copying an angle would be used. To construct a reflection, constructing a perpendicular segment from a point to a line would be used.

Practice and Apply

18.

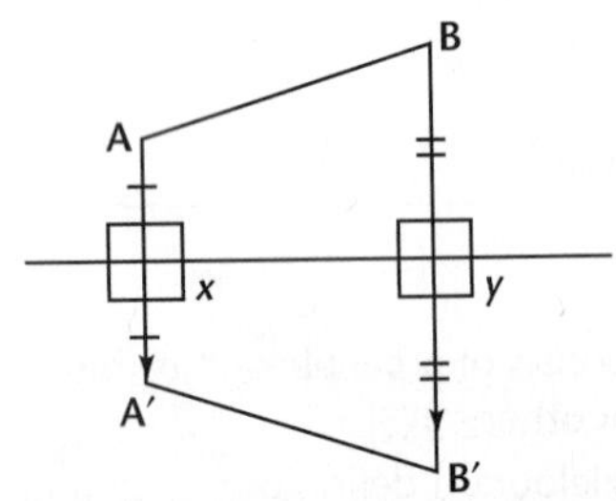

(also, Exercises 19, 20, 21)

19. $\overline{AA'} \perp \overline{XY}$; $\overline{BB'} \perp \overline{XY}$

Every segment connecting a point and its preimage is cut at right angles by the mirror of the reflection.

20. $AX = A'X$; $BY = B'Y$

Every segment connecting a point and its preimage is bisected by the mirror of the reflection.

21. $\overline{AA'} \parallel \overline{BB'}$; $\overline{AA'}$ and $\overline{BB'}$ are both perpendicular to line l.

28–29. Check student drawings. Each side of the triangle must be rotated by the same angle, using a protractor in 28 and compass and straightedge in 29.

Look Back

30. Set your compass equal to a distance greater than XY. Place your compass on X and draw an arc above and below. Using the same compass setting, place your compass on Y and draw arcs that intersect the arcs you drew from X. Label the points of intersection A and B. Draw $\overline{AB}$. $\overline{AB}$ is the perpendicular bisector of $\overline{XY}$

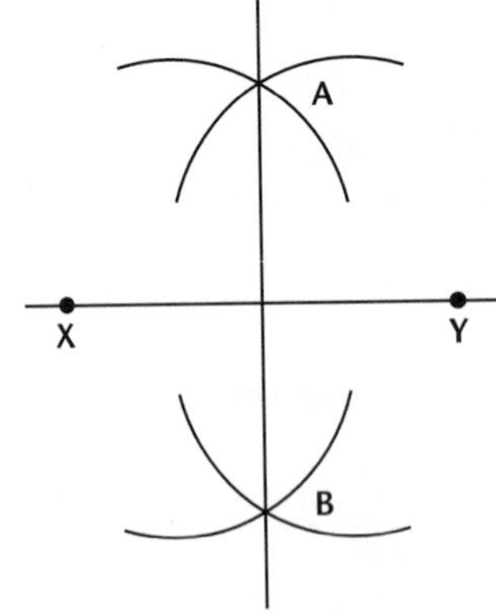

31. Label the angle to be copied $\angle ABC$. Draw a line l and mark a point on the line Y to be the vertex of the copy. Draw a circle with center B and a congruent circle with center Y. Mark the points where the circle intersects $\angle ABC$ as D and E. Mark the point where the circle intersects line l as Z. Set your compass equal to DE. Make an arc with center Z and radius DE. Mark the point where the circle with center Y intersects the arc with center Z as X. Draw $\overline{YX}$. $\angle ABC \cong \angle XYZ$.

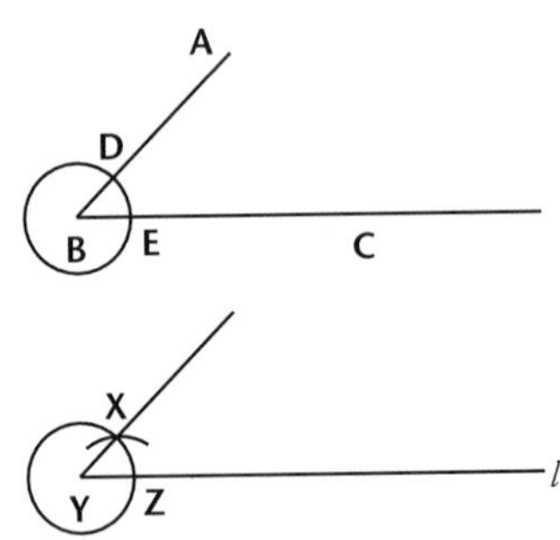

32. Label the triangle to be copied $\triangle ABC$ as shown. Draw line l and mark a point on the line D. Set your compass equal to AC. Place the point on your compass on D and draw an arc that intersects line l. Mark the point F. Set your compass equal to AB. Place the point of your compass on D and draw an arc in the area above. Set your compass equal to CB. Place the point of your compass on F and draw an arc the intersects the arc you drew using D. Label the point of intersection E. Draw $\triangle DEF$. $\triangle DEF \cong \triangle ABC$.

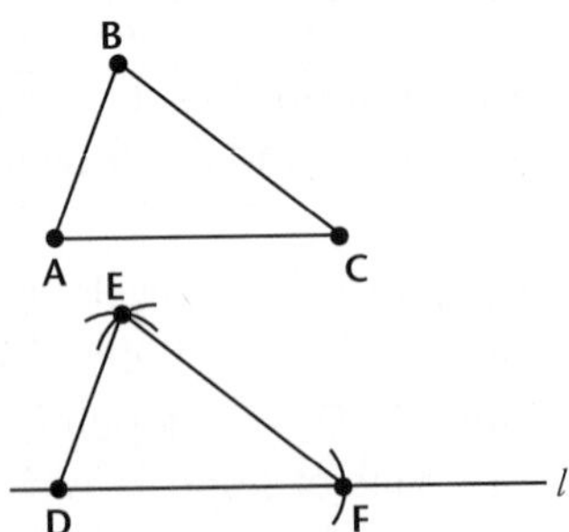

Lesson 5.1, Pages 244–251

Exploration 1

1. Check student drawings. For example, possible length, width pairs are (2,10), (3,9), and (6,6).
2. $h = \frac{24 - 2b}{2} = \frac{1}{2}(24 - 2b) = 12 - b$

3.

b	$h = 12 - b$	$A = bh$
1	11	11
2	10	20
3	9	27
4	8	32
5	7	35
6	6	36
7	5	35

The area values increase until $b = 6$ and then decrease.

4.

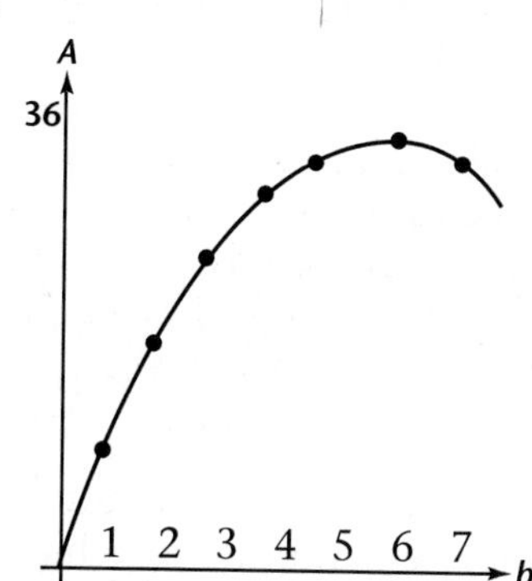

5. 6; 6
6. A square

Exploration 2

1. $h = \frac{3600}{b}$

2.

b	$h = 3600 \div b$	$P = 2b + 2h$
10	360	740
20	180	400
30	120	300
40	90	60
50	72	244
60	60	240
80	45	250

3.

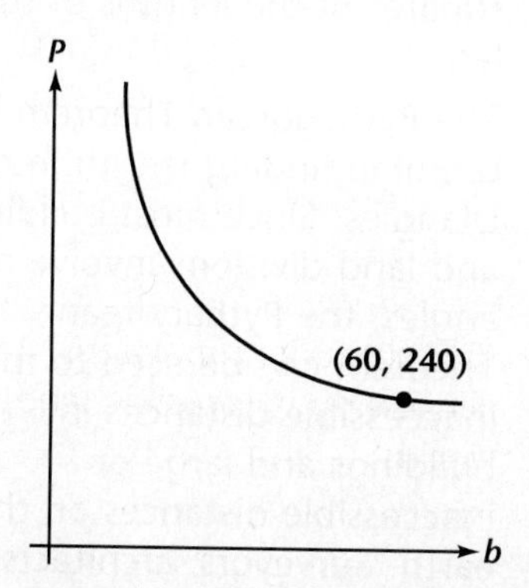

4. 60; 60
5. A square, this is the same rectangle that has maximum area for a given perimeter.

Communicate

1. Draw a polygon inside the pool so that its edges are nearly flat and find the perimeter of the polygon.
2. Divide the surface area into non overlapping rectangles, find the area of each rectangle and add the area of the rectangles.
3. If boundary pieces overlapped, then when finding the sum of the piece lengths some boundary parts would be counted twice.
4. Divide the area into squares or rectangles and find the sum of their areas. The more squares or rectangles used, the better the estimate.

Lesson 5.2, Pages 254–260

Exploration 1, Part 1

2. Right triangles
3. The triangles are congruent. The area of each triangle is half the area of the rectangle.
4. Yes. Match the diagonals of the two triangles.
5. $A = \frac{1}{2}bh$

Exploration 1, Part 2

2. Yes. If two lines are both perpendicular to a third line, then the lines are parallel.
3. A diagonal of a rectangle divides it into two congruent triangles by SSS, SAS or ASA.
4. They are equal because the shaded part is made up of two triangles, each of which is congruent to one of the two triangles that make up the unshaded part.
5. $A = \frac{1}{2}bh$

Exploration 2

2. Right triangle
3. Rectangle; $A = bh$
4. $A = bh$; A parallelogram can be made into a rectangle by removing a right triangle from one side of the parallelogram and matching it onto the other side of the parallelogram.

5.

Statements	*Reasons*
1. $ABCD$ is a parallelogram	1. Given
2. $\overline{AD} \cong \overline{BC}$	2. Opposite sides of a parallelogram $\cong$.
3. $m\angle AEB = m\angle AEF = 90°$	3. Definition of altitude.
4. $\overline{AB} \parallel \overline{DC}$	4. Parallelogram definition
5. $\angle ABE \cong \angle DCF$	5. If lines $\parallel$, then corresponding $\angle$s $\cong$.
6. $\triangle ABE \cong \triangle DCF$	6. AAS
7. $\overline{AE} \cong \overline{DF}$	7. CPCTC
8. $\overline{AE} \parallel \overline{DF}$	8. If two lines are perpendicular to the same line, then the lines are parallel.
9. $ADFE$ is a parallelogram	9. If one pair of opposite sides of a quadrilateral are both congruent and parallel, then it is a parallelogram.
10. $ADFE$ is a rectangle	10. If one angle of a parallelogram is a right angle, then it is a rectangle.

Exploration 3

2.

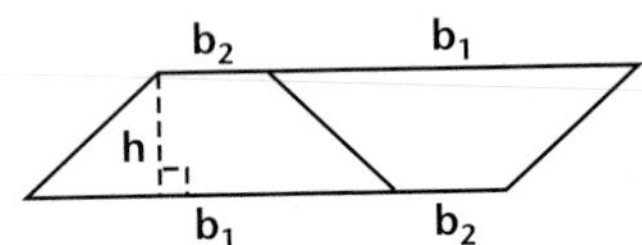

3. One method: Using the diagram above. Two copies of a trapezoid can be put together to form a parallelogram with base $= b_1 + b_2$, height $= h$, and Area $= (b_1 + b_2)h$. Since there are two trapezoids in the parallelogram, the area of one of the trapezoids will be $A = \frac{(b_1+b_2)h}{2}$.

Communicate

1. 7 in^2; The diagonal of a parallelogram divides the parallelogram into two congruent triangles, each of which has half the area of the parallelogram.
2. Consider a parallelogram with $b = 10$ and $h = 2$ and a second parallelogram with $b = 5$ and $h = 4$. The parallelograms are not congruent, but they have the same area.

   Consider a triangle with $b = 10$ and $h = 2$ and a second triangle with $b = 5$ and $h = 4$. The triangles are not congruent, but they have the same area.
3. No. To find the area of a trapezoid, only the lengths of the bases and the height of the trapezoid must be known.

Practice and Apply

8. 150 units^2
9. 70 units^2
10. 64 units^2
11. 270 units^2
12. 143.5 units^2
13. 108 units^2
14. 255 units^2
15. 375 units^2

Look Back

27.

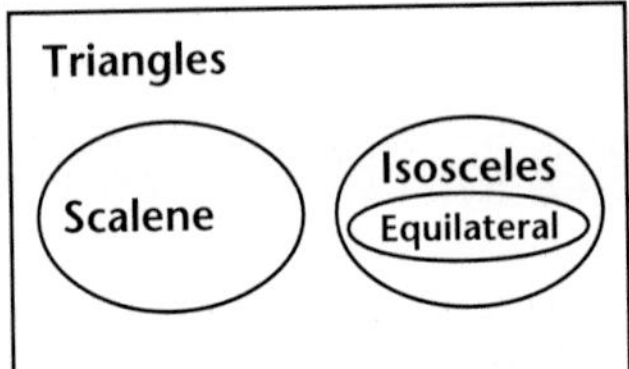

28.

Statements	Reasons
1. $\angle A \cong \angle D$, $\angle EFC \cong \angle BCF$, $\overline{AF} \cong \overline{DC}$	1. Given
2. $\overline{FC} \cong \overline{FC}$	2. Reflexive
3. $\overline{AC} \cong \overline{DF}$	3. Overlapping Segments Theorem
4. $\triangle ABC \cong \triangle DEF$	4. ASA

Lesson 5.3, Pages 261–266

Exploration 1

4. $C = \pi d$; $C = 2\pi r$

Exploration 2

2. Parallelogram
3. Yes.
4. Triangles; h becomes closer and closer to r.
5. $b = \frac{C}{2} = \frac{2\pi r}{2} = \pi r$
6. $A = \pi r^2$; As the number of sectors increases, the assembled sectors will more closely resemble a rectangle whose area is that of the circle.

Communicate

1. Circular yard: $100 = 2\pi r \Rightarrow r = \frac{100}{2\pi} = 15.92$ ft, so $A = \pi(15.92^2)^a$ 796 ft^2

   Square yard: $100 = 4s \Rightarrow s = 25$; so $A = 25^2 = 625$ ft^2

   The circular yard provides more area.
2. *B*. *A* and *B* are traveling at the same revolutions per minute; however, since the circumference that *B* travels is longer than the circumference that *A* travels, *B* moves faster.
3. The decimal form of $\frac{22}{7} \approx 3.1429$ is slightly larger than π but it is easier to use in some contexts. The value of π cannot be found exactly because π is an irrational number.

   About 3.141592654 depending on how many decimal places the calculator holds.
4. Triangle: $P = 15.71$ units; Square: $P = 12$ units; Rectangle: $P = 13$ units; Circle: $P = 10.63$ units

   The triangle has the largest perimeter, and the circle has the smallest perimeter.

Lesson 5.4, Pages 267–274

Exploration

3. Every time.
4. Yes. It is unlikely that the numbers with the property would be listed by chance.

Communicate

1. For any right triangle, the square of the length of the hypotenuse is equal to the sum of the squares of the lengths of the legs.
2. The Pythagorean Theorem is useful in finding length in right triangles. Since most buildings and land divisions involve right angles, the Pythagorean Theorem can be used to measure inaccessible distances in buildings and large or inaccessible distances on the earth. Surveyors, architects,

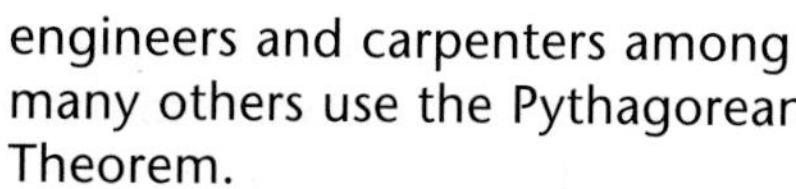

engineers and carpenters among many others use the Pythagorean Theorem.

3. They would make use of the fact that a 3-4-5 triangle is a right triangle, since $3^2 + 4^2 = 5^2$.

4. $h^2 = 1^2 + 1^2 \Rightarrow h = \sqrt{2}$ units

Practice and Apply

19. 3,4,5; 5,12,13; 7,24,25; 9,40,41; 11,60,61; 13,84,85

Let $a = m$, $b = \frac{m^2-1}{2}$ and $c = \frac{m^2+1}{2}$

$a^2+b^2 = m^2 + \left(\frac{m^2-1}{2}\right)^2 = m^2 + \frac{m^4-2m^2+1}{4} = \frac{4m^2}{4} + \frac{m^4-2m^2+1}{4} = \frac{m^4+2m^2-1}{4}$

$c^2 = \left(\frac{m^2+1}{2}\right)^2 = \frac{m^4+2m^2+1}{4}$

$a^2 + b^2 = c^2$, so any triple generated will work.

20. 4,3,5; 6,8,10; 8,15,17; 10,24,26; 12,35,37; 14,48,50

Let $a = 2m$, $b = m^2 = 1$ and $c = m^2+1$

$a^2+b^2 = (2m)^2 + (m^2-1)^2 = 4m^2 + m^4 - 2m^2 + 1 = m^4 + 2m^2+1$

$c^2 = (m^2+1)^2 = m^4 + 2m^2+1$

$a^2 + b^2 = c^2$, so any triple generated will work.

26. Each of the outer triangles has area $= \frac{1}{2}ab$, so the total area of the 4 outer triangles is $2ab$. Each side of square $ABCD$ is $a+b$ and the area of a square with sides $a+b$ is $(a+b)^2$. So the area of $EFGH$ is $(a+b)^2 - 2ab$. Each of the inner triangles has area $= \frac{1}{2}ab$, so the total area of the 4 inner triangles is $2ab$. Each side of square $IJKL$ is b-a and the area of a square with sides $a-b$ is $(b-a)^2$. So the area of $EFGH$ is $(a-b)^2 + 2ab$. Since the area of $EFGH$ is $2c^2$, adding gives $2c^2 = (a-b)^2 + (a+b)^2 \Rightarrow 2c^2 = a^2 - 2ab + b^2 + a^2 + 2ab + b^2 \Rightarrow 2c^2 = 2a^2 + 2b^2 \Rightarrow c^2 = a^2 + b^2$

Lesson 5.5, Pages 275–282

Exploration

1.

Shorter leg	Hypotenuse	Longer leg
1	2	$\sqrt{3}$
2	4	$2\sqrt{3}$
3	6	$3\sqrt{3}$
4	8	$4\sqrt{3}$
5	10	$5\sqrt{3}$

2. The longer leg is $\sqrt{3}$ times the shorter leg.

3. a. $2x$ b. $x\sqrt{3}$

4.

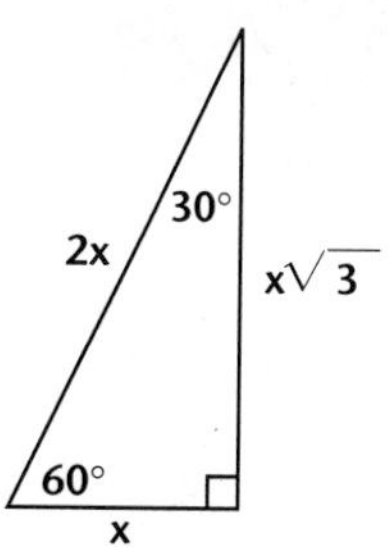

Communicate

1. $6\sqrt{2}$; The length of the hypotenuse is the length of a leg times $\sqrt{2}$.

2. The length of the longer leg is the length of the shorter leg times $\sqrt{3}$: $4\sqrt{3}$.

The length of the hypotenuse is twice the length of the shorter leg : 8.

3. Find the length of the segment from the center of the polygon to the midpoint of a side.

4. Find one half of the product of the apothem and the perimeter of the figure.

5.

Hypotenuse
$\sqrt{2}$
$2\sqrt{2}$
$3\sqrt{2}$
$4\sqrt{2}$
$5\sqrt{2}$

Lesson 5.6, Pages 283–288

Exploration: Part 1

2. $A(0,5)$; $B(1,\sqrt{24})$; $C(2,\sqrt{21})$; $D(3,\sqrt{16})$; $E(4,\sqrt{9})$

3. $\sqrt{25} = 5$; $\sqrt{24}$; $\sqrt{21}$; $\sqrt{16} = 4$; $\sqrt{9} = 3$

4. $\sqrt{25} + \sqrt{24} + \sqrt{21} + \sqrt{16} + \sqrt{9} \approx 21.48$

5. 85.93; It overestimates because the rectangles include area outside the circle.

6. $A = 25\pi \approx 78.5398$ $E = .094$

Part 2

$A \approx 4(\sqrt{24} + \sqrt{21} + \sqrt{16} + \sqrt{9}) \approx 65.9262$. This method underestimates because parts of the circle are not included.
$E = .161$

Part 3

$\frac{85.9262 + 65.9262}{2} = 75.9262$.

This is closer to the true value.
$E = .033$

Communicate

1. The differences give the lengths of the legs of the right triangle used to compute the distance.

2. The distance formula uses the Pythagorean Theorem to find distance on the coordinate plane.

3. Divide the area into rectangles. First use rectangles that extend beyond the area, then use rectangles that are completely within the area. Find the areas of both sets of rectangles, then find the average of the results.

4. As the widths of the rectangles get smaller they more closely approximate the arcs of the circle. There is less difference between the areas of the rectangles and the portions of the circle they approximate.

Practice and Apply

15. Sample answer shown.

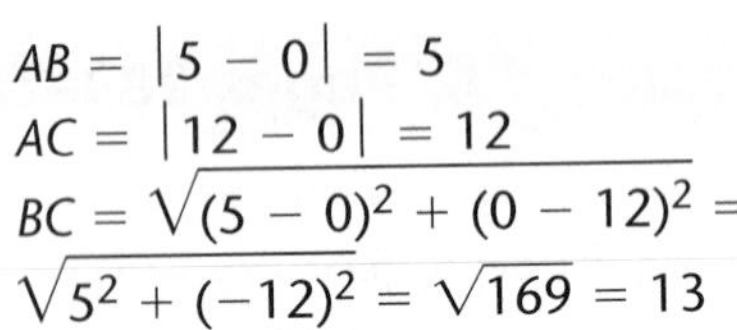

$AB = |5 - 0| = 5$
$AC = |12 - 0| = 12$
$BC = \sqrt{(5 - 0)^2 + (0 - 12)^2} = \sqrt{5^2 + (-12)^2} = \sqrt{169} = 13$

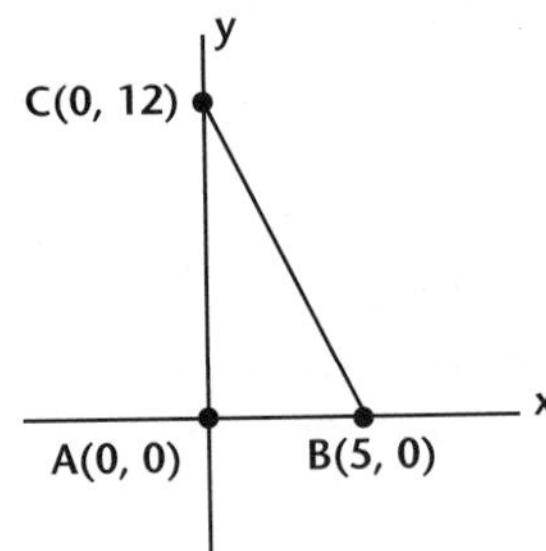

16. Sample answer shown.
$AC = \sqrt{(0 - -3)^2 + (4 - 0)^2} = \sqrt{3^2 + 4^2} = \sqrt{25} = 5$
$BC = \sqrt{(3 - 0)^2 + (0 - 4)^2} = \sqrt{3^2 + (-4)^2} = \sqrt{25} = 5$

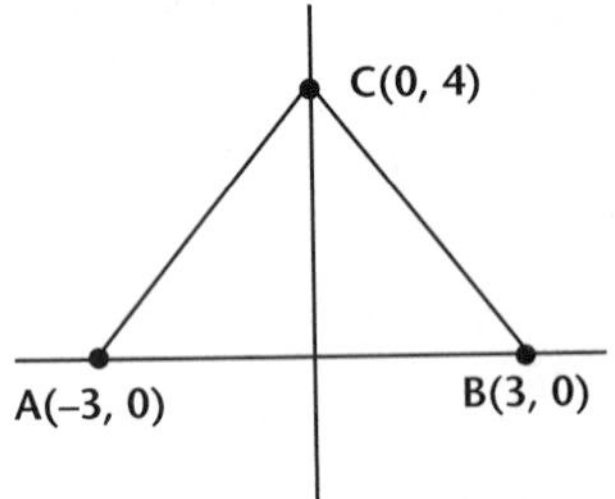

17. Sample answer when rectangles of width 1 are used follows.
From the left: $(1)(2) + (1)(3) = 5$ (high estimate)
From the right: $(1)(6) + (1)(3) = 9$ (low estimate)
Average $= \frac{5+9}{2} = 7$

18. Sample answer when rectangles of width $\frac{1}{2}$ are used follows.
From the left: $\left(\frac{1}{2}\right)(2) + \left(\frac{1}{2}\right)(1.75) + \left(\frac{1}{2}\right)(1) = 2.375$ (high estimate)
From the right: $\left(\frac{1}{2}\right)(1) + \left(\frac{1}{2}\right)(1.75) = 1.375$ (low estimate)
Average $= \frac{2.375+1.375}{2} = 1.875$

Look Back

19. Slope of $\overline{WX} = \frac{-6}{8} = \frac{-3}{4}$; Slope of $\overline{XY} = \frac{4}{3}$
Slope of $\overline{YZ} = \frac{6}{-8} = \frac{-3}{4}$; Slope of $\overline{ZW} = \frac{4}{3}$
Since the slopes of the opposite sides are equal, the opposite sides are parallel and the figure is a parallelogram.
Since the slopes of consecutive sides are negative reciprocals, the sides are perpendicular and the figure is a rectangle.

Lesson 5.7, Pages 289–295

Exploration

4. The relative error should be 0.1 or lower.

Communicate

1. $\frac{1}{2}$; Since M is the midpoint of $\overline{AB}$, $AM = MB$. The distance from A to M is half the distance from A to B. $\frac{AM}{AB} = \frac{1}{2}$.

2. $\frac{1}{4}$; Since C is the midpoint of $\overline{AM}$, the distance from A to C is half the distance from A to M. So the distance from A to C is one fourth the distance from A to B. $\frac{AC}{AB} = \frac{1}{4}$.

3. 1. $AB = AB$, $\frac{AB}{AB} = 1$

4. $\frac{3}{4}$; The probability of landing on the segment is 1 and the probability of landing on $\overline{AC}$ is $\frac{1}{4}$. So the probability that the point is not on $\overline{AC} = 1 - \frac{1}{4} = \frac{3}{4}$.

Lesson 6.1, Pages 304–310

Exploration 1

1.

1) 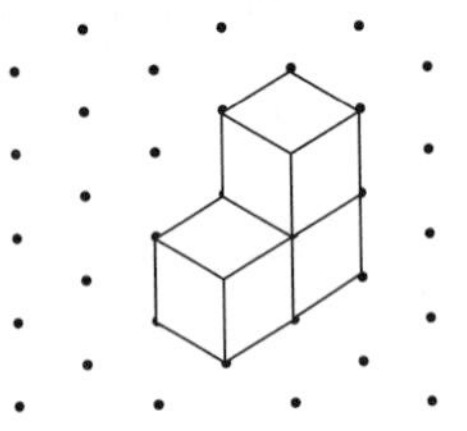

2) 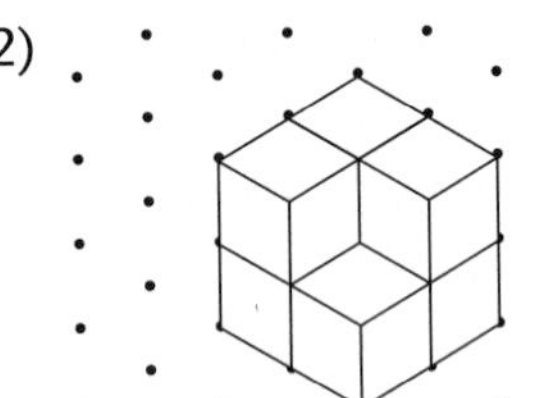

3) 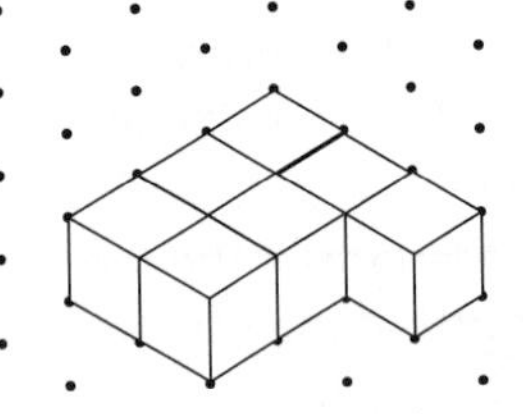

4) 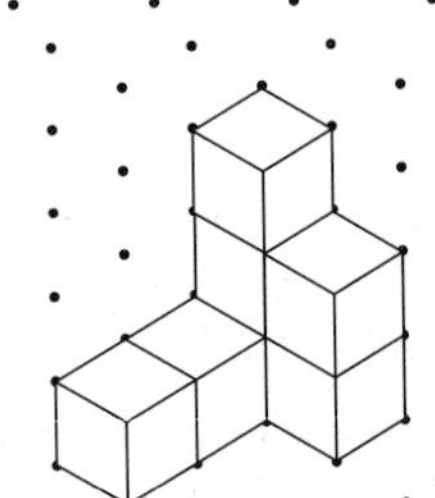

2.

1) 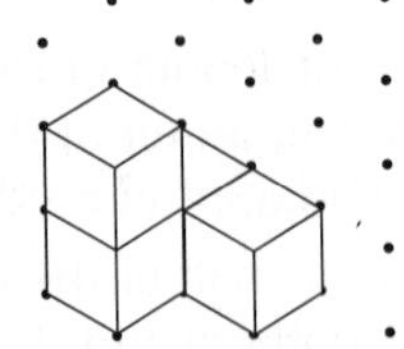

2) 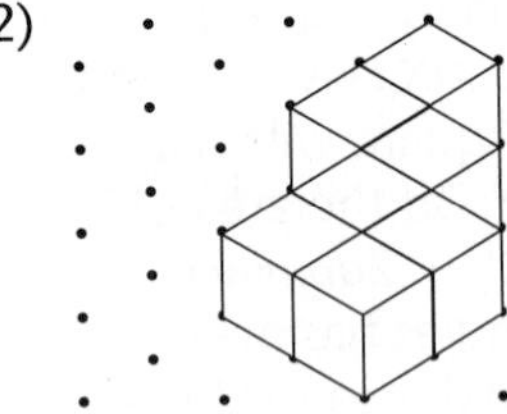

3) 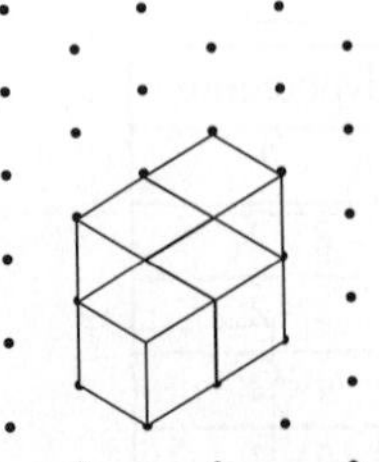

4) 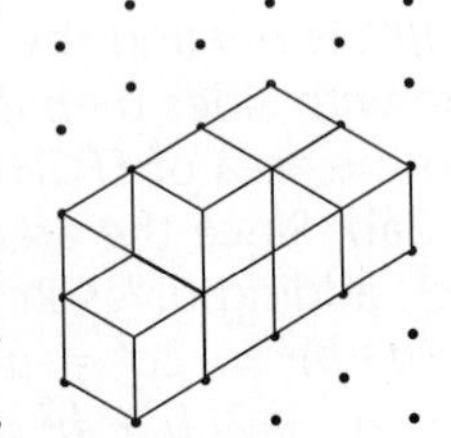

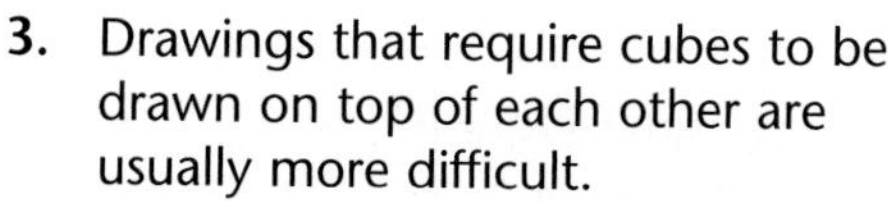

3. Drawings that require cubes to be drawn on top of each other are usually more difficult.

Exploration 2

3. Yes. Parts of the figure might be hidden by the visible cubes, such as a cubic hole in the last figure.

Exploration 3

1. Assume the "front" view faces the reader.

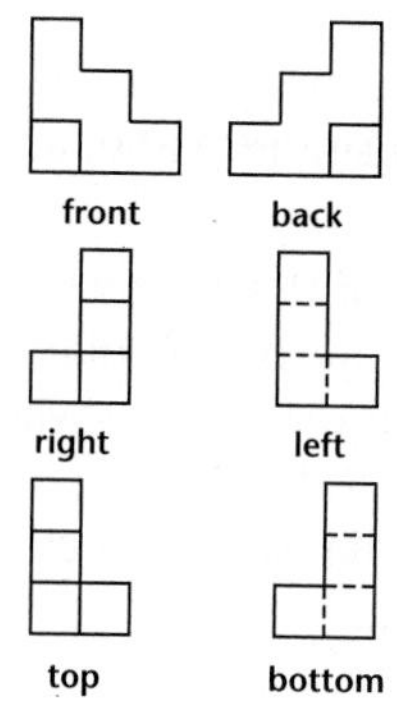

2. Count the cubes used in constructing the figure.
3. Find the area of each of the 6 views and add them. This counts the area of each face once, and hence the gives the total surface area.

Communicate

1. Represent the hidden parts of the cube using dashed lines for the hidden edges.
2. a
3. d
4. c
5. b
6. 5
7. 8
8. 20
9.

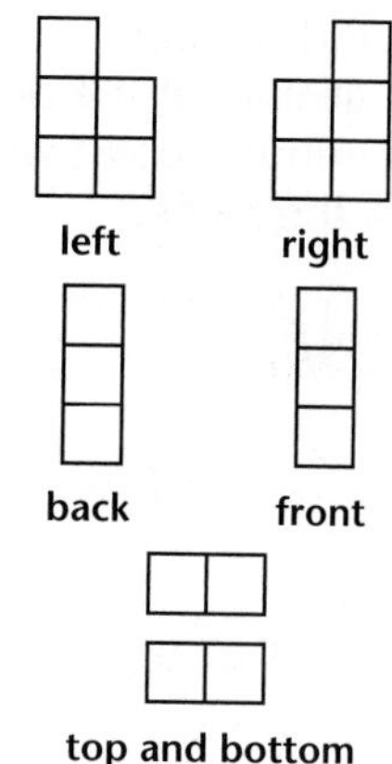

Practice and Apply

10–15. Answers will vary. Check student drawings.

18.

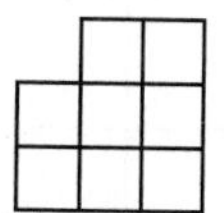

20. Yes, there are 4 faces that have the shape and size shown below: two from the top, one from the back and one from the left.

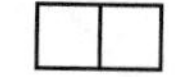

23. Answers will vary. Check student drawings.
24. Count the exposed cube faces.
25. 5 cubes
26. Answers will vary. In general, for a given volume (number of cubes), more symmetrical arrangements tend to give a smaller surface area.

30.

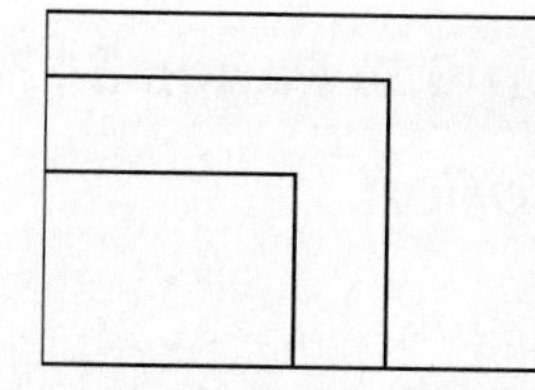

31.

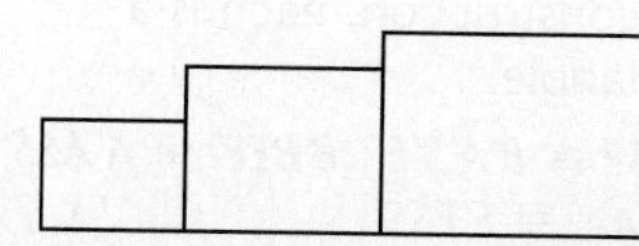

Lesson 6.2, Pages 311–316

Exploration 1—Part 1

1. $\overline{AE}$; $\overline{DH}$; $\overline{CG}$; $\overline{BF}$

   Yes; No; Yes
2. $\overline{AE} \parallel \overline{DH}$; $\overline{AE} \parallel \overline{CG}$; $\overline{AE} \parallel \overline{BF}$; $\overline{DH} \parallel \overline{CG}$; $\overline{DH} \parallel \overline{BF}$; $\overline{CG} \parallel \overline{BF}$

   $\overline{AB} \parallel \overline{CD}$; $\overline{AB} \parallel \overline{GH}$; $\overline{AB} \parallel \overline{EF}$; $\overline{CD} \parallel \overline{GH}$; $\overline{CD} \parallel \overline{EF}$; $\overline{GH} \parallel \overline{EF}$

   $\overline{AD} \parallel \overline{BC}$; $\overline{AD} \parallel \overline{FG}$; $\overline{AD} \parallel \overline{EH}$; $\overline{BC} \parallel \overline{FG}$; $\overline{BC} \parallel \overline{EH}$; $\overline{FG} \parallel \overline{EH}$
3. Yes, the segments are in different planes.

   Examples of skew segments are: $\overline{AE}$ and $\overline{GH}$; $\overline{DH}$ and $\overline{GF}$; $\overline{CG}$ and $\overline{EF}$; $\overline{BF}$ and $\overline{EH}$

Exploration 1—Part 2

1. 6 : $ABCD \parallel EFGH$; $CDHG \parallel BAEF$; $AEHD \parallel BFGC$
2. Two planes are parallel if and only if they do not intersect.
3. 3 : $ABCD \parallel EFGH$; $CDHG \parallel BAEF$; $AEHD \parallel BFGC$

Exploration 2—Part 1

$\overline{AB} \perp AEHD$ and $\overline{AB} \perp BFGC$; $\overline{CD} \perp AEHD$ and $\overline{CD} \perp BFGC$

$\overline{GH} \perp AEHD$ and $\overline{GH} \perp BFGC$; $\overline{EF} \perp AEHD$ and $\overline{EF} \perp BFGC$

$\overline{AE} \perp ABCD$ and $\overline{AE} \perp EFGH$; $\overline{DH} \perp ABCD$ and $\overline{DH} \perp EFGH$

$\overline{CG} \perp ABCD$ and $\overline{CG} \perp EFGH$; $\overline{BF} \perp ABCD$ and $\overline{BF} \perp EFGH$

$\overline{AD} \perp ABFE$ and $\overline{AD} \perp DCGH$; $\overline{BC} \perp ABFE$ and $\overline{BC} \perp DCGH$

$\overline{FG} \perp ABFE$ and $\overline{FG} \perp DCGH$; $\overline{EH} \perp ABFE$ and $\overline{EH} \perp DCGH$

2. Answers will vary but should contain some of the elements of the definition in Step 4 of Part 2 of this exploration. Yes.
3. Yes, the sketch should show that the pencil may revolve around the line in a plane perpendicular to the line.
4. It is perpendicular to the plane of the paper.

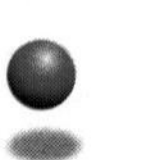

5. Yes.
6. ... the segment is perpendicular to all lines in the plane that pass through the point where the line (or segment or ray) intersects the plane.

Exploration 2—Part 2

1. $\overline{AB} \parallel CDHG$ and $\overline{AB} \parallel EFGH$; $\overline{BC} \parallel AEHD$ and $\overline{BC} \parallel EFGH$
   $\overline{CD} \parallel ABFE$ and $\overline{CD} \parallel EFGH$; $\overline{AD} \parallel BCGF$ and $\overline{AD} \parallel EFGH$
   $\overline{AE} \parallel CDHG$ and $\overline{AE} \parallel BFGC$; $\overline{DH} \parallel ABFE$ and $\overline{DH} \parallel BFGC$
   $\overline{CG} \parallel ABFE$ and $\overline{CG} \parallel AEDH$; $\overline{BF} \parallel AEHD$ and $\overline{BF} \parallel CDHG$
   $\overline{EF} \parallel ABCD$ and $\overline{EF} \parallel CDHG$; $\overline{FG} \parallel ABCD$ and $\overline{FG} \parallel AEHD$
   $\overline{GH} \parallel ABCD$ and $\overline{GH} \parallel ABFE$; $\overline{EH} \parallel ABCD$ and $\overline{EH} \parallel BCGF$
   ... it does not intersect the plane.
2. Yes.
3. Yes.
4. ... it does not intersect the plane. Or
   ... it is parallel to a line contained in the plane.

Exploration 3

1. 4
2. They are perpendicular.
4. ... two rays that are perpendicular to the angle edge at the same point.
6. The measures are different unless the rays were drawn perpendicular to the line of the crease.

Communicate

1. Answers will vary.
2. Answers will vary.
3. Answers will vary.
4. Not necessarily. Consider for example the corner of a room where three lines meet that are all perpendicular to each other but are not parallel.

5. No. The line must be perpendicular to all lines in the plane that pass through the point where the line meets the plane.

Practice and Apply

24.

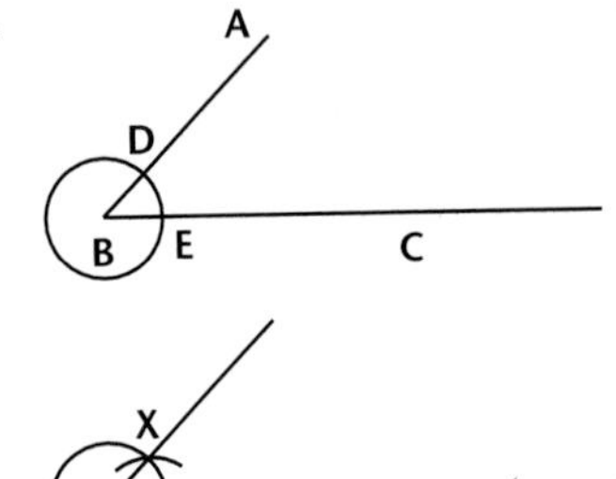

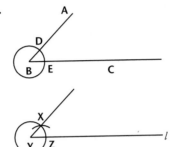

25.

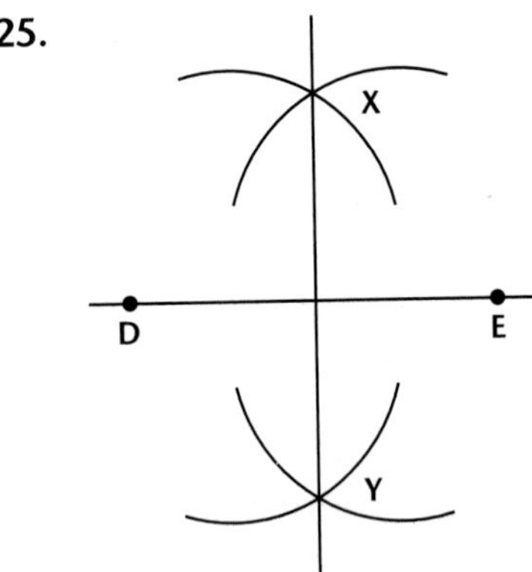

26.

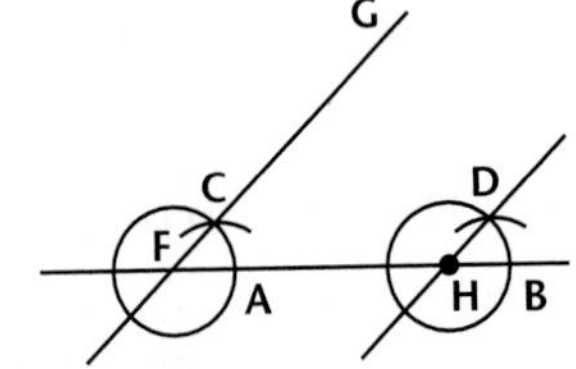

27.

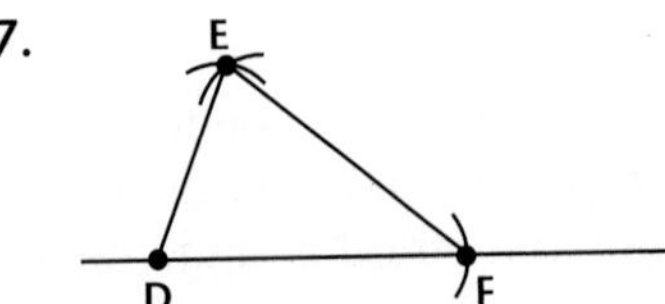

Lesson 6.3, Pages 317–322

Exploration

Part 1

1. By construction, each is a rectangle.
2. $BASE \cong B'A'S'E'$; $B'BEE' \cong A'ASS'$; $B'BAA' \cong E'ESS'$
3. a. Rectangle
   b. Parallelogram
   c. Rectangle
   For b, the deformation leaves the sides parallel but not at right angles to the adjacent sides.
4. All are still true.

Part 2

1. No. The upper base is not the translation image of the lower base.
2. No
3. So that the lateral faces are each contained in a single plane, and hence flat.
4. Yes. Give one of the bases two translations in perpendicular directions. Opposite sides will still be both parallel and congruent, making the faces parallelograms.

Communicate

1. hexagonal right prism
2. right triangular prism
3. right pentagonal prism
4. oblique hexagonal prism
5. *ABC* and *DEF* are triangular bases. *BCDF* and *ABFE* are parallelograms and *ACDE* is a rectangle.

Lesson 6.4, Pages 326–330

Communicate

1. *z*-axis
2. *y*-axis
3. *x*-axis
4. *xy*-plane
5. *xz*-plane
6. *xy*-plane
7. From the given information, the area of the base is $2 \cdot 3 = 6$, and the height is 5, for a volume of 30 cubic units.

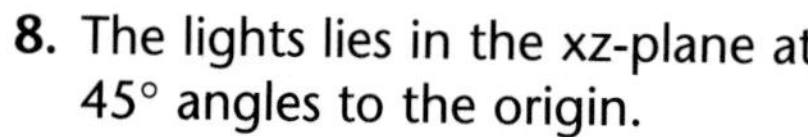

8. The lights lies in the xz-plane at 45° angles to the origin.

Practice and Apply

9.

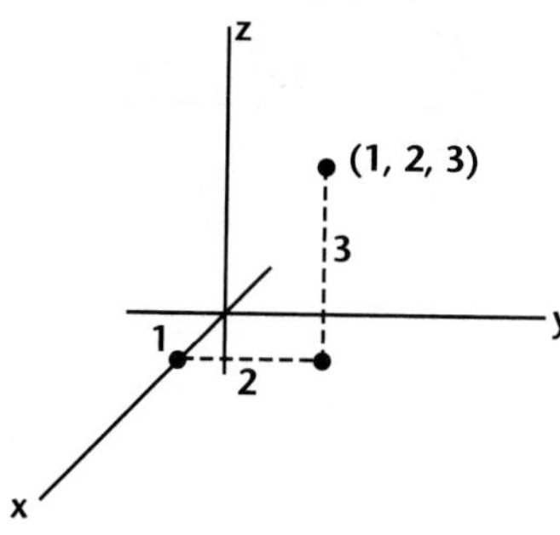

10.

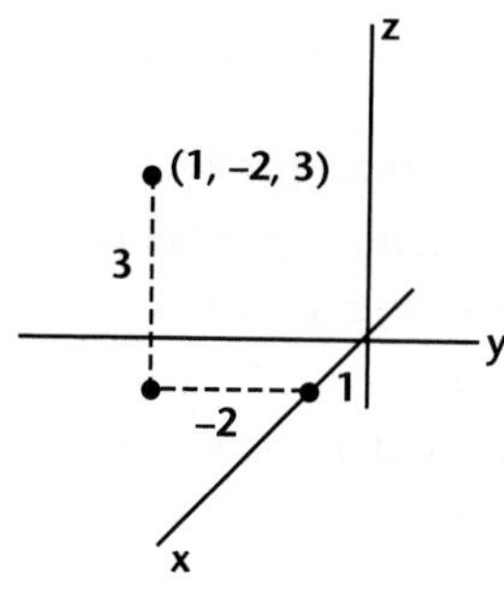

11.

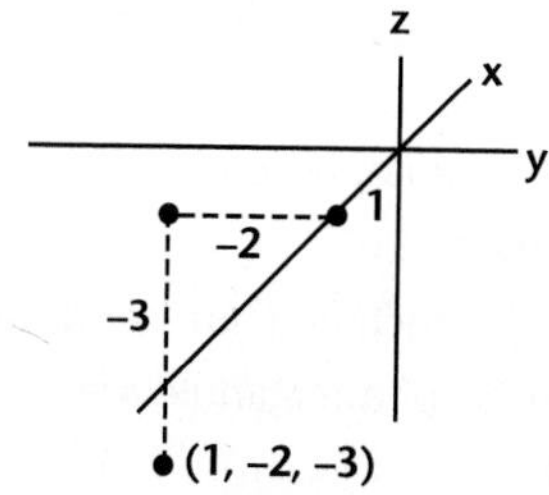

12.

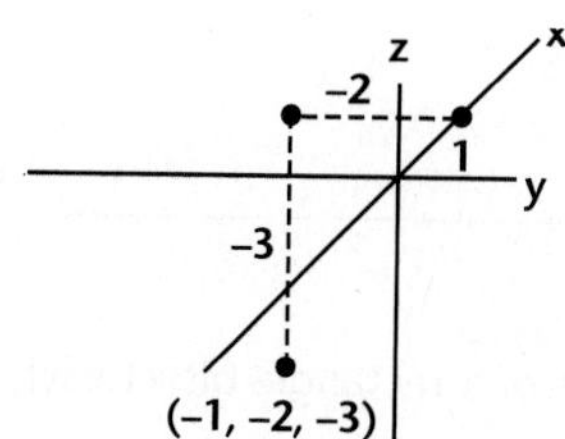

17.

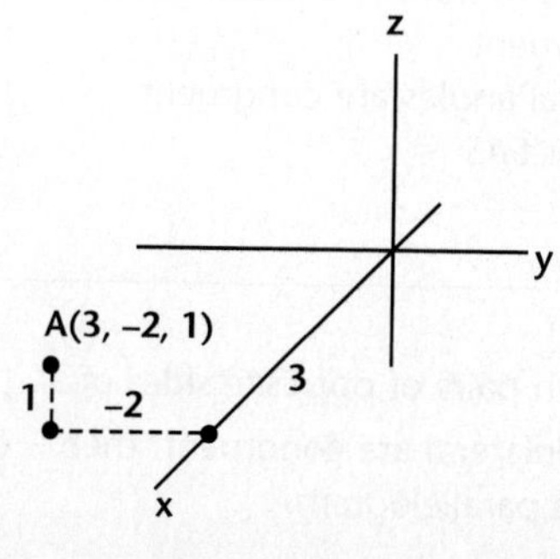

18.

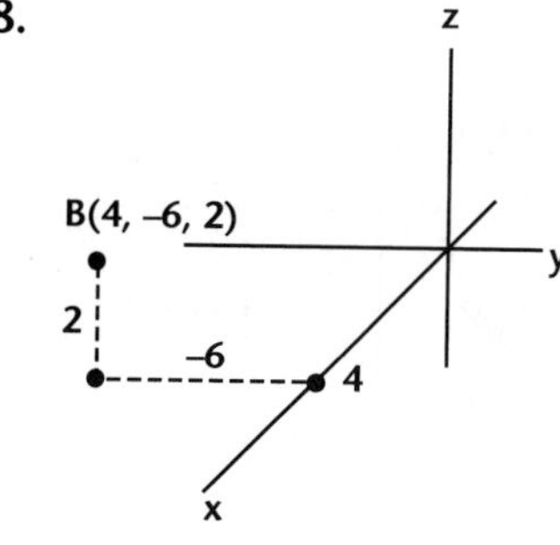

19.

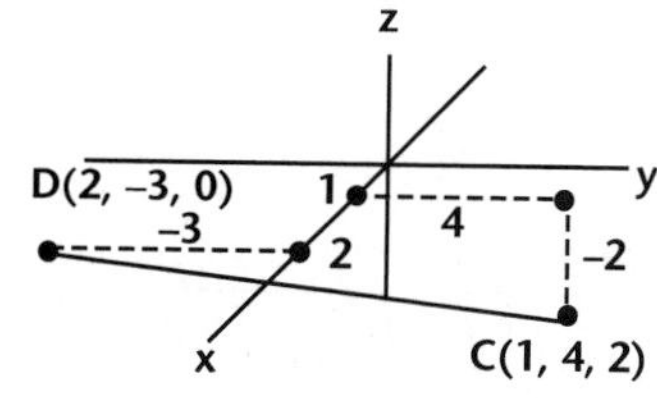

20.

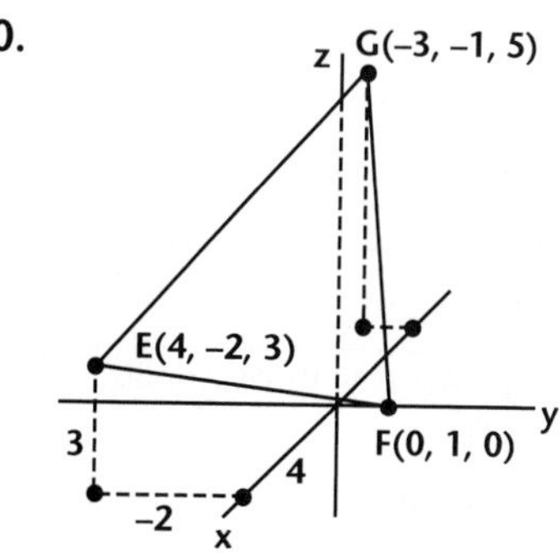

Lesson 6.5, Pages 331–335

Communicate

1. $Ax + By + Cz = D$, where A, B, C, and D are constants.
2. For any values of x and y that satisfy the equation of the plane, all values of z will also work. The plane is parallel to the z-axis.
3. For any values of y and z that satisfy the equation of the plane, all values of x will also work. The plane is parallel to the x-axis.
4. For any value of x that satisfies the equation of the plane, all values of y and z will also work. The plane is parallel to both the y-axis and the z-axis.

Practice and Apply

5.

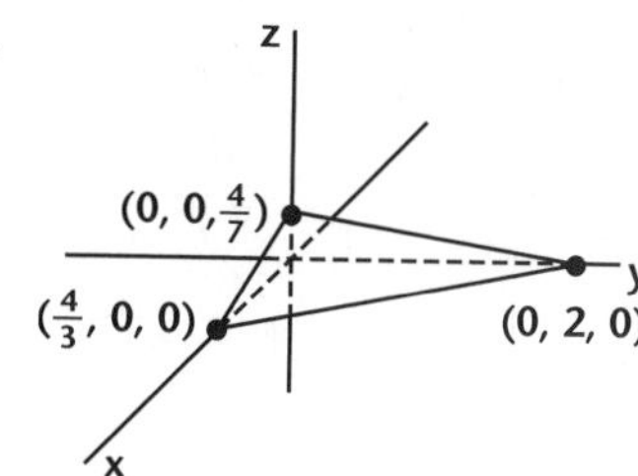

6.

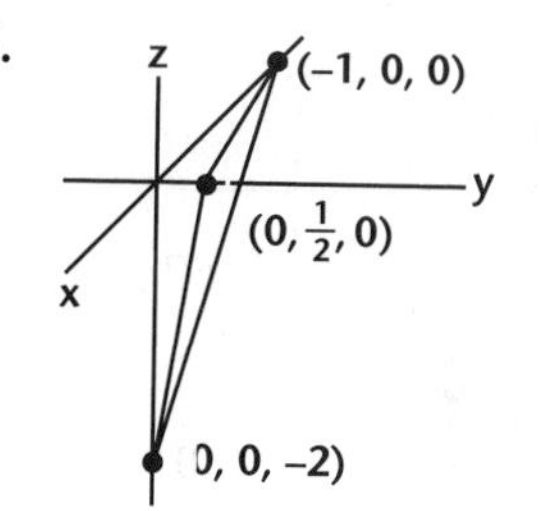

7.

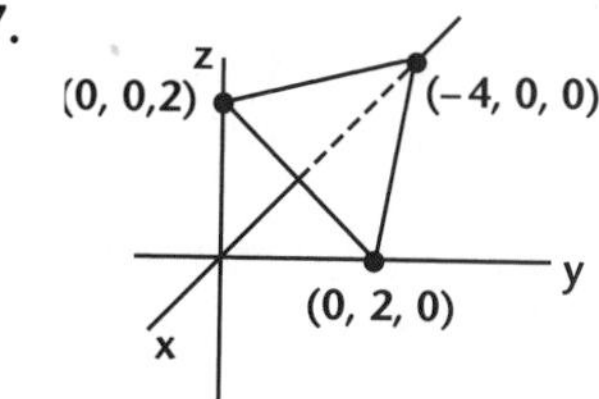

8.

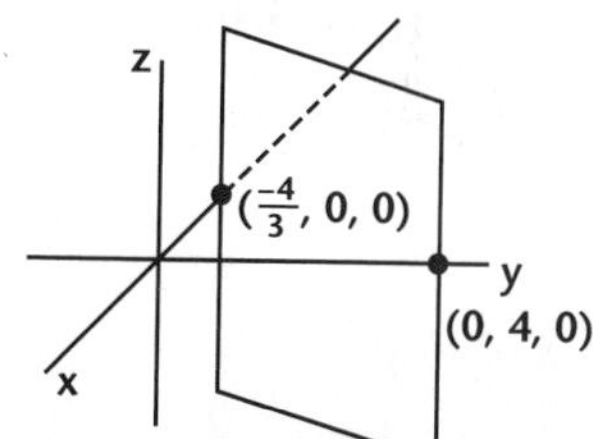

Parallel to z-axis

9.

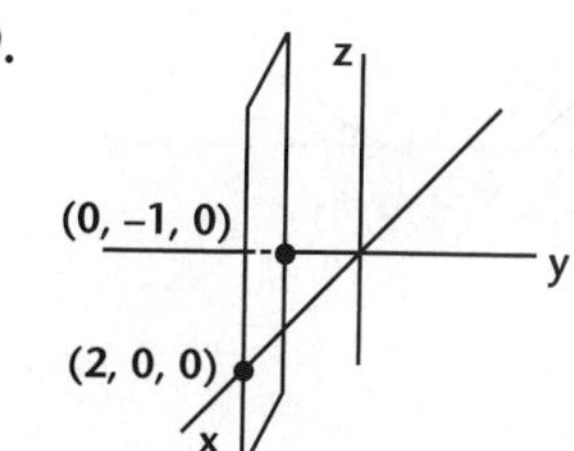

Parallel to z-axis.

10.

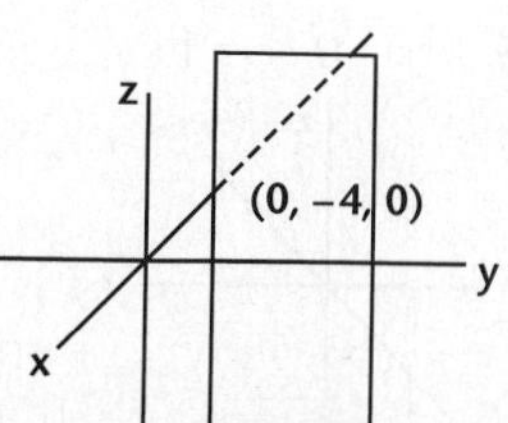

Parallel to both y-axis and z-axis.

11.

t	x	y	z
1	1	2	3
2	2	4	6

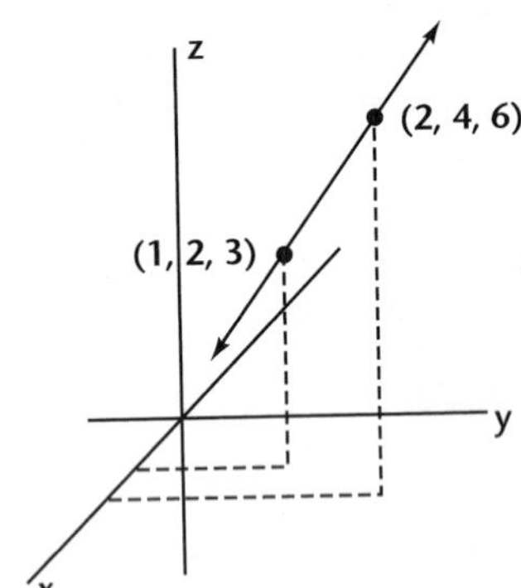

12.

t	x	y	z
1	2	3	0
2	3	5	0

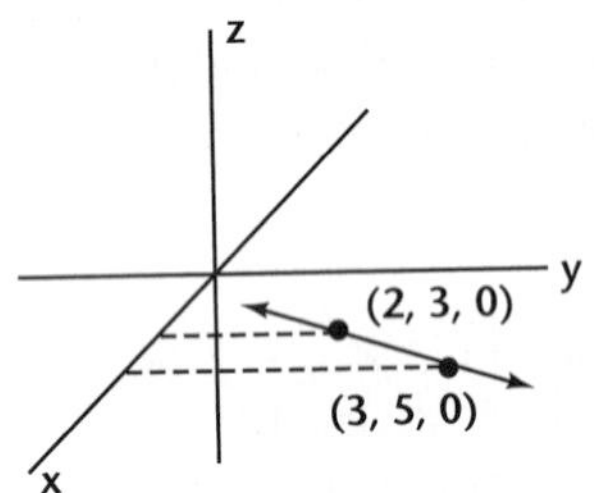

13.

t	x	y	z
1	$\frac{2}{3}$	1	0
2	$\frac{4}{3}$	2	1

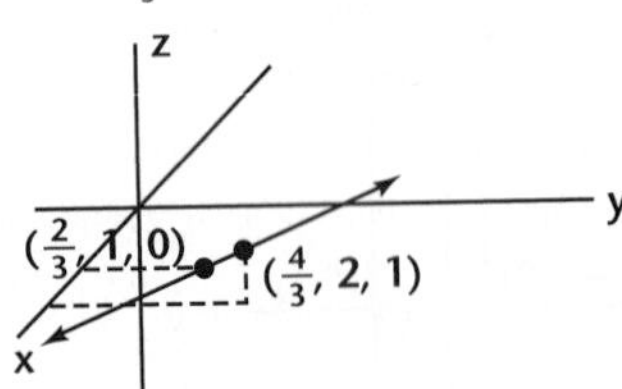

14.

t	x	y	z
1	3	0	−3
2	6	0	1

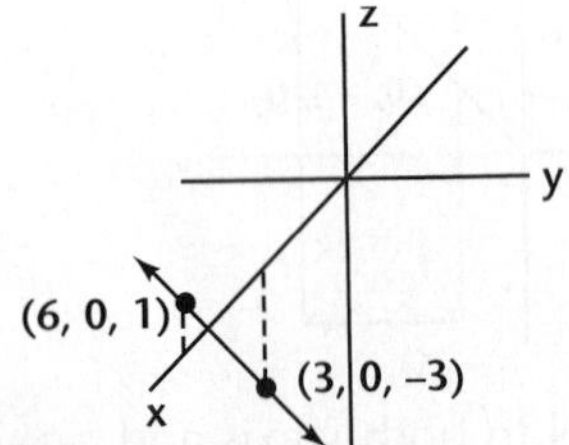

17.

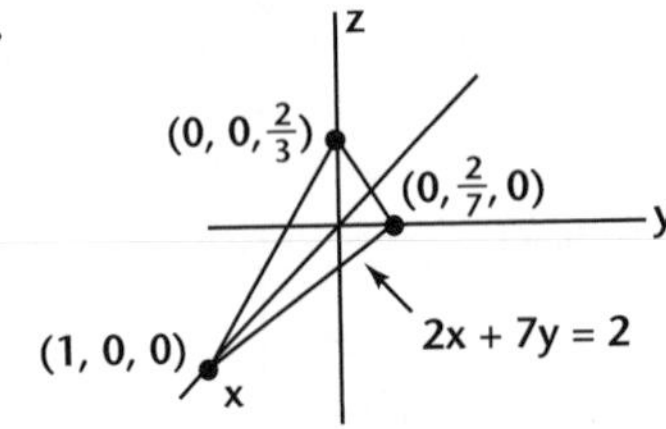

18.

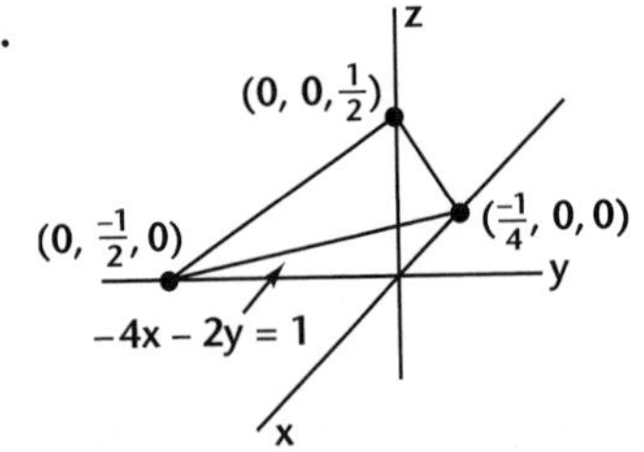

Look Back

20.

Given: $\angle 1 \cong \angle 2$; $\overline{EH} \cong \overline{FG}$

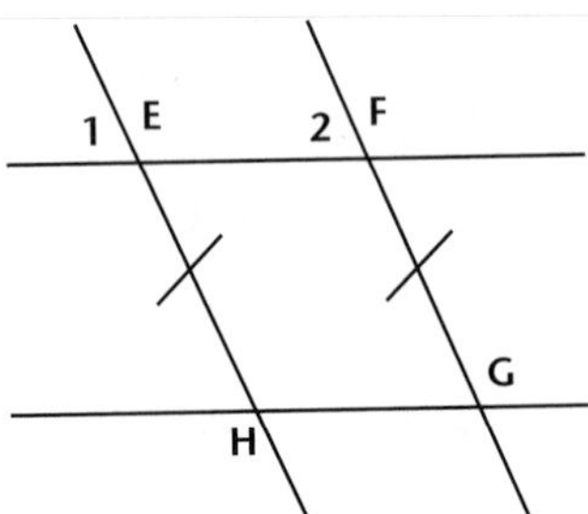

Prove: $\overline{EF} \cong \overline{HG}$

Statements	*Reasons*
1. $\angle 1 \cong \angle 2$; $\overline{EH} \cong \overline{FG}$	1. Given
2. $\overline{EH} \parallel \overline{FG}$	2. Corr angles ≅ imply parallel lines
3. *EFGH* is a parallelogram	3. If one pair of sides of a quadrilateral is both // and ≅, then it is a parallelogram.
4. $\overline{EF} \cong \overline{HG}$	4. Opposite sides of a parallelogram are congruent.

21.

Statements	*Reasons*
1. Rectangle *ABCD* with diagonals $\overline{AC}$ and $\overline{BD}$	1. Given
2. $\angle ADC \cong \angle BCD$	2. Measures of the angles of a rectangle are all 90°
3. $\overline{AC} \cong \overline{BD}$	3. Diagonals of a rectangle are congruent
4. $\overline{AD} \cong \overline{BC}$	4. Opposite sides of a rectangle are congruent
5. $\overline{DC} \cong \overline{DC}$	5. Reflexive
6. $\triangle ADC \cong \triangle BCD$	6. SSS or SAS

22.

Statements	*Reasons*
1. Rectangle *ABCD* with diagonals $\overline{AC}$ and $\overline{BD}$ intersecting at *E*	1. Given
2. $\overline{AE} \cong \overline{CE}$ and $\overline{DE} \cong \overline{BE}$	2. Diagonals of a rectangle bisect each other
3. $\overline{AD} \cong \overline{BC}$	3. Opposite sides of a rectangle are congruent
4. $\angle AED \cong \angle BEC$	4. Vertical angles are congruent
5. $\triangle AED \cong \triangle BEC$	5. SSS or SAS

23.

Statements	*Reasons*
1. $\overline{DA} \cong \overline{RT}$ and $\overline{DT} \cong \overline{AR}$	1. Given
2. *DART* is a parallelogram	2. If both pairs of opposite sides of a quadrilateral are congruent, then it is a parallelogram.

Lesson 6.6, Pages 336–343

Communicate

1. … will seem to meet at the same point.
2. … will meet at a point on the horizon of the drawing.
3. … intersect the parallel lines. They become shorter the closer they get to the vanishing point.
4. Consider railroad tracks or roads. Parallel lines appear to get closer together as they get farther away from a viewer. When lines that we think should be parallel (such as the rails of a railroad track) are drawn as converging, the drawing gives the illusion of depth.
5. The vanishing point in perspective drawings is the point at which parallel lines seem to meet.

Look Back

19.

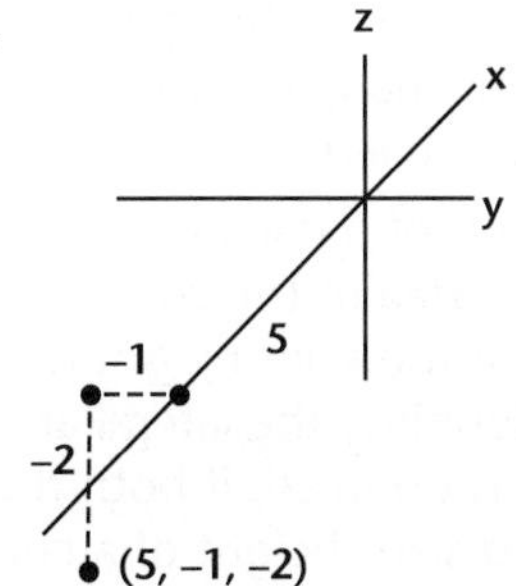

20.

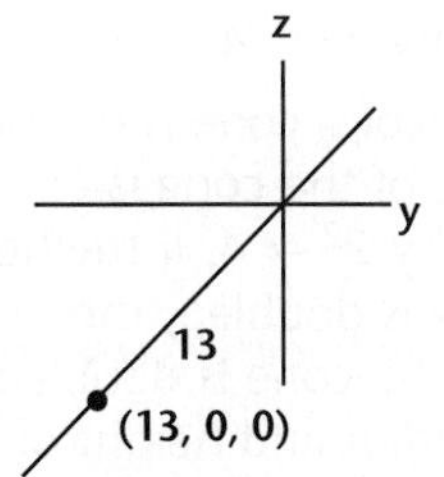

21.

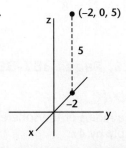

Lesson 7.1, Pages 354–358

Exploration 1; Part 1

1.

Length	Surfacel Area	Volume	SA/Vol
1 6	1	6	
2	10	2	5
3	14	3	$\frac{14}{3}$
4	18	4	$\frac{9}{2}$
5	22	5	$\frac{22}{5}$
n	$4n+2$	n	$\frac{4n+2}{n}$

2. $\frac{402}{100}$; The ratio of surface area to volume decreases as n increases. The ratio approaches 4.

Exploration 1; Part 2

1.

Length	Surface Area	Volume	SA/Vol
1	6	1	6
2	24	8	3
3	54	27	2
4	96	64	$\frac{3}{2}$
5	150	125	$\frac{6}{5}$
n	$6n^2$	n^3	$\frac{6}{n}$

2. As the length of a side increases, the ratio of surface area to volume decreases.

Exploration 2

1.

Side	Length	Width	Height	Volume
1	12	9	1	108
2	10	7	2	140
3	8	5	3	120
4	6	3	4	72
5	4	1	5	20
x	$14-2x$	$11-2x$	x	$x(14-2x)(11-2x)$

3. a. $x = 2$ b. $V = 140$

Communicate

1. Maximum volume. The builder would want to maximize the amount that could be stored in a building using a limited amount of wood and minimize storage costs.
2. Maximum volume. The producer would want to maximize the amount of soup in a container that uses a limited amount of material and minimize packaging cost.
3. Maximum surface area. The designer would want to maximize the area that could absorb energy and minimize the cost per unit of energy produced.
4. Maximum surface area. The engineers design the shuttle to maximize the area covered by the tiles relative to the volume of the shuttle. Then heat is absorbed on re-entry in the most efficient manner.
5. Answers will vary. Sample answer: Maximum surface area. The engineers want the maximum area of tire on the surface to maximize traction and stability.

Practice and Apply

11. The flatworm's shape maximizes its surface area and thus maximizes the amount of oxygen it can absorb.
12. The air sacks maximize the surface area and allows oxygen to be absorbed efficiently.
13. The volume of water the cactus can hold is maximized while the surface area through which water can evaporate and heat can be absorbed is minimized. When water is available, the cactus can store the maximum amount and retain it more easily.
14. Tall trees with broad leaves grow where there is competition for light and maximum surface area is important. In the desert there is competition for water, not light, and minimum surface area is important for conserving water.
15. For an $n \times 1 \times 1$ prism, the ratio approaches 4. For an $n \times n \times n$ prism, the ratio is inversely proportional to n.

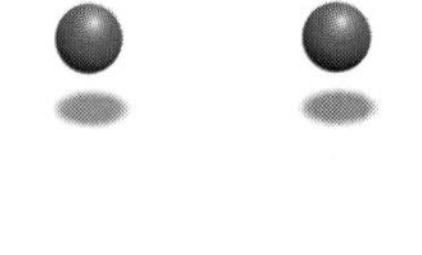

16. It is natural to make the conjecture that the rectangular prism that has the smallest surface area for a given volume is the cube.

Lesson 7.2, Pages 359–365

Communicate

1. Find the areas of all the lateral faces and add twice the area of a base.
2. A right prism is a polygon that has been extended straight up. So, to find the volume of a right prism, find the area of the base and multiply by the height.
3. Cavalieri's Principle states that if two solids have equal heights and if the areas of their equal cross sections are always equal, then the solids have the same volume. To find the volume of an oblique prism, find the area of its base and multiply by its height—its volume is the same as that of a right prism with the same base and height.
4. Yes, provided the bases have the same area. Then the cross sections have the same area and the heights are the same, Cavalieri's principle applies.
5. The volume is doubled.

Lesson 7.3, Pages 366–372

Exploration

2. 3 times
3. $V = \frac{1}{3}Bh$

Communicate

1. Multiply the slant height by the perimeter of the base and the result by one half. Add the area of the base to the result.
2. Multiply the area of the base by the height and divide the result by three.
3. The slant height is longer because the slant height is the hypotenuse of the right triangle with the height as one leg.
4. The volume of a pyramid is one third the volume of a rectangular prism with the same base and height.
5. The altitude of a regular right pyramid intersects the base at the center of the base, the point equidistant from the vertices.

Lesson 7.4, Pages 373–378

Exploration

1. A rectangular prism
2. $V = (\pi r)(r)(h) = \pi r^2 h$
3. $V = \pi r^2 h$, where r is the radius and h is the height of the cylinder.

Communicate

1. $V = \pi r^2 h$
2. To find the lateral area, multiply the circumference of the base by the height of the cylinder. Add twice the area of the base to the result.

   The result is $SA = 2\pi rh + 2\pi r^2 = 2\pi r(h + r)$
3. If the cylinder is cut into wedges, it can be assembled into a shape that is nearly a rectangular prism.
4. When the radius of a cylinder is doubled, the volume of the cylinder is multiplied by $2^2 = 4$.
5. When the height of a cylinder is doubled, the lateral area of the cylinder is doubled. There is no change in the area of the base.

Lesson 7.5, Pages 379–386

Exploration 1

1. $c = 2\pi(7) = 14\pi$
2. $C = 2\pi(15) = 30\pi$
3. $\frac{c}{C} = \frac{14\pi}{30\pi} = \frac{7}{15} = .47$
4. $A = \pi(15^2) = 225\pi$; $225\pi(\frac{7}{15}) = 105\pi = L$
5. $B = \pi(7^2) = 49\pi$; $105\pi + 49\pi = 154\pi$; The surface area of the cone.

Exploration 2

1–3. $c = 2\pi r$; $C = 2\pi l$; $\frac{c}{C} = \frac{2\pi r}{2\pi l} = \frac{r}{l}$

$LA = \frac{r}{l}\pi l^2 = \pi rl$; $B = \pi r^2$

$SA = LA + B = \pi rl + \pi r^2$

Yes, this is a proof of the formula below.

Communicate

1. Multiply the product of the radius and the slant height of the cone by π. Add the result to the area of the base.
2. Find the product of the area of the base and the height, then divide by three.
3. The slant height is longer because the slant height is the hypotenuse of the right triangle with the height as one leg.
4. If the radius of a cone is doubled, the lateral area of the cone is doubled. If the slant height of a cone is doubled, the lateral area of the cone is doubled. If both the radius and slant height of a cone are doubled, the lateral area is multiplied by $2^2 = 4$.
5. If the radius of a cone is doubled, the volume of the cone is multiplied by $2^2 = 4$. If the height of the cone is doubled, the volume of the cone is doubled. If both the radius and height of a cone are doubled, the volume is multiplied by $2^3 = 8$.

Lesson 7.6, Pages 387–393

Communicate

1. Square the radius of the sphere and multiply by 4π.

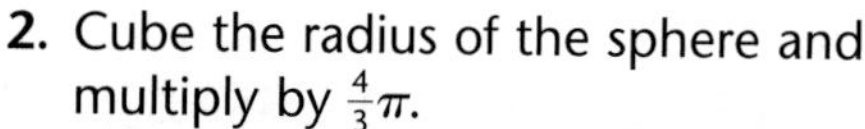

2. Cube the radius of the sphere and multiply by $\frac{4}{3}\pi$.
3. When the radius of a sphere is doubled, the surface area of the sphere is multiplied by $2^2 = 4$.
4. When the radius of a sphere is doubled, the volume of the sphere is multiplied by $2^3 = 8$.
5. The "Pythagorean" Right-Triangle Theorem is used to find the radii of the circles that are the intersections when planes cut through a sphere at various heights from the center of the sphere.

   A right triangle is set up with the radius of the sphere as the hypotenuse and the height from the sphere center to the plane of circle as one leg. The radius of the circle is the other leg. The areas of those circles are compared to the areas of the annuli that form when the same planes intersect a cylinder with a double cone cut out of it.

Lesson 7.7, Pages 396–403

Exploration 1, Part 1

2.

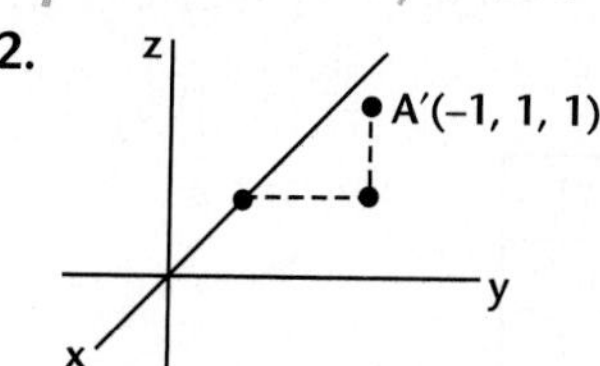

3. *zy*-plane.

   The *zy*-plane is the perpendicular bisector of $\overline{AA'}$.
4. When a point is reflected through a plane, the reflection plane is the perpendicular bisector of the line segment connecting the point to its image. A point and its image under a reflection through a plane are on opposite sides of the plane and equidistant from the plane.
5. Reflection in the *xz*-plane.

   Reflection in the *xy*-plane.
7. When a figure is reflected through a plane, the reflection plane is the perpendicular bisector of all the line segments connecting the points of the figure to their images. Every point of a figure and its image under a reflection through a plane are on opposite sides of the plane and equidistant from the plane.

Exploration 1, Part 2

1.

Octant of image	Coordinates of image
front-right-bottom	(2,3,−4)
front-left-top	(2,−3,4)
back-right-top	(−2,3,4)
back-right-bottom	(−4,5,−6)
back-left-top	(−4,−5,6)
front-right-top	(4,5,6)

2. a. (x,y,−z) b. (x,−y,z) c. (−x,y,z)

Exploration 2

2. A figure in space has reflectional symmetry if one side of the figure is the image of the other side when reflected through a plane drawn through the figure. If a plane can be drawn through a figure so that every point in the figure has an image point on the other side of the plane such that the plane is the perpendicular bisector of the line segments connecting the points and their images, the figure has reflectional symmetry.

Exploration 3

2.

z
A(−1, 1, 1)
A′(1, −1, 1)
B(−1, 1, 0)
y
B′(1, −1, 0)
x

3. The plane containing the line y = x and perpendicular to the xy plane.
4. (0,0,1); (0,0,0)
5. Yes; Yes; They are perpendicular to the z-axis.
6. When a figure is rotated around an axis in space, each point in the figure moves around the axis of rotation in a circle contained in a plane that is perpendicular to the axis of rotation. The radius of the circle is the distance from the point to the axis of rotation and the center of the circle is on the axis of rotation.

Exploration 4

2. A figure in space has rotational symmetry if one side of the figure is the image of the other side when rotated around a line drawn through the figure. If a line can be drawn through a figure so that every point in the figure has a coplanar images point in the figure that is a rotation of the same amount on a circle with radius equal to the distance to the line and center on the line, the figure has rotational symmetry and the line is the axis of rotation.

Communicate

1. In three-dimensional symmetry the "mirror" is a plane, while in two-dimensional symmetry the "mirror" is a line. In three-dimensional symmetry, the line segment connecting a point to its image is perpendicular to a plane, while in two-dimensional symmetry the line segment connecting a point to its image is perpendicular to a line.

   In three-dimensional symmetry, every point and its image are equidistant from a plane, while in two-dimensional symmetry every point and its image are equidistant from a line.
2. Cylinder
3. Cone
4. The segment is three times a long and reflected through the xy-plane, the xz-plane and the yz-plane. This is equivalent to a dilation and a reflection through the origin.

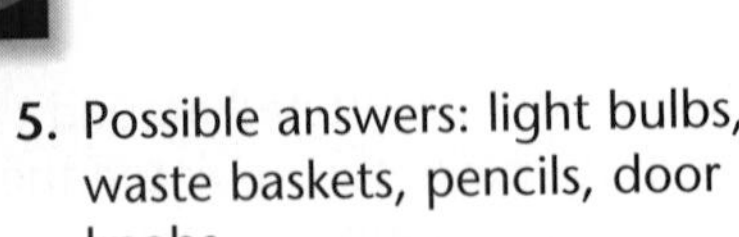

5. Possible answers: light bulbs, waste baskets, pencils, door knobs.

Practice and Apply

6.

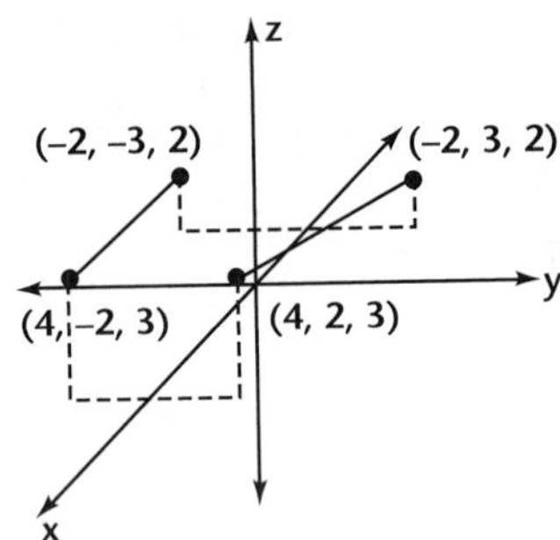

7.

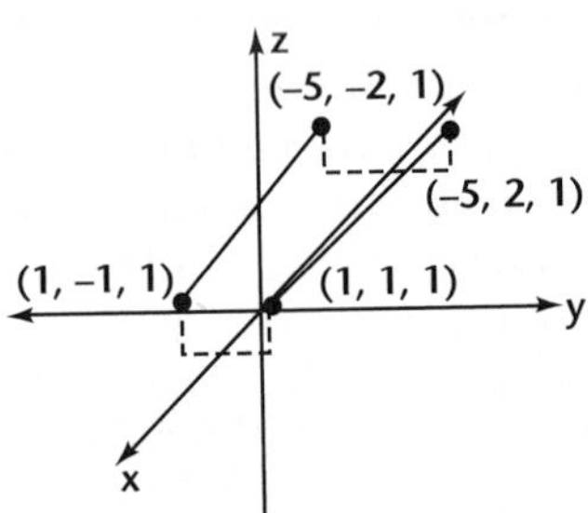

8.

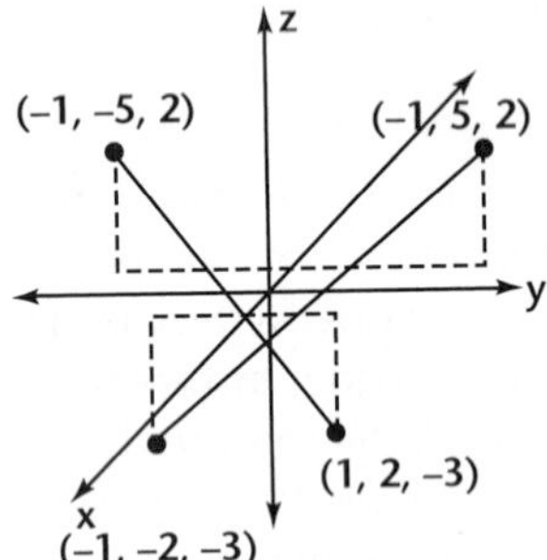

16.

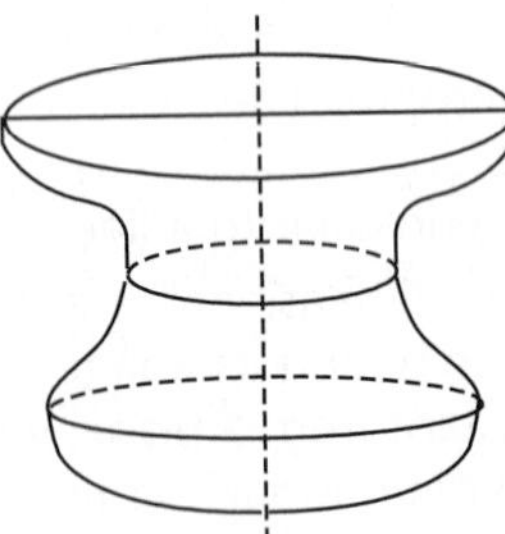

17.

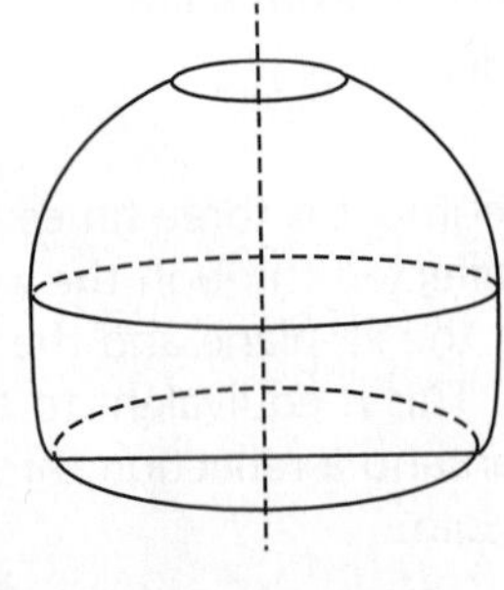

18.

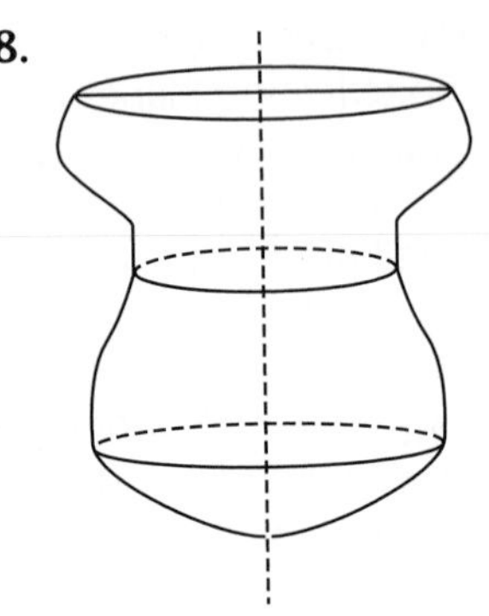

19.

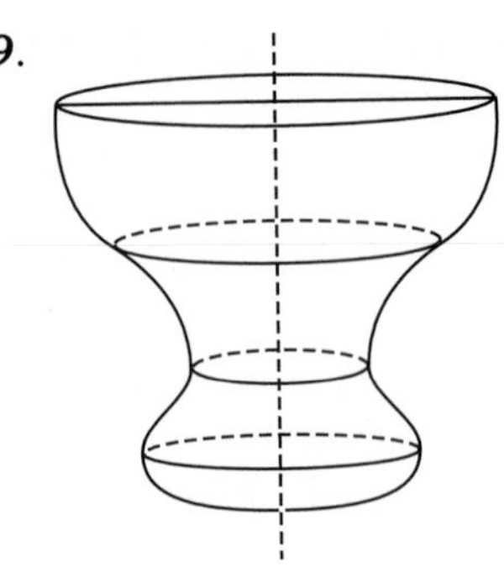

Lesson 8.1, Pages 412–418

Exploration 1, Part 2

1. a line
2.

A	Scale Factor	A′	OA	OA′	$\frac{OA'}{OA}$
(2,4)	0.5	(1,2)	$\sqrt{20} = 2\sqrt{5}$	$\sqrt{5}$	0.5
(2,4)	2	(4,8)	$\sqrt{20} = 2\sqrt{5}$	$4\sqrt{5}$	2
(2,4)	3	(6,12)	$\sqrt{20} = 2\sqrt{5}$	$6\sqrt{5}$	3
(2,4)	4	(8,16)	$\sqrt{20} = 2\sqrt{5}$	$8\sqrt{5}$	4
(2,4)	−1	(−2,−4)	$\sqrt{20} = 2\sqrt{5}$	$2\sqrt{5}$	1
(2,4)	n	(2n,4n)	$\sqrt{20} = 2\sqrt{5}$	$2n\sqrt{5}$	n

3. The image of a point under a scale factor transformation on n is $|n|$ times as far from (0,0).
$OP = \sqrt{x^2 + y^2}$;
$OP' = \sqrt{(xn)^2 + (yn)^2} = \sqrt{x^2n^2 + y^2n^2} = \sqrt{n^2(x^2 + y^2)} = n\sqrt{x^2 + y^2}$, therefore $\frac{OP'}{OP} = n$

Exploration 1, Part 3

1.

A	B	SF	A′	B′	AB	A′B′	$\frac{A'B'}{AB}$
(2,4)	(6,1)	0.5	(1,2)	(3,0.5)	5	2.5	0.5
(2,4)	(6,1)	2	(4,8)	(12,2)	5	10	2
(2,4)	(6,1)	3	(6,12)	(18,3)	5	15	3
(2,4)	(6,1)	4	(8,16)	(24,4)	5	20	4
(2,4)	(6,1)	−1	(−2,−4)	(−6,−1)	5	5	1
(2,4)	(6,1)	n	(2n,4n)	(6n,1n)	5	5n	n

2. All segments have slope $= \frac{-3}{4}$.
3. The image of a segment under a scale factor transformation of n is $|n|$ times as long as the pre-image. The image has the same slope as the pre-image and is parallel to the pre-image.
4. $PQ = \sqrt{(x_2 - x_1)^2 + (y_2 - y_1)^2}$;
$P'Q' = \sqrt{(nx_2 - nx_1)^2 + (ny_2 - ny_1)^2} =$
$\sqrt{n^2(x_2 - x_1)^2 + n^2(y_2 - y_1)^2} = \sqrt{n^2((x_2 - x_1)^2 + (y_2 - y_1)^2)} = \sqrt{n(x_2 - x_1)^2 + (y_2 - y_1)^2}$;
$P'Q' = n(PQ)$
Both segments have slope $= \frac{y_2 - y_1}{x_2 - x_1}$.

Exploration 3

1. $y = \frac{4}{3}x$

   (6,8) is on the line because it works in the equation of the line: $8 = \frac{4}{3}(6)$.

   (9,12) is on the line because it works in the equation of the line: $12 = \frac{4}{3}(9)$.

2. $y = \frac{b}{a}x$

   (2a,2b) is on the line because it works in the equation of the line: $2b = \frac{b}{a}(2a)$.

   (3a,3b) is on the line because it works in the equation of the line: $3b = \frac{b}{a}(3a)$.

   (na,nb) is on the line because it works in the equation of the line: $nb = \frac{b}{a}(na)$.

Communicate

1. A scale factor is the number by which both coordinates of a point are multiplied. The image of a point under a scale factor transformation by n is $|n|$ times as far from (0,0).
2. A dilation is a transformation in which a scale factor and a point called the center of dilation are given. In a dilation with scale factor n and center P, each point on the image is collinear with P and with the corresponding point on the pre-image. The image point is $|n|$ times as far from P as the corresponding point on the pre-image. The size of the image under a dilation can be different than the size of the pre-image.
3. When the scale factor is 1, the image is the same as the pre-image. When the scale factor is 0, the image is the center of dilation. When the scale factor is negative, the image is on the opposite side of the center of dilation as the pre-image. When the scale factor is a fraction, points on the image are closer to the center of dilation than the corresponding points on the pre-image and the image is smaller than the pre-image.
4. The scale factor would give the ratio between the diameters of the images and the pre-image, and it would give the ratio between the heights of the images and the pre-image.

Practice and Apply

18.

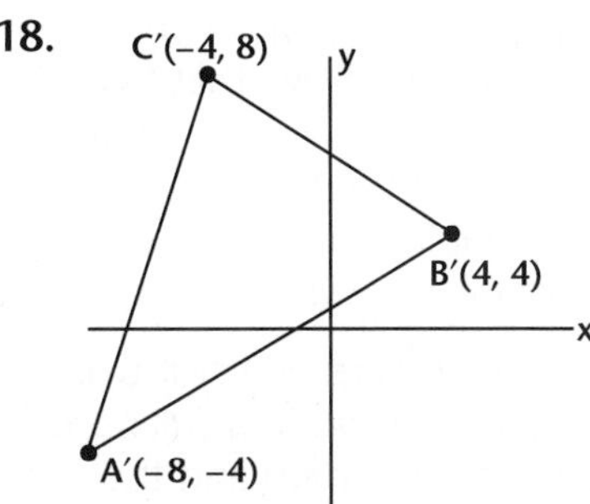

19.

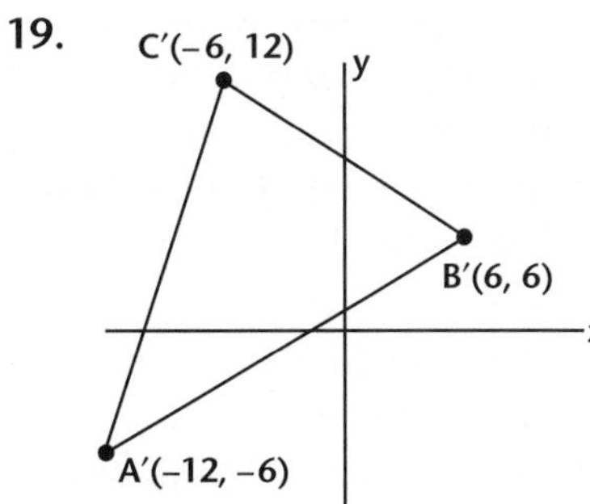

20.

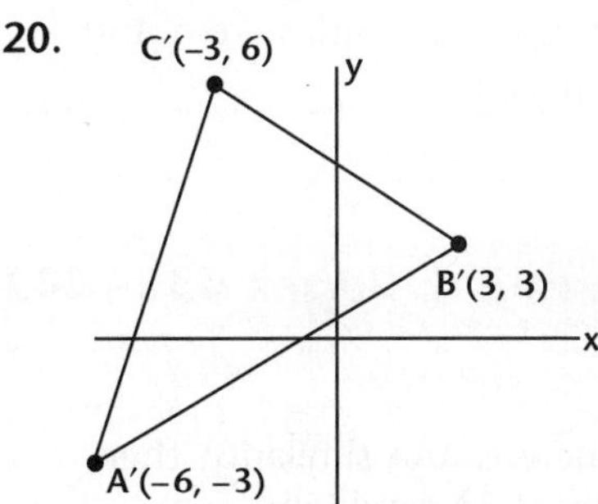

21.

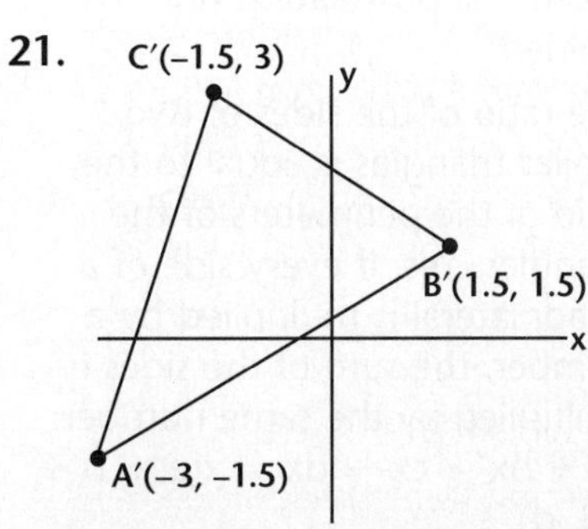

Lesson 8.2, Pages 419–428

Exploration

2. $\angle A \cong \angle D$; $\angle B \cong \angle E$; $\angle C \cong \angle F$
3. $\frac{AB}{DE} = \frac{BC}{EF} = \frac{AC}{DF}$
4. Yes, the ratios of corresponding sides are equal.
5. Both are needed.

   Figures at left: Lengths of the corresponding sides are in proportion, but the figures aren't similar because the angles are different.

   Figures at right: Angles are congruent, but the figures aren't similar because the sides aren't in proportion.
6. Two polygons are similar if and only if their corresponding angles are congruent and their corresponding sides are proportional.

Communicate

1. $\overline{RT}$ and $\overline{IO}$; $\overline{TW}$ and $\overline{OU}$; $\overline{RW}$ and $\overline{IU}$
2. $\frac{KM}{RT} = \frac{MP}{TA} = \frac{PQ}{AW} = \frac{QK}{WR}$
3. $\triangle FRG \sim \triangle BYN$
4. $\triangle CXZ \sim \triangle MLV$
5. Yes, corresponding angles are congruent and the corresponding sides are proportional, with the ratio between the sides being 1.
6. $GHJK \sim KLMN$. All corresponding angles are congruent and all corresponding sides are in proportion.
7. $\frac{a+b}{b} = \frac{a}{b} + \frac{b}{b}$ and $\frac{c+d}{d} = \frac{c}{d} + \frac{d}{d'}$; and $\frac{b}{b} = \frac{d}{d} = 1$

Practice and Apply

24. Estimates should vary from 70,000 to 75,000 acres. Methods for determining the area will vary.
25. Estimates should vary from 10,500 to 11,250.

Lesson 8.3, Pages 429–435

Exploration 1

3. $m\angle F = 70°$; $m\angle C = 70°$; $\frac{DE}{AB} = \frac{EF}{BC} = \frac{DF}{AC}$
4. Corresponding angles are congruent.
5. Corresponding sides are proportional.
6. Yes
7. Two triangles are similar if two pairs of corresponding angles are congruent.

Exploration 2

3.

$\triangle DEF$	$\triangle ABC$
$m\angle D \approx 46.5°$	$m\angle A \approx 46.5°$
$m\angle E \approx 104.5°$	$m\angle B \approx 104.5°$
$m\angle F \approx 29°$	$m\angle C \approx 29°$

4. Corresponding angles are congruent.
5. Corresponding sides are proportional.
6. Yes
7. Two triangles are similar if all three pairs of corresponding sides are in the same proportion to each other.

Exploration 3

3.

$\triangle DEF$	$\triangle ABC$	$\frac{\triangle DEF}{\triangle ABC}$
$DF = 10$	$AC = 5$	$\frac{2}{1}$
$m\angle D = 53°$	$m\angle A = 53°$	$\frac{1}{1}$
$m\angle E = 90°$	$m\angle B = 90°$	$\frac{1}{1}$
$m\angle F = 37°$	$m\angle C = 37°$	$\frac{1}{1}$

4. Corresponding angles are congruent.
5. Corresponding sides are proportional.
6. Yes
7. Two triangles are similar if two pairs of corresponding sides are in the same proportion and the included angles are congruent.

Communicate

1. Two triangles are similar if two pairs of corresponding angles are congruent. To prove that two triangles are similar, it is enough to show that two pairs of corresponding angles are congruent. ; AA-Similarity
2. Two triangles are similar if all three pairs of corresponding sides are in the same proportion to each other. To prove that two triangles are similar, it is enough to show that the three pairs of corresponding sides are proportional. ; SSS-Similarity
3. Two triangles are similar if two pairs of corresponding sides are in the same proportion and the included angles are congruent. To prove two triangles congruent, it is enough to show that two pairs of corresponding sides are in the same proportion and the included angles are congruent. ; SAS-Similarity
4. Consider a square with sides 5 cm and a rectangle with sides 5 cm and 10 cm. The corresponding angles are congruent, but the corresponding sides are not in proportion.

Lesson 8.4, Pages 436–442

Communicate

1. If there is AAA similarity, then there is AA similarity, so an additional postulate is not needed.
2. The ratio of the sides of two similar triangles is equal to the ratio of the perimeters of the triangles. Yes. If every side of a quadrilateral is multiplied by a number, the sum of the sides is multiplied by the same number: $ax + bx + cx + dx = x(a + b + c + d)$.
3. No. Consider the isosceles triangle with angles 45°, 45°, and 90° and the isosceles triangle with angles 50°, 50°, and 80°.
4. Yes. The measures of all the angles of all equilateral triangle are 60°.
5. No. Consider a 45°-45°-90° triangle and a 30°-60°-90° triangle.

Lesson 8.5, Pages 443–451

Communicate

1. It is impossible to measure the height of the mountain directly, so the height of the mountain must be found indirectly using similar triangles.
2. Yes; Proportional Medians Theorem
3. Yes; Proportional Altitudes Theorem; The theorem holds for all similar triangles, including isosceles and equilateral ones.
4. Eratosthenes measured the shadow cast by a stick in Alexandria on the summer solstice when a rod in Aswan wouldn't cast a shadow. Eratosthenes found the angle that the sun's rays formed with the stick to be 7.5°. Since the sun's rays are parallel, Eratosthenes knew the angle at the center of the earth that corresponded to the 500 miles between Aswan and Alexandria was also 7.5°. Eratosthenes then used the ratio between 7.5° and 360°, the measure of an entire circle with center at the earth's center, to find the circumference of the earth.

   If parallel lines are cut by a transversal, then alternate interior angles are congruent.

Lesson 8.6, Pages 452–459

Exploration 1

1. $\frac{9}{1}$
2. $\frac{4}{1}$
3. $\frac{9}{4}$

4. $\frac{16}{9}$
5. $(\frac{a}{b})^2 = \frac{a^2}{b^2}$

Exploration 2

1. $\frac{27}{1}$
2. $\frac{27}{8}$
3. $\frac{8}{1}$
4. $\frac{8}{1}$
5. $(\frac{a}{b})^3 = \frac{a^3}{b^3}$

Exploration 3

1.

Height scale factor	Volume scale factor	Cross-sectional radius scale factor
4	4	$\sqrt{64} = 8$
5	125	$\sqrt{125} = 11.2$
100	1,000,000	$\sqrt{1000000} = 1000$

2. $\sqrt{20^3} = (89.44)(2.5) =$ 223.6 cm = 2.2 meters
$\sqrt{50^3} = (353.55)(2.5) = 883.9$ cm = 8.8 meter
$(1000)(2.5) = 2500$ cm = 25 meters
3. The legs would be proportionally thicker because their thickness must increase by more than the increase in height.

Communicate

1. The ratio of the volumes is the ratio of the sides cubed.
2. All cubes are similar because they have the same shape and, since all sides of a cube have the same length, all corresponding sides have the same ratio. All spheres are similar because they have the same shape and, since lengths on a sphere are determined only by the sphere radius, all corresponding lengths in any two spheres will have the same ratio as the ratio between the radii of the spheres. This is also true for other three dimensional figures which have faces that are regular polygons. For example, a tetrahedron is a pyramid in which all faces are equilateral triangles. All tetrahedrons are similar.
3. First find the ratio of the sides by finding the square root of the ratio of the areas. Cube the ratio of the sides to find the ratio of the volumes.
4. Volume is multiplied by $2^2 = 4$.
5. The amount of weight a bone can support is proportional to its cross-sectional area. If an animal gets proportionally bigger, its volume and weight will increase as the cube of the increased height. To support the increased weight, the bone radius must thus increase as the square root of the increased weight.

Lesson 9.1, Pages 470–477

Exploration 1

4. Equilateral triangle. All angles measure 60° because all sides are congruent.
8. Yes. All the sides are the same length as the radius of the circle. All the interior angles measure 120°, twice the measure of an angle in an equilateral triangle.
9. Yes. All the central angles measure 60°.; Yes. In the construction, a rotation was divided into 6 equal parts of 60°.

Exploration 2

5. All six of the outer circles meet at the center of the original circle. A regular hexagon would be formed if the points where the petals of the flower meet the circle were connected. All the petals of the flower are congruent.

Exploration 3

1. Yes
2.

Statements	*Reasons*
$AB = CD$	Given
$AP = CP$, $BP = DP$	Circle definition -radii of a circle are congruent
$\triangle APB \cong \triangle CPD$	SSS
$m\angle APB = m\angle CPD$	CPCTC
$m\overset{\frown}{AB} = m\overset{\frown}{CD}$	Definition of arc measure.

3. In a circle, congruent chords intercept congruent arcs.

Communicate

1. True. A diameter is a chord that passes through the center of the circle.
2. False. A radius has one endpoint at the center of the circle, not both endpoints on the circle.
3. True. The two endpoints of the chord are points on the circle.
4. True. The triangles formed by joining the endpoints of the chords to the center of the circle are congruent by SSS. The central angles for the two chords are therefore congruent and the arcs have the same measure. See Exploration 3.
5. True. The two arcs will have congruent central angles. The triangles formed by connecting the endpoints of the chords to the center of the circle are

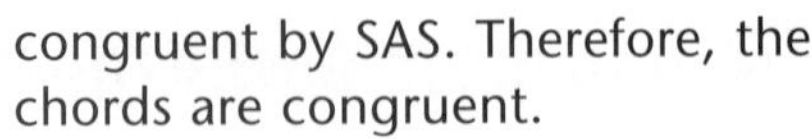

congruent by SAS. Therefore, the chords are congruent.

Practice and Apply

23. In a circle, if two arcs are congruent, then their chords are congruent.
Given: O is the center of the circle, $m\overset{\frown}{AB} = m\overset{\frown}{CD}$

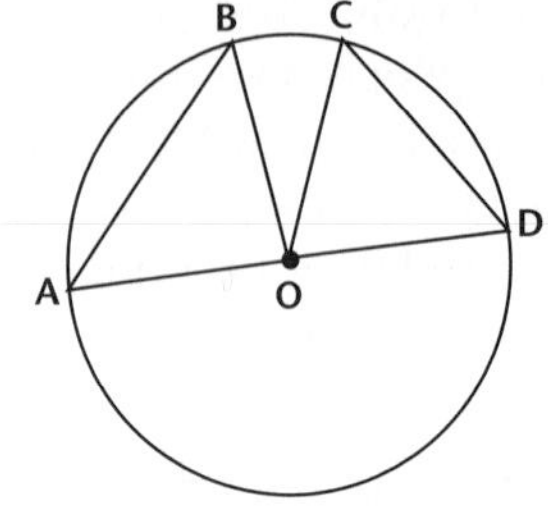

Prove: $\overline{AB} = \overline{CD}$

Statements	*Reasons*
1. O is the center of the circle, $m\overset{\frown}{AB} = m\overset{\frown}{CD}$	1. Given
2. $\angle AOB \cong \angle DOC$	2. Definition of arc measure
3. $\overline{OA} \cong \overline{OD}$, $\overline{OB} \cong \overline{OC}$	3. Definition of circle – radii of a circle are congruent
4. $\triangle AOB \cong \triangle DOC$	4. SAS
5. $\overline{AB} = \overline{CD}$	5. CPCTC

Lesson 9.2, Pages 478–483

Exploration 1

3. $AX = BX$

4. In a circle, if a radius is perpendicular to a chord of the circle, then the radius bisects the chord.

5. Because $\overline{PX} \perp \overline{AB}$, ;an$AXP \cong \angle BXP$. $\overline{PA} \cong \overline{PB}$ because the radii of a circle are congruent. $\overline{PX} \cong \overline{PX}$ by the Reflexive Property. $\triangle APX \cong \triangle BPX$ by HL.

Exploration 2

3. $m\angle PQR$ approaches 90°.

4. If a radius is drawn to the point of tangency, then the radius is perpendicular to that line.

Communicate

1. A line and a circle can intersect in 0 points, in 1 point, or in 2 points.

2. two points

3. One point

4. The point of tangency is the point where a tangent line intersects a circle.

Practice and Apply

17. Given: Chord AB in circle with center O

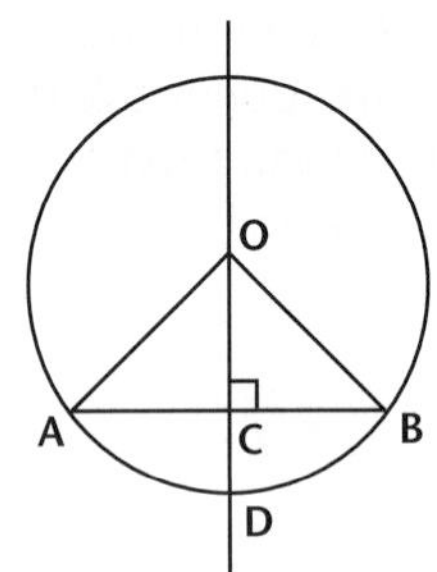

Prove: The perpendicular bisector of $\overline{AB}$ passes through the center of the circle

Statements	*Reasons*
1. Chord $\overline{AB}$ in circle with center O	1. Given
2. Draw $\overline{OC}$ such that $\overline{OC} \perp \overline{AB}$ and D on the line and circle	2. There is exactly one perpendicular from a point to a line.
3. $\overline{OC} \cong \overline{OC}$	3. Reflexive
4. $OA \cong OB$	4. Circle definition – radii of a circle are congruent.
5. $m\angle ACO = m\angle BCO = 90°$	5. Definition of perpendicular.
6. $\triangle ACO \cong \triangle BCO$	6. HL
7. $\overline{AC} \cong \overline{BC}$	7. CPCTC
8. C is the midpoint of $\overline{AB}$	8. Midpoint definition
9. $\overline{OD}$ is the perpendicular bisector of $\overline{AB}$	9. Perpendicular bisector definition
10. $\overline{OD}$ is a radius	10. Definition of radius
11. $\overline{OD}$ passes through O	11. Definition of radius

Lesson 9.3, Pages 484–490

Exploration 1

4. $m\angle AVC = 74°$; $m\overset{\frown}{AC} = 106°$

5. The measure of an inscribed angle is half the measure of its intercepted arc.

Exploration 2, Part 1

1. $m\angle 1 = m\angle 2$

2. $m\angle 3 = m\angle 1 + m\angle 2$

3.

$m\angle 1$	$m\angle 2$	$m\angle 3$	$m\overset{\frown}{AC}$
20°	20°	40°	40°
30°	30°	60°	60°
40°	40°	80°	80°
x°	x°	2x°	2x°

$m\overset{\frown}{AC} = m\angle 3$ because $\angle 3$ is the central angle that intercepts $\overset{\frown}{AC}$.

4. $m\angle 1 = \frac{1}{2} m\overset{\frown}{AC}$

Exploration 2, Part 2

1. $m\angle 1 = \frac{1}{2} m\overset{\frown}{AX}$; $m\angle 4 = \frac{1}{2} m\overset{\frown}{CX}$

2.

$m\angle 1$	$m\overset{\frown}{AX}$	$m\angle 4$	$m\overset{\frown}{CX}$	$m\angle AVC$	$m\overset{\frown}{AXC}$
20°	40°	20°	40°	40°	80°
30°	60°	20°	40°	50°	100°
40°	80°	50°	100°	90°	180°
x°	(2x)°	y°	(2y)°	x° + y°	2(x° + y°)

$m\overset{\frown}{AXC} = m\overset{\frown}{AX} + m\overset{\frown}{CX}$

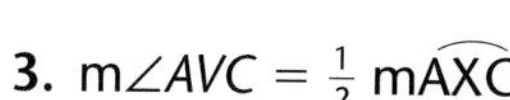

3. $m\angle AVC = \frac{1}{2} m\overset{\frown}{AXC}$

Communicate

1. $m\angle 2 = m\angle BAO = 22°$ because $\triangle AOB$ is isosceles. $m\angle 1 = m\angle 2 + m\angle BAO = 44°$ because $\angle 1$ is an exterior angle for $\triangle AOB$.
2. Since $m\angle 1 = 44°$ and $\angle 1$ is the central angle that intercepts $\overset{\frown}{AC}$, $m\overset{\frown}{AC} = 44°$.
3. $\angle DAB : \angle CBA$
4. $m\angle 2 = m\angle BAO = 22°$ because $\triangle AOB$ is isosceles.
   Or
   $m\angle 2 = \frac{1}{2} m\overset{\frown}{AC} = 22°$
5. $m\overset{\frown}{BD} = 2m\angle DAB = 44°$
6. Both angles intercept the same arc and both angles measure half the measure of the arc they intercept.

Lesson 9.4, Pages 491–499

Exploration 1.

Case 1

1. $m\angle AVC = 90°$; $m\overset{\frown}{AV} = 180°$
2. The relation ship is the same as between an in scribed angle and its intercepted arc: $m\angle AVC = \frac{1}{2} m\overset{\frown}{AV}$.

Case 2

a.

$m\overset{\frown}{AV}$	$m\angle 1$	$m\angle 2$	$m\angle PVC$	$m\angle AVC$
120°	120°	30°	90°	60°
100°	100°	40°	90°	50°
80°	80°	50°	90°	40°
60°	60°	60°	90°	30°
$x°$	$x°$	$\frac{180° - x°}{2}$	90°	$\frac{x°}{2}$

b. $\frac{1}{2}$

Case 3

a.

$mAVC$	$m\angle 1$	$m\angle 2$	$m\angle PVC$	$m\angle AVC$
200°	160°	10°	90°	100°
220°	140°	20°	90°	110°
240°	120°	30°	90°	120°
260°	100°	40°	90°	130°
$x°$	$360° - x°$	$\frac{x° - 180°}{2}$	90°	$\frac{x°}{2}$

b. $\frac{1}{2}$; The relationship is the same as between an inscribed angle and its intercepted arc: the measure of the angle is half the measure of the intercepted arc. The measure of a secant-tangent (or chord) angle with its vertex on the circle is equal to half the measure of its intercepted arc.

Exploration 2

1. $m\angle AVC = m\angle 1 + m\angle 2$
2.

$m\overset{\frown}{AC}$	$m\overset{\frown}{BD}$	$m\angle 1$	$m\angle 2$	$m\angle AVC$	$m\angle DVB$
160°	40°	80°	20°	100°	100°
180°	70°	90°	35°	125°	125°
$x_1°$	$x_2°$	$\frac{x_1°}{2}$	$\frac{x_2°}{2}$	$\frac{x_1° + x_2°}{2}$	$\frac{x_1° + x_2°}{2}$

3. The measure of an angle formed by two secants or chords is equal to half the sum of the measures of the two intercepted arcs.

Exploration 3

1. $m\angle 1 = m\angle AVC + m\angle 2 \Rightarrow$
   $m\angle AVC = m\angle 1 - m\angle 2$
2.

$m\overset{\frown}{BD}$	$m\overset{\frown}{AC}$	$m\angle 1$	$m\angle 2$	$m\angle AVC$
200°	40°	100°	20°	80°
250°	60°	125°	30°	95°
100°	50°	50°	25°	25°
80°	20°	40°	10°	30°
$x_1°$	$x_2°$	$\frac{x_1°}{2}$	$\frac{x_2°}{2}$	$\frac{x_1° - x_2°}{2}$

3. The measure of an angle formed by two secants with vertex outside the circle is equal to half the difference between the measures of the two intercepted arcs.

Communicate

1. half of the measure of its intercepted arc.
2. half of the measure of its intercepted arc.
3. half of the sum of the measures of the two intercepted arcs.
4. half of the difference between the measures of the two intercepted arcs.

Look Beyond

38.

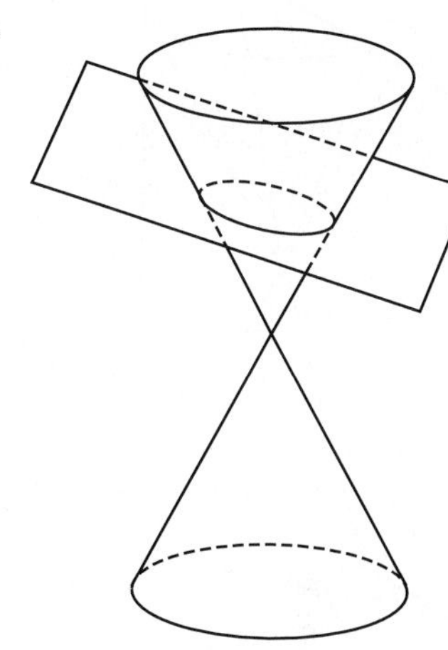

39.

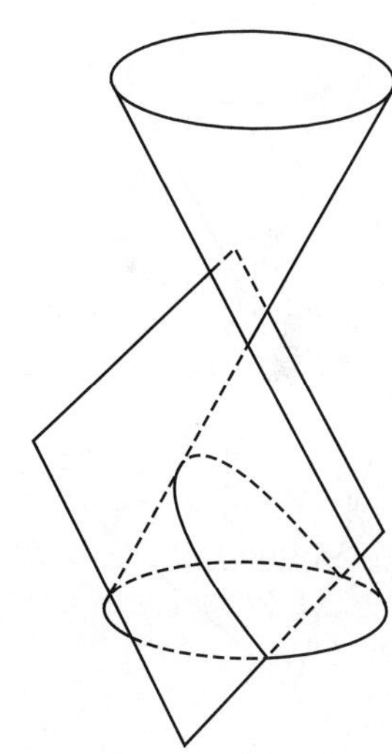

40.

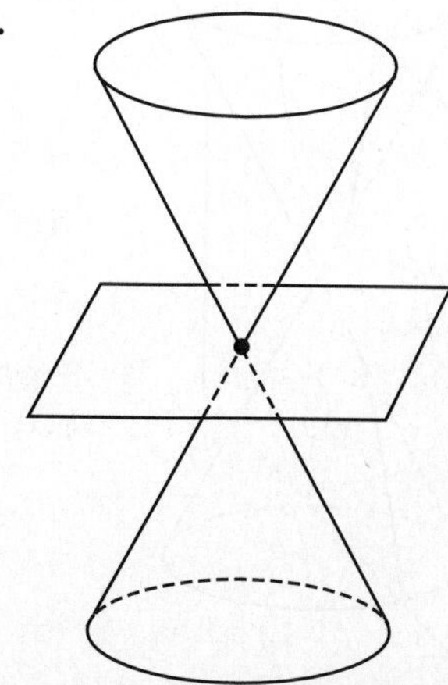

41.

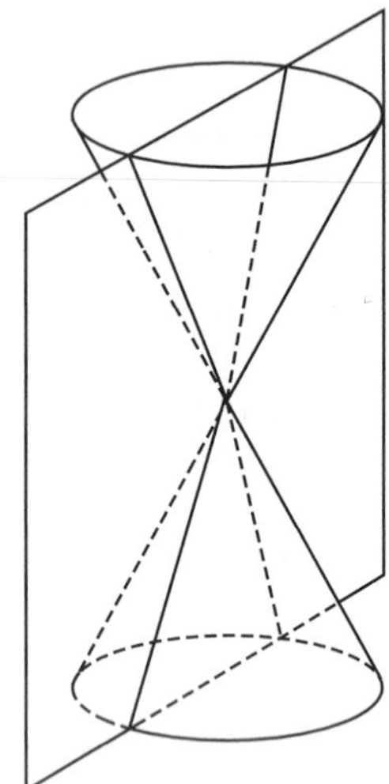

42.

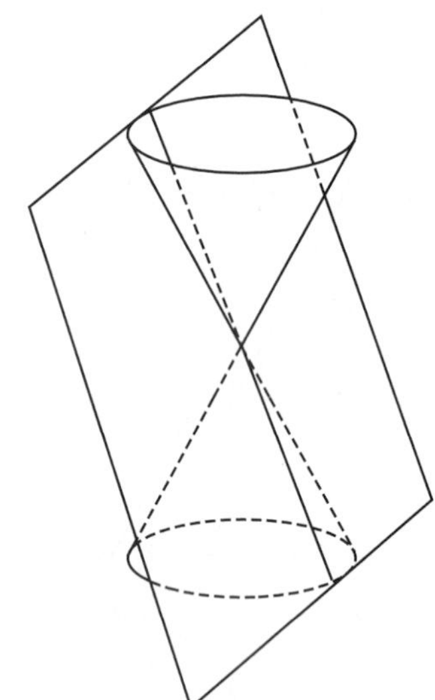

43.

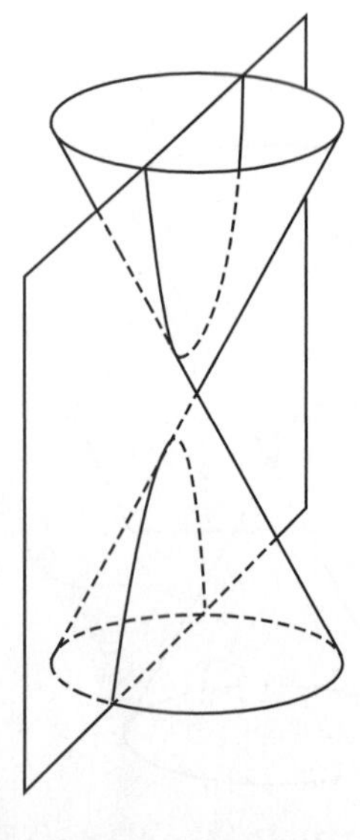

Lesson 9.5, Pages 502–508

Exploration 1

3. The lengths of the two segments that are tangent to a circle from the same external point are equal.
4. Given: $\overline{XA}$ and $\overline{XB}$ are tangent to the circle with center P at A and B
   Prove: $\overline{XA} \cong \overline{XB}$

Statement	*Reason*
1. $\overline{XA}$ and $\overline{XB}$ are tangent to the circle with center P at A and B.	1. Given
2. $\overline{PA} \cong \overline{PB}$	2. Definition of circle.
3. $\overline{PX} \cong \overline{PX}$	3. Reflexive
4. $\overline{PA} \perp \overline{XA}$; $\overline{PB} \perp \overline{XB}$	4. Tangent lines are $\perp$ to the radius drawn to the point of tangency.
5. $\triangle PAX \cong \triangle PBX$	5. HL
6. $\overline{XA} \cong \overline{XB}$	6. CPCTC

Exploration 2, Part 1

2. $\triangle AOC$; $\triangle BOD$
3. $\angle ADB$ and $\angle ACB$ They are congruent.
4. $\angle AOC$ and $\angle BOD$; Vertical angles $\triangle AOC \sim \triangle BOD$; AA-similarity
5. $\triangle BXC$; They are similar by AA-similarity.
6. $\frac{AX}{BX} = \frac{XD}{XC}$
7. <u>the product of the lengths of the other secant segment and its external segment.</u>

Exploration 2, Part 2

1. $\overline{XB} \cong \overline{XD}$
2. $\frac{AX}{BX} = \frac{XD}{XC}$; $(AX)(XC) = (BX)(BX) = (BX)^2$
   The length of the tangent segment squared.

Exploration 3

2. $\triangle XAD$ and $\triangle XBC$
3. $\angle DAC$ and $\angle CBD$; $\angle ADB$ and $\angle BCA$; They are congruent.
4. $\angle AXD$ and $\angle BXC$; Vertical angles; $\triangle AXD \sim \triangle BXC$ by AA-similarity.
5. $\frac{DX}{AX} = \frac{CX}{BX}$; $(AX)(CX) = (BX)(DX)$
6. <u>the product of the lengths of the other segment</u>

Communicate

1. $\overline{AC}$; $\overline{BC}$
2. $\overline{DF}$
3. $\overline{DE}$
4. $\triangle AXC$ and $\triangle DXB$ or $\triangle AEB$ and $\triangle DEC$
5. $\angle ABD$ and $\angle DCA$, $\angle ABE$ and $\angle DEC$, $\angle BEC$ and $\angle DEA$, $\angle XAC$ and $\angle XDB$, $\angle ACX$ and $\angle DBX$
6. 6 units
7. 5 units

Lesson 9.6, Pages 509–515

Communicate

1. (10,0), (−10,0), (0,10), (0,−10)
2. (8,0), (−8,0), (0,8), (0,−8)
3. $(\sqrt{50},0)$, $(-\sqrt{50},0)$, $(0,\sqrt{50})$, $(0,-\sqrt{50})$
4. $(x - h)^2 + (y - k)^2 = r^2$,
   The center of the circle is the point (h,k)
5. $(x - 4)^2 + (y + 5)^2 = 49$
6. $x^2 + y^2 = 6.25$
7. $(x + 1)^2 + (y + 7)^2 = 100$
8. Circle, Center = (0,0); $r = \sqrt{10}$ units

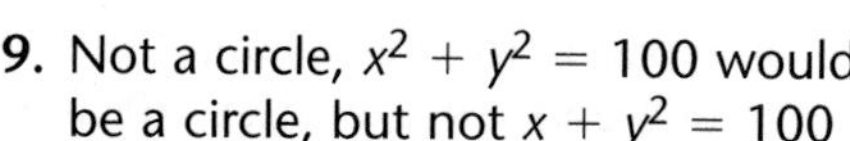

9. Not a circle, $x^2 + y^2 = 100$ would be a circle, but not $x + y^2 = 100$
10. Not circle, $x^2 + y^2 = 64$ would be a circle, not $x^3 + y^3 = 64$.
11. Not circle, $x^2 + y^2 = 9$ would be a circle, not $x^2 - y^2 = 9$.

Practice and Apply

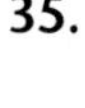

35.

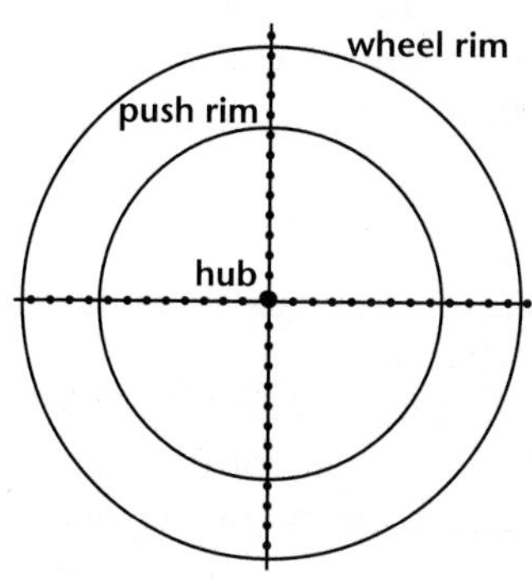

wheel rim: $x^2 + y^2 = 150.0625$
push rim: $x^2 + y^2 = 78.77$

Lesson 10.1, Pages 524–532

Exploration 1

5. The ratio of the side opposite to the acute angle of a right triangle to the side adjacent to the acute angle is constant in a right triangle if the measure of the acute angle is 30°.

Exploration 2

5. The graph is increasing at an increasing rate. While the difference between the measures of the angles used are the same, the differences between the tangent ratios for the angles are increasing.

Exploration 3

4. Specific answers will vary because of differing student choices for the length of YZ. The following formula will work for all choices: $\tan \angle ZXY = \frac{YZ}{XY} \Rightarrow (XY)(\tan \angle ZXY) = YZ \Rightarrow XY = \frac{YZ}{\tan \angle ZXY}$.

5. Ruler measurement uncertainties are the chief source of error (i.e., rulers are accurate only to their smallest unit of measure), but the two methods should give fairly consistent results.

Communicate

1. The tangent ratio measure the quotient of the side opposite to the adjacent side, while the seked, or contangent, is the reciprocal of this ratio.
2. The tangent ratio increases. As the measure of an angle increases, the measure of the side opposite the angle increases relative to the side adjacent to the angle.
3. $\tan \angle A = \frac{1}{1} = 1$
4. $\tan \angle A = \frac{2}{1} = 2$
5. $\tan \angle A = \frac{5}{12}$

Lesson 10.2, Pages 533–540

Exploration 1

1. The value of sin θ is increasing. ; $\sin 0° = 0$; $\sin 90° = 1$
2. The value of cos θ is decreasing. ; $\cos 0° = 1$; $\cos 90° = 0$
3. As an angle θ increases from 0° to 90°, the value of sin θ increases, while the value of cos θ decreases.
4.

θ	$\sin \theta$	$\cos \theta$
0°	0	1
10°	0.174	0.985
20°	0.342	0.940
30°	0.5	0.866
40°	0.643	0.766
50°	0.766	0.643
60°	0.866	0.5
70°	0.940	0.342
80°	0.985	0.174
90°	1	0

5.

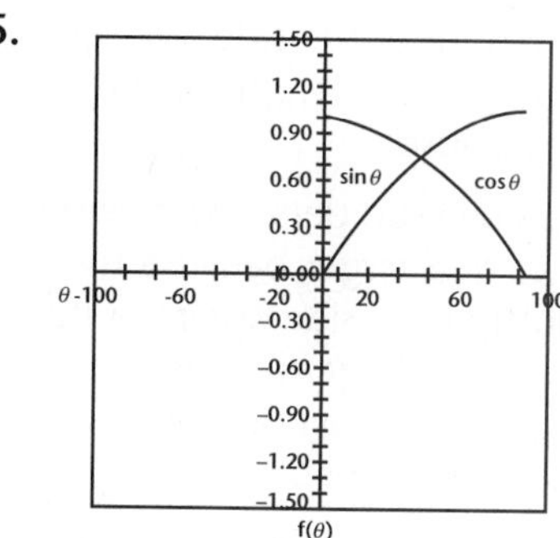

Exploration 2, Part 1

1.

θ	$\sin \theta$	$\cos \theta$	$\tan \theta$	$\frac{\sin \theta}{\cos \theta}$
20°	0.3420	0.9397	0.3640	0.3640
40°	0.6428	0.7660	0.8391	0.8391
60°	0.8660	0.5	1.7321	1.7321

2. $\tan \theta = \frac{\sin \theta}{\cos \theta}$
3. $\frac{\text{sine of } \angle A}{\text{cosine of } \angle A} = \frac{\text{length of side opposite } \angle A}{\text{length of side adjacent to } \angle A} = \tan \angle A$

   Yes, the equation uses the definitions of the sine and cosine ratios and algebra to prove the conjecture.

Exploration 2, Part 2

1.

θ	$\sin \theta$	$\cos \theta$	$(\sin \theta)^2 + (\cos \theta)^2$
20°	0.3420	0.9397	1
40°	0.6428	0.7660	1
60°	0.8660	0.5	1

2. The graph is the horizontal line $y = 1$.
3. The value of y is always 1. ; $(\sin \theta)^2 + (\cos \theta)^2 = 1$
4. Definition of sine and cosine ratios.
   Laws of Exponents and adding with a common denominator.
   By the Pythagorean Theorem, $a^2 + b^2 = c^2$. So, $\frac{a^2 + b^2}{c^2} = 1$
   Therefore, $(\sin \theta)^2 + (\cos \theta)^2 = 1$.

Communicate

1. the length of the side opposite the angle; the length of the hypotenuse

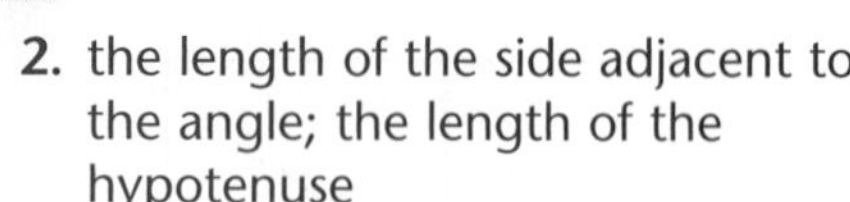

2. the length of the side adjacent to the angle; the length of the hypotenuse
3. The sine ratio gets larger as the angle gets larger because the length of the side opposite the angle gets larger relative to the length of the hypotenuse as the angle gets larger.
4. The cosine ratio gets smaller as the angle gets larger because the length of the side adjacent to the angle gets smaller relative to the length of the hypotenuse as the angle gets larger.
5. By definition, the tangent is the ratio of the lengths of the opposite side to the adjacent side. Also the sine and the cosine of the angles could be computed. The tangent is the ratio of these.
6. Consider a right triangle with sides *a*, *b*, and hypotenuse, *c*, with angles *A*, *B*, and *C* opposite their respective sides. Then:

   $\sin A^2 + \cos A^2 = (\frac{a}{b})^2 + (\frac{b}{c})^2 = \frac{(a^2 + b^2)}{c^2}$.

   The "Pythagorean" Right-Triangle Theorem states that $a^2 + b^2 = c^2$ for any right triangle, therefore $\frac{(a^2 + b^2)}{c^2} = 1$.

Lesson 10.3, Pages 541–547

Exploration 1

1. sin 0° = 0; sin 90° = 1
2. cos 0° = 1; cos 90° = 0; tan 0° = 0; tan 90° is undefined because $\tan \theta = \frac{\sin \theta}{\cos \theta}$ and cos 90° = 0, so the calculator shows "error".
3. The calculator gives values.
4. The calculator gives values.
5. sin 65° = .9063; sin 115° = .9063; sin 70° = .9397; sin 110° = .9397

   cos 65° = .4226; cos 115° = −.4226; cos 70° = 3420; cos 110° = −.3420

   tan 65° = 2.1445; tan 115° = −2,1445; tan 70° = 2.7475; tan 110° = −2.7475

   The sine of an angle and its supplement are equal. The both the cosine and tangent of an angle and its supplement are opposites.

Exploration 2

1. $P'(\frac{\sqrt{3}}{2}, \frac{1}{2}) = (.8660, .5)$
2. $P'(\frac{-\sqrt{3}}{2}, \frac{-1}{2}) = -.8660, -.5)$
3. $P'(\frac{\sqrt{3}}{2}, \frac{-1}{2}) = (.8660, -.5)$
4.

θ	x-coordinate	y-coordinate	cos θ	sin θ
30°	$\frac{\sqrt{3}}{2} = 0.8660$	$\frac{1}{2} = 0.5$	0.8660	0.5
150°	$\frac{-\sqrt{3}}{2} = -0.8660$	$\frac{1}{2} = 0.5$	−0.8660	0.5
210°	$\frac{-\sqrt{3}}{2} = 0.8660$	$\frac{-1}{2} = -0.5$	−0.8660	−0.5
330°	$\frac{\sqrt{3}}{2} = 0.8660$	$\frac{-1}{2} = -0.5$	0.8660	−0.5

5. cos θ is the x-coordinate of the rotation image of (1,0) under a rotation of θ about the origin. sin θ is the *y*-coordinate of the rotation image of (1,0) under a rotation of θ about the origin.
6. Yes. The image of (1,0) under a rotation of 30° is in the first quadrant, so the values of sin θ and cos θ are positive and can be found using a triangle.

Exploration 3

1.

Quadrant	sin θ	cos θ	tan θ
I	+	+	+
II	+	−	−
III	−	−	+
IV	−	+	−

3. sin θ positive in Quadrants I and II.

   cos θ positive in Quadrants I and IV.

   tan θ positive in Quadrants I and III.

   These agree with Part 1
4. Any pair in the form θ and (180° − θ). Examples: 20° and 160° , 45° and 135°.
5. Any pair in the form θ and (360° − θ) or θ and −θ. Examples: 20° and 340°, 50° and −50°.
6. Any pair in the form θ and (180° + θ). Examples: 40° and 220°, 70° and 250°.
7. Yes. sin θ is positive in Quadrants I and II.
8. No, cos 60° = .5 and cos 120° = −.5; Modify as cos θ = −cos(180° − θ).

Communicate

1. The unit provides a way to use the rotation of a unit segment about the origin in order to extend the possible angles for the trig functions up to and beyond 360°.
2. On the unit circle, the *x*-value at 90° is 0 and the *y*-value is 1. Thus, sin 90° = 1, cos 90° = 0, and tan 90° is undefined
3. On the unit circle, the *x*-value at 180° is 1 and the *y*-value is 0. Thus, sin 180° = 0, cos 180° = −1, and tan 180° = 0
4. In Quadrant III, both the *x*- and *y*-values are negative. Therefore, cosine is negative.
5. In Quadrant II, both the *x*-values are negative and the *y*-values are positive. Therefore, sine is positive.
6. In Quadrant III, both the *x*- and y-values are negative.
7. The position of a point on the pump rises and falls repeatedly, passing through the same set of positions each cycle. The vertical motion can be represented the y values of the point. Thus the sine function represents the vertical motion.

Practice and Apply

In 8–11 Check student sketches.

8. sin 45° = 0.71; cos 45° = 0.71; tan 45° = 1.00
9. sin 110° = 0.94; cos 110° = −.34; tan 110° = −2.75

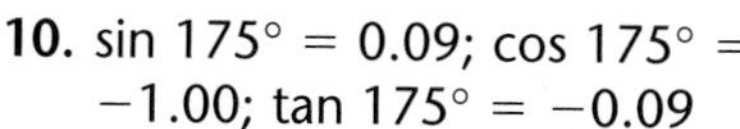

10. sin 175° = 0.09; cos 175° = −1.00; tan 175° = −0.09

11. sin 450° = 1; cos 450° = 0; tan 450° is undefined

Lesson 10.4, Pages 548–554

Exploration 1, Part 2

1.

θ	0°	90°	180°	270°	360°
Sine	0	1	0	−1	0
Cosine	1	0	−1	0	1

2. 0°: $x' = x$, $y' = y$

   90°: $x' = -y$; $y' = x$

   180°: $x' = -x$; $y' = -y$

   270°: $x' = y$; $y' = -x$

   360°: $x' = x$, $y' = y$

   A rotation of 0° or 360° has no effect on the coordinates of the figure being rotated. Under a rotation of 90°, the *x*-coordinate of the image is the opposite of the *y*-coordinate of the pre-image and the *y*-coordinate of the image is equal to the *x*-coordinate of the pre-image. Under a rotation of 180°, the *x*-coordinate of the image is the opposite of the *x*-coordinate of the pre-image and the *y*-coordinate of the image is the opposite of the *y*-coordinate of the pre-image.

   Under a rotation of 270°, the *x*-coordinate of the image is equal to the *y*-coordinate of the pre-image and the *y*-coordinate of the image is the opposite of the *x*-coordinate of the pre-image.

3. A rotation of −90° has the same effect as a rotation of 270°.

   A rotation of 540° has the same effect as a rotation of 180°.

Exploration 2

3. A single rotation of 70° is equal to a 30° rotation followed by a 40° rotation. Rotations in a plane can be added.

Communicate

1. $\begin{bmatrix}1 & 0\\0 & 1\end{bmatrix}$ The result is the same matrix because by definition the identity matrix is designed that way.

2. 0° or any multiple of 360°

3. Examples: 90° and −270°, 120° and −240°

Practice and Apply

24. $\begin{bmatrix}0 & -1\\1 & 0\end{bmatrix}$ 25. $\begin{bmatrix}0 & 1\\-1 & 0\end{bmatrix}$

26. $\begin{bmatrix}.707 & -.707\\.707 & .707\end{bmatrix}$ 27. $\begin{bmatrix}.5 & -.866\\.866 & .5\end{bmatrix}$

28. $A'(1.414,1.414)$, $B'(0,5.657)$, $C'(2.121,4.950)$

29. $A''(0,2)$, $B''(-4,4)$, $C''(-2,5)$

28–29.

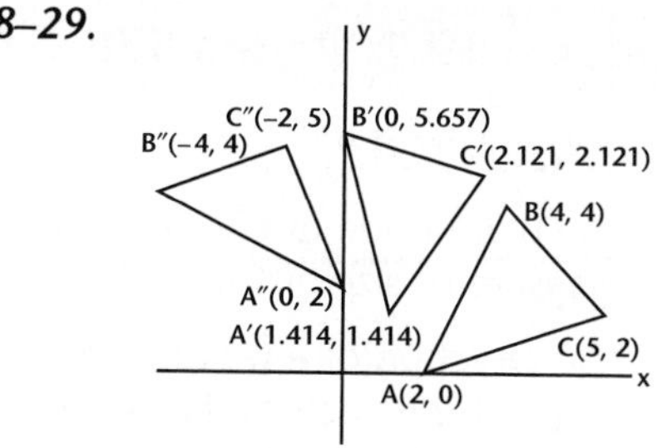

30. The triangle is rotated by 90° about the origin. $\begin{bmatrix}0 & -1\\1 & 0\end{bmatrix}$

   $[R_{60}]$n $[R_{60}]$n $\begin{bmatrix}4\\0\end{bmatrix}$

31. $\begin{bmatrix}.5 & -.866\\.866 & .5\end{bmatrix} \cdot \begin{bmatrix}2\\3.464\end{bmatrix}$

   $\begin{bmatrix}-.5 & -.866\\.866 & -.5\end{bmatrix}\begin{bmatrix}-2\\3.464\end{bmatrix}$

   $\begin{bmatrix}-1 & 0\\0 & -1\end{bmatrix}\begin{bmatrix}-4\\0\end{bmatrix}$

   $\begin{bmatrix}-.5 & .866\\-.866 & -.5\end{bmatrix}\begin{bmatrix}-2\\-3.464\end{bmatrix}$

   $\begin{bmatrix}.5 & .866\\-.866 & .5\end{bmatrix}\begin{bmatrix}2\\-3.464\end{bmatrix}$

   $\begin{bmatrix}1 & 0\\0 & 1\end{bmatrix}\begin{bmatrix}4\\0\end{bmatrix}$

32. $\begin{bmatrix}4\\0\end{bmatrix}$

33. $\begin{bmatrix}4\\0\end{bmatrix}$ They are all the same

35. Yes. Rotation matrices have unique qualities which allow them to be commutative under multiplication. Generally, matrices are not commutative under multiplication. For example:

$$\begin{bmatrix}2 & 3\\1 & 4\end{bmatrix} \times \begin{bmatrix}1 & 1\\1 & 1\end{bmatrix} = \begin{bmatrix}5 & 5\\5 & 5\end{bmatrix}$$

but

$$\begin{bmatrix}1 & 1\\1 & 1\end{bmatrix} \times \begin{bmatrix}2 & 3\\1 & 4\end{bmatrix} = \begin{bmatrix}3 & 7\\4 & 7\end{bmatrix}.$$

Lesson 10.5, Pages 555–561

Exploration

3. $\frac{\sin A}{a} = \frac{\sin B}{b} = \frac{\sin C}{c}$

Communicate

1. $\frac{\sin A}{a} = \frac{\sin B}{b} = \frac{\sin C}{c}$. The Law of Sines relates the sides of a triangle to the angles opposite those sides. Using this property may allow unknown distances or angles to be computed.

2. If an acute angle is computed using the Law of Sines, it is the only one possible. If the angle is obtuse, more than one angle may be consistent with the Law of Sines.

3. Two angles and a side of a triangle are known, or two sides and an angle opposite one of the sides of the triangle are known.

4. Otherwise there would be two variables in any proportion from the Law of Sines.

5. An example is shown. Any proportion from the Law of Sines would have two variables because an angle and the side opposite the angle are not known.

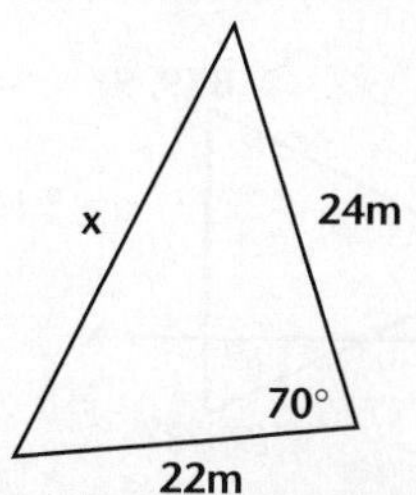

Lesson 10.6, Pages 564–569

Communicate

1. The square of a side of any triangle is equal to the sum of the squares of the other two sides of the triangle minus the product of two times the other two sides and the cosine of the angle opposite the side.
2. Any proportion in the Law of Sines would have contained two variables. No.
3. Two sides and an included angle of a triangle, or all three sides of a triangle.
4. First an altitude of the oblique triangle is drawn so that each angle in the oblique triangle is also an angle in a right triangle. Right triangle trigonometry and the Pythagorean Theorem are applied in the right triangles to prove the Law of Cosines for the oblique triangle.

Practice and Apply

17. By the Pythagorean Theorem, $a^2 = h^2 + (c_1 + c)^2 = h^2 + c_1^2 + 2c_1c + c^2$.
18. From right angle trigonometry, $h = b \sin(180 - A) = b \sin A$. Also, $c_1 = b \cos(180 - A) = -b \cos A$.
19. Substituting, $a^2 = (b \sin A)^2 + (-b \cos A)^2 + 2(-b \cos A)(c) + c^2$. Simplifying and using the identity $(\sin A)^2 + (\cos A)^2 = 1$ gives the Law of Cosines:

$$a^2 = b^2 + c^2 - 2bc \cos A$$

Look Back

20.

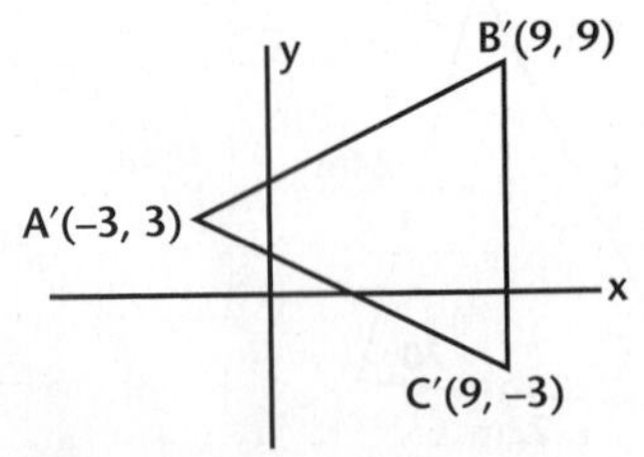

21.

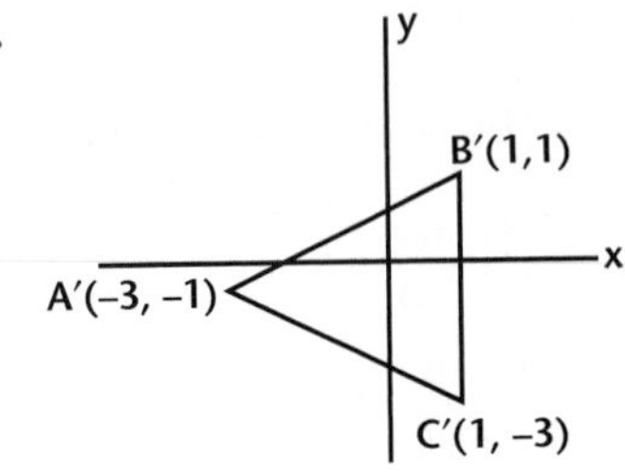

22.

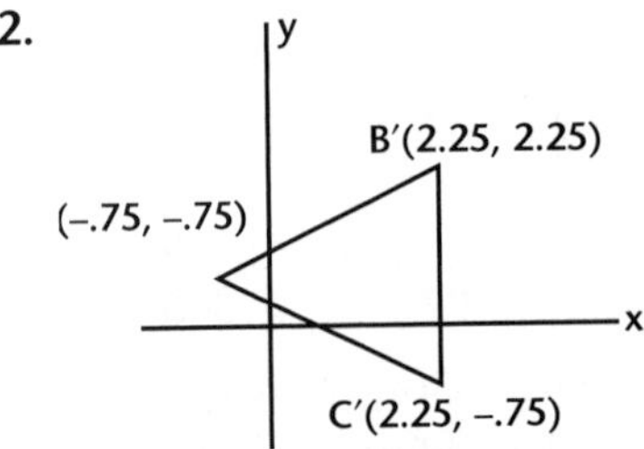

Lesson 10.7, Pages 570–575

Communicate

1. The magnitude of a vector is the measure or size of the vector.
2. The direction of a vector is the angle of the vector relative to a reference direction.
3. magnitude = 175 knots; direction = Northwest or 45° North of West
4. Magnitude of resultant = 12 knots = still water speed-current speed.

   Direction of resultant is upstream.
5. Magnitude of resultant = 0. ; Resultant has no direction.

Practice and Apply

6.

7. .

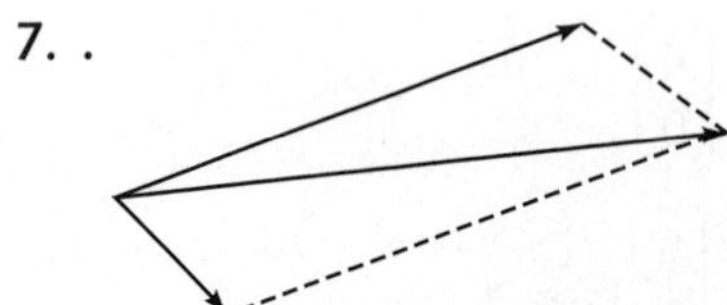

8.

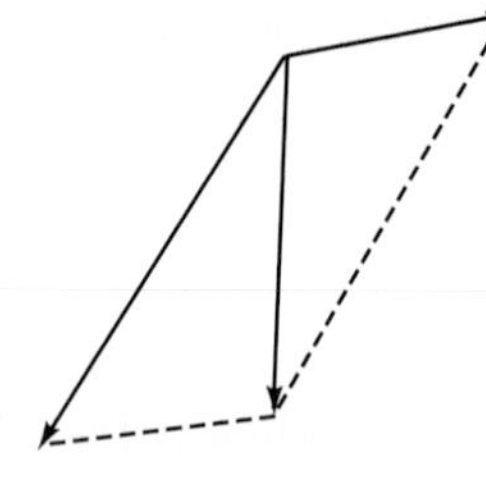

9.

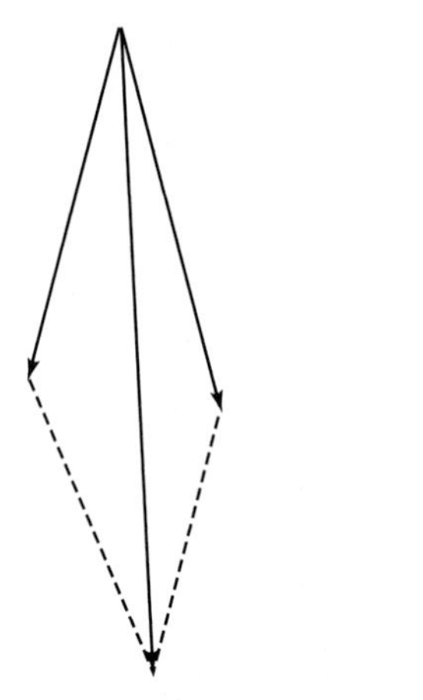

10.

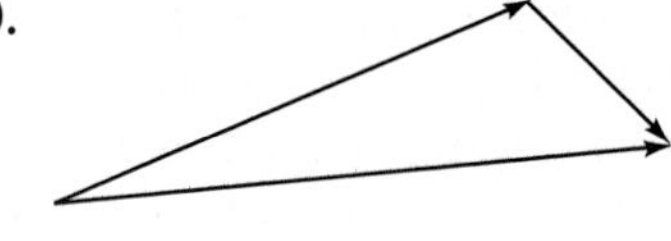

11.

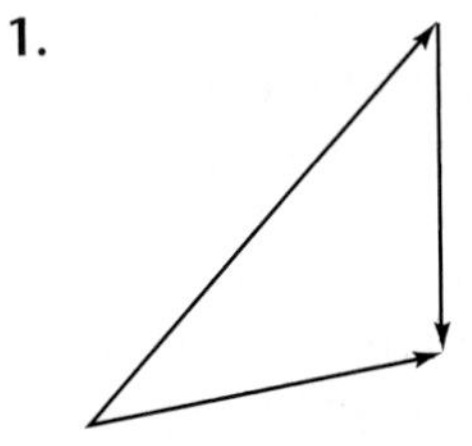

12.

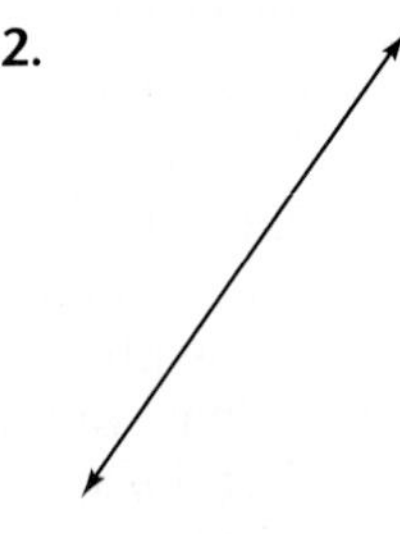

13.

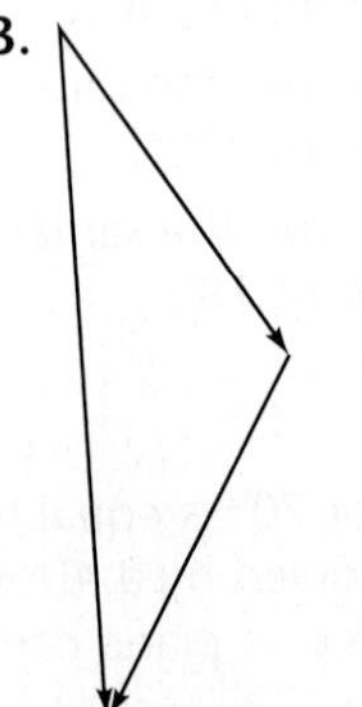

14.

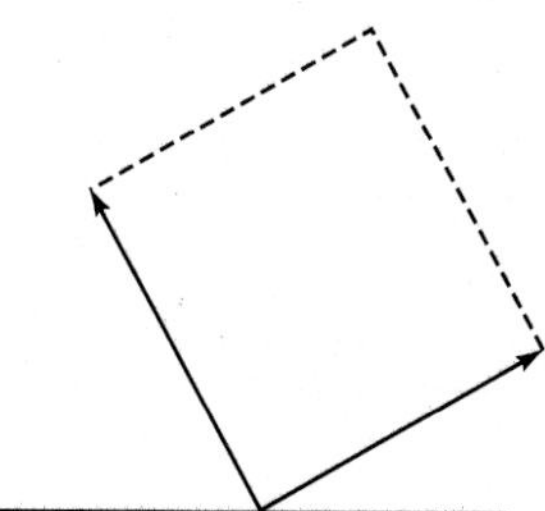

15. The angles of the parallelogram are $180° - 60° - 25° = 95°$ and $180° - 95° = 85°$

16.

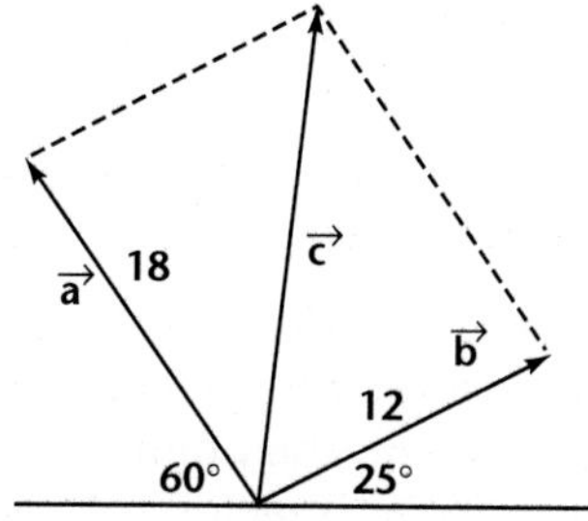

19. $|a| = \sqrt{89} = |b|$
The vectors have the same magnitude, but they are not equal because they are not parallel.

Lesson 11.1, Pages 586–593

Exploration 1

2. l = 25.9 cm
4. Approximately 1.62

Exploration 2

4. The rectangles appear golden. ; Sample data: $\frac{25.9}{16} \approx 1.62$; $\frac{16}{9.9} \approx 1.62$; $\frac{9.9}{6.1} \approx 1.62$; $\frac{6.1}{3.8} \approx 1.61$; $\frac{3.8}{2.3} \approx 1.65$

Exploration 3

3. $BC = \sqrt{2^2 + 1^2} = 5$
6. The ratio between the sides of the rectangle is $\frac{l}{w} = \frac{1+\sqrt{5}}{2}$, which is the Golden Ratio.

Communicate

1. Construct a vertical segment of length 2 units. Then mark a point one unit along the perpendicular to the segment through the top endpoint of the segment. Mark an additional $\sqrt{5}$ units along the perpendicular segment. Draw the rectangle whose sides are the horizontal and vertical segments. This is a golden rectangle, since $\frac{l}{w} = \frac{1+\sqrt{5}}{2}$ which must be true for a golden rectangle by the Pythagorean Theorem.
2. Many possible answers. Some examples are the Parthenon, seashells, and the leaf patterns on a variety of plants.
3. In a Golden Rectangle, $\frac{\text{length}}{\text{width}} = \frac{\text{width}}{\text{length} - \text{width}}$. When a square is constructed inside a Golden Rectangle, the large and small rectangles in the figure that results are similar.
4. When a square is constructed inside a Golden Rectangle, the large and small rectangles in the figure that results are similar, so the small rectangle is also a golden rectangle. More golden rectangles could be formed by constructing squares inside the smaller golden rectangles. Repeating the process replicates a sequence of similar golden rectangles.
5. Answers will vary

Look Back

25. sin 17° ≈ .29; cos 17° ≈ .96; tan 17° ≈ .31
26. sin 387° ≈ .45; cos 387° ≈ .89; tan 387° ≈ .51
27. sin 65° ≈ .91; cos 65° ≈ .42; tan 65° ≈ 2.14
28. sin 122° ≈ .85; cos 122° ≈ −.53; tan 122° ≈ −1.60

Lesson 11.2, Pages 594–599

Exploration 1, Part 1

2.

Destination point	Distance from 0
A(1,1)	2
B(3,4)	7
C(4,0)	4
D(−4,4)	8
E(−3,−2)	5
F(3,−4)	7

3. The taxidistance between (0,0) and $(x,y) = |x| + |y|$.

Exploration 1, Part 2

Sample:

1.

(x_1,y_1)	(x_2,y_2)	x_1	x_2	y_1	y_2	Distance
A(1,1)	C(4,0)	1	4	1	0	4
B(3,4)	C(4,0)	3	4	4	0	5
D(−4,4)	B(3,4)	−4	3	4	4	7
D(−4,4)	F(3,−4)	−4	3	4	−4	15
A(1,1)	E(−3,−2)	1	−3	1	−2	7
E(−3,−2)	F(3,−4)	−3	3	−2	−4	8

2. Taxidistance between (x_1,y_1) and $(x_2,y_2) = |x_2 - x_1| + |y_2 - y_1|$.

3.

Radius	Number of Points in Circle	Circumference
1	4	8
2	8	16
3	12	24
4	16	32
5	20	40
r	4r	8r

Exploration 2

2. Student drawings should resemble the pattern shown in text.

3. *Continued*

If the radius is r, the number of points is 4r. The circumference is 8r.

4. $\pi = \frac{C}{d} = \frac{8r}{2r} = 4$

Communicate

1. In taxicab geometry points are located at the intersection points of horizontal and vertical lines forming a grid. To travel from point to point one may only move along the lines.
2. Horizontal and vertical lines represent streets. Points on a taxicab grid may only be at the intersection of two "streets". Coordinates of taxicab points are always integers.
3. The distance between two points is the smallest number of "blocks" a taxi must travel to go from one point to another. Taxidistance between (x_1,y_1) and (x_2,y_2) = $|x_2 - x_1| + |y_2 - y_1|$. Find the number of blocks in the vertical direction and in the horizontal direction by finding he absolute values of the differences between the x-coordinates and the y-coordinates. Add the absolute values.
4. Many possible answers. Samples: Finding the shortest driving distance between two points in a city, finding alternate routes from one point to another if one route is closed.

Practice and Apply

5.

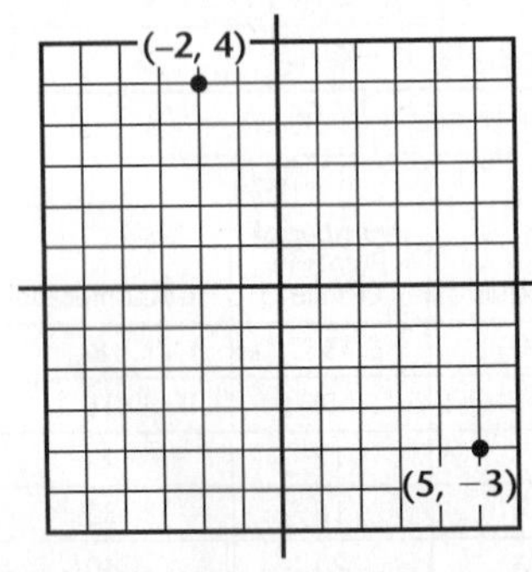

$AB = |-2 - 5| + |4 - -3| = 14$

Find the distance between the x-coordinates and add this to the distance between the y-coordinates.

8.

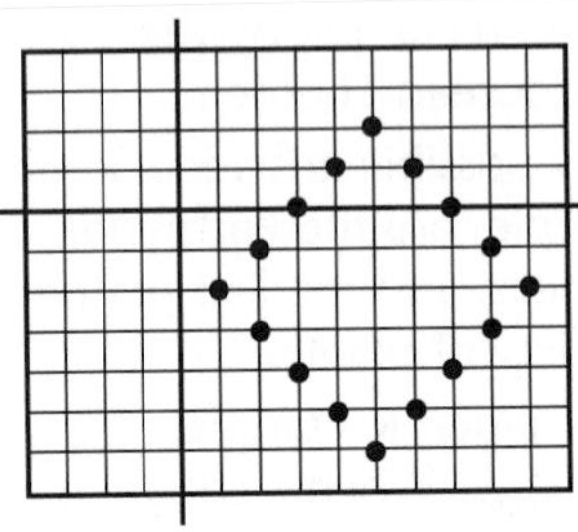

11.

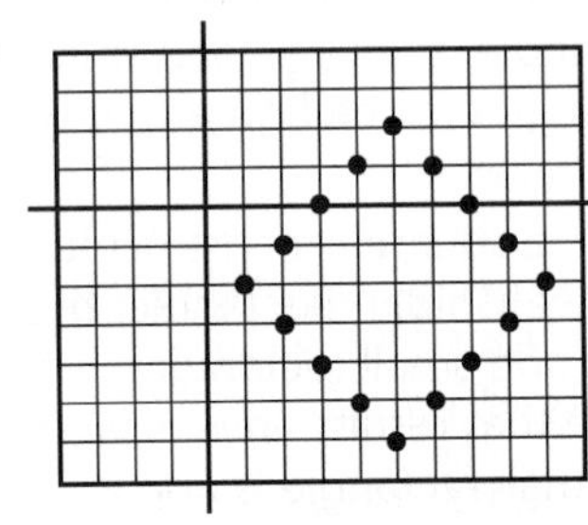

22. a–c. Taxidistance 3 is shown.

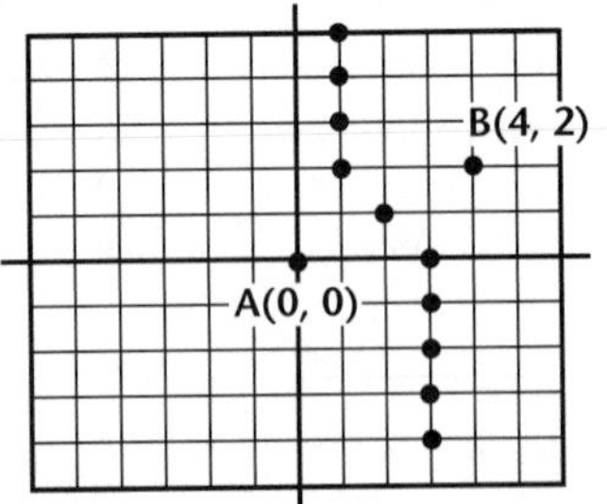

23. Points on the perpendicular bisector of a segment are equidistant from the endpoints of the segment in both Euclidean and Taxicab geometry.

However, in Taxicab geometry, the perpendicular bisector is not "straight" and the points on the perpendicular bisector are not "connected".

Lesson 11.3, Pages 600–606

Exploration

All are traversable except *F* and *H*.

Figure	Number of Vertices	Number of Odd Vertices	Number of Even Vertices	Can it be Traversed?
A	4	0	4	Yes
B	4	2	2	Yes
C	4	2	2	Yes
D	3	0	3	Yes
E	6	2	4	Yes
F	7	6	1	No
G	7	2	5	Yes
H	8	4	4	No

Communicate

1. People wondered if it were possible to take a walk and cross each of the seven bridges of Konigsberg exactly once.
2. A network can be traversed if it has 2 or fewer odd vertices. An odd vertex may only be the starting or end point of a traversal.
3. Either the player on the far right or the player on the far left must start. If one starts the other will be the end point. The network has exactly two odd vertices.
4. No. The network representing the

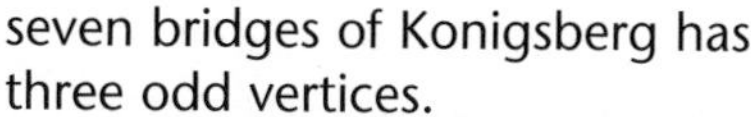

seven bridges of Konigsberg has three odd vertices.

5. Sample answers: Finding the best route for making a large number of deliveries. Designing a highway system to serve several communities. Designing an office complex for a company headquarters. Organizing a complex production process.

Practice and Apply

14. .

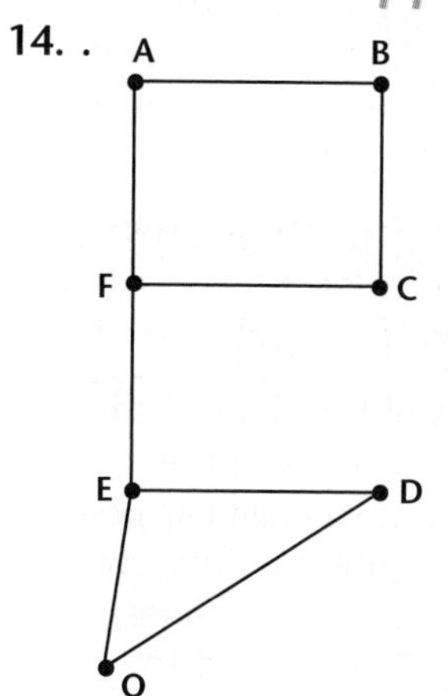

17. .

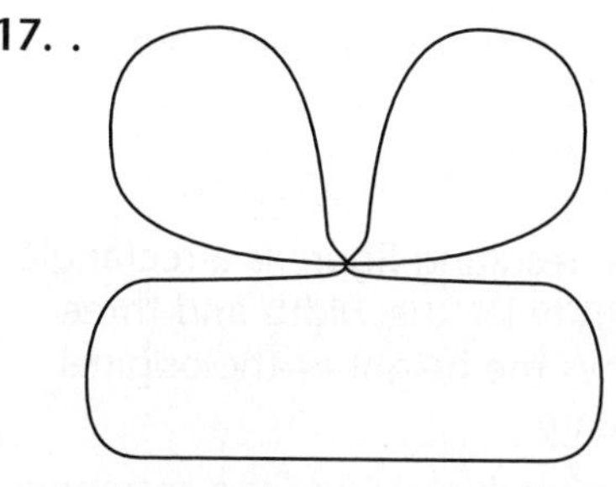

18. The network is traversable and is an Euler Circuit because the one vertex is even.

19. The officer can park anywhere along the network.

Look Back

23.

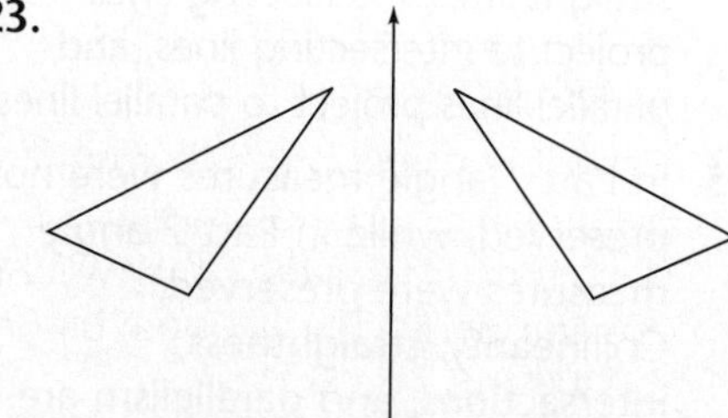

24.

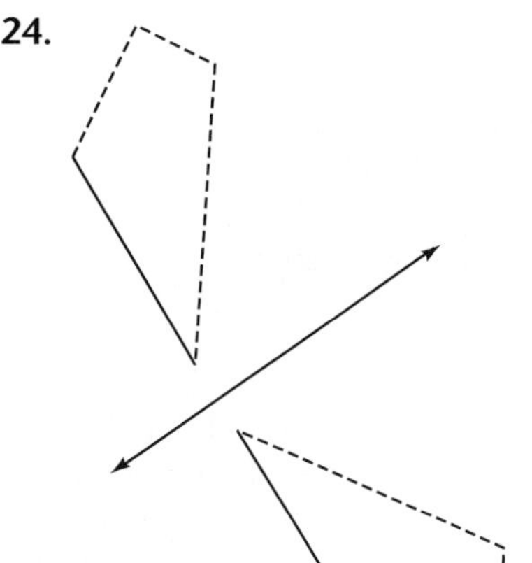

Lesson 11.4, Pages 607–613

Communicate

1. Two shapes are topologically equivalent if one of them can be distorted into the other without cutting or tearing.
2. Yes. They both contain three regions and one intersection.
3. No. The figure on the right has two intersection points, while the one on the left has none. The figure on the right would have to be cut to make it equivalent to the figure on the right.
4. An invariant is a property that stays the same no matter how a figure is distorted.
5. Every simple closed curve divides the plane into two distinct regions. To connect a point in the inside region with a point on the outside region, you must cut the curve.

Practice and Apply

8. Sample answer:

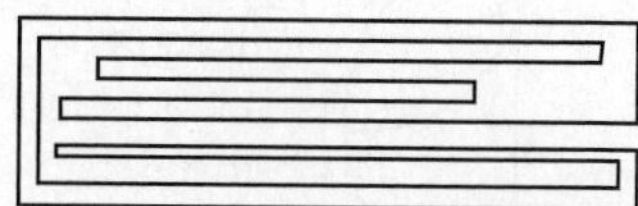

9. Assume the top and bottom faces of the original cube are *ABCD* and *EFGH*.

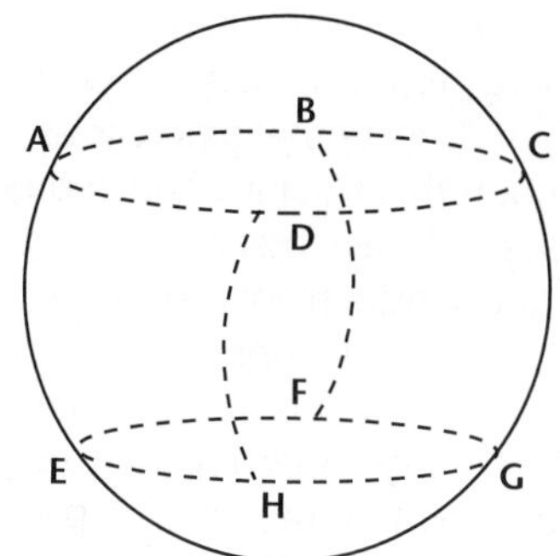

10.

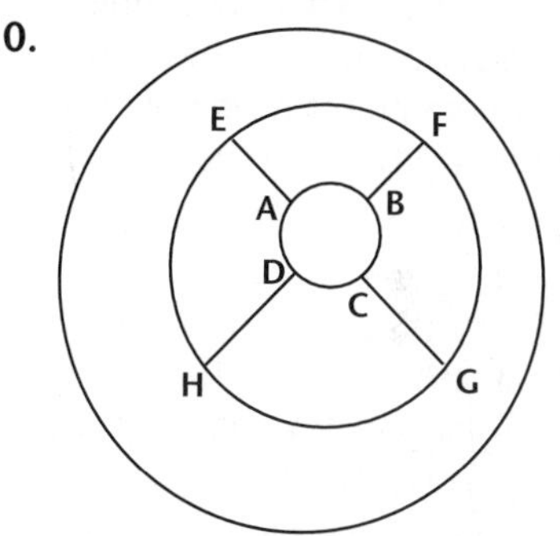

Lesson 11.5, Pages 614–620

Communicate

1. Euclid's Fifth Postulate. If two straight lines in a plane are cut by a third line such that the sum of the interior angles on one side is less than 180°, then the lines will meet if they are extended.
2. None; An infinite number.
3. Early, this was because of our sense intuition of parallel lines. Later on, people tried to prove that the Fifth Postulate was a theorem by proving it from the other four postulates. When it was shown that the Fifth Postulate could not be proven, mathematicians challenged the Fifth Postulate in other ways.
4. No parallels and an infinite number of parallels.
5. A line is a great circle. You can go around the circle an infinite number of times. There is no specific starting point or ending point.

Practice and Apply

7. Consider polar opposite sphere points a single "point." Two

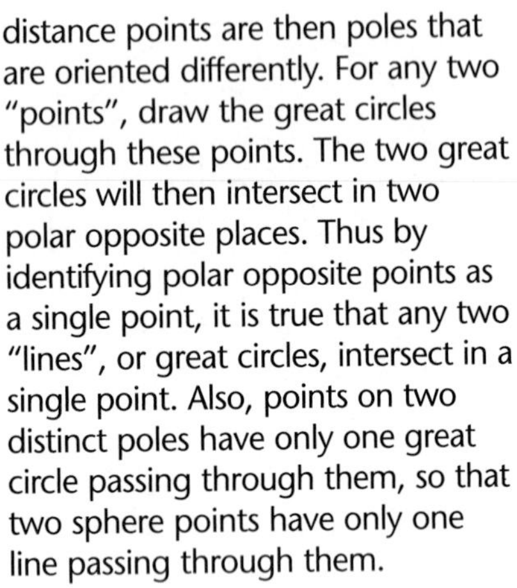

distance points are then poles that are oriented differently. For any two "points", draw the great circles through these points. The two great circles will then intersect in two polar opposite places. Thus by identifying polar opposite points as a single point, it is true that any two "lines", or great circles, intersect in a single point. Also, points on two distinct poles have only one great circle passing through them, so that two sphere points have only one line passing through them.

8. 2-gon *AB*.

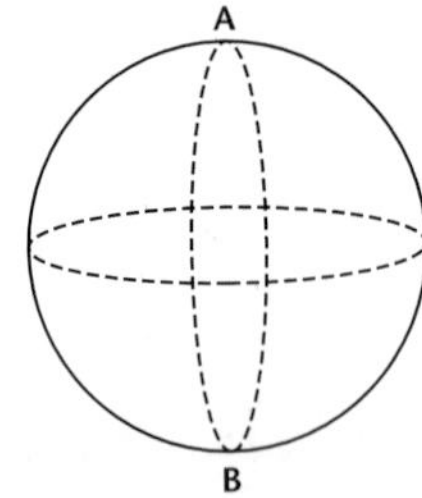

The vertices must be at the poles because the points of intersection of two lines are at the two poles.

10. 4-gon *ABCD* with diagonal $\overline{BD}$

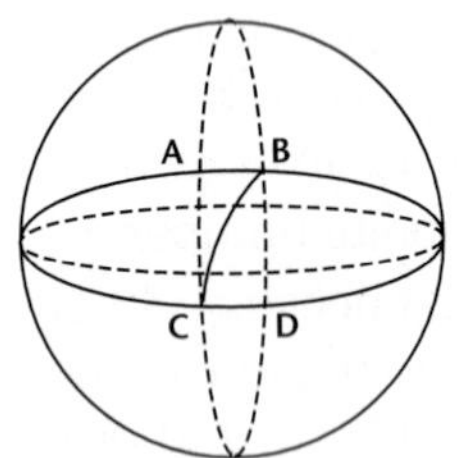

12. $\triangle ABC$

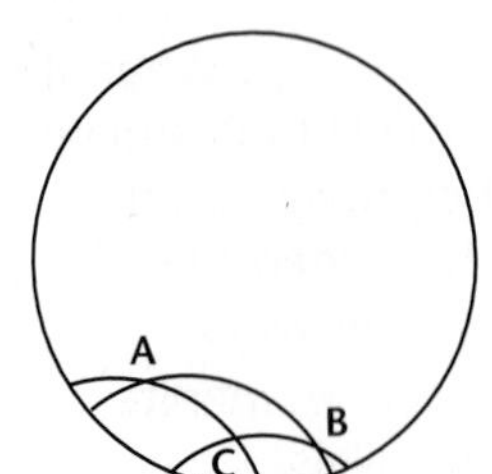

14.

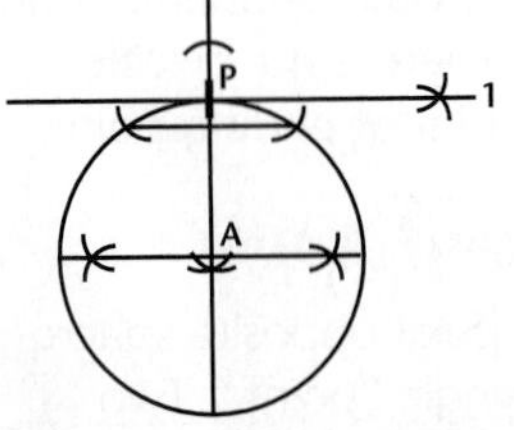

16. An infinite number.

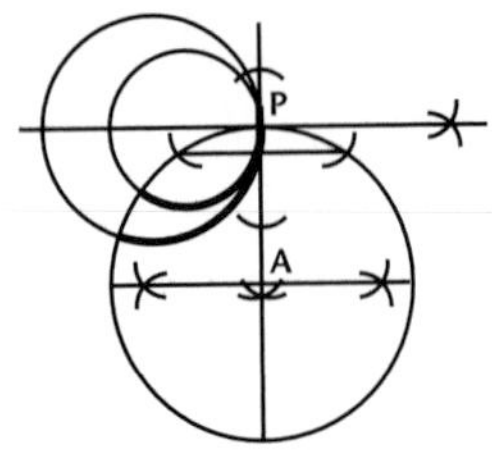

18. An infinite number. ; In Euclid's Parallel Postulate, there is exactly one line in a plane parallel to a given line through a point not on the line. Euclid's Parallel Postulate does not hold in Poincare's model. Poincare's model is a model of a Non-Euclidean geometry.

19.

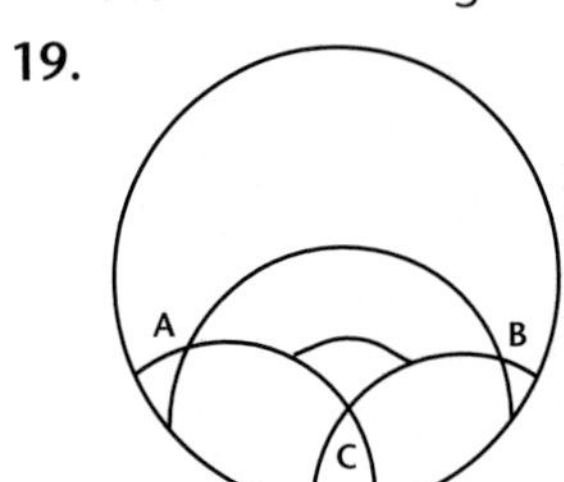

23. Quadrilateral *ABCD*.

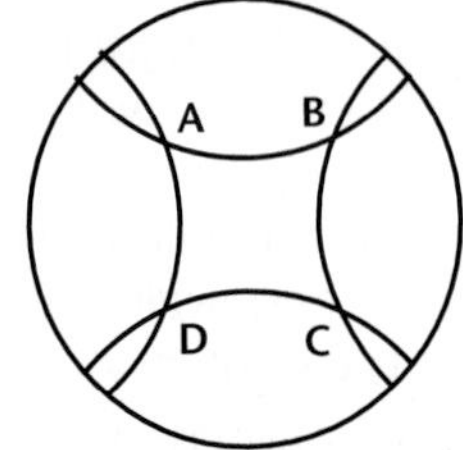

24. The sum of the measures of the angles of the quadrilateral is less than 360°. A diagonal of the quadrilateral will divide the quadrilateral into two triangles. Since the sum of the measures of the angles of each triangle is less than 180°, the sum of the measures of the angles of the quadrilateral will be less than 360°.

25. .

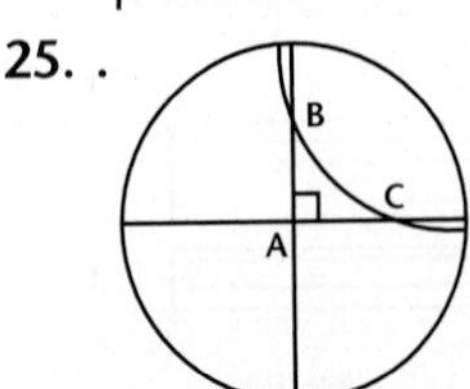

The sum of the acute angles in the triangle is less than 90°.

Look Beyond

32. Proof approach:

$m\angle 1 + m\angle 3 = 180°$; $\angle 2 + \angle 4 = 180°$; $m\angle D = m\angle 1$; $m\angle E = m\angle 2$; $m\angle D + m\angle 3 = 180°$; $m\angle E + m\angle 4 = 180°$; $m\angle D + m\angle 3 + m\angle E + m\angle 4 = 360°$;
However, the sum of the measures of the angles of a quadrilateral is less than 360° in the geometry of Lobachevski. (See Exercise 24.) Therefore, $\triangle ABC \cong \triangle DEF$.

Lesson 11.6, Pages 621–628

Exploration 1, Part 1

The images of the sides of the square are the sides of the parallelogram, so collinear points project to collinear points, straight lines project to straight lines, intersecting lines project to intersecting lines, and parallel lines project to parallel lines.

Exploration 1, Part 2

2. The resulting figure is a rectangle with twice the width and three times the height of the original square.

3. The resulting figure is a rectangle with three times the width and twice the height of the original square.

4. The image of each side of the square is a side of a rectangle, so collinear points project to collinear points, straight lines project to straight lines, intersecting lines project to intersecting lines, and parallel lines project to parallel lines.

5. In Part 1 angle measures were not preserved, while in Part 2 angle measures were preserved. Collinearity, straightness, intersections, and parallelism are preserved in both.

Exploration 2, Part 1

6. X, Y and Z are collinear. If A, B, and C are three distinct points on one line and A', B', and C' are three distinct points on a second line, then the intersections of $\overline{AB'}$ and $\overline{BA'}$, $\overline{AC'}$ and $\overline{CA'}$, and $\overline{BC'}$ and $\overline{CB'}$ are collinear.

Exploration 2, Part 2

9. X, Y and Z are collinear. If one triangle is a projection of another triangle, then the lines containing the corresponding sides of the two triangles meet at points on a line.

Communicate

1. An affine transformation distorts a figure without regard for size or shape. Collinearity, intersection of lines, straightness of lines, and parallelism are preserved.
2. All are affine transformations. The transformation in part c is also a dilation and the transformation in part b is also a reflection.
3. It is the study of the properties of figures that do not change under a projection.
There is no concept of size, measurement, or congruence.
Its theorems state facts about the positions of points and the intersections of lines.
An unmarked straightedge is the only tool allowed for drawing figures.
4. The Theorem of Pappus: If A, B, and C are three distinct points on one line and A', B', and C' are three distinct points on a second line, then the intersections of $\overline{AB'}$ and $\overline{BA'}$, $\overline{AC'}$ and $\overline{CA'}$, and $\overline{BC'}$ and $\overline{CB'}$ are collinear.
The Theorem of Desargues: If one triangle is a projection of another triangle, then the lines containing the corresponding sides of the two triangles meet at points on a line.

Practice and Apply

5. .

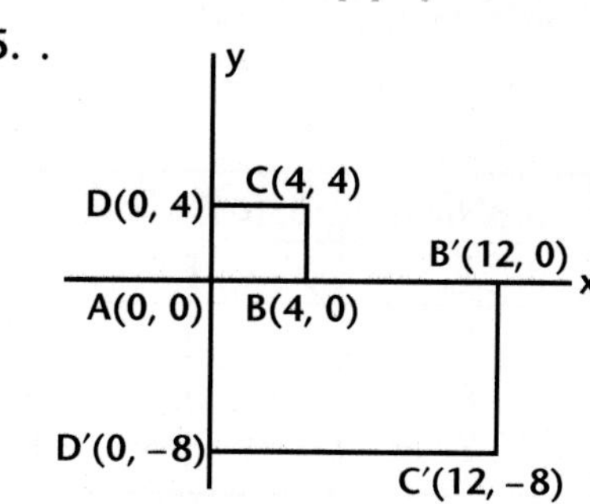

6. .

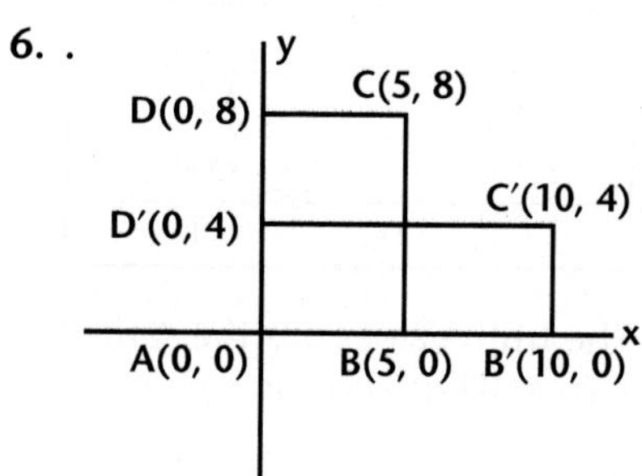

7. .

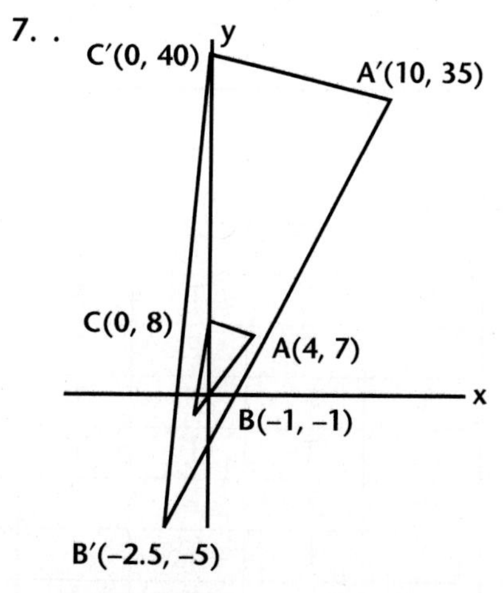

14.

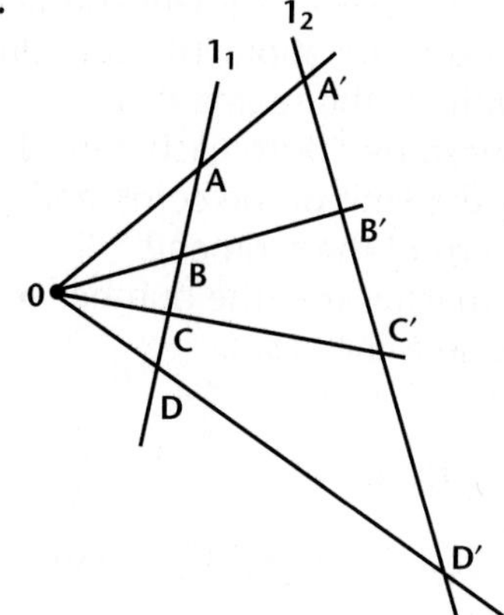

15.

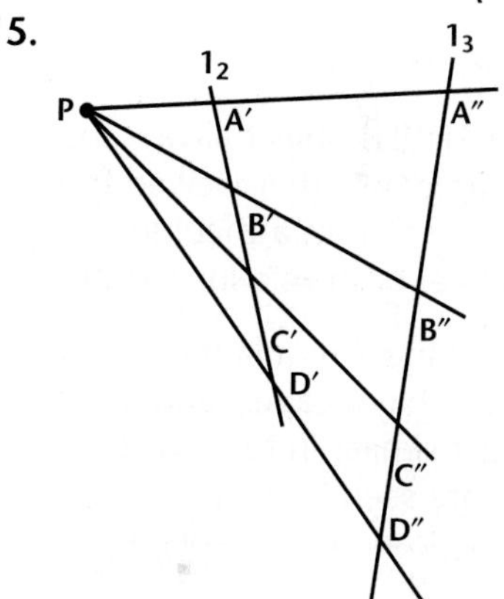

16.

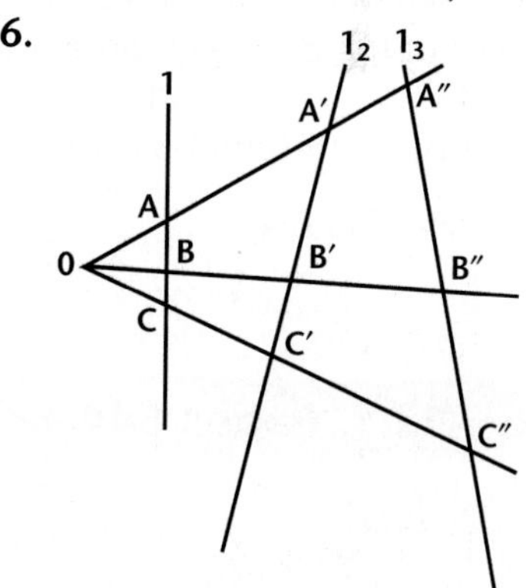

Lesson 11.7, Pages 629–635

Exploration 1

5.

Iteration	0	1	2	3	4	5	n
Number of Segments	1	2	4	8	16	32	2^n
Combined Length	27	18	12	8	$\frac{16}{3}$	$\frac{32}{9}$	$27\left(\frac{2}{3}\right)^n$

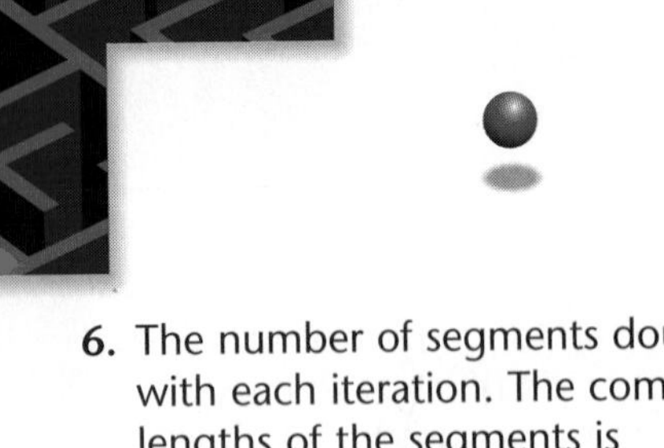

6. The number of segments doubles with each iteration. The combined lengths of the segments is reduced by $\frac{1}{3}$ with each iteration. No, the limit of the combined lengths of the segments approaches 0 as the number of iterations increases.

Exploration 2

2. The result is a Sierpinski Gasket.

Communicate

1. A self-similar object always looks the same, whether viewed from very close, from a medium distance, or from a long distance.
2. A coastline. Irregularities in the form of bays, capes, and inlets appear similar in form whether they are seen from space, or from much closer in an airplane.
3. After many iterations, all that remains of the original segment is a set of very short, pointlike segments that resembles dust.
4. An infinite number.

Lesson 12.1, Pages 644–649

Communicate

1. Valid. The argument is in the form *modus ponens.*
2. Not valid. The converse of a premise may not be true even if the premise is .
3. Valid. The argument is in the form *modus tollens.*
4. Valid. The argument is in the form *modus ponens.*
5. Yes. If one of the premises is false or faulty, a valid argument may lead to a conclusion that is not true.

Lesson 12.2, Pages 650–655

Exploration

1.

p AND q	$\sim(p$ AND $q)$
T	F
F	T
F	T
F	T

2.

$\sim p$	$\sim q$	$\sim p$ OR $\sim q$
F	T	T
T	F	T
T	T	T

3. The truth values are the same for $\sim(p$ AND $q)$ and for $\sim p$ OR $\sim q$.

4. $\sim p$ OR $\sim q$

Communicate

1. A conjunction is true when both of its statements are true.
2. A disjunction is true when either or both of its statements are true.
3. False. One of the statements, $(4)(5) = 9$, is false.
4. True. One of the statements, `All triangles have three angles', is true.
5. True. Both statements are true.
6. False. Both statements are false.
7. The state is not Alaska.
8. The weather is rainy.

Practice and Apply

28.

p	q	r	s	p AND q	r AND s	$(p$ AND $q)$ OR $(r$ AND $s)$
T	T	T	T	T	T	T
T	T	T	F	T	F	T
T	T	F	T	T	F	T
T	F	T	T	F	T	T
F	T	T	T	F	T	T
T	T	F	F	T	F	T
T	F	T	F	F	F	F
F	T	T	F	F	F	F
T	F	F	T	F	F	F
F	T	F	T	F	F	F
F	F	T	T	F	T	T
T	F	F	F	F	F	F
F	T	F	F	F	F	F
F	F	T	F	F	F	F
F	F	F	T	F	F	F
F	F	F	F	F	F	F

For p AND q OR r AND s to be false, p or q and r or s must be false. At least one statement in both conjunctions must be false.

29.

p	q	r	p OR q	(p OR q) OR r
T	T	T	T	T
T	T	F	T	T
T	F	T	T	T
F	T	T	T	T
T	F	F	T	T
F	T	F	T	T
F	F	T	F	T
F	F	F	F	F

p OR q OR r is false when all three statements are false.

30.

p	q	r	p AND q	p AND q AND r
T	T	T	T	T
T	T	F	T	F
T	F	T	F	F
F	T	T	F	F
T	F	F	F	F
F	T	F	F	F
F	F	T	F	F
F	F	F	F	F

p AND q AND r is true when all three statements are true.

Look Beyond

38. Suppose the statement is true. If the statement is true, then there is a contradiction. Therefore, the statement is false.

Lesson 12.3, Pages 658–664

Communicate

1. Converse: If the moon is made of green cheese, then today is February 30th.
   Inverse: If today is not February 30th, then the moon is not made of green cheese.
   Contrapositive: If the moon is not made of green cheese, then today is not February 30th.
2. Converse: If a triangle is equilateral, then all the sides of the triangle are congruent.
   Inverse: If all the sides of a triangle are not congruent, then it is not equilateral.
   Contrapositive: If a triangle is not equilateral, then all the sides of the triangle are not congruent.
3. Converse: If I did not buy cereal, then I did not go to the market.
   Inverse: If I go to the market, then I buy cereal.
   Contrapositive: If I bought cereal, then I went to the market.
4. Converse: If I am not late to school, then the car started.
   Inverse: If the car does not start, then I will be late to school.
   Contrapositive: If I am late for school, then the car did not start.
5. If a is true and b is false.

Practice and Apply

8. The conditional is false. AAA is not enough to prove two triangles congruent. Converse: If two triangles are congruent, then the three angles of one triangle are congruent to the three angles of the other triangle. ; True, by the definition of congruent figures.
   Inverse: If the three angles of one triangle are not congruent to the three angles of another triangle, then the triangles are not congruent. ; True, by the definition of congruent figures.
   Contrapositive: If two triangles are not congruent, then the three angles of one triangle are not congruent to the three angles of the other triangle. ; False, because the original statement is false.
9. Conditional is true. The sum of two even numbers is an even number. Converse: If $p + q$ is an even number, then p and q are even numbers. ; False. 3 + 5 is an even number. Inverse: If p and q are not even numbers, then $p + q$ is not an even number. ; False. 3 + 5 is an even number. Contrapositive: If $p + q$ is not an even number, then p and q are not even numbers; True, because the original conditional is true.
10. Conditional is false. Water will freeze at higher temperatures at higher altitudes. Converse: If its temperature is less than or equal to 32°F, then water freezes. ; True Inverse: If water does not freeze, then its temperature is greater than 32°F. ; True Contrapositive: If its temperature is greater than 32°F, then water does not freeze. ; False, because the original conditional is false.
11. If $\sim q$, then $\sim p$. ; If $\sim(\sim p)$, then $\sim(\sim q)$. = If p, then q. ; The contrapositive of the contrapositive of a statement is the same as the original statement.
12. b is the contrapositive of the statement. Since the statement is

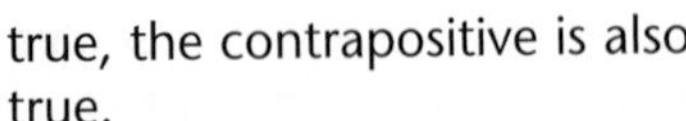

true, the contrapositive is also true.

13. If a triangle is a right triangle with hypotenuse c and legs a and b, then $a^2 + b^2 = c^2$. Converse: If $a^2 + b^2 = c^2$, then the triangle is a right triangle with hypotenuse c and legs a and b. ; True. Proved earlier.

    Inverse: If a triangle is not a right triangle with hypotenuse *c* and legs *a* and *b*, then $a^2 + b^2 \neq c^2$. ; True, because the inverse is logically equivalent to the converse.

    Contrapositive: If $a^2 + b^2 \neq c^2$, then the triangle is not a right triangle with hypotenuse c and legs a and b. ; True, because the contrapositive is logically equivalent to the Pythagorean Theorem.

14.–15. Student answers will depend on whether the converse of the chosen theorem or postulate is true.

16. If a person is a senior, then the person must report to the auditorium.

17. If a point is on the perpendicular bisector of a segment, then the point is equidistant from the endpoints of the segment.

18. If you expect to be late, then call me.

Lesson 12.4, Pages 665–670

Communicate

1. A contradiction is a claim that both a statement and its negation are true.

2. Sample answer:

   The defendant committed the crime. The defendant did not commit the crime. They are contradictory because one is the negation of the other.

3. a = b and a ≠ b, for example.

4. The proof starts by assuming the negation of the statement to be proved. Then a contradiction is reached by logical arguments. The conclusion is then made that the original statement must be true.

5. In an indirect proof a conditional If p then q is assumed to be true. Then ~q is proven by logical arguments. Then ~p must be true. Proving alibis in court is often done in this manner.

Practice and Apply

6. Yes. The proof starts by assuming the opposite of the statement to be proved. Then a contradiction is reached after logical arguments. Finally, the opposite of the assumption is stated to be true because the assumption must be false.

7. Yes. The proof starts by assuming the opposite of the statement to be proved. Then a contradiction is reached after logical arguments. Finally, the opposite of the assumption is stated to be true because the assumption must be false.

8. No. The proof does not start by assuming the opposite of the statement to be proved.

9. No. The proof does not start by assuming the opposite of the statement to be proved.

10. Yes. The proof starts by assuming the opposite of the statement to be proved. Then a contradiction is reached after logical arguments. Finally, the opposite of the assumption is stated to be true because the assumption must be false.

26.

Statements	*Reasons*
1. $\overline{BD}$ bisects $\angle ABC$, $\overline{BD}$ is not a median	1. Given
2. Assume $\overline{AB} \cong \overline{BC}$	2. Negate proof statement
3. $\angle A \cong \angle C$	3. Isosceles Triangle Theorem
4. $\overline{BD} \cong \overline{BD}$	4. Reflexive
5. $\angle ABD \cong \angle CBD$	5. Definition of bisector
6. $\triangle ABD \cong \triangle CBD$	6. AAS
7. $\overline{AD} \cong \overline{CD}$	7. CPCTC
8. D is the midpoint of $\overline{AC}$	8. Midpoint definition
9. $\overline{BD}$ is a median of $\triangle ABC$	9. Median definition

The last step contradicts the given statement that $\overline{BD}$ is not a median. The assumption must therefore be false. So the conclusion is $\overline{AB} \ncong \overline{BC}$.

27.

Statements	*Reasons*
1. $\overline{CT} \ncong \overline{BK}$	1. Given
2. Assume $\overline{BC}$ and $\overline{KT}$ bisect	2. Negate proof statement each other.
3. $\overline{CW} \cong \overline{BW}$, $\overline{TW} \cong \overline{KW}$	3. Definition of bisection
4. $\angle CWT \cong \angle BWK$	4. Vertical $\angle$s $\cong$
5. $\triangle CWT \cong \triangle BWK$	5. SAS
6. $\overline{CT} \cong \overline{BK}$	6. CPCTC

The last statement is a contradiction of the given statement that $\overline{CT} \ncong \overline{BK}$. The assumption must therefore be false. So the conclusion is $\overline{BC}$ and $\overline{KT}$ do not bisect each other.

28.

	Statements		*Reasons*
1.	$\angle 1 \ncong \angle 2$	1.	Given
2.	Assume $m\angle 1 = 90°$	2.	Negate proof statement
3.	$\angle 1$ and $\angle 2$ are a linear pair	3.	Definition of linear pair
4.	$m\angle 1 + m\angle 2 = 180°$	4.	The sum of the measures of a linear pair is 180°.
5.	$m\angle 2 = 180° - m\angle 1$	5.	Subtraction Property of =.
6.	$m\angle 2 = 180° - 90° = 90°$	6.	Substitution

The last step is a contradiction because $m\angle 1 = m\angle 2 = 90°$ contradicts the given statement that $\angle 1 \ncong \angle 2$. The assumption must therefore be false. So the conclusion is $m-1 \neq 90°$.

29. Assume that my client is guilty. If my client is guilty, then my client was at the scene of the crime when it happened. But that contradicts the fact that my client was miles away from the scene of the crime when it happened. The assumption must therefore be false and my client must be innocent.

30. Assume that this is Elm Street. If this is Elm Street, then there is a brick house on the corner. But that contradicts the fact that all the houses on this street are made of wood. The assumption must therefore be false and this is not Elm Street.

31. Assume that the temperature is below 32°F. If the temperature is below 32°F, then the water on the sidewalk would freeze. But that contradicts the fact that the water on the sidewalk is not frozen. The assumption must therefore be false and the temperature must be 32°F or higher.

33.

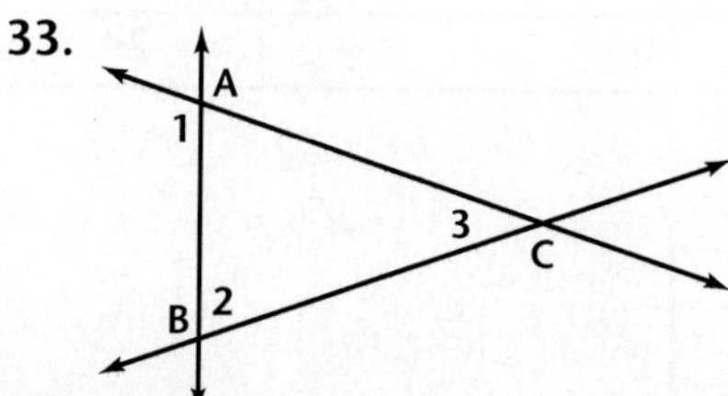

Assume that the lines are not parallel. If the lines are not parallel, then they meet at a some point C. Since $\angle 1$ is an exterior angle of $\triangle ABC$, $m\angle 1 = m-2 + m-3$. But this means that $m\angle 1 > m\angle 2$ because $m\angle 3 > 0°$. So $\angle 1$ is not congruent to $\angle 2$ which contradicts the given statement that the alternate interior angles are congruent. Therefore, the assumption that the lines are not parallel must be false and the lines are parallel.

Look Back

37.

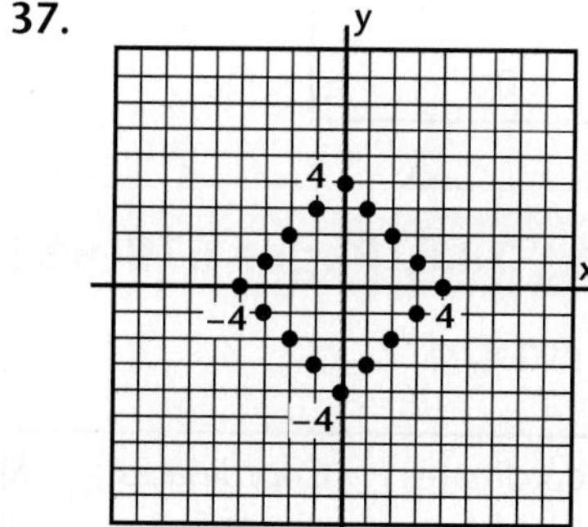

38.

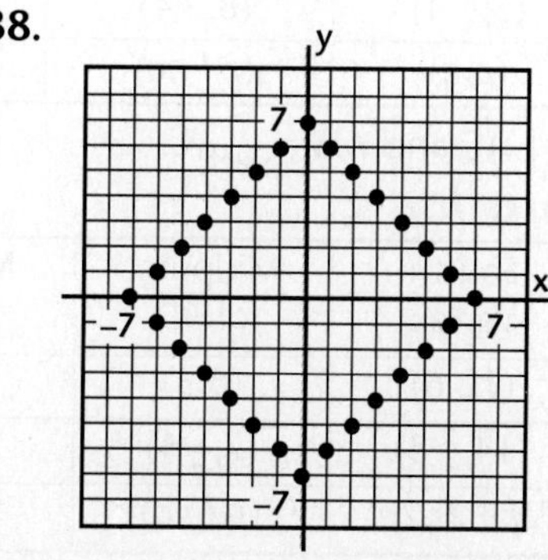

Lesson 12.5, Pages 671–678

Exploration

1.

TV	TV after pressing POWER button
1	0
0	1

2.

Play button	Record button	Tape Recorder
1	1	1
1	0	0
0	1	0
0	0	0

3.

Student pedal	Instructor pedal	Brake
1	1	1
1	0	1
0	1	1
0	0	0

4. The combination of the values in the first two columns were considered. Both of the buttons must be pushed for the tape recorder to record in step 2, while only one of the brakes must be pushed to stop the car in step 3.

5. Both the student pedal and the instructor pedal would have to be pushed for the car to stop. Neither the student or the instructor has complete control of the car.

Communicate

1.

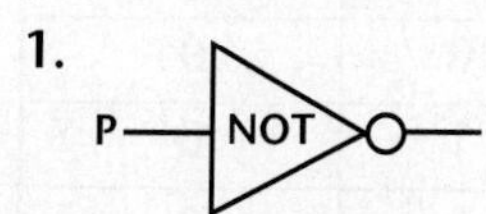

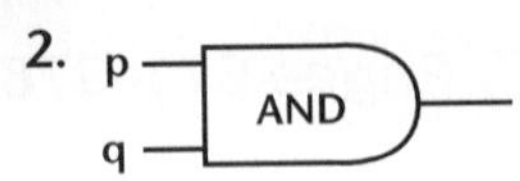

2. p, q → AND

3. p, q → OR

Practice and Apply

26. (p AND q) OR r

p	q	r	(p AND q)	(p AND q) OR r
1	1	1	1	1
1	1	0	1	1
1	0	1	0	1
1	0	0	0	0
0	1	1	0	1
0	1	0	0	0
0	0	1	0	1
0	0	0	0	0

27. p AND (q OR r)

p	q	r	(q OR r)	p AND (q OR r)
1	1	1	1	1
1	1	0	1	1
1	0	1	1	1
1	0	0	0	0
0	1	1	1	0
0	1	0	1	0
0	0	1	1	0
0	0	0	0	0

4. 0

5. 1

6. 0

7. 1

29. AND gate; Door Locked AND Power Button;

Door Locked, Power Button → AND

Door Locked	Power Button	Dishwasher
1	1	1
1	0	0
0	1	0
0	0	0

30. OR gate; Living Room Switch OR Hall Switch;

Living Room Switch, Hall Switch → OR

Livingroom Switch	Hall Switch	Light
1	1	1
1	0	1
0	1	1
0	0	0

31. AND gate; Power Button AND Play Button

Power Button, Play Button → AND

Power Button	Play Button	CD Player
1	1	1
1	0	0
0	1	0
0	0	0

Look Beyond

39. Use the people and equipment in the picture to estimate the area inside the grid. Then use the grid to estimate the area of the spill.

40. Answers will vary. About 4800 ft^2.

Lesson 12.6, Pages 679–684

Exploration 1

3.

Coordinates of M	Coordinates of S	Slope of $\overline{MS}$	Slope of $\overline{AC}$	Length of $\overline{MS}$	Length of $\overline{AC}$
(1,3)	(5,3)	0	0	4	8
(3,−4)	(8,−4)	0	0	5	10
(p,q)	$(p+r,q)$	0	0	r	$2r$

4. $\overline{MS} \parallel \overline{AC}$ and $MS = \frac{1}{2}AC$.

Exploration 2

Fourth Vertex	Midpoint of $\overline{BD}$	Midpoint of $\overline{AC}$
(12,6)	(6,3)	(6,3)
(14,−8)	(7,−4)	(7,−4)
$(2p+2r,2q)$	$(p+r,q)$	$(p+r,q)$

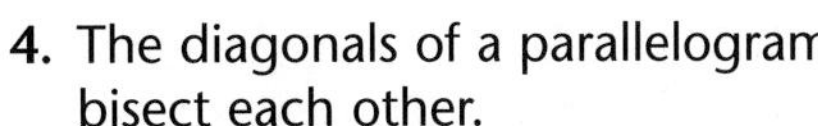

4. The diagonals of a parallelogram bisect each other.

Exploration 3

1. The y-coordinate is the same as x-coordinate.
2. $(\frac{a+b}{2}, \frac{a+b}{2})$; Yes, the coordinates are the same, so the point lies on the line y = x.
3. Slope of y = x is 1.
4. Slope $= \frac{a-b}{b-a} = \frac{a-b}{-(a-b)} = -1$
5. The lines are perpendicular.
6. The line y = x is the perpendicular bisector of the line through (a,b) and (b,a). Therefore, (a,b) and (b,a) are equidistant from the line and on opposite sides of the line. By definition of reflection through a line, (a,b) and (b,a) are the images of each other when reflected through $y = x$.

Communicate

1. Finding the remaining points is made easier. Calculations are made easier.
2. Set up a general parallelogram using variables for coordinates. Use the distance formula to find the lengths of the sides of the parallelogram.
3. Set up a general rhombus using variables for coordinates. Use the slope formula to find the slopes of the diagonals of the rhombus. If the slopes of the diagonals are opposite reciprocals, then the diagonals are perpendicular.
4. Prove that the diagonals of a rhombus divide the rhombus into congruent triangles. This creates congruent adjacent angles where the diagonals intersect, showing that the diagonals are perpendicular.

Practice and Apply

16. $AD = \sqrt{(2s-0)^2 + (0-0)^2} = 2s$;
$BC = \sqrt{(2r-2p)^2 + (2q-2q)^2}$
$= 2r - 2p$;
$MS = \sqrt{(r+s-p)^2 + (q-q)^2}$
$= r + s - p$
$\frac{AD + BC}{2} = \frac{2s + 2r - 2p}{2} = s + r - p = MS$

17. Slope of $\overline{AD} = \frac{0-0}{0} - 12 = 0$;
Slope of $\overline{BC} = \frac{2q-2q}{2r-2p} = 0$;
Slope of $\overline{MS} = \frac{q-q}{r+s-p} = 0$;
Since the slopes of the bases and the midsegment are equal, they are parallel.

21. Given: Right triangle ABC with vertices $A(0,0)$, $B(0,2b)$, and $C(2a,0)$ and hypotenuse BC.

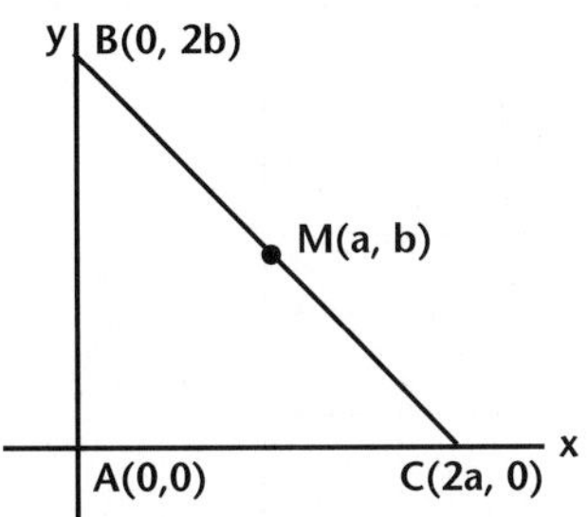

Prove: The midpoint of BC is equidistant from $A, B,$ and C.

The midpoint M of
$BC = (\frac{2a+0}{2}, \frac{0+2b}{2}) = (a,b)$.
$AM = BM = CM = \sqrt{a^2 + b^2}$.
Therefore the midpoint of the hypotenuse is equidistant from the vertices of a right triangle.

24. Given: Quadrilateral $ABCD$ with midpoints of its sides $W, X, Y,$ and Z.

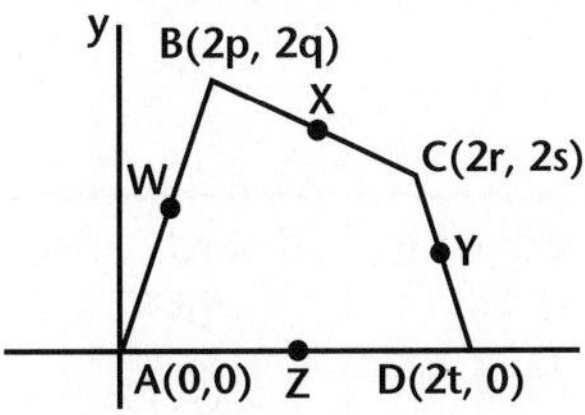

Prove: $WXYZ$ is a parallelogram.
Midpoint of $\overline{AB}$: $W = (p,q)$;
Midpoint of $\overline{BC}$; $X = (p+r,q+s)$;
Midpoint of $\overline{CD}$: $Y = (r+t,s)$;
Midpoint of $\overline{DA}$: $Z = (t,0)$.
Slope of $\overline{WX} = \frac{s}{r}$;
Slope of $\overline{YZ} = \frac{s}{r}$;
Slope of $\overline{XY} = \frac{-q}{t-p}$; Slope of $\overline{WZ} = \frac{-q}{t-p}$.
Since the slopes of opposite sides are equal, the opposite sides of ABCD are parallel and ABCD is a parallelogram by definition.

Look Beyond

30. Given: Cicle with center (0,0) and radius a.

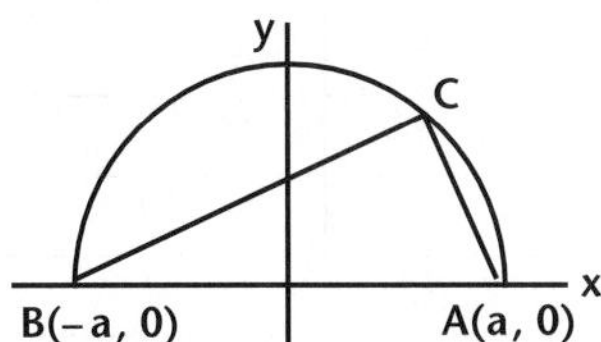

Prove: $\angle ACB$ is a right angle.
Let the x-coordinate of C be b. Then the y-coordinate c of C can be found using the Pythagorean Theorem: $b^2 + c^2 = a^2$ since a is the radius of the circle. So, $c = \sqrt{a^2 - b^2}$.
Slope of \o(AC,\s\up8(—)) = $\frac{\sqrt{a^2-b^2}-0}{b-a} = \frac{\sqrt{a^2-b^2}}{b-a}$;
Slope of \o(BC,\s\up8(—)) = $\frac{\sqrt{a^2-b^2}-0}{b+a} = \frac{\sqrt{a^2-b^2}}{b+a}$;
The product of the slopes = \f(\r($a^2 - b^2$), $b - a$) × \f(\r($a^2 - b^2$), $b + a$) $= \frac{a^2-b^2}{b^2-a^2} = \frac{a^2-b^2}{-(a^2-b^2)} = -1$
Since the product of the slopes is -1, $\overline{AC} \perp \overline{BC}$ and $\angle ACB$ is a right angle.
If the y-coordinate is selected first a similar argument may be used.

31. $AG = BD = CH = \sqrt{48}$; The diagonals of a cube are congruent.
Midpoint of $\overline{AG} = (2,2,2)$;
Midpoint of $\overline{BD} = (2,2,2)$;
Midpoint of $\overline{CH} = (2,2,2)$; The diagonals of a cube all bisect each other.